D0203260

Ecology and Classification of North American Freshwater Invertebrates

Second Edition

Ecology and Classification of North American Freshwater Invertebrates

Second Edition

Edited by

JAMES H. THORP

Bayard and Virginia Clarkson Professor of Biology
Department of Biology
Clarkson University
Potsdam, New York

and

ALAN P. COVICH

Department of Fishery and Wildlife Biology
Colorado State University
Fort Collins, Colorado

San Diego San Francisco New York Boston London Sydney Tokyo

Front cover images: From Jurine, L. (1820). Histoire des Monocles, qui se trouvent aux environs de Geneve. J. J. Paschoud, Paris, France.

This book is printed on acid-free paper. 

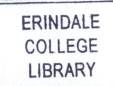

Academic Press
A Harcourt Science and Technology Company
525 B Street, Suite 1900, San Diego, California 92101-4495, USA
http://www.academicpress.com

Academic Press
Harcourt Place, 32 Jamestown Road, London NW1 7BY, UK
http://www.academicpress.com

Library of Congress Catalog Card Number: 00-109117

International Standard Book Number: 0-12-690647-5

PRINTED IN THE UNITED STATES OF AMERICA
01 02 03 04 05 06 07 EB 9 8 7 6 5 4 3 2 1

"For aminals and callapiters and other wonders of nature seen through our children's eyes."

Dr. James Thorp

"To my family and to Professor G. E. Hutchinson for their inspiration and patience."

Dr. Alan Covich

Contents

Contributors

Numbers in parentheses indicate the pages on which the authors' contributions begin.

Arthur E. Bogan (331) North Carolina State Museum of Natural Sciences, Research Laboratory, Raleigh, North Carolina 27607.

Patricia E. Bossert (135) Science Department, Northport High School, Northport, New York 11768.

Ralph O. Brinkhurst (431) Hermitage, Tennessee 37076.

Kenneth M. Brown (297) Department of Biological Sciences, Louisiana State University, Baton Rouge, Louisiana 70803.

David R. Cook (551) 7836 North Invergordon Place, Paradise Valley, Arizona 85253.

Alan P. Covich (1, 19, 777) Department of Fishery and Wildlife Biology, Colorado State University, Fort Collins, Colorado 80523.

Ronald W. Davies (465) Faculty of Science, Monash University, Clayton, Victoria 3168, Australia.

L. Denis Delorme (811) National Water Research Institute, Canada Centre for Inland Waters, Burlington, Ontario, L7R 4A6.

Stanley I. Dodson (849) Department of Zoology, University of Wisconsin, Madison, Wisconsin 53706.

David G. Frey† (849).

Thomas M. Frost† (97).

Stuart R. Gelder (431) Department of Biology, University of Maine at Presque Isle, Presque Isle, Maine 04769.

Fredric R. Govedich (465) Department of Biological Sciences, Monash University, Clayton, Victoria 3168, Australia.

† Deceased.

Anne E. Hershey (733) Department of Biology, University of North Carolina–Greensboro, Greensboro, North Carolina 27402.

William L. Hilsenhoff (661) Department of Entomology, University of Wisconsin, Madison, Wisconsin 53706.

H. H. Hobbs III (955) Department of Biology, Wittenberg University, Springfield, Ohio 45501.

William D. Hummon (181) Department of Biological Sciences, Ohio University, Athens, Ohio 45701.

Jurek Kolasa (155) Department of Biology, McMaster University, Hamilton, Ontario L8S 4K1 Canada.

Gary A. Lamberti (733) Department of Biological Sciences, University of Notre Dame, Notre Dame, Indiana 46556.

Robert F. McMahon (331) Department of Biology, The University of Texas at Arlington, Arlington, Texas 76019.

Diane R. Nelson (527) Department of Biological Sciences, East Tennessee State University, Johnson City, Tennessee 37614.

George O. Poinar, Jr. (255) Department of Entomology, Oregon State University, Corvallis, Oregon 97331.

Janet W. Reid (915) Department of Invertebrate Zoology, National Museum of Natural History, Smithsonian Institution, Washington, DC 20560.

Henry M. Reiswig (97) Redpath Museum and Department of Biology, McGill University, Montreal, Quebec H3A 2K6.

Anthony Ricciardi (97) Department of Biology, Dalhousie University, Halifax, Nova Scotia B3H 4J1.

Robert W. Sanders (43) Department of Biology, Temple University, Philadelphia, Pennsylvania 191ZZ.

Lawrence B. Slobodkin (135) Department of Ecology and Evolution, State University of New York, Stony Brook, New York 11794.

Bruce P. Smith (551) Biology Department, Ithaca College, Ithaca, New York 14850.

Ian M. Smith (551) Biodiversity Section, ECORC, Central Experimental Farm Agriculture and Agri-Food Canada, Ottawa, Ontario, Canada K1A 0C6.

Terry W. Snell (195) School of Biology, Georgia Institute of Technology, Atlanta, Georgia 30332.

David Strayer (181) Institute of Ecosystem Studies, Millbrook, New York 12545.

William D. Taylor (43) Department of Biology, University of Waterloo, Waterloo, Ontario N2L 3G1, Canada.

James H. Thorp (1, 19, 777) Department of Biology, Clarkson University, Potsdam, New York 13699.

Robert Lee Wallace (195) Department of Biology, Ripon College, Ripon, Wisconsin 54971.

Craig E. Williamson (915) Department of Earth and Environmental Sciences, Lehigh University, Bethlehem, Pennsylvania 18015.

Timothy S. Wood (505) Department of Biological Sciences, Wright State University, Dayton, Ohio 45435.

Preface

It is a pleasure to write the preface to the second edition of this book because it indicates that our authors produced a first edition that met the needs of our audience. To justify having those same professionals find the second edition useful, as well as attracting a new generation of readers, it was incumbent upon us to develop a text that improves and updates the first edition. We hope that you will find our authors have done so!

The reader will discover that almost every chapter has undergone major changes, with updated literature references, new coverage of invertebrate biology and ecology, and extensive revisions to taxonomy and classification. In a few cases, authors have made important improvements in figures and have introduced more detailed taxonomic keys. We have expanded the number of authors (for Chapters 4, 11, 13, 16, and 22) and added a new chapter (Chapter 18 by Drs. Anne Hershey and Gary Lamberti). As in the first edition, we limited our coverage of aquatic insects to save space and avoid excessive duplication with current textbooks on aquatic insect ecology and classification. However, we opted to add a chapter on ecology to provide an expanded coverage of aquatic insects at the introductory level to accompany a more taxonomically centered chapter on insects (Chapter 17 by Dr. William Hilsenhoff).

This edition is strongly focused on freshwater taxa found in surface waters of the United States and Canada. This encompasses at least 10,000–15,000 named species while avoiding the broad diversity of tropical aquatic invertebrates in North America. Groundwater and estuarine species are typically ignored or barely mentioned. We have also severely limited coverage of parasitic species and taxa that are only occasionally associated with aquatic habitats.

As in the first edition, we have asked the authors to set the taxonomic keys in their chapters at a level where the average reader could reasonably be assured of reaching an accurate organismal identification without being an expert on that group. This approach is counter to a tendency in some texts to write dichotomous keys to the species level for the nonexpert. In many highly specific keys, nonspecialists use the keys to obtain both a species name and, sometimes, a false sense of security that accuracy has been achieved. To supplement information at the generic or higher level in our text, the chapters provide information on where the reader can find more detailed, regional taxonomic keys.

As in our first book, many people contributed to this edition besides the authors. At Academic Press, we appreciate, in particular, efforts of our editor, Dr. Charles R. Crumly, and his able colleague and assistant, Ms. Donna James. At Clarkson University, we especially thank Linda Delosh and Bobbi Baillargeon for help in managing aspects of the book's production and chapter distribution. Each chapter was reviewed by one to three outside experts; their help is acknowledged in some chapters and through individual letters from editors and authors. We also express our appreciation again to the following reviewers of the second edition, while sincerely apologizing in advance for any we have overlooked: Drs. Thomas A. Langen, Denis H. Lynn, Michael L. Pace, Henry M. Reiswig, Anthony Ricciardi, David Strayer, Mitchell J. Weiss, John W. Fleeger, W. T. Edmondson, Jeffrey D. Jack, Elizabeth J. Walsh, Andreas Schmidt-Rhaesa, Dreux J. Watermolen, Eileen H. Jokinen, Charles Lydeard, Caryn C. Vaughn, Mark J. Wetzel, Dean W. Blinn, Byron T. Backus, Judith E. Winston, Clark W. Beasley, William R. Miller, John C. Conroy, Thomas Simmons, Sharon K. Jasper, Manuel L. Pescador, Arthur C. Benke, Timothy B. Mihuc, David C. Culver, B. Brandon Curry, Alison J. Smith, John E. Havel, William R. DeMott, Anna M. Hill, and David M. Lodge. We continue to appreciate suggestions from readers on how individual chapters or the book as a whole could be improved. Future comments and suggestions can be directed to authors of individual chapters or to the editors.

One of the sad tasks not faced in the first edition is announcing the demise of past authors. We greatly regret the passing of Professors David G. Frey (a coauthor of Chapter 21: Cladocera and Other Branchiopoda) and, at a relatively young age, Thomas M. Frost (senior author of Chapter 4: Porifera). Dr. Stanley I. Dodson handled all revisions to Chapter 21, and Tom Frost had finished coauthoring Chapter 4 before he died in a tragic accident in the summer of 2000. They will be sorely missed by the two of us and by their many other colleagues and friends.

As premature as it may sound, we have already begun thinking about our third, and probably last edition of this book with the current editors. While we expect another ten years will have elapsed before the third edition is completed, we realize it will require new authors in a number of areas as more senior people retire, shift research focal areas, or move into academic administration. If you are a reader who has the expertise, interest, and drive to write such a chapter, please feel free to contact one of us . . . but wait a few years so we have time to relax and handle those other tasks that have piled up on our desks while we completed work on the second edition!

James H. Thorp and Alan P. Covich

1 INTRODUCTION TO FRESHWATER INVERTEBRATES

James H. Thorp
Department of Biology
Clarkson University
Potsdam, New York 13699

Alan P. Covich
Department of Fishery and Wildlife Biology
Colorado State University
Fort Collins, Colorado 80523

I. INTRODUCTION

Next to their more vibrant, and often larger, marine relatives, freshwater invertebrates may initially seem drab and uninspiring. Yet once students of freshwater biology penetrate beyond superficial appearances and thoroughly examine the diverse structural, physiological, behavioral, and general ecological adaptations of freshwater invertebrates, few fail to be impressed by the enticing nature of these fascinating animals.

This chapter serves as a very brief introduction to freshwater invertebrates of Canada and the United States. Following a short discussion of procedures for naming and classifying organisms, we briefly describe the taxonomic groups covered in Chapters 3–23. We also include a taxonomic key to help you begin the process of classifying freshwater invertebrates. This dichotomous key will lead you to the appropriate chapter for more detailed biological and taxonomic information. Additional information on freshwater invertebrates can be found in books by Hutchinson (1993) and Merritt and Cummins (1996) as well as in various texts on limnology and stream ecology (e.g., Wetzel, 2001).

II. APPROACHES TO TAXONOMIC CLASSIFICATION

No phylum is totally immune to an occasional reshuffling of species, renaming of constituent taxa, and recalculation of total species richness (number of species). This dynamic feature of the discipline is, in part, a necessary response to the acquisition of new knowledge, but it also results from the ubiquitous and almost inherent disagreements that usually typify groups of two or more scientists! Taxonomists are frequently labeled as "splitters" or "lumpers" according to whether, respectively, they tend to acknowledge more or fewer species within an assemblage. Classifications they produce based on diverse scientific approaches, such as cladistics and numerical (= phenetics) taxonomy, can produce dramatically different views of relationships among taxa. Moreover, molecular systematics has sometimes shaken the traditional taxonomic schemes based on classical phenotypic characters. Although arguments at the species level are especially rampant, debates occasionally break out regarding the categorization of higher tiers, including classes and phyla. While reading through this book, you will often

Ecology and Classification of North American Freshwater Invertebrates. 2nd Edition

encounter references to current taxonomic debates. The final classification of taxa has been left to the authors of individual chapters. As editors, we have allowed the authors a great amount of leeway in this area, requesting only that they inform the reader about the principal areas of dispute.

Species are named today using a system of binominal nomenclature (= names composed of two parts) based on the 18th-century proposal of the brilliant Swedish naturalist Carl von Linné (or Carolus Linnaeus, the latinized form he preferred to use). This is the "genus and species" designation familiar to most biology students (e.g., *Homo sapiens* for the human species). You will occasionally see a species designated with three names (a trinomen), which is permissible under guidelines of an international code described below. For example, *Bosmina (Sinobosmina) freyi* is the genus, subgenus, and species name of a water flea, or cladoceran, common in rivers (it is one of a species complex formerly called *Bosmina longirostris*). In contrast, *Pheidole xerophila tucsonica* is the genus, species, and subspecies name of a harvester ant living in the desert near Tucson, Arizona. Scientific names are either in Latin or in a form that has been latinized (e.g., *B.(S.) freyi* was formed by latinizing the last name of Dr. David G. Frey, to honor this now-deceased coauthor of Chapter 21).

The final arbiter for naming taxa according to this system is the International Commission on Zoological Nomenclature, a judicial body elected periodically by the International Congress of Zoology to evaluate taxonomic names proposed by scientists. They use a set of agreed-upon rules, termed the International Code of Zoological Nomenclature (ICZN). The commission bases its decisions on rules covering the priority of names and designation of "type" taxa. As the name implies, "priority" specifies that the first name assigned in a publication is generally the official name of that taxon. The commission has, however, plenary powers to set aside the "rightful" name of a species if doing so would further the cause of science. For example, alteration of a familiar, long-standing name could cause major confusion in the scientific literature. Occasionally, two distinct taxa are inadvertently assigned the same binominal name by scientists working in different institutions. In that case, the species described last loses its name as a junior homonym. To lessen the state of confusion and diminish possibilities for future arguments, one individual of each species is designated as the prime example of that taxon. Such "type" specimens are deposited in one of many internationally recognized museums and made available for study by responsible scientists. Type specimens can be replaced only if the original is inadvertently lost or destroyed.

Taxonomic arguments among scientists fall into two broad categories: "What is a species?" and "How do you properly recognize evolutionary relationships among species?" In the first case, phylogenetic (including are they interfertile), evolutionary, and recognition definitions are employed. In the second instance, scientists use phenetics, cladistics, and phylogenetic taxonomy to establish relationships, as described below.

Systematists, which include many authors of our text, are involved in naming species (alpha taxonomy) and establishing phylogenetic relationships among groups (beta and gamma taxonomy). Most of these scientists can be divided in modern times into those employing numerical taxonomy (= numerical phenetics) and those using the more popular cladistics (= phylogenetic systematics), although there is unacknowledged blending of the two approaches in some cases (especially when gene sequencing data are being used). Each approach analyzes attributes of a species, or "characters," (e.g., morphological features or genetic sequences), to determine relationships among groups. In a very simplified definition, numerical taxonomists sum a large number of randomly selected characters possessed by multiple species and determine phylogenetic relationships based on the degree of similarity among species (i.e., shared characters), without placing relative evolutionary values on different characters. In contrast, cladists emphasize the recency of common descent (shared "derived" characters), a genealogical technique that places relative evolutionary values on different characters. Their approach depends significantly on identification of primitive (= ancestral) vs derived (i.e., more recently evolved) characters.

Information on systematic zoology is published in various journals, such as the *Zoological Record, Zeitschrift für Wissenschaftliche Zoologie* (Abteilung B), and *Bulletin Signaletique* (CNRS), as well as in numerous society publications and journals dealing with a broad range of subjects including classification. Readers interested in knowing where best to search for papers on a given taxon should consult the Literature Cited section in the appropriate chapter of this book, or see more specialized texts, such as Merritt and Cummins (1996) on genera of freshwater insects.

The International Code of Zoological Nomenclature passes judgment on names of genera and species but does not strictly control higher classification. For that reason, some disagreements exist about the numbers and names of higher classification tiers, especially at the ordinal level and above (e.g., the phyla Cnidaria vs Coelenterata, Bryozoa vs Ectoprocta, and disagreements on whether Crustacea is a phylum, subphylum, or class). The system for adding suffixes to taxonomic names as a means of designating classification level has

TABLE I Principal Zoological Ranks and Suffixes Where Designated

Kingdom
Phylum
Subphylum
Superclass
Class
Subclass
Cohort
Superorder
Order
Superfamily (-oidea)[a]
Family (-idae)
Subfamily (-inae)
Tribe (-ini)[a]
Subtribe (-ina)[b]
Genus
Subgenus
Species
Subspecies

[a] An ending recommended but not mandatory according to the International Code of Zoological Nomenclature.
[b] An ending customary but not cited in the code.

been confusing in the past but is beginning to change for the better. A nonexclusive list of the principal zoological ranks and suffixes is shown in Table I.

The importance of correct identification and consistent approaches to classification soon becomes apparent to all people interested in learning about general biotic relationships. Vast amounts of information are catalogued in research libraries and museums that use the Linnean system of binominal classification. Once a specimen is correctly identified, a large number of pertinent references in specialized journals can be accessed, not only in systematics but also in ecology, evolution, animal behavior, and medical sciences. Furthermore, the global revolution known as the Internet has opened a tremendous resource for taxonomists and other biologists. By using computer-based information retrieval systems present in most libraries or your personal access to the World Wide Web, you can rapidly track down biological and taxonomic research on an incredible number of species. The ease with which student and advanced scientists can now obtain information on many groups was almost unimaginable 10 years ago. To help you get started, we have included in Table II some of the favorite sites recommended by our authors. Keep in mind, however, that these are only a small minority of available sites and that the life span of a particular Web address is unpredictable and relatively fleeting compared to many of the sources used by investigators in the past (unless the Web site is directly maintained by a museum).

TABLE II Brief Selection of Web Sites for Freshwater Invertebrates

Taxonomic group and site	*Site maintained by*
General sites[a]	
Aquatic Sciences and Fisheries Abstracts	Corporate
Cambridge Biological Abstracts	Corporate
Current Contents	Corporate
Science Citation Index	Corporate
Web of Science	Corporate
Zoological Record	Corporate
Invertebrate Classification and Ecology in General[b]	
benthos.org/	North American Benthological Society
nysfola.org/links.html	Individuals
ucmp.berkeley.edu/help/taxaform.html	Paleontology Museum, UC Berkeley
york.biosis.org/zrdocs/zoolinfo/grp_dipt.htm	BIOSIS (not-for-profit organization) and the Zoological Society of London
Protozoa	
uga.edu/protozoa/home.htm	Society of Protozoologists
130.158.208.53/WWW/Servers/Protistologists.html	Corporate/University
megasun.bch.umontreal.ca/protists/protists.html	University of Montreal
Porifera	
mailbase.ac.uk/lists/porifera/files/	Individuals
ucmp.berkeley.edu/porifera/porifera.html	Paleontology Museum, UC Berkeley
Rotifera	
member.aol.com/bdelloid1/deloid.htm	Individuals
nazca.mmtlc.utep.edu/~rotifer/main.html	Individuals
Nematoda and Nematomorpha	
ianr.unl.edu/son/	Society of Nematologists

(Continues)

TABLE II (Continued)

Taxonomic group and site	*Site maintained by*
Mollusca: Gastropoda	
flmnh.ufl.edu/natsci/malacology/malacology.htm#Top	Florida Museum of Natural History
ummz.lsa.umich.edu/mollusks/index.html	Museum of Zoology, University Michigan
Mollusca: Bivalvia	
courses.smsu.edu/mcb095f/gallery/	Individuals
inhs.uiuc.edu/cbd/collections/mollusk.html	Illinois Natural History Survey
Annelida	
inhs.uiuc.edu/~mjwetzel/mjw.inhsCAR.html	Illinois Natural History Survey
keil.ukans.edu/~worms/annelid.html	Individuals/University of Kansas
Ectoprocta (Bryozoa)	
civgeo.rmit.edu.au/bryozoa/default.html	Individuals/Royal Melbourne Institute of Technology
Arthropoda: Insecta and Collembola	
afn.org/~iori/oinlinks.html	Individuals
entweb.clemson.edu/database/trichopt/	International Trichopteran Committee
coleopsoc.org/	Coleopterists Society
famu.org/mayfly/	Florida Agriculture & Mechanical University
mc.edu/~stark/stonefly.html	Individuals
ouc.bc.ca/eesc/iwalker/intpanis/	Individuals
Arthropoda: Crustacea: Crustacea: General and Peracarida	
odu.edu/~jrh100f/amphome/	Individuals
york.biosis.org/zrdocs/zoolinfo/arthrop.htm#crustaceans	BIOSIS
nmnh.si.edu/gopher-menus/Isopods.html	National Museum of Natural History
Arthropoda: Crustacea: Ostracoda	
uh.edu/~rmaddock/IRGO/irgohome.html	International Research Group on Ostracoda, University Houston
Arthropoda: Crustacea: Branchiopoda (Cladocera)	
cladocera.uoguelph.ca/	Individuals
Arthropoda: Crustacea: Copepoda	
nmnh.si.edu/iz/copepod/	National Museum of Natural History
Arthropoda: Crustacea: Decapoda	
bioag.byu.edu/mlbean/crayfish/crayhome.htm	Monte Bean Life Science Museum, BYU
nmnh.si.edu/gopher-menus/Crayfish.html	National Museum of Natural History

Note. Most of these sites have links to many other sites.

[a] We do not guarantee the accuracy or quality of any of these sites. All of these sites were open in the spring of 2000, but we have no idea how long they will remain accessible. Sites maintained by museums are likely to be long-lasting. Sites designated as being maintained by "individuals" are usually maintained by someone at a university, but the commitment of that university to the long-term survival of the site is usually unknown.

[b] These require that your institution subscribe to the service (i.e., someone pays!).

[c] Many of these Internet Web sites begin with http://www. Delete the www if those letters do not work.

III. SYNOPSES OF THE NORTH AMERICAN FRESHWATER INVERTEBRATES

All major phyla of invertebrates, other than Echinodermata, have some freshwater representatives, but only a few are more diverse in freshwaters than in the world's oceans. Some salient features of these freshwater taxa, as discussed in Chapters 3–23, are summarized in the remainder of this section.

A. Protozoa

Protozoa are unicellular (or acellular) eukaryotes existing either as individuals or as members of a loose-knit colony (Fig. 1). The term protozoa is a taxon of convenience, having no significance in classification schemes within the kingdom Protista (or Protoctista) other than as a traditional functional grouping of all heterotrophic and motile species. Modern taxonomic treatments divide the kingdom Protista into 27 phyla (more or less), 2 of which contain significant numbers of free-living, phagotrophic species. These two phyla, Ciliophora (ciliates) and Sarcomastigophora (the majority of amoeboid protozoa), are the subject of Chapter 3, by William D. Taylor and Robert W. Sanders.

Protozoa play important, though often overlooked, roles in freshwater ecosystems. For example, flagellate protozoans are the predominant predators of picoplankton (primarily bacteria and small cyanobacteria)

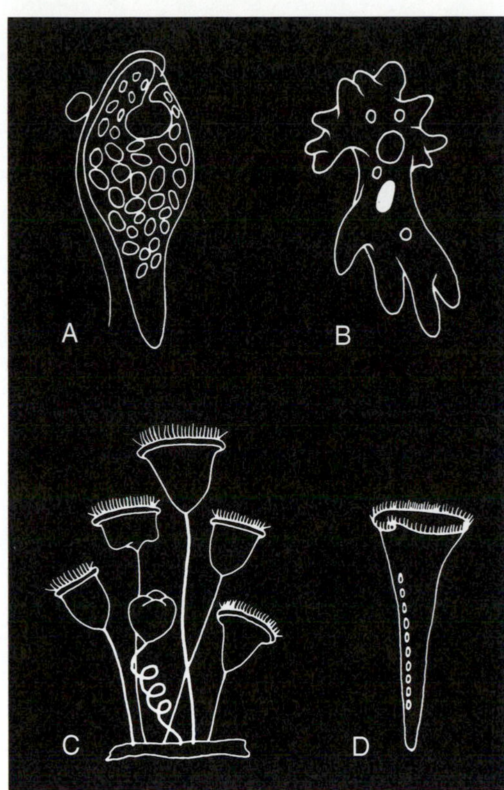

FIGURE 1 Solitary and colonial protozoa: (A) The zooflagellate *Khawkinea*; (B) *Amoeba*; (C) the colonial ciliate *Vorticella*; and (D) the solitary ciliate *Stentor*. See Chapter 3. [These and subsequent figures in this chapter were redrawn and modified from illustration in Chapters 3–22.]

and thus are essential components of the "microbial loop" within lakes and rivers. Many genera consume bacteria within the surficial layer of sediments and in the water column, and larger taxa capture and eat algae or other protozoa. They, in turn, are preyed upon by some species of oligochaetes, chironomids, and rotifers.

B. Porifera

About 27 species of sponges colonize freshwater habitats in Canada and the United States, with a general east-to-west and north-to-south decline in species richness. Although less diverse than protozoa, freshwater members of the phylum Porifera—the sponges—nonetheless are commonly found on hard substrates in many lakes and streams (Fig. 2). Except for the marine mesozoa and placozoa, sponges have the simplest structure of all multicellular phyla, showing little variation in external form and having a cellular-level (or incipient, tissue-level) organization. The most prominent feature of their anatomy is the progressively finer series of canals permeating the entire sponge body. Water is propelled through this canal system both passively, as a result of its basic hydrodynamic structure, and actively by choanocytes (collar cells). These cells also extract bacteria and algae from the water column by suspension (filter) feeding. Many species also contain symbiotic algae (green algae, except for one sponge that shelters a yellow-green algal symbiont), which provide supplemental nutrition.

The annual life cycle of a freshwater sponge is characterized by shifts between periods of active growth and dormancy in the gemmule stage. Availability of suitable hard substrate and adequate food seem to be the critical factors regulating the spread and growth of sponges, as described by Thomas M. Frost, Henry M. Reiswig, and Anthony Ricciardi in Chapter 4. Few animals prey on sponges, and most of those graze only a small portion of the sponge's tissues, leaving the sponge alive to grow and reproduce.

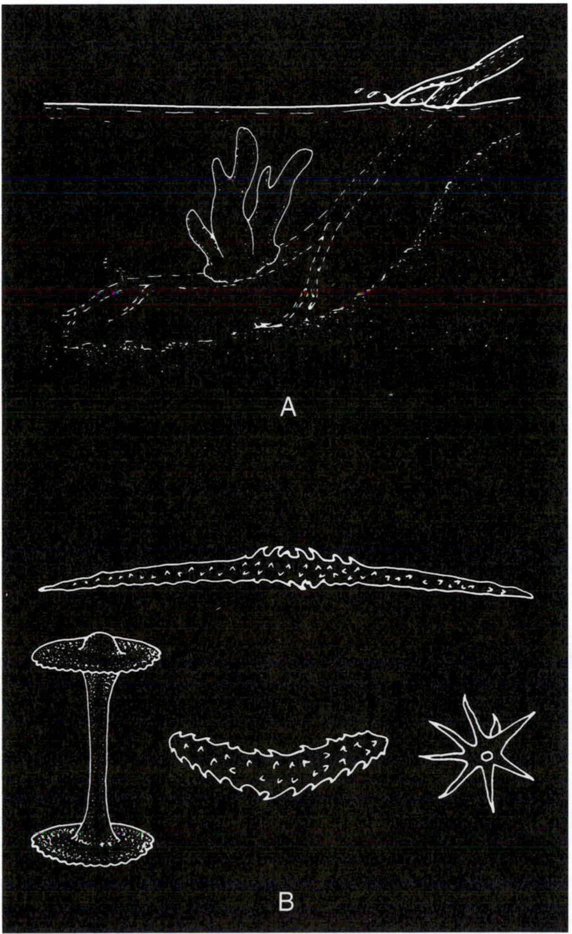

FIGURE 2 Freshwater sponges (phylum Porifera): (A) Drawing of a living sponge on a stick; (B) representative sponge spicules (all 25–150 μm in length). See Chapter 4.

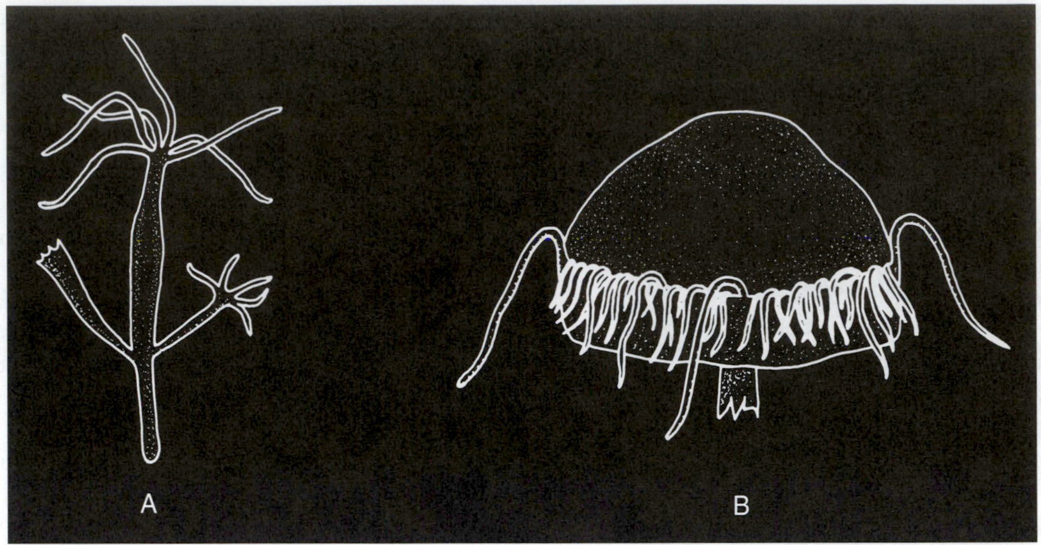

FIGURE 3 Freshwater coelenterates (phylum Cnidaria): (A) A polypoid Hydra; (B) a medusoid *Craspedacusta*. See Chapter 5.

C. Cnidaria

Cnidarians, or coelenterates, have been relatively unsuccessful in adapting to freshwater, being represented only by one class (Hydrozoa) with four or five purely freshwater genera: the common hydras (brown *Hydra* and green *Chlorohydra*, or *H. viridissima*) and the rare medusoid *Craspedacusta* and polypoid *Calposoma* and *Polypodium* (Fig. 3). Cordylophora, a mostly estuarine taxon, occasionally penetrates relatively freshwaters, where it lives in a colonial, sessile form. Despite their low diversity, cnidarians—or at least hydras—are routinely present on almost any reasonably hard substrate in permanent ponds and lakes and in streams ranging from headwater streams to large rivers.

The cnidarian body consists of a generally thin, acellular layer of mesoglea sandwiched between two thicker cellular layers (ecto- and endoderm), all surrounding a central body cavity (the coelenteron). Coelenterates dispatch their mostly invertebrate prey (both zooplankton and some benthos) with ectodermal batteries of stinging or sticky nematocysts. They are, in turn, fed upon by a few flatworms and sometimes crayfish. Green hydra are named for the presence of endosymbiotic *Chlorella*, algae that provide a significant source of nutrition for cnidarians. The biology of freshwater Cnidaria is described in Chapter 5, by Lawrence B. Slobodkin and Patricia E. Bossert.

D. Flatworms: Turbellaria and Nemertea

In Chapter 6 on unsegmented flatworms, Jurek Kolasa describes the biology and classification of the more than 200 species of turbellarians (phylum Platyhelminthes) and three species of ribbon worms (phylum Nemertea, or Rhynchocoela) found in North America (Fig. 4). He discusses the acoelomate Turbellaria, the simplest bilaterally symmetrical eumetazoan with three embryonic cell layers, and the more complex, coelomate Nemertea, which has an anus and a closed circulatory system. These latter flatworms possess an

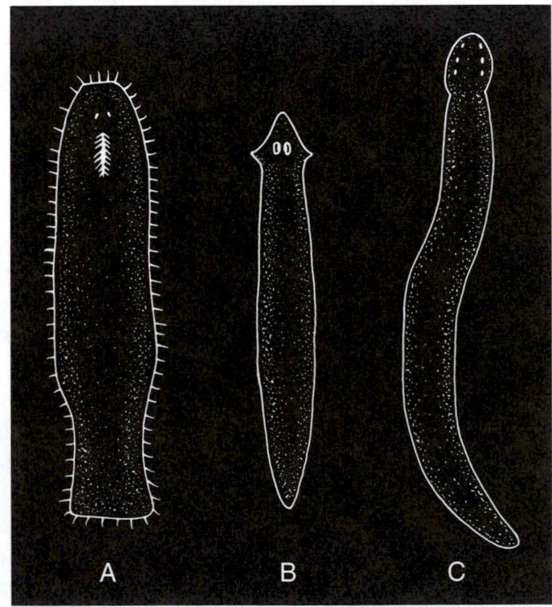

FIGURE 4 Turbellarian flatworms (phylum Platyhelminthes) and a ribbon worm (phylum Nemertea): (A) *Macrostomum*; (B) *Dugesia*; and (C) the numertean *Prosotoma*. See Chapter 6.

eversible, muscular proboscis resting in a rhynchocoel. Despite their common name, flatworms are usually cylindrical in cross section!

Flatworms are well represented in the benthos of lakes and streams. Ribbon worms of the single North American genus *Prostoma* are rarely reported but can be common in their typical shallow, littoral zone habitats; all are carnivores that capture their prey with a sticky proboscis. The much more diverse turbellarians are distributed more widely than ribbon worms; but they are not much better known, in part because they are very difficult to identify with the collection and preservation techniques normally employed in field studies. The great majority of these flattened or cylindrical worms are free-living denizens of lakes and streams from shallow to deep waters. Most are small (<1 mm long), especially the more diverse microturbellaria (~150 species in North America). Better known, however, are the larger triclads (most >10 mm long), which are often called planaria. Turbellarians consume prey in accordance with their size, variously eating bacteria, algae, protozoa, and small or large invertebrates.

E. Gastrotricha

The pseudocoelomate gastrotrichs (Fig. 5) are common residents of epigean waters throughout North America, with densities ranging from 100,000 to 1,000,000 animals per square meter. Although fewer than 100 freshwater species from North America have been reported, the actual diversity of the phylum Gastrotricha is probably much greater in freshwater (and many marine species are also present). Possibly because gastrotrichs are so small (50–800 μm), our ecological knowledge of this group has not advanced very far. They colonize both aerobic and anoxic surface sediments, thrive as members of the aufwuchs assemblage on aquatic plants, and occasionally venture into the plankton. Oddly enough, given their size, they are very rare in groundwater (other than the hyporheic zone of streams). Gastrotrichs thrive on a diet of bacteria, algae, protozoa, and detritus. Not much is known about their enemies, but these certainly include amoeba, cnidarians, and predaceous midges. Their ecology and classification are examined in Chapter 7 by David L. Strayer and William D. Hummon.

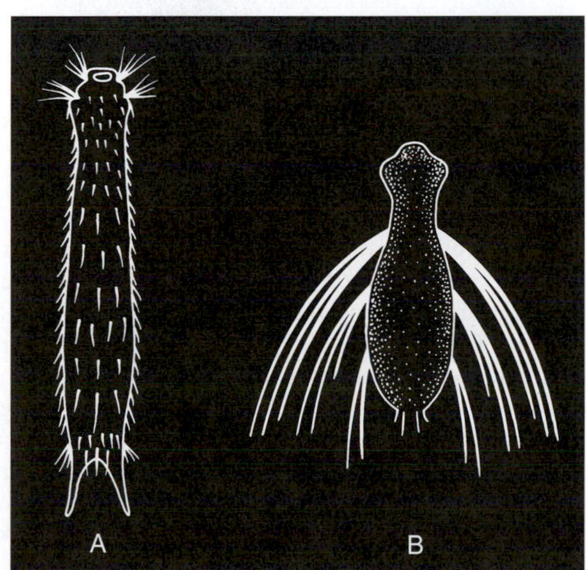

FIGURE 5 Representative freshwater Gastrotricha: (A) *Chaetonotus*; (B) Stylochaeta. See Chapter 7.

F. Rotifera

Perhaps no other phylum is as clearly associated with freshwater as is Rotifera (Fig. 6). Nearly 2000 species of rotifers, or "wheel animals," inhabit freshwaters throughout the world, whereas only about 50 species are exclusively marine. These unsegmented, pseudocoelomates are distinguished by two principal anatomical features: an apical, ciliated region known as the corona and a muscular pharynx, termed the mastax, with its complex set of hard jaws. The ciliated corona is employed for both locomotion and food-gathering. Rotifers range in size from minute creatures barely 100 μm long to giants of 2 mm or more!

As Robert L. Wallace and Terry W. Snell point out in Chapter 8, rotifers are one of the three principal animal taxa in the plankton (along with protozoa and microcrustaceans). Because they are more efficient than cladocera when feeding on minute algae, rotifers can exert a greater grazing pressure on the small picoplankton. Other rotifers are important predators on bacteria, protozoa, and small metazoa in the plankton. Rotifers play a critical role in the microbial (nutrient) loop within freshwater lakes and rivers. Rotifers are the numerical dominants of most large river zooplankton communities. Many species are also benthic or nearly so. Larval fish, some protozoa, insect larvae, microcrustaceans, and other rotifers are numbered among their predators.

G. Nematoda and Nematomorpha

Nematodes, which are best known to the average person for damaging crops, are also abundant in freshwater lakes and streams where they spend all or a large part of their life cycle (Fig. 7A). Chapter 9 by George O. Poinar covers 66 genera (encompassing 300–500 species), representing over 95% of the

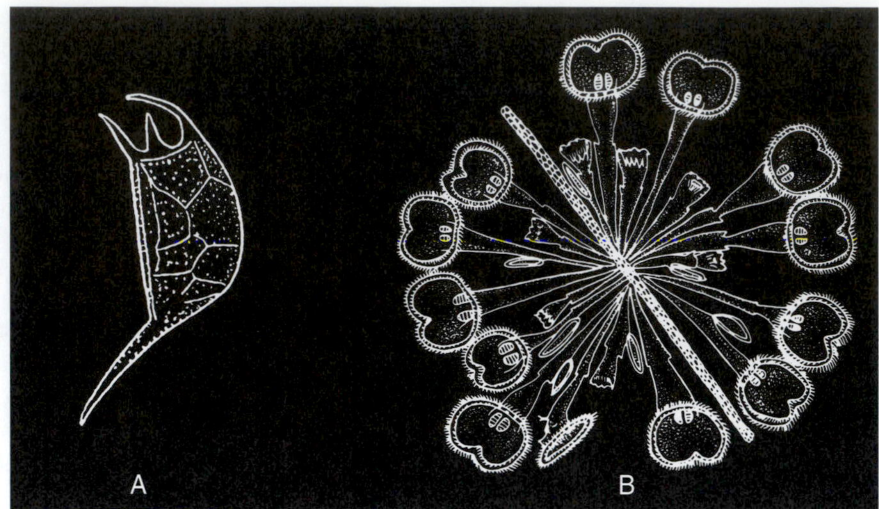

FIGURE 6 "Wheel animals" of the phylum Rotifera: (A) A Solitary *Keratella*; (B) a colony of *Sinantherina*. See Chapter 8.

known species of roundworms. Nematodes are unsegmented worms with a body cavity (pseudocoel) and complete alimentary tract; they lack jointed appendages and specialized circulatory and respiratory systems. Roundworms are significant components of the micro- and macrobenthos of many aquatic ecosystems. For example, in Mirror Lake, New Hampshire, they constitute 60% of all individuals in the benthic metazoa, although only about 1% of the total biomass (Strayer, 1985, as cited in Chapter 9). Many nematodes are parasitic in plants and animals, but these are only briefly mentioned in Chapter 9 by George O. Poinar. Free-living roundworms, the focus of that chapter, obtain nourishment from bacteria, algae, and protozoa. While flatworms and crayfish feed occasionally on roundworms, their greatest threat comes from other, predaceous nematodes.

Another phylum of worms that, in a very rough way, resembles nematodes is Nematomorpha (Fig. 7B). These hairworms, or horsehair worms, are free-living as adults but parasitic as larvae in various invertebrates. Like nematodes, they are unsegmented pseudocoelomates, but their intestine is nonfunctional, and they are generally much larger than roundworms. The evolutionary position of nematomorphs remains controversial, but they probably have a distant relationship with nematodes. All but one genus occurs in freshwater, and seven genera from North American lakes and streams are reported.

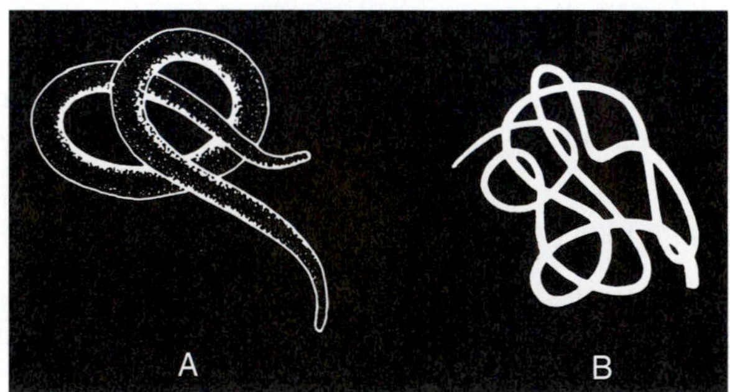

FIGURE 7 Illustrated is a common roundworm (A; phylum Nematoda) and a less frequently observed, adult horsehair worm (B; phylum Nematomorpha). See Chapter 9.

H. Mollusca

Two classes within the phylum Mollusca contain large numbers of freshwater species: the classes Gastropoda (snails) and Bivalvia (mussels and clams). These ubiquitous and abundant molluscs are discussed separately in Chapters 10 (snails) and 11 (mussels and clams).

1. Gastropoda

Snails are one of the more diverse freshwater groups (having about 659 species in North America) and are certainly among the most easily recognized by the public. These soft-bodied, unsegmented coelomates have a body organized into a muscular foot, distinct head region, and visceral mass: a fleshy mantle covers the viscera and secretes a spiraled, univalve, calcareous shell (Fig. 8). They are commonly found in littoral regions of lentic and lotic habitats, where they collect bottom detritus, graze on the algal film covering rocks, wood snags, and plants or even float upside down at the water surface, consuming trapped and neustonic algae. Recent ecological work on freshwater gastropods,

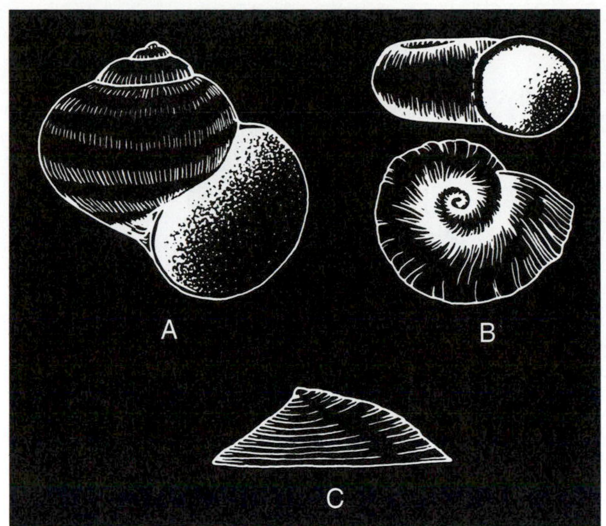

FIGURE 8 Three types of shell morphologies in freshwater snails and limpets (phylum Mollusca, class Gastropoda): (A) The most common, spiral shell, as shown by *Pomacea*: (B) two views of the planospiral snail *Planorbella* (*Helisoma*); and (C) the conical freshwater limpet *Ferrissia*. See Chapter 10.

as described by Kenneth M. Brown in Chapter 10, has focused on their important roles as grazers on periphyton and as prey of benthic-feeding fish and invertebrates, especially crayfish but also others like leeches and predaceous water bugs (Belostomatidae). Freshwater limpets are included in Gastropoda, although they are not related closely to their marine relatives.

2. Bivalvia

The North American fauna of freshwater bivalves is the richest in the world, with over 250 species of mussels and clams (Fig. 9; see also Chapter 11, by Robert F. McMahon and Arthur E. Bogan). Bivalves, or pelecypods, are major deposit and suspension feeders within permanent lakes, streams, and large rivers, where they are often the largest invertebrates in body mass. Except for the smaller pea clams (Sphaeriidae), native mussels (Unionacea) are uniquely capable of growing large enough to escape predation by almost all aquatic vertebrates and invertebrates. Bivalves are much less motile than gastropods, and their body is enclosed in two hinged, calcareous shell valves. A radula, which is employed by snails for grazing, is absent in clams and mussels. Unlike the vast majority of North American freshwater invertebrates, bivalves have been of direct economic importance to humans; they were formerly used for food (by Native Americans) and button production and are now harvested for freshwater pearls or for the core elements needed to grow marine-cultured pearls. Many native species are listed on state and federal lists of "threatened and endangered species" as a result of human disruption of their aquatic habitats, loss of fish hosts for larval stage, and invasion of Asian clams (*Corbicula fluminea*), zebra mussels (*Dreissena polymorpha*), and quagga mussels (*D. bugensis*). These exotic species have become the numerical and/or biomass dominants in many North American waters, especially east of the great plains, and they are exerting substantial ecological and economic effects.

I. Annelida

The phylum Annelida consists primarily of three relatively diverse freshwater groups (oligochaete worms, branchiobdellidans, and leeches) and a few minor freshwater taxa. Annelids are legless, segmented coelomates with serially arranged organs; they occur in both lentic and lotic environments, where they tend to be either detritivores (oligochaetes) or predators (leeches and branchiobdellidans). The taxonomy of Annelida is in great dispute—an observation reflected by the diverse views of authors in Chapters 12 and 13. A strong area of controversy involves the rank of higher groups, beginning at whether Clitellata (which encompasses all annelids except polychaetes) is a superclass, class, or subclass; this then affects ranks of lower groups like the oligochaetes and leeches. Unless the reader is a student of annelid biology and evolution,

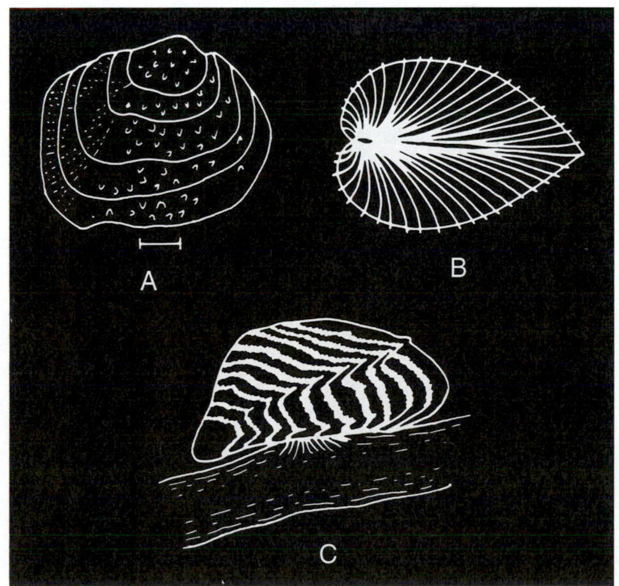

FIGURE 9 One native and two invadere species of freshwater bivalves (phylum Mollusca,class Bivalvia): (A) The native mussel *Quadrula*; (B) the ubiquitous Asiatic clam, *Corbicula*; and (C) a newcomer to North America, the rapidly spreading zebra mussel, *Dreissena*. See Chaper 11.

one can set aside these arguments for the time being and concentrate on learning other details of annelid biology, ecology, and classification.

1. Oligochaeta and Branchiobdellida

In Chapter 12, Ralph O. Brinkhurst and Stuart R. Gelder discuss the free-living oligochaete worms (~150 species) and the related branchiobdellidans, a less-well-studied group of 107 species in North America (Fig. 10). Oligochaetes are among the more widely distributed freshwater taxa and one of the very few invertebrates to colonize moist terrestrial environments. In addition to the characteristics of the phylum as a whole, oligochaetes lack suckers and generally have four bundles of setae on every segment but the first. Oligochaetes are best known for their ingestion of sediments in the littoral and profundal zones (a process prominent in nutrient recycling), but some species feed on benthic algae and other epiphytic organisms. Their great abundance and ease of consumption and assimilation make them favorite prey items for many invertebrate and vertebrate predators. Branchiobdellidans are ectocommensals of mostly astacoidean crayfish, where they consume epibionts living on crayfish, spare food lost by the host in the feeding process, tissue from host wounds, and other branchiobdellidans. They have a posterior sucker for attaching to their hosts.

2. Leeches and Acanthobdellids

Another ecologically important group of annelids is represented by 73 North American species of leeches,

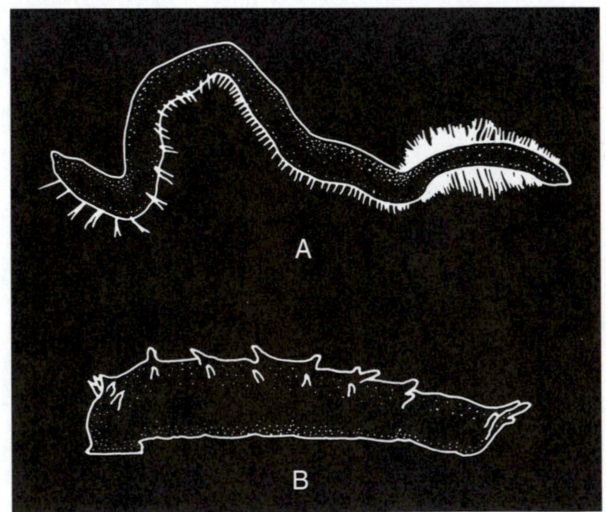

FIGURE 10 Two freshwater annelids with distinctive lifestyles: (A) The free-living oligochaete *Branchiura*, showing its posterior gill filaments; (B) *Ceratodrilus*, a branchiobdellidan which is an ectocommensal of crayfish. See Chapter 12.

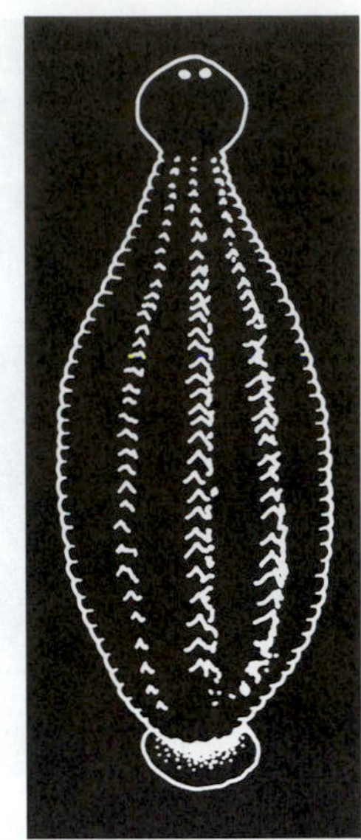

FIGURE 11 Schematic view of the leech Placobdella. See Chapter 13.

as described in Chapter 13, by Ronald W. Davies and Fredric R. Govedich. A single species of leech-like Acanthobdellidae occurs in North America; it is a temporary ectoparasite of salmonid fish. Unlike most oligochaetes, leeches have suckers (anterior and posterior) and lack setae (Fig. 11). Another major distinction between these two annelids is their feeding niches. Most leeches are predators of invertebrates, such as chironomid midges, oligochaetes, amphipods, and molluscs; but nonscientists best know this group for the minority that are ectoparasites, feeding on the blood of vertebrates (sanguivory), including incautious humans. Predators of leeches include fish, birds, snakes, salamanders, and a large variety of freshwater invertebrates.

J. Bryozoa

Bryozoans, or "moss animals," are generally sessile, colonial invertebrates with ciliated tentacles (called "lophophore") for capturing suspended solitary algae, protozoa, and organic particles (Fig. 12). All but one North American species are in the phylum Ectoprocta, or Bryozoa, and are composed of animals whose mouth, but not anus, is located within the whorl of feeding tentacles. Although the diversity of freshwater Bryozoa is not high (24 species in North America), they are actually quite common occupants of hard, relatively stationary, and biologically inactive substrates. Most live in relatively warm water (18–28°C). They occur in lakes but thrive better in flowing water, where

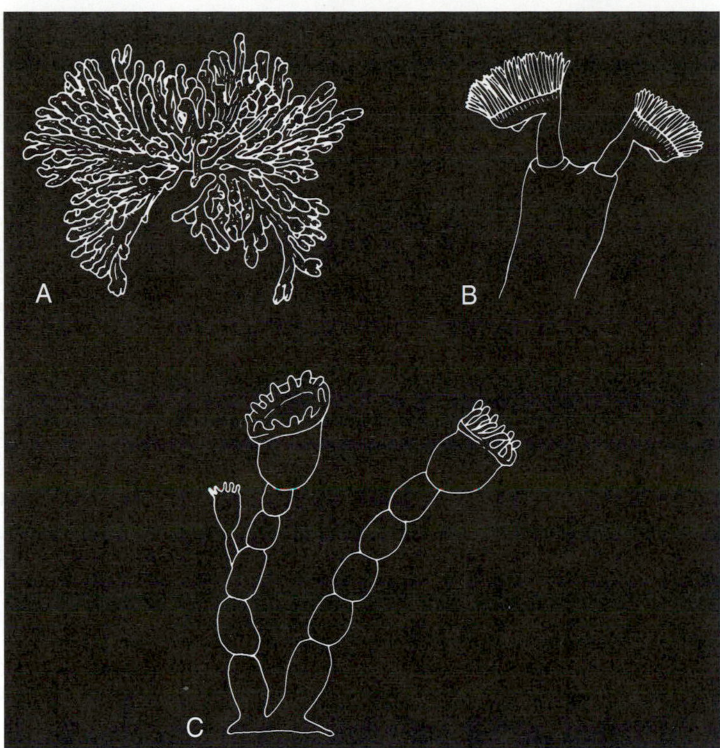

FIGURE 12 Freshwater bryozoans: (A) *Plumatella* colony (phylum Bryozoa or Ectoprocta;) (B) close-up of two individuals in an ecotoproct colony; and (C) the ubiquitous entoproct *Urnatella*. See Chapter 14.

a constant source of suspended food encounters their tentacles. Our ecological knowledge of Bryozoa is relatively sparse despite the long-standing familiarity of scientists with this group. The only nonectoproct bryozoan is the cosmopolitan, suspension-feeding *Urnatella gracilis*. It is classified in the phylum Entoprocta, a taxon with very dubious relation to the ectoprocts. Both the mouth and anus of entoprocts are placed within the whorl of feeding tentacles; other important structural differences from the ectoprocts also exist. The biology, ecology, and classification of bryozoans are described in Chapter 14, by Timothy S. Wood.

K. Tardigrada

Tardigrades, or "water bears," are easy to overlook because of their diminutive nature (adults average 250–500 μm), but they are widely dispersed in freshwater and terrestrial environments, and a few of the ~800 species worldwide occur in the ocean. Although the taxonomy of Tardigrada needs to be revised, the diversity of freshwater species in North America is low compared to most other phyla with freshwater members: 3 genera with 5 species are strictly freshwater, while 6 genera with 13–14 species are typically freshwater. The actual number of water bear species in North America is probably considerably higher than this low value. Tardigrades are segmented micrometazoa with four fleshy legs terminating in claws (Fig. 13). They have piercing/sucking mouthparts and obtain nutrition from plants, animals, and detritus. An important part of the water bear life cycle is the latent stage (cryptobiosis) entered into when environmental conditions become inhospitable. The biology and systematics of these little-known creatures are described by Diane R. Nelson in Chapter 15.

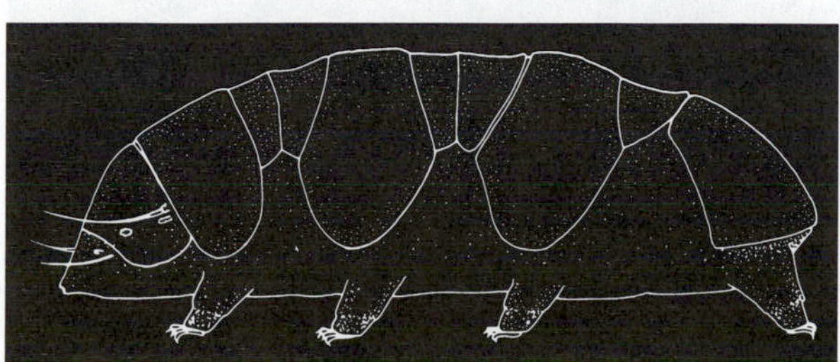

FIGURE 13 Lateral view of a heterotardigrade (phylum Tardigrada). See Chapter 15.

L. Arthropoda

The most successful terrestrial phylum and one of the most prominent freshwater taxa is Arthropoda. Its three subphyla with freshwater members—Chelicerata (water mites and aquatic spiders), Uniramia (aquatic insects), and Crustacea (crayfish, fairy shrimp, copepods, etc.)—are all diverse and important components of lakes and streams. (Some authors prefer to group the insects, myriapods, and crustaceans as classes within a single subphylum, Mandibulata.) Arthropods occupy every heterotrophic niche in benthic and pelagic habitats of most permanent and temporary aquatic systems and are well represented in groundwater habitats. As adults, these metameric coelomates are characterized by a chitinous exoskeleton and stiff, jointed appendages modified as legs, mouth parts, and antennae (except in water mites).

1. Water Mites

The class Arachnida (subphylum Chelicerata) is represented in freshwater by a single genus of semiaquatic, true spiders (*Dolomedes*) and a colossal number of water mites (subclass Acari; Fig. 14). In Chapter 16, Ian M. Smith, David R. Cook, and Bruce P. Smith, discuss the biology and classification of several distantly related groups of mites, the most diverse of which is Hydrachnida. As they point out, a single square meter of substrate from a littoral weed bed in a eutrophic lake may contain 2000 mites representing up to 75 species and a comparable area of rocky stream bed may yield 5000 individuals in 50 or more species! Over 1500 species are estimated to be present in Canada and the United States. Because of their prodigious numbers and wide distribution, mites probably play a major, albeit often ignored, role in aquatic ecosystems as predators (e.g., on ostracodes, early instars of insects, various zooplankton, and insect eggs), ectoparasites (mostly on insects), and prey of fish, other mites, and a few other invertebrates. For example, mites interact intimately with all life stages of aquatic insects, parasitizing 20–50% of adults in natural populations. Mites differ from members of other arthropod subphyla in lacking antennae; adults generally have four pairs of jointed walking legs and segmentation is greatly reduced or not immediately apparent.

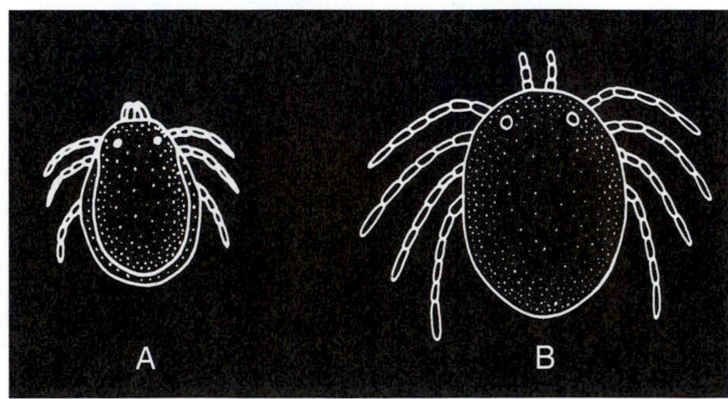

FIGURE 14 Generalized larval (A) and adult (B) water mites (phylum Arthropoda, Acari). See Chapter 16.

2. Aquatic Insects and Collembola

Perhaps the most diverse and best studied groups of freshwater invertebrates are the 10 insect orders containing the roughly 6000 species of aquatic insects (Fig. 15). William L. Hilsenhoff introduces the classification and general characteristics of aquatic insects in Chapter 17, and Anne E. Hershey and Gary A. Lamberti tackle the daunting task of summarizing some basic ecological aspects of aquatic insects in Chapter 18, leaving the more detailed discussions to books devoted to all or specific groups of aquatic insects (e.g., McCafferty, 1981; Ward, 1992; Merritt and Cummins, 1996). Like the water mites, insects occupy every freshwater habitat and heterotrophic niche and are present with tremendous densities and diversities. The majority spend most of their lives as larvae, only briefly departing the freshwater milieu to mate. Aquatic insects feature mandibles and one pair of antennae; adults have three pairs of legs and generally possess functional wings. Chapter 17 also covers the semiaquatic springtails, or Collembola. These wingless arthropods were once classified as insects but are now generally regarded as members of the separate class Entognatha within the superclass Hexapoda, which also includes insects. They are usually associated with the water surface (neuston) or riparian habitats. Although a number of species are sometimes collected in or on water, only about 15 species are regarded as semiaquatic or riparian.

3. Crustaceans

Nearly 4000 species of crustaceans inhabit freshwaters around the world, occupying a great diversity of habitats and feeding niches. Within pelagic and littoral zones, water fleas and copepods are the principal macrozooplankton, and benthic littoral areas shelter vast numbers of seed shrimps, scuds, and other crustaceans. An omnivorous feeding habit is typical of crustaceans, although there are many strict herbivores, carnivores, and detritivores. Members of the subphylum Crustacea are characterized by a head with paired mandibular jaws, a pair of maxillae, and two pairs of antennae. Their appendages are often biramous.

a. Peracarida and Branchiura Several groups of Crustacea, most in the class Malacostraca, superorder Peracarida, have a relatively small number of freshwater representatives. Two of these, Amphipoda and

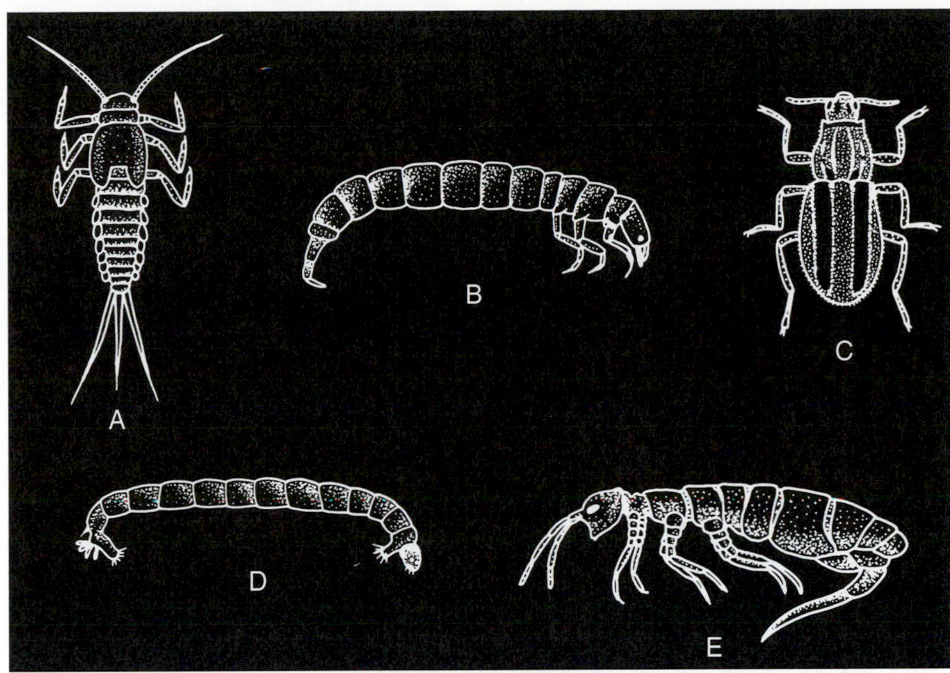

FIGURE 15 Five arthropods in the superclass Hexapoda: (A) *Baetis*, a mayfly nymph (animals drawn in a-d are all in the class Insecta); (B) the caddisfly larva *Polycentropus*; (C) and adult elmid beetle *Stenelmis*; (D) the midge larva *Chironomus*; and (E) a semi-aquatic springtail (class Entognatha, order Collembola). See Chapter 17.

Isopoda, are important taxa within marine systems. Chapter 19 (by Alan P. Covich and James H. Thorp) includes both an introduction to the subphylum Crustacea and coverage of the ecology and systematics of three malacostracan orders representing less than 200 species: Amphipoda (scuds), Isopoda (aquatic sow bugs), and Mysidacea (opossum shrimp) (Fig. 16). Of these three taxa, the scuds are the most widely distributed and diverse within North American freshwaters, for example. The peracaridans are omnivorous and opportunistic, consuming items ranging from algae and organic matter to small, living invertebrates and flesh of dead bivalve molluscs. They occur in ecosystems ranging in size from large rivers to small ponds and are also well represented among subterranean fauna. The subclass Branchiura, or "fishlike," attach themselves to the gills of fish; only one genus occurs in North American freshwaters (*Argulus*).

b. Ostracoda Approximately 420 species of mussel and seed shrimps (class Ostracoda) found on this continent are easily distinguished from other crustaceans by the typical, bivalved carapace enveloping their bodies (Fig. 17). This protective covering is composed of chitin and calcite. Ostracodes occur within the benthos of nearly every conceivable aquatic system from temporary ponds to large rivers to groundwater habitats, but they rarely enter the plankton even though many species are good swimmers. Almost all are free living and most are herbivores on attached algae or detritivores. They are, in turn, consumed by fish, waterfowl, and various benthic and planktonic invertebrate predators, as discussed by L. Denis Delorme in Chapter 20.

c. Cladocera and Other Branchiopoda The class Branchiopoda includes the common *Daphnia* and other cladocera frequently studied in the laboratory component of introductory biology courses, along with lesser known representatives of the class such as the elusive fairy shrimp of ephemeral pools and saline lakes (Fig. 18). Branchiopods are a heterogeneous group linked by similar mouth parts and leaf-like thoracic legs (phyllopods). Virtually all species within the eight extant orders of Branchiopoda are limited to freshwaters (mostly lentic systems). As pointed out by Stanley I. Dodson in Chapter 21, branchiopods occupy key positions in aquatic communities. As consumers, they are algivorous herbivores, detritivores (often assimilating bacteria on benthic or suspended organic matter), and occasionally predators of small invertebrates. They are important prey items in the diets of many fish, waterfowl, and certain other vertebrate and invertebrate predators. Many species are planktonic, while others live in shallow-water benthic habitats. Some, like the fairy

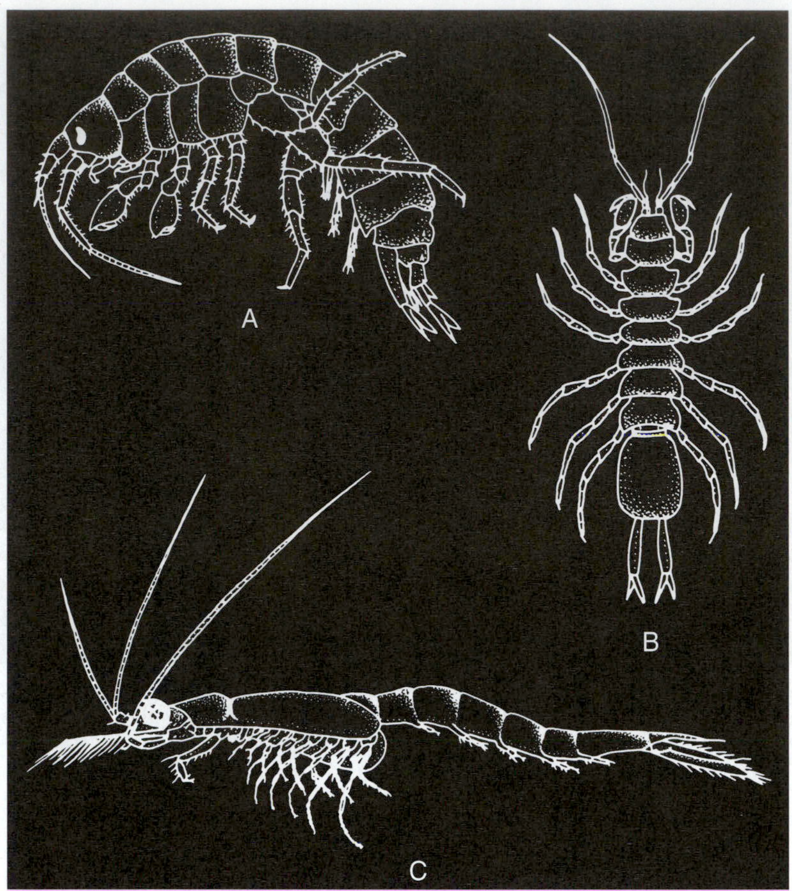

FIGURE 16 Representative freshwater crustaceans: (A) *Gammarus*, a scud (order Amphipoda); (B) *Caecidotea*, an aquatic sow bug (order Isopoda); and (C) *Mysis*, an opossum shrimp (order Mysidacea). See Chapter 19.

shrimp *Artemia,* are adapted to temporary ponds and to hypersaline ponds and lakes. Scientists estimate that the United States and Canada contain ~400 species of cladocera and 78 species of noncladoceran branchiopods.

d. Copepoda Copepods are very common in aquatic and semiaquatic habitats, spanning the gamut from ephemeral pools, groundwaters, and wetlands to the benthos and plankton of large lakes and rivers. They are also major constituents of marine zooplankton and can be abundant in moist terrestrial environments. Most are free-living, but two of the seven orders contain species parasitic on freshwater fish. Slightly over 300 species from North American fresh waters are described, 10 of which are considered to be introduced. They typically represent more than 50% of the zooplankton biomass in rivers and lakes. As was the case with cladocera, copepods have an integral role in aquatic food webs both as primary and secondary consumers (most are omnivorous) and as a major source of food for many young and adult fish, other copepods, and other invertebrates, such as larval phantom midges (*Chaoborus*). Copepods can be distinguished from other small (0.5—2.0 mm) crustaceans by their possession of: (1) a cylindrical, segmented body; (2) two setose, caudal rami on the posterior end of the abdomen (which are used in swimming); and (3) a single, simple anterior eye (Fig. 19). The biology and classification of this important crustacean are

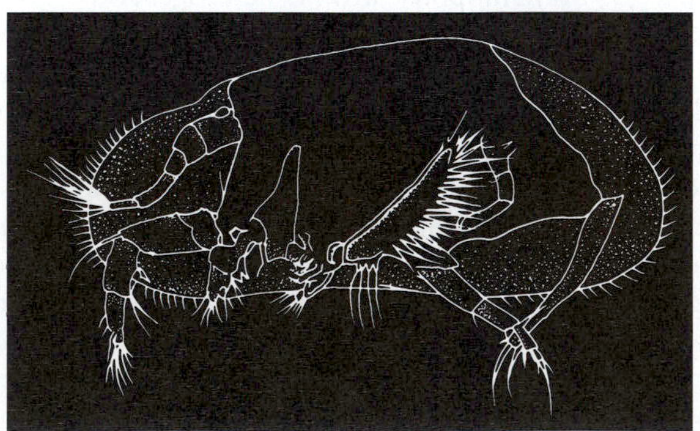

FIGURE 17 Female ostracode *Candoma* (crustacean class Ostracoda) with left value of carepace removed. See Chapter 20.

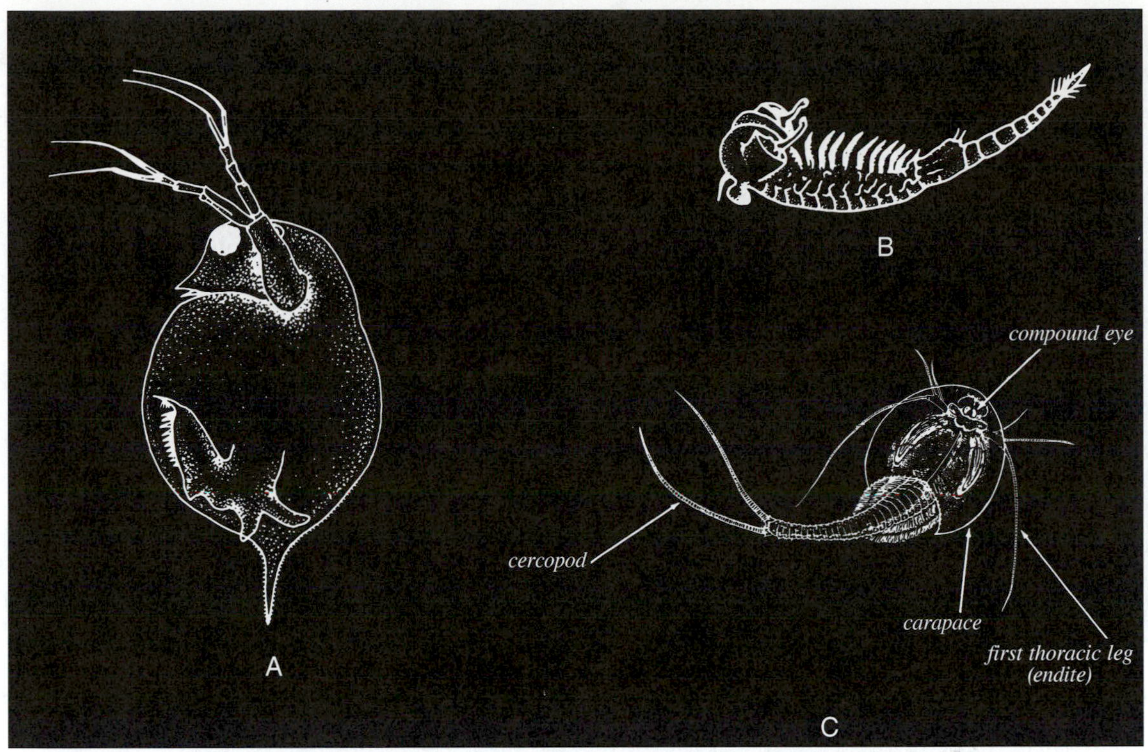

FIGURE 18 Tow branchiopod crustaceans: (A) The widely distributed "water flea" *Daphnia*; (B) the more localized fairly shrimp *Eubranchipus*, showing an adult male (C) *Triops*, a notostracan tadpole shrimp. See Chapter 21.

discussed by Craig E. Williamson and Janet W. Reid in Chapter 22.

e. Decapoda Decapoda is the most diverse order of the class Malacostraca in marine and freshwater ecosystems. It encompasses 342 species of crayfish and 17 species of shrimp in epigean and subterranean freshwaters north of Mexico (Fig. 20). All are characterized by terminal claws on the first three of five pairs of thoracic appendages and a branchial chamber

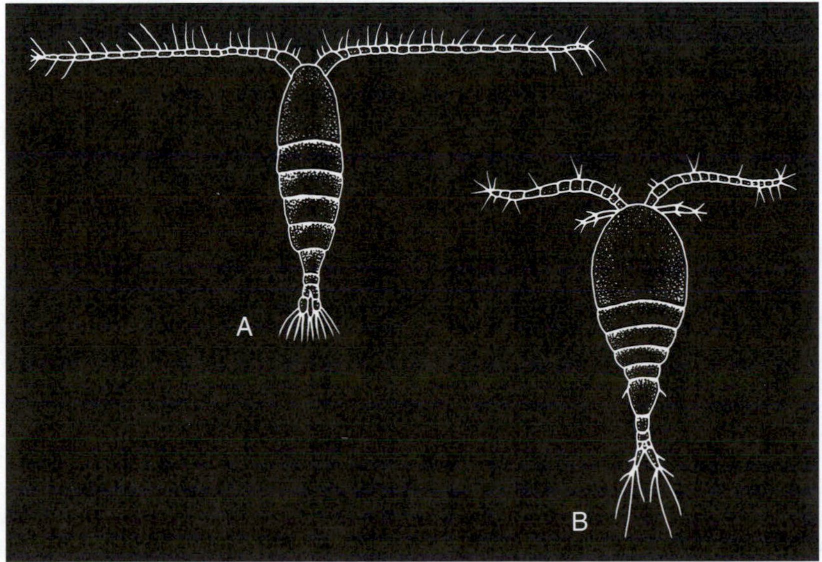

FIGURE 19 Two copepod crustaceans: (A) A generalized calanoid species; (B) a representative cyclopoid copepod. See Chapter 22.

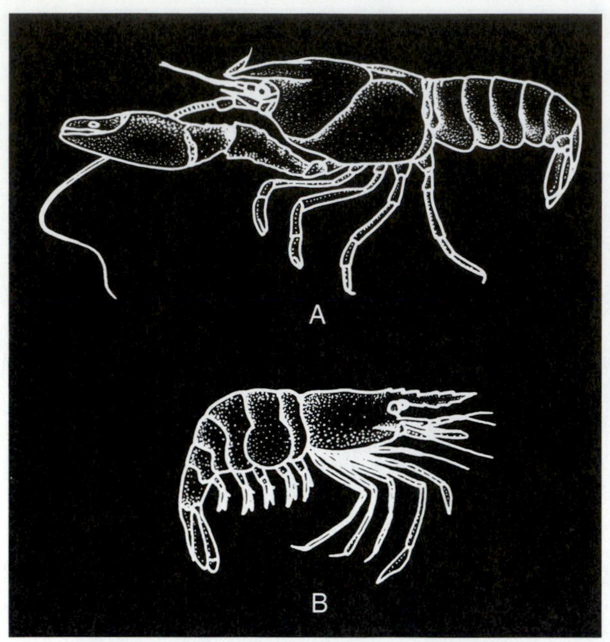

FIGURE 20 Crayfish (A) are extremely common decapod crustaceans, while palaemonid shrimp (B) have a more restricted distribution (although occasionally abundant). See Chapter 23.

enclosed within the carapace (see Chapter 22 by Horton H. Hobbs). Decapods are the largest motile invertebrates in North American freshwaters. Their size, activity, and omnivorous diet make them key players in the benthos of many ecosystems. For example, they can influence species distribution of snails within a community by selectively preying on certain size classes or shell morphologies, and they can be voracious herbivores that devastate vascular plants in ponds. Decapods, unlike practically all other North American invertebrates, are increasingly consumed by humans, especially in the southern United States.

M. Taxonomic Key to Major Taxa of Freshwater Invertebrates

The following dichotomous, taxonomic key is designed specifically to lead the reader to appropriate chapters within this book (i.e., it should not be employed in other contexts). It can be used for the larval and adult stages of aquatic insects, but it generally applies to the adults of other taxa.

1a. Unicellular (or acellular) organisms present as individuals or colonies; heterotrophic and/or autotrophic; protozoans (Fig. 1) .... kingdom Protista (Chapter 3)

1b. Multicellular, heterotrophic organisms (sometimes with symbiotic autotrophs) 2

2a(1b). Multicellular animals without discrete organs and with a cellular-level (or incipient tissue-level) construction; sponges (Fig. 2) .... phylum Porifera (Chapter 4)

2b. Multicellular animals with organ or organ-system construction and two or three embryonic cell layers 3

3a(2b). Two embryonic cell layers; adults with a central body cavity opening to the exterior and surrounded by cellular endoderm, and acellular mesoglea, and a cellular ectoderm; polyps (e.g., hydra) and medusae (e.g., *Craspedacusta*) (Fig. 3) .... phylum Cnidaria (Chapter 5)

3b. Three embryonic cell layers; adults with cellular ectoderm, mesoderm, and endoderm 4

4a(3b). Flattened or cylindrical, bilateral, acoelomate worms with only one opening to the digestive tract; flatworms (some occasionally called planaria) (Fig. 4) .... phylum Platyhelminthes: class Turbellaria (Chapter 6)

4b. Pseudocoelomate or coelomate eumetazoa with a complete digestive tract 5

5a(4b). Flattened, unsegmented worms with an eversible proboscis in a rhynchocoel and a closed circulatory system; ribbon worms (Fig. 5) .... phylum Nemertea (Chapter 6)

5b. Not with above characteristics 6

6a(5b). Pseudocoelomates; legs, lophophorate tentacles, and segmentation absent; gastrotrichs (Fig. 5), rotifers (Fig. 6), roundworms (Nematoda; Fig. 7a), or hairworms (Nematomorpha; Fig. 7b) 7

6b. Coelomates; legs, lophophorate tentacles, and/or metameres present; snails and mussels (Mollusca; Figs. 8 and 9), segmented worms (Annelida; Figs. 10 and 11), bryozoans (Fig. 12), water bears (Tardigrada; Fig. 13), and arthropods (mites, insects, and crustaceans; Figs. 14–20) 10

7a(6a). Small (50–800 μm), spindle- or tenpin-shaped, ventrally flattened pseudocoelomates; more or less distinct head bearing sensory cilia; cuticle usually ornamented with spines or scales of various shapes; posterior of body normally formed into a furca with distal adhesive tubes; gastrotrichs (Fig. 5) .... phylum Gastrotricha (Chapter 7)

7b. Not with above characteristics 8

8a(7b). Nonvermiform pseudocoelomates with apical (and usually ciliated) corona and muscular pharynx (mastax) with complex set of jaws; wheel animals, or rotifers (Fig. 6) .... phylum Rotifera (Chapter 8)

8b. Not with above characteristics; nonsegmented worms 9

9a(8b). Alimentary tract present; cylindrical body usually tapering at both ends; noncellular cuticle without cilia, often with striations, punctuations, minute bristles, etc.; most less than 1 cm long (except family Mermithidae); roundworms (Fig. 7a) .... phylum Nematoda (Chapter 9)

9b. Degenerate intestine; anterior and posterior tips normally are obtusely rounded or blunt; cuticle opaque and epicuticle usually criss-crossed by minute grooves; pseudocoel frequently filled with mesenchymal cells with large clear vacuoles, giving tissue a foamy appearance; length several centimeters to 1 m, width 0.25–3 mm; only adults with free-living stage; hairworms or horsehair worms (Fig. 7b) .... phylum Nematomorpha (Chapter 9)

10a(6b). Soft-bodied coelomates, usually with a hard calcareous shell; most with ciliated gills, a ventral muscular foot, and a fleshy mantle covering internal organs; snails and mussels (Figs. 8 and 9) .... phylum Mollusca 11

10b. Body not enclosed in a single, spiraled shell or in a hinged, bivalved shell, or if a bivalved shell is present, then animal has jointed legs. 12

11a(10a). Body enclosed in a single shell, which usually has obvious spiral coils (limpets are the exception); body basically partitioned into head, foot, and visceral mass; radula present; snails and limpets (Fig. 8) .... phylum Mollusca: class Gastropoda (Chapter 10)

11b. Body enclosed in two hinged shells; head and radula absent; mussels and clams (Fig. 9) .... phylum Mollusca: class Bivalvia (Chapter 11)

12a(10b). Legs absent in all life stages 13

12b. Adults and most larval stages with legs; if larvae without legs or prolegs (some insects), then cephalic region with paired mandibles .... phylum Arthropoda 16

13a(12a). Lophophorate tentacles absent; segmented worms 14

13b. Ciliated, lophophorate feeding tentacles present 15

14a(13a). No suckers and four bundles of chaetae on each segment except the first (Oligochaeta and Aphanoneura; Fig. 10A) "or" posterior, disk-shaped sucker on segment 11 and head composed of peristomium and three segments, ectoparasites on crayfish (Branchiobdellida; (Fig. 10B) .... phylum Annelida (in part) (Chapter 12)

14b. Body with anterior and posterior suckers and no chaetae; leeches (Euhirudinea; Fig. 11) "or" posterior sucker on four segments, ectoparasites on salmonid fish (*Acanthobdella peledina*) .... phylum Annelida (in part) (Chapter 13)

15a(13b). Anus opens outside a generally U-shaped lophophore; bryozoans (Fig. 12A) .... phylum Ectoprocta (Bryozoa) (Chapter 14)

15b. Both mouth and anus open within lophophore; calyx with single whorl of 8–16 ciliated tentacles (Fig. 12b) .... phylum Entoprocta (Chapter 14)

16a(12b). Minute adults with four pairs of clawed, nonjointed legs; water bears (Fig. 13) .... phylum Tardigarda (Chapter 15)

16b. Adults and most larvae with jointed legs; arthropods 17

17a(16b). Antennae absent; body divided into cephalothorax (prosoma) and abdomen (opisthosoma); adults generally with four pairs of legs; water mites or the pseudo-aquatic water spiders (Fig. 14) .... subphylum Chelicerata, class Arachnida .... (Chapter 16)

17b. Antennae present; insects, springtails, and crustaceans 18

18a(17b). Mandibles and one pair of antennae; adults (most of which are terrestrial) generally with functional wings, larvae with wing pads or without any indication of wings; larvae and adults normally with three pairs of legs (some larvae without legs); wingless adults in class Entognatha (order Collembola, the springtails) and winged adults are in class Insecta (aquatic insects) (Fig. 15) .... subphylum Uniramia (Chapters 17 and 18)

18b. Two pairs of antennae and mandibles (at some life stage); adults with variable numbers of legs, never with wings .... subphylum Crustacea 19

19a(18b). Body with 19 segments (head: 5; thorax: 8; abdomen: 6); abdomen with pleopods; carapace covers head and all or part of thorax .... class Malacostraca 20

19b. Body enveloped in a bivalved carapace (ostracodes), "or" thoracic legs flattened and leaf-like (water fleas and other branchiopods), "or" 11 postcephalic segments and a single cephalic eye (copepods) 21

20a(19a). Malacostracans with first three of five pairs of thoracic appendages modified as maxillipeds (including usually large chelae on first pair in crayfish) and carapace encloses branchial chamber; crayfish and atyid and palaemonid shrimp (Fig. 20) .... order Decapoda (Chapter 23)

20b. Malacostracan crustaceans with other characteristics, including scuds (order Amphipoda), aquatic sow bugs (order Isopoda), and opossum shrimp (order Mysidacea) (Fig. 16) .... (Chapter 19)

21a(19b). Body enveloped in laterally compressed, bivalved carapace; mussel and seed shrimp (Fig. 17) .... class Ostracoda (Chapter 20)

21b. Body not enclosed within a bivalved carapace 22

22a(21b). Thoracic legs flattened and leaf-like (phyllopod), mandibles are simple unsegmented rods with inner corrugated grinding surfaces; last body segment with a pair of spines or claws; water fleas (cladocera) and other branchiopods (Fig. 8) .... class Branchiopoda (Chapter 21)

22b. Cylindrical, segmented body, two setose caudal rami on posterior end of abdomen, and a single, simple anterior eye; copepods (Fig. 19) .... class Copepoda (Chapter 22)

LITERATURE CITED

Hutchinson, G. E. 1993. A treatise on limnology, Vol. IV, The zoobenthos. Wiley, New York.

McCafferty, W. P. 1981. Aquatic entomology. Science Books International, Boston, MA.

Merritt, R. W., Cummins K. W., Eds. 1996. An introduction to the aquatic insects of North America. 3rd ed. Kendall/Hunt, Dubuque, IA.

Ward, J. V. 1992. Aquatic insect ecology. 1. Biology and habitat. Wiley, New York.

Wetzel, R. G. 2001. Limnology, 3rd ed. Academic Press, San Diego, CA.

2 AN OVERVIEW OF FRESHWATER HABITATS

James H. Thorp

Department of Biology
Clarkson University
Potsdam, New York 13699

Alan P. Covich

Department of Fishery and Wildlife Biology
Colorado State University
Fort Collins, Colorado 80523

I. INTRODUCTION

The contribution of inland waters to the total biospheric water content is insignificant in terms of percentage (< 1% according to Wetzel, 2001) but absolutely crucial from the perspective of terrestrial and freshwater life. Although inland lakes contain 100 times as much water as surface rivers, most lake water is held within massive basins, such as the Laurentian Great Lakes of North America, Lake Baikal of Siberia, and Lake Tanganyika of East Africa. Because most freshwater invertebrates are clustered within shallow, well-lighted zones of lakes, the relative importance of small ponds, creeks, and rivers as habitats for aquatic invertebrates is much greater than their volume percentages would otherwise indicate. Composition, species richness, and total density of invertebrates vary considerably among inland water habitats, as discussed in this chapter.

Distributions of invertebrates are influenced by interactions among physical, chemical, and biological characteristics. The general importance of these abiotic and biotic factors is examined in this chapter. Detailed information on the ecology of individual taxa in inland water ecosystems can be gleaned from Chapters 3–23.

II. LOTIC ENVIRONMENTS

Flowing waters, or lotic environments, were a principal pathway for evolutionary movement of animals from the sea to lakes and land. Even today, many taxa of freshwater invertebrates are restricted to headwater streams and rivers by unique environmental characteristics of these ecosystems. Compared to nonflowing waters, or lentic ecosystems, streams are generally more turbulent than lakes and, therefore, stratification of the water mass with a thermocline is rare. High turbulence generally maintains high oxygen concentrations, reduces within-stream temperature differences, and more evenly distributes plankton and suspended or dissolved nutrients. Temperatures in streams fluctuate over a smaller range than is typical of shallow littoral zones of lentic ecosystems, where most lake-dwelling animals reside (Hynes, 1970). Except in the northernmost rivers, ice is less commonly encountered and is generally not as thick as in lakes. Flowing water habitats frequently possess more habitat heterogeneity, and the food web in forested drainage basins is more dependent on allochthonous production (externally produced plant matter), even though instream production can be important (Thorp and Delong, 1994). Lotic

Ecology and Classification of North American Freshwater Invertebrates. 2nd Edition

ecosystems are also more permanent on both evolutionary and ecological time frames than most lentic habitats. Both heterogeneity and permanence are thought to increase diversity within these ecosystems.

In this section, we review features of epigean, or surface waters. Characteristics of aquatic environments where water flows underground (hyporheic, phreatic, and cave environments) are discussed in Section III.

A. The Physical–Chemical Milieu

1. Basin Morphometry and Characteristics of Flowing Water

Although stream invertebrates are influenced directly by the physical features of the local habitat and only indirectly by geological forces shaping the morphometry of the entire ecosystem, ecologists and invertebrate zoologists must consider all those factors when attempting to understand what controls community composition. Some elements of stream morphometry that influence aquatic communities are stream order, channel patterns (e.g., meandering and braiding), erosion, deposition, and formation and characteristics of pools and riffles.

Classifying streams by their order, or pattern of tributary connections, is a simple technique that aids analysis of longitudinal (i.e., upstream–downstream) changes in stream characteristics within a single catchment; however, it is less useful, and indeed often misleading for comparing streams in different catchments. For example, communities within streams in the eastern coastal plain bear no close resemblance to communities within similar-order streams in the southwestern deserts (when flooded) (cf. Figs. 1C and 1D respectively). According to this scheme, as refined by Strahler (1952) and illustrated in Figure 2, a continuously running, headwater creek with no permanent tributaries is classified as a first-order stream (permanency is defined here as not being ephemeral on a seasonal or other regular basis). Two first-order streams combine to form a second-order segment; two second-order creeks join as a third-order stream, etc. Stream order increases only when two streams of equivalent rank come together; hence, a third-order stream that flows into a fourth-order system has no effect on the subsequent numbering scale of the larger stream. The Mississippi River, the largest river in North America, attains a stream order of 11–13 in its lower reaches (depending on which maps are used as the basis for analysis). Unfortunately, this scheme is not as useful for many streams in arid portions of the continent where a dry stream bed (Fig. 1D) may be replaced by a raging torrent in a matter of minutes or a few hours.

Viewed from the air, streams clearly do not flow in a simple, straight direction for significant distances (Fig. 3A), nor do stream beds maintain constant depths for long stretches. Instead, the channel traces a meandering pattern of gentle and sharp bends through the basin with alternating bars, pools, and riffles (Fig. 4) as the water travels downstream along a roughly helical path. The distance between meanders is a function of several factors, especially stream width. The river channel also has a tendency to undergo braiding (i.e., the division of the channel into a network of branches) in response to the presence of erodible banks, sediment transport, and large, rapid, and frequent variations in stream discharge (Gordon *et al.*, 1992).

Water flows in a somewhat helical manner along the surface toward an undercutting bank where the deepest pool is usually located and then passes along the bottom toward a "point" or "alternate" sediment bar on the opposite bank from the pool. Sediment is continually being eroded from some areas (e.g., undercut banks and the head of islands) and deposited in other regions (such as on bars and the foot of islands). Successive deepwater pools are separated by shallow riffle regions near the midpoint of two pools. This sequence of bars, pools, and riffles is characteristic of both straight and meandering segments of most rivers, especially those with poorly sorted loads. (See Calow and Petts, 1992; Gordon *et al.*, 1992; and references cited therein for additional details on fluvial geomorphology).

Scientists have appreciated the influence of alternating pools, riffles, and (to a much lesser extent) sediment bars for some time. In comparison to pools, riffles are characterized by greater habitat heterogeneity, sediment size, stream velocity, and sometimes oxygen content. In general, well-flushed, stony riffles support more species and individuals than silty reaches and pools (Hynes, 1970). This richness is a consequence, in part, of the different adaptations required to live in these physically distinct areas, especially with regard to dissolved oxygen.

Our initial separation of aquatic ecosystems into flowing- and standing-water environments recognizes the fundamental importance of this character to the biotic composition of inland water environments. Scientists measure water movement in streams in several ways, including stream discharge, turbulence, and velocity. Mean and peak values of stream discharge influence various ecosystem characteristics such as substrate particle size, bed movement, nutrient spiraling, and rate of animal drift. Stream discharge is calculated by the equation

$$Q = wdv$$

where "Q" is the discharge in cubic meters per second (cms) or cubic feet per second (cfs), "w" the width, "d"

FIGURE 1 Lower-order streams: (A) Rocky-bed mountain stream in Glacier National Park, Montana; (B) rocky-bed stream from the Smoky Mountains of North Carolina; (C) sandy, coastal plain stream from South Carolina; (D) intermittent, dry-bed stream from Arizona (photographs by J. H. Thorp).

FIGURE 1 (*Continued*)

the depth, and "v" the velocity. As stream flow increases, shear stress is magnified. When the stress reaches a critical point, the substrate is picked up, rolled, or generally moved downstream. Various aspects of stream discharge, including mean annual discharge and yearly and historical extremes, are described annually for most rivers in the United States in the Water Data Reports published by the U.S. Geological Survey. This information is derived from multiple gaging stations located near the mouth of each river and often extending into the headwaters; water-quality data are also available for a few gaging stations in some of these rivers. These reports are available in the government documents section of most major libraries and are accessible through the USGS Web site. For more details on hydrologic processes, consult

FIGURE 2 Bifurcation of streams illustrating the system for classiifying lotic ecosystems according to stream order. The maximum order of the hypothetical stream shown here is sixth.

articles or books on this subject, such as Allan (1995), Calow and Petts (1992, 1994), Gordon *et al.* (1992), Newbury (1984), and Statzner *et al.* (1988).

In addition to being affected by the mean and range of stream discharge, freshwater invertebrates are strongly influenced by water velocity. An invertebrate's ability to maintain its position on a rock and the amount of food passing its way (especially for suspension feeders) are both affected by current velocity. The mean current velocity of a stream at any given time usually occurs at a depth of about 0.6 of the maximum depth, as measured downward from the water surface (or 0.4 of the maximum depth as measured upward from the stream bottom). The current velocity diminishes progressively toward the bottom from that point and drops dramatically near the substrate in a 1- to 3-mm-thick area known as the boundary layer. For this reason, velocities experienced by two invertebrates living within boundary layers of separate streams—one with high and the other with low average current velocity—may not differ appreciably. Many stream invertebrates are partially protected from dislodgment by having a low profile that keeps their bodies mostly within the boundary layer. Other possible adaptations for life in fast flow include streamlining, reduction of projecting structures, development of suckers, friction pads, hooks or grapples, and secretions of adhesive products.

Under normal flow conditions, the average current velocity varies from the headwaters to the mouth of a river. Although many people believe that headwater streams flow faster than large-order rivers, this popular idea has been repeatedly disproved (e.g., Leopold, 1953; Ledger, 1981). This erroneous view of streams may have arisen because the waters of a turbulent headwater stream (Figs. 1A and 1B) can "appear" to casual observers to be moving faster than the waters of a higher order, laminar flow river (Fig. 3A), unless the large river is shallow and thus more turbulent (Fig. 3B). It is important to note, however, that flow within microhabitats in any river order can substantially exceed the average current velocity of the stream reach.

Fluctuating current velocity rather than constancy is typical of all lotic systems. Except in the largest rivers, zero or near-zero flow rates have been recorded at least once in most lotic environments since the U.S. Geological Survey began recording stream discharges in the 19th century. The smaller the stream and the drier the surrounding area (e.g., desert streams), the more likely that a lotic system will be intermittent on at least a seasonal basis. At the other extreme, all streams have peak discharges on seasonal and multiyear bases. Flood events, or spates, may be very predictable, such as those produced by seasonal snow melt in mountainous streams, or they may occur dramatically over unpredictable time periods (Resh *et al.*, 1988). Current velocities during floods and normal periods are related to the width, depth, and roughness of the stream bed, but rarely exceed 3 m/s even during floods (falling water seldom surpasses 6 m/s) (Hynes, 1970). Above 2 m/s, most rivers begin to enlarge their beds by erosion, unless constrained by rock banks or human construction.

Invertebrates are often adapted morphologically or in life-history traits to moderate, seasonally predictable spates, but large unpredictable floods may eliminate all or most of the fauna of a stream by causing bed movement and severely scouring the bottom. When these scouring events occur, the sources for recolonization may be assemblages found in the adjacent hyporheic zone, in surface waters located upstream or downstream, or in separate, nearby stream catchments (Gibert *et al.*, 1994; Jones and Mulholland, 2000).

2. The Importance of Substrate Type

The nature and provision of substrates are of prime importance to survival because the overwhelming majority of lotic invertebrates are benthic (see references in Hynes, 1970; Allan 1995). Substrates provide sites for resting, food acquisition, reproduction, and development (e.g., places for pupal case attachment), as well as refuges from predators and inhospitable physical conditions. Hard substrates are generally valuable only for their surface qualities (except for species that bore into wood), whereas relatively soft substrates

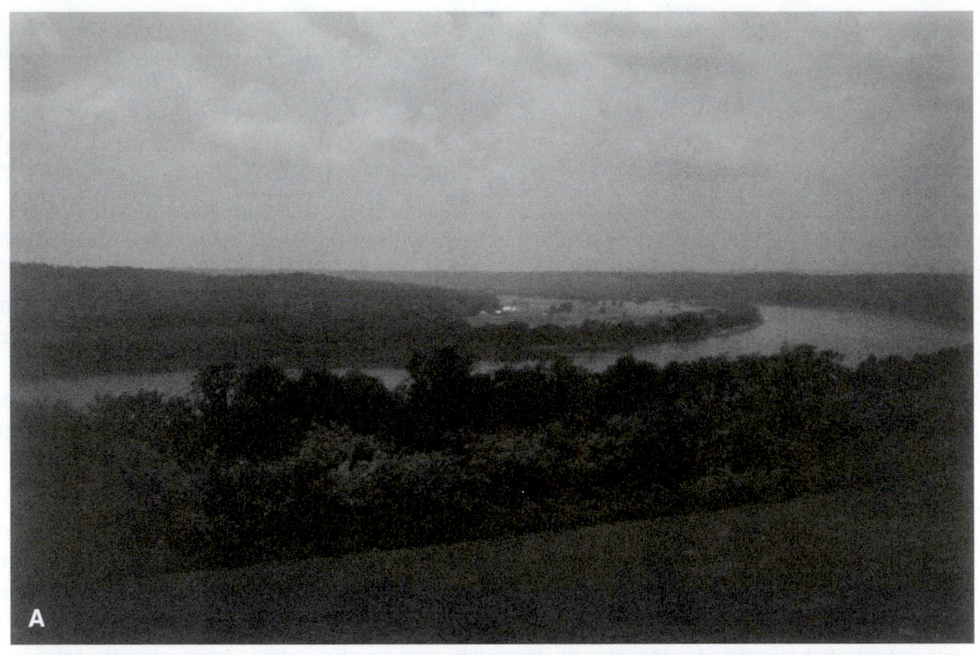

FIGURE 3 (A) Ohio River above Louisville, Kentucky; here the river is meandering with a sandy-gravel bottom and a more laminar flow; (B) Rio Grande near Santa Fe, New Mexico; here this large river is more turbulent than the Ohio River (photographs by J. H. Thorp).

(ranging from sand to silt) provide a three-dimensional environment. The desirability of a given substrate type can be evaluated for a specific taxon using a combination of at least the following interrelated characteristics (not necessarily in order of importance):

1. Whether it is mineral (ranging from large boulders to fine clay particles), living (algae, mosses, vascular plant stems and roots, and, rarely, animal hosts), or formerly living (e.g., abscised leaves and snags from tree limbs and trunks).
2. The silt and organic content of, or on, the substrate, including whether it is bare or covered by significant amounts of algae, mosses, or vascular plants.
3. Its particle size, surface texture, and porosity.
4. Its stability, as judged by its retention in the

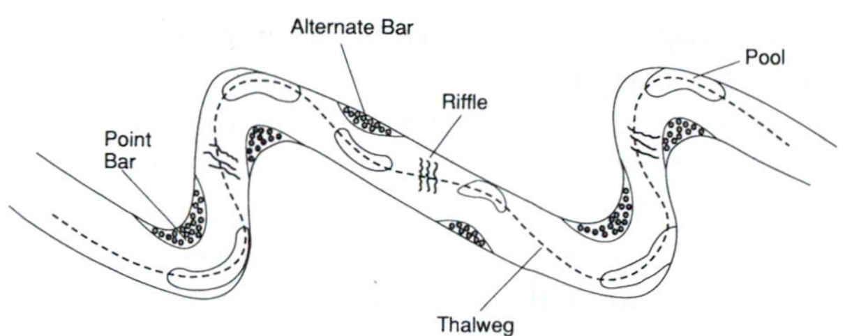

FIGURE 4 Schematic drawing of a stream showing straight reaches and meanders. In both sections of a river, a sequence of deep (pools) and shallow (riffle) areas is manifested. Point bars and alternate bars develop as bed material is deposited in areas opposite from the pools. The thalweg, a line connecting the deepest part of the channel, migrates back and forth across the river.

system (e.g., leaves are a short-lived substrate, whereas boulders are retained almost indefinitely) and its tendency to shift position (ranging from almost constantly shifting sand to boulders that rarely turn).

5. The physical-chemical milieu in which the substrate resides, such as the water depth at which it is located, the ambient oxygen content, and the range of current velocities surrounding it.

Even if a particular site meets all the physical and chemical criteria for a desirable substrate, it may not be appropriate for an individual invertebrate if a competitor has preempted the space or if it is located in an area that would make a resident especially susceptible to predators or abiotic disturbances. For these reasons, the distribution of a species across a spectrum of substrate types may not reflect so much the selectivity of the invertebrate species as the effects of substrate limitation. While a particular taxon may not benefit greatly from substrate heterogeneity, the species diversity of the entire community tends to rise with an increase in habitat complexity. In a like manner, an individual may benefit from maximum substrate, but the community as a whole may flourish better with some disturbances. This latter concept, embodied in part by the "intermediate disturbance hypothesis" (Connell, 1978; Ward and Stanford, 1983; Thorp and Cothran, 1984), suggests that intermediate levels of biotic or abiotic disturbances (e.g., frequency of substrate shifting) can promote maximum species diversity under certain circumstances. For a more detailed discussion of the general effects of substrate characteristics, the reader should initially consult Allan (1995).

3. Thermal and Chemical Constraints on Distribution

a. The Thermal Environment Because temperatures vary on a seasonal basis and a gradient exists from headwaters to mouth, it is not surprising that the distribution and abundance of freshwater invertebrates are correlated with the annual mean and range of temperatures (e.g., Hynes, 1970; Ward, 1986). A general stream temperature pattern on an annual basis is a progressive increase from headwaters to mouth, especially in streams originating in the mountains and fed primarily by terrestrial runoff. [This general pattern may not hold for low-gradient, north-flowing streams.] Streams originating from coldwater springs, however, with their relatively constant temperatures in the headwaters, can be warmer in winter near their source than somewhat farther downstream; in the summer, the opposite pattern exists (Minckley, 1963). In many temperate zone rivers, annual fluctuation in temperature increases downstream, while diel temperature changes peaks in mid-order reaches (Statzner and Higler, 1985). In contrast, diel and annual temperature amplitudes are very low in middle and lower reaches of tropical streams, although high-elevation, tropical streams have significant seasonal temperature gradients (Covich, 1988).

Predictions of thermal patterns must be modified to account for effects of tributaries (often colder) and dams on downstream reaches. Likely differences in thermal regimes between natural and regulated rivers are described by the "serial discontinuity concept" (Stanford *et al.*, 1988), which predicts that the thermal character of the stream below a dam will be "reset" toward that typical of reaches above the dam. This pattern is evident if you calculate the number of degree-days (sum of the average daily temperatures over a period) in stream reaches above and below the dam. The timing of insect emergence, an invertebrate's tolerance of specific habitats, and annual productivity of individual species and communities are all influenced directly or indirectly by the number of degree-days in those habitats (Sweeney and Vannote, 1978).

Species richness and the nature of invertebrate fauna vary along a continuum from headwaters to the mouth. Crustaceans (benthic and planktonic), molluscs, and fish increase in abundance and diversity with stream order, whereas insects generally decrease downstream. Because of previously noted patterns for temperature means and fluctuations with stream order, it is not surprising that aquatic ecologists saw a link between thermal regimes and species diversity. Temperature may influence organisms directly or have an indirect effect through changes in oxygen saturation levels (the oxygen-carrying capacity of water decreases at higher temperatures). While it is true that a great many lotic invertebrates are more stenothermal (i.e., surviving only in a narrow temperature range) than their lentic counterparts (Hynes, 1970), it has not been demonstrated that the relationship between invertebrate diversity and temperature is more than coincidental. The problem in distinguishing causative factors is that many biotic and abiotic parameters other than temperature also change along river gradients. For example, habitat complexity, which has often been demonstrated to significantly influence diversity, varies downstream, often reaching a maximum value in mid-order sections of rivers.

b. The Chemical Environment Certain natural chemical features of streams and lakes significantly affect invertebrate abundance and diversity by influencing habitat quality. Of these, the most crucial are probably dissolved oxygen and salinity (water hardness). Hydrogen ion concentration, or pH, is occasionally important in natural systems. Anthropogenic pollution has had a severe impact on the integrity of aquatic ecosystems by affecting these chemical parameters as well as by introducing organic and inorganic toxicants.

As a general rule, dissolved oxygen decreases downstream. This pattern occurs because the upper reaches of a river are usually more turbulent, the surface to volume ratio is more favorable for diffusion, and temperatures are lower. In the last case, oxygen solubility is related in an inverse, nonlinear manner to temperature; this gas is markedly more abundant in cold waters. Pure water in equilibrium with air at standard pressure holds 12.770, 10.084, 8.263, and 6.949 mg O_2/L at 5, 15, 25, and 35°C, respectively (see Mortimer, 1981, in Wetzel, 1983, p. 158). Because oxygen is produced and respired by plants, it may be locally more abundant or even supersaturated during the day in regions with heavy plant growth. However, oxygen potentially decreases at night to lethal levels in weed-choked streams because of high oxygen demands for decomposition and plant respiration. The probability of lethal oxygen concentrations occurring for the invertebrate species typical of a stream reach rises as currents diminish; hence, the problem is more serious in lakes and seasonally sluggish streams than in fast-flowing lotic habitats.

Invertebrates differ both in their respiratory needs and in their anatomical and behavioral adaptations for obtaining oxygen. Respiratory rates of lotic species are generally dependent on external oxygen concentration (tension) at medium to low levels, while oxygen consumption rates of animals in lentic or typically sluggish lotic environments tend to be independent of ambient tensions, at least at medium-to-high concentrations (Hynes, 1970). Currents provide a steady renewal of oxygen to the respiratory surfaces of invertebrates. Where velocities are too low, the animal may need to ventilate its gills or obtain atmospheric oxygen. For example, crayfish normally move water over their respiratory surfaces with "gill balers" but some species climb partially or completely out of water if oxygen tensions drop too low (returning when gills need moistening). Because invertebrates have different oxygen requirements, it is tempting to assume that interspecific differences in distribution are partially caused by oxygen limitations (e.g., contrasting assemblages of caddisflies in mountain streams and floodplain rivers). Certainly many stream animals, such as larvae of the blackfly *Simulium*, cannot obtain adequate oxygen in still water (Hynes, 1970), but this explanation for restriction from lentic habitats is less effective in resolving species differences in distribution along the entire stream reach. Even if one demonstrated in the laboratory that the oxygen tensions in high-order rivers were inadequate to support a caddisfly from a mountain stream, this is not evidence that oxygen was the principal factor leading to the present pattern of distribution. Other factors, such as interspecific competition, resource type (e.g., seston particle size), substrate form, or predators, may have been initially responsible, with fine tuning of respiratory adaptations coming later.

The ionic content of inland waters is usually extremely dilute in comparison to seawater, with four cations (Ca^{2+}, Mg^{2+}, Na^+, and K^+) and four anions (HCO_3^-, CO_3^{2-}, SO_4^{2-}, and Cl^-) representing the total ionic salinity in most freshwater systems, for all practical purposes (Wetzel, 2001). Soft waters are generally found in catchments with acidic igneous rocks, whereas hard waters (relatively high salinities for freshwater) occur in drainage regions with alkaline earths, usually derived from calcareous rocks. The global mean salinity of river water is 120 mg/L; North American waters are slightly harder, averaging 142 mg/L [from Livingstone (1963) and Benoit (1969) in Wetzel (1983, p. 180)]. Most soft waters in North America have salinities <50 mg/L. Salinities are slightly elevated during periods of low flow, and they generally increase downstream because of bedrock erosion (Hynes, 1970) and human influences on watersheds.

Except for a few hypersaline environments, such as mineral springs and alkaline and brine lakes, inland waters are hypotonic to the internal fluids of freshwater species; therefore, all freshwater animals need some ability either to osmoregulate (almost all species) or, at the minimum, to control the relative abundances of certain ions (necessary even for freshwater Cnidaria; see Chapter 5). Because animals with either partially calcareous exoskeletons (e.g., crayfish, ostracodes, and cladocera) or hard calcareous shells (snails and mussels) have an added need for calcium ions, one might expect them to be particularly excluded from soft-water habitats (see Chapter 18). Indeed, molluscs are rare in soft-water lakes and less abundant in soft-water streams, but this limitation is not as severe as in lentic ecosystems. Molluscs can live in some soft-water streams, because they can tolerate extremely dilute tissue fluids (the lowest of any metazoa; see Chapter 11). The minimum concentration of Ca^{2+} tolerated by mussels (2–2.5 mg Ca/L) is just adequate to prevent the rate of shell dissolution from exceeding the rate of shell deposition. Despite their heightened need for calcium, crayfish can be abundant in many soft-water streams of the Southeast where these ions are difficult to obtain. The problem of ionic loss at the critical period when the old exoskeleton is molted is partially overcome by the crayfish habit of consuming its shed exuvia as a way of extracting additional calcium ions. This behavior and the efficient osmoregulatory system of the crayfish (with gastrolith formation) allow this decapod to colonize a greater chemical range of habitats (see Chapter 23).

In the last decade, the catastrophic effects of acid precipitation and mine drainage on freshwater ecosystems have drawn great attention from scientists and politicians alike, but relatively unpolluted rainwater and runoff from thick forest litter are also slightly acidic (though not at lethal levels). Bogs and the streams draining them are among the most naturally acidic ecosystems; sphagnum bogs usually have a pH below 4.5 and occasionally below 3.0 (Wetzel, 2001). Streams draining limestone catchments are well buffered by carbonates and therefore have moderate to high pH values as well as plentiful supplies of calcium. In contrast, acidic waters are normally low in calcium, causing potential osmotic problems for their residents. Despite probable ionic and pH difficulties which may have eliminated certain taxa, naturally acidic ecosystems often support a surprisingly diverse invertebrate assemblage.

B. Ecosystem Changes along the Stream Course

1. *Patterns along the Continuum*

Physical and biological characteristics clearly change from headwaters to the river mouths. The river widens, deepens, and becomes more turbid downstream, often removing part of the bottom from the photic zone. Stream shading from riparian trees lessens, with a corresponding increase in percentage of the surface receiving direct solar radiation. Concurrently, the relative importance values for autochthonous (internally produced) and allochthonous carbon to energy budgets alter along with the ratio of production to respiration. As a result of these and other habitat changes, the invertebrate fauna undergoes dramatic shifts in species composition, relative abundances, and functional feeding groups.

This holistic view of the river was not synthesized until formation of the river continuum concept (RCC) by a group of aquatic ecologists working in temperate deciduous streams of North America (Vannote *et al.*, 1980; Sedell *et al.*, 1989). This important theory has proved to be one of the most hotly debated in recent decades, and that fact alone contributes to the pivotal role it has played in aquatic ecology. Individual hypotheses in the RCC have received both support and criticism (as examples of the latter, see Winterbourn, 1982; Statzner and Higler, 1985; Junk *et al.*, 1989; Thorp and Deong, 1994). It is not our purpose to defend or assail this holistic, ecosystem concept; instead, we wish to acquaint you with some of its predictions as they could influence your understanding of the habitat and ecology of lotic invertebrates. Because the RCC was originally developed from studies of relatively undisturbed, North American streams with forested headwaters (cf. Fig. 1B), the following comments may not be as applicable to stream invertebrate communities in other biomes or regions of the world.

The river continuum concept relies on three theoretical pillars of support. The first principle is that stream communities originate in response to a continuous gradient of physical variables present from headwaters to the mouth; this view contrasts with a previous portrayal of natural streams as a series of disjunct communities proceeding downstream in a stepwise fashion. The second canon is that biotic communities within the "wetted channel" of the river cannot be divorced from either the adjacent riparian zone or the surrounding biotic and geomorphic catchment area that funnels water, nutrients, and other material into the aquatic ecosystem. The final tenet is that the nature of a downstream assemblage is inextricably linked with processes occurring upstream (this has been challenged for floodplain rivers by Junk *et al.*, 1989).

Of the many physical, chemical, and biological predictions made by the RCC, the one most relevant to understanding invertebrate ecology concerns longitudinal changes in the relative abundance of functional feeding groups and their food resources. The proposed relationships between stream size and progressive alterations in community composition are illustrated in Figure 6. As heterotrophs, invertebrates consume living and/or dead organic matter, processing it in ways characteristic of several "functional feeding groups." These groups include "shredders," which break up coarse particulate organic matter (CPOM) such as abscised leaves; "collectors," which gather or filter CPOM, fine particulate organic matter (FPOM), and/or live, drifting organisms; "grazers" or "scrapers," which remove algae or other aufwuchs growing on rocks or other substrates; and "predators," which may include carnivores and some herbivores.

The traditional means of assigning a functional feeding label to a life stage of a particular species has been either to observe the organism's feeding behavior (preferably in the field) or to analyze its gut contents on a seasonal basis (e.g., Benke and Wallace, 1997). This task is time consuming and may be impractical for some taxa. References to the food resources and feeding behavior of species or larger taxonomic categories are reported in many research papers (for examples, see chapters in this book and the tables and references on aquatic insects in Merritt and Cummins, 1996). The danger in relying on general sources, however, is the well-established observation that functional feeding classifications can vary significantly with life-history stage, season, and other ecological features; it is also difficult to assign omnivores to any single or consistent functional feeding group (see summary of some limitations of this method in Mihuc, 1997). An approach that overcomes some of these problems is to construct food webs using data on stable isotope and/or fatty acid content of producers (dissolved organic matter through living terrestrial and aquatic plants) and aquatic consumers; these data reflect the cumulative feeding patterns of freshwater invertebrates over periods of up to 3 months (e.g., Thorp *et al.*, 1998).

The RCC predicts that the invertebrate assemblages of headwater streams, with their heavy canopy of riparian trees, will be dominated by shredders and collectors. As the stream widens and more light reaches the water surface, autochthonous production from benthic algae and rooted macrophytes increases, allowing the development of a significant grazer fauna; shredders should diminish dramatically in these mid-order reaches of the river. Farther downstream, the river deepens so that much of the bottom is below the photic zone (supposedly limiting autotrophic production in the river), and the channel is hypothesized to be sufficiently wide to preclude any significant input of allochthonous production from the riparian zone. In this high-order section of the river, collector–filterers are expected to dominate the ecosystem. These conclusions on primary sources of organic matter have been strongly challenged for large rivers by both proponents of the flood pulse concept (Junk *et al.*, 1989), who emphasize the importance of allochthonous production generated on river floodplains, and proponents of the riverine productivity model (Thorp and Delong, 1994), who stress the greater importance of instream (autochthonous) primary production to secondary (animal) production, especially at higher trophic levels.

2. Nature of the River's Source

One weakness in the original formulation of the RCC was its reliance on data collected from a narrow range of stream sizes (mostly stream orders 1–3). Even today, our knowledge of assemblages living at opposite ends of the continuum—springs and intermittent streams versus large rivers—is woefully inadequate (Meyer, 1990).

Research on intermittent, or ephemeral, streams (Fig. 1D) is more common in arid regions of North America (e.g., Matthews, 1988; Forrester *et al.*, 1999), but has expanded into headwater streams in more humid regions of eastern North America (e.g., Delucchi and Peckarsky, 1989). Almost all streams are subject to seasonal and/or aperiodic disturbances from unusually high current velocities and stream discharges, but only intermittent streams fluctuate regularly from dry to wetted basin. When a stream stops flowing, a resident invertebrate may have several options for its survival or that of its future progeny: it may move downstream, seek shelter in deep pools, enter the hyporheic zone (i.e., the wetted substrate below the surface waters, as described in Section III.A), emerge as an adult (some aquatic insects), develop a resistant stage, or lay eggs and then die. Although the composition of the invertebrate fauna in permanent and intermittent streams may differ, life-history patterns of invertebrates in ephemeral streams are not particularly unique (Delucchi and Peckarsky, 1989), perhaps because options for migrating downstream or entering the hyporheic zone are readily available for many species. The ecology of that portion of the ecosystem, however, may be significantly different from that of a community functioning within permanent streams (e.g., Forester *et al.*, 1999).

There is a tendency to think of streams as originating exclusively from terrestrial runoff, but many headwater systems are derived from springs. Near their source, spring-fed streams often vary in physical, chemical, and biological features from other types of first-order streams (e.g., Covich, 1988). Temperatures within spring-fed streams are relatively more constant, and oxygen concentrations are occasionally lower, sometimes to near-lethal levels for metazoa. In other cases, hot springs may provide habitats that range into the upper lethal limits for eukaryotes (see Chapters 19 and 20) and even prokaryotes. Metazoan life near the upwelling region is rarely possible; but, as ambient temperatures approach 40°C or lower in side channels and pools and in areas farther downstream, conditions become suitable for some invertebrates (e.g., Barnby and Resh, 1988). These include a very restricted number of species of nematodes, oligochaetes, mites, dipterans (such as brine flies), and ostracodes and a few other crustaceans. Geothermal springs are often extremely vigorous environments not only because temperatures may approach the boiling point but also because their waters are frequently laden with high concentrations of sulfur. The ionic content of springs can differ from that of from nearby streams supplied by runoff from land. Springs in limestone regions are well buffered and are frequently highly charged with calcium bicarbonate (Hynes, 1970). Saline springs exist where the ionic content is too high for many invertebrates.

The biota of springs often vary from other first-order streams as a result of their connection to underground rivers and their aquatic constancy (Gibert *et al.*, 1994; Jones and Mulholland, 2000). A net placed near the spring source can capture unusual subsurface species that have been washed out of their subterranean environments. In this way, blind troglobitic shrimp have been retrieved from surface waters in the springs of the San Marcos River of Texas (Longley, 1986; Strenth *et al.*, 1988). Because springs are relatively uniform environments over long ecological periods, it is not unusual to find relict species in these environments that have survived the retreat of the glaciers only in these limited habitats (Hynes, 1970). For discussion on endemic groundwater fauna in unglaciated eastern North America, see Strayer *et al.* (1995).

3. Large Rivers

At the opposite extreme of the continuum, the invertebrate biota and ecological processes in large rivers (e.g., Figs. 3A, 3B, and 5) have rarely been studied in the past (although this is beginning to change) despite their importance to humans. Hynes (1989) pointed out that only 4% of all ecological publications of flowing waters have involved large rivers. Consequently, certain misconceptions about the nature of large rivers have arisen, such as descriptions of the "typical" large river as a slow, meandering, soft-bottomed stream. In contrast, observations by one of us (JHT) in the Mississippi, Ohio, and Tennessee Rivers suggest that the bottom is more likely to be sand, gravel, or cobble except in the shallow, slow-moving water near banks and in the mouths of tributaries where silt often accumulates. Furthermore, evidence for differences in meandering patterns along a river continuum has not yet been generated, and in comparison to upstream reaches, the current velocity is generally higher in large rivers. Because governments have been removing fallen trees and regulating both flow and channel depth in navigable rivers since the mid-1800s, the present nature of these ecosystems may be considerably different from the original state (Minshall, 1988). For example, removal of snags has greatly reduced the abundance of shallow-water hard substrates, thereby influencing composition of both the insect assemblage and the fish that normally fed upon these insects (cf. Benke *et al.* 1994).

Many aquatic ecologists lacking experience in large rivers (other than reservoirs) often think of them as "lakelike," perhaps expecting similar physical and chemical conditions. Unlike lentic environments, however, the helical flow of riverine currents prevents formation of thermoclines and major zones of oxygen depletion within the water column; this has considerable influence on processes of nutrient cycling (or spiraling in rivers). The greater depth and turbidity of sediment-laden rivers reduce the amount of benthos receiving light so that only a narrow photic zone exists. The resulting restriction on phytoplankton net production, along with microbial breakdown of allochthonous

FIGURE 5 Photograph of a large river (the Ohio River upstream of Louisville, Kentucky), showing the laminar flow nature of its waters (photograph by J. H. Thorp).

detritus, is thought to be a primary factor producing a heterotrophic metabolic state in most large rivers (Howarth *et al.*, 1996). The potential importance of physical factors in an advective environment (Pace *et al.*, 1992) and their influence on the role of biotic interactions (Thorp and Casper, unpublished data) are currently being investigated.

The RCC appears generally correct in predicting the important role for suspension feeders in large rivers, but it underestimates the role of some other groups, such as grazers. As predicted by the RCC, suspension feeders predominate in both the benthos (e.g., mussels and midges) and pelagic zones (various crustacean and rotifer zooplankton) of large rivers, especially since the invasion of zebra and quagga mussels into eastern rivers (Thorp *et al.*, 1998). However, the predicted absence of grazers, as would be anticipated in the Ohio River from Figure 6, is certainly incorrect. Abundant populations of snails in the Ohio consume various types of benthic algae, other aufwuchs, and detritus (depending on the depth distribution of snails; K. Greenwood and J. H. Thorp, unpublished data). Invertebrates in general tend to be clustered in nearshore (especially on rocks, as in Fig. 5), shallow-water regions of rivers (Thorp and Delong, 1994), where they have access to both benthic algae and allochthonous material entering from the riparian zone.

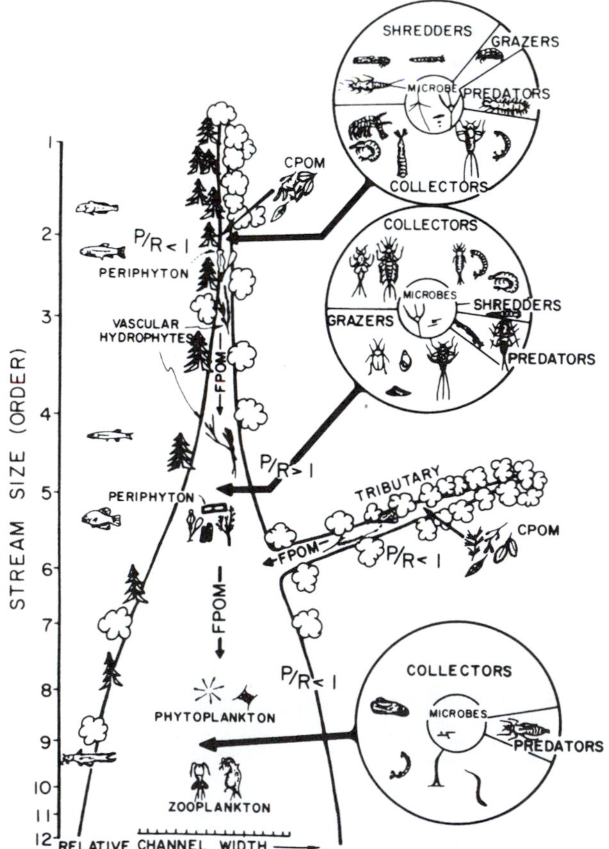

FIGURE 6 Predicted relationship between stream size and structural and functional attributes of a freshwater ecosystem as embodied in the river continuum concept. P/R refers to the ratio of production to respiration (from Fig. 1 in Vannote *et al.*, 1980).

Because the North American Benthological Society focuses on nonpelagic ecology and members of the American Society of Limnology and Oceanography are primarily oriented to lentic and oceanic environments, studies of plankton ecology in rivers are relatively rare in North America, though more common in Europe. However, the increasing importance of zebra mussels in rivers has contributed to studies of these potamoplankton (from the Greek *potamos*, or river) and benthic–pelagic coupling processes (e.g., Pace *et al.*, 1998; Jack and Thorp, 2000; Thorp and Casper, unpublished data).

One of the more interesting trends in large-river research, that seems to be developing includes freshwater zooplankton. This trend revolves around the importance of lateral slack-water areas, or storage zones (Reckendorfer *et al.*, 1999, Thorp and Casper, unpublished data), to community and ecosystem processes in the river as a whole.

III. UNDERGROUND AQUATIC HABITATS

A. Hyporheic and Phreatic Zones

Until recently, aquatic ecologists considered the boundaries of rivers to extend only from the air-water interface down to the silty, sandy, or rocky "bottom" of the river. We now realize that this definition is overly limiting and view streams as four-dimensional ecosystems with longitudinal (upstream-downstream), vertical (deeper into the sediments), lateral (floodplain), and temporal components (cf. Ward, 1989).

Early this century, scientists recognized the potential importance of interstitial spaces within sediments of lakes and streams as refuges for small animals. This interstitial habitat is termed the "hyporheic zone" within lotic systems and the "psammon zone" in lentic environments. The depth of the hyporheic zone varies greatly according to bed topography, substrate porosity, and water velocity (Boulton *et al.*, 1998), being greater, for example, in rivers having a gravel bed rather than a mud bottom (the habitable interstitial space is influenced by, but is not directly related to, sediment porosity). The overlying surface, or epigean, waters strongly influence the nutrient content, oxygen tension, and other physical and chemical characteristics of this zone and, consequently, the nature of its fauna, or "hyporheos." Likewise, upwelling from the hyporheic zone can provide nutrients to stream organisms. While many organisms dwell much of their lives in the hyporheic, it is not clear whether this zone serves as the refuge from stream spates that it was once thought to be (Gayraud *et al.*, 2000).

Almost all freshwater invertebrate phyla have representatives in hyporheos, with the probable exceptions of Cnidaria and Porifera. They include several orders of insects (e.g., midges, elmid beetles, and stoneflies), mites, crustaceans (including cladocera, ostracodes, and amphipods), oligochaetes, turbellarians, rotifers, snails, tardigrades, and protozoa.

Hyporheic organisms are usually abundant in streams unless the substrate is bedrock, clay, or other material containing pore sizes under 50 μm (Williams, 1984), but their densities reflect differences in organic content and oxygen, among other factors. Below a few centimeters, the density of hyporheos is generally inversely related to substrate depth, with very few individuals occurring much below 1 m into the sediment. Because the hyporheic food web is based on detrital inputs from surface waters (Boulton *et al.*, 1998), animal density is strongly related to the amount of organic carbon present in the interstitial spaces (Brunke and Gonser, 1999). Organisms in the hyporheic, and presumably the phreatic, are more resistant to low oxygen conditions (hypoxia) than are epigean species, but they cannot permanently tolerate suboxic conditions (Malard and Hervant, 1999). The distribution and local diversity of hyporheic invertebrates are also influenced by historical patterns of glaciation (Strayer and Reid, 1999).

The hyporheic zone also extends laterally into the stream bank, where often the distance it penetrates horizontally from the river proper, is similar to that it penetrates vertically into the river bed. Data from the Flathead River, a gravelly stream of the northern Rocky Mountains, have shown, however, that the hyporheos permeates groundwater as far as 2 km from the surface water of the river (Stanford and Ward, 1988). It has been known for some time that a unique fauna can occupy the phreatic zone—a saturated underground area where the water moves slowly laterally in the absence of direct influence from surface water rivers. The study by Stanford and Ward, however, demonstrated that the groundwater community within up to 2 km from the Flathead River was hyporheic rather than phreatic because there were physical (flow), nutrient, and biotic connections between the two areas. The biota of the adjacent phreatic zone was dominated by subterranean crustaceans, while the hyporheic zone supported stoneflies and other typically riverine taxa.

B. Aquatic Habitats within Caves and Other Karst Topography

Regions of the world containing large amounts of soluble limestone and adequate underground water often form complex subterranean cavities; such karst topography is widespread in North America (Fig. 7). Karsts develop when rocks are fractured and are especially susceptible to solution (e.g., crystalline, high calcite limestone). Over time, small cavities can enlarge to produce huge underground caverns. The public generally becomes aware of solution processes only when an above-ground entrance exists or a sinkhole develops (also called sinks or, more properly, dolines).

Caves can be defined as karsts with an entrance, passageways, rooms, and a terminal blockage to all but water and material it carries. Cave environments provide habitats (Fig. 8) for many types of invertebrates which differ in several significant ways from surface, or epigean, aquatic habitats. They have very constant temperatures, that are equal to the average yearly surface temperature of the local geographic region, light is dim or absent, and abiotic disturbances are probably less frequent and severe (although floods occur in caves).

Food is quite scarce in caves, as it usually arrives only through groundwater or periodic movement of animals that seek refuge within an open cave after having fed outside (e.g., bats which deposit organic guano in the cave). The principal forms of organic matter supporting the food web are particulate (POM) and dissolved organic matter (DOM) along with heterotrophic bacteria growing in water or on suspended and benthic detritus and some chemoautotrophic bacteria. Consequently, food webs are relatively simple, consisting mostly of detritivores and their predators. The density and species diversity of a single cave community are usually quite low.

In contrast to the simplicity of a cave community, the species composition of one cave can be markedly

FIGURE 7 Karst topography in the 48 contiguous states of the United States (from Ritter, 1986).

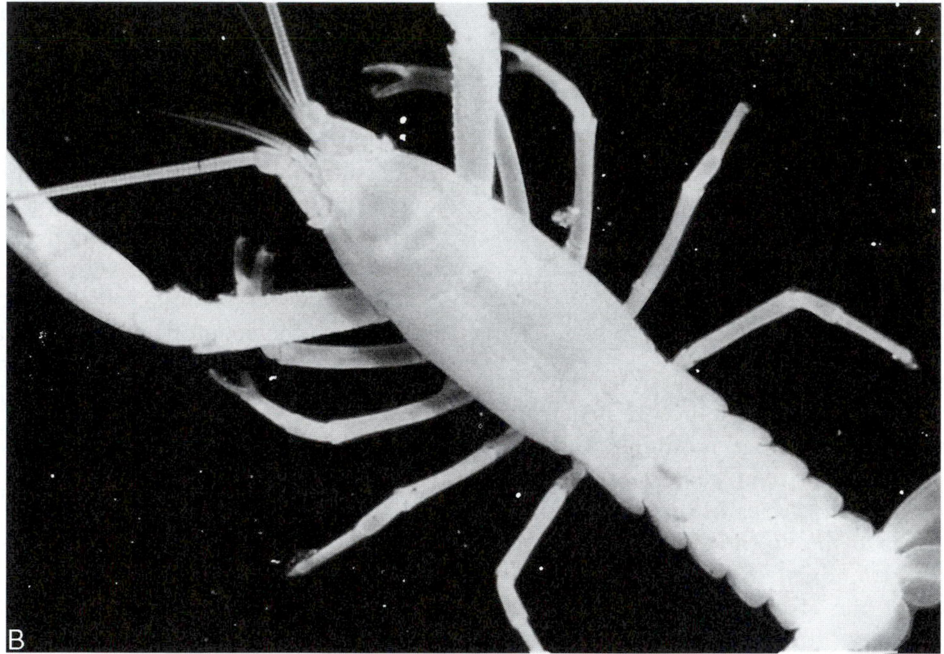

FIGURE 8 (A) A cave habitat in southern Indiana; (B) The blind, troglobitic crayfish *Orconectes testii* (both photographs courtesy of H. H Hobbs, III).

different from that of an adjacent but unconnected neighboring cave. Endemism is especially prominent in caves because of inherent geographic isolation. The species radiation of amphipods in North America has been much greater in karsts than in surface waters (see Chapter 19). A total of 927 species of invertebrates that live exclusively in caves or associated subterranean habitats in the contiguous 48 states in the United States, have been described so far, with more than 50% of the aquatic species and subspecies derived from only 18 counties representing less than 1% of the land area (Culver *et al.*, 2000)! Just over 50% of these subterranean species are from only a single county, and 95% of the cave species are listed by the Nature Conservancy as vulnerable or imperiled. It should be evident, therefore, that habitat protection at those sites is absolutely critical to species survival.

Obligate cave species, or troglobites (as opposed to the facultative troglophiles), usually have low metabolic rates and a much longer lifespan than their surface relatives. They are also frequently blind and lack body pigmentation. Reproductive yield is spread over longer periods with fewer but more yolky eggs formed during each reproductive bout. For example, a female crayfish of the troglobitic species *Orconectes inermis* in Indiana caves may carry only a few dozen large eggs, while nearby epigean, congeneric individuals can bear hundreds of small eggs (J. H. Thorp, personal observation, and H. H. Hobbs, III, personal communication; see also ecologically similar species in Fig. 8).

Information on karst communities can be gleaned from journals such as the *National Speleological Society Bulletin*. See also Chapters 19 and 23 in this book for details about cave amphipod and decapod crustaceans, respectively. For other information on the ecology of groundwater communities, see specialty texts such as Gibert *et al.* (1994) and Jones and Mulholland (2000).

IV. LENTIC ECOSYSTEMS

Unlike the situation for lotic habitats, the chemistry, physics, geology, and biology of lentic (or lacustrine) environments are well covered in several standard limnology textbooks (e.g., Cole, 1994; Lampert and Sommer, 1997; Wetzel, 2001), most of which devote very little space to stream studies. Lake limnology is also discussed in more depth in various specialty texts (e.g., Taub, 1984; Carpenter, 1988) and in four volumes on limnology by Hutchinson (1957, 1967, 1975, 1993). For this reason, our discussion of lake habitats is briefer than that accorded to stream ecosystems.

A. Geomorphology and Abiotic Zonation of Lakes

Although rivers and streams are important biologically, more than 99% of the inland surface waters are in fresh and saline lakes. These standing water, or lentic, ecosystems occur on every continent but are concentrated in the formerly glaciated regions of the Northern Hemisphere. Except for ~20 lakes with depths >400 m, most natural and human constructed lentic systems have average depths of <20 m (Wetzel, 2001). Most are geologically young, dating from the last glacial period in the case of natural lakes or from the last century for human-made lakes. According to Hutchinson (1957), natural lakes are formed by over 70 distinct processes. Most result from catastrophic phenomena (landslides and glacial, tectonic, and volcanic activities), but some form less violently from the action of rivers (e.g., oxbow lakes), waves, and rock solution. Before the advent of extensive, commercial fur trapping in North America, ponds built by beavers were a pervasive feature of the headwaters of many North American streams (Naiman *et al.*, 1986).

Lentic ecosystems can be divided into several abiotic zones, primarily based on distance from shore, light penetration, and temperature change (Fig. 9). The photic zone extends downward to the depth of 1% light penetration; all primary producers and most heterotrophic animals live within this zone. Areas below these depths are variously called the aphotic or profundal zone (the latter often in reference to benthos). The shallow, nearshore region of the photic zone where rooted macrophytes exist is termed the littoral zone. The entire mass of open water located away from both the shore and the littoral zone is known as the limnetic or pelagic zone.

On a seasonal basis, most lentic ecosystems in North America become stratified with a layer of lighter water, the epilimnion, floating over the denser hypolimnion (Fig. 9). At least during the summer, the epilimnion is considerably warmer than the hypolimnion. These two zones are separated by a layer of rapid temperature change called the metalimnion. The boundary between the epilimnion and the hypolimnion, where temperature changes occur most rapidly with depth, is defined as the thermocline (it is typically within the metalimnion). Because there is minimal exchange of water between upper and lower zones during stratification, the hypolimnion is frequently lower in oxygen, higher in nutrients, and different in pH and other chemical concentrations. When lake stratification breaks down for a short period during one or more seasons, much or all of the water mass recirculates in a process referred to as lake turnover. If a lake mixes completely twice a year, it is referred to as dimictic; if it turns over only once a year it is called monomictic. In other lentic systems, complete mixing either rarely takes place (in oligomictic lakes) or occurs more than twice per year (polymictic lakes). In addition to thermal zones, some lakes are chemically stratified (commonly with a heavier layer of saline water in the hypolimnion). Some of these lakes never mix completely and are termed meromictic; they only circulate in the upper zones. As a consequence, they are typically devoid of oxygen in deeper zones, where nutrients can accumulate over time.

The benthic zone can be further divided into somewhat arbitrary categories based on other ecosystem characteristics, such as substrate type, wave action, and

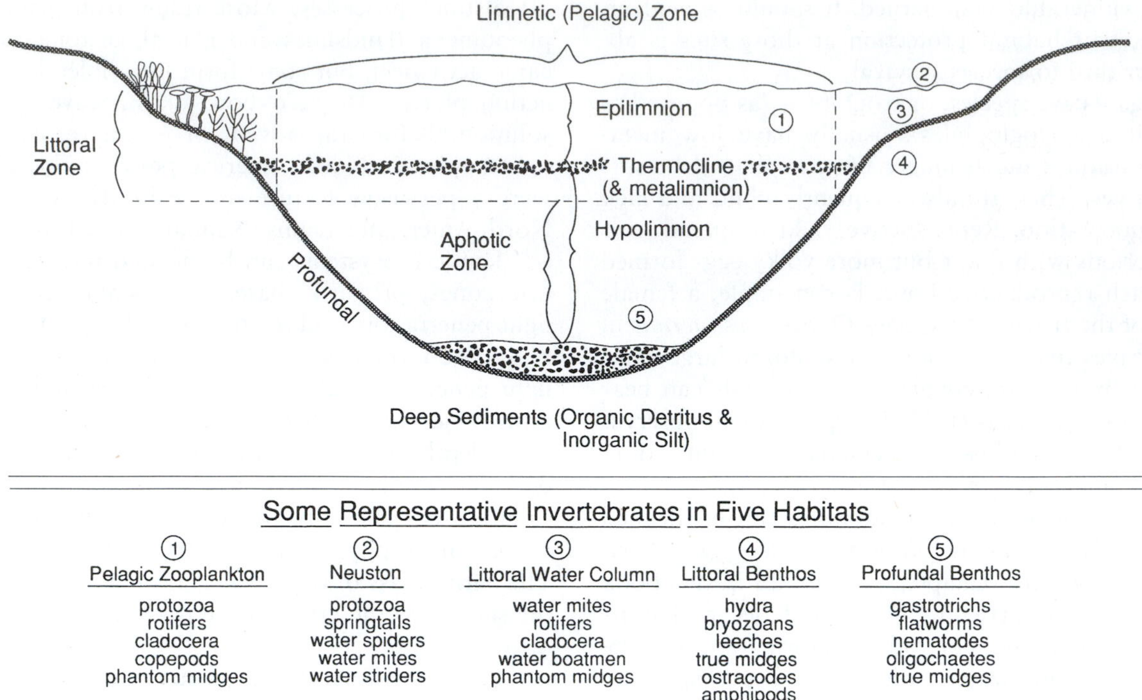

FIGURE 9 Biotic and abiotic zones within a lake along with a list of some representative freshwater invertebrates found within these zones.

vegetation type. The importance of substrate type for lentic environments is similar to that described earlier for lotic ecosystems.

B. Biotic Zonation of Lakes

Lake invertebrate assemblages can be subdivided primarily into "zooplankton", within the water column, and "benthos," which live in, on, or just above the bottom (commonly in association with macrophytes). A third, smaller group encompasses the "neuston," which live at the air–water interface. These first two categories remain quite distinct except in three general cases. A few species, such as the phantom midge *Chaoborus*, migrate upward at dusk into the plankton and return to the benthos near dawn. Meroplankton, in contrast to the much more diverse holoplankton, spend only a portion of their life cycle in the water column. For example, chironomid midges are benthic for most of the larval period (the short-lived adults are aerial), but swim within the water column during the earliest stages of their larval existence. Dreissenid mussels, which invaded North America in the late 1980s, have a veliger stage that is planktonic for up to 3 weeks (Garton and Haag, 1993; see also Chapter 11). Although not true members of the plankton, many nearshore species, such as littoral zone water fleas (cladocera; Chapter 20), alternate periods of resting or foraging on the bottom with intervals of swimming and foraging in the water column.

1. Neuston

The term neuston refers to the assemblage of organisms associated with the surface film of lakes, oceans, and slow-moving portions of streams. It generally includes species that live just underneath the water surface (hyponeuston), individuals that are above but immersed in the water (epineuston), and taxa that travel over the surface on hydrophobic structures (superneuston or, more properly, a form of epineuston). This name is similar to, or a subset of the older name, pleuston (sometimes neuston is used in reference to the microscopic components of the more encompassing pleuston). The density of neustonic organisms decreases with increasing turbulence.

The neustonic food web is primarily supported by a thin bacterial film on the upper surface of the water, a concentration of phytoplankton near the surface, and allochthonous inputs from trapped terrestrial and aquatic organisms. Protozoa are common in this assemblage (which also includes algae and floating macrophytes), whereas other typically planktonic taxa are rare (one exception is the cladoceran *Scapholeberis*). Moving over the water surface are springtails (Collembola), some arachnids (mites and water spiders), and various families of heteropteran bugs (e.g., water striders, Gerridae) (refer to Chapter 16 for information on arachnids and to Chapters 17 and 18 for details about insects and springtails). Larger neustonic species are especially vulnerable to predation from both aquatic and terrestrial predators, and all species must be adapted to the higher ultraviolet radiation present near the water surface (see Williamson *et al.*, 1999, for recent perspectives on UV and lake plankton).

2. Zooplankton of Pelagic and Littoral Habitats

Zooplankton include all animals in the water column that float, drift, or swim weakly (i.e., they are at the mercy of currents). Fish, as powerful swimmers, are the principal members of freshwater "nekton" in littoral and pelagic environments. Some predatory zooplankton are active swimmers and comprise the "nektoplankton"; examples are the large cladoceran *Leptodora kindtii* (Chapter 21) and mysid "shrimp" *Mysis relicta* (Chapter 19). Zooplankton thrive within two distinct habitats: open-water epilimnion and the nearshore littoral zone. The littoral habitat is the more heterogeneous of the two in terms of spatial and temporal complexity. The seasonal presence of macrophytic plants (algae and vascular plants) and proximity of planktonic and benthic habitats provide important aspects of physical complexity to the littoral zone. Additionally, temperatures fluctuate more rapidly in shallow water, and wave turbulence is greater. Oxygen is rarely limiting nearshore (except, sometimes in eutrophic lakes), but the effects of ice formation can be more severe. The majority of lentic research has dealt with the pelagic zone.

Biological interactions in the two planktonic habitats are different in response to those spatial and temporal differences and to variance in bases of the two food webs. Pelagic plankton rely on nutrients regenerated by lake turnover, offshore transport, and internal recycling; photosynthesis is exclusively by phytoplankton. Littoral plankton derive nutrients from the same three sources, but they have first access to allochthonous carbon and nutrients. They can also gain energy and recycled nutrients from the benthos on a continual basis (not possible for pelagic zooplankton except during lake turnover). Phytoplankton, periphyton, and macrophytes provide photosynthetic products to support the invertebrate assemblage of the productive littoral zone.

Predator–prey relationships and zooplankton behaviors vary between pelagic and littoral zones. Open-water copepods, cladocera, mysids, and a few rotifers migrate vertically on a diel basis. Fewer littoral zooplankton migrate, and among those that do, a typical pattern is for offshore movement at dusk and inshore migration at dawn. Ecologists believe that an important cause for both horizontal and vertical migration is predator avoidance. The types of predators in the two habitats are considerably different. Aquatic insects in general are of minor importance in the pelagic zone, but insect predators, such as dragonflies and predaceous beetles, are extremely abundant nearshore. Although predaceous adult and larval fish are usually more numerous in shallow, vegetated regions, the greater habitat complexity of the littoral habitat reduces foraging efficiency relative to open-water regions. Microcrustacean predators are present in both habitats.

3. Littoral and Profundal Benthos

Benthic invertebrates are often divided into three arbitrary size classes based on the mesh dimensions of the sieve used to process the samples. Macroinvertebrates are retained by coarse sieves (≥ 200 μm mesh) and are usually sorted with No. 60 or 35 U.S. Geological Sieves (250 and 500 μm mesh, respectively). Meiofauna, which are plentiful but rarely identified, need to be collected with fine sieves ranging from 40 to 200 μm mesh (usually ≤ 100 μm); microbenthos pass even these minute openings. The effort required to process benthic samples, especially from vegetated and/or silty habitats, increases tremendously with diminishing mesh size; consequently, species smaller than macrofauna are often ignored. While these size categories in themselves have no natural ecological or taxonomic significance, the frequent exclusion of the smaller forms from ecological analyses may have severely influenced our perception of how benthic assemblages function. For example, Strayer (1985) estimated that the micro- and meiobenthic animals contributed 68% of the species, 98% of the individuals, 25% of the biomass, and 35% of the production of the zoobenthos in Mirror Lake, New Hampshire. Studies in Mirror Lake revealed that over half of the benthic micro- and meiofauna live in the top centimeter of sediments (Strayer, 1985, 1986).

Most freshwater benthos reach their maximum densities and diversity in shallow water and decline perceptibly with increasing depth in the profundal zone. Few macroinvertebrates tolerate conditions in the profundal zone beneath the seasonal thermocline, but micro- and meiofauna can be abundant in deep water. This pattern probably reflects gradients of oxygen availability, habitat heterogeneity, and food resources—all of which are greater in the littoral zone. Studies of vegetated and nonvegetated regions of the littoral zone demonstrate the great value of macrophytes in reducing predation rates on benthic macrofauna (e.g., Hershey, 1985). This refuge is especially crucial because benthic animals are generally poor swimmers and have difficulty escaping highly motile predators.

A variety of distinctive habitats are available for benthic invertebrates. Nematodes, gastrotrichs, flatworms, and other meiobenthos frequent infaunal habitats where they move between sediment particles (Chapters 6–9). Those living on or among sand grains are called psammon or interstitial organisms. Macroinvertebrates, such as freshwater mussels, oligochaetes, and some crayfish, regularly burrow in the substrate. Many motile and sedentary species live on the surface of the mud, rocks, or submerged plant debris. Smaller animals and algae that live on firmer substrates are sometimes called aufwuchs. In addition to the attached aufwuchs, macrophytes support some larger, motile invertebrates which either feed on the aufwuchs (e.g., grazing snails; Chapter 10) or use the plant stems and leaves as perching sites for resting or foraging (e.g., sprawling dragonfly nymphs; Chapters 17 and 18).

C. Wetlands, Ephemeral Ponds, and Swamps

For many years the public has ignored or labeled as undesirable the vast acreage of wetlands, ephemeral ponds, and swamps in North America, considering them breeding grounds for mosquitoes and snakes. Many have been drained, filled, and bulldozed for housing and commercial development without regard to their intrinsic value to wildlife and the environment. Our knowledge of these fascinating aquatic ecosystems is relatively limited in comparison to information available about lakes and streams. Interestingly enough, one factor that has begun reversing this trend has been the implementation of strict environmental laws to protect the broad category of wetlands. Such laws have also led to an influx of research funds to study these ecosystems which have spawned a huge variety of journal articles (e.g., Zimmer *et al.*, 2000), books (e.g., Batzer *et al.*, 1999; Mitsch and Gosselink, 2000), and even a new journal (*Wetlands*).

In its broader definition, the term freshwater wetlands refers to nontidal ecosystems whose soils are saturated with water on a permanent or seasonal basis. Emergent aquatic vegetation is prominent and may alternate with annual terrestrial plants in these marshy, or palustrine, habitats. The extensive biomass of trees, herbaceous vegetation, grasses, and other plants in many wetlands is responsible for their recently recognized abilities to help filter pollutants from the environment. Wetlands vary from shallow, seasonally ephemeral ponds (such as Carolina bays and other pocosins, Sharitz and Gibbons, 1982) to relatively permanent vegetation-choked marshes to semi-lotic alluvial swamps (Fig. 10; see also Mitsch and Gosselink, 2000).

Because wetlands are generally more ephemeral than other lentic ecosystems, their biotic communities are strongly influenced by water cycles. Wetlands invertebrates are affected by ecosystem permanency (e.g., years since last exposure), predictability of drying, and season of drying (if at all). The nature of the community also reflects volume and depth of open water and sometimes current velocity (for alluvial swamps). See Batzer *et al.* (1999) for habitat specific information on invertebrate ecology from most types of freshwater wetlands in North America.

Invertebrates of seasonal wetlands, such as the Carolina bays of the southeastern United States, have evolved various characteristics that have directly or indirectly adapted them for ecosystems that alternate between aquatic and terrestrial states. Wiggins *et al.* (1980) divided animals of temporary pools into four groups based on their adaptations for tolerating or avoiding drought and their period of recruitment. "Group 1 are year-round residents incapable of active dispersal, which avoid desiccation either as resistant stages or by burrowing into pool sediments. Group 2 are spring recruits which must oviposit on water but subsequently aestivate and overwinter in the dry basin in various stages. Group 3 are summer recruits ovipositing in the dry basin and overwintering as eggs or larvae. Group 4 are nonwintering migrants leaving the pool before the dry phase, which is spent in permanent water; they return in spring to breed." When a temporary pool refills after a drought, the first colonizers are usually detritivores which exploit the abundant biomass of recently dead and decaying terrestrial vegetation; these species provide the animal tissue that supports the later arrival of predaceous invertebrates (Wiggins *et al.*, 1980). Many species in ephemeral environments are ecological generalists, because they live in both temporary and permanent aquatic ecosystems.

Alluvial swamps are at an opposite continuum from temporary wetland pools. Contrasting with the typical picture of a swamp as a stagnant marsh, alluvial (or riverine) swamps contain many areas of slowly to rapidly moving waters; portions of these wetlands somewhat resemble a highly braided stream. While it is true that southeastern alluvial swamps, such as those bordering the Savannah River between South Carolina and Georgia, are infested with alligators and poisonous snakes, they are otherwise extremely beautiful ecosystems with very few aerial insect pests once you have moved inward from the surrounding brush and terrestrial forest. Alluvial swamps are extremely heterogeneous, organically rich environments; oxygen does not appear to be limiting, at least in the shaded, flowing water regions.

FIGURE 10 (A) Photograph of a low-flow portion of a bald cypress—water tupelo, alluvial swamp of the Savannah River. Floating platforms in center are suspending wood snags as part of a colonization study by Thorp *et al.* (1995). (B) Small wetland area in northern New York created by a beaver dam. Shown are cattails (foreground), open water covered by duckweed (middle), and various emergent vegetation and woody plants (photographs by J. H. Thorp).

The many soft- and hard-sediment habitats of swamps are home to a great diversity and density of aquatic invertebrates. In a study of an alluvial swamp in South Carolina, Thorp *et al.* (1985) showed that invertebrates rapidly colonized wood snags, reaching a rough steady state in numbers within the first week of an 8-week study. Peak densities were equivalent to nearly 18,000 animals/m^2! "Filter-feeding taxa were numerically dominant early but soon were subordinate to gatherer and scraper functional feeding groups. Current velocity, seston particle size [as a food resource], dispersal capacity and competition for space may be important factors affecting community structure and colonization patterns in these aquatic ecosystems" (p. 56).

D. Hypersaline Lakes

Natural saline lakes are common worldwide but generally have neither the average size nor abundance of freshwater lakes (Eugster and Hardie, 1978; Hammer, 1984). They are frequently encountered in the prairies and the Great Basin of the western United States and in the provinces of Alberta, Saskatchewan, and British Columbia of western Canada (Hammer, 1984). The Great Salt Lake of Utah is the largest saline lake in North America. Saline lakes usually occur in more arid regions where closed basins promote high concentrations of salts (Fig. 11). These lentic ecosystems include saltern lakes, which are high in sodium chloride; lakes with large concentrations of sulfates and borates; and soda lakes characterized by abundant sodium carbonates and bicarbonates. Salinities in the Great Salt Lake have been recorded as high as 200 g/L (versus an average of 35 g/L in the world's oceans). In addition to high salinities, hypersaline lakes differ in pH, metal concentrations, and alkalinity from more typical inland lakes.

The abiotic characteristics of saline lakes make them extremely rigorous environments. Cyanobacteria colonize saline lakes as do some sedges, but the submerged macrophyte flora is relatively sparse in comparison to that in freshwater lakes. Because invertebrate density and diversity are enhanced by abundant littoral plants, the depauperate flora of saline lakes and the lack of certain other microhabitats undoubtedly depress the invertebrate fauna of briny pools. Few invertebrates of inland waters can regulate osmotic and ionic concentrations of their tissues in a hypertonic environment. As a result, hypersaline environments contain very few species, although their densities may be high. For example, the only permanent metazoa in Mono Lake, California (Fig. 11), are the ubiquitous brine shrimp, *Artemia salina*, and the brine fly genus *Ephydra* (see also Herbst, 1999). High densities of brine shrimp in Mono Lake are crucial to survival of

FIGURE 11 Saline Mono Lake in arid central California. The light "rocks" in the foreground are actually calcium carbonate mounds known as "tufa"; many of these arise within the lake proper. Visible on the island in the background is a small volcanic cone (photograph by J.H. Thorp).

migratory eared grebes, *Podiceps nigricollis,* which molt flight feathers while at Mono Lake, depending on these grazing crustaceans as their sole food source (Cooper *et al.*, 1984). This simple but important food chain is threatened by the possible demise of the brine shrimp as a result of rising salinities brought on by removal of freshwater from the Mono Lake catchment to provide water for the human population in Southern California. Other saline lakes support more species, including the rotifer *Brachionus plicatilis,* the calanoid copepod *Diaptomus nevadensis,* and several families of flies, beetles, and true bugs.

As saline lakes become less salty, the diversity and density of their flora and fauna are enhanced. For example, the meromictic Waldsea Lake of Saskatchewan, which is one of the better studied saline (≥ 3 ppt total dissolved solids) inland lakes in Canada, has a much higher species diversity than the much saltier Mono Lake. It contains two macrophyte species, some filamentous algae, a sparse phytoplankton population, and a dense but rather uniform population of zooplankton, dominated by the calanoid copepod *Diaptomus connexus,* the rotifers *Hexarthra fennica* and *Brachionus plicatilis,* and the water flea *Daphnia similis* (Hammer, 1984). The littoral zone supports high numbers of relatively few species, including several species of true bugs, nine genera of beetles, midges (mostly *Cricotopus*), a few caddisflies, the damselfly *Enallagma clausum,* one snail species (*Lymnaea stagnalis*), and several genera of crustaceans (see references in Hammer, 1984). Although Waldsea Lake is certainly diverse compared to the hypersaline Mono Lake, its community is depauperate in comparison to most permanent, freshwater lentic environments.

The inland waters of North America are sufficiently diverse and abundant to prevent the valid formation of any simple, all-encompassing theory on how biotic communities of these aquatic ecosystems are regulated. This does not imply, however, that we should abandon the task of assimilating and evaluating data from a diversity of ecosystems in order to understand better the effects of a few identifiable factors, such as the influence of habitat characteristics. A comparative approach, combined with intensive descriptive and experimental studies, will in the long run enhance our ability to develop a holistic view of stream and lake ecosystems. As we have shown in this chapter and as you will learn in Chapters 3–23, habitat preferences of an organism are extremely crucial in determining its distribution, survival, and reproductive output. The nature of the habitat, however, is only one of a large suite of factors influencing distribution and density of freshwater invertebrates.

LITERATURE CITED

Allan, J. D. 1995. Stream ecology: Structure and function of running waters. Chapman and Hall, New York.

Barnby, M. A., Resh, V. H. 1988. Factors affecting the distribution of an endemic and a widespread species of brinefly (Diptera: Ephydridae) in a northern California thermal saline spring. Annals of the Entomological Society of America 81:437–446.

Batzer, D. P., Rader, R. B., Wissinger, S. A. 1999. Invertebrates in freshwater wetlands of North America: ecology and management, Wiley, New York, 1100 p.

Benke, A. C., Jacobi, D. I. 1994. Production dynamics and resource utilization of snag-dwelling mayflies in a blackwater stream. Ecology 75(5):1219–1232.

Benke, A. C., Wallace, J. B. 1997. Trophic basis of production among riverine caddisflies: implications for food web analysis. Ecology 78:1132–1145.

Boulton, A. J., Findlay, S., Marmonier, P., Stanley, E. H., Valett,. H. M. 1998. The functional significance of the hyporheic zone in streams and rivers. Annual Review of Ecology and Systematics 29:59–81.

Brunke, M., Gonser, T. 1999. Hyporheic invertebrates—the clinal nature of the interstitial communities structured by hydrological exchange and environmental gradients. Journal of the North American Benthological Society 18:344–362.

Calow, P., Petts, G. E. Eds. 1992. The rivers handbook: hydrological and ecological principles, Vol. 1, Blackwell Science, Oxford, England, 526 p.

Carpenter, S. R. Ed. 1988. Complex interactions in lake communities, Springer-Verlag, New York. 283 p.

Cole, G. A. 1994. Textbook of limnology, 4th ed. Mosby, St. Louis, MO, 412 p.

Connell, J. H. 1978. Diversity in tropical rain forests and coral reefs. Science 199:1302–1310.

Cooper, S. D., Winkler, D. W., Lenz, P. H. 1984. The effects of grebe predation on a brine shrimp population. Journal of Animal Ecology 53:51–64.

Covich, A. P. 1988. Geographical and historical comparisons of neotropical streams: biotic diversity and detrital processing in highly variable habitats. Journal of the North American Benthological Society 7:361–386.

Culver, D. C., Master, L. L., Christman, M. C., Hobbs, H. H. 2000. Obligate cave fauna of the 48 contiguous Unites States. Conservation Biology 14:386–401.

Delucchi, C. M., Peckarsky, B. A. 1989. Life history patterns of insects in an intermittent and a permanent stream. Journal of the North American Benthological Society 8:308–321.

Eugster, H. P., Hardie, L. A. 1978. Saline lakes, *in*: A. Lerman, Ed., Lakes: chemistry, geology, and physics. Springer-Verlag, New York, pp. 237–293.

Forester, G. E., Dudley, T. L., Grimm, N. B. 1999. Trophic interactions in open systems: effects of predators and nutrients on stream food chains. Limnology and Oceanography 44:1187–1197.

Garton, D. W., Haag, W. R. 1993. Seasonal reproductive cycles and settlement patterns of *Dreissena polymorpha* in western Lake Erie. Chapter 6, *in*: Nalepa, T.F. and D.W. Schloesser, Eds., Zebra mussels: biology, impacts, and control. Lewis Publishers, Boca Raton, FL, pp. 111–128.

Gayraud, S., Philippe, M., Maridet, L. 2000. The response of benthic macroinvertebrates to artificial disturbance: drift or vertical movement in the gravel bed of two sub-alpine streams? Archiv für Hydrobiologie 147:431–446.

Gibert, J., Danielopol, D. L., Stanford, J. A. Eds. 1994. Groundwater ecology. Academic Press, San Diego, 571 p.

Gordon, N. D., McMahon, T. A., Finlayson, B. L. 1992. Stream hydrology: an introduction for ecologists. Wiley, New York, 526 p.

Hammer, U. T. 1984. The saline lakes of canada, *in*: F.B. Taub, Ed., Ecosystems of the world, No. 23: lakes and reservoirs. Elsevier, New York, pp. 521–540.

Herbst, D. B. 1999. Biogeography and physiological adaptations of the brine fly genus *Ephydra* (Diptera: Ephydridae) in saline waters of the Great Basin. Great Basin Naturalist 59:127–135.

Hershey, A. E. 1985. Effects of predatory sculpin on the chironomid communities in an arctic lake. Ecology 66:1131–1138.

Howarth, R. W., Schneider, R., Swaney, D. 1996. Metabolism and organic carbon fluxes in the tidal freshwater Hudson River. Estuaries 19:848–865.

Hutchinson, G. E. 1957. A treatise on limnology. I. Geography, physics, and chemistry. Wiley, New York, 1015 p.

Hutchinson, G. E. 1967. A treatise on limnology. II. Introduction to lake biology and limnoplankton. Wiley, New York, 1115 p.

Hutchinson, G. E. 1975. A treatise on limnology. III. Limnological botany. John Wiley, New York, 660 p.

Hutchinson, G. E. 1993. A treatise on limnology. IV. The zoobenthos. Wiley, New York, 944 p.

Hynes, H. B. N. 1970. The ecology of running waters. University of Toronto Press, Toronto, Ont. 555 p.

Hynes, H. B. N. 1989. Keynote address, *in*: Dodge, D. P., Ed., Proceedings of the International Large River Symposium, Can. Sp. Publ. Fish. Aquat. Sci. 106, pp. 5–10.

Jack, J. D., Thorp, J. H. 2000. Field experiments on regulation of potamoplankton in a large river by a benthic suspension-feeding mussel, *Dreissena polymorpha*. Freshwater Biology (in press).

Jones, J. B., Mulholland, P. J. Eds. 2000. Streams and ground waters. Academic Press, San Diego, 425 p.

Junk, W. J., Bayley, P. B., Sparks, R. E. 1989. The flood pulse concept in river-floodplain systems, *in*: Dodge, D. P., Ed., Proceedings of the International Large River Symposium, Can. Sp. Publ. Fish. Aquat. Sci. 106, pp. 110–127.

Lampert, W., Sommer, U. 1997. Oxford University Press, New York, 382 p. [Translation of a German text published in 1993.]

Ledger, D. C. 1981. The velocity of the River Tweed and its tributaries. Freshwater Biology 11:1–10.

Leopold, L. B. 1953. Downstream change of velocity in rivers. American Journal of Science 251: 606–624.

Longley, G. 1986. The biota of the Edwards aquifer and the implications for paleozoogeography. *in*: Abbott P. L., Woodruff C.M., Jr., Eds., The Balcones Escarpmernt, Geological Society of America, Central Texas, pp. 51–54.

Malard, F., Hervant, F. 1999. Oxygen supply and the adaptations of animals in groundwater. Freshwater Biology 41:1–30.

Matthews, W. J. 1988. North American prairie streams as systems for ecological study. Journal of the North American Benthological Society 7:387–409.

Merritt, R. W., Cummins, K. W. Eds. 1996. An introduction to the aquatic insects of North America. 3rd ed., Kendall/Hunt, Dubuque, Iowa.

Meyer, J. L. 1990. A blackwater perspective on riverine ecosystems. BioScience 40:643–651.

Mihuc, T. B. 1997. The functional trophic role of lotic primary consumers: generalist versus specialist strategies. Freshwater Biology 37:455–462.

Minckley, W. L. 1963. The ecology of a spring stream Doe Run, Meade County, Kentucky. Wildlife Monographs No. 11 (A publication of the Wildlife Society), 124 p.

Minshall, G. W. 1988. Stream ecosystem theory: a global perspective. Journal of the North American Benthological Society. 7:263–288.

Mitsch, W. J., Gosselink, J. G. Eds. 2000. Wetlands, 3rd ed., Wiley, New York, 750 p.

Naiman, R. J., Melillo, J. M., Hobbie, J. E. 1986. Ecosystem alteration of boreal forest streams by beaver (*Castor canadensis*). Ecology 67:1254–1269.

Newbury, R. W. 1984. Hydrologic determinants of aquatic insect habitats, *in*: Resh, V. H., Rosenberg D. M., Eds., The ecology of aquatic insects. Praeger, New York, pp. 323–357.

Pace, M. L., Findlay, S. E. G., Fischer, D. 1998. Effects of an invasive bivalve on the zooplankton community of the Hudson River. Freshwater Biology 39:103–116.

Pace, M. L., Findlay, S. E. G., Lints, D. 1992. Zooplankton in advective environments: The Hudson River community and a comparative analysis. Canadian Journal of Fisheries and Aquatic Sciences 49:1060–1069.

Reckendorfer, W., Keckeis, H., Winkler, G., Schiemer F. 1999. Zooplankton abundance in the River Danube, Austria: the significance of inshore retention. Freshwater Biology 41:583–591.

Resh, V. H., Brown, A. V., Covich, A. P., Gurtz, M. E., Li, H. W., Minshall, G. W., Reice, S. R., Sheldon, A. L., Wallace, J. B., Wissmar, R. C. 1988. The role of disturbance in stream ecology. Journal of the North American Benthological Society 7:433–455.

Ritter, D. F. 1986. Process geomorphology, 2nd ed., Wm. C. Brown, Debugue, Iowa.

Sedell, J. R., Richey, J. E., Swanson, F. J. 1989. The river continuum concept: a basis for the expected behavior of very large rivers? *in*: Dodge, D.P., Ed., Proceedings of the International Large River Symposium, Can. Sp. Publ. Fish. Aquat. Sci. 106, pp. 49–55.

Sharitz, R. R., Gibbons, J. W. 1982. The ecology of southeastern shrub bogs (pocosins) and Carolina bays: a community profile. FWS/OBS-82/04. U.S. Fish and Wildlife Service, Division of Biological Services, Washington, D. C, 93 p.

Stanford, J. A., Hauer, F. R., Ward, J. V. 1988. Serial discontinuity in a large river system. Verhandlungen Internationale Vereinigung fur theoretische und angewandte Limnologie 23:114–118.

Stanford, J. A., Ward, J. V. 1988. The hyporheic habitat of river ecosystems. Nature 335:64–66.

Statzner, B., Gore, J. A. Resh, V. H. 1988. Hydraulic stream ecology: observed patterns and potential applications. Journal of the North American Benthological Society 7:307–360.

Statzner, B., Higler, B. 1985. Questions and comments on the River Continuum Concept. Canadian Journal of Fisheries and Aquatic Sciences 42:1038–1044.

Strayer, D. 1985. The benthic micrometazoans of Mirror Lake, New Hampshire. Archive für Hydrobiologie/Suppl. 72(3):287–426.

Strayer, D. 1986. The size structure of a lacustrine zoo-benthic community. Oecologia 69:513–516.

Strayer, D., May, S. E., Nielsen, P., Wollheim, W., Hausam, S. 1995. An endemic groundwater fauna in unglaciated eastern North America. Canadian Journal of Zoology 73:502–508.

Strayer, D. L., Reid, J. W. 1999. Distribution of hyporheic cyclopoids (Crustacea: Copepoda) in the eastern United States. Archiv für Hydrobiologie 145:79–92.

Strahler, A. N. 1952. Dynamic basis of geomorphology. Geological Society of America Bulletin 63:1117–1142.

Strenth, N. E., Norton, J. D., Longley, G. 1988. The larval development of the subterranean shrimp *Palaemonetes antrorum* Benedict (Decapoda, Palaemonidae) from central Texas. Stygologia 4:363–370.

Sweeney, B. W., Vannote, R. L. 1978. Size variation and the distribution of hemimetabolous aquatic insects: two thermal equilibrial hypotheses. Science 200:444–446.

Taub, F. B. Ed. 1984. Lakes and reservoirs. Ecosystems of the world, Vol. 23, Elsevier, New York, 643 p.

Thorp, J. H., Cothran, M. L. 1984. Regulation of freshwater community structure at multiple intensities of dragonfly predation. Ecology 65:1546–1555.

Thorp, J. H., Delong, M. D. 1994. The riverine productivity model: an heuristic view of carbon sources and organic processing in large river ecosystems. Oikos 70(2):305–308.

Thorp, J. H., Delong, M. D., Casper, A. F. 1998. *In situ* experiments on predatory regulation of a bivalve mollusc (*Dreissena polymorpha*) in the Mississippi and Ohio Rivers. Freshwater Biology 39:649–661.

Thorp, J. H., Delong, M. D., Greenwood, K. S., Casper, A. F. 1998. Isotopic analysis of three food web theories in constricted and floodplain regions of a large river. Oecologia 117:551–563.

Thorp, J. H., McEwan, E. M., Flynn, M. F., Hauer, F. R. 1985. Invertebrate colonization of submerged wood in a cypress-tupelo swamp and blackwater stream. American Midland Naturalist 113:56–68.

Vannote, R. L., Minshall, G. W., Cummins, K. W., Sedell, J. R., Cushing, C. E. 1980. The river continuum concept. Canadian Journal of Fisheries and Aquatic Sciences 37:130–137.

Ward, J. V. 1986. Altitudinal zonation in a Rocky Mountain stream. Archiv fur Hydrobiologie, Suppl. 74:133–199.

Ward, J. V. 1989. The four-dimensional nature of lotic ecosystems. Journal of the North American Benthological Society 8:2–8.

Ward, J. E., Standford, J. A. 1983. The intermediate disturbance hypothesis: an explanation for biotic diversity patterns in lotic ecosystems, *in*: Fontaine T.D., Bartell, S.M. Eds., Dynamics of lotic ecosystems. Ann Arbor Science Publishers, Ann Arbor, MI, pp. 347–356.

Wetzel, R. G. 1983. Limnology, 2nd ed., Saunders, Philadelphia, 857 p.

Wetzel, R. G. 2001. Limnology, 3rd ed., Academic Press, San Diego (in press).

Wiggins, G. B., Mackay, R. J., Smith, I. M. 1980. Evolutionary and ecological strategies of animals in annual temporary pools. Archive fur Hydrobiologie/Suppl. 58(1):97–206.

Williams, D. D. 1984. The hyporheic zone as a habitat for aquatic insects and associated arthropods. *in*: Resh, V.H., Rosenberg, D.M. Eds., The ecology of aquatic insects. Praeger, New York, pp. 430–455.

Williamson, C. E., Hargreaves, B., Orr, P. S., Lovera, P. A. 1999. Does UV play a role in changes in predation and zooplankton community structure in acidified lakes? Limnol. Oceanogr. 44(3, part 2):774–783.

Winterbourn, M. J. 1982. The River Continuum Concept-reply to Barmuta and Lake. New Zealand Journal of Marine and Freshwater Research 16:229–231.

Zimmer, K. D., Hanson, M. A., Butler, M. G. 2000. Factors influencing invertebrate communities in prairie wetlands: a multivariate approach. Canadian Journal of Fisheries and Aquatic Sciences 57:76–85.

3 PROTOZOA

William D. Taylor
Department of Biology
University of Waterloo
Waterloo, Ontario N2L 3G1, Canada

Robert W. Sanders
Department of Biology
Temple University
Philadelphia, Pennsylvania 191ZZ

I. INTRODUCTION

Protozoa are ubiquitous; they are present in an active state in all aquatic or moist environments, and cysts are present everywhere in the biosphere, ready to give rise to active populations. They play an important role in many natural communities, although they are often overlooked. In the laboratory, protozoa often play the role of model organisms; cell biologists, physiologists, geneticists, and even developmental biologists frequently turn to protozoa to address questions that would be more difficult to answer with metazoa, but in the hope that the answers are relevant to metazoa. The result is a surprising amount of information about a relatively few species. Accordingly, detailed monographs are available on several taxa, including *Blepharisma* (Giese, 1973), *Paramecium* (Wichterman, 1986), *Tetrahymena* (Elliott, 1973) and *Amoeba* (Jeon, 1973). Each monograph contains thousands of references. Ecologists also have used protozoan populations and communities as model systems to investigate processes, such as competition and predation, or to test models, such as island biogeography, that would be much more difficult to study using larger organisms. Again, the result has been an extensive body of literature that is largely restricted to a few species and/or to unusual or artificial habitats.

In the last 20 years, there has been an explosion of information about the ecology of free-living protozoa, especially the planktonic species that were virtually ignored in the earlier literature. This increase in interest followed the development of direct-count methodologies for aquatic bacteria, and an appreciation of the importance of bacteria and their predators in aquatic systems. The first focus was on the heterotrophic flagellates and ciliates that consumed bacterial production and passed it to larger consumers, the "microbial loop." As aquatic ecologists came to realize that protozoa are not only present, but play a major quantitative role in material and energy flow, this interest expanded to other feeding guilds and freshwater environments.

Ecology and Classification of North American Freshwater Invertebrates. 2nd Edition

The goal of this chapter is to provide an introduction to the ecology and classification of protozoa for ecologists who wish to include them in studies of aquatic environments. The chapter focuses on the biology of the animal-like protozoa, largely ignoring the related pigmented organisms which are covered in phycological sources. The chapter is oriented toward the organisms, not the freshwater habitats themselves. Readers wishing to learn about the fauna of particular habitats might consult the bibliography on freshwater protozoa prepared by Finlay and Ochsenbein-Gattlen (1982) or some more recent works on particular habitats: lake plankton (Beaver and Crisman, 1989; Laybourn-Parry, 1992; Riemann and Christoffersen, 1993; Carrias *et al.*, 1996 Foissner et al. 1999), streams (Bott and Kaplan, 1990; Carlough and Meyers, 1991; Sleigh *et al.*, 1993), and wetlands (Pratt and Cairns, 1985; Decamp and Warren, 1999).

What are protozoa?

Protozoa are unicellular or colonial eukaryotes, including all of the heterotrophic and motile ones. It is a taxon of convenience. Although the protozoa were considered to be a single phylum of eukaryotic animals in early classifications, modern treatments distribute the protozoa among many phyla. This reflects the immense diversity of this assemblage, and the fact that differences among these unicellular organisms are at least as profound as those which separate plant and animal phyla (see Section III.D). Margulis and Schwartz (1988), in their five-kingdom classification of living organisms, place the unicellular eukaryotes and some of their multicellular descendants, 27 phyla in all, in the kingdom Protoctista. Other modern classifications partition the protists into more or fewer phyla. The classification of unicellular eukaryotes is in a state of flux, with much work to be done. In any event, it is among these groups that the organisms we know as protozoa are scattered.

For the purpose of this chapter, we define the term protozoa as those unicellular or colonial eukaryotes that are heterotrophic, and will restrict ourselves to a discussion of free-living, phagotrophic forms in freshwater. Among the protozoan phyla we will not discuss here are the marine Foraminifera and Radiolaria, the soil-dwelling Labyrinthulamycota, Plasmodiophoromycota and Acrasida, the saprophytic Hyphochytridiomycota, Oomycota and Chytridiomycota, and the parasitic Apicomplexa, Microspora, and Myxospora. The classification we use, and citations to figures illustrating the major taxa, are outlined in Table I. It is designed to be as current as possible, but still retain the older names. Therefore, it emphasizes classes, subclasses, or orders, largely according to which retains the most familiar name for the group in question. Our major

TABLE I Major Taxomomic Categories of Protozoa as Used in This Chapter

Functional Group / *Phylum* / *Class* / *subclass* / *Order*	*Figure reference*
Phytoflagellates (largely autotrophic, but some mixotrophic and heterotrophic)	
Cryptophyta	13J
Dinoflagellata	13A–C
Euglenida	13D, F–I
Chrysophyta	13K–N; 14A
Chlorophyta	13E
Zooflagellates (heterotrophic)	
Zoomastigina	
Choanoflagellida	14B–H
Kinetoplastida	14J, L–P
bicoecids	14I
diplomonads	14M
cercomonads	14K
Ameboid Protozoa	
Karyoblastea	16V
Rhizopoda	
Heterolobosea	
Schizopyrenida	15A–C
Lobosea	
Gymnamoebia	
Euamoebida	15D–M,O,P
Leptomyxida	
Acanthopodida	15N
Testacealobosia	
Arcellinida	16B–Q
Himatismenida	16A
Filosea	
Aconchulinida	16R, 17A
Testaceafilosida	16S–W; 17B–K
Granuloreticulosida	
Athalamea	17P–R
Monothalamea	17L–O
Foraminiferea (marine only)	
Actinopoda	
Heliozoa	
Actinophryida	18B,C
Desmothoracida	18A
Ciliophryida	18E
Centrohelida	18D,F,H
Rotosphaerida	18G
Ciliated Protozoa	
Ciliophora	
Karyorelictea	23J
Heterotrichea	22A,B,D–F
Spirotrichea	
Hypotrichia	20X,Y
Choreotrichia	21S,T,U,X
Stichotrichia	21A–I,N–Q
Oligotrichia	21V,W
Odontostomatida	21,J,M,L
Armophorida	21K,R
Litostomatea	22G,I,K,L,N,P–W; 23D–I

(Continues)

TABLE I (Continued)

Functional Group / *Phylum* / *Class* / *subclass* / *Order*	*Figure reference*
Phyllopharyngea	
Phyllopharyngia	24T
Suctoria	19A–P
Nassophore	
Nassulida	20Z
Microthoracida	20V,W
Colpodea	22C; 23K; 24N,P–S
Prostomatea	
Prostomatida	22H,J,O
Prorodontidae	23A–C
Plagiopylea	24M
Oligohymenophorea	
Peniculia	24A,B,D,E,G,I,J
Scuticociliatia	23L–X
Hymenostomatia	24 C,F,H,O
Peritrichia	20 A–U

points of reference were Page (1998), Lynn and Small (1997), and various chapters in Lee *et al.* (1985) and Margulis *et al.* (1989).

Despite the fact that the term protozoa is perhaps not a proper taxonomic group, it remains a useful functional one. For ecologists conducting studies involving aquatic animals, including those of microscopic proportions, it is natural to include the animal-like protozoa in those studies. After all, it is functional similarity that generally concerns the practising zoologist or ecologist.

II. ANATOMY AND PHYSIOLOGY

A. Protozoa as Cells and Organisms

Large protozoa that have many nuclei, or large, amitotic, polyploid nuclei, may be best thought of as acellular rather than unicellular creatures. Nonetheless, some species have been studied extensively by cell biologists as model cells, and ecologists interested in protozoa can glean much useful information from their labors. This section will rely heavily on information from those few taxa. We will first describe some of the organelles which are important and widely distributed among protozoa, then discuss major groups of protozoa with respect to their more unique organelles. Lastly, we will revert to an overall view in discussing environmental physiology of protozoa. In all sections, we will emphasize those organelles that relate to

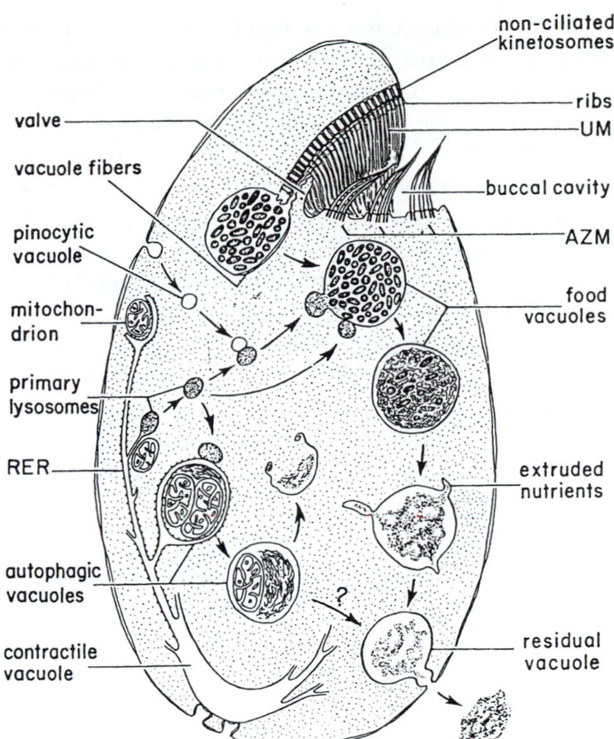

FIGURE 1 The formation and fate of food vacuoles, autophagic vacuoles, and pinocytotic vacuoles in *Tetrahymena pyriformis*. (From Elliott and Clemmons, 1966.)

feeding, locomotion, ecology, and morphology at the light-microscope level.

B. Protozoan Organelles

1. Food Vacuoles

Phagocytosis in protozoa leads to the formation of food vacuoles (Fig. 1). Digestion is, as one would expect, primarily intracellular in free-living protozoa. The formation and behavior of food vacuoles is best known in the ciliates *Tetrahymena* (see Rasmussen, 1976; Nilsson, 1976, 1987) and *Paramecium* (Allen, 1984). A more general review is provided by Nisbet (1984).

The formation of food vacuoles is stimulated by the presence of particulate food. In protein broth medium, *Tetrahymena* shows very slow rates of food vacuole formation, ingestion, and growth unless particles are present (Ricketts, 1972), even though it can grow without forming food vacuoles in complex media via carrier-mediated uptake of dissolved compounds across the plasma membrane. During starvation, vacuoles resembling food vacuoles may be created to digest cellular components. Formation of such "cytolosomes" is also known from *Amoeba* (Chapman-Andresen, 1973).

Many protozoa have a fixed site of food-vacuole formation; a cytostome with accompanying structures to aid in ingestion. In others, for example, Rhizopoda and Heliozoa, the site of ingestion is flexible, perhaps only limited to a characteristic region of the cell. The Suctoria have many sites for ingestion. The contents of a single vacuole may vary from a single large food item in macrophagous species to thousands of small items, such as bacteria, in microphagous forms. Similarly, individual protozoa may have one-to-many food vacuoles. The vacuoles of microphagous forms may be perfectly spherical, while larger vacuoles containing single prey may be quite irregular, conforming to the contours of the prey. Macrophagous protozoa may distend themselves to surround large food items, leave them protruding from their cytoplasm, or even share a prey item among several individuals. Large food items may be digested by cytoplasmic extensions in dinoflagellates and possibly other protozoa (Fig. 2).

Digestion may be initiated by the toxicysts of some predatory ciliates. In others, it begins when granules (acidosomes) fuse with the developing food vacuole. The pH of the new vacuole drops precipitously, then increases minutes later. Hydrolytic enzymes, such as acid phosphatase, are introduced as the food vacuole fuses with lysosomes. The size of the food vacuole may diminish, reflecting the absorption of water and nutrients, and vacuoles may coalesce. Some protozoa, such as ciliates, have a fixed site of egestion or cytoproct. When the contents of the food vacuole are released, the membrane surrounding the vacuole is retained and recycled (Allen, 1984). It is unlikely that a cellular equivalent of a digestive track exists in protozoa. Nevertheless, it has been possible to estimate the feeding rate of protozoa in the field from the rate at which food vacuoles are eliminated (Goulder, 1972).

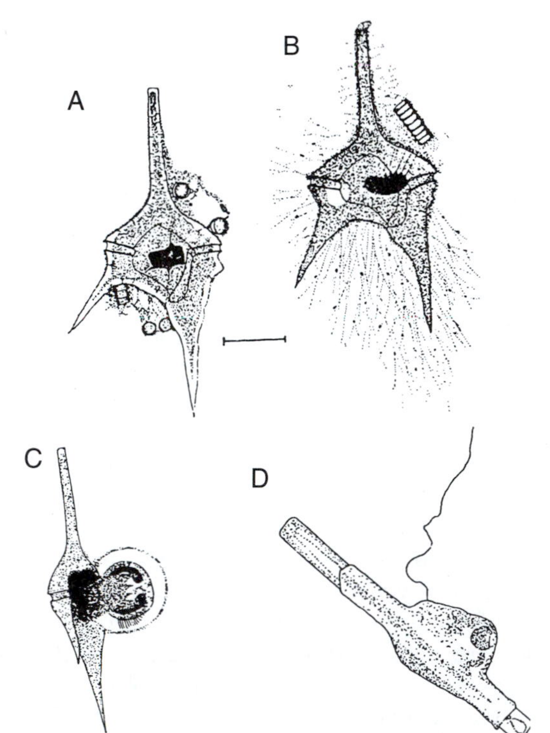

FIGURE 2 Phagocytosis of large food items by protoplasmic extensions in the dinoflagellate *Ceratium hirudinella* (A–C), and the consumption of a large diatom by a *Paraphysomonas*-like chrysophyte (D). (A–C from Hofeneder, 1930: D after Suttle *et al.* 1986.)

2. Contractile Vacuoles

These organelles are widely distributed in protists and are almost universal in freshwater protozoa and sponges. They are absent in dinoflagellates and are generally lacking in those protists with rigid cell walls and/or from marine or endosymbiotic habitats. A large body of evidence suggests that contractile vacuoles eliminate water and, thereby, counteract the tendency of naked protists to swell in freshwater (Patterson, 1980). The activity of the contractile vacuole is correlated with changes in the osmotic balance between the cell and its medium, and a loss of contractile function results in swelling of the cell. Of course, the occurrence of contractile vacuoles in naked, freshwater protists and their absence in most rigid-walled, marine or endosymbiotic ones also support this theory.

Associated with the contractile vacuole is a specialized cytoplasm, the spongiome, which collects the solute to be expelled by contractile vacuole. The spongiome may contain vesicles or canals visible with the light microscope, and therefore be of diagnostic value (Fig. 3). Patterson (1980) recognized four types of contractile vacuole complexes in protozoa, with six different patterns of behavior. Permanent contractile vacuole pores are present in ciliates and are readily visible in silver-stained specimens. The contractile vacuole may have a fixed location in some other groups, for example, in the gullet or reservoir of cryptomonad or euglenoid flagellates, but the fixed pores, if they exist, are not visible.

Several important functional aspects of the contractile vacuole complex remain unknown; for example, the composition of the fluid expelled, and, therefore, the role of the contractile vacuole complex in salt balance and excretion. It is also debatable whether expulsion of contractile vacuole contents is active or passive, and how the activity of the contractile vacuole is controlled.

Dinoflagellates possess a functionally homologous organelle, the pusule. This organelle consists of a single

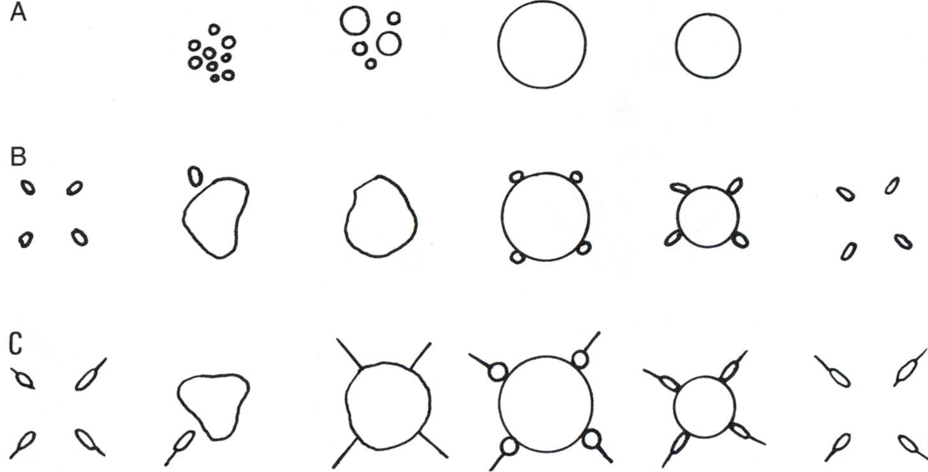

FIGURE 3 Top views of three morphologies of contractile vacuoles found in protozoa, showing their cycles of activity from left to right as seen with a light microscope. (A) A vacuole which forms by the coalescing of smaller vesicles, as is common in Rhizopoda. Vacuole (B) has ampullae in fixed locations, while vacuole (C) also has visible canals. (After Patterson, 1980.)

chamber or tubule associated with a flagellar opening, and it is lined with vesicles (Dodge and Gruet, 1987).

3. Cilia and Flagella

Flagella, cilia, and derived structures are widely distributed in eukaryotes, including protozoa. Their ultrastructural similarity, including the basic 9 pairs +2 pattern of microtubules in the shaft, or 9 triplets in the base, indicates homology despite the great diversity in form and function. Many animal sense organs contain cilia, which emphasizes that they are probably intrinsically sensory as well as motile. Cilia or flagella which are thought to be sensory occur in many protozoa, especially the "brosse" kinetids of prostomateans, and the flagellum of phototactic euglenids.

It is well established that the motile force in cilia and flagella is the sliding of microtubules in the shaft against each other. Therefore, the active movement is the bending of shaft and not the movement of shaft relative to the base. The bending of shaft takes on the appearance of a beat or stroke in short cilia or a wave in long flagella.

Flagella usually occur in pairs and are relatively long (often longer than the cell that carries them, and reaching lengths of up to 50 μm). A major functional and morphological dichotomy among flagella is between those with and without mastigonemes (Fig. 4). Mastigonemes are thin, hairlike projections perpendicular to the shaft of the flagellum. They are not visible with light microscopy, but are functionally important. A distally directed wave of bending in a smooth flagellum will push the cell in the opposite direction to which the flagellum is pointed, as in animal sperm. The same wave form in a flagellum with mastigonemes will pull the cell in the direction in which the flagellum is pointed. To understand the motility of protozoa, especially those using cilia and flagella, it is necessary to appreciate that the apparent viscosity of water increases as the size of organisms moving through it decreases (Holwill, 1974; Fenchel, 1987). To flagellates, water is a liquid so viscous that they sink only very slowly and they not glide from inertia.

The difference between cilia and flagella is one of form and function. Cilia are generally short and densely packed. They are organized in parallel rows (kineties) to produce fields of cilia whose activities are coordinated (Fig. 5A). The movement of adjacent somatic cilia is not quite synchronous; rather, each is slightly out of phase with its neighbors, or metachronous. A field of cilia, therefore, moves in waves that may or may not correspond in direction to the actual power stroke of the cilia. Compound ciliary organelles or polykinetids are restricted fields of cilia acting as a single unit in locomotion, as in the cirri of hypotrichs (Fig. 20Y), or feeding, as in the oral apparatus of hymenostomes (Fig. 5B). They may be relatively long compared to other cilia and fused so that their compound nature is not obvious under light microscopy, except as indicated by their diameter.

Ciliary movement is faster than flagellar movement, even accounting for ciliates generally being larger than flagellates (Sleigh and Blake, 1977). Indeed, within both these groups there is relatively little increase in absolute swimming speed with body size;

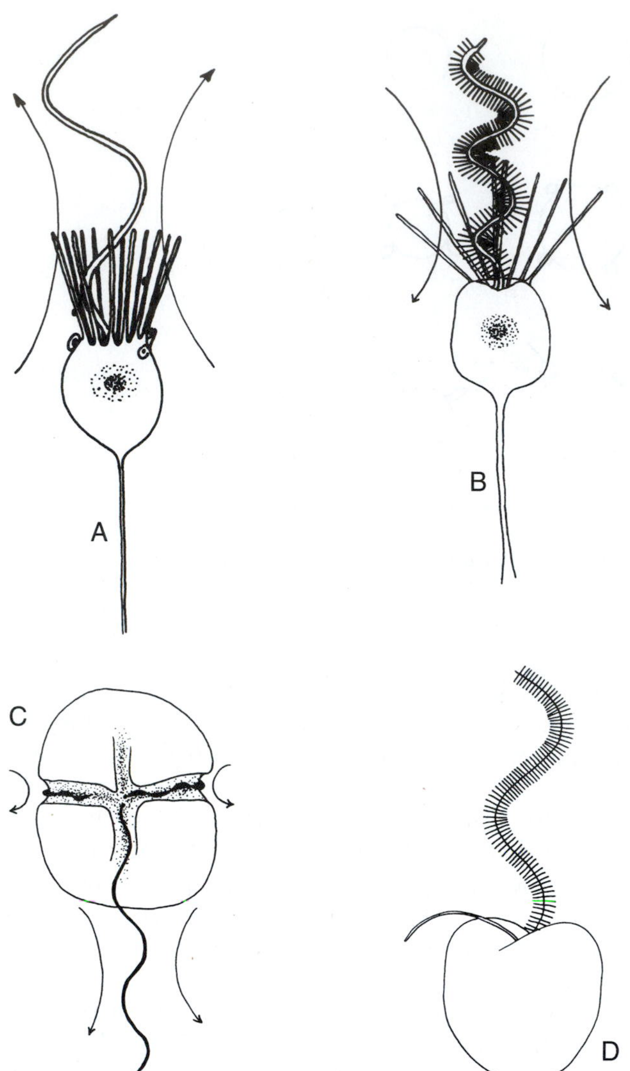

FIGURE 4 Different uses of flagella in two sessile bacterivores, and two swimming forms. Arrows indicate movement of water relative to the cells. Note that distally directed waves of beating produce opposite forces in smooth flagella and flagella with mastigonemes. (A) The choanoflagellate *Monosiga,* modified from Fenchel (1982a); (B) *Actinomonas,* which is probably a heliozoan, also modified from Fenchel; (C) a dinoflagellate; (D) a chrysophyte.

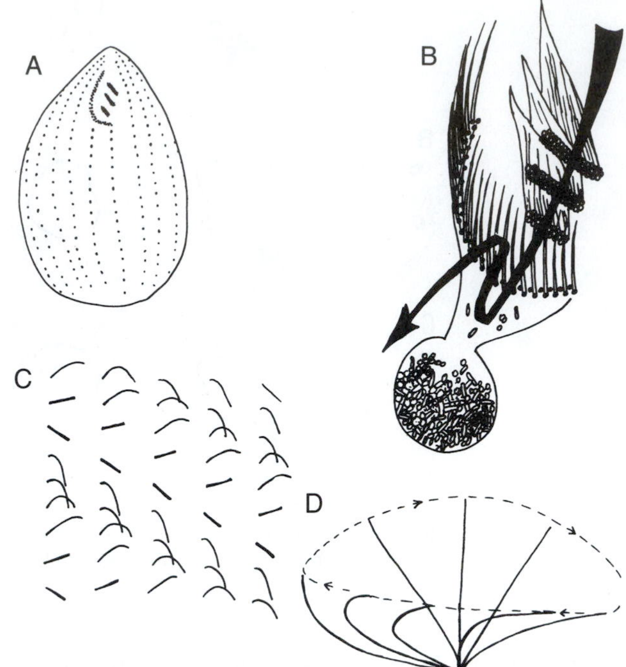

FIGURE 5 The distribution of cilia on *Tetrahymena* (A). The compound ciliary organelles associated with the mouth (B) are used to collect bacteria into the forming food vacuole. Somatic cilia beat metachronously to propel the cell (C). (D) The path of a single cilium viewed from the side, showing the power stroke and the recovery stroke in different planes.

flagellates average about 0.2 mm/s, while ciliates average about 1 mm/s.

4. Extrusomes

Extrusive organelles or extrusomes are membrane-bound organelles associated with the pellicle of protists and containing material which can be ejected or extruded from the cell. Most are visible at the light-microscope level, at least in the extruded state. They are widespread in occurrence, and diverse in structure and function; they are probably not homologous. The function of many is in doubt (Haackbell *et al.*, 1990). However, following Hausman (1978) we will group them according to their functional and structural similarity for discussion, omitting those restricted to marine groups.

Spindle trichocysts of nassophorean ciliates, especially those of *Paramecium*, are well known. They are abundantly distributed over the cortex (e.g., 8000 per cell in *Paramecium*) and each may discharge a pointed projectile on a proteinaceous shaft in response to mechanical or physical stimulation. These are probably effective deterrents to some potential predators (Harumoto, 1994) although specialists, such as the predatory ciliate *Didinium*, are undaunted (Fig. 6). Expulsion of the trichocyst is extremely rapid and is driven by a change in the paracrystalline structure of protein in the shaft of the trichocyst. The increase in length of the shaft following expulsion is approximately eightfold.

Extrusomes are found in several flagellate groups; most dinoflagellates have spindle trichocysts similar to those of ciliates. Until recently, these groups were not considered to be even remotely related. However, recent studies of ribosomal RNA suggest they are less distant than one might suppose based on gross morphology (Gunderson *et al.*, 1987). Ejectisomes are found in most

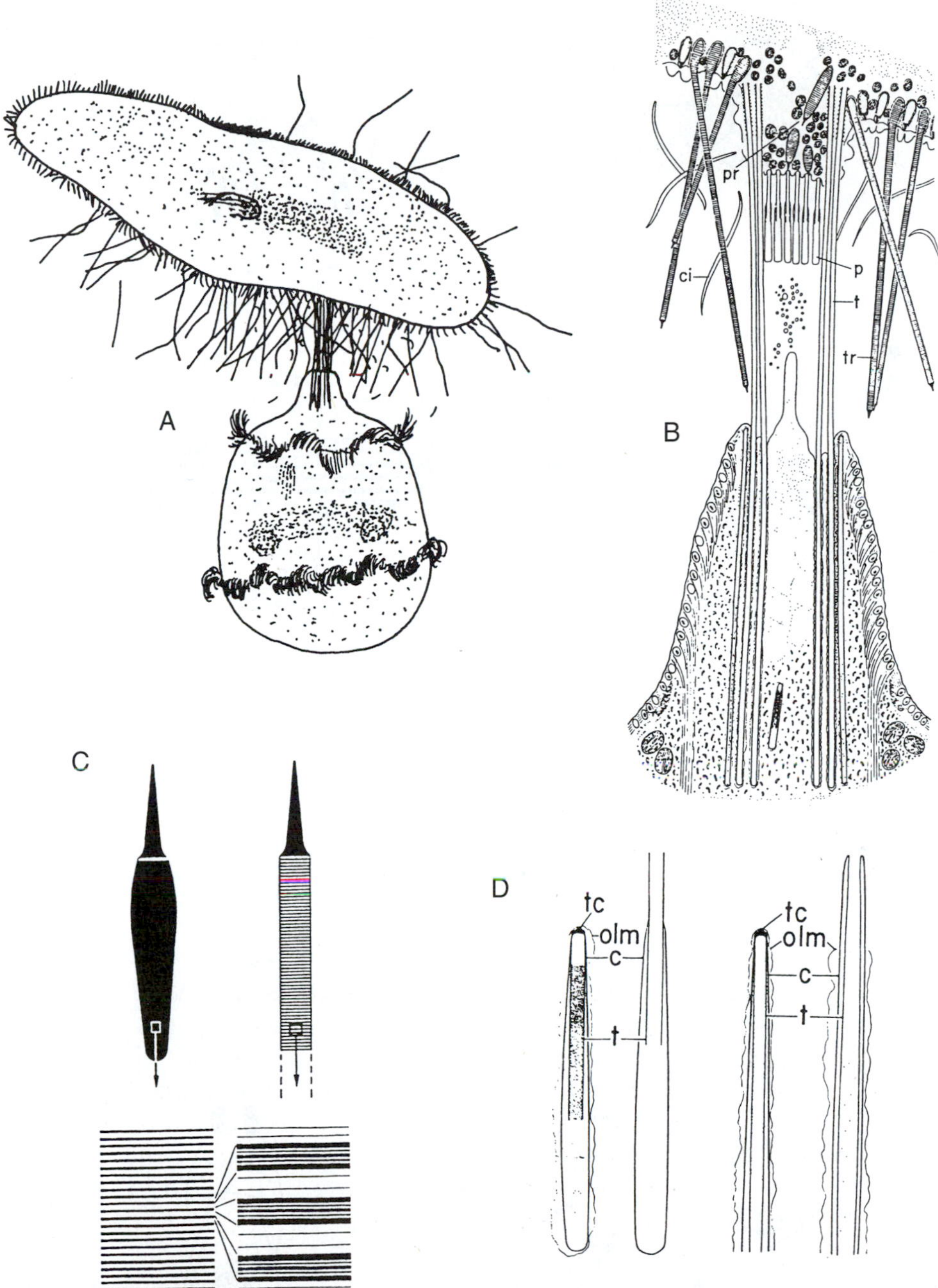

FIGURE 6 (A) Capture of *Paramecium* by *Didinium.* Note the use of toxicysts by *Didinium,* and the spindle trichocysts released by *Paramecium* in response. (B) The oral apparatus and zone of contact, as revealed by electron microscopy. Note that there are two varieties of extrusomes involved in *Didinium's* attack, and that the deciliation of *Paramecium* is part of the result. (C) The ultrastructure of spindle trichocysts of *Paramecium* in the resting (left) and discharged (right) state. (D)Ultrastructure of *Didinium* pexicysts (left) and toxicysts (right). Both are shown in resting and discharged states. c; Capsule. ci; cilia. olm; outer limiting membrane p; discharged pexicysts. pr; protuberance of *Paramecium* cytoplasm. t; discharged toxicyst. tc; terminal cap. tr; discharged trichocyst. (B and D are from Wessenberg and Antipa, 1970; C is from Hausman, 1978.)

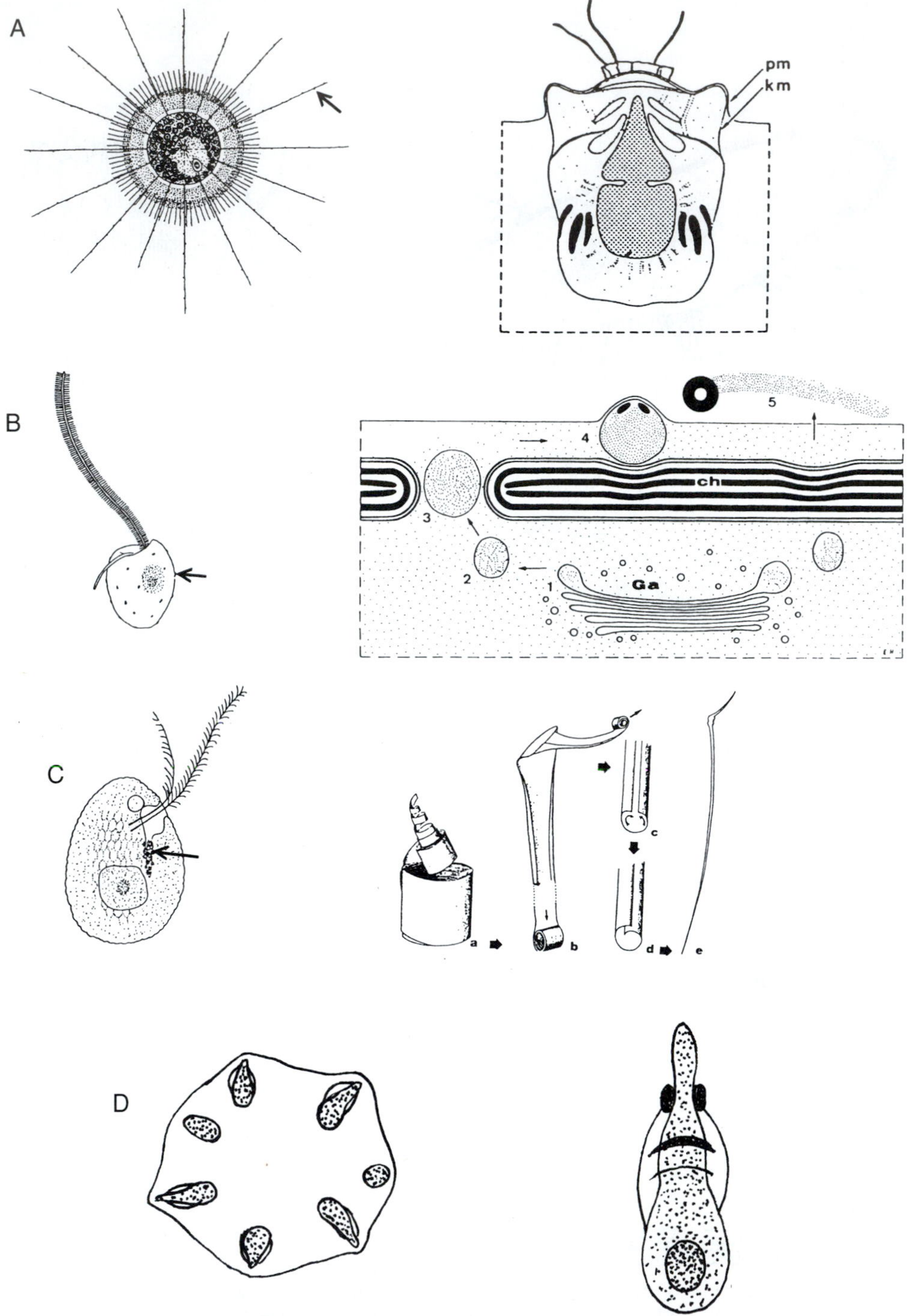

FIGURE 7 Examples of extrusomes in other protozoa. (A) Kinetocysts from the axopodia of the heliozoan *Hetrophrys* (from Lee *et al.*, 1985). (B) Discobolocysts from the cell surface of the chrysophyte *Ochromonas*, and the steps from formation (1) to discharge (5). (C) Ejectisomes from the gullet of *Cryptomonas*. a; undischarged. b-d; discharging, showing uncoiling in two directions and rolling of the uncoiled ejectisome into a tube. e; discharged ejectisome. (D) Haptocysts from the tentacles of a suctorian. Arrows indicate the location of the extrusomes to the right. Extrusomes in A–C are from Hausman (1978). Haptocysts in D are drawn from Rudzinska (1965).

cryptomonads and in a few other phytoflagellates (Fig 7). They are typically dimorphic, although of similar morphology; larger ejectisomes are associated with the gullet, while smaller ones are external to the gullet and near the anterior of the cell. Because some cryptophytes are heterotrophic, it is tempting to speculate that the ejectisomes are used in food capture. However, the gullet is not the site of phagotrophy, so an offensive role of these extrusomes is doubtful. The mechanism of expulsion of ejectisomes is very different from that of spindle trichocysts; the nonexpelled ejectisome is a tight coil of ribbonlike material which uncoils and rolls into a tube on expulsion. A similarly coiled secondary structure forms a pointed tip.

Discobolocysts are also extrusomes of flagellates, in this case primarily of chrysophytes. They are almost spherical, rather than spindle-shaped, but their expulsion is reminiscent of spindle trichocysts; the shaft of the discobolocyst elongates, apparently through a conformational change in the material in the shaft (Fig. 7). Although their function is unknown, it is possible that they discourage phagotrophs.

Extrusomes are used for capturing food by predatory ciliates in the classes Prostomatea and Litostomatea, which have toxicysts, and Phyllopharyngea, which have haptocysts. Toxicysts are extruded by eversion, and in doing so release material which appears to be toxic to the prey. Several types of toxicysts are known, and up to three kinds may be found within a species. The use of toxicysts by *Didinium* in capturing *Paramecium* is illustrated in Figure 6. Although the toxicysts of *Didinium* are concentrated on the proboscis, they may be elsewhere in other ciliates. *Loxophyllum* (Fig. 23I) has them in clusters along its lateral margin, while *Actinobolina* (Fig. 22G) has toxicysts on tentacles that are widely distributed over the body.

Haptocysts are found in the tips of the feeding tentacles of Suctoria (Fig. 19). Prey (mostly other ciliates) which touch the feeding tentacles are caught and held by the extrusion of the haptocysts (Fig. 7) which penetrate the prey, inject substances (probably including hydrolytic enzymes) and aid in the fusion of membranes of the predator and prey. The cytoplasm of the prey is then drawn into the suctorian through the feeding tentacles.

Some Heliozoa possess extrusomes of various types, which may give the elongate pseudopodia (axopodia) a bumpy appearance (Fig. 7). Kinetocysts resemble haptocysts or small trichocysts. Extrusomes in Heliozoa also include mucocysts (see below) and electron-dense bodies that also expel amorphous material onto the cell surface. All probably facilitate prey adhesion and capture.

Mucocysts are widespread in ciliates and flagellates. They secrete an amorphous material onto the cell surface that, at least in many species, is involved in cyst formation. *Tetrahymena* can use mucocysts to create a temporary capsule in response to introduction of a dye; this is probably a defensive reaction, although it is not known which other stimuli cause this response. Wolfe (1988) suggested that mucocysts may be the protozoan's first line of defense. Mucocysts may also be involved in nutrition; the mucous secreted may be reingested after attracting dissolved compounds or flocculating small food items (see Section III.C). Euglenoid flagellates have larger muciferous bodies in addition to mucocysts, although some authors do not differentiate between the two. These provide a continuous coat of mucilage on the surface of the cell and, in some species, participate in cyst formation, adhesion to the substratum, or stalk formation.

C. Anatomy of Flagellates

The term flagellates is applied to protists with one-to-many flagella. In free-living taxa, as opposed to parasitic species, the number of flagella is limited; *Paramastix* has two rows of 8–12 flagella, but most others have 1–4 (usually 2). There are several groups of heterotrophic flagellates in freshwater: choanoflagellates, kinetoplastids, diplomonads, cercomonads, and bicoecids. These are raised to phyla by some authors, while bicoecids are occasionally classified with chrysophytes. Some Rhizopoda, the schizopyrenids or amoeboflagellates, also have flagella.

Other groups of flagellates contain mostly or entirely autotrophic forms with chloroplasts. However, many of the pigmented, autotrophic taxa are capable of phagotrophy, a condition referred to as mixotrophy (Sanders, 1991). Among these autotrophic and mixotrophic groups there are nonpigmented, wholly heterotrophic species. The groups with many mixotrophic or heterotrophic taxa include cryptophytes, chrysophytes, dinoflagellates, and euglenoid flagellates. These groups are usually considered as phyla.

The functional significance of mixotrophy varies widely. Different species use the ability to phagocytose other cells to supplement photosynthesis in low light, to gain access to nutrients, or to survive long periods in the dark (Jones, 1997). In keeping with our desire to cover the animal-like protists, we will survey heterotrophic and mixotrophic flagellates, ignoring primarily autotrophic groups. Pigmentation and chloroplast morphology are important taxonomic characters for some groups, but here we will emphasize features found in heterotrophs.

Choanoflagellates, or collared flagellates, are distinctive for the collar that surrounds the single flagellum (Figs. 4, 14C–H). They bear a strong resemblance to choanocytes of Porifera. The collar may be difficult to distinguish with light microscopy; but when examined with electron microscopy, it is evidently composed of microvilli. The microvilli intercept bacteria drawn past the cell by the flagellum. Because most choanoflagellates are attached or colonial, the distally directed waves of the smooth flagellum serve to push water past the protozoan. Bacteria caught on microvilli are transported to the cell and ingested by pseudopodia at the base of the collar (Leadbeater and Morton, 1974). Many choanoflagellates attach to the substrate, and many have an external, loose-fitting covering or lorica (although, again, it may be difficult to see with the light microscope). In one marine family (Acanthoecidae), the lorica is basketlike.

Bicoecids (Fig. 14I) resemble choanoflagellates, although they lack a collar. Like choanoflagellates, they are enclosed in a lorica and have a flagellum that is used to create a feeding current. A second flagellum lies along the cell and continues posteriorly to become an attachment to the base of the lorica.

Kinetoplastids (Fig. 14J,N–P) are known mostly as parasites, especially *Trypanosoma* and its relatives, but many members of the suborder Bodina live in freshwater (Vickerman, 1976). They have a unique single mitochondrion which runs the length of the cell and one or more DNA-rich organelles, kinetoplasts, associated with this mitochondrion near the flagella. The best known genus is *Bodo*, which, like other bodonids, has two flagella (Fig. 14J). One trails, often in contact with the substrate, while the other extends ahead. The trailing flagellum may attach temporarily to the substrate. In the related, but sessile and colonial, genus *Cephalothamnium* (Fig. 14L), the attachment is permanent.

Diplomonads and cercomonads are relatively unimportant groups to the freshwater ecologist; diplomonads are largely parasitic, with a few free-living forms that may be found in organically enriched water (e.g., *Hexamita*, Fig. 14M). There is one freshwater cercomonad, *Cercomonas* (Fig. 14K), and a common soil flagellate, *Heteromita*. *Cercomonas*, like the diplomonadida, tends to be found in organically enriched environments.

The cryptomonads include many common heterotrophs and autotrophs; mixotrophy has been reported, but is not common. The two flagella are unequal in length and different in appearance (Fig. 7). The longer one has two rows of mastigonemes, while the shorter has only one. These flagella arise from a subapical invagination commonly referred to as a "gullet," although it does not appear to be the site of ingestion in heterotrophic forms. The contractile vacuole empties into this vestibule. Ejectisomes (see above) may not be visible to the light microscopist unless they are discharged, which they sometimes are after fixation. The pellicle is covered with plates, although these also are not generally visible.

The dinoflagellates (Fig. 13A–C) are a very large and unique group; they are probably more important in marine than in freshwater environments. Certainly, our knowledge of this group is mostly based on marine studies. Their unique arrangement of flagella—one spirals around the cell in a groove (girdle) and the second is distally directed in another groove (sulcus)—makes them distinctive. Again, heterotrophy and mixotrophy are common (Fig. 2). Because dinoflagellates sometimes ingest large prey, their phagotrophic habits have been more often recognized. A covering of plates may or may not be present (hence "armored" and "naked" dinoflagellates). Ingestion in some species is assisted by an appendage, the peduncle, which emerges from the sulcus and allows them to subdue prey many times their size (Gaines and Elbrächter, 1987; Jacobson and Anderson, 1986). The trichocysts also assist in prey capture in some forms. Very large prey may be digested by protoplasmic extensions, and this tendency intergrades with ectoparasitism, which is common in this group. Marine and brackish species are noted for the production of toxins.

Chrysophytes also contain both colorless heterotrophs and pigmented mixotrophs. Even among the mixotrophs, the degree to which different taxa rely on phagotrophy varies greatly. Chrysophytes are generally small, and their prey are bacteria. They have two unequal flagella: one long and directed anteriorly with rows of mastigonemes; and the other short, smooth, and directed laterally (Fig. 4D, Fig. 13K–M). They are naked or covered with fine siliceous scales which are not always visible with light microscopy; many are amoeboid. Their carbohydrate storage product, chrysolaminarin, occurs in liquid globules and may be useful in recognizing members of this group.

Euglenids are generally large flagellates with two flagella, although in many taxa only one flagellum emerges from the gullet (Fig. 13D). Several heterotrophic species creep over the substrate with the second flagellum trailing and hidden beneath the cell (Fig. 15F–H) as in some bodonids and *Cercomonas*. Indeed, molecular genetic studies indicate that the euglenids may be related to the kinetoplastids, and both may be rather remote from other flagellates (Gunderson *et al.*, 1987). Heterotrophic species may have an ingestatory apparatus consisting of two rods, allowing them to swallow relatively large items. Again, the gullet is not used in ingestion.

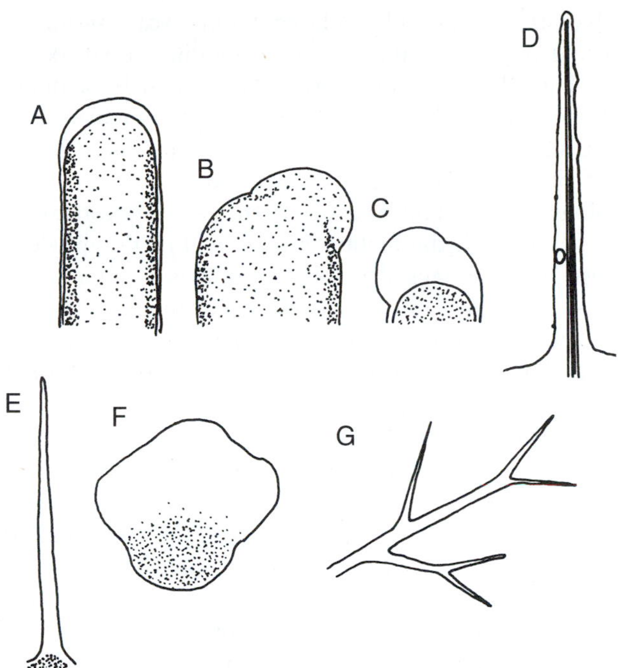

FIGURE 8 Types of pseudopodia. (A) The cylindrical pseudopodia of Amoebidae. (B) The cylindrical, granular, eruptive pseudopodia of Pelomyxidae. (C) The flattened, hyaline, eruptive pseudopodia of some Entamoebidae. (D) The axopodia of Heliozoa. (E) The filiform pseudopodia of some Mayorellidae. (F) The broad, flattened pharopodia of some Mayorellidae. (G) The filopodia of athalamids. (Modified from Bovee and Jahn, 1973.)

D. Anatomy of Amoeboid Protozoa

These are the protozoa we generally refer to as amoebas, in reference to the well-known *Amoeba proteus*. Locomotion is mostly by pseudopodia. Flagella are restricted to temporary swimming stages in the order Schizopyrenida. Rhizopoda and Karyoblastea mostly have blunt pseudopodia called lobopodia, where as the Filosea and Granuloreticulosida have pseudopodia ending in fine points (Fig. 8). Amoebas range in size from only a few micrometers to 2 mm in diameter. Although many lack a fixed external morphology, the characteristic morphologies shown by the various taxa are surprisingly distinctive even if difficult to quantify (Fig. 15). By using also the number, size, and structure of organelles and characteristics of tests (where present), identification is not as difficult for living specimens as might be imagined.

The morphology of amoebas is plastic. Many adopt a stellate morphology (Fig. 9) if suspended in water. Few are truly planktonic, but rather they live on surfaces or in sediments. Most show an inner granular cytoplasm and an outer hyaline cytoplasm, or hyaloplasm, with a characteristic thickness and distribution around the cell. Locomotion may be achieved by extending many pseudopodia simultaneously, as in *Amoeba* (Fig. 15F), or by moving as a single mass on a broad front (e.g., Figs. 8F, 15E,P), or as a cylinder (limax amoebas, Figs. 8A, 15I,K,L). Not only do pseudopodia have characteristic shapes, but the tail end or a uroid may be distinctive (Fig. 15J,L), and the cell surface may be distinctly sculptured, as in *Thecamoeba* (Fig. 15D). The dynamics of movement also varies; some amoebas move continuously and smoothly, while the cytoplasm of others seems to burst forward intermittently; this motion is called eruptive.

Members of the subclass Testacealobosia are distinguished by their external coverings, called tests (Fig. 16A–Q). These are generally vase-shaped, with a single opening through which pseudopodia emerge. The test may be proteinaceous, but is often covered with secreted plates or inorganic particles. Many are terrestrial, but benthic forms are common and a few are planktonic. Planktonic species have oil droplets or gas-filled vacuoles to compensate for the weight of the test (Meisterfeld, 1991). Others use the gas vacuoles, which can appear in a matter of minutes, to leave the substrate and rise to the surface film where pseudopodia are extended at the air-water interface (Ogden, 1991).

Schizopyrenida are small amoebas, many of which are capable of transforming into flagellates (Fig. 9C).

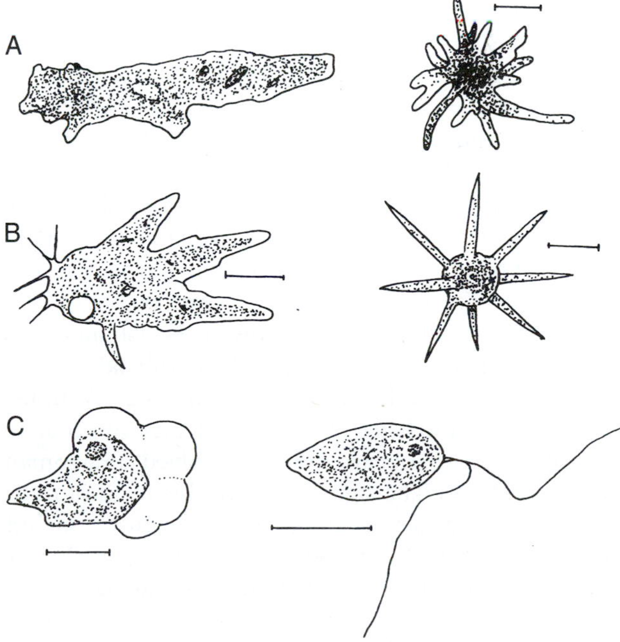

FIGURE 9 Some amoebas and their "floating forms" on the right. (A) *Ameoba proteus*, scale = 100 μm. (B) *Polychaos timidum*, scale = 20 μm. (C) *Naegleria gruberi*, scale = 10 μm. (After Page, 1976.)

The flagellated form may have two or four flagella. As amoebas, they are small, uninucleate, and usually simple in form. The amoeboid form may be a soil-dweller, transforming into a flagellate on being suspended in water. The best known amoeboflagellate of this kind is *Naegleria*. *Naegleria gruberi* was considered by Page (1976) as possibly the most common amoeba in freshwater. At one time, *Naegleria* was best known as a model system for microtubule assembly, since it could be induced to assemble its flagellum in the laboratory. The similar *Naegleria fowleri* was discovered as a cause of fatal encephalitis in humans in 1958, and the genus is now better known in infamy as an unwanted resident of air conditioners and swimming pools. Pathogenic strains are able to survive at 37°C, which, fortunately, most free-living amoebas cannot. Increased abundance of pathogenic *N. fowleri* is a dangerous consequence of thermal pollution (Tyndal *et al.*, 1989). The interesting history of this discovery, as well as other aspects of this group, can be found in Carter (1968) and Singh (1975).

Filosea and Granuloreticulosida are amoebas with fine, pointed pseudopodia. Filosea (Figs. 16S–U,W, 19A–K) have single or branched filopodia with fine points (Fig. 8G), while Granuloreticulosida have a network of branching and anastomosing reticulopodia (Fig. 17L–R). Most members of these two classes have tests. The freshwater granuloreticulosids are among the least studied protozoa, although their marine relatives are very well known.

Naked amoebas feed by phagocytosis, the smaller being bacterivores and the larger being predators of algae, other protozoa, and small metazoa. Particulate food is surrounded by ectoplasm, sometimes engulfed by a pseudopod, and enveloped in a food vacuole. In contrast to most other phagotrophic protozoa, there is no fixed site of ingestion or cytostome. Testacealobosia are largely algivores, and many feed on filamentous algae, drawing in filaments through the opening in their test.

Pinocytosis or "cell drinking" can be induced in *Amoeba* by dissolved organic substances. Channels appear in the ectoplasm through which substances concentrated on the cell surface are imbibed. Although *Amoeba* does not have mucocysts, a mucous coat on the cell surface appears to be involved (Chapman-Andresen, 1973). The significance of this form of nutrition in the field is unknown, but one must conclude that *Amoeba* uses this ability somewhere other than on microscope slides in undergraduate laboratories!

E. Anatomy of Heliozoa

The Heliozoa are primarily freshwater counterparts of the better known marine Radiolaria (classes Polycystinea and Phaeodarea) and Acantharia. All these groups have unique pseudopodia, called axopodia (Fig. 8D). Axopodia are strengthened by a microtubular array called an axoneme or stereoplasm. [The term axoneme is also used to describe the microtubular core of cilia and flagella, but this does not imply homology.] The origin and ultrastructure of axonemes is diverse; indeed, the Heliozoa may well be polyphyletic. Most Heliozoa lack the skeleton that is so characteristic of Radiolaria and Acantharia, although some are covered in siliceous or organic scales (Fig. 18H,F), and some have a perforated shell or capsule (order Desmothoracida, Fig. 18A).

As diverse as the evolutionary origin of the stereoplasm may be, the axopodia supported by the microtubles are similar in function. Their surface is sticky, possibly because of extrusomes on the axopodia or near their bases (Fig. 7). The outside cytoplasmic layer of the axopods appears to be in constant motion, streaming out and returning on opposite sides of each axopod. Food items adhere and are transported with the flow of cytoplasm until they reach the surface of the cell body and are engulfed. In this way, they are functionally analogous to the microvilli of choanoflagellates. Food items may range from picoplankton to mesozooplankton in different Heliozoa. Large prey items, such as rotifers and copepods in the case of *Actinosphaerium*, may be entangled in several axopodia and engulfed by pseudopods. Several individuals may participate in the capture of one prey.

Although Heliozoa are frequently planktonic, they are found primarily on or near the benthos. Some Heliozoa traverse the bottom with a unique tumbling motion resulting from controlled changes in the length of the axopods. Many sessile forms with stalks are known. In sessile forms, cell division is likely to be unequal, producing a dispersal stage that may be flagellated or amoeboid.

F. Anatomy of Ciliates

Unlike many of the other protozoan groups, the phylum Ciliophora is certainly monophyletic; despite their diversity, they share many features. After discussing these common characteristics, we will consider the important differences among subphyla and classes, emphasizing organelles involved in locomotion and feeding.

The presence of cilia is an obvious and distinctive feature. The cilia may be reduced in number, especially in sessile forms, or organized into larger compound ciliary organelles, such as cirri. The only large group that does not always possess cilia is the Suctoria, sessile predators whose dispersal stages are, however, ciliated.

This distinctive group is easily recognized by its feeding tentacles. The novice should take care not to confuse small, ciliated animals with ciliates; the size range of ciliates overlaps that of several metazoan groups, such as turbellarians, rotifers, and gastrotrichs.

The cortex of Ciliophora is a wonderful example of the complex cell ultrastructure to be found among protozoa. We refer the reader to Lynn (1981) and Lynn and Corliss (1990). This ultrastructure is important in the definition of higher taxa, while the arrangement of both somatic cilia and the compound ciliary organelles, including those associated with the cytostome, are important at all taxonomic levels.

The other distinctive characteristic of ciliates is their unique nuclear organization or nuclear dualism. One or more large amitotic, highly polyploid macronuclei are functional in RNA synthesis, while one or more smaller, diploid micronuclei function as gametic nuclei during conjugation (see Section III.B).

The ciliates are divisible into ten classes (Small and Lynn, 1985). Class Karyorelictea is thought primitive for the group, with numerous macronuclei that are not highly polyploid. They are largely benthic. The freshwater genus *Loxodes* (Fig. 23J) is perhaps better known in its environmental physiology and field ecology than any other heterotrophic protozoan (reviewed in Goulder, 1980; Finlay 1990). Compound ciliary organelles associated with the cytostome are prominent in the classes Heterotrichea and Spirotrichea. Large heterotrichs, such as *Stentor* and *Spirostomum* (Fig. 22F,A) are familiar as teaching material. Spirotrichs are abundant in many freshwater habitats, from plankton (choreotrichs and oligotrichs, Fig. 21S–W) to anaerobic benthos (odontostomes and armophorids, Fig. 21J–N,R). Most consume algae and other nanoplankton-sized food particles. Stichotrichs and hypotrichs (Fig. 21A–H,N–Q, 20X–Z) are mostly dorsoventrally flattened crawlers, whose somatic cilia are all compound cirri.

Like spirotrichs and heterotrichs, Nassophorea tend to be algivores except for one small and poorly known group, the microthoracids. The Nassophorea are named for their basket-like "nasse" or cyrtos supporting the cytopharynx (Fig. 12).

Classes Prostomatea and Litostomatea are largely predators, often of other ciliates(Fig. 22G-W, 23A-I). Prostomes generally have apical cytostomes, while many litostomes have subapical, sometimes slitlike cytostomes. The mouth is encircled by a crown of cilia from whose bases (kinetosomes) arise the "rhabdos," a cylinder of microtubules surrounding, and supporting the cytopharynx. Toxicysts are found in most species and are used to subdue active prey. Toxicysts may be found around the cytostome, on a proboscis, on tentacles, or elsewhere on the body. A number of short, specialized kineties (rows of kinetosomes) are often found near the anterior. This "brosse" (brush) probably assists in prey recognition.

Class Phyllopharyngea contains the distinctive Suctoria (Fig. 19), sessile or free-floating predators of other ciliates. Suctoria are unusual in that most have several feeding tentacles rather than a single mouth. Prey ciliates stick to these tentacles because of the firing of haptocysts (see extrusomes, above), and their cytoplasm is withdrawn through the tentacles. Suctoria reproduce by unequal binary fission (budding), which yields a ciliated dispersal stage or "swarmer." The other subclass, the Phyllopharyngia, contains surface-associated algivores (e.g., *Chilodonella*, Fig. 24T), plus a diverse array of marine epizooic forms.

Colpodea (Fig. 24K,L,N,P–S) are not as frequent in freshwater environments, most being terrestrial bacterivores. They are more likely to be encountered in small, temporary waters, and can be readily obtained by incubating terrestrial vegetation or soil in water. Plagiopylea (Fig. 24M) were formerly placed in the Colpodea, and they do resemble colpodids in form. Unlike colpodids, they are strict anaerobes.

The Oligohymenophorea are mostly microphagous. These are named for the compound ciliary organelles that are found in a buccal cavity surrounding the cytostome. The most common pattern (subclasses Hymenostomatia, Scuticociliatia, Peniculia; Figs. 24C,H, 23L–X, 24A–E,G,I,J) comprises three polykinetids on the left side of the buccal cavity and an undulating membrane on the right. The net result is three brushes, the polykinetids, working against a curved wall, the undulating membrane, to deliver small particles to the cytostome. Most are effective bacterivores, although some are histophagous (consuming tissue of injured or recently killed animals), parasitic, or algivorous. The large subclass Peritrichia (Fig. 20A–U) contains sessile bacterivores in which the buccal cavity is deepened as an infundibulum, and the polykinetids wind down it to the cytostome after encircling a prominent peristome. Somatic ciliature is absent in most species. Many are attached to the substrate by a contractile stalk, as in the common *Vorticella*. Many are colonial, and a few secondarily free-swimming.

G. Environmental Physiology

1. Factors Affecting the Distribution of Protozoa

Protozoa are found over an extremely broad range of environmental conditions, but individual species, especially those adapted to unusual circumstances, may have relatively narrow limits to their distribution. Several microbial communities with characteristic fauna have been described. Biological, physical, and chemical factors must interact to govern the distribution of these assemblages in a manner that defies simple description. Nonetheless, several physical–chemical factors are important to the growth and/or distribution of protozoa. Among these temperature and oxygen are probably the most important and best known. Light is obviously important to autotrophic forms, and the negative influence of UV light on protozoa is probably an important factor that has received too little attention (Sommaruga *et al.*, 1996).

An extensive and still rapidly expanding literature concerns the importance of environmental contaminants. The distribution of protozoa with respect to organic pollution and its associated environmental conditions has been used as part of a system for classifying the state of aquatic habitats according to their fauna (Sládecek, 1973). Foissner (1988) and Foissner and Berger (1996) have clarified the taxonomy of ciliates used as indicators and provided a summary of the "saprobity values" of ciliates. Similarly, Bick (1972) provided a guide to ciliates useful as biological indicators in freshwater, along with their ecology and distribution with respect to several parameters.

A list of species occurrences versus environmental parameters was assembled by Noland (1925). His conclusion that "the wide range of tolerance that most ciliates show toward the physico-chemical factors studied, together with the evident correlation of several of these factors with the food habits of the species, suggests that the nature and amount of available food has more to do with distribution of the freshwater ciliates than any other one factor" still appears to be warranted. The physical nature of the substrate, including the biological substrate of periphyton, is also important to the community of surface-oriented protozoans that develops on it (Picken, 1937; Ricci, 1989). Indeed, the relative importance of biotic components of the environment to the distribution of species versus the structural and chemical–physical components, is difficult to ascertain from survey data.

2. Temperature

As in all organisms, metabolic processes and growth in protozoa are dependent on temperature. Because of the difficulty of measuring growth rates *in situ* for protozoa, several workers have investigated the relationship of temperature and growth rate in the laboratory to facilitate the extrapolation of laboratory data on growth rate to field situations (Finlay, 1977; Baldock *et al.*, 1980). Generally, Q_{10} values for protozoa are about 2.0, as they are for many biological processes, but important deviations occur (Baldock and Berger, 1984; Müller and Geller, 1993).

Active protozoa are found in freshwater environments at temperatures between 0 and 50°C. For example, protozoa contribute to the plankton community of almost permanently ice-covered lakes in Antarctica (Laybourn-Parry *et al.*, 1995). Some protozoa can be cultured in the laboratory at or near 0°C, but it is doubtful whether all protozoa can function at such low temperatures, or whether those that can perform any better than one would expect from extrapolating the metabolic rate versus temperature curve for species collected from more moderate temperatures to 0°C. Lee and Fenchel (1972) found that *Euplotes* collected from Antarctic, temperate, and subtropical waters had similar growth–temperature relationships; but the Antarctic strains did not multiply above about 12°C, while the subtropical strain multiplied even above 35°C, but could not reproduce below 6°C.

A similar situation is to be expected with respect to high temperatures; most species continue to increase their rate of cell division as temperature increases to 25°C or more and tolerate at least 30°C. But there are numerous records of species from hot environments, up to 50°C (Noland and Goldjics, 1967). Tolerance to extremes of temperature and some other variables have been induced by gradual acclimation (see Fauré-Fremiet, 1967).

3. Oxygen

Protozoa are sufficiently small that they do not need organelles to increase the surface area for respiration. Aerobic forms appear to function at extremely low oxygen tensions and, thereby, avoid competition and predation from larger animals. It is in these circumstances, especially when they are generated by high densities of bacteria in response to organic enrichment, that protozoa are most abundant. Aerobic protozoa that are typically found under such conditions are called microaerophilic. The succession of species following organic enrichments and the resulting changes in oxygen tension were a focus for some of the earliest research in protozoan ecology (Noland and Goldjics, 1967).

A great deal of research has been done on the autecology of *Loxodes*, a microaerophilic ciliate living in the benthos and plankton of hypereutrophic waters where it consumes nanoplanktonic algae (Fenchel and Finlay, 1991). One of the most surprising outcomes of this research is the discovery that *Loxodes* can use nitrate rather than oxygen as an electron acceptor. It undergoes diel vertical migrations between the anoxic hypolimnion and oxygenated epilimnion, but cannot survive in either if constrained. It aggregates at low O_2 concentrations guided, in part, by unique organelles (Müller's vesicles) now known to be responsible for geotaxis. The direction of the geotactic response is determined by the oxygen tension, which is probably detected by cytochromes, and by light, which is detected by pigment granules (pigmentocysts) in the pellicle. Although other ciliates have been intensively studied in the laboratory, the work on *Loxodes* is unusual in the degree to which laboratory studies have been guided by field observations. This suggests that autecological work on other species could be very rewarding.

Some microaerophilic ciliates contain endosymbiotic green algae (see Section III). Despite their photosynthetic symbionts, they aggregate at depths where O_2 and light are low (Finlay *et al.*, 1987). At very low O_2 tensions, they move into dim light suggesting a role for the symbionts not only as producers of carbon for their hosts but as a means for maintaining a low intracellular level of oxygen.

Many free-living protozoa are tolerant of anaerobic conditions (Fenchel and Finlay, 1991). Among the ciliates, this diverse assemblage of species have been referred to as "sapropelobionts" (e.g., Jankowski, 1964a,b), although some of the species embraced by this term are microaerophilic rather than anaerobic. Fenchel (1969) suggested that the term "sulfide ciliates" is more appropriate for the truly anaerobic ciliates of reducing sediments. These are intolerant of oxygen and lack mitochondria and cytochrome oxidase. Instead of mitochondria, some anaerobic protozoa have hydrogenosomes, organelles that resemble mitochondria, but reduce pyruvate and generate hydrogen. Epi- and endozooic bacteria are commonly associated with anaerobic protozoa, and methanogenic endosymbionts, in particular, may be closely associated with the hydrogenosomes (Fenchel and Finlay, 1991). Ecological studies of freshwater species have appeared in recent years (Wagener *et al.*, 1990, Massana and Pedrós-Alió, 1994; Guhl *et al.*, 1996).

Other protozoa which can probably exist under anaerobic conditions include the giant amoeba *Pelomyxa palustris*, which has been said to consume at least some oxygen (Bovee and Jahn, 1973) although it has no mitochondria (Daniels, 1973). Oxygen consumption in some anaerobes may be a protective mechanism (Fenchel and Finlay, 1991). The diplomonads, which have two free-living genera (*Hexamita* and *Trepomonas*) also lack mitichondria. Flagellated protozoa in the anaerobic hypolimnia of eutrophic lakes may reach abundances as great or greater than those in the surface waters (Bennett *et al.*, 1990). Many true anaerobes among the protozoa belong to taxa that include gut-dwelling endosymbionts.

4. Environmental Contaminants

The relative abundance and diversity of protozoa has also been proposed as a useful indicator of organic and toxic pollution (Cairns *et al.*, 1972), and the response of protozoa to contaminants has been reviewed in that light (Cairns, 1974). Others have advocated the use of protozoa in single-species bioassays (Nilsson, 1989). Relatively quick respiration bioassays may reveal both decreases and transient increases in respiration (Slabbert and Morgan, 1982). Because tests with protozoa are relatively rapid, they are particularly useful for studies of quantitative structure–activity relationships (QSAR) where the toxicity of many compounds must be compared. The response of *Tetrahymena* to a wide variety of compounds is similar to that of a cyprinid, the fathead minnow (Bearden and Schultz, 1997). Protozoan predators and prey have been used to explore complex effects of contaminants on communities (Doucet and Maly, 1990, Fernandez-Leborans and Novillo-Vajos, 1993).

Among environmental contaminants, the response of protozoa to oil and other hydrocarbons is relatively well known. They are generally resistant to hydrocarbons (Rogerson and Berger, 1981); the ciliate *Colpidium colpoda* accumulated hydrocarbons in intracellular inclusions, up to about 20% of the cell volume. The number of mucocysts increased, implicating these organelles as having a protective function (see Section II.B.4). Scott *et al.* (1984) found that freshwater ponds treated with oil alone or oil and oil-dispersant mixtures had numbers and biomass of protozoa similar to those in control ponds, although some taxa were reduced while others increased.

The toxicity of various hydrocarbons to *Tetrahymena* was similar when differences in solubility and partitioning between the water and ciliates were considered, indicating a similar mode of action once they were in the cell (Rogerson *et al.*, 1983). Biomagnification of chlorinated hydrocarbons may be substantial. For the pesticide Mirex, the concentration in *Tetrahymena* was 193 times that present in the medium, and 82% of the available contaminant was in the cells (Cooley *et al.*, 1972). For Arochlor™ 1254, the biomagnification was 60-fold. A wide variety of insecticides have been shown to cause dose-dependent inhibition of growth and biomagnification in *Tetrahymena* (Dhanaraj *et al.*, 1989; Lal *et al.*, 1987). Some were toxic at higher concentrations.

The toxicity of heavy metals to protozoa also has been explored in some detail (Ruthven and Cairns, 1973; Doucet and Maly, 1990; Fernandez-Leborans and Novillo-Vajos, 1993). Intraspecific variation in heavy metal tolerance appears to be related to environmental variation in contamination (Nyberg and Bishop, 1983). To some degree, tolerance to metals is specific, but is also generally greater in outbreeding species (Nyberg, 1974; see Section III.B for discussion of breeding systems). Free-ion concentration, rather than the total amount of metal, appears to determine toxicity (Stoecker *et al.*, 1986).

5. Energetics of Protozoa

Because protozoa are fast-growing and easily cultivated in the laboratory, their energetics have been studied extensively. Ingestion, growth, respiration, egestion, and other parameters have been measured or estimated, and quotients cataloged. We deal with some of these parameters elsewhere; for example, maximum growth rates are discussed in Section II.B, some field estimates of growth rate are mentioned in Section III.B and egestion in Section III.C.10). Laybourn-Parry (1984) summarized much of the existing information on protozoan energetics, and Calow (1977) integrated it into his wider review of animal energetics.

Fenchel's (1974) classic paper drew attention to the relationship between body size and rate of population increase for organisms, and the more rapid growth rate of the smaller as compared to the larger ones. His analysis also suggested a cost associated with multicellularity; both net growth efficiency (growth/assimilation) and growth are generally larger for protozoa than for metazoa of similar size. Calow's summary of conversion efficiencies confirms that protozoa are generally efficient converters of energy, with many species showing growth efficiencies of 50% or more. However, efficiencies for some taxa are much lower (e.g., Scott, 1985, and references therein) and it is difficult to generalize. Given the high growth rates of protozoa, we expect that field populations of protozoa will be found to have high production-to-biomass (*P*/*B*) ratios relative to metazoa, on the order of one per day. However, few field estimates of growth rate are available. Field estimates of growth rates include Schönborn (1977) and Taylor (1983b) for benthic populations, and Taylor and Johannsson (1991) and Carrick *et al.* (1992) for planktonic populations. It appears that planktonic ciliates undergo periods of consumer and resource control, and these may be different for different species (Havens and Beaver, 1997; Wang and Heath, 1997).

III. ECOLOGY AND EVOLUTION

A. Diversity and Distribution

The concept of species within the protozoa has always been problematic. The definition of species as interbreeding or potentially interbreeding populations is irrelevant to the many asexual forms. On the other hand, studies of the genetics of what were once believed to be species, such as *Paramecium aurelia*, *Tetrahymena pyriformis*, and *Euplotes patella*, have revealed groups of interfertile mating types (Corliss and Daggett, 1983). These groups of mating types are called syngens by protozoologists, but are essentially sibling species. New species names have been created to recognize them in some cases (e.g., *P. monaurelia*, *P. biaurelia*, and so on). Careful morphological analysis has uncovered differences among some of these species, but in practice, they are not morphologically separable. They must be identified by mating reactions with known lines, or by using electrophoretic differences in some cytoplasmic enzymes. At the same time, the phenotypic plasticity of protozoa has also led to the description of invalid species names (Gates, 1978). Finlay *et al.* (1996) provided a carefully reasoned analysis of this problem, and concluded that "the biological species concept is neither appropriate nor practicable" for ciliated protozoa. They concluded that we must instead be pragmatic and consider morphological species. The same would be more true of phyla where sex is less common or absent. It remains that some protozoan groups have as many species as the more familiar animal taxa. For example, ciliates comprise some 7000 species (Corliss, 1979), with about 3000 known free-living species (Finlay *et al.*, 1996), of the tens of thousands of protozoan species.

At this point it should be admitted that the taxonomy of free-living protozoa, at least as practiced by ecologists, is in a primitive state. A textbook on protozoa will advise the student to collect protozoa at the margin of a pond or in some other small and protected waters rich in organic material. Collections from these environments are likely to yield rich assortment of large protozoa, readily identifiable at least to genus under the dissecting or compound phase microscope. The working freshwater ecologist will more likely be interested in lake plankton, littoral or profundal benthos, or stream aufwuchs. Samples from these situations will more likely yield minute forms too small to be seen under a dissecting microscope, and too few to be examined under a compound microscope without concentration. They may not be figured in the textbooks. Many will be undescribed species, or species descriptions will be available only in specialist journals, not to be found in most libraries, or so outdated as to be of dubious value. Largely because of these problems, most ecologists who include protozoa in their studies of aquatic habitats do not identify protozoa, even if they do count them and measure them for biomass estimates. Techniques of sample preservation, concentration, and enumeration (see Section IV), if they exist for a particular biotope, may not allow for identification. Molecular techniques such as species-specific oligonucleotide probes have been developed and may prove to be useful tools for identification of particular protozoan groups in natural samples (Lim, 1996).

The distribution of most free-living protozoa is probably broad. Certainly, some morphological species are worldwide in distribution and are essentially ubiquitous. Experiments using enriched and incubated field samples led Fenchel *et al.* (1997) to conclude that although local diversity of protozoa may be high, global diversity may not be; morphological species are usually present in appropriate habitats, even if not detected by initial sampling. Protozoa quickly colonize new environments, as was shown by Maguire's (1977) study of new volcanic islands. On the other hand, sibling species of the *Paramecium aurelia* complex have limited distributions on a world wide scale (Sonneborn, 1975) even though the species complex is ubiquitous.

B. Reproduction and Life History

Most protozoa reproduce by equal binary fission, and therefore their life histories are simple. Multiple fission is more common among endosymbionts but does also occur in free-living forms. Among free-living protozoa, multiple fission appears to be related to intense temporal heterogeneity in the environment. For example, it is found in some terrestrial forms which are active only for brief periods when their environment is moist, or in histophages whose food resources are very patchy and ephemeral (Fig. 10). Therefore, some analogy can be drawn between the significance of multiple fission in certain free-living protozoa and some endosymbionts using the common theme of dispersal (Taylor, 1981).

A phenomenon perhaps related to multiple fission, in that it is related to dispersal, is unequal fission or budding. Binary fission is only approximately equal, and differences in the partitioning of cytoplasm occur and are adjusted in the next cell cycle. But a consistent asymmetry of daughter cells is widely distributed among sessile forms; the larger daughter cell remains attached, while the smaller is motile. This pattern undoubtedly has evolved independently a number of times. It is perhaps a reflection of the probability of survival of the two daughter cells not being equal, and resources being partitioned according to the odds. The smaller daughter cell is usually a propagule, capable of establishing itself in a new location. In some cases, (e.g., *Zoothamnium*, Fig. 20H,I), it may function as a gamete.

Generation times are highly variable among protozoa, but the maximum rates of population increase attainable when food is not a constraint generally increase with temperature and decrease with body size (see Section II.5). Multiple regressions of growth rate on cell size and temperature are sometimes used to estimate production despite shortcomings (Müller and Geller, 1993). However, even for protozoa of the same size and growing at the same temperature, maximum rates of cell division vary widely among taxa. Some of this variation is correlated with genome size and may reflect a causative relationship. Among most eukaryotic cells with diploid, mitotic nuclei, there is a positive relationship between DNA content and cell size, and a negative relationship between DNA content and division rate. It appears that, among such cells, a large cell must have a proportionately large DNA content and nucleus, and a slow division rate. This relationship holds among most plant and animal cells and among most protozoa. However, it has also been established in plants that polyploidy allows a large DNA content and large cell size, but without a concomitant decrease in division rate. Division rate is apparently related to the size of the genome, not the number of copies.

The diverse nuclear arrangements of protozoa can be viewed as various solutions to the genome size/cell

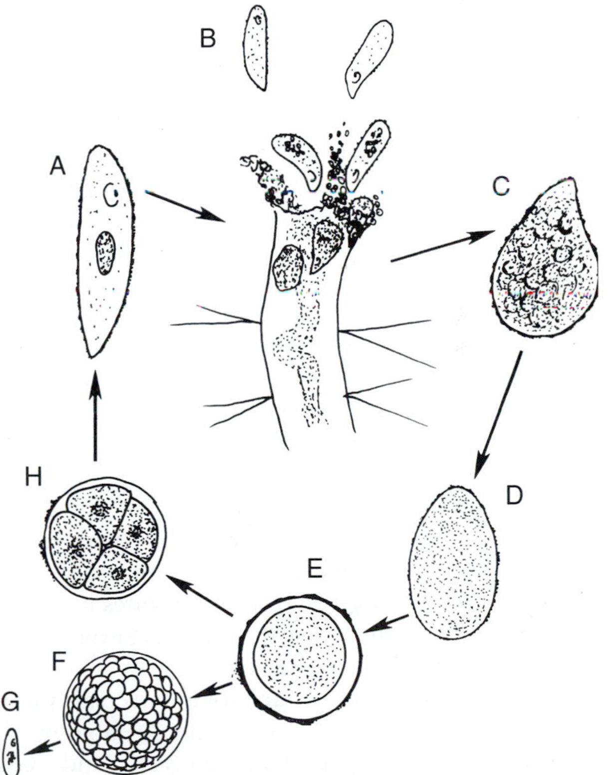

FIGURE 10 Life cycle of the histophagous ciliate *Ophryoglena*. Change from the theront (hunting morphology) (A) to the trophont (feeding) morphology (B,C) occurs when food is found. The protomont stage (D) seeks a suitable place to settle, and leads to the cyst or tomont stage (E,H). The tomont usually produces 4–8 new theronts, or may produce a much larger number of microtheronts (F,G). The function of microtheronts is unknown. Theronts may encyst if they do not find food, and re-emerge as small secondary theronts. (Modified from Canella and Rocchi-Canella, 1976.)

size/fission rate constraint (Cavalier-Smith, 1980). Multinucleate cells, nuclear dualism, amitotic nuclei, and colony formation may represent ways to get larger without slowing growth. For example, in ciliates, cell size is correlated with macronuclear DNA content, but cells divide as frequently as one would expect from the size of their micronucleus (Shuter *et al.*, 1983). In this context, both multinuclearity and multicellularity can be seen as means of avoiding the constraints imposed by the basic eukaryotic plan.

Encystment is a feature of many protozoan life cycles, especially those that live in environments that dry out (Corliss and Esser, 1974). Many protozoa also produce cysts at times when environmental conditions are unsuitable for other reasons, such as lack of food. Encystment under these conditions can be considered as a means of coping with temporal heterogeneity. Still other protozoa incorporate encystment into the life cycle following feeding (the "digestive cyst") or during cell division (the "reproductive cyst"). Digestive cysts are common in ciliates that rapidly consume large amounts of food, possibly after a long period of searching, e.g., histophages and predators. Some species enter a cyst after feeding and emerge after cell division, combining these functions. The encysted stage may reduce energy expenditure and/or the risk of predation when the acquisition of food is either not possible or unnecessary. Nonencysting protozoa may persist days or weeks in the absence of food; this survival time is not closely related to cell size, is density-dependent, and probably reflects differences in energy costs associated with locomotory behavior (Jackson and Berger, 1985). Survival time was found to be negatively correlated with intrinsic rate of population increase (see Section III.C.2) among eleven species, but there were exceptions to this trend. Crawford and Stoecker (1996) examined survival and respiration during starvation in a marine mixotrophic ciliate *Strombidum capitatum*. There was no decrease in dark respiration with starvation, and the starving population comprised both larger cells with low specific respiration rates and small cells with high specific respiration rates. Survival time was longest for the large cells.

Sex is widespread, but far from universal among protozoa. For example, it is common among the Ciliophora, Chlorophyta, Chrysophyta, and Granureticulosida, limited among Dinoflagellata and Heliozoa, and absent in most if not all Zoomastigina and Rhizopoda. Grell (1973) and Margulis and Sagan (1986) discussed the occurrence and variety of sexual phenomena among the protists. Grell organized sexual phenomena into three basic types: (1) gametogamy, in which gametes fuse as free-swimming cells; (2) autogamy, where gametes or gametic nuclei of the same individual fuse; and (3) gamontogamy, where mates (gamonts) unite to exchange gametes or gametic nuclei. Gametogamy is found in flagellates where, typically, isogametes are produced by multiple fission. However, anisogamy also occurs. Autogamy is present in Heliozoa and some Ciliophora. Gamontogamy is the usual form of sex in Ciliophora.

As in most small metazoa, sexuality interrupts asexual reproduction, and is associated with environmental change. It should be emphasized that sexuality is not necessarily associated with reproduction in protozoa and is best understood as a separate phenomenon (Margulis and Sagan, 1986). Indeed, sex usually marks the end of the existence of an individual protist; it becomes the gamete or gametes, and its genome ceases to exist. The restricted occurrence of sex among protozoa, and the separation of sex from reproduction in many taxa, make protozoa a promising group for further study concerning the evolution and significance of sex.

The relationship between sex and asexual reproduction has been carefully examined in some Ciliophora, especially *Paramecium* (see Wichterman, 1986, for a review of this topic). Exchange of haploid nuclei between two conjugants (gamontogamy) leads to two individuals with new and different genotypes. Each is the founding member of a new clone, expanding by asexual binary fission. Individuals of a new clone will undergo a period of "sexual immaturity" when sex does not occur. The duration of this immature period varies among species; species with short immature periods are considered to be "inbreeders," as mating with close relatives is favored, whereas species with long immature periods are considered to be "outbreeders."

The immature period is followed by a period of sexual maturity when conjugation will occur under the following circumstances: (1) a partner of a complementary mating type is available; (2) both are at the right stage of the cell cycle; and (3) both are slightly starved. If conjugation does not occur before the end of the period of maturity, a period of sexual senescence follows. The rate of binary fission declines, cells are less likely to survive conjugation if a mate is found, and autogamy (self-fertilization) may occur. Although autogamy may prolong vigor, the clones ultimately die out. An analogy can be drawn between the life of a clone of ciliates and the development of a metazoan from a zygote—a large population of cells develops asexually from the single cell resulting from sexual fusion, but ultimately the size and life span of that population are limited by senescence which must be circumvented by another sexual episode.

As some ciliates do not undergo conjugation and others vary greatly in the frequency necessary to prevent senescence (a strain of *Tetrahymena* without a micronucleus and, therefore, not capable of sex was isolated in 1923 and is still going strong), one is tempted to speculate on the function of both sex and senescence. Is sex required to postpone senescence by replacing irreparably damaged DNA by recombination? Or, is senescence a mechanism to eliminate micronuclear or gametic genes that have not been expressed for many generations? Or, is the need for sex a function of the unique amitosis of the ciliate macronucleus? The finite life span of animal cells in tissue culture versus the immortality of cultures of cancer cells begs similar questions. That autogamy in ciliates is a stop-gap measure to postpone senescence contrasts with sex in the heliozoan *Actinophrys sol*, where fusion of the products of meiosis within the same cell is the normal pattern. Bell (1988) reviewed this fascinating subject and concluded that senescence of both sexual and asexual protists can be viewed as a common phenomenon, explicable in terms of the genetic deterioration of finite populations.

Although sex is a necessary event in the life cycle of some ciliates, it is a dangerous undertaking; genetic load is high in many ciliates, as is the mortality following conjugation. As noted previously, some species avoid close matings by having a long immaturity period. As if conjugation was not dangerous enough, prokaryotic endosymbionts render some ciliates "mate-killers"; their partners in conjugation pass on their genes but are lethally affected (see Section III.C.4).

C. Ecological Interactions

1. *Intraspecific Interactions among Protozoa*

Protozoa have been used extensively in the study of competition and predator–prey interactions. Their obvious advantages for this type of study include small size, short generation time, and, for some species, ease of maintenance in the laboratory. These advantages do not carry over to field experiments, and little work has been done *in situ*.

Intraspecific competition in protozoa is evident in the dynamics of batch cultures, where a few individuals are introduced into a large, but finite environment. If inoculated into batch cultures as starved individuals, some ciliates will respond to this situation by growing rapidly and increasing in size past the biovolume at which the cell might be expected to divide. This has been interpreted as a strategy to sequester the maximum amount of food should the supply become exhausted. However, if food is still available, these cells enter into a phase of exponential growth where they divide at a similar size in each cycle. When the food supply is depleted, some individuals will continue to divide at the expense of cell size. Again, this can be interpreted as an adaptation for an r-selected or colonizing species to produce as many propagules as possible (Taylor, 1981; Fenchel, 1987).

Curds and Cockburn (1971) found that, in order to model the population dynamics of *Tetrahymena* in batch cultures, it was necessary to include interference among individuals in their equations. This was not necessary for continuous cultures. The exhaustion of food in some species induces cannibalism, sometimes with abrupt changes in morphology for the cannibals (see Giese, 1973, for a thorough discussion of this phenomonen in the ciliate *Blepharisma*). Another form of interference is the "killer" phenomenon. Some protozoa harbor prokaryotic endosymbionts that render them lethal to conspecifics (mate-killers were mentioned above).

2. *Interspecific Competition among Protozoa*

Interspecific competition in laboratory cultures of ciliates has been studied by Gause and numerous authors since. Gause (1934) provided clear evidence that environmental factors could influence the outcome of competition.

The Lotka–Volterra equations predict that the winner of exploitative competition for resources in stable environments should be the species with the greater K or carrying capacity, that is, the more efficient user of the resource. However, K is usually measured as numbers, not biomass, so smaller species will tend to have a higher K. In his studies of competitive interactions of bacterivorous ciliates, Luckinbill (1979) confirmed that K did not predict the outcome of competition. But, in accordance with part of the r/K selection hypothesis, the intrinsic rate of increase, r, was negatively related to competitive ability. [The r/K selection hypothesis holds that r, which measures fitness in density-independent or stochastic environments, is negatively correlated among species with K, which measures fitness in stable environments.] Thus the theorem seems to hold for ciliates, except that K, being highly correlated to body size, does not measure competitive ability in stable environments. As noted above, r is also correlated with body size.

To factor out this correlation, Taylor (1978a,b) expressed intrinsic rate of increase as r/r_e, where r_e is the expected r for a ciliate based on a regression of r on body volume. Therefore, r/r_e is a measure of how fast-growing or r-selected a ciliate is for its body size. This measure was found to be negatively correlated with frequency of occurrence in the field, and positively correlated with macronuclear ploidy (Taylor and Shuter, 1981). This research suggests that protozoan communities contain fast-growing but ephemeral opportunists, as well as more conservative, specialist species that have more modest growth rates, but stable populations. A similar picture emerged from a comparison of the population dynamics of a relatively fast and a relatively slow-growing species in the laboratory (Luckinbill and Fenton, 1978).

Field studies of exploitative competition in protozoa are few, as are field measurements of growth rates for comparison to laboratory estimates under conditions of excess food. Accordingly, we know little about the relative importance of density-dependence or independence. Work on planktonic protists suggests that field populations may experience periods of rapid growth and periods of little or no growth (Taylor and Johannsson, 1991; Carrick *et al.*, 1992).

Maguire (1963a,b) found strong evidence for the exclusion of colonizing species in the genus *Colpoda* by *Paramecium* and associated species. Hairston and Kellerman (1965) and Hairston (1967) found that one member of the *Paramecium aurelia* species complex dominated sympatric sibling species in competition experiments and predominated in field collections. The field population they studied appeared to be food-limited. In contrast, Taylor (1983b) found that sessile ciliates on an artificial substrate in a stream had high growth and mortality rates, implying little competition. The extensive studies on colonization by Cairns and co-workers (reviewed in Cairns and Yongue, 1977) provide indirect evidence for species interactions, including competition.

There is also competition between protozoa and metazoa consuming similar resources. Potential metazoan competitors include some rotifers and crustaceans, although it is difficult in some cases to distinguish between the effects of competition and predation (Wickham and Gilbert, 1993). Likewise, the presence of predatory protozoa may affect the outcome of competition between other protozoan species (Lawler, 1993).

3. Predatory Interactions among Protozoa

Many protozoa are predators of other protozoa. For example, Chardez (1985) listed a diverse assortment of protozoa consuming testaceans. Protozoa have been used extensively in laboratory studies of predator–prey interactions, including such topics as selectivity, stability of predator–prey interactions, and the behavioral and energetic determinants of functional and numerical response. Much of this research has been done with *Didinium nasutum*. Hewett (1980a,b) and Berger (1979) showed that *Didinium* changes its body size, feeding behavior, and feeding and growth rate when fed different prey and may encounter some difficulty handling large prey after adjusting to small ones. The extrusomes which enable *Didinium* and other predatory ciliates, dinoflagellates, and Heliozoa to handle their prey were mentioned in Section II. Although extrusomes are important weapons of many other ciliates, especially in the Litostomatea and Suctoria, other predators are simply swallowers, including the giant cannibal forms of *Blepharisma* and *Tetrahymena*, large ciliates such as *Bursaria*, and some other predators such as *Peranema* among the Euglenida and the large amoebas.

Interest in the "microbial loop" (the route by which picoplankton production reaches larger consumers) has focused attention on the contribution of protozoa, as consumers of picoplankton, to production at higher trophic levels. A microbial loop leading to fish via crustacean zooplankton is likely to be inefficient because of the number of steps involved, and even more inefficient if predatory protozoa process some of that production. Nevertheless, protozoa are ingested by many metazoa and often appear to be a high-quality food (Sanders and Wickham, 1993).

Relatively little is known about the quantitative significance of predatory protozoa in planktonic food webs. In some instances ciliates appear to be the major predators of heterotrophic nanoflagellates (Weisse *et al.*, 1990), but the relative predation impact on flagellates by ciliates versus metazoa appears to vary seasonally (Weisse, 1991; Sanders *et al.*, 1994). Known predators of other protozoa, such as *Dileptus* and *Actinosphaerium*, are commonly observed in the plankton. The prey of many abundant litostomes is not known, but they are likely to feed on other protozoa. Use of fluorescently labeled prey has identified small ciliates, including *Strobilidium*, *Halteria*, and tintinnids, as predators of heterotrophic nanoflagellates in lakes (Cleven, 1996). In some estuarine systems, dinoflagellates may be important predators of ciliates (Bockstahler and Coats, 1993) but this is not documented in freshwaters. Presence of predatory ciliates, amoebas, and turbellaria elicits morphological changes in the ciliate *Euplotes* that reduce predation (Kusch, 1993). Predation within the protozoan microplankton is still an unknown loss in the transfer of energy to higher trophic levels and therefore bears on the debate concerning the importance of the microbial loop.

4. Symbionts of Protozoa

Symbionts of protozoa include both ecto- and endosymbiotic prokaryotes and eukaryotes, including a wide variety of parasites (Curds, 1977; Jeon, 1983; Lee and Corliss, 1985). The most obvious eukaryote symbionts of freshwater protozoa are algae of the genus *Chlorella*, often called zoochlorellae. These are most widely studied in the ciliate *Paramecium bursaria*, but are also known from a wide variety of other ciliates as well as flagellates, amoebas, and many metazoa. In some cases, the symbiosis appears to benefit the protozoan because of the photosynthates released by the algae; *P. bursaria*, for example, probably receives maltose from its symbionts and can grow better at low food concentrations when they are present.

Although some alga-bearing protozoa display positive phototaxis and aggregate at the water surface (Berninger *et al.*, 1986), others aggregate deeper, in dim light. Some of these microaerophilic forms are negatively phototactic, unless the medium is anaerobic (Finlay *et al.*, 1987), and zoochlorellae may provide low internal oxygen tensions in anaerobic external environments. Therefore, there may be two functional types of protozoan–algal symbioses: one in which the primary benefit for the host is to receive photosynthate, characterized by positive phototaxis, and one in which the primary benefit for the host is oxygen production, characterized by oxygen-dependent phototaxis.

Some choreotrichs (*Strombidum* spp.) harbor isolated chloroplasts as symbionts (Rogerson *et al.*, 1989; Stoecker *et al.*, 1989). The chloroplasts may be from more than one species of alga. In some cases, whole cells may be retained as symbionts. The host ciliate becomes functionally mixotrophic, able to photosynthetically fix significant amounts of carbon. The haptorid ciliate *Perispira ovum* ingests the flagellate *Euglena*, incorporating its chloroplasts, mitochondria and paramylon bodies (Johnson *et al.*, 1995). It probably uses oxygen produced by the chloroplasts to live in low-oxygen environments.

Prokaryotic endosymbionts are widely distributed in protozoa. For example, several species of anaerobic protozoa have endosymbiotic methanogenic bacteria (Section II.H.3). Some endosymbionts are infectious parasites which may kill their hosts, while some are necessary for the survival of their hosts. In some cases, the function of the symbionts is unknown. Perhaps the most interesting endosymbionts from an ecological point of view are the intensively studied "killer" particles of ciliates. Some transform their hosts into "killers" that cause the death of conspecific "sensitives" simply by their proximity. Others transform their hosts into "mate-killers," whose partners in conjugation are lethally affected. The significance of these endosymbionts to field populations of *Paramecium* has been investigated by Landis (1981). He concluded that the infection must be maintained by natural selection, because killers rarely have a chance to mate and transmit their infection to new hosts. Mating and cell division are the only known means of transmission.

5. Epizooic and Epiphytic Protozoa

Many freshwater protozoa have developed epizooic associations with animals. Some associations are likely to be rather nonspecific, involving species that can also be found attached to, or crawling on, nonliving substrates as part of the aufwuchs community. Other associations (e.g., peritrichs of the order Mobilida that are epizooic on fish) are perhaps better left to the parasitologists. However, some cases deserve brief mention here because they are common and likely to be noticed.

One common association is between planktonic, colonial cyanobacteria and peritrich ciliates (Pratt and Rosen, 1983; Kerr, 1983). Peritrichs are frequently so abundant as to rival the biomass of their host colony, and they likely affect its nutrient environment and motility. Smaller peritrichs, choanoflagellates, and bicoecids are common epizooics on diatoms and filamentous cyanobacteria (Fig. 11A).

Peritrichs and Suctoria colonize the exoskeleton of planktonic crustacea. Kankaala and Eloranta (1987)

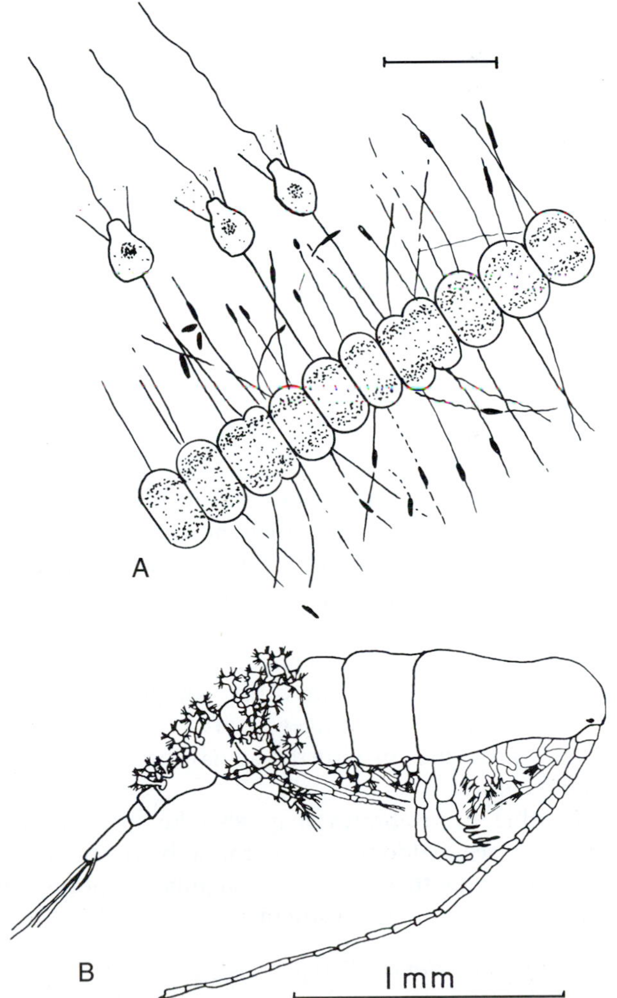

FIGURE 11 (A) Choanoflagellates epiphytic on filamentous cyanobacteria from Georgian Bay, Lake Huron. Scale bar = 10 μm. (B) *Tokophrya quadripartita* (Suctoria) epizooic on *Limnocalanus* from Lake Michigan. (From Evans *et al.*, 1979.)

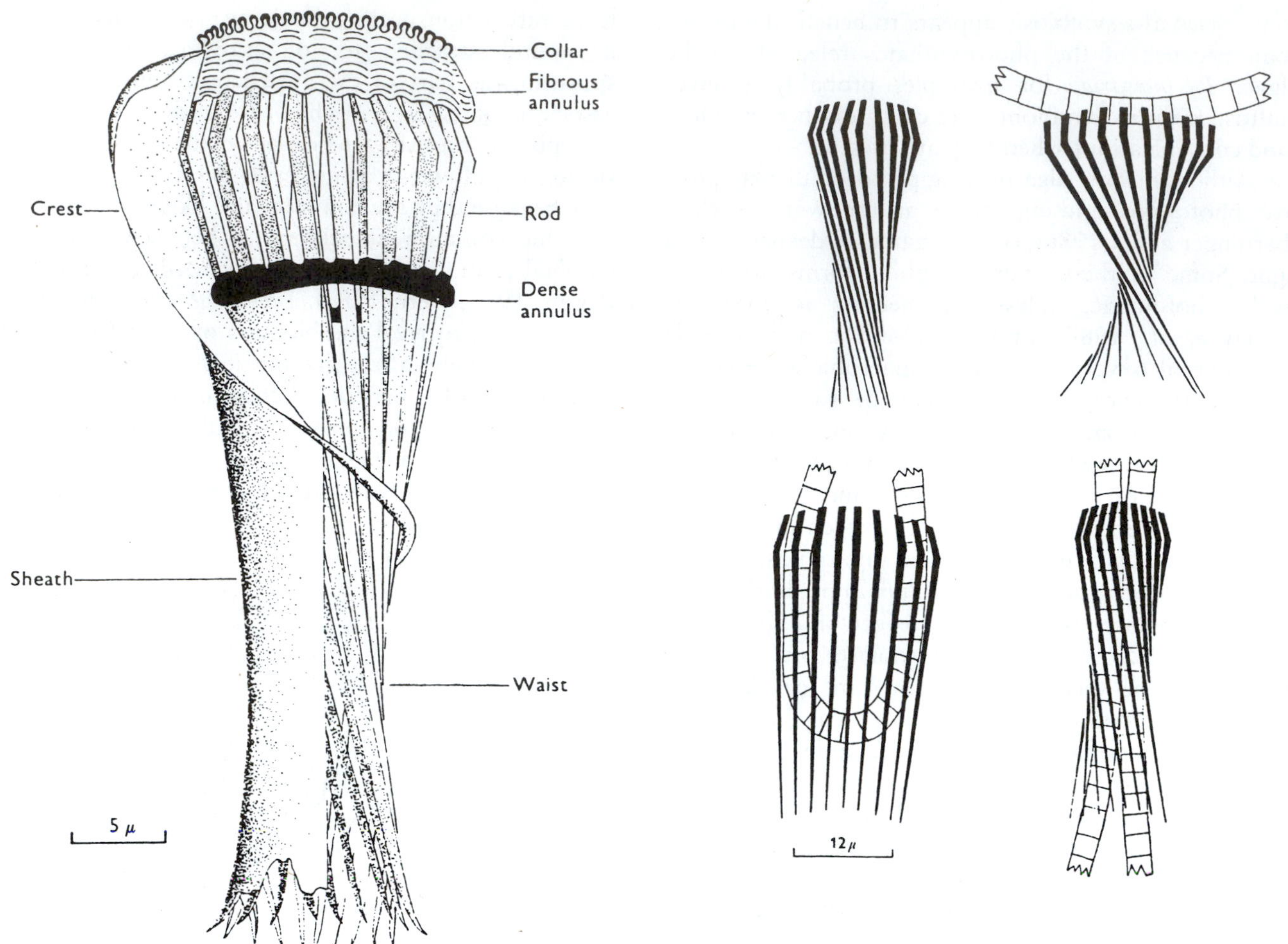

FIGURE 12 The cyrtos of *Nassula,* and how it is used to ingest filamentous cyanobacteria. (From Tucker, 1968.)

found that the clearance rates of *Vorticella* epizooic on *Daphnia* can rival those of their host, suggesting that competition for food with *Vorticella* could be significant to the *Daphnia*. Epizooic suctoria (*Tokophrya* spp., Fig. 12B) on mysid and calanoid Crustacea in the Great Lakes may be heavy enough to impair their hosts (Evans *et al.*, 1981) and may also compete with them for food. Henebry and Ridgeway (1980) suggested that epizooic protozoa on Crustacea may be useful pollution indicators.

Benthic invertebrates may also be infested with peritrichs (see Baldock, 1986, for a brief summary). The biomass of these may be a significant portion of the protozoan biomass in streams.

6. Protozoa as Bacterivores

Protozoa are most widely known as bacterivores. Bacterivorous protozoa are abundant in activated-sludge sewage treatment plants, and, indeed, are necessary for their proper functioning (Curds and Cockburn, 1970). At the opposite extreme, there is considerable evidence that in plankton communities, especially oligotrophic ones, heterotrophic flagellates are the dominant bacterivores and that much of the carbon flow passes through them (e.g., Sanders *et al.*, 1989; Nagata, 1988; Carrias *et al.*, 1996). Although ciliates may be abundant bacterivores in enriched environments (hence, the old name "infusorians" or animals from organic infusions), they are thought to be quantitatively unimportant in many planktonic ones. However, in some planktonic environments, especially those where cladoceran crustacea can seasonally reduce flagellate abundances significantly, ciliates may become the dominant bacterivores (Simek and Straskrabová, 1992). Heterotrophic flagellates, and to a lesser degree ciliates, also may be important bacterivores in the water column of rivers (Carlough and Meyers, 1991), although bacterivory by protists in this environment has barely been addressed.

The relative importance of flagellates and ciliates as bacterivores in different habitats and during different seasons is, therefore, an open question. The question should probably be phrased in terms of the size distribution of grazers of picoplankton, and the answer probably involves species interactions and food web structure as well as trophic status.

The small Heliozoa are an overlooked group of largely planktonic bacterivores, and small amoebas are bacterivorous also. Mixotrophic flagellates may play a significant role as bacterivores in some freshwater plankton (Bird and Kalff, 1987; Sanders and Porter, 1988; Berninger *et al.*, 1992). The importance of mixotrophs as bacterivores varies considerably both temporally and spatially, and the factors determining their abundance and feeding rates in natural systems are not entirely clear. However, light, temperature, dissolved nutrient concentrations, and food levels are likely to affect both the abundance and the relative importance of phagotrophy versus photosynthesis in these organisms (Bird and Kalff, 1987; Sanders *et al.*, 1989, 1990; Jones and Rees, 1994). Many protozoa glean bacteria from surfaces or attached to particles (Albright *et al.*, 1987; Caron, 1987), but the quantitative role of bacterivorous protozoa in benthic habitats is relatively unknown. Pioneering studies in this area include Baldock and Sleigh (1988) and Bott and Kaplan (1989). In experiments with stained lake sediment, Starink *et al.* (1994) found over 90% of protozoa ingested bacteria at rates up to 73 bacteria protist^{-1} hour^{-1}.

Small protozoa or "heterotrophic nanoplankton" (flagellates, amoebas, and some small ciliates) capture and consume bacteria individually, with rates of ingestion of tens to hundreds per hour (e.g., Fenchel, 1982b). Most ciliates, in contrast, use their complex oral ciliature to collect hundreds into each food vacuole, producing feeding rates up to thousands of bacteria per hour (e.g., Taylor, 1986). Although the former are likely to be selective in their feeding, the larger ciliates are probably very limited in their ability to discriminate. The use of fluorescent microspheres to measure bacterivory (McManus and Fuhrman, 1986) resulted in renewed interest in the ability of protozoa to discriminate against inorganic particles in favor of bacteria (Nygaard *et al.*, 1988) and in the influence of chemical coatings and surface charge (Sanders, 1988). Both ciliate and flagellate bacterivores discriminate among bacteria on the basis of size (Sanders, 1988; Simek and Chrzanowski, 1992; Simek *et al.*, 1994). Several studies have documented that ciliates differ in their ability to use different bacteria as food. Some may be unable to consume filamentous bacteria, while other bacteria, especially pigmented ones, produce toxins that render them unsuitable to support the growth of some ciliates (Dive, 1973; Taylor and Berger, 1976).

A poorly understood and usually ignored facet of bacterivory in protozoa is the role of extracellular products. The presence of some ciliates leads to the flocculation of bacteria (Hardin, 1943; Curds, 1963; Nilsson, 1976). Mucous-like extracellular materials are involved in uptake of dissolved material and possibly small particles by pinocytosis in *Amoeba*. Small amoebas secrete various hydrolases into their medium (Schuster, 1979) as do flagellates. Porter (1988) suggested that flocculation and external digestion may be important in bacterivorous flagellates. The involvement of mucocysts or other extrusomes is likely, but not confirmed.

As one might expect, feeding rate (functional response) and growth rate (numerical response) of bacterivores increase with bacterial density, and the nature of these response curves has been studied because of their relevance to interspecific competition and the control of bacterial numbers. In addition to the expected hyperbolic functions, thresholds or sigmoid responses have also been observed, and attributed to changes from feeding to searching behavior at low particle concentrations (e.g., Taylor, 1978a; Sanders, 1988). Accordingly, low concentrations of bacteria (10^4–10^7 per milliliter, usually 10^5–10^6) can persist even in stationary-phase cultures of bacterivorous protozoa. The behavioral and physiological mechanisms responsible for this phenomenon are still in doubt (Sambanis and Frederickson, 1988); however, it is probably relevant to the densities of picoplankton observed in field situations.

7. Algivorous Protozoa

Larger protozoa are less likely to be bacterivores, and most consume algae or other protozoa. Suspension-feeding ciliates each ingest a distinct size-spectrum of particles, related to the form and function of the oral apparatus (Fenchel, 1980), and this spectrum generally shifts in larger ciliates toward larger particles, i.e., algae and other protozoa. However, even small prostomes can be important predators of flagellates (Sommaruga and Psenner, 1993). Larger flagellates, amoebas, and Heliozoa also consume other algae and protozoa.

Protozoa feeding on relatively large items are more likely to be selective, as the rate at which particles are consumed is much reduced. For example, Rapport *et al.* (1972) documented selective feeding by the heterotrich ciliate *Stentor* consuming other protozoa at up to about five prey per minute. Stentors consistently biased their feeding toward *Tetrahymena* relative to

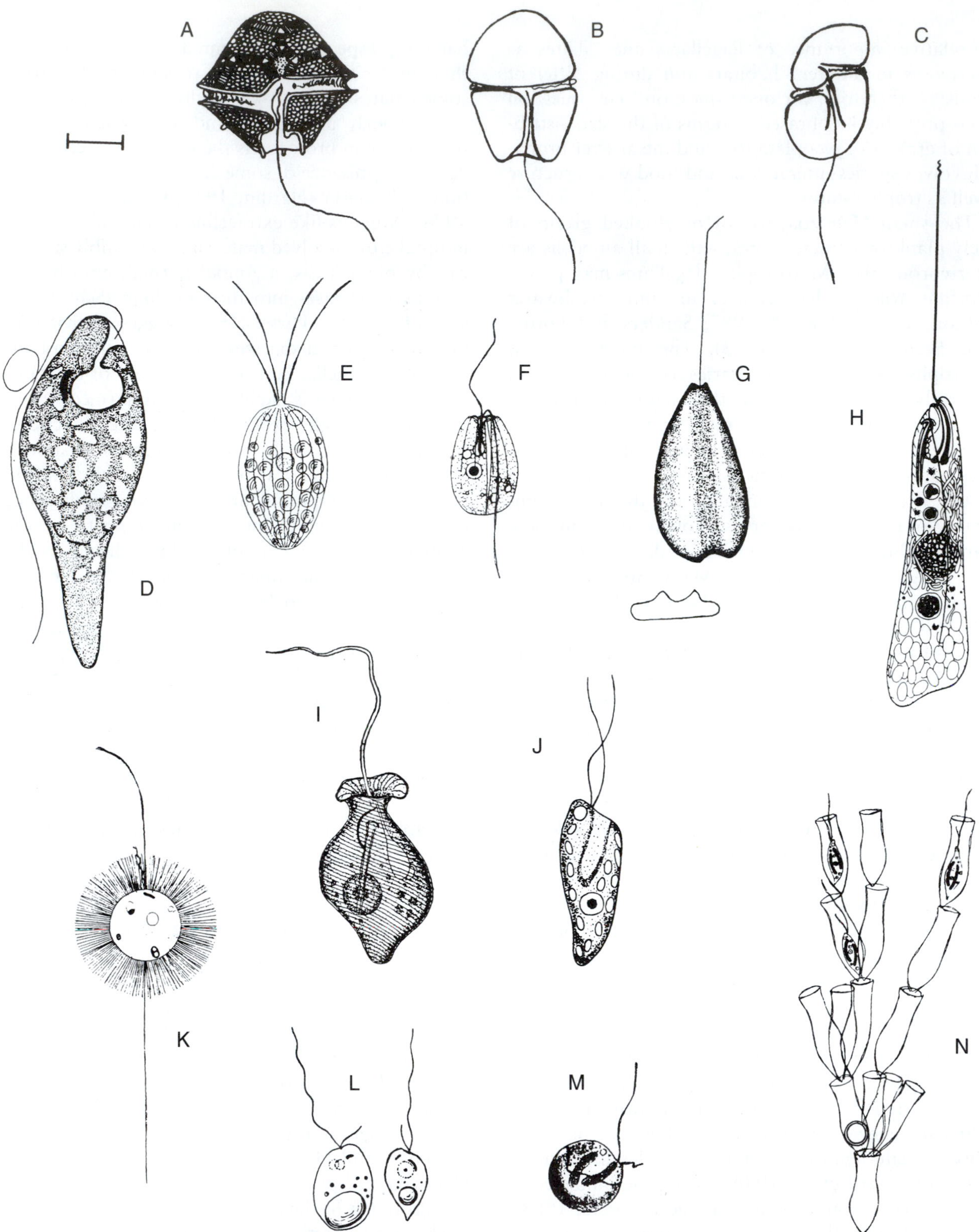

FIGURE 13 (A) *Peridinium;* (B) *Gymnodinium;* (C) *Gyrodinium;* (D) *Khawkinea halli;* (E) *Polytomella citri;* (F) *Entosiphon sulcatum;* (G) *Petalomonas abcissa;* (H) *Peranema trichophorum;* (I) *Urceolus cyclostomas;* (J) *Chilomonas paramecium;* (K) *Paraphysomonas vestita;* (L) *Spumella (=* Monas) vivapara, two cell shapes; (M) *Ochromonas variabilis*[a]; (N) *Dinobryon sertularia*(mixotrophic). (After: Bourrelly, 1968, L; Calaway and Lackey, 1962, E, F, J, N; Eddy, 1930, A; Jahn and McKibben, 1937, D; Leedale, 1985, H; Pascher, 1913, M; Lemmermann, 1914, K; Shawhan and Jahn, 1947, G; Smith, 1950, I.) Scale = 5 μm for E; 10 μm for A, B, C, G, I, J, K, L, M; 20 μm for D, F, H, N.

three autotrophic flagellates when mixed versus single species feeding suspensions were compared. The behavioral response of algivorous choreotrich ciliates appears to involve both mechanical and chemical stimuli, and suggests that although ciliates may ingest inert particles, they may discriminate among them as well as among food organisms (Buskey and Stoecker, 1989). Selective feeding has been noted for *Amoeba* (Mast and Hahnert, 1935) and *Thecamoeba* (Page, 1977). Ciliates may have relatively large ecological impacts on phytoplankton populations; in Lake Constance ciliates consumed about 14% of the daily primary production during a phytoplankton spring bloom, which was equivalent to the amount removed by metazoan zooplankton (Weisse *et al.*, 1990).

Although most phagotrophic flagellates are bacterivores, larger species may be effective algivores. Dinoflagellates, for example, may consume diatoms and other large phytoplankton (see Section II.B.1). Zoomastigina are probably under-rated as consumers of algae (Canter, 1973; Goldman and Caron, 1985; Suttle *et al.*, 1986). The colorless chrysophyte *Paraphysomonas* (Fig. 2D) was observed to feed selectively on diatoms, phagocytosing cells that were nearly their own volume (Grover, 1989). The phagotrophic flagellate *Peranema* (Fig. 13H) feeds on related autotrophic euglenids.

The consumption of filamentous algae (including cyanobacteria) would appear to be mechanically difficult for protozoa, but many have specialized to this role. Large amoebas, such as *Pelomyxa* include filamentous algae in their omnivorous diets, and some Testacealobosia, such as *Arcella* and *Difflugia*, are readily cultured on filamentous algae which they engulf through the aperture of their tests. The nassophorine ciliates contain many forms which feed on filamentous algae, including *Nassula*, *Pseudomicrothorax*, and *Frontonia*. *Nassula* spp. (Fig. 11) appear to limit their diet to *Oscillatoria*. Detailed ultrastructural studies have revealed the role of the cyrtos in feeding, especially how it is used to break the algal filament so that ingestion can begin (Tucker, 1968; Peck, 1985).

Abundant epibenthic diatoms, especially noncolonial and unattached pennate diatoms such as *Navicula*, usually attract abundant herbivorous ciliates and amoebas. Many dorsoventrally flattened and thigmotactic ciliates in the Phyllopharyngia and Nassophorea are diatom feeders. Their functional response has been described by Balczon and Pratt (1996).

8. Histophagous Ciliates

Several genera and species of ciliates feed on metazoan tissue. Some, like *Coleps hirtus* and *Prorodon* spp., are omnivores and only opportunistic histophages. Strict histophages, feeding only on fresh tissue, include all hymenostomes of the genus *Ophryoglena* (Fig. 10) and some *Tetrahymena* spp. (e.g., Corliss, 1973; Lynn, 1975). *Deltopylum* is also a histophage; but in our experience, it is relatively rare. *Philaster* appears to be marine only. These forms feature complex life cycles with reproductive cysts and multiple fission, which are probably adaptations to an extremely patchy food resource (Section III.A). Like other carrion feeders, they increase the efficiency of food webs by bypassing the detrital pathway. *Ophryoglena*, and the related fish parasite *Ichthyophthirius*, share the unique organelle of Lieberkühn (Canella and Rocchi-Canella, 1976), or watchglass organelle, whose function is unknown but probably related to their feeding mode.

Related hymenostomes are insect parasites. *Lambornella* can be a parasite of the cuticle of mosquito larvae (Washburn *et al.*, 1988), and, it and some *Tetrahymena* spp. may have significant impacts on aquatic Diptera (Golini and Corliss, 1981; Egerter *et al.*, 1986). *Espejoia* feeds on the egg masses of aquatic insects.

9. Metazoan Predators of Protozoa

Protozoa, as quantitatively important components of plankton, aufwuchs, or sediment communities, probably contribute to the nutrition of many animal-feeding guilds, such as filter-feeders, scrapers, and deposit-feeders. However, because most protozoa are fragile and without structures that resist digestion, they are seldom recorded among the gut contents of animals despite their importance (McCormick and Cairns, 1991). Among benthic animals, minute predators, such as Naididae (Oligochaeta), Orthocladinae (Diptera, Chironomidae), Turbellaria, and Copepoda consume protozoa selectively (Taylor, 1983b; Archbold and Berger, 1985; Kusuoka and Watanbe, 1989).

More is known about the feeding of planktonic animals on protozoa (Porter *et al.*, 1985; Sanders and Wickham, 1993; Jack and Gilbert, 1997; Stoecker and Capuzzo, 1990), reflecting the recent interest in the microbial loop (Section III.C.3) and how picoplankton production reaches higher trophic levels (e.g., Pomeroy, 1974; Sherr *et al.*, 1987). Most planktonic rotifers consume heterotrophic flagellates, and many prey on ciliates (Arndt, 1993; Gilbert and Jack, 1993). Copepoda especially are known to be consumers of protozoa, and recent studies suggest that copepods may actively select protozoa from among similar-sized plankton (Burns and Gilbert, 1993; Wickham, 1995a,b). Cladocera can ingest ciliates as well as both photosynthetic and nonphotosynthetic flagellates (Sanders and Porter, 1988; Wickham and Gilbert, 1993). There is evidence, however, that ciliates are not always suitable as food for Cladocera (Debiase *et al.*, 1990). Fish would seem to be unlikely predators of protozoa, but protozoa are of important food for some larval fishes (Kornienko, 1971; Kentouri and Divanach, 1986).

Many planktonic protozoa are capable of rapid or saltatory movements that reduce their probability of being captured. Speeds attained during these movements exceed 100 body lengths per second, and they reduce the frequency of capture compared to species that lack such movements (Gilbert, 1994). Morphological adaptations that reduce probablity of ingestion can be induced by the presence of predators in only a few hours in some ciliates (Wicklow, 1997). The ciliate *Lambornella* reacts to the presence of predatory mosquito larvae by shifting morphologically and behaviorally to become a mosquito parasite (Washburn *et al.*, 1988).

10. Protozoa and Nutrient Cycling

Regeneration of the inorganic constituents of living organisms occurs through a process involving ingestion, digestion and egestion by consumers, catabolism and autolysis of all organisms, and the enzyme-mediated breakdown of nonliving organic matter by decomposers. Autotrophs are net consumers of these inorganic constituents, whereas phagotrophs regenerate them. Decomposers, the heterotrophic bacteria and fungi, regenerate carbon; but they may or may not be net producers of other inorganic constituents, depending on the composition of the organic matter they decompose. If it is poor in other constituents, such as organic nitrogen and phosphorus (as when it is mostly carbohydrate), then decomposers will compete with autotrophs for these other constituents. In the final analysis, the incorporation and regeneration of inorganic constituents will be in balance unless organic matter is stored or exported.

Heterotrophic protozoa fit into this picture as typical consumers, eating other organisms and releasing both unassimilated organic material and end products of catabolism, and as major consumers of the decomposers. The latter role is manifested in the ability of bacterivorous protozoa to accelerate the breakdown of organic material by bacteria (Stout, 1980). This property of protozoa is mostly due to their keeping the bacteria in an actively growing state at moderate abundances, rather than directly due to their contribution to the remineralization process.

Protozoa also contribute directly to mineralization, and some early studies suggested protozoa were particularly active in releasing inorganic nutrients. However, the basis of this view has been challenged (Taylor, 1986). Indeed, the relatively independent literature on energetics of protozoa (Section II) suggests that protozoa are more rather than less efficient in converting the food they ingest into cytoplasm. Recent studies on nutrient recycling by flagellates suggest that mineralization of nitrogen and phosphorus by protozoa may be particularly low when it is most needed, that is, when their food organisms are themselves N- or P-limited (Goldman *et al.*, 1985; Caron *et al.*, 1988). *In situ* measurements of nutrient release relative to ingestion are few, but release of P for planktonic ciliates appears to be close to 50%, or so we expect based on conversion efficiencies (Taylor and Lean, 1991). Protozoa probably make an important contribution to the regeneration of nutrients in many environments, but no more so than one would expect from their contribution to consumption and production.

D. Evolutionary Relationships

The relationship of protozoa to plants and animals, as well as the evolutionary relationships among the protozoa, are controversial areas of current research. New tools for the study of phylogeny coming from molecular biology are contradicting morphology-based classifications, but the number of organisms examined is still small. Here, we will briefly introduce some of the major theories and controversies.

One change in view which has occurred over recent years is that the major dichotomy among living things is not between plants and animals, but between prokaryotes and eukaryotes. The evolution of eukaryotes from the prokaryotes is perhaps the major event in organic evolution on earth, and may have occurred by essentially Lamarckian means: the inheritance of acquired characteristics (Taylor, 1987). That is, the eukaryotic cell represents the symbiotic amalgamation of several distantly related prokaryotes (Margulis, 1981). Although this concept is controversial, the controversy is largely one of degree in that the symbiotic origin of some organelles, such as double-membraned mitochondria and chloroplasts, is well supported by molecular techniques and is now widely accepted. Gupta *et al.* (1994) argued on the basis of a 70-kDa heatshock protein that the eukaryotic nucleus (also double-membraned) may also have been derived by engulfing a bacterium.

Within the unicellular eukaryotes, or Protista, one of the most primitive forms is the giant amoeba *Pelomyxa* (Karyoblastea); it lacks mitochondria and has amitotic nuclear fission. Many consider the flagellum to be as old as the eukaryote cell, but disagree whether colorless flagellates are primitive or derived from pigmented forms. Analyses of ribosomal RNA strongly support the former argument for most Zoomastigina. Some protists without mitochondria, including the colorless (and parasitic) flagellate *Giardia,* are clearly very close to the prokaryotes relative to other protozoa. This strongly suggests that their features are primitive rather than derived (Sogin *et al.*, 1989, 1996). The same analysis reveals that the chlorophyte *Chlamydomonas* is more closely aligned with higher plants than other protozoa, including other autotrophic phyla (Sogin *et al.*, 1989).

Although our view of the relationship among unicellular eukaryotes is in a state of flux, the molecular techniques currently being applied reveal that these groups have had long and independent evolutionary histories. For example, the genetic distances within ciliates or flagellates are greater than the distances between chlorophytes and higher plants, or between plants and animals (Sogin and Elwood, 1986; Gunderson *et al.*, 1987).

There are two major schools of thought with respect to the relationship between protozoa and metazoa (Hanson, 1977). One considers that the protozoan ancestor of metazoa is a colonial flagellate, probably a choanoflagellate, and that Porifera, Placozoa, Cnidaria, some gastrotrichs, and other animals with monociliated epidermal cells are among the most primitive animals. Others believe that primitive animals were derived by the compartmentalization of a ciliated, multinucleate protozoan. The early development of animals best supports the former hypothesis, as do the data collected thus far on ribosomal RNA; the protozoa least distant from animals are flagellates, not ciliates, which seem to have had a very long independent history.

Perhaps the most important conclusions an ecologist can draw from the dynamic field of protist systematics is that although the protozoa are all unicellular or colonial and comprise one to many phyla in different classification schemes, they are more distant among themselves than multicellular animals from the evolution point of view, and at least as morphologically and physiologically diverse.

IV. COLLECTING, REARING, AND PREPARATION FOR IDENTIFICATION

The small size, delicacy, and lack of hard parts in most protozoa makes the quantitative collection and enumeration of protozoa particularly difficult. The methods for sample collection and examination vary with the group of interest and the habitat, and there is room for improvement in almost all the procedures mentioned below. General techniques for the beginner are provided in Lee *et al.* (1985) and Finlay *et al.* (1988).

For some groups—especially amoebas and heterotrophic flagellates, when identification is required or when the sample contains sediment or some other material—examination of live organisms in fresh collections under a dissecting and/or compound microscope is the only method available. With care, this can be done in a quantitative manner (e.g., Bott and Kaplan, 1989). The "most probable number" method of enumeration, based on inoculating culture medium with small aliquots of sample and scoring the fraction of such inocula producing cultures, is common in the study of soil protozoa and has been applied to freshwater. However, the resultant estimates sometimes have been very low relative to direct counts (Bott and Kaplan, 1989). Given the difficulty in culturing many freshwater protozoa, this method should be reserved for estimating species known to be able to grow on the medium used from small inocula. Extraction procedures have been devised for benthic protozoa, but are not widely used (Hairston and Kellerman, 1964; Fenchel, 1967). Fixed samples are not useful in benthic studies, as the task of enumerating fixed but motionless protozoa in sediment is virtually impossible.

Studies on epibenthic or aufwuchs protozoa have generally resorted to artificial substrates. Those that allow for microscopic examination are particularly useful: glass slides (Schönborn, 1977), petri dishes (Spoon and Burbank, 1967; Taylor, 1983b), and strips of nylon mesh (Taylor, 1983a). Polyurethane foam has been extensively used as a substrate in studying protozoan community interactions and response to pollution (Cairns *et al.*, 1979).

Planktonic samples present fewer problems. Large protozoa can be settled from samples and counted by inverted microscopy as for phytoplankton. Useful fixatives are Lugol's Iodine (1% v/v, or higher concentration in saline or hard water, e.g., Taylor and Heynen, 1987) or mercuric chloride (5% saturated mercuric chloride plus a drop of 0.04% bromothymol blue, Pace and Orcutt, 1981). These procedures do not lend themselves to the identification of ciliates, nor to the detection of chromatophores in small flagellates. Filtration methods may aid with both of these problems. The quantitative protargol or QPS method of Montagnes and Lynn (1987a,b) and Skibbe (1994) produces permanent, quantitative, stained preparations for identification of ciliates and flagellates, although it requires that samples be fixed in a concentrated Bouin's fixative. The use of epifluorescence microscopy on filtered samples (Caron, 1983) can facilitate the separation of heterotrophic and phototrophic protists and is generally the method of choice for enumerating very small planktonic protozoa, even though it does not lend itself to identification.

Culture of most protozoa is best begun by maintaining samples collected in the field. Samples kept in the laboratory will usually undergo succession, with different species waxing and waning in abundance. Screening out metazoan predators, such as copepods, may encourage protozoa to increase in number. Supplementing the organic or nutrient content of the sample may help. Cereal grains (rice or wheat, boiled for quicker effect or plain), broth from boiled vegetable matter (hay, lettuce), or nutrient broth are commonly used. Remember that adding too much is easy, and results in foul conditions that may encourage species different from those important in the original sample; or one may simply kill all the aerobic protozoa. Generally, additions of organic matter to cultures should not be more than about 0.1% w/v for whole vegetable matter, and no more than a tenth as much (100 mg/L) for more labile material.

The first step in isolating species of interest is to prepare some protozoa-free medium by filtering or boiling water from the site where the protozoa were collected, and inoculating individuals of the desired species back in. Mineral water is frequently used for culturing heterotrophic forms. Again, rice or wheat grains or organic infusions are useful in promoting the moderate growth of bacteria for bactivorous species. Algae or other protozoa may be required as food for others. White worms (enchytreids) are convenient food for histophagous forms.

Ultimately, you may wish to establish monoxenic (the organism of interest and its prey only) or axenic cultures, possibly cloned from a single individual. Artificial freshwater media are described in Finlay *et al.* (1988) and Thompson *et al.* (1988). Although involving marine protozoa, the recent efforts by Gifford (1985) might prove a useful template for attempts to culture difficult freshwater forms. A manual of elementary techniques is found in Finlay *et al.* (1988). General and specific techniques for studying the ecology of free-living protozoa are also compiled in Kemp *et al.* (1993).

V. IDENTIFICATION OF PROTOZOA

A. Notes about the Key

It is often difficult and time consuming to identify protozoa to the level of species. In many cases, unambiguous identification requires specialized staining techniques or the use of electron microscopy. The taxonomic grouping used in this key is an amalgamation of publications by specialists on the different groups (e.g., Page, 1998; Lynn and Small, 1997; Lee *et al.*, 1985; Margulis *et al.*, 1989). While the taxonomy is based on ultrastructural features and morphometric attributes that in many cases are not readily discerned with light microscopy, this key is designed to make possible the identification of many protozoa to a family or genus level using light microscopy alone. Although observation of living organisms is important for identification, the key should still be useful for many fixed samples. Most of the specialized terms used in the key are defined in the general glossary. If one becomes more interested in identifying protozoa to a species level, specialized techniques and keys in the publications cited earlier can be used.

The genera listed in the key are representative of the families and are not complete. Likewise, the illustrations used as examples here are of one or more species considered typical of a genus, but do not necessarily represent all of the species in the genus.

Advances in the systematics of the protozoa are continuous. Due to the heterogeneity of the group, it is often difficult to remain abreast of the changes. A new edition of *Lee et al.* (1985), *An Illustrated Guide to the Protozoa*, is expected to be published by the Society of Protozoologists at any time, and will likely have changes in systematic organization.

Classification below the class level in the Zoomastigina and the ordinal level in Heliozoa are not conclusively agreed upon, or are under revision. Thus, the use of familial names was avoided for these groups in the key.

Phyla of flagellates marked with an asterisk (*) contain many or mostly photosynthetic members; genera listed have one or more species that lack chloroplasts and are thus heterotrophic, or contain chloroplasts, but also exhibit heterotrophic nutrition.

B. Taxonomic Key to Functional Groups of Protozoa

1a. Locomotion via pseudopodia as in Figure 8. Rarely, flagella may be present as well (e.g., Fig. 9 C) .... Amoeboid Protozoa (Section V.D)

1b. Simple cilia or compound ciliary organelles characteristic (but see Fig. 19A–P) and present in at least one part of life cycle; subpellicular infraciliature present even when cilia are not; two types of nuclei (macronucleus and micronucleus) with rare exceptions (Fig. 23J). Ciliated Protozoa, Phylum Ciliophora (see Section V.E)

1c. Commonly 1 to 4 flagella, 16 to 24 in one free-living genus (*Paramastix*); binary fission.......... Flagellated Protozoa (see Section V.C)

C. Taxonomic Key to Major Groups and Common Genera of Flagellated Protozoa

1a. With median groove (annulus and sulcus); two flagella, one extending transversely around the cell.......... Phylum Dinoflagellata*
(*Peridinium, Gymnodinium, Gyrodinium;* Fig. 13A–C)

1b. Without a median groove.......... 2

2a (1b). Often large flagellates with pharyngeal rods (sometimes difficult to see) or containing the reserve material paramylon; one or usually two flagella arise from within an anterior invagination (reservoir); contractile vacuole associated with reservoir; elongate; body shape often plastic when living. Phylum Euglenida* (*Khawkinea, Entosiphon, Petalomonas, Peranema, Urceolus;* Fig. 13D,F–I)

2b. Cells not as above.......... 3

3a (2b). Flagellates with two or four equal flagella. Phylum Chlorophyta* (*Polytoma, Polytomella*; Fig. 13E)

3b. Cells with two unequal flagella.......... 4

4a (3b). Small cells with a deep, subapical gullet; two nearly equal length flagella arise from gullet.......... Phylum Cryptophyta* (*Chilomonas*; Fig. 7C, Fig. 13J)

4b. Cells without deep gullet.......... 5

5a (4b). Typically with two unequal flagella; long flagellum usually directed forward during swimming, with short flagellum directed backwards if emergent; compact Golgi body frequently visible anterior to nucleus of larger cells; chrysolaminarin vescicle(s) often fill posterior of cell; contractile vacuole usually present in extreme anterior end of cell; many species capable of simultaneous photosynthesis and phagocytosis. Phylum Chrysophyta* (*Dinobryon, Monas, Ochromonas, Paraphysomonas, Uroglena*; Fig 7B, 13K–N, 14A)

5b. Cell not as above. Phylum Zoomastigina.......... 6

6a (5b). Single anterior flagellum encircled laterally by a tentacular, funnel-shaped collar; solitary or colonial; with or without theca. Class Choanoflagellida (*Codosiga, Desmarella, Sphaeroeca, Diploeca, Monosiga, Salpingoeca*; Fig. 4A, 14B–H)

6b. Without collar, or with indistinct collar.......... 7

7a (6b). Like Choanoflagellida above, but with second trailing flagella attached to the base of lorica; protoplasmic collar rudimentory. Bicoecid flagellates, affinities uncertain (*Bicoeca*; Fig. 14I)

7b. Without protoplasmic collar.......... 8

8a (7b). Flagellates with two unequal length flagella, one trailing; ingestion by pseudopodia. (*Cercomonas*; Fig. 14K)

8b. Not as above.......... 9

9a (8b). With one or usually two flagella arising from a flagellar pocket (depression); characteristically elongate or bean shaped. Unique organelle, the kinetoplast, usually associated with the flagella. Class Kinetoplastida (*Bodo, Cephalothamnium, Pleuromonas, Rhynchomonas*; Fig. 14J,L,N–P)

9b Cells with one or two nucleus-flagella complexes (karyomastigonts) each with 1–4 flagella; no Golgi apparatus; when two karyomastigonts, mirrored symmetry of nuclei and flagella. Diplomonads (*Hexamita*; Fig. 14M)

D. Taxonomic Key to the Major Groups and Common Genera of Protozoa with Pseudopodia

1a. Large, cylindrical or ovoid multinucleate amoeba; bacterial symbionts; usually containing mineral particles; nonmotile flagella-like extensions; oxygen poor habitats Phylum Karyoblastea. (one genus, *Pelomyxa*; Fig. 8B, Fig. 16V)

1b. Pseudopodia hyaline, usually blunt, eruptive (Fig. 8A,C,F) but filiform process may occur. May be flagellated stages.......... 2

1c. Hyaline, filiform pseudopodia (filopodia) sometimes branching, not anastomosing (Fig. 8E); no known flagellate stages or spores. Class Filosea.......... 21

1d. Threadlike and delicate pseudopodia with finely granular appearance, often branching and anastomosing to form complex reticulum (reticulopodia, Fig. 8G). Test often present. Phylum Granuloreticulosea.......... 27

1f. Long slender axopods (Fig. 8D) radiating from cell; with or without skeletal elements. Phylum Heliozoa.......... 32

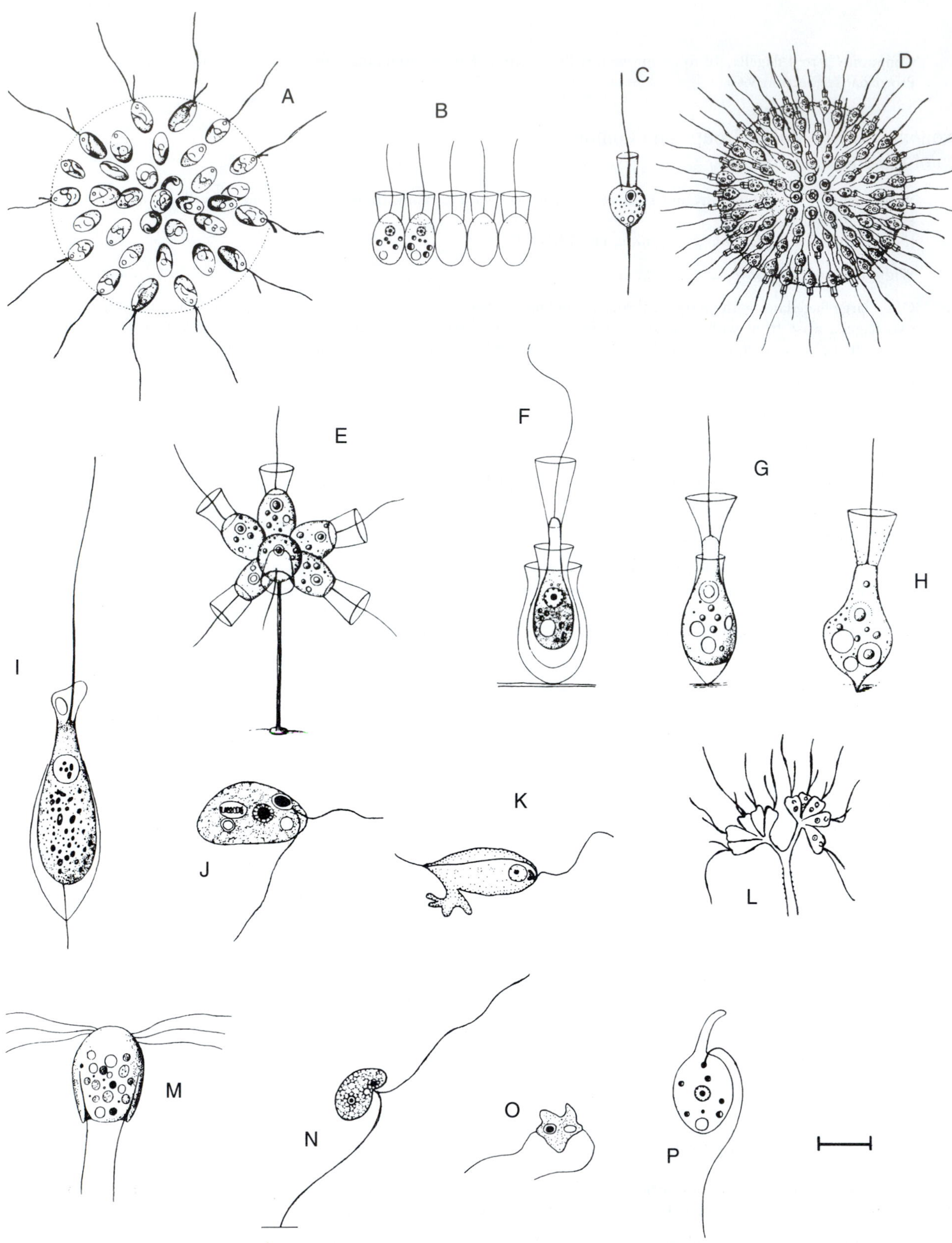

FIGURE 14 (A) *Uroglena americana*(mixotrophic); (B) *Desmarella moniliformis;* (C,D) *Sphaeroeca volvox,* individual cell and colony; (E) *Codosiga botrys;* (F) *Diploeca placita;* (G) *Salpingoeca fusiformis;* (H) *Monosiga ovata;* (I) *Bicoeca lacustris;* (J) *Bodo caudatus;* (K) *Cercomonas sp.;* (L) *Cephalothamnion cyclopum;* (M) *Hexamita inflata;* (N,O) *Pleuromonas jaculans,* attached and amoeboflagellate forms; (P) *Rhynchomonas nasuta.* (After: Bourrelly, 1968, L; Calaway and Lackey, 1962, N, O, P; Lackey, 1959, B, F; Lee, 1985, K; Pascher, 1913, A; Pascher, 1914, C, D, E, G, H, I, J, M.) Scale = 2.5 μm for P; 5 μm for F, G, H, I, K, L; 10 μm for A, B, C, J, M, N, O; and 20 μm for E.

2a (1b). Small, usually cylindrical amoebae, many with a flagellated stage (Fig. 9C). Phylum Rhizopoda, Class Heterolobosea, Order Schizopyrenida 3

2b No flagellated stages. Class Lobosea 4

3a (2a). Amoeboid form cylindrical, monopodial, eruptive; temporary flagellate stages common; nucleolus divides to form polar masses in mitosis....................Vahlkampfiidae (*Valkampfia, Naegleria*; Fig. 15A–B)

3b Nucleolus disintegrates during mitosis, nuclear membrane does not; single known freshwater species; commonly flattened; no flagellate stage known....................Gruberellidae (*Stachyamoeba*; Fig. 15C)

4a (1b). Lacking external test. Subclass Gymnamoebia 5

4b. Incompletely enclosed in a test or other flexible cuticle of microscales. Subclass Testacealobosia 13

5a (4a). Cylindrical or flattened; flattened forms with regular outline; no trailing uroidal filaments, with rare exceptions; not strikingly eruptive. Order Euamoebida 6

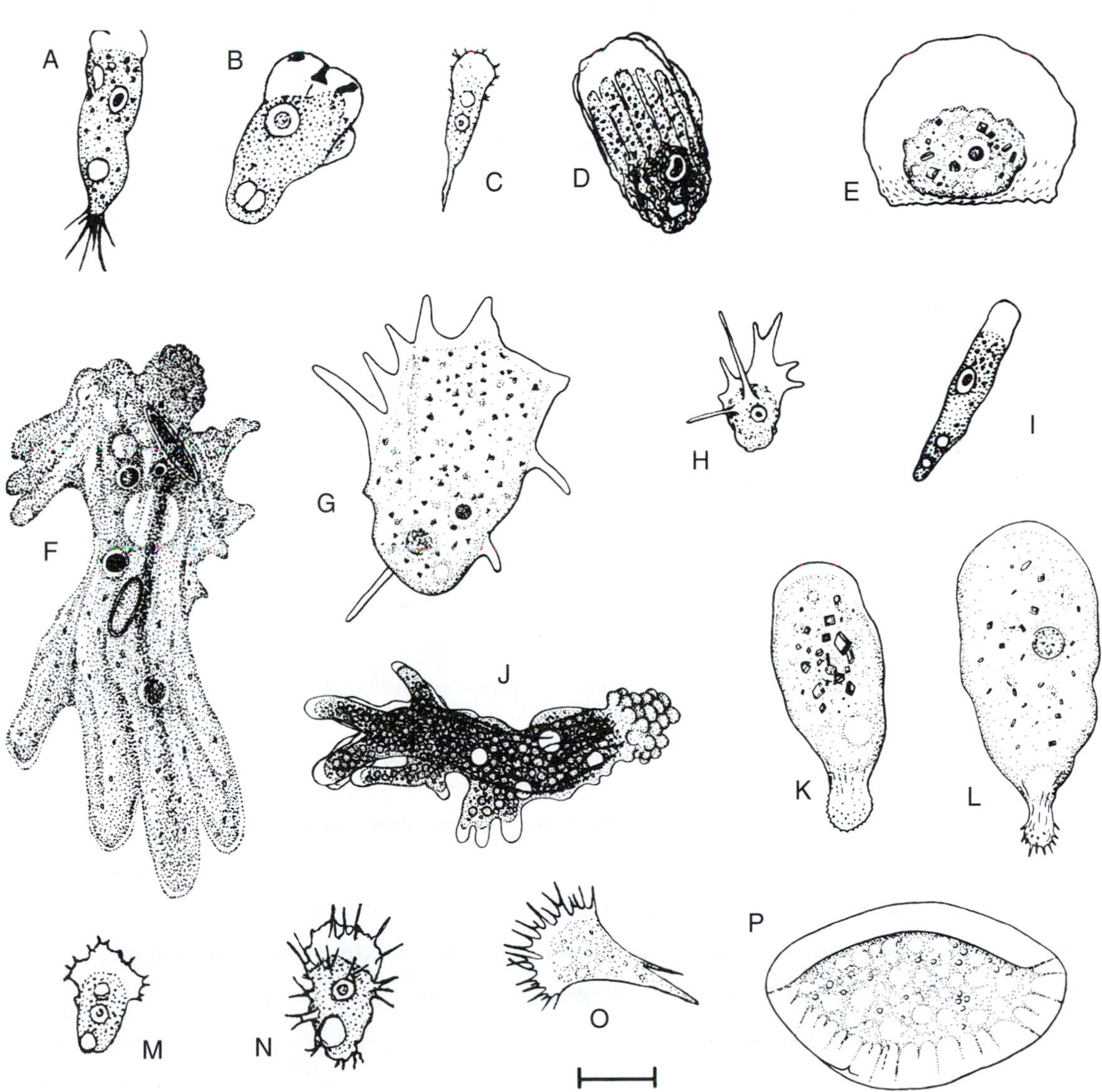

FIGURE 15 (A) *Vahlkampfia avaria;* (B) *Naegleria;* (C) *Stachyamoeba lipophora;* (D) *Thecamoeba sphaeronucleolus;* (E) *Vanella miroides;* (F) *Amoeba proteus;* (G) *Mayorella bigomma;* (H) *Vexillifera telemathalassa;* (I) *Hartmanella vermiformis;* (J) *Chaos illinoisense;* (K) *Saccamoeba lucens;* (L) *Trichamoeba cloaca;* (M) *Echinamoeba exudans;* (N) *Acanthamoeba;* (O) *Filamoeba nolandi* (P) *Hylodiscus rubicundus.* (After: Bovee, 1985, A, B, C, D, H, I, J, M, N, O; Kudo, 1966, F; Page, 1988, E, G, K, L, P.) Scale = 10 μm for A, B, C, E, I, M; 15 μm for H, N, O, P; 30 μm for D, G, K, L; 50 μm for F; and 100 μm for J.

5b. Usually flattened; frequent changes in shape typical; sometimes eruptive; subpseudopodia usually present, often furcate 11

6a (5a). Without subpseudopodia 7

6b. With subpseudopodia 9

7a. (6a). Cell body flattened 8

7b. Cell body subcylindrical 10

8a (7a). Cell usually oblong; cresent-shaped hyaline margin at anterior end; pellicle-like layer, with dorsum often wrinkled and/or ridged; usually uninucleate; no cytoplasmic crystals. Thecamoebidae (*Thecamoeba*; Fig. 15D)

8b. Body usually fan-shaped, oval, or spoon-shaped, with hyaline margin occupying up to half of length. Vannellidae (*Vannella,* Fig. 15E)

9a (6b). Subpseudopodia hyaline, blunt, digitiform, usually from anterior hyaline margin; uninucleate; nucleolar material in central body. Paramoebidae (*Mayorella,* Fig. 15G)

9b. Few slender, conical, or linear subpseudopodia, from anterior hyaline margin or cell surface; uninucleate. Vexilliferidae (*Vexillifera,* Fig. 15H)

10a (7b). Most species polypodial; length usually more than 75 μm; uni- or multinucleate; numerous cytoplasmic crystals. Amoebidae (*Amoeba, Chaos, Trichamoeba*; Fig. 15F,J,L)

10b. Cell monopodial, pseudopods rare; uninucleate with central nucleolus; cytoplasmic crystals in some; cysts common. Hartmannellidae (*Hartmannella, Saccamoeba*; Fig. 15I,K)

11a (5b). Cell regularly discoid, flattened ovoid, or fan-shaped; usually broader than wide; postcentral granular mass, usually surrounded, sometimes completely, by hyaline border with short subpseudopodia. Hyalodiscidae (*Hyalodiscus*; Fig. 15P)

11b. Not as above 12

12a (11b). Cell flattened, broad and irregular in outline, though sometimes elongate during locomotion; slender tapering subpseudopodia, sometimes furcate, produced from broad, hyaline lobopodium; often with small lipid globules; uninucleate. Order Acanthopodida, Acanthamoebidae (*Acanthamoeba*; Fig. 15N)

12b. Several to many fine, sometimes furcate subpseudopodia, finer than in Acanthamoebidae. Echinamoebidae (*Echinamoeba, Filamoeba*; Fig. 15M,O)

13a (4b). Test more or less rigid with distinct aperture. Order Arcellinida 11

13b. Discoid or sometimes globose amoeba incompletely closed in a flexible tectum, no well defined aperature. Order Himatismenida, Cochliopodiidae (*Cochliopodium,* Fig. 16A)

14a (13a). Pseudopods conical, clear, sometimes anastomosing. Suborder Phryganellina, Phryganellidae (*Phryganella*; Fig. 16B–C)

14b. Pseudopods digitate and finely granular 15

15a (14b). Test membranous or chitinoid, pliable or rigid; no plates or scales, but may have attached debris. Suborder Arcellina 16

15b. Test chitinoid or not, rigid, with embedded and/or attached plates, scales, siliceous granules. Suborder Difflugina 18

16a (15a). Test flexible to semi-rigid; aperture ventral with variable shape. Microcoryciidae (*Penardochlamys*; Fig. 16H–I)

16b. Test rigid, chitinoid 17

17a (16b). Test often areolar, smooth; aperture ventral, round. Arcellidae (*Arcella, Pyxidicula*; Fig. 16D,F–G)

17b. Test oval to flask-shaped, nonareolar, clear; aperture terminal. Hyalospheniidae (*Hyalosphenia*; Fig. 16L–M)

18a (15b). Aperture round, broadly oval or wavy. Difflugiidae (*Difflugia, Lesquereusia*; Fig. 16J–K,N)

18b. Aperture slit-like or narrowly oval 19

19a (18b). Aperture terminal 20

19b. Aperture antero-ventral, invaginated, slit-like with overhanging lip. Plagiopyxidae (*Plagiopyxis*; Fig. 16E)

20a (19a). Test particles rectangular. Paraquadrulidae (*Quadrulella*; Fig. 16O–P)

20b. Test particles not rectangular. Nebelidae (*Nebela*; Fig. 16Q)

21a (1d). Without distinct test, sometimes with scales. Order Aconchulinida 22

21b. With test. Order Testaceafilosida 23

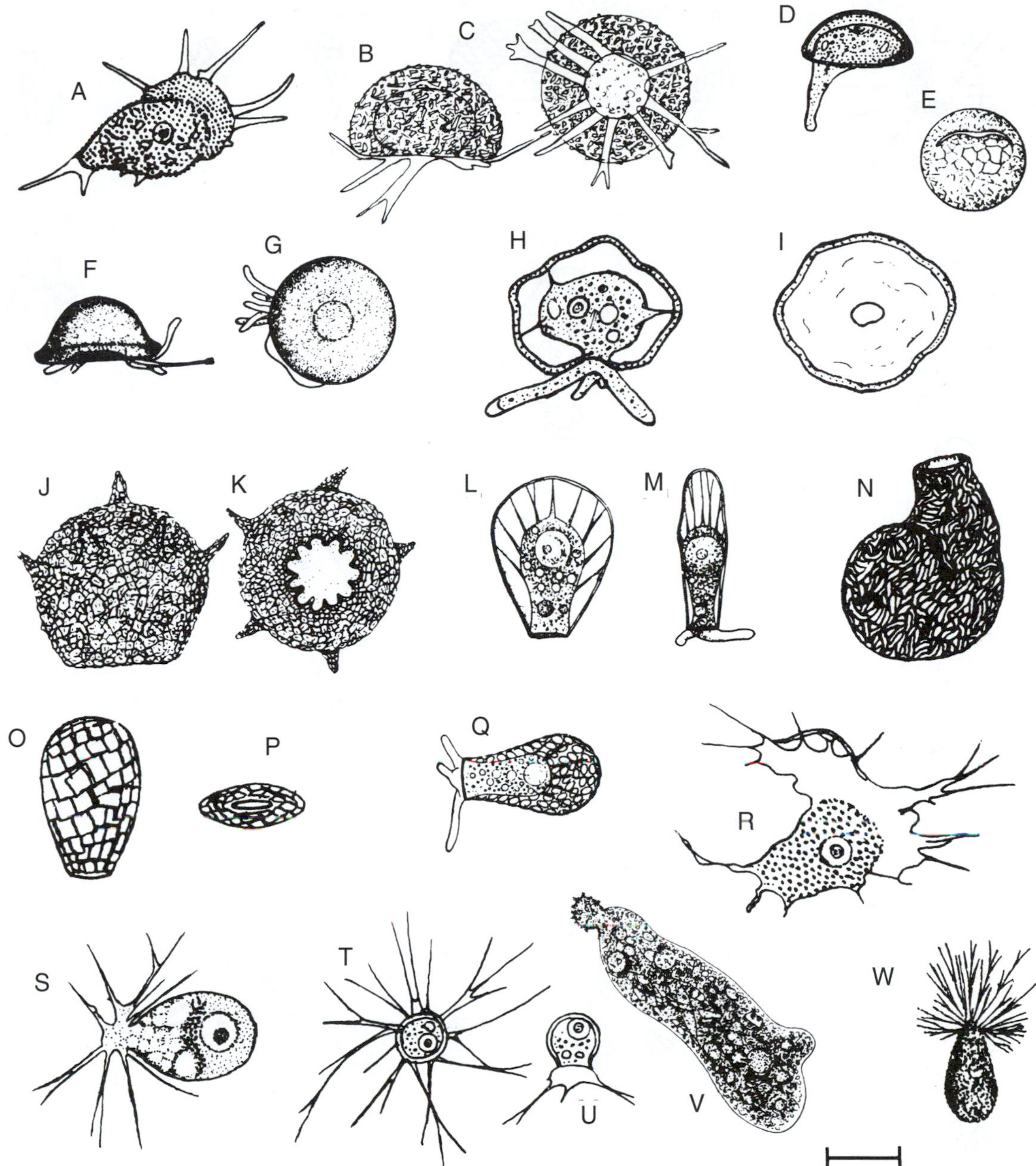

FIGURE 16 (A) *Cochliopodium bilimbosum;* (B, C) *Phryganella nidulus;* side and oral views; (D) *Pyxidicula operculata;* (E) *Plagiopyxis callida;* (F, G) *Arcella vulgaris,* side and dorsal views; (H, I) *Penardochlamys arcelloides,* side and oral views; (J, K) *Difflugia corona,* side and oral views; (L, M) *Hyalosphenia cuneata;* (N) *Lesquereusia spiralis;* (O, P) *Quadrulella symmetrica;* (Q) *Nebela collaris;* (R) *Penardia granulosa;* (S) *Chlamydophrys minor;* (T, U) *Lecythium hyalinum,* dorsal and side; (V) *Pelomyxa palustrus;* (W) Pseudo-difflugia gracilis; (After: Bovee, 1985, A, B, C, H, I, J, K, N, O, P, R, T, U; Deflandre, 1959, D, E, F, G, L, M, Q, S, W; Kudo, 1966, V.) Scale = 10 μm for C, R, S; 25 μm for A, H; 30 μm for L, T, W; 45 μm for N, O; 60 μm for Q; 90 μm for B, E, F, J; and 500 μm for V.

22a (21a). Filopodia nonanastomosing and more or less radiate. Vampyrellidae (*Vampyrella*; Fig. 17A)

22b. Filopodia distally branching and anastomosing. Reticulosidae (*Penardia*; Fig. 16R)

23a (21b). Test without scales; may have spines, attached debris........24

23b. Test with secreted, siliceous scales arranged in definitive patterns........25

24a(23a). Test round, thin; may have spines or spicules. Chlamydophryidae (*Chlamydophrys, Lecythium*; Fig. 16S–U)

24b. Test not round. Pseudodifflugiidae (*Pseudodifflugia*; Fig. 16W)

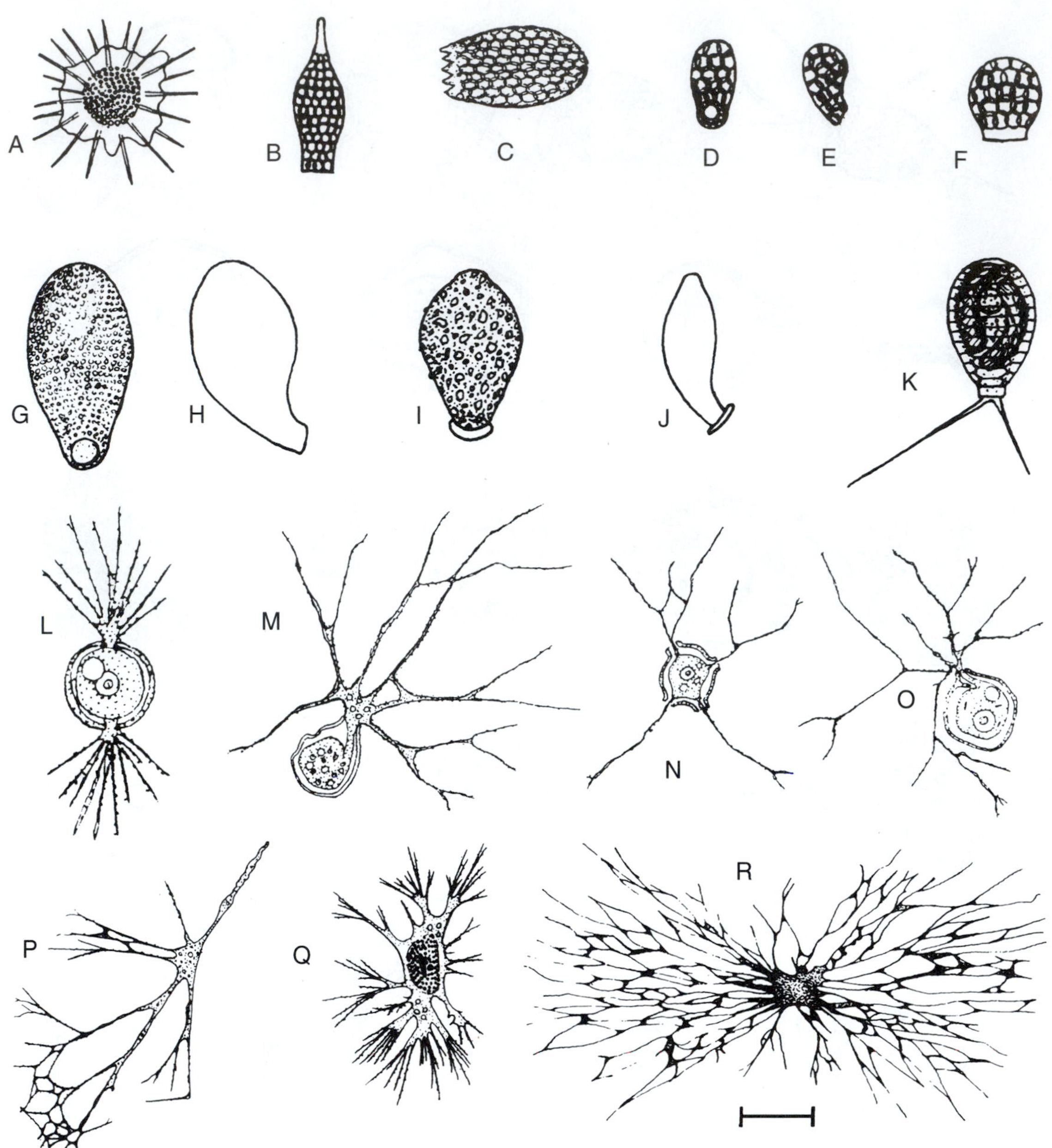

FIGURE 17 (A) *Vampyrella lateritia;* (B) *Paraeuglypha reticulata;* (C) *Euglypha tuberculata;* (D, E) *Trinema enchelys,* oral and side views; (F) *Sphenoderia lenta;* (G, H) *Cyphoderia ampulla;* (I, J) *Campascus triqueter;* (K) *Paulinella chromataphora;* (L) *Diplophyrys archeri;* (M) *Lieberkuehnia wagnerella;* (N) *Microcometes paludosa;* (O) *Microgromia haeckeliana;* (P) *Biomyxa vagans;* (Q) *Chlamydomyxa montana;* (R) *Reticulomyxa filosa.* (After: Bovee, 1985, B, D, E, F, G, H, I, J, K, L, M, N, O, P, Q, R; Deflandre, 1959, A, C.) Scale = 10 μm for L, N, O; 15 μm for K; 25 μm for A, B, C; 40 μm for D, F, G; 50 μm for I, M, Q; 80 μm for P; and 10,000 μm for R.

25a (23b). Scales round to elliptical, thin, overlapping, adjacent or scattered. Euglyphidae (*Paraeuglypha, Euglypha, Trinema, Sphenoderia;* Fig. 17B–F)

25b. Scales not as above 26

26a (25b). Scales thick, cylindrical; test usually with aperture at end of a neck bent to one side. Cyphoderiidae (*Cyphoderia, Campascus;* Fig. 17G–I)

26b. Scales long, with long axes perpendicular to pseudostome; with two symbiotic cyanobacterial filaments. Paulinellidae (*Paulinella;* Fig. 17K)

27a (1e).	Phylum Granuloreticulosea. With a test. Class Monothalamea	28
27b.	Without a test. Class Athalamea	30
28a (27a).	Test with one pseudostome (opening)	29
28b.	With more than one pseudostomes. Amphitremidae (*Diplophyrys, Microcometes*; Fig. 17L,N)	
29a (28a).	Test flattened on one side. Microgromiidae (*Microgromia*; Fig. 17O)	
29b.	Test not flattened on one side. Lieberkuehnidae (*Lieberkuehnia*; Fig. 17M)	
30a (27b).	Naked amoebae with or without anastomosing pseudopodia; one or a few nuclei Order Athalamida	31
30b.	Naked, multinucleate and highly reticulate plasmodia. Order Promycetozoida, Family Reticulomyxidae (*Reticulomyxa*; Fig. 17R)	
31a (30a).	Body mass changeable in form. Biomyxidae (*Biomyxa*; Fig. 17P)	
31b.	Body mass more or less round and central. Chlamydomyxidae (*Chlamydomyxa*; Fig. 17Q)	
32a (1f).	Class Heliozoa; Axopodia without visible axonemes	33
32b.	Axonemes usually visible with light microscopy	34
33a (32a).	Cell enclosed in latticed organic capsule (skeleton); generally stalked; no centroplast; cell body spherical in adults. Order Desmothoracida (*Clathrulina*; Fig. 18A)	

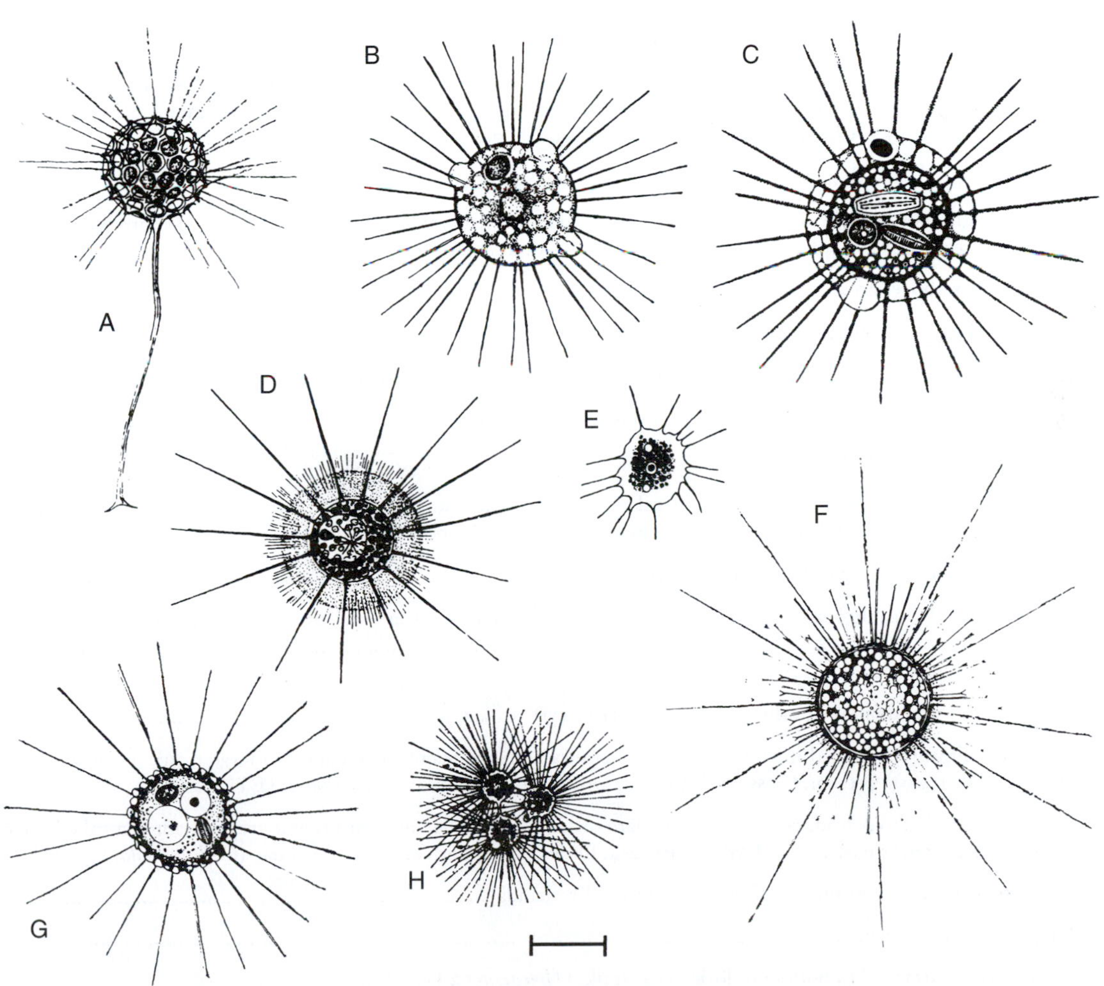

FIGURE 18 (A) *Clathrulina elegans;* (B) *Actinophrys sol;* (C) *Actinosphaerium eichhorni;* (D) *Heterophrys myriopoda;* (E) *Ciliophrys infusionum;* (F) *Acanthocystis turfacea;* (G) *Lithocolla globosa;* (H) *Raphidiophrys elegans.* (After: Deflandre, 1959, H; Kudo, 1966, B, C, D, E; Rainer, 1968, A, F, G.) Scale = 15 μm for E; 30 μm for B, D, G; 50 μm for A; 75 μm for F, H; and 160 μm for C.

33b. Lack centroplast and apparently lack axonemes, the lack of microtubules (axonemes) in axopods has not been confirmed and this order may be abandoned. Order Rotosphaerida (*Lithocolla*; Fig. 18G)

34a (32b). Stalked or not; long thin granule-studded axopods that usually arise from centroplast; axonemes frequently discernible with light microscopy; usually with skeleton of siliceous or organic plates and/or spicules; highly contractile axopods or stalk. Order Centrohelida. (*Heterophrys, Acanthocystis, Raphidiophrys*; Figs. 7, 18D, F,H)

34b. No skeleton 35

35a (34b). No centroplast or axoplast; axopods granual-studded and thicker at bases; large central nucleus surrounded by lacunar ectoplasm or several nuclei at periphery of central area with vesicular ectoplasm. Order Actinophryida (*Actinophrys, Actinosphaerium*; Fig. 18B–C)

35b. Microtubule organizing center of dense plaques from the nuclear membrane or from centroplast; may be confused with Actiniphryida without knowledge of fine structure (origin and pattern of axopod microtubules). Order Ciliophyrida (*Ciliophrys*; Fig. 18E)

E. Taxonomic Key to Major Groups and Common Genera of Ciliates

1a. Suctorial tentacles present; no true cytostome or cytopharynx; adults (trophonts) usually sessile, many species ectosymbiotic, some planktonic species; cilia absent except in free-swimming dispersal stages; with or without lorica. Class Phyllopharygea, Subclass Suctoria 9

(Note: Patterns of division and release of the larvae form the basis for dividing the subclass into orders. Other characters are useful in identification, but knowledge of the full life cycle is often required before suctorians can be confidently assigned to a genus.)

1b. Cilia present in unencysted stages; no suctorial tentacles 2

2a (1b). Conspicuous buccal ciliature at apical pole; buccal ciliature winds clockwise toward the center when viewed from oral end; somatic cilature reduced or absent; mobile or sessile; solitary, gregarious or colonial; some species loricate; oral region can contract and withdraw in most species. Class Oligohymenophora, Subclass Peritrichia 17

2b. Not as above 3

3a (2b). Oral area usually bordered by a well-developed adoral zone of membranelles (AZM) consisting of more than three membranelles (polykineties); with or without ventral cirri 26

3b. Oral area not as above; no ventral cirri 4

4a (3b). Cytostome at or near surface; buccal cavity, if present, without cilia or paroral membranes 5

4b. Cytostome at end of a buccal cavity with cilia or paroral membrane(s) associated 6

5a (4a). Cytostome at or near anterior end, or continuing down side as a slit 52

5b. Cytostome lateral or ventral 7

6a (4b). Paroral membrane(s) within or leading into buccal cavity 65
(Sometimes difficult to see; follow both alternatives in key when in doubt)

6b. No paroral membrane; simple cilia in buccal cavity 82

7a (5b). Cytostome a barely visible lateral slit on the convex side of a tapering front end, or a lateral opening at the base of an anterior proboscis 63
(see also Spathidiidae, 55)

7b. Circular mouth located mid-ventrally; no proboscis; large cyrtos 8

8a (7b). Body nearly ellipsoid, rounded in cross section, sometimes flattened ventrally; medium to large (some $>100\ \mu m$); densely ciliated all over; oral depression present. Class Nassophorea, Order Nassulida, Nassulidae (*Nassula*; Fig. 20Z)

8b. Body usually flattened; ventrum ciliated, dorsum bare or with a few cilia; anterior preoral arcs of right ventral ciliary rows continuous with more posterior parts. Class Phyllopharyngea, Order Cyrtophorida, Chilodonellidae (*Chilodonella*; Fig. 24T)

9a (1a). Budding and cytokinesis on surface of trophont. Order Exogenida 10

9b. Budding begins in a pouch 12

10a (9a). Trophont basally attached to bottom of lorica near stalk. (*Thecacineta*; Fig. 19A)

10b. Not as above 11

11a (10b). With vase-like lorica, sometimes with stalk-like base; tentacles extend through one or more slits. Urnulidae (*Metacineta, Paracineta*; Fig. 19B–D)

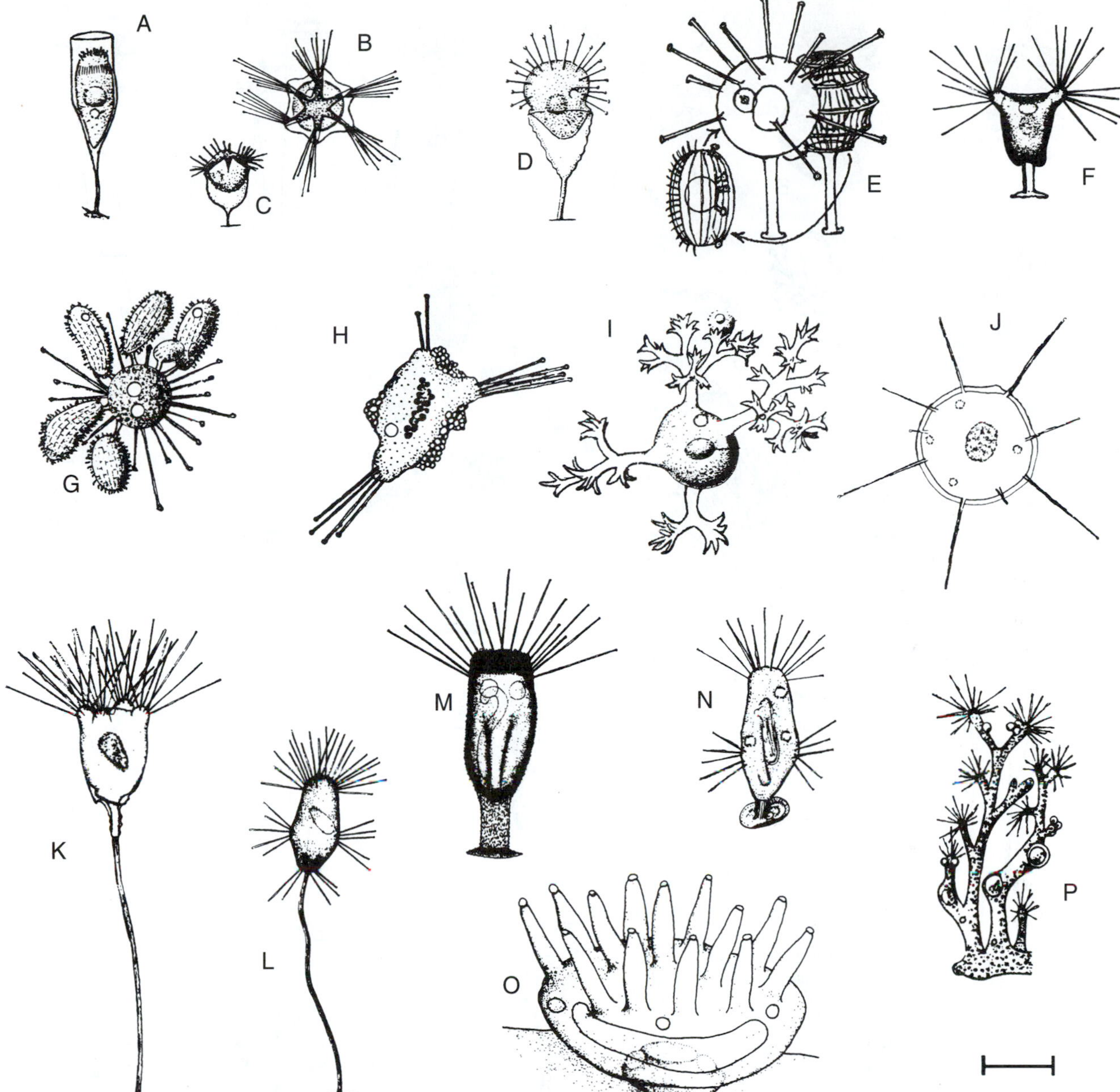

FIGURE 19 (A) *Thecacineta cothurniodes;* (B, C) *Metacineta mystacina,* top and side views; (D) *Paracineta crenata;* (E) *Podophyra fixa,* showing trophont, encysted form, and swarmer; (F) *Acineta limnetis;* (G) *Sphaerophyra magna;* (H) *Trichophyra epistylidis;* (I) *Dendrocometes paradoxus;* (J) *Heliophyta reideri;* (K) *Tokophyra quadripartita;* (L) *Multifasciculatum elegans;* (M) *Squalorophyra macrostyla;* (N) *Discophyra elongata;* (O) *Stylocometes digitalis;* (P) *Dendrosoma radians.* (After: Corliss, 1979, P; Goodrich and Jahn, 1943, F, K, L, M; Kent, 1880–1882, G, I; Matthes, 1972, J, O; Noland, 1959, A, B, C, D, N; Small and Lynn, 1985, E, H.) Scale = 15 μm for E, H, J, O; 30 μm for A, D, F, G; 50 μm for I, L, M, N; 75 μm for B, K; and 200 μm for P.

11b. Trophont small, pyriform to spherical; usually stalked; tentacles apical or evenly distributed; often attached to other ciliates. Podophryidae (*Podophyra, Sphaerophyra*; Fig. 19E,G)

12a (9b). Budding and cytokinesis completed in brood pouch. Order Endogenida 13

12b. Cytokinesis begun in pouch, completed exogenously. Order Evaginogenida 15

13a (12a). Trophont without stalk or stalk-like process; attached to substrate by broad part of body. Dendrosomatidae (*Dendrosoma, Trichophyra*; Fig. 19H,P)

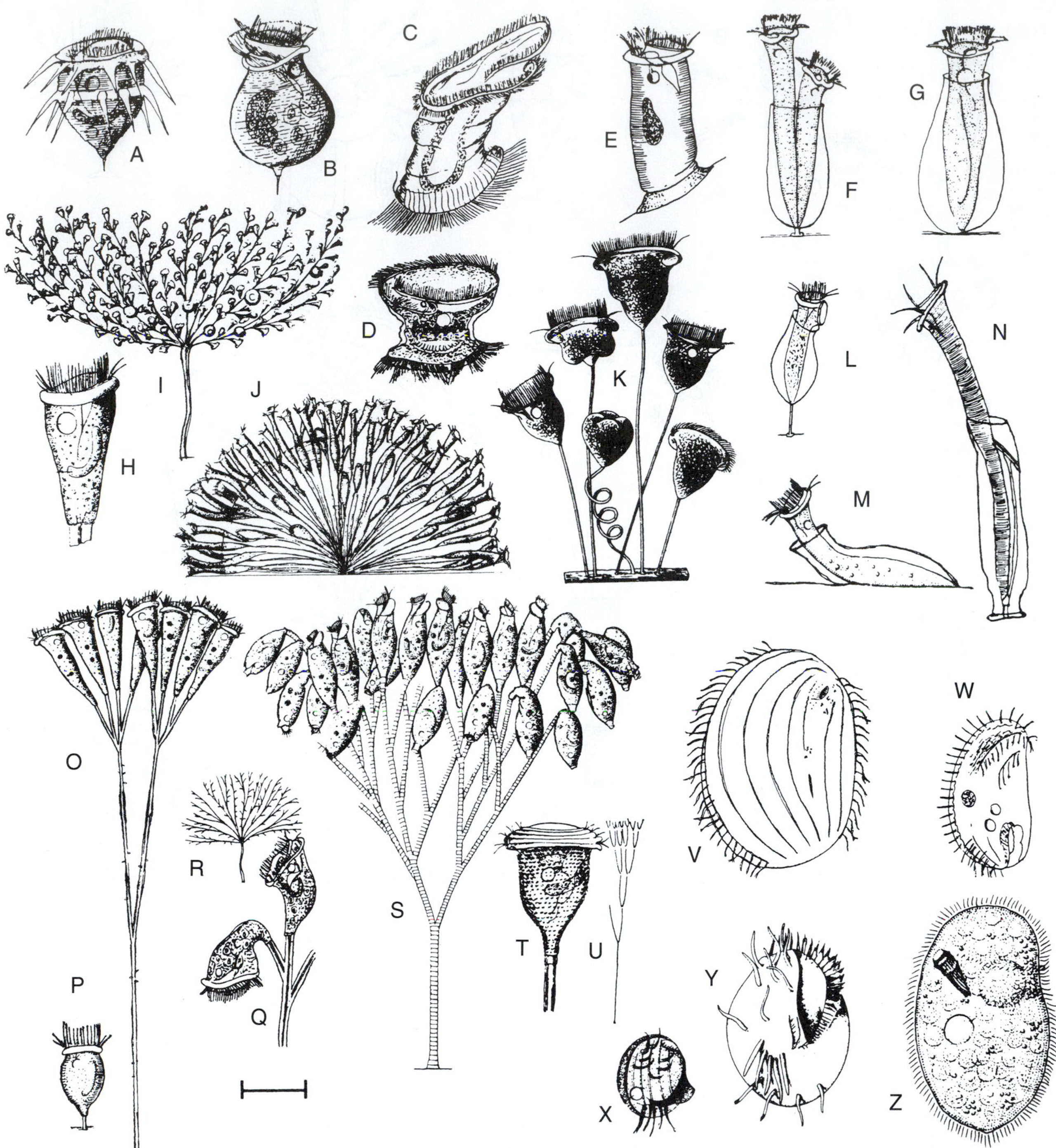

FIGURE 20 (A) *Hastatella radians;* (B) *Astylozoon faurei;* (C) *Urceolaria mitra;* (D) *Trichodina pediculis;* (E) *Scyphidia physarum;* (F) *Cothurnia imberbis;* (G) *Vaginicola ingenita;* (H, I) *Zoothamnium arbuscula,* individual and colony; (J) *Ophrydium eichhorni;* (K) *Vorticella campanula;* (L) *Pyxicola affinis;* (M) *Platycola longicollis;* (N) *Thuricola folliculata;* (O) *Epistylis plicatilis;* (P) *Rhabdostyla pyriformis;* (Q, R) *Carchesium polypinum,* individual and colony; (S) *Opercularia nutans;* (T, U) *Campanella umbellaria,* individual and colony; (V) *Pseudomicrothorax agilis;* (W) *Microthorax pusillus;* (X) *Aspidisca costata;* (Y) *Euplotes patella;* (Z) *Nassula ornata.* (After: Corliss, 1959, V, Y; Kahl, 1930–1935, A, B, C, D, E, H, L, N, Q, R, T, U, W; Kent, 1880–1882, I, J, K, O, S, X; Noland, 1959, F, G, M, P.) Scale = 15 μm for V, W; 20 μm for A, B, G, P; 25 μm for D, H, F, X; 30 μm for C, Z; 40 μm for E, L, M, N, S; 50 μm for O, Y; 75 μm for K, Q, U; and 200 μm for J.

13b. Not as above 14

14a (13b). With lorica, usually stalked; tentacles in a few fascicles or distributed over body. Acinetidae (*Acineta*; Fig. 19F)

14b. Lacks lorica; tentacles single, fascicles, or evenly distributed. Tokophryidae (*Tokophyra, Multifasciculatum*; Fig. 12 and 19K,L)

15a (12b). Conical tentacles; body often hemispherical . Dendrocometidae (*Dendrocometes, Stylocometes*; Fig. 19I,O)

15b. Not as above, capitate tentacles 16

16a (15b). Body discoidal, flattened against substrate, with peripheral attachment ring. Heliophryidae (*Heliophrya,* Fig. 19J)

16b. Thin tentacles; body spherical to ovoid; some species loricate and/or stalked. Discophryidae (*Discophrya, Squalorophyra*; Fig. 19M,N)

17a (2a). Trophont mobile; symbiotic on other organisms; with complex, aboral holdfast. Order Mobilida 18

17b. Trophont only rarely motile, attached with stalk; often attached to other organisms; many gregarious or colonial species. Order Sessilida 19

18a (17a). Denticles of holdfast with hooks and spines; ectosymbionts on *Hydra,* fishes, Amphibia. Trichodinidae (*Trichodina*; Fig. 20D)

18b. Denticles of holdfast simple, toothed; elongated macronucleus, ectosymbionts on turbellarians. Urceolariidae (*Urceolaria;* Fig. 20C)

19a (17b). Free-swimming; aboral end drawn to point with one or two rigid bristles or cilia; swim with oral end forward. Astylozoidae (*Astylozoon, Hastatella*; Fig. 20A,B)

19b. Aborally attached to substratum directly or by stalk 20

20a (19b). Colonial, with zooids in gelatinous matrix. Ophrydiidae (*Ophrydium;* Fig. 20J)

20b. Not in gelatinous matrix 21

21a (20b). Stalk absent, tapering body may resemble short stalk. Scyphidiidae (*Scyphidia*; Fig. 20E)

21b. Stalked, or, if stalkless, in lorica 22

22a (21b). With contractile stalk; solitary or colonial 23

22b. If stalked, stalk not contractile; solitary or colonial 24

23a (22a). Each zooid independently contractile; if solitary, entire stalk contractile. Vorticellidae (*Carchesium, Vorticella*; Fig. 20K,Q–R)

23b. Entire colony contracts in unison due to shared, continuous myonemes. Zoothamniidae (*Zoothamnium*; Fig. 20H–I)

24a (22b). With vase-shaped lorica, stalkless or short stalked; slender body attached to lorica aborally. Vaginicolidae (*Cothurnia, Platycola, Pyxicola, Thuricola, Vaginicola*; Fig. 20F,G,L–N)

24b. Not as above 25

25a (24b). Stalked; peristomal area raised on short neck with furrow separating it from margin of zooid. Operculariidae (*Opercularia;* Fig. 20S)

25b. Stalked; solitary or colonial; peristomal area not on neck, no furrow. Epistylididae (*Campanella, Epistylis, Rhabdostyla;* Fig. 20O, P, T, U)

26a (3a). Locomotor cirri present on ventral surface; generally dorso-ventrally flattened. Class Spirotrichea, subclass Hypotrichia 27

26b. Not as above 36

27a (26a). Adoral zone of membranelles well developed, prominent 28

27b. AZM reduced; no marginal or caudal cirri, ventral cirri conspicuous. Class Spirotrichea, Subclass Hypotrichia, Order Euplotida, Aspidiscidae (*Aspidisca*; Fig. 20X)

28a (27a). Right marginal row of cirri absent, one or more left marginal cirri; large caudal cirri; paroral membrane only as a field on right of oral area. Class Spirotrichea, Subclass Hypotrichia, Order Euplotida (*Euplotes*; Fig. 20Y)

28b. Left oral cilia as a "collar" and "lapel", right oral cilia variable with one or more paroral membranes. Class Spirotrichea, Subclass Stichotrichia 29

29a (28b). Ventral cirri not in distinct file from anterior to posterior (exception: Gastrostyla) Oxytrichidae (*Gastrostyla, Oxytricha, Stylonychia*; Fig. 21A,C,E)

29b. Ventral cirri in distinct linear or zig-zag file(s) 30

30a (29b). Frontoventral cirri run length of ventrum as zig-zag files. Urostylidae (*Urostyla, Uroleptus*; Fig. 21B,D)

30b.	Ventral cirri in one or more linear files of varied length	31
31a (30b).	Body in lorica; without marginal cirri; anterior end long, narrow, contractile, Chaetospiridae (*Chaetospira*; Fig. 21N,O)	
31b.	Not as above	32
32a (31b).	Files of cirri curved or spiralled along body	33
32b.	Files of cirri straight or oblique, ventral only	35
33a (32a).	Files of cirri end on dorsum; oral region extends over one-fourth of body length. Strongylidiidae (*Strongylidium*; Fig. 21P)	
33b.	Not as above	34

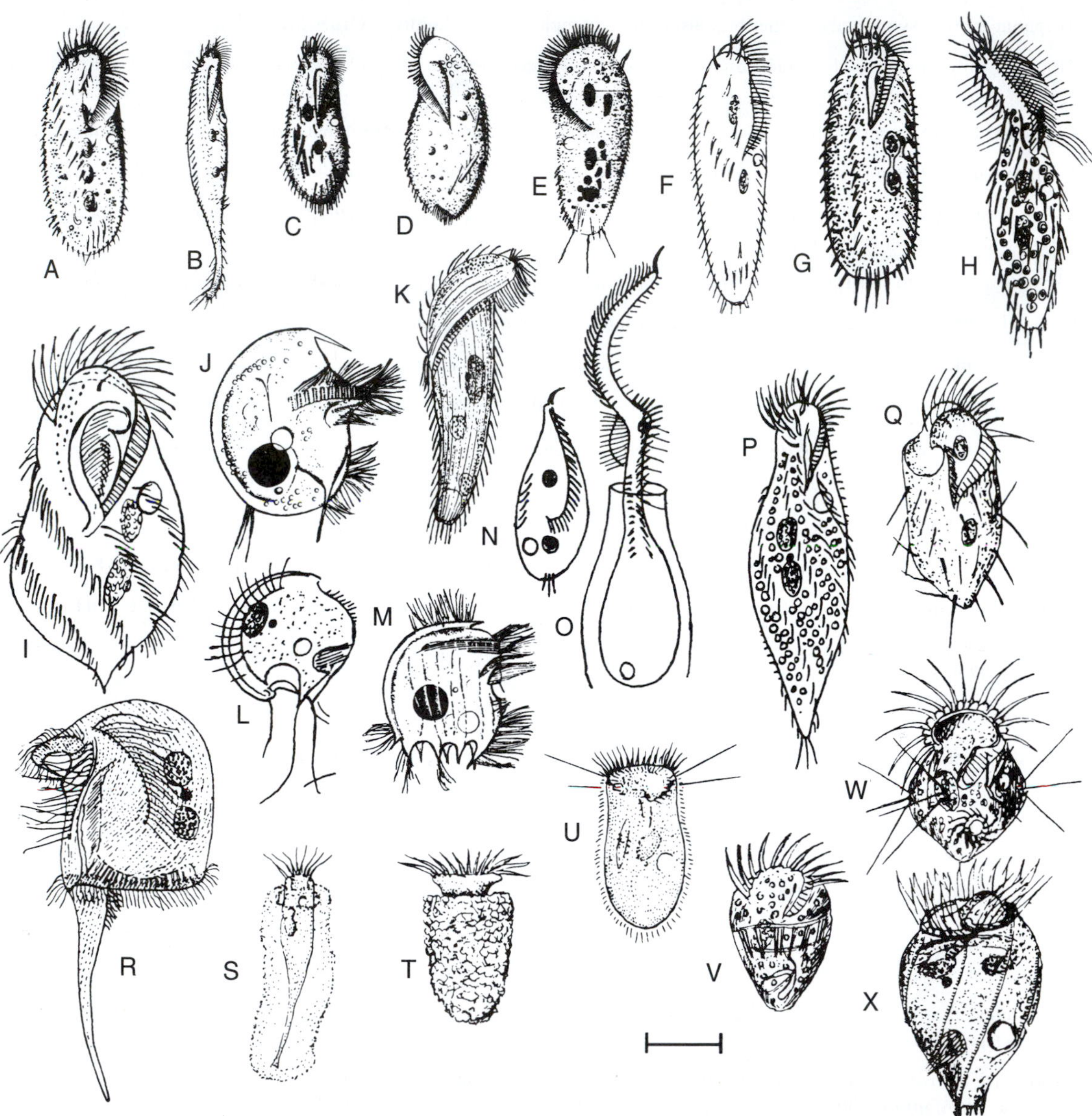

FIGURE 21 (A) *Gastrostyla steini;* (B) *Uroleptus piscis;* (C) *Oxytricha fallax;* (D) *Urostyla grandis;* (E) *Stylonychia mytilus;* (F) *Gonostomum affine;* (G) *Amphisiella oblonga;* (H) *Stichotricha aculeata;* (I) *Hypotrichidium conicum;* (J) *Discomorphella pectinata;* (K) *Metopus es;* (L) *Myelostoma flagellatum;* (M) *Saprodinium dentatum;* (N, O) *Chaetospira mülleri,* contracted and extended forms; (P) *Strongylidium crassum;* (Q) *Psilotricha acuminata;* (R) *Caenomorpha medusula;* (S) *Tintinnidium fluviatile;* (T) *Tintinnopsis cylindricum;* (U) *Strombidinopsis setigera;* (V) *Strombidium viride;* (W) *Halteria grandinella;* (X) *Strombilidium gyrans.* (After: Jankowski, 1964a, b, J, M; Kahl, 1930–1935, F, G, H, I, K, L, N, O, P, Q, R, V, W, X; Kent, 1880–1882, A, B, C, D, E; Noland, 1959, S, T, U.) Scale = 15 μm for L; 25 μm for H, W, X; 30 μm for F, I, J, P,Q, R, T; 40 μm for G, K, M, N, O, S, U, V; 60 μm for B, E; 80 μm for A, C; and 140 μm for D.

34a (33b). Ventral files of cirri curve to left rear. Psilotrichidae (*Psilotricha*; Fig. 21Q)

34b. Ventral files of cirri spiral obliquely around body. Spirofilidae (*Hypotrichidium, Stichotricha*; Fig. 21H,I)

35a (32b). Files of cirri parallel to long axis of oral region along its right border. Gonostomatidae (*Gonostomum*; Fig. 21F)

35b. Files of cirri not parallel to oral region, one or more of the ventral cirral files extends from anterior to well past mid-ventrum. Amphisiellidae (*Amphisiella*; Fig. 21G)

36a (26b). Body surface ciliated 46

36b. Somatic ciliature sparse or absent 37

37a (36b). Body flattened, rigid, often with spines; body ciliature present as short, generally obvious rows; oral ciliature relatively inconspicuous; mainly anaerobic. Class Spirotrichea, Order Odontostomatida 38

37b. Not as above 40

38a (37a). Body box-shaped, generally with short posterior spines; short rows of body cilia at front and rear, those in front parallel to and forward of oral opening. Epalxellidae (*Saprodinium*; Fig. 21M)

38b. Body round to discoidal 39

39a (38b). Distinct anterior spine; band of cilia on ridge overhanging buccal cavity. Discomorphellidae (*Discomorphella*; Fig. 21J)

39b. Body ciliature very sparse; lacks spines; pair of cirri at posterior end. Myelostomatidae (*Myelostoma*; Fig. 21L)

40a (37b). Membranelles numerous in complete or almost complete circle at oral end; body generally conical or bell-shaped; mostly planktonic. Class Spirotrichea, Subclasses Choreotrichia and Oligotrichia 41

40b. Body rounded or conical, spirally twisted to left; twisting of body less prominent than Metopidae; often with long caudal spine; oral region spiralled; dense preoral cilia along anterior border of oral cavity; body cilia absent except for caudal tuft and anterior cirrus-like tufts; anaerobic. Class Spirotrichea, Order Armophorida (*Metopus, Caenomorpha*; Fig. 21M,R)

41a (40a). Oral ciliature forms a closed circle. Subclass Choreotrichia 42

41b. Oral cilature forms an open anterior collar with an anterioventral "lapel". Subclass Oligotrichia 45

42a (41a). Loricate, attached to inner wall of lorica 43

42b. Not loricate 44

43a (42a). Delicate, gelatinous or mucoid tubular lorica, with attached debris, often translucent . Tintinnididae (*Tintinnidium*; Fig. 21S)

43b. Lorica rigid, agglomerate of mineral grains and diatom fragments. Codonellidae (*Tintinnopsis*; Fig. 21T)

44a (42b). Rows of somatic ciliature equally distributed around body, extending length of body; body cilia may be long. Strombidinopsidae (*Strombidinopsis*; Fig. 21U)

44b. One or more usually short rows of body cilia; body ciliature short. Suborder Strobilidiina, Strobilidiidae (*Strobilidium*, Fig. 21X)

45a (41b). Somatic ciliature reduced to a few bristle-like cirri arranged circumferentially around the equator effecting a jumping motion. Halteriidae (*Halteria*; Fig. 21W)

45b. Somatic ciliature greatly reduced or absent, no bristles; sometimes with prominent girdle close to middle of cell. Strombidiidae (*Strombidium*; Fig. 21V)

46a (36a). Uniform somatic ciliation; often with conspicuous tuft of caudal cilia; buccal membranelles large, but not always obvious; anterior end of body twisted left; oral region spiralled with dense preoral ciliation; in richly organic environments with low oxygen. Class Spirotrichea, Order Armophorida, Metopidae (*Metopus;* Fig. 21K)

46b. Not as above 47

47a (46b). Somatic ciliature well developed, uniform, usually inserts on oral region; body size large; many contractile species; rarely loricate; some species pigmented. Class Heterotrichea, Order Heterotrichida 48

47b. Somatic ciliature well developed; body large (500–1,000 μm), broad; buccal cavity prominent, funnel-like; macronucleus elongate; carnivorous. Class Colpodea, Order Bursariomorphida, Bursariidae (*Bursaria*; Fig. 22C)

48a (47a). Body trumpet-shaped; highly contractile; oral ciliature spirals clockwise around flared anterior end; often pigmented and/or with algal endosymbionts. Stentoridae (*Stentor*; Fig. 22F)

48b. Body not trumpet-shaped 49

49a (48b). Body laterally compressed, pyriform to elipsoid; often pigmented pink, red, or purple; long, narrow peristome on left margin; noncontractile. Blepharismidae (*Blepharisma*; Fig. 22B)

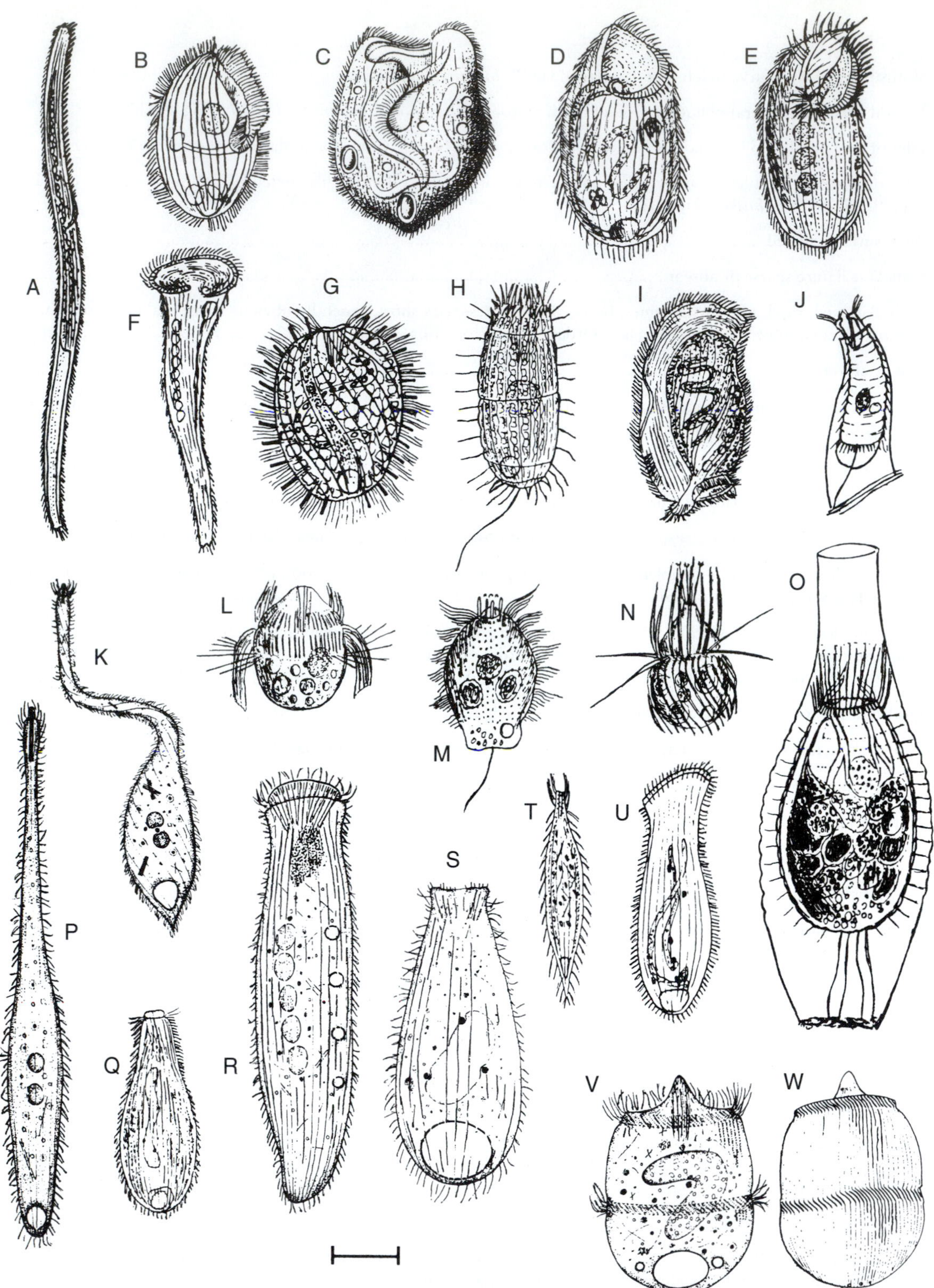

FIGURE 22 (A) *Spirostomum minus;* (B) *Blepharisma lateritium;* (C) *Bursaria truncatella;* (D) *Climacostomum virens;* (E) *Condylostoma tardum;* (F) *Stentor polymorphus,* half extended; (G) *Actinobolina radians;* (H) *Coleps hirtus;* (I) *Bryophyllum lieberkühni;* (J) *Metacystis recurva;* (K) *Lacrymaria olor;* (L) *Askensia volvox;* (M) *Urotricha farcta;* (N) *Mesodinium pulex;* (O) *Vasicola ciliata;* (P) *Trachelophyllum apiculatum;* (Q) *Enchelyodon elegans;* (R) *Homalozoon vermiculare;* (S) *Enchelys simplex;* (T) *Chaena teres;* (U) *Spathidium spathula;* (V, W) *Didinium nasutum,* live and silver stained. (After: Dragesco, 1966a, K, S, V, W; Dragesco, 1966b, P, R; Kahl, 1930–1935, A, B, D, E, F, G, H, I, J, L, M, N, O, Q, T, U; Kent, 1880–1882, C.) Scale = 10 μm for M, N; 20 μm for H, J, L, P, S; 30 μm for G, O, U; 40 μm for B, K, T; 60 μm for E, Q, R; 80 μm for D, V, W; 100 μm for A, F, I; and 200 μm for C.

49b. Body shape not as above 50

50a (49b). Body large, very long; oral cavity shallow; peristomal area long, narrow, oral ciliature sometimes inconspicuous; contractile. Spirostomidae (*Spirostomum*; Fig. 22A)

50b. Buccal ciliature prominent; oral cavity deeper than above 51

51a (50b). Buccal cavity occupying much of anterior part of body; paroral membrane inconspicuous; somatic ciliation dense; with or without endosymbiotic algae. Climacostomidae (*Climacostomum*; Fig. 22D)

51b. Paroral membrane prominent; contractile; somatic ciliation dense. Condylostomatidae (*Condylostoma*; Fig. 22E)

52a (5a). Body ovoid; with tentacles extending from the cell in all directions when at rest, but retracted and hardly visible when swimming. Class Litostomatea, Order Pharyngophorida, Actinobolinidae (*Actinobolina*; Fig. 22G)

52b. Tentacles absent 53

53a (52b). Translucent $CaCO_3$ plates (armor) in cortex; typically barrel-shaped; body frequently spiny, often with prominent anterior and caudal thorns; long caudal cilium common; brosse present, but inconspicuous. Class Prostomatea, Order Prorodontida, Colepidae (*Coleps*; Fig. 22H)

53b. No such armor, not as above 54

54a (53b). Sessile in pseudochitinous lorica; one or more caudal cilia; paratenes obvious. Class Prostomatea, Order Prostomatida, Metacystidae (*Metacystis, Vasicola*; Fig. 22J,O)

54b. Not as above 55

55a (54b). Mouth at anterior end, rounded or only slightly elongated 56

55b. Apex fan-shaped to varying degree; mouth a long slit, beginning at anterior end; slit may extend down side of cell. Litostomatea, Order Haptorida, Spathidiidae (*Byrophyllum, Spathidium*; Fig. 22I,U)

56a (55a). Mouth at anterior pole, surrounded by an unciliated area and often raised as a blunt cone; circumferential ciliary girdle of closely apposed cilia, otherwise naked or with short cilia. Class Litostomatea, Order Haptorida 57

56b. Uniform ciliation around mouth, and typically over the entire cell 58

57a (56a). Girdle ciliature of one type, one or two girdles. Didiniidae (*Didinium*; Figs. 6, 22V,W)

57b. Body ciliature of two types, one cirrus-like, as a single girdle. Mesodiniidae (*Askenasia, Mesodinium*; Figs. 22L,N)

58a (56b). Rear one-third to one-fifth of body unciliated, except for one or more long caudal cilia; other somatic ciliation evenly distributed. Class Prostomatea, Order Prorodontida, Urotrichidae (*Urotricha*; Fig. 22M) ....

58b. Not as above 59

59a (58b). Body uniformly ciliated at rear; mouth with oral dome or bulb; cytopharynx not permanently inverted, but inverts during ingestion. Class Litostomatea, Order Haptorida 60

59b. Not as above; lack oral dome 62

60a (59a). Anterior tapering to a ciliated neck; neck set off by groove and a circle of longer cilia; contractile . Lacrymariidae (*Lacrymaria*; Fig. 22K)

60b. Not as above 61

61a (60b). Body long, generally >4 times width; ovoid or flask-shaped; oral region simple dome, sometimes pointed. Trachelophyllidae (*Chaenea, Trachelophyllum*; Fig. 22P,T)

61b. Oral region flattened at apex; cytostome circular to ovoid; body usually shorter than four times width. Enchelyidae (*Homalozoon, Enchelys, Enchelyodon*; Fig. 22Q,R,S)

62a (59b). Short "dorsal brush" (brosse) extends backward from apical mouth on one side (may not be obvious without silver staining); body ellipsoid. Class Prostomatea, Order Prorodontida, Prorodontidae (*Prorodon, Pseudoprorodon*; Fig. 23A,B)

62b. Dorsal brush lacking, may have reduced brosse; similar to, and easily confused with Prorodon. Class Prostomatea, Order Prorodontida, Holophryidae (*Holophyra*; Fig. 23C)

63a (7a). Body long, flattened; mouth on convex side of tapering anterior end; ciliated only on right side of body. Class Karyorelictea, Order Loxodida, Loxodidae (*Loxodes*; Fig. 23J)

63b. Not as above 64

64a (63b). Cytopharynx permanent, round, located at base of proboscis or tapering front end; ciliature along proboscis; toxicysts on ventrum of proboscis. Class Litostomatea, Order Pharyngophorida, Tracheliidae (*Dileptus, Paradileptus, Trachelius*; Fig. 23D,E,H)

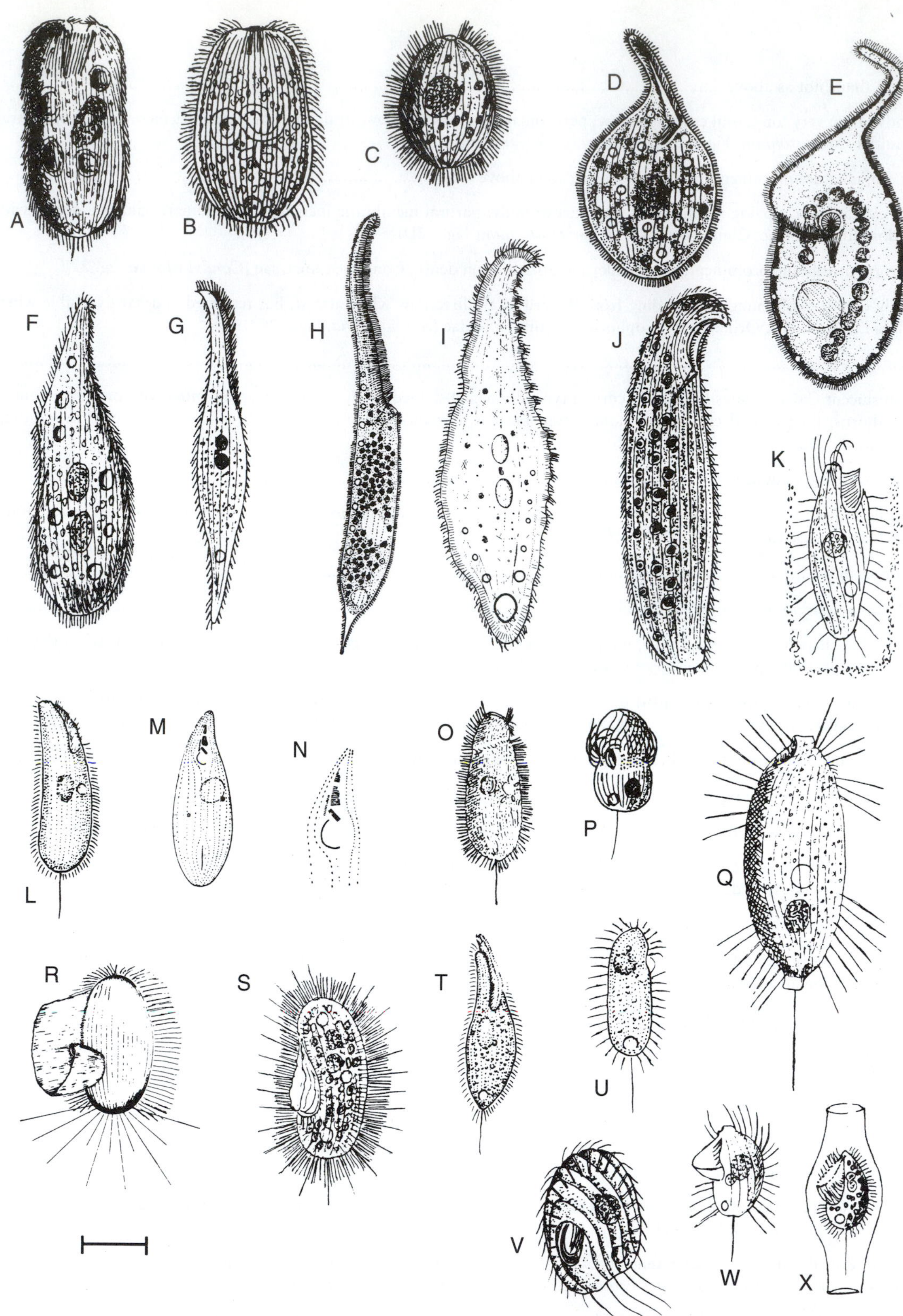

FIGURE 23 (A) *Prorodon teres;* (B) *Pseudoprorodon ellipticus;* (C) *Holophyra simplex;* (D) *Trachelius ovum;* (E) *Paradileptus robustus;* (F) *Amphileptus claparedi;* (G) *Litonotus fasciola;* (H) *Dileptus anser;* (I) *Loxophyllum helus;* (J) *Loxodes magnus;* (K) *Cyrtolophosis mucicola;* (L, M, N) *Philasterides armata,* live, silver stained, and oral detail of silver-stained specimen; (O) *Loxocephalus plagius;* (P) *Urozona bütschlii;* (Q) *Balanonema biceps;* (R) *Pleuronema coronatum;* (S) *Histiobalantium natans;* (T) *Cohnilembus pusillus;* (U) *Uronema griseolum;* (V) *Cinetochilum margaritaceum;* (W) *Cyclidium glaucoma;* (X) *Calyptotricha pleuronemodies.* (After: Corliss, 1959, R; Dragesco, 1966a, I; Grolière, 1980, M, N; Kahl, 1930–1935, A, B, C, F, G, J, K, O, P, Q, S, V, W, X; Kudo, 1966, D, E, H; Noland, 1959, L, T, U.) Scale = 10 μm for K, Q; 15 μm for P, V; 20 μm for T, U, W, X; 25 μm for G, H, L, M; 30 μm for C, I, S; 40 μm for B, R; 50 μm for F; 60 μm for A, O; and 75 μm for D, E, J.

64b. Mouth slit-like, located along convex side of tapering anterior end, evident only when feeding; body laterally compressed; both sides ciliated, cilia may be only bristles on 1 side. Class Litostomatea, Order Pleurostomatida, Amphileptidae (*Amphileptus, Litonotus, Loxophyllum;* Fig. 23F,G,I)

65a (6a). With linear oral furrow or groove leading from anterior end to mouth, bordered on right by paroral membrane 66

65b. Without linear preoral groove bordered by membranes, but generally with membranes inside oral cavity 71

66a (65a). With an apparent double paroral membrane on the right side of the furrow (actually the paroral membrane plus a row of somatic cilia). Class Oligohymenophora, Order Scuticociliatida, Cohnilembidae (*Cohnilembus*; Fig. 23T)
(Many genera of the Order Scuticociliatida, e.g., lines 66–70 and lines 75–76, are difficult to identify without special preparation techniques such as Protargol or other silver staining)

66b. Paroral membrane along furrow not "double" 67

67a (66b). Preoral furrow long, shallow, ciliated with three ciliary fields; generally a single caudal cilium; body long, ovoid. Class Oligohymenophora, Order Scuticociliatida, Philasteridae (*Philasterides*; Fig. 23L–N)

67b. Bottom of preoral furrow not ciliated, paroral membrane curves around rear of mouth 68

68a (67b). Dwells in lorica that is open at both ends; resembles *Pleuronema* or *Cylidium.* Class Oligohymenophora, Order Scuticociliatida, Calyptotrichidae (*Calyptotricha*; Fig. 23X)

68b. Not in a lorica as above 69

69a (68b). Small (15–60 μm long); with few somatic cilia; distinct caudal cilium. Oligohymenophora, Order Scuticociliatida, Cyclidiidae (*Cyclidium*; Fig. 23W)

69b. Larger than above 70

70a (69b). Paroral membrane a prominent velum, extends from anterior to well past equator of cell; one to many stiff, long caudal cilia; a single contractile vacuole. Class Oligohymenophora, Order Scuticociliatida, Pleuronematidae (*Pleuronema*; Fig. 23R)

70b. Somatic ciliation uniform; long stiff cilia distributed over body surface; paroral membrane less prominent than above. Class Oligohymenophora, Order Scuticociliatida, Histiobalantiidae (*Histiobalantium*; Fig. 23S)

71a (65b). Oral cavity huge, covers most of ventrum; dorsal side convex; cilia of paroral membrane long. Class Nassophorea, Order Peniculida, Lembadionidae (*Lembadion*; Fig. 24G)

71b. Mouth smaller than above, not over half of body length 72

72a (71b). With one or more long caudal cilia 73

72b. Without caudal cilia 77

73a (72a). With one or two girdles of cilia 74

73b. Cilia not in girdles, usually in longitudinal rows 75

74a (73a). Oral cavity equatorial; distinct constriction in the ciliated girdle; ends bare except for caudal cilium. Class Oligohymenophora, Order Scuticociliatida, Urozonidae (*Urozona*; Fig. 23P)

74b. Oral cavity deep, equatorial; two distinct ciliary girdles; ciliary tuft on posterior end. Class Nassophorea, Order Peniculida, Urocentridae (*Urocentrum*; Fig. 24D)

75a (73b). Mouth at, or anterior to, middle of cell 76

75b. Oral area large, toward posterior half of cell ($\approx$mid-ventral); body usually flattened; cilia denser ventrally. Class Oligohymenophora, Order Scuticociliatida, Cinetochilidae (*Cinetochilum*; Fig. 23V)

76a (75a). Small, <50 μm; anterior pole flat, unciliated. Class Oligohymenophora, Order Scuticociliatida, Uronematidae (*Uronema*; Fig. 23U)

76b. Somatic ciliation even; body long-ovoid; oral area small, closer to anterior end. Class Oligohymenophora, Order Scuticociliatida, Loxocephalidae (*Balanonema, Loxocephalus*; Fig. 23O,Q)

77a (72b). Body small, ovoid to ellipsoid; in mucilaginous case from which it can emerge freely; cytostome near anterior end. Class Colpodea, Order Cyrtolophosidida, Cyrtolophosidae (*Cyrtolophosis*; Fig. 23K)

77b. Not associated with mucilaginous envelope 78

78a (77b). Body distinctly heart- or cone-shaped; oral region relatively large, covers most of ventral surface. Class Nassophorea, Order Peniculida, Stokesiidae (*Stokesia*; Fig. 24B)

78b. Body shaped otherwise 79

79a (78b). Body ovoid to ellipsoid; mouth in anterior half of body, pointed in front, with pre- and/or post-oral sutures, sometimes hard to see;

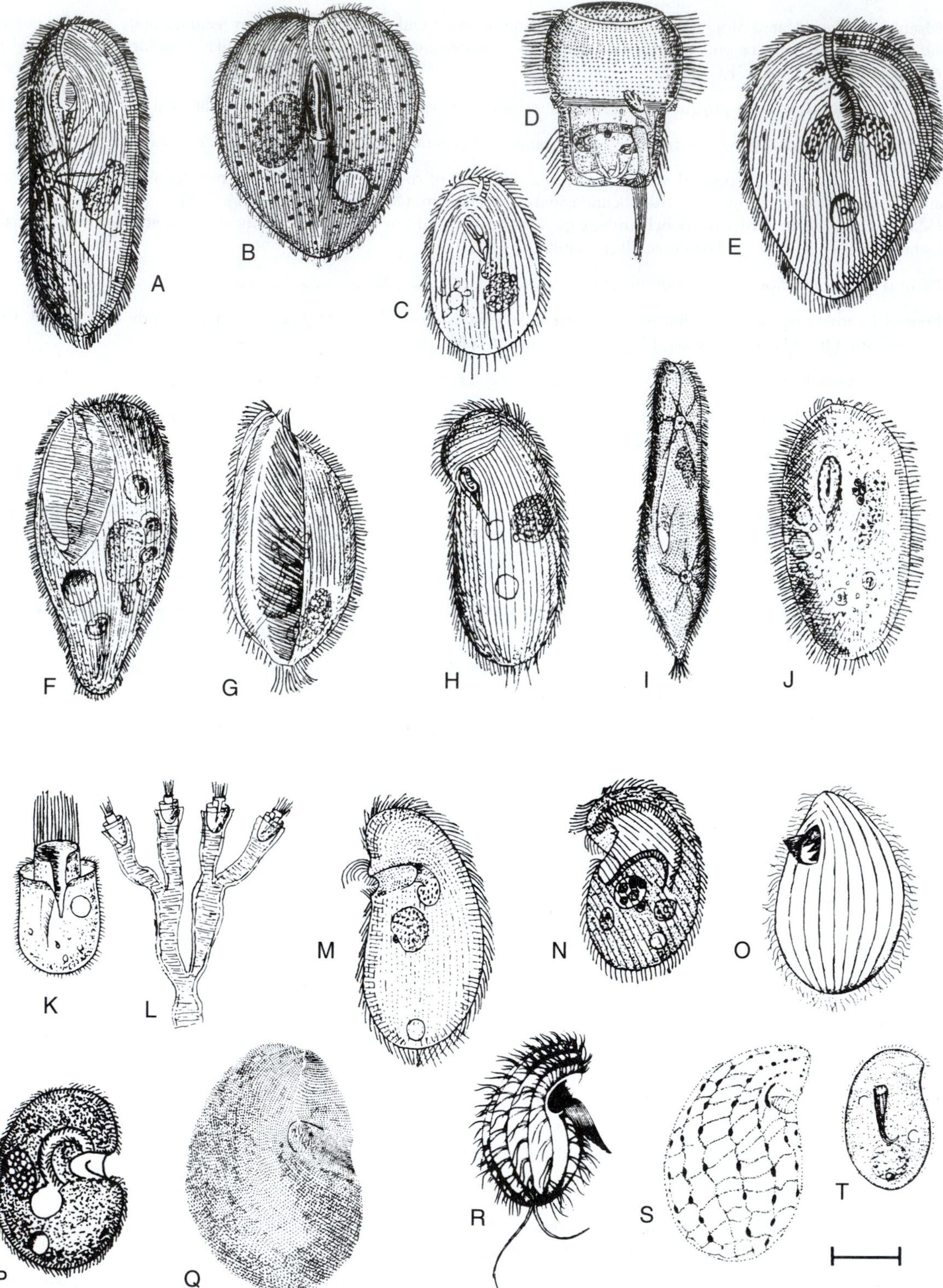

FIGURE 24 (A) *Frontonia leucas;* (B) *Stokesia vernalis;* (C) *Glaucoma scintillans;* (D) *Urocentrum turbo;* (E) *Disematostoma bütschlii;* (F) *Turaniella vitrea;* (G) *Lemabadion magnum;* (H) *Colpidium colpoda;* (I) *Paramecium caudatum;* (J) *Clathrostoma viminale;* (K, L) *Maryana socialis,* individual and colony; (M) *Plagiopyla nasuta;* (N) *Bresslaua vorax;* (O) *Tetrahymena pyriformis;* (P, Q) *Tillina magna,* live and line drawing of silver-stained specimen; (R, S) *Colpoda steinii,* live and silver stained; (T) *Chilodonella uncinata.* (T) (After: Corliss, 1959, O, R; Dragesco, 1966b, B; Kahl, 1930–1935, A, C, D, E, F, G, H, I, J, K, L, M, P; Kudo, 1966, N; Lynn, 1976, S; Lynn, 1977, Q; Noland, 1959, T.) Scale = 15 μm for G. O. R; 25 μm for C, H; 30 μm for D, F; 40 μm for B, E, J, M; 60 μm for I, N; and 75 μm for A, K, Q.

nematodesmata prominent to side and rear of mouth; contractile vacuoles with long collecting canals. Class Nassophorea, Order Peniculida, Frontoniidae (*Disematostoma, Frontonia*; Fig. 24A)

79b. Mouth typically rounded or blunt in front; body ciliature with preoral suture, no postoral suture; one paroral membrane and three membranelles, which are typically inconspicuous. Order Hymenostomatida, Suborder Tetrahymenina 80

(Genera of this suborder, lines 80 and 81, are difficult to identify without special preparation techniques, i.e., silver staining)

80a (79b). Body pyriform to cylindrical; bases of membranelles of uniform width; mouth roughly triangular. Tetrahymenidae (*Tetrahymena*; Fig. 1, 24O)

80b. Not as above 81

81a (80b). Right ventral cilia curve left and twist forward parallel to suture. Turaniellidae (*Colpidium, Turaniella*; Fig. 24F,H)

81b. Right ventral cilia curve left, but do not twist forward toward mouth. Glaucomidae (*Glaucoma*; Fig. 24C)

82a (6b). In a gelatinous tube-like lorica, which may have dichotomous branches; with a terminal cone-like protuberance bearing long cilia and surrounded by a circular adoral groove leading to the buccal cavity. Class Colpodea, Order Colpodida, Marynidae (*Maryna*; Fig. 24K,L)

82b. No gelatinous case; mouth area not as described above 83

83a (82b). Body small (<100 μm), laterally flattened; ciliation relatively sparse, on pellicular ridges 84

83b. Not as above 85

84a (83a). Cytostome in rear half of body; small cyrtos hidden by ventral pellicular fold. Class Nassophoria, Order Microthoracida, Microthoracidae (*Microthorax*; Fig. 20W)

84b. Cytostome opens laterally in anterior one-third of body, long tubular cyrtos; body nearly oval. Class Nassophoria, Order Propeniculida, Leptophrygidae (*Pseudomicrothorax*; Fig. 20V)

85a (83b). Body ciliation uniform; ventral cytostome in an oral vestibule; nematodesmata form a ring or "basket" around mouth; similar to Frontoniidae. Class Nassophorea, Order Peniculida, Clathrostomatidae (*Clathrostoma*; Fig. 24J)

85b. Mouth area sunken in, with oral groove leading to it, or with ciliated pharyngeal tube, or both 86

86a (85b). Medium to large ($\geq$150 μm in some species); body foot-shaped or ellipsoidal, with pointed or rounded caudal end; uniform body ciliation; long, broad, ciliated oral groove leads into buccal cavity; two contractile vacuoles; one species with algal symbiont. Class Nassophorea, Order Peniculida, Parameciidae (*Paramecium*; Fig. 6 and 24I)

86b. Not as above; with single contractile vacuole 87

87a (86b). Oral groove transverse, in anterior part of body; cytostome at base of deep tubular pocket; densely ciliated; common in anaerobic habitats. Class Plagiopylea (*Plagiopyla*; Fig. 24M)

87b. Body reniform, indented in middle on one side; oral groove passes around groove toward mouth; division in a cyst. Class Colpodea, Order Colpodida, Colpodidae (*Bresslaua, Tillina, Colpoda*; Fig. 24N,P–S.)

LITERATURE CITED

Albright, L. J., Sherr, E. B., Sherr, B. F., Fallon, R. D. 1987. Grazing of ciliated protozoa on free and particle-associated bacteria. Marine Ecology Progress Series 38:125–129.

Allen, R. D. 1984. *Paramecium* phagosome membrane: from oral region to cytoproct and back again. Journal of Protozoology 31:1–6.

Archbold, J. H. G., Berger, J. 1985. A qualitative assessment of some metazoan predators of *Halteria grandinella*, a common freshwater ciliate. Hydrobiologia 126:97–102.

Arndt, H. 1993. Rotifers as predators on components of the microbial food web (bacteria, heterotrophic flagellates, ciliates)—a review. Hydrobiologia 255:193–204.

Balczon, J. M., Pratt, J. R. 1996. The functional response of two benthic algivorous ciliated protozoa with differing feeding strategies. Microbial Ecology 31:209–224.

Baldock, B. M. 1986. Peritrich ciliates epizoic on larvae of *Brachycentrus subnilus* (Trichoptera): importance in relation to the total protozoan population in streams. Hydrobiologia 131:125–131.

Baldock, B. M., Baker, J. H., Sleigh, M. A. 1980. Laboratory growth rates of six species of freshwater Gymnamoebia. Oecologia (Berlin) 47:156–159.

Baldock, B. M., Berger, J. 1984. Effects of low temperature on the growth of four fresh-water amoebae (Protozoa: Gymnamoebia). Transactions of the American Microscopical Society 103: 233–239.

Baldock, B. M., Sleigh, M. A. 1988. The ecology of benthic protozoa in rivers—seasonal variation in numerical abundance in fine sediments. Archive für Hydrobiologie 111:409–422.

Bearden, A. P., Schulz, T. W. 1997. Structure-activity relationships for *Pimephales* and *Tetrahymena*: a mechanism of action approach. Environmental Toxicology and Chemistry 16:1311–1317.

Beaver, J. R., Crisman, T. L. 1989. The role of ciliated protozoa in pelagic freshwater ecosystems. Microbial Ecology 17:111–136.

Bell, G. 1988. Sex and death in Protozoa, Cambridge University Press, Cambridge, MA.

Bennett, S. J., Sanders, R. W., Porter, K. G. 1990. Heterotrophic, autotrophic and mixotrophic nanoflagellates: seasonal abundances and bacterivory in a eutrophic lake. Limnology and Oceanography 35:1821–1832.

Berger, J. 1979. The feeding behavior of *Didinium nasutum* on an atypical prey ciliate (*Colpidium campylum*). Transactions of the American Microscopical Society 141:261–275.

Berninger, U.-G., Finlay, B. J., Canter, H. M. 1986. The spatial distribution and ecology of zoochlorellae-bearing ciliates in a productive pond. Journal of Protozoology 33:557–563.

Berninger, U.-G., Caron, D. A., Sanders, R. W. 1992. Mixotrophic algae in three ice-covered lakes of the Pocono Mountains, USA. Freshwater Biology 28:263–272.

Bick, H. 1972. Ciliated protozoa, An illustrated guide to the species used as biological indicators in freshwater biology, World Health Organization, Geneva. 198 pp.

Bird, D. F., Kalff, J. 1987. Algal phagotrophy: regulating factors and importance relative to photosynthesis in *Dinobryon* (Chrysophyceae). Limnology and Oceanography 32:277–284.

Bockstahler, K. R., Coats, D. W. 1993. Grazing of the mixotrophic dinoflagellate *Gymnodinium sanguineum* on ciliate populations of Chesapeake Bay. Marine Biology 116:477–487.

Bott, T. L., Kaplan, L. A. 1989. Densities of benthic protozoa and nematodes in a Piedmont stream. Journal of the North American Benthological Society 8:187–196.

Bott, T. L., Kaplan, L. A. 1990. Potential for protozoan grazing of bacteria in streambed sediments. Journal of the North American Benthological Society 9:336–345.

Bourelly, P. 1968. Les algues d'eau douce, Vol. 2. N. Boubée, Paris.

Bovee, E. C. 1985. Classes Lobosea Carpenter; Filosea Leidy; Granuloreticulosa DeSaedeleer; Order Athalamida Haeckel, *in*: Lee, J. J. Hutner, S. H., Bovee, E. C. Eds. An illustrated guide to the Protozoa, Society of Protozoologists, Lawrence, Kansas pp. 158–252.

Bovee, E. C., Jahn, T. L. 1973. Taxonomy and phylogeny, *in*: Jeon, K. W. Ed. The biology of *Amoeba,* Academic Press, New York, pp. 37–82.

Burns, C. W., Gilbert, J. J. 1993. Predation on ciliates by freshwater calanoid copepods: rates of predation and relative vulnerabilities of prey. Freshwater Biology 30:377–393.

Buskey, E. J., Stoecker, D. K. 1989. Behavioral responses of the marine tintinnid *Favella* sp. to phytoplankton: influence of chemical, mechanical and photic stimuli. Journal of Exerimental Marine Biology and Ecology 132:1–16.

Cairns, J. Jr. 1974. Protozoans (Protozoa). *in*: Hart C. W. Jr., Fuller, S. L. H. Eds. Pollution ecology of freshwater invertebrates. Academic Press, New York, pp. 1–18.

Cairns, J. Jr., Kuhn, D. L., Plafkin, J. L. 1979. Protozoan colonization of artificial substrates, *in*: Wetzel, R. L. Ed. Methods and measurements of periphyton communities: a review, Special Technical Publication 690, American Society for Testing and Materials, Philadelphia, Pennsylvania, pp. 34–57.

Cairns, J. Jr., Lanza, G. R., Parker, B. C. 1972. Pollution related structural and functional changes in aquatic communities with emphasis on freshwater algae and protozoa. Proceedings of the Academy of Natural Sciences of Philadelphia 124:79–127.

Cairns, J. Jr., Yongue, W. H. 1977. Factors affecting the number of species in freshwater protozoan communities, *in*: Cairns J. Jr., Eds. Aquatic microbial communities. Garland Publishing, New York, pp. 257–303.

Calaway, W. T., Lackey, J. B. 1962. Waste treatment protozoa, Florida Engineering and Industrial Experiment Station, Gainesville, FL.

Calow, P. 1977. Conversion efficiencies in heterotrophic organisms. Biological Reviews of the Cambridge Philosophical Society 52:385–409.

Canella, M. F., Rocchi-Canella, I. 1976. Biologie des Ophryoglenina. Contributions à la connaissance des Ciliés VIII. Annali dell Universitià Università di Ferrara. (Nuova Serie) 3:suppl. 2.

Canter, H. M. 1973. A new primitive protozoan devouring centric diatoms in the plankton. Zoological Journal of the Linnean Society 52:63–83.

Carlough, L. A., Meyers, J. L. 1991. Bacterivory by sestonic protists in a southeastern blackwater river. Limnology and Oceanography 36:873–883.

Caron, D. A. 1983. Technique for enumeration of heterotrophic and phototrophic nannoplankton, using epifluorescence microscopy, and comparison with other procedures. Applied and Environmental Microbiology 46:491–498.

Caron, D. A. 1987. Grazing of attached bacteria by heterotrophic microflagellates. Microbial Ecology 13:203–218.

Caron, D. A., Goldman, J. C., Dennett, M. R. 1988. Experimental demonstration of the roles of bacteria and bacterivorous protozoa in plankton nutrient cycles. Hydrobiologia 159:27–40.

Carrias, J.-F., Amblard, C., Bourdier, G. 1996. Protistan bacterivory in an oligomesotrophic lake: importance of attached ciliates and flagellates. Microbial Ecology 31:249–268.

Carrick, H. J., Fahnenestiel, G. L., Taylor, W. D. 1992. Growth and production of planktonic protozoa in Lake Michigan: in situ versus in vitro comparisons and importance to food web dynamics. Limnology and Oceanography 37:1221–1235.

Carter, R. F. 1968. Primary amoebic meningoencephalitis: clinical, pathological and epidemiological features of six fatal cases. Journal of Pathology and Bacteriology 96:1–25.

Cavalier-Smith, T. 1980. *r*- and *K*-tactics in the evolution of protist developmental systems: genome size, phenotype diversifying selection, and cell cycle patterns. BioSystems 12:43–59.

Chapman-Andresen, C. 1973. Endocytic processes. *in*: Jeon, K. W. Ed. The biology of *Amoeba*. Academic Press, New York, pp. 319–348.

Chardez, D. 1985. Protozoaires prédateurs de Thécamoebiens. Protistologica 21:187–194.

Cleven, E.-J. 1996. Indirectly fluorecently labelled flagellates (IFLF): a tool to estimate the predation on free-living heterotrophic flagellates. Journal of Plankton Research 18:429–442.

Cooley, N. R., Keltner, J. M. Jr., Forester, J. 1972. Mirex and Aroclor® 1254: effect on and accumulation by *Tetrahymena pyriformis*. Journal of Protozoology 19:636–638.

Corliss, J. O. 1959. An illustrated guide to the higher groups of the ciliated protozoa, with a definition of terms. Journal of Protozoology 6:265–284.

Corliss, J. O. 1973. History, taxonomy, and evolution of species of *Tetrahymena*, *in* Elliott, A. M. Ed. Biology of *Tetrahymena*, Dowden, Hutchinson and Ross, Stroudsburg, PA, pp. 1–55.

Corliss, J. O. 1979. The ciliated protozoa, Characterization, classification and guide to the literature, Pergamon Press, Oxford.

Corliss, J. O., Daggett, P.-M. 1983. "*Paramecium aurelia*" and "*Tetrahymena pyriformis*": current status of the taxonomy and nomenclature of these popularly known and widely used ciliates. Protistologica 19:307–322.

Corliss, J. O., Esser, S. C. 1974. Comments on the role of the cyst in the life cycle and survival of free-living protozoa, Transactions of the American Microscopical Society 93:578–593.

Crawford, D. W., Stoecker, D. K. 1996. Carbon content, dark respiration and mortality of the mixotrophic planktonic ciliate *Strombidium capitatum*, Marine Biology 126:415–422.

Curds, C. R. 1963. The flocculation of suspended matter by *Paramecium caudatum*. Journal of General Microbiology 33:357–363.

Curds, C. R. 1977. Microbial interactions involving protozoa, *in*: Skinner, F. A., Shewan, J. M. Eds. Aquatic Microbiology, Academic Press, New York, pp. 69–105.

Curds, C. R., Cockburn, A. 1970. Protozoa in biological sewage-treatment processes. II. Protozoa as indicators in the activated sludge process. Water Research 4:237–249.

Curds, C. R., Cockburn, A. 1971. Continuous monoxenic culture of *Tetrahymena pyriformis*. Journal of General Microbiology 66:95–108.

Daniels, E. W. 1973. Ultrastructure. *in*: Jeon, K. W. Ed., The biology of Amoeba. Academic Press, New York, pp. 125–169.

Debiase, A. E., Sanders, R. W., Porter, K. G. 1990. Relative nutritional value of ciliate protozoa and algae as food for *Daphnia*. Microbial Ecology 19:199–210.

Decamp, O., Warren, A. 1999. Bacterivory in ciliates isolated from constructed wetlands (reed beds) used for wastewater treatment. Water Research 32:1989–1996

Deflandre, G. 1959. Rhizopoda and Actinopoda, *in:* W. T. Edmondson, editor. Freshwater Biology. Wiley, New York, pp. 232–264

Dhanarj, P. S., Caushal, B. R., Lal, R. 1989. Effects on uptake by the metabolism of aldrin and phorate in a protozoan, *Tetrahymena pyriformis*. Acta Protozoologica 28:157–163.

Dive, D. 1973. Nutrition holozoique de *Colpidium campylum* aux dépens de bactéries pigmentées ou synthétisant des toxines. Protistologica 10:517–525.

Dodge, J. D., Gruet, C. 1987. Dinoflagellate ultrastructure and complex organelles. *in*: Taylor, F. J. R. Ed., The biology of Dinoflagellates. Academic Press, New York, pp. 92–142.

Doucet, C. M., Maly, E. J. 1990. Effects of copper on the interaction between the predator *Didinium nasutum* and its prey *Paramecium caudatum*. Canadian Journal of Fisheries and Aquatic Sciences 47:1122–1127.

Dragesco, J. 1966a. Observations sur quelques ciliés libres. Archiv für Protistenkunde 109:155–206.

Dragesco, J. 1966b. Ciliés libres de thonon et sus environs. Protistologica 2:59–95.

Eddy, S. 1930. The freshwater armored or thecate dinoflagellates. Transactions of the American Microscopical Society 49: 277–321.

Egerter, D. E., Anderson, J. R., Washburn, J. O. 1986. Dispersal of the parasitic ciliate *Lambornella clarkii*: implications for ciliates in the biological control of mosquitoes. Proceedings of the National Academy of Sciences 83:7335–7339.

Elliott, A. M., Ed. 1973. Biology of *Tetrahymena*. Dowden, Hutchinson and Ross, Stroudsburg, PA.

Evans, M. S., Sell, D. W., Beeton, A. M. 1981. *Tokophrya quadripartita* and *Tokophrya* sp. (Suctoria) associations with crustacean zooplankton in the Great Lakes region. Transactions of the American Microscopical Society 100:384–391.

Fauré-Fremiet, E. 1967. Chemical aspects of ecology, *in*: Florkin, M. Scheer, B. T., Eds. Chemical ecology, Vol. 1, Protozoa. Academic Press, NewYork, pp. 21–54.

Fenchel, T. 1967. The ecology of the marine microbenthos. I. The quantitative importance of ciliates as compared with metazoans in various types of sediments. Ophelia 4:121–137.

Fenchel, T. 1969. The ecology of the marine microbenthos. IV. Structure and function of the benthic ecosystem, its chemical and physical factors and the microfauna communities with special reference to the ciliated protozoa. Ophelia 6:1–182.

Fenchel, T. 1974. Intrinsic rate of increase: the relationship with body size. Oecologia (Berlin) 14:317–326.

Fenchel, T. 1980. Suspension feeding in ciliated protozoa: functional response and particle size selection. Microbial Ecology 6:1–11.

Fenchel, T. 1982a. Ecology of heterotrophic microflagellates. I. Some important forms and their functional morphology. Marine Ecology Progress Series 8:211–223.

Fenchel, T. 1982b. Ecology of heterotrophic microflagellates. II. Bioenergetics and growth. Marine Ecology Progress Series 8:225–231.

Fenchel, T. 1987. Ecology of Protozoa. Springer-Verlag, Berlin.

Fenchel, T., Perry, T., Thane, A. 1977. Anaerobiosis and symbiosis with bacteria in free-living ciliates. Journal of Protozoology 24:154–163.

Fenchel, T., Finlay, B. J. 1991. The biology of free-living anaerobic ciliates. European Journal of Protistology 26:201–215.

Fenchel, T., Estaban, G. E., Finlay, B. J. 1997. Local versus global diversity of microorganisms: cryptic diversity of ciliated protozoa. Oikos 80:220–225.

Fernandez-Leborans, G., Novillo-Vajos, A. 1993. Changes in the structure of a freshwater protozoan community subjected to cadmium. Ecotoxicology and Environmental Safety 25:271–279.

Finlay, B. J. 1977. The dependence of reproductive rate on cell size and temperature in freshwater ciliated protozoa. Oecologia (Berlin) 30:75–81.

Finlay, B. J. 1990. Physiological ecology of free-living protozoa. Advances in Microbial Ecology 12:1–35.

Finlay, B. J., Berninger, U.-G., Stewart, L. J., Hindle, R. M. Davison, W. 1987. Some factors controlling the distribution of two pond-dwelling ciliates with algal symbionts (*Frontonia vernalis* and *Euplotes daideleos*). Journal of Protozoology 34:349–356.

Finlay, B. J., Corliss, J. O., Esteban, G., Fenchel, T. 1996. Biodiversity at the microbial level: the number of free-living ciliates in the biosphere. Quarterly Review of Biology 71:221–237.

Finlay, B. J., Ochsenbein-Gattlen, C. 1982. Ecology of free-living protozoa. A bibliography. Occasional paper No. 17, Freshwater Biological Association, Ambleside, Cumbria, UK.

Finlay, B. J., Rogerson, A., Cowling, A. J. 1988. A beginners guide to the collection, isolation, cultivation and identification of freshwater protozoa. Culture Collection of Algae and Protozoa, Freshwater Biological Association, Ambleside, England.

Foissner, W. 1988. Taxonomic and nomenclatural revision of Sládecek's list of ciliates (Protozoa: Ciliophora) as indicators of water quality. Hydrobiologia 166:1–64.

Foissner, W., Berger, H. 1996. A user-friendly guide to the ciliates (Protozoa: Ciliophora) commonly used by hydrobiologists as bioindicators in rivers, lakes, and waste waters, with notes on their ecology. Freshwater Biology 35:375–482.

Foissner, W., Berger, H., Schaumburg, J. 1999. Identification and ecology of limnetic plankton ciliates. Informations berichte des Bayer. Landesamtes für Wasserwirtschaft, Heft 3/99.

Gaines, G., Elbrächter, M. 1987. Heterotrophic nutrition, *in*: Taylor, F. J. R. Ed. The biology of Dinoflagellates, Academic Press, New York, pp. 224–268.

Gates, M. A. 1978. An essay on the principles of ciliate systematics. Transactions of the American Microscopical Society 97:221–235

Gause, G. F. 1934. The struggle for existence. Williams and Wilkins, Baltimore, MD.

Giese, A. C. 1973. *Blepharisma*. Stanford University Press, Stanford, CA.

Gifford, D. J. 1985. Laboratory culture of marine planktonic oligotrichs (Ciliophora; Oligotrichida). Marine Ecology Progress Series 23:257–267.

Gilbert, J. J. 1994. Jumping behavior in the oligotrich ciliates *Strobilidium velox* and *Halteria grandinella*, and its significance as a defense against rotifer predators. Microbial Ecology 27:189–200.

Gilbert, J. J., Jack, J. D. 1993. Rotifers and predators on small ciliates. Hydrobiologia 255/256:247–253.

Goldman, J. C., Caron, D. A. 1985. Experimental studies on an omnivorous microflagellate: implications for grazing and nutrient recycling in the marine microbial food chain. Deep-Sea Research 32:899–915.

Goldman, J. C., Caron, D. A., Andersen, O. K., Dennett, M. R. 1985. Nutrient cycling in a microflagellate food chain. Marine Ecology Progress Series 24:231–242.

Golini, V. I., Corliss, J. O. 1981. A note on the occurrence of the hymenostome ciliate *Tetrahymena* in chironomid larvae. Transactions of the American Microscopical Society 100:89–93.

Goodrich, J. P., Jahn, T. L. 1943. Epizoic suctoria (Protozoa) from turtles. Transactions of the American Microscopical Society 62:245–253.

Goulder, R. 1972. Grazing by the ciliated protozoon *Loxodes magnus* on the alga *Scenedesmus* in a eutrophic pond. Oecologia (Berlin) 13:177–82.

Goulder, R. 1980. The ecology of two species of primitive ciliated protozoa commonly found in standing freshwaters (*Loxodes magnus* Stokes and *L. striatus* Penard). Hydrobiologia 72: 131–158.

Grell, K. G. 1973. Protozoology. Springer-Verlag, Berlin.

Grolière, C.-A. 1980. Morphologie et stomatogenèse chez deux ciliés Scuticociliatida des genres *Philasterides* Kahl, 1926 et *Cyclidium* O.F. Müller, 1786. Acta Protozoologica 19: 195–206.

Grover, J. P. 1989. Effects of Si:P supply ratio, supply variability, and selective grazing in the plankton: an experiment with a natural algal and protistan assemblage. Limnology and Oceanography 34:349–367.

Gunderson, J. H., Elwood, H., Ingold, A., Kindle, K., Sogin, M. L. 1987. Phylogenetic relationships between chlorophytes, chrysophytes, and oomycetes. Proceedings of the National Academy of Sciences 84:5823–5827.

Guhl, B. E., Finlay, B. J., Schink, B. 1996. Comparison of ciliate communities in the anoxic hypolimnia of three lakes: general features and the influence of lake characteristics. Journal of Plankton Research 18:335–353.

Gupta, R. S., Golding, G. B. 1996. The origin of the eukaryotic cell. Trends in Biochemical Science 21:166–171.

Haack-bell, B., Hohenberger, R., Plattner, H. 1990. Trichocysts of *Paramecium*: secretory organelles in search of their function. European Journal of Protistology 25:289–305.

Hairston, N. G. 1967. Studies on the limitation of a natural population of *Paramecium aurelia*. Ecology 48:904–910.

Hairston, N. G., Kellerman, S. L. 1964. *Paramecium* ecology: electromigration of field samples and observations on density. Ecology 45:373–376.

Hairston, N. G., Kellerman, S. L. 1965. Competition between varieties 2 and 3 of *Paramecium aurelia*: the influence of temperature in a food-limited system. Ecology 46:134–139.

Hanson, E. D. 1977. The origin and early evolution of animals, Wesleyan University Press, Middletown, CT.

Hardin, G. 1943. Flocculation of bacteria by protozoa. Nature (London) 151:642

Harumoto, T. 1994. The role of trichocyst discharge and backward swimming behavior of *Paramecium* from *Dileptus margaritifer*. Journal of Eukaryotic Microbiology 41:560–564.

Hausmann, K. 1978. Extrusive organelles of protozoa. International Review of Cytology 52:197–276.

Havens, K. E., Beaver, J. R. 1997. Consumer vs. resource control of ciliate protozoa in a copepod-dominated lake. Archiv für Hydrobiologie 140:491–511.

Henebry, M. S., Ridgeway, R. T. 1980. Epizoic ciliated protozoa of planktonic copepods and cladocerans and their possible use as indicators of organic water pollution. Transactions of the American Microscopical Society 98:495–508.

Hewett, S. W. 1980a. Prey-dependent cell size in a protozoan predator. Journal of Protozoology 27:311–313.

Hewett, S. W. 1980b. The effect of prey size on the functional response and numerical response of a protozoan predator to its prey. Ecology 61:1075–1081.

Hofeneder, H. 1930. Über die animalische Ernährung con *Ceratium hirudinella* O.F. Müller and über die Rolle des Kernes bei dieser Zellfunkton. Archiv für Protistenkunde 71:1–32.

Holwill, M. E. J. 1974. Hydrodynamic aspects of ciliary and flagellar movement, *in*: Sleigh, M. A. Ed. Cilia and flagella, Academic Press, New York, pp. 143–175.

Jack, J. D., Gilbert, J. J. 1997. Effects of metazoan predators on ciliates in freshwater plankton communities. Journal of Eukaryotic Microbiology 44:194–199.

Jackson, K. M., Berger, J. 1985. Life history attributes of some ciliated protozoa. Transactions of the American Microscopical Society 104:53–62.

Jacobson, D. M., Anderson, D. M. 1986. Thecate heterotrophic dinoflagellates: feeding behavior and mechanisms. Journal of Phycology 22:249–258.

Jahn, T. L., McKibben, W. R. 1937. A colorless euglenoid flagellate, *Khawkkinea halli* n. gen., n. sp. Transactions of the American Microscopical Society 56:48–54.

Jankowski, A. W. 1964a. Morphology and evolution of Ciliophora. III. Diagnoses and phylogenesis of 53 sapropelebionts, mainly of the order Heterotrichida. Archive für Protistenkunde 107:185–294.

Jankowski, A. W. 1964b. Morphology and evolution of Ciliophora. IV. Sapropelebionts of the family Loxocephalidae fam. nova, their taxonomy and evolutionary history. Acta Protozoologica 2:33–58.

Jeon, K. W., Ed. 1973. The biology of *Amoeba*. Academic Press, New York.

Jeon, K. W., Ed. 1983. Intracellular symbiosis. International Review of Cytology. Supplement 14.

Johnson, P. W., Donaghay, P. L., Small, E. B., Sieburth, J. McN: 1995. Ultrastructure and ecology of *Perispira ovum* (Ciliophora: Litostomatea): and aerobic, planktonic ciliate that sequesters the chloroplasts, mitochondria, and paramylon of *Euglena proxima* in a microoxic habitat. Journal of Eukaryotic Microbiology 57:323–335.

Jones, H. L. J. 1997. A classification of mixotrophic protists based on their behaviour. Freshwater Biology 37:35–43.

Jones, R. I., Rees, S. 1994. Influence of temperature and light on particle ingestion by the freshwater phytoflagellate *Dinobryon*. Archiv für Hydrobiologie 132:203–211.

Kahl, A. 1930–1935. Urtiere oder Protozoa. I: Wimpertiere oder Ciliata (Infusoria), *in:* Dahl, F. Ed. Die Tierwelt Deutschlands 18 (1930), 21 (1931), 25 (1932), 30 (1935). Fischer-Verlag, Jena, pp. 1–886.

Kankaala, P., Eloranta, P. 1987. Epizooic ciliates (*Vorticella* sp.) compete for food with their host *Daphnia longispina* in a small polyhumic lake. Oecologia (Berlin) 73:203–206.

Kemp, P. F., Sherr, B. F., Sherr, E. B., Cole, J. J. 1993. Handbook of methods in aquatic microbial ecology, Lewis Publishers, Boca Raton.

Kent, W. S. 1881–1882. A manual of the infusoria, Vol. 2, David Bogue, London.

Kentouri, M., Divanah, P. 1986. Sur l'mportance des ciliés pélagiques dans l'alimentation des stades larvaires de poissons. (On the importance of pelagic ciliates in the feeding of larval stages of fish.) Année Biologique 25:307–318.

Kerr, S. J. 1983. Colonization of blue-green algae by *Vorticella* (Ciliata, Peritrichida). Transactions of the American Microscopical Society 102:38–47.

Kornienko, G. S. 1971. The role of infusoria in the feeding of larvae of herbivorous fish. Journal of Ichthyology 11:241–246.

Kudo, R. R. 1966. Protozoology, 5th ed., C. C. Thomas, Springfield, IL.

Kusch, J. 1993. Induction of defense morphological changes in ciliates. Oecologia (Berlin) 94:571–575.

Kusuoka, Y., Watanabe, Y. 1989. Distinction of emigration by telotroch formation and death by predation in peritrich ciliates: SEM observations on the remaining stalk ends. FEMS Microbiology Ecology 62:7–12.

Lackey, J. B. 1959. Zooflagellates, *in:* Edmondson, W. T. Ed., Freshwater Biology. Wiley, New York, pp. 190–231.

Lal, S., Saxena, B. M., Lal. R. 1987. Uptake, metabolism and effects of DDT, Fenthrothion and chlorphyrifos on *Tetrahymena pyriformis*. Pesticide Science 21:181–191.

Landis, W. G. 1981. The ecology of the killer trait, and interactions of five species of the *Paramecium aurelia* complex inhabiting the littoral zone. Canadian Journal of Zoology 59:1734–1743.

Lawler, S. P. 1993. Direct and indirect effects in microcosm communities of protists. Oecologia. 93:184–190.

Laybourn-Parry, J. 1984. A functional biology of free-living Protozoa. University of California Press, Berkley, CA.

Laybourn-Parry, J. 1992. Protozoan plankton ecology. Chapman & Hall, London/New York.

Laybourn-Parry, J., Bayliss, P., Ellis-Evans, J. C. 1995. The dynamics of heterotrophic nanoflagellates and bacterioplankton in a large ultra-oligotrophic Antarctic lake. Journal of Plankton Research 17:1835–1850.

Leadbeater, B. S. C., Morton, C. 1974. A microscopical study of a marine species of *Codosiga* James-Clark (Choanoflagellata) with special reference to the ingestion of bacteria. Biological Journal of the Linnean Society 6:337–347.

Lee, J. J., Corliss, J. O. 1985. Symposium on "Symbiosis in Protozoa": introductory remarks. Journal of Protozoology 32:371–372.

Lee, J. J., Small, E. B., Lynn, D. H., Bovee, E. C. 1985. Some techniques for collecting, cultivating, and observing protozoa, *in* Lee, J. J., Hutner, S. H., Bovee, E. C. Eds. An illustrated guide to the Protozoa, Society of Protozoologists, Lawrence, KS, pp. 1–7.

Lee, C. C., Fenchel, T. 1972. Studies on ciliates associated with sea ice from Antarctica. II. Temperature responses and tolerances in ciliates from Antarctic, temperate and tropical habitats. Archive für Protistenkunde 114:237–244.

Leedale, G. F. 1985. Order 3. Euglenida Bütschli, 1984, *in*: Lee, J. J., Hutner, S. H., Bovee, E. C., Eds., An illustrated guide to the Protozoa. Society of Protozoologists. Lawrence, KS, pp. 41–54.

Lemmerman, E. 1914. Protomastiginae, *in:* Pascher, A. Ed. Die süsswasser-flora deutschlands, österreichs und der schweiz, I. Flagellatae I. Fischer, Jena, pp. 30–121.

Lim, E. L. 1996. Molecular identification of nanoplanktonic protists based on small subunit ribosomal RNA gene sequences for ecological studies. Journal of Eukaryotic Microbiology 43: 101–106.

Luckinbill, L. S. 1979. Selection and the *r*/*K* continuum in experimental populations of protozoa. American Naturalist 113: 427–437.

Luckinbill, L. S., Fenton, M. M. 1978. Regulation and environmental variability in experimental populations of protozoa. Ecology 59:1271–1276.

Lynn, D. H. 1975. The life cycle of the histophagous ciliate *Tetrahymena corlissi* Thompson, 1955. Journal of Protozoology 22:188–195.

Lynn, 1976. Comparative ultrastructure and systematics of the Colpodida. Structural conservatism hypothesis and a description of *Colpoda steini* Maupas. Journal of Protozoology 23:302–314.

Lynn, 1977. Comparative ultrastructure and systematics of the Colpodida. Fine structure specializations associated with large body size in *Tillina magna* Gruber, 1880. Protistologica 12:629–648.

Lynn, D. H. 1981. The organization and evolution of the microtubular organelles in ciliated protozoa. Biological Reviews of the Cambridge Philosophical Society 56:243–292.

Lynn, D. H., Corliss, J. O. 1990. Phylum Ciliophora (Ciliates), *in*: Harrison, F. W., Corliss, J. O., Eds., Microscopic anatomy of invertebrates, Vol. 1: The Protozoa. Wiley-Liss, New York, pp. 331–465.

Lynn, D. H., Small, E. B. 1997. A revised classification of the phylum Ciliophora Doflein, 1901. Rev. Soc. Mex. Hist. Nat. 47:65–78.

Maguire, B. Jr. 1963a. The passive dispersal of small aquatic organisms and their colonization of isolated bodies of water. Ecological Monographs 33:161–185.

Maguire, B. Jr. 1963b. The exclusion of *Colpoda* (Ciliata) from superficially favorable habitats. Ecology 44:781–784.

Maguire, B. Jr. 1977. Community structure of protozoans and algae with particular emphasis on recently colonized bodies of water, *in*: Cairns, J. Jr., Ed. Aquatic microbial communities, Garland Publishing, New York, pp. 358–397.

Margulis, L. 1981. Symbiosis in cell evolution. W.H. Freeman, San Francisco, CA.

Margulis, L., Corliss, J. O., Melkonian, M., Chapman, D. J., Eds. 1989. Handbook of Protoctista, Jones and Bartlett Publishers, Boston.

Margulis, L., Schwartz, K. V. 1988. Five kingdoms: an illustrated guide to life on earth. 2nd ed., Freeman, W. H. San Francisco, CA.

Margulis, L., Sagan, D. 1986. Origins of sex, Yale University Press, New Haven.

Massana, R., Pedrós-Alió, C. 1994. Role of anaerobic ciliates in planktonic food webs: abundance, feeding and impact on bacteria in the field. Applied and Environmental Microbiology 60:1325–1334.

Mast, S. O., Hahnert, W. F. 1935. Feeding, digestion, and starvation in *Amoeba proteus* Leidy. Physiological Zoology 8:255–272.

Matthes, D. 1954. Suktorienstudien. VI. Die gattung Heliophrya Saedeleer and Tellier 1929. Archiv für Protistenkunde 100: 143–152.

McCormick, P. V., Cairns, J. Jr. 1991. Effects of micrometazoa on the protistan assemblage of a littoral food web. Freshwater Biology 26:111–119.

McManus, G. B., Fuhrman, J. A. 1986. Bacterivory in seawater studied with the use of inert fluorescent particles. Limnology and Oceanography 31:420–426.

Meisterfeld, R. 1991. Vertical distribution of *Difflugia hydrostatica* (Protozoa, Rhizopoda). Internationale Vereinigung für Theoretische und Angewandte Limnologie Verhandlungen 24:2726–2728.

Montagnes, D. J. S., Lynn, D. H. 1987a. A quantitative protargol stain (QPS) for ciliates: a method description and test of its quantitative nature. Marine Microbial Food Webs 2:83–93.

Montagnes, D. J. S., Lynn, D. H. 1987b. Agar embedding on cellulose filters: an improved method of mounting protists for protargol and Chatton-Lwoff staining. Transactions of the American Microscopical Society 106:183–186.

Müller, H., Geller, W. 1993. Maximum growth rates of aquatic ciliated protozoa: the dependence on body size and temperature reconsidered. Archiv für Hydrobiologie 126:315–327.

Nagata, T. 1988. The microflagellate-picoplankton food linkage in the water column of Lake Biwa. Limnology and Oceanography 33:504–517.

Nilsson, J. R. 1976. Phagotrophy in *Tetrahymena*. *in* Levandowsky, M.

and Hutner, S. H., Eds. Biochemistry and Physiology of Protozoa. Vol. 2., Academic Press, New York, pp. 339–379.

Nilsson, J. R. 1989. *Tetrahymena* in cytotoxicology: with special reference to effects of heavy metals and selected drugs. European Journal of Protistology 25:2–25.

Nilsson, J. R. 1987. Structural aspects of digestion of *Escherichia coli* in *Tetrahymena*. Journal of Protozoology 34:1–6.

Nisbet, B. 1984. Nutrition and feeding strategies in Protozoa. Croom Helm, London.

Noland, L. E. 1925. The factors affecting the distribution of fresh water ciliates. Ecology 6:437–452.

Noland, L. E. 1959. Ciliophora, *in:* Edmondson, W. T. Ed., Freshwater Biology, Wiley, New York, pp. 265–297.

Noland, L. E., Goldjics, M. 1967. Ecology of free-living Protozoa, *in:* Chen, T.-T. Ed. Research in protozoology. Vol. 2, Pegamon, Oxford, pp. 215–266.

Nyberg, D. 1974. Breeding systems and resistance to environmental stress in ciliates. Evolution 28:367–380.

Nyberg, D., Bishop, P. 1983. High levels of phenotypic variability of metal and temperature tolerance in *Paramecium*. Evolution 37:341–357.

Nygaard, K., Borsheim, K. Y., Thingstad, T. F. 1988. Grazing rates on bacteria by marine heterotrophic microflagellates compared to uptake rates of bacteria-sized monodisperse fluorescent latex beads. Marine Ecology Progress Series 44:159–165.

Ogden, C. G. 1991. Gas vacuoles and flotation in the testate amoeba *Arcella discoides*. Journal of Protozoology 38:217–220.

Pace, M. L., Orcutt, J. D. Jr. 1981. The relative importance of protozoans, rotifers, and crustaceans in a freshwater zooplankton community. Limnology and Oceanography 26:822–830.

Page, F. C. 1976. An illustrated key to freshwater and soil amoebae, with notes on cultivation and ecology. Scientific Publication Number 34, Freshwater Biological Association, Ambleside, Cumbria, England.

Page, F. C. 1977. The genus *Thecamoeba* (Protozoa, Gymnamoebia). Species distinctions, locomotive morphology, and protozoan prey. Journal of Natural History 11:25–63.

Page, F. C. 1987. The classification of 'naked' amoebae (Phylum Rhizopoda). Archive für Protistenkunde 133:199–217.

Page, F. C. 1988. A new key to freshwater and soil Gymnamoebae, Freshwater Biological Association, Ambleside, England.

Pascher, A. 1913. Chrysomonadidae, *in:* Pascher, A. Ed., Die süsswasser-flora deutschlands, österreichs und der schweiz, I. Flagellatae II. Fischer, Jena, pp. 7–95.

Patterson, D. J. 1980. Contractile vacuoles and associated structures: their organization and function. Biological Reviews of the Cambridge Philosophical Society 55:1–46.

Patterson, D. J. 1993. The current status of the free-living heterotrophic flagellates. Journal of Eukaryotic Microbiology 40:606–608.

Peck, R. K. 1985. Feeding behavior in the ciliate *Pseudomicrothorax dubius* is a series of morphologically and physiologically distinct events. Journal of Protozoology 32:492–501.

Picken, L. E. R. 1937. The structure of some protozoan communities. Journal of Ecology 25:368–384.

Pomeroy, L. R. 1974. The ocean's foodweb: a changing paradigm. BioScience 24:499–504.

Porter, K. G. 1988. Phagotrophic flagellates in microbial food webs. Hydrobiologia 159:89–97.

Porter, K. G., Sherr, E. B., Sherr, B. F., Pace, M. C., Sanders, R. S. 1985. Protozoa in planktonic food webs. Journal of Protozoology 32:409–415.

Pratt, J. R., Cairns, J. Jr. 1985. Functional groups in the protozoa: roles in differing ecosystems. Journal of Protozoology 32: 415–423.

Pratt, J. R., Rosen, B. H. 1983. Associations of species of *Vorticella* (Peritrichida) and planktonic algae. Transactions of the American Microscopical Society 102:48–54.

Rainier, H. 1968. Heliozoa, *in:* Dahl, F. Ed. Die tierwelt deutschlands, Vol. 56, Fischer-Verlag, Jena, pp. 3–174.

Rapport, D., Berger, J., Reid, D. B. 1972. Determination of food preference of *Stentor coeruleus*. Biological Bulletin 142:103–109.

Rasmussen, L. 1976. Nutrient uptake in *Tetrahymena*. Carlsberg Research Communications 41:143–167.

Ricci, N. 1989. Microhabitats of ciliates: specific adaptations to different substrates. Limnology and Oceanography 34: 1089–1097.

Ricketts, T. R. 1972. The interaction of particulate material and dissolved foodstuffs in food uptake by *Tetrahymena pyriformis*. Archive für Mikrobiologie 81:344–349.

Riemann, B., Christoffersen, K. 1993. Microbial trophodynamics in temperate lakes. Marine Microbial Food Webs 7:69–100.

Rogerson, A., Berger, J. 1981. Effect of crude oil and petroleum-degrading microorganisms on the growth of freshwater and soil protozoa. Journal of General Microbiology 124:53–59.

Rogerson, A., Finlay, B. J., Berninger, U.-G. 1989. Sequestered chloroplasts in the freshwater ciliate *Strombidium viride* (Ciliophora: Oligotrichina). Transactions of the American Microscopical Society 108:117–126.

Rogerson, A., Shiu, W. Y., Huang, G. L., Mackay, D., Berger, J. 1983. Determination and interpretation of hydrocarbon toxicity to ciliate protozoa. Aquatic Toxicology 3:215–228.

Rudzinska, M. A. 1965. The fine structure and function of the tentacles of *Tokophrya infusionum* and ultrastructural changes in food vacuoles during digestion. Journal of Cell Biology 25: 459–477.

Ruthven, J. A., Cairns, J. Jr. 1973. Response of fresh-water protozoan artificial communities to metals. Journal of Protozoology 20:127–135.

Sambanis, A., Fredrickson, A. G. 1988. Persistence of bacteria in the presence of viable, nonencysting bacterivorous ciliates. Microbial Ecology 16:197–212.

Sanders, R. W. 1988. Feeding by *Cyclidium* sp. (Ciliophora, Scuticociliatida) on particles of different sizes and surface properties. Bulletin of Marine Science 43:446–457.

Sanders, R. W. 1991. Mixotrophic protists in marine and freshwater ecosystems. Journal of Protozoology 38:76–80.

Sanders, R. W., Porter, K. G. 1988. Phagotrophic flagellates. Advances in Microbial Ecology 10:167–192.

Sanders, R. W., Porter, K. G., Bennett, S. J., DeBiase, A. E. 1989. Seasonal patterns of bacterivory by flagellates, ciliates, rotifers and cladocerans in a freshwater planktonic community. Limnology and Oceanography 34:673–687.

Sanders, R. W., Porter, K. G., Caron, D. A. 1990. Relationship between phototrophy and phagotrophy in the mixotrophic chrysophyte *Poterioochromonas mahlhamensis*. Microbial Ecology 19:97–109.

Sanders, R. W., Leeper, D. A., King, C. H., Porter, K. G. 1994. Food web structure and zooplankton grazing impacts on photosynthetic and heterotrophic nanoplankton. Hydrobiologia 288: 167–181.

Sanders, R. W., Wickham, S. A. 1993. Planktonic protozoa and metazoa: predation, food quality and population control. Marine Microbial Food Webs 7:197–223.

Schönborn, W. 1977. Production studies on Protozoa. Oecologia (Berlin) 27:171–184

Schuster, F. L. 1979. Small amebas and ameboflagellates. *in* Levandowsky M., Hutner, S. H. Eds. Biochemistry and physiology of Protozoa, Vol. 1, Academic Press, New York, pp. 215–285.

Scott, J. M. 1985. The feeding rates and efficiencies of a marine ciliate, *Strombidium* sp., grown under chemostat steady-state conditions. Journal of Experimental Marine Biology and Ecology 90:81–95.

Scott, B. F., Wade, P. J., Taylor, W. D. 1984. Impact of oil and oil-dispersant mixtures on the fauna of freshwater ponds. The Science of the Total Environment 35:191–206.

Shawhan, F. M., Jahn, T. L. 1947. A survey of the genus *Petalomonas* Stein (Protozoa: Euglenida). Transactions of the American Microscopical Society 66:182–189.

Sherr, E. B., Sherr, B. F., Albright, L. J. 1987. Bacteria: Link or sink? Science 235:88–89.

Shuter, B. J., Thomas, J. E., Taylor, W. D., Zimmerman, A. M. 1983. Phenotypic correlates of genomic DNA content in unicellular eukaryotes and other cells. American Naturalist 122:26–44.

Simek, K., Straskrabová, V. 1992. Bacterioplankton production and protozoan bacterivory in a mesotrophic reservoir. Journal of Plankton Research 14:773–787.

Simek, K., Chrzanowski, T. H. 1992. Direct and indirect evidence of size-selective grazing on pelagic bacteria by freshwater nanoflagellates. Applied and Environmental Microbiology 58:3715–3720.

Simek, K., Vrba, J., Hartman, P. 1994. Size-selective feeding by *Cyclidium* sp. on bacterioplankton and various sizes of cultured bacteria. FEMS Microbial Ecology 14:157–168.

Singh, B. N. 1975. Pathogenic and nonpathogenic amoebae, MacMillan, London.

Skibbe, O. 1994. An improved quantitative protargol stain for ciliates and other planktonic protists, Archiv für Hydrobiologie 130:339–348.

Slabbert, J. L., Morgan, W. S. G. 1982. A bioassay technique using *Tetrahymena pyriformis* for the rapid assessment of toxicants in water. Water Research 16:517–523.

Sládecek, V. 1973. System of water quality from the biological point of view. Ergebnisse der Limnologie 7:1–218.

Sleigh, M. A., Blake, J. R. 1977. Methods of ciliary propulsion and their size limitations, *in*: Pedley, T. J. Ed. Scale effects in animal locomotion. Academic press, New York, pp. 243–256.

Sleigh, M. A, Baldock, B. A., Baker., J. H. 1993. Protozoan communities in chalk streams. Hydrobiologia 248:53–64.

Small, E. B., Lynn, D. H. 1985. Phylum Ciliophora Doflein, 1901. *in*: Lee, J. J., Hutner, S. H., Bovee, E. C., Eds., An illustrated guide to the Protozoa, Society of Protozoologists. Lawrence, KS, pp. 393–575.

Smith, G. M. 1950. The fresh-water algae of the United States, 2nd ed., McGraw-Hill, New York.

Sogin, M. L., Elwood, H. J. 1986. Primary structure of the *Paramecium aurelia* small-subunit rRNA coding region: phylogenetic relationships within the Ciliophora. Journal of Molecular Evolution 23:53–60.

Sogin, M. L., Gunderson, J. H., Elwood, H. J., Olonso, R. A., Peattie, D. A. 1989. Phylogenetic meaning of the Kingdom concept: an unusual ribosomal RNA from *Giardia lamblia*. Science 243:75–77.

Sogin, M. L., Silberman, J. D., Hinkle, G., Morrison, H. D. 1996. Problems with molecular diversity in the Eukarya. *in Roberts*, D. McC., Sharp, P., Alderson, G., Collins, M. Eds., Evolution of microbial life. Cambridge University Press, Cambridge, pp. 167–184.

Sommaruga, R., Oberleiter, A., Psenner, R. 1996. Effect of UV radiation on the bacterivory of a heterotrophic nanoflagellate. Applied and Environmental Microbiology 62:4395–4400

Sommaruga, R., Psenner, R. 1993. Nanociliates of the order Prostomatida: their relevance in the microbial food web of a mesotrophic lake. Aquatic Sciences 55:179–187.

Sonneborn, T. M. 1975. The *Paramecium aurelia* complex of fourteen sibling species. Transactions of the American Microscopical Society 94:155–178.

Spoon, D. M., Burbanck, W. D. 1967. A new method for collecting sessile ciliates in plastic petri dishes with tight-fitting lids. Journal of Protozoology 14:735–744.

Starink, M., Krylova, I. N., Bär-Gilissen, M.-J., Bak, R. P. M., Cappenberg, T. E. 1994. Rates of benthic protozoan grazing on free and attached sediment bacteria measured with fluorescently stained sediment. Applied and Environmental Microbiology 60:2259–2264.

Stoecker, D. K., Capuzzo, J. M. 1990. Predation on protozoa: its importance to zooplankton. Journal of Plankton Research 12:891–908.

Stoecker, D. K., Sunda, W. G., Davis, L. H. 1986. Effects of copper and zinc on two planktonic ciliates. Marine Biology 92:21–29.

Stoecker, D. K., Silver, M. W., Michaels, A. E., Davis, L. H. 1989. Enslavement of algal chloroplasts by four *Strombidium* spp. (Ciliophora, Oligotrichida). Marine Microbial Food Webs 3:79–100.

Stout, J. D. 1980. The role of protozoa in nutrient cycling and energy flow. Advances in Microbial Ecology 4:1–50.

Suttle, C. A., Chan, A. M., Taylor, W. D., Harrison, P. J. 1986. Grazing of planktonic diatoms by microflagellates. Journal of Plankton Research 8:393–398.

Taylor, F. J. R. 1987. Some eco-evolutionary aspects of intracellular symbiosis, *in*: Jeon, K. W. Ed., Intracellular symbiosis. International Review of Cytology. Suppliment 14, pp. 1–28.

Taylor, W. D. 1978a. Growth responses of ciliate protozoa to the abundance of their bacterial prey. Microbial Ecology 4: 207–214.

Taylor, W. D. 1978b. Maximum growth rate, size and commonness in a community of bacterivorous ciliates. Oecologia (Berlin) 36:263–272.

Taylor, W. D. 1981. Temporal heterogeneity and the ecology of lotic ciliates. *in*: Locke, M. A. Williams, D. D. Eds., Perspectives in running water ecology, Plenum, New York, pp. 209–224.

Taylor, W. D. 1983a. A comparative study of the sessile ciliates of several streams, Hydrobiologia 98:125–133.

Taylor, W. D. 1983b. Rate of population increase and mortality for sessile ciliates in a stream riffle. Canadian Journal of Zoology 61:2023–2028.

Taylor, W. D. 1986. The effect of grazing by a ciliated protozoan on phosphorus limitation in batch culture. Journal of Protozoology 33:47–52.

Taylor, W. D., Berger, J. 1976. Growth responses of cohabiting ciliate protozoa to various prey bacteria. Canadian Journal of Zoology 54:1111–1114.

Taylor, W. D., Heynen, M. L. 1987. Seasonal and vertical distribution of Ciliophora in Lake Ontario. Canadian Journal of Fisheries and Aquatic Sciences 44:2185–2191.

Taylor, W. D., Johannsson, O. 1991. A comparison of estimates of productivity and consumption by zooplankton for planktonic ciliates in Lake Ontario. Journal of Plankton Research 13:363–372.

Taylor, W. D., Lean, D. R. S. 1991. Phosphorus pool sizes and fluxes in the epilimnion of a mesotrophic lake. Canadian Journal of Fisheries and Aquatic Sciences 48:1293–1301.

Taylor, W. D., Shuter, B. J. 1981. Body size, genome size, and intrinsic rate of increase in ciliated protozoa. The American Naturalist 118:160–172.

Thompson, A. S., Rhodes, J. C., Pettman, I. 1988. Catalogue of strains. Culture collection of Algae and Protozoa, Freshwater Biological Association, Ambleside, Cumbria, England.

Tucker, J. B. 1968. Fine structure and function of the pharyngeal basket in the ciliate *Nassula*. Journal of Cell Science 3: 493–514.

Tyndall, R. L., Ironside, K. S., Metler, P. L., Tan, E. L., Hazen, T. C., Fliermanns, C. B. 1989. Effect of thermal additions on the density and distribution of thermophilic amoebae and pathogenic *Naegleria fowleri* in a newly created cooling lake. Applied and Environmental Microbiology 55:722–732.

Vickerman, K. 1976. The diversity of kinetoplastid flagellates, *in*: Lumsden, W. H. R., Evans, D. A. Eds., Biology of Kinetoplastida. Academic Press, New York, pp. 1–34.

Wagener, S., Schulz, S., Hanselmann, K. 1990. Abundance and distribution of anaerobic protozoa and their contribution to methane production in Lake Cadagno (Switzerland). FEMS Microbial Ecology 74:39–48.

Wang, S.-J., Heath, R. T. 1997. The distribution of protozoa across a trophic gradient, factors controlling their abundance and importance in the plankton food web. Journal of Plankton Research 19:491–518.

Washburn, J. O., Gross, M. E., Mercer, D. R., Mercer, J. R. 1988. Predator-induced trophic shift of a free-living ciliates: parasitism of mosquito larvae by their prey. Science 240:1193–1195.

Weisse, T. 1991. The annual cycle of heterotrophic freshwater nanoflagellates: role of bottom-up versus top-down control. Journal of Plankton Research 13:167–185.

Weisse, T., Müller, H., Pinto-Coelho, R. M., Schweizer, A., Springmann, D., Baldringer, G. 1990. Response of the microbial loop to the phytoplankton spring bloom in a large prealpine lake. Limnol. Oceanogr. 35:781–794.

Wessenberg, H., Antipa, G. 1970. Capture and digestion of *Paramecium* by *Didinium nasutum*. Journal of Protozoology 17: 250–270.

Wickham, S. A. 1995a. Cyclops predation on ciliates: species-specific differences and functional responses. Journal of Plankton Research 17:1633–1646.

Wickham, S. A. 1995b. Trophic relationships between cyclopoid copepods and ciliated protists: complex interactions link the microbial and classic food webs. Limnology and Oceanography 40:1173–1181.

Wickham, S. A., Gilbert, J. J. 1993. The comparative importance of competition and predation by Daphnia on ciliated protozoa. Archive für Hydrobiologie 126:289–314.

Wicklow, B. J. 1997. Signal-induced defensive phenotypic changes in ciliated protists: morphological and ecological implications for predator and prey. Journal of Eukaryotic Microbiology 44:176–188.

Wichterman, R. 1986. The biology of *Paramecium*, 2nd ed., Plenum Press, New York.

Wolfe, J. 1988. Analysis of *Tetrahymena* mucocyst material with lectins and Alcian blue. Journal of Protozoology 35:46–51.

PORIFERA[1]

Thomas M. Frost[†]
Henry M. Reiswig
Redpath Museum and
Department of Biology
McGill University
Montreal, Quebec H3A 2K6

Anthony Ricciardi
Department of Biology
Dalhousie University
Halifax, Nova Scotia B3H 4J1

I. INTRODUCTION

Sponges in fresh water? Many people react this way when they are told that there might be sponges in a lake or stream. Most people have seen sponges from the ocean but are unaware that they also occur in freshwater. Yet sponges are common and sometimes abundant inhabitants of a wide variety of freshwater habitats. In some situations they comprise a major component of the benthic fauna and may play important roles in ecosystem processes in freshwater.

Sponges are the simplest of the multicellular phyla. They lack organs and tissues are their highest level of organization. Specialized cells accomplish many basic biological functions in sponges. Despite their simplicity, however, sponges display a variety of elegant adaptations to freshwater habitats including a strong capacity for osmoregulation, complex life cycles, a capability to feed selectively on a broad range of particulate resources, and, in many species, an intimate association with symbiotic algae. This chapter will introduce you to the structure, function, and diversity of freshwater sponges. It will emphasize their taxonomy and their basic ecology. For those whose interests are whetted by this introduction, Bergquist (1978) and Simpson (1984) provide detailed treatments of the biology of the phylum Porifera in general.

II. STRUCTURE AND PHYSIOLOGY

The simplicity of a sponge's external features contrasts sharply with the complexity of its internal structure and function. Viewed macroscopically, freshwater sponges exhibit a limited range of non-distinct body forms, but a microscopic examination reveals a variety of internal features. This contrast derives from the basic organization of sponges in which tissues rather than organs represent the highest level of

[1]The preparation of this chapter was supported by several grants from the National Science Foundation. We thank Janet Blair, Joan Elias, Susan Knight, and Yolanda Lukaziewski for their assistance in its preparation.

[†]It is with profound regret that we learned in the late summer of 2000 of the death of Tom Frost who gave his life saving another person from drowning. He will be missed by his many friends and colleagues.

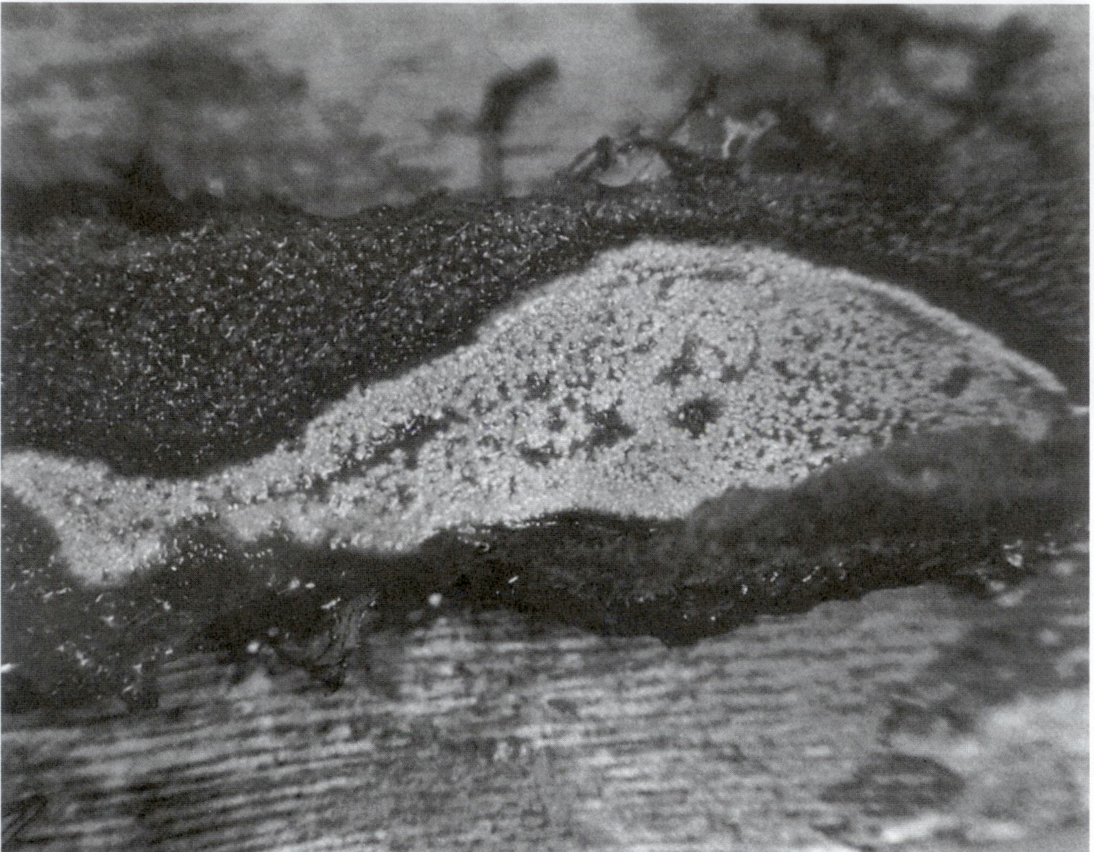

FIGURE 1 Gemmulated colonies of *Spongilla lacustris* and *Duosclera mackayi* exhibiting an encrusted growth form. The distinct zone between the two species illustrates the typical nonmerging front that occurs when two different freshwater species grow into contact with each other on a substratum.

morphological complexity. In fact, specialized cells operating independently or in association with related cells accomplish such primary functions as food gathering, digestion, and reproduction. An appreciation of the utility of sponge structure, therefore, is possible only by including microscopic examinations. Likewise, it is difficult to discuss the physiological functions of sponges independently of their microscopic structure and these topics are presented jointly, in a general overview.

There has been substantial debate as to whether sponges should be considered as colonies or individuals (Hartman and Reiswig, 1973; Simpson, 1973). This reflects the continuing difficulties that ecologists, evolutionary biologists, and zoologists in general encounter in deciding on what an individual is (Santelices, 1999). From an ecological perspective, it is most important to consider that sponges are much more like colonies, such as corals or bryozoans, in terms of their ecological function than they are like individual arthropods or vertebrates. Overall, however, concepts of colony or individual are not really appropriate for sponges, because of their level of biological organization.

A. External Morphology

Freshwater sponges display a variety of morphologies that range from encrusting (Fig. 1), to rounded (Fig. 2) and finger-like (Fig. 3) growth forms. Because form can vary substantially within a species (Penney and Racek, 1968), external morphology is of very limited use in sponge taxonomy. Variation in growth form is influenced by environmental conditions such as water movement (Jewell, 1935; Palumbi, 1984) and the availability of light (Frost and Williamson, 1980). Differences can be particularly dramatic between lentic and lotic habitats (Neidhoefer, 1940).

The surface structure of sponges varies from flat (Fig. 1) to strongly convoluted (Fig. 2). Although there are some species-associated differences in surface characteristics, this feature, too, is strongly influenced by habitat conditions. Finer-scale characteristics of external sponge anatomy are related to internal features.

FIGURE 2 A colony of *Ephydatia muelleri* exhibiting a rounded growth form encrusted on a branch from a fallen tree (from Neidhoefer, 1940).

FIGURE 3 A colony of the common freshwater sponge *Spongilla lacustris* exhibiting a characteristic, fingerlike growth form.

Examples, discussed below, include the incurrent and excurrent portions of the feeding canals and extensions of the skeletal system.

B. Internal Anatomy and Physiology

The basic organization of a sponge consists of a surface epithelium, composed of pinacocytes, surrounding an organic matrix, termed the mesohyl, which contains a broad variety of specialized cells. Cells within the mesohyl interact with the epidermal cells to accomplish basic functions including the processing of water for food uptake and gas exchange, digestion, structural support, and reproduction. Some sponge cells are highly plastic in their behavior and can shift their function with changing environmental conditions. De Vos *et al.* (1991) provided a detailed atlas of the overall general morphology of sponges illustrating most cell types with scanning and transmission electron micrographs.

All sponge cells, including those in the epidermis and in the mesohyl, appear to be involved in osmoregulation (Weissenfels, 1975). This capacity has been critical in allowing the invasion of freshwater by sponges from their original marine habitats.

1. Water Processing System

A sponge can be considered, essentially, as a series of progressively finer filters and a mechanism that circulates water through them. The sponge filters are capable of removing particles ranging from large algae to bacterial cells less than 1 μm in diameter and, possibly, to colloidal organic matter (Frost, 1987). In addition, the movement of water through the canal system, with its extensive surface area, fosters the transport of gases and excreted materials. The main features of the sponge water-processing system (Figs. 4 and 5) have been identified through the efforts of numerous investigators (reviewed in Frost, 1976b, 1987) with several papers by Weissenfels and his co-workers (Weissenfels, 1975, 1976, 1980, 198l, 1982,1992; Langenbruch and Weissenfels, 1987) providing particularly important details on structure. Although there are some subtle differences in structures among species, the summary below provides a general overview of the structures associated with sponge water processing.

Water first enters a sponge through a large number of 50-μm diameter openings termed ostia (single: ostium) that are spread across its surface epithelium. Passing through the surface, inflowing water next enters a large subdermal cavity. From there, water flows

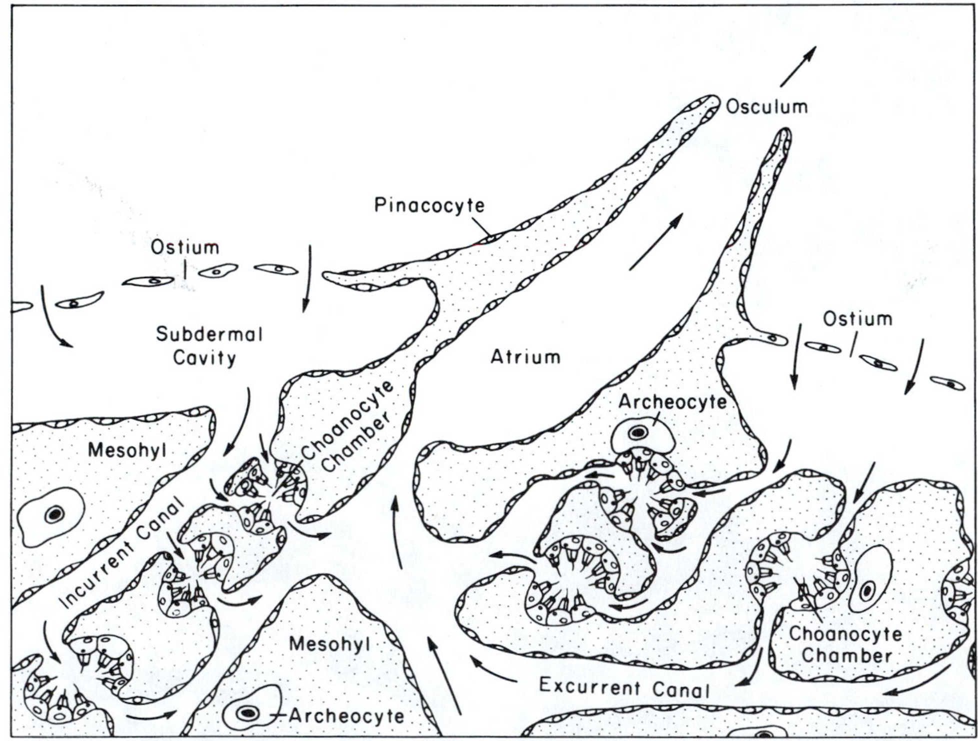

FIGURE 4 A schematic diagram indicating the primary components of the sponge feeding canal system. The diagram is not drawn to scale, but Figure 5 provides an indication of the size and arrangement of the feeding system in an actual sponge. Water flow patterns are indicated by arrows.

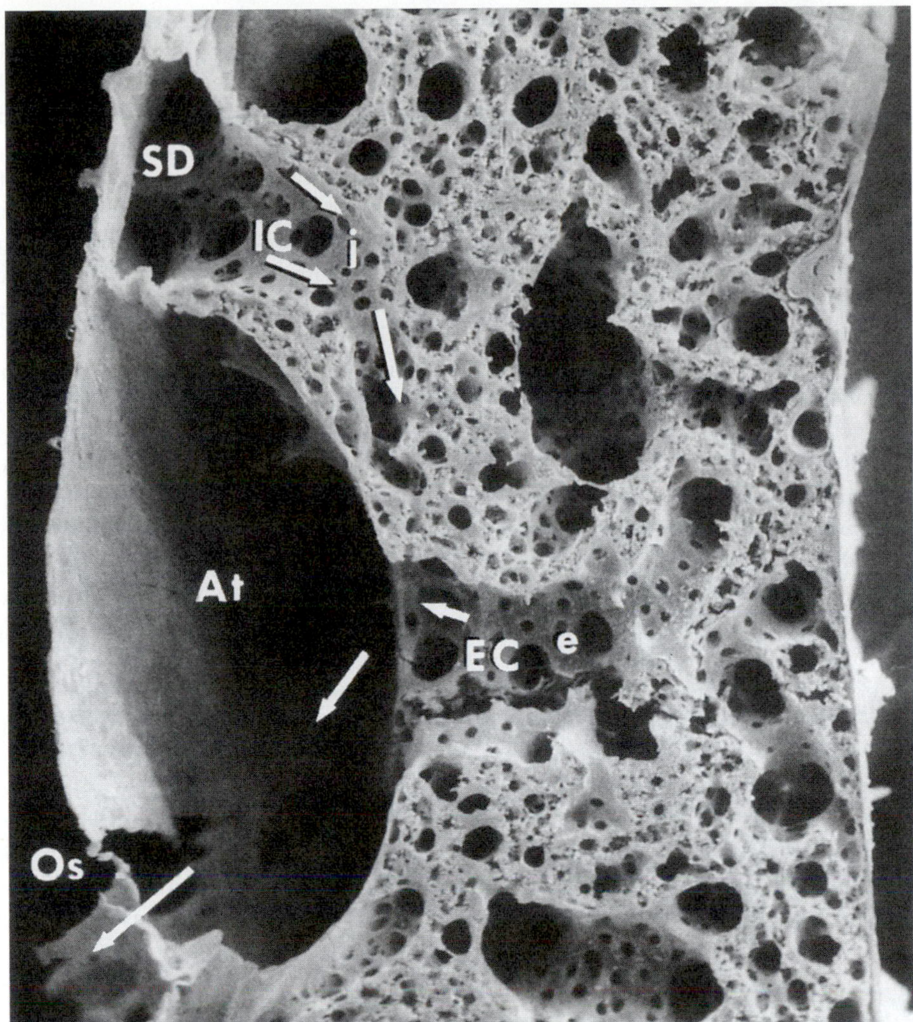

FIGURE 5 A scanning electron micrograph of a cross section of the freshwater sponge *Ephydatia fluviatilis* indicating the form, size, and arrangement of the feeding canal system. Illustrated are a sudbdermal cavity (SD), an incurrent canal (IC) with lateral branches into the sponge (i), and excurrent canal (EC) and its lateral branches (e), an atrium (At), and an osculum (Os). Arrows indicate the flow path of water and numerous choanocyte chambers are visible throughout the sponge. Magnification, approximately 100× (from Weissenfels, 1982).

into incurrent canals which branch repeatedly into canals of progressively narrower diameters, until they reach a choanocyte chamber (Fig. 4).

Choanocyte chambers are the keystones of the sponge feeding system. Each chamber consists of numerous choanocytes that are characterized by the presence of a flagellum and a collar of microvilli. Beating by the flagella within these chambers is primarily responsible for setting up the flow of water through the sponge. Water flow is also facilitated by a sponge's basic hydrodynamic structure. Vogel (1974) has demonstrated that water will actually flow passively, albeit at a reduced rate, through the feeding canals of killed sponges because of this structure. Choanocytes also serve as the final filters in the feeding system. The microvilli that make up the collars of the choanocytes are spaced to form openings less than 0.1 μm in diameter.

Once water has passed the choanocyte chambers, it enters the excurrent portion of the water processing system. Exiting the choanocyte chambers, the canals join together in an anastomosing fashion forming larger and larger diameter vessels, finally entering a broad chamber termed an atrium. From this chamber water exits the sponge body through a specialized structure termed an osculum. Such oscula channel water from large portions of a sponge and each sponge is likely to have several oscula. Water is forced from the oscula with sufficient force so as to enable materials that have passed through the sponge to be transported far enough away from the sponge that these materials are unlikely to be recaptured by the flow of water entering that sponge.

Although the description above may suggest that the structure of the sponge canals was fixed, the system is actually dynamic. The positions of both the incurrent and excurrent canals, as well as their components, can shift as the sponge grows or as portions of the feeding canals are occluded by ingested materials (Mergner, 1966). Similarly, the mesohyl of freshwater sponges exhibits rhythmic condensation that assists the choanocyte chamber in pumping water (Weissenfels, 1990).

2. Digestion

A second key feature of sponge feeding involves the capacity of the cells within the canal network to engulf particles through phagocytosis. Cells ranging from the porocytes that form the ostia on the sponge surface to the pinacocytes lining the incurrent canals and the choanocytes exhibit this capacity for phagocytosis. Digestion and the transport of nutrients within the sponge body also involve phagocytic activities in which materials are exchanged between cells. Information on activities by cells in sponge digestion has been gained by several detailed studies employing light and electron microscopy (van Weel, 1949; Schmidt, 1970; Willenz, 1980).

Phagocytosis can occur when a particle makes contact with a surface within the canal system. Food particles large enough to occlude a sponge's ostia are taken up directly at the epithelium. Smaller particles are taken up after contact with the cells lining the canal walls or with the collars of the choanocytes. Following initial uptake, food particles are transferred to digestive cells, termed archaeocytes, which move freely within the mesohyl. As digestion proceeds, archaeocytes bearing ingested materials move through the sponge interior and transfer nutritional materials to other cells, such as those involved with reproduction or skeleton formation.

After the digestion of ingested materials is completed, archaeocytes move to the excurrent portion of the canal network. Nondigested material is released into the water flowing out of the sponge by a reverse phagocytic action. Egested particles are then carried by excurrent flow out through oscula at the sponge surface.

Detailed analyses of cell activities in sponge digestion have revealed a surprising degree of complexity. The time necessary for a particle to traverse the digestive system completely, from initial uptake to release into an excurrent canal, can vary markedly with the nature of the food particle and with environmental conditions. Ordinarily, digestible food particles may be held within archaeocytes for more than 12 h prior to release into the sponge excurrent. Under certain circumstances, sponges can sense indigestible particles and release them after much shorter time periods (Frost, 1980a). The capacity to speed the processing time for ingested materials is most evident when a sponge is faced with a dense suspension of particles. At such times, food particles may be taken up, cycled, and released almost immediately (Schmidt, 1970).

The sophistication of the sponge feeding system is particularly evident in digestive activities because the rapid cycling of ingested particles can be highly selective. While overabundant particles are being quickly cycled through the sponge, other food particles, present in lower concentrations in the feeding suspension, are held within the sponge for periods that are sufficiently long to allow normal digestion (Frost, 1980a, 1981). Sponges, therefore, are selective feeders, not in their initial uptake of particles as selective feeding is typically defined, but in their ultimate use of these resources (Frost, 1980b).

A sponge's capacity to vary particle transit time is linked to the large number of cells that are operating within it. Digestive cells are sufficiently numerous so that, under normal feeding conditions, each cell handles only one ingested particle at a time (Willenz, 1980). Each individual archeocyte then, can be considered as a separately functioning digestive unit with the capacity to tune its actions to the characteristics of the food particle that it is processing.

3. Skeleton

The gross structure of freshwater sponges derives from an interplay between two fundamentally different components—a mineral skeleton made up of siliceous structures termed spicules and an organic skeleton made up of collagen. Collagen binds spicules together into rigid structures yielding the sponge's basic framework. All freshwater sponges exhibit skeletal systems comprised of siliceous spicules and collagen. Marine sponges exhibit a broader variety of supporting structures (Berquist, 1978).

a. Siliceous Spicules Some major groups of sponges are unique among multicellular organisms in their use of silica as a primary component of their skeleton. Freshwater sponge spicules are composed of opalescent silica that has been laid down along an axial, organic filament by specialized cells termed sclerocytes. Considerable diversity exists in spicule morphology, and, because much of this variability is species-specific, their structure plays a critical role in sponge taxonomy.

Three general classes of spicules are recognized: (1) megascleres, which make up the main framework of

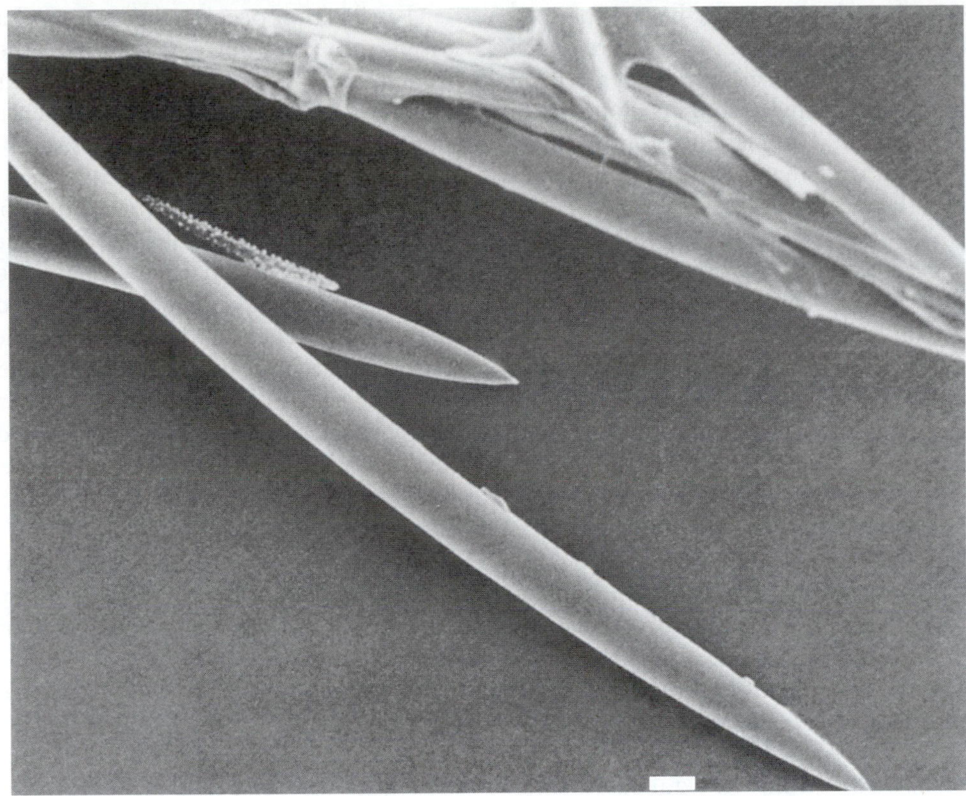

FIGURE 6 A scanning electron micrograph of megascleres from *Spongilla lacustris* collected in a lake with a high concentration of dissolved silica. A spined microsclere is also present. The bar is 10 μm in length.

a sponge; (2) microscleres, which appear to add structural reinforcement for sponge tissues; and (3) gemmoscleres, which form part of the resistant coat of gemmules (see below). Megascleres are needle-shaped structures (Fig. 6 and Fig. 7) that range in length from 150 to 450 μm. Microscleres are similar in form to megascleres but are usually less than one-fifth their length (Fig. 6). Gemmoscleres are similar in size to microscleres and both exhibit a variety of forms ranging from needlelike to dumbbell or star-shaped. Any of these three spicule forms may be smooth or spined depending on the species. An elaborate terminology has been developed to describe the varied shapes of spicules. These terms have been summarized in Boury-Esnault

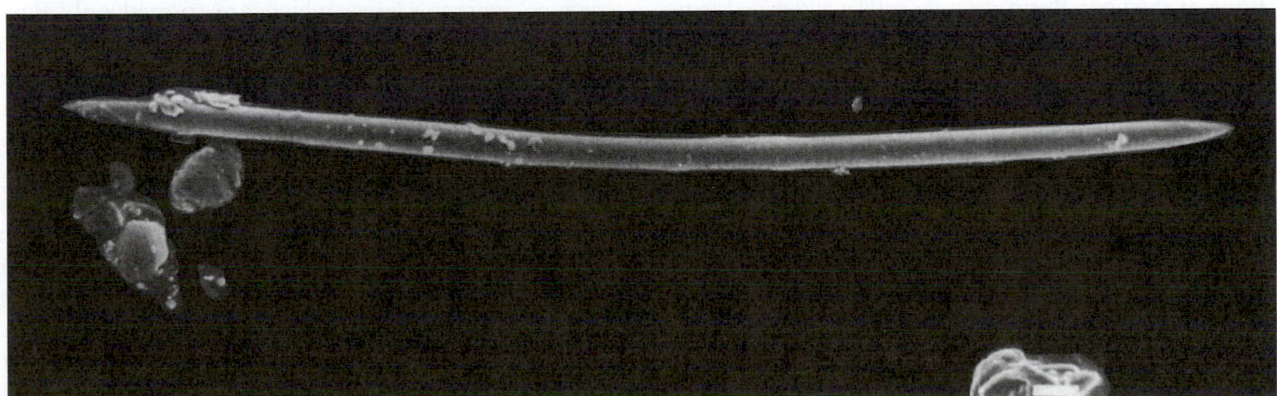

FIGURE 7 A scanning electron micrograph of a megasclere of *Spongilla lacustris* collected in a lake with low concentration of dissolved silica. The bar is 10 μm in length.

and Rützler (1997), but their use has been kept to a minimum in this chapter. The fine-scale structure of the spicules, particularly the gemmoscleres, plays a critical role in the taxonomy of sponges, and its importance cannot be overemphasized. The classification of freshwater sponges is based fundamentally on the structure of gemmule spicules (Penney and Racek, 1868).

b. Collagen The predominant form of collagen in sponges is spongin, which serves primarily to bind the spicules of the inorganic skeleton together. A second form of collagen provides smaller-scale structure within the sponge mesohyl. Finally, collagen, often combined with gemmoscleres, forms the resistant coat of gemmules.

4. Reproduction

Both sexual and asexual processes can play major roles in sponge reproduction. Sexual reproduction involves the activities of numerous, isolated reproductive cells functioning throughout the body of an active sponge. Asexual reproduction often involves major changes in all of the cells within a sponge.

a. Sexual Reproduction Although detailed observations have been made for only a few freshwater species, it appears that the mode of sexual reproduction is usually gonochoristic, with each separate sponge being entirely male or female. In at least some cases, however, the gender of a particular sponge is not fixed but appears to depend on environmental conditions. In at least one species, *S. lacustris,* sponges have been shown to switch sex from one year to the next (Gilbert and Simpson, 1976a). Bisbee (1992) did report one hermaphrodite specimen in a population of *S. lacustris* in South Carolina where most sponges exhibited only male or female elements.

Sexual reproduction is accomplished by specialized cells that develop within the mesohyl during limited portions of the year. Reproductive cells are derived from other, more common sponge cells. Spermatogenesis occurs within distinct spermatic cysts which appear to be formed from choanocytes. Egg cells (oocytes) seem to develop either from archeocytes or choanocytes. Oocyte growth is dependent upon a variety of nurse cells that are phagocytized as the egg develops (Gilbert, 1974). Because these events do not occur for extensive periods, successful sexual reproduction requires synchronous timing of egg and sperm development across sponge populations in a habitat. This has been documented for some freshwater species (Simpson and Gilbert, 1973). This requirement leads to spectacularly synchronized events of extensive sperm production in some marine species (Reiswig, 1970).

Fertilization occurs when sperm that have been released into the open water by other sponges are brought into the canal systems of female sponges by their feeding currents. Sperm are then taken up by the sponge, probably by choanocytes, and conveyed to an egg cell where fertilization takes place.

As is the general case within the Porifera, freshwater sponges are viviparous. Larvae undergo extensive development prior to their release from the mother sponge. Material for growth is provided by nurse cells. In the final stages of development, the sponge larva, termed a parenchymula, contains archeocytes, pinacocytes, sclerocytes, and choanocyte chambers. In addition, its surface is covered by flagellated epithelial cells which allow the larva to swim upon release.

Once development is completed, larvae are released through the excurrent portion of the feeding canal system. The larvae swim until they settle onto a substratum, where they undergo metamorphosis and quickly develop the structures typical of adult sponges.

b. Asexual Reproduction Asexual reproduction in sponges may be simple or complex. On one extreme, clonal development of separate, independently functioning sponges can take place through fragmentation. The plastic nature of sponge cells makes it possible for even very small pieces to develop into active, fully functional sponges. Propagation in this fashion occurs commonly in some habitats (Frost *et al.,* 1982).

On the other extreme, most freshwater sponges typically form gemmules in a complex process of dedifferentiation in which the structures of the active sponge regress into masses of cells surrounded by resistant coats (Figs. 1 and 8). This process, which often occurs prior to the onset of rigorous environmental conditions, permits sponges to weather periods of stress and to disperse into other habitats.

The coat of a mature gemmule usually consists of three collagen layers. In most species, specialized spicules, gemmoscleres, which are formed only during gemmule formation, are also embedded within the gemmule coat. The coat of each gemmule is continuous with a much thinner layer at a single specialized structure, termed a micropyle or foraminal aperture, through which cells emerge during germination. Within the gemmule are a large number of specialized cells, thesocytes, which contain numerous yolk inclusions. Thesocytes can store energy-rich materials which are used for growth following germination. The aggregation of cells to form gemmules takes place within the mesohyl. Gemmules are often present in large numbers within the vestigial skeletons of previously active sponges.

Completely formed gemmules exhibit low metabolic rates until germination is induced, and are

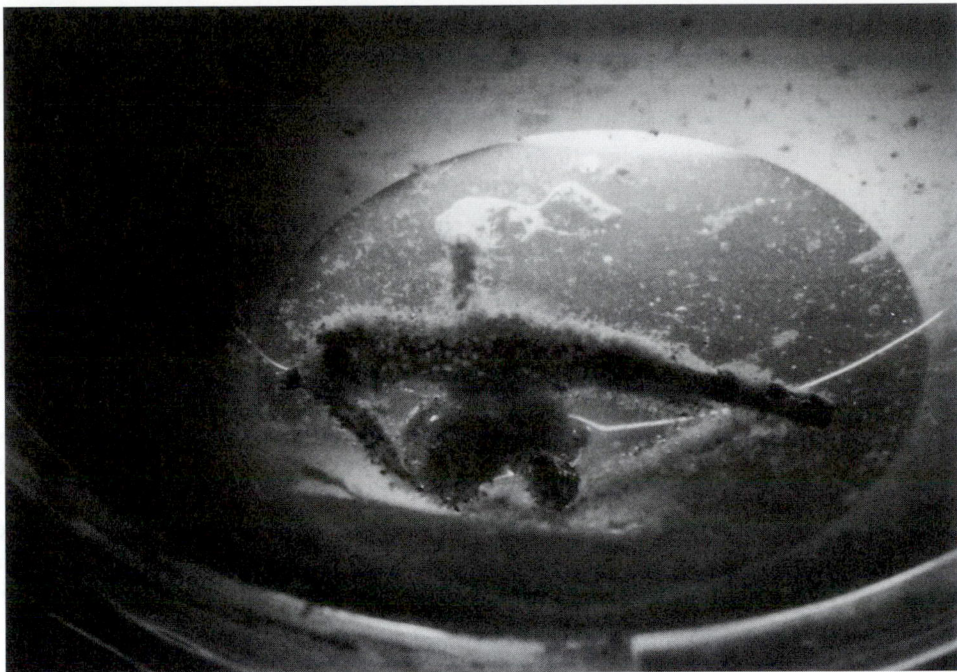

FIGURE 8 Numerous gemmules (the small whitish spheres) of *Spongilla lacustris* in a formerly active fingerlike growth of the sponge.

extremely resistant to environmental stress. Gemmules of a common freshwater species *Ephydatia muelleri* survived exposure to −80°C for more than 9 weeks (Barbeau *et al.*, 1989) although cold tolerance has been shown to vary substantially among species (Ungemach *et al.*, 1997). Gemmules can also survive anoxic conditions for several months (Reiswig and Miller, 1998). Upon germination, the thesocytes increase their metabolic rate, exit from the gemmule coat through the micropyle, and differentiate into the varied cells typical of an active sponge.

At least one freshwater sponge species is also capable of another form of asexual reproduction, budding. Buds exhibit many of the structures of active sponges, but in a smaller form that is capable of dispersal within water (Saller, 1990). They do not, however, exhibit the same resistance to adverse environmental conditions as gemmules.

5. Biochemistry

Sponges exhibit a variety of unusual chemical constituents. Some, particularly those identified as secondary metabolites, have been linked with an ability to deter predation (Pennings *et al.*, 1994). Particularly unusual among chemical features of sponges has been a major portion of fluorine in the body of a marine species (Gregson *et al.*, 1979). Most unusual chemicals have been documented in marine species, but some are also likely to occur in freshwater species. It has been suggested that they may provide some defenses against predation and overgrowth by epizooic organisms for freshwater sponges (Frost, 1976a).

III. ECOLOGY AND EVOLUTION

A. Diversity and Distribution

1. Diversity

The diversity of sponges in freshwater is low in comparison with that in the marine habitats from which they have evolved. Although the entire phylum Porifera consists of more than 5000 species (Bergquist, 1978), there are probably fewer than 300 freshwater species worldwide. In the most recent taxonomic revision of the Spongillidae, the largest family of freshwater sponges, Penney and Racek (1968) list a total of 95 species in 18 genera . While it is certain that there are additional, undescribed species, particularly in tropical regions that have not been thoroughly investigated (e.g., the Amazon Basin; Volkmer-Ribeiro, 1981), it is clear that freshwater forms represent a small group within the sponge phylum.

In turn, North American freshwater sponges represent only a restricted subset of the world's freshwater species. Overall, within both the Spongillidae and Metaniidae, we have generated a list of 27 clearly distinct species in 11 genera that appear to be valid reports for the portion of North America north of Mexico and the Caribbean.

2. Distribution

From the broadest perspective, the distribution of sponge species is influenced by biogeographic factors. On an intermediate level, the general environmental conditions that exist within a lake or stream will control the presence of a particular sponge species there. Finally, conditions such as light, substrata, and wave action will determine sponge distribution within a particular water body.

The freshwater sponge species in North America exhibit a variety of distribution patterns (Penney, 1960; Penney and Racek, 1968). Several species have been reported throughout the United States and Canada, while others have been observed in only one or a few habitats. In some cases, a broad distribution in North America represents only a subset of a more extensive range. *Ephydatia fluviatilis* and *Eunapius fragilis* are distributed worldwide, while *Spongilla lacustris, Ephydatia muelleri,* and *Trochospongilla horrida* occur throughout temperate regions of the Northern Hemisphere. Some species exhibit more restricted cross-continental distributions. *Anheteromeyenia ryderi* occurs primarily in eastern North America but has also been reported once in Central America and from several locations in Western Europe (Økland and Økland, 1989). Other species are broadly distributed throughout North America, but have not been observed on other continents (e.g., *Duosclera mackayi* and *Trochospongilla pennsylvanica*). Of the remaining North American species, most appear to be restricted to limited regions within Canada and the United States; and, in the most extreme cases, several species have been reported from single locations.

It is important to note that this chapter provides limited coverage of North America. It considers only sponges in Canada and the United States. Limited information on Mexican sponges is reported in Rioja (1953). Smith (1994) presented the first report of a freshwater sponge from the West Indies. Abundant sponges have been observed in a wetland in Costa Rica (T. Frost, personal observation), and it is certain that there are sponges throughout Mexico and Central America.

Within Canada and the United States, there have been numerous reports on the occurrence of sponges in provinces and states or regions within them. Some recent examples include reports for Alberta (Clifford, 1991), Arizona (Sowka, 1999), Connecticut (De Santo and Fell, 1996), eastern Canada (Ricciardi and Reiswig, 1993), western Montana (Barton and Addis, 1997), and southern Lake Michigan (Lauer and Spacie, 1996). Penney (1960) presented an extensive review of earlier reports on freshwater sponges. Despite these reports, information available on the occurrence of sponges does not match their overall distribution.

The apparently limited ranges of some species may result more from a lack of observations rather than actual restrictions in sponge distributions. Accepting this limitation, however, there appear to be some general trends in the distribution of sponges across North America. For example, relatively few species of sponges have been reported in western regions, and there appears to be a general east-to-west decrease in sponge diversity. A similar trend occurs between northern and southern regions, with more species reported from the northern United States and southern Canada than from the more southern regions of the continent. In some situations, the limited distribution of a particular species may involve adaptations to climatic conditions. In many other cases, biogeographic factors, particularly limitations on dispersal, are likely to control regional distribution patterns. There are some notable exceptions to limitations on dispersal, however (e.g., Jones and Rützler, 1975, Økland and Økland, 1989).

Within regional distribution patterns, the factors that regulate the occurrence of a species in a particular lake or stream are less well understood. While *Spongilla alba* appears to be restricted to brackish water habitats (Poirrier, 1976), other species are influenced by a variety of environmental factors. Jewell (1935, 1939) provided the most detailed quantitative information available on this topic. In comparing species across lakes in northern Wisconsin, she found considerable variation in each species' habitat requirements. Some species were restricted to a narrow subset of environmental conditions, while others occurred throughout a broad range of habitats. Chemical factors were correlated with the distributions of some species; but for other species, more recent analyses indicate that there is wide tolerance of chemical conditions (Colby *et al.,* 1999). In addition, some species were favored by flowing water, while others were more typically associated with standing water. In general, it is difficult to pinpoint any specific factors that control the distribution of sponges. Many species are sufficiently broad in their distribution so that many lakes can be expected to contain at least some small specimens. This is particularly true in some regions, e.g., northern Wisconsin, where sponges occur in nearly every lake. In some cases, particularly in smaller lakes, large populations of several sponge species may be common. Aside from cases of geographic isolation, sponges would be excluded only from habitats with frequent physical disturbances, with high pollution levels, or with high loadings of silt or particulates that can clog their feeding systems (Harrison, 1974). Not all pollutants affect sponges, however. They appear to be tolerant of high levels of some contaminants, particularly heavy metals (Richelle *et al.,* 1995).

Within a lake or stream that is suitable for growth, sponge distribution is controlled by finer scale environmental features. For most sponges, a hard substratum is essential for growth, and the absence of a surface on which to grow will limit the distribution of sponges even where other conditions are highly favorable. Only a few species, notably *Spongilla lacustris* and, to a lesser extent, *Ephydatia muelleri*, can grow in soft sediments. In rivers and streams, sponges are excluded from regions with high flow due to physical disruption. In lakes, sponges can be limited in shallow waters by wave action or ice scour (Bader, 1984). In deeper waters, sponges can be limited by low oxygen or by colder temperatures. Some species with symbiotic algae may be limited by light availability, particularly where water is darkly stained by dissolved organic materials, but species with less dependence on symbionts may thrive in such habitats.

B. Reproduction and Life History

The annual life-history pattern for most freshwater sponges involves periods of active growth alternating with dormancy and includes both asexual and sexual reproductive processes. Transitions to, and from, dormant stages usually involve the asexual processes of gemmule formation and hatching. Asexual reproduction also occurs during active growth periods, both in the formation of gemmules and in the generation of separately functioning sponges through fragmentation. Sexual reproduction occurs only for a limited time during periods of active growth. Key features of the freshwater sponge life cycle are summarized in Figure 9 and discussed in detail below. Overall, reproduction in sponges involves responses to two distinct problems, propagation within a single habitat and dispersal among habitats. Within-habitat processes are discussed separately below prior to a consideration of dispersal.

1. Dormancy

In most cases, dormant periods for freshwater sponges are characterized by the complete transformation of all active tissue in a sponge into gemmules (Simpson and Fell, 1974; Simpson, 1984). More rarely, tissue regression yields a growth form that functions as a gemmule without its specialized structures. In either situation, sponge feeding ceases and respiratory activities are greatly reduced. Most studies of dormancy in sponges have focused on transitions involving gemmules.

Despite the common role of gemmules in dormancy, their presence does not in itself necessarily signify a transition to a dormant life-cycle period. In many cases, particularly where sponges are large, a few gemmules are produced during most of the active season.

Most freshwater sponges undergo a period of dormancy at some time during the year, typically during periods of environmental stress. In temperate habitats, dormant periods occur most commonly during winter. This is the case for the majority of sponges in North America. However, at lower latitudes this pattern can be reversed with dormancy occurring during hot, summer periods and active growth taking place during cooler seasons (Poirrier, 1974; Ilan *et al.,* 1996). Dormant periods have also been observed in response to drying in ephemeral habitats. While dormancy is a common feature of freshwater sponge life cycles, it does not always occur; some sponges maintain active growth even under winter ice cover (Frost, personal observation).

The timing of dormancy varies both among species within particular habitats and among habitats for individual species. In Lake Pontchartrain, Louisiana, *Spongilla alba* exhibits a typical pattern of gemmule formation during winter months (Poirrier, 1976), while a co-occurring population of *Ephydatia fluviatilis* exhibits a reciprocal pattern of summer dormancy (Harsha *et al.,* 1983). At the same time, populations

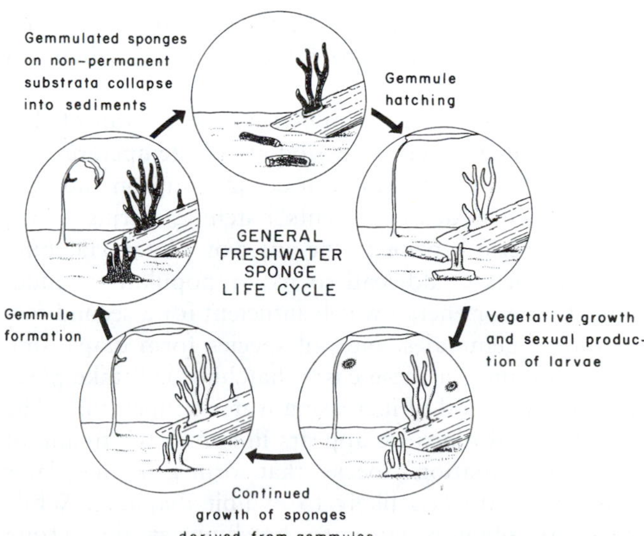

FIGURE 9 General features of the annual life history of freshwater sponges (see text for detailed explanations).

of *E. fluviatilis* in more northern habitats form dormant stages in winter. *S. lacustris* populations in New Hampshire gemmulated during winter and were active during summer (Simpson and Gilbert, 1973), while population in South Carolina were inactive during summer periods (Bisbee, 1992). Sexual reproduction occurred at different times during the seasonal cycle in these areas as well. In Little Rock Lake, Wisconsin, *S. lacustris* undergoes a winter dormancy period while co-occurring populations of *Corvomeyenia everetti* and *Ephydatia muelleri* maintain active growth during the entire winter. Populations of *E. muelleri* in nearby Wisconsin lakes undergo typical winter dormancy.

The transition from an active to a dormant state for sponges usually appears to be cued by environmental factors, such as changing temperatures or declining water levels. In some species this transition may also involve a complex response to previous growth conditions (Gilbert, 1975). Within a region, the timing of gemmulation is nearly synchronous across habitats for all populations of a species that exhibit dormancy (Frost, personal observation). In contrast, different species within a region can vary substantially in the seasonal timing of dormancy and in the rate at which the transition from active tissue to gemmules proceeds. In Mud Pond, New Hampshire, *Trochospongilla pennsylvanica* begins forming gemmules in mid-August when water temperatures are at their maximum while *S. lacustris* does not form gemmules until October when water temperatures reach 10°C (Simpson and Gilbert, 1973).

Sponge species also vary in the factors that induce their release from dormancy (Simpson and Fell, 1974). Temperature appears to be the primary environmental cue that induces gemmule hatching in temperate habitats. Gemmules of some sponges hatch while water temperatures are at near-winter levels suggesting that another cue such as photoperiod or the availability of light is operating.

In some species, gemmules exhibit a form of diapause in which a certain period at cold temperatures is necessary before hatching can occur (Simpson and Fell, 1974). Such a delay prevents hatching during short, warming periods prior to the onset of winter. Hatching at such times could lead to major population reductions if time or energy was insufficient for a second formation of gemmules. Not all species form diapausing gemmules and, in these cases, hatching will take place during any period of increasing water temperature. The occurrence of diapause appears linked to the timing of gemmule formation. Species that form gemmules later in the year are less likely to exhibit diapause. While there are obvious costs to hatching at the wrong time, there may also be advantages to being able to hatch quickly without diapause under some circumstances. Studies of dormancy in freshwater sponges have generally focused on cold-water populations, and less is known about the factors that control the transitions to, and from, dormant stages when they occur in response to hot or dry conditions.

In general, while it is clear that the formation of dormant stages plays a crucial role in the overall life history of most sponges, substantial variability is apparent in: (1) the morphological forms involved in dormant stages; (2) the timing of transitions to, and from, dormant periods; and (3) the occurrence of diapause. This variability suggests that the overall value of dormancy and the successful strategies by which it can be employed differ markedly among habitats and that sponge responses to such differences involve a diversity of adaptations.

2. Growth, Reproduction, and Dispersal within Habitats

Nondormant periods in the freshwater sponge life cycle appear to be characterized by continuous growth. Although quantitative studies have been limited, sponge growth in some habitats can be extremely prolific. In a pond in New Hampshire, 10 mg dry mass of gemmule tissue of *S. lacustris* that hatched in spring produced an average of 18 g of active sponge tissue just prior to gemmule formation the following fall (Frost *et al.*, 1982). *Ephydatia fluviatilis* exhibited comparable prolific growth rates in a European stream (Pronzato and Manconi, 1995). Growth is largely indeterminate for individual sponges, which may reach surprisingly large sizes. In some optimal habitats, for example, a single specimen of *S. lacustris*, consisting of intertwined fingers (as in Fig. 1), can occupy more than a cubic meter of space (Frost, personal observation).

Sexual reproduction for most freshwater sponge populations occurs synchronously throughout a habitat in a 1–3 month period following hatching from dormant conditions (Simpson and Gilbert, 1973). Synchrony across a population is particularly important, because otherwise the sperm released by male sponges would have a low probability of successfully fertilizing an egg (Gilbert, 1974).

The motile larvae produced by sexual reproduction in freshwater sponges can play an important role in the establishment of new sponges. In particular, the colonization of new substrata will often depend on such motile forms. While settling on an appropriate substratum is clearly critical for the successful growth of a new sponge, little information is available on substrate choice by settling sponge larvae. Studies of a variety of marine organisms suggest, however, that substrate selection by freshwater sponge larvae may be possible.

Fragmentation of intact specimens into separately functioning units may also play an important role in freshwater sponge life cycles. In some cases, such fragmentation occurs during periods of tissue regression and may not lead to growth on new substrata. In other habitats, however, particularly those with large amounts of aquatic vegetation to which sponges can attach, fragmentation and growth may occur repeatedly during an active season leading to dispersal throughout a habitat (Frost *et al.*, 1982).

Although gemmules are frequently associated with dormant periods, they may also be present in freshwater species during periods of growth. Many sponge species routinely contain a few gemmules within otherwise active tissue, particularly in areas of thick growth. In an extreme case, *S. lacustris* develops specialized, summer gemmules which have much thicker coats than the gemmules formed for overwinter periods (Gilbert and Simpson, 1976b). These investigators suggested that summer gemmules may serve in a "bet-hedging" strategy against some environmental catastrophe which could occur prior to more extensive gemmule formation. It seems likely that summer gemmules may also serve in dispersal among lakes or streams.

3. Dispersal Among Habitats

As is the case for most freshwater organisms, an ability to disperse among different habitats is critical for the continued success of sponges in freshwater environments. Most lakes and streams are ephemeral from an evolutionary perspective. Nearly all lakes in North America are less than 20,000 years old (Frey, 1963), a brief period compared to the 100 million year history of freshwater sponges (Racek and Harrison, 1975).

While there has been little documentation of dispersal by freshwater sponges among habitats, it seems likely that gemmules are the primary agent for such movement. The two other life-cycle stages that might function in dispersal among habitats, larvae and sponge fragments, possess neither the structural nor the physiological mechanisms necessary for movement out of water or across significant distances. It is conceivable that these forms, particularly the larvae, could function in short-range dispersal among connected water bodies, but their fragile structure would preclude any out-of-water transport. Gemmules, on the other hand, with their resistant coats and ability to withstand harsh environmental conditions appear to be well suited to dispersal among habitats. Ricciardi and Reiswig (1993) have reported rafting of *S. lacustris* gemmules during flooding in the Ottawa River.

C. Ecological Interactions

As is the case with all organisms, the size of a freshwater sponge population, and its subsequent impact on an ecosystem, results from an interplay of growth and loss processes. Growth is regulated by the availability of nutrients and suitable habitat. Losses are influenced by physical environmental conditions and, potentially, by interactions with other organisms. Although quantitative studies are few, it appears that the dynamics of freshwater sponge populations are mediated largely by the availability of suitable microhabitats within a lake or stream. Where substrates for sponge growth are short lived or where physical disturbances are frequent, growth and loss rates can be extremely rapid and may operate primarily in a density independent fashion. Under such circumstances, sponge populations would be regulated by an interplay between the physical environment and their own growth and life history processes. Concomitantly, interactions with other organisms would exert only minimal influences on sponge population processes. In contrast, where substrates are permanent, sponge populations appear more likely to be regulated by intra- or interspecific biotic interactions (Jackson and Buss, 1975). In many sponge habitats where growth on permanent substrates is rare, such biotic interactions could be expected to exert only a secondary role in their population dynamics.

1. Availability of Substrates

The development of freshwater sponge populations usually depends on the availability of hard substrates. Most species must be attached to a substrate in order to grow. In addition, all species seem to depend on the presence of some solid surface for the successful settling and growth of their free-swimming larvae. A variety of surfaces are suitable. Examples range from boulders or exposed bedrock to the branches of fallen trees to the leaves and stems of aquatic macrophytes. Human built structures in water, such as bridge foundations, dams, spillways, and floats, can also serve as suitable surfaces for sponge growth. Potential substrates vary markedly in their permanence relative to the longevity of sponges, and these differences, in turn, have important consequences for sponge population dynamics.

Among North American species, only *S. lacustris* has been routinely observed growing out of soft sediments. A quantitative investigation of its population dynamics (Frost *et al.*, 1982), however, illustrated a crucial role for substrates even where sponges did not depend upon such substrates for their growth. In Mud Pond, New Hampshire, *S. lacustris* grows either directly from soft bottom sediments or attaches to any of several species of aquatic macrophytes which exhibit large summer populations, but which die back completely before winter. Due to the near absence of any permanent substrates in Mud Pond, gemmulated sponges there overwinter almost exclusively in sediments on the pond bottom. Successful hatching in spring depends upon the presence of enough gemmulated tissue to grow out of these sediments, and many sponges are not large enough to make this transition. Most sponges are unsuccessful in hatching, and spring populations are sparse. Hatching from sediments, therefore, is the limiting step in the Mud Pond population's dynamics. Despite remarkable growth during summer (a 1,000-fold increase in biomass), the population of *S. lacustris* in Mud Pond was maintained at nearly constant levels throughout five years of observations. Essentially, the population is maintained in a density-independent fashion by a combination of the lack of permanent substrates and the difficulty of gemmule hatching from soft sediments.

Sponge population dynamics appear to be substantially different in habitats where more permanent substrates are available. Populations of *Ephydatia fluviatilis* growing on hard substrates in a river on Sardinia, Italy exhibited growth rates that were still high, but were substantially lower than those reported for *S. lacustris* (Pronzato and Manconi, 1995). Populations were persistent on rocks over several years making transitions from active to gemmulated tissue. They were substantially disrupted, however, by a major flood but did recolonize habitats fairly quickly (Pronzato and Manconi, 1995). In general, there appear to be periods of substantial population growth with substantial degeneration during other parts of the year, a process that can be categorized as density-independent (Pronzato and Manconi, 1995). Finally, in Mary Lake, Wisconsin, *S. lacustris* grows attached to trees that have fallen into the lake and sponges repeatedly make the transition from active tissue to gemmules and back to active tissue within the same skeleton over the course of several years. In contrast with the large fluctuations in sponge density observed in Mud Pond, it seems that the populations of *S. lacustris* in Mary Lake exhibit higher densities throughout the year and lower growth rates during the course of the summer than those seen in Mud Pond (Frost, personal observation). Under such circumstances, population dynamics can be characterized primarily as density-dependent.

Overall, the availability of substrates exerts a major control over sponge population dynamics. For many species, the lack of substrates limits growth completely. But, even where species are not wholly dependent on substrates, their nature and availability can exert a major influence over sponge population dynamics. The longevity of an individual sponge, including repeated transitions between active and gemmulated states, would appear to be directly related to the permanence of its substrate. In addition, it seems likely that the importance of both intra- and interspecific biotic interactions in sponge population dynamics will be influenced by the time that a particular substrate has been available for colonization. Only on long-lived substrates would sponges be likely to reach sufficient densities to interact with each other.

2. Nutritional Ecology

When suitable habitats are available, sponges are capable of substantial growth. In some cases, these high growth rates appear to be linked primarily to a sponge's efficient particle-gathering mechanisms. In other situations, however, sponges have an alternative means of obtaining resources. Many freshwater sponges contain large numbers of intracellular algae. They form symbiotic associations that combine autotrophy and heterotropy. In addition, because sponges require substantial amounts of silica for their spicules, population growth and maintenance have the potential to be influenced by silica availability.

a. Sponge Feeding Although there have been several detailed studies of the mechanisms by which sponges feed and of the resources they consume (Frost, 1987), much less is known about the influence of food availability on sponge growth under natural conditions. The rates at which sponges filter water can be extremely high. During summer, specimens of *S. lacustris* typically filtered more than 6 ml/h per miligram dry mass of tissue (Frost, 1980b). At this rate, a finger-sized sponge filters more than 125 L per day . Because a sponge's feeding effectively removes all particles ranging in size from bacteria to large algae from the water it filters, these rates suggest that sponges may exhaust available food under some circumstances. Food limitation would be particularly likely in still water or where particulate resources are sparse. Pile *et al.* (1997) demonstrated that such depletion occurred in Lake Baikal, Russia, where sponges consumed a variety of picoplankton (cells <2 μm in diameter) in substantial quantities.

b. Algal Symbionts In many cases, freshwater sponges are bright green due to the presence of large quantities of chlorophyll contained within extensive populations of algal symbionts (Gilbert and Allen, 1973a, b). This situation is so common that sponges are frequently mistaken for plants in freshwater habitats. The algal symbiosis in freshwater sponges is representative of a large number of algal–invertebrate associations that are common in marine and freshwater habitats (Reisser, 1984). In these mutualistic associations, the nutritional processes of algae and invertebrates are closely coupled. In freshwater sponges, and in many other invertebrates, algae are maintained endosymbiotically within the cells of their host.

Algal–invertebrate symbioses combine autotrophic processes with normal animal heterotrophy. In the resulting, mixotrophic nutrition, symbiotic algae provide photosynthetically fixed carbon to the invertebrate host which, in turn, supplies nutrients such as nitrogen or phosphorus or carbon dioxide to the algae. This combination has proved particularly successful in nutrient poor conditions, such as coral reefs (Muscatine and Porter, 1977) but occurs over a broad range of habitats.

Algal symbionts play a major role in the growth of some freshwater sponges. Contributions from symbiotic algae accounted for 50–80% of the growth of *S. lacustris* in Mud Pond, New Hampshire (Frost and Williamson, 1980). The contribution of algae was determined by maintaining sponges *in situ* under darkened conditions, which led to the loss of their algae. Comparisons were then made between the growth of such aposymbiotic sponges and that of normal green sponges maintained under lighted but otherwise identical conditions. In similar experiments in Little Rock Lake, Wisconsin, *S. lacustris* and two other sponge species were incapable of any growth and died after a few weeks under darkened conditions (Frost, personal observation). Sand-Jensen and Pedersen (1994) also demonstrated a substantial contribution by symbiotic algae to the growth of *S. lacustris* in a Danish stream. Not all sponges depend on algal symbionts, however. Some species can live in darkness, where their nutrition is based solely on heterotrophic processes.

The factors that control the proportional contribution of autotrophic and heterotrophic processes to overall sponge nutrition are less clear. That sponges failed to grow under darkened conditions in only some habitats indicates that the relationship between a sponge and its algal symbionts can range from facultative to obligate depending on environmental conditions. Even for species that sometimes depend to a large extent upon their algae, the amount of algal chlorophyll within a sponge (a rough measure of the potential contribution of its algal symbionts) varies markedly, both among habitats for a species and among species within a single habitat. In a survey of northern Wisconsin lakes, the average concentration of algal chlorophyll in *S. lacustris* varied among lakes by more than a factor of three (Frost, personal observation). Similarly, in Little Rock Lake, Wisconsin, three sponges exhibited consistent species-specific differences in their chlorophyll content that were coupled with significantly different rates of both photosynthesis and feeding (Frost and Elias, 1990). These differences occurred even though the three sponge species shared identical environmental conditions.

The symbiotic relationship between sponges and algae differs fundamentally among sponge species. Some species depend, to a large extent, on algae for growth, while others exhibit little dependence. These differences among species may involve specializations to particular environmental conditions, in that sponges that contain large amounts of algal chlorophyll may be capable of growth only where light is readily available. Comparing habitats, it seems likely that sponges would vary the density of symbionts that they contain in response to the relative availability of light and particulate resources. However, this relationship is not simple. Detailed examinations of several sponge species indicate that increasing particulate resources may favor higher rather than lower concentrations of algae (Frost and Elias, 1990).

Until recently, it had been thought that all symbiotic associations of algae and invertebrates in freshwater involved the presence of one group of green alga termed zoochlorellae (Reisser, 1984). Invertebrates hosting this alga include protozoans, hydra, flatworms, clams, and sponges. It has now been determined, however, that at least one freshwater sponge species, *Corvomeyenia everetti*, and likely its congener *C. carolinensis*, contain a yellow-green alga as their symbiont (Frost *et al.*, 1997). Volkmer-Ribeiro (1986) has suggested that the *Corvomeyenia* species have an evolutionary history that is long distinct from sponges with green-algal symbionts. These distinct histories combined with numerous overall similarities between sponges with green and yellow-green symbionts indicate that two, closely convergent symbiotic associations have evolved separately in the freshwater sponges. Morphological affinities between marine algae and the yellow-green symbiont suggest that the symbiotic association in *Corvomeyenia* may have developed in an ancestral marine form and been maintained during a subsequent invasion of freshwater. It is important to note that differences between zoochlorellae and yellow-green algae cannot explain the observed differences in algal contributions to overall sponge nutrition described above. Substantial differences in the contribution of algae have been observed in comparisons of sponge species that contain only green algal symbionts.

The variety of ecological interactions between sponges and algae and the complexity of their evolutionary history indicate that algal symbioses have been important in the diversity and distribution of freshwater sponges.

c. Potential Alternative Nutritional Modes Additional means of gathering food resources have been documented recently for some marine sponges. Vacelet and Boury-Esnault (1995) have described a deep-sea sponge that lacks a typical sponge water-processing system and feeds on small animals with a mechanism that can be considered as carnivorous. However, some carnivorous sponges near deep-sea hydrothermal vents exhibit a symbiotic relationship with methanotrophic bacteria (Vacelet *et al.*, 1995). Neither nutritional mode has been documented in freshwater species, but their existence in marine habitat raises interesting possibilities.

d. Influence of Silica on Skeleton Structure The availability of silica within a habitat may exert both direct and indirect effects on sponge growth. As discussed previously, some sponges are restricted to high or low silica habitats. However, some species are distributed across habitats exhibiting a broad range of silica conditions, where they exhibit major responses to its availability. For example, in populations of *S. lacustris* in different lakes in northern Wisconsin, the total amount of biogenic silica and both the number and morphology of megascleres vary directly with silica concentrations in the lake (Frost *et al.*, in preparation). Total biogenic silica and spicule width (Fig. 6) decrease as less silica is available. In contrast, however, the number of megascleres increases as their width decreases (Fig. 7). With this combination of changes, the total surface area of megascleres remains constant as silica decreases—a situation that increases the structural support gained per unit of available silica. This suggests a direct and complex response by sponges to different availabilities of silica.

Responses to low silica conditions are related to changes in the stiffness and growth form of sponges. Specimens of *S. lacustris* from high silica habitats are much stronger and more resistant to breakage than those growing where silica is low. Sponges in low silica habitats, therefore, may be limited to growing closely attached to substrates or in waters with little wave action. Also, because spicules may play a role in the resistance of sponges to predation (see below), populations in low silica habitats may be more vulnerable to consumption.

3. Biotic Interactions

a. Competition for Substrates The availability of substrate plays a crucial role in the population dynamics of freshwater sponges. As such, competition for surfaces would seem to have a potentially important effect on sponge distribution and abundance. A sponge has the potential to interact with a substrate any time that its growth leads to contact with another sponge or any other organisms. There are reports of significant interactions for a substrate within, and between, sponge species, as well as between sponges and other organisms. Such interactions do not appear to occur commonly, however, and the role of competition for a substrate in the regulation of freshwater sponge populations remains largely unknown. In part, this is due to the minimal attention that interactions for substrates have received in freshwater in general. In addition, because freshwater sponge populations are often limited by density-independent processes, the abundance of sponges within a habitat is usually sparse with little potential for interactions on substrates.

When specimens of same sponge species grow into contact with each other on a substrate, two distinctly different results have been observed (Van de Vyver, 1970). In some cases, the separate sponges will form a single, functionally integrated unit. In other instances, a distinct structural barrier is erected in areas of contact and the two sponges continue to function as completely separate units. The genetic relationship between sponges appears to control whether they will merge or develop a barrier. Separate sponges grown from fragments or gemmules taken from the same original sponge will always merge with each other. In contrast, experimental manipulations have indicated the existence of distinct strains within a sponge species which will not merge with each other (Van de Vyver and Willenz, 1975). Studies of marine sponges originally suggested that only sponges that are genetically identical would merge with each other (Niegel and Avise, 1983). However, subsequent analyses have indicated that, while the probability of merging appears to be a function of the genetic similarity of sponges, specimens need not be genetically identical in order to join into a single unit (Stoddart *et al.*, 1985). Such fusion interactions in sponges operate at a cellular level (Van de Vyver, 1979) and have attracted the attention of researchers interested in the basic process of cell–cell recognition. These studies have direct implications for the understanding of immune-response systems in more complex invertebrates and in vertebrates (Smith and Hildemann, 1984).

Interactions for substrates among different sponge species have been thoroughly documented in marine systems while they have received little attention in freshwater. Marine studies have demonstrated a complex hierarchy in which certain sponge species can completely overgrow and kill other species (Jackson and Buss, 1975). These interactions have a major influence on overall community ecology on substrates (Jackson, 1977). Observations of interspecies interactions in freshwater sponges have been limited to instances where sponges have been shown to develop a nonmerging front similar to that described above for intra-species interactions (Fig. 1). In habitats where substrates are extremely long-lived, it seems likely that some sponge species may be able to completely overgrow other sponges, but the significance of such interactions in freshwater habitats remains to be demonstrated.

Interactions for substrates between sponges and other organisms had likewise received little attention in freshwater. Recently, however, sponges have been shown to overgrow and kill the zebra mussel, *Dreissena polymorpha*, a species whose recent invasion of North America via the Laurentian Great Lakes has been extensive and has raised substantial concerns (Ricciardi *et al.*, 1995). Unfortunately, sponge overgrowth does not appear to be a solution to zebra-mussel related problems. More generally, for attached organisms overgrowth by sponges must be a factor in other freshwater habitats. The spreading of a sponge across any surface must involve the overgrowth of microscopic algae and bacteria. Sponges often encrust on aquatic macrophytes, but do not seem to harm the plants. In a few cases, freshwater sponges have been observed to overgrow and kill colonies of bryozoans (Frost, personal observation). The overall importance of sponge overgrowth for other organisms remains to be assessed.

b. Predation and Infaunal Organisms Conspicuous predation upon sponges, in which major portions are consumed by a larger organism, is rare. A variety of smaller organisms consume portions of sponges, but they appear to act as grazers or parasites, leaving the sponge that they consume largely intact. Such species that feed on sponges represent a subset of the diverse infaunal community that typically occurs in freshwater sponges. Other infaunal organisms appear only to make use of the structure provided by a sponge in a commensal relationship.

Resistance to predation by larger organisms appears to be characteristic of the Porifera in general (Randall and Hartman 1968; Resh, 1976). Mechanical and chemical defenses combine to make sponges resistant to predation. Spicules provide mechanical deterrence to predators; these sharp spines are thought to act as an irritant to its mouth and digestive system. The effectiveness of spicules in deterring predation was evident in experimental tests in which snails would not consume *S. lacustris* from habitats where it contained robust spicules, but would consume specimens from habitats in which low silica availability limited the size of spicules (T. Frost and A. Covich, personal observation). Chemical deterrence to predation has been attributed to a variety of toxic and pharmacologically active compounds that are well documented in marine sponges (Cecil *et al.* 1976) and are also likely to occur in freshwater species.

Despite the general effectiveness of sponge defense mechanisms, a few organisms prey effectively upon sponges. In some cases sponge predators have developed a nearly complete dependence on sponges. The few large predators on freshwater sponges include fishes, crayfishes, ring-necked ducks and, possibly, snails. Extensive predation by fish on sponges has been documented in Africa (Dominey, 1987) and South America (Volkmer-Ribiero and Grosser, 1981), but is not common in North America. Crayfish were implicated as causing major reductions in a sponge population in a Massachusetts stream (Williamson, 1979), but widespread consumption of sponges by crayfish has not been reported. Juvenile ring-necked ducks have been reported to sometimes depend on sponges as a food source (McAuley and Longcore, 1988). Snails will consume and can subsist on freshwater sponges under laboratory conditions (T. Frost and A. Covich, personal observation); but here too, their significance as predators under natural conditions remains unknown.

Numerous small invertebrates reside within, or attached to, freshwater sponges. Examples include protozoans, oligochaetes, nematodes, rotifers, bivalves, water mites, and aquatic insects (Berzins, 1950; Resh, 1976; Volkmer-Ribiero and De Rosa-Barbosa, 1979). A variety of relationships exist between these infaunal organisms and their sponge hosts. Some infaunal species feed upon sponges while others use the sponge body as a habitat. In some cases the relationships are obligate, at least for certain life cycle stages, and such species may show a striking degree of specialization to the sponge host. Other organisms are only occasionally associated with sponges.

Several groups of aquatic insects typify organisms with obligate, specialized feeding on freshwater sponges. Numerous species within three insect orders, the Diptera, Neuroptera, and Trichoptera, depend on freshwater sponges during major portions of their life cycle (Resh, 1976). One neuropteran family, the Sisyridae, is so commonly associated with sponges that its members are referred to as Spongilla Flies. Not all relationships between insects and sponges are obligate. Some insect genera that contain obligately, sponge-feeding species also contains species that are only occasionally or never associated with sponges (Resh *et al.*, 1976). The variety of interactions evident between sponges and related insect species indicates that the evolutionary history of insect–sponge relationships is diverse and complex.

Obligate relationships with freshwater sponges have also been documented for water mites (Proctor and Pritchard, 1990) that apparently depend on the sponge primarily for structure. The water mites live and lay their eggs within the feeding canal system of the sponge. These and other organisms that live within sponges (e.g., clams, insects and protozoans) may be taking advantage of the lack of predation on sponges. Spicules and chemical defenses may provide a general refuge against predators on infauna as well as sponges.

The community ecology of the infaunal organisms themselves is also complex. The presence and abundance of a particular infaunal species results from an interplay of its own life cycle, the life cycle of the host sponge and potentially, the abundance of other infaunal organisms. Data from northern Wisconsin sponge populations indicate that most other infaunal species are rare or absent during periods when water mites are abundant within sponges (J. Elias and T. Frost, personal observation). Again, this is an area that has received little attention.

4. Functional Role in Ecosystems

In most lakes and streams, populations of freshwater sponges are sufficiently sparse that they are unlikely to play a major role in ecosystem function. However, there are marked exceptions to this general pattern. In some habitats, sponges are the dominant component of the benthic community and exert a substantial influence on total nutrient cycling and primary production. Quantitative analyses of freshwater sponge populations are rare, but a detailed study of one population illustrates the potentially major impact of sponges on ecosystems (Frost, 1978; Frost *et al.*, 1982). At the peak of its population size in early October, *S. lacustris* in Mud Pond, New Hampshire reached an average biomass of greater than 3.5 g dry mass per square meter. At this density, the total sponge population in the pond would filter more than ten million liters of pond water per day, a rate that would cycle a volume equivalent to that of the entire pond every 7 days. Since the sponges are capable of removing most phytoplankton and bacteria from the water that they filter, their potential impact on Mud Pond is substantial. Because of their algal symbionts, sponges can also account for a substantial portion of the primary production in Mud Pond. During late summer and fall, the chlorophyll within the sponges is approximately equivalent to that in the entire phytoplankton community (Frost, 1978; Frost and Williamson, 1980).

Sponges also play a major role in flowing-water ecosystems. Mann *et al.* (1972), in a detailed study of the River Thames in England, found that sponges accounted for nearly 40% of the total production by benthic animals.

Clearly, sponges can be a major component of the biota of some ecosystems. While detailed quantitative studies are rare, observations indicate that the populations in Mud Pond and the River Thames are representative of sponges in a variety of other lakes and streams. Sponge populations in some northern Wisconsin lakes, for example, appear to be substantially larger than those in Mud Pond (T. Frost, personal observation). At the same time, sponges in many ecosystems represent only a minor component when compared with other organisms.

Even when sponges account for substantial portions of an ecosystem's primary and secondary production, they may not interact directly with higher trophic levels. The lack of predation on sponges can lead to an accumulation of material in sponge biomass that is not passed directly up the food chain. Interaction of sponge with other ecosystem components may be indirect, operating primarily to limit the availability of materials to other organisms within the food web.

5. Paleolimnology

The siliceous nature of sponge spicules makes them resistant to decomposition under most circumstances. As such they are usually well preserved in lake sediments and have the potential to serve as a useful tool in indicating historic lake conditions. In situations where specific habitat requirements can be defined for a species (Jewell, 1935), the presence of its spicules in the sediments of a lake from which it is currently absent can indicate changes over time in the environmental conditions of that lake. Likewise, even when a sponge species is present throughout an extended period within a lake, quantitative changes in spicule morphology over time can indicate shifts in the availability of silica within that lake. For example, lakes throughout northern Wisconsin appear to have exhibited a more than 50% reduction in silica availability over the last 12,000 years (Kratz *et al.*, 1991). Paleolimnological investigations employing sponge spicules are a potentially fruitful area for study (Harrison, 1986). Studies in 28 Connecticut water bodies revealed that most lakes had maintained the same sponge populations over the past century, but that a few lakes had exhibited changes in sponge populations perhaps indicating a decline in water quality (Paduano and Fell, 1997). Spicule presence and distribution has been used to infer the history of selected Florida soils (Schwandes and Collins, 1994). Further studies have the potential to reveal other important features of lake history.

D. Evolutionary Relationships

All freshwater sponges belong to the class Demospongiae, the most diverse and morphologically complex of the four major groups in the sponge phylum. Within the demosponges, the evolutionary history of the freshwater sponges is long and polyphyletic. Fossil freshwater sponges are reported at least 100 million years before (Racek and Harrison, 1975); and, although there is still debate in this area, there appear to have been at least three, and possibly four, separate, successful invasions of freshwater from marine habitats. The separate invasions are presently classified into distinct families among the freshwater sponges (Volkmer-Ribeiro and De Rosa-Barbosa, 1979, Volkmer-Ribeiro, 1986).

The phylogenetic diversity of freshwater sponges is not well represented in North America where species are confined primarily to the most cosmopolitan family, the Spongillidae (Penney and Racek, 1968). Volkmer-Ribeiro (1986) proposed that species in the genus *Corvomeyenia* should be grouped, along with several South American species, in the family Metaniidae with an evolutionary history in freshwater that is distinct from the spongillids. Only two North American species, *Corvomeyenia everetti* and *C. carolinensis*, occur in this proposed family which has a worldwide, primarily tropical distribution (Volkmer-Ribeiro, 1986). Not all sponge biologists have recognized the Metaniidae as a separate family (e.g., Ricciardi and Reiswig, 1993), however.

The two other freshwater sponge families are absent from North America. Species in the Lumbomerskiidae appear to be confined to a few large and ancient lakes in Europe and Asia (e.g., Lakes Baikal and Ochrid), and the family is characterized by a large degree of endemism. Species in the Potamolepidae are widely distributed throughout Africa and South America (Volkmer-Ribeiro and De Rosa-Barbosa, 1979).

IV. COLLECTING, REARING, AND PREPARING SPONGES FOR IDENTIFICATION

Collecting sponges is usually straightforward. In many cases, sponges grow in shallow water and can be obtained simply by hand or with a long-handled rake. Where sponges are rare, it may be easiest to collect them while snorkeling. This may be the case particularly when collecting in areas with numerous hard substrates where small stones must be overturned and the undersides of fallen trees and large boulders must be examined. Where sponges grow in deeper waters, scuba techniques afford the most efficient means of collection. Dredges or nets dragged across substrates in deeper waters are likely to miss most sponges and, at the same time, are likely to disrupt substantial bottom areas. In very deep waters, however, these may be the only practical means of collecting.

When searching a lake or stream for sponges, it is best to examine as broad a variety of substrates as possible. Aquatic macrophytes, rocks, and fallen logs are common sites for sponge growth. In bog habitats, the roots of vegetation growing at the edges of bog mats or on the undersides of the mats themselves are likely sites for sponges. Where sponges are rare, it may be necessary to examine a large number of different substrates before finding any specimens. In other habitats, however, sponges will be conspicuous.

Growing or maintaining sponges under controlled environmental conditions is much more difficult than collecting them. Some investigators have grown small sponges from gemmules in the laboratory for cytological investigations (Rasmont, 1962; Fell, 1974). Such specimens can also be particularly useful in examining smaller scale structural details of sponges. Poirrier *et al.* (1981) have developed an effective, continuous-flow system for sponges using bacteria as a food source. Laboratory investigations of this sort can provide important insights into sponge ecology (e.g., Harsha *et al.*, 1983), particularly when controlled conditions are essential. Sponges appear to be highly sensitive to environmental conditions, however, and the responses of sponges obtained in laboratory studies must be compared carefully with their behavior under field situations. As such, to evaluate sponge behavior under natural conditions, it is often best to work with freshly collected sponges and to conduct experiments *in situ* whenever possible.

The identification of freshwater sponges depends primarily on characteristics of spicules and on features of intact gemmules. Successful species identifications are dependent on obtaining all of the spicules (megascleres, gemmoscleres, and, if present in a species, microscleres) that can occur in a species. Thus, it is absolutely essential to obtain samples of sponges that include all spicule types. Gemmoscleres are particularly important, and this can present a problem because gemmules (Figs. 1 and 8) may only occur during certain times of the year. Some gemmoscleres may remain in active sponge tissue after gemmules have hatched, so there is some chance of identifying sponges without gemmules visible in them. Gemmules should be sought carefully in any survey, however.

As the primary step in species identification, samples of all types of spicules that occur in a species should be prepared for microscopic examination. To obtain spicule preparations, organic portions of a sponge are digested with acid and the remaining spicules are mounted on a glass microscope slide. Sponge tissue can be digested with nitric acid in a centrifuge tube that is immersed in boiling water for 1 h. The spicules can then be concentrated by gentle centrifugation, after which the acid is poured off and the spicules are then washed with ethanol or methanol.

Washing and centrifugation should be repeated at least two times after which the spicules are mounted on a microscope slide using a cover slip and a permanent medium (e.g., Permount). Reiswig and Browman (1987) described a similar, but more precise technique for a quantitative method of processing sponge spicules.

Intact gemmules can be removed from sponge tissue, dried, and mounted directly on microscope slides. Specimens of entire sponges are necessary for museum collections. Whole sponges can be dried or preserved in alcohol. Simpson (1963) describes methods that can be used for preparing freshwater sponges for cytological observations, such as are necessary for examining reproductive cycles.

V. CLASSIFICATION

The classification scheme reported here is derived primarily from Penney and Racek (1968), who have completed the most recent taxonomic revision of the family Spongillidae, including the genus *Corvomeyenia*. Their monograph provides more detailed descriptions of most of the species reported here and should be consulted for rigorous taxonomic investigations. Information for several species in this chapter is based on reports other than those Penney and Racek (1968), most of which have been published since 1968. In the cases where substantial information about a species has been obtained from a source other than, or in addition to, Penney and Racek, these sources have been cited along with the species description in the list of species below. In addition, there are a number of cases where species reported for Canada and the United States by Penney and Racek (1968) have been judged subsequently as invalid (e.g., Poirrier, 1974, 1977; Harrison and Harrison, 1977; Harrison, *et al.*, 1977; Harrison, 1979), and these species have not been included here. Ricciardi and Reiswig (1993) provide another valuable source of information on 15 sponge species that occur in eastern Canada, and this report should be consulted for more detailed information on these species.

In general, the freshwater sponge species in Canada and the United States can be distinguished by characteristics of their spicules. Taxa are separated primarily by the presence, shape, and spination of microscleres and/or gemmoscleres, although the presence or absence of spines on megascleres can be a useful characteristic in some cases in distinguishing species within a genus. Spicules are usually either needle or rodlike (Fig. 6) or dumbell-shaped (Fig. 10). The latter are termed birotulate, and their rounded ends are called rotules. In the descriptions here, a spicule should be assumed to be needle or rodlike, unless it is described as otherwise. Also, the descriptions reported here are based on examinations with light microscopes. More detailed examinations with scanning electron microscopes can reveal fine structures, primarily spines, that are not apparent in light microscope observations. Such finer scale observations should be interpreted with caution when employing the descriptions and key that are provided here.

Gemmoscleres are critical in the classification of freshwater sponges, and it is important to include gemmules when preparing sponge specimens for identification. Obtaining gemmoscleres may be a problem, particularly in young sponges derived from larvae, where gemmules and their spicules may not be present during certain times of the year. Examine a collected specimen closely to insure that gemmules are present (Figs. 1 and 8). It is possible, particularly in larger specimens, that some gemmoscleres may have been left in a sponge body from a previous year's gemmules. However, it is also possible that a few stray gemmoscleres or

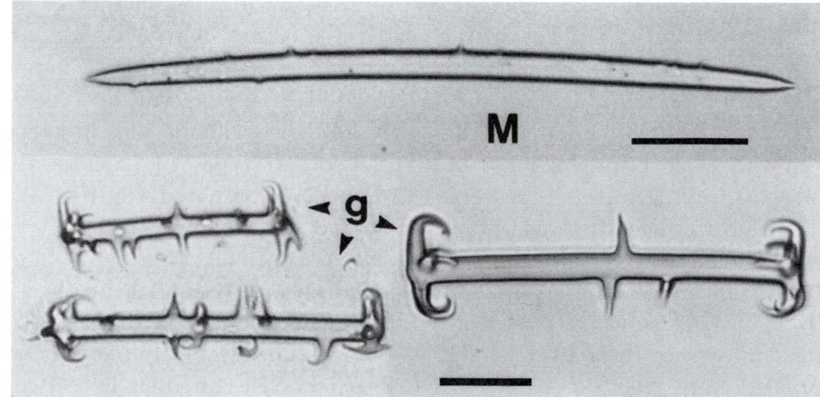

FIGURE 10 Megasclere (M, scale bar = 50 μm) and gemmoscleres (g, scale bar = 25 μm;) of *Anheteromeyenia argyrosperma* (from Ricciardi and Reiswig, 1993).

even entire gemmules from another sponge species may have been incorporated into a sponge specimen. The growth of the tissue of one sponge species over the gemmules of another species has lead to some erroneous species identifications (Ricciardi and Reiswig, 1992). Thus, identifications based on only a few gemmoscleres should be made with caution.

In some species with microscleres, it may be difficult to distinguish between microscleres and gemmoscleres on the basis of structure alone. By separating gemmules from intact sponges and making separate spicule preparations of them, it is possible to obtain samples in which gemmoscleres predominate. Alternatively, by carefully sampling tissue from the surface of a sponge in areas of new growth it may be possible to obtain material in which gemmoscleres are rare and in which microscleres and megascleres predominate. Because of the important role that microscleres play in separating species, identifications based on only a few microscleres should also be made with caution.

It is important to consider that ecomorphic variation can be substantial for some sponge species, particularly *Anheteromeyenia ryderi, Ephydatia fluviatilis,* and *Spongilla lacustris*. Cautions about particularly troublesome situations are raised in the species descriptions below. Ricciardi and Reiswig (1993) provide more detailed information on the eastern Canadian species.

In using the key presented here, there are only two cases in which characteristics other than the structure of spicules are required to distinguish among taxa. In contrasting the genera *Ephydatia* and *Radiospongilla*, it is necessary to obtain a cross section of intact gemmules in order to view the arrangement of gemmoscleres within the gemmule coat. Similarly, within the genus *Heteromeyenia*, it is necessary to employ the size, shape, and structures of the foraminal aperture of entire gemmules to distinguish among species. Particularly important for *Heteromeyenia* are the forms of structures on the end of the foraminal tubule termed cirrous projections.

Ricciardi and Reiswig (1993) reported an extensive survey of sponges in eastern Canada in which they reported 15 species. The report is extremely useful and should be consulted for detailed information on the species that they report. It even includes a key for distinguishing, to a certain extent, among sponges for which gemmoscleres are not available—a critical requirement of the key here and for using Penney and Racek (1968).

A. Freshwater Sponge Species of Canada and the United States

Summarized below are the spicular characteristics and distribution patterns of freshwater sponges reported from the United States and Canada. Dimensions presented for spicules are the ranges that have been reported for them; average values to be expected are very close to the mean of the numbers listed below. In some cases, mean dimension values are reported as xx-(mean)-xx, where xx represents ranges. It is important to note that these mean values are from Ricciardi and Reiswig (1993). They may be specific to sponges from eastern Canada and may not be representative of sponges growing in other regions. Values for ranges are the widest values that have been published in the papers reported here. Some sponge spicule features may vary from habitat to habitat. For a few species information is also provided on features of gemmules, such as the distribution of gemmoscleres within them, their opening structures (foraminal apertures), and their overall distribution within a sponge body. Geographic distribution patterns emphasize the occurrence of sponges in the United States and Canada, but also describe the overall distributions reported for each species.

Anheteromeyenia argyrosperma (Potts)

Spicules: megascleres slender, 240-(284)-304 μm in length, and sparsely covered with small, sharply pointed spines; microscleres absent; gemmoscleres birotulates of two distinct length groups, 65-(81)-89 or 110-(130)-160 μm, with both size classes similar in form exhibiting spines on their entire shaft and conspicuous recurved, clawlike hooks on their ends.

Distribution: reported from the eastern half of North America from Florida to Canada but confined to this region.

Anheteromeyenia ryderi (Potts)

Spicules: megascleres are extremely variable from habitat to habitat with lengths ranging from 141-(220)-279 μm; variable shape and coverage with broadly conical spines; microscleres absent; gemmoscleres birotulates of two distinct length groups with distinct differences in their shapes, 28-(34)-41 or 45-(64)-75 μm in length; shorter forms have shafts with only one or a few spines and flattened rotules with a large number of small teeth; larger forms are robust with numerous recurved spines on their shaft and with strongly recurved hooks on their ends.

Distribution: reported from the eastern half of North American, from Louisiana to Canada, but confined to this region.

Corvomeyenia carolinensis (Harrison)

Spicules: megascleres slender, straight to slightly curved and entirely smooth, ranging 194–280 μm in length; microscleres birotulate with straight to strongly

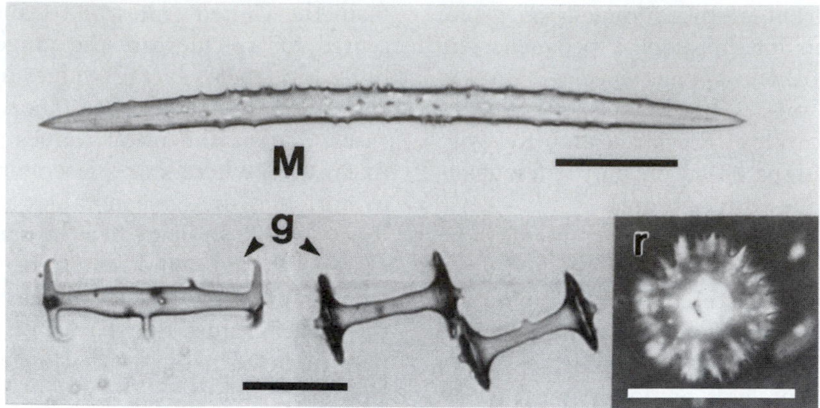

FIGURE 11 Megasclere (M, scale bar = 50 μm) and gemmoscleres (g, scale bar = 25 μm;) including a rotule (r) of *Anheteromeyenia ryderi* (from Ricciardi and Reiswig, 1993).

curved (>80%), smooth shafts, 15–25 μm in length terminating in rotules 4–7 μm in diameter with 4–6 recurved hooks; gemmoscleres birotulates with straight to slightly curved, smooth shafts 60 to 158 μm in length and terminating in rotules of 13–22 μm diameter with 5–8 recurved hooks (Harrison, 1971).

Distribution: reported from one pond in South Carolina and one lake in Connecticut.

Corvomeyenia everetti (Mills)

Spicules: megascleres slender, slightly curved and entirely smooth, 143-(218)-285 μm in length (in rare cases, a variable number of megascleres may be sparsely spined); microscleres birotulates, 14-(18)-26 μm in length, terminating in rotules 3-(5)-7 μm in diameter with 3–6 small, distinctly recurved spines;

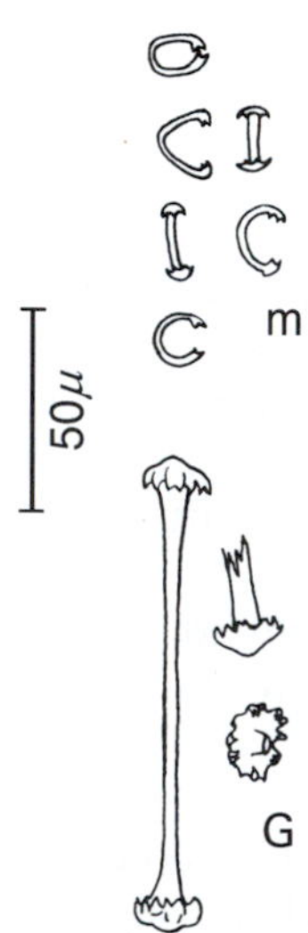

FIGURE 12 Microscleres (m) and gemmoscleres (G) of *Corvomeyenia carolinensis* (after Harrison, 1971).

gemmoscleres birotulate with straight to slightly curved, smooth shafts, 33-(59)-78 μm in length and terminating in rotules 10-(20)-26 μm in diameter bearing 5–7 recurved hooks. See information on *Corvospongilla novaeterrae* below and in Ricciardi and Reiswig (1993).

Distribution: reported only from the eastern half of the northern United States and Canada.

Corvospongilla becki (Poirrier)

Spicules: megascleres stout, 130–218 μm in length, usually curved, covered with spines, the spines being larger near the ends of a spicule; microscleres birotulate, 25–44 μm in length, rotules 9–17 μm in diameter, usually with four recurved hooks; gemmoscleres of two distinct length classes, the smaller class is 28–56 μm in length, slightly to strongly curved and spined, except in the inner curved region, the larger class 71–139 μm in length, straight to slightly curved and completely spined (somewhat similar in form to the gemmoscleres of *S. lacustris*—Fig. 29) (Poirrier, 1978). See information on *Corvospongilla novaeterrae* below.

Distribution: reported from one lake in Louisiana.

Corvospongilla novaeterrae (Potts)

Spicules: megascleres stout, smooth, 112-(154)-170 μm in length and relatively scarce; microscleres are abundant birotulates, 13-(21)-32 μm in length with smooth shafts, rotules dome shaped with 3–6 spines; gemmoscleres highly variable, bearing numerous large, recurved spines which tend to be concentrated near the ends of the shaft, sometimes making spicules appear birotulate with lengths 21-(39)-63 μm and widths, excluding spines, 3-(6)-9 μm. This species is remarkably similar to *Corvomeyenia everetti* in its basic growth form, except that *C. novaeterrae* gemmoscleres are not thin, elongate birotulates and *C. novaeterrae* gemmules

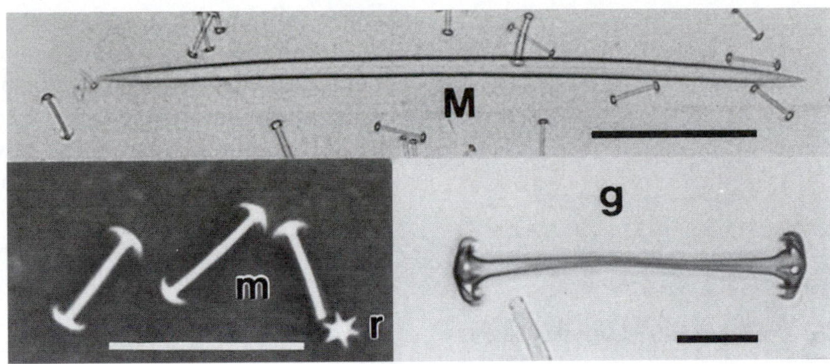

FIGURE 13 Megasclere (M, scale bar = 50 μm), microscleres (m, scale bar = 25 μm) including a rotule (r) and gemmosclere (g, scale bar = 25 μm) of *Corvomeyenia everetti* (from Ricciardi and Reiswig, 1993).

have a very poorly developed outer layer. See detailed information on *C. novaeterrae* in Ricciardi and Reiswig (1993).

Distribution: reported from a few lakes in the maritime provinces of Canada and one lake in Connecticut (De Santo and Fell, 1996).

Dosilia palmeri (Potts)

Spicules: megascleres slender, 370–450 μm in length, slightly curved to nearly straight, covered with sparse spines in their central portion; microscleres stellate with 8–12 usually smooth rays arising from a central nodule, length is extremely variable; gemmoscleres birotulates, 55–85 μm in length, occurring in two subtly distinct size classes, with strong spines on their central shaft and with equally sized rotules, 23–25 μm in diameter, bearing numerous blunt recurved teeth (Penney, 1960).

Distribution: reported from Florida in North America north of Mexico and from other locations in Central America.

Dosilia radiospiculata (Mills)

Spicules: megascleres slender, 290–400 μm in length, entirely smooth or covered with minute spines; microscleres stellate with 6–8 microspined rays projecting from their center, length is extremely variable; gemmoscleres birotulates of two distinctly different size classes with longer forms exhibiting nonspined shafts 120–230 μm in length and shorter forms exhibiting strongly spined shafts 45–82 μm in length. [There is some question as to whether the two species of *Dosilia* are distinct or simply ecomorphic variants of a single species.]

Distribution: reported from the Canadian border south to Mexico, but only from this region.

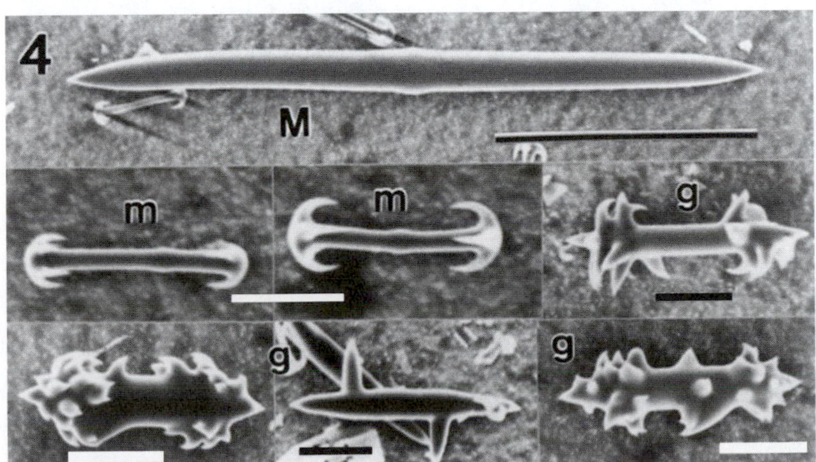

FIGURE 14 Megasclere (M, scale bar = 50 μm), microscleres (m, scale bar = 10 μm) and gemmoscleres (g, scale bar = 10 μm) of *Corvospongilla novaeterrae* (from Ricciardi and Reiswig, 1993).

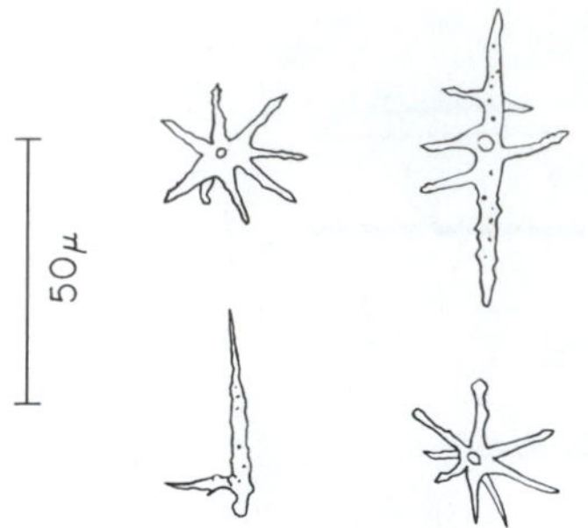

FIGURE 15 Microscleres of *Dosilia radiospiculata* (after Penney and Racek, 1968).

Duosclera mackayi (Carter)

Spicules: two distinct classes of megascleres, the first is relatively scarce, straight or slightly curved, 177-(200)-302 μm in length, 7-(12)-18 μm in width (excluding spines), covered with coarse spines and the second is somewhat shorter, 79-(156)-267 μm in length and generally broader, 2-(8)-20 μm in width (excluding spines) and densely covered with spines that are long, pointed, strongly recurved near the tips of the spicule and perpendicular near its center; microscleres absent; gemmoscleres are the same as the second class of megascleres, which are always present in a specimen, even in the absence of gemmules. When intact gemmules are present, gemmoscleres are arrayed tangentially to the surface of the gemmule except near the foraminal aperture. The overall orientation of gemmules can help distinguish *Duosclera mackayi* from a similar species, *E. fragilis*. In *D. Mackayi* the foraminal apertures are always oriented inward or towards a substrate while those in *E. fragilis* are always directed outward or upward from a pavement layers. This species has recently been classified as a new genus (Reiswig and Ricciardi, 1993). Previously reported as *Eunapius igloviformis* in Penney and Racek (1968) and as *Eunapius mackayi* in Ricciardi and Reiswig (1993).

Distribution: throughout the United States and Canada, but confined to these regions.

Ephydatia fluviatilis (Linneaus)

Spicules: megascleres usually slightly curved, 210-(343)-439 μm in length, and usually entirely smooth, although in some cases, some sparsely spined megascleres co-occur with smooth forms; microscleres absent; gemmoscleres birotulates of one class, 20-(23)-30 μm in length with a slender, smooth shaft, which sometimes has 1–4 large spines, with flat irregularly shaped rotules of equal, 13-(18)-24 μm diameters and usually more than 20 teeth that are not deeply incised. *E. fluviatilis* can sometimes be confused with *E. muelleri*. These species can be most surely distinguished by comparing gemmosclere length to rotule diameter.

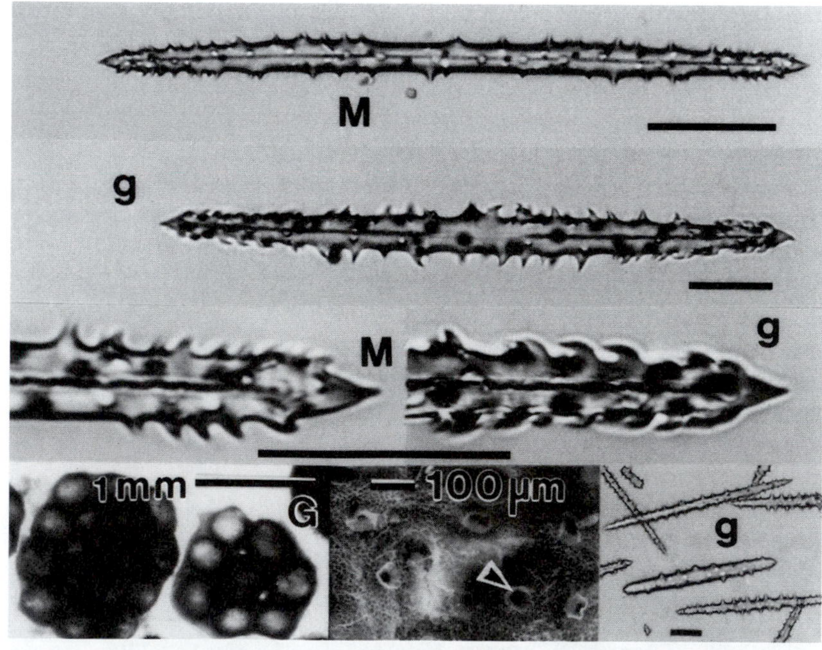

FIGURE 16 Megascleres (M, scale bars = 50 and 25 μm for the magnified section), gemmoscleres, which are actually a second size category of megasclere (g, scale bars = 25 μm) and gemmules of *Duosclera mackayi*, the arrowhead indicates the foraminal aperture on the underside of a gemmule cluster (from Ricciardi and Reiswig, 1993).

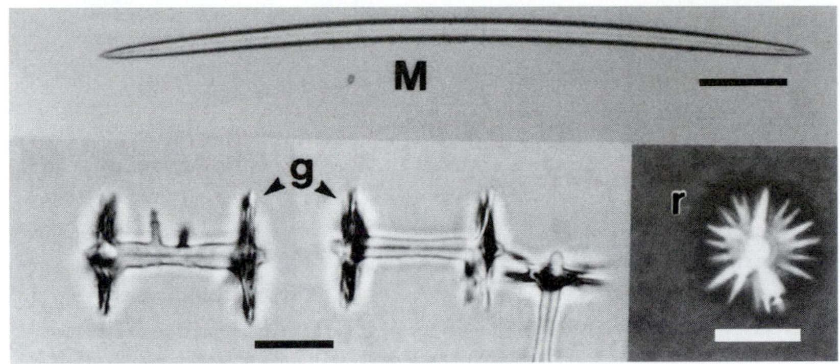

FIGURE 17 Megasclere (M, scale bar = 50 μm) and gemmoscleres (g, scale bar = 25 μm) and a rotule (r, scale bar = 10 μm) of *Ephydatia fluviatilis* (from Ricciardi and Reiswig, 1993).

Gemmosclere length is always greater than rotule diameter in *E. fluviatilis* while gemmosclere length is less than or equal to rotule diameter in *E. muelleri* (Ricciardi and Reiswig, 1993).

Distribution: appears to be truly cosmopolitan with more frequent occurrence in temperate than in tropical zones.

Ephydatia millsii (Potts)

Spicules: megascleres slightly curved to nearly straight, 180–270 μm in length, with numerous small spines except at the tips; microscleres absent; gemmoscleres birotulates of one size class, 36–48 μm in length, with smooth shafts that are clearly broader near the rotules and with distinctly flat, disk-shaped rotules of equal, 22–28 μm diameters and only very small incisions at their margins.

Distribution: reported only from Florida.

Ephydatia muelleri (Lieberkühn)

Spicules: megascleres straight to slightly curved, 171-(245)-350 μm in length, usually with small spines, typically lacking at the tips, but smooth in rare cases; microscleres absent; gemmoscleres birotulate of one class, 8-(17)-28 μm in length, with thick smooth shafts and with flat irregularly shaped rotules of equal, 8-(15)-27 μm diameters and usually have fewer than 12 teeth that are deeply incised into long rays. *E. muelleri* can sometimes be confused with *E. fluviatilis*. These species can be most surely distinguished by comparing gemmosclere length to rotule diameter. Gemmosclere length is always greater than rotule diameter in *E. fluviatilis* while gemmosclere length is less than or equal to rotule diameter in *E. muelleri* (Ricciardi and Reiswig, 1993).

Distribution: widely distributed throughout the Northern Hemisphere with a preference for temperate regions.

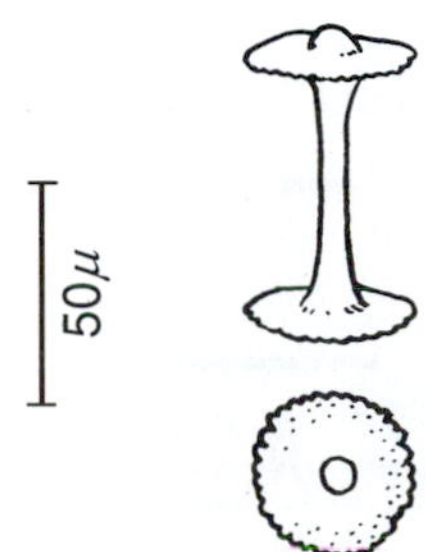

FIGURE 18 Gemmoscleres of *Ephydatia millsii* (after Penney and Racek, 1968).

Eunapius fragilis (Leidy)

Spicules: megascleres entirely smooth, 165-(189)-271 μm in length; microscleres absent; gemmoscleres straight to slightly curved, covered with conspicuous spines often more dense near the tips, 32-(57)-140 μm. Mature gemmules enclosed in a common brown coat forming a pavement layer cemented to the substrate (Gp in Fig. 20) or in individual clusters of 2–4 gemmules. The overall orientation of gemmules can help distinguish *E. fragilis* from a similar species, *Duosclera mackayi*. In *E. fragilis*, the foraminal apertures are always directed outward from a cluster or upward from a pavement layer while those in *D. mackayi* are always oriented inward or toward a substrate (Ricciardi and Reiswig, 1993).

Distribution: truly cosmopolitan.

Heteromeyenia baileyi (Bowerbank)

Spicules: megascleres slender, 216-(247)-320 μm in length, smooth or with sparse microspines except near the tips; microscleres 53-(67)-85 μm in length, delicate,

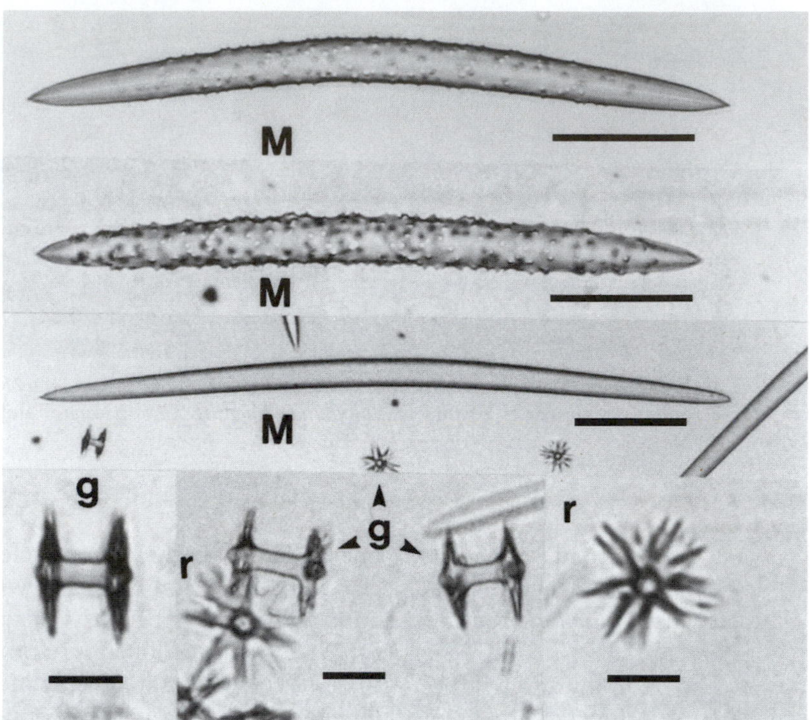

FIGURE 19 Spined and smooth megascleres (M, scale bar = 50 μm) and gemmoscleres (g, scale bar = 10 μm), and a rotule (r, scale bar = 10 μm) of *Ephydatia mülleri* (from Ricciardi and Reiswig, 1993).

slightly curved to almost straight with spines that occur throughout their length, but which increase in length towards the central region where they are often knobbed and distinctly perpendicular to the main axis; gemmoscleres birotulate of two intergrading length classes: (1) short birotules with flat, serrated rotules (13-(18)-22 μm in diameter), 38-(51)-60 μm in length, and (2) long birotules 49-(70)-86 μm in length with rotules (18-(22)-28 μm in diameter) composed of long recurved hooks giving an umbrellalike appearance (hooks often have knobbed tips). Foraminal apertures of gemmules do not have terminal cirrous projections like those for *H. tubisperma* (Fig. 24), *H. latitenta* (Fig. 22), and *H. tentasperma* (Fig. 23). The long

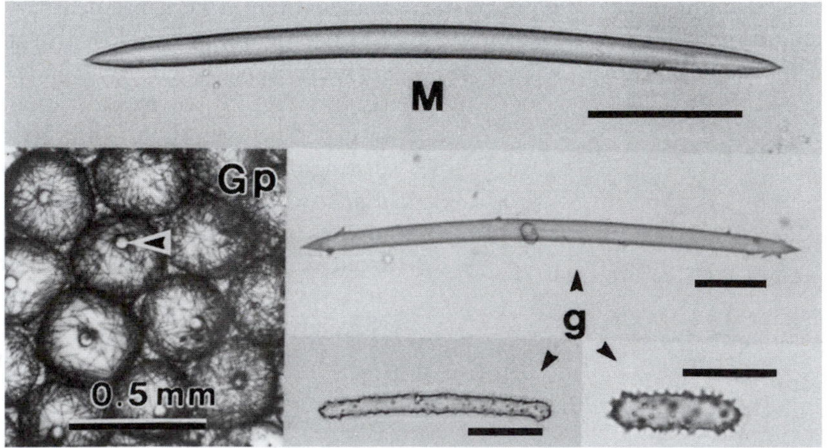

FIGURE 20 Megasclere (M, scale bar = 50 μm), gemmoscleres (g, scale bars = 25 μm), and a gemmule pavement of *Eunapius fragilis*, the arrowhead indicates a foraminal aperture on the dorsal surface of the gemmule pavement (from Ricciardi and Reiswig, 1993).

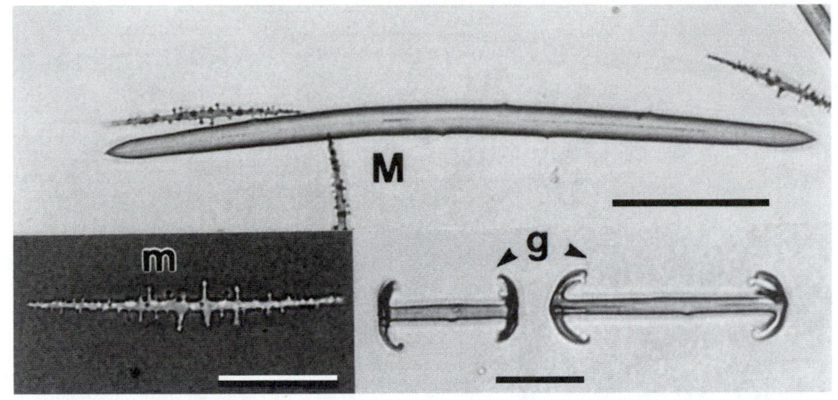

FIGURE 21 Megasclere (M, scale bar = 50 μm), microsclere (m, scale bar = 25 μm) and gemmoscleres (g, scale bar = 25 μm) of *Heteromeyenia baileyi* (from Ricciardi and Reiswig, 1993).

middle spines on the microscleres of *H. baileyi* are also useful in distinguishing it from other species in the genus. Their maximum length is on average greater than the width of the microsclere.

Distribution: widely distributed throughout the eastern United States and Canada with a few reports from Europe.

Heteromeyenia latitenta (Potts)

Spicules: megascleres straight, 265–285 μm in length, smooth to sparsely microspined; microscleres slender and entirely spined with those in the central region only slightly larger than those on the ends, 85–100 μm; gemmoscleres birotulates of two length groups, 50–55 or 60–78 μm with shafts bearing numerous stout and pointed spines, both rotules of equal, 16–18 μm diameters, margins forming numerous conspicuous, recurved teeth. Foraminal apertures of gemmules with one or two, very long, cirrous projections starting from a disk that is initially flat but rounded in its terminal regions.

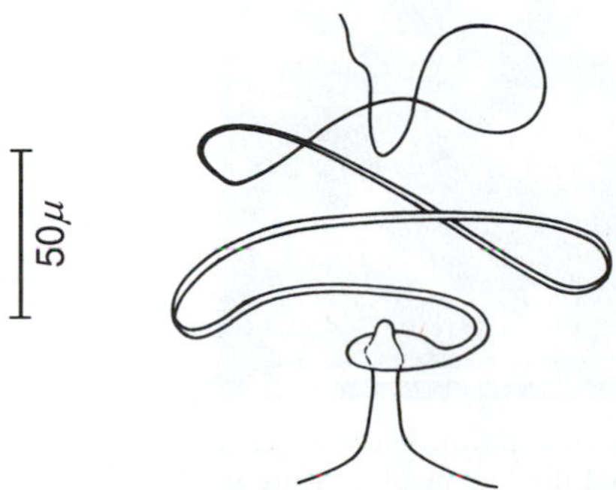

FIGURE 22 Foraminal aperture of *Heteromeyenia latitenta* (after Neidhoefer, 1940).

Distribution: reported only from northeastern United States.

Heteromeyenia tentasperma (Potts)

Spicules: megascleres very slender, 260–280 μm in length, with sparse microspines; microscleres slender with sparse microspines, 75–80 μm; gemmoscleres birotulates of two length groups, 50–55 or 65–72 μm with stout shafts bearing a small number of acute spines and with both rotules of equal, 15-18 μm diameters and consisting of an arrangement of lateral spines. Foraminal apertures of gemmules distinctly tubular and relatively short with 3–6 long and irregular cirrous projections.

Distribution: reported only from northeastern to midwestern United States.

Heteromeyenia tubisperma (Potts)

Spicules: megascleres slender, 190-(290)-337 μm in length, smooth to sparsely microspined; microscleres, 73-(100)-118 μm in length, slender and entirely spined with spines near the tips small and recurved and those near the central portion distinctly larger, straight, and with knobs; gemmoscleres birotulates of two roughly defined size classes, with both classes 33-(44)-70 μm in length with stout shafts bearing a small number of acute spines and with both rotules of equal, 12-(19)-25 μm

FIGURE 23 Foraminal aperture of *Heteromeyenia tentasperma* (after Neidhoefer, 1940).

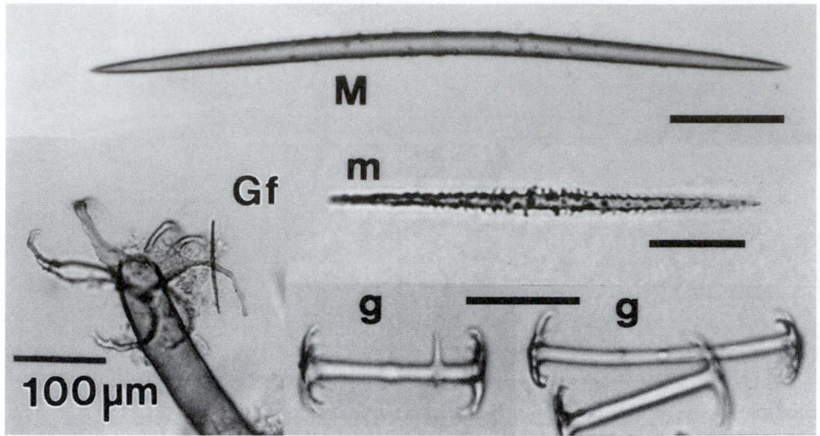

FIGURE 24 Megasclere (M, scale bar = 50 μm), microsclere (m, scale bar = 25 μm), and gemmoscleres (g, scale bar = 25 μm) and a foraminal aperture (GF) of *Heteromeyenia tubisperma* (from Ricciardi and Reiswig, 1993).

diameters, consisting of an arrangement of lateral spines. Foraminal apertures of gemmules distinctly tubular, slender and very long (0.5–0.9 times the diameter of the gemmule) with 5 or 6 cylindrical cirrous projections. It is important to note that developing gemmules may have stunted foraminal development such that they resemble *H. latitenta* and *H. tentasperma*. Specimens should be thoroughly inspected for fully developed gemmules. *H. tentasperma* can be distinguished from *H. baileyi* which has shorter microscleres with longer spines.

Distribution: reported only from the eastern half of North America.

Radiospongilla cerebellata (Bowerbank)

Spicules: megascleres straight to slightly curved, 240–330 μm in length, entirely without spines; microscleres absent although immature gemmoscleres may be abundant in some portions of the dermal membrane; gemmoscleres usually distinctly curved and rarely straight, 72–110 μm in length, covered with abundant spines that are pronouncedly recurved toward the terminal ends (in intact gemmules, gemmoscleres are arrayed in two distinct layers with those in the inner layer arranged in a radial fashion and those in the outer layer arranged tangentially to the inner layer) (Poirrier, 1972).

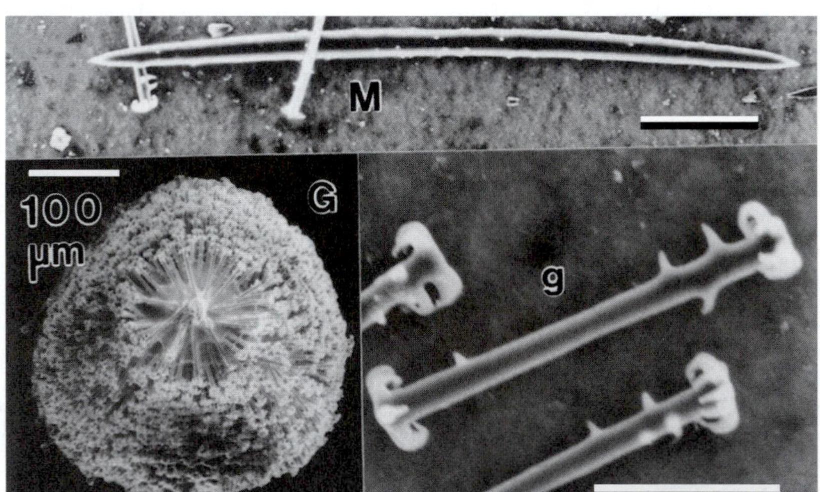

FIGURE 25 Megasclere (M, scale bar = 50 μm), gemmoscleres (g, scale bar = 25 μm), and a gemmule (G) showing a crater like depression around the foraminal aperture of *Radiospongilla crateriformis* (from Ricciardi and Reiswig, 1993).

Distribution: reported only from Texas in the United States, but widely distributed in tropical and subtropical Asia and Africa.

Radiospongilla crateriformis (Potts)

Spicules: megascleres slender and slightly curved, 240-(278)-300 μm in length, sparsely covered by very small spines, except at their tips; microscleres absent; gemmoscleres slender and covered with small conical spines, except at the tips where slightly recurved spines are sufficiently dense to form pseudorotules, 60-(71)-80 μm in length. Gemmoscleres in intact gemmules are arranged radially around the gemmule coat except in the immediate vicinity of the foraminal aperture where they form a craterlike depression leaning away from the aperture.

Distribution: reported primarily from the eastern half of the United States but as far west as Wisconsin and Texas; eastern Canada (Ricciardi and Reiswig, 1993), also reported in China, Japan, Southeast Asia, and Australia.

Spongilla alba (Carter)

Spicules: megascleres entirely smooth, 144–420 μm in length; microscleres slender and slightly curved with erect spines that are longer in the central region, 49–124 μm; gemmoscleres slightly to moderately curved with stout, sharp, recurved spines that are more dense at the ends, forming distinct heads, 48–130 μm (Poirrier, 1976).

Distribution: occurs in warmer regions worldwide with a strong preference for brackish water, reported from the southeast coastal regions of the United States.

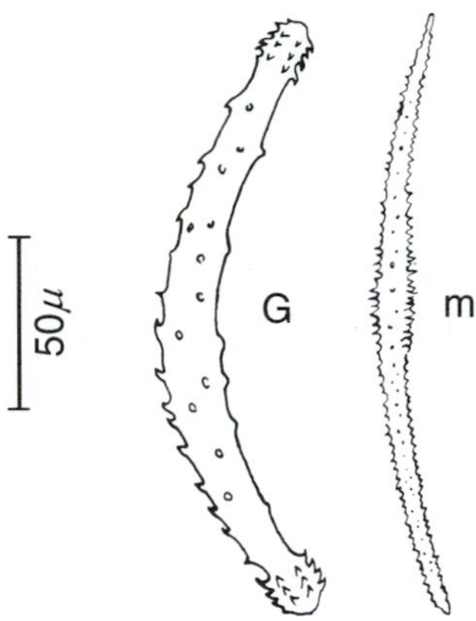

FIGURE 26 Microsclere (m) and gemmosclere (G) of *Spongilla alba* (after Penney and Racek, 1968).

Spongilla aspinosa (Potts)

Spicules: megascleres slender and entirely smooth, 155-(274)-338 μm in length; microscleres range from rare to abundant in number, smooth or very sparsely microspined, 21-(50)-78 μm in length; gemmoscleres smooth, resembling small megascleres, 129-(274)-306 μm in length.

Distribution: reported from the United States from Florida to Michigan, and from eastern Canada.

Spongilla cenota (Penney and Racek)

Spicules: megascleres stout and completely smooth, 310–410 μm in length; microscleres numerous and

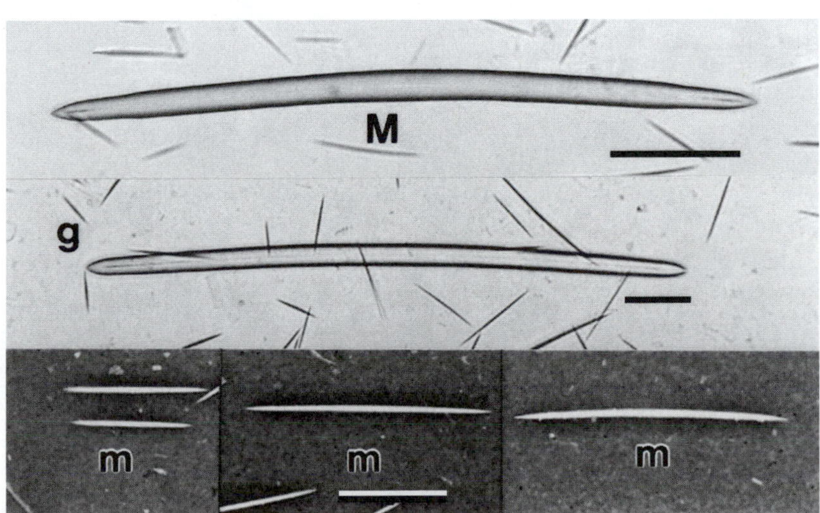

FIGURE 27 Megasclere (M, scale bar = 50 μm), microscleres showing a range of forms (m, scale bar = 25 μm) and gemmosclere (g, scale bar = 25 μm) of *Spongilla aspinosa* (from Ricciardi and Reiswig, 1993).

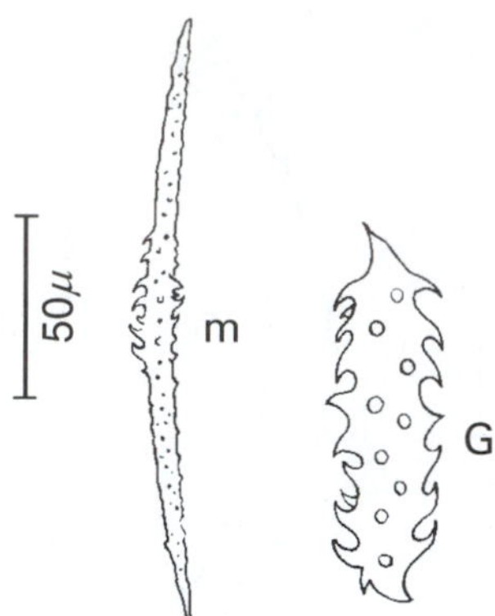

FIGURE 28 Microsclere (m) and gemmosclere (G) of *Spongilla cenota* (after Penney and Racek, 1968).

slender, covered with small spines at their tip and with a group of clearly larger spines in their center, 68–123 μm; gemmoscleres extremely stout, entirely covered with stout, sharp, recurved spines 65–86 μm (Poirrier, 1976).

Distribution: reported only from Florida, and the Yucatan in Mexico.

Spongilla lacustris (Linneaus)

Spicules: megascleres entirely smooth, 158-(254)-362 μm in length; microscleres densely covered with small spines, 30-(61)-130 μm in length; gemmoscleres absent from winter gemmules and slightly curved, usually covered with strong, curved spines that tend to be concentrated near the tips, 18-(32)-130 μm long in thick-coated summer gemmules (Poirrier et al., 1987).

Distribution: throughout the Northern Hemisphere including frequent reports from throughout the United States and Canada.

Stratospongilla penneyi (Harrison)

Spicules: megascleres slightly curved, 215–296 μm in length, smooth to very delicately microspined, microscleres slender, 38–75 μm in length, covered with very small spines, gemmoscleres curved to distinctly bent, 48–123 μm in length, smooth to delicately microspined with sharply pointed apices.

Distribution: reported from only one location, a canal in southern Florida (Harrison, 1979).

Trochospongilla horrida (Weltner)

Spicules: megascleres straight to slightly curved, 155-(187)-250 μm in length, covered with stout,

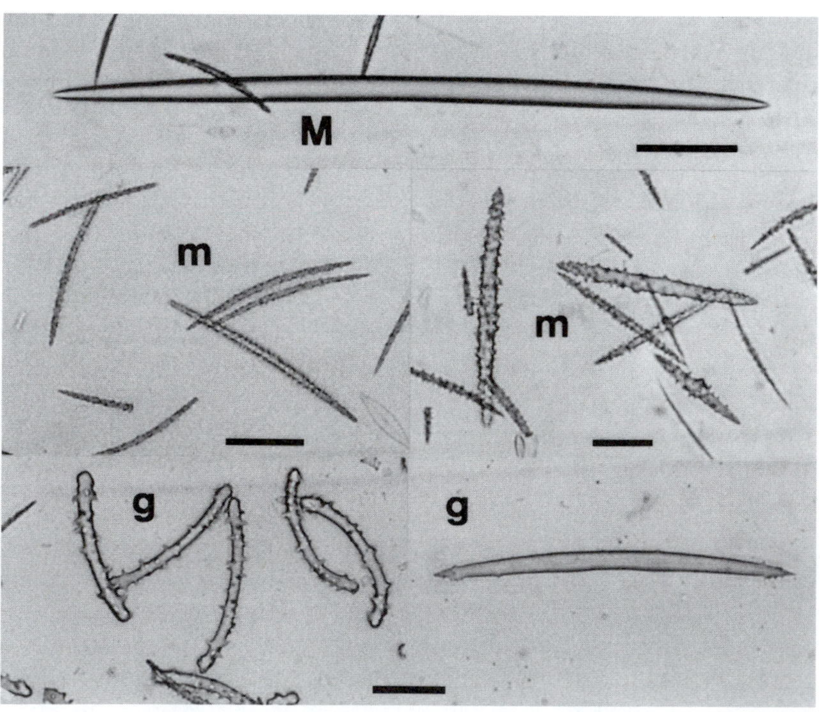

FIGURE 29 Megasclere (M, scale bar = 50 μm), microscleres showing a range of aberrant forms in the right plate (m, scale bar = 25 μm) and gemmosclere (g, scale bar = 25 μm) of *Spongilla lacustris* (from Ricciardi and Reiswig, 1993).

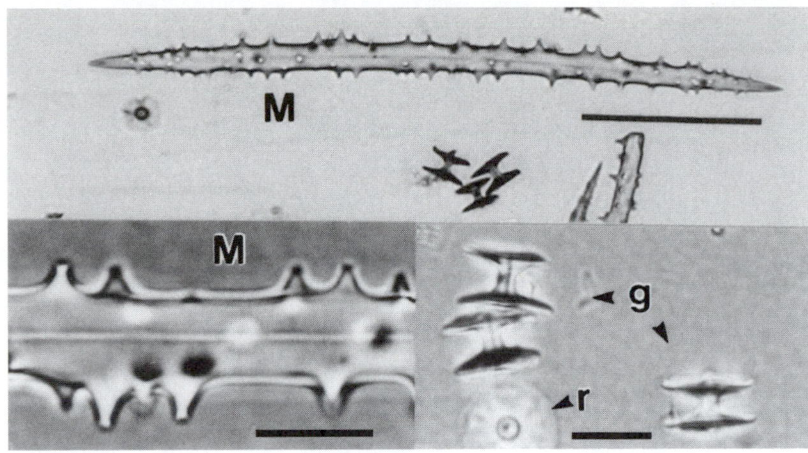

FIGURE 30 Megascleres (M, scale bars = 50 and 25 μm for magnified portion which shows the truncated spines typical of the genus) and gemmoscleres (g, scale bar = 10 μm) with rotule (r) of *Trochospongilla horrida* (from Ricciardi and Reiswig, 1993).

blunt, truncated spines; microscleres absent; gemmoscleres small birotulates, 8–10 μm in length, with stout smooth shafts and with rotules of nearly equal size, with the smaller between 9 and 13 μm in diameter and the larger between 13 and 16 μm with circular margins (overall length of gemmoscleres is never greater than the diameter of the smaller rotule). Compare with information for *T. pennsylvanica* to confirm an identification.

Distribution: widely dispersed throughout the Northern Hemisphere although with few reports in eastern Canada.

Trochospongilla leidii (Bowerbank)

Spicules: megascleres straight to slightly curved and entirely smooth, 150–170 μm in length; microscleres absent; gemmoscleres small birotulates, 11 μm in length, terminating in rotules with circular margins and equal, 12–14 μm diameters.

Distribution: reported from limited regions of the eastern United States, with one report of a population from the Panama Canal (Jones and Rützler, 1975).

Trochospongilla pennsylvanica (Potts)

Spicules: megascleres slender and slightly curved, 100-(253)-432 μm in length, entirely covered with blunt, truncated spines; microscleres absent; gemmoscleres small birotulates with slender shafts, 11-(17)-41 μm in length, terminating in rotules with circular margins and usually with two distinctly different diameters, 3-(9)-23 μm and 13-(24)-41 μm. In some cases both rotules of the gemmoscleres of *T. pennsylvanica* appear to have the same diameters leading them to be identified incorrectly as *T. horrida* using the key to this chapter. In such cases, the two species can be distinguished by the length of the gemmosclere shaft relative to the rotule diameter. The shaft length is longer than, or in rare cases, equal to, the diameter for the smaller rotule's diameter in *T. pennsylvanica* and shorter than the rotule's diameter for *T. horrida* (Ricciardi and Reiswig, 1993).

Distribution: reported throughout, but apparently restricted to, the North American continent.

B. Key to the Freshwater Sponges in Canada and the United States

Please note that the successful use of this key depends on obtaining a fully representative sample of all

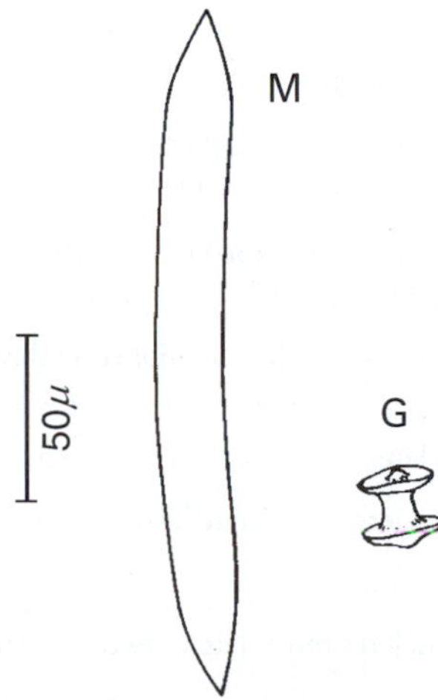

FIGURE 31 Megasclere (M) and gemmosclere (G) of *Trochospongilla leidii* (after Penney and Racek, 1968).

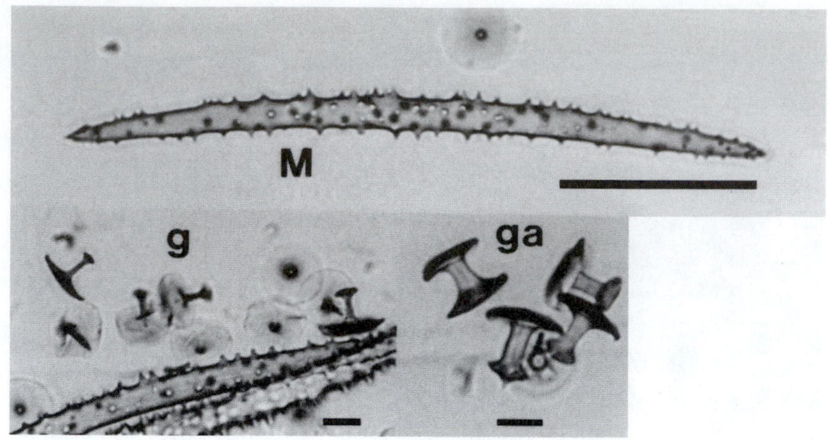

FIGURE 32 Megasclere (M, scale bars = 50 μm) and gemmoscleres (g, scale bar = 10 μm) with rotules (r) or aberrant forms (ga) of *Trochospongilla pennsylvanica* (from Ricciardi and Reiswig, 1993).

of the types of spicules that can occur within a sponge species (megascleres, gemmoscleres, and, if they occur in a species, microscleres). If a species ordinarily exhibits gemmoscleres or microscleres and they are not contained within a specimen, it will not be possible to identify even its genus. The one exception to this rule occurs for *Spongilla lacustris* which is unique among the species in this key in its lack of gemmoscleres in its winter gemmules. If gemmules are clearly present in a specimen but gemmoscleres are absent, it is most likely *S. lacustris*. [Caution is necessary here in that some foreign bodies in sponges may resemble gemmules superficially, for example, the eggs of water mites. Gemmules can be distinguished by their highly resistant coats.] It is important to note that *S. lacustris.* may also have gemmoscleres; they occur when summer gemmules are or have been present.

Also note that this key is intended only for use with the species listed above from Canada and the United States. The couplets in this key have been designed only for this particular subset of the world's freshwater sponges. Those examining specimens from other regions will need to consult Penney and Racek (1968) and more recent taxonomic work (e.g., Volkmer-Ribiero, 1986, and the references cited in the species list). The key is arranged systematically so that genera are separated prior to species within a genus. In most cases, therefore, a valid identification to genus may be possible even if a species has not previously been reported in Canada or the United States. Even to separate genera, however, it is necessary to have included all the possible spicule types from a specimen.

1a. Microscleres present. 2

1b. Microscleres absent. 16

2a(1a). Microscleres star-shaped (Fig. 15) or birotulate (Figs. 10 and 11). 3

2b. Microscleres rod-shaped to needlelike in structure. (Fig. 21 and Fig. 24). 8

3a(2a). Microscleres star-shaped with several rays extending from central region of the spicules (Fig. 15) .... Genus *Dosilia* 4

3b. Microscleres birotulate (Figs. 12 and 13) 5

4a(3a). Gemmoscleres birotulate in two distinctly different size categories (45–82 μm in length and 120–230 μm in length). .... *Dosilia radiospiculata*

4b. Gemmoscleres birotulate in two size categories that are nearly equal in length (55–85 mm). .... *Dosilia palmeri*

5a(3b). Gemmoscleres birotulate (Figs. 12 and 13). .... Genus *Corvomeyenia* 6

5b Gemmoscleres rod- or needle-shaped (Fig. 14).
.... Genus *Corvospongilla* 7

6a(5a). Microscleres straight to slightly curved birotules with a predominance of straight forms (Fig. 12).
.... *Corvomeyenia everetti*

6b. Microscleres a mixture of straight to strongly curved birotules with a predominance of curved forms (Fig. 11).
.... *Corvomeyenia carolinensis.*

7a(5b). Megascleres covered with spines
.... *Corvospongilla becki*

7b. Megascleres smooth
.... *Corvospongilla novaeterrae*

8a(2b). Gemmoscleres birotulate (as in Fig. 21).
.... Genus *Heteromeyenia* 9

8b. Gemmoscleres needlelike (as in Fig. 17) or absent. 12

9a(8a). Foraminal aperture of gemmules without terminal cirrous projections (contrast with Figs. 22, 23, and 24).
.... *Heteromeyenia baileyi*

9b. Foraminal aperture of gemmules with distinct, terminal cirrous projections (Figs. 22, 23, and 24). 10

10a(9b). Foraminal aperture of gemmules with one to two very long cirrous projections starting from a flat disk, flat and ribbonlike at the base and cylindrical at the end. (Fig. 22) .... *Heteromeyenia latitenta*

10b. Foraminal aperture of gemmules with 3–6 cirrous projections that are short when compared with those of *H. latitenta.* 11

11a(10b). Foraminal aperture of gemmules distinctly tubular and very long ranging from 0.5 to 0.9 times the diameter of the gemmule (Fig. 23) .... *Heteromeyenia tubisperma*

11b. Foraminal aperture of gemmules distinctly tubular, but short, less than 0.4 times the diameter of the gemmule (Fig. 24).
.... *Heteromeyenia tentasperma*

12a(8b). Gemmoscleres smooth or covered with very fine spines 13

12b. Gemmoscleres absent or covered with robust, conspicuous spines (Figs. 28 and 29).
.... Genus *Spongilla*, except for *S. aspinosa* 14

13a(12a). Microscleres smooth or very sparsely microspined (Fig. 27). .... *Spongilla aspinosa*

13b. Microscleres covered with very small spines. .... *Stratospongilla penneyi*

14a(12a). Gemmoscleres clublike with spines conspicuously more dense on the ends than in the center (Fig. 26). .... *Spongilla alba*

14b. Gemmoscleres absent or with spines distributed evenly across the length of the spicule (Figs. 28 and 29). 15

15a(14b). Microscleres with conspicuously denser and longer spines in the central region (Fig. 28). .... *Spongilla cenota*

15b. Microscleres with spines distributed evenly across the length of the spicule (Fig. 29). .... *Spongilla lacustris*

16a(1b). Megascleres of a single size category with distinctly different forms of gemmoscleres 17

16b. Megascleres in two distinctly different size categories (Fig. 16) with the smaller category serving also as gemmoscleres although both sizes of megascleres are always present even when gemmules are absent. .... *Duosclera mackayi*

17a(16a). Gemmoscleres birotulate (Fig. 30) 18

17b. Gemmoscleres rod shaped or needlelike (Fig. 20) 25

18a(17a). Margins of rotules completely smooth (Figs. 30 and 31).
.... Genus *Trochospongilla* 19

18b. Margins of rotules distinctly spined or serrated (Figs. 17 and 19) 21

19a(18a). Megascleres conspicuously spined (Fig. 30) 20

19b. Megascleres smooth or with very fine spines (Fig. 31) .... *Trochospongilla leidii*

20a(19a). Gemmoscleres with rotules of two distinctly different diameters (Fig. 32) .... *Trochospongilla pennsylvanica*

20b. Gemmoscleres with rotules of nearly equivalent diameters. (Fig. 30) .... *Trochospongilla horrida*

21a(18b). Gemmoscleres occurring in two distinct size classes (Fig. 10) .... Genus *Anheteromeyenia* 22

21b. Gemmoscleres occurring in only one size class (Fig. 17) .... Genus *Ephydatia* 23

22a(21a). Gemmoscleres of two classes that are similar in shape, but clearly different in length (Fig. 10) .... *Anheteromeyenia argyrosperma*

22b. Gemmoscleres of two classes that are distinct in both shape and size (Fig. 11). .... *Anheteromeyenia ryderi*

23a(21b). Rotules of gemmoscleres with margins that are nearly smooth bearing numerous, very small incisions (Fig. 18). .... *Ephydatia millsii*

23b. Rotules of gemmoscleres with clearly indented margins (Fig. 17). 24

24a(23b). Gemmoscleres less than 20 μm in length with rotules bearing fewer that 12 teeth deeply incised into long rays, gemmosclere length less than or equal to rotule diameter (Fig. 19). .... *Ephydatia muelleri*

24b. Gemmoscleres greater than 25 μm in length with rotules bearing greater than 20 teeth that are not deeply incised (Fig. 17). .... *Ephydatia fluviatilis*

25a(17b). Most gemmoscleres in intact gemmules arrayed in a distinctly radial fashion within the gemmule coat (Fig. 25). .... Genus *Radiospongilla* 26

25b. Gemmoscleres in intact gemmules arrayed only tangentially to the surface of the gemmule. .... *Eunapius fragilis*

26a(25a). Megascleres sparsely covered with small spines (Fig. 25). .... *Radiospongilla crateriformis*

26b. Megascleres entirely smooth. .... *Radiospongilla cerebellata*

LITERATURE CITED

Bader, R. B. 1984. Factors affecting the distribution of a freshwater sponge. Freshwater Invertebrate Biology 3:86–95.

Barbeau, M. A., Reiswig, H. M., Rath, L. C. 1989. Hatching of freshwater sponge gemmules after low temperature exposure: *Ephydatia mülleri* (Porifera: Spongillidae). Journal of Thermal Biology 14:225–231.

Barton, S. H., Addis, J. S. 1997. Freshwater sponges (Porifera: Spongillidae) of western Montana. Great Basin Naturalist 57:93–103.

Bergquist, P. R. 1978. Sponges, University of California Press, Berkeley, CA.

Berzins, B. 1950. Observations on rotifers on sponges. Transactions of the American Microscopical Society 69:189–193.

Bisbee, J. W. 1992. Life cycle, reproduction and ecology of freshwater sponges in a South Carolina pond. I. Life cycle and reproduction of *Spongilla lacustris*. Transactions of the American Microscopical Society 111:77–88.

Boury-Esnault, N., Rützler, K. 1997. Thesaurus of sponge morphology, Smithsonian Contributions to Zoology Number 596, Smithsonian Institution Press, Washington, DC.

Cecil, J. T., Stempien, M. F. Jr., Ruggieri, G. D. Nigrelli, R. F. 1976. Cytological abnormalities in a variety of normal and transformed cell lines treated with extracts from sponges, *in*: Harrison, F. W., Cowden, R. R. Eds., Aspects of sponge biology, Academic Press, New York, NY, pp. 171–182.

Clifford, H. F. 1991. Aquatic invertebrates of Alberta, The University of Alberta Press, Edmonton, Alberta, Canada.

Colby, A. C. C., Frost, T. M., Fischer, J. M. 1999. Sponge distribution and lake chemistry in northern Wisconsin lakes: Minna Jewell's survey revisited. Memoirs of the Queensland Museum 44:93–99.

De Santo, E. M., Fell, P. E. 1996. Distribution and ecology of freshwater sponges in Connecticut. Hydrobiologia 341:81–89.

De Vos, L., Rützler, K., Boury-Esnault, N., Donadey C., Vacelet. J. 1991. Atlas of sponge morphlogy, Smithsonian Institution Press, Washington, DC.

Dominey, W. 1987. Sponge-eating by *Pungu maclareni*, an endemic Cichlid fish from Lake Barombi Mbo, Cameroon. National Geographic Research 3:389–393.

Fell, P. E. 1974. Porifera, *in*: Giese, A. C., Pearse, J. S. Eds., Reproduction of marine invertebrates, Vol. I, Academic Press, New York, NY, pp. 51–132.

Frey, D. G. 1963. Limnology in North America. University of Wisconsin Press, Madison, WI.

Frost, T. M. 1976. Investigations of the aufwuchs of freshwater sponges. I. A quantitative comparison between the surfaces of *Spongilla lacustris* and three aquatic macrophytes. Hydrobiologia 50:145–149.

Frost, T. M. 1976. Sponge feeding: A review with a discussion of some continuing research, *in*: Harrison, F. W., Cowden, R. R. Eds., Aspects of sponge biology, Academic Press, New York, NY, pp. 283–298.

Frost, T. M. 1978. The impact of the freshwater sponge *Spongilla lacustris* on a sphagnum bog pond. Verhandlungen Internationale Vereinigung für Theoretische und Angewandte Limnologie 20:2368–2371.

Frost, T. M. 1980a. Clearance rate determinations for the freshwater sponge *Spongila lacustris*: effect of temperature, particle type and concentration, and sponge size. Archiv für Hydrobiologie 90:330–356.

Frost, T. M. 1980b. Selection in sponge feeding processes, *in*: Smith, D. C., Tiffon, Y. Eds., Nutrition in lower Metazoa, Pergammon Press, Oxford, England, pp. 33–44.

Frost, T. M. 1981. Analysis of ingested materials within a freshwater sponge. Transaction of the American Microscopical Society 100:271–277.

Frost, T. M. 1987. Porifera *in*: Pandian, T. J., Vernberg, J. F. Eds., Animal energetics, Vol. 1, Academic Press, New York, NY, pp. 27–53.

Frost, T. M., de Nagy, G. S., Gilbert, J. J. 1982. Population dynamics and standing biomass of the freshwater sponge, *Spongilla lacustris*. Ecology 63:1203–1210.

Frost, T. M., Elias, J. E. 1990. The balance of autotrophy and heterotrophy in three freshwater sponges with algal symbionts, *in*: Rützler, K. Ed., Proceedings of the Third International Conference on Sponge Biology, Smithsonian Press, Washington, DC, pp. 478–484.

Frost, T. M., Graham, L. E., Elias, J. E., Haase, M. J., Kretchmer, D. W., Kranzfelder, J. A. 1997. A yellow-green algal symbiont in the freshwater sponge *Corvomeyenia everetti*: convergent evolution of symbiotic associations. Freshwater Biology 38:395–399.

Frost, T. M., Williamson, C. E. 1980. In situ determination of the effect of symbiotic algae on the growth of the freshwater sponge *Spongilla lacustris*. Ecology 61:1361–1370.

Gilbert, J. J. 1974. Field experiments on sexuality in the freshwater sponge *Spongilla lacustris*. Control of oocyte production and the fate of unfertilized oocytes. The Journal of Experimental Zoology 188:165–178.

Gilbert, J. J. 1975. Field experiments on gemmulation in the freshwater sponge *Spongilla lacustris*. Transactions of the American Microscopical Society 94:347–356.

Gilbert, J. J., Allen, H. L. 1973a. Studies on the physiology of the green freshwater sponge *Spongill lacustris*: primary productivity, organic matter, and chlorophyll content. Verhandlungen Internationale Vereinigung für Theoretische und Angewandte Limnlogie 20:2368–2371.

Gilbert, J. J., Allen, H. L. 1973b. Chlorophyll and primary productivity of some green, freshwater sponges. Internationale Revue der Gesamten Hydrobiologie 58:633–658.

Gilbert, J. J., Simpson, T. L. 1976a. Sex reversal in a freshwater sponge. The Journal of Experimental Zoology 195:145–151.

Gilbert, J. J., Simpson, T. L. 1976b. Gemmule polymorphism in the freshwater sponge *Spongilla lacustris*. Archiv für Hydrobiologie 78:268–277.

Gregson, R. P., Baldo, B. A., Thomas, P. G., Quinn, R. J., Berquist, P. R., Stephens, J. F., Horne, A. R., 1979. Flourine is a major constituent of the marine sponge *Halichondria moorei*. Science 206:1108–1109.

Harrison, F. W. 197l. A taxonomical investigation of the genus *Corvomeyenia* Weltner (Spongillidae) with an introduction of *Corvomeyenia carolinensis* sp. nov. Hydrobiologia 38:123–140.

Harrison, F. W. 1974. Sponges (Porifera: Spongillidae). *in*: Hart, C. W. Jr., Fuller, S. L. H. Eds., Pollution ecology of freshwater invertebrates, Academic Press, New York, NY, pp. 29–66.

Harrison, F. W. 1979. The taxonomic and ecological status of the environmentally restricted spongillid species of North America. V. *Ephydatia subtilis* (Weltner) and *Stratospongilla penneyi* sp. nov. Hydrobiologia 62:99–105.

Harrison, F. W., Warner, B. G. 1986. Fossil fresh-water sponges (Porifera: Spongillidae) from Western Canada: An overlooked group of Quaternery paleoecological indicators. Transactions of the American Microscopical Society 105:110–120.

Harrison, F. W., Harrison, M. B. 1977. The taxonomic and ecological status of the environmentally restricted spongillid species of North America. II. *Anheteromeyenia biceps* (Lindenschmidt, 1950). Hydrobiologia 62:107–111.

Harrison, F. W., Harrison, M. B. 1979. The taxonomic and ecological status of the environmentally restricted spongillid species of North America. IV. *Spongilla heterosclerifera* Smith 1918. Hydrobiologia 62:99–105.

Harrison, F. W., Johnston, L., Stansell, D. B., McAndrew, W. 1977. The taxonomic and ecological status of the environmentally restricted spongillid species of North America. I. *Spongilla sponginosa* Penney 1957. Hydrobiologia 53:199–202.

Harsha, R. E., Francis, J. C., Poirrier, M. A. 1983. Water temperature: A factor in the seasonality of two freshwater sponge species, *Ephydatia fluviatilis* and *Spongilla alba*. Hydrobiologia 102:145–150.

Hartman, W. D., Reiswig, H. M. l973. The individuality of sponges, *in*: Boardmann, R. S., Cheetham, A. H., Oliver, W. A. Eds., Animal colonies. Dowden, Hutchinson and Ross, Stroudsburg, PA, pp. 567–584.

Ilan, M., Dembo, M. G., Gasith, A. 1996. Gemmules of sponges from a warm lake. Freshwater Biology 35:165–172.

Jackson, J. B. C. 1977. Competition on marine hard substrata: the adaptive significance of solitary and colonial strategies. American Naturalist 111:743–767.

Jackson, J. B. C., Buss, L. W. 1975. Allelopathy and spatial competition among coral reef invertebrates. Proceedings of the National Academy of Science, USA. 72:5160–5163.

Jewell, M. E. 1935. An ecological study of the fresh-water sponges of northern Wisconsin. Ecological Monographs 5:461–504.

Jewell, M. E. 1939. An ecological study of the fresh-water sponges of northern Wisconsin, II. The influence of calcium. Ecology 20:11–28.

Jones, M. L., Rützler, K. 1975. Invertebrates of the Upper Chamber, Gaún Locks, Panama Canal, with emphasis on *Trochospongilla leidii* (Porifera). Marine Biology 33:57–66.

Kratz, T. K., Frost, T.M., Elias, J. E., Cook, R. B. 1991. Reconstruction of a regional, 12,000-year silica decline in lakes using fossil sponge spicules. Limnology and Oceanography 36:1244–1249.

Langenbruch, P-F., Weissenfels, N. 1987. Canal systems and choanocyte chambers in freshwater sponges (Porifera, Spongillidae). Zoomorphology 107:11–16.

Lauer, T. E., Spacie, A. 1996. New records of freshwater sponges (Porifera) for southern Lake Michigan. Journal of Great Lakes Research 22:77–82.

Mann, K. H., Britton, R. H., Kowalczewski, A., Lack, T. J., Mathews, C. P., McDonald, I. 1972. Productivity and energy flow at all trophic levels in the River Thames, England, *in*: Kajak, Z., Hillbricht-IlKowska, A. Eds., Productivity problems of freshwaters. PWN, Polish Scientific Publishers, Warsaw, Poland, pp. 579–596

McAuley, D. G., Longcore, J. R. 1988. Foods of juvenile ring-necked ducks: relationships to wetland pH. Journal of Wildlife Management 52:177–185.

Mergner, H. 1966. Zum Nachweis der Artspezifität des Induktionsstoffes bei Osculatrohrneubildungen von Spongilliden. Verhandlunger der Deutschen Zoologischen Gesselschaft in Göttingen 30:522–564.

Muscatine, L., Porteri, J. W. 1977. Reef corals: mutualistic symbioses adapted to nutrient-poor environments. Bioscience 27:454–460.

Neidhoefer, J. R. 1940. The fresh-water sponges of Wisconsin. Transactions of the Wisconsin Academy of Sciences, Arts, and Letters 32:177–197.

Niegel, J. E., Avise, J. C. 1983. Histocompatability bioassays of population structure in marine sponges. Journal of Heredity 74:134–140.

Økland, K. A., Økland, J. 1989. The amphiatlantic freshwater sponge *Anheteromeyenia ryderi* (Porifera: Spongillidae): taxonomic-geographic implications of records from Norway. Hydrobiologia 171:177–188.

Paduano, G. M., Fell., P. E. 1997. Spatial and temporal distribution of freshwater sponges in Connecticut lakes based upon analysis of siliceous spicules in dated sediment cores. Hydrobiologia 350:105–121.

Palumbi, S. R. 1984. Tactics of acclimation: morphological changes of sponges in an unpredictable environment. Science 225:1478–1480.

Penney, J. T. 1960. Distribution and bibliography (1892–1957) of the freshwater sponges. University of South Carolina Publications, Series 3, 3:1–97.

Penney, J. T., Racek, A. A. 1968. Comprehensive revision of a worldwide collection of freshwater sponges (Porifera: Spongillidae). United States National Museum Bulletin 272.

Pennings, S. C., Pablo, S. R., Paul, V. J. Duffy, J. E. 1994. Effects of sponge secondary metabolites in different diets on feeding by three groups of consumers. Journal of Experimental Marine Biology and Ecology 180:137–149.

Pile, A. J., Patterson, M. R., Savarese, M., Chernykh, V. I., Fialkov, V. A. 1997. Trophic effects of sponge feeding within Lake Baikal's littoral zone. 2. Sponge abundance, diet, feeding efficiency, and carbon flux. Limnology and Oceanography 42:178–184.

Poirrier, M. A. 1972. Additional records of Texas freshwater sponges (Spongillidae) with the first record of *Radiospongilla cerebellata* (Bowerbank, 1863) from the Western Hemisphere. The Southwestern Naturalist 16:434–435.

Poirrier, M. A. 1974. Ecomorphic variation in gemmoscleres of *Ephydatia fluviatilis* Linneaus (Porifera: Spongillidae) with comments upon its systematics and ecology. Hydrobiologia 44:337–347.

Poirrier, M. A. 1976. A taxonomic study of the *Spongilla alba, S. cenota, S. wagneri* species group (Porifera: Spongillidae) with ecological observations of *S. Alba*, *in*: Harrison, F. W., Cowden, R. R. Eds., Aspects of sponge biology. Academic Press, New York, NY, pp. 203–214.

Poirrier, M. A. 1977. Systematic and ecological studies of *Anheteromeyenia ryderi* (Porifera: Spongillidae) in Louisiana. Transactions of the American Microscopical Society 96:62–67.

Poirrier, M. A. 1978. *Corvospongilla becki* N. Sp., A fresh-water sponge from Louisiana. Transactions of the American Microscopical Society 97:240–243.

Poirrier, M. A., Francis, J. C., Labiche, R. A. 1981. A continuous-flow system for growing fresh-water sponges in the laboratory. Hydrobiologia 79:225–259.

Poirrier, M. A., Martin, P. S., Baerwald, R. J. 1987. Comparative morphology of microsclere structure in *Spongilla alba, S. cenota*, and *S. lacustris* (Porifera: Spongillidae). Transactions of the American Microscopical Society 106:302–310.

Proctor, H. C., Pritchard, G. 1990. Variability in the life history of *Unionicola crassipes*, a sponge-associated water mite (Acari: Unionicolidae) Canadian Journal of Zoology 68:1227–1232.

Pronzato, R., Manconi, R. 1995. Long-term dynamics of a freshwater sponge population. Freshwater Biology 33:485–495.

Racek, A. A., Harrison, F. W. 1975. The systematic and phylogenetic position of *Palaelospongilla chubutensis* (Porifera: Spongillidae). Proceedings of the Linnean Society of New South Wales 99:157–165.

Randall, J. E., Hartman, W. D. 1968. Sponge-feeding fishes of the West Indies. Marine Biology 1:216–225.

Rasmont, R. 1962. The physiology of gemmulation in fresh-water sponge. Sympsoium of the Society for the Study of Development and Growth 20:3–25.

Reisser, W. 1984. The taxonomy of green algae endosymbiotic in ciliates and a sponge. British Phycological Journal 19:309–318.

Reiswig, H. M. 1970. Porifera: sudden sperm release by tropical Demospongiae. Science 170: 538–539.

Reiswig, H. M., Browman, H. I. 1987. Use of membrane filters for Microscopid preparations of sponge spicules. Transactions of the American Microscopical Society 106:10–20.

Reiswig, H. M., Ricciardi, A. 1993. Resolution of the taxonomic status of the freshwater sponges *Eunapius mackayi*, *E. Inglov-iformis*, and *Spongila johanseni* (Porifera: Spongillidae). Transactions of the American Microscopical Society 112: 262–279.

Reiswig, H. M., Miller, T. L. 1998. Freshwater sponge gemmules survive months of anoxia. Invertebrate Biology 117:1–8.

Resh, V. H. 1976. Life cycles of invertebrate predators of freshwater sponge, *in*: Harrison, F. W., Cowden, R. R. Eds., Aspects of sponge biology, Academic Press, New York, NY, pp. 299–314.

Resh, V. H., Morse, J. C., Wallace, I. D. 1976. The evolution of the sponge feeding habit in the caddisfly genus*Ceraclea* (Trichoptera: Leptoceridae). Annals of the Entomological Society of America 69:937–941.

Ricciardi, A., Reiswig, H. M. 1992. *Spongilla heterosclerifia* Smith, 1918 is an interspecific freshwater sponge mixture (Porifera, Spongillidae). Canadian Journal of Zoology 70:352–354.

Ricciardi, A., Reiswig, H. M. 1993. Freshwater sponges (Porifera, Spongillidae) of eastern Canada: taxonomy, distribution, and ecology. Canadian Journal of Zoology 71:665–682.

Ricciardi, A., Snyder, F. L., Kelch, D. O., Reiswig, H. M. 1995. Lethal and sublethal effects of sponge overgrowth on introduced dreissenid mussels in the Great Lakes—St. Lawrence River System. Canadian Journal of Fisheries and Aquatic Sciences 52:2695–2703.

Richelle, E., Degoudenne, Y., Dejonghe L., Van de Vyver, G. 1995. Experimental and field studies on the effect of selected heavy metals on three freshwater songe species: *Ephydatia fluviatilis*, *Ephydatia mueleri* and *Spongilla lacustris*. Archv für Hydrobiologie 135:209–231.

Rioja, E. 1953. Datos historicos acerca de las esponjas de agua dulce de Mexico. Revista de La Sociedad Mexicana de Historia Natural 14:51–57.

Saller, U. 1990. Formation and construction of asexual buds of the freshwater sponge *Radiospongilla cerebellata* (Porifera, Spongillidae). Zoomorphology 109:295–301.

Sand-Jensen, K., Pedersen, M. F. 1994. Photosynthesis by symbiotic algae in the freshwater sponge, *Spongilla lacustris*. Limnology and Oceanography 39:551–561.

Santelices, B. 1999. How many kinds of individual are there. Trends in Ecology and Evolution 14:152–155.

Schmidt, I. 1970. Phagocytose et pinocytose chez les Spongillidae. Zeitschrift für vergleichende Physiologie 66:398–420.

Schwandes, L. P., Collins, M. E. 1994. Distribution and significance of freshwater sponge spicules in selected Florida Soils. Transactions of the American Microscopical Society 113:242–257.

Simpson, T. L. 1963. The biology of the marine sponge *Microciona prolifera* (Ellis and Solander): I. A study of cellular function and differentiation. The Journal of Experimental Zoology 154: 135–151.

Simpson, T. L. 1973. Coloniality among the Porifera, *in*: Boardmann, R. S., Cheetham, A. H., Oliver, W. A. Eds., Animal colonies. Dowden, Hutchinson and Ross, Stroudsburg, PA, pp. 549–565.

Simpson, T. L. 1984. The cell biology of sponges. Springer-Verlag, New York, NY.

Simpson, T. L., Fell, P. E. 1974. Dormancy among the Porifera: gemmule formation and hatching in freshwater and marine sponges. Transactions of the American Microscopy Society 93:544–577.

Simpson, T. L., Gilbert, J. J. 1973. Gemmulation, gemmule hatching, and sexual reproduction in fresh-water sponges: I. The life cycle of *Spongilla lacustris* and *Tubella pennsylvanica*. Transactions of the American Microscopy Society 92:422–433.

Smith, D. G. 1994. First report of a fresh-water sponge (Porifera: Spongillidae) from the West Indies. Journal of Natural History 28:981–986.

Smith, L. C., Hildemann, W. H. 1984. Alloimmune memory is absent in *Hymeniacidon sinapium*, a marine sponge. The Journal of Immunology 133:2351–2355.

Sowka, P. A. 1997. Occurrence of two species of freshwater sponges (*Dosilia radiospiculata* and *Ephydatia muelleri*) in Arizona. Southwestern Naturalist 44:211–212.

Stoddart, J. A., Ayre, D. J., Willis, G., Heyward, A. J. 1985. Self-recognition in sponges and corals? Evolution 39:461–463.

Ungemach, L. F., Souza, K., Fell, P. E., Loomis, S. H. 1997. Possession and loss of cold tolerance by sponge gemmules: a comparative study. Invertebrate Biology 116:1–5.

Vacelet, J., Boury-Esnaault, N. 1995. Carnivorous sponges. Nature 373:333–335.

Vacelet, J., Boury-Esnaault, N., Fiala-Medioni, A., Fisher, C. R. 1995. A methanotrophic carnivorous sponge. Nature 377:296.

van Weel, P. B. 1949. On the physiology of the tropical fresh-water sponge *Spongilla proliferens* Annand. I. Ingestion, digestion, and excretion. Physiolgia Comparata et Oecologia 1:110–128.

Van de Vyver, G. 1970. La non confluence intraspécifique chez les spongiaires et la notion d'individu. Extrait des Annales d'Embryologie et de Morphogenèse 3:251–262.

Van de Vyver, G. 1979. Cellular mechanisms of recognitions and rejection among sponges. *in*: Levi, C., Boury-Esnault, N. Eds., Biologie des Spongiaires. Colloques Internationaux du Centre National De La Recherche Scientifique, No. 291, Paris, France, pp. 195–204.

Van de Vyver, G., Willenz, Ph. 1975. An experimental study of the life cycle of the fresh-water sponge *Ephydatia fluviatilis* in its natural surroundings. Wihelm Roux' Archiv 177:41–52.

Vogel, S. 1974. Current induced flow through the sponge, *Halichondria*. Biological Bulletin (Woods Hole, Massachusetts) 147: 443–456.

Volkmer-Ribeiro, C. 1981. Porifera. *in*: Hurlbert, S. H., Rodriquesz, G., Santos, N. D. Eds., Aquatic biota of tropical South America, Part 2: Anarthropoda. San Diego State University, San Diego, CA, pp. 86–95.

Volkmer-Ribeiro, C. 1986. Evolutionary study of the freshwater sponge genus *Metania* GRAY, 1867: III. Metaniidae, new family. Amazoniana 9:493–509.

Volkmer-Ribeiro, C., Rosa-Barbosa, R. De. 1979. Neotropical freshwater sponges of the Family Potamolepidae Brien, 1967. *in*: Levi, C., Boury-Esnault, N. Eds., Biologie des Spongiaires. Colloques Internationaux du Centre National De La Recherche Scientifique, No. 291, Paris, France., pp. 503–511.

Volkmer-Ribeiro, C., Grosser, K. M. 1981. Gut contents of *Leporinus obtusidens* "sensu" Von Ihering (Pisces, Characoidei) used in a survey for freshwater sponges. Revista Brasileira de Biologia 41:175–183.

Weissenfels, N. 1975. Bau und Funktion des Süsswasserschwamms *Ephydatia fluviatilis* L. (Porifera). II. Anmerkungen zum Körperbau. Zeitschrift für Morphologie der Tiere 81:241–256.

Weissenfels, N. 1976. Bau und Funktion des Süsswasserschwamms *Ephydatia fluviatilis* L. (Porifera). III. Nahrungsaufnahme, Verdauung und Defäkation. Zoomorphologie 85:73–88.

Weissenfels, N. 1980. Bau und Funktion des Süsswasserschwamms *Ephydatia fluviatilis* L. (Porifera). VII. Die Porocyten. Zoomorphologie 95:27–40.

Weissenfels, N. 1981. Bau und Funktion des Süsswasserschwamms *Ephydatia fluviatilis* L. (Porifera). VIII. Die Entstehung und Entwicklung der Kragengeissselkammern und ihre Verbindung mit dem ausführenden Kanalsystem. Zoomorphology 98:35–45.

Weissenfels, N. 1982. Bau und Funktion des Süsswasserschwamms *Ephydatia fluviatilis* L. (Porifera). IX. Rasterelektronmikrosckpische Histologie und Cytologie. Zoomorphology 100:75–87.

Weissenfels, N. 1990. Condensation rhythm of fresh-water sponges (Spongillidae, Proifera). European Journal of Cell Biology 53:373–383.

Weissenfels, N. 1992. The filtrations apparatus for food collection in freshwater sponges (Porifera, Spongilidae). Zoomorphology 112:51–55.

Willenz, P. 1980. Kinetic and morphological aspects of particle ingestion by the freshwater sponge *Ephydatia fluviatilis* L., *in*: Smith, D. C., Tiffon, Y. Eds., Nutrition in lower Metazoa, Pergammon Press, Oxford, England, pp. 163–187.

Williamson, C.W. 1979. Crayfish predation on freshwater sponges. American Midland Naturalist 101:245–246.

5 CNIDARIA

Lawrence B. Slobodkin
Department of Ecology and Evolution
State University of New York
Stony Brook, New York 11794

Patricia E. Bossert
Science Department
Northport High School
Northport, New York 11768

I. INTRODUCTION

Medusae, anemones, corals, and other polyps compose the ancient and remarkably successful phylum, Cnidaria. They occur as fossils in the lithographic stone of the Mid-Cambrian Burgess shale, and have not changed very dramatically since then. Fossilized cnidarian embryos are reported from the lower Cambrian (Kouchinsky *et al.,* 1999). The phylum name, Cnidaria, is derived from the Greek term for "nail", based on their possession of nematocysts, which look like rods attached to a round capsule. The other name for the phylum is Coelenterata, a term alluding to their saclike internal space, the coelenteron. A general overview of the phylum and survey of older literature is provided by Hyman (1940) and the papers in various conference proceedings (e.g., Mackie, 1976).

The freshwater representatives of Cnidaria are small animals belonging to the class Hydrozoa, with relatively few species and somewhat monotonous anatomy. They consist of the following taxa:

1. The common and familiar hydras, a group of secondarily simple, solitary polyps;
2. The sporadically common *Craspedacusta* and *Limnocodium,* jellyfish with minute, polypoid larvae (Boulenger and Flower, 1928; Fuhrman, 1984).
3. *Calposoma,* a tiny, colonial polyp, which is so small and inconspicuous that it is probably more common than it appears to be (Rahat and Campbell, 1974; Fuhrman, 1984).
4. *Polypodium,* which spends part of its life cycle as a parasite in the eggs of sturgeon and part as an ambulatory, predaceous polyp (Lipin, 1926; Raikova, 1973, 1980); this species has been described primarily from eastern European rivers, but should also be found in North American sturgeon.
5. Various estuarine coelenterates, which may occasionally occur in relatively fresh water; colonial, sessile animals of the genus *Cordylophora* will serve as examples of these (Hubschman and Kishler, 1972; Roos, 1979).

Hydra and *Cordylophora* belong to the order Hydroida of the class Hydrozoa. *Craspedacusta, Limnocnida,* and *Calposoma* are classified as members of the order Limnomedusae. The parasitic *Polypodium*

Ecology and Classification of North American Freshwater Invertebrates. 2nd Edition

has been assigned to the order Trachylina of the same class (cf. Hyman, 1940). The fact that the other three orders of the class Hydrozoa (Actinulida, Siphonophora, and Hydrocorallina) and the other three classes of coelenterates (Scyphozoa, Cubozoa, and Anthozoa) are not represented outside of the sea is curious. There is not even any serious speculation as to why freshwater invasion by coelenterates has been so severely limited. There is no special osmoregulatory organ in the phylum, but this is not an explanation since its absence has not stopped the successful invaders. Because of their small size, soft bodies, and often sessile habits, freshwater Cnidaria are either not collected or not well preserved in most routine collecting procedures. They are, however, widely distributed and can be found in most ponds and streams when a specific search is made. When they are abundant, they can be major predators of small invertebrates and even of tiny fish. In turn, *Hydra* is fed upon by flatworms, and crayfish eat *Craspedacusta*. Probably, other animals prey on freshwater Cnidaria, but this has not been carefully studied.

II. GENERAL BIOLOGY OF CNIDARIA

A. Body Plan

All cnidarians share a simple body plan of a central cavity surrounded by two cellular layers (Fig. 1). The endoderm lines the interior cavity, the coelenteron. Between the endoderm and the ectoderm is an intermediate, noncellular mesoglea. Nematocysts aid in feeding and in repelling predators, and are present in all freshwater and marine cnidarians.

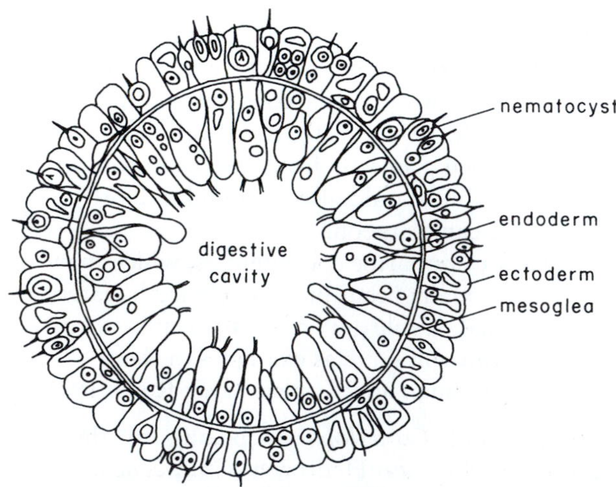

FIGURE 1 Cross section through digestive cavity (coelenteron) of generalized cnidarian.

The feeding aperture or mouth leads into the coelenteron, which functions as a gut. By convention, the end of the animal with the mouth is termed "oral", the opposite end "aboral." The oral aperture serves at different times as mouth and anus. When the mouth is closed, the pressure of fluid in the coelenteron can stiffen the body, even in the complete absence of any hard tissue. The coelenteron, therefore, functions also as a hydrostatic skeleton.

The mesoglea varies enormously in thickness among different members of the group. At its most meager, as in *Hydra*, it is not more than 200 μm thick, containing only wandering cells, nonliving fibrous components, and fibers from neuromuscular cells. It is more fully developed in medusae, such as *Craspedacusta,* and may contain a great deal of collagenous or gelatinous material. The ectodermal body wall has neural and contractile properties and is also the location for ripe nematocysts.

The presence of definite cell layers with differentiated functions distinguishes these animals as true metazoa. The absence of mesoderm implies that they do not have organs like higher metazoa; therefore, such terms as "tentacles" and "gut" are used in a functional sense.

The basic body plan can be manifested as either a polyp or a medusa (Fig. 2). Polyps are typically elongated along the oral–aboral axis. Medusae are approximately bell-shaped and usually have their greatest body dimension perpendicular to the oral–aboral axis. The coelenteron of a polyp is usually deeper than its body is wide, while a medusa is usually wider than its coelenteron is deep. Also, medusae generally have relatively thicker mesoglea. In some species, different generations, and, in some cases, the same individual organism at different stages of its development, can adopt the form of either a polyp or a medusa.

Around the feeding aperture of polyps or the edge of the bell of medusae is typically a ring of tentacles. Tentacles are extensions of the two cellular layers into more or less elongated projections. The coelenteron

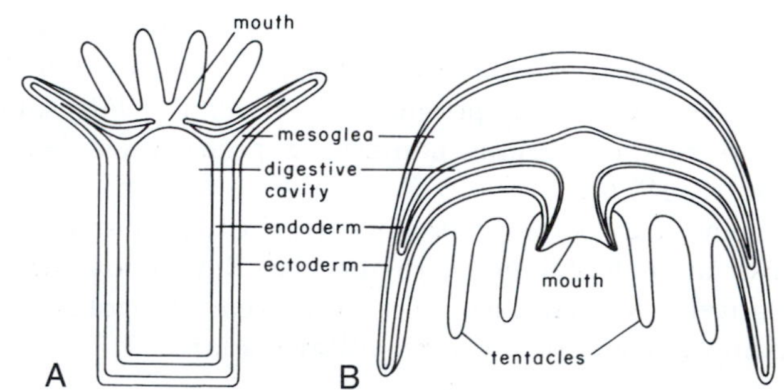

FIGURE 2 Basic body plan of Cnidaria showing (A) polyp and (B) medusa.

may or may not extend into the tentacles. Tentacular ectoderm is especially rich in nematocysts, which may be arranged in rosette or ring-shaped batteries. Tentacles are used in food capture, defense, and in some cases, locomotion.

B. Nematocysts

The phylum is characterized by the presence of cnidoblasts, ectodermal cells that produce the cell products called cnida or nematocysts. Because there are many kinds of elaborately spined nematocysts, these structures are valuable characters for the classification of coelenterates, particularly in such morphologically monotonous groups as *Hydra*.

Nematocysts consist of a capsule containing a threadlike tube (Fig. 3A). Near the base of the tube is a projection reminiscent of a trigger. Remnants of the living cnidoblast are absent from mature nematocysts. Many nematocysts come to lie on the tentacles, often after having been moved through the mesoglea from some other region.

After firing, a nematocyst consists of a long thread with a capsule at its base (Fig. 3B). Some nematocysts (the stenoteles or penetrants) are open at the thread tip, giving the appearance of a hypodermic syringe with a long needle. Stenoteles may have a complex of thornlike structures around the base of the thread. Stenoteles eject a neurotoxin, which partially paralyzes the prey. Desmonemes (or volvonts), another form of nematocyst, seem to lack poison, but rather eject sticky threads that wind about the spines and hairs on the body of the prey, interfering with movement. The capsules of some volvonts remain fixed to the tentacles after firing so that their exploded threads fasten prey to the tentacles as if by many tiny ropes or grappling hooks.

At least 17 morphologically distinct forms of nematocysts are present in the phylum (Fig. 4). The penetrants and volvonts of *Hydra* come in several forms, classified in terms of capsule size, spination of the threads, and shape and distribution of the basal spines. A single animal may have five or more types of nematocysts. Nematocysts of basically the same type may differ among species in how the thread is coiled inside the capsule prior to eversion. Some may appear like a coiled spring, with gyres at right angles to the longest dimension of the capsule, while others are coiled parallel to

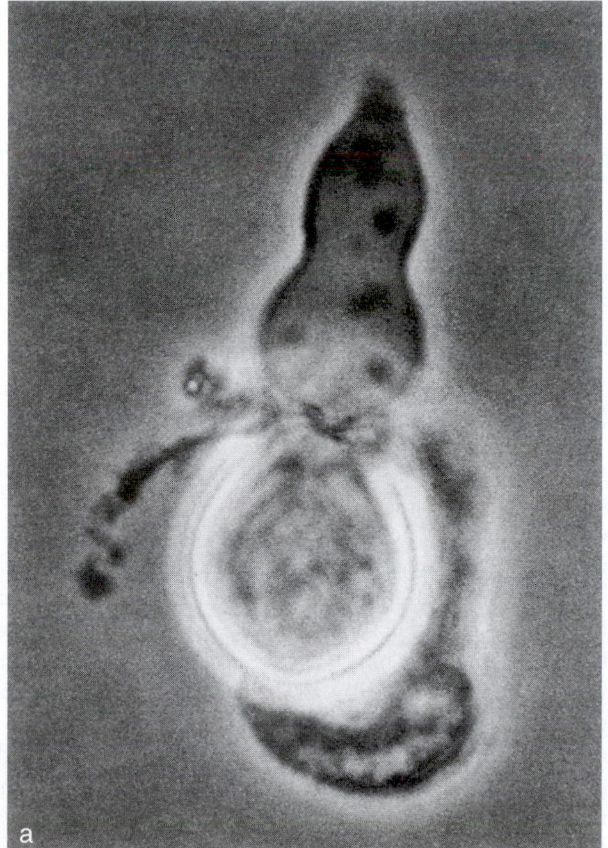

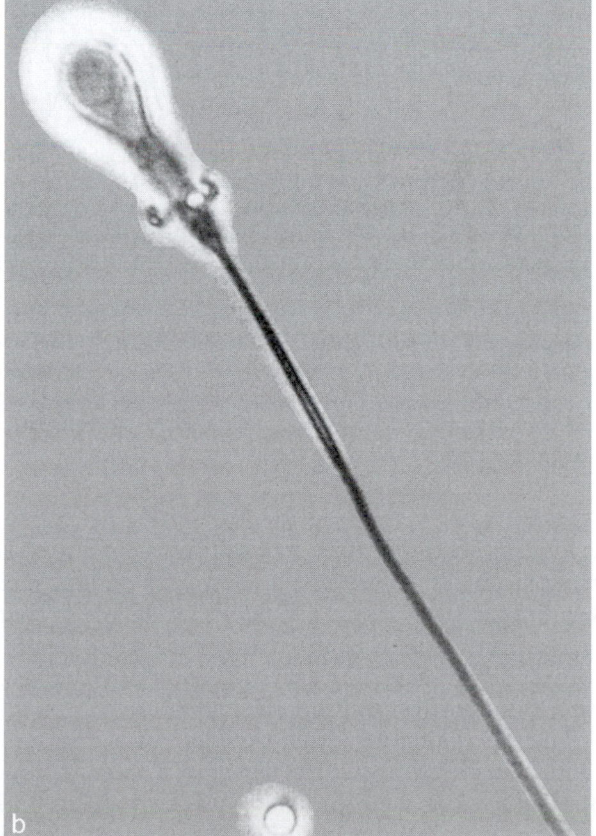

FIGURE 3 Discharged nematocyst from *Hydra*.

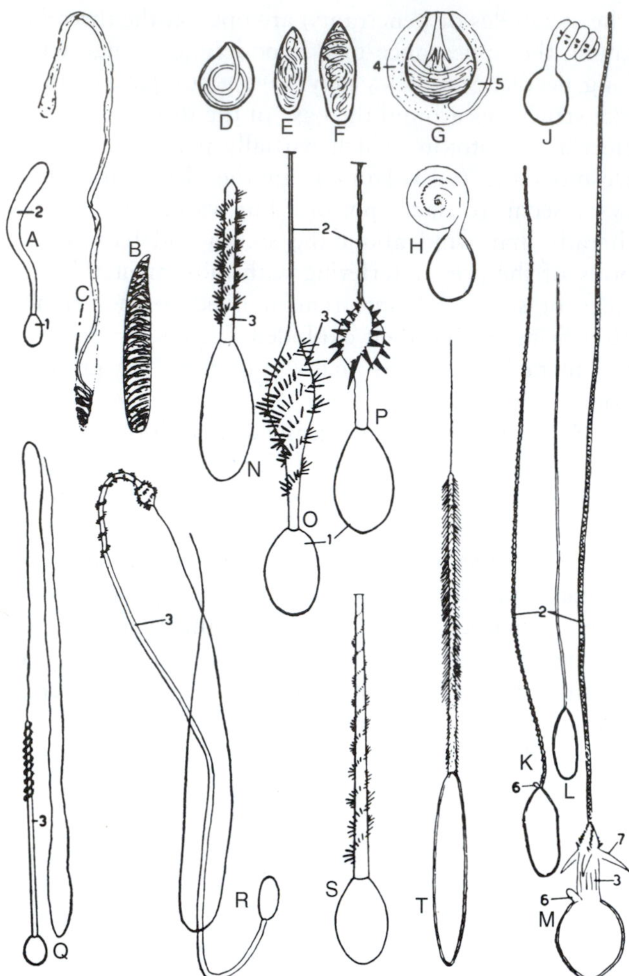

FIGURE 4 Some of the types of nematocysts present in Cnidaria (redrawn from Hyman, 1940); (A) rhopaloneme (only in Siphonophora); (B) spirocyst; (C) same as B, unraveling (not discharged); (D) desmoneme [(H) and (J) same as D but not discharged]; (E) atrichous hydrorhiza [(L) same as E, but discharged]; (F) holotrichous isorhiza [(K) same as F but discharged]; (G) stenotele inside its cnidoblast [(M) same as G but discharged]; (N) microbasic amastigophore; (O) homotrichous microbasic eurytele; (P) heterotrichous microbasic eurytele; (Q) macrobasic mastigophore; (R) teleotrichous macrobasic eurytele; (S) heterotrichous anisorhiza; and (T) microbasic mastigophore. 1, Capsule; 2, tube; 3, butt; 4, cnidoblast; 5, its nucleus; 6, lid; and 7, stylet.

the long dimension of the capsule. Nematocyst structure was initially considered a central taxonomic character; but recent evidence indicates that details of shape and coiling, and also the relative abundance of nematocysts of different types, are somewhat variable even within clones of *Hydra* (Campbell, 1987).

The microanatomy and function of nematocysts is a rich research area. The enormous interest about nematocysts and their production arises in part from the following observations. On the level of electron microscopy, they are extremely elaborate structures making them of interest as examples of complex cell differentiation. Also, the mechanism by which the thin thread of the nematocyst everts from its coiled state within the capsule (like an enormously elongated, inverted finger of a glove suddenly turning itself inside out) is a difficult problem in fluid pressures. The poison that is secreted by some nematocysts through the open tip of the thread is of medical interest. Furthermore, the enormous diversity of shapes, sizes, and spination of nematocysts within single organisms and among the different coelenterates, poses a problem in cell differentiation and genetics. Readers interested in more details on nematocysts should consult the large review volume by Hessinger and Lenhoff (1988), which is too lengthy to summarize here.

C. Feeding

After a prey has been stung and encumbered by the nematocysts, the tentacles move the victim to the mouth, which opens to admit it into the coelenteron. Often the prey is still alive and active, but this ceases as soon as the gastric cells lining the coelenteron secrete digestive juices.

Since there is no anus, food cannot be passed along the gut while digestion continues (as in higher metazoa). Instead, feeding stops until the digestion process is completed and the indigestible remnants have been regurgitated. In hydra, ingested food decomposes within an hour into a slurry of particles. The role of food vacuoles is reminiscent of the feeding process in protozoa and sponges (Barnes, 1966). The free borders of the digestive cells ingest particles by pinocytosis. Individual food particles are enclosed in vacuoles that are moved through endodermal cells. Eventually, their indigestible residues are ejected by the endodermal cells and returned to the coelenteron to leave ultimately through the mouth.

D. Reproduction and Metamorphoses

In the coelenterates, many kinds of reproduction exist (Hyman, 1940). Like all metazoa, they can reproduce sexually. The fertilized eggs may produce larvae differing anatomically and ecologically from the adult sexually reproducing stage. In marine coelenterates, there may be an elaborate succession of larval stages, some of which may form colonies or reproduce vegetatively by budding or fragmentation. In any particular species, one or more of these stages may be missing. There are medusae that produce eggs that go through various larval stages to produce new medusae (Fig. 5A). Some polyps produce gonads (Fig. 5A), and others bud off medusae that either swim away to become sexual (Fig. 5C) or remain attached to

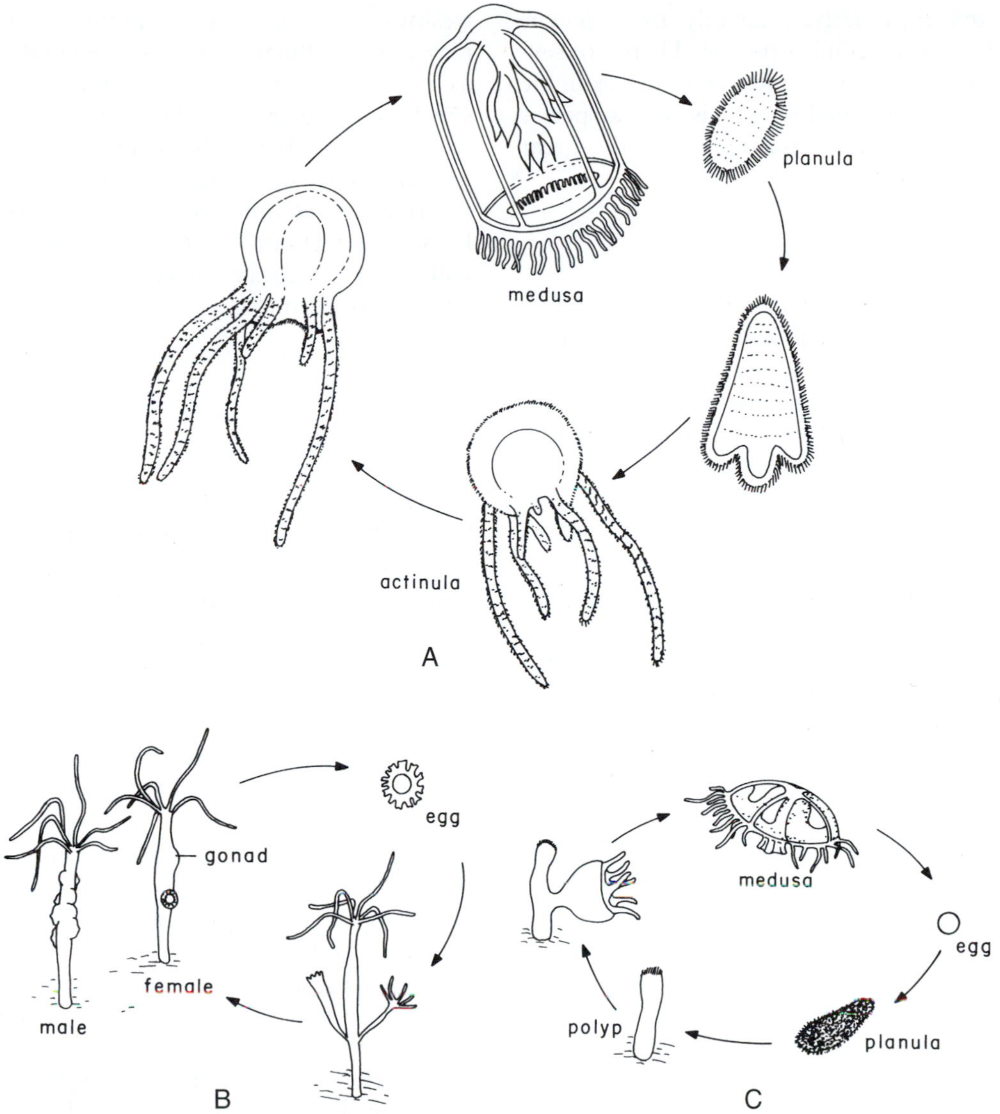

FIGURE 5 (A) Sexual medusa produces larva; larva forms new medusa; (B) sexual polyp generates new polyp; and (C) polyp produces medusa which becomes sexual and forms new polyp [redrawn from Barnes, 1966.]

their parents and become sexual without ever feeding independently.

Polyps or medusae may produce new individuals that may or may not resemble their "parent." A new individual may separate from its parent; but in many coelenterates, the asexually produced individuals stay attached and form a colony. The development of colonies is absent in hydra, but occurs among all other freshwater Cnidaria.

In the anatomically simplest coelenterate colonies, such as the colonial microhydra, "larvae" of the freshwater *Craspedacusta*, all attached individuals are essentially similar in both form and function (Payne, 1924). Each has its own mouth and coelenteron and each is capable of budding new polyps. Microhydra larvae can produce medusae by budding. Colonies of many coelenterates are much more elaborate.

In a common modification of this process, certain members of the colony become "gonozooids" that neither feed nor have tentacles; instead, they consist of a stalk rising from the common stolon that buds off medusae. These medusae, in turn, either form gonads while still attached to their colony or leave the colony and then produce gonads.

The best known freshwater coelenterates, the hydra, have lost the medusa stage entirely. In hydra, polyps may produce new polyps by asexual budding or may temporarily switch to sexual reproduction using gonads. Fertilized eggs derived directly from polyps may then hatch to produce new polyps. There are several marine examples, but only one freshwater example of medusae budding new medusae. This is in a species presumed endemic to Lake Tanganyika, in the great African rift (Thiel, 1973).

E. Ecological Interactions

Although some species may gain part of their nourishment from intracellular algal symbionts, all cnidarians are carnivores. Their prey consists primarily of small coelomate animals, organisms that evolved long after the appearance of coelenterates. They generally do not feed on protozoa, nematodes, or sponges, nor will these animals trigger the nematocysts. This raises a curious question, for which there is no evident answer: "What did the first coelenterates eat?"

Hydra are extremely effective predators. Prey includes crustaceans, worms, and fish and insect larvae. They have been seriously considered as a biological control organism for mosquitoes. *Hydra* and *Craspedacusta* can both sting fishes very badly. Fish that are too big to swallow often cannot survive being stung. For this reason, hydra are sometimes serious pests of fish hatcheries.

Nematocysts are sufficiently unpleasant that relatively few predators attack coelenterates. Turtles, fish, crabs, worms, echinoderms, and flatworms are among the predators on marine coelenterates. Predation on the freshwater coelenterates is not well studied. Crayfish eat *Craspedacusta* (Dodson and Cooper, 1983). Flatworms are reported to eat *Hydra* (Hyman, 1940; Kanaev, 1969), but we have seen small flatworms withdraw from contact with hydras.

Some marine coelomates are not affected by nematocysts and are commensal with coelenterates. On coral reefs, clown fish live among tentacles of large anemones and other fish occur only in close association with the Portuguese man-of-war. In freshwater, *Anchistropus,* a chydorid cladoceran, has been observed clinging to, and apparently feeding on, the body wall of hydras (Griffing, 1965; Slobodkin, personal observation). The amoeba *Hydramoeba hydroxena* feeds on hydra (Stiven, 1976), and the hypotrichous ciliate *Kerona* lives on the surface of hydras (Hyman, 1940). *Polypodium hydriforme* is parasitic on a fish for part of its life (Raikova, 1980).

III. DESCRIPTIVE ECOLOGY OF FRESHWATER CNIDARIA

Their anatomical simplicity and the ease with which hydra can be cultured in the laboratory make freshwater Cnidaria very important as experimental material in cellular and developmental biology, as well as in neurobiology, biochemistry (Ciernichiari *et al.,* 1969), and genetics (Sabine *et al.,* 1987; Schummer *et al.,* 1992). Hydra also lends itself to studies in cell growth, morphogenesis, microanatomy, and symbiosis (Ciernichiari *et al.,* 1969; Dunn, 1986, 1987, 1988; Bossert and Dunn, 1996). In the present chapter, we will focus on cnidarian ecology and natural history, which have been less well studied.

A. Hydra

1. General Biology of Hydra

Hydra are small polyps from 1 to 20 mm in body length. The body is crowned by up to 10 or 12 tentacles. Usually the tentacles are of approximately the same length as the body, but may be somewhat shorter, particularly in the green hydra, and can exceed 20 cm in length in hungry brown hydra in quiet water. There are no medusae. Reproduction is by budding or by gametes produced directly from ectoderm. The fertilized eggs may enter a resting stage. When development proceeds, the egg immediately develops into a polyp (see Fig. 5).

Hydra are found attached to almost any reasonably hard surface. Slight bacterial films may make surfaces more attractive, but heavy growth of microalgae may be avoided. Water lily stems, charophytes, dead leaves, sticks, and stones are favored substrates. They may appear as single animals or as dense furlike aggregations. There are reports of fishing nets becoming completely covered with brown hydra, resulting in rashes on the hands and arms of the fishermen (Batha, 1974).

If the wait for food is longer than approximately 12 h, the hydra begin to change their locations on the substrate (Ritte, 1969; Lenhoff and Lenhoff, 1986). They have two different methods of movement. Small-scale movements on a substrate may occur by attaching the stretched tentacles to the substrate or, in shallow water, to the surface film, releasing the pedal attachment and contracting the tentacles and reattaching. Sufficiently crowded or hungry hydras float off their substrate (Lomnicki and Slobodkin, 1966). They may appear in the plankton or be found floating upside down with the pedal disk in the water surface film and their tentacles trailing.

Batha (1974) extensively documented the existence of planktonic hydra in Lake Michigan through use of divers and of suitable attachment surfaces suspended in midwater. In addition, the reports of fishnets covered by hydra and observations by Griffing (1965) of sudden relocations of hydras within a single pond suggest that hydras are probably much more important components of lake plankton than has been generally realized.

Floating animals will sink and settle either from wave and current action or from having just fed. This behavior keeps hydra in areas of abundant food. It is also evolutionarily important as a dispersal mechanism. From the perspective of a naturalist, it has the effect of making it relatively easy to collect hydra at the outflow of lakes and ponds.

They are not tolerant of heavy metals, but they can thrive in even highly eutrophic water. They can survive at temperatures from near freezing to 25°C.

Several *Hydra* species may coexist in a pond; often a small green hydra and, at least, one large brown species will co-occur. Also, strains of *Hydra* may replace each other seasonally. Bossert (1988) showed that green hydra collected several months apart from the same small pond had very different size and growth characteristics when maintained under very similar conditions in the laboratory.

Despite many years of collection and observation, we have never found a species of green hydra that was consistently larger than any strain of asymbiotic brown hydra, nor have we found any contradictory account in the scientific literature. If it occurs, it certainly seems rare. Some large brown hydra can be caused to become green in the laboratory (Rahat and Sugiyama, 1993), but it is not clear that this is significant in nature.

Hydra occur in freshwater from the Amazon to Alaska and from Siberia to Africa at depths from shallow water to 60 m or more. This cosmopolitan distribution may be a result of the portability of the thecate eggs; or perhaps, as has been suggested by Campbell (1987), it might be due to the four species groups (see Section V) having differentiated before the primeval continental masses separated in the Mesozoic era. A recent study of the hydras of Madagascar (Campbell, 1999), which has been an island for more than 100 million years, shows that they differ very little from mainland hydra, indicating that either the rate of evolution of the genus *Hydra* is very slow or that they are surprisingly well dispersed. One species group (*oligactis*) is missing from Madagascar. Since there is no obvious reason why hydra of this group should be less mobile than the others, their absence strengthens the interpretation of slow evolution. This implies that hydras are evolutionarily very old.

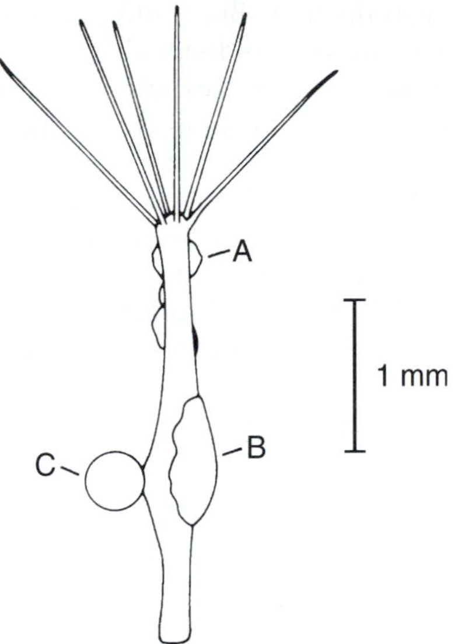

FIGURE 6 Hydra. (A) Male gonad; (B) female gonad; and (C) egg [redrawn from Campbell, 1987.]

2. Reproduction and Mortality

Asexual budding is the primary reproductive mechanism of hydra during periods of population increase. Under optimal conditions of food supply, temperature, and water quality, each adult hydra polyp can produce two buds per day and each bud can mature and begin reproducing in a week or less. Generally, the smaller strains of hydra bud more rapidly than the larger ones, at equal feeding rates. Buds are from 12 to 20% of the size of the mother, varying with the strain.

While Hydra usually reproduce by producing free-living buds, under deleterious environmental conditions they may develop testes or single egg ovaries and engage in sexual reproduction. External fertilization by free-swimming sperm occurs while the egg is attached to the body wall of the mother. The embryo may enter a resting period of days or even months before proceeding with direct development into a new polyp.

During sexual reproduction, gonads develop along the stalk in place of buds (Fig. 6). Mature male gonads are a mound of tissue with a distinct apical nipple from which sperm extrude. Mature female gonads consist of a single large egg cell resting on a cushion of smaller cells. The fertilized eggs are surrounded by a theca, which may be smooth or ornamented or may consist of polyhedral plates. The features of the theca are of taxonomic importance.

Sexuality in hydra seems to occur only under deleterious environmental conditions, although the precise cues are unknown. Chemical changes due to temperature fluctuations, and, perhaps, sudden nutritional variation can all induce sexuality at certain times (Lenhoff and Loomis, 1961; Kanaev, 1969). Individual hydra have been reported as unisexual and some bisexual individuals have been noted (Hyman, 1940; Kanaev, 1969). If conditions improve, sexual hydra can return to asexual reproduction. Buds and gonads may occur concurrently in the same individual.

Most studies of hydra sexuality are made on animals that have been maintained in the laboratory. Batha (1974) has suggested that sexual reproduction is very rare in nature, based on the complete absence of gonads among thousands of brown hydra collected in Lake Michigan. Although we have found eggs in green hydra collected from a small pond and other field reports of gonads do exist (Kanaev, 1969), it seems clear that in hydra the numerically most important reproductive process is budding.

In most plants and animals, population size and genetic recombination are associated through the process of sexual reproduction, but sexual reproduction in hydra is not significant to population dynamics. It seems important for maintaining genetic heterogeneity and permitting escape from temporarily unsuitable conditions (cf. the chapter on "Escape in Time and Space" in Slobodkin, 1980).

In a single pond or stream, most of the animals are likely to belong to a rather small number of vegetative reproductive lines, being genetically identical except for occasional mutations. Even those that have emerged from eggs are likely to be the result of relatively close inbreeding, since sexuality is usually found in crowded local populations whose members are descended from a very small number of clonal lines. There is no direct evidence on sperm survival in nature, but it seems unlikely that sperm can travel great distances.

Hydra that are very small, either because of youth or starvation, will not produce buds. If nutrition is limited, animals may be kept indefinitely in a condition of neither growing nor budding. Both body size and budding rate are proportional to feeding rate up to a point of food saturation.

The capacity of hydra to reduce its body size is of ecological interest. Animals with hard skeletons are committed to a particular body size in the sense that they cannot shrink below a given point during periods of food shortage. The inability to reduce size may contribute to death by starvation. By contrast, hydras shrink in size when starved, but can be restored to full reproductive size by increasing their food supply. At a given temperature, hydras will come to a steady-state body size and budding rate if food supply is sufficient and constant.

The time for complete size adjustment to either temperature change or feeding level change on the part of an individual brown hydra is approximately 3 weeks. Reduction of temperature decreases growth rate, but increases body size. These size differences seem to be due to changes in cell number rather than cell size (Fig. 7) (Hecker and Slobodkin, 1976). Because

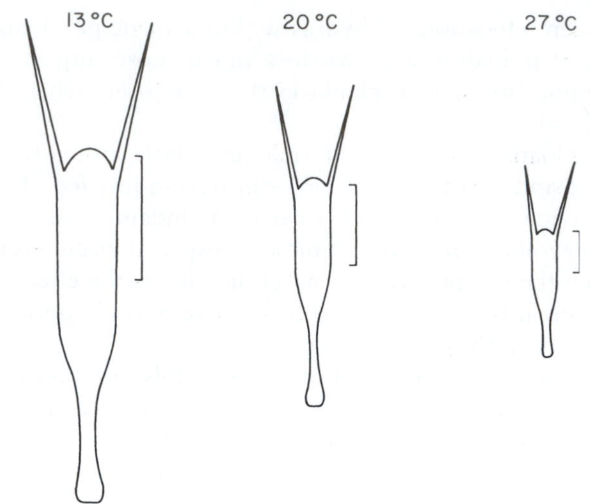

FIGURE 7 Relative sizes of hydra raised at different temperatures. Scale shown is the same size at each temperature [redrawn from Hecker and Slobodkin, 1976.]

larger animals have greater food requirements, the smaller green hydra can produce buds when fed one or two *Artemia* salina nauplii per day while the larger brown hydra require 5–10 per day before they will bud.

3. Feeding

The tentacles extend above and lateral to the body. They are generally motionless except in the presence of potential food or during locomotion. In healthy animals, the tentacles are cylindrical or tapered, but never clubbed. The length of the tentacles varies somewhat with species, but considerably more with environmental circumstances. In quiet water, animals with bodies no longer than 2 cm have been observed to constrict their tentacles into thin threads extending at least 20 cm (L. B. Slobodkin, personal observation). This is unusual however, and tentacles equal to three body lengths or less are more typical.

Tentacles are quickly retracted if the animal is disturbed or if organisms brush against them. Whole prey or extracts of their body fluids will initiate active waving movements of the tentacles (Lenhoff, 1983). If a prey organism brushes against the tentacles, nematocysts will discharge, poisoning the prey and attaching it to the tentacle. Other tentacles and their nematocysts then join the attack. In a matter of 1–4 min, the prey will have been pressed against the mouth surface by the tentacles and will have entered the coelenteron. Meanwhile, the tentacles for subsequent swallowing may have caught other prey.

The number of prey swallowed in one feeding encounter varies with the size of the hydra, the size of the prey, and the previous feeding condition of the hydra. One large *Daphnia magna* may fill the coelenteron completely, stretching its walls to give the appearance of a *Daphnia* stuffed into a thin expandable sack of hydra tissue.

During the digestion process, the body may become rounded as the swallowed prey are reduced to a slurry. After approximately 1 h, material is suddenly discharged from the coelenteron through the mouth, the hydra momentarily appearing like a punctured balloon. The columnar shape is then restored, tentacles regain their virulence, and the animal waits for its next meal. There is evidence that several days of starvation will increase the appetite of a brown hydra. Brown hydra can survive more than 40 days without food, and green hydra can live without food 4 months or longer.

In general, hydra eat small, open-water plankters, but are less effective at capturing animals that normally inhabit underwater surfaces. The common cladoceran genera *Simocephalus, Scapholebris,* and *Chydoris* and at least some ostracods are immune to the activities of hydra (Schwartz *et al.,* 1983). Hydra can eat very small fish and insects, but sufficiently large animals with hard skeletons and strong swimming force can escape after being stung. The long bristles on small midge larvae have been found to impede predation by hydra (Hershey and Dodson, 1987).

Apparently, hydra primarily feed on the kinds of prey that they are least likely to encounter. This suggests that hydras are sufficiently important as natural predators that only those crustaceans that have evolved immunity to hydra can coexist closely with them.

4. Immortality and Regeneration

Trembley, in the eighteenth century, had already demonstrated the capacity of hydra to regenerate perfectly, even after severe mutilations. The regenerative powers of hydra make them favorite classroom objects. They can be decapitated, bisected, have their tentacles amputated, or even be turned inside out and in a matter of a few days regenerate missing parts or regain their proper organization. Within a clone, rings of hydra stalk can be threaded on hairlike quoits and may fuse to form a single tube (Burnett, 1973). Even without operations, different accidental conditions in the field or laboratory will produce hydra with various mutilations: more or fewer tentacles, missing heads and so on. All of these hydras will reorganize themselves neatly if water chemistry is not deleterious and if they have been reasonably well-nourished before the mutilations occurred.

Since hydras can regenerate so well, are they potentially immortal? Immortality in any organism is impossible to demonstrate during a research program of finite duration, and therefore the question cannot be unequivocally settled. However, what we generally mean by immortality in organisms is the absence of any signs whatsoever of senescence or permanent scars of the past. Martinez (1997) maintained laboratory cohorts of hydra for more than 3 years and demonstrated the absence of any symptoms of senescence. Single polyps have been maintained in our laboratory without budding for at least a year, their lives terminating only from human error. There is no clear evidence for senescence of any kind in hydra. If hydra are starved they will become smaller; but do not die until they are too small to feed. Newborn hydra are approximately at this minimum size (Slobodkin *et al.,* 1991).

It seems likely that hydra polyps are potentially immortal. All deaths in hydra can be assigned to such things as alterations in water quality or food supply, temperature shocks, excessively severe starvation, or predation.

5. Symbiosis

While species of *Hydra* may be difficult to distinguish, there is a very clear distinction between the brilliant green color of some species of *Hydra* and the yellow, brown, and gray colors of those that do not have symbiotic algae. The green hydra are accepted as a taxonomically distinct group and have been assigned their own generic status as *Chlorohydra,* which we use here interchangeably with "green hydra." No certainty exists as to the monophyletic character of this genus.

The green color arises from *Chlorella*-like cells, unicellular algae each occupying a vacuole in the endodermal cells of their hosts. Each endodermal cell contains 10–35 algae-laden vacuoles (Fig. 8). The precise number varies with species, but is nearly constant within a hydra strain and a set of environmental circumstances (Bossert, 1986). Algae are also found in the central cells of the tentacles, but in somewhat smaller numbers. The endoderm of the buds contains algae like those of the mother.

Ample evidence exists showing that green hydra gain nourishment from their symbionts. Radioactive-tracer experiments have demonstrated that maltose, a secondary photosynthetic compound, leaks from the algal symbionts to their hosts (Muscatine, 1965; Ciernichiari *et al.,* 1969; Lenhoff, 1974). Also, there is microscopic evidence that algal cells can be attacked by host lysozymes (Dunn, 1986, 1987, 1988). In competition experiments between brown hydra and green hydra populations, light provides a significant advantage to the green hydra (Slobodkin, 1964).

The algae can be removed from some strains of green hydra by use of a variety of techniques, including photosynthetic poisons, prolonged darkness or extremely strong light, and dilute glycerin solutions

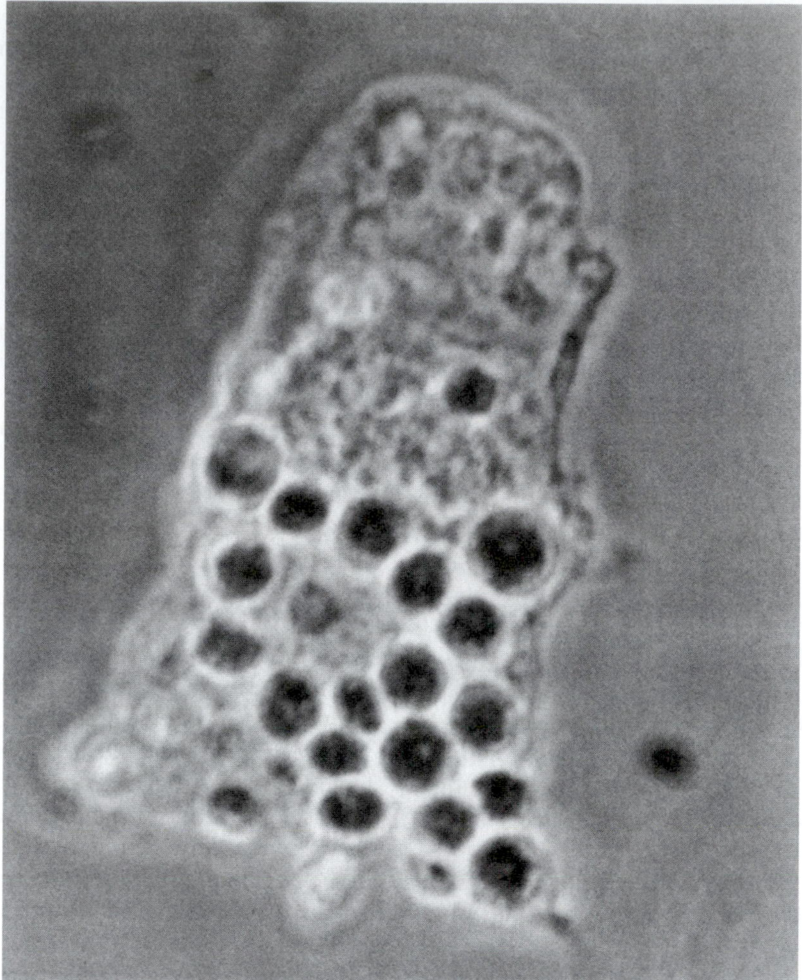

FIGURE 8 An algae-laden endodermal cell from *Hydra* (*viridis*) (1000×).

(Pardy, 1983). Such hydras are referred to as aposymbiotic and are susceptible to reinvasion by algae. It is possible to develop green patches in brown hydra by injecting their coelenteron with algae or by feeding green hydra to brown ones (L. B. Slobodkin, personal observation), but this coloration fades with time.

If the association with algae is advantageous, why are big hydras not green? Symbiotic relationships involve a delicate interaction between the two partners, particularly when one partner lives within the cells of the other. A primary requirement for stable endosymbiosis is a mechanism providing balanced rates of mitosis of the host and symbiont cells. If there were no control of the algal rate of increase, they would be expected to kill their host cells by filling them with algae; and if the hydra cells excessively limited the algae, they would be eliminated and the hydra would be brown.

While the number of algae per host cell stays constant in all green hydra, maintaining this constancy seems more difficult in the larger green hydra strains. Bossert and Dunn (1996) have shown that algal cells are increasing faster than host cells in green hydra strains of all sizes, but the disparity between algal and animal mitotic rates is greatest in the largest strains. Dunn (1986, 1988) suggested that algal cells are being actively expelled or digested by all green hydra, but especially by larger strains.

Size in green and brown hydra is apparently regulated by the relative amounts of several hormones, some of which activate the formation of buds while others inhibit budding (Sabine *et al.*, 1987). Bossert (1988) found that one of the hormones implicated in producing smaller hydra size also inhibits algal mitosis within green hydra cells. Apparently, the same hormonal mechanism that produces small size aids in maintaining the balance between algal cell and hydra cell increases, while a hormonal balance that permits larger size makes control of algae more difficult and, in the largest strains, impossible. More investigation of this problem is needed.

Green hydras are never found in nature as aposymbionts and seem to have evolved a dependence on algae. However, these algae have probably not evolved a dependence on hydra. In fact, how the algae benefit from the association is not at all clear. Certainly, the *Chlorella* in a green hydra are immune to being eaten by filter feeders, have an assured source of mineral nutrients, CO_2, and nitrogen, and are moved into light as the hydra moves. However, the actual rate of increase of algal cells inside hydra is probably lower than those of algal cells outside, and it is not obvious that the number of algal cells contained in the entire green hydra population is a significant fraction of the natural algal population in the lake or pond.

6. Toxicology

With the development of modern techniques, there has been an impressive increase in the number and significance of research programs focused on hydra. For example, like any small, easily cultured animal, hydra has been attractive for use in ecotoxicology. In these tests, the survival or death rate of organisms is typically related to concentrations of particular toxins.

Toxicity tests using specially prepared hydra can provide evidence as to whether or not a particular toxin is likely to cause cancer. The procedure involves mass culture of *Hydra attenuata,* and disassociation of the hydra into individual cells that are aggregated into "artificial embryos." The small, mixed-cell hydra patties organize themselves into complete and perfect animals in less than a week. It has been found that toxins which do not produce cancers are approximately equally toxic to intact hydra and to the reconstituting patties while known carcinogens are much more toxic to the regenerating cell mass than to the intact hydra (Johnson *et al.*, 1982).

7. Molecular Biology

The regeneration studies, and the toxicological studies as well as mathematical analyses of hydra population dynamics and evolution (Slobodkin *et al.*, 1991), are all possible because of the extreme anatomical and behavioral simplicity of hydra.

Loewenhoek first described hydra in 1704. They became an object of intense scientific concern with the work of Trembley (Lenhoff and Lenhoff, 1986). Trembley collected and described green and brown hydra, feeding them on small invertebrates and dissecting them with a razor. He established their remarkable capacity to recover from damage that would kill most animals. Among the ingenious dissections were removal and splitting of the head, and producing multiheaded animals. He also sliced hydra into rings. In those cases, a new head developed if the slice had been nearer the original head, while a foot developed if the slice had been closer to the original foot. He even threaded whole hydra onto bristles and watched as they turned themselves inside out, leaving the bristle behind. There existed in the 18th century a belief that animals at birth were endowed with a kind of uniqueness, an animal soul. Transforming a single hydra into 10 others identical with itself by slicing it into pieces effectively eliminated the idea of a unique animal soul (Burnett, 1973).

It was also generally assumed in those days that green organisms that benefited from light were plants. Trembley demonstrated that green hydra were animals that in certain respects acted as if they were plants.

Studies of hydra continued to contribute to our understanding of basic biological questions through the next two hundred years. *Hydra* research has gained in importance in the past two decades since it has been found that the molecular and genetic apparatus of hydra is also amenable to analysis. The current volume of work is very large; hence, only a brief indication of what is being done will be included here.

The biochemical and genetic properties of Cnidaria are of significance in studies of the evolution of metazoa. Some early workers felt that hydra, arguably the simplest metazoa, must be very close to the ancestor of all metazoa. It is now clear that neither hydra itself nor other cnidarians, such as *Scyphozoa, Cubozoa,* and *Hydrozoa* (to which hydra belongs), are on the main branch that led from Cnidaria to other metazoa. This conclusion is based on the fact that they are the only metazoa in which mitochondrial DNA is arranged in a linear, rather than circular pattern. However, another major group of Cnidaria, the anthozoans, have circular molecules of mitochondrial DNA like all other metazoa and would, therefore, seem closer to the main line of metazoan evolution than hydra (Bridge *et al.*, 1992; Warrior and Gall, 1985).

Despite this finding, the extreme simplicity of hydra anatomy suggests that it may provide a useful molecular model for higher metazoa. Signal transduction pathways mediated by cAMP or inositol triphosphate are present in hydra as they are in virtually all higher organisms.

The control of basic body form in hydra involves an interaction of homeotic genes, some of which are largely identical with body form genes in mice and fruit flies (Bernfield, 1993; Aerne, 1995; Finnerty and Martindale, 1999; Galliot *et al.*, 1999). The implications of this are that the basic developmental processes of the first metazoa (or perhaps advanced premetazoan ancestors of the first metazoa) evolved a developmental control mechanism which has remained

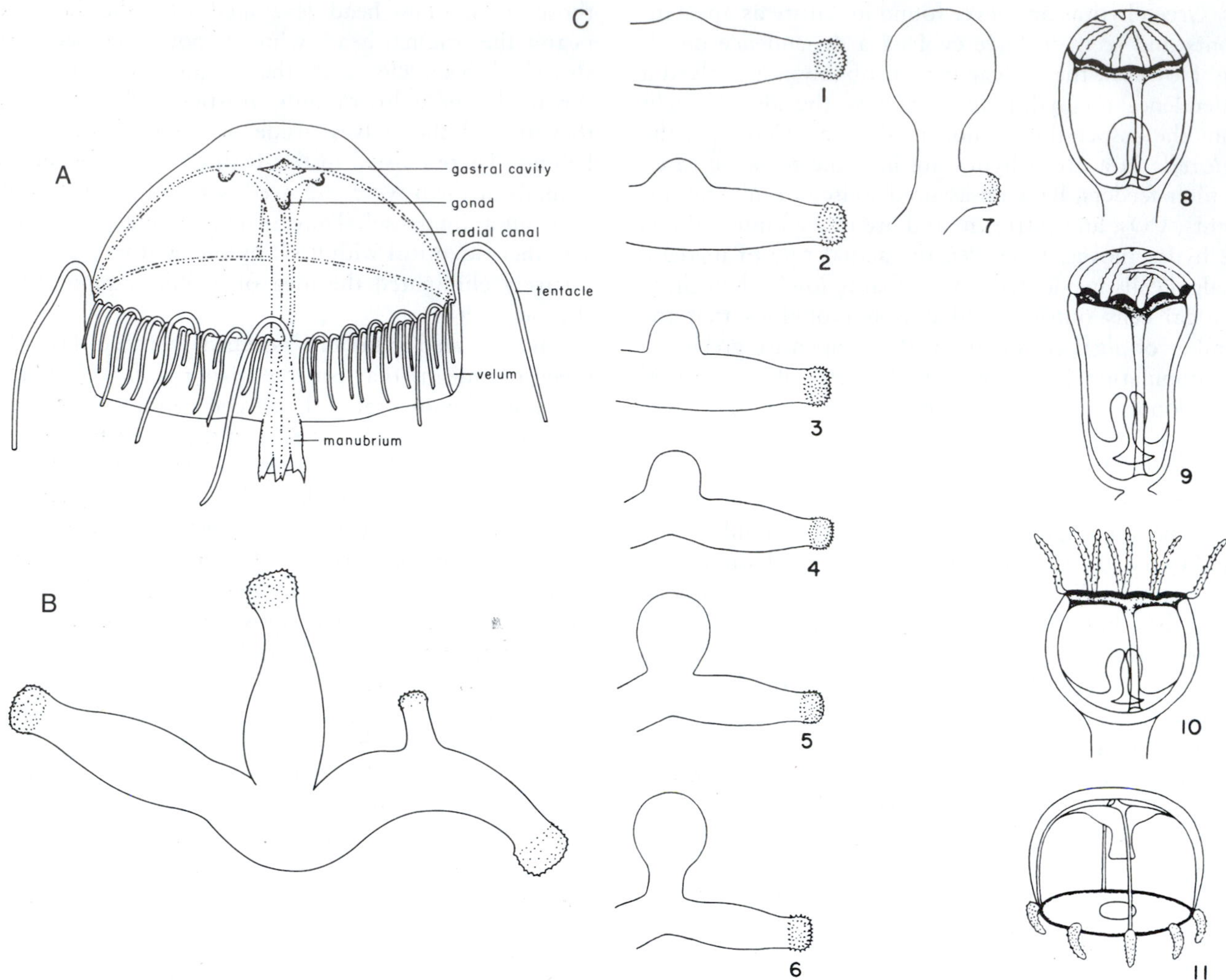

FIGURE 9 (A) *Craspedacusta sowberii,* also named *Limnocodium victoria* (see text); (B) larval stage of A, also named *Microhydra ryderi* (see text); and (C) stages in development of *Craspedacusta* [redrawn from Payne, 1924.]

essentially intact throughout all subsequent metazoan evolution!

B. Craspedacusta

The first scientific accounts of freshwater medusae were based on specimens found in the giant water lily tank of Regents Park in London in 1880 (Fig. 9A). Two authors described these separately. Lankester named them *Craspedacusta sowberii* after the discoverer, Sowber, and Allman named the same organisms *Limnocodium victoria* after the lily. In the same year and in the same tanks, a tiny colonial hydroid was found. This animal had no tentacles; each polyp terminated in a bulbous "capitulum" studded with nematocysts. In the center of the capitulum was the mouth. These little polyp colonies (Fig. 9B) were initially, and correctly, assumed to be the larval stage of the medusae. However, very similar polyps were discovered in a water tank in Philadelphia and were described and named as a separate species. It was not until 1928 that it became once again obvious that the polyp named *Microhydra ryderi* was the larva of *Craspedacusta* (Boulenger and Flower, 1928).

Since their initial description, *Craspedacusta* medusae have been found in many locations around the world, apparently transported with ornamental aquatic plants and with the water hyacinth, a recently spread pest species. They have also been found in locations free of imported plants; for example, Lake Gatun in the Panama Canal zone (Slobodkin, personal observation). Since first documented in the United States in 1908, these unpredictable and sporadic organisms have been reported in 35 states. The jellyfish have been found in lakes, farm ponds, quiet coves of rivers, and old water-filled quarries. When they occur they are locally abundant but sufficiently rare to call to public attention (Anonymous, 1999).

1. Life Cycle

The most conspicuous stage of the life cycle is the small medusa (0.5–1.5 cm). Except in the Yang-tse river system of China, the occurrence of noticeable populations of the medusae of *Craspedacusta* is sporadic and often surprising to local naturalists. As the water warms up in a pond or in the slow current of a stream backwater, a swarm of medusae appears, often where it has never been seen before or at least has not been apparent for many years. These medusae feed on zooplankton. As they grow, the number of tentacles increases from 8–12 up to as many as 100. After several weeks of growth, gonads develop in pouches of the radial canals. Fertilized eggs produce a small crawling planula, which then differentiates into a microhydra (Payne, 1924). The planula consists of two cell layers forming a double-walled, sausage-shaped sac. The differentiation of the planula into a microhydra is rather simple. It stands on end and both a mouth and a capitulum develop on the unattached end, thereby forming a polyp.

The new polyps may continue to bud off new microhydra or medusae. Buds may remain attached after they have developed a capitulum, producing colonies of up to 12 polyps attached to a common stolon. Starvation and severely abnormal temperatures cause the polyps to shrink to a cellular ball, surrounded by a chitinlike membrane. This can persist through the severe conditions and then redifferentiate as a polyp.

The buds of medusae appear initially as rounded swellings of the polyp wall (Fig. 9C). Over a period of several weeks, they enlarge and develop a central manubrium. An endoderm-lined circular canal with four radial canals leads to the base of the manubrium. Eight tentacles emerge from the circular canal. The medusa bud is initially covered by a layer of tissue that eventually perforates centrally, remaining as the vellum of the adult. The mouth opens into the manubrium. The medusa is by now at least as large as the polyp, to which it is still attached at the aboral end. Eventually, it begins locomotory pulsation and separates from the polyp, completing the life cycle (Payne, 1924). Medusae usually reproduce sexually; fertilized eggs develop into planulae, which transform into microhydra.

2. Ecology

The microhydra polyps feed in essentially the same fashion as hydra. Their small size limits their prey, but this is partially compensated for by their colonial growth pattern, which permits several polyps to make a simultaneous attack. Free-living medusae feed on various crustacean zooplankters. They can even kill, but apparently not swallow, the large (0.5 cm), predaceous cladoceran *Leptodora* (Dodson and Cooper, 1983).

As a medusa feeds, it grows larger, adding additional tentacles at both the vellum margin and its inner edge. Several hundred tentacles are found on mature animals. Medusae appear sporadically in shallow ponds, natural lakes, and artificial reservoirs throughout the north and south temperate zones. Often entire medusa populations are unisexual. The polyps are so inconspicuous that they probably have a much broader distribution than the literature reports.

Many locations containing polyps are reported not to have medusae. This bewildering picture has been clarified by Acker (1976) and Kramp (1950) . Acker suggested that medusa production by the polyps requires temperatures greater than approximately 20°C during a period of increasing temperature and adequate, but not enormous, food levels. Other environmental alterations may be significant but have not been tested. These conditions are reminiscent of those that produce sexuality in hydra.

Acker and Kramp maintain that the original natural habitat of *Craspedacusta sowerbii* is the Yang-tse-Kiang region of China. In the upper river valley, two *Craspedacusta* species, *C. sowerbii* and *C. sinensis*, coexist, whereas only *C. sowerbii* reaches the downstream areas. From this habitat, *C. sowerbii* traveled with water hyacinths and other plants to its present worldwide distribution.

Shallow pools exist along the lower Yang-tse valley which are subject to large temperature changes and to sudden flooding from the main river during high water. Plankton populations in these ponds fluctuate strongly. In these ponds, medusae are a recurring phenomenon throughout the year. The occurrence of medusae in the spring is so regular that they are given a common name that translates to "peach blossom fish."

The generic name *Limnocodium* is usually applied to freshwater medusae of the Old World and *Craspedacusta* to those of the New; but it is not clear if this is more than a geographic distinction nor has there been enough investigation to determine how many species of freshwater medusae actually exist. Due to their sporadic occurrence, the changes in size and tentacle number with developmental stage, the simplicity of larval anatomy, and the difficulties of preserving specimens, morphologic studies of freshwater medusae are difficult. As in hydra, many species have been described, but perhaps some of these are based on nutritional or developmental history or preservation artifacts. The strongest evidence for multiple species is that sympatric populations of two species of medusae are known from the Yang-tse and that a medusa from Lake Tanganyika has an extraordinarily different life cycle. In this latter case, a mature medusa buds new medusae from its manubrium in addition to reproducing sexually (Bouillon, 1957).

C. Calposoma

This is a colonial polyp not much bigger than a paramecium and similar in general appearance to the microhydra except that it is considerably smaller and has tentacles rather than a capitulum. As far as is known, there is only one species, *Calposoma dactyloptera* (Fuhrman, 1984). Some of its properties have been described by Rahat and Campbell (1974). These organisms are so small that their tentacles consist of a single cell, a tentaculocyte, which contains a row of miniscule nematocysts. The full natural history of these animals and the details of their life cycle are not known. Their general anatomy is reminiscent of a miniaturized hydra with stiff tentacles, but this does not necessarily indicate taxonomic or evolutionary proximity.

D. Cordylophora

The genus *Cordylophora* is an athecate member of the primarily brackish water and marine hydrozoan family Clavidae. It grows as a branching colony up to 5 cm high. The feeding polyps have a conical hypostome on which filiform tentacles are irregularly arranged. Colonies also include gonophores, which produce gonads. There is a chitinous periderm (Fig. 10). During periods of stress, the animals regress to "metanonts", masses of resting tissue in the hydrorhizae (Naumov and Stepan'iants, 1980). The metanonts appear when the plant stalks on which they grow begin decomposing in the fall (Roos, 1979).

Cordylophora was first found in the Caspian and Black seas. It has been suggested that it evolved during the time that these two bodies of water were connected. It now occurs in rivers of Europe and America, apparently having spread during the last century attached to ship bottoms. The recent expansion of range may be attributed to the relatively greater speed of seagoing vessels during the last hundred years which would shorten the time of salt water immersion between brackish water ports (Roos, 1979).

Industrial and navigational developments have extended the region of brackish water in some estuaries, and the increasing pollution of rivers may have duplicated estuarine conditions. This may be expected to encourage the geographic spread and the upriver

FIGURE 10 *Cordylophora* [Redrawn from Roos, 1979].

movement of not only *Cordylophora* but also other estuarine Cnidaria (Hubschman and Kishter, 1972). The possibility of foreign Cnidaria appearing in habitats where they have never been seen is also enhanced by rapid vessels and the use of water ballast discharged in ports of arrival. *Cordylophora* is one example of an expanding range. We expect that there are many others, which we have not attempted to survey.

E. Polypodium Hydriforme

The remaining freshwater coelenterate that we will consider is the strange and poorly understood *Polypodium hydriforme,* originally described by (Lipin, 1911, 1926) and later examined in detail by Raikova (1973, 1980). This tiny *Hydrozoan* was first discovered in Eastern Europe as a parasite inside the eggs of the European sterlet (*Acipenser ruthenus*), a small member of the sturgeon family. It is particularly interesting as it is perhaps the only endoparasitic coelenterate.

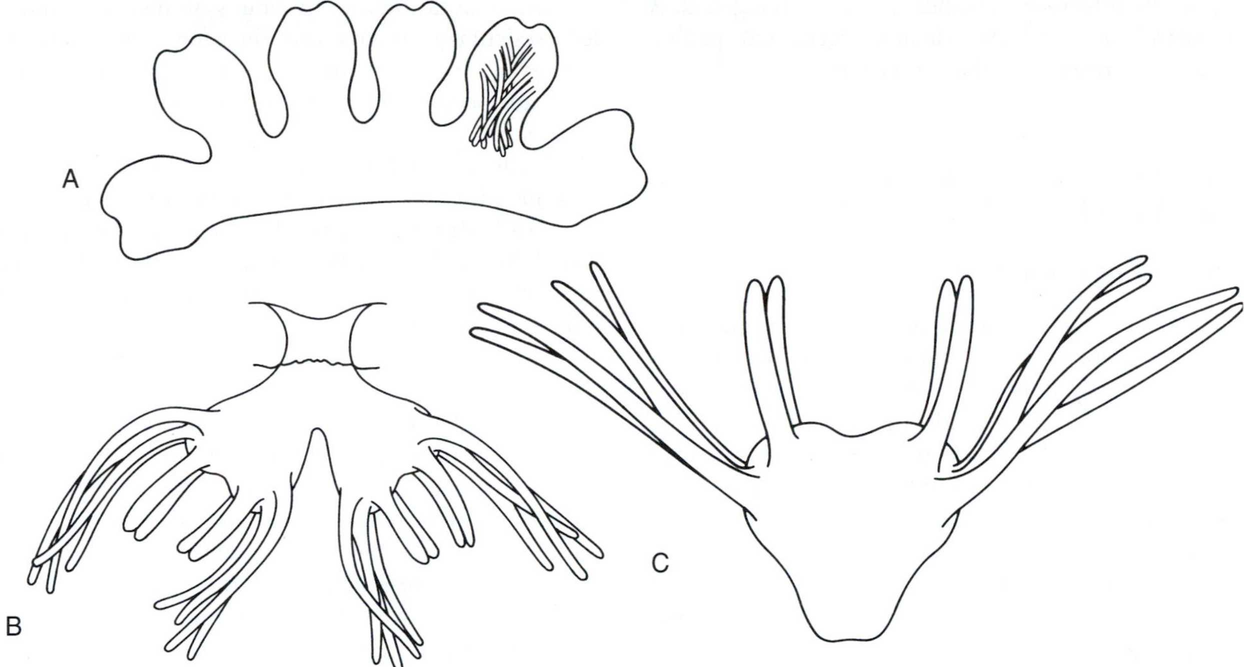

FIGURE 11 (A) *Polypodium* stolon with apex of knobs directed toward center of egg; (B) emerging polyp; and (C) mature polyp [redrawn from Lipin, 1911.]

Obviously there are modifications associated with its parasitic habit that make classification difficult, but it is reported to have only one type of nematocyst. These match the type found in the narcomedusae, an extremely toxic group of hydrozoans.

The development of the larval *Polypodium* (Fig. 11) is closely coordinated with that of the sturgeon egg. A binucleate, single cell stage is known from immature sterlet oocytes. The two nuclei are unequal in size and chromosome number. The smaller, reportedly haploid nucleus is surrounded by the large polyploid nucleus, which develops into a trophic envelope around the embryo formed by the division of the small nucleus and its surrounding cytoplasm. By the time the host oocyte has started to accumulate yolk, the *Polypodium* is a two-layered planula, approximately 1-mm long. It has a flagellated external layer, which eventually will become the endoderm and an internal layer of ultimate ectoderm, all surrounded by a capsule that serves as a digestive organ for consuming yolk (Raikova, 1980).

After a month, the planula has developed into a stolon with internally directed buds and tentacles. At this stage, the *Polypodium* is a colony consisting of a straight stolon from which project as many as a dozen knobs, arranged linearly with their apices toward the center of the egg (Fig. 11A). Each knob develops two indentations. From each indentation, 12 tentacles, four of which are short and stubby, project into the stolon itself (Fig. 11B).

As the fish eggs ripen and are released from the fish, the tentacles evert through a slit in the stolon. Simultaneously, the knobs invert, developing a coelenteron lined by what had been the surface exposed to the egg yolk. The stolon breaks up and the knobs now appear as somewhat bifurcated polyps with 12 tentacles on each head and a coelenteron full of fish egg yolk (Fig. 11C).

The free-living polyps subdivide by longitudinal fission. They crawl on the bottom using tentacles as walking legs, aided by nematocysts (isotrichous isorhiza) that hold the substrate. They feed on turbellaria and oligochaetes. Gonads form on the polyps: both single sex and hermaphroditic individuals are known. The genital anatomy is considerably more complex than that of hydra. These gonads and their accessory structures arise from endoderm. The presumed ovule is diploid and is released into the gastric cavity. The presumed male gonads become filled with binucleate cells. One nucleus remains haploid, while the other becomes polyploid. These may fall out of the coelenteron, but there are observations of polyps crawling onto young sterlets and placing these gonads on the fish. The transition from fish surface to immature oocyte has not been observed.

Polypodium parasitizes at least five species of *Acipenser* in all of the major rivers of the former Soviet Union. Eighty percent of sterlet (*Acipenser ruthenus*) and 20% of the sturgeon (*A. guldenstadti*) are infested, with sporadic infections in other species of the genus. A careful search of North American sturgeon will probably reveal the presence of this cnidarian.

IV. COLLECTION AND MAINTENANCE OF FRESHWATER CNIDARIA

A. Collecting Techniques

Hydras are so ubiquitous that their presence should be expected in any reasonably unpolluted body of water. Unfortunately, since they are usually sedentary, they are not easily found in plankton tows. Also, lacking hard body parts, they are often badly damaged by preservatives. A careful examination of suitable substrates is required. If rocks, leaves, submerged vegetation (e.g., *Myriophyllum, Elodea*) are collected and placed overnight in a glass or enamel pan, and the pan is carefully examined under a low-power dissection microscope, hydras are usually found. Abundance will depend on the seasonal distribution of zooplankton and may vary among lakes.

Since hydras float when hungry and attach again after they have been fed, the best place to took for them is often the downstream end of a lake or the pools in the stream immediately below the lake. If there is a dam, this should be examined with particular care. In a typical body of water, there may be at least three species of *Hydra*: a green one, a large brown one, and an intermediate-sized brown one.

Collecting of the medusae of *Craspedacusta* must be done with buckets rather than nets. The greater the sample volume the better (to prevent anoxia and excess temperature change). A compromise must be found between excessive agitation of the water, which will damage the animals, and insufficient stirring, which will permit them to settle or to become anoxic. The microhydra larvae can be found on fragmentary organic debris, much like hydra, but we have not collected them.

B. Maintenance Procedures

Craspedacusta and *Cordylophora* can be kept in natural water and fed on field-collected zooplankton. Even the free-living stage of *Polypodium* can be kept in the laboratory and fed on oligochaetes and turbellaria (Raikova, 1973). Dodson and Cooper (1983) have fed a range of foods to *Craspedacusta,* including large rotifers and copepods.

Generally, it is extremely difficult to maintain jellyfish in aquaria for any length of time. Their tissues are so fragile that they are battered to pieces by most aquarium aerating or stirring systems. The more or less sedentary polyps are therefore easier to study. However, the microhydra larvae of *Craspedacusta* have not been extensively studied, perhaps due to their small size.

Hydras are laboratory animals par excellence. General directions for most of the things one might want to do with hydra are discussed in detail in the papers collected by Lenhoff (1983). They can be maintained in artificial pond water or bottled spring water and fed on either live natural foods, such as cladocerans (e.g., *Daphnia*), copepods, or larvae of midges and mosquitoes. They will not eat protozoa.

Although the brine shrimp, *Artemia,* do not normally occur in freshwater, many pond animals eat them avidly in the laboratory. Many investigators use *Artemia* nauplii as food for hydra, *Craspedacusta,* planarians, and other small freshwater invertebrates. The *Artemia* eggs are collected at the edges of salt ponds and lakes. They are sold in vacuum-packed cans for use by aquarists and may be stored for long periods at cool temperatures.

The procedure for feeding hydra with *Artemia* consists of following the package directions for hatching the brine shrimp (a process requiring about 1 day), draining and rinsing the hatched nauplii free of the salt water, and adding them to the hydra. The nauplii must be alive and vigorous or the hydra will not eat them. Dead *Artemia* can be fed to flatworms! *Artemia* are best drained in a net that can be made by inserting a taut sheet of bolting silk in an embroidery hoop. The *Artemia* and their salt solution are poured into the net, which should be wet on both sides to speed drainage. The shimmering mass of *Artemia* and eggs are then rinsed into a new container of hydra culture water. If this is permitted to stand for 15 min, the active nauplii will swim toward a light, leaving the unhatched eggs behind. These active swimmers can be taken with a medicine dropper and fed to the hydra. This keeps salt and hydra separate and also permits elimination of most of the unhatched *Artemia* eggs.

These small prey may be eaten in great numbers. Large brown hydra can eat as many as 100 *Artemia* at a single meal, and even the smallest green hydra can consume 1 or 2. No more than 10 min after the first prey is swallowed, the swallowing process stops even if some prey remain on the tentacles. We have seen that organisms may brush against the tentacles with impunity for some time after a hydra has been satiated. This suggests either that nematocysts are no longer discharging or that the nematocyst supply has been temporarily exhausted.

The primary danger in this feeding process is that water contaminated with either dead *Artemia* or with the regurgitation products of previous feedings is highly detrimental. The live food must therefore be added to the hydra; and after approximately 1 h, the hydra must be placed in clean water, either by moving them or by discarding the tainted water. Since hydra usually stick to the substrate, the old medium can be simply poured off and a new medium added. The glass surface will eventually become coated with bacteria. This can be prevented somewhat by scraping the glass with a rubber spatula before discarding the old medium, but this is not usually effective for more than about 1 week, at which time it is best to provide new, clean containers. *Hydras* are remarkably sensitive to heavy metals and to detergents. Even a very short length of copper tubing in a water supply will kill hydra. Therefore, in cleaning the dishes, physical dirt is often less dangerous than the detergents. Dishes must be very thoroughly rinsed in metal-free water. To obtain nontoxic water, one may either use water from the collecting site, pretest the water with hydra to make sure that the water is innocuous, or use glass distilled or deionized water. When rinsing dishes containing hydra, appropriate salts must be added to the distilled water.

Various recipes for culture media are available. We use two stock solutions adapted by K. Dunn (personal communication) from solutions developed by Loomis, Lenhoff, and others. Solution A contains 81 g $NaHCO_3$ in 1 L of distilled water. Solution B contains 7.46 g KCl, 20.33 g $MgCl_2 \cdot 6H_2O$, and 147.02 g $CaCl_2 \cdot 2H_2O$ in 1 L of distilled water. These stock solutions are added at the rate of 1 mL/L to distilled water to make artificial pond water. Hydras die at temperatures above approximately 30°C, but most seem healthy at temperatures as low as 5°C. The growth and reproductive rates are proportional to temperature, as are the food demands. Stocks are therefore best maintained at low temperatures except when rapid growth is desired.

V. CLASSIFICATION OF FRESHWATER CNIDARIA

The five groups of freshwater Cnidaria are clearly distinct from each other. Except for the hydras, they are relatively rare and sporadic in their distribution. In North America there is no evidence, at present, for more than one species of *Craspedacusta, Polypodium,* or *Calposoma*. There may be several species of *Cordylophora*, but we are considering *Cordylophora* as one example of a brackish water cnidarian rather than a completely freshwater coelenterate. Our taxonomic key, therefore, consists of five very coarse divisions, one of which, that for the hydras, is then subdivided in somewhat greater detail.

A. Species Groups of Hydras

There are two closely related genera of hydras: the brown genus known as *Hydra* and the green hydras assigned to the genus *Chlorohydra*. By convention, the term hydra refers to members of both genera, unless otherwise specified. Hydras tend to look superficially similar, except for the distinction between the asymbiotic and green species. On more careful examination, differences become apparent, not only among apparently unrelated species but even within a clone. Asymbiotic animals from the same clone change color depending on the color of their food. All hydra will also change body size as a function of the amount of food and of temperature (Hecker and Slobodkin, 1976). Despite the within-clone plasticity, unrelated hydra maintained under identical conditions, side by side in the laboratory, retain consistent differences. Batha (1974) noticed that field-collected hydra differ in appearance from their own clonal descendants maintained for long periods in the laboratory. Even animals collected from the same pond at different seasons and kept in the laboratory sometimes show consistent differences in laboratory cultures, despite apparent similarities of the field-collected specimens. The effect of all this is to make taxonomy extremely difficult. Some investigators have tended to describe new species on the basis of rather unstable, variant appearances, while others have despaired of making any precise identifications.

Campbell (1987) has provided a compromise position, in which he has classified the hydra into clearly distinct "species groups," each composed of an uncertain number of more closely related species whose precise identity may have to await new techniques of examination. Color, presence or absence of a body stalk, nematocyst shape, and the order of appearance of tentacles on the new buds are the characters used. The structure of the theca around the fertilized eggs, the appearance of gonads, proportions of different kinds of nematocysts, microscopic details of symbiotic algae, and certain physiological characteristics are all likely to be important in subdividing the various species groups. Unfortunately, these subdivisions are difficult for all but the most serious students and professionals. Since the species group is adequate for most purposes, it seems advisable to quote extensively from Campbell's descriptions. The four species groups are given as follows.

1. The easily recognized group *Hydra viridissima* (also known as *Chlorohydra*). In addition to the brilliant green color, Campbell lists the following characteristics: "small to moderate body size and tentacles, stalk not distinct from the rest of the column; hermaphroditic;

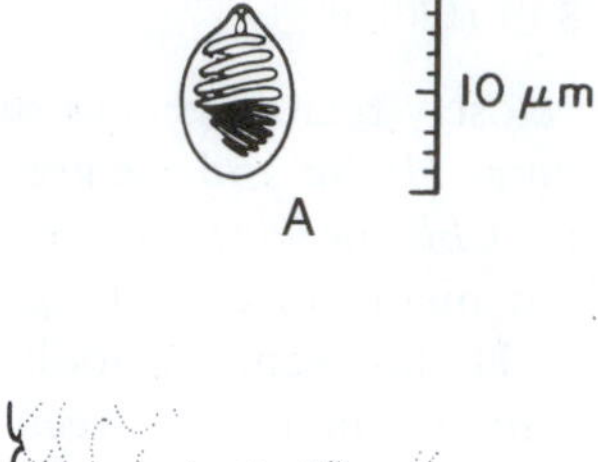

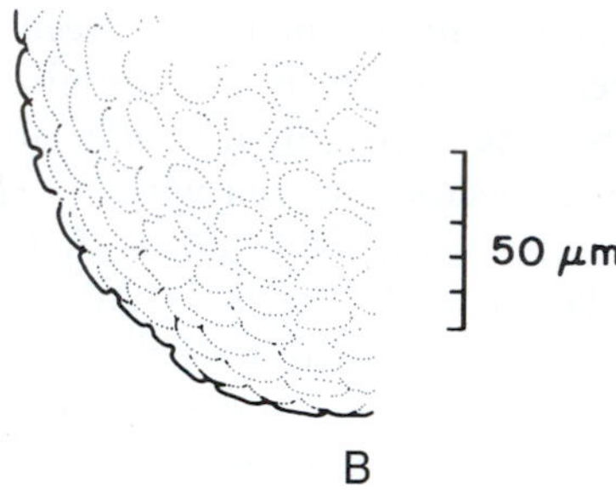

FIGURE 12 (A) Holotrichous isorhiza and (B) nonspiny embryotheca. [Redrawn from Campbell, 1987.]

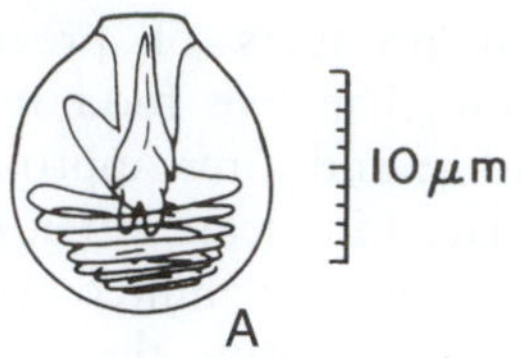

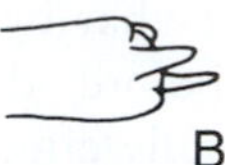

FIGURE 13 (A) Stenotele and (B) tentacle formation in *Hydra oligactis* group. [Redrawn from Campbell, 1987.]

embryotheca spherical, made of polygonal plates; nematocysts very small; tentacles arising simultaneously on buds."

2. The stalked *Hydra* or *Oligactis* group: "large size; pronounced translucent stalk in large individuals; long tentacles; dioecious; spherical embryo with simple theca; slender stenotele (capsules) and large, blunt, cylindrical holotrichous isorhiza (capsules); distinct golden color in culture due to yellow crystals in the ectoderm; two lateral tentacles arising before the others in the bud" (see Fig. 12).
3. The *vulgaris* group or common hydra "11 moderate size tentacles of moderate length; dioecious or monoecious, sometimes switching between the two conditions; spherical embryotheca ornamented with spines; broad stenotele (capsules); slender holotrichous isorhiza; often slipper-shaped, with the anterior end broadly pointed; tentacles arising on the buds simultaneously or nearly so" (see Fig. 13).
4. The "gracile *Hydra*" or *braueri* group: "small to medium size; tentacles moderately short; hermaphroditic; embryotheca and egg flattened, with the embryotheca adherent to the substrate; embryotheca smooth or papillate; tentacles arising simultaneously on the buds; holotrichous isorhiza (capsule) broader than half its length; and stenotele (capsule) plump; body often palecolored during laboratory cultivation."

No single character, not even the nematocysts, can by itself distinguish one group from another. Every group probably contains a large number of species, each of which has been described under a large number of names. Perhaps biochemical procedures may eventually sort them out, but for the moment and for most purposes, strains collected from the field should be classified to a species group and then supplied with enough ancillary description so that other workers can at least know if they have the same or a similar organism. Campbell (1987) provided a key to the four groups of *Hydra*, which we have incorporated in a key to the other freshwater Cnidaria.

B. Taxonomic Key to Genera of Freshwater Cnidaria

The polyp and medusa stages of a single species of coelenterate can differ greatly, both morphologically and ecologically. The key has been arranged to give the same results whether one starts with polyps or medusae. Also, the key refers to "species groups" of *Hydra* as names in parentheses.

1a.	Parasitic in fish eggs	*Polypodium*
1b.	Nonparasitic	2
2a(1b).	Medusae	*Craspedacusta*
2b.	Polyps	3
3a(2b).	Polyp without basal attachment, with oral surface downward	*Polypodium*
3b.	Polyps with basal disk	4
4a(3b).	Solitary polyps with single circle of multicellular tentacles; no medusa, gonads are body stalk	7

4b. Colonial polyps 5

5a(4b). Filiform tentacles, irregularly arranged on conical hypostome; branching colony with gonophores .... *Cordylophora*

5b. Colonial; atentacular or tentacles unicellular 6

6a(5b). Atentacular, oral capitulum; polyps from common stolon; may bud medusa from polyp; "microhydra" .... *Craspedacusta*

6b. Minute with unicellular tentacles .... *Calposoma*

7a(4a). Bright green .... *Hydra* (*viridis*)

7b. Not green 8

8a(7b). Holotrichous isorhiza nematocysts at least half as broad as long; embryotheca without spines .... *Hydra* (*braueri*)

8b. Holotrichous isorhiza nematocysts slender or if plump, then embryotheca spined 9

9a(8b). Buds acquire two lateral tentacles before others appear; otherwise stenotele nematocysts at least 1.5 times as long as broad .... *Hydra* (*oligactis*)

9b. Buds acquire tentacles in some other order; otherwise stenotele nematocysts less than 1.5 times as long as broad .... *Hydra* (*vulgaris*)

LITERATURE CITED AND FURTHER READINGS

Acker, T. S. 1976. *Craspedacusta sowerbii:* an analysis of an introduced species, *in:* Mackie, G. Ed., Coelenterate biology and behavior, Plenum, New York, pp. 219–226.

Aerne, B. 1995. Cnidarian hox genes. Developmental Biology 169:547–56.

Angradi, T. R. 1998. Observations of freshwater jellyfish, *Craspedacusta sowerbii* Lankester (Trachylina: Petasidae), in a West Virginia reservoir. Brimleyana 25:34–42.

Anonymous. 1999. Fresh water jellyfish in Rhode Island, URI Watershed Watch Home Page.

Barnes, R. D. 1966. Invertebrate Zoology, Saunders, Philadelphia, PA.

Batha, J. V. 1974. The distribution and ecology of the genus *Hydra* in the Milwaukee area of Lake Michigan. Ph. D. thesis, Zoology Department, University of Wisconsin, Milwaukee.

Bernfield, M. 1993. Molecular basis of morphogenesis. [Symposium with the Society for Developmental Biology] Wiley-Liss, New York.

Bossert, P. 1986. Regulation of intracellular algae by various strains of the symbiotic *Hydra viridissima.* J. Cell Sci. 85:187–195.

Bossert, P. 1988. The effect of *Hydra* strain size on growth of endosymbiotic alga. Ph.D. thesis, Department of Biology, State University of New York, Stony Brook, NY.

Bossert, P., Dunn, K. 1996. Regulation of intracellular algae by various strains of the symbiotic *Hydra viridissima.* Journal of Cell Science 85:187–195.

Bouillon, J. 1957. Etude monographique du genre *Limnocnida* (Limnomedusae). Annales de la Societe Royale Zoologique de Belgique 87:254–500.

Boulenger, C. L., Flower, W. U. 1928. The Regent's Park medusa, *Craspedacusta* sowerbii and its identity with C. (*Microhydra*) *ryderi.* Proceedings of the Zoological Society of London 66:1005–1015.

Bridge, D., Cunningham, C., Schierwater, B., DeSalle, R., Buss, L. 1992. Class level relationships in the phylum Cnidaria: Evidence from mitochondrial genome structure. Proceedings National Academy of Science USA 89:8750–8753.

Burnett, A. L. 1973. Biology of Hydra, Academic Press, New York.

Campbell, R. D. 1987. A new species of *Hydra* (Cnidaria: Hydrozoa) from North America with comments on species clusters within the genus. Zoological Journal of the Linnean Society 91:243–263.

Campbell, R. D. 1999. The *Hydra* of Madagascar (Cnidaria: Hydrozoa). Annales de Limnologie-International Journal of Limnology 35(2):95–104.

Ciernichiari, E., Muscatine, L., Smith, D. C. 1969. Maltose excretion by the symbiotic algae of *Hydra viridis.* Proceedings of the Royal Society of London Series B 173:557–576.

Colasanti, M. 1995. NO in hydra feeding response. Nature 374(6522):505.

Devries, D. R. 1992. The freshwater jellyfish *Craspedacusta sowerbyi:* a summary of its life history, ecology, and distribution. Journal of Freshwater Ecology 7(1):7–16.

Dodson, S. I., Cooper, S. D. 1983. Trophic relationships of the freshwater jellyfish *Craspedacusta* sowerbii Lankester 1880. Limnology and Oceanography 28:345–351.

Dunn, K. W. 1986. Adaptations to endosymbiosis in the green *Hydra, Hydra viridissima.* Ph. D. thesis, Department of Biology, State University of New York, Stony Brook.

Dunn, K. W. 1987. Growth of endosymbiotic algae in the green hydra, *Hydra viridissima.* Journal of Cell Science 88:571–578.

Dunn, K. W. 1988. The effect of host feeding on the contribution of endosymbiotic algae to the growth of green hydra. Biology Bulletin 175:193–201.

Elliott, J. K., Elliott, J. M., Leggett, W. C. 1997. Predation by hydra on larval fish: field and laboratory experiments with bluegill (*Lepomis macrochirus*). Limnology and Oceanography 42(6): 1416–1423.

Finnerty, J. R., Martindale, M. Q. 1999. Early evolution of Hox and ParaHox genes: evidence from the Cnidaria. Developmental Biology 210(1):41.

Fuhrman 1984. *Craspedacusta* sowerbii Lank. et un noveau coelentere d'eau douce, *Calpasoma* dacryloprera. n.g., n.sp. Revue Suisse de Zoologie 46:363–368.

Galliot, B., de Vargas, C., Miller, D. 1999. Evolution of homeobox genes: Q(50) Paired-like genes founded the Paired class. Development Genes and Evolution 209(3):186–197.

Griffing, T. C. 1965. Dynamics and energetics of populations of brown hydra. Ph.D. thesis, Zoology Department, University of Michigan, Ann Arbor, MI.

Grosvenor, W., Rhoads, D. E., Kassimon, G. 1996. Chemoreceptive control of feeding processes in hydra. Chemical Senses 21(3):313–321.

Hecker, B., Slobodkin, L. B. 1976. Responses of *Hydra oligactis* to temperature and feeding rate, *in:* Mackie, G. O. Ed., Coelenterate ecology and behavior, Plenum, New York, pp. 175–186.

Hershey, A., Dodson, S. 1987. Predator avoidance by *Cricotopus:* cyclomorphosis and the importance of being big and hairy. Ecology 68:913–920.

Hessinger, D. Lenhoff, H. M. Eds. 1988. The biology of nematocysts, Academic Press, London.

Hubschman, J. H., Kishler, W. J. 1972. *Craspedacusta sowerbii* Lankester 1880 and *Cordylophora lacustris* Allman 1871 in Western Lake Eric (Coelenterata). Ohio Journal of Science 72:318–332.

Hyman, L. H. 1940. The invertebrates: Protozoa through Ctenophora, McGraw-Hill, New York.

Johnson, E. M., Gorman, R. M., Gabel, B., George, M. E. 1982. The *Hydra attenuata* system for detection of teratogenic hazards. Teratogenesis, Carcinogenesis, and Mutagenesis 2:263–276.

Kanaev, I. I. 1969. Hydra: essays on the biology of freshwater polyps. Lenhoff, H. M. Ed. [Translated by Burrows, E. T. and Lenhoff, H. M.] Published by H. T. Burrows and H. M. Lenhoff, University of Miami, Coral Gables, FL. [Originally published by the Soviet Academy of Sciences, Moscow, 1952.] 453 pp.

Kouchinsky, A., Bengtson, S., Gershwin, L. A. 1999. Cnidarian-like embryos associated with the first shelly fossils in Siberia. Geology 27(7):609–612.

Kramp, P. L. 1950. Freshwater medusae in China. Proceedings of the Zoological Society of London 120: 165–184.

Lenhoff, H. M. 1974. On the mechanism of action and evolution of receptors associated with feeding and digestion, *in:* Muscatine, L. Lenhoff, H. M. Eds., Coelenterate biology: reviews and new perspectives, Academic Press, New York, pp. 211–214.

Lenhoff, H. M. 1983. Hydra: research methods, Plenum Press, New York, 463 pp.

Lenhoff, H. M., Loomis, W. F. 1961. The biology of *Hydra* and of some other coelenterates, University of Miami Press, Coral Gables, FL.

Lenhoff, S. G., Lenhoff, H. M. 1986. *Hydra* and the birth of experimental biology, 1744: Abraham Trembley's Memoires concerning the polyps, Boxwood Press, Pacific Grove, CA.

Link, J., Keen, R. 1995. Prey of deep-water hydra in Lake Superior. Journal of Great Lakes Research 21(3):319–323.

Lipin, A. 1911. Morphologie und biologie von *Polypodium.* Zoologische Jahrbuecher, Abteilung fuer Anatomie und Ontogenie der Tiere 31.

Lipin, A. 1926. Polypodium. Zoologische Jahrbuecher, Abteilung fuer Anatomic und Ontogenie der Tiere.

Lomnicki, A., Slobodkin, L. B. 1966. Floating in *Hydra* littoralis. Ecology 47:881–889.

Mackie, G. O. 1976. Coelenterate ecology and behavior, Plenum, New York.

Martinez, D. 1997. On senescence in asexual metazoans. Ph.D. thesis, Department of Ecology and Evolution, State University of New York at Stony Brook, Stony Brook, NY. 181 pp.

Muscatine, L. 1965. Symbiosis of *Hydra* and algae. III Extracellular products of the algae. Comparative Biochemistry and Physiology 16:177–192.

Naumov, D. V., Stepan'iants, S. D. 1980. Teoreticheskoe i prakticheskoe znachenie kishechnopolostnykh: sbornik. Leningrad, Zool. in-t.

Pardy, R. 1983. Preparing aposymbiotic *Hydra* and introducing symbiotic algae into aposymbiotic *Hydra, in:* Lenhoff, H. M. Ed., *Hydra:* Research methods, Plenum, New York, pp. 393–401.

Payne, F. 1924. A study of the freshwater medusae, *Craspedacusta ryderi.* Journal of Morphology 38:397–430.

Rahat, M., Campbell, R. 1974. Nematocyst migration in the polyp and 1 celled tentacles of the minute freshwater coelenterate *Calposoma* dactyoptgera. Transactions of the American Microscopical Society 93:379–385.

Rahat, M., Sugiyama, T. 1993. The endodermal cells of some brown *Hydra* are autonomous in their ability to host or not to host symbiotic algae: an analysis of Chimera. Endocytobiosis and Cell Research 9(2–3):223–231.

Raikova, E. 1973. Life cycle and systematic position of *Polypodium hydriforme* Ussov (Coelenterata, a cnidarian parasite of the eggs of Acipenseridae. Publications of the Seto Marine Biological Laboratory 20:165–174.

Raikova, E. 1980. Morphology, ultrastructure and development of the parasitic larva and its surrounding trophamnion of *Polypodium hydriforme* Coelenterata. Cell Research 206:487–500.

Ritte, U. Z. 1969. Floating and sexuality in laboratory populations of *Hydra littoralis.* Ph. D. thesis Zoology Department. University of Michigan, Ann Arbor, MI, 350 pp.

Roos, P. 1979. Two stage life cycle of a *Cordylophora* population in the Netherlands. Hydrobiologia 62:231–239.

Sabine, A., Hoffmeister, H., Chica, H., Schaller, H. C. 1987. Head activator and head inhibitor are signals for nerve cell and differentiation in hydra. Developmental Biology 122:72–77.

Schummer, M., Scheurlen, I., Schaller, C., Galliot, B. 1992. Hom Hox Homeobox genes are present in hydra (*Chlorohydra viridissima*) and are differentially expressed during regeneration. Embo Journal 11(5):1815–1823.

Schwartz, S. S., Hann, B. C., Herbert, D. N. 1983. The feeding ecology of *Hydra* and possible implications in the structuring of pond plankton communities. Biological Bulletin (Woods Hole, Mass) 164:136–142.

Slobodkin, L. 1964. Experimental populations of Hydrida. British Ecological Society Jubilee Symposium. Journal of Animal Ecology 33 (Supplement):131–148.

Slobodkin, L. 1980. Growth and regulation of animal populations, Dover, New York.

Slobodkin, L. Bossert, P., Mattessi, C., Gatto, M. 1991. A review of some physiological and evolutionary aspects of body size and bud size of *Hydra.* Hydrobiologia 216:377–382.

Spadinger, R., Maier, G. 1999. Prey selection and diel feeding of the freshwater jellyfish, *Craspedacusta sowerbyi.* Freshwater Biology 41(3):567–573.

Stiven, A. 1976. The quantitative analysis of the hydra-hydramoeba host–parasite system, *in:* Mackie, G.O. Ed., Coelenterate ecology and behavior, Plenum, New York, pp. 389–400.

Thiel, H. 1973. *Limnocnida indica* in Africa. Publications of Seto Marine Biological Laboratory 20:73–79.

Thorp, J. H., Barthalmus, G. T. 1975. Effects of crowding on growth rate and symbiosis in green hydra. Ecology 56:206–212.

Walsh, E. J. 1995. Habitat-specific predation susceptibilities of a littoral rotifer to two invertebrate predators. Hydrobiologia 313:205–211.

Warrior, R., Gall, J. 1985. The mitochondrial DNA of *Hydra attenuata* consists of two linear molecules. Archives des Sciences (Geneva) 38:439–445.

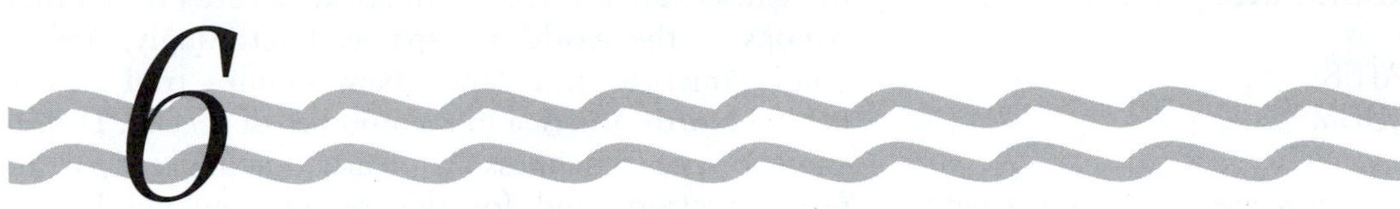

6 FLATWORMS: TURBELLARIA AND NEMERTEA

Jurek Kolasa

Department of Biology
McMaster University
Hamilton, Ontario L8S 4K1 Canada

Turbellaria and Nemertea are common and often very numerous inhabitants of freshwaters. Even though more than 200 species of Turbellaria and three species of Nemertea live in North America, their ecology and systematics have been less studied than that of many other common aquatic invertebrates. An obvious reason for this limited attention is the difficulty posed by preservation. Most turbellarians become unrecognizable after a routine preservation of field samples in alcohol or formalin. Study of live specimens is the best method and many interesting discoveries lie ahead in the areas of reproductive biology, dispersal, endosymbiosis, and community structure.

Ecology and Classification of North American Freshwater Invertebrates. 2nd Edition

TURBELLARIA

I. INTRODUCTION: STATUS IN THE ANIMAL KINGDOM

The turbellarian flatworms are the lowest acoelomate Bilateria with only a single opening to the digestive tract; that is, they lack a definitive anus. Cestoidea (tapeworms), Trematoda (flukes), and Turbellaria constitute the phylum Platyhelminthes. Because of tradition and practical considerations, turbellarians are divided into microturbellarians and macroturbellarians. This division does not reflect phylogenetic relationships, but rather superficial morphological similarities. Also, depending on the methodology chosen, the taxa discussed may have a rank different from the one used here (see Ehlers, 1986). Several orders and suborders of Turbellaria are known from freshwaters (Table I, Fig. 1).

One order, Acoela, has been recently excluded from Turbellaria on the basis of DNA sequence analysis (Ruiz-Trillo *et al.*, 1999) but we retained it in this chapter for practical reasons.

Of the approximately 400 species of freshwater microturbellaria, about 350 have been recorded in Europe (Lanfranchi and Papi, 1978), and approximately 150 occur in North America (Appendix 6.1), many of which are the same as those found in Europe. Many species and genera of microturbellarians remain to be discovered in North America. Records from other regions of the world are sparse. Interestingly, Africa, South America, and Papua New Guinea trail closely behind North America in the number of species. Differences in species richness are clearly due in great part to research effort; and for this reason, the number of species in North America must be highly underestimated. Triclads (often called planarians) are more thoroughly studied. The number of recorded species in North America is approximately 40, but the actual number of taxa is likely to be substantially higher.

II. ANATOMY AND PHYSIOLOGY

A. General External and Internal Anatomical Features

Most freshwater microturbellaria are less than 1 mm in length, although some can reach several millimeters. Triclads are distinctly larger, with most species exceeding 10 mm and the largest being several centimeters long.

The turbellarian body is elongated, relatively soft, and usually tapered at the ends. Sometimes lateral flaps near the cerebral region or a short tail like section are present. With the exception of triclads, flatworms are generally not flat—despite their common name. Most

TABLE I Selected Characteristics of Commonly Recognized Turbellarian Orders and Suborders Occurring in Freshwaters

Taxon[a] *order suborder superfamily*	*Approx. number of species*	*Average body size (mm)*	*Comments and special features*
Microturbellarians			
Catenulida	60	0.5–1	Thin chains of zooids; common in various habitats
Acoela	3	0.5–1	Most species marine
Macrostomida	50	1.0–3	Common in various habitats
Prolecithophora	5	5.0–10	Many marine species
Lecithoepitheliata	10	3.0–10	Many marine species; some are terrestrial or semiaquatic
Proseriata	4	2.0–5	Many marine species
Rhabdocoela			
Dalyellioida	100	0.8–1	Common in various habitats
Temnocephalida		1.0–14	Commensals on crust aceans, snails, turtles, one parasitie
Typhloplanoida			
Typhioplanida	150	0.5–6	Common in various habitats including terrestrial and semiaquatic
Kalyptorhynchia	15	1.0–2	Rare; most species marine
Macroturbellarians			
Tricladida	100	5–20	Greatest diversity associated with karst habitats

[a] Taxonomical rank is indicated by indentation.

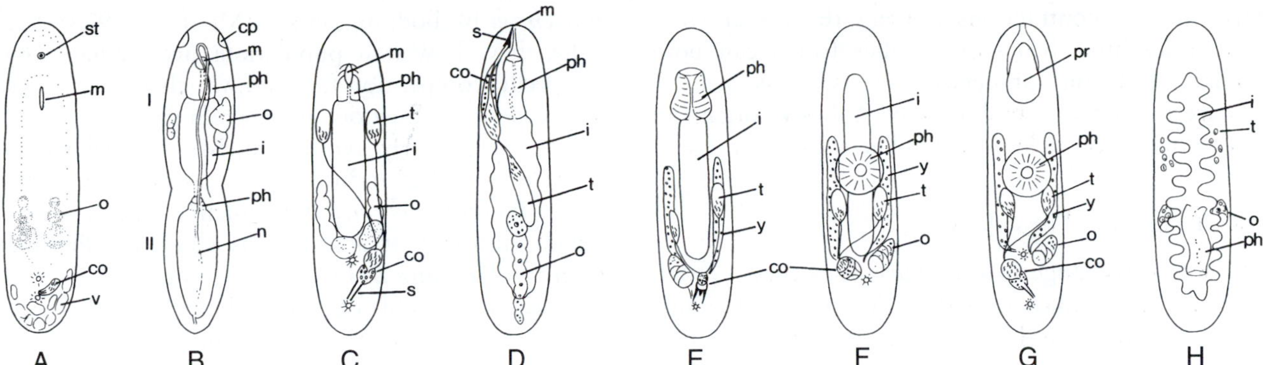

FIGURE 1 Schematic representation of higher turbellarian taxa found in freshwaters with an emphasis on the most useful diagnostic features. (A) Acoela (one unidentified species); (B) Catenulida (e.g., *Catenula, Stenostomum*, I and II represent two successive zooids); (C) Macrostomida (*Macrostomum, Microstomum*); (D) Lecithoepitheliata (e.g., *Prorhynchus, Geocentrophora*); (E) Dalyellioida (*Gieysztoria, Microdalyellia*); (F) Typhloplanoida (e.g., *Mesostoma, Olisthanella, Castrada*); (G) Kalyptophynchia (e.g., *Gyratrix*; intestine omitted for clarity); and (H) Proseriata and Tricladida (e.g., *Bothrioplana, Dendrocoelopsis*). Co, Copulatory organ; cp, ciliated pits; i, intestine, m, mouth; n, protonephridial duct; o, ovary; ph, pharynx; pr, proboscis; s, stylet, st, statocyst; t, testes; v, vacuolized tissue; and y, yolk glands.

microturbellaria are circular in cross section, with some differentiation of shape of the dorsal and ventral surfaces. Moreover, species that reproduce asexually may be composed of several zooids (Figs. 3B, 4A,B, 12A), giving a chainlike appearance to an animal. Turbellarians may be colorless, white, red, bluish, green, black, brown, or yellowish depending on epidermal and parenchymal pigments, gut content, and symbiotic algae. Some species, such as *Dugesia tigrina* and *Hydrolimax grisea*, develop characteristic patchy patterns of pigments.

Anatomically, the most prominent turbellarian features are a ciliated epidermis, rhabdoids, an intestine without anus, ventral mouth, and a complex reproductive system (Figs. 1, 9, 12D). The ciliated epidermis covers the entire body surface. Rhabdoids are rodlike, light-refracting structures produced in the epidermis (dermal rhabdoids) or in the parenchyma (adenal rhabdoids). It is often difficult to determine the type of rhabdoids in live specimens. Adenal rhabdoids are most abundant at or near the front extremity. They often aggregate in juxtaposed groups that are known in the taxonomic literature as rod tracts. Wrona (1986) reviewed hypotheses on the role of rhabdoids in triclads and has provided experimental evidence that rhabdoids contribute to mucus production. Rhabdoids are also poisonous and used in prey immobilization or in deterring predators. Other components of the body walls include a basal membrane and muscle fibers. The fibers are longitudinal, circular, or in larger forms, diagonal. Some fibers go through the parenchyma. The parenchyma is a loose tissue filling the body cavity and containing various cell types, including secretory, excretory, reproductive, and formative cells and their insoluble excretions, that fulfill important physiological functions (Hyman, 1951). In large turbellarians, the parenchyma is usually more compact than in smaller forms, where it can be quite loose and vacuolated. Turbellarians possess numerous glands, of which the frontal, pharyngeal, shell-producing, and rhabdoid-producing glands are most common.

A vast array of sensory organs enables turbellarians to react to environmental stimuli. These organs include sensory hairs, eyes, statocysts, and chemoreceptors of various kinds. Some species possess aggregations of sensory cells, or ciliated pits, in lateral depressions situated in front of the brain ganglion. Exact functions of the ciliated pits and structures associated with them are unknown, but at least some of these structures (i.e., the refractive bodies of *Stenostomum*) have been implicated in phototaxis (Marcus, 1951). Some evidence exists that these refractive bodies function as lenses (Tyler and Burt, 1988). Thigmotaxis is common and plays a role in a range of behaviors such as choosing substrate, hunting prey, and avoiding predators. Among North American freshwater Turbellaria, a statocyst is present only in *Catenula, Rhynchoscolex, Otomesostoma*, and an unidentified representative of Acoela (Strayer, 1985; Kolasa *et al.*, 1987). Turbellarians have two types of simple eyes: inverted and direct cup eyes (Bedini *et al.*, 1973). The number of eyes is variable, but one pair of eyes is most frequently present.

Locomotion in the Turbellaria is based on the movement of cilia. Some triclads, however, are capable of fast muscular contractions allowing them to crawl in bursts of leechlike behavior. Smaller forms can both swim and glide on submersed objects, while heavier forms are restricted to gliding only. Adhesive organs in the form of pseudosuckers or papillae are often present.

B. Environmental Physiology

Gaseous exchange in turbellarians occurs through their body walls. Such exchange depends on the surface/volume ratio and is not very efficient. Consequently, flatworms may have a limited ability to adapt to low-oxygen conditions. This conclusion appears to be true for most flatworms and this limitation is often exploited in extraction of specimens from samples (e.g., Young, 1970; Schwank, 1981). However, the ability to cope with reduced oxygen levels does not seem to differ substantially from that of other aquatic invertebrates. Oxygen adaptability differs more among species adapted to diverse habitats than among major taxa (cf. Heitkamp, 1979b). For example, species of *Macrostomum* normally found in stream headwaters are more sensitive to low oxygen levels, and perhaps high temperature, than are representatives of the same genus from physically more fluctuating habitats (J. Kolasa, personal observation). Some species, particularly in the genus *Phaenocora*, have developed adaptations to cope with anaerobic conditions (Young and Eaton, 1975). They showed that the presence of algae in *Phaenocora typhlops* enhances its ability to survive low-oxygen conditions. This same species, as well as a related *P. unipunctata* (Öersted) produces hemoglobin, allowing them to store oxygen during burrowing in the anoxic mud. Many marine species, however, live in low-oxygen sediments, but have neither hemoglobin nor symbiotic algae.

Heitkamp (1979a, b) studied respiration rates of *Opistomum pallidum*, two species of *Mesostoma*, and a species hosting symbiotic zoochlorellae—*Dalyellia viridis*. He showed the respiration to be related to the body size and shape (surface in a very flat *Mesostoma ehrenbergi*, or weight in cylindrical *Mesostoma lingua* and other species). Interestingly, *Dalyellia viridis* that are symbiotic with zoochlorellae had the highest respiration rate of all studied species; possibly their symbiotic algae contribute significantly to the observed rate. Heitkamp (1979b) found the respiration rate of microturbellarians to vary depending on the temperature and diurnal rhythm.

A considerable effort has been devoted to studying nutrition and digestion in Turbellaria, particularly by Jennings (1977). In this context, the lipid storage capacity of flatworms and its importance to population dynamics and competition is discussed for *Dugesia polychroa* by Boddington and Mettrick (1977). Specifically, they show that populations are regulated by intraspecific competition for food, which affects individual size, cocoon production, and cocoon viability.

Triclads clearly avoid light. Microturbellarians appear to be either indifferent or positively phototactic (e.g., *Typhloplana viridata* which contains symbiotic algae), although if a sample is suddenly illuminated, many will attempt to hide in sediments.

Desiccation of eggs can be tolerated by some species and it may be an obligatory factor in releasing them from diapause and stimulating embryonic development. Similarly, some triclads such as *Phagocata* may undergo encystment as entire or fragmented animals. Temperature may also have an important effect on reproduction. Dumont and Schorreels (1990) showed that clutch size, total offspring, and daily offspring all depended on temperature rather than food in *Mesostoma lingua*.

Although the majority of Turbellaria appear to have some habitat preferences, within a limited range of physical conditions many species have adaptations conferring considerable flexibility. Some members of the Catenulida are found in habitats as different as fast-flowing streams and temporary pools (e.g., *Stenostomum leucops, Catenula lemnae*). *Mesostoma* species have been reported from both warm rice fields in the Sacramento Valley in California as well as from snowmelt pools at elevations above 2400 m (Collins and Washino, 1979).

Even though microturbellarians are delicate and fragile organisms, they appear to have a significant potential for dispersal. Results of an experiment conducted by Jan Ciborowski (University of Windsor) were surprising. Ciborowski studied colonization of artifical ponds (flexible swimming pools) situated on the roof of a university building for several weeks. About 50% of these ponds were successfully colonized by six different species of microturbellaria (unpublished data). With only one exception, these species are not known to produce resistant disseminules such as eggs or cocoons.

III. ECOLOGY

A. Life History

In most species, miniature replicas of the adult hatch directly from eggs; these juveniles differ from adults chiefly by the absence of reproductive systems. Some freshwater turbellarians, however, may be ovoviviparous (*Mesostoma*) or may have a larval stage distinctly different from the adult (*Rhynchoscolex*). Numerous modifications of the life cycle have evolved among turbellarians. Many seasonally occurring species are univoltine, particularly those associated with temporary habitats or at the extremes of their geographical ranges. Other species are multivoltine, with the number of generations depending on habitat availability (see Heitkamp, 1982, for an excellent account). A similar diversity of life cycles is observed in triclads, but the ecological reasons for this diversity remain hidden. Reproductive patterns may sometimes coincide with major taxonomic subdivisions. Calow and Read (1986) studied European representatives of the three families that are also represented in North America: Dugesiidae, Planariidae, and Dendrocoelidae. These investigators have found that species breeding over several seasons (Dugesiidae, Planariidae) invest less in reproduction than members of Dendroceolidae, which reproduce once and die.

As a rule, turbellarians are hermaphroditic. Various modifications of both egg formation and development are found and may include self-fertilization, as in *Mesostoma ehrenbergi* (Fiore and Ioalé, 1973; Göltenboth and Heitkamp, 1977), protandry or progyny in many other species, or loss of either the male or female gonad as in Catenulida (Rieger, 1986). All of these modifications have implications for habitat adaptability, clonal diversity, rate of population growth, and ability to colonize.

Asexual reproduction by means of paratomy, that is transverse division of the body, is common in several genera of microturbellaria, particularly in Catenulida and Macrostomida. In macroturbellarians, architomy (transversal fission of the caudal region) occurs frequently in Dugesiidae and in some Planariidae (Calow and Read, 1986; Rieger, 1986). In some species of *Stenostomum*, and in some populations of the common triclad *Dugesia tigrina*, sexual organs have never been observed.

B. Distribution

Freshwater turbellarians are largely free-living animals, although a few European freshwater species such as *Varsoviella kozminski* (Gieysztor and Wiszniewski) and *Phaenocora beauchampi* (Sekera) are ectoparasitic on crustaceans. Several other freshwater forms, including triclads, occurring in Europe and Australia are commensal on crustaceans and turtles (e.g., Jennings, 1985). The great majority of freshwater turbellarians are free-living and live in various aquatic systems such as ponds, lakes, streams, hyporheic water, ditches, and temporary puddles. However, some may be found in aquatic habitats normally excluded from the domain of freshwater ecology, such as water films among fallen leaves in a mesic forest or in capillary soil water of a grassy meadow (Sayre and Powers, 1966).

In North America, microturbellarians are known as far north as Igloolik, a small island at 68°N latitude off Melville Peninsula. Two species, *Mesostoma* and an unidentified mesostomid, were collected by P.D.N. Hebert and associates (unpublished). In Alaska and in the area of Hudson Bay in northern Canada, as many as six different species are known (Holmquist, 1967; Schwartz and Hebert, 1986). Further south, in temperate climatic zones, the species diversity increases and can easily reach 20 to perhaps 60 species in a single lake or pond. For example, there are 23 species of flatworms in the oligotrophic Mirror Lakes, New Hampshire (Strayer, 1985) and as many as 57 species occur in the polytrophic Lake Zbechy in Poland (Kolasa, 1979). Streams also have rich flatworm fauna. In a short section of Wappinger Creek, an eastern tributary of the Hudson River in New York, 15 species were found (Kolasa *et al.*, 1987). A survey of several streams, including springs and underground water in the same area revealed 32 species of microturbellarians, some of which are still undescribed. Because of a dearth of systematic studies, few microturbellarians are known from the West Coast. Case and Washino (1979), Collins and Washino (1979), and other related papers report *Mesotoma lingua* from California, together with some other common species in rice fields.

Many species, particularly among the Catenulida and the Macrostomida, are widely distributed worldwide, while species of the "higher" Turbellaria, such as the Typhloplanida and Tricladida, are geographically more restricted. About 3% of all species known from Europe are designated as cosmopolitan in distribution (Lanfranchi and Papi, 1978), but studies in the tropics and Southern Hemisphere (Mead and Kolasa, 1984) suggest that this figure is too low.

Caves and underground waters hold many unique and endemic species, particularly of triclads (e.g., Kenk, 1987). Some troglobiotic triclads and microturbellarians have recent or remote marine origins (e.g., Kawakatsu and Mitchell, 1984a; Kolasa, 1977b). An extensive study of European underground triclads (Gourbault, 1972) revealed a suit of distinct distributional, reproductive, and physiological characteristics. These often include specialized underground microhabitat selection, greater frequency of sexual reproduction, and slower metabolism as compared to epigean (or surface-dwelling) triclads.

In contrast, a number of interstitial and underground water microturbellarians, for example *Bothrioplana semperi, Limnoruanis romanae*, and *Stenostomum pegephilum*, occur on more than one continent. Distributions of most genera appear to be worldwide among microturbellarians, but not triclads, many genera of which are known exclusively from either North

America, Eurasia, South America, or Australia (Ball, 1974).

The majority of triclads from North America are described from the eastern and southern United States and Canada (Kenk, 1972). Four triclad species inhabit Alaskan waters and 10 live in Tennessee (Chandler and Darlington, 1986). Several species of triclads with recent marine origins occur in freshwater caves of Mexico (Kawakatsu and Mitchell, 1984b; Benazzi and Giannini, 1971).

Boreo-alpine distribution of many cold-water species has been tied to glaciations in Europe and similar patterns also seem to occur in North America (Hampton, 1988). At lower latitudes, similarities between South American and North American microturbellian composition (cf. Marcus, 1945) suggest considerable interchanges of species.

The ecological distribution of both microturbellarians and triclads has been studied more intensively in Europe, but most of the results are directly relevant to North American fauna. The distribution of microturbellarians in lakes has been investigated by Strayer (1985), Chodorowski (1959), Kolasa (1977a, 1979), Rixen (1968), Schwank (1976), and Young (1970, 1973a). These studies show that microturbellarians differentiate among substrate types and depth zones. The flatworm assemblages of sand and gravel bottoms are distinctly different from assemblages in habitats characterized by bottoms of silt, coarse organic matter, or vegetation. Plant successional zones also differ in their species assemblages. Although the littoral zone has the greatest density and diversity of flatworms, there is apparently no limit to how deep some species may be found as long as oxygen is available. For example, *Otomesostoma auditivum* lives at a depth of 15–45 m in Echo Lake in the Sierra Nevada range, but it occurs at a depth of 145 m in some European alpine lakes (Hyman, 1955).

Ecological differentiation of microturbellarians in running waters is more pronounced than in lakes. European studies showed distinct assemblages of species associated with springs, headwaters, and lower stretches of streams and rivers (epirhithron, metarhithron, hyporhithron, and potamon) (Schwank, 1981; Kolasa, 1983). These regularities in zonation of flatworm assemblages are consistent with patterns observed among other lotic taxa. These patterns, however, still need to be related to newer ecological approaches to running waters, i.e., to the river continuum concept (Vannote *et al.*, 1980). Among those classical zones, epirhithron shows the greatest species richness. No comparable studies have been published on North American microturbellarians. Zonation of triclads in streams was studied in a greater detail, including experimentation (Reynoldson, 1982). Competition, as well as thermal requirements were implicated as important factors separating several species of flatworms along a water course and in standing waters (Claussen and Walters, 1982; Pattee, 1980; Reynoldson, 1982).

Both microturbellarians and triclads live in habitats where the physical conditions require additional adaptations in comparison to an average pond or stream environment. Wet mosses and leaves on the water edge, water containers at the base of bromeliad leaves, the high Arctic, ice melt ponds at elevations over 4000 m in Central Asia, and hot springs are habitats for one or more species of flatworms. Representatives of freshwater families are also commonly found in humid terrestrial habitats.

C. Behavioral Ecology

Laboratory studies reveal that turbellarians have a limited ability to learn simple tasks such as choosing white over dark branches of a maze (McConnell, 1967). Observations of prey handling (personal observation), however, indicate that their learning ability may be much greater for tasks that might be advantageous under natural conditions, such as prey or predator recognition.

Possible behavior modification or behavior determination associated with symbiosis is a potentially rewarding research area. *Typhloplana viridata*, a common spring pond species, is unique in being positively phototactic. This behavior may be associated with the presence of symbiotic algae. It might be of general interest to know control mechanisms of this behavior.

D. Foraging Relationships, Predators, and Parasites

Microturbellarians eat bacteria, algae, protozoa, and small invertebrates, while triclads feed predominantly on larger invertebrates. Scavenging is common in both morphological groups. Various *Mesostoma* species have been studied in detail and illustrate the diversity of feeding habits and techniques within one genus. Leaf-shaped *M. ehrenbergi* usually suspend themselves in the water column on a mucous filament attached to the surface film (Göltenboth and Heitkamp, 1977). In that position, the worm waits for a cladoceran or an insect larva to come into contact. Using its front end, *M. ehrenbergi* traps the prey with sticky mucus, wraps around it, and proceeds to feed. According to Dumont and Carels (1987), *M. lingua* uses neurotoxins to immobilize its prey in addition to mucus. *Mesostoma* can also glide on the surface of underwater objects or swim directly in the water column. Both activities enhance its chances of encountering prey. Unlike *M. ehrenbergi*, many other species are spindle-shaped (e.g., *M. arctica*). These species hunt more often on the bottom or among submersed objects. Although several *Mesostoma* studied spend more than 50% of their time resting on various objects, they can catch prey as soon as it comes close enough to create vibrations in the water (J. Kolasa, personal observation). At that point, *Mesostoma* can erect the front portion of its body and actively move it around in search for contact with the prey. Several species of *Mesostoma*, such as *M. vernale* (Hyman) and *M. californicum* (Hyman), as well as species of *Bothromesostoma* have a flat ventral surface with which they cling to the surface film. These species are more agile and much faster than other species of *Mesostoma* and they prefer hunting at the water surface.

Immobilizing the prey and sucking its body fluids is a common feeding behavior in microturbellarians equipped with a strong pharynx rosulatus, or in triclads with a pharynx plicatus. Many other turbellarians, however, swallow the whole victim or bite chunks of prey. For example, species of *Stenostomum* and *Macrostomum* that feed on algae, protozoa, rotifers, and other meiofauna usually swallow the prey whole. Other, such as *Stenostomum predatorium* and various *Phaenocora* species may attack and partly consume larger turbellarians (Kepner and Carter, 1931), oligochaetes, or other soft invertebrates. Finally, a few microturbellarians (e.g., *Gyratrix hermaphroditus* and *Prorhynchus stagnalis*) use their copulatory stylets to stab prey.

Chemical detection of prey undoubtedly plays an important role in feeding. Injured invertebrates attract species of both triclads and microturbellarians under laboratory conditions. The presence of injured or recently killed organisms allows scavenging by many individuals that played no role in immobilizing the prey.

Turbellarians themselves are subject to predation and parasitism. Ciliates and flagellates are frequent parasites of Catenulida and Typhloplanida, and nematodes have been found in Lecithoepitheliata (*Prorhynchus*; J. Kolasa, unpublished). Fish occasionally eat triclads (Davies and Reynoldson, 1971). Microturbellarians (*Phaenocora typhlops*) may occasionally be eaten by invertebrates, such as a chironomid *Anatopynia* (Young, 1973b) or other Turbellaria (e.g., *Stenostomum predatorium*).

E. Population Regulation and Density

It is not clear whether there is any single mode of population regulation that applies to flatworms. Asexual reproduction of *Dugesia tahitiensis* was found to be density-dependent under laboratory conditions (Peter, 1995). The few studies conducted in Britain indicate that intraspecific competition for food may be very important for *Phaenocora typhlops* (Young, 1975), while interspecific competition may play a role in regulating several planarians (Reynoldson, 1982). In European *Polycelis tenuis* and, perhaps, in American *Cura foremanii* (Girard), this regulation is expressed by a negative relationship between the worm body size and density. In fact, long-term changes in the population size of *Polycelis* suggest that limit cycles or cyclical fluctuations of the density due to intraspecific competition are involved (Reynoldson, 1982, and references therein).

Densities of microturbellaria vary according to the habitat, season, and species. No major generalizations have been made, but some patterns are beginning to emerge, particularly along seasonal and successional axes. In small, second- or third-order streams, the greatest densities of microturbellaria are found in the lower reaches, i.e., in the metarhithron and hyporhithron, while the highest richness is found in the epirhithron zone (Kolasa, 1983). For example, a mean density of 1280 individuals/m^2 (of all species) was recorded in Wappinger Creek, New York, in an artifical substrate after a week of colonization.

Schwank (1981), using volume instead of surface to quantify turbellarian abundance in submontane streams in Germany, found densities of about 80 microturbellarians/L. If one assumes an average sample depth into the substrate of 5 cm, this translates into over 4000 individuals/m^2. His data suggest a corresponding density of triclads of 32 individuals/L and 1600 individuals/m^2. In oligotrophic Mirror Lake, turbellarian densities varied from 40,000 individuals/m^2 in shallow littoral regions to almost zero in profundal zones; the lakewide mean was 27,000 individuals/m^2 (Strayer, 1985). Other studies reported densities of 800 individuals/m^2 of Lake Michigan (Nalepa and Quigley, 1983), 3500 individuals/m^2 (annual mean) in shallow littoral areas of Zbechy Lake in Poland (Kolasa, 1979), 3100 individuals/m^2 in Lake Paajarvi in Finland (Holopainen and Paasivirta, 1977), to 9500 individuals/m^2 in ponds in Germany (Heitkamp, 1982). These values have to be interpreted cautiously, however, in the light of seasonal succession and abundance changes, which ordinarily produce the greatest richness and abundance in early summer (Bauchhenss, 1971; Strayer, 1985; Chodorowski, 1960).

F. Functional Role in the Ecosystem

The functional role of species is typically discussed in terms of their trophic interactions. All triclads are predatory. Several invertebrates and vertebrates may consume triclads and are a significant source of their mortality. However, triclads constitute a relatively minor component of the diet of their predators (Davies and Reynoldson, 1971). Triclads themselves consume isopods, midges, oligochaetes, caddisflies, mayflies, ostracodes, and cladocerans. The ability to exploit various resources differs among species and may result in strong competition with a simultaneous expansion of invading species (e.g., Gee and Young, 1993).

Heitkamp (1982) reports intense predation on *Mesostoma* and *Rhynchomesostoma* by *Dalyellia* in temporary pools. He has not determined the frequency of such predation nor established whether this predation results in population regulation.

High densities of both triclads and microturbellarians suggest that their role in biotic interactions of benthic communities may be greater than their contribution to the diet of other organisms. Densities of *Mesostoma* species (*lingua*) in excess of 1000 individuals/m^2 were observed in more than 30% of the rice fields studied by Collins and Washino (1979) in California. In some cases, microturbellaria may regulate population dynamics of zooplankton in ponds, as demonstrated by Maly *et al.* (1980). They found that the feeding rates of *Mesostoma ehrenbergi* were two zooplankton per day (estimated from Fig. 1). However, Schwartz and Hebert (1982) reported that this species can consume about ten cladocerans per day. Blaustein (1990) and Blaustein and Dumont (1990) found that *Mesostoma lingua* might have major numerical impact on mosquito larvae, cladocerans, and ostracodes in rice fields, but little effect on copepods. More important, perhaps, is the functional role of microturbellaria as consumers of protozoa, rotifers, and algae, especially by the usually abundant *Stenostomum leucops* and similar forms. Unfortunately, no quantitative data on this subject are available.

Although there are no explict studies of the energy flow through lake meiobenthos, the detailed study of the energy budget of Mirror Lake (Strayer and Likens, 1986) indicated this flow to be equivalent to that to zooplankton. Furthermore, Strayer (1985) showed that Turbellaria, and *Rhynchoscolex* in particular, were a substantial component of meiobenthos in Mirror Lake. This statement might be somewhat misleading because meiobenthos itself is a minor component of the whole budget.

IV. CURRENT AND FUTURE RESEARCH PROBLEMS

A. Dispersal

Understanding the geographical and ecological distribution of Turbellaria is important because their distribution has evolutionary and ecological implications. Yet this understanding will suffer seriously unless the dispersal capabilities of flatworms are assessed in terms of distance and rates. Hebert and Payne (1985) found low levels of gene flow in Arctic *Mesostoma*. Their calculated dispersal rates are inconsistent with both the unpublished data from Ciborowski's experiment mentioned earlier and with common observations that even newly created water bodies have a rich complement of species. A series of questions may be posed. Is dispersal different at various latitudes? Do dispersal strategies and efficiency differ along major taxonomic lines? Are habitat generalists better dispersers than specialists? Are species associated with a particular habitat type (e.g., temporary ponds) better dispersers as opposed to cave dwellers? If yes, what in their life cycle makes them so? These questions point out the limits in factual knowledge and ecological interpretation of the many aspects of turbellarian dispersal.

B. Rhabdoids

Wrona's (1986) study demonstrated a lack of support for hypotheses that rhabdoids play a major role in contacts with prey or predators in triclads. In many microturbellarians however, rhabdoids, and adenal products in particular are channeled toward the anterior of the body. The front end usually plays a crucial role in exploration of the environment and in initiation of attack against prey. This special position is highly suggestive of more active applications of rhabdoids than the mere production of mucus. Therefore, the role of rhabdoids warrants further research.

C. Endosymbiosis

Green algae occur in *Phaenocora, Dalyellia, Typhloplana, Castrada*, and some other microturbellarians. The nature of the association between the algae and turbellarians is not fully understood. However, it could be used as a convenient general model for laboratory manipulation.

D. Community Ecology

Comparative research on the density and richness of turbellarian assemblages in various water types might suggest which factors have relatively greater influence in structuring flatworm communities. Quantification and standardization of sampling procedures may be a major hurdle for comparative ecologists. Different density estimates obtained in quantification of lake Turbellaria may be due to minor differences in techniques (e.g., Strayer, 1985).

V. COLLECTING, REARING, AND IDENTIFICATION TECHNIQUES

A variety of techniques can be used to obtain qualitative and quantitative samples of Turbellaria. These techniques may involve: scooping with a plankton net over aquatic vegetation or sediments; collecting bottom sediments by means of bottom samplers, such as Ekman, Ponar, or multiple corer (Schwank, 1981; Strayer, 1985); filtering water taken from wells, springs, hyporheic interstitial, or ponds or puddles using a plankton net (Kolasa *et al.*, 1987); or band baiting (e.g., *Mesostoma*, triclads) into traps using pieces of liver or injured aquatic insects (Case and Washino, 1979; Kenk, 1972). In streams, a Surber sampler or similar devices can be applied.

Samples should be cooled, topped with water, and transported to the laboratory as quickly as possible. In the laboratory, samples can be transferred to jars or beakers and left for a couple of hours for the water to stagnate. When the water clears, animals swim toward the surface or glide on the glass where they can be picked up with a pipette for further examination. Alternatively, turbellarians can be directly separated from the sediments under a stereomicroscope, but this procedure is more time consuming. Turbellaria can also be removed from mineral substrates having a low organic matter content by gently heating the sample on a hot plate so that the surface of the sediment reaches 30–32°C (Kolasa, 1983). Other methods of extracting flatworms from sediments were developed for marine sands (Martens, 1984; Noldt and Wehrenberg, 1984). These methods are based on various combinations of anaesthetizing worms (with $MgCl_2$ or ethanol), stirring water to separate mineral particles, and sieving to catch worms. Some of these methods may be adaptable to freshwater fauna.

Many triclads and microturbellarians can be maintained in the laboratory in small glass containers as long as the water is regularly changed and appropriate food provided. Synthetic pond water is best for common species. Temperatures between 17 and 25°C may be adequate for most species, except for cold-water forms. Food requirements vary with the species and variety of easily available aquatic invertebrates must be tested to ensure success. These may include cladocerans, mosquito larvae (for *Mesostoma, Dugesia*), oligochaetes (for *Phaenocora*), protozoa, or other small turbellarians (*Stenostomum*) (e.g., Kolasa, 1987; Yu and Legner 1976). Often, pieces of larger invertebrates, such as *Asellus, Gammarus*, and *Tubifex*, can be successfully used. McConnell (1967) provided many useful hints on rearing triclads.

Most turbellarians can be identified by squash mounts of live animals. Identification is difficult for a novice. Squash mounts are prepared by placing a live individual in a drop of water under a microscope coverslip. Next, excess water is removed with a fine pipette or a strip of filter paper until the specimen is immobilized. Pressure on the specimen can be varied by removing or adding water. Such mounts, with a little bit of practice, reveal the arrangement and appearance of internal organs—information usually necessary for species determination. Whenever further study is desirable, animals can be fixed and preserved as in the following steps.

1. Anaesthetizing with 7% ethanol, 0.1% chloretone, 1% hydroxylamine hydrochloride, or slowing their locomotion by placing them in a small volume of water on ice (this step is optional);
2. Killing with hot fixatives, such as Stieve's, Gilson's, Bouin's, room temperature 70% ethanol, or (optional) killing cooled worms with glutaraldehyde (see electron microscopic techniques for details);
3. Rinsing mercury residues of the above fixatives with 50% ethanol solution of iodine (except Bouin's);
4. Storing specimens in 70% ethanol.

Animals so prepared can be sectioned and analyzed with standard histological methods, including staining with Mallory's stain, Delafield's hematoxylin, and other common stains.

VI. IDENTIFICATION OF NORTH AMERICAN GENERA OF FRESHWATER TURBELLARIA

The key provided uses a combination of both phylogenetic and superficial morphological and anatomical characters. Although the phylogenetically important characters permit greater confidence, they are often very difficult to use by an inexperienced researcher. It is strongly recommended that identification be confirmed by comparing the specimen at hand with exact taxonomic descriptions of the species available in other publications. Monographs by Luther (1955, 1960, 1963, for Dalyelliidae, Macrostomidae, Typhloplanidae), Nuttycombe (1956), Nuttycombe and Waters (1938, for *Catenula, Stenostomum*), Ferguson (1939, for *Macrostomum*), and Gilbert (1938, for *Phaenocora*) are particularly useful.

A. Taxonomic Key to Orders and Suborders of Turbellaria

As some characters may be difficult to determine, it may be advisable to match the general body plan of the individual being identified with one of the pictures in Figure 1A–H. For additional schematics one should consult a guide to world Turbellaria (Cannon, 1986).

1a. Simple female gonad; egg entolecithal (Acoela, Macrostomida, Catenulida; freshwater members of the latter two often reproduce asexually and ovary is absent) (Figs. 1A–C, 12A, B) 2

1b. Heterocellular female gonad, with yolk-producing part separate from the oocyte-producing part (all other Turbellaria) (Figs. 1D–H, 2) 4

2a(1a). Mouth opens directly or through a pharynx simplex into the body; no protonephridia; no distinct intestine; gonads without clear walls (Fig. 1A); rare in freshwater .... Acoela
[Only one species (unidentified) collected in North America (from Mirror Lake in New Hampshire; Strayer, 1985).]

2b. Mouth opens to a pharynx simplex; protonephridia present; epithelial, ciliated intestine present 3

3a(2b). Protonephridia with a single, central excretory duct, often two or more zooids; male gonopore, if present, situated on the dorsal side and in the front of the body; ovary without oviducts and supplementary organs; statocyst present in some species (Figs. 1B, 12A, B) .... Catenulida (see Section VI.B)

3b. Protonephridia with a pair of main excretory ducts; male gonopore on the ventral side of the posterior part of body; female gonopore usually separate and in front of the male one, more than one zooid in the family Microstomidae only; no statocyst (Figs. 1C, 5A, B) .... Macrostomida (see Section VI.C)

4a(1b). Single germovitellarium where ova are surrounded by yolk cell epithelium; pharynx variabilis; copulatory organ armed with a sclerotized stylet near the pharynx (Figs. 1D, 2A) .... Lecithoepitheliata (see Section VI.D)

4b. Ova not surrounded by yolk epithelium; ovary separate or combined with vitellarium (Figs. 1E–H, 2B, C) 5

5a(4b). Pharynx plicatus or variabilis; testicles and ovaries dispersed; germ cells dispersed or aggregated (Fig. 1H) .... Proseriata, Prolecithophora, and Tricladida (see Section VI.E)

5b. Pharynx bulbosus (doliiformis or rosulatus, Fig. 1E–G) 6

6a(5b). Pharynx doliiformis, directed forward (Fig. 1E) .... Dalyellioida (see Section VI.H)

6b. Pharynx rosulatus directed ventrally or, exceptionally as in *Phaenocora*, ventrally and forward 7

7a(6b). Without a proboscis (Fig. 1F) .... Typhloplanoida (see Section VI.F)

7b. Proboscis present (Fig. 1G) .... Kalyptorhynchia (see Section VI.G)

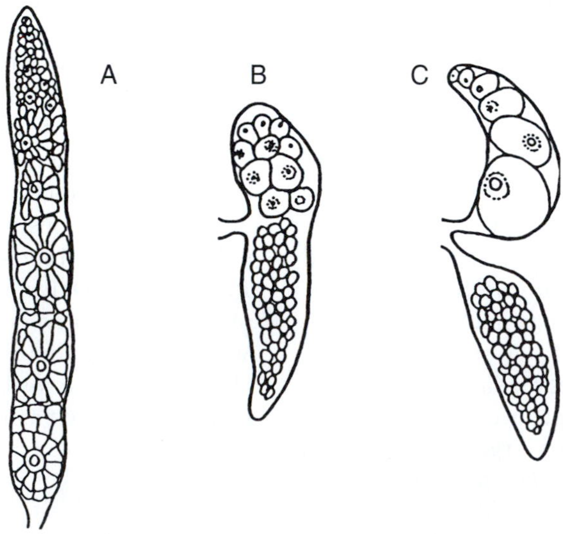

FIGURE 2 Examples of different types of heterocellular female gonads. (A) *Prorhynchus, Geocentrophora*; (B) *Bothrioplana*; and (C) *Typhloplana, Dalyellia.*

B. Taxonomic Key to Catenulida

1a. Brain simple, compact, oval 2

1b. Brain composed of paired frontal and posterior lobes; a group of sensory cells in front of the brain, often arranged pseudometamerically, i.e., in parallel rows .... family Stenostomidae 5

2a(1a). Brain situated usually at the base of, but not further than in the middle of prostomium (frontal body section without the intestine); with ciliated ventral or lateroventral furrow separating prostomium from the rest of body .... family Catenulidae 3

2b. Brain situated in the first 1/3 of prostomium .... family Chordariidae *Chordarium*

3a(2a). A ring of strongly ciliated longitudinal grooves developed on the prostomium (Fig. 3A) .... *Suomina*

3b. No such grooves present 4

4a(3b). Clearly open intestine extends, approximately, to the middle of the postpharyngeal section of body (Figs. 3B,C, 12A) .... *Catenula*

4b. Open intestine extends much further, leaving no more than 1/5 to 1/4 of the postpharyngeal section closed; usually the intestinal lumen of individual zooids is connected .... *Dasyhormus*

5a(1b). Adult individuals composed of two or more zooids; gonads often absent 6

5b. Adult individuals (with gonads) without signs of asexual divisions; with long prostomium, ciliated pits absent (Figs. 3D,F, 12B) .... *Rhynchoscolex*

6a(5). A section of the intestine near the pharynx with a strong ring of muscles (Fig. 4E) .... *Myostenostomum*

6b. No such structure, intestine uniformly ciliated (Fig. 4A–D) .... *Stenostomum*

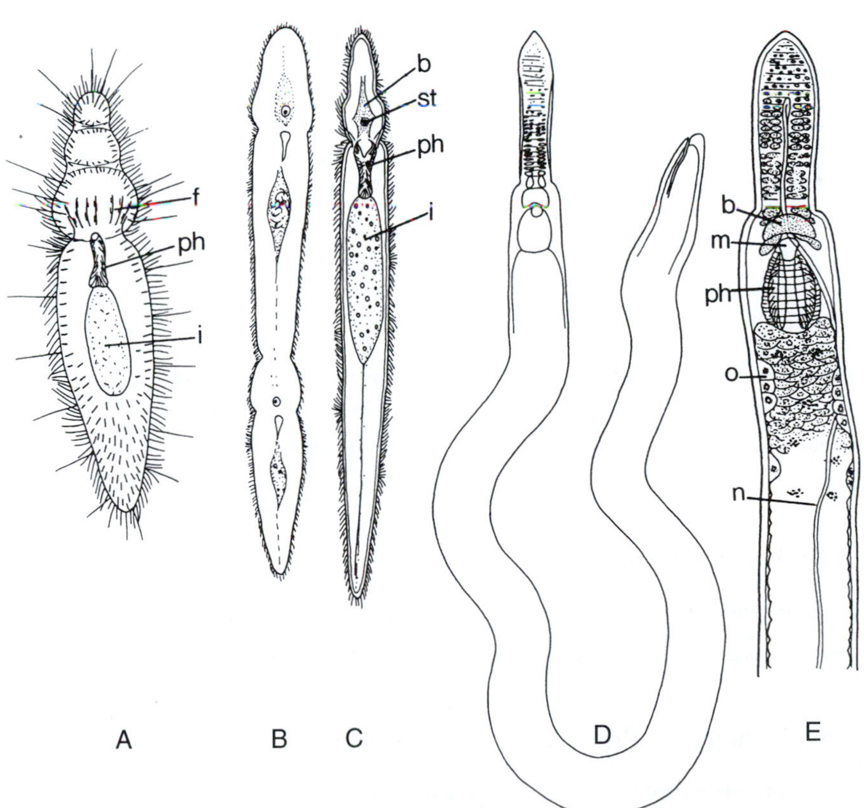

FIGURE 3 Some representatives of Catenulida. (A) *Suomina turgida*; (B) *Catenula lemnae*; (C) *Catenula leptocephala*; and (D) and (E) *Rhynchoscolex simplex*. Length of species varies between 0.35 and 5.00 mm. B, Brain; f, furrow; i, intestine; m, mouth; n, protonephridial duct; o, ovary; ph, pharynx; and st, statocyst.

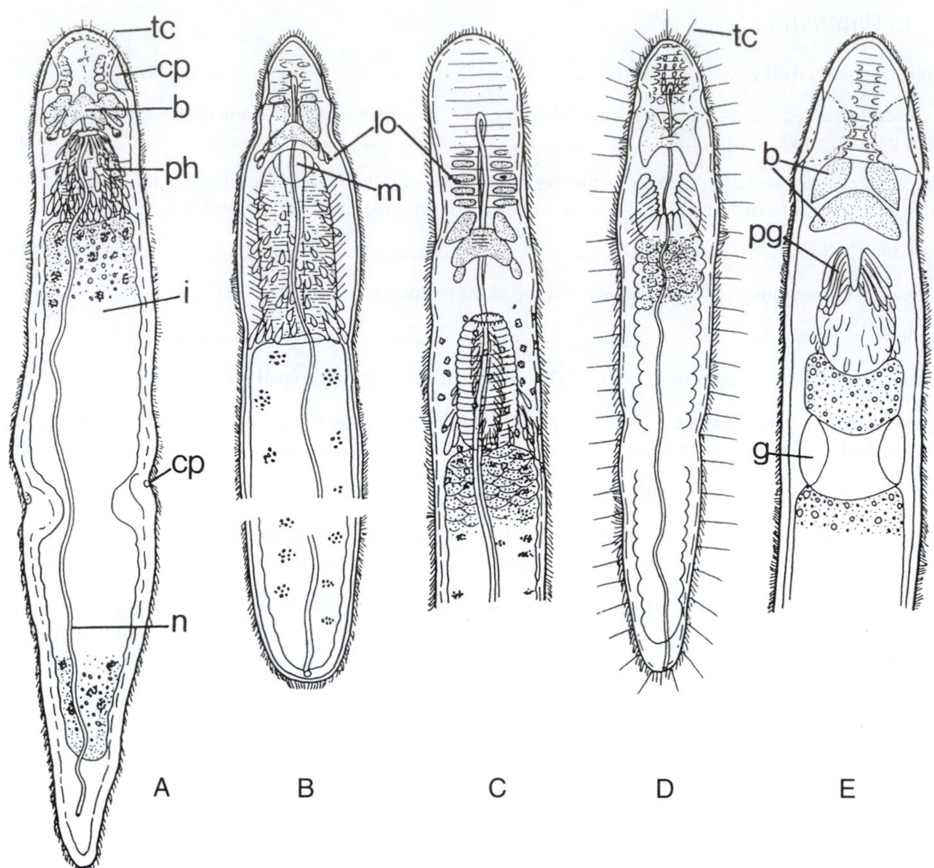

FIGURE 4 Some representatives of Catenulida. (A) *Stenostomum leucops*; (B) *S. beauchampi*; (C) *S. glandulosum*; (D) *S. brevipharyngium*; and (E) *Myostenostomum tauricum*. B, Brain; cp, ciliated sensory pits; g, muscular ring of intestine or gizzard; i, intestine; lo, light-refracting organs; m, mouth; n, protonephridial duct; pg, pharyngeal glands; ph, pharynx; and tc, tactile cilia.

C. Taxonomic Key to Macrostomida

1a. Ciliated pits present; intestine extends in front of mouth; numerous zooids separated by division planes (Fig. 5B,C) family Microstomidae .... *Microstomum*

1b. No ciliated pits; intestine entirely behind the pharynx; no signs of asexual division (Fig. 5A, D, E) ....family Macrostomidae *Macrostomum*

D. Taxonomic Key to Lecithoepitheliata

1a. Stylet of the copulatory organ straight; no eyes .... *Prorhynchus*

1b. Stylet of the copulatory organ curved; eyes present except in some cave forms .... *Geocentrophora*

E. Taxonomic Key to Prolecithophora, Proseriata, and Tricladida

1a. Mouth situated in front and directed forward .... *Hydrolimax*

1b. Mouth situated elsewhere and directed backward or ventrally 2

2a(1b). Mouth situated near the middle and directed ventrally; statocysts present .... *Otomesostoma*

2b. Mouth directed backward; no statocyst 3

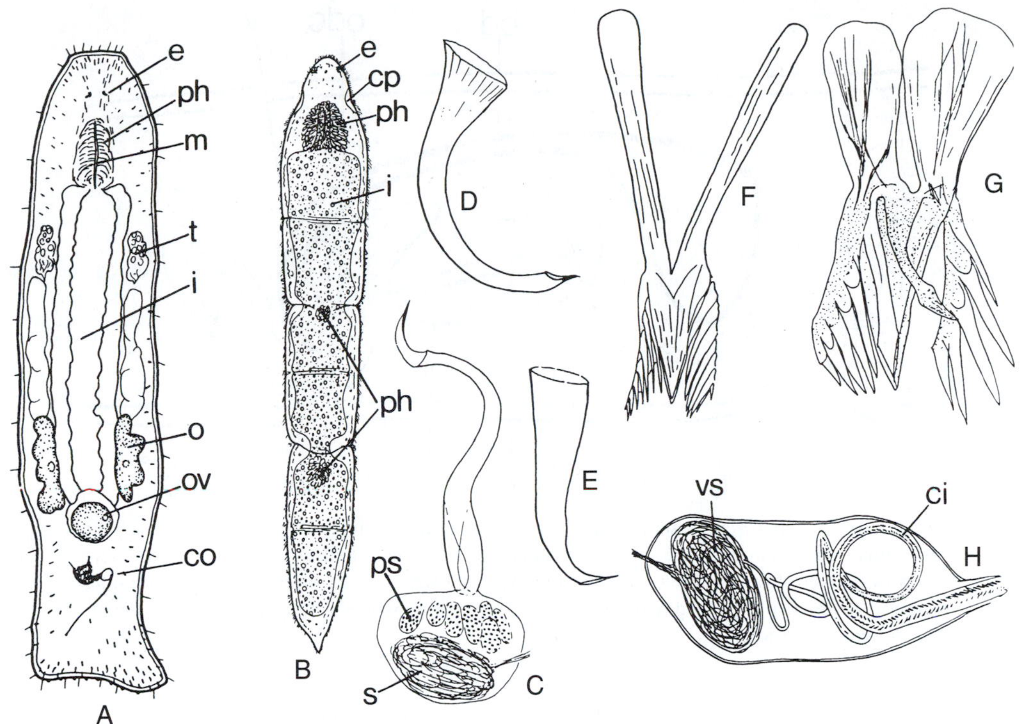

FIGURE 5 Representatives of Macrostomida and copulatory organs of various families. (A) *Macrostomum tuba*; (B) *Microstomum lineare*; (C) copulatory organ of *M. lineare*; (D) stylet of *Macrostomum gilberti*; (E) stylet of an unidentified *Macrostomum* from New Haven, CT; (F) sclerotized copulatory apparatus of *Microdalyellia rossi* (Dalyelliidae); (G) sclerotized copulatory apparatus of *M. tennesseensis*; and (H) copulatory organ of *Opistomum pallidum* containing a spiny reversible cirrus (Typhloplanidae). Ci, Cirrus; co, copulatory organ; cp, ciliated pits; e, eye; i, intestine; m, mouth; o, ovary; ov, egg; ph, pharynx; ps, prostatic secretions; s, sperm; t, testes; and vs, seminal vesicle.

3a(2b).	Intestine split into two branches behind the pharynx (Figs. 6–7)	Tricladida[1]
3b.	Intestine not composed of three branches (Fig. 8A)	
4a(3a).	Internal pharyngeal muscles in two distinct layers (Planariidae) (Fig. 6)	5
4b.	Internal pharyngeal muscles in one layer of mixed longitudinal and circular fibers (Dendrocoelidae) (Fig. 6)	12
5a(4a).	Oviducts, separate or united, open into end part of the bursa stalk	6
5b.	Oviducts united, the common oviduct opens in the genital atrium (Fig. 6)	7
6a(5a).	Testes extend to the level of pharynx	*Cura*
6b.	Testes extend to the posterior end of body	*Dugesia*
7a(5b).	Numerous eyes arranged in a band around the anterior end of body (Fig. 7G)	*Polycelis*
7b.	Eyes absent or one pair only; if numerous, then not along the head margin	8
8a(7b).	Adenodactyl present (Fig. 6)	*Planaria*
8b.	Adenodactyl absent	9
9a(8b).	Anterior end with an adhesive organ	10
9b.	Anterior end without an adhesive organ	11
10a(9a).	Body elongated, flat, with a well-developed postpharyngeal section	*Sphalloplana*

[1] Substantial parts of the key to Tricladida follow Kenk (1972). Separation of *Hymanella* from *Phagocata* is based on Ball *et al.* (1981).

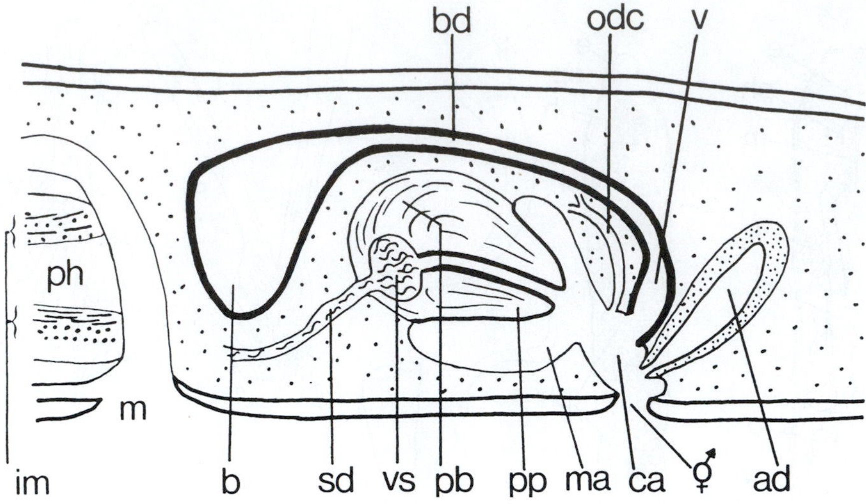

FIGURE 6 Schematic representation of the major anatomic features used in taxonomy of Tricladida. Ad, Adenodactyl; b, bursa; bd, bursal duct; ca, common genital atrium; im, internal muscle layer of the pharynx (two possible types shown); m, mouth; ma, male atrium; odc, common oviduct; ph, pharynx; pb, penis bulb; pp, penis papilla; sd, sperm duct; v, vagina; and vs, seminal vesicle.

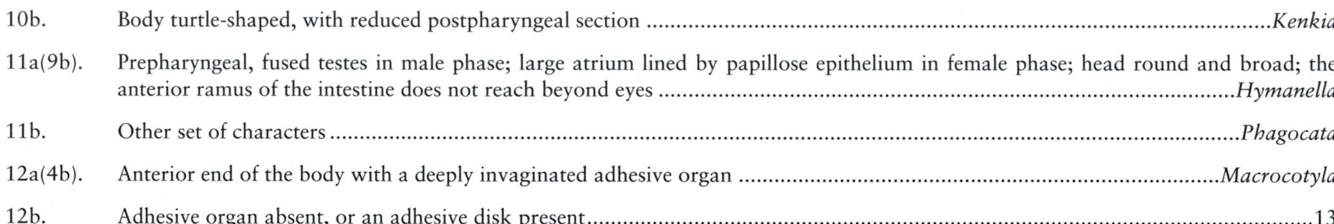

10b. Body turtle-shaped, with reduced postpharyngeal section .... *Kenkia*

11a(9b). Prepharyngeal, fused testes in male phase; large atrium lined by papillose epithelium in female phase; head round and broad; the anterior ramus of the intestine does not reach beyond eyes .... *Hymanella*

11b. Other set of characters .... *Phagocata*

12a(4b). Anterior end of the body with a deeply invaginated adhesive organ .... *Macrocotyla*

12b. Adhesive organ absent, or an adhesive disk present 13

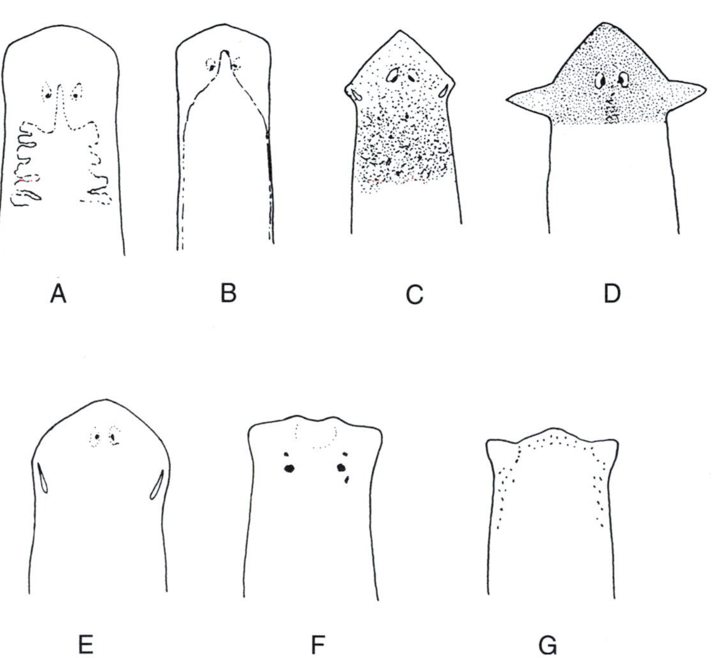

FIGURE 7 Some common representatives of Tricladida. (A) *Hymanella retenuova*; (B) *Phagocata velata*; (C) *Dugesia tigrina*; (D) *D. dorotocephala*; (E) *D. polychroa*; (F) *Procotyla fluviatilis*; and (G) *Polycelis coronata* (A and B modified from Ball *et al.*, 1981).

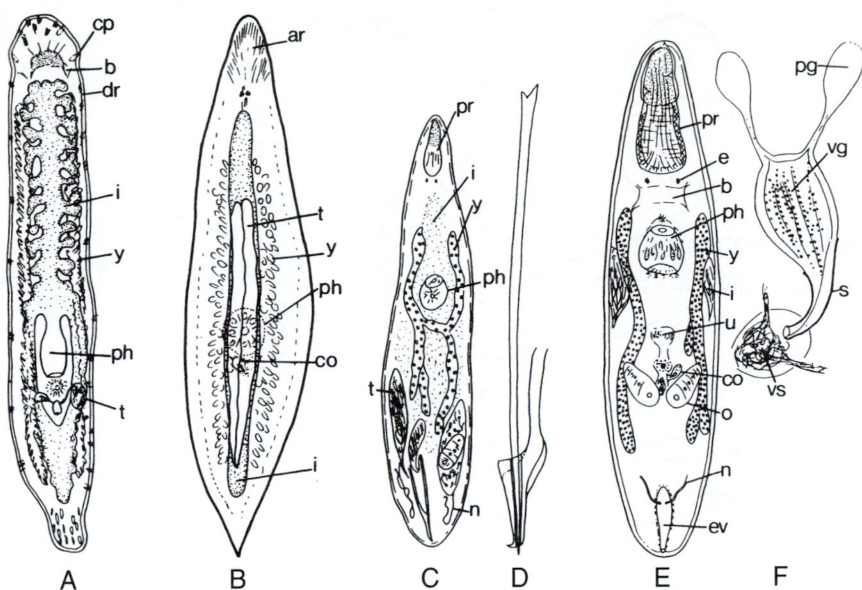

FIGURE 8 Representatives of (A) Proseriata (*Bothrioplana semperi*); (B) Typhloplanoida (*Mesosomta craci*); and (C–F) Kalyptohynchia; (C) *Gyratrix hermaphroditus* and (D) its stylet; (E) *Opisthocystis goettei* and (F) its copulatory organs. Ar, Adenal rhabdoids; b, brain; co, copulatory organ; cp, ciliated pits; dr, dermal rhabdoids; e, eye; ev, excretory vesicle; i, intestine; n, protonephridial duct; o, ovary; pg, pharyngeal glands; ph, pharynx; pr, proboscis; s, stylet; t, testes; vg, granular vesicle; vs, seminal vesicle; and y, yolk glands.

13a(12b). Penis papilla present 14

13b. Penis papilla absent, bulb large .... *Rectocephala*

14a(13a). Penis bulb rounded, containing a seminal vesicle .... *Dendrocoelopsis*

14b. Penis bulb elongated, contains a prostatic vesicle .... *Procotyla*

F. Taxonomic Key to Typhloplanoida

The basic plan of the freshwater typhloplanoid reproductive system is given in Fig. 9.

1a. Yolk glands paired, situated on both sides of the intestine 2

1b. Yolk glands developed as single strand of cells above the intestine (Fig. 10E, F) .... *Limnoruanis*

2a(1a). Pharynx developed as a typical, round or slightly elongated pharynx rosulatus and oriented ventrally or somewhat forward 3

2b. Pharynx elongated or cylindrical, oriented backward; copulatory organ with a spiny cirrus (Fig. 5H) .... *Opistomum*

3a(2a). Testes under yolk glands 4

3b. Testes above yolk glands 11

4a(3a). Protonephridial ducts open separately on the body surface (Protoplanellinae) (as in Fig. 10C) 5

4b. Protonephridial ducts combined with mouth or gonopore (as in Fig. 10A) 7

5a(4a). Front extremity with a central depression and set slightly apart by two lateral, ciliated pits; pharynx in the first third of body (Fig. 10C) *Prorhynchella*

5b. Front extremity not set apart, rounded; pharynx in the second to third portion of body 6

6a(5b). Reproductive complex with a bursa copulatrix; ductus ejaculatorius usually surrounded by gelatinous matrix (Fig. 11E, F; cf., Fig. 9) .... *Krumbachia*

6b. Bursa copulatrix absent .... *Amphibolella*

7a(4b). Protonephridial ducts joined and combined with mouth; no retractable proboscis (Typhloplaninae) 8

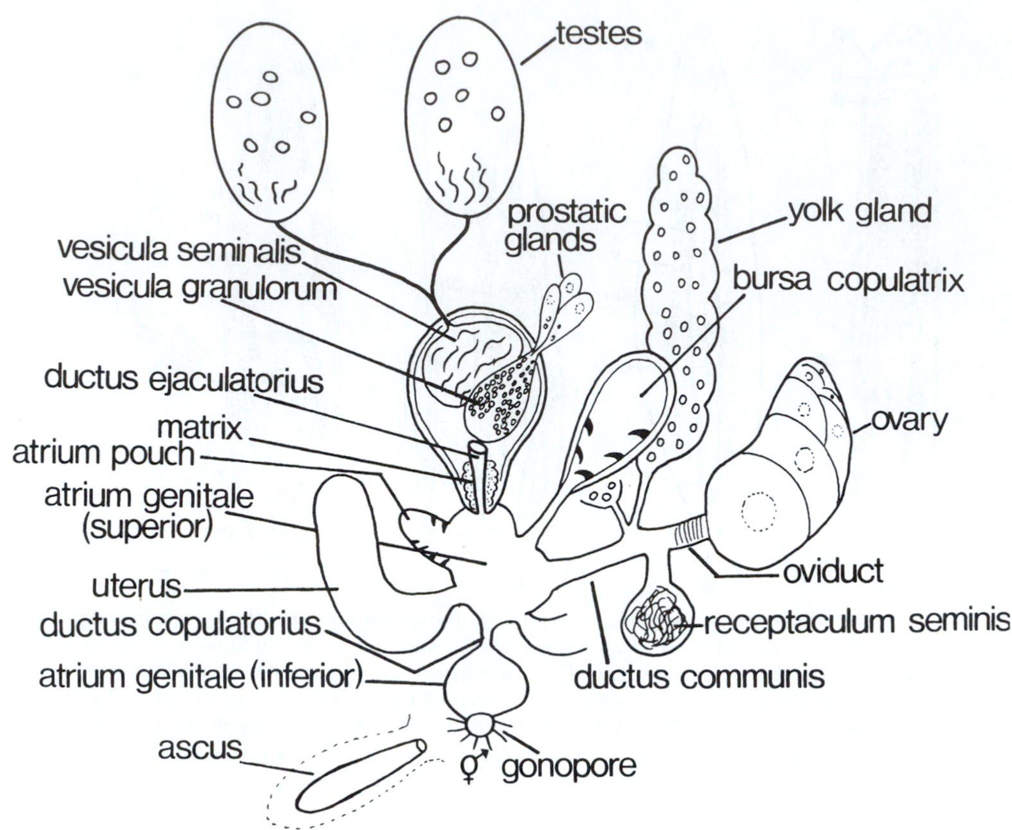

FIGURE 9 A schematic representation of the main reproductive organs and structures of freshwater Typhloplanoida. In some species, additional structures may be present or the relative arrangement among the organs may be slightly different.

7b.	Protonephridial ducts open into the gonopore; retractable front extremity (Fig. 10D)	*Rhynchomesostoma*
8a(7a).	With a bursa copulatrix and an atrium copulatorium (Fig. 11D)	*Castrada*
8b.	One of the above organs missing	9
9a(8b).	With eyes (Figs. 10A, 12C)	*Strongylostoma*
9b.	No eyes	10
10a(9b).	Atrium copulatorium absent; bursa copulatrix present	*Typhloplanella*
10b.	Both bursa copulatrix and atrium copulatorium absent; always with green algae; pharynx in the first half of body; copulatory organ pear-shaped	*Typhloplana*
11a(3b).	Pharynx developed as a typical, round pharynx rosulatus	13
11b.	Pharynx barrel-shaped, directed clearly forward, and situated near the front extremity (Fig. 11B)	12
12a(11b).	Ductus copulatorius absent; inferior genital atrium inconspicuous (Fig. 11B)	*Phaenocora*
12b.	With ductus copulatorius; exceptionally large inferior atrium; occurs in sulfuric springs	*Pseudophaenocora*
13a(11a).	An elongated supplementary organ, the ascus, present in the proximity of the gonopore; or with a well-defined gland complex in its place (Ascophorinae) (cf. Fig. 9)	14
13b.	Ascus absent	16
14a(13a).	Pharynx in the middle of the first half of body	*Protoascus*
14b.	Pharynx approximately in the middle of body or slightly behind	15
15a(14b).	Gonopore near the middle of body	*Ascophora*

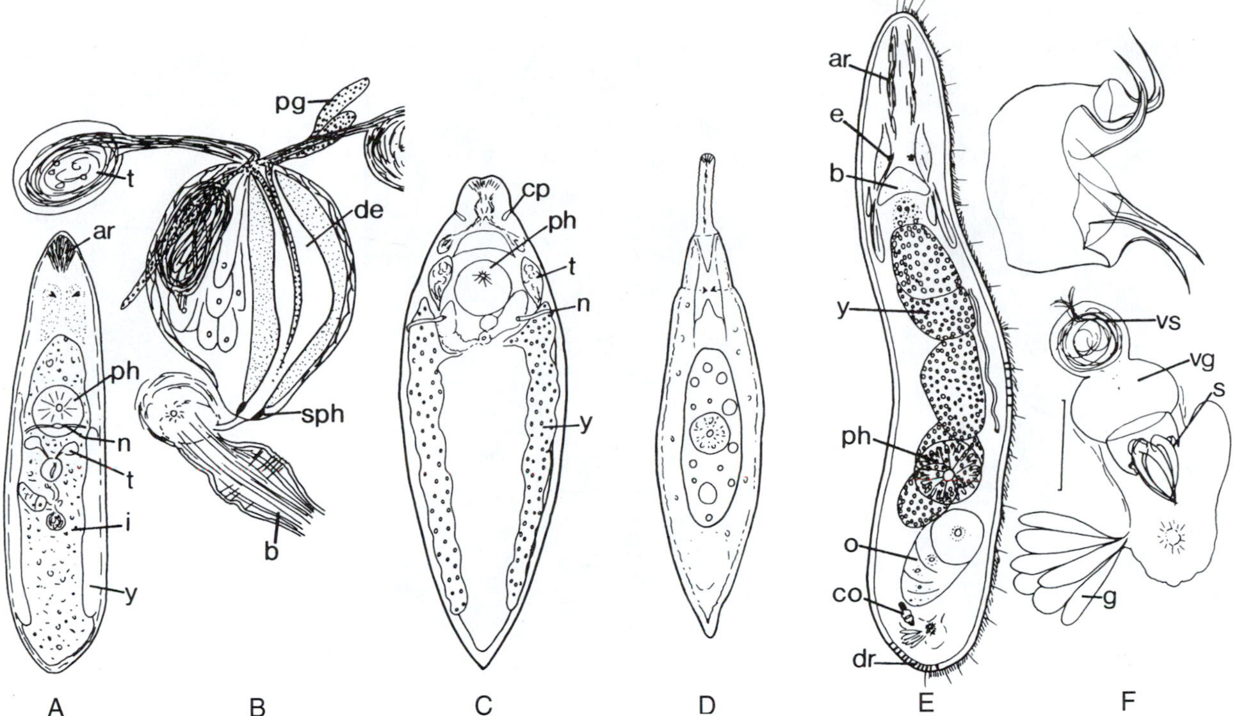

FIGURE 10 Some representatives of Typhloplanoida. (A) *Strongylostoma simplex*; (B) copulatory organ with testes of *S. simplex*; (C) *Prorhynchella minuta*; (D) a juvenile of *Rhynchomesostoma rostratum*; (E) *Limnoruanis romanae*; and (F) sclerotized part of the copulatory organ, partly reverted (upper picture), and the copulatory organ with the gonopore and glands (lower picture) in *L. romanae* (C modified from Rusebush, 1939). Ar, Adenal rhabdoids; e, eye; g, glands; i, intestine; n, protonephridial duct; o, ovary; pg, pharyngeal glands; ph, pharynx; s, stylet,; sph, sphincter; t, testes; vg, granular vesicle; vs, seminal vesicle; and y, yolk glands.

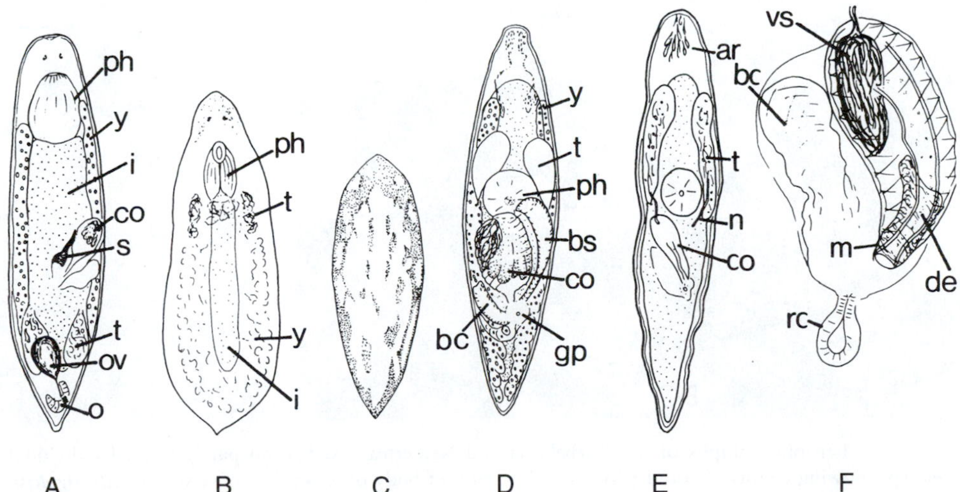

FIGURE 11 Representatives of the Dalyellioida (*Castrella*) and Typhloplanida (others); (A) *Castrella pinguis*; (B) *Phaenocora* sp. (from Churchill, Manitoba); (C) *Mesostoma vernale* (the habitus of this species is also characteristic of various *Bothromestoma*); (D) *Castrada virginiana*; (E) *Krumbachia hiemalis*; and (F) copulatory complex of *Krumbachia hiemalis*. ar, Adenal rhabdoids; bc, copulatory bursa; bs, blind sac of the atrium; co, copulatory organ; de, ductus ejaculatorius; gp, gonopore; i, intestine; m, matrix; n, protonephridial duct; o, ovary; ov, egg capsule; ph, pharynx; r, seminal receptacle; s, stylet; t, testes; vs, seminal vesicle; and y, yolk glands.

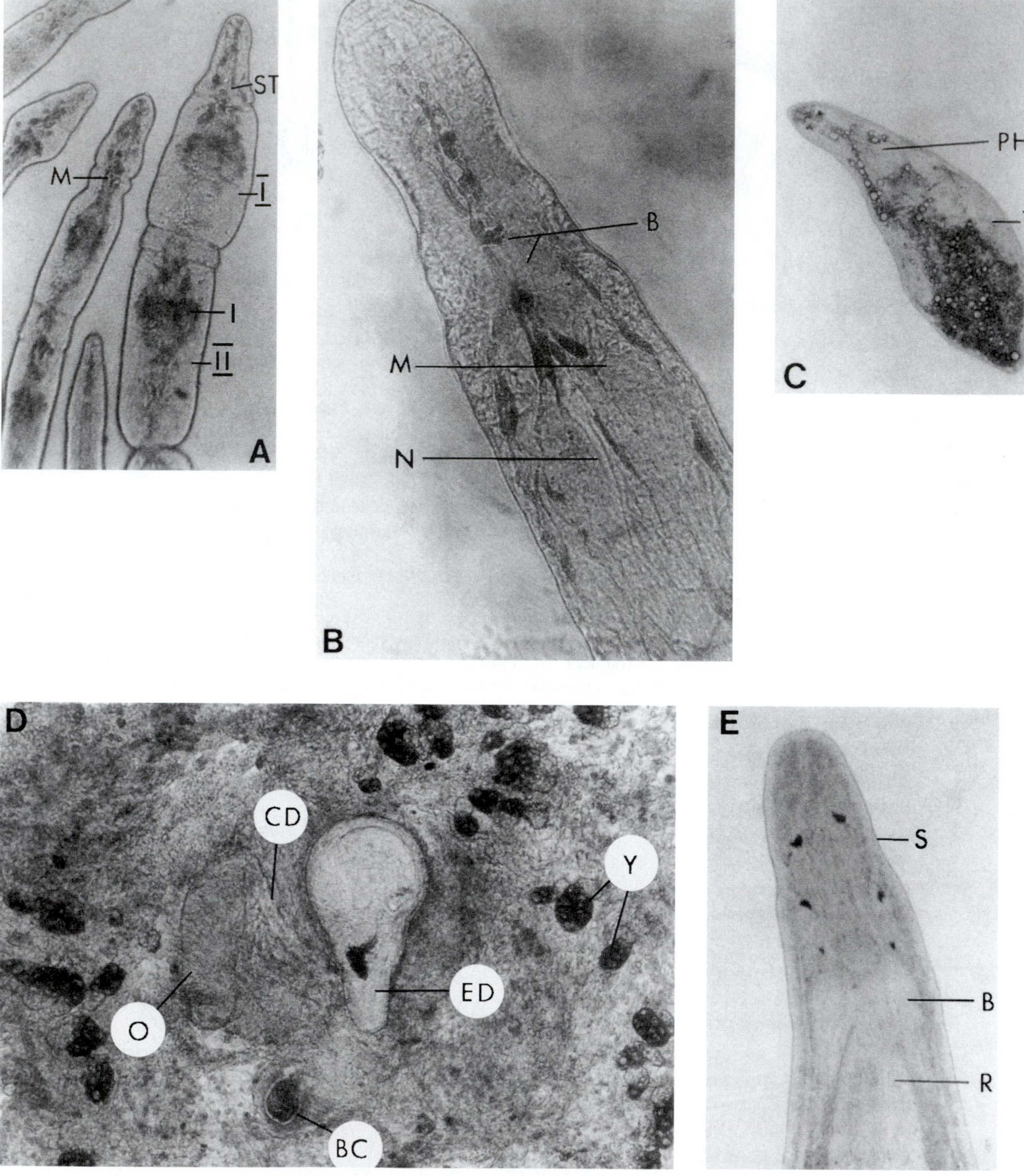

FIGURE 12 Microphotographs of live Turbellaria and Nemertea. (A) Several partly framed individuals of *Catenula* sp. showing chains of zooids; (B) frontal portion of body of *Rhynchoscolex* sp.; (C) *Strongylostoma* sp., a representative of Typhloplanoida; (D) squash preparation of the portions of the reproductive system of *Mesostoma craci*, with the copulatory organ in the middle; and (E) a frontal portion of body of a nemertean *Prostoma* sp. B, Brain; BC, bursa copulatrix; CD, common duct; ED, ejaculatory duct; prostatic granulations are visible in its proximity; I, intestine; M, mouth; N, protonephridium; O, ovary; PH, pharynx; R, rhynchocoel; S, sensory pit; ST, statocyst; T, testicle; Y, yolk glands; I, first zooid; and II, second zooid.

15b. Gonopore in the third posterior part of body .... *Dochmiotrema*

16a(13b). Protonephridial ducts open separately on the body surface; eyes often present........ *Olisthanella*

16b. Protonephridial ducts open into mouth 17

17a(16b). With a ventral pit and a canal connecting the bursa copulatrix to the ovovitellin duct; often flat on the ventral side and strongly convex on the dorsal........ *Bothromesostoma*

17b. Without the above characters (Figs. 11C, 8B, 12D)........ *Mesostoma*

G. Taxonomic Key to Kalyptorhynchia and Similar Forms

1a. Excretory vesicle absent; protonephridial ducts open directly at the surface of body; single ovary (Fig. 8C)........ 2

1b. Protonephridial ducts open into an excretory vesicle in the rear end of body; paired ovary (Fig. 8E, F)........ *Opisthocystis*[2]

2a(1a). Copulatory organ with an almost straight stylet (Fig. 8C, D) .... *Gyratrix*

2b. Copulatory organ with a curved stylet........ *Microkalyptorhynchus*[3]

H. Taxonomic Key to Dalyellioida

1a. With a paired ovary; copulatory organ with a straight, funnel-like tube........ *Pilgramilla*

1b. With a single ovary; copulatory organ with a more of less complex sclerotized structure (e.g., Figs. 5F, G, 11A) 2

2a(1b). Sclerotized structure directly attached to the rest of the copulatory organ........ 3

2b. Sclerotized structure in a separate pocket adjacent to the bulb of the copulatory organ (Fig. 11A)........ *Castrella*

3a(2a). Sclerotized structure typically with two handles and two lateral branches carrying spines, auxiliary parts may be present (Fig. 5F, G) 4

3b. Sclerotized structure different; usually in the form of a spiny ring .... *Gieysztoria*

4a(3a). One to several eggs in the parenchyme; normally, older individuals green due to symbiotic zoochlorellae........ *Dalyellia*

4b. Never more than one egg; no zoochlorellae .... *Microdalyellia*

NEMERTEA

VII. GENERAL CHARACTERISTICS, EXTERNAL AND INTERNAL ANATOMICAL FEATURES

Nemerteans are coelomate (Turbeville and Ruppert, 1985) and unsegmented worms possessing an eversible muscular proboscis resting in the rhynchocoel, a walled, longitudinal dorsal cavity connected with the mouth (Fig. 13 and 14, see also Figure 12E in Section VI.F.[4] Unlike the Turbellaria, Nemertea have an anus as well as a closed-blood circulatory system. A monolayered, ciliated epidermis and several layers of muscles enclose the body. The excretory system is protonephridial. Sensory structures include eyes, statocysts (absent in freshwater nemerteans), and cerebral, frontal, and lateral neural organs. The reproductive system is much simpler than in Turbellaria. Freshwater nemerteans are hermaphroditic and often protandric. Spermatozoa and ova are produced in numerous gonads situated in diverticulae of the intestine. As a rule, fertilization is external and self-fertilization can also occur. Development is direct. Most information provided in this section on nemerteans is based on an excellent review of freshwater and terrestrial nemerteans by Moore and Gibson (1985). That review will be an invaluable source of data and references for anyone working on the ecology and biology of freshwater nemerteans.

[2] An unidentified kalyptorhynchid occurring in a tributary of the Hudson River (Kolasa *et al.*, 1987) also meets this criterion, although it may belong to a family other than that containing *Opisthocystis*. *Klattia* is a synonym of *Opisthocystis* (Karling T.G. 1956. Ark. Zool. 9:187–279).

[3] Taxonomic placement within Kalyptorhynchia is provisional. Name discussed by W.E. Hazen (1953. Morphology, taxonomy, and distribution of Michigan rhabdocoeles. Ph.D. Thesis, Univ. of Michigan, Ann Arbor. 122 pp.).

[4] Until 1985, Nemertea were considered to be acoelomate. With the new finding (Turbeville and Ruppert, 1985), they can no longer be associated with other flatworms. Our presentation here is more for convenience than systematic reasons.

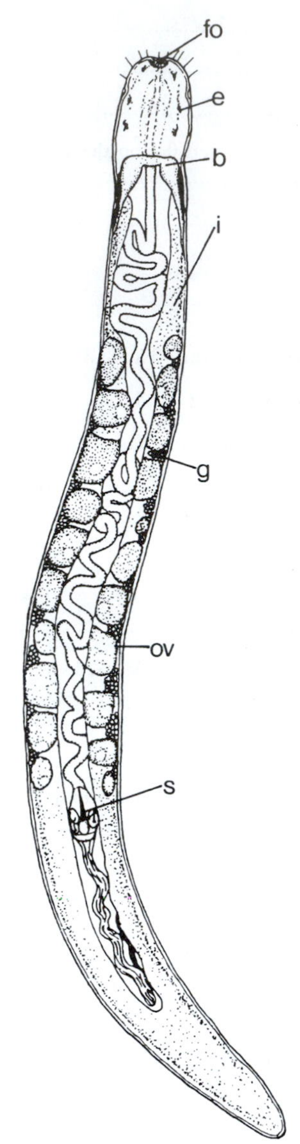

FIGURE 13 A semi-schematic view of *Prostoma* sp. b, Brain; e, eye; fo, frontal organ; g, gonad; i, intestine; ov, egg; s, stylet.

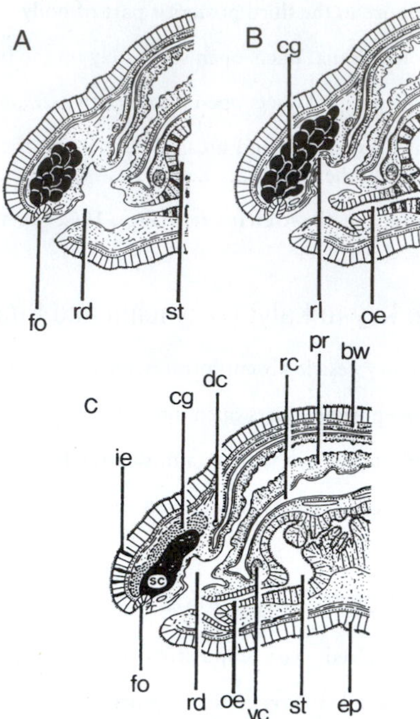

FIGURE 14 Sections through the anterior extremity of three North American nemerteans. (A) *Prostoma eilhardi*; (B) *P. graecense*; (C) *P. canadiensis* (after Gibson and Moore, 1976, 1978). bw, Body wall musculature; cg, cephalic gland lobular region; dc, dorsal cerebral commissure; ep, epidermis; fo, frontal organ; id, improvised ducts from cephalic gland lobules; oe, oesophagus; pr, proboscis; rc, rhynchocoel; rd, rhynchodaeum; rl, rhychodaeal longitudinal muscle fibers; st, stomach; and vc, ventral cerebral commisure.

Twelve species of freshwater nemerteans are classified into two, possibly polyphyletic families and six genera (Moore and Gibson, 1985). In the family Heteronemertea, three monotypic genera are recognized: *Planolineus* (Beauchamp, 1928), *Siolineus* (du Bois-Reymond Marcus, 1948), and *Apartronemertes* (Wilfert and Gibson, 1974). In the family Hoplonemertea, the genus *Prostoma* (Duges, 1828) contains most of the species. Two other genera, *Campbellonemertes* (Moore and Gibson, 1972) and *Potamonemertes* (Moore and Gibson, 1973) are known from islands in the southern Pacific, Campbell Island and South Island, New Zealand, respectively.

VIII. ECOLOGY

A. Life History and General Ecology

Most nemerteans are marine, although a small minority are terrestrial or freshwater. The freshwater forms are benthic, predatory worms 10–40 mm in length and of pink coloration. *Prostoma* feed readily on oligochaete worms and, occasionally, on crustaceans, nematodes, turbellarians, midge larvae, and other small invertebrates (McDermott and Roe, 1985). Feeding activity is most intense at night and the sticky proboscis captures prey. Little is known about habitat selection by freshwater nemerteans. North American *Prostoma* species have been found in both streams and lakes, including Lake Huron. In lakes, *Prostoma* species appear to be associated with filamentous algae.

Even though some nemerteans have strong regenerative capabilities, reproduction is exclusively sexual, with both male and female organs present at the same time. Since only limited quantitative studies have been conducted on the biology of freshwater nemerteans in the field, mechanisms of population regulation remain unknown. Clearly, *Prostoma* can reproduce rapidly with up to 210 eggs per reproductive episode throughout the year as long as the temperature is above 10°C (Young and Gibson, 1975). A study of one population in England suggests that *Prostoma* population density increases during the cold months of the year and declines during the summer, with smaller individuals recorded during the winter (Gibson and Young, 1976). It is not clear, however, if this reproduction pattern is general or due to cattle trampling of the shore vegetation where the study was conducted.

Distribution of nemerteans in North America is not well known. (Strayer, 1985) reported *Prostoma rubrum* from Mirror Lakes, New Hampshire. However, *P. rubrum* is no longer a valid species (Gibson and Moore, 1976), and this observation has to be treated as *Prostoma* sp. *Prostoma canadiensis* is known from one site in Lake Huron (Gibson and Moore, 1978), where is occurs to a depth of 20 m. *Prostoma* species are probably much more common in freshwaters than the infrequent records from eastern North American might indicate. A *Prostoma* species (*P. rubra*) was also reported in vegetation of slow coastal streams, marshes, and estuarine pools in St. Vincent, Caribbean (Harrison and Rankin, 1976). So far, no *Prostoma* have been found in temporary water bodies.

Three established species of nemerteans are known from North America (Fig. 14); a fourth, *P. asensoriatum*, is a questionable species. Distribution of these species in North America is summarized in Figure 15, but they are probably present in most standing and running waters. With the exception of the dubious *P. asensoriatum*, all three remaining species have been

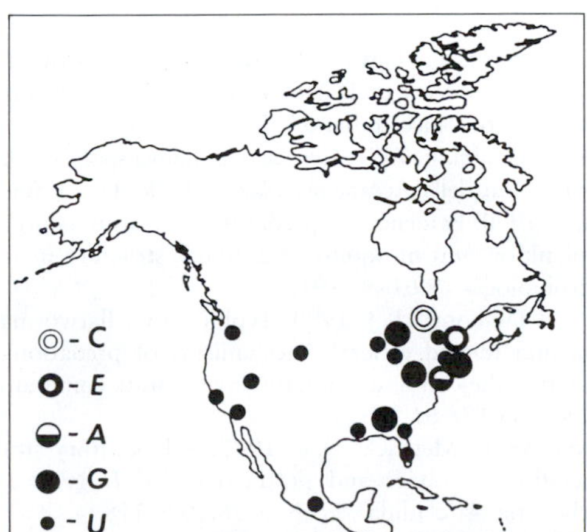

FIGURE 15 Known distribution of *Prostoma* species in North America (after Gibson, 1985; modified and updated). C, *P. canadiensis*; E, *P. eilhardi*; A, *P. asensoriatum*; G, *P. graecense*; and U, unidentified *Prostoma*.

recorded in other parts of the world, including South America, Europe, Africa, and Australia. Zoologists do not know whether the worldwide pattern of distribution is natural or, instead, a result of recent human-mediated dispersal. Analysis of genetic diversity of various populations might shed some light on this question.

B. Physiological Adaptations

Osmoregulation is an ecologically important function in nemerteans as in all other freshwater invertebrates with permeable body walls. It is controlled by the cerebral organs and involves several organs and enzymatic systems associated with blood vessels (Moore and Gibson, 1985). Well-developed nephridia may play a role in osmoregulation as well as in the removal of nitrogenous wastes (but the latter has not yet been demonstrated). Gases, particularly oxygen, are exchanged across the surface of the body. The role of blood vessels in oxygen transport is unclear.

Nemerteans, like turbellarians, release copious mucus whose various functions probably include defense, locomotion, physiological barrier, and encystment.

C. Behavioral Ecology

There are no studies devoted exclusively to the behavior of freshwater nemerteans. Nevertheless, some observations on feeding behavior, escape reactions, and locomotion are available. Adult nemerteans can only crawl, while small juveniles also swim. Forward locomotion of *Prostoma* may involve ciliary movement alone, ciliary movement in combination with muscular waves, sinusoidal curves when the animal pushes through the vegetation, or peristalsis when the animal moves backward. The proboscis provides the fastest movement forward. This long organ is rapidly everted and attached to substrate, and the body is then pulled toward the point of attachment.

Typically, prey are captured by a sticky proboscis and then swallowed whole by the widely distended mouth (McDermott and Roe, 1985). The mechanisms of prey detection are poorly understood. In terrestrial nemerteans, these mechanisms are unusual in that they involve ambushing strategy and reliance on mechanical stimuli.

D. Functional Role in the Ecosystem

Necessarily, the functional importance of nemerteans in the ecosystem can only be indirectly inferred from their trophic position and relative densities. I have found densities of *Prostoma* species in Wappinger Creek, a tributary of the Hudson River, to vary between 50 and 590 individuals/m^2, which constitutes a small fraction of all predatory invertebrates identified at the study site. Similarly, low densities of *P. canadiensis* (up to 140 individuals/m^2) were observed on gravelly and muddy substrates in Lake Huron.

IX. CURRENT AND FUTURE RESEARCH PROBLEMS

In view of the limited knowledge of nemertean ecology, almost any area of research will provide valuable information. New information on phenology, population dynamics, and habitat selectivity would permit a better evaluation of the role of nemerteans in freshwater communities. Specific studies in the biology of nemerteans appear particularly promising in the areas of reproductive biology and evolutionary ecology. The ecological role of femaleness later in the life cycle, the production of resting eggs, and the environmental versus genetic cues of life stages all may offer exciting and general models for evolutionary ecology.

X. COLLECTION, CULTURING, AND PRESERVATION

Collection of nemerteans is similar to collection of any soft-bodied, benthic invertebrates such as turbellarians or oligochaetes living in the substrate or on aquatic plants. Sieving and vigorous washing may damage specimens. Placing samples in beakers or jars and topping with water allows nemerteans to crawl toward the edges of water where they can be collected with a pipette.

Freshwater nemerteans are relatively easy to culture. They may be maintained and will reproduce in aquaria or even in petri dishes in clean, oxygenated water and on a diet of live aquatic invertebrates. Chopped tubificid oligochaetes provide a good diet.

Worms can be preserved in 80% ethanol. Before preservation, however, worms should be narcotized using 7% ethanol or chloretone until their movement ceases. According to Moore and Gibson (1985), gin can be used in the field if pure ethanol is unavailable, but only as a last resort!

XI. TAXONOMIC KEY TO SPECIES OF FRESHWATER NEMERTEA IN NORTH AMERICA

It should be noted that Gibson and Moore (1976, 1978) provided keys to freshwater nemerteans of the world. Those keys are more suitable if you encounter a species that has not yet been recorded in North America. The present key is an adaptation of the keys by Gibson and Moore (1976, 1978) pertinent to established North American *Prostoma*.

1a. Cephalic glands open via frontal organs and by improvised ducts (Fig. 14C); esophagus distinct but unciliated; rhynchodaeum with isolated longitudinal muscle strands; proboscis with twelve nerves. *P. canadiensis* (Gibson and Moore, 1978)

1b. Cephalic glands open only via frontal organ, no improvised ducts 2

2a(1b). With a distinctive ciliated esophagus; cephalic glands reach back to the brain (Fig. 14B); rhynchodaeum with a well-developed layer of longitudinal muscle .... *P. graecense* (Bohmig, 1892)

2b. Different combination of characters; indistinct and unciliated esophagus (Fig. 14A); rhynchodaeum without specifically associated layer of longitudinal muscle fibers; proboscis with 9–10 nerves .... *P. eilhardi* (Montgomery, 1894)

ACKNOWLEDGMENTS

I am very grateful to L.R.G. Cannon, J. Moore, R. Gibson, A.P. Mead, R. Kent, S. Schwartz, D. Strayer, and S. Tyler for helpful comments on the manuscript, additional references, and permission to reproduce figures, glossary entries, and fragments of keys.

LITERATURE CITED

Ball, I. R. 1974. A contribution to the phylogeny and biogeography of the freshwater triclads (Platyhelminthes: Turbellaria), *in*: Riser, N.W., Morse, M.P. Eds., Biology of the Turbellaria. McGraw-Hill, New York, pp. 339–401.

Bauchhenss, J. 1971. Die Kleinturbellarien Frankens. Ein Beitrag zur Systematik und Ökologie der Turbellaria excl. Tricladida in Süddeutschland. Internationale Revue des gesamten Hydrobiologie 56:609–666.

Bedini, C., Ferrero, E., Lanfranchi, A. 1973. Fine structure of the eyes in two species of Dalyelliidae (Turbellaria Rhabdocoela). Monitore Zoologico Italiano 7:51–70.

Benazzi, M., Giannini, E. 1971. *Cura azteca*, nuova specie di planaria del Messico. Atti della Accademia Nazionale dei Lincei 50:477–481.

Blaustein, L. 1990. Evidence for predatory flatworms as organizers of zooplankton and mosquito community structure in rice fields. Hydrobiologia 199:179–191.

Blaustein, L., Dumont, H. J. 1990. Typhloplanid flatworms (*Mesostoma* and related genera): mechanisms of predation and evidence that they structure aquatic invertebrate communities. Hydrobiologia 198:61–77.

Boddington, M. J., Mettrick, D. F. 1977. A laboratory study of the population dynamics and productivity of *Dugesia polychroa* (Turbellaria: Tricladida). Ecology 58:109–118.

Calow, P., Read, D. A. 1986. Ontogenetic patterns and phylogenetic trends in freshwater flatworms (Tricladida); constraint or selection? Hydrobiologia 132:263–272.

Cannon, L. R. G. 1986. Turbellaria of the world. A guide to families and genera, Queensland Museum, Brisbane, QLd., Australia.

Case, T. J., Washino, R. K. 1979. Flatworm control of mosquito larvae in rice fields. Science 206:1412–1414.

Chandler, C. M., Darlington, J. T. 1986. Further field studies on freshwater planarians of Tennessee (Turbellaria: Tricladida): III. Western Tennessee. Journal of Freshwater Ecology 3:493–501.

Chodorowski, A. 1959. Ecological differentiation of turbellarians in Harsz Lake. Polskie Archiwum Hydrobiologii 6:33–73.

Chodorowski, A. 1960. Vertical stratification of Turbellaria species in some littoral habitats of Harsz Lake. Polskie Archiwum Hydrobiologii 8:153–163.

Claussen, D. L., Walters, L. M. 1982. Thermal acclimation in the freshwater Planarians, *Dugesia tigrina* and *D. dorotocephala*. Hydrobiologia 94:231–236.

Collins, F. H., Washino, R. K. 1979. Factors affecting the density of *Culextarsalis* and *Anopheles freeborni* in northern California rice fields. Proceedings of California Mosquito Control Association 46:97–98.

Davies, R. W., Reynoldson, T. B. 1971. The incidence and intensity of predation on lake-dwelling triclads in the field. Journal of Animal Ecology 40:191–214.

Dumont, H. J., Carels, I. 1987. Flatworm predator (*Mesostoma* cf. *lingua*) releases a toxin to catch planktonic prey (*Daphnia magna*). Limnology and Oceanography 32:699–702.

Dumont, H. J., Schorreels, S. 1990. A laboratory study of the feeding of *Mesostoma lingua* (Schmidt) (Turbellaria: Neorhabdocoela) on *Daphnia magna* Straus at four different temperatures. Hydrobiologia 198:79–89.

Ehlers, U. 1986. Comments on a phylogenetic system of the Platyhelminthes. Hydrobiologia 132:1–12.

Ferguson, F. F. 1939. A monograph of the genus *Macrostomum* O. Schmidt 1848. Part II. Zoologischer Anzeiger 126:131–144.

Fiore, L., Ioalé, P. 1973. Regulation of the production of subitaneous and dormant eggs in the turbellarian *Mesostoma ehrenbergii* (Focke). Monitore Zoologico Italiano 7:203–224.

Gee, H., Young, J. O. 1993. The food niches of the invasive *Dugesia tigrina* (Girard) and indigeneous *Polycelis tenuis* Ijima and *P. nigra* (Müller) (Turbellaria; Tricladida). Hydrobiologia 254: 99–106.

Gibson, R., Moore, J. 1976. Freshwater nemerteans. Zoological Journal of the Linnean Society 58:177–218.

Gibson, R., Moore, J. 1978. Freshwater nemerteans: new records of *Prostoma* and a description of *Prostoma canadiensis* sp. nov. Zoologischer Anzeiger 201:77–85.

Gibson, R., Young, J. O. 1976. Ecological observations on a population of the freshwater hoplonemertean *Prostoma jenningsi* Gibson and Young 1971. Archiv für Hydrobiologie 78:42–50.

Gilbert, C. M. 1938. Two new North American Rhabdocoeles—*Phaenocora falciodenticulata* nov. spec. and *Phaenocora kepneri* adenticulata nov. subspec. Zoologischer Anzeiger 122:208–223.

Göltenboth, F., Heitkamp, U. 1977. *Mesostoma ehrenbergi* (Focke, 1836) Plattwürmer (Strudelwürmer). Biologie, mikroskopische Anatomie und Cytogenetik, *in*: Siewing, R., Ed., Grosses Zoologisches Praktikum Heft 6a:pp. 1–60.

Gourbault, N. 1972. Recherches sur les triclades paludicoles hypogés. Memories du Muséum. National d'Histoire Naturelle 73:1–249.

Hampton, A. M. 1988. Altitudinal range and habitat of triclads in streams of the Lake Tahoe basin. American Midland Naturalist 120:302–312.

Harrison, A. D., Rankin, J. J. 1976. Hydrobiological studies of Eastern Lesser Antillean Islands. II. St. Vincent: freshwater fauna—its distribution, tropical river zonation and biogeography. Archiv für Hydrobiologie Suppl 50:275–311.

Hebert, P. D. N., Payne, W. 1985. Genetic variation in populations of the hermaphroditic flatworm *Mesostoma lingua* (Turbellaria, Rhabdocoela). Biological Bulletin (Woods Hole, MA) 169:143–151.

Heitkamp, U. 1979a. Abundanzdynamik und reproduktionsphasen limnophiler tricladen (Turbellaria) aus stehenden kleingewassern. Hydrobiologia 65:49–57.

Heitkamp, U. 1979b. Die Respirationsrate neorhabdocoeler Turbellarien mit unterschiedlicher Temperaturvalenz. Archiv für Hydrobiologie 87:95–111.

Heitkamp, U. 1982. Untersuchungen zur Biologie, Ökologie und Systematik limnischer Turbellarien periodischer und perennierender Kleingewässer Südniedersachsens. Archiv für Hydrobiologie, Supplement 64:65–188.

Holmquist, Ch. 1967. Turbellaria of northern Alaska and northwestern Canada. Internationale Revue des gesamten Hydrobiologie 52:123–139.

Holopainen, I. J., Paasivirta, L. 1977. Abundance and biomass of the meiozoobenthos in the oligotrophic and mesohumic lake Paajarvi, southern Finland. Annales Zoologici Fennici 14:124–134.

Hyman, L. H. 1951. The invertebrates: Platyhelminthes and Rhynchocoela, the acoelomate Bilateralia. McGraw-Hill, New York.

Hyman, L. H. 1955. Descriptions and records of freshwater Turbellaria from the United States. American Museum Novitates 1714:1–36.

Jennings, J. B. 1977. Patterns of nutrition in free-living and symbiotic Turbellaria and their implications for the evolution of parasitism in the phylum Platyhelmithes. Acta Zoologica Fennica 154:63–79.

Jennings, J. B. 1985. Feeding and digestion in the aberrant planarian *Bdellasimilis barwicki* (Turbellaria:Tricladidae:Procerodidae): an ectosymbiote of freshwater turtles in Queensland and New South Wales. Australian Journal of Zoology 33:317–327.

Kawakatsu, M., Mitchell, R. W. 1984a. A list of retrobursal triclads inhabiting freshwaters and aquatic probursal triclads regarded as marine relicts, with corrective remarks on our 1984 paper published in Zoological Science, Tokyo. Occasional Publications, Biological Laboratory of Fuji Women's College, Sapporo 11:1–8.

Kawakatsu, M., Mitchell, R. W. 1984b. Redescription of *Dugesia azteca* (Benazzi et Giannini, 1971) based upon the material collected from the type locality in Mexico, with corrective remarks. Bulletin of the National Science Museum, Ser. A 10:37–50.

Kenk, R. 1972. Freshwater planarians (Turbellaria) of North America, *in*: EPA Biota of Freshwater Ecosystems, Identification manual 1:1–81.

Kenk, R. 1987. Freshwater triclads (Turbellaria) of North America. XVI. More on subterranean species of Phagocata of the Eastern United States. Proceedings of the Biological Society of Washington 100:664–673.

Kepner, W. A., Carter, J. S. 1931. Ten well-defined new species of *Stenostomum*. Zoologischer Anzeiger 98:108–123.

Kolasa, J. 1977a. Bottom fauna of the heated Konin Lakes. Turbellaria and Nemertini. Monografie Fauny Polski 7:29–48.

Kolasa, J. 1977b. Remarks on the marine-originated Turbellaria in the freshwater fauna. Acta Zoologica Fennica 1954:81–87.

Kolasa, J. 1979. Ecological and faunistical characteristics of Turbellaria in the eutrophic Lake Zbechy. Acta Hydrobiologica 21:435–459.

Kolasa, J. 1983. Formation of the turbellarian fauna in a submontane stream. Acta Zoologica Cracoviensia 26:57–107.

Kolasa, J. 1987. Population growth in some *Mesostoma* species (Turbellaria), predatory on mosquitoes. Freshwater Biology 18:205–212.

Kolasa, J., Strayer, D., Bannon-O'Donnell, E. 1987. Microturbellarians from interstitial waters, streams, and springs in southeastern

New York. Journal of the North American Benthological Society 6:125–132.

Lanfranchi, A., Papi, F. 1978. Turbellaria (excl. Tricladida), *in*: Illies, J. Ed., Limnofauna Europaea, G. Fisher, Stuttgart.

Luther, A. 1955. Die Dalyelliiden. Acta Zoologica Fennica 87:1–337.

Luther, A. 1960. Die Turbellarien Ostfennoskandiens. I. Acoela, Catenulida, Macrostomida, Lecithoepitheliata, Prolecithophora, und Proseriata. Fauna Fennica 7:1–155.

Luther, A. 1963. Die Turbellarien Ostfennoskandiens IV. Neorhabdocoela 2. Typhloplanoida: Typhloplanidae, Solenopharyngidae und Carcharodopharyngidae. Fauna Fennica 16:1–163.

Maly, E. J., Schoenholtz, S., Arts, M. T. 1980. The influence of flatworm predation on zooplankton inhabiting small ponds. Hydrobiologia 76:233–240.

Marcus, E. 1945. Sobre microturbellarios do Brasil. Comunicaciones Zoologicas del Museo de Historia Natural de Montevideo 25:1–97.

Marcus, E. 1951. Contributions to the natural history of Brazilian Turbellaria. Comunicaciones Zoologicas del Museo de Historia Natural de Montevideo 63:1–25.

Martens, P. M. 1984. Comparison of three different extraction methods for Turbellaria. Marine Ecology Progress Series 14: 229–234.

McConnell, J. V. 1967. A manual of psychological experimentation on planarians. A special publication of the Worm Runner's Digest. J. V. McConnell, Ann Arbor, MI.

McDermott, J. J., Roe, P. 1985. Food, feeding behavior and feeding ecology of nemerteans. American Zoologist 25:113–125.

Mead, A. P., Kolasa, J. 1984. New records of freshwater Microturbellaria from Nigeria, West Africa. Zoologischer Anzeiger 212:257–271.

Moore, J., Gibson, R. 1985. The evolution and comparative physiology of terrestrial and freshwater nemerteans. Biological Reviews 60:257–312.

Nalepa, T. F., Quigley, M. A. 1983. Abundance and biomass of meiobenthos in nearshore Lake Michigan with comparisons to macrozoobenthos. Journal of the Great Lakes Research 9:523–529.

Noldt, U., Wehrenberg, C. 1984. Quantitative extraction of living Platyhelminthes from marine sands. Marine Ecology Progress Series 20:193–201.

Nuttycombe, J. W. 1956. The *Catenula* of the eastern United States. American Midland Naturalist 55:419–433.

Nuttycombe, J. W., Waters, A. J. 1938. The American species of the genus *Stenostomum*. Proceedings of the American Philosophical Society 79:213–301.

Pattee, E. 1980. Coefficients thermiques et ecologie de quelques planaires d'eau douce. VII: leur zonation naturelle. Annales de Limnologie 16:21–41.

Peter, R. 1995. Regenerative and reproductive capacities of the fissiparous planarian *Dugesia tahitiensis*. Hydrobiologia 305:261.

Reynoldson, T. B. 1982. The population biology of Turbellaria with special reference to the freshwater triclads of the British Isles. Advances in Ecological Research 13:235–365.

Rieger, R. M. 1986. Asexual reproduction and the turbellarian archetype. Hydrobiologia 132:35–45.

Rixen, J.-U. 1968. Beitrag zur Kenntnis der Turbellarienfauna des Bodenseegebietes. Archiv für Hydrobiologie 64:335–365.

Ruiz-Trillo, I., Riutort, M., Littlewood, D. T. J., Herniou, E. A., Baguñà, J. 1999. Acoel flatworms: earliest extant bilateralian metazoans, not members of Platyhelminthes. Science 283:1919–1923.

Sayre, R. M., Powers, E. M. 1966. A predacious soil turbellarian that feeds on free-living and plant-parasitic nematodes. Nematologica 12:619–629.

Schwank, P. 1976. Quantitative Untersuchungen an litoralen Turbellarien des Bodensees. Jh. Ges. Naturkde. Württemberg 131: 163–181.

Schwank, P. 1981. Turbellarien, Oligochaeten and Archianneliden des Breitenbachs und anderer oberchessischer Mittelgebirgsbäche. II. Die Systematik und Autökologie der einzelnen Arten. Archiv für Hydrobiologie, Supplement 62:1–85.

Schwartz, S. S., Hebert, P. D. N. 1982. A laboratory study of the feeding behaviour of the rhabdocoel *Mesostoma ehrenbergii* on pond Cladocera. Canadian Journal of Zoology 60:1305–1307.

Schwartz, S. S., Hebert, P. D. N. 1986. Prey preference and utilization by *Mesostoma lingua* (Turbellaria, Rhabdocoela) at a low arctic site. Hydrobiologia 135:251–257.

Strayer, D. 1985. The benthic micrometazoans of Mirror Lake, New Hampshire. Archiv für Hydrobiologie, Supplement 72: 287–426.

Strayer, D., Likens, G. E. 1986. An energy budget for the zoobenthos of Mirror Lake, New Hampshire. Ecology 67:303–313.

Turbeville, J. M., Ruppert, E. E. 1985. Comparative ultrastructure and evolution of Nemertines. American Zoologist 25:53–71.

Tyler, S., Burt, M. D. B. 1988. Lensing by a mitochondrial derivative in the eye of *Urastoma cyprinae* (Turbellaria, Prolecithophora). Fortschritte der Zoologie 36:229–234.

Vannote, R. L., Minshall, G. W., Cummins, K. W., Sedell, J. R., Cushing, C. E. 1980. The river continuum concept. Canadian Journal of Fisheries and Aquatic Sciences. 37:130–137.

Wrona, F. 1986. Distribution, abundance, and size of rhabdoids in *Dugesia polychroa* (Turbellaria: Tricladida). Hydrobiologia 132:287–293.

Young, J. O. 1970. British and Irish freshwater Microturbellaria: historical records, new records and a key for their identification. Archiv für Hydrobiologie 67:210–241.

Young, J. O. 1973a. The occurence of microturbellaria in some British lakes. Archiv für Hydrobiologie 72(2):207–208.

Young, J. O. 1973b. The prey and predators of *Phaenocora typhlops* (Vejdovsky) (Turbellaria: Neorhabdocoela) living in a small pond. Journal of Animal Ecology 42:637–643.

Young, J. O. 1975. The population dynamics of *Phaenocora typhlops* (Vejdovsky) (Turbellaria: Neorhabdocoela) living in a pond. Journal of Animal Ecology 44:251–262.

Young, J. O., Eaton, J. W. 1975. Studies on the symbiosis of *Phaenocora typhlops* (Vejdovsky) (Turbellaria; Neorhabdocoela) and *Chlorella vulgaris* var. *vulgaris*, Fott & Novakova (Chloroccales). II. An experimental investigation into the survival value of the relationship to host and symbiont. Archiv für Hydrobiologie 75:225–239.

Young, J. O., Gibson, R. 1975. Some ecological studies on two populations of the freshwater hoplonemertean *Prostoma eilhardi* (Montgomery, 1894) from Kenya. Verhandlungender internationale Vereinigung für. Limnologie 19:2803–2810.

Yu, H.-S., Legner, E. F. 1976. Regulation of aquatic diptera by planaria. Entomophaga 21:3–12.

APPENDIX 6.1 List of North American Species of Microturbellaria

---

The arrangement of orders and families follows with some modifications from that of Lanfranchi and Papi (1978).

Catenulida
- Family Catenulidae
  - *Catenula confusa*
  - *Catenula lemnae*
  - *Catenula leptocephala*
  - *Catenula sekerai*
  - *Catenula virginia*
  - *Suomina turgida*
- Family Chordariidae
  - *Chordarium europaeum*
- Family Stenostomidae
  - *Myostenostomum tauricum*
  - *Rhynchoscolex platypus*
  - *Rhynchoscolex simplex*
  - *Rhynchoscolex* sp. 1
  - *Rhynchoscolex* sp. 2
  - *Stenostomum anatirostrum*
  - *Stenostomum anops*
  - *Stenostomum arevaloi*
  - *Stenostomum beauchampi*
  - *Stenostomum brevipharyngium*
  - *Stenostomum ciliatum*
  - *Stenostomum cryptops*
  - *Stenostomum glandulosum*
  - *Stenostomum grande*
  - *Stenostomum kepneri*
  - *Stenostomum leucops*
  - *Stenostomum mandibulatum*
  - *Stenostomum membranosum*
  - *Stenostomum occultum*
  - *Stenostomum pegephilum*
  - *Stenostomum predatorium*
  - *Stenostomum pseudoacetabulum*
  - *Stenostomum simplex*
  - *Stenostomum temporaneum*
  - *Stenostomum tuberculosum*
  - *Stenostomum unicolor*
  - *Stenostomum uronephrium*
  - *Stenostomum ventronephrium*
  - *Stenostomum virginianum* (probably synonymous with *S. unicolor*)
- Macrostomida
  - Family Macrostomidae
  - *Macrostomum collistylum*
  - *Macrostomum curvistylum*
  - *Macrostomum gilberti*
  - *Macrostomum glochostylum*
  - *Macrostomum lewisi*
  - *Macrostomum norfolkense*
  - *Macrostomum ontarioense*
  - *Macrostomum orthostylum*
  - *Macrostomum phillipsi*
  - *Macrostomum reynoldsi*
  - *Macrostomum riedeli*
  - *Macrostomum ruebushi*
  - *Macrostomum sensitivum*
  - *Macrostomum sillimani*
  - *Macrostomum* sp. 1, indeterm.
  - *Macrostomum* sp. 2, indeterm.
  - *Macrostomum* sp. 3, indeterm.
  - *Macrostomum stirewalti*
  - *Macrostomum tenneseense*
  - *Macrostomum tuba*
  - *Macrostomum lineare*
  - *Macrostomum virginianum*
  - Family Microstomidae
  - *Microstomum lineare*

Lecithoepitheliata
- Family Prorhynchidae
- *Geocentrophora cavernicola*
- *Geocentrophora sphyrocephala*
- *Prorhynchus stagnalis*

Proseriata

Family Plagiostomidae
- *Hydrolimax grisea*

Prolecithophora

Family Bothrioplanidae
- *Bothrioplana semperi*

Family Otomesostomidae
- *Otomesostoma auditivum*

Dalyellioida

Family Provorticidae
- Genus indet. species indet
- *Pilgramilla virginiensis*

Family Dalyelliidae
- *Castrella pinguis*
- *Castrella graffi*
- *Dalyellia viridis*
- *Microdalyella abursalis*
- *Microdalyella circobursalis*
- *Microdalyella fairchildi*
- *Microdalyella gilesi*
- *Microdalyella rheesi*
- *Microdalyella rochesteriana*
- *Microdalyella rossi*
- *Microdalyella ruebushi*
- *Microdalyella sillimani*
- *Microdalyella tennesseensis*
- *Microdalyella virginiana*

Typhloplanida

Family Typhloplanidae
- *Amphibolella spinulosa*

Ascophora elegantissima
Bothromesostoma personatum
Castrada hofmanni
Castrada lutheri
Castrada virginiana
Castrada sp. indet
Krumbachia cf. hiemalis
Krumbachia minuta
Krumbachia virginiana
Limnoruanis romanae
Mesostoma arctica
Mesostoma californicum
Mesostoma columbianum
Mesostoma craci
Mesostoma curvipenis
Mesostoma ehrenbergii
Mesostoma macroprostatum
Mesostoma platygastricum
Mesostoma vernale
Mesostoma virginianum
Olisthanella truncula
Opistomum pallidum
Phaenocora agassizi
Phaenocora falciodenticulata
Phaenocora highlandense
Phaenocora kepneri
Phaenocora lutheri
Phaenocora virginiana
Prorhynchella minuta
Protoascus wisconsinensis
Pseudophaenocora sulflophila
Rhynchomesostoma rostratum
Strongylostoma cf., *elongatum*
Strongylostoma radiatum
Typhloplanella halleziana

Kalyptorhynchia

Family Polycystidae
Gyratrix hermaphroditus
Opistocystis goettei

Family undetermined
Microrhynchus virginianus

7 GASTROTRICHA

David Strayer
Institute of Ecosystem Studies
Millbrook, New York 12545

William D. Hummon
Department of Biological Sciences
Ohio University
Athens, Ohio 45701

I. INTRODUCTION

The gastrotrichs are among the most abundant and poorly known of the freshwater invertebrates. Gastrotrichs are nearly ubiquitous in the benthos and periphyton of freshwater habitats, with densities typically in the range of 100,000–1,000,000 individuals/m^2. Nonetheless, the remarkable life cycle of freshwater gastrotrichs was worked out only recently, and we know almost nothing about how the distribution and abundance of these animals are controlled in nature. The impact of freshwater gastrotrichs on their food resources and on freshwater ecosystems has not yet been investigated. Twelve genera and fewer than 100 species of freshwater gastrotrichs are now known from North America. Because the North American gastrotrich fauna has received so little study, these numbers understate the real diversity of the fauna.

Formerly placed in the Aschelminthes or Nemathelminthes, gastrotrichs usually are now considered to constitute a phylum of their own. The evolutionary placement of gastrotrichs is unclear; some authors (e.g., Lorenzen, 1985; Wallace *et al.*, 1996) believe that they are most closely related to nematodes, kinorhynchs, loriciferans, nematomorphs, and priaulids, while others (e.g., Conway Morris, 1993) think that gastrotrichs are more closely related to rotifers and acanthocephalans. Important general references on gastrotrichs include Remane (1935–1936), Hyman (1951), Voigt (1958), d'Hondt (1971a), Hummon (1982), Ruppert (1988), Schwank (1990), and Kisielewski (1991, 1998). The phylum contains two orders: the Macrodasyida, which consists almost entirely of marine species, and the Chaetonotida, which contains marine, freshwater, and semiterrestrial species. Unless noted otherwise, the information included in this chapter refers to freshwater members of the Chaetonotida.

Macrodasyids usually are distinguished from chaetonotids by the presence of pharyngeal pores and more numerous adhesive tubules (Fig. 1). Macrodasyids are common in marine and estuarine sands, but are barely represented in inland freshwaters. Two species of freshwater gastrotrichs have been placed in the Macrodasyida.

Ruttner-Kolisko (1955) described an aberrant gastrotrich, *Marinellina flagellata* (Fig. 1A), from the hyporheic zone of an Austrian river. Unfortunately, she was able to find only two specimens, both of them apparently immature. Because *Marinellina* has a pair of anterior lateral structures that Ruttner-Kolisko interpreted as adhesive tubules, she placed this species in the Macrodasyida. Remane (1961) rejected the assignment of *Marinellina* to the macrodasyids and placed it instead in the chaetonotid family Dichaeturidae. Kisielewski (1987) reaffirmed Ruttner-Kolisko's

Ecology and Classification of North American Freshwater Invertebrates. 2nd Edition

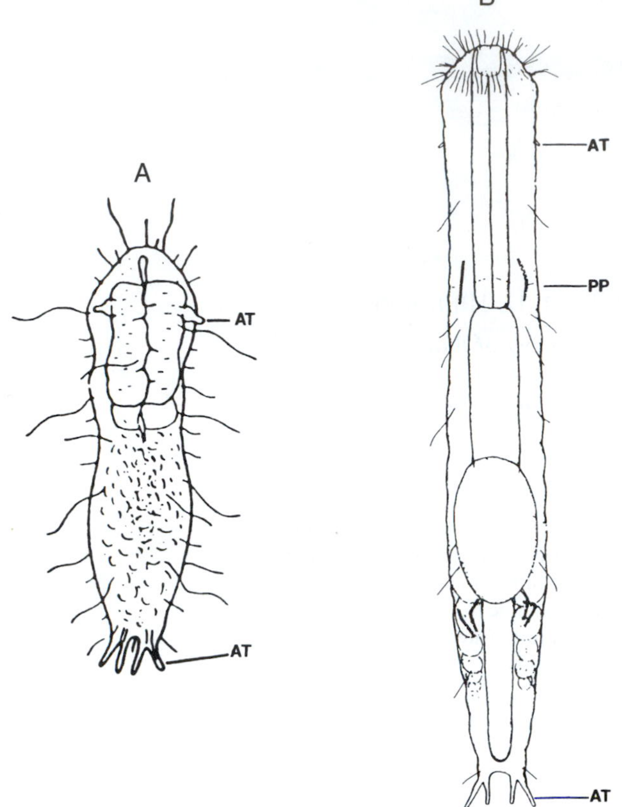

FIGURE 1 Freshwater macrodasyid gastrotrichs: (A) *Marinellina flagellata*, and (B) *Redudasys fornerise*. AT, adhesive tube; and PP, pharyngeal pore. From Ruttner-Kolisko (1955) and Kisielewski (1987).

original placement of the species. An animal similar to *Marinellina* was just discovered in the hyporheic zone of another Austrian stream (Schmid-Araya and Schmid, 1995); but until it is studied critically, the systematic placement of these enigmatic Austrian gastrotrichs will remain unclear.

Kisielewski (1987) discovered an undoubted macrodasyid, *Redudasys fornerise* (Fig. 1B), from the psammon of a Brazilian reservoir. Additional freshwater macrodasyids probably will be found when appropriate habitats (psammon, hyporheic zone) are explored. The distribution, biology, and evolutionary relationships of any such species will be of great interest.

II. Anatomy and Physiology

The following brief account of gastrotrich anatomy is summarized chiefly from Remane (1935–1936), Hyman (1951), Hummon (1982), and Boaden (1985), which should be consulted for greater detail.

A. External Morphology

Gastrotrichs are colorless animals, spindle- or tenpin-shaped, and ventrally flattened (Fig. 2). Freshwater gastrotrichs are 50–800 μm long. Conspicuous external features include a more or less distinct head, which bears sensory cilia, and a cuticle which, in most species, is ornamented with spines or scales of various shapes. In the most common freshwater family, the Chaetonotidae (as well as in the rare Dichaeturidae and Proichthydiidae), the posterior end of the body is formed into a furca, which contains distal adhesive tubes that allow the animal to attach itself tenaciously to surfaces. In other families, these structures are absent, but the posterior end of the body may bear long spines or sensory bristles. The ventral side of the animal bears longitudinal rows or patches of cilia, which provide the forward-gliding locomotion of the animal.

B. Organ System Function

The digestive system begins with a subterminal mouth, which may be surrounded by a ring of short bristles. Between the mouth and the pharynx lies a cuticular buccal capsule, which is often somewhat protrusible. The muscular pharynx is similar to the nematode pharynx, with a triradiate, Y-shaped lumen. Often, there are anterior and posterior swellings, but these lack the valves characteristic of the pharyngeal bulbs of nematodes. Posterior to the pharynx is an undifferentiated gut, which empties into the anus.

The paired reproductive organs lie lateral to the gut in the posterior half of the body. In young animals, large, developing, parthenogenetic eggs are present. As there are no oviducts, the egg is released through a rupture in the ventral body wall. Older animals become hermaphrodites (see below) and bear both sperm sacs and developing sexual (i.e., meiotic) eggs lateral to the gut. An unpaired organ of unknown function, the X-organ, lies posterior to the gonads near the anus.

The brain is bilobed, straddling the pharynx. Sensory organs include long cilia and bristles, which presumably are tactile. Balsamo (1980) demonstrated that *Lepidodermella squamata* is photosensitive, although it does not appear to have distinct multicellular photoreceptors. Gray and Johnson (1970) found evidence of a tactile chemical sense in a marine macrodasyid; it seems probable that freshwater chaetonotids possess similar chemosensory abilities.

The excretory system consists of a pair of protonephridia (Brandenburg, 1962) in the anterior mid-

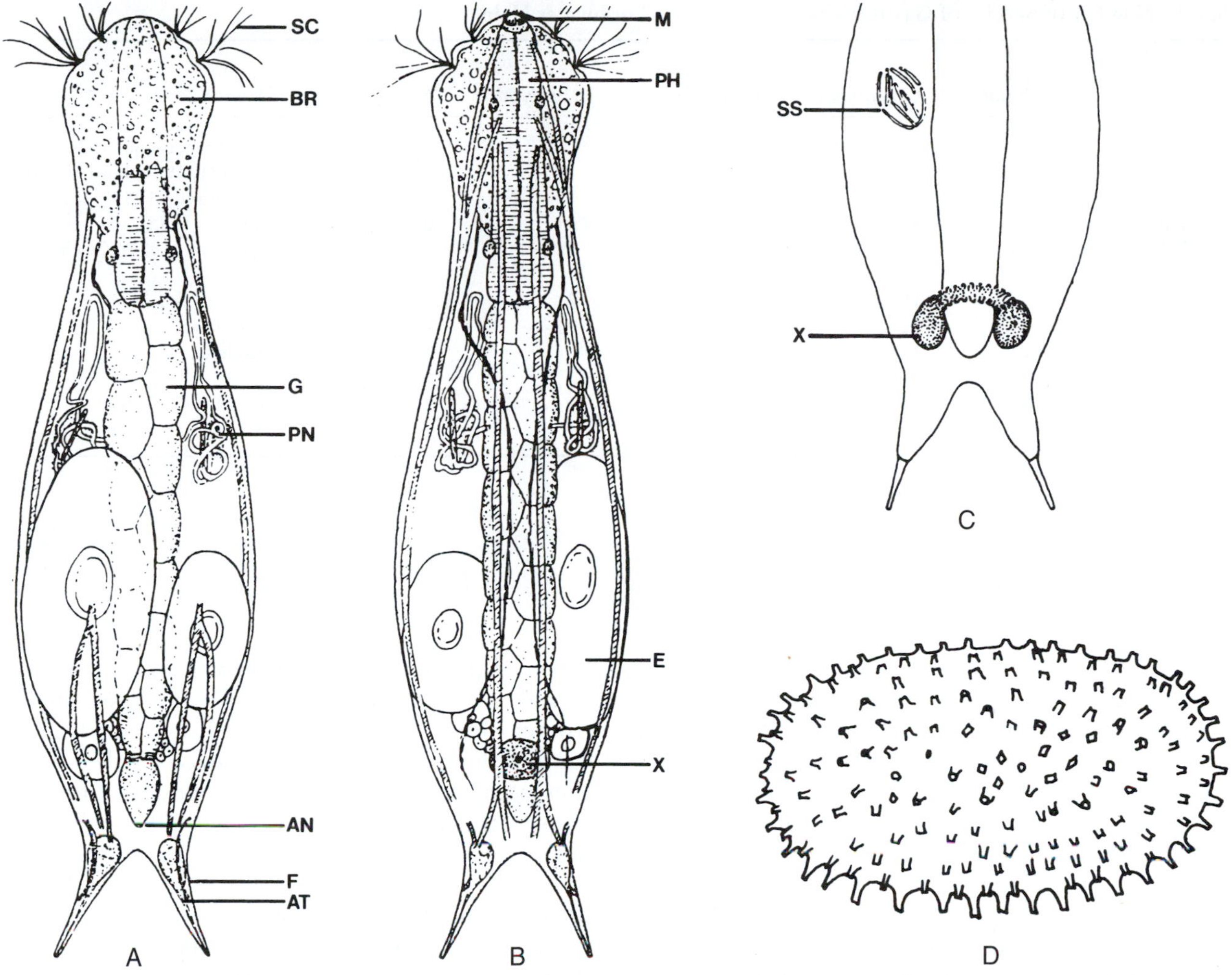

FIGURE 2 Schematic illustration of a typical chaetonotid gastrotrich showing (A) dorsal view, (B) ventral view, (C) posterior end of a hermaphrodite, showing sexual organs, and (D) egg. AT, adhesive tubes; AN, anus; BR, brain; E, egg; F, furca; G, gut; M, mouth; PH, pharynx; PN, protonephridium; SC, sensory cilia; SS, sperm sac; and X, X-organ. Modified from Remane (1935–1936), Voigt (1958), and Kisielewska (1981).

body, which empty through pores on the ventral body surface. There is no circulatory or respiratory system, per se.

III. ECOLOGY AND EVOLUTION

A. Diversity and Distribution

Gastrotrichs are widely distributed in freshwaters in surface sediments and among vegetation. Some species of the Neogosseidae and Dasydytidae are good swimmers and are occasionally reported from the plankton of shallow, weedy lakes (e.g., Hutchinson, 1967; Green, 1986; Kisielewski, 1991). However, none of the gastrotrichs has become as truly planktonic as the daphnid cladocerans or ploimate rotifers.

Gastrotrichs have been found in a wide range of freshwater and semiterrestrial habitats. They are abundant in most lakes, ponds, and wetlands (Tables I and II). Kisielewski (1981, 1986a) showed that gastrotrich density and species richness are positively correlated with the productivity of the habitat, and several workers have found that gastrotrich density is highest in highly organic sediments (e.g., Strayer, 1985). Apparently, most gastrotrichs live very near the sediment

TABLE I Number of Species of Gastrotrichs Found in Some Freshwater Habitats.

Habitat	*Total*	*Chaetonotus*	*Other Chaetonotidae*	*Dasydytidae*	*Neogosseidae*	*Source*
European lakes[a]	18	12	5	<1	0	Preobrajenskaja, 1926; Balsamo, 1981, 1990, Bertolani and Balsamo, 1989
Mirror Lake, NH	20–32	8–20	12	0	0	Strayer, 1985
Ponds, Poland[b]	21	10	7	3	<1	Nesteruk, 1986; Kisielewski, 1986a
Peat bogs, Poland[c]	27	16	9	2	0	Kisielewski, 1981
Bog pools, Poland[d]	24	12	8	4	0	Kisielewska, 1982
Phragmites mats, Romania	28	16	7	5	0	Rudescu, 1968
Ponds, Brazil[e]	38	14	12	10	2	Kisielewski, 1991
Rivers, Brazil[e]	22	5	12	4	1	Kisieleswki, 1991

[a] Mean of four lakes.
[b] Mean of seven ponds.
[c] Mean of four bogs.
[d] Mean of two pools.
[e] Sum of several collecting sites.

surface in lakes (Fig. 3). Gastrotrichs also are abundant in unpolluted streams, where they inhabit sand bars, sometimes in great numbers (Hummon *et al.*, 1978; Hummon, 1987). Apparently, they are scarce in groundwaters other than the hyporheic zone (Renaud-Mornant, 1986). Their rarity in underground waters is surprising, because many marine gastrotrichs are interstitial in habit (d'Hondt, 1971a; Renaud-Mornant, 1986), and because the small size and bacterial diet of gastrotrichs would seem to preadapt them to the groundwater habitat.

Gastrotrichs are among the few animals commonly found in anaerobic environments (Moore, 1939; Cole, 1955; Strayer, 1985), remaining abundant even during extended periods (i.e., months) of anoxia. The physiological basis of the anaerobiosis of freshwater gastrotrichs has not yet been studied. It seems likely that some freshwater gastrotrichs possess a sulfide detoxification mechanism similar to that demonstrated for marine gastrotrichs (Powell *et al.*, 1979, 1980) and freshwater nematodes (Nuss, 1984; Nuss and Trimkowski, 1984) to deal with the elevated concentrations of H_2S that often accompany extended anoxia.

Little is known of the factors that control the distribution of individual species of gastrotrichs in freshwater. Arguing largely from analogy with marine work (d'Hondt, 1971a), we might expect factors of primary importance to include the granulometry, stability,

TABLE II Density and Biomass of Gastrotrichs in Some Freshwater Habitats.

Site	*Density* (No./m^2)	*Biomass* (mg DM/m^2)	*Source*
Lake Bikcze, Poland	1,160,000[a]	23[a,b]	Nesteruk, 1996
Lake Brzeziczno, Poland	920,000[a]	30[a,b]	Nesteruk, 1996
Lake Piaseczno, Poland	910,000[c]	23[b,c]	Nesteruk, 1996
Mirror Lake, NH	130,000[d]	1[d]	Strayer, 1985
Lake Suviana, Italy	57,000[e]	—	Madoni, 1989
Lake Erie, OH	50,000[f]	—	Evans, 1982
Three small ponds, Poland	1,600,000–2,600,000	25–78[b]	Nesteruk, 1996
Mississippi River, MN	130,000–230,000[g]	—	Hummon, 1987, and unpublished

[a] Littoral zone.
[b] Converted from wet mass by multiplying by 0.15.
[c] Mean of three stations.
[d] Lakewide mean.
[e] Mean of two deepwater stations.
[f] Beaches.
[g] Sand bars.

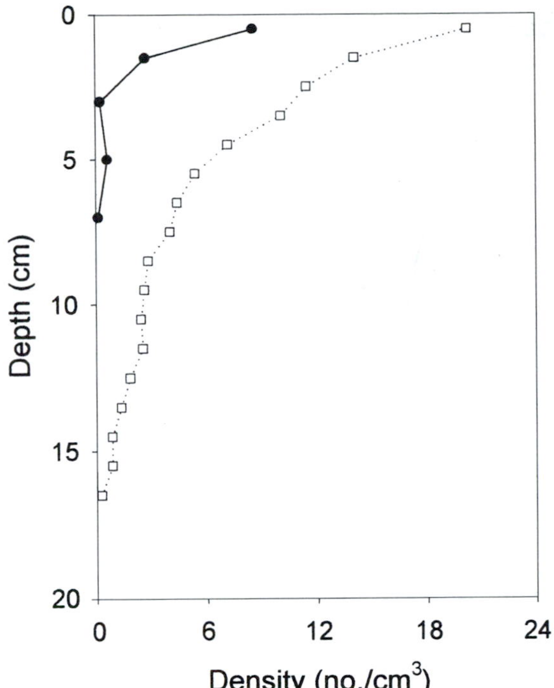

FIGURE 3 Vertical distribution of gastrotrichs within the sediments of Lake Brzeziczno, Poland, (dashed line) and Mirror Lake, New Hampshire, (solid line) as a function of depth from the sediment surface. From data of Strayer (1985) and Nesteruk (1991).

packing, and organic content of the sediment, the amount of dissolved oxygen, and the density and composition of communities of microbes and predators. Also, culture work suggests that the inorganic chemistry of the water and the presence of anthropogenic contaminants can exert a strong influence on gastrotrich populations (Hummon, 1974; Faucon and Hummon, 1976; Hummon and Hummon, 1979).

Many genera of freshwater gastrotrichs are known to have intercontinental or cosmopolitan distributions. Exceptions include several genera so far known only from Brazil (the macrodasyid *Redudasys*, the chaetonotids *Arenotus* and *Undula*, and the dasydytid *Ornamentula*); *Dichaetura* and *Marinellina*, rare species known only from Europe, and the proichthydiids *Proichthydium* and *Proichthyioides*, which have been found only at their type localities in Argentina and Japan, respectively. The geographic distribution of individual species is not well known, because of the primitive state of the species-level taxonomy of freshwater gastrotrichs and the paucity of field studies throughout most of the world. Even in North America, only Arkansas (Davis, 1937), Illinois (Robbins, 1965, 1973; Horlick, 1975), northern Indiana (Pfaltzgraff, 1967), Michigan (Brunson, 1947, 1948, 1950), and eastern Ohio (Craft, 1993) have received even cursory surveys of freshwater gastrotrichs. Thus, the 76 species that Schwank (1990) reported from North America probably are a small fraction of North America's freshwater gastrotrich fauna. Some species have been reported to occur over broad ranges, including in some cases more than one continent (e.g., Kisielewski, 1991). Until further studies are made, such reported intercontinental distributions should be regarded with caution (cf. Frey, 1982; Todaro *et al.*, 1996), especially because "conspecifics" collected from different continents often exhibit marked morphological differences from one another (e.g., Robbins, 1973, Emberton, 1981).

Chaetonotus typically dominates gastrotrich faunas everywhere in freshwater, both in terms of numbers of species and numbers of individuals (Table I). Other genera of the Chaetonotidae are common in all kinds of freshwaters, but are usually less abundant than *Chaetonotus*. The Dasydytidae and Neogosseidae are less widespread than the Chaetonotidae, usually living in weedy, productive waters, where they may, however, become numerically abundant (e.g., Blinn and Green, 1986; Kisielewski, 1981, 1986a; Nesteruk, 1986). Dasydytids and neogosseids also are more abundant and speciose in the tropics than in temperate waters (Table I; Kisielewski 1991). The very rarely seen Dichaeturidae and Proichthydiidae have been collected from cisterns, underground waters, tree holes, and among moss (Remane, 1935–1936, Ruttner-Kolisko, 1955; Sudzuki, 1971).

B. Reproduction and Life History

Until recently, populations of freshwater gastrotrichs were thought to consist entirely of parthenogenetic females (e.g., Hyman; 1951; Pennak, 1978). However, more detailed recent studies have dispelled this notion, and have revealed a remarkable life cycle among freshwater gastrotrichs (Fig. 4) (Weiss and Levy, 1979; Hummon, 1984a–c, 1986; Levy, 1984).

Newly hatched gastrotrichs are relatively large (approximately two-thirds the length of adults—Brunson, 1949) and already contain developing parthenogenetic eggs. These eggs develop rapidly under favorable conditions. At 20°C, the first egg may be laid within a day after the mother hatches. Typically, a total of four parthenogenetic eggs is laid over a 4-day period. Apparently, parthenogenesis is apomictic, so that offspring are genetically identical to their mother.

There are two kinds of parthenogenetic eggs. The more common kind, tachyblastic eggs, develop immediately and hatch quickly (within a day of being laid, at 20°C). Occasionally, the final parthenogenetic egg laid by a female is not a tachyblastic egg, but a resting, or opsiblastic, egg. Opsiblastic eggs are thick-shelled, a

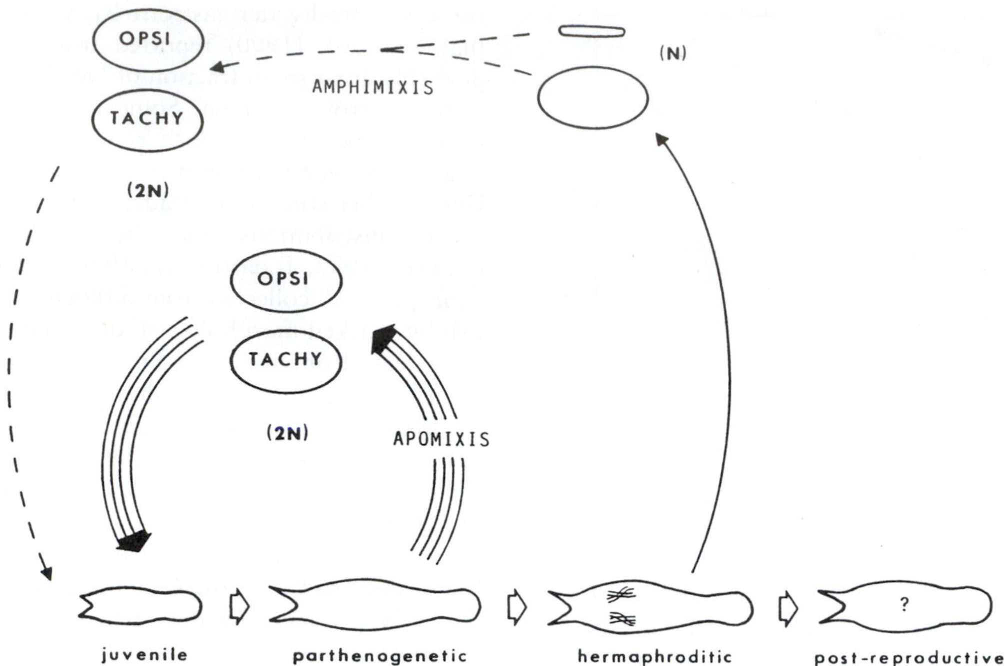

FIGURE 4 Schematic diagram of the proposed generalized life cycle for freshwater chaetonotid gastrotrichs, based predominately on study of *Lepidodermella squamata*. Dashed lines show hypothetical events which have not yet been demonstrated. From Levy (1984), after ideas presented by Levy and Weiss (1980), with permission of the authors. OPSI, opsiblastic egg; and TACHY, tachyblastic egg.

little larger than tachyblastic eggs, and are very resistant to freezing and drying (Brunson, 1949). The factors that induce the production of opsiblastic eggs are not well known, although such eggs are often produced by animals in crowded cultures. Opsiblastic eggs are almost always the final egg produced by an animal, even if the total number of eggs is fewer than four.

Following the production of parthenogenetic eggs, animals develop into hermaphrodites (Fig. 2C) (Hummon and Hummon, 1983; 1988). During this time, sperm and meiotic sexual eggs are produced, and the X-body grows. These changes occur slowly, over the period of a week after the last parthenogenetic egg is laid. No one has yet observed sperm transfer or fertilization, although Levy and Weiss (Levy and Weiss, 1980; Levy, 1984) reported finding a third kind of egg (the "plaque-bearing egg") in cultures of *Lepidodermella squamata*. They suggested that plaque-bearing eggs may be the product of sexual reproduction. Because the sperms are few in number (32–64 per animal) and nonmotile, fertilization probably is internal. Animals reared in isolation do not appear to produce fertilized sexual eggs, so cross-fertilization probably is the rule. The absence of ducts associated with the male or female reproductive system makes it difficult to suggest a mechanism of sperm transfer (Hummon 1986, described one bizarre possibility). Probably the enigmatic X-organ is involved. Much remains to be learned about the postparthenogenetic sexual phase and its importance in nature.

The life cycle just described is unique among invertebrates and offers considerable ecological flexibility to gastrotrich populations. The initial parthenogenetic phase allows for explosive population growth under favorable conditions: workers have commonly reported growth rates (r) of 0.1–0.5 per day in laboratory cultures (e.g., Hummon, 1974; Faucon and Hummon, 1976; Hummon and Hummon, 1979; Hummon, 1986). Production of parthenogenetic resting eggs (opsiblastic eggs) buffers the population against unfavorable conditions and presumably allows for dispersal among habitats. Finally, the subsequent sexual phase introduces genetic recombination. Sexual reproduction is most likely to occur in populations in which rates of mortality are low enough to allow some gastrotrichs to reach the age required for sexual development.

C. Ecological Interactions

Gastrotrichs feed on bacteria, algae, protozoans, detritus, and small inorganic particles. Bacteria probably are of primary importance. Bennett (1979) demonstrated that *Lepidodermella squamata* readily digested bacteria and found that this gastrotrich would not survive in laboratory cultures in the absence of bacteria. He reported that *L. squamata* could digest the green alga *Chlorella* as well, but suggested that algae were of secondary importance in gastrotrich diets. Gray and Johnson (1970) showed that the marine macrodasyid *Turbanella hyalina* could choose among various strains of natural bacteria, apparently on the basis of a tactile chemical sense. Thus, the quality as well as quantity of bacterial populations was important to the gastrotrichs. Freshwater gastrotrichs are most likely capable of similar fine discrimination among bacteria and other prey.

Reported predators of gastrotrichs include heliozoan and sarcodine amoebae, cnidarians, and tanypodine midges (Brunson, 1949; Bovee and Cordell, 1971; Moore, 1979), but many other benthic predators presumably feed on gastrotrichs. Nothing is known about the importance of predation in regulating populations of freshwater gastrotrichs or about the quantitative importance of gastrotrichs as a food item for various predators.

We have no direct information on what regulates gastrotrich populations in nature. Gastrotrichs are capable of enormous population growth (10–50% per day) in laboratory cultures. If potential growth rates are anywhere near this high in nature, as seems likely in some circumstances, then there must be an equally high counterbalancing mortality, perhaps from predation. There are no detailed studies of the population dynamics of freshwater gastrotrichs in nature. The few quantitative studies of the seasonal dynamics of freshwater gastrotrich populations (Fig. 5; see also Nesteruk, 1986) have shown that population densities are usually (but not always) lowest during the winter. We do not know what drives these seasonal dynamics, but seasonal changes in water temperature, food supply, and predation pressures are obvious possibilities.

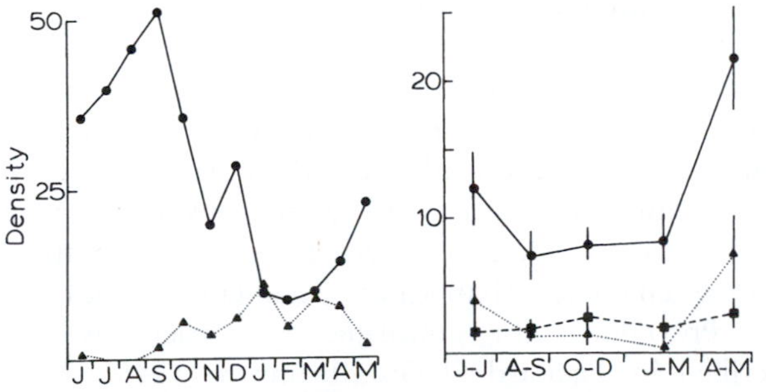

FIGURE 5 Seasonal trends in gastrotrich abundance. Left: density (number/cm^3) of two species of dasydytids in the surface sediments of a boggy pool ("Complex B") in Poland: *Setopus dubius* (●) and *Dasydytes ornatus* (▲). From data of Kisielewska (1982). Right: density (number/cm^2) of all gastrotrichs (●), an unidentified species (probably of *Heterolepidoderma*) (■), and *Lepidodermella triloba* (▲) on the gyttja sediments of Mirror Lake, New Hampshire (original). Plotted points are means ± s.e. (n = 16).

Gastrotrichs, along with nematodes and rotifers, are among the most abundant animals in the freshwater benthos, often having densities on the order of 10–100 individuals/cm^2 (= 100,000–1,000,000 individuals/m^2) (Table II). However, because there have been no direct measurements of the roles of gastrotrichs in freshwater ecosystems, their importance can only be guessed at. There is only a single, tentative estimate of gastrotrich metabolism in freshwater: Strayer (1985) estimated secondary production and respiration of gastrotrichs in Mirror Lake, New Hampshire, each to be 50–100 mg dry mass per square meter per year, which is less than 1% of total production or respiration of the zoobenthic community. This estimate suggests that gastrotrich metabolism, and processes such as nutrient regeneration that are correlated with metabolism, are of minor importance in freshwater ecosystems. It would be imprudent, however, to dismiss gastrotrichs as quantitatively unimportant to ecosystem functioning without actually measuring gastrotrich activities under defined conditions. The extraordinarily high rates of population turnover and potentially highly selective feeding behavior of gastrotrichs suggest that they may exert a considerable influence on the composition of natural bacterial communities.

D. Evolutionary Relationships

The phylogenetic relationships among the various aschelminth groups are not yet clear. Several points of morphological similarity between nematodes and gastrotrichs (summarized by Lorenzen, 1985, and Wallace *et al.*, 1996) suggest that these two groups are allied, but other analyses (Conway Morris, 1993; Garey *et al.*, 1996) have placed gastrotrichs closer to rotifers and even platyhelminthes than to nematodes. Within the Gastrotricha, only the broad outlines of phylogeny have been sketched out. Three major groups of gastrotrichs are widely recognized: the order Macrodasyida and the suborders Multitubulata and Paucitubulata of the order Chaetonotida. The Multitubulata (which includes only the marine *Neodasys*) exhibit many primitive characteristics and are in some ways intermediate between the Macrodasyida and the Paucitubulata (which contains almost all of the freshwater gastrotrichs). Therefore, Boaden (1985) suggested that modern Macrodasyida and Paucitubulata are descended from a *Neodasys*-like ancestor. Nevertheless, on the basis of digestive tract ciliature and other characters, we believe that the ancestral gastrotrich was more likely similar to the dactylopodolid macrodasyids.

Evolutionary relationships among the families, genera, and species of freshwater gastrotrichs are still largely unknown because of a paucity of basic morphological, biochemical, and zoogeographical information about most species. An enormous number of species have not even been described, let alone studied. Even in North America, perhaps 75–90% of the species of freshwater gastrotrichs are undescribed. The correct systematic placement of some genera (e.g., *Dichaetura, Marinellina*) will require much additional study. Furthermore, there is growing concern (e.g., Remane, 1935–1936; Ruppert, 1977; Kisielewski, 1981, 1987, 1991, 1997; Schwank, 1990) that some of the traditionally defined genera of the Chaetonotidae may be inadequate. Better morphological information and the discovery of species having characters intermediate between genera is leading to a blurring of distinctions between genera such as *Chaetonotus, Heterolepidoderma*, and *Lepidodermella*. Further, it is becoming increasingly clear that some of the characters traditionally used to define gastrotrich genera (e.g., adhesive tubes, cuticle ornamentation) are subject to convergent evolution (e.g., Kisielewski, 1991; Balsamo and Fregni, 1995), and so might not reflect common lines of descent. Finally, it seems likely that *Ichthydium*, which is defined by an absence of cuticular ornamentation, is polyphyletic. In response to these concerns, Schwank (1990) and Kisielewski (1991) erected new genera of chaetontotids and redefined existing genera, but the classification of this group may need to be redone once more complete information becomes available.

IV. COLLECTING, REARING, AND PREPARATION FOR IDENTIFICATION

Gastrotrichs may be collected by taking samples of sediments or vegetation. For quantitative work on sediment-dwelling species, small-diameter (2–5 cm) cores are preferable. For plant-dwelling forms, quantitative samples probably could be obtained by modifying sampling methods developed for macroinvertebrates (e.g., Downing, 1984; Jackson, 1997) to use very fine mesh and small sample volumes or subsampling.

Because living animals are preferable to preserved animals for many purposes, it often is desirable to extract the animals from the sample prior to preservation. If the sample must be preserved immediately, workers recommend narcotizing the animals with 1% $MgCl_2$ for 10 min, then fixing them in 10% formalin with rose bengal (e.g., Hummon, 1981, 1987).

It is difficult to extract or count the gastrotrichs from a sample. Some workers have handpicked or counted animals under a dissecting microscope (e.g., Hummon, 1981, 1987; Evans, 1982; Strayer, 1985), but this procedure is tedious. Density gradient centrifugation (Nichols, 1979; Schwinghamer, 1981) ought to be useful in extracting gastrotrichs from sediment, but has not yet been tested on freshwater gastrotrichs. Nesteruk (1987) found that a modified Baermann funnel was useful for extracting chaetonotids, but not dasydytids, from pond sediments. Sieves should be avoided or used cautiously, because gastrotrichs are too small to be retained quantitatively on even very fine mesh sieves. For example, Hummon (1981) found that a 37-μm mesh sieve retained only 31% of the gastrotrichs in a series of samples taken from the upper Mississippi River. For quantitative work, it is important to check the efficiency of whatever extraction or counting method is used, because gastrotrichs are so small and can be easily overlooked (cf. Strayer, 1985, pp. 295–296).

Living gastrotrichs are preferable to dead gastrotrichs for taxonomic work. Living gastrotrichs often are too active for critical observations to be made, so they must be slowed down. This can be accomplished by gently squeezing the animal (either with a rotocompressor—Spoon, 1978—or by removing some of the water from beneath a cover slip with a tissue), by placing it in a viscous medium such as methylcellulose, or by narcotizing it. Cocaine was the traditional narcotic of choice (Brunson, 1959), but it is now difficult to obtain for laboratory use. d'Hondt (1967) recommended using MS 222, and we have had very good success with narcotizing gastrotrichs by bleeding 1.8% neosynephrine (available at pharmacies) under the cover slip. Animals may be killed with formalin or fumes of osmium tetroxide (e.g., Brunson, 1950) following narcotization. Osmium tetroxide is a superior fixative, but is dangerous and should be used with extreme care. It is sometimes necessary to examine individual scales, which can be isolated from an animal by bleeding 2% acetic acid under the cover slip. For serious taxonomic work, videomicroscopy of living animals may provide better permanent documentation of species characters than killed specimens or slides.

Procedures for culturing gastrotrichs were described by Packard (1936), Brunson (1949), Townes (1968), Hummon (1974), and Bennett (1979). Gastrotrichs have been cultured on 0.1% malted milk, raw egg yolk, wheat grain infusion, baked lettuce infusion, and baker's yeast. Brunson (1949) recommended that animals be acclimated gradually to culture media when collected from the wild. Hummon (1974) described a procedure for starting individual cultures of known-age animals from eggs that may be especially useful for bioassay work.

V. TAXONOMIC KEY

It is relatively easy to identify most North American freshwater gastrotrichs to genus and very difficult to identify them to species. Species identification requires a keen eye, careful observation, a cooperative gastrotrich, and some luck, because most of the freshwater gastrotrichs of North America undoubtedly are undescribed. The following works are helpful in species identification: Brunson (1950, 1959), who keyed and illustrated species then known from North American freshwaters; Robbins (1965, 1973), who gave additional information and drawings of North American species; d'Hondt (1971b), who gave a key to the species of *Lepidodermella* and defined three subgenera of *Ichthydium*; Kisielewski (1981), who made a critical evaluation of the morphological characters that must be measured to describe (or identify) a species; Kisielewski (1986b), who gave a recent treatment of *Aspidiophorus*; Schwank (1990), who provided illustrated keys (in German) for all known freshwater gastrotrichs worldwide; and Kisielewski (1991), who described many Brazilian species and addressed several important issues in gastrotrich systematics. The following key includes genera of freshwater gastrotrichs known from or likely to be found in North America.

1a. Animal with at least three pairs of adhesive tubules (one anterior and two posterior) and a pair of pharyngeal pores, body strap-shaped (Fig. 1A,B); an almost entirely marine group not yet reported from North American freshwaters............order Macrodasyida

1b. Animal lacking adhesive tubules (Fig. 7A–H) or with one pair (very rarely two pairs) of adhesive tubules posteriorly (Fig. 6A–L), and no pharyngeal pores, body strap-shaped or tenpin-shaped; common and widespread in freshwater ....order Chaetonotida 2

2a (1). Posterior end of body usually with furca and adhesive tubules; body usually strap-shaped or tenpin-shaped (Fig. 6A-L)..................3

2b. Posterior end of body without furca or adhesive tubules, although sometimes bearing spines or pegs; body usually tenpin or bottle-shaped (Fig. 7A–H)..............12

3a (2). Furca doubly branched; scales and spines sparse or absent (Fig. 6A); a rare genus not yet reported from North America .... Dichaeturidae, *Dichaetura*

3b. Furca singly branched; scales and spines present or absent (Fig. 6B–L); common and widespread 4

4a (3). Branches of furca heavy, sickle-shaped, curved, and tapered, not distinctly divided into a cone-shaped basal part and a distal duct; head with long cilia that are not arranged in bundles; head plates absent (Fig. 6B); a rare family not yet reported from North America ....Proichthydiidae

4b. Branches of furca usually with a cone-shaped base and a distal adhesive duct; body often with numerous spines or scales; head with cilia that are arranged in bundles; cephalic plates present (Fig. 6C–L); common and widespread ....Chaetonotidae 5

5a (4). Branches of furca ringed and often very long; body often large and without a distinct neck (Fig. 6C) ....*Polymerurus*

5b. Branches of furca not ringed (Fig. 6D–L)..............6

6a (5). Dorsal surface of body without spines or scales (except for dorsal sensory bristles or a few scales at the base of the furca (Fig. 6D) ... .*Ichthydium*

6b. Dorsal surface of body with numerous scales or spines (Fig. 6E–L) 7

7a (6). Spines or spined scales present and often numerous (Fig. 6E–G, J) 8

7b. Spines absent (occasionally a few thin spines are present at the base of the furca) (Fig. 6H,I,K,L) 9

8a. (7). Ventral scales different from dorsal scales; spines of various types (Fig. 6E–G); common and widespread..............*Chaetonotus*

8b. Ventral scales similar to dorsal scales; posterior part of body with several long spines that reach beyond the end of the furca (Fig. 6J); not yet reported from North America ....*Lepidochaetus*

9a (7). Furca without adhesive tubes; body markedly tenpin-shaped, with groups of long cilia on the head and posterior part of the body (Fig. 6H); a semiplanktonic genus known only from Brazil ....*Undula*

9b. Furca with adhesive tubes; body strap-shaped or tenpin-shaped; without groups of long cilia on head and posterior body (Fig. 6I,K,L) 10

10a (9). Dorsal surface of body covered with stalked scales (Fig. 6I) ....*Aspidiophorus*

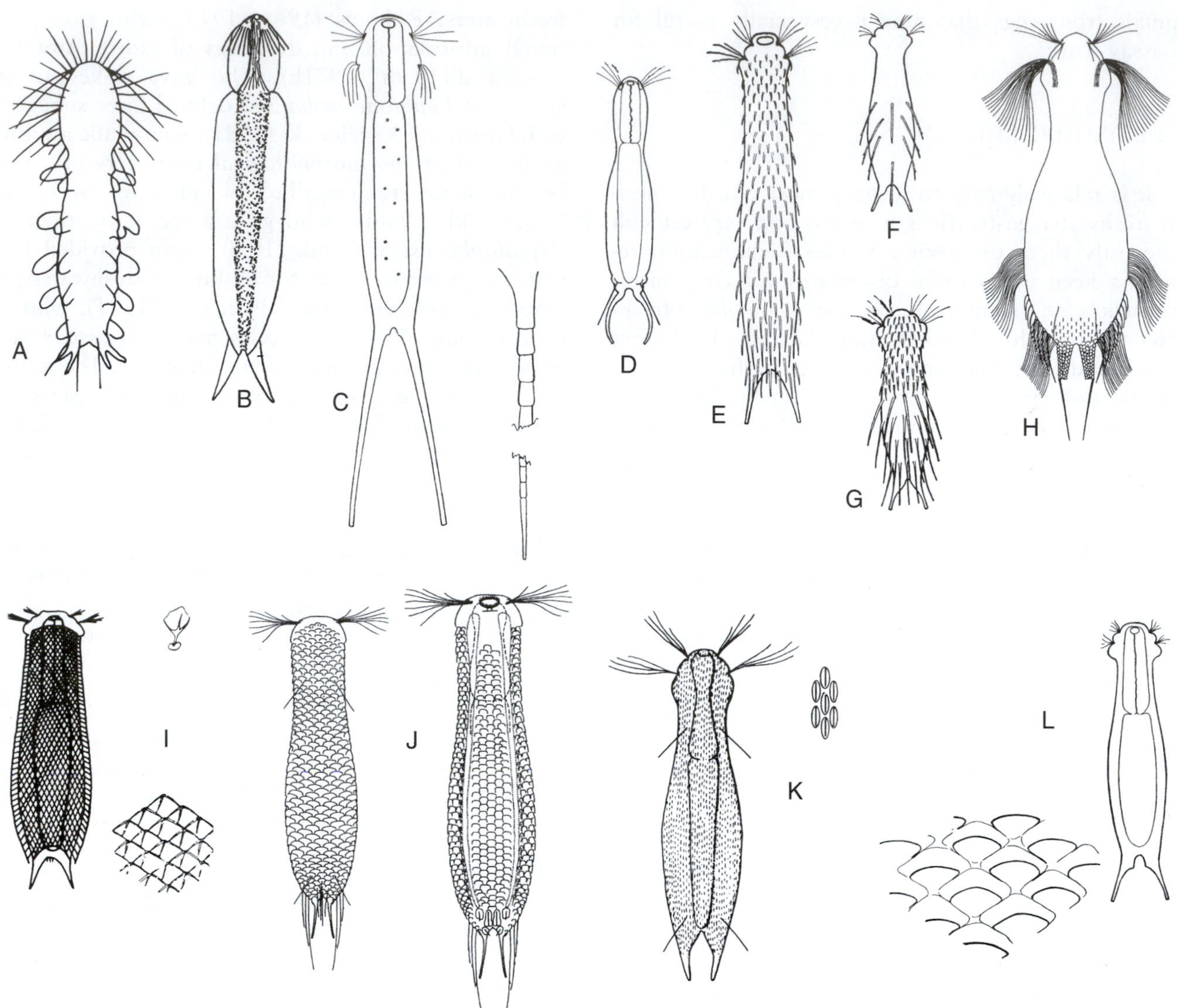

FIGURE 6 Genera of freshwater Dichaeturidae, Proichthydiidae, and Chaetonotidae. (A) *Dichaetura*; (B) *Proichthydium; (C) Polymerurus*, showing detail of ringed branches of furca; (D) *Ichthydium*, (E–G) *Chaetonotus*, showing examples of spination; (H) *Undula*; (I) *Aspidiophorus*, with detail of coat of scales and a single scale; (J) *Lepidochaetus*, in dorsal (left) and ventral (right) views, (K) *Heterolepidoderma*, with detail of scales; and (L) *Lepidodermella*, with detail of scales. From Remane (1935–1936), Brunson (1950), Hyman (1951), Voigt (1958), Robbins (1965), and Kisielewski (1991).

10b. Scales on dorsal surface of body unstalked (Fig. 6K,L) 11

11a (10). Scales elongate, with longitudinal keels (Fig. 6K) .... *Heterolepidoderma*

11b. Scales not keeled (Fig. 6L) .... *Lepidodermella*

12a (2). Head with club-shaped tentacles (Fig. 7A,B) .... Neogosseidae 13

12b. Head without club-shaped tentacles (Fig. 7C–H) .... Dasydytidae 14

13a (12). Posterior end of body with two groups of long spines (Fig. 7A); elements of mouth-ring jointed .... *Neogossea*

13b. Posterior end of body with single medial group of spines (Fig. 7B); elements of mouth-ring unjointed .... *Kijanebalola*

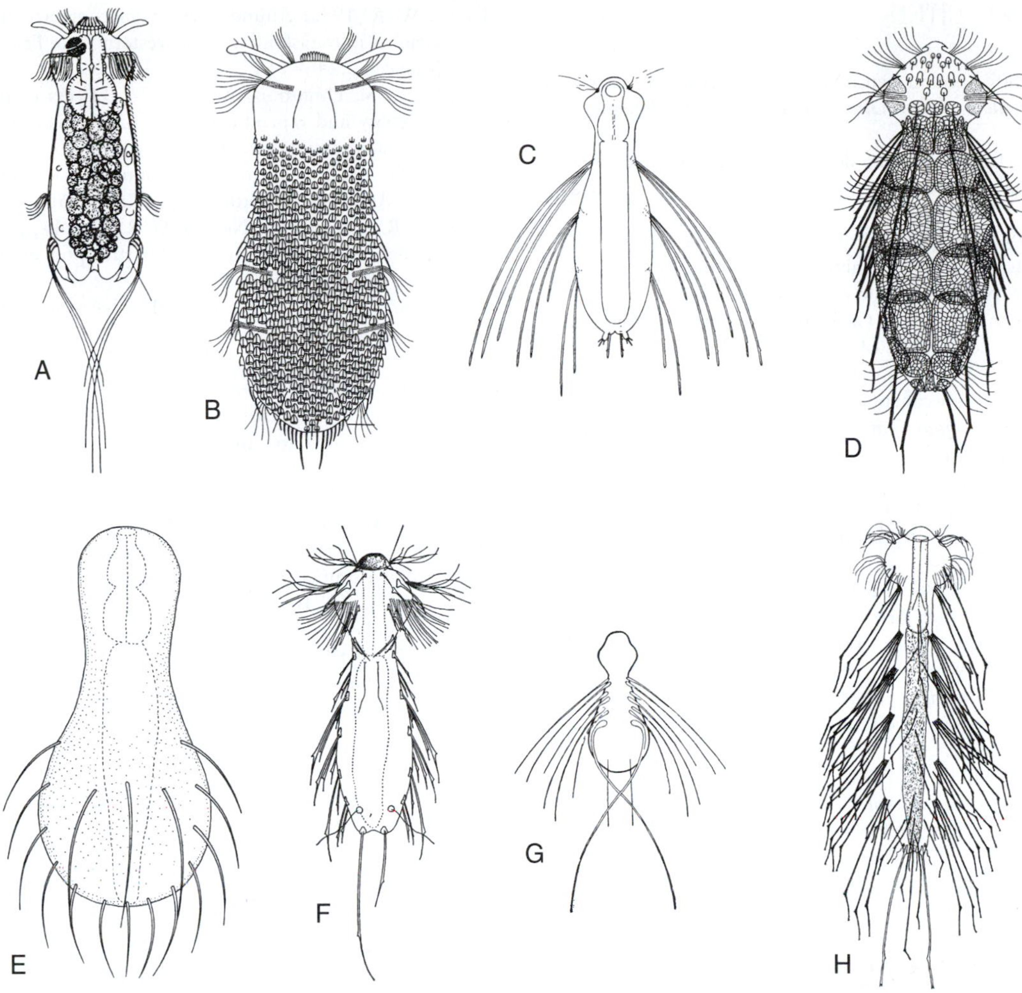

FIGURE 7 Genera of freshwater Neogosseidae and Dasydytidae. (A) *Neogossea*; (B) *Kijanebalola*; (C) *Stylochaeta;* (D) *Ornamentula*; (E) *Anacanthoderma*; (F) *Setopus*; (G) *Haltidytes*, and (H) *Dasydytes*. From Brunson (1950), Krivanek and Krivanek (1958), Balsamo (1983), Schwank (1990), and Kisielewski (1991).

14a (12). Posterior end of body with pair of peglike protuberances (Fig. 7C) .... *Stylochaeta*

14b. Posterior end of body without peglike protuberances (Fig. 7D–H) 15

15a (14). Body enclosed in a "lorica" of large, thick, ornamented scales (Fig. 7D); known only from Brazil .... *Ornamentula*

15b. Scales absent or small and inconspicuous (Fig. 7E–H) 16

16a (15). Head much narrower than body and scarcely wider than neck; lateral spines absent or identical to dorsal spines; pharynx with two bulbs (Fig. 7E); not yet reported from North America .... *Anacanthoderma*

16b. Head distinctly wider than neck; body with long lateral spines; pharynx with one bulb or no bulb (Fig. 7F–H) 17

17a (16). Some lateral spines movable and sharply bent basally; posterior end of body rounded, without rear spines (Fig. 7G) .... *Haltidytes*

17b. All spines fixed, straight or bent distally; posterior end of body usually with spines (Fig. 7F,H) 18

18a (17). Lateral spines with 1–3 lateral denticles and often terminally bifurcated; spines of uniform thickness from their base to the last lateral denticle; pharynx usually with a distinct posterior bulb (Fig. 7H) .... *Dasydytes*

18b. Lateral spines tapered, with at most one weak lateral denticle and never terminally bifurcated; pharynx without posterior bulb (Fig. 7F) .... *Setopus*

VI. LITERATURE CITED

Balsamo, M. 1980. Spectral sensitivity in a freshwater gastrotrich (*Lepidodermella squammatum* Dujardin). Experientia 36: 830–831.

Balsamo, M. 1981. Gastrotrichi della Toscana: il lago di Sibolla. Bolletino del Museo Civico di Storia Naturale Verona 7: 547–571.

Balsamo, M. 1983. Gastrotrichi (Gastrotricha). Guide per il Riconoscimento delle Specie Animali delle Acque Interne Italiane 20. Consiglio Nazionale delle Ricerche.

Balsamo, M., Fregni, E. 1995. Gastrotrichs from interstitial freshwater, with a description of four new species. Hydrobiologia 302:163–175.

Bennett, L. W. 1979. Experimental analysis of the trophic ecology of *Lepidodermella squammata* (Gastrotricha:Chaetonotida) in mixed culture. Transactions of the American Microscopical Society 98:254–260.

Bertolani, R., Balsamo, M. 1989. Tardigradi e Gastrotrichi del Trentino: il Lago di Tovel. Studi Trentini di Scienze Naturali 65:83–93.

Blinn, D. W., Green, J. 1986. A pump sampler study of microdistribution in Walker Lake, Arizona, USA.: a senescent crater lake. Freshwater Biology 16:175–185.

Boaden, P. J. S. 1985. Why is a gastrotrich?, *in*: Conway Morris, S., George, J. D., Gibson, R., Platt, H. M. Eds., The origins and relationships of lower invertebrates. Clarendon Press, Oxford, UK, pp. 248–260.

Bovee, E. C., Cordell, D. L. 1971. Feeding on gastrotrichs by the heliozoon *Actinophrys sol*. Transactions of the American Microscopical Society 90:365–369.

Brandenburg, J. 1962. Elektronenmikroscopische Untersuchungen des Terminalapparates von *Chaetonotus* sp. (Gastrotrichen) als Beispeil einer Cyrtocyte bei Aschelminthen. Zeitschrift für Zellforschung 57:136–144.

Brunson, R. B. 1947. Gastrotricha of North America. II. Four new species of *Ichthydium* from Michigan. Transactions of the Michigan Academy of Science, Arts, and Letters 33:59–62.

Brunson, R. B. 1948. *Chaetonotus tachyneusticus*: a new species of gastrotrich from Michigan. Transactions of the American Microscopical Society 67:350–351.

Brunson, R. B. 1949. The life history and ecology of two North American gastrotrichs. Transactions of the American Microscopical Society 68:1–20.

Brunson, R. B. 1950. An introduction to the taxonomy of the Gastrotricha with a study of eighteen species from Michigan. Transactions of the American Microscopical Society 69:325–352.

Brunson, R. B. 1959. Gastrotricha, *in*: Edmondson, W. T. Ed., Freshwater biology. 2nd ed., Wiley, New York, pp. 406–419.

Cole, G. A. 1955. An ecological study of the microbenthic fauna of two Minnesota lakes. American Midland Naturalist 53:213–230.

Conway Morris, S. 1993. The fossil record and the early evolution of the Metazoa. Nature 361:219–225.

Craft, L. 1993. A taxonomic survey of freshwater gastrotrichs from eastern Ohio. MS thesis, Ohio University, 169 pp.

Davis, K. B. 1937. The Gastrotricha (Chaetonotoidea) of Washington County, Arkansas. MA thesis, University of Arkansas, 94 pp.

Downing, J. A. 1984. Sampling the benthos of standing waters, *in*: Downing, J. A., Rigler, F. H. Eds., A manual on methods for the assessment of secondary productivity in freshwaters. 2nd ed., Blackwell Scientific Publications, Oxford, UK, pp. 87–130

Emberton, K. C. 1981. First record of *Chaetonotus heideri* (Gastrotricha: Chaetonotidae) in North America. Ohio Journal of Science 81:95–96.

Evans, W. A. 1982. Abundances of micrometazoans in three sandy beaches in the island area of western Lake Erie. Ohio Journal of Science 82:246–251.

Faucon, A. S., Hummon, W. D. 1976. Effects of mine acid on the longevity and reproductive rate of the Gastrotricha *Lepidodermella squammata* (Dujardin). Hydrobiologia 50:265–269.

Frey, D. G. 1982. Questions concerning cosmopolitanism in Cladocera. Archiv für Hydrobiologie 93:484–502.

Garey, J. R., Krotec, M., Nelson, D. R., Brooks, J. 1996. Molecular analysis supports a tardigrade-arthropod association. Invertebrate Biology 115:79–88.

Gray, J. S., Johnson, R. M. 1970. The bacteria of a sandy beach as an ecological factor affecting the interstitial gastrotrich *Turbanella hyalina* Schultze. Journal of Experimental Marine Biology and Ecology 4:119–133.

Green, J. 1986. Associations of zooplankton in six crater lakes in Arizona, Mexico and New Mexico. Journal of Zoology (A)208:135–159.

d'Hondt, J.-L. 1967. Effets de quelques anesthétiques sur les Gastrotriches. Experientia 23:1025–1026.

d'Hondt, J.-L. 1971a. Gastrotricha. Oceanography and Marine Biology: Annual Review 9:141–192.

d'Hondt, J.-L. 1971b. Note sur quelques Gastrotriches Chaetonotidae. Bulletin de la Société Zoologique de France 96:215–235.

Horlick, R. G. 1975. *Dasydytes monile*, a new species of gastrotrich from Illinois. Transactions of the Illinois State Academy of Science 68:61–64.

Hummon, M. R. 1984a. Reproduction and sexual development in a freshwater gastrotrich. 1. Oogenesis of parthenogenic eggs (Gastrotricha). Zoomorphology 104:33–41.

Hummon, M. R. 1984b. Reproduction and sexual development in a freshwater gastrotrich. 2. Kinetics and fine structure of postparthenogenic sperm formation. Cell and Tissue Research 236:619–628.

Hummon, M. R. 1984c. Reproduction and sexual development in a freshwater gastrotrich. 3. Postparthenogenic development of primary oocytes and the X-body. Cell and Tissue Research 236:629–636.

Hummon, M. R. 1986. Reproduction and sexual development in a freshwater gastrotrich. 4. Life history traits and the possibility of sexual reproduction. Transactions of the American Microscopical Society 105:97–109.

Hummon, M. R., Hummon, W. D. 1979. Reduction in fitness of the gastrotrich *Lepidodermella squammata* by dilute mine acid water and amelioration of the effect by carbonates. International Journal of Invertebrate Reproduction 1:297–306.

Hummon, M. R., Hummon, W. D. 1983. Gastrotricha, *in*: Adiyodi K. G., Adiyodi, R. G. Eds., Reproductive biology of invertebrates. Vol II: Spermatogenesis and sperm function. Wiley, London, pp. 195–205.

Hummon, W. D. 1974. Effects of DDT on longevity and reproductive rate in *Lepidodermella squammata* (Gastrotricha, Chaetonotida). American Midland Naturalist 92:327–339.

Hummon, W. D. 1981. Extraction by sieving: a biased procedure in studies of stream meiobenthos. Transactions of the American Microscopical Society 100:278–284.

Hummon, W. D. 1982. Gastrotricha, *in*: Parker, S. P. Ed., Synopsis and classification of living organisms. Vol. 1. McGraw–Hill, New York, pp. 857–863.

Hummon, W. D. 1987. Meiobenthos of the Mississippi headwaters, *in*: Bertolani, R. Ed., Biology of tardigrades. Selected Symposia and Monographs UZI., 1, Mucchi, Modena, Italy, pp. 125–140.

Hummon, W. D., Hummon, M. R. 1983. Gastrotricha, *in*: Adiyodi, K. G., Adiyodi, R. G. Eds., Reproductive biology of invertebrates. Vol. I: Oogenesis, oviposition, and oosorption. Wiley, London, pp. 211–221.

Hummon, W. D., Hummon, M. R. 1988. Gastrotricha, *in*: Adiyodi, K. G., Adiyodi, R. G. Eds., Reproductive biology of invertebrates. Vol. III: Accessory sex glands. Oxford and IBH Publishing Company, New Delhi, pp. 81–85.

Hummon, W. D., Evans, W. A., Hummon, M. R., Doherty, F. G., Wainberg, R. H., Stanley, W. S. 1978. Meiofaunal abundance in sandbars of acid mine polluted, reclaimed, and unpolluted streams in southeastern Ohio, *in*: Thorp, J. H., Gibbons, J. W. Eds., Energy and environmental stress in aquatic ecosystems. DOE Symposium Series (CONF–771114). National Technical Information Service, Springfield, VA, pp.188–203.

Hutchinson, G. E. 1967. A treatise on limnology, Vol 2: Introduction to lake biology and the limnoplankton. Wiley, New York.

Hyman, L. H. 1951. The invertebrates: Acanthocephala, Aschelminthes, and Entoprocta: the pseudocoelomate Bilateria. Vol. 3. McGraw-Hill, New York.

Jackson, M. J. 1997. Sampling methods for studying macroinvertebrates in the littoral vegetation of shallow lakes. The Broads Authority, Norwich, UK.

Kisielewska, G. 1981. Hermaphroditism of freshwater gastrotrichs in natural conditions. Bulletin de l'Académie Polonaise des Sciences, Série des Sciences Biologiques 29:167–172.

Kisielewska, G. 1982. Gastrotricha of two complexes of peat hags near Siedlce. Fragmenta Faunistica 27:39–57.

Kisielewski, J. 1981. Gastrotricha from raised and transitional peat bogs in Poland. Monografie Fauny Polski 11:1–143.

Kisielewski, J. 1986a. Freshwater Gastrotricha of Poland. VII. Gastrotricha of extremely eutrophicated water bodies. Fragmenta Faunistica 30:267–295.

Kisielewski, J. 1986b. Taxonomic notes on freshwater gastrotrichs of the genus *Aspidiophorus* Voigt (Gastrotricha: Chaetonotidae), with descriptions of four new species. Fragmenta Faunistica 30:139–156.

Kisielewski, J. 1987. Two new interesting genera of Gastrotricha (Macrodasyida and Chaetonotida) from the Brazilian freshwater psammon. Hydrobiologia 153:23–30.

Kisielewski, J. 1991. Inland-water Gastrotricha from Brazil. Annales Zoologici 43 (Suppl 2):1–168.

Kisielewski, J. 1997. On the subgeneric division of the genus *Chaetonotus* Ehrenberg (Gastrotricha). Annales Zoologici 46:145–151.

Kisielewski, J. 1998. Brzuchorzeski (Gastrotricha). Fauna Slodkowodna Polski 31. Polskie Towarzystwo Hydrobiologiczne, Uniwersytet Lodzki.

Krivanek, R. C., Krivanek, J. O. 1958. A new and a redescribed species of *Neogossea* (Gastrotricha) from Louisiana. Transactions of the American Microscopical Society 77:423–428.

Levy, D. P. 1984. Obligate postparthenogenetic hermaphroditism and other evidence for sexuality in the life cycles of freshwater Gastrotricha. Dissertation, Rutgers University, New Brunswick, NJ. 257 pp.

Levy, D. P., Weiss, M. J. 1980. Sperm in the life cycle of the freshwater gastrotrich, *Lepidodermella squammata*. American Zoologist 20:749 (abstract).

Lorenzen, S. 1985. Phylogenetic aspects of pseudocoelomate evolution, *in*: Conway Morris, S., George, J. D., Gibson, R., Platt, H.M. Eds., The origins and relationships of lower invertebrates. Clarendon Press, Oxford, pp. 210–223.

Madoni, P. 1989. Community structure of the microzoobenthos in Lake Suviana (Tusco-Emilian Apennines). Bolletino di Zoologia 56:159–165.

Moore, G. M. 1939. A limnological investigation of the microscopic benthic fauna of Douglas Lake, Michigan. Ecological Monographs 9:537–582.

Moore, J. W. 1979. Some factors influencing the distribution, seasonal abundance and feeding of subarctic Chironomidae (Diptera). Archiv für Hydrobiologie 85:302–325.

Nesteruk, T. 1986. Freshwater Gastrotricha of Poland. IV. Gastrotricha from fish ponds in the vicinity of Siedlce. Fragmenta Faunistica 30:215–233.

Nesteruk, T. 1987. Assessing the efficiency of three methods of extracting freshwater Gastrotricha from bottom mud. Acta Hydrobiologica 29:219–226.

Nesteruk, T. 1991. Vertical distribution of Gastrotricha in organic bottom sediment of inland water bodies. Acta Hydrobiologica 33:253–264.

Nesteruk, T. 1996. Density and biomass of Gastrotricha in sediments of different types of standing waters. Hydrobiologia 324: 205–208.

Nichols, J. A. 1979. A simple flotation technique for separating meiobenthic nematodes from fine-grained sediments. Transactions of the American Microscopical Society 98:127–130.

Nuss, B. 1984. Ultrastrukturelle und ökophysiologische Untersuchungen an kristalloiden Einschlüssen der Muskeln eines sulfidtoleranten limnischen Nematoden (*Tobrilus gracilis*). Veröffentlichungen des Instituts für Meeresforschung in Bremerhaven 20:3–15.

Nuss, B., Trimkowski, V. 1984. Physikalische Mikroanalysen an kristalloiden Einschlüssen bei *Tobrilus gracilis* (Nematoda, Enoplida). Veröffentlichungen des Instituts für Meeresforschung in Bremerhaven 20:17–27.

Packard, C. E. 1936. Observations on the Gastrotricha indigenous to New Hampshire. Transactions of the American Microscopical Society 55:422–427.

Pennak, R. W. 1978. Freshwater invertebrates of the United States. 2nd ed., Wiley-Interscience, New York.

Pfaltzgraff, G. H. 1967. A preliminary study of the Gastrotricha of northern Indiana. Proceedings of the Indiana Academy of Sciences 76:400–404.

Powell, E. N., Crenshaw, M. A., Rieger, R. M. 1979. Adaptations to sulfide in the meiofauna of the sulfide system. I. ^{35}Sulfide accumulation and the presence of a sulfide detoxification system. Journal of Experimental Marine Biology and Ecology 37:57–76.

Powell, E. N., Crenshaw, M. A., Rieger, R. M. 1980. Adaptations to sulfide in the sulfide-system meiofauna. Endproducts of sulfide detoxification in three turbellarians and a gastrotrich. Marine Ecology Progress Series 2:169–177.

Preobrajenskaja, E. N. 1926. Zur Verbreitung der Gastrotrichen in den Gewässern der Umgebung zu Kossino (in Russian, with a German summary). Arbeiten der biologischen Station zu Kossino (bei Moskau) 4:3–14.

Remane, A. 1935–1936. Gastrotricha und Kinorhyncha. Klassen und Ordnungen des Tierreichs 4 (Abteilung 2, Buch 1, Teil 2, Lieferung 1–2):1–242, 373–385.

Remane, A. 1961. *Neodasys uchidai* nov. spec., eine zweite *Neodasys*-Art (Gastrotricha Chaetonotida). Kieler Meeresforschungen 17:85–88.

Renaud-Mornant, J. 1986. Gastrotricha, *in*: Botosaneanu, L. Ed., Stygofauna mundi. E. J. Brill, Leiden, The Netherlands, pp. 86–109.

Robbins, C. E. 1965. Two new species of Gastrotricha (Aschelminthes) from Illinois. Transactions of the American Microscopical Society 84:260–263.

Robbins, C. E. 1973. Gastrotricha from Illinois. Transactions of the Illinois State Academy of Science 66:124–126.

Rudescu, L. 1968. Die Rotatorien, Gastrotrichen und Tardigraden der Schilfrohrgebiete des Donaudeltas. Hidrobiologia (Bucharest) 9:195–202.

Ruppert, E. E. 1977. *Ichthydium hummoni* n. sp., a new marine chaetonotid gastrotrich with a male reproductive system. Cahiers de Biologie Marine 18:1–5.

Ruppert, E. E. 1988. Gastrotricha, *in*: Higgins, R. P., Thiel, H. Eds., Introduction to the study of meiofauna. Smithsonian Institution Press, Washington, pp. 302–311.

Ruttner-Kolisko, A. 1955. *Rheomorpha neiswestnovae* und *Marinellina flagellata*, zwei phylogenetisch interessante Wurmtypen aus dem Süsswasserpsammon. Österreichische Zoologische Zeitschrift 6:55–69.

Schmid-Araya, J. M., Schmid, P. E. 1995. The invertebrate species of a gravel stream. Jahresberichte der Biologischen Station Lunz 15:11–21.

Schwank, P. 1990. Gastrotricha, *in*: Schwoerbel, J., Zwick, P. Eds., Süsswasserfauna von Mitteleuropa, Band 3. Gustav Fischer Verlag, Stuttgart, pp. 1–252.

Schwinghamer, P. 1981. Extraction of living meiofauna from marine sediments by centrifugation in a silica sol–sorbitol mixture. Canadian Journal of Fisheries and Aquatic Sciences 38:476–478.

Spoon, D. M. 1978. A new rotary microcompressor. Transactions of the American Microscopical Society 97:412–416.

Strayer, D. 1985. The benthic micrometazoans of Mirror Lake, New Hampshire. Archiv für Hydrobiologie Supplementband 72: 287–426.

Sudzuki, M. 1971. Die das Kapillarwasser des Lückensystems bewohnenden Gastrotrichen Japans I. Zoological Magazine 80: 256–257.

Todaro, M. A., Fleeger, J. W., Hu, Y. P., Hrincevich, A.W., Foltz, D. W. 1996. Are meiofaunal species cosmopolitan? Morphological and molecular analysis of *Xenotrichula intermedia* (Gastrotricha: Chaetonotida). Marine Biology 125:735–742.

Townes, M. M. 1968. The collection, identification, and cultivation of gastrotrichs. Turtox News 46:99–101.

Voigt, M. 1958. Gastrotricha. Die Tierwelt Mitteleuropas, Band I, Lieferung 4a, pp. 1–45.

Wallace, R. L., Ricci, C., Melone, G., 1996. A cladistic analysis of pseudocoelomate (aschelminth) morphology. Invertebrate Biology 115:104–112.

Weiss, M. J., Levy, D. P. 1979. Sperm in "parthenogenetic" freshwater gastrotrichs. Science 205:302–303.

PHYLUM ROTIFERA

Robert Lee Wallace
Department of Biology
Ripon College
Ripon, Wisconsin 54971

Terry W. Snell
School of Biology
Georgia Institute of Technology
Atlanta, Georgia 30332

I. INTRODUCTION

The phylum Rotifera or Rotatoria comprises of approximately 2000 species of unsegmented, bilaterally symmetrical, pseudocoelomates, possessing two distinctive features (Fig. 1). First, at the apical end (head) is a ciliated region called the corona, which is used in locomotion and food gathering. In adults of some forms, ciliation is lacking and the corona is a funnel or bowl-shaped structure at the bottom of which is the mouth. Second, a muscular pharynx, the mastax, possessing a complex set of hard jaws, called trophi, is present in all rotifers.

When viewing the anterior end of most rotifers one is struck with the idea of a rotating wheel. This is due to the metachronal beat of cilia on the corona, a structure usually composed of two concentric rings: trochus and cingulum (Fig. 2). This same image provided early microscopists with the name for the phylum: the etymon is Latin, *rota*, "wheel" and Latin, *ferre*, "to bear" equals "wheel bearers." Although rotifers are often confused with ciliated protozoans and gastrotrichs by beginning students, those organisms do not possess trophi and their ciliation is not distributed in the same way as in rotifers. Rotifers are small organisms, generally ranging from 100–1,000 μm long, although a few elongate species may surpass 2,000 μm or more. Very few rotifers are parasitic (May, 1989); nearly all are free-living herbivores or predators.

Collectively this phylum is widely distributed, being found in all freshwater habitats at densities generally ranging up to about 1,000 individuals/L. However, rotifers occasionally become abundant if sufficient food is available, and can attain population densities of >5,000 individuals/L. In some rather unusual water bodies, exceedingly large populations can develop; sewage ponds may contain about 12,000 individuals/L (Seaman *et al.*, 1986), and at certain times in soda water bodies in Chad, much more than 100,000 individuals/L may occur (Iltis and Riou-Duvat, 1971)! Although most inhabit freshwaters, some genera also have members that occur in brackish and marine waters. For example, about 20 of the 32 species comprising the genus *Synchaeta* are described as marine (Nogrady, 1982). However, only about 50 species of rotifers are exclusively marine. In general, rotifers are not as diverse or as abundant in marine environments as microcrustaceans, but they occur in many nearshore

Ecology and Classification of North American Freshwater Invertebrates. 2nd Edition

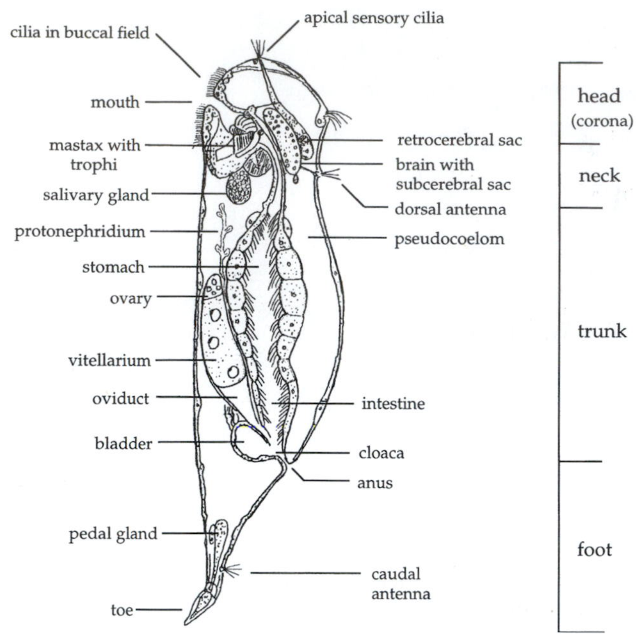

FIGURE 1 Lateral view of a generalized rotifer. (Modified from Koste and Shiel, 1987, with permission.)

marine communities (Egloff, 1988) and occasionally comprise the dominant portion of the biomass (Schnese, 1973; Johansson, 1983). One unusual group of rotifers, the bdelloids (Fig. 3), may be found inhabiting the film of water covering mosses, lichens, and liverworts. Additionally, they are often abundant in soils (Pourriot, 1979); estimates of their densities range from about 32,000 to more than 2 million individuals/m^2, depending on soil moisture levels. Because of their feeding habits, and the fact that they are sometimes more numerous than nematodes, rotifers play an important role in nutrient cycling in soils (Pourriot, 1979).

Most rotifers are free-moving, either swimming as members of the plankton or crawling over plants or within the sediments; however, some sessile species live permanently attached to freshwater plants (Wallace, 1980). The vast majority of rotifers are solitary, but about 25 species form colonies of various sizes (Wallace, 1987). All freshwater rotifers are either exclusively parthenogenetic or produce males for a limited time each year. Therefore, unless collections are made

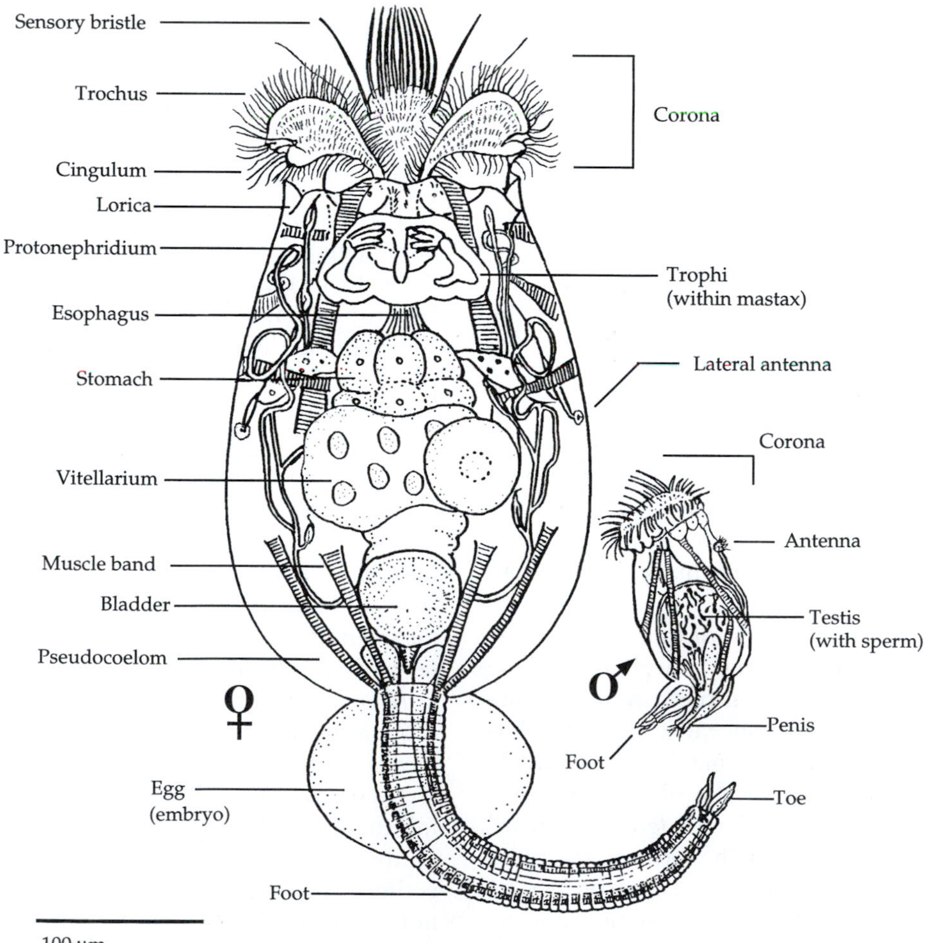

FIGURE 2 Female and male *Brachionus plicalitis*. (Modified from Pourriot, 1986, with permission.)

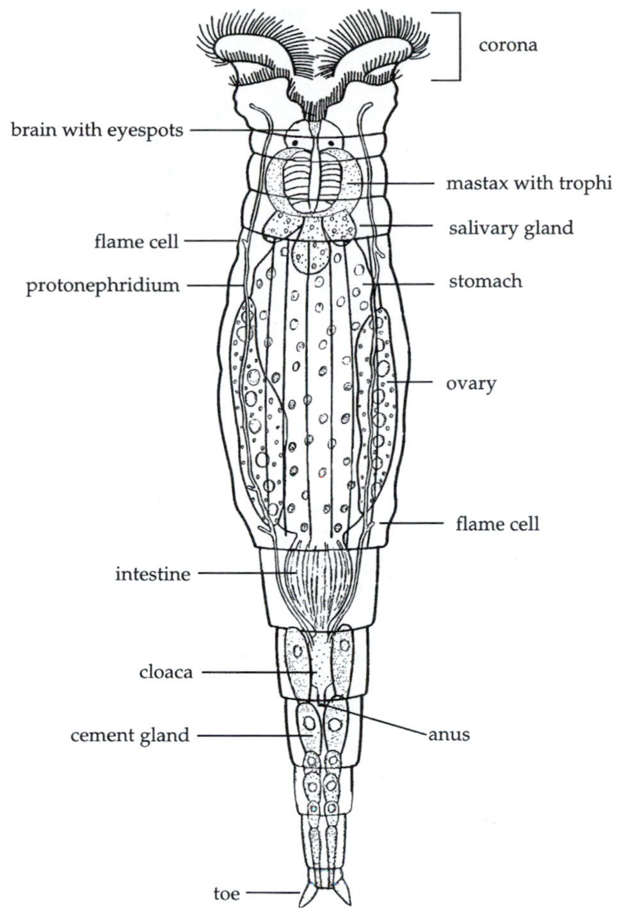

FIGURE 3 Typical bdelloid rotifer (*Philodina*). (Modified from several sources.)

frequently, male rotifers may never be seen. Three very different classes of rotifers are commonly recognized (Seisonidea, Bdelloidea, Monogononta).

Additional accounts of this phylum may be found in most texts of general and invertebrate zoology, and in some specialized books about freshwaters (Edmondson, 1959, pp. 420–494; Hutchinson, 1967, pp. 506–551; Pennak, 1989, pp. 169–225). For detailed reviews of the biology of rotifers consult the works of de Beauchamp (1965), Hyman (1951, pp. 59–151), Koste (1978), Ruttner-Kolisko (1974), and Nogrady *et al.* (1993). In the 1800s there were some beautifully illustrated works that still offer an excellent view of these animals (e.g., Hudson and Gosse, 1886). There is no single scientific journal or set of journals in which researchers publish their work on rotifers; the field is simply too diverse. However, every three years, since 1976, a small group of workers (approximately 50–100) have gathered to hold the International Rotifer Symposium. To date, nine such meetings have been held and most of the proceedings have been published as a special volume of the journal *Hydrobiologia*. Some of the papers discussed in this chapter were presented at those meetings.

II. ANATOMY AND PHYSIOLOGY

A. External Morphology

Rotifers are saccate-to-cylindrical in shape, sometimes appearing wormlike (e.g., many bdelloids, Fig. 3). Typically, the body is comprised of four regions: head (with corona); neck; body; and foot (Fig. 1). Although these regions may be marked by folds in the body wall, which function like joints, rotifers are not metameric (segmented). Unfortunately, the generalizations noted here do not represent all rotifers well. In some forms, the neck and foot may be quite prominent, while in others, they are absent (Fig. 4). In the class Monogononta, male rotifers (Fig. 2) and juveniles (larval females) of sessile rotifers are usually much smaller than adult females (Fig. 5). In addition, male rotifers are often structurally simpler (e.g., the gut commonly functions only as a food reserve) and the larvae of sessile forms generally have a different morphology from that of the adult (Wallace, 1980). When either males or larvae are found in plankton tows, their strikingly different morphologies may lead to improper identification.

The foot is an appendage that extends from the body ventrally (Fig. 2). It usually possesses two toes, but the number may vary from 0 to 4. The foot also may possess pedal glands whose ducts exit near the toes. These glands secrete a sticky cement for temporarily attaching the rotifer to substrates. In larvae of sessile rotifers, the cement forms a bond with the substrate that is not easily detached; if dislodged, sessile forms do not reattach (Wallace, 1980).

Rotifers possess a syncytial integument or body wall containing a filament layer of varying thickness called the intracytoplasmic lamina (Clément, 1985). This feature apparently is shared with the parasitic phylum, Acanthocephala, indicating a phylogenetic relationship (see Section III.D.). The integument of a halophile rotifer, *Brachionus plicatilis*, was examined biochemically by Bender and Kleinow (1988) and their work indicates that it contains two filamentous, keratinlike proteins (M_r 39,000 and 47,000) cross-linked by disulfide bonds. Species in which major portions of the integument are thickened are termed "loricate;" those in which the integument is thin and very flexible are termed "illoricate." In general, thickness of the body wall is of little taxonomic significance since extremes may be found within a single family or genus (e.g., *Cephalodella*). Further, even in loricate forms portions of the integument (lorica) have a less well-developed intracytoplasmic lamina, making the lorica

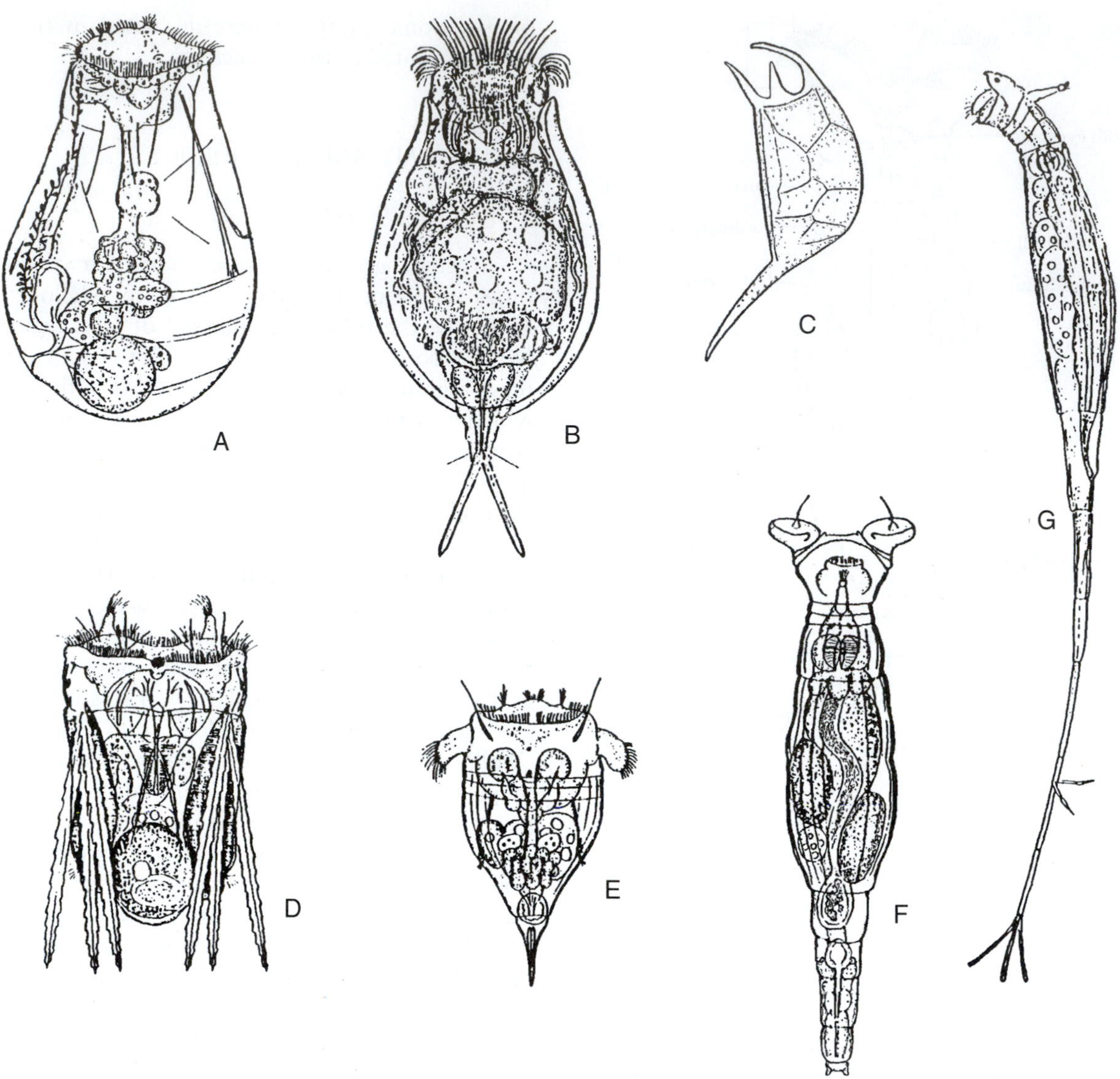

FIGURE 4 Representative rotifers. Monogononts: (A) *Asplanchna*; (B) *Euchlanis*; (C) *Keratella* (lateral view); (D) *Polyarthra*; and (E) *Synchaeta*. Bdelloids: (F) *Philodina*; and (G) *Rotaria*. (From Koste, 1976, with permission.)

flexible in that region. Flexibility is found in the region of the corona, often in the foot, and at articulations between movable spines and the body. In some rotifers, the integument has various projections (e.g., bumps and fixed or movable spines) that serve various functions, chief among these being protection from some predators (see Sections III.A.1 and III.C.4).

B. Organ-System Function

Internally, rotifers possess a perivisceral cavity (pseudocoelom) in which are suspended muscles, nerves, and digestive, reproductive, and protonephridial organs (Figs. 1, 2, and 3). Respiratory and circulatory systems are absent. Rotifers possess two remarkable features. First, all postembryonic tissues are syncytial. Second, all individuals of a species have a consistent number of nuclei in each organ throughout life. [N.B.: There are approximately 900 nuclei per female (Hyman, 1951).] This feature, called "eutely," is also seen in a few other invertebrates (e.g., nematodes, Chapter 9). Most nuclei may be seen using a standard light microscope, but special optics such as differential interference contrast (Nomarski) enhance their visibility.

1. Corona

There is considerable variation in the shape of the anterior ends of rotifers, and at least seven different

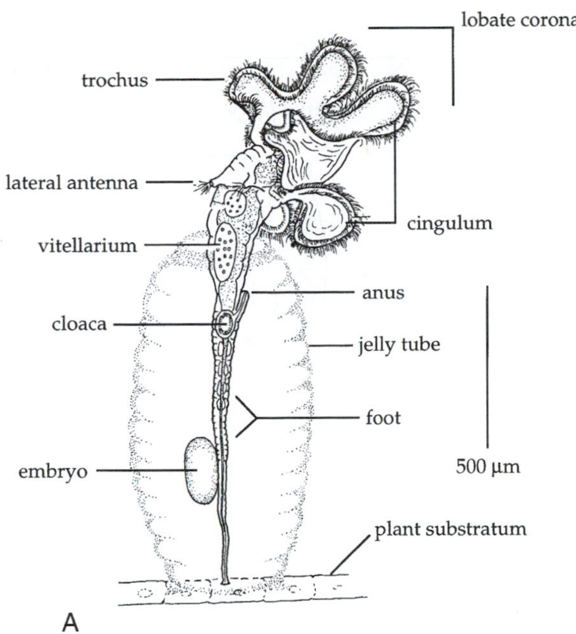

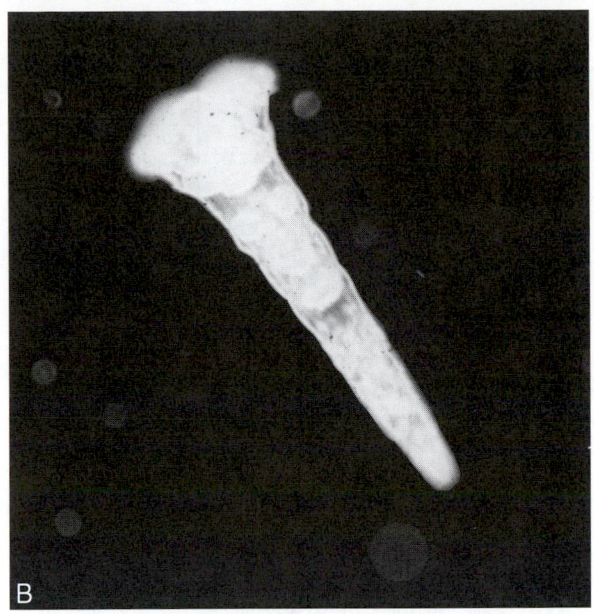

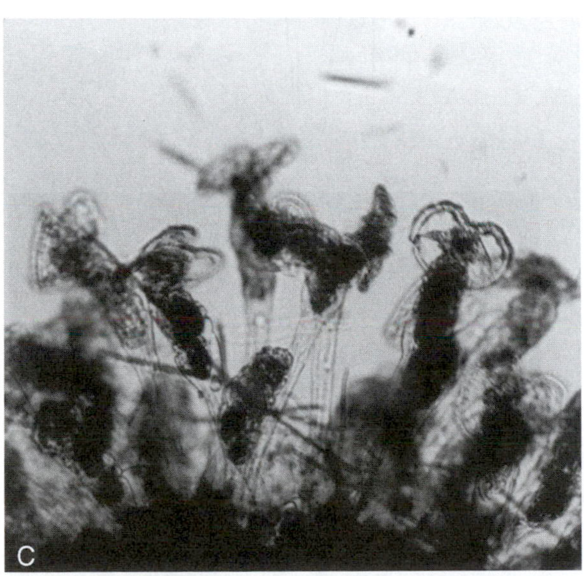

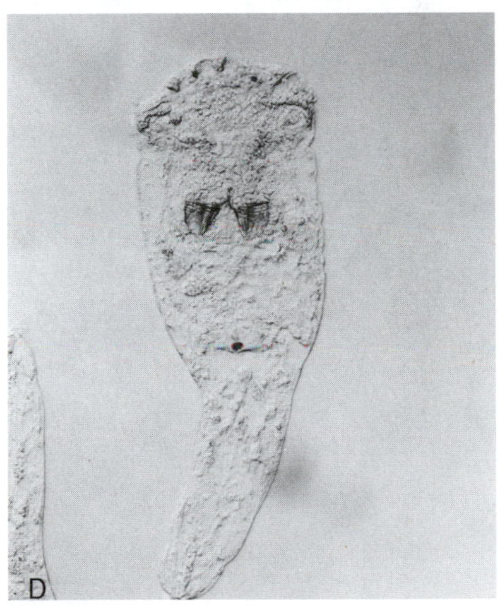

FIGURE 5 Adults and larvae of some sessile species (family Flosculariidae). (A) Adult *Octotrocha speciosa* (from Koste, 1989, with permission); (B) larval *Octrotrocha speciosa* (dark field illumination, length = 1200 μm); (C) colony of adult *Lacinularia flosculosa* (bright field illumination, individual length = 1200 μm); and (D) larval *Lacinularia flosculosa* (bright field illumination, larvae somewhat compressed by the cover glass, length = 450 μm).

types of coronae have been described based on the placement of the mouth and the distribution of cilia (Koste and Sheil, 1987). The corona form of many rotifers is comprised of two ciliated rings called the trochus and cingulum (Figs. 2 and 5). These structures are responsible for the production of water currents that are used in locomotion and feeding. However, not all rotifers possess this coronal ciliation. Adults of the family Collothecidae exhibit the most extreme variation from the typical plan; in most collothecids, cilia are nearly or completely lacking from the corona and long setae surround the rim of a funnel-shaped structure known as the infundibulum (Latin, a funnel) (Fig. 6). These setae prevent escape of prey when the edges of the infundibulum fold over the victim, capturing it in a fashion similar to the Venus flytrap. However, some adult atrochidae possess no setae on their corona (e.g. *Cupelopagis*). In other groups, ciliation

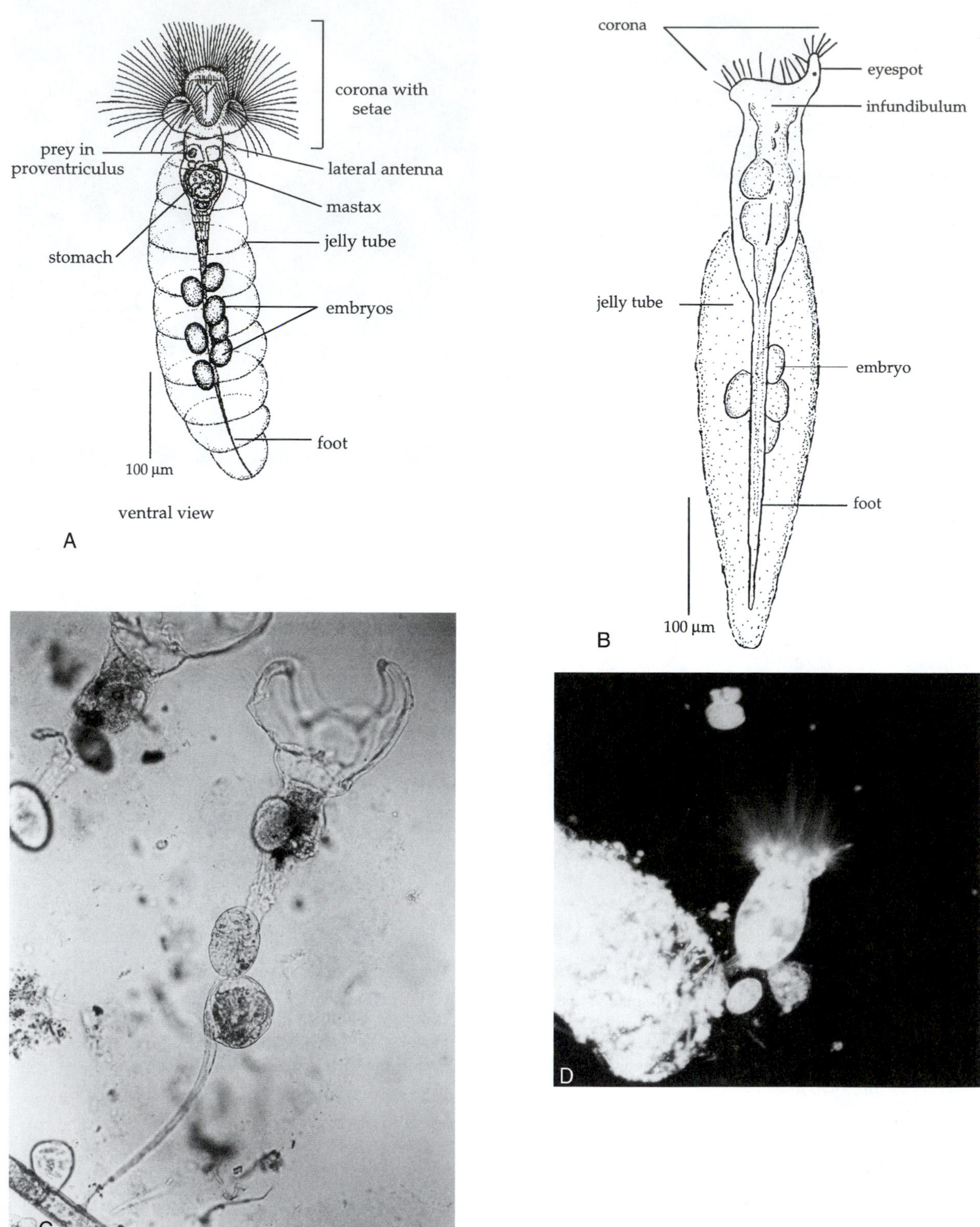

FIGURE 6 Three representatives of the order Collothecacea. (A) *Collotheca trilobata*, a sessile form; (B) *Collotheca mutabilis*, a planktonic form; (C) *Collotheca* sp. with two developing embryos (width of corona = 100 μm); and (D) *Collotheca cornuta*, dark field illumination (length of setae = 200 μm). (A, modified from Wallace *et al.*, 1989, with permission; and B, modified after several sources.)

may be limited to a ventral field or several lobes, as is seen in some creeping forms and some bdelloids. Other structures that may be present on the corona include cirri, sensory antennae, and palps.

2. Trophi and Gut

Once food is captured by the corona, it enters a ventral mouth by passing through a short, ciliated tube into a muscular pharynx, termed the "mastax." The mastax possesses a chitinous lining on the inside, developed as a set of translucent jaws called trophi. The trophi may work the food in various ways (e.g., grinding) before it is passed to the esophagus and swallowed (Figs. 1 and 2). In the family Collothecidae, a portion of the mastax is enlarged into a food-storage organ known as the proventriculus (Fig. 6). The mastax leads posteriorly to an esophagus, stomach, and in most species, intestine and anus, but the gut ends in a blind stomach in some genera (i.e., *Asplanchna*, *Asplanchnopus*). The posterior portion of the intestine (cloaca) receives eggs from the oviduct and fluid from either a bladder or directly from paired protonephridia (Fig. 1). Often the gut is pigmented, depending on the nature of recently ingested material. Different species found in the same sample may possess guts that vary in color due to differences in diet.

Rotifer trophi are composed of several hard parts and associated musculature, which articulate in a specific spatial arrangement. In their basic form, trophi consist of three functional units: an incus (Latin, anvil) and paired mallei (Latin, hammer) (Fig. 7). The incus is composed of three pieces: a fulcrum and a pair of rami (Latin, branch) that move like forceps and articulate with the fulcrum at their bases. Each malleus consists of two parts: manubrium and uncus. The manubrium (Latin, handle) resembles a club-shaped structure extended at one end into a cauda (Latin, tail) and flared at the other end (head). The manubrium articulates with a toothed structure called the uncus (Latin, hook). The plane of movement of the pieces comprising the malleus is at right angles to that of the rami. In some species, the trophi may be modified by reduction of the basic parts, addition of accessory structures, or by asymmetrical development of one or more of the pieces.

The trophi of rotifers have been recognized as important taxonomic features: classes, orders, and families, and even species may be determined based on the details of the trophi alone (e.g., Salt *et al.*, 1978). Nine different types of trophi are recognized based on the size and shape of the seven pieces and the presence of any accessory parts (Figs. 7–10), transitional and aberrant types also are known. In "Malleate" trophi (Fig. 8A–C, E, F, P), all parts of the incus and mallei are well developed and functional, but the rami are characteristically massive and may possess teeth along the inner margin. Further, the unci have 4–7 large teeth. This form works by grasping food and grinding it before pumping the crushed material into the esophagus. Malleate trophi are present in such common rotifers as *Keratella* and *Kellicottia*. "Malleoramate" trophi (Fig. 8I) are found only in the order Flosculariacea and resemble the malleate form except that, in the malleoramate form, the rami are strongly toothed and the unci possess many thin teeth. Similar to the malleoramate form, "ramate" trophi (Fig. 9) have large, semicircular shaped rami and unci with many teeth. Ramate trophi are generally considered to be limited to the Bdelloidae (Melone *et al.*, 1998a). Only members of the order Collothecacea (family Collothecidae) possess "uncinate" trophi (Fig. 8H). These trophi are characterized by unci possessing few teeth, usually with one large and a few small ones. "Virgate" trophi (Fig. 8D, K–O, Q, S) are modified for piercing and pumping and generally can be recognized by the long fulcrum and manubria and the presence of a powerful hypopharyngeal muscle. Some trophi of this form are asymmetrical. Virgate trophi are found in the common genera *Notommata*, *Polyarthra*, and *Synchaeta*. "Forcipate" trophi (Fig. 8J) as the name implies, have an action like forceps, whereby the trophi are projected from the mouth to grasp prey, which are then brought into the mouth and swallowed. Forcipate trophi are limited to the family Dicranophoridae. "Incudate" trophi (Fig 8G, R) function by grasping prey with a forceps-like action, but this form has a different morphology than the forcipate type; the rami are quite large and the mallei very small. The mastax actually initiates prey capture by creating a suction, drawing prey into the mouth, which is then stuffed into the stomach with the aid of the trophi. *Asplanchna*, which has no anus, uses its trophi to extract undigestable materials from its

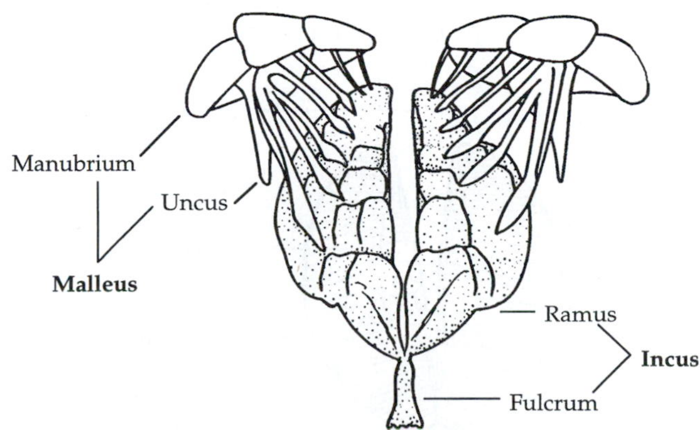

FIGURE 7 Generalized rotifer trophi. (Modified from Wallace *et al.*, 1989, with permission.)

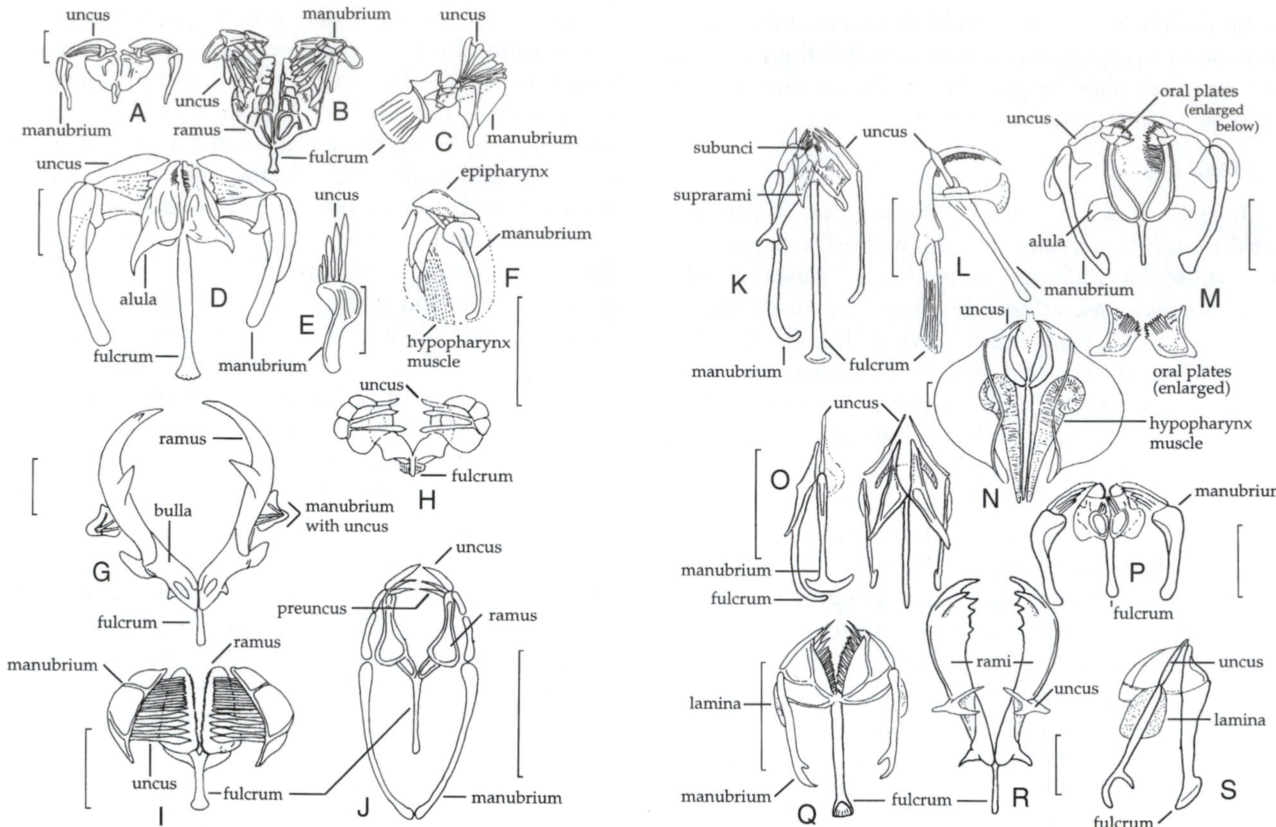

FIGURE 8 Rotifer trophi types. (A) malleate trophi of *Epiphanes*, ventral; (B) malleate trophi (*Epiphanes*), ventral elevated; (C) malleate trophi (*Epiphanes*), lateral; (D) virgate trophi of *Notommata*, dorsal; (E) malleate trophi of *Proales*; (F) showing hypopharynx (lateral); (G) incudate trophi of *Asplanchna*; (H) uncinate trophi of *Collotheca*; (I) malleoramate of *Ptygura*; (J) forcipate trophi of *Dicranophorus*; (K) asymmetrical virgate trophi of *Trichocera rattus*; (L) virgate trophi of *Eothinia*; (M) virgate trophi of *Itura*, oral plates enlarged; (N) virgate trophi of *Synchaeta* with the powerful hypopharynx muscle; (O) virgate trophi of *Ascomorpha*; (P) malleate trophi of *Proales gigantea*, an intermediate form between malleate and virgate types; (Q) virgate trophi of *Cephalodella*, dorsal; (R) incudate trophi of *Asplanchna*; and (S) virgate trophi of *Cephalodella*, lateral. Bars = 20 μm. (From Koste and Shiel, 1987, with permission.)

stomach. Incudate trophi are limited to the family Asplanchnidae. “Cardate” trophi (Fig. 10) are found only in the family Lindiidae and function by producing a pumping action, without the hypopharyngeal muscle. The “fulcrate” type of trophi have been described as an aberrant form and are incompletely understood (Edmondson, 1959, p. 432). This form is found only in the class Seisonidea, a very small group of marine rotifers.

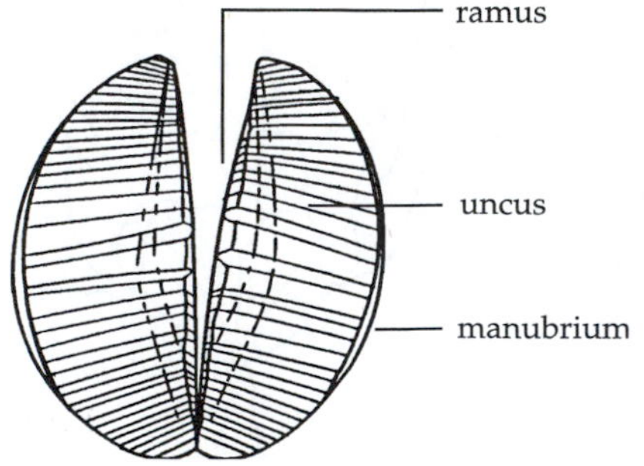

FIGURE 9 Ramate trophi of bdelloid rotifers. (Modified from several sources.)

3. Organ Systems

Research on the ultrastructure of rotifers has proceeded rapidly during the past decade and promises to continue to do so. For a comprehensive overview on the topic of ultrastructural research on rotifers, consult the works of Amsellem and Ricci (1982), Clément (1977, 1980, 1987), Clément *et al.* (1983), Clément and Wurdak (1991), Wurdak (1987), Wurdak and Gilbert (1976), and Wurdak *et al.* (1977, 1983).

a. Muscular System The muscular system consists of small groups of longitudinal and circular muscles inserted at various points on the integument or between the integument and viscera (Fig. 2). In loricate species,

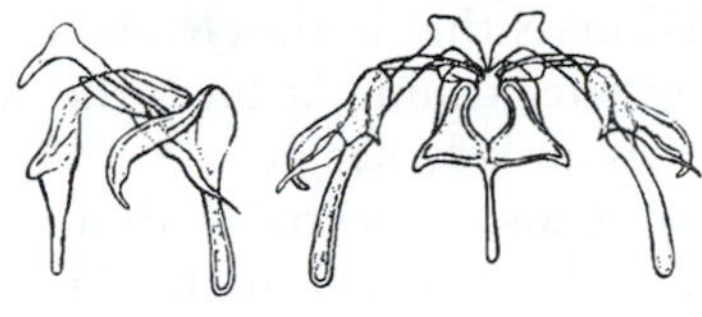

FIGURE 10 Cardate trophi of the family Lindiidae. (From Harring and Myers, 1922, with permission.)

the integument provides a firm structure against which the muscles work. Muscle contraction can increase the pressure of the pseudocoel, which then acts as a hydrostatic skeleton. In some species, muscles retract the corona, which increases pressure within the pseudocoel, thus expanding flaccid portions of the integument, known as body-wall outgrowths (*Asplanchna*, Fig. 11). In others, such as *Brachionus* (Fig. 12), retracting the corona makes spines, that articulate with the body in the posterolateral region, swing forward. When in this extended position these spines interfer with predators like *Asplanchna* which would otherwise consume the *Brachionus* (Gilbert, 1966, 1967). Muscles also are present in the viscera, particularly in the mastax and stomach. Striated, longitudinal muscles are responsible for retracting the corona and foot and for moving certain articulating spines that are not positioned by hydrostatic pressure. Some species possess powerful muscles that control movement of certain locomotory appendages (e.g., *Hexarthra*, Fig. 13). Contractions of these muscles cause a swift downward sweep of the

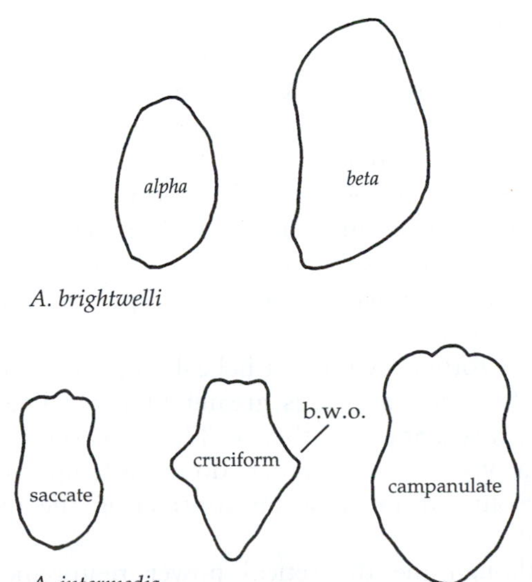

FIGURE 11 Body form variability in the genus *Asplanchna*. (b.w.o. = body-wall outgrowths). (Modified from Gilbert, 1980a, with permission.)

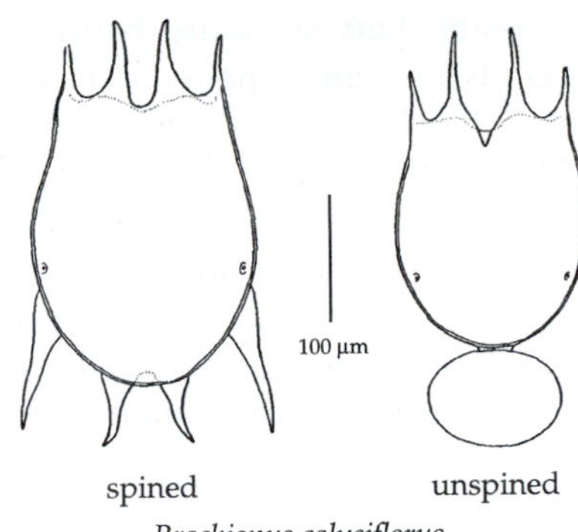

FIGURE 12 Spined and unspined *Brachionus calyciflorus*. (From Koste, 1976, with permission.)

appendages which results in a rapid displacement or jump of the rotifer (see Section II.C.1).

b. Nervous System The nervous system is simple, consisting of only a cerebral ganglion or brain (Fig. 3) located dorsally on the mastax, a few other ganglia present in the mastax and foot, and three types of sensory organs, namely mechano-, chemo-, and photoreceptors. Mechanoreceptive bristles are situated on the corona, while several antennae are located elsewhere on the body surface, usually laterally and caudally (Fig. 2). Chemoreceptive pores are also present on the corona. Many species possess one or more photoreceptive eyespots, sometimes accompanied by a pigmented spot (Fig. 3). When present, eyespots are located in the anterior end, usually near the brain (Clément, 1980; Clément *et al.*, 1983). While most rotifers retain the eyespots throughout life, larvae of sessile rotifers often lose them during metamorphosis. Paired, ventral nerve cords proceed from the brain along the length of the body into the foot. Several other ganglia are usually found in the nerve cords at the exit points for lateral nerves.

One interesting structure found in the apical region of many bdelloid and monogonont rotifers is the

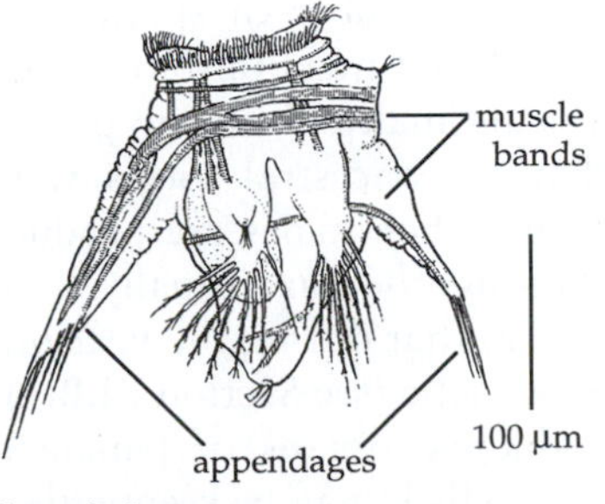

FIGURE 13 *Hexarthra* showing positioning of muscles that initiate jumps. (Modified from several sources.)

retrocerebral organ. This structure consists of paired subcerebral glands and an unpaired retrocerebral sac, both with ducts that lead to the surface of the corona (Fig. 8.1). While the function of the retrocerebral organ is unknown, Edmondson (1959, p. 422) has noted that there are fewer protonephridia in rotifers that possess well-developed retrocerebral organs. However, Clément (1977) suggests that it may function as an exocrine gland, perhaps lubricating the anterior part of the body.

Although information on neurobiochemistry is very limited, research has shown that acetylcholine functions as a neurotransmitter (i.e., a cholinergic system) in twelve species of rotifers from six families (Nogrady and Alai, 1983, Nogrady and Keshmirian, 1986a, b). In addition, norepinephrine neuroreceptor sites (i.e., an adrenergic system) have been reported in *Brachionus plicatilis*, and a widespread catecholaminergic neuronal systems have been observed in species of *Asplanchna* and *Brachionus* (see Keshmirian and Nogrady, 1988).

c. Protonephridium A paired protonephridial system comprised of tubules and flame cells functions in excretion and osmoregulation in all rotifers (Figs. 1 and 2). Usually there are only a small number (<6) of flame cells, but large rotifers may possess many more. For example, *Asplanchna sieboldi* may have up to 100 flame cells (Ruttner-Kolisko, 1974, p. 11). Normally, the tubules drain into a urinary bladder which leads to a cloaca, but the bladder is absent in some species and a contractile cloaca assumes its function.

d. Reproductive System Besides their separation by anatomical details of the trophi, the three classes of rotifers are differentiated based on the anatomy of their reproductive systems (Wallace, 1998). The gonads are paired in both the Seisonidea and the Bdelloidea, but in the latter class, males are completely unknown and reproduction is always asexual. Members of the third class (Monogononta) have only one gonad. Although males are present in this group, they have not been described for a large number of species. However, it is generally assumed that most (all) monogononts are capable of producing males given the proper conditions, or at least that the ancestral forms were capable of male production. When males are produced by monogonont populations, they are usually limited to a few days or a week, so that during the year most reproduction is parthenogenetic (see Section III.B.1).

The reproductive organs of female rotifers comprise three units, which may be seen with a light microscope: ovary, vitellarium, and follicular layer (Amsellem and Ricci, 1982). The ovary of a rotifer is a small, syncytial mass that is closely associated with the yolk-producing vitellarium. At birth the adult complement of ovocytes already has been formed in the ovary. The vitellarium is also syncytial with a constant number of nuclei, a characteristic useful in the taxonomy of some species. The follicular layer surrounds both the ovary and the vitellarium; in some species, this layer forms the oviduct, which connects with the posterior portion of the gut forming a joint exit, the cloaca (Fig. 1).

In general, monogonont males are much smaller (about 100 μm long) than females. In nearly all forms, the gut of males is reduced to a rudiment or is absent entirely (Fig. 2). In those forms with a rudimentary gut (e.g., *Asplanchna*), it serves as an energy source for the fast swimming, nonfeeding male. The single testis is large and saccate, usually containing <50 freely floating mature sperm. A ciliated vas deferens leads from the testis to the penis, usually with one or rarely two pairs of accessory (prostate) glands that discharge into it (see Section III.B.2).

Commonly, rotifers are oviparous, that is, they release their eggs outside the body where the embryos develop. Many planktonic rotifers carry their eggs attached to the body of the mother by a thin thread (*Brachionus*), while others fix them to a substrate (*Euchlanis*) or release them into the plankton (*Notholca*). A few species retain the embryo in the body until the offspring hatches: ovoviviparous (e.g., *Asplanchna* and *Cupelopagis*).

C. Environmental Physiology

1. Locomotion

All rotifers swim during at least a portion of their life cycle and some swim continuously, never attaching even temporarily to surfaces. Swimming influences the acquistion of food and mates, increases the number of encounters with predators, and promotes the dispersal of the larvae of sessile forms. Some rotifers are unusual in that they can swim, but routinely remain in mucus tubes attached to a substrate (e. g., certain *Cephalodella* species), while others are free-floating mucus sheaths (*Ascomorpha*).

Most rotifers swim in a helical pattern, so that the actual distance traveled is greater than the linear displacement (see Starkweather, 1987). However, for practical reasons, most researchers do not attempt to calculate absolute distance when considering the distance traveled.

Although the theoretical power requirements of swimming (i.e., the theoretical energy required to overcome water resistance) is <1% of the total metabolism, the actual energetic cost appears to be much greater.

In *Brachionus plicatilis*, Epp and Lewis (1984) calculated this cost to be approximately 38% of the total metabolism (See also Epp and Lewis, 1979).

Swimming speeds of male and female *Brachionus plicatilis* have been measured by Luciani *et al.* (1983), Epp and Lewis (1984), and Snell and Garman (1986). Swimming speeds also have been measured for *Asplanchna brightwelli* (Coulon *et al.*, 1983), two species of *Keratella* (Gilbert and Kirk, 1988), and for *Anuraeopsis fissa*, *Asplanchna girodi*, *Brachionus calyciflorus*, *Euchlanis dilatata*, *K. americana*, *Lecane quadridentata*, *Brachionus patulus*, and *Trichocerca pusilla* (Rico-Martinez and Snell, 1998). These studies have shown that swimming speed is temperature–dependent and varies among strains. Values recorded for *B. plicatilis* at 25°C range from 0.6 to 0.9 mm/sec for females and from 1.3 to 1.5 mm/sec for males, with young and old females swimming about 30% slower than mature females. *Asplanchna brightwelli* and *Keratella* females normally swim at 0.9 and 0.5 mm/sec, respectively. *Euchlanis dilatata* is one of the fastest species, with males and females both swimming at 0.98 mm/sec at 25°C.

Little is known about the swimming speeds of the larvae of sessile rotifers, but the swimming speeds of the larvae of *Ptygura beauchampi* are know to vary with age. New born (0–2 h) through mid-aged (ca., 3 h) larvae swam about 2–2.5 mm/sec, faster than the rates reported for *Brachionus* males. However, older larvae (>4.5 h) swam at approximately 1.0 mm/sec. Concomitant with these changes in swimming speed was an increase in turning frequency (Wallace, 1980).

Correlations of rotifer ultrastructure with turning frequency, locomotion in general, and other behaviors have been a field of systematic investigation by Clément and his co-workers (Clément, 1987; Clément *et al.*, 1983; Chauoy, 1995). Their efforts have demonstrated that rotifers are excellent models for comparative neurobehavioral studies, but much more work remains to be done before a complete synthesis of structure, function, and behavior will be possible.

Spines and other appendages influence both swimming speed and sinking rates in rotifers. For example, Stemberger (1988) has found that unspined *Keratella testudo* generally swim faster and sink more slowly than spined forms. On the other hand, *Polyarthra* normally swims at a much slower velocity (0.24 mm/sec) than any species previously discussed, but is capable of very short bursts (about 0.065s long) of rapid movements called jumps. During a jump, *Polyarthra* may attain velocities greater than 50 mm/sec, with a mean velocity of 35 mm/sec, or >100 times its normal swimming speed (Gilbert, 1985a, 1987; Starkweather, 1987). Jumps are produced by movements of 12 appendages, called paddles, which articulate with the body near the head. When *Polyarthra* detect a local disturbance in the water a few body lengths away, powerful striated muscles rapidly flex some of the paddles upward and then during the jump they return to their original positions. During this process the rotifer is displaced an average of 1.25 mm (ca. 12 body lengths). Jumping helps this rotifer escape invertebrate predators, such as *Asplanchna* (Gilbert, 1980b; Gilbert and Williamson, 1978) and first instar *Chaoborus* (Moore and Gilbert, 1987) as well as the filtering currents of microcrustaceans, such as *Daphnia* (Gilbert, 1985b, 1987). They are also effective against naive workers attempting to remove *Polyarthra* from plankton samples using micropipets! The genera *Filinia* and *Hexarthra* also can jump rapidly, but movement in these forms has not been well studied (see Section III.C.4). In contrast, *Keratella* cannot jump to avoid the filtering currents of daphnids and, as a result, are severely damaged or killed when swept into the branchial chambers of cladocerans (Burns and Gilbert, 1986a,b; Gilbert and Stemberger, 1985b). However, *Keratella* is not without some escape abilities; it is capable of increasing its swimming speed by a factor of about 3.5 when it encounters inhalant currents of *Daphnia* or the predatory rotifer *Asplanchna* (Gilbert and Kirk, 1988).

Another source of phenotypic variation in body size is polyploidy. Substantial intrapopulational variation in body size was reported for *Euchlanis dilatata* (Walsh and Zhang, 1992). Two distinct body sizes were observed in Devils Lake, Oregon over 12 months of intensive sampling. The smaller morphotype was about 225 μm in length and the larger averaged about 275 μm. The larger morphotype had a chromosome number of 21 and the smaller morphotype had 14. The haploid chromosome number found in males of this species is $n = 7$. This is the first case of polyploidy reported for rotifers, but new techniques for staining and counting rotifer chromosomes may make more systematic explorations possible.

2. Physiological Ecology

Physiological tolerances of organisms prescribe environments where survival and reproduction are possible. Thus, an environmental tolerance curve for a species summarizes the range of environments where reproduction occurs and culminates at the upper and lower lethal limits for the species (e.g., temperature range). Within the tolerance curve, the environmental optimum for a species is that environment where survival and reproduction are maximal. Therefore, a set of tolerance curves (including temperature, pH, etc.) indicate the breadth of adaptation of a species and its niche width. Environmental tolerances and niche widths have been characterized for a few rotifer species. For example, Epp and Lewis (1980) described the response of *Brachionus plicatilis* to temperature. That study determined respiration rate over temperature ranging from 15 to 32°C and recorded Q_{10s} of 1.9–2.4 for broad temperature intervals. Respiration levels off between 20 and 28°C, indicating that *B. plicatilis* can maintain a constant metabolic rate over this temperature range. At higher and lower temperatures, respiration rate increased, presumably because of thermal stress which was beyond the homeostatic capability of this rotifer.

Snell (1986) showed that amictic and mictic females have similar temperature tolerance curves. Amictic *B. plicatilis* females reproduced at 20 and 40°C, whereas mictic females did not. In general, amictic females reproduced over a broader environmental range of temperatures salinity, and food level than did mictic females.

The effect of pH on the distribution and abundance of rotifers is a topic that has received a good deal of attention. However, since hydrogen-ion concentration is related to other important chemical parameters in freshwaters, studies of rotifer occurrence as a function of pH alone are of limited value. Nevertheless, some extensive early work by Myers in the 1930s demonstrated that rotifer species can be classified into a few broad groups based on pH alone: alkaline species, acid species; and those with a broad range (see also Edmondson, 1944). More recently, Berzins and Pejler (1987) concluded that species found in oligotrophic waters had pH optima at or below neutrality (pH of 7.0) and those species common to eutrophic waters had optima at or above neutrality. Further, they noted that acid-water species were often nonplanktonic or semiplanktonic.

Much less is known of the specific metabolic responses of rotifers to pH. The influence of pH on *B. plicatilis* was described by Epp and Winston (1978). They found that swimming activity and respiration rate were not significantly different at pH values of 6.5–8.5. Snell *et al.* (1987) examined the pH from 4.0 to 9.9 and found swimming activity depressed below pH 5.6 and above pH 8.7. Alkaline waters depressed swimming activity more than acidic conditions. Increasing acidification of lakes as a result of acid rain has been widespread in North America. Acidification leads to predicable changes in the rotifer zooplankton, such as the increasing dominance of *Keratella tauracephala* (Brett, 1989). This observation has been repeated in the experimental acidification of lakes (Gonzalez and Frost, 1994). In general, it appears that the importance of rotifers within lakes increases with acidity (Frost *et al.*, 1998).

Osmoregulation has been investigated by Epp and Winston (1977) for *B. plicatilis*, a species that is common in highly alkaline waters and salt lakes. This species was an osmoconformer capable of tolerating osmolarities exceeding 957 mOsm/L (33.5 ppt). Ito (1960) suggested that the upper osmotic limit for *B. plicatilis* may be as high as 2860 mOsm/L (97 ppt). Transfer of this rotifer directly from 41 to 957 mOsm/L caused considerable mortality, but acclimation to high osmolarities was achieved by gradually increasing ionic concentrations. *Brachionus plicatilis* does not tolerate low osmolarities well, and this probably accounts for its restriction to alkaline and brackish waters.

Although most rotifers require oxygen concentrations significantly above 1.0 mg/L, some can tolerate anaerobic or near-anaerobic conditions for short periods. Other species routinely live in oxygen-poor regions, such as the hypolimnion of eutrophic lakes or in sewage ponds. The physical and chemical factors described above, along with food (Bogdan and Gilbert, 1987) and predation (Williamson, 1983), define the niche boundaries for rotifer species. Niche boundaries for field populations have been described by Edmondson (1944), Makarewicz and Likens (1975, 1979), Miracle (1974), and Miracle *et al.* (1987). These investigations have generally revealed that rotifer species extensively partition the environment, thus avoiding negative interactions.

3. Environmental Toxicology

Because rotifers fill an important ecological role and are relatively easy to culture, interest is growing in their use for aquatic toxicity testing. The response of rotifers to a variety of toxicants has been characterized in both natural and laboratory populations. For example, effects of various insecticides, herbicides, and wastewater on natural rotifer populations have been investigated (Snell and Janssen, 1995). In general, rotifers seem to serve as good indicators of environmental water quality (Sladecek, 1983), and the use of rotifer population dynamics as sensitive indicators of toxicity has been promoted (Halbach, 1984; Halbach *et al.*, 1981, 1983). A multispecies approach to toxicity assessment with a laboratory microcosm consisting primarily of rotifers has been attempted by Jenkins and Buikema (1985). Many short-term, acute toxicity tests with a variety of substances have been conducted using rotifers, e.g., insecticides, heavy metals, free ammonia, sodium dodecyl sulfate, crude oil and petrochemicals (Snell and Janssen, 1998).

Unfortunately, the results of these studies are not directly comparable, as their methodologies varied considerably. However, testing protocols are comparable from several studies of the halophile *B. plicatilis* and a number of median lethal concentrations have been reported. (LC_{50} is the concentration of a compound which kills 50% of a cohort within a specified exposure time.) Capuzzo (1979a, b) found *B. plicatilis* to be very sensitive to free chlorine and chloramine. After a 30-min exposure at 25°C, LC_{50} values of 0.09 and <0.01 mg/L were recorded for chlorine and chloramine, respectively. *Brachionus* is more sensitive to mercury (LC_{50} of 0.045 mg/L; Gvozdov, 1986) than to chromium ($LC_{50} > 500$ mg/L; Persoone *et al.*, 1989). Free ammonia is also toxic, with LC_{50} values of 20.4 and 17.7 mg/L at salinities of 15 and 30 ppt, respectively (Snell and Janssen, 1995).

Pesticides, in contrast, are not very toxic to *B. plicatilis*. The 24-h LC_{50} values for five organophosphate pesticides and one organochlorine pesticide ranged from 0.9 to 150 mg/L for three different strains of *B. plicatilis* (Serrano *et al.*, 1986). These same investigators observed that pesticide resistance of this rotifer is about 1000 times greater than that reported for other aquatic organisms. Since pesticides are generally targeted to arthropods, the low sensitivity of rotifers to these compounds is not surprising.

Several median lethal concentrations for a variety of compounds have been reported for freshwater rotifers, mainly in the genus *Brachionus*. Dad and Kant Pandya (1982) recorded 24-h LC_{50} values for *B. calyciflorus* exposed to two insecticides, and Couillard *et al.* (1989) found the following relative metal toxicities: Hg > Cu > Cd > Zn > Fe > Mn. LC_{50} values of 600 and 0.16 mg/L for phenol and pentachlorophenol, respectively, were reported by Halbach *et al.* (1983) for *B. rubens*.

A standardized rotifer toxicity test for marine and freshwater that employs test animals derived from hatching *B. plicatilis* or *B. calyciflorus* resting eggs has been described (ASTM, 1991). This test is less expensive because stock cultures are not necessary to obtain test animals, and it is simple, rapid, and sensitive, so the use of rotifers in aquatic toxicology is expected to expand.

4. Anhydrobiosis

The ability of rotifers to tolerate desiccation and then be revived sometime later has been known since the early 1700s, when Leeuwenhoek observed rehydration of rotifers found in dry sediments of rain gutters. Anhydrobiosis, which is also termed cryptobiosis and osmobiosis, is limited to bdelloids. Unlike monogononts, bdelloids cannot produce cysts. The term cryptobiosis refers to the slow or hidden metabolism which is characteristic of these animals, while anhydrobiosis and osmobiosis emphasize the processes whereby loss of water is accomplished through evaporation and external osmotic pressure, respectively. While in the desiccated state, these rotifers resemble a wrinkled barrel (or tun, as in the tardigrades, see Chapter 15), with the head and foot retracted into the animal's trunk (Fig. 14). The significance of this phenomenon is clear, as many bdelloids inhabit environments that dry completely at irregular intervals. A few strictly aquatic bdelloids cannot withstand desiccation (Ricci, 1998b).

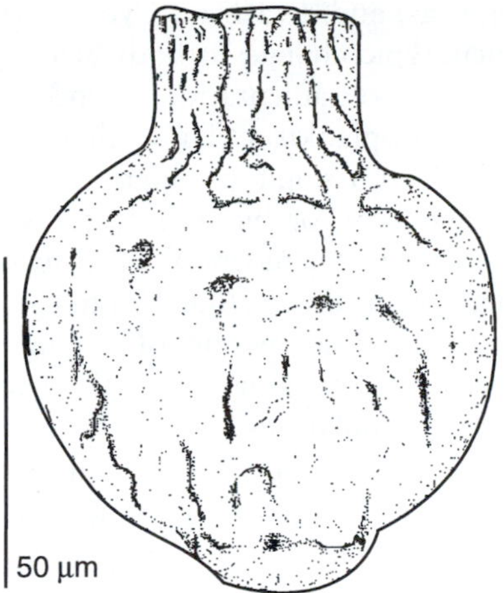

FIGURE 14 Body of a desiccated bdelloid rotifer (*Habrotrocha rosa*) with the corona withdrawn into the trunk. (Bar = 50 μm.)

Anhydrobiosis involves more than simple drying; unless loss of metabolic water proceeds slowly, the rotifer usually dies. During anhydrobiosis, the fine structure of cells is retained, but in a greatly modified state (Ricci, 1987; Schramm and Becker, 1987). Changes that occur internally include a 50% reduction in the volume of the pseudocoel, a condensation of cells and organs, and a decrease in cytoplasmic volume, so that the entire animal is only about 25–30% of its original size. Nuclei, mitochondria, endoplasmic reticula, and other organelles form a compact mass within cells of the anhydrobiotic animal (Schramm and Becker, 1987).

Desiccated bdelloids have been reported to be viable even after more than two decades in the anhydrobiotic state. Recovery from anhydrobiosis may require as little as 10 min, or it may take several hours, according to prevailing environmental conditions. Survival is negatively affected by starvation before desiccation and by moist environments and high temperatures during the desiccation period (Ricci 1987; Ricci, *et al.*, 1987, Schramm and Becker, 1987).

III. ECOLOGY AND EVOLUTION

A. Diversity and Distribution

1. Phenotypic Variation

Phenotypic variation is an important adaptive mechanism in rotifers, but has posed difficult problems for systematists. This variation arises by several mechanisms including cyclomorphosis, dietary- and predator-induced polymorphisms, polymorphisms in hatchlings from resting eggs, and dwarfism. Cyclomorphosis is the seasonal phenotypic change in body size, spine length, pigmentation, or ornamentation found in successive generations of zooplankton. These changes are phenotypic alterations in a single population that are related to physical, chemical, or biologic features of the environment. Each different morphological form is called a morphotype. Specifically excluded from cyclomorphotic change are seasonal succession of sibling species and clonal replacements of genotypes, both of which are genetic changes in populations.

A striking phenotypic change in morphology that is associated with a dietary polymorphism was described for three *Asplanchna* species (*brightwelli*, *intermedia*, *sieboldi*) by Gilbert (1980a). Diets that include the plant product *a*-tocopherol (vitamin E) induce saccate females, the smallest morphotype, to produce cruciform daughters. Cruciforms have lateral outgrowths of the body wall (Fig. 11) that protect them from cannibalism by conspecifics by making them larger and, thus, more difficult to ingest if captured. In the presence of *a*-tocopherol and certain prey types, cruciforms can produce a third morphotype called campanulates (more prevalent in *A. sieboldi* and *A. intermedia*). Campanulates are very large females (>2000 μm), which heavily cannibalize saccate females. Female polymorphism is much less pronounced in *A. brightwelli* where there is a 50–60% increase in body size, but no campanulates are produced and body wall outgrowths are slight. Dietary polymorphism in *Asplanchna* (gigantism) may have evolved originally as a generalized growth response to larger prey typical of eutrophic waters (Gilbert, 1980a; Gilbert and Stemberger, 1985c). The tocopherol response probably is adaptive, because it signals the availability of nutritious rotifer and microcrustacean prey.

Another source of phenotypic variation is predator-induced polymorphisms. Spined and unspined forms had been recognized in several rotifer species for many years, but the cause(s) and significance(s) of these variations remained an enigma (Fig. 12). However, Gilbert (1966, 1967) was the first to show that spine production could be induced in the offspring of female *B. calyciflorus* if adults were exposed to culture medium which had previously held the predatory species *Asplanchna*. Gilbert (1967) also demonstrated that such spines were strong deterrents to predation by *Asplanchna* (see Section III.C.4). However, Stemberger (1990) has shown that food concentration can dramatically modify the development of spines in *B. calyciflorus*.

Two additional sources of phenotypic variation are polymorphisms called "aptera generations" in the hatchlings of resting eggs and dwarfism, both of which have been reported in rotifers of some tropical crater lakes (Green, 1977). Aptera morphotypes were initially thought to be different species of *Polyarthra*, but later were shown to be forms lacking the paddles that are characteristic of this genus. Only the generation hatching from resting eggs lacks paddles; their parthenogenetic offspring develop into typical morphotypes. Similar polymorphisms between resting egg hatchlings and parthenogenetic generations were described for *Keratella quadrata* and are suspected for *Notholca acuminata* (Amrén, 1964). Dwarfism in *Brachionus caudatus* in Cameroon crater lakes was described by Green (1977) and is characterized by reduced body size and spination as compared to normal morphotypes. Green speculated that high temperature combined with reduced food supply may cause this condition.

2. Distribution and Population Movements

Water bodies are not uniform habitats with respect to concentrations of food and predators and abiotic factors such as dissolved oxygen concentration, light intensity, temperature, and water movements. Therefore, it is not surprising to find that rotifers are not evenly distributed in lakes and ponds; often there is considerable variability with respect to their horizontal and vertical distribution. For example, Hoffman (1982) showed that two *Filinia* species had very different vertical distribution patterns over the course of a year in a small lake in Germany (Fig. 15). Striking horizontal variabilty in rotifer distribution has been reported by a variety of researchers: for example, by Green (1985) for a tropical lake. Thus, we find that some rotifers are strictly littoral, being found in open waters only as occasional migrants, whereas others are pelagic; however, the depth at which they are found is a function of season. Consequently, experienced researchers can determine to some degree where and when a water sample has been taken, merely by the composition of the rotifer species it contains. Although these major distribution patterns are the result of differential population growth and water currents and other large-scale water movements, within-lake distribution patterns can be influenced to a lesser degree by locomotory behaviors. One commonly recognized behavior of marine and freshwater zooplankton is a daily (diel) vertical migration in the water column in which the animals usually come to the surface only during the night. During the day, zooplankton avoid visual predators (fish) that occupy near-surface waters, but when they return to the surface at night they can exploit the rich algal resources present there. In rotifers, diel migrations are never as dramatic as those typical of microcrustaceans; the population maximum usually changes only about 1–2 m

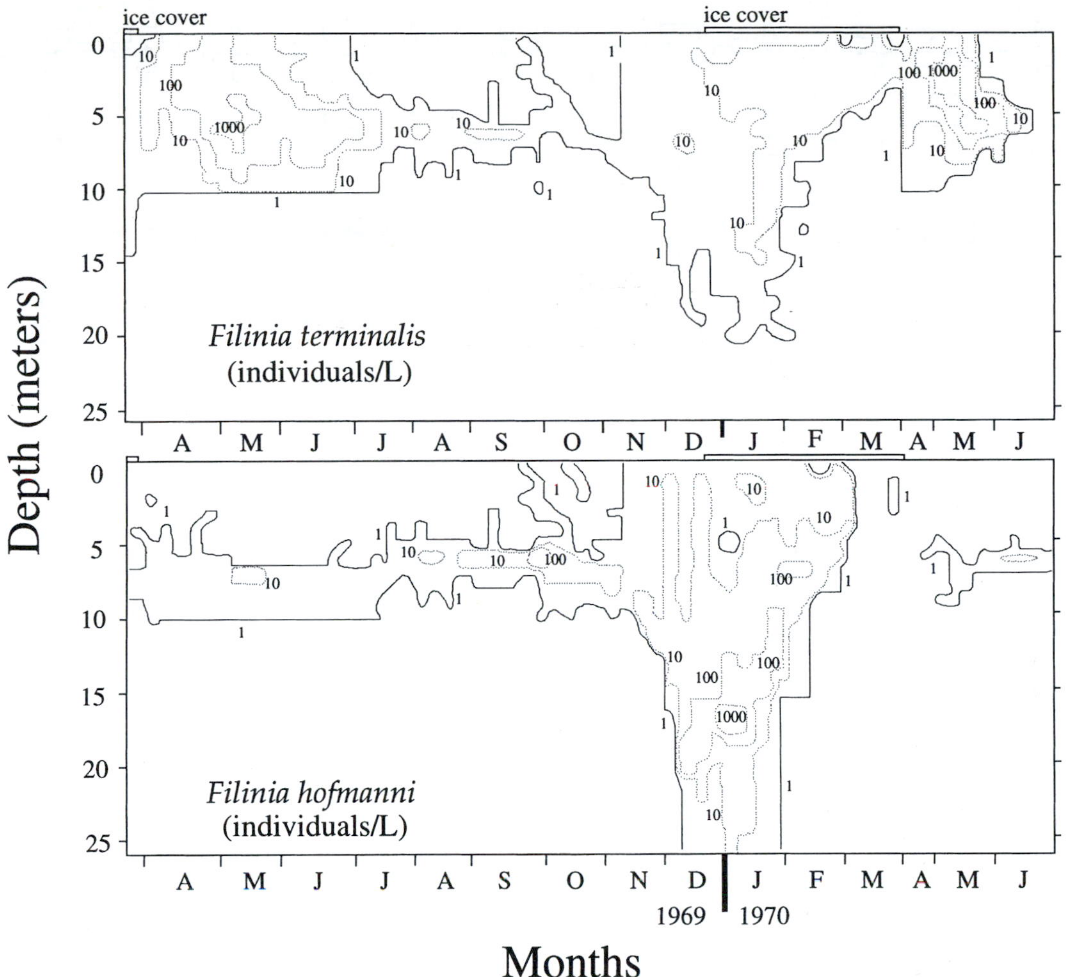

FIGURE 15 Temporal and spatial distribution of two species of *Filinia* in Lake Plußee, Germany. Solid line indicates limits of the population at one individual per liter; dotted line indicated boundaries of higher population levels (numbers of individuals per liter). (Modified from data in Hofmann, 1982, with permission.)

over a daily cycle (Fig. 16). However, ovigerous (egg-bearing) and nonovigerous females may have different migration patterns, and this can cause serious errors in the calculations of birth rates. Errors of nearly an order of magnitude depending on the sampling protocol may occur, if the population is sampled at only one depth or at different times during the day (Magnien and Gilbert, 1983).

Horizontal movement of rotifers has been investigated by Preissler (1980) who showed that some pelagic rotifers apparently avoid the inshore region. Preissler demonstrated this phenomenon, known as "avoidance of the shore," by using a circular plexiglass arena in which he simultaneously monitored the swimming direction of several zooplankton species. When he artifically altered the shadow produced by the natural elevation of the shoreline by adding a black collar around the arena, both *Asplanchna priodonta* and *Synchaeta pectinata* swam away from the shadow. The littoral rotifer *Euchlanis dilatata* apparently showed no preference based on light intensity.

Research on the abundance and distribution of rotifers in lotic systems (flowing waters) has lagged far behind that of lentic ones, but we do know that both rotifer diversity and abundance are usually much lower in rivers than in lakes. Further, most of what is present seems to be derived from upstream lakes or secondary channels with minimal reproduction within the channel (Saunders and Lewis, 1989). In some instances, the number of species is very high, but usually only a few of these are numerically significant (<10). The density of rotifers encountered in lotic (flowing) systems is usually much smaller than that commonly present in lentic (lakes) ones (ca. 5–400 individuals/L), but in some instances it may be quite high (ca. >3,000 individuals/L).

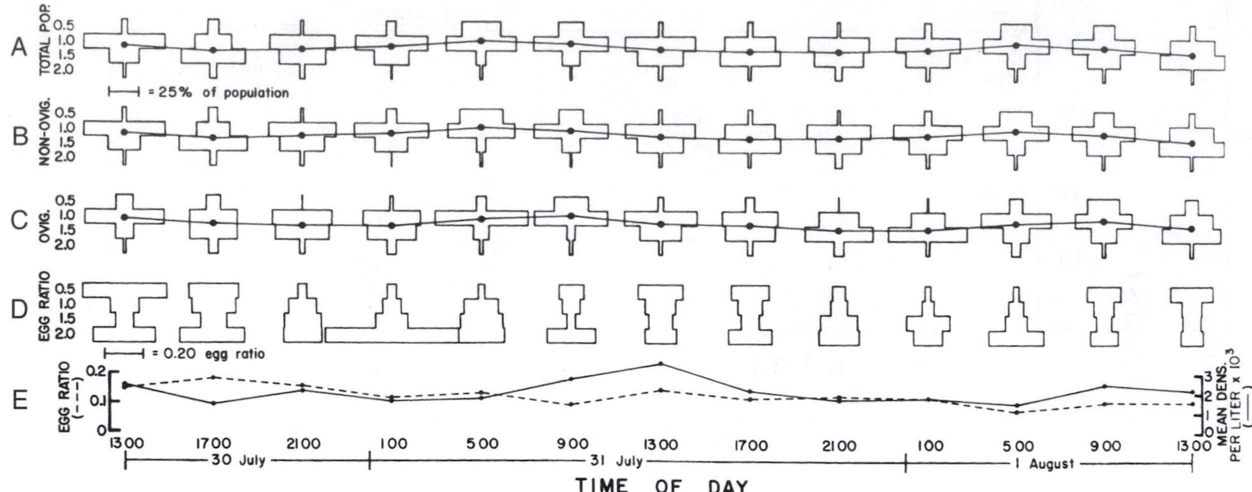

FIGURE 16 Depth distribution of *Keratella crassa* in a small, shallow lake during a 48 h period in the summer of 1980. (A) Total population; (B) nonovigerous population; (C) ovigerous population; (D) egg ratios calculated for each sampling point; (E) egg ratio and mean density of the entire population throughout the water column. Kite diagrams express the population as relative percentages for each sampling point, while the line indicates the mean depth of the entire population. (From Magnien and Gilbert, 1983, with permission.)

3. Biogeography

Early studies of rotifer distributions were dominated by the belief that all rotifers are cosmopolitan, an idea supported by the fact that the resting eggs of monogononts and the anhydrobiotic stages of bdelloids are transported by birds and other mechanisms. Such passive dispersal, it was argued, is very effective at ensuring that most species become globally distributed. However, data has accumulated suggesting that this conclusion was premature. Rotifers may not be as easily dispersed as previously thought (Jenkins and Underwood, 1998). Green (1972) showed a latitudinal zonation in certain planktonic rotifers, and Pejler (1977b) provided data demonstrating that some species have restricted distributions. In general, endemism seems to be an important theme in several genera, including *Keratella*, *Notholca*, and *Synchaeta* (Dumont, 1983). Exhaustive inventories of rotifer fauna in a few lakes has suggested that 150–160 rotifer species reside in a typical temperate lake as compared to more than 210 species for a tropical lake (Dumont and Segers, 1996). These numbers are likely to go higher as sessile and contractile species become more well known. These authors argue that the theory of cosmopolitanism must be abandoned for rotifers and cladocerans.

In comparison to the substantial research effort developed elsewhere, there has been relatively little work on rotifer biogeography in the United States and most of that work was done prior to the 1950s or is limited to specific geographic areas (e.g., the Great Lakes). Nevertheless, in North America there are several examples of species with restricted distributions in the genera *Anuraeopsis*, *Brachionus*, *Lecane*, and *Lepadella*. Central America and the southern United States have close affinities to tropical South American fauna, while further north, affinities with Europe are apparent and endemism is strong in *Keratella*, *Notholca*, and *Synchaeta* (Chengalath and Koste, 1987). Dumont (1983) suggested that continental drift and Pleistocene glaciations best explain the current biogeography of rotifers.

Most biogeographical analyses of rotifers have been limited to collating information on the location of populations from descriptions found in the literature. This problem has additional difficulty because many species undergo periodic polymorphism (cyclomorphosis) (see Section III.A.1), with each population having a slightly different morphology. In the past, this phenomenon confused some workers who considered each morphotype to be a new species. Furthermore, the question of what constitutes a rotifer species becomes problematic as males have never been described for many taxa. Future efforts to clarify these matters probably will be greatly aided by modern techniques in gel electrophoresis (e.g., King, 1977; King and Zhao, 1987; Snell and Winkler, 1984) and by using mating behavior to define species boundaries (Snell and Hawkinson, 1983; Snell *et al.*, 1988; Snell, 1989).

In the future, biogeographic research may play an important role in monitoring environmental conditions by examining the structure of rotifer communities in a variety of lakes subject to environmental pertubations:

e.g., heavy metal input (MacIsaac *et al.*, 1987) and acid precipitation (Frost *et al.*, 1998).

4. Colonial Rotifers

Most rotifers are solitary and interact only as potential prey or mates, but about 25 species (in eight genera) of the class Monogononta form permanent colonies (Wallace, 1987). All colonial forms are members of two families (Flosculariidae and Conochilidae) of the order Flosculariacea (Fig. 17). None of these taxa are predators and all reproduce in a fashion typical of monogononts. There appears to be a relationship between coloniality and sessile existence. About 70% of colonial forms (18 species) are sessile, but even more striking is the fact that all seven genera of the family Flosculariidae have colonial species. This has led some workers to suggest that sessile species are somehow preadapted for the evolution of coloniality (see Wallace, 1987 for a review). The form of colony production has significant implications. Because colonial rotifers do not reproduce by budding or the formation of specialized zooids, colony members are not intimately connected, as are the colonial bryozoans (Chapter 14); therefore, energy resources cannot be shared among colony members. Some colonial rotifers produce tubes from either hardened secretions (e.g., *Limnias*), pellets (e.g., *Floscularia conifera*), or gelatinous secretions (e.g., *Lacinularia* and *Conochilus*). These tubes are

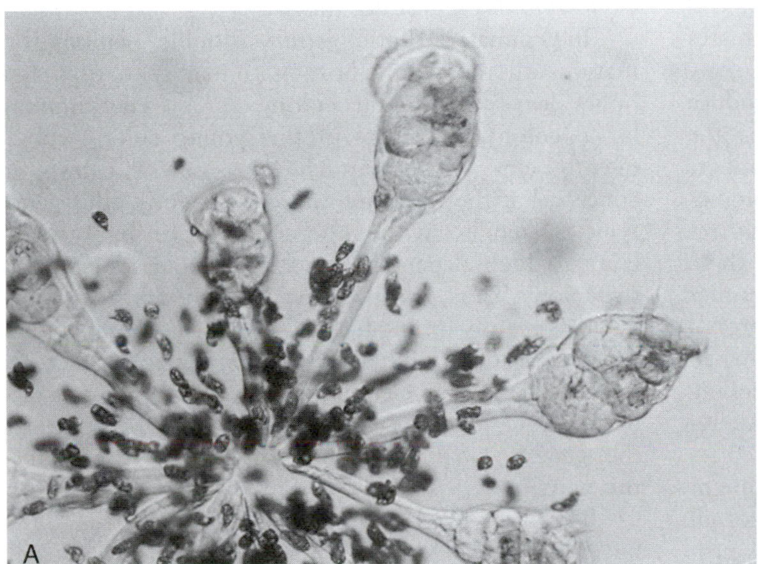

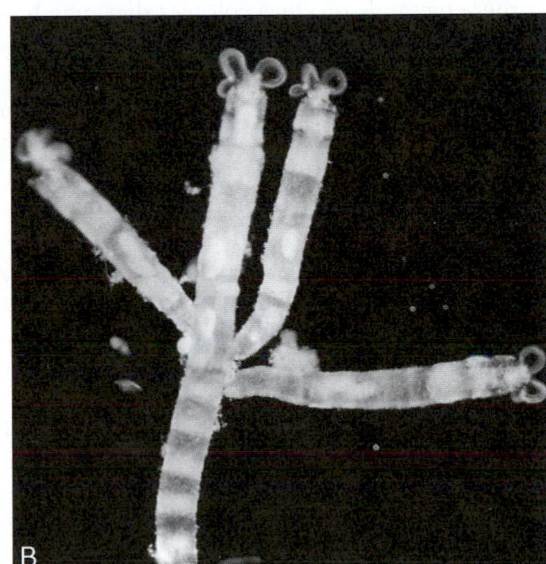

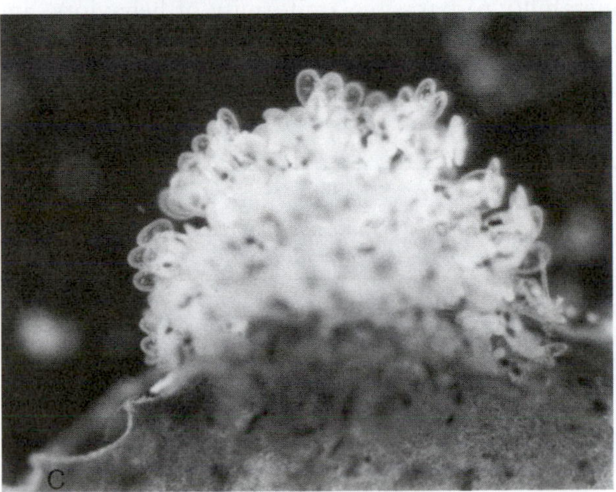

FIGURE 17 Colonial rotifers. (A) A portion of a small, planktonic colony of *Conochilus unicornis*: animals live within a gelatinous mass which they secrete, also present are many small euglenoids; (B) a small, sessile colony of *Floscularia*: adults reside within tubes they construct of tiny pellets of compacted detritus; and (C) a large sessile colony of *Sinantherina*.

important in the overall structure of the colony as they provide a substrate or matrix for the addition of new members to the colony (see Section III.A.5).

The number of individuals in a colony varies greatly among genera. *Floscularia ringens* usually builds colonies with fewer than five individuals, as do some members of the family Conochilidae. In this family, small colonies, composed of only an adult and one-to-several young, are produced by some species, while others may have more than 250 individuals within a colony. Although most colonial species construct small colonies of fewer than 5–24 individuals, some routinely produce large colonies of 50–200 individuals (e.g., *Floscularia conifera*, *Sinantherina socialis*, *Conochilus hippocrepis*), and a few produce colonies of truly gargantuan size (>1000; *Lacinularia*).

Colonies form by one of three different methods, but these methods produce colonies that may have a different degree of genetic relatedness among colony mates (Table I). In allorecruitive (Greek, *allo*, "other") colony formation, free-swimming young (larvae) produce colonies by settling on tubes of sessile adults. Thus, the colonies of allorecruitive species grow in size (numbers of individuals per colony) and in the number of colonies within the habitat (colony density) by intercolonial recruitment of young. Because the larvae joining these colonies may come from females belonging to another colony, genotypic relatedness within the colony is probably low. These colonies are transitory, beginning when larvae attach to a previously settled adult and ending when larval recruitment to an old colony ceases. Late recruits to an allorecruitive colony may suffer reduced fecundity, because death of the founding individual might lead to dislodgment from the substrate and death of all members of the colony. A few allorecruitive species produce interspecific colonies of two or more species. Some species of the genus *Floscularia* reproduce this way.

In autorecruitive (Greek, *auto*, "self") colony formation, the young remain with their mother in the colony. Thus, the colonies of autorecruitive species grow in size (numbers of individuals per colony) by intracolonial recruitment of young back into the parental colony and in the number of colonies within the habitat (colony density) by colony fission. Because the young joining these colonies come from the parent colony, genotypic relatedness within the colony is probably high. Here autorecruitive colonies are long-lived and develop continuously throughout the season, increasing in size as new individuals are added and decreasing only if the colony divides. Late recruits to these colonies probably do not suffer diminished fecundity because the colony continues though individuals die. Autorecruitive species rarely produce interspecific colonies. Species of the genera *Conochilus* and *Conochiloides* reproduce this way.

In geminative (Latin, *gemin*, "double") colony formation, all the young born within a span of a few hours leave the parent colony as a free-swimming larval colony. Members of this young colony subsequently explore and attach to a new substrate in concert, but they may join an older colony, thus genotypic relatedness within the colony is probably moderate to high depending on the particular history of the colony. Thus, because the size of the larval colonies depends mainly on the fecundity of the parental colony, the number of individuals in a colony does not increase over its life time. Geminative species rarely produce interspecific colonies, but it is possible. Species of the genera *Lacinularia* and *Sinantherina* reproduce this way.

The consequences of high genetic relatedness have not been explored, but may include increased vulnerability to parasites, uniformity of behaviors, and decreased genetic diversity of resting eggs.

TABLE I Colony Formation in Rotifers

Colony formation category	*Increase in colony size (individuals/colony)*	*Mode of increase of colony density in the habitat*	*Predicted genotypic relatedness within colonies*	*Frequency of interspecific colony formation*	*Examples*
Allorecruitive (intercolonial)	Young are recruited to established colonies or remain solitary	Young leave the parental colony and may establish a new colony	Low	Common	*Floscularia* (Fig. 17B & 43A)
Autorecruitive (intracolonial)	Young stay within the parental colony	Fission: adult colonies fragment producing smaller colonies	High	Absent	*Conochilus* (Fig. 17A)
Geminative	Rare-to-absent	Cohorts of larvae collectively establish new colonies	Moderate-to-high	Rare-to-absent	*Lacinularia* (Fig. 5C) *Sinantherina* (Figs. 17C)

Two hypotheses have been offered to explain the adaptive significance of coloniality in rotifers. One hypothesis suggests that colonial animals possess an energetic advantage over solitary individuals of the same species; for example, colonial *F. conifera* apparently live longer and mature faster than the solitary individuals (Edmondson, 1945). Juxtaposition of filtering currents produced by two individuals may permit increased filtering rate and/or enhanced filtering efficiency. Although experiments have not supported the idea that coloniality affects filtration rates, Wallace (1987) has provided some information that supports the view that coloniality increases filtering efficiency. A second hypothesis argues that colonial existence can protect individuals from certain predators. Gilbert (1980b) and Diéguez and Balseiro (1998) showed that large *Conochilus* colonies are less vulnerable to attack by certain invertebrate predators, because they are too large to be engulfed whole and because individual rotifers can retract into the refuge of the gelatinous matrix of the colony.

5. Sessile Rotifers

Sessile species are found in two families of rotifers: Flosculariidae (7 genera) and Collothecidae (5 genera). Although they are often overlooked because their habitats generally are not examined thoroughly, these forms are actually quite common in lakes and rivers in North America, occasionally reaching very high densities on plant surfaces (>6 individuals/mm^2) especially in bogs and small eutrophic ponds (Wallace, 1980; Wallace and Edmondson, 1986).

The juvenile motile stages of sessile rotifers (Fig. 5) are not considered to be larvae by some researchers, but others have used this term and there is a conceptual parallel to other sessile invertebrates that undergo an extensive metamorphosis after settlement (Wallace, 1980). Not surprisingly, the behavior of larvae changes dramatically once they come into contact with a potential attachment site. These new behaviors have been described using terms such as selection, choice, and preference. However, it should be recognized that the use of such words does not imply cognition by the rotifer; they are merely convenient terms to describe this phenomenon.

Several workers have demonstrated that larvae can select a particular substrate from among those available for settlement (Wallace, 1980). For example, larvae of *F. conifera* settle with a greater frequency on the tubes of conspecifics than on aquatic plants, although there is substantially more plant surface available. This propensity leads to the formation of intraspecific colonies, each with 50 or more individuals. During the growing season, about 75% of the entire population may be colonial (Wallace, 1977).

Some species apparently attach to a surface based on the chemistry of the water in the vicinity of the substrate. Wallace and Edmondson (1986) have shown that greater than 90% of *Collotheca gracilipes* larvae prefer the undersurface of *Elodea canadensis* leaves to the upper surface (Fig. 18). Choice of the undersurface is apparently in response to the way *Elodea* can alter the concentration of calcium ions in the water immediately around the leaf. In water having a pH in the

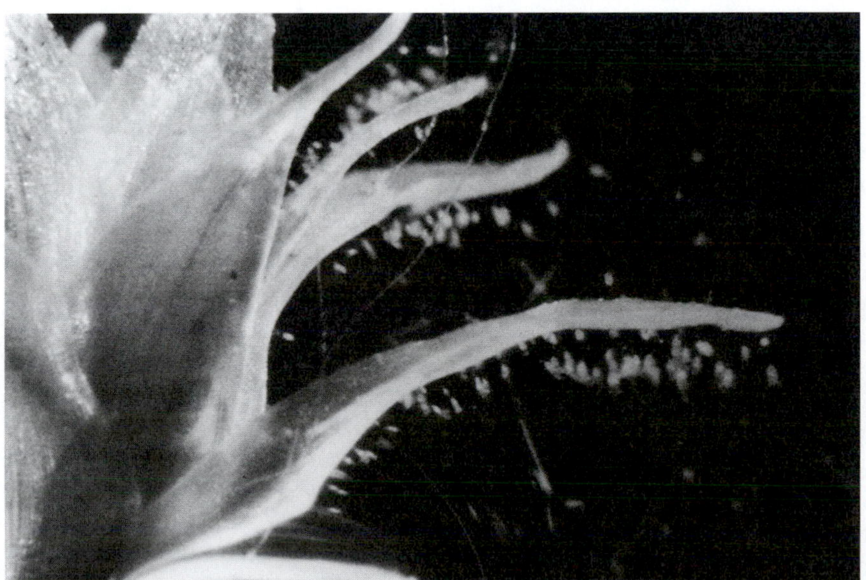

FIGURE 18 *Collotheca gracilipes* colonizing the apical meristern of the aquatic macrophyte *Elodea canadensis*. Many (>25) individuals may be seen attached to the under surfaces of the leaves. (Length of the lowest leaf is ≈ 1 cm.) (From Wallace and Edmondson, 1986, with permission.)

neutral-to-alkaline range, *Elodea* acts as a polar plant, removing Ca^{2+} from beneath the leaf and releasing it above. This choice provides a superior habitat in comparison to the upper surface. In short-term, laboratory experiments, young *C. gracilipes* that attached to undersurfaces of *Elodea* leaves grew significantly taller and produced more eggs per female than those attached to the upper surfaces of the same leaves. Several other species of sessile rotifers exhibit strong preferences for particular substrates, but the significance of these associations has not been fully elucidated (Wallace, 1980).

Larval substrate selection behaviors have been described for three species: *Ptygura beauchampi*, *Sinantherina socialis*, and *C. gracilipes*. In general, larvae appear to react to potential surfaces in similar ways. Very young larvae avoid settling for periods up to several hours after hatching. Once this refractory period is past, all surfaces are explored, but some (i.e., the preferred substrates) receive much more attention and elicit different behaviors, including some reminiscent of male mating behavior (see Section III.B.2). A larva will traverse these surfaces with both its corona and foot in contact with the surface, occasionally stopping in this slightly bent position. The larva may continue exploration of the surface for several minutes, but eventually it attaches to the surface using cement from glands in the foot and then undergoes metamorphosis.

Immediately after attachment and metamorphosis have occurred, the young of most sessile rotifers begin to secrete a protective tube. Often this secretion is in the form of a clear, gelatinous material (e.g., *Collotheca* and *Stephanoceros*), but in others the tube becomes somewhat opaque by the adherence of debris and colonizing microorganisms (e.g., several *Ptygura* species and *Lacinularia*). One species of *Limnias* forms a cement tube that looks like a series of rings, placed one on top of another. Perhaps the most fascinating example of tube construction is found in the genus *Floscularia*. Some species in this genus possess a small ciliated cup on the ventral side of the head to which some tiny mineral particles and other small debris collected by the corona are shunted. The ciliated cup appears to be in constant motion, mixing gelatinous secretions with the particles to form small pellets, either in the shape of bullets or balls. Once a pellet is fully formed, the rotifer places it on top of the tube in an action resembling the movements of a bricklayer. In this way the tube is constantly elongated as the animal grows. Because pellets are manufactured from particles collected from the water, they may have a greenish tint, but they are usually brown. When a heavy rainstorm temporarily suspends soil in the water, the pellets that are produced by *Floscularia* usually turn out to be very dark. Thus, the event of the storm is marked as a dark ring in the tubes of all the animals alive at that time. Edmondson (1945) noted this and used suspensions of powdered carmine and carbon black to mark the tubes of *F. conifera*, so that he could study the dynamics of population growth and various aspects of rotifer life history in a field population (See Fig. 17B).

B. Reproduction and Life History

1. Reproduction

The type of reproduction varies among the three classes of rotifers (Wallace, 1998). Species in the class Seisonidea reproduce exclusively through bisexual means, with gametogenesis occurring via classical meiosis and the production of two polar bodies. At the other extreme, members of the class Bdelloidea reproduce entirely by asexual parthenogenesis, a process which includes two equational divisions producing two polar bodies. No males have been observed in bdelloids. Species in the class Monogononta exhibit cyclical parthenogenesis where asexual reproduction predominates, but sexual reproduction also occurs occasionally.

a. Cyclical Parthenogenesis Cyclical parthenogenesis in monogononts (Fig. 19) takes place in the absence of males (amictic phase), but periodically males are

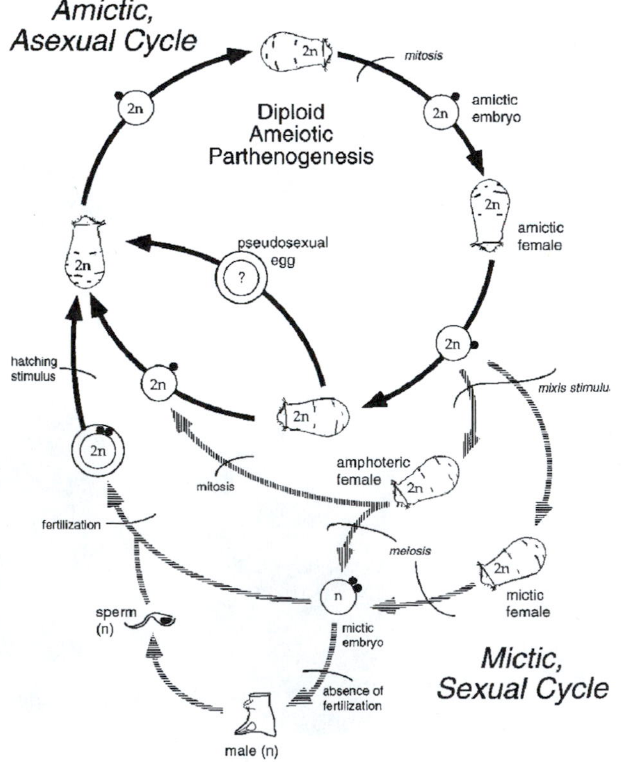

FIGURE 19 Generalized life cycle for monogonont rotifers. The life cycle of bdelloids consists of only the parthenogenetic portion.

produced and sexual reproduction takes place (mictic phase). Amictic females are diploid and produce diploid eggs (called amictic eggs) that develop mitotically via a single equational division into females (Wallace, 1998). The term "summer egg" has been applied to the eggs produced by amictic females, but this term is misleading because amictic eggs can be produced at any time of the year, depending on the species in question.

Most of the monogonont rotifer life cycle is spent in the amictic phase, but in certain conditions, sexual reproduction occurs concurrently within the population, presumably initiated by specific environmental stimuli. Upon receiving the mictic stimulus, amictic females begin producing both mictic and amictic daughters. The proportion of mictic daughters and the duration of their production, depends on the strength of the mictic stimulus. Mictic female production is only the initial step in mictic reproduction and is followed by male production, fertilization, and finally formation of a diapausing embryo called a resting egg or cyst (also winter or fertilized egg) (Fig. 19). Mictic females produce haploid eggs via meiosis that, if unfertilized, develop into haploid males. Males are typically smaller, faster swimming, and shorter lived than females. The males of some species are structually reduced, but extremes exist from fully developed males (*Rhinoglena*) to a condition in which certain organs are degenerate (*Conochilus*). Among sessile species the male is free-swimming while the female is permanently attached to a substrate.

Fertilized mictic eggs are diploid and develop into resting eggs possessing thick, often sculptured, walls (Fig. 20). These dormant stages are very resistant to harsh environmental conditions and may be dispersed over wide areas by the wind, water, or migrating animals. After a period of dormancy, which varies among species, resting eggs respond to species-specific environmental cues and hatch, releasing diploid, amictic females that enter into the asexual phase of the life cycle (Fig. 19). The stimuli that induce hatching may include changes in light, temperature, salinity, and oxygen concentration (Wallace, 1998).

With a few notable exceptions, the stimulus for initiating sexual reproduction is still poorly understood. Dietary α-tocopherol (vitamin E) controls the shift from amictic to mictic reproduction in most *Asplanchna* species (Gilbert, 1980a), and photoperiod plays a similar regulatory role in *Notommata* (Pourriot and Clément, 1981). In the genus *Brachionus*, population density has been most frequently cited as a stimulus for mictic female production (Wallace, 1998; cf, Pourriot *et al.*, 1986). In addition to environmental factors, genetic factors also play a major role in determining the sensitivity of particular strains to mictic stimuli (e.g., Snell and Hoff, 1985; Lubzens *et al.*, 1985).

An unusual variation of the standard monogonont life cycle has been observed in three genera: *Asplanchna*, *Conochilus*, and *Sinantherina*. In some populations amphoteric females (individuals that produce both female and male offspring) have been recorded (e.g., Ruttner-Kolisko, 1977b; King and Snell, 1977). Amphoteric rotifers have the ability to produce both diploid (female) and haploid (male) eggs, with some producing both females and resting eggs, or males and resting eggs (Fig. 17). The presence of amphoteric reproduction in other rotifers has not been fully investigated, and its significance in the life history of those genera that possess it remains to be determined. There have been reports of the production of pseudosexual eggs; that is, the production of resting eggs via parthenogenesis in the absence of males. However, little in known of this phenomenon.

Two different types of parthenogenetic amictic eggs have been described in *Synchaeta pectinata* (Gilbert, 1995). The traditional type is thin shelled (1.4 μm) that develops without diapause. The second type has a thick-shelled (9 μm) amictic egg that enters an obligatory diapause after 1–3 cleavage divisions. The mean duration of diapause is about 14 days, and it is not greatly extended by low temperature. These eggs are induced immediately after starvation and seem to be produced at no additional energetic cost compared to traditional amictic eggs. The adaptive significance of diapausing amicitic eggs seems to be to increase the ability of populations to survive short-term food limitation. *Synchaeta pectinata* also produce resting eggs following sexual reproduction and these have the capacity for dormancy extending up to several years.

b. Resting Eggs The density of resting eggs in sediments has not been routinely examined and the few published studies have used different methods for estimating density. Because of their capacity for extended dormancy, resting eggs theoretically could accumulate to high densities in sediments. Nipkow (1961) found resting egg densities for seven species to range from 100 to 600 eggs/cm^2 in lake sediments, but *Synchaeta oblonga* densities ranged from 3,000 to 4,000 eggs/cm^2. Snell *et al.*, (1983) found an average of 200 resting eggs/cm^2 for *Brachionus plicatilis* in the surface sediments of a subtropical brackish pond. These workers showed that the highest densities were on the sediment surface, and numbers declined exponentially to a depth of 7 cm. May (1987) examined resting egg densities in sediments from Loch Leven, Scotland, by recording the number of animals hatching from sediments incubated at temperatures of 5, 10, and 15°C. This technique is

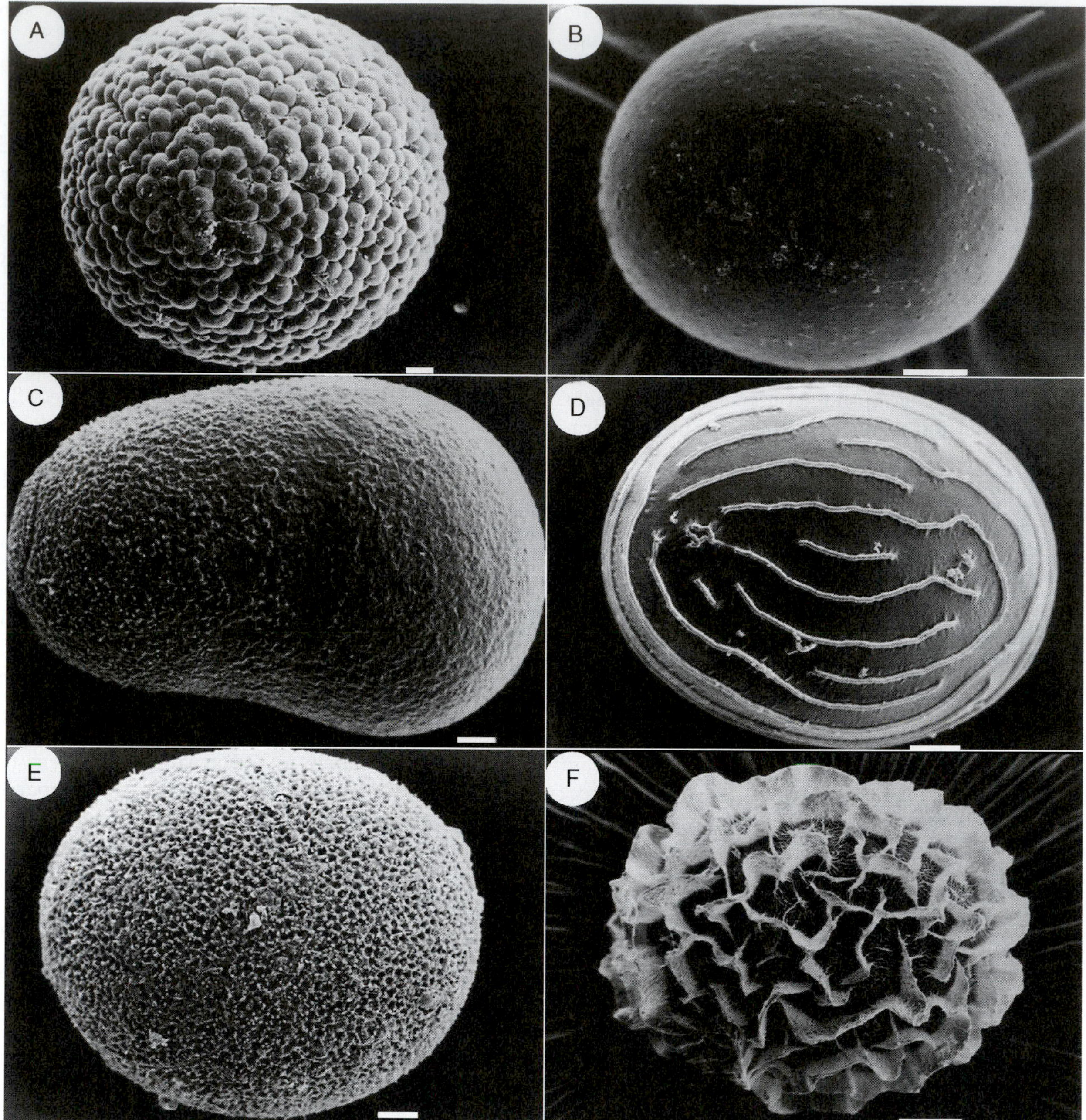

FIGURE 20 Resting eggs of several monogonont rotifers. (A) *Asplanchna girodi*; (B) *Hexarthra fennica*; (C) *Brachionus calyciflorus*; (D) *Conochiloides natans*; (E) *Sinantherina socialis*; and (F) *Kellicottia bostoniensis*. (Bars = 10 μm; the wrinkled backgrounds in panels B and F are the supporting membranes.) (Original scanning electron photomicrographs by Hendrik Segers.)

not directly comparable to the others just noted, and probably yields lower estimates of resting egg density (see also May, 1986). Marcus *et al.* (1994) observed *B. plicatilis* resting eggs at densities up to 1532 individuals/cm^2 in anoxic and sulfidic sediments of Pettaquamscutt estuary in Rhode Island. Rotifer resting eggs were abundant as deep as 12 cm into these sediments and were the most abundant among any of the zooplankton taxa present. *B. plicatilis* resting eggs older than 40 years could be hatched and used to initiate viable populations.

The number of resting eggs in sediments is a function of previous resting egg production, mortality in the sediments, and hatching rates. Certain abiotic features such as sedimentation rates and the uneven deposition of sediments in the benthos (sediment focusing) may affect the final density at particular sites in a lake or estuary.

2. Reproductive Behavior

Detailed descriptions of mating behaviors are available for *Asplanchna brightwelli* (Aloia and Moretti, 1973), three species of *Brachionus* (e.g., Gilbert, 1963b; Snell and Hawkinson, 1983), and *Anuraeopsis fissa, Asplanchna girodi, Euchlanis dilatata, Keratella americana, Lecane quadridentata, Brachionus patulus*, and *Trichocerca pusilla* (Rico-Martinez and Snell, 1998). Males and females show pronounced sexual dimorphism, males being smaller and faster swimmers (Fig. 2). Lacking a functional foot, males swim constantly without attaching. Because males and females swim randomly, the probability of male–female encounters in planktonic species can be modeled mathematically (Snell and Garman, 1986). In the colonial, sessile rotifer *Sinantherina socialis*, males may copulate with several females of one colony in succession. Females take no active role in locating a mate or reacting to the male once an encounter occurs. Males, in contrast, display a distinct mating behavior upon encountering conspecific females.

Mating behavior begins only when the corona of the male squarely contacts the female. However, not all head-on encounters result in mating; indeed, the probability of copulation in laboratory cultures generally varies from 10 to 75% in *Brachionus plicatilis*, depending on the strain (Snell and Hawkinson, 1983). This requirement for head-on contact by the male is due to the presence of chemoreceptors in his coronal region, which apparently respond only to a species-specific glycoprotein on the surface of the female (Snell *et al.*, 1995). Mating begins with the male swimming circles around the female, skimming over the surface of her lorica (Fig. 21). During this phase, the male maintains contact with the female with both his corona and penis (this requires the male to remain in a slightly bent position). After several seconds of circling, the male attaches his penis to the female, usually in the region of her corona, and loses coronal contact. After about 1.2 min of copulation in *B. plicatilis* (Snell and Hoff, 1987), sperm transfer is completed and copulation is terminated when the male and female break apart and swim away. Newborn *B. plicatilis* males only have about 30 sperm and transfer two or three at each insemination (Snell and Childress, 1987).

3. Aging and Senescence

Rotifers have long been used as models of aging and senescence. Jennings and Lynch (1928) conducted

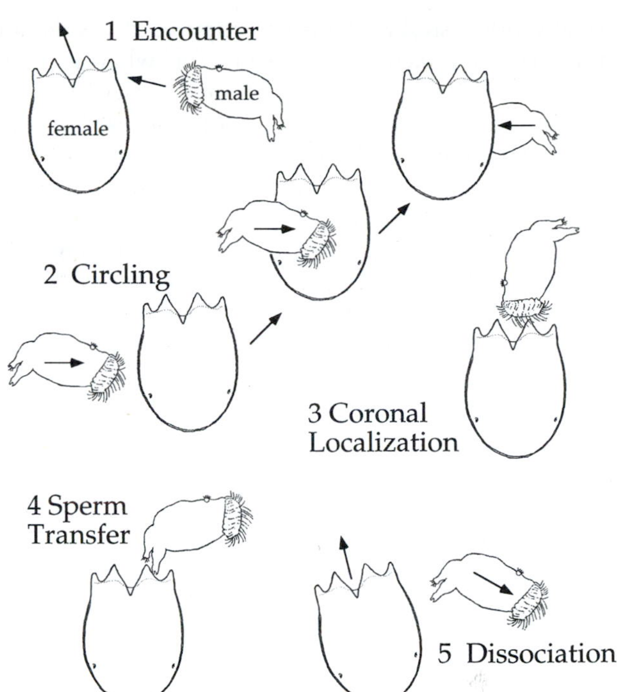

FIGURE 21 Male mating behavior in brachionid rotifers. Arrows indicate general swimming movements by male and female rotifers. (Male not drawn to scale.)

the first major study of aging in rotifers using the monogonont *Proales sordida*. They found the mean lifespan to be 8 days at 20°C, but a maternal effect was noted. Offspring from older mothers had shorter lifespans and more variability in their developmental rates and fecundities. Maternal effects were a theme expanded upon by Lansing who, in a series of papers during the 1940s and 1950s, showed that parental age also influenced longevity in the bdelloid *Philodina citrina* (see King, 1969 for a review). Lansing established "orthoclones" by isolating the first offspring of parental females, then the first offspring of F1 females, and so forth, yielding a clone derived exclusively of the first-born females. Orthoclones derived from the offspring of older females also were developed. This protocol is a powerful tool for investigating effects on aging because the only difference among orthoclones is the age of their mothers. In *P. citrina*, mean lifespans of 5-, 11-, and 16-day orthoclones were 23.8, 18.1, and 16.6 days, respectively. Lansing's general conclusion was that young orthoclones have increasingly longer lifespans, while older orthoclones have progressively shorter lifespans, resulting eventually in clonal extinction. Rate of senescence, Lansing hypothesized, was controlled by a maternal effect that was transmissible and cumulative. However, the effect was reversible since the first-born females from an old orthoclone outlived their parents. Based on these and other studies, Lansing proposed that calcium is the maternal factor responsible for what is now known as the "Lansing Effect." Rate of calcium accumulation and its importance in rotifer senescence became the reigning hypothesis of the day. Since Lansing's work, researchers have reported similar results in some species, but not in others.

King (1983) re-examined Lansing's data using life table analysis and concluded that Lansing probably did not observe an aging effect. King showed that short-lived lines derived from old parents reproduced earlier and at higher rates in succeeding generations. Long-lived lines derived from young parents delayed initial reproduction to later age classes. King argued that the Lansing effect actually resulted from shifts in fecundity patterns that concentrated reproduction in a few age classes in short-lived lines. Snell and King (1977) demonstrated that concentrated reproduction in rotifers could shorten lifespan. Until more is known about the physiological basis for the "Lansing Effect", King suggested that it be viewed with skepticism.

The pattern of reproduction also modifies longevity (Snell and King, 1977). Female *Asplanchna brightwelli* with high rates of reproduction concentrated early in life have shorter mean lifespans (2.5 days at 25°C) than females with slower reproduction distributed over a greater portion of the life- span (mean lifespan = 4.5 days). The length of prereproductive and postreproductive periods were similar for short- and long-lived individuals. However, length of the reproductive period was about twice as long in long-lived individuals. These data suggest that reproduction influences individual survival negatively, but deleterious effects are minimized when the reproductive period is spread over a greater portion of the lifespan. Further, Rougier and Pourriot (1977) found that maternal age is negatively correlated with mictic female production. When *Brachionus calyciflorus* females began reproducing at age 2 days at 15°C, 80–90% of their daughters were mictic. By age 8 days, only 25% of daughters were mictic, and reproduction by 12 day old females yielded no mictic daughters. This decline in mictic female production with age was linear in 2-, 6-, and 10-day orthoclones.

Male lifespan has been examined far less thoroughly than that of females, but available data indicates that males live only about half as long as females (Fig. 22) (King and Miracle, 1980; Snell, 1977).

Other experiments on aging in rotifers have indicated that vitamin E (a-tocopherol) can extend mean lifespan about 10% at 23°C in the bdelloid *Philodina* sp., but maximum lifespan is unaffected (Enesco and Verdone-Smith, 1980). Vitamin E, at a concentration of 20 μg/mL, also was found to extend the lifespan

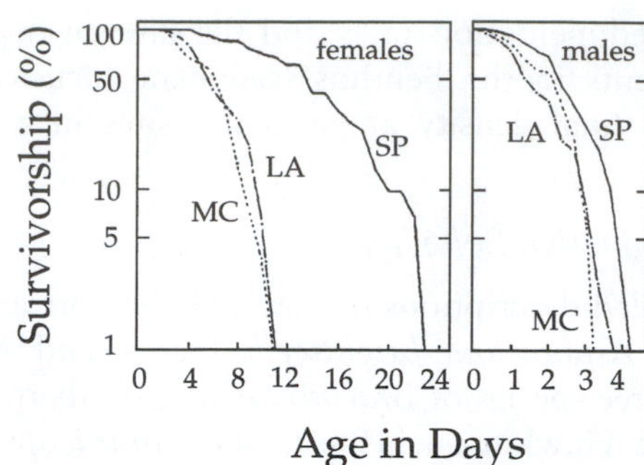

FIGURE 22 Female and male survival of three clones of *Brachionus plicatilis* at 25°C. Clone LA is from LaJolla, CA; clone MC is from McKay Bay near Tampa, FL; clone SP is from Castellon, Spain. Note that the Y axis (survivorship %) is a log scale. (From King and Miracle, 1980, with permission.)

in *A. brightwelli* (Sawada and Enesco, 1984). Of several other antioxidants tested, however, only thiazolidine-4-carboxylic acid (TCA) extended the mean life span significantly, presumably by quenching free radical reactions.

Other factors that affect mean life span in *A. brightwelli* are photoperiod (light:dark cycle, L:D), dietary restriction (Sawada and Enesco, 1984), and ultraviolet (UV) radiation (Enesco, 1993). Rotifers exposed to a 6:18 or 0:24 L:D photoperiod showed an 18% and a 22% increased life span, respectively, over a control photoperiod of 12:12 Dietary restriction has life-extending effects in a wide variety of animals. In *A. brightwelli*, two different modes of dietary restriction yielded longer life spans. Reduction of food intake increased life span by 14.2%, whereas intermittent feeding produced an 11.8% longer life span. Ultraviolet irradiation in the 200–4800-J/m^2 range shortened life span significantly, and life span declined logarithmically as UV dose increased.

4. Population Dynamics

Studies of rotifer population dynamics attempt to explain the causes of changes in population size. By separating and quantifing the relative contributions of reproduction, mortality, and dispersal, ecologists are able to identify the factors regulating population size and the determinants of average abundance. Because some species are easily cultured and experimentally manipulated, rotifers have proven to be useful models for investigating the dynamics of animal populations. Techniques also exist for investigating the dynamics of natural rotifer populations, further broadening their usefulness in ecological studies.

a. Life Tables The most rigorous approach to studying population dynamics is lifetable analysis. Life tables are age-specific records of all reproduction and mortality occurring in a population. From these observations, vital population statistics can be calculated such as intrinsic rate of increase (r_m), net reproductive rate (R_o), life span, generation time, reproductive value, and stable age distribution. The general protocol for gathering lifetable data on rotifers is to collect a cohort of newborn females, isolate them in small volumes of culture medium (ca. 1 mL), and make daily observations of their survival and reproduction. Maternal females usually are transferred daily to fresh medium with a specific food level, and offspring are counted and removed. A table is then constructed of age, number surviving in each age class, and number of offspring produced in each age class. At least 10 replicate females are grouped into each population and probability of surviving to age x (l_x) and age-specific fecundity (m_x) are calculated.

The detailed observations required for life-table analysis are usually only possible in laboratory populations and a number of genera have been examined in that way, including several species of *Brachionus* (Halbach, 1970; Pourriot and Rougier, 1975; King and Miracle, 1980; Walz, 1987), *Asplanchna* (Gilbert, 1977; Snell and King, 1977), and bdelloids (Ricci, 1983). In certain cases, it is possible to perform life table analyses on field populations (i.e., *Floscularia conifera*, Edmondson, 1945). Although these and other studies have examined survival and reproduction of parthenogenetic females, survivorship curves for male rotifers also are available for *A. brightwelli* (Snell, 1977) and *B. plicatilis* (King and Miracle, 1980).

The major environmental factors affecting age-specific survival and reproduction are temperature, food quantity and quality, genetic strain, and reproductive type (either amictic, or fertilized or unfertilized mictic females). Increased temperature shortens survivorship. In *B. calyciflorus*, for example, 50% of a cohort survive to age 16 days at 15°C, 11 days at 20°C, and 5 days at 25°C (Fig. 23). Age-specific fecundity also is compressed at higher temperatures. At 15°C, fecundity occurs at a low rate over 20 days. In contrast, at 25°C reproduction occurs at a high rate and is compressed into 10 days. Females produce a total of 13 offspring at 15°C, 16.6 at 20°C, and 12.9 at 25°C. The intrinsic rates of increase (r_m) calculated from these l_x and m_x values are 0.34, 0.48, and 0.82 offspring per female per day at the respective temperatures (Halbach, 1970).

The effect of food quantity on l_x and m_x schedules is illustrated in Figure 24. *Brachionus calyciflorus* was fed the green alga *Chlorella* at densities of $0.05-5 \times 10^6$ cells/mL at 20°C. The best survival occurred at algal concentrations 0.5×10^6 and 1.0×10^6 cells/mL, where mean life span was 9 days (Halbach and Halbach-Keup, 1974). Life span decreased at lower food levels, reaching 2.5 days at 0.05×10^6 cells/mL. Lower life spans also were recorded at food concentrations above 1.0×10^6 cells/mL, probably the result of accumulation of algal metabolic products that are toxic to rotifers. Fecundity also peaked at 1.0×10^6 cells/mL, with a mean of 17 offspring per female. In contrast, at 0.05×10^6 cells/mL, lifetime fecundity was only 0.5 offspring per female whereas 3 offspring per female were produced at 5.0×10^6 cells/mL.

Intraspecific differences in survival among strains is illustrated in Figure 22. King and Miracle (1980) examined three strains of *B. plicatilis* and found that their mean life spans at 25°C ranged from 6 to 13.5 days. Because these strains were acclimated to, and tested in, a common environment, the observed differences must be genetic. Also shown in the same figure are male survivorship curves; interstrain differences are again apparent with male life spans.

The three different female types (amictic, and unfertilized and fertilized mictic) differ markedly in their age-specific survival and fecundity. In *Brachionus urceolaris*, for example, amictic females live at 20°C an average of 9 days and produce 20 female offspring; unfertilized mictic females live 9.5 days and produce 25 male offspring; and fertilized mictic females live 10 days and

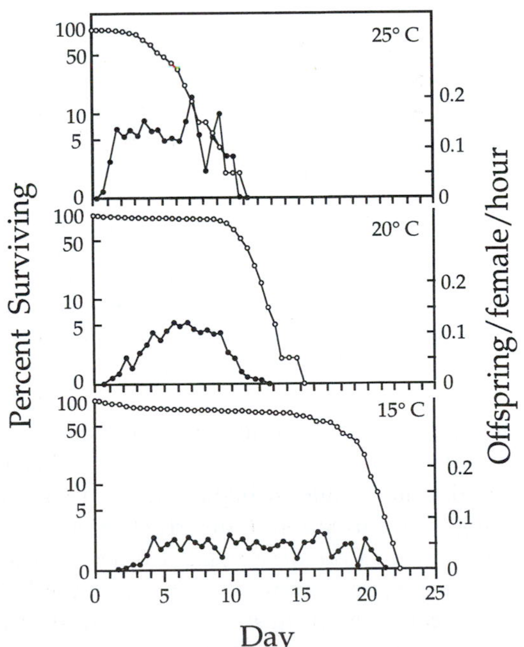

FIGURE 23 Effect of temperature on age-specific survival and fecundity of *Brachionus calyciflorus*. The left Y axis is the percent of the cohort surviving (open circles) and the right Y axis is the offspring per female per hour (closed circles). (From Halbach, 1970, with permission.)

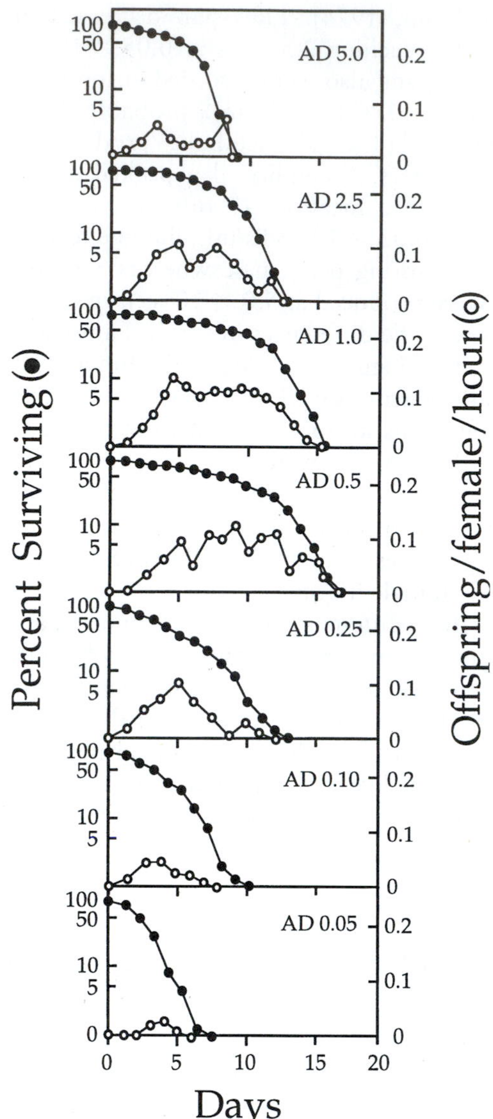

FIGURE 24 Effect of food level on age-specific survival and fecundity of *Brachionus calyciflorus*. The left Y axis is the percent of the cohort surviving (closed circles) and the right Y axis is the offspring per female per hour (open circles). The food is *Chlorella pyrenoidosa* provided at algal doses (AD) in the (0.05–5) × 10^6 per ml/range. (From Halbach and Halbach-Keup, 1974, with permission.)

produce four resting eggs (Fig. 25). The greatly reduced fecundity of fertilized females is typical of other rotifer species and is likely due to higher energy requirements for resting egg formation (Gilbert, 1980a). Also reported in Figure 25 are development times for each egg (time from egg extrusion to hatching) and the birth intervals between eggs. Birth intervals are shortest in the middle of the reproductive period.

Individuals within a single rotifer population differing in their growth forms (e.g., spined and unspined) can possess very different intrinsic rates of increase under identical culture conditions. Stemberger (1988) has shown that the r_m value for spined and unspined

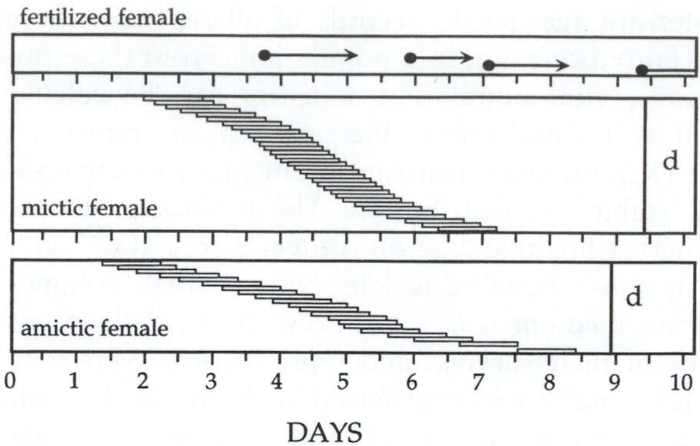

FIGURE 25 The fecundity schedule and egg developmental times of three types of *Brachionus urceolaris* females at 20°C with excess food. Bottom panel: amictic females producing diploid female eggs; middle panel: unfertilized mictic females producing haploid male eggs; top panel: fertilized females producing diploid resting eggs (solid circles). Arrows in the top panel represent the time of resting egg attachment to fertilized mictic females(hrs). Each bar in the middle and lower panels represents one egg and the length of the bar indicates the developmental time for that egg(days). (Modified from Ruttner-Kolisko, 1974, with permission.)

forms of *Keratella testudo* can differ significantly, but the difference depends on food concentration (Fig. 26). At low food levels, there is no significant difference in r_m for the two forms of this rotifer. At higher food concentrations, however, the unspined form had much greater values of r_m. Presumably, a similar phenomena will be found in other species in which there are spined and unspined forms (see Sections III.A.1, and III.C.4).

When life-history patterns of zooplankton are compared, rotifers have higher r_m values than either cladocerans or copepods (Allan, 1976). The r_m values of these groups range from 0.2 to 1.6, 0.2 to 0.6, and 0.1 to 0.4 offspring per female per day, respectively. Rotifers achieve their high population growth rates because of short development times, which more than compensate for their small clutch sizes. Rotifers also

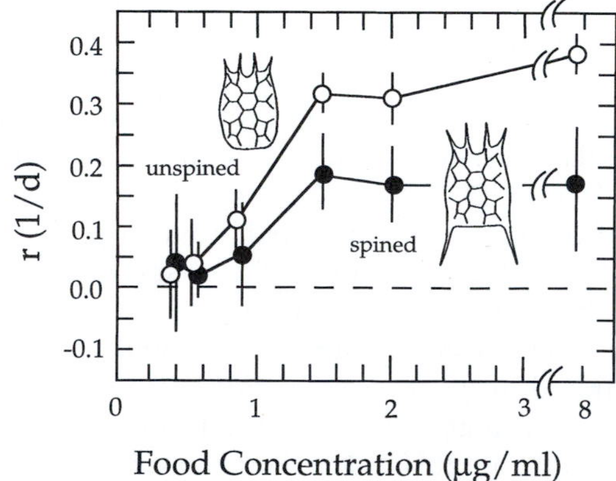

FIGURE 26 Intrinsic rates of population growth of spined (open Figures) and unspined (closed Figures) *Keratella testudo*. Dashed line, r = 0. (Modified from Stemberger, 1988, with permission.)

show the greatest response to increased temperatures. The high population-growth rates of rotifers may be an adaptive response to predation, against which most rotifer species have few defenses (Stemberger and Gilbert, 1987a). As a result of these characteristics, rotifers are regarded as being able to quickly exploit new conditions.

b. Dynamics of Field Populations Annual cycles of natural rotifer populations have been characterized for several species. One particularly thorough multiyear study was conducted by Herzig (1987) in Neusiedlersee, a shallow, well-mixed Austrian lake. Figure 27 shows 12 years of data for the abundance of 13 rotifer species. Some species, like *Keratella quadrata,* have relatively stable population sizes and occupy the lake nearly continuously. In contrast, abundances of other species, such as *Rhinoglena fertoensis*, *Filinia longiseta*, and *Hexarthra fennica*, fluctuate much more. Fluctuation of rotifer abundance over two or three orders of magnitude during a seasonal cycle is typical of many natural populations.

Techniques have been developed to characterize the dynamics of natural zooplankton populations (Edmondson, 1993). The instantaneous rate of increase of a population (r) is the difference between instantaneous birth (b) and death (d) rates

$$r = b - d \tag{1}$$

The value of r for a given population may be estimated from the equation

$$r = \frac{\ln N_t - \ln N_0}{T} \tag{2}$$

This procedure requires two successive estimates of population size (N_o and N_t) separated by time interval T. The estimate of r is based on several assumptions about the population, including that it is growing exponentially with a stable age distribution. One can also estimate the birth rate in this population by counting the number of eggs carried per female. The finite hatching rate is

$$B = \frac{E}{D} \tag{3}$$

where E is the number of eggs per female and D the developmental rate at a specific temperature. The value E is determined directly from samples of the population and values of D by observing hatching of eggs from a sample brought back to the laboratory. For example, Herzig (1983) reports values of D for several species. When E and D have been calculated, the instantaneous birth rate, b, can be estimated from the following equation by Palohiemo (1974)

$$b = \frac{\ln(E + 1)}{D} \tag{4}$$

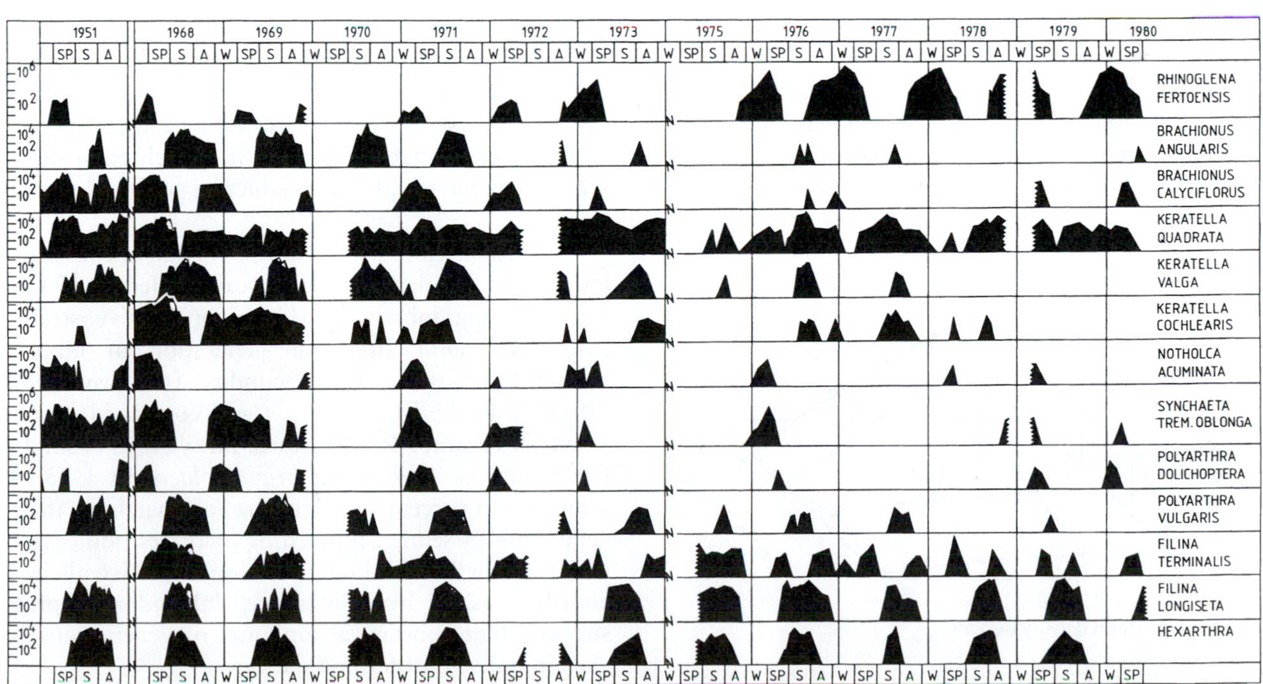

FIGURE 27 Phenologies of several important rotifer species in Neusiedlersee, Austria. The X axis is the season (winter, spring, summer, and autumn) and the Y-axis the rotifer abundance per cubic meter. (From Herzig, 1987, with permission.)

With these data, it is possible to estimate the death rate, d, by subtraction

$$d = b - r \qquad (5)$$

The egg ratio technique makes it possible to predict growth of natural rotifer populations where reproduction by amictic females is parthenogenetic. A few simple measurements allow ecologists to estimate important population parameters that summarize processes of birth and death occurring in the population. This technique has proven useful for investigating the population dynamics and secondary production of zooplankton (e.g. Edmondson and Litt, 1987). However, before applying the egg ratio method, the works of Edmondson (1993) and Palohiemo (1974) should be consulted.

The dynamics of natural rotifer populations are affected by a number of environmental factors including temperature, food quantity and quality, exploitative and interference competition, predation, and parasitism. Temperature is a major factor affecting fertility, mortality, and developmental rates. In general, higher temperatures do not increase the number of offspring produced per female, but instead, shorten birth intervals by decreasing developmental time (Edmondson, 1993; Pourriot, 1965). Life span also is reduced, but the net effect of higher temperatures is an increased population growth rate. In laboratory studies, higher temperatures and elevated food levels combine synergistically to increase significantly mean rate of population increase (r) of *Brachionus calyciflorus* (Fig. 28). Other abiotic environmental factors, such as oxygen concentration, light intensity, and pH, also influence rotifer population dynamics (see Hofmann, 1977 for a review).

Food availability is a major biotic factor regulating rotifer population growth. Species have markedly different food requirements for reproduction. Threshold food concentration, the food level where population growth is zero, was determined for eight species of planktonic rotifers by Stemberger and Gilbert (1985). Small species like *Keratella cochlearis* had lower threshold food concentrations than larger species like *Asplanchna priodonta* or *Synchaeta pectinata* (Fig. 29). The logarithm of the threshold concentration was positively related to the logarithm of rotifer body mass, so the smallest species had the lowest food thresholds. The food concentration required to support 50% of the maximum population growth rate ($r_{max}/2$) varied 35-fold among the eight rotifer species and also was positively related to body mass. Small rotifer species, therefore, appear best adapted to food-poor environments because those species possess low food thresholds. Larger species, in contrast, are better adapted to food-rich environments, where they have higher reproductive potentials.

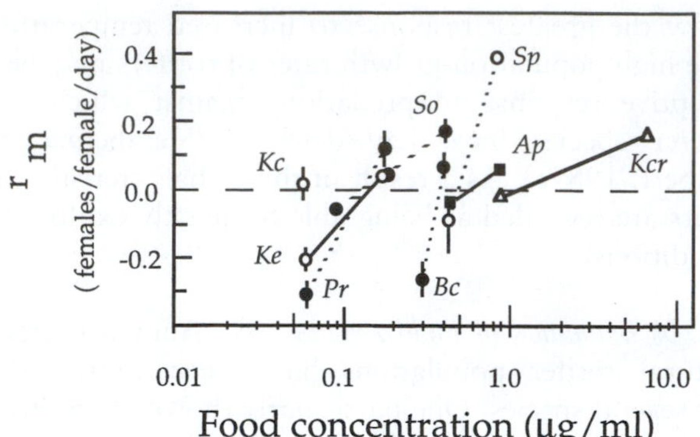

FIGURE 29 Threshold food levels where population growth rate (r_m) = 0. Food concentration is algal dry mass/mL; r_m is offspring per female per day. The solid line indicates food concentrations required to maintain a population growth of zero. Values above the dashed line (positive r_m) indicates increasing population densities; values below the solid line (negative r_m) indicate decreasing population densities. Species code: Kc = *Keratella cochlearis*, Ke = *Keratella earlinae*; Kcr, Keratella crassa; So, *Synchaeta oblonga*; Sp, *Synchaeta pectinata*; Bc, *Brachionus calyciflorus*; Pr, *Polyarthra remata*; and Ap, *Asplanchna priodonta*. (From Stemberger and Gilbert, 1985b, with permission.)

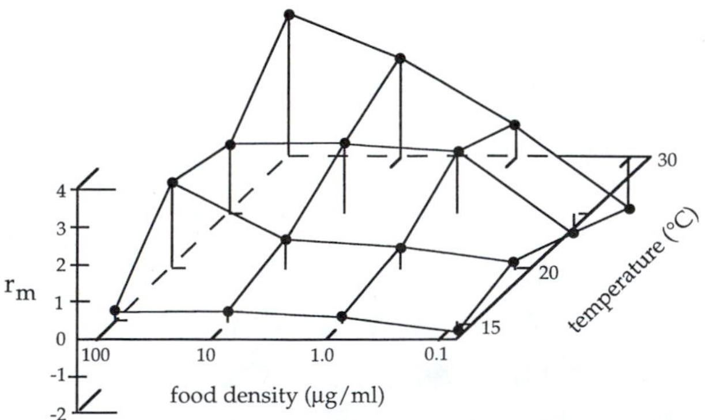

FIGURE 28 Synergistic effects of temperature and food density on mean rate of population increase (r_m) of *Brachionus calyciflorus*. (From Starkweather, 1987, with permission.)

c. Genetic Variation Our knowledge of rotifer genetics is in its infancy. An early worker, Hertel (1942) observed strong inbreeding depression in five strains of *Epiphanes senta* after one generation of inbreeding by self-fertilization. Net fecundity (R_0) was reduced 70–80% in the F1 generation and continued to decline slowly in F2 and F3 generations. Birky (1967) found similar results in *Asplanchna brightwelli*, with a 25–45% reduction in percentage of resting eggs hatching after one generation of selfing. Inbreeding increases homozygosity, thus potentially revealing deleterious recessive alleles previously masked by dominance. Inbreeding depression suggests that substantial amounts of genetic variability exist in natural rotifer populations.

Electrophoretic analysis of allozymes may be used to characterize genetic variation in natural populations.

Although application of these techniques to rotifers has been very limited, allozymes have been used to distinguish strains. Snell and Winkler (1984) analyzed five enzymes representing seven loci in 17 strains of *Brachionus plicatilis* from all over the world. This work showed no relationship between geographic proximity and genetic similarity. In the same species, Serra and Miracle (1985) found low allozyme variation, with only one or two electrophoretic patterns present in each population. These researchers suggested that this low variability resulted from intense inbreeding and interclonal competition, both of which lead to stabilization of just a few highly adapted genotypes. Although genetic variance within populations was low, Serra and Miracle found substantial variance between populations. In contrast, King and Zhao (1987) demonstrated that several electromorphs were present in each collection of *B. plicatilis* from Soda Lake in Nevada, rather than a few dominant types as reported by Serra and Miracle (1985). King and Zhao also observed no seasonal succession of genotypes and concluded that *B. plicatilis* in Soda Lake follow a pattern of incomplete genetic discontinuity. In this pattern, several genotypes persist throughout the year, but their frequencies increase or decrease in response to environmental changes (King, 1972). Allozyme variation was analyzed in five populations of the bdelloid *Macrotrachela quadricornifera* using six enzymes (Pagani, *et al.*, 1991). Relatively high polymorphism was found, with only 27% of alleles shared by all strains.

A variety of perspectives on the genetic structure of rotifer populations have been offered. Ruttner-Kolisko (1963) argued that most rotifer populations are highly homogeneous due to founder effects and strong selection eliminating all but a few genotypes. As a result, bisexual reproduction in rotifers is characterized by extreme inbreeding. Such a genetic structure produces fragmented gene pools where geographically separate populations also are isolated genetically. Birky (1967) disagreed with this view, stating that the substantial heterozygosity revealed by inbreeding experiments suggests that rotifer populations are not genetically homogeneous. The great morphological similarity among geographically separated populations (Pejler, 1977a; Dumont, 1983) further supports the idea that rotifer gene pools are better integrated than Ruttner-Kolisko proposed.

Seasonal changes in rotifer population structure have been characterized by King (1972) who proposed three models; one based on physiological adaptation; and two based on genetic changes. Working with *Asplanchna girodi* in a small pond in Florida, King (1977, 1980) concluded that the complete genetic discontinuity model best described his observations. *Asplanchna girodi* actually had a succession of genetically distinct populations, each hatching from resting eggs, completing their life cycle, and returning to dormancy as the next population was emerging. Little overlap occurred between populations, thus making them completely discontinuous. Competition (Snell, 1979) and the presence of cyanobacteria (Snell, 1980) were major ecological factors promoting succession among these rotifer populations. The population structure differs from the incomplete genetic discontinuity structure that King and Zhao (1987) reported for *B. plicatilis* in Soda Lake.

Experimental data from interpopulation crosses also illuminate the genetic structure of rotifer populations. Some strains of rotifers can mate successfully with strains from widely separated habitats, while others such as *Asplanchna brightwelli* (Birky, 1967; King, 1977) are reproductively isolated.

Behavioral reproductive isolating mechanisms are well developed in rotifers and can be used to define species boundaries (Snell, 1989). Gilbert (1963b) investigated this phenomenon in *Brachionus calyciflorus* and found that mating behavior was strictly species-specific; *B. calyciflorus* males failed to mate with females of *B. angularis*, *B. quadridentatus*, *Synchaeta* or *Euchlanis*. Although Snell and Hawkinson (1983) reported no behavioral reproductive isolation among temporally separated populations of *B. plicatilis* from the same bay, reduced probabilities of copulation among spatially and geographically isolated populations revealed incipient behavioral isolation.

Reproductive isolating mechanisms are not always effective barriers to gene flow. Pejler (1956) argued that the morphological intermediates of *Polyarthra vulgaris* and *P. dolichoptera* he found in several Swedish tarns were the result of interspecific hybridization. Ruttner-Kolisko (1969) provided some direct evidence of interspecific hybridization when she successfully hybridized *B. urceolaris* and *B. quadridentatus*. Resting eggs were formed from this cross (in both directions), but they could be hatched only when female *B. urceolaris* were crossed with male *B. quadridentatus*. Asymmetry in hybridization is typical of closely related species and suggests that reproductive isolation between *B. urceolaris* and *B. quadridentatus* has yet to be completed.

C. Ecological Interactions

1. Foraging Behavior

Although the diets of some rotifers are highly specialized (Bogdan and Gilbert, 1987), many species consume a wide variety of both plant and animal prey and may be described as generalist suspension feeders (Walz, 1997). For example, *Asplanchna* species generally are considered predatory, yet their main food often can be large algae. In contrast, the "herbivorous" *Brachionus* and *Ptygura* will also eat small ciliates.

Thus, distinctions between predator and herbivore in rotifers often are not clear; it is more constructive to consider the ways in which rotifers encounter and process food items rather than what they eat.

Research on food selection by rotifers began by correlating the density of natural populations of rotifers to the food available (Edmondson, 1965) and by determining food selectivities using laboratory cultures and/or natural populations (Bogdan and Gilbert, 1982, 1987; Salt, 1987). In the latter studies, natural foods (i.e., algae, bacteria, yeast) or artificial materials (e.g., latex microspheres) are used to observe rotifer feeding behavior and to calculate feeding rates. Natural foods have been labeled using radioisotopes, but loss of label from the food can result in significant errors. However, even the simple procedure of adding a few drops of food suspension to a Petri dish with rotifers may lead to interesting observations concerning how rotifers process food. For example, using powdered carmine, Wallace (1987) demonstrated that adult *Sinantherina socialis* do not act independently of one another when in a colony. Instead, the coronae of the animals in a large section of the colony all face in the same direction for several minutes. Animals exhibiting this behavior are called an array. Arrays determine where the dominant water currents flow to the colony, how the water currents flow around the colony, and where the currents leave the colony.

Rotifers feed in ways that are directly related to their general life history. Although most planktonic rotifers, such as *Asplanchna*, *Brachionus*, *Polyarthra*, and *Rhinoglena*, swim at similar rates through comparable areas of the water column, the ways in which they encounter and capture food can be quite different. *Brachionus* uses its swimming currents to sweep small food particles into the buccal region for processing; this is often termed filter or suspension feeding. *Asplanchna*, on the other hand, does not use its swimming currents to gather food. Once this rotifer contacts a potential food item with its corona, it may or may not attempt to ingest the item based on such factors as hunger level and the size and type of prey (Gilbert and Stemberger, 1985b; Salt *et al.*, 1978). Some benthic rotifers creep along algal filaments and feed by piercing the filament and sucking the cytoplasm from the cells; examples are *Notommata copeus* (Clément *et al.*, 1983) and *Trichocerca rattus* (Clément, 1987). Sessile rotifers capture food in one of two ways, depending on the family. Members of the family Flosculariidae (e.g., *Floscularia*, *Ptygura*, *Sinantherina*) create feeding currents in a manner similar to the planktonic suspension feeders. In contrast, all collothecid rotifers are ambush raptors (e.g., *Collotheca*, *Stephanoceros*). Once a prey enters the infundibular region of the corona of these rotifers, long setae (*Collotheca*) or arms (*Stephanoceros*) fold over the prey, capturing it much like the action of a Venus flytrap. All members of the family Atrochidae lack setae (e.g., *Cupelopagis*). In these forms, an enlarged, sometimes sheetlike corona folds over prey to capture them. In both cases, prey are pushed through the rotifer's mouth and into the proventriculus where they are stored until the mastax transfers them into the stomach. When prey are particularly abundant, several live prey may be seen in the proventriculus.

Once food is captured, the process of ingestion begins, but not all captured food items are consumed. By observing individual *Brachionus calyciflorus* in various densities of suspended food particles, Gilbert and Starkweather (1977, 1978) showed that this rotifer regulated its ingestion of food particles by three different mechanisms. First, rotifers used pseudotrochal cirri to screen certain large particles away from the mouth. Second, particles collected by the corona may be rejected later by cilia within the buccal field. Finally, particles that gain entrance to the oral cavity may be returned to the buccal field by action of oral cilia for subsequent rejection. The mastax also may actively reject particles. Microphagus rotifers, such as *Brachionus* and *Ptygura*, use the mastax to push food (usually larger particles) out of the oral cavity. Members of the genera *Asplanchna* and *Asplanchnopus* also use their trophi to remove materials from their stomachs, but the materials removed by these raptors are empty carapaces of hard-bodied prey such as cladocerans. Occasionally, predatory rotifers such as *Cupelopagis*, may become so engorged that attempting to ingest another prey item results in the loss, from the proventriculus, of one previously captured.

2. Functional Role in the Ecosystem

Freshwater zooplankton are dominated by three taxonomic groups (protozoans, rotifers, and microcrustaceans), yet it is the microcrustaceans that usually receive the most attention from researchers. Such a disparity is probably to be expected for two reasons, one significant and one trivial. First, microcrustaceans commonly account for a greater proportion of the total zooplankton biomass than either protozoans or rotifers. Second, microcrustaceans are easier organisms to manipulate and observe in laboratory situations. However, although rotifers usually have a smaller standing biomass than microcrustaceans, their high turnover rate makes them very important to the trophic dynamics of freshwater planktonic communities (e.g., Bogdan and Gilbert, 1982, Neil, 1984).

Feeding rates of zooplankton generally are referred to as filtration or clearance rates and are measured as microliters of water cleared of a certain food type per animal per unit time (i.e., μL/animal/h). Usually rotifer clearance rates are lower than those of cladocerans and copepods, although the rates depend heavily on food type, temperature, and animal size (Bogdan and Gilbert, 1982). For most rotifers, clearance rates are commonly between 1 and 10 μL/animal/h, whether determined in the laboratory or in the field. However, a few species can achieve levels exceeding 50 μL/animal/h (e.g., Bogdan and Gilbert, 1987).

Using estimates of clearance rates, Starkweather (1987) noted that even moderately sized rotifers with body volumes of about 10^{-3} μL process enormous amounts of water with respect to their size: $>10^3$ times their own body volume each hour! Ingestion rates (biomass consumed per animal per unit time) also are very high for rotifers. An adult rotifer may consume food resources equal to 10 times its own dry weight per day. Therefore, by having assimilation efficiences (i.e., assimilation divided by ingestion) of between 20 and 80%, rotifers convert a good deal of their food to animal biomass, which may be passed on to the next trophic level (Starkweather, 1980, 1987).

Although microcrustaceans generally have higher clearance rates than rotifers (about 10–150 μL/animal/h for cladocerans and 100–800 μL/animal/h for copepods) rotifers can exert greater grazing pressure on phytoplankton than some small cladocerans. Bogdan and Gilbert (1982) report that in a small eutrophic lake, *Keratella cochlearis* accounted for about 80% of the community grazing pressure on small algae during the year. These two workers also showed that *K. cochlearis* had filtering rates about 5–13 times higher than that for the cladoceran *Bosmina longirostris* (Gilbert and Bogdan, 1984). Therefore, under certain conditions, rotifers may be important competitors to small, filter-feeding microcrustaceans and are important in nutrient recycling in aquatic systems (Makarewicz and Likens, 1979). Furthermore, rotifers can alter the species composition of algae in certain systems. Schlüter *et al.* (1987) showed that intense feeding by *Brachionus rubens* can cause a shift in the dominant algal species from *Scenedesmus* to a spined algae, *Micractinium*. Apparently, this shift is based on the inability of *B. rubens* to consume algae with protective spines.

Although generally not ingested, cyanobacteria are increasingly recognized as playing an important role in determining zooplankton species composition. Large cladocera are more sensitive to the cyanobacteria *Anabena affinis* than rotifers (Gilbert, 1990). The mechanism for this differential sensitivity to cyanobacteria toxicity is based on different tendencies to ingest cyanobacterial filaments and different physiological tolerances of the toxin (Kirk and Gilbert, 1992).

Having the ability to reproduce rapidly, rotifers may account for 50% or more of the zooplankton production, depending of the prevailing conditions (Herzig, 1987). This production, in turn, can be an important food source for other rotifers, *Asplanchna* (Gilbert and Stemberger, 1985b), cyclopoid and calanoid copepods (Williamson, 1983), malacostracans (*Mysis*, Threlkeld *et al.*, 1980), insect larvae (*Chaoborus*, Moore, 1988; Moore and Gilbert, 1987), and fish (Hewitt and George, 1987).

Abundance and species composition of rotifers often reflect the trophic status of lakes. For example, Hillbricht-Ilkowska (1983), Walz *et al.* (1987), and others have reported changes in the maximal total-population density of several orders of magnitude when lakes were subject to intense eutrophication. Individual species sometimes undergo dramatic population changes during those periods. Walz *et al.* (1987) showed that the density of *Asplanchna* in Lake Constance increased its maximum population levels 280-fold over a period of 28 years. Population declines also have been seen. Edmondson and Litt (1982) indicated that *Keratella cochlearis* was abundant during the years when Lake Washington had elevated concentrations of dissolved phosphorus, low water transparency, and high algal densities. However, as these waterquality parameters improved, the population of this rotifer declined dramatically. Overall, there was at least a 20-fold increase, followed by a decline during a period of 15 years.

Studies of the interactions of rotifers with other organisms will probably continue to receive attention in the future, especially predator–prey interactions (Gilbert, 1966 1967; Salt, 1987; Williamson, 1983), interference competition between other herbivorous zooplankton and rotifers (Burns and Gilbert, 1986a, 1986b; Gilbert and Stemberger, 1985a), the toxic effects of cyanobacteria (Snell, 1980; Starkweather and Kellar, 1987), mate recognition (Snell *et al.*, 1995), and life-history strategies (Walz, 1995). One underutilized tool for the study of energetics and trophic interactions is the chemostat, which, unlike batch cultures, maintains constant experimental conditions (Boraas and Bennet, 1988; Walz, 1993).

3. Competition with Other Zooplankton

Rotifers, cladocerans, and copepods often compete for limited food resources and, in general, rotifers are relatively poor exploitative competitors because their clearance rates are usually many times lower than those of daphnids (see Section III.C.2). Rotifers also have a more limited size range of particles they can ingest compared to cladocerans and are less resistant to starvation (Kirk, 1997). Most rotifers eat algal cells in the

4–17 μm range and are much less efficient at ingesting smaller or larger cells (Gilbert, 1985b). In contrast, most cladocerans eat algal and bacterial cells in the 1–17 μm range, and some can ingest much larger cells (Hall *et al.*, 1976; Bogdan and Gilbert, 1982). Thus, cladocerans generally have broader food niches than rotifers in terms of food type and size, and through direct competition may suppress rotifer population growth. This outcome may be reversed, however, when a sufficient quantity of suspended sediments is present in a lake (Kirk, 1991).

Exploitative competition between rotifers and daphnids is readily demonstrated when these zooplankton are grown in single and mixed cultures. *Brachionus calyciflorus* and *Daphnia pulex* both grow well on the alga *Nannochloris oculata* in single species cultures. When both species are present, however, *Daphnia* removes an increasingly larger proportion of algal cells until the rotifers gradually starve to extinction after 2–3 weeks (Fig. 30). *Daphnia* is unaffected by the presence of rotifers.

Population growth of certain rotifers also is inhibited by *Daphnia* through interference competition (Burns and Gilbert, 1986a, b; Gilbert and Stemberger, 1985b; Gilbert, 1989a). For example, in the presence of any one of four *Daphnia* species, *Keratella cochlearis* suffers mechanical damage (i.e., killed, wounded, or loss of eggs) when swept into the branchial chamber of the daphnids. No species differences were detected among the *Daphnia* used, but the rate at which rotifers were killed was directly proportional to daphnid body length. *Keratella* also was found in the guts of some *Daphnia*, documenting a new pathway for trophic interactions. *Daphnia* have a major impact on *Keratella* populations when the cladocerans are larger than 2 mm and present in densities of 1–5 individuals/L. Similar effects have been reported for short-term bottle experiments involving the cladoceran *Scapholeberis kingi* and the rotifer *Synchaeta oblongata* (Gilbert, 1989b). Of course, interference and exploitative competition may occur simultaneously (Gilbert 1985b, 1988).

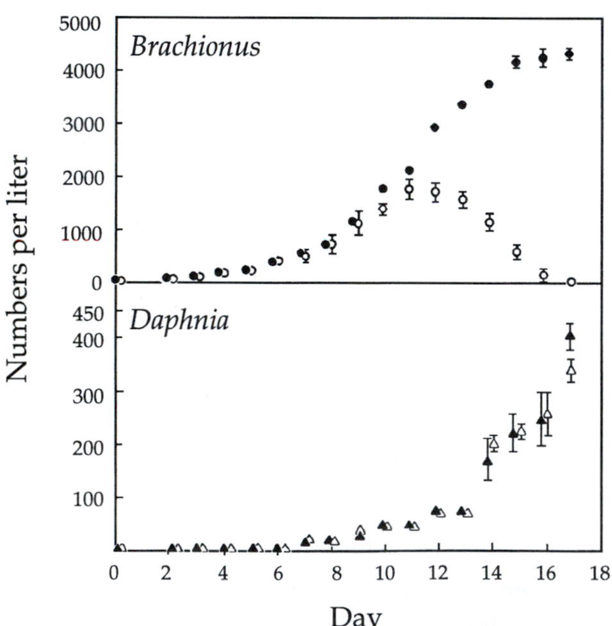

FIGURE 30 Competition between *Brachionus calyciflorus* and *Daphnia pulex*. *Brachionus* and *Daphnia* were grown in single species (closed symbols) and mixed-species (open symbols) batch cultures at 20°C, daily renewed with 5×10^6 *Nannochloris* cells/mL. Population size (Y axis) is the number of individuals in the 80 mL culture. Error bars (±1 se) are visible when they exceed the size of the symbols. (From Gilbert, 1985, with permission.)

Predation is another important regulatory factor in rotifer population dynamics, as rotifers are prey for several aquatic predators including protozoans, other rotifers, insect larvae, cladocerans, copepods, and planktivorous fish. From many studies we know that predation affects rotifer population dynamics both directly (by contributing to mortality) and indirectly as a selective force shaping rotifer morphology, physiology, and behavior (see the next section). However, one area that deserves more study is the predatory interactions of protozoans and rotifers (e.g., protozoans as predators—Finlay *et al.*, 1987 and Nogrady *et al.*, 1993; rotifers as predators—Arndt, 1993; and Gilbert and Jack, 1993).

4. Predator–Prey Interactions

Rotifers possess several types of defensive mechanisms that have apparently evolved in response to predation. These mechanisms include morphological features, escape movements, and other ways of avoiding potential predators (see Stemberger and Gilbert, 1987a; Williamson, 1983, 1987).

Most rotifers are transparent and quite small, some comparable in size to protozoans (ca. 60–250 μm long). While these features benefit limnoplanktonic rotifers by reducing their visibilty to fish, a small body size renders rotifers more vulnerable to invertebrate predators, which feed by touch rather than by sight. Many rotifers produce a thickened integument (lorica) and/or spines and other projections, or carry their eggs, all of which have been shown to reduce the ability of predatory zooplankton to prey upon them (e.g., Gilbert, 1966, 1967, 1980b; Stemberger and Gilbert, 1984b). Spines are produced in some rotifers (e.g., *Brachionus calyciflorus*, Figs. 12 and 31; *Keratella cochlearis*; *Keratella testudo*, Fig. 26) in response to a build-up of soluble substances released by invertebrate predators such as *Asplanchna* and copepods (Gilbert, 1967; Stemberger and Gilbert, 1984b; Stemberger,

1984). We now know that exposure to a water-soluble factor secreted by the predatory rotifer *Asplanchna* or the copepods *Epischura*, *Mesocyclops*, and *Tropocyclops* induces *de novo* spine formation or spine lengthening in several planktonic rotifers (Stemberger and Gilbert, 1987a, 1987b) (see Section III.A.1). Rotifers for which such polymorphisms have been observed include *Brachionus calyciforus*, *B. bidentata*, *B. urceolaris*, *Filinia longiseta passa*, *Keratella cochlearis*, *K. slacki*, and *K. testudo*. The adaptive significance of spined morphotypes is a significant reduction in capture and ingestion by invertebrate predators by making the rotifer more difficult to manipulate and swallow. The presence of spines on *B. calyciflorus* is a good example of the phenomenon (Fig. 12). The way this works is as follows. When disturbed by a potential predator, *B. calyciflorus* will retract its corona and by doing so increases the hydrostatic pressure within its pseudocoelom. The elevated pressure causes the posterolateral spines to swing forward and outward (Fig. 31). Thus, the apparent body volume is larger. For *Asplanchna*, this increase in the size of its prey is sufficient to prevent ingestion after capture. After a period of time during which *Asplanchna* attempts to swallow *B. calyciflorus* (usually >60 sec), the predator will release (reject) the prey which then swims away unharmed. Thus, the invertebrate predator releases a biochemical cue (an allelochemic) that initiates a developmental change in subsequent generations of the rotifer (spine production) reducing the effect of predation. Some forms of *K. cochlearis* also possess posterior spines, which make them four times more likely to be rejected after capture by *A. girodi* than unspined forms (Stemberger and Gilbert, 1984b). Most rejected prey swim away unharmed. However, occasionally spined forms are swallowed by *Asplanchna* or a similar genus (*Asplanchnopus*) and the spines become lodged in the pharynx or esophagus of the predator.

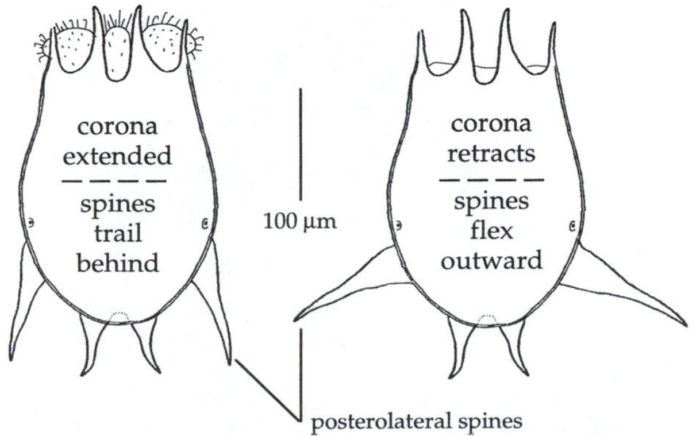

FIGURE 31 Extension of the postero-lateral spines in *Brachionus calyciflorus*. (Modified from Koste, 1976, with permission.)

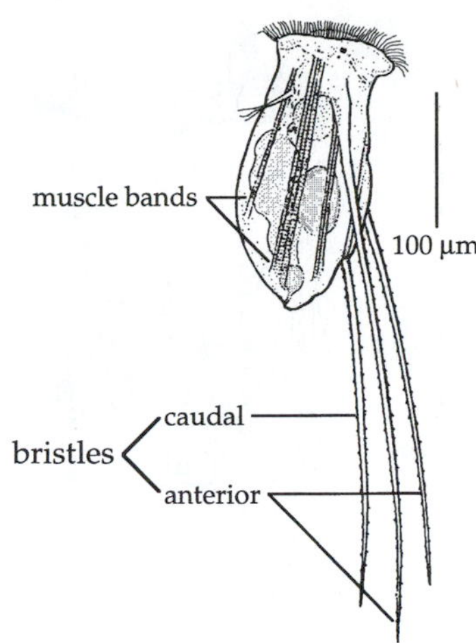

FIGURE 32 *Filinia* (lateral view). This rotifer possesses long spines which are used in making rapid jumps to escape predators. (Modified from several sources.)

Presumably, both predator and prey die when that happens.

Spine production has additional consequences beyond predatory avoidance. Stemberger (1988) has demonstrated that, in the absence of predators, reproductive effort (survivorship and fecundity) and hydrodynamic characteristics (sinking rates and swimming speeds) vary in spined and unspined *Keratella testudo*. These relationships were further complicated by food concentration (see Section III.B.4.a, and Fig. 26).

Transparent mucus sheaths produced by some planktonic rotifers (e.g., *Conochilus* and *Lacinularia* species) appear to function in a similar way to spines. The sheaths make the rotifers effectively larger and, therefore, less vulnerable to invertebrate predators (e.g., *Asplanchna* and predatory copepods) without making them more visible to fish (Gilbert, 1980b; Stemberger and Gilbert, 1987a, 1987b; Wallace, 1987).

Three rotifer genera can often escape predators by making rapid jumps using a variety of appendages: long spines (*Filinia*, Fig. 32), setous armlike appendages (*Hexarthra*, Fig. 13), and paddles (*Polyarthra*, Fig. 33) (see Section II.C.1). Many rotifers assume a passive posture, displaying what has become known as the "dead-man response," rather than fleeing when predators attack (e.g., *Asplanchna*, *Brachionus*, *Keratella*, *Sinantherina*, and *Synchaeta*). This simple behavior is merely a retraction of the corona into the body and passive sinking. Contraction of the corona stops the animal from swimming, thus eliminating the vibrations it produces in the water that may be detected by the predator. In addition, this behavior may make the rotifer more difficult to grasp in its turgid state. In *Sinantherina spinosa*, this passive posture exposes a group of

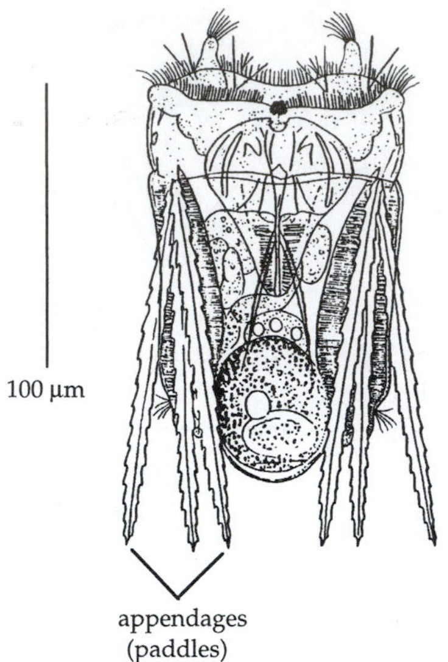

FIGURE 33 *Polyarthra.* This rotifer possesses long paddlelike appendages which are used in making rapid jumps to escape predators. (Modified from several sources.)

small spines on its ventral body surface, which may function in defense against planktivorous fish (see Wallace, 1987).

To date, only one species of rotifers, *Sinantherina socialis*, has been shown as unpalatable to small zooplanktivorous fishes (Felix *et al.*, 1995). While neither the nature nor the location of the unpalatability factor(s) is known, these authors speculate that this colonial species possesses an unidentified chemical held in the glandlike "warts" which are located at the anterior end of the animals.

Defenses against predators like spines, mucus sheaths, thickened loricas, and escape movements are energetically demanding. Some species, like *Synchaeta pectinata*, are not well defended against predators, but have evolved very high, maximal population growth rates that offset mortality from predation. Avoiding potential predators in space and/or in time is another simple yet effective defense mechanism against predators. Some rotifers occupy the habitat at a different time of year than an important predator, migrate vertically or horizontally in the habitat, or live in zones with low oxygen concentration, thereby missing predatory pressures altogether.

5. Parasitism on Rotifers

The importance of parasites in controlling population density in rotifers has not been examined thoroughly, although a few studies have correlated parasitic infection with a decrease in population density of planktonic species (Ruttner-Kolisko, 1977a). In certain cases, it appears that parasites can cause the demise of an entire population within a few days (Edmondson, 1965).

The sporozoan parasite, *Microsporidium* (*Plistophora aerospora*) frequently infects planktonic rotifers possessing thin loricas (Ruttner-Kolisko, 1977a). For example, members of the genera *Asplanchna*, *Brachionus*, *Conochilus*, *Epiphanes*, *Polyarthra*, and *Synchaeta* have been reported to be parasitized. Apparently, water temperature is an important mediating factor in the spread of the parasite as infection rate drops off at water temperatures below 20°C. In infected rotifers, the pseudocoel of the animal becomes nearly filled with sausage-shaped cysts of *Microsporidium* (Fig. 34).

Several workers have described endoparasitic fungi which attack soil rotifers of the genera *Adineta* and *Philodina*. These reports describe three avenues of parasitic attack: adhesion, ingestion, and injection. Tzean and Barron (1983) have described one fungus that forms peglike, adhesive appendages on both small conidia (spores, ca. 30 μm long) and long vegetative hyphae.

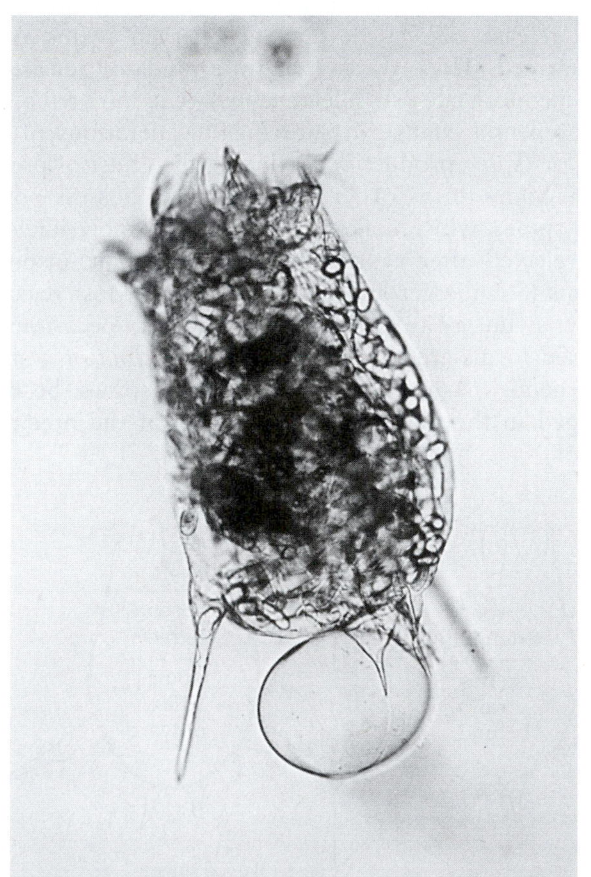

FIGURE 34 Photomicrograph of *Brachionus* infected with *Microsporidium* within its body cavity (pseudocoel).

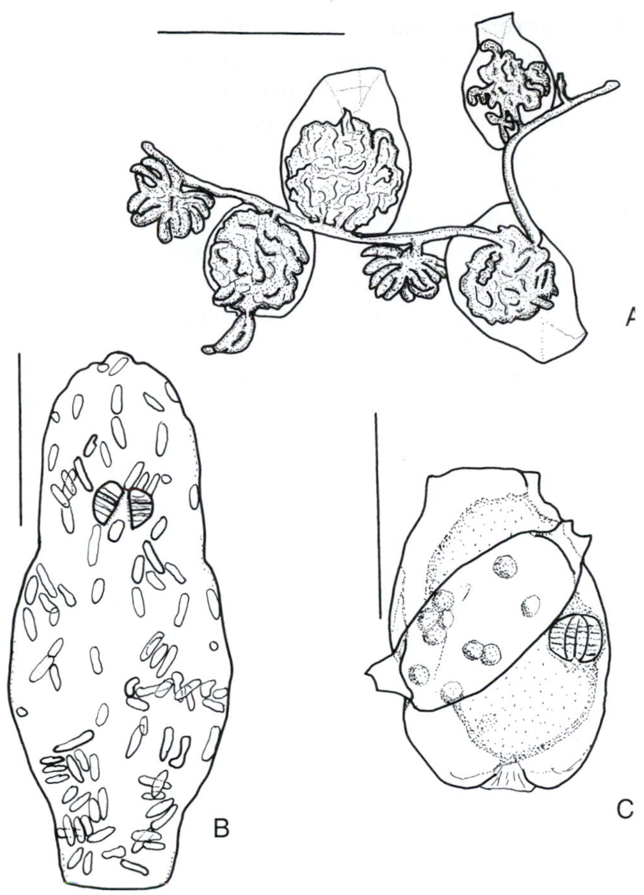

FIGURE 35 Fungal parasites of rotifers. (A) Four rotifers trapped by adhesive pegs on the vegetative hyphae of *Cephaliophora*; (B) unicellular, assimilative hyphae of *Triaculus*; and (C) zoosporangium of *Haptoglossa* with two evacuation tubes and ten encysted zoospores. (Bars ≈ 100 μm.)

Once the adhesive pegs attach to a rotifer, they germinate and rapidly colonize the pseudocoel (Fig. 35A). At least three different genera of fungi produce spores (Fig. 35B) that initiate parasitic attack when ingested (e.g., Barron, 1980a; Barron and Tzean, 1981). A third avenue of infection occurs via hypodermic injection of a vegetative cell into the rotifer (Barron, 1980b; Robb and Barron, 1982). Once inside the host, these fungal cells grow into assimilative hyphae, producing more infective cells either inside or outside the rotifer (Fig. 35C).

6. Aquaculture

In freshwater communitites, rotifers serve as food for several invertebrate predators which, in turn, are consumed by many economically important fish. They also are used directly as food by many planktivorous adult fish (O'Brien, 1987). Perhaps the most important ecological link between rotifers and fish is the consumption of rotifers by larval fish. At an early age, most fish larvae feed on microzooplankton, rotifers being one of the most important components of the diet. In general, rotifers are highly nutritious and their biochemical composition can be further improved by specialized diets (Watanabe *et al.*, 1983). Many fish are well adapted to locate, pursue, capture, and ingest rotifers, and rotifers are easy prey because they swim slowly and frequently lack predator defenses (Stemberger and Gilbert, 1987a).

Aquaculturists have exploited this important relationship between planktivorous fish and their rotifer prey in intensive aquaculture systems. This field has developed into a major technical discipline in several countries, most notably Israel, Japan, and China. Most of this work has the practical goal of determining the correct biotic and abiotic factors necessary to maintain mass cultures of rotifers (usually *Brachionus plicatilis*) as food organisms for larval shrimp and fish (Lubzens, 1987; Lubzens *et al.*, 1989).

Most systems for mass culturing of rotifers are simple batch cultures capable of producing kilogram quantities of rotifer biomass each day (Hagiwara *et al.*, 1997). A substantial amount of effort has been devoted to identifying optimal culture conditions for the most commonly grown species (*Brachionus rubens* and *B. calyciflorus* in freshwater and *B. plicatilis* in saltwater). For mass cultures of *B. rubens* raised at 20°C, the optimal pH is 6.0–8.0, with upper and lower lethal limits of 9.5 and 4.5, respectively (Schlüter, 1980). Oxygen levels as low as 1.15 mg/L sustained good growth, but 0.72 mg oxygen/L was strongly limiting (Schlüter and Groeneweg, 1981). Schlüter (1980) also tested eight algal species and found that *Scenedesmus costato-granulatus*, *Kirchneriella contorta*, and *Chlorella fusca* produced the best rotifer growth. Algal concentrations of 70 mg/L dry weight of *Scenedesmus* were optimal. *Brachionus rubens* reproduction in mass cultures is inhibited by ammonia levels of 3–5 mg NH_3-N/L (Schlüter and Groeneweg, 1985). Because about 50% of the nitrogenous wastes excreted by *Brachionus* is ammonia, waste product removal is a serious consideration in mass culture design (Hirata and Nagata, 1982).

Although rotifers have been important in marine fish aquaculture for many years (Lubzens, 1987), freshwater rotifers are just beginning to be used. Nauplii of brine shrimp (*Artemia*) have been used extensively as larval feed, but these marine organisms do not live more than about 6 h in freshwater, and their decomposition contributes to the organic load in culture tanks. A freshwater organism is preferred for freshwater aquaculture, and rotifers fill this role well. Rotifers also are being used as food for young fish larvae too small to begin feeding on *Artemia*. For these reasons the use of rotifers in freshwater aquaculture is expected to grow.

D. Evolutionary Relationships

The phylogenetic position of the eight tradition "pseudocoelomate" phyla represents a problem that continues to generate debate among workers. Within this larger question several opinions have been expressed concerning the evolution of rotifers, including some that propose a polyphyletic origin of the phylum. One theory suggests that rotifers evolved from an ancestor possessing a fluid-filled body cavity (Lorenzen, 1985), perhaps a regressed coelom, while another proposes that rotifers were derived from an ancestral acoel turbellarian (phylum Platyhelminthes, Chapter 6) (Hyman, 1951). We believe that the evidence favors the view that rotifers and platyhelminthes are related; these phyla share several common features including protonephridia, a ciliated integument which possesses mucus glands, and a cerebral eyespot with a pigment cup (Wallace *et al.*, 1996; Garey *et al.*, 1998).

The most critical issue regarding the origin of the Rotifera is the question of its relationship to phylum Acanthocephala. Presence of the intracytoplasmic lamina within a syncytial epidermis as well as other morphological features in both phyla suggests that they are closely related even though acanthocephalans are 5–1500 times larger in size than the largest rotifer (Wallace *et al.*, 1996). Lorenzen (1985) argues that structures identical to the rostrum and lemnisci of acanthocephalans are found in bdelloids, thus making the two monophyletic. However, Ricci's (1998a) recent critique argues against homology of these structures and a recent cladistic analysis of some 60 morphological characters rejected Lorenzen's thesis that acanthocephalans and bdelloid rotifers are related (Melone *et al.*, 1998b). Nevertheless, molecular analyses of 18S rRNA genes from a few well-known species place the acanthocephalans firmly within the Rotifera. One view places the acanthocephalans as a highly modified sister group to the bdelloids (Garey *et al.*, 1996, 1998), while another places them as highly modified rotifers related to the sister group of bdelloids and monogononts (Mark Welch, unpublished observations). If additional genetic analysis confirms these studies, the definition of what constitutes the phylum Rotifera will need to be completely re-interpreted. A composite overview of these hypotheses is illustrated in Figure 36.

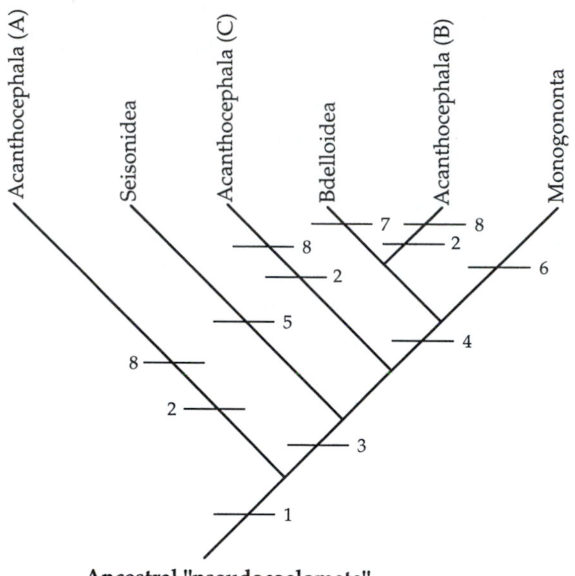

FIGURE 36 Phylogenetic relationships within phylum Rotifera at the class level. Illustrated here are the three different hypotheses regarding the relationship of acanthocephalans and rotifers. (A) The traditional hypothesis based on an analysis of morphological data concludes that acanthocephalans are a sister group to rotifers (Melone *et al.*, 1998). (B) A recent hypothesis based on an analysis of molecular data (nuclear 18S rRNA and mitochondrial 16 S rRNA) concludes that acanthocephalans are a highly modified sister group to bdelloid rotifers (Garey *et al.*, 1998). (C) Results of recent research of molecular data (nuclear gene: 82 kd, heat-shock protein) concludes at acanthocephalans are highly modified rotifers related to the sister group of bdelloids plus monogononts (Mark Welch, 2000). Numbers refer to changes in character state as gains or losses. Character states are numbers as follows: (1) intracytoplasmic lamina; (2) females with uterine bell, ciliated epidermal cells absent, cephalic gland absent; (3) ciliated corona present, mastax with trophi present, retrocerebral apparatus present and paired, dorsal antenna present; (4) Eurotatoria: urinary collection bladder present, manubrium present in trophi, paired retrocerebral apparatus fused into a single unit; (5) male genital ducts U-shaped, penis absent; (6) male haploidy, single gonads; (7) adult capable of a type of diapause, loss of male, stomach syncytial, fulcrum absent in trophi; and (8) specialized parasitic features.

IV. COLLECTING, REARING, AND PREPARATION FOR IDENTIFICATION

A. Collections

Collecting rotifers does not require complex or expensive equipment. One can almost always collect several species of planktonic rotifers by towing a fine-mesh (25–50 μm) net through any body of water. [N.B.: Nets with greater mesh sizes tend to miss small-bodied forms.] Productive lakes and ponds usually provide especially good sampling sites. More elaborate equipment such as closing nets and Clarke–Bumpus samplers work well, but are not necessary unless required by a specific experimental protocol. Water collected by discrete sampling devices (e.g., Van Dorn sampler, Kemmerer bottle, plankton trap, pumps) is then filtered through a net of appropriate size. In weedy areas, a dip net or flexible collecting tube (Pennak, 1962) are very useful. Another simple method to collect rotifers is to submerge a 3–4 L (1 gallon) glass jar in a weedy region and to arrange loosely a few aquatic plants in it. If using this technique, fill the jar to about 2–5 cm from the top and place it near a subdued light source, such as a north window. Rotifers which swim to the surface on the lighted side may be removed using a transfer pipette.

Certain aquatic plants, such as *Elodea*, *Myriophyllum*, and filamentous algae are good substrata to examine for the presence of sessile rotifers, but *Utricularia* usually provides the richest diversity (Wallace, 1980). Plants with highly dissected leaves may be examined in small dishes using a dissecting microscope. Broad-leaved plants must be cut into strips and examined on edge.

The upper few centimeters of moist sand taken just above the water line along a lake or a marine shoreline or the hyporheic interstitial zone of a stream bed usually provides several species of rotifers. Unfortunately, very few studies have been conducted on rotifers from this habitat termed "psammon," in part because of the difficulty in separating the organisms from the sand. Recent studies have revealed a remarkable diversity (35–85% of the fauna) and abundance (up to 10^5 individuals/L) of rotifers in this habitat. Under certain conditions rotifers may be found at depths of up to 60 cm into the interstitial. For a review of these topics consult Schmid-Araya (1995).

Sediments collected from the bottom using a dredge, coring apparatus, or suction device usually provide several bdelloid and monogonont species. The upper few centimeters of sediments from a core collected in the winter or early spring normally contain resting eggs, which can be induced to hatch within a few days when several milliliters of sediment are incubated at ambient spring or summer temperatures (May, 1986, 1987).

Do not overlook laboratory aquaria or holding tanks as potential sources of material. We have found some unusual species in aquaria that had remained almost unattended for months. One might try adding a small amount of sediments from several sites as a way of adding variety to the rotifer community within the aquarium. Sessile rotifers may be present if the aquarium contains aquatic plants recently collected from the field. However, if you attempt to keep sessile forms, be sure first to remove all snails from the aquarium. Populations of rotifers may be maintained or even increased by adding a small amount of food once or twice a week.

B. Culture

King and Snell (1977), Ricci (1984), and Stemberger (1981) have summarized the general procedures for culturing rotifers. Some species of rotifers may be cultured in the laboratory quite easily, obtaining densities of $>10^5$ individuals/L in a few weeks. Others seem impossible to keep even for short periods. Several species are commonly raised under a variety of conditions (Table II). Most of these need regular care a few times per week (changing the medium, feeding, etc.). However, some species require little care. For example, *Habrotrocha rosa* is easily cultured in dilute broths of decaying powdered baby food and may be nearly ignored for weeks at a time without extinction of the culture (see Bateman, 1987). Techniques used to culture protozoans (e.g., making extracts and infusions of various grains, manure, soil, etc.) have been adopted for the culture of some rotifer species with great success. Most rotifer cultures can be maintained xenically (i.e., containing several contaminating organisms) without much problem. In fact, sometimes rotifers are found contaminating cultures of protozoans, microcrustaceans, etc., that have been provided by commerical biological supply companies. Rotifers have been maintained under axenic (Dougherty, 1963) or monoxenic culture (Gilbert, 1970) conditions. Sophisticated culture techniques using single- and two-stage chemostats have been described by several workers (Starkweather, 1987; Walz, 1993).

C. Preparation for Identification

Three preservatives are commonly used to preserve rotifers: formalin and Lugol's iodine at concentrations of 5% or less, and ethanol at about 30–50%. Lugol's has two advantages over formalin. It is less toxic and it stains the specimens slightly; the staining makes the animals more visible during sorting procedures. Unfortunately, Lugol's iodine makes the lorica skrink and deforms the specimens. Other stains such as rose bengal are commonly used to stain specimens preserved in either formalin or ethanol.

Unless special precautions are taken before fixation, illoricate rotifers (especially bdelloids) will contract into a completely unidentifiable lump, making identification difficult if not impossible. Such specimens have lost all value for taxonomic purposes, although the trophi still may be useful (see later discussion). Histological fixatives (e.g., Bouins) or sugar-formalin solutions (used to prevent osmotic shock in microcrustaceans) are not particularly helpful in preventing contraction in rotifers. Whenever possible, live specimens should be identified first, then fixed to determine the effect of a particular fixative on body shape.

Fortunately, several anesthetics have been found to work well on rotifers (Edmondson, 1959; May, 1985; Nogrady and Keshmirian, 1986a, b; Pennak, 1989 p.

TABLE II Examples of Rotifers That Have Been Cultured

Species	*Culture conditions*	*Reference*
Bdelloids		
Habrotrocha rosa	Batch cultures of tap water and powdered baby food, dog food, etc.	Bateman (1987), Schramm and Becker (1987), Wallace personal observations
Philodina acuticornis	Monoxenic culture	Meadow and Barrow (1971)
several bdelloids	Batch cultures	Ricci (1984), Pourriot (1958)
Monogononts		
Asplanchna spp.	Batch cultures	Aloia and Moretti (1973), Gilbert (1967), Stemberger (1981), Stemberger and Gilbert (1984a)
Brachionus angularis	Continuous culture	Walz (1983a)
B. calyciflorus	Batch culture	Gilbert (1963a)
	Monoxenic cultures	Gilbert (1970)
	Chemostat culture	Boraas and Bennet (1988)
B. patulus	Batch culture	Rao and Sarma (1986)
B. plicatilis	Batch culture	Hirayama and Ogawa (1972)
	Mass culture	Lubzens (1987)
	Chemostat culture	Droop and Scott (1978)
B. rubens	Batch culture	Pilarska (1972)
	Mass culture	Schlüter and Groeneweg (1981)
	Chemostat culture	Rothhaupt (1985)
Cephalodella forficula	Batch culture	Dodson (1984)
Encentrum linnhei	Monoxenic and Chemostat cultures	Scott (1983), Scott (1988)
Euchlanis dilatata	Batch culture	King (1967)
Keratella cochlearis	Continuous and batch cultures	Walz (1983a,b)
Keratella testudo	Batch culture	Stemberger and Gilbert (1987b)
Lecane inermis	Batch culture	Dougherty (1963)
Notommata copeus	Batch culture	Pourriot (1977)
Polyarthra major	Batch culture	Stemberger (1981)
Polyarthra vulgaris	Batch culture	Buikema et al. (1977)
Synchaeta cecelia	Batch culture	Egloff (1988)

(For additional references see *Hydrobiologia* volumes 255/256 and 313/314; Pourriot (1965: 122) provides a list of species that have been cultured prior to about 1959.)

195), but no single anesthetic has been shown to be universally effective. Care should be taken when working with anesthetic agents; some are controlled substances and/or are toxic. An alternative approach is to use the simple hot-water fixiation technique described by Edmondson (1959, p. 433) which generally gives excellent results with a number of species. It is generally a good idea to place live samples in jars over ice for the return trip to the laboratory, although we have found that a few species suffer when cooled (e.g., *Sinantherina socialis*).

Techniques for preservation and mounting of rotifers for examination using light microscopy (Edmondson, 1959, p. 433 Pennak, 1989, p. 196; De Smet, 1998) and transmission and scanning electron microscopy (Amsellem and Clément, 1980; Melone, 1998; Kleinow, 1998) have been described. However, the striking beauty of the living rotifer is utterly lost during almost any fixation and mounting procedure.

Observations of live rotifers are not always easily accomplished. The goal is to retard their movements without crushing or distorting the specimen with the cover glass. Both objectives may be accomplished with a compression microscope slide (Taylor, 1993). If a compressor is unavailable, then tiny pieces of broken cover glass (not recommended) or clay supports (recommended) work well as corner supports. If the clay supports are too high, a slight pressure from a pencil on each corner will reduce the height of the cover glass to the desired level. With some practice, one can trap a planktonic rotifer sufficiently to prevent swimming without undue constriction. Sessile rotifers are handled more easily. Plant material with attached rotifers can be trimmed with iridectomy scissors to a size suitable for placement on a microscope slide. The animal will remain in place without the need of compression as long as the plant material is large enough to act as an anchor.

Methylcellulose or other viscous agents and fibrous material, such as glass wool or shredded filter paper, also may be used to impede swimming species. Unfortunately, methylcellulose interferes with ciliary function, and fibers reduce observation to a game of hide-and-go-seek. Any lighting conditions may be used, as long as you are careful not to overheat the specimens. Strobe lighting provides a marvelous view of ciliary movements, and darkfield illumination is often spectacular!

The final identification of many rotifer species requires examination of the trophi which, in certain species, may be done by compressing an intact animal. However, Myers (1937) described a method to extract trophi from surrounding soft tissues using a small volume of bleach (sodium hypochlorite) in a depression slide (see also Edmondson, 1959, pp. 432–433, Pennak, 1989, pp. 195–196; Stemberger, 1979). A regular microscope slide may be used, if the cover glass is supported so the arrangements of the pieces of the trophi are not disturbed by compression. Because the trophi are liberated rather quickly when the bleach comes into contact with the rotifer, it is necessary to find the animal rapidly; otherwise it becomes necessary to scan the entire slide for the small trophi. Be aware that bubbles may form and obscure your view when bleach comes into contact with some biological materials.

Methods for preparing trophi for scanning electron microscope (SEM) work have been described by several workers including a recent methodology by Kleinow (1998) that gives stereoscopic pictures. In all work with trophi, it is important to remember that these structures are very small (<50 μm long) and that they are three dimensional objects with a particular spatial arrangement among all seven pieces comprising the trophi.

V. CLASSIFICATION AND SYSTEMATICS

A. Classification

Classification schemes differ slightly in the ways that they treat the three basic groups of rotifers. In this key, three classes (Seisonidea, Bdelloidea, Monogononta) are recognized. The first two are comprised of rotifers with two gonads and are sometimes considered orders within the class Digononta, leaving class Monogononta separate (e.g., Pennak, 1989 p. 196). Another method of classification emphasizes the unique anatomy and obligatory sexual reproduction of the Seisonidea, separating it from Bdelloidea and Monogononta, which are then placed in the class Eurotatoria (Koste, 1978, p. 48; Melone *et al.*, 1998b). European workers tend to rank rotifers as a class within the phylum Aschelminthes, while North American authors treat rotifers as a phylum (in which case the group known as aschelminths does not receive taxon status).

B. Systematics

1. Class Seisonidea

This monogeneric taxon (*Seison*) comprises only two recognized species of large (2–3 mm) dioecious, marine rotifers that are epizoic on the gills of a leptostracan crustacean, *Nebalia*. The corona is very reduced and is not used in food gathering or locomotion. Both species have paired gonads and a functional gut in both sexes. The trophi are fulcrate. Females have ovaries without vitellaria. Sexes are of similar size and morphology. These unusual marine rotifers are not included in the following key. For a detailed review of this class consult the work of Ricci *et al.*, (1993).

2. Class Bdelloidea

Comprising 18 genera and just over 360 species (Ricci 1987), bdelloids possess a rather uniform morphology and are exclusively females, reproducing by parthenogenesis. Bdelloids are characterized by having paired ovaries with vitellaria, more than two pedal glands, and ramate trophi. Most bdelloids are microphagous with a corona of either two trochal disks or a modified ciliated field. Bdelloids often have a vermiform body with a pseudosegmentation consisting of annuli, which permit shortening and lengthening of the body by telescoping. One order (Bdelloida) comprising four families (Adinetidae, Habrotrochidae, Philodinavidae, Philodinidae) is recognized. Bdelloids generally are not caught in plankton tows, although they may be found in waters with dense vegetation. Bdelloids often occur in sediments or among plant debris or crawling on the surfaces of aquatic plants. Some forms inhabit the capillary water films formed in soils (Pourriot 1979) or covering mosses. Many species are capable of becoming desiccated and then rehydrated (see Section II.C.4). Unfortunately, there has been no systematic review of North American bdelloids. The works of Bartos (1951), Burger (1948), Donner (1965), Koste and Shiel (1986), and Pourriot (1979) should be consulted for more detail on bdelloids.

3. Class Monogononta

Class Monogononta comprises the largest group of rotifers with more than 1600 species in about 95 genera of benthic, free swimming, and sessile forms. All are assumed to be dioecious with one gonad. In many species, males have never been observed. Females possess one ovary with a vitellarium. Males usually are structurally reduced with a vestigal gut that functions only in energy storage. Males generally are ephemeral, usually being present in the plankton for only a few days or weeks each year. Monogononts are microphagous or raptorial, but a few are parasitic. The corona is modified as broad-to-narrow disks or earlike lobes or is vase-shaped with reduced ciliation and long setae used in prey capture. In this key, three orders are recognized (Collothecacea, Flosculariacea, Ploimida), collectively comprising 24 families. Other workers

TABLE III List of the Rotifer Families Recognized in the Key Presented Here

Order Bdelloidea
- Adinetidae (4')
- Habrotrochidae (2)
- Philodinavidae (4)
- Philodinidae (3)

Order Collothecacea
- Collothecidae (5)

Order Flosculariacea
- Conochilidae (8)
- Filiniidae (11)
- Flosculariidae (9)
- Hexarthridae (7)
- Testudinellidae (11')
- Trochosphaeridae (10)

Order Ploimida
- Asplanchnidae (15)
- Birgeidae (14')
- Brachionidae (23')
- Colurellidae (19)
- Dicranophoridae (12) [De Smet, 1997]
- Epiphanidae (22)
- Euchlanidae (21)
- Gastropidae (27')
- Lecanidae (21') [Segers, 1995a]
- Lindiidae (14)
- Microcodonidae (26)
- Mytilinidae (20)
- Notommatidae (26') [Nogrady and Pourriot, 1995] includes Ituridae [Pourriot, 1997] and Scaridiidae [Segers, 1995b]
- Proalidae (22') [De Smet, 1996]
- Synchaetidae (27)
- Trichocercidae (24)
- Trichotriidae (23)

References in square brackets indicate recent publications done as a part of an ongoing series of guides that are updating rotifer taxonomy.

include the first two orders as suborders within order Gnesiotrocha (de Beauchamp, 1965; Koste, 1978; Koste and Shiel, 1987). Table III lists the families recognized in the subsequent key and their relevant couplet number.

C. Taxonomic Keys

There are a variety of keys to rotifers including those by Edmondson (1959) and Pennak (1989), covering the North American fauna, that may be followed to the level of genus. However, the specialized keys of Koste and Shiel (1987) for Australian waters, Pontin (1978) for the British Isles, Stemberger (1979) for the Great lakes, Ruttner-Kolisko (1974) for planktonic rotifers in general, and Koste's (1978) revision of Voigt for the rotifers of central Europe are also important. This second group of works deals nearly exclusively with monogonont rotifers. While these keys may not be comprehensive enough to cover all of the variations in size and morphology that are sometimes found within a species, they will probably prove to be more than adequate for most work where detailed descriptions are needed. A comprehensive work of providing a worldwide guide to the rotifers is currently underway under the auspices of H.J.F. Dumont as coordinating editor. Volume 1 in the series (Nogrady *et al.*, 1993) provides a general key to the level of family. To date four subsequent volumes have been published by SPB Academic Publishers. These volumes collectively cover the following taxa: Volume 2, Lecanidae; Volume 3, Notommatidae and Scaridiidae; Volume 4, Proalidae; and Volume 5. Dicranophoridae. A very simple key is available from Ward's Natural Science.

One of the fundamental differences among the higher taxonomic levels in phylum Rotifera is the structure of the trophi, with characteristic types of jaws being recognized in each family (see Section II.B.2). [Transitional forms are also known (Koste, 1978; Koste and Shiel, 1987).] Although the following key is designed with the nonspecialist in mind, commensurate with the central importance of the trophi, we have used both the structure of the trophi and other obvious characters of anatomy and morphology as priniciple points of separation. In keeping with the nature of this text, the key is taken only to the level of family following, for the monogononts, the recent taxonomy of Koste (1978), Koste and Shiel (1987), and Nogrady *et al.* (1993).

D. Taxonomic Key to Families of Freshwater Rotifera

1a. Rotifers with paired ovaries (Digononta) and ramate trophi (Fig. 9) .... order Bdelloidea 2

1b. Rotifers with a single ovary and trophi other than ramate .... class Monogononta 5
[Although this step obviously is very important, special care is generally not necessary to resolve this couplet. The ramate trophi of bdelloids are usually identifable in whole animals without resorting to their isolation (see Melone *et al.*, 1998b and Section IV.C). If the type of trophi and number of ovaries cannot be determined, there are other clues that may be useful for live organisms. Upon contacting a substrate, many bdelloids will crawl on the surface in a manner reminiscent of a leech, hence the etymon of the name (Greek, *bdella*, leech). Further, the corona of some bdelloids has the appearance of two separate wheels, while only a few sessile monogononts of the family Flosculariidae give this impression.]

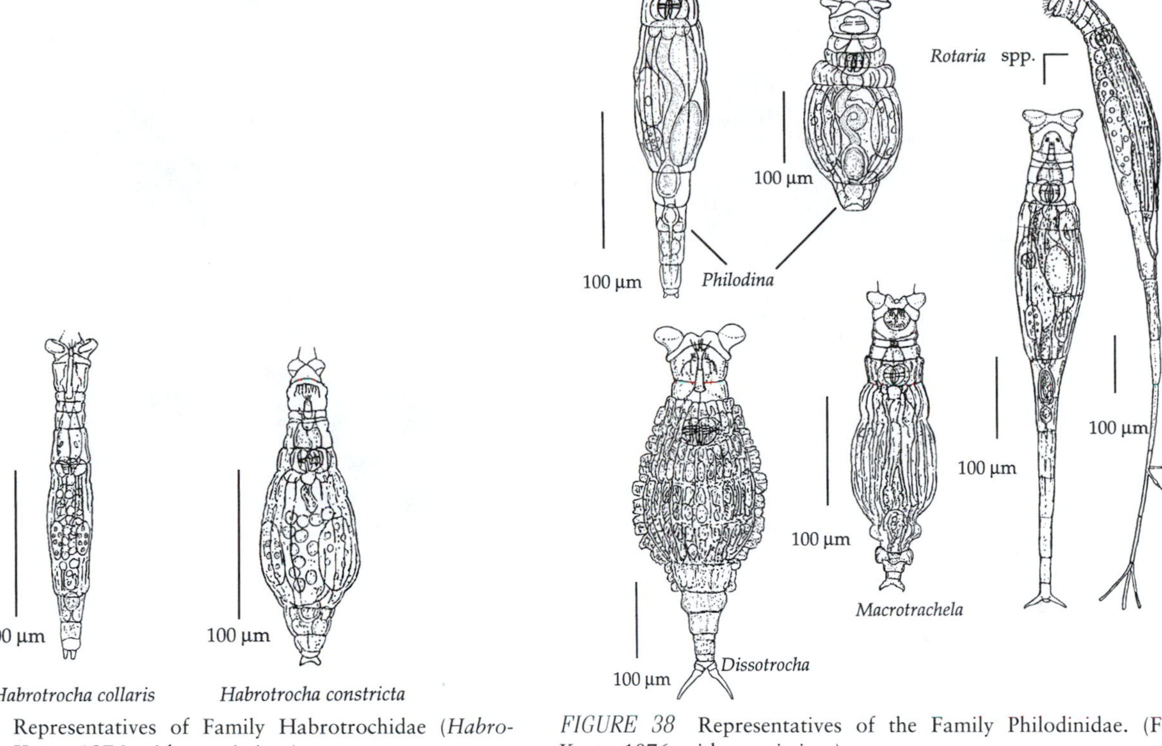

FIGURE 37 Representatives of Family Habrotrochidae (*Habrotrocha*). (From Koste, 1976, with permission.)

FIGURE 38 Representatives of the Family Philodinidae. (From Koste, 1976, with permission.)

2a(1a) Stomach without lumen, as a syncitial mass of food vacuoles that gives the gut a frothy appearance .... family Habrotrochidae
[Three genera (e.g., *Habrotrocha*) reported in North America; mainly on moss or benthic, includes about 120 species (e.g., Fig. 37.]

2b Stomach with lumen 3

3a2b. Corona with the appearance of two separate wheels (trochus) extended on pedicels .... family Philodinidae
[About 10 genera and >150 species, of which *Philodina* and *Rotaria* are common; also *Dissotrocha* and *Macrotrachella* (Fig. 38).]

3b Corona not as above 4

4a(3b). Corona ciliated lobes or regions near mouth; rostrum ciliated .... family Philodinavidae
[One rare species reported from North America (*Abrochta*, Fig. 39).]

4b. Corona as flat ciliated fields on ventral side (no cingulum) .... family Adinetidae
[Two genera (*Adineta*, *Bradyscela*) of about 15 species (Fig. 39).]

5a(1b). Trophi uncinate (Fig. 8H); corona lacking typical ciliated regions, but is open as an infundibulum with or without elongate setae around the margin .... order Collothecacea comprising 2 families: Atrochidae and Collothecidae
[Collothecidae (*Collotheca* and *Stephanocerous*) possess elongate setae (Figs. 6 and 40); fewer than 50 species; mostly sessile, ca. five species are free swimming, many produce clear gelatinous tubes; none are colonial, but may colonize substrata forming dense groupings (cf. Flosculariidae). Atrochidae (*Acyclus*, *Atrochus*, *Cupelopagis*) lack setae; each genus monospecific (Fig. 41).]

5b. Trophi other than uncinate; corona with typical ciliated lobes or fields, or ciliated bands 6

6a(5b). Rotifers possessing malleoramate trophi (Fig.8I), order Flosculariacea 7
[Malleoramate trophi possess unci with numerous fine teeth nearly completely overlying the rami (unlike the malleate type; see the description in couplet 16); teeth close to the fulcrum are usually larger than those more distant. In live animals, the nearly constant movement of the trophi in a grinding or poundinglike action is characteristic.]

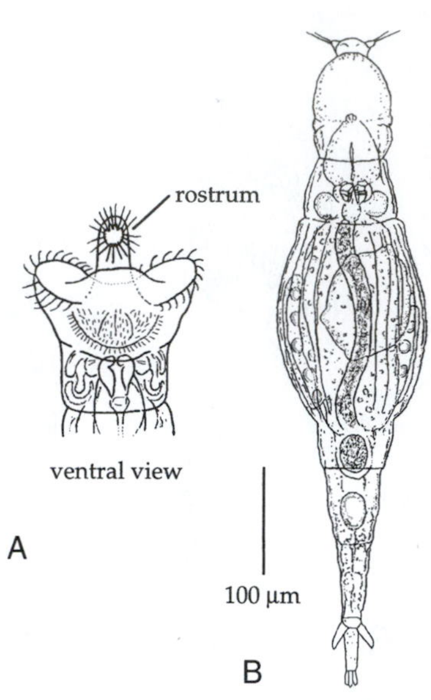

FIGURE 39 Representatives of two bdelloid families. (A) Family Philodinavidae, *Abrochta*, ventral view of corona (adapted from several sources); and (B) family Adinetidae, *Adineta*, dorsal view of adult (from Koste, 1976, with permission).

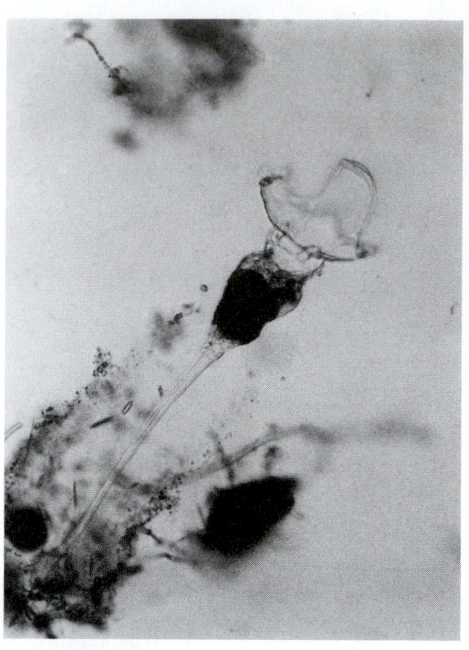

FIGURE 40 Representative of Family Collothecidae (*Collotheca* sp. within its gelatinous tube which is embedded with detritus and diatoms).

6b Rotifers with trophi other than malleoramate ....order Ploimida 12

7a(6a). Conical body with six hollow, armlike, setose appendages that are outgrowths of the integument; appendages inserted with powerful muscles ....family Hexarthridae
[Monogeneric family, genus *Hexarthra (Pedalia)* with about eight species (Fig. 13), some of which inhabit salt or brackish waters. Although illoricate, *Hexarthra* spp. preserve well in formalin without serious contraction.]

7b. Body lacking hollow, armlike, setose appendages 8

8a(7b). Rotifers with a corona of horseshoe or U-shaped ciliated bands; possessing a ventral gap in the coronal ciliation; free-swimming; solitary or small to very large colonies (>150 individuals) within a gelatinous matrix ....family Conochilidae

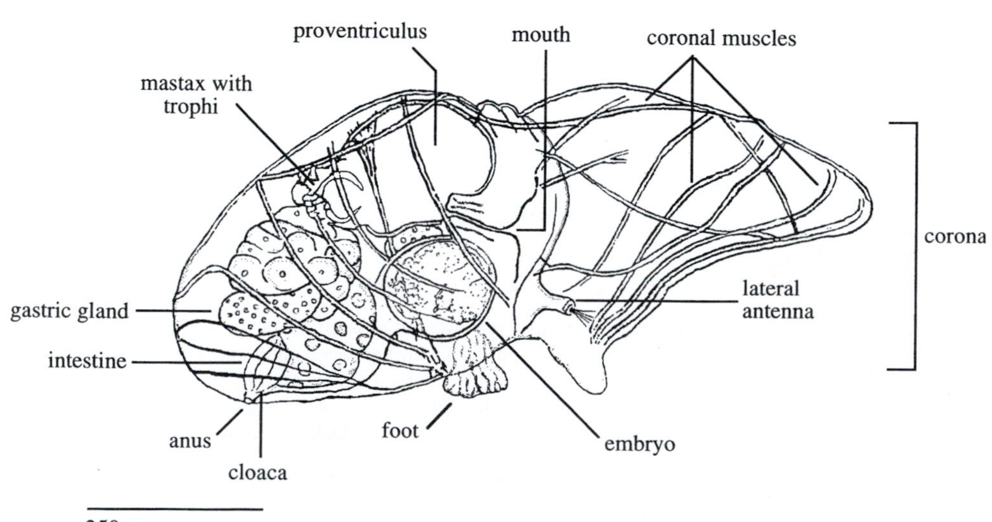

FIGURE 41 Representatives of family Atrochidae (*Cupelopagis vorax*). (Adapted from several sources.)

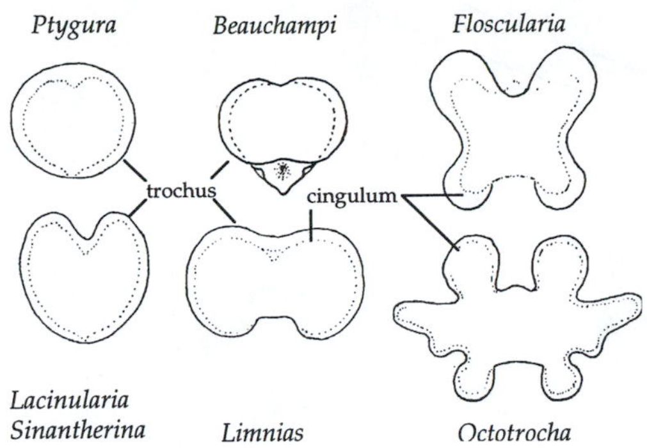

FIGURE 42 Coronae of the family Flosculariidae. (Modified from several sources.)

[Two genera recognized, based on the position of antennae (*Conochilus* and *Conochiloides*), but Koste (1978) and Ruttner-Kolisko (1974) do not consider antennal position to be a significant feature and subsume *Conochiloides* within *Conochilus* (Fig. 17).]

8a. Rotifers with corona other than a horseshoe or U-shaped ciliated band; solitary or colonial; with or without a gelatinous matrix 9

9a(8b). Rotifers typically with elongate bodies and large, circular to lobate, earlike corona (Fig. 42); solitary or colonial; with or without a tube or gelatinous matrix; mostly sessile, but some free swimming .... family Flosculariidae
[Seven genera of mainly sessile species (Figs. 5, 17C, and 43), a few are free-swimming; includes the genera *Floscularia*, *Lacinularia*, *Ptygura*, and *Sinantherina*.]

9b. Rotifers lacking a large, circular to lobate, earlike corona 10

10a(9b). Spherical rotifers with a corona as a circular band around the equator or toward one end .... family Trochosphaeridae
[Genus *Trochosphaera* (Fig. 44) with two species in eutrophic waters, rare, but when present may be very abundant. A second rare genus (*Horaëlla*) apparently not reported in the United States.]

10b. Rotifers not spherical in shape 11

11a(10b). Rotifers with two movable anterior spines (bristles) below the corona, of varying lengths often much longer than the body and one (rarely two) rigid caudal spines; foot absent .... family Filiniidae
[Genus *Filinia* (Fig. 32) with two North American species and apparently many hybrids. The long spines point posteriorly when the animal is swimming, but the anterior ones point forward when the corona is withdrawn.]

11b. Rotifers without long spines as described above .... family Testudinellidae
[Two dissimilar genera. (1) Genus *Testudinella*, lorica greatly flattened dorsoventrally, with dorsal and ventral plates fused along the lateral margin; foot annulated and retractile ending in a ciliated cup, which is difficult to see; approximately 12 littoral species up to 250 μm long (Fig. 45). (2) Genus *Pompholyx*, body having 4 nearly equal lobes in cross section; with a pair of frontal eyespots, (Fig. 45) common in lakes and ponds.]

12a(6b). Rotifers with forcipate trophi (Fig. 8J) .... family Dicranophoridae
[About 12 genera of creeping, littoral rotifers (Fig. 46) with symmetrical or asymmetrical forcipate trophi, a feature easily determined by gentle compression of the specimen; no planktonic or semiplanktonic species are present in this family.]

12b. Rotifers with trophi other than forcipate 13

13a(12b). Mostly littoral rotifers possessing cardate trophi (Fig. 10) having a sucking action or trophi highly modified and stomach with zoochlorellae 14

13b. Rotifers with trophi other than cardate 15

14a(13a). Manubria of the trophi with hooks that may be determined in lateral view of the animal (Fig. 10); body spindle-shaped .... family Lindiidae
[Monogenetic family, Genus *Lindia* (Fig. 47); mostly littoral.]

14b. Highly modified trophi, possessing pseudunci; stomach with zoochlorellae, gastric glands absent (Fig. 48) .... family Birgeidae
[Monospecific family (*Birgea enantia*, a rare littoral species).]

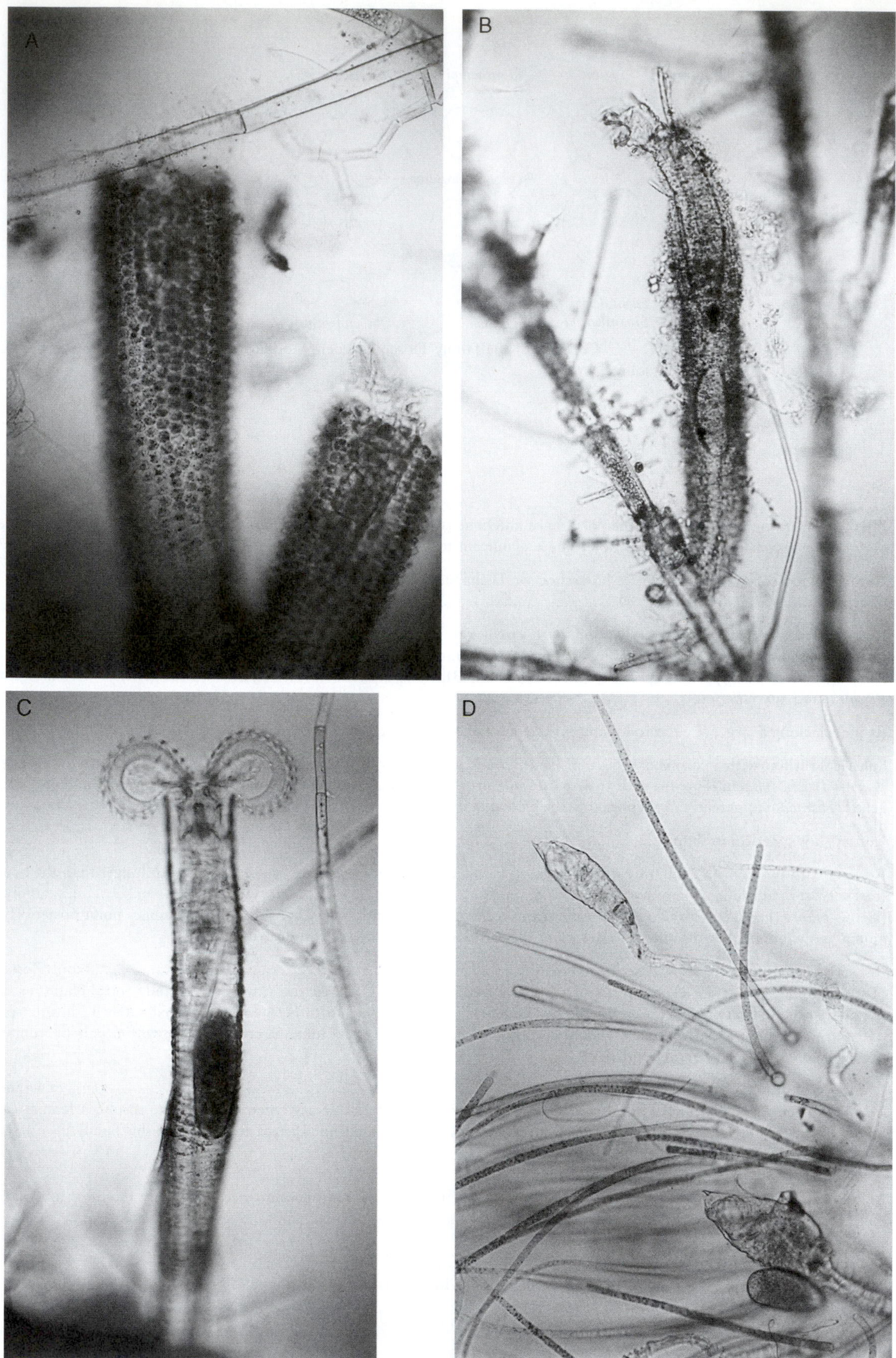

FIGURE 43 Representatives of family Flosculariidae. (A) *Floscularia conifera*, portion of a sessile colony, with two adults inside a tube constructed of pellets made of detritus (tube length may reach 2000 μm); (B) *Ptygura barbata* attached to a filamentous alga (total animal length ≈ 300 μm); (C) *Limnias melicerta*, sessile in its clear cement tube (individuals ≈ 900 μm); and (D) *Ptygura mucicola*, sessile within the colonial cyanobacterium, *Gloeotrichia* (body length ≈ 100 μm).

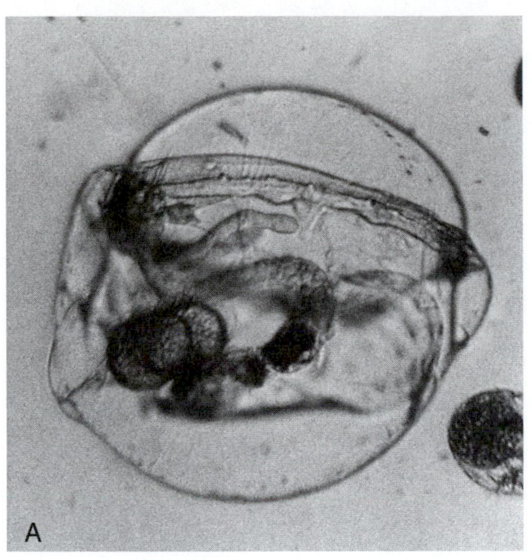

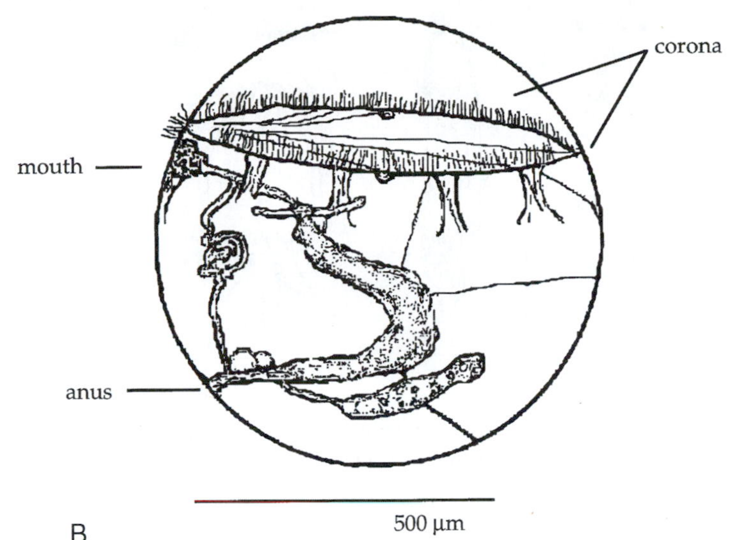

FIGURE 44 Representative of family Trochosphaeridae. (A) Photomicrograph of *Trochosphaera* (specimen slightly distorted); and (B) line drawing interpretation. (Diameter of the individuals ≈ 800 μm; original photomicrograph by Tom Nogrady.)

15a(13b). Saccate rotifers, with incudate or modified incudate trophi, (Fig. 8G,R) some lacking intestine and anus..........family Asplanchnidae
[Three genera. Two lacking an intestine: *Asplanchna*, foot absent, with six species (Figs. 11 and 49) and *Asplanchnopus*, foot present, with three species. One rare, benthic genus with intestine and foot, *Harringia*, (Fig. 50) with two species.]

15b. Rotifers not possessing incudate trophi 16

16a(15b). Rotifers possessing malleate trophi 17
[In malleate trophi (Fig. 8A,B,C,E,F,P), each uncus has only 4–7 teeth (unlike the condition found in malleoramate trophi; see description couplet 6). The fulcrum may be short (malleate) or long (submalleate).]

16b. Rotifers possessing virgate trophi 24
[Virgate trophi are often asymmetrical (Fig. 8D,K,L,M,N,O,Q,S), with a long fulcrum and manubria and generally small rami.]

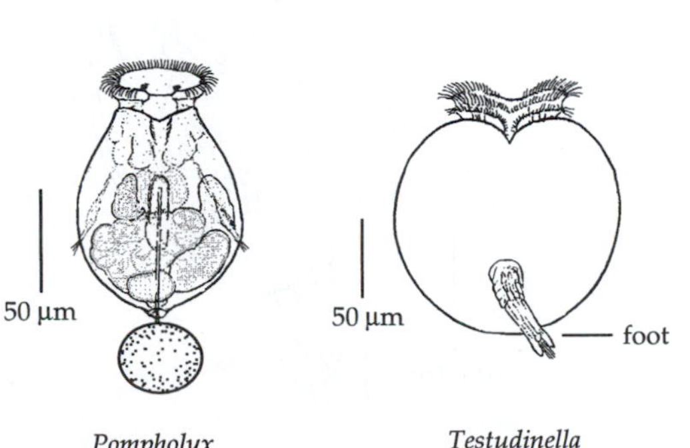

FIGURE 45 Representatives of family Testudinellidae. (Ventral views of *Testudinella* and *Pompholyx*.) (Modified from several sources.)

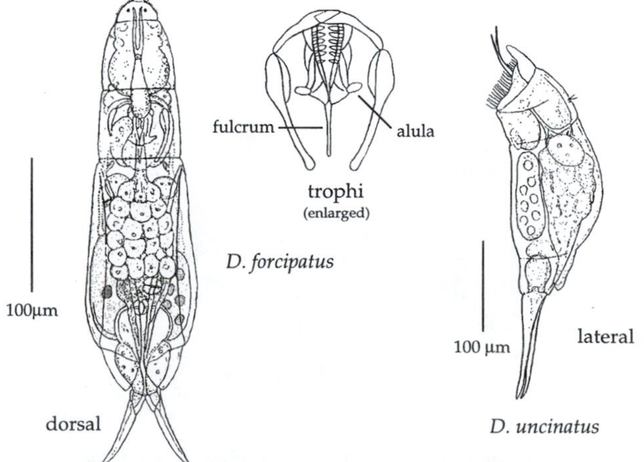

FIGURE 46 Representative of family Dicranophoridae (*Dicranophorus*). (From Koste, 1976, with permission.)

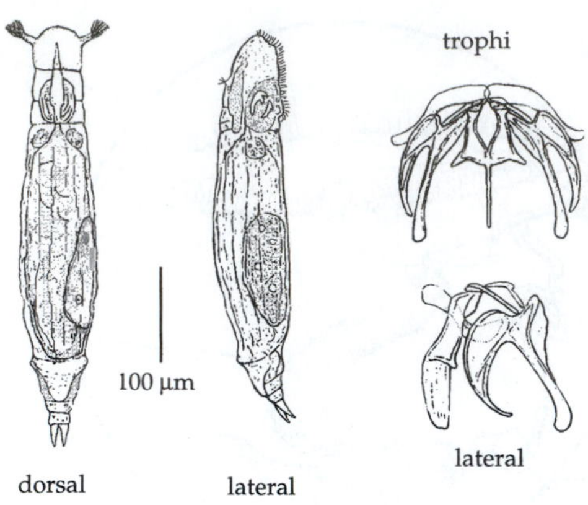

FIGURE 47 Representative of family Lindiidae (*Lindia*). (From Harring and Myers, 1922, with permission.)

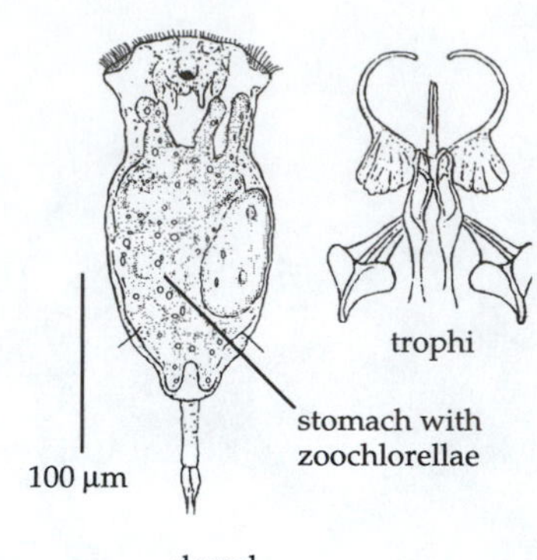

FIGURE 48 Representative of family Birgeidae (*Birgea*). (From Harring and Myers, 1922, with permission.)

17a(16a). Loricate rotifers: body wall thickened and firm 18

17b. Illoricate rotifers: body wall not thickened or firm 22
[Interpreting whether the body wall is thickened (loricate) or not (illoricate) can be difficult and it does take some experience to judge this characteristic. To get an indication as to how firm the lorica is, follow this procedure. When working with fresh material gently, apply pressure from the point of a pencil or probe onto the cover glass while observing the specimen. If the body puffs out under pressure and returns to its original shape when the pressure is released, the specimen is illoricate. Loricate forms will exhibit much less flexibility of the body wall. In preserved materials, illoricate forms tend to shrivel up, while the body wall of loricate rotifers will retain its shape even if the inner organs separate from the body wall, collapsing into a central mass of tissue.]

18a(17a). Lorica possessing furrows, grooves, or sulci; or with a dorsal, semicircular head shield covering the corona; or with a very strongly developed lorica and a dorsal transverse ridge 19

18b. Lorica lacking furrows, grooves, sulci or dorsal head shield; without a strongly developed lorica and dorsal transverse ridge 23

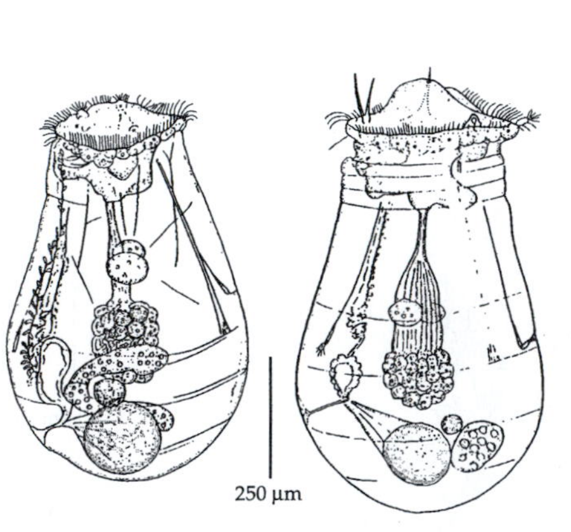

FIGURE 49 Representative of family Asplanchnidae (*Asplanchna*). [From Koste, 1976, with permission.]

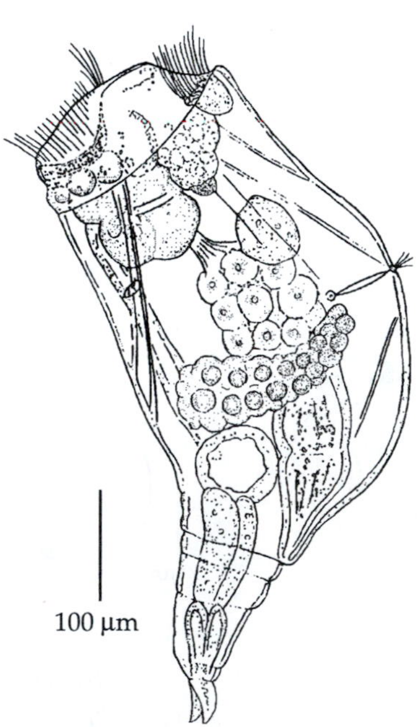

FIGURE 50 Representative of family Asplanchnidae (*Harringia*). [From Koste, 1976, with permission.]

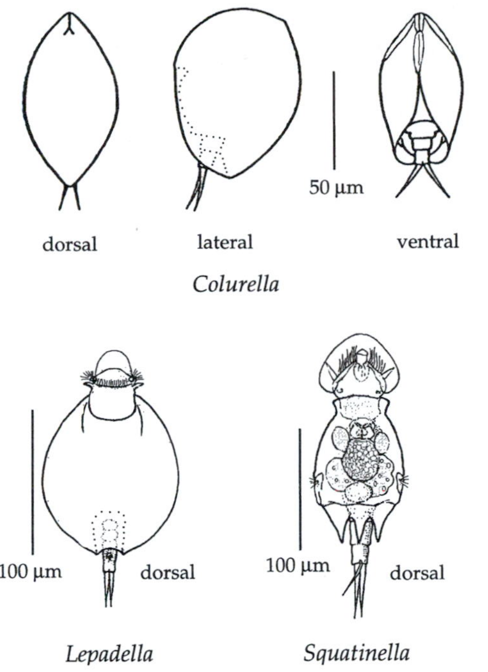

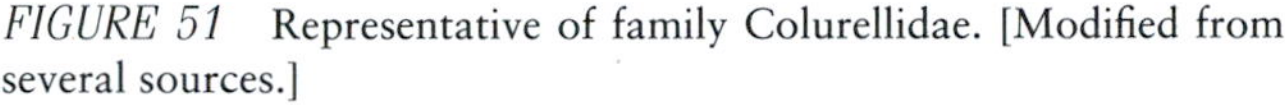

FIGURE 51 Representative of family Colurellidae. [Modified from several sources.]

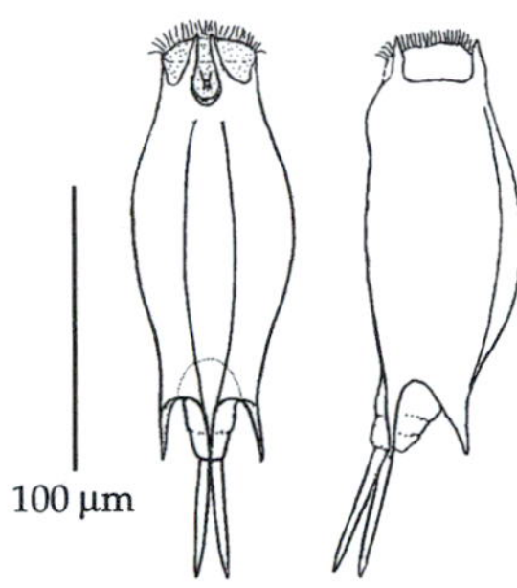

FIGURE 52 Representative of family Mytilinidae (*Mytilina*). [Modified from several sources.]

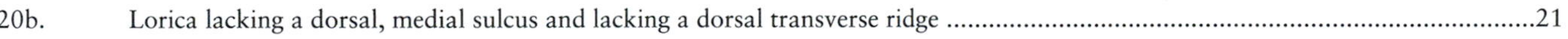

19a(18a). Lorica with medial, ventral furrow (but no lateral furrow or grooves) extending the full length of the animal or with a ventral notch in which the foot lies, or with a dorsal, semicircular head shield covering the corona ....family Colurellidae
[Four genera found in the littoral zone including *Colurella* (possessing a medial ventral furrow, Fig. 51A), a very common genus of some 15 species, which browse on epiphytic microorganisms; *Lepadella* (possessing a ventral notch in which the foot lies, Fig. 51B), very common with many species; *Squatinella* (possessing a dorsal semicircular head shield covering the corona, Fig. 51C). [N.B.: *Diplois daviesiae* (family Euchlanidae), a rare monospecific genus present in *Sphagnum* bogs will key to family Colurellidae, if the lateral furrows of this species are missed.]

19b. Lorica lacking medial, ventral furrow or notch, and lacking a head shield 20

20a(19b). Lorica with dorsal, medial sulcus (double keel) or with a strongly developed lorica usually possessing a dorsal transverse ridge .... family Mytilinidae
[Two littoral genera: *Mytilina* (Fig. 52) possessing a laterally flattened lorica with a dorsal longitidutinal sulcus, spines on all four anterior corners, about 10 species common, and *Lophocharis* with a strongly developed lorica, usually having a dorsal transverse ridge.]

20b. Lorica lacking a dorsal, medial sulcus and lacking a dorsal transverse ridge 21

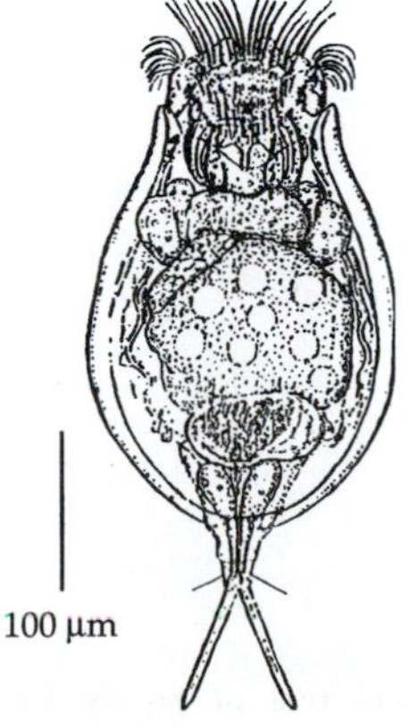

◀ *FIGURE 53* Representative of family Euchlanidae (*Euchlanis*). [Modified from several sources.]

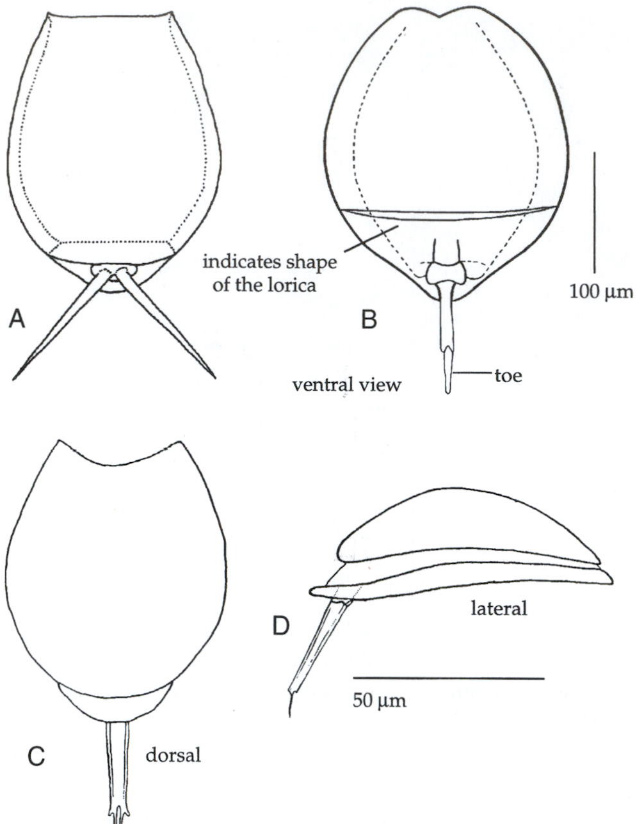

FIGURE 54 Representatives of the family Lecanidae. (A & B) *Lecane* spp.; (C) *Lecane lunaris* (dorsal); and (D) lateral view. (A and B from several sources; C and D from Dartnall and Hollowday, 1985, with permission.)

21a(20b) Foot projecting from between dorsal and ventral plates at the posterior end of the lorica; dorsal and ventral plates separated by a deep furrow or groove .... family Euchlanidae
[Four genera common in the littoral rotifers, including *Euchlanis* (Fig. 53).]

21b. Foot projecting through a hole in the ventral plate at the posterior end of the lorica; dorsal and ventral plates connected by a weak furrow or groove .... family Lecanidae

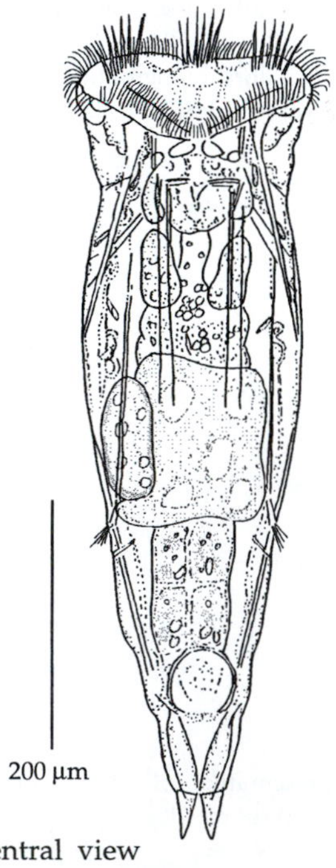

FIGURE 55 Representative of family Epiphanidae (*Epiphanes*). (From Dartnall and Hollowday, 1985, with permission.)

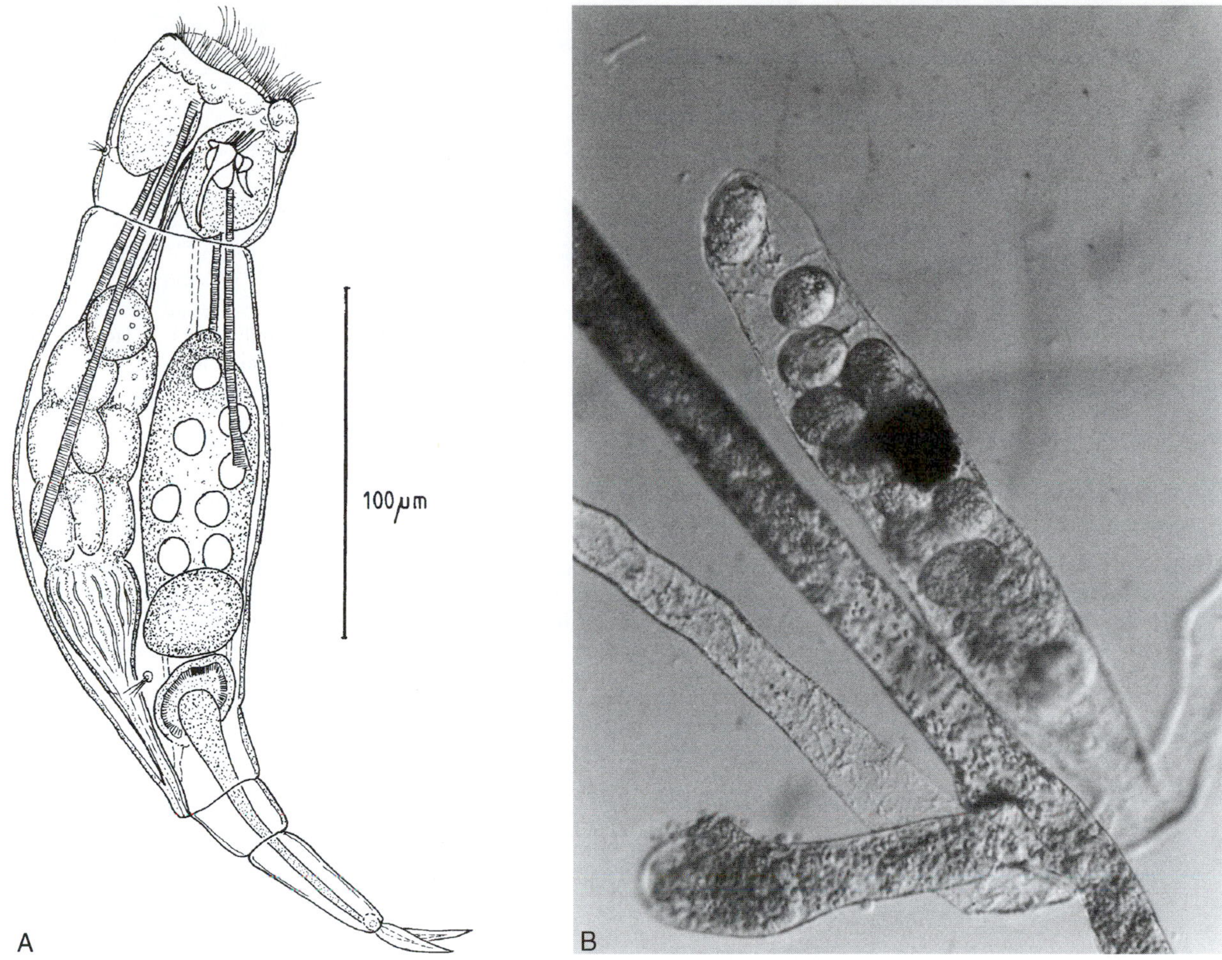

FIGURE 56 Representatives of family Proalidae. (A) *Proales* (from Chengalath, 1985, with permission); and (B) photomicrograph of adult of *Proales werneckii* (dark body) with numerous eggs within a gall of *Vaucheria* sp. [Gall ≈ 600 μm long.]

[A large number of littoral species in two genera (Fig. 54): *Lecane* (with two toes which may be partially fused at the base) and *Monostyla* (one toe or toe divided distally); Koste (1978) has subsumed *Monostyla* within *Lecane*.]

22a(17b). Mouth set in a funnel-shaped buccal field .... family Epiphanidae
[Six genera including *Epiphanes* (Fig. 55), *Cyrtonia*, *Mikrocodides*, and *Rhinoglena* all littoral species or in small, shallow lakes. *Epiphanes*, sometimes incorrectly described as a 'typical rotifer' in textbooks, is not common, but may be found in ponds receiving animal wastes.]

22b. Mouth set in an oblique, ciliated field on ventral side; body wormlike to fusiform .... family Proalidae
[Four genera, including the large genus *Proales*; generally found in littoral zones and sandy beaches (Fig. 56A). *Proales daphnicola* is planktonic and may be attached to cladocerans; *Proales werneckii* is parasitic in galls that it induces in filaments of aquatic *Vaucheria* spp (Fig. 56B).]

23a(18b). Lorica extending beyond body to head, foot, and toes .... family Trichotriidae
[Three genera, including *Macrochaetus* (Fig. 57A), with numerous bilaterally placed spines (about seven littoral species) and *Trichotria* (Fig. 57B), with a pair of heavy spines on the dorsal side of the foot (about 10 littoral species).]

23b. Lorica not extending beyond body .... Family Brachionidae
[A large family comprising 6 genera of common rotifers. *Brachionus*, a very common genus with about 25 species of littoral and planktonic rotifers (Fig. 58). *B. plicatilis* is frequently found in salt and brackish waters. *Kellicottia*, with two common planktonic species, possessing long anterior spines of unequal length (Fig. 59). *Keratella*, with more than 15 species, having the dorsal

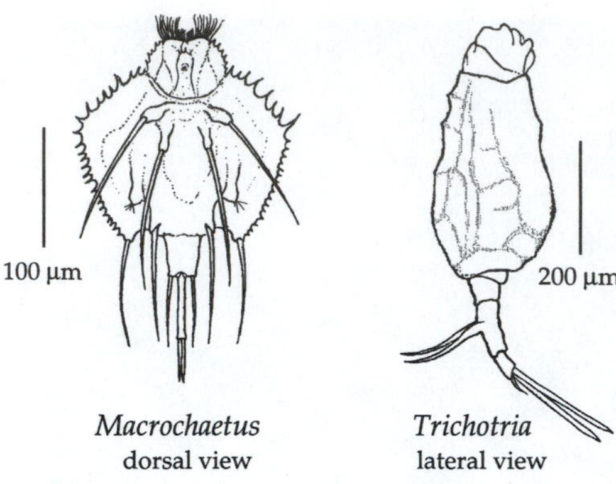

FIGURE 57 Representatives of family Trichotriidae (*Macrochaetus* and *Trichotria*). [Modified from several sources.]

plate decorated with a characteristic facet pattern (Fig. 60). *K. cochlearis* is very common. *Notholca* a genus common in cool waters (Fig. 61).]

24a(16b). Body twisted as a partial helix (asymmetrical) and/or trophi asymmetrical; or small, saccate animals, predatory within the colonial alga, *Volvox* ....family Trichocercidae
[Three genera: (1) genus *Trichocera* (body twisted as a partical helix; unequal toes; asymmetrical trophi) with some 90 species, may be important in the plankton (Fig. 62); (2) genus *Elosa* (body showing in cross section three lobes, two lateral and one dorsal, plus ventrally an inconspicous lobe; asymmetrical trophi) with two species common in the psammon and with *Sphagnum* moss; (3) *Ascomorphella volvocicola*, predatory within *Volvox* (Fig. 63).]

24b. Body and trophi not as described above 25

25a(24b). Free-swimming rotifers 27

25b. Crawling or creeping rotifers in the littoral (ocassional species in the plankton) 26

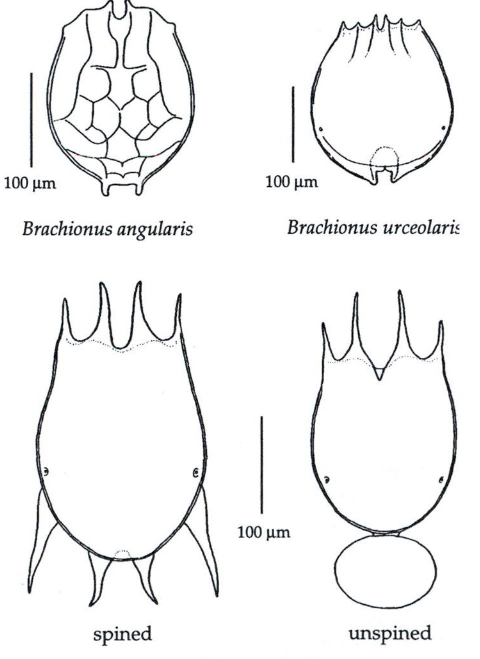

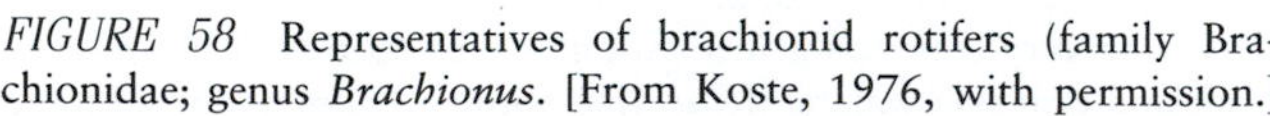

FIGURE 58 Representatives of brachionid rotifers (family Brachionidae; genus *Brachionus*. [From Koste, 1976, with permission.]

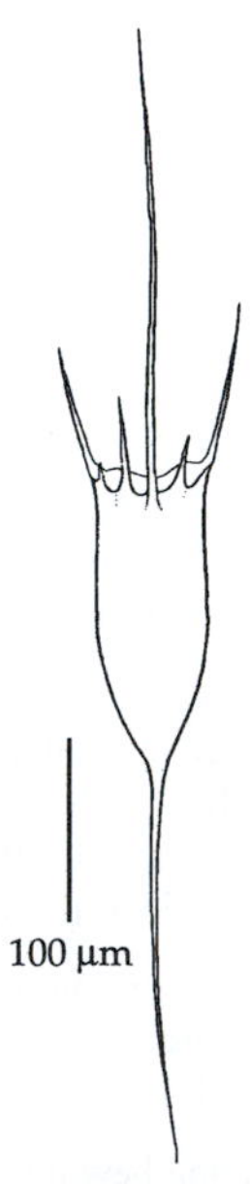

FIGURE 59 *Kellicottia* (dorsal view) a common genus of the family Brachionidae. (From several sources.)

A

lateral

dorsal

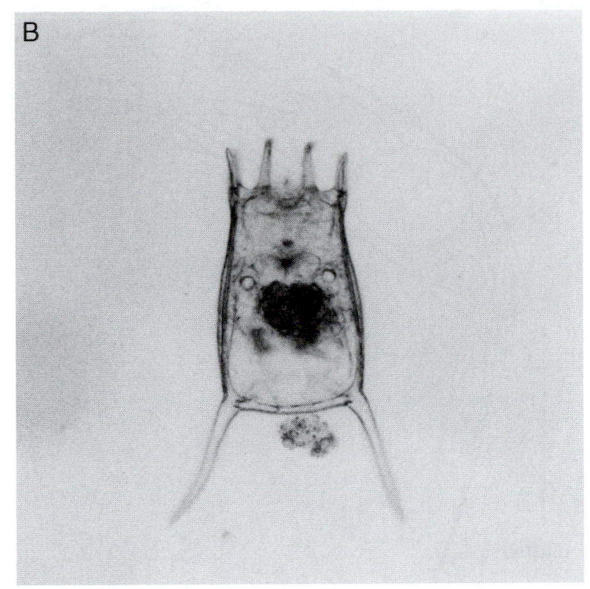

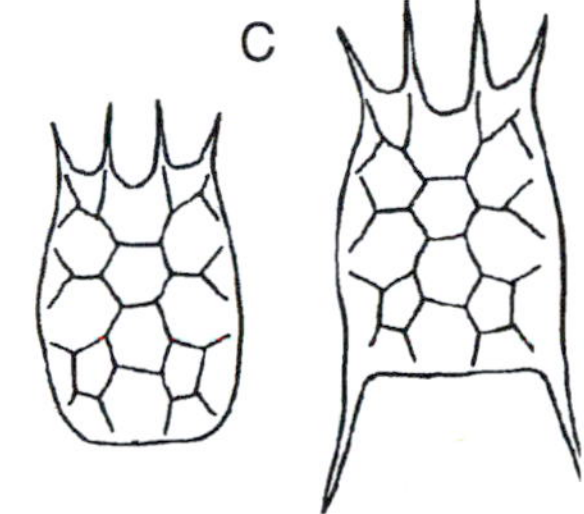

FIGURE 60 Variation in the genus *Keratella*, a common genus of the family Brachionidae. (Individuals ≈ 100–200 μm in length.) (A) *K. cochlearis* (from Koste, 1976, with permission); (B) *K. quadrata*; and (C) *K. testudo* (from Stemberger, 1988, with permission).

26a(25b). Purple plates positioned anterior to the mastax; corona wide, flat, and somewhat circular; foot jointed and long, about 50% of the total length of the animal, with a single toe ....family Microcodonidae
[Monospecific family of one uncommon species, *Microcodon clavus*, mostly littoral, but may be found in the plankton.]

26b. Lacking purple plates as described earlier, corona ventral and not circular ....family Notommatidae
[A varied family with many species in 13 genera including *Cephalodella* (Fig. 64), *Notommata* (Fig. 65), and *Pleurotrocha*. Also included at this point in the key are the families Ituridae and Scaridiidae.]

27a(25a). Corona with four prominant setae (sensory bristles) and auricles (earlike structures); or possessing 12 movable, flattened, sword-to-paddle- or feather-shaped appendages (paddles); or with a sculptured lorica having ridges, grooves, or areolations .... family Synchaetidae
[Four genera: with sensory bristles and auricles, *Synchaeta* (Fig. 66), comprising some 12 freshwater species plus others of marine and brackish waters, may be important in the plankton; with paddles *Polyarthra* (Fig. 33), fewer than 10 species separated based on paddle morphology, nuclei of vitellarium, and lateral antennae; *Pseudoploesoma* (Fig. 67A) and *Ploesoma* (Fig. 67B).]

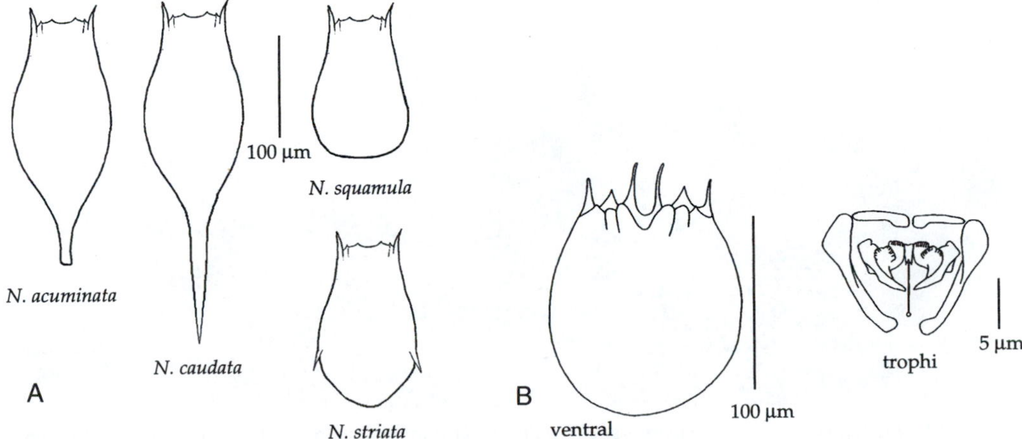

FIGURE 61 *Notholca*, a common genus of the family Brachionidae, usually found in cool waters. (A) Variation within the genus (modified from several sources); and (B) *N. squamula*. (Modified from May, 1980, with permission).

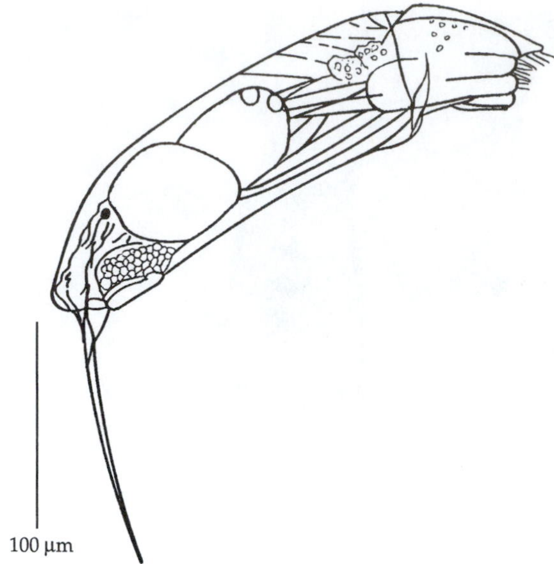

FIGURE 62 *Trichocera* (lateral view) an important genus of the family Trichocercidae. Lateral view. [From Wallace *et al.*, 1989, with permission.]

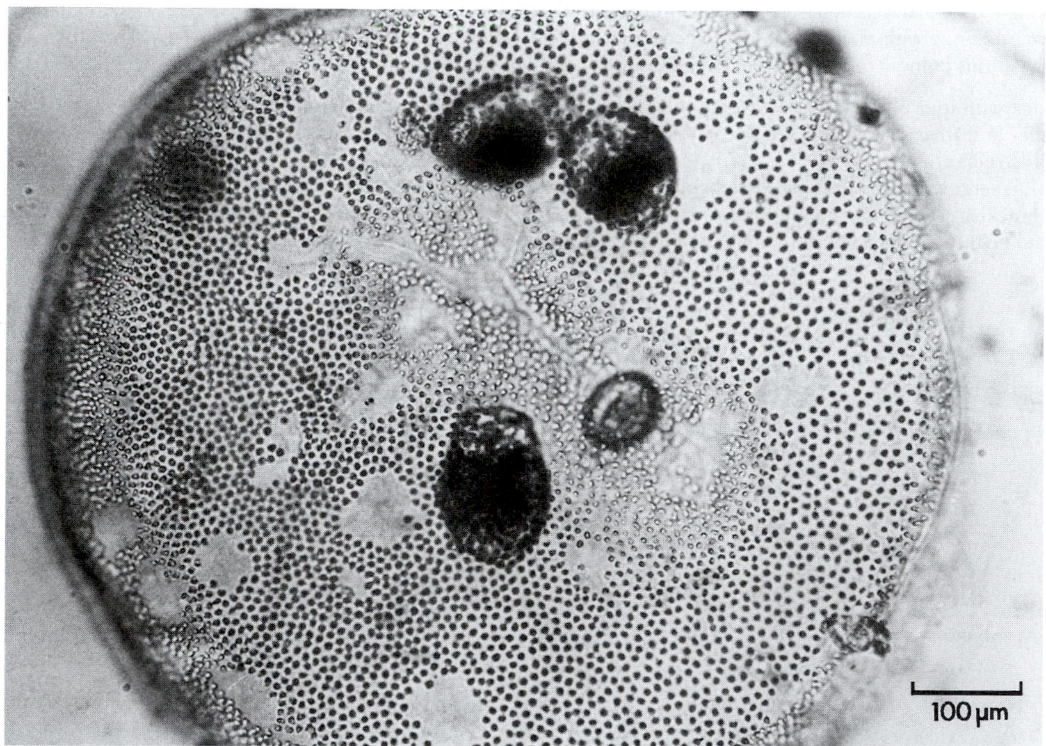

FIGURE 63 Predatory activity of *Ascomorphella volvocicola* (family Trichocercidae) on its prey, the colonial alga *Volvox*. Seen here are at least three rotifers feeding on the cells of a colony. Gaps in the cells of the alga represent the damage done by the rotifer; an egg is seen just to the right of center. (Photomicrograph by courtesy of Russ Shiel.)

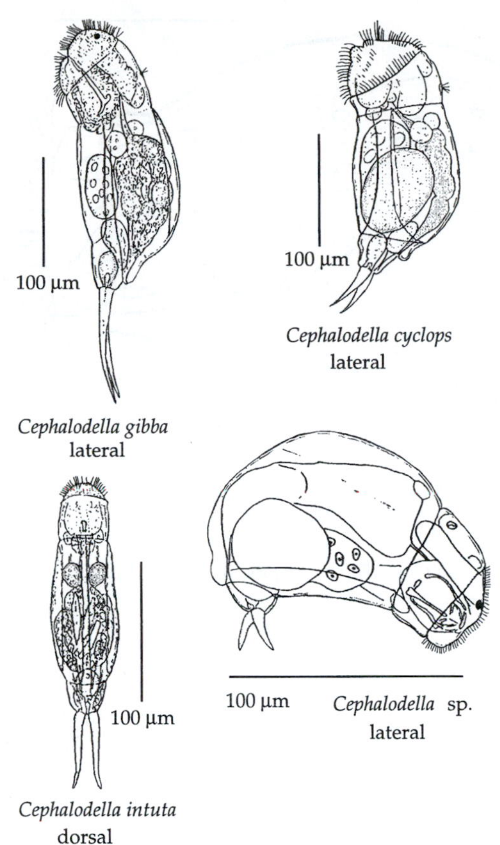

FIGURE 64 Representatives of family Notommatidae (*Cephalodella*). (*Cephalodella* sp. from Dartnall and Hollowday, 1985; others from Koste, 1976, with permission.)

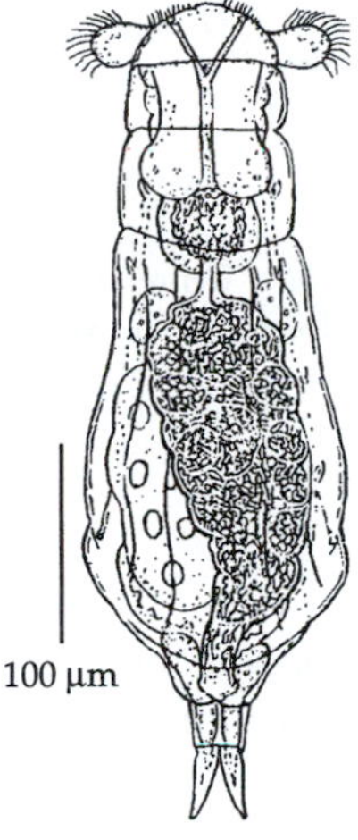

FIGURE 65 Representative of family Notommatidae (*Notommata*, dorsal view). (From Koste, 1976, with permission.)

27b. Lacking prominant setae, auricles, and paddles, and lorica not sculptured ....family Gastropodidae
[Two genera: *Ascomorpha* (Fig. 8.68A) (incorporating *Chromogaster*) comprising 6 species, possessing a fingerlike projection (palp) from the corona; *Gastropus* (Fig. 8.68B) with three species having a laterally compressed body, may be important in the plankton.]

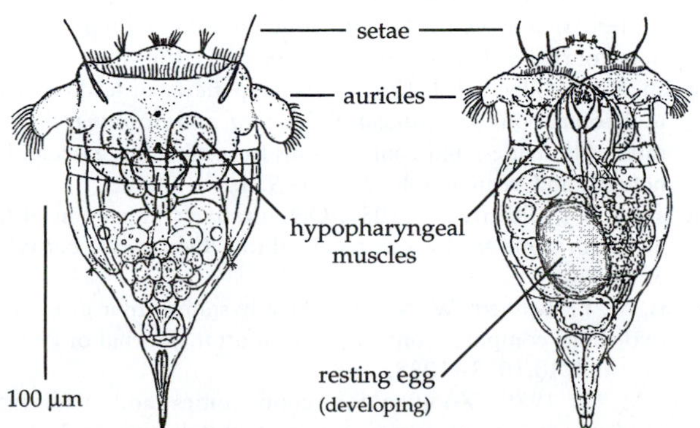

FIGURE 66 Representatives of family Synchaetidae (*Synchaeta*). [From Koste, 1976, with permission.]

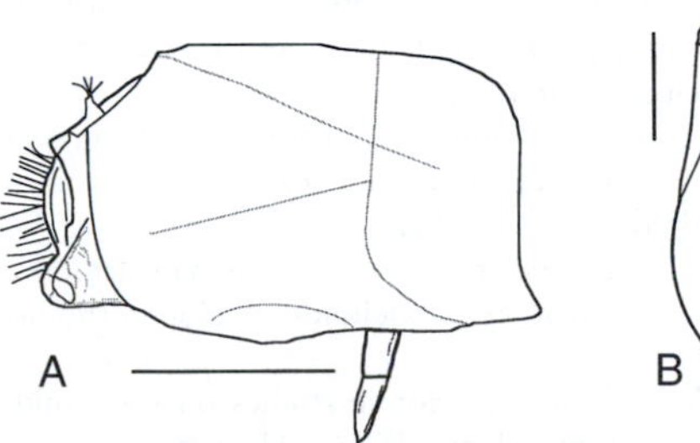

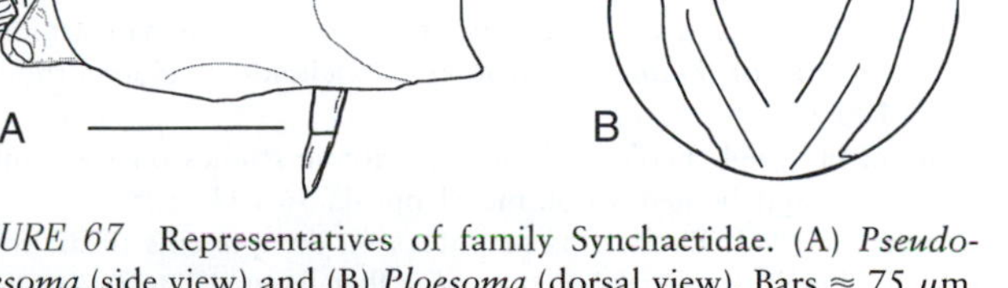

FIGURE 67 Representatives of family Synchaetidae. (A) *Pseudoploesoma* (side view) and (B) *Ploesoma* (dorsal view). Bars ≈ 75 μm. (Modified from several sources.)

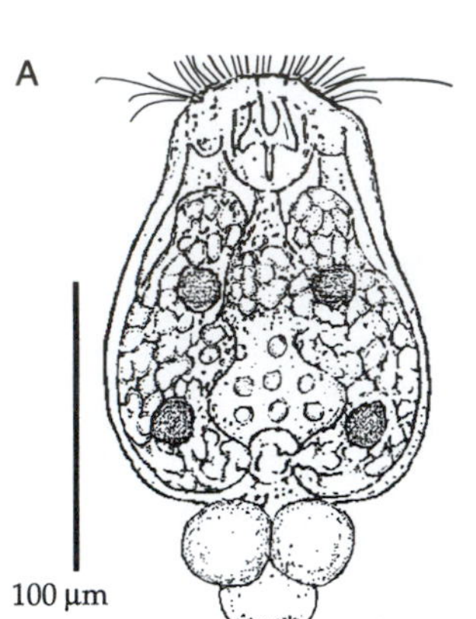

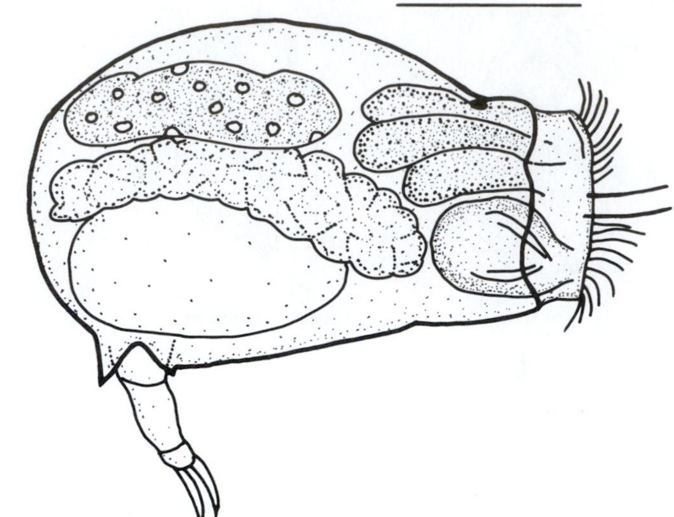

FIGURE 68 Representatives of family Gastropodidae. (A) *Ascomorpha* sp.; (B) *Gastropus*, (bar = 100 μm). (A from Koste, 1976, with permission; and B modified from several sources.)

ACKNOWLEDGMENTS

We wish to thank our mentors J. J. Gilbert and C. E. King for all the help and encouragement they have given us over the years. We also acknowledge Dr. W. Koste, a good friend for advice and inspiration; a worker who knows more about rotifers than we ever thought possible. Drs. W. T. Edmondson, J. J. Gilbert, T. Nogrady, P. L. Starkweather, R. S. Stemberger, L. J. Walsh, and G. H. Wittler reviewed the manuscript for this chapter, a task very much appreciated by the authors. However, we remain responsible for any errors which may still be present. We thank H. L. "Chico" Taylor for his insight in preserving and mounting rotifers for taxonomic work.

LITERATURE CITED

Allan, J. D. 1976. Life history patterns in zooplankton. American Naturalist 110:165–180.

Aloia, R. C., Moretti, R. L. 1973. Sterile culture techniques for some species of the rotifer *Asplanchna*. Transactions of the American Microscopical Society 92:364–371.

Amsellem, J., Clément, P. 1980. A simplified method for the preparation of rotifers for transmission and scanning electron microscopy. Hydrobiology 73:119–122.

Amsellem, J., Ricci, C. 1982. Fine structure of the female genital apparatus of *Philodina* (Rotifera, Bdelloidea). Zoomorphology 100:89–105.

Amrén, H. 1964. Ecological and taxonomic studies on zooplankton from Spitsbergen. Zool. Bidr. Uppsala 36:193–208.

Arndt, H. 1993. Rotifers as predators on components of the microbial web (bacteria, heterotrophic flagellates, ciliates)—a review. Hydrobiologia 255/256:231–246.

ASTM (American Society of Testing and Materials). 1991. Standard guide for acute toxicity test with the rotifer *Brachionus*. Vol. 11.05, E 1440, West Conshohocken, PA.

Barron, G. L. 1980a. Fungal parasites of rotifers: *Harposporium*. Canadian Journal of Botany 58:443–446.

Barron, G. L. 1980b. A new *Haptoglossa* attacking rotifers by rapid injection of an infective sporidium. Mycologia 72:1186–1194.

Barron, G. L., Tzean, S. S. 1981. A subcuticular endoparasite impaling bdelloid rotifers using three-pronged spores. Canadian Journal of Botany 59:1207–1212.

Bartos, E. 1951. The Czechoslovak Rotatoria of the order Bdelloidea. Vestník Ceskoslovenské Zoologické Spolecnosti 15:241–500.

Bateman, L. 1987. A bdelloid rotifer living as an inquiline in leaves of the pitcher plant, *Sarracenia purpurea*. Hydrobiologia 147: 129–133.

de Beauchamp, P. 1965. Classe des Rotifères, *in*: Grassé, P. P. Ed., Traité de Zoologie IV, 3, Masson, Paris, pp. 1225–1379.

Bender, K., Kleinow, W. 1988. Chemical properties of the lorica and related parts from the integument of *Brachionus plicatilis*. Comparative Biochemistry and Physiology 89B:483–487.

Berzins, B., Pejler, B. 1987. Rotifer occurrence in relation to pH. Hydrobiologia 147:107–116.

Birky, Jr., C. W., 1967. Studies on the physiology and genetics of the rotifer *Asplanchna* III. Results of outcrossing, selfing, and selection. Journal of Experimental Zoology 165:104–116.

Bogdan, K. G., Gilbert, J. J. 1982. Seasonal patterns of feeding by natural populations of *Keratella*, *Polyarthra*, and *Bosmina*: clearance rates, selectivities, and contributions to community grazing. Limnology and Oceanography 27:918–934.

Bogdan, K. G., Gilbert, J. J. 1987. Quantitative comparison of food niches in some freshwater zooplankton. Oecologia (Berlin) 72: 331–340.

Boraas, M. E., Bennett, W. N. 1988. Steady-state rotifer growth in a two-stage, computer-controlled turbidostat. Journal of Plankton Research 10:1023–1038.

Brett, M. T. 1989. Zooplankton communities and acidification processes (a review). Water, Air and Soil Pollution 44:387–414.

Buikema, A. L. Jr., Cairns, Jr., J., Edmunds, P. C., Krakauer, T. H. 1977. Culturing and ecology studies of the rotifer, *Polyarthra vulgaris*. EPA-600/3-77-051. Environmental Research Laboratory, Office of Research and Development, U.S. Environmental Protection Agency, Duluth, Minnesota.

Burger, A. 1948. Studies on the moss dwelling bdelloids (Rotifera) of eastern Massachusetts. Transactions of the American Microscopical Society 67:111–142.

Burns, C. W., Gilbert, J. J. 1986a. Effects of daphnid size and density on interference between *Daphnia* and *Keratella cochlearis*. Limnology and Oceanography 31:848–858.

Burns, C. W., Gilbert, J. J. 1986b. Direct observations of the mechanisms of interference between *Daphnia* and *Keratella cochlearis*. Limnology and Oceanography 31:859–866.

Capuzzo, J. 1979a. The effect of temperature on the toxicity of chlorinated cooling waters to marine animals—a preliminary review. Marine Pollution Bulletin 10:45–47.

Capuzzo, J. 1979b. The effect of halogen toxicants on survival, feeding and egg production of the rotifer *Brachionus plicatilis*. Estuarine and Coastal Marine Science 8:307–316.

Charoy, C. 1995. Modification of the swimming behaviour of *Brachionus calyciflorus* (Pallas) according to food environment and individual nutritive state. Hydrobiologia 313/314: 197–204.

Chengalath, R., Koste, W. 1987. Rotifera from northwestern Canada. Hydrobiologia 147:49–56.

Clément, P. 1977. Ultrastructural research on rotifers. Archiv für Hydrobiologie Beiheft 8:270–297.

Clément, P. 1980. Phylogenetic relationships of rotifers, as derived from photoreceptor morphology and other ultrastructural analysis. Hydrobiologia 73:93–117.

Clément, P. 1985. The relationships of rotifers, *in*: Conway Morris, S., George, J. D., Platt, H. M., Eds. The origins and relationships of lower invertebrates. Systematics Association, Vol. 28, Clarendon Press, Oxford, pp. 224–247.

Clément, P. 1987. Movements in rotifers: correlations of ultrastructure and behavior. Hydrobiologia 147:339–359.

Clément, P., Wurdak, E., Amsellem, J. 1983. Behavior and ultrastructure of sensory organs in rotifers. Hydrobiologia 104:89–130.

Clément, P., Wurdak, E. 1991. Rotifera. *in*: Harrison, F. W., Ruppert, E. E. Eds. Microscopic Anatomy of Invertebrates, Vol. 4: Aschelminthes, Wiley-Liss, New York, pp. 219–297.

Couillard, Y., Ross, P., Pinel-Alloul, B. 1989. Acute toxicity of six metals to the rotifer *Brachionus calyciflorus*. Toxicity Assessment 4:451–462.

Coulon, P. Y., Charras, J. P., Chasse, J. L., Clément, P., Cornillac, A., Luciani, A., Wurdak, E. 1983. An experimental system for automatic tracking and analysis of rotifer swimming behavior. Hydrobiologia 104:197–202.

Dad, N. K., Kant Pandya, V. 1982. Acute toxicity of two insecticides to rotifer *Brachionus calyciflorus*. International Journal of Environmental Studies 18:245–246.

Dartnall, H. J. G., Hollowday, E. D. 1985. Anarctic rotifers. British Antarctic Survey Scientific Reports 100:1–46.

De Smet, W. 1996. *in*: Dumont, H. J. (Ed.) Rotifera, Vol. 4: The Proalidae (Monogononta). Guide to the identification of the microinvertebrates of the continental waters of the world. SPB Academic Publishers, The Hague, The Netherlands. 102 pp.

De Smet, W. 1997. Rotifera *in*: Dumont, H. J., Vol. 5: The Dicranophoridae (Monogononta). Guide to the identification of the microinvertebrates of the continental waters of the world. SPB Academic Publishers, The Hague, The Netherlands. 344 pp.

De Smet, W. 1998. Preparation of rotifer trophi for light microscopy. Hydrobiologia 387/388:117–121.

Diéguez, M., Balseiro, E. 1998. Colony size in *Conochilus hippocrepis*: defense adaptation to predator stage sizes. Hydrobiologia 387/388:421–425.

Dodson, S. I. 1984. Ecology and behaviour of a free-swimming tube-dwelling rotifer *Cephalodella forficula*. Freshwater Biology 14: 329–334.

Donner, J. 1965. Ordnung Bdelloidea. Bestimmungsbücher zur Bodenfauna Europas. Vol. 6. Akademie Verlag, Berlin. 267 pp.

Dougherty, E. C. 1963. Cultivation and nutrition of micrometazoa III. The minute rotifer *Lecane inermis*. Journal of Experimental Zoology 153:183–186.

Droop, M. R., Scott, J. M. 1978. Steady-state energetics of a planktonic herbivore. Journal of the Marine Biological Association of the United Kingdom 58:749–772.

Dumont, H. 1983. Biogeography of rotifers. Hydrobiologia 104: 19–30.

Dumont, H., Segers, H. 1996. Estimating lacustrine zooplankton species richness and complimentarity. Hydrobiologia 341: 125–132.

Edmondson, W. T. 1944. Ecological studies of sessile Rotatoria, Part I. Factors affecting distribution. Ecological Monographs 14:32–66.

Edmondson, W. T. 1945. Ecological studies of sessile Rotatoria, Part II. Dynamics of populations and social structure. Ecological Monographs 15:141–172.

Edmondson, W. T. 1959. Rotifera, *in*: W. T. Edmondson, Ed. Freshwater Biology, 2nd Ed. Wiley, New York, pp. 420–494.

Edmondson, W. T. 1965. Reproductive rate of planktonic rotifers as related to food and temperature in nature. Ecological Monographs 35:61–111.

Edmondson, W. T. 1993. Experiments and quasi-experiments in limnology. Bulletin of Marine Science 53:65–83.

Edmondson, W. T., Litt, A. H. 1982. *Daphnia* in Lake Washington. Limnology and Oceanography 30:180–188.

Edmondson, W. T., Litt, A. H. 1987. *Conochilus* in Lake Washington. Hydrobiologia 147:157–162.

Egloff, D. A. 1988. Food and growth relations of the marine microzooplankter, *Synchaeta cecelia* (Rotifera). Hydrobiologia 157: 129–141.

Enesco, H. E. 1993. Rotifers in aging research: use of rotifers to test various theories of aging. Hydrobiologia 255/256:59–70.

Enesco, H. E., Verdone-Smith, C. 1980. *a*-Tocopherol increases lifespan in the rotifer *Philodina*. Experimental Gerontology 15: 335–338.

Epp, R. W., Lewis, W. M. 1979. Sexual dimorphism in *Brachionus plicatilis* (Rotifera): evolutionary and adaptive significance. Evolution 33:919–928.

Epp, R. W., Lewis, W. M. 1980. Metabolic uniformity over the environmental temperature range in *Brachionus plicatilis* (Rotifera). Hydrobiologia 73:145–147.

Epp, R. W., Lewis, W. M. 1984. Cost and speed of locomotion for rotifers. Oecologia 61:289–292.

Epp, R. W., Winston, P. W. 1977. Osmotic regulation in the brackish-water rotifer *Brachionus plicatilis* (Muller). Journal of Experimental Biology 68:151–156.

Epp, R. W., Winston, P. W. 1978. The effects of salinity and pH on the activity and oxygen consumption of *Brachionus plicatilis* (Rotatoria). Comparative Biochemistry and Physiology 59A:9–12.

Felix, A., Stevens, M. E., Wallace, R. L. 1995. Unpalatability of a colonial rotifer, *Sinantherina socialis*, to small zooplanktivorous fishes. Invertebr. Biol. 114:139–144.

Finlay, B. J., Curds, C. R., Bamforth, S. S., Bafort, J. M. 1987. Cilated Protozoa and other microorganisms from two African soda lakes (Lake Nakuru and Lake Simbi, Kenya). Arch für Protistenkunde 133:81–91.

Frost, T. M., Montz, P. K., Gonzalez, M. J., Sanderson, B., Arnott, S. 1998. Rotifer responses to increased acidity: long-term patterns during the experimental and two manipulation of Little Rock Lake. Hydrobiologia 387/388:141–152.

Garey, J. R., Near, T. J., Nonnemacher, M. R., Nadler, S. A. 1996. Molecular evidence for Acanthocephala as a subtaxon of Rotifera. Journal of Molecular Evolution 43:287–292.

Garey, J. R., Schmidt-Rhaesa, A. Near, T. J. Nadler, S. A. 1998. The evolutionary relationships of rotifers and acanthocephalans. Hydrobiologia 387/388:83–91.

Gilbert, J. J. 1963a. Mictic female production in the rotifer *Brachionus calyciflorus*. Journal of Experimental Zoology 153:113–123.

Gilbert, J. J. 1963b. Contact chemoreceptors, mating behavior and reproductive isolation in the rotifer genus *Brachionus*. Journal of Experimental Biology 40:625–641.

Gilbert, J. J. 1966. Rotifer ecology and embryological induction. Science 151:1234–1237.

Gilbert, J. J. 1967. *Asplanchna* and postero-lateral spine production in *Brachionus calyciflorus*. Archiv für Hydrobiologie 64:1–62.

Gilbert, J. J. 1970. Monoxenic cultivation of the rotifer *Brachionus calyciflorus* in a defined medium. Oecologia (Berlin) 4:89–101.

Gilbert, J. J. 1977. Effect of the non-tocopherol component of the diet on polymorphism, sexuality and reproductive rate of the rotifer *Asplanchna sieboldi*. Archiv für Hydrobiologie 80:375–397.

Gilbert, J. J. 1980a. Female polymorphisms and sexual reproduction in the rotifer *Asplanchna*: evolution of their relationship and control by dietary tocopherol. American Naturalist 116:409–431.

Gilbert, J. J. 1980b. Feeding in the rotifer *Asplanchna*: behavior, cannibalism, selectivity, prey defenses, and impact on rotifer communities, *in:* W. C. Kerfoot, Ed., Evolution and ecology of zooplankton communities. Univ. Press of New England, Hanover, NH, pp. 158–172.

Gilbert, J. J. 1985a. Escape response of the rotifer *Polyarthra*: a high-speed cinematographic analysis. Oecologia (Berlin) 66:322–331.

Gilbert, J. J. 1985b. Competition between rotifers and *Daphnia*. Ecology 66:1943–1950.

Gilbert, J. J. 1987. The *Polyarthra* escape response: defense against interference from *Daphnia*. Hydrobiologia 147:235–238.

Gilbert, J. J. 1988. Suppression of rotifer populations by *Daphnia*: a review of the evidence, the mechanisms, and the effects on zooplankton community structure. Limnology and Oceanography 33(6, part 1):1286–1303.

Gilbert, J. J. 1989a. The effect of *Daphnia* interference on a natural rotifer and ciliate community: short–term bottle experiments. Limnology and Oceanography 34(3):606–617.

Gilbert, J. J. 1989b. Competitive interactions between the rotifer *Synchaeta oblonga* and the cladoceran *Scapholeberis kingi* Sars. Hydrobiologia 186/187:75:80.

Gilbert, J. J. 1990. Differential effects of *Anabena affinis* on cladocerans and rotifers: mechanisms and implications. Ecology 71: 1727–1740.

Gilbert, J. J. 1995. Structure, development and induction of a new diapause stage in rotifers. Freshwater Biology 34:263–270.

Gilbert, J. J., Bogdan, K. G. 1984. Rotifer grazing: *in situ* studies on selectivity and rates, *in*: Meyers, D. G., Strickler, J. R. Eds., Trophic Interactions within Aquatic Ecosystems. American Association for the Advancement of Science, Selected Symposium 85. Westview, Boulder, CO, USA pp. 97–133.

Gilbert, J. J., Jack, J. D. 1993. Rotifers as predators on ciliates. Hydrobiologia 255/256:247–253.

Gilbert, J. J., Kirk, K. L. 1988. Escape response of the rotifer *Keratella*: description, stimulation, fluid dynamics, and ecological significance. Limnology and Oceanography 33(6, Part 2):1440–1450.

Gilbert, J. J., Starkweather, P. L. 1977. Feeding in the rotifer *Brachionus calyciflorus* I. Regulatory mechanisms. Oecologia (Berlin) 28:125–131.

Gilbert, J. J., Starkweather, P. L. 1978. Feeding in the rotifer *Brachionus calyciflorus* III. Direct observations on the effects of food type, food density, change in food type, and starvation on the incidence of pseudotrochal screening. Internationale Vereinigung für theoretische und angewandte Limnologie, Verhandlungen 20:2382–2388.

Gilbert, J. J., Stemberger, R. S. 1985a. Control of *Keratella* populations by interference competition from *Daphnia*. Limnology and Oceanography 30:180–188.

Gilbert, J. J., Stemberger, R. S. 1985b. Prey capture in the rotifer *Asplanchna girodi*. Internationale Vereinigung für theoretische und angewandte Limnologie, Verhandlungen 22:2997–3000.

Gilbert, J. J., Stemberger, R. S. 1985c. The costs and benefits of gigantism in polymorphic species of the rotifer *Asplanchna*. Archiv Für Hydrobiologie, Beiheft 21:185–192.

Gilbert, J. J., Williamson, C. E. 1978. Predator–prey behavior and its effect on rotifer survival in associations of *Mesocyclops edax*, *Asplanchna girodi*, *Polyarthra vulgaris*, and *Keratella cochlearis*. Oecologia (Berlin) 37:13–22.

Gonzalez, M. J., Frost. T. M. 1994. Comparisons of laboratory bioassays and a whole lake experiment: Rotifer responses to experimental acidification. Ecological Applications 4:69–80.

Green, J. 1972. Latitudinal variations in associations of planktonic rotifers. Journal of Zoology, London 167:31–39.

Green, J. 1977. Dwarfing of rotifers in tropical crater lakes. Archiv für Hydrobiologie Beiheft 8:232–236.

Green, J. 1985. Horizontal variations in associations of zooplankton in Lake Kariba. Journal of Zoology, London 206:225–239.

Gvozdov, A. O. 1986. Phototaxis as a test function in bioassay. Gidrobiologichesky Zhurnal 22:65–68.

Hagiwara, A., Snell, T. W., Lubzens, E., Tamaru, C. (Eds.). 1997. Live food in Aquaculture., Vol. 358, Kluwer Academic Publishers, Belgium, Hydrobiologia.

Halbach, U. 1970. Die Ursachen der Temporal Variation von *Brachionus calyciflorus* Pallas (Rotatoria). Oecologia 4:262–318.

Halbach, U. 1984. Population dynamics of rotifers and its consequences for ecotoxicology. Hydrobiologia 109:79–96.

Halbach, U., Halbach–Keup, G. 1974. Quantitative Beiehungen zwischen Phytoplankton und der Populationdynamik des Rotators *Brachionus calyciflorus* Pallas. Befunde aus Laboratoriums–experimenten und Freilanduntersuchungen. Archiv für Hydrobiologie 73:273–309.

Halbach, U., Wiebert, M., Westermayer, M., Wissel, C. 1983. Population ecology of rotifers as a bioassay tool for ecotoxicological tests in aquatic environments. Ecotoxicology and Environmental Safety 7:484–513.

Halbach, U., Wiebert, M., Wissel, C., Kalus, J., Beuter, K., Delion, M. 1981. Population dynamics of rotifers as bioassay tools for toxic effects of organic pollutants. Internationale Vereinigung für theoretische und angewandte Limnologie, Verhandlungen 21:1141–1146.

Hall, D. J., Threlkeld, S. T., Burns, C. W., Crowley, P. H. 1976. The size–efficiency hypothesis and the size structure of zooplankton communities. Annual Review of Ecology and Systematics 7: 177–208.

Harring, H. K., Myers, F. J. 1922. The rotifers of Wisconsin. Transactions of the Wisconsin Academy of Sciences, Arts, and Letters 20:553–662.

Hertel, E. W. 1942. Studies on vigor in the rotifer *Hydatina senta*. Physiological Zoology 15:304–324.

Herzig, A. 1983. Comparative studies on the relationship between temperature and the duration of embryonic development of rotifers. Hydrobiologia 104:237–246.

Herzig, A. 1987. The analysis of planktonic rotifer populations: a plea for long–term investigations. Hydrobiologia 147:163–180.

Hewitt, D. P., George, D. G. 1987. The population dynamics of *Keratella cochlearis* in a hypereutrophic tarn and the possible impact of predation by young roach. Hydrobiologia 147:221–227.

Hillbricht–Ilkowska, A. 1983. Response of planktonic rotifers to the eutrophication process and to the autumnal shift of blooms in Lake Biwa, Japan. I. Changes in abundances and composition of rotifers. The Japanese Journal of Limnology 44:93–106.

Hirata, H., Nagata, W. 1982. Excretion rates and excreted compounds of the rotifer *Brachionus plicatilis* O.F. Müller in culture. Memoirs of the Faculty of Fisheries Kagoshima University 31:161–174.

Hirayama, K., and S. Ogawa. 1972. Fundamental studies on the physiology of the rotifer for its mass culture I. Filter feeding of the rotifer. Bulletin of the Japanese Society of Scientific Fisheries 38:1207–1214.

Hofmann, W. 1977. The influence of abiotic factors on population dynamics in planktonic rotifers. Archiv für Hydrobiologie Beiheft 8:77–83.

Hofmann, W. 1982. On the coexistence of two pelagic *Filinia* species (Rotatoria) in Lake Plußee I. Dynamics of abundance and dispersion. Archiv für Hydrobiologie 95:125–137).

Hudson, C. T., Gosse, P. H. 1886. The Rotifera; or wheel-animalcules, both British and foreign. Vol. I & II. Longmans, Green, London.

Hutchinson, G. E. 1967. A treatise on limnology. Vol. 2. Introduction to lake biology and the limnoplankton. Wiley, New York.

Hyman, L. H. 1951. The Invertebrates: Acanthocephala, Aschelminthes, and Entoprocta. The pseudocoelomate Bilateria. Vol. 3. McGraw–Hill, New York.

Iltis, A., Riou–Duvat, S. 1971. Variations saisonniéres du peuplement en rotifères des eaux natronées du Kanem (Tchad). Cah. O.S.T.R.O.M. ser. Hydrobiol. 5(2):101–112.

Ito, T. 1960. On the culture of the mixohaline rotifer *Brachionus plicatilis* O.F. Müller, in seawater. Report of the Faculty of Fisheries, Prefectural University of Mie 3:708–740.

Jenkins, D. G., Buikema, A. L. 1985. Plankton rotifer responses to herbicide stress in *in situ* microcosms. Virginia Journal of Science 36:144.

Jenkins, D.G., Underwood, M. O. 1998. Zooplankton may not disperse readily in wind, rain, or waterfowl. Hydrobiologia 387/388:15–2.

Jennings, H. S., Lynch, R. S. 1928. Age, mortality, fertility and individual diversities in the rotifer *Proales sordida* Gosse. I. Effects of the age of the parent on characteristics of the offspring. Journal of Experimental Zoology. 50:345–407.

Johansson, S. 1983. Annual dynamics and production of rotifers in an eutrophication gradient in the Baltic Sea. Hydrobiologia 104:335–340.

Keshmirian, J., Nogrady, T. 1988. Histofluorescent labelling of catecholaminergic structures in rotifers (Aschelminthes) II. Males of *Brachionus plicatilis* and structures from sectioned females. Histochemistry 89:189–192.

King, C. 1967. Food, age and the dynamics of a laboratory population of rotifers. Ecology 48:111–128.

King, C. E. 1969. Experimental studies of aging in rotifers. Experimental Gerontology 4:63–79.

King, C. E. 1972. Adaptation of rotifers to seasonal variation. Ecology 53:408–418.

King, C. E. 1977. Genetics of reproduction, variation, and adaptation in rotifers. Archiv für Hydrobiologie Beiheft 8:187–201.

King, C. E. 1980. The genetic structure of zooplankton populations. *in:* Kerfoot, W. C. Ed., Evolution and ecology of zooplankton communities. Univ. Press of New England, Hanover, NH, pp. 315–328.

King, C. E. 1983. A re–examination of the Lansing effect. Hydrobiologia 104:135–139.

King, C. E., Miracle, M. R. 1980. A perspective on aging in rotifers. Hydrobiologia 73:13–19.

King, C. E., Snell, T. W. 1977. Culture media (natural and synthetic): Rotifera *in:* M. Rechcigal, Jr., Ed. CRC Handbook Series in Nutrition and Food. Sect. G: Diets, Culture Media, Food Supplements. Vol II: Food habits of, and diets for invertebrates and vertebrates—zoo diets. CRC Press, Cleveland, Oh, pp. 71–75.

King, C. E., Zhao, Y. 1987. Coexistence of rotifer (*Brachionus plicatilis*) clones in Soda Lake, Nevada. Hydrobiologia 147:57–64.

Kirk, K.L. 1991. Inorganic particles alter competition in grazing plankton: the role of selective feeding. Ecology 72:915–923.

Kirk, K.L., Gilbert, J. J. 1992. Variation in herbivore response to chemical defenses: zooplankton foraging on toxic cyanobacteria. Ecology 73:2208–2217.

Kirk, K. L. 1997. Life-history responses to variable environments: starvation and reproduction in planktonic rotifers. Ecology 78: 434–441.

Kleinow, W. 1998. Stereopictures of internal structures and rotifer trophi. Hydrobiologia 123–129.

Koste, W. 1976. Über die Rädertierbestände (Rotatoria) der oberen und mittleren Hase in de Jahren 1966–1969. Osnabrücker Naturwissenschaftliche. Mitteilungen 4:191–263.

Koste, W. 1978. Rotatoria. Die Rädertiere Mitteleuropas. 2 vols, Gébrüder Borntraeger, Berlin.

Koste, W. 1989. *Octotrocha speciosa*, eine seltene, sessile Art mit einem merkwürdigen Räderorgan. Mikrokosomos 78:115–121.

Koste, W., Shiel, R. J. 1986. Rotifera from Australian inland waters. I. Bdelloidea (Rotifera: Digononta). Australian Journal of Marine and Freshwater Research 37:765–792.

Koste, W., Shiel, R. J. 1987. Rotifera from Australian inland waters. II. Epiphanidae and Brachionidae (Rotifera: Monogononta). Invertebrate Taxonomy 7:949–1021.

Lorenzen, S. 1985. Phylogenetic aspects of pseudocoelomate evolution, *in:* Conway Morris, S., George, J. D., Platt, H. M. Eds., The origins and relationships of lower invertebrates. Systematics Association, Vol. 28. Clarendon Press, Oxford, pp. 210–223.

Lubzens, E. 1987. Raising rotifers for use in aquaculture. Hydrobiologia 147:245–255.

Lubzens, E., Minkoff, G., Maron, S. 1985. Salinity dependence of sexual and asexual reproduction in the rotifer *Brachionus plicatilis*. Marine Biology 85:123–126.

Lubzens, E., Tandler, A., Minkoff, G. 1989. Rotifers as food in aquaculture. Hydrobiologia 186/187:387–400.

Luciani, A., Chasse, J.–L., Clément, P. 1983. Aging in *Brachionus plicatilis*: the evolution of swimming as a function of age at two different calcium concentrations. Hydrobiologia 104:141–146.

MacIsaac, H. J., Hutchinson, T. C., Keller, W. 1987. Analysis of planktonic rotifer assemblages from Sudbury, Ontario, area lakes of varying chemical composition. Canadian Journal of Fisheries and Aquatic Sciences 44:1692–1701.

Magnien, R. E., Gilbert, J. J. 1983. Diel cycles of reproduction and vertical migration in the rotifer *Keratella crassa* and their influence on the estimation of population dynamics. Limnology and Oceanography 28:957–969.

Makarewicz, J. C., Likens, G. E. 1975. Niche analysis of a zooplankton community. Science 190:1000–1003.

Makarewicz, J. C., Likens, G. E. 1979. Structure and function of the zooplankton community of Mirror Lake, New Hampshire. Ecological Monographs 49:109–127.

Marcus, N. H., Lutz, R., Burnett, W., Cable, P. 1994. Age, viability, and vertical distribution of zooplankton resting eggs from an anoxic basin: Evidence of an egg bank. Limnology and Oceanography 39:154–158.

Mark Welch, D. (2000). Evidence from a protein-coding gene that acanthocephalans are rotifers. Invertebrate Biology:17–26.

May, L. 1980. On the ecology of *Notholca squamula* Müller in Loch Leven, Kinross, Scotland. Hydrobiologia 73:177–180.

May, L. 1985. The use of procaine hydrochloride in the preparation of rotifer samples for counting. Internationale Vereinigung für theoretische und angewandte Limnologie, Verhandlungen 22: 2987–2990.

May, L. 1986. Rotifer sampling—a complete species list from one visit? Hydrobiologia 134:117–120.

May, L. 1987. Effect of incubation temperature on the hatching of rotifer resting eggs collected from sediments. Hydrobiologia 147:335–338.

May, L. 1989. Epizoic and parasitic rotifers. Hydrobiologia 186/187: 59–67.

Meadow, N. D., Barrows, Jr., C. H. 1971. Studies on aging in a bdelloid rotifer. Journal of Experimental Zoology 176:303–314.

Melone, G. 1998. The rotifer corona by SEM. Hydrobiologia 387/388:131–134.

Melone, G., Ricci, C., Segers, H. 1998a. The trophi of Bdelloidea (Rotifera): a comparative study across the class. Canadian Journal of Zoology 76:1755–1765.

Melone, G., Ricci, C., Segers, H., Wallace, R. L. 1998b. Phylogenetic relationships of phylum Rotifera. Hydrobiologia 387/388: 101–107.

Miracle, M. R. 1974. Niche structure in freshwater zooplankton: a principal components approach. Ecology 55:1306–1316.

Miracle, M. R., Serra, M., Vincente, E., Blanco, C. 1987. Distribution of *Brachionus* species in Spanish mediterranean wetlands. Hydrobiologia 147:75–81.

Moore, M. V. 1988. Density-dependent predation of early instar *Chaoborus* feeding on multispecies prey assemblages. Limnology and Oceanography 33:256–269.

Moore, M. V., Gilbert, J. J. 1987. Age-specific *Chaoborus* predation on rotifer prey. Freshwater Biology 17:223–236.

Myers, F. J. 1937. A method of mounting rotifer jaws for study. Transactions of the American Microscopical Society 56:256–257.

Neill, W. E. 1984. Regulation of rotifer densities by crustacean zooplankton in an oligotrophic montane lake in British Columbia. Oecologia 61:175–181.

Nipkow, F. 1961. Die Radertiere im Plankton des Zürchsee und ihre Entwisklungsphasen. Schweizer Journal Hydrobiologie 23: 398–461.

Nogrady, T. 1982. Rotifera, *in:* Parker, S. P. Ed. Synopsis and classification of living organisms. McGraw–Hill., New York, pp. 865–872.

Nogrady, T., Alai. M.1983. Cholinergic neurotransmission in rotifers. Hydrobiologia 104:149–153.

Nogrady, T., Keshmirian, J. 1986a. Rotifer neuropharmacology I. Cholinergic drug effects on oviposition of *Philodina acuticornis* (Rotifera, Aschelminthes). Comparative Biochemistry and Physiology 83C:335–338.

Nogrady, T., Keshmirian, J. 1986b. Rotifer neuropharmacology II. Synergistic effect of acetyl–choline on local anesthetic activity in *Brachionus calyciflorus* (Rotifera, Aschelminthes). Comparative Biochemistry and Physiology 83C:339–344.

Nogrady, T., Pourriot, R. 1995. Rotifera, Vol. 3: The Notommatidae. *in:* Dumont, H. J., Ed., Guide to the identification of the microinvertebrates of the continental waters of the world. SPB Academic Publishers, The Hague, The Netherlands, 248 pp.

Nogrady, T., Wallace, R. L., Snell, T. W. 1993. Rotifera: Vol. 1: Biology, Ecology and Systematics. *in:* Dumont, H. J. Ed., Guide to the identification of the microinvertebrates of the continental waters of the world. SPB Academic Publishers bv, The Hague, The Netherlands, 142 pp.

O'Brien, W. J. 1987. Planktivory by freshwater fish: thrust and parry in pelagia, *in:* Kerfoot, C. W., Sih, A. Eds. Predation: direct and indirect impacts on aquatic communities. Univ. Press of New England, Hanover, NH, pp. 3–16.

Pagani, M., Ricci, C., Bolzern, A. M. 1991. Comparison of five strains of a parthenogenetic species, *Macrotrachela quadricornifera* (Rotifera, Bdelloidea) I!. Isoenzymatice patterns. Hydrobiologia 211:157–163.

Paloheimo, J. E. 1974. Calculation of instantaneous birth rates. Limnology and Oceanography 19:692–694.

Pejler, B. 1956. Introgression in planktonic Rotatoria with some points of view on its causes and results. Evolution 10:246–261.

Pejler, B. 1977a. General problems on rotifers taxonomy and global distribution. Archiv für Hydrobiologie Beiheft 8:212–220.

Pejler, B. 1977b. On the global distribution of the family Brachionidae (Rotatoria). Archiv für Hydrobiologie (Suppl.) 53: 255–306.

Pennak, R. W. 1962. Quantitative zooplankton sampling in littoral vegetation areas. Limnology and Oceanography 7:487–489.

Pennak, R. W. 1989. Freshwater invertebrates of the United States, 3rd ed. Wiley, New York.

Persoone, G., Van De Vel, A. Van Steertegem, M., De Nayer, B. 1989. Predictive value of laboratory tests with aquatic invertebrates: influence of experimental conditions. Aquatic Toxicology 14: 149–66.

Pilarska, J. 1972. The dynamics of growth of experimental populations of the rotifer *Brachionus rubens* Ehrbg. Polish Archives for Hydrobiology 19:265–277.

Pontin, R. S. 1978. A key to the British freshwater planktonic Rotifera. Scientific Publications No. 38. Freshwater Biological Association, Cumbria, UK.

Pourriot, R. 1958. Sur l' élevage des Rotifères au laboratoire. Hydrobiologia 11:189–197.

Pourriot, R. 1965. Recherches sur l'ecologie des Rotifères. Vie et Milieu (Suppl.) 21:1–224.

Pourriot, R. 1977. Food and feeding habits of Rotifera. Archiv für Hydrobiologie Beiheft. 8:243–260.

Pourriot, R. 1979. Rotifères du sol. Revue d'Ecologie et de Biologie du Sol. 16:279–312.

Pourriot, R. 1986. Les rotifères—Biologie. *in*: G. Barnabé, Ed. Aquaculture Vol. 1. Technique et documentation, Lavoisier, Paris.

Pourriot, R. 1997. Rotifera, Vol. 5: The Ituridae (Monogononta). *in:* Dumont, H. J. Ed., Guides to the identification of the microinvertebrates of the continental waters of the world. SPB Academic Publishers, The Hague, The Netherlands. 344 pp.

Pourriot, R., Clément, P. 1981. Action de facteurs externes sur la reproduction et le cycle reproducteur des Rotifers. Acta Oecologia Generale 2:135–151.

Pourriot, R., Rougier, C. 1975. Dynamique d'une population experimentale de *Brachionus dimidatus* (Bryce) (Rotifère) en fonction de la nourriture et de la temperature. Annals Limnology 11: 125–143.

Pourriot, R., Rougier, C., Benest, D. 1986. Food quality and mictic female control in the rotifer *Brachionus rubens* Ehr. Bulletin Society Zoologie Frances 111:105–112.

Preissler, K. 1980. Field experiments on the optical orientation of pelagic rotifers. Hydrobiologia 73:199–203.

Rao, T. R., Sarma, S .S. S. 1986. Demographic parameters of *Brachionus patulus* Muller (Rotifera) exposed to sublethal DDT concentrations at low and high food levels. Hydrobiologia 139: 193–200.

Ricci, C. 1983. Life histories of some species of Rotifera Bdelloida. Hydrobiologia 104:175–180.

Ricci, C. 1984. Culturing of some bdelloid rotifers. Hydrobiologia 112:45–51.

Ricci, C. 1987. Ecology of bdelloids: how to be successful. Hydrobiologia 147:117–127.

Ricci, C. 1998a. Are lemnisci and proboscis present in the Bdelloidea? Hydrobiologia 387/388:93–96.

Ricci, C. 1998b. Anhydrobiotic capabilities of bdelloid rotifers. Hydrobiologia 387/388:321–326.

Ricci, C., Vaghi, L., Manzini, M. L. 1987. Desiccation of rotifers (*Macrotrachela quadricornifera*): survival and reproduction. Ecology 68:1488–1494.

Ricci, C., Melone, G., Sotgia, C. 1993. Old and new data on Seisonidea (Rotifera). Hydrobiologia 255/256:495–511.

Rico-Martinez, R., Snell, T. W. 1998. Mating behavior in eight rotifer species: using cross-mating tests to study species boundaries. Hydrobiologia 356:165–173.

Robb, E. J., Barron, G. L. 1982. Nature's ballistic missile. Science 218:1221–1222.

Rothhaupt, K. O. 1985. A model approach to the population dynamics of the rotifer *Brachionus rubens* in two–stage chemostat culture. Oecologia 65:252–259.

Rougier, C., Pourriot, R. 1977. Aging and control of the reproduction in *Brachionus calyciflorus* (Pallas) (Rotatoria). Experimental Gerontology 12:137–151.

Ruttner–Kolisko, A. 1963. The interrelationships of the Rotatoria, *in*: E. C. Dougherty, Ed., The Lower Metazoa. Univ. of California Press, Berkeley., CA, pp. 263–272.

Ruttner–Kolisko, A. 1969. Kreuzungexperimente zwischen *Brachionus urceolaris* and *Brachionus quadridentatus*, ein Beitrag zur Fortpflanzungbiologie der heterogonen Rotatoria. Archiv für Hydrobiologie 65:397–412.

Ruttner–Kolisko, A. 1974. Planktonic rotifers biology and taxonomy. Die Binnengewässer (Supplement) 26:1–146.

Ruttner–Kolisko, A. 1977a. The effect of the microsporid *Plistophora asperospora* on *Conochilus unicornis* in Lunzer Untersee (LUS). Archiv für Hydrobiologie Beiheft 8:135–137.

Ruttner–Kolisko, A. 1977b. Amphoteric reproduction in a population of *Asplanchna priodonta*. Archiv für Hydrobiologie Beiheft 8:178–181.

Salt, G. W. 1987. The components of feeding behavior in rotifers. Hydrobiologia 147:271–281.

Salt, G. W., Sabbadini, G. F., Commins, M. L. 1978. Trophi morphology relative to food habits in six species of rotifers (Asplanchnidae). Transactions of the American Microscopical Society 97:469–485.

Saunders, J. F. III, Lewis, Jr, W. M. 1989. Zooplankton abundance in the lower Orinoco River, Venezuela. Limnol. Oceanogr. 34: 397–409.

Sawada, M., Enesco, H. E. 1984. A study of dietary restriction and lifespan in the rotifer *Asplanchna brightwelli* monitored by chronic neutral red exposure. Experimental Gerontology 19:3 29–334.

Schlüter, M. 1980. Mass culture experiments with *Brachionus rubens*. Hydrobiologia 73:45–50.

Schlüter, M., Groeneweg, J. 1981. Mass production of freshwater rotifers on liquid wastes. I. The influence of some environmental factors on population growth on *Brachionus rubens* Ehrenberg 1838. Aquaculture 25:17–24.

Schlüter, M., Groeneweg, J. 1985. The inhibition by ammonia of population growth of the rotifer, *Brachionus rubens*, in continuous culture. Aquaculture 46:215–220.

Schlüter, M., Groeneweg, J., Soeder, C. J. 1987. Impact of rotifer grazing on population dynamics of green microalgae in high–rate ponds. Water Research 10:1293–1297.

Schmid-Araya, J. M. 1998. Rotifers in interstitial sediments. Hydrobiologia 231–240.

Schnese, W. 1973. Relations between phytoplankton and zooplankton in brackish coastal water. Oikos (Suppl.) 15:28–33.

Schramm, U., Becker, W. 1987. Anhydrobiosis of the bdelloid rotifer *Habrotrocha rosa* (Aschelminthes). Zeitschrift fuer Mikroskopisch–Anatomische Forschung 101:1–17.

Scott, J. M. 1983. Rotifer nutrition using supplemented monoxenic cultures. Hydrobiologia 104:155–166.

Scott, J. M. 1988. Effect of growth rate on the physiological rates of a chemostat–grown rotifer *Encentrum linnhei*. Journal of the Marine Biological Association of the United Kingdom 68: 165–177.

Seaman, M. T., Gophen, M., Cavari, B. Z., Azoulay, B. 1986. *Brachionus calyciflorus* Pallas as agent for removal of *E. coli* in sewage ponds. Hydrobiologia 135:55–60.

Segers, H. 1995a. Rotifera, Volume 2: The Lecanidae (Monogononta)., *in:* Dumont, H. J., Ed. Guides to the identification of the microinvertebrates of the continental waters of the world. SPB Academic Publishers, The Hague, The Netherlands. 226 pp.

Segers, H. 1995b. Rotifera, Volume 3: The Scaridiidae. *in:* Dumont, H. J., Ed., Guides to the identification of the microinvertebrates of the continental waters of the world. SPB Academic Publishers, The Hague, The Netherlands. 248 pp.

Serra, M., Miracle, M. 1985. Enzyme polymorphisms in *Brachionus plicatilis* populations from several Spanish lagoons. Internationale Vereinigung für theoretische und angewandte Limnologie, Verhandlungen 22:2991–2996.

Serrano, L., Miracle, M. R., Serra, M. 1986. Differential response of *Brachionus plicatilis* (Rotifera) ecotypes to various insecticides. Journal of Environmental Biology 7:259–275.

Sladecek, V. 1983. Rotifers as indicators of water quality. Hydrobiologia 100:169–201.

Snell, T. W. 1977. Lifespan of male rotifers. Archiv für Hydrobiologie Beiheft 8:65–66.

Snell, T. W. 1979. Intraspecific competition and population structure in rotifers. Ecology 60:494–502.

Snell, T. W. 1980. Blue–green algae and selection in rotifer populations. Oecologia 46:343–346.

Snell, T. W. 1986. Effects of temperature, salinity and food level on sexual and asexual reproduction in *Brachionus plicatilis* (Rotifera). Marine Biology 92:157–162.

Snell, T. W. 1989. Systematics, reproductive isolation and species boundaries in monogonont rotifers. Hydrobiologia 186/187: 299–310.

Snell, T. W., Childress, M. J. 1987. Aging and loss of fertility in male and female *Brachionus plicatilis* (Rotifera). International Journal of Invertebrate Reproduction and Development 12:103–110.

Snell, T. W., Garman, B. L. 1986. Encounter probabilities between male and female rotifers. Journal of Experimental Marine Biology and Ecology 97:221–230.

Snell, T. W., Hawkinson, C. A. 1983. Behavioral reproductive isolation among populations of the rotifer *Brachionus plicatilis*. Evolution 37:1294–1305.

Snell, T. W., Hoff, F. H. 1985. The effect of environmental factors on resting egg production in the rotifer *Brachionus plicatilis*. Journal of the World Mariculture 16:484–497.

Snell, T. W., Hoff, F. H. 1987. Fertilization and male fertility in the rotifer *Brachionus plicatilis*. Hydrobiologia 147:329–334.

Snell, T. W., King, C. E. 1977. Lifespan and fecundity patterns in rotifers: the cost of reproduction. Evolution 31:882–890.

Snell, T. W., Winkler, B. C. 1984. Isozyme analysis of rotifer protein. Biochemical Systematics and Ecology 12:199–202.

Snell, T. W., Burke, B. E., Messur, S. D. 1983. Size and distribution of resting eggs in a natural population of the rotifer *Brachionus plicatilis*. Gulf Research Reports 7:285–288.

Snell, T. W., Childress, M. J., Boyer, E. M., Hoff, F. H. 1987. Assessing the status of rotifer mass cultures. Journal of the World Aquaculture Society 18:270–277.

Snell, T. W., Childress, M. J., Winkler, B. C. 1988. Characteristics of the mate recognition factor in the rotifer *Brachionus plicatilis*. Comparative Biochemistry and Physiology 89A:481-485.

Snell, T. W., Janssen, C. R. 1995. Rotifers in ecotoxicology: a review. Hydrobiologia 313/314:231–247.

Snell, T. W., Janssen, C. R. 1998. Microscale toxicity testing with rotifers, *in*: Wells, P. G., Lee, K., Blaise, C. Eds., Microscale Toxicity Testing in Aquatic Toxicology, CRC Press, Boca Raton, FL, pp. 409–422.

Snell, T. W., Rico-Martinez, R., Kelly, L., N. Battle, T. E. 1995. Identification of a sex pheromone from a rotifer. Marine Biology 123: 347–353.

Starkweather, P. L. 1980. Aspects of the feeding behavior and trophic ecology of suspension feeding rotifers. Hydrobiologia 73:63–72.

Starkweather, P. L. 1987. Rotifera. Pages 159–183 *in*: T. J. Pandian and F. J. Vernberg, Eds., Animal energetics. Vol. 1: Protozoa through Insecta. Academic Press, Orlando, FlO.

Starkweather, P. L., Kellar, P. E. 1987. Combined influences of particulate and dissolved factors in the toxicity of *Microcyctis aeruginosa* (NRC–SS–17) to the rotifer *Brachionus calyciflorus*. Hydrobiologia 147:375–378.

Stemberger, R. S. 1979. A guide to rotifers of the Laurentian Great Lakes. U.S. Environmental Protection Agency, Cincinnati, Oh. (Available from National Technical Information Service, Springfield, VA, PB80–101280.)

Stemberger, R. S. 1981. A general approach to the culture of planktonic rotifers. Canadian Journal of Fisheries and Aquatic Sciences 38:721–724.

Stemberger, R. S. 1984. Spine development in the rotifer *Keratella cochlearis*: induction by cyclopoid copepods and *Asplanchna*. Freshwater Biology 14:639–647.

Stemberger, R. S. 1988. Reproductive costs and hydrodynamic benefits of chemically induced defenses in *Keratella testudo*. Limnology and Oceanography 33:593–606.

Stemberger, R. S. 1990. Food limitation, spination, and reproduction in *Brachionus calyciflorus*. Limnology and Oceanography 35: 33–44.

Stemberger, R. S., Gilbert, J. J. 1984a. Body size, ration level, and population growth in *Asplanchna*. Oecologia (Berlin) 64:355–359.

Stemberger, R. S., Gilbert, J. J. 1984b. Spine development in the rotifer *Keratella cochlearis*: induction by cyclopoid copepods and *Asplanchna*. Freshwater Biology 14:639–647.

Stemberger, R. S., Gilbert, J. J. 1985. Body size, food concentration and population growth in planktonic rotifers. Ecology 66:1151–1159.

Stemberger, R. S., Gilbert, J. J. 1987a. Defenses of planktonic rotifers against predators. *in*: Kerfoot, W. C., Sih, A. Eds. Predation: direct and indirect impacts on aquatic communities. Univ. Press of New England, Hanover, NH, pp. 227–239.

Stemberger, R. S., Gilbert, J. J. 1987b. Multiple species induction of morphological defenses in the rotifer *Keratella testudo*. Ecology 68:370–378.

Taylor, H. L. 1993. The Taylor microcompressor Mark II. Microscope 41:19-20.

Threlkeld, S. T., Rybock, J. T., Morgan, M. D., Folt, C. L., Goldman, C. R. 1980. The effects of an introduced invertebrate predator and food resource variation on zooplankton dynamics in an ultraoligotrophic lake, *in*: Kerfoot, W. C. Ed., Evolution and Ecology of Zooplankton Communities. Univ. Press of New England, Hanover, NH, pp. 555–568.

Tzean, S. S., Barron, G. L. 1983. A new predatory hypomycete capturing bdelloid rotifers in soil. Canadian Journal of Botany 61:1345–1348.

Wallace, R. L. 1977. Distribution of sessile rotifers in an acid bog pond. Archiv für Hydrobiologie 79:478–505.

Wallace, R. L. 1980. Ecology of sessile rotifers. Hydrobiologia 73: 181–193.

Wallace, R. L. 1987. Coloniality in the phylum Rotifera. Hydrobiologia 147:141–155.

Wallace, R. L. 1998. Rotifera, *in*: Knobil, E., Neil. J. D., Eds., Encyclopedia of reproduction. Vol. 4. Academic Press, San Diego, CA, pp. 118–129.

Wallace, R. L., Edmondson, W. T. 1986. Mechanism and adaptive significance of substrate selection by a sessile rotifer. Ecology 67:314–323.

Wallace, R. L., Taylor, W. K. Litton, J. R. 1989. Invertebrate zoology. Macmillan, New York. 337 pp.

Wallace, R. L., Ricci, C., Melone, G. 1996. A cladistic analysis of pseudocoelomate (aschelminth) morphology. Invertebrate Biology 115:104–112.

Walsh, E. J. Zhang, L. 1992. Polyploidy and body size variation in a natural population of the rotifer *Euchlanis dilatata*. Journal Evolutionary Biology 5:345–353.

Walz, N. 1983a. Continuous culture of the pelagic rotifers *Keratella cochlearis* and *Brachionus angularis*. Archiv für Hydrobiologie 98:70–92.

Walz, N. 1983b. Individual culture and experimental population dynamics of *Keratella cochlearis* (Rotatoria). Hydrobiologia 107: 35–45.

Walz, N. 1987. Comparative population dynamics of the rotifers *Brachionus angularis* and *Keratella cochlearis*. Hydrobiologie 147:209–213.

Walz, N. (Ed.) 1993. Plankton regulation dynamics. Springer, New York, 308 pp.

Walz, N. 1995. Rotifer populations in plankton communities: energetics and life history strategies. Experentia 51:437–453.

Walz, N. 1997. Rotifer life history strategies and evolution in freshwater plankton communities, *in*: Streit, B., Städler, T., Lively, C.M. Eds., Evolutionary Ecology of Freshwater Animals. Birkhäuser Verlag, Basel, pp. 119–149.

Walz, N., Elster, H.-J. Mezger, M. 1987. The development of the rotifer community structure in Lake Constance during its eutrophication. Archiv für Hydrobiologie (Suppl) 74:452–487.

Watanabe, T., Kitajima, C., Fujita, S. 1983. Nutritional values of live food organisms used in Japan for mass propagation of fish: a review. Aquaculture 34:115–143.

Williamson, C. E. 1983. Invertebrate predation on planktonic rotifers. Hydrobiologia 104:385–396.

Williamson, C. E. 1987. Predator–prey interactions between omnivorous diaptomid copepods and rotifers: the role of prey morphology and behavior. Limnology and Oceanography 32:167–177.

Wurdak, E. 1987. Ultrastructure and histochemistry of the stomach of *Asplanchna sieboldi*. Hydrobiologia 147:361–371.

Wurdak, E., Clément, P., Amselem, J. 1983. Sensory receptors involved in the feeding behavior of the rotifer *Asplanchna brightwelli*. Hydrobiologia 104:203–212.

Wurdak, E., Gilbert, J. J. 1976. Polymorphism in the rotifer *Asplanchna sieboldi*: fine structure of saccate, cruciform and campanulate females. Cell Tissue Research 169:435–448.

Wurdak, E., Gilbert, J. J., Jagels, R. 1977. Resting egg ultrastructure and formation of the shell in *Asplanchna sieboldi* and *Brachionus calyciflorus*. Archiv für Hydrobiologie Beiheft 8:298–302.

9 NEMATODA AND NEMATOMORPHA

George O. Poinar, Jr.
Department of Entomology
Oregon State University
Corvallis, Oregon 97331

NEMATODA

I. INTRODUCTION

The phylum Nematoda comprises a wide range of roundworms that can be grouped either in nutritional (microbotrophic, predaceous, and parasitic in plants, invertebrates, or vertebrates) or ecological (soil, marine, freshwater, ectoparasitic, endoparasitic, or free-living) categories (Poinar, 1983). Relatively little attention has been devoted to this important group of invertebrates by freshwater biologists because of difficulties associated with sampling, extraction, and identification.

Nematode representatives of essentially all of the nutritional categories mentioned above occur in freshwater during one or more life stages. However nematodes are treated here only if all or a large part of their life cycle occurs in freshwater habitats (these habitats are discussed elsewhere). Nematode parasites of vertebrates that live in or frequent freshwater usually occur in the freshwater habitat only as eggs or within intermediate hosts and, therefore, are not discussed in this chapter. Mermithid parasites of insects are included since the eggs, infective stages, postparasitic juveniles, and adults occur in freshwater habitats.

Only nematode genera containing obligate aquatic species (those which live nowhere else) that have been reported from North America are included in the keys in the present work. The systematic arrangement of these genera is listed in Appendix 9.1. As more studies are undertaken, it is likely that genera described from other continents will also be represented in North America. In addition, the reader should note that

Ecology and Classification of North American Freshwater Invertebrates. 2nd Edition

undescribed or unreported genera may be encountered in routine sampling.

The related group, Nematomorpha, or hairworms, are frequently encountered by freshwater biologists. Superficially, they resemble mermithid nematodes, and share similar life history strategies with certain species of the latter group.

II. MORPHOLOGY AND PHYSIOLOGY

Nematodes are nonsegmented, wormlike invertebrates lacking jointed appendages but possessing a body cavity and a complete alimentary tract (Fig. 1). While they lack specialized respiratory and circulatory systems, nematodes possess a well-developed nervous system, an excretory system, and a set of longitudinal muscles.

With the exception of the family Mermithidae, most freshwater nematodes are under 1 cm in length. Although the basic body shape and anatomic plan of all nematodes are similar, there are some characteristics of most freshwater nematodes that are lacking in many terrestrial forms. One of these is the presence of three unicellular hypodermal glands commonly located in the upper part of the tail, immediately behind the anus (Figs. 1 and 2D). These caudal glands produce a secretion that is carried by a canal to the tip of the tail. At this point, the canal is attached to a valve (spinneret), which passes through a pore in the tail terminus (Figs. 1 and 6d). The secretion produces an adhesive deposit by which the nematodes can attach themselves to the surfaces of submerged rocks, plants, invertebrates, and debris.

Mucus may also be produced by pharyngeal glands of aquatic nematodes. The deposits from the caudal and pharyngeal glands can be used to form slimy traces over the substrate. Microorganisms that become entrapped in these deposits serve as food when the nematodes ingest the mucus together with the attached microorganisms (Riemann and Schrage, 1978).

A. Cuticle

The exterior body covering on all nematodes is a noncellular, flexible, multilayered structure called

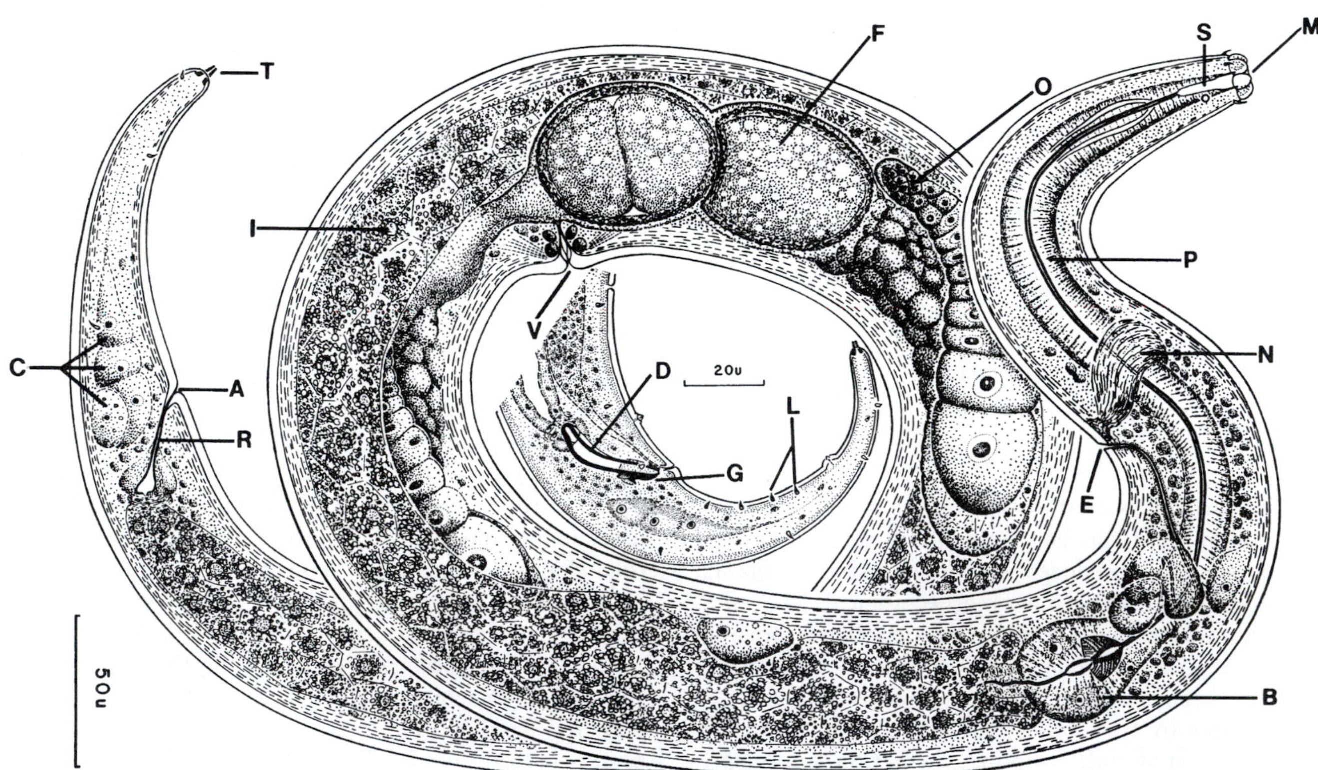

FIGURE 1 Adult female of *Plectus* (Plectidae) [center insert shows tail of male]. M, mouth; S, stoma; P, pharynx; N, nerve ring; E, excretory pore; B, valvated basal bulb of pharynx; I, intestine; R, rectum; A, anus; C, caudal glands; T, spinneret; V, vulva; O, ovary; F, egg; D, spicule; G, gubernaculum; L, genital papillae. [Drawing by A. Maggenti, Univeristy of California, Davis; labeling by G. Poinar.]

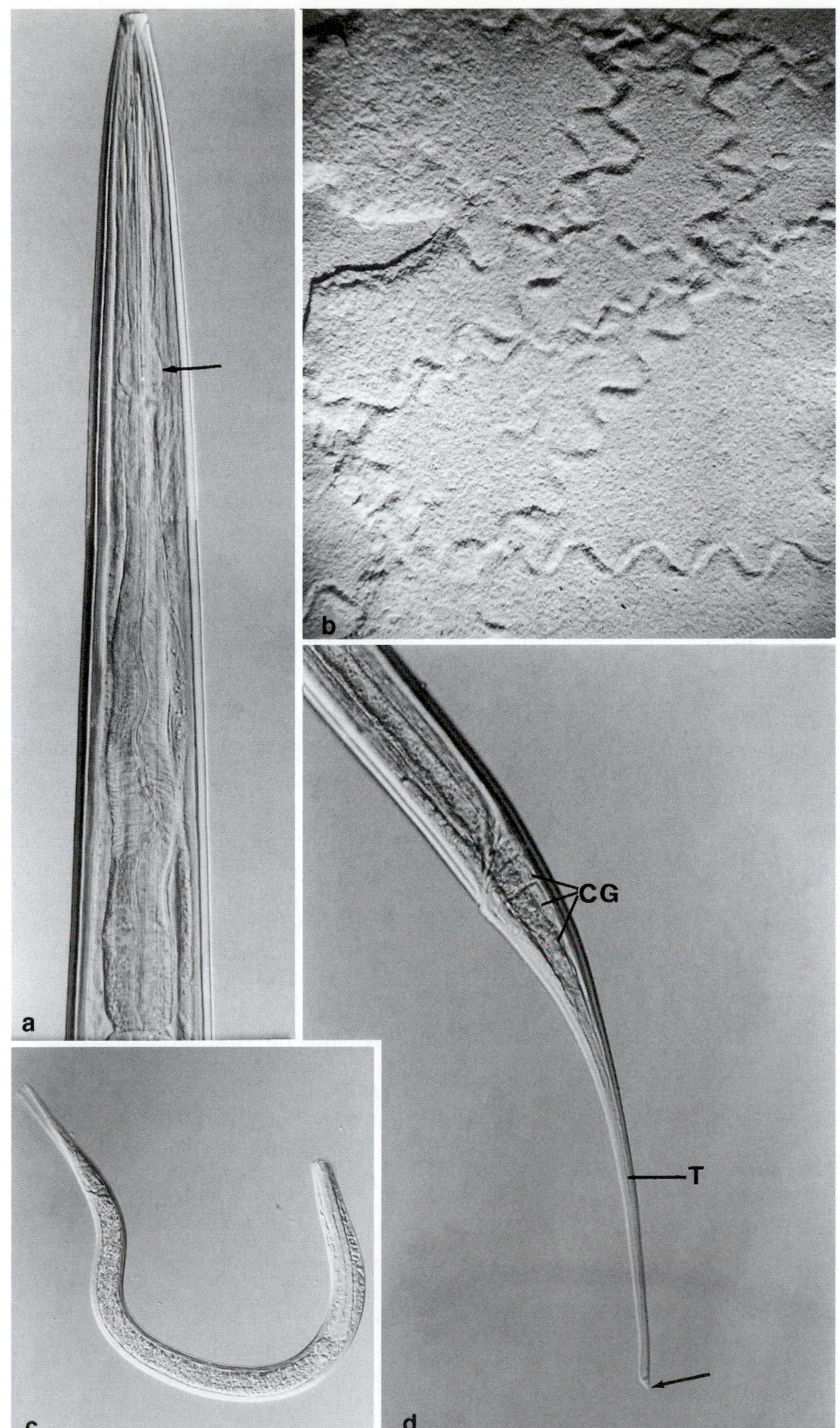

FIGURE 2 (a) Gradually tapering pharynx of a freshwater *Dorylaimus* sp. Note absence of distinct valve in basal portion of pharynx. Arrow shows the nerve ring. (b) Fossil tracts thought to be made by freshwater nematodes. From the 50 million-year-old (Middle Eocene) Green River sediments of Lake Uinta in the Soldier summit area near Provo, Utah. (c) A juvenile freshwater nematode capable of swimming by rapid vibratory body movements. (d) Tail of a freshwater nematode showing three caudal glands (CG), the caudal gland tube (T) and spinneret (arrow).

the cuticle. It is secreted by a layer of underlying hypodermal cells and covers the entire external surface of the nematode, with portions entering various external openings such as the stoma, rectum, vagina, amphids (lateral sense organs), and sometimes the excretory pore. The cuticle is composed of an outer cortex, a middle matrix, and an inner fiber or basal zone. All these regions are constructed of sublayers, which vary in thickness according to the species and stage. The outermost lipoidal layer of the cortex is important because it serves as a semipermeable membrane, allowing water and solutes to pass in and out of the body cavity.

The cuticle of freshwater nematodes may bear longitudinal striations, punctations, bristles, setae, or somatic alae (Figs. 6b and 7a). When present, the bursa is attached to the male tail and is composed of flattened flaps of cuticle which guide and hold the male in place during mating (Fig. 6c). All these cuticular structures are useful characters for identifying nematode groups or species. The cuticle is normally shed four times during development. At each molt (ecdysis), the old cuticle splits and is discarded. In a few forms, the old cuticle is retained, producing a double cuticle or sheath around the nematode (Fig. 8b).

B. Hypodermis

The hypodermis is a layer of tissue that is located beneath the cuticle and is responsible for the formation of the cuticle. It is generally a thin layer of tissue that expands into the coelom to form longitudinal cords between the muscle fields. Most nematodes possess lateral, dorsal, and ventral cords that contain the nuclei and other cytoplasmic inclusions of the hypodermis. The hypodermal cords originate as distinct cells but later lose their identity and form a syncytium.

C. Muscular System

Nematode movement is controlled by longitudinal muscles because these organisms do not possess circular muscles. These somatic muscles are composed of many adjacent cells which are innervated by nerves lodged in the ventral and dorsal hypodermal cords. Most nematode movement superficially resembles that of many other elongate, stiff-bodied, appendageless animals, whether they be vertebrates (snakes) or invertebrates (biting midge larvae). In locomotion, the side-to-side bending of the body is produced by alternating contractions and relaxations of opposite muscle sectors that are controlled by the central nervous system. When placed on flat surfaces in the laboratory, nematodes normally crawl on their sides. Some aquatic nematodes can swim by rapid vibratory or thrashing side-to-side movements of the body.

D. Digestive System

All freshwater nematodes, except the Mermithidae, contain a continuous alimentary tract composed of a stoma, pharynx, and intestine (Fig. 1). The stoma (mouth) is the most anterior part of the digestive tract and is usually lined with cuticle. The stomal walls are formed by a fusion of various smaller segments or rings (rhabdions), which may be separated as is found in the Cephalobidae, or fused, as in the Rhabditidae (Fig. 8a). The structure of the stoma generally reflects the food selection of the bearer. For example, predaceous nematodes tend to have a wide stoma armed with teeth (Fig. 7b), a stylet, or a spear (Fig. 9b), whereas microbotrophic forms generally have a small tubular stoma (Fig. 8a), and omnivores a funnel-shaped mouth cavity (Fig. 10a). Plant parasites have protrusible stylets which can be inserted into plant tissues (Fig. 8b), and the infective stages of invertebrate parasites often have a small stylet used in penetrating the cuticle or intestinal wall of the host.

The pharynx (or esophagus) lies between the stoma and intestine and passes food from the mouth into the intestine. It is usually composed of radial muscles, which upon dilation allow the pharynx to function as a pump or vacuum. The pharynx also contains glands that produce secretions used in digestion, escape from the egg shell, molting, and penetration of host tissue. There are normally three portions to the pharynx: (1) the corpus (sometimes enlarged to form a distinct metacorpus or median bulb); (2) the isthmus (usually surrounded by the nerve ring); and (3) the basal bulb (often containing valves to keep food particles from reversing their direction of flow) (Fig. 8a). When nematodes feed, material is sucked up into the mouth by reduced pressure resulting from contraction of the radial muscles of the pharynx. Food is passed along the lumen by waves of alternating contractions and relaxations of the pharyugeal muscles. The pharyngeal lumen opens behind the food, thus creating suction pressure, and closes in front of the food, forcing it through.

The intestine is a single-cell-thick, cylindrical tube that runs from the pharynx to the anal opening (cloacal opening in males). It is composed of an anterior ventricular region, a midintestine or intestine proper, and a posterior rectal region (not always obvious). The inner surfaces of the cells facing the lumen are lined with microvilli which seem to be responsible for the major uptake of nutrients. A pharyngeal-intestinal valve is located in the anterior portion, behind the basal bulb of the pharynx. In some freshwater forms, this valve is quite distinct and elongated (10c). An intestinal–rectal valve is located at the junction of the intestine and rectum. This unicellular sphincter muscle controls the amount of water passing out of the body. Many rhabditid-type nematodes have three rectal glands, which empty into the rectum. These glands are not to be confused with the caudal glands, which produce secretions from the tail terminus (Fig. 2d).

The Mermithidae are unusual in having the intestine detached from the remainder of the alimentary tract and modified into a trophosome or food storage organ. During development, the trophosome separates from the pharynx and rectum, thus remaining as a separate, isolated organ.

E. Excretory System

Although variable in structure, the excretory system of nematodes falls into two basic types. The first is frequently found in freshwater forms and consists of a ventral excretory gland (renette cell) connected by an excretory duct to the excretory pore on the surface of the cuticle. The ventral pore usually opens in the pharyngeal or anterior intestinal region of the body. The second type is a tubular system consisting of a series of longitudinal excretory canals; pooled contents of these canals pass into a joining canal which connects with the excretory duct and pore. Most freshwater nematodes excrete nitrogen in the form of ammonia or urea. The excretory system also has an osmoregulatory function of removing water that diffuses into the pseudocoelom and so regulates the amount of turgor pressure in the body cavity; some scientists believe that this is the primary function of the "excretory" system.

F. Respiration

Most nematodes are aerobic organisms, at least during their developmental period. Many nematodes can survive short periods of anaerobic conditions but only a few forms can survive anoxia indefinitely. Strayer (1985) listed representatives of three nematode genera that he considered as benthic species found under anoxic conditions in Mirror Lake. A specialized respiratory system is lacking in nematodes. Instead, they obtain and lose gases by simple diffusion. Body cavity cells, when present, appear to play no role in oxygen transport.

G. Nervous System

The nervous system of nematodes centers around a central anterior mass of ganglia or "brain," which connects with nerve cords extending anteriorly and posteriorly through the body. The central ganglionic mass is closely associated with the circumpharyngeal commissure or nerve ring (Fig. 2a). Nerves extending anteriorly from the nerve ring innervate the amphids and cephalic sense organs, while those running posteriorly control the male tail papillae.

H. Sense Organs

The size, shape, position, and spatial arrangement of sense organs are all characteristic of nematode species or groups and are therefore frequently used as taxonomic characters. The anterior sensory papillae are found on the nematode head. The basic number is 16, but usually some have become vestigial. There is normally an inner circle of six labial papillae around the mouth, a second ring of six labial papillae slightly behind the first set, and a third circle of four cephalic papillae still further back on the head. In mermithids, the cephalic papillae may also include adjacent enlargements of the hypodermal tissue.

The amphids are paired, lateral sense organs that open to the exterior on the nematode cuticle. They may be located on the same circle as the labial or cephalic papillae or further back in the neck region. The term amphid generally refers to the exterior configuration of the amphidial opening, although the amphid also consists of an amphidial pouch, amphidial gland, and amphidial nerves. The structure of the amphidial openings and their location are very important taxonomic characters. They can vary from small pores to large circles (Fig. 7c), pockets (Fig. 10b), or spirals (Fig. 7d). Sexual dimorphism may be evident with either the male or female possessing a considerably larger amphid usually of the same general shape. Amphids may be multifunctional and evidence suggests that they serve as photoreceptors, olfactors, or sensory glands. Some are sensitive to pH and ions.

Other innervated papillae on nematodes are deirids (paired papillae located laterally near the nerve ring), postdeirids (similar structures located at the midbody), and genital papillae (found on the ventral surface around the cloacal opening of males). When elongate, the genital papillae are usually attached to thin membranous outgrowths of cuticle (bursa) (Fig. 6c). The presence or absence of a bursa and its structure are important taxonomic characters.

A few freshwater nematodes possess what are thought to be light receptor organs. These receptors normally consist of pigmented areas in the neck region, which may or may not be associated with a cuticular-structured lens. If a lensatic body is present, then the organ is sometimes referred to as an ocellus or pseudocellus.

I. Reproductive System

The majority of nematodes, including the freshwater forms, reproduce by amphimixis (sperm and eggs come from separate individuals). However, some forms have developed uniparental reproduction (i.e., autotoky), and this condition also occurs in aquatic genera. In the latter, autotoky usually expresses itself as parthenogenesis, where progeny arise from unfertilized eggs. Hermaphroditism seems quite rare among freshwater nematodes (Poinar and Hansen, 1983), apparently occurring in *Chronogaster troglodytes* (Poinar and Sarbu, 1994) and some others. Asexual reproduction does not occur in the Nematoda.

In amphimictic reproduction, the male nematode always places sperm inside the female; external fertilization has never been demonstrated. Some aquatic species exhibit "traumatic insemination" when the male penetrates the female cuticle with his spicules and releases sperm into her body cavity.

Female nematodes contain one or two gonads, which open to the exterior on the ventral side of the body at the vulva (Figs. 1 and 6a). The vulva is connected to a muscular tube, the vagina, which in turn leads into the uterus followed by the oviduct and ovary. A spermatheca (sperm-collecting area) is usually present between the uterus and oviduct. With double-ovary species (didelphic), the gonads are usually opposite and join at a common vagina (Fig. 6a). Single-ovary forms (monodelphic) have retained the anterior ovary, which is often reflexed into the posterior part of the body (Fig. 8d).

The male may have a single testis (monorchic) or two testes (diorchic), which lead to a common seminal vesicle and vas deferens before entering the cloacal chamber, a common opening for the reproductive and digestive systems. Males of freshwater nematodes all possess one or two sclerotized structures termed spicules (Figs. 1 and 6c). These are inserted into the vagina of the female at insemination (except in traumatic insemination when they penetrate the cuticle of the female). Some males also have another sclerotized structure, the gubernaculum. The spicules are contained within the walls of the cloacal chamber while the gubernaculum is attached to the floor of the cloacal chamber. The latter structure supports the spicules during their movement in and out of the cloacal opening (Fig. 6c). On the ventral surface of the male tail is usually some type of genital papillae, which may or may not be supported by a bursa (Fig. 6c).

Through the release and detection of sex attractants, male and female nematodes are able to locate each other. Once together, the nematodes intertwine and the posterior region of the male contacts the female vulva. Sometimes an adhesive compound is produced to seal the union. Spicules are then inserted and sperm is transferred from the vas deferens of the male into the vagina and uterus of the female. The pair may remain joined for only several minutes or may be united for days. It is common to find "balls" of mermithids in stream bottoms all coiled together in continuous copulation. Individual sperm as well as secondary spermatocytes may be transferred during mating. In many nematodes, sperm maturation is completed in the female. Nematode sperm is variable in size and shape, but is always nonflagellated and usually amoeboid.

Parthenogenetic forms can reproduce by mitotic or meiotic parthenogenesis. In the former, the diploid somatic number of chromosomes is retained, whereas in the latter, two maturation divisions occur which are similar to oogenesis in amphimictic species. A diploid chromosome number is established by fusion of the nonextruded polar nucleus with the egg pronucleus or by doubling of the chromosome before the first cleavage division occurs.

III. DEVELOPMENT AND LIFE HISTORY

The eggs of nematodes differ from all other stages in containing chitin in their shells. During embryonic development, nematodes show predeterminate cleavage, which means that cells destined to form specific tissues appear early in the embryo. Freshwater forms like the mermithids may deposit eggs with fully formed juveniles ready to hatch upon receiving the right stimuli. Other species deposit eggs in the single-cell stage, with embryonic development occurring in the environment. Postembryonic development in nematodes (development after hatching) is similar to the gradual type of metamorphosis in insects. Aside from an increase in size, proportional changes of various organs, and the development of the gonads and second sexual characteristics, there are few differences between juvenile and adult nematodes (Fig. 11c). For this reason, the immature stages of nematodes should be called juveniles and not larvae, because the latter usually implies complete metamorphosis where the immature stages differ radically from the adults. [Despite these differences, however, the term larva is widely used today in referring to the immature, postembryonic stages of nematodes.]

Hatching is influenced by temperature and external stimuli. The eggs of the mermithid *Pheromermis pachysoma* hatch when ingested by aquatic insects.

Eggs of other freshwater nematodes hatch when the water reaches a certain temperature. Eggs that undergo anabiosis as a result of desiccation will not hatch until the area is flooded. This probably enhances survival for nematodes in temporary water sources, although this point still requires elucidation. In most cases, it is the first-stage juvenile that emerges from the egg; however, in mermithids, juveniles molt once in the egg and then emerge as second-stage juveniles. If equipped with a spear or tooth, the young nematode will use these to break out of the shell, otherwise it simply employs pressure and pharyngeal secretions to soften the shell wall.

All nematodes undergo four molts and have six stages during their development (i.e., egg, first-stage juvenile, second-stage juvenile, third-stage juvenile, fourth-stage juvenile, and adult). During each molt, the cuticle is shed and replaced by another secreted by the hypodermis. Sometimes two molts may occur almost simultaneously, as in postparasitic juvenile Mermithidae (Fig. 9a), and occasionally the shed skin is retained around the body of the next stage (Fig. 8b). Free-living nematodes generally complete their development rapidly in comparison to other metazoans. This can be 3–5 days for rhabditids (under optimum conditions) and generally from 1 to 6 weeks for most other freshwater forms. Therefore, nematode populations can turn over very quickly (see Schiemer, 1983, 1987 for energetic studies on nematodes).

Aside from premolt and quiescent stages, most nematodes feed continuously throughout the growth period. Most freshwater nematodes take food in through the mouth and absorb nutrients into the intestinal cells. During their parasitic development in the hemolymph of invertebrates, mermithids absorb nutrients directly through their body walls. Nematode growth entails both enlargement and multiplication of cells. Growth is frequently nonproportional from stage to stage. For example, the intestine often will grow faster than the pharynx, so the proportion of the two organs differs in each stage. This information can be used to determine nematode stages, along with gonad development.

The types of food vary among the freshwater nematode groups. Microbotrophic forms ingest bacteria, algae, single-celled fungi, and protozoa. Predaceous species attack small metazoans such as nematodes, annelids, early insect stages, and molluscs. Many are omnivorous and will consume microbial life as well as metazoans. Plant parasites feed on the cytoplasm of higher plants, algae and filamentous fungi. Parasites of invertebrates absorb nutrients from the hemolymph of various taxa, especially insects in the case of the Mermithidae and snails in the case of *Daubaylia*. Vertebrate parasites may occur in the guts or body cavities of fish, aquatic birds, or amphibians, where they feed on host fluids and tissues. Only the egg stages are free-living and these are not likely to be encountered or recognized during routine collecting activities.

The dependence of nematodes on moisture has brought a tremendous selection pressure to withstand periods of desiccation. Some nematodes form resistant stages, while in others, all life stages can tolerate drought conditions for various periods. Resistant stages include the egg and the second- through fourth-stage juveniles. Some nematodes have been maintained for years (up to 25 years) in anabiosis and then revived in water.

Very little is known about the resistant stages, dispersal, and survival of freshwater nematodes. Jacobs (1984) mentions the "ability for passive dispersal" of resistant forms, but does not discuss how this dispersal occurs. It is very likely that the eggs and, possibly, some juvenile stages of forms adapted to semitemporary water sources can enter a resistant, quiescent stage capable of surviving desiccation and possibly temperature extremes; these have been demonstrated in some soil and plant nematodes. Passive dispersal may occur when these resistant stages are blown by heavy winds or washed to new areas by flash floods. The possibility of freshwater nematodes being carried in mud attached to the body parts of various water-frequenting animals, such as wading and swimming birds, is also feasible. The presence of freshwater nematodes in temporary ponds has not been investigated.

IV. ECOLOGY

A. Fossil Record

The oldest known possible fossil record of aquatic nematodes consists of tracks dating from the Upper Precambrian of Australia and Europe (Hantzschel, 1975). Tracks similar to these in the Green River sediments of Lake Uinta (Middle Eocene) (Fig. 2b) were considered to be made by nematodes (Moussa, 1969). These more recent tracks occur in the Soldier summit area near Provo, Utah, and were made during the second lacustrine phase of the Green River sediments. Lake Uinta was a large freshwater lake that occupied the Uinta and Piceance Creek Basins in Utah and Colorado. These tracks could have marked the passage of fairly large freshwater nematodes moving near the shore. The early Precambrian tracks could well be nematodes since it is highly probable that the phylum dates back to that period when representatives of the now predominantly aquatic order Araeolaimida probably existed (Poinar, 1983) and biting midges

(Ceratopogonidae: Diptera) were not present, which are a group of insects whose legless larvae move in a fashion similar to nematodes.

The aquatic mermithids represent the earliest and most frequenty encoutered undisputed nematode fossils. The earliest of these occurs in 120–135 million-year-old amber from Lebanon (Fig. 12a). It was originally described as occurring in a biting midge (Ceratopogonidae: Diptera) (Poinar *et al.*, 1994), but according to Art Borkent (personal communication) the host is actually a member of the Chironomidae. Since the generic placement of this specimen was based in part on its host family, it is now prudent to transfer the species *libani* from *Heleidomermis* Rubstov into a new genus *Cretacimermis* Poinar. Thus, this nematode, which is the oldest definite fossil nematode, should now be called *Cretacimermis libani* Poinar, Acra and Acra (1994). Another aquatic mermithid fossil species is *Heydenius dominicanus* from Dominican amber (Poinar, 1984). Two specimens of this species were found in close association with an adult cranefly (Tipulidae: Diptera) and an adult mosquito (Culicidae: Diptera) (Fig. 12b). Based on the record of present-day mermithid infections of these two host groups, and the discovery of a female mosquito with a mermithid still coiled in its abdomen in Dominican amber, it can be assumed that *H. dominicanus* was a parasite of the adult mosquito. Such rare fossils show the antiquity of some parasitic associations.

B. Habitats of Freshwater Nematodes

Nematodes constitute an important and significant portion of the zoobenthic community of freshwater habitats. This community has been separated by benthic ecologists into three size categories (Strayer, 1985). Animals retained on sieves with a mesh ranging from 200 to 2000 μm compose the macrofauna, those that are retained on sieves with a mesh ranging from 40 to 200 μm construct the meiofauna, and those that pass through a 40-μm mesh sieve make up the microfauna. Freshwater nematodes covered in the present chapter could be considered as belonging to all groups; the large free-living stages of the Mermithidae would be considered macrofauna, the majority of the adults and juveniles of the other groups would fall under meiofauna, and the young juveniles and eggs of many forms would fall into the microfaunal category. Thus, for nematodes in general, these size categories are little used, although they are helpful if only certain stages are being studied.

Nematodes constituted 60% of all the benthic metazoans in Mirror Lake, New Hampshire, and are probably the most abundant benthic animals in freshwater (Strayer, 1985). In Mirror Lake, nematode abundance showed a distinct depth distribution, with a strong minimum at 7.5 m. More than 90% of the nematode biomass at 7.5 m was composed of species of the large predatory nematodes belonging to the genus *Mononchus*. Strayer (1985) also reported that 63% of the nematodes in Mirror Lake lived in the top 2 cm of sediment, while only 18% penetrated to a depth greater than 4 cm. Species of the genera *Monhystera* and *Ethmolaimus*, which constituted 55% of the nematodes in Mirror Lake, were found in the benthic zone. Although the nematodes outranked all other animals in abundance (constituting 59% of all metazoan individuals in the benthos), they still represented only 1% of the zoobenthic biomass. Nematodes normally contribute between 1 and 15% of the zoobenthic biomass in lakes, and other studies show that they constitute from 40 to 80% of all meiobenthic animals (Holopainen and Paasivirta, 1977; Oden, 1979; Nalepa and Quigley, 1983).

Nematodes that occur in freshwater have traditionally been called free-living nematodes. The term free-living is an ecological ranking, which means that these nematodes have no symbiotic association with multicellular plants or animals. Such nematodes obtain nourishment from bacteria, algae, protozoa, and other unicellular organisms. Nematodes with this type of nutrition are best referred to as microbotrophic, although they have been called saprophytic, saprophagous, bacteriophagous, microbivorous, and microphagous. Such nematodes act as secondary consumers, feeding on bacteria and fungi. They may serve to maintain these microbial populations, although many of the free-living forms will die if bacterial populations reach high numbers, which is why most of the normal aquatic nematodes are absent from areas high in sewage.

The term free-living, however, is also used to refer to a particular stage in the life cycle of a nematode—often a stage in the life cycle of a plant-parasitic or animal-parasitic nematode—when it has either ended or not yet begun its parasitic relationship. Examples in the freshwater habitat are the mermithids. The egg, infective second-stage juvenile, postparasitic juvenile, and adult mermithids are free-living and nonfeeding. Nourishment is obtained from the hemocoel of insects only during a portion of their juvenile development. Thus, free-living in the present work refers to those nematodes that have some developmental stages occurring in freshwater, irrespective of their nutritional requirements or their status at other developmental stages.

The habitats of freshwater nematodes vary as do the categories of freshwater resources. Two important requirements for freshwater nematodes to complete their development are continuation of the water source and availability of food and oxygen. Many freshwater nematodes feed on microorganisms that flourish in the gyttja, an organic deposit composed of excretory material from benthic animals. It is most obvious in lakes and ponds, where it is often combined with decomposing plant debris. This sediment increases in density with increasing depth. Core samples are used for quantitative measurements of zoobenthos in the gyttja (Strayer, 1985).

Annual productivity of freshwater nematodes in an Austrian alpine lake was studied by Bretschko (1973). In the deeper benthos, below 20 m, there were 2–4 annual generations with the total yield of nematode biomass estimated at 66 kg per year. It is interesting that the production was higher in shallow water and during the winter when the lake was covered with ice. For example, *Tobrilus grandipapellatus* was collected at populations of 235,000 individuals/m^2 in the winter, but only 60,000 individuals/m^2 in the summer. This could be related to the availability of oxygen and density of micribial populations.

Nematodes of all environments, including freshwater, are among the most numerous animals feeding on both primary decomposers such as bacteria and fungi, as well as primary producers such as algae and higher plants (Nicholas, 1984). The significance of nematodes in the general economy of the ecosystem is not known, and attempts to estimate their contributions to energy flow have been made only with terrestrial and marine forms. Such estimates have used the equation:

$$C = P + R + E + U$$

where C is the consumption, P the production, R the respiration, E the ejecta (feces), and U is excretion.

Although freshwater nematodes have been little studied, estuarine forms have been examined in detail (Warwick and Price, 1979). Some 40 species of nematodes occur on a mud flat in the Lynher Estuary in southern Britain. The population varies from 8–9 × 10^6 individuals/m^2 in winter to nearly 23 × 10^6 individuals/m^2 in late spring (15% of the macrofauna). Warwick and Price (1979) calculated a total annual oxygen (O_2) consumption of 28 liter O_2/m^2 per year, which is equivalent to some 11 g of metabolized carbon (C). The annual production was calculated at about 6.6 g C/m^2 per year. Previous calculations with the soil nematode *Caenorhabditis briggsae* showed that the conversion of bacteria into nematode tissue varied from 13 to 20% (Nicholas, 1984).

The question of nematodes living under anoxic conditions has been addressed by several workers. In deep lakes, the hypolimnion may become anaerobic for certain periods, resulting in a reduction of nematode fauna, but not eradication. In Lake Tiberias, in Israel, Por and Moary (1968) found large numbers of *Eudorylaimus andrassyi* at a depth of 43 m during the winter oxygen-free period.

In the Neusiedlersee in Austria, *Tobrilus gracilus* occurs in muddy anaerobic zones beyond the limits of emergent vegetation (Schiemer, 1978). In his study of the nematodes in Mirror Lake, Strayer (1985) found most species in the littoral zone, but species of *Ethmolaimus, Monhystera,* and unidentified Tylenchidae were highly abundant in the anaerobic sediments at a depth of 10.5 m.

Extreme habitats for freshwater nematodes include high-temperature hot springs. Table I lists some freshwater nematodes collected at unusually high temperatures. The listing of 61.3°C for *Aphelenchoides* sp. is not only a record for nematodes, but for all metazoan life forms!

TABLE I Survival of Freshwater Nematodes in Hot Water Springs at High Temperatures

Nematode species	*Temperature* (°C)	*Location*	*Reference*
Aphelenchoides sp.	61.3	New Zealand	Rahm (1937)
Aphelenchoides sp.	35.0	New Zealand	Winterbourn and Brown (1967)
Aphelenchus sp.	57.6	Chile	Rahm (1937)
Doryalimus atratus Linstow	47.0	Italy	Issel (1906)
Dorylaimus atratus Linstow [*D. thermae* Cobb]	40.0	Wyoming, USA	Hoeppli (1926)
Dorylaimus atratus Linstow [*D. thermae* Cobb]	53.0	Wyoming, USA	Cobb, in Hoeppli (1926)
Euchromadora striata (Eberth)	52.0	Italy	Meyl (1954)
Monhystera ocellata Bütschli	52.0	Italy	Meyl (1954)
Monhystera gerlachii Meyl	52.0	Italy	Meyl (1954)
Plectus sp.	57.6	Chile	Rahm (1937)
Rhabdolaimus brachyuris Meyl	52.0	Italy	Meyl (1954)
Theristus pertenuis Bresslau and Schuurmans Stekhoven	52.0	Italy	Meyl (1954)
Tylocephalus sp.	45.3	New Zealand	Winterbourn and Brown (1967)

Other specialized habitats encompass brackish and estuarine waters, inland saline lakes, cave streams, and associations with specific aquatic plants. "Artificial water sources" include canals, waste water, tap water, wells, filter beds of sewage treatment plants, and temporary water sources. Nematodes found in these habitats often do not have any relation to true freshwater forms or a freshwater habitat, and the majority are not treated in this work.

A unique cave habitat has resulted in what appears to be the first aquatic nematode specialized for survival in hydrogen sulfide-rich thermo-mineral waters. These conditions are found in the Movile cave, located in a limestone plateau in Southern Dobrogea, Romania (Poinar and Sarbu, 1994). The nematode, *Chronogaster troglodytes,* lives in floating microbial mats comprised primarily of mycelia of fungi belonging to the class Oomycetes and various bacterial populations. Selection for this habitat could have undergone millions of years since a Miocene age for the origin of Movile cave has been suggested. Also, while members of the genus *Chronogaster* are mostly aquatic, they, like many freshwater nematodes, are quite broad in their selection of habitat. Other species of *Chronogaster* have been collected from sandy or sandy loam soils, clay soils, pasture soils and even forest soils. In fact one species, *C. africana* is mentioned as occurring in multiple aquatic and terrestrial habitats (Heynes and Coomans, 1980). Such plasticity was probably an important feature in the specialization of *C. trogoldytes* in its unique underground cave habitat. A list of nematodes recovered from caves and subterranean waters is presented by Poinar and Sarbu (1994).

Previous workers have attempted to separate the freshwater-dependent nematodes from the facultative forms. Micoletzky (1922) distinguished between completely aquatic, principally aquatic, amphibious, and principally terrestrial species; while more recently, Jacobs (1984) presented an ecological classification with stenohygrophilic nematodes representing the strictly aquatic forms and enhygrophilic nematodes representing the semi-aquatic forms. The strictly aquatic or stenohygrophilic taxa covered here occur in many microhabitats (Fig. 3). As examples, periphytic forms live among plant roots, bryophilic taxa associate with moss and liverworts, endobenthic species thrive within the sediment and bottom debris, epibenthic groups live on the surface of the bottom, haptobenthic forms exist on

FIGURE 3 Scene of the bottom of a typical fast-flowing stream showing several nematode microhabitats. (A) Stream bed containing endobenthic forms; (B) bed surface containing epibenthic groups; (C) rock surface containing haptobenthic groups; (D) *Nostoc* algal pads containing mermithids that parasitize *Cricotopus* midges (Chironomidae) living inside the algae.

the surface of submerged rocks, debris, and aquatic plants, and planktonic taxa live continuously in the water column. The latter group is poorly known and mainly found in turbid waters where their occurrence may represent more of a dispersal than a habitat mode.

Many genera of freshwater stenobygrophilic nematodes are widely dispersed, occurring on more than one continent. Some species are eurytopic, occurring in diverse habitats, while others are stenotopic and restricted to a few specialized habitats.

Freshwater nematodes have been considered as indicators of water pollution (Zullini, 1976). Heavy metals and other pollutants settle and are taken up by organic matter in the sediments. Nematodes ingest this material and those that are sensitive to the pollutants may die. Representatives of the Chromodorida are apparently more sensitive to pollution that those of the Rhabditida (Zullini, 1976). In extremely polluted waters, only the Rhabditoidea may be quite abundant. A two-year study sampling nematodes along two streams in Indiana indicated that an analysis of the benthic nematode community structure could be useful in evaluating disturbance to aquatic habitats (Ferris and Ferris, 1972).

C. Enemies

Perhaps the greatest enemies of freshwater nematodes are other predaceous nematodes. Representatives of the Mononchida, Dorylaimida, and Enoplida are known to attack and devour a range of small invertebrates in their surroundings. The crayfish *Pacifastacus leniusculus* was observed to feed occasionally on nematodes (Flint, 1976). The freshwater, rhabdocoel turbellarian, *Microstomum* feeds on nematodes (Stirewalt, 1937), as will the freshwater nemertean worm, *Prostoma* (Coe, 1937). The role of disease organisms in regulating populations of freshwater nematodes is not known, but the author has frequently collected specimens of *Tobrilus* infested with what appeared to be microsporidian spores (Fig. 9c). Other records of probable protozoan diseases in freshwater nematodes have been reported in *Chromadora, Dorylaimus, Ironus, Monhystera, Plectus, Theristus, Trilobus, Tripyla, Prodesmodora, Achromadora, Paraphanolaimus,* and *Desmolaimus* (Poinar and Hess, 1988).

V. COLLECTING AND REARING TECHNIQUES

A. Sampling Methods

There are two basic types of sampling: qualitative and quantitative. The former method is used in preliminary studies to determine which kinds of nematodes are present in the ecosystem. The latter method is used to evaluate the components of a population, the proportion of each in the environment, and the dynamics of the population over time.

Qualitative methods for sampling freshwater nematodes depend on the water flow and the zone to be sampled. In sampling nematodes from beds of fast-flowing streams, a net, an instrument to turn over rocks and debris on the stream bed, and a set of sieves, are necessary (Fig. 4a). The net is first anchored with the opening facing upstream. A pick or geologist's hammer is used to disturb the stream bed slightly upstream from the net (Fig. 4b). The stream washes nematodes and other debris from the disturbed area into the net. The contents of the net are then placed in enamel collecting pans with water and set aside for extraction. Nets can be pulled over aquatic plants to obtain nematodes living on their surfaces. The type of nematode being collected will determine the mesh size of the net selected. It is recommended that, in order to obtain a fair sample of all nematodes, a mesh size no larger than 40 μm be used.

A number of devices have been devised to collect freshwater nematodes (Filipjev and Schuurmans Stekhoven, 1959). These include dredges, wormnets, bottom catchers, mudsuckers, swab or sledge trawls, and plankton nets. Their use depends on the type of material desired. For sluggish or stationary water sources, samples of the bottom (mud, sand) can be placed in a water-filled container (Fig. 4c) which is then shaken, the suspension allowed to settle for 3–4 sec, and the supernatant then poured into a second container for extraction. Rocks and other submerged material can be placed in buckets and the nematodes washed off with a water spray or by vigorous shaking or brushing.

Quantitative sampling methods involve the use of augers or probes to remove core samples of measurable sizes from the beds (Strayer, 1985). Other quantitative methods can be devised on the basis of need and the desired biotype to be investigated. Techniques applied to marine nematodes may also be suitable for freshwater forms. For further information, see Hulings and Gray (1971), Holme and McIntyre (1971), and Downing and Rigler (1984). It should be remembered that any estimate of the nematode population depends on the type of sample taken. Thus, a biased sample will produce a biased estimate of the nematode population. Also, it should be noted that nematodes are sensitive animals and all sampling and extraction techniques should be as gentle as possible.

B. Extraction

The process of extraction involves removing nematodes from samples collected in various freshwater

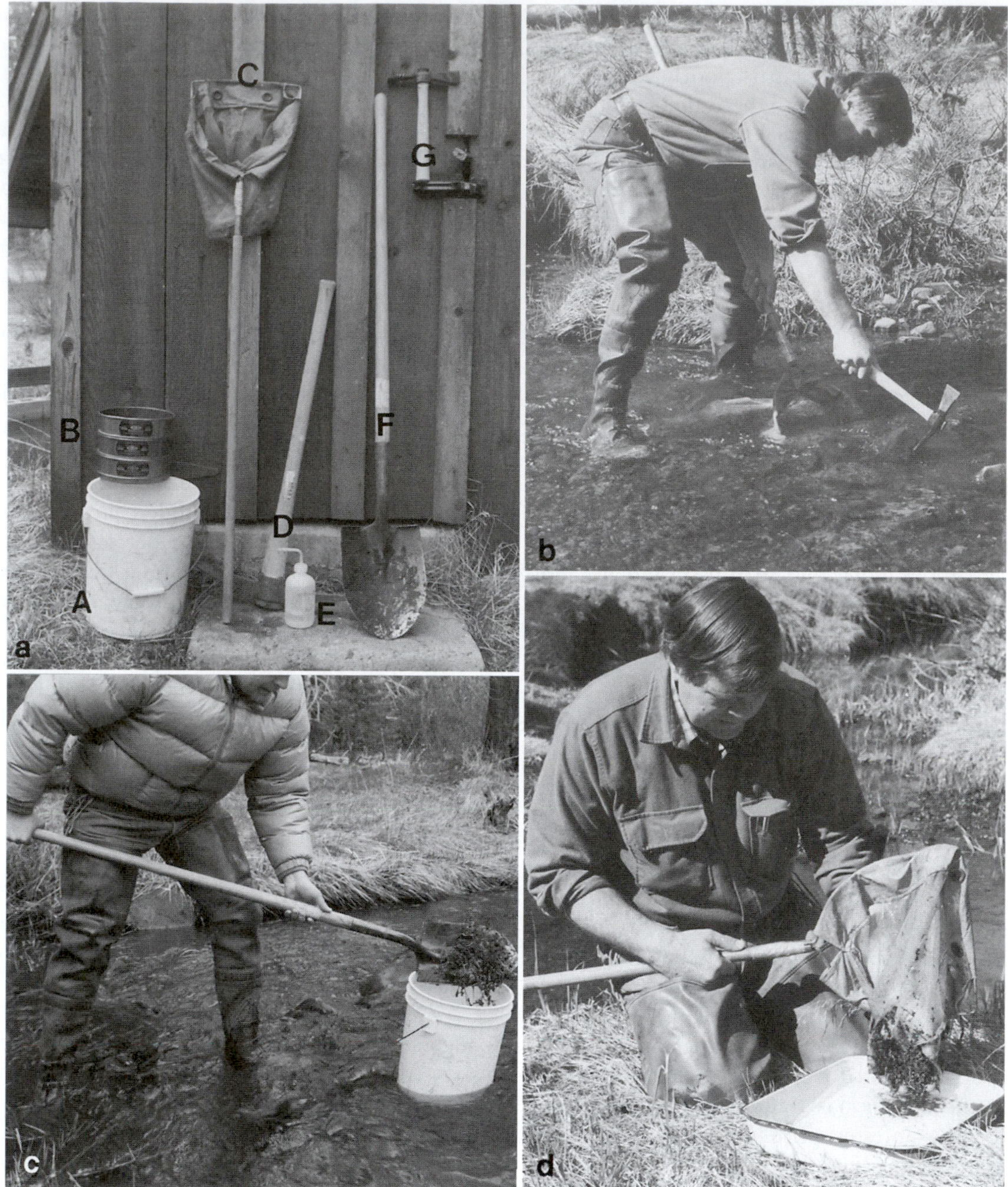

FIGURE 4 (a) Equipment needed to sample freshwater nematodes. A, pail; B, set of gradated sieves; C, net; D, pick; E, squeeze bottle; F, shovel, G, geologist's hammer. (Hip boots are not shown.) (b) With the geologist's hammer, the stream bed is disturbed upstream from the positioned net. (c) With a shovel, samples of the stream bed are removed together with roots of aquatic plants. (d) Large nematodes (mermithids and dorylaimids) and debris are transferred from the net into an enamel examination pan.

habitats (Figs. 4d, 5a–d). The basic principles of nematode extraction are based on the fact that most nematodes have a specific gravity between 1.10 and 1.14, thus they will slowly sink in freshwater. They are denser than fine clay particles, but lighter than sand. The extraction methods used for soil and plant-parasitic nematodes are wholly adequate for freshwater forms.

Although the majority of freshwater nematodes are too small to view with the naked eye, the free-living stages of many aquatic mermithids are large enough to be handpicked from the samples (Fig. 5c). They can be carefully lifted with a fine needle containing an L-shaped bend at the tip or with a pair of fine forceps (Fig. 5d). Other mermithids that are small enough to

FIGURE 5 (a) Nematodes and debris collected from the stream bed are passed through a series of three sieves with openings ranging from 0.4–0.04 mm across. (b) Nematodes are washed off the surface of the final screen with a spray of water from the plastic squeeze bottle. (c) Free-living stages of mermithids are large enough to be seen with the naked eye and can be picked up with forceps. (d) Larger nematodes can be transferred directly to a small vial for later killing and fixing.

escape visual spotting can be collected with the extraction processes.

One of the oldest and simplest methods of extracting nematodes from a wide range of samples is the Cobb sieving and gravity method, or a modification of the latter. This consists of making a water suspension of the nematodes and pouring it through a series of screens of different mesh sizes to collect the nematodes. The suspension is made by placing the samples in a pail or other suitably large container (Fig. 4c), adding 2–3 L of water (preferably from the original source), allowing the mixture to set for 10–20 sec to allow the heavier particles to sink, and then pouring the suspension containing the nematodes through a series of screens (Fig. 5a). If extraneous large debris (wood particles, rocks) is present, the samples should first be poured through a 20–40 mesh sieve (openings are 0.840–0.350 mm). The sizes of sieves selected will, of course, influence the size range of nematodes extracted. If a sample of all available nematodes is desired, then a final fine screen of 400–500 mesh (openings are 0.035–0.026 mm) can be used. The problem with the fine screens is that they become easily clogged with debris. Such screens should be tapped rapidly and gently on the under surface with the fingers to assist passage of the water. The nematodes, together with debris, will be trapped on the surface of the 400- or 500-mesh sieve. Both can be removed by directing a small jet of water on the back of the sieve and washing the contents into a separate container (Fig. 5b). Certain biases are inherent in sieving extraction methods (Hummon, 1981). Since they may be unsuitable for quantitative studies (Viglierchio and Schmidt, 1983), the more tedious method of sorting through unsieved samples under a dissecting microscope may be desirable (Strayer, 1985).

The final step consists of separating the nematodes from the fine debris. For this, the Baermann funnel method, or a modification thereof, can be used. All that is needed is a funnel with a piece of rubber tubing attached to the stem. The tube should be closed with a screw clamp. The nematode–debris mixture is placed on a fine cloth or facial tissue, which is supported by a piece of screen held in the top, wide portion of the funnel. Water is slowly added to the funnel until the nematode–debris mixture is just covered. The clamp should be opened to eliminate the air trapped in the funnel stem and to allow a continuous column of water to flow into the funnel. The nematodes crawl through the debris and facial tissue, drop through the screen, and settle at the base of the funnel stem, where they can be drawn off by releasing the clamp. The nematodes should be drawn off every 6 h; the apparatus can be operated in this fashion for several days. Then the nematodes can be counted or handpicked under a dissecting microscope.

A more recently devised method of isolating nematodes from samples involves the principle of increasing the specific gravity of the solution to make the nematodes float to the surface. Substances such as sugar, salt, or Ludox can be added to the nematode–water mixture in the centrifugal-flotation and sugar-flotation techniques (Ayoub, 1977). If sugar is used, then 673 g added to 1 L of water will result in a specific gravity of 1.18, which is greater than that of nematodes (causing them to float on the surface). Detailed instructions for these and other methods of extracting nematodes from soil and plant samples are presented by Ayoub (1977) and Nichols (1979).

Quantitative methods of extraction for monitoring nematode populations over a period of time often involve the use of elutriators, which use an upcurrent of water to separate nematodes from the medium. Such methods are rather elaborate and are discussed by Southey (1978).

C. Fixing and Mounting

After the nematodes have been extracted, they should be held in water (preferably water from their original collecting source) prior to fixation. The nematodes should be killed before being placed in fixative; otherwise they become distorted and difficult to examine. They are most easily killed with heat, which also tends to relax and extend them. Too much heat will destroy the internal tissues, but good results can be obtained by pouring hot water (60–70°C) over the nematodes. They should then be transferred to the fixative as soon as possible. The best fixative for nematodes is TAF (7 mL 40% formalin, 2 mL triethanolamine, and 91 mL distilled water). However, 3–5% formalin or 70% ethanol can also be used if TAF is not available. Some distortion may occur with ethanol, however.

The nematodes should be fixed for at least 2–3 days before transfer to glycerin for mounting on microscope slides. Nematodes are most easily conveyed to glycerin by the evaporation method, which requires little handling. Fixed specimens are transferred to a dish containing a solution of 70 mL ethanol (95%), 5 mL glycerol, and 25 mL water. The dish is partly covered for the first 3 days to allow the alcohol and water to evaporate at a slow rate; then the cover is removed for the next 14 days. Finally, the containers (now with mostly glycerin) are placed in a desiccator or an oven (35°C) for another 2 weeks to drive out the remaining water. The nematodes will then be in a relatively pure solution of glycerin and can be mounted directly on microscope slides. Such slides are termed permanent since they can be kept for years for continuous study. Eventually, the nematode tissues tend to become transparent. Temporary slides are made by mounting the nematodes directly after fixation in the fixing solution (formalin or alcohol). They may last several months if the ringing seal is tight.

The following steps can be taken to make permanent slides with specimens transferred to glycerin (modified from Poinar, 1983):

1. With a small pointed instrument (dental pulp canal file, small insect pin mounted on a wooden splint, needle), transfer the nematode(s) to a small drop of glycerin placed in the center of a microscope slide.
2. Push the nematode to the bottom of the drop. Then add at three equidistant points around the nematode, small supports (coverslip pieces, wire) with a width equal to or slightly wider than that of the nematode. Push these also to the bottom of the drop.
3. Place a cover slip over the drop and lower it slowly at a 30° angle so that air bubbles will not become entrapped in the glycerin as the slide touches the mounting medium.
4. Add more glycerin, if needed, by placing a small drop at the edge of the coverslip and allowing it to move under the glass and spread throughout. If the coverslip is floating on glycerin, then remove the excess by blotting it with a moistened (water) piece of tissue.
5. Carefully seal the edges of the coverslip with a ringing compound, such as nail polish or Turtox slide ringing cement (General Biological Supply House, Chicago, Il).
6. Label the slide with information regarding the date, locality, and collector.

D. Culturing

Some free-living, microbotrophic freshwater nematodes can be cultured in the laboratory if they are presented with a growing colony of their preferred food (i.e., bacterial or algal species) and maintained under environmental conditions (temperature, oxygen) comparable to those present at the collecting site. Nuttycombe (1937) successfully cultured several species of freshwater nematodes using a wheat grain infusion method. Between 200 and 300 grains of wheat seeds were added to 250 mL of spring water in a flask. This mixture and a separate flask of pure spring water were heated to a boil and allowed to cool. The spring water was poured into 200 mL Petri dishes with 3–4 grains of the boiled wheat. The dishes with the wheat seeds were allowed to stand for 2–3 days before the nematodes were added. Microorganisms brought in with the nematodes grew on the wheat seeds, and the microbotrophic nematodes fed on the microbes.

Postparasitic juveniles of freshwater mermithids can be held in the laboratory until they mature for identification. Care should be taken so that both the temperature and the amount of dissolved oxygen parallel those of the collection site. Periodic transfers to new containers with freshwater may be necessary to prevent the buildup of fungi which will destroy the nematodes. Petersen (1972) devised a method of culturing a mosquito mermithid parasite *(Romanomermis culicivorax)* through its entire life cycle. The mosquito chosen as a laboratory host was *Culex pipiens quinquefasciatus* (Say), because it could be reared continuously in crowded conditions, was easily maintained in colony, and was highly susceptible to the nematode. Postparasitic juvenile mermithids were placed in wet sand within plastic trays (36 × 25 × 100 cm). The postparasites molted to the adult stage, mated, and oviposited in the sand. The eggs, which embryonated in the moist sand, could be maintained in their resting state for several months. When the trays were flooded with water, the infective stages emerged from the sand and were capable of infecting newly hatched mosquito larvae. The infected insect larvae were kept in a separate container and fed. By the time the mosquitoes were ready to pupate, the nematodes had completed their development and started to emerge. The emerging postparasitic juveniles were then transferred to wet sand to continue the cycle. Adequate numbers of mermithids were available for both scientific investigations and biologic control studies with this rearing method.

VI. IDENTIFICATION

Once the specimens are mounted, they can be examined and identified with a compound microscope. Because the identification of nematodes involves the use of many fine details, a microscope equipped with oil immersion should be employed (1000×). The use of dioscopic illumination after Nomarski or differential interference contrast better reveals minute features of the cuticle (such as the amphid structure). If possible, living or just-killed nematodes should also be examined under the microscope. Their movements can be reduced by removing water from under the coverslip and compressing the nematode between the microscope slide and coverslip. Certain characters are much clearer on living specimens and there is a certain thrill in watching these creatures move under a microscope that is absent with preserved material. However, measurements can best be made on fixed nematodes.

In preparing the following taxonomic key, the following works dealing with freshwater nematodes were consulted: Cobb (1914); Chitwood and Allen (1959); Ferris *et al.* (1973); Tarjan *et al.* (1977); Pennak (1978, 1989); Esser *et al.* (1985); Esser and Buckingham (1987); Jacobs (1984); Tsalolikhin (1983); Gerber and Smart (1987). The first five references also have keys which can be consulted by the interested reader for cross-reference. For descriptions of various genera and a listing of species, Goodey (1963) is still very helpful, although now out of date.

Some 66 genera of freshwater nematodes are included in the present key. Between 300 and 500 species of nematodes are probably included within these 66 genera. Only those genera that contain species that are obligately aquatic (live nowhere else) and have been reported from North America are included. Over 95% of the aquatic nematodes encountered should be included in the following key.

A. Taxonomic Key to Families and Selected Genera of Freshwater Nematoda

The following key pertains to the families and selected genera of freshwater nematodes and freeliving stages of parasitic nematodes found in freshwater habitats of North America. Following this is a separate key to the genera of aquatic Mermithidae.

1a. Elongated and threadlike, generally over 1 cm in length and usually observable with the naked eye; found on the bottom surface or several centimeters within the bed (Figs. 5c and 10a, b) 2

lb. Not elongate, generally under I cm in length and observable only with a lens (Figs. 2c and 11c); found on all exposed surfaces (rocks, on or in plants) as well as on the bottom surface and within the bed 3

2a (1a). Forms usually long (6 cm or more in length), leathery body wall, range in color from light brown to black (Fig. 13a) .... Hairworms (phylum Nematomorpha)

2b. Forms usually smaller (under 6 cm in length), body wall fragile, range in color from white to rose, green and yellow (Fig. 5c) .... Mermithidae (see Section VI.B.)

3a (1b). Head of nematode bearing a stylet (Figs. 8b and 9b) 4

3b. Teeth and other armature may be present, but a stylet is absent (Figs. 7a, b and 8a) 16

4a (3a). Stylet knobs usually absent (Fig. 9b); if present, then pharynx lacking a valvated metacorpus; pharynx narrow anterior and thickened posterior at junction with intestine (Fig. 2a); valvated metacorpus absent .... Dorylaimida 11

4b. Stylet knobs usually present (Fig. 8b); pharynx composed of a valvated metacorpus followed by a slender isthmus and basal glandular bulb leading into the intestine (Figs. 8b, c) 5

5a (4b). Dorsal gland outlet in precorpus; metacorpus moderate in size (Fig. 8b) (less than three-fourths of body width .... Tylenchida 8

5b. Dorsal gland outlet in metacorpus; metacorpus large (three-fourths of body width or more) (Fig. 8c) .... Aphelenchida 6

6a(5b). Stylet usually with faint or inconspicuous knobs (Fig. 8c); tail tip usually pointed; males lacking bursa and gubernaculum .... Aphelenchoididae 7

6b. Stylet without knobs; tail tip usually rounded; males with a bursa (Fig. 6c) and gubernaculum .... Aphelenchidae *Aphelenchus* Bastian, 1865

7a (6a). Tail shape elongate, filiform .... *Seinura* Fuchs, 1931

7b Tail shape short, conical .... *Aphelenchoides* Fischer, 1894

8a(5b). Head bearing distinct setae (Fig. 7a) .... Atylenchidae *Atylenchus* Cobb, 1913

8b Head without setae 9

9a (8b). Procorpus fused with large, oval metacorpus; cuticle strongly annulated; adult female with cuticular sheath and well-developed stylet (Fig. 8b) .... Criconematidae *Hemicycliophora* de Man, 1921

9b. Procorpus distinct and narrow before reaching the expanded metacorpus, cuticle not strongly annulated; adult female without cuticular sheath; styles may or may not be well developed .... Tylenchidae 10

10a (9b). Large distinct stylet, ovaries paired, amphidelphic (Fig. 6a); tail tapering but not filiform .... *Hirschmanniella* Luc and Goodey, 1963

10b. Stylet small and inconspicuous, ovary single, prodelphic; tail filiform .... *Tylenchus* Bastian, 1865

11a (4a). Pharynx gradually widens toward basal bulb (Fig. 2a) 12

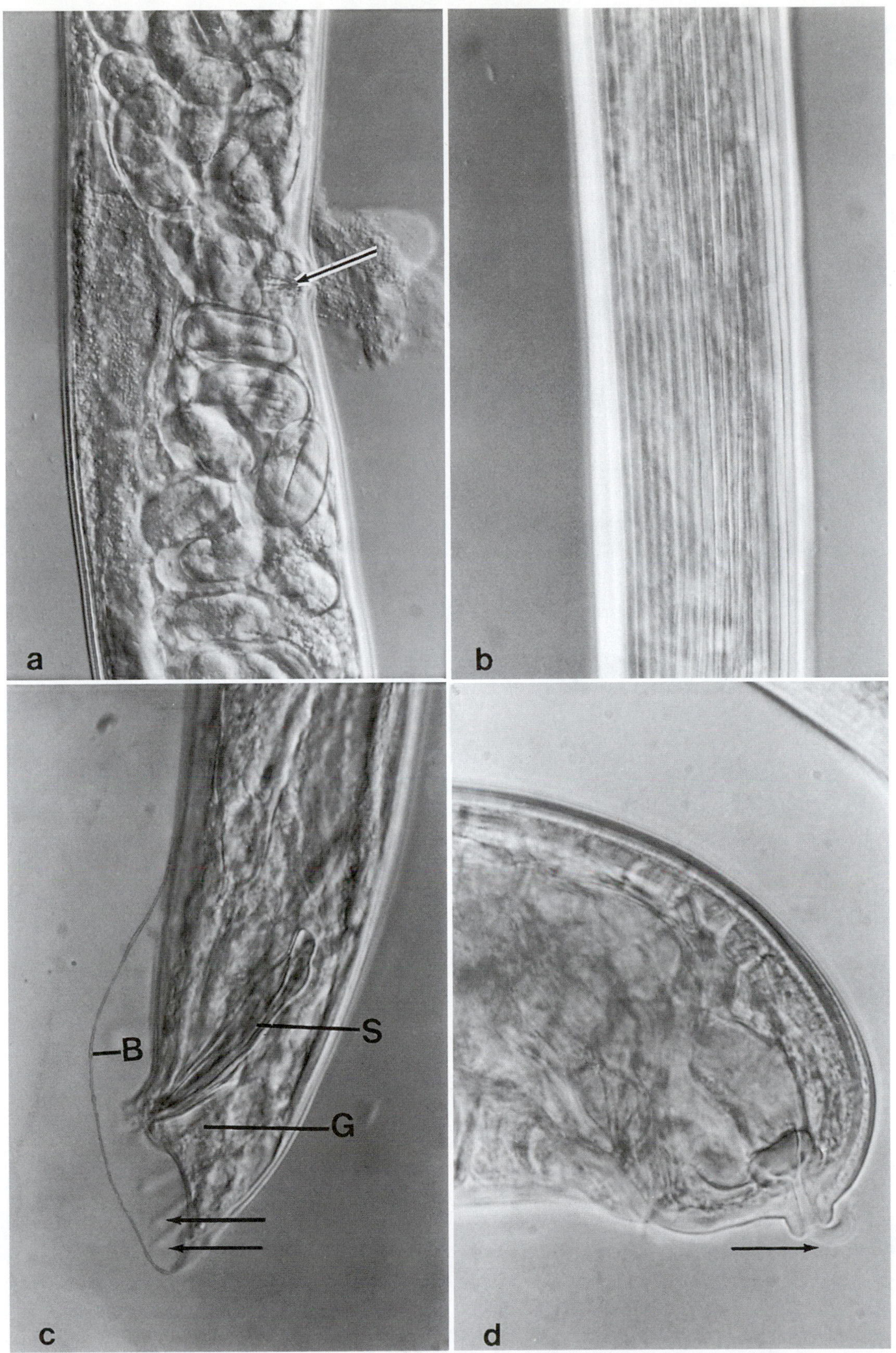

FIGURE 6 (a) Midbody region of a female *Pellioditis* sp. (Rhabditidae) showing the vagina (arrow) and paired opposite uteri filled with developing eggs. Note copulation deposit surrounding the vulva opening. (b) Longitudinal striations lining the cuticle of *Dorylaimus* sp. (Dorylaimidae). (c) Tail of male *Pellioditis* (Rhabditidae) showing spicule (S), gubernaculum (G), bursa (B), and bursal papillae or rays (arrows). (d) Tail of an aquatic nematode showing caudal gland mucous being emitted from the terminus (arrow).

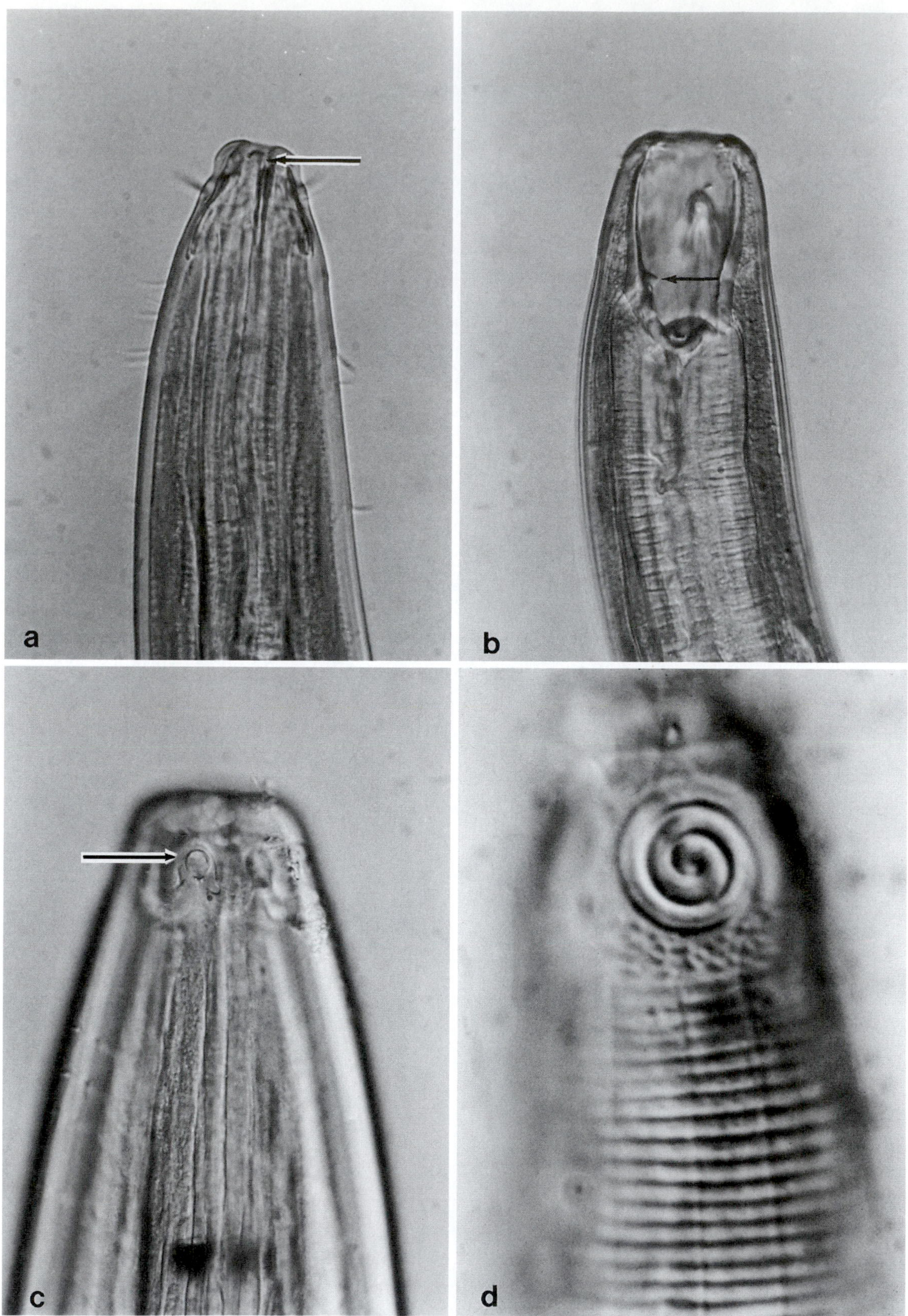

FIGURE 7 (a) Anterior end of an aquatic nematode showing a narrow tube-shaped stoma (arrow) and cuticular bristles or setae. (b) Large oval, cup-shaped stoma of a predatory aquatic nematode. Note dorsal tooth (arrow) on stomal wall. (c) Nematode bearing a medium-sized, circular amphid. (d) Nematode bearing a large, spiral-shaped amphid.

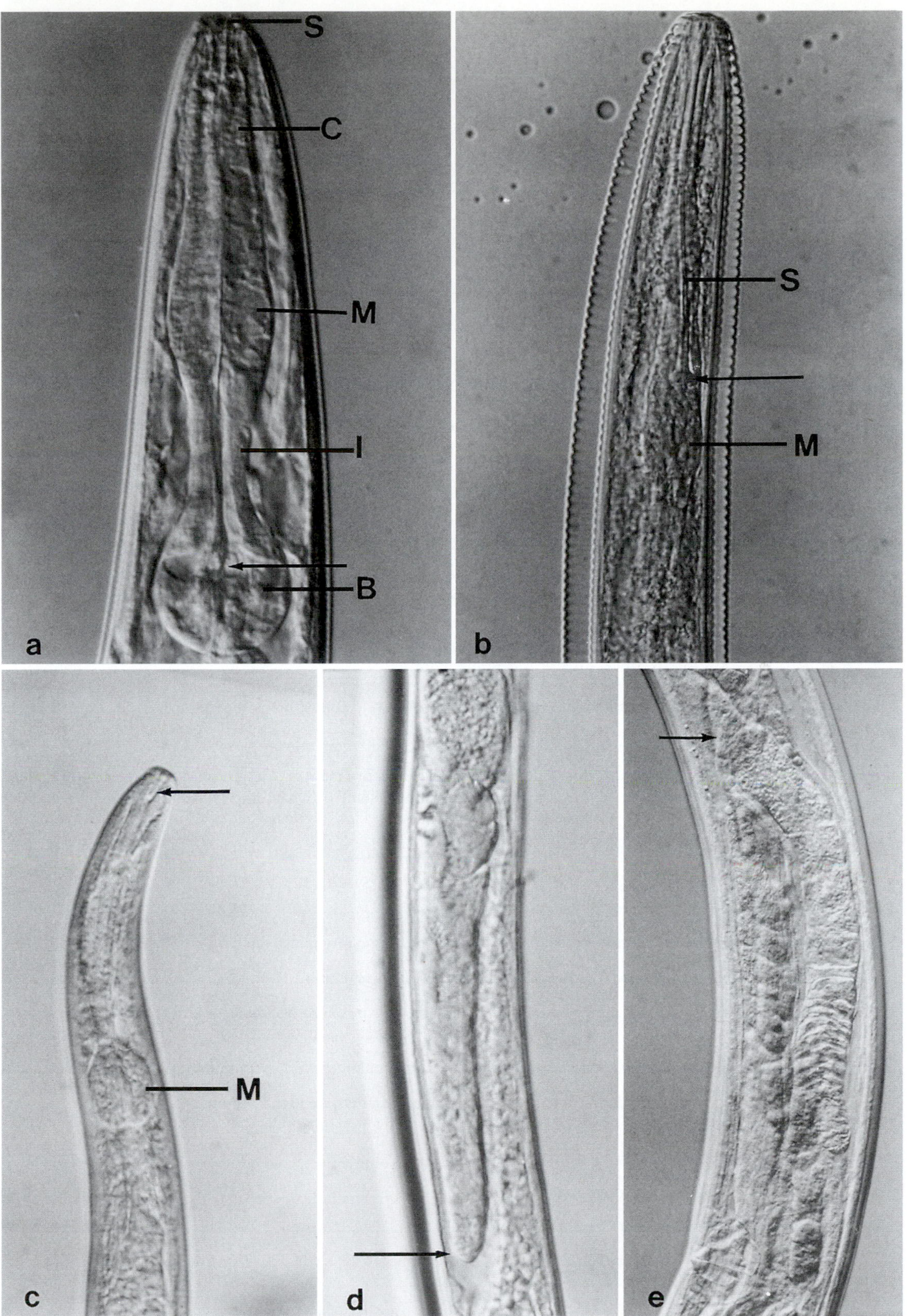

FIGURE 8 (a) Pharyngeal region of *Pellioditis* (Rhabditidae) showing tubular stoma (S), corpus (C), metacorpus (M), isthmus (I), and basal bulb (B). Arrow shows valve in basal bulb. (b) Outer ensheathing cuticle of *Hemicycliophora* (Hemicycliophoridae). Note long stylet (S) with prominent basal knobs (arrow) and valvated metacorpus (M). (c) Anterior of *Aphelenchoides* sp. (Aphelenchoididae) showing small stylet lacking knobs (arrow) and large metacorpus (M) with valve. (d) Outstretched ovary (arrow shows tip). (e) Reflexed ovary (arrow shows point of reflexion).

11b. Pharynx abruptly widens to form the basal bulb which contains a valvular chamber (Fig. 8a) ....Campydoridae *Aulolaimoides* Micoletzky, 1915

12a (11a). Head bearing a narrow protrusible mural spear that is symmetrically pointed at tip ....Nygolaimidae *Nygolaimus* Cobb, 1913

12b. Head bearing a thick protrusible axial spear that is asymmetrically pointed at tip (sloped on one side) (Fig. 9b) .... Dorylaimidae 13

13a (12b). Stomal area bearing distinct teeth or denticles around the tip of the stylet 14

13b. Stomal area lacking distinct teeth or denticles in the stylet tip region 15

14a (13a). Four large teeth together with mural denticles present ....*Paractinolaimus* Meyl, 1957.

14b. Four large teeth only present...*Actinolaimus* Cobb, 1913

15a (13b). Cuticle thick and longitudinally ridged (Fig. 9b) ....*Dorylaimus* Dujardin, 1845

15b. Cuticle smooth ....*Mesodorylaimus* Andrassy, 1959

16a (3b). Amphids minute and borne on the lateral lips, excretory duct cuticularized (canal lined with a fine layer of cuticle); caudal glands absent, conspicuous cephalic setae absent, bursa present or absent 17

16b. Amphids usually enlarged and located on the neck region, excretory duct rarely cuticularized (except in Plectidae); head often with conspicuous setae; bursa usually absent. 21

17a (16a). Pharynx with an expanded median bulb (metacorpus) containing a longitudinal valve and a glandular nonvalvated basal bulb Diplogasteridae .... *Rhabditolaimus* Fuchs, 1915

17b. Pharynx may or may not contain a metacorpus, but if it does, it does not contain a valve; basal pharyngeal bulb with or without a basal valve..18

18a(17b). Pharynx elongate and lacking a valve in the basal bulb or bulb area, slender; slow moving forms (Fig. 9d) ....Daubayliidae *Daubaylia* Chitwood and Chitwood, 1934

18b. Pharynx normal in length, with a basal bulb containing a valve (Fig. 8a); stouter, quick-moving forms 19

19a(18b). Stoma with rhabdions separate, female with a single ovary, bursa absent ....Cephalobidae 20

19b. Stoma with rhabdions fused, cylindrical (Fig. 8a), female with a single or paired ovaries, bursa usually present (Fig. 6c) .... Rhabditidae [Most of the genera in this large family are soil forms, some of which appear from time to time in aquatic habitats. *Mesorhabditis* (Osche, 1952), which has a single ovary and a bursa, occurs ;n sewage bed and run-off water. *Pellioditis* (Doughtery, 1953), which has paired ovaries also occurs in aquatic habitats.]

20a (19a). Lateral fields reach to tail tip; female tail rounded..*Cephalobus* Bastian, 1865

20b. Lateral fields end at the phasmids; female tail usually pointed...*Eucephalobus* Steiner, 1936

21a (16b). Amphids spiral (Fig. 7d), loop-like, stirrup-shaped or circular (Fig. 7c); pharynx usually with a basal bulb (Fig. 8a).......................22

21b. Amphids pore-like to pocket-like, pharynx usually without a distinct basal bulb 41

22a (21a). Ovary usually outstretched (reflexed in *Prismatolaimus*) (Fig. 8d), amphids usually circular (Fig. 7c) but may be slitlike .... Monhysterida 23

22b. Ovaries reflexed (Fig. 8e) (except in *Odontolaimus* which also has circular amphids), amphids variable (including circular)...........25

23a (22a). Amphids slit-like, faint; stoma with denticulated cushions (padded areas covered with minute projections).. *Prismatolaimus* de Man, 1880

23b. Amphids circular, distinct, stoma lacking denticulated cushions 24

24a (23b). Stoma shallow; pharynx without distinct basal bulb (as in Fig. 2a) ....*Monhystera* Bastian, 1865

24b. Stoma elongate; pharynx with distinct basal bulb (as in Fig. 8) ....*Monhystrella* Cobb, 1918

25a (22b). Caudal glands and spinneret present (Fig. 2d); stoma usually armed with teeth (Fig. 7b); knobs, bristles (Fig. 7a), or punctations usually present on cuticle ....Chromadorida 26

25b. Caudal glands and spinneret present or absent; stoma may or may not be armed with teeth; bristles sometimes present on cuticle,especially around head ....Araeolaimida 33

26a (25a). Amphids circular (Fig. 7c); cuticular punctations minute ....Microlaimidae *Microlaimus* de Man, 1880

26b. Amphids spiral (Fig. 7d), kidney shaped or circular; cuticular punctations coarse 27

27a (26b). Amphids circular (as in Fig. 7c) or spiral (as in Fig. 7d), pharyngeal–intestinal junction large, tri-radiate.............Cyatholaimidae 29

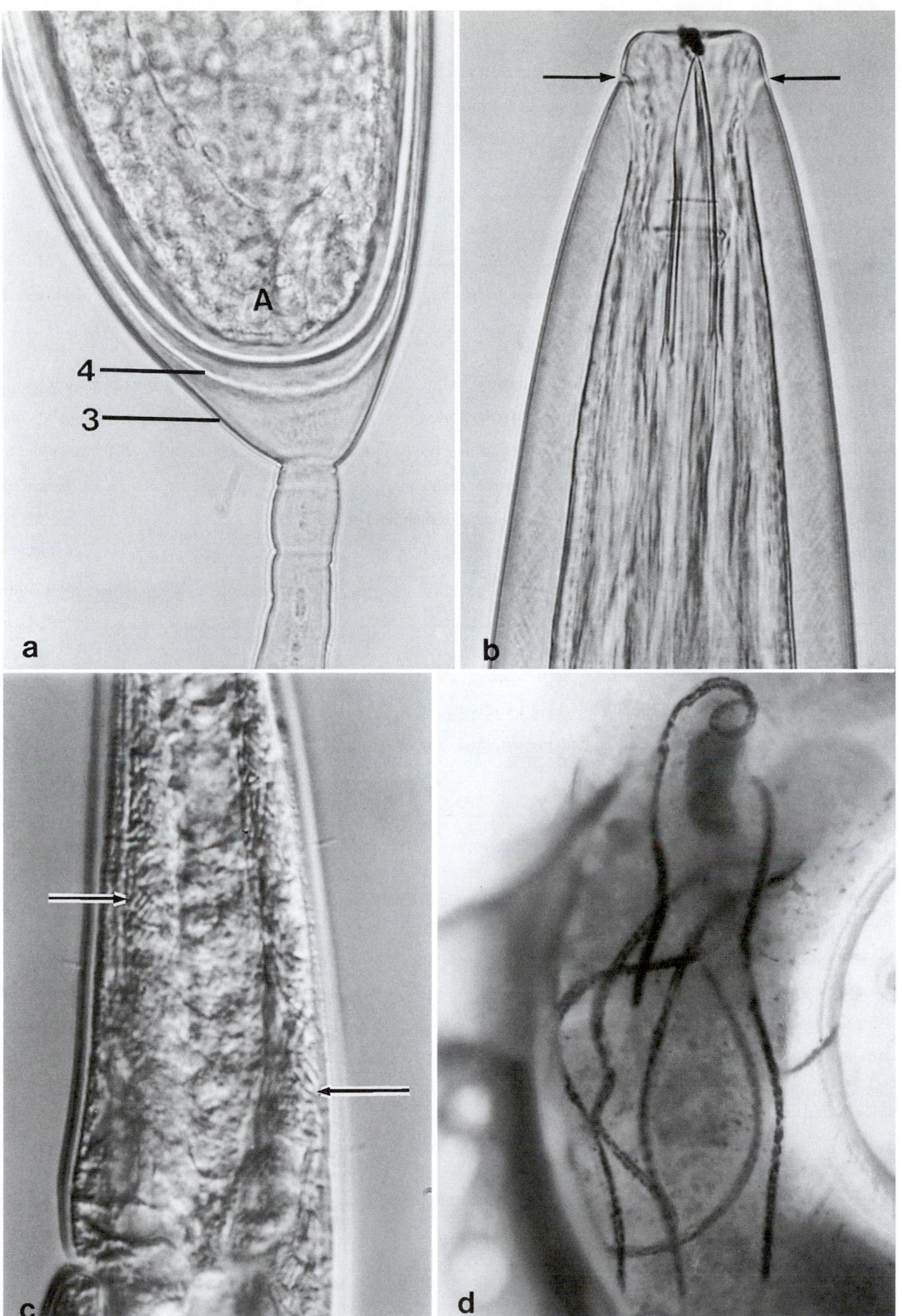

FIGURE 9 (a) Posterior portion of a postparasitic juvenile mermithid (Mermithidae) in the process of molting. Note shedding of third- (3) and fourth-stage (4) cuticles. A, adult. (b) Head of *Dorylaimus* sp. (Dorylaimidae) showing base of lip region (arrows) and stylet lacking basal knobs. (c) Tail region of a *Tobrilus* sp. (Tobrilidae) showing spores of a microsporidian parasite filling the hypodermal tissue (arrows). (d) Adults of *Daubaylia* (Daubayliidae) inside an aquatic snail.

27b. Amphids spiral or kidney shaped, pharyngeal–intestinal junction small, not tri-radiate .... Chromadoridae 28

28a (27b). Amphids represented as a broad transverse slit; five or more preanal genital papillae in male .... *Chromadorita* Filipjev, 1922

28b. Amphids represented as a flattened, broken ring; no more than three preanal genital papillae in male .... *Punctodora* Filipjev, 1930

29a (27a). Amphids circular (Fig. 7c) 30

29b. Amphids spiral (Fig. 7d) 31

30a (29a). Stoma cuplike; length shorter than neck width .... *Prodesmodora* Micoletzky, 1923

30b. Stoma tubular (Fig. 7a); length two or more times neck width .... *Odontolaimus* de Man, 1880

31a (29b). Stoma tubular (Fig. 7a), there may be three anterior equal teeth in the stoma, but single large dorsal tooth is lacking .... *Ethmolaimus* de Man, 1880

31b. Stoma cup- or funnel-shaped, armed with a prominent dorsal tooth (as in Fig. 7b) much larger than any other teeth 32

32a (31b). Amphids located at level of stoma, small subventral teeth absent .... *Paracyatholaimus* Micoletzky, 1924

32b. Amphids located posterior to stoma, one or two small subventral teeth present .... *Achromadora* Cobb, 1913

33a (25b). Caudal glands absent; stoma funnel-shaped with separate rhabdions .... Teratocephalidae 34

33b. Caudal glands present (Fig. 2a) or absent; stoma tubular with rhabdions fused 35

34a (33a). Amphids porelike; cuticular annulation strong (Fig. 8b) .... *Teratocephalus* de Man, 1876

34b. Amphids spiral (Fig. 7d); cuticular annulation weak .... *Euteratocephalus* Andrassy, 1958

35a (33b). Ovary outstretched (Fig. 8d); amphids circular .... Axonolaimidae *Cylindrolaimus* de Man, 1880

35b. Ovary reflexed (Fig. 8e) 36

36a (35b). Pharynx usually with a basal valvated bulb (Figs. 1 and 8a) 37

36b. Pharynx with or without a basal bulb, if bulb present, then valve absent 39

37a (36a). Basal bulb of pharynx lacking valves .... *Anonchus* Cobb, 1913

37b. Basal bulb of pharynx with valves (Fig. 8a) 38

38a (37b). Pharyngeal basal bulb with postbulbar extension (Fig. 10c) connecting it with the intestine; caudal glands absent .... *Chronogaster* Cobb, 1913

38b. Phayrngeal basal bulb without postbulbar extension, abutting the intestine; caudal galnds present (Fig. 1) .... *Plectus* Bastian, 1865

39a (36b). Basal bulb lacking .... Bastianiidae *Bastiania* de Man, 1876

39b. Basal bulb present .... Camacolaimidae 40

40a (39b). Amphids circular (Fig. 7c), basal pharyngeal bulb absent .... *Aphanolaimus* de Man, 1880

40b. Amphids spiral (Fig. 7d), a slight basal pharyngeal bulb present .... *Paraphanolaimus* Micoletzky, 1923

41a (21b). Stoma usually oval in outline (Fig. 7b), heavily sclerotized and armed with one or more teeth, setae and bristles absent .... Mononchidae

41b. Stoma not in the form of a heavily sclerotized oval cavity, teeth or denticles may be present, setae and bristles may be present (Fig. 7a) 42

42a (41b). Cuticle of head double .... Oncholaimidae 43

42b. Cuticle of head normal 44

43a (42a). Setae or bristles present on tail .... *Oncholaimus* Dujardin, 1845

43b. Setae and bristles usually lacking on tail .... *Mononchulus* Cobb, 1918

44a (42b). Stoma heavily sclerotized, cylindrical .... Ironidae 49

44b. Stoma not heavily sclerotized, funnel-shaped or tubular 45

45a (44b). Stoma vestigial and unarmed; pharynx base expanded and set off to form a slight, elongate bulb .... Alaimidae 46

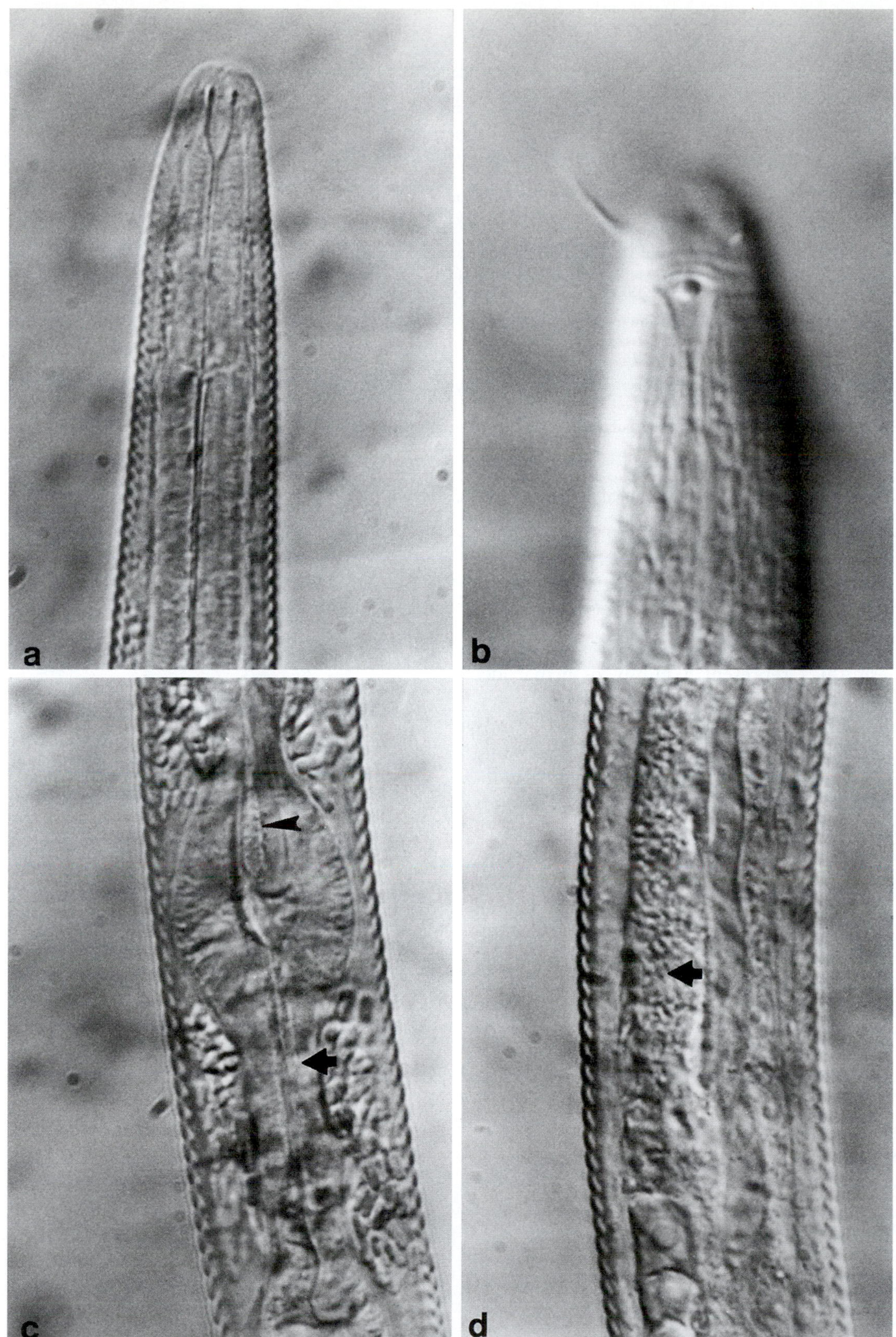

FIGURE 10 *Chronogaster troglodytes* Poinar and Sarbu, a specialized aquatic cave nematode from Movile Cave in Romania. (a) Anterior end with funnel-shaped mouth cavity. (b) Head showing stirrup-shaped amphid with a circular opening. (c) Basal pharyngeal region showing valve (fleche) in basal bulb and postbulbar extension (arrow) between the basal bulb and intestine. (d) Spermlike bodies (arrow) which occur near the proximal portion of the female gonad, and the absence of males, suggests that *C. troglodytes* is hermaphroditic.

45b. Stoma distinct, or if vestigial then armed with an inconspicuous median tooth; pharynx may be expanded at base, but rarely set off from the remainder in the form of a bulb 47

46a(45a). Amphids porelike, minute .... *Alaimus* de Man, 1880

46b. Amphids cup-shaped, distinct .... *Amphidelus* Thorne, 1939

47a(45b). Base of pharynx expanded to form a valvated bulb; three rodlike thickenings compose posterior part of stoma .... Leptolaimidae .... *Rhabdolaimus* de Man, 1880

47b. Pharynx cylindrical, not with basal valvated bulb; stoma a simple tube without rodlike thickenings .... Tripylidae 48

48a(47b). Three lips; amphids porelike, minute; stoma cylindrical .... *Tripyla* Bastian, 1865

48b. Six lips; amphids cup-shaped, distinct; stoma funnel-shaped .... *Tobrilus* Andrassy, 1959

49a(44a). Stoma tubular, with two minute teeth at the base; excretory pore opening posterior to head .... *Cryptonchus* Cobb, 1913

49b. Stoma tubular, with three anterior hooklike teeth; excretory pore opening in head area .... *Ironus* Bastian, 1865

50a(41a). Pharyngeal-intestinal junction tuberculate; dorsal tooth on stoma pointing posteriorally .... *Anatonchus* Cobb, 1916

50b. Pharyngeal-intestinal junction not tuberculate; dorsal tooth on stoma pointing anteriorally 51

51a (50b). Ventral stomatal ridge with longitudinal row of denticles .... *Prionchulus* Cobb, 1916

51b. Ventral stomatal ridge absent, or if present then unarmed .... *Mononchus* Bastian, 1865

B. Taxonomic Key to Extant Genera of Freshwater Mermithidae

The mermithids represent a family of nematodes that have become parasitic on other invertebrates. Over half of the described genera parasitize only aquatic insects. Two genera, *Aranimermis* and *Pheromermis,* enter the immature stages of aquatic insects which are later consumed by terrestrial arthropods. However, the free-living stages of these and all other aquatic mermithids (postparasitic juvenile, adults, eggs, and infective juveniles) are found in the aquatic habitat. Only the developing second- and third-stage juveniles parasitize the aquatic stages of insects and other invertebrates; the other stages do not take nourishment. Eggs and infective stage juveniles are microscopic and normally not taken in samples, but the postparasitic juveniles and adults are easily seen and collected. Since the keys are based on adult characters, postparasitic juveniles should be held in water until they molt to the adult stage. One important character in identifying mermithids is the number of hypodermal cords present. These can be determined by cutting thin cross sections with a razor blade by hand, and examining them under the microscope. The hypodermal cords protrude through the muscle fields in dorsal, ventral, lateral, and often the subdorsal and subventral regions. Only genera known to occur in North America are included.

1a. Adult cuticle with cross-fibers; thick, robust white nematodes usually found along the edges of bogs, springs, and streams (parasites of wasps, ants, and horseflies) (Fig. 11b) .... *Pheromermis* Poinar, Lane, and Thomas, 1976

1b. Adult cuticle without cross-fibers, robust or slender, white, green, pink, brown, or yellow nematodes found in lake, stream, and pond beds 2

2a. With four cephalic papillae .... *Pseudomermis* de Man, 1903 (syn. *Tetramermis* Steiner, 1925)

2b. With six cephalic papillae 3

3a (2b). With single or fused (rare) spicules 4

3b. With paired, separate spicules 8

4a (3a). Spicule medium to long, more than twice anal body width 5

4b. Spicule short, less than twice anal body width 6

5a (4a). Mouth terminal; spicules J-shaped; vulval flap present; [parasites of midges (Chironomidae)] .... *Lanceimermis* Artyukhovsky, 1969

5b. Mouth normally shifted ventrally; spicule curved but not J-shaped; vulva flap absent; [parasites of blackflies (Simuliidae), midges (Chironomidae), and mayflies (Ephemeroptera)] .... *Gastromermis* Micoletzky, 1923

6a(4b). Spicule shorter than anal body width; amphids small; [parasites of mosquitoes (Culicidae)] .... *Perutilimermis* Nickle, 1972

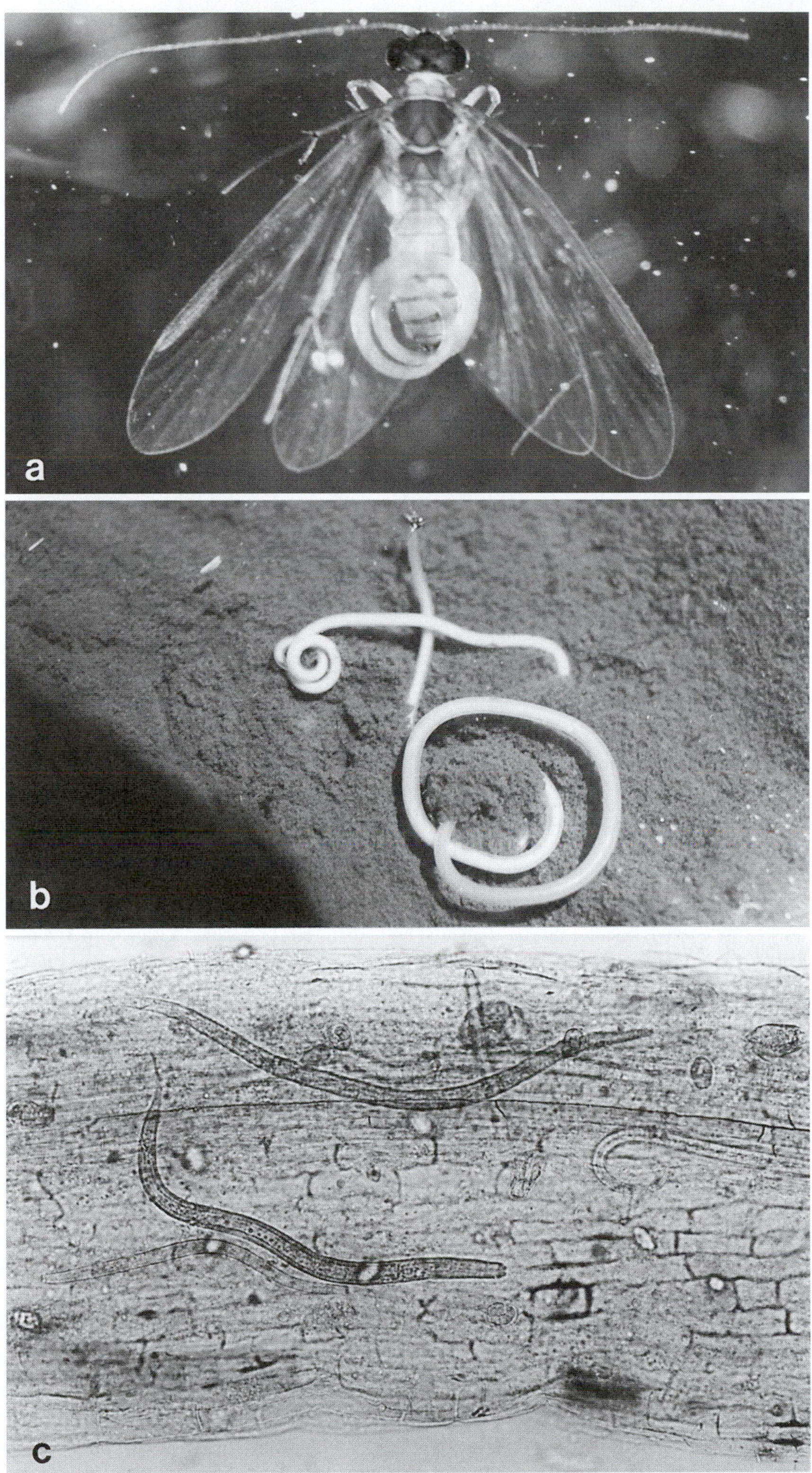

FIGURE 11 (a) A postparasitic juvenile aquatic mermithid emerging from an adult caddis fly, Trichoptera (collected by H. Wolda and submitted by R. Schuster). (b) Adults of *Pheromermis pachysoma* (Mermithidae) in a California spring bed. These nematodes have emerged from parasitized queen wasps visiting the spring and will undergo molting, mating and oviposition in the spring. (c) Various eggs and juvenile stages of freshwater nematodes developing in a decaying, submerged leaf.

6b. Spicule longer than anal body width; amphids medium to large 7

7a(6b). Tail pointed; eight hypodermal cords; [parasites of midges (Chironomidae) and mosquitoes (Culicidae)] ...*Hydromermis* Corti, 1902

7b. Tail rounded; six hypodermal cords; [parasites of midges (Chironomidae) and blackflies (Simuliidae)] .....*Limnomermis* Daday, 1911

8a (3b). Vagina straight or nearly so, barrel- or pear-shaped 9

8b. Vagina S-shaped, U-shaped, or elongate with both ends curved..13

9a (8a). Spicules shorter than cloacal body width..10

9b. Spicules equal to or longer than cloacal body width..11

10a (9a). Head expanded into a bulblike shape with very thick cuticle; male lacking lateral–dorsal genital papillae; [parasites of midges (Chironomidae)] ....*Capitomermis* Rubtsov, 1968

10b. Head not expanded into a bulblike shape with thick cuticle; male with genital papillae on lateral-dorsal surface; [parasites of biting midges (Ceratopogonidae)] ....*Heleidomermis* Rubtsov, 1970

11a(9b). Six hypodermal cords; [parasites of blackflies (Simuliidae)] ....*Mesomermis* Daday, 1911

11b. Eight hypodermal cords 12

12a(11b). Spicules 2–4 times anal body width; [parasites of mosquitoes (Culicidae)] ....*Romanomermis* Coman,1961

12b. Spicules 1-2 times anal body width; [parasites of mosquitoes (Culicidae) and midges (Chironomidae)] .... *Octomyomermis* Johnson, 1963

13a (8b). Eight hypodermal cords 16

13b. Six hypodermal cords..14

14a(13b). Spicule length less than two times anal diameter; [parasites of midges (Chironomidae) and mosquitoes (Culicidae)] .... *Strelkovimermis* Rubtsov, 1969

14b. Spicule length three or more times cloacal diameter 15

15a(14b). Spicules ten or more times body width at cloaca; vagina elongate with bends at both ends; postparasitic juveniles with distinct tail appendage;[parasites of diving beetles (Dytiscidae)]..*Drilomermis* Poinar and Petersen, 1978

15b. Spicules 3–10 times body width at cloaca; vagina elongate with 3–6 irregular bends; postparasitic juveniles with indistinct or no tail appendage; [parasites of spiders] ....*Aranimermis* Poinar and Benton, 1986

16a(13a). Vagina elongate, slightly curved; spicules shorter than anal body diameter; [parasites of mosquitoes (Culicidae)].. *Culicimermis* Rubtsov and Isaeva, 1975

16b. Vagina S-shaped, ends distinctly curved; spicules range from shorter than anal body diameter to about 10 times tail diameter 17

17a(16b). Spicule length equal to or shorter than tail diameter; amphids small; cephalic crown well developed; [parasites of mosquitoes (Culicidae)] ....*Empidomermis* Poinar, 1977

17b. Spicule length between one and ten times anal diameter; amphids medium–large, cephalic crown absent or only slightly developed; [parasites of blackflies (Simuliidae)] ....*Isomermis* Coman, 1953

NEMATOMORPHA

VII. INTRODUCTION

Members of the poorly known phylum Nematomorpha (Gordiacea) constitute a relict group that has no clear or close relationship with any other living forms. Fossil hairworms are rare and controversial.The earlier determination of *Gordius tenuifibrosus* (Voigt, 1938) as a fossil hairworm (identified from a 15-mm long fragment of subcuticular tissue in the Eocene brown coals of the Geisel Valley near Halle, Germany) is now considered questionable in light of similar characters found on members of the Mermithidae (Poinar, 1999). And while it has been proposed that fossil palaeoscolecid worms are larval hairworms (Xianguang and Bergstrom, 1994), similarities between these two taxa are only superifical and provide no evidence for phylogenetic relationship (Poinar, 1999). The only unequivocal record of fossil Nematomorpha are two hairworms, *Paleochordodes protus* Poinar (1999), emerging from a cockroach in Dominican amber dated at 15–45 million years (Fig 20a). However, the Nematomorpha certainly originated considerably earlier (probably in the Early Paleozoic) as an offshoot of now extinct lines. Chitwood (1950) considered that Nematomorpha had the closest ties (albeit distant) with nematodes and rotifers (the Acanthocephala and

Echinodermata were tied for the next closest kin). Using cladistic methods, Schmidt-Rhaesa (1998b) also concluded that nematomorphs are closest to nematodes; however, it is difficult to establish lineages when such basic characters as the number of molts during the growth phase has not been confirmed. The body plan of the Nematomorpha is unique in the animal kingdom and even the sperm are considered aberrant and support the contention that this group separated quite early from the other members of the Aschelminthes and

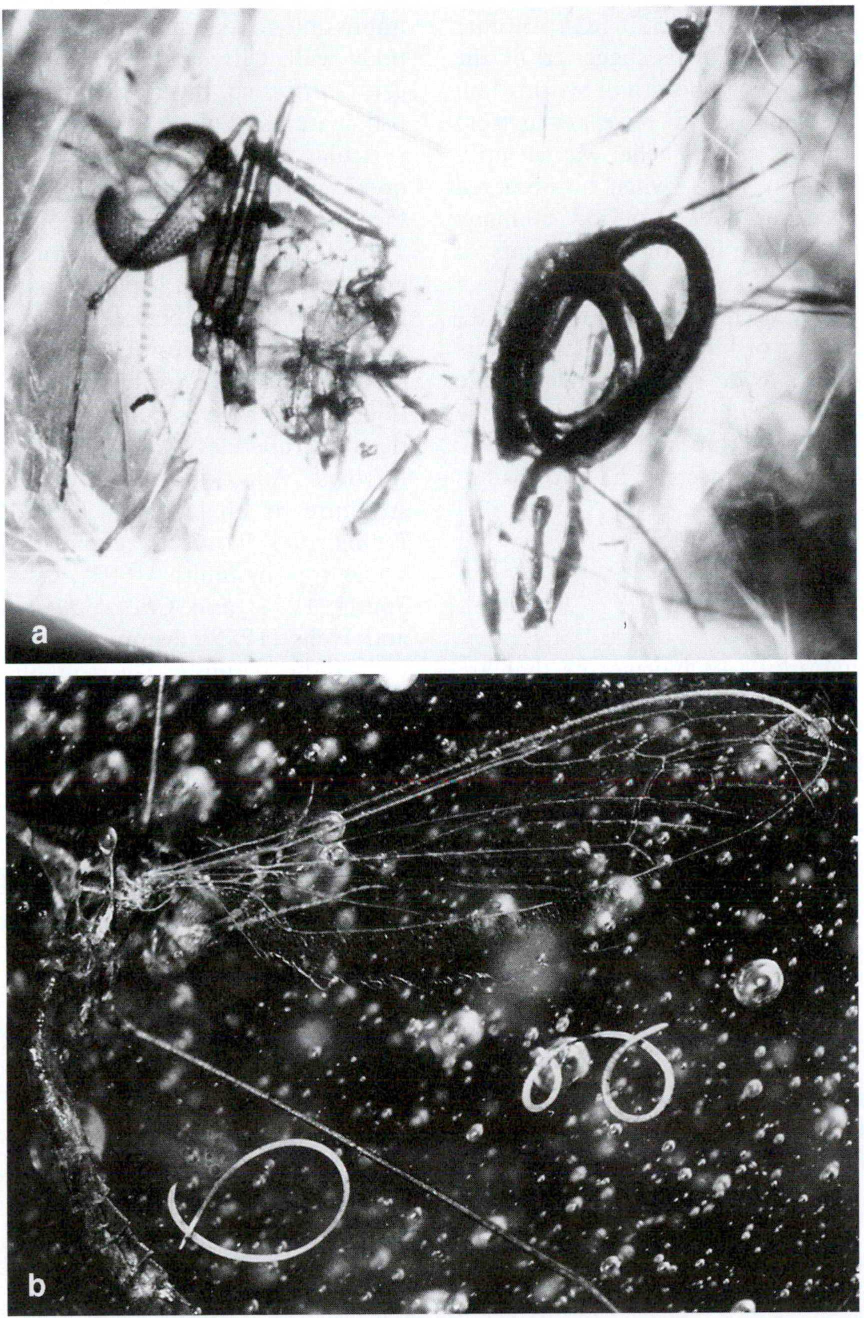

FIGURE 12 (a) Oldest known undisputed fossil nematode, *Cretacimermis libani* (Poinar Acra and Acra) from 120–135 million-year-old Lower Cretaceous Lebanese amber. This mermithid infected and developed in the aquatic larval stage of a midge (Chironomidae) and was carried over into the adult host. (b) Two mermithid nematodes, *Heydenius dominicanus* Poinar, in 20–40 million-year-old Dominican amber. These nematodes probably emerged from an adult mosquito also enclosed in the amber (figure shows an adjacent cranefly adult in the amber; the mosquito could not be included in the photo at this magnification).

evolved independently thereafter (Lora Lami Donin and Cotelli, 1977) (Fig. 15).

Some myths surround this group, perhaps associated with the proverbial "Gordian knot" (representing a problem solvable only by drastic action as demonstrated by Alexander the Great cutting the knot that could not be untied, which bound a chariot to a pole at Gordium, the capital of Phrygia, in 333 BC). Another myth, still believed by some today, is suggested by the common name "hairworms" or "horsehair worms" indicating that they were supposed to have arisen from horse hairs that fell into water. This belief was scientifically disproven by Leidy in 1870, when he observed horse hairs placed in water over a period of many months without " . . . having had the opportunity of seeing their vivification."

Adult hairworms have been associated with the digestive and urogenital tract of humans and larval hairworms will burrow into a wide range of invertebrate and vertebrate tissue, including human facial tissue sometimes resulting in orbital tumors (Watson, 1960). A review of the public health aspects of hairworms has been provided by Cappucci (1982).

VIII. MORPHOLOGY AND PHYSIOLOGY

It is the free-living adults of hairworms that are normally described because these stages are most frequently encountered in sampling. They are dark (rarely white), slender worms ranging in length from several centimeters to 1 m and in width from 0.25 to 3 mm (Fig. 13a). The color of most adults varies from yellowish to black; and although the anterior end is generally attenuated, both tips tend to be obtusely rounded or blunt (divided in some males and females). Because the cuticle is normally opaque, it is impossible to examine the internal organs through the body wall. Ultrastructural studies of a North American *Gordius* sp. revealed some interesting morphological features (Eakin and Brandenburger, 1974). The cuticular structures that are such important taxonomic characters are actually sculpturing on the surface of the thin, superficial epicuticle (Figs. 13b, 16a, 19). This epicuticle is normally criss-crossed by grooves or furrows, leaving small elevations of irregular areas (areoles) between them. The surfaces of the areoles may be smooth (Fig. 19a) or may bear setae (bristles) (Fig. 19b) or cuticular projections (tubercles) (Fig. 19c,d), arranged singly or in clusters. These bristles and tubercles may also occur in the interareolar furrows. Additional observations on the cuticular structure of hairworms include investigations of *Paragordius varius* by Zapotosky (1971), *Parachordodes* spp. by Smith (1991), *G. aquaticus difficilis* by Smith (1994) and *Chordodes morgani* by Chandler and Wells (1989). Many of these studies involve the use of the scanning electron microscope which is a useful tool for examining details of the surface sculpturing on the epicuticle.

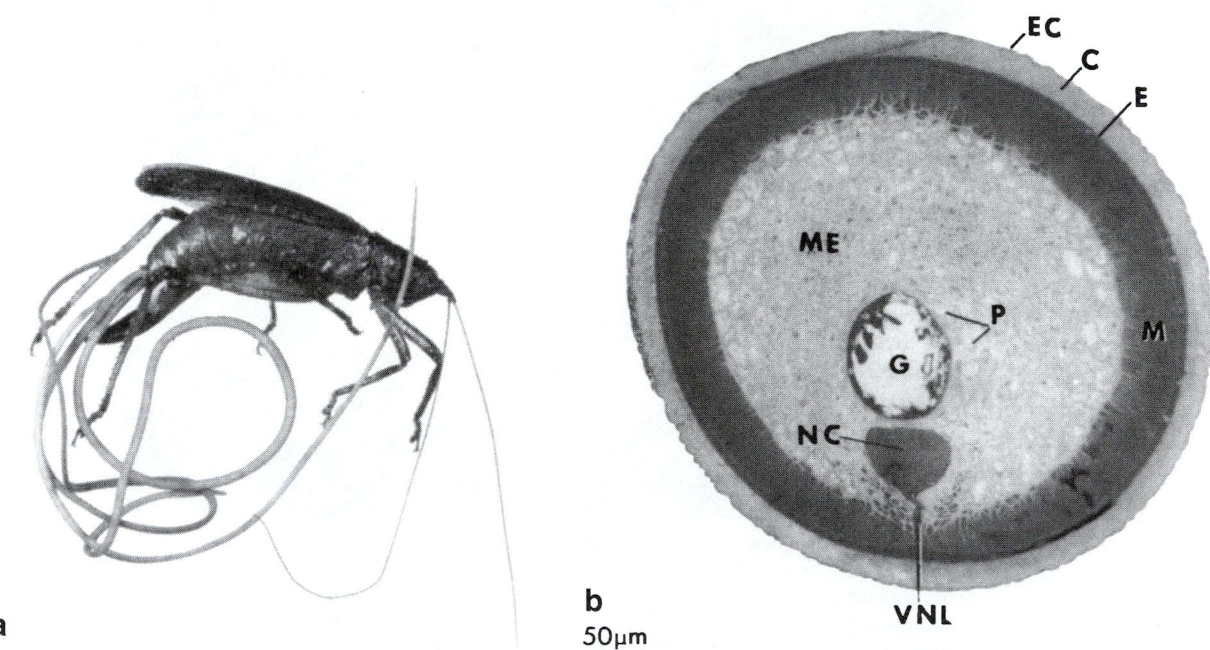

FIGURE 13 (a) A hairworm (Nematomorpha) emerging from its orthopteran host. (b) Transverse section of *Gordius* sp. C, cuticle; E, epidermis; EC, epicuticle; G, gut; M, muscles; ME, mesenchyme; NC, nerve cord; P, pseudocoel; VNL, ventral neural lamella. [Courtesy of R. M. Eakin and J. L. Brandenburger.]

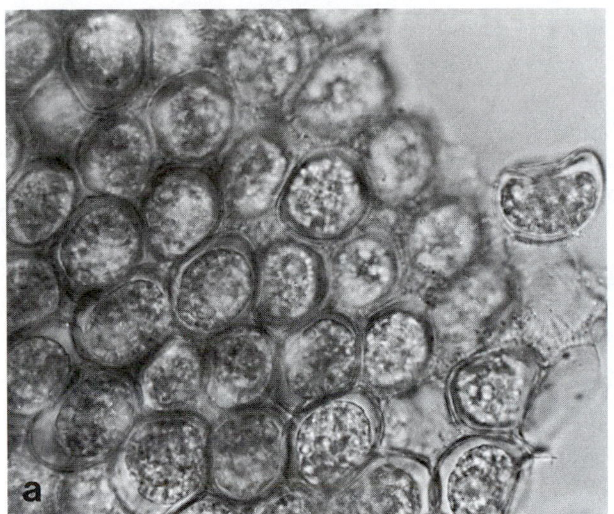

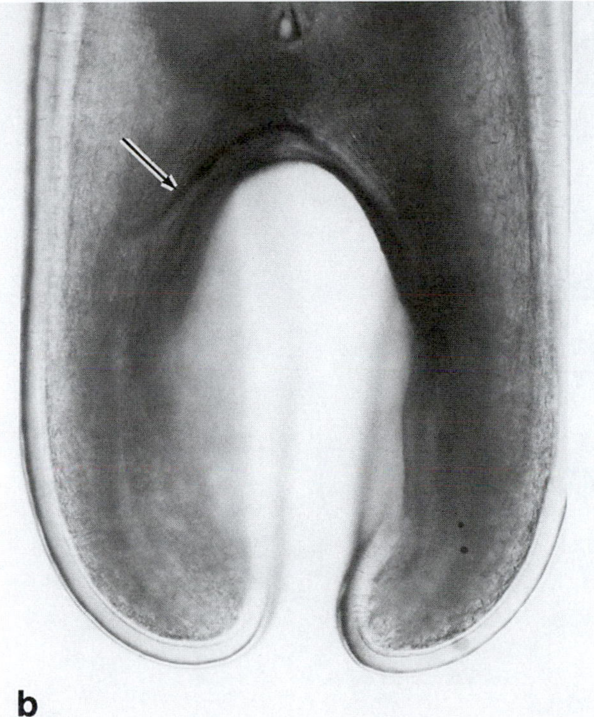

FIGURE 14 (a) Layer of eggs of *Euchordodes nigromaculatus* Poinar scraped off a boulder in Cave Stream, Craigieburn forest, New Zealand. Adhering the eggs to submerged substrates is an adaptation of hairworms living in fast flowing streams. (b) Ventral view of tail of *Gordius dimorphus* Poinar showing bilobed tail with lobes longer than wide, spherical cloacal opening located ventrally and a semicircular postcloacal fold (arrow).

The cuticle, beneath the epicuticle, is composed of many crossing layers of cylindrical, nonperiodic collagenous fibers, spirally coiled along the length of the adult (Fig. 16). Beneath the cuticle is the epidermal layer, constructed of a single layer of interdigitated cells. The musculature consists only of overlapping, longitudinal, flat cells; circular muscles are absent. Ultrastructural details of the musculature of *G. aquaticus* and *Nectomena munidae* are provided by Schmidt-Rhaesa (1998a). The pseudocoel is nearly filled with mesenchymal cells containing large clear vacuoles, which give the tissue a foamy appearance. These cells are embedded in a supporting collagenous matrix (Fig. 16b).

The mouth is located at the calotte or anterior tip of the body (Fig. 18a). This region is often lighter in color than the rest of the body and contains the nonfunctional pharynx, which, in turn, leads into a degenerate intestine lined with epithelium one cell in thickness. The gut epithelial cells of *Gordius* sp. contain numerous microvilli that project into the lumen (Eakin and Brandenburger, 1974).

In both sexes, the genital ducts empty into the intestine, forming a cloaca lined with cuticle (Fig. 18b). Because adults do not feed, the cloaca is probably used solely for reproductive purposes. All nematomorphs are amphimictic. Males have paired cylindrical testes, each of which connects with the cloaca via a separate sperm duct. Females possess paired ovaries, and the eggs, after passing through separate oviducts, enter the cloaca independently.

Adults become sexually mature soon after emerging from their hosts and copulation occurs after the male coils its posterior end around the terminus of the female. The spermatozoa are either deposited as a spermatophore on the female terminus, from where they migrate into the cloaca, or are deposited directly into the female cloaca. Ultrastructural studies of the sperm show that they are composed of a main cell body attached to a nuclear portion containing an elongated nucleus with a helical configuration (Fig. 15a,b) (Lora Lami Donin and Cotelli, 1977) (Schmidt-Rhaesa, 1997). The sperm structure is similar in many aspects to that of mermithid nematodes (Poinar and Hess-Poinar, 1993). There are unique features, one of which is what has been called a multivesicular complex, actually a number of flattened pockets, each filled with mucopolysaccharide appearing substances which surround the nuclear portion of the spermatozoan (Fig. 15a,b). The eggs are deposited singly or in clusters which often form elongate strings held together by secretions produced by the antrum (anterior portion of the female cloaca). In certain species that occur in fast flowing waters, ovipositional modifications occur. While conducting studies on *Euchordodes nigromaculatus* in a stream in New Zealand, the present author noted that the eggs were attached as a coating to rocks and other submerged debris (Fig. 14a) (Poinar, 1989). This unique manner of oviposition explains how certain hairworms can maintain populations in fast-flowing streams. Other investigators have noted eggs of Euchordodes attached to twigs and other stationary objects in the water (Bohall *et al.,* 1997) (D.J. Watermolen, personal observation).

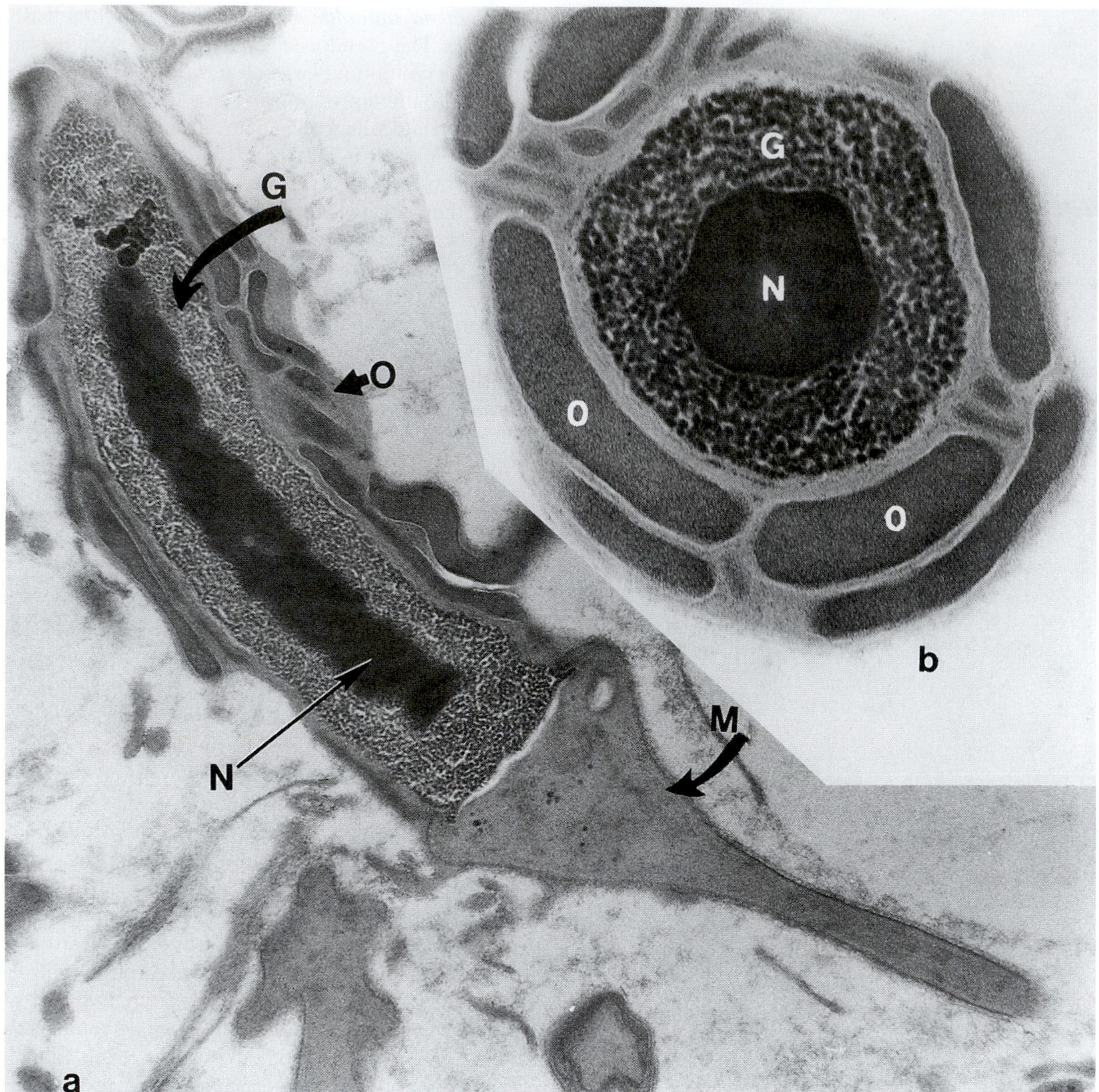

FIGURE 15 Electron micrographs of the lateral (a) and transverse (b) views of sperm cells of *Neochordodes occidentalis* (Montgomery). M, main cell body; N, nucleus; G, glycogen deposit surrounding nucleus; O, membraneous organelles forming the multivesicular complex. (Photos courtesy of Roberta Poinar.)

The preparasitic stage that hatches from the egg is morphologically quite different from the adult worm and can, therefore, properly be called a larva. [Recall that the term larva implies that some type of metamorphosis occurs before the adult stage.] The larva consists of a presoma or preseptum (featuring an evaginable proboscis armed with cuticular spines) and a body or postseptum containing adult tissue primordia (Figs. 17, 18c,d). Ultrastructural studies of the larvae of *Paragordius varius* demonstrated quite clearly the nature of the proboscis and spines (Zapotosky, 1974, 1975). It is unfortunate that so few larvae of described nematomorph species are known, since they possess their own specific characters which could be of taxonomic value, as noted by Poinar and Doelman (1974) and Bohall *et al.* (1997).

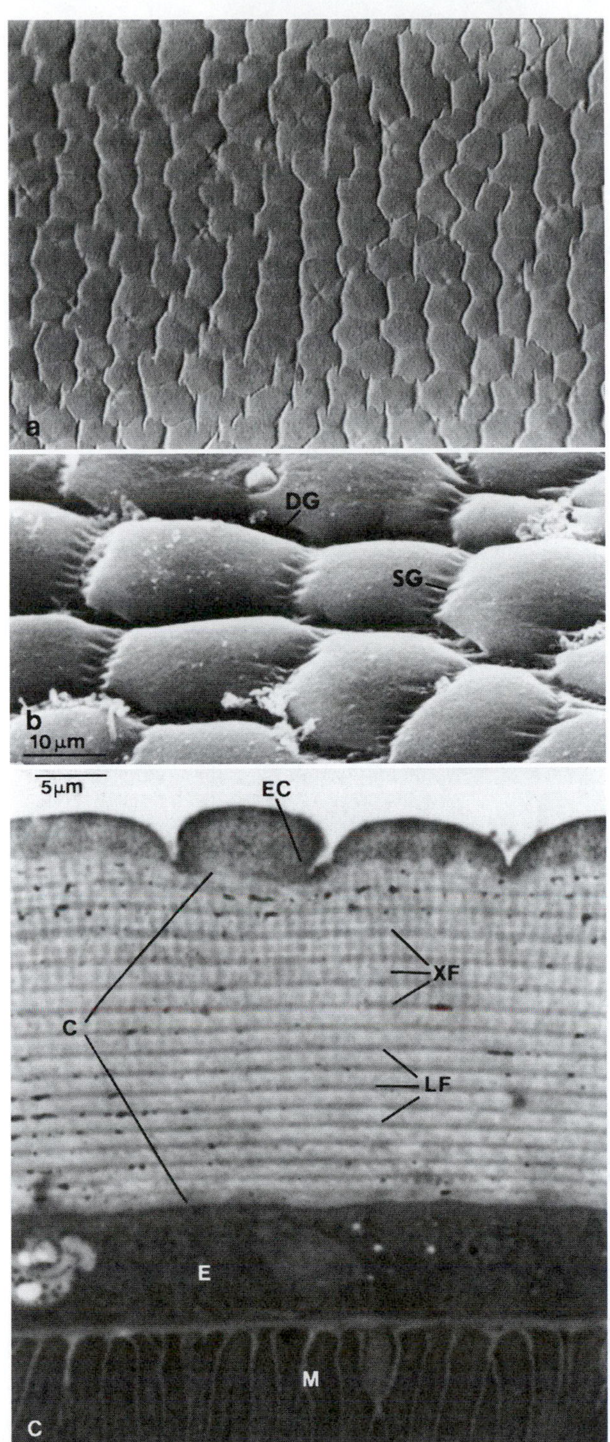

FIGURE 16 (a) Surface view of the areolar pattern on the epicuticle of *Neochordodes occidentalis* (Montgomery). (b) Raised areoles and intra-areolae furrows on the epicuticle of a *Gordius* sp. DG, deep groove separating areoles; SG, shallow grove separating areoles. (Courtesy of R. M. Eaken and J. L. Brandenburger.) (c) Cross section through the body wall of a *Gordius* sp. C, cuticle; E, epidermis; EC. epicuticle; LF, cuticular fibers cut longitudinally; M, muscle plates; XF, cuticular fibers cut crosswise. [Courtesy of R. M. Eaken and J. L. Brandenbruger.]

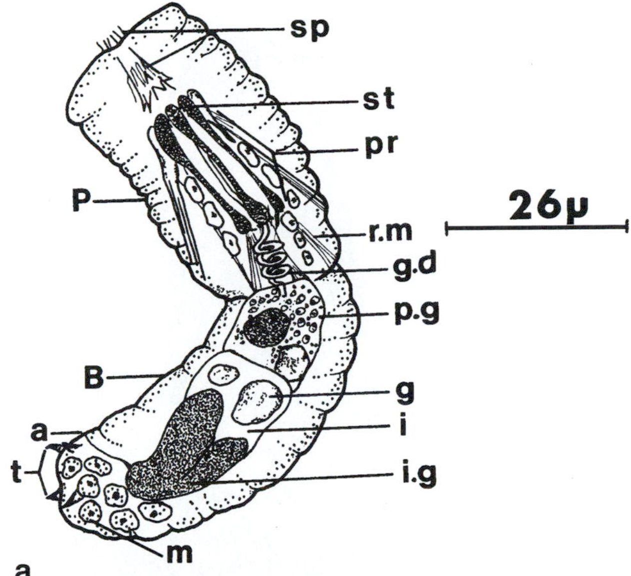

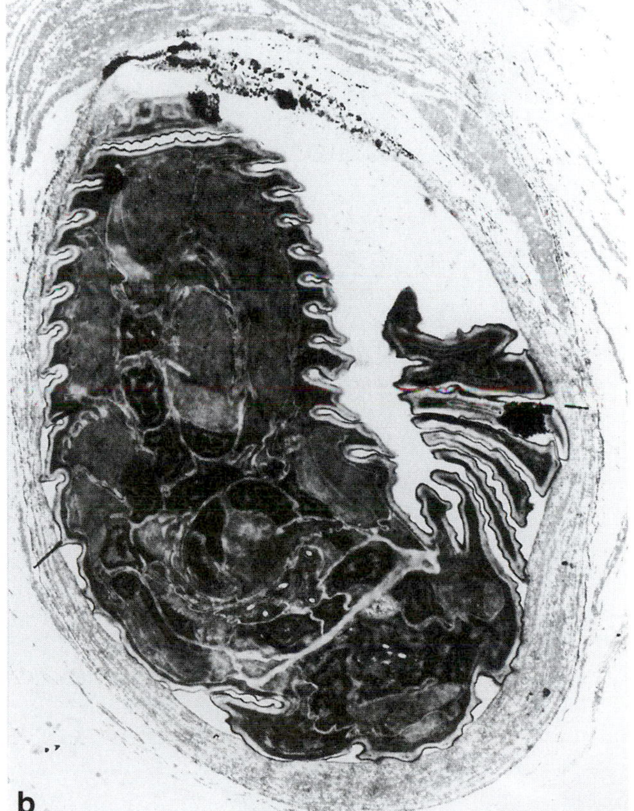

FIGURE 17 (a) Figure of the preparasitic larva of *Neochordodes occidentalis* (Montgomery). P, presoma (preseptum); B, body (postseptum); a, anus; g, globule. g,d, gland duct; i, intestine; i.g, intestinal gland; m, mesenchyme cells; p.g, preintestinal gland; pr, protrusible proboscis; r.m, retractor muscles; st, stylets; sp, circulet of spines; t, tail spines (modified from Poinar and Doelman, 1974.). (b) Electron micrograph of a larva of *N. occidentalis* encysted in the gut wall of a tadpole of the tree frog *Hyla regilla* Baird and Girard. Note that the surrounding cyst is composed of several layers. [Photo courtesy of Roberta Poinar.]

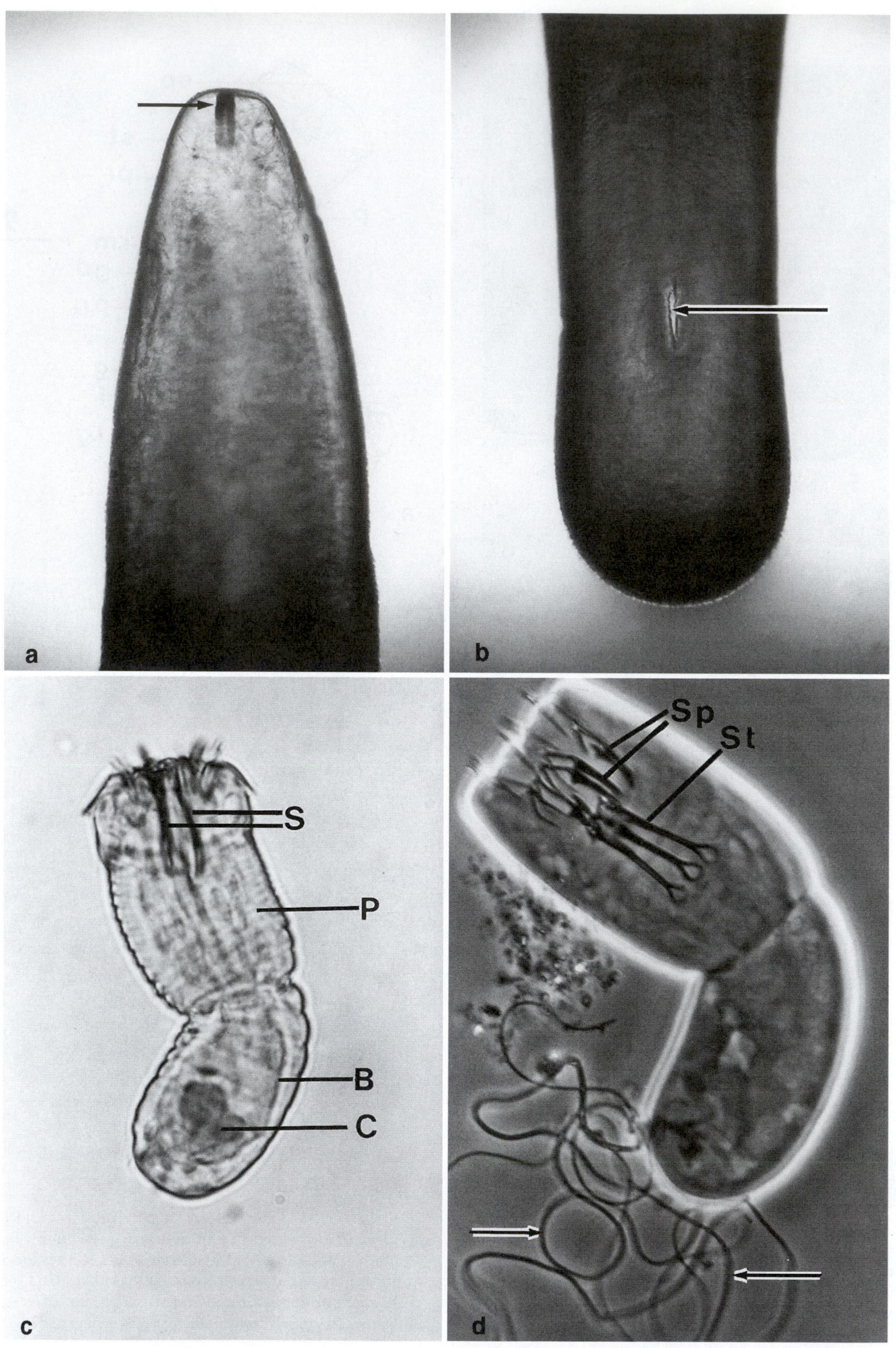

FIGURE 18 (a) Anterior end of an adult *N. occidentalis*. Note the degenerate pharynx (arrow). (b) Ventral view of male tail of *N. occidentalis* showing the elongate cloacal aperature (arrow). (c) Preparasitic larva of *N. occidentalis* showing presoma (P), body (B), intestinal gland (C), and stylets (S). (d) Preparasitic larva of *N. occidentalis* showing extruded secretions from the intestinal gland (arrows), stylet (ST), and spines (SP).

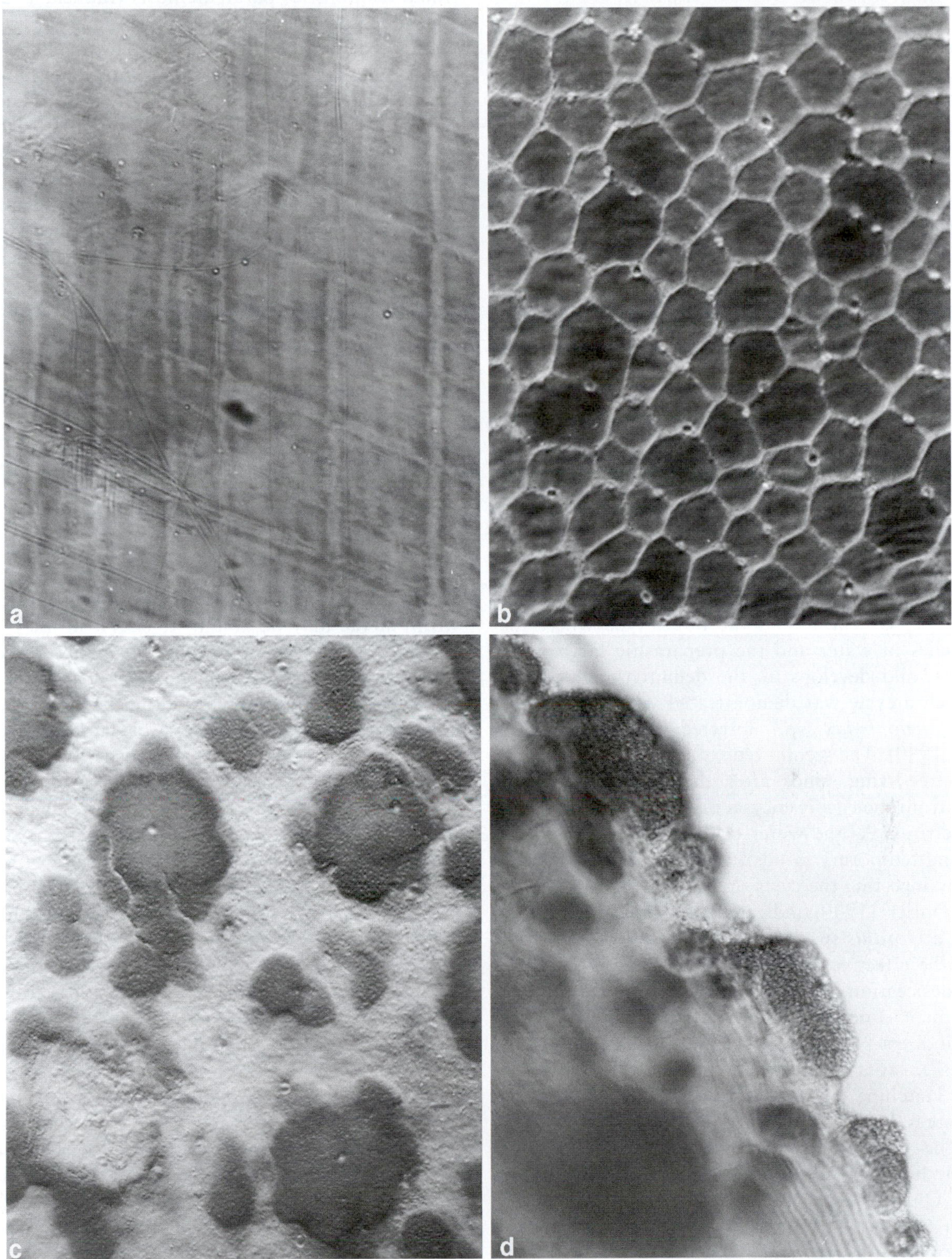

FIGURE 19 (a) Surface view of the body wall of *Gordius dimorphus* Poinar showing the absence of areoles and the presence of parallel criss-cross fibers. (b) Surface view of the areoles of *Euchordodes nigromaculatus* Poinar with minute spines present, especially in the furrows between the flattened areoles. (c) Surface view of the areoles of *Gordionus diblastus* (Orley) showing variation in areole size and distribution. (d) Lateral view of the raised areoles of *G. diblastus* showing size variability.

IX. DEVELOPMENT AND LIFE HISTORY

The four stages in the life history of nematomorphs are the egg, the preparasitic larva that hatches from the egg, the parasitic larva that develops within an invertebrate, and the free-living, aquatic adult. Parasitic larvae of all Nematomorpha mature only within invertebrates, which can be called the developmental or definitive hosts. A second type of host involved in the life cycle of probably the great majority of hairworms is a transport or paratenic host. The preparasitic larva enters the hemocoel and internal tissues of this host, but does not develop further until the paratenic host is eaten by a scavenger or predator. The paratenic host is usually an invertebrate, but can also be a vertebrate (e.g., tadpoles, fish, etc.). Poinar and Doelman (1974) noted that when larvae of *Neochordodes occidentalis* encysted in tadpoles (*Hyla regilla*), many of the cysts were composed of several overlapping layers as is shown in Figure 17b, in contrast to those normally formed in insects. However, such encysted larvae of *N. occidentalis* in the gut of tadpole paratenic hosts showed no abnormal features.

Three types of life cycles have been described for nematomorphs. The first is the direct type where the egg hatches in water and the preparasitic larva is ingested by, and develops in, the definitive invertebrate host. Such a cycle was demonstrated with Trichoptera larvae *(Stenophylax* sp.) infected by *Gordius* sp. (Dorier, 1930). The second type of cycle could be called indirect-free-living, since after the preparasitic larva hatches from the egg, it encysts on submerged leaves or other detritus. As the water dries up (found in forms inhabiting temporary ponds), the definitive host is infected by ingesting the cysts on the now exposed vegetation. Dorier (1930) indicated this type of development in a *Gordius* sp. that infected millipedes. For this cycle to be effective, the cysts would have to be resistant to desiccation. Further confirmation of this type of development is needed. The third type of cycle can be called indirect-paratenic where, after hatching, the preparasitic larva is ingested by an invertebrate or vertebrate. Hatching occurs when the mature preparasitic larva forces itself through the wall of the egg (Fig. 20b). The parasite burrows into the tissues of this paratenic host and then encysts, but does not develop further. Only when the paratenic host is eaten by a predator (carabid beetle, praying mantis, dragonfly) or omnivore (various Orthoptera) does the parasite initiate development. This type of cycle occurs in the great majority of investigated hairworms, including a *Chordodes* developing in praying mantids that ingest infected mayfly paratenic hosts (Inoue, 1962), a *Gordius* developing in *Dytiscus* that feed on infected tadpole (*Rana temporaria*) paratenic hosts (Blunck, 1922), *Euchordodes nigromaculatus* developing in stenopelmatids that fed on infected stoneflies, mayflies or caddisflies (Poinar, 1991), *Gordius robustus* developing in crickets that were fed infected meal worms (Hanelt and Janovy, 1999) and probably *Neochordodes occidentalis* which readily infects and encysts in mosquitoes and other aquatic insects (Poinar and Doelman, 1974). In fact, it is apparent that this indirect-paratenic type of cycle is the most typical type of development for hairwroms. The first two types of development mentioned above need confirmation.

There is no evidence that the preparasitic larvae can burrow directly through the outer body wall of either the paratenic or definitive host. In all cases, they enter by way of mouth and encyst in the midgut (Fig. 20e) or burrow through the midgut and encyst in the tissues of the body cavity. In cases of mass infection as demonstrated in laboratory studies with larvae of *N. occidentalis* invading mosquito larvae, the tissues may be damaged to such a degree that the host dies (Poinar and Doelman, 1974). In cases where the primary paratenic host is eaten by an insect predator which is not suitable for further parasitic development, the hairworm larvae may be able to re-encyst in the predator. Encysted larvae in dobsonflies probably represented cases of secondary paratenic hosts (Poinar, 1991).

Host immunity to hairworms has thus far been documented only in the case of larvae entering or remaining in paratenic hosts. The usual type of host reaction to such parasitic larvae is humoral melanization which may occur immediately after entry as in the case of *N. occidentalis* invading mosquito larvae (Fig. 20c,d) (Poinar and Doelman, 1974) or after the larvae have encysted in the internal tissues of the paratenic host as with hairworm larvae in caddis-flies, mayflies, stoneflies (Fig. 20e) and especially chironomid larvae (Poinar, 1991). Such melanization reactions often result in the death of invading larvae (Fig. 20d), although in cases where the host reaction occurred after encystment, it is possible that the parasites died of other causes and were then melanized.

When parasitic development has been completed and the hairworm is ready to emerge, the host must come into contact with water. This poses no problem for parasites in aquatic hosts, but can present obstacles when the host is a terrestrial arthropod. Although entry into water may occur accidentally, it is more likely that the hosts suddenly have a desire to reach water. This may result from partial desiccation due to the actions of the parasite or be a consequence of some abnormal stimuli from parasite products that affect physiological centers of the host. When the host enters water, the hairworms emerge and either slowly sink or initiate

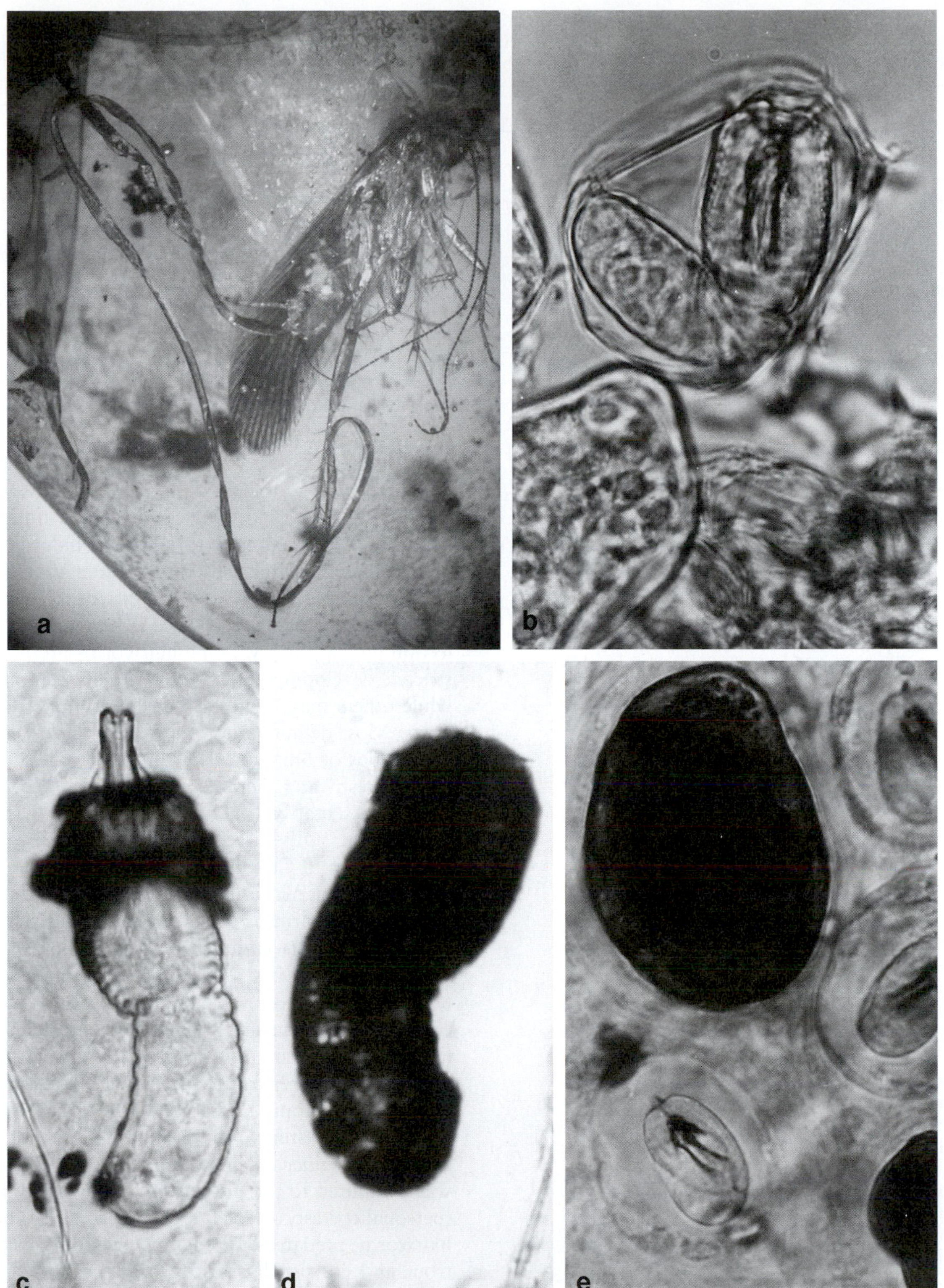

FIGURE 20 (a) Two specimens of *Paleochordodes protus* Poinar (1999) that were in the process of emerging from their cockroach host in 15–45 million-year-old Dominican amber. These represent the first unequivocal fossil Nematomorpha. (b) Pre-parasitic larva of *Neochordodes occidentalis* ready to emerge from its egg. (c) Host reaction showing initial stages of melanization of an invading larva of *N. occidentalis* that just penetrated into the hemocoel of a fourth stage mosquito larva (experimental infection; see Poinar and Doelman, 1974). (d) Melanized larva of *N. occidentalis* killed from the host reaction of a fourth stage mosquito larva (experimental infection; see Poinar and Doelman, 1974). (e) Dead, melanized encysted larvae of probably *Euchordes nigromaculatus* in the intestinal wall of the stonefly, *Stenoperla prasina* (natural infection; see Poinar, 1991).

TABLE II List of Host families Reported to contain the Developing Stages of Nematomorpha[a]

Taxon	Reference
Phylum Annelida	
Class Hirudinea	
Order Rhynchobdellida	
Family Glossiphoniidae	Leidy (1878)
Order Pharyngobdellida	
Family Erpobdellidae	Sawyer (1971)
Phylum Mollusca	
Class Gastropoda	
Subclass Pulmonata	
Family Lymnaeidae	Chitwood and Chitwood (1937)
Phylum Arthropoda	
Class Arachnida	
Order Scorpionida	
Family Vaejovidae	S. Williams (personal communication)
Order Ambylpygi	
Family Phrynidae	Sciacchitano (1958)
Class Crustacea	
Order Notostraca	
Family Apodidae	Linstow (1878)
Order Decapoda	
Family Caridae	Linstow (1878)
Class Chilopoda	
Order Lithobiomorpha	
Family Lithobiidae	Dorier (1930)
Order Scolopendromorpha	
Family Scolopendridae	Dorier (1930)
Class Diplopoda	
Order Glomerida	
Family Glomeridae	Dorier (1930)
Order Spirobolida	
Family Spirobolidae	Cooper and Storck (1973)
Order Julida	
Family Julidae	Sahli (1972)
Class Insecta	
Order Odonata	
Family Libellulidae	Heinze (1937)
Order Orthoptera	
Family Acrididae	Blunck (1922)
Family Blattidae	Sciacchitano(1958)
Family Gryllacrididae	Poinar (1989)
Family Gryllidae	Montogomery (1907)
Family Rhaphidophoridae	Poulin (1996)
Family Stenopelmatidae	Poinar (1991)
Family Tettigonidae	Thorne (1940)
Order Mantodea	
Family Mantidae	Sciacchitano (1958)
Order Phamsatida	
Family Phasmatidae	Römer (1895)
Order Dermaptera	
Family Forficulidae	Sciacchitano (1958)
Order Hemiptera	
Family Notonectidae	Zalmon (1977)
Order Neuroptera	
Family Sialidae	Mellanby (1951)
Order Lepidoptera	
Family Saturnidae	Camerano (1897)
Order Coleoptera	
Family Carabidae	Blunck (1922)
Family Chrysomelidae	Dorier (1930)
Family Dytiscidae	Blunck (1922)
Family Silphidae	Blunck (1922)
Family Tenebrionidae	Baylis (1944)
Order Hymenoptera	
Family Apidae	Cury (1946)

[a] It is often impossible to determine whether authors are referring to paratenic or developmental hosts in their reports. Questionable references have been omitted.

undulating body movements. Males are claimed to be more active than females. As the parasites mature in the host, they change from a light, cream color to a yellowish brown or dark brown color. At the time of emergence, most have already turned their natural color, but in some cases, further darkening occurs after emergence.

Since there are so few records of known gordiids from identified hosts, very little is known about host selection and specificity. Some hairworms are probably restricted developmentally to certain invertebrate genera, while others may be able to develop in a wide range of hosts. Most definitive hosts are medium-to-large bodied predaceous or omnivorous arthropods. The great majority of freshwater nematomorphs have been collected from representatives of the insect orders Coleoptera and Orthoptera. Other host groups include myriapods (Diplopoda and Chilopoda), crustaceans, and leeches. A list of invertebrate families known to contain developing stages of hairworms is presented in Table II. The diversity of paratenic hosts is quite great, extending from trematodes (Cort, 1915) to vertebrates. Preparasites will attempt to burrow and encyst into the soft tissues of any cold blooded organism that ingests them. The natural enemies of hairworms are poorly known. Fish ingest the adults and the preparasites are probably fed upon by invertebrate micropredators. A *Gordius* adult that was brought to nestlings by a parent bird (a tapaculo, *Scelorchilus rubecula*) in a temperate rain forest in Chile was submitted to the present author by Mary F. Willson (personal correspondence, 1994), so birds can be listed as hairworm predators. Predation by rock bass, brown trout and crayfish has also been documented (Cochran et. al., 1999).

X. SAMPLING

Except for the marine genus *Nectonema,* all known representatives of the Nematomorpha occur in freshwater. Their habitats range from watering troughs and puddles to rivers, lakes and subterranean streams. Several researchers have commented on their occurrence in the Great Lakes (Watermolen and Haen, 1994; Cochran *et al.*, 1990). No special sampling technique has been developed for hairworms. They are large enough to be seen with the naked eye and can be netted or lifted out of the water by hand. Stream forms often accumulate in slower moving water off to the side of the main current. Holes dug into small streamlets will often catch and hold adults that are being carried downstream and nets can be placed across small streams to collect the unattached adults. Late summer and spring are the best times to look for the free-living adults. Although the long, whitish egg strands of some species can sometimes be spotted in water, the preinfectives are too small to be found and are usually noted only by chance when examining the paratenic hosts. Perhaps dissecting paratenic hosts and searching for encysted larvae is the easiest way to determine if hairworms are present in an aquatic system. The eggs of those species adapted to fast-flowing streams can be noted stuck to the surfaces of rocks and other debris. Growing larvae in developmental hosts can be found throughout the year, but are much less frequent than are encysted larvae in paratenic hosts and are nearly impossible to keep alive if they are removed from their host prior to completion of their development. No methods are available for *in vitro* development or culture of nematomorphs. May (1919) was able to monitor development by injecting preparasitic larvae of *Gordius* into the body cavity of long-horned grasshopper hosts (Tettigonidae). However, because of the fragile character of the preparasitic forms and the relatively long period of parasitic development (several months) coupled with the problem of maintaining the hosts, few workers have even bothered to culture hairworms *in vivo*.

XI. IDENTIFICATION

Adult hairworms can be preserved in 5‰ formalin or 70% alcohol for identification purposes. All diagnostic characters are external features associated with the head, tail, and epicuticular surfaces. The extremities can be removed and mounted in lactophenol or glycerin after dehydration in an alcohol series. For cuticular examination, small slivers of epicuticle can be removed from the body regions with a razor blade and placed in lactophenol or dehydrated in glycerin. The underlying epidermis and muscle tissue can then be scraped away and the cuticular slice mounted (outer surface up) in lactophenol or glycerin. This will expose the areoles and their ornamentation. Some workers prefer to clear the sections in xylene before mounting them. The scanning electron microscope has been used to reveal cuticular details and, although it can be a useful tool, workers should not base descriptions on characters only visible with the SEM since it is not practical for many, including field biologists. Besides, adequate diagnostic characters can normally be observed at least with oil immersion objectives (1000 ×). In a study comparing the light and scanning electron microscope in differentiating cuticular characters separating *Parachordodes lineatus* and *P. violaceus*, Smith (1991) noted that while details were clearer with SEM, the light microscope revealed the essential characters needed to separate the species. Success in large part with the light microscope depends greatly on skill in preparing the sample. Both males and females should be available for accurate identification; however, some males can be identified alone using the following key.

The classification used here follows that presented by Dorier (1965) in the Traité de Zoologie. It also incorporates the contributions of Heinze (1952) as well as studies on North American forms by May (1919), Leidy (1851), Carvalho (1942), Montgomery (1898a, b, 1899, 1907), Poinar and Doelman (1974), Redlich (1980), Chandler (1985), Chandler and Wells (1989), Smith (1991; 1994), Pennak (1989) and others.

A. Taxonomic Key to Extant Genera of Nematomorpha

Although some of the genera included have not been reported from North America, it is highly probable that they as well as other still undescribed genera occur in the Nearctic. Those genera with reported North American representatives are marked with an asterisk.

1a. Adults with natatory hairs; marine forms .... Nectonematoidea (Rauther, 1930)
.... **Nectonema* Verrill, 1879

1b. Adults lacking natatory hairs; freshwater forms .... Gordioidea Rauther, 1930 2

2a (1b). Cuticle covered with ridges; male cloaca terminal .... *Chordodiolus* Heinze, 1935

2b. Cuticle not covered with ridges, male cloaca subterminal 3

3a (2b). Cuticle smooth, lacking aeroles or with flat, smooth areoles; male tail bilobed (Fig. 14b), with a postcloacal fold (Fig. 14b); female tail entire .... Gordiidae May, 1919 4

3b. Cuticle with distinct areoles usually ornamented with bristles or tubercles; male tail entire or if bilobed, then without a postcloacal cresent; female tail entire or trilobed .... Chordodidae May, 1919 5

4a (3a). Lobes of male tail rounded (Fig. 14b); head region not attenuated .... *Gordius Linneaus, 1766*

4b. Lobes of male tail pointed; head region attenuated .... *Acutogordius* Heinze, 1952

5a (3b). Male tail entire; female tail entire 6

5b. Male tail bilobed; female tail entire or trilobed 9

6a (5a). Epicuticle with one type of areole, containing short bristles and tubercles * *Neochordodes* Carvallo, 1942 ....

6b. Epicuticle with two or more types of areoles 7

7a (6b). Areoles containing prominent tubercles, crowns or papillae; female cloaca terminal .... * *Chordodes* Creplin, 1847

7b. Areoles without prominent tubercles, crowns or papillae; female cloaca subterminal 8

8a (7b). Minute spines present on cuticle, especially in furrows between areoles (Fig. 19b); slender, small forms .... *Euchordodes* Heinze, 1937

8b. Interareolar furrows without spines; normal-sized forms .... **Pseudochordodes* Carvalho, 1942

9a (5b). Female tail trilobed .... * *Paragordius* Camerano, 1897

9b. Female tail entire 10

10a (9b). Areoles arranged in rows separated by ridges .... *Beatogordius* Heinze, 1934

10b. Areoles randomly arranged 11

11a (10b). Interareolar furrows large, forming a network around the areoles .... *Semigordionus* Heinze, 1952

11b. Interareolar furrows normal 12

12a(11b). A single type of areole present although they may vary in size and be single, bifid or trifid (Fig. 19c,d) .... *Gordionus* G. W. Muller, 1927

12b. Two different types of areoles present 13

13a(12b). Lobes on male tail longer than wide; adult head clearly flattened; pore canals present on larger areoles in males .... **Parachordodes* Camerano, 1897

13b. Lobes on male tail approximately as long as wide; adult head slightly flattened; pore canals absent or if present, then only in interareolar areas in males .... *Paragordionus* Heinze, 1935

LITERATURE CITED

Ayoub, S. M. 1977. Plant Nematology. An agricultural training aid. California Department of Food and Agriculture, Sacramento. 157 pp.

Baylis, H. A. 1944. Notes on the distribution of hairworms (Nematomorpha: Gordiidae) in the British Isles. Proceedings of the Zoological Society of London 113: 193–197.

Blunck, H. 1922. Die Lebensgeschichte der im Gelbrand schmarotzenden Saitenewürmer. Zoologischer Anzeiger 54:111–132, 145–162.

Bohall, P. J., Wells, M. R., Chandler, C. M. 1997. External morphology of larvae of *Chordodes morgani* (Nematomorpha). Invertebrate Biology 116:26–29.

Bretschko, G. 1973. Benthos production of a highmountain lake: Nematoda. Verhandlungen des Instituts für Vereinforschung in Limnologica 18:1421–1428.

Camerano, L. 1897. Monografia dei Gordii. Memorie della Reale Accademia delle Scienze di Torino 47:339–419.

Cappucci, Jr., D. T. 1982. Gordian worms (Hairworms): biological and public health aspects, *in:* Steele, J. H. Editor-in-Chief, CRC Handbook Series in Zoonoses, Section C: Parasitic Zoonoses, Vol. II, CRC Press, Inc., Boca Raton, FL, pp. 193–203.

Carvalho, J. M. C. 1942. Studies on some Gordiaceae of North and South America. Journal of Parasitology 28:213–222.

Chandler, C. M. 1985. Horsehair worms (Nematomorpha, Gordioidea) from Tennessee, with a review of taxonomy and distribution in the United States. Journal of the Tennessee Academy of Science 60:59–62.

Chandler, C. M., Wells, M. R. 1989. Cutlcular features of *Chordodes morgani* (Nematomorpha) using scanning electron microscopy. Transactions of the American microscopy Society 108: 152–158.

Chitwood, B. G. 1950. Nemic relationships, *in:* Chitwood, B. G. Chitwood, M. B., Eds., Introduction to Nematology. University Park Press, Baltimore, MD, pp. 227–241.

Chitwood, B. G., Allen, M. W. 1959. Nemata, *in:* Edmondson, W. T., Ed., Fresh-water biology. 2nd ed., Wiley, New York, pp. 368–401.

Chitwood, B. G., Chitwood, M. B. 1937. Snails as hosts and carriers of nematodes and Nematomorpha. The Nautilus 50: 130–135.

Cobb, N. A. 1914. North American fresh-water nematodes. Transactions of the American Microscopical Society 33:35–100.

Cochran, P. A., Watermolen, D. J., Haen, G. L. 1990. Horsehair worms (Phylum Nematomorpha) in Lake Michigan. Journal of Great Lakes Research 16:485–487.

Cochran, P. A., Kinziger, A. P., Poly, W. J. 1999. Predation on Horsehair worms (Phylum Nematomorpha). Journal of Freshwater Ecology 14:211–218.

Coe, W. R. 1937. Methods for the laboratory culture of nemerteans, *in:* Needham, J. G., Ed., Culture methods for invertebrate animals. Dover, New York, pp. 162–165.

Cooper, C. L., Storck, T. W. 1973. *Gordius* sp., a new host record. Ohio Journal of Science 73:228.

Corbel, J.-C. 1967. Les parasites des Orthopteres. Annales de Biologie 6:391–426.

Cort, W. W. 1915. *Gordius* larvae parasitic in a trematode. Journal of Parasitology 1:198–199.

Cury, R. 1946. Molestias das abelhas. Biologico 12:241–254.

Dorier, A. 1930. Recherches biologiques et systematiques sur les Gordiaces. Travaux du laboratoire d'hydrobiologie et de pisciculture de l' Universite de Grenoble, 22 année:1–183.

Dorier, A. 1965. Classe Gordiaces, *in:* Grasse, P.-P., Ed., Traite de Zoologie. Vol. 4. Masson, Paris, pp. 201–1222.

Downing, J. A., Rigler, F. H., Eds. 1984. A manual on methods for the assessment of secondary production in freshwaters. Blackwell, Oxford.

Eakin, R. M., Brandenburger, J. L. 1974. Ultrastructural features of a Gordian worm (Nematomorpha). Journal of Ultrastructure Research 46:351–374.

Esser, R. P., Buckingham, G. R 1987. Genera and species of free-living nematodes occupying fresh-water habitats in North America, *in:* Veech, J. A., Dickson, D. W. Eds. Vistas on Nematology. Society of Nematologists, Hyaltsville, MD, pp. 477–487.

Esser, R. P., Buckingham, G. R., Bennett, C. A., Harkcom, K. J. 1985. A survey of phytoparasitic and free-living nematodes associated with aquatic macrophytes in Florida. Proceedings of the Florida Soil and Crop Science Society 44:150–155.

Ferris, V. R., Ferris, J. M. 1972. Nematode community structure: a tool for evaluating water resource environments. Technical report No. 30. Water Resources Research Center, Purdue University, Lafayette, IN.

Ferris, V. R., Ferris, J. M., Tjepkema, J. P. 1973. Genera of freshwater nematodes (Nematode) of Eastern North America. U.S. Environmental Protection Agency Identification Manual No. 10. 38 pp.

Filipjev, I. N., Schuurmans Stekhoven, Jr., J. H. 1959. Agricultural Helminthology. Brill, Leiden, The Netherlands.

Flint, R. W. 1976. The natural history, ecology and proauction of the crayfish, *Pacifastacus leniusculus,* in a subalpine lacustrine environment. Ph.D. Thesis, University of California, Davis.

Gerber, K., Smart, Jr., G. C. 1987. Plant-parasitic nematodes associated with aquatic vascular plants, *in:* Veech, J. A., Dickson, D. W., Eds. Vistas on Nematology. Society of Nematologists, Hyaltsville, MD, pp. 488–501.

Goodey, J. B. 1963. Soil and Freshwater Nematodes. Methuen, London.

Hanelt, B. Janovy, Jr., J. 1999. The life cycle of a horsehair worm, *Gordius robustus* (Nematomorpha: Gordioidea). Journal of Parasitology 85:139–141.

Hantzschel, W. 1975. Trace fossils and problematics, *in:* Moore, R. C., Ed., Treatise on Invertebrate Paleontology. Part W, Supplement 1. The Geology Society of America. Boulder, CO, pp. 1–269 .

Heinze, K. 1937. Die Saitenwürmer (Gordioidea) Deutschlands. Eine systematisch-faunistische Studie über insektenparasiten aus der Gruppe der Nematomorpha. Zeitschrift für Parasitenkunde 9:263–344.

Heinze, K. 1952. Über Gordioidea, eine systematische Studie über Insektenparasiten aus der Gruppe der Nematomorpha. Zeitschrift für Parasitenkunde 7:657–678.

Heyns, J., Coomans, A. 1980. Freshwater nematodes from South Africa. 5. *Chronogaster* Cobb, 1913. Nematologica 26: 245–265.

Hoeppli, R. J. C. 1926. Studies of free-living nematodes from the thermal waters of Yellowstone Park. Transactions of the American Microscopical Society 45:234–255.

Holme, N. A., McIntyre, A. O., Eds. 1971. Methods for the study of marine benthos. IBP Handbook No. 16. Blackwell, Oxford.

Holopainen, I. J., Paasivirta, L. 1977. Abundance and biomass of the meozoobenthos in the oligotrophic and mesohumic lake Paajarvi, southern Finland. Annales Zoologica Fennland 14:124–134.

Hulings, N. C., Gray, J. S., Eds. 1971. A manual for the study of meiofauna. Smithsonian Contributions to Zoology 78:84 pp.

Hummon, W. D. 1981. Extraction by sieving: a biased procedure in studies of stream meiofauna. Transactions of the American Microscopical Society 100:278–284.

Inoue, I. 1962. Studies on the life history of *Chordodes japonensis,* a species of Gordiaceae. III. The mode of infection. Annotationes Zoological Japonenses 35:12–19.

Issel, R. 1906. Sulla termobiosi negli animal) acquatici. Ricerche faunistiche e biologiche. Attidella Societa Ligustica di Scienze natural) e geografiche 17:3–72.

Jacobs, L. J. 1984. The free-living inland aquatic nematodes of Africa—a review. Hydrobiologia 113:259–291.

Kirjanova, E. S. 1958. On the structure of the copulative organs of males of the freshwater hairworms (Nematomorpha, Gordioidea). Zoological Zhurnal 37:359–372. (In Russian.)

Leidy, J. 1870. The gordius, or hairworm. The American Entomologist and Botanist 2:193–197.

Leidy, J. 1878. On *Gordius* infesting the cockroach and leech. Proceedings of the Academy of Natural Sciences of Philadelphia 30:383.

Linstow, O. von. 1878. Compendium der Helminthologie, Hanover.

Lora Lami Donin, C., Cotelli, F. 1977. The rod-shaped sperm of Gordioidea (Aschelminthes, Nematomorpha). Journal of Ultrastructural Research 61:193–200.

May, H. G. 1919. Contributions to the life histories of *Gordius robustus* Leidy and *Paragordius varius* (Leidy). Illinois Biological Monographs 5:1–119.

Mellanby, H. 1951. Animal life in freshwater. A guide to freshwater invertebrates. 4th ed., Methuen, London.

Meyl, A. H. 1954. Beiträge zur Kenntnis der Nematodenfauna vulkanisch erhitzter Biotope. Zeitschrift für Morphologie und Ökologie der Tiere 42:421–448.

Micoletzky, H. 1922. Die freilebenden Erd-Nematoden. Archiv für Naturgeschichte A87: 1–650.

Montgomery, Jr., T. H. 1907. The distribution of the North American Gordiacea, with descriptions of a new species. Proceedings of the Academy of Natural Sciences of Philadelphia, 59: 270–272.

Montgomery, Jr., T. H. 1899. Synopses of North American invertebrates. II. Gordiaces (hairworms). American Naturalist 33: 647–652.

Montgomery, Jr., T. H. 1898a. The Gordiacea of certain American collections with particular reference to the North American fauna. Bulletin of the Museum of Comparative Zoology 32: 1–59.

Montgomery, Jr., T. H. 1898b. The Gordiacea of certain American collections, with particular reference to the North American fauna. II. Proceedings of the California Academy of Science 1:333–344.

Moussa, M. T. 1969. Nematode fossil tracks of Eocene age from Utah. Nematologica 15:376–380.

Nalepa, T. F., Quigley, M. A. 1983. Abundance and biomass of the meiobenthos in nearshore Lake Michigan with comparisons to the macrobenthos. Journal of Great Lakes Research 9:530–547.

Nicholas, W. L. 1984. The biology of free-living nematodes. Clarendon Press, Oxford.

Nichols, J. A. 1979. A simple flotation technique for separating meiobenthic nematodes from fine-grained sediments. Transactions of the American Microscopical Society 98:127–130.

Nuttycombe, J. W. 1937. Wheat-grain infusion, *in:* Needham, J. G. Ed. Culture methods for invertebrate animals. Dover, New York, pp. 135–136.

Oden, B. J. 1979. The freshwater littoral meiofauna in a South Carolina reservoir thermal effluents. Freshwater Biology 9:291–304.

Pennak, R. W. 1978. Fresh-water invertebrates of the United States. 2nd ed., Wiley, New York.

Pennak, R. W. 1989. Freshwater invertebrates of the United States. 3rd ed., Wiley, New York.

Petersen, J. J. 1972. Procedures for the mass rearing of a mermithid parasite of mosquitoes. Mosquito News 32:226–230.

Poinar, Jr., G. O. 1983. The Natural History of Nematodes. Prentice-Hall. Englewood Cliffs, N J.

Poinar, Jr., G. O. 1984. *Heydenius dominicanus* n. sp. (Nematoda: Mermithidae), a fossil parasite from the Dominican Republic. Journal of Nematology 16:371–375.

Poinar, Jr., G. O. 1989. Unpublished observations in Australia and New Zealand.

Poinar, Jr., G. O. 1991. Hairworm (Nematomorpha: Gordioidea) parasites of New Zealand wetas (Orthoptera: Stenopelmatidae) Canadian Journal of Zoology 69:1592–1599.

Poinar, Jr., G. O. 1999. *Paleochordodes protus* n.g., n.sp. (Chordodidae:Nematomorpha), parasites of a fossil cockroach, with a critical examination of other fossil hairworms and helminthes of extant cockroaches (Blattaria: Insecta). Invertebate Biology (118:109–115.).

Poinar, Jr., G. O., Doelman, J. J. 1974. A reexamination of *Neochordodes occidentalis* (Mont.) comb. n. (Chordodidae: Gordioidea): larval penetration and defense reaction in *Culex pipiens* L. Journal of Parasitology 60:327–335.

Poinar, Jr., G. O., Hansen, E. 1983. Sex and reproductive modifications in nematodes. Helminthological Abstract, Series B 52:145–163.

Poinar, Jr., G. O., Hess, R. 1988. Protozoan diseases of nematodes, *in:* Poinar, G. O. Jr. Jansson, H.-B., Eds., Diseases of Nematodes. Vol. 1, pp. 103–131, CRC Press, Boca Raton, FL.

Poinar, Jr., G. O., Hess-Poinar, R. 1993. The fine structure of *Gastromermis* sp. (Nematoda: Mermithidae) sperm. Journal of Submicroscopic Cytology and Pathology 25:417–431.

Poinar, Jr., G. O., Sarbu, S.M. 1994. *Chronogaster troglodytes* sp.n. (Nemata: Chronogasteridae) from Movile Cave, with a review of carvernicolous nematodes. Fundamental and applied Nematology 17:231–237.

Poinar, Jr., G.O., Acra, A., Acra, F. 1994. Earliest fossil neam tode (Mermithidae) in Cretaceous Lebanese amber. Fundamental and applied Nematology 17:475–477.

Por, F. D., Moary, D. 1968. Survival of a nematode and an oligochaete species in the anaeorbic benthal of Lake Tiberios, Oikos 19:388–391.

Poulin, R. 1996. Observations on the free-living adult stage of *Gordius dimorphus* (Nematomorpha: Gordioidea), Journal of Parasitology 82:845–846.

Rahm, G. 1937. Grenzen des Lebens? Studien in heissen Quellen. Forschungen und Fortschritte 13:381–387.

Redlich, A. 1980. Description of *Gordius attoni* sp.n. (Nematomorpha, Gordiidae) from Northern Canada. Canadian Journal of Zoology 58:382–385.

Riemann, F., Schrage, M. 1978. The mucus-trap hypothesis on feeding of aquatic nematodes and implications of biodegradation and sediment texture. Oecologia 34:75–88.

Römer, F. 1895. Die Gordiiden des Naturhistorischen Museums in Hamburg. Zoologische Jahrbücher, Abtheilung fur Systematik, Geographie und Biologie der Thiere 8:790–803.

Sahli, F. 1972. Modifications des caracteres sexuel secondaires males chez les Julidae (Myriapods, Diplopoda) sous l'influence de Gordiaces parasites. Comptes Rendus de l'Academie des Sciences 274:900–903.

Sawyer, R. T. 1971. Erpobdellid leeches as new hosts for the Nematomorpha, *Gordius*. Journal of Parasitology 57:285.

Schiemer, F. 1978. Verteilung und Systematik der freilebenden Nematoden des Neusiedlersee. Hydrobiologie 58:167–194.

Schiemer, F. 1983. Comparative aspects of food dependence and energetics of freeliving nematodes. Oikos 41:32–42.

Schiemer, F. 1987. Nematoda, *in:* Pandian, T. J., Vernberg, F. J. Eds., Animal energetics. Vol. 1: Protozoa through Insecta. Academic Press, New York, pp. 185–215.

Schmidt-Rhaesa, A. 1997. Ultrastructural observations of the male reproductive system and spermatozoa of *Gordius aquaticus* L. 1758 (Nematomorpha). Invertebrate Reproduction and Development 32: 31–40.

Schmidt-Rhaesa, A. 1998a. Muscular ultrastructure in *Nectonema munidae* and *Gordius aquaticus* (Nematomorpha). Invertebrate Biology 117:38–45.

Schmidt-Rhaesa, A. 1998b. Phylogenetic relationships of the Nematomorpha- a discussion of current hypothesis. Zoologischer Anzeiger 236:203–216.

Sciacchitano, I. 1958. Gordioidea del Congo Belga. Annales Museum de la Republic de Congo Belge 67:7–111.

Smith, D. G. 1991. Observations on the morphology and taxonomy of two *Parachordodes* species (Nematomorpha, Gordioidea, Chordidae) in southern New England (USA). Journal of Zoology (London) 225: 469–480.

Smith, D. G. 1994. A reevaluation of *Gordius aquaticus difficilis* Montgomery, 1898 (Nematomorpha, Gordioidea, Gordiidae). Proceedings of the Academy of Natural Sciences of Philadelphia 145:29–34.

Southey, J. F. 1978. Laboratory methods for work with plant and soil nematodes. Tech. Bull. No. 2. Ministry of Agriculture, Fisheries and Food, London.

Stirewalt, M. A. 1937. The culture of *Microstomum, in:* Needham, J. G., Ed., Culture methods for invertebrate animals. Dover, New York, pp. 149–150.

Strayer, D. 1985. The benthic micrometazoans of Mirrow Lake, New Hamphire. Archiv für Hydrobiologie, Supplement 72:287–426.

Thorne, G. 1940. The hairworm, *Gordius robustus* Leidy, as a parasite of the Mormon cricket, *Anabrus simplex* Haldeman. Journal of the Washington Academy of Science 30:219–231.

Tarjan, A. C., Esser, R. P., Chang, S. L. 1977. An illustrated key to nematodes found in freshwater. Journal of Water Pollution Control Federation 49:23182–337.

Tsalolikhin, S. Y. 1983. The nematode families Tobrilidae and Tripylidae: World fauna. Nauka, Leningrad. (In Russian.)

Viglierchio, D. R., Schmidt, R. V. 1983. On the methodology of nematode extraction from field samples: Comparison of the methods for soil extraction. Journal of Nematology 15:450–454.

Voigt, E. 1938. Ein fossiler Saitenwurm *(Gordius tenuifibrosus* n.sp.) aus der Eozänen Braunkohle des Geiseltales. Nova Acta Leopoldina, new series 5:351–360.

Warwick, R. M., Price, R. 1979. Ecological and metabolic studies on free-living nematodes from an estuarine mudflat. Estuary Coastal Marine Science 9:257–271.

Watermolen, D. J., Haen, G. L. 1994. Horsehair worms (Phylum Nematomorpha) in Wisconsin, with notes on their occurrence in the Great Lakes. Journal of Freshwater Ecology 9:7–11.

Watson, J. M. 1960. Medical Helminthology. Bailliere, London.

Winterbourn, M. J., Brown, T. J. 1967. Observations on the faunas of two warm streams in the Taupo Thermal Region. New Zealand Journal of marine and freshwater Resources 1:38–50.

Xianguang, H., Bergstrom, J. 1994. Palaeoscolecid worms may be nematomorphs rather than annelids. Lethaia 27:11–17.

Zalmon, F. G. 1977. A nematomorph parasitizing a back swimmer (Hemiptera: Notonectidae). The American Midland Naturalist 97:229–230.

Zapotosky, J. E. 1971. The cuticular ultrastructure of *Paragordius varius* (Leidy, 1851) (Gordioidea, Chordodidae). Proceedings of the Helminthological Society of Washington 38:228–236.

Zapotosky, J. E. 1974. Fine structure of the larval stage of *Paragordius varius* (Leidy, 1851) (Gordioidea: Paragordiidae). 1. The Preseptum. Proceedings of the Helminthological Society of Washington 41:209–221.

Zapotosky, J. E. 1975. Fine structure of the larval stage of *Paragordius varius* (Leidy, 1851) (Gordioidea: Paragordiidae). 2. The postseptum. Proceedings of the Helminthological Society of Washington 42:103–111.

Zullini, A. 1976. Nematodes as indicators of river pollution. Nematology Mediterranean 4:13–22.

APPENDIX 9.1 Systematic Arrangement of the Nematode Genera[a]

Phylum Nematoda
- Class Adenophorea
  - Subclass Chromadorida
  - Order Araeolaimida
    - Suborder Araeolaimina
    - Superfamily Axonolaimoidea
      - Family Axonolaimidae *(Cylindrolaimus)*
    - Superfamily Leptolaimoidea
      - Family Bastianiidae *(Bastiania)*
      - Family Leptolaimidae *(Rhabdolaimus)*
    - Superfamily Camacolaimidae
      - Family Camacolaimidae *(Aphanolaimus, Paraphanolaimus)*
    - Superfamily Plectoidea
      - Family Chronogasteridae *(Chronogaster)*
      - Family Plectidae *(Anonchus, Plectus)*
      - Family Teratocephalidae *(Euteratocephalus, Teratocephalus)*
  - Order Monhysterida
    - Superfamily Monhysteroidea
      - Family Monhysteridae *(Monhystera, Monhystrella, Prismatolaimus)*
  - Order Chromadorida
    - Suborder Chromadorina
    - Superfamily Chromadoridea
      - Family Chromadoridae *(Chromadorita, Punctodora)*
    - Suborder Cyatholaimina
    - Superfamily Cyatholaimoidea
      - Family Cyatholaimidae *(Achromadora, Ethmolaimus, Odontolaimus, Paracyatholaimus, Prodesmodora*
      - Family Microlaimidae *(Microlaimus)*
  - Order Enoplida
  - Suborder Enoplina
    - Superfamily Tripyloidea
      - Family Tripylidae *(Tobrilus, Tripyla)*
        - Family Alaimidae *(Alaimus, Amphidelus)*
        - Family Ironidae *(Cryptonchus, Ironus)*
  - Suborder Oncholaimina
    - Superfamily Oncholaimoidea
        - Family Oncholaimidae *(Mononchulus, Oncholaimus)*
  - Order Dorylaimida
    - Suborder Dorylaimina
        - Superfamily Dorylaimoidea
        - Family Dorylaimidae *(Actinolaimus, Dorylaimus, Eudorylaimus, Mesodorylaimus, Paractinolaimus)*
        - Family Nygolaimidae *(Nygolaimus)*
      - Superfamily Leptonchoidea

(Continues)

APPENDIX 9.1 (Continued)

- Family Campydoridae *(Aulolaimoides)*
- Order Mononchida
  - Suborder Mononchina
    - Superfamily Mononchoidea
      - Family Mononchidae *(Anatonchus, Mononchus, Prionchulus)*
- Order Mermithida
  - Suborder Mermithina
    - Superfamily Mermithoidea
      - Family Mermithidae *(Aranimermis, Capitomermis, Cretacimermis, Culicimermis, Drilomermis, Empidomermis, Gastromermis, Heleidomermis, Heydenius, Hydromermis, Isomermis, Lanceimermis, Limnomermis, Mesomermis, Octomyomermis, Perutilimermis, Pheromermis, Pseudomermis, Romanomermis, Strelkovimermis)*
- Class Secernentea
  - Order Tylenchida
    - Suborder Tylenchina
      - Superfamily Hoplolaimoidea
      - Family Hoplolaimidae *(Hirschmaniella)*
    - Superfamily Tylenchoidea
      - Family Tylenchidae *(Tylenchus)*
    - Superfamily Atylenchoidea
      - Family Atylenchidae *(Atylenchus)*
- Suborder Criconematina
  - Superfamily Hemicyclcophoroidea
    - Family Hemicycliophoridae *(Hemicycliophora)*
- Order Aphelenchida
  - Suborder Aphelenchina
    - Superfamily Aphelenchoidoidea
      - Family Aphelenchoididae *(Aphelenchoides, Seinura)*
  - Superfamily Aphelenchidae
    - Family Aphelenchidae *(Aphelenchus)*
- Order Rhabditida
  - Suborder Diplogasteroidea
- Superfamily Diplogasteroidea
  - Family Diplogasteridae *(Rhabditolaimus)*
- Superfamily Rhabditoidea
  - Family Cephalobidae *(Cephalobus, Eucephalobus)*
    - Family Rhabditidae *(Caenorhabditis, Mesorhabditis, Pellioditis)*
    - Family Daubayliidae *(Daubaylia)*[a]

[a] The above include all genera cited in this chapter.

10 MOLLUSCA: GASTROPODA

Kenneth M. Brown

Department of Biological Sciences
Louisiana State University
Baton Rouge, Louisiana 70803

I. INTRODUCTION

Gastropods are the most diverse class of the phylum Mollusca, with anywhere from 40,000 to 100,000 species, depending on the authority (Bieler, 1992; Ponder and Lindberg, 1997). In North America, there are 14 families, 88 genera and 659 species of freshwater snails (Bogan, 1999). Snails are common organisms along the margins of lakes and streams. They feed on detritus, graze on the periphyton covering of macrophytes or cobble, or even float upside down at the water surface, supported by the surface tension, and feed on algae trapped in the same fashion (Fig. 1). In fact, gastropods often control the amount and composition of periphyton in both lotic and lentic environments. They are also the basis of food chains dominated by sport fish. Predators, by controlling snail populations, may indirectly facilitate algal producers. Freshwater snails also have extensive intraspecific variation in life histories, productivity, morphology and feeding habits that adapt them to live in uncertain freshwater habitats. There is increasing concern, however, that many riverine gastropod populations in the southeastern United States are currently endangered because of habitat alterations such as impoundments.

Snails are soft bodied, unsegmented animals, with a body organized into a muscular foot, a head, a visceral mass containing most of the organ systems, and a fleshy mantle which secretes the calcareous shell. Gastropods have a univalve shell and possess a filelike radula used in feeding on the periphyton coverings of rocks or plants. Traditionally, gastropods were divided into three subclasses: Prosobranchia (including 53% of modern species), Opistobranchia (4%), and Pulmonata (43%). Prosobranch snails possess a gill (ctenidium) and a horny (flexible) or calcareous operculum, or "trap door," which is pulled in after the foot to protect the animal. Pulmonates have secondarily re-invaded freshwaters from the terrestrial habitats used by their ancestors, use a modified portion of the mantle cavity as a lung, and lack an operculum.

Later authors divided the prosobranchs into the ancestral archeogastropods, the mesogastropods, and most-derived neogastropods, with a trend toward reduction in number of gills, auricles, and kidneys. Cox (1960) combined the meso- and neogastropods into the Caenogastropoda. More recent authors have also considered the valvatids to be more closely allied to the opistobranchs and pulmonates than to the prosobranch groups, and they have suggested combining the three

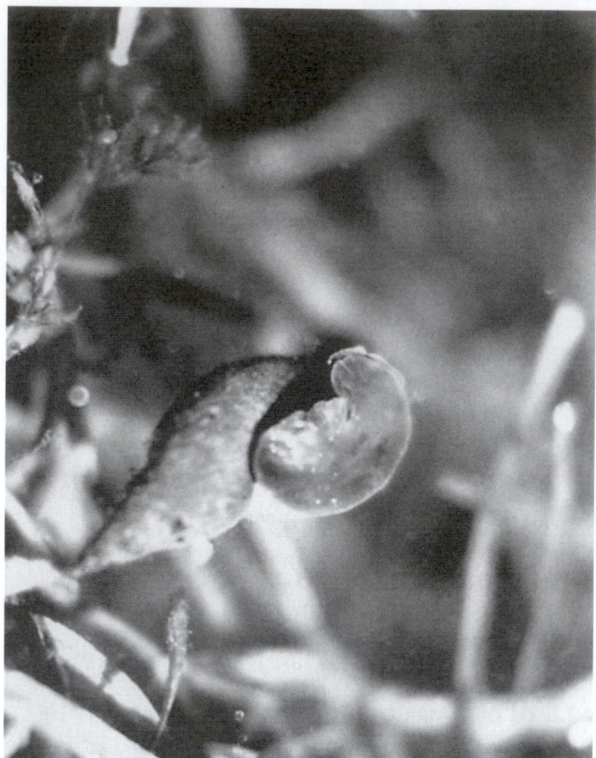

FIGURE 1 An adult *Lymnaea stagnalis* (length about 5 cm), supported by the surface tension, is foraging upside down at the water surface.

former groups into the Heterogastropoda (Kosuge, 1966). Cephalopods are thought to be the closest sister group to the gastropods, and patellid limpets (a marine, intertidal group) are thought to be the most primitive gastropods (Ponder and Lindberg, 1997). More details on the time line of gastropod evolution are given by Bandel (1997).

The first gastropods are thought to have had coiled shells, and several trends have occurred throughout gastropod evolution (Ponder and Lindberg, 1997). Torsion (a 180° rotation of the gut so that the anus and mantle cavity lie anterior) is a functional homology of all gastropods, although it is secondarily reduced in opistobranchs and pulmonates. Anatomical complexity has also become secondarily simplified, including the radula, the circulatory system, and digestive anatomy. External fertilization and pelagic larvae are also considered ancestral traits, with internal fertilization, direct development and encapsulated eggs being derived traits. In contrast, respiratory systems, neuro-secretory structures, and life histories have increased in complexity through evolutionary time. A detailed introduction to gastropod biology can be found in Fretter and Graham (1962), and in volumes on the general biology of molluscs edited by Wilbur (1983) or the biology of pulmonates edited by Fretter and Peake (1975, 1978).

II. ANATOMY AND PHYSIOLOGY

A. External and Internal Morphology

1. Shell

The structure of the shell is important in systematics. Shells can have a simple conical shape (Fig. 2A), as in the limpets, family Ancylidae, with new shell material secreted at the margin. Second, the shell can be planospiral (with the whorls all in one plane), as in the pulmonate family Planorbidae (Fig. 2B). Whorls may be elevated into a "spire," as in the pulmonate families Physidae and Lymnaeidae, and in various prosobranch families (Fig. 2 C, D). Russell-Hunter (1983) discusses how new shell material is secreted by the mantle for each of these three basic shapes. If shells are placed with the spire facing away from the observer and the aperture (opening from which the foot extends) upward, shells with the aperture on the left are sinistral (e.g., the physids), whereas those on the right are dextral (e.g., the lymnaeids and prosobranchs). Spiral shells have a central supporting member, the columella, similar to the center support of a spiral staircase. The columella adds to the strength of the shell, and provides an attachment point for the soft parts via the columellar muscle.

A spiral shell (Fig. 2D) is convenient for illustrating terminology used in systematics. The pointed end of the shell above the aperture is the apex. Shell length in spiral shells is measured from the apex to the lower tip of the aperture, while the greatest diameter is used in planospiral shells. The spire is separated into a number

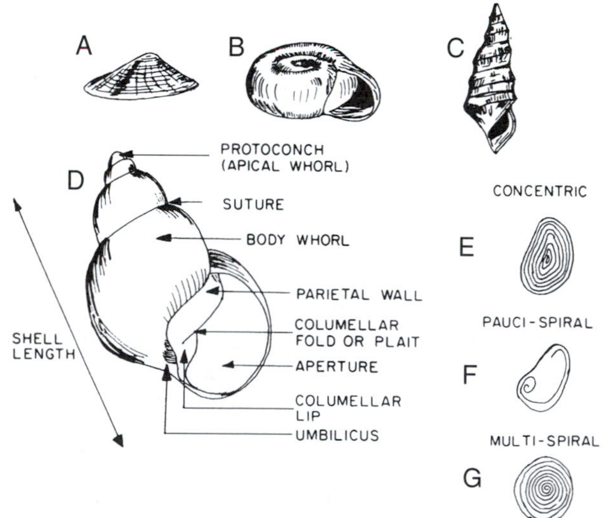

FIGURE 2 Basic anatomy of the shell, including shell architecture (A, conical; B, planospiral; C and D, spiral, with major shell features indicated in D) and three types of opercula (E, concentric, F, paucispiral, and G multispiral).

of whorls by sutures (Fig. 2). The most apical whorl is the nuclear whorl or protoconch, the initial shell of the newly hatched snail or "spat." The final and biggest whorl (representing the most recent shell growth) is the body whorl, ending in the aperture. Whorls may be rounded and sutures deep and well defined (as in a typical lymnaeid shell, Fig. 2D), or they may be flattened and sutures shallow, as in the prosobranch family Pleuroceridae (Fig. 2C). Spiral shells thus have a variety of shapes. If the whorls are flat, the shell is termed cone-shaped, while moderately inflated whorls produce a "subglobose" shell, and shells that are almost circular are globose; there is a continuum between these three shapes. Shell thickness varies from thin and fragile (as in many pulmonates) to thick and resistant to crushing (as in the prosobranchs).

Part of the aperture is often reflected over the body whorl at the columella (Fig. 2D) to form an inner lip. If there is a channel between the inner lip and the body whorl, the shell is umbilicate or perforate (the opening of the channel is called the umbilicus and leads up and inward into the columella). Imperforate shells lack an umbilicus. While freshwater shells are not as ornate as their marine relatives (Vermeij and Covich, 1978), shells may have spines, ridges running along the margins of the whorls (called carina), colored bands, or small malleations (hammerings) on the surface. Ridges at right angles to the whorls are called costae, while smaller ridges running spirally along the whorls are lirae. The operculum is useful in classifying prosobranchs. If the growth lines lie completely within each other, the operculum is concentric; whereas, lines arranged in a spiral are termed multispiral or paucispriral (see examples in Fig. 2 E–G). Further discussion on shell sculpture is given in Fretter and Graham (1962) and Burch (1989).

The shell is composed of an outer periostracum of organic (mostly protein) composition which may limit shell abrasion or dissolution of shell calcium carbonate by acid waters. Beneath the periostracum is a thick layer of crystalline calcium carbonate with some protein material as a matrix for the calcium crystals (Russell-Hunter, 1978). Calcium carbonate is either absorbed directly from water or is sequestered from food (McMahon, 1983). Snails in calcium-poor waters (<5 mg/L, Lodge *et al.*, 1987) expend energy to absorb calcium against a gradient, and many species are therefore limited to calcium-rich habitats.

2. Soft Anatomy

The soft parts are separated into: (1) head, (2) foot, (3) visceral mass, and (4) the mantle (Fig. 3). Aquatic pulmonates and prosobranchs possess eyes at the base of their tentacles, unlike terrestrial pulmonates

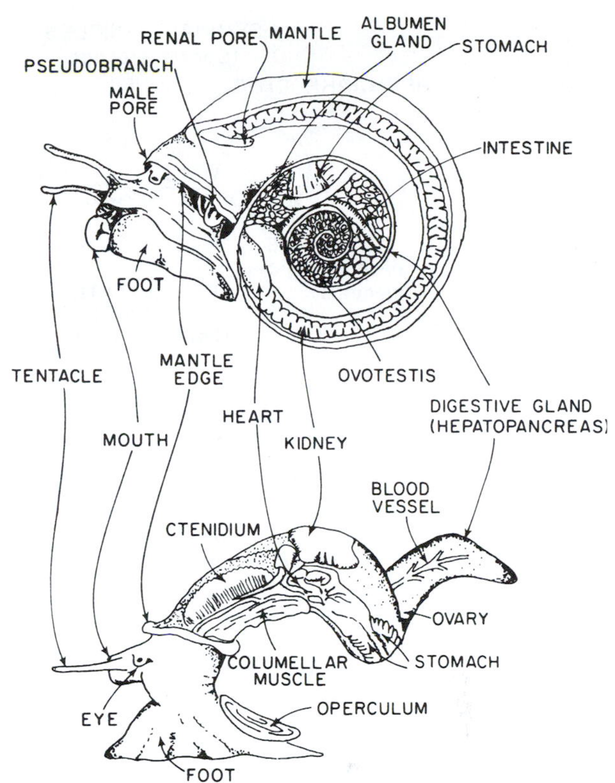

FIGURE 3 Basic internal anatomy (with shell removed) of a planorbid pulmonate (above, after Burch, 1989), and a pleurocerid prosobranch (below, after Pechenik, 1985).

whose eyes are at the tips of the tentacles. The muscular foot is provided with both cilia and secretory epithelium to secrete mucus for locomotion, as well as pedal muscles which produce waves of contraction to push the animal forward. The visceral hump includes most of the organs of digestion and reproduction. The mantle covers the visceral mass, and underlays the shell, which it secretes. The anterior mantle, over the head, possesses a mantle cavity, where the gill or ctenidium is located in prosobranchs. Gastropods have ganglia innervating each of these areas. Further information on internal anatomy can be found in Fretter and Graham (1962), Barnes (1987), Pechenik (1985), or Hyman (1967).

B. Organ System Function

1. Reproductive System

Prosobranchs are usually dioecious, and males use the enlarged right tentacle as a copulatory organ (in the viviparids), or possess a specialized penis or verge (in the hydrobiids, pomatiopsids, and valvatids) or have no copulatory organ (thiarids and pleurocerids). The penis is usually located behind the male's right tentacle. Many pleurocerids lay clutches of a few eggs, whereas

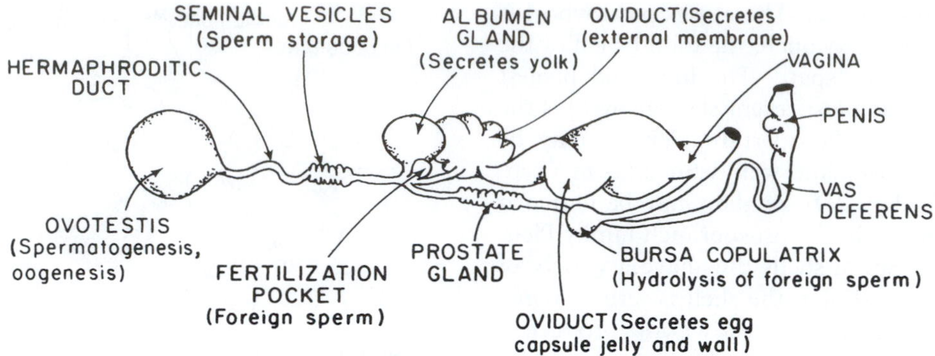

FIGURE 4 Anatomy of the reproductive system of the monoecious pulmonate *Physella* (after Duncan, 1975). Sperm are produced in the ovotestis, stored in the seminal vesicle, exit via vas deferens, and are inserted into the vagina of a second individual via the introvertable penis; or sperm may fertilize the same individual's eggs in the hermaphroditic duct. Most foreign sperm are hydrolyzed in *Physa* in the bursa copulatrix, but some do fertilize eggs in the fertilization pocket. The albumen gland secretes yolk around the egg, and the oviduct wall secretes both the egg membrane and the jelly and external wall of the egg case.

viviparids produce eggs which hatch and develop in a fold of the anterior mantle (the pallial oviduct), and are born free-living. Some viviparids in the genus *Campeloma,* especially more northern populations, are parthenogenetic (Van Cleave and Altringer, 1937; Vail, 1978; Johnson, 1992).

Pulmonates, in contrast, are all monoecious. The basic components of the pulmonate reproductive system are shown in Figure 4. Sperm and eggs are produced in the ovotestis and exit via a common hermaphroditic duct in all pulmonates except ancylids, which have two openings and are thus obligate cross-fertilizers. The albumen gland adds protein and nutrients to the egg. Eggs are either fertilized in the hermaphroditic duct by the same individual's sperm, or by sperm from another individual near the junction of the hermaphroditic duct and the oviduct. The external egg membranes are then secreted in the oviduct. Eggs are laid in gelatinous egg cases and attached to plants, submersed wood, other snails or rocks. Males have an introvertible penis, and fertilization is internal. Duncan (1975) gives a more detailed discussion of egg and sperm formation, copulation and fertilization, and egg capsule deposition.

Although pulmonates are simultaneous hermaphrodites, most species out-cross when possible. Pulmonates that "self" usually mature at later ages and have lower fecundity. For example, *Physella heterostropha* suffers a reduction of 65% in fecundity when not allowed to out-cross (Wethington and Dillon, 1993). Snails act as males or females based on the amount of stored sperm that they possess (Wethington and Dillon, 1996). Similarly, the African snail *Bulinus globosus* suffers a 50% reduction in fecundity, and an 18% reduction in hatching rate, under obligate self-fertilization, indicating a significant genetic load of recessive lethal genes (Jarne *et al.,* 1991). In other physids (Eileen Jokinen, personal communication), the male system is larger in juveniles than in adults, and protandry (a form of sequential hermaphroditism where individuals switch sex from male to female as they age and increase in size) may occur.

With the costs to fitness indicated in the preceding paragraph, one might wonder why hermaphroditism occurs in pulmonates. Most pulmonates are slow-moving and go through seasonal "bottlenecks" (precipitous declines in density). The chances of finding a mate in such situations are small, providing a selective advantage for monoecy. Pulmonates are dispersed passively as spat trapped in mud on birds feet (Boag, 1986 and references therein), and solitary, monoecious immigrants have an obvious advantage. The cost of inbreeding depression is evidently less than not being able to reproduce at all. Parthenogenesis (being able to reproduce without males) may have a similar adaptive value for prosobranch snails that are isolated in small headwater streams (Vail, 1978).

2. Digestive System

Food is brought into the mouth by rasping movements of the radula, a filelike structure (Fig. 5) resting on a cartilage (the odontophore) to which muscles which extend and retract the radula are attached. When the radula is extended, it contacts the substratum, and algal particles are scraped off when retractors pull the radula back into the mouth. The radula may also pulverize food particles by grinding them against the roof of the mouth. A long esophagus leads to the

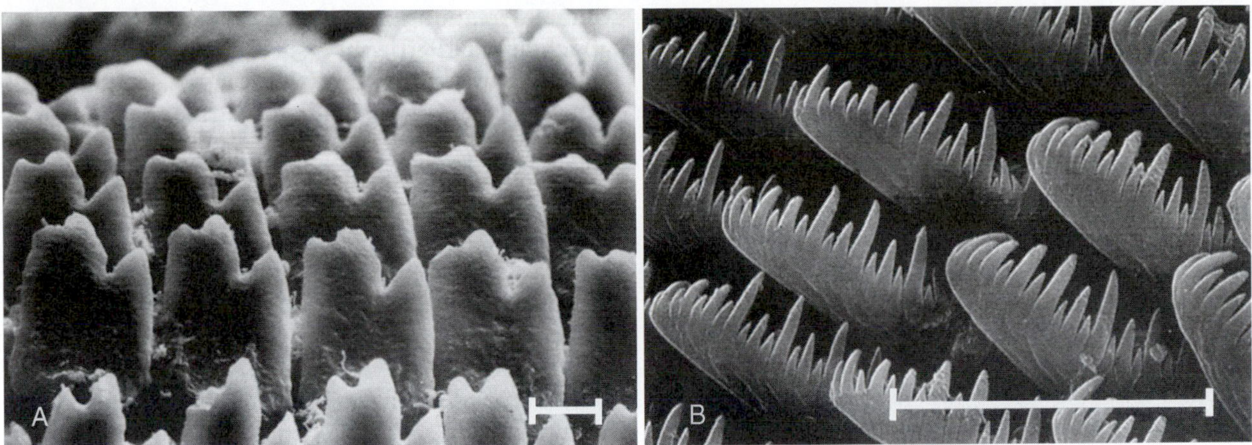

FIGURE 5 Scanning electron microscope photographs of radular teeth of (A) the lymnaeid, *Pseudosuccinea columella* and (B) the physid, *Physella vernalis* (from Kesler *et al.*, 1986). Note the lymnaeid has fairly large teeth with few cusps. These teeth, along with large, cropping jaws in the buccal mass, a grinding gizzard equipped with sand, and high cellulase levels allow it to specialize on filamentous green algae (Kesler *et al.*, 1986). *Physella* has many, small teeth with long cusps, arranged in chevron-shaped rows, useful for piercing detritus and attached bacteria.

stomach, located in the visceral mass. Some gastropods possess a specialized crop where sand grains further abrade food particles. Digestive enzymes are produced by the digestive gland, the hepatopancreas. Considerable digestion also occurs intracellularly in the hepatopancreas. Snails are one of the few animal groups to possess and possibly synthesize cellulases (Kesler, 1983; Kesler *et al.*, 1986), that degrade the algal cell walls.

3. Respiratory and Circulatory Systems

The respiratory system differs radically between prosobranchs and pulmonates. Freshwater prosobranchs have a single ctenidium or gill (Aldridge, 1983). The ctenidium, usually in the mantle cavity, has leaflike triangular plates richly supplied with blood vessels. Oxygen poor blood passes across the plates in the opposite direction to oxygen-rich water currents (generated by ctenidal cilia). This countercurrent mechanism assures diffusion of oxygen from water into the blood. Pulmonates lost their gill during their intermediate terrestrial phase, and have a vascularized pocket in the mantle used as a lung (hence their name). The opening of the lung is the pneumostome. Pulmonates either rely on surface breathing or have a limited capacity for oxygen transfer across their epithelial tissues (McMahon, 1983). Some physids and lymnaeids fill the mantle pocket with water and use it as a derived gill (Russell-Hunter, 1978). Ancylids and planorbids have re-adapted further to aquatic conditions by using a conical extension of epithelium as a gill, and planorbids also have a respiratory pigment, hemoglobin, which increases the efficiency of oxygen transport (McMahon, 1983).

4. The Excretory System

Gastropods have a permeable epidermis and are subject to osmotic inflow of water from their hypo-osmotic surroundings. Thus, they must pump out excess water in their urine. The gastropod coelom is little more than a small cavity (pericardium) surrounding the heart. The coelomic fluid is largely a filtrate of the blood, containing waste molecules, such as ammonia, that are filtered across the wall of the heart. Additional wastes are actively secreted into the coelom by the walls of the pericardium. The coelomic fluid then enters a metanephridial tubule (the coelomoduct) where selective resorption of salts and further secretion of wastes occurs. The urine is then discharged into the mantle cavity. Freshwater gastropods excrete nitrogen both as ammonia and as urea. Ammonia is adaptive in aquatic environments because, although it is toxic, it is extremely soluble and readily diffuses away. Pulmonates often produce urea which is better in terrestrial situations, or during hibernation or estivation, because it is relatively nontoxic and can be stored in the blood until able to be excreted (McMahon, 1983). Further discussion of excretion can be found in Martin (1983) for gastropods in general and in Machin (1975) for pulmonates.

C. Environmental Physiology

Pulmonates usually face wider extremes of temperature variation than do most prosobranchs (McMahon, 1983). Most temperate pulmonates, for example, can withstand temperatures near 0°C for several months, whereas tropical pulmonates can withstand temperatures near 40°C for extended periods. This is undoubtedly adaptive because of the seasonal and diurnal variation in temperature in many pond habitats. Pulmonates better regulate changes in metabolic rate with changing temperatures than do prosobranchs. For example, Q_{10} values for 18 pulmonate species were less (averaging 2.2) than in 13 prosobranch species (averaging 2.8) (McMahon, 1983). As the average temperature increases, snails grow faster and reproduce at an earlier age, with more generations per year. For example, the subtropical pulmonate *Physella cubensis* grows slowly and will not mature at 10°C (Thomas and McClintock, 1990), but hatching times and ages at maturity decrease, and growth rates increase as laboratory temperatures are increased to 30°C.

Increasing water temperature is also the cue for onset of reproduction in many temperate pulmonates. The ability of pulmonates to reproduce in cold water allows adults to breed early in the spring and juveniles to grow rapidly to adult size before the end of the summer. Pulmonates can secrete a mucus covering over the aperture, called an epiphragm, to retard moisture loss during dry periods (Boss, 1974; Jokinen, 1978).

In terms of adaptation to hypoxia, pulmonates: (1) tolerate greater variation in dissolved oxygen than do prosobranchs (possibly because prosobranchs rarely experience hypoxia in the littoral zones of lakes or the fast flowing rivers they are common in); and (2) also regulate oxygen consumption at varying levels of dissolved oxygen better than prosobranchs, which are "oxy-conformers" (Brown *et al.,* 1998). Pulmonates apparently withstand lower oxygen tensions by either surface breathing or reliance on anaerobic metabolism (McMahon, 1983).

Approximately 45% of freshwater gastropods are restricted to waters with calcium concentrations greater than 25 mg/L, and 95% to levels greater than 3 mg/L. Although it may not take much energy to absorb calcium from water, assuming adequate water hardness, it may be energetically costly to secrete calcium into the shell against an electrochemical gradient. For example, in the northeastern United States (Jokinen, 1987), *Aplexa elongata, Helisoma (Planorbella) trivolvis,* and all lymnaeids except *Fossaria* are limited to calcium-rich waters. McMahon (1983) and Lodge *et al.* (1987) discuss the degree of calcium regulation, and the relationship of external calcium level to shell thickness and growth. Shell accretion may lag behind rapid tissue growth in eutrophic habitats, producing thinner shelled individuals (McMahon, 1983).

III. ECOLOGY AND EVOLUTION

A. Diversity and Distribution

Although widespread in Asia, pleurocerids reach their greatest abundance in the rivers and streams of the southeastern United States (Table I). Most are fairly large snails that are quite common in rocky riffles or shoals in both headwaters and large rivers. However, many species have been lost to stream alteration by impoundments, and present-day diversity has decreased dramatically (see Section III.E). Females possess an egg laying sinus on the right side of the foot (Dazo, 1965).

TABLE I The Diversity of Freshwater Snail Families in North America[a]

Subclass	*Family*	*Number of genera*	*Number of species*	*Area of greatest diversity*[b]
Prosobranchia	Ampullaridae	2	4	SE
	Bithyniidae[c]	1	1	NE
	Hydrobiidae	28	152	U
	Micromelaniidae	1	1	SE
	Neritinidae	1	1	SE
	Pleuroceridae	7	156	SE
	Pomatiopsidae	1	6	SE, W
	Thiaridae[c]	2	3	S
	Valvatidae	1	11	U
	Viviparidae	5	29	E
Pulmonata	Acroloxidae	1	1	W, NE
	Ancylidae	4	13	U
	Lymnaeidae	9	58	N
	Physidae	4	43	U
	Planorbidae	11	47	U

[a] Data compiled from Burch (1989) and Neves *et al.* (1998).
[b] N, North; E, East; W, West; S, South and U, ubiquitous.
[c] Introduced.

Shell anatomy is used to classify genera, although many species show considerable variation in shell characters. The shells of pleurocerids are solid and the aperture may bear a canal anteriorally. The operculum is corneous and paucispiral.

Viviparids are worldwide in distribution, and these large snails are fairly diverse (Table I) throughout the eastern states and Canadian provinces. *Campeloma, Lioplax,* and *Tulotoma* are endemic to North America. Although the first two genera are widespread, *Tulotoma* was once considered extinct and has only recently been rediscovered in the Coosa River in Alabama, where this filter feeding, brooding snail is common under rocks in shallow water shoals in large tributaries (Hershler *et al.,* 1990). *Viviparus* is quite common in rivers and lakes throughout eastern North America. Another large viviparid *Cipangopaludina chinensis* was introduced to the United States in the 1890's and has now spread throughout the United States, especially in hard-water lakes with sandy or muddy substrates (Jokinen, 1982).

Ampullarids are a tropical, mostly amphibious family with a mantle cavity provided both with a gill and a lung. The two genera present in Florida, *Pomacea* and *Marisa*, are quite large snails (50–60 mm). *Marisa* has also been introduced to rivers in central Texas.

The Neritinidae are a marine, tropical group. A few species have invaded estuarine and freshwater habitats, for example *Neritina reclivata* in Florida, Georgia, Alabama, and Mississippi. The paucispiral, calcareous operculum has a pair of projections which lock the operculum against the teeth on the aperture, providing a stronger defense.

Of the 20 species of valvatids in the northern hemisphere, 11 are found in North America (Table I), mostly in lakes in northern states. Valvatids are egg-laying hermaphrodites, with a single, featherlike gill carried on the left side, and a pallial tentacle carried on the right side of the shell as the animal crawls. Valvatids have small (approximately 5 mm diameter) dextral shells, with a corneous, slightly concave and thin, multispiral operculum. The spire is only slightly elevated, and the shells are sometimes carinated.

Hydrobiids are extremely diverse and exist in freshwater, brackish, and marine habitats. They are small but have a thick shell which protects them from fish predation, and they dominate the gastropod assemblages of many northern lakes (Brown, 1997). There are 103 genera worldwide (Burch, 1989), and they are diverse in North America (Table I). In fact, hydrobiids may be the most diverse group of freshwater snails, as 58 new species in the genus *Pyrgulopsis* were recently described from small springs in the southwestern United States (Hershler, 1998). Although these populations often reach high densities, many species are endemic to only a few springs, increasing chances of extinction. The genus *Fluminicola*, which is common in rivers in the northwestern United States, is also quite diverse (Hershler and Frest, 1996). The hydrobiids have small, dextral shells with a paucispiral operculum. Because of the similarity of their shells, the structure of the verge (penis) is used in classification.

The six North American pomatiopsids are similar in general anatomy to the hydrobiids. Pomatiopsids are, however, amphibious and are often found inhabiting stream banks several meters above the water's surface, while hydrobiids are truly aquatic (Burch, 1989). Pomatiopsid systematics are discussed by Davis (1979).

Thiarid females are parthenogenetic, brooding eggs in a pouch in the neck region, which opens on the right side. The similar-shelled pleurocerids, discussed above, are on the contrary dioecious and oviparous.

Ancylids have a worldwide distribution, and all possess a simple cone-shaped shell. In North America, they have reached moderate diversity (Table I). Ancylids have sinistral shells, with the apex inclined slightly to the right, and the gill (pseudobranch) and many of the internal organs opening on the left side of the body. Their streamlined shape allows them to colonize fast-flowing streams, where they are common on rocks or macrophytes, although some species are also common in lentic environments.

The family Acroloxidae occurs mainly in Eurasian lakes and ponds. Only one species of *Acroloxus* occurs in the United States, in Colorado and southeastern Canada. Since the apex in *Acroloxus* is tipped to the left, the aperture is considered to be dextral, unlike the ancylid limpets.

Lymnaeids are worldwide in distribution, and are the most diverse pulmonate group in the northern United States and Canada (Table I). Lymnaeids have broad triangular tentacles, and lay long, sausage-shaped egg masses. Lymnaeids have fairly uncomplicated, large teeth on their radula (Fig. 5), which are useful for cropping long strands of filamentous algae(Kesler *et al.,* 1986). One group of lymnaeids, found along the Pacific coast of North America, has limpet-shaped shells, but they are larger than ancylids.

Physids also have a worldwide distribution and are ubiquitous in North America (Table I); few aquatic environments lack physids, because of their ease of introduction. Their shells are small, sinistral, with raised spires. Their tentacles and foot are slender and they have fingerlike mantle extensions (except for the genus *Aplexa*) and lay soft, crescent-shaped egg masses. Their radular teeth are smaller and more complicated in shape than in the lymnaeids (Fig. 5), and they are better at harvesting more tightly attached periphyton species like diatoms, or in feeding on detritus (Kesler *et al.*, 1986). Their rate of crawling is much more rapid than most gastropods, and this along with their early age at maturity and high fecundity explains why they are so wide-spread (Brown *et al.*, 1998).

Planorbids are widespread and fairly diverse snails (Table I), and range in size from minute (1 mm) to quite large (30 mm) in North America. They possess hemoglobin as a respiratory pigment, sometimes giving the tissue a red hue. Their egg cases are flat and circular, with harder membranes than those of the lymnaeids or physids. Planorbids are considered closely related to the ancylids (McMahon, 1983).

B. Reproduction and Life History

Freshwater snails are extremely interesting because of the variety of observed life-history patterns. For example, freshwater pulmonates are oviparous (egg-laying) hermaphrodites, and are usually annual and semelparous (i.e., they reproduce once and die). On the other hand, almost all prosobranchs are dioecious and are often iteroparous, with perennial life cycles. Prosobranchs can be oviparous or ovoviviparous (Russell-Hunter, 1978; Calow, 1978, 1983; Brown, 1983). Annual species have essentially a 1 year life cycle, whereas perennials often live and reproduce for 5 years or more.

Russell-Hunter (1978) and Calow (1978) have classified life histories for freshwater gastropods (Fig. 6). At one end are annual adults that reproduce in the spring and die (e.g., there is complete replacement of generations). Most pulmonates belong to this group, including species from the genera *Lymnaea*, *Physella*, and *Aplexa* (the original studies are listed in Calow, 1978). In the second category (Fig. 6B), reproduction occurs in both spring and late summer with both cohorts surviving the winter, or (Fig. 6C) where there again is complete replacement of generations. In Figure 6D–F, there are three reproductive intervals, with varying degrees of replacement of generations. These would predominantly be populations in subtropical or tropical environments. Finally, there are populations that can truly be considered perennial and iteroparous (Fig. 6G); most are prosobranchs. For example, the pleurocerid genus *Elimia* has life cycles lasting from 6 to 11 years in Alabama steams, with several cohorts overlapping at any one time (Richardson *et al.*, 1988; Huryn *et al.*, 1994). Adults reproduce in the spring and summer and are more common in areas of low current velocity, while juveniles are common in the fall in high flow areas (Huryn *et al.*, 1994; Johnson and Brown, 1997).

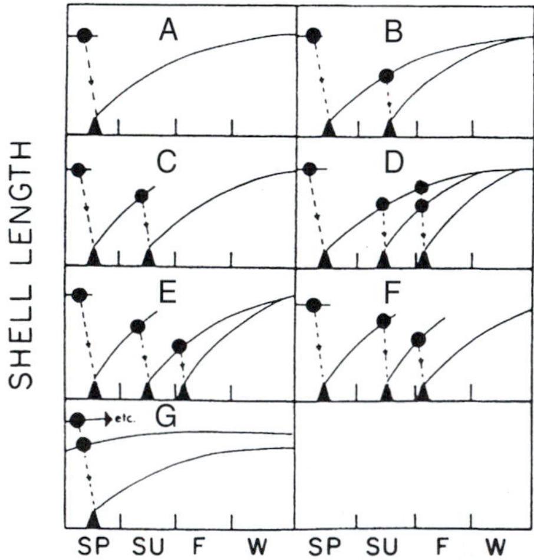

FIGURE 6 The variety of life cycles in freshwater snails. Curves represent the growth of individual cohorts (in millimeters of shell length), circles represent size at maturity while triangles represent appearance of egg cases in samples. Panel A represents annual semelparity, and B–F represent increasing numbers of cohorts per year, while panel G represents perennial iteroparity (from Calow, 1978). SP, spring; S, summer; F, fall and W, winter.

Jokinen (1985) also found Russell-Hunter and Calow's life-cycle categories useful in describing lifecycle patterns in twelve species of gastropods inhabiting a small Connecticut lake. Four species, *Amnicola limosa*, *Helisoma anceps*, *Helisoma* (*Planorbella*) *companulatum* and *Laevapex fuscus*, had simple, annual patterns. Four species had at least two breeding seasons a year, with varying replacement of cohorts, including *Pseudosuccinea collumella*, *Planorbula armigera*, *Promenetus exacuous*, and *Gyraulus deflectus*. Three species, *Fossaria modicella*, *Gyraulus circumstriatus*, and *Physella ancillaria*, had continuous breeding throughout the field season. The prosobranch *Campeloma decisum* was the only species with a perennial reproductive pattern. Jokinen (1985) suggested that divergence in life history patterns between taxonomically similar species lessened competition, by allowing some species to specialize on detritus (common in the spring), and others periphyton (common in the summer).

Marine snails have enormous fecundity, but individual eggs are extremely small. Most marine snails are prosobranchs, and freshwater prosobranchs (their descendants) probably have relatively small eggs as a result. In fact, viviparids and thiarids may have evolved ovoviviparity (and in some cases the production of larger embryos) to cope with the much more unpredictable conditions in freshwaters (Calow, 1978, 1983). Similarly, the loss of the planktonic veliger, and shortening of the developmental period in oviparous freshwater snails has been attributed to the more variable physicochemical conditions (Calow, 1978).

Reproductive effort (percent of energy devoted to reproduction) is lower in iteroparous freshwater snails than in semelparous ones (Browne and Russell-Hunter, 1978). For example, pulmonate families are semelparous with relatively high reproductive output (Brown, 1983). Pulmonates also reproduce at smaller sizes and earlier ages, produce more eggs, have larger clutch sizes, greater shell growth rates, shorter life cycles and smaller final shell sizes than viviparid prosobranchs.

The reduced fecundity but increased parental care found in viviparids (vs semelparous pulmonates) increases offspring survival. Life tables for *Viviparus georgianus* (Jokinen *et al.,* 1982) and for the pulmonate *Lymnaea elodes* (Brown *et al.,* 1988) indicate that survival to maturity is indeed much less than 1% in all of the pulmonate populations, but over 40% in the ovoviviparous prosobranch. Prosobranchs are also sexually dimorphic in life-history patterns. Female viviparids live longer (males usually survive for only one reproductive season) and reach larger sizes (Van Cleave and Lederer, 1932; Browne, 1978; Jokinen, 1982; Jokinen *et al.,* 1982. Pace and Szuch, 1985; Brown *et al.,* 1989).

Numerous studies have implicated periphyton productivity (Eisenberg, 1966, 1970; Burky, 1971; Hunter, 1975; Browne, 1978; Aldridge, 1982; Brown, 1985) in determining voltinism patterns, growth rates, fecundity, and gastropod secondary production (see review in Russell-Hunter, 1983). Other important factors include water hardness and temperature. For example, populations of *Elimia* in limestone-substrate streams in Alabama have greater annual production than populations in relatively impermeable slate or sandstone-substrate streams (Huyrn *et al.,* 1995), both because of the higher alkalinity and greater buffering of low temperatures in the winter by ground water. Similarly, populations of *Lymnaea stagnalis* in Canada often take several seasons to complete their life cycle (Boag and Pearlstone, 1979) while populations are annual in the warmer waters of Iowa (Brown, 1979).

Lymnaea peregra in wave-swept habitats in English lakes have *r*-selected life-history traits (e.g., early reproduction and high reproductive output) in comparison to populations in less-harsh habitats (Calow, 1981). However, other studies of life history variation in molluscs do not agree as well with the predictions of *r*- and *K*-theory (see discussion in Burky, 1983).Transplant studies, where individuals from separate populations are reared in a common environment, have usually indicated that environmental effects on life histories are much more important than genetic differences between populations. For example, populations of *Lymnaea elodes* reared in more productive ponds lay nine times as many eggs, have an annual versus a biennial reproductive cycle, and reach larger individual sizes (Brown, 1985).

Although genetic polymorphism has been studied using gel electrophoresis in terrestrial pulmonates and freshwater prosobranchs more than in aquatic pulmonates, freshwater snails still appear to have levels of genetic polymorphism intermediate to terrestrial and marine species (Brown and Richardson, 1988). Terrestrial snails inhabit patchily-distributed microclimates which increase chances for low population densities and self fertilization, resulting in little genetic polymorphism within populations. Freshwater snails decline to low densities because of seasonal bottlenecks and thus may also self. Marine environments are less seasonal, and many marine snails have planktonic larvae, facilitating gene flow and increasing polymorphism.

C. Ecological Interactions

1. Habitat and Food Selection, Effects on Producers

Slow moving, silty habitats are occupied by pulmonates or detritivorous prosobranchs such as viviparids, whereas fast current areas are dominated by limpets or prosobranch grazers like pleurocerids (Harman, 1972). There is also evidence for habitat selection on the species level. *Campeloma decisum* is positively rheotactic (moves upstream) and aggregates at any barrier (e.g., logs, riffle zones, etc., Bovbjerg, 1952). *Physella integra* and *Lymnaea emarginata* prefer cobble substrates with attached periphyton, while *Helisoma anceps* and *Campeloma rufrum* prefer sand (Clampitt, 1973; Brown and Lodge, 1993). Most gastropods in northern Wisconsin lakes, however, prefer periphyton covered cobble over sand or macrophytes (Brown and Lodge, 1993). Cobble, since it offers greater surface area and is present year-round, develops a richer periphyton coating. Substrate selection may even occur on a finer level: gastropods from an English pond prefer periphyton from the macrophytes the snails were found on in the field (Lodge, 1986). Snails often move among habitats as well. Migrations occur to deeper water in the fall in lakes and back to the littoral zone in the spring (Cheatum, 1934; Clampitt, 1974; Boag, 1981). Pond pulmonates on the other hand burrow into the substrate with declining temperatures (Boerger, 1975).

Freshwater gastropods are herbivores or detritivores, but occasionally ingest carrion (Bovbjerg, 1968) or small invertebrates associated with periphyton (Cuker, 1983). Periphyton is easier to scrape, and contains higher concentrations of nitrogen and other limiting nutrients than macrophyte tissue (Russell-Hunter, 1978; Aldridge, 1983). For example, carbon to nitrogen (e.g., carbohydrate to protein) ratios are below 10.1:1, while macrophytes have ratios of 24.1:1 (McMahon *et al.*, 1974). Algal and diatom remains therefore dominate in the guts of snails (Calow, 1970;

TABLE II Feeding Preferences of Freshwater Snails

Feeding type	*Family*	*References*
Algivores (scrapers)	Ancylidae	Calow (1973a, b, 1975)
	Lymnaeidae	Bovbjerg (1968, 1975), Brown (1982), Calow (1970), Cuker (1983), Hunter (1980), Kairesalo and Koskimies (1987), Kesler *et al.* (1986), Lodge, (1986)
	Neritinidae	Jacoby (1985)
	Pleuroceridae	Aldridge (1982, 1983), Dazo (1965), Goodrich, (1945)
	Viviparidae	Duch (1976), Jokinen *et al.*, (1982)
Detritivores or bacterial feeders	Physidae	Brown (1982), Kesler *et al.*, (1986), Townsend (1975)
	Planorbidae	Calow (1973b, 1974a, b)
	Viviparidae	Chamberlain (1958), Pace and Szuch, (1985), Reavell (1980)
Filter feeders	Bithyniidae	Tashiro (1982), Tashiro and Colman (1982), Brendelberger and Jurgens, (1993)
	Viviparidae	Brown *et al.* (1989)

1973a, 1975; Calow and Calow, 1975; Reavell, 1980; Kesler *et al.,* 1986; Lodge, 1986; Madsen, 1992). However, gastropods at high densities can exhaust periphyton and then consume macrophytes, suppressing macrophyte species richness (Sheldon, 1987).

Feeding preferences for the freshwater gastropod families are summarized in Table II. Lymnaeids are "microherbivores", scraping algae and diatoms from rocks or macrophytes, but grow more rapidly when animal tissue is in their diet (Bovbjerg, 1968). For example, although *Pseudosuccinea columella* (a lymnaeid) is an omnivore, it still consumes more algae than the sympatric *Physa vernalis* (Kesler *et al.,* 1986). The lymnaeid also possesses higher levels of cellulases, as well as a radula and jaws well adapted for cropping algae (Fig. 5), and a gizzard filled with sand that can macerate food. The physid, a detritivore, lacks these adaptations.

Both the limpet family Ancylidae and the prosobranch family Pleuroceridae are also considered to graze on periphyton (Table II). *Ancylus fluviatilis* selectively grazes diatoms, but the limpet has little effect on periphyton communities, due either to adaptations of algal and diatom species to grazing or to relatively low limpet abundances (Calow, 1973a, b). Aldridge (1983) concluded that pleurocerid grazers feed on periphyton rather than macrophyte tissue again because of higher levels of nitrogen.

The prosobranch *Bithynia tentaculata* both grazes on periphyton and uses its ctenidium to capture phytoplankton that is consolidated into a mucous string which loops from the mantle cavity to the mouth (Brendelberger and Jurgens, 1993). Indeed, filter feeding may be more efficient than scraping as increasing levels of phytoplankton shade out periphyton, explaining why this species has become so abundant in nutrient rich, eutrophic lakes in New York (Tashiro, 1982; Tashiro and Colman, 1982).

Viviparus georgianus is a micro-algivore (Duch, 1976; Jokinen *et al.,* 1982) or a detritivore (Pace and Szuch, 1985). Most viviparids, however, are probably detritivores or utilize bacteria associated with detritus (Table II). As macrophytes decompose, nitrogen levels increase, again increasing their value as food resources. For example, *Viviparus* reaches extremely high densities (from 151 to 608 individuals m^2) in wooded streams in Michigan (Pace and Szuch, 1985) and detritus-rich bayous in Louisiana (up to 1,700 individuals m^2, Brown *et al.,* 1989).

Both physids (Kesler *et al.,* 1986) and planorbids (Calow, 1973b, 1974a) prefer detritus (Table II). For example, although widely-spread species like *Physella gyrina* and *Helisoma trivolvis* did not prefer detritus over periphyton in laboratory experiments, another physid, *Aplexa elongata*, which is much more common in wooded ponds with a rich detritus food base, preferred detritus (Brown, 1982).

Originally, snail algivores were considered indiscriminant grazers, taking all components of the periphyton (Hunter, 1980; Hunter and Russell-Hunter, 1983). However, limpets and planorbids are selective (Calow, 1973a, b), and *Lymnaea peregra* grazes selectively on filamentous green algae (Lodge, 1985). *Planorbis vortex* ingests diatoms in greater quantities than found in the periphyton, but is still predominantly a detritivore.

Gastropods have the ability to locate macrophytes through distant chemoreception (Croll, 1983). For example, *Lymnaea peregra* is positively attracted to *Ceratophylum demersum* because of dissolved organic materials excreted by the macrophyte (Bronmark, 1985b). Similarly, *Potamopyrgus jenkinsi* orients toward both plant and animal extracts (Haynes and Taylor, 1984), while *Biomphalaria glabrata* either orients toward or away from specific macrophytes (Bousefield, 1979).

Almost all experimental manipulations have indicated snail grazers can decrease periphyton standing crops (Table III, see also review in Bronmark, 1989). For example, *Physella* at high densities reduces algal

TABLE III The Effects of Experimental Manipulations of Gastropods on Periphyton Biomass, Production and Assemblage Structure

Group	*Genus*	*Decreased algal biomass?*	*Increased algal production?*	*Favored adnate species?*	*Studies*
Prosobranchs	*Elimia* or *Juga* (N = 11)	73%	17%	100%	Mulholland *et al.* (1983), Steinman *et al.* (1987), Lamberti *et al.* (1987), Marks and Lowe (1989), McCormick and Stevenson (1989), Hill and Harvey (1990), Mulholland *et al.* (1991), Tuchman and Stevenson (1991), Hill *et al.* (1992), Rosemond *et al.* (1993)
	Theodoxus (N = 1)	Y	NO	Y	Jacoby (1985)
	Amnicola (N = 1)	Y	N.M.	Y	Kesler (1981)
Pulmonates	*Physella* (N = 3)	100%	N.M.	100%	Doremus and Harman (1977), Lowe and Hunter (1988), Swamikannu and Hoagland (1989)
	Promenetus (N = 1)	NO	NO	NO	Doremus and Harman (1977)
	Lymnaea, *Physella*, *Helisoma*	Y	Y	Y	Hunter (1980)

Percentages refer to the number of studies noting an affect. For single studies: Y, effect; NO, no effect, N.M., not measured.

biomass by 97%, and richness by 66% (Lowe and Hunter, 1988). In some cases, snails also increase algal production, perhaps by decreasing total biomass and lowering algal competition for light or nutrients, removing senescent cells, or increasing rates of nutrient cycling. Pulmonate gastropods can also alter the quality (e.g., nitrogen-to-carbon ratios and chlorophyll a levels) of periphyton (Hunter, 1980), as can prosobranchs in streams (Steinman *et al.*, 1987). Although low levels of grazing may stimulate production, higher snail densities decrease both biomass and production (McCormick and Stevenson, 1989; Swamikannu and Hoagland, 1989). Grazing may not, however, have as much of an impact on shaded streams, where light may be the primary limiting factor (Hill and Harvey, 1990). Even in unshaded steams there may be little overall effect of grazers, because the loss of the algal over-story to grazing is compensated for by the competitive release and increased growth of adnate algae (Hill and Harvey, 1990).

Snail grazers selectively remove larger filamentous green algae, and leave smaller, adnate species behind (Table III). Under slight gastropod grazing pressure, periphyton assemblages are dominated by filamentous green algae, but more intensely grazed assemblages are dominated by more tightly adhering or toxic species such as cyanobacteria. Snail grazers may, in fact, indirectly facilitate macrophytes, if periphyton coverings shade and limit macrophyte growth, and snails prefer periphyton. For example, the growth of *Ceratophylum demersum* increased when gastropod grazers were present (Bronmark, 1985b), although increased growth also occurred when plants were exposed only to snail-conditioned water, indicating increased nutrient recycling was a cause (Underwood, 1991). Similarly, when sunfish depress snail abundances and increase periphyton abundance in fish enclosures, they may indirectly depress macrophytes (Martin *et al.*, 1992). The molluscivorous tench (a fish) can have similar cascading effects on European gastropods, periphyton, and macrophytes (Fig. 7). Thus, interactions between gastropod predators, snails, periphyton, and macrophytes may be very complex in natural systems.

2. Factors Regulating Population Size

One of the first demonstrations of population regulation under field conditions was with *Lymnaea elodes* (Eisenberg, 1966, 1970). When the density of adult snails in pens in a small pond was increased, adult fecundity declined, as did juvenile survival. With addition of a high quality resource, spinach, an increase in the number of eggs per mass occurred. Evidently, the availability of micronutrients in periphyton was the crucial variable (Eisenberg, 1970). Brown (1985) provided additional evidence by transferring juvenile *Lymnaea elodes* among a series of ponds differing in periphyton productivity. There was an exponential increase in juvenile growth rates with increasing pond productivity, and snails in the most productive pond laid nine times as many eggs as snails in the two less productive ponds. A number of field studies also provide indirect evidence for the importance of resource abundance: populations in more eutrophic habitats have more generations a year, more rapid shell growth, and lay more eggs (Burky, 1971; Hunter, 1975; McMahon, 1975; Eversole, 1978). Highly eutrophic sites may however

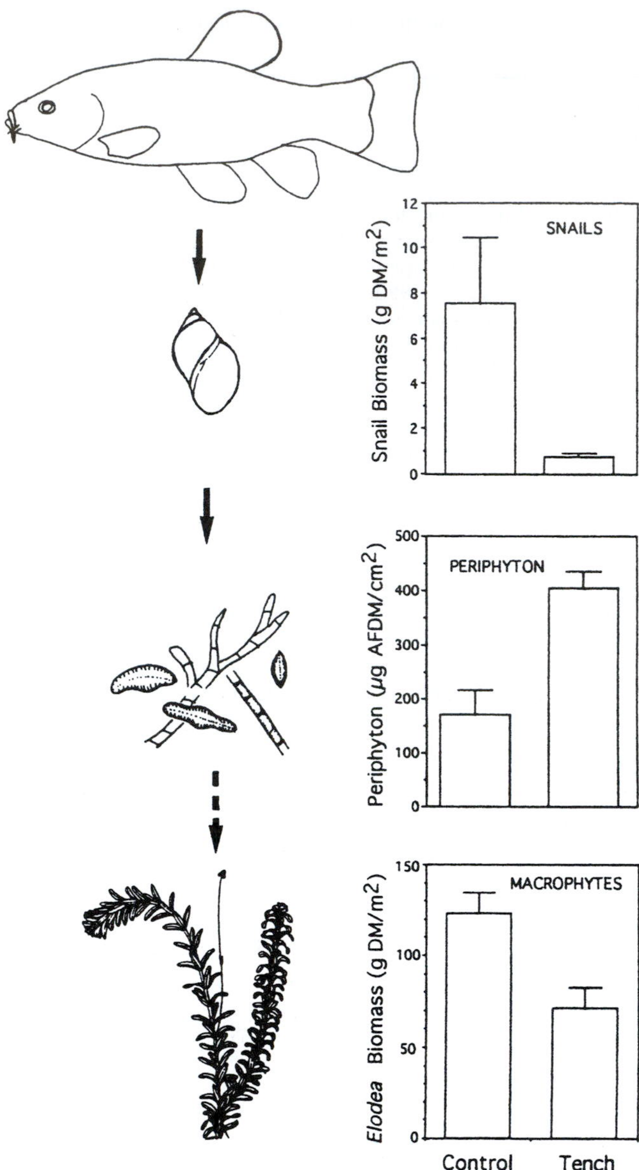

FIGURE 7 Cascading (also called top-down) effects of a molluscivorous fish (the tench) on gastropods, periphyton and macrophytes, based on a field manipulation (from Bronmark and Vermaat, 1998). Note how snail numbers are depressed by fish predation (top), and reduced gastropod grazing results in periphyton increases, and macrophytes are shaded by increasing periphyton cover, and decrease.

be detrimental to gastropods. For example, gastropod diversity declined over a fifty year period as Lake Oneida, New York, became increasingly eutrophic (Harman and Forney, 1970). In lotic systems, manipulation of periphyton resources and the abundances of pleurocerid grazers have indicated both that food resources can control snail density and size distributions, and that the snails can, in turn, control their food resources (Hill *et al.*, 1992; Rosemond *et al.*, 1993).

Although less studied than the role of periphyton, the parasitic larvae of trematode worms may also impact snail population dynamics and evolution (Holmes, 1983). The adult worm produces eggs that are expelled in the feces of the final host (a vertebrate) and hatch into an infectious larval stage called a miracidium. Depending on the parasite group, the miracidium either penetrates the epithelium of the snail or is consumed as an egg and hatches within the snail. Once inside the snail host, miracidia asexually produce several stages (redia and sporocysts) which eventually produce thousands of cercaria that infect the final host or another intermediate host. Infections either accelerate or decelerate snail growth (Brown, 1978; Anderson and May, 1979; Holmes, 1983, see discussion in Minchella *et al.*, 1985). Immediately after infection or exposure, snail egg production rates increase dramatically (Minchella and Loverde, 1981), evidently reducing the eventual costs to snail fitness. As parasites consume the ovotestis and hepatopancreas, however, both growth and egg production of "patent" snails (those infections in the final stage where cercaria are emerging from snails) drop below that of uninfected snails (Minchella *et al.*, 1985).

The role of trematodes in controlling snail populations is unclear, as prevalence (percentage of population infected) varies considerably (Brown, 1978; Holmes, 1983). For example, prevalence in *Lymnaea elodes* in Indiana ponds varied from as low as 4% to as high as 49% (Brown *et al.*, 1988). Prevalences were higher in less productive ponds, evidently because food limitation caused longer snail life cycles that increased chances for snails to be located by miracidia. Life-table models predicted that the number of offspring produced per adult in the next generation declined by 14 to 21% in parasitized populations of *L. elodes*.

Assemblages of trematode larval species within a snail host can themselves be considered communities, with relative abundances and types of biotic interactions dependent on the snail species, the type of final host, and the microhabitat the snail occupies. For example, multispecies infections are common and interspecific competition rare in *Physella gyrina*, evidently because it is quite mobile and reproduces continually throughout the field season, traits which increase chance of infection (Snyder and Esch, 1993). *Helisoma anceps* is more sedentary and has a simple annual life-history pattern; its parasitic trematode community is characterized by fewer multiple infections and a competitive dominance hierarchy among parasites. Regarding microhabitats, trematode larvae that infect snails when eggs are consumed (and that have frogs as final hosts) are more common in snails collected from shallow habitats in ponds (Sapp and Esch, 1994); in contrast, while larvae that infect snails as miracidia (and

TABLE IV Comparison of Average Standing Stocks, Productivity, and Turn-over Times for Populations of Pulmonate and Prosobranch Snails Reported in Russell-Hunter and Buckley (1983)

Subclass	*Mean biomass*[a] + *s.e.*	*Mean production*[b] + *s.e.*	*Mean* turnover time[c] + *s.e.*
Pulmonata	0.98 ± 0.50 (6)	5.71 ± 2.44 (10)	98.0 ± 9.46 (10)
Prosobranchia	4.64 ± 1.80 (4)	3.56 ± 0.51 (5)	385.3 ± 33.5 (4)

(*N*) Number of species averaged.
[a] g C/m^2.
[b] mg C/m^2 per day.
[c] days.

have waterfowl as final hosts) occur in all microhabitats, but only when the final hosts are common.

Trematode parasites and their snail hosts are also extremely interesting from a co-evolutionary viewpoint (Holmes, 1983; Minchella *et al.,* 1985). Because invertebrates cannot easily acquire resistance to parasites, frequency-dependent selection may operate to ensure the fitness of any genotype less vulnerable to a particular trematode (Holmes, 1983). Parasites may also cause a shift in investment of resources from costly reproduction to growth and maintenance and, thus, even increase survivorship of infected snails (see also Baudoin, 1975; Minchella *et al.,* 1985). Because trematode larvae evolve surface antigens that mimic those of the snail host, application of one evolutionary model, called the "red queen" hypothesis, suggests that gastropods are constantly evolving new antigen phenotypes merely to stay ahead of their trematode parasites. For example, the New Zealand prosobranch *Potamopyrgus* has both sexual morphs (able to produce genetically more variable offspring through recombination) and parthenogenetic morphs. The frequency of sexual reproduction is positively correlated with trematode prevalence, as one would expect if sexual morphs have an advantage (Lively, 1987). Sexual morphs are also more common in shallow water, where waterfowl (the final hosts) occur (Lively and Jokela, 1995). Infected snails are also more likely to forage during the day than uninfected snails or gravid females, suggesting parasites may modify snail behavior so they are more likely to be consumed by waterfowl, to complete the life cycle (Levri and Lively, 1996). Experimental studies in New Zealand have suggested trematode populations are locally adapted to better infect their own host populations than snails from different lakes (Lively, 1989). Gene flow has not swamped such local adaptation, even though rates of gene flow are greater among parasite populations than among their snail hosts (Dybdhal and Lively, 1996), However, sexual reproduction may not be favored in all snail–trematode systems. In *Campeloma*, certain trematode metacercaria actually feed on sperm, and parthenogenetic morphs thus have an advantage (Johnson, 1992).

3. Production Ecology

Average standing crop biomass, productivity, and turnover times (these terms are explained in the glossary) differ between prosobranchs and pulmonates (Table IV, summarized from Russell-Hunter and Buckley, 1983). Standing crops are greater on the average for prosobranchs than for pulmonates, although prosobranch populations studied to date have not included genera with smaller individuals, such as *Amnicola* and *Valvata*.

Even if pulmonates have lower standing stocks, their rapid growth and short life cycles still result in higher average production rates and shorter turnover times (Table IV). For example, two pleurocerid grazers in an Alabama stream, *Elimia cahawbensis* and *E. clara*, have considerable biomasses of 2–5 g ash-free dry mass (=AFDM) per square meter, but slow growth rates and long life cycles result in relatively low rates of secondary production (0.5–1.5 g AFDM/m^2). The combination of high biomass and low production results in a low production to biomass ratio of 0.3 (Richardson *et al.,* 1988). Prosobranch detritivores may, however, have higher production rates. *Viviparus subpurpureus* and *Campeloma decisum*, with high densities and short life cycles in Louisiana bayous, have high standing crop biomasses (10–20 g AFDM/m^2) and production rates (20–40 g AFDM/m^{2}per year) among the highest known for freshwater molluscs (Richardson and Brown, 1989).

4. Ecological Determinants of Distribution

Water hardness and pH are often considered major factors determining the distributions of freshwater snails (Boycott, 1936; Macan, 1950; Russell-Hunter, 1978; Okland, 1983; Pip, 1986). For example, in the acid-rain affected lakes of the Adirondack mountains of New York, Jokinen (1991) found that snail diversity declined with declining pH and increasing altitude.

However, in lake districts with adequate calcium (above about 5 mg/L $CaCO_3$), or in the normal (nonacidified) range of pH, relationships between physico-chemical parameters and gastropod diversity are less clear (Lodge *et al.,* 1987; Jokinen, 1987). For example, species in New York lakes overlap broadly in the ranges of physico-chemical variables in which they occur (Harman and Berg, 1971). Physico-chemical parameters, therefore, set the limits for gastropod distributions, but are not as important in explaining the relative abundance patterns and densities of gastropods in most hard-water, circum-neutral lakes.

On a biogeographic scale, a factor determining gastropod distributions is dispersal ability. Studies have indicated diversity increases with the area of lakes and ponds (Lassen, 1975; Browne, 1981; Bronmark, 1985a; Jokinen, 1987). Since immigration rates generally increase and extinction rates decrease with increasing habitat size, larger habitats, all else being equal, usually support more species (MacArthur and Wilson, 1967). Jokinen (1987), following the ideas of Diamond (1975), found that snail species were not distributed randomly with increasing snail diversity. Five snails were "high-S" species (e.g., those that occurred only at the most diverse sites) including *Valvata tricarinata*, *Gyraulus deflectus*, *Laevapex fuscus*, and two species of *Lyogyrus*. Most gastropods were widely distributed across sites ("C–D" tramps in Diamond's terminology). The species with the greatest dispersal ability (called "super tramps" by Diamond) were found only at low diversity sites; these included the lymnaeid *Pseudosuccinea collumella* and the limpet *Ferrissia fragilis*.

After dispersal, successful colonization depends on the presence of suitable substrates. For example, a significant relationship exists between the number of gastropod species and the number of substrates in lakes and streams in New York, USA (Harman, 1972). In fact, substrate preferences determined in the laboratory are good predictors of the types of ponds where the snails are common (e.g., algivores in open ponds, detritivores in wooded ponds; Brown, 1982). Gastropod diversity is also positively related to macrophyte biomass, probably because macrophytes increase surface area for periphyton colonization (Brown and Lodge, 1993).

Disturbance may also determine the assemblage of snails. In temporary ponds, diversity is lowered by frequent drying; and in habitats that go hypoxic, diebacks of macrophyte and gastropod populations will also occur (Lodge and Kelly, 1985; Lodge *et al.,* 1987). Disturbance may also limit some species from areas such as wave-swept shores, littoral zones of reservoirs, etc. Similarly, in streams, current velocity may affect distribution and growth. Adult *Elimia* avoid high-flow areas, perhaps because increased flow makes movement more difficult and lowers grazing rates or increases metabolic rate, thus lowering adult size (Johnson and Brown, 1997). Juveniles, because they are smaller and can exploit the boundary layer, are actually more common in fast-flowing areas.

In large lakes, biotic interactions such as interspecific competition or predation, are important in determining gastropod diversity and abundance (Lodge *et al.,* 1987). In regards to competition, some evidence suggests it is rare in freshwater snails. For example, pulmonate pond snails in the midwestern USA overlap little on food and habitat dimensions of the niche, although some species do compete (Brown, 1982). Other studies also indicate differences in resource utilization. For example, the European ancylid *Ancylus fluviatilis* prefers diatoms and is found on the top of cobble where periphyton is abundant; whereas the co-occurring planorbid *Planorbis contortus* is a detritivore and occurs more often under stones, where detritus accumulates (Calow, 1973a, b, 1974a, b). Similarly, *Pseudosuccinea columella* and *Physa vernalis* possess differences in their gut and radulae allowing coexistence in New England ponds (Kesler *et al.,* 1986). Furthermore, niche partitioning may also be facilitated by differences among snails in enzymatic activity (Calow and Calow, 1975; Brendelberger, 1997) or radular morphology (Barnese *et al.,* 1990).

However, there is some indirect evidence for competition. First, there are often fewer coexisting congeners in field samples than would be expected, based on mathematical simulations (Dillon, 1981, 1987). Second, competition has been inferred from changes in relative abundance of gastropod species through time, that is, by apparent competitive exclusion (Harman, 1968). Third, niche overlap can, in fact, be higher in more diverse snail assemblages found in large lakes. For example, the abundances of gastropod species are often positively associated with high-diversity macrophyte beds in lakes, and experiments show most species have similar preferences for macrophytes (Brown, 1997). Finally, pulmonates are common in ponds or in vegetated areas of lakes, while prosobranchs are rare in ponds and common in lakes and rivers. One explanation could be competitive exclusion of pulmonates from lakes by prosobranchs, or exclusion of prosobranchs from ponds by the same mechanism. However, additional explanations include greater vulnerability of prosobranchs to hypoxic conditions in ponds, poorer dispersal abilities of prosobranchs, or the fact that thin-shelled pulmonates are more vulnerable to shell-crushing fish common in lakes or rivers (Brown *et al.,* 1998).

Indeed, Lodge *et al.* (1987) argue that predators determine the composition of gastropod assemblages in lakes. Indirect evidence includes the fact that gastropods have a number of anti-predator adaptations. These include thick shells to protect against shell-crushing predators (Vermeij and Covich, 1978; Stein *et al.,* 1984, Brown and DeVries, 1985; Saffran and Barton, 1993; Brown, 1998), as well as escape behaviors such as shaking the shell or crawling above the water to protect against shell-invading invertebrate predators (Townsend and McCarthy, 1980; Bronmark and Malmqvist, 1986; Brown and Strouse, 1988; Alexander and Covich, 1991 a, b; Covich *et al.,* 1994). Molluscivores are common in lakes and rivers. For example, the pumpkinseed sunfish, *Lepomis gibbosus*, and the redear or shell-cracker sunfish, *Lepomis microlophus*, specialize on gastropod prey and have pharyngeal teeth adapted to crush shells. Crayfish will select snails instead of grazing on macrophytes if given a choice (Covich, 1977), using their mandibles to chip shells back from the aperture and selecting species with thinner shells (Saffran and Barton, 1993; Brown, 1998).

Although Osenberg (1989) argued that lentic snail assemblages are limited by food resources, most experimental evidence supports a strong role for predators in determining snail diversity and abundance (Table V). For example, the central mud minnow can significantly lower the density of relatively thin-shelled snails in permanent ponds. When fish densities were manipulated in pens, the number of eggs and juveniles of *Lymnaea elodes* was significantly less in the presence of fish. The small, gape-limited fish feed on eggs and juveniles and may restrict *Lymnaea elodes* from permanent habitats such as large marshes or lakes. Pumpkinseed sunfish also strongly prefer large, weak-shelled gastropod species in laboratory experiments, and thin-shelled species also decline dramatically in pumpkinseed enclosures in lakes (see references in Table V). Thus, most thin-shelled pulmonates should occur in lakes only in macrophyte beds, where they have a refuge from visual predators; and sandy areas should be dominated by thicker shelled pulmonates like *Helisoma* or by prosobranchs. Indeed, gastropod distributions among habitats within Indiana and Wisconsin lakes do follow these patterns (Lodge *et al.,* 1987; Brown and Lodge, 1993; Lodge *et al.,* 1998). In fact, fish predators may indirectly benefit poorer gastropod competitors by preferentially removing dominant species. Lake trout, for example, preferentially consume larger lymnaeids, releasing valvatids from competition in arctic lakes (Hershey, 1990; Merrick *et al.,* 1991). However, experimental manipulations suggest such indirect effects do not extend through four trophic levels. Piscivores (top predators feeding on fish) in Swedish lakes cannot depress predators of snails to the extent that interactions cascade down the food web to facilitate snails which negatively impact periphyton (Bronmark and Weisner, 1996).

TABLE V Summary of Experimental Field Manipulations Testing the Role of Snail Predators in Controlling Their Prey

Study	*Prey*	*Predator*	*Conclusions*
Brown and Devries 1985	*Lymnaea elodes*	mud minnows	decreased abundance of eggs and juveniles of thin-shelled *L. elodes*
Sheldon 1987	assemblage	sunfish	removal of fish increased snails which decreased macrophytes through herbivory
Kesler and Munns 1989	assemblage	belastomatid bugs	decreased snail abundances in New England pond
Osenberg 1989	assemblage	littoral sunfish	fish controlled only rare, large snail spp., competition between snails considered more important
Hershey (1990), Merrick *et al.* (1991)	assemblage	lake trout	trout removed better competitor (*L. elodes*), which favored poorer competitor, *Valvata*
Martin *et al.* 1992	assemblage	*L. microlophus* *L. macrochiras*	fish decreased snails 10-fold, increased periphyton biomass 2-fold, decreased algal cell size, decreased macrophytes.
Bronmark *et al.* 1992	assemblage	pumpkinseed (*L. gibbosus*)	fish decreased snails, increased periphyton, favored adnate algae
Bronmark 1994	assemblage	Tench	fish decreased snails, increased periphyton, decreased macrophytes
Lodge *et al.* 1994	assemblage	*O. rusticus*	crayfish decreased snails and macrophytes, had no effect on periphyton
Daldorph and Thomas 1995	assemblage	sticklebacks	fish decreased thin-shelled snail spp., increased periphyton
Bronmark and Weisner 1996	assemblage	piscivores & molluscivores	molluscivores did decrease snails, but piscivores did not control molluscivores and increase snails indirectly

Assemblage refers to a natural community of gastropods.

Invertebrate, shell-invading predators may also limit snail populations or cause shifts in gastropod relative abundances. Although some leeches, such as *Nephelopsis obscura*, have fairly low feeding rates (Brown and Strouse, 1988), crayfish can consume up to or over a hundred snails per night. The crayfish *Orconectes rusticus* significantly reduced snail abundances in enclosure experiments in Wisconsin lakes (Table V), and an interlake survey indicated that snail abundances were negatively correlated with crayfish catches. Crayfish also affect gastropod habitat selection: even though cobble habitat is preferred by many snails (see earlier discussion), it also provides refugia from fish predators for crayfish. Crayfish predation overrides the effects of rich food resources (Weber and Lodge, 1990).

Crayfish are also quite selective molluscivores. They prefer thinner-shelled gastropods (Saffran and Barton, 1993; Brown, 1998), and often prefer juvenile snails which have shorter handling times and higher profitability unless these more vulnerable prey are protected by macrophytes (Nystrom and Perez, 1998). Crayfish predators also shift size distributions of *Physella virgata* upwards in Oklahoma streams. This may be due both to size-selective predation and to the diversion of energy of the snail from reproduction to growth in order to reach a size-based refuge from predation (Crowl, 1990; Crowl and Covich, 1990). Interestingly, only snails in vulnerable size classes respond by crawling out of the water after predation by crayfish has occurred on conspecifics (Alexander and Covich, 1991b). Other invertebrate predators may also be important. For example, belostomatid bugs can eat as many as five snails per bug per day in the laboratory (Kesler and Munns, 1989).

Fish predators may alter snail foraging behavior as well. In experiments, *Physella* selects refugia in the presence of sunfish predators, only leaving when sufficiently starved (Turner, 1997). This behavioral interaction may provide an additional explanation for why periphyton biomass increases when fish predators are present. For example, use of open habitats by snails decreases with increasing predation risk from sunfish (Fig. 8), with the result that periphyton biomass also increases (Turner, 1997). Interestingly, such indirect facilitation of periphyton does not occur upon experimental addition of crayfish to enclosures, probably because crayfish are also omnivorous and graze on periphyton (Lodge *et al.*, 1994).

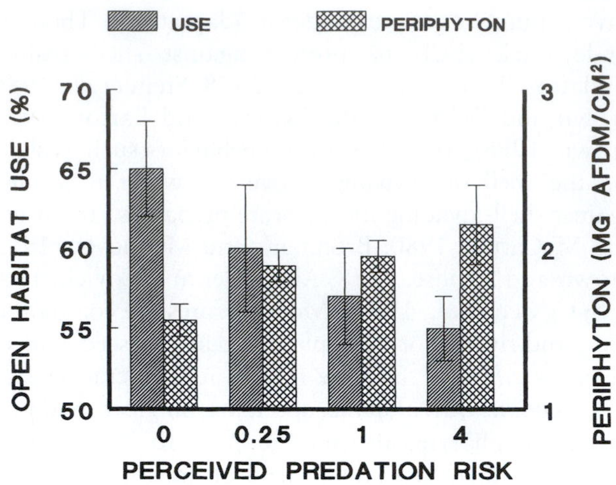

FIGURE 8 Indirect effect of mortality from fish predation on snail grazing rates. As more crushed snails were added to experimental pools (predation risk perceived by snails, *x*-axis), snails used refuges more, resulting in reduced use of open areas (left *y*-axis). Note how periphyton biomass (ashfree dry mass or AFDM, right *y*-axis) increased in open area as a result of reduced snail grazing (after Turner, 1997).

5. Suggestions for Further Work

Studies of the production ecology of smaller prosobranchs, as well as snail populations in other areas besides the northeastern United States are still needed. Questions still exist on subjects such as:

1. How pulmonates are physiologically adapted to relatively ephemeral or hypoxic aquatic habitats, including studies of differences in metabolic pathways and nitrogen excretion (McMahon, 1983).
2. The relative roles of the physiological and ecological constraints to shell versus tissue growth, given the remarkable intraspecific variation in shell structure and composition in freshwater gastropods.
3. Physiological bases of feeding preferences (such as relative cellulase activities, etc.).
4. Micro-preferences (e.g., preferences for certain algae or diatom species), for example, whether detritivores are consuming leaves or more nutritious bacteria or fungi.
5. The nutritional quality or toxicity of different food types and their role in preferences by gastropods.

Further research on abiotic and biotic factors influencing snail assemblages is also necessary. Abiotic factors such as pond drying, anoxia, floods, or macrophyte die-offs could explain much about gastropod population dynamics or species composition. Further work is also necessary to determine whether the allopatric distributions of pulmonates and prosobranchs are due to competition (see discussion in Brown *et al.*, 1998). Further work is needed as well on the relative role of invertebrate predators like leeches, belostomatid bugs, and scyomyzid fly larvae (Eckblad, 1973) in structuring snail assemblages.

D. Evolutionary Relationships

Marine prosobranchs are ancestral to freshwater prosobranchs and terrestrial and freshwater pulmonates. Freshwater prosobranchs adapted to the dilute osmotic conditions of estuaries and then rivers, and most modern families are widespread because their adaptive radiations predated continental breakups and drift. Davis (1979, 1982) and Clarke (1981) pointed out that prosobranch dispersal is limited in most cases to slow movement of adults along streams and rivers. Such populations are more likely to become isolated, promoting chances of speciation and adaptive radiation. Examples include the dramatic adaptive radiation of hydrobiids (Davis, 1982) and pleurocerids in lotic habitats (Burch, 1989; Lydeard *et al.*, 1997).

In contrast, pulmonates, due to their greater rates of passive dispersal as spat on birds and insects, have more widespread distributions. For example, they are widespread in the northern and northeastern states, as well as in Canada (Harman and Berg, 1971; Clarke, 1981). Immigration rates to ponds average 0.8 and can be as high as 9 species per year (Lassen, 1975; Davis, 1982). Pulmonates also predominate in shallow, more ephemeral habitats less than 100 km^2 in area and with durations less than 10^3 years. Populations probably do not exist long enough in such habitats for speciation to occur, explaining why pulmonates are less speciose than prosobranchs (Russell-Hunter, 1983; Clarke, 1981; Davis, 1982).

Pulmonates apparently evolved from intertidal prosobranchs that relied less and less on aquatic respiration (Morton, 1955; McMahon, 1983). Modern estuarine pulmonates like *Melampus* may resemble these ancestral species. The intermediate, terrestrial pulmonates lost the ctenidium, and gave rise both to modern terrestrial pulmonates (order Stylommatophora) and the aquatic pulmonates (order Basommatophora). Figure 9 illustrates a hypothetical phylogeny of aquatic pulmonates based on progressive re-adaptation to aquatic habitats (see discussion in McMahon, 1983). Lymnaeids are in some cases still amphibious (e.g., occurring in temporary wetlands), while physids are intermediate in their adaptation to aquatic habitats (some species come to the surface to breathe, while others have filled the lung with water and can be found in deeper lakes). Ancylids and planorbids have secondarily acquired more aquatic specializations such as the evolution of secondary gills, and in the planorbids, hemoglobin.

Several phylogenetic trees for freshwater snails have been published recently, using both classical morphological and recently developed molecular methods. Jung (1992) used the more traditional characters of shell and soft anatomy to propose a cladogram (Fig. 10) for certain species of planorbids, using 65 characters. He pointed out that systematic studies should not rely on preserved specimens, since fixation alters soft anatomy. In his phylogeny, *Biomphalaria* is considered more ancestral, and *Planorbella trivolvis* and *Helisoma anceps* the most derived species.

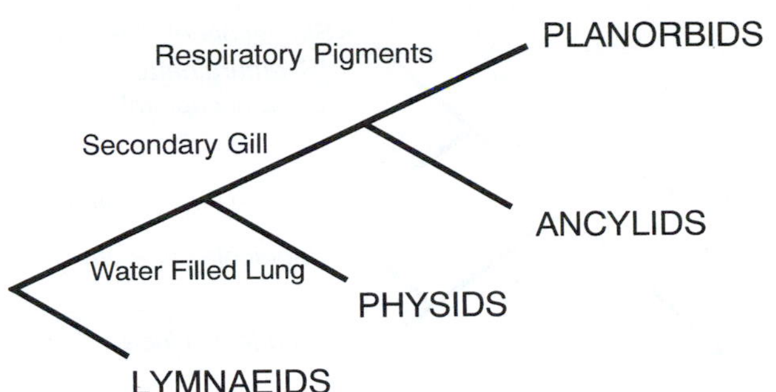

FIGURE 9 Hypothetical phylogeny for freshwater pulmonates assuming increased adaptation through time to secondarily-invaded aquatic habitats. The argument is that air-breathing lymnaeids are limited to shallow water, while physids, which in some cases can fill their lung with water and use it as a gill, are found in deeper water, and ancylids and planorbids are the most specialized for an aquatic life style, with secondarily derived epithelial gills and in the planorbids, hemoglobin.

Remigio and Blair (1997) published a phylogeny for certain lymnaeid species, using populations from both North America and Europe (Fig. 11). They considered molecular methods, such as sequencing ribosomal DNA in mitochondria, superior techniques in comparison to other characters such as shell morphology, which often converge among genera. For example, based on molecular information, they considered karyotypes with 18 chromosomes to be ancestral, and 16 chromosomes to be a derived character, unlike

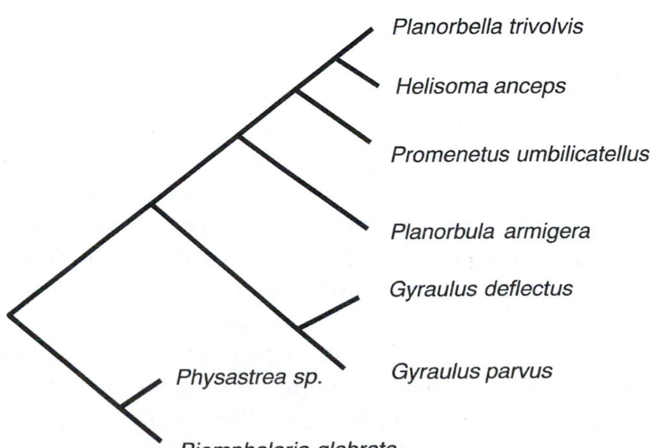

FIGURE 10 Simplified cladogram for several planorbid species studied by Jung (1992), based on shell and soft anatomy.

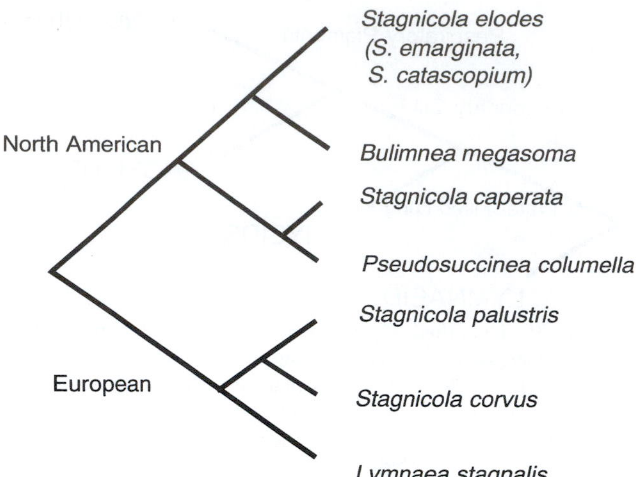

FIGURE 11 Cladogram for a subgroup of the lymnaeid species found in both North America and Europe, based on molecular data. Simplified from Remigio and Blair (1997).

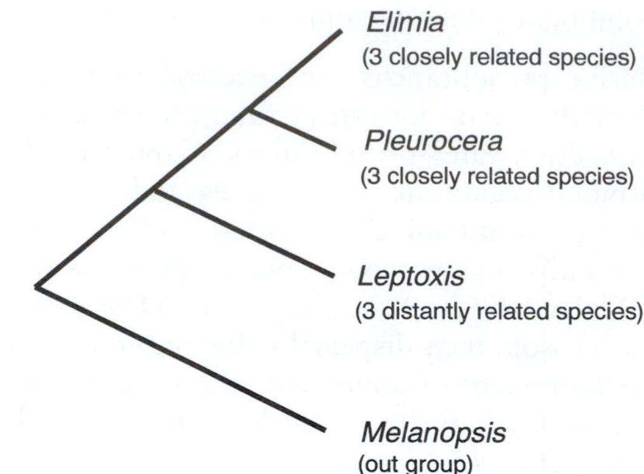

FIGURE 12 Cladogram for a subgroup of species in several prosobranch genera in the Mobile River Basin, based on molecular data. The greater divergence of leptoxid species suggests a polyphyletic origin for this genus. Simplified from Lydeard *et al.* (1997).

earlier studies. European and North American stagnicolids (*S. elodes* and *S. palustris*) were not considered each other's closest relatives, contrary to interpretations based on morphology. *Stagnicola catascopium*, *S. emarginata*, and *S. elodes*, based on genetic divergence, could actually be conspecific taxa. On the other hand, two populations of *Lymnaea stagnalis* (one in Germany and the other in Italy) were genetically quite divergent, indicating isolation by the geographic barrier of the Alps. The authors propose that lymnaeids evolved in the late Jurassic to early Cretaceous on the common continent Laurasia, and were separated into distinct Palearctic and Nearctic groups at the start of the Cenozoic (65 million years ago) when ancestral species still had 18 chromosomes. The genus *Stagnicola* was probably extant at that time, explaining why North American and European species are so divergent.

Lydeard *et al.* (1997) used mitochondrial rDNA sequences to study the evolutionary relationships of representative prosobranch snails in the Mobile River Basin in the southeastern United States (Fig. 12). This basin may have once had the most diverse gastropod fauna in the world, with 118 species and nine families. Pleurocerids make up almost two thirds of the surviving fauna. Because of impoundments, the genus *Gyrotoma*, once found in shallow riffles in the Coosa River, has gone extinct, along with 14 species of *Elimia*, and 11 species of *Leptoxis*. Lydeard's phylogeny considers 14 prosobranch species (with *Melanopsis* as an out group) and suggests *Elimia* and *Pleurocera* are monophyletic genera, but that *Leptoxis* is paraphyletic. The remaining *Elimia* species in the Mobile Basin are genetically fairly similar, while the remaining three species of *Leptoxis* are quite divergent in their mitochondrial 16S rDNA sequence.

Further studies of genetic relationships among pleurocerids found in the Mobile River Basin suggest that earlier species descriptions, relying solely on shell morphology, have overestimated the true diversity of the system. For example, Lydeard *et al.* (1998), again looking at 16S rDNA sequences, considered *Elimia hydei* basal to all other elimiads, and *Elimia crenatella*, *E. showalteri*, *E. fascians*, and *E. E. caelatura* to be true species. The low genetic divergence within another clade led them to question the status of six speces: *Elimia carinocostata*, *E. gerhardtii*, *E. haysiana*, *E. alabamiensis*, *E. olivula*, and *E. cylindracea*. Dillon and Lydeard (1998) used allozyme polymorphism to study relationships among Mobile Basin leptoxids. They argued that *Leptoxis praerosa* and *L. picta* were deserving of species status, but that *L. taeniata* and *L. ampla* should be subsumed within *L. picta*. On the other hand, Dillon and Ahlstedt (1997) used allozyme variation to show that *Athernia anthonyi* and *Leptoxis praerosa* were different species, although adult shell morphology is similar.

Several caveats about the evolution of freshwater snails should be kept in mind (Davis, 1982). First, phylogenetic analyses, and therefore biogeographic studies, are often limited by the poor systematic information available on many groups, as well as the poor fossil record (only shells are preserved, which often show convergence). Second, a number of factors can explain current distributions, as illustrated for hydrobiids and pomatiopsids by Davis (1982): (1) phylogenetic events such as centers of origin and adaptive radiation; (2) past history events such as continental drift and geological alteration of stream and river flow; (3) dispersal powers, and (4) ecological factors (see earlier discussion).

E. Conservation Biology of Freshwater Snails

Freshwater snails, especially prosobranch species in the southeastern United States, are facing rates of extinction similar to the more widely known case of North American unionid bivalves (Lydeard and Mayden, 1995; Neves *et al.,* 1998; Jenkinson and Todd, 1998). Freshwater snails account for 60% of all North American molluscan species, and 61% of these snails are in the Southeast. Historically, there were 118 species of snails in the Mobile River Basin, 96 species in the Tennessee River, and 36 species each in the Cumberland and Apalachicola River basins. Thirty-eight of the 118 species in the Mobile River Basin are known to be extinct. The decline in diversity in the Mobile River Basin varies from 33 to 84% in different drainages.

Nine species of gastropods are currently listed as federally endangered (Bogan, 1999), although over 200 species from 11 families are candidates (69% of which are in the Southeast). In the southeastern states alone, two viviparids, 50 hydrobiids, 83 pleurocerids, nine pulmonates, three ancylids and six planorbid species are considered at risk (Neves *et al.* 1998). Many of the same factors that have imperiled unionid bivalves are responsible: impoundments which inundate shallow riffles and decrease oxygen and alter water temperature, increased turbidity from agricultural activities, improper sewage treatment, and other pollution sources, to name but a few. Organic enrichment, pesticides and pathogens endanger about a quarter of the impaired river-miles in North American rivers, while excess nutrients and siltation occur, respectively, in 37 and 45% of the impacted river-miles (Neves *et al.,* 1998). At the current time there is little governmental interest in managing imperiled snail populations, although the large, knobbed-shell species, *Io fluvialis*, has been successfully reintroduced to one river in its original range.

IV. COLLECTING AND CULTURING FRESHWATER GASTROPODS

A number of sampling techniques exist (reviewed in Russell-Hunter and Buckley, 1983), although some are not quantitative. The least quantitative technique, but one that often gives large numbers of individuals and a good idea of species composition, is sweep netting with a net of 1-mm mesh (Brown, 1979, 1997).

In soft sediments or sand, quantitative samples can be collected with an Ekman grab or a corer. When sampling macrophytes, the sampler must collect both plants with attached snails and the substrate with any bottom-dwelling species. Examples of such samplers are described in Gerking (1957), Savino and Stein (1982) and Lodge *et al.* (1994). In coarse sand, Ponar dredges are the best alternative. In cobble, little recourse to direct counts of given areas by visual search is available. One can estimate densities by individually covering the rocks with aluminum foil, and then estimating the area from weight-to-area regressions for the foil.

The best sorting technique for gastropods is hand-sorting. Large adults can be removed visually, while samples are washed through a graded series of sieves (the smallest having a mesh of approximately 0.5 mm) to remove mud but retain smaller snails. Samples are then carefully backwashed from the sieves into shallow water on flat white trays, the vegetation teased apart, and the whole tray examined in a systematic fashion to remove small gastropods and egg cases.

Temperate gastropods will grow well at 15–20°C, while subtropical species grow better at 20–25°C. Any hard substrate with a dense periphyton covering can be added for food, although artificial foods such as lettuce, cereal, or spinach are sometimes used. Provide food only as needed to avoid fouling containers. Detritivores should be fed leaf litter colonized with bacteria and fungi (i.e., held for at least 2 weeks in a pond or stream).

Avoid crowding of snails, as growth and reproduction are sensitive to density. An approximate rule is one snail per liter. Water should be re-circulated through a gravel or charcoal filter, or at least be changed weekly. Conspecifics should be paired so that mating can occur. Culturing of "weedy" species like physids is best done at low temperatures to retard egg production, and constant removal of egg cases is necessary to prevent population "explosions". Adequate lighting (with a 12L:12D cycle) is necessary to promote periphyton growth in aquaria (gro-lites®work well).

V. IDENTIFICATION OF THE FRESHWATER GASTROPODS OF NORTH AMERICA

The following key is modified from Burch's (1989) key for North American freshwater snails. Taxa are keyed down to genera in all families except the diverse hydrobiids. Readers interested in keying specimens down to the species level should consult Burch (1989) or keys addressing regional snail faunas (Harman and Berg, 1971; Jokinen, 1983; Thompson, 1984; Jokinen, 1992).

After collecting specimens, first relax them with tricaine methane sulfonate, menthol crystals, polypropylene phenoxytol, or sodium nembutol and then preserve them in 70% ethyl alcohol. Preserved specimens are better than dried shells, as possession of an operculum will be one of the determinations necessary.

Collect a number of different-sized specimens, because closely related genera sometime differ only in adult shell size. Sizes reported in the key, following Burch (1989), are small shells less than 10 mm, medium between 11 and 29, and large greater than 30 mm in length. Size bars in keys are in millimeters. Refer to the glossary for shell sculpture terminology like costae, lirae, etc. In some cases, soft tissue characters are used in the key (for example the structure of the verge, or penis, located behind the male's right tentacle, is important in hydrobiids). The shells of preserved individuals can be carefully crushed and removed, after living snails are relaxed with any of the chemicals mentioned above. Examine the verge under a dissecting microscope.

If examination of the radula is necessary, dissect out the buccal mass (a muscular mass behind the mouth that surrounds the radula) from a relaxed specimen and place it in a 10% potassium hydroxide solution to digest the tissue. After several hours, remove any remaining tissue and rinse the radula in 70% alcohol, stain it and mount it on a slide for examination under a compound microscope. However, digestion with strong bases may damage the radula. An alternative technique (Holznagel, 1998) is more involved, but also provides a method to extract DNA for phylogenetic work, and can be used with frozen or alcohol-fixed specimens (but not those preserved in formalin). First prepare stock solutions of NET buffer and Proteinase-K. For the former, add 1 mL pH 8.0 Tris buffer, 2 mL 0.5 M EDTA (ethylene diamine tetra-acetic acid), 1 mL 5M NaCl, and 20 mL of 10% SDS (sodium dodecyl sulfate) solution to 76 mL of de-ionized water. For the latter, add 20 mg Proteinase-K to 1 mL de-ionized water and store at −20°C until use. Next remove the body from the shell. This is fairly easy in recently thawed or relaxed specimens, but for nonrelaxed specimens the shell may have to be gently cracked and removed from the body. Dissect away the head, and rinse with water. Place the tissue in a 1.5 mL microcentrifuge tube with 500 μL of NET buffer and 10 μL of Proteinase-K. Cap the tube and place it on a 37°C mixing table for 3 h (for fixed specimens) or 4 days (for dried specimens). The buffer and enzyme may need daily replacement. Observe the radula under a dissecting scope to determine if tissue digestion is complete, and rinse it with de-ionized water several times. The supernatant can be used for extraction of DNA with phenol/chloroform techniques (see Sambrook *et al.*, 1989). Store the radula in 25% ethanol until ready for mounting on a microscope slide.

A. Taxonomic Key to Families and Selected Genera of Freshwater Gastropods

1a. Shell an uncoiled cone (limpet, or cap shaped) (Fig. 2) 2

1b. Shell coiled 8

2a(1a). Apex nearly central; adult up to 12 mm; Pacific drainage. . . Family Lymnaeidae 3

2b. Apex to the right or left of the median line; adult 7 mm or less; dextral or sinistral 4

3a(2a). Apex central (Fig. 13A) .... *Lanx*

3b. Apex anterior; Columbia river (Fig. 13B) .... *Fisherola*

4a(2b). Shell dextral (apex tips to left when viewed with aperture facing up); Rocky Mountain lakes, northeastern Ontario and north-central Quebec (Fig. 13C) .... Family Acroloxidae .... *Acroloxus*

4b. Shell sinistral (apex tips to right when viewed with aperture facing up); throughout North America .... Family Ancylidae 5

5a(4b). Shell elevated, apex in midline, tinged with pink or red inside and out, radially striate, with a notch-shaped depression in unworn specimens; apertural lip broad and flat. In southeastern rivers (Fig. 13D) .... *Rhodacmea*

5b. Shell height variable; apex in midline or to right, the same color as the rest of the shell, finely radially striate or smooth; widely distributed in lotic or lentic habitats 6

6a(5b). Apex with fine radial striae, eroded in older specimens; aperture width variable, open or with a horizontal shelf in the posterior; one lobed, flat pseudobranch; widely distributed in lotic and lentic habitats (Fig. 13E) .... *Ferrissia*

6b. Shell more depressed; apex without radial striae; aperture ovate to subcircular, always open; two-lobed pseudobranch, the lower elaborately folded; in eastern states and south in lentic habitats 7

7a(6b). Apex tipped to right, tentacles colorless; In canals in southern Florida and Texas (Fig. 13F) .... *Hebetancylus*

7b. Apex in midline; Tentacles with a black core; East of the Mississippi in lentic habitats, occasionally in south-central streams (Fig. 13G) .... *Laevapex*

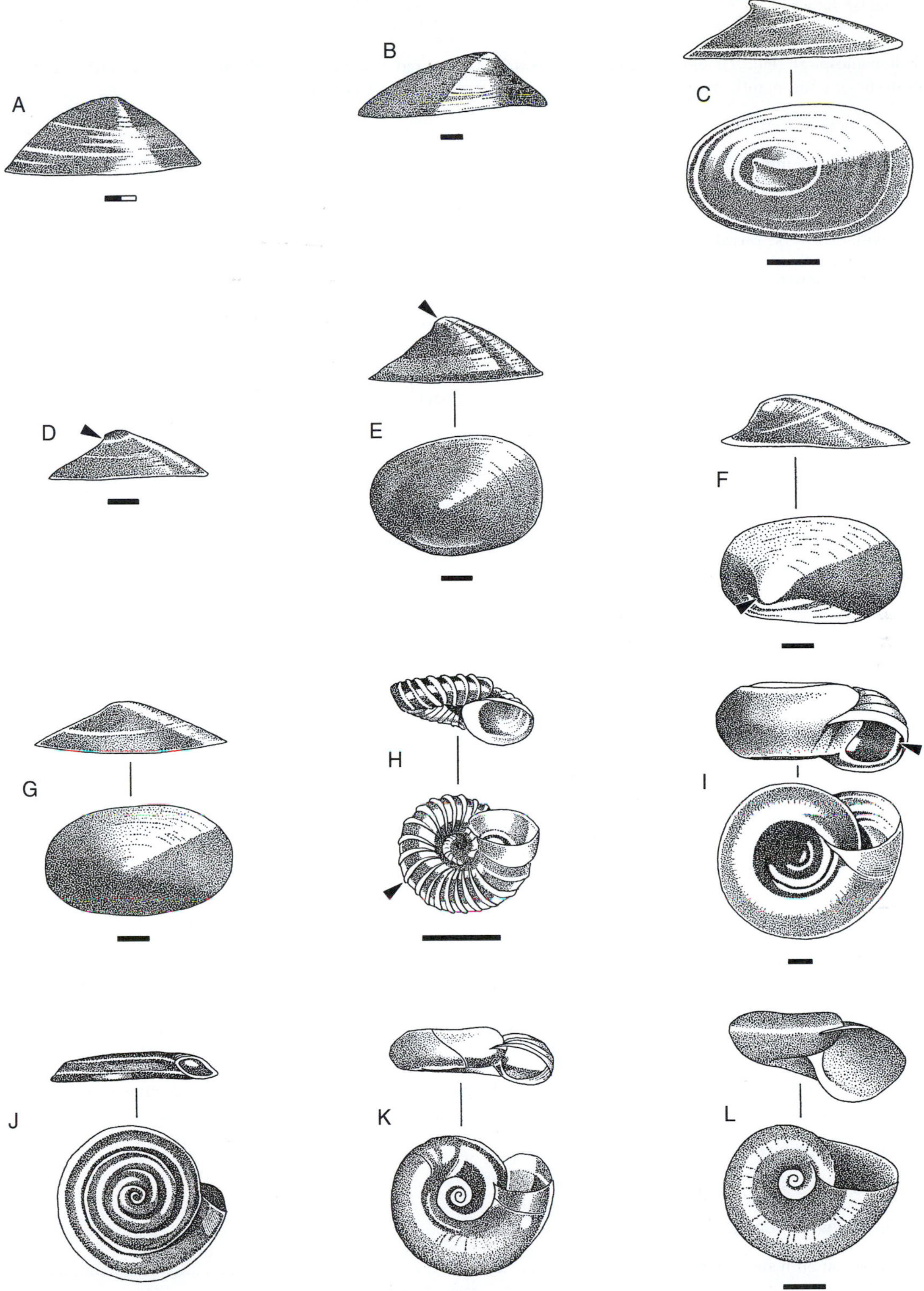

FIGURE 13 Representative limpets and planorbids. (A) The lymnaeid limpet *Lanx patelloides* (note large size, west coast distribution); (B) lymnaeid limpet *Fisherola nutalli*; (C) *Acoloxus;* (D) Ancylid limpet *Rhodacmea rhodacme* (= *hinkezi*) (note notched depression and southeastern distribution); (E) ancylid limpet *Ferrissia rivularis* (note elevated shell, some species possess posterior "shelf" in shell); (F) ancylid limpet *Hebetancylus excentricus* (note depressed apex to the right of midline, colorless tentacles, and southern distribution); (G) *Laevapex fuscus* (note obtuse apex near midline of shell, black pigmented tentacles, and widespread distribution in eastern backwaters and southern, slow-flowing streams). (H) *Armiger crista* (note costae); (I) *Planorbula armigera* (= *jenksii*) (note teeth in aperture); (J) *Drepanotrema kermatoides*; (K) *Gyraulus deflectus*; and (L) *Menetus dilatatus* (A–L after Burch, 1982).

8a(1b). Shell planospiral (Fig. 2); blood (hemolymph) nearly always red (contains hemoglobin); pseudobranch (false gill) near pneumostome or anus; mantle margin simple .... Family Planorbidae 9

8b. Shell with raised spire 20

9a(8a). Adult less than 8 mm in diameter 10

9b. Adult more than 10 mm in diameter 16

10a(9a). Shell with transverse raised ridges (costae). Canada and northern United States. (Fig. 13H) .... *Armiger*

10b. Shell without costae 11

11a(10b). Adults 2 mm or less in diameter; Coosa River, Alabama .... *Neoplanorbis*

11b. Adults more than 2 mm in diameter 12

12a(11b). Aperture or body whorl with internal "teeth" or lamellae set back one quarter whorl from aperture (Fig. 13I .... *Planorbula*

12b. Aperture or body whorl without teeth 13

13a(12b). Shell extremely flat, multiwhorled, or with numerous, low, close-set spiral ridges (lirae). Florida, Texas and southern Arizona (Fig. 13J) .... *Drepanotrema*

13b. Shell not extremely flat or multi whorled, lacking lirae 14

14a(13b). Body whorl height equal from one side to the other (Fig. 13K) .... *Gyraulus*

14b. Body whorl rapidly increases in height toward aperture 15

15a(14b) Body whorl rounded (Fig. 13L) .... *Menetus*

15b. Body whorl with angular edge (carina) (Fig. 14a) .... *Promenetus*

16a(9b). Adult greater than 30 mm; operculum present; penis to right of mantle; gelatinous eggs; In Florida, and introduced to rivers in central Texas .... Family Pilidae .... (Ampullaridae) (Fig. 18a) .... *Marisa*

16b. Adults less than 30 mm, operculum absent 17

17a(16b). Shell thin, fragile, body whorl relatively depressed; Florida, Texas, Arizona (Fig. 14B) .... *Biomphalaria*

17b. Shell thicker, solid; body whorl often high 18

18a(17b). Body whorl extremely large; Western in distribution (Fig. 14C) .... *Vorticifex*

18b. Body whorl not extremely large 19

19a(18b). Shell spire (left side) with deep conical depression, with strong carina (Fig. 14D) .... *Helisoma*

19b. Shell spire (left side) with shallow depression, whorls rounded on at least one side, without carina; aperture lip sometimes flared (Fig. 14E,F) .... *Planorbella*

20a(8b). Shell sinistral (aperture to left when shell viewed with spire pointing away from observer); mantle margin may be digitate or lobed .... Family Physidae 21

20b. Shell dextral (aperture to right) 24

21a(20a) Mantle edge with fingerlike projections 22

21b. Mantle edge without projections, but may be serrated 23

22a(21a). Projections on both sides of mantle .... *Physa*

22b. Projections on parietal side of mantle (Fig. 14G–I) .... *Physella*

23a(21b). Serrations extend beyond apertural lip, partly overlap shell; Texas (Fig. 14J) .... *Stenophysa*

23b. Shell elongate, surface black and glossy, sutures smooth, in wooded ponds in Canada and northern United States (Fig. 14K) .... *Aplexa*

24a(20b). Aperture with teeth on the parietal inner margin; operculum present and calcareous, paucispiral (Fig. 2); adult length about 20 mm; gill featherlike. Florida and southern Georgia .... Family Neritinidae .... (Fig. 14I) *Neritina*

24b. Shell without teeth on parietal wall, operculum present or absent. 25

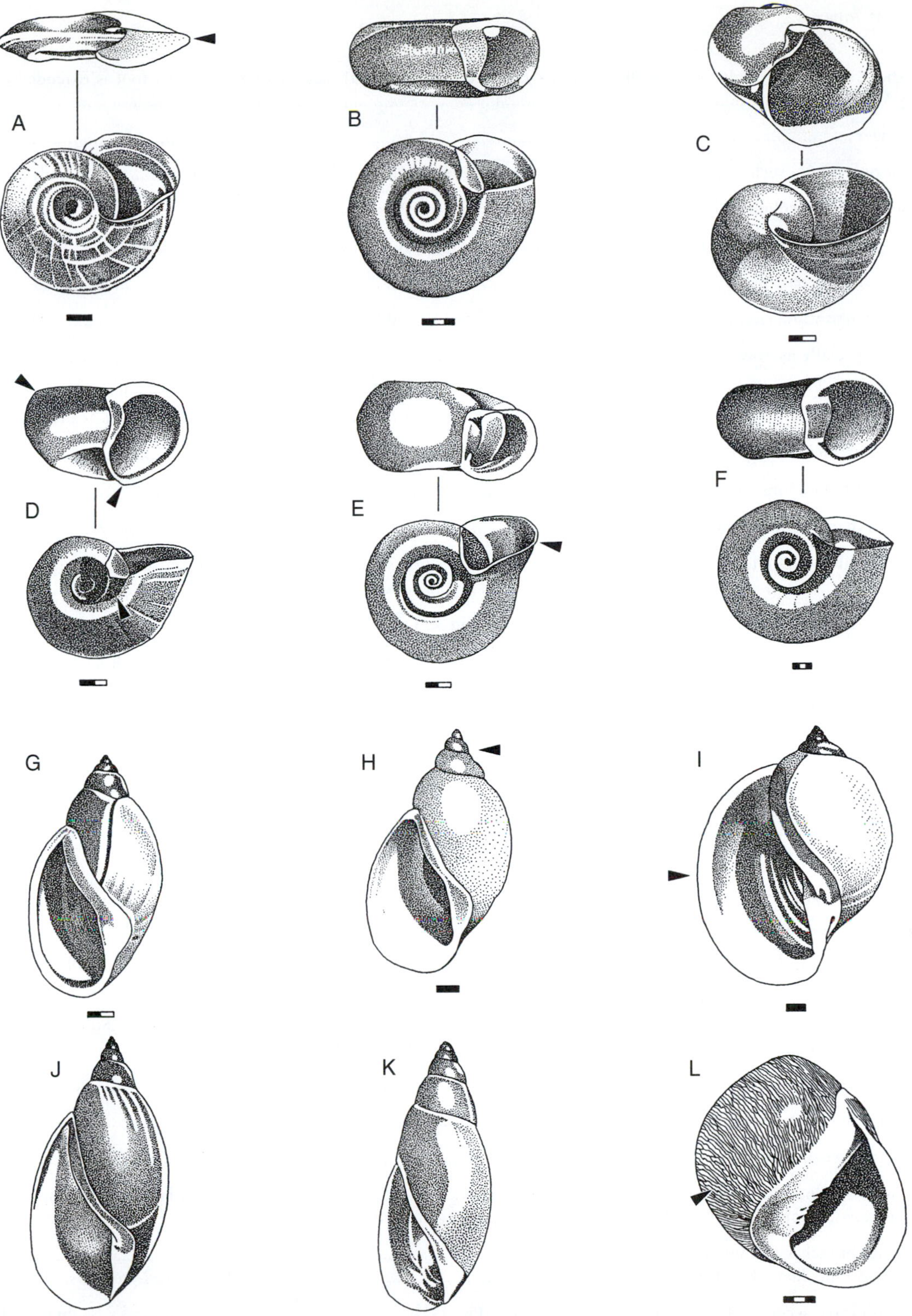

FIGURE 14 Representative planorbids, physids, and neritinids. (A) *Promenetus exacuous* (note flaired body whorl and carina); (B) *Biomphalaria glabrata*; (C) *Vorticifex* (=*Parapholyx*) *effusa*; (D) *Helisoma anceps* (note strong growth lines and carina); (E) *Planorbella* (=*Helisoma*) *companulata* (sometimes called *companulatum*, note flaired lip of aperture); (F) *Planorbella* (=*Helisoma*) *trivolvis* (this species reaches 20 mm in diameter and is extremely common); (G) *Physella* (=*Physa*) *gyrina* (very similar and possibly subspecies are *P. anatina* and *P. virgata* of the midwest and south, the only physid to reach 20 mm shell length); (H) *Physella* (=*Physa*) *integra* (note the more elevated spire than *P. gyrina*, rarely reaches 10 mm shell length); (I) *Physella* (=*Petrophysa*) *zionis* (note the large aperture); (J) *Stenophya*: (K) *Aplexa elongata* (note the spindle shape and lustrous almost-black shell); and (I) *Neritina reclivata* (note markings on shell and teeth on parietal wall). [A–I after Burch, 1982.]

25a. Spire short, operculum multispiral; shell sometimes carinate; external, featherlike gill visible when foot is extended (Fig. 15A, B) Family Valvatidae .... *Valvata*

25b. Spire longer 26

26a. Without operculum; shell relatively fragile and easily crushed; respiration by pouch in mantle cavity.......... Family Lymnaeidae 27

26b. Shell stronger; operculum and gills present.......... Subclass Prosobranchia 33

27(26a). With large, globose body whorl and extremely large aperture (Fig. 15C).......... *Radix*

27b. Body whorl more narrow 28

28a(27b). Shell very narrow and elongate; southern Ontario; north-central United States to Vermont (Fig. 15D) .... *Acella haldemani*

28b. Shell not especially narrow 29

29a(28b) Shell tear-shaped, transparent and fragile, with large, oval aperture and body whorl, small spire; amphibious. Eastern North America (Fig. 15E).......... *Pseudosuccinea*

29b. Shell not extremely thin or fragile 30

30a(29b) Adults more than 35 mm in length.......... 31

30b. Adults less than 35 mm in length 32

31a(30a). Shell with narrow, pointed spire, moderately fragile; in marshes or streams in north-central and eastern United States, Canada (Fig. 15F).......... *Lymnaea stagnalis*

31b. Shell globose, thick, with relatively wide spire. Great Lakes and St. Lawrence River drainages (Fig. 15G).......... *Bulimnea megasoma*

32a(30b). Adult shell more than 13 mm, with microscopic spiral striations or malleations; columella with a well-developed twist or plait. Widely distributed in ponds and marshes in northern states or in alpine regions of western states (Fig. 15H) .... *Stagnicola*

32b. Adult shell less than 13 mm, spiral sculpture usually absent; columella without a twist or plait; (Fig. 15I).......... *Fossaria*

33a(26b). Operculum multispiral or paucispiral (Fig. 2), outer margins not concentric.......... 34

33b. Operculum concentric (center may be paucispiral) 47

34a(33a). Adult less than 7 mm; males possess verge (penis) behind right tentacle.......... 35

34b. Adults more than 15 mm in length, males lack verge.......... 40

35a(34a). Spire high, head–foot region divided on each side by longitudinal groove; eyes in prominent swellings on the outer bases of the tentacles; amphibious or terrestrial; crawls with steplike movement; (Fig. 16A) .... Family Pomatiopsidae.......... *Pomatiopsis*

35b. Shell length variable; head–foot region not divided; eyes at same location but not on prominent swellings; totally aquatic.......... Family Hydrobiidae 36

36a(35b). Verge simple, no accessory lobes (Fig. 17A).......... Subfamily Lithoglyphinae .... *Lepyrium*, *Cochliopina*, *Fluminicola*, *Antrobia*, *Clappia*, *Somatogyrus*

36b. Verge with accessory lobes (Fig. 17 B-E).......... 37

37a(36b). Verge with two arms (mitten shaped, Fig. 17B, C).......... 38

37b. Verge with three or more divisions (Fig. 17D, E) 39

38a. Verges with one small and one large lobe with glandular areas .... Subfamily Nymphophilinae (Fig 17B) .....*Orygocerus, Birgella, Striobia, Marstonia, Rhapinema, Notogilla, Spilochlamys, Cincinnatia, Pyrgulopsis, Fontelicella, Natricola*

38b. Verges with branches of approximately similar diameter.......... Subfamily Amnicolinae (Fig. 17C) .... *Amnicola*, *Lyrogyrus*, *Hauffenia*, *Horatia*

39a. Verges with three accessory lobes branching off main lobe (Fig. 17D).......... Subfamily Hydrobiinae .... *Probythinella, Hoyia, Tryonia, Pyrgophorus, Littoridinops, Aphaostracon, Hyalopyrgus*

39b. Verge ending in three similar-sized lobes (Fig. 17E).......... Subfamily Fontigentinae .... *Fontigens*

40a(34b). Mantle edge papillate; parthenogenetic (males absent); females brood young in a pouch posterior to head; Introduced from Florida to Texas .... Family Thiaridae 41

40b. Mantle edge smooth; dioecious (males present), females oviparous, with egg-laying sinus on the right side of the foot.......... Family Pleuroceridae 42

41a(40a). Rounded whorls with spiral grooves and transverse raised lines (costae) (Fig. 16B) .... Introduced in Florida to Arizona .... *Melanoides*

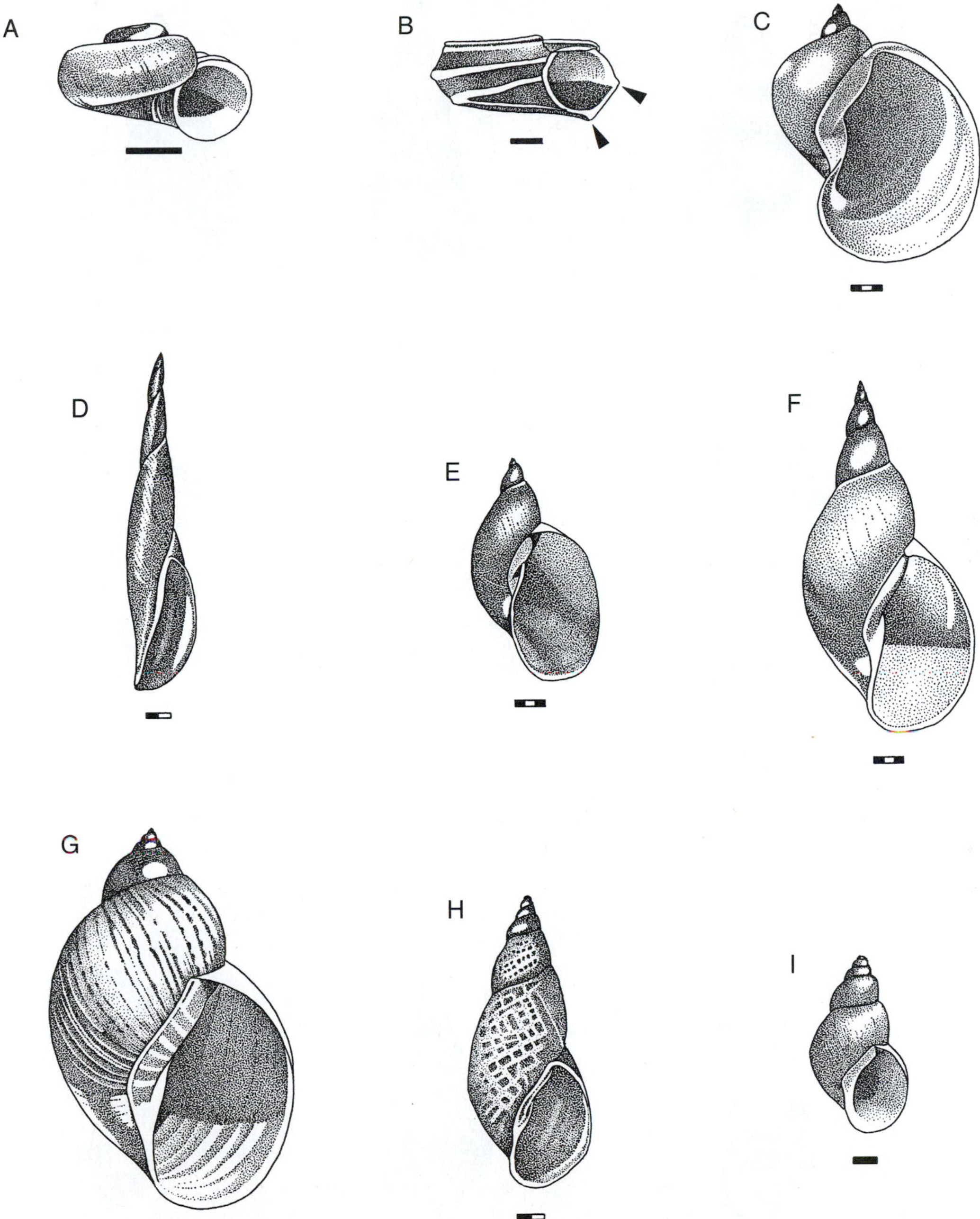

FIGURE 15 Representative valvatids and lymnaeids. (A) *Valvata sincera*; (B) *Valvata tricarinata* (note carina); (C) *Radix* (=*Lymnaea*) *auricularia* (note expanded body whorl); (D) *Acella haldemani* (note extremely narrow shell); (E) *Pseudosuccinea columella* (note thin, transparent shell and amphibious habit); (F) *Lymnaea stagnalis* (note large and fragile shell); (G) *Bulimnea megasoma* (note thick, large shell); (H) *Stagnicola* (=*Lymnaea*) *elodes* (note that malleations sometimes present, and this species is common only in ponds and marshes, rarely lakes); and (I) *Fossaria* (=*Lymnaea*) *humilis* (note small size and amphibious habit). [Figures A–I after Burch, 1982.]

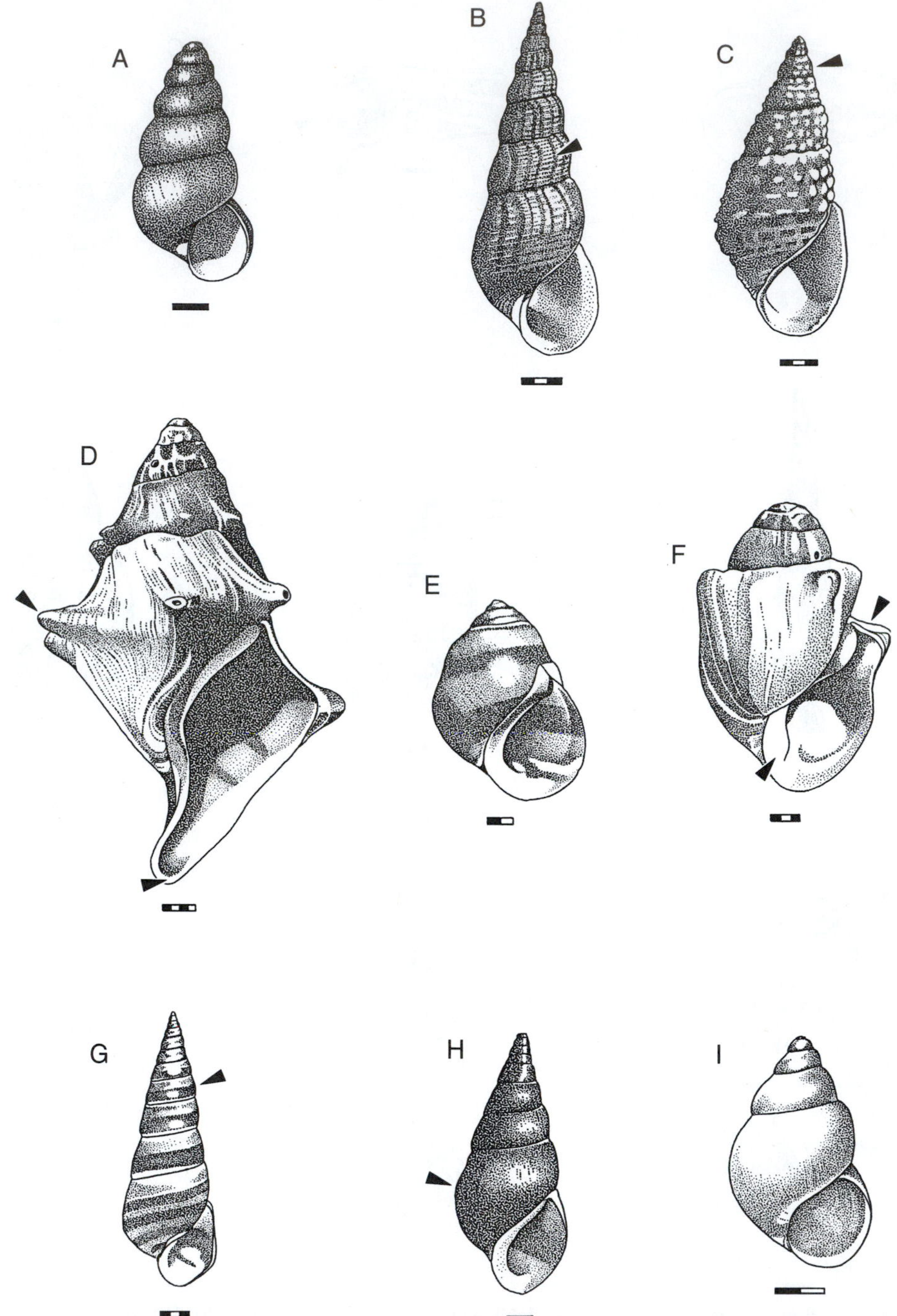

FIGURE 16 Representative potamiopsids, pleurocerids and bithyniids. (A) *Pomatiopsis lapidaria*; (B) *Melanoides tuberculata* (note costae and lirae); (C) *Thiara granifera* (note tubercules and flattened whorls near apex); (D) *Io fluvialis* (length of spines variable); (E) *Leptoxis praerosa*; (F) *Lithasia geniculata* (note thickened anterior aperture lip and shoulders on whorls); (G) *Pleurocera acuta* (note acute angle on anterior aperture and relatively flat whorls); (H) *Elimia* (= *Goniobasis*, = *Oxytrema*) *livescens* (note more rounded whorls, although there is tremendous variation in shell sculpture in this genus); and (I) *Bithynia tentaculata*. [A–I after Burch, 1982.]

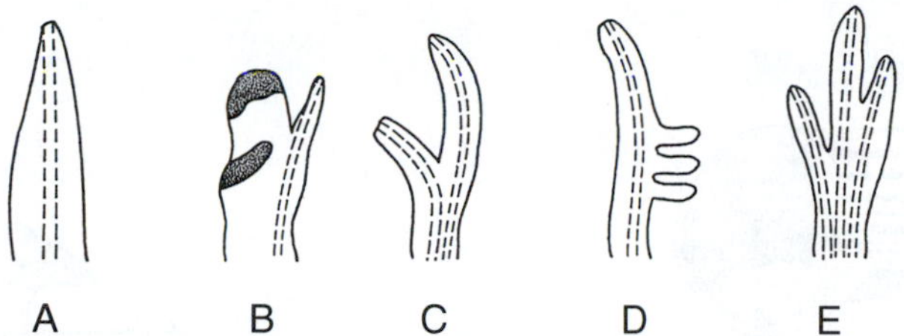

FIGURE 17 Structure of the verge (penis) in various subfamilies of hydrobiids. (A) Simple, single axis verge of subfamily Lithoglyphinae; (B) glandular crests on verge of subfamily Nymphophilinae; (C) two-ducted verge of subfamily Amnicolinae; (D) verge with accessory lobes in subfamily Hydrobiinae; and (E) threeducted verge of subfamily Fontigentinae [A–E after Burch, 1982].

41b. Whorls flat near spire; spiral rows of nodules on shell (Fig. 16C)................Introduced to Florida and Texas..........................*Thiara*

42a(40b). Large, often with elongated spines and a long, anterior canal on aperture (Fig. 16D); rivers in eastern Tennessee and western Virginia..*Io*

42b. Shell size and sculpture quite variable; aperture sometimes ends with a short canal 43

43a(42b) Shell medium-to-small, a subglobose, or globose cone (Fig. 16E); lateral radular teeth with broad, bluntly rounded median cusps; ....*Leptoxis*

43b. Shell usually a narrow cone; lateral radular teeth with narrow, triangular median cusps..44

44a(43b). Shell length medium, either cone-shaped or sub-globose, often with spines or nodules; inside margin of the aperture thickened, aperture meets body whorl nearly at right angle (Fig. 16F)..*Lithasia*

44b. Shell length and sculpture variable, base of aperture either rounded or with a short canal; inner margin of the aperture without thickening 45

45a. Anterior (basal tip of) aperture usually pointed; whorls flat-sided....................Mississippi, Great Lakes and Hudson River drainages (Fig. 16G)..*Pleurocera*

45b. Anterior aperture and whorls more rounded..46

46a(45b). In river drainages east of the Mississippi River (Fig. 16H) ....*Elimia*

46b. In river drainages west of the Mississippi ....*Juga*

47a(33b). Adults less than 15 mm; operculum calcareous; from Wisconsin to Pennsylvania and New York (Fig. 16I)...........Family Bithyniidae *Bithynia tentaculatum*

47b. Adults more than 20 mm (some up to 50 or 60 mm); operculum corneous..48

48a(47b). Globose, length up to 60 mm; penis on right side of mantle; calcareous eggs. Florida..........Family Pilidae (Ampullariidae) (Fig. 18B) ....*Pomacea*

48b. Subglobose, right tentacle thicker in males, used as a penis; females ovoviviparous...........Throughout the United States and Canada Family Viviparidae 49

49a(48b). Adults over 35 and up to 50 mm; shell relatively thin, whorls not shouldered (Fig. 18C); introduced throughout United States.......... ....*Cipangopaludina*

49b. Shell thick and less than 35 mm 50

50a(49b). Shell sometimes with one or two spiral rows of nodules; outer margin of aperture concave (when observed from the side); inside (columellar) margin of operculum folded inward; Alabama rivers (Fig. 18D)..*Tulotoma*

50b. Shell without nodules; aperture not as above 51

51a(50b). Operculum with spiral center and concentric edge; whorls with a median angle or low ridge (Fig. 18E)................................*Lioplax*

51b. Operculum entirely concentric; whorls without angles or ridges 52

51a(51b). Shell often with colored bands or hirsute in juveniles; aperture nearly circular (Fig. 18F)..*Viviparus*

51b. Colored bands absent; aperture longer than wide, forms shoulder at junction with body whorl (Fig. 18G)......................*Campeloma*

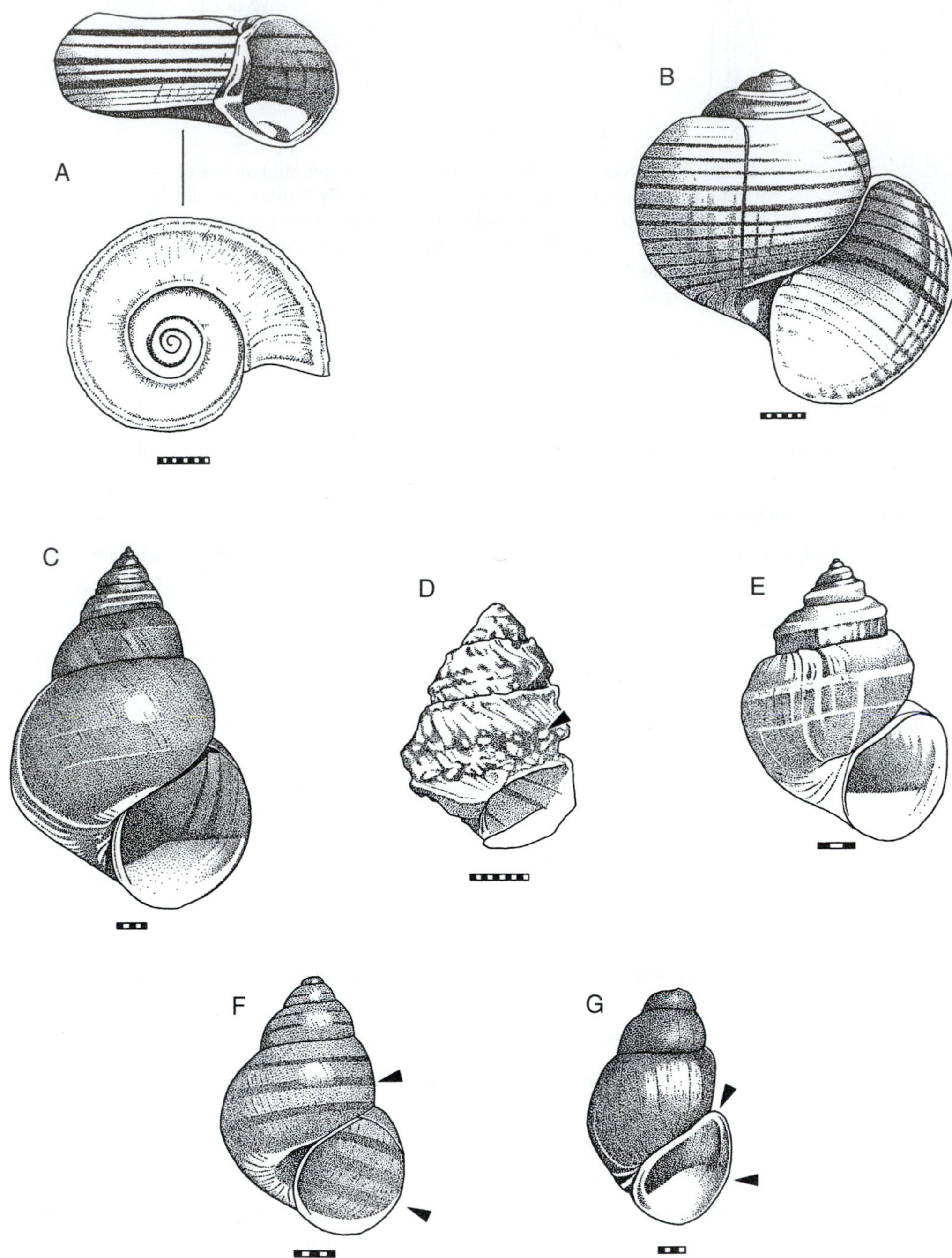

FIGURE 18 Representative ampullarids and viviparids. (A) *Marisa cornuarietis* (note large, planospiral shell, occurs in southern Florida and introduced to rivers in central Texas); (B) *Pomacea paludosa* (large, 'Apple' snail common in southern Florida); (C) *Cipangopaludina japonica* (a large, introduced, but now common species in North America); (D) *Tulotoma magnifica* (in some southeastern rivers, note tubercules which may be absent in some morphs or species); (E) *Lioplax subcarinata*; (F) *Viviparus georgianus* (common, note circular operculum, and bands which may disappear in adults); (G) *Campeloma decisum* (some times referred to as *decisa*; note operculum is longer than it is wide, and shouldered junction of aperture and body whorl). [Figures A–G after Burch, 1982.]

ACKNOWLEDGMENTS

I would like to thank Dr. Jeffery Tamplin for the considerable effort involved in drafting the excellent figures for the key, and Mr. Sean Keenan for his invaluable help in preparing the chapter.

LITERATURE CITED

Aldridge, D. W. 1982. Reproductive tactics in relation to life cycle bioenergetics in three natural populations of the freshwater snail, *Leptoxis carinata*. Ecology 63:196–208.

Aldridge, D. W. 1983. Physiological ecology of freshwater prosobranchs, *in*, Russell-Hunter, W. D. Ed., The Mollusca, Vol. 6, pp. 329–358, Academic Press, Orlando, FL.

Alexander, J. E., Jr., Covich, A. P. 1991a. Predator avoidance by the freshwater snail *Physella virgata* in response to the crayfish *Procambarus simulans*. Oecologia 87:435–442.

Alexander, J. E., Jr., Covich, A. P. 1991b. Predation risk and avoidance behavior by two freshwater snails. Biological Bulletin 180:387–393.

Anderson, R. M., May, R. M. 1979. Prevalence of schistosome infections within molluscan populations: observed patterns and theoretical predictions. Parasitology 79:63–94.

Bandel, K. 1997. Higher classification and pattern of evolution of the gastropoda. Cour. Forsch. Inst. Senckenberg 201:57–81.

Barnes, R. D. 1987. Invertebrate Zoology, 5th ed., Saunders, Philadelphia, PA.

Barnese, L. E., Lowe, R. L., Hunter, R. D. 1990. Comparative grazing efficiency of pulmonate and prosobranch snails. Journal of the North American Benthological Society 9:35–44.

Baudoin, M. 1975. Host castration as a parasite strategy. Evolution 29:335–352.

Bieler, R. 1992. Gastropod phylogeny and systematics. Annual Review of Ecology and Systematics 23:311–338.

Boag, D. A. 1981. Differential depth distribution among freshwater pulmonate snails subjected to cold temperatures. Canadian Journal of Zoology 9:733–737.

Boag, D. A. 1986. Dispersal in pond snails: potential role of waterfowl. Canadian Journal of Zoology 64:904–909.

Boag, D. A., Pearlstone, P. S. M. 1979. On the life cycle of *Lymnaea stagnalis* (Pulmonate:Gastropoda) in Southwestern Alberta. Canadian Journal of Zoology 52:353–362.

Boerger, H. 1975. Movement and burrowing of *Helisoma trivolvis* (Say) (Gastropoda: Planorbidae) in a small pond. Canadian Journal of Zoology 53:456–464.

Bogan, A.E. 1999. North American freshwater gastropod diversity and conservation. *in*: The First Symposium of the Freshwater Mollusk Conservation Society. 17–19, March, 1999 Chattanooga, TN (Abstract).

Boss, K. J. 1974. Oblomovism in the Mollusca. Transactions of the American Microscopical Society 93:460–481.

Bousefield, J. D. 1979. Plant extracts and chemically triggered positive rheotaxis in *Biomphalaria grabrata* (Say), snail intermediate host of *Schistosoma mansoni*. Journal of Applied Ecology 16:681–690.

Bovbjerg, R. V. 1952. Ecological aspects of dispersal of the snail *Campeloma decisum*. Ecology 33:169–176.

Bovbjerg, R. V. 1968. Responses to food in lymnaeid snails. Physiological Zoology 41:412–423.

Boycott, A. E. 1936. The habitats of freshwater Mollusca in Britain. Journal of Animal Ecology 5:116–186.

Brendelberger, H. 1997. Determination of digestive enzyme kinetics: a new method to define trophic niches in freshwater snails. Oecologia 109:34–40.

Brendelberger, H., Jurgens, S. 1993. Suspension feeding in *Bithynia tentaculata* (Prosobranchia: Bithyniidae) as affected by body size, food and temperature. Oecologia 94:36–42.

Bronmark, C. 1985a. Freshwater snail diversity: effects of pond area, habitat heterogeneity and isolation. Oecologia 67:127–131.

Bronmark, C. 1985b. Interactions between macrophytes, epiphytes, and herbivores: an experimental approach. Oikos 45:26–30.

Bronmark, C. 1989. Interactions between epiphytes, macrophytes and freshwater snails: a review. Journal of Molluscan Studies 55:299–311.

Bronmark, C. 1994. Effects of tench and perch on interactions in a freshwater benthic food chain. Ecology 75:1818–1824.

Bronmark, C., Malmqvist, B. 1986. Interactions between the leech *Glossiphonia complanata* and its gastropod prey. Oecologia 69:268–276.

Bronmark, C., Klosiewski, S. P., Stein, R. A. 1992. Indirect effects of predation in a freshwater, benthic food chain. Ecology 73: 1662–1674.

Bronmark, C., Weisner, S.E.B. 1996. Decoupling of cascading trophic interactions in a freshwater, benthic food chain. Oecologia 108:534–541.

Bronmark, C., Vermaat, J. E. 1998. Chapter 3. Complex-fish-snail-epiphyton interactions and their effects on submerged freshwater macrophytes *in*, Jepperson, E., Sondergaard, M., Cristofferson, K. Eds., The structuring role of submerged macrophytes in lakes, Springer, New York, pp. 47–68.

Brown, D. S. 1978. Pulmonate molluscs as intermediate hosts for digenetic trematodes, *in*, Fretter, V., Peake, J. Eds. Pulmonates, Vol. 2A, Systematics, evolution, and ecology. Academic Press, New York, pp. 287–333.

Brown, K. M. 1979. The adaptive demography of four freshwater pulmonate snails. Evolution 33:417–432.

Brown, K. M. 1982. Resource overlap and competition in pond snails: an experimental analysis. Ecology 63:412–422.

Brown, K. M. 1983. Do life history tactics exist at the intraspecific level? Data from freshwater snails. American Naturalist 121: 871–879.

Brown, K. M. 1985. Intraspecific life history variation in a pond snail: The roles of population divergence and phenotypic plasticity. Evolution 39: 387–395.

Brown, K.M. 1997. Temporal and spatial patterns of abundance in the gastropod assemblage of a macrophyte bed. American Malacological Bulletin 14:27–33.

Brown, K. M. 1998. The role of shell strength in the foraging of crayfish for gastropod prey. Freshwater Biology 40: 255–260.

Brown, K. M., DeVries, D. R. 1985. Predation and the distribution and abundance of a pulmonate pond snail. Oecologia 66: 93–99.

Brown, K. M., Richardson, T. D. 1988. Genetic polymorphism in gastropods: a comparison of methods and habitat scales. American Malacological Bulletin 6:9–17.

Brown, K. M., Strouse, B. H. 1988. Relative vulnerability of six freshwater gastropods to the leech *Nephelopsis obscura* (Verrill). Freshwater Biology 19:157–166.

Brown, K. M., Leathers, B. K., Minchella, D. J. 1988. Trematode prevalence and the population dynamics of freshwater pond snails. American Midland Naturalist 120:289–301.

Brown, K. M., Varza, D. E., Richardson, T. D. 1989. Life cycles and population dynamics of two subtropical viviparid snails (Prosobranchia: Viviparidae). Journal of the North American Benthological Society 8:222–228.

Brown, K.M., Lodge, D.M. 1993. The importance of specifying null models: are invertebrates really more abundant in vegetated habitats? Limnology and Oceanography 38: 217–275.

Brown, K. M., Alexander, J. E., Thorp, J. H. 1998. Differences in the ecology and distribution of lotic pulmonate and prosobranch gastropods. American Malacological Bulletin 14:91–101.

Browne, R. A. 1978. Growth, mortality, fecundity, and productivity of four lake populations of the prosobranch snail, *Viviparus georgianus*. Ecology 59: 742–750.

Browne, R. A. 1981. Lakes as islands: biogeographic distribution, turnover rates, and species composition in the lakes of central New York. Journal of Biogeography 8: 75–83.

Browne, R. A., Russell-Hunter, W. D. 1978. Reproductive effort in molluscs. Oecologia 27:23–27.

Burch, J. B. 1989. North American Freshwater Snails. Malacological Publications, Hamburg, MI, 365 pp.

Burky, A. J. 1971. Biomass turnover, respiration, and interpopulation variation in the stream limpet *Ferrissia rivularis*. Ecological Monographs 41: 235–251.

Burky, A. J. 1983. Physiological ecology of freshwater bivalves, *in*, Russell-Hunter, W. D., Ed., The Mollusca, Vol. 6, Ecology, pp. 281–327, Academic Press, Orlando, FL.

Calow, P. 1970. Studies on the natural diet of *Lymnaea peregra obtusa* (Kobelt) and its possible ecological implications. Proceedings of the Malacological Society of London 39:203–215.

Calow, P. 1973a. Field observations and laboratory experiments on the general food requirements of two species of freshwater snail, *Planorbis contortus* Linn. and *Ancylus fluviatilis*. Proceedings of the Malacological Society of London 40:483–489.

Calow, P. 1973b. The food of *Ancylus fluviatilis* (Mull.), a littoral, stone-dwelling herbivore. Oecologia 13:113–133.

Calow, P. 1974a. Evidence for bacterial feeding in *Planorbis contortus* Linn. (Gastropoda: Pulmonata) Proceedings of the Malacological Society of London 41:145–156.

Calow, P. 1974b. Some observations on the dispersion patterns of two species populations of littoral, stone-dwelling gastropods (Pulmonata). Freshwater Biology 4:557–576.

Calow, P. 1975. The respiratory strategies of two species of freshwater gastropods (*Ancylus Fluviatilis* Mull. and *Planorbis contortus* Linn.) in relation to temperature, oxygen concentration, body size, and season. Physiological Zoology 48:114–129.

Calow, P. 1978. The evolution of life-cycle strategies in fresh-water gastropods. Malacologia 17:351–364.

Calow, P. 1981. Adaptational aspects of growth and reproduction in *Lymnaea peregra* from exposed and sheltered aquatic habitats. Malacologia 21:5–13.

Calow, P. 1983. Life-cycle patterns and evolution, *in*, Russell-Hunter, W. D. Ed., The Mollusca, Vol. 6, Ecology., pp. 649–680. Academic Press, Orlando, FL.

Calow, P., Calow, L. J. 1975. Cellulase activity and niche separation in freshwater gastropods. Nature 255:478–480.

Chamberlain, N. A. 1958. Life history studies of *Campeloma decisum*. Nautilus 72: 22–29.

Cheatum, E. P. 1934. Limnological investigations on respiration, annual migratory cycle, and other related phenomena in freshwater pulmonate snails. Transactions of the American-Microscopical Society 53:348–407.

Clampitt, P. T. 1973. Substratum as a factor in the distribution of pulmonate snails in Douglas Lake, Michigan. Malacologia 12: 379–399.

Clampitt, P. T. 1974. Seasonal migratory cycle and related movements of the fresh-water pulmonate snail, *Physa integra*. American Midland Naturalist 92:275–300.

Clarke, P. H. 1981. The freshwater molluscs of Canada. National Museum of Natural Sciences, Ottawa, Canada.

Covich, A. P. 1977. How do crayfish respond to plants and Mollusca as alternate food resources? Freshwater Crayfish 3:165–169.

Covich, A.P., Crowl, T.A., Alexander, J.E., Jr., Vaughn, C.C. 1994. Predator-avoidance responses in freshwater decapod-gastropod interactions mediated by chemical stimuli. Journal of the North American Benthological Society 13:283–290.

Cox, L.R. 1960. Gastropoda. General characteristics of Gastropoda, *in*:, Treatise on Invertebrate Paleontology, Part I. Mollusca 1, Moore, R.C. Ed., pp. 84–169, Lawrence: Univ. of Kansas.

Croll, R. P. 1983. Chemoreception. Biological Reviews 58:293–319.

Crowl, T. A. 1990. Life-history strategies of a freshwater snail in response to stream permanence and predation: balancing conflicting demands. Oecologia 84:238–243.

Crowl, T. A., Covich, A. P. 1990. Predator-induced life-history shifts in a freshwater snail. Science 247:949–951.

Cuker, B. E. 1983. Competition and coexistence among the grazing snail *Lymnaea*, Chironomidae, and microcrustacea in an arctic epilithic lacustrine community. Ecology 64:10–15.

Daldorph, P. W. G., Thomas, J. D. 1995. Factors influencing the stability of nutrient-enriched freshwater macrophyte communities: the role of stickelbacks *Pungitus pungitius* and freshwater snails. Freshwater Biology 33:271–289.

Davis, G. M. 1979. The origin and evolution of the Pomatiopsidae, with emphasis on the Mekong River hydrobiid gastropods. Monographs of the Academy of Natural Sciences, Philadelphia, Number 20.

Davis, G. M. 1982. Historical and ecological factors in the evolution, adaptive radiation, and biogeography of freshwater molluscs. American Zoologist 22:375–395.

Dazo, B. Z. 1965. The morphology and natural history of *Pleurocera acuta* and *Goniobasis livescens*. Malacologia 2:1–80.

Diamond, J. M. 1975. Assembly of species communities, *in*: Cody, M., Diamond, J. M. Eds., Ecology and Evolution of Communities. pp. 342–344.

Dillon, R. T. 1981. Patterns in the morphology and distribution of gastropods in Oneida Lake, New York, detected using computer-generated null hypotheses. American Naturalist 118:83–101.

Dillon, R. T. 1987. A new Monte Carlo method for assessing taxonomic similarity within faunal samples: reanalysis of the gastropod community of Oneida Lake, New York. American Malacological Bulletin 5:101–104.

Dillon, R.T., Jr., Ahlstedt, S.A. 1997. Verification of the specific status of the endangered Anthony's river snail, *Athearnia anthonyi*, using allozyme electrophoresis. Nautilus 110:97–101.

Dillon, R.T., Jr., Lydeard, C. 1998. Divergence among Mobile Basin populations of the pleurocerid snail genus, *Leptoxis*, estimated by allozyme electrophoresis. Malacologia 39:113–121.

Doremus, C. M., Harman, W. N. 1977. The effects of grazing by physid and planorbid freshwater snails on periphyton. Nautilus 91:92–96.

Duch, T. M. 1976. Aspects of the feeding habits of *Viviparus georgianus*. Nautilus 90:7–10.

Duncan, C. J. 1975. Reproduction, *in*: Fretter and Peake, Eds., Pulmonates, Vol. 1, Functional anatomy and physiology. Academic Press, Orlando, FL, pp. 309–366.

Dybdahl, M. F., Lively, C. M. 1996. The geography of coevolution: comparative population structures for a snail and its trematode parasite. Evolution 50:2264–2275.

Eckblad, J. W. 1973. Experimental predation studies of malacophagous larvae of *Sepedon fuscipennis* (Diptera: Sciomyzidae) and aquatic snails. Experimental Parasitology 33:331–342.

Eisenberg, R. M. 1966. The regulation of density in a natural population of the pond snail, *Lymnaea elodes*. Ecology 47:889–906.

Eisenberg, R. M. 1970. The role of food in the regulation of the pond snail, *Lymnaea elodes*. Ecology 51:680–684.

Eversole, A. G. 1978. Life cycles, growth and population bioenergetics in the snail *Helisoma trivolvis* (Say). Journal of Molluscan Studies 44:209–222.

Fretter, V.,Graham, A. 1962. British prosobranch molluscs. Ray Society, London.

Fretter, V., Peake, J. 1975. Pulmonates, Vol. 1, Functional anatomy and physiology. Academic Press, Orlando, FL.

Fretter, V., Peake, J. 1978. Pulmonates, Vol. 2A, Systematics, Evolution, and Ecology. Academic Press, Orlando, FL.

Gerking, S. D. 1957. A method of sampling the littoral macrofauna and its application. Ecology 38:219–225.

Goodrich, B. 1945. *Goniobasis livescens* of Michigan. Miscellaneous Publications of the Museum of Zoology of the University of Michigan 64:1–26.

Harman, W. N. 1968. Replacement of pleurocerids by *Bithynia* in polluted waters of central New York. Nautilus 81:77–83.

Harman, W. N. 1972. Benthic substrates: their effect on fresh-water mollusca. Ecology 53:271–277.

Harman, W. N., Forney, J. L. 1970. Fifty years of change in the molluscan fauna of Oneida Lake, New York. Limnology and Oceanography 15:454–460.

Harman, W. N., Berg, C. O. 1971. The freshwater snails of central New York. Search (Agriculture) 1:1–68.

Haynes, A., Taylor, B. J. R. 1984. Food finding and food preference in *Potamopyrgus jenkensi* (E. A. Smith)(Gastropoda: Prosobranchia) Archives fur Hydrobiologie 100:479–491.

Hershey, A. E. 1990. Snail populations in arctic lakes: competition mediated by predation? Oecologia 82:26–32.

Hershler, R. 1998. A systematic review of the hydrobiid snails (Gastropoda:Rissooidea) of the Great Basin, Western United States. Part I. Genus *Pyrgulopsis*. Veliger 41:1–132.

Hershler, R., Frest, T.J. 1996. A review of the North American freshwater snail genus *Fluminicola* (Hydrobiidae). Smithsonian Contributions to Zoology. 583:1–41.

Hershler, R., Pierson, J.M., Krotzer, R.S. 1990. Rediscovery of *Tulotoma magnifica* (Conrad) (Gastropoda:Viviparidae). Proceedings of the Biological Society of Washington 103:815–824.

Hill, W. R., Harvey, B. C. 1990. Periphyton responses to higher trophic levels and light in a shaded stream. Canadian Journal of Fisheries and Aquatic Sciences 47:2307–2314.

Hill, W. R., Boston, H. L., Steinman, A. D. 1992. Grazers and nutrients simultaneously limit lotic primary productivity. Canadian Journal of Fisheries and Aquatic Sciences 49:504–512.

Holmes, J. C. 1983. Evolutionary relationships between parasitic helminths and their hosts. *in*: Futuyma, D. J., Slatkin, M. Eds., *Coevolution*, pp. 161–185. Sinnauer, Sunderland, MA.

Holznagel, W.E. 1998. Research note: a non-destructive method for cleaning gastropod radulae from frozen, alcohol-fixed, or dried material. American Malacological Bulletin 14:181–183.

Hunter, R. D. 1975. Growth, fecundity, and bioenergetics in three populations of *Lymnaea palustris* in upstate New York. Ecology 56:50–63.

Hunter, R. D. 1980. Effects of grazing on the quantity and quality of freshwater aufwuchs. Hydrobiologia 69:251–259.

Hunter, R. D., Russell-Hunter, W. D. 1983. Bioenergetic and community changes in intertidal aufwuchs grazed by *Littorina littorea*. Ecology 64:761–769.

Huryn, A. D., Koebel, J. W., Benke, A. C. 1994. Life history and longevity of the pleurocerid snail *Elimia*: a comparative study of eight populations. Journal of the North American Benthological Society 13:540–556.

Huryn, A. D., Benke, A. C., Ward, G. M. 1995. Direct and indirect effects of geology on the distribution, biomass, and production of the freshwater snail *Elimia*. Journal of the North American Benthological Society 14:519–534.

Hyman, L. H. 1967. The invertebrates, Vol. VI, Mollusca I. McGraw-Hill, New York.

Jacoby, J. M. 1985. Grazing effects on periphyton by *Theodoxus fluviatilis* (Gastropoda) in a lowland stream. Journal of Freshwater Biology. 3:265–274.

Jarne, P., Finet, L., Delay, B., Thaler, L. 1991. Self-fertilization versus cross-fertilization in the hemaphroditic freshwater snail *Bulinus globosus*. Evolution 45:1136–1146.

Jenkinson, J. J., Todd, R. M. 1998. Management of native molluscan resources, *in*: Benz, G. W., Collins, D. E. Eds., Aquatic fauna in peril: the southeastern perspective. Southeast Aquatic Research Institute, Special Publication 1.

Johnson, S. G. 1992. Parasite-induced parthenogenesis in a freshwater snail: stable, persistent patterns of parasitism. Oceologia 89:533–541.

Johnson, P. D., Brown, K. M., 1997. The role of current and light in explaining the habitat distribution of the lotic snail *Elimia semicarinata*. Journal of the North American Benthological Society. 15:344–369.

Jokinen, E. H. 1978. The aestivation pattern of a population of *Lymnaea elodes*. American Midland Naturalist 100:43–53.

Jokinen, E. H. 1982. *Cipangopaludina chinensis* (Gastropods:Viviparidae) in North America, review and update. Nautilus 96:89–95.

Jokinen, E. H. 1983. The freshwater snails of Connecticut. State Geological and Natural History Survey of Connecticut, Department of Environmental Protection, Publication 109.

Jokinen, E. H. 1985. Comparative life history patterns within a littoral zone snail community. Verh. Internat. Verein. Limnol. 22:3292–3299.

Jokinen, E. H. 1987. Structure of freshwater snail communities: species–area relationships and incidence categories. American Malacological Bulletin 5:9–19.

Jokinen, E. H. 1991. The malacofauna of the acid and non-acid lakes and rivers of the Adirondack mountains and surrounding lowlands, New York state, U.S.A. Verh. Internat. Verein. Limnol. 24:2940–2946.

Jokinen, E. H. 1992. The freshwater snails (Mollusca: Gastropoda) of New York state. New York State Museum Bulletin 482, 112 pp.

Jokinen, E. H., Guerette, J., Kortmann, R. W. 1982. The natural history of an ovoviviparous snail, *Viviparus georgianus* (Lea), in a soft-water eutrophic lake. Freshwater Invertebrate Biology 1:2–17.

Jung, Y. 1992. Phylogenetic relationships of some planorbid genera (Gastropoda: Lymnophila). Malacological Review 25:73–102.

Kairesalo, T., Koskimies, I. 1987. Grazing by oligochaetes and snails on epiphytes. Freshwater Biology 17:317–324.

Kesler, D. H. 1981. Periphyton grazing by *Amnicola limosa*: an enclosure-exclosure experiment. Journal of Freshwater Ecology 1:51–59.

Kesler, D. H. 1983. Cellulase activity in gastropods: should it be used in niche separation? Freshwater Invertebrate Biology 2:173–179.

Kesler, D. H., Jokinen, E. H., Munns, W. R., Jr. 1986. Trophic preferences and feeding morphology of two pulmonate snail species from a small New England pond, U.S.A.Canadian Journal of Zoology 64:2570–2575.

Kesler, D. H., Munns, W. R., Jr. 1989. Predation by *Belostoma flumineum* (Hemiptera): an important cause of mortality in freshwater snails. Journal of the North American Benthological Society 8:342–350.

Kosuge, S. 1966. The family Triphoridae and its systematic position. Malacologia 4:297–324.

Lamberti, G. A., Ashkenas, L. R., Gregory, S. V., 1987. Effects of three herbivores on periphyton communities in laboratory streams. Journal of North American Benthological Society 6:92–104.

Lassen, H. H. 1975. The diversity of freshwater snails in view of the equilibrium theory of island biogeography. Oecologia 19:1–8.

Levri, E. P., Lively, C. M. 1996. The effects of size, reproductive condition and parasitism on foraging behavior in a freshwater snail, *Potamopyrgus antipodarum*. Animal Behavior 51:891–901.

Lively, C. M. 1987. Evidence from a New Zealand snail for the maintenance of sex by parasitism. Nature 328:519–521.

Lively, C. M. 1989. Adaptation by a parasitic trematode to local populations of its host. Evolution 43:1663–1671.

Lively, C. M., Jokela, J. 1995. Spatial variation in infection by digenetic trematodes in a population of freshwater snails (*Potamopyrgus antipodarum*). Oecologia 103:509–517.

Lodge, D. M. 1985. Macrophyte–gastropod associations: observations and experiments on macrophyte choice by gastropods. Freshwater Biology 15:695–708.

Lodge, D. M. 1986. Selective grazing on periphyton: a determinant of freshwater gastropod microdistributions. Freshwater Biology 16:831–841.

Lodge, D. M., Kelly, P. 1985. Habitat disturbance and the stability of freshwater gastropod populations. Oecologia 68:111–117.

Lodge, D. M., Brown, K. M., Klosiewski, S. P., Stein, R. A., Covich, A. P., Leathers, B. K., Bronmark, C. 1987. Distribution of freshwater snails: spatial scale and the relative importance of physicochemical and biotic factors. American Malacological Bulletin 5:73–84.

Lodge, D. M., Kershner, M. W., Aloi, J. E., Covich, A. P. 1994. Effects of an omnivorous crayfish (*Orconectes rusticus*) on a freshwater littoral food web. Ecology 75:1265–1281.

Lodge, D. M., Stein, R. A., Brown, K. M., Covich, A. P., Bronmark, C., Garvey, J. E., Klosiewski, S. P. 1998. Predicting impact of freshwater exotic species on native biodiversity: challenges in spatial scaling. Australian Journal of Ecology 23:53–67.

Lowe, R. L., Hunter, R. D. 1988. Effects of grazing by *Physa integra* on periphyton community structure. Journal of the North American Benthological Society 7:29–36.

Lydeard, C., Mayden, R. L. 1995. A diverse and endangered aquatic ecosystem of the Southeast United States. Conservation Biology 9:800–805.

Lydeard, C., Holznagel, W. E., Garner, J., Hartfield, P., Pierson, M. 1997. A molecular phylogeny of Mobile River drainage basin pleurocerid snails (Caenogastropoda: Cerithioidea) Molecular Phylogenetics and Evolution 7:117–128.

Lydeard, C., Yoder, J.H., Holznagel, W.E., Thompson, F.G., Hartfield, P. 1998. Phylogenetic utility of the 5'-half of mitochondrial 16S rDNA gene sequences for inferring relationships of *Elimia* (Cerithioidea:Pleuroceridae) Malacologia 39:183–193.

Macan, T. T. 1950. Ecology of freshwater Mollusca in the English Lake District. Journal of Animal Ecology 19:124–146.

MacArthur, R. H., Wilson, E. O., 1967. The theory of island biogeography. Princeton University Press, Princeton, NJ. USA.

Machin, J. 1975. Water relationships, *in*: Fretter, V., Peake, J. Eds., Pulmonates, Vol. 1, Functional anatomy and physiology, pp. 105–164. Academic Press, Orlando, FL.

Madsen, H. 1992. Food selection by freshwater snails in the Gezira irrigation canals, Sudan. Hydrobiologia 228:203–217.

Marks, J. C., Lowe, R. L. 1989. The independent and interactive effects of snail grazing and nutrient enrichment on structuring periphyton communities. Hydrobiologia 185:9–17.

Martin, A. W. 1983. Excretion., *in*: Wilbur, K. M. Ed., The molluscs, Vol. 5, Physiology, pp. 353–407. Academic Press, Orlando, FL.

Martin, T. H., Crowder, L. B., Dumas, C. F., Burkholder, J. M. 1992. Indirect effects of fish on macrophytes in Bayes Mountain Lake: Evidence for a littoral trophic cascade. Oecologia 89:476–481.

McCormick, P. V., Stevenson, R. J. 1989. Effects of snail grazing on benthic algal community structure in different nutrient environments. Journal of the North American Benthological Society 8:162–172.

McMahon, R. F. 1975. Growth, reproduction, and bioenergetic variation in three natural populations of a freshwater limpet *Laevapex fuscus*. Proceedings of the Malacological Society of London 41:331–352.

McMahon, R. F. 1983. Physiological ecology of freshwater pulmonates, *in*: Russell-Hunter, W. D. Ed. The mollusca, Vol. 6, Ecology, pp. 359–430, Academic Press, Orlando, FL.

McMahon, R. F., Hunter, R. D., Russell-Hunter, W. D. 1974. Variation in aufwuchs at six freshwater habitats in terms of carbon biomass and of carbon: nitrogen ratio. Hydrobiologia 45:391–404.

Merrick, G. W., Hershey, A. E., McDonald, M. E. 1991. Lake trout (*Salvelinus namaycush*) control of snail density and size distribution in an arctic lake. Canadian Journal of Fisheries and Aquatic Sciences. 48:498–502.

Minchella, D. J., Loverde, P. T. 1981. A cost of increased early reproductive effort in the snail *Biomphalaria glabrata*. American Naturalist 118:876–881.

Minchella, D. J., Leathers, B. K., Brown, K. M., McNair, J. K. 1985. Host and parasite counter-adaptations: An example from a freshwater snail. American Naturalist 126:843–854.

Morton, J. E. 1955. The evolution of the Ellobiidae with a discussion on the origin of the pulmonata. Proceedings of the Zoological Society of London 125:127–168.

Mulholland, P. J., Newbold, J. D., Elwood, J. W., Horn, C. L. 1983. The effect of grazing intensity on phosphorus spiralling in autotrophic streams. Oecologia 58:358–366.

Mulholland, P. J., Steinman, A. D., Palumbo, A. V., Elwood, J. W., Kirschtel, D. B. 1991. Role of nutrient cycling and herbivory in regulating periphyton communities in laboratory streams. Ecology 72:966–982.

Neves, R. J., Bogan, A. E., Williams, J. D., Ahlstedt, S. A., Hartfield, P. W. 1998. Status of aquatic mollusks in the southeastern United States: A downward spiral of diversity, *in*: Benz, G.W., Collins, D. E. Eds. Aquatic fauna in peril: The southeastern perspective. Southeast Aquatic Research Institute, Special Publication 1.

Nystrom, P., Perez, J. R. 1998. Crayfish predation and the common pond snail (*Lymnaea stagnalis*): the effect of habitat complexity and snail size on foraging efficiency. Hydrobiologia 368:201–208.

Okland, J. 1983. Factors regulating the distribution of freshwater snails (Gastropoda) in Norway. Malacologia 24:277–288.

Osenberg, C. W. 1989. Resource limitation, competition and the influence of life history in a freshwater snail community. Oecologia 79:512–519.

Pace, G. L., Szuch, E. J. 1985. An exceptional stream population of the banded apple snail, *Viviparus georgianus*, in Michigan. Nautilus 99:48–53.

Pechenik, J. A. 1985. Biology of the invertebrates. Prindle, Weber & Schmidt, Boston.

Pip, E. 1986. The ecology of freshwater gastropods in the central Canadian region. Nautilus 100:56–66.

Ponder, W.F., Lindberg, D.R. 1997. Towards a phylogeny of gastropod molluscs: an analysis using morphological characters. Zoological Journal of Linnaean Society 119:83–265.

Reavell, P. E. 1980. A study of the diets of some British freshwater gastropods. Journal of Conchology 30:253–271.

Remigio, E. A., Blair, D. 1997. Molecular systematics of the freshwater snail family Lymnaeidae (Pulmonata: Basommatophora) utilizing mitochondrial ribosomal DNA sequences. Journal of Molluscan Studies 63:173–185.

Richardson, T. D., Scheiring, J. F., Brown, K. M. 1988. Secondary production of two lotic snails (Pleuroceridae: *Elimia*). Journal of the North American Benthological Society 7: 234–245.

Richardson, T. D., Brown, K. M. 1989. Secondary production of two subtropical viviparid prosobranchs. Journal of the North American Benthological Society 8:229–236.

Rosemond, A. D., Mulholland, P. J., Elwood, J. W. 1993. Top–down and bottom–up control of stream periphyton: effects of nutrients and herbivores. Ecology 74:1264–1280.

Russell-Hunter, W. D. 1978. Ecology of freshwater pulmonates, *in*: Fretter, V., Peake, J., Eds., The pulmonates, Vol. 2A, Systematics, evolution and ecology , pp. 335–383, Academic Press, Orlando, FL.

Russell-Hunter, W. D. 1983. Ecology of freshwater pulmonates, *in*: Russell-Hunter, W. D. Ed., The Mollusca, Vol. 6, Ecology. pp. 335–383, Academic Press, Orlando, FL.

Russell-Hunter, W. D., Buckley, D. E. 1983. Actuarial bioenergetics of nonmarine molluscan productivity. *in*: Russell-Hunter, W. D., Ed., The Mollusca, Vol. 6, Ecology. pp. 463–503, Academic Press, Orlando, FL.

Saffran, K. A., Barton, D. R. 1993. Trophic ecology of *Orconectes propinquus* (Girard) in Georgian Bay (Ontario, Canada). Freshwater Crayfish 9:350–358.

Sambrook, J., Fritch, E.F., Maniatis, T. 1989. Molecular cloning, a laboratory manual, 2nd ed., Cold Spring Habor Laboratory Press, Cold Spring Harbor, NY, (3 vols).

Sapp, K. H., Esch, G. W. 1994. The effects of spatial and temporal heterogeneity as structuring forces for parasite communities in *Helisoma anceps* and *Physa gyrina*. American Midland Naturalist 132:91–103.

Savino, J. F., Stein, R. A. 1982. Predator-prey interaction between large-mouth bass and bluegills as influenced by simulated submerged vegetation. Transactions of the American Fisheries Society 111:255–266.

Sheldon, S. P. 1987. The effects of herbivorous snails on submerged macrophyte communities in Minnesota lakes. Ecology 68: 1920–1931.

Snyder, S. C., Esch, G. W. 1993. Trematode community structure in the pulmonate snail *Physa gyrina*. Journal of Parisitology 79:205–215.

Stein, R. A., Goodman, C. G., Marschall, E. A. 1984. Using time and energetic measures of cost in estimating prey value for fish predators. Ecology 65:702–715.

Steinman, A. D., McIntire, C. D., Lowry, R. R. 1987. Effects of herbivore type and density on chemical composition of algal assemblages in laboratory streams. Journal of North American Benthological Society 6:189–197.

Swamikannu, X., Hoagland, K. D. 1989. Effects of snail grazing on the diversity and structure of a periphyton community in an eutrophic pond. Canadian Journal of Fisheries & Aquatic Sciences 46:1698–1704.

Tashiro, J. S. 1982. Grazing in *Bithynia tentaculata*: age specific bioenergetic patterns in reproductive partitioning of ingested carbon and nitrogen. American Midland Naturalist 107: 133–150.

Tashiro, J. S., Colman, S. D. 1982. Filter feeding in the freshwater prosobranch snail *Bithynia tentaculata*: bioenergetic partitioning of ingested carbon and nitrogen. American Midland Naturalist 107:114–132.

Thomas, D. L., McClintock, J. B. 1990. Embryogenesis and the effects of temperature on embryonic development, juvenile growth rates, and the onset of oviposition in the freshwater pulmonate gastropod *Physella cubensis*. Invertebrate Reproduction and Development 17:65–71.

Thompson, F. G. 1984. The freshwater snails of Florida: A manual for identification. Univ. of Florida Press, Gainesville, FL.

Townsend, C. R. 1975. Strategic aspects of time allocation in the ecology of a freshwater pulmonate snail. Oecologia 19:105–115.

Townsend, C. R., McCarthy, T. K. 1980. On the defense strategy of *Physa fontinalis* (L.), a freshwater pulmonate snail. Oecologia 46:75–79.

Tuchman, N. C., Stevenson, P. J. 1991. Effects of selective grazing by snails on benthic algal succession. Journal of North American Benthological Society 10:430–443.

Turner, A. M. 1997. Contrasting short-term and long-term effects of predation risk on consumer-habitat use and resources. Behavioral Ecology 8:120–125.

Underwood, G. J. C. 1991. Growth enhancement of the macrophyte *Ceratophylum demersum* by the presence of the snail *Planorbus planorbus*: the effects of grazing and chemical conditioning. Freshwater Biology 26:325–334.

Vail, V. A. 1978. Seasonal reproductive patterns in three viviparid gastropods. Malacologia 17:73–97.

Van Cleave, H. J., Altringer, D. A. 1937. Studies on the life cycle of *Campeloma rufrum*, a freshwater snail. American Naturalist 71:167–184.

Van Cleave, H. J., Lederer, L. G. 1932. Studies on the life cycle of the snail, *Viviparus contectoides*. Journal of Morphology 53: 499–522.

Vermeij, G. J., Covich, A. P. 1978. Co-evolution of freshwater gastropods and their predators. American Naturalist 112: 833–843.

Weber, L. M., Lodge, D. M. 1990. Periphytic food and predatory crayfish: relative roles in determining snail distribution. Oecologia 82:33–39.

Wethington, A. R., Dillon, R. T. Jr. 1993. Selfing, outcrossing, and mixed mating in the freshwater snail *Physa heterostropha*: lifetime fitness and inbreeding depression. Invertebrate Biology 116: 192–199.

Wethington, A. R., Dillion, R. T., Jr. 1996. Gender choice and gender conflict in a non-reciprocally mating simultaneous hermaphrodite, the freshwater snail, *Physa*. Animal Behavior 51: 1107–1118.

Wilbur, K. M. 1983. The Mollusca, Vol. 1–6. Academic Press, Orlando, FL.

11 MOLLUSCA: BIVALVIA

Robert F. McMahon
Department of Biology
Box 19498
The University of Texas at Arlington
Arlington, TX 76019

Arthur E. Bogan
North Carolina State Museum
of Natural Sciences
Research Laboratory
4301 Ready Creek Road
Raleigh, NC 27607

I. INTRODUCTION

North American (NA) freshwater bivalve molluscs (class Bivalvia) fall in the subclasses Paleoheterodonta (Superfamily Unionoidea) and Heterodonta (Superfamilies Corbiculoidea and Dreissenoidea). They have enlarged gills with elongated, ciliated filaments for suspension feeding on plankton, algae, bacteria, and microdetritus. The mantle tissue underlying and secreting the shell forms a pair of lateral, dorsally connected lobes. Mantle and shell are both single entities. During development, the right and left mantle lobes extend ventrally from the dorsal visceral mass to enfold the body. Each lobe secrets a calcareous shell valve which remains connected by a mid-dorsal isthmus (Allen, 1985). Like all molluscs, the shell valves consist of outer proteinaceous and inner crystalline calcium carbonate elements (Wilbur and Saleuddin, 1983). The lateral mantle lobes secrete shell material marked by a high proportion of crystalline calcium carbonate making them thick, strong and inflexible, while the mantle isthmus secretes primarily protein, forming a dorsal elastic hinge ligament uniting the calcareous valves (Fig. 1). The hinge ligament is external in all freshwater bivalves. Its elasticity opens the valves while the anterior and posterior shell adductor muscles (Fig. 2) run between the valves and close them in opposition to the hinge ligament which opens them on adductor muscle relaxation.

The mantle lobes and shell completely enclose the bivalve body, resulting in cephalic sensory structures becoming vestigial or lost. Instead, external sensory structures are concentrated on the mantle margins where they are exposed to the external environment when the valves open. Compared to other molluscs, the bivalve body is laterally compressed and dorso-ventrally expanded, adapting them for burrowing in sediments, enclosure by shell valves and mantle protecting their soft tissues from abrasion and preventing fine sediments from entering the mantle cavity where they could interfere with gill suspension feeding. These adaptations along with a highly protrusile, muscular, spadelike foot used for burrowing, have made bivalves the most successful infaunal suspension feeders in marine and freshwater habitats.

Ecology and Classification of North American Freshwater Invertebrates. 2nd Edition

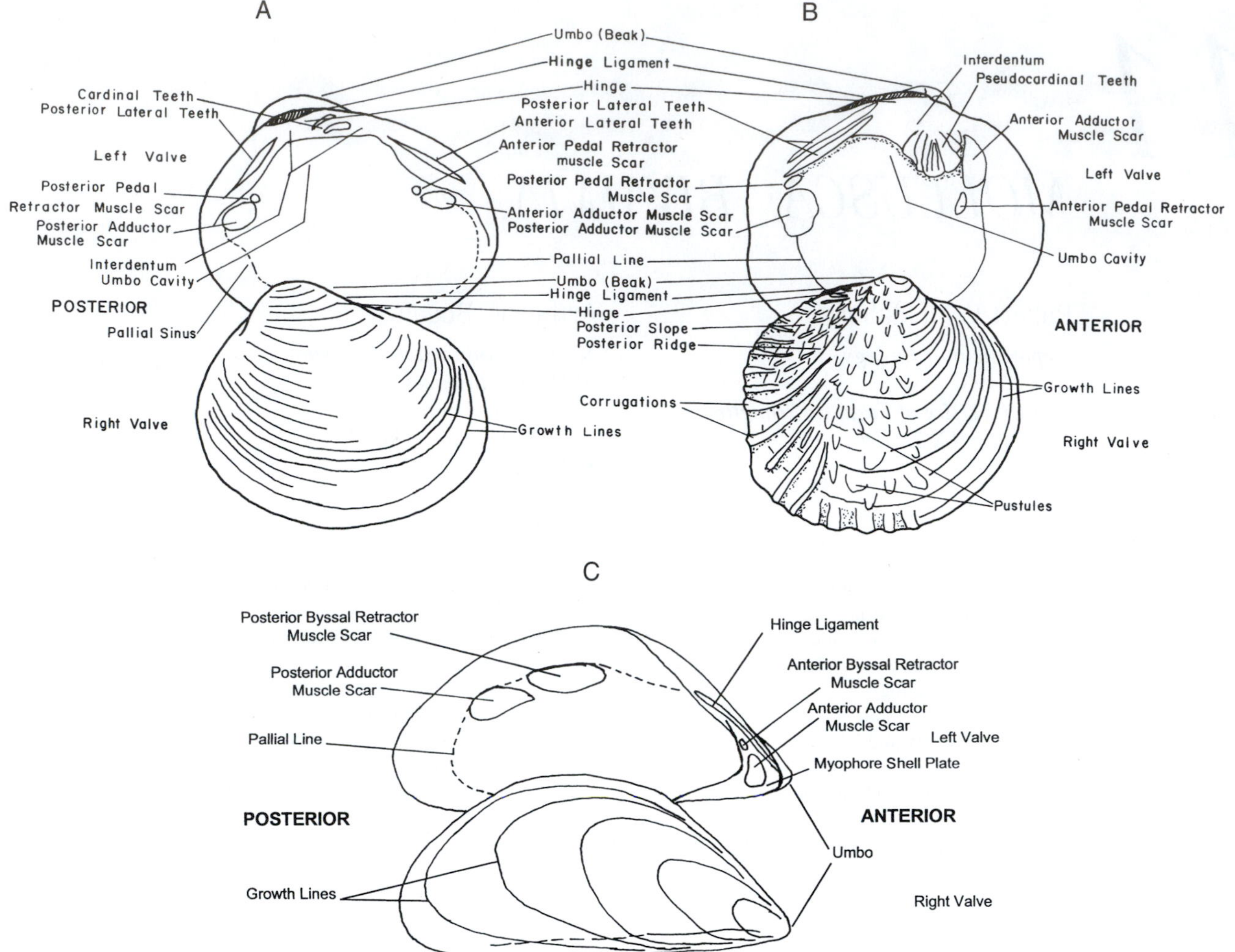

FIGURE 1 General morphological features of the shells of (A) corbiculoidean, (B) unionoidean, and (C) dreissenoidean freshwater bivalves.

The NA bivalve fauna is the most diverse in the world, consisting of approximately 308 extent native and seven introduced taxa (Turgeon *et al.*, 1998). The majority of species fall in the superfamily Unionoidea with 278 native NA and 13 recognized subspecies in 49 genera in the family, Unionidae and five species in two genera in the family, Margaritiferidae. Many species in this group have unique morphological adaptations and highly endemic, often endangered populations (Neves *et al.*, 1997). In the superfamily, Corbiculoidea, the family Sphaeriidae, has 36 native and four introduced NA species. While falling into only four genera, the Sphaeriidae are more cosmopolitan than unionoideans (note "unionoideans" as used here refers to all species within the superfamily, Unionoidea while "unionid" used later in the chapter refers only to those unionoidean species in the family, Unionidae), several genera and species having pandemic distributions. *Corbicula fluminea* falls within the Corbiculacea (family Corbiculidae). This species invaded NA freshwaters in the early 1900s and now extends throughout the coastal and southern United States and Mexico, becoming the dominant benthic species in many habitats (McMahon, 1983a,1999). More recently, two dreissenid species, *Dreissena polymorpha* (zebra mussel) and *Dreissena bugensis* (quagga mussel) have invaded North America. *D. polymorpha* was discovered in Lake St. Clair and Lake Erie in 1988 after introduction in 1986. It now extends throughout the Great Lakes, St. Lawrence River, and the Mississippi River and most of its major tributaries (Mackie and Schloesser, 1996). *D. bugensis* was first found in the Erie-Barge Canal and eastern Lake Erie, NY, in 1991 and has since spread through Lakes Erie and Ontario and the St. Lawrence River (Mills *et al.*, 1996). Both species are likely to have been simultaneously introduced as planktonic veliger larvae released with ballast water from ships entering the Great Lakes from ports on the Bug and

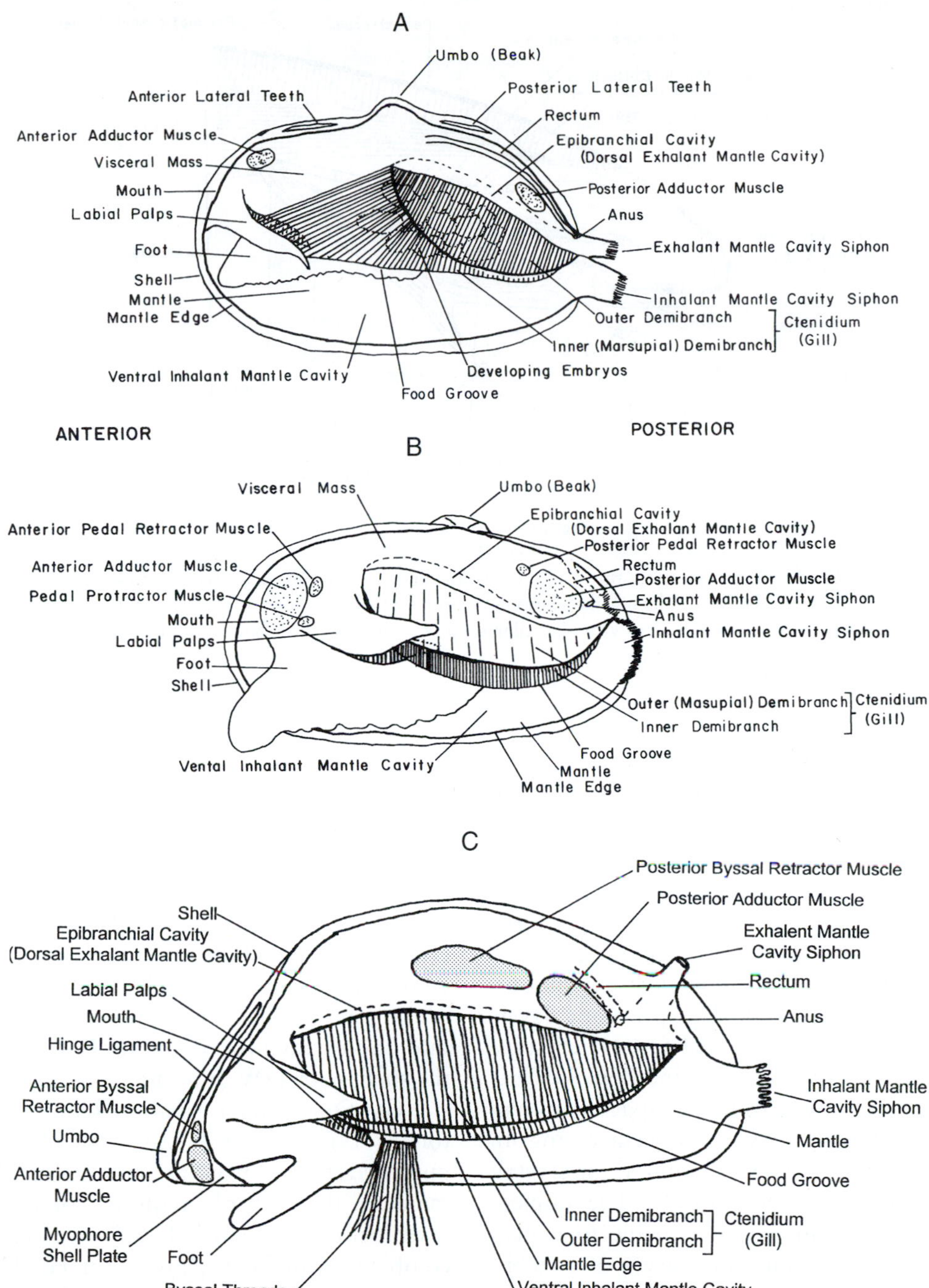

FIGURE 2 General external anatomy of the soft tissues of (A) corbiculoidean, (B) unionoidean, and (C) dreissenoidean freshwater bivalves.

Dnieper rivers in the Black Sea, Ukraine (Marsden *et al.*, 1996, Mills *et al.*, 1996).

II. ANATOMY AND PHYSIOLOGY

NA freshwater bivalves fall into three superfamilies: Unionoidea, Corbiculoidea, and Dreissenoidea. External shell morphology among these groups vary (Fig. 1), but their soft tissue morphologies are relatively similar and, thus, will be discussed in general terms below.

A. External Morphology

1. Shell

Bivalve shells consist of calcium carbonate ($CaCO_3$) crystals embedded in a proteinaceous matrix, both secreted by underlying mantle tissue. In most

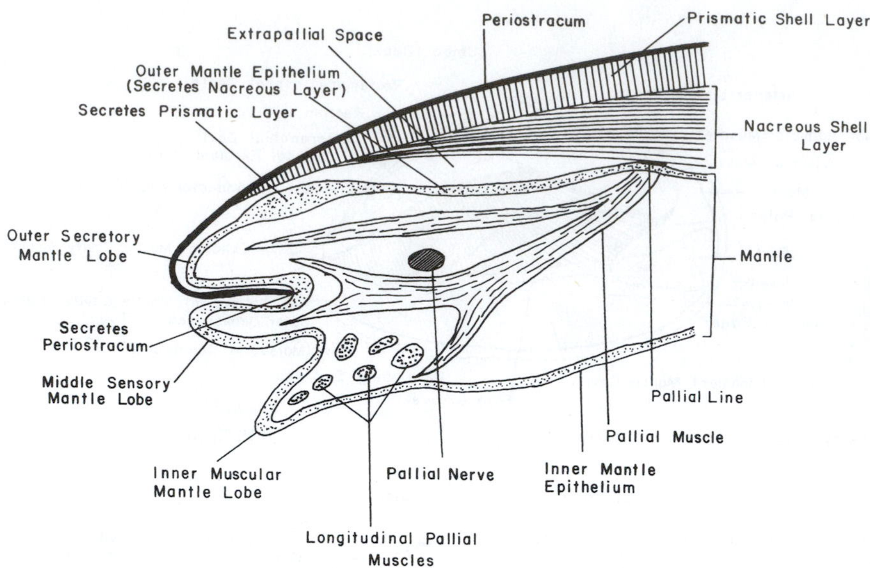

FIGURE 3 A crosssection through the mantle and shell edges of a typical freshwater unionoidean bivalve displaying the anatomic features of the shell, mantle, and mantle edge. Sphaeriids have a complexed cross-lamellar shell structure and lack nacre, but their mantle edge has a similar structure.

bivalves, the shell consists of three parts: an outer proteinaceous periostracum secreted from the periostracal groove in the mantle edge, and underlying, calcareous prismatic and nacre layers in which $CaCO_3$ crystals are embedded within an organic matrix (Fig. 3). The periostracum is initially secreted free of the other layers, but soon fuses with the underlying prismatic layer secreted by a portion of the mantle edge just external to the periostracal groove (Fig. 3). The prismatic layer consists of a single layer of elongated $CaCO_3$ crystals oriented 90° to the horizontal shell plane (Fig. 3). The periostracum edge seals the extrapallial space between mantle and shell, allowing $CaCO_3$ concentrations in that space to reach saturation levels required for crystal deposition (Saleuddin and Petit, 1983). The tanned protein of the periostracum is impermeable to water, preventing shell $CaCO_3$ dissolution. Layers of conchiolin (i.e., tanned protein) occurring within unionoidean shells, are suggested to retard shell dissolution in calcium poor freshwaters and are not found in the Corbiculoidea or Dreissenoidea (Kat, 1985). The nacreous layer is continuously secreted by underlying mantle epithelium. It consists of consecutive layers of small $CaCO_3$ crystals deposited parallel to the shell plane embedded in an organic matrix of chitinlike mucopolysaccharide and protein (Machado *et al.*, 1994) (Fig. 3). Microstructure and texture patterns of $CaCO_3$ crystals differ among major molluscan classes and interspecifically within groups (Hedegaard and Wenk, 1998). Continual secretion of the nacreous layer thickens and strengthens the shell, thus accounting for most of its mass.

Calcium (Ca^{2+}) and bicarbonate (HCO_3^-) necessary for shell $CaCO_3$ crystal deposition are transported from the external medium across the body epithelium into the hemolymph (blood). HCO_3^- ions are also generated from metabolically released CO_2 reacting with hemolymph water ($CO_2 + H_2O \rightleftarrows H^+ + HCO_3^-$). These ions are actively transported across the mantle into the extrapallial fluid for $CaCO_3$ deposition into the shell matrix (Wilbur and Saleuddin, 1983).

$CaCO_3$ crystal formation requires release of protons (H^+) ($Ca^{2+} + HCO_3^- \rightleftarrows CaCO_3 + H^+$) which must be removed from the extrapallial fluid in order to maintain the high pH required for $CaCO_3$ deposition (extrapallial fluid pH 7.4–8.3). It is proposed that H^+ combines with HCO_3^- to form H_2CO_3, followed by dissociation into CO_2 and H_2O which diffuse from the extrapallial fluid into the hemolymph. The enzyme, carbonic anhydrase, which catalyzes this reaction, is present in mantle tissue (Wilbur and Saleuddin, 1983). Extrapallial fluid pH is higher in freshwater than marine bivalves, favoring shell $CaCO_3$ deposition at the lower Ca^{2+} concentrations of the dilute hemolymph of freshwater species (Wilbur and Saleuddin, 1983).

$CaCO_3$ and shell matrix precipitate from the extrapallial fluid. Ca^{2+} and HCO_3^- are actively concentrated in the extrapallial fluid, thus favoring $CaCO_3$ deposition (Wilbur and Saleuddin, 1983). The organic shell matrix separates individual calcareous crystals and binds them with the crystal layers into a unified structure. It also has crystal-nucleating sites (possibly composed of Ca^{2+}-binding polypeptides) on which $CaCO_3$ crystals initially form. $CaCO_3$ crystals grow on these sites to produce a new layer of nacre (Wilbur and Saleuddin, 1983).

A minimum of one ATP appears to be required for deposition of two Ca^{2+}ions, with additional ATP required for active HCO_3^- transport (Wilbur and Saleuddin, 1983). Thus, fast-growing bivalves or those with massive shells must devote increased proportions of maintenance energy to shell mineral deposition. Deposition of shell organic matrix and periostracum also requires metabolic energy in the form of four ATP for every peptide bond. Although rarely more than 10% of shell dry mass, the shell organic matrix can account for 30–50% of the total dry organic matter (i.e., shell + tissue organic matter), requiring up to one-third of the total energy allocated to organic growth (Wilbur and Saleuddin, 1983). Fast-growing, thin-shelled species may devote proportionately less assimilated energy to shell production than slower growing, thick-shelled species, allowing greater energy allocation to tissue growth and reproduction, thereby increasing fitness. However, a thinner, more fragile shell increases probability of predation, lethal desiccation or damage and/or dislodgment from the substrate, reducing fitness. Shell sculpture and shape appear highly correlated with habitat among unionoideans. Species adapted to soft substrates in low-flow habitats have thinner shells, with smaller hinge teeth and greater lateral compression preventing sinking into the substrate while those adapted to faster flowing habitats with harder substrates have, thicker, more inflated, more sculptured shells which anchor them more firmly in the substrate, preventing dislodgment during high water flow (Watters, 1994a). Similarly, concentric sulcations on the shells of *Corbicula fluminea* increase anchoring capacity in their lotic habitats (McMahon, 1999). Therefore, the ratio of energy allocation to shell versus tissue growth appears to be an adaptive strategy, energetic trade-offs between these two processes being evolved under species-specific niche selection pressures (see Section III.B).

Shell external morphology varies among species. Externally, shells may have concentric or radial corrugations of varying sizes and densities, ridges, pigmented rays or blotches in the periostracum. These features, along with shape of the posterior shell ridge and overall shell shape are major taxonomic characters for distinguishing species (see Section V). Another major external shell feature is the umbo or beak, an anteriorly curving, dorsally projecting structure on each valve, which is the oldest portion of the shell (Fig. 1).

Major internal shell features include the hinge and projecting hinge teeth, which interlock to hold the valves in exact juxtaposition, forming the fulcrum on which they open and close, and various mussel scars. These can also be major taxonomic characters (see Section V). In corbiculids, massive conical cardinal teeth form just below the umbos (one in the right and two in the left valve), anterior and posterior of which lie lateral teeth, usually in the form of elongate lamellae (Fig. 1A). In contrast, unionoideans have no true cardinal teeth. Instead, massive, raised, pseudocardinal teeth develop near the umbonal end of the anterior lateral teeth, serving a function similar to cardinal teeth. Elongated, lamellar, posterior lateral teeth extend posteriorly from the umbos (Fig. 1A). Among the unionoideans, hinge teeth are vestigial or lost in the tribe Anodontini. The superfamily, Dreissenoidea, is characterized by a mussel-type shell without teeth in which the anterior end is greatly reduced and posterior end expanded. Thus, the umbos lie at the anterior end making the shell decidedly, anteriorly pointed (Fig. 1C). The nacre of freshwater bivalve shells may have species-specific colors. Internal-shell muscle scars mark the insertion points of anterior and posterior adductor muscles, anterior and posterior pedal (foot) retractor muscles, pedal protractor muscles and pallial line muscles which attach the mantle margin to the shell (Figs. 2 and 3). Posteriorly, the pallial line may be indented marking the pallial sinus into which inhalant and exhalant siphons are withdrawn during valve closure. The extent of the pallial sinus is directly correlated with size and length of the siphons.

2. Locomotory Structures and Burrowing

With the exception of epibenthic dreissenoids which attach to hard substrates with proteinaceous byssal threads, the vast majority of NA freshwater bivalves burrow in sediments. Some species lie on a hard substrate, using the foot to wedge into crevices or under rocks. All species locomote with a muscular, motile, anterio-ventrally directed, protrusile foot (Fig. 2).

Bivalve burrowing has been detailed by Trueman (1983) and described for the unionidean, *Anodonta cygnea* (Wu, 1987). The burrowing cycle begins with relaxation of the adductor muscles, allowing shell valves to open against the substratum by hinge ligament expansion, thus anchoring the shell in place. Contraction of transverse and circular muscles around pedal hemolymph sinuses (i.e., open blood spaces) causes the foot to narrow and lengthen, forcing it anteriorly into the substrate. Once extended, its distal tip is anchored either by expansion with hemolymph or lateral curving into the substrate. Shell adductor muscles then contract, rapidly closing the valves, releasing their hold on the substrate. Rapid valve closure expels a jet of water from the mantle cavity, anteriorly through the pedal gape, loosening compacted sediments at the anterior shell margin. Thereafter, alternating contraction of anterior and posterior pedal retractor muscles simultaneously rock the anterior shell margin dorsoventrally and pull it forward into the loosened sediments against the anchored foot tip. Once a new position is achieved, shell adductor muscles relax, re-anchoring the open valves in the substrate, reinitiating the burrowing cycle. NA freshwater bivalves have relatively short siphons (the mantle edges are not fused to form true siphons in unionoideans and siphons are highly reduced or absent in some species of *Pisidium*), thus, they generally are found either just beneath the sediment surface or with the posterior shell margins just above it. Living near the sediment surface can lead to dislodgment, thus a number of riverine unionoidean taxa have shell and pedal morphologies adapted for rapid reburial (Watters, 1994a).

Many juvenile freshwater bivalves crawl considerable distances over the substrate before settlement, holding the shell valves upright on an extended, dorsoventrally flattened foot. Such surface locomotion also occurs among adult sphaeriids (Wu and Trueman, 1984) and in juvenile *C. fluminea* (observed by the author). Crawling involves extension of the foot, anchoring its tip with mucus and/or a muscular attachment sucker, followed by pedal retractor muscle contraction which pulls the body forward. Pedal surface locomotion is reduced or lost in most adult unionoideans, but is retained to varying degrees in adult sphaeriids, *C. fluminea* and dreissenids. Adult dreissenids may spontaneously release from the byssus and crawl long distances before re-attachment. Single byssal threads are produced during crawling to prevent dislodgment. Pedal locomotion is particularly common in juvenile and immature dreissenids and allows their escape from locally poor conditions or highly dense mussel clumps (Clarke and McMahon, 1996a).

B. Organ-System Function

1. Circulation

Bivalves have an 'open' circulatory system in which hemolymph is not always enclosed in vessels. Rather, circulatory fluid is carried by vessels from the heart to various parts of the body where it passes into open, spongy hemocoels (i.e., blood sinuses). In the hemocoels, it bathes tissues directly, percolating through them before returning to the heart via the gills. Circulatory fluid in open systems is called "hemolymph." Bivalves have large hemolymph volumes, accounting for 49–55% of total body water (Jones, 1983). The bivalve heart ventricle uniquely surrounds the rectum and pumps oxygenated hemolymph from the gills and mantle via the kidney through anterior and posterior vessels (Jones, 1983, Narain and Singh, 1990) (Fig. 4) which subdivide into smaller vessels to various parts of the body, including pallial arteries to the mantle and visceral arteries to the visceral mass and foot. These secondary arteries further subdivide into many tiny vessels that open into the hemocoels where cellular exchange of nutrients, gases, and wastes occurs. Thereafter, deoxygenated hemolymph is carried from the body tissues and organs to the mantle and gills to be reoxygenated and, thence, to the heart. Evolution of an open circulatory system in molluscs, including bivalves, is associated with coelom reduction. The heart ventricle is surrounded by a coelomic remnant, the "pericardial cavity," enclosed by the pericardial epithelium or "pericardium." Other coelomic remnants include spaces comprising the kidneys ('coelomoducts') and gonads (Jones, 1983).

Like most bivalves, the hemolymph of all freshwater species has no specialized respiratory pigments for O_2 transport (Bonaventura and Bonaventura, 1983). Instead, O_2 is transported dissolved directly in the hemolymph fluid which has an O_2-carrying capacity essentially equivalent to water. The very low metabolic demands and extensive gas exchange surfaces (mantle and gills) of bivalves allow maintenance of a primarily aerobic metabolism in spite of a reduced hemolymph O_2-carrying capacity. As proteinaceous respiratory pigments are the primary blood pH buffer in most animals, bivalve blood acid–base balance is dependent on other mechanisms (see Section II.C.3).

2. Gills and Gas Exchange

In lamellibranch bivalves, including all NA freshwater species, the gills are expanded beyond the requirements for gas exchange as they are also used for suspension (filter) feeding (Fig. 2), the main mode of food acquisition for the majority of species (see Section III.C.2). The left and right gills, or ctenidia, consist of an axis which extends anterolaterally along the visceral mass. Many long, thin, inner and outer filaments extend laterally from the axis. In all NA freshwater bivalves, filaments are fused together (an evolutionarily advanced condition) and penetrated by a series of pores or ostia (i.e., eulamellibranch condition). From the axis, the filaments first extend ventrally (descending filament limbs) and then reflect dorsally (ascending filament limbs) to attach distally to the dorsal mantle wall (outer filaments) or the dorsal side of the visceral mass (inner filaments), forming two V-shaped, porous curtains called the outer and inner "demibranchs" (Fig. 5).

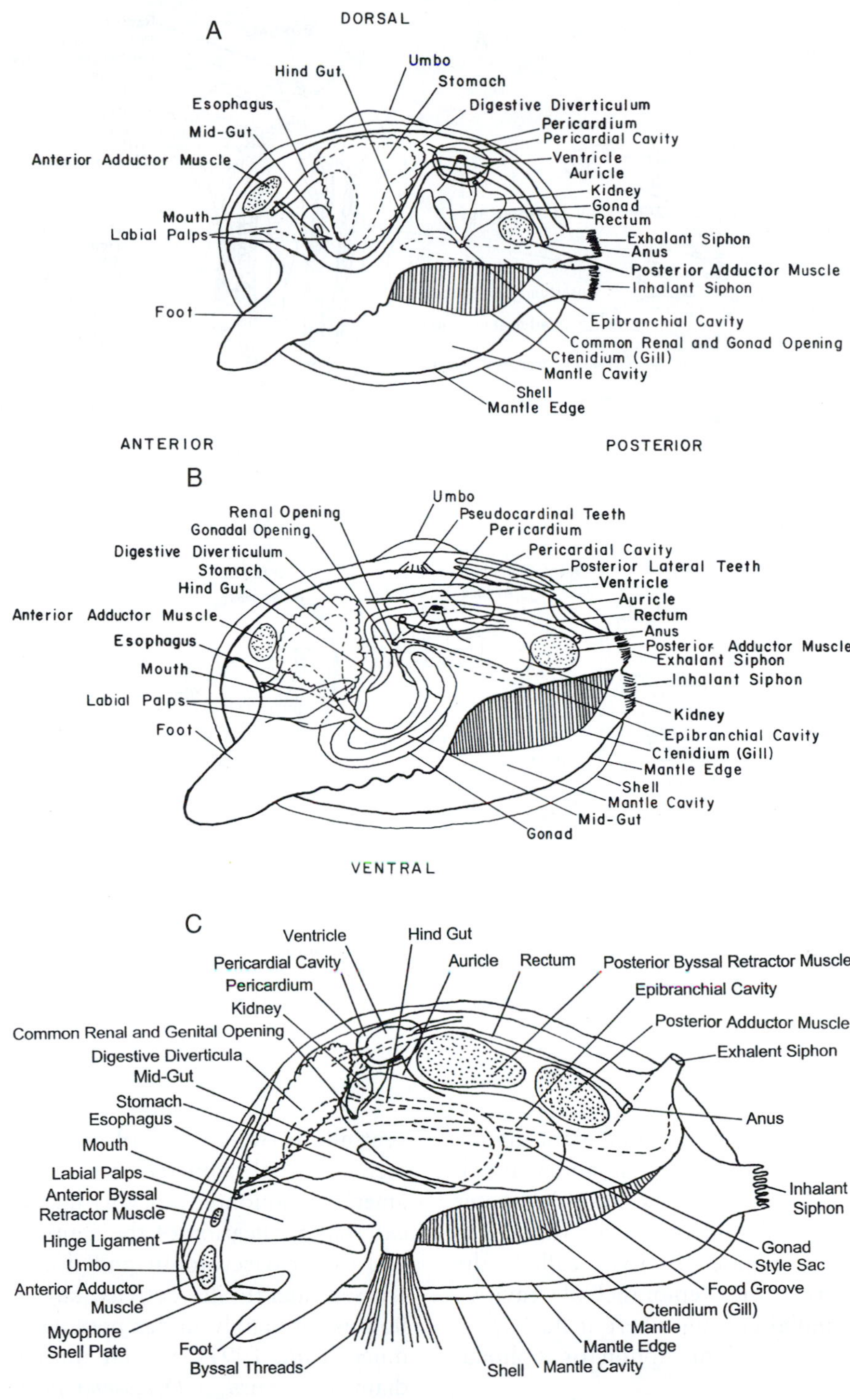

FIGURE 4 General internal anatomy, organs, and organ systems of the soft tissues of (A) corbiculoidean, (B) unionoidean, and (C) dreissenoidean freshwater bivalves.

The demibranchs completely separate the mantle cavity into ventral inhalant and dorsal exhalant portions. The descending and ascending filament limbs are held closely adjacent by tissue bridges called "interlamellar junctions" and enclose an area called a "water tube" or interlamellar space (Fig. 5).

Feeding and respiratory currents are sustained by 'lateral cilia' on the adjacent external filament surfaces

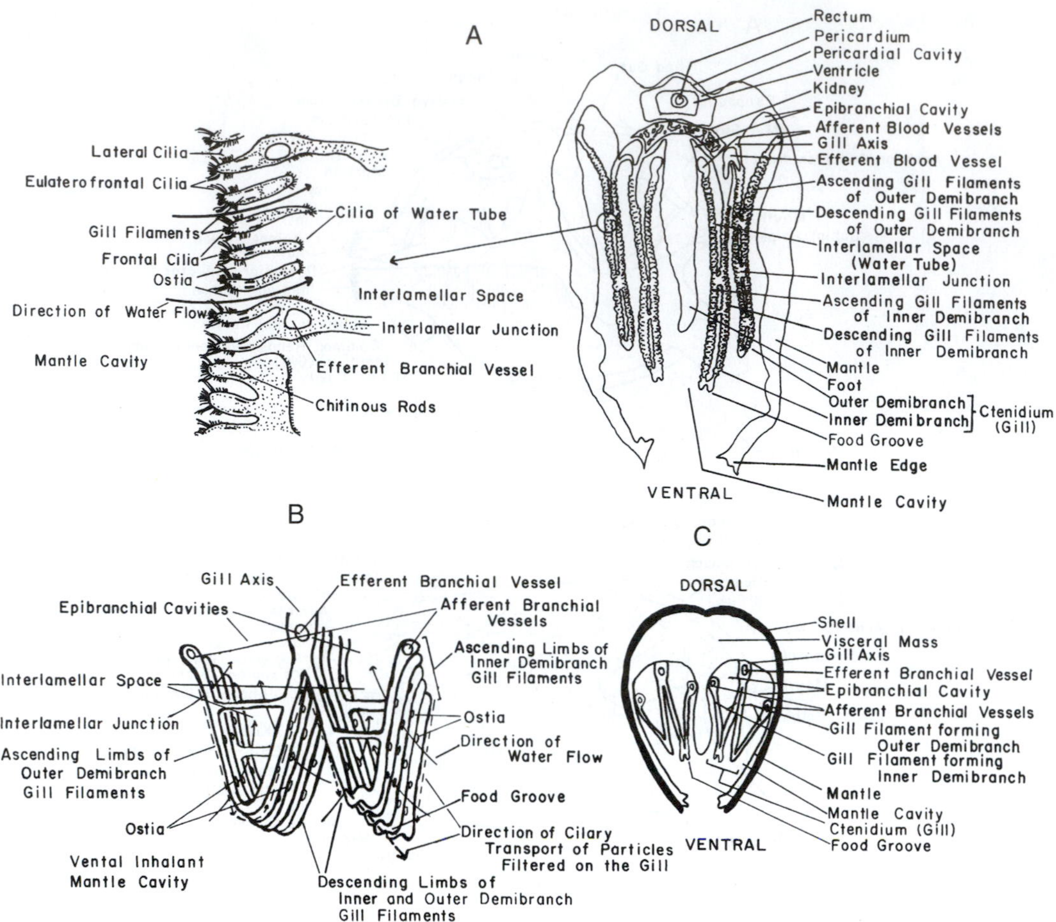

FIGURE 5 The structural features of the gills (ctenidia) of freshwater bivalves. (A) Cross section through the central visceral mass, ctenidia, and mantle of a typical freshwater unionoidean, with high magnification view showing details of filaments, ostia, and ciliation. (B) Diagrammatic representation of the respiratory and feeding currents across the ctenidium. (C) Diagrammatic cross-sectional representation of lammellibranch bivalve ctenidia.

(Fig. 5). They drive water entering the inhalant mantle cavity via the inhalant siphon or siphonal notch through the ostia between filaments into the water tubes. Water passes dorsally up the water tubes into the dorsal 'exhalant mantle cavity,' called the "suprabranchial cavity" or "epibranchial cavity," formed above the ctenidial curtain where it flows posteriorly, exiting via the exhalant siphon or siphonal notch (Figs. 2 and 5).

While the lateral cilia drive water across the gill, rate of water flow is partially controlled by size of the ostia. In unionoideans, two muscle bands control ostial pore size. Antero-posterior oriented muscle bands are attached to vertically oriented chitinous filament supporting rods which alternate with rows of ostia. Horizontal muscle contraction pulls adjacent supporting rods together to reduce ostial pore diameter, slowing water flow. A second set of muscle bands run dorsoventrally between rows of ostia and, when contracted, increase ostial diameter, increasing flow rate. Ostial diameter of unionoidean gills increased 2–3 times with external application of the neurotransmitter, serotonin, and external serotonin and dopamine application increased lateral cilia activity, suggesting that gill water flow is ultimately under nervous system control (Gardiner *et al.*, 1991). Similar muscular control of ostial diameter occurs in *Dreissena polymorpha* where contraction of horizontally oriented muscle bands and spincterlike ostial muscles reduce ostial diameter, slowing water flow. External application of serotonin relaxes these muscles increasing ostial diameter and gill water flow (Medler and Silverman, 1997).

3. Excretion and Osmoregulation

Freshwater bivalves, like all freshwater animals, have hemolymph and tissue osmotic concentrations greater than the very dilute freshwater medium, resulting in constant ion loss and water gain. Osmoregulatory problems are compounded by the extensive gill and mantle surfaces of bivalves over which ion and water fluxes occur (Dietz, 1985). To reduce these fluxes, freshwater bivalves have evolved the lowest hemolymph and cell osmotic concentrations of any metazoan, being 20–50% of that found in most other freshwater species (Dietz, 1985). In spite of their reduced osmotic concentration, the extensive epithelial surfaces of bivalves still cause water and ion fluxes to be greater than those recorded in other freshwater species, leading to urine clearance rates of 20–50 ml/k per hour (Dietz, 1985).

In unionoideans, Na^+ is taken up in exchange for outward transport of the metabolic waste cations, H^+ and NH_4^+, and, perhaps, Ca^{2+}. Chloride (Cl^-) ion uptake is in exchange for metabolically produced HCO_3^- or OH^-. Active Ca^{2+} uptake also occurs (Burton, 1983). In freshwater snails, the major source of Ca^{2+} is ingested food (McMahon, 1983b), while the relative roles of food and transport in the Ca^{2+} balance of freshwater bivalves are unknown. Sodium ion (Na^+) uptake does not require presence of Cl^- which is indicative of separate transport systems for these ions (Burton, 1983); however, freshwater dreissenids appear to co-transport Na^+ and Cl^- (Horohov *et al.,* 1992). While active transport is the major route by which unionoideans gain ions, exchange diffusion (i.e., transport of an ion linked with diffusion of a second ion species down its concentration gradient) accounts for 67% of Na^+ uptake in *C. fluminea. C. fluminea* and *D. polymorpha* have higher ion-transport rates than unionoideans and sphaeriids, reflecting their geologically recent penetration of freshwaters (Dietz, 1985, Horohov *et al.,* 1992, Wilcox and Dietz, 1995). *C. fluminea* has a hemolymph solute concentration that is higher (Dietz, 1985) and *D. polymorpha,* that is lower than most other freshwater bivalves (Dietz *et al.,* 1996a). The low hemolymph ion-concentration of *D. polymorpha* occurs in spite of its high rates of active ion uptake, indicating that it is much more ion-permeable than other freshwater bivalves, perhaps due to its recent evolutionary invasion of freshwaters (Dietz *et al.,* 1994, 1995, 1996a, b). In contrast, epithelial ion permeability in *C. fluminea* is low (Zheng and Dietz, 1998a) and active ion uptake rate high (Zheng and Dietz, 1998b), allowing maintenance of higher hemolymph ion concentrations and tolerance greater ambient osmolarity variation than other freshwater bivalves (Zheng and Dietz, 1998b). Interestingly, exchange diffusion may account for up to 90% of Cl^- turnover in unionoideans in pond water, but when salt-depleted, active transport dominates Cl^-uptake (Dietz, 1985). Na^+ and Cl^- uptake can occur over the body epithelia in unionoideans, but the majority occurs over the gills (Dietz, 1985), with the epithelial cells of water canals connecting gill ostia to water tubes being a major site for active ion and osmotic regulation (Kays *et al.,* 1990). Active ion uptake and hemolymph-ion concentrations in unionoideans and, by inference, other freshwater bivalves, may be regulated by neurotransmitter substances released by gill neurons (Dietz *et al.,* 1992).

Excess water is eliminated via coelomoducts or kidneys. The heart auricle walls initially ultrafilter the blood. Hydrostatic pressure generated by auricle contraction forces blood fluid, ions and small organic molecules through the auricle walls into the pericardial cavity surrounding the heart. Filtration occurs through podocyte cells of the pericardial gland lining the inner auricular surface and, perhaps, through the efferent branchial vein running from the longitudinal kidney vein to the auricles (Martin, 1983) (Fig. 6). Only larger hemolymph protein, lipid and carbohydrate molecules cannot pass the pericardial gland filter. The filtrate exits the pericardial cavity via left and right "renopericardial openings" in the pericardial wall to enter the "reopericardial canals" leading to the left and right kidneys. Larger organic waste molecules are actively transported into the filtrate by the kidney. In the Unionoidea, the rectal wall is extremely thin where surrounded by the ventricle, suggesting that larger organic waste molecules may be transported through it directly into the rectum in this region (Narain and Singh, 1990). Competed excretory fluid passes via "nephridiopores" into the epibranchial cavity to be carried on exhalant water flow out the exhalant siphon (Figs. 2 and 4A) (Martin, 1983).

While little studied in freshwater bivalves, the kidney is the presumed site of major active ion resorption from the filtrate into the hemolymph. Dietz and Byrne (1999) have demonstrated reabsorption of filtrate sulfate ion (SO_4^{-2}) in the kidney of *D. polymorpha.* As kidney walls appear relatively impermeable to water, active filtrate-ion resorption leads to formation of a dilute excretory fluid facilitating excess water excretion with filtrate osmolarity being 50% that of hemolymph in the unionid, *Anodonta cygnea* (Martin, 1983). Filtrate-ion reabsorption is less energetically expensive than active-ion uptake from the freshwater medium because ion-concentration gradients between coelomoduct fluid and hemolymph are far less than those between hemolymph and freshwater.

Due to rapid water influx, an elevated excretory fluid production is required in freshwater bivalves to maintain osmotic balance, being approximately 0.03 mL/g wet tissue per day in *A. cygnea* (Martin, 1983). In *D. polymorpha* excretory fluid production was 2–3 mL/g dry

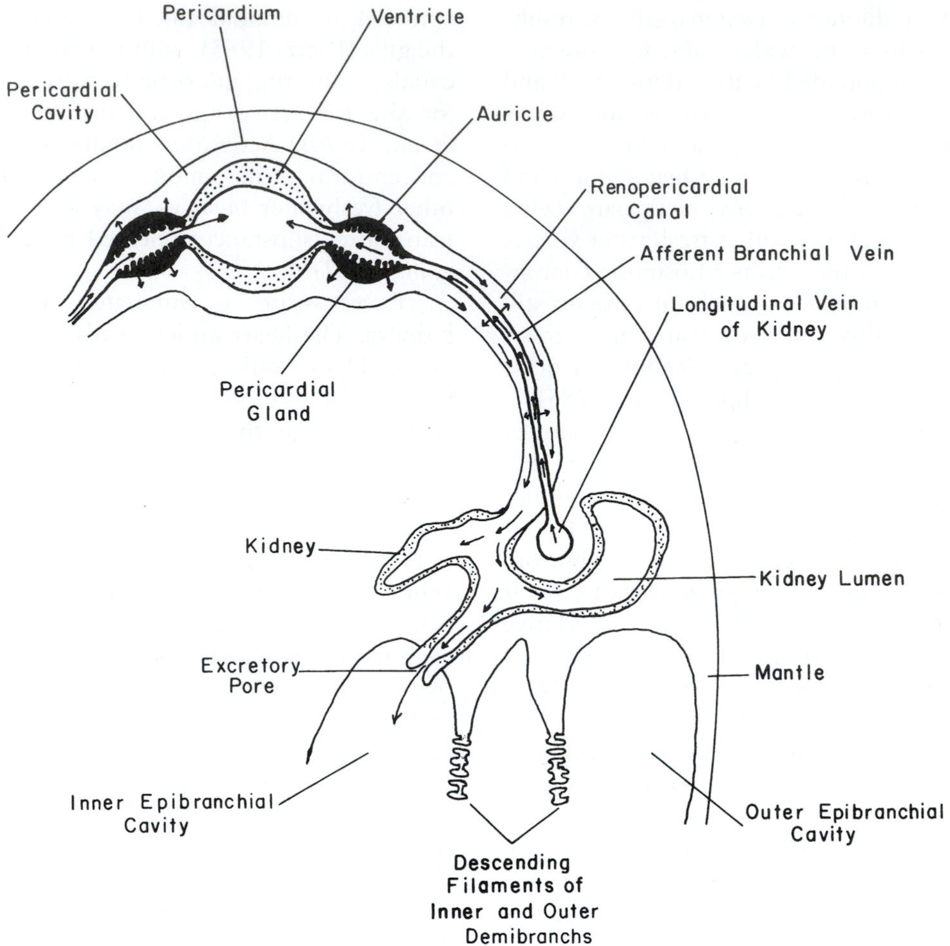

FIGURE 6 Cross-sectional representation of the anatomic features of the excretory system of a typical freshwater bivalve. Arrows indicate pathways for the excretion of excess water in the hemolymph through the excretory system to be eliminated at the excretory pore (redrawn from Martin, 1983).

tissue per day (Dietz and Byrne, 1999) which, when converted to a wet tissue weight value of 0.1–0.15 mL/g dry tissue weight per day (based on dry tissue weight being 0.05 wet tissue weight) is 3.3–5.0 times greater than that of *A. cygnea* (Martin, 1983), reflecting the high epithelial osmotic permeability of *D. polymorpha* compared to other freshwater bivalves (Dietz *et al.*, 1996a, b). In spite of renal ion absorption, the excretory fluid of freshwater bivalves has a considerably higher ion concentration than freshwater, making excretion a major avenue of ion loss. Ions lost by excretion and epithelial diffusion are recovered by active transport from the medium across epithelial surfaces, particularly that of the gills (Dietz, 1985, also see Section II.C.6).

The main nitrogenous waste product of freshwater bivalves is ammonia (NH_3) diffused across external epithelia as dissolved NH_3 or ammonium ion (NH_4^+). NH_4^+ can also be actively transported to the external medium. In marine bivalves, 65–72% of nitrogen excretion is as ammonia/ammonium ion and 13–28% as urea (Florkin and Bricteux-Gregoire, 1972), values likely to be similar in freshwater species. Some freshwater unionoideans appear to be able to detoxify ammonia/ammonium ions by conversion to urea and, perhaps, amino acids and proteins (Mani *et al.*, 1993).

4. Digestion and Assimilation

Freshwater bivalves are suspension (filter) feeders, filtering algae, bacteria and suspended microdetrital particles from water flowing through the gill. Filtered material is transported by gill ciliary tracks to the "labial palps" where it is sorted on corrugated ciliated surfaces into food and nonfood particles before food particles are carried by cilia to the mouth. Some freshwater species may also utilize the foot to feed on sediment organic detrital particles (Reid *et al.*, 1992). Filter and pedal detritus feeding are described in Section III.C.2, while this section is devoted to food digestion and assimilation.

The bivalve mouth is a simple opening flanked laterally by left and right pairs of labial palps, whose ciliary tracks deliver filtered food from the gills to the mouth as a constant stream of fine particles. After ingestion, food particles pass through a short, ciliated esophagus where mucus secretion and ciliary action bind them into a mucus string before entering the stomach. The stomach, lying in the anterio-dorsal portion of the visceral mass (Fig. 4), is a complex structure with ciliated sorting surfaces and openings to a number of digestive organs and structures. On its ventral floor, posterior to the opening of the midgut, is an elongated, evaginated, blind-ending tube, the "style sac." The distal end of the style sac secretes the "crystalline style," a long mucopolysaccharide rod that projects from the style sac into the stomach. Style-sac cells secrete digestive enzymes into the style matrix and have cilia which slowly rotate the style. Stomach and style pH ranges from 6.0–6.9, acidity depending on phase of digestion (Morton, 1983).

The rotating style tip projects against a chitinous plate or "gastric shield" on the dorsal roof of the stomach. The gastric shield is penetrated by microvilli from underlying epithelial cells, believed to release digestive enzymes onto the shield surface (Morton, 1983). Style rotation mixes stomach contents and winds the esophageal mucus food string on to the style tip where slow release of its embedded enzymes begins extracellular digestion. Abrasion of the style tip against the gastric shield, triturates food particles and allows their further digestion by enzymes released from the eroding style matrix and from digestive epithelial cells underlying the shield.

After initial trituration and digestion at the gastric shield, food particles released into the stomach may have several fates. Particles are size-sorted. A ciliated ridge, the "typhlosole," running the length of the midgut returns large particles to the stomach for further digestion and trituration and eventually selects and passes indigestible particles to the hindgut/rectum for egestion. Ciliated surfaces on the posterior "sorting caecum" of the stomach direct sufficiently small food particles into tubules of the "digestive diverticulum" for final intracellular digestion (Fig. 4). The diverticula have the lowest fluid pH of the gut (Morton, 1983). Larger food particles falling on the caecum are recycled into the stomach for further trituration and enzymatic digestion, thus particles may pass over its ciliated sorting surfaces several times before acceptance for intracellular digestion or rejection into the rectum to form feces.

Digestive cells lining the lumina of the tiny, blind-ending, terminal tubules of the digestive diverticula take up fine food particles by endocytosis into food vacuoles for final digestion and assimilation. After completion of intracellular digestion/assimilation, the apical portions of the digestive cells, which contain food vacuoles with undigested wastes and digestive enzymes, are shed into the tubule lumina as "fragmentation spherules" which are returned to the stomach on ciliated rejection pathways. Breakdown of fragmentation spherules in the stomach releases their food vacuole contents, hypothesized to be a major source of stomach acidity and extracellular digestive enzymes (Morton, 1983).

Rejected indigestible matter passes through the short hindgut into the rectum for egestion from the anus, opening into the epibranchial cavity on the posterior face of the posterior shell adductor muscle just upstream from the exhalant siphon or opening, allowing feces expulsion on the exhalant current. Undigested particles are consolidated into discrete, dense, fecal pellets by mucus secreted by the hindgut/rectum, preventing their uptake on the inhalant current.

The cerebropleural and visceral ganglia release neurohormones which influence glycogenesis (Joosse and Geraerts, 1983). The vertebrate glycogenic hormones, insulin and adrenalin, have similar effects on unionoideans. Insulin injected into the Indian unionid, *Lamellidens corrianus,* resulted in declining hemolymph glucose concentration and increase in pedal and digestive diverticular glycogen stores, while adrenaline injection induced glycogen store breakdown and increased hemolymph sugar concentration (Jadhav and Lomte, 1982b). Elevated hemolymph glucose concentrations stimulates gut epithelial cells to produce an insulin-like substance in the unionoideans, *Unio pictorum* and *A. cygnea,* which stimulated activity of glucose synthetase, increasing uptake of hemolymph glucose into glycogen stores (Joose and Geraerts, 1983). Cerebropleural ganglionic neurosecretory hormones regulate pedal and digestive diverticular accumulation and release of protein and nonprotein stores in *L. corrianus* (Jadhav and Lomte, 1983).

5. Reproductive Structures

The paired gonads of freshwater bivalves lie close to the digestive diverticula. Among unionoideans, their paired condition is difficult to discern. Unionoidean gonads envelop the lower portions of the intestinal tract and sometimes extend into the proximal portions of the foot (Fig. 4B), while those of sphaeriids, *C. fluminea,* and freshwater dreissenids lie more dorsally in the visceral mass, extending on either side of the stomach, intestine and digestive diverticula (Fig. 4A, and C; Mackie, 1984; Claudi and Mackie, 1993). A short gonoduct leading from each gonad opens into the epibranchial cavity allowing gamete release on exhalant

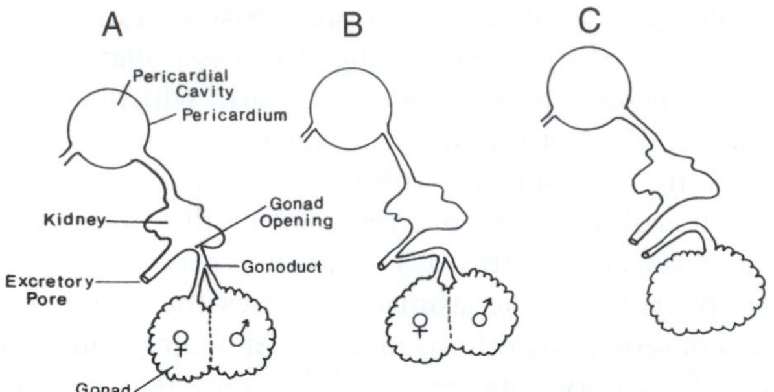

FIGURE 7 Schematic representations of typical reproductive systems of bivalves. (A) The primitive condition in some marine bivalve species: male or female gametes are released from the gonoduct opening into the kidney and passed externally through the excretory pore (shown here is a primitive marine hermaphroditic bivalve; the anatomy is essentially similar in gonochoristic species). (B) Hermaphroditic freshwater corbiculoidean bivalves (*Corbicula fluminea* and sphaeriids): male and female gametes are passed through the gonoducts into the kidney duct close to the excretory pore from which gametes are shed. (C) Gonochoristic freshwater unionoideans: gametes pass to the outside through a gonoduct and gonopore totally separate from the kidney duct and excretory pore. (After Mackie, 1984.)

currents (Fig. 4). Among unionoideans, which are generally gonochoristic except for a few hermaphroditic species of Anodontinae and Ambleminae (Hoeh *et al.*, 1995), the tracts and openings of the renal and reproductive systems are entirely separate, an advanced condition (Fig. 7C). In sphaeriids and *C. fluminea,* which are hermaphroditic, the gonads have distinct regions in which either male or female acini produce sperm or eggs. In these latter groups and the gonochoristic freshwater dreissenids, ducts carrying eggs or sperm unite into a single gonoduct carrying gametes from each gonad into the distal end of the kidney—allowing gamete discharge into the epibranchial cavity via the nephridial canal and nephridiopore (Fig. 7A)—or directly into the nephridial canal, discharging through a common pore on a papilla in the epibranchial cavity (Fig. 7B; Mackie, 1984; Kraemer *et al.,* 1986).

Almost all NA freshwater bivalves are ovoviviparous, brooding developing embryos in specialized gill marsupia. Only the dreisssenids release sperm and eggs externally, leading to development of free swimming larvae (Nichols, 1996). In brooding species, interlamellar spaces are modified to form gill marsupia (brood chambers). Fully formed juveniles are released from the marsupia via the exhalant siphon in sphaeriids and *C. fluminea.* In unionoideans, bivalved "glochidia larvae" are released from either the exhalant siphon, specialized gill pores, or ruptures in the ventral margin of the marsupial gills. Unionoidean glochidia parasitize fish hosts before metamorphosing into free—living juveniles. Interlamellar spaces of the outer demibranch form marsupia in unionoideans except in the subfamily, Ambleminae, which form marsupia in both demibranchs (Burch, 1975b). In contrast, marsupia form in the inner demibranchs of sphaeriids and *C. fluminea.* Among species of the unionid genus, *Lampsilis,* marsupia develop only in the posterior portions of the outer demibranch from which glochidia are released through small pores directly into the inhalant siphon (for further details, see Section III.B.1).

In *C. fluminea,* interlamellar marsupial brooding spaces have no specialized structures. In sphaeriids, embryos are enclosed in specialized brood chambers evaginated from gill filaments into the interlamellar space. Among anodontid unionids, the marsupial interlamellar space is divided by septa into separate chambers associated with each filament. Each space is further subdivided by lateral septa into a central marsupium and inner and outer water tubes transport water from the gill ostia to the epibranchial cavity, an advanced structure not found in other unionoidean groups. In the primitive unionoidean families, Margartiferidae and Ambleminae, the entire outer demibranch forms the marsupium, while among the more advanced tribes, Pleurobemini and Lampsilini, the marsupia are limited to specific portion of the outer demibranch (Mackie, 1984).

Some unionoidean species display sexual dimorphism, generally rare in other freshwater bivalves. In sexually dimorphic species, glochidial incubation distends the female outer marsupial demibranch (Mackie, 1984). Thus, posterior portions of the valves of female lampsilines are inflated compared to males, accommodating the expanded posterior marsupia. Less obvious general inflation of the female shell occurs in the anodontids (Mackie, 1984).

Monoecious unionoideans (some species Anodontinae and Ambleminae) and all sphaeriids are generally simultaneous hermaphrodites, concurrently producing mature eggs and sperm. *C. fluminea* has an unusual pattern of producing only eggs early after maturity (shell length $\approx$6 mm) followed later by sperm production, thereafter, remaining simultaneously hermaphroditic throughout life (Kraemer and Galloway, 1986).

Bivalves lack copulatory organs. Except for dreissenids which release both sperm and eggs externally (Nichols, 1996), freshwater bivalves release only sperm externally which is transported on inhalant currents of other individuals to unfertilized eggs in the gill marsupia. Among corbiculids, self-fertilization appears common in sphaeriids, occurring near the conjunction of male and female gonoducts (Mackie, 1984), while, in *C. fluminea,* embryos frequently occur in the lumina of gametogenic follicles and gonoducts, indicative of self fertilization within the gonad (Kraemer *et al.,* 1986). Self-fertilization makes hermaphroditic species highly invasive (see Section III.B).

Temperature appears to be the main reproductive stimulus in most freshwater bivalves. Gametogenesis and fertilization are initiated above and maintained within critical ambient water temperature thresholds. Other environmental factors which may influence reproduction include: neurosecretory hormones, density dependent factors, diurnal rhythms, food availability and parasites. Evidence for neurosecretory and density controls has been demonstrated in sphaeriids (Mackie, 1984). Zebra mussels spawn in response to the neurosecretory hormone, serotonin, and presence of algal extracts (Ram *et al.,* 1996). Evidence of increasing activity and metabolic rate during dark hours in unionoideans (McCorkle *et al.,* 1979; Englund and Heino, 1994a), *C. fluminea* (McCorkle-Shiley, 1982) and *D. polymorpha* (Borcherding, 1992) suggests that spawning activity and gamete/glochidial/juvenile release rates may also display diurnal rhythmicity.

Sperm of freshwater bivalves may have ellipsoid or conical nuclei and acrosomes of variable complexity (Mackie, 1984). Unionoidean sperm have rounded or short cylindrical heads considered a primitive condition (Rocha and Azevedo, 1990; Lynn, 1994) while that of *Dreissena* and *Corbicula* have elongate heads (Kraemer *et al.,* 1986; Ram *et al.,* 1996). The sperm of *C. fluminea* is uniquely biflagellate (Kraemer *et al.,* 1986). Sperm with elongate heads may be adapted for swimming in gonadal and oviductal fluids more viscous than water and are associated with internal fertilization in gonadal ducts rather in marsupia (Mackie, 1984). However, the elongate-headed sperm of *D. polymorpha* is an exception to this rule, because it externally fertilizes the egg. *D. polymorpha* has sperm with a straight head and an oval, bulbous acrosome while *D. bugensis* sperm has a curved head with a conically shaped acrosome (Denson and Wang 1994). There appears to be specific site for sperm recognition and entry on the vegetal pole of unionoidean eggs (*Truncilla truncata*) where the vitelline coat forms a corrugated surface surrounding a truncated cone (Focarelli *et al.,* 1990).

The eggs of freshwater bivalves are round and have greater yolk volumes than marine species with planktonic larvae. Only freshwater dreissenids have relatively small eggs (40–70 μm) associated with their external fertilization and small, free-swimming veliger larva, which grows considerably in the plankton before settlement and metamorphoses to a juvenile (219–365 μm) (Nichols, 1996). The larger, yolky eggs of all other species contain nourishment supporting development to a more advanced juvenile/glochidium stage. The Sphaeriidae, the smallest adult freshwater bivalves, produce the largest eggs, resulting in very small brood sizes ranging from 6–24 per adult in the genus *Sphaerium,* 1–135 per adult in the genus *Musculium,* and 3.3–6.7 per adult in the genus *Pisidium* (Burky, 1983). Adult *Musculium partumeium,* only 4.0 mm in shell length (SL), release juveniles of 1.4 mm SL (Hornbach *et al.,* 1980; Way *et al.,* 1980). In contrast, unionoideans and *C. fluminea* have smaller eggs and release smaller glochidia/juveniles (generally $<$0.3 mm SL) and have much larger brood sizes (10^3–10^6 per adult) (Burky, 1983; McMahon, 1999). The small eggs of zebra mussels allow them to have massive fecundities of $>10^6$ per adult female (Nichols, 1996) required for successful external fertilization and planktonic larval development. Evolutionary implications of bivalve fecundities are discussed in Section III.B.

The bivalve egg is surrounded by a vitelline membrane that is relatively thin in sphaeriids and thicker in unionoideans and *C. fluminea* (Mackie, 1984). In unionoideans, it remains intact throughout most of embryonic development. It disintegrates during early development in sphaeriids (Heard, 1977), allowing developing embryos to absorb nutrients from brood sacs without embryos and/or from nutrient cells lining the interlamellar spaces of marsupial gills. In dreissenids, the vitelline membrane is lost early to release the free-swimming trochophore larvae which metamorphoses into a planktonic veliger (Nichols, 1996). The vitelline membrane is also lost early to release a free-swimming trochophore retained through development of a juvenile clam in the marsupial gills of *C. fluminea* (Kraemer and Galloway, 1986).

6. Nervous System and Sense Organs

The bivalve head, entirely enclosed within the mantle and shell valves, is not in direct contact with the external environment. Thus, cephalic structures including sense organs have been lost, the head having only a mouth and associated labial palps (Fig. 2). Loss of cephalic sense organs has lead to bivalve nervous system being far less centralized than other advanced molluscan species. A pair of cerebropleural ganglia lateral to the esophagus near the mouth, are interconnected by dorsal, superesophageal commissures (Fig. 8). From these extend two pairs of nerve chords. Paired dorsal nerve chords extend posteriorly through the visceral mass to a pair of visceral ganglia on the anterio-ventral surface of the posterior shell adductor muscle while paired cerebro-pedal nerve chords innervate a pair of pedal ganglia in the foot (Ruppert and Barnes, 1994).

The pedal and cerebropedal ganglia exert motor control over the pedal and anterior shell adductor muscles, while motor control of the siphons and posterior shell adductor muscle is affected by the visceral ganglia. Coordination of pedal and valve movements

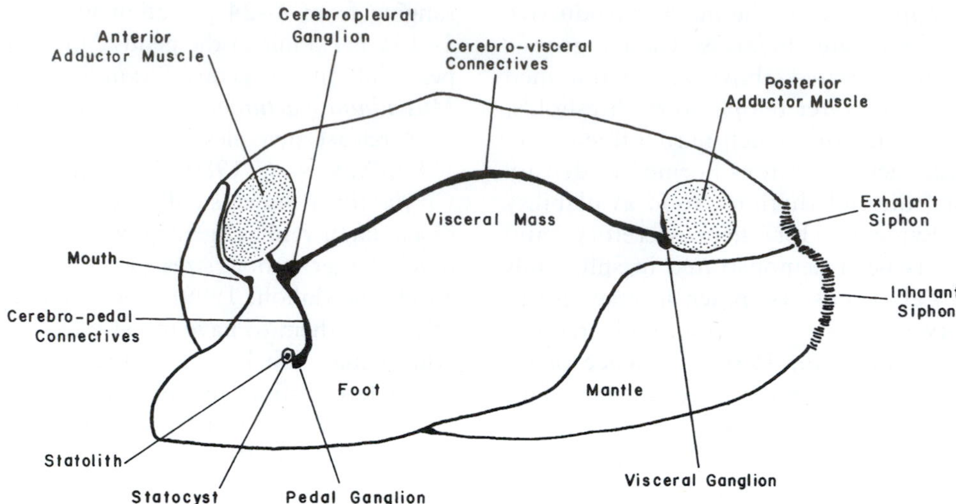

FIGURE 8 The anatomic features of the central nervous system of a typical unionoidean freshwater bivalve (central nervous system anatomy is essentially similar in freshwater corbiculoidean and dreissenoidean bivalves).

during burrowing and locomotion (see Section II.A.2) resides in the cerebropedal ganglia.

Sense organs are concentrated on the mantle edge and in siphonal tissues most directly exposed to the external environment. Mantle edge sense organs are most concentrated on the middle sensory mantle lobe (Fig. 3). Photoreceptor cells detect changes in light intensity associated with shadow reflexes, phototaxis and diurnal rhythms; while tentacles and stiffened, immobile stereocilia are associated with tactile mechanoreceptor organs perceiving direct contact displacement (touch) or vibrations. On the siphon margins, such tactile receptors prevent drawing of large particles into the mantle cavity. When these mechanoreceptors are impinged by large particles, the siphons are closed by sphincter muscles and/or rapidly withdraw to prevent particle entry. Stronger mechanical stimuli to the siphons cause the valves to be rapidly adducted, forcing water from the siphons, ejecting any impinging material. Under intense mantle or siphon stimulation, siphons are withdrawn and the valves tightly closed, a common predator defense behavior in all bivalve species.

A pair of statocysts lying near or within the pedal ganglia are enervated by commissures from the cerebropedal ganglion (Fig. 8). Statocysts are greatly reduced in sessile marine species (i.e., cemented oysters), suggesting their importance to locomotion and burrowing in free-living freshwater species. They are lined with ciliary mechanoreceptors responding to pressure exerted by a calcareous statolith or series of smaller, granular statoconia held within the statocyst vesicle (Kraemer, 1978). As gravity sensing organs, statocysts detect body orientation and, thus function in geotactic and positioning responses during burrowing and pedal locomotory behavior.

Bivalves have a pair of organs called osphradia on the dorsal wall of epibranchial cavity, underlying the visceral ganglion in unionoideans and *C. fluminea* (Kraemer, 1981). Because they are profusely enervated, and neuronally connected to the visceral ganglion, shell adductor muscles and kidney, they have been considered sense organs, but their sensory function(s) are debated. In gastropods, where osphradia are located on the incurrent side of the ctenidium, they have been considered to be mechanoreceptors, sensing suspended particles in inhalant water flow, or chemoreceptors. However, in bivalves, the epibranchial (exhalant) mantle cavity position of osphradia where they receive only filtered water and their lack of extensive neuronal gill connections appear to preclude either a mechano- or chemoreceptor function. Rather, bivalve osphradia are hypothesized to be light sensing organs that control seasonal activities such as spawning, or which provide sensory input to kidney function, and/or diurnal patterns of shell valve adduction (Kraemer, 1981).

C. Environmental and Comparative Physiology

1. Seasonal Cycles

Freshwater bivalves display seasonal physiological responses associated with temperature and reproductive cycles. Metabolic rates vary seasonally in freshwater bivalves (for reviews see Burky, 1983; Hornbach, 1985; McMahon, 1996, 1999) with rates being generally greatest in summer and least in winter due to

TABLE I Seasonal Variation in the Oxygen Consumption Rates ($\dot{V}_{O_2}$) of Selected Species of Freshwater Bivalves

Species and source	*mg dry tissue weight*	*Maximum* $\dot{V}_{O_2}$ *μl* O_2*/h*	*Minimum* $\dot{V}_{O_2}$ *μl* O_2*/h*	*Ratio min:max* $\dot{V}_{O_2}$	$Q_{10(Acc.)}$[a]	*Seasonal temperature range* (°C)
Sphaerium striatinum	25 mg	31.5	1.40	22.5:1	4.7	2–22
Hornbach *et al.* (1983)	7 mg	15.8	0.78	20.3:1	4.5	2–22
	2 mg	8.7	0.24	33.6:1	5.8	2–22
Pisidium compressum	3 mg	0.71	0.04	17.5:1	—	—
Way and Wissing (1984)	0.9 mg	0.27	0.03	9.0:1	—	—
	0.05 mg	0.13	0.02	6.5:1	—	—
Pisidium variabile	3 mg	1.13	0.22	5.14:1	—	—
Way and Wissing (1984)	0.9 mg	0.42	0.10	4.20:1	—	—
	0.05 mg	0.15	0.02	7.5:1	—	—
Pisidium walkeri	1.3 mg	1.11	0.85	1.3:1	1.1	1–26
Burky and Burky (1976)	0.02 mg	0.017	0.0074	2.3:1	1.4	1–26
Corbicula fluminea	348 mg	430.6	34.4	12.5:1	3.2	7–29
Williams (1985)	204 mg	308.2	31.5	9.8:1	2.8	7–29
	60 mg	143.3	25.8	5.6:1	2.2	7–29
Dreissena polymorpha	—	7.2	2.1	3.6:1	2.3	6–21
Quigley *et al.* (1992)						
Pyganodon grandis	10 g	4690	400	11.7:1	2.7	6–31
Huebner (1982)	5 g	2690	250	10.8:1	2.6	6–31
Lampsilis radiata	5 g	2090	170	12.3:1	2.8	6–31
Huebner (1982)	2 g	810	80	10.1:1	2.6	6–31

[a] $Q_{10(Acc.)}$ is the respiratory Q_{10} value computed from a change in the $\dot{V}_{O_2}$ value recorded for individuals field acclimated to their respective seasonal maximum and minimum temperatures.

temperature effects (Burky, 1983; Hornbach, 1985). Annual variation in metabolic rate can be extensive (Table I). Maximal summer O_2 consumption rates ($\dot{V}_{O_2}$) may be 20–33 times minimal winter rates over a seasonal range of 2–22°C in *Sphaerium stratinum* (Hornbach *et al.*, 1983) or as little as 1.3 times minimal winter rates in *Pisidium walkeri* (1–26°C) (Burky and Burky, 1976) (Table I).

Immediate temperature increases induce a corresponding metabolic rate increase in ectothermic animals such as bivalves. Acute, temperature-induced variations in metabolic rate or $\dot{V}_{O_2}$ (or in any rate function) are represented by Q_{10} values, the factor by which a rate function changes with a 10°C temperature increase, computed over any temperature range as:

$$Q_{10} = \left(\frac{\text{Rate}_2}{\text{Rate}_1}\right)^{10/[T_2 - T_1]}$$

where Rate_1 is the rate at lower temperature, Rate_2 the rate at higher temperature, T_1 the lower temperature (°C), and T_2 the higher temperature (°C). Q_{10} for metabolic rate in the majority of ectotherms is 2–3, essentially that of chemical reactions. Thus, Q_{10} values outside this range indicate active metabolic regulation, values <1.5 suggesting active metabolic suppression and, >3.5, active metabolic stimulation with temperature change. Freshwater bivalve Q_{10} values are highly variable within and among species, ranging from 0.2 to 14.8 among 20 species of sphaeriids (Hornbach, 1985) and from 1.2 to 3.72 among three species of sphaeriids (Alimov, 1975). Q_{10} ranged from 1.5 to 4.1 for *D. polymorpha* over 10–30°C depending on prior acclimation temperature and temperature range of determination (Alimov, 1975; McMahon, 1996). For 10°C acclimated individuals of *C. fluminea,* respiratory Q_{10} was 2.1 over 5–30°C (McMahon, 1979a). Among unionoideans, in *Lampsilis siliquoidea,* Q_{10} ranged from 1.88 to 4.98 and in *Pyganodon grandis,* from 1.27 to 10.35 (Huebner, 1982). While numerous environmental and physiological factors may affect the Q_{10} values of freshwater bivalves, no general patterns emerge. Rather, metabolic response to temperature appears to have evolved under species-specific microhabitat selection pressures (Hornbach, 1985).

Temperature effects could lead to massive seasonal metabolic fluctuations; rates being suboptimal during colder months and supraoptimal during warmer months. Thus, many ectothermic species display metabolic temperature acclimation, involving compensatory adjustment of metabolic rate to a new temperature regime over periods of a few days to several weeks. Typically, metabolic rates are adjusted upward upon acclimation to colder temperatures and downward upon acclimation to warmer temperatures which dampens metabolic fluctuation with seasonal temperature change, allowing year-round maintenance of near optimal metabolic rates. For most species, such "typical" seasonal metabolic acclimation is partial, with metabolic rates not returning to an absolute optimal level in a new temperature regime. The degree of metabolic temperature compensation can be detected by comparing Q_{10} values of $\dot{V}_{O_2}$ in instantaneous response to acute temperature change with those measured after acclimation [Acclimation Q_{10} or $Q_{10(\mathrm{Acc.})}$]. If $Q_{10(\mathrm{Acc.})}$ approximates 1.0, metabolic temperature compensation is nearly perfect with $\dot{V}_{O_2}$ regulated near an optimal level throughout the year. If $Q_{10(\mathrm{Acc.})}$ is less than 2.0 or considerably less than acute Q_{10}, acclimation is partial, with the metabolic rate approaching, but not reaching, the optimal level. If $Q_{10(\mathrm{Acc.})}$ is 2–3 or equivalent to the acute Q_{10}, the species is incapable of temperature acclimation. If $Q_{10(\mathrm{Acc.})}$ is greater than 3.0 or acute Q_{10}, inverse (reverse) acclimation is displayed in which acclimation to a colder temperature further depresses metabolic rate, and further stimulates it on acclimation to a warmer temperature.

Three of the above acclimation patterns occur in freshwater bivalves: (1) no capacity for acclimation (i.e., $Q_{10(\mathrm{Acc.})}$ equivalent to acute Q_{10}) is displayed by the unionids *P. grandis* and *L. radiata* (Huebner, 1982) and *D. polymorpha* (Quigley *et al.*, 1992; McMahon, 1996); (2) partial acclimation, by *Pisidium walkeri* where $Q_{10(\mathrm{Acc.})}$ is considerably less than acute Q_{10} (Burky and Burky, 1976); and (3) reverse acclimation (i.e., $Q_{10(\mathrm{Acc.})}$ is greater than acute Q_{10}) displayed by *Sphaerium striatinum* (Hornbach *et al.*, 1983) and *C. fluminea* (Williams and McMahon, unpublished data) (Table I). The adaptive advantage of reverse acclimation has been questioned because it results in massive seasonal swings in metabolic rates, but among freshwater bivalves it may reduce utilization of energy stores during nonfeeding, overwintering periods (Burky, 1983).

$\dot{V}_{O_2}$ is also related to individual size or biomass in all animals as follows:

$$\dot{V}_{O_2} = aM^b$$

where M is the individual biomass, and a and b are constants. It can be rewritten as a linear regression in which $\dot{V}_{O_2}$ and M are transformed into logarithmic values:

$$\mathrm{Log}_{10}\,\dot{V}_{O_2} = a + b(\mathrm{Log}_{10}\,M)$$

where a and b are the Y-intercept (i.e., $\dot{V}_{O_2}$ at $\mathrm{Log}_{10}\,M = 0$ or $M = 1$) and slope (i.e., increase in $\mathrm{Log}_{10}\,\dot{V}_{O_2}$ per unit increase in $\mathrm{Log}_{10}\,M$), respectively. Thus, a measures the relative magnitude of $\dot{V}_{O_2}$ and b, the rate of increase in $\dot{V}_{O_2}$ with increasing biomass. If $b = 1$, $\dot{V}_{O_2}$ increases in direct proportion to M. If b is >1, $\dot{V}_{O_2}$ increases at a proportionately greater rate than M, and if b is <1, $\dot{V}_{O_2}$ increases at a proportionately slower rate than M. Thus, b values <1 indicate that weight-specific $\dot{V}_{O_2}$ (i.e., $\dot{V}_{O_2}$ per unit body mass) decreases with increasing body mass while $b > 1$ indicates that it increases with increasing mass. Conventional wisdom suggests that b values range from 0.5 to 0.8 such that weight specific $\dot{V}_{O_2}$ decreases with increasing body mass. While generally true for vertebrates, it is less characteristic of invertebrates, and particularly of molluscs, including freshwater bivalves. Among 14 species of sphaeriids, b ranged from 0.12 to 1.45 (Hornbach, 1985). Limited data suggest that unionoideans have more typical b values, being 0.9 for *L. radiata,* and 0.77 for *P. grandis* (Huebner, 1982), with a b value range of 0.71–0.75 being reported for a number of unionoideans (Alimov, 1975). A b value of 0.63 is reported for *D. polymorpha* (Alimov, 1975). For *C. fluminea,* b ranged from 1.64–1.66 depending on season, temperature and individual condition (Williams and McMahon, unpublished data). Alimov (1975) estimated an average b of 0.73 for this species. The b value changes with season in some species (Hornbach *et al.*, 1983; Way and Wissing, 1984), but remains constant in others (Burky and Burky, 1976; Huebner, 1982) and can also vary with reproductive condition, reported to increase in some brooding adult sphaeriids (Way and Wissing, 1982) but not in others (Burky, 1983; Hornbach, 1985). Metabolic rate may also vary with physiologic state, increasing in starved individuals of *C. fluminea* (Williams and McMahon, unpublished data) and declining in *Musculium partumieum* during estivation or habitat drying (Way *et al.*, 1981).

Comparison of a values among species of *Pisidium* indicated that weight-specific $\dot{V}_{O_2}$ in this group (mean $a = 0.399$) is about 1/3 that of species of *Musculium* (mean $a = 1.605$) or *Sphaerium* (mean $a = 1.439$) (Hornbach, 1985). Reduced metabolic rate in pisidiids may reflect their reduced relative gill surface areas (Hornbach, 1985) and hypoxic interstitial suspension feeding habitats (Lopez and Holopainen, 1987), reduced metabolic demand of profundal pisidiids perhaps accounting for their high tolerance of hypoxia (Burky, 1983; Holopainen, 1987).

Annually, a in *C. fluminea* varied from −0.12 to 1.43 (mean = 0.72) (Williams and McMahon, unpublished data) while overall a for *D. polymorpha* was 0.140 (Alimov, 1975). The annual range in a for *P. grandis* was −0.13 to −1.098 (mean = −0.563) and, for *L. radiata*, −0.403 to −1.331 (mean = −0.800) (Huebner, 1982) while Alimov (1975) reported a to range from 0.016 to 0.096 for a number of unionoideans. The elevated a of *C. fluminea* and *D. polymorpha* relative to unionideans and sphaeriids (range = 0.399–1.605) indicates that both species have higher $\dot{V}_{O_2}$ and metabolic rates than other freshwater bivalves. In contrast, the low a values among unionideans reflect their relatively low metabolic rates compared to other freshwater bivalve groups.

In bivalves, shell production and tissue growth account for a large proportion of metabolic demand, requiring greater than 20% of total metabolic expenditure in young marine blue mussels, *Mytilus edulis* (Hawkins *et al.,* 1989). Thus, unionoideans with the slowest growth rates among freshwater bivalves (see section III.B.2) also have the lowest metabolic rates, while the fast-growing *C. fluminea* and *D. polymorpha* have the highest metabolic rates (McMahon, 1996, 1999).

In some sphaeriids, $\dot{V}_{O_2}$ is influenced by growth and reproductive cycles: maximal metabolic rates occurring during periods of peak adult and brooded juvenile growth (Burky and Burky 1976; Hornbach *et al.,* 1983; Way *et al.,* 1981; Way and Wissing, 1984), perhaps as a result of the elevated metabolic demands associated with tissue growth (Hawkins *et al.,* 1989), elevated $\dot{V}_{O_2}$ of brooded developmental stages, and the energetic costs of providing maternal metabolites to brooded juveniles (Mackie, 1984). In contrast, metabolic rates in *C. fluminea* (Williams and McMahon, unpublished data) and unionoideans (Huebner, 1982) were unaffected by embryo brooding, perhaps because these species do not provide maternal nourishment to brooded embryos.

Filtration rates also vary seasonally and with abiotic conditions in sphaeriids. In *S. striatinum* (Hornbach *et al.,* 1984b) and *M. partumeium* (Burky *et al.,* 1985a), filtration rates decreased with increased particle concentration and decreased temperature. Maximal filtration rates occurred during warmer summer months and peaked during reproduction. In *M. partumeium,* filtration rate and $\dot{V}_{O_2}$ declined in aestivating individuals prior to summer habitat drying (Way *et al.,* 1981; Burky, 1983). Filtration rate is directly correlated with temperature in *D. polymorpha,* being maximal in summer (MacIsaac, 1996) while it appears to be relatively temperature independent in field-acclimated specimens of *C. fluminea* (Long and McMahon, unpublished data).

Freshwater bivalves also display distinct seasonal cycles in tissue biochemical content, primarily related to reproductive cycle. Protein, glycogen and lipid contents are maximal during gonad development and gametogenesis in the freshwater unionid, *Lamellidens corrianus,* and are minimal during glochidial release (Fig. 9A); a pattern repeated in protein and lipid contents of individual tissues (Fig. 9B–D) (Jadhav and Lomte, 1982a). Similarly, overwintering, nonreproductive individuals of *C. fluminea* had twice the biomass and greater nonproteinaceous energy stores than

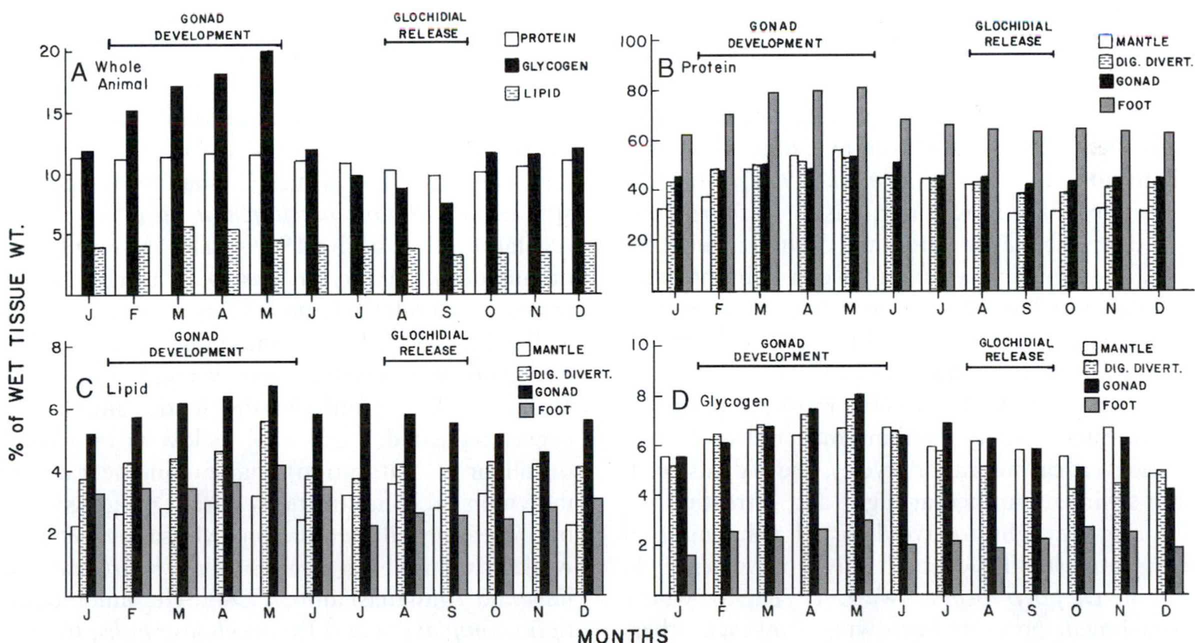

FIGURE 9 Seasonal variation in the protein, lipid, and glycogen contents of the wholebody and various tissues of the freshwater unionoidean mussel, *Lamellidens corrianus,* relative to the reproductive cycle. All organic contents are expressed as percentages of wet tissue weight. (A) Annual vartiation in whole-body contents of proteins (open histgrams), glycogen (solid histograms), and lipid (cross-hatched histograms). Remaining figures represent levels of protein (B), lipid (C) or glycogen (D) in the mantle (open histograms), digestive diverticulum (cross-hatched histograms), gonad (solid histograms), and foot (stippled histograms). Horizontal bars at the top of each figure represent reproductive cycles, indicating periods during which either gonads develop or glochidia are released. Gonad development is associated with increases in organic content, and glochidial release, with decreases in organic content of the whole body and various tissues. (From the data of Jadhav and Lomte, 1982a.)

summer reproductive individuals (Williams and McMahon, 1989). Thus, reproductive effort appears to require massive mobilization of organic energy stores from somatic tissues to support gonad development and gametogenesis in freshwater bivalves.

In the sphaeriids, *Sphaerium corneum* and *Pisidium amnicum,* increase in tissue glycogen content after fall reproduction was associated with a two-to-three-fold increase in anoxia tolerance of winter- over summer-conditioned individuals, the low glycogen contents of summer-conditioned individuals apparently reducing their capacity for anaerobic metabolism. Thus, fall accumulation of glycogen stores not only supported spring gametogenesis, but also provided anaerobic substrate for survival of winter anoxia induced by ice cover (Holopainen, 1987).

2. Diurnal Cycles

Freshwater bivalves display diurnal cycles of metabolic activity. Active Na^+ uptake peaked during dark hours in *C. fluminea* (McCorkle and Shiley, 1982) and the unionid, *Toxolasma texasiensis* (Graves and Dietz, 1980). This rhythmicity was lost in constant light, suggesting that ion uptake rates were driven by exogenous changes in light intensity. Such diurnal ion transport rhythms appear closely linked to activity rhythms. In the unioniodeans, *Ligumia subrostrata,* (McCorkle *et al.,* 1979), and European *Anodonta anatina* and *Unio tumidus* (Englund and Heino, 1994a) valve gaping activity peaks during dark periods. Rhythmic gaping in *L. subrostrata* was lost in constant light, suggesting this behavior is also driven by exogenous variation in light intensity (McCorkle *et al.,* 1979). Similarly, diurnal O_2 consumption rate rhythms in *L. subrostrata* appeared driven by light intensity, declining with increase in, and increasing with decrease in light intensity, however, it also had an endogenous component, persisting for 14 days in constant light (McCorkle *et al.,* 1979). Thus, at least some freshwater bivalves may have diurnal activity rhythms, being most active during dark hours. Such activity rhythms may reflect diurnal feeding and vertical migration cycles, individuals feeding at the sediment surface at night and retreating below it during daylight to avoid visual fish and bird predators. Interestingly, diurnal valve movements do not occur in *D. polymorpha* (Walz, 1978a, b), whose byssal attachment prevents burrowing; however, others have reported diurnal valve movement in this species (Borcherding, 1992).

3. Other Factors Affecting Metabolic Rates

Bivalve $\dot{V}_{O_2}$ can be suppressed by pollutants such as heavy metals, ammonia, and cyanide, which degrade metabolic functioning (Lomte and Jadhav, 1982a) leading to extirpation (Starrett, 1971; Williams *et al.,* 1992) or reduced growth rates (Grapentine, 1992). Increased levels of suspended solids impaired $\dot{V}_{O_2}$ and induced starvation in three unionoidean species, indicating interference with gill respiratory and filter-feeding currents (Aldridge *et al.,* 1987). It also suppresses $\dot{V}_{O_2}$ (Alexander *et al.,* 1994) and filtration rate in *D. polymorpha* (Lei *et al.,* 1996). Similary, high particle concentrations depress filtration rates in *C. fluminea* (Way *et al.,* 1990a). Bivalve metabolic rates may also be density dependent. $\dot{V}_{O_2}$ declined with increased density in the unionid, *Elliptio complanata.* $\dot{V}_{O_2}$ was three times greater in singly held individuals relative to groups of seven or more, suggesting release of a pheromone suppressing metabolic rates of nearby individuals (Paterson, 1983).

Unionoideans and sphaeriids display varying degrees of respiratory regulation when subjected to progressive hypoxia (Burky, 1983). Both *D. polymorpha* (McMahon, 1996) and *C. fluminea* (McMahon, 1979a) are highly O_2 dependent, their $\dot{V}_{O_2}$ declining proportionately with declining partial pressure of O_2 (P_{O_2}). Such species are generally relatively intolerant of prolonged hypoxia and restricted to well-oxygenated habitats. In contrast, other freshwater bivalve species are oxygen independent and regulate $\dot{V}_{O_2}$ at relatively constant levels with progressive hypoxia until a critical $\dot{V}_{O_2}$ is reached below which $\dot{V}_{O_2}$ declines proportionately with further decline in P_{O_2}. Such species can inhabit waters periodically subjected to prolonged hypoxia. Thus, the sphaeriids, *Sphaerium simile* and *Pisidium casertanum,* from hypoxic profundal habitats are relatively O_2 independent (Burky, 1983) while the shallow temporary pond species, *Musculium partumeium,* is relatively poor O_2 regulator (Hornbach, 1991). The Austrailian riverine unionoidean, *Alathyria jacksoni,* rarely experiencing hypoxia, is a relatively poor O_2 regulator while a periodically hypoxic, Australian, pond species, *Velesunio ambiguus,* was a strong oxygen regulator (Sheldon and Walker, 1989). Similarly, the lentic unionioidean, *A. cygnea* regulated $\dot{V}_{O_2}$ at a P_{O_2} as low as 14.3 Torr (0.9% of full air O_2 staturation), maintaining near constant hemolymph O_2 concentrations during progressive hypoxia by increasing gill ventilation (Massabuau *et al.,* 1991) while maintaining a near-constant heart rate (Michaelidis and Anthanasiadou, 1994). The unioniodeans, *Elliptio complanata* and *Pyganodon grandis,* from a small, Canadian, euthrophic lake, were extreme O_2 regulators, maintaining near constant $\dot{V}_{O_2}$ down to 1 mg O_2/L (P_{O_2} $\approx$18 Torr, 11.3% of full air O_2 saturation) (Fig. 10) (Lewis, 1984). Their capacity for extreme $\dot{V}_{O_2}$ regulation is adaptive, as winter ice cover made the lake severely hypoxic and overwintering individuals burrow deeply into hypoxic sediments (Lewis, 1984). In contrast, the

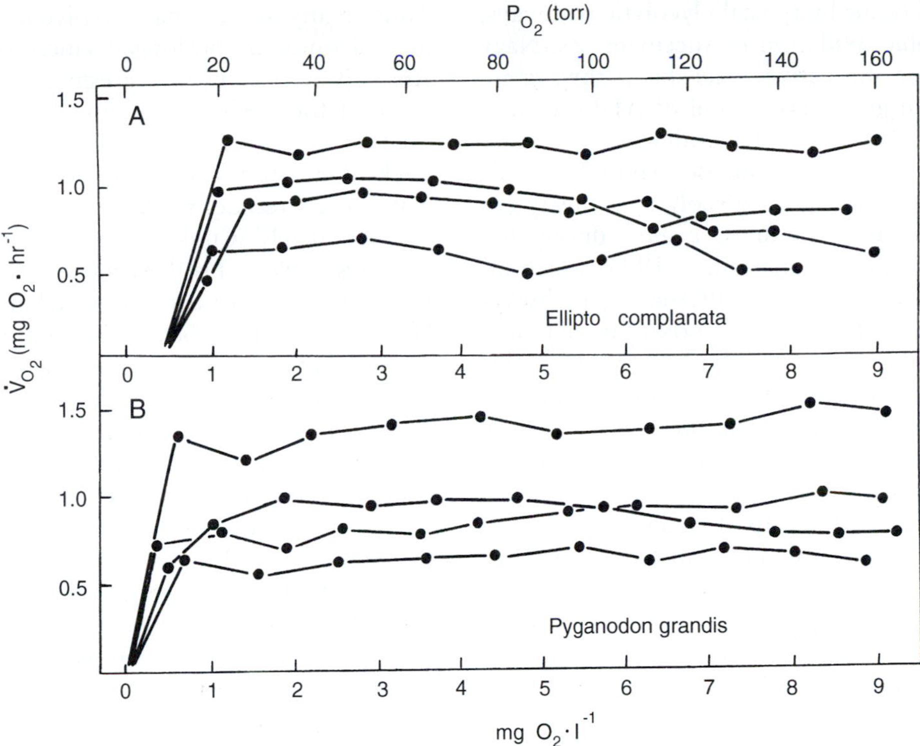

FIGURE 10 Respiratory responses of the freshwater unionoidean mussels, *Elliptio complanata* and *Pyganodon grandis* to ambient O_2 concentrations declining from near full-air saturation (8–9 mg O_2/l, lower horizontal axis, P_{O_2} = 140–160 torr or mg Hg, upper horizontal axis) to the concentration at which O_2 uptake ceases. Respiratory responses of four individuals of (A) *E. complanata*, and (B) four individuals of *P. grandis*. Both species maintained normal O_2 uptake rates at a P_{O_2} as low as 15–20 torr (9–13 percent of full-air O_2 saturation) suggesting elevated capacity for O_2 regulation of oxygen uptake. (Redrawn from Lewis, 1984.)

$\dot{V}_{O_2}$ of tropical and semitropical species not experiencing winter hypoxia tends to be more O_2 dependent (McMahon, 1979a; Das and Venkatachari, 1984), the $\dot{V}_{O_2}$ of the subtropical species, *C. fluminea,* declining to very low levels with just a 30% decline in P_{O_2} below full air O_2 saturation levels (McMahon, 1979a).

Some freshwater bivalves are very tolerant of extreme hypoxia or even anoxia allowing survival in hypoxic conditions below the thermocline of stratified lakes or above reducing substrates (Butts and Sparks, 1982; Belanger, 1991). Profundal sphaeriid species tolerate extreme hypoxia and anoxia throughout summer lake stratification (Holopainen and Jonasson, 1983), surviving 4.5 to >200 days of anoxia, depending on season and temperature (Holopainen, 1987). The unionoidean *A. cygnea* survived 22 days of anoxia (Zs.-Nagy *et al.,* 1982). In contrast, other freshwater bivalves are highly intolerant of hypoxia/anoxia, including *D. polymorpha* and *C. fluminea,* restricting them to well-oxygenated waters (Effler and Siegfried, 1994; Johnson and McMahon, 1998; Matthews and McMahon, 1999). Byssal thread production was inhibited in *D. polymorpha* below a P_{O_2} of 40 Torr (25% of full air O_2 saturation) (Clarke and McMahon, 1996b).

Individual size also affects tolerance of anoxia/hypoxia. Juveniles of the unionoidean, *Elliptio complanata,* are less hypoxia-tolerant than adults (Sparks and Strayer, 1998). In contrast, anoxia and hypoxia tolerance decreases with increased size in *C. fluminea* and *D. polymorpha* (Johnson and McMahon, 1998, Matthews and McMahon, 1999). The general trend for decreased juvenile hypoxia tolerance among freshwater bivalves may restrict some species to well-oxygenated habitats even though the adults may be hypoxia-tolerant (Sparks and Strayer, 1998).

Under anoxia, bivalves rely on anaerobic metabolic pathways which are not those of glycolysis; but, instead, involve simultaneous catabolism of glycogen and aspartate or other amino acids to yield the end products, alanine and succinate. Succinate can be further degraded into volatile fatty acids such as proprionate or acetate (Zs.-Nagy *et al.,* 1982; de Zwaan, 1983; van den Thillart and de Vries, 1985). Anoxic for 6 days, *A. cygnea* maintained 52–94% of aerobic ATP levels, higher than could occur by typical glycolytic pathways, due to its anaerobic oxidation of succinate (Zs.-Nagy *et al.*, 1982). These alternative anaerobic pathways are more efficient than glycolysis (2 mol of ATP preduced per mole of glucose catabolized), producing 4.71–6.43 mol of ATP per mole of glucose catabolized (de Zwaan, 1983). Indeed, typical glycolytic pathway enzyme activities are reduced in *A. cygnea* during hypoxia (Michaelidis and Athanasiadou, 1994) which favors metabolite catabolism by alternative pathways and allows greater tolerance of hypoxia/anoxia than less efficient glycolysis. Further, the higher ATP yields of the alternative pathways allow excretion of anaerobic endproducts, rather than retention of acidic lactate to deleterious levels as occurs in species dependent on glycolysis (de Zwaan, 1983).

During anerobiosis, buildup of acidic end products in bivalve tissues and hemolymph can cause considerable tissue and hemolymph acidosis (i.e., decrease in pH). Without respiratory pigments, freshwater bivalve hemolymph has little inherent buffering capacity, thus bivalves mobilize shell calcium carbonate into the hemolymph through the mantle epithelium (Machado *et al.*, 1990) to buffer acidosis (Heming *et al.*, 1988; Byrne and McMahon, 1994; Burnett, 1997). Thus, when anaerobic, pallial fluid pH of the unionoidean *Margaritifera margaritifera* remained highly constant, but its Ca^{2+} concentration increased (Heming *et al.*, 1988). Similarly, blood Ca^{2+} levels rose eightfold in *L. subrostrata* (Dietz, 1974) and nearly fivefold in *C. fluminea* (Byrne *et al.*, 1991a) under anoxia induced by emersion in a pure N_2 atmosphere. A three-fold increase in hemolymph Ca^{2+} concentration occurred in *Pyganodon grandis* made anoxic by 6 days air emersion (Byrne and McMahon, 1991).

The gills of unionoideans (Steffens *et al.*1985) harbor extensive extracellular calcium phosphate concretions that could buffer hemolymph pH. However, their mass increases during prolonged hypoxia and is inversely related to hemolymph pH and directly to blood Ca^{2+} concentration, suggesting that Ca^{2+} released from the shell during hypoxia is sequestered in gill concretions to prevent its loss by excretion or diffusion. On return to normoxia, release and redeposition in the shell of concretion Ca^{2+} is likely to be much less energetically demanding than replacement of lost shell Ca^{2+} from the dilute freshwater medium (Silverman *et al.*, 1983). The enzyme, carbonic anhydrase, is bound to these concretions, facilitating exchange of concretion Ca^{2+} (uptake or release) with the hemolymph (Istin and Girard, 1970a, b).

4. Desiccation Resistance

Body water content in freshwater bivalves can vary greatly seasonally and during drought periods. Thus, many species have evolved adaptations allowing tolerance of prolonged emersion and desiccation stress (Cáceres, 1997). Freshwater bivalves may be emersed for weeks or months during seasonal or unpredictable droughts (Byrne and McMahon, 1994). Lack of mobility leaves individuals stranded in air as water levels recede, while others occur in habitats that dry completely. Unlike other freshwater invertebrates (Cáceres, 1997), bivalves have no obvious structures for maintenance of aerial gas exchange when emerged, However, they have behavioral and physiological adaptations that maintain aerobic metabolism while minimizing water loss rates (Byrne and McMahon, 1994).

Survival of emersion in freshwater bivalves is correlated with capacity to control water loss. Among freshwater species, *C. fluminea* and *D. polymorpha* are highly emersion intolerant, tolerating only 8–36 days (McMahon, 1979b) or 3–27 days emersion (McMahon *et al.*, 1993), respectively, depending on temperature and relative humidity. *Dreissena bugensis* is even less emersion tolerant than *D. polymorpha* (Ussary and McMahon, 1994; Ricciardi *et al.*,1995). In these species and unionoideans, death occurs in most cases at a critical threshold of water loss regardless of temperature or relative humidity; thus, rate of water loss dictates emersion tolerance times (McMahon, 1979b; McMahon *et al.*, 1994; Byrne and McMahon, 1994; Ricciardi *et al.*, 1995). In contrast, sphaeriids inhabiting temporary ponds survive emersion for several months during summer drying (Burky, 1983). In some species, juveniles and adults survive emersion (Collins, 1967; McKee and Mackie, 1980), while in others, only juveniles survive (McKee and Mackie, 1980; Way *et al.*, 1981). Sphaeriids burrow into sediments prior to emersion (Burky, 1983) as do some unionoideans (Cáceres, 1997).

$\dot{V}_{O_2}$ declines in both emersed, aestivating *M. partumeium* (Way *et al.*, 1981) and emersed individuals of *C. fluminea* (McMahon and Williams, 1984; Byrne *et al.*, 1990). Reduction in metabolic demand in emersed individuals allows long-term maintenance on limited energy reserves.

Freshwater bivalves have a number of mechanisms for maintaining gas exchange while emersed. *Sphaerium occidentale* is emersed for several months in its ephemeral pond habitats. In air, its $\dot{V}_{O_2}$ is 20% that of aquatic rates. Gas exchange occurs across specialized pyramidal cells extending through shell punctae which allows continuous valve closure, minimizing water loss (Collins, 1967). In *C. fluminea,* aerial $\dot{V}_{O_2}$ is 21% that of aquatic rates (McMahon and Williams, 1984). Emersed specimens periodically gape and expose mantle tissue margins cemented together with mucus (Byrne *et al.*, 1988) during which high rates of aerial $\dot{V}_{O_2}$ are sustained while no O_2 uptake occurs during valve closure (McMahon and Williams, 1984).

Bursts of metabolic heat production occur during mantle edge exposure, associated with aerial oxygen consumption and apparent repayment of an O_2 debt accumulated during intervening anaerobic periods of valve closure (Byrne *et al.,* 1990). Thus, periodic mantle margin exposure allows maintenance of aerial gas exchange while greatly reducing the tissue surface area exposed and duration of exposure, greatly reducing water loss rates (Byrne and McMahon, 1994). Frequency and duration of mantle edge exposure is reduced in *C. fluminea* with increased temperature, decreased relative humidity and increasing duration of emersion, suggesting that increased desiccation pressure leads to a greater reliance on anaerobic metabolism to slow evaporative water loss (Byrne *et al.,* 1988).

Emersed unionoideans periodically expose their mantle edges (Byrne and McMahon, 1994). Emersed specimens of *Ligumia subrostata* periodically expose mantle edges and maintain aerial $\dot{V}_{O_2}$ at 21–23% of aquatic rate (Dietz, 1974). Periodic mantle edge exposure also occurs in emersed specimens of *M. margaritifera* (Heming *et al.,* 1988), *Pyganodon grandis* (Byrne and McMahon, 1991), *Pyganodon grandis, Toxolasma parvus* and *Uniomersus tetralasmus* (Byrne and McMahon, 1994). *P. grandis* has a thin shell whose margins do not completely seal when closed and a high frequency of mantle edge exposure when emersed (>25–90% of emersion time), resulting in rapid water loss and poor emersion tolerance (2–32 days). It avoids desiccation by rapid down-shore migration during reductions in habitat water levels (Byrne and McMahon, 1994). More emersion-tolerant species like *T. parvus* and *U. tetralasmus* have thicker, tightly sealing shells and their frequency and duration of mantle edge exposure is reduced with increasing duration of emersion and decreasing relative humidity, conserving water as desiccation pressure increases. *T. parvus* continuously closes the valves in latter stages of emersion, allowing it to survive emersion up to 145 days (Byrne and McMahon, 1994). *U. tetralasmus* is found in small, variable-level, lentic habitats and is highly emersion tolerant. In air, it occludes the siphons with viscous mucus, preventing direct atmospheric exposure of inner mantle tissues. Its frequency and duration of mantle edge exposure is quite low compared to other species and it ceases mantle edge exposure in the latter stages of emersion, making water loss rates lower than recorded for other freshwater bivalves and allowing it to survive emersion for up to 578 days (Byrne and McMahon, 1994).

In air, unionoideans and *C. fluminea* utilize shell Ca^{2+} to buffer accumulating HCO_3^- (Byrne and McMahon, 1994) manifested by accumulation of Ca^{2+} and HCO_3^- in mantle cavity fluids of emersed *M. margaritifera* (Hemming *et al.,* 1998) and hemolymph of emersed *C. fluminea* (Byrne *et al.,* 1991a) and *P. grandis* (Byrne and McMahon, 1991). Mantle edge exposure allows release of CO_2 generated by metabolic and shell-buffering processes in both *C. fluminea* (Byrne *et al.,* 1991a) and unionoideans (Byrne and McMahon, 1991).

In emersed *C. fluminea* there is little evidence of O_2 debt payment after re-immersion, suggesting that this and other freshwater bivalves may remain primarily aerobic while emersed (Byrne *et al.,* 1990).

The Unionoidea include the most emersion-tolerant NA freshwater bivalve species (for a review, see Byrne and McMahon, 1994). Their capacity to tolerate prolonged emersion and/or migrate vertically with changing water levels (White, 1979) may partially account for their dominance in larger NA river drainages subject to extensive seasonal water-level fluctuations. Unionoidean growth, reproduction and other life-history phenomena may be partially driven by seasonal water-level variation. Thus, anthropomorphic impoundment/regulation of flow rates and levels within these drainages could be contributing to the present decline of their unionoidean populations and species diversity (Williams *et al.,* 1992).

The mode of nitrogen excretion or detoxification during emersion is unresolved for freshwater bivalves. Ammonium ion, NH_4^+, is the major nitrogenous excretory product of aquatic molluscs (Bishop *et al.,* 1983). Due to its toxicity, NH_4^+ is generally not accumulated in emersed molluscs even though its high solubility precludes release as ammonia gas (NH_3). When emersed, aquatic snails detoxify NH_4^+ by conversion to urea or uric acid, to be excreted on re-immersion. However, most bivalves cannot convert NH_4^+ to urea or uric acid (Bishop *et al.,* 1984). Without the capacity to detoxifiy NH_4^+, how do freshwater bivalves tolerate emersion? *Corbicula fluminea,* unlike intertidal bivalves (Bishop *et al.,* 1983), does not catabolize amino acids during emersion, precluding NH_4^+ formation (Byrne *et al.* 1991b). Similarly, the unionideans, *Lamellidens corrianus* and *L. marginalis,* depend almost exclusively on carbohydrate metabolism while emersed (Lomte and Jadhav, 1982c; Sahib *et al.,* 1983). Interestingly, some unionoideans may be capable of converting ammonia to urea or other protective metabolites (Summathi and Chetty, 1990; Mani *et al.,* 1993) which may partially account for their extensive emersion tolerance.

Tolerance of emersion may also be associated with physiological and biochemical alterations in emersed individuals. Thus, individuals of *Sphaerium occidentale* and *Musculium securis* from an emersed population were more emersion tolerant than individuals from an immersed, active population (McKee and Mackie, 1980), suggesting that gradual emersion may induce emersion resistant biochemical and physiological alterations such as a shift to carbohydrate dominated catabolism and reduced metabolic demand. Such biochemical and physiological compensation may be mediated by neurosecretory hormones (Lomte and Jadhav, 1981a).

There have been almost no studies of freeze tolerance in freshwater bivalves even though many species occupy shallow, temperate habitats in which winter water-level reduction could emerse individuals in subfreezing air. Marine intertidal mytilid mussels tolerate exposure to subfreezing conditions by allowing freezing of hemolymph and interstitial fluids while preventing cell freezing, an adaptation manifested by their lack of hemolymph supercooling during freezing (Aarset, 1982). In contrast, *D. polymorpha* is intolerant of emersion below −3°C (Clarke *et al.,* 1993) and displays hemolymph supercooling (Paulkstis *et al.,* 1996). Poor freeze tolerance in *D. polymorpha* may reflect its tendency to settle at depths of >1 m, preventing winter emersion (Clarke *et al.,* 1993). Freeze sensitivity/tolerance in other freshwater bivalves (particularly shallow water species) is ripe for further study.

5. Gill Calcium Phosphate Concretions in Unionoideans

Dense calcium phosphate [$Ca_3(PO_4)_2$] concretions occur in unionoidean tissues. Mantel concretions may provide Ca^{2+} for shell deposition (Davis *et al.,* 1982; Jones and Davis, 1982; Istin and Girard, 1970a). The mantle contains high concentrations of the enzyme, carbonic anhydrase which may be bound to the concretions (Istin and Girard, 1970b). This enzyme catalyzes the reaction of CO_2 and H_2O to form bicarbonate ion (HCO_3^-), suggesting that it facilitates mobilization of concretion Ca^{2+}.

Dense extracellular calcium phosphate concretions (diameter = 1–3 μm) also occur in unionoidean gills (Silverman *et al.,* 1983, 1988; Steffens *et al.,* 1985). They develop from amorphous material initially concentrated in membrane bound vesicles in gill connective tissue cells (Silverman *et al.,* 1989). Concretions account for up to 60% of gill dry weight in some species (Silverman *et al.,* 1985). They are most dense along parallel nerve tracts oriented at 90° to the gill filaments (Silverman *et al.,* 1983; Steffens *et al.,* 1985). They are 25% protein by dry weight. One of the proteins is similar to vertebrate, calcium-binding calmodulin. This protein is most abundant in protein granules of concretion-forming cells prior to concretion mineral deposition, suggesting that it acts as site for calcium phosphate deposition (Silverman *et al.,* 1988).

Besides storing shell Ca^{2+} released to the hemolymph to buffer respiratory acidosis (see Section II.C.3), gill concretions provide a ready source of maternal Ca for shell deposition in brooded glochidia (Silverman *et al.,* 1985, 1987). Thus, gill concretion mass in brooding individuals of *L. subrostrata* and *P. grandis* was only 47 and 70% that of nonbrooding individuals, respectively (Silverman *et al.,* 1985). ^{45}Ca tracer studies indicate that 90% of glochidial shell Ca was of maternal origin in *P. grandis,* the most likely source being gill concretions, with nonmineralized Ca accounting for only 8% of glochidial Ca in *L. subrostrata* (Silverman *et al.,* 1987).

6. Water and Salt Balance

In hypo osmotic freshwater, bivalves lose ions and gain water from their hyperosmotic tissues and hemolymph. Excess body water is excreted as a fluid hypo-osmotic to tissues and hemolymph via the kidneys while ions lost in excretion and over surface epithelia are actively recovered over the gills and epithelial tissues as described in Section II.B.3 (reviewed by Deaton and Greenberg, 1991). Relatively salinity tolerant *C. fluminea* survives exposure to a salinity of 10–14 ppt (Morton and Tong, 1985), above which it remains isosmotic to the medium (Gainey and Greenberg, 1977). Unionoideans and *D. polymorpha* have lower hemolymph osmolarities than *C. fluminea* (Dietz *et al.,* 1996a) and generally lose capacity for osmotic and volume regulation above 3–4 ppt (Hiscock, 1953; Wilcox and Dietz, 1998). *C. fluminea* regulates fluid volume by actively increasing hemolymph free amino acid concentration in hyperosmotic media (Gainey and Greenberg, 1977; Matsushima *et al.,* 1982), preventing water loss by active equilibration of hemolymph and medium osmolarity. Such volume regulation does not occur in other freshwater bivalves because their low tissue osmolarities limit the free amino acid pool (Yamada and Matsushima, 1992; Dietz *et al.,* 1996a,b, 1998). Volume regulation may have been lost in unionoideans and *D. polymorpha,* because they are not exposed to hyperosmotic conditions in freshwater. Instead, both groups retain only limited cell volume regulation through regulation of intracellular inorganic ion concentrations (Dietz *et al.,* 1996a, b, 1998). The elevated osmotic concentration and free amino acid pool volume regulation of *C. fluminea* reflects its recent evolution from an estuarine ancestor (Dietz, 1985) in which such adaptations are common (Deaton and Greenberg, 1991).

Both unionoideans and *C. fluminea* respond to maintenance in very dilute media by increasing active Na^+ uptake rate to maintain hemolymph ion concentration (Dietz, 1985). The activity of (Na^+ and K^+)-activated ATPase, an enzyme required for active Na^+/K^+ transport, increased in mantle and kidney tissues of salt-depleted *C. fluminea,* suggesting activation of Na^+ transport. This response did not occur in the salt-depleted

Lampsilis straminca claibornesis, suggesting that active ion uptake regulation has been lost in unionoideans with a longer freshwater fossil history than *C. fluminea* (Deaton, 1982). In *C. fluminea* Cl^- is the major blood ion, maintained by elevated active epithelial Cl^- uptake. In contrast, in the unionoidean, *Toxolasma texasiensis,* Cl^- and HCO_3^- are equivalent hemolymph anionic components. Because HCO_3^- can be readily mobilized by respiratory and metabolic processes, unionoideans may depend to a greater extent on active Na^+ relative to Cl^- uptake to maintain hemolymph osmotic balance (Byrne and Dietz, 1997). The anionic hemolymph component of *D. polymorpha,* like that of *C. fluminea,* is dominated by Cl^-, but this species actively absorbs equal levels of Na^+ and Cl^- to maintain ionic stasis, suggesting co-transport of these ions (Horohov *et al.,* 1992). Retention of Cl^- as the main hemolymph anion in *C. fluminea* and *D. polymorpha* reflects their recent evolution from estuarine ancestors (Deaton and Greenberg, 1991; Byrne and Dietz, 1997; Horohov *et al.,* 1992), while the greater dependence of *C. fluminea* on Cl^- uptake for ionic regulation and its increased hemolymph osmotic concentration (Dietz *et al.,* 1994) suggests that it entered freshwater more recently than *D. polymorpha.* Interestingly, although *D. polymorpha* can hyperosmotically regulate hemolymph ion concentrations in dilute freshwaters, it is less tolerant of elevated medium ion concentrations than other bivalves restricting its penetration of estuarine habitats (Horohov *et al.,* 1992; Dietz *et al.,* 1994, 1998; Wilcox and Dietz, 1998).

In the Asian unionoidean, *Anodonta woodiana,* mantle cavity water osmolarity at 34 mosmol/L was 76% that of the hemolymph (45 mosmol/L) with pallial fluid Na^+, K^+, and Cl^- concentrations being 71, 76 and 72% those of the hemolymph, respectively. Maintenance of elevated mantle water ion concentrations suggests that it acts as an osmotic buffer, reducing the gradient for, and, thus, the rate of diffusive ion loss to the dilute freshwater medium (Matsushima and Kado, 1982). However, loss of a hyperosmotic mantle fluid to the external medium through the exhalant siphon could also be a major route for ion loss.

In unionoideans and *C. fluminea,* the enzyme, carbonic anhydrase (CA), which catalyzes formation of carbonic acid (H_2CO_3) from water and carbon dioxide, occurs in gill and mantle tissues (Henery and Saintsing, 1983). H_2CO_3 degrades into H^+ and CO_3^- (bicarbonate ion) which are counter ions for active Na^+ and Cl^- uptake (Horohov *et al.,* 1992; Dietz *et al.,* 1994; Byrne and Dietz, 1997). Gill and mantle CA activity increases when freshwater bivalves are held in extremely dilute media while CA inhibition by actazolamide results in reduced Na^+ and Cl^- uptake rates, strong evidence of the role of CA in ion regulation (Henery and Saintsing, 1983).

Freshwater bivalve hemolymph osmotic concentrations are the lowest recorded for multicellular invertebrates (Dietz, 1985; Deaton and Greenberg, 1991). Among unionoideans, hemolymph osmolarity for the European species, *Anodonta cygnea,* was 40–50 mosmol/L or 4–5% that of seawater. *D. polymorpha* has the lowest hemolymph concentration of all freshwater bivalves (30–36 mosmol/L or 3.0–3.6% that of seawater) (Dietz *et al.,* 1994), while *C. fluminea* has the highest value at 65 mosmol/L (6.5% that of seawater) (Zheng and Dietz, 1998a). That for other freshwater invertebrates ranges from 100 to 400 mosmol/L or 10–40% that of seawater (Burky, 1983). Low hemolymph osmotic concentrations in freshwater bivalves reduce the gradient for transepithelial osmosis and ion diffusion across their extensive mantle and gill epithelial surfaces to that which can be balanced by water excretion and active ion uptake at energetically feasible levels (Burton, 1983). Hydrostatic pressure generated by heart beat in *A. cygena* allows filtration of enough hemolymph plasma into the pericardial space to account for urine production and is twice that of the marine clam, *Mya arenaria,* suggesting that freshwater bivalves can excrete water at higher rates than less hypoosmotically stressed estuarine species (Jones and Peggs, 1983). Exposure of *Pyganodon* sp. to a very dilute medium induced formation of extensive extracellular membrane spaces in the deep infoldings of kidney epithelial cells, perhaps increasing surface area for active ion uptake from excretory fluids producing a more dilute urine and/or allowing increased excretion of excess water (Khan *et al.,* 1986).

Increasing osmotic gradients for water uptake from marine through estuarine into freshwater habitats required that the external epithelia of bivalves became increasing impermeable to water with their evolutionary transition through these environments. Thus, increased epithelial osmo-resistance characterizes all freshwater bivalves (Deaton and Greenberg, 1991; Dietz, 1985; Dietz *et al.,* 1994; Byrne and Dietz, 1997; Zheng and Dietz, 1998a). Among freshwater bivalves, *C. fluminea* is least osmotically permeable (Zheng and Dietz, 1998a) and *D. polymorpha* most permeable (Dietz *et al.,* 1995), while unionoideans are of intermediate permeability (Dietz *et al.,* 1996a, b; Zheng and Dietz, 1998a). The "osmotically tight" epithelium of *C. fluminea* allows it to maintain higher hemolymph osmolarity than other freshwater bivalves while producing normal excretory fluid volumes (Zheng and Dietz, 1998a). In contrast, "osmotically leaky" epithelium in *D. polymorpha* results in very elevated excretory fluid production (Dietz *et al.,* 1995), requiring extremely high rates of epithelial ion uptake to replace ions lost in voluminous excretory fluids (Dietz and Byrne, 1997). Thus, the main osmoregulatory adaptations of *C. fluminea* and *D. polymorpha* are quite different, the former being reduced epithelial permeability, and that of the latter, increased water excretion and active epithelial ion uptake.

Hormones regulate osmotic control in freshwater bivalves. Cyclic AMP (cAMP) stimulates active Na^+ uptake by unionoideans, while prostglandins inhibit it. In contrast, prostglandin inhibitors stimulate N^+ uptake (Dietz *et al.,* 1982; Graves and Dietz, 1982; Saintsing and Dietz, 1983). Serotonin stimulates tissue accumulation of cAMP, increasing active Na^+ uptake. Thus, an antogonistic relationship between serotonin and prostaglandins modulates adenylate cyclase-catalyzed cAMP stimulation of active Na^+ uptake. Not surprisingly, high concentrations of serotonin occur in unionoidean gill nerve tracts (Dietz, 1985; Dietz *et al.,* 1992). Gill concentrations of the neurotransmitters, dopamine and norepinephrine, greatly declined in unionoideans salt-depleted in extremely dilute mediums, while serotonin was regulated at near-normal levels, suggesting that serotonin regulates Na^+ uptake for ionic balance. The circadian rhythms of Na^+ uptake in freshwater clams (Graves and Dietz, 1980; McCorkle-Shiley, 1982) may be mediated by an antagonistic serotonin/prostaglandin hormonal system (Dietz, 1985).

When cerebropleural or visceral ganglia were ablated, individuals of the Indian unionoidean, *Lamellidens corrianus,* rapidly lost osmoregulatory capacity which was restored by injection of ganglia extracts, indicating that ganglion neurosectory hormones are involved in water balance (Lomte and Jadhav, 1981b). The affinity of the pedal ganglion of *A. cygena* for monoamines controlling ion/water balance is temperature dependent indicative of a seasonal component to osmoregulation (Hiripi *et al.,* 1982).

III. ECOLOGY AND EVOLUTION

A. Diversity and Distribution

North American freshwater bivalves distributions, particularly for unionoideans, have been well described. Species distribution maps for unionoideans and sphaeriids exist for Canada (Clarke, 1973), the United States (LaRocque, 1967a) and for the whole of North American (Burch, 1975a, b; Parmalee and Bogan, 1998). LaRocque (1967b) describes living and Pleistocene fossil assemblages at specific NA localities. There is also a massive literature, too numerous to cite here, describing species occurrences or species assemblages in various NA drainage systems.

Native (non-introduced) NA sphaeriid species have broad distributions, often extending from the Atlantic to Pacific coasts. Introduced to NA from southeast Asia in early 1900s (McMahon, 1999), *Corbicula fluminea* (i.e., light-colored shell morph of *Corbicula*) has a similarily widespread NA distribution, inhabiting drainages on the west coast of the United States, the southern tier of states, and throughout states east of the Mississippi River, with the exception of the most northern states, and into northern Mexico (McMahon, 1999) (Fig. 11). A second, unidentified species of *Corbicula* (i.e., the dark-colored shell morph) is restricted to isolated, spring-fed drainages in southcentral Texas and southern California and Arizona (Fig. 11) (Britton and Morton, 1986). In contrast, NA unionoidean species generally have more restricted distributions. Few species range on both sides of the continental divide and a large number are limited to single drainage systems (Burch, 1975b; LaRocque, 1967a).

1. Dispersal

Widespread NA distributions of sphaeriids and *Corbicula* relative to unionoideans reflect fundamental differences in their dispersal capacities. Unionoideans depend primarily on host fish glochidial transport for dispersal (Kat, 1984), their ranges reflecting those of their host fish species (Haag and Warren, 1998). While host fish glochidial transport increases probability of dispersal into favorable habitats, as host fish and adult unionoidean habitat preferences generally coincide (Kat, 1984), it limits the extent of dispersal, leading to highly endemic species. For example, electrophoretic studies of peripheral Nova Scotian populations of unionoideans suggest that they invade new habitats mostly by host fish dispersal (Kat and Davis, 1984), barriers to host fish dispersal being barriers to unionid dispersal. Thus, distributions of modern and fossil NA interior basin unionoidean assemblages are limited to areas below major waterfall barriers to fish host upstream migration in the Lake Champlain drainage system of New York, Vermont, and Quebec (Smith, 1985a) and re-establishment of *Anodonta implicata* populations in the upper Connecticut River drainage closely followed restoration of its anadromous glochidial clupeid fish host populations, by construction of fishways past numerous impoundments preventing upstream fish host dispersal (Smith, 1985b). Vaughn and Taylor (2000) have found that >50% of the variation in unionoidean assemblages is associated with regional distribution and abundances of fishes, indicating that fish community structure is a determinant of mussel community structure.

Sphaeriids and *C. fluminea* have evolved dispersal mechanisms that make them more invasive than unionoideans, accounting for their more cosmopolitan distributions. Juvenile sphaeriids disperse between

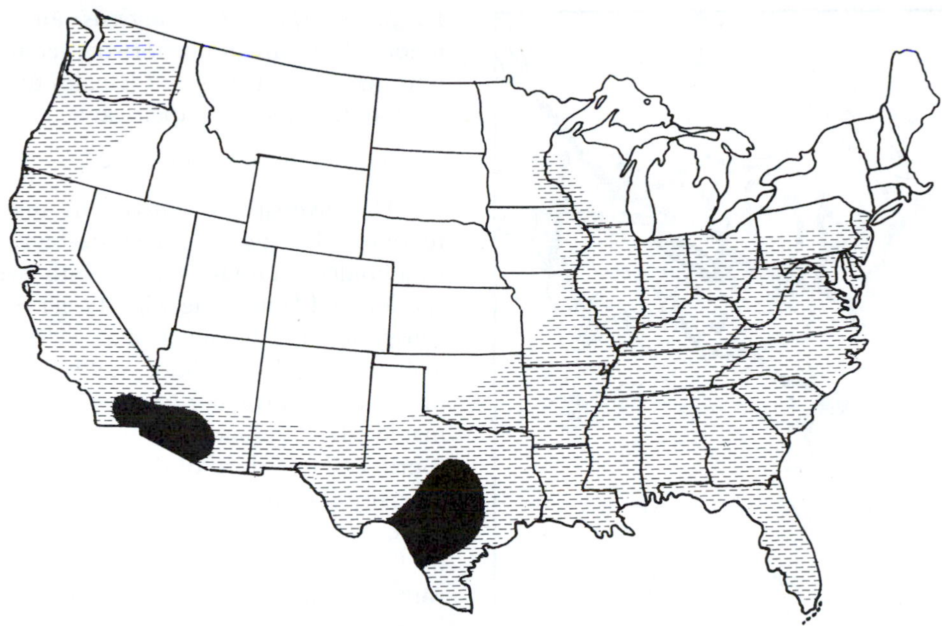

FIGURE 11 Distribution of *Corbicula* in the United States. Hatched area is the distribution of the light-colored morph of *Corbicula, Corbicula fluminea*. The solid areas are the distribution of the dark-colored morph of *Corbicula*, yet to be assigned a species designation.

drainage systems by clamping shell valves onto limbs of aquatic insects, feathers of water fowl (Burky, 1983), or even limbs of salamanders (Davis and Gilhen, 1982) first noted by Darwin (1882). Some sphaeriids survive ingestion and regurgitation by ducks, which commonly feed on them, allowing long-distance dispersal (Burky, 1983). The rapid spread of *C. fluminea* through NA drainage systems (McMahon, 1999), while partly anthropomorphically mediated, also resulted from its natural dispersal capacities. The long juvenile mucilaginous byssal thread or filamentous algae on which juveniles settle becomes entangled in the feet or feathers of shore birds or water fowl, making them transport vectors (McMahon, 1999). Its natural dispersal capacity has allowed it to spread into areas of northern Mexico where human-mediated transport is highly unlikely (Hillis and Mayden, 1985), and into southern Britain during interglacial periods (McMahon, 1999). *Dreissena polymorpha,* byssally attaches to floating wood or boat and barge hulls, facilitating long distance transport (Mackie and Schloesser, 1996), and to macrophytic vegetation that can be transported across drainages by nesting shore birds and water fowl.

Juveniles of *C. fluminea* can be passively transported long distances downstream suspended in water currents (McMahon, 1999). Water currents also disperse the planktonic veliger of *D. polymorpha* (Mackie and Schloesser, 1996). Adult *C. fluminea* can be carried downstream over substratum by water currents (Williams and McMahon, 1986), a process facilitated by production of a mucus dragline from the exhalant siphon (Prezant and Charlermwat, 1984). Passive hydraulic dispersal of juvenile and adult *C. fluminea* not only accounts for its extraordinary ability to invade downstream portions of drainages after introduction (1999), but also leads to its impingement and fouling of industrial, agricultural, and municipal raw-water systems (1999). Similarly, current-mediated transport of free-swimming veliger and passively suspended juvenile *D. polymorpha,* and of adults attached to floating substrates or carried as clumps of individuals over the bottom accounts for its dispersal in European drainage systems after escape from the Caspian Sea (Mackie and Schloesser, 1996). This species rapidly spread downstream throughout the Great Lakes and St. Lawrence River from its original upstream introduction into Lake St. Clair in 1985–1986 (Mackie and Schloesser, 1996). It has also been carried throughout most of the Mississippi River and adjoining tributaries both by downstream hydrological transport and upstream by attachment to the hulls of commercial barges (Mackie and Schloesser, 1996) (Fig. 12). Zebra mussels invaded the lower Great Lakes, St. Lawrence River and Erie-Barge Canal by 1990, later invading portions of the upper Great Lakes, the Hudson River, and the Finger Lakes. *D. polymorpha* entered the Mississippi Drainage from Lake Michigan through the Illinois River and spread

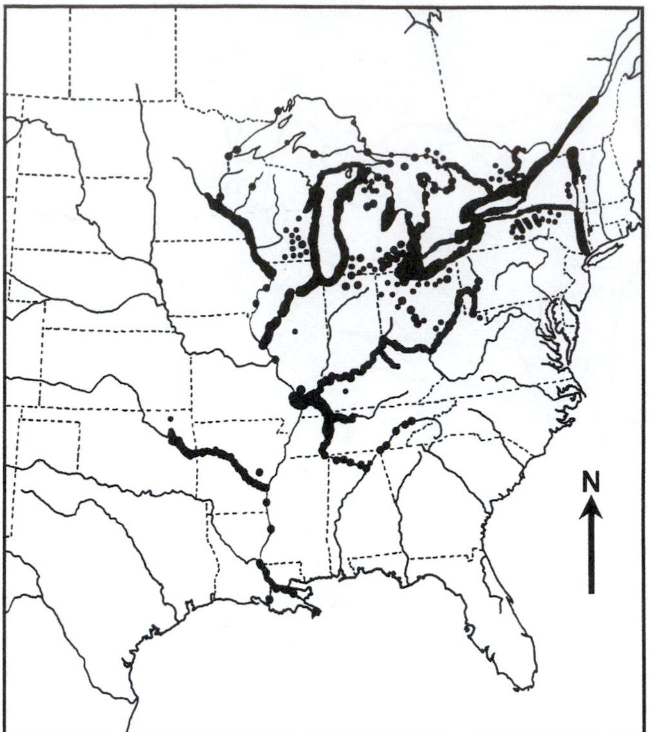

FIGURE 12 Distribution of the zebra mussel, *Dreissena polymorpha*, in North America as of Spring, 2000.

downstream to the Mississippi Delta and upstream to La Crosse, Wisconsin. It had invaded all major tributaries of the Mississippi River by 1998 with the exception of the Missouri River, where the first specimens were reported upstream of its confluence with the Platte River in 1999. As of early 2000, *D. polymorpha* occupied drainages in 21 eastern and midwestern U.S. states and in the Canadian provinces of Ontario and Quebec, including 62 confirmed populations in small, isolated inland lakes (Fig. 12). *Dreissena bugensis,* the second dreissenid species occupying NA inland waters is sympatric with *D. polymorpha* in Lakes Huron, Erie, and Ontario and the western end of the Erie-Barge Canal (New York Sea Grant, 1999).

Capacity for downstream transport and byssal attachment has made *D. polymorpha* a major NA biofouling pest species, recapitulating its history in Europe (Claudi and Mackie, 1993). Juvenile sphaerids are also passively, hydrologically transported downstream (McKillop and Harrison, 1982) which may be an important dispersal mode for many species in this family. In contrast, hydrological transport is extremely rare in unionoideans (Imlay, 1982).

As both sphaeriids and *C. fluminea* are self-fertilizing hermaphrodites (see Section II.B.5), a single individual can found a new population. In contrast, *D. polymorpha, D. bugensis* and the majority of unionoideans are gonochoristic, requiring simultaneous introduction of males and females to found a new population, thus reducing their capacity as invaders.

2. Anthropomorphic Impacts

The diversity of native NA unionoid bivalves is represented by two of the six recognized families of the Unionoidea: Margaritiferidae with two genera and five species; and Unionidae with 49 genera, 278 species, 13 subspecies (Turgeon *et al.,* 1998; Johnson, 1998; Williams and Fradkin, 1999). North American unionoidean diversity represents 51 of the approximately 165 unionoidean genera, and between one-fourth and one-third of the world's unionoidean species diversity (Bogan, 1993; Bogan and Woodward, 1992). Neves *et al.* (1997) documented that 261 of these taxa or 91% of NA unionoidean diversity occurs in the southeastern United States where it is focused in the Mobile Bay, Tennessee and Cumberland River basins. Alabama, with parts of the Mobile Bay and Tennessee River basins, has the greatest unionoidean diversity with 175 taxa followed by Tennessee with 129 taxa.

This great diversity of unionoidean bivalves began to be impacted as it began to be described, with European expansion across North America. It was noticed quite early on that the unionoidean fauna was declining (Higgins, 1858). By the turn of the 20th century, people were observing that the unionoidean fauna of entire regions was being decimated or had disappeared. Rhoads (1899) described extirpation of mussels from the Monongahela River at Pittsburgh due to damming and pollution. Ortmann (1909) noted the loss of the freshwater mussels, crayfish and fish fauna from the upper Ohio River Basin in western Pennsylvania due to acid run-off from coal mines and the complete destruction of the Pigeon River unionoidean fauna in east Tennessee by pollution (Ortmann, 1918). Van der Schalie (1938) warned that the Tennessee Valley Authority's construction of dams on the Tennessee River could lead to long-term negative impacts on, and the eventual destruction of, its freshwater fauna.

Extinction of NA freshwater bivalve species was not reported until Stansbery (1970, 1971) listed 11 presumed extinct taxa. Turgeon *et al.,* (1988) listed 13 taxa, Williams *et al.,* (1993) listed 12% of the unionoid taxa as extinct, and most recently Turgeon *et al.,* (1998) listed 35 taxa as presumed extinct (Table II). As of 1997, 77 NA unionid taxa were endangered, 43 threatened and 73 of special concern with only 70 species listed as currently stable and additional species being added to federal and state lists yearly (Williams *et al.,* 1993; Neves *et al.,* 1997). Massive historical losses of unionoideans have been revealed by comparison of

TABLE II Extinct Freshwater Unionoidean Bivalves

Species	*Common name*	*States of occurrence*
Alasmidonta mccordi Athearn, 1964	Coosa elktoe	AL
Alasmidonta robusta Clarke, 1981	Carolina elktoe	NC, SC
Alasmidonta wrightiana (Walker, 1901)	Ochlockonee arcmussel	FL
Elliptio nigella (Lea, 1852)	Winged spike	AL, GA
Epioblasma arcaeformis (Lea, 1831)	Sugarspoon	AL, KY, TN
Epioblasma biemarginata (Lea, 1857)	Angled riffleshell	AL, KY, TN
Epioblasma flexuosa (Rafinesque, 1820)	Leafshell	AL, IL, IN, KY, OH,TN
Epioblasma florentina florentina (Lea, 1857)	Yellow blossom	AL, KY, TN
Epioblasma haysiana (Lea, 1834)	Acornshell	AL, KY, TN, VA
Epioblasma lenior (Lea, 1842)	Narrow catspaw	AL, TN
Epioblasma lewisii (Walker, 1910)	Forkshell	AL, KY, TN
Epioblasma obliquata obliquata (Rafinesque, 1820)	Catspaw	AL, IL, IN, KY, OH, TN
Epioblasma personata (Say, 1829)	Round combshell	IL, IN, KY, OH
Epioblasma propinqua (Lea, 1857)	Tennessee riffleshell	AL, IL, IN, KY, OH,TN
Epioblasma sampsonii (Lea, 1861)	Wabash riffleshell	IL, IN, KY
Epioblasma stewardsonii (Lea, 1852)	Cumberland leafshell	AL, KY, TN
Epioblasma torulosa gubernaculum (Reeve, 1865)	Green blossom	TN, VA
Epioblasma torulosa torulosa (Rafinesque, 1820)	Tubercled blossom	AL, IL, IN, KY, OH,TN, WV
Epioblasma turgidula (Lea, 1858)	Turgid blossom	AL, AR, TN
Lampsilis binominata Simpson, 1900	Lined pocketbook	AL, GA
Medionidus mcglameriae van der Schalie, 1939	Tombigbee moccasinshell	AL
Pleurobema altum (Conrad, 1854)	Highnut	AL, GA
Pleurobema avellanum Simpson, 1900	Hazel pigtoe	AL
Pleurobema bournianum (Lea, 1840)	Scioto pigtoe	OH
Pleurobema chattanoogaense (Lea, 1858)	Painted clubshell	AL, GA, TN
Pleurobema flavidulum (Lea, 1861)	Yellow pigtoe	AL
Pleurobema hagleri (Frierson, 1900)	Brown pigtoe	AL
Pleurobema hanleyianum (Lea, 1852)	Georgia pigtoe	AL, GA, TN
Pleurobema johannis (Lea, 1859)	Alabama pigtoe	AL
Pleurobema murrayense (Lea, 1868)	Coosa pigtoe	AL, GA, TN
Pleurobema nucleopsis (Conrad, 1849)	Longnut	AL, GA
Pleurobema rubellum (Conrad, 1834)	Warrior pigtoe	AL, GA, TN
Pleurobema troschelianum (Lea, 1852)	Alabama clubshell	AL, GA, TN
Pleurobema verum (Lea, 1861)	True pigtoe	AL
Quadrula tuberosa (Lea, 1840)	Rough rockshell	TN, VA

(From Williams *et al.*, 1998b.)

present species assemblages with those of earlier surveys or with recent fossil assemblages (Neves and Zale, 1982; Parmalee and Klippel, 1982, 1984; Parmalee *et al.*, 1982; Ahlstedt, 1983; Havlik, 1983; Stern, 1983; Hoeh and Trdan, 1984; Miller *et al.*, 1984; Hartfield and Rummel, 1985; Taylor, 1985; Mackie and Topping, 1988; Nalepa and Gauvin, 1988; Starnes and Bogan, 1988; Baily and Green, 1989; Bogan, 1990; Anderson *et al.*, 1991; Counts *et al.*, 1991; Hornbach *et al.*, 1992; Parmalee and Hughes, 1993; Hoke, 1994; Blalock and Sickel, 1996). Extirpations of sphaeriid faunae are far less common, but have occurred (Paloumpnis and Starrett, 1960; Mills *et al.*, 1966). Unionoidean assemblages in Indian middens near the upper Ohio River yielded at least 32 species, while a 1921 survey yielded only 25 of the midden species, and a 1979 survey, only 13 of the midden species in the same area, indicative of massive species extirpation (Taylor and Spurlock, 1982). Similar historical loss of Indian midden species has occurred in the Tennessee River Drainage (Parmalee, 1988). Further evidence of environmental change in the upper Ohio River includes recent establishment of 15 unionid species previously unreported on Indian middens or earlier surveys (Taylor and Spurlock, 1982).

Such historical data clearly indicate the toll that anthropomorphic activities are taking on the NA unionoidean fauna (Neves, 1993; Neves *et al.*, 1997). The United States Fish and Wildlife Service began listing unionoideans as threatened and endangered after passage of the Endangered Species Act of 1973. By 1998, 56 taxa were listed as endangered (Turgeon *et al.*, 1998), with about 70% of NA unionoideans at some level of imperilment (Williams *et al.*, 1993). Bogan (1998) reviewed the causes for decline of NA freshwater bivalve diversity and attributed it, and species extinctions, to habitat destruction (loss of both unionoidean and host fish habitat), pollution including acid mine runoff, pesticides, heavy metals, commercial exploitation and introduced species. Bogan (1997) summarized this:

> "The central cause of this decline and decimation of the freshwater molluscan fauna is the modification and destruction of their aquatic habitat, with sedimentation as a leading major factor. Sources of sedimentation include poor agricultural and timbering practices. Damming of major rivers has also had a dramatic impact on this fauna with the loss of unionid obligate host fish due to changes in local water quality and loss of habitat. In-stream gravel mining, dredging, and canalization have further eliminated stable aquatic habitat. Acidic mine drainage and various point and non-point pollution sources also continue to decimate local aquatic mollusk populations."

Evidence of negative anthropomorphic impacts on unionoidean populations is extensive. The freshwater pearling industry can extirpate entire populations (Laycock, 1983), overfishing for pearls being a major factor in the decline of the pearl mussel, *Margaritifera magaritifera* in Great Britain (Young and Williams, 1983a). Commercial unionid shell fisheries that provide seed pearls for the marine cultured pearl industry negatively impact NA populations (Williams *et al.,* 1993). Impoundments of rivers slow flow and allow accumulation of silt, leading to mussel fauna reductions (Duncan and Thiel, 1983; Parmalee and Klippel, 1984; Stern, 1983; Starnes and Bogan, 1988; Blay, 1990; Williams *et al.,* 1992; Houp, 1993; Parmalee and Hughes, 1993). Impoundments also eliminate glochidial fish hosts (Mathiak, 1979) or prevent fish host dispersal of glochidia. Release of cold, hypolimnetic water from impoundments negatively impacts downstream unionoidean populations (Ahlstedt, 1983; Clarke, 1983; Vaughn and Taylor, 1999). Controlled water releases from impoundments lead to major flow-rate oscillations, either scouring the bottom of suitable mussel substrates during high flows or causing lethal aerial exposure during low flows (Miller *et al.,* 1984; Vaughn and Taylor, 1999). Channelization of drainage systems for navigation or flood control is detrimental to unionoideans. Increased flow velocity and propeller wash elevate suspended solids, which interfere with mussel filter feeding and O_2 consumption (Aldridge *et al.,* 1987; Payne and Miller, 1987). It reduces availability of stabilized sediments, sand bars, and low-flow areas, all preferred unionoidean habitats (Payne and Miller, 1989; Strayer and Ralley, 1993; Strayer, 1999a).

Pollution adversely affects bivalves. Unionoidean faunas can be negatively impacted or extirpated by industrial pollution (Zeto *et al.,* 1987; Wade *et al.,* 1993), urban wastewater effluents (sewage, silt, pesticides) and resultant eutrophication and hypoxia (Gunning and Suttkus, 1985; St. John, 1982; Neves and Zale, 1982; Arter, 1989; Strayer, 1993), silt and acid discharges from mines (Taylor, 1985; Warren *et al.,* 1984; Anderson *et al.,* 1991) siltation from bank erosion due to deforestation, destruction of riparian zones, and poor agricultural practice (Hartfield, 1993; Williams *et al.,* 1993) or disturbance and silt from river-bed gravel mining (Brown and Curole, 1997). Advent of modern sewage treatment on the Pearl River, Louisiana, allowed re-establishment of five previously absent unionid species (Gunning and Suttkus, 1985).

Nonindigenous species (i.e., "biological pollution") present a new threat to NA unionoidean taxa. The nonindigenous zebra mussel, *D. polymorpha,* byssally attaches in great numbers to the exposed, posterior, siphonal shell regions of unionoideans, leading to their slow starvation as the zebra mussels strip suspended food particles from their inhalant current. Thus, zebra mussel infestations have extirpated a number of unionoidean species populations from the lower Great Lakes (Schloesser *et al.,* 1996).

3. Physical Factors

Physical factors influence bivalve distributions. While environmental requirements are species specific, a number of generalities appear warranted. Sediment type clearly affects distribution patterns. Unionoideans are generally most successful in areas where flow is moderate with a stable substrate of coarse sand or sand–gravel mixtures, and are generally absent from substrates with heavy silt loads and very low water flow (Strayer and Ralley, 1993; Strayer, 1999a). In the Wisconsin and St. Croix rivers, only 7 of 28 unionid species occurred in sand–mud sediments, the majority preferring sand–gravel mixtures. Only three species, *Pyganodon grandis, Lampsilis teres,* and *L. siliquoidea,* preferred sand–mud substrates (Stern, 1983). Some unionoidean species select preferred sediment types (Baily, 1989). In rivers subjected to periodic high flows, unionoideans oriented with siphons facing upstream to a greater extent than those in stable flow rivers, a position which presents the narrowest profile to the current, reducing chances of flow-induced dislodgment (Di Maio and Corkum, 1997). The relatively specific substrate and flow requirements of many unionoidean species may account for their patchy and clumped distributions (Strayer, 1999a). Courser sediments allowing free exchange of interstitial and surface waters may be a requirement for early survival of recently excysted juvenile unionoideans with low pH, hypoxia, and ammonia accumulation in interstitial water being correlated with juvenile mortality in sediments where free water exchange is reduced (Buddensiek *et al.,* 1993). Thus, sediment limitations for juvenile development may result in patchy or clumped adult unionoidean distributions. In contrast, *C. fluminea* is able to colonize

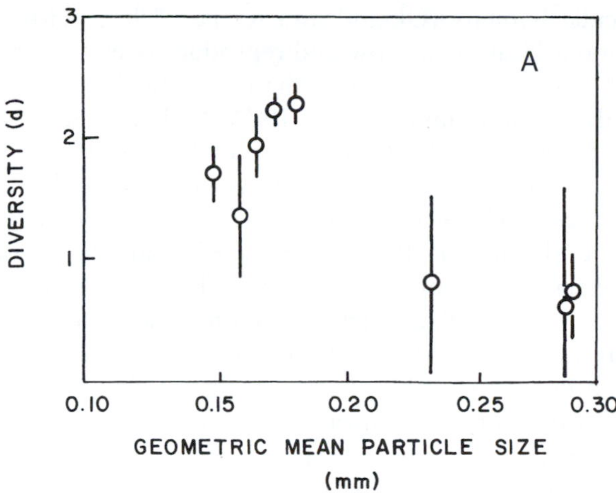

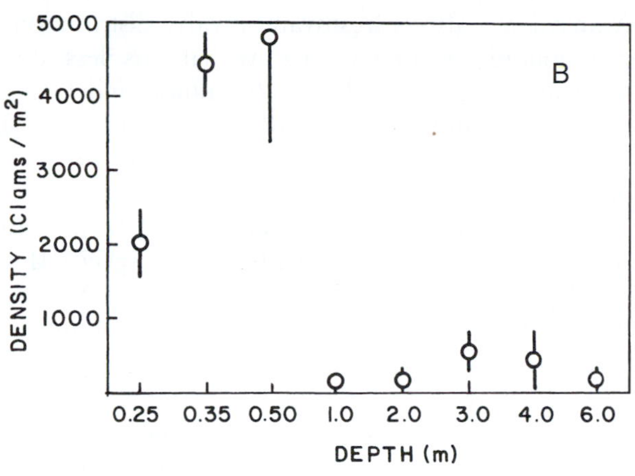

FIGURE 13 Sediment relationships in sphaeriid clam communities from sites along a depth transect in Britannia Bay, Ottawa River, Canada. (A) Mean sphaeriid density (Shannon–Weaver *d*) values for various sphaeriid communities in relation to mean sediment geometric particle size. (B) Mean sphaeriid density values at various depths. Vertical bars about points are 95 percent confidence limits. Note increase in diversity with decrease in mean sediment particle size and maximization of density at depths less than one meter. (Redrawn from data of Kilgour and Mackie, 1988.)

habitats ranging from bare rock through gravel/sand to silt (McMahon, 1999), which has allowed it to invade a wide variety of NA drainage systems. Its optimal habitat is oxygenated sand or gravel–sand (Belanger *et al.*, 1985).

In contrast to unionoideans, species diversity in *Pisidium* increases with decreasing particle size (Fig. 13A), becoming maximal at a mean particle diameter of 0.18 mm (Kilgour and Mackie, 1988). In southeastern Lake Michigan, *Pisidium* density and diversity were maximal in very fine sand–clay and silt–clay sediments, while peak *Sphaerium* diversity occurred at somewhat larger particle sizes (Zbeda and White, 1985). Thus, there are differences in substrate preferences among sphaeriids, perhaps associated with sediment organic detritus feeding mechanisms in *Pisidium* (see Section III.C.2).

Apparent differences in substrate preferences may be associated with species-specific differences in optimal water velocities. Unionoideans are most successful where velocities are low enough to allow sediment stability, but high enough to prevent excessive siltation (Strayer and Ralley, 1993; Strayer, 1999a), making well-oxygenated, coarse sand and sand–gravel beds optimal habitats for riverine species with such habitats being more critical for survival of juvenile unionoideans than adults (Buddensiek *et al.*, 1993). Low or variable velocities allow silt accumulations that make sediments too soft for maintenance of position in adult unionoideans (Lewis and Riebel, 1984; Salmon and Green, 1983) or which interfere with their filter feeding and gas exchange (Aldridge *et al.*, 1987). In contrast, periodic scouring of substrates during high flows removes substrate and unionoideans, and prevents their successful resettlement (Young and Williams, 1983b). Indeed, unionoidean densities decline in areas of high flow (Way *et al.*, 1990b). Sediment type did not affect burrowing in three lotic unionid species (*P. grandis, Elliptio complanata,* and *Lampsilis siliquoidea)* (Lewis and Riebel, 1984), suggesting that it is not involved in substrate preferences, but unionoideans may move to preferred sediments (Baily, 1989). *C. fluminea,* with its relatively heavy, ridged shell and rapid burrowing ability, is better adapted for life in high current velocities and unstable substrates than most unionoideans (McMahon, 1999). In the Tangipahoa River, Mississippi, it colonizes unstable substrates from which unionoideans are excluded (Miller *et al.*, 1986). In contrast to the majority of unionoideans and *C. fluminea,* many sphaeriid species occur in small ponds and the profundal portions of large lakes, where water flow is negligible and the substrate has high silt contents and heavy organic loads (Burky, 1983). Preference of some sphaeriids for low-flow, silty habitats may reflect interstitial sediment detritus feeding, particularly in *Pisidium* (Lopez and Holopainen, 1987) (see Section III.C.2). Byssal attachment allows *D. polymorpha* to inhabit relatively high flow areas compared to other bivalves (pediveliger larvae settle in flows up to 1.5m/sec). It also makes it successful epibenthic species in lentic habitats with a preponderance of hard substrates from which native NA bivalves are generally eliminated (Claudi and Mackie, 1993).

Water depth affects freshwater bivalve distributions. Most unionoideans prefer shallow habitats less than 4–10 m deep (Stone *et al.,* 1982; Salmon and Green, 1983; Machena and Kautsky, 1988; Way *et al.,* 1990b), although some species occur in deeper lotic waters if well oxygenated. *C. fluminea* is also restricted to shallow, near-shore lentic habitats (McMahon, 1999), as are the majority of *Sphaerium* and *Musculium* species (Fig. 13B) (Zbeda and White, 1985; Kilgour and Mackie, 1988). In contrast, some species of *Pisidium* inhabit profundal regions of lakes (Holopainen and Jonasson, 1983; Kilgour and Mackie, 1988). Depth distributions of *D. polymorpha* vary between habitats; however, adults rarely occur above 2 m and dense populations extend to 4–60 m, but are restricted to well-oxygenated waters above the epilimnion. The quagga mussel, *D. bugensis,* extends to greater, oxygenated depths than *D. polymorpha* in the Great Lakes (Mills *et al.,* 1996), but is unlikely to penetrate hypoxic hypolimnetic waters as it is less hypoxia tolerant than *D. polymorpha* (P. D. Johnson and R. F. McMahon, unpublished data). Recently settled juveniles of *D. polymorpha* migrate to deeper water (Mackie and Schloesser, 1996), perhaps avoiding wave-induced agitation (Clarke and McMahon, 1996b) or ice scour.

Limitation of most lentic bivalves to shallow habitats may reflect their poor hypoxia tolerance. In lentic habitats, hypolimnetic waters are often hypoxic. As many species of unionoideans, *Sphaerium* and *Musculium* (Burky, 1983), *Dreissena* (McMahon, 1996) and *C. fluminea* (McMahon, 1999) cannot maintain normal O_2 uptake under severely hypoxic conditions, they are mostly restricted to shallow, well-oxygenated habitats. Juvenile unionids are less hypoxia-tolerant than adults (Dimock and Wright, 1993), preventing colonization of hypoxic habitats. In contrast, many species of *Pisidium* are extreme $\dot{V}_{O_2}$ regulators (Burky, 1983) allowing them to inhabit highly hypoxic hypolimnetic habitats (Jonasson, 1984a, b). However, summer hypoxia retards growth and reproduction in profundal *Pisidium* populations, indicating that hypoxia can deleteriously impact even hypoxia-tolerant species (Halopainen and Jonasson, 1983). Hypoxia-intolerant *C. fluminea* invaded the profundal regions of a small lake only after artificial aeration of its hypoxic hypolimnetic waters (McMahon, 1999). Sewage-induced hypoxia in the Pearl River, Louisiana, extirpated its unionoidean fauna (Gunning and Suttkus, 1985). Even hypoxia-tolerant profundal *Pisidium* communities are extirpated by extreme hypoxia (Jonasson, 1984a).

Ambient pH does not greatly limit the distribution of freshwater bivalves. The majority of species prefer waters of pH above 7.0; species diversity declines in acidic habitats (Okland and Kuiper, 1982). However, unionoideans can grow and reproduce over a pH range of 5.6–8.3, a pH of less than 4.7–5.0 being the absolute lower limit (Fuller, 1974; Okland and Kuiper, 1982; Kat, 1982; Hornbach and Childers, 1987). Low pH may have sublethal effects on unionoideans. It reduces shell thickness (Hinch *et al.,* 1989), tissue cholesterol content (Rao *et al.,* 1987), and hemolymph concentrations of Na^+, K^+, and Cl^- (Pynnönen, 1991; Mäkelä and Oikari, 1992). Unionoidean glochida and juveniles are less low pH tolerant than adults (Huebner and Pynnönen, 1992; Dimock and Wright, 1993), perhaps preventing colonization of mildly acidic waters. In contrast, *Pyganodon grandis* showed no change in hemolymph Na^+, K^+, and Cl^- concentrations after transfer into an acidic lake (pH 5.9) from an alkaline lake (Malley *et al.,* 1988). Some sphaeriids are relatively insensitive to pH or alkalinity. Species richness, growth and reproduction in sphaeriid faunas of six low-alkalinity lakes was similar to those in higher alkalinity lakes (Rooke and Mackie, 1984a, b; Servos *et al.,* 1985). *Musculium partumeium* and *Pisidium casertanum* inhabited acid lakes in New York State (pH $<$6.0) (Jokinen, 1991). Indeed, maximal laboratory growth and reproduction in *Musculium partumeium* occurred at pH 5.0, suggesting adaptation to moderately acidic habitats (Hornbach and Childers, 1987).

Low-pH habitats generally have low calcium concentrations. Low pH leads to shell dissolution and eventual mortality if shell penetration occurs (Kat, 1982). Sphaeriids occur in waters with Ca concentrations as low as 2 mg Ca/L, while the unionid, *Elliptio complanata,* occurs at 2.5 mg Ca/L (Rooke and Mackie, 1984a). Among freshwater bivalves, *D. polymorpha* is the most calciphilous and pH intolerant; adults requiring pH $>$ 6.5 and $>$12 mg Ca/L and veliger larvae requiring pH $>$7.4 and $>$24 mg Ca/L for successful development (McMahon, 1996). Freshwater bivalves actively take up Ca^{2+} at medium concentrations as low as 0.5 mM Ca/L (0.02 mg Ca/L, see Section II.A.1), which is far below their minimal ambient Ca^{2+} concentration of 2–2.5 mg/L. Thus, the minimum ambient calcium concentration appears to be that at which the rate of calcium uptake and deposition to the shell exceeds that of calcium loss from shell dissolution and diffusion, allowing maintenance of shell integrity and growth. As many factors affect shell deposition and dissolution rates (e.g., temperature, pH, and calcium concentration), the minimal Ca concentration and/or pH tolerated by a species may vary between habitats dependent on interacting biotic and abiotic parameters and are often species-specific. Low calcium waters usually have low concentrations of other biologically important ions, making them inhospitable to bivalves even if Ca concentrations are suitable for shell growth.

Temperature influences bivalve distributions; species have specific upper and lower limits for survival and reproduction (Burky, 1983). For example, intolerance of <2°C prevents *C. fluminea* from colonizing drainages in the northcentral United States, which reach 0°C in winter (Fig. 11) (McMahon, 1999), leading to low-temperature winter-population kills on the northern edge of its range (Sickel, 1986). Thus, many northern U.S. *C. fluminea* populations are restricted to areas receiving heated effluents (McMahon, 1999). In contrast, the maximal temperature for development of *D. polymorpha* eggs and larva is 24°C and adults do not tolerate temperatures >30°C, preventing this species from colonizing southern and southwestern U.S. waters which exceed 30°C in summer (McMahon, 1996). Indeed, in the most southern U.S. states, *D. polymorpha* only occurs in the lower Mississippi River which rarely exceeds 30°C (Hernandez *et al.*, 1995).

Water-level variation affects bivalve distributions. Declining water levels during droughts or dry periods expose relatively immotile bivalves for weeks or months to air. Restriction of many bivalve populations to shallow waters makes them susceptible to emersion. Many sphaeriid species and some unionoidean taxa are highly tolerant of air exposure, surviving prolonged seasonal emersion in ephemeral or variable level habitats (Burky, 1983; White, 1979; Byrne and McMahon, 1994). These species display unique emersion adaptations (Byrne and McMahon, 1994 and Section II.C.4).

Freshwater bivalve distribution is also related to stream size or order. Unionoidean species diversity increased with distance downstream in the Sydenhan River, Ontario (Mackie and Topping, 1988). Similarly, stream size proved to be the only useful predictor of unionoidean species richness of six tested factors (i.e., stream size, stream gradient, hydrologic variablity, Ca concentration, physiographic province and presence/absence of tides) in the Susquehanna, Delaware and Hudson River drainages (Strayer, 1993). The number of unionoidean species in Michigan drainage systems increased proportionately with system size; mainly by species additions (Strayer, 1983) (Fig. 14), but species richness could not be completely accounted for by drainage area size (note the high degree of variation in species richness versus drainage area values in Fig. 14), suggesting that other environmental variables affect unionoidean distribution patterns (Strayer, 1993). In the Ohio River drainage, the number of fish species was directly related to drainage area, and the number of unionoidean species directly related to the number fish species, suggesting that unionoidean diversity is partially dependent on fish glochidial host species availability (Watters, 1993). Similarly, *Margaritifera hembeli* populations with the highest juvenile recruitment rates occurred in stream reaches with the highest densities of its fish host (Johnson and Brown, 1998). Among other variables affecting species richness are stream hydrology, affected by surface geology and soil porosity. Porous soils retain water, buffering runoff, so that streams draining them have constant flows and rarely dry, allowing them to support greater mussel diversity than streams draining soils of poor water-infiltration capacity that are prone to flooding–drying cycles. In variable-flow streams, emersion or low oxygen in stagnant pools during dry periods, and bottom scouring and high silt loads during floods reduce species richness (Strayer, 1983). Indeed, flow stability, high O_2, reduced flood risk, and low silt loads of larger streams appear to account for their increased bivalve species richness (Fig. 14) (Strayer, 1983, 1993; Way *et al.*, 1990b; Strayer and Ralley, 1993). However, some unionoidean species, such as *Amblema plicata, Pyganodon grandis* and *Fusconaia flava,* are adapted to small, variable flow streams (Strayer, 1983; Di Maio and Corkum, 1995) while other species favor tidally influenced portions of rivers (Strayer, 1993).

The above physical factors may interact to determine distributions of bivalves. The density of the unionoidean, *M. hembeli,* was affected by stream order, hardness, depth, substrate size and compaction, and

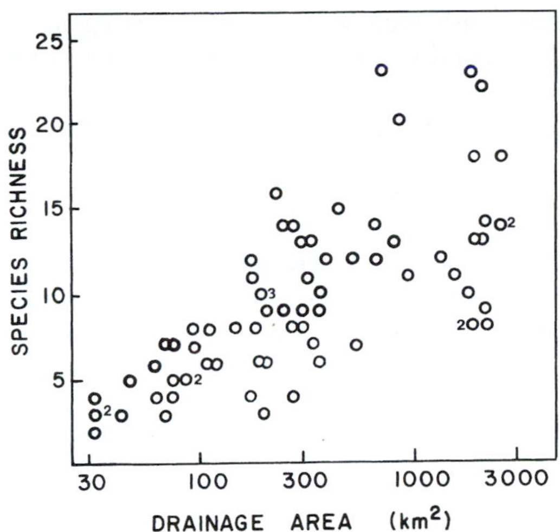

FIGURE 14 Unionoidean mussels species richness (total number of species present at a particular site) as a function of stream size measured by its total drainage area in km^2 in southeastern Michigan, United States. Numbers next to points indicate the number of observations falling on that point. The relationship between drainage area and mussel species richness was statistically significant $r = 0.68$, $P < 0.001$). (Redrawn from Strayer, 1983.)

flow rate, with highest densities occurring in shallow areas of second-order streams, of hardness >8 mg/L, with higher flow rates, and stable sediments with larger particle sizes (Johnson and Brown, 2000).

B. Reproduction and Life History

North American freshwater bivalves display extraordinary variation in life history and reproductive adaptations. Life-history traits (e.g., those affecting reproduction and survival, including growth, fecundity, life span, age to maturity, and population energetics) have been reviewed for freshwater molluscs (Calow, 1983; Russell-Hunter and Buckley, 1983) and specifically for freshwater bivalves (Burky, 1983; Mackie, 1984; Mackie and Schloesser, 1996; McMahon, 1999). Most research involves the Sphaeriidae, with less information for unionoideans. Sphaeriids are good subjects for life-history studies, because of their greater abundances, ease of collection and laboratory maintenance, relatively simple hermaphroditic life cycles, ovoviviparity, semelparity, release of completely formed miniature adults, and relatively short life spans. In contrast, unionoideans are more difficult subjects because they are gonochoristic, long-lived, interoparous, often rare and difficult to collect, and have life cycles complicated by the parasitic glochidial stage. The life-history traits of *C. fluminea* and *D. polymorpha* have been intensely studied due to their invasive nature and economic importance as fouling organisms (Mackie and Schloesser, 1996; Nichols, 1996; McMahon, 1999).

1. Unionoidea

The life-history characteristics of freshwater unionoideans are clearly different from those of sphaeriids, *C. fluminea* or *D. polymorpha* (Table III). The majority of unionids live in large, stable aquatic habitats which buffer them from periodic catastrophic population reductions, typical of smaller, unstable aquatic environments (see Section III.A). In stable habitats, long-lived adults accumulate in large numbers (Payne and Miller, 1989). A few species inhabit ponds

TABLE III Summary of the Life History Characteristics of North American Freshwater Bivalves, Unionoidea, Sphaeriidae, *Corbicula fluminea, Driessena polymorpha*

Life history trait[a]	*Unionoidea*	*Sphaeriidae*	*Corbicula fluminea*	*Dreissena polymorpha*
Life span (years)	<6–>100 (Species dependent)	<1–>5 (species dependent)	1–4	4–7
Age at maturity (years)	6–12	>0.17–<1.0 (1 year in some species)	025–0.75	0.5–2
Reproductive mode	Gonochoristic (few hermaphroditic species)	Hermaphroditic	Hermaphrodictic self-fertilizing	Gonochoristic
Growth rate	Rapid prior to maturity, slower thereafter	Slow relative to Unionoideans, *C. fluminea* or *D. polymorpha*	Rapid throughout life	Rapid throughout life
Fecundity (young per avg. adult per breeding season)	200,000–17,000,000 per female	3–24 (*Sphaerium*) 2–136 (*Musculium*) 3–7 (*Pisidium*)	35,000	30,000–40,000 per female
Juvenile size at release	Very small, 50–450 μm	Large 600–4150 μm	Very small 250 μm	Extremely small, 40 μm
Relative juvenile survivorship	Extremely low	High	Extremely low	Extremely low
Relative adult survivorship	High	Intermediate	Low 2–41% per year	Intermediate 26–88% per year
Semelparous or iteroparous	Highly iteroparous	Semelparous or Iteroparous (species dependent)	Moderately iteroparous	Moderately iteroparous
Number of reproductive efforts per year	One	1–3 (continuous in some species)	Two (spring and fall)	One (2–8 months long)
Assimilated energy respired (%)	—	21–91 (Avg. = 45%)	11–42	—
Nonrespired energy allocated to growth (%)	85.2–97.5	65–96 (Avg. = 81%)	58–71	96.1
Nonrespired energy allocated to reproduction (%)	2.8–14.8	4–35 (Avg. = 19%)	15	4.9
Turnover time in days (= mean standing crop biomass : biomass per day ratio)	1790–2849	27–1972 (generally <80)	73–91	53–869 (habitat dependent)
Habitat stability	Stable	Intermediately stable	Unstable	Moderately unstable

[a] See text for literature citations to data on which this table was based.

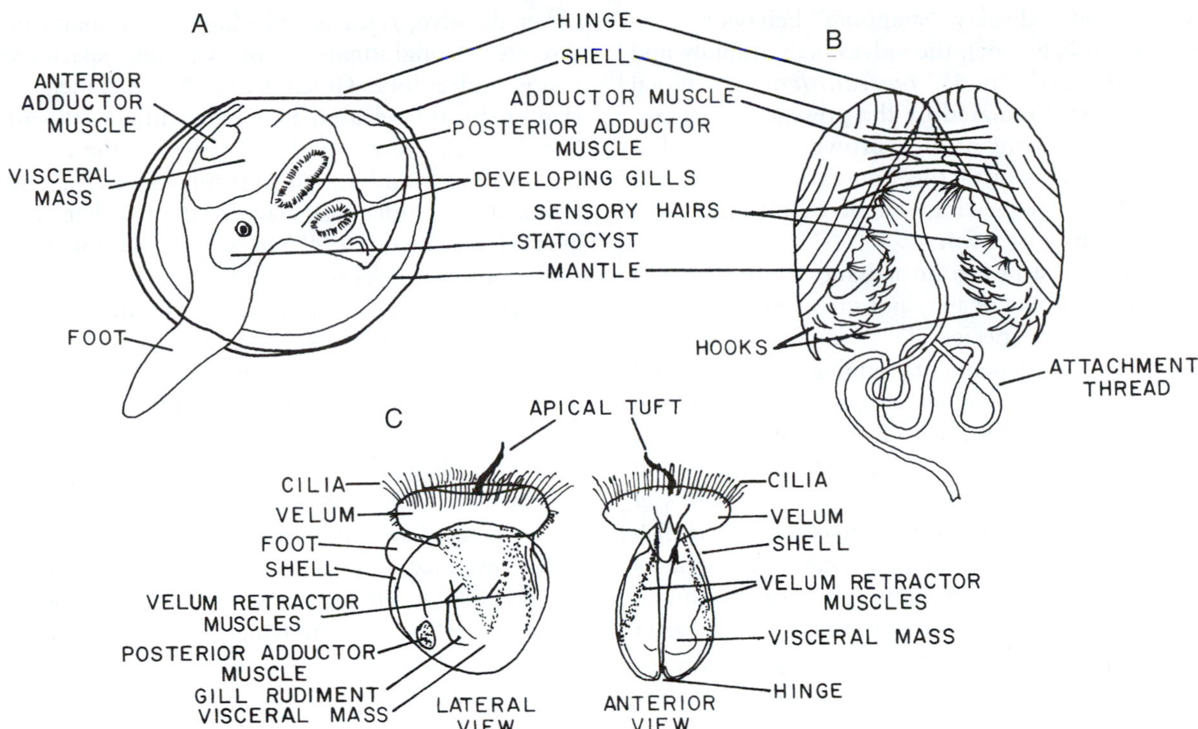

FIGURE 15 Anatomic features of freshwater bivalve larval stages. (A) the D-shaped juvenile of *Corbicula fluminea*, the freshwater Asian clam (shell length = 200 μm). (B) The glochidium larva of unionoideans, which is parasitic on fish. Depicted is the glochidium of *Pyganodon* characterized by the presence of paired spined hooks projecting medially from the ventral edges of the shell valves and an attachment thread; not all unionoidean species have glochidia with these characteristics (50–400 μm in diameter depending on species). (C) Lateral and anterior views of the free-swimming, planktonic veliger stage of *Dreissena polymorpha*, the zebra mussel; the veliger is 40–290 μm in diameter and uses the ciliated velum to swim and feed on phyto- and bacterioplankton. Presence of a foot as shown in this diagram indicates development to the settlement competent pediveliger stage. The juveniles of freshwater sphaeriid species are large and highly developed, having essentially adult features (Fig. 2A) at birth.

(Burch, 1975b), but their life-history traits have not been studied.

An important aspect of the unionoidean life-history traits is their parasitic glochidium larval stage (Fig. 15B). With the exception of *Simpsonaias ambigua*, whose glochidial host is the aquatic salamander, *Necturus maculosus*, all other known NA unionoideans have glochidial fish hosts (Watters, 1994b). The significance of glochidia unionoidean reproduction has been reviewed by Kat (1984) and details of glochidial incubation and development in marsupial brood pouches were described in Section II.B.5. The glochidium has a bivalved shell adducted by a single muscle. Its mantle edge contains sensory hairs. In the genera *Unio, Anodonta, Pygauodon, Megalonaias,* and *Quadrula,* a long threadlike structure projects from the mantle beyond the ventral valve margin which may be involved with detection of and/or attachment to, fish hosts.

There are three general forms of glochidia. In the subfamily Anodontinae, "hooked glochidia" occur, with triangular valves from whose ventral edges project an inward curving hinged hook covered with smaller spines (Fig. 15B). On valve closure, the hooks penetrate the skin, scales, or fins of fish hosts, allowing glochidial attachment and encystment on host external surfaces. The majority of NA unionoideans produce "hookless glochidia" with more rounded valves bearing reinforcing structures and/or small spines or stylets on their ventral margins which generally attach and encyst on fish gills. "Axe-head glochidia" of the genus *Potamilus* have a flared ventral valve margin, and near-rectangular valves which may have hooklike structures on each corner. Their host attachment sites are unknown (Kat, 1984).

Glochidia initially attach to fish by clamping (snapping) the valves onto fins, scales, and/or gill filaments. They are not host-specific in attachment, attaching to any contacted fish (Kat, 1984). In contrast, gravid adult unionids (European *Anodonta anatina*) appear to release glochidia in response to chemicals released by host fish (Jokela and Palokamgas, 1993). Once released, glochidia display 'snapping' behavior' (i.e., host attachment behavior), the valves being rapidly and repeatedly adducted. In *M. margaritifera,* glochidial valve snapping is stimulated by the presence of mucus, blood, gill tissue, or fins of their brown trout host, but not by water currents or tactile stimulation (Young and Williams, 1984a), suggesting use of chemical cues to detect and attach to host fish.

Glochidia encyst in fish host tissues within 2–36 h of attachment and may or may not grow during encystment, depending on species. Time to juvenile excystment is also species-dependent, ranging from 6 to 160 days, but is reduced at higher temperatures (Zale and Neves, 1982; Kat, 1984). Unsuitable host fish reject glochidia after encystment (Kat, 1984), fish blood serum components dictating host suitability (Neves *et al.,* 1985). Since glochidia nonselectively attach to fish (Kat, 1984; Neves *et al.,* 1985), host suitability appears more dependent on fish immunity than on glochidial host recognition. Indeed, even suitable host fish reject glochidia, rejection rate being host fish species-dependent (Haag *et al.,* 1999). Numbers of *M. margaritifera* glochidia encysted in a brown trout population declined with time (Young and Williams, 1984b), with laboratory studies revealing that only 5–12% of *M. margaritifera* glochidia successfully excyst from infected host fish, indicating host rejection of most encysted individuals (Young and Williams, 1984a).

Unionoideans have adaptations that increase likelihood of glochidia contact with fish hosts. Glochidial release occurs annually, but duration of release is species-dependent with some species having more than one annual release period (Fukuhara and Nagata, 1988; Gordon and Smith, 1990). Cycles of gametogenesis and glochidial release may be controlled by neurosecretory hormones (Nagabhushanam and Lomte, 1981). Tachytictic mussels are short-term breeders, whose glochidial development and release occurs between April and August; shedding of glochidia often corresponding with either migratory periods of anadromous fish hosts, or nesting periods of their host fish. Fish hosts of tachytictic unionoideans often construct nests in areas harboring dense unionoidean populations, where fish host nest construction by fanning away substrates, and fish host fanning of developing embryos provide optimal conditions for glochidial– host contact. Thus, a high proportion of nest-building fish species, such as centrarchids, are glochidial hosts for NA unionoideans (Watters, 1994b). In contrast, bradytictic unionoidean species are long-term breeders, retaining developing glochidia in gill marsupia throughout the year, releasing them in summer (Kat, 1984).

When released from adult mussels, glochidia are generally bound by mucus into discrete packets, which either dissolve, releasing glochidia, or remain intact as discrete "conglutinates" of various species-specific forms and colors. Glochidia with attachment threads (Fig. 15B) are released in tangled mucus threads that dissolve relatively rapidly. Many of these glochidia possess hooks and attach to fish-host external surfaces. Hooked glochidia are larger than other types of glochidia (Bauer, 1994). In some unionoideans, mucilaginous networks of glochidia persist, enhancing host contact by suspending glochidia above the substrate.

Unionids with hookless glochidia that attach to fish gills may release conglutinates that mimic the food items of their fish hosts. They resemble brightly colored oligochaetes, flatworms, or leeches. Some species hold their wormlike conglutinates partially extruded from the exhalant siphon, making them more obvious to fish hosts. Some *Lamsilis* species, such as, *Lampsilis perovalis,* produce a "supercongultinate" that contains all the glochidia in the marsupial gill and resembles a small fish. It is tethered to the female's exhalant siphon by a long, transparent mucus strand (Fig. 16A). In flowing water, it mimics the darting motions of a small fish, eliciting attacks by host fish (Haag *et al.,* 1995). Consumption of such conglutinates releases glochidia within the buccal cavity of the fish followed by transport to gill filament attachment sites on ventilatory currents.

The most unusual unionoidean host food mimicry involves pigmented muscular, posterior mantle edge extensions in female *Lampsilis* and *Villosa.* These "mantle flaps' (Fig. 16B) resemble the small fish or macroinvertebrate prey of their fish hosts (Haag and Warren, 1999). Gravid females extend the posterior shell margins well above the substrate and pulsate the projecting mantle flaps to mimic a small, actively swimming fish or moving macroinvertebrate. When fish strike these mantle lures, glochidia are forcibly released through posterior pores in the marsupial gill (often projected between the mantle flaps, Fig. 16B) assuring glochidial ingestion and contact with the fish host gills (Kat, 1984; Haag and Warren, 1999).

As the glochidia of some unionoideans do not grow while encysted, their parasitic nature has been questioned. However, *in vitro* glochidial culture experiments suggest that they absorb organic molecules from fish tissues and require fish plasma for development and metamorphosis (Isom and Hudson, 1982), indicative of a true host–parasite relationship. Indeed, glochidial infection damages fish hosts, especially juvenile fish (Cunjak and McGladdery, 1991; Panha, 1993).

Like other parasites, glochidia are shed in huge numbers to ensure the maximum host contact and attachment. Unionoidean fecundity ranges from 10,000 to 17,000,000 glochidia per female per breeding season (Parker *et al.,* 1984; Young and Williams, 1984b;

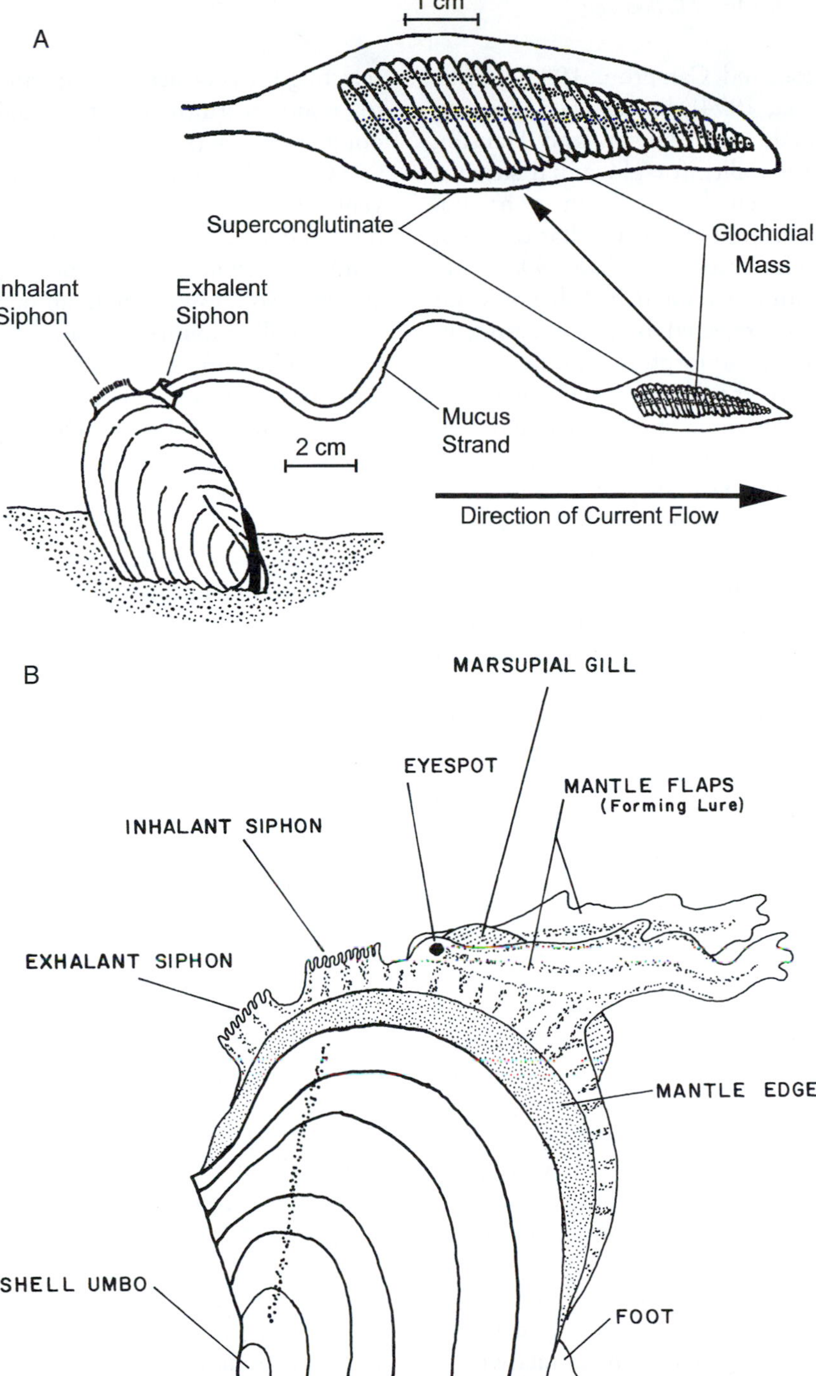

FIGURE 16 Reproductive adaptations in female Lamsilid unionoideans. (A) Superconglutinate tethered on a mucus strand extending from the exhalant siphon of *Lampsilis perovalis.* Glochidia are encapsulated in the flattened terminal lure (detailed in the upper portion of the figure) at the end of the mucus strand which makes darting fishlike movements in water currents, enticing host fish to strike the lure, thus assuring release of glochidia into its buccal cavity for attachment to the fishes' gills. (Redrawn from Haag *et al.*, 1995) (B) The modified, mantle flaps of a female specimen of *Lampsilis ovata*, which mimic the small fish prey (note eye spot and lateral linelike pigmentation) of its predatory host fish species. The posterior portions of the marsupial outer demibranchs are projected from the mantle cavity to lie between the flap lures. When the mantle flap lures are struck by an attacking host fish, glochidia are released through pores in the marsupial gill ensuring their entry into the Figh's buccal cavity and attachment to its gills. Superconglutinates and mantle flap lures are characteristic of the unionoidean genus, *Lampsilis*.

Paterson, 1985; Paterson and Cameron, 1985; Jansen and Hanson, 1991; Bauer, 1994), brood size increasing exponentially with female size (Janson and Hanson, 1991; Bauer, 1994). Probability for glochidial survival to metamorphosis is extremely small. In a natural population of *M. margaritifera,* a species that does not produce glochidial conglutinates, only 0.0004% of released glochidia successfully encysted in fish hosts. Of these, only 5% were not rejected before excystment; and, of those successfully metamorphosing, only 5% established as juveniles in the substrate (Young and Williams, 1984b). Thus, only one in every 100,000,000 shed glochidia became settled juveniles. High glochidial mortality makes the effective fecundity of unionoideans extremely low, which is not unusual for a species adapted to stable habitats. Similarly, only 0.007% of released glochida of *Pyganodon grandis* encysted in their host fish (*Perca flavescens*), however, once encysted, they had relatively high survival, 27% surviving to, and excysting after, two years (Jansen and Hanson, 1991). Based on these data and the fecundity ranges listed in Table III, only 0.002–0.17 of the glochidia produced by a female unionoidean would successfully settle as juveniles in sediment. Therefore, the main advantages of the glochidial stage appear to be a directed dispersal by fish hosts into favorable habitats (Kat, 1984) and utilization of fish-host energy resources to complete development to a juvenile large enough to effectively compete after settlement in adult habitats (Bauer, 1994).

There is a trade-off between glochidial size and number of glochidia released. Species with large glochidia (including toothed glochidia) produce fewer glochidia while species with small glochidia release them in great numbers (Bauer, 1994). Species releasing fewer, larger glochidia, tend to have a greater range of host-fish species and prefer lower flow habitats (Bauer, 1994). Production of many small glochidia occurs in species with limited fish hosts and may increase chances of glochidial–host contact, while having a wide range of host-fish species allows production of fewer, larger glochidia, because chances of glochidium contact with a suitable host are higher. In the highly K-selected, stable habitats of most unionoideans, excysting at a large size makes juvenile mussels more competitive, increasing likelihood of survival to maturity. Thus, small glochidia of unionoideans, such as *Margaritifera,* remain encysted for longer periods (>1000 days), allowing growth to a more competitive size before excystment. In contrast, among species with larger glochidia, the encystment period is much shorter (200–300 days) before attainment of competitive juvenile size. The long encystment periods of small glochidia reduce their chances for successful excystment due to increased chances of host mortality and/or rejection by the host's immune system, selecting for production of greater numbers of glochidia to assure recruitment of enough juveniles to sustain population size (Bauer, 1994).

Glochidial utilization of fish-host energy stores prevents their direct competition with adults for limited food and space resources (Bauer, 1994), as occurs in juvenile *C. fluminea* and *D. polymorpha.* Glochidial parasitism also allows female unionoideans to devote relatively small amounts of nonrespired, assimilated energy to reproduction (2.8–14.5% of total nonrespired, assimilated energy), leaving the majority for somatic tissue growth (Table V) (Negus, 1966; James, 1985; Paterson, 1985). Allocation of a high proportion of energy to tissue growth is characteristic of species adapted to stable habitats (i.e., K-selected species). As unionoideans are often very long-lived (*Margaritifera margaritifera* up to 130–200 years) and highly iteroparous (>6–10 reproductive periods throughout life), allocation of the majority of nonrespired energy to growth increases probability of adult survival to future reproductive efforts. Increased growth rate and reduction of reproductive effort increases competitiveness and lowers probability of predation and/or mortality associated with reproductive effort or dislodgment from the substrate during floods. All these characteristics increase unionoidean fitness in stable habitats (Sibly and Calow, 1986; Bauer, 1994).

In the majority of unionoideans, greatest shell growth occurs in immature individuals during the first 4–6 years of life (Hanson *et al.,* 1988; Payne and Miller, 1989; Harmon and Joy, 1990) (Fig. 17A). Indeed, relative shell growth in young unionids is greater than in sphaeriids or *C. fluminea* and *D. polymorpha* (Table III). Unionoidean shell-growth rate declines exponentially with age, but tissue biomass accumulation rate remains constant or increases with age (Fig. 17A, B; see also Haukioja and Hakala, 1978). Thus, early in life, increases in shell size and biomass occur preferentially over tissue accumulation; whereas after maturity (>6 years), shell growth slows and tissue accumulates at proportionately higher rates. Delayed maturity in unionoideans (6–12 years, Table III) allows allocation of all nonrespired assimilation to growth early in life and, due to an exponential relationship between size and fecundity, leads to reproductive effort being primarily sustained by the oldest, largest individuals (Downing *et al.,* 1993).

Growth rates in unionoideans are affected by abiotic conditions. Shell growth of field-enclosed specimens *P. grandis* declined with depth, being greatest >3 m (Hanson *et al.,* 1988). However, in natural populations, depth did not influence shell growth in *P. grandis* (Hanson *et al.,* 1988) or *A. woodiana* (Kiss and Pekli, 1988), suggesting that vertical migration

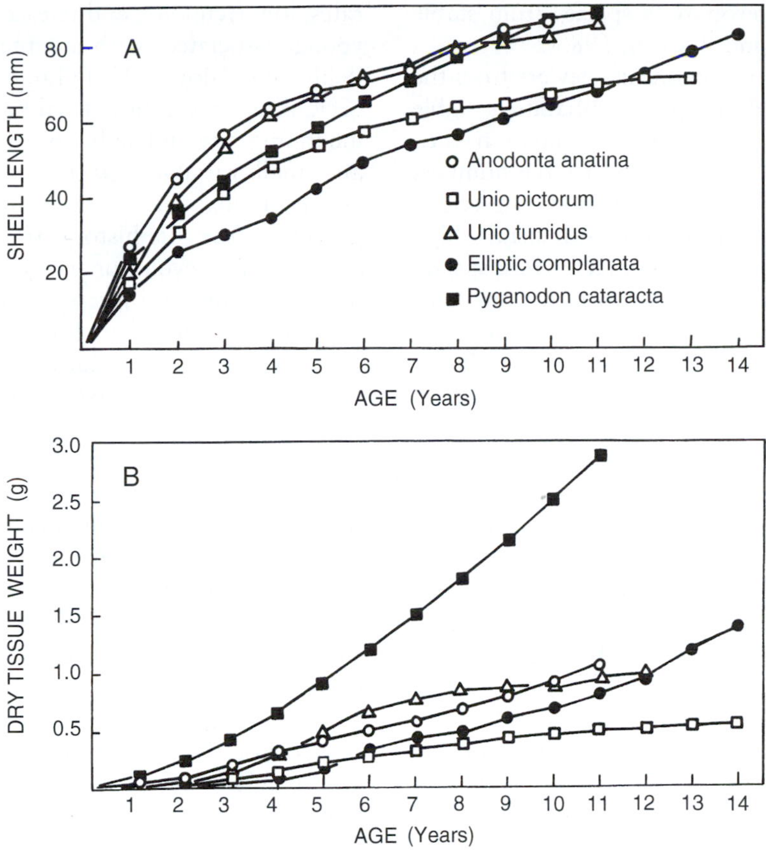

FIGURE 17 The shell and tissue growth of selected species of unionids (*Anodonta anatina*, open circles; *Unio pictorum*, open squares; *Unio tumidus*, open tiangles; *Elliptio complanata*, solid circles; and *Pyganodon cataracta*, solid squares). (A) Mean shell-length increase with increasing age over the entire life span of each species. (B) Mean dry tissue-weight increase with increasing age over the entire life span of each species. Note that increase in shell length declines with age, while dry tissue weight increases either linearly or exponentially with age. (Data from Negus, 1966; Paterson, 1985; and Paterson, and Cameron, 1985.)

may mask depth influences on growth in free-living individuals (Hanson *et al.*, 1988). Unionoidean growth rates also vary seasonally with level of precipitation (Timm and Mutvei, 1993) and can be depressed by pollutants (Harman and Joy, 1990).

Once mature, large adult unionoideans display high age-specific survivorship between annual reproductive efforts, being 81–86% in 5–7 year-old *Anodonta anatina* (Negus, 1966), greater than 80% in mature *M. margaritifera* (12–90 years) (Bauer, 1983), with a lower value of 15.6% in *Pleurobema collina* (Hove and Neves, 1994). High adult survivorship, long life spans, and low juvenile survivorship account for the preponderance of large adult individuals in natural unionoidean populations (Bauer, 1983; James, 1985; Negus, 1966; Paterson, 1985; Paterson and Cameron, 1985; Tevesz *et al.*, 1985; Hansen *et al.*, 1988; Huebner *et al.*, 1990; Woody and Holland-Bartels, 1993; Hove and Neves, 1994; Vaughn and Pyron, 1995). Populations dominated by adults are characteristic of stable, highly competitive habitats (Sibly and Calow, 1986).

The preponderance of large, long-lived adults in unionoidean populations causes them to have high proportions of standing crop biomass relative to biomass production. The relationship between standing crop biomass and biomass production rate can be expressed as turnover time, computed in days as the average standing crop of a population divided by its mean daily productivity rate (Russell-Hunter and Buckley, 1983). Long-lived unionids, with populations dominated by large adults, have extremely long turnover times ranging from 1790–2849 days (computed from data in James, 1985; Negus, 1966; Paterson, 1985; Huebner *et al.*, 1990) compared with sphaeriids (27–1972 days), *C. fluminea* (73–91 days), or *D. polymorpha* (53–869 days) (Table III). Such extended turnover times characterize long-lived, iteroparous species from stable habitats (Russell-Hunter and Buckley, 1983).

In only one aspect do unionoideans deviate from the life-history traits expected of species inhabiting stable habitats and experiencing extensive competition (i.e., K-selected). That is in the production of large numbers of small young (glochidia). However, as described above, this is an adaptation ensuring sufficiently high probability of glochidial–host fish contact. Thus, species producing conglutinates resembling fish host prey items have greater glochidial–host contact and produce fewer (200,000–400,000 glochidia per female) and larger glochidia (Kat, 1984). In contrast, *M. margaritifera,* which releases very small, dispersed glochidia, has extraordinarily high fecundities (up to 17,000,000 glochidia per female) (Young and Williams, 1984b).

Extended life spans, delayed maturity, low effective fecundities, reduced dispersal, high habitat selectivity, poor juvenile survival and extraordinarily long turnover times make unionoidean populations highly susceptible to human perturbations. Because of their life-history traits (particularly long life spans and low effective fecundities), they do not recover rapidly once decimated by pollution or other human- or naturally-mediated habitat disturbances (see Section III.A). Successful juvenile settlement appears particularly affected by disturbance, with population structures indicating periods when entire annual generations are not recruited (Bauer, 1983; Negus, 1966; Payne and Miller, 1989). Reduction in population densities may reduce fertilization success in unionoideans. Complete failure of fertilization in *Elliptio complanata* occurred at densities <10 mussels/m^2 with densities >40 mussels/m^2 required for 100% of females to be fertilized (Downing *et al.,* 1993). Thus, anthropomorphically influenced reduction of unionoidean population densities to low levels could prevent further recruitment. Such disturbance-induced lack of juvenile recruitment raises the specter of many NA unionoidean populations being composed of dwindling numbers of long-lived adults destined for extirpation as anthropomorphic disturbances prevent reproduction and/or juvenile recruitment in aging adult populations (for an example, see Vaughn and Pyron, 1995).

2. Sphaeriidae

The Sphaeriidae display great intra- and interspecific life history variation (Holopainen and Hanski, 1986; Mackie, 1984; Way, 1988). Like unionoideans, their life-history traits do not fall into suits associated with stable or unstable habitats. Instead, they include the short life spans, early maturity, small adult size, and increased energetic allocation to reproduction associated with adaptation to unstable habitats, and the slow growth rates, low fecundity, and release of large, fully developed young associated with adaptation to stable habitats (Sibly and Calow, 1986) (Table III). Sphaeriids are euryoecic (i.e., have a broad habitat range); some members inhabiting stressful habitats, such as ephemeral ponds, and small, variable flow streams, while others live in stable, profundal lake habitats (Burky, 1983). Here, we generalize the life-history traits of sphaeriids within adaptive and evolutionary frameworks. However, the degree of inter- and intraspecific life-history variation in this group is such that, for each generality, specific exceptions can be cited (Mackie, 1979).

Central to understanding sphaeriid life-history traits is their viviparous reproduction (Mackie, 1978). All species brood embryos in specialized chambers formed from evaginations of the exhalant side of the inner demibranch gill filaments. Maternal nutrient material is supplied to embryos supporting their growth and development to release as fully formed miniature adults (see Section II.B.5). Thus, even though adult sphaeriids are the smallest of all NA freshwater bivalves, they release the largest young (Mackie, 1984). Among 13 sphaeriid species, average birth shell length ranged from 0.6 to 4.15 mm (Burky, 1983; Holopainen and Hanski, 1986; Hornbach and Childers, 1986; Hornbach *et al.,* 1982; Mackie and Flippance, 1983a), much larger than unionoidean glochidia (0.05–0.45 mm, Bauer, 1994), *C. fluminea* juveniles (0.25 mm, McMahon, 1999), or *D. polymorpha* veliger larvae (0.04–0.07 mm, Nichols, 1996) (Table III). Based on shell lengths, newborn sphaeriids have (3.4–2.1) $\times$ 10^5 times greater biomass than newborns of other groups. Ratios of maximum adult shell length : birth shell length in sphaeriids range from 2.8:1 to 5.4:1, suggesting that newborn juveniles have biomass 0.6–4.6% of the maximum biomass of adults. There is a significant direct relationship between these two parameters, shown in Figure 18.

Large offspring size reduces sphaeriid fecundity. Average clutch sizes range from 3 to 24 per young per adult for *Sphaerium,* 2 to 136 per young per adult for *Musculium,* and 1.3 to 16 per young per adult for *Pisidium* (Burky, 1983; Holopainen and Hanski, 1986). Even with reduced fecundity, large juvenile biomass requires allocation of large amounts of nonrespired energy to reproduction ($\bar{x}$ = 19%, Burky, 1983) compared to unionoideans (<14.8%), *C. fluminea* (15%) or *D. polymorpha* (4.9%) (Table III). As developing juveniles have high metabolic rates (Burky, 1983; Hornbach, 1985; Hornbach *et al.,* 1982) and are supported by adult energy stores (Mackie, 1984), estimates of reproductive costs based on released juvenile biomass may grossly underestimate actual reproductive costs in this group.

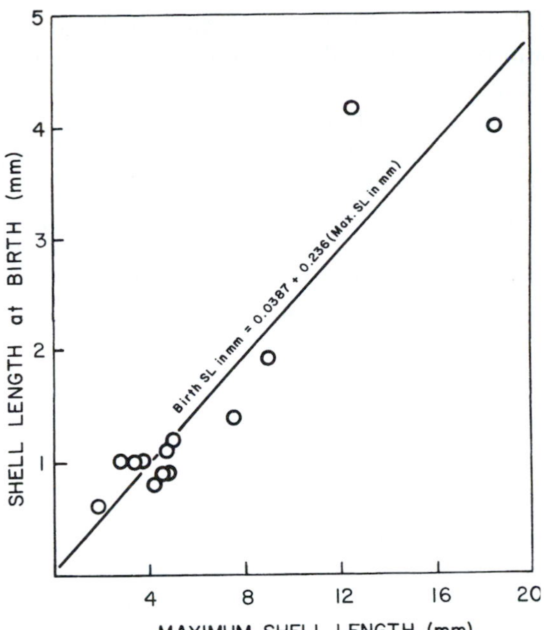

FIGURE 18 The relationship between shell length (SL) of juveniles at birth and maximal SL of adults for 13 species of sphaeriid freshwater clams. Note that juvenile birth length increases linearly with maximal adult size, suggesting that adult size may limit juvenile birth size in sphaeriids. The solid line represents the best fit of a linear regression relating birth size to maximal adult size as follows: Birth SL = 0.0387 + 0.236 mm (maximal adult SL in mm) ($n = 13$, $r = 0.935$, $F = 76.3$, $P < 0.0001$). (Data from Holopainen and Hanski, 1986; Hornbach and Childers, 1986; Hornbach *et al.*, 1982; and Mackie and Flippance, 1983a.)

Life-history hypotheses predict that the low fecundity and large birth size of sphaeriids would maximize fitness in stable habitats. However, the majority of sphaeriids inhabit variable ponds and streams or profundal habitats subject to hypoxia (Burky, 1983; Holopainen and Hanski, 1986). Burky (1983) and Way (1988) have argued that the harsh conditions of such habitats are seasonally predictable, making them, in reality, stable. Thus, sphaeriid species inhabiting them have adaptations that prevent catastrophic population reductions during seasonal episodes of environmental stress (see Section II.C). Thus, sphaeriid populations approach carrying capacity, leading to selection for life-history adaptations associated with the intense intraspecific competition characteristic of life in stable habitats.

Seasonal water-level fluctuation and hypoxia often more severely impact smaller individuals. Thus, production of large, well-developed juveniles by sphaeriids may increase their probability of surviving predictable environmental stress. As such stress can cause high adult mortality (Burky *et al.*, 1985b; Holopainen and Jonasson, 1983; Hornbach *et al.*, 1982; Jonasson 1984b), fitness would also be increased by devoting greater proportions of nonrespired assimilation to any one reproductive effort, as chances of adult survival to the next reproduction are low. Thus, sphaeriids devote relatively higher proportions of energy to reproduction than other freshwater bivalves (Burky, 1983) (Table III). This combination of *r*- selected and K-selected traits, optimizing sphaeriid fitness in habitats with periodic, predictable stress, is a life-history strategy called "bet hedging" (Stearns, 1980).

The early maturation, characteristic of many sphaeriid species (Burky, 1983; Holopainen and Hanski, 1986) (Table III), is due to the advanced development of newborn juveniles, and allows reproduction before onset of seasonal environmental stress. This trait is particulary displayed by sphaeriid species inhabiting ephemeral ponds and streams. They have rapid growth, early maturity, and reduced numbers of reproductive efforts. Thus, *Musculium lacustre, Pisidium clarkeanum,* and *P. annandalei* from temporary drainage furrows in Hong Kong live less than 1 year, and have only one or two life-time reproductive efforts (Morton, 1985, 1986). Similarly, *Musculium lacustre, Pisidium casertanum, P. variable, Spharium fabule,* and *Musculium securis* had only 1–3 reproductive efforts within life spans of 1 to 2 years (Mackie, 1979) and a lake species, *S. japonicum,* is univoltine and semelparous (Park and Okino, 1994). An ephemeral pond population of *Musculium partumeium* lived 1 year, was semelparous, reproduced just prior to pond-drying, and devoted a large proportion of nonrespired assimilation (18%) to reproduction (Burky *et al.*, 1985b). Ephemeral pond populations of *M. lacustre* were similarly univoltine and semelparous, devoting 19% of nonrespired energy to reproduction (Burky, 1983). A *Sphaerium striatinum* population, subject to flooding, lived 1 year or less and reproduced biannually (Hornbach *et al.*, 1982), devoting 16.1% of nonrespired assimilation to reproduction (Hornbach *et al.*, 1984b). In contrast, when populations of these species occur in more permanent habitats, they become bivoltine, reproducing in both spring and fall. Spring generations are iteroparous, reproducing in the fall and following spring; whereas fall generations are semelparous, reproducing only in the spring (Mackie, 1979; Burky, 1983; Burky *et al.*, 1985b).

Some *Pisidium* species live in the profundal portions of large, more permanent lentic habitats where they tolerate periodic hypoxia after summer formation of the hypolimnion (see Section II.C.3). As these profundal species are tolerant of hypoxia, such environments are more stable than small ponds and streams, although far less productive, reducing food availability (Holopainen and Hanski, 1986). Thus, profundal sphaeriids display life-history strategies differing from those in small ponds and streams

(Burky, 1983; Holopainen and Hanski, 1986). Profundal populations have reduced growth rates, longer life spans of 4 to 5 years, delayed maturation often exceeding 1 year, higher levels of iteroparity, and univoltine reproductive patterns (Holopainen and Hanski, 1986), all life-history traits characteristic of more stable habitats (Sibly and Calow, 1986). Interestingly, shallow-water and profundal populations of the same species of *Pisidium* may display different life-history tactics. Compared to profundal populations, shallow-water populations grow more rapidly, mature earlier, have shorter life spans, and tend toward semelparity (Holopainen and Hanski, 1986), life-history traits associated with unstable habitats. As pisidiids have well-developed dispersal capacities, this variation in life-history tactics among species' populations occupying different habitats in unlikely to result from genetic adaptation. Rather, much of this variation appears to be environmentally induced plasticity and is reflected in the highly variable turnover times reported for sphaeriids (range = 27–1972 days, Table III).

Freshwater bivalve growth rates are highly dependent on ecosystem productivity. Populations from productive habitats have greater assimilation, allowing allocation of greater absolute amounts of nonassimilated energy to growth (Burky, 1983). Shallow, freshwater habitats are usually highly productive, warm, and rarely O_2-limited and, thus, support higher growth rates. Conversely, profundal environments are less productive, cooler, and often O_2-limited, leading to lower growth rates. Since sphaeriids mature at a species-specific size irrespective of growth rate (Burky, 1983; Holopainen and Hanski, 1986), rapid growth leads to early maturity in shallow-water habitats and slow growth to delayed maturity in profundal habitats. Sphaeriids also die at species-specific terminal sizes whether terminal size is attained rapidly or slowly. As growth rate determines the time required to reach terminal size, fast-growing, shallow-water individuals reach terminal sizes more rapidly (often within less than 1 year) allowing participation in only one or two reproductive efforts, while slow-growing, profundal individuals (Holopainen and Hanski, 1986) reach terminal size more slowly, allowing participation in a greater number of reproductive efforts. Similar interpopulation growth rate and reproductive variation has been recorded in several NA sphaeriid species (Mackie, 1979). The fundamentally ecophenotypic nature this variation has been demonstrated for *Pisidium casertanum*. When individuals were reciprocally transferred between two populations with different life-history traits or co-reared in the laboratory, the majority of life-history trait differences proved to be environmentally induced. However, electrophoresis revealed genetic differences between populations and transfer and laboratory corearing demonstrated a small portion of observed variation to be genetically based (Hornbach and Cox, 1987). Extensive ecophenotypic plasticity may account for the euryoecic nature and cosmopolitan distributions of sphaeriids (see Section III.A). Certainly, capacity to adjust growth rates, maturity, reproductive cycles, life cycles, and energetic allocation patterns to compensate for habitat biotic and abiotic variation enables sphaeriids to have relatively wide niches and, thus broad distributions.

3. *Corbicula fluminea*

The introduced Asian freshwater clam, *Corbicula fluminea,* unlike unionoideans and sphaeriids, has life-history traits adapting it to unstable, unpredictable habitats (McMahon, 1999), making it the most invasive of all NA freshwater bivalves. *C. fluminea* grows rapidly, in part because it has higher filtration and assimilation rates than other bivalves (Foe and Knight, 1986a; Lauritsen, 1986a; Way *et al.,* 1990a). In a natural population, a relatively small proportion of assimilation (29%) was devoted to respiration (Table V), the majority (71%) being allocated to growth and reproduction (Aldridge and McMahon, 1978). Laboratory studies have also showed that 59–78% (Lauritsen, 1986a) or 58–89% of assimilation (Foe and Knight, 1986a) was allocated to tissue production. Thus, *C. fluminea* has among the highest net production efficiencies recorded for any freshwater bivalve, reflected by its low turnover times, ranging from 73 to 91 days (Table III).

The very high proportion of nonrespired assimilation (85–95%) devoted to growth in *C. fluminea* (Aldridge and McMahon, 1978; Long and McMahon, unpublished data), sustains rapid shell growth (shell length = 15–30 mm in the first year of life, 35–50 mm in the terminal third-to-fourth year) (McMahon, 1983a). High growth rates decrease probability of predation, as fish and bird predators feed only on small individuals (Section III.C.3) thus, increasing probability of survival to next reproduction in this moderately iteroparous species. Increase in shell size occurs at the expense of tissue production during summer maximal growth (Long and McMahon, unpublished data), suggesting that larger shells optimize fitness. High growth rates sustain the highest population production rates (10.4–14.6 g organic carbon/m^2 per year, Aldridge and McMahon, 1978) reported for any freshwater bivalve (Burky, 1983), with dense populations producing 1000–4500 g C/m^2 per year (McMahon, 1999).

Newly released juveniles are small (shell length $\approx$250 μm), but completely formed, with a characteristically D-shaped shell, adductor muscles, foot, statocysts, gills, and digestive system (Kraemer and Galloway, 1986) (Fig. 15A). Juveniles are denser than water and settle and anchor to sediments or hard surfaces with a single mucilaginous byssal thread. However, they are small enough to be suspended in turbulent water currents and hydrologically transported (McMahon, 1999). A relatively low amount of nonrespired assimilation is allocated to reproduction (5–15%; Aldridge and McMahon, 1978; Long and McMahon, unpublished data), equivalent to that of unionids, but less than sphaeriids (19%, Table III). However, the species' elevated assimilation rates allow higher actual allocation to reproduction than in other species.

As the juvenile of *C. fluminea* is small (organic carbon biomass = 0.136 μg), fecundity is large, ranging from 97 to 570 juveniles per adult per day during reproductive efforts, yielding an average annual fecundity of 68,678 juveniles per adult per year (McMahon, 1999). Early juvenile survivorship is extremely low and mortality rates remain high throughout life (74–98% in the first, 59–69% in the second, and 93–97% in the third year) (McMahon, 1999), causing populations to be dominated by juveniles and immatures. High adult mortality and population dominance by immature individuals characterizes species adapted to unstable habitats (Stearns, 1980).

The majority of NA *C. fluminea* populations have two annual reproductive periods (i.e., are bivoltine) (McMahon, 1999). It is hermaphroditic and capable of self-fertilization (Kraemer and Galloway, 1986; Kraemer *et al.*, 1986) such that single individuals can found new populations. Spermiogenesis occurs only during reproductive periods, but gonads contain mature eggs throughout the year (Kraemer and Galloway, 1986).

C. fluminea matures within 3–6 months at a shell length of only 6–10 mm (Kraemer and Galloway, 1986). Thus, juveniles born in spring can grow to maturity and participate in reproduction the following fall (McMahon, 1999) (Table III). Maximum life span is variable between populations and within populations, ranging from 1 to 4 years (McMahon and Williams, 1986a; McMahon, 1999). Early maturity allows this iteroparous species to participate in 2–7 reproductive efforts, depending on life span.

A relatively short life span, early maturity, high fecundity, bivoltine reproduction, high growth rates, small juvenile size, and capacity for downstream dispersal make *C. fluminea* both highly invasive and adapted for life in truly unstable, disturbed lotic habitats subject to unpredictable catastrophic faunal reductions. Its high reproductive potential and growth rates allow it to rapidly attain high densities after invading a new habitat or after catastrophic population declines. Thus, it is successful in NA drainage systems subject to periodic anthropomorphic interference such as channelization, navigational dredging, sand and gravel dredging, commercial and/or recreational boating, and organic and/or chemical pollution, compared to less resilient unionoideans or sphaeriids (McMahon, 1999).

Surprisingly, *C. fluminea* is more susceptible to environmental stresses, such as temperature extremes, hypoxia, emersion, and low pH than most sphaeriids and unionoideans (McMahon, 1999), making it highly susceptible to human disturbance. With only a limited capacity to tolerate unpredictable environmental stress, why is *C. fluminea* so successful in disturbed habitats? The answer lies in its ability to recover from disturbance-induced catastrophic population crashes much more rapidly than either sphaeriids or unionids. *C. fluminea* rapidly re-establishes populations even if disturbance has reduced them to a few widely separated individuals, as all individuals are hermaphrodites capable of self-fertilization and have high fecundities. Downstream dispersal of juveniles from viable upstream populations allows rapid re-establishment of decimated populations, the accelerated growth, high fecundity, and relatively short life spans of this species allowing recovery to normal age–size distributions and densities within 2–4 years (McMahon, 1999). Biannual (bivoltine) reproduction also increases its probability of surviving catastrophic density reductions, as it prevents loss of an entire generation to chance environmental disturbance ("bet hedging," Stearns, 1980). Capacity for rapid recolonization allows *C. fluminea* to sustain populations in substrates subject to periodic flood scouring from which slower growing and late maturing unionoideans are eliminated (Way *et al.*, 1990b, Section III.B.1).

Like sphaeriids, NA *C. fluminea* populations display high interpopulation variation in life-history traits (McMahon, 1999). As there is little or no genetic variation among NA populations (McLeod 1986), this variation must be ecophenotypic. Growth rates increase and time to maturity and life spans decrease in populations from more productive habitats (McMahon, 1999). On the northern border of its NA range, low temperatures reduce growth and reproductive periods, such that populations are univoltine rather than bivoltine. A slow growing *C. fluminea* population was univoltine and semelparous in an oligotrophic Texas lake (McMahon, unpublished data). Even within populations, life-history tactics vary, dependent on year-to-year variations in temperature and primary productivity (McMahon and Williams, 1986a; Williams and McMahon, 1986). As in sphaeriids, ecophenotypic life-history trait variation is adaptive in *C. fluminea,* allowing colonization of a variety of habitats. Thus, this species is highly euryoecic and highly invasive, making it the single most successful and economically costly aquatic animal introduced to North America (McMahon, 1999).

4. Dreissena Polymorpha

The zebra mussel, *Dreissena polymorpha,* is the most recently introduced NA bivalve species (Mackie and Schloesser, 1996). Like C. *fluminea,* its life-history characteristics (McMahon, 1996; Nichols, 1996) make it highly invasive. Unlike other NA bivalves, it releases sperm and eggs making fertilization completely external. Developments leads to a free-swimming planktonic veliger larva (Fig. 15C) which feeds and grows in the plankton for 8–10 days before settlement (Nichols, 1996). The planktonic veliger enhances the zebra mussel dispersal ability. Veligers in the Illinois River traveled at least 306 km downstream before settlement (Stoeckel *et al.,* 1997). Numbers of dispersing veligers can be astounding. Stoeckel *et al.* (1997) estimated that veligers in the Illinois River reached densitiess >250 veligers/L which resulted in veligers passing a fixed point at a rate as 75×10^6 per second. Total annual veliger flux was estimated to be 1.935×10^{14} and 2.131×10^{14} per year in 1994 and 1995, respectively. Adults byssally attached to floating objects are also transported downstream.

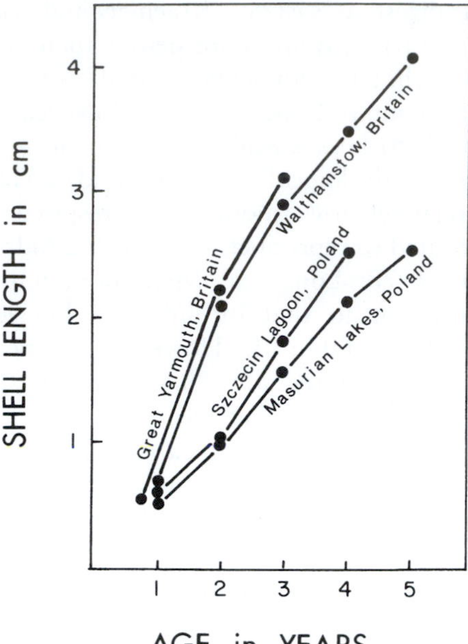

FIGURE 19 Shell-growth rates in European populations of the zebra mussel, *Dreissena polymorpha*. Growth rates for North American populations in Lake Erie are similar to or greater than faster growing British populations depicted in this figure.

Zebra mussels sexually mature in the first or second year of life (first year in most NA populations) have life spans of 3–5 years in Europe and 2–3 years in North America (Mackie and Schloesser, 1996) and, like *C. fluminea,* sustain high growth rates throughout life (Fig. 19). It is iteroparous and univoltine; an individual participating in 3–4 annual reproductive periods within its life span. Reproduction is initiated above 10–12°C, but is maximized above 18°C. Spawning is dependent on abiotic and biotic conditions, producing peaks of veliger density within a reproductive season (McMahon, 1996; Nichols, 1996; Ram *et al.,* 1996; Stoeckel, *et. al.* 1997). The freshly hatched veliger is small (diameter = 40–70 μm), but the pediveliger grows to 180–290 μm just prior to settlement (Nichols, 1996), leading to a 100 to 400-fold biomass increase during the 8–10 day planktonic period.

Maximal *D. polymorpha* adult size ranges from 3.5 to 5 cm depending on growth rate, which, like terminal size, is dependent on habitat primary productivity and temperature. Growth rate of caged mussels in Lake St. Clair, Michigan, initially enclosed at an average shell length (SL) of 4.2 m was 0.095 mm per day over the first 150 days reaching a mean SL of 14.3 mm and, thereafter, slowing down to a mean of 0.015 mm per day for the next 182 days reaching a mean SL of 17.0 mm (Bitterman *et al.,* 1994). Growth rates of young mussels in the lower Mississippi River were temperature-dependent, peaking at 0.06–0.08 mm per day at ≈16–17°C and declining at higher temperatures (Allen *et al.,* 1999), a result supporting laboratory studies showing that increasing metabolic demands reduce assimilated energy allocation to growth above 15°C (Walz, 1978a).

Like *C. fluminea, D. polymorpha* allocates a very high percentage (96.1%) of nonrespired assimilation to somatic growth, leaving only 3.9% for reproduction (Mackie *et al.,* 1989; Table III); however, Sprung (1991) reports a 30% postspawning reduction in female body mass, indicating higher reproductive energy allocation in some populations. Allocation of a large proportion of nonrespired assimilation to growth allows individuals to rapidly increase in size, making them more competitive and less subject to predation. Zebra mussel veligers settle on the shells of established individuals, forming thick mats or clusters many shells deep, in which competition for space and food must be intense. Thus, rapid postsettlement juvenile growth is highly adaptive as it allows stable byssal attachment to the substrate and positioning of siphons at the mat surface, where food and O_2 are most available. In spite of the low levels of energy devoted to reproductive effort by *D. polymorpha,* its very small eggs make individual fecundity large, ranging from 30,000 to $>10^6$ eggs per female (Sprung, 1991; Mackie and Schloesser, 1996).

Dreissena polymorpha population densities range from 7000 to 114,000 individuals/m^2 and standing crop biomasses from 0.05–15 kg/m^2 (Claudi and

Mackie, 1993). These high densities and biomasses result from the tendency of juveniles to settle on substrates inhabited by adults forming dense mats or clumps many individuals thick. High individual growth rates and population densities lead to very high productivities, estimated to be 0.05–15 g C/m^2 per year in European populations, and ≈75 g C/m^2 per year in NA Great Lakes populations based on dry tissue mass productivity values in Mackie and Schloesser (1996). The NA productivity value for *D. polymorpha* is still relatively low compared to values of 1000–4500 g C/m^2 per year estimated for dense NA *C. fluminea* populations (McMahon, 1999). Population growth and productivity are highly habitat-dependent in *D. polymorpha,* making turnover times variable, ranging from low values of 53 days (highly productive) to high values of 869 days (low productivity; Table III).

A high growth rate throughout life, elevated fecundity, short life span, and a capacity for adult and larval down-stream dispersal make *Dreissena* (like *Corbicula*) a highly invasive species. However, unlike *C. fluminea, D. polymorpha* populations are restricted to more stable habitats such as larger, permanent lakes and rivers. Its apparent preference for stable habitats is reflected by its endemic range in the Caspian Sea and Ural River (large stable habitats), avoidance of shallow, near-shore lentic- and low-flow lotic habitats, relatively long age to maturity (generally at least 1 year of life), iteroparity, gonochorism, and relatively high adult survivorship (26–88% per year; Mackie and Schloesser, 1996).

Restriction of the original range of *D. polymorpha* to the Caspian Sea and Ural River also suggests a limited dispersal capacity between drainage systems. Its dispersal through Europe occurred only in the nineteenth century (and continues in western Asia today) as interconnenting canal systems transported it between drainages. Dispersal of *D. polymorpha* between drainages is primarily by human vectors, including transport of adults attached to boat/barge hulls, ballast water dumping, and transport of veligers through canal systems interconnecting catchments. Adults are poorly tolerant of emersion (McMahon, 1996), precluding extensive natural overland dispersal unless human-mediated. Thus, dispersal of this species between NA catchments has been primarily by human vectors, with larger, permanent, navigable bodies of water being most susceptible (Mackie and Schloesser, 1996). Without anthropomorphic vectors, dispersal of *D. polymorpha* through NA drainages would almost certainly have proceeded at a much slower pace than recorded for *C. fluminea,* but human activities have already lead to this species being widely distributed in major NA drainages east of the Mississippi River (Ram and McMahon, 1996; New York Sea Grant, 1999).

C. Ecological Interactions

1. Behavioral Ecology

Other than borrowing (see Section II.A.2), information on bivalve behavior is sparse. Reviews of molluscan neurobiology and behavior (Willows, 1985, 1986) have no bivalve references. Lack of information, rather than lack of complex and intriguing behaviors, reflects difficulties in making behavioral observations on predominantly sessile bivalves surrounded by a shell.

There are a number of reproductive behaviors in freshwater bivalves, including those ensuring unionioidean glochidial contact with fish hosts (Section II.B). Adult *C. fluminea* display downstream dispersal behavior during reproductive periods. While juvenile clams (SL <2 mm) are hydrologically transported throughout the year, immatures (SL = 2–7 mm) and adults (SL >7 mm) leave the substrate to be passively hydrologically transported over the sediment surface ("rolling") only prior to reproductive periods (Fig. 20). Dispersing adults have lower dry tissue weights, lower tissue organic carbon-to-nitrogen ratios (Williams and McMahon, 1989), higher levels of ammonia excretion, and reduced molar O_2 consumption-to-N_2 excretion ratios than those remaining in the substrate (Williams and McMahon, 1985), all indicative of poor nutritional condition. Thus, downstream dispersal allows starving individuals to disperse from areas of low food availability and high intraspecific competition into areas more nutritionally favorable for reproductive efforts (Williams and McMahon, 1986, 1989).

Some adult unionoidean bivalves and *C. fluminea* display surface locomotory behavior involving the same movements of foot and valves described in Section II.B.2 for burrowing. Unionoidean surface locomotion is fairly common (Imlay, 1982), leaving tracts in sediments 3–10-m long (Golightly, 1982). The adaptive significance of bivalve surface locomotion is not well understood as it could attract potential predators. However, it may be involved with pedal feeding on organic sediment deposits (see Section III.C.2).

Some species, such as *Pyganodon grandis,* migrate vertically with seasonal changes in water level (White, 1979) to avoid emersion. Other species, such as *Uniomerus tetralasmus, C. fluminea,* and some sphaeriids, remain in position and suffer prolonged emersion (see Section II.C.4). In Texas, fire ants, *Solenopsis invicta,* kill emersed bivalves (McMahon, unpublished), suggesting that this introduced insect may be a new threat to unionoideans and sphaeriids as it expands its range in the southeastern U.S.

Some sphaeriids, *C. fluminea,* and *D. polymorpha* also crawl on hard substrates or macrophytes. *Sphaerium corneum* holds the valves erect and crawls

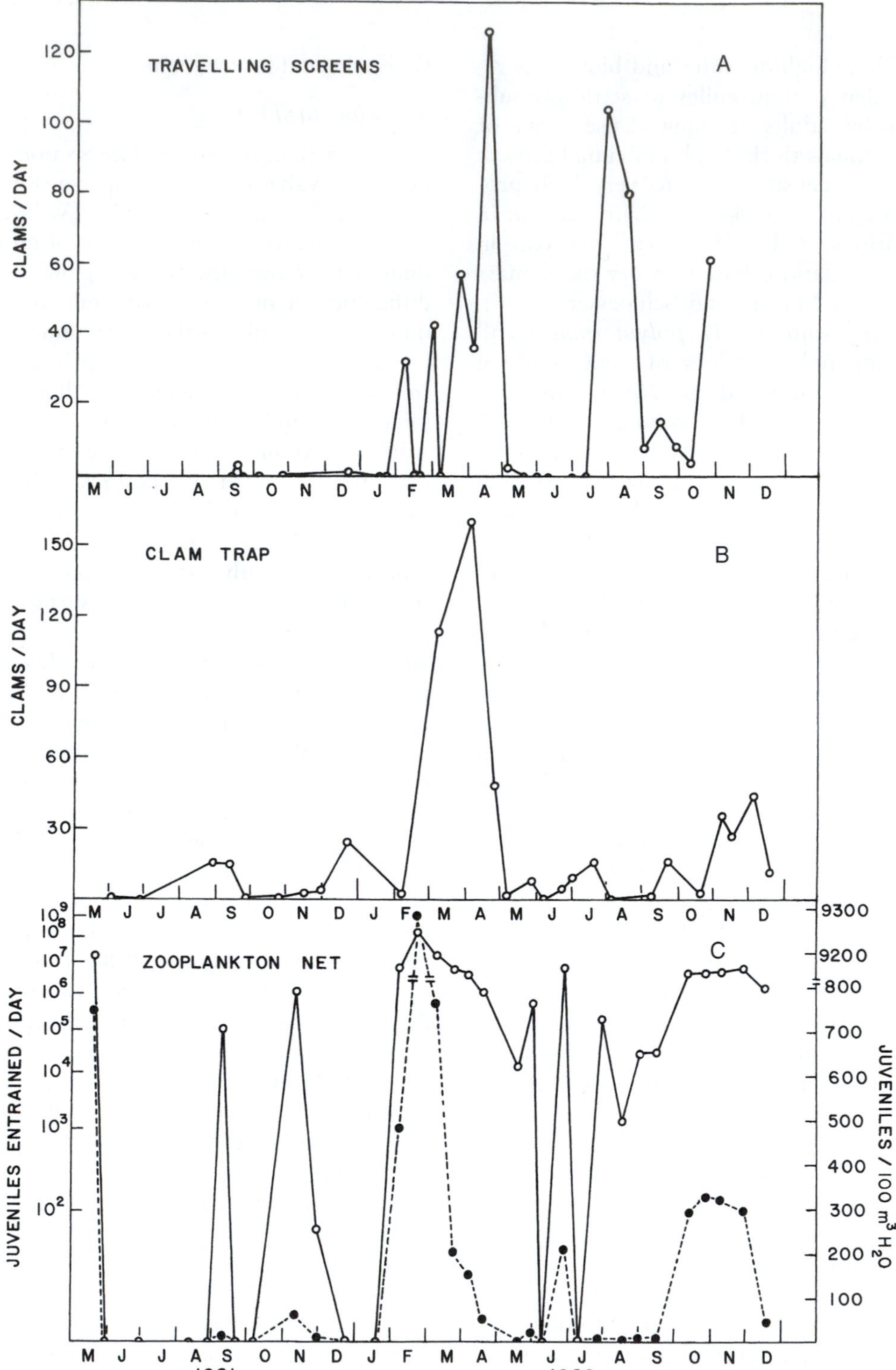

FIGURE 20 Seasonal variation in downstream dispersal behavior by juvenile, subadult and adult *Corbicula fluminea* in the water intake canal of a power station. (A) Rate of impingement of adults dispersing downstream onto traveling screens in front of water intake embayments (shell length >10 mm). (B) Rate of retention of subadults (shell length = 1–7 mm) in a clam trap held on the substrate surface of the intake canal. (C) Juveniles suspended in the water-column (shell length <2 mm). Right vertical axis represents density of juveniles in the intake canal water column (solid circles connected by dashed lines). Left vertical axis is number of juveniles entrained daily with intake water (open circles connected by solid lines). Note that the adult and subadult clams display maximal downstream dispersal behavior during only two periods, March–May and September–November). Juveniles occurred in the water column throughout the year; peak juvenile water column densities occurred during reproductive periods and midwinter periods of low ambient water temperature. (From Williams and McMahon, 1986.)

TABLE IV The Effects of Temperature on the Percentage of Time Spent in Various Valve Movement Behaviors in Emersed Specimens of *Corbiucla fluminea*[a]

Temperature (°C)	*Time with valves closed* (%)	*Time with mantle edge exposed* (%)	*Time with mantle edge parted or attempting to burrow* (%)
15	29.5	65.8	4.7
25	51.2	43.5	5.3
35	90.5	9.1	0.4

[a] From Byrne *et al.* (1988).

by extending the foot tip, anchoring it with mucus, and pulling the body forward by pedal muscle contraction (Wu and Trueman, 1984). Juvenile *C. fluminea* crawl on hard subtrates in the same manner, while adults similarly crawl while lying on one of the valves (Cleland and McMahon, unpublished). Small specimens of *D. polymorpha* are active crawlers and can climb smooth vertical surfaces. They routinely discard their byssus and migrate to a new position for reattachment. Thus, juveniles migrate to deeper waters in the winter to avoid ice scour (Mackie *et al.*, 1989).

Freshwater bivalves also respond to external cues. Chief among these is valve closure in response to irritating external stimuli when detected by mantle edge and siphonal sense organs (Section II.B.6). Valve closure seals internal tissues from damage by external irritants. Almost all freshwater bivalves tolerate some degree of anaerobiosis (Section II.C.3). Thus, individuals exposed to irritants may keep the valves shut for relatively long periods until normal external conditions return. Valve closure occurs in response to heavy metals (Doherty *et al.,* 1987), chlorine, and other biocides (Mattice *et al.,* 1982; McMahon and Lutey, 1988), suspended solids (Aldridge *et al.,* 1987), and pesticides (Borcherding, 1992). This ability allows *C. fluminea* to avoid intermittent exposure to chlorination or other biocides, making chemical control of their fouling difficult (Mattice *et al.,* 1982). Valve closure in immediate response to mantle edge or siphonal tactile stimulation is a predator defense mechanism. Valve closure response in *D. polymorpha* is used to monitor water pollution in Europe (Borcherding and Volpers, 1994).

Freshwater bivalves have adaptive behavioral responses to prolonged emersion. Physiological adaptations were discussed in Section II.C.4. Here, behavioral responses are described in greater detail. When emersed, *C. fluminea* displays four major behaviors: (1) escape responses involving attempts to burrow; (2) valves gaped widely with mantle edges parted, opening the mantle cavity to the atmosphere; (3) valves narrowly gaped with mantle edges exposed, but cemented together with mucus; and (4) valves clamped shut. Behaviors (1) and (2) are never displayed more than 6% of time in air. Exposure of sealed mantle edges allows aerial gas exchange (Byrne *et al.,* 1988), but results in evaporative water loss. When valves are closed, water loss is minimized; but O_2 uptake ceases (Byrne *et al.,* 1988; McMahon and Williams, 1984). As temperature increases (Table IV) or relative humidity decreases, duration of mantle edge exposure decreases (McMahon, 1979b; Byrne *et al.,* 1988). Hence, behaviors associated with water loss are reduced in response to increased desiccation pressure. Indeed, relative humidities near zero or temperatures above 30–35°C induce continual valve closure (Byrne *et al.,* 1988) (Table IV), reducing water loss, but making individuals anaerobic. Similar behaviors occur in emersed unionoideans (Byrne and McMahon, 1994). Less emersion-tolerant unionoideans, like *P. grandis,* display high levels of mantle edge exposure (25–90% of time emersed) and high rates of water loss. More emersion-tolerant species, like *Uniomerus tetralasmus* and *Toxolasma parvus,* have reduced levels of mantle edge exposure behavior when emerged (<25% of time emersed). The extremely emersion-tolerant *U. tetralasmus* plugs the siphon openings with mucus to reduce water loss and spend little time with mantle edges exposed (Byrne and McMahon, 1994). Ability of emersed bivalves to adjust aerial respiratory responses to external dessication pressures is a complex behavior, requiring capacity to sense, integrate, and respond to external temperature and relative humidity levels and internal osmotic concentration.

Freshwater bivalves may also have circadian patterns of behavior. Both ion-uptake and O_2-consumption rates are greater during dark than light hours in unionoideans and *C. fluminea* (McCorkle *et al.,* 1979; Graves and Dietz, 1980; McCorkle-Shiley, 1982; Section II.C.2) indicative of circadian-activity patterns in feeding, reproduction, and burrowing. The density of juvenile *C. fluminea* in the water column increases during dark hours (McMahon, unpublished) suggesting that adults preferentially release juveniles at night.

Freshwater bivalves may have circadian burrowing cycles, retreating into sediments during light hours to avoid visual predators like fish and birds. The unionoideans, *Anodonta anatina* and *Unio tumidis*, displayed a diurnal activity rhythm with valves open for longer periods at night than during the day (Englund and Heino, 1994a). Similar diurnal valve movement behavior occurs in *D. polymorpha* (Borcherding, 1992).

2. Filter Feeding

a. Particle Capture Most freshwater bivalves feed by filtering suspended material from gill water flow (Section II.B.2) driven by lateral cilia located on each side of the filaments. Projecting laterally from each filament's leading edge are "cirri" made up of partially fused tufts of eulaterofrontal cilia occurring at intervals of 2–3.5 μm (Figs. 5A and 21). Eulaterofrontal cirri on adjacent filaments project toward each other forming a stiffened grid on which suspended seston (phytoplankton, bacteria, and fine detritus) is retained from water flow driven between the filaments by the lateral cilia. Filtering may also involve creation of eddies by cirri in which seston particles settle. Filtered particles range from 1 to 10 μm, with *D. polymorpha* and *C. fluminea* both able to filter smaller particles ($\leq$1.0 μm) than most other freshwater species (Fig. 22) (Jørgensen *et al.*, 1984; Paterson, 1984; Way *et al.*, 1990a; Silverman *et al.*, 1995). It has been claimed that a mucus net covering the gill is the primary filter. While the role of mucus in bivalve filtering is debated, there is little doubt that the dense mesh of the cirri could form an effective filter (Morton, 1983; Way, 1989). Particles are captured at the level of the cirri (Jørgensen, 1996), suggesting their involvement in filtration. However, adjacent cirri are separated by 2–3.5 μm, which until recently, made their mode of particle capture an enigma because filtered particles were much smaller than the intercirral distance. Two-particle capture mechanisms have been hypothesized: direct physical cirral filtration; or capture by water currents created by gill cilia (Silverman *et al.*, 1996a). Recent confocal microscopic examinations of living gill tissue of *D. polymorpha* (Silverman *et al.*, 1996a, 1996b) and marine mytilid mussels (Silverman *et al.*, 1999) support direct cirral particle capture as the main feeding mode.

The cirrus is composed of a pair of adjacent fused ciliary sheets aligned with inhalant water flow, extending from a single gill epithelial cell (Fig. 21D.b). Thirty-eight-to-42 cilia form the cirrus. Each cilium has a distinct basal "hinge" region (Fig. 21D.c) and a free distal end characterized by reduced numbers of microtubules. Cilia increase in length from the inhalant to the trailing edge of the cirrus (Fig. 21D.b). Cirral cilia move in unison on the hinge. When they reflex into the space between adjacent filaments, the free cirral cilia tips splay out into the inhalant current forming a filter of submicron dimensions (Fig. 21D.a) on which previously free-moving particles ($<$0.75 μm diameter) are trapped. When the cirrus is then flexed back over the leading edge of the filament, trapped particles are released (Silverman *et al.*, 1996a, 1996b, 1999) (Fig 21D.a), allowing them to entrain in mucus transported either ventrally or dorsally by filament frontal cilia (Beninger *et al.*, 1997) to specialized, ciliated food grooves on the ventral edges or bases of the demibranches depending on species (Figs. 2 and 5A–C). In the unionoidean, *Pyganodon cataracta*, particles and mucus transported by filament frontal cilia become entrained and concentrated in a mucus thread on reaching the ventral food grooves where they are carried anteriorly to the labial palps (Tankersley, 1996) (Figs. 5A and 21). The role of mucus in bivalve food particle transport is being debated Other investigators feel that ciliary-based, fluid-mechanical forces are primarily responsible for particle transport with mucus playing a minor role if any (Jørgensen, 1996). Frontal cirri similar to those of *D. polymorpha* occur in sphaeriid, unionoidean and corbiculacean bivalves (Fig. 21) (Way *et al.*, 1989) and marine bivalves (Beninger *et al.*, 1997) where they are presumed to similarly function in particle capture.

b. Sorting of Filtered Particles Mucus-entrained, filtered particles are carried anteriorly to the labial palps, paired triangular flaps on each side of the mouth (Figs. 2 and 4). The outer labial palp lies against the outer ventral side of the outer demibranch, and the inner palp, against the inner ventral side of the inner demibranch (Fig. 2). Palp epithelial surfaces facing away from the demibranches are smooth, while those adjacent to the demibranch have a series of parallel ridges lying obliquely to a ciliated oral groove formed in the fused dorsal junction of the inner and outer labial palps leading to the mouth (Morton, 1983). In the unionoidean, *P. cataracta*, the palps transfer the food-laden mucus thread from the ctenidial food groove directly to the mouth for ingestion (Tankersley, 1996). In marine eulammelibranch bivalves, the mucus thread becomes mechanically fluidized on the palp surface by a combination of grinding movements of the opposing palp surfaces, transport across the corrugated palp surface and chemical reduction in mucus viscosity. Fluidization, breaks the mucus thread into a fragmented, flocculent mucus–particle slurry prior to ingestion (Beninger *et al.*, 1997). After fluidization, the corrugated palp surface sorts filtered particles into those accepted for ingestion and those rejected. Generally, smaller particles are carried over the tops of palp corrugations by cilia to be deposited in the oral groove for ingestion. Denser, larger

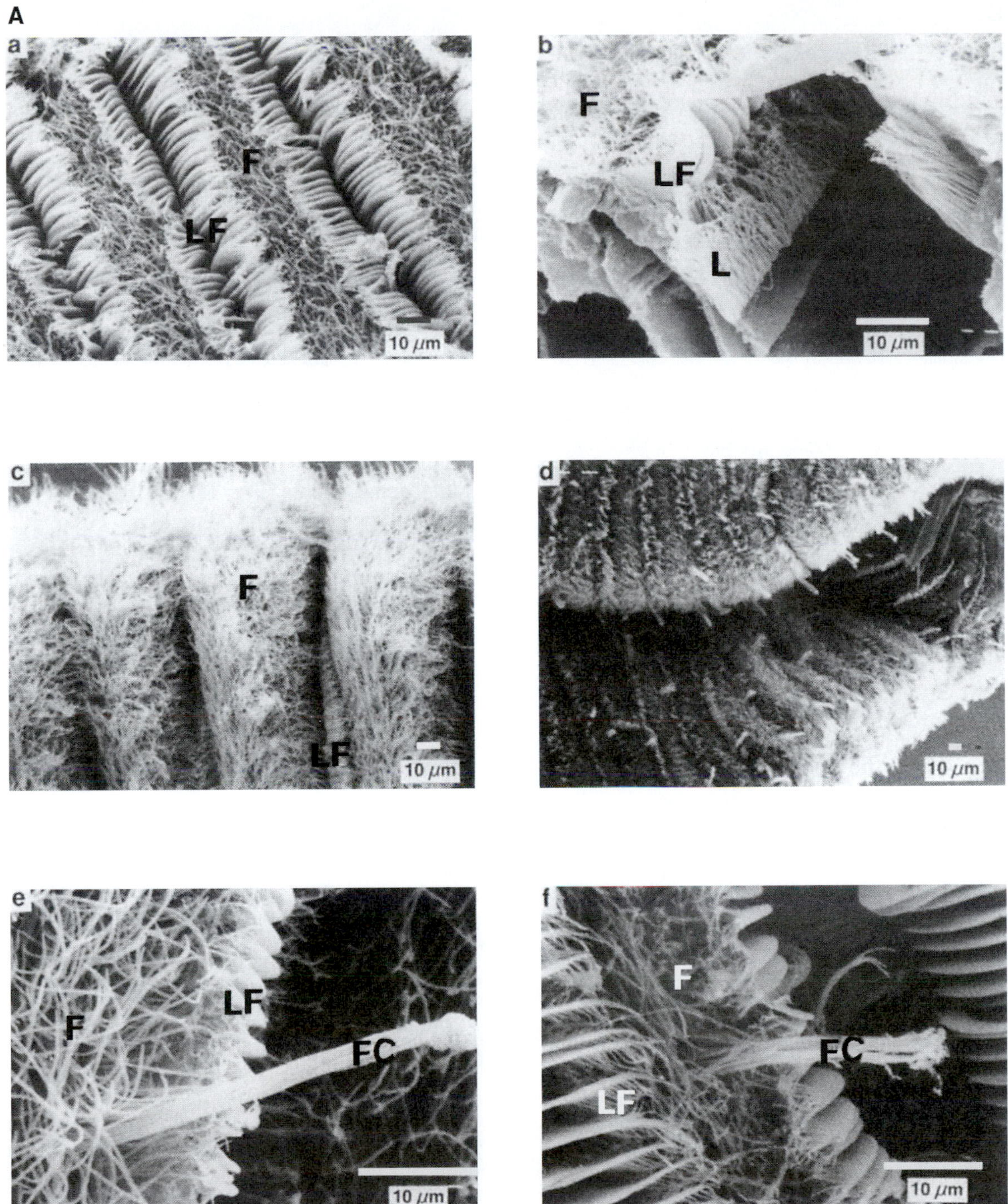

FIGURE 21 Scanning electron micrographs of the gill ciliation of representative freshwater bivalves. (A) *Musculium transversum* (Sphaeriidae): (a) arrangement of frontal and eulaterofrontal ciliation forming cirri on the leading edge of the gill filaments in the midgill region; (b) oblique view of a cross-sectional fracture of the gill showing frontal and eulataterofontal cilia forming cirri on the leading edge of the filament and lateral cilia on the side of the filament; (c) ciliation of the food groove on the ventral edge of the gill; (d) posterior portion of the outer (foreground) and inner (background) demibranchs of the gill; (e) a frontal cirrus formed from fused cilia emerging between the eulaterofrontal cirri and frontal cilia on the leading edge of a filament; (f) a frontal cirrus emerging from a band of frontal cilia. (B) *Corbicula fluminea* (Corbiculidae): (a) leading edge of a gill filament showing frontal cilia, eluatorofrontal cilia forming cirri, lateral cilia and frontal cirri; (b) high-magnification view of frontal cilia and eulateralfrontal cirri; (c) oblique view of a longitudinal fracture through the midgill region showing all four types of ciliation, including lateral cilia lining the sides of the gill filiament;

(Continues)

B

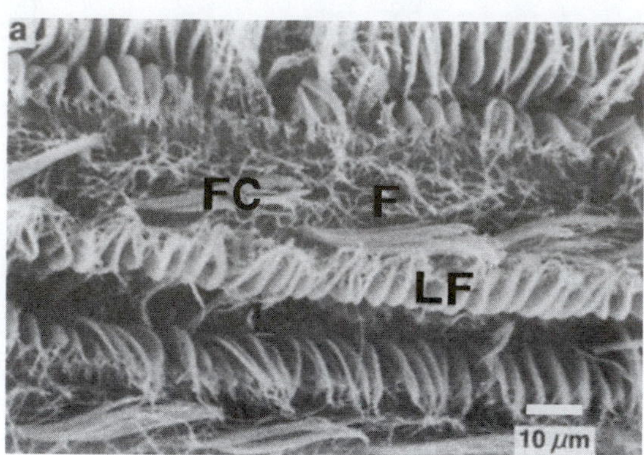

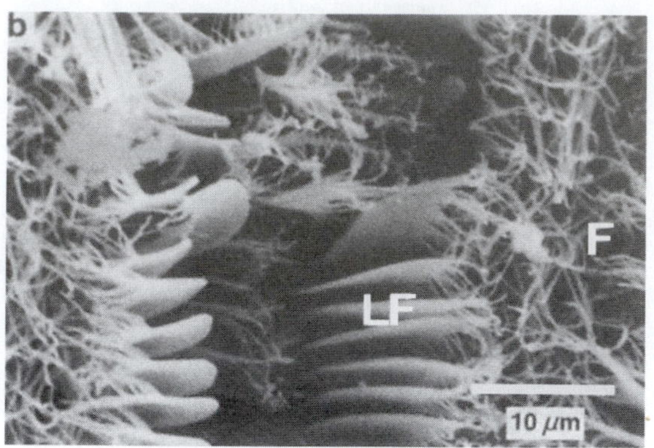

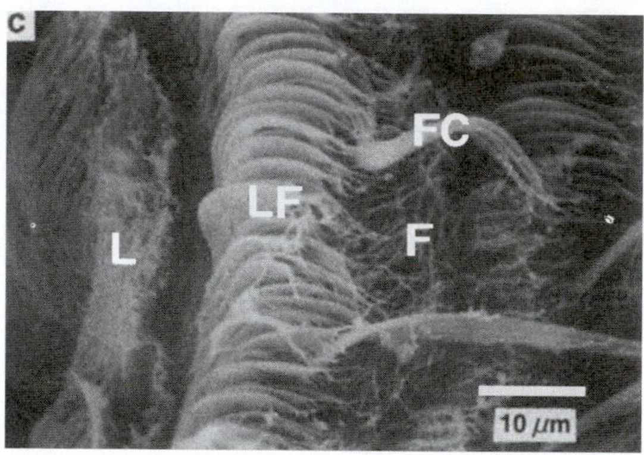

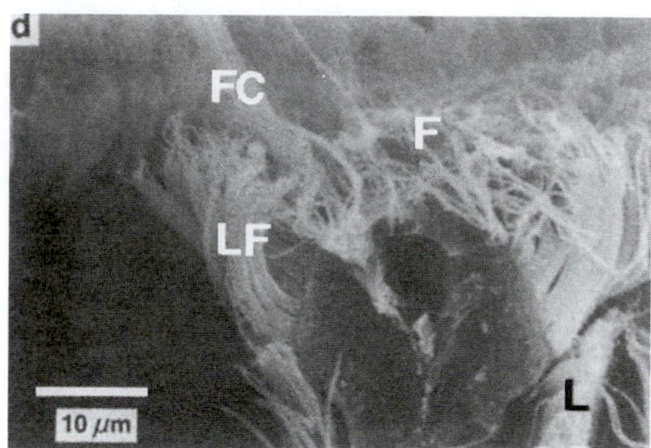

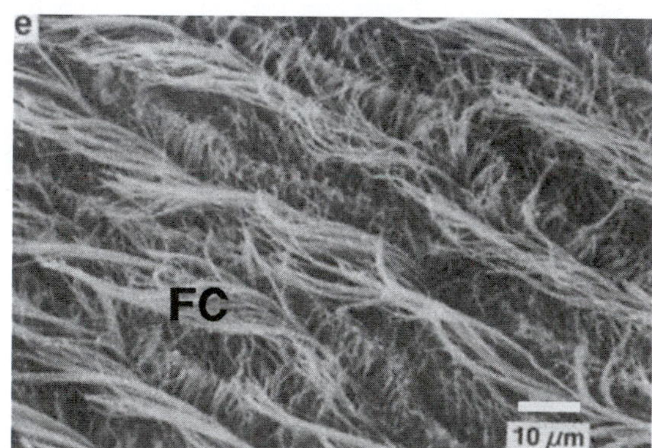

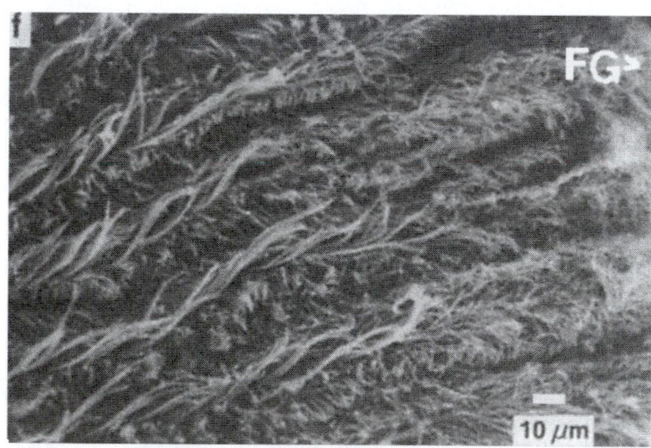

FIGURE 21 (Continued)
(d) crosssection of the midgill region showing the origin of the frontal cirrus; (e) lower region of the gill showing the presence of less well organized frontal cilia; (f) food groove region at the ventral edge of the inner demibranch; note loss of ciliary organization in food groove, including lack of frontal cirri. (C) *Obliquaria reflexa* (Unionoidea): (a) frontal cilia and eulaterofrontal cilia forming cirri on the leading edges of several gill filaments (500X); (b) high-power view showing lateral cilia on sides of the gill filaments between tracts of frontal cilia and eulaterofrontal cilia forming cirri on adjacent filaments. (D) *Dreissena polymorpha* (Dreissenoidea): (a) frontal cilia on the leading edge of a gill filament and some eulaterofrontal cirri reflexed into the space between adjacent filaments with the free, distal ends of their cilia splayed out to form the primary gill filtering mechanism and other cirri reflexed over the frontal cilia in a position in which they release particles captured from water flow between adjacent filaments to frontal cilia of the food groove; (b) detailed structure of a eulaterofrontal cirrus formed from two rows of opposed cilia fused basally, free at their distal tips, with a distinct

(Continues)

C

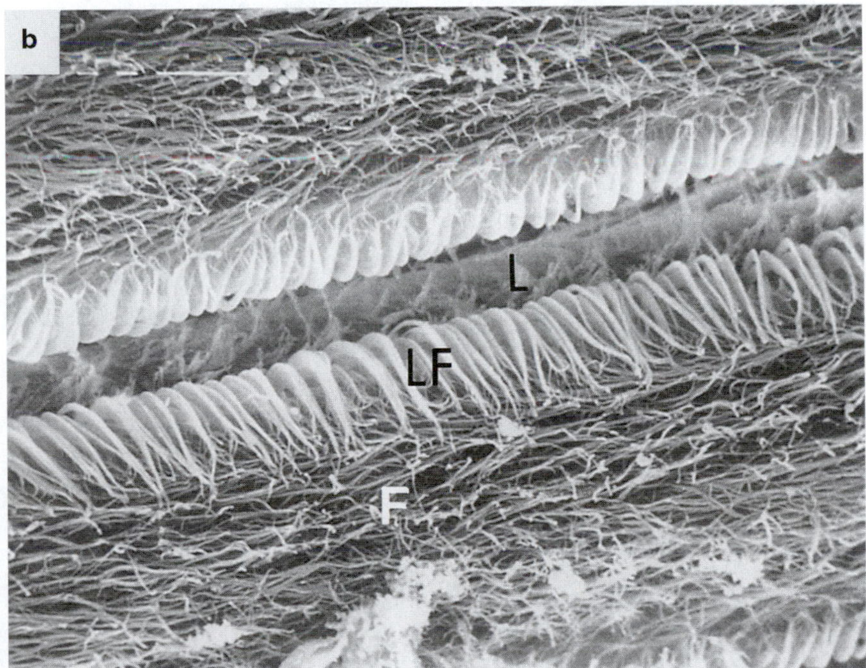

FIGURE 21 (Continued)
basal hinge on which the cirral cilia simultaneously flex between the interfilament space (food capture) and the frontal cilia on the leading edge of the filament (food release); (c) transmission electron micrograph showing details of the unique hinge on which cirral cilia flex between the interfilament space and the frontal cilia. Note lack of frontal cirri characteristic of the (A) Sphaeriidae, and (B) *Corbicula fluminea* in (C) unionoideans and (D) *Dreissena* Label key for figures A–C : F, frontal cilia; FC, frontal cirrus; L, lateral cilia, LF eulaterofrontal cirri; and FG, food groove. [Photomicrographs supplied by Tony Deneka and Daniel Hornbach (Macalester

(Continues)

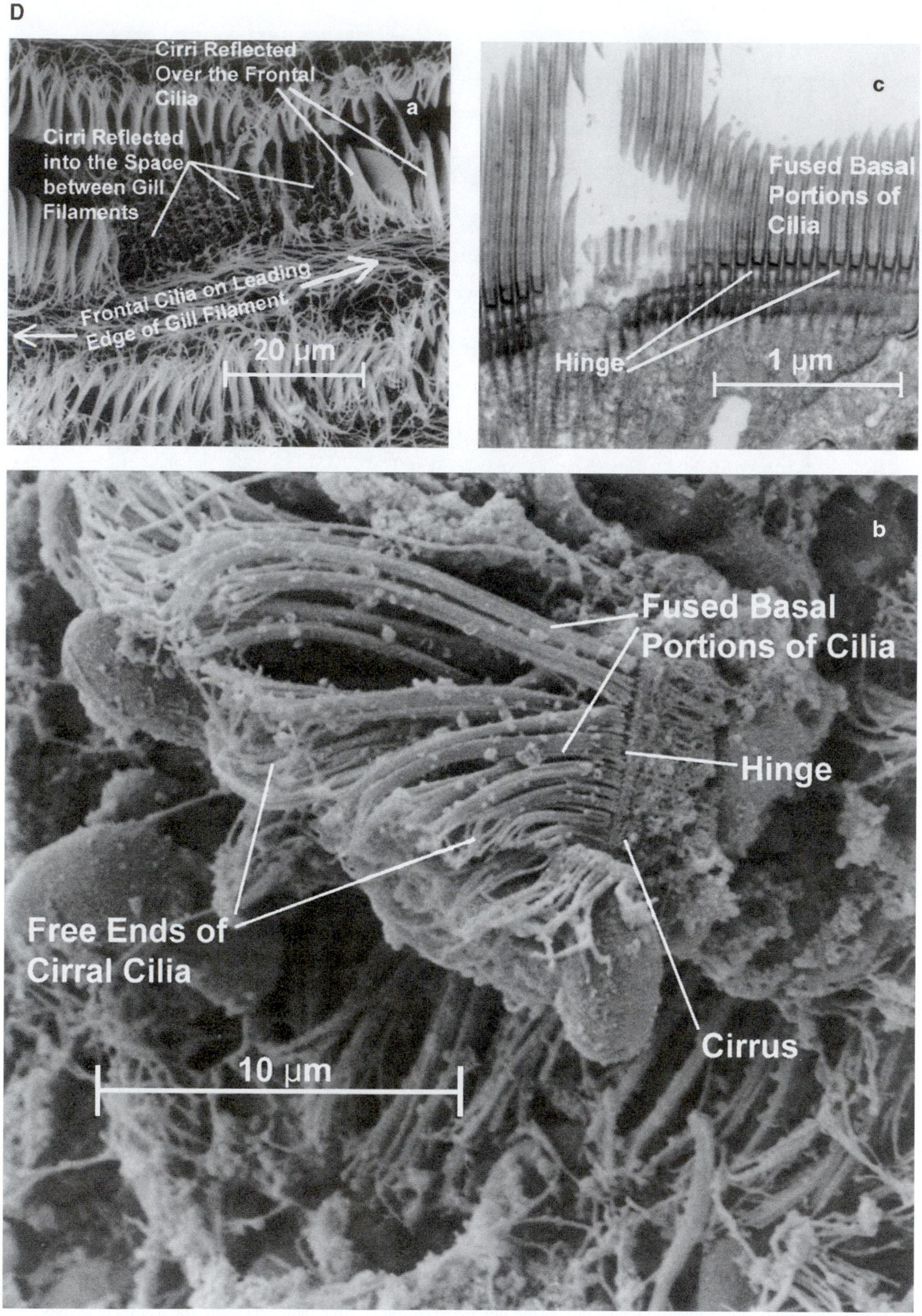

D

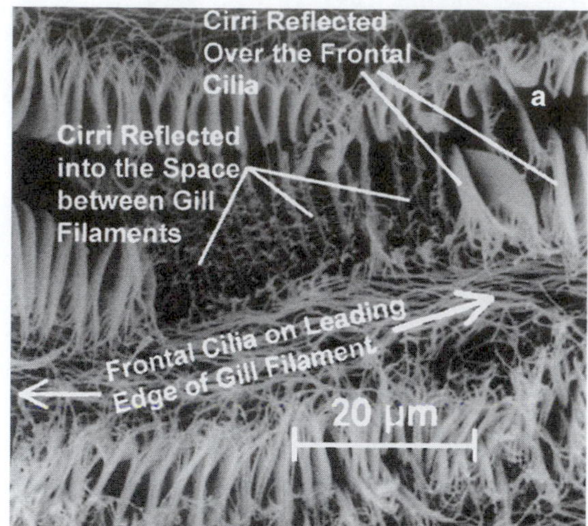

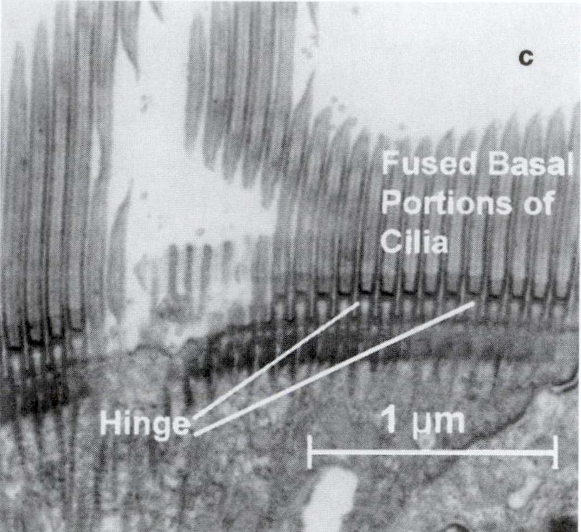

FIGURE 21 (Continued)

College), Carl M. Way (U.S. Army Corps of Engineers), and Harold Silverman and Thomas H. Dietz (Louisiana State University, Baton Rouge).]

particles fall between the corrugations where cilia carry them to the ventral edge of the palp to be bound in mucus and released onto the mantle as pseudofeces (pseudofeces are mucus-bound particles that have not been ingested) (Morton, 1983; Beninger *et al.,* 1997). Pseudofeces are carried by mantle cilia to the base of the inhalant siphon (Fig. 2), where they are periodically expelled with water forced from the inhalant siphon by rapid valve adduction (i.e., valve clapping) (Morton, 1983). The palp may also sort medium-sized particles. Smaller medium-sized particles tend to remain longer on corrugation crests than in intervening grooves, thus, reaching the oral groove for ingestion, while larger, medium-sized particles remain longer in the grooves leading to eventual rejection as pseudofeces. Thus, labial palp particle sorting may be primarily by particle size and, perhaps, density. The function of labial palps in particle sorting is being debated, palps appearing to have little if any sorting function in some species (Tankersley, 1996) and extensive sorting functions in others (Beninger *et al.,* 1997). Labial palp function in freshwater bivalves requires further study.

The width of palp corrugations and gill filaments can be adjusted by both muscular activity and distention of gill and palp blood sinuses, thus allowing active particle-size selection (Morton, 1983). When fed natural seston, the percentage of particles retained by the unionoidean, *Elliptio complanata,* declined linearly over a particle size range of 5 (nearly 100%) to $>10\mu m$ (less than 10% retention), suggesting size selection (Paterson, 1984). Similar size selection for particles occurs in the sphaeriid, *Musculium transversum* (Way, 1989), and *C. fluminea* (Way *et al.,* 1990a). Some unionoidean species may select different algal types. In the same river pool, *Amblema plicata* ingested a greater proportion of green algae, a lower diversity of algal species, and different algal species than *Ligumia recta* (Bisbee, 1984). Interspecific differences in particle selection may allow some sympatric species to divide the feeding niche, avoiding competition, however, other unionoideans are nonselective in feeding and have high diet overlap between species.

c. Filtering Rate The filtering rate of bivalves can be computed as follows:

$$C = \frac{M}{t}\left(In\frac{C_0}{C_t}\right)$$

where M is the volume of suspension filtered, t the time, and Co and Ct the concentrations of filtered particles at time o and t, respectively (Jørgensen *et al.,* 1984). Filtration rates are elevated in *C. fluminea* compared to other freshwater bivalves. For a 25-mm shell length (SL) specimen, rate ranges from 300 to 2500 L/h depending on temperature and seston concentration (Foe and Knight, 1986b; Lauritsen, 1986b; Way *et al.,* 1990a). The filtering rate of *C. fluminea* fed natural seston concentrations was not significantly correlated with field water temperature (Long and McMahon, unpublished) and was independent of temperature between 20 and 30°C in the laboratory (Lauritsen, 1986b), suggesting capacity for temperature compensation of gill ciliary filtering activity. The filtration rate is also somewhat temperature independent in *Sphaerium striatinum* and *Musculium partumeium,* reaching maximal values during reproductive periods when water temperatures, were below midsummer maxima (Burky *et al.,* 1981, 1985a; Way *et al.,* 1981). Filtering rates in *D. polymorpha* are similar to those in *C. fluminea* (Lei *et al.,* 1996) and 2–4 times higher than those in sphaeriids and unionoideans (Kryger and Riisgård, 1988).

High particle concentrations reduce filtering rates (Burky *et al.,* 1985a; Way, 1989; Way *et al.,* 1990a; Lei *et al.,* 1996). Filtering rates in *S. striatinum* (Hornbach *et al.,* 1984a) and *M. partumeium* (Burky *et al.,* 1985a) were depressed at concentrations below natural seston, making particle ingestion nearly constant above a critical particle concentration. Thus, filtering may have be regulated to control ingestion rate. Particle concentrations within ambient seston concentrations do not affect filtering rate in *C. fluminea* (Mattice, 1979), ingestion rates increasing directly with concentration. However, filtration rate in *C. fluminea* declined three-fold with increasing particle concentration between 0.33 and 2.67 μl algal volume/L suggesting inhibition at very high concentrations, but increasing algal concentration over this range still resulted in a 3.5-fold rise in ingestion rate (Lauritsen, 1986b). Similar reduction in filtration rate at particle concentrations above natural seston levels occurs in the sphaeriid, *M. transversum* (Way, 1989). In contrast to *C. fluminea,* filtration rate of the unionoidean, *E. complanta,* was depressed by increasing natural seston concentrations (Paterson, 1984). Filtration rate differed among populations of *C. fluminea,* but ingestion rates were similar, suggesting regulation of the filtration rate to control ingestion rates (Way *et al.,* 1990a). In *D. polymorpha,* filtration rate was reduced when feeding on artificial plastic microspheres compared to natural seston. When filtering particles <1.5 μm in diameter, filtration rate decreased with decreasing particle concentration below a critical upper limit, and with decreasing temperature. Filtration rate was also affected by prior temperature acclimation (Lei *et al.,* 1996). Such data indicate that the seasonal and environmental filter-feeding responses of freshwater bivalves are species-specific and laboratory measurements of filtration rates are likely to be invalid unless carried out at natural seston concentrations and sizes and with natural seston rather than artificial particles. Filter feeding and particle retention rates were depressed in gravid females of the unionoidean, *Pyganodon cataracta,* in which particle transport was slowed down and the ability to retain small particles (<6 μm) reduced, when the marsupial gills were distended with glochidia (Tankersley, 1996).

Suspended silt can inhibit filtering and consumption rates in freshwater bivalves, perhaps by overwhelming ciliary filtering and sorting mechanisms. High silt concentrations reduced filtering and metabolic rates in three unionoidean species whether exposure was infrequent (once every 3 h) or frequent (once every 0.5 h), suggesting interference with gill ciliary activity. Furthermore, individuals experiencing frequent exposure increased reliance on carbohydrate catabolism (Aldridge *et al.,* 1987), symptomatic of short-term starvation; perhaps, the basis of exclusion of unionoideans from habitats with high silt loads (Adam, 1986). In contrast, natural suspended sediment concentrations did not affect shell or tissue growth in *C. fluminea* (Foe and Knight, 1985), perhaps, accounting for its ability to colonize high-flow lotic habitats with greater suspended solids than tolerated by most unionoideans (Payne and Miller, 1987; Section III.A.). In contrast, growth in juvenile unionoideans is stimulated by small amounts of suspended silt in their algal food (Hudson and Isom, 1984), perhaps because juveniles (*Villosa iris*) may feed primarily on sediment interstitial water through the pedal gape, leading to ingestion of silt, microdetritus and bacteria, with little consumption of algae (Yeager *et al.,* 1994).

In smaller streams, where phytoplankton productivity may be low, most suspended organic matter is particulate organic detritus or heterotrophic bacteria and fungi (Nelson and Scott, 1962). While experiments are few, at least some freshwater bivalves efficiently filter suspended bacteria. *D. polymorpha* appears to extract particles of bacteria size (<1.0 μm in diameter) (Sprung and Rose, 1988, Fig. 22) as can *C. fluminea* (Silvermann *et al.,* 1995) with *D. polymorpha* and *C. fuminea* able to clear the bacterium, *Escherichia coli,* 30 and 3 times faster than the lotic unoinid, *Toxolasma texasiensis,* respectively. Increased capacity for bacterial filtration in *D. polymorpha* appears associated with its enlarged gill with a 100-fold greater density of cirri relative to *C. fluminea,* that of *T. texasiensis* has even fewer and smaller cirri than *D. polymorpha* or *C. fluminea* (Silverman *et al.,* 1995). Filter-feeding by *D. polymorpha* may significantly reduce bacterioplankton densities in the inner portion of Saginaw Bay, Lake Huron (Cotner *et al.,*). *Musculium transversum* (Way, 1989a) and some unionoidean species (Fig. 22) do not efficiently filter bacterial-size particles. Unionoideans appear specialized to filter larger particles, such as phytoplankton, detritus, protozoa, and even small zooplankton, such as rotifers (Singh *et al.,* 1991). In contrast, six unionoidean species adapted to lotic habitats where seston was dominated by bacteria and detritus could filter bacteria at significantly higher rates than three species adapted to lentic habitats where seston was dominated by phytoplankton. Lotic species had more complex cirri (>25 vs <16 cilia per cirral plate) and greater gill cirral density than lentic species, indicative of a filtering system better adapted for fine particle filtration (Silverman *et al.,* 1997) . Because bacteria and detrital particles may comprise most of the organic matter in some, particularly lotic, habitats, bacterial feeding warrants greater study in freshwater bivalves.

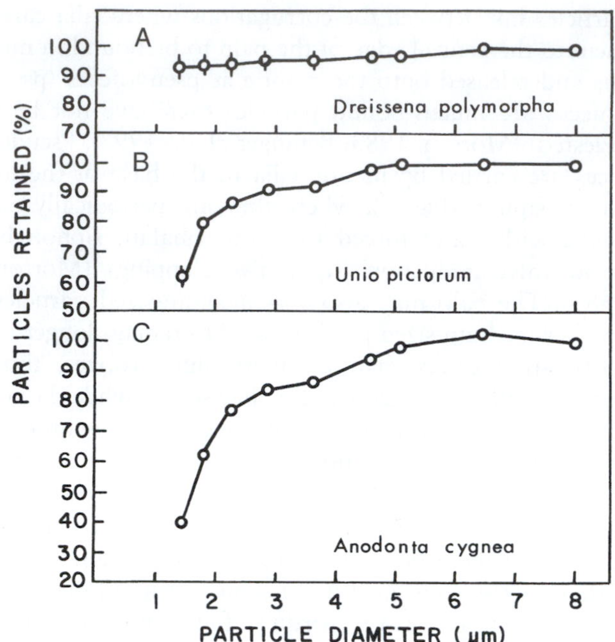

FIGURE 22 Relationship between the percentage of suspended particles retained by the gill filtering mechanism of freshwater bivalves and particle size, as particle diameter in micrometers. Horizontal axis for all figures is the percentage of total particles retained from suspensions passing over clam gill filtering systems in (A) *Dreissena polymorpha*, the unionoideans, (B) *Unio pictorum*, and (C) *Anodonta cygnea*. Vertical lines about points are standard deviations. Note that all three species efficiently retained particles of 2.5–8 μm diameter, equivalent to most unicellular algae, but only *D. polymorpha* efficiently retained bacteria-sized particles of 1 μm diameter. (Redrawn from Jørgensen *et al.*, 1984.)

d. Sediment Feeding At least some freshwater bivalves may supplement phytoplankton filter feeding by consuming organic detritus or interstitial bacteria from sediments. The infaunal habit of many *Pisidium* species suggests dependence on nonplanktonic food sources. Many pisidiids can efficiently filter small interstitial bacteria. They burrow continually through sediments with the shell hinge downward, drawing dense interstitial water and bacteria through parted ventral mantle edges into the mantle cavity to be filtered on the gills (Lopez and Holopainen, 1987). Both *Pisidium casertanum* and *P. conventus* feed in this manner, filtering bacteria of smaller diameter (<1 μm) than most bivalves (Lopez and Holopainen, 1987) (Fig. 22). Filtering of rich interstitial bacterial flora may be associated with the reduced ctenidial surface areas and low filtration and metabolic rates of *Pisidium* species compared to *Sphaerium* or *Musculium* (Hornbach, 1985; Lopez and Holopainen, 1987). Similar interstitial bacteria and detritus feeding occurs in juvenile unionids (Yeager *et al.,* 1994). *M. transversum* uses its long inhalant siphon to vacuum detrital particles from the sediment surface (Way, 1989).

Pedal feeding on sediment organic detritus may be another important food source in some freshwater bivalves. Pedal feeding occurs in marine and estuarine bivalves (Morton, 1983; Reid *et al.,* 1992), but is little studied in freshwater species. *Corbicula fluminea* draws sediment detrital particles over the ciliated foot epithelium into the mantle cavity, where they accumulate in the food groove of the inner demibranch and are carried to the palps for sorting and ingestion (Way *et al.,*1990b; Reid *et al.,* 1992). Cellulose particles (40–120 μm in diameter) soaked in various amino acid solutions and mixed with sediments accumulated in the stomach of pedal-feeding *C. fluminea* (McMahon, unpublished). Thus, *C. fluminea* removed sediment organic matter at a rate of 50 mg/clam per day and doubled growth rates when allowed to pedal feed in addition to filter feeding (Hakenkamp and Palmer, 1999). Similar pedal feeding has been reported in *M. transversum* (Way, 1989) and in juvenile *Villosa iris* (Yeager *et al.,* 1994). Thus, pedal deposit feeding may be an important auxiliary feeding mode in freshwater and marine bivalves and may have important impacts on sediment organic dynamics (Hakenkamp and Palmer, 1999).

Pedal feeding may be more universal in freshwater species than previously suspected. In a stream *S. striatinum* population, only 35% of organic carbon assimilation occurred by filter feeding, leaving 65% to come from sediment detrital sources (Hornbach *et al.,* 1984b), presumably by pedal feeding. Other sphaeriid species reach highest densities in sediments of high organic content (Zbeda and White, 1985) or in habitats receiving organic sewage effluents, suggesting dependence on pedal deposit-feeding (Burky, 1983). Pedal deposit-feeding mechanisms may explain the extensive horizontal sediment locomotion of a number of freshwater bivalves (Section II.D), as it allows pedal contact with organic detritus.

3. Population Regulation

a. Abiotic Population Regulation Most of the information regarding regulation of freshwater bivalve populations is anecdotal. Fuller (1974) reviews factors affecting population density and reproductive success. Catastrophic abiotic factors causing reductions in bivalve populations are reviewed in Section III.B. Among these are silt accumulation in sediments of impounded rivers and in the water column during flooding. Silt interferes with gill filter feeding and gas exchange (Aldridge *et al.,* 1987; Payne and Miller, 1987) causing massive mortality in unionoidean populations (Adam, 1986, Section III.A). In contrast, many sphaeriids thrive in silty sediments (Burky, 1983; Kilgour and Mackie, 1988) associated with their utilization of sediment detrital food sources (Section III.C.2). Shell growth in *C. fluminea,* a silt tolerant species, can be temporarily inhibited by high turbidity (Fritz and Luz, 1986).

Temperature extremes also affect bivalve populations. *C. fluminea* tolerates 2°C (Mattice, 1979) to 36°C (McMahon and Williams, 1986b). Both cold-induced winter (Blye *et al.,* 1985; Sickel, 1986) and heat-induced summer kills (McMahon and Williams, 1986b) are reported for this species. The incipient upper lethal limit for *D. polymorpha* is 30°C, preventing its invadsion of many southern U.S. drainage systems (McMahon, 1996). Temperature limits for other NA bivalves are not as well known, but, on average, appear to be broader than either *C. fluminea* or *D. polymorpha* (Burky, 1983). Juvenile unionoideans are less temperature tolerant than adults (Dimock and Wright, 1993), suggesting that temperature limitation operates at the juvenile level in this group. Within the tolerated range, temperature may have detrimental effects on reproductive success. Sudden temperature decreases cause abortive glochidial release in unionoideans (Fuller, 1974), and temperatures above 30–33°C inhibit reproduction in *C. fluminea* (McMahon, 1999). Sphaeriid densities were reduced in midsummer by heated effluents (Winnell and Jude, 1982). Heated effluents may also stimulate bivalve growth, inducing early maturity and increasing reproductive effort. They can provide heated refugia in which cold-sensitive species may overwinter as in northernmost NA populations of *C. fluminea* (McMahon, 1999).

Low ambient pH can depauperate or extirpate bivalve populations. Bivalve populations have been extirpated by acid mine drainage (Taylor, 1985; Warren *et al.,* 1984). Highly acidic waters cause shell erosion and eventual death (Kat, 1982). Perhaps of more importance than pH in regulating freshwater bivalve populations are water hardness and alkalinity. Unionoideans do not occur in New York drainage systems with calcium concentrations <8.4 mg Ca/L (Fuller, 1974). In waters of low alkalinity, Ca concentrations may be too low for shell deposition; the lower limit for most freshwater bivalves being 2–2.5 mg Ca/L (Okland and Kuiper, 1982; Rooke and Mackie,

1984a). Growth and fecundity were suppressed in a *Pisidium casertanum* population in a low Ca concentration relative to that a higher Ca concentration (Hornbach and Cox, 1987). Waters of low alkalinity have little pH-buffering capacity, subjecting them to seasonal pH variation. This makes them particularly sensitive to acid rain, which detrimentally impacts bivalve faunae (Rooke and Mackie, 1984a). Shell Ca content is related to water Ca concentration. Of 10 Canadian freshwater bivalve species, two sphaeriids and one unionoidean displayed no relationship between shell Ca content and water Ca concentration, shell Ca content decreased with increased water Ca concentration in two sphaeriid species and increased with water Ca concentration in four sphaeriid and two unionoidean species (Mackie and Flippance, 1983b).

Lowering of water levels during dry periods induce bivalve mortality by exposing individuals to air. Such emersion caused near 100% mortality in *C. fluminea* populations (White, 1979) due to its poor desiccation tolerance (Bryne *et al.*, 1988). Many unionoidean species are relatively tolerant of emersion. When sympatric populations of nine unionoidean species, the sphaeriid *Musculium transversum,* and *C. fluminea* were emersed for several months, *M. transversum* and *C. fluminea* suffered 100% mortality while the unionoideans experienced only 50% mortality, because they either migrated downshore or resisted desiccation (White, 1979). Like *C. fluminea, D. polymorpha* is relatively emersion intolerant, particularly above 20°C (McMahon, 1996), restricting it to habitats with relatively stable water levels and making it unlikely to develop dense populations in variable level reservoirs.

Low environmental O_2 concentrations can be detrimental to freshwater bivalves. Continual eutrophication of Lake Estrom, Denmark, so reduced ambient profundal O_2 concentrations (0–0.2 mg O_2/L for three months) that massive reductions of sphaeriid populations ensued (Jonasson, 1984a, b), including hypoxia-tolerant species, such as *Pisidium casertanum* and *P. subtruncatum* in which the lower limit for aerobic respiration was 1.7 mg O_2/L (Jonasson, 1984b). Adult unionoideans are relatively hypoxia tolerant. Juveniles are more hypoxia sensitive, those of *Utterbackia imbecillis* and *Pyganodon cataracta* tolerating anoxia for less than 24 h (Dimock and Wright, 1993). Thus, restriction of unionoidean species to well oxygenated waters may be a juvenile rather than adult constraint.

Pollution is also detrimental to freshwater bivalves (reviewed by Fuller, 1974) including chemical wastes (Zeto *et al.*,1987), asbestos (Belanger *et al.*,1986a), organic sewage effluents (Gunning and Sutkkus, 1985; Neves and Zale, 1982; St, John, 1982), heavy metals (Belanger *et al.*, 1986b; Fuller, 1974; Lomte and Jabhav, 1982a; Grapentine, 1992), chlorine and paper mill effluents (Fuller, 1974), acid mine drainage (Taylor, 1985; Warren *et al.*, 1984) and PCBs (Grapentine, 1992). Potassium ions, even in low concentration (>4–7 mg K^+/liter), can be lethal to freshwater bivalves including *D. polymorpha* (Claudi and Mackie, 1993), excluding bivalves from watersheds where potassium is naturally abundant (Fuller, 1974).

b. Biotic Population Regulation There have been a few studies of biotic regulation of freshwater bivalve populations. Freshwater bivalves are host for a number of parasites. They are intermediate hosts for digenetic trematodes (Fuller, 1974). Such infections cause sterility in gastropods, and have been reported to induce sterility in the unionoidean, *Anodonta anatina,* allowing diversion of the mussel's energy stores from production of its gametes to production of trematode cercariae (Jokela *et al.*, 1993), reducing the mussel population's reproductive capacity. Parasitic nematode worms inhabit the guts of unionoideans (Fuller, 1974). The external oligochaete parasite *Chaetogaster limnaei* resides in the mantle cavities of unionoideans (Fuller, 1974), *C. fluminea* (Sickel, 1986) and *D. polymorpha* (Conn *et al.*, 1996). *Dreissena polymorpha* is host to a wide diversity of parasites (Malloy *et al.*, 1997), but their impacts on this species are unknown. All of these parasites probably contribute to the regulation of freshwater bivalve population densities, but the degree to which they do has received little experimental attention.

Water mites of the family Unionicolidae, including *Unionicola* and *Najadicola,* are external parasites of unionoideans (Vidrine, 1990). Both mature and preadult mites are parasitic, attaching to gills, mantle, and the visceral epithelium (depending on species) (Mitchell, 1955; Chapter 16). Heavy mite infestations cause portions of the gills to be shed, abortion of developing glochidia, or even death (Fuller, 1974), perhaps making unionicolid mites a major regulator of unionoidean populations. The larval stage of the chironomid, *Ablabesmyia janta,* parasitizes the unionoidean gill filaments and may exclude unionicolid mites from the gills they infest (Vidrine, 1990).

Disease has been little studied. There is little evidence of viral or bacterial involvement in die-offs of *C. fluminea* (Sickel, 1986) or unionoideans (Fuller, 1974).

In some *C. fluminea* populations, massive die-offs (particularly of older individuals) occur after reproductive efforts (Aldridge and McMahon, 1978; McMahon and Williams, 1986a; Williams and McMahon, 1986), probably due to reductions in tissue energy reserves of postreproductive individuals (Williams and McMahon, 1989). Such postreproductive morality also occurs in sphaeriids (Burky, 1983).

TABLE V List of the Major Fish Predators of Freshwater Bivalves[a]

Family	Genus and species	Common name
Clupeidae	*Alosa sapidissima*	American shad
Cyprinidae	*Cyrinus carpio*	Common carp
Catostomidae	*Ictiobus bubalus*	Smallmouth buffalo
	Ictiobus niger	Black buffalo
	Minytrema melanops	Spotted sucker
	Moxostoma carinatum	River redhorse
Moronide	*Morone saxatilis*	Striped bass
Ictaluridae	*Ictalurus furcatus*	Blue catfish
	Ictalurus punctatus	Channel catfish
Centrarchidae	*Lepomis gulosus*	Warmouth
	Lepomis macrochirus	Bluegill
	Lepomis microlophus	Red-ear sunfish
Sciaenidae	*Aplodinotus grunniens*	Freshwater drum
Acipenseridae	*Acipenser fulvesens*	Lake sturgeon

[a] Data from Fuller (1974), McMahon (1983a), Robinson and Wellborn (1988), and Sickel (1986).

Predation may be the most important regulator of freshwater bivalve populations. Shore birds and ducks feed on sphaeriids and *C. fluminea* (Dreier, 1977; Paloumpis and Starrett, 1960; Smith *et al.*, 1986; Thompson and Sparks, 1977, 1978) and *D. polymorpha* (Mazak *et al.*, 1997). *C. fluminea* densities were 3–5 times greater in enclosures excluding diving ducks (Smith *et al.*, 1986). Water fowl feeding also significantly reduces *D. polymorpha* populations (Mackie and Schloesser, 1996). Crayfish feed on small bivalves including *C. fluminea* (Covich *et al.*,1981) and *D. polymorpha* (Mackie *et al.*, 1989), and, in some habitats, may regulate bivalve population densities. Fire ants (*Solenopis invicta*) prey on clams emersed by receding water levels. In additions, turtles, frogs, and the mudpuppy salamander *Necturus maculosus* all feed to a limited extent on small or juvenile bivalves (Fuller, 1974). Free-living oligochaetes prey on newly released glochidia (Fuller, 1974).

Perhaps the major predator of freshwater bivalves are fish. A number of NA fish are molluscivores (Table V). While most molluscivorous fish feed on small bivalve species or juveniles of larger bivalves (SL <7 mm), several routinely take large adults, including carp (*Cyprinus carpio*), channel catfish (*Ictalurus punctatus*) and freshwater drum (*Aplodinotus grunniens*). The latter species crushes shells with three massive, highly muscular pharyngeal plates, as do carp to a lesser extent. Bivalve prey size may be limited by the dimensions of the shell-crushing apparatus. Thus, 273–542-mm long *A. grunniens* did not take *D. polymorpha* with SL >11.4 mm, but, below this size, fish size and mussel size taken were not related. Thus, predation by *A. grunniens* is unlikely to control *D. polymorpha* populations (French and Love, 1995). In contrast, channel catfish swallow bivalves intact (J. C. Britton, personal communication). The majority of molluscivorous fish may limit predation to smaller bivalves because shells of larger individuals are too strong to crack or crush. The shell strength of *C. fluminea* increases exponentially with size [$\log_{10}$ force in newtons to crack the shell $= -0.76 + 2.31$ ($\log_{10}$ SL in mm)]; it is 6.5 times stronger than the very thick-shelled estuarine bivalve, *Rangia cuneata* (Fig. 23) (Kennedy and Blundon, 1983). Shell strength is directly correlated with shell thickness. In specimens with an SL of 20 mm, the 3.0 mm thick shell of the unionoidean, *Fusconaia ebena,* was more crush resistant than the 1.3 mm thick shell of *C. fluminea* which, in turn, was more crush resistant than the fragile 0.31 mm thick shell of *D. polymorpha.* The fragile shell of *D. polymorpha* makes it prone to predation by diving ducks, crayfish and fish relative other freshwater bivalves (Miller *et al.,* 1994).

Fish predation can deplete freshwater bivalve populations. Sphaeriid diversity and density increased in habitats from which molluscivorous fish were excluded (Dyduch-Falniowska, 1982). Robinson and Wellborn (1988) reported that, 11 months after settlement, densities of *C. fluminea* were 29 times greater in enclosures excluding fish. In contrast, exclusion of fish and turtles did not increase density or species richness of a combined unoinoidean and sphaeriid community in a cooling reservoir (Thorp and Bergey, 1981).

As fish are intermediate hosts for unionoidean glochidia, the size of fish host population can influence mussel reproductive success. Absence of appropriate fish hosts have caused extirpation of unionoideans in a number of NA aquatic habitats (Fuller, 1974; Kat and Davis, 1984; Smith, 1985a, b; Neves *et al.,* 1997; Vaughn, 1997; Section III. A.) and availability of fish hosts may regulate recruitment (Johnson and Brown, 1998). Indeed, restoration of host fish populations produced remarkable recoveries of some endangered unionoidean populations (Smith, 1985b). Thus, anthropomorphic activities reducing fish host population densities can result in destruction of unionoidean populations, making relationships between unionoideans and their glochidial host fish an important future management consideration for drainage systems and their fisheries to ensure continued health of their remaining unionoidean fauna (Watters, 1993; Neves, 1997).

Mammals prey on freshwater bivalves. Otters, minks, muskrats, and raccoons eat bivalves and may regulate populations of some species (Fuller, 1974). Raccoons and muskrats feed on *C. fluminea* and may account for reductions of adult clam densities in the shallow, near-shore waters of some Texas rivers, where

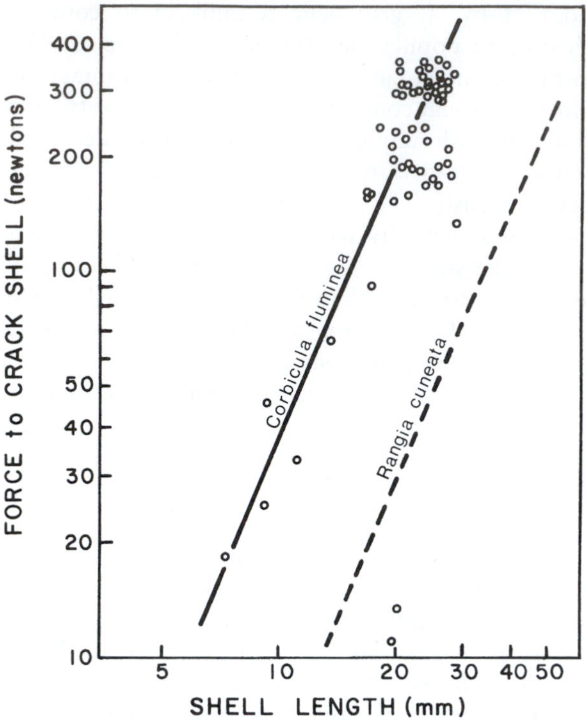

FIGURE 23 Shell strength of *Corbicula fluminea*. Solid line is the best fit of a geometric mean estimate of a least squares log–log linear regression, relating shell length (SL) to force required to crack the shell (open circles) as follows: Log_{10} force to crack the shell in newtons = −0.76 + 2.31 ($\log_{10}$ mm SL). Dashed line is the best fit of a similar log–log linear regression for the estuarine wedge calm, *Rangia cuneata*, as follows: $\log_{10}$ force to crack the shell in newtons = −0.736 + 2.42 ($\log_{10}$ SL in mm). *R. cuneata*, with the strongest shell of eight tested estuarine bivalve species, has a thicker shell than does *C. fluminea*. However, as the regression slope values for the two species are nearly equal, the elevated y-intercept for *C. fluinea* (−0.76 compared to −1.73 for *R. cuneata*) indicates that its shell is approximately an order of magnitude stronger. (Redrawn from Kennedy and Blundon, 1983.)

feeding sites are marked by thousands of opened shells (McMahon, unpublished observations). Even wild hogs root out and destroy shallow-water unionoidean beds (Fuller, 1974). Muskrats annually consummed 3% of individuals of *Pyganodon grandis* in a small Canadian lake, equivalent to 31% of the mussel's annual tissue production. They preferentially consumed larger mussels (shell length >55 mm), strongly affecting mussel population size–age structure, resulting in a biomass decline of large individuals, reducing the population's annual reproductive effort (Hanson *et al.*, 1989).

c. Competitive Interactions There have been no experimental evaluations of interspecific or intraspecific competition among freshwater bivalves; only anecdotal observations. Unionoidean or sphaeriid population declines coincident with establishment of *C. fluminea* have occurred in NA habitats (McMahon, 1999; Sickel, 1986). *C. fluminea* may detrimentally impact sphaeriid populations after invasion, but there is little evidence of major impacts of *C. fluminea* on NA unionoideans (Strayer, 1999b). Thus, *C. fluminea* had little impact on a Tennessee River unionoidean community, the *C. fuminea* population declining over time while unionid density remained stable (Miller *et al.*, 1992). Rather, habitats are first made inhospitable to indigenous unionoideans by channelization, dredging, over-fishing or other water management practices. *C. fluminea* is more tolerant of high flows, suspended solids, and silting caused by human activities in drainage systems, allowing it to rapidly colonize such disturbed habitants in which unionoideans have been reduced or extirpated (McMahon, 1999). In contrast, establishment of *C. fluminea* in drainage systems receiving minimal human interference has little effect on native bivalve populations (McMahon, 1999; Sickel, 1986), suggesting an inability of *Corbicula* to outcompete native unionoideans in undisturbed habitats. The impacts of *C. fluminea* on native, NA bivalves remains subject to debate (reviewed by Strayer, 1999b) and should be experimentally addressed.

In contrast to *C. fluminea,* there is clear evidence of direct negative impacts of *D. polymorpha* on NA unionoidean populations. *Dreissena polymorpha* byssally attaches to the posterior portions of unionoidean valves projecting above the substrate where they can reach densities which strip seston from the unionoidean's inhalant current, leading to its slow starvation. Extensive *D. polymorpha* infestations can cause unionoideans to be dislodged from the substrate, leading to death. *D. polymorpha* infestation has caused near complete extirpation of unionoideans from the lower Great Lakes and St. Lawrence River (Schloesser *et al.*, 1996; Strayer, 1999b). Interestingly, short-term brooding unionoideans (tribes Amblemini and Pleurobemini) appear less sensitive to zebra mussel impacts than long-term brooders (tribe Lampsilini and subfamily Anodontinae), perhaps due to their greater tissue energy reserves (Strayer, 1999b). Unionoidean populations in shallow (1 m), near-shore waters of western Lake Erie were less subject to *D. polymorpha* infestation than those at greater depths (Schloesser *et al.*, 1997), perhaps because wave-induced agitation inhibits byssal attachment (Clarke and McMahon, 1996b).

Dreissena polymorpha can have indirect impacts on N.A. bivalve communities. After *it* colonized the lower Hudson River in 1992, sphaeriid and unionoidean populations declined even though *D. polymorpha* did not attach to their shells. This decline resulted from a massive reduction in phytoplankton density by *D. polymorpha* feeding, causing starvation of indigenous bivalves (Strayer, 1999b).

d. Density-Dependent Mechanisms of Population Regulation Factors directly impacting freshwater bivalve population densities may have secondary impacts associated with reduction in gamete fertilization through reduction of sperm concentration in the water column. Thus, egg fertilization in the marsupial gills of female *Elliptio complanata* was directly correlated with population density, fertilization failing completely at <10 adults/m^2 (Downing *et al.*, 1993).

Published experimental investigations of intraspecific, density-dependent, population regulation mechanisms in freshwater bivalves are also lacking. However, some evidence is available from descriptive field studies of *C. fluminea* and sphaeriids. In *C. fluminea*, extremely successful recruitment of newly settled juveniles occurred after adult population decimation (McMahon and Williams, 1986a, b); whereas, juvenile settlement is generally far less successful in habitats harboring dense adult populations. High adult density may prevent successful juvenile recruitment, thus preventing extensive juvenile–adult competition. This hypothesis was tested by enclosing adult *C. fluminea* at different densities in a lake harboring a dense *Corbicula* population. While adult density had no effects on adult or juvenile growth rates, it affected juvenile settlement. Settlement in enclosures without adults was 8702 juveniles/m^2. Maximal settlement (26,996 juveniles/m^2) occurred at 329 adults/m^2; and was least at the highest adult densities of 659 and 1976 clams/m^2, being only 8642 and 3827 juveniles/m^2, respectively (McMahon, unpublished data). Thus, adults of *C. fluminea* may inhibit juvenile settlement. Mackie *et al.*, (1978) found that high adult densities in *Musculium securis* caused significant reduction in the number of juveniles incubated, implying that intraspecific adult competition regulates reproductive effort. In addition, interspecific competition between *M. securis* and *M. transversum* caused reduction in reproductive capacity in the subdominant species, with species dominance dependent on habitat. While intraspecific density effects and interspecific competition may regulate *Corbicula* and sphaeriid populations, results may not be the same for highly K-selected unionoidean populations which require further study before the effects of thinning of mussel beds by commercial harvesters on juvenile recruitment and adult growth can be assessed.

4. Functional Role in the Ecosystem

Bivalves play important roles in freshwater ecosystems (reviewed by Strayer *et al.*, 1999a, b). Because they can achieve high densities and are filter feeders, they are important consumers of phytoplankton primary productivity. Unionoideans, with filtering rates on the order of 300 mL/g dry tissue/h (Paterson, 1984), can account for a large proportion of total consumption of phytoplankton productivity. As unionoidean densities range from 15 individuals/m^2 (Paterson, 1985) to 28 individuals/m^2 (Negus, 1966) and dry tissue standing crop biomasses from 2.14 g/m^2 (Paterson, 1985) to 17.07 g/m^2 (Negus, 1966), filtering by unionoidean communities could be as high as 15–122.9 L/m^2 per day. Thus, on an annual basis, a eutrophic lake unionoidean community filtered 79% of total lake volume, removing 92–100% of all suspended seston. However, even with such high cropping rates, high phytoplankton reproductive rates resulted in unionoideans reducing seston by only 0.44% (Kasprzak, 1986). While the unionoidean community accounted for only 0.46% of seston removed by all phytoplanktivorous animals in the lake, they made up 85% of its standing crop biomass. Their biomass contained a high proportion of the phosphorus load of the lake, limiting the quantity of phosphorus available for phytoplankton production. Similarly, the unionoidean community of Lake St. Clair ingested 13.5% of the total phosphorus load of the lake from May through October and deposited 64% of consumed phosphorus in the sediments, providing nutrients for rooted macrophytic vegetation and invertebrate deposit feeders (Nalepa *et al.*, 1991). *D. polymorphia* populations in five European lakes had an average dry weight biomass 38 times that of submerged macrophytes and average standing crop phosphorus and nitrogen contents 2.94 (range, 0.39–7) and 4.21 (range, 0.59–9.4) times that of submerged macrophytes, respectively, greatly reducing availability of these growth-limiting inorganic nutrients to the aquatic plant community (Stanczykowska, 1984). In the western basin of Lake Erie, *D. polymorpha* has become a major factor in phosphorus flux; its uptake of phosphorus and nitrogen potentially limiting phytoplankton growth and its low N:P excretion ratios (<20:1) potentially shifting phytoplankton community structure toward cyanobacteria (Arnott and Vanni, 1996).

Sphaeriids also have important impacts on lentic phytoplankton communities. The sphaeriid community of an oligotrophic lake comprised only 0.12% of total planktivore biomass and 0.2% of the bivalve biomass, but accounted for 51% of total seston consumption (Kasprzak, 1986). This probably resulted from their relatively high filtration rates and population productivity (Burky *et al.*, 1985a; Hornbach *et al.*, 1984a, b). In smaller lotic and lentic habitats, sphaeriids can consume a major portion of primary productivity. In a canal community dominated by sphaeriids (98% of planktivore biomass), they accounted for 96% of total seston consumption (Krasprzak, 1986). A stream population of *S. straitinum* was estimated to filter 3.67 g organic C/m^2 per year or 0.0004% of seston organic carbon flowing over them (Hornbach *et al.,* 1984b). In contrast, filtration by a pond population of *M. partumeium* removed 13.8 g C/m^2 per year as seston (Burky *et al.,* 1985a). Based on the data of Burky *et al.,* (1985b), standing crop seston levels for the entire pond ranged annually between 46 and 550 g carbon. Average annual seston organic carbon consumption by the *M. partumeium* community was estimated to be roughly 3.9 g C per day, or 0.7–8.5% of total standing crop seston carbon per day. These data indicate that sphaeriid communities crop a significant portion of the primary productivity of their small lentic habitats.

With extremely high filtering rates and dense populations (McMahon, 1999), *C. fluminea* is potentially a major consumer of phytoplankton productivity. In the Potomac River, phytoplankton densities and chlorophyll a concentrations declined as it flowed over dense beds of *C. fluminea,* both measures falling 20–75% lower than upstream values; current phytoplankton levels in this river section are considerably lower than those prior to *C. fluminea* invasion. It was estimated that the *C. fluminea* population filtered the entire water column in the 3–4 day period required for it to pass the river reach where the population was most dense (Cohen *et al.,* 1984). Similar reduction in phytoplankton density of a lotic habitat by *C. fluminea* has been reported by Leff *et al.* (1990). Reductions in phytoplankton density and chlorophyll a concentration must have been due to clam filter feeding, particularly as discharge volume variation, zooplankton feeding, toxic substances, and nutrient limitations were not different from other river sections (Cohen *et al.* 1984). Similarly, Lauritsen (1986b) estimated that a *C. fluminea* population of 350 clams/m^2 in the Chowan River, NC, at an average depth of 5.25 m and clam filtering rate of 564–1010 mL/h, filtered the equivalent of the overlying water column every 1–1.6 days. At an average depth of 0.25 m, current flow of 18.5 m/min, and average clam filtering rate of 750 mL/h (Lauritsen, 1986b), the entire water column of the Clear Fork of the Trinity River flowing over a *C. fluminea* population with an average adult density of 3750 clams/m^2 was filtered every 304 m of river reach, or every 16 min (McMahon, unpublished observations). Such massive filtering of the water column by dense bivalve populations could keep seston concentrations at low levels and limit the energy available to other seston-feeding species. Consumption of the majority of primary productivity by dense bivalve populations and the accumulation of that production in large, relatively long-lived, predator-resistant adult clams may not only limit the inorganic nutrients available for primary productivity and the energy available to other primary consumers, but may also divert energy flow from higher trophic levels.

Bivalves, by filtering suspended seston, are water clarifiers and organic-nutrient sinks. When co-cultured with channel catfish, *C. fluminea* increased ambient water O_2 concentrations by reducing seston and turbidity levels (Buttner, 1986). Dense bivalve populations, by removing phytoplankton and other suspended material, increase water clarity, and, by binding suspended sediments into pseudofeces, accelerate sediment deposition rates (Prokopovich, 1969). Dense *D. polymorpha* populations clarify water which increases light penetration, stimulating the growth of rooted aquatic macrophytes (Mackie and Schloesser, 1996).

Such major ecological impacts by *D. polymorpha* have been observed in several locations in the lower Great Lakes, St. Lawrence River and Hudson River harboring dense mussel populations (reviewed by MacIsaac, 1996). In these habitats, *D. polymorpha* filtering increases water clarity. In the Hudson River, *D. polymorpha* filtered the equivalent of the entire water column every 1.2–3.6 days (Strayer *et al.,* 1999). Phytoplankton densities are most negatively impacted by *D. polymorpha* filtering in waters <1.85 m above mussel beds (MacIsaac *et al.,* 1999). Besides phytoplankton and small zooplankton, *D. polymorpha* filters suspended clay and silt particles and binds them into pseudofeces deposited in sediments (MacIsaac, 1996; Strayer *et al.,* 1999). Phytoplankton reductions by *D. polymorpha* have been extensive, ranging from 59 to 91% in portions of the Great Lakes (MacIsaac, 1996) and from 80 to 90% in the lower Hudson River, with mussels less efficiently removing diatoms than other smaller algae (Strayer *et al.,* 1999). Filter feeding by *D. polymorpha* can reduce bacterioplankton densities, but can also stimulate bacterioplankton growth through nutrient or organic carbon excretion (Cotner *et. al.,* 1995). Feeding by dense *D. polymorpha* populations appears to favor development of cyanobacteria blooms in some NA waters, perhaps due to reduction in N:P ratio, enhanced light penetration, greater buoyancy of cyanobacteria filaments, and/or chemical or mechanical inhibition of mussel filtering by cyanobacteria (MacIsaac *et al.,* 1996).

Zooplankton populations are reduced by *D. polymorpha* filtering. Small zooplankton such as rotifers and even *Dreissena* veligers can be directly filtered; rotifer density in western Lake Erie declining by 74% within 5 years of *D. polymorpha* colonization, while larger zooplankton species that avoid entrainment on the mussel's inhalant currents were relatively unaffected (MacIsaac, 1996). Similarly, zebra mussel invasion brought about 80–90% reductions in rotifers, tintinnids, and copepod naupli densities in the lower Hudson River (Strayer *et al.,* 1999). Suppression of small zooplankton could reduce fish populations whose larvae utilize them as food and result in density increases of bacterioplankton on which small zooplankton feed (Strayer *et al.,* 1999). Increased water clarity and sediment nutrient levels appear to stimulate increased aquatic macrophyte growth in many areas colonized by *D. polymorpha,* favoring fish species associated with them over open-water species (MacIsaac, 1996). Increased macrophyte growth also appeared to favor associated invertebrate species, whose densities increased dramatically in the Hudson River after *D. polymorpha* colonization while deep-water benthic invertebrates concurrently declined (Strayer *et al.,* 1999). Thus, overall, *D. polymorpha* populations appear to divert resources from the pelagic zone and deep-water sediments to vegetated shallows and zebra mussel beds (Strayer *et al.,* 1999), leading to increases in benthic invertebrate abundance and diversity due to habitat structure enhancement among accumulated mussel shells and increased food supply, enhancement of physical habitat appearing to have the greatest impact (Botts *et al.,* 1996). Increased bentic invertebrate densities support density increases in fish populations feeding on them (MacIsaac, 1996; Ricciardi *et al.,* 1997; Stewart *et al.,* 1998; Strayer *et al.,* 1999), These postcolonization impacts of *D. polymorpha* are extensive, with almost all of 11 measured ecological variables being shifted by >50% after establishment of mussel populations in the lower Hudson River (Strayer *et al.,* 1999). Increased food availability, associated with feeding on *D. polymorpha,* may also be responsible for postzebra mussel colonization increases in diving-duck flock sizes and staging durations in several portions of Lake Erie (MacIsaac, 1996).

As freshwater clams filter suspended organic detritus and bacteria and consume sediment interstitial bacteria and organic detritus (Section III.D.2), they may be significant members of the aquatic decomposer assemblage. *Pisidium casertanum* and *P. corneum* feed primarily by filtering sediment bacteria (Lopez and Holopainen, 1987). Hornbach *et al.* (1982, 1984b) estimated that only 24–35% of energy needs of a population of *Sphaerium striatinum* were met by filter feeding, the remaining coming from sediment organic detritus. *C. fluminea* pedally feeds on sediment detritus (Reid *et al.,* 1992), impacting the decomposer community (Hakenkamp and Palmer, 1999). *Musculium transversum* utilizes its long inhalant siphon to vacuum organic detritus from the sediment surface (Way, 1989). The detritivorous habit of many freshwater bivalve species may divert primary productivity ordinarily lost to respiration of the detritivorous community back into bivalve tissue where it is made re-available to higher trophic levels.

The activity of freshwater bivalves may also directly affect habitat physical characteristics. Deposition of calcium in shells may reduce ambient Ca concentrations. In a *C. fluminea* population of 32 clams/m^2, annual fixation of shell $CaCO_3$ was 0.32 kg $CaCO_3$/m^2 per year (Aldridge and McMahon, 1978). At this rate, shell $CaCO_3$ fixation in a dense population of 100,000 clams/m^2 (Eng, 1979) could be 50–60 kg Ca CO_3/m^2 per year. Such rapid removal of calcium to shells accumulating in the sediments could reduce water hardness, particularly in lentic habitats (Huebner *et al.,* 1990). Seasonal cycles of shell growth could induce seasonal cycles in water Ca concentration, being greatest in winter when shell growth is minimal and least in summer when shell growth is maximal (Rooke and Mackie, 1984c). Bivalves can also affect solute flux between sediments and the water column. Unionoidean mussels enhanced release of nitrate and chloride and inhibited $CaCO_3$ release from sediments (Matisoff *et al.,* 1985). Dense sphaeriid populations can be the principal affectors of sediment dissolved O_2 demand, in cases reaching levels characteristic of semipolluted and polluted streams (Butts and Sparks, 1982). *D. polymorpha* and *C. fluminea* increase sedimentation rates, negatively or positively impact phosphate and nitrate fluxes between sediments and the water column, change the N:P ratio of the water column, clarify water, and change sediment physical make up by accumulation of their shells (Arnott and Vanni, 1996; Strayer *et al.,* 1999).

Small bivalves (sphaeriids, juvenile unionoideans, *D. polymorpha,* and *C. fluminea*) can be a major food source for third trophic level carnivores, including fish and crayfish (Section III.C.3). Bivalve flesh has a low caloric content, 3.53–5.76 kcal/g ash-free dry weight (Wissing *et al.,* 1982), reflecting low lipid:protein ratios. *C. fluminea* has tissue organic C:N ratios ranging from 4.9–6.1:1, indicative of a 51–63% of dry weight protein content (Williams and McMahon, 1989). High protein content makes bivalve flesh an excellent food source to sustain predator tissue growth.

Many freshwater bivalve populations are highly productive (particularly sphaerids, *C. fluminea,* and *D. polymorpha*), rapidly converting primary productivity into tissue energy available to third trophic level carnivores. Productivity values range from 0.019 g C/m^2 per year for a *Pisidium crassum* population to 10.3 g C/m^2 per year for a *C. fluminea* population (Aldridge and McMahon, 1978) and 75 g C/m^2 per year for a *D. polymorpha* population (Mackie and Schloesser, 1996); the average for 11 sphaeriid species and two species of *Corbicula* was 2.5 g C/m^2 per year (Burky, 1983). Other values include 12.8–14.6 g C/m^2 per year for an Arizona *C. fluminea* canal population (computed from data of Marsh, 1985) and 1.07 g C/m^2 per year for a New Brunswick lake unionoidean population of *Elliptio complanata* (computed from data of Paterson, 1985).

Bivalves are generally more efficient at converting assimilated food energy into new tissue than most second trophic level aquatic animals because their sessile, filter-feeding habits minimize energy expended in food acquisition, allowing efficient transfer of energy from primary production to bivalve predators. Thus, bivalves can be important conduits of energy fixed by phytoplankton photosynthesis to higher trophic levels. However, such trophic energy transfer is mainly through smaller species and juvenile specimens, as large adults are relatively immune to predation (Section 3.C).

The measure of efficiency of conversion of assimilated energy absorbed across the gut wall into energy fixed in new tissue growth is net growth efficiency:

$$\% \text{ Net Growth Efficiency} = \frac{P}{A(100)}$$

where P is the productivity rate or rate of energy or organic carbon fixation into new tissue growth by an individual or population and A the assimilation rate or rate of energy or organic carbon assimilated by an individual or population. The greater the net growth efficiency of a second trophic level species, the more efficient its conversion of assimilated energy into flesh, and, therefore, the greater the potential for energy flow though it to third trophic level predators (reviewed by Russell-Hunter and Buckley, 1983; Holopainen and Hanski, 1986; Hornbach, 1985). Net growth efficiencies (Table III), are 9–79% in sphaeriids (average = 55%) and 58–89% in *C. fluminea,* indicating efficient conversion of assimilated energy into flesh compared to other aquatic second trophic level animals.

5. Bivalves as Biomonitors

Freshwater bivalves, particularly unionoideans *C. fluminea* and *D. polymorpha,* have characteristics such as long life spans, and growth and reproductive rates sensitive to environmental perturbation (Burky, 1983) that make them good biomonitors (Imlay, 1982). They can be held in field enclosures without excessive maintenance. Shell growth is sensitive to environmental variation and/or disturbance (Way and Wissing, 1982; Fritz and Lutz, 1986; Belanger *et al.,* 1986a, b; Way, 1988), and the valves remain as evidence of death, allowing mortality rates to be estimated. Shells can be marked by tags (Rosenthal, 1969; Young and Williams, 1983c), including floating tags tethered to the shell (Englund and Heino, 1994b), paint, or shell-etching (McMahon and Williams, 1986a).

Bivalves can be collected throughout the year and easily shipped alive long distances. Because large species stay in place (exception is *C. fluminea*), they experience conditions in the monitored environment throughout life (Imlay, 1982). Annual shell-growth increments allows determination of variation in heavy-metal pollutant levels over long periods by analysis of levels in successively secreted shell layers (Imlay, 1982). However, there is controversy regarding use of shell-growth rings to age unionoideans. Growth rings are produced annually in some species (Neves and Moyer, 1988), but not in others (Downing *et al.,* 1992). Large size allows analysis of pollutant levels in single individuals, and the wide geographical ranges of some species (Imlay, 1982) allow comparisons across drainage systems. Table VI indicates the types of pollutants and environmental perturbations monitored by freshwater bivalves. The broad distributions of *C. fluminea* and *D. polymorpha* in North America and their dense, readily collectible populations make them excellent biomonitors compared to unionoideans, many species of which have restricted distributions and/or are threatened or endangered (Section III.A).

D. Evolutionary Relationships

Sphaeriid, dreissenids, corbiculids, and unionoideans represent separate evolutionary invasions of freshwaters. Sphaeriids, dreissenids, and corbiculids fall in the order Veneroida, and unionoideans in the order Unionoida (Allen, 1985). The Veneroida became successful infaunal filter feeders though evolution of inhalant and exhalant siphons, allowing maintenance of respiratory and feeding currents while burrowed. Their eulamellibranch gills with fused filaments (Fig. 5) were a clear advancement over the primitive (i.e., filibranch) condition with separate gill filaments attached by interlocking cilia (Allen, 1985).

The family, Sphaeriidae, in the superfamily, Corbiculacea, with a long freshwater fossil history extending from the Cretaceous (Keen and Dance, 1969) has evolved along two major lines. The first, represented by *Pisidium,* involves adaptation to life in organically rich sediments by filtering interstitial bacteria (Lopez and Holopainen, 1987), making them dominant in profundal habitats. The second, represented by *Sphaerium* and *Musculium,* involves adaptation to life in small, shallow, lentic or lotic habitats subject to predictable seasonal perturbation, such as habitat-drying. Their adaptations include estivation during prolonged emergence (Section III.A). Like many of the Veneroida, the byssus is absent in adults of most species.

The superfamily Unionoidea, like the Sphaeriidae, has a long freshwater fossil history, extending from at

TABLE VI List of some Investigations Involving Utilization of Bivalves to Monitor Effects or Levels of Pollutants in Freshwater Habitats

Pollutant monitored	*Species utilized*	*Literature citations*
Arsenic	*Corbicula fluminea*	Elder and Mattraw (1984), Price and Knight (1978), Tatem, (1986)
Cadmium	Eight unionoidean species,	Price and Knight (1978)
	Anodonta anatina	Hemelraad *et al.* (1985)
	Anodonta cygnea	Hemelraad *et al.* (1985)
	Corbicula fluminea	Elder and Mattraw (1984), Graney *et al.* (1984), Price and Knight (1978), Tatem (1986)
Chromium	Eight unionoidean species	Price and Knight (1978)
	Corbicula fluminea	Elder and Mattraw (1984), Tatem (1986)
Copper	*Corbicula fluminea*	Annis and Belanger (1986), Elder and Mattraw (1984), Price and Knight (1978), Tatem (1986)
	Lamellidens marginalis	Hameed and Raj (1990)
Iron	*Corbicula fluminea*	Tatem (1986)
Lead	*Corbicula fluminea*	Annis and Belanger (1986), Elder and Mattraw (1984), Price and Knight (1978), Tatem (1986)
	Lamellidens marginalis	Hameed and Raj (1990)
Manganese	Eight unionoidean species	Price and Knight (1978)
	Corbicula fluminea	Elder and Mattraw (1984), Tatem (1986)
	Lamellidens marginalis	Hameed and Raj (1990)
Mercury	*Corbicula fluminea*	Elder and Mattraw (1984), Price and Knight (1978)
Nickel	*Lamellidens marginalis*	Hameed and Raj (1990)
Tin	Eight unionoidean species	Price and Knight (1978)
Zinc	*Anodonta* sp.	Herwig *et al.* (1985)
	Corbicula fluminea	Belanger *et al.* (1986b), Elder and Mattraw (1984), Foe and Knight (1986c)
	Lamellidens marginalis	Hameed and Raj (1990)
Asbestos	*Corbicula fluminea*	Belanger *et al.* (1986a, 1987)
Octachlorostyrene	*Lampsilis siliquiodea*	Elder and Mattraw (1984), Tatem (1986)
Polychlorinated	*Corbicula fluminea*	Elder and Mattraw (1984), Tatem (1986)
Biphenols (PCBs)	*Lampsilis siliquoidea*	Pugsely *et al.* (1985)
	Dreissena polymorpha	Fisher *et al.*1993
Pentachlorophenol (PCP)	*Dreissena. Polymorpha*	Borcherding (1992), Fisher *et al.* (1993)
Pesticides	*Corbicula fluminea*	Elder and Mattraw (1984), Hartley and Johnston (1983), Tatem (1986)
Sewage effluents	*Corbicula fluminea*	Foe and Knight (1986c), Horne and MacIntosh (1979), Weber (1973)
Power station effluents	*Corbicula fluminea*	Dreier and Tranquilli (1981), Ferris *et al.* (1988), Foe and Knight (1987), McMahon and Williams (1986b)
Natural waters	*Dreissena polymorpha*	Borcherding and Volpers (1994)
	Unio tumidus	Englund and Heino (1994a)
	Anodonta anatina	Englund and Heino (1994a)

least the Triassic (Haas, 1969). A long fossil history and tendency for reproductive isolation due to gonochorism and the parasitic glochidial stage (Section III.A) which allows sympatric speciation via new glochidial fish host acquisition (Graf, 1997) has led to extensive radiation, the Unionoidea being represented worldwide by 150 genera and a great number of species (Allen, 1985). Their origin remains unclear. Similarity of shell structure suggests a relationship to marine fossil species in the order Trigonoida (Allen, 1985). Like the majority of sphaeriids, adult unionoideans lack a byssus, which would be of no use in their infaunal habitats. They dominant the shallow regions of larger, relatively stable lentic and lotic habitats, particularly in stable sand–gravel sediments (Section III.A). The division of the superfamily, Unionoidea, into the presumed more primitive family, Margaritiferidae, and the more derived family, Unionidae, has been partially supported by mitochondrial DNA analysis of NA species. This study also supported the monophyly of the NA unionoidean subfamilies Anodontinae and Ambleminae and tribe Pleuorblemini, but provided only weak support for the tribe Lampsilini (Lydeard *et al.*, 1996). Further molecular studies are needed to elucidate the complex phylogeny of the Unionoidea, including analysis of species outside of North America.

The Sphaeriidae and Unionoidea, with long freshwater fossil histories, have evolved distinct niches. While both occur in stable habitats, the majority of sphaeriids inhabit smaller ponds and streams or the profundal regions of lakes, and the majority of unionoideans inhabit stable sediments of shallow portions of larger rivers and lakes (Sections III.A and III.B).

C. fluminea entered freshwater only in recent (Pleistocene) times (Keen and Casey, 1969), making it unlikely to complete effectively with sphaeriids or unionoideans whose long fossil histories have allowed them to become highly adapted to their preferred stable freshwater habitats. Indeed, *C. fluminea* does not appear to seriously impact native NA bivalves in undisturbed freshwaters (McMahon, 1999; Section C.3). Rather, it is adapted for life in unstable habitats—subject to periodic catastrophic perturbation, particularly flood-induced sediment disturbance—which are generally unsuitable for sphaeriids or unionoideans (Section III.A). Among its adaptations to high-flow habitats are: (1) capacity for rapid burrowing; (2) a strong, heavy, concentrically sculptured, inflated shell; and (3) juvenile retention of a byssal thread—all of which allow maintenance of position in sediments (Vermeij and Dudley, 1985). Additionally, its high fecundity, elevated growth rate, early maturity, and extensive capacity for dispersal (Section III.B) allow it to colonize habitats from which other bivalves have been extirpated and rapidly recover from catastrophic population reductions. Thus, *C. fluminea* evolved a niche in unstable, disturbed habitats not utilized by sphaeriids or unionoideans, having evolved from an estuarine *Corbicula* ancestor inhabiting similar environments in the upper, near-freshwater portions of estuaries (Morton, 1982).

The superfamily Dreissenacea, containing *D. polymorpha,* while superficially resembling marine mytilacean mussels (order Mytiloida), is placed in the Veneroida because of its advanced eulamellibranch ctenidium. Adult dreissenaceans retain an attachment byssus, considered a primitive condition. In contrast, the mytiliform shell with its reduction of the anterior valves and anterior adductor muscle, is a derived adaptation to an epibenthic niche characterized by byssal attachment to hard surfaces (Allen, 1985; Mackie and Schloesser, 1996). *D. polymorpha* probably evolved from an estuarine dreissenacean ancestor of the genus *Mytilopsis.* As with *C. fluminea,* its fossil record indicates a recent, Pleistocene introduction to freshwaters (Mackie *et al.*, 1989). Also, like *C. fluminea, D. polymorpha* appears to have successfully colonized freshwaters because its niche (i.e., an epibenthic species attached to hard substrates) is not occupied by infaunal, burrowing sphaeriids or unionoideans, thus minimizing competition with these more advanced groups.

Freshwater bivalves have characteristics that are uncommon in comparable marine species. Most shallow-burrowing marine species are gonochoristic with external fertilization and free-swimming planktonic veligers. A planktonic veliger is nonadaptive in lotic freshwater habitats, as it would be carried downstream before settlement, eliminating upstream populations. Thus, the planktonic veliger stage is suppressed in sphaeriids, which brood embryos in gill marsupia and release large, fully formed juveniles ready to take up life in the sediment. In unionids, eggs are retained in gill marsupia and hatch into the glochidium, whose parasitism of fish hosts allows upstream as well as downstream dispersal and permits release of well-developed juveniles into favorable habitats (Section III.B). Even the recently evolved *C. fluminea* has suppressed the planktonic veliger stage of its immediate estuarine ancestors (Morton, 1982), producing a small, but fully formed juvenile whose byssal thread allows immediate settlement and attachment to the subtrate. Retention of external fertilization and a free-swimming veliger in *D. polymorpha* are primitive characteristics reflecting its recent evolution from an estuarine ancestor.

Downstream veliger dispersal would preclude *D. polymorpha* from establishing populations in high-flow lotic habitats, unless upstream impoundments provide replacement stock. In fact, in Europe and Asia, this species is most successful in large, lentic or low-flow lotic habitats such as canals or large rivers (Mackie *et al.*, 1989). Serial impoundment of rivers has provided lentic refugia for this species in otherwise lotic drainages, facilitating its invasion of NA waters (Mackie and Schloesser, 1996).

Another common adaptation in freshwater species is hermapharoditism with a capacity for self-fertilization. Hermaphroditism makes all individuals in a population reproductive, allowing rapid population expansion during favorable conditions. Hermaphroditism allows sphaeriids to re-establish populations after predictable seasonal perturbation and *C. fluminea,* along with its high fecundity, to rapidly re-establish populations after catastrophic habitat disturbance. It makes both sphaeriids and *C. fluminea* invasive, as the introduction of a single individual can found a new population. In contrast, the gonochorism of most unionoideans and *D. polymorpha* limits their invasive capacity, reflected by their stable, relatively undisturbed habitats. Hermaphroditism occurs in only seven species within three subfamilies of NA Unionoidea (Hoeh *et al.*, 1995) with widely different habitats and reproductive traits. Selection pressures for evolution of hermaphroditism within these species have not been elucidated.

Finally, unionoideans and sphaeriids have different shell morphologies relative to comparable shallow-burrowing marine bivalves. Freshwater species generally do not have denticulated or crenulated inner-valve margins, extensive radial or concentric shell ridges, tightly sealing valve margins, well-developed hinge teeth, overlapping shell margins, or uniformally thick, inflated, strong shells to the degree displayed by shallow-burrowing marine species (Vermeij and Dudley, 1985). These shell characteristics allow shallow-burrowing marine species to maintain position in unstable substrates and/or resist shell cracking or boring by large predators common in marine habitats. Lack of such structures in unionoideans and sphaeriids reflects their preference for stable sediments and the general absence of effective predators on adult freshwater bivalves; certainly there are no shell-boring predators of NA freshwater species. Indeed, the shells of freshwater unionoideans display considerably less nonlethal, predator-induced damage than shallow-burrowing marine species (Vermeij and Dudley, 1985). While retention by *C. fluminea* of a thick, extremely strong shell (Kennedy and Blundon, 1983; Miller *et al.*, 1992), lacking pedal and siphonal gapes, with a well-developed hinge, shell inflation, and concentric ornamentation, may represent a primitive condition, it allows this species to inhabit sediments too unstable to support unionoideans or sphaeriids. Such primitive characteristics reflect similar adaptations in the immediate estuarine ancestor of this recently evolved freshwater species. In contrast, the thin, fragile shell of *D. polymorpha* (Miller *et al.*, 1992) may have evolved due to byssal attachment of the species to hard substrates where the shell is not subject to damage from unstable sediments.

IV. COLLECTING, PREPARATION FOR IDENTIFICATION, AND REARING

A. Collecting

Small sphaeriid clams are best collected by removing sediments containing clams with: (1) a trowel or shovel (shallow water); and (2) a long-handled dip net or shell scoop with a mesh of <0.35 mm (moderate depths) or by Ekman, Ponar, or Peterson dredges (deeper water), the heavier Ponar and Peterson dredges being better in lotic habitats. Drag dredges must have sediment catch bags of small mesh (1 mm or less). Use of scuba is also an effective means of collecting sphaeriids. Larger sphaeriids species can be separated from fine sediments with a 1-mm mesh sieve, but a 0.35-mesh is required for smaller specimens. Fragile sphaeriids should be separated from sediments by gentle vertical agitation of the sieve at the water surface to prevent shell damage. Coarse material should be removed prior to sieve agitation to avoid shell breakage. Exceptionally small, fragile species can be collected by washing small quantities of sediment into settlement pans, specimens being revealed after sediments settle and the water clears.

Ekman, Ponar, or Peterson dredges are best for quantitative samples of sphaeriids, because they remove sediments from under specific sediment surface areas. Quadrat frames can be placed on the bottom (Miller and Payne, 1988) and all surface sediments within the frame collected for sorting. Such frames can be utilized in shallow or deeper waters (the latter in conjunction with scuba equipment). Drag dredging at specified speeds for known intervals allows partial sample quantification. Core sampling devices consisting of tubes driven into the substrate to specific depths can yield both accurate estimates of sphaeriid density and sediment depth distributions. Core samplers sample only a small surface area, thus are best utilized with dense populations and/or repetitive sampling.

Low densities of some unionoidean populations make their collection difficult. Where populations are dense, shoveling sediments through a sieve allows collection of a broad range of sizes and age classes (Miller and Payne, 1988). As unionoideans are larger than sphaeriids, sieve mesh sizes of 0.5–1 cm may be appropriate unless recently settled juveniles must be collected. In less dense, shallow-water populations, hand-picking using a glass-bottomed bucket to locate specimens can often suffice. In turbid waters and/or soft sediments (sand or mud), shell rakes may be utilized. Hand-picking and shell rakes can select larger, older individuals. In shallow, turbid water, using hands and fingers to systematically feel for mussels buried in soft sediments can be effective, but selects larger individuals and can lead to cut fingers. On rocky or boulder bottoms, unionoideans generally accumulate in crevices or near downstream bases of large rocks. Here, only hand-picking (in conjunction with scuba techniques in deeper water) is effective. Unionoideans may also be collected by drag or brail dredges. Mussel brails (often used by commercial shellers) consist of a bar with attached lines terminating in blunt-tipped gang hooks. When dragged over unionid beds, mussels clamp the valves onto the hooks allowing them to be brought to the surface.

Quantitative sampling of unionoideans can be accomplished with quadrats, along with scuba in deeper waters (Isom and Gooch, 1986; Miller and Payne, 1988; Waller *et al.*, 1993). Only heavy Ponar and Peterson dredges bite deeply enough into the substrate to take large unionoidean species. In sparse populations, errors in density estimates can result from the small surface areas that these dredges sample, requiring numerous samples to improve accuracy. Drag dredges sample larger areas and may be partially quantitative, but generally do not provide accurate estimates. A skimmer dredge collected 62.3% of mussels in its path, but resulted in 10% mortality of thin-shelled species (Miller *et al.,* 1989) Unionoideans may also be collected during natural or planned water level drawdowns (White, 1979) by collecting either emersed individuals or those migrating downshore and accumulating at the water's edge.

In a comparative study of sampling, quadrat sampling provided the most accurate analysis of a unionoidean community, but was difficult and expensive to carry out in deeper waters, where systematic sampling by scuba produced the best results (Isom and Gooch, 1986). In a comparison of the accuracy of quadrat and qualitative timed searches for unionoideans, it was found that timed searches overestimated large species, highly sculptured species and those whose shells protruded from the substrate, while underestimating buried, small, smooth-shelled species. In contrast, quadrat sampling tended to underestimate rare species and total number of species unless a large number of quadrats were sampled (Vaughn *et al.,* 1997). Extensive sampling may be required to find rare species, particularly if their population densities need to be assessed; thus, time and funding limitations may prevent accurate assessment of presence/absence and population densities of rare unionoideans (Kovalak *et al.,* 1986; Vaughn *et al.,* 1997). Accurate sampling of rare unionoidean species is required for their effective management and conservation. As many NA unionoidean species are endangered, one should ascertain the status of any species before permanently removing specimens from their natural habitats. Take only dead shells or, if absolutely necessary, a few living specimens for identification.

C. fluminea is easily collected because it occurs in high densities, prefers shallow waters, and is easily identified. Individuals can be separated from sediments with a 1-mm mesh sieve. Qualitative samples are best obtained by shovel or drag dredge (Williams and McMahon, 1986) and quantitative samples by quadrat frame (Miller and Payne, 1988), or Peterson (Aldridge and McMahon, 1978) or Ekman dredges (Williams and McMahon, 1986). In rock or gravel substrates, *C. fluminea* are best taken by hand, hand trowel, or shovel; collected sediments should be passed though a coarse mesh sieve to separate specimens. In fast-flowing streams, specimens of *C. fluminea* accumulate in crevices or behind the downstream sides of large rocks.

D. polymorpha is best collected by scuba with manual removal, using quadrat frames for quantification, because of its byssal attachment to hard surfaces and its preference for deeper waters ($<$1–2 m). Ripping of individuals from the byssus damages their tissues, thus the byssus should be cut with a knife or sharp trowel before removal. Juveniles can be collected after they have settled on settlement blocks or submerged buoys set out during the reproductive season. Ekman, Ponar, Peterson, and drag dredges are generally not suitable for collection of attached epibenthic species such as *D. polymorpha,* but they can be used for populations on soft or semisoft benthic substrates (Hunter and Baily, 1992).

Juveniles of sphaeriids and *C. fluminea* can be surgically removed from brood sacs or collected from sediments after release with a fine mesh sieve (mesh size $<$0.35 mm). Juvenile *C. fluminea* can also be obtained by release from freshly collected, gravid adults left in water for 12–24 h (Aldridge and McMahon, 1978). Planktonic *C. fluminea* juveniles and *D. polymorpha* veligers may be taken with a zooplankton net towed behind a boat or held in current flow. Their densities may be assessed by passing known volumes of water through a zooplankton net. Plankton net mesh size for collection of *C. fluminea* juveniles should be $<$200 μm and $\lesssim$40 μm for all planktonic stages of *D. polymorpha,* including the trochophore. Recently settled juveniles of unionoideans may be sieved from sediments. Glochidia can be surgically removed from demibranchs of gravid females or encysted glochicia taken from the fins, pharyngeal cavity, or gills of their fish hosts (for a list of unionoidean fish hosts, see Watters, 1994b).

B. Preparation for Identification

To preserve bivalves, larger individuals should first be narcotized or relaxed, allowing tissues to be preserved in a lifelike state and preservatives to penetrate tissues through gaped shell valves. Live bivalves placed directly in fixatives clamp the valves, which prevents preservative penetration. There are a number of bivalve relaxing agents (Coney, 1993; Araujo *et al.,* 1995), including alcoholized water (either 3% ethyl alcohol by volume with water or 70% ethyl alcohol added slowly, drop by drop, to the medium until bivalves gape), chloroform added slowly to the medium, methol crystals (one level teaspoon per liter, scattered on the water surface), propylene phenoxetol and phenoxetol BPC (5 mL of product emulsed with 15–20 mL of water, added to water containing bivalves or introduction of a droplet equal to 1% of holding water volume), phenobarbitol added in small amounts to the holding medium, magnesium sulfate (introduced into holding medium over a period of several hours to form a 20–30% solution by weight), magnesium chloride (7.5% solution by weight), and urethane.

None of these agents works equally well with all species. Laboratory tests of narcotizing agents against

C. fluminea revealed propylene phenoxetol to be the only agent capable of relaxing this species for experimental surgery, and allowing recovery on return to fresh medium (Kropf-Gomez and McMahon, unpublished). Heating bivalves to 50°C for 30–60 min causes most species to relax and gape widely, but is lethal. In larger specimens, wooden pegs or portions of matchsticks forced between the valves prior to fixation allows preservative penetration of tissues.

The best long-term preservative for freshwater bivalves is 70% ethyl alcohol (by volume with water). Specimens may be initially fixed in 5–10% formaldehyde solutions (by volume with 40% formaldehyde solutions) for 3–7 days. But formaldehyde is acidic and dissolves the calcareous portion of shells unless pH-neutralized by the addition of powdered calcium carbonate ($CaCO_3$) to make a saturated solution, 5 g of powdered sodium bicarbonate ($NaHCO_3$) per liter, or 1.65 g of potassium dihydrogen orthophosphate and 7.75 g disodium hydrogen orthophosphate per liter (Smith and Kershaw, 1979). After 3–7 days in formaldehyde, specimens should be transferred to 70% ethyl alcohol for permanent preservation. Addition of 1–3% glycerin (by volume) to alcohol preservatives keeps tissues soft and pliable (Smith and Kershaw, 1979). Smaller species (shell length <15 mm) generally do not require relaxation. Tissues to be utilized in microscopy should be preserved in gluteraldehyde or Bouin's solution.

Shells can be cleaned with a mild soap solution and soft brush. Organic material can be digested from shell surfaces by immersion in a dilute (3% by weight with water) solution of sodium or potassium hydroxide at 70–80°C, thereafter, removing remaining organic matter with a soft brush. For dry-keeping, the shell periostracal surface should be varnished or covered with petroleum jelly to prevent drying, cracking, and/or peeling. Numbers identifying collection and specimen can be marked on the inner shell surface with India ink.

In order to remove soft parts from living bivalves, immerse them in water at >60°C and remove tissues after valves fully gape. Separated flesh can be fixed in 70% alcohol. For fragile sphaeriids, flesh is best removed with the tip of a fine needle, manipulating specimens with a fine brush. Shells should be dried in air at room temperature, not in an oven; heat causes shells to crack and their periostracum to crack and peel.

Both soft tissue and shell characteristics are utilized in the identification of freshwater bivalves, so both must be preserved for species identification. For unionoideans, *C. fluminea,* and *D. polymorpha,* most diagnostic taxonomic characteristics can be seen by eye or with a 10 × hand lens. For sphaeriids, a dissecting microscope with at least 10–30 × power or a compound microscope is required (Ellis, 1978). Anatomical details are best observed on dry shells or soft tissues immersed in water.

Identification of recently released juvenile bivalves is difficult and may require preparation of stained slide whole mounts. Glochidia are best identified by removal from a gravid adult of an identified species as are juvenile sphaeriids. Only the juveniles of *C. fluminea* and larval stages of *D. polymorpha* are routinely found in the plankton. The juvenile of *C. fluminea* (Fig. 15A) is easily recognizable, and can be readily separated from the pediveliger of *D. polymorpha* by the presence of a fully formed foot, lack of a velum and a D-shaped shell while the foot is formed only in the pediveliger stage of *D. polymorpha* which has an umbonal shell (Nichols and Black, 1994). The planktonic veliger of *D. polymorpha* is clearly distinguishable by its ciliated velum, lack of a foot and D-shaped shell (Nichols and Black, 1994; Fig. 15C). The glochidia of many unionoidean species have specific fish species hosts (Watters, 1994b), whose identification will assist glochidial identification. Identification of glochidia without knowledge of the host fish or species of origin is extremely difficult, characters such as presence/absence of attachment hooks or threads and glochidial size may allow assignment to a family or subfamily, but may be unreliable even at these higher taxonomic levels (Bauer, 1994).

C. Rearing Freshwater Bivalves

For artificial rearing of freshwater bivalves, water in holding tanks should be temperature-regulated. Adequate aeration, filtration, and ammonia removal systems are required, as some species have low hypoxia and ammonia tolerances (Byrne *et al.* 1991a, 1991b).

As unionoideans and *C. fluminea* are filter phytoplankton, they require a constant supply of filterable food to remain healthy and growing. The best artificial food appears to be algal cultures. When fed monoalgal cultures of the green algae, *Ankistrodesmus* and *Chlorella vulgaris* or the cyanobacteruim, *Anabaena oscillariodes,* assimilation efficiencies in *C. fluminea* were 47–57% and net growth efficiencies, 59–78%, making them excellent food sources for this species (Lauritsen, 1986a).

Artificial diets do not appear to be as successful as algal cultures in maintaining *C. fluminea* growth. When fed either ground nine-grain cereal, rice flour, rye, bran, brewers' yeast, or artificial trout food, small *C. fluminea* (5–8 mm SL) starved on all but nine-grain cereal, the latter supporting little tissue growth. Supplementing these grain diets with live green algae (*Ankistrodesmus* sp.) greatly enhanced tissue growth, but the greatest growth occurred on pure *Ankistrodesmus* cultures (Foe and Knight, 1986b). *C. fluminea* fed mixed cultures of the green algae *Pedinomonas* sp., *Ankistrodesmus* sp., *Chlamydomonas* sp., *Chorella* sp., *Scenedesmus* sp., and *Selenastrum* sp. had maximal growth when mixtures did not include *Selenastrum,* which was toxic to this species. Greatest tissue growth occurred in clams fed mixed cultures of all five remaining algal species. Growth declined with number of algal species in feeding cultures; feeding with two algal species resulting in starvation (Foe and Knight, 1986b). Thus, artificial bivalve culture appears to be best supported on mixed algal diets, but certain toxic algal species must be avoided.

Temperature also affects bivalve growth rate. When 5–8 mm SL specimens of *C. fluminea* were fed mixed algal cultures of *Chamydomonas, Chlorella,* and *Ankistrodesmus* at 10^5 cells/mL, assimilation efficiencies were maximal at 16 and 20°C (48–51%). Tissue growth was maximal at 18–20°C and became negative (tissue loss) at ≥30°C (Foe and Knight, 1986a), suggesting 18–20°C to be an ideal culture temperature for this species and, perhaps, other NA bivalves. In algae-fed cultures of *D. polymorpha,* growth was maximal at 15°C and suppressed at 20°C (Walz, 1978b). However, tissue growth in a natural population of *C. fluminea* increased up to 30°C (McMahon and Williams, 1986a), indicating that artificial culture systems are not equivalent to field conditions in supporting bivalve growth. In this regard, excellent tissue growth in *C. fluminea* was supported by algal cultures produced by several days' exposure to sunlight of water taken from the natural habitat of the clam to increase its algal concentration (Foe and Knight, 1985). Thus, ideal culture conditions for unionoideans and *C. fluminea* would appear to be a 20°C holding temperature and feeding with natural algal assemblages whose growth has been promoted with inorganic nutrients and exposure to sunlight; maximal growth being achieved in *D. polymorpha* under the same conditions at 15°C.

Specimens of *C. fluminea, D. polymorpha* and unionoideans may be held for long periods in the laboratory without feeding. *C. fluminea* survived 154 days of starvation at room temperature (22–24°C) while sustaining tissue weight losses ranging from 41 to 71% (Cleland *et al.,* 1986). Similarly, I have held unionoideans in the laboratory for many months without feeding. Samples of *D. polymorpha* experienced 100% mortality after being held in the laboratory without feeding for 166, 514, and 945 days at 25, 15, and 5°C, respectively (Chase-Off and McMahon, unpublished data). Thus, maintenance at low temperature (<10°C) greatly prolongs the time bivalves may be held in good condition without feeding.

C. fluminea has never been reared successfully to maturity or carried through a reproductive cycle in artificial culture, although field-collected, nongravid adults released juveniles after four months in laboratory culture (King *et al.,* 1986). The glochidium stage makes unionoideans difficult to rear in the laboratory as it requires encystment in a fish host for successful juvenile metamorphosis, but it can be accomplished (Young and Williams, 1984a). Application of an immunosuppressant agent has allowed glochidial transformation to juveniles on nonhost fish where they would otherwise be rejected by the immune system of the host fish (Kirk and Layzer, 1997). Glochidia of several unionoidean species have been transformed into juveniles *in vitro* in a culture medium containing physiologic salts, amino acids, glucose, vitamins, antibiotics, and host fish plasma (Isom and Hudson, 1982). Juvenile *utterbackia imbecillis* and *Epioblasma triquetra* have been successfully cultured in a medium of river water exposed to sunlight for 1–4 days to enhance algal concentration. Addition of silt enhanced juvenile growth in both species, while feeding artificial, mixed cultures of three algal species resulted in starvation (Hudson and Isom, 1984), an observation supported by recent studies suggesting that juvenile unionoideans feed primarily on interstitial water drawn from sediments through the pedal gape, such that silt and associated microdetritus and bacteria may be their main ingested food source (Yeager *et al.,* 1994).

Many sphaeriid species can be easily maintained in simple artificial culture systems. Ease of artificial culture in this group may relate to their feeding on sediment organic detritus (Burky *et al.,* 1985b; Hornbach *et al.,* 1984b) and interstitial bacteria (Lopez and Holopainen, 1987), making use of algal cultures as a food source unnecessary. Hornbach and Childers (1987) maintained *Musculium partumeium* through successful reproduction in beakers with 325 mL of filtered river water, without sediments, on a diet of 0.1 mg of finely ground Tetra Min© fish food/clam per day. The first generation in this simple culture system survived 380–500 days, a life span equivalent to the natural population (Hornbach *et al.,* 1980). Similarly, Rooke and Mackie (1984c) maintained 20 adult *Pisidium casertanum* in a 6-L aquaria with sediments for 35 weeks without feeding, suggesting that individuals fed on sediment bacteria or organic deposits.

Live molluscs rapidly remove dissolved calcium from culture media (Rooke and Mackie, 1984c), thus calcium levels in holding media should be augmented by the addition of $CaCO_3$. The ease with which sphaeriids can be artificially cultured makes them ideal for laboratory microcosm experiments. Ideal culture conditions appear to include provision of natural sediments, a source of calcium (i.e., ground $CaCO_3$), and finely ground food of reasonable protein content, such as aquarium fish food or brewers' yeast, which may be directly assimilated by clams or support the growth of interstitial bacteria upon which they feed (Lopez and Holopainen, 1987).

V. IDENTIFICATION OF THE FRESHWATER BIVALVES OF NORTH AMERICA

A. Taxonomic Key to the Superfamilies of Freshwater Bivalvia

There are five bivalve superfamilies with freshwater representatives in North America. Of these the Unionoidea, Corbiculoidea, and Dreissenoidea contain the true freshwater species and comprise the vast majority of freshwater bivalve fauna. The remaining two superfamilies, Cyrenoidea and Mactroidea, each contain one brackish water species that can extend into freshwater coastal drainages and so are included here. In North America, the Dreissenacea is represented by two introduced freshwater species and an estuarine species. The Corbiculoidea includes 36 native and five introduced species in six genera; the Unionoidea are composed of 278 native species and 13 subspecies in 49 genera (Turgeon *et al.*, 1998). Separate taxonomic keys are provided here for the latter two supperfamilies.

1a. Shell hinge ligament is external 2

1b. Shell hinge ligament is internal 4

2a(1a). Shell with lateral teeth extending anterior and posterior of true cardinal teeth (Fig. 1), shells of adults generally small (<25 mm in shell length, shell thin and fragile; exceptions are the genera *Polymesoda* and *Corbicula*) .... superfamily Corbiculoidea [See Section V.B.]

2b. Shell without lateral teeth extending anterior and posterior of cardinal teeth 3

3a(2b). Shell hinge with two cardinal teeth and without lateral teeth; shell thin and fragile, 12–15 mm long with small umbos; *Cyrenoida floridana* (Dall) (extends from brackish into coastal freshwater drainages in Florida) .... superfamily Cyrenoidea

3b. Shell without true cardinal teeth, when present, lateral teeth only occur posterior to usually well-developed pseudocardinal teeth (Fig. 1), pseudocardial teeth absent or vestigial in some species; Shells of adults are generally large (>25 mm in shell length) .... superfamily Unionoidea [See Section V.C.]

4a(1b). Hinge with anterior and posterior latoral teeth on either side of cardinals; Shell massive, adults 25–60 mm shell length, obliquely ovate; *Rangia cuneata* (Gray) (extends from brackish into coastal freshwater drainages from Delaware to Florida to Veracruz, Mexico) .... superfamily Mactroidea

4b. Hinge without teeth; shell mytiloid in shape, anterior end reduced and pointed, hinge at anterior end, posterior portion of shell expanded, anterior adductor muscle attached to internal apical shell septum, attached to hard substrates by byssal threads .... superfamily Dreissenoidea 5

5a(4b). Periostracum bluish brown to tan without a series of dorsoventrally oriented black zigzag markings, anterior end hooked sharply ventrally, ventral shell margins not distinctly flattened over entire ventral side of valves; restricted to brackish water habitats. *Mytilopsis leucophaeata* (Conrad) (extends into coastal freshwater drainages from New York to Florida to Texas and Mexico) .... *Mytilopsis*

5b. Periostracum light tan and often marked with a distinct series of black vertical zigzag markings; anterior portion of shell not ventrally hooked, ventral shell margins can be flattened, restricted to freshwaters, *Dreissena polymorpha* (Pallas) (Fig. 24); (a European species introduced into the Great Lakes in Lake St. Clair and present by December 2000 throughout the Great Lakes, the St. Lawrence River and the Mississippi Drainage. and Dreissona bugensis Andrusov now in lakes Erie and Ontario, the Erie-Barge Canal and the St. Lawrence River .... *Dreissena*

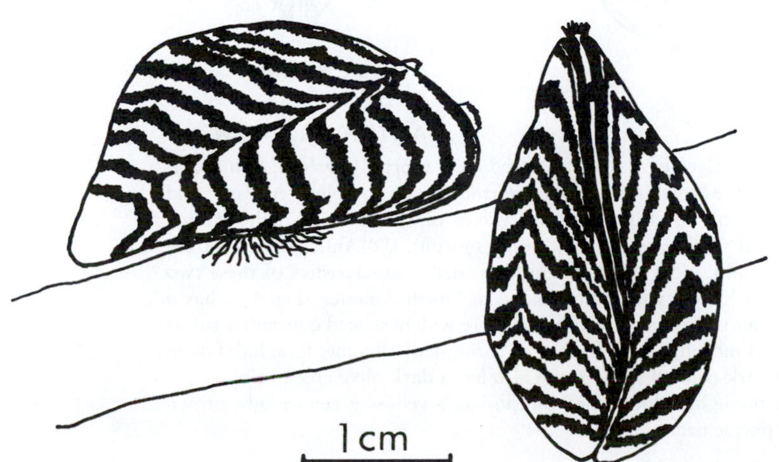

FIGURE 24 The external morphology of the shell of the zebra mussel, *Dreissena polymorpha*.

B. Taxonomic Key to the Genera of Freshwater Corbiculacea

This key is based on the excellent species key for North American freshwater Corbiculoidea by Burch (1975a) with additional material from Clarke (1973) for the Canadian Interior Basin, and Mackie *et. al.* (1980) for the Great Lakes. The Corbiculoidea have ovate, subovate, or trigonal shells with lateral hinge teeth anterior and posterior to the cardinal teeth. All North American species expect *Corbicula fluminea* are in the family Sphaeriidae. The family designation Pisidiidae has also been commonly applied to this group, but the International Commission of Zoological Nomenclature (ICZN) placed the Spheariidae (Name number 573) on the Offical List of Family Names (Opinion 1331) in 1985; hence, Sphaeriidae is used as the family designation in this key and the rest of the chapter. In North America, the Sphaeriidae comprise the dominant bivalve fauna in small, often ephemeral ponds, lakes, and streams, the profundal portions of lakes and in silty substrates. Identification is generally based on shell morphology, but requires, in some cases, soft tissue morphology.

1a. Shells large (maximum adult shell length >25 mm), thick, lateral teeth serrated ....family Corbiculida 2

1b. Shells generally small (maximum shell length <25 mm), thin, lateral teeth smooth ....family Sphaeriidae 4

2a(1a). Maximum adult shell length generally <50 mm, shell ornamented by distinct, concentric sulcations, anterior and posterior lateral teeth with many fine serrations, simultaneous hermaphrodites, massive numbers of small (length <0.3 mm) developmental stages (>1000) incubated directly in inner demibranchs, released juveniles (<5 mm SL) anchor to substratum with a single mucilaginous byssal thread (Fig. 25)..*Corbicula* 3

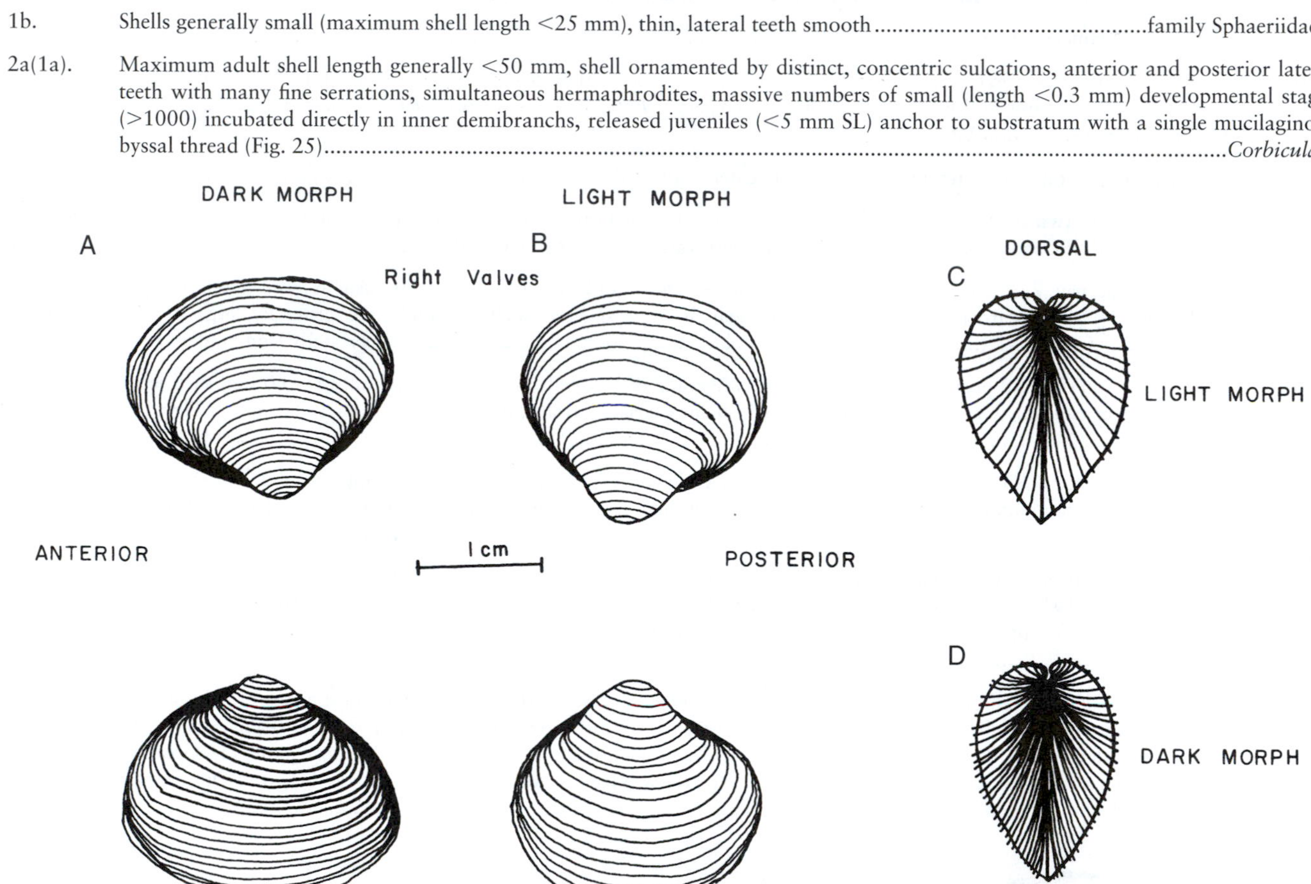

FIGURE 25 External morphology of the shell valves of the light-colored shell morph (*Corbicula fluminea*) and dark-colored shell morph (*Corbicula* sp.) of the North American *Corbicula* species complex. (A) Right and left valves of *Corbicula* sp. (dark-colored morph). (B) Right and left valves of *C. fluminea* (light-colored morph). (C) Anterior view of the shell valves of *C. fluminea* (light-colored morph). (D) Anterior view of the shell valves of *Corbicula* sp. (dark-colored morph). Note the distinguishing shell characteristics of these two species. *C. fluminea*, the light-colored morph, which is widely distributed in North America (Fig. 11), has a more nearly trigonal shell, taller umbos, a greater relative shell width and more widely spaced concentric sulcations than does *Corbicula* sp., the dark-colored morph, that is limited to spring-fed, alkaline, lotic habitats in the southwestern United States (Fig. 11). The dark-colored shell morph also has a dark olive green-to-black periostracum and deep royal blue nacre, while the light-colored shell morph has a yellow green-to-light brown periostracum and white-to-light blue or light purple nacre.

2b. Maximum adult shell length generally >50 mm, shell ornamentation of many fine, closely spaced concentric striations, embryos not incubated in demibranchs, dioecious, periostracum deep brown in color, three cardinal teeth, estuarine, restricted to brackish waters in the tidal portions of rivers. *Polymesoda caroliniana* (Bosc) (Virginia to northern Florida to Texas)..............................*Polymesoda*

3a(2a). Shell nacre white with light blue, rose, or purple highlights, particularly at shell margin, muscle scars of same color intensity as rest of nacre, periostracum yellow to yellow-green or brown with outer margins always yellow or yellow-green in healthy, growing specimens, shell trigonal to ovate, umbos inflated and distinctly raised above dorsal shell margin, shell length : shell height ratio ≈1.06, shell length : shell width ratio ≈1.47, shell height : shell width ratio ≈ 1.38, concentric shell sulcations widely spaced, 1.5 sulcations/mm shell height (Hillis and Patton, 1982); introduced in the early 1900s, it has spread throughout drainage systems of the United States and coastal northern Mexico (Fig. 11), the "light-colored shell morph" of *Corbicula* or Asian clam (Fig. 25B, C).*Corbicula fluminea* (Müller)

3b. Shell nacre uniformly royal blue to deep purple over entire internal surface, muscle scars more darkly pigmented than rest of nacre, periostracum dark olive green to black, edges of valves in healthy, growing specimens not yellow or yellow-green, shell more ovate and laterally compressed with umbos less inflated and less distinctly raised above the dorsal shell margin than in *C. fluminea*, shell length : height ratio ≈1.15, shell length : width ratio ≈ 1.65, shell height : width ratio ≈ 1.43, concentric shell sulcations narrowly spaced, particularly at umbos, 2.1 sulcations/mm shell height (Hillis and Patton, 1982); introduced, distribution limited to highly oligotrophic, permanent, spring-fed, calcium carbonate-rich streams in the southwestern United States (Britton and Morton, 1986); (Fig. 11). Called the dark-colored shell morph of *Corbicula*, its taxonomic status is uncertain, electrophoretic (Hills and Patton, 1982; McLeod, 1986) and physiological evidence (Cleland *et al.*, 1986) suggest it to be distinct from *C. fluminea* (Fig. 25A, D) ....*Corbicula* sp.

4a(1b). Both inhalant and exhalant mantle cavity siphons present and well developed, umbos lie anterior of center..............................5

4b. Only exhalant mantle cavity siphon present, inhalant siphon either absent or formed as a slit in the posterior-ventral mantle edges, umbos posterior of center, generally small, shell length 0.5–12 mm, embryos in inner demibranch held in thick-walled sacs, each with individual chambers for embryos, no byssal gland, 24 species widely distributed in North Amercia; for species identifications and distributions see Burch (1975a) (Fig. 26A)..............................subfamily Pisidiinae *Pisidium*

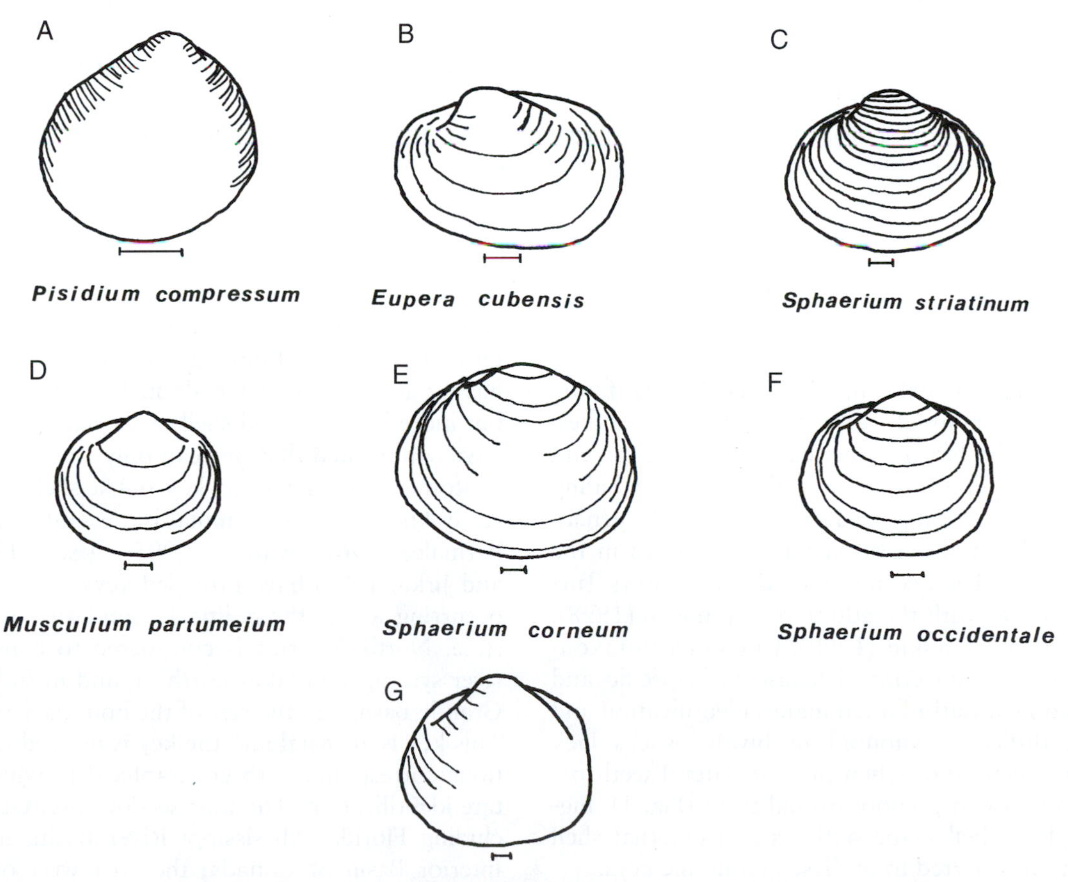

FIGURE 26 Diagrams of the external morphology of the left shell valve of species representative of the North American genera of the freshwater bivalve family, Sphaeriidae. Size scaling bar is 1-mm long.

5a(4a). Inhalant and exhalant mantle cavity siphons partially fused, embryos incubated in inner demibranchs in thin-walled longitudinal pouches, no byssal gland, shell with two cardinal teeth in each valve, without external mottling....................subfamily Sphaeriinae 6

5b. Inhalant and exhalant siphons not fused, embryos develop in individual chambers formed between inner and outer lamellae of inner demibranchs, functional byssal gland present, only one cardinal tooth in each shell valve, with external, mottled pigmentation, *Eupera cubensis* (Prime) (Atlantic coastal plain drainages from southern Texas to North Carolina, Caribbean Islands) (Fig. 26B) ....subfamily Euperinae *Eupera*

6a(5a). Shell sculptured with relatively coarse or widely spaced striae (≤8 striae/mm in middle of shell), shell relatively massive and strong, *Sphaerium simile* (Say) (Southern Canada from New Brunswick to British Columbia, south from Virginia to Wyoming, *S. striatinum* (Lamarck) (Canada from New Brunswick to the upper Yukon River, throughout the United States, Mexico, and Central America), *S. fabale* (Prime) (Southern Ontario to Georgia and Alabama) (Fig. 26C) ....*Sphaerium*

6b. Shell relatively thin, often fragile, with many fine, narrowly spaced striae (≥12 striae/mm in middle of shell) 7

7a(6b). Shell of adults <8 mm in length 8

7b. Shell of adults >8 mm in length 9

8a(7a). Posterior valve margin at near right angle to dorsal margin, shells roughly rhomboidal, umbos large and distinctly elevated above dorsal shell margin, *Musculium partumeium* (Say) (United States and southern Canada), *M. transversum* (Say) (Canada and the United States east of the continental divide, extending into Mexico), *M. securis* (Prime) (Nova Scotia to British Columbia southwestern Northwest Territories in Canada, United States except for southwest) (Fig. 26D) ....*Musculium*

8b. Posterior and dorsal margins rounded or forming an obtuse angle, shells ovate, *Sphaerium corneum* (Linnaeus) (introduced from Europe, localities in southern Ontario and Lakes Champlain and Erie), *S. nitidum* Westerlund (distribution is holarctic, northern Canada to northern United States), *S. occidentale* (Prime) (Canada from New Brunswick to southeastern Manitoba, northern United States south to Florida, west to Utah and Colorado) (Fig. 26 E, F) ....*Sphaerium*

9a(7b). Umbos large, distinctly elevated above the dorsal shell margin ....10.

9b. Umbos small, indistinctly elevated above the dorsal shell margin, *Sphaerium corneum* (Linnaeus), (introduced, localities in Ontario, Lakes Champlain and Erie) (Fig. 26E) ....*Sphaerium*

10a(9a). Shell rounded, umbos not prominent, *Sphaerium occidentale* (Prime) (New Brunswick to southeastern Manitoba, northern United States south to Florida in the east and Utah and Colorado in the west) (Fig. 26F) ....*Sphaerium*

10b. Posterior end of shell truncate, shell rhomboidal, umbos prominent, *Musculium lacustre* (Müller) (From treeline in Canada south throughout all but southwestern United States into central America) (Fig. 26G)..*Musculium*

C. Taxonomic Key to the Genera of Freshwater Unionoidea

The Unionoidea make up the large bivalve fauna (shell length >25mm) of permanent freshwater lakes, rivers, and ponds. North America has the richest and most diverse unionoidean fauna in the world, including 278 species and 13 recognized subspecies in 49 genera in the Unionidae and five species in two genera in the Margaritiferidae. The taxonomy used here follows Turgeon *et al.* (1998) with the addition of Johnson (1998), and Williams and Fradkin (1999). Unionoidean taxonomy remains very uncertain because intraspecific and interpopulation variation often makes identification and systematics difficult. Unionoidean bivalve shells lack true cardinal teeth and, when present, lateral teeth occur only posterior to pseudocardinal teeth (Fig. 1). Figure 27 displays shell-shape outlines and external shell ornamentations referred to in these taxonomic keys.

The key in the first edition of this chapter, as well as many of the major keys to freshwater bivalves (Walker, 1918; Clench, 1959; Burch, 1975b; Clarke, 1973), relied on the integrated use of important anatomical structures and shell characters to identify unionoidean bivalves. However, most field biologists do not have the luxury of having a fully gravid female in hand for identification, but more likely, a dead shell or valve. Several colleagues have commented that one can only use a key containing anatomical characters to key out a shell when you already know what the animal is. Several state keys (e.g., Parmalee, 1967; Watters, 1995; Oesch, 1984; Strayer and Jirka, 1997) have provided keys to shells of species occurring within the political boundaries of a particular state. North America is considered to consist of those river systems and lakes north of, and including, the Rio Grande Basin and the rest of the boundary with Mexico. This key is artificial and the key is divided into four sections corresponding to geographical provinces to facilitate identification. The four sections are: Gulf Coast, including Florida; Mississippi River Basin, including the Interior Basin of Canada; the area west of the Rocky Mountain divide; and the Atlantic Coast. These subdivisions are a simplification of the 12 faunal provinces recognized by Parmalee and Bogan (1998). Table VII lists

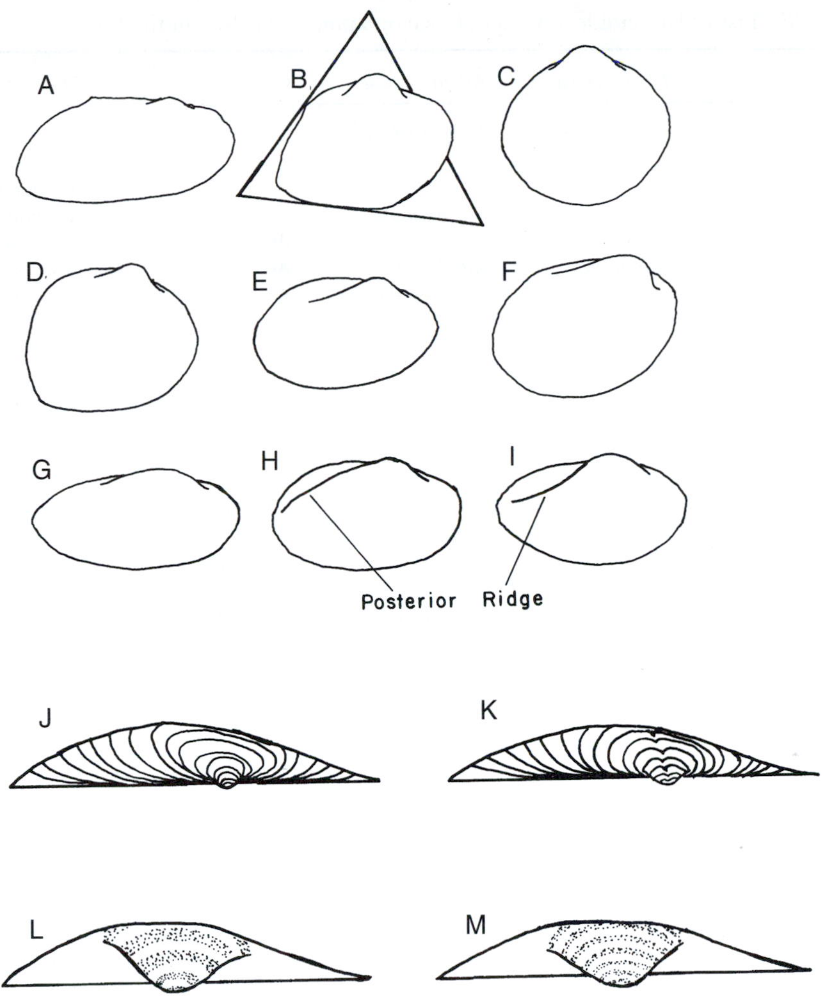

FIGURE 27 Illustrations of the diagnostic shell features or characters used for taxonomic identification in Section V.C. Shell-shape descriptions: (A) rhomboidal; (B) triangular or trigonal; (C) round; (D) quadrate; (E and F) oval or ovoid; and (G) elliptical. Posterior shell-ridge morphology: (H) posterior ridge convex; and (I) posterior ridge concave. Concentric ridge structures of umbos: (J) single-looped concentric ridges; (K) double-looped concentric ridges; (L); coarse concentric ridges; and (M) fine concentric ridges. (Redrawn from Burch, 1975b.)

the genera found in each of these four sections. This geographical approach allows a biologist to focus only on those genera actually occurring in their region and ignores the rest of the diversity (*e.g.*, west of the Rocky Mountains has only three genera: one Margaritiferidae and two Unionidae). We have attempted to provide a logical and clear key to the unionoidean genera of North America based solely on shell characters. The valve or pair of valves to be identified is assumed for the purposes of this key to be adults. Identification of very small juvenile shells and small adult specimens can be difficult. Landmarks, structures and shell shape are based on complete shells. We have attempted to use obvious characters, but many of them require familiarity with unionoidean shell morphology and the range of variation in each of these characters. Strayer and Jirka (1997; p. 28) stated the problem most clearly when they said, "Keys based on shell characters are inevitably filled with vague, subjective terms, are frustrating for beginners to use, and misidentify many shells." They have finally put into print what most malacologists have said: **"Users should know that if they rely solely on this key, they will misidentify many shells."** (Strayer and Jirka, 1997; p. 28). There is no substitute for comparing the specimen to be identified with other identified specimens in a museum collection or specimens that have had their identification verified by a specialist. This key is an introduction to the genera of North American unionoideans. Each couplet which contains a genus, also lists some of the species in that genus which should key out to that cou-

TABLE VII List of Unionoidean Genera by Geographic Area Used in the Key

	Pacific coast	*Atlantic coast*	*Gulf coast*	*Interior basin*
Margaritiferidae	*Margaritifera*	*Margaritifera*	*Margaritifera*	
				Cumberlandia
Unionidae				*Actinonaias*
		Alasmidonta	*Alasmidonta*	*Alasmidonta*
			Amblema	*Amblema*
	Anodonta	*Anodonta*	*Anodonta*	*Anodonta*
		Anodontoides	*Anodontoides*	*Anodontoides*
			Arcidens	*Arcidens*
				Arkansia
				Cyclonaias
				Cyprogenia
			Cyrtonaias	
			Disconaias	
				Dromus
			Ellipsaria	*Ellipsaria*
		Elliptio	*Elliptio*	*Elliptio*
			Elliptoideus	
			Epioblasma	*Epioblasma*
		Fusconaia	*Fusconaia*	*Fusconaia*
			Glebula	
	Gonidea			
				Hemistena
		Lampsilis	*Lampsilis*	*Lampsilis*
		Lasmigona	*Lasmigona*	*Lasmigona*
				Lemiox
		Leptodea	*Leptodea*	*Leptodea*
		Lexingtonia		*Lexingtonia*
		Ligumia	*Ligumia*	*Ligumia*
			Medionidus	*Medionidus*
			Megalonaias	*Megalonaias*
			Obliquaria	*Obliquaria*
			Obovaria	*Obovaria*
				Pegias
			Plectomerus	*Plectomerus*
				Plethobasus
		Pleurobema	*Pleurobema*	*Pleurobema*
			Popenaias	
			Potamilus	*Potamilus*
			Ptychobranchus	*Ptychobranchus*
		Pyganodon	*Pyganodon*	*Pyganodon*
			Quadrula	*Quadrula*
			Quincuncina	*Simpsonaias*
		Strophitus	*Strophitus*	*Strophitus*
		Toxolasma	*Toxolasma*	*Toxolasma*
			Tritogonia	*Tritogonia*
			Truncilla	*Truncilla*
		Uniomerus	*Uniomerus*	*Uniomerus*
		Utterbackia	*Utterbackia*	*Utterbackia*
				Venustaconcha
		Villosa	*Villosa*	*Villosa*
Total	3	18	37	43

plet. This list of species is not complete, but is representative of the species in the genus with the listed set of characters. Two genera are a source of much confusion, *Elliptio* and *Pleurobema*. The South Atlantic Slope *Elliptio* species complex has yet to be worked out. Johnson (1970) lumped a considerable number of taxa, which may, in fact, be valid species. The Gulf Coast *Pleurobema* species complex is the other major area of confusion. There are a large number of named taxa, but it is not clear at the present time which is a valid species and which is part of a cline. This group is still in need of further work. We would recommend that, to identify specimens to the species level, the user move to a key for the state or province in which he/she is working. These volumes will give detailed descriptions, keys and figures of the species occurring in your local area (e.g., Florida: Clench and Turner, 1956; Johnson, 1972; Tennessee: Parmalee and Bogan, 1998; Louisiana: Vidrine, 1993; New York: Strayer and Jirka, 1997; Illinois: Parmalee, 1967; Missouri: Oesch, 1984; and Ohio: Watters, 1995). A detailed list by state of freshwater bivalve literature can be found in Williams *et al.* (1993).

1a. Origin of shell is from west of the Rocky Mountains 4

1b. Shell is not from this area 2

2a. (1b) Origin of shell is from the Atlantic Coast of North America 6

2b. Shell is not from this area 3

3a. (2b) Origin of shell is from the Mississippi River Basin 26

3b. Origin of shell is from the Gulf Coast of North America 89

West of the Rocky Mountains

4a. (1a) Shell has lateral muscle scars [small pits for the attachment of mantle muscles to the shell], typically dark periostracum, relatively thick shell with pseudocardinal and lateral teeth .... *Margaritifera falcata* (Fig. 11.28B)

4b. Shell lacks lateral muscle scars, periostracum typically not dark and hinge teeth reduced or absent 5

5a. (4b) Shell has a very strongly developed posterior ridge, reduced hinge teeth, occurs from British Columbia to central California, east to Nevada and Idaho .... *Gonidea angulata* (Fig. 28U)

5b. Shell lacks the sharp posterior ridge, lacks any evidence of hinge teeth .... *Anodonta* [in part] [*A. beringiana, A. californiensis, A. dejecta, A. kennerlyi, A. nuttalliana, A. oregonensis*] (Fig. 28F)

Atlantic Coast

6a. (2a) Shell has lateral muscle scars, typically dark periostracum, relatively thick shell with pseudocardinal teeth and lateral teeth .... *Margaritifera margaritifera* (Fig. 28B)

6b. Shell lacks lateral muscle scars, periostracum may not be dark, hinge teeth present or absent 7

7a. (6b) Shell with hinge teeth absent or greatly reduced 8

7b. Shell with pseudocardinal teeth present, with or without lateral teeth 12

8a. (7a) Umbo not projecting above the hinge-line .... *Utterbackia* [in part] [*U. imbecillis*] (Fig. 28AW)

8b. Umbo projecting above the hinge-line 9

9a. (8b) Beak sculpture double looped (Fig. 27), shell uniformly thin .... *Pyganodon* [in part] [*P. cataracta*] (Fig. 28AN)

9b. Beak sculpture consists of concentric bars 10

10a. (9b) Nacre usually orange in the beak cavity, pseudocardinal tooth area represented by a thickening near the umbo, ventral shell margin uniform thickness .... *Strophitus* [in part] [*S. undulatus*] (Fig. 28AR)

10b. Nacre bluish or white, hinge plate uniformly thin, teeth or swellings absent, ... 11

11a. (10b) Ventral margin with a prominent thickened area along the anterior ventral magin below the pallial line, *Anodonta* [in part] [*A. implicata*] (Fig. 28F)

11b. Ventral margin not as above, shell elongate, Anodontoides [in part] [A. ferussacianus, restricted to the Susquehanna and Hudson River basins] (Fig. 28G)

12a. (7b). Shell with lateral teeth absent or reduced, neither functional nor interlocking .... *Alasmidonta* [in part] [*A. varicosa, A. marginata susquehannae, A. undulata*] (Fig. 28D)

12b. Shell truncated, with well-developed lateral teeth 13

13a. (12b) Right valve with two lateral teeth, small, rare .... *Alasmidonta* [in part] [*A. heterodon*] (Fig. 28D)

13b. Right valve with one lateral tooth 14

14a. (13b) Shell with spines on the umbo and down on to the disk of the shell 15

14b. Shell lacks any evidence of spines 16

15a. (14a) Shell thick and may be large .... *Elliptio* [in part] [shell from the Altamaha River Basin, Georgia, *Elliptio spinosa*, or the Tar/Neuse River Basin, North Carolina *Elliptio steinstansana*] (Fig. 28P)

15b. Shell small and from the James River Basin, Virginia .... *Pleurobema collina* (Fig. 28AJ)

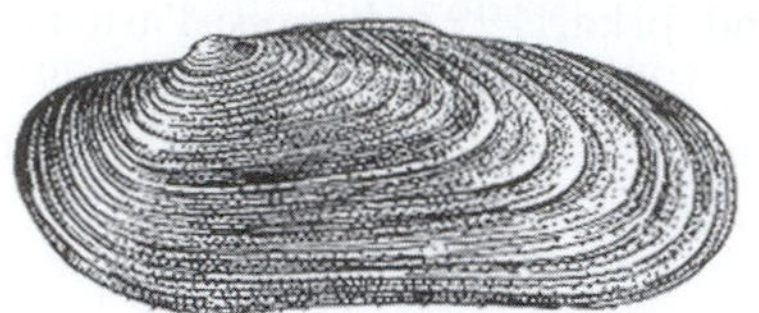

A. *Cumberlandia monodonta*

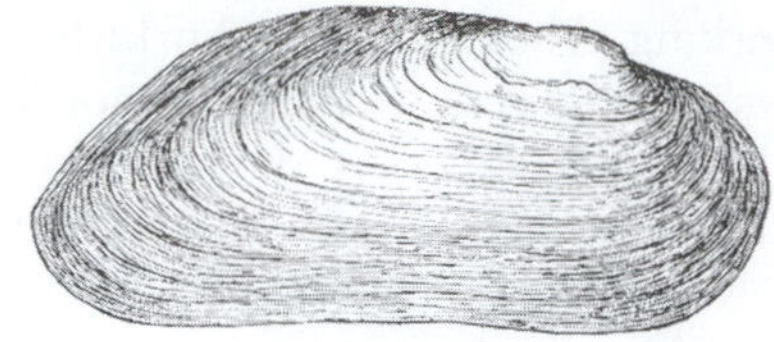

B. *Margaritifera margaritifera*

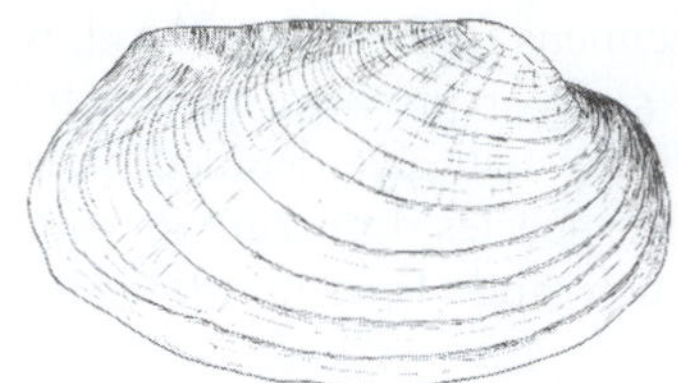

C. *Actinonaias ligamentina*

D. *Alasmidonta undulata*

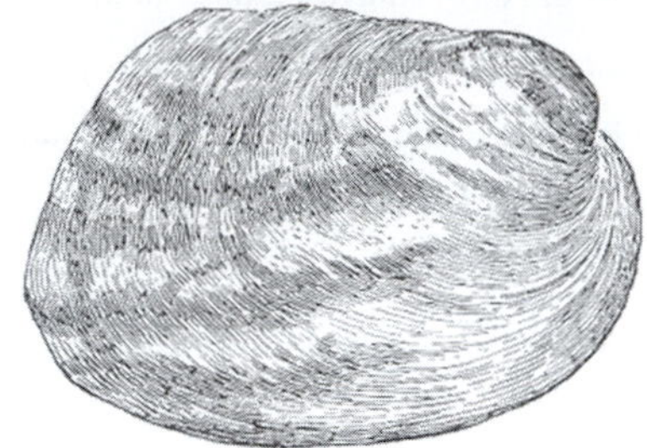

E. *Amblema plicata*

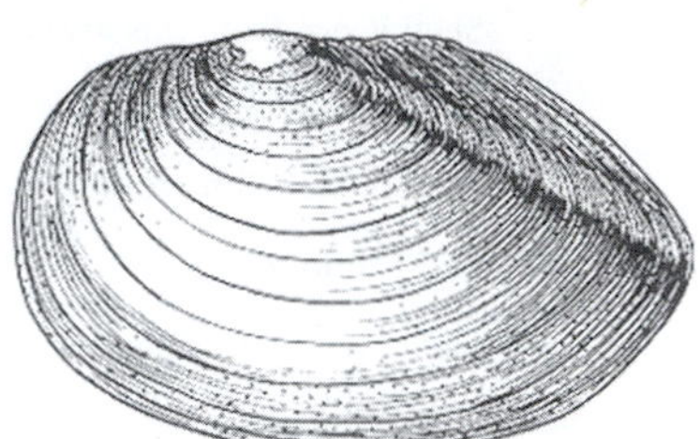

F. *Anodonta* sp.

G. *Anodontoides ferussacianus*

H. *Arcidens confragosa*

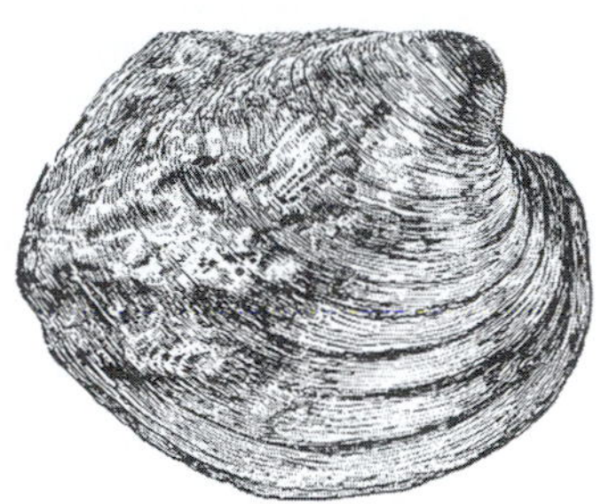

I. *Arkansia wheeleri*

J. *Cyclonaias tuberculata*

K. *Cyprogenia stegaria*

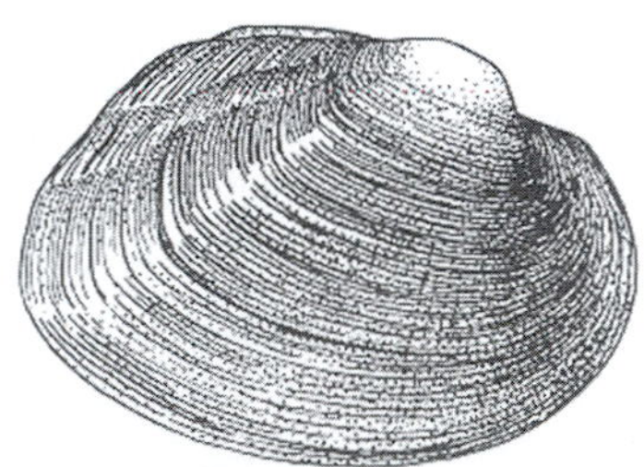

L. *Cyrtonaias tampicoensis*

M. *Disconaias discus*

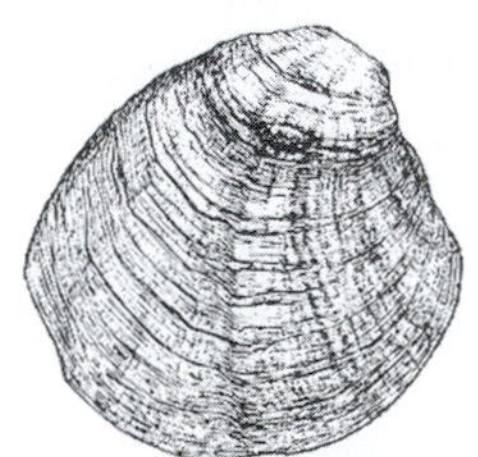

N. *Dromus dromas*

O. *Ellipsaria lineolata*

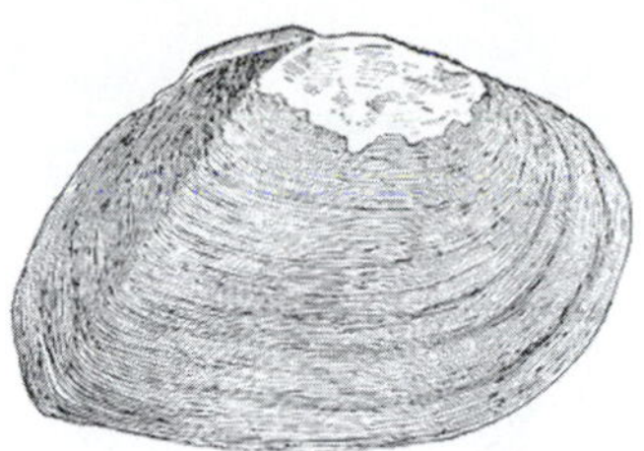

P. Elliptio crassidens

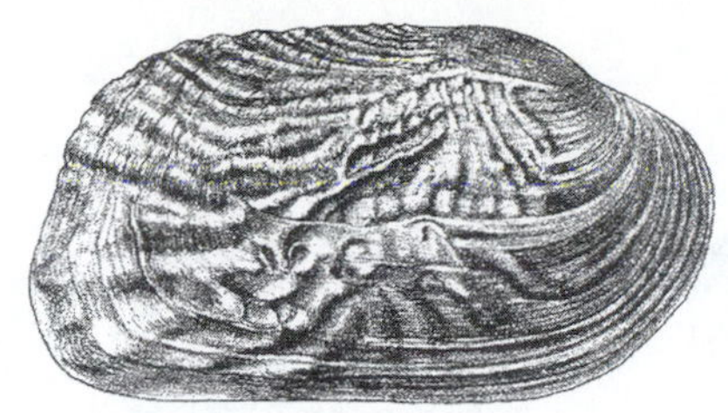

Q. Elliptoideus sloatianus

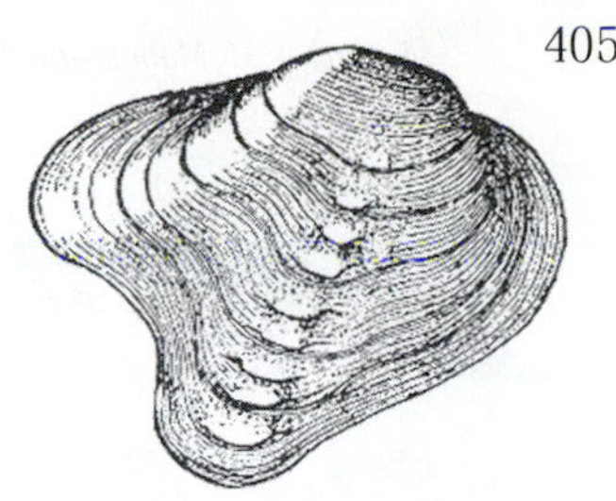

R. Epioblasma flexuosa

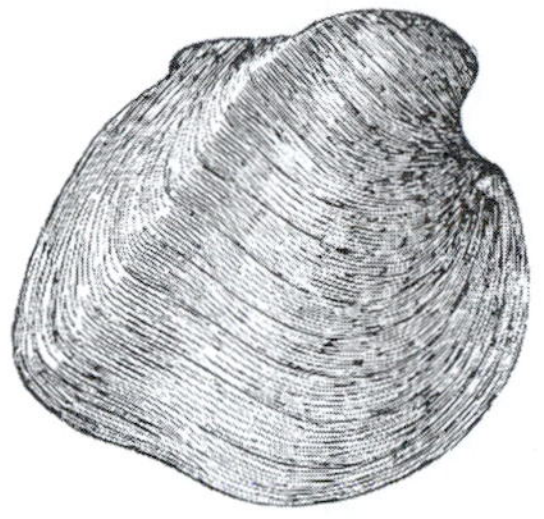

S. Fusconaia flava

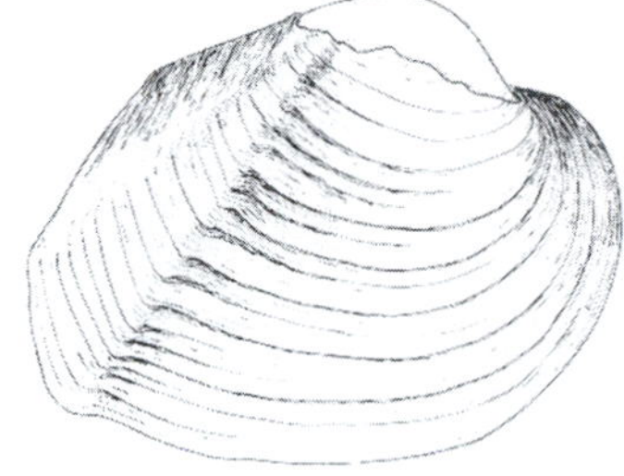

T. Glebula rotundata

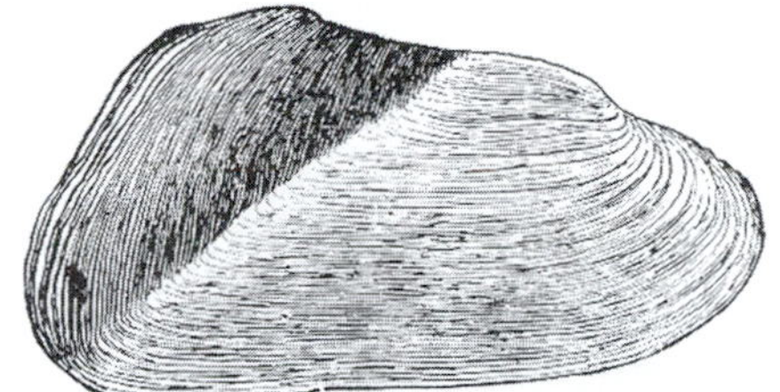

U. Gonidea angulata

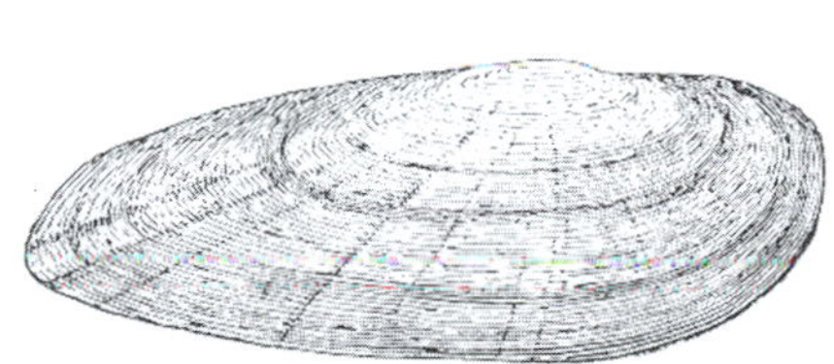

V. Hemistena lata

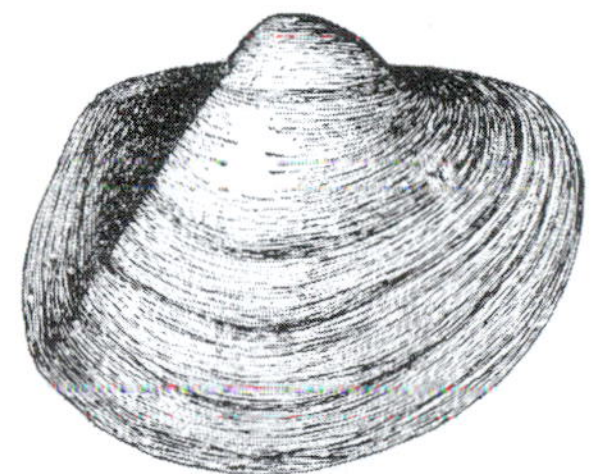

W. Lampsilis ovata

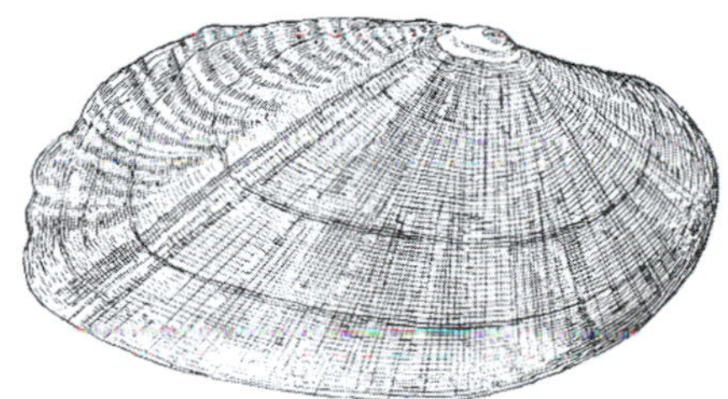

X. Lasmigona costata

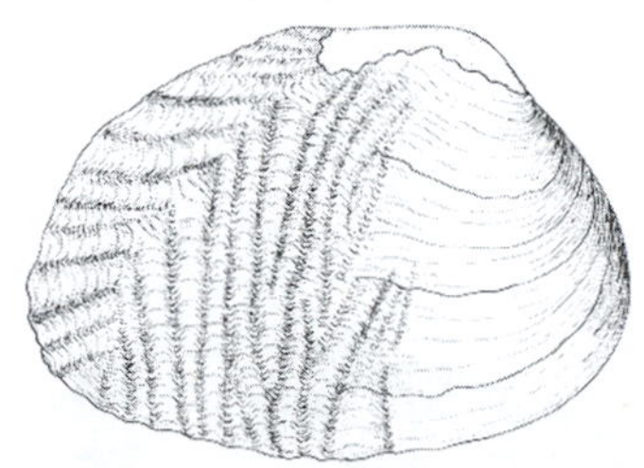

Y. Lemiox rimosus

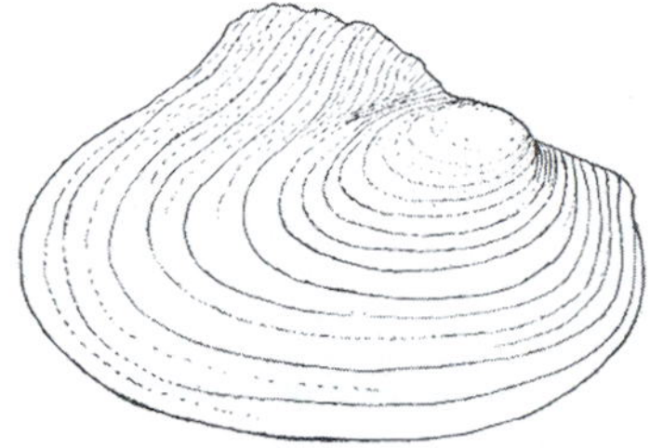

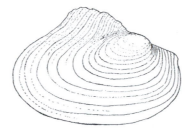

Z. *Leptodea fragilis*

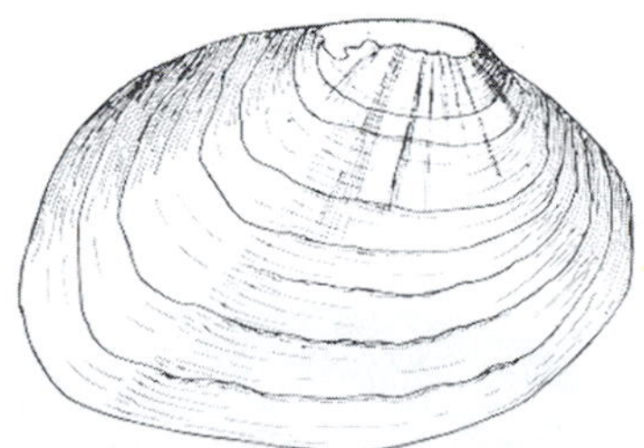

AA. *Lexingtonia dolabelloides*

FIGURE 28 Diagrams of the external morphology mostly of right valves of the shell of species representative of the North American genera of the freshwater bivalve superfamily, Unionoidea (Figs. 28A–AY). Figures B–E, G–K, N, P, R–AC, AE–AG, AI, AJ, AL–AQ, AR–AU, reprinted with permission of Mrs. W.T. Edmondson from W.T. Edmondson (1959); Figures Q, AW, reprinted with permission of J.B. Burch from Burch (1973); Figures A, F, L, M, AD, AH, AK, AX, AY, reprinted with permission from the *Treatise on Invertebrate Paleontology*, courtesy of The Geological Society of America and the University of Kansas, © 1969.

(Continues)

AB. *Ligumia recta*

AC. *Medionidus conradicus*

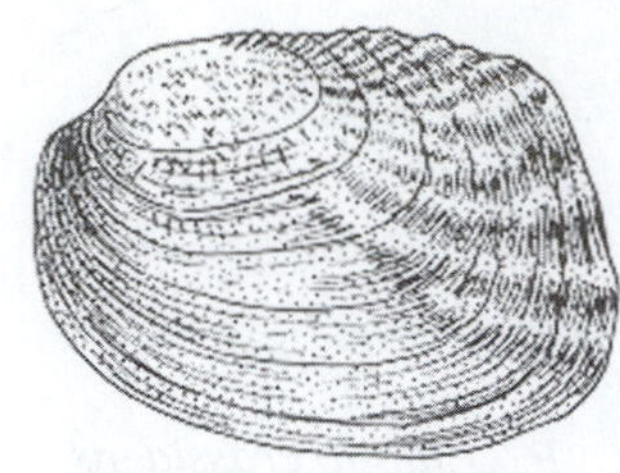
AD. *Megalonaias nervosa*

AE. *Obliquaria reflexa*

AF. *Obovaria retusa*

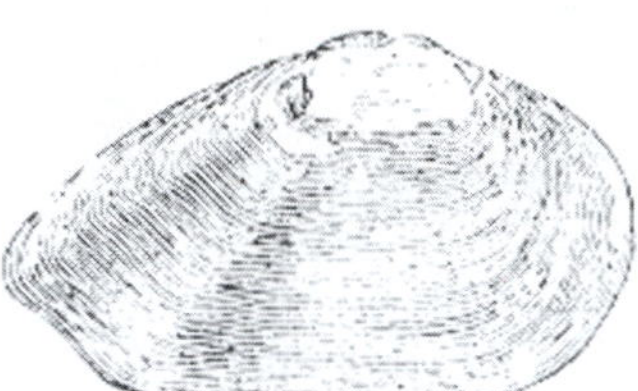
AG. *Pegias fabula*

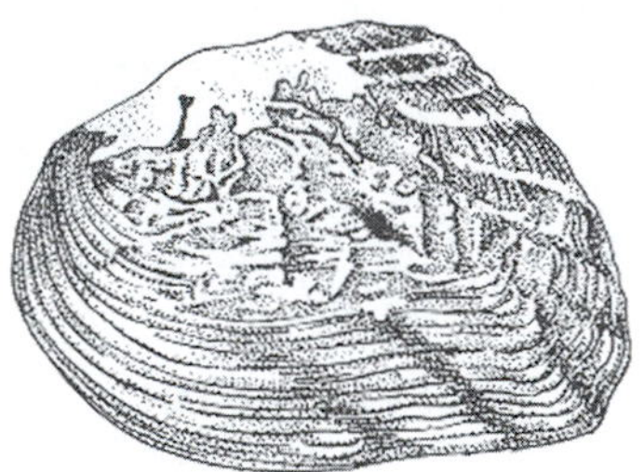
AH. *Plectomerus dombeyanus*

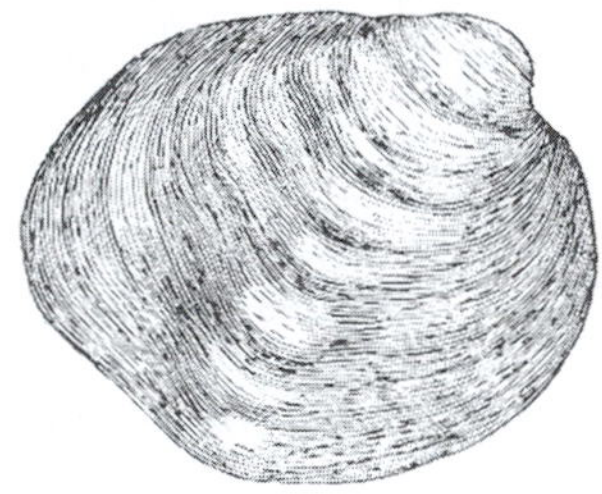
AI. *Plethobasus cyphyus*

AJ. *Pleurobema clava*

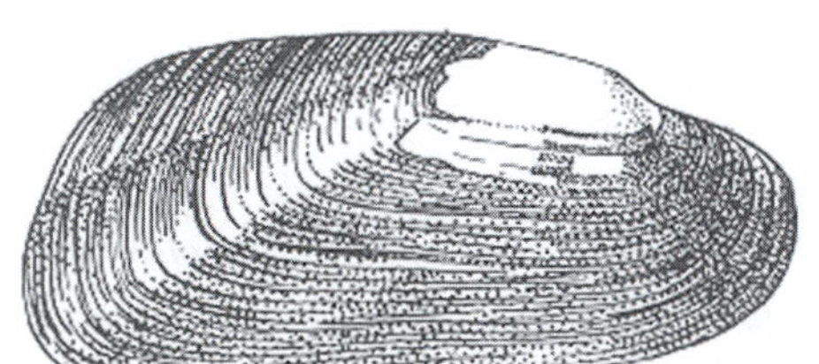
AK. *Popenaias popei*

AL. *Potamilus alatus*

AM. *Ptychobranchus fasciolaris*

FIGURE 28 (*Continued*)

AN. *Pyganodon grandis*

AO. *Quadrula quadrula*

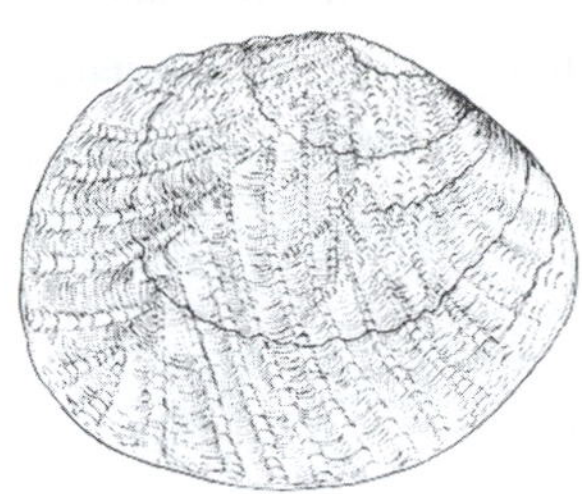

AP. *Quincuncina infucata*

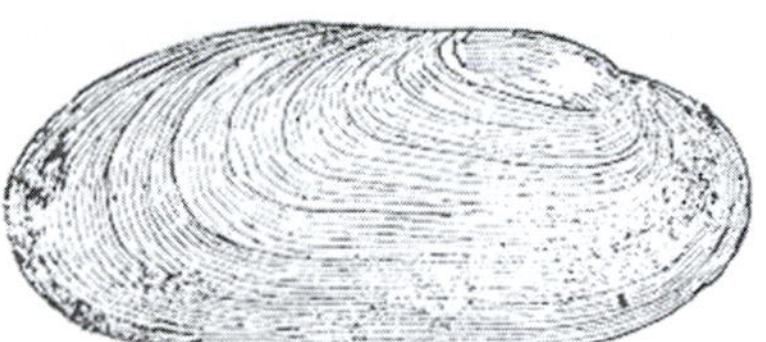

AQ. *Simpsonaias ambigua*

AR. *Strophitus undulatus*

AS. *Toxolasma texasiensis*

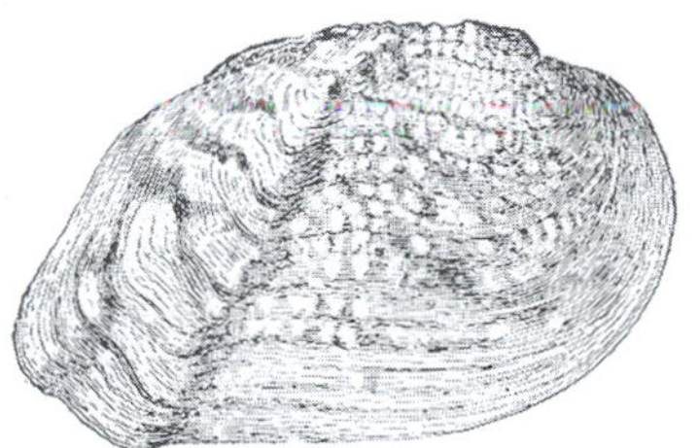

AT. *Tritogonia verrucosa*

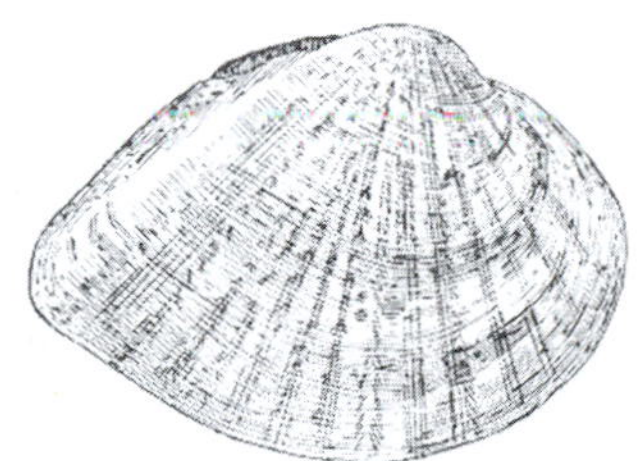

AU. *Truncilla truncata*

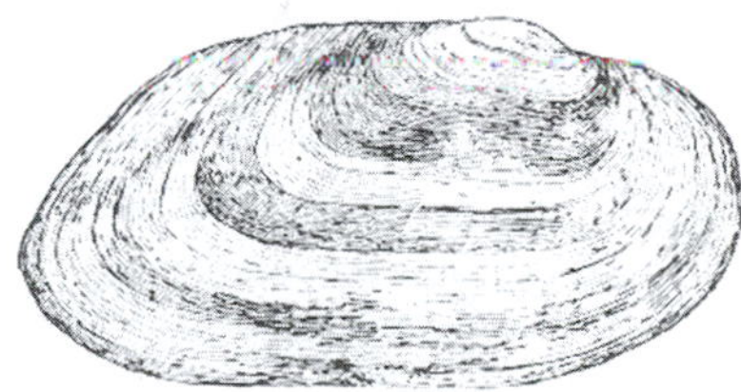

AV. *Uniomerus tetralasmus*

AW. *Utterbackia imbecillis*

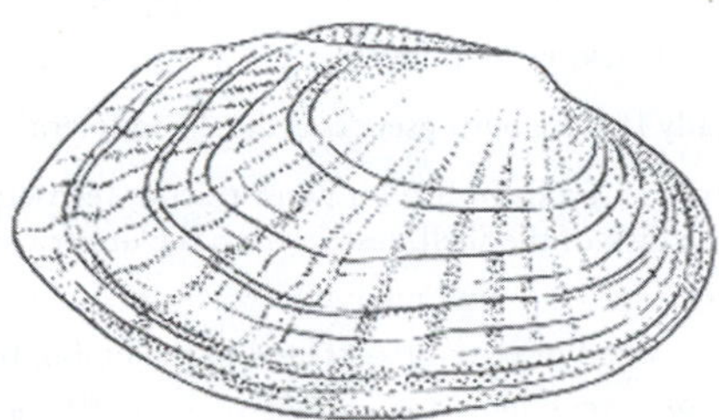

AX. *Venustaconcha ellipsiformis*

AY. *Villosa villosa*

FIGURE 28 (*Continued*)

16a. (14b) Hinge line in left valve with an additional small interdental or accessory tooth, giving the appearance of three pseudocardinal teeth, shell more or less compressed, shell shape rhomboid, periostracum dark green with numerous green rays, beak sculpture consists of prominent bars .... *Lasmigona* [in part][*L. decorata, L. robusta, L. subviridis*](Fig. 28X)

16b. Left valve without extra interdental tooth 17

17a. (16b) Shell shape rectangular to broadly triangular 18

17b. Shell shape oval, round or rhomboid 19

18a. (17a) Shell is from the James River Basin, Virginia .... *Lexingtonia subplana* (Fig. 28AA)

18b. Shell is from an area extending from the Roanoke River Basin south to the headwaters of the Savannah River Basin .... *Fusconaia masoni* (Fig. 28S)

19a. (17b) Shell shape rhomboid or rectangular 20

19b. Shell shape oval or round 23

20a. (19a) Shell usually more than twice as long as high 21

20b. Shell usually less than twice as long as high 25

21a. (20a) Nacre color white, shell inflated, ... 22

21b. Nacre color typically some shade of purple, but ranges from white to salmon to purple .... *Elliptio* [in part] [this genus contains basically three shell shapes, narrow and elongate: the *Elliptio lanceolata* complex; rectangular with various degrees of inflation: the *Elliptio complanata* complex; those shells with short shell length, not too tall and inflated: the *Elliptio icterina* complex] (Fig. 28P)

22a. (21a) Periostracum unrayed, shell thick, posterior end angled, periostracum mat or fuzzy, rectangular in shell shape, *Uniomerus caroliniana* (Fig. 28 AV)

22b. Periostracum rayed in juveniles, posterior end tapered to a point in middle of posterior margin, periostracum not mat, *Ligumia nasuta* (Fig. 28AB)

23a. (19b) Adult shell typically <40 mm in length, with a fuzzy or mat textured dark periostracum .... *Toxolasma* [in part] [*T. pullus*] (Fig. 28AS)

23b. Adult shell >40 mm in length, lacking the pronounced fuzzy periostracum 24

24a. (23b) Shell shape oval to elongate oval, periostracum very shiny to mat with rays .... *Lampsilis* [in part] [*L. cariosa, L. dolabraeformis, L. radiata, L. splendida*] (Fig. 28W)

24b. Shell shape oval, periostracum dull yellow, without rays or with fine rays all over the shell, found in or near tidewater, nacre often a salmon color .... *Leptodea* [in part] [*L. ochracea*] (Fig. 28Z)

25a. (20b) Periostracum light colored with numerous green rays, shell relatively thin, oval to elongate oval, bladelike pseudocardinal teeth .... *Villosa* [in part] [*V. delumbis, V. villosa, V. vibex*] (Fig. 28AY)

25b. Periostracum dark to black, shell thick, no green rays, shell shape oval to round .... *Villosa* [in part] [*V. constricta*] (Fig. 28AY)

Mississippi River Basin

26a. (3a) Shell with lateral muscle scars on the inside center of the valve .... *Cumberlandia monodonta* (Fig. 28A)

26b. Shell lacking any evidence of lateral muscle scars 27

27a. (26b) Shell not as above, lacking lateral teeth or all evidence of any hinge teeth 28

27b. Shell with both pseudocardinal and lateral teeth present 34

28a. (27a) Shell with pseudocardinal teeth, but lacking lateral teeth 29

28b. Shell with greatly reduced or totally lacking both pseudocardinal and lateral teeth 30

29a. (28a) Shell <25 mm, typically periostracum eroded off with periostracum restricted to narrow strip on shell margin, beak sculpture heavy bars extending down on the disk of the shell, restricted to the upper Cumberland and upper Tennessee River basins .... *Pegias fabula* (Fig. 28AG)

29b. Shell not small, periostracum not eroded or erosion restricted to the umbo, beak sculpture consists of heavy bars, widespread in the Mississippi River Basin .... *Lasmigona* [in part] [*L. costata, L. complanata*] (Fig. 28X)

30a. (28b) Shell with pseudocardinal tooth consisting of a slight swelling or knob, lateral tooth consisting of a rounded ridge, beak cavity copper colored, umbo centrally located .... *Strophitus undulatus* (Fig. 28AR)

30b. Hinge teeth completely absent 31

31a. (30b) Umbonal area of shell projects above the plane of the hinge line 32

31b. Umbo level with hinge line or below the level of the hinge line 33

32a. (31a) Umbo elevated well above the hinge line, shell shape variable, ranging from oval, elliptical to rhomboid, beak sculpture consists of double loopedridges (Fig. 27) with projections on the bottom of the loop .... *Pyganodon* [in part] [*P. grandis*] (Fig. 28AN)

32b. Umbo projecting only slightly above the hinge line, shell elongated, inflated, beak sculpture consists of 3 or 4 fine sharp concentric ridges .... *Anodontoides ferussacianus* (Fig. 28G)

33a. (31b) Shell shape round to circular, compressed, beak sculpture consists of 4–5 pairs of broken irregular double looped ridges .... *Anodonta* [in part] [*A. suborbiculata*] (Fig. 28F)

33b. Shell shape elongate, inflated, dorsal and ventral margins nearly straight and parallel, beak sculpture consists of 5–6 fine irregular concentric ridges .... *Utterbackia* [in part] [*U. imbecillis*] (Fig. 28AW)

34a. (27b) Shell with sculpture on the external shell surface 35

34b. Shell with no plications, undulations, ridges, pustules or nodules 55

35a. (34a) Shell with prominent pustules or nodules and/or undulations, plications or ridges 36

35b. Shell with only prominent, well-defined pustules, knobs or nodules 46

36a. (35a) Shell with undulations, ridges, or plications, no pustules or nodules on the surface 37

36b. Shell with undulations, ridges or plications and pustules on the disk or umbonal area 44

37a. (36a) Shell with undulations, plications, ridges restricted to the posterior slope 38

37b. Shell with undulations, plications, ridges not restricted to the posterior slope 41

38a. (37a) Shell shape rectangular to oval, slablike, compressed, left valve with an accessory dentacle posterior to pseudocardinal teeth .... *Lasmigona* [in part] [*L. costata*](Fig. 28X)

38b. Shell shape more elongate, lacks the accessory dentacle 39

39a. (38b) Shell elongate, inflated with a blue to blue-green nacre color, restricted to the Tennessee and Cumberland River basins .... *Medionidus* [in part] [*M. conradicus*] (Fig. 28AC)

39b. Shell elongate, inflated, nacre not as above 40

40a. (39b) Shell thin, posterior slope steeply angled to moderately angled .... *Alasmidonta* [in part] [*A. marginata*] (Fig. 28D)

40b. Shell thicker, posterior slope rounded, posterior ridge rounded, restricted to the Tennessee and Cumberland River Basins .... *Ptychobranchus* [in part] [*P. subtentum*] (Fig. 28AM)

41a. (37b) Shell shape rectangular, with a distinct posterior ridge, purple nacre, open and shallow beak cavity .... *Plectomerus dombeyanus* (Fig. 28AH)

41b. Shell without a distinct posterior ridge, typically white nacre 42

42a. (41b) Shell thick, compressed, maximum shell length <50 mm, low plications or wrinkles restricted to the posterior slope and posterior ventral margin of the shell, this species restricted to the Tennessee River Basin .... *Lemiox rimosus* (Fig. 28Y)

42b. Shell not as above 43

43a. (42b) Shell variably inflated, shell shape round to quadrate, heavy pseudocardinal teeth, deep compressed beak cavity, sculpture consists of typically three well-developed ridges or undulations running across the shell from anterior of the umbo to the posterior ridge, widespread in the Mississippi River Basin .... *Amblema plicata* (Fig. 28E)

43b. Shell somewhat inflated, shell shape quadrate, or approaching oval, pseudocardinal teeth curved, heavy sculpturing on the posterior half of the shell, species is restricted to the Red River Drainage in Oklahoma and the Ouachita River Drainage in Arkansas .... *Arkansia wheeleri* (Fig. 28I)

44a. (36b) Shell elongate with a well-developed diagonal posterior ridge .... *Tritogonia verrucosa* (Fig. 28AT)

44b. Shell shape not elongate, lacking well-defined posterior ridge 45

45a. (44b) Well-developed plications running across the shell and pustules on the disk and umbonal area .... *Megalonaias nervosa* (Fig. 28AD)

45b. Plications not well developed and wavy, not straight, plications on posterior slope, two rows of pustules or knobs on umbo .... *Arcidens confragosa* (Fig. 28H)

46a. (35b) Pustules distributed across most of the shell 47

46b. Pustules restricted to a single or double row 50

47a. (46a) Pustules covering most of the shell, purple nacre, beak cavity deep and compressed .... *Cyclonaias tuberculata* (Fig. 28J)

47b. Pustules covering most of the shell, nacre color white to pink not purple 48

48a. (47b) Pustules consistently small, with a shallow sulcus down the center of the disk, nacre white .... *Cyprogenia* [in part] [*C. stegaria* in the Ohio, Tennessee and Cumberland rivers; *C. aberti* in Arkansas, Oklahoma] (Fig. 28K)

48b. Pustules generally larger, and often of variable size and shape 49

49a. (48b) Disk of the shell usually with a broad green stripe running down the disk of the shell with a variable number of pustules, varies from a very few to covering most of the shell, nacre color mostly white, beak cavity deep and open not compressed ....*Quadrula* [in part] [*Q. pustulosa*] (Fig. 28AO)

49b. Disk of shell without a broad green stripe, consistently covered with a large number of pustules, nacre color often pink, salmon or white, beak cavity open and shallow..................*Plethobasus* [in part] [*P. cooperianus*] (Fig. 28AI)

50a. (46b) Pustules forming two distinct rows..................51

50b. Pustules or nodules forming a single row..................52

51a. (50a) Two rows of pustules with one row on the posterior ridge, shell thick ......*Quadrula* [in part], [*Q apiculata,* this species has small pustules in the sulcus between the two rows of pustules; *Q. nodulata*, this species lacks a sulcus between the rows of pustules and may only have very few pustules; *Q. quadrula*; *Q. fragosa,* this species has a well-developed wing posterior to the umbo] (Fig. 28AO)

51b. Radiating row of knobs, shell inflated, thin to relatively thick, umbo high and full, pseudocardinal teeth compressed, not curved; beak sculpture consists of irregular nodules forming two loops which continue down and across most of the shell in two radiating rows, [mostly in small and juvenile shells] pustules fuse, become elongate and form plications, plications on the posterior slope of adults..................*Arcidens confragosus* (Fig. 28H)

52a. (50b) The single row of pustules or knobs restricted to the posterior ridge..................*Quadrula* [in part] [*Q. cylindrica*, with pustules on the posterior ridge while *Q. cylindrica strigillata* of the upper Tennessee River Basin has small pustules over most of the shell as well as pustules on the posterior ridge.] (Fig. 28AO)

52b. The single row of pustules or knobs forming a single row down the center of the disk..................53

53a. (52b) Pustules large, located in a row down the center of the shell and limited to about three pustules on each valve ....*Obliquaria reflexa* (Fig. 28AE)

53b. Pustules more numerous, but still in a centrally located row 54

54a. (53b) Pustules usually wider than tall, shell periostracum a waxy yellow color ....*Plethobasus* [in part] [*P. cicatricosus, P. cyphyus*] (Fig. 28AI)

54b. Pustules rather small, running down the centerline of the disk, but interrupted by a well-defined, raised ridge running around the shell like a growth line, restricted to the Tennessee and Cumberland River drainages ....*Dromus dromas* (Fig. 28N)

55a. (34b) Shell with a well-developed dorsal wing projecting above the hinge line, the wing usually posterior of the umbo, but some species may also have an anterior wing [some older shells, such as in *Leptodea fragilis*, maybe lacking the dorsal wing]..................56

55b. Shell without a well-developed dorsal wing..................58

56a. (55a) Hinge teeth thin pseudocardinal teeth compressed, thin and not projecting perpendicular to the axis of the hinge line ....*Leptodea* [in part] [*L. fragilis, L. leptodon*] (Fig. 28Z)

56b. Well-developed dorsal wing..................57

57a. (56b) White nacre, thick-shelled ....*Lasmigona* [in part] [*L. complanata*] (Fig. 28X)

57b. Purple nacre, thin-to-thick shelled, may be inflated..................*Potamilus* [in part] [*P. alatus, P. purpuratus*] (Fig. 28AL)

58a. (55b) Shell shape round 59

58b. Shell shape not round 68

59a. (58a) Shell compressed, shell shape round 60

59b. Shell not compressed, but more inflated, shell shape round, oval..................61

60a. (59a) Shell compressed, shape nearly circular, without lateral teeth ....*Lasmigona* [in part] [*L. complanata*] (Fig. 28X)

60b. Shell compressed, with lateral teeth, steep posterior slope, shell thick with well-developed hinge teeth, periostracum light yellow with interrupted rows of green chevrons ....*Ellipsaria lineolata* (Fig. 28O)

61a. (59b) Umbo central or nearly central in the dorsal margin..................62

61b. Umbo anterior of the center of the shell 65

62a. (61a) Broad green ray extending from the umbo down only a short distance onto the disk of the shell ....*Quadrula* [in part] [*Q. pustulosa* without pustules] (Fig. 28AO)

62b. No broad green ray 63

63a. (62b) Shell with a raised ridge one quarter of the distance from the umbo running around the shell parallel with the growth lines. Restricted to the Tennessee and Cumberland River basins ....*Dromus dromas* (Fig. 28N)

63b. Shell without a raised ridge 64

64a. (63b) Shell surface smooth, without a broad central sulcus .... *Obovaria* [in part] [*O. jacksoniana*, from the Mississippi River embayment shell somewhat more oval than round; *O. retusa*, nacre color white; nacre color purple inside the pallial line, white outside of the pallial line; *O. subrotunda*, shell round, thick shelled, nacre typically white but some populations with deep purple.] (Fig. 28AF)

64b. Shell with broad shallow-to-pronounced sulcus, fine green rays, shallow beak cavity, well-developed hinge teeth, female shells with a swollen, extended or expanded portion of the posterior slope and posterior ventral margin of the shell............................ *Epioblasma* [in part] [*E. capsaeformis*, *E. florentina florentina*, *E. florentina walkeri*, *E. lewisi*, *E. flexuosa*, *E. biemarginata*, *E. stewardsoni*, *E. propinqua*, *E. torulosa torulosa*, *E. torulosa gubernaculum*, *E. torulosa biloba, E. sampsoni*] (Fig. 28R)

65a. (61b) Beak cavity deep compressed .... *Fusconaia* [in part][*F. subrotunda*, *F. ebena*] (Fig. 28S)

65b. Beak cavity shallow and open..66

66a. (65b) Restricted to the Tennessee River Basin, shell with green rays at least on the umbo area with a thick shell, heavy lateral teeth, very shallow sulcus or missing, shallow-to-no beak cavity ....*Lexingtonia dolabelloides* (Fig. 28AA)

66b. Not restricted to the Tennessee River Basin, but widespread 67

67a. (66b) Long axis of the main pseudocardinal teeth parallel with the lateral teeth, umbo anterior ....*Obovaria* [in part, *O. olivaria*] (Fig. 28AF)

67b. Teeth not as above, shell wedge shaped to triangular or approaching square with or without a central sulcus, nacre white to deep pink ....*Pleurobema* [in part] [*P. rubrum*, *P. sintoxia*] (Fig. 28AJ)

68a. (58b) Shell shape not oval...69

68b. Shell shape oval to oblong..85

69a. (68a) Shell shape rectangular, triangular, square with sulcus...70

69b. Shell shape elongate, two-to-four times longer than high...75

70a. (69a) Beak cavity deep...71

70b. Beak cavity shallow or broad and open 72

71a. (70a) Shell shape rectangular, with green rays on the umbo ....*Fusconaia* [in part] [*F. flava*, *F. cuneolus*, *F. cor*, *F. ozarkensis*, *F. barnesiana* has a shallow beak cavity but belongs with this group] (Fig. 28S)

71b. Shell shape triangular to broadly triangular.......................................*Pleurobema* [in part] [*P. cordatum*, *P. plenum*] (Fig. 28AJ)

72a. (70b) Posterior slope not very sharp or rounded, covered with lines of green chevrons ....*Truncilla* [in part; *T. truncata*, *T. donaciformis*] (Fig. 28AU)

72b. Posterior slope very sharp or steep 73

73a. (72b) Shell thin, inflated, abruptly truncate posteriorly, posterior slope lighter colored than the rest of the shell .... *Alasmidonta* [in part] [*A. marginata*; shell thin, shell shape elongate, posterior slope not a steep sharp angle, headwaters of the Little Tennessee River in western North Carolina, *A. raveneliana*; streams of the upper Cumberland River Basin, *A. atropurpurea*] (Fig. 28D)

73b. Shell thin to thick, inflated posterior slope steep shell thick, shell shape rectangular..74

74a. (73b) Females with posterior slope inflated into a marsupial swelling, males evenly rounded posterior ridge and not inflated posterior ridge, nacre white ....*Epioblasma* [in part] [*E. triquetra* wide spread, *E. arcaeformis* extinct and formerly restricted to the Tennessee and Cumberland River basins.] (Fig. 28R)

74b. Posterior margin not modified into a marsupial swelling, nacre purple.........................*Elliptio* [in part] [*E. crassidens*] (Fig. 28P)

75a. (69b). Thin shelled..76

75b. Thick shelled 79

76a. (75a) Shell compressed, very thin, periostracum rayed, nacre iridescent except for a purple wash in the open beak cavity ....*Hemistena lata* (Fig. 28V)

76b. Shell inflated...77

77a. (76b) Hinge teeth thin, periostracum rayless, adult shell length <50 mm..*Simpsonaias ambigua* (Fig. 28AQ)

77b. Periostracum usually rayed, adult shell length >50 mm 78

78a. (77b) Shell hinge thin, usually heavily rayed...*Villosa* [in part] [*V. iris* complex, *V. taeniata*] (Fig. 28AY)

78b. Shell yellow, plain or rayed.......................*Lampsilis* [in part] [*L. teres*, *L. rafinesqueana*, *L. siliquoidea*, *L. virescens*] (Fig. 28W)

79a. (75b) Shell compressed 80

79b. Shell inflated...82

80a. (79a) Adult shell length <40 mm, rayed, compressed, widespread, ... *Villosa* [in part] [*V. fabalis*] (Fig. 28AY)

80b. Adult shell length >40 mm 81

81a. (80b) Shell hinge massive, curved, nacre white, periostracum with interrupted green rays .... *Ptychobranchus* [in part] [*P. fasciolaris, P. occidentalis*] (Fig. 28AM)

81b. Hinge lighter, straight, nacre purple to white, periostracum typically unrayed .... *Elliptio* [in part] [*E. dilatata*] (Fig. 28P)

82a. (79b). Adult shell length <55mm

82b. Adult shell length >55mm

83a. (82a). Periostracum olive green with numerous rays *Villosa* [in part] [*V. trabalis,* restricted to the Tennessee River Basin above Muscle Shoals and the upper Cumberland River basin, inflated, white nacre, *V. perpurpurea*, upper Tennessee River Basin, inflated, purple nacre] (Fig. 28AY)

83b. Periostracum ranges from greenish to black without rays *Toxolasma* [in part] [T. lividus, purple nacre, *T. parvus* white nacre, *T. texasiensis* white to salmon nacre] (Fig. 28AS)

84a. (82b) Shell inflated, periostracum rayless, posterior shell margin comes to a blunt point ventrally, coarse concentric beak sculpture opening posteriorly .... *Uniomerus* [*U. tetralasmus, U. declivus*] (Fig. 28AV)

84b. Shell inflated, periostracum rayed as a juvenile, becoming dark brown to black as an adult, posterior shell margin comes to a point in middle of posterior margin, nacre white to white with purple wash in umbo area .... *Ligumia* [*L. nasuta, L. recta, L. subrostrata*] (Fig. 28AB)

85a. (68b) Shell shape oval 86

85b. Shell shape oblong 87

86a. (85a) Shell shape oval, inflated, rayed, thin to thick shell, posterior ridge rounded or sharp, nacre variously white to pink or salmon .... *Lampsilis* [in part] [*L. cardium, L. fasciola, L. ovata, L. reeviana,* L. *rafinesqueana, L. higginsi, L. abrupta*] (Fig. 28W)

86b. Shell shape elongate oval .... *Villosa* [in part] [*V. taeniata*, interrupted broad green rays, restricted to the Tennessee and Cumberland River basins; in the following three species males are oval while female shells have a truncate posterior margin, *V. vanuxemensis*, purple nacre, restricted to the Tennessee River Basin and the central Cumberland River Basin, *V. ortmanni*, restricted to Kentucky, *V. lienosa*, variously colored nacre] (Fig. 28AY)

87a. (85b) Shell shape oblong, inflated to globose, unrayed .... *Potamilus* [in part] [S-shaped, very compressed hinge line, very inflated and thin shelled *P. capax*; inflated, black shiny periostracum and purple nacre *P. purpuratus*] (Fig. 28AL)

87b. Shell shape oblong, inflated but not globose 88

88a. (87b) Adult shell length <70 mm, shell shape elliptic to oblong, thick shelled, nacre white with a tinge of salmon to coppery color in center of the shell, shallow beak cavity, rayed, with those rays on the posterior slope thick and wavy .... *Venustaconcha ellipsiformis* (Fig. 28AX)

88b. Adult shell length >70 mm, shell shape oblong to oval, thin to thick shelled, nacre white to iridescent, beak cavity broad and open, moderately deep, usually rayed but rays on posterior slope not thin or wavy .... *Actinonaias* [*A. pectorosa*, thin shelled, restricted to the Tennessee and Cumberland River basins; *A. ligamentina*, widespread and typically thick shelled.] (Fig. 28C)

Gulf Coast

89a (3b) Shell with lateral muscle scars in the interior center of the valve .... *Margaritifera* [in part] [*M. hembeli, M. marrianae*] (Fig. 28B)

89b. Shell without lateral muscle scars in the interior center of the valve 90

90a. (89b) Shell with pseudocardinal and lateral teeth greatly reduced or absent 91

90b. Shell with both pseudocardinal and lateral teeth present 95

91a. (90a) Pseudocardinal and lateral teeth present, but reduced 92

91b. Pseudocardinal and lateral teeth completely absent 93

92a. (91a) Shell with pseudocardinal tooth consisting of a slight swelling or knob, lateral tooth consisting of a rounded ridge, beak cavity copper colored, umbo centrally located .... *Strophitus* [in part] [*S. subvexus, S. connasaugaensis*] (Fig. 28AR)

92b. Shell with pseudocardinal tooth thin, reduced and compressed, umbo projecting only slightly above the hinge line, shell elongated, inflated, beak sculpture consists of 3–4 fine, sharp, concentric ridges .... *Anodontoides* [in part] [*A. radiatus*] (Fig. 28G)

93a. (91b) Umbo elevated well above the hinge line, shell shape variable, ranging from oval, elliptic or rhomboid, beak sculpture consists of double looped with projections on the bottom of the loop .... *Pyganodon* [in part] [*P. grandis*] (Fig. 28AN)

93b. Umbo level ridges with hinge line or below the level of the hinge line 94

94a. (93b) Shell shape round to circular, compressed, beak sculpture consists of 4–5 pairs of broken irregular double looped ridges.. *Anodonta* [in part] [*A. suborbiculata, A. heardi*] (Fig. 28F)

94b. Shell shape elongate, inflated dorsal and ventral margins nearly straight and parallel, beak sculpture consists of 5–6 fine irregular concentric ridges .... *Utterbackia* [in part] [*U. imbecillis, U. peggyae, U. peninsularis*] (Fig. 28AW)

95a. (90b) Shell with sculpture on the external shell surface 96

95b. Shell without plications, undulations, ridges, pustules or nodules 109

96a. (95a) Shell with pustules or nodules and/or undulations, plications or ridges 97

96b. Shell with only pustules, knobs or nodules 105

97a. (96a) Shell with undulations, ridges, or plications, but not pustules or nodules on the surface 98

97b. Shell with undulations, ridges or plications and pustules on the disc or umbonal area 103

98a. (97a) Shell surface with undulations, plications, ridges restricted to the posterior slope 99

98b. Shell surface with undulations, plications, ridges not restricted to the posterior slope 101

99a. (98a) Shell shape round to oval, slablike, compressed, left valve with an accessory dentacle posterior to pseudocardinal teeth plications occur on dorsal wing .... *Lasmigona* [in part] [*L. complanata*] (Fig. 28X)

99b. Shell shape more elongate, lacks the accessory dentacle 100

100a. (99b) Shell inflated with a blue to blue-green nacre color .... *Medionidus* [in part] [*M. accutissimus, M. mcglameriae, M. parvulus, M. penicillatus, M. simpsonianus, M. walkeri*] (Fig. 28AC)

100b. Shell inflated, nacre not as above, shell thin, posterior slope steeply angled to moderately angled .... *Alasmidonta* [in part] [*A. triangulata, A. wrightiana*] (Fig. 28D)

101a. (98b) Shell shape rectangular, with a distinct posterior ridge, purple nacre, open and shallow beak cavity .... *Plectomerus dombeyanus* (Fig. 28AH)

101b. Shell shape round to rectangular, without a distinct posterior ridge 102

102a. (101b) Shell thick, compressed, low plications or wrinkles on posterior slope and undulations and ridges on the disk of the shell, nacre with a central white area and a peripheral purple band, restricted to the Apalachicola and Ochlockonee River basins .... *Elliptoideus sloatianus* (Fig. 28Q)

102b. Shell variably inflated, shell shape round to quadrate, heavy pseudocardinal teeth, deep compressed beak cavity, sculpture consists of three to five well-developed ridges or undulations running across the shell from anterior of the umbo to the posterior ridge, nacre color white to iridescent .... *Amblema* [in part] [*A. plicata, A. neislerii, A. elliotti*] (Fig. 28E)

103a. (97b) Shell shape elongate, with a well-developed diagonal posterior ridge .... *Tritogonia verrucosa* (Fig. 28AT)

103b. Shell shape not elongate, lacking well-defined posterior ridge 104

104a (103b). Shell shape round to quadrate, lacking well-defined posterior ridge, has well developed plications running across the shell and pustules on the disk or umbonal area .... *Megalonaias nervosa* (Fig. 28AD)

104b. Plications not well developed and wavy, not straight, plications on posterior slope, two rows of pustules or knobs on umbo .... *Arcidens confragosa* (Fig. 28H)

105a. (96b) Pustules distributed across most of the shell 106

105b. Pustules restricted to a single or double row 107

106a. (105a). Adult shell length <60 mm, sculpture fine, shell oval to elongate oval, pustules and chevrons .... *Quincuncina* [*Q. infucata, Q. burkei* restricted to the Choctawhatchee River Basin; *Q. mitchelli* from the Rio Grand, Guadalupe, Colarado and Brazos rivers in Texas] (Fig. 28AP)

106b. Adult shell length not restricted to <60 mm, pustules covering most of the shell, nacre color white to pink, not purple, with a variable number of pustules, from a very few to pustules covering most of the shell, nacre color mostly white, beak cavity deep and open, not compressed .... *Quadrula* [in part] [*Q. petrina, Q. houstonensis, Q. couchiana, Q. aurea, Q. asperata*] (Fig. 28AO)

107a. (105b) Pustules or nodules large, forming a single row down the center of the disk, and limited to about three pustules on each valve .... *Obliquaria reflexa* (Fig. 28AE)

107b. Pustules forming two distinct rows especially on the umbonal area 108

108a. (107b) Two rows of pustules, with one row on the posterior ridge, shell thick .... *Quadrula* [in part] [*Q apiculata,* has small pustules in the sulcus between the two rows of pustules; *Q. nodulata* lacks a sulcus between the rows of pustules and may only have very few pustules; *Q. quadrula, Q. rumphiana*] (Fig. 28AO)

108b. Radiating rows of knobs, shell inflated, thin to relatively thick, umbo high and full, pseudocardinal teeth compressed, not curved, beak sculpture consists of irregular nodules forming two loops which continue down and across most of the shell in two radiating rows .... *Arcidens confragosus* (Fig. 28H)

109a. (95b) Shell with a well-developed dorsal wing projecting above the hinge line, the wing usually posterior of the umbo [see Fig. 30] but some species may also have an anterior wing [some older shells such as in *Leptodea fragilis* may be lacking the dorsal wing] 110

109b. *Shell without a well-developed dorsal wing* 111

110a. (109a) *Pseudocardinal teeth moderately heavy, projecting perpendicular to the axis of the hinge line, purple nacre thin to thick shelled, may be inflated* .... *Potamilus* [*P. inflatus, P. purpuratus*] (Fig. 28AL)

110b. Pseudocardinal teeth compressed, thin and not projecting perpendicular to the axis of the hinge line .... *Leptodea* [*L. fragilis, L. amphichaenus*] (Fig. 28Z)

111a. (109b) Shell shape round 112

111b. Shell shape not round 117

112a. (111a) Shell compressed, shell shape round, with lateral teeth, steep posterior slope, shell thick with well-developed teeth, periostracum light yellow with interrupted rows of green chevrons .... *Ellipsaria lineolata* (Fig. 28O)

112b. Shell not compressed, round to oval 113

113a. (112b) Umbo central or nearly central in the dorsal margin 114

113b. Umbo anterior of the center of the dorsal margin of the shell 116

114a. (113a) Broad green ray extending around the umbo down only a short distance onto the disk of the shell .... *Quadrula* [in part] [*Q. asperata* without pustules] (Fig. 28AO)

114b. No broad green ray 115

115a. (114b) Shell surface smooth, without a broad central sulcus .... *Obovaria* [in part] [*O. jacksoniana, O. unicolor*] (Fig. 28AF)

115b. Shell with broad, shallow to pronounced sulcus, fine green rays, shallow beak cavity, well developed hinge teeth, female shells with a swollen, extended or expanded portion of the posterior slope and posterior ventral margin of the shell .... *Epioblasma* [in part] [*E. penita, E. metastriata, E. othcaloogensis*] (Fig. 28R)

116a. (113b) Beak cavity deep compressed .... *Fusconaia* [in part][*F. ebena, F. succissa, F. escambia*] (Fig. 28S)

116b. Beak cavity shallow and open, shell shape wedge shaped to triangular or approaching square with or without a central sulcus, nacre white to deep pink .... *Pleurobema* [in part] [*P. marshalli*] (Fig. 28AJ)

117a. (111b) Shell shape not oval 118

117b. Shell shape oval 130

118a. (117a) Shell shape triangular, square to rectangular, with sulcus 119

118b. Shell shape elongate, tapered posteriorly, two to four times longer than high 123

119a. (118a) Beak cavity deep 120

119b. Beak cavity shallow or broad and open 121

120a. (119a) Shell shape rectangular with green rays on the umbo .... *Fusconaia* [in part] [*F. cerina*] (Fig. 28S)

120b. Shell shape triangular to broadly triangular .... *Pleurobema* [in part] [*P. taitianum*] (Fig. 28AJ)

121a. (119b) Posterior slope rounded, not very steep, covered with lines of green chevrons .... *Truncilla* [in part] [*T. cognata, T. macrodon, T. truncata, T. donaciformis*] (Fig. 28AU)

121b. Posterior ridge very sharp, posterior slope steep 122

122a. (121b) Shell thin, inflated, abruptly truncate posteriorly, posterior slope lighter colored than the rest of the shell .... *Alasmidonta* [in part] [*A. triangulata*] (Fig. 28D)

122b. Shell thin to thick, inflated posterior slope steep shell thick, shell shape rectangular, posterior margin not modified into a marsupial swelling, often tapering posteriorly, nacre varies pale salmon to purple .... *Elliptio* [in part] [*E. crassidens*] (Fig. 28P)

123a. (118b). Thin shelled 124

123b. Thick shelled 126

124a. (123a) Shell compressed, shell shape elongate to almost rectangular, very thin, periostracum rayed, nacre iridescent, white to dull purple, beak cavity shallow, lateral teeth long and curved to straight, pseudocardinal very small, periostracum brown with faint rays as a juvenile, restricted to the Rio Grande Basin, Texas .... *Popenaias popei* (Fig. 28AK)

124b. Shell inflated 125

125a. (124b) Hinge thin, periostracum usually heavily rayed .... *Villosa* [in part] [*V. nebulosa* complex, *V. villosa, V. vibex*] (Fig. 28AY)

125b. Hinge not thin, periostracum yellow, plain or rayed .... *Lampsilis* [in part] [*L. teres, L. straminea, L. hydiana, L. bracteata*] (Fig. 28W)

126a. (123b) Shell compressed 127

126b. Shell inflated 128

127a. (126a) Hinge massive, curved, nacre white, periostracum with interrupted green rays .... *Ptychobranchus* [in part] [*P. greeni*] (Fig. 28AM)

127b. Hinge thinner, straight, nacre purple to white, periostracum typically unrayed .... *Elliptio* [in part] [*E. arctata, E. arca*] (Fig. 28P)

128a. (126b) Rayed as a juvenile becoming dark brown to black as an adult, tapered to a point in middle of posterior margin, nacre white to white with purple wash in umbo area, *Ligumia* [*L. recta, L. subrostrata*] (Fig. 28AB)

128b. Shell unrayed

129a. (128b) Adult shell length >50mm, bluntly pointed posterior ventrally, coarse concentric beak sculpture seeming to radiate from a single point, *Uniomerus* [in part] [*U. tetralasmus, U. declivus*] (Fig. 28AV)

129b. Adult shell length 50 mm, bluntly rounded or squared off posteriorly, concentric beak sculpture in form of bars or ridges, angled up posteriorly *Toxolasma* [in part] [T. parvus, T. texasiensis] (Fig. 28AS)

130a. (117b) Shell shape oval 131

130b. Shell shape oblong 133

131a. (130a) Posterior half of the pseudocardinal tooth divided into several parallel vertical ridges (Fig. 29), thick shelled, inflated, periostracum dark brown to dark olive .... *Glebula rotundata* (Fig. 28T)

131b. Pseudocardinal tooth not divided into vertical ridges as above 132

132a. (131b) Shell shape oval, inflated, rayed, thin-to-thick shell, posterior ridge rounded or sharp, nacre variously white to pink or salmon .... *Lampsilis* [in part] [*L. ornata, L. binominata*] (Fig. 28W)

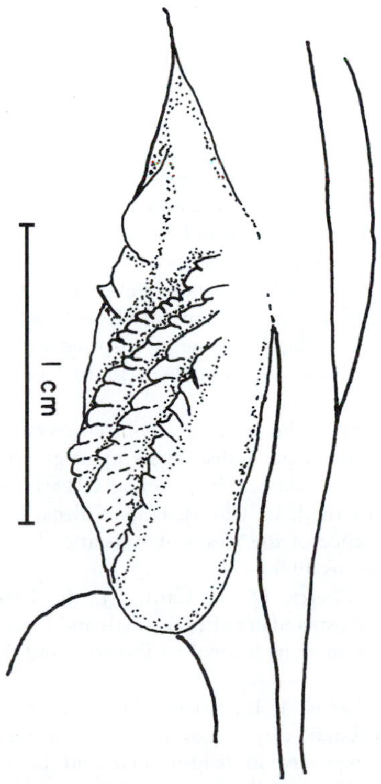

FIGURE 29 Structure of the posterior pseudocardinal teeth of the right valve of *Gelbula rotundata*. Note that the posterior pseudocardinal teeth are deeply divided into parallel, vertical, plicate lamellae, a tooth arrangement uniquely characteristic of this species. (Redrawn from Burch, 1975b.)

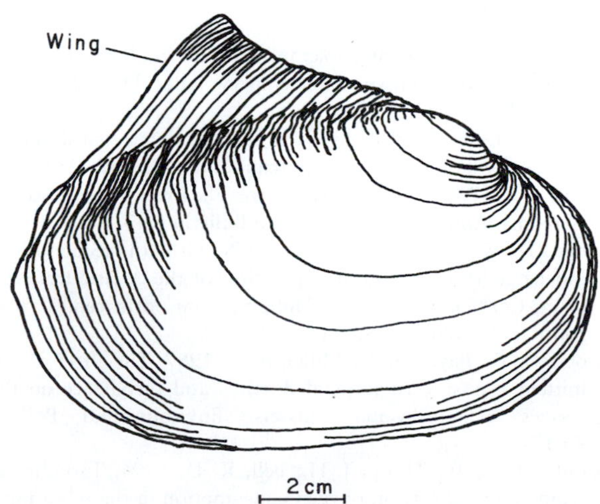

FIGURE 30 Right shell valve of the unionid, *Potamilus alatus*, showing the thin, extensive dorsal projection of the shell posterior to the umbos forming a "wing" whose presence or absence is a diagnostic characteristic valuable for identification of a number of unionid species. (Redrawn from Burch, 1975b.)

132b. Shell shape elongate, oval .... *Villosa* [in the following two species male shells are oval while female shells have a truncate posterior margin, *V. vanuxemensis umbrans*, purple nacre, restricted to the upper Coosa River Basin, V. *lienosa*, variously colored nacre] (Fig. 28AY)

133a. (130b) Shell shape oblong, inflated to globose, unrayed .... *Potamilus* [in part] [*P. amphichaenus*, P. *purpuratus*] (Fig. 28AL)

133b. Shell shape oblong, inflated, but not globose 134

134a. (133b) Shell shape elongate oval, shell relatively thin to thick, compressed, pseudocardinal teeth compressed, thin, lateral teeth long, beak cavity shallow, periostracum tan to brown, with dark brown rays posteriorly, restricted to the Rio Grande Basin in Texas .... *Disconaias* [in part] [*D. salinasensis, D. conchos*] (Fig. 28M)

134b. Shell shape oval, moderately thick shell, inflated, beak cavity relatively deep, periostracum dull/shiny, red-brown to brown, juveniles with faint rays, Brazos River to Rio Grande Basin, Texas .... *Cyrtonaias tampicoensis* (Fig. 28L)

ACKNOWLEDGMENTS

Alan P. Covich and Caryn C. Vaugh critically reviewed the chapter and made many valuable suggestions for its improvement. Thomas H. Dietz, Daniel Hornbach, Carl M. Way, and Harold Silverman provided the scanning and transmission electron micrographs of bivalve gill ciliation presented in Figure 21. Paul W. Parmalee, Caryn C. Vaughn, Dan Spooner, Valeria Rice and Jeff Garner kindly reviewed and commented on the unionoidean key. Jonathon A. Raine and Chad Rubins provided technical support with the preparation of the illustrations used in Figure 28. Cindy Bogan provided suggestions and encouragement during the drafting of the unionoidean key. Mrs. W. T. Edmondson kindly granted her permission for us to use the figures from her late husband's second edition of Ward and Wipple (Edmondson, 1959).

LITERATURE CITED

Aarset, A. V. 1982. Freezing tolerance in intertidal invertebrates (a review). Comparative Biochemistry and Physiology *A*73: 571–580.

Adam, M. E. 1986. The Nile bivalves: how do they avoid silting during the flood? Journal of Molluscan Studies 52:248–252.

Ahlstedt, S. A. 1983. The Mollusca of the Elk River in Tennessee and Alabama. American Malacological Bulletin 1:43–50.

Aldridge, D. W., McMahon, R. F. 1978. Growth, fecundity, and bioenergetics in a natural population of the Asiatic freshwater clam, *Corbicula manilensis* Philippi, from north central Texas. Journal of Molluscan Studies 44:49–70.

Aldridge, D. W., Payne, B. S., Miller, A. C. 1987. The effect of intermittent exposure to suspended solids and turbulence on three species of freshwater mussels. Environmental Pollution 45:17–28.

Alexander, J. E., Jr., Thorp, J. H., Fell, R. D. 1994. Turbidity and temperature effects on oxygen consumption in the zebra mussel (*Dreissena polymorpha*). Canadian Journal of Fisheries and Aquatic Science 51:179–184.

Alimov, A. F. 1975. The rate of metabolism of freshwater bivalve molluscs. Soviet Journal of Ecology 6:6–13.

Allen, A. J. 1985. Recent Bivalva: their form and evolution. *in:* Trueman, E. R., Clarke, M. R., Eds. The Mollusca, Vol. 10, Evolution. Academic Press, New York, pp. 337–403.

Allen, Y. C., Thompson, B. A., Ramcharan, C. W. 1999. Growth and mortality rates of the zebra mussel, *Dreissena polymorpha*, in the lower Mississippi River. Canadian Journal of Fisheries and Aquatic Science 56:748–759.

Anderson, R. M., Layzer, J. B., Gordon, M. E. 1991. Recent catastrophic decline of mussels (Bivalvia: Unionidae) in the Little South Fork Cumberland River, Kentucky. Brimleyana 17:1–8.

Annis, C. G., Belanger, T. V. 1986. *Corbicula manilensis,* potential bio–indicator of lead and copper pollution. Florida Scientist 49:30.

Araujo, R., Remón, J. M., Moreno, D., Ramos, M. A. 1995. Relaxing techniques for freshwater molluscs: trials for evaluation of different methods. Malacologia 36:29–41.

Arnott, D. L., Vanni, M. J. 1996. Nitrogen and phosphorous recycling by the zebra mussel (*Dreissena polymorpha*) in the Western Basin of Lake Erie. Canadian Journal of Fisheries and Aquatic Science 53:646–659.

Arter, H. E. 1989. Effect of eutrophication on species composition and growth of freshwater mussels (Mollusca, Unionidae) in Lake Hallwil (Aargau, Switzerland). Aquatic Sciences 51:87–99.

Baily, R. C., 1989. Habitat selection by a freshwater mussel: an experimental test. Malacologia 31:205–210.

Baily, R. C., Green, R. H. 1989. Spatial and temporal variation in a population of freshwater mussels in Shell Lake, N.W.T. Canadian Journal of Fisheries and Aquatic Science 46:1392–1395.

Bauer, G. 1983. Age structure, age specific mortality rates and population trend of the freshwater pearl mussel (*Margaritifera margaritifera*) in north Bavaria. Archiv für Hydrobiologie 98: 523–532.

Bauer, G. 1994. The adaptive value of offspring size among freshwater mussels (Bivalvia; Unionidae). Journal of Animal Ecology 63:933–944.

Belanger, S. E. 1991. The effect of dissolved oxygen, sediment and sewage treatment plant discharges upon growth, survival and density of Asiatic clams. Hydrobiologia 218:113–126.

Belanger, S. E., Farris, J. L., Cherry, D. S., Cairns, J., Jr., 1985. Sediment preference of the freshwater Asiatic clam, *Corbicula fluminea.* Nautilus 99:66–73.

Belanger, S. E., Cherry, D. S., Cairns, J., Jr. 1986a. Uptake of chrysotile asbestos fibers alters growth and reproduction of Asiatic clams. Canadian Journal of Fisheries and Aquatic Sciences 43:43–52.

Belanger, S. E., Farris, J. L., Cherry, D. S., Cairns, J., Jr. 1986b. Growth of Asiatic clams (*Corbicula* sp.) during and after long-term zinc exposure in field-located and laboratory artificial streams. Archives of Environmental Contamination and Toxicology 15:427–434.

Belanger, S. E., Cherry, D. S., Cairns, J., Jr., McGuire, M. J. 1987. Using Asiatic clams as a biomonitor for chrysotile asbestos in public water supplies. Journal of the American Water Works Association 79:69–74.

Beninger, P. G., Lynn, J. W., Dietz, T. H., Silverman, H. 1997. Mucociliary transport in living tissue: two-layer model confirmed in the mussel, *Mytilus edulis* L. Biological Bulletin (Woods Hole, Mass.) 193:4–7.

Bisbee, G. D. 1984. Ingestion of phytoplankton by two species of freshwater mussels, the black sandshell, *Ligumia recta*, and the three ridger, *Amblema plicata*, from the Wisconsin River in Oneida County, Wisconsin. Bios 58:219–225.

Bishop, S. H., Ellis, L. L., Burcham, J. M. 1983. Amino acid metabolism in molluscs. *in:* Hochachka, P. W., Ed. The Mollusca. Vol. 1: Metabolic biochemistry and molecular biomechanics. Academic Press, New York, pp. 243–327.

Bitterman, A. M., Hunter, R. D., Haas, R. C. 1994. Allometry of shell growth of caged and uncaged mussels (*Dreissena polymorpha*) in Lake St. Clair. American Malacological Bulletin 11:41–49.

Blalock, H. N., Sickel, J. B. 1996. Changes in mussel (Bivalvia: Unionidae) fauna within the Kentucky portion of Lake Barkley since impoundment of the lower Cumberland River. American Malacological Bulletin 13:111–116.

Blay, J., Jr. 1990. Fluctuations in some hydrological factors and the condition index of *Aspatharia sinuata* (Bivalvia, Unionacea) in a small Nigerian Reservoir. Archiv für Hydrobiologie 117:357–363.

Blye, R. W., Ettinger, W. S., Nebane, W. N. 1985. Status of Asiatic clam (*Corbicula fluminea*) in south eastern Pennsylvania-1984. Proceedings of the Pennsylvania Academy of Science 59:74.

Bogan, A. E. 1990. Stability of recent unionid (Mollusca: Bivalvia) communities over the past 6000 years. *in:* Miller, W., III, Ed. Paleocommunity temporal dynamics: the long-term development of multispecies assemblages. The Paleontological Society, Special Publication No. 5, Washington, DC, pp. 112–136.

Bogan, A. E. 1993. Freshwater bivalve extinctions: search for a cause. American Zologist 33:599–609.

Bogan, A. E. 1997. The silent extinction. American Paleontologist 15:2–4.

Bogan, A. E. 1998. Freshwater molluscan conservation in North America: problems and practices. *in:* Seddon, M. B., Holmes, A. M., Eds. Molluscan conservation: a strategy for the 21st Century. I. J. Killeen, Journal of Conchology, Special Publication No. 2, pp. 223–230.

Bogan, A. E., Woodward, F. R. 1992. A review of the higher classification of Unionoida (Mollusca: Bivalvia). Unitas Malacologica Abstracts of the Eleventh International Malacological Congress, Siena, Italy, pp. 17–18.

Bonaventura, C., Bonaventura, J. 1983. Respiratory pigments: structure and function. *in:* Hochachka, P. W., Ed. The Mollusca. Vol. 2: Environmental chemistry and physiology. Academic Press, New York, pp. 1–50.

Borcherding, J. 1992. Another early warning system for the detection of toxic discharges in aquatic environments based on valve movements of the freshwater mussel, *Dreissena polymorpha*. Limnologie Aktuell 4:127–146.

Borcherding, J., Volpers, M. 1994. The "*Dreissena*-monitor"—first results on the application of this biological early warning system in the continuous monitoring of water quality. Water Science Technology 29:199–201.

Botts, P. S, Patterson, B. A., Schloesser, D. W. 1996. Zebra mussel effects on benthic invertebrates: physical or biotic? Journal of the North American Benthological Society 15:179–184.

Britton, J. C., Morton, B. 1986. Polymorphism in *Corbicula Fluminea* (Bivalvia: Corbiculidae) from North America. Malacological Review 19:1–43.

Brown, K. M., Curole, J. P. 1997. Longitudinal changes in the mussels of the Amite River: endangered species, effects of gravel mining and shell morphology. *in:* Cummings, K. S., Buchanan, A. C., Mayer, C. A., Naimo, T. J., Eds. Conservation and management of freshwater mussels II: Initiatives for the future, Proceedings of a Symposium, 1995, Upper Mississippi River Conservation Committee, Rock Island, IL, pp. 236–246.

Buddensiek, V., Engel, H., Fleischauer-Rössing, S., Wächtler, K. 1993. Studies on the chemistry of interstitial water taken from defined horizons in the fine sediments of bivalve habitats in several northern German lowland waters. II: Microhabitats of *Margaritifera margaritifera* L., *Unio crassus* (Philipsson), and *Unio tumidus* Philipsson. Archive feur Hydrobiologie 127:151–166.

Burch, J. B. 1975a. Freshwater sphaeriacean clams (Mollusca: Pelecypoda) of North America. Malacological Publications, Hamburg, MI.

Burch, J. B. 1975b. Freshwater unionacean clams (Mollusca: Pelecypoda) of North America. Malacological Publications, Hamburg, MI.

Burky, A. J. 1983. Physiological ecology of freshwater bivalves. *in:* Russell-Hunter, W. D., Ed. The Mollusca. Vol. 6 : Ecology. Academic Press, New York, pp. 281–327.

Burky, A. J., Burky, K. A. 1976. Seasonal respiratory variation and acclimation in the pea clam, *Pisidium walkeri* Sterki. Comparative Biochemistry and Physiology. *A*55:109–114.

Burky, A. J., Hornbach, D. J., Way, C. M. 1981. Growth of *Pisidium casertanum* (Poli) in west central Ohio. Ohio Journal of Science 81:41–44.

Burky, A. J., Benjamin, R. B., Conover, D. G., Detrick, J. R. 1985a. Seasonal responses of filtration rates to temperature, oxygen availability, and particle concentration of the freshwater clam *Musculium partumeium* (Say). American Malacological Bulletin 3:201–212.

Burky, A. J., Hornbach, D. J., Way, C. M. 1985b. Comparative bioenergetics of permanent and temporary pond populations of the freshwater clam, *Musculium partumeium* (Say). Hydrobiologia 126:35–48.

Burnett, L. E. 1997. The challenges of living in hypoxic and hypercapnic aquatic environments. American Zoologist 37:633–640.

Burton, R. F. 1983. Ionic regulation and water balance. *in:* Saleuddin, A. S. M., Wilbur, K. M., Ed. The Mollusca. Vol. 5: Physiology, Part 2. Academic Press, Orlando, FL, pp. 292–352.

Buttner, J. K. 1986. *Corbicula* as a biological filter and polyculture organism in catfish rearing ponds. Progessive Fish-Culturist 48:136–139.

Butts, T. A., Sparks, R. E. 1982. Sediment oxygen demand-fingernail clam relationship in the Mississippi River Keokuk Pool. Transactions of the Illinois Academy of Sciences 75:29–39.

Byrne, R. A., Dietz, T. H. 1997. Ion transport and acid-base balance in freshwater bivalves. Journal of Experimental Biology 200:457–465.

Byrne, R. A., McMahon, B. R. 1991. Acid–base and ionic regulation, during and following emersion, in the freshwater bivalve, *Anodonta grandis simpsoniana* (Bivalvia: Unionidae). Biological Bulletin (Woods Hole, Mass.) 181:289–297.

Byrne, R. A., McMahon, R. F. 1994. Behavioral and physiological responses to emersion in freshwater bivalves. American Zoologist 34:194–204.

Byrne, R. A., McMahon, R. F., Dietz, T. H. 1988. Temperature and relative humidity effects on aerial exposure tolerance in the freshwater bivalve *Corbicula fluminea*. Biological Bulletin (Woods Hole, Mass.) 175:253–260.

Byrne, R. A., Gnaiger, E., McMahon, R. F., Dietz, T. H. 1990. Behavioral and metabolic responses to emersion and subsequent reimmersion in the freshwater bivalve, *Corbicula fluminea*. Biological Bulletin (Woods Hole, Mass.) 178:251–259.

Byrne, R. A., Shipman, B. N., Smatresk, N. J., Dietz, T. H., McMahon, R. F. 1991a. Acid–base balance during emergence in the freshwater bivalve *Corbicula fluminea*. Physiological Zoology 64:748–766.

Byrne, R. A., Dietz, T. H., McMahon, R. F. 1991b. Ammonia dynamics during and after prolonged emersion in the freshwater clam *Corbicula fluminea* (Müller) (Bivalvia: Corbiculacea). Canadian Journal of Zoology 69:676–680.

Cáceres, C. E. 1997. Dormancy in invertebrates. Invertebrate Biology 116:371–383.

Cairns, J., Jr., Cherry, D. S. 1983. A site-specific field and laboratory evaluation of fish and Asiatic clam population responses to coal fired power plant discharges. Water Science and Technology 15:31–58.

Calow, P. 1983. Life-cycle patterns and evolution. *in:* Russell-Hunter, W. D., Ed. The Mollusca. Vol. 6: Ecology. Academic Press, New York, pp. 649–678.

Clarke, A. H. 1973. The freshwater molluscs of the Canadian interior basin. Malacologia 13:1–509.

Clarke, A. H. 1983. The distribution and relative abundance of *Lithasia pinguis* (Lea), *Pleurobema plenum* (Lea), *Villosa trabalis* (Conrad), and *Epioblasma sampsoi* (Lea). American Malacological Bulletin 1:27–30.

Clarke, M., McMahon, R. F., Miller, A. C., Payne, B. S. 1993. Tissue freezing points and time for complete mortality on exposure to freezing air temperatures in the zebra mussel (*Dreissena polymorpha*) with special reference to dewatering during freezing conditions as a mitigation strategy, *in*: Tsou, J. L., Muussalli, Y. G., Eds. Proceedings: Third International Zebra Mussel Conference 1993, EPRI TR-102077, Electric Power Research Institute, Palo Alto, CA, pp. 1–145.

Clarke, M., McMahon, R. F. 1996a. Comparison of byssal attachment in dreissenid and mytilid mussels: mechanisms, morphometry, secretion, biochemistry, mechanics and environmental influences. Malacological Review 29:1–16.

Clarke, M., McMahon, R. F. 1996b. Effects of hypoxia and low-frequency agitation on byssogenesis in the freshwater mussel *Dreissena polymorpha*. Biological Bulletin (Woods Hole, Mass.) 191:413–420.

Claudi, R., Mackie, G. L. 1993. Practical manual for zebra mussel monitoring and control. Lewis Publishers, Boca Raton, FL.

Cleland, J. D., McMahon, R. F., Elick, G. 1986. Physiological differences between two morphotypes of the Asian clam. *Corbicula*. American Zoologist 26:103A.

Clench, W. J. 1959. Mollusca. *in*: Edmondson, W. T., Ed. Freshwater biology. 2nd ed., Wiley, New York, pp. 1117–1160.

Clench, W. J., Turner, R. D. 1956. Freshwater mollusks of Alabama, Georgia, and Florida from the Escambia to the Suwannee River. Bulletin of the Florida State Museum, Biological Sciences 1:97–239.

Cohen, R. R. H., Dressler, P. V., Phillips, E. P. J., Cory, R. L. 1984. The effect of the Asiatic clam, *Corbicula fluminea,* on phytoplankton of the Potomac River, Maryland. Limnology and Oceanography 29:170–180.

Collins, T. W. 1967. Oxygen-uptake, shell morphology and dessication of the fingernail clam, *Sphaerium occidentale* Prime. Ph.D. Thesis, University of Minnesota, Minneapolis.

Coney, C. C. 1993. An empirical evaluation of various techniques for anesthetization and tissue fixation of freshwater Unionoida (Mollusca: Bivalvia), with a brief history of experimentation in molluscan anesthetization. Veliger 36:413–424.

Conn, D. B., Ricciardi, A., Babapulle, M. N., Klien, K. A., Rosen, D. A. 1996. *Chaetogaster limnaei* as a parasite of the zebra mussel *Dreissena polymorpha*, and the quagga mussel *Dreissena bugensis* (Mollusca: Bivalvia). Parasitology Research 82:1–7.

Cotner, J. B., Gardner, W. S., Johnson, J. R., Sada, R. H., Cavaletto, J. F., Heath, R. T. 1995. Effects of zebra mussels (*Dreissena polymorpha*) on bacterioplankton: evidence for both size-selective consumption and growth stimulation. Journal of Great Lakes Research 21:517–528.

Counts, C. L., III, Handwerker, T. S., Jesien, R. V. 1991. The Naiades (Bivalvia: Unionoidea) of the Delmarva Peninsula. American Malacological Bulletin 9:27–37

Covich, A. P., Dye, L. L., Mattice, J. S. 1981. Crayfish predation on *Corbicula* under laboratory conditions. American Midland Naturalist 105:181–188.

Cunjak, R. A., McGladdery, S. E. 1991. The parasite–host relationship of glochidia (Mollusca: Margaritiferidae) on the gills of young of the year Atlantic salmon (*Salmo salar*). Canadian Journal of Zoology 69:353–358.

Darwin, C. 1982. On the dispersal of freshwater bivalves. Nature 25: 529.

Das, V. M. M., Venkatachari, S. A. T. 1984. Influence of varying oxygen tension on the oxygen consumption of the freshwater mussel *Lamellidens marginalis* (Lamarck) and its relation to body size. Veliger 26:305–310.

Davis, D. S., Gilhen, J. 1982. An observation of the transportation of pea clams, *Pisidium adamsi,* by bluespotted salamanders, *Ambystoma laterale.* The Canadian Naturalist 96:213–215.

Davis, W. L., Jones, R. G., Knight, J. P., Hagler, H. K. 1982. An electron microscopic histochemical and X-ray microprobe study of spherites in a mussel. Tissue and Cell 14:61–67.

Deaton, L. E. 1982. Tissue (NA + K)-activated adenosinetriphosphatase activities in freshwater and brackish water bivalve molluscs. Marine Biology Letters 3:107–112.

Deaton, L. E., Greenberg, M. J. 1991. The adaptation of bivalve molluscs to oligohaline and fresh waters: phylogenetic and physiological aspects. Malacological Review 24:1–18.

Denson, D. R., Wang, S. Y. 1994. Morphological differences between zebra and quagga mussel spermatozoa. American Malacological Bulletin 11:79–81.

de Zwaan, A. 1983. Carbohydrate catabolism in bivalves. *in:* Hochachka, P. W., Ed. The Mollusca. Vol. 1: Metabolic biochemistry and molecular biomechanics. Academic Press, New York, pp. 137–175.

Dietz, T. H. 1974. Body fluid composition and aerial oxygen consumption in the freshwater mussel, *Ligumia subrostrata* (Say): effects of dehydration and anoxic stress. Biological Bulletin (Woods Hole, Mass.) 147:560–572.

Dietz, T. H. 1985. Ionic regulation in freshwater mussels: a brief review. American Malacological Bulletin 3:233–242.

Dietz, T. H., Byrne, R. A. 1997. Effects of salinity on solute clearance from the freshwater bivalve, *Dreissena polymorpha* Pallas. Experimental Biology Online 2:11–20.

Dietz, T. H., Byrne, R. A. 1999. Measurement of sulfate uptake and loss in the freshwater bivalve *Dreissena polymorpha* using semimicroassay. Canadian Journal of Zoology 77:331–336.

Dietz, T. H., Scheide, J. I., Saintsing, D. G. 1982. Monoamine transmitters and cAMP stimulation of Na transport in freshwater mussels. Canadian Journal of Zoology 60: 1408–1411.

Dietz, T. H., Wilson, J. M., Silverman, H. 1992. Changes in monoamine transmitter concentration in freshwater mussel tissues. Journal of Experimental Zoology 261:355–358.

Dietz, T. H., Lessard, D., Silverman, H., Lynn, J. W. 1994. Osmoregulation in *Dreissena polymorpha*: the importance of Na, Cl, K, and particularly Mg. Biological Bulletin (Woods Hole, Mass.) 187:76–83.

Dietz, T. H., Byrne, R. A., Lynn, J. W., Silverman, H. 1995. Paracellular solute uptake by the freshwater zebra mussel, *Dreissena polymorpha*. American Journal of Physiology 269 (Regulatory and Integrative Comparative Physiology 38):R300–R307.

Dietz, T. H., Wilcox, S. J., Byrne, R. A., Lynn, J. W., Silverman, H. 1996a. Osmotic and ionic regulation of North American zebra mussels (*Dreissena polymorpha*). American Zoologist 36: 364–372.

Dietz, T. H., Wilcox, S. J., Byrne, R. A., Silverman, H. 1996b. Effects of hyperosmotic challenge on the freshwater bivalve *Dreissena polymorpha*: Importance of K^+. Canadian Journal of Zoology 75:697–705.

Dietz, T. H., Neufeld, D. H., Silverman, H., Wright, S. H. 1998. Cellular volume regulation in freshwater bivalves. Journal of Comparative Physiology *B* 168:87–95.

Di Maio, J., Corkum, L. D. 1995. Relationship between the spatial distribution of freshwater mussels (Bivalvia: Unionidae) and the hydrological variability of rivers. Canadian Journal of Zoology 73:663–671

Di Maio, J., Corkum, L. D. 1997. Patterns of orientation in unionids as a function of rivers with differing hydrological variability. Journal of Molluscan Studies 63:531–539.

Dimock, R.V., Jr., Wright, A. H. 1993. Sensitivity of juvenile freshwater mussels to hypoxic, thermal and acid stress. The Journal of the Elisha Mitchell Scientific Society 109:183–192.

Doherty, F. G., Cherry, D. S., Cairns, J., Jr. 1987. Valve closure responses of the Asiatic clam *Corbicula fluminea* exposed to cadmium and zinc. Hydrobiologia 153:159–167.

Downing, W. L., Shostell, J., Downing, J. A. 1992. Non–annual external annuli in the freshwater mussels *Anodonta grandis grandis* and *Lampsilis radiata siliquiodea*. Freshwater Biology 28:309–317.

Downing, J. A., Rochon, Y., Pérusse, M., Harvey, H. 1993. Spatial aggregation, body size, and reproductive success in the freshwater mussel, *Elliptio complanata*. Journal of the North American Benthological Society 12:148–156.

Dreier, H. 1977. Study of *Corbicula* in Lake Sangchris. *in:* The annual report for fiscal year 1976, Lake Sangchris project, section 7. Illinois Natural History Survey, Urbana, IL, pp. 7.1–7.52.

Dreier, H., Tranquilli, J. A. 1981. Reproduction, growth, distribution, and abundance of *Corbicula* in an Illinois cooling lake. Illinois Natural History Survey Bulletin 32:378–393.

Duncan, R. E., Thiel, P. A. 1983. A survey of the mussel densities in pool 10 of the upper Mississippi River. Technical Bullletin No. 139. Wisconsin Department of Natural Resources, Madison. 14 pp.

Dyduch-Falniowska, A. 1982. Oscillations in density and diversity of *Pisidium* communities in two biotopes in southern Poland. Hydrobiological Bulletin 16:123–132.

Edmondson, W. T. (Ed.). 1959. Freshwater biology (second edition). Wiley, New York.

Effler, S. W., Siegfried, C. 1994. Zebra mussel (*Dreissena polymorpha*) populations in the Seneca River, New York: impact on oxygen resources. Environmental Science and Technology 28:2216–2221.

Elder, J. F., Mattraw, H. C., Jr. 1984. Accumulation of trace elements, pesticides, and polychlorinted biphenyls in sediments and the clam *Corbicula manilensis* of the Apalachicola River, Florida. Archives of Environmental Contamination and Toxicology 13:453–469.

Ellis, A. E. 1978. British freshwater bivalve Mollusca. Keys and notes for identification of the species. Academic Press, New York.

Eng, L. L. 1979. Population dynamics of the Asiatic clam, *Corbicula fluminea* (Müller), in the concrete-lined Delta–Mendota Canal of central California. *in:* Britton, J. C., Ed. Proceedings, first international *Corbicula* symposium. Texas Christian University Research Foundation, Fort Worth, TX, pp. 39–68.

Englund, V., Heino, M. 1994a. Valve movement of *Anodonta anatina* and *Unio tumidus* (Bivalvia, Unionidae) in a eutrophic lake. Annales Zoologici. Fennici 31:257–262.

Englund, V. P. M., Heino, M. P. 1994b. A new method for the identification marking of bivalves in still water. Malacological Review 27: 111–112.

Ferris, J. L., van Hassel, J. H., Belanger, S. E., Cherry, D. S., Cairns, J., Jr. 1988. Application of cellulolytic activity of Asiatic clams (*Corbicula* sp.) to instream monitoring of power plant effluents. Environmental Toxicology and Chemistry 7:701–713.

Fisher, S. W., Gossiaux, D. G., Bruner, K. A., Landrum, P. F. 1993. Investigation of the toxicokinetics of hydrophobic contaminants in zebra mussel. *in*: Nalepa, T. F., Schloesser, D., Eds. Zebra mussels: biology, impacts, and control. Lewis Publishers, Boca Raton, FL, pp. 465–488.

Florkin, M., Bricteux-Gregoire, S. 1972. Nitrogen metabolism in molluscs. *in*: Florkin, M., Scheer, B. T., Eds. Chemical zoology, Vol. VII, Mollusca. Academic Press, San Diego, CA, pp. 301–348.

Focarelli, R., Rosa, D., Rosati, F. 1990. Differentiation of the vitelline coat and the polarized site of sperm entrance in the egg of *Unio elongatulus* (Mollusca, Bivalvia). Journal of Experimental Zoology 254:88–96.

Foe, C., Knight, A. 1985. The effect of phytoplankton and suspended sediment on the growth of *Corbicula fluminea* (Bivalvia). Hydrobiologia 127:105–115.

Foe, C., Knight, A. 1986a. A thermal energy budget for juvenile *Corbicula fluminea*. American Malacological Bulletin, Special ed. No. 2:143–150.

Foe, C., Knight, A. 1986b. Growth of *Corbicula fluminea* (Bivalvia) fed on artificial and algal diets. Hydrobiologia 133:155–164.

Foe, C., Knight, A. 1986c. A method for evaluating the sublethal impact of stress employing *Corbicula fluminea*. American Malacological Bulletin, Special ed. No. 2:133–142.

Foe, C., Knight, A. 1987. Assessment of the biological impact of point source discharges employing Asiatic clam. Archives of Environmental Contamination and Toxicology 16:39–51.

French, J. R. P., III, Love, J. G. 1995. Size limitation on zebra mussels consumed by freshwater drum may preclude the effectiveness of drum as a biological controller. Journal of Freshwater Ecology 10:379–383.

Fritz, L. W., Lutz, R. A. 1986. Environmental perturbations reflected in internal shell growth patters of *Corbicula fluminea* (Mollusca: Bivalvia). Veliger 28:401–417.

Fukuhara, S, Nagata, Y. 1988. Frequency of incubation of *Anodonta woodiana* in small pond. Venus 47:271–277.

Fuller, S. L. H. 1974. Clam and mussels (Mollusca: Bivalvia). *in:* Hart, C. W., Jr., Fuller, S. L. H., Eds. Pollution ecology of freshwater invertebrates. Academic Press, New York, pp. 215–273.

Gainey, L. F., Jr., Greenberg, M. J. 1977. Physiological basis of the species abundance-salinity relationship in molluscs: a speculation. Marine Biology (Berlin) 40:41–49.

Gardiner, D. B., Silverman, H., Dietz, T. H. 1991. Musculature associated with the water canals in freshwater mussels and response to monoamines *in vitro*. Biological Bulletin (Woods Hole, Mass.) 180:453–465

Golightly, C. G., Jr. 1982. Movement and growth of Unionidae (Mollusca: Bivalvia) in the Little Brazos River, Texas. Ph.D. Thesis, Texas A&M University, College Station, Tx.

Graf, D. L. 1997. Sympatric speciation of freshwater mussels (Bivalvia: Unionoidea): a model. American Malacological Bulletin 14:35–40.

Graney, R. L., Jr., Cherry, D. S., Cairns, J., Jr. 1984. The influence of substrate, pH, diet and temperature upon cadmium accumulation in the Asiatic clam (*Corbicula fluminea*) in laboratory streams. Water Research 18:833–842.

Grapentine, L. C. 1992. Growth rates of unionid bivalves upstream and downstream of industrial outfalls contaminated with trace metals. Canadian Technical Report of Fisheries and Aquatic Sciences 1863:141–143.

Graves, S. Y., Dietz, T. H. 1980. Diurnal rhythms of sodium transport in the freshwater mussel. Canadian Journal of Zoology 58:1626–1630.

Graves, S. Y., Dietz, T. H. 1982. Cyclic AMP stimulation and prostaglandin inhibition of Na transport in freshwater mussels. Comparative Biochemistry and Physiology *A* 71:65–70.

Gordon, M. E., Smith, D. G. 1990. Autumnal reproduction in *Cumberlandia monodonta* (Unionoidea: Margaritiferidae). Transactions of the American Microscopical Society 109:407–411.

Gunning, G. E., Suttkus, R. D. 1985. Reclamation of the Pearl River. A perspective of unpolluted versus polluted waters. Fisheries 10:14–16.

Haag, W. R., Warren, M. L., Jr. 1998. Role of ecological factors and reproductive strategies in structuring freshwater mussel communities. Canadian Journal of Fisheries and Aquatic Science 55:297–306.

Haag, W. R., Warren, M. L., Jr. 1999. Mantle displays of freshwater mussels elicit attacks from fish. Freshwater Biology 42:35–40.

Haag, W. R., Butler, R. S., Hartfield, P. D. 1995. An extraordinary reproductive strategy in freshwater bivalves: prey mimicry to facilitate larval dispersal. Freshwater Biology 34:471–476.

Haag, W. R., Warren, M. L., Jr., Shillingsford, M. 1999. Host fishes and host-attracting behavior of *Lampsilis altilis* and *Villosa vibex* (Bivalvia: Unioniodae). American Midland Naturalist 141:149–157.

Haas, F. 1969. Superfamily Unionacea Fleming, 1828. *in:* Moore, R. C., Ed. Treatise on invertebrate paleontology. Part N: Mollusca. The Geological Society of America, Boulder, CO, pp. 411–467.

Hakenkamp, C. C., Palmer, M. A. 1999. Introduced bivalves in freshwater ecosystems: the impact of *Corbicula* on organic matter dynamics in a sandy bottom stream. Oecologia 119:445–451.

Hameed, P. S., Raj, A. I. M. 1990. Freshwater mussel, *Lamellidens marginalis* (Lamark) (Mollusca: Bivalvia: Unionidae) as an indicator of river pollution. Chemistry and Ecology 4:57–64.

Hanson, J. M., McKay, W. C., Prepas, E. E. 1988. The effects of water depth and density on the growth of a unionid clam. Freshwater Biology 19:345–355.

Hanson, J. M., MacKay, W. C., Prepas, E. E. 1989. Effect of size-selective predation by muskrats (*Ondatra zebithicus*) on a population of unionid clams (*Anodonta gradis simpsoniana*). Journal of Animal Ecology 58:15–28.

Harmon, J. L., Joy, J. E. 1990. Growth rates of the freshwater mussel, *Anodonta imbecillis* Say 1829, in five West Virginia wildlife station ponds. American Midland Naturalist 124:372–378.

Hartfield, P. 1993. Headcuts and their effects on freshwater mussels. *in*: Cummings, K. S., Buchanan, A. C., Koch, L. M., Eds. Conservation and Management of Freshwater Mussels, Proceedings of a Symposium, 1992, Upper Mississippi River Conservation Committee, Rock Island, IL, pp. 131–141.

Hartfield, P., Rummel, R. G. 1985. Freshwater mussels (Unionidae) of the Big Black River, Mississippi. Nautilus 99:116–119.

Hartley, D. M., Johnston, J. B. 1983. Use of the freshwater clam *Corbicula manilensis* as a monitor for organochlorine pesticides. Bulletin of Environmental Contamination and Toxicology 31:33–40.

Haukioja, E., Hakala, T. 1978. Life-history evolution in *Anodonta piscinails* (Mollusca, Pelecypoda). Oecologia 35:253–266.

Havlik, M. E. 1983. Naiad mollusk populations (Bivalvia: Unionidae) in Pools 7 and 8 of the Mississippi River near La Crosse, Wisconsin. American Malacological Bulletin 1:51–60.

Hawkins, A. J. S., Widdows, J., Bayne, B. L. 1989. The relevance of whole-body protein metabolism to measured costs of maintenance and growth in *Mytilus edulis*. Physiological Zoology 62:745–763.

Heard, W. H. 1977. Reproduction of fingernail clams (Sphaeriidae: *Sphaerium* and *Musculium*). Malacologia 16:421–455.

Hedegaard, C., Wenk, H-R. 1998. Microstructure and texture patterns of mollusc shells. Journal of Molluscan Studies 64:133–136.

Hemelraad, J., Herwig, H. J., Holwerda, D. A., Zandee, D. I. 1985. Accumulation, distribution and localization of cadmium in the freshwater clam *Anodonta* sp. Marine Environmental Research 17:196.

Heming, T. A., Vinogradov, G. A., Klerman, A. K., Komov, V. T. 1988. Acid–base regulation in the freshwater pearl mussel *Margaritifera margaritifera:* effect of emersion and low water pH. Journal of Experimental Biology 137:501–511.

Henery, R. P., Saintsing, D. G. 1983. Carbonic anhydrase activity and ion regulation in three species of osmoregulating bivalve molluscs. Physiological Zoology 56:274–280.

Hernandez, M. R., McMahon, R. F., Dietz, T. H. 1995. Investigation of geographic variation in the thermal tolerance of zebra mussels, *Dreissena polymorpha. in*: Proceedings of the fifth zebra mussel and other aquatic nuisance organisms conference 1995. Ontario Hydro, Ont., Canada, pp. 195–209.

Herwig, H. J., Hemelraad, J., Howerda, D. A., Zandee, D. I. 1985. Cytochemical localization of cadmium and tin in bivalves. Marine Environmental Research 17:196–197.

Higgins, F. 1858. A catalogue of the shell-bearing species, inhabiting the vicinity of Columbus, Ohio, with some remarks thereon. Twelfth Annual Report, Ohio Sate Board of Agriculture for 1857:548–555.

Hillis, D. M., Mayden, R. L. 1985. Spread of the Asiatic clam, *Corbicula* (Bivalvia: Corbiculacea), into the new world tropics. Southwestern Naturalist 30:454–456.

Hillis, D. M., Patton, J. C. 1982. Morphological and electrophoretic evidence for two species of *Corbicula* (Bivalvia: Corbiculacea) in north central America. American Midland Naturalist 108:74–80.

Hinch, S. G., Kelly, L. J., Green, R. H. 1989. Morphological variation of *Elliptio complanata* (Bivalvia: Unionidae) in differing sediments of soft-water lakes exposed to acidic deposition. Canadian Journal of Zoology 67:1895–1899.

Hiripi, L., Burrell, D. E., Brown, M., Assanah, P., Stanec, A., Stefano, G. B. 1982. Analysis of monoamine accumulations in the neuronal tissues of *Mytilus edulis* and *Anodonta cygnea* (Bivalvia)-III. Temperature and seasonal influences. Comparative Biochemistry and Physiology, *C* 71:209–213.

Hiscock, I. D. 1953. Osmoregulation in Australian freshwater mussels (Lamellibranchiata). I. Water and chloride exchange in *Hyridella australis* (Lam.). Australian Journal of Marine and Freshwater Research 4:317–329.

Hoeh, W. R., Trdan, R. J. 1984. The freshwater mussels (Pelecypoda: Unionidae) of the upper Tittabawassee River drainage, Michigan. Malacological Review 17:97–98.

Hoeh W. R., Frazer, K. F., Naranjo-García, E., Trdan, R. J. 1995. A phylogenetic perspective on the evolution of simultaneous hermaphroditism in a freshwater mussel clade (Bivalvia: Unionidae: *Utterbackia*). Malacological Review 28:25–42.

Hoke, E. 1994. A survey and analysis of the unionid mollusks of the Elkhorn River Basin, Nebraska. Transactions of the Nebraska Academy of Sciences 21:31–54.

Holopainen, I. J. 1987. Seasonal variation of survival time in anoxic water and the glycogen content of *Sphaerium corneum* and *Pisidium amnicum* (Bivalvia: Pisidiidae). American Malacological Bulletin 5:41–48.

Holopainen, I. J., Hanski, I. 1986. Life history variation in *Pisidium*. Holarctic Ecology 9:85–98.

Holopainen, I. J., Jonasson, P. M. 1983. Long-term population dynamics and production of *Pisidium* (Bivalvia) in the profundal of Lake Esrom, Denmark. Oikos 41:99–117.

Hornbach, D. J. 1985. A review of metabolism in the Pisidiidae with new data on its relationship with life history traits in *Pisidium casertanum*. American Malacological Bulletin 3:187–200.

Hornbach, D. J. 1991. The influence of oxygen availability on oxygen consumption in the freshwater clam *Musculium partumeium* (Say) (Bivalvia: Sphaeriidae). American Malacological Bulletin 9:39–42.

Hornbach, D. J., Childers, D. L. 1986. Life-history variation in a stream population of *Musculium partumeium* (Bivalvia: Pisidiidae). Journal of the North American Benthological Society 5:263–271.

Hornbach, D. J., Childers, D. L. 1987. The effect of acidification on life-history traits of the freshwater clam *Musculium partumeium* (Say, 1822) (Bivalvia: Pisidiidae). Canadian Journal of Zoology 65:113–121.

Hornbach, D. J., Cox, C. 1987. Environmental influences on life history traits in *Pisidium casertanum* (Bivalvia: Pisidiidae): field and laboratory experimentation. American Malacological Bulletin 5:49–64.

Hornbach, D. J., Way, C. M., Burky, A. J. 1980. Reproductive strategies in the freshwater sphaeriid clam, *Musculium partumeium* (Say), from a permanent and a temporary pond. Oecologia 44:164–170.

Hornbach, D. J., Wissing, T. E., Burky, A. J. 1982. Life-history characteristics of a stream population of the freshwater clam *Sphaerium striatinum* Lamarck (Bivalvia: Pisidiidae). Canadian Journal of Zoology 60:249–260.

Hornbach, D. J., Wissing, T. E., Burky, A. J. 1983. Seasonal variation in the metabolic rates and Q_{10-} values of the fingernail clam, *Sphaerium strianum* Lamark. Comparative Biochemistry and Physiology. *A*76:783–790.

Hornbach, D. J., Way, C. M., Wissing, T. E., Burky, A. J. 1984a. Effect of particle concentration and season on the filtration rates of the freshwater clam, *Sphaerium striatinum* Lamark (Bivalvia: Pisidiidae). Hydrobiologia 108:83–96.

Hornbach, D. J., Wissing, T. E., Burky, A. J. 1984b. Energy budget for a stream population of the freshwater clam. *Sphaerium striatium* Lamark (Bivalvia: Pisidiidae). Canadian Journal of Zoology 62:2410–2417.

Hornbach, D. J., Miller, A. C., Payne, B. S. 1992. Species composition of the mussel assemblages in the upper Mississippi River. Malacological Review 25:119–128.

Horne, F. R., MacIntosh, S. 1979. Factors influencing distribution of mussels in the Blanco River of central Texas. Nautilus 94:119–133.

Horohov, J., Silverman, H., Lynn, J. W., Dietz, T. H. 1992. Ion transport in the freshwater zebra mussel, *Dreissena polymorpha*. Biological Bulletin (Woods Hole, MA) 183:297–303.

Houp, R. E. 1993. Observations on long-term effects of sedimentation on freshwater mussels (Mollusca: Unionidae) in the North Fork of Red River, Kentucky. Transactions of the Kentucky Academy of Sciences 54:93–97.

Hove, M. C., Neves, R. J. 1994. Life history of the endangered James spiny mussel *Pleurobema collina* (Conrad, 1837) (Mollusca: Unionidae). American Malacological Bulletin 11:29–40.

Hudson, R. G., Isom, B. G. 1984. Rearing of juveniles of the freshwater mussels (Unionidae) in a laboratory setting. Nautilus 98:129–135.

Huebner, J. D. 1982. Seasonal variation in two species of unionid clams from Manitoba, Canada: respiration. Canadian Journal of Zoology 60:650–564.

Huebner, J. D., Pynnönen, K. S. 1992. Viability of glochidia of two species of *Anodonta* exposed to low pH and selected metals. Canadian Journal of Zoology 70:2348–2355.

Huebner, J. D., Malley, D. F., Donkersloot, K. 1990. Population ecology of the freshwater mussel *Anodonta grandis grandis* in a Precambrian Shield lake. Canadian Journal of Zoology 68:1931–1941.

Hunter, D. R., Baily, J. F. 1992. *Dreissena polymorpha* (zebra mussel): colonization of soft substrata and some effects on unionid bivalves. Nautilus 106:60–67.

Imlay, M. J. 1982. Use of shells of freshwater mussels in monitoring heavy metals and environmental stresses: a review. Malacological Review 15:1–14.

Isom, B. G., Gooch, C. 1986. Rational and sampling designs for freshwater mussels Unionidae in streams, large rivers, impoundments and lakes. *in*: Isom, B. G., Ed., Rationale for sampling and interpretation of ecological data in the assessment of freshwater ecosystems, ASTM STP 894. American Society for Testing and Materials, Philadelphia, PA, pp. 46–59.

Isom, B. G., Hudson, R. G. 1982. *In vitro* culture of parasitic freshwater mussel glochidia. Nautilus 96:147–151.

Istin, M., Girard, J. P. 1970a. Dynamic state of calcium reserves in freshwater clam mantle. Calcified Tissue Research 5:196–206.

Istin, M., Girard, J. P. 1970b. Carbonic anhydrase and mobilization of calcium reserves in the mantle of lamellibranchs. Calcified Tissue Research 5:247–260.

Jadhav, M. L., Lomte, V. S. 1982a. Seasonal variation in biochemical composition of the freshwater bivalve, *Lamellidens corrianus*. Rivista di Idrobiologia 21:1–17.

Jadhav, M. L., Lomte, V. S. 1982b. Hormonal control of carbohydrate metabolism in the freshwater bivalve, *Lamellidens corrianus*. Rivista di Idrobiologia. 21:27–36.

Jadhav, M. L., Lomte, V. S. 1983. Neuroendocrine control of midgut gland in the bivalve, *Lamellidens corrianus* (Prasad) (Mollusca: Lamellibranchiata). Journal of Advanced Zoology 4:97–104.

James, M. R. 1985. Distribution, biomass and production of the freshwater mussel, *Hyridella menziesi* (Gray), in Lake Taupo, New Zealand. Freshwater Biology 15:307–314.

Jansen, W. A., Hanson, J. M. 1991. Estimates of the number of glochidia produced by clams (*Anodonta grandis simpsoniana* Lea), attaching to yellow perch (*Perca flavescens*), and surviving to various ages in Narrow Lake, Alberta. Canadian Journal of Zoology 69:973–977.

Johnson, P. D., Brown, K. M. 1998. Intraspecfic life history variation in the threatened Louisiana pearlshell mussel, *Margaritifera hembeli*. Freshwater Biology 40:317–329.

Johnson, P. D., Brown, K. M. 2000. The importance of microhabitat factors and habitat stability to the threatened Louisiana pearl shell, *Margaritifera hebeli* (Conrad). Canadian Journal of Zoology 78:1–7.

Johnson, P. D., McMahon, R. F. 1998. Effects of temperature and chronic hypoxia on survivorship of the zebra mussel (*Dreissena polymorpha*) and Asian clam (*Corbicula fluminea*). Canadian Journal of Fisheries and Aquatic Science 55:1564–1572.

Johnson, R. I. 1970. The systematics and zoogeography of the Unionidae (Mollusca: Bivalvia) of the southern Atlantic Slope Region. Bulletin of the Museum of Comparative Zoology 140:263–450.

Johnson, R. I. 1972. The Unionidae (Mollusca: Bivalvia) of Peninsular Florida. *Bulletin of the Florida State Museum, Biological Sciences* 16:181–249 + addendum.

Johnson, R.I. 1998. A new mussel *Potamilus metnecktayi* (Bivalvia: Unionidae) from the Rio Grande system, Mexico and Texas with notes on Mexican *Disconaias*. Occasional papers on Mollusks 5(76):427–455.

Jokinen, E. H. 1991. The malacofauna of the acid and non-acid lakes and rivers of the Adirondack Mountains and surrounding lowlands, New York State, USA. Verhandlungen-Internationale Vereinigung fur Limnologie 24:2940–2946.

Jokela, J., Palokamgas, P. 1993. Reproductive tactics in *Anodonta* clams: parental host recognition. Animal Behavior 1993:618–620.

Jokela, J., Uotila, L., Taskinen, J. 1993. Effect of the castrating trematode parasite *Rhipidocotyle fennica* on energy allocation of freshwater clam *Anodonta piscinalis*. Functional Ecology 7:332–338.

Jonasson, P. M. 1984a. Decline of zoobenthos through five decades of euthrophication in Lake Esrom. Verhandlungen Internationale Vereinigung fur Theoretische and Angewandte Limnologie 22:800–804.

Jonasson, P. M. 1984b. Oxygen demand and long term changes in zoobenthos. Hydrobiologia 115:121–126.

Jones, H. D. 1983. Circulatory systems of gastropods and bivalves. *in:* Saleuddin, A. S. M., Wilbur, K. M., Eds. The Mollusca. Vol. 5: Physiology, Part 2. Academic Press, New York, pp. 189–238.

Jones, H. D., Peggs, D. 1983. Hydrostatic and osmotic pressure in the heart and pericardium of *Mya arenaria* and *Anodonta cygnea*. Comparative Biochemistry and Physiology 76A:381–385.

Jones, R. G., Davis, W. L. 1982. Calcium containing lysosomes in the outer mantle epithelial cells of *Amblema*, a freshwater mollusc. Anatomic Record 203:337–343.

Joosse, J., Geraerts, W. P. M. 1983. Endocrinology. *in:* Saleuddin, A. S. M., Wilbur, K. M., Eds. The Mollusca, Vol. 4: Physiology, Part 1. Academic Press, New York, pp. 318–406.

Jørgensen, C. B. 1996. Bivalve filter feeding revisited. Marine Ecology Progress Series 142:287–302.

Jørgensen, C. B., Kiorboe, T., Mohlenberg, F., Riisgard, H. U. 1984. Ciliary and mucus-net filter feeding, with special reference to fluid mechanical characteristics. Marine Ecology Progress Series 15:283–292.

Kasprzak, K. 1986. Role of the Unionidae and Sphaeriidae (Mollusca, Bivalvia) in the eutrophic Lake Zbechy and its outflow. Internationale Revue der Gesamten Hydrobiologie 71:315–334.

Kat, P. W. 1982. Shell dissolution as a significant cause of mortality for *Corbicula fluminea* (Bivalvia: Corbiculidae) inhabiting acidic waters. Malacological Review 15:129–134.

Kat, P. W. 1984. Parasitism and the Unionacea (Bivalvia). Biological Review 59:189–207.

Kat, P. W. 1985. Convergence in bivalve conchiolin layer microstructure. Malacological Review 18:97–106.

Kat, P. W., Davis, G. M. 1984. Molecular genetics of peripheral population of Nova Scotian Unionidae (Mollusca: Bivalvia). Biological Journal of the Linnean Society 22:157–185.

Kays, W. T., Silverman, H., Dietz, T. H. 1990. Water channels and water canals in the gill of the freshwater mussel, *Ligumia subrostrata*: ultrastructure and histochermistry. Journal of Experimental Zoology 254:256–269.

Keen, M., Casey, R. 1969. Family Corbiculidae Gray, 1847. *in:* Moore, R. C., Ed. Treatise on invertebrate paleontology. Part N: Mollusca. The Geological Society of America., Boulder, CO, pp. 664–669.

Keen, M., Dance, P. 1969. Family Pisidiidae, Gray, 1857. *in:* Moore, R. C., Ed. Treatise on invertebrate paleontology. Part N: Mollusca. The Geological Society of America, Boulder, CO, pp. 669–670.

Kennedy, V. S., Blundon, J. A. 1983. Shell strength in *Corbicula* sp. (Bivalvia: Corbiculacea) from the Potomac River, Maryland. Veliger 26:22–25.

Khan, H. R., Aston, M. L., Saleuddin, A. S. M. 1986. Fine structure of the kidneys of osmotically stressed *Mytilus, Mercenaria,* and *Anodonta.* Canadian Journal of Zoology 64:2779–2787.

Kilgour, B. W., Mackie, G. L. 1988. Factors affecting the distribution of sphaeriid bivalves in Britannia Bay of the Ottawa River. Nautilus 102:73–77.

King, C. A., Langdon, C. J., Counts, C. L., III. 1986. Spawning and early development of *Corbicula fluminea* (Bivalvia: Corbiculidae) in laboratory culture. American Malacological Bulletin 4:81–88.

Kirk, S. G., Layzer, J. B. 1997. Induced metamorphosis of freshwater mussel glochidia on nonfish host. Nautilus 110:102–106.

Kiss, A., Pekli, J. 1988. On the growth of *Anodonta woodiana* (Lea 1834) (Bivalvia: Unionacea). Bulletin of the University of Agricultural Sciences, Gödöllö 1:119–124.

Kovolack, W. P., Dennis, S. D., Bates, J. M. 1986. Sampling effort required to find rare species of freshwater mussels. *in*: Isom, B. G., Ed. Rationale for sampling and interpretation of ecological data in the assessment of freshwater ecosystems, ASTM STP 894. American Society for Testing and Materials, Philadelphia, PA, pp. 34–45.

Kraemer, L. R. 1978. Discovery of two distinct kinds of statocysts in freshwater bivalved mollusks: some behavioral implications. Bulletin of the American Malacological Union pp. 24–28.

Kraemer, L. R. 1981. The osphradial complex of two freshwater bivalves: Histological evaluation and functional context. Malacologia 20:205–216.

Kraemer, L. R., Galloway, M. L. 1986. Larval development of *Corbicula fluminea* (Müller) (Bivalvia: Corbiculacea): an appraisal of its heterochrony. American Malacological Bulletin 4:61–79.

Kraemer, L. R., Swanson, C., Galloway, M., Kraemer, R. 1986. Biological basis of behavior in *Corbicula fluminea,* II. Functional morphology of reproduction and development and review of evidence for self-fertilization. American Malacological Bulletin, Special Edition. No. 2:193–201.

Kryger, J., Riisgård, H. U. 1988. Filtration rate capacities in 6 species of European freshwater bivalves. Oecologia 77:34–38.

LaRocque, A. L. 1967a. Pleistocene Mollusca of Ohio. Part 2. State of Ohio Division of Geological Survey, Department of Natural Resources, Columbus. pp. 113–356.

LaRocque, A. L. 1967b. Pleistocene Mollusca of Ohio. Part1. State of Ohio Department of Natural Resources, Division of Geological Survey, Columbus. pp. 1–111.

Lauritsen, D. D. 1986a. Assimilation of radiolabeled algae by *Corbicula.* American Malacological Bulletin, Special Edition No. 2:219–222.

Lauritsen, D. D. 1986b. Filter-feeling in *Corbicula fluminea* and its effect on seston removal. Journal of the North American Bentholoigical Society 5:165–172.

Laycock, G. 1983. Vanishing naiads. Audubon 85:26–28.

Leff, D. S., Burch, J. L., McArthur, J. V. 1990. Spatial distribution, seston removal, and potential competitive interactions of the bivalves *Corbicula fluminea* and *Elliptio complanata* in a coastal plain stream. Freshwater Biology 24:409–416.

Lei, J., Payne, B. S., Wang, S. Y. 1996. Filtration dynamics of the zebra mussel, *Dreissena polymorpha*. Canadian Journal of Fisheries and Aquatic Science 53:29–37.

Lewis, J. B. 1984. Comparative respiration in two species of freshwater unionid mussels (Bivalvia). Malacological Review 17:101–102.

Lewis, J. B., Riebel, P. N. 1984. The effect of substrate on burrowing in freshwater mussels (Unionidae). Canadian Journal of Zoology 62:2023–2025.

Lomte, V. S., Jadhav, M. L. 1981a. Effect of desiccation on the neurosecretory activity of the freshwater bivalve, *Parreysia corrugata.* Science and Culture 47:437–438.

Lomte, V. S., Jadhav, M. L. 1981b. Neuroendocrine control of osmoregulation in the freshwater bivalve, *Lamellidens corrianus.* Journal of Advanced Zoology 2:102–108.

Lomte, V. S., Jadhav, M. L. 1982a. Effects of toxic compounds on oxygen consumption in the freshwater bivalve, *Corbicula regularis* (Prime, 1860). Comparative Physiology and Ecology 7:31–33.

Lomte, V. S., Jadhav, M. L. 1982b. Respiratory metabolism in the freshwater mussel *Lamellidens corrianus*. Marathwada University Journal of Science 21:87–90.

Lomte, V. S., Jadhav, M. L. 1982c. Biochemical composition of the freshwater bivalve, *Lamellidens corrianus* (Prasad, 1922). Rivista di Idrobiologia 21:19–25.

Lopez, G. R., Holopainen, I. J. 1987. Interstitial suspension-feeding by *Pisidium* sp. (Pisididae: Bivalvia): a new guild in the lentic benthos? American Malacological Bulletin 5:21–30.

Lynn, J. W. 1994. The ultrastructure of the sperm and motile spermatozeugmata released from the freshwater mussel *Anodonta grandis* (Mollusca: Bivalvia, Unionoidae). Canadian Journal of Zoology 72:1452–1461.

Lydeard, C., Mulvey, M., Davis, G. M. 1996. Molecular systematics and evolution of reproductive traits of North American freshwater unionacean mussels (Mollusca: Bivalvia) as inferred from 16S rRNA gene sequences. Philosophical Transactions of the Royal Society of London *B*, 351:1593–1603.

Machodo, J. F., Ferreira, K. G., Ferreira, H. G., Fernandes, P. L. 1990. The acid–base balance of the outer mantle epithelium of *Anodonta cygnea*. Journal of Experimental Biology 150:159–169.

Machado, J., Marvo, R., Ferreira, C., Moura, G., Ries, M., Coimbra, C. 1994. Study on mucopolysaccharides as a shell component of *Anodonta cygnea*. Bulletin de l'Institut Océanographique, Monaco 14:141–150.

Machena, C., Kautsky, N. 1988. A quantitative diving survey of benthic vegetation and fauna in Lake Kariba, a tropical man-made lake. Freshwater Biology 19:1–14.

MacIsaac, H. J. 1996. Potential abiotic and biotic impacts of zebra mussels on the inland waters of North America. American Zoologist 36:287–299.

MacIsaac, H. J., Johannsson, O. E., Ye, J., Sprules, W. G., Leach, J. H., McCorquodale, J. A., Grigorovich, I. A. 1999. Filtering impacts of an introduced bivalve (*Dreissena polymorpha*) in a shallow lake: application of a hydrodynamic model. Ecosystems 2:338–350.

Mackie, G. L. 1978. Are sphaeriid clams ovoviparous or viviparous? Nautilus 92:145–147.

Mackie, G.L. 1979. Growth dynamics in natural populations of Sphaeriidae clams (*Sphaerium*, *Musculium*, *Pisidium*). Canadian Journal of Zoology 57:441–456.

Mackie, G. L. 1984. Bivalves. *in:* Tompa, A. S., Verdonk, N. H., van der Biggelaar, J. A. M. Eds. The Mollusca. Vol. 7: Reproduction. Academic Press, New York, pp. 351–418.

Mackie, G. L., Flippance, L. A. 1983a. Life history variations in two populations of *Sphaerium rhombiodeum* (Bivalvia: Pisidiidae). Canadian Journal of Zoology 61:860–867.

Mackie, G. L., Flippance, L. A. 1983b. Intra and interspecific variation in calcium content of freshwater Mollusca in relation to calcium content of the water. Journal of Molluscan Studies 49:204–212.

Mackie, G. L., Schloesser, D. W. 1996. Comparative biology of zebra mussels in Europe and North America: an overview. American Zoologist 36:244–258.

Mackie, G. L., Topping, J. M. 1988. Historical changes in the unionid fauna of the Sydenham River Watershed and downstream changes in shell morphometrics of three common species. Canadian Field Naturalist 102:617–626.

Mackie, G. L., Quadri, S. U., Reed, R. M. 1978. Significance of litter size in *Musculium securis* (Bivalvia: Sphaeriidae). Ecology 59:1069–1074.

Mackie, G. L., White, D. S., Zdeba, T. W. 1980. A guide to the freshwater mollusks of the Laurentian Great Lakes with special emphasis on the genus, *Pisidium*. EPA-600/3-80-068. United States Environmental Protection Agency, Duluth, MN.

Mackie, G. L., Gibbons, W. N., Muncaster, B. W., Gray, I. M. 1989. The zebra mussel, *Dreissena polymorpha:* a synthesis of European experience and a preview for North America. 0-7729-5647-2. Water Resources Branch, Ontario Ministry of the Environment, Ont., Canada.

Mäkelä, T. P., Oikari, A. O. J. 1992. The effects of low water pH on the ionic balance in freshwater mussel *Anodonta anatina* L. Annales Zoologici Fennici 29:169–175.

Malley, D. F., Huebner, J. D., Donkersloot, K. 1988. Effects on ionic composition of blood and tissues of *Anodonta grandis grandis* (Bivalvia) of an addition of aluminum and acid to a lake. Archives of Environmental Contamination and Toxicology 17:479–491.

Malloy, D. P., Karatayev, A. Y., Burlakova, L. E., Kurandina, D. P., Laruelle, F. 1997. Natural enemies of zebra mussels, and ecological competitors. Reviews in Fisheries Science 5:27–97.

Mani, V. U., Rao, P. V., Reddy, K. S., Indira, K., Rajendra, W. 1993. Bioaccumulation and clearance of ambient ammonia by the freshwater water mussel, *Lamellidens marginalis* (L.). Indian Journal of Comparative Animal Physiology 11:83–88.

Marsden, J. E., Spidle, A. P., May, B. 1996. Review of genetic studies of *Dreissena* sp. American Zoologist 36:259–270.

Marsh, P. C. 1985. Secondary production of introduced Asiatic clam, *Corbicula fluminea,* in a central Arizona canal. Hydrobiologia 124:103–110.

Martin, W. A. 1983. Execretion. *in:* Saleuddin, A. S. M., Wilbur, K. M. Eds. The Mollusca. Vol. 5: Physiology, Part 2. Academic Press. New York, pp. 353–405.

Massabuau, J-C., Burtin, B., Wheatly, M. 1991. How is O_2 consumption maintained independent of ambient oxygen in mussel, *Anodonta cygnea*? Respiration Physiology 83:103–114.

Mathiak, H. A. 1979. A river survey of the unionid mussels of Wisconsin 1973–1977. Sand Shell Press. Horicon, WIS.

Matisoff, G., Fisher, J. B., Matis, S. 1985. Effect of microinvertebrates on the exchange of solutes between sediments and freshwater. Hydrobiologia 122:19–33.

Matsushima, O., Kado, Y. 1982. Hyperosmoticity of the mantle fluid in the freshwater bivalve, *Anodonta woodiana*. Journal of Experimental Biology 221:379-381.

Matsushima, O., Sakka, F. and Kado, Y. 1982. Free amino acid involved in intracellular osmoregulaion in the clam, *Corbicula*. Journal of Science of Hiroshima University, Series B, Division *1* 30:213–219.

Mattice, L. S. 1979. Interactions of *Corbicula* sp. with power plants. *in;* J. C. Britton, Ed. Proceedings, first international *Corbicula* symposium. Texas Christian University Research Foundation, Fort Worth, TX, pp. 119–138.

Mattice, J. S., McLean, R. B., Burch, M. B. 1982. Evaluation of short-term exposure to heated water and chlorine for control of the Asiatic clam (*Corbicula fluminea*). Oak Ridge National Laboratory, Environmental Science Division, Publication No. 1748. U.S. National Technical Information Service, Department of Commerce, Springfield, VA, 33 pp.

Matthews, M. A., McMahon, R. F. 1999. Effects of temperature and temperature acclimation on survival of zebra mussels *(Dreissena polymorpha*) and Asian clams (*Corbicula fluminea*) under extreme hypoxia. Journal of Molluscan Studies 65: 317–325.

Mazak, E. J., MacIsaac, H. J., Servos, M. R., Hesslein, R. 1997. Influence of feeding habits on organochlorine contaminant accumulation in waterfowl of the Great Lakes. Ecological Applications 7:1133–1143.

McCorkle, S., Shirley, T. C., Dietz, T. H. 1979. Rhythms of activity and oxygen consumption in the common pond clam, *Ligumia subrostrata* (Say). Canadian Journal of Zoology 57: 1960–1964.

McCorkle-Shiley, S. 1982. Effects of photoperiod on sodium flux in *Corbicula fluminea* (Mollusca: Bivalvia). Comparative Biochemistry and Physiology. *A*71:325–327.

McKee, P. M., Mackie, G. L. 1980. Dessication resistance in *Sphaerium occidentale* and *Musculium securis* (Bivalvia: Sphaeriidae) from a temporary pond. Canadian Journal of Zoology 58:1693–1696.

McKillop, W. B., Harrison, A. D. 1982. Hydrobiological studies of the eastern Lesser Antillean Islands. VII. St. Lucia: behavioural drift and other movements of freshwater marsh molluscs. Archiv für Hydrobiologie 94:53–69.

McLeod, M. J. 1986. Electrophoretic variation in North America *Corbicula*. American Malacological Bulletin, Special ed. No. 2:125–132.

McMahon, R. F. 1979a. Response to temperature and hypoxia in the oxygen consumption of the introduced Asiatic freshwater clam *Corbicula fluminea* (Müller). Comparative Biochemistry and Physiology *A*63:383–388.

McMahon, R. F. 1979b. Tolerance of aerial exposure in the Asiatic freshwater clam, *Corbicula fluminea* (Müller). *in:* Britton, J. C. Ed. Proceedings, first international *Corbicula* symposium. Texas Christian University Research Foundation, Fort Worth, TX, pp. 227–241.

McMahon, R. F. 1983a. Ecology of an invasive pest bivalve, *Corbicula. in:* Russell-Hunter, W. D. Editor. The Mollusca. Vol. 6: Ecology. Academic Press, New York, pp. 505–561.

McMahon, R. F. 1983b. Physiological ecology of freshwater pulmonates. *in:* Russell-Hunter, W. D. Ed. The Mollusca. Vol. 6: Ecology. Academic Press, New York, pp. 360–430.

McMahon, R. F. 1996. The physiological ecology of the zebra mussel, *Dreissena polymorpha*, in North America. American Zoologist 36:339–363.

McMahon, R. F. 1999. Invasive characteristics of the freshwater bivalve, *Corbicula fluminea. in:* Claudi, R., Leach, J. H. Eds. Nonindigenous freshwater organisms: vectors, biology and impacts. Lewis Publishers, Boca Raton, FL, pp. 315–343.

McMahon, R. F., Lutey, R. W. 1988. Field and laboratory studies of the efficacy of poly[oxyethylene(dimethyliminio)ethylene(dimenthyliminio)ethylene dichloride] as a biocide against the Asian clam, *Corbicula fluminea. in:* Proceedings: service water reliability improvement seminar. Electric Power Research Institute, Palo Alto, CA, pp. 61–72.

McMahon, R. F., Williams, C. J. 1984. A unique respiratory response to emersion in the introduced Asian freshwater clam *Corbicula fluminea* (Müller) (Lamellibranchia: Corbiculacea). Physiological Zoology 57:274–279.

McMahon, R. F., Williams, C. J. 1986a. A reassessment of growth rate, life span, life cycles and population dynamics in a natural population and field caged individuals of *Corbicula fluminea* (Müller). American Malacological Bulletin, Special ed. No. 2:151–166.

McMahon, R. F., Williams, C. J. 1986b. Growth, life cycle, upper thermal limit and downstream colonization rates in a natural population of the freshwater bivalve mollusc, *Corbicula fluminea* (Müller) receiving thermal effluents. American Malacological Bulletin, Special ed. No. 2:231–239.

McMahon, R. F., Ussary, T. A., Clarke, M. 1993. Use of emersion as a zebra mussel control method, Technical Report EL-93-1, U.S. Army Corps of Engineers, Washington, DC.

Medler, S., Silverman, H. 1997. Functional organization of intrinsic gill muscles in zebra mussels, *Dreissena polymorpha* (Mollusca: Bivalvia), and response to transmitters *in vivo*. Invertebrate Biology 116:200–212.

Michaelidis, B., Anthanasiadou, P. 1994. Effect of reduced oxygen tension on the heart rate and the kinetic properties of glycolytic key enzymes PFK, PK and glycogen phosphorylase from the freshwater mussel, *Anodonta cygnea*. Comparative Biochemistry and Physiology *B*108:165–172.

Miller, A. C., Payne, B. S. 1988. The need for quantitative sampling to characterize demography and density of freshwater mussel communities. American Malacological Bulletin 6:49–54.

Miller, A. C., Rhodes, L., Tipit, R. 1984. Changes in the naiad fauna of the Cumberland River below Lake Cumberland in central Kentucky. Nautilus 98:107–110.

Miller, A. C., Payne, B. S., Aldridge, D. W. 1986. Characterization of a bivalve community in the Tangipahoa River, Mississippi. Nautilus 100:18–23.

Miller, A. C., Whiting, R., Wilcox, D. B. 1989. An evaluation of a skimmer dredge for collecting freshwater mussels. Journal of Freshwater Ecology 5:151–154.

Miller, A. C., Payne, B. S., Tippet, R. 1992. Characterization of a freshwater mussel (Unionidae) community immediately downriver of Kentucky Lock and Dam in the Tennessee River. Transactions of the Kentucky Academy of Sciences 53:154–161.

Miller, A. C., Lei, J., Tom J., 1994. Shell strength of the non-indigenous zebra mussel *Dreissena polymorpha* (Pallas) in comparison to two other freshwater bivalve species. Veliger 37:319–321.

Mills, E. L., Rosenberg, G., Spidle, A. P., Ludyanskiy, M., Pligin, Y., May, B. 1996. A review of the biology and ecology of the quagga mussel (*Dreissena bugensis*), a second species of freshwater dreissenid introduced to North America. American Zoologist 36:271–286.

Mills, H. B., Starrett, W. C., Belrose, F. C. 1966. Man's effect on the fish and wildlife of the Illinois River. Biological Notes (Illinois Natural History Survey) No. 57:24 pp.

Mitchell, R. D. 1955. Anatomy, life history, and evolution of the mites parasitizing freshwater mussels. Miscellaneous Publications Museum of Zoology University of Michigan 89:1–28.

Morton, B. 1982. Some aspects of the population structure and sexual strategy of *Corbicula* cf. *fluminails* (Bivalvia: Corbiculacea) from the Pearl River. Peoples Republic of China. Journal of Mollusca Studies 48:1–23.

Morton, B. 1983. Feeding and digestion in Bivalvia. *in:* Saleuddin, A. S. M., Wilbur, K. M. Eds. The Mollusca. Vol. 5: Physiology, Part 2. Academic Press, New York, pp. 65–147.

Morton, B. 1985. The population dynamics, reproductive strategy and life history tactics of *Musculium lacustre* (Bivalvia: Pisidiidae) in Hong Kong. Journal of Zoology (London) (*A*) 207:581–603.

Morton, B. 1986. The population dynamics and life history tactics of *Pisiduim clarkeanum* and *P. annandalei* (Bivalvia: Pisidiidae) sympatric in Hong Kong. Journal of Zoology (London) (*A*) 210:427–449.

Morton, B., Tong, K. Y. 1985. The salinity tolerance of *Corbicula fluminea* (Bivalvia: Corbiculoidea) from Hong Kong. Malacological Review 18:91–95.

Nagabhushanam, R., Lomte, V. S. 1981. Observations on the relationship between neurosecretion and periodic activity in the mussel, *Parreysia corrugata*. Indian Journal of Fisheries 28:261–265.

Nalepa, T. F., Gauvin, J. M. 1988. Distribution, abundance, and biomass of freshwater mussels (Bivalvia: Unionidae) in Lake St. Clair. Journal of Great Lakes Research 14:411–419.

Nalepa, T. F., Gardner, W. S., Malczyk, J. M. 1991. Phosphorus cycling by mussels (Unionidae: Bivalvia) in Lake St. Clair. Hydrobiologia 219:239–250.

Narain, A. S., Singh, K. 1990. Unionid rectal structure in relation to function and with particular reference to the piercing of heart by gut. Proceedings of the National Academy of Science, India 60:245–252.

Negus, C . 1966. A quantitative study of growth and production of unionid mussels in the River Thames at Reading. Journal of Animal Ecology 35:513–532.

Nelson, D. J., Scott, D. C. 1962. Role of detritus in the productivity of a rock–outcrop community in a Piedmont stream. Limnology and Oceanography 7:396–413.

Neves, R. J. 1993. A state-of-the-unionids address. *in*: Cummings, K. S., Buchanan, A. C., Koch, L. M. Eds. Conservation and management of freshwater mussels. Proceedings of a Symposium, 1992, Upper Mississippi River Conservation Committee, Rock Island, IL, pp. 2–9.

Neves, R. J. 1997. A national strategy for the conservation of native freshwater mussels. *in*: Cummings, K. S., Buchanan, A. C., Mayer, C. A., Naimo, T. J. Eds. Conservation and management of freshwater mussels II: initiatives for the future. Proceedings of a Symposium, 1995, Upper Mississippi River Conservation Committee, Rock Island, IL, pp. 1–10.

Neves, R. J., Moyer, S. N. 1988. Evaluation of techniques for age determination of freshwater mussels (Unionidae). American Malacological Bulletin 6:179–188.

Neves, R. J., Zale, A. V. 1982. Freshwater mussels (Unionidae) of Big Moccasin Creek, Southwestern Virginia. Nautilus 96:52–54.

Neves, R. J., Weater, L. R., Zale, A. V. 1985. An evaluation of host fish suitability for glochidia of *Villosa vanuxemi* and *V. nebulosa* (Pelecypoda: Unionidae). American Midland Naturalist 113:13–18.

Neves, R. J., Bogan, A. E., Williams, J. D., Ahlstedt, A. S., Hartfield, P. W. 1997. Status of aquatic mollusks in the southeastern United States: a downward spiral of diversity. *in:* Benz, G. W., Collins, D. E. Eds. Aquatic fauna in peril: the southeastern perspective, Special Publication 1, Southeast Aquatic Research Institute, Chattanooga, TN, pp. 43–85.

New York Sea Grant. 1999. North American range of the zebra mussel. Dreissena! 10:8–9.

Nichols, S. J. 1996. Variations in the reproductive cycle of *Dreissena polymorpha* in Europe, Russia, and North America. American Zoologist 36:311–325.

Nichols, S. J. and Black, M.G. 1994. Identification of larvae: the zebra mussel (*Dreissena polymorpha*), quagga mussel (*Dreissena rosteriformis bugensis*), and Asian clam (*Corbicula fluminea*). Canadian Journal of Zoology 72:406–417.

Oesch, R. D. 1984. Missouri naiades. A guide to the mussels of Missouri. Missouri Department of Conservation, Jefferson City, MO.

Okland, K. A., Kuiper, J. G. J. 1982. Distribution of small mussels (Sphariidae) in Norway, with notes on their ecology. Malacologia 22:469–477.

Ortmann, A. E. 1909. The destruction of the freshwater fauna in western Pennsylvania. Proceedings of the American Philosophical Society 48:90–110.

Ortmann, A. E. 1918. The nayades (freshwater mussels) of the Upper Tennessee drainage. With notes on synonymy and distribution. Proceedings of the American Philosophical Society 57:521–626.

Paloumphis, A. A., Starrett, W. C. 1960. An ecological study of benthic organisms in three Illinois River flood plain lakes. American Midland Naturalist 64:406–435.

Panha, S. 1993. Glochidiosis and juvenile production in a freshwater pearl mussel, *Chamberlainia hainesiana*. Invertebrate Reproduction and Development 24:157–160.

Park, J., Okino, T. 1994. Seasonal distribution and growth of *Sphaerium* (*Musculium*) *japonicum* (Bivalvia; Sphaeriidae) in Lake Suwa, Japan. Journal of the Faculty of Science, Shinsiu University 29:49–56.

Parker, R. S., Hackney, C. T., Vidrine, M. F. 1984. Ecology and reproductive strategy of a south Louisiana freshwater mussel, *Glebula rotundata* (Lamarck) (Unionidae: Lampsilini). Freshwater Invertebrate Biology 3:53–58.

Parmalee, P. W. 1967. The freshwater mussels of Illinois. Illinois State Museum Popular Science Series 8:1–108.

Parmalee, P. W. 1988. A comparative study of late prehistoric and modern molluscan faunas of the Little Pigeon River System, Tennessee. American Malacological Bulletin 6:165–178.

Parmalee, P. W., Bogan, A. E. 1998. The freshwater mussels of Tennessee. The University of Tennessee Press, Knoxville, TN.

Parmalee, P. W., Hughes, M. H. 1993. Freshwater mussels (Mollusca: Pelecypoda: Unionidae) of Tellico Lake: Twelve years after impoundment of the Little Tennessee River. Annals of the Carnegie Museum 62:81–93.

Parmalee, P. W., Klippel, W. E. 1982. A relic population of *Obovaria retusa* in the middle Cumberland River, Tennessee. Nautilus 96:30–32.

Parmalee, P. W., Klippel, W. E. 1984. The naiad fauna of the Tellico River, Monroe County, Tennessee. American Malacological Bulletin 3:41–44.

Parmalee, P. W., Klippel, W. E., Bogan, A. E. 1982. Aboriginal and modern freshwater mussel assemblages (Pelecypoda: Unionidae) from the Chickamauga Reservoir, Tennessee. Brimleyana 8: 75–90.

Paterson, C. G. 1983. Effect of aggregation on the respiration rate of the freshwater unionid bivalve, *Elliptio complanata* (Solander). Freshwater Invertebrate Biology 2:139–146.

Paterson, C. G. 1984. A technique for determining apparent selective filtration in the fresh-water bivalve *Elliptio companata* (Lightfoot). Veliger 27:238–241.

Paterson, C. G. 1985. Biomass and production of the unionid, *Elliptio companata* (Lightfoot) in an old reservoir in New Brunswick, Canada. Freshwater Invertebrate Biology 4: 201–207.

Paterson C. G., Cameron, I. F. 1985. Comparative energetics of two populations of the unionid, *Anodonta cataracta* (Say). Freshwater Invertebrate Biology 4:79–90.

Paulkstis, G. L., Janzen, F. J., Tucker, J. K. 1996. Response of aerially-exposed zebra mussels (*Dreissena polymorpha*) to subfreezing temperatures. Journal of Freshwater Ecology 11:513–519.

Payne, B. S., Miller, A. C. 1987. Effects of current velocity in the freshwater bivalve *Fusconaia ebena*. American Malacological Bulletin 5:177–179.

Payne, B. S., Miller, A. C. 1989. Growth and survival of recent recruits to a population of *Fusconaia ebena* (Bivalvia: Unionidae) in the lower Ohio River. American Midland Naturalist 121:99–104.

Prezant, R. S., Charlermwat, K. 1984. Flotation of the bivalve *Corbicula fluminea* as means of dispersal. Science 225:1491–1493.

Price, R. E., Knight, L. A., Jr. 1978. Mercury, cadmium, lead and arsenic in sediments, plankton, and clams from Lake Washington and Sardis Reservoir, Mississippi, October 1975–May 1976. Pesticides Monitoring Journal 11:182–189.

Prokopovich, N. P. 1969. Deposition of clastic sediments by clams. Sedimentary Petrology 39:891–901.

Pugsley, C. W., Hebert, P. D. N., Wood, G. W., Brotea, G., Obal, T. W. 1985. Distribution of contaminants in clams and sediments from the Huron–Erie corridor. I-PCBs and octachlorostyrene. Journal of Great Lakes Research 11:275–289.

Pynnönen, K. 1991. Influence of aluminum and H^+ on the electrolyte homeostatsis in the Unionidae, *Anodonta anatina* L. and *Unio pictorum* L. Archives of Environmental Contamination and Toxicology 20:218–255.

Quigley, M. A., Gardiner, W. S., Gordon, W. M. 1992. Metabolism in the zebra mussel (*Dreissena polymorpha*) in Lake St. Clair of the Great Lakes. *in*: Nalepa, T. F., Schloesser, D. W. Eds. Lewis Publishers, Boca Raton, FL, pp. 295–306.

Ram, J. L., McMahon, R. F. 1996. Introduction: the biology, ecology, and physiology of zebra mussels. American Zoologist 36:239–243.

Ram, J. L., Fong, P. F., Garton, D. W. 1996. Physiological aspects of zebra mussel reproduction: maturation, spawning and fertilization. American Zoologist 36:326–338.

Rao, K. R., Muley, S. D., Mane, U. H. 1987. Impact of some ecological factors on cholesterol content from soft tissues of freshwater bivalve *Indonaia caeruleus*. Journal of Hydrobiology 3:67–69.

Reid, R. G. B., McMahon, R. F., Foighil, D. O., Finnigan, R. 1992. Anterior inhalant currents and pedal feeding in bivalves. Veliger 35:93–104.

Rhoads, S. N. 1899. On a recent collection of Pennsylvanian mollusks from the Ohio River system below Pittsburgh. Nautilus 12:133–138.

Ricciardi, A., Serrouya, R., Whoriskey, F. G. 1995. Aerial exposure tolerance of zebra and quagga mussels (Bivalvia: Dreissenidae): implications for overland dispersal. Canadian Journal of Fisheries and Aquatic Science 52:470–477.

Ricciardi, A., Whoriskey, F. G., Rasmussen, J. B. 1997. The role of zebra mussel (*Dreissena polymorpha*) in structuring macroinvertebrate communities on hard substrata. Canadian Journal of Fisheries and Aquatic Science 54:2596–2608.

Robinson, J. V., Wellborn, G. A. 1988. Ecological resistance to the invasion of a freshwater clam, *Corbicula fluminea:* fish predation effects. Oecologia 77:445–452.

Rocha, E., Azevedo, C. 1990. Ultrastructural study of the spermatogenesis of *Anodonta cygnea* L. (Bivalvia, Unionidae). Invertebrate Reproduction and Development 18:169–176.

Rooke, J. B., Mackie, G. L. 1984a. Mollusca of six low-alkalinity lakes in Ontario. Canadian Journal of Fisheries and Aquatic Science 41:777–782.

Rooke, J. B., Mackie, G. L. 1984b. Growth and production of three species of molluscs in six low alkalinity lakes in Ontario, Canada. Canadian Journal of Zoology 62:1474–4178.

Rooke, J. B., Mackie, G. L. 1984c. Laboratory studies of the effects of Mollusca on alkalinity of their freshwater environment. Canadian Journal of Zoology 62:793–797.

Rosenthal, R. J. 1969. A method of tagging mollusks underwater. Veliger 11:288–289.

Ruppert, E. E., Barnes, R. D. 1994. Invertebrate zoology, 6th ed., Saunders College Publishing, Philadelphia, PA.

Russell-Hunter, W. D., Buckley, D. F. 1983. Actuarial bioenergetics of nonmarine molluscan productivity. *in:* Russell-Hunter, W. D. Ed. The Mollusca. Vol. 6: Ecology. Academic Press, New York, pp. 464–503.

Sahib, I. K. A., Narasimhamurthy, B., Sailatha, D., Begum, M. R., Prasad, K. S., Roa, K. V. R. 1983. Orientation in the glycolytic potentials of carbohydrate metabolism in the selected tissues of the aestivating freshwater pelecypod, *Lamellidens marginalis* (Lamarck). Comparative Physiology and Ecology 8:180–184.

Saintsing, D. G., Dietz, T. H. 1983. Modification of sodium transport in freshwater in mussels by prostaglandins, cyclic AMP and 5-hydroxytryptamine: effects of inhibitors of prostaglandin synthesis. Comparative Biochemistry and Physiology C76: 285–290.

Saleuddin, A. S. M., Petit, H. P. 1983. The mode of formation and the structure of the periostracum. *in:* Saleuddin, A. S. M., Wilbur, K. M. Eds. The Mollusca. Vol. 4: Physiology, Part 1: Academic Press, New York, pp. 199–234.

Salmon, A., Green, R. H. 1983. Environment determinants of unionid clam distribution in the Middle Thames River, Ontario. Canadian Journal of Zoology 61:832–838.

Schloesser, D. W., Nalepa, T. F., Mackie, G. L. 1996. Zebra mussel infestation of unionid bivalves (Unionidae) in North America. American Zoologist 36:300–310.

Schloesser, D. W., Smithee, R. D., Longton, G. D., Kovalak, W. P. 1997. Zebra mussel induced mortality of unionids in firm substrata of western Lake Erie and a habitat for survival. American Malacological Bulletin 14:67–74.

Servos, M. R., Rooke, J. B., Mackie, G. L. 1985. Reproduction of selected Mollusca in some low alkalinity lakes in south-central Ontario. Canadian Journal of Zoology 63:511–515.

Sheldon, F., Walker, K. F. 1989. Effects of hypoxia on oxygen consumption by two species of freshwater mussel (Unionacea; Hyriidae) from the River Murray. Australian Journal of Marine and Freshwater Research 40:491–499.

Sibly, R. M., Calow, P. 1986. Physiological ecology of animals: an evolutionary approach. Blackwell, London.

Sickel, J. B. 1986. *Corbicula* population mortalities: factors influencing population control. American Malacological Bulletin, Special Edition. No. 2:89–94.

Silverman, H., Steffens, W. L., Dietz, T. H. 1983. Calcium concretions in the gills of a freshwater mussel serve as a calcium reservoir during periods of hypoxia. Journal of Experimental Zoology 227:177–189.

Silverman, H., Steffens, W. L., Dietz, T. H. 1985. Calcium from extracellular concretions in the gills of freshwater unionid mussels is mobilized during reproduction. Journal of Experimental Zoology 236:137–147.

Silverman, H., Kays, W. T., Dietz, T. H. 1987. Maternal calcium contribution of glochidial shells in freshwater mussels (Eulamellibranchia: Unionidae). Journal of Experimental Biology 242: 137–146.

Silverman, H., Silby, L. D., Steffens, W. L. 1988. Calmodulin-like calcium-binding protein identified in the calcium-rich mineral deposits from freshwater mussel gills. Journal of Experimental Zoology 247:224–231.

Silverman, H., Richard, P. E., Goddard, R. H., Dietz, T. H. 1989. Intracellular formation of calcium concretions by phagocytic cells in freshwater mussels. Canadian Journal of Zoology 67: 198–207.

Silverman, H., Achberger, E. C., Lynn, J. W,. Dietz, T. H. 1995. Filtration and utilization of laboratory-cultured bacteria by *Dreissena polymorpha*, *Corbicula fluminea* and *Carunculina texasensis*. Biological Bulletin (Woods Hole, Mass.) 189:308–319.

Silverman, H., Lynn, J. W., Achberger, E. C., Dietz, T. H., 1996a. Gill structure in zebra mussels: bacterial-sized particle filtration. American Zoologist 36:373–384.

Silverman, H., Lynn, J. W., Achberger, E. C., Dietz, T. H. 1996b. Particle capture by the gills of *Dreissena polymorpha*: structure and function of the latero-frontal cirri. Biol. Bull (Woods Hole, Mass.) 191:42–54.

Silverman, H., Nichols, S. J., Cherry, J. S., Achberger, E. C., Lynn, J. W., Dietz, T. H. 1997. Clearance of laboratory-cultured bacteria by freshwater bivalves: differences between lentic and lotic unionids. Canadian Journal of Zoology 75:1857–1866.

Silverman, H., Lynn, J. W., Beninger, P. G., Dietz, T. H. 1999. The role of latrero-frontal cirri in particle capture by the gills of *Mytilus edulis*. Biological Bulletin (Woods Hole, MA) 197:368–376.

Singh, D K., Thakur, P. K., Munshi, J. S. D. 1991. Food and feeding habits of a freshwater bivalve, *Parreysia favidens* (Benson) from the Kosi River System. Journal of Freshwater Biology 3:287–293.

Smith, B. J., Kershaw, R. C. 1979. Field guide to the non-marine molluscs of south eastern Australia. Australian National University Press, Canaberra.

Smith, D. G. 1985a. A study of the distribution of freshwater mussels (Mollusca: Pelecypoda: Unionidae) of the Lake Champlain drainage in northwestern New England. American Midland Naturalist 114:19–29.

Smith, D. G. 1985b. Recent range expansion of the freshwater mussel, *Anadonta implicata* and its relationship to clupeid fish restoration in the Connecticut River system. Freshwater Invertebrate Biology 4:105–108.

Smith, L. M., Vangilder, L. D., Hoppe, R. T., Morreale, S. J., Brisbin, I. L., Jr. 1986. Effect of diving ducks on benthic food resources during winter in South Carolina, U.S.A. Wildfowl 37:136–141.

Sparks, B. L., Strayer, D. L. 1998. Effects of low dissolved oxygen on juvenile *Elliptio complanata* (Bivalvia: Unionidae). Journal of the North American Benthological Society 17:129–134.

Sprung, M. 1991. Costs of reproduction: a study on the metabolic requirements of the gonads and fecundity of the bivalve *Dreissena polymorpha*. Malacologia 33:63–70.

Sprung, M., Rose, U. 1988. Influence of food size and food quantity on the feeding of the mussel *Dreissena polymorpha*. Oecologia 77:526–532.

Stanczykowska, A. 1984. Role of bivalves in the phosphorus and nitrogen budget in lakes. Verhandlungen-Internationale Vereinigung fur Theoretische und Angewandte Limnologie 22:982–985.

Stansbery, D. H. 1970. 2. Eastern freshwater mollusks. (I.) The Mississippi and St. Lawrence River systems. American Malacological Union Symposium on Rare and Endangered Mollusks. Malacologia 10:9–22.

Stansbery, D. H. 1971. Rare and endangered freshwater mollusks in eastern United States. *in*: Jorgensen, S. E., Sharp, R. E., Eds. Proceedings of a symposium on rare and endangered mollusks (naiads) of the U.S. Region 3, Bureau Sport Fisheries and Wildlife, U.S. Fish Wildlife Service. Twin Cities, MN, pp. 5–18f.

Starrett, W. C. 1971. Mussels of the Illinois River. Illinois Natural History Survey Bulletin 30:264–403.

Starnes, L. B., Bogan, A. E. 1988. The mussels (Mollusca: Unionidae) of Tennessee. American Malacological Bulletin 6:9–37.

Stearns, S. C. 1980. A new view of life-history evolution. Oikos 35:266–281.

Steffens, W. L., Silverman, H., Dietz, T. H. 1985. Localization and distribution of antigens related to calcium-rich deposits in the gills of several freshwater bivalves. Canadian Journal of Zoology 63:348–354.

Stern, E. M. 1983. Depth distribution and density of freshwater mussels (Unionidae) collected with scuba from the lower Wisconsin and St. Croix Rivers. Nautilus 97:36–42.

Stewart, T. W., Miner, J. G., Lowe, R. L. 1998. Quantifying mechanisms for zebra mussel effects on benthic macroinvertebrates: organic matter production and shell-generated habitat. Journal of the North American Benthological Society 17:81–94.

St. John, F. L. 1982. Crayfish and bivalve distribution in a valley in southwestern Ohio. Ohio Journal of Science 82:242–246.

Stoeckel, J. A, Schneider, D. W., Soeken, L. A., Blodgett, K. D., Sparks, R. E. 1997. Larval dynamics of a riverine metapopulation: implications for zebra mussel recruitment, dispersal, and control in a large river system. Journal of the North American Benthological Society 16:586–601.

Stone, N. M., Earl, R., Hodgson, A., Mather, J. G., Parker, J., Woodward, F. R. 1982. The distributions of three sympatric mussel species (Bivalvia: Unionidae) in Budworth Mere, Cheshire. Journal of Molluscan Studies 48:266–274.

Strayer, D. 1983. The effects of surface geology and stream size on freshwater mussel (Bivalvia: Unionidae) distribution in southeastern Michigan, U.S.A. Freshwater Biology 13:253–264.

Strayer, D. L. 1993. Macrohabitats of freshwater mussels (Bivalvia: Unionacea) in streams of the northern Atlantic Slope. Journal of the North American Benthological Society 12:236–246.

Strayer, D. L. 1999a. Use of flow refuges by unionid mussels in rivers. Journal of the North American Benthological Society 18:468–476.

Strayer, D. L. 1999b. Effects of alien species on freshwater mollusks in North America. Journal of the North American Benthological Society 18:74–98.

Strayer, D. L., Jirka, K. J. 1997. The pearly mussels of New York State. The New York State Museum Memoir 25, The University of the State of New York, The State Education Department, pp. xiii, 1–113.

Strayer, D. L., Ralley, J. 1993. Microhabitat use by assemblage of stream-dewelling unionaceans (Bivalvia), including two rare species of *Alasmidonta*. Journal of the North American Benthological Society 12:247–258.

Strayer, D. L., Caraco, N. F., Cole, J. J., Findlay, S., Pace, M. L. 1999. Transformation of freshwater ecosystems by bivalves. Bioscience 49:19–27.

Summathi, V. P., Chetty, A. N. 1990. Ambient ammonia clearance by bivalve mollusc *Lamellidens corrianus* (Lea). Environment and Ecology 8:1333–1334.

Tankersley, R. A. 1996. Multipurpose gills: effect of larval brooding on the feeding physiology of freshwater unionid mussels. Invertebrate Biology 11:243–255.

Tatem, H. E. 1986. Bioaccumulation of polychlorinated biphenyls and metals from contaminated sediment by freshwater prawns, *Macrobrachium rosenbergii* and clams, *Corbicula fluminea*. Archives of Environmental Contamination and Toxicology 15:171–183.

Taylor, R. W. 1985. Comments on the distribution of freshwater mussels (Unionacea) of the Potomac River headwaters in West Virginia. Nautilus 99:84–87.

Taylor, R. W., Spurlock, B. D. 1982. The changing Ohio River naiad fauna: a comparison of early Indian Middens with today. Nautilus 96:49–51.

Tevesz, M. J. S., Cornelius, D. W., Fisher, J. B. 1985. Life habits and distribution of riverine *Lampsilis radiata luteola* (Mollusca: Bivalvia). Kirtlandia 41:27–34.

Thompson, C. M., Sparks, R. E. 1977. Improbability of dispersal of adult Asiatic clams, *Corbicula manilensis,* via the intestinal tract of migratory water fowl. American Midland Naturalist 98: 219–223.

Thompson, C. M., Sparks, R. E. 1978. Comparative nutritional value of a native fingernail clam and the introduced Asiatic clam. Journal of Wildlife Management 42:391–396.

Thorp, J. H., Bergey, E. A. 1981. Field experiments on responses of a freshwater, benthic macroinvertebrate community to vertebrate predators. Ecology 62:365–375.

Timm, H., Mutvei, H. 1993. Shell growth of the freshwater unionid *Unio crassus* from Estonian rivers. Proceedings of the Estonian Academy of Sciences Biology 42:55–67.

Trueman, E. R. 1983. Locomotion in molluscs. *in:* Saleuddin, A. S. M., Wilbur, K. M., Eds. The Mollusca. Vol. 4: Physiology. Part 1. Academic Press, New York, pp. 155–198.

Turgeon, D. D., Bogan, A. E., Coan, E. V., Emerson, W. K., Lyons, W. G., Pratt, W. L., Roper, C. F. E., Scheltema, A., Thompson, F. G., Williams, J. D. 1988. Common and scientific names of aquatic invertebrates from the United States and Canada: Mollusks. American Fisheries Society Special Publication 16, Bethesda MD.

Turgeon, D. D., Quinn, J. F., Jr., Bogan, A. E., Coan, E. V., Hochberg, F. G., Lyons, W. G., Mikkelsen, P., Neves, R. J., Roper, C. F. E., Rosenberg, G., Roth, B., Scheltema, A., Sweeney, M. J., Thompson, F. G., Vecchione, M., Williams, J. D. 1998. Common and scientific names of aquatic invertebrates from the United States and Canada: Mollusks, *American Fisheries Society Special Publication 26,* 2nd ed., pp. 536 [Also on CD-ROM].

Ussary, T. A., McMahon, R. F. 1994. Comparative study of the dessication resistance of zebra mussels (*Dreissena polymorpha*) and quagga mussels (*Dreissena bugensis*). *in*: Proceedings: Fourth international zebra mussel conference '94. Wisconsin Sea Grant Institute, Madison, WI, pp. 351–369.

van der Schalie, H. 1938. Contributing factors in the depletion of naiads in eastern United States. Basteria 3:51–57.

van den Thillart, G., de Vries, I. 1985. Excretion of volatile fatty acids by anoxic *Mytilus edulis* and *Anodonta cygnea*. Comparative Biochemistry and Physiology *B*80:299–301.

Vaughn, C. C. 1997. Regional patterns of mussel species distributions in North American rivers. Ecography 20:107–115.

Vaughn, C. C., Pyron, M. 1995. Population ecology of the endangered Ouachita rock–pocketbook mussel, *Arkansia wheeleri* (Bivalvia: Unionidae), in the Kiamichi River, Oklahoma. American Malacological Bulletin 11:145–151.

Vaughn, C. C., Taylor, C. M. 1999. Impoundments and the decline of freshwater mussels: a case study of an extinction gradient. Conservation Biology 13:912–920.

Vaughn, C. C., Taylor, C. M. 2000. Macroecology of a host–parasite relationship. Ecography 23:11–20.

Vaughn, C. C., Taylor, C. M., Eberhard, K. J. 1997. A comparison of the effectivenes of timed searches vs. quadrat sampling in mussel surveys. *in*: Cummings, K. S., Buchanan, A. C., Mayer, C. A., Naimo, T. J., Eds. Conservation and management of freshwater mussels II: initiatives for the future. Proceedings of a Symposium, 1995, Upper Mississippi River Conservation Committee, Rock Island, IL, pp. 157–162.

Vermeij, G. J., Dudley, E. C. 1985. Distribution of adaptations: a comparison between the functional shell morphology of freshwater and marine pelecypods. *in:* Trueman, E. R., Clarke, M. R., Eds. The Mollusca. Vol. 10: Evolution. Academic Press, New York, pp. 461–478.

Vidrine, M. F. 1990. Fresh-water mussel–mite and mussel–*Ablabesmyia* associations in Village creek, Hardin County, Texas. Proceedings of the Louisiana Academy of Sciences 53:1–4.

Vidrine, M. F. 1993. The historical distributions of freshwater mussels in Louisiana. Gail Q. Vidrine Collectibles. Eunice, LA.

Wade, W. C., Hudson, R. G., McKinney, A. D. 1993. Comparative response of *Ceriodaphnia dubia* and juvenile *Anodonta imbecillis* to selected complex industrial whole effluents. *in*: Cummings, K. S., Buchanan, A. C., Koch, L. M., Eds. Conservation and management of freshwater mussels. Proceedings of a Symposium, 1992, Upper Mississippi River Conservation Committee, Rock Island, IL, pp. 109–112.

Walker, B. 1918. The Mollusca. *In*: Ward, H. B., Whipple, G. C., Eds. Fresh-water Biology, New York, New York, pp. 957–1020.

Waller, D. L., Rach, J. J., Cope, W. G., Luoma, J. A. 1993. A sampling method for conducting relocation studies with freshwater mussels. Journal of Freshwater Ecology 8:397–399.

Walz, N. 1978a. The energy balance of the freshwater mussel *Dreissena polymorpha* Pallas in laboratory experiments and in Lake Constance. I.. Patterns of activity, feeding and assimilation efficiency. Archiv fur Hydrobiologie 55(Supp.):83–105.

Walz, N. 1978b. The energy balance of the freshwater mussel *Dreissena polymorpha* Pallas in laboratory experiments and in Lake Constance. III. Growth under standard conditions. Archiv fur Hydrobiologie 55(Supp.):121–141.

Warren, M. L., Jr., Cicerello, D. R., Camburn, K. E., Fallo, G. J. 1984. The longitudinal distribution of the freshwater mussels (Unionidae) of Kinniconick Creek, northeastern Kentucky. American Malacological Bulletin 3:47–53.

Watters, G. T. 1992. Unionids, fishes, and the species area curve. Journal of Biogeography 19:481–490.

Watters, G. T. 1993. Mussel diversity as a function of drainage area and fish diversity: management implications. *in*: Cummings, K. S., Buchanan, A. C., Koch, L. M. Eds. Conservation and management of freshwater mussels. Proceedings of a Symposium, 1992, Upper Mississippi River Conservation Committee, Rock Island, IL, pp. 113–116.

Watters, G. T. 1994a. Form and function of unionoidean shell scupture and shape (Bivalvia). American Malacological Bulletin 11:1–20.

Watters, G. T. 1994b. Annotated bibliography of the reproduction and propagation of the unionoidea (primarily of North America). Ohio Biological Survey Miscellaneous Contributions 1:1–159.

Watters, G. T. 1995. A guide to the freshwater mussels of Ohio. Division of Wildlife, The Ohio Department of Natural Resources.

Way, C. M. 1988. An analysis of life histories in freshwater bivalves (Mollusca: Pisidiidae). Canadian Journal of Zoology 66: 1179–1183.

Way, C. M. 1989. Dynamics of filter-feeding in *Musculium transversum* (Bivalvia: Sphaeriidae). Journal of the North American Benthological Society 8:243–249.

Way, C. M., Wissing, T. E. 1982. Environmental heterogeneity and life history variability in the freshwater clams, *Pisidium variable* (Prime) and *Pisidium compressum* (Prime) (Bivalvia: Pisidiidae). Canadian Journal of Zoology 60:2841–2851.

Way, C. M., Wissing, T. E. 1984. Seasonal variability in the respiration of the freshwater clams, *Pisidium variabile* (Prime) and *P. compressum* (Prime) (Bivalvia: Pisidiidae). Comparative Biochemistry and Physiology *A*78:453–457.

Way, C. M., Hornbach, D. J., Burky, A. J. 1980. Comparative life history tactics of the sphaeriid clam, *Masculium partumeium* (Say), from a permanent and temporary pond. American Midland Naturalist 104:319–327.

Way, C. M., Hornbach, D. J., Burky, A. J. 1981. Seasonal metabolism of the sphaeriid clam, *Musculium partumeium,* from a permanent and temporary pond. Nautilus 95:55–58.

Way, C. M., Hornbach, D. J., Deneka, T., Whitehead, R. A. 1989. A description of the ultrastructure of the gills of freshwater bivalves, including a new structure, the frontal cirrus. Canadian Journal of Zoology 67:357–362.

Way, C. M., Hornbach, D. J., Miller-Way, C. A., Payne, B. S., Miller, A. C. 1990a. Dynamics of filter-feeding in *Corbicula fluminea* (Bivalvia: Corbiculidae). Canadian Journal of Zoology. 68:115–120.

Way, C. M., Miller, A. C., Payne, B. S. 1990b. The influence of physical factors on the distribution and abundance of freshwater mussels (Bivalvia: Unionacea) in the lower Tennessee River. Nautilus 103:96–98.

Weber, C. I. 1973. Biological field and laboratory methods for measuring the quality of surface waters and effluents (macroinvertebrate section). EPA-670/4-73-001. National Environment Research Center, Office of Research and Development, United States Environment Protection Agency, Cincinnati, OH.

White, D. S. 1979. The effect of lake-level fluctuations on *Corbicula* and other pelecypods in Lake Texoma, Texas and Oklahoma. *in:* Britton, J. C., Ed. Proceedings, first international *Corbicula* symposium. Texas Christian University Research Foundation, Fort Worth, Tx, pp. 82–88.

Wilbur, K. M., Saleuddin, A. S. M. 1983. Shell formation. *in:* Saleuddin, A. S. M., Wilbur, K. M., Eds. The Mollusca. Vol. 4: Physiology, Part 1. Academic Press, New York, pp. 236–287.

Wilcox, S. J., Dietz, T. H. 1995. Potassium transport in the freshwater bivalve, *Dreissena polymorpha.* Journal of Experimental Biology 198:861–868.

Wilcox, S. J., Dietz, T. H. 1998. Salinity tolerance of the freshwater bivalve *Dreissena polymorpha* (Pallas, 1771) (Bivalvia: Dreissenidae). Nautilus 111:143–148.

Williams, C. J. 1985. The population biology and physiological ecology of *Corbicula fluminea* (Müller) in relation to downstream dispersal and clam impingement on power station raw water systems. Master's Thesis, The University of Texas at Arlington, Arlington, Tx.

Williams, C. J., McMahon, R. F. 1985. Seasonal variation in oxygen consumption rates, nitrogen excretion rates and tissue organic carbon: nitrogen ratios in the introduced Asian freshwater bivalve, *Corbicula fluminea* (Müller) (Lamellibranchia: Cobiculacea). American Malacological Bulletin 3:267–268.

Williams, C. J., McMahon, R. F. 1986. Power station entrainment of *Corbicula fluminea* (Müller) in relation to population dynamics, reproductive cycle and biotic and abiotic variables. American Malacological Bulletin Special Edition No. 2:99–111.

Williams, C. J., McMahon, R. F. 1989. Annual variation of tissue biomass and carbon and nitrogen content in the freshwater bivalve, *Corbicula fluminea,* relative to downstream dispersal. Canadian Journal of Zoology 67:82–90.

Williams, J. D., Fradkin, A. 1999. *Fusconaia apalachicola*, a new species of freshwater mussel (Bivalvia: Unionidae) from precolumbian archaeological sites in the Apalachicola Basin of Alabama, Florida, and Georgia. Tulane Studies in Zoology and Botany 31:51–62]

Williams, J. D., Fuller, S. L. H., Grace, R. 1992. Effects of impoundments on freshwater mussels (Mollusca: Bivalvia: Unionidae) in the main channel of the Black Warrior and Tombigbee Rivers in western Alabama. Bulletin of the Alabama Museum of Natural History 13:1–10.

Williams, J. D., Warren, M. L., Jr., Cummings, K. S., Harris, J. L., Neves, R. J. 1993. Conservation status of freshwater mussels of the United States and Canada. Fisheries 18:6–22.

Willows, A. O., Ed. 1985. The Mollusca Vol. 8: Neurobiology and behavior, Part 1. Academic Press, New York.

Willows, A. O., Ed. 1986. The Mollusca Vol. 9: Neurobiology and behavior, Part 2. Academic Press, New York.

Winnell, M. H., Jude, D. J. 1982. Effect of heated discharge and entrainment on benthos in the vicinity of the J. H. Campbell Plant, Eastern Lake Michigan, 1978–1981. Special Report No. 94. Great Lakes Research Division, University of Michigan, Ann Arbor, MI, 202 pp.

Wissing, T. E., Hornbach, D. J., Smith, M. S., Way, C. M., Alexander, J. P. 1982. Caloric contents of corbiculacean clams (Bivalvia: Heterodonta) from freshwater habitats in the United States and Canada. Journal of Molluscan Studies 48:80–83.

Woody, C. A., Holland-Bartels, L. 1993. Reproductive characteristics of a population of the washboard mussel *Megalonaias nervosa* (Rafinesque 1820) in the upper Mississippi River. Journal of Freshwater Ecology 8:57–66.

Wu, W-L., Trueman. 1984. Observation on surface locomotion of *Sphaerium corneum* (Linneaus) (Bivalvia: Sphariidae). Journal of Molluscan Studies 50:125–128.

Wu, W-L. 1987. The burrowing of *Anodonta cygnea* (Linnaeus, 1758) (Bivalvia: Unionidae). Bulletin of Malacology Republic of China 13:29–41.

Yamada, A., Matsushima, O. 1992. The relation of d-alanine and alanine racemase activity in molluscs. Comparative Biochemistry and Physiology *B*103:617–621.

Yeager, M. M., Cherry, D. S., Neves, R. J. 1994. Feeding and burrowing behaviors of juvenile rainbow mussels, *Villosa iris* (Bivalvia: Unionidae). Journal of the North American Benthological Society 13:217–222.

Young, M. R., Williams, J. 1983a. The status and conservation of the freshwater pearl mussel *Margaritifera margaritifera* Linn. in Great Britain. Biological Conservation 25:35–52.

Young, M. R., Williams, J. 1983b. Redistribution and local recolonization by the freshwater pearl mussel *Margaritifera margaritifera* (L.). Journal of Conchology 31:225–234.

Young, M. R., Williams, J. C. 1983c. A quick secure way of making freshwater pearl mussels. Journal of Conchology 31:190.

Young, M. R., Williams, J. 1984a. The reproductive biology of the freshwater pearl mussel *Margaritifera margaritifera* (Linn.) in Scotland II. Laboratory studies. Achiv für Hydrobiologie 100: 29–43.

Young, M. R., Williams, J. 1984b. The reproductive biology of the freshwater pearl mussel *Margaritifera margaritifera* (Linn.) in Scotland I. Field studies. Archiv für Hydrobiologie 99: 405–422.

Zale, A. V., Neves, R. J. 1982. Fish hosts of four species of lampsiline mussels (Mollusca: Unionidae) in Big Moccasin Creek, Virginia. Canadian Journal of Zoology 60:2535–2542.

Zbeda, T. W., White, D. S. 1985. Ecology of the zoobenthos of southeastern Lake Michigan near the D. C. Cook Nuclear Plant. Part 4: Pisidiidae. Special Report No. 13. Great lakes Research Division, Institute of Science and Technology, University of Michigan, Ann Arbor, MI.

Zeto, M. A., Tolin, W. A., Smith, J. E. 1987. The freshwater mussels of the upper Ohio River, Greenup and Belleville Pools, West Virginia. Nautilus 101:182–185.

Zheng, H., Dietz, T. H. 1998a. Paracellular solute uptake in the freshwater bivalves *Corbicula fluminea* and *Toxolasma texasensis*. Biological Bulletin (Woods Hole, MA) 194:170–177.

Zheng, H., Dietz, T. H. 1998b. Ion transport in the freshwater bivalve *Corbicula fluminea*. Biological Bulletin (Woods Hole, MA) 194:161–169.

Zs.-Nagy, I., Holwerda, D. A., Zandee, D. I. 1982. The cytosomal energy production of molluscan tissues during anaerobiosis. Basic and Applied Histochemistry 26 (Suppl.):8–10.

12

ANNELIDA: OLIGOCHAETA, INCLUDING BRANCHIOBDELLIDAE

Ralph O. Brinkhurst
205 Cameron Court
Hermitage, TN 37076

Stuart R. Gelder
Department of Biology
University of Maine at Presque Isle
Presque Isle, ME 04769

OLIGOCHAETA

I. INTRODUCTION TO THE OLIGOCHAETA

One family of aquatic Oligochaeta[1], the Tubificidae (sludge worms), has been recognized since the time of Aristotle for its ability to develop dense colonies in organically polluted waters (Fig. 1). Beginning with the concepts that all sludge worms are pollution-tolerant indicators, recent work has made biologists aware of the wide range of ecological niches occupied by aquatic annelids in general, and oligochaetes and branchiobdellids in particular. They are found in every kind of freshwater and estuarine habitat, not only in organic mud. Specialized forms can be found in groundwater, in oligotrophic lakes and streams, and as commensals or symbionts on crayfish, molluscs, and even tree frogs.

[1] Oligochaeta may be a parahyletic group. The term is used here in the absence of a resolution into monophyletic taxa.

Ecology and Classification of North American Freshwater Invertebrates. 2nd Edition

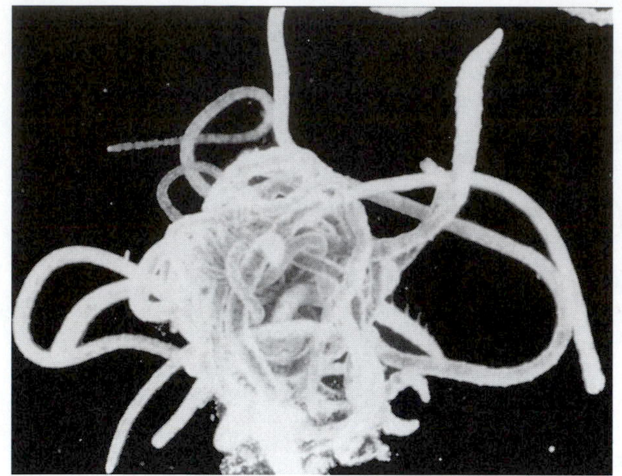

FIGURE 1 Clump of living tubificids (photograph by P. M. Chapman).

Several North American genera prey on other worms or a variety of small invertebrates. The most surprising recent discovery has been the description of a rich diversity of tubificids at all depths in the oceans of the world.

While most students are familiar with the terms Oligochaeta and Clitellata, there is no general agreement about the equivalent rankings of the various annelid groups. Some authors place the arthropods and annelids in a single phylum, but most texts separate them. Within the Annelida, it is possible to classify the Aclitellata (including Polychaeta) and the Clitellata as superclasses or classes. The former arrangement allowed us to refer to the Polychaeta, Oligochaeta, Hirudinea, and Branchiobdellida as classes, whereas the latter would demote these to orders. Relationships among the last three taxa, as well as the monotypic Acanthobdellida, are presently under investigation using cladistic methods, which greatly assist the process of determining sister-group relationships. Results of recent cladistic analyses provide evidence of a sister-group relationship between lumbriculids and branchiobdellids, which share a unique arrangement of genital organs. As a result, we have returned to an earlier classification of branchiobdellids as a taxon of family rank. The Aeolosomatidae are now recognized as a non-clitellate taxon Aphanoneura.

Biologists who reject Hennigian methods may well classify the various groups in a different way from those who do not. The various terms will be used here based on the concept of superclass Clitellata with a series of classes, although we strongly suspect that the ranking of the Acanthobdellida will shortly be revised.

II. OLIGOCHAETE ANATOMY AND PHYSIOLOGY

A. Anatomy

Oligochaete worms are bilaterally symmetrical, segmented coelomates with four bundles of chaetae on every segment except the first. In earthworms, the chaetal "bundles" usually consist of two chaetae each; but in aquatic species there are often several, and there may even be more than a dozen per bundle. In a few exceptional forms, the chaetae are absent in some or even all segments. Aquatic species (superorder Microdrili) are smaller than earthworms (Megadrili) as the name suggest, and they have simple body walls and none of the specialized regions of the digestive tract found in the earthworms. The clitellum is single-layer thick, and is limited to the genital regions in microdriles, and the eggs are large and yolky. In earthworms, secretions from the multilayered clitellum provide an alternative source of food for the eggs, which are small and less yolky. The clitellum is located well behind the gonadal segments. In aquatic species, the male and female ducts are connected to external genital pores located either in the same segment as the gonads they serve, or in the segment immediately behind them. In earthworms, the male ducts are extended back to open well behind the gonadal segments. It is generally thought that there were originally four gonadal segments, two with paired testes, and two with paired ovaries, in X–XIII (Roman numerals are used for segment number by convention). These have been reduced in number in various ways, as gamete storage and sperm transmission through copulation have evolved (see Section III.D). In some entire families and in a few individual species in other families, the gonadal segments may be found much nearer to the anterior end than usual. This may be due to a tendency to regenerate fewer than normal number of anterior segments in the predominantly asexual-reproducing forms in which this happens (family Naididae). This explanation will not suffice for sexually reproducing forms in which the forward shift involves just one segment (family Lumbriculidae). In the family Opistocystidae, the gonads are located behind segment XX in all but the one species discussed here.

The arrangement of the various reproductive organs of some typical aquatic oligochaetes is illustrated in Figure 2. The reproductive system and chaetae of one side are illustrated, but in life these animals are bilaterally symmetrical. The anatomy of the reproductive system is easy to understand if the stages in reproduction from sperm development to egg hatching are described. The mother cells of the sperm leave the testes to float free in the coelom. The amount of space

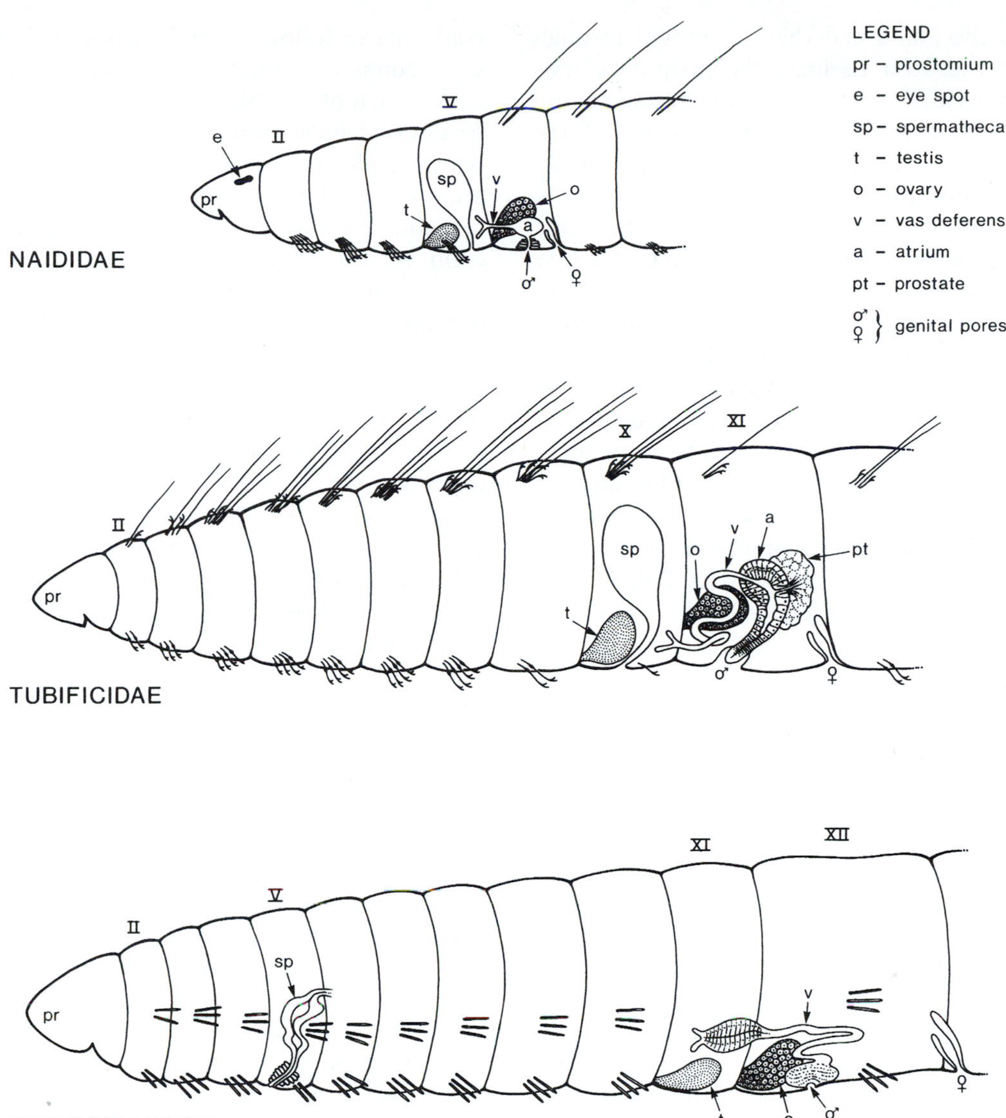

FIGURE 2 Reproductive systems and external characteristics of three major families of Oligochaeta. The worms are shown with the anterior to the left, and only the structures of the left side are visible. The dorsal and ventral chaetae are illustrated. In the Naididae, an example with dorsal chaetae beginning in VI is shown. Segments are identified with Roman numerals by convention (septa would be VII/VIII, VIII/IX, etc.). a, Atrium; e, eye spot; o, ovary; pr, prostomium; pt, prostate; sp, spermatheca; t, testis; v, vas deferens; genital pores.

available to the developing sperm masses, or morulae, is increased by the development of outpushings of the testicular segments, slightly anteriad but more extensively posteriad through several segments. These are called sperm sacs. The sperm eventually cluster around large funnels suspended on the posterior walls of the testicular segment, sometimes well within the sperm sacs. These sperm funnels lead into vasa deferentia that originally drained via the male pores, located near the ventral chaetae of the post-testicular segment. Various structures have evolved, presumably at the point where the vasa deferentia contact the original body wall. Inversion of the body wall has led to the evolution of atria. These muscular bodies were originally lined with ectodermal cells, many of which had a glandular function. They presumably fed and lubricated the sperm stored in the atria ready for copulation. The glandular cells are thought to have migrated inward, through the muscle layers, so that their cell bodies now lie in the coelom. This resulted in both an increase in the space available to the glandular cell bodies and an increase in the space available for sperm within the atria. The ways in which the gland cells have migrated through the muscles of the atria to form the prostate glands vary, and this provides a very useful taxonomic feature (though not one that can be seen without some effort). The term prostate gland is used for very different structures in earthworms and branchiobdellids. In a further evolutionary development, another section of the body wall becomes part of the male reproductive structure between the atria and the male pores. Various types of penes are found in what are considered to be the more advanced groups (family Tubificidae, family Lumbriculidae) and have evolved quite independently several times. In some penes, the normal cuticular layer over the epidermis has become hypertrophied to form cuticular penis sheaths. Sometimes this cuticular layer is hardly noticeable, but the cuticular tubes may become thick and rigid with soft penes free within them. Cuticular penis sheaths are said to be present if the cuticle over the penes can be seen in cleared, whole-mounted worms and has a consistent shape. In some species, basement membranes or cuticular lining of eversible penes appear on whole mounts, but these are usually irregular in form. These are called "apparent penis sheaths."

When worms copulate, sperm is passed to the partner and is usually deposited in saclike spermathecae. There may be one or more pairs of these structures located in or near the gonadal segments, commonly in front of the testicular segments, or in them. The number and position of the spermathecae may vary within the Lumbriculidae, but there is usually one pair in the single testicular segment (families Tubificidae, and Naididae), or separated from the other genitalia in V (family Enchytraeidae). In a few species, spermathacae may be present in some specimens but not others, or they may be missing altogether. In *Bothrioneurum* (Tubificidae), the sperm are placed in spermatophores that are attached externally to the body wall of the partner, usually near the gonopores, during copulation. In members of the tubificid subfamily Tubificinae, two types of sperm are present, which come together in organized masses called spermatozeugmata. These can be recognized within the spermathecae of mated worms. Other forms of spermatozeugmata are found in other subfamilies.

After a worm has copulated and has sperm in its spermathecae, the clitellum secretes a cocoon. The eggs have already developed in the coelomic space of the ovarian segment. This segment is extended posteriad into egg sacs. As the sperm sacs already displace the septa of several segments behind the testes, the egg sacs and their contained eggs are only detectable in segments beyond the sperm sacs. The egg and sperm sacs have been omitted from Figure 2 because it becomes confusing to follow a whole series of fingerlike protrusions running through several segments, all contained within each other. [Note that egg and sperm sacs of the Acanthobdellida and Hirudinea are not contained within each other, but lie parallel to each other. They are paired, bilateral structures just like the rest of the reproductive system.] Eggs are deposited into the cocoon via the female ducts. These consist of nothing more than ciliated funnels on the posterior walls of the ovarian segment that open into the intersegmental furrow. In a few primitive forms, the female ducts actually penetrate the postovarian segment and open on its anterior surface. It is assumed that fertilization occurs in the cocoon. The basic position for the spermathecal, male, and female pores is in the longitudinal line of the ventral (ventrolateral) chaetal bundles, but there is some variation in this due to midventral fusion, the development of median copulatory bursae, or the migration of the spermathecal pores to a more dorsal position. While the spermathecal pores may open near the anterior furrow of the segment they occupy, all the genital pores commonly lie close to the ventral chaetae.

The worm finally releases the cocoon, which formed as a ring of material around it, secreted (presumably) by the clitellum. The ends of the cocoon close, and the fertilized eggs develop inside. Small worms eventually escape through the openings at the ends of the cocoons and are recognizable as immature specimens, which can be identified if they happen to belong to one of the species with characteristic chaetae.

The various reproductive organs and the modified chaetae that may develop close to the genital pores can best be found by looking carefully in the correct segments. When the keys mention genital structures, these will only be found on mature specimens. An immature aquatic oligochaete appears to consist of very little more than the body wall with its chaetae and a simple tubular gut running through the coelom. Very few specialized anatomical features have been added to the basic vermiform shape, unlike the situation in the Polychaeta. That is why it is generally useless to illustrate the general habit of oligochaetes; useful illustrations are those of the chaetae and reproductive systems. In contrast to immature specimens, a mature worm will have an external clitellum, though this is not as prominent as that of an earthworm, and the egg and sperm sacs are full and, therefore, visible. The worm is often somewhat distended in the genital region, and the reproductive organs obscure the normal view of the gut.

As most worms burrow through soft substrates and extract their food by ingesting them, the range of externally visible anatomical adaptations are few because this habit imposes severe limitations on body shape. The prostomium lies in front of the mouth, and while its body cavity is clearly coelomic, it is considered to be presegmental. The prostomium may bear a median anterior prolongation, the proboscis. In *Bothrioneurum*, there is a median dorsal pocket containing sensory cells, the prostomial pit (Fig. 3). Segment numbering begins with the peristomium, the small segment around the mouth. In aquatic species, the peristomium is completely separate from the prostomium. It is also presegmental, and bears no chaetae. This is important to remember when using cleared, whole-mounted specimens as it is often easier to locate specific segments by counting chaetal bundles (which begin in Segment II).

The chaetae vary in form as well as number. This variation provides very useful key characters, but there is increasing evidence that the fine details of chaetal form, or even the presence or absence of a particular type of chaetae, such as hairs, can be affected by quite simple environmental variables such as pH or conductivity (Chapman and Brinkhurst, 1987). The Enchytraeidae have simple-pointed chaetae, which may be straight or sigmoid. The chaetae in a bundle may vary in size (Fig. 4). Hair chaetae are elongate, simple shafts. They are found only in the dorsal bundles of some species of Naididae, Tubificidae, and in *Crustipellis* (Opistocystidae). They are present in both dorsal and ventral bundles in the Aeolosomatidae (Aphanoneura, Aclitellata), this being one of a series of fundamental differences between these minute freshwater worms and true oligochaetes. Hair chaetae are termed hispid or serrate when they bear a series of fine lateral hairs. This condition may vary within a

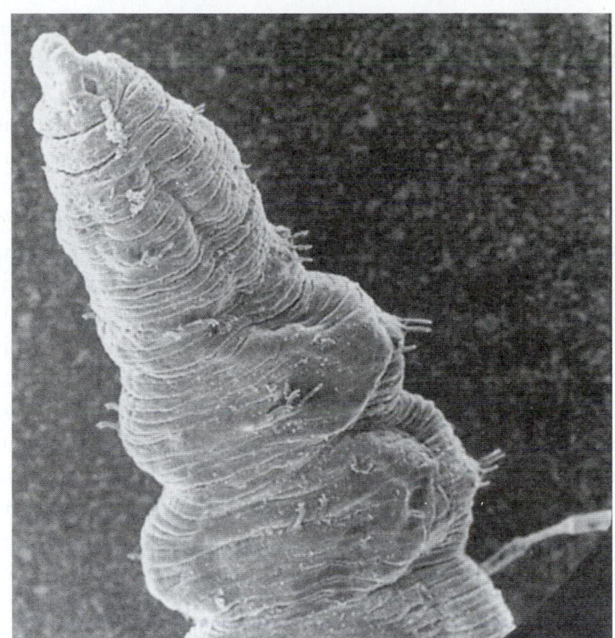

FIGURE 3 Prostomial pit, *Bothrioneurum* (photograph by P. M. Chapman).

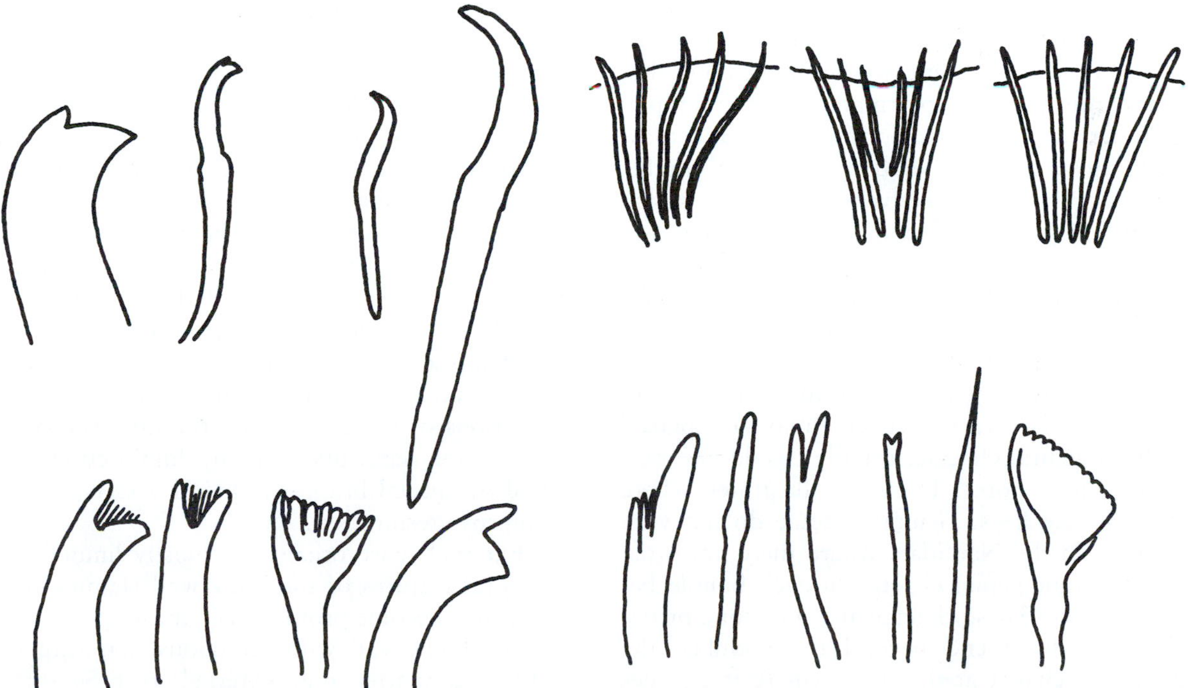

FIGURE 4 Representative chaetae. From the left, top row: lumbriculid bifid with reduced upper tooth, head end, and entire; small dorsal and large ventral sickle-shaped chaetae of *Haplotaxis;* three types of bundles from the Enchytraeidae. Bottom row: two pectinates, a palmate, and a bifid dorsal chaeta, Tubificidae; six types of dorsal needles, Naididae.

species depending on seasonal or environmental variables, but it is still used as a taxonomic character for many naidids for which the limits of intraspecific variability have yet to be established. In a few naidids, the lateral hairs may be especially prominent along one side of the hair chaeta, and these are called plumose hairs from their featherlike appearance. In one or two instances, minute bifid tips have been observed at the ectal end of hair chaetae, indicating their probable origin from the much more ubiquitous bifid chaetae. This bifid form, with the ectal end divided into two teeth, is characteristic of aquatic as opposed to terrestrial oligochaetes, although the functional reason for this is unknown. Bifid chaetae may be the only type present in all bundles, and the ventral bundles usually contain nothing else. Some simple-pointed chaetae may be found in a few anterior ventral bundles (in the tubificid genus *Spirosperma,* for instance), or they may accompany the hair chaetae in the dorsal bundles in some Naididae. A common combination in the dorsal bundles is hair chaetae accompanied by pectinate chaetae in, at least, the preclitellar bundles. Pectinate chaetae are usually of the same general form as the bifids for the particular taxon, but the space between the two teeth is filled with either a comb of thin teeth or, more rarely, a ridged web. In a few naidids and tubificids, these webbed chaetae may be expanded as palmate chaetae. In general, the dorsal chaetae accompanying the hair chaetae in the majority of naidid species differ markedly from the ventral chaetae. For this reason, they have come to be termed needles, whether they are palmate, pectinate, bifid, or simple-pointed. Some authors like to substitute the terms fascicle, capilliform, bifurcate, and crotchet for bundle, hair, bifid, and chaeta, respectively. They also use the terms distal and proximal for upper and lower tooth, the designations used in describing the variations in the length and thickness of the teeth, often a useful characteristic. [*Anglo-Saxon* terms are more succinct, and are used internationally, but the longer Latin terms are more readily translated into French and Spanish.] The normal ventral chaetae may be replaced by specialized genital chaetae, usually beside the spermathecal or penial pores. These are illustrated where appropriate in the keys. Genital chaetae do not vary much in form in the Naididae, where there are often two or three blunt penial chaetae in each bundle beside the male pore. These chaetae are quite long proximally, but the distal end beyond the nodulus (the swelling on the chaeta at the point where it emerges from the chaetal sac) is short. Penial chaetae are usually somewhat long and thicker than the normal ventral chaetae. Penial chaetae, like those of the Naididae, may be found in the Rhyacodrilinae, a subfamily of Tubificidae, though there are usually several to a bundle. They are usually arranged with the outer ends close together, the inner ends spread out. In Tubificinae, the ventral chaetae in the penial segment are usually lost but the spermathecal chaetae are often modified, with elaborate distal ends shaped like trowels. They may be used to introduce the modified spermatozeugmata into the spermathecae during copulation. The functions of both these and the penial chaetae as claspers are based on speculation because aquatic oligochaetes have rarely been observed mating. Spermathecal chaetae are often associated with special glands. Where these are large, and there is a great deal of competition for space in the genital segments, the spermathecal chaetae and glands lie immediately in front of the spermathecal segment (*Rhizodrilus,* Tubificidae). In general, the number and diversity of chaetae is reduced toward the posterior end of the worm, and modified teeth on the bifids commonly become "normal." There are a few exceptions. The upper teeth of the chaetae of *Amphichaeta* (Naididae) get longer posteriorly, and the posterior segments of *Telmatodrilus vejdovskyi* (Tubificidae) bear chaetae with almost brushlike tips (see illustrations in the key).

A few other anatomical features are used in aquatic oligochaete taxonomy. External gills are found in a few taxa. The posterior segments of the tubificid *Branchiura* bear single dorsal and ventral gill filaments. These are longest in the median segments of the gill-bearing region, very short at each end of the row. *Dero* (Naididae) has gills at the posterior end of the body, usually enclosed in a gill chamber (branchial fossa). These are best observed in living specimens. All aquatic oligochaetes have red blood pigments which aid oxygen uptake and transport. In most species, respiration takes place through the posterior body wall or even within the pre-anal part of the intestine. Water may be pumped in and out of the anus for this purpose. Most aquatic oligochaetes protrude the tail end out of the sediment into the water and may wave the tail to create enough disturbance to maintain high concentrations of oxygen around the tail. The body wall of posterior segments in many lumbriculids is penetrated by special branching blood vessels, presumably to improve respiration.

External sense organs are mostly limited to ultrastructural features of the body wall (Jamieson, 1981). These may become prominent in those species that protect the body wall with secretions that trap bacteria and foreign matter in an external crust. Sensory papillae can be seen protruding through such layers in living specimens, but they are usually retracted in fixed material. This has led to arguments about their presence or absence in certain species, but we may assume that any worm with its normal body-wall sensory system denied access to information by being encrusted has developed localized retractile sensory papillae. It is also easy to understand why the necessarily sensitive prostomium and first two or three segments may be naked in taxa that have protected body walls, and why the anterior end is often retractile within the rest of the protected body. Simple pigmented eye spots are located on the lateral surface of segment I in a number of naidids, but may be present or absent in various specimens of the same species.

The majority of aquatic oligochaetes feed by ingesting sediment, extracting nutrition from the organic matter and especially the bacteria, which comprise less than 10% of it. They feed continuously in a conveyor-belt fashion, often ingesting recolonized feces. Food is obtained by everting the roof of the pharynx through the mouth. Glandular cells associated with the pharynx have evolved into pharyngeal or septal glands, in which the cell bodies lie in the coelomic cavities of several anterior segments, attached to the posterior septa. Intercellular canals run forward from these cell bodies along the lateral sides of the gut to open dorsally on the pharyngeal pad, which is one of the characteristics of worms of this group; this feature is absent in the Aeolosomatidae. The pharyngeal glands are used in taxonomy of the Enchytraeidae, but not in other families. The pharyngeal glands are lost in predatory oligochaetes, but traces of the dorsal pad can still be found in some (*Chaetogaster*, Naididae) although it has been lost in others (*Phagodrilus*, Lumbriculidae, and *Haplotaxis*, Haplotaxidae).

B. Physiology

The primary emphasis of studies on worm physiology has been on respiration, as it provides a simple model of the value of red blood pigment function and anaerobic metabolism. A good review of respiration studies with reference to their use in determining production in worms was provided by Bagheri and McLusky (1984), who studied the estuarine tubificid *Tubificoides benedii* (*benedeni*). Oxygen consumption varied from >0.01 to <0.5 μL/mg/h, but while there were small, apparent linear increases in rate with temperature, these were not statistically significant. These results are among the lowest reported for worms, suggesting that the experimental technique is not creating stress in the test animals. The response of respiration levels to stress has been suggested as a tool in pollution biology (Brinkhurst *et al.*, 1983). Most tubificids respire at a more or less constant rate as the external oxygen concentration is lowered. Once a critical value is reached, respiration rate falls dramatically. Stress can elevate or depress the constant respiration rate or it may shift the critical value. It may also cause a reduction in, or a total loss of, respiratory control above the critical value. In some instances, respiratory control actually increased in the face of supposed stress factors. A unique phenomenon is that mixing certain tubificid species together will actually reduce their respiration rates by a significant amount (Chapman *et al.*, 1982a). This result was anticipated by earlier studies of worm production using monocultures and mixed cultures of three tubificid species. In those studies with mixed populations (summarized in Brinkhurst and Austin, 1978), growth rates increased when respiration decreased, as did assimilation efficiencies. These experiments seemed to provide an explanation for the way in which tubificids live in small clumps of mixed species. Some worms preferred the recolonized feces of another species as food and sought the company of that species. Even when they were only in water from monocultures of the appropriate second species, worms moved less, ate more, and (in the presence of living worms) ingested more of their preferred food. This caused a shift in all of the variables measured in production work based on estimating caloric or carbon budgets. The work provided an unusual example of the value of multi-species flocks.

Freshwater oligochaetes, such as the pollution-tolerant tubificids *Tubifex tubifex* and *Limnodrilus hoffmeisteri*, can withstand exposure to 10 ppt salinity, but others can only survive at 5 ppt (Chapman *et al.*, 1982b, c). A low level of salinity actually seems to improve the ability of freshwater worms to withstand stress factors. It is less surprising that the provision of sediment improves stress resistance. Estuarine species may withstand a very wide range of environmental stress, including extreme salinity values (*Paranais*, Naididae; *Monopylephorus*, Tubificidae). It would be a mistake to believe that these animals are exposed to a wide range of salinities due to daily tidal excursions, however. In muddy situations in rivers such as the Fraser in British Columbia, the "interstitial" salinity goes through a seasonal cycle related to the capture and release of snow in the headwaters. As temperatures increase, the flow of freshwater increases and tidal seawater incursions are limited to the outer estuary. All of the benthic communities move slowly up and down the estuary in response to this annual rhythm of interstitial salinity, maintaining themselves at what are presumably preferred salinity ranges for each species. It is selection for these optimal salinities (that remain quite constant within the mud on short-term basis) that causes the sequence of species observed along the salinity gradient in estuaries (Chapman, 1981; Chapman and Brinkhurst, 1981).

While worms in cocoons may be able to survive desiccation, there is no experimental evidence of this. Adult *Tubifex tubifex* (Tubificidae) are able to form cysts for adult worms that can survive unmoistened for 14 days, or for as long as 70 days with occasional dampening. Once the worms have been moistened, they may emerge in 20 h (Kaster and Bushnell, 1981). The lumbriculid *Lumbriculus variegatus* can also encyst, but it only survived for 4 days under experimental conditions. In both instances, the covering of foreign matter, which helps to disguise the cyst, is also essential to the survival of the worms. There are records of a handful of other lumbriculids undergoing fission in cysts, but no physiological work has been done. *Lumbriculus* also forms a different type of cyst to withstand freezing (Olsson, 1981). Specimens of the tubificid *Limnodrilus profundicola* were collected by R. Brinkhurst at a height of ca. 4000 m on Mount Evans, Colorado. After being accidentally frozen into a solid block of ice, the worms emerged quite unscathed when thawed the next day. As so many aquatic oligochaetes, especially the Naididae, are cosmopolitan or at least very widely distributed, one would expect more evidence of adaptations to counter desiccation. Dispersal mechanisms of oligochaetes are still a matter for speculation.

The development of eye spots is unique to the Naididae, a family of small, delicate worms, some of which can make rudimentary swimming movements. Other naidid species, while not free-living, occupy tubes attached to aquatic plants and may protrude the anterior end. The majority of aquatic oligochaetes avoid the light by keeping the head end buried in their food supply, the sediment. Most aquatic worms have quite sensitive tails and have a rapid withdrawal reflex based on a mechanosensory detection system that is responsive to the approach of potential predators (Zoran and Drewes, 1987). *Branchiura* has the fastest escape reaction reported for any invertebrate. This speed is associated with the development of the lateral giant nerve fibers in the ventral nerve cord. These are responsible for rapid forward transmission of sensory information. The median giant nerve fiber transmits posteriad from the head end. In *Lumbriculus,* which lives in shallow littoral habitats, the body is extended up to the air–water interface, with the tail lying horizontally in it. Not surprisingly, these worms have what Drewes and Fourtner (1988) termed hindsight or light-sensitive tails, and they respond to passing shadows. Light can be damaging to worms.

A great deal of work has been done in the laboratory on the effect of toxic substances on worms, summarized by Chapman and Brinkhurst (1984) and Wiederholm *et al.* (1987).

III. ECOLOGY AND EVOLUTION OF OLIGOCHAETA

A. Diversity and Distribution

Many aquatic oligochaetes are cosmopolitan, or at least widely distributed. This is especially true of the Naididae, but even here significant patterns are beginning to emerge. The supposedly primitive naidid subfamily Stylarinae is scarce in the southern United States, such genera as *Stylaria, Arcteonais, Ripistes,* and *Vejdovskyella* being common in northern locations. *Dero and Allonais* species (members of the Naidinae) are well represented in subtropical and tropical regions. There are other north–south distribution patterns. The tubificid *Arctodrilus wulikensis* is found in Alaska, as are several lumbriculid species that are part of the Pacific Rim fauna mentioned later. *Phallodrilus hallae,* a tubificid limited to Lake Superior, is related to a large number of marine taxa in the subfamily Phallodrilinae. Unlike the northern forms, tubificid species restricted to southern or southeastern localities belong to genera that are otherwise more widely distributed.

Among the tubificids and lumbriculids, some taxa are limited to the western United States and Canada. Some of these are related to or are the same as Asian Pacific Rim counterparts, suggesting an evolutionary distribution pattern. Several tubificids are known only from Lake Tahoe, which may prove to be home to a number of endemic species, though some forms first described there have since been found elsewhere, mostly still in California.

The apparent limitation of some species to the eastern United States may simply reflect a greater frequency of caves or more interest in cave faunae in those areas. At least six lumbriculid species fall into this category, but there are also some eastern surface water tubificid species. The groundwater tubificid *Rhyacodrilus falciformis* is a European species now known to be widespread in such habitats on Vancouver Island, British Columbia, and in Illinois, and there is a lumbriculid known from a well in California; so there may be as rich a western groundwater fauna (awaiting description) as there is in eastern North America. Many groundwater species have very limited distributions, although the distinctions between species in genera such as *Trichodrilus* are often based on minute differences. This may mean that biologists look more carefully for differences where they anticipate that they may exist.

The reasons for these geographic patterns have not been studied, partly because they have only been recognized for a few years. There may be ecological reasons for the distribution of some taxa. Species of *Aulodrilus* (Tubificidae) may be widely distributed, but they appear to reproduce asexually in the more temperate zones . This may suggest a recent spread from warmer areas where they do breed sexually. Some species can be shown to be recent introductions by humans. The North American *Limnodrilus cervix* has been introduced to Britain and Sweden via ports and canals (Milbrink, 1983). It is possible that the diverse collection of *Potamothrix* (Tubificidae) species in the St. Lawrence–Great Lakes represents a series of introductions of common European species into North America. This is more certain for *Psammoryctides barbatus* (Tubificidae), a very recent discovery in the St. Lawrence River in Quebec (Vincent *et al.*, 1978), as well as the estuarine and coastal species *Tubificoides benedii* found recently by the author in Vancouver Harbor, British Columbia. One of the most common tropical tubificids, *Branchiura sowerbyi,* was first described from a Victorian greenhouse in Britain, where it had been introduced along with the tropical plants on display. It has now become adapted to temperate waters, though it still thrives best in reservoirs and canals, especially where the temperature is elevated. It has apparently spread all across North America since it was first recorded in Ohio in 1931; but, this could simply reflect a gradual spread of awareness of this easily recognized species (Brinkhurst, 1965). Some of these regional distributions are mentioned in the keys, but the reader should always be prepared for the unusual.

Niche discrimination in aquatic oligochaetes is less obvious than zoogeographic patterns. The majority of these worms are adapted to live in sediments ranging from sand to mud. They can be found in pockets of such sediments in stony habitats as well as in lowland rivers, lakes, and ponds where soft substrates are the norm. There is none of the usual distinction between lacustrine and riverine species. The best guess that can be made about niche discrimination is that the precise nature of the organic matter in sediments, and the microflora it supports, determines the outcome of interspecific competition. The ecophysiological work on multispecific flocks described earlier is probably applicable to most situations and not limited to the *Tubifex–Limnodrilus–Quistadrilus* association in which it was initially described. Milbrink (1993) obtained similar results with *T. tubifex* and *Potamothrix moldaviensis.* In most habitats, between 5 and 15 species can be found. There are some special adaptations, such as the groundwater fauna already mentioned. Several genera of the Naididae are no longer restricted to life in the sediment, but may be found among aquatic weeds in quiet rivers and ponds. *Lumbriculus* lives among vegetation in pond and stream margins. *Dero* (*Allodero*) species are symbionts of frogs, which, once released from their host come to resemble specimens of *Dero (Aulophorus),* rendering the taxonomic distinction suspect. The predators *Haplotaxis* and *Phagodrilus* are found in springs, seepages, and groundwater, but *Chaetogaster* is more ubiquitous. Estuarine species form another group, but a number of species are restricted to tidal freshwater where there is little, if any, detectable saltwater. None of these exhibit any marked anatomical adaptations to their way of life, with the exception of the pharyngeal structures of the predators and loss of chaetae and gills in *Allodero.*

The significance of food in the ecological specialization of the aquatic oligochaetes, especially the tubificids, is demonstrated by the sequence of species groups that inhabit progressively more organically polluted stretches of rivers or more eutrophic lakes. The Lumbriculidae and some naidid genera such as *Pristina* and *Pristinella* are found in streams with low organic matter in the sediments, and they may be accompanied by tubificids belonging to the Rhyacodrilinae in particular (*Rhyacodrilus, Bothrioneurum, and Rhizodrilus*). Tubificinae, such as *Potamothrix* and *Aulodrilus,* can be found in situations with a reasonable supply of organic matter but also a good oxygen supply, although *Tubifex tubifex,* several *Limnodrilus* species, and *Quistadrilus* are found in areas where the oxygen supply can become reduced. This basic pattern has been expanded, quantified, and recognized in Europe as well as North America (e.g., Lang, 1984). At first, this type of work followed the much earlier pattern of establishing lists of chironomid midge species characteristic of lake types (reviewed in Brinkhurst, 1974). The maximum value in identifying aquatic oligochaetes in pollution studies today is to add information for use in data matrices to be analyzed by multivariate statistics, such as cluster analyses. There are relatively few invertebrate species in freshwater muddy habitats, in contrast to stony streams (where there are far more insects) or marine habitats (where there are many more mud-dwelling taxa). The taxonomic identites of the communities from each sample site are retained in such analyses, whereas they are lost in simpler diversity models.

B. Reproduction and Life History

Field studies on reproduction in the Naididae were reviewed by Loden (1981). Asexual reproduction, (usually by paratomy), predominates, but sexual reproduction apparently enables populations to persist through periods of unsuitable conditions. Abundance varies seasonally depending on the geographic region, with spring, summer, or winter maxima being produced asexually. Most species apparently cease feeding as they mature because the gut degenerates, and the mouth may even become sealed. According to Loden, some species incorporate the anterior part of the worm in the cocoon, but Lochhead and Learner (1984) disputed this in describing a life history for *Nais*. They reported that worms from this and other naidid genera are able to shed their cocoons. According to these investigators, the adult worms are more resistant to environmental stress than the cocoons. All cocoons of aquatic oligochaetes may be disguised with mud particles, or may be placed out of reach of predators (Newrlka and Mutayoba, 1987).

Most tubificids reproduce sexually, but there are several parthenogenetic species. The latter often lack spermathecae and the testes may also be reduced or absent. The vasa deferentia may be reduced to solid strands in these species, but the atria are usually retained. A few other species seem able to reproduce asexually by fission. The results of various field studies produce conflicting interpretations of the life histories of such common genera as *Limnodrilus*. There is less disagreement about events in less adaptable species found in more restricted habitats. The worms may mature a few weeks after hatching if temperature and food supplies are adequate, and they may breed more than once in a season. Alternative studies suggest that many species mature in their second year. After breeding, these forms resorb the gonads and gonoducts and would be recognized as immature specimens. The next year they develop a new set of reproductive organs and breed again, after which most species die. Some published examples are the studies of Block *et al.* (1982) and Lazim and Learner (1986a, b) for *Limnodrilus* species, Christensen (1984) on asexual reproduction, and Poddubnaya (1984) on parthenogenesis. Because of the problems encountered in interpreting the results of field studies on life histories, considerable effort has been put into the development of laboratory cohort studies (Bonacina *et al.*, 1987). In general, oligochaete life histories are probably not as precisely tuned to environmental variables like day length as they are in more highly evolved forms such as insects. This may be attributable to simpler neurosecretory systems in worms, but there is no evidence to support this idea. Food and temperature may cause considerable local variation in life cycles, especially in the opportunistic species like *Limnodrilus hoffmeisteri*. Unfortunately, these abundant, cosmopolitan species are the most convenient laboratory animals, but they are so adaptable and perhaps atypical that comparable work should be done with species found in more restricted habitats.

C. Ecological Interactions

There is some published work on the food and feeding behavior of the Naididae, but a much larger literature exists on the relationship between tubificids (and one lubriculid species *Stylodrilus heringianus*) and the sediments that they ingest (and therefore defecate) continuously.

Algae and other epiphytic material are the main food of many naidids, but *Nais* feeds on heterotrophic, aerobic bacteria like the tubificids. There is no specialized gut flora in either group (Brinkhurst and Chua, 1969; Harper *et al.*, 1981). There is no evidence of selective ingestion or digestion of specific bacteria in *Nais*, but selective digestion is important in tubificids. *Nais elinguis* will select algal filaments in preference to monofilament nylon thread and prefers some algae to others or to glass beads (Bowker *et al.*, 1985). Algae also form part of the food supply of the tubificids *Limnodrilus claparedeianus* and *Rhyacodrilus sodalis* and the lumbriculid *Lumbriculus variegatus*.

Studies of defecation rate and the chemical and bacterial content of feces in tubificids have usually exploited the fact that these worms will continue to feed with their heads in the sediment and their tails in the water column, even when such colonies in small containers are inverted. If a layer of cotton or fine netting is used to prevent the sediment drifting down past the worms into the collecting funnel placed beneath the culture, it is easy to determine the amount and the chemical content of the feces. If the feces are collected over very short time periods, the bacterial flora of the feces can be studied. Similar analyses of the sediment provided as food make it possible to study the effect of passage through the gut on the organic matter content and bacterial flora (Brinkhurst and Austin, 1978). Fecal samples can also be collected from cultures oriented normally, but the feces then have to be collected from the artificial sediment surface instead of being allowed to drop into a container. Both methods are potentially liable to error and cultures must be handled very carefully. The behavior of the worms is affected by changes in temperature and light and by physical disturbance. The results derived from cultures maintained in both orientations have usually shown that worms are not affected by being inverted, and that, if either method is subject to error in determining the amount of fecal material produced, these errors cancel out. Only Kaster *et al.* (1984) suggested that worms in the normal position produce more feces (45–110%) than those in inverted cultures. Such obvious differences would have been detectable in earlier studies had they existed. Defecation rate seems to be very difficult to determine. The nutritional value of the food provided may decline as the experiment proceeds, and this may cause an increase in the amount of sediment processed in order to feed the colony. Experimental results are affected by periods of acclimation to the artificial sediments, which may be as long as a month. Two supposedly identical cultures maintained side by side may behave quite differently (Appleby and Brinkhurst, 1970). Defecation rates may be used in studies of worm production when associated with estimates of organic matter content and caloric value of the food and feces. They may also be used in studies of sedimentation and bioperturbation of sedimentary layers in freshwater ecosystems. The lumbriculid *Stylodrilus heringianus* builds up large populations in many regions of the Great Lakes, where it behaves in the same way as a tubificid. Krezoski *et al.* (1984) used gamma spectroscopy methods to measure the rates at which a layer of sediment (labeled with radioactive cesium) and overlying water (labeled with radioactive sodium) were transported by the burrowing activity of worms and other benthos. The results of these and similar studies all support the idea that the superficial sediment layers in lakes are thoroughly mixed by benthic organisms, and that solute transport across the mud–water interface is greatly enhanced by them. Worms are probably the most important animals involved in this because of the depths to which they penetrate the sediment, the large populations that they build up, and their ceaseless activity. Gardner *et al.* (1981) showed that invertebrates could be responsible for most of the phosphorus released from aerobic sediments in Lake Michigan. Chatarpaul *et al.* (1979) showed that worms accelerated nitrogen loss from coarse stream sediments, indicating their importance in rivers as well as lakes. There is now a growing realization that the sediments are not just the final resting place for organic matter and contaminants, but that they may become a source for them (Karlckhoff and Morris, 1985).

D. Evolutionary Relationships

There are contradictory ideas about the evolution of any biological group, many of which are due to fundamental differences in the methods used. There are strongly held opinions about the way in which relationships should be determined, even among those who employ cladistic methods and those who agree with at least the basic concepts of Hennig. The following account is a personal one, based on experience with most families of aquatic oligochaetes, an exposure to marine and freshwater worms worldwide, and some familiarity with cladistics. This is not said in order to validate this particular view, but to help the reader identify the biases it inevitably includes. At least it should not suffer from the geographic or taxonomic parochialism that tends to cause taxonomic inflation—the tendency to elevate the taxa in which we are most interested to the highest level possible.

We should perhaps consider what the original worm looked like, rather than asking whether polychaetes or oligochaetes came first. It seems quite reasonable to suppose that an animal looking a lot like a very simple earthworm, but with a reproductive system that released eggs and sperm freely into the environment, provides a common ancestor to both groups (Fig. 5). There has been some support for this idea among polychaete biologists, but, in general, they seem to have difficulty deciding on relationships above the family level, and on what constitutes a primitive worm. The "simplified earthworm" with two chaetae per bundle and no parapodia is adopted following the ideas of Clark (1964) regarding the evolution of segments and coelom. There is, of course, no reason to suppose that every group evolved in the ocean. The teleost fish provide a good example of a group that may well have undergone a very large diversification in freshwater before re-invading the ocean, so that the existence of a large number of marine Tubificidae is not *prima facie* evidence for their antiquity, even within that family, though the possibility should not be excluded. All possible combinations of the reproductive system segmental arrangement within the oligochaetes (and other Clitellata) can be derived from a worm with four genital segments, two with paired testes in each, followed by two with paired ovaries. The original position of this series, termed GI-GIV (Brinkhurst, 1982), seems to have been in X–XIII. The organisms with this arrangement that exit today are found in the Haplotaxidae, with only one or two exceptions that are thought to be primitive members of groups descended directly from that family (Fig. 6). There are some exceptions that have been explained elsewhere, but only the barest outline will be presented here. [Challenges to this largely Austral group as being ancestral come from the fact that *Haplotaxis* itself is a highly modified genus of predatory worms (see the taxonomic key).] It is then thought that a number of families have quite independently lost the anterior testes (GI) and the posterior ovaries (GIV), making copulation easier (there are fewer pores to bring into contact with each other). This happens because the development of egg and sperm sacs enables more gametes to be stored in the coelom from a single pair of gonads than could be achieved in a single segment when the sacs were small or undeveloped. Similarly, the evolution of multiple chaetae per bundle seems to have arisen independently a number of times, most notably in the Enchytraeidae, Tubificidae, and Naididae. In the former, most species have two chaetae per bundle, a few "specialists" have no chaetae in some or all bundles (another common but independent evolutionary step found in several families), but some have several chaetae. These are all simple-pointed, however, which seems to be correlated

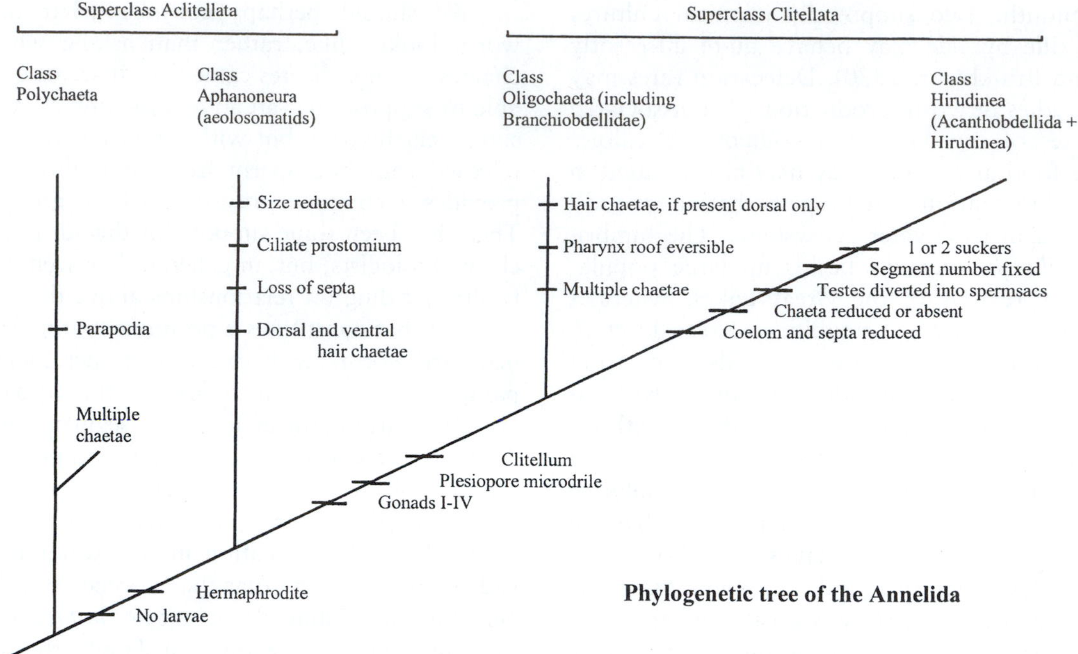

FIGURE 5 Phylogenetic tree of the Annelida. This diagram represents an outline of the major (potential) monophylies and the apomorphic character states that define them (solid bars across lines). Note that the Branchiobdellidae is shown as included within the Oligochaeta, which may also be true for the Hirudinea according to recent molecular evidence. Aphanoneura is expected to be eliminated as the Aeolosomatidae become placed within the Polychaeta in future publication.

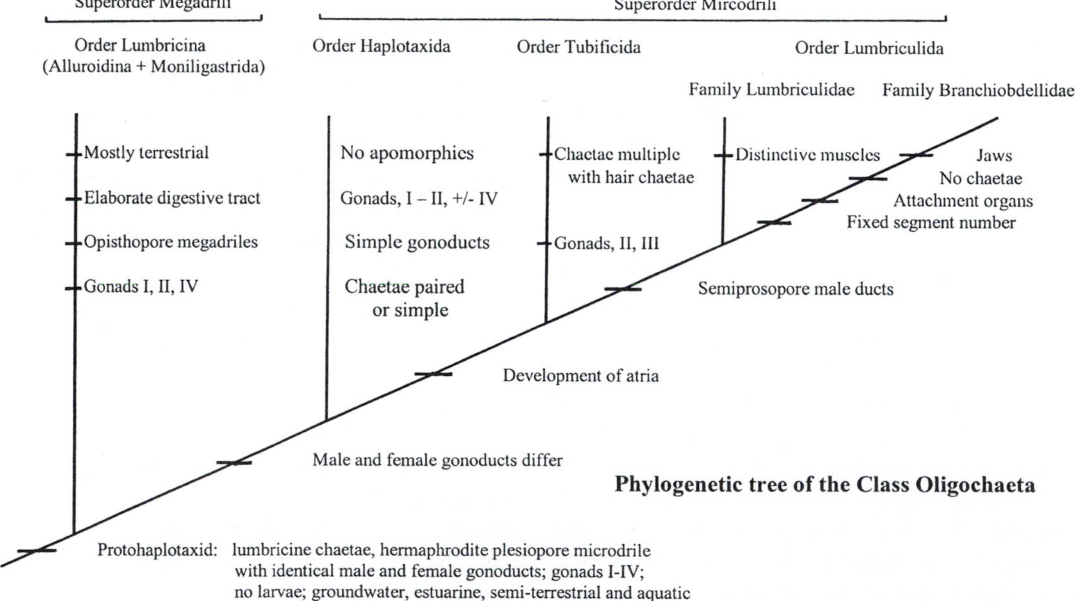

FIGURE 6 Phylogenetic tree of the Oligochaeta. While the Hirudinea (Acanthobdellida + Euhirudinea) are said to be related to the Lumbriculida according to molecular analyses, they do not have semiprosoporous male ducts (possibly plesioporous and prosoporus respectively, a single pair of male ducts in *Acanthobdella*, testes within sperm sacs confluent with the male pores with no male funnels in Euhirudinea). The form of the male ducts in *Acanthobdella* and Euhirudinea could have arisen independently from a semiprosoporous condition, in which case they could form a monophyly or monophylies within the Lumbriculida, but the molecular and anatomical schemes need to be resolved and hence the Hirudinea are excluded from this figure.

with the fact that most members of this family are at best only semiaquatic if not totally terrestrial. Chaetae with bifid tips are a hallmark of aquatic oligochaetes, as are more complex pectinate and palmate chaetae and their derivatives, hair chaetae. The hair chaetae of the Phreodrilidae (a small, Southern Hemisphere family) seem to have evolved independently of those of the Naididae and Tubificidae. These are the two most closely related families of aquatic oligochaetes, and they are apparently the most highly evolved in many respects. The Naididae exhibit several adaptations to an aquatic life as opposed to one restricted to burrowing in moist sediments. They have simple eyes, a few have specialized chaetae in a few anterior segments, and some can swim in a very simple way. The predominance of asexual reproduction enables them to exploit a more changeable environment than that available in the sediments, and in some as yet unidentified way has enabled most of them to become very widely distributed. Their diet now includes algae as well as bacteria, and we can assume that they can absorb macromolecules through the body wall as food (as do the freshwater and marine tubificids). The male reproductive system has become progressively more complex in these groups. In the Enchytraeidae, the male ducts may terminate in what appear to be more glandular than muscular structures, which are not yet identifiable as atria. The tubificids have muscular atria and the higher forms have stalked prostate glands, elaborate penes, and either penis sheaths, penial chaetae, or spermathecal chaetae.

The Lumbriculidae are more of a puzzle. They have usually retained both testicular segments, but have lost the second ovaries (GIV). Many advanced forms secondarily loose GI, and a few parthenogenetic forms have an irregular, asymmetrical, or even exaggerated number of gonads. The chaetae are restricted to two per bundle, but these are usually bifid rather than simple-pointed as postulated in the plesiomorphic condition. Some biologists see the paired chaetae as evidence of a close relationship between the lumbriculids and the earthworms, suggesting that the latter are the derivatives of these aquatic forms. This puts an enormous weight on just one character, but these people also regard hair chaetae as being retained from a polychaete ancestor. The difficulty with relating the lumbriculids to the earthworms is that the lumbriculids (and branchiobdellidans) have a unique way of reducing the number of male pores. They have their atria in GII and both pairs of male ducts drain into them. In Lumbriculids, the prostate glands consist of a diffuse covering of gland cells and none of them have developed stalked prostates. Cuticular penis sheaths are very rare; genital chaetae are limited to a single species. Too much has been made of the apparent diversity of reproductive segments in this family. While the variation is greater than that of other families, the most common form does seem to be the ancestral one (Brinkhurst, 1989). While the aquatic oligochaetes closest to their presumed ancestors seem to be found in Australia and New Zealand, not unlike many other freshwater invertebrate groups, other families are widely distributed. According to a cladistic analysis of some Enchytraeidae (Coates, 1989), South America seemed to contain the basal forms. In the Naididae, it is the northern Stylarinae that appear to the oldest (Nemec and Brinkhurst, 1987). The Lumbriculidae inhabit the Northern Hemisphere, with a very high proportion of them being limited to, or extending eastward from, Lake Baikal in Siberia. The European *Cookidrilus* seems to most resemble the presumed ancestor, with the widespread genera *Stylodrilus* and *Trichodrilus* being quite generalized (Brinkhurst, 1989). There has yet to be a detailed analysis of the more cosmopolitan Tubificidae, but there is a greater diversity of genera in the Northern Hemisphere, with very few restricted to the Southern Hemisphere (as *Antipodrilus*) and very few adapted to the subtropical and tropical regions, apart from *Branchiura* and, perhaps, *Aulodrilus*.

We are currently trying to assess the relationships between the clitellate taxa usually termed Oligochaeta, Hirudinea, and Acanthobdellida. In a preliminary revision (Brinkhurst and Gelder, 1989), it was shown that the supposed link between the predatory lumbriculid worm *Agriodrilus* from Lake Baikal (Russia) and the non oligochaete taxa in this group is purely circumstantial, made even clearer by the recent discovery of a parallel evolution in the North American *Phagodrilus*. Without presenting all of the data here, it now looks as if the branchiobdellids and Lumbriculidae share a common ancestor, both having two pairs of male ducts draining through atria in GII, unless this character proves to be a convergence. Hence, the former taxon is referred to as Branchiobdellidae in this chapter. It remains to be seen if the male ducts of the Hirudinea and the Acanthobdellida are at all related to what we might call the semiprosopore state of the male ducts in the lumbriculids (only one pair of male ducts are now thought to be in their testicular segment in the ancestor of the family according to a cladistic analysis). The extreme modification of the male ducts in the leeches, in which the testes lie in sacs behind the two successive segments bearing male and female pores, is considered to be derived from a system with the testes in front of the ovaries, as in the oligochaetes. With the loss of the coelom, the testes are supposed to be transferred into the surviving sperm sacs as a series of small gonads replacing a single original pair. Complex male ducts have been developed, but these are in the form of a continuous tube in which the sperm are shed internally. These ducts need to be reexamined for any trace of relict sperm funnels, to see if there is any evidence that they were ever prosoporous like the lumbriculids. The acanthobdellids, represented by one species with one probable synonym, have four pairs of chaetae in a few anterior segments including the oral segment. There is no trace of a prostomium, or peristomium, and perhaps some anterior segments (Brinkhurst, 1999). *Acanthobdella* also retains or possibly recovers the coelom (which is lost in leeches), but the testes still lie within the sperm ducts. Two intriguing aspects are the pattern of eyes and the unique sperm-receiving device shared by this genus and the piscicolid leeches. Other revised information about the branchiobdellids, discussed later, also weakens the argument for a relationship between them and the leech groups. That argument was based on a very superficial examination of the detailed facts available. Textbook accounts in this and many other areas of supposed evolutionary relationships should be read with caution, especially if they give no clear indication of the theoretical basis on which the opinions are based.

IV. COLLECTING, REARING, AND PREPARATION OF OLIGOCHAETES FOR IDENTIFICATION

A. Collection

Oligochaetes may be collected using any of the standard array of samplers from dip nets to grabs and core tubes. Collectors specifically seeking worms can use refined histological fixatives in the field. A great deal of material is collected in the course of ecological surveys where time does not permit such care. Strictly quantitative samples from soft sediments are best obtained with narrow core tubes without core retainers. Using core tube lengths appropriate to the type of substrate can save a great deal of frustration, but most people seem to try to work with a single set of tubes. Soft sediments can be washed through screens and sorted in the laboratory, preferable after being preserved in their entirety in the field. If cores are used, the volume to be carried back from the field may be quite small. Field sorting is relatively inaccurate, and small specimens may well be missed. The best preservative for bulk samples is 10% buffered formalin (4% formaldehyde solution), but is easier to carry 40% in the field and add this to an aqueous sample to approximate 10% in the whole sample. Addition of rose bengal or phloxine B in very small amounts aids sorting, and the stain can be washed out of the worms when they are transferred to 70% alcohol for storage. The use of larger amounts of stain creates problems in identifying worms. Sorting should be done with stereomicroscopes under good lighting. Care should be taken to avoid diluting the alcohol when transferring worms into the storage vial. Many North American collections are sent for identification with the worms half decomposed because bulk samples were preserved in alcohol, not formalin, or the original alcohol has not been drained and replaced with 70%. Storage can be quite permanent in patent lip vials with neoprene stoppers. The stoppers tend to swell and seal the vial against alcohol loss. We have some in our laboratories that have not needed attention for 20 years. If these vials and stoppers are not available, vials must be sealed with tape, wax, or parafilm. Some neoprene stoppers dry and crack with age, and all collections should be curated to maintain their integrity. Museum quality material should be donated to institutions that can provide curation.

If the collector is not taking quantitative samples for ecological or environmental studies and is working in a remote or unusual location, then it may be worth the effort of sorting, fixing and preserving some specimens while in the field if they are readily visible. Putting the sample in an enamel dish and allowing everything to settle will reveal even small worms as they begin to move. Carrying whole samples back to the laboratory to sort them while the worms are alive can cause selective mortalities, as some species surviving elevated temperatures and lowered oxygen concentration better than others (and the latter always seem to be the more interesting species from a taxonomic viewpoint). Live worms can be sorted from sediment and detritus by spreading clean sand over the sample or putting the sample on a screen set over clean water. The worms will then actively migrate into the sand or water. This can be useful in samples full of plant fragments.

Samples taken with dip nets or other methods that include coarse substrates can be processed in the field to reduce the volume and weight of the sample. Repeated, careful elutriation into a screen, screen bucket, or back through the original dip net can be used, imitating a gold-panning technique. The amount of vigor used can be selected in order to separate enough of the fine fraction including all the specimens from the heavy material. The coarse residues should, of course, be carefully checked for large, heavy, or attached organisms, or specimens trapped in empty shells and other unusual situations.

B. Rearing Oligochaetes

Culture of worms can be as simple as placing the substrate and the worms in an aerated aquarium, preferable where temperatures can be kept at appropriate levels, usually between 4–10°C depending on the original habitat. Small cultures used for laboratory experiments can be maintained in shallow vessels half filled with sterilized lake sand. Filtered lake water is added every two weeks and frozen lettuce strips are buried in the sand to provide bacteria as food. Mass culture to provide fish food can utilize manure to feed the worms. Marian and Pandian (1984) described raising 7.5-mg *Tubifex tubifex* every 42 days on a mixture of 75% cow dung and 25% fine sand, maintaining the oxygen level at 3 mg/L. Fresh cow dung is added every four days, and harvesting is supposedly best done at night. Aston (1984) cultured *Branchiura sowerbyi* at 25°C. Leitz (1987) reviewed attempts to mass culture *Lumbriculus variegatus*. Production rates of 15 kg/m^2 and population doubling times of 11–42 days are reported. The use of live worms as fish food has many advantages over other diets. Most commercially available live worms are now *Lumbriculus* (called 'black worms" in commerce) rather than *Limnodrilus*.

C. Preparation for Identification

Many biologists believe that these animals are hard to identify. It is worth indicating that, as they are hermaphrodites, there is no difficulty with keys for males and females, and there are no larval forms. About 60% of the species can be identified from superficial somatic characters, but the rest require mature specimens. It is, of course, always preferable to confirm identities from mature specimens wherever possible, but it is not essential. Those used to identifying organisms with stereomicroscopes may be bothered by the need to make slide mounts, but simple methods can be used in most instances. The serious taxonomist will want to make more elaborate preparations and may want to section specimens to check details, but this is not necessary for routine identification. There are only about 150 freshwater oligochaetes and a small number of branchiobdellidans in North America, and many of these are rare and local. Recent keys have been written with the beginner in mind, and so they emphasize external or readily visible internal characters. Such keys work well if they proceed directly to the specific level, but it is a little more difficult to write keys to the genera, as preferred in this volume. Generic characteristics are mostly detailed aspects of the soft parts of the male reproductive structures. These may be difficult to see on rapidly made preparations in which the soft parts are dissolved. The keys presented here do not always proceed directly to genera for this reason, but we have managed to write keys involving externally visible characteristics. Some reference to geographic distribution is made in the keys, but this has been done only where the information available seems reliable. Fully illustrated keys with supporting descriptions of all the species are available elsewhere (Kathman and Brinkhurst, 1998).

Once the worms are preserved in alcohol, they must be prepared for study with a compound microscope. Examination of specimens from remote or unusual locations should begin with a study of the external characteristics, such as position of the gonopores and other obvious features. However, for routine identifications of specimens from ordinary localities, very little benefit can be derived from this step in the process. In this case, the worms can either be immersed in a temporary mounting medium by replacing the alcohol with Amman's lactophenol, or they may be washed with water and then placed in separate drops of lactophenol on slides. It is usually possible to mount live worms under a single coverslip (with two coverslips per slide) unless the worms are very large. These wet mounts should be set aside for 1–2 days on stacking trays. Amman's medium is quite corrosive, so it should be handled with care and any spillage should be wiped off microscope stages immediately. It consists of 20% phenol, 20% lactic acid, 20% water, and 40% glycerine. It is best stored in dark bottles; however, it is hygroscopic and does not have a good shelf life. Somewhat more permanent mounts can be made with media such as CMCP and polyvinyl lactophenol. Such slides should usually be sealed with a ring of material (Glyceel, nail polish, etc.) around the edges of the coverslips to prevent the media from drying up.

None of these methods produce satisfactory permanent slides, nor are they adequate for working with the Enchytraeidae or the marine Tubificidae, in which internal anatomy must be seen. For this type of material, permanent whole mounts using stained and dehydrated worms are mounted in Canada Balsam or one of its more recent substitutes. To prevent worms from becoming brittle, they should be cleared in methyl salicylate rather than xylene. This will enable advanced students to try bisecting (sagitally) the separated genital region of each worm or even dissecting out the male ducts when necessary. This should be limited to identification of new taxa to the generic level, when details of the male reproductive system must be established. Serial sections (usually sagital longitudinal) are also employed in taxonomic work, but not in routine identification of well-known species. It is important to be knowledgeable about the expected locations of the various organs before attempting dissection or interpretation of serial sections.

The procedure for examining a worm on a slide is as follows. Check the prostomium for a proboscis or sense organ. Next, examine the ventral chaetae of the first two or three bundles and determine the form of the dorsal bundles and where they begin (usually in II or between IV and VI). Establish the number and form of these chaetae (bifid, pectinate, hair chaetae, etc.). Determine the relative lengths of the teeth. As these are illustrated by taxonomists with the ectal or outer end of the whole chaeta toward the top edge or corner of the figure, the upper tooth is always drawn lying above the lower tooth. When viewing a whole-mounted specimen, the chaetae often lie flat on the body wall and so no distinction of position can be made; therefore, always visualize a chaeta in the upright position. Care should be taken to examine several chaetae from an exactly lateral aspect, because slight deviations can produce apparent distortion of the relative lengths of the teeth. Measurements often betray the bias of the eye of the observer and are recommended where an eyepiece micrometer scale is available. The worm should then be searched for genital characters. Carefully check the appropriate segments (usually X–XI in a tubificid, for example) to see if the ventral chaetae are modified. If the dorsal and ventral chaetae are alike, make sure that you have examined at least three bundles in those segments to ensure that a ventral bundle has been examined. Check the penial segment to see if there is a penis sheath (they should be paired, of course) even if it is thin and inconspicuous. Check the posterior chaetae and the body itself for gills or any other special features. Then proceed to check the key. If the worm has distinct features, [like a single dorsal hair with single simple needles (these beginning in V) and the anterior ventral (II–V) and posterior ventral (VI onward) bundles have chaetae that differ in length, width, and shape], it will rapidly become possible to skip the key altogether and proceed straight to *Nais* or *Dero* depending on the presence of gills. Many such easy identifications to major groups or to genera can be made with a little practice. Identifications will require a properly aligned microscope, as chaetae and penis sheaths have refractive indices that do not differ much from the background, and too much light makes them impossible to see. Oil immersion lenses must be used to see pectinations on chaetae in tubificids (although with practice they can be identified without), and must also be used to determine the form of the needles (dorsal chaetae) in naidids.

V. TAXONOMIC KEYS FOR OLIGOCHAETA

A. Taxonomic Key to Families of Freshwater Oligochaeta and Aphanoneura

1a. Minute worms, 1–2 mm or chains of animals up to 10 mm; hair chaetae in both dorsal and ventral bundles; worms move in a gliding motion using cilia; body wall with colored or refractile epidermal glands; no eversible pharyngeal pad; no clitellum, only ventral copulatory glands in rare mature specimens; nervous system ladderlike (Fig. 7)........Aphanoneura ....Aeolosomatidae [Key to world species in Brinkhurst and Jamieson (1971)]

1b. Larger worms; hair chaetae restricted to dorsal bundles or absent; worms move by alternate contraction of longitudinal and circular muscles; no colored epidermal glands; eversible pharyngeal roof present (except in aquatic earthworms); clitellum present in mature worms; nervous system fused ventrally ....Oligochaeta 2

2a(1b). Ventral chaetae large, sickle-shaped, single (i.e., two per segment) dorsal chaetae small, straight, often missing from some or all segments; prostomium very long, furrowed, no proboscis; mouth large, muscular pharynx present; body elongate, slender, resembling a gordian worm (Fig. 8)........Haplotaxidae........*Haplotaxis*
[Worms in this genus resemble the European *H. gordioides* (Hartman, 1821), but mature specimens have yet to be described from North America. There may be several species (now regarded as synonyms) differing by the number and distribution of dorsal chaetae.]

2b. All chaetae paired, or multiple in a bundle, unless partially or totally absent; prostomium short, often conical, without a transverse furrow, but with or without a proboscis; body form very rarely elongate and slender........3

3a(2b). Chaetae all paired from II or, if more numerous, all simple-pointed........4

3b. Chaetae more than two per bundle, usually bifid, sometimes with pectinate and hair chaetae dorsally, simple-pointed chaetae rare (limited to a few anterior ventral bundles, or single needles with the hairs dorsally)........7

4a(3a). Chaetae paired, bifid, with small to rudimentary upper teeth (Fig. 9........Lumbriculidae (in part)........(Section V.B)

4b. Chaetae simple-pointed........5

5a(4b). Thick-bodied worms with clitellum several segments behind the gonopores (XII–XIV)................*Megadrili* [Mostly terrestrial, but some aquatic species. Most aquatic species are members of the Sparganophilidae (*Sparganophilus*) or the Lumbricidae (*Eiseniella*).]

5b. Thin-bodied worms with clitellum one cell layer thick and in region of gonopores (X–XII or further forward)........6

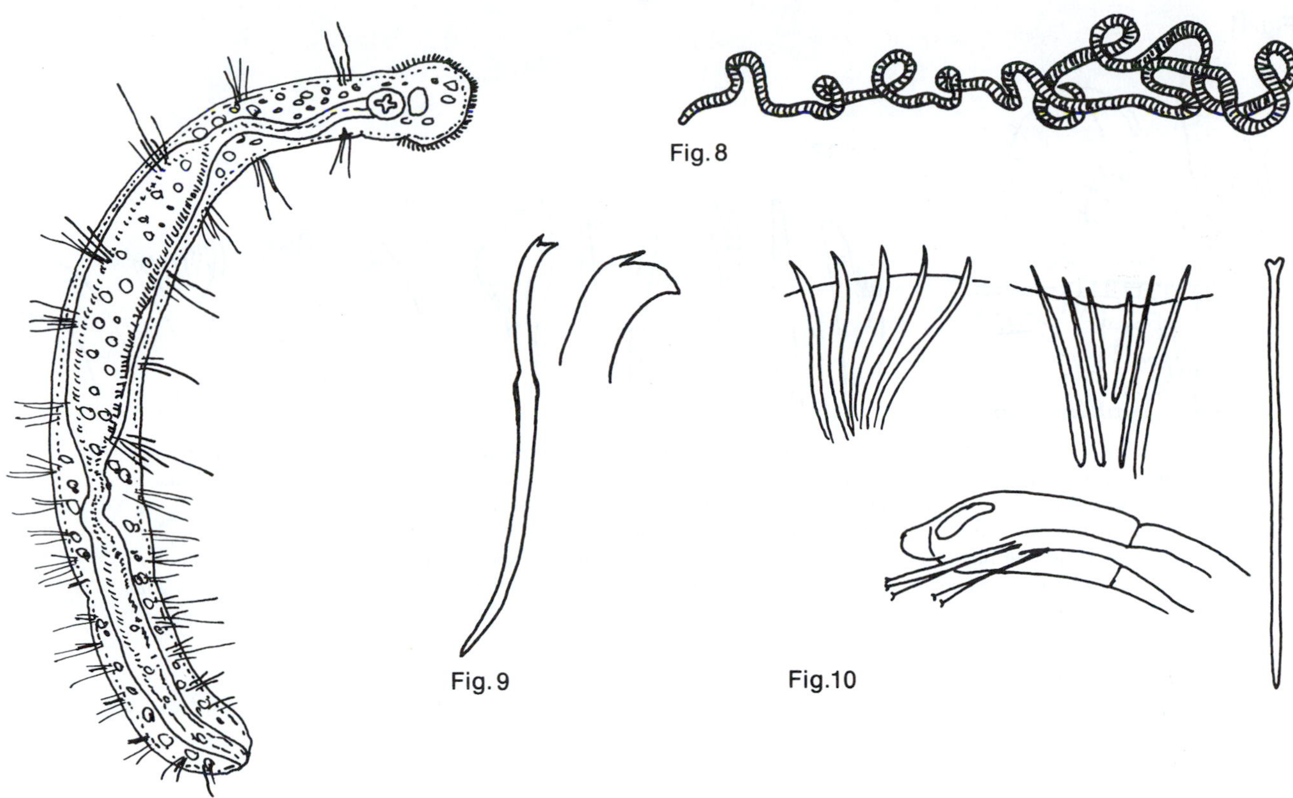

FIGURE 7 *Aeolosoma* sp. Note ciliated prostomium, lack of septa, simple pharynx, dorsal and ventral hair chaetae, and colored body wall inclusions (irregular circles). *FIGURE 8* *Haplotaxis* sp. Whole specimens can be even longer in relation to the breadth. *FIGURE 9* Bifid chaetae, Lumbriculidae. *FIGURE 10* Enchytraeid chaetae. These may be straight, curved, with or without noduale, and may differ in length within a bundle. All chaetae are usually identical. The anterior end of *Barbidrilus* (below) have ventral bundles of strange rodlike chaetae with bifid tips only in II and III, shown on the right.

6a(5b). Larger worms (difficult to mount under a coverslip) with sigmoid, nodulate chaetae; proboscis present or absent; spermathecae in or adjacent to the gonadal segments, male pores between VIII and X .... Lumbriculidae (in part) .... (Section V.B)

6b. Smaller worms with chaetae often straight, commonly not nodulate, sometimes missing from some if not all segments (in *Barbidrilus* Loden and Locy, 1981 with uniquely forked chaetae in II–III ventrally, others missing). No proboscis. Spermathecae open in V, male pores in XII (Figs. 2 and 10) .... Enchytraeidae

7a(3b). Worms (1.7–3.0 mm) with posterior end bearing one median and two lateral processes. Genital region with testes in XI, ovaries and atria in XII, spermathecae in XIII (Fig. 11) .... Opistocystidae
[The only definite North American record is for *Crustipellis tribranchiata* (Harman, 1970), restricted to Louisiana, Mississippi, Florida, and North Carolina. The tail processes make this obvious (Harman and Loden, 1978).]

7b. Posterior end naked, or with gills, or with two lateral processes plus gills, no median process (Fig. 26); spermathecae in the testicular segment, atria in the ovarian segment, these being in X–XI or further forward (Fig. 2) 8

8a(7b). Length usually <1 cm; hair chaetae usually present in dorsal bundles but absent in some genera; hair chaetae commonly associated with simple-pointed or bifid chaetae rather than the rarer palmate or pectinate chaetae, these dorsal chaetae (needles) very often differ from the ventrals in form (Fig. 12); dorsal chaetae may begin behind II, often in V or VI, sometimes elsewhere; dorsal chaetae often limited to one or two needles, and one or two hairs per bundle; ventral bundles with numerous bifid chaetae, those of II or even II–V often differ in form and thickness from the rest; asexual reproduction by budding forms chains of individuals; mature specimens have spermathecae in IV, V or VI with the male pores one segment behind them (Fig. 2); eye spots may be present; gills may surround the anus (Fig. 26); prostomium may bear a proboscis (Fig. 23 .... Naididae .... (Section V.C)

8b. Length usually >1 cm, width usually 0.5–1.0 mm; when hair chaetae are present dorsally, they are usually accompanied by pectinate chaetae that closely resemble the ventral chaetae, apart from having intermediate teeth (Fig. 13); when hair chaetae are absent dorsally, all bundles usually contain similar bifid chaetae; dorsal chaetae begin in II, normally several per bundle; reproduction

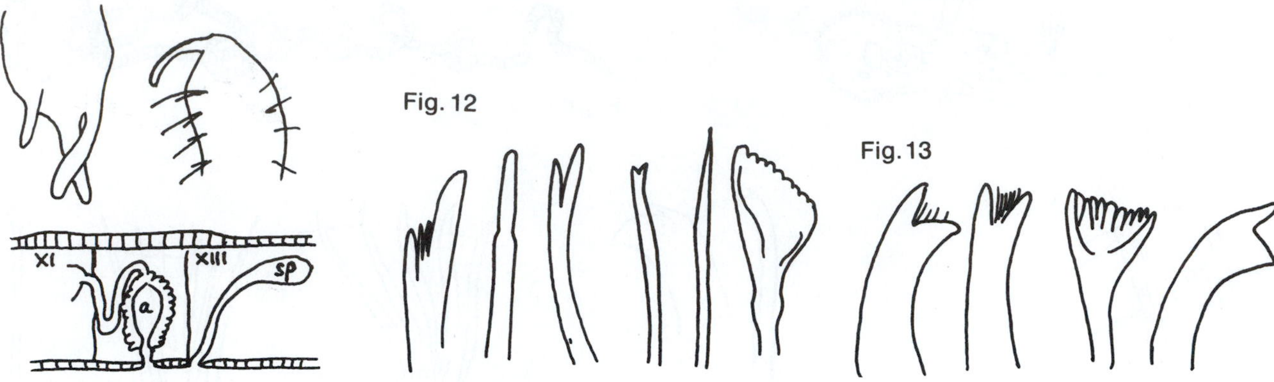

FIGURE 11 *Crustipellis* (Opistocystidae). Tail end to left, head end with proboscis to right, genital segments below (a, atrium, sp, spermatheca, both of left side). *FIGURE 12* Various forms of dorsal chaetae, needles, of Naididae. These usually differ from the more normally bifid ventrals. *FIGURE 13* Dorsal chaetae of Tubificidae. Pectinates, on the left, differ relatively little from a bifid, shown on the right. Ventral bundles usually consist of bifids, sometimes with minute pectinations in those species with pectinate dorsals. Pectinates are usually accompanied by hairs in the dorsal bundles, rarely are bifid dorsals accompanied by hairs. All chaetae may be bifid.

normally sexual, rarely by fragmentation; spermathecal pores normally on X, male pores on XI (Fig. 2); no eyes or proboscis; no gills around anus, single dorsal and ventral gill filaments on some posterior segments of one species (Fig. 30)....................Tubificidae ....(Section V.D)

B. Taxonomic Key to Genera and Selected Species of Freshwater Lumbriculidae

1a. Prostomium with a proboscis (Figs. 14 and 15)....................2

1b. Prostomium without a proboscis....................6

2a(1a). Chaetae bifid, at least in anterior segments....................3

2b. Chaetae all simple-pointed....................4

3a(2a). Proboscis elongate; chaetae bifid in anterior segments (Fig. 14)....................*Kincaidiana hexatheca* Altman, 1936
[This species is limited to the Pacific Northwest.]

3b. Proboscis short; all chaetae bifid, but with colateral teeth (set side by side) (Fig. 15).................... *Rhynchelmis brooksi* Holmquist, 1976....................[Known only from northern Alaska.]

4a(2b). Cave-dwelling species with short proboscis....................*Trichodrilus allegheniensis* Cook, 1971
[Known only from Tennessee.]

4b. Surface water species with long proboscis....................5

5a(4b). Atria of male reproductive system with spiral muscles (Fig. 16)....................*Eclipidrilus* (in part)
[This genus was reviewed by Brinkhurst (1988). The species with probosces are known mostly from the southeastern United States, with one record from Montana.]

5b. Atria without spiral muscles....................*Rhynchelmis* (in part)
[Several species distributed from California–Nevada to Alaska and Northwest Territories of Canada. See Fend and Brinkhurst (2000)]

6a(1b). Chaetae bifid....................7

6b. Chaetae simple-pointed....................8

7a(6a). Elongate (to 100 mm or more) slender worms, the front end often green in life, the rest dark red to black; elaborately branched blood vessels laterally in the body wall of posterior segments (Fig. 17); reproduces sexually and asexually (fragmentation with or without encystment); no permanent everted penes....................*Lumbriculus*
[One widely distributed species, *Lumbriculus variegatus* (Mller, 1774), inhabits shallow water, where it floats the tail in the air–water interface. Several other possible taxa, mostly from Alaska, are almost certainly variants of this (see *Lumbriculus* and *Thinodrilus,* in Holmquist 1976).]

7b. Short (25–40 mm), tapering worms, pale to white color; blood vessels in the posterior lateral body wall with short, unbranched

lateral diverticulae; reproduction sexual; permanently everted soft penes on X close together near the midline on mature specimens .. .. *Stylodrilus heringianus* Claparéde, 1862
[Widespread in the Great Lakes and Canada, possibly a European introduction.]

8a.(6b). Groundwater species 9

8b. Surface water species 11

9a(8a). Four pairs of male pores on X, two pairs of spermathecal pores on IX (Fig. 18) .... *Spelaedrilus multiporus* Cook, 1975
[Known only from a cave in Virginia.]

9b. Male pores paired on X, spermathecal pores paired in either IX, XI, or XI and XII 10

10a.(9b). Spermathecal pores paired on IX .... *Stylodrilus*
[*S. californianus* Rodriguez, 1996, California; *S. wahkeenensis* Rodriguez and Coates, 1996, Oregon, and possibly Tennessee and Alabama.]

10b. Spermathecal pores paired in XI or XI and XII .... *Trichodrilus* (in part)
[From caves in West Virginia, Illinois, Washington.]

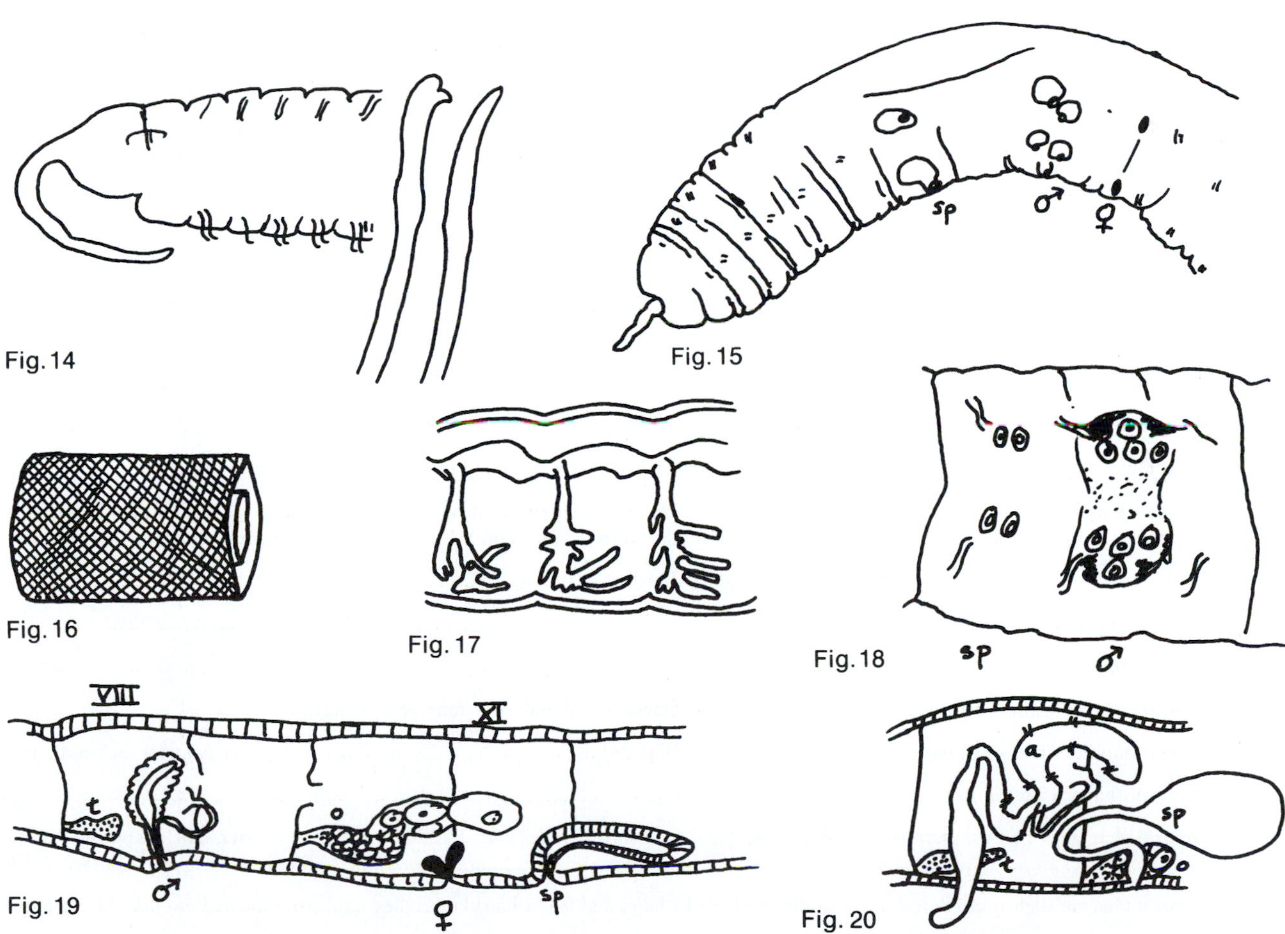

FIGURE 14 Proboscis and chaetae of *Kincaidiana hexatheca. FIGURE 15* Proboscis of *Rhynchelmis brooksi. FIGURE 16* Spiral atrial muscles, *Eclipidrilus. FIGURE 17* Blood vessels of the lateral body wall of posterior segments, *Lumbriculus. FIGURE 18* Ventral view of segments IX (to left) and X showing unique gonopores of *Spelaedrilus* (sp, spermathecal pores; male symbol, male pores, 4 and 8 of each, respectively.) *FIGURE 19* Reproductive systems of *Styloscolex,* anterior to left, left side shown. Note testis (t) and atrium in VIII, ovary (o) and female funnel in X, and spermatheca (sp) in XI. *FIGURE 20* Reproductive system, *Kincaidiana freidris,* anterior to left, organs of left side shown. Note testis (t) and atrium (a) in VIII, prostate gland stalks shown as well as male duct, ovary (o), and spermatheca (sp) in IX.

11a(8b). Male pores paired in VIII 12

11b. Male pores paired in X....13

12a(11a). Spermathecal pores in XI, ovaries in X (Fig. 19) ....*Styloscolex opisthothecus* Sokol'skaya, 1969
[This mostly Asian species is restricted to northern Alaska.]

12b. Spermathecal pores paired in IX, ovaries in IX (Fig. 20) ....*Kincaidiana freidris* Cook, 1966
[This species may not be congeneric with *K. hexatheca,* but cladistic analyses suggest they are quite closely related. Both are Pacific Northwest species, this one from California.]

13a(11b). Spermathecal pores paired in IX ....*Stylodrilus sovaliki* (Holmquist, 1976) and ....*Tenagodrilus musculus* Eckroth and Brinkhurst, 1996
[*S. sovaliki* is from Alaska, but it closely resembles its European congener *S. absoloni* (Hrabe, 1970). *T. musculus* is from Alabama.]

13b. Spermathecal pores either single median in VIII–IX or in IX, or paired in IX ....*Eclipidrilus* (in part)
[One species, (*E. frigidus* Eisen, 1881) is from Idaho and California; the other three (*E. lacustris* Verrill, 1871, *E. ithys* Brinkhurst, 1998, and *E. fontanus* Wassell, 1984) are from northeastern North America].

C. Taxonomic Key to Genera and Selected Species of Freshwater Naididae

1a. No dorsal chaetae present (in some rare specimens the dorsal chaetae are recovered, but these can be recognized because of the absence of ventral chaetae in III–V and the presence of an enlarged pharynx and reduced prostomium associated with a predatory habit) ....*Chaetogaster*

1b. Dorsal chaetae present (rare in *Ophidonais*) 2

2a(1b). Dorsal chaetal bundles without hair chaetae 3

2b. Dorsal chaetal bundles with hair chaetae (Fig. 2) 7

3a(2a). Dorsal chaetae begin in II or III....4

3b. Dorsal chaetae begin in V or VI 5

4a(3a). Dorsal chaetae begin in II; no gap between chaetal bundles of III and IV....*Homochaeta naidina* Bretscher, 1896

4b. Dorsal chaetae begin in III; gap between chaetal bundles of III and IV (Fig. 21) ....*Amphichaeta*

5a(3b). Dorsal chaetae begin in V; estuarine species ....*Paranais*

5b. Dorsal chaetae begin in VI; freshwater species 6

6a(5b). Dorsal chaetae stout, solitary, bluntly simple-pointed or notched at the outer end (Fig. 22)....*Ophidonais serpentina* (Müller, 1773)

6b. Dorsal chaetae curved, with bifid tips, 2–4 per bundle. [See also *Piguetiella,* in which hair chaetae may be missing in most, if not all, dorsal bundles; some *Uncinais* specimens are said to have dorsal chaetae in V.] ....*Uncinais uncinata* (Orsted, 1842)

7a(2b). With a proboscis on the prostomium (visible as a stump if broken off) (Fig. 23) 8

7b. Without a proboscis on the prostomium 11

8a(7a). Dorsal chaetae begin in II....*Pristina*

8b. Dorsal chaetae begin in VI (observe with care as ventral chaetae of IV and/or V may be missing)....9

9a(8b). Dorsal bundles of VI–VIII with 2–16 giant hair chaetae (Fig. 24) ....*Ripistes parasita* (Schmidt, 1874)

9b. No giant hair chaetae....10

10a(9b). Ventral chaetae with a characteristic double bend (Fig. 25); dorsal bundles with 1–3 hairs and 3–4 shorter, hairlike needles ....*Stylaria lacustris* (Linnaeus, 1767)

10b. Ventral chaetae slightly curved; dorsal bundles with 8–18 hairs and 9–12 hairlike needles ....*Arcteonais lomondi* (Martin, 1907)

11a(7b). Dorsal chaetae begin in II or III....12

11b. Dorsal chaetae begin in IV or further back....14

12a(11a). Body wall covered with foreign matter adhering to glandular secretions ....*Stephensoniana*

12b. Body wall naked....13

13a(12b). Dorsal chaetae begin in II; hair chaetae in all bundles, none especially thin, either none especially elongate "or" elongate hair chaetae in III....*Pristinella*

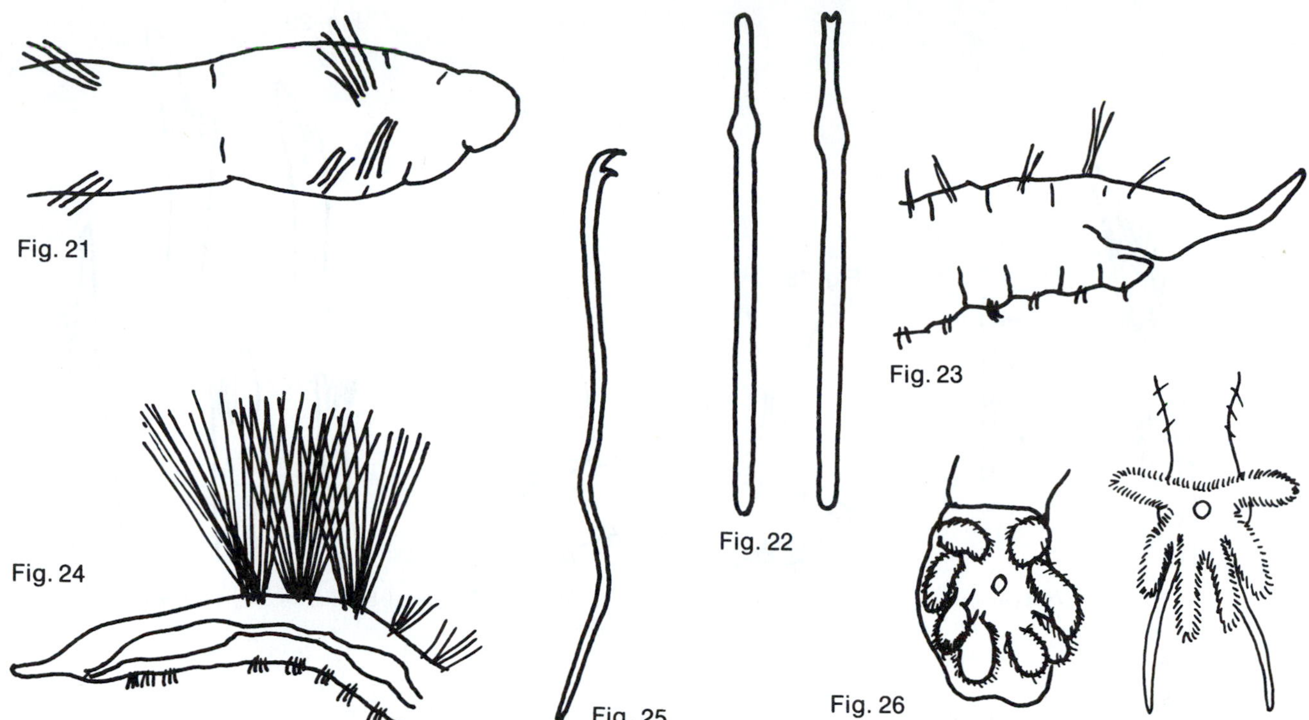

FIGURE 21 Anterior end (to right), *Amphichaeta,* showing dorsal chaetae beginning in III, and gap between chaetae of III and IV, particularly pronounced in this species, but noticeable throughout the genus. *FIGURE 22* Dorsal chaetae, *Ophidonais.* These are single, or even absent in many segments. *FIGURE 23* Proboscis on the prostomium of a naidid, this one with dorsal chaetae from II, but this varies (see key). *FIGURE 24* Dorsal chaetae of *Ripistes* begin in VI, with giant hairs in three segments. *FIGURE 25* Ventral chaeta of *Stylaria lacustris. FIGURE 26* Perianal gills of *Dero. Dero (Aulophorus)* with long palps (right) and *Dero (Dero)* without palps (left).

13b. Dorsal chaetae either begin in III, hairs of III elongate "or" begin in II, the hairs being longest in midbody, and often missing from a number of segments, and exceptionally thin when present .... *Bratislavia*

14a(11b). Posterior end of the body with a branchial fossa, normally with gills (Fig. 26); some species symbionts in tree frogs, these develop gills once outside their host. .... *Dero*
[There are three subgenera, *Allodero* containing the symbiotic species, *Aulophorus* with long palps on each side of the branchial fossa, and *Dero* that lacks palps. In a strict cladistic sense, *Allodero* would not be recognized.]

14b. No gills; free-living species 15

15a(14b). Dorsal chaetae begin in XVIII–XX, each bundle with a single short hair and a robust needle .... *Haemonais waldvogeli* Bretscher, 1900

15b. Dorsal chaetae begin in V, VI, or VII (note that ventral chaetae of IV and/or V may be missing) 16

16a(15b). Hair chaetae up to 9 per bundle, thick, with long lateral hairs (Fig. 27) .... *Vejdovskyella*

16b. Hair chaetae 1–3 per bundle at most, thin with or without fine lateral hairs 17

17a(16b). One to three especially long hair chaetae in each dorsal bundle of VI, 1 or 2 ordinary hair chaetae in the remainder (Fig. 28) .... *Slavina appendiculata* (d'Udekem, 1855)

17b. No elongate hair chaetae 18

18a(17b). Hair chaetae very short (84–120 μm), present in only one or two bundles, some with no hair chaetae at all (see 6) .... *Piguetiella*

18b. Hair chaetae mostly 1 or 2 per bundle (two species with up to 5), long (180–200 μm or more), present in all bundles from V or VI. 19

19a(18b). Characteristic needle chaetae with thick, unequal teeth, sometimes clearly pectinate (Fig. 29); ventral chaetae change form and length slightly between V and VI .... *Allonais*

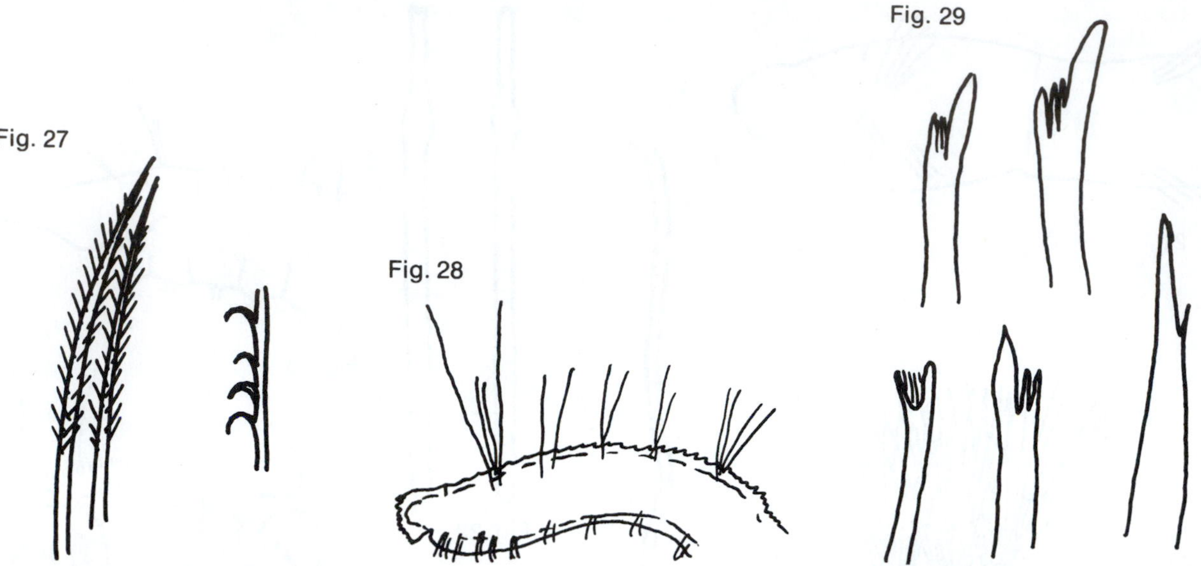

FIGURE 27 Hair chaetae of *Vejdovskyella,* detail to right. *FIGURE 28* Dorsal bundles of VI in *Slavina* with elongate hairs. *FIGURE 29* Needle chaetae of various *Allonais* species.

19b. Needle chaetae simple-pointed, bifid, or faintly pectinate; either ventral chaetae of II differ from the rest "or" those of II–V differ markedly in length, width, and form from the rest 20

20a(19b). Needle chaetae bifid or faintly pectinate; ventral chaetae of II may differ from the rest .... *Specaria*

20b. Needle chaetae simple-pointed, bifid, or pectinate; ventral chaetae of II–V differ strongly in length, thickness, and form from the rest in the majority of species. .... *Nais*
[Species of *Dero* with the posterior end of the body missing may key out here because there will be no gills. Compare specimens to those with gills; the anterior ventral chaetae in *Dero* species have very long upper teeth.]

D. Taxonomic Key to Genera and Selected Species of Freshwater Tubificidae

This key is designed to maximize the use of external characters. As it is primarily a key to genera, it is necessary to refer mostly to characters found on mature specimens. Immature specimens can often be identified on the basis of chaetal form alone, but then the key must proceed directly to the specific level (Kathman and Brinkhurst, 1998). In this key, there are two genus groups that are difficult to separate on external characters; and where this is true, evidence from distribution patterns is introduced as an additional aid. Keys for worms of this family are traditionally written at the species level.

1a. Single dorsal and ventral gill filaments on posterior segments (Fig. 30) .... *Branchiura sowerbyi* Beddard, 1892
[Broken anterior fragments may be mistaken for *Aulodrilus pluriseta* (see 18b).]

1b. No gill filaments 2

2a(1b). Body wall papillate (with projections usually covered with foreign matter attached by secretions) (Fig. 31) .... *Telmatodrilus* (in part), *Spirosperma, Quistadrilus*
[*Telmatodrilus (Alexandrovia) onegensis* Hrabe, 1962 is recorded from Alaska; the subgenus is often elevated to generic status, which may prove correct.]

2b. Body wall naked 3

3a(2b). Spermathecal chaetae replace normal ventral chaetae in the spermathecal segment, very rarely in adjacent segments also (Fig. 34); with or without modified penial chaetae (Figs. 39, 41, 42) and cuticular penis sheaths (Fig. 35) 4

3b. No modified spermathecal chaetae 8

4a(3a). Thin, hollow-tipped spermathecal chaetae in X, similar chaetae in XI together with apparent cuticular penis sheaths (Fig. 32) .... *Haber speciosus* (Hrabe, 1931)

4b. Spermathecal chaetae not duplicated (or, if so, on VII or VIII and very rarely scattered from VI–XII); with or without cuticular penis sheaths) 5

5a(4b). Spermathecal chaetae broad, spatulate, one each side of IX associated with long tubular glands; ventral chaetae of X similar or absent; several penial chaetae in each ventral bundle of XI with short, knobbed distal ends; male ducts open into large median inversion of the body wall of XI (Fig. 33) .... *Rhizodrilus lacteus* Smith, 1900

5b. Without this unique set of characters combined 6

6a(5b). Spermathecal chaetae much broader and longer than the normal ventral chaetae, distal ends hollow, located on X (or in one instance on VII or VIII or even irregularly between VI and XII); no cuticular penis sheaths .... *Potamothrix*

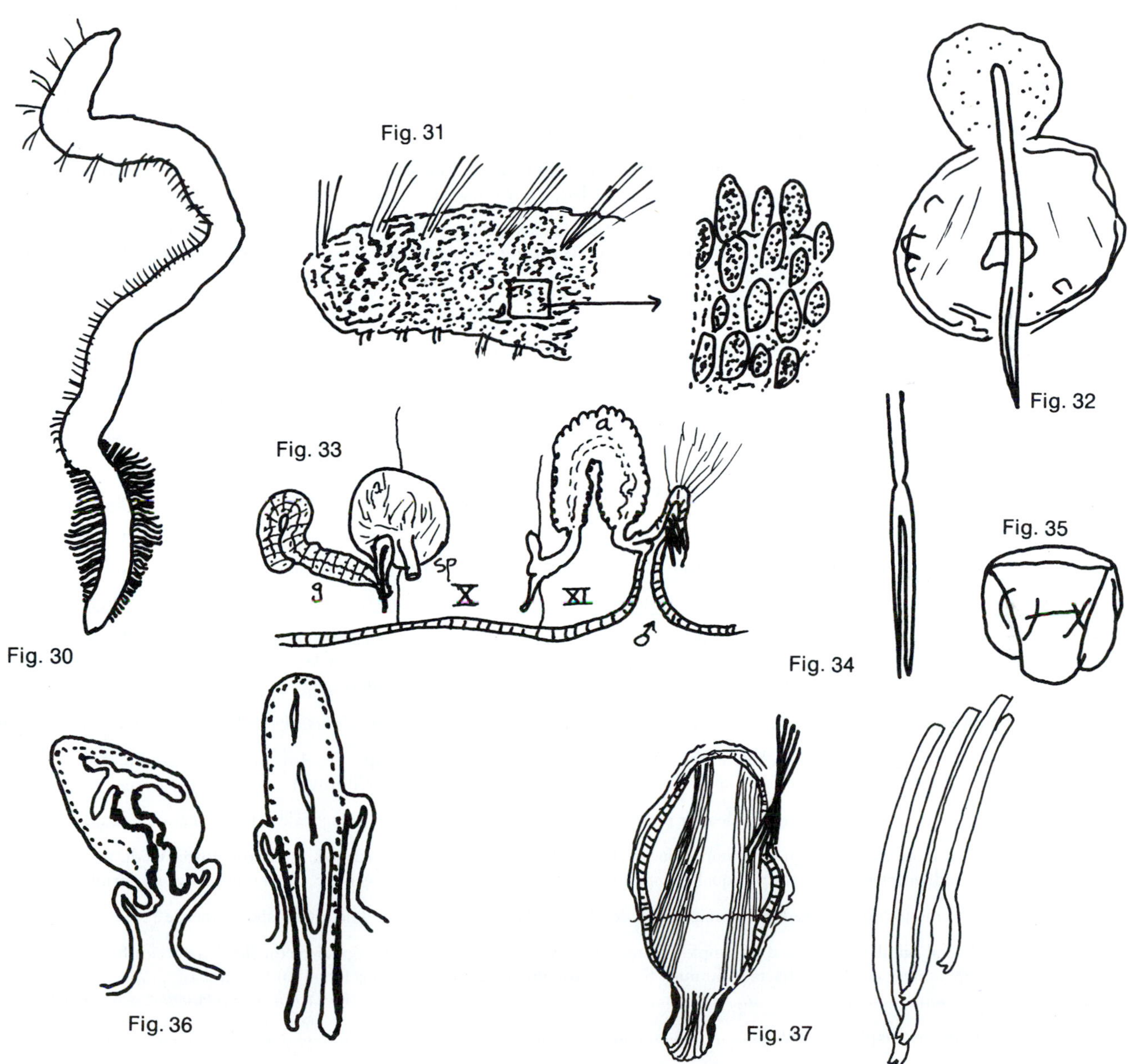

FIGURE 30 *Branchiura,* showing posterior fan of gill filaments. *FIGURE 31* Papillate body wall of *Spirosperma.* *FIGURE 32* Spermathecal chaeta of *Haber* (in which penial chaetae similar.) *FIGURE 33* Reproductive organs, left side (anterior to left) of *Rhizodrilus lacteus.* Note spermathecal chaeta and gland (g) in IX, spermatheca (sp) in X, atrium (a) in XI discharging into median chamber that also receives penial chaetae (on right, opposite atrial pore). *FIGURE 34* Spermathecal chaeta of *Isochaetides.* *FIGURE 35* Penis sheath of *Isochaetides.* *FIGURE 36* Inverted (left) and everted penis of *Psammoryctides.* *FIGURE 37* *Varichaetadrilus,* penis (left) with penial sheath on tip (shown pointing down) and penial chaetae, top right and (enlarged) to right.

[Most species are restricted to the Great Lakes basin or extend into adjacent states; only one is widespread. These may be introduced from Europe, where they are characteristic of productive waters.]

6b. Spermathecal chaetae longer than the normal ventral chaetae, slender distally, somewhat like a hypodermic needle (Fig. 34); true or apparent cuticular penis sheaths present....7

7a(6b). True cuticular penis sheaths present, but not much thicker than the normal cuticular layer of the body wall (Fig. 35)......*Isochaetides*

7b. Apparent penis sheaths present (probably the cuticular linings of eversible penes), look like crumpled cylinders in whole-mounts of most specimens (Fig. 36)....*Psammoryctides*
[One species is a recent introduction to the St. Lawrence River (Quebec) from Europe, another is known only from coastal swamps of the Gulf of Mexico, and a third is known from California and the Great Lakes.]

8a(3b). Penial chaetae present replacing the normal ventral chaetae of XI (Figs. 37, 39, 41, 42)....9

8b. Penial chaetae absent, ventral chaetae of XI usually missing in mature specimens....12

9a(8a). Penial chaetae bifid but with shortened distal ends and elongate proximal ends, somewhat thicker than the normal ventral chaetae; with short penis sheaths on the ends of erectile penes (Fig. 37)....*Varichaetadrilus* (in part)
[Distributed form Alaska to California and Alberta (but see also 14).]

9b. Penial chaetae strongly modified, with knobbed or hooked tops to short distal ends, elongate proximal ends, arranged fanwise with heads close together (as if they function as claspers) "or" large, single and sickle-shaped (Figs. 39, 41, 42)....10

10a(9b). Male pores open into an eversible chamber in XI that contains so-called paratria (Fig. 38) and penial chaetae when present (Fig. 39); sensory pit on the prostomium (Fig. 3); sperm attached to the body wall close to the male pore in external spermatophores (Fig. 40)....*Bothrioneurum vejdovskyanum* Stolc, 1888

10b. No median copulatory chamber or sensory pit; sperm loose in spermathecae after copulation....11

11a(10b). Penial chaetae in XI, 3–6 simple-pointed chaetae with hooked tips (Fig. 41); somatic chaetae 3–5 bifids from II–XX, simple-pointed chaetae from XXV–XXX posteriad, no dorsal hair chaetae ....*Phallodrilus hallae* Cook and Hiltunen, 1975

11b. Penial chaetae with knobbed tips, or single, sickle-shaped (Fig. 42); no simple-pointed somatic chaetae (except in the ventral bundles of one species, when accompanied by bifid chaetae, and the dorsal bundles of that species possess hair chaetae).... *Rhyacodrilus*

12a(8b). Cuticular penis sheaths present in XI (Figs. 42–45) 13

12b. Cuticular penis sheaths absent....15

13a(12a). Penis sheaths more or less elongate, cylindrical, with penes free within them (Fig. 43); all chaetae bifid ....*Limnodrilus*

13b. Penis sheaths annular to conical, attached to the surface of the penis (Figs. 44 and 45) 14

14a(13b). Short cuticular penis sheaths, annular to tub-shaped (Fig. 44)....*Varichaetadrilus (*in part), *Tubifex, Ilyodrilus* (in part)
[The *Varichaetadrilus* species included here are reported from the southeastern United States; *I. frantzi* Brinkhurst, 1965 is restricted to freshwater at the head of estuaries from British Columbia (Canada) to California.]

14b. Penis sheaths conical (Fig. 45)....*Tasserkidrilus, Ilyodrilus templetoni* (Southern, 1904)

15a(12b). Chaetae simple-pointed anteriorly, behind the clitellum the chaetae have brushlike tips (Fig. 46) ....*Telmatodrilus vejdovskyi* Eisen, 1879 [This species is known from British Columbia (Canada) to California.]

15b. Chaetae bifid, or with upper tooth divided, or pectinate to palmate, all with the teeth in a single plane....16

16a(15b). Pharynx and mouth enlarged, prostomium reduced (Fig. 47); chaetae of III broad, curved, with large lower teeth, the rest thinner, straighter, and with teeth more nearly equal; no spermathecae, presumably parthenogenetic ..*Teneridrilus mastix* (Brinkhurst, 1978)

16b. Pharynx and mouth not enlarged, prostomium well developed; chaetae of III not modified; with spermathecae 17

17a(16b). Ventral chaetal bundles with up to 4 simple-pointed or bifid chaetae with reduced upper teeth. [In this species from Lake Tahoe there are apparently no penial chaetae, although these are usually present in this genus (see 11b).] ....*Rhyacodrilus brevidentatus* Brinkhurst, 1965

17b. No simple-pointed ventral chaetae 18

18a(17b). Dorsal chaetal bundles with 2–4 pectinate-to-palmate chaetae from II–XIV, from there posteriad all dorsal chaetae bifid; anterior ventral chaetae 3–5 per bundle, bifid with the upper teeth thinner than but only a little longer than the lower, but beyond XIV the upper teeth become much longer than the lower....*Arctodrilus wulikensis* Brinkhurst and Kathman, 1983
[This species is recorded from the Brooks Range, Alaska.]

18b. Chaetae progressively enlarged from III posteriad, becoming very broad with large lower teeth and recurved distal ends, "or" no hair chaetae dorsally and large numbers of bifid chaetae with thin, short upper teeth, becoming palmate beyond VI in one species, "or" hair chaetae present but all ventral chaetae again characteristically numerous and with short upper teeth (Fig. 48 .....*Aulodrilus*

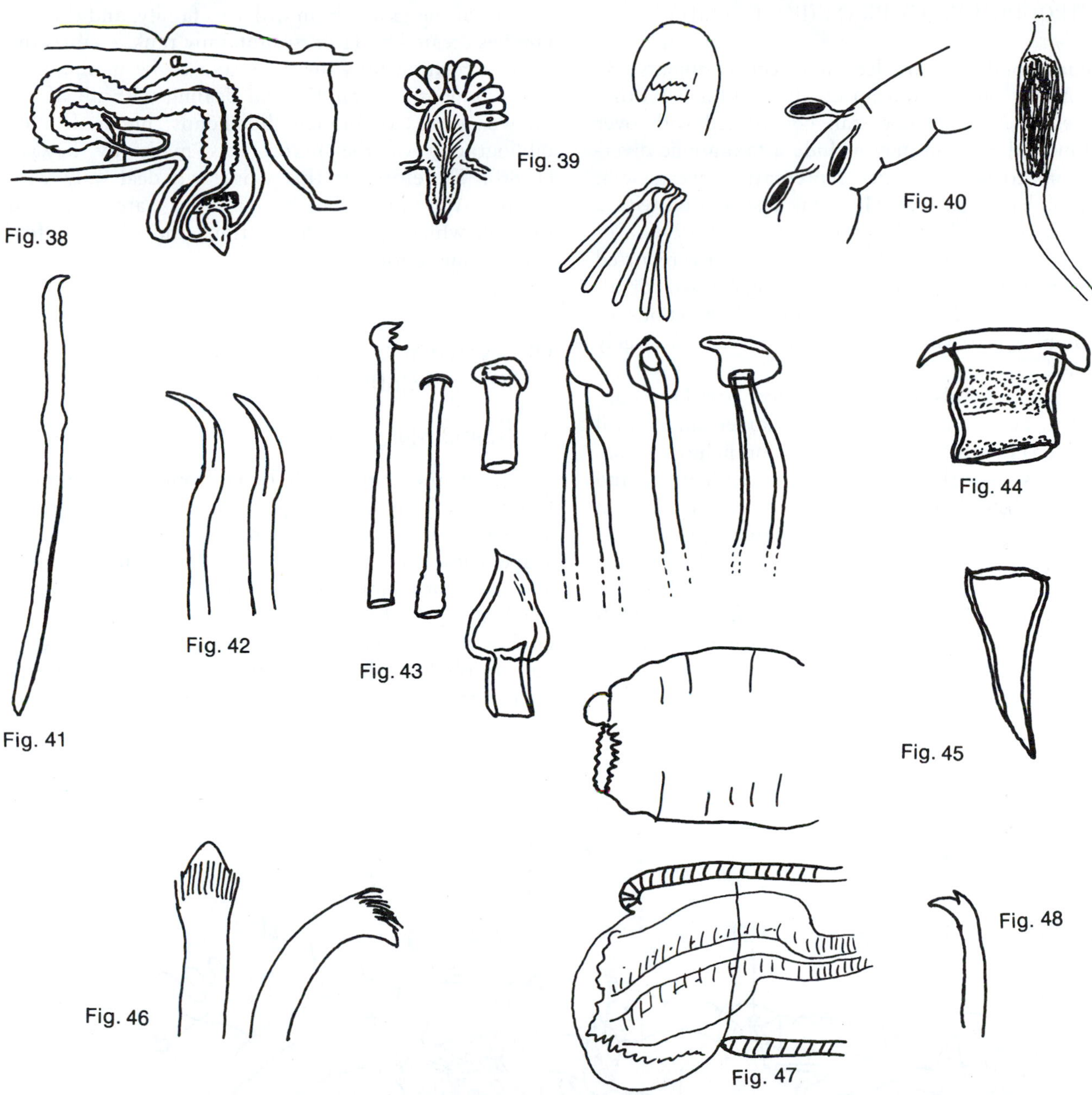

FIGURE 38 Paratrium of *Bothrioneurum,* shown left *in situ* in everted terminal portion of male duct, detail to right. *FIGURE 39* Penial chaetae, *Bothrioneurum,* detail of tip above, bundle below. *FIGURE 40* Spermatophores of *Bothrioneurum,* attached to body wall around gonopores (left) and detail (right). Note sperm enclosed, with no duct through attachment stalk. *FIGURE 41* Penial chaeta, *Phallodrilus hallae. FIGURE 42* Penial chaetae, *Rhyacodrilus falciformis. FIGURE 43* Penes of various *Limnodrilus* species, three to right much longer than shown (40 times the length when fully developed). *FIGURE 44* Penis sheath of *Tubifex tubifex,* often thin and difficult to see, but granular surface often a clue. *FIGURE 45* Penis sheath of *Ilyodrilus,* short conical form. The terminal opening often looks torn, and individuals with a second portion as long as that illustrated here may be seen, this second part presumably shed before mating. *FIGURE 46* Posterior brushlike chaetae, *Telmatodrilus. FIGURE 47* Inverted (above) and everted pharynx of *Teneridrilus mastix. FIGURE 48* Ventral chaeta, *Aulodrilus.*

BRANCHIOBDELLIDAE

VI. INTRODUCTION TO BRANCHIOBDELLIDAE

Branchiobdellids are leechlike, ectosymbionts living primarily on astacoidean crayfish and a few other crustaceans. A single host can be infested with over 1,000 individuals and may include a taxonomic diversity of up to seven species representing five genera. Branchiobdellids have an Holarctic distribution with a total of 21 genera and 149 species (Gelder, 1996a), of which 15 genera containing 107 species are reported from North America. In spite of being known for almost 250 years, the ecophysiological interactions of this complex ectosymbiotic association remain largely unstudied.

Branchiobdellids have for the most part been considered to be a family, or at best a superfamily, until Holt (1965) raised them to an order, Branchiobdellida. He did this to make the taxon independent of, and equal in rank to, both the oligochaetes and leeches. In a review of the various ranks and names given to the branchiobdellids (Gelder, 1996a), it was found that the taxon has reached class rank, Branchiobdellae, but this was an overly inflated rank for the group. Holt (1986) arranged the branchiobdellid genera into five families within the order. This gave rise to the term, used in the first edition, "branchiobdellid" as the common epithet for members of the order, thus leaving branchiobdellid for members of that family. The demotion of the taxon from order to family, and the five families created by Holt to subfamily rank, enables the relative relationships of these taxa to be maintained. There is no single publication dealing with all of the known species of branchiobdellidans, but a detailed bibliography of the taxon was made by Sawyer (1986). Therefore, as this work will deal only with North American genera, citations are included through which species descriptions and further information can be found.

VII. ANATOMY AND PHYSIOLOGY OF BRANCHIOBDELLIDS

A. External Morphology

Adult branchiobdellids range from 0.8–10 mm in length. The body in most species is rod-shaped (terete), although some have a characteristic pyriform or flask-shape with either ventral or dorso-ventral flattening. The segment number is constant and, based on the number of paired ganglia, is accepted to be fifteen (XV) as shown by Roman numerals in Figure 49. The external morphology shows a peristomium and three segments forming a head, and the remaining obvious body

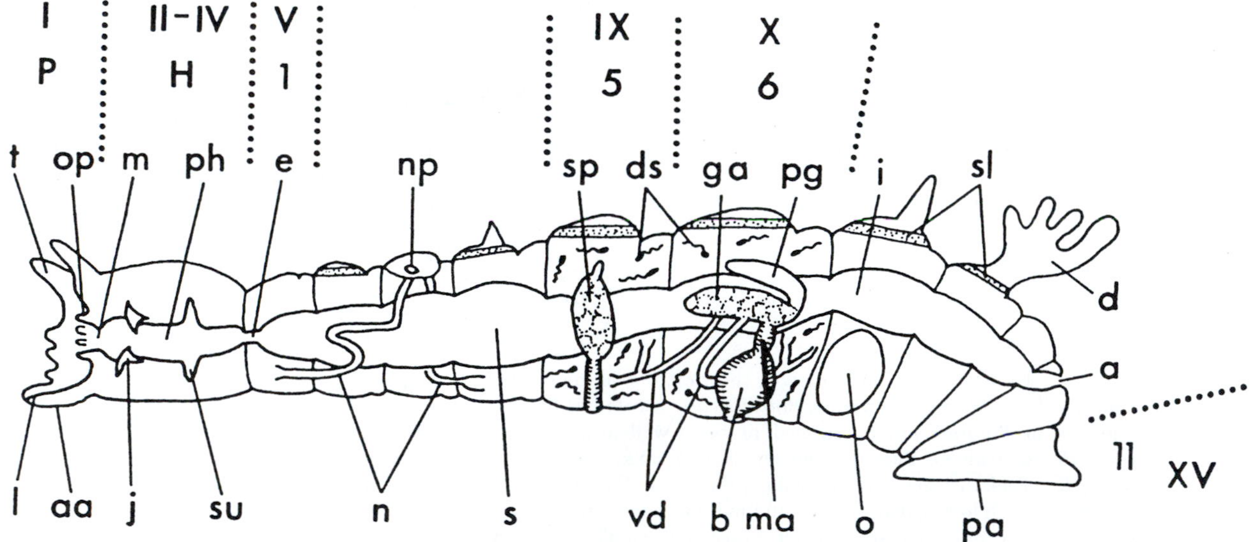

FIGURE 49 Diagram of the lateral aspect of a hypothetical branchiobdellidan showing anatomic characters used in the taxonomy of this group: a, anus; aa, anterior attachment surface; b, bursa; d, digitiform dorsal appendages; ds, developing sperm; e, esophagus; ga, glandular atrium; i, intestine; j, jaw; l, lip; m, mouth; ma, muscular atrium; n, nephridia; np, nephridial pore; o, ovum; op, oral papillae; pa, posterior attachment surface; pg, prostate gland; ph, pharynx; s, stomach; sl, supernumerary longitudinal muscles; sp, spermatheca; su, sulcus; t, tentacle; vd, vasa deferentia; H, head; P, peristomium; 1-11, body segments; I-XV, segments based on paired ganglia.

segments are numbered in Arabic numerals, 1 to 11. The latter nomenclature is the one used in the descriptive literature on branchiobdellids. Segment 11 is modified into the posterior, disk-shaped attachment organ (sucker), and contains a single pair of ganglia, XV. Peristomial tentacles or lobes may be present on the dorsal lip, and the mouth is surrounded usually by 16 oral papillae. Each body segment is divided into a major anterior and a minor posterior annulus, although a few species have the major annulus further subdivided into two. Dorsal transverse ridges across the major annuli are produced by supernumerary longitudinal muscles (sl) and sometimes support dorsal digitiform or multibranched appendages (d) (Gelder, 1996b).

B. Organ System Function

The alimentary canal consists of a pair of sclerotized jaws (j) situated in the anterior pharynx followed by one to three pairs of sulci (su), a short esophagus (e), stomach (s), intestine (i), and an anus (a) opening dorso-medially onto segment 10 (Fig. 49). Locomotory movements are leechlike and involve an anterior and a posterior attachment organ (Gelder and Rowe, 1988). The anterior attachment site is on the ventral surface of the ventral lip and is not the mouth as assumed generally. The muscle cells have the nucleus and cytoplasm in the medulla with the contractile fibers in the cortex. The ultrastructure of the cells in three species from the Branchiobdellinae are all round circomyarian in their contractile fiber arrangement. A species from each of the Bdellodrilinae and Cambarincolinae have a few round circomyarian fibered cells, but most are "polyplatymyarian" (Valvassori *et al.,* 1994). The pore(s) of the asymmetrically arranged anterior pair of nephridia open to the exterior on segment 3 (np) and the posterior pair are located in segment 8, but open on segment 9.

A pair of testes are located in segments 5 and 6, and the sperm develop (ds) in the coelom of those segments (Fig. 49). The ultrastructure of spermatozoa from four species representing the Branchiobdellinae, Bdellodrilinae, and Cambarincolinae showed an unexpected variation in morphology (Ferraguti and Gelder, 1990). Additional studies have confirmed that morphological variations in the spermatozoa occur between genera, but that interspecific variations are dimensional only (Ferraguti, 2000). Two vasa efferentia, both with a ciliated funnel, merge into a vas deferens and the two vasa deferentia, one from each segment, enter the glandular atrium in segment 6. When present, the prostate gland forms a protuberance or a diverticulum from the glandular atrium. The connecting muscular atrium passes into the penis which is surrounded by a muscular bursa (b). Most penes are either eversible or protrusible, although a few species have a penis that shows varying degrees of reduction. The bursal atrium opens mesad on the ventral surface of segment 6. The terms, "spermiducal gland" (= glandular atrium) and "ejaculatory duct" (= muscular atrium) as used in the First Edition (Brinkhurst and Gelder, 1991) followed Holt (1986). These have been replaced by the terms in parentheses to promote consistency in anatomical terminology with oligochaetologists in general.

The spermatheca (sp), when present, is located in segment 5 and consists of a glandular ental region connected to a muscular duct that opens medially onto the ventral surface of the segment. A pair of ovaries in segment 7 produce eggs that mature in the coelom, and are released through a pair of short ducts in the body wall of that segment.

The vascular system consists of a dorsal and ventral longitudinal vessels that are connected by four paired, lateral branches in the head and a single pair of branches in segments 1, 7, and 10, respectively. The dorsal vessel becomes a sinus surrounding the gut epithelium from segments 3–7.

The central nervous system consists of a dorsal brain, circumesophageal connectives and a paired ventral cord extending into the last segment. The head contains four pairs of ganglia with each body segment containing a single pair of ganglia. The nerve cord is displaced laterally in segments 5 and 6 where the spermatheca and male genitalia pass mesad through the ventral body wall.

C. Environmental Physiology

Most branchiobdellids browse on the epibionts, such as diatoms and ciliates, attached to the host and are opportunistic commensals when fragments of the host's food are available (Jennings and Gelder, 1979). Other opportunistic food sources include cannibalism and exposed host tissues at injury sites. Contrary to the reputation of these worms, a parasitic regime has been demonstrated only in *Branchiobdella hexodonta,* and inferred in a few other species. Food is sucked into the mouth and then the jaws close to crop microorganisms from the substrate or fragment large particles. Various mucoid secretions assist the ingestion of food, but the term "salivary gland" as quoted by Sawyer (1986) is inappropriate and was corrected by Gelder and Rowe (1988). Digestion is extracellular in the stomach and intestine with uptake occurring in the cells of the latter region. Low oxygen tensions do not appear to affect branchiobdellids adversely, although some species tolerate higher water temperatures better than others.

VIII. ECOLOGY AND EVOLUTION

A. Diversity and Distribution

The branchiobdellids are characterized by a lack of anatomical and ecological diversity in comparison with equivalent clitellate taxa. The range of the branchiobdellids in North America extends from southern Canada to Costa Rica (Gelder, 1999). A few species are found distributed throughout this region, but the others have a more limited range (see Section IX.B). The zoogeography of species is becoming more confused and obscure as crayfish and their symbionts are introduced into new areas for aquaculture, live food, research, and teaching purposes.

B. Reproduction and Life History

Information on chromosome numbers is restricted to three species, *Branchiobdella astaci* $2n = 14$, *B. kozarovi* $2n = 12$, and *B. parasita* $2n = 10$ (Mihailova and Subchev, 1981), and they refute previous reports of $n = 4$ and 8 for the first species. Germ cells from the testes are released into the coelom of segments 5 and 6, where spermatogenesis occurs. The mature spermatozoa then congregate at the mouth of the sperm funnels prior to their passage to the penis. Prior to copulation, the muscular bursa is partially or completely everted (Figs. 50–53), and comes close to, or contacts with, the epidermis surrounding the recipient's spermathecal pore. This is followed by the extended penis entering the pore for insemination. Single inseminations have been observed; however, it is not known if reciprocal insemination occurs. Spermatophores on the epidermis of *Xironogiton victoriensis* have been examined by Gelder (1996c). Members of *Ellisodrilus* and *Caridinophila* do not have spermatheca so their method for sperm reception and storage is open to speculation. Although the use of spermatophores has been proposed as a solution to their problem, examinations of preserved specimens of *Ellisodrilus* have found no such packets.

The secretions from the clitellar gland cells have been characterized histochemically and compared with those in other clitellate annelids (Gelder and Rowe, 1988). These secretions produce the pedunculate cocoon and nourishment to sustain the embryo while it develops. Branchiobdellid cocoons deposited on live hosts will develop to maturity and hatch out. The hatching rate as observed on live hosts is temperature dependent, and occurs after 10–12 days at 20–22°C. The embryo is lecithotrophic and then becomes an albuminotrophic cryptolarva. In this latter phase, precocious feeding of the surrounding nutrient fluid starts with ingestion by a transitory pharynx and the digestion yolk-sac epithelium. There are no reports of growth rates or the time a juvenile requires to reach sexual maturity. A few reports have noted that cocoons were deposited on cast exoskeletons of their hosts and the glass walls of their container, but embryos did not develop and emerge. However, Woodhead (1945) is the only person to report that branchiobdellids mated, attached their cocoons to the walls of the stock dishes, and young hatched a few days later.

C. Ecological Interactions

Branchiobdellids have been reported on freshwater crustaceans, primarily astacoidean crayfish. Although all

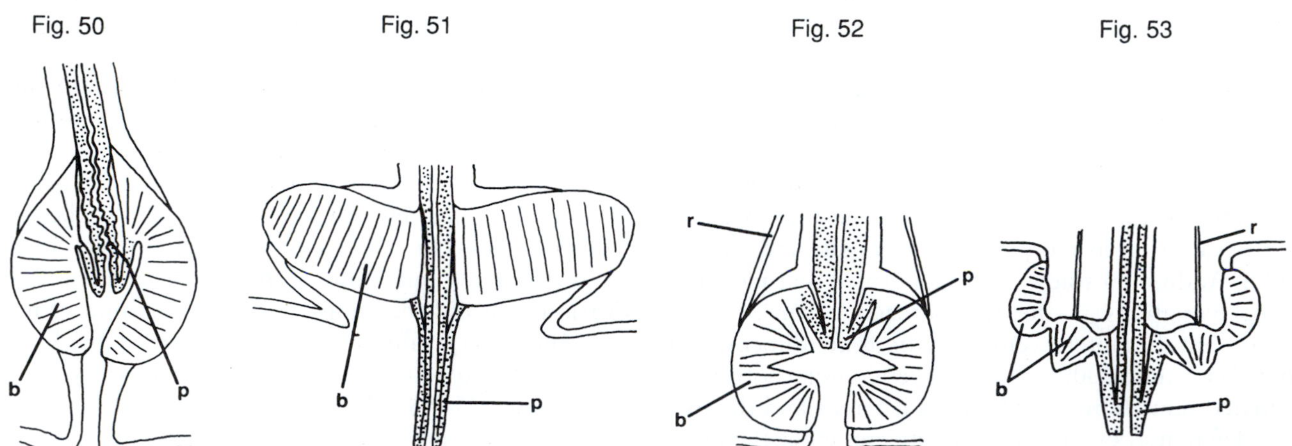

FIGURE 50 Diagrammatic section of a withdrawn, eversible penis: b, bursa, p, penis. *FIGURE 51* Diagrammatic section of an extended, eversible penis; b, bursa; p, penis. *FIGURE 52* Diagrammatic section of a withdrawn, protrusible penis; b, bursa; p, penis; r, retractor muscle. *FIGURE 53* Diagrammatic section of an extended, protrusible penis: b, bursa, p, penis; r, retractor muscle.

crayfish appear to be potential hosts, some species actively remove any branchiobdellids from their exoskeleton (Gelder and Smith, 1987). Noncrayfish hosts were reported to be acceptable in troglodytic habitats (Holt, 1973b) and in Central America at and beyond the southern limits of the geographical range of the astacoidean crayfish hosts (Gelder, 1999). Branchiobdellids have been reported on the blue crab, *Callinectes* sp., in low salinity waters adjacent to the Gulf of Mexico (Overstreet, 1983), and so the possibility exists that a marine branchiobdellid will be found.

Branchiobdellids also harbor ectosymbionts and parasites. Externally, these include stalked, solitary and colonial ciliate protozoans, and rotifers. Internally, branchiobdellids serve as host to the third juvenile stage of the nematode *Dioctophyme renale* (giant kidney worm) (Schmidt and Roberts, 1989), and the trematode mesocercaria of *Pharyngostomoides procyonis* (reviewed in Schell, 1985).

D. Evolutionary Relationships

The branchiobdellids form a monophyly (Gelder and Brinkhurst, 1990). The leechlike appearance of branchiobdellids has caused some researchers to included them with the leeches. However, the presence of a semiprosopore male ducts (Fig. 6) in both the branchiobdellids and lumbriculid oligochaetes, strongly supports the branchiobdellids returning to family rank as a sister group to the lumbriculids (Brinkhurst, 1999). The anatomical modifications shown by branchiobdellids, such as constant segment number, loss of chaetae, development of anterior and posterior attachment organs, jaws, and other features, are not considered homologous to similar states in leeches but, instead, reflect a specialization to an ectosymbiotic lifestyle as proposed by Stephenson (1930). The relocation of the branchiobdellids back into the oligochaetes is the result of the best interpretation of the available data. Phylogenetic analyses will soon be available using matrices of nucleotide sequences in selected genes from a significant number of clitellate species. Whether the new methodologies will provide new phylogenies or confirm some of those already established remains to be seen.

IX. IDENTIFICATION OF BRANCHIOBDELLIDS

A. Collecting, Rearing, and Preparation for Identification

Crayfish and other crustacean hosts are collected by netting or baiting a trap (see Chapter 22) and then placed directly into a container with preservative. If branchiobdellids are to be examined it is recommended that the preservative formalin–alcohol–acetic acid (FAA) be used (70 mL ethanol + 5 mL formaldehyde (full strength) + 1 mL glacial acetic acid + 30 mL water). The volume of preservative should be at least three times that of the crustaceans to ensure thorough preservation of the symbionts and host. Branchiobdellids removed from a live host have been maintained *in vitro* for several months.

Microscopical examination of a living branchiobdellid—in water slightly squeezed between slide and cover-glass—will enable all of the anatomical characters to be observed. First examine the specimen from a dorsal or ventral aspect as this enables one to observe the peristomium, jaws and location of the anterior nephridial pore(s). The specimen can then be rolled to show a lateral aspect, allowing the spermatheca and male genitalia to be studied.

Live specimens can be relaxed using carbonated drinks (such as club soda) or a 1–2% magnesium chloride solution (Delly, 1985), straightened or flattened, and then immediately preserved. Unstained specimens are dehydrated in graded ethanol solutions, cleared in clove oil or oil of wintergreen (methyl salicylate), and mounted in Canada balsam for permanent preparations. Specimens usually curl when preserved on the host, and so these worms can only be examined from the lateral aspect. Large branchiobdellids (4–10 mm long) will require microdissection to expose their genitalia. The necessary instruments can be made from fragments of a razor blade and insect (very small) pins glued to the end of cocktail sticks. For techniques to prepare serial sections and to stain specimens, refer to a book on animal microtechiques, such as Humason (1979).

Gelder (1996a) compiled a checklist of branchiobdellids and included citations to all papers containing type descriptions.

B. Taxonomic Key to Genera of Freshwater Branchiobdellidae

1a. Two anterior nephridial pores 2

1b. One anterior nephridial pore 4

2a(1a). Ventral flattening of body 3

2b. Dorso-ventral flattening of body (Fig. 54) .... *Xironogiton* [British Columbia, Oregon, Washington, California, Idaho, Wyoming, and the Appalachian Region from Maine to Georgia.]

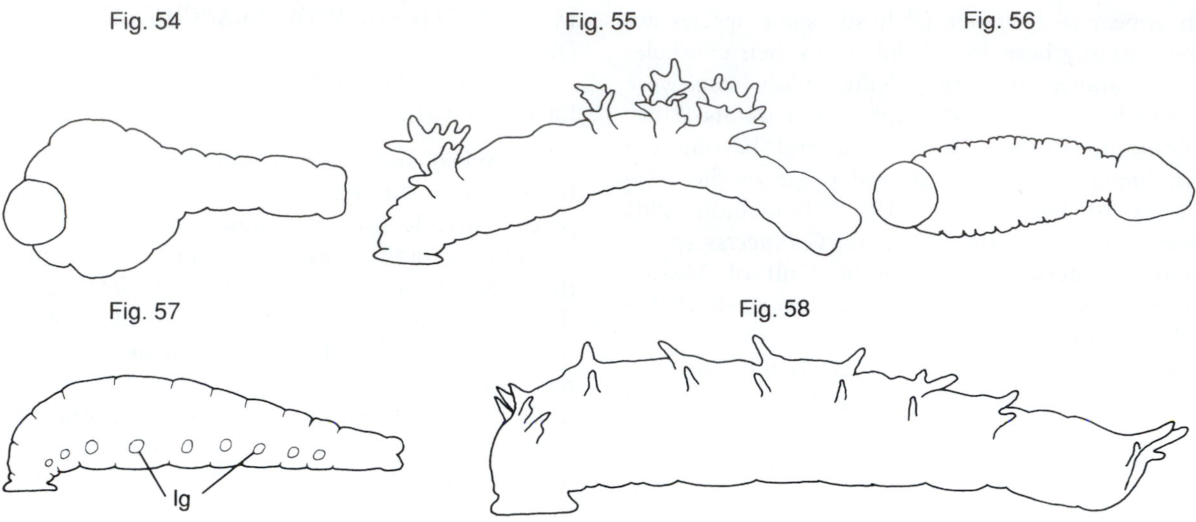

FIGURE 54 Ventral view of *Xironogiton instabilis,* length about 2 mm. *FIGURE 55* Lateral view of *Pterodrilus alcicornis,* length about 1.5 mm. (Redrawn from Holt, 1986b.) *FIGURE 56* Ventral view of *Triannulata magna,* length about 4.5 mm. (Redrawn from Holt, 1974.) *FIGURE 57* Lateral view of *Bdellodrilus illuminatus,* length about 3 mm, lg, lateral glands. *FIGURE 58* Lateral view of *Ceratodrilus thysanosomus,* length about 3 mm. (Redrawn and modified from Holt, 1960.)

3a(2a). Vasa deferentia entering glandular atrium mesad .... *Ankyrodrilus* [Tennessee, Virginia]

3b. Vasa deferentia entering glandular atrium entad .... *Xironodrilus* [East-central USA]

4a(1b). Vasa deferentia entering glandular atrium mesad 5

4b. Vasa deferentia entering glandular atrium entad 7

5a(4a). Body surface with indistinct segments 6

5b. Body surface with distinct segments (Fig. 59) .... *Cronodrilus* [Georgia]

6a(5a). Lateral segmental glands, segments 1–9 (Fig. 55) .... *Bdellodrilus* [Southern Canada to southern Mexico]

6b. Lateral segmental glands not in segments 1–9 .... *Uglukodrilus* [California, Oregon, Washington]

7a(4b). Spermatheca present 8

7b. Spermatheca absent .... *Ellisodrilus* [Tennessee, Kentucky, Indiana, Michigan]

8a(7a). Penis protrusible (Figs. 52 and 53) 9

8b. Penis eversible (Figs. 50 and 51) 11

9a(8a). Prostate body present 10

9b. Prostate body absent (Fig. 60) .... *Magmatodrilus* [California]

10a(8b). Dorsal appendages present, range from paired digitiform unit to single transverse ridge on segment 8 (Fig. 56) .... *Pterodrilus* [Eastern United States]

10b. Dorsal appendages absent (Fig. 61) .... *Cambarincola* [Southern Canada to Costa Rica]

11a(9a). Prostate body absent 12

11b. Prostate body present 13

12a(11a). Two annuli per segment (Fig. 62) .... *Sathodrilus* (in part) [Florida, California, Mexico]

12b. Three annuli per segment (Fig. 57) .... *Triannulata* [Oregon, Washington]

13a(11b). Dorsal appendages present (Fig. 58) .... *Ceratodrilus* [Utah, Wyoming, Idaho, Montana, Oregon]

13b. Dorsal appendages absent. [This grouping cannot be resolved further without detailed histological preparation.](see Fig. 63) .... *Sathodrilus* (in part) [South Carolina, Georgia, Mexico] (see Fig. 64) .... *Sathodrilus* (in part) [Southern Canada to Mexico] *Oedipodrilus* [Tennessee, Kentucky, Ohio, Pennsylvania, Indiana, Illinois] .... *Tettodrilus* [Tennessee]

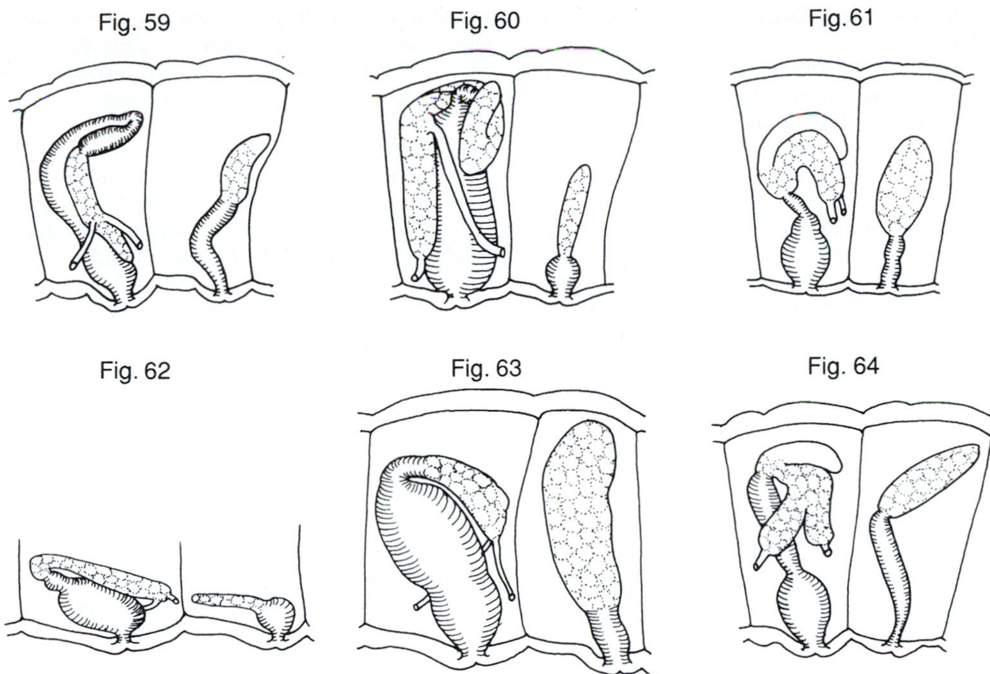

FIGURE 59 Lateral view of the male genitalia and spermatheca in *Cronodrilus ogyguis*. (Redrawn and modified from Holt, 1968a.) *FIGURE 60* Lateral view of the male genitalia and spermatheca in *Magmatodrilus obscurus*. (Redrawn and modified from Holt, 1967.) *FIGURE 61* Lateral view of the male genitalia and spermatheca in *Cambarincola fallax*. *FIGURE 62* Lateral view of the male genitalia and spermatheca in *Sathodrilus hortoni*. (Redrawn from Holt, 1973a.) *FIGURE 63* Lateral view of the male genitalia and spermatheca in *Sathodrilus carolinensis*. (Redrawn from Holt, 1968a.) *FIGURE 64* Lateral view of the male genitalia and spermatheca in *Sathodrilus attenuatus*. (Redrawn from Holt, 1981.)

LITERATURE CITED

Appleby, A. G., Brinkhurst, R. O. 1970. Defecation rate of three tubificid oligochaetes found in the sediment of Toronto Harbour, Ontario. Journal of the Fisheries Research Board of Canada 27:1971–1982.

Aston, R. J. 1984. The culture of *Branchiura sowerbyi* (Tubificidae, Oligochaeta) using cellulose substrate. Aquaculture 40:89–94.

Bagheri, E. A., McLusky, D. S. 1984. The oxygen consumption of *Tubificoides benedeni* (Udekem) in relation to temperature, and its application to production biology. Journal of Experimental Marine Biology 78:187–197.

Block, E. M., Moreno, G., Goodnight, C. J. 1982. Observations on the life history of *Limnodrilus hoffmeisteri* (Annelida, Tubificidae) from the Little Calumet River in temperate North America. International Journal of Invertebrate Reproduction 4:239–247.

Bonacina, C., Bonomi, G., Monti, C. 1987. Progress in cohort cultures of aquatic Oligochaeta. Hydrobiologia 155:163–169.

Bowker, D. W., Wareham, M. T., Learner, M. A. 1985. A choice chamber experiment on the selection of algae as food and substrata by *Nais elinguis* (Oligochaeta, Naididae). Freshwater Biology 15:547–557.

Brinkhurst, R. O. 1965. Studies on the North American aquatic Oligochaeta II. Tubificidae. Proceedings of the Academy of Natural Sciences, Philadelphia 117:117–172.

Brinkhurst, R. O. 1974. The benthos of lakes. Macmillan, London.

Brinkhurst, R. O. 1982. Evolution in the Annelida. Canadian Journal of Zoology 60:1043–1059.

Brinkhurst, R. O. 1989. A phylogenetic analysis of the Lumbriculidae (Annelida, Oligochaeta). Can. J. Zool. 67:2731–2739.

Brinkhurst, R. O. 1998. On the genus *Eclipidrilus*, including a description of *E. ithys* n. sp. Canadian Journal of Zoology 76:644–659.

Brinkhurst, R. O. 1999. Lumbriculids, branchiobdellidans and leeches, a review of progress. Hydrobiologia. (in press).

Brinkhurst, R. O., Austin, M. J. 1978. Assimilation by aquatic oligochaetes. Internationale Revue der gesamten Hydrobiologie 63:863–868.

Brinkhurst, R. O., Chua, K. E. 1969. Preliminary investigation of the exploitation of some potential nutritional resources by three sympatric tubificid oligochaetes. Journal of the Fisheries Research Board of Canada 26:2659–2668.

Brinkhurst, R. O., Gelder, S. R. 1989. Did the lumbriculids provide the ancestors of the branchiobdellids, acanthobdellids and leeches? Hydrobiologia 180:7–15.

Brinkhurst, R. O., Jamieson, B. G. M. 1971. Aquatic Oligochaeta of the world. Oliver and Boyd, Edinburgh, UK.

Brinkhurst, R. O., Chapman, P. M., Farrell, M. J. 1983. A comparative study of respiration rates of some aquatic oligochaetes in relation to sublethal stress. Internationale Revue der gesamten Hydrobiologie 68:683–699.

Chapman, P. M. 1981. Measurement of the short-term stability of interstitial salinities in subtidal estuarine sediments. Estuarine Coastal and Shelf Science 12:67–81.

Chapman, P. M., Brinkhurst, R. O. 1981. Seasonal changes in interstitial salinities and seasonal movements of subtidal benthic invertebrates in the Fraser River Estuary. Estuarine and Coastal Shelf Science 12:49–66.

Chapman, P. M., Brinkhurst, R. O. 1984. Lethal and sublethal tolerances of aquatic oligochaetes with reference to their use as a biotic index of pollution. Hydrobiologia 115:139–144.

Chapman, P. M., Brinkhurst, R. O. 1987. Hair today, gone tomorrow: induced chaetal changes in tubificid oligochaetes. Hydrobiologia 155:45–55.

Chapman, P. M., Farrell, M. A., Brinkhurst, R. O. 1982a. Effects of species interactions on the survival and respiration of *Limnodrilus hoffmeisteri* and *Tubifex tubifex* (Oligochaeta, Tubificidae) exposed to various pollutants and environmental factors. Water Research 16:1405–1408.

Chapman, P. M., Farrell, M. A., Brinkhurst, R. O. 1982b. Relative tolerance of selected aquatic oligochaetes to individual pollutants and environmental factors. Aquatic Toxicology 2:47–67.

Chapman, P. M., Farrell, M. A., Brinkhurst, R. O. 1982c. Relative tolerances of selected aquatic oligochaetes to combinations of pollutants and environmental factors. Aquatic Toxicology 2:69–78.

Chatarpaul, L., Robinson, J. B., Kaushik, N. K. 1979. Role of tubificid worms on nitrogen transformations in stream sediment. Journal of the Fisheries Research Board of Canada 36:673–678.

Christensen, B. 1984. Asexual propagation and reproductive strategies in aquatic oligochaetes. Hydrobiologia 115:91–95.

Clark, R. B. 1964. Dynamics in metazoan evolution. Clarendon Press, Oxford.

Coates, K. A. 1989. Phylogeny and origins of the Enchytraeidae. Hydrobiologia 180:17–33.

Delly, J. G. 1985. Narcosis and preservation of freshwater animals. American Laboratory pp. 31–40.

Drewes, C. D., Fourtner, C. R. 1988. Hindsight and rapid escape in an aquatic oligochaete. Page 278.5, *in:* Society for Neuroscience, Abstracts of the 18th Annual Meeting, Toronto, Ontario, November 13–18 1988. 14(1).

Ferraguti, M. 2000. Clitellata, *in* Adyiodi, K. G., Adyiodi, R. G., (Series Eds.), Reproductive biology of invertebrates. Jamieson, B. G. M., (Vol. Ed.), Progress in male gamete ultrastructure and phylogeny, Vol. IX, Pt B. Oxford & IBH Pub. Co., New Delhi, India, pp. 125–182.

Ferraguti, M., Gelder, S. R. 1990. The comparative ultrastructure of spermatozoa from five branchiobdellidans (Annelida: Clitellata). Canadian Journal of Zoology 69:1945–1956.

Fend, S. V, Brinkhurst, R. O. Hydrobiologia 428:1–59. A revision of *Rhynchelmis*. (2000).

Gardner, W. S., Nalepa, T. F., Quigley, M. A., Malczyk, J. M. 1981. Release of phosphorus by certain benthic invertebrates. Canadian Journal of Fisheries and Aquatic Science 38;978–981.

Gelder, S. R. 1996a. A review of the taxonomic nomenclature and a checklist of the species of the Branchiobdellae (Annelida, Clitellata). Proceedings of the Biological Society of Washington 109:653–663.

Gelder, S. R. 1996b. Description of a new branchiobdellidan species, with observations on three other species, and a key to the genus *Pterodrilus* (Annelida, Clitellata). Proceedings of the Biological Society of Washington 109:256–263.

Gelder, S. R. 1996c. Histochemical characterization of secretions in the reproductive systems of two species of branchiobdellidans (Annelida, Clitellata): a new character for the phylogenetic matrix? Hydrobiologia 334:219–227.

Gelder, S. R. 1999. Zoogeography of branchiobdellidans (Annelida) and temnocephalidans (Platyhelminthes) ectosymbiotic on freshwater crustaceans, and their reactions to one another *in vitro*. Hydrobiologia (406:21–31.).

Gelder, S. R., Brinkhurst, R. O. 1990. An assessment of the phylogeny of the Branchiobdellida (Annelida: Clitellata) using PAUP. Canadian Journal of Zoology 68:1318–1326.

Gelder, S. R., Hall, L. A. 1990. Description of *Xironogiton victoriensis* n.sp. from British Columbia, Canada with some remarks on other species and a Wagner analysis of *Xironogiton* (Clitellata: Branchiobdellida). Canadian Journal of Zoology 68:2352–2359.

Gelder, S. R., Rowe, J. P. 1988. Light microscopical and cytochemical study on the adhesive and epidermal gland cell secretions of the branchiobdellid *Cambarincola fallax* (Annelida: Clitellata). Canadian Journal of Zoology 66:2057–2064.

Gelder, S. R., Smith, R. C. 1987. Distribution of branchiobdellids (Annelida: Clitellata) in northen Maine, U.S.A. Transactions of the American Microscopical Society 106:85–88.

Harman, W. J., Loden, M. S. 1978. A re-evaluation of the Opitocystidae (Oligochaeta) with descriptions of two new genera. Proceedings of the Biological Society of Washington 91:453–462.

Harper, R. M., Fray, J. C., Learner, M. A. 1981. A bacteriological investigation to elucidate the feeding biology of *Nais variabilis* (Oligochaeta, Naididae). Freshwater Biology 11:227–236.

Holmquist, C. 1976. Lumbriculids (Oligochaeta) of Northern Alaska and Northwestern Canada. Zoologische Jahrbucher (Systematik) 103:377–431.

Holt, P. C. 1960. The genus *Ceratodrilus* hall (Branchiobdellidae, Oligochaeta) with the description of a new species. Virginia Journal of Science 11:53–77.

Holt, P. C. 1965. The systematic position of the Branchiobdellidae (Annelida: Clitellata). Systematic Zoology, 14:25–32.

Holt, P. C. 1967. Status of the general *Branchiobdella* and *Stephanodrilus* in North America with description of a new genus (Clitellata: Branchiobdellida). Proceedings of the United States National Museum 124:1–10.

Holt, P. C. 1968. New genera and species of branchiobdellid worms (Annelida: Clitellata). Proceedings of the Biological Society of Washington 81:291–318.

Holt, P. C. 1973a. A summary of the branchiobdellid (Annelida: Clitellata) fauna of Mesoamerica. Smithsonian Contributions to Zoology, No. 142:1–40.

Holt, P. C. 1973b. Branchiobdellids (Annelida: Clitellata) from some eastern North American caves, with descriptions of new species of the genus *Cambarincola*. International Journal of Speleology 5:219–256.

Holt, P. C. 1973c. Epigean branchiobdellids (Annelida: Clitellata) from Florida. Proceedings of the Biological Society of Washington 86:79–104.

Holt, P. C. 1974. An emendation of the genus *Triannulata* Goodnight 1940, with the assignment of *Triannulata montana* to *Cambrincola* Ellis 1912 (Clitellata: Branchiobdellida). Proceedings of the Biological Society of Washington 87:57–72.

Holt, P. C. 1981. New species of *Sathodrilus* Holt 1968, (Clitellata: Branchiobdellida) from the Pacific drainage of the United States, with the synonymy of *Sathodrilus virgiliae* Holt 1977. Proceedings of the Biological Society of Washington 94:848–862.

Holt, P. C. 1986. Newly established families of the order Branchiobdellida (Annelida: Clitellata) with a synopsis of the genera. Proceedings of the Biological Society of Washington 99:676–702.

Humason, G. L. 1979. Animal Tissue Techniques. 4th edition Freeman, San Francisco, CA.

Jamieson, B. G. M. 1981. The ultrastructure of the Oligochaeta. Academic Press, London.

Jennings, J. B., Gelder, S. R. 1979. Gut structure, feeding and digestion in the branchiobdellid oligochaete *Cambarincola macrodonta* Ellis 1912, an ectosymbiote of the freshwater crayfish *Procambarus clarkii*. Biological Bulletin 156:300–314.

Karlckhoff, S. W., Morris, K. R. 1985. Impact of tubificid oligochaetes on pollution transport in bottom sediments. Environmental Science and Technology 19:51–56.

Kaster, J. L., Bushnell, J. H. 1981. Cyst formation by *Tubifex tubifex* (Tubificidae). Transactions of the American Microscopical Society 100:34–41.

Kaster, J. L., Val Klump, J., Meyer, J., Krezoski, J., Smith, M. E. 1984. Comparison of defecation rates of *Limnodrilus hoffmeisteri* Claparéde (Tubificidae) using two different methods. Hydrobiologia 111:181–184.

Kathman, R. D., Brinkhurst, R. O. 1998. Guide to the freshwater oligochaetes of North America. Aquatic Resources Center, College Grove, TN. 264 pp.

Krezoski, J. R., Robins, J. A., White, D. S. 1984. Dual radiotracer measurement of zoobenthos-mediated solute and particle transport in freshwater sediments. Journal of Geophysical Research 89:7937–7947.

Lang, C. 1984. Eutrophication of Lakes Leman and Neuchatel (Switzerland) indicated by oligochaete communities. Hydrobiologia 114:131–138.

Lazim, M. N., Learner, M. A. 1986a. The life-cycle and production of *Limnodrilus hoffmeisteri* and *L. udekemianus* (Oligochaeta, Tubificidae) in the organically enriched Moat-Feeder stream, Cardiff, South Wales. Archiv fur Hydrobiologie (Supplement) 4:200–225.

Lazim, M. N., Learner, M. A. 1986b. The life-cycle and productivity of *Tubifex tubifex* (Oligochaeta, Tubificidae) in the Moat-Feeder stream, Cardiff, South Wales. Holarctic Ecology 9:185–192.

Leitz, D. M. 1987. Potential for aquatic oligochaetes as live food in commercial aquaculture. Hydrobiologia: 155:309–310.

Lochhead, G., Learner, M. A. 1984. The cocoon and hatchling of *Nais variabilis* (Naididae, Oligochaeta). Freshwater Biology 14:189–193.

Loden, M. S. 1981. Reproductive ecology of Naididae. Hydrobiologia 83:115–123.

Marian, M. P., Pandian, T. J. 1984. Culture and harvesting techniques for *Tubifex tubifex*. Aquaculture 42:303–315.

Mihailova, P. V., Subchev, M. A. 1981. On the karyotype of three species of the family Branchiobdellidae (Annelida: Oligochaeta). Comptes rendus l'Academie bulgare des Sciences 34:265–267.

Milbrink, G. 1983. An improved environmental index based on the relative abundance of oligochaete species. Hydrobiologia 102:89–97.

Milbrink, G. 1993. Evidence for mutualistic interactions in freshwater oligochaete communities. Oikos 68: 317–322.

Moore, J. P. 1895. *Pterodrilus,* a remarkable discodrilid. Proceedings of the Academy of Natural Sciences of Philadelphia pp. 449–454.

Nemec, A. F. L., Brinkhurst, R. O. 1987. A comparison of methodological approaches to the subfamilial classification of the Naididae (Oligochaeta). Canadian Journal of Zoology 65:691–707.

Newrlka, P., Mutayoba, S. 1987. Why and where do oligochaetes hide their cocoons? Hydrobiologia 155:171–178.

Olsson, T. I. 1981. Overwintering of benthic macroinvertebrates in ice and frozen sediment in a North Swedish River. Holarctic Ecology 4:161–166.

Overstreet, R. M. 1983. Metazoan symbionts of crustaceans. *in:* Provenzano A. J., Jr., Ed. The biology of Crustacea, pathobiology. Vol. 6, Academic Press, New York, pp. 155–250.

Poddubnaya, T. L. 1984. Parthenogenesis in Tubificidae. Hydrobiologia 115:97–99.

Sawyer, R. T. 1986. Leech biology and behaviour. Clarendon, Oxford. pp. 1065.

Schell, S. C. 1985. Handbook of trematodes of North America, North of Mexico. University of Idaho Press, Moscow, ID.

Schmidt, G. D., Roberts, L. S. 1989. Foundations of parasitology. Times Mirror/Morsby, Boston, MA.

Stephenson, J. 1930. The Oligochaeta. Clarendon Press, Oxford.

Valvassori, R., de Eguileor, M., Lanzavecchia, G., Gelder, S. R. 1994. Comparative body wall musculature and muscle fibre ultrastructure in branchiobdellidans (Annelida: Clitellata), and their phylogenetic significance. Hydrobiologia 278:189–199.

Vincent, V., Vaillancourt, G., McMurray, S. 1978. Premiere mention de *Psammoryctides barbatus* (Grube) (Annelida, Oligochaeta) en Amerique du Nord et note sur sa distribution dans le haut estuaire du Saint-Laurent. Le Naturaliste Canadien 105:77–80.

Wassell, J. T. 1984. Revision of the lumbriculid oligochaete *Eclipidrilus* Eisen, 1881 with description of three subgenera and *Eclipidrilus (Leptodrilus) fontanus* n. subgen. n. sp. from Pennsylvania. Proceedings of the Biological Society of Washington 97:78–85.

Wiederholm, T., Wiederholm, A. -M., Milbrink, G. 1987. Bulk sediment bioassays with five species of fresh–water oligochaetes. Water, Air and Soil Pollution 36:131–154.

Woodhead, A. E. 1945. The life-history cycle of *Dioctophyma renale,* the giant kidney worm of man and many other mammals. Journal of Parasitology, supplement 31:12.

Zoran, M. J., Drewes, C. D. 1987. Rapid escape reflexes in aquatic oligochaetes: variations in design and function of evolutionary conserved giant fiber systems. Journal of Comparative Physiology (A) 161:729–738.

ADDITIONAL SUGGESTED READINGS

In addition to the literature actually cited in the text, the following publications will provide more specific information on taxonomy (Brinkhurst and Wetzel, 1984) and data derived from a series of international conferences on aquatic oligochaete biology.

Brinkhurst, R. O., Cook, D. G., Eds. 1980. Aquatic Oligochaete biology. Plenum Press, New York.

Brinkhurst, R. O., Wetzel, M. J. 1984. Aquatic Oligochaeta of the World: Supplement. A Catalogue of New Freshwater Species, Descriptions, and Revisions. Canadian Technical Report of Hydrography and Ocean Sciences 44:i-v + 101 pp.

Bonomi, G., Erseus, C., Eds. 1984. Aquatic Oligochaeta. Developments in hydrobiology, No. 24. Junk, Dordrecht, Netherlands.

Brinkhurst, R. O., Diaz, R. J., Eds. 1987. Aquatic Oligochaeta. Developments in hydrobiology, No. 40. Junk, Dordrecht, Netherlands.

Coates, K. A., Reynoldson, T. B., Reynoldson, T. B., Eds. 1996. Aquatic oligochaete biology. VI. Developments in hydrobiology, No. 115. Junk, Dordrecht, Netherlands.

Kaster, J. L. (Ed.). 1989. Aquatic Oligochaeta. Developments in Hydrobiology, No. 51. Junk, Dordrecht, Netherlands.

Healy, B., Reynoldson, T. B., Coates, K. A., Eds. 1999. Aquatic oligochaete biology. VII. Developments in hydrobiology. Junk, Dordrecht, Netherlands (in press.).

Reynoldson, T. B., Coates, K. A., Eds. 1994. Aquatic oligochaete biology. VI. Developments in hydrobiology, No. 95. Junk, Dordrecht, Netherlands.

13

ANNELIDA: EUHIRUDINEA AND ACANTHOBDELLIDAE

Ronald W. Davies*
Faculty of Science
Monash University
Clayton, Victoria 3168
Australia

Fredric R. Govedich
Department of Biological Sciences
Monash University
Clayton, Victoria 3168
Australia

I. INTRODUCTION TO EUHIRUDINOIDEA AND ACANTHOBDELLIDA

Leeches form an important component of the benthos of most lakes and ponds, and some species commonly occur in the quieter flowing sections of streams and rivers. There are a total of 73 species presently recorded from North America, of which the majority are predators feeding on chironomids, oligochaetes, amphipods, and molluscs. The young of several species feed on zooplankton, but *Motobdella montezuma* is unique in specializing on planktonic amphipods using mechanoreception to detect its prey. Other species of leeches are temporary sanguivorous (blood feeding) ectoparasites of fish, turtles, amphibians, crocodilians, water birds, or occasionally humans. The majority of predaceous leeches have an annual or biannual life cycle, breeding once and then dying (semelparity). However, at least two species have been shown to be genetically capable of breeding several times (iteroparity) although exhibiting phenotypic semelparity in the field. The majority of sanguivorous leeches are iteroparous showing saltatory growth after the three or more blood meals required to reach mature size.

Because of the diversity of physico-chemical conditions encountered in different aquatic ecosystems and temporally within a single habitat, leeches show considerable physiological plasticity. They are able to live in waters with very low salt concentrations as well as waters with salinities that exceed the salinity of seawater. Similarly, some species are capable of surviving in

* Present address: Department of Biological Sciences, University of Calgary, Calgary, Alberta, Canada T2N 1N4.

Ecology and Classification of North American Freshwater Invertebrates. 2nd Edition

the absence of oxygen (anoxia) for more than 60 days and in the presence of supersaturated water (hyperoxia) for shorter periods.

In many small ponds and lakes, leeches are the top predators and are ideal for studying intra-and interspecific resource competition, niche overlap, and predator–prey interactions. In larger lakes and rivers, leeches may form an important component of the diet of fish; hence, they are grown commercially for fish bait. Sanguivorous leeches are being investigated extensively in relation to the pharmacological properties of their salivary secretions, especially their anticoagulants.

The leech-like *Acanthobdella peledina*, a temporary ectoparasite of salmonids, is the only species of Acanthobdellidae in North America. Its ecology and feeding have not been extensively studied, but are likely to be similar to the Piscicolidae—a family of leeches also ectoparasitic on fish.

II. TAXONOMIC STATUS AND CHARACTERISTICS

There is an undisputed close taxonomic affinity between leeches and oligochaetes, although there are differences of opinion about the exact nature of the relationship. This has resulted in leeches being variously classified as Hirudinea, Hirudinoidea, or Euhirudinea. Purschke *et al.* (1993) showed that not only did the Branchiobdellae (see Chapter 12), Acanthobdellidae, and Euhirudinea resemble each other, but they were also closely related. Their evolutionary relationship remains unanswered, but the present consensus is to consider them all superorders of the subclass Clitellata. The Clitellata are thought to have evolved in the interstitial environment of lagoons near the tributary outlets and entrance to channels opening into the sea in the Palaeozoic, 500 million years ago. The first documented fossil of a leech dates back 150 million years to the Jurassic. However, a large annulate with a circular structure similar to a leech sucker discovered at Waukesha, Wisconsin (Briggs, 1991), confirmed as a leech, will extend their history back 400 million years to the Silurian. Cocoons of extant leeches have been recorded from Holocene sediments but Manum *et al.* (1991) suggest that ovoid fossils 1.0–10.0 mm long with a netted or a feltlike wall fabric from the Lower Cretaceous (150 million years) were very probably leech cocoons.

Only a few families of microdrilid oligochaetes can be considered as possible close relatives of the leeches. The naiids are the best candidates as most live in aerated freshwater, which seems to be the original environment of leeches, many of them are predators, they have a reduced and sometimes fixed number of

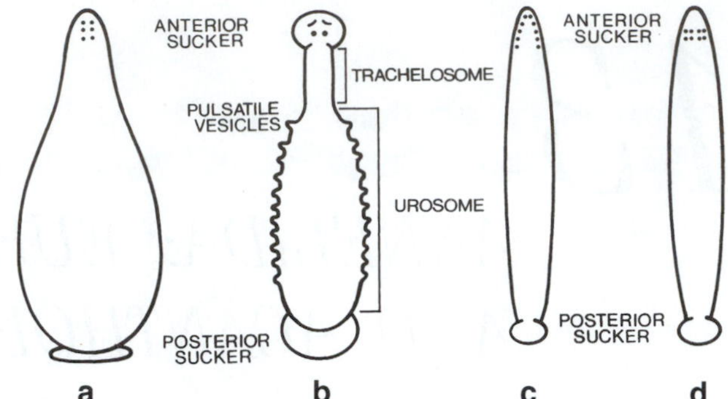

FIGURE 1 Dorsal view of the general body shape of a member of each of the families: (a) Glossiphoniidae; (b) Piscicolidae; (c) Hirudinidae; and (e) Erpobdellidae.

segments, they have eyes of very similar structure to those of leeches and at least one genus (*Chaetogaster*) has epizoic species which move by looping and which like leeches lack a prostomium (Omodeo, 1998).

The diagnostic features of leeches are: (1) 32 postoral metameres (Fernandez, 1980, Weisblat *et al.*, 1980) plus a nonsegmental prostomium; (2) a reduced or obliterated coelom; (3) absence in the adult of chaetae and septa; (4) the presence of an anterior (oral) sucker and a usually larger posterior (anal) sucker (Fig. 1); (5) superficial subdivision of the metameres into annuli (Figs. 2 and 29a, b); and (6) a median, unpaired, male genital pore placed anterior to the single median ventral, female genital pore (Figs. 3 and 4a).

The characteristics delimiting the Euhirudinea are primarily related to their feeding and feeding behavior indicative that leeches are oligochaetes which at first become predators and then, subsequently, on at least two separate occasions some became sanguivorous (Siddall and Burreson, 1995).

The phylogenetic position of the Acanthobdellidae has been debated. Because they share some morphological, behavioral, and ecological characteristics more typically attributed to leeches some researchers have included them within the same taxonomic group as the

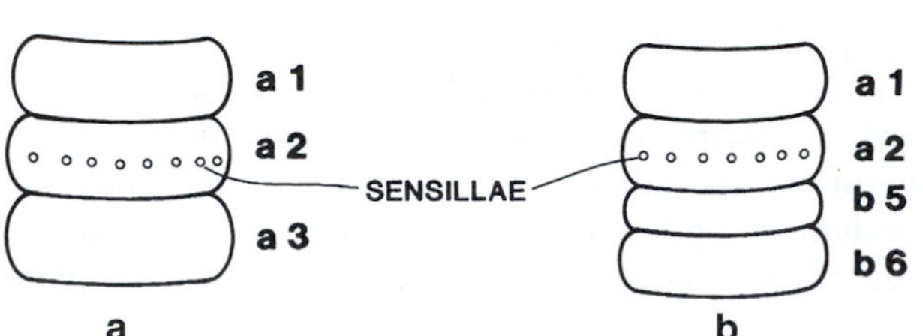

FIGURE 2 Annulation showing (a) leech 3-annulate condition and (b) *Acanthobdella peledina* 4-annulate condition.

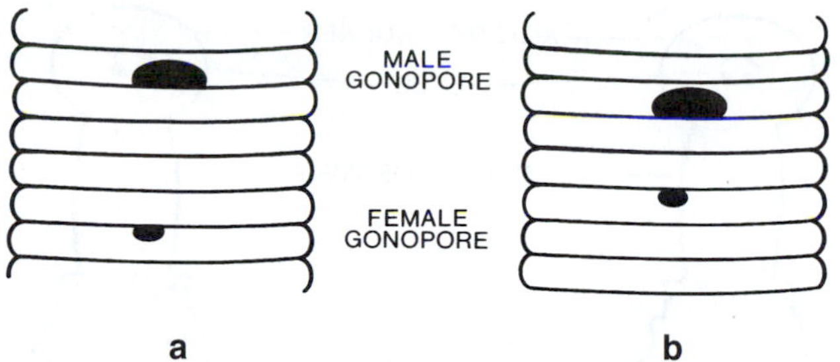

FIGURE 3 Ventral view of the anterior male and posterior female gonopores separated by (a) four annuli and (b) two annuli.

leeches (Mann, 1961, Elliott and Mann, 1979; Klemm, 1982, 1985; Sawyer, 1972, 1986). However, recent research shows that the Acanthobdellidae, along with the Branchiobdellidae, are sister groups to the leeches sharing a common ancestor (Siddall and Burreson, 1995, 1996). Acanthobdellidae are temporary ectoparasites of salmonid fishes but, unlike leeches, with five anterior coelomic compartments separated by distinct septa, each anterior segments bearing two pairs of chaetae (setae) (Fig. 4b). The Acanthobdellidae do not have an anterior sucker, but a small posterior sucker is formed from four of the 30 segments forming the body. Each segment is superficially subdivided into four annuli (Fig. 2b).

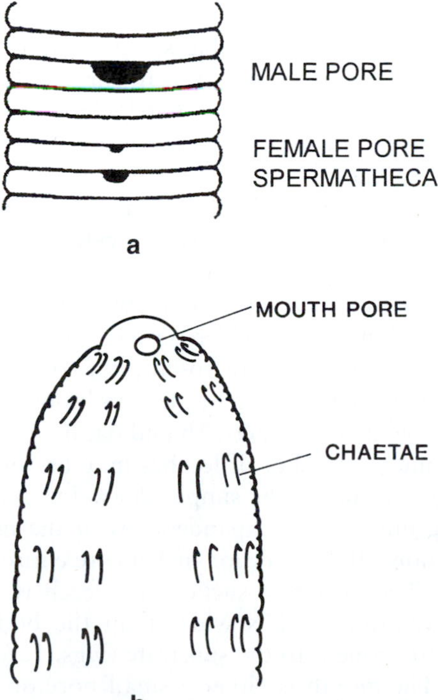

FIGURE 4 *Acanthobdella peledina*, ventral view of (a) the male and female gonopores with a spermatheca and (b) the anterior view showing the absence of an anterior sucker, the presence of a mouth pore, and five pairs of hooked chaetae.

Classically, oligochaetes and leeches are placed within the phylum Annelida either in the order Hirudinea, class Clitellata, or in the class Euhirudinea. Soös (1965, 1966a, b, c, 1967, 1968, 1969a, b) placed leeches in the taxonomically undefined Hirudinoidea, but subsequently both Soös (1969b) and Richardson (1969) referred them to the class Hirudinoidea. Correctly, Klemm (1985) and Sawyer (1986) indicated that in accordance with the Zoological Code the suffix "-oidea" should only be added to the stem name of a superfamily and recommended the suppression of the class Hirudinoidea. Davies (1971, and many subsequent references) classified leeches as Hirudinoidea implicitly as a superfamily of the class Oligochaeta.

Until recently, these differences were primarily academic because everyone agreed that as Hirudinea, Euhirudinea or Hirudinoidea, leeches were in the phylum Annelida and closely related to oligochaetes. However, Sawyer (1986) defined the class Hirudinea in a significantly different way, with Euhirudinea, Acanthobdellidae, Branchiobdellidae, and Agriodrilidae as subclasses within the phylum Uniramia. Thus, the class Euhirudinea (sensu Sawyer, 1986) required the removal of this class from the phylum Annelida, and its placement into a phylum with the Onychophora, Myriapoda, and Hexapoda. The evidence to support this is scant (Sawyer, 1984, 1986) and contentious (Davies, 1987), and the majority of authors did not follow this recommendation and retained the leeches as Euhirudinea, Hirudinea or Hirudinoidea as a class of Annelida, a view supported by Zrzavy *et al.* (1998). The term Euhirudinea used in this chapter refers to a class of Annelida containing the leeches and not the Uniramian class defined by Sawyer (1986).

Sawyer (1986) combined many of the families and genera described and accepted by previous researchers, including those described in detail by Richardson (1969, 1971, 1975). In addition, he placed the North American members of the genus *Dina* into *Erpobdella* leading to further confusion about the relationships within the Erpobdellidae and, particularly, between the genus *Erpobdella* and *Mooreobdella*. These changes were not justified scientifically nor were complete descriptions of the modified families and genera given. This has resulted in taxonomic confusion. Leech families and genera described prior to Sawyer (1986) have been retained in this chapter and until a formal revision, including a re-analysis of the higher level relationships of leeches, has been published it is recommended that these names be retained.

III. EUHIRUDINEA AND ACANTHOBDELLIDAE

A. General Anatomy

The body of a leech consists of two preoral, nonmetameric segments and 32 postoral metameres designated I through XXXIV, while the Acanthobdellidae have only 29 postoral metameres (designated I through XXIX). Each segment is usually divided externally by superficial furrows into 2–16 annuli (Figs. 2 and 29). Segments with the full number of annuli (complete segments) are found in the middle of the body; because this number is generally characteristic of the genus or species, it is referred to in the keys. Annuli features can be most easily seen in the lateral margins of the ventral surface.

Moore (1898) numbered the three primary annuli a1, a2, and a3 from the anterior, with annulus a2 (neural annulus) containing the nerve cord ganglion. The neural annulus (a2) is marked externally by transverse rows of cutaneous sensillae (Figs. 2 and 29). Repeated bisection of the primary annuli gives more complex annulation.

The number and position of the eyes are also important diagnostic features. In some genera, coalescence of eyes sometimes occurs, but the lobed nature of the eyes usually indicates the original condition (Fig. 5b). Eyespots (oculiform spots) can also occur on the lateral margins of the body and on the posterior sucker (Figs. 5c, 6b, and 19 b, c, d).

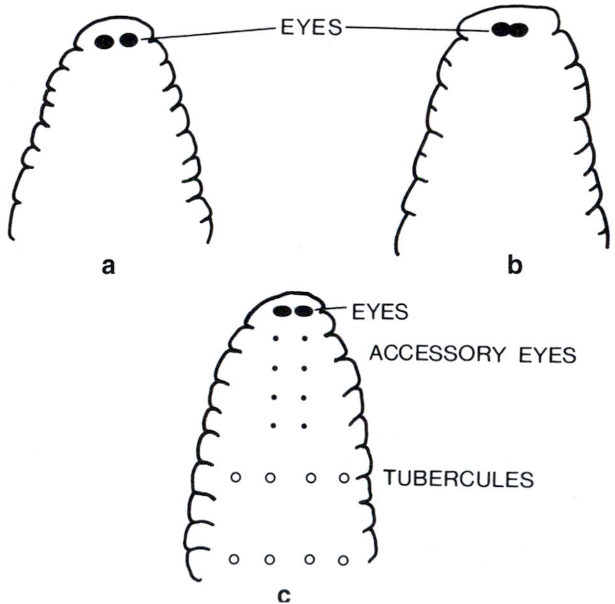

FIGURE 5 Dorsal view of the head of *Batracobdella* with a single pair of eyes (a) close together; (b) lobed; and (c) *Placobdella hollensis* with one pair of eyes followed by pairs of accessory eyes.

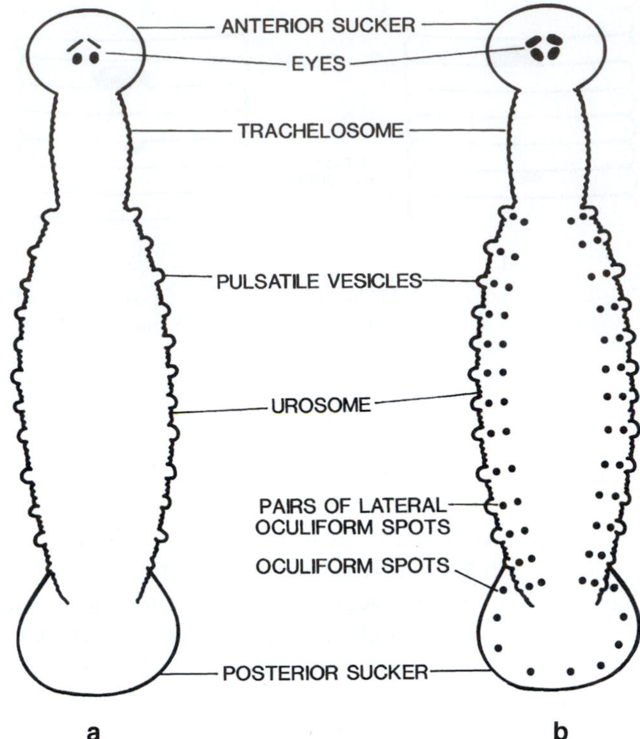

FIGURE 6 (a) *Cystobranchus verrilli*; and (b) *Cystobranchus meyeri* showing the pulsatile vesicles on the urosome and the presence on *C. meyeri* of oculiform spots on the posterior sucker and the paired lateral oculiform spots on the urosome.

Papillae, small protrusible sense organs, are often widely scattered over the dorsal surface. Tubercles, which are larger protrusions that include some of the dermal tissues and muscles, may also be present and can bear papillae.

The body of most leeches and acanthobdellids is not divisible into regions. With the genera *Illinobdella* and *Piscicolaria* as exceptions, the body of the family Piscicolidae is divided into a narrow anterior trachelosome and a longer and wider posterior urosome (Figs. 1b and 6). Lateral gills are absent, but some genera of Piscicolidae have paired pulsatile vesicles on the neural annuli of the urosome (Figs. 1b and 6a, b).

The anterior sucker of leeches may be very prominent (Figs. 1b, 6, 7b) or simple (Figs. 1a, c, d, 7a, c), consisting only of the expanded lips of the mouth. In the acanthobdellids, anterior suckers are entirely absent (Fig. 4b). The posterior sucker of a leech is generally directed ventrally and is wider than the body at the point of attachment to the substrate (Figs. 1, 6, 19, and 22–24). The mouth is either a small pore on the edge (Fig. 7a) or center (Fig. 7b) of the ventral surface of the anterior sucker, or is large and occupies the entire cavity (Fig. 7c) of the anterior sucker. In the family Hirudinidae, the buccal cavity (which may or may not

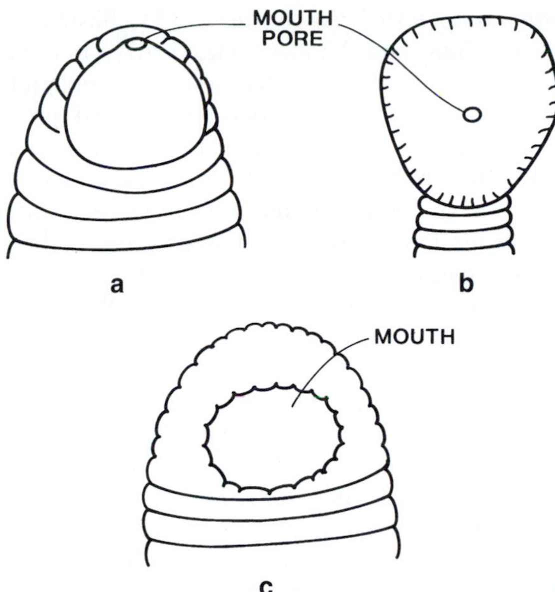

FIGURE 7 Ventral view of the anterior sucker of: (a) the mouth pore on the anterior rim of the sucker (e.g., *Marvinmeyeria, Placobdella, Oligobdella*); (b) near the center of the sucker (Piscicolidae, *Batracobdella, Helobdella*); and (c) the mouth occupying the entire cavity (e.g., Hirudinidae, Erpobdellidae).

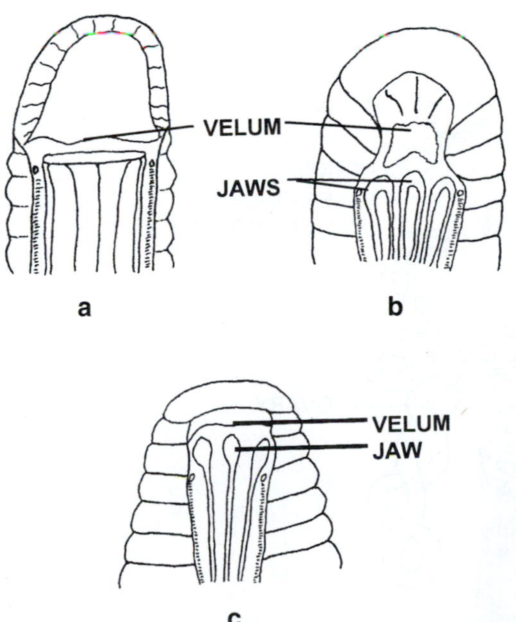

FIGURE 8 Ventral view of the dissection of the mouth and buccal cavity of: (a) *Mollibdella grandis*; (b) *Percymoorensis marmorata*; and (c) *Macrobdella decora* showing the velum, and the relative size of the jaws in *Macrobdella decora* and *P. marmorata* and the absence of jaws in *Mollibdella grandis*.

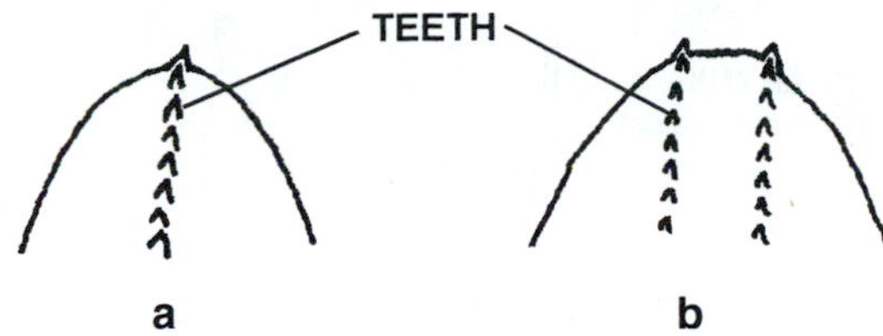

FIGURE 9 Surface view of the jaws of Hirudinidae showing the teeth arranged in: (a) one (monostichodont) or (b) two (distichodont) rows.

contain jaws) is separated from the cavity of the anterior sucker by a flap of skin (the velum) (Fig. 8). When present, there are three muscular jaws (two ventrolateral and one dorso-medial) (Fig. 8b, c) bearing teeth arranged in one (monostichodont) (Fig. 9a) or two (distichodont) rows (Fig. 9b).

In the Erpobdellidae, the buccal cavity contains three muscular ridges, but in the Glossiphoniidae and Piscicolidae, the pharynx is modified to form an eversible muscular proboscis. Opening into the pharynx are the salivary glands which secrete some digestive enzymes. In bloodsucking leeches the salivary glands are primarily related to the process of bloodsucking rather than digestion. The components of the salivary secretions and their functions can be summarized as lubrication, proteolytic inhibition, spreading (hyaluronidase), vasodilation, and anticoagulation (e.g., hirudin or hementin) (Sawyer, 1986). The anticoagulants produced by North American sanguivores have not been investigated extensively with work so far concentrating on the European *Hirudo medicinalis* and the South American *Haementaria ghilianii*. Seymour *et al.* (1990) and Krezel *et al.* (1994) reported the presence and structure of decorsin, a potent inhibitor of platelet aggregation from *Macrobdella decora*.

The pharynx in most erpobdellid leeches leads to a simple tubelike crop that lacks lateral ceca. However, this is not the case in the erpobdellid genus *Motobdella* which has one or two pairs of simple ceca or postceca located in the posterior portions of the crop (Davies *et al.* 1985; Govedich *et al.* 1998). The crop in most leech families is adapted for the storage of fluids and usually has 1–11 pairs of lateral ceca. Between the last pair of crop ceca, the intestine leads to the anus (Fig. 10a). Only in the Glossiphoniidae does the anterior portion of the intestine give off four pairs of simple ceca (Fig. 10b). The anus usually opens on the dorsal surface on or near segment XXVII, just anterior to the posterior sucker. In a few species, the anus is displaced anteriorly. *Motobdella* has one or two pairs of post ceca (Fig. 10c).

Digestion and absorption mainly occur in the intestine although the crop is also used in predatory species.

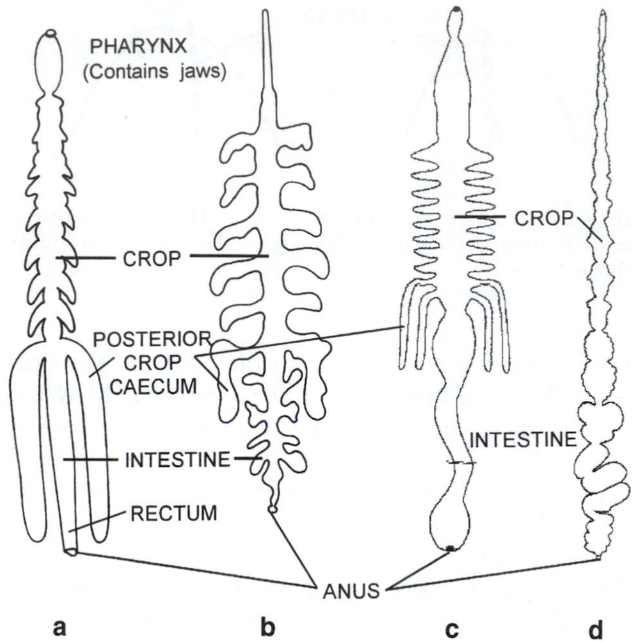

FIGURE 10 Stylized ventral view of the alimentary canal of (a) Hirudinidae; (b) Glossiphoniidae; (c) *Motobdella*; and (d) *Acanthobdella peledina*.

In sanguivores, digestion is primarily by enzymes produced by endosymbiotic bacteria. The bacteria involved are species specific for each species of sanguivorous leech (Jennings and Vanderlande, 1967), e.g. *Hirudo medicinalis* utilizes *Aeromonas hydrophila* (*Pseudomonas hirudinis*) (Büsing, 1951, Büsing *et al.*, 1953, Jennings and Vanderlande, 1967). Endopeptidases are absent or rare in leeches, but exopeptidases occur abundantly in predatory species and protease (Zebe *et al.*, 1986) and protease inhibitors have been recorded (Roters and Zebe, 1992a, b). Because of their dependence on endosymbiotic bacteria, digestion in sanguivorous leeches takes weeks or months; but for predatory species, digestion requires just a few days (Davies *et al.*, 1977, 1978, 1981, 1982a; Wrona *et al.*, 1979, 1981).

The alimentary canal of acanthobdellids (Fig. 10c) is similar to that of the glossiphoniid leeches. A small eversible proboscis leads into the muscular pharynx, which, in turn, leads into a large tubular crop lacking ceca. The crop leads into the intestine, which bears six pairs of lateral ceca; the anus is located dorsally between segments XXV and XXVI. Digestion is probably similar to sanguivorous leeches with the use of endosymbiotic bacteria.

The male and female gonopores are visible in mature leeches on the midline of the ventral surface of segments XI and XII, respectively, and are separated by a species-specific number of annuli (Figs. 3, 4, and 27). The anterior male gonopore is larger and more easily seen than the female gonopore. Leeches do not have true testes or ovaries, but instead possess testisacs and ovisacs. The testisacs are located posterior to segment XI and are either discrete, paired spherical structures (Fig. 11a) (e.g., Hirudinidae, Glossiphoniidae, and

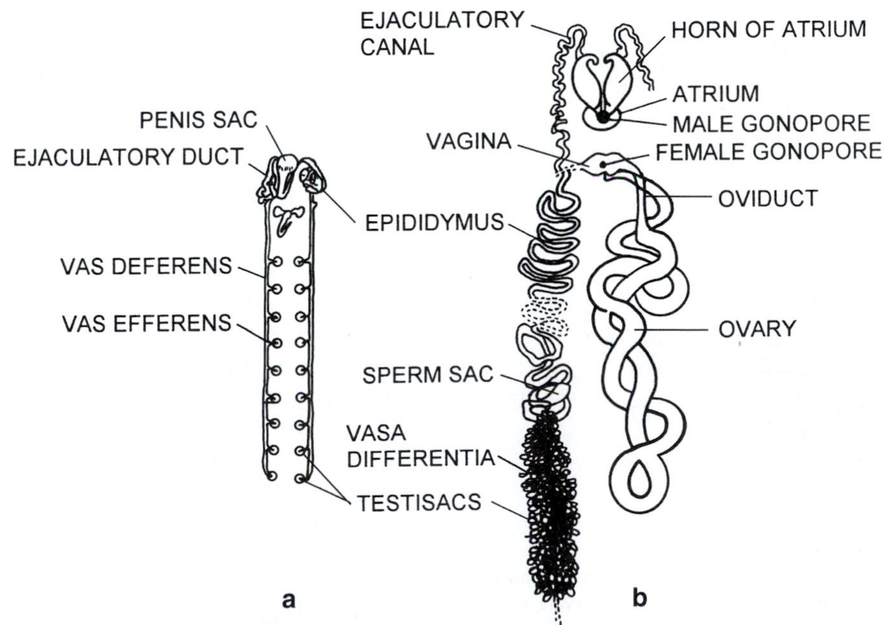

FIGURE 11 Reproductive systems of: (a) *Hirudo medicinalis* (Hirudinidae); and (b) *Nephelopsis obscura* (Erpobdellidae).

Piscicolidae) located intersegmentally or multifollicular columns resembling bunches of grapes lying on either side of the ventral nerve cord (Fig. 11b) (e.g., Erpobdellidae; Singhal and Davies, 1985; Davies *et al.*, 1985, Govedich *et al.*, 1998). Short vasa efferentia connect the testisacs to the vasa deferentia on each side of the body, which run anteriorly to form large coiled epididymis (sperm vesicles). Paired ejaculatory ducts run from the epididymis through the atrial cornua and unite to form a medial atrium consisting of a bulb and eversible penis. There is a single pair of ovisacs which are either small, spherical organs confined to segment XII or elongate, coiled tubes sometimes bent back on themselves such that their blind ends lie close to the female gonopore (Singhal and Davies, 1985; Davies *et al.*, 1985). Two oviducts run from the ovisacs to converge and form the common oviduct leading to the female gonopore (Fig. 11).

Leeches are frequently recorded as sequential protandrous hermaphrodites (Van Damme, 1974; Lasserre, 1975). *Nephelopsis obscura* shows sequential protandry during its first bout of gametogenesis, although there is a short transition period when gametogenic stages from both sexes co-occur. In the second cycle of gametogenesis, mature sperm and ova occur simultaneously (Davies and Singhal, 1988) and thus this cycle demonstrates simultaneous hermaphroditism. At the time of egg laying, a cocoon is secreted by the clitellum, a specialized region of epidermis usually posterior to the gonopores. As the cocoon moves toward and passes over the head, fertilized eggs pass into the cocoon from the female gonopore.

The reproductive system of *Acanthobdellida peledina* is basically similar with a pair of elongate testisacs and ovisacs and median unpaired male and female gonopores.

The excretory system of leeches and acanthobdellids consists of a maximum of 17 pairs of highly modified metanephridia opening along the body between segments VII and XXII. Many species do not exhibit the full number of metanephridia, showing reductions in the more anterior metameres and/or the segments containing gonads. In addition to excreting nitrogenous wastes (mainly ammonia), the nephridia help maintain water and salt balance in the body. With the osmotic pressure of the body higher than the surrounding freshwater, there is a tendency for an inward flow of water through the epidermis. The nephridia pass out excess water and help retain inorganic ions. Only about 6% of the salts found in the primary urine are excreted, the rest being reabsorbed (Zerbst-Boroffka, 1975). During acclimation to hypersaline media, several different active and passive processes are responsible for the changes of osmolarity and electrolytic concentrations in the blood of the leech (Nieczaj and Zerbst-Boroffka, 1993).

In the leeches and acanthobdellids, the large coelom, which in Annelida is typically divided by intersegmental septa, has to varying degrees been obliterated (Fig. 12). In the Glossiphoniidae (Fig. 12a), Piscicolidae (Fig. 12d), and acanthobdellids, the blood

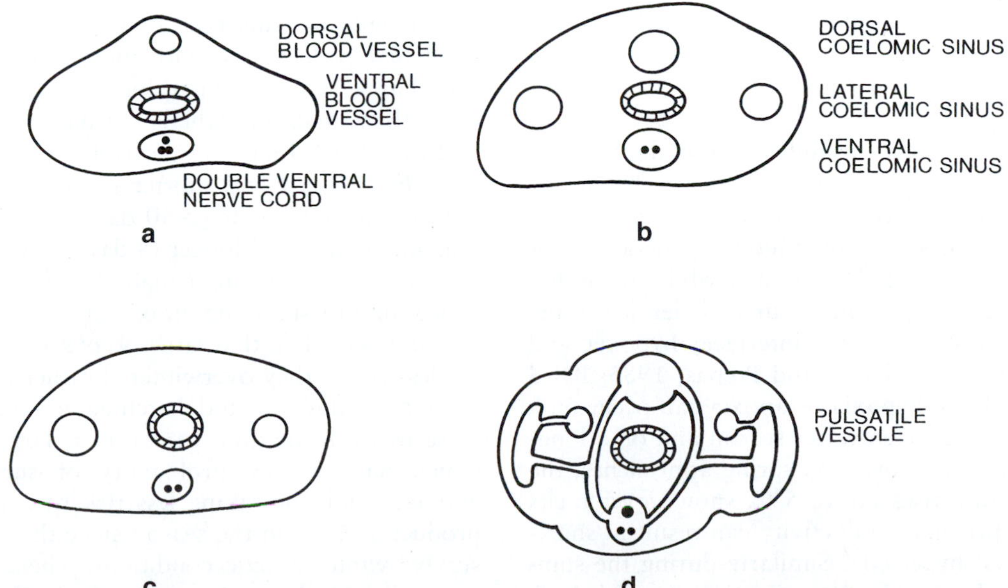

FIGURE 12 Stylized transverse sections of: (a) Glossiphoniidae; (b) Hirudinidae; (c) Erpobdellidae; and (d) Piscicolidae showing the coelomic sinuses, dorsal and ventral blood vessels, and the double ventral nerve cord.

vascular system consists of: (1) a ventral vessel enclosed with the nerve cord in the ventral lacuna; (2) a dorsal blood vessel enclosed in the dorsal lacuna; and (3) transverse blood vessels connecting the dorsal and ventral vessels in the anterior and posterior parts of the body. In the Hirudinidae (Fig. 12b) and Erpobdellidae (Fig. 12c), the true blood vascular system is completely lost; instead, the blood circulates in the coelomic lacuna system, forming a hemocoel.

Oxygen uptake in freshwater leeches and acanthobdellids is through the general body surface, although some species of Piscicolidae have small pulsatile vesicles filled with coelomic fluid that function as accessory respiratory organs. Most leeches can also ventilate their body surfaces by dorsoventral undulations along the body reminiscent of swimming, but with the posterior sucker firmly attached. However, *Acanthobdella peledina* does not ventilate (Dahm, 1962).

All leeches and acanthobdellids show typical looping; locomotory movements consisting of body elongation and shortening, with the anterior sucker (leeches only) and posterior sucker serving alternately as points of attachment. Many leech species are also able to swim using dorsoventral undulations. All Erpobdellidae and Hirudinidae are good swimmers, but only a few species of Glossiphoniidae and Piscicolidae are efficient swimmers. Some species (*Placobdella hollensis*) swim readily as juveniles and adults, but others (*Placobdella ornata* and *Placobdella parasitica*) swim only as juveniles (Sawyer, 1981). The erpobdellid genus *Motobdella* is unique in its swimming ability and specializes in feeding almost exclusively on pelagic amphipods by using mechanoreception to identify and track its prey in the water column (Blinn and Davies, 1990; Blinn *et al.*, 1987, 1988, 1990; Davies *et al.*, 1985).

B. Physiology

Perhaps the greatest variability in the abiotic environment experienced by leeches occurs in reference to temperature and dissolved oxygen concentrations

In stratified lakes, the epilimnion is generally at or above saturation during the summer, while in the hypolimnion oxygen depletion occurs. Under ice, especially at the substrate–water interface, hypoxia and possibly anoxia occur (Babin and Prepas, 1985; Baird *et al.*, 1987a). Thus, hypoxia or anoxia can occur on a diel basis during the open-water season and on a long-term basis during ice cover. As spring approaches, the water temperature rises above 5°C, snow cover melts and under-ice primary productivity can result in short-term (1–2 days) hyperoxia. Similarly during the summer, long-term hyperoxia can persist for several days. Both anoxia and hyperoxia can cause mortality of vertebrates and invertebrates (Casselman and Harvey, 1975). In Europe, Mann (1956) showed that oxygen uptake by some leech species was proportional to the oxygen concentration of the water (i.e., they were conformers), while other species (*Glossiphonia complanata*, *Helobdella stagnalis*) showed a comparatively constant oxygen uptake over a range of dissolved oxygen concentrations of the water (i.e., they were regulators). *Nephelopsis obscura* and *Erpobdella punctata* exposed to short-term hypoxia both showed a decline in oxygen uptake with reduced oxygen tension in the water (Wrona and Davies, 1984). When exposed to hypoxic conditions, most leeches, but not *Acanthobdella peledina*, show ventilation movements with the posterior sucker attached to the substrate and the body thrown into dorsoventral flexions (Dahm, 1962). It is assumed that the primary function of ventilation is respiratory since ventilation increases under conditions of low oxygen availability. While *N. obscura* and *E. punctata* ventilate in hypoxic conditions, ventilations also occur under saturated (normoxic) conditions and always cease when oxygen decreases below 20% saturation. This suggests ventilation may have additional roles.

Many leeches can withstand anaerobic conditions (Von Brand, 1944), but little research has been done on the ecological effect of anoxia (or hypoxia) on the leeches, with the majority of work related to use of leeches as indicator species in the European saprobic system (reviewed in Sladacek and Kosel, 1984).

In flow-through experiments, Davies *et al.* (1987) showed inter- and intraspecific differences in the survivorship of *N. obscura* and *E. punctata* to anoxic conditions and demonstrated the errors of using static experiments for this type of study. For both species, survivorship decreased with increased water temperature. Large *N. obscura* had longer survival times (>50 days) than small individuals (12 days) at 5°C and also at 20°C (8 and 4 days, respectively). Conversely, at 5°C large *E. punctata* had shorter survival times (25 days) than small individuals (>50 days), but at 20°C, large individuals survived longer (8 days) than small individuals (2 days). As young (small) *N. obscura* have a low probability of surviving anoxia at winter temperatures (5°C), it is evident that large *N. obscura* would be taking less risk if they overwintered rather than hazarding cocoon production and hatching of young up to or close to ice formation. In contrast, large *E. punctata*, which have a low probability of surviving winter anoxia, would be taking less risk to continue cocoon production later in the season since their offspring can survive winter anoxic conditions. These results correlate well with the patterns of reproduction observed in the field (Davies, 1978; Davies *et al.*, 1977).

The metabolic and physiological adaptations of *N. obscura* to long-term anoxia have been studied under laboratory conditions mimicking winter conditions (Reddy and Davies, 1993a, b; Dratnal *et al.*, 1993). Lactic acid did not accumulate in either large or small individuals over a 40-day period of anoxia; however, succinate and alanine did. Both glycogen and amino acids were utilized as energy reserves during the anoxic period along with stored lipids. Following the 40-day period of anoxia, glycolytic flux and adenosine triphosphate diphosphate (ATP/ADP) were significantly reduced in both small and large individuals. Upon a return to normoxia, preanoxic levels of glycogen, malate, succinate, alanine, and aspartate are restored within 24 hs. In addition, the preanoxic energy charge is restored within the first 6 hs of normoxia. All small and 98% of large individuals survived the 40-day anoxic period without significant losses in body weight; however, activity was reduced when compared to *N. obscura* not exposed to anoxic conditions (Reddy and Davies, 1993a, b). Physiological adaptations to the effects of severe hypoxia are similar to those observed for anoxic conditions and in *N. obscura* include decreased growth, enhanced energy allocation of glycogen, triclyglycerols and total lipids. Body weight and size at maturity were not affected by hypoxia; however, the time to maturity was longer for *N. obscura* under hypoxic conditions (Dratnal *et al.*, 1993).

Even less work has been done on the effect of hyperoxia on leeches. Survival of *N. obscura* and *E. punctata* exposed to hyperoxia (200, 300%) increased with leech size and decreased with increased temperature. The survivorship time for all size classes of *N. obscura* and *E. punctata* greatly exceeded the maximum recorded duration (1–2 days) of hyperoxia in the field during the spring. However, during the summer (20°C), if hyperoxiain the 200–300% range persisted for more than 20 days, medium and large *N. obscura* would show higher mortality than comparable sizes of *E. punctata* (Davies and Gates 1991a, b).

With very low or undetectable levels of superoxide dismutase in *N. obscura*, Singhal and Davies (1987) concluded that *N. obscura* had two possible defenses against hyperoxia. These were to move away from areas of hyperoxia or, as recorded by Singhal and Davies (1987), to secrete large amounts of mucus over the epidermis, which reduces oxygen diffusion into the body and thereby the effects of oxygen toxicity. Davies and Everett (1977) and Davies *et al.* (1977) recorded intra- and interspecific differences in the timing of seasonal movements between lentic microhabitats of *N. obscura* and *E. punctata*. The results from these investigations on hyperoxia show that small leeches are highly intolerant of hyperoxia, especially at higher temperatures, and suggest that their seasonally earlier movement to deeper water is related to avoidance of potentially lethal oxygen conditions. Similarly, the presence of only large leeches in the macrophyte zone, where hyperoxia is more common and more persistent, is compatible with the greater tolerance of the large size classes to hyperoxic conditions.

During winter, leeches frequently have to rely on energy stored during periods favorable to energy acquisition although energetic demands can be mitigated by decreased activity. During the winter *N. obscura* exhibits degrowth (Dratnal and Davies, 1990) and posthatchling-sized individuals that have over wintered are different, from similarly sized individuals hatched in the summer with respect to ecophysiological process, life-history and fitness characteristics (Dratnal and Davies, 1990). The cryoprotective glycerol concentration in winter leeches are significantly higher (Reddy and Davies, 1993b), but they have had only half the total lipid concentration energy reserves of similarly sized individuals hatched in the summer. Under optimal feeding conditions, lipid depletion at maturity is fully compensated for in over wintered leeches through increased food ingestion, higher absorption efficiency, and the allocation of more energy to lipid accumulation than other growth functions (Reddy *et al.*, 1992). At, or just prior to sexual maturity, both winter and summer leeches show decreases in total protein, glycogen, and total lipid as a result of the increased energetic demand of gametogenesis and maturation.

Ultrastructural investigations of the neurons of *N. obscura* exposed to anoxia and hyperoxia (Singhal *et al.*, 1988) showed a marked reduction in numbers of mitochondria, ribosomes, and neurotransmitter vesicles with swollen and fragmented cristae. These changes are more severe in anoxia than hyperoxia.

In most lakes and rivers inhabited by leeches, ionic content and total dissolved salts (TDS) in the water show little temporal variability. However, significant geographic variability occurs between lakes and rivers primarily related to the edaphic features of the underlying geological formations. It has been shown that the distribution of leeches is, at least in part, related to TDS of the habitat waters (Mann, 1968; Herrmann, 1970a, b; Scudder and Reynoldson and Davies, 1976). Similarly, in Europe, Mann (1955) showed that the composition of the leech fauna in lentic habitats was a function of water hardness. It has been demonstrated that the relative concentrations of ions in the medium modify the mortality of aquatic animals (Beadle, 1939; Croghan, 1958; Linton *et al.*, 1983a, b).

The blood of leeches is relatively concentrated compared to oligochaetes and is hyperosmotic to the water in which they live. *Nephelopsis obscura* and *Helobdella stagnalis* were able to maintain themselves hyperosmotically in a medium between 15.6 and 59.5 mOsm/L and 47 and 112.7 mOsm/L, respectively, but were conformers at higher medium concentrations. *Theromyzon trizonare* was a regulator in both hypo- and hyperosmotic media. Leeches are also capable of regulating body volume when exposed to varying salinities (Rosca, 1950; Madanmohanrao, 1960; Smiley and Sawyer, 1976; Reynoldson and Davies, 1980; Linton *et al.* 1982). *Nephelopsis obscura* was able to maintain its weight in medium from 100–207 mosmol/L while *E. punctata* could only regulate between 58 and 200 mOsm/L. Weight regulation and volume regulation are probably not regulated per se, although these variables change in a predictable fashion with change in the osmotic concentration of the environment. It is more likely that leeches regulate the osmotic pressure of the body fluids; hence, with influxes and effluxes of water and salts, weight and volume change.

Using a three-way factorial experimental design, the effect of water temperature, ionic content, and TDS on the mortality and reproduction of *N. obscura* and *E. punctata* were examined (Linton *et al.*, 1983a, b). A strong interaction among temperature, ionic content, and TDS was observed on the mortality of *E. punctata*, but *N. obscura* showed uniformly low mortality unaffected by temperature, ionic content, or TDS. For both species, temperature showed the greatest influence on cocoon production and ionic content showed the least. Cocoon production by *E. punctata* was reduced in waters with low TDS and the proportion of nonviable cocoons was higher in low TDS than in high TDS.

IV. ECOLOGY

A. Diversity

Leeches are represented in North America by four families. Glossiphoniidae are either predators of macroinvertebrates or temporary ectoparasites of freshwater fish, turtles, amphibians, or water birds; Piscicolidae are parasites of fishes as well as some Crustacea; Erpobdellidae are primarily predators of macroinvertebrates and zooplankton; and, Hirudinidae are either predators of macroinvertebrates (e.g., oligochaetes, snails) or bloodsucking ectoparasites of freshwater and terrestrial Amphibia and mammals. Sanguivorous (bloodsucking) leeches spend a relatively short period of time taking a blood meal on the host and are more frequently found free living in the benthos. For this reason, host specificity is not often a very useful feature for identification of leeches.

While the majority of leeches parasitic on fish belong to the family Piscicolidae, two species of Glossiphoniidae (*Actinobdella inequiannulata* and *Placobdella pediculata*) show a strong preference for fish hosts. *Placobdella pediculata* recorded mainly from the midwestern United States has a high degree of host specificity for the freshwater drum *Apolodinotus grunniens* (Bur, 1994). However, other species of glossiphoniids (e.g., *P. montifera* and *Desserobdella phalera*) are also occasionally found on fishes. Members of the Erpobdellidae and Hirudinidae have been recorded on fishes, but it is doubtful that any species in these families regularly feeds on these hosts. All leeches require a firm substrate for attachment and are capable of feeding on body fluids. If a healthy or injured fish can provide either of these requirements, freshwater leeches will attach themselves. Parasitic leeches generally attach periodically to the fishes, take a blood meal, and then move back into the benthos. Acanthobdellids feed similarly on coldwater fishes (Bur, 1994).

B. Foraging Relationships

Leeches are either predators, consuming macroinvertebrates in the benthos and plankton or ectoparasitic sanguivores, feeding on vertebrates. The effects of predatory fish on benthic communities have been widely studied some showing weak or no effects whereas other studies have shown strong effects. Dahl and Greenberg (1997) compared prey consumption by brown trout (*Salmo trutta*) and the leech *Erpobdella octoculata* and showed that while one leech consumes much less than one trout, the role of leeches in structuring benthic communities is very similar.

Leeches and acanthobdellids are primarily fluid feeders; because, even though some predators ingest whole prey, they quickly evacuate the hard skeletal parts through the mouth or anus after extracting body fluids. Sanguivory occurs in the Hirudinidae, which have three-toothed jaws used to bite and penetrate the host's skin and in the Piscicolidae, and some Glossiphoniidae, which in contrast, lack jaws and penetrate the host with a proboscis. The acanthobdellid similarly has a short proboscislike specialization of the anterior foregut.

Theromyzon are sanguivorous glossiphoniids feeding on waterfowl and require a minimum of three blood meals to reach maturity (Davies, 1984; Wilkialis and Davies, 1980b). In comparison, *Placobdella papillifera* and *Haementaria ghilianii* require a minimum of four (Davies and Wilkialis, 1982; Sawyer *et al.*, 1981). The glossiphoniids generally appear to require significantly fewer blood meals than *Hirudo medicinalis*, which takes 10 or more meals to reach maturity (Blair, 1927; Pütter 1907, 1908), although in the laboratory at 26°C, only 4–5 blood meals are required. For *T. trizonare*,

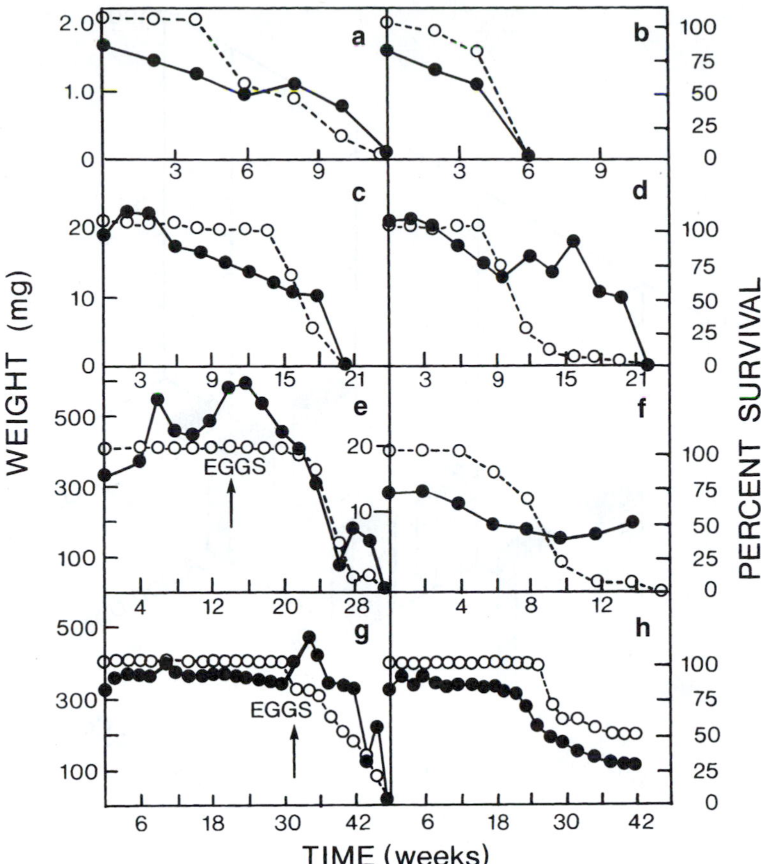

FIGURE 13 Changes in mean weight (0–0) and the percentage of survival (0–0) for populations of *Theromyzon rude* exposed to different water temperatures and feeding regimes: (a) fed once, maintained at 20°C; (b) fed once, maintained at 5°C; (c) fed twice, maintained at 20°C; (d) fed twice, maintained at 5°C; (e) fed three times, maintained at 20°C; (f) immature (14.7 mg) collected from the field, fed once, maintained at 20°C; and (g) immature (14.7 mg) collected from the field, fed once, maintained at 5°C.

over 80% of the population takes three meals in the first 6 months after hatching (Fig. 13). The remainder of the population overwinters after two meals and take the third in the spring so that all the population reproduces approximately 12 months after hatching. Some individuals that overwinter after three meals decline in weight before spring and require a fourth meal before reproduction commences. When ingesting blood, pathogens such as syphilis, erypsipelas, tetanus, cholera, hepatitis B and human immune deficiency virus (HIV) enter the gut and can survive for long periods (Nehili *et al.*, 1994). However, as these pathogens do not penetrate the unicellular leech salivary glands, direct transmission to another host is unlikely.

Predatory leeches either suck fluids with a proboscis (Glossiphoniidae) or have a suctorial mouth (Erpobdellidae and Hirudinidae). The range of prey species is normally quite diverse and changes seasonally with prey availability. In some instances, prey selection varies intraspecifically with body size and/or age class (Davies *et al.*, 1978, 1981, 1988; Wrona *et al.*, 1979, 1981). Because leeches either do not ingest identifiable hard parts or quickly evacuate them, the only accurate and efficient method of examining feeding is with serologic techniques. Antisera against prey are produced (Davies, 1969) and then used for a serologic test of the gut contents.

The feeding of *Nephelopsis obscura* and *Erpobdella punctata* was investigated using specific rabbit antisera against Cladocera/Copepoda, Chironomidae, Oligochaeta, Amphipoda, and Gastropoda. Apart from the absence of Gastropoda in the diet of *E. punctata*, there were no significant differences at the species level in prey utilization, niche breadth, and evenness (Fig. 14). However, temporal differences in prey utilization were evident, and both intra- and interspecific resource

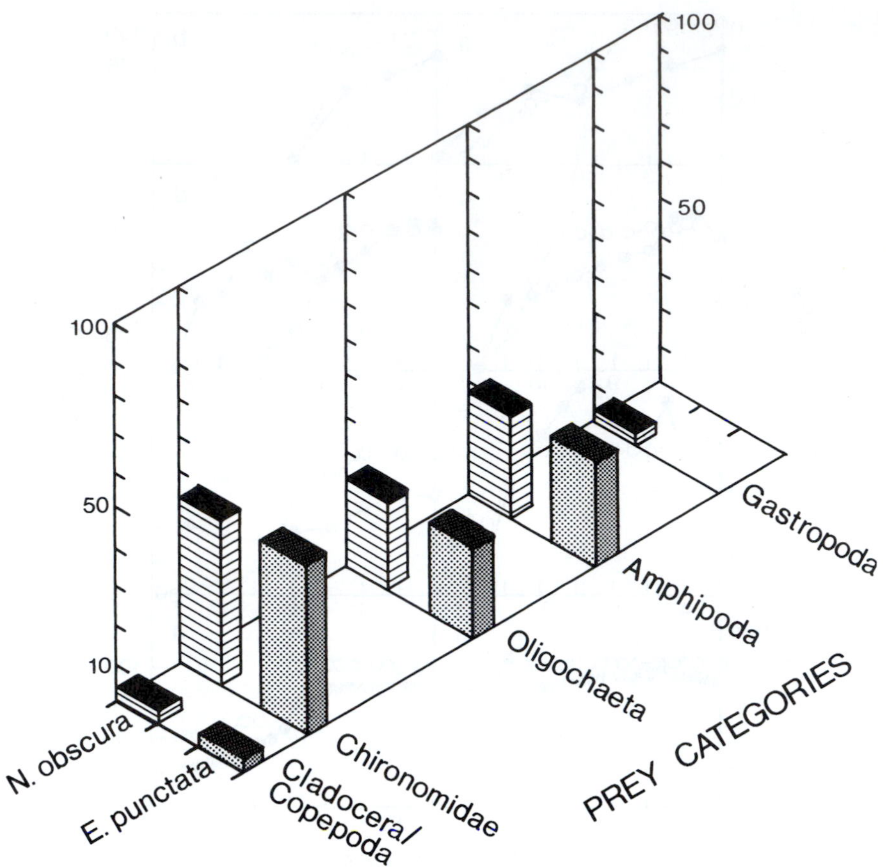

FIGURE 14 Frequency of prey utilization histograms (%) for *Nephelopsis obscura* and *Erpobdella punctata*.

partitioning occurred as a function of different weight class utilization of prey (Davies *et al.*, 1981; Martin *et al.*, 1994c). *Erpobdella punctata* has also been recorded as a phoront (Khan and Frick, 1997) on salamanders, with the salamander providing the leech with food particles from its feeding. Similarly, temporal differences in feeding and intraspecific weight class differences in prey utilization were found in sympatric *Glossiphonia complanata* and *Helobdella stagnalis* (Wrona *et al.*, 1981) (Fig. 15). The size of both predators significantly affected feeding success on prey (Martin *et al.*, 1994a, b).

In freshwater ecosystems, benthic species have a highly heterogeneous spatial and temporal distribution and invertebrate predators, including leeches, experience variable interfeeding intervals. Smith and Davies (1996a,b) measured the acquisition of energy and its allocation to components of bioenergetic balance throughout the life cycle in three groups of *N. obscura* fed at low, medium and high frequency. As feeding frequency increased, the total amounts of energy ingested, feces plus mucus produced, somatic and reproductive growth, energy storage (total lipid), and respiration all

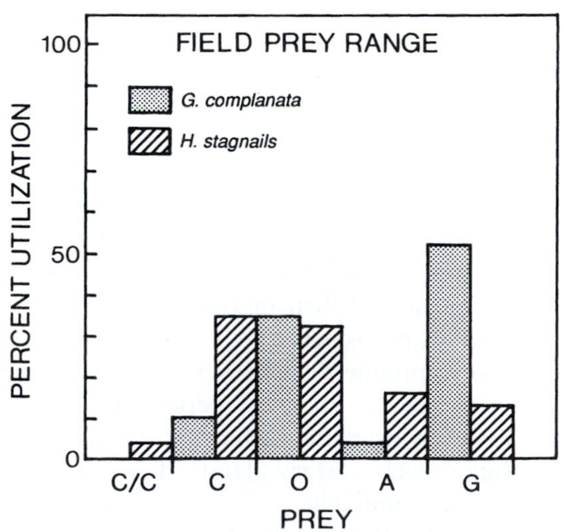

FIGURE 15 Frequency of prey utilization histograms (%) for *Glossiphonia complanata* and *Helobdella stagnalis* (C/C, Cladocera/ Copepoda; C, Chironomidae; O, Oligochaeta; A, Amphipoda; G, Gastropoda).

increased. In the medium and high frequency feeding treatments, the relative proportion of growth energy allocated to storage was constant, i.e. the ratio of reserves to structural tissue remains the same.

While most predatory leech species utilize a variety of prey, *Motobdella montezuma* is an exception. The pelagic endemic amphipod *Hyalella montezuma* comprises nearly 90% of the diet of the endemic *M. montezuma* in the thermally constant environment of Montezuma Well (Arizona), even though numerous other potential prey are abundant throughout the year (Blinn *et al.*, 1987; Davies *et al.*, 1988). This restricted diet was confirmed by both gut content and serological analyses. The highly stable conditions in Montezuma Well have contributed to the very close relationship between the endemic predator and prey. Using a bioenergetic model McLoughlin *et al.* (1999) showed that the specialized foraging utilized by *M. montezuma* is more efficient to the benthic opportunistic foraging strategy typical of most other erpobdellids. However, rather than being an obligate forager on *H. montezuma*, which is abundant during most of the year, *Motobdella montezuma* exploits other prey types during the winter, thereby achieving an overall higher rate of energy gain.

C. Reproduction and Life History

All leeches are hermaphrodites showing protandry or cosexuality (Davies and Singhal, 1988), with reciprocal cross-fertilization as the general rule. Fertilization, which is internal, is accomplished in the majority, of the Glossiphoniidae and all of the Piscicolidae and Erpobdellidae by attaching a spermatophore to the body of the partner. The spermatozoa penetrate the body wall and make their way to the ovisacs via the coelomic sinuses. The clitellar region is the most frequent site for the deposition of spermatophores, but they can be attached to any part of the body of the partner. In some species of Piscicolidae, there is a specialized region for the reception of spermatophores; fertilization is affected only by spermatophores deposited there. In the Hirudinidae, reciprocal internal fertilization is brought about by the insertion of an eversible penis into the vagina of the partner.

Once fertilization occurs, the eggs are deposited into a cocoon secreted by the clitellum. It has been widely assumed that the presence of a visible clitellum closely parallels maturation of the female reproductive system, and in nonreproductive leeches the clitellar glands are barely distinguishable from the epithelial cells (Fernandez *et al.*, 1992). The clitellum is especially prominent in Erpobdellidae and has been used by numerous authors to determine sexual maturity. Biernacka and Davies (1995) showed that sexual maturity of *Nephelopsis obscura* cannot be judged by the presence or absence of an externally visible clitellum. Although a high proportion of the population exhibited a visible clitellum at some time, it was not persistent (lasting no longer than 7 days) and thus could easily be missed in studies with weekly or longer sampling periods. Feeding regime affects the timing of the clitellum development but, regardless of feeding regime, all the leeches exhibited two periods with a visible clitellum. The first appearance of a clitellum did not coincide with either spermatogenic or oogenic maturity, but at the second appearance of a clitellum fully mature ova were present. The presence of a visible clitellum is thus not a good indicator of maturity in *Nephelopsis obscura* and probably not a good indicator in most or all species of Erpobdellidae. The presence of a spermatophore is also not a good indicator of maturity. Singhal *et al.* (1985) found only 5% of mature *N. obscura* collected from the field had observable spermatophores, and Biernacka and Davies (1995) found that only 4% of the mature size range animals had a spermatophore attached.

There can be a considerable delay between copulation and cocoon deposition, e.g. in field populations of *Helobdella stagnalis* copulation occurs in the fall and cocoon deposition takes place in the spring (Davies and Reynoldson, 1976). Erpobdellid cocoons are thick walled, oval, and attached to a firm substrate (stone, leaf, stem, wood). The cocoons of Piscicolidae and Hirudinidae are only loosely attached to the substrate and are usually spherical. In species of Hirudinids depositing their cocoons in moist habitats out of water, the outer wall of the cocoon is spongy, which is thought to reduce water loss. The cocoons of Glossiphoniidae are very thin-walled (sometimes called egg cases) and are either deposited on the substrate and immediately covered by the body of the parent or are attached to the ventral surface of the parent. In both cases, the hatchlings are attached to the ventral body wall and are carried around by the parent for a long period, which for *Theromyzon tessulatum* can be 5 months and for *T. trizonare,* 1 month (Wilkialis and Davies, 1980a). *Glossiphonia complanata* adults have been found to provide their young with nutrients that are passed through the body wall of the parent into the posterior sucker of the young (De Eguileor *et al.*, 1994). Reductions in body weight and size have been recorded for brooding glossiphoniids; however, these losses have been attributed to reduced feeding potentials and to the energy expended ventilating the young (Calow and Riley, 1982; Milne and Calow, 1990). It is possible that other species which brood attached young may provide nutrients to the developing young; however, further studies need to be conducted on a variety of glossiphoniid leech species. A comparison on reproductive output for glossiphoniid that brood eggs and young, and for erpobdellids, that leave their eggs in an encapsulated cocoon, shows that brooding is not metabolically more expensive than encapsulation. There are no extra energy costs incurred in terms of carrying or ventilating broods, but weight loss due to reduction in feeding does occur. These costs appear to be energetically similar to the cost of encapsulation for erpobdellids, and it is suggested that brooding and encapsulation represent alternative evolutionary paths for brood protection.

The life cycles of all leeches and acanthobdellids consist of egg, juvenile, and mature hermaphrodite adult, which reproduces and produces more eggs. Adults of predatory species usually reproduce only once before death (i.e., semelparity), although some sanguivorous species reproduce several times over a few years (i.e., iteroparity). Semelparity and iteroparity are generally considered genetically different reproductive strategies determined through evolutionary processes. Studies on the reproductive biology of leeches have shown considerable variation in life history with annual or biennial life cycles in predatory species. Some sanguivorous species (e.g., *T. trizonare*) live for several years with the duration of the life cycle dependent on the interval between blood meals (Davies, 1984). Baird *et al.* (1986, 1987b) showed that changes in water temperature, size at reproduction, energy loss during reproduction, and postreproductive feeding of the erpobdellid *Nephelopsis obscura* significantly affected postreproductive mortality. Although genetically iteroparous, *N. obscura is* like most predatory leeches and some sanguivorous species in being almost always semelparous in the field because of high postreproductive mortality. Thus, it is quite possible that the range of life histories reported for various species of leeches falls within the range of environmentally induced variability. Indeed, it is conceivable that all leeches are genetically iteroparous but phenotypically exhibit semelparity under most conditions. This flexibility would ensure the long-term persistence of genotypes in a variable environment, where a more rigid strategy might have a low or zero fitness in some years. Freshwater ecosystems in North America are highly variable in terms of both abiotic (experience oxygen concentration and temperature) and biotic (prey availability) environmental parameters, and Bulmer (1985) suggested that iteroparity is found in variable environments as a bet-hedging or risk-avoidance strategy.

The life cycle and life history of comparatively few freshwater leech species in North America have been examined. In Alberta, two generations of *N. obscura* are produced annually, one in the spring and one in the late summer. The spring generation is the progeny of the heavier individuals from the spring generation of the previous year and from the late-summer generation produced two years previously. The late-summer generation is produced by portions of the previous year's spring and late summer generations. Thus, each generation produces young after either 12 and 15 months or 12 and 19 months, although each individual reproduced only once (Davies and Everett, 1977) (Fig. 16). Reddy *et al.* (1992) showed that *N. obscura* which had overwintered differed physiologically, throughout their life history, from individuals which had not. The first generation of *N. obscura* in early summer is the progeny of the heavier individuals from the early summer generation of the previous year and the second late summer generation is produced by portions of the previous year's early and late summer generations. The late summer generation has a faster growth rate, higher lipid storage, higher proportion of energy available for growth and metabolism, and earlier valuation of larger maximum size than the early summer generation. However, there were no differences in consumption of prey or respiration between the two generations (Qian and Davies, 1994). As there are no significant genetic differences between the early and late summer generations, the differences in physiological ecology and life history traits are attributed to differences in prehistory of the parents. In Minnesota, *N. obscura* delays its age of reproduction to two years and in so doing attains a larger

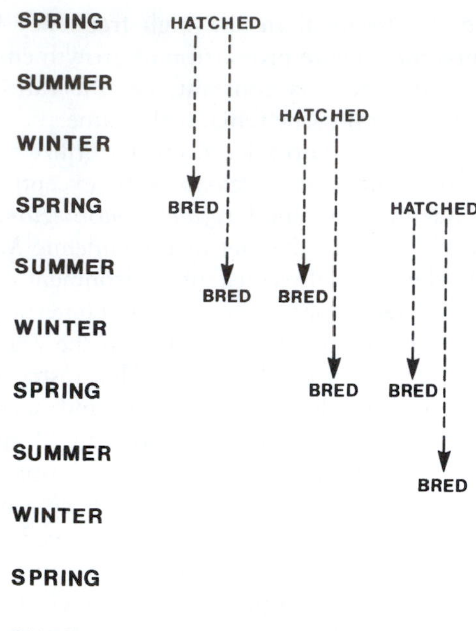

FIGURE 16 Life cycles of spring and summer hatching *Nephelopsis obscura* in Alberta, Canada.

body size (Peterson, 1982, 1983). *Erpobdella punctata* is usually subdominant to *N. obscura* and reproduces after one year. When *E. punctata* is dominant, it has a more complex cycle with reproduction after one or two years, at a larger size, over a shorter season (Davies *et al.*, 1977). *Helobdella stagnalis* can produce either one or two generations per year depending on the water temperature regime (Davies and Reynoldson, 1976). In warmer conditions, the overwintering population reproduces in the spring and dies on completion of brooding. The spring generation grows rapidly and a second generation of young is produced in the summer, which forms the overwintering population. In colder temperature regimes, members of the overwintering population reproduce in the late spring/early summer and die after brooding. After summer growth, this generation forms the next overwintering population. Davies (1978) showed that populations of *H. stagnalis* from cold water regimes are able to produce two generations per year in appropriate conditions.

Several environmental factors have been implicated in affecting the reproduction of leeches, including food availability and water temperature. Water temperature regime can affect the relative reproductive success and growth of species (Davies and Reynoldson, 1976; Davies, 1978; Wrona *et al.*, 1987), as well as depth distribution and seasonal movement (Gates and Davies, 1987). A bioenergetics simulation model of the growth and life history of *N. obscura* (Linton and Davies, 1987) showed that its growth is more sensitive to prey variation between years than to temperature variation. Davies and Singhal (1988) showed that, in the field, *N. obscura* exhibits sequential protandry in the first bout of gametogenesis and simultaneous hermaphroditism in the second. In the laboratory maintained under constant conditions, oogenesis initially follows spermatogenesis but mature ova first appeared shortly after Stage 4 of spermatogenesis with the final stages of spermatogenesis lagging behind oogenesis (sequential protogyny). However, for about 3 weeks mature spermatozoa and ova co-occur, i.e., simultaneous hermaphroditism occurs. This is quite different from what occurs in the field. Clearly, *N. obscura* shows plasticity in the phenology of the development of hermaphroditism, the differences between the field and laboratory resulting from the absence of an environmental cue. This was subsequently shown to be a period of low temperature (3–5°C) for a minimum of 7 days between stages 3 and 4 of spermatogenesis.

Kalarani *et al.* (1995) demonstrated the presence of ecdysone in *N. obscura* and showed correlations between ecdysone concentrations and gonad maturation and between ecdysone concentration and utilization of energy reserves during gametogenesis. Responses to ecdysone varied with both the ecdysone concentration and the duration of exposure, indicating the importance of the sequential changes in ecdysone to the target tissues. Hormonal control of reproduction in leeches using brain homogenates has been shown by a number of authors (Kulkarni *et al.*, 1980; Hagadorn 1969; Webb, 1980; Webb and Omar, 1981).

The life cycle of *Acanthobdella peledina* is basically annual with young immature individuals first occurring on their host in June. These individuals leave the fish in October on reaching sexual maturity. Reproduction with egg and cocoon production occur off the fish and have not been observed (Holmquist, 1974).

D. Dispersal

The dispersal of freshwater leeches and acanthobdellids is generally assumed to be the result of passive transfer from one water body to another by other animals. This seems probable for parasitic sanguivorous species, but less probable for the majority of freshwater leeches which are predators of benthic invertebrates. However, some examples have been recorded with *Marrinmeyeria lucida* transported by birds (Daborn, 1976), and Khan and Frick (1997) showed that a high proportion of salamanders (*Ambystoma maculosum*) in South Carolina served as phoretic hosts to *Erpobdella punctata*. Platt *et al.* (1993) similarly recorded *Helobdella stagnalis*, previously found on fish, phoretic on *Ambystoma tigrinum* in Indiana.

Davies *et al.* (1982b) examined passive transfer by ducks of two parasitic species (*Theromyzon trizonare* and *Placobdella papillifera*) and two predatory species (*Helobdella stagnalis* and *Nephelopsis obscura*). *Theromyzon trizonare* adults were transferred both in the nares, while taking a blood meal, and on the duck's body under the feathers. *Placobdella papillifera* were also conveyed on the body of the duck, but adult *H. stagnalis* and *N. obscura* were not transported. Adult leeches ingested by ducks were not recovered from the duck feces, but viable *N. obscura* cocoons were recovered from the feces of fed ducks.

Passive dispersal by wind has never been recorded, but the transport of cocoons attached to macrophytes is possible. Davies (1979) showed that *Helobdella stagnalis, H. triserialis, Glossiphonia complanata*, and *Mooreobdella fervida* were dispersed to Anticosti Island by sea currents from the Quebec north shore of the Gulf of St. Lawrence. Dispersal by currents in large lakes or through rivers is also highly probable, and leeches have been recorded in the drift (Elliott, 1973). Accidental transport of leeches by humans has not yet been shown in North America, but has been demonstrated in other areas of the world (Sawyer,

1986). Active directional dispersion up rivers was recorded for *Percymoorensis marmorata* by Richardson (1942) and for *E. punctata* by Sawyer (1970). *Erpobdella punctata* has been found to be a phoront on the salamander *Ambystoma maculatum*, and it is likely that as the salamanders move from pond to pond the leeches are transported (Khan and Frick, 1997).

Passive dispersal, while unlikely, does appear to be the mechanism by which leeches colonize new ecosystems. The passive transport of individuals between systems can be estimated using genetic markers and the level of genetic similarity between populations can be used to estimate the amount of migration between populations. In a recent study of *Erpobdella punctata*, the estimated migration rates between populations were found to be extremely low, regardless of the distance between neighboring populations. However, some populations are more closely related to each other than to others indicating the potential for either migrants or founding individuals to be transported between ecosystems (Govedich *et al.*, 1999).

Differences in intra- and inter-population genetic structure is primarily determined by gene flow with poorly dispersing species exhibiting high inter- and low intra-population variation. In general, there is greater genetic divergence with increased physical isolation restricting gene flow. Qian and Davies (1996) determined whether genetic differences among three populations of *Nephelopsis obscura* were the result of differential dispersal or environmental heterogeneity. Two of the populations examined were from sites in Alberta, Canada, geographically close (80 km) to each other while the third population in Utah, was geographically distant (1200 km) from the other sites. Despite the high probability of dispersal between the close sites and the lower probability of dispersal between these two sites and the distant site, one of the sites in Alberta differed at only one locus from the Utah population and both these populations differed from the other Alberta population at five loci. These genetic differences were correlated with the dissolved oxygen regimes acting as a selective force rather than geographic distance between populations.

E. Predation

In North America, predators of leeches include fish, birds, garter snakes, newts, and salamanders (Pearse, 1932; Bartonek and Hickey, 1969a; Bartonek and Murthy, 1970; Bartonek and Trauger, 1975; Bartonek, 1972; Able, 1976; Arnold, 1981; Davies *et al.*, 1982b; Kephart, 1982; Kephart and Arnold, 1982), as well as insects, gastropods, and amphipods (Hobbs and Figueroa, 1958; Pritchard, 1964; Sawyer, 1970; Anderson and Raasveldt, 1974; Blinn *et al.*, 1988). Young and Spelling (1986), Spelling and Young (1987) and Young (1987), reviewing predation on three lentic leech species in Great Britain, recorded a similar range of predators as well as interspecific leech predation and cannibalism.

Leeches lack hard body parts and are thus difficult, if not impossible, to identify from the gut contents of predators. Serology is the most appropriate method for analyzing predation on soft-bodied animals (Davies, 1969), but this approach requires prior knowledge of the range and size of potential predators to be collected from the field and tested. In the laboratory, Cywinska and Davies (1989) found that four species of dytiscids, one species of Odonata, two species of hemipterans, and two species of amphipods fed on one or more size classes of *N. obscura*. In Great Britain, Young and Spelling (1986) also noted that species within these taxa fed on *E. octoculata, Glossiphonia complanata*, and *Helobdella stagnalis* in addition to triclads, trichopterans, and megalopterans. It would thus appear that larval and adult coleopterans and nymphal odonates and hemipterans are potentially voracious predators of leeches throughout the Holarctic.

In experiments with predation on cocoons, Cywinska and Davies (1989) recorded two methods of cocoon consumption; either the embryos and eggs within the cocoons were fed upon, by piercing holes through the cocoon wall (e.g., Coleoptera larvae and adult Hemiptera), or else the whole cocoon (sometimes, plus attached plant leaf) was consumed (e.g., amphipods and adult Coleoptera). Predation has also been recorded by gastropods on *E. punctata* cocoons (Sawyer, 1970) and by a mite on *Haemopsis sanguisuga* cocoons (Bennike, 1943). The only known vertebrate predators of leech cocoons are salamanders (Pearse, 1932) and waterfowl (Davies *et al.*, 1982b).

Predation rates on *N. obscura* in the laboratory were inversely related to size, with individuals >30 mg being consumed only at very low rates by certain coleopterans. Young and Spelling (1986) found a similar trend of increased consumption of leeches by most predators with declining leech weight. This can be explained by the inability of some potential predators to capture and overcome the more powerful larger leeches, rapid satiation of predators with larger prey, or the ability of larger leeches to produce more mucus and so impede the movements of the potential predators.

A caveat of these feeding experiments was that predators were starved, had no alternative food available, and had no choice in size of *N. obscura* prey. In such conditions, the results of the experiments probably showed the maximum potential predation rates and suggested that predation on cocoons and size classes <10 mg might be substantial. Considering the very high densities of potential predators in many lakes and ponds and the ability of some predators to consume large numbers of leeches and/or their cocoons, the effects of predation in the field are potentially significant in macrophyte zones.

The rapid decline in density of hatchling and small (< 10 mg) *N. obscura*, observed in the field by Davies and Everett (1977), could be explained by predation in the shallow macrophyte zones where cocoons are deposited and hatch. Placement of cocoons by *M. montezuma* on deeply submerged *Potamogeton* stems at least 3.5 m below the water surface at the interface of the littoral-pelagic zone in Montezuma Well may be a strategy to reduce predation (Davies *et al.*, 1988).

Predation in the water column is almost entirely restricted to fish, amphibia, and birds. The presence of large numbers of leeches in the gut contents of fish (particularly from rivers) and the commercial development of leeches for bait in some areas indicate predation of fish can be at least occasionally heavy.

F. Behavior

Sawyer (1981) suggested that the behavior of leeches and acanthobdellids could be divided into a number of elementary responses: crawling (vermiform, inchworm); swimming; shortening (fast, slow); searching (head movement, body waving); and an alert posture when the body is extended and held motionless for some time. However, species differ markedly in the frequency of expression of these different responses. Behavior of leeches can be modified by size (age) and physiological state (starved or fed, and mature or immature) as well as by extrinsic environmental parameters monitored by eyes, sensillae, chemoreceptors, and mechanoreceptors (T-cells).

The foraging behaviors of sanguivores and predators are usually different. Sanguivorous leeches in the proximity of a potential host respond to changes in illumination (shadows), vibrations in the water, chemical stimuli, and possibly differences in temperature between the host and the surrounding water. Compared to predatory leeches, the sanguivores are less negatively phototactic and nocturnal, especially when hungry. *Theromyzon tessulatum, T. trizonare, Piscicola geometra*, and *Calliobdella vivida*, when hungry, have all been found in the open water ready to attach to a host (Herter, 1929a, b; Meyer and Moore, 1954; Sawyer and Hammond, 1973). *Hirudo medicinalis* and *Theromyzon tessulatum* are stimulated to attach to and bite hosts with temperatures of 37–40°C (Dickinson and Lent, 1984), and *Placobdella costata is* attracted from as far as 15 cm to hosts with temperatures of 33–35°C (Mannsfeld, 1934). *Haementaria ghilianii* and *Hirudo medicinalis* are both sensitive to water disturbances (traveling surface waves) (Sawyer, 1981; Young *et al.*, 1981), and it is likely that most sanguivores respond in a similar manner.

Water disturbance is frequently the first indication of the presence of a potential host; the leech moves toward the disturbance using cues detected by its sensilla (Friesen and Dedwylder, 1978; Friesen, 1981; Derosa and Friesen 1981). *Theromyzon tessulatum* and *T. trizonare* respond to water disturbance in a similar manner, but adult brooding *T. tessulatum* are unique in that they move toward a potential host even though they never take a blood meal. If the parent *T. tessulatum* locates a host, it attaches to it and the brooding young leave to seek their first blood meal (Wilkialis and Davies, 1980a).

Although movements toward a potential host by sanguivorous leeches are not specific, the next step of determining whether or not the potential host is an appropriate source of blood involves some selectivity. After attachment to the potential host with the posterior sucker, the leech makes a number of searching movements with the anterior portion of the body. If the host is not appropriate, the leech detaches. Attachment to a host does not always occur at an appropriate location. For example, *T. trizonare* primarily feed in the nares; but if one attaches to the feathers rather than the beak it will remain on the bird for up to 30 min, moving about in search of a suitable location to feed. Host recognition by sanguivores presumably results from chemoreception and/or mechanoreception.

Predatory leeches are generally more nocturnal and negatively phototactic than sanguivores (Elliott, 1973; Anholt and Davies, 1987; Davies and Kassera, 1989; Davies *et al.*, 1996). Angstadt and Moore (1997) demonstrated that *Nephelopsis obscura* swimming is entrained to a 12-h light/12-h dark cycle and that a circadian rhythm of swimming persists in constant darkness. Davies *et al.* (1996) tested the hypotheses that foraging behavior responses such as numbers of meals consumed, the number of encounters with prey and feeding time as well as activity (swimming) were higher in the dark, and increased with the size of *N. obscura* and the length of the starvation period. While swimming activity was longer in the dark, not all the foraging behavior variables followed this pattern. There was an inverse relationship between swimming time and the number of meals consumed and the poor correspondence between swimming time and the number of prey encounters indicates that swimming is not always directly associated with increased foraging and is sometimes related to movements between microhabitats. Thus, increased nocturnal activity is not always indicative of increased foraging activity.

Anholt (1986) showed that *N. obscura* reduced its foraging activity when satiated and this was confirmed through direct observations of encounters and feeding rates as a function of satiation. Increasing satiation in *N. obscura* produces a threshold level above which prey with natural escape responses cannot be successfully captured. Only a small fraction of encountered tubificids were ingested by starved *N. obscura* when the prey had a substrate refuge. The loss of prey capture ability with increasing satiation is a behavioral foraging constraint. When its threshold level is reached, further successful foraging on prey with a substrate refuge is impossible. This threshold level varies for different prey types, e.g., Anholt (1986) showed that the tube-dwelling chironomids *Chironomus riparius* and *Glypotendipes paripes* were superior prey types compared to tubificids in promoting leech growth, and concluded that *T. tubifex* could evade capture by *N. obscura* more efficiently than chironomids. This was confirmed by direct observations which showed escape times by chironomids longer than the escape times of tubificids.

Groups of *N. obscura* were provided with sufficient prey with either a substrate (MS) or without a substrate (NS) to allow asymptotic food ingestion without prey patch depletion. In both groups of leeches, feeding was not reduced by low prey encounters, but by the predator's ability to capture more prey. Thus, the significantly different feeding rates exhibited by MS and NS leeches resulted from different factors determining the asymptote of food ingestion. In MS leeches this was primarily the foraging constraint threshold, but in NS group it was primarily the gut capacity determining maximal feeding potential (Dratnal *et al.*, 1992).

As MS leeches showed a higher growth efficiency, the lower food acquisition was compensated for by allocating a greater fraction of energy to growth and/or by higher energy absorption efficiency, a result of longer food retention time in a less packed alimentary canal. Thus, the foraging constraint significantly affected the internal status of *N. obscura* with a feedback system between the two. Foraging results in increasing satiation, which reduces and eventually stops foraging, until some level of hunger (decreased satiation) is regained. Oxygen uptake (respiration) increased in fed *N. obscura* suggesting an increased energetic requirement for the processes directly or indirectly associated with digestion with the resulting energetic trade-off producing a lower overall activity in fed leeches.

Although several modifying, behavioral factors apply to *N. obscura* in the field they do not preclude the constraint on foraging by *N. obscura* affecting its size-specific rate of feeding and life-history traits, especially as chironomids and tubificids form major components of the diet in the field (Davies *et al.*, 1978).

The existence of mechanisms based on individual size-dependent limitation in food ingestion are of great importance not only for the indeterminate growth of *N. obscura* but must also strongly modify size-dependent food partitioning within the predator population—an important factor in population regulation.

To behave adaptively, animals often give higher priority to some behaviors over others. Misell *et al.* (1998) looked at the effect of feeding behavior on other behaviors, such as swimming, crawling, and shortening, in *Hirudo medicinalis*. The priority feeding had, relative to other behaviors, showed that as expected feeding occupies a high position in the hierarchy of behaviors and dominated all the other behaviors tested. A feeding *H. medicinalis* will ignore strong mechanical stimuli which normally cause vigorous escape responses to continue feeding. Although little is know about the neural basis of feeding in *H. medicinalis*, the role of serotonin has been studied. Increases in serotonin result in an increase in feeding and vice versa, leading to the hypothesis that serotonin is effectively a feeding hormone.

While some predators also show increased activity in response to disturbances in the water, they are generally opportunistic, feeding on prey encountered through either their own locomotion or the movements of the prey. Prey detection by predator leeches has received little attention since Gee (1913) noted that *Mooreobdella microstoma* responded to vibrations in the water and chemical extracts of its prey by unidirectional swimming and random movements of the anterior part of its body. Davies *et al.* (1982a) examined the chemosensory detection of prey by *N. obscura* and showed that without direct tactile contact, *N. obscura* were unable to detect and react to any of the common prey types tested (i.e., they did not show clinotaxis). Whether or not a particular prey appears in the diet of a predatory leech depends on the probability that it will be encountered, the probability that it will be attacked once encountered, and the probability that an attack will result in capture. Simon and Barnes (1996) concluded that *Percymoorensis marmorata* use olfactory rather than mechano sensory cues to detect their prey (oligochaetes). Olfactory responses of the medicinal leech (*Hirudo medicinalis*) are very specific although arousal and searching for prey are primarily initiated by water movement. The response of *Bdellerogatus plumbeus* to its main prey *Erpobdella punctata* are also chemosensory (Riggs, 1980). However, Young *et al.* (1995) concluded that *Helobdella stagnalis*, *Glossiphonia complanata* and *Erpobdella octoculata* do not depend on chemosensory cues to detect prey and that chemosensory cues are only important when the leech is in contact with the prey as shown by Davies *et al.* (1982a). They also concluded that the inability of leeches to detect damaged prey in heterogenous environments was the reason for very high juvenile mortality.

An exception to the general rule that the diet of predatory leeches is usually nonspecific and contains a wide variety of prey is demonstrated by *Motobdella montezuma*. Blinn *et al.* (1988) showed that *M. montezuma* forages selectively on the pelagic amphipod *Hyalella montezuma* despite the presence of a wide variety of alternate prey. *Motobdella montezuma* discriminates between two congeneric amphipod prey by mechanoreception, exhibiting a high response to the acoustic vibration signals of the endemic *H. montezuma* but not to *H. azteca*. Not only does *M. montezuma* discriminate between species by mechanoreception, but it also selectively discriminates between size classes, feeding principally on juveniles. Although amphipods form part of the diet of both *N. obscura* and *E. punctata*, neither species shows a significant response to their acoustic vibration signals (Blinn and Davies, 1989). The unusual feeding behavior exhibited by *Motobdella* can be linked to the highly specialized sensillae that ring the mouth (Fig. 17). These sensillae are composed of cilia that are in contact with the water where vibrations from potential prey can be detected. The number of cilia and the size of the sensillae have been found to be significantly greater in *Motobdella* than in the related erpobdellid genera *Erpobdella* and *Nephelopsis* (Govedich *et al.*, 1998).

Substrate type affects leech foraging with *Erpobdella octoculata* foraging more effectively over cobble substrates than over fine gravel with differences in prey type consumed (Dahl and Greenberg, 1997). When leeches and trout were together, trout foraging was unaffected by leeches but leech foraging was affected by trout. Although one leech may consume much less than one trout at field densities, leeches have similar effects as fish in structuring lotic benthic communities (Dahl, 1998).

Some animals form temporary or permanent aggregations or groups which may reduce an individual's risk of predation, enhance mating or food intake, ameliorate physiological stress, coordinate and enhance dispersal or synchronize development. Leeches commonly form aggregations in the field, especially the benthic predatory species. Smith and Davies (1996a, b) examined the effects of different group sizes on acquisition and allocation of energy in *Nephelopsis obscura* and evaluated the physiological cost and benefits of group living. In terms of growth, asymptotic biomass, ingestion and respiration group sizes larger than one and less than 10 were found to be optimal—a group size within the typical range of groups sizes found on natural stony shores.

Oxygen concentrations affect both the growth of *N. obscura* and its feeding behavior (Davies *et al.*, 1992; Davies and Gates, 1991a, b). Higher rates of feeding occurs in 100 and 300% oxygen saturation regions compared to hypoxic regimes. This is related to the greater proportion of time spent on or in the mud substrate and thus increased prey encounter rate. Since hypoxia and anoxia cause high mortality at summer temperatures, differences in vertical distribution of *N. obscura* are behavioral adaptations to minimize stress. Seasonal migrations of *N. obscura* from deep water to shallow water in the spring and to deep water from shallow water in the fall cannot be solely attributed to temperature preferences and it is likely that avoidance of oxygen stress also plays a role.

G. Population Regulation

On a geological time scale, only a very few freshwater ecosystems are long-lived and even on a shorter time scale of 1–20 years, many lentic ecosystems are ephemeral. In ephemeral ecosystems, leeches may not have sufficient time to reach carrying capacity. Evidence for competition between leech species has been presented by Davies *et al.* (1982c) for *Nephelopsis obscura* and *Erpobdella punctata* and by Wrona *et al.* (1981) for *Glossiphonia complanata* and *Helobdella stagnalis*. However, only the studies on competition between *N. obscura* and *E. punctata* are comprehensive in relation to the established criteria (Reynoldson and Bellamy, 1970; Williamson, 1972).

Reynoldson and Davies (1976, 1980) showed a considerable overlap in the range of water conductances in which *N. obscura* and *E. punctata* can regulate their weights. The sympatric distributions of the species, therefore, appear to be related to common salinity tolerances. However, among sympatric populations, there is a strong inverse relationship between the numerical abundances of *N. obscura* and *E. punctata* in lentic ecosystems (Davies *et al.*, 1977). Dominance of *E. punctata* over *N. obscura* is a rare temporary event and the switch to dominance by *N. obscura* is rapid. The reproductive strategies of *E. punctata* vary depending on whether it is dominant or subordinate (Davies *et al.*, 1977). Thus, the biological interactions between *N. obscura* and *E. punctata* are amenable to explanation based on competition.

The utilization of common food resources by *N. obscura* and *E. punctata* has been demonstrated by numerous laboratory and field studies (Davies and Everett, 1977; Davies *et al.*, 1978, 1981, 1982c). At the species level, there were no significant differences in the

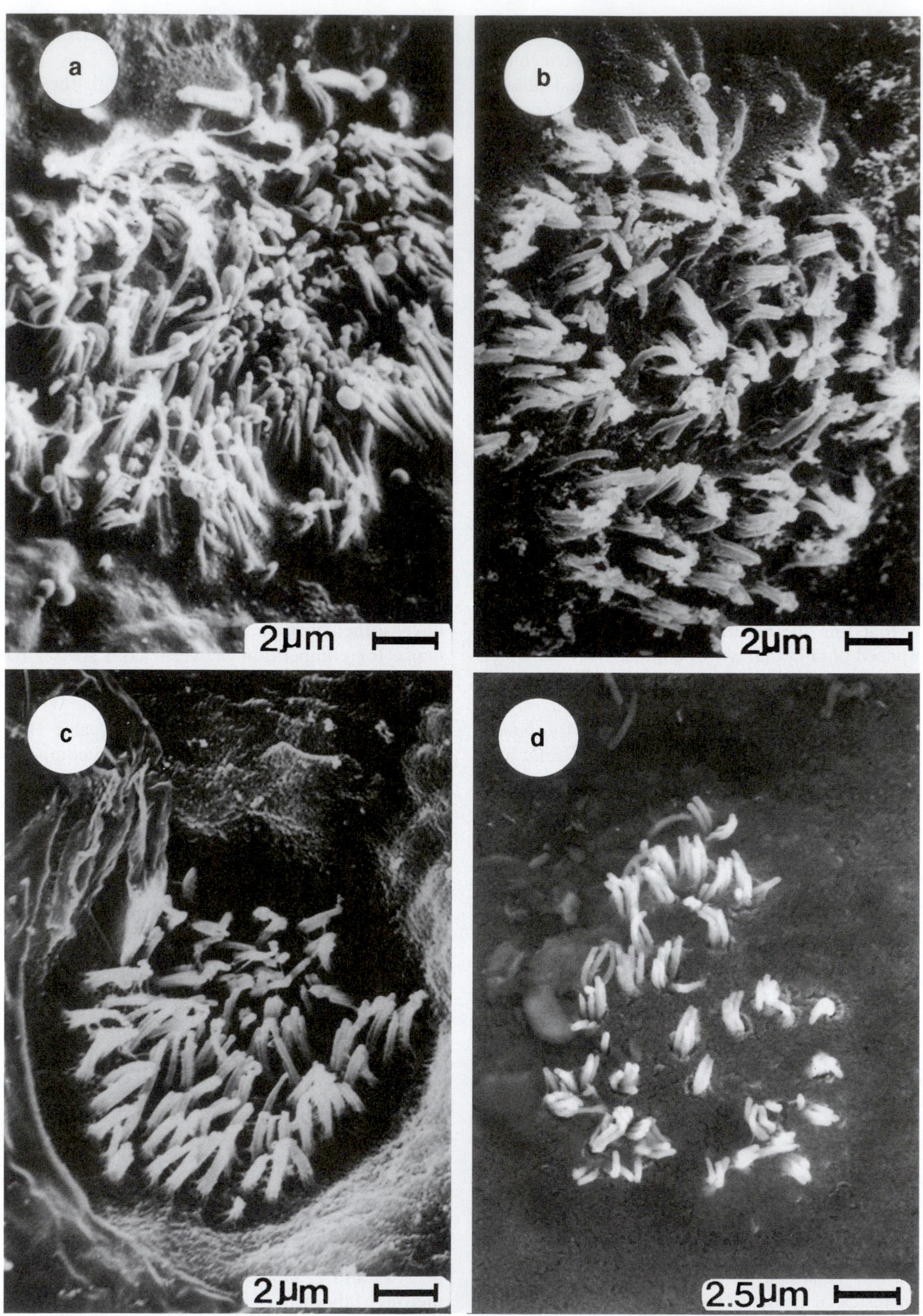

FIGURE 17 Photomicrographs (5000 ×) showing differences in the size of sensilla and the number of cilia within sensilla of: (a) *Motobdella sedonensis*; (b) *M. montezuma*; (c) *Erpobdella punctata*; (d) *Nephelopsis obscura* (Govedich *et al.*, 1998).

diets of either species between years and the trends in prey utilization were similar each year (Fig. 14). Intraspecific food resource partitioning, as a function of differential weight class utilization of prey was shown by both species. Such resource partitioning minimizing intraspecific competition, increases population fitness (Giesel, 1974). The change in feeding ecology of *E. punctata* with changes in numerical dominance provided further evidence for interspecific competition. Furthermore, a significant decrease in mean niche overlap and niche breadth was found for both species when *N. obscura* became dominant. The majority of criteria established to determine interspecific competition have been supported by the data collected on *N. obscura* and *E. punctata*; however, the manipulation of food resources has not yet been completed. In British lakes, semelparous leech species show very high (95–98%) mortality of juveniles (Young *et al.*, 1995). Many young leeches are inept at capturing prey, and feeding success increases when adult leeches are also present suggesting that mortality of young leeches is high because of their inability to locate damaged prey in an heterogenous environment (Dahl, 1998).

H. Ecotoxicology

Aquatic ecosystems seem to be sinks for many environmental contaminants such as heavy metals, pesticides and other synthetic or natural pollutants. The impacts of these pollutants are generally only seen when the problems have reached critical end points such as fish kills or algal blooms. However, more subtle effects, such as changes in growth, reproduction, behavior, physiology or survival of freshwater species, may act as indicators for potential larger scale problems. In the longer term, these subtle chronic effects play a role on health and efficiency of freshwater ecosystems. Biomonitoring and ecotoxicological studies utilizing freshwater leeches, especially benthic predatory species, have recently become increasingly more common.

Changes in the community are frequently assessed by using bioindicators (species or groups of species for which changes in abundance reflects the quality of the environment. Leeches were used by Bendell and McNichol (1991) as indicators of acidification on 40 small lakes, many of which had been acidified by nickel smelting works at Sudbury, Ontario, Canada. Nine species of leeches were recorded from the 20 lakes with pH > 5.5. Several species of leech disappeared over a narrow range of pH which suggests that acidity or a single chemical or biological factor correlated with pH is responsible. However, Grantham and Hann (1994) showed that lake pH in the Experimental Lakes Area of Ontario was not a dominant variable, describing only a small amount of the variance in the species–environment relationship.

Leeches have shown superior chlorophenol bioconcentrating ability over many other aquatic organisms. Prahacs and Hall (1996) evaluated *N. obscura* as a potential biomonitor of chlorinated phenolic compounds discharged from bleach pulp mills on the Lower Fraser River, British Columbia, Canada. They found a strong linear correlation between water contaminant concentration (0.1–10.0 μg/L) and bioconcentration of chloroguaiacols. Bioconcentration was inversely related to pH. As *in situ* biomonitors, Prahacs *et al.* (1996) showed that and increase in chlorine dioxide substitution in leeches reduced amounts of chlorinated phenolics accumulated, with a sharp decrease observed at chlorine dioxide levels >90%.

An alternative approach to bioindicators is the use of biomarkers, defined as measurable variations in cellular or biochemical components or processes, structures or functions, that provides information on a stress and its effects at the population level. The effects of exposure of adult sexually mature *Nephelopsis obscura* for 4–24 days to 0.0–580 μg/L cadmium showed that the numbers of ova and spermatozoa per unit biomass were significantly reduced with increasing cadmium concentration and exposure time, as were the masses of the testisacs and ovisacs (Davies *et al.*, 1995). Exposure of cocoons to cadmium (0.0–4000 μg/L) had a significant effect on post hatchling survivorship (Wicklum *et al.*, 1997). The effects of chronic exposure to cadmium (0.0–50 μg/L) were examined in terms of changes in energy acquisition and allocation (somatic growth, reproductive growth, feces plus mucus, active respiration, resting respiration and activity time) (Wicklum and Davies, 1996). Leeches in the high cadmium concentration group had significantly lower biomass production, survivorship, and ingestion rate. Assimilation efficiency, reproductive investment, resting respiration and active respiration were unaffected by cadmium, but leeches in the high concentration group were significantly less active and produced more mucus. Decreased ingestion and increased mucus production in the high concentration of cadmium contributed to a lower growth rate. Although cadmium had no effect on resting or active respiration, there were significant effects on total respiration as leeches exposed to high cadmium doses spent less time in active and, thus, in total respiration. Macroinvertebrates are generally good indicators of toxic stress in wetlands; but while phorate (a commonly used pesticide) killed all amphipods and chironomids, snails and leeches were tolerant and could survive heavy doses by behavioral adaptations and/or physiological tolerances (Dieter *et al.*, 1996). Leeches are also not good monitors of mercury (usually as methyl mercury) biomagnification and are not a dietary risk to waterfowl or fish (McNicol *et al.*, 1997). Studying polychlorinated biphenyls discharged to a lotic system in Ontario, fish and leeches occupying the top of the food web accumulated more PCBs than organisms lower in the food web, thus indicating that biomagnification (uptake through ingestion) through trophic transfer and not bioconcentration is the primary mechanism governing contamination (i.e., uptake as a chemical from the water) (Zaranko *et al.*, 1997).

V. COLLECTION AND REARING

A. Collection

Leech and acanthobdellid ectoparasites are sometimes found attached to hosts and are usually easily removed. Most leech species (but not *Acanthobdella peledina*) are free-living for the majority of their life cycle and are normally attached to a firm substrate or free swimming, as are the predatory leech species. They can be collected with bottom samplers and dip or plankton nets.

Free-living leeches can be sampled quantitatively using colonization chambers, grab samplers, or timed collections from rocks and stones. As a general rule, the plan area of the largest leech collected, or its area of escape response, should not exceed 5% of the sampler area (Green, 1979). The only benthic samplers that even approximately meet these criteria are grab samplers such as the Ekman grab (Ekman, 1911). Another problem faced when quantitatively sampling free-living leeches is that they occur in both the sediment and the vegetation. The Gerking box sampler (Gerking, 1957) is one of the few methods of quantitatively estimating population density, separable into these two components. This technique has, however, a number of technical problems including the difficulty of clipping macrophytes close to the substrate, the entrainment of sediments and associated fauna in the macrophyte sample, the invasive influence of the operator, and the difficulty of integrating the macrophyte and sediment samples because the benthic samples are taken from inside the area from which the macrophyte samples are collected. These problems are overcome by using a sampler consisting of a bottom mounted detachable grab connected to a box sampler with levered, spring-loaded jaws for cutting macrophytes (Gates *et al.* 1987). The sampler is triggered from above the water and simultaneously compartmentalizes the soft sediments in the grab and the phytomacrofauna in the water column above.

B. Rearing

Leeches are generally easy to maintain in the laboratory provided they are maintained under approximately similar conditions encountered in the field. Appropriate dissolved oxygen and temperature regimes are particularly important. Light regime is less critical, but if an ambient light regime cannot be provided, darkened conditions are preferable. Leeches are very susceptible to rapid changes in ionic content, temperature, and dissolved oxygen concentrations. If field conditions cannot be provided, they should be slowly acclimated to the new conditions (e.g., temperature changes exceeding 5°C per day will usually cause very high mortality, but a temperature change of 1°C per day within the normal field range will usually result in little mortality).

Predatory leech species are easily fed in the laboratory. If suitable prey at high densities are provided, most leeches will feed avidly when hungry. For maintenance, *ad libitum* feeding 1 day per week is sufficient.

Most predatory species brought into the laboratory and provided with ambient field conditions will feed, grow, and reproduce. The viability of the cocoons produced is sometimes decreased if the parent has been maintained in the laboratory for a long period. It is much more difficult to raise hatchlings produced in the laboratory through to mature size, capable of producing viable cocoons. Each species requires special conditions and there is much research to be done in this area.

Conversely, while some sanguivorous leeches are more difficult to feed in the laboratory, reproduction, cocoon production, and hatching are generally easier. Some sanguivores (e.g., *Placobdella papillifera*) will eat if immersed directly into blood. Many others (e.g., *Hirudo medicinalis*) will feed through a membrane if blood at an appropriate temperature (30–40°C) is on the other side. For these species, the simplest technique is to provide the hungry sanguivore with a sausage made from intestinal epithelium filled with warmed blood. Other species (e.g., *Theromyzon trizonare*) have not yet been induced to feed artificially and must be provided with a live host and given the opportunity to reach an appropriate feeding site (the nares of waterbirds for *T. trizonare*). However, when suitable sites are available, cocoon production occurs readily.

VI. IDENTIFICATION

A. Preparation of Specimens

Some species can be studied live, but permanently stained slides and serial sections are required for identification of a few species. Clearing, without permanent staining, is often required for determining the ocular number and arrangement.

Most leeches contract strongly when placed in cold fixative and should be narcotized first. Before narcotizing the specimens, the colors of the dorsal and ventral surfaces should be recorded, as chromatophores are dissolved or altered by most narcotizing agents. Leeches are generally difficult to relax properly, and any of the following three methods are recommended:

1. Add drops of 95% methanol slowly to the water containing the leech, gradually increasing the concentration for about 30 min until movement ceases. When the leech is limp and no longer responds to touch, pass it between the fingers to straighten it and remove excess mucous.
2. Add carbonated water or bubble in CO_2 until movement of the leech stops. Straighten the leech out on a slide and slowly add warm 70% ethanol and a few drops of glacial acetic acid until it is covered.
3. Add a drop or two of 6% nembutal until movements stop and then straighten the leech on a slide.

For some specimens, slight flattening is occasionally desirable. This is best done between two glass slides with weights added if necessary. After relaxation, the leeches should be fixed for 24 h (depending on size) in 5 or 10% formalin. For histological preparations, Fleming's or Bouin's fixatives should be used. Large leeches should be injected with 70% ethanol to ensure preservation of the internal organs.

To preserve a leech, wash the fixed specimen with deionized water to remove the formalin and then add either 70% ethanol or 5% buffered formalin. Although formalin preserves the colors longer, ethanol is recommended since it is safer to use and less destructive to soft tissues. Leeches fixed in formalin or Bouin's or Fleming's fixative can be stained in Mayer's paracarmine, borax carmine, or Hams' hematoxylin for 12–94 h, destained in a 1% HCl–70% ethanol solution until the leech epidermis is free of stain, and then neutralized in a 1% NH_4OH–70% ethanol solution. Fast green or eosin are routinely used as counterstains. Stained specimens should be dehydrated in progressively higher concentrations of ethanol, cleared in methyl salicylate, and mounted in a neutral pH mounting medium.

B. Important Features for Identification

Identification of leeches can be difficult, if not impossible, with poorly preserved specimens. External features such as the following are used whenever possible: annulation; number of eyes and their arrangement; presence or absence of papillae, tubercles, and pulsatile vesicles; relative size of anterior (oral) and posterior (anal) suckers; size and position of the mouth; and, the relative positions of the male and female gonopores. The only notes on coloration included in the following keys are those known to persist after preservation. In a few instances, reference is made to internal anatomical features. These have been kept to a minimum and are used only when positive identification is impossible without them. To distinguish between *Nephelopsis obscura* and the *Dina-Mooreobdella* complex, the atrial cornua must be examined (Fig. 18). The preserved specimen should be pinned out with the ventral surface up and a transverse incision made across the body three or four annuli) posterior to the male gonopore. Cuts should be made anteriorly up the lateral margins of the body for about 20 annuli) and the posterior edge of the flap lifted forward exposing the inner tissues, which must be cleared to expose the atrium. To help

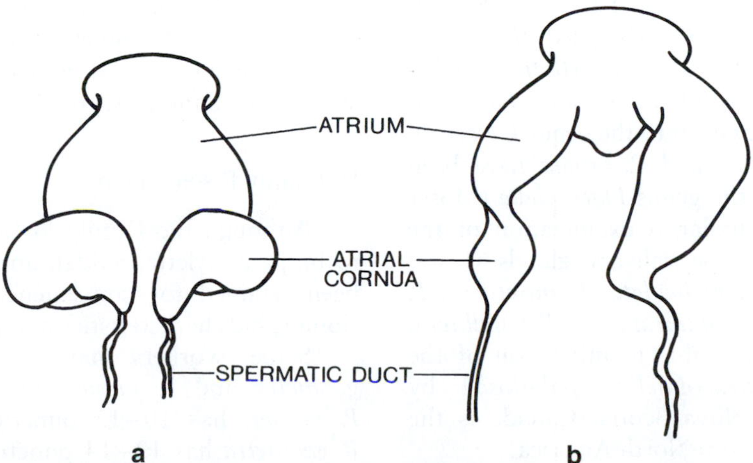

FIGURE 18 (a) Spirally curved atrial cornua of *Nephelopsis obscura*; and (b) the simply curved atrial cornua of *Dina/Mooreobdella*.

identify *Desserobdella* the salivary gland can be examined in specimens cleared with clove oil.

To examine the velum and jaws, the specimen should be pinned out and a median ventral incision made from the lower lip of the anterior sucker back far enough for the margins to be pinned out to expose the inner surface of the pharynx (Fig. 8). Details of the teeth (Fig. 9) can only be seen by removal of a jaw and making a temporary mount on a microscope slide.

The length of leeches and acanthobdellids is dependent on species, age, physiological state, and the degree of relaxation. The measurements presented in the keys are maximum total lengths of relaxed individuals.

C. Family Glossiphoniidae

Confusion exists in the systematics of these genera, and many species have undergone considerable synonymy. Although some taxonomists consider that the genus *Haementeria* (De Filippi, 1849) has priority over *Placobdella* (Blanchard, 1893), the latter name is widely recognized and accepted as a genus in North America and is well represented with nine species. Lukin (1976) recognized important differences between *Glossiphonia complanata* and *G. heteroclita*, and suggested that the latter species be placed in the subgenus *Alboglossiphonia* elevated to generic level by Klemm (1982). There has been considerable confusion among the species assigned to the genera *Batracobdella* Viguier and *Placobdella* Blanchard. Barta and Sawyer (1990) erected a new genus *Desserobdella*, differentiated by the presence of two pairs of coalesced eyes and one pair of diffuse salivary glands compared to the two pairs of pairs of compact salivary glands in *Placobdella* and a single pair of eyes in *Batracobdella*. *Placobdella* species primarily specialize on reptilian and crocodilian hosts whereas species of *Desserobdella* utilize amphibian or piscine hosts. Reexamination of type specimens by Barta and Sawyer (1990) and Jones and Woo (1990) assigned *Placobdella* (*Batracobdella*) *picta*, *Placobdella* (*Batracobdella*) *phalera*, *Batracobdella cryptobranchii* and *Batracobdella* (*Placobdella*) *michiganensis* to the genus *Desserobdella*. *Placobdella parasitica* and *P. ornata* have been confirmed as members of the genus *Placobdella* (Moser and Desser, 1996), but similar re-examination of the number and structure of the salivary glands is still needed for *P. hollensis*, *P. pediculata*, *P. montifera*, *P. nuchalis*, *P. translucens*, *P. mulilineata*, and *P. papillifera*.

The single and questionable identification of the European species *Batracobdella paludosa* by Pawlowski (1948) from Nova Scotia, Canada is the only record of this genus from North America.

Theromyzon tessulatum is characteristically a European species, but its presence in North America has been well documented (Davies, 1971, 1973, 1991; Klemm, 1982; Sawyer, 1986; Oosthuizen and Davies, 1993). The first species of *Theromyzon* reported from North America, described by Baird (1869) as *Glossiphonia rudis*, was based on a brief nondiagnostic description. Because Moore and Meyer (1951) and Meyer and Moore (1954) collected specimens from the same locality (Great Bear Lake) they assumed their specimens with the male and female gonopores separated by three annuli were *Theromyzon rude*. Thereafter, all specimens of *Theromyzon* collected in North America with three annuli between the gonopores have been identified as *T. rude*. However, examination and dissection of the syntypes (Oosthuizen and Davies, 1992) showed that one specimen was *Placobdella ornata* and the second, which became the holotype of *Glossiphonia rudis*, had only two annuli separating the gonopores. Specimens with three annuli between the gonopores were subsequently shown to belong to a new species *T. trizonare* and all specimens previously identified as *T. rude* are, in fact, *T. trizonare*. In North America, there are three species with two annuli between the gonopores (Oosthuizen and Davies, 1993): *T. maculosum*, *T. bifarium*, and *T. rude*. *Theromyzon maculosum*, characteristically a European species, is easily distinguished from *T. bifarium* and *T. rude* by the presence of two independent oviducts with two female pores compared to one female pore in the latter two species. *Theromyzon bifarium* is distinguished from *T. rude* by the presence of a cylindrical male atrium, paired dorso-lateral male ducts which enter the atrium separately. Although *T. sexoculatum* (Moore, 1898) and *T. biannulatum* (Klemm, 1977) have been described from North America with two annuli between the gonopores they are not valid species.

The genus *Actinobdella* was at first erroneously placed in the family Piscicolidae by Moore (1901), who was misled by the unusual annulation of *Actinobdella inequiannulata*. Complete segments of *Actinobdella* are 3-annulate (a1, a2, a3), which are further divided into six unequal annuli (b1, b2, b3, b4, b5, b6).

D. Family Piscicolidae

Although the Family Piscicolidae has been shown to be paraphyletic (Siddall and Burreson, 1995) it has been retained for convenience and is based on plesiomorphic characteristics and parasitism of fish.

Some workers have suggested that *Piscicola geometra* and *P. milneri* are synonymous, although *P. milneri* has 10–12 punctiform eyespots, whereas *P. geometra* has 12–14 punctiform eyespots and dark pigmented rays on the posterior sucker (Fig. 19c, d). Klemm (1977) indicated that the number of punctiform

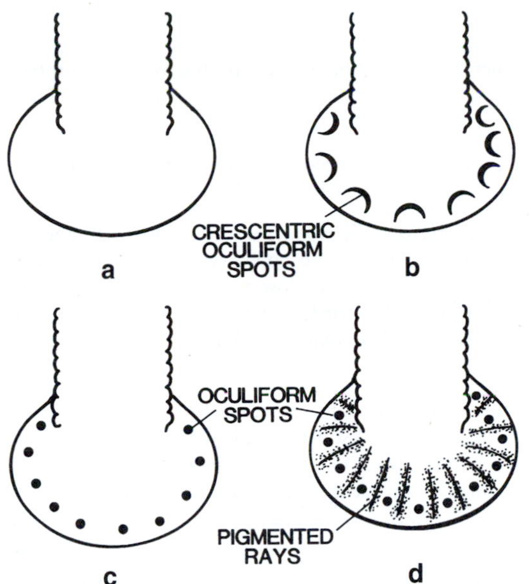

FIGURE 19 Posterior suckers of: (a) *Piscicola punctata* without oculiform spots; (b) *Piscicola salmositica* with 8–10 crescentiform oculiform spots; *Piscicola geometra* with 12–14 oculiform spots separated by pigmented rays; and (c) *Piscicola Milnera* with 10–12 punctiform oculiform spots on posterior sucker.

eyespots on the posterior sucker and the presence or absence of the pigmented rays varies with the age of the specimens. As the gonopores of *Piscicola milneri* are separated by two annuli, and by three annuli in *P. geometra*, they are here considered distinct species.

The genera *Myzobdella* and *Illinobdella* are very similar (Meyer, 1940; Daniels and Sawyer, 1973), but their relationship needs to be thoroughly investigated before the synonymy suggested by Daniels and Sawyer (1973) and Sawyer (1986) can be fully accepted. In the present chapter, a conservative approach has been used. If later evidence supports the propositions of Daniels and Sawyer (1973) and Sawyer (1986), then it will be relatively simple to combine the records for *Illinobdella* with those for *Myzobdella*. However, if supporting data are not forthcoming, subsuming *Illinobdella* with *Myzobdella* will result in an irretrievable loss of information.

E. Family Hirudinidae

Richardson (1969, 1971) divided the North American Haemopisoid leeches into the genera *Mollibdella, Bdellarogatis*, and *Percymoorensis*, rather than placing them all in the genus *Haemopis*. This was accepted by Soös (1969b), Davies (1971–1973), and Klemm (1972a), and is accepted here. Sawyer and Shelley (1976) and Sawyer (1986), for unspecified reasons, recombined them all back into the genus *Haemopis*, a view also adopted by Klemm (1977, 1982, 1985).

F. Family Erpobdellidae

There is some disagreement as to whether *Dina* and *Mooreobdella* are distinct genera (Moore, 1959) or subgenera of the genus *Dina* (Soös 1968). The diagnostic feature is whether the vasa deferentia form preatrial loops extending to ganglion XI (*Dina*) or do not have preatrial loops (*Mooreobdella*). This is difficult to determine, and thus in this key, the genera (or subgenera) are lumped together in a *Dina–Mooreobdella* complex, separated on external features. Although placed in the genus *Erpobdella* by Davies *et al.* (1985), *Motobdella montezuma* has several morphological traits which distinguish it from *Erpobdella* including posterior and crop ceca, large sensilla and the location of the anus next to the posterior sucker. *Motobdella sedonensis* can readily be distinguished from *M. montezuma* in having fewer papillae (1.0–14 vs 14–18) and in having the papillae on the dorsal surface not extending round the body as *M. montezuma*.

VII. TAXONOMIC KEYS

A. Euhirudinea and Acanthobdella

1a. Body divided into 29 postoral segments; segments subdivided superficially into four annuli (a1, a2, b5, b6) (Fig. 2b): no anterior sucker, mouth on ventral surface of segment III; no jaws; posterior suckering of four segments: two pairs of chaetae on five consecutive anterior segments; distal ends of chaetae bent to form hooks (Fig. 4b); chaetae absent from remainder of the body; paired ventral male gonopores: single median ventral male gonopore; (Fig. 4a) length 22 mm; ectoparasite of salmonids.................................. *Acanthobdella peledina* Grube, 1851

1b. Body divided into 32 postoral segments; segments subdivided superficially into 3–16 annuli (Figs. 2a and 31), anterior sucker present consists of four segments: mouth on ventral surface of anterior sucker (Fig. 7); jaws present or absent; posterior sucker consists of seven segments; chaetae absent from entire body; median ventral unpaired male and female gonopores (Figs. 3 and 29) ....Euhirudinea

B. Species of Euhirudinea

1a. Mouth a small pore on ventral surface of anterior sucker through which a muscular pharyngeal proboscis can be protruded (Fig. 7a, b); no jaws or teeth .... Rhynchobdellida 2

1b. Mouth large, occupying the entire cavity of anterior sucker (Fig. 7c); no protrusible proboscis; jaws with teeth either present or absent (Fig. 8) .... Arhynchobdellida 3

2a(1a) Body flattened dorso-ventrally and much wider than head (Fig. 1a) (except *Placobdella montifera* and *P. nuchalis*); body not cylindrical (except for *Helobdella elongata* which is subcylindrical); not differentiated into two body regions: anterior sucker ventral, more or less fused to body and narrower than body; body never divided into anterior trachelosome and posterior urosome; eggs in membranous cocoons and young brooded on ventral surface of parent; one, two, three or four pairs of eyes; no oculiform eye spots on posterior sucker; segments 3-annulate (a1, a2, a3) (Fig. 2a) (except *Oligobdella biannulata* which is 2-annulate) .... Family Glossiphoniidae 4

2b. Body cylindrical and usually long and narrow; body sometimes divided into a narrow anterior trachelosome and a wider posterior urosome (*Illinobdella* and *Piscicolaria*) (Figs. 1b, 6); anterior sucker expanded and distinct from body: zero, one, or two pairs of eyes; pulsatile vesicles along the lateral margins present (*Piscicola* and *Cystobranchus*) or absent; seven or more annuli per segment (except *Piscicolaria reducta* which is 3-annulate); oculiform eye spots sometimes present on posterior sucker (Figs. 6b, 19b–d); no brooding of cocoons or young .... Family Piscicolidae 37

3a(1b) Five pairs of eyes arranged in an arch on segments II–VI with the third and fourth pairs of eyes separated by one annulus (Fig. 20a); body elongate (Fig. 1c); jaws with teeth either present or absent; nine or ten pairs of testisacs arranged metamerically (Fig. 11a); pharynx short .... Family Hirudinidae 49

3b. Zero, three, or four pairs of eyes in separate labial and buccal groups (Fig. 25b); body elongate (Fig. 1d); no jaws; testisacs small and numerous (Fig. 11b); pharynx about one-third of body length .... Family Erpobdellidae 62

4a(2a) Posterior sucker conspicuous with a marginal circle of 30–60 glands and retractile papillae, their positions being indicated dorsally by faint radiating ridges (Fig. 21) .... *Actinobdella* 5

4b. Posterior sucker without a marginal circle of glands or retractile papillae 6

5a(4a) Posterior sucker on short, distinct pedicel with 29–31 digitate processes on rim (Fig. 21); somites 3- or 6-annulate; dorsal papillae in 1–5 longitudinal rows; length 22 mm .... *Actinobdella inequiannulata* Moore, 1901

5b. Posterior sucker on short distinct pedicel with about 60 digitate processes on rim; somites six-annulate with b_3 and b_5 the largest and most conspicuous; length 11 mm; dorsal papillae in five longitudinal rows .... *Actinobdella annectens* Moore, 1906

6a(4b) Zero, one, or two pairs of eyes; a series of paired accessory eyes sometimes present along body (Fig. 5c) 7

6b. Three or four pairs of eyes 30

7a(6a) Mouth apical or subapical on rim of anterior sucker (Fig. 7a); zero ,one or two pairs of eyes 8

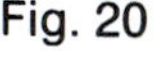

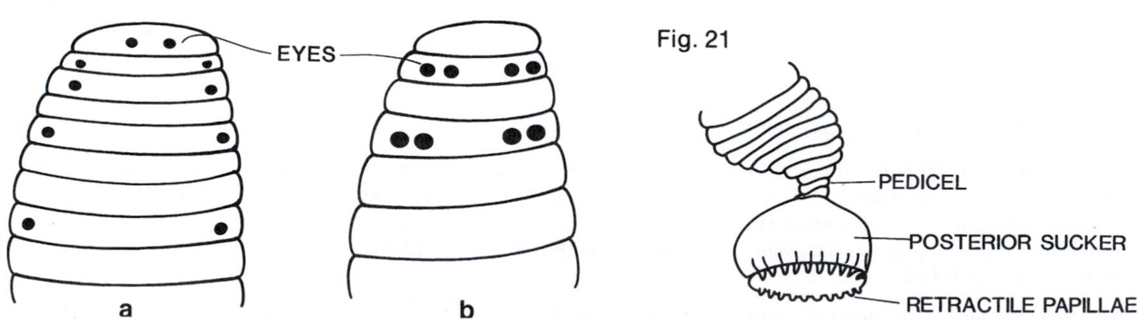

FIGURE 20 Arrangement of eyes in: (a) Hirudinidae; and (b) Erpobdellidae. *FIGURE 21* Lateral view of the posterior sucker of *Actinobdella inequiannulata* showing the pedicel and the retractile papillae around the margin.

7b. Mouth within anterior sucker and clearly not on rim (Fig. 7b); one or two pair of eyes; gonopores separated by one or two annuli (Fig. 3b) 18

8a(7a) Male and female gonopores united in common bursal pore; one pair of eyes well separated; body smooth without papillae; six pairs of crop ceca; length 22 m.................... *Marvinmeyeria lucida* Moore, 1954

8b. Male and female gonopores separated by two annuli (Fig. 3b); zero or one pair of eyes (Fig. 5a) [*Placobdella hollensis* has several pairs of accessory eyes on the neck (Fig. 5c)]; eyes close together or confluent (Fig. 5b) (except *Placobdella montifera* and *P. nuchalis* which have eyes well separated); body usually papillated; seven pairs of crop ceca .... *Placobdella/Oligobdella* 9

9a(8b) One small pair of eyes on segment II and one pair of larger sometimes coalesced eyes; no supplementary eyes 10

9b. One pair of small eyes on segment II and a larger sometimes coalesced pair of eyes on segment III followed by an indefinite number of pairs of accessory eyes (Fig. 5c); tubercles large and rough; length 30 mm.................... *Placobdella hollensis* (Whitman, 1892)

10a(9a) Anus between segments XXIII and XXIV with the 16 postanal annuli forming a slender stalk (pedicel) which bears the posterior sucker; no papillae; length 35 mm .... *Placobdella pediculata* (Hemingway, 1908)

10b. Anus close to the posterior sucker; no pedicel 11

11a(10b) Margins of posterior sucker denticulate (Fig. 22) head expanded and discoid and set off from body by a narrow neck (Fig. 23).....12

11b. Margins of posterior sucker not denticulate 13

12a(11a) Dorsum with three prominent tuberculate keels or ridges (Fig. 23); length 16 mm.................... *Placobdella montifera* Moore, 1906

12b. Dorsum smooth; no keels or ridges; length 25 mm .... *Placobdella nuchalis* Sawyer and Shelley, 1976

13a(11b) Medium longitudinal row of tubercles on dorsal surface are large and conspicuous; tubercles bear several papillae giving a rough, warty appearance; ventrum unstriped; length 40 mm.................... *Placobdella ornata* (Verrill, 1872)

13b. Dorsal papillae, if present, simple cones with smooth domes 14

14a(13b) Dorsum smooth; dorsal papillae inconspicuous or absent 15

14b. Dorsal papillae small and conical but distinct 17

15a(14a) Dorsum smooth; no papillae or tubercles; dorsum with conspicuous white genital and anal patches; one or more medial white patches; white bar on neck 16

15b. Dorsal papillae inconspicuous or absent; when present, papillae arranged in five longitudinal rows; ventral surface with 11–12 blue, brown, or green stripes; annulus a_3 in middle of body without distinct cross furrow; length 65 mm .... *Placobdella parasitica* (Say, 1824)

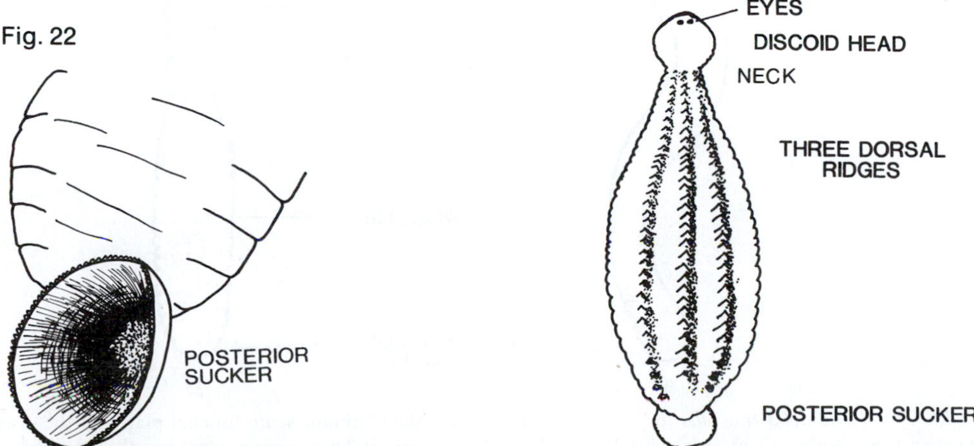

FIGURE 22 Posterior sucker of *Placobdella montifera* showing the denticulate margins. *FIGURE 23* *Placobdella montifera* showing the expanded discoid head and dorsum with three prominent tuberculate dorsal ridges (keels).

16a(15a) 3-annulate (Fig. 2a); posterior sucker not conspicuous; length 11 mm .... *Placobdella translucens* Sawyer and Shelley, 1976

16b. 2-annulate; posterior sucker large; length 7 mm .... *Oligobdella biannulata* (Moore, 1900)

17a(14b) Dorsum with 5–7 longitudinal rows of small papillae; papillae on posterior sucker; annulus a_3 in middle of body with distinct cross furrow; length 45 mm .... *Placobdella papillifera* (Verrill, 1872)

17b. Dorsum with few or numerous small papillae in five longitudinal rows; length 50 mm .... *Placobdella multilineata* Moore, 1953

18a(7b) One or two pairs of eyes; if only one part of eyes present, these are close together or lobed indicating coalescence (Fig. 5a, b); gonopores separated by two annuli (Fig. 3b) 19

18b. One pair of eyes which are well separated (Fig. 24); gonopores separated by one annulus .... *Helobdella* 23

19a(18a) One large pair of eyes on segment IV sometimes coalesced (Fig. 5a, b); seven pairs of crop ceca; papillae absent; length 20 mm .... *Batracobdella paludosa* (Carena, 1824)

19b. One pair of eyes on somite II and a larger pair of eyes on segment III eyes sometimes confluent (Fig. 5b); one pair of diffuse salivary glands; two annuli separating gonopores (Fig. 3b) .... *Desserobdella* 20

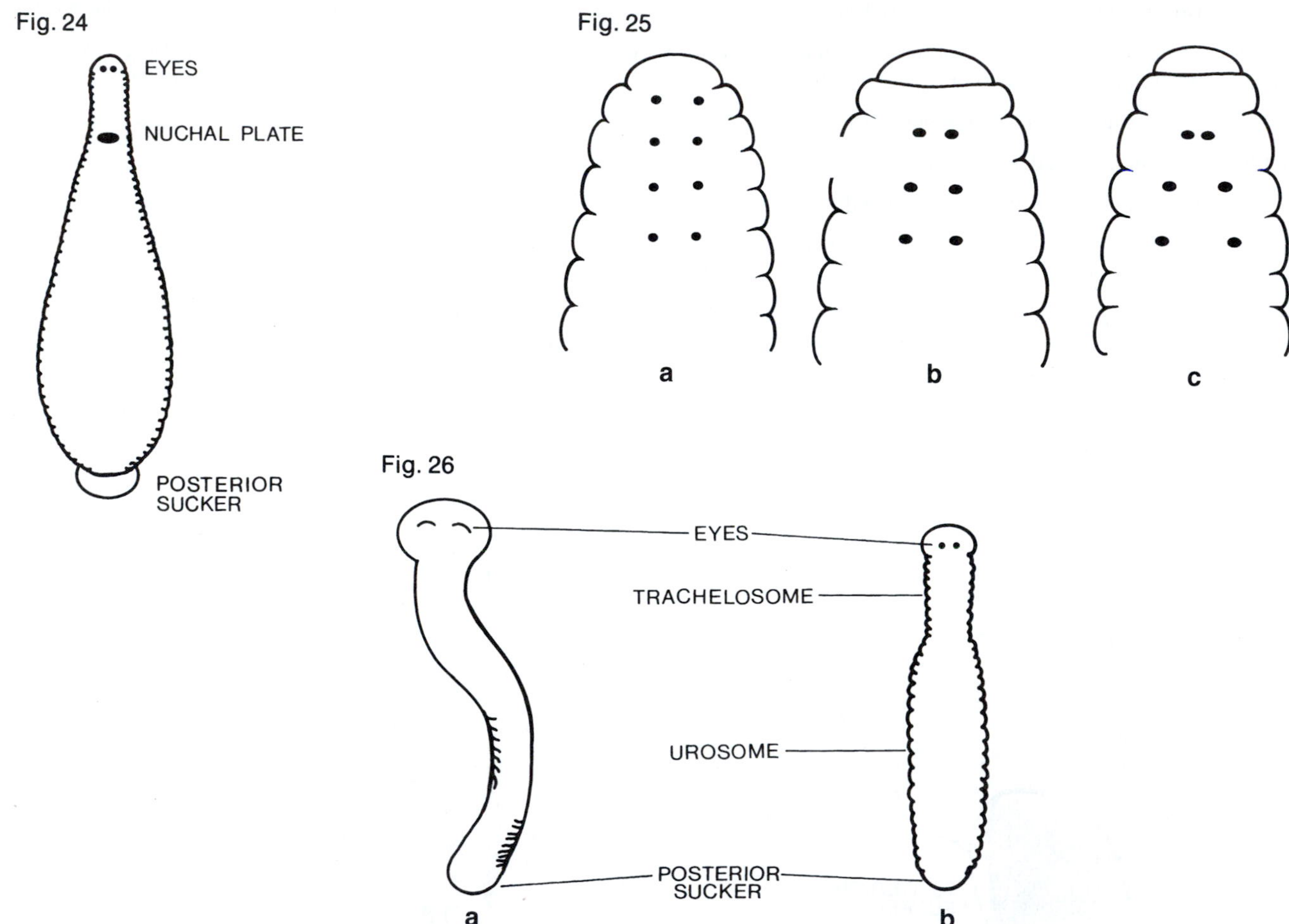

FIGURE 24 *Helobdella stagnalis* (or *H. californ*ia) showing the chitinous scute (nuchal plate) on the dorsal surface and the single pair of eyes. *FIGURE 25* Dorsal view of: (a) *Theromyzon trizonare*; (b) *Glossiphonia complanata*; and (c) *Alboglossiphonia heteroclita* showing the arrangement of the eyes. *FIGURE 26* (a) *Myzobdella lugubris*; and (b) *Piscicolaria reducta* showing the trachelosome and urosome without pulsatile vesicles, the relative sizes of the eyes, and the weakly developed posterior sucker.

20a(19b) Dorsum with white genital and anal patches, one or more medial white patches and a white bar on neck; no papillae on dorsum; five longitudinal rows of white prominences surrounded by yellowish dots, equidistant longitudinally and transversely; body very flattened; length 10 mm .... *Desserobdella michiganensis Sawyer, 1972*

20b. Dorsum without white patches and bar on neck 21

21a(20b) Posterior sucker separated from body on a short pedicel; dorsum smooth; eight rows of inconspicuous sensillae on dorsal annuli; only recorded parasitism on *Cryptobranchus alleganiensis*; length 17 mm .... *Batracobdella (Desserobdella) cryptobranchii* Johnson and Klemm, 1977

21b. Posterior sucker not on pedicel; posterior sucker small; low small papillae present; 3-annulate (Fig. 3a); length 25 mm 22

22a(21b) Dorsal papillae present; lateral metameric and dorsal color markings; 60 radiating papillaeon rim of posterior sucker; feeds mainly on fish; length 10 mm .... *Deserobdella phalera* (Graf, 1899)

22b. No dorsal papillae; no lateral metameric and dorsal colour markings; no papillae on rim of posterior sucker; feeds mainly on amphibians; length 5 mm .... *Desserobdella picta* (Verrill, 1872)

23a(19b) Brown, horny, chitinous scute (nuchal plate) on the dorsal surface of segment VIII (Fig. 24) 24

23b. Without a nuchal plate 25

24a(23a) Pigment pattern arranged in longitudinal stripes on dorsum; six branched pairs of crop ceca; posterior sucker pigmented on dorsum; length 18 mm .... *Helobdella californía* Kutschera, 1988

24b. No longitudinal stripes on dorsum; six unbranched pairs of crop ceca, the last pair posteriorly; posterior sucker unpigmented; length 14 mm .... *Helobdella stagnalis* (*Linnaeus*, 1758)

25a(23b) Dorsal surface smooth 26

25b. Dorsal surface with 3–7 longitudinal series of papillae, or with scattered papillae 28

26a(25a) Body unpigmented and translucent; body rounded and subcylindrical; lateral margins almost parallel; posterior sucker small and terminal; one pair of crop ceca; length 25 mm .... *Helobdella elongata* (Castle, 1900)

26b. Body pigmented, with or without longitudinal or transverse bands; body flat with posterior wider than tapering anterior; six pairs of crop ceca 27

27a(26b) Dorsum without transverse pigmentation; dorsum with six prominent longitudinal white stripes alternating with brown stripes; length 14 mm .... *Helobdella fusca* (Castle, 1900)

27b. Dorsum with transverse brown interrupted stripes alternating with irregular white banks; length 10 mm .... *Helobdella transversa* Sawyer, 1972

28a(25b) Dorsal surface with 5–9 longitudinal rows of papillae; papillae large, conspicuous, and rounded; lightly or unpigmented; length 14 mm .... *Helobdella papillata* (Moore, 1906)

28b. Dorsal surface with three or fewer incomplete series of small papillae or with scattered papillae; length 25–30 mm 29

29a(28b) Ratio of body width to body length 0.26 ± 0.02 SD. Dorsal surface with longitudinal strips many of which are interrupted by circular zones of unpigmented skin. Five pairs of crop caeca which have numerous secondary diverticula with a bumpy outline .... *Helobdella triserialis* (Blanchard, 1849)

29b. Ratio of body width to body length 0.37 ± 0.01 SD. Dorsal surface with five narrow longitudinal stripes which extend unbroken along the body. Five pairs of crop ceca which are bilobed and smooth in outline .... *Helobdella robusta* Shankland, Martindale, Nardelli-Haefinger, Baxter and Price, 1992

30a(6b) Four pairs of eyes on paramedian lines of segments II–V (Fig. 25a); body very soft .... *Theromyzon* 31

30b. Three pairs of eyes (Fig. 25b, c) (coalesced eyes sometimes occur, but the lobed nature indicates the original condition); body firm 35

31a(30a) Gonopores separated by two annuli 33

31b. Gonopores separated by three or four annuli 32

32a(31b) Gonopores separated by three annuli cocoons attached to ventral body wall, two female pores .... *Theromyzon trizonare* Davies and Oosthuizen, 1992

32b. Gonopores separated by four annuli cocoons attached directly to substrate, single female pore .... *Theromyzon tessulatum* (Müller, 1776)

33a(31a) Two female pores .... *Theromyzon maculosum* (Rathke, 1862)

33b. One female pore 34

34a(33b) Female atrium cylindrical .... *Theromyzon bifarium* Oosthuizen and Davies, 1993

34b. Female atrium spherical .... *Theromyzon rude* (Baird, 1869)

35a(30b) First pair of eyes closer together than succeeding two pairs, i.e., eyes arranged in triangular pattern (Fig. 25c); no papillae; male and female ducts open into a common gonopore; little pigmentation; generally amber colored; length 10 mm .... *Alboglossiphonia heteroclita* (Linnaeus, 1761)

35b. Eyes equidistant in two paramedian rows (Fig. 25b) 36

36a(35b) Dorsum sometimes with papillae on annulus a_2 in six longitudinal rows; pair of paramedial stripes on dorsum and ventrum; six pairs of crop ceca; gonopores separated by two annuli (Fig. 3b); length 25 mm .... *Glossiphonia complanata* (Linnaeus, 1758)

36b. Dorsum with large, distinct papillae on annuli a_2 and a_3; dorsum with numerous, irregularly shaped whitish spots; seven pairs of crop ceca; length 25 mm .... *Boreobdella verrucata* (Müller, 1844)

37a(2b) Posterior sucker flattened, as wide or wider than the widest part of body (Fig. 1b); pulsatile vesicles on lateral margins of neural annuli of urosome (Fig. 6a, b); zero or two pairs of eyes 38

37b. Posterior sucker concave, weakly developed, and narrower than widest part of the body; no pulsatile vesicles (Fig. 26); zero or one pair of eyes 45

38a(37a) Body divided into anterior trachelosome and posterior urosome; 11 pairs of pulsatile vesicles not very apparent in preserved specimens; each pulsatile vesicle covers two annuli; body cylindrical or sometimes slightly flattened; two pairs of eyes; both anterior and posterior suckers wider than body; oculiform spots present on posterior sucker (Fig. 19b, c, d))(except *Piscicola punctata*) (Fig. 19a): 14-annulate (except *Piscicola punctata* which is 3-annulate) .... *Piscicola* 39

38b. Body divided into distinct anterior trachelosome and posterior urosome; 11 pairs of large and distinct pulsatile vesicles easily seen in preserved and live specimens (Figs. 1b and 6); each pulsatile vesicle covers four annuli; well-developed anterior and posterior suckers; no papillae; zero or two pairs of eyes (Fig. 6); oculiform spots on posterior sucker present or absent (Fig. 9a); 7-annulate; length 80 mm .... *Cystobranchus* 42

39a(38a) Posterior sucker with 8–14 oculiform spots (Fig. 19b, c, d) 40

39b. Posterior sucker without oculiform spots (Fig. 19a); 3-annulate (Fig. 2a); two pairs of crescent-shaped eyes: gonopores separated by three or four annuli; length 16 mm .... *Piscicola punctata* (Verrill, 1872)

40a(39a) Posterior sucker with 10–14 punctiform oculiform spots (Fig. 19c, d) 41

40b. Eight to ten crescent-shaped oculiform spots on the posterior sucker (Fig. 19b); gonopores separated by two annuli; 14-annulate length 31 mm .... *Piscicola salmositica* Meyer, 1946

41a(40a) With 10–12 (usually 10) punctiform oculiform spots on posterior sucker; dark rays absent from posterior sucker (Fig. 19c): gonopores separated by two annuli (Fig. 3a); anterior pair of eyes like heavy dashes twice as long as wide; length 24 mm *Piscicola milneri* (Verrill, 1874)

41b. With 12–14 punctiform, oculiform spots on posterior sucker, separated by an equal number of dark pigmented rays (Fig. 19d); gonopores separated by three annuli; anterior pair of eyes like fine dashes five times as long as wide; length 30 mm ..*Piscicola geometra* (Linnaeus, 1758)

42a(38b) With oculiform spots on posterior sucker 43

42b. Without oculiform spots on posterior sucker 44

43a(42a) Eight oculiform spots on posterior sucker; two rows of 12 lateral oculiform spots on each side of body (Fig. 6b); two pairs of eyes; length 7 mm .... *Cystobranchus meyeri* Hayunga and Grey, 1976

43b. Ten oculiform spots on posterior sucker; no lateral ocelli; two pairs of eyes; length 15 mm .... *Cystobranchus virginicus* Hoffman, 1964

44a(42b) Two pairs of eyes; the first pair forming conspicuous dashes at 45° to the longitudinal axis; the second part of eyes ovoid (Fig. 6a); gonopores separated by two annuli (Fig. 3b); length 30 m .... *Cystobranchus verrilli* Meyer, 1940

44b. No eyes; length 30 mm .... *Cystobranchus mammillatus* (Malm, 1863)

45a(37b) Body divided into small trachelosome and larger urosome (Fig. 26b); length-to-width ratio 3–6:1; one pair of eyes; 12- or 14-annulate .... *Myzobdella lugubris* Leidy, 1851

45b. Body not divided into trachelosome and urosome (Fig. 26a); zero or one pair of eyes 46

46a(45b) Length-to-width ratio $\geqq$ 10:1; zero or one pair of eyes in posterior half of the head; 12- or 14-annulate .... *Illinobdella* 47

46b. Length-to-width ratio 4–5:1 (Fig. 26b); one pair of eyes; 3-annulate (Fig. 2a); dorsum with six black/brown longitudinal stripes; length 8 mm .... *Piscicolaria reducta* Meyer, 1940

47a(46a) Length-to-width ratio about 10:1; 14-annulate .... *Illinobdella alba* Meyer, 1940

47b. Length-to-width ratio ≥15:1 48

48a(47b) Length-to-width ratio 15:1; one pair of eyes (sometimes absent); anus 15 annuli anterior to posterior sucker; 14-annulate; posterior sucker little more than concavity of the posterior body .... *Illinobdella richardsoni* Meyer, 1940

48b. Length-to-width-ratio 18:1; one pair of eyes; anus ten annuli or less anterior to posterior sucker; 12-annulate; posterior sucker half as wide as posterior body .... *Illinobdella elongata* Meyer, 1940

49a(3a) With copulatory gland pores on the ventral surface, 10 or 11 annuli posterior to the male gonopores (Figs. 27 and 28) .... *Macrobdella* 50

49b Ventral copulatory gland pairs absent 53

50a(49a) With 24 copulatory gland pores (four rows with six gland pores each) (Fig. 28a) on raised pads; dorsum with red/orange spots; two to two-and-a-half annuli between gonopores; length 150 mm .... *Macrobdella sestertia* Whitman, 1886

50b. With four, six, or eight copulatory gland pores (Fig. 28); with or without red/orange spots on dorsum 51

51a(50b) Eight copulatory gland pores (Fig. 28b) (2 rows of 4); two annuli between gonopores; dorsum without red/orange spots; length 150 mm .... *Macrobdella ditetra* Moore, 1953

51b. Four or six copulatory gland pores; dorsum with red/orange spots 52

52a(51b) Four copulatory gland pores (Fig. 27) (two rows of two); five to five and-a-half annuli between gonopores; jaws well developed with about 65 monostichodont acute teeth (Fig. 9a); 10 pairs of testisacs; ventrum red/orange; length 150 mm .... *Macrobdella decora* (Say, 1824)

52b. Six copulatory gland pores (Fig. 28c) (three rows of two); four or five annuli between gonopores; ventrum yellow/grey; length 150 mm .... *Macrobdella diplotertia* Meyer, 1975

53a(49b) Glandular area around gonopores; gonopores separated by three or four annuli .... *Philobdella* 54

53b. No glandular area around gonopores; gonopores separated by 5–7 annuli 55

54a(53a) 20–26 distichodont teeth per jaw (Fig. 9b); lateral margins of dorsum without discrete spots; length 85 mm .... *Philobdella floridana* (Verrill, 1874)

54b. 35–48 distichodont teeth per jaw; seven pairs of testisacs; irregular black spots on margins of dorsum; length 85 mm .... *Philobdella gracilis* Moore, 1901

55a(53b) Jaws absent (Fig. 8a) 56

55b. Jaws present (Fig. 8b, c) and denticulate (Fig. 9) 57

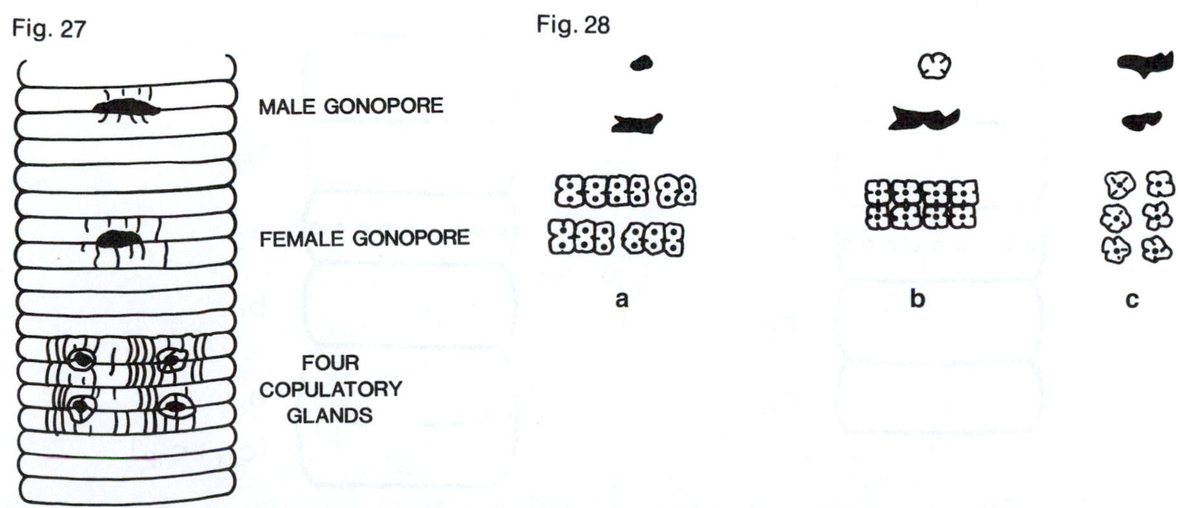

FIGURE 27 Ventral view of the male and female gonopores of *Macrobdella decora*. *FIGURE 28* Diagrammatic representation of the relative slopes and positions of the male and female gonopores and copulatory glands of: (a) *Macrobdella sestertia*; (b) *Macrobdella ditetra*; and (c) *Macrobdella diplotertia*.

56a(55a) Lower surface of velum smooth (Fig. 8a); gonopores in the furrows between the annuli and separated by five annuli; pharynx with 12 internal ridges; length 300 mm .... *Mollibdella grandis* (Verrill, 1874)

56b. Lower surface of velum papillate; gonopores in the middle of annuli and separated by five annuli; pharynx with 15 internal ridges .... *Bdellarogatis plumbeus* (Moore, 1912)

57a(55b) Jaws small and retractable into narrow-mouthed tubular pits; 9–25 distichodont (Fig. 9b) teeth per jaw .... *Percymoorensis* 58

57b. Jaws large; 35–100 acute monostichodont (Fig. 9a) teeth; length 100 mm .... *Hirudo medicinalis* Linnaeus, 1758

58a(57a) Gonopores separated by five to five-and-a-half annuli; female gonopore small 59

58b. Gonopores separated by six and-a-half to seven annuli; female gonopore large and conical length 200 mm .... *Percymoorensis septagon* Sawyer and Shelley, 1976

59a(58a) Dorsum with median black stripe 60

59b. Dorsum with irregular, scattered black spots 61

60a(59a) Dorsum uniformly black or slate gray; posterior sucker narrower than body; jaws with 20–25 pairs of teeth; length 250 mm .... *Percymoorensis lateralis* (Say, 1824)

60b. Dorsum brown/green; posterior sucker as wide as body; 9–14 pairs teeth .... *Percymoorensis kingi* (Mather, 1954)

61a(59b) Jaws with 10–12 teeth; posterior sucker about 75% of maximal body width and attached by very short pedicel; length 75 mm .... *Percymoorensis lateromaculata* (Mather, 1963)

61b. Jaws with 12–16 pairs of teeth; posterior sucker about half maximum body width; length 150 mm .... *Percymoorensis marmorata* (Say, 1824)

62a(3b) Somites 5-annulate (b_1, b_2, a_2, b_5, b_6) (Fig. 29a) with all annuli equal in length; three pairs of eyes; gonopores separated by two annuli (Fig. 3b); length 100 mm 63

62b. Somites six or seven-annulate (Fig. 29b); annuli of unequal length units will be either subdivided or longer than the others; in any group of six consecutive annuli at least one annulus narrower or wider than the others; three or four pairs of eyes 65

63a(62a) Eyes all similar in size; annuli not raised on dorsum and without papillae; anus located three or four segments anterior to posterior sucker; mouth small; gonopores in furrows separated by 2–5 annuli .... *Erpobdella punctata* (Leidy, 1870)

63b. Eyes differ in size with the second and third pairs smaller than the anterior pair; each annulus raised on dorsum with 10–18 small white-tipped papillae; anus located at base of posterior sucker; one or two pairs of crop ceca gonopores in furrows separate by two annuli .... *Motobdella* 64

64a(63b) 14–18 small, white-tipped papillae in a ring about each annulus; mouth very large (4.0 mm); two pairs of crop ceca nephridiopores not visible, internal surface of posterior sucker pigmented .... *Motobdella montezuma* Davies, Singhal and Blinn, 1985

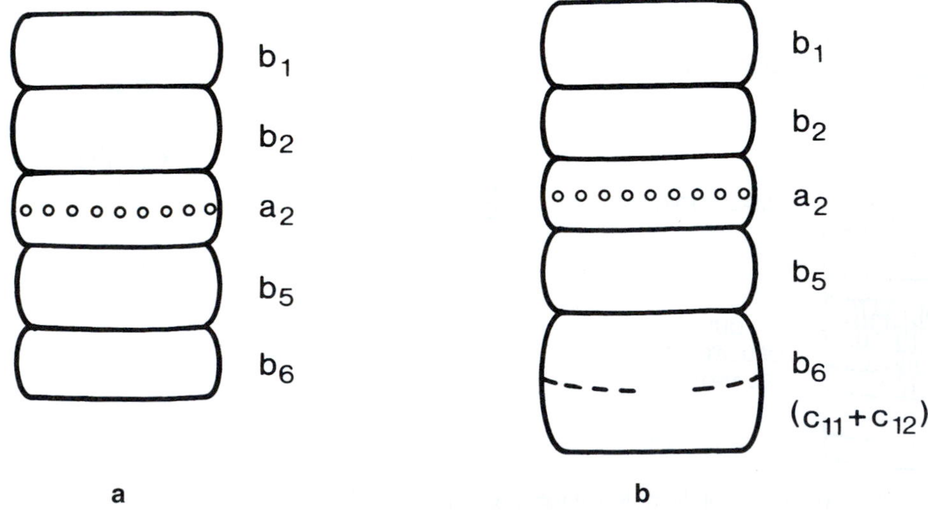

FIGURE 29 (a) Annulation of *Erpobdella* with all annuli (b_1, b_2, a_2, b_5, b_6) equal in length; and (b) *Nephelopsis, Dina*, or *Mooreobdella* with the annuli (b_1, b_2, a_2, b_5, c_{11}, c_{12}) of unequal lengths.

64b. 10–14 small, white-tipped papillae on dorsal surface only of each annulus; mouth 1.1 mm in diameter; nephridiopores visible as small white spots; internal surface of posterior sucker not pigmented; one or two pars of crop ceca................*Motobdella sedonensis* Govedich, Blinn, Keim and Davies, 1998

65a(62b) Four pairs of eyes, two labial and two buccal of similar size 66

65b. Zero or three pairs of eyes ....Dina/Mooreobdella 68

66a(65a) Two annuli between gonopores (Fig. 3a); anus one segment anterior to poster sucker; atrial cornua spirally coiled like the horn of a ram (Fig. 18a); length 100 mm..................*Nephelopsis obscura* Verrill, 1872

66b. Three or more annuli between gonopores; atrial cornua simply curved (Fig.18b)..................*Dina* 67

67a(66b) Three-and-a-half to four annuli between gonopores; body heavily blotched with a median stripe; anus large, opening on a conical tubercle; length 60 mm..................*Dina dubia* (Moore and Meyer, 1951)

67b. Three to three-and-a-half annuli between gonopores; nearly pigmentless or with a few dark spots; anus small, not on tubercle; length 30 mm..................*Dina parva* Moore, 1912

68a(65b) Dorsum lacking scattered black pigment..................69

68b. Dorsum with scattered black pigment; two annuli between gonopores (Fig. 3a); length 55 mm ....*Mooreobdella melanostoma* Sawyer and Shelley, 1976

69a(68a) Two to two-and-a-half annuli between gonopores..................70

69b. Three to four-and-a-half annuli between gonopores 72

70a(69a) Male gonopore surrounded by circle of papillae; zero or three pairs of eyes; two annuli between gonopores (Fig. 3a); length 15 mm *Dina anoculata* Moore, 1898

70b. No papillae around male gonopore; three (rarely four) pairs of eyes 71

71a(70b) Two annuli between gonopores (Fig. 3a); three (rarely four) pairs of eyes; length 50mm..................*Mooreobdella fervida* Verrill, 1872

71b. Either 2 or 2 1/2 annuli between gonopores; three pairs of eyes; length 30 mm..................*Mooreobdella bucera* Moore, 1949

72a(69b) Three pairs of eyes; gonopores separated by *three* annuli; length 50 mm ....*Mooreobdella microstoma* (Moore, 1901)

72b. Gonopores separated by 4–4 1/2 annuli; three pairs of eyes; length 40 mm............*Mooreobdella tetragon* Sawyer and Shelley, 1976

ACKNOWLEDGMENTS

It is a pleasure to acknowledge the technical assistance and patience of Mrs Thelma Thompson in the preparation of this chapter. We would also like to acknowledge Dr. Bonnie Bain for her assistance in proofreading and reviewing this chapter and key.

VIII. LITERATURE CITED

Able, K. W. 1976. Cleaning behaviour in the cyprinodont fishes: *Fundus majalis*, *Cyprinodon variegatus* and *Lucania parva*. Chesapeake Science 17:35–39.

Anderson, R. S., Raasveldt. L.G 1974. *Gammarus* predation and the possible effects of *Gammarus* and *Chaoborus* feeding on the zooplankton composition in some small lakes and ponds in western Canada. Canadian Wildlife Service Occasional Paper No. 18.

Angstadt, J. D., Moore. W.H. 1997. A circadian rhythm of swimming behavior in a predatory leech of the Family Erpobdellidae. The American Midland Naturalist 137:165–172.

Anholt, B. R. 1986. Prey selection by the predatory leech *Nephelopsis obscura* in relation to three alternative models of foraging. Canadian Journal of Zoology 64:649–555.

Anholt, B. R., Davies, R. W. 1987. Effects of hunger level on the activity of the predatory leech *Nephelopsis obscura* Verrill (Hirudinoidea: Erpobdellidae). American Midland Naturalist 177:307–311.

Arnold, S. J. 1981. Behavioural variation in natural populations. 1. Phenotypic, genetic and environmental correlations between chemoreceptive responses to prey in the garter snake *Thamnophis elegans*. Evolution 35:489–509.

Babin, J., Prepas. E. E. 1985. Modeling winter oxygen depletion rates in ice-covered temperate zone lakes in Canada. Canadian Journal of Fisheries and Aquatic Sciences 42:239–249.

Baird, D. J., Linton, L. R., Davies. R. W. 1986. Life-history evolution and post-reproductive mortality risk. Journal of Animal Ecology 55:295–302.

Baird, D. J., Gates, T. E., Davies. R. W. 1987a. Oxygen conditions in two prairie pothole lakes during winter ice cover. Canadian Journal of Fisheries and Aquatic Sciences 44:1092–1095.

Baird, D. J., Linton, L. R., Davies. R. W. 1987b. Life-history flexibility as a strategy for survival in a variable environment. Functional Ecology 1:45–48.

Baird, W. 1869. Descriptions of some new suctorial annelids in the collection of the British Museum. Proceedings of the Zoological Society of London. pp. 310–318.

Barta, J. R., Sawyer, R. T. 1990. Definition of a new genus of glossiphoniid leech and a redescription of the type species,

Clepsine picta Verrill, 1872. Canadian Journal of Zoology 68:1942–1950.

Bartonek, J. C. 1972. Summer foods of American Widgeon, Mallards and Green-winged Teal near Great Slave Lake, Northwest Territories. Canadian Field Naturalist 86:373–376.

Bartonek, J. C., Hickey, J. J. 1969a. Selective feeding by juvenile diving ducks in summer. The Auk 86:457493.

Bartonek, J. C., Hickey, J. J. 1969b. Food habits of canvasbacks, redheads and lesser scaup in Montana. Condor 71:280–290.

Bartonek, J. C., Murdy, H. W. 1970. Summer foods of lesser scaup in subarctic taiga. Arctic 23:35–44.

Bartonek, J. C., Trauger, D. L. 1975. Leeches (Hirudinea) infestations among waterfowl near Yellowknife, Northwest Territories. Canadian Field Naturalist 89:234–243.

Beadle, L. D. 1939. Regulation of the haemolymph in the saline water mosquito larva *Aedes detritus* (Edw.). Journal of Experimental Biology 16:346–362.

Bendell, B. E., McNicol, D. K. 1991. An assessment of leeches (Hirudinea) as indicators of lake acidification. Canadian Journal of Zoology 69:130–133.

Bennike, S. A. B. 1943. Contributions to the ecology and biology of the Danish freshwater leeches. Folia Limnologica Scandinavica 2:1–109.

Biernacka, B., Davies, R. W. 1994. The effects of temperature and feeding regime on hermaphroditism and gametogenesis of *Nephelopsis obscura*, a freshwater predatory leech. Canadian Journal of Zoology 72:1639–1642.

Biernacka, B., Davies, R. W. 1995. Is a visible clitellum an index of sexual maturity in erpobdellid leeches? Invertebrate Reproduction and Development 28:97–101.

Blair, W. N. 1927. Notes on *Hirudo medicinalis*, the medicinal leech, as a British species. Proceedings of the Zoological Society of London 11:999–1002.

Blanchard, E. 1849. Annelides. Hirudineanos. Gay's historia fisca y politico de Chile. Zoologia, Paris 3:43–50.

Blanchard, R. 1893. Courtes notices sur les Hirudinees: X. Hirudinees de l'Europe boreale. Bulletin Societe Zoologique de France 18:93–94.

Blinn, D. W., Davies, R. W. 1989. The evolutionary importance of mechanoreception in three erpobdellid leech species. Oecologia 79:6–9.

Blinn, D. W., Davies, R. W. 1990. Concomitant diel vertical migration of a predatory leech and its amphipod prey. Freshwater Biology 24:401–407.

Blinn, D. W., Davies, R. W., Dehdashti, B. 1987. Specialized pelagic feeding by *Erpobdella montezuma* (Hirudinea). Holarctic Ecology 10:235–340.

Blinn. D. W., Pinney, C. Wagner, V. T. 1988. Intraspecific discrimination of amphipod prey by a freshwater leech through mechanoreception. Canadian Journal of Zoology 66:427–430.

Blinn, D. W., Dehdashti, B. Runck, C. Davies, R. W. 1990. The importance of prey size and density in an endemic predator–prey couple (leech *Erpobdella montezuma*–amphipod *Hyalella montezuma*). Journal of Animal Ecology 59:187–192.

Briggs, D. E. 1991. Extraordinary fossils. American Scientist 79:130–141.

Bulmer, M. G. 1985. Selection for iteroparity in a variable environment. American Naturalist 136:63–71.

Bur, M.T. 1994. Incidence of the leech *Actinobdella pediculata* on freshwater drum in Lake Erie. Journal of Great Lakes Research 20:768–770.

Büsing, K. H. 1951. *Pseudomonas hirudinis*, ein bakterieller Darmsybiont des Blutegels (*Hirudo offcinalis*). Zentralblatt fuer Bakterienologie Parasitenkunde 157:478–484.

Büsing, K. H., Döll, W., Freytag, K. 1953. Die Bakterienflora der medizinischen Blutegel. Archiv fur Mikrobiologie 19:52–86.

Calow, P., Riley, H. 1982. Observations on reproductive effort in British erpobdellid and glossiphoniid leeches with different life cycles. Journal of Animal Ecology 51:697–712.

Carena, H. 1824. Monographic due genre *Hirudo*. Supplement. Memoire Accademia Science, Torino 28:331–337.

Casselman, J. M., Harvey, H. H. 1975. Selective fish mortality resulting from low winter oxygen. Internationale Vereinigung für Theoretische und Angewandte Limnologie 19:2418–2429.

Castle, W. E. 1900. Some North American fresh-water Rhynchobdellidae, and their parasites. Bulletin of the Museum of Comparative Zoology. Harvard 36:17–64.

Croghan, P. C. 1958. The survival of *Artemia salina* (Linn.) in various media. Journal of Experimental Biology 35:213–216.

Cywinska, A., Davies, R. W. 1989. Predation on the erpobdellid leech *Nephelopsis obscura* in the laboratory. Canadian Journal of Zoology 67:2689–2693.

Daborn, G. R. 1976. Colonization of isolated aquatic habitats. Canadian Field-Naturalist 90:56–57.

Dahl, J. 1998. The impact of vertebrate and invertebrate predators on a stream benthic community. Oecologia 117:217–226.

Dahl, J., Greenberg, L. 1997. Foraging rates of a vertebrate and an invertebrate predator in stream enclosures Oikos 78:459–466.

Dahm, A. G. 1962. Distribution and biological patterns of *Acanthobdella peledina* Grube from Sweden (Hirudinea, Acanthobdellidae). Acta Universitatis Lundensis Arrskrift 58:1–35.

Daniels, B. A., Sawyer R. T. 1973. Host-parasite relationship of the fish leech *Illinobdella moorei* Meyer and the white catfish *lctalurus catus* (Linnaeus). Bulletin of the Association of Southeastern Biologists 20:48.

Davies, R. W. 1969. The production of antisera for detecting specific triclad antigens in the gut contents of predators. Oikos 20:248–260.

Davies, R. W. 1971. A key to the freshwater Hirudinoidea of Canada. Journal of the Fisheries Research Board of Canada 28:543–552.

Davies, R. W. 1972. Annotated bibliography to the freshwater leeches (Hirudinoidea) of Canada. Fisheries Research Board of Canada., Technical Report No. 306.15 pp.

Davies, R. W. 1973. The geographic distribution of freshwater Hirudinoidea in Canada. Canadian Journal of Zoology 51:531–545.

Davies, R. W. 1978. Reproductive strategies shown by freshwater Hirudinoidea. Internationale Vereinigung für Theoretische und Angewandte Limnologie 20: 2378–9381.

Davies, R. W. 1979. Dispersion of freshwater leeches (Hirudinoidea) to Anticosti Island, Quebec. Canadian Field-Naturalist 93: 310–313.

Davies, R. W. 1984. Sanguivory in leeches and its effects on growth, survivorship and reproduction of *Theromyzon rude*. Canadian Journal of Zoology 62:589–593.

Davies, R. W. 1987. All about leeches. *Nature (London)* 325:585.

Davies, R. W. 1991. Annelida: Leeches, Polychaetes, and Acanthobdellids, *in*: Thorp, J. H. Covich, A. P. Eds. Ecology and Classification of North American Freshwater Invertebrates. Academic Press, San Diego.

Davies, R. W., Everett, R. P. 1977. The life history, growth and age structure of *Nephelopsis obscura* Verrill, 1872 (Hirudinoidea) in Alberta. Canadian Journal of Zoology 55:620–627.

Davies, R. W., Gates, T. E. 1991a. Intra- and interspecific differences in the response of two lentic species of leeches to seasonal hyperoxia. Canadian Journal of Fisheries and Aquatic Science 48:1124–1127.

Davies, R. W., Gates, T. E. 1991b. The effects of different oxygen regimes on the feeding and vertical distribution of *Nephelopsis obscura* (Hirudinoidea). Hydrobiologia 211: 51–56.

Davies, R. W., Kassera, C. E. 1989. Foraging activity of two specis of predatory leeches exposed to active and sedentary prey. Oecologia 81:329–334.

Davies, R. W., Oosthuizen, J. H. 1992. A new species of duck leech from North America formerly confused with *Theromyzon rude* (Rhynchobdellida: Glossiphoniidae. Canadian Journal of Zoology 71:770–775.

Davies, R. W., Reynoldson, T. B. 1976. A comparison of the life-cycle of *Helobdella stagnalis* (Li. 1758) (Hirudinoidea) in two different geographic areas in Canada. Journal of Animal Ecology 45:457–470.

Davies, R. W., Singhal, R. N. 1988. Cosexuality in the leech, *Nephelopsis obscura* (Erpobdellidae). International Journal of Invertebrate Reproduction and Development 13:55–64.

Davies, R. W., Wilkialis, J. 1982. Observations on the ecology and morphology of *Placobdella papillifera* (Verrill) (Hirudinoidea: Glossiphoniidae) in Alberta, Canada. American Midland Naturalist 107:316–324.

Davies, R. W., Reynoldson, T. B., Everett, R. P. 1977. Reproductive strategies of *Erpobdella punctata* (Hirudinoidea) in two temporary ponds. Oikos 29:313–319.

Davies, R. W., Wrona, F. J., Everett, R. P. 1978. A serological study of prey selection by *Helobdella stagnalis* (Hirudinoidea). Journal of Animal Ecology 48:181–194.

Davies, R. W., Wrona, F. J., Linton, L., Wilkialis, J. 1981. Inter- and intra-specific analyses of the food niches of two sympatric species of Erpobdellidae (Hirudinoidea) in Alberta, Canada. Oikos 37:105–111.

Davies, R. W., Linton, L. R., Parsons, W., Edgington, E. S. 1982a. Chemosensory detection of prey by *Nephelopsis obscura* (Hirudinoidea: Erpobdellidae). Hydrobiologia 97:157–161.

Davies, R. W., Linton, L. R. Wrona, F. J. 1982b. Passive dispersal of four species of freshwater leeches (Hirudinoidea) by ducks. Freshwater Invertebrate Biology 1:40–44

Davies, R. W., Wrona, F. J. Linton, L. 1982c. Changes in numerical dominance and its effects on prey utilization and interspecific competition between *Erpobdella punctata* and *Nephelopsis obscura* (Hirudinoidea): an assessment. Oikos 39:92–99.

Davies, R. W., Singhal, R. N. Blinn, D. W. 1985. *Erpobdella montezuma* (Hirudinoidea: Erpobdellidae), a new species of freshwater leech from North America. Canadian Journal of Zoology 63:965–969.

Davies, R. W., Yang, T., Wrona, F. J. 1987. Inter- and intra-specific differences in the effects of anoxia on erpobdellid leeches using static and flow-through systems. Holarctic Ecology 10:149–153.

Davies, R. W., Blinn, D. W., Dehdashti, B., Singhal, R. N. 1988. The comparative ecology of three species of Erpobdellidae (Annelida-Hirudinoidea). Archiv fur Hydrobiologie 111:601–614.

Davies, R. W., Monita, D. M. A. Dratnal, E. Linton, L. R. 1992. The effects of different oxygen regimes on the growth of a freshwater leech. Ecography 15:190–194.

Davies, R .W., Singhal, R. N., Wicklum D. D. 1995. Changes in reproductive potential of the leech *Nephelopsis obscura* (Erpobdellidae) as biomarkers for cadmium stress. Canadian Journal of Zoology 73:2192–2196.

Davies, R. W., Dratnal, E., Linton, L. R. 1996. Activity and foraging behaviour in the predatory freshwater leech *Nephelopsis obscura* (Erpobdellidae). Functional Ecology 10:51–54.

De Eguileor, M., Daniel, S., Giordana, B., Lanzavecchia, G., Valvassori, R. 1994. Trophic exchanges between parent and young during development of *Glossiphonia complanata* (Annelida, Hirudinea). Journal of Experimental Zoology 269:389–402.

De Filippi, F. 1849. Sopra un nuovo genere (*Hlaementeria*) di Annelidi della famigliadelle Sanguisughe. Memoires de Academie de Science., Tonno 10:1–14.

Derosa, X. S., Friesen, W. O. 1981. Morphology of leech sensilla: observations with the scanning electron microscope. Biological Bulletin 160:383–393.

Dickinson, M. H., Lent, C. M. 1984. Feeding behaviour of the medicinal leech, *Hirudo medicinalis L.* Journal of Comparative Physiology 154:449–455.

Dieter, C. D., Duffy, W. G., Flake, L. D. 1996. The effect of phorate on wetland macroinvertebrates. Environmental Toxicology and Chemistry 15:308–312.

Dratnal, E., Davies, R. W. 1990. Feeding and energy allocation in *Nephelopsis obscura* following exposure to winter stress. Journal of Animal Ecology 59:763–773.

Dratnal, E., Dratnal, P. A., Davies, R. W. 1992. The effect of food availability and foraging constraints on the life history of a predatory leech, *Nephelopsis obscura*. Journal of Animal Ecology 61:373–379.

Dratnal, E., Reddy, D. C., Biernacka, B., Davies, R. W. 1993. Facultative physiological adaptation and compensation to winter stresses in the predatory leech *Nephelopsis obscura*. Functional Ecology 7:91–96.

Ekman, S. 1911. Die Bodenfauna des Vattern, Qualitative und Quantitativ Untersucht. Internationale Revue der gesamten Hydrobiologie 7:146–204.

Elliott, J. M. 1973. The diel activity pattern, drifting and food of the leech *Erpobdella octoculata* (L.) (Hirudinea: Erpobdellidae) in a Lake District stream. Journal of Animal Ecology 42:449–459.

Elliott, J. M., Mann, K. H. 1979. A key to the British freshwater leeches. Freshwater Biological Association Scientific Publication No. 40,72 pp.

Fauvel, P. 1922. Un nouveau serpulien d'eau Saumatre *Mercierella enigmatica*. Bulletinde la Societe Zoologique de France 47: 425–431.

Fernandez, J. 1980. Embryonic development of the Glossiphoniid leech *Theromyzon rude*: characterization of developmental stages. Developmental Biology 76:245–262.

Fernandez, J., Tellez, V., Olea, N. 1992. Hirudinea. Microscopic Anatomy of Invertebrates 7:323–394.

Frey, H., Leuckart, R. 1847. Verzeichnis der zur Fauna Helgolands gehörenden wirbellosen seethiere. Beiträgen zur Kenntniss wirbelloser Thiere, pp. 136–168.

Friesen, W. O. 1981. Physiology of water motion detection in the medicinal leech. Journal of Experimental Biology 92:255–275.

Friesen, W. O., Dedwylder, R. D. 1978. Detection of low-amplitude water movements: a new sensory modality in the medicinal leech. Neuroscience Abstracts 4:380.

Gates, T. E., Davies, R. W. 1987. The influence of temperature on the depth distribution of sympatric Erpobdellidae (Hirudinoidea). Canadian Journal of Zoology 65:1243–1246.

Gates, T. E., Baird, D. J., Wrona, F. J., Davies, R. W. 1987. A device for sampling macroinvertebrates in weedy ponds. Journal of the North American Benthological Society 6:133–139.

Gee, W. 1913. The behavior of leeches with special reference to its modifiability. University of California Publications in Zoology 11:197–305.

Gerking, S. D. 1957. A method of sampling the littoral macrofauna and its application. Ecology 38:219–266.

Giesel, J. T. 1974. The biology and adaptability of natural populations. Mosley, St. Louis, Mio.

Govedich, F. R., Blinn, D. W. Hevley, R. H., Keim, P. S. 1999. Cryptic radiation in erpobdellid leeches in xevic landscapes: a molecular analysis of population differentiation. Canadian Journal of Zoology 77:52–57.

Govedich, F. R., Blinn, D. W., Keim, P., Davies, R. W. 1998. Phylogenetic relationships of three genera of Erpobdellidae (Hirudinoidea), with a description of a new genus, *Motobdella*, and species, *Motobdella sedonensis*. Canadian Journal of Zoology 76:2164–2171.

Graf, A. 1899. Hirudineenstudien. Acta Acadamemiae Leopoldensia, Halle 72:215–404.

Grantham, B. A., Hann, B. J. 1994. Leeches (Annelida: Hirudinea) in the Experimental Lakes Area, Northwestern Ontario, Canada: Patterns of species composition in relation to environment. Canadian Journal of Fisheries and Aquatic Science 51:1600–1607.

Green, R. H. 1979. Sampling design and statistical methods for environmental biologists. Wiley, New York.

Grube, A. E. 1851. Annulaten reise in den Aussersten Norden und Ostens Sibiriens. Herausgegeben von Dr. A. Th. Middendorff, St. Petersburg 1:1–254.

Hagadorn, I. R. 1969. Hormonal control of spermatogenesis in *Hirudo medicinalis*. II. Testicular response to brain removal during the phase of testicular maturity. Genetics and Comparitive Endocrolology 12:469–478.

Hartman, O. 1951. The literature of the polychaetous annelids. Vol 1. Allan Hancock Foundation. University of Southern California, Los Angeles, 290 pp.

Hartman, O. 1952. *Iphitime* and *Ceratocephala* (polychaetous annelids) from California. Bulletin of the Southern California Academy of Science 51:9–20.

Hayunga, E. G., Grey, A. J. 1976. *Cystobranchus meyeri* sp. n. (Hirudinea: Piscicolidae) from *Catostomus commersoni* Lacepede in North America. Journal of Parasitology 62:621–627.

Hemingway, E. E. 1908. The anatomy of *Placobdella pediculata in*: The leeches of Minnesota. Geological and Natural History of Minnesota, Zoology Series No. 5, pp. 29–63.

Herrmann, S. J. 1970a. Systematics, distribution and ecology of Colorado Hirudinea. American Midland Naturalist 83:1–37.

Herrmann, S. J. 1970b. Total residual tolerances of Colorado Hirudinea. Southwestern Naturalist 15:261–273.

Herter, K. 1929a. Temperaturversuche mit Egeln. Zeitschrift fuer Vergleichende Physiologie 10:248–271.

Herter, K. 1929b. Reizphysiologisches Verhalten und Parasitismus des Entenegels *Protoclepsis tesselata* O. F. Müller. Zeitschrift fuer Vergleichende Physiologie10:272–308.

Hobbs, H. H., Figueroa, H. V. 1958. The exoskeleton of a freshwater crab as a microhabitat for several invertebrates. Virginia Journal of Science 9:395–396.

Hoffman, R. L. 1964. A new species of *Cystobranchus* from southwestern Virginia (Hirudinea: Piscicolidae). American Midland Naturalist 72:390–395.

Holmquist, C. 1974. A fish leech of the genus *Acanthobdella* found in North America. Hydrobiologia 44:241–245.

Holt, P. C. 1969. The systematic position of the Branchiobdellidae (Annelida: Clitellata). Systematic Zoology 14:25–32.

Jennings, J. B., Vanderlande, V. M. 1967. Histochemical and bacteriological studies on digestion in nine species of leeches (Annelida: Hirudinea). Biological Bulletin 133:166–183.

Johnson, G. M., Klemm, D. J. 1977. A new species of leech, *Batracobdella cryptobranchii* n. sp. (Annelida: Hirudinea) parasitic on the Ozark hillbender. Transactions of the American Microscopical Society 90:331–377

Johnson, H. P. 1903. Freshwater nereids from the Pacific coast and Hawaii, with remarks on freshwater Polychaeta in general. Mark Anniversary Volume. Holt. New York, pp. 205–223.

Jones, S. R .M., Woo, P. T. K. 1990. Redescription of the leech *Desserobdella phalera* (Graf, 1899) n. comb. (Rhynchobdellida: Glossiphoniidae), with notes on its biology and occurrence in fishes. Canadian Journal of Zoology 68:1951–1955.

Kalarani, V., Reddy, D. C., Habibi, H. R., El-Shimy, N., Davies, R. W. 1995. Occurrence and hormonal action of ecdysone on gametogenesis and energy utilization in the leech *Nephelopsis obscura* (Erpobdellidae). The Journal of Experimental Zoology 273:511–518.

Kephart D. G. 1982. Microgeographic variation in the diets of garter snakes. Oecologia (Berlin) 52:287–291.

Kephart, D. G., Arnold, S. J. 1982. Garter snake diets in a fluctuating environment: a seven-year study. Ecology 63:1232–1236.

Khan, R. N., Frick, M. G. 1997. *Erpobdella punctata* (Hirudinea: Erpobdellidae) as phoronts on *Ambystoma maculatum* (Amphibia: Ambystomatidae). Journal of Natural History 31: 157–161.

Klemm, D. J. 1972a. Freshwater Leeches (Annelida: Hirudinea) of North America. United States Environmental Protection Agency Identification Manual No. 8,53 pp.

Klemm, D. J. 1972b. Freshwater polychaetes (Annelida) of North America. United States Environmental Protection Agency Identification Manual No. 4, 15 pp.

Klemm, D. J. 1977. A review of the leeches (Annelida: Hirudinea) in the Great Lakes region. Michigan Academician 9:397–418.

Klemm, D. J. 1982. Leeches (Annelida: Hirudinea) of North America. United States Environmental Protection Agency 600/3–82–025, 177, pp.

Klemm, D. J. 1985. A Guide to the Freshwater Annelida (Polychaeta, Naidid and Tubificid Oligochaeta, and Hirudinea) of North America. Kendall/Hunt, Dubuque, IA. 198 pp.

Krezel, A. M., Wagner, G., Seymour-Ulmer, J., Lazarus, R. A. 1994. Structure of the RGD protein Decorsin: conserved motif and distinct function in leech proteins that affect blood clotting. Science 264:1944–1947.

Kulkarni, G. K., Nagabhushanen, R., Hanumarte, M. M. 1980. Impact of some neurohumors on the brain neurosecretory profile of the Indian leech, *Poecilobdella viridis* (Blanchard). Biological Journal 2:37–41.

Kutschera, U. 1988. A new leech species from North America, *Helobdella californica* nov. sp. (Hirudinea: Glossiphoniidae). Zoologischer Anzeiger 220:173–178.

Lasserre, P. 1975. Clitellata, *in*: Giese A. C. Pearse, J. S. Eds. Reproduction of marine invertebrates. Vol. 3; Annelids: Echiurans. Academic Press, New York, pp. 215–275.

Leidy, J. 1851. Description of *Myzobdella*. Proceedings of the Academy of Natural Sciences, Philadelphia 1851–1853:243.

Leidy, J. 1858. *Manavankia speciosa*. Proceedings of the Academy of Natural Sciences, Philadelphia 10:90.

Leidy, J. 1870. Description of *Nephelis punctata*. Proceedings of the Academy of Natural Sciences, Philadelphia 22:89.

Light, J. E., Siddall, M. E. 1999. Phylogeny of the leech family Glossiphoniidae based on mitochondrial gene sequences and morphological data. Parasitology 85 (in press).

Linnaeus, C. 1758. Systema naturae. Lipsiae, 10th ed.

Linnaeus, C. 1761. Fauna suecia. Stockholm. 578 pp.

Linton, L. R., Davies, R. W. 1987. An energetics model of an aquatic predator and its application to life-history optima. Oecologia 71:552–559.

Linton, L. R., Davies, R. W., Wrona, F. J. 1982. Osmotic and respirometric responses of two species of Hirudinoidea to changes in water chemistry. Comparative Biochemistry and Physiology 71A:243–247.

Linton, L. R., Davies, R. W., Wrona, F. J. 1983a. The effects of water temperature, ionic content and total dissolved solids on

Nephelopsis obscura and *Erpobdella punctata* (Hirudinoidea, Erpobdellidae). I. Mortality. Holarctic Ecology 6:59–63.

Linton, L. R., Davies, R. W., Wrona, F. J. 1983b. The effect of water temperature, ionic content and total dissolved solids on *Nephelopsis obscura* and *Erpobdella punctata* (Hirudinoidea, Erpobdellidae). II. Reproduction. Holarctic Ecology 6:64–68.

Lukin, E. I. 1976. Leeches of fresh and brackish water bodies. Fauna of the Union of Soviet Socialist Republics. Leeches. Vol. 1. Academy of Science, USSR, Moscow. 484 pp.

Madanmohanrao, G. 1960. Salinity tolerances and oxygen consumption of the cattle leech, *Hirudinaria granulosa*. Proceedings of the Indian Academy of Science 11:211–218.

Malm, C. W. 1863. Svenska Iglar, Disciferae. Gotheborg. Vetenskapsakademien Arhandlingar i Naturskosarendon 8:153–263.

Mann, K. H. 1955. Some factors influencing the distribution of freshwater leeches in Britain. Internationale Vereinigung für Theoretische und Angewandte Limnologie 12:582–587.

Mann, K. H. 1956. A study of the oxygen consumption of five species of leech. Journal of Experimental Biology 33:615–626.

Mann, K. H. 1961. Leeches (Hirudinea) their structure, physiology, ecology and embryology. Pergamon Oxford, 201 pp.

Mannsfeld, W. 1934, *in*: Literature uber Hirudineen bis zum Jahre 1938. Bronn's Klassen und Ordnungen des Tierreichs. Bd. 4, Abt. III: Buch 4. T.2, pp. 539–642.

Manum, S. B., Bose, M. N., Sawyer, R. T. 1991. Clitellate cocoons in freshwater deposits since the Triassic. Zoologica Scripta 20:347–366.

Martin, A. J., Sealy, R. M. H., Young, J. O. 1994a. Does body size differences in the leeches *Glossiphonia complanata* (L.) and *Helobdella stagnalis* (L.) contribute to co-existence? Hydrobiologia 273:67–75.

Martin, A. J., Sealy, R. M. H., Young, J. O. 1994b. Food limitations in lake-dwelling leeches: field experiments. Journal of Animal Ecology 63:93–100.

Martin, A. J., Sealy, R. M. H., Young, J. O. 1994c. The consequences of a food refuge collapse on a guild of lake-dwelling triclads and leeches. Hydrobiologia 277:187–195.

Mather, C. K. 1954. *Haemopis kingi*, new species (Annelida: Hirudinea). American Midland Naturalist 52:460–468.

Mather, C. K. 1963. *Haemopis latero-maculatum*, new species (Annelida: Hirudinea). American Midland Naturalist 70:168–174.

McLoughlin, N. J., Blinn, D. W., Davies, R. W. 1999. An energetic evaluation of a predator–prey (leech-amphipod) couple in Montezuma Well, Arizona, USA. Functional Ecology 13:45–50.

McNicol, D. K., Bendell, B. E., Mallory, M. L. 1995. Evaluating macroinvertebrate responses to recovery from acidification in small lakes in Ontario, Canada. Water, Air and Soil Pollution 85:451–456.

McNicol, D. K., Mallory, M. L., Mierle, G., Scheuhammer, A. M., Wong, A. K. H. 1997. Leeches as indicators of dietary mercury exposure in non-piscivorous waterfowl in Central Ontario, Canada. Environmental Pollution 95:177–181.

Meyer, M. C. 1940. A revision of the leeches (Piscicolidae) living on freshwater fishes of North America. Transactions of the American Microscopical Society 59:354–376.

Meyer, M. C. 1946. A new leech *Piscicola salmositica*, n. sp. (Piscicolidae), from steelhead trout (*Salmo gairdneri gairdneri* Richardson, 1838). Journal of Parasitology 32:467–476.

Meyer, M. C. 1975. A new leech, *Macrobdella diplotertia* sp. n. (Hirudinea: Hirudinidae), from Missouri. Proceedings of the Helminthological Society of Washington 42:83–85.

Meyer, M. C., Moore, J. P. 1954. Notes on Canadian leeches (Hirudinea), with the description of a new species. Wasmann Journal of Biology 12:63–96.

Milne, I. S., Calow, P. 1990. Costs and benefits of brooding in glossiphoniid leeches with special reference to hypoxia as a selection pressure. Journal of Animal Ecology 59:41–56.

Misell, L. M., Shaw, B. K., Kristan W. B. Jr. 1998. Behavioral hierarchy in the medicinal leech, *Hirudo medicinalis*: feeding as a dominant behavior. Behavioural Brain Research 90:13–21.

Moore, J. P. 1898. The leeches of the U.S. National Museum. Proceedings of the United States National Museum 21:543–563.

Moore, J. P. 1900. A description of *Microbdella biannulata* with especial regard to the constitution of the leech somite. Proceedings of the Academy of Natural Sciences, Philadelphia 52:50–73.

Moore, J. P. 1901. The Hirudinea of Illinois. Illinois State Laboratory of Natural History Bulletin 5:479–547.

Moore, J. P. 1906. Hirudinea and Oligochaeta collected in the Great Lakes region. Bulletin of the Bureau of Fisheries 25:153–171.

Moore, J. P. 1912. Classification of the leeches of Minnesota, *in*: The leeches of Minnesota. Geology and Natural History of Minnesota, Zoology Series No. 5:63150.

Moore, J. P. 1949. Hirudinea, *in*: R. Kenk, Ed. The animal life of temporary and permanent ponds in southern Michigan. Miscellaneous Publications of the Museum of Zoology, University of Michigan No. 71, pp. 38–39.

Moore, J. P. 1953. Three undescribed North American leeches (Hirudinea: Notulae). National Academy of Natural Science, Philadelphia No. 250, pp. 1–13.

Moore, J. P. 1954. Cited in Meyer and Moore (1954).

Moore, J. P. 1959. Hirudinea, *in*: Edmonson, W. T. Ed. Freshwater Biology. 2nd ed. Wiley, New York, pp. 542–557.

Moore, J. P., Meyer, M. C. 1951. Leeches (Hirudinea) from Alaskan and adjacent waters. Wasmann Journal of Biology 9:11–77.

Moser, W. E., Desser, S. S. 1996. Morphological, histochemical, and ultrastructural characterization of the salivary glands and proboscises of three species of glossiphoniid leeches (Hirudinea: Rhynchobdellida). Journal of Morphology 225:1–18.

Müller, F. 1844. De Hirudinea circa Berolinum hueusque observatis. Dissitation Inaug., Berlin. pp 23–25.

Müller, O. F. 1776. Verrnium terrestrium et fluviatilum, seu animalium infusonorum helminthicorum, et testaceorum, non marinorum succincta historia. Hauniae et Lipsiae (Helminthica) 1:37–51.

Nehili, M., Ilk, C., Melhorn, H., Ryhnau, K., Dick, W., Njajou, M. 1994. Experiment on the possible role of leeches as vectors of animal and human pathogens:a light and electron microscopy study. Parasitology Research 80:277–290.

Nieczaj, R., Zerbst-Boroffka, I. 1993. Hyperosmotic acclamation in the leech, *Hirudo medicinalis* L.: energy, metabolisms, osmotic, conic and volume regulation. Comparative Biochemistry and Physiology 106A:595–602.

Omodeo, P. 1998. History of *Clitellata*. Italian Journal of Zoology 65:51–73.

Oosthuizen, J. H., Davies, R. W. 1992. Redescription of the duck leech, *Theromyzon rude* (Baird, 1869) (Rhynchobdellida: Glossiphoniidae). Canadian Journal of Zoology 70:2028–2033.

Oosthuizen, J. H., Davies, R. W. 1993. A new species of *Theromyzon* (Rhynchobdellida: Glossiphoniidae), with a review of the genus in North America. Canadian Journal of Zoology 71:1311–1318.

Pawlowski, L. K. 1948. Contribution a la connaissance des sangsus (Hirudinea) de la Nouvelle-Ecosse, de Terre-Neuve et des iles francais Saint-Pierre et Miquelon. Fragmenta faunistica Musei Zoologici Polonici 5:317–353.

Pearse, A. S. 1932. Parasites of Japanese salamanders. Ecology 13:135–152.

Peterson, D. L. 1982. Management of ponds for bait leeches in

Minnesota. Minnesota Department of Natural Resources, Section of Fisheries Investigation. Report No. 357:43.

Peterson, D. L. 1983. Life cycle and reproduction of *Nephelopsis obscura* Verrill (Hirudinea; Erpobdellidae) in permanent ponds of northwestern Minnesota. Freshwater Invertebrate Biology 2:165–172.

Platt, T. R., Sever, D. M. Gonzalez, V. L. 1993. First report of the predacious leech *Helobdella stagnalis* (Rhynchobdellida:Glossiphoniidae) as a parasite of an Amphibian, *Ambystoma tigrinum* (Amphibia: Caudata). The American Midland Naturalist 129:208–210.

Prahacs, S. M., Hall, K. J. 1996. Leeches as *in situ* biomonitors of chlorinated phenolic compounds. Part 1: Laboratory investigations. Water Research 30:2293–2300.

Prahacs, S. M., Hall, K. H. Duncan, W. 1996. Leeches as *in situ* biomonitors of chlorinated phenolic compounds. Part 2: Pulp mill investigations. Water Research 30:2301–2308.

Pritchard, G. 1964. The prey of dragonfly larvae (Odonata: Anisoptera) in ponds in northern Alberta. Canadian Journal of Zoology 42:785–800.

Purschke, G., Westheide, W., Rohde, D. Brinkhurst, R. O. 1993. Morphological reinvestigation and phylogenetic relationships of *Acanthobdella peledina* (Annelida: Clitellata). Zoomorphology (Berline) 113:91–101.

Pütter, A. 1907. Der Stoffwechsel der Blutegels (*Hirudo medicinalis* L.). I. Zeitschriftfur Allgemeine Physiologie 6:217–286.

Pütter, A. 1908. Der Stoffwechsel der Blutegels (*Hirudo medicinalis*) II. Zeitschrift für Allgemeine Physiologie 7: 16–61.

Qian, Y., Davies, R. W. 1994. Differences in bioenergetic and life-history traits between two generations of *Nephelopsis obscura* due to the prehistory of their parents. Freshwater Ecology 8:102–109.

Qian, Y., Davies, R. W. 1996. Inter-population genetic variation in the freshwater leech *Nephelopsis obscura* in relation to environmental heterogeneity. Hydrobiologia 325:131–136.

Rathke, H. 1862. Beitrage zur entwicklungsgeschicte der Hirudineen. Leipzig, Verlag von Wihelm Engelmann 116 pp.

Reddy, D. C., Dratnal, E., Davies, R. W. 1992. Dynamics of energy storage in a freshwater predator (*Nephelopsis obscura*) following winter stresses. Canadian Journal of Fisheries and Aquatic Sciences 49:1583–1587.

Reddy, D. C., Davies, R. W. 1993a. Metabolic adaptations by the leech *Nephelopsis obscura* during long-term anoxia and recovery. Journal of Experimental Zoology 265:224–230.

Reddy, D. C., Davies, R. W. 1993b. Bioenergetic responses to the initiation of winter and glycerol formation in the leech *Nephelopsis obscura*. Canadian Journal of Zoology 71:2036–2041.

Reynoldson, T. B., Bellamy, L. S. 1970. The establishment of interspecific competition in action between *Polycelis nigra* (Müll.) and *P. tenuis* (Ijima) (Turbellaria, Tricladida). *in*: Proceedings of the Advanced Study Institute, Dynamics Number Population, Oosterbruck, pp. 282–297.

Reynoldson, T. B., Davies, R. W. 1976. A comparative study of the osmoregulatory ability of three species of leech (Hirudinoidea) and its relationship to their distribution in Alberta. Canadian Journal of Zoology 54:1908–1911.

Reynoldson, T. B., Davies, R. W. 1980. A comparative study of weight regulation in *Nephelopsis obscura* and *Erpobdella punctata* (Hirudinoidea). Comparative Biochemistry and Physiology 66:711–714.

Richardson, L. R. 1942. Observations on migratory behaviour of leeches. Canadian Field Naturalist 56:67–70.

Richardson, L. R. 1969. A contribution to the systematics of the hirudinid leeches with descriptions of new families genera and species. Acta Zoologica Academiae Scientiarum Hungaricae 15:97–149.

Richardson, L. R. 1971. A new species from Mexico of the Nearctic genus *Percymoorensis* and remarks on the family Haemopidae (Hirudinoidea). Canadian Journal of Zoology 49:1095–1103.

Richardson, L. R. 1975. A contribution to the general zoology of the land-leeches (Hirudinoidea: Haemadipsoidea Superfam. nov.). Acta Zoologica Academiae Scientarium Hungaricae. 21: 119–152.

Riggs, M. R. 1980. Helminth parasites of leeches of the genus *Haemopis*. Master's thesis, Iowa State University.

Rosca, D. I. 1950. Duration of survival and variations in body weight in *Hirudo medicinalis* placed in solutions of increasing salinity. Studii si Cercetari de Biologie Cluj. 1:211–222.

Roters, F. J., Zebe, E. 1992a. Protease inhibitors in the alimentary tract of the medicinal leech *Hirudo medicinalis* : *in vivo* and *in vitro* studies. Journal of Comparative Physiology 162B:85–92.

Roters, F. J. Zebe, E. 1992b. Proteinors of the medicinal leech *Hirudo medicinalis*:purification and partial characterisation of three enzymes from the digestive tract. Comparative Biochemistry and Physiology 102B:627–634.

Sawyer, R. T. 1970. Observations on the natural history and behaviour of *Erpobdella punctata* (Leidy) (Annelida: Hirudinea). American Midland Naturalist 83: 65–80.

Sawyer, R. T. 1972. North American freshwater leeches, exclusive of the Piscicolidae, with a key to all species. Illinois Biological Monograph No. 46,154 pp.

Sawyer, R. T. 1981. Leech biology and behavior. *in*: Muller, K. J., Nicholls, I. G., Stent, G. S. Eds. The neurobiology of the leech. Cold Spring Harbor Lab., Cold Spring Harbor. New York, pp. 7–26.

Sawyer, R. T. 1984. Arthropodization in the Hirudinea: evidence for a phylogenetic link with insects and other Uniramia. Zoological Journal of the Linnean Society, London 80:303–322.

Sawyer, R. T. 1986. Leech biology and behaviour. Clarendon, Oxford. 1065 pp.

Sawyer, R. T., Hammond, D. L. 1973. Distribution, ecology and behavior of the marine leech *Calliobdella carolinensis* (Annelida: Hirudinea), parasitic on the Atlantic menhaden in epizootic proportions. Biological Bulletin 145:373–388.

Sawyer, R. T., Shelley, R. H. 1976. New records and species of leeches (Annelida: Hirudinea) from North and South Carolina. Journal of Natural History 10: 65–97.

Sawyer, R. T., LePont, F., Stuart, D. K., Kramer, A. P. 1981. Growth and reproduction of the giant glossiphoniid leech *Haementaria ghilianii*. Biological Bulletin (*Woods Hole, Mass.*) 160: 322–331.

Say, T. 1824. Sur les *Hirudo parasitica, laterialis, marmorata* et *decora* dans le voyage du Major Long. Natural History, Zoology Philadelphia 2:253–387.

Scudder, G. G. E., Mann, K. H. 1968. The leeches of some lakes in the southern interior plateau region of British Columbia. Syesis 1:203–209.

Seymour, J. L., Henzel, W. J., Nevins, B., Stults, J. T., Lazarus, R. A. 1990. A potent glycoprotein IIb–IIIa antagonist and platelet aggregation inhibitor from the leech *Macrobdella decora*. The Journal of Biological Chemistry 265:10143–10147.

Shankland, M., Martindale, M. Q., Nardelli-Haefinger, D., Baxter, E., Price D. J. 1992. Origin of segmental identity in the development of the leech nervious system. Development Supplement 2:29–38.

Siddall, M. E., Burreson, E. M. 1995. Phylogeny of the Euhirudinea: independent evolution of blood feeding by leeches? Canadian Journal of Zoology 73:1048–1064.

Siddall, M. E., Burreson, E. M. 1996. Leeches (Oligochaeta Euhirudinea), their phylogeny and the evolution of life-history strategies. Hydrobiologia 334:277–285.

Simon, T. W., Barnes, K. 1996. Olfaction and prey search in the carnivorous leech *Haemopis marmorata*. The Journal of Experimental Biology 199:2041–2051.

Singhal, R. N., Davies, R. W. 1985. Descriptions of the reproductive organs of *Nephelopsis obscura* and *Erpobdella punctata* (Hirudinoidea: Erpobdellidae). Freshwater Invertebrate Biology 5:91–97.

Singhal, R. N., Davies, R. W. 1987. Histopathology of hyperoxia in *Nephelopsis obscura* (Hirudinoidea: Erpobdellidae). Journal of Invertebrate Pathology 50:33–39.

Singhal, R. N., Davies, R. W., Shah, K. L. 1985. The taxonomy and morphology of the leeches (Hirudinoidea: Glossiphoniidae) parasitic on turtles from the Beas River (India) including descriptions of two new species and one redescription. Zoologischer Anzeiger 215:147–155.

Singhal, R. N., Sarnat, H. B., Davies, R. W. 1988. Effect of anoxia and hyperoxia on the neurons in the leech *Nephelopsis obscura* (Erpobdellidae): ultrastructural studies. Journal of Invertebrate Pathology 52:409–418.

Sladacek, V., Kosel, V. 1984. Indicator value of freshwater leeches (Hirudinea) with a key to the determination of European species. Acta Hydrochimie und Hydrobiologie 12:451–461.

Smiley, J. W., Sawyer, R. T. 1976. Osmoregulation in the leeches *Macrobdella dietra* and *Haemopsis marmorata*. Bulletin of the Association of Southeastern Biologists 23:96.

Smith, D. E. C., Davies, R. W. 1996a. Changes in energy allocation by the predator *Nephelopsis obscura* exposed to differences in prey availability. Canadian Journal of Zoology 75:606–612.

Smith, D. E. C. Davies, R. W. 1996b. Effects of aggregative behavior on the bioenergetics of the freshwater predatory leech *Nephelopsis obscura* (Erpobdellidae). Freshwater Biology 36:647–659.

Soös, A. 1965. Identification key to the leech (Hirudinoidea) genera of the world, with a catalogue of the species. I. Family: Piscicolidae. Acta Zoologica Academiae Scientarium Hungaricae 2:417–463.

Soös, A. 1966a. Identification key to the leech (Hirudinoidea) genera of the world, with a catalogue of the species. II. Families: Semiscolecidae, Trematobdellidae, Americobdellidae, Diestecostomatidae. Acta Zoologica Academiae Scientarium Hungaricae 12:145–160.

Soös, A. 1966b. Identification key to the leech (Hirudinoidea) genera of the world, with a catalogue of the species. III. Family: Erpobdellidae. Acta Zoologica Academiae Scientarium Hungaricae 12:371–407.

Soös, A. 1966c. On the genus *Glossiphonia* Johnson, 1816, with a key and catalogue to the species (Hirudinoidea: Glossiphoniidae). Annales Historico-Naturales Musei Nationalis Hungarici 58:271–279.

Soös, A. 1967. Identification key to the leech (Hirudinoidea) genera of the world, with a catalogue of the species. IV. Family: Haemodipsidae. Acta Zoologica Academiae Scientiarum Hungaricae 13:417–432.

Soös, A. 1968. Identification key to the species of the genus *Erpobdella* de Blainville,1818 (Hirudinoidea: Erpobdellidae). Annales Historico-Naturales Musei Nationalis Hungarici 60:141–145.

Soös, A. 1969a. Identification key to the leech (Hirudinoidea) genera of the world, with a catalogue of the species. V. Family: Hirudinidae. Acta Zoologica Academiae Scientiarum Hungaricae 15:151–201.

Soös, A. 1969b. Identification key to the leech (Hirudinoidea) genera of the world, with a catalogue of the species. VI. Family: Glossiphoniidae. Acta Zoologica Academiae Scientiarum Hungaricae 15:397–454.

Spelling, S. M., Young, J. O. 1987. Predation on lake-dwelling leeches (Annelida: Hirudinea): An evaluation by field experiment. Journal of Animal Ecology 56:131–146.

Treadwell, A. L. 1914. Polychaetous annelids of the Pacific coast in the collections of the zoological museum of the University of California. University of California Publications in Zoology 13:175–234.

Van Damme, N. 1974. Organogénèse de l'appareil genital chez la sangsue *Erpobdella octoculata* L. (Hirudinée Pharyngobdelle). Archives de Biologie 18:373–377.

Verrill, A. E. 1872. Descriptions of North American freshwater leeches. American Journal of Science 3:126–139.

Verrill, A. E. 1874. Synopsis of the North American freshwater leeches. Report to the United States Fisheries Commission 1872/73:666–689.

Von Brand, T. 1944. Occurrence of anaerobiosis among invertebrates. Biodynamica 4:185–328.

Webb, R.A. 1980. Spermatogenesis in leeches I. Evidence for a gonadotropic peptide hormone produced by the supraoesophageal ganglion of *Erpobdella octoculata*. General and Comparative Endocrinology 42:401–412.

Webb, R. A., Omar, F. E. 1981. Spermatogenesis in leeches. II. The effect of the supraoesophageal ganglion and ventral nerve cord ganglia on spermatogenesis in the North American medicinal leech *Macrobdella decora*. Genetics and Comparative Endocronology 44:54–63.

Webster, H. 1879. Annelida Chaetopoda of New Jersey. New York State Museum of Natural History Report 32:101–125.

Weisblat, D. A., Harper, G., Stent, G. S., Sawyer, R. T. 1980. Embryonic cell lineages in the nervous system of the glossiphoniid leech *Helobdella triserialis*. Developmental Biology 76:58–78.

Whitman, C. O. 1886. The leeches of Japan. Quarterly Journal of Microscopical Science 26:317–416.

Whitman, C. O. 1892. The metamerism of *Clepsine*. *in*: Festschrift 70 sten Geburtstag R. Leuckarts. 285–295.

Wicklum, D., Davies, R. W. 1996. The effects of chronic cadmium stress on energy acquisition and allocation in a freshwater benthic invertebrate predator. Aquatic Toxicology 35:237–252.

Wicklum, D., Smith, D. E. C. Davies, R. W. 1997. Mortality, preference, avoidance, and activity of a predatory leech exposed to cadmium. Archives of Environmental Contamination Toxicology 32:178–183.

Wilkialis, J., Davies, R. W. 1980a. The reproductive biology of *Theromyzon tessulatum* (Glossiphoniidae: Hirudinoidea), with comments on *Theromyzon rude*. Journal of Zoology (London) 192:421–429.

Wilkialis, J., Davies, R. W. 1980b. The population ecology of the leech (Hirudinoidea: Glossiphoniidae) *Theromyzon tessulatum*. Canadian Journal of Zoology 58:906–912.

Williamson, M. 1972. The analysis of biological populations. Arnold, London.

Wrona, F. J., Davies, R. W. 1984. An improved flow-through respirometer for aquatic macroinvertebrate bioenergetic research. Canadian Journal of Fisheries and Aquatic Sciences 41:380–385.

Wrona, F. J., Davies, R. W., Linton, L. 1979. Analysis of the food niche of *Glossiphonia complanata* (Hirudinoidea: Glossiphoniidae). Canadian Journal of Zoology 57:2136–2142.

Wrona, F. J., Davies, R. W., Linton, L. Wilkialis, J. 1981. Competition and co-existence between *Glossiphonia complanata* and *Helobdella stagnalis* (Glossiphoniidae: Hirudinoidea). Oecologia 48:133–137.

Wrona, F. I., Linton, L. R., Davies, R. W. 1987. Reproductive success and growth of two species of Erpobdellidae: the effects of water temperature. Canadian Journal of Zoology 65:1253–1256.

Young, J. O. 1987. Predation on leeches in a weedy pond. Freshwater Biology 17: 161–167.

Young, J. O., Spelling, S. M. 1986. The incidence of predation on lake-dwelling leeches. Freshwater Biology 16:465–477.

Young, S. R., Dedwylder, R., Friesen, W. O. 1981. Response of the medicinal leech to water waves. Journal of Comparative Physiology 144: 111–116.

Young, J. O., Seaby, R. M. H., Martin, A. J. 1995. Contrasting mortality in young freshwater leeches and triclads. Oecologia 101:317–323.

Zaranko, D. T., Griffiths, R. W., Kaushik, N. K. 1997. Biomagnification of polychlorinated biphenyls through a riverine food web. Environmental Toxicology and Chemistry 16:1463–1471.

Zebe, E., Roters, F. J., Kaiping, B. 1986. Metabolic changes in the medicinal leech *Hirudo medicinalis*, following feeding. Comparative Biochemistry and Physiology 84A:49–55.

Zrzavy, J, Mihulka, S., Kepka, P., Bezdek, A., Tietz, D. 1998. Phylogeny of the Metazoa based on morphological and 18S ribosomal DNA evidence. Cladistics 14:249–285.

Zerbst-Boroffka, I. 1975. Function and ultrastructure of the nephridium in *Hirudo medicinalis L.* III. Mechanisms of the formation of primary and final urine. Journal of Comparative Physiology B 100:307–316.

14 BRYOZOANS

Timothy S. Wood

Department of Biological Sciences
Wright State University
Dayton, Ohio 45435

I. INTRODUCTION

Bryozoans are among the most commonly encountered animals that attach to submerged surfaces in fresh water. During warm months of the year they are found in almost any lake or stream where there are suitable attachment sites. The variety of forms ranges from wisps of stringy material to massive growths weighing several kilograms. In the days before sand filtration, bryozoans were notorious for clogging the distribution pipelines of public water systems. Today they still foul water intake lines, irrigation nozzles, and wastewater treatment systems.

In general, bryozoans are sessile, modular invertebrates with ciliated tentacles that capture suspended food particles. Historically they have included what are now recognized as two very distinct and unrelated phyla: Ectoprocta and Entoprocta. Within the ectoproct bryozoans are two major classes: Phylactolaemata is an exclusively freshwater group and the main focus of this chapter; Gymnolaemata is a vastly larger, polyphyletic collection of species, mostly marine except for a few species in the subclass Ctenostomata which occur in fresh or brackish water. In this chapter ctenostomes are included in the general discussion. Entoproct bryozoans, phylogenetically distant from ectoprocts, are treated in a separate section, but still included in this chapter more for reasons of tradition than systematics.

II. ANATOMY AND PHYSIOLOGY

A. External Morphology

1. Zooids and Lophophores

The phylactolaemate bryozoan colony is composed of identical zooids fused seamlessly into a single structure (Fig. 1). Major anatomical features are shown in Figure 2. Each zooid in the colony has two basic parts: an organ system, or polypide, which can be partially protruded as a unit; and a body wall which can enclose the entire polypide and which separates the colony's interior from the surrounding water. These parts are joined in different places by three structures: a tentacular sheath, prominent retractor muscles, and a slender funiculus loosely running from the gut caecum to a point on the interior of the body wall.

The polypide bears a prominent lophophore composed of ciliated tentacles arranged around a central mouth. In *Fredericella* species, the tentacles simply encircle the mouth (Fig. 3). In all other phylactolaemates the lophophore extends dorsally in a bilateral pair of arms, forming a U-shaped structure with an outer row of long tentacles and an inner row of shorter ones (Fig. 4). In most species, each tentacle bears one medial and two lateral tracts of cilia which beat in metachronal waves. Tentacles are loosely joined near their base by an intertentacular membrane, effectively forming a groove

Ecology and Classification of North American Freshwater Invertebrates. 2nd Edition

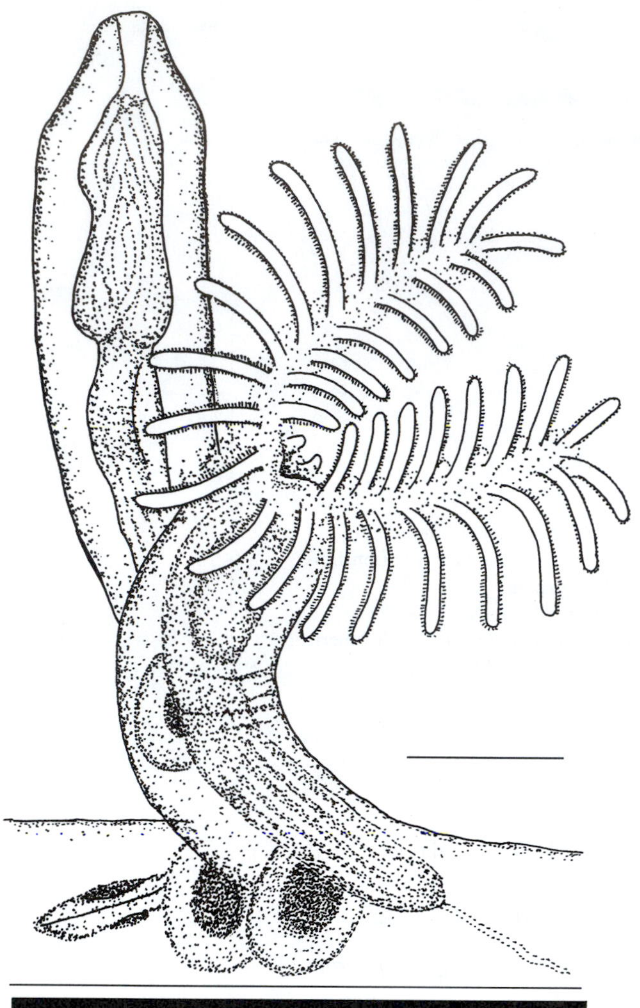

FIGURE 1 Two zooids of a tubular bryozoan colony, one retracted and the other extended in a feeding position. Scale bar = 0.5 mm.

between rows of tentacles. Two such grooves converge from the lophophore arms toward the mouth. Adjacent to mouth is the epistome, a small, heavily ciliated lobe which may function in food selection.

In gymnolaemate (ctenostome) bryozoans the zooids are tubular and rather delicate looking, with small, conical lophophores (Fig. 8a–d). The body wall is thin and transparent. Colonies are diffuse, consisting of creeping chains of zooids sometimes separated by narrow pseudostolons.

2. Statoblasts

A conspicuous feature of phylactolaemate bryozoans is the asexual production of encapsulated dormant buds, called statoblasts. The various distinctive types are widely used for species identification. Most species produce free statoblast that can be released from the colony as disseminule, each bearing a peripheral band of sclerotized chambers normally filled with a gas for buoyancy. In most species, the free statoblasts (often called floatoblasts) acquire their buoyant gas in a late developmental stage within the colony, while in a few other species the floatoblast must be dried or frozen before becoming buoyant. Many species with self-inflating floatoblasts also produce specialized sessile statoblasts ("sessoblasts") cemented to the substrate upon which colonies grow, and bearing a distinctive lamella around the rim. A third type of statoblast, the piptoblast, is a simple, beanlike structure known only in *Fredericella* species, lacking specialized external structures and held firmly within the tubular branches of the colony. Mukai (1982) provides an excellent overview of statoblast development.

All statoblasts are composed of paired valves joined at an equatorial suture. In most floatoblasts the valves have two structural layers (Fig. 5a). An inner capsule contains germinal tissue and food reserves; the outer periblast covers the capsule completely and includes the peripheral annulus and a central fenestra. In most species, the so-called dorsal (or cystigenic) valve generally has a smaller fenestra and correspondingly wider annulus than the ventral (deutoplasmic) valve. At least one species forms leptoblasts in which the capsule is lacking and a fully formed zooid is enclosed by the periblast only

Sessoblasts have a similar arrangement of inner capsule and outer periblast. The thin peripheral lamella is considered homologous with the floatoblast annulus, although it is never inflated. The basal valve provides a large, irregular ring which adheres tightly to the substrate.

Gymnolaemate bryozoans do not produce statoblasts. Instead, many of the species restricted to fresh or brackish water form special thick-walled hibernacula. These irregularly shaped bodies are integrated into the colony structure and attached firmly to the substrate, remaining long after the colony disintegrates. Like statoblasts, hibernacula survive desiccation, changes of salinity, and other suboptimal conditions, although the limits of their resistance have not been documented.

B. Organ-System Function

1. Coelom, Neural System, and Body Wall

Most ectoproct colonies have a spacious coelom shared by all zooids. A clear coelomic fluid is circulated by cilia on the peritoneum. The main gastric coelom communicates through an incomplete diaphragm with the coelom of the lophophore, which extends fully into every tentacle.

A small nerve ganglion is located on the diaphragm at the base of the epistome between the mouth and the

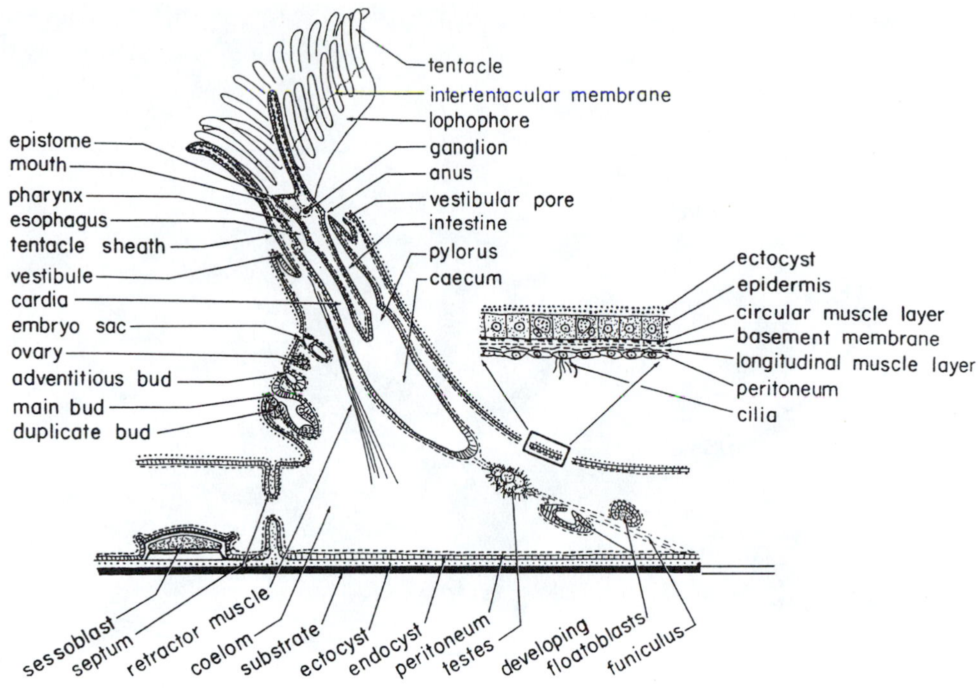

FIGURE 2 Schematic longitudinal section of a generalized colony (based on Mukai, 1982).

anus. A single nerve tract extends to each of the paired arms of the lophophore, with nerves branching off to each tentacle. Other nerves from the ganglion innervate the epistome, the tentacle sheath, and the digestive tract. There is no nervous communication between zooids.

The body wall is composed of living tissues, collectively called the endocyst, together with a nonliving outer ectocyst (Fig. 2). Lining the main coelom is a peritoneum bearing scattered tracts of cilia. Behind the peritoneum lies a thin, longitudinal muscle layer followed by a basement membrane. Overlying this is a thin layer of circular muscles and a single layer of epidermis. Near the apex of the zooid, the endocyst folds inward to form a flat pocket, the vestibule, then joins the

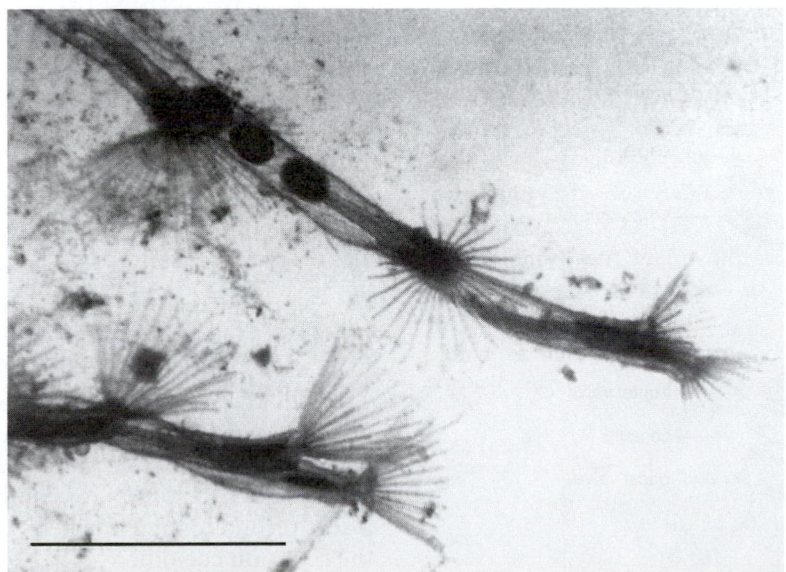

FIGURE 3 Portion of a *Fredericella* colony showing characteristic circular outline of the lophophore. Scale bar = 2 mm.

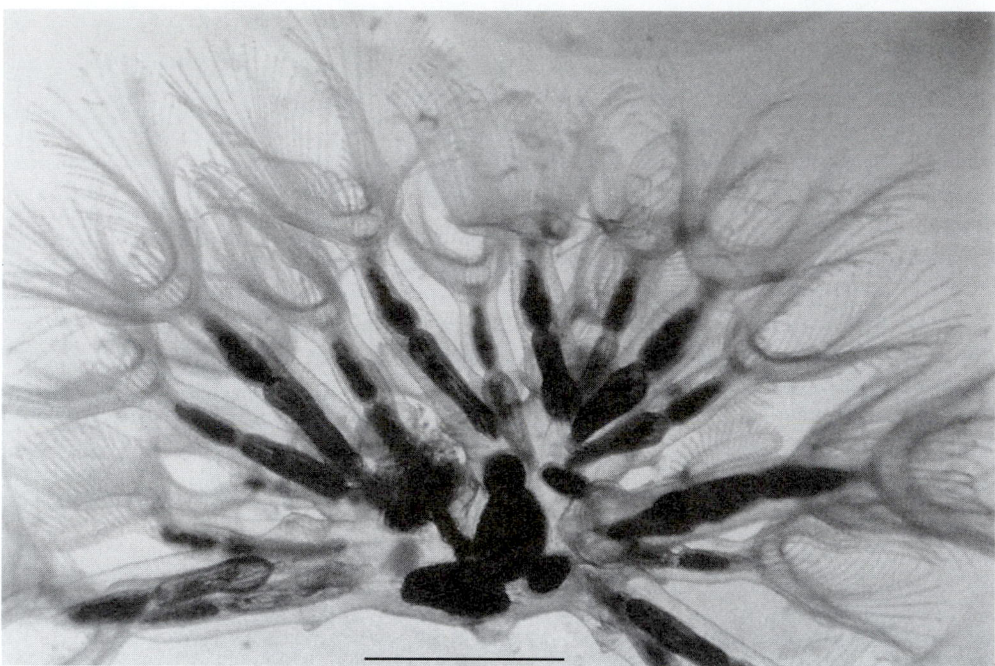

FIGURE 4 Portion of a *Lophopodella* colony showing the typical horseshoe-shaped lophophore. Scale bar = 2 mm.

lophophore as an eversible tentacle sheath. When the polypide is fully retracted, a well-defined opening, or orifice, remains at the zooid tip. A pore in the vestibular body wall of many species is normally closed except to permit the release of floatoblasts from living colonies.

In some tubular species, a portion of the body wall grows inwardly to form a kind of sclerotized, reinforcing ring or incomplete septum. Those branches of tubular colonies that are attached to a substrate may have a raised line, or keel, running along the outer surface.

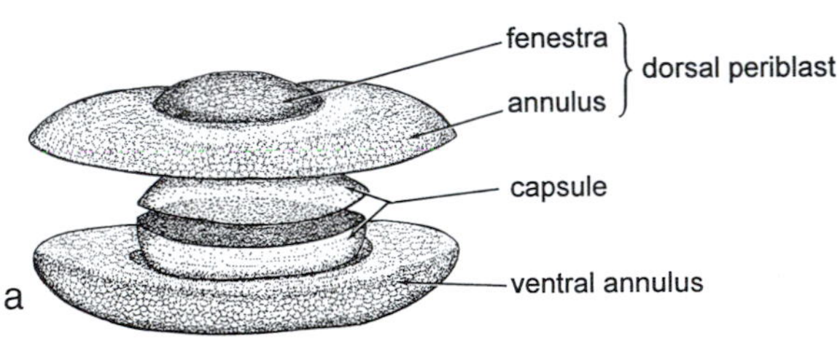

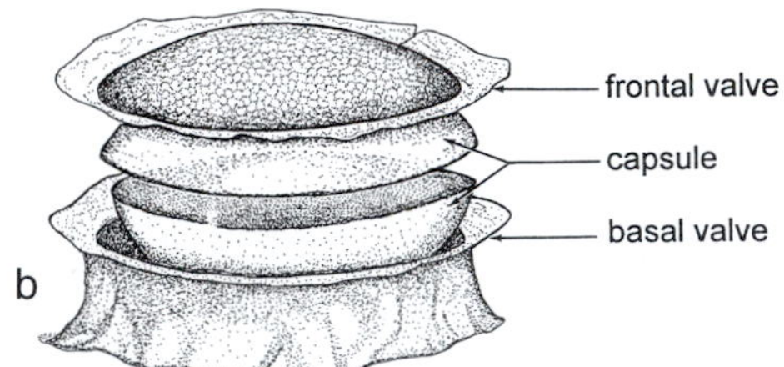

FIGURE 5 Sclerotized components of generalized phylactolaemate statoblasts in exploded view. (a) Floatoblast; (b) sessoblast.

The composition of the ectocyst is highly variable among species. In most tubular colonies it is sclerotized and remains intact for a time after living parts have died. In at least two species, the ectocyst is composed largely of chitinous microfibrils with random orientation. A thin, proximal electron-dense layer is overlain by a thicker stratum of looser fibrils (Goethals *et al.*, 1984). Many organic and inorganic particles adhere to this outer coat, including bacteria which also penetrate the interior of the loose layer. The ectocyst in these species varies from thin and flexible to leathery, opaque, and brittle. With age the ectocyst may darken as the the chitin is thickened and tanned. Thus, the growing tips of the zooids are often distinctly lighter in color than the rest of the colony. Rao *et al.* (1978) found that with increasing age a second, separate sclerotized layer can form in the inner body wall of *Plumatella casmiana*.

Mukai *et al.* (1984) have shown that the body wall breaks down when contact is made between adhesive pads from germinating statoblasts of the same species. The resulting colonies are confluent, possibly allowing a sort of outcrossing when sperm are circulated within the common coelom.

In nontubular phylactolaemates the ectocyst is gelatinous and often restricted to certain parts of the colony. Mukai and Oda (1980) reported many vacuolar cells, apparently with a secretory function, in the epidermis of these species. The massive jellylike substance produced by *Pectinatella magnifica* is more than 99% water and contains partially denatured protein along with some chitin, calcium, and sodium chloride (Morse, 1930).

2. Feeding Mechanics

The lophophore is a complex feeding organ capable of capturing a wide variety of food particles. In contrast with marine bryozoans, the lophophore of most freshwater species is large and powerful, bearing many closely positioned tentacles. Such an organ may be particularly well suited to lentic habitats; certainly those species with the largest lophophores seem to grow best in quiet water. In species where lophophores are tightly crowded (e.g., *Cristatella* or *Pectinatella*) the processed water exits along defined excurrent channels. This results in a significant feeding efficiency, according to the model proposed by Eckman and Okamura (1998). In a study using *Plumatella repens*, the ingestion rate per zooid was higher for large colonies than for small ones, and also varied directly with ambient water current (Okamura and Doolan, 1993).

Gilmour (1978) showed that differential beating of the lateral and frontal cilia sort particles by density, rejecting heavier, inedible materials before they reach the epistome. Kaminski (1991a) made a distinction between ciliary feeding, which brings in nannoplankton (5–10 μm), and tentacular feeding in which the tentacles orient segments of filamentous algae for ingestion. Flicking movements of individual tentacles sometimes throw particles toward the mouth; at other times the tips of tentacles are brought together to prevent the escape of an active protozoan or rotifer.

Exactly how the lophophore captures minute particles is not entirely clear. In U-shaped lophophores the membrane joining the base of all tentacles creates an intertentacular groove of relatively still water. Bullivant (1968) suggested that particles accelerating toward the lophophore are impelled the final short distance into this groove by their own momentum, while the water is deflected abruptly to the sides. This would be consistent with the observation that species with the most powerful lophophores are able to capture the smallest particles (Kaminski, 1984).

Food particles entering the mouth collect momentarily in a short, ciliated pharynx, then are swallowed through a narrow esophagus to the Y-shaped stomach. A short unciliated cardia leads to the long, cylindrical cecum, where particles are churned by rhythmic peristaltic contractions. In well-fed colonies waste products collect in brownish, longitudinal bands, giving the cecum a striped appearance. A short proximal part of the cecum, the pylorus, leads through a narrow sphincter to a straight intestine. Here processed particles are consolidated, infused with mucus, and ejected as a fecal pellet through the anus, which lies just outside the whorl of tentacles. The bundle of eliminated material is too large to be re-ingested by neighboring zooids, and it falls away from the colony.

3. Reproductive Systems and Larvae

Phylactolaemate bryozoans are sexually active during a single brief period of the year. Sperm develop in conspicuous masses on the funiculus of certain zooids, from which they later detach to circulate passively throughout the colony coelom. Egg clusters develop on the peritoneum ventrally within the zooid. There is now direct evidence for outcrossing in populations of *Cristatella mucedo* based on RADP assays (Jones *et al.*, 1994), but the mechanism for gamete transfer is unknown. Sperm motility has never been reported, nor have sperm even been observed leaving one colony or entering another. Self-fertilization thus appears likely in most species except in instances where two colonies have become confluent. Assuming that fertilization occurs in the coelom, the zygote somehow migrates to a special sac formed as an ingrowth of the body wall. The embryo develops into a specialized free-swimming structure usually called a "larva," although technically a motile colony. Embryology of the planktonic larva and the steps leading to its metamorphosis are summarized by Hyman (1959). Fine structure is described by Franzén and Sensenbaugh (1983). The larva is composed of a heavily ciliated outer mantle and an inner pear-shaped mass (Fig. 6a). The inner parts include one to four fully formed polypides together with their funiculi and associated musculature. Released from the colony usually after dark, the larva swims with its aboral pole forward, which contains special gland cells and a neural center. Settling normally occurs within an hour (Fig. 6b–d).

Among most Gymnolaemata, sexual reproduction leads to the development of small larvae totally unlike those of Phylactolaemata. Settling and metamorphosis occur shortly after their release, since the larvae have no feeding or digestive organs. The larva of *Paludicella* and its embryological development were described briefly by Braem (1896). However, virtually nothing is known of larval ecology or behavior among freshwater ctenostome bryozoans.

All bryozoan zooids are capable of budding new zooids. In the Phylactolaemata, buds arise from a specific site on the midventral wall of the parental zooid (Figs. 2 and 6a). The development of a new zooid is accompanied by the appearance of new bud primordia,

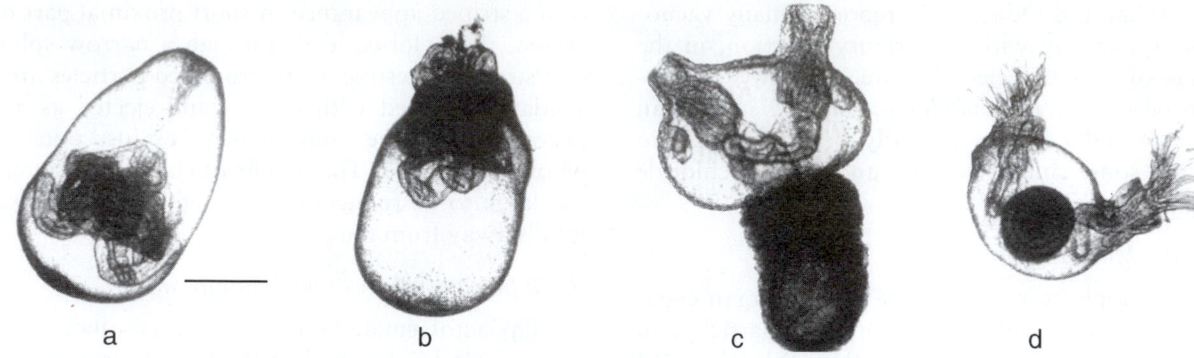

FIGURE 6 Settling and transformation of a larva. Scale bar for all figures = 500 μm. (a) Swimming larva containing two fully developed zooids; the more tapered end is the oral pole; (b) the mantle peels back immediately after the larva attaches to a glass substrate; (c) constricted mantle appears as a dark mass; the two zooids are now clearly visible, each with its own duplicate bud; (d) the mantle is drawn into the body cavity, lophophores are extended, and the two zooids begin feeding (from Zimmerman, 1979).

which may or may not develop further. Brien (1953) showed that every zooid bears two bud primordia: a "main bud," which forms the first daughter zooid, and an "adventitious bud" between the main bud and parental zooid, which becomes the second daughter zooid. In addition, each main bud itself has a small "duplicate bud" primordium on its ventral side. In the process of budding, duplicate and adventitious buds become main buds, and new duplicate and adventitious buds are formed. These relations have been nicely clarified by Mukai *et al.* (1987).

In ctenostomes the new zooidal bud originates from an outswelling of the body wall of a parental zooid. An interior wall then grows across the base of the bud to separate it from the parental body cavity. Most often, a new bud develops at the distal end of a lineal series of zooids, and as it grows a wall forms at the tip to separate it from the body cavity of the next distal bud in the series.

In addition to asexual budding, certain phylactolaemate species increase their numbers significantly by an active fission of the colony. This is a common occurrence among the globular colonies of *Cristatella, Pectinatella, Lophopodella, Lophopus*, and *Asajirella*. The products of fission glide slowly apart along the substrate by a means that is not fully understood.

C. Environmental Physiology

Little is known about the digestive physiology of phylactolaemate bryozoans, except that the esophagus environment is alkaline, the stomach is acidic, and the intestinal pH is neutral (Marcus, 1926). Passage time through the gut varies from one to 24 hs depending on the ingestion rate. Zones of basophilic and acidophilic cells line the cecum, but there is no agreement among investigators on their function.

Statoblasts have been the subject of considerable study. The period of obligate dormancy imposed on most statoblasts serves to maintain populations through periods of unfavorable environmental conditions. Those statoblasts not fixed to immobile substrates may also function in passive dispersal, carried either by water currents or by migrating animals. Brown (1933) demonstrated viability of *Plumatella* statoblasts after they had passed in separate trials through the digestive tracts of a salamander, frog, turtle, and duck.

Like statoblasts, the hibernacula of gymnolaemates maintain the populations during periods of unfavorable conditions. Unlike most statoblasts, however, they are an integral part of the colony, can never be released freely into the water and thus would not normally function as independent disseminules.

The resistance of statoblasts to environmental stress has been reviewed by Bushnell and Rao (1974). Most *Plumatella* and *Fredericella* statoblasts survive desiccation or freezing for periods of 1 to 2 years. Some *Plumatella* floatoblasts can germinate after more than 4 years in cold water. *Lophopodella carteri* statoblasts have remained viable during dry storage for over 6 years. Other less hardy statoblasts include those of *Pectinatella magnifica*, which do not withstand desiccation and survive only brief periods of freezing temperatures.

Statoblasts stored in water at a favorable temperature eventually germinate after a variable dormant period. However, when that period is prolonged by continuous exposure to cold, dry, anaerobic, or other unfavorable conditions, a near simultaneous germination can be achieved when a suitable environment is restored. In some species, light is an important factor for germination (Oda, 1980; Richelle *et al.*, 1994). Ultrastructure of the statoblast suture and its possible relation to germination has been studied by Bushnell and Rao (1974) and Rao and Bushnell (1979).

The budgeting of energy resources for colony growth and reproduction is poorly understood. In many cases, gametogenesis precedes statoblast production, but the two processes may also operate concurrently. The first zooid to germinate from a statoblast never forms statoblasts itself, but it may participate in gametogenesis. In laboratory colonies of *Plumatella emarginata*, Mukai and Kobayashi (1988) found sexual activity only in certain colonies, which varied in size from as few as 10 zooids. Only those colonies with the most vigorous development of testes eventually formed mature larvae.

The regulation of sessoblast development in Plumatellidae has not yet been clarified. Apparently, statoblast primordia can differentiate into either floatoblasts or sessoblasts, and this differentiation becomes evident during the late epidermal-disk stage (Mukai and Kobayashi, 1988). In natural populations, Wood (1973) noted that most sessoblasts were formed either early or late in the season, leading him to suppose that this was a response to suboptimal conditions. However, laboratory-reared *Plumatella* colonies frequently form both floatoblasts and sessoblasts under apparently favorable growing conditions.

III. ECOLOGY AND EVOLUTION

A. Diversity and Distribution

Freshwater bryozoans are generally restricted to relatively warm water, flourishing at 15–28°C. However, *Cristatella mucedo* has been found at 6°C (Bushnell, 1966), and *Lophopus crystallinus* survives at 0°C for brief periods (Marcus, 1934). *Fredericella indica* lives through the winter in most of North America, budding new zooids at 3°C and producing piptoblasts above 8°C. All of these species also grow well at higher temperatures. Only *Stephanella hina* has never been found above 17°C and is probably restricted to cool water (Smith, 1989a). Temperatures as high as 37°C were recorded for living *Plumatella nitens* and *P. fruticosa* (Bushnell, 1966). Shrivastava and Rao (1985) noted that *Plumatella emarginata* in India shows only meager growth at 34°C.

Phylactolaemates tolerate a wide range of pH, but they favor slightly alkaline water. This was the conclusion of Tenney and Woolcott (1966) in a North Carolina study where pH ranged from 6.3 to 8.0. A *Fredericella* species (probably *F. indica*) has been collected in acidic conditions as low as pH 4.9 (Everitt, 1975), *Plumatella fruticosa* at pH 5.7 (Bushnell, 1966), and both *P. repens* and *Hyalinella punctata* at pH 6.3 (Tenney and Woolcott, 1966).

Most species occur in both still and running water, but *Plumatella emarginata* and *Fredericella indica* grow especially well in lotic habitats. Among those species associated with still waters are *Plumatella nitens, P. rugosa, P. recluse, Lophopus crystallinus, Hyalinella punctata*, and *P. orbisperma*. Many common species tolerate turbid water, but *Pectinatella magnifica* does not, possibly because its large colonies are necessarily more exposed to settling particles (Cooper and Burris, 1984).

The reputation of bryozoans as clean water species is not entirely warranted. Henry *et al.* (1989) found *Fredericella sultana, Plumatella emarginata*, and *P. fungosa* thriving in high concentrations of heavy metals and PCBs, although none of these had accumulated in the tissues. Bushnell (1974) also recognized these species to be particularly tolerant of extreme contamination from sewage and industrial wastes. More recently, *Plumatella vaihiriae* has been noted as an important fouling species in secondary clarifiers and tertiary sand filters of wastewater treatment plants (Wood and Marsh, 1999). According to Sládeček (1980), both *Plumatella repens* and *P. fungosa* survive at only 30% saturation of dissolved oxygen. Many species, such as *Lophopodella carteri* flourish both in the laboratory and the field when supplied with large quantities of suspended organic particles. In Europe, the distribution of *Plumatella fungosa* is correlated with nutrient-enriched water (Job, 1976). *Pectinatella magnifica* in the United States, Japan, and Korea become luxuriant in areas that are visibly eutrophic.

One clearly limiting factor for all species is the availability of suitable surface on which to grow. Almost any solid, biologically inactive material is acceptable, including rocks, glass, plastics, automobile tires, and aged wood. Surfaces that are seldom successfully colonized include newly dead wood, corroded metal, and oily or tarred materials. Aquatic plants are common substrates, although in a detailed survey of bryozoan–plant relations, Bushnell (1966) found little evidence for species specificity. Among large, floating leafed plants, the lotus lily, *Nelumbo lutea*, supports heavy growths of bryozoans, while other lilies, such as *Nymphaea* or *Nuphar*, are seldom colonized.

Free floatoblasts, of course, are passive carriers of living material to new substrates. Swimming larvae, on the other hand, actively select attachment sites. Hubschman (1970) demonstrated particle size discrimination in larvae of *Pectinatella magnifica*, concluding that rock particles smaller than 1-mm diameter were avoided. Additional selection criteria are likely. Curry *et al.* (1981) found evidence for selective settling of *Paludicella articulata* larvae on mussel shells.

In general, freshwater bryozoan species are widely distributed but variable in local abundance. Some species appear to be distributed as metapopulations with regular cycles of local extinction followed by new immigration (Okamura, 1997). Two of our most common species, *Fredericella indica* and *Plumatella reticulata*, are rare in Europe (Massard *et al.*, 1992; Massard and Geimer, 1996); similarly, the abundant European *Plumatella fungosa* is seldom found on our continent. A near worldwide distribution is enjoyed by *Plumatella casmiana, Lophopodella carteri*, and *Paludicella articulata*. However, the old notion that many other species have a cosmopolitan distribution was apparently based on faulty identification. *Cristatella mucedo* is restricted to holarctic regions, while *Fredericella browni* and *Pottsiella erecta* are most common where winters are mild. About 25% of North American species are considered rare, and at least one, *Lophopus crystallinus*, is seriously endangered. Some species are known to be expanding their range: *Pectinatella magnifica* jumped from North America to Europe and Japan and has recently invaded Korea; *Lophopodella carteri* is increasingly common since its introduction to North America in the 1930s; and the once scarce *Plumatella vaihiriae* is popping up in lakes and wastewater treatment facilities across the continent. Species believed to be endemic to North America now include *Plumatella nitens* (Wood, 1996), *P. orbisperma* (Ricciardi and Wood, 1992), *P. recluse* (Smith, 1992) and the tiny ctenostome, *Sineportella forbesi* (Wood and Marsh, 1996).

All known North American species have been reported east of the Mississippi River and north of the 39th parallel. Nineteen of the 24 species occur in states bordering the Great Lakes and an additional two species are described only from New England. Only the brackish *Victorella pavida* has not been found further north than Chesapeake Bay.

Relatively little is known about freshwater bryozoans west of Ontario and the Mississippi River. Most states and provinces have no published records of bryozoans. However, various reports show Colorado and Utah with natural populations of "*Plumatella repens*", *P. casmiana, P. emarginata, P. fruticosa, P. fungosa, Fredericella indica*, and *Cristatella mucedo*. Other western sightings show *Cristatella mucedo* in British Columbia (Reynolds, 1976), and "*Plumatella repens*" in Arizona and Nevada (Wilde *et al.*, 1981; Dehdashti and Blinn, 1986). *Pectinatella magnifica* appears to be well established in eastern Texas (Neck and Fullington, 1983) and in the Pacific Northwest (personal communications). In Nebraska, the most abundant species appears to be *Plumatella vaihiriae* (Wood, unpublished).

The most recent regional surveys are those of Bushnell (1965a–c) in Michigan, Wood (1989) in Ohio, Smith (1989b) in Massachusetts, Ricciardi and Reiswig (1994) in eastern Canada, and Marsh and Wood in Illinois (in preparation). Each of these has revealed significant new species records for North America.

B. Reproduction and Life History

In most parts of the United States, bryozoan populations have two or more generations during the growing season. Only *Fredericella indica* typically is found throughout the year living even under a cover of ice. Populations of other species are normally maintained during unfavorable seasons by dormant statoblasts. An unusual exception to this pattern was reported by Dehdashti and Blinn (1986) in a population of "*Plumatella repens*" inhabiting an Arizona cave, where nearly constant conditions permitted uninterrupted, active growth.

Winter dormancy in statoblasts is broken by conditions favorable to colony growth. In temperate climates the principal triggering factor appears to be temperature. Viable statoblasts often germinate within a few days of each other when the water temperature rises above 8°C. In those regions where winter water temperature never approaches freezing, the time of germination may be more variable.

The single zooid that emerges from a statoblast is termed the ancestrula. It eventually buds one-to-five new zooids, each similarly capable of multiple zooid budding. Colonies grow rapidly as water temperature rises. Bushnell (1966) described naturally occurring colonies of *Plumatella nitens* (identified at the time as *P. repens*) which tripled and quadrupled in size within one week. Doubling times of five days have been reported for both *Plumatella casmiana* and *Fredericella indica* (Wood, 1973).

In species of *Plumatella, Fredericella*, and *Lophopodella* statoblasts may appear in colonies having fewer than five zooids; in most other species they are not formed until colonies are much larger. Among natural populations, a second generation usually arises either from the statoblasts or larvae of early spring colonies, or even from fragments of colonies surviving from the first period of growth. Third generation colonies also are known, but any statoblasts they produce normally remain in diapause until the following spring. *Plumatella casmiana* is unique in releasing thin-walled statoblasts which germinate immediately, so there may be many overlapping generations throughout the growing season.

Gametogenesis, when it occurs, is normally encountered in the first spring generation of colonies, with larvae released 2–4 weeks after fertilization. Sperm masses develop on the funiculus as ova appear on the peritoneum. The sperm later break free to circulate in the coelom for 1 to 2 weeks. This is followed by larval development and eventual release. There appears to be little overlap in larval release dates among closely related species living in the same area. In two Ohio lakes, free-swimming larvae of *Plumatella rugosa* appeared throughout June; *P. casmiana* larvae were released during July; then came *Hyalinella punctata* in late July–early August, followed by *Plumatella emarginata* in late August–early September (Zimmerman, 1979).

Sexual activity in phylactolaemates is often reported as rare even in such careful studies as that of Wöss (1996). Among the Plumatellidae, with a prolific production of asexual statoblasts, sexual reproduction would seem a relatively unimportant means for recruitment. However, in *Pectinatella magnifica* and *Hyalinella punctata*, with only one statoblast generation per year, sexually produced larvae play a significant role in establishing new colonies (Hubschman, 1970; Wöss, 1999). Likewise, in *Fredericella* species, because of their fixed statoblasts, larvae are assumed to be instrumental for both population growth and dispersal.

C. Ecological Interactions

The constant flow of water through bryozoan lophophores creates a favorable environment for the microscopic aufwuchs community. Protozoans, rotifers, gastrotrichs, microcrustaceans, and other small animals congregate especially on and around branching tubular colonies of bryozoans. Flatworms, oligochaetes, snails, orbatid mites, crayfish, and such insect larvae as caddisflies and midges occasionally graze on the living zooids. Predation is seldom extensive.

Chironomid larvae enter old tubes of *Plumatella repens*, hastening their disintegration, and occasionally damaging living portions of the colony. Some workers consider such damage to be accidental rather than the result of direct feeding, while others regard chironomid larvae as important predators. Colonies of *P. casmiana* escape such harm, possibly because larvae cannot fit themselves inside the narrow, branching tubules, but instead build detritus tubes alongside the exterior colony wall where they cause little harm. *Pectinatella magnifica* is often a host to chironomid larvae, which find shelter by burrowing into the gelatinous base close to the substrate and causing no apparent damage. One of these, *Tendipes pectinatellae*, is reported to be specifically commensal on *P. magnifica* (Dendy and Sublette, 1959).

Myxozoan and microsporidian parasites have long been identified with freshwater bryozoan hosts. These have been recently confirmed and further described from *Cristatella mucedo* by Okamura (1996) and Canning *et al.* (1996, 1997). Myxozoans are easily seen with light microscopy when they circulate in the coelomic fluid. Infected colonies lose their normal vigor and soon die. *Buddenbrockia*, an active, wormlike parasite, has been described from the coelom of several plumatellids (e.g., Schröder, 1912; Marcus, 1941). Easily detected through the clear body wall of *Hyalinella punctata*, for example, it seems to cause little distress to the hosts as long as food remains abundant to the host. Life cycle details of any of these parasites are so far unknown.

Extensive predation by fish has not been verified. Walburg *et al.* (1971) lists "bryozoans" among the stomach contents of six reservoir fish species, but provides no details. When Dendy (1963) observed bluegills biting off pieces of *Plumatella* sp., he believed the fish were doing so only to obtain insect larvae associated with the colonies. A homogenate of *Lophopodella carteri* tissues is highly toxic to fish (Meacham and Woolcott, 1968), and attempted feeding by fish on this or any other gelatinous species has never been observed. Freshwater prawns will graze on colonies of *Plumatella vaihiriae*, but only when no other food is available (Bailey-Brock and Hayward, 1984), suggesting either a lack of nutritional value or a repellant chemistry of these colonies.

As sessile suspension feeders, bryozoans handle a wide variety of food. Richelle *et al.* (1994) clearly demonstrated the ability of *Plumatella fungosa* to thrive on a diet of suspended bacteria. Other ingested particles include diatoms, desmids, green algae, cyanobacteria, dinoflagellates, rotifers, small nematodes, protozoa, and even microcrustaceans, along with bits of detritus and inorganic materials. In a comparative study of three species, Kaminski (1984) found that 95% of ingested particles were under 5 μm in diameter. *Plumatella repens*, with its wide mouth and strong gut musculature, ingested larger organisms than did either *Cristatella mucedo* or *Plumatella fruticosa*. These included rotifers (*Keratella* sp.), colonial green algae, and cyanobacteria as large as 75 μm. In general, organisms with long body extensions were able to avoid ingestion by bryozoans, while small, rounded shapes were taken most easily.

Analysis of stomach contents alone does not reveal the important sources of bryozoan nutrition. Rotifers and green algae have been known to pass through the gut completely unharmed (Hyman, 1959) although there are reports that a large portion of ingested organisms are variously damaged (Rüsche, 1938; Kaminski, 1991a).

The quantity of seston removed from a 460-ha eutrophic lake by *Plumatella fungosa* is estimated at 15 metric tons per year (Kaminski, 1991b). At the same time, 8.8 tons of fecal pellets are deposited in the sediments, about twice the amount contributed by fish or waterfowl, but less than that deposited by molluscs.

D. Evolutionary Relationships

Ectoprocta is one of three animal phyla collectively known as lophophorates. The other two, Brachiopoda and Phoronida, are represented by a small number of solitary, mostly sessile, marine animals. For all of these, the main unifying feature is a lophophore with hollow tentacles containing an extension of the coelom, and with cilia beating in metachronal waves to bring water and suspended food toward the mouth.

As a group, the lophophorates show no clear affinity with any other invertebrates. The semblance of radial cleavage, mesoderm formation, and other elements of morphology and development in branchiopods appears distinctly deuterostome (Zimmer, 1973; Emig, 1984). However, recent analysis of 18S rDNA sequence data suggest that lophophorates are protostomes (Halanych *et al.*, 1995), a conclusion supported by other biochemical and morphological studies (e.g., Willmer, 1990; Gutmann, 1978).

Structural similarities have been noted between ectoprocts and the wormlike sipunculids. The anterior introvert of a sipunculid is protruded and withdrawn in a manner very similar to that of the ectoproct polypide, using coelomic pressure and retractor muscles. The sipunculid introvert ends in a mouth surrounded by hollow, ciliated, tentacular outgrowths. The gut is U-shaped, and the anus is situated near the base of the introvert.

In fact, nucleotide sequence data suggest that sipunculids and gymnolaemate bryozoans may be closely related, as are the brachiopods and phoronids (Mackey *et al.*, 1996; Cohen and Gawthrop, 1996). However, the same data indicate that lophophorates, in general, and ectoprocts, in particular, are not monophyletic assemblages. The gymnolaemate, *Alcyonidium gelatinosum* and the phylactolaemate, *Plumatella repens* appear to be neither closely related to each other nor to the brachiopod–phoronid grouping.

Nevertheless, it is reasonable so far to propose an ancestral line that led directly to phoronids and phylactolaemate ectoprocts. Only these two groups have a crescentic lophophore and an epistome. They also both produce new buds from a region on the oral (ventral) side of the adult, while in other ectoprocts budding is in an anal direction (Jebram, 1973). Many phoronids produce special, fat bodies for energy storage, which are strikingly similar to early developmental stages of phylactolaemate statoblasts. While the larvae, in general, are quite different, the phoronid, *Phoronis ovalis*, has a unique, ciliated actinotroch (Silén, 1954) that appears similar to the phylactolaemate "larva." Significantly, *P. ovalis* also forms colonies which resemble somewhat those of the phylactolaemate family Fredericellidae.

Phoronids and gymnolaemate ectoprocts are probably more distantly related, as reflected especially by their different embryology and larval metamorphosis. The phoronid actinotroch larva, for example, has a distinct coelom, the larval gut is retained in the adult, and the lophophore develops from the metatroch ring of cilia. By contrast, the gymnolaemate cyphonautes larvae has no larval coelom, the larval gut is not retained, and tentacles develop from the episphere region.

Within the class Phylactolaemata, similarities in the mode of colony growth and statoblast morphology have long been regarded as the basis for evolutionary relationships. *Fredericella* species are thought to exhibit primitive features with their simple statoblasts and an open, dendritic pattern of colony branching. The circular outline of the lophophore, associated with a relatively small number of tentacles, may also be a primitive character. The evolutionary trends include a shift toward greater compactness of the colony and the accomodation of increasing numbers of tentacles on the lophophore. Statoblasts serve the two seemingly conflicting roles of dispersal and of retaining a position on proven favorable substrate.

Among the Plumatellidae and Stephanellidae all species retain a basically tubular design. The two statoblast functions are served by distinctly different statoblast types: floatoblasts and sessoblasts.

In Lophopodidae, Pectinatellidae, and Cristatellidae the colonies are gelatinous and globular, with no trace of branching. This development is coupled with a larger lophophore and a single remarkable statoblast design that incorporates a buoyant ring with marginal hooks. These statoblasts function both as an anchor and a means for dispersal. Radiating spines and hooks were apparently acquired independently by all three families in an interesting case of parallel evolution.

Hyalinella punctata, normally classified with the Plumatellidae, shares many features with the Lophopodidae, and may actually link the two groups. Such features include a gelatinous colony wall, the absence of a sessoblast, and a large floatoblast which, like those of lophopodids, is initially nonbuoyant upon release.

Fossil statoblasts resembling those of recent *Plumatella, Hyalinella*, and *Stephanella* species appear in rock strata from the Late Permian (Vinogradov, 1996), demonstrating a remarkable stability in these taxa. Future fossil discovery and analysis could well provide key information on phylactolaemate systematics.

The freshwater Gymnolaemata, all ctenostomes, are members of an ancient marine group. They share with the Cheilostome bryozoans many features expressed in the development of both larvae and colonies. At this point, however, phylogenetic relationships are still speculative.

IV. ENTOPROCTA

Entoprocta is a small group of about 60 species distinct in almost every way from the Ectoprocta but historically included with them under the name "bryozoan." Both groups are sessile with ciliated tentacles and an incomplete separation of budded zooids, but the similarities stop there. The only known freshwater species, *Urnatella gracilis*, was discovered in North America in 1851. Two other species have been proposed elsewhere based mainly on number of stalk segments and morphology of the basal plate. Emschermann (1965) considers all of these synonymous with *U. gracilis*. Although normally classified as Urnatellidae, the genus is similar to the marine species *Barentsia* and may be united with it in the family Pedicellinidae.

The zooid is a bulbous calyx borne on a flexible, segmented stalk measuring up to 5-mm long (Fig. 7). Several such stalks, either solitary or sparingly branched, may arise from a basal plate. The calyx bears a single whorl of 8–16 uniformly short, ciliated tentacles. The area of the calyx enclosed by the tentacles, called an atrium or vestibule, includes both the mouth and the anus. When the zooid is disturbed, the tentacles fold over the vestibule and are covered by a tentacular membrane.

Internal organs include a fully ciliated digestive tract and a large medial ganglion. A number of excretory flame bulbs communicate through short ducts to a common nephridiopore; additional flame bulbs occur in the stalk. The body cavity is a pseudocoel filled with loose mesenchyme and extending into each tentacle.

The calyx is deciduous, dropping off at the onset of cold temperatures or other unfavorable conditions. The

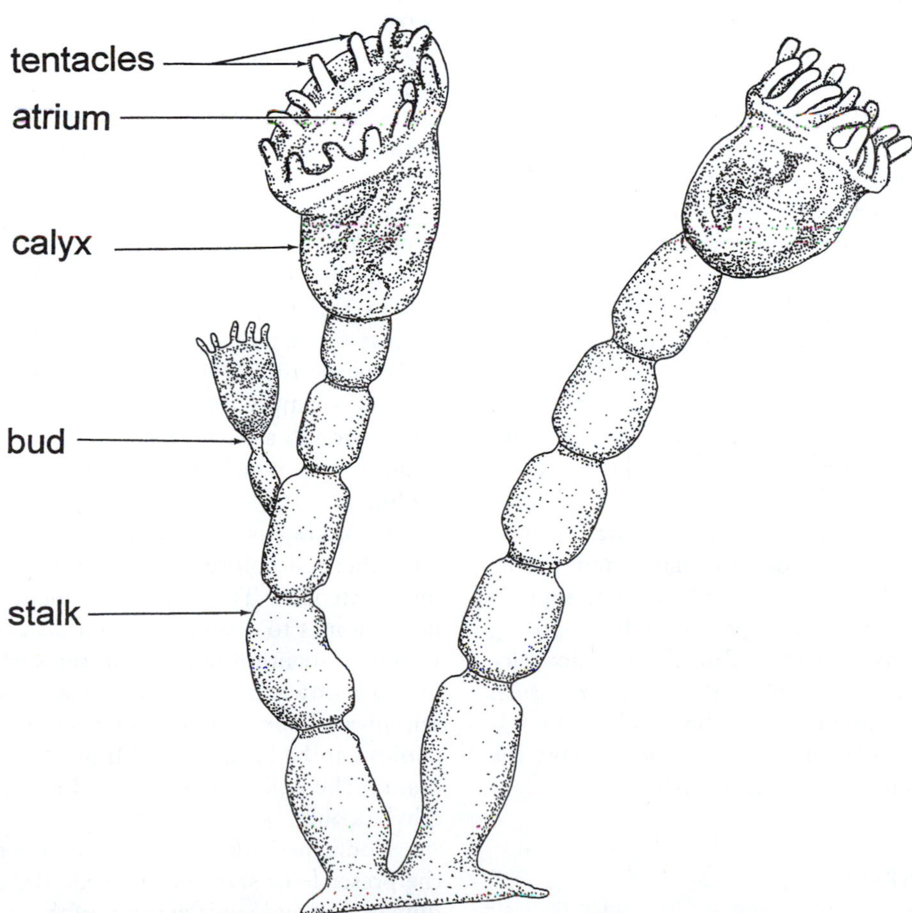

FIGURE 7 Small entoproct colony (*Urnatella gracilis*) showing external structures.

basal segments of stalks, containing food and germinal tissue, can survive the winter much like ectoproct statoblasts, forming new calyxes when the water temperature returns to around 15°C. New zooids develop by budding from the stalk, as illustrated particularly well by Oda (1982). It is reported that young colonies can disperse locally by crawling over the substrate (Protosov, 1980). Sexual reproduction leads to the development of swimming larvae, presumably similar to those of other Pedicellinidae; details are unknown.

Like other bryozoans, *Urnatella gracilis* is a suspension feeder, consuming organic particles, unicellular algae, and protozoans. Two especially important foods are the diatom *Melosira* and two green algae species, *Pediastrum duplex* and *P. simplex* (Weise, 1961).

Urnatella gracilis is known from every continent except Antarctica and Australia. In North America its reported distribution ranges from the east to the west coast and from Florida, Louisiana, and Texas to as far north as Michigan. Zooids attach to almost any substrate, including rocks, sticks, aquatic plants, bivalve shells, ectoprocts, and such debris as nails, beverage cans, and lead fishing weights (Eng, 1977). Individuals occur most frequently in flowing water or in shallow areas of large lakes where there is extensive water movement. Tolerance to a wide range of chemical and physical conditions has been noted most recently by Hull *et al.* (1980), and Cusak and McCullough (1985).

In some areas, especially on new substrate, entoprocts may comprise a sizeable fraction of the macroinvertebrate community. Stalk densities of over $225{,}000/m^2$ and an average biomass of $868\ mg/m^2$ (dry weight) were reported from a seven month old artificial stream in Mississippi (King *et al.*, 1988).

Entoproct phylogeny is even less clear than that of the Ectoprocta. Any similarities to ectoproct bryozoans are clearly the result of convergent evolution rather than common ancestry. A close comparison of entoproct tentacles and the ectoproct lophophore reveals little in common despite their outwardly similar structure and function. Body cavities of the two groups are likewise irreconcilable: a pseudocoel on one hand and a true coelom on the other. Analysis of 18S r-RNA data support a wide separation between entoprocts and ectoprocts, placing the entoprocts instead among the pseudocoelomates (Mackey *et al.*, 1996). Nevertheless, puzzles such as this are expected in animals so highly modified by sessile life style and modular architecture, and for now the question of entoproct origins remains open.

V. STUDY METHODS

Bryozoans are normally found in areas of shallow water where there is a suitable firm substrate. Although most species are plainly visible to the unaided eye, a 10× Coddington-type magnifier is useful for field identification. Many species can be detected by examining the edges of floating objects for their free statoblasts. Some gelatinous colonies can be nudged gently from the substrate with little damage, but with other species it is better to collect pieces of substrate with colonies attached. Basic equipment for this includes a sturdy knife for organic substrates and a cold chisel and hammer for rocks. In lakes, where colonies are elusive, statoblasts can often be retrieved from sediments using a stack of standard seives with mesh openings of 1.0 mm, 500, and 150 μm (Jones *et al.*, 1998).

Swimming larvae are most easily detected by placing large colonies in a shallow tray of water. If larvae are present, they will generally be released with gentle prodding, or will emerge spontaneously during the nighttime hours. They are best seen against a dark background. Before their release from the colony, zooids with larvae appear swollen at the tip.

Bryozoans are not difficult to rear in the laboratory, but the substrate on which they attach must always be inverted so that fecal material and other settling debris will not accumulate around the zooids. Most species can be kept at room temperature. At lower temperatures there are fewer problems with fouling organisms, but colony growth is slow. Food appears to be the most important variable for success with laboratory-reared bryozoans. Oda (1980) maintained colonies of *Lophopodella carteri* on pure cultures of *Chlamydomonas reinhardtii*. Wayss (1968) kept *Plumatella repens* for three years using a variety of unicellular green algae in 1:1 Knop's solution and soil extract. Others have achieved limited short-term success using mixed protozoan cultures, pet fish food, and fine detritus. One reliable and trouble-free source of food is simply the suspended organic particles circulated through a dark rearing tank from a large, well-lit aquarium in which active fish are maintained (Wood, 1996).

If specimens are to be preserved, they should be anaesthetized before fixing so the lophophores will remain extended. The most convenient method is to confine colonies to a small covered dish of water with thin wafers of menthol floating on the surface. The menthol diffuses slowly into the water and relaxes the zooids within an hour or two. Lophophores of anaesthetized zooids can be hardened with drops of full-strength formalin. The colony is then fixed and preserved in 70% ethyl alcohol.

Confirmed identification of most species requires the presence of statoblasts inside the colony. Statoblast dimensions and surface topography are species-specific. In specimens stored for an appreciable time, any gas in the floatoblast annulus is replaced by liquid, making the entire capsule clearly visible through the periblast. In this case, it is important not to mistake capsule dimensions for those of the fenestra. Length and width of sessoblasts should be measured from the base of the lamella, not from its outer edge.

Surface features of the floatoblast fenestra can often be seen in isolated valves using ordinary light microscopy. However, features of the annulus require the use of a scanning electron microscope. The most common surface pattern in the floatoblast annulus shows the sclerotized cells in relief, like cobblestones, a condition called "paved" (Fig. 12a). The annulus may also be reticulated (Fig. 12b) or adorned with tiny nodules (Fig. 12d). In various species the floatoblast fenestra may be smooth (Fig. 12a), reticulated (Fig. 12b), tuberculated, or bearing a reticulum in which each cell contains a single interstitial tubercle (Fig. 12d). Other diagnostic features of the floatoblast include polar grooves on the dorsal valve (Fig. 11b) and appearance of the intact suture (Fig. 12c). All recent species descriptions include statoblast details such as these.

For detailed examination of a statoblast using light microscopy, it is useful to first separate the component parts by placing it for about 1 min in a hot solution of potassium hydroxide. Then transfer it to distilled water where the valves should separate spontaneously. Use fine insect pins (size 000) to assist in this process and to tease away the yolky contents. I find it convenient to arrange these parts under the microscope and photograph them for my working records.

To prepare statoblasts for scanning electron microscopy it is first necessary to remove manually the thin membrane that often adheres tightly to the surface. If subsequent cleaning is necessary, ultrasonic treatment is too harsh. Instead, place the statoblasts in a small Eppendorf vial with a 0.05 M sodium hexametaphosphate, then hold the vial against the shaft of an electric vibrating blade shaver. Wash the statoblasts in several changes of distilled water. Freeze-drying helps prevents distortion of the fenestra. Sputtering is advisable but not essential.

Well over half the recognized phylactolaemate species, are represented by type specimens in major museums, including all species named since the 1940s (Wood, 1999). Even long neglected and desiccated material remains valuable for morphological study of statoblasts.

Certain techniques for subcellular study of freshwater bryozoans have proven particularly useful. These include analysis by karyotype (Backus and Mukai, 1987), by randomly amplified polymorphic DNA (Okamura *et al.*, 1993), 18S ribosomal DNA (Halanych *et al.*, 1995), and the identification of polymorphic microsatillite loci (Freeland *et al.*, 1999).

VI. TAXONOMIC KEY TO NORTH AMERICAN FRESHWATER BRYOZOANS

1. Zooid composed of bulbous body on externally segmented stalk; tentacles folding individually toward center when zooid is disturbed; statoblasts absent (Fig. 7). Phylum Entoprocta . . . *Urnatella gracilis* Leidy, 1851.

1′. Zooid without externally segmented stalk; tentacles withdrawn together when zooid is disturbed; statoblasts may be present. Phylum Ectoprocta 2

2(1′). Colony composed of branching tubules, body wall transparent-to-opaque; statoblasts (if present) with smooth margins; tentacles fewer than 65 3

2′. Colony globular and either lobed or entire in outline; body wall transparent; statoblasts with peripheral spines, hooks, or pointed extensions; sessoblasts absent; tentacles more than 65 25

3(2). Extended lophophore circular in outline; statoblasts (if present) piptoblasts only (Fig. 9a, b); tentacles fewer than 25 4

3′. Extended lophophore U-shaped in outline; statoblasts either floatoblasts or sessoblasts (or both); tentacles more than 25 10

4(3). Statoblasts never formed; ectocyst stiff, shiny, and transparent; tentacles fewer than 20; individual zooids clearly demarcated by internal septa; orifice appears quadrangular when lophophore is withdrawn. Gymnolaemata, Ctenostomata 5

4′. Statoblasts formed, especially in zooids attached to substrate, but possibly missing in some specimens; ectocyst not stiff and shiny; tentacles more than 19. Phylactolaemata, Fredericellidae, *Fredericella* 8

5(4). Each zooid includes a uniformly narrow, sinuous, stolonlike tubule by which it is joined to a parental zooid 6

5′. Stolonlike tubules absent, zooids branch from each other at nearly right angles; widely distributed in North America and worldwide. (Fig. 8b). Paludicellidae. *Paludicella articulata* (Ehrenberg, 1831).

6(5). Tentacles numbering exactly 8. Victorellidae 7

6′ Tentacles more than 15; zooids ranging from straight and erect-to-bulbous and recumbant. (Fig. 8c). Known throughout eastern North America, especially south of the 40th parallel. Pottsiellidae. *Pottsiella erecta* (Potts, 1884).

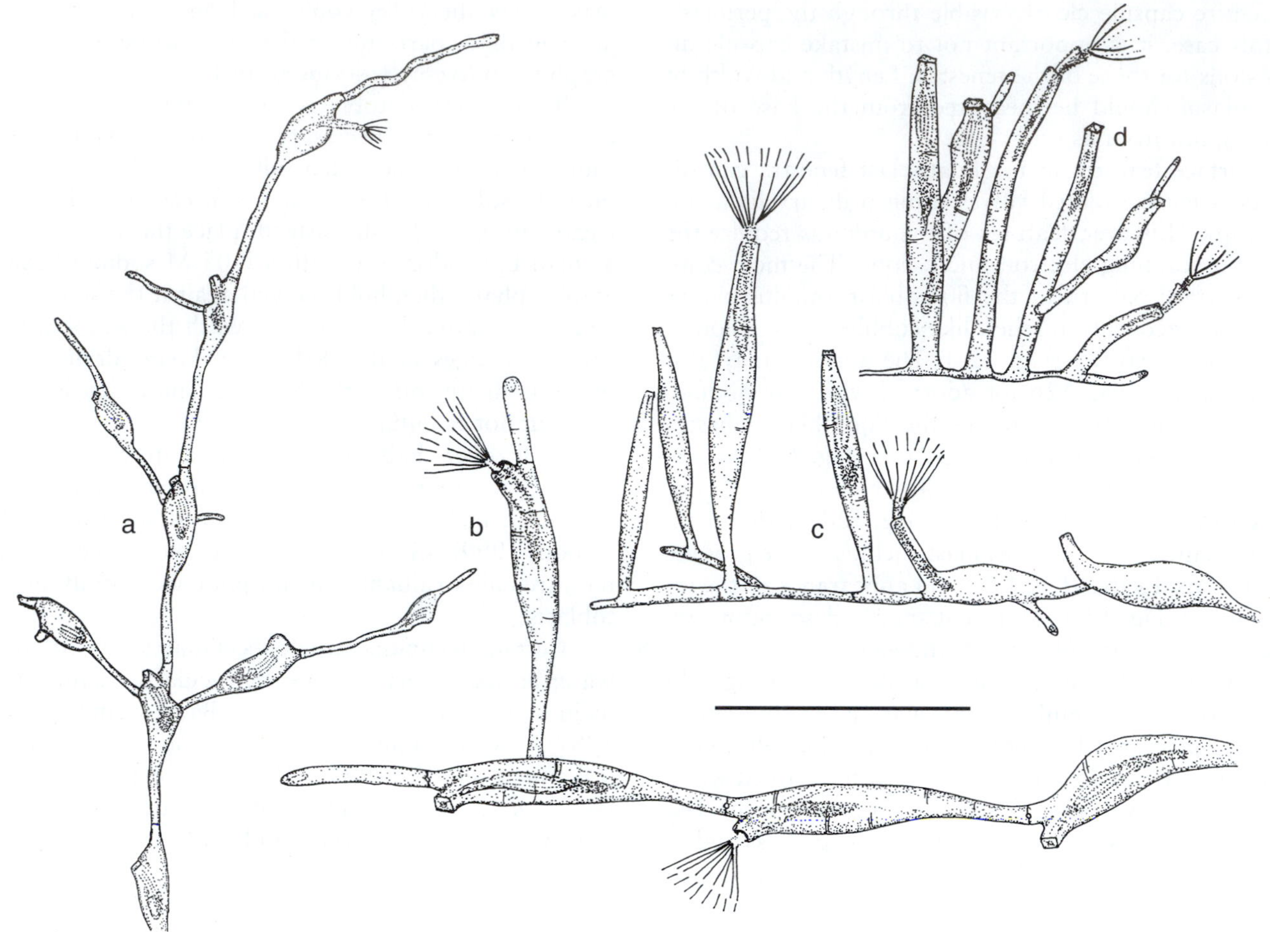

FIGURE 8 Gymnolaemate colonies occurring in fresh water. Scale bar for all figures = 1 mm. (a) *Sineportella forbesi*; (b) *Paludicella articulata*; (c) *Pottsiella erecta*; and (d) *Victorella pavida*.

7(6) Erect portion of zooid 0.6–1.6 mm long and only slightly contracile (Fig. 8d); occurring mainly in brackish water; reported in North America south from Chesapeake Bay and in southeastern Louisiana. *Victorella pavida* (Kent, 1870).

7′ Erect portion of zooid never more than 0.3 mm long and highly contractile (Fig. 8a), known only from a single site in east central Illinois. *Sineportella forbesi* Wood and Marsh, 1996.

8(4′) Piptoblast surface densely pitted or minutely roughened; appearing dull and granular when dry..9

8′ Piptoblast surface appearing mirror-smooth and shiny when dry. Common in Europe, confirmed in North America only from a single site in Curry County, Oregon. *Fredericella sultana* (Blumenbach, 1779).

9(8) Piptoblast broadly oval to round (Fig. 9b), often more than one per zooid; valves covered by a minutely wrinkled mantle which is easily removed by 5-min immersion in concentrated KOH solution to reveal smooth sclerotized surface. Reported throughout North, Central and South America. *Fredericella browni* (Rogick, 1945).

9′ Piptoblast oval-to-elongate (Fig. 9a), seldom more than one per zooid; surface uniformly pitted and unaffected by KOH. Common throughout North America, also scattered sites in Europe and Asia; probably includes several species not yet distinguished. (Fig. 13a). *Fredericella indica* Annandale, 1909.

10(3′) Floatoblasts circular or nearly so, with little difference in appearance between dorsal and ventral sides; colony wall thick, soft, transparent. Uncommon..11

10′ Floatoblasts oval-or-oblong; fenestrae of two valves distinctly different in size; colony wall not necessarily thick or transparent. Abundant and widespread. Plumatellidae 13

11(10) Zooids erect, extending up to 4 mm. but collapsing when removed from water; polypide and lophophore relatively small; floatoblast fully reticulate and lacking tubercles (Fig 9m). Confirmed in North America only from several sites in Massachusetts, although floatoblasts reported from Oregon. Stephanellidae. *Stephanella hina* Oka 1908.

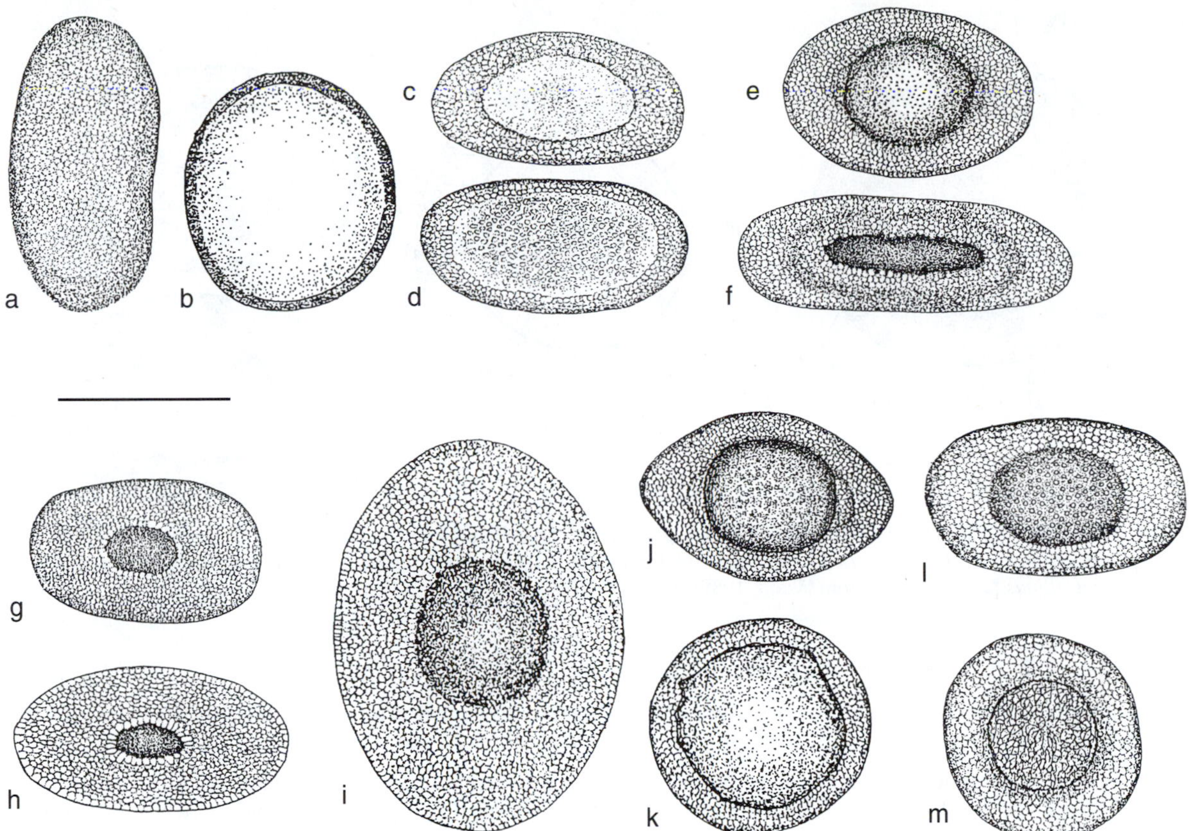

FIGURE 9 Free statoblasts from the families Plumatellidae and Fredericellidae. Scale bar for all figures = 250 μm. (a) *Fredericella indica*; (b) *Fredericella browni*; (c) *Plumatella casmiana* (capsuled floatoblast); (d) *Plumatella casmiana* (leptoblast); (e) *Plumatella repens*; (f) *Plumatella fruticosa*; (g) *Plumatella reticulata*; (h) *Plumatella emarginata*; (i) *Hyalinella punctata*; (j) *Plumatella vaihiriae*; (k) *Plumatella orbisperma*; (l) *Plumatella bushnelli*; and (m) *Stephanella hina* [a–e and g–i modified from Wood, 1989; j based on Rogick and Brown, 1942; k and m based on Bushnell, 1965b].

11′ Zoids not as above, floatoblast surface variable. Plumatellidae 12

12(11′) Floatoblast fenestra tuberculated, annulus adorned with minute nodules (Figs. 12d and 9k), sessoblast circular or nearly so; thick walls of large colonies fusing into a firm, transparent matrix. Known only from a few sites in Michigan and the northern Great Lakes. *Plumatella orbisperma* (Kellicott, 1882).

12′ Floatoblast fully reticulated and without tubercles; sessoblast long oval-to-rectangular. Known only from small ponds in forested sites of New England. *Plumatella recluse* Smith, 1992.

13(10′) Colony wall thick, soft, and transparent; colony entirely adherant to substrate; zooids forming low mounds when polypides are retracted (Fig. 13d); floatoblast width greater than 300 μm, length over 400 μm (Fig. 9i), dark and seldom buoyant upon release from colony, sessoblasts never formed. Widely distributed. *Hyalinella punctata* (Hancock, 1850).

13′ Floatoblast and colony not as above, or sessoblasts present 14

14(13′) Width of floatoblast dorsal annulus at poles never less than length of dorsal fenestra (Figs. 14.9g, h); internal septa frequent......... 15

14′. Width of floatoblast dorsal annulus at poles less than length of dorsal fenestra (Fig. 9e, f, j, l); internal septa present or absent...... 16

15(14). Floatoblast dorsal valve nearly flat, long edge curved; suture between valves visible in dorsal view (Figs. 5a, 9h, and 12a); sessoblast surface uniformly granular, tentacles about 40. Internal septa perpendicular to colony wall; mouth region often red. Common and widely distributed, especially in flowing water. *Plumatella emarginata* Allman, 1844.

15′. Floatoblast valves almost equally convex, long edge relatively straight (Fig. 9g), sessoblast surface roughened by network of raised lines; tentacles about 32; internal septa slightly oblique; mouth region never red. Common throughout North America and south at least to Panama, also reported from Israel. *Plumatella reticulata* Wood, 1988.

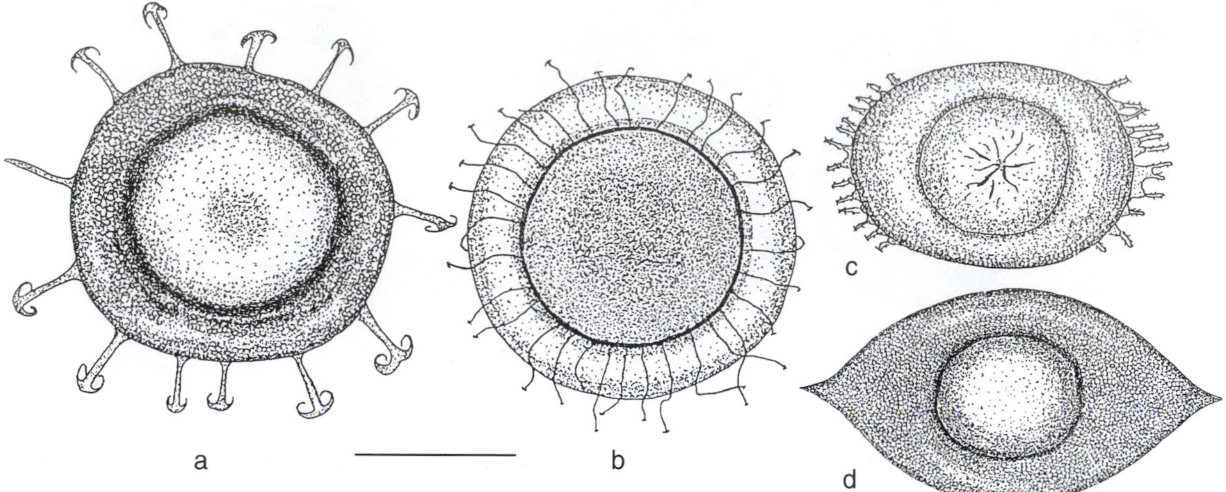

FIGURE 10 Statoblasts in the families Lophopodidae, Pectinatellidae, and Cristatellidae. Scale bar for all figures = 500 μm. (a) *Pectinatella magnifica*; (b) *Cristatella magnifica*; (c) *Lophopodella carteri*; (d) *Lophopus crystallinus* [a–c modified from Wood, 1989].

16(14′). Floatolast and sessoblast length not more than twice width 17

16′. Floatoblast and sessoblast length more than twice width (Fig. 9f); colonies with thin branches largely free of substrate. Widely distributed, especially in northern, oligotrophic habitats. *Plumatella fruticosa* Allman, 1844.

17(16) Floatoblast length-to-width greater than 1.6 (Fig. 9c, d); colony composed of short, profusely branched tubes always closely adherent to substrate (Fig. 13b), tubes becoming erect and fused when crowded; floatoblast dorsal fenestra lacking prominent tubercles;

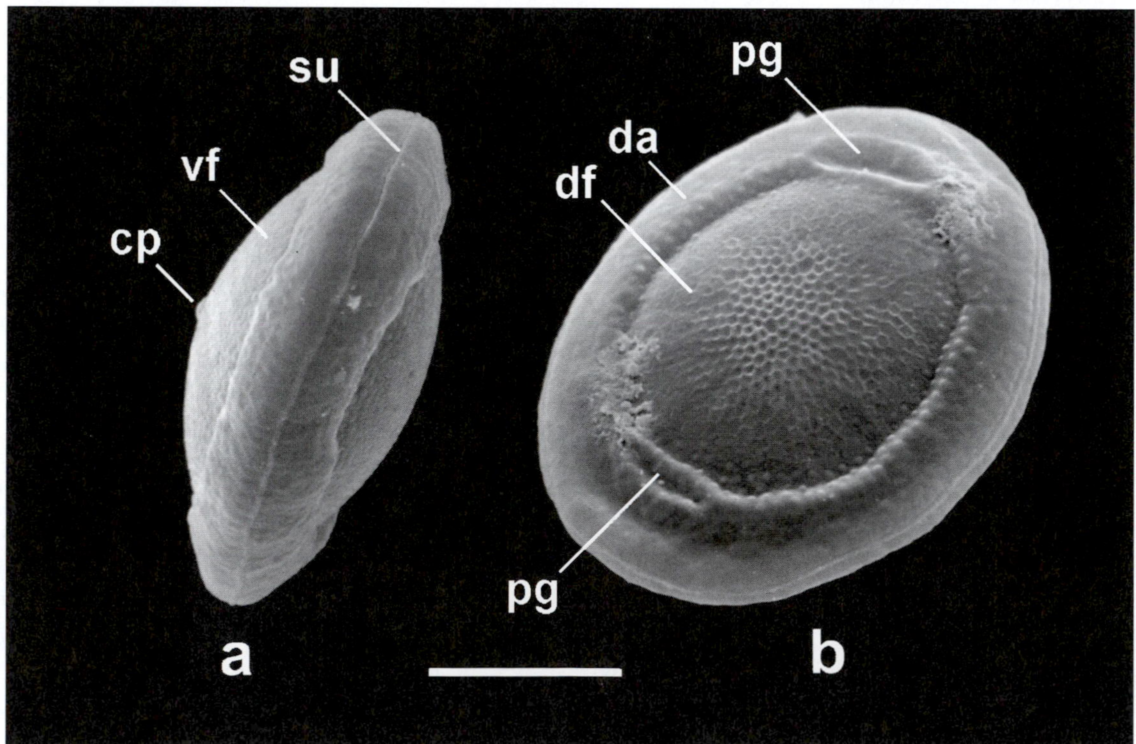

FIGURE 11 Scanning electron micrographs showing floatoblast features. Scale bar = mm. (a) Lateral view; (b) frontal view of dorsal valve. cp, central prominance; da, dorsal annulus; df, dorsal fenestra; pg, polar groove; su, suture; and vf, ventral fenestra. Scale bar = 100 μm.

delicate thin-walled floatoblast sometimes present (Fig. 9d); sessoblast lamella relatively narrow. Common and widely distributed. *Plumatella casmiana* (Oka, 1907).

17′ Floatoblast length to width less than 1.6 (Fig. 9e,j,l); colony branches not particularly short, colony or statoblasts not as above 18

18(17′) Floatoblast distinctly tapered at the poles, with strongly reticulated fenestrae (Figs. 9j, 12b); dorsal annulus bulging around inner capsule to form a distinct shoulder; sessoblast surface densely pitted and without tubercles. Uncommon but often locally abundant, ranging throughout the United States. *Plumatella vaihiriae* (Hastings, 1929).

18′ Floatoblast rounded at poles, fenestra not strongly reticulated, annulus extending evenly from suture to fenestra without bulging around capsule; sessoblast surface densely tuberculate 19

19(18′) Floatoblast lateral edges relatively straight (Fig. 9l); dorsal fenestra with prominent tubercles annulus with dense nodules visible by SEM; sessoblast lamella relatively wide. Known only from a single site on a North Carolina barrier island. *Plumatella bushnelli* Wood, 2001.

19′ Floatoblast lateral edges evenly curved (Fig. 9e); dorsal fenestra with weak tubercles and slight reticulation; common and widespread 20

20(19′) Floatoblast dorsal fenestra small, not more than half the diameter of ventral fenestra; ventral annulus markedly wider at poles than at sides 21

20′ Floatoblast dorsal fenestra large, considerably more than half the diameter of ventral fenestra (Fig. 11b); ventral annulus uniformly narrow. Known from Massachusetts to Wisconsin, mostly north of the 41st parallel. *Plumatella nitens* Wood, 1996.

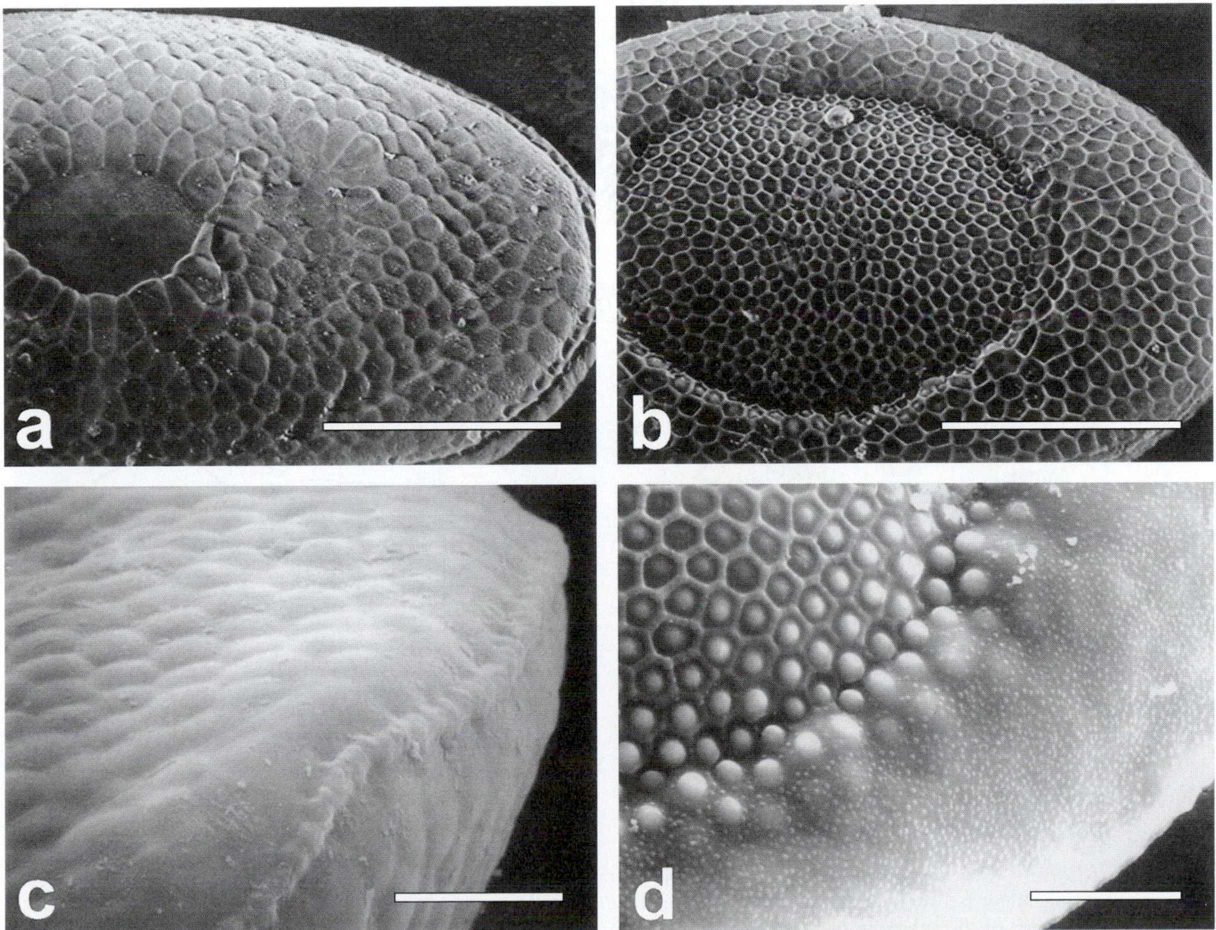

FIGURE 12 Scanning electron micrographs showing floatoblast surface morphology. Scale bar = mm. (a) *Plumatella emarginata* dorsal valve with "paved" annulus and smooth fenestra; (b) *Plumatella vaihiriae* with reticulated fenestra and annulus; (c) *Hyalinella punctata* with paved annulus and suture as a raised cord; and (d) *Plumatella orbisperma* with reticulation and "interstitial" tuibercles on the annulus, small nodules on fenestra.

21(20) Colony tubules seldom fused together; scanning electron microscopy (SEM) reveals no tubercles on floatoblast annulus, although tiny nodules may be present, internal septa are uncommon. 22

21′ Large colony forming tight masses of fused tubules without free branches; SEM reveals tubercles on floatoblast annulus; many internal septa; occurring in highly eutrophic water; abundant in Europe, but generally uncommon in North America. *Plumatella fungosa* (Pallas, 1768).

22(21) SEM shows floatoblast annulus surface smooth except for tiny, rash-like nodules 23

22′ SEM shows floatoblast annulus surface roughened by concave cell walls, nodules absent; species common and widespread in North America, especially in lentic habitats (Fig. 13c). *Plumatella rugosa* (Wood, Geimer and Massard 1998).

23(22) SEM shows nodules only on floatoblast annulus 24

23′ SEM shows nodules abundant on both floatoblast annulus and fenestra; known from Illinois, Ohio, and western New York. *Plumatella nodulosa* Wood, 2001.

24(23) Floatoblast suture defined by a lumpy, double cord; known only from Illinois. *Plumatella similirepens* Wood, 2001.

24′ Floatoblast suture marked by two crooked rows of large, toothlike tubercles; known from Europe and New Zealand, but not yet documented from North America. *Plumatella repens* (Linnaeus, 1758)

25(2′) Mouth region with red pigmentation; prominent pair of white spots at end of each arm of lophophore; statoblasts round with hooked spines radiating from outer margin of annulus (Fig. 10a). Colony gelatinous and slimy, ranging from a flat sheet to football-sized mass. (Fig. 13f). Common and widely distributed. Pectinatellidae. *Pectinatella magnifica* (Leidy, 1851)

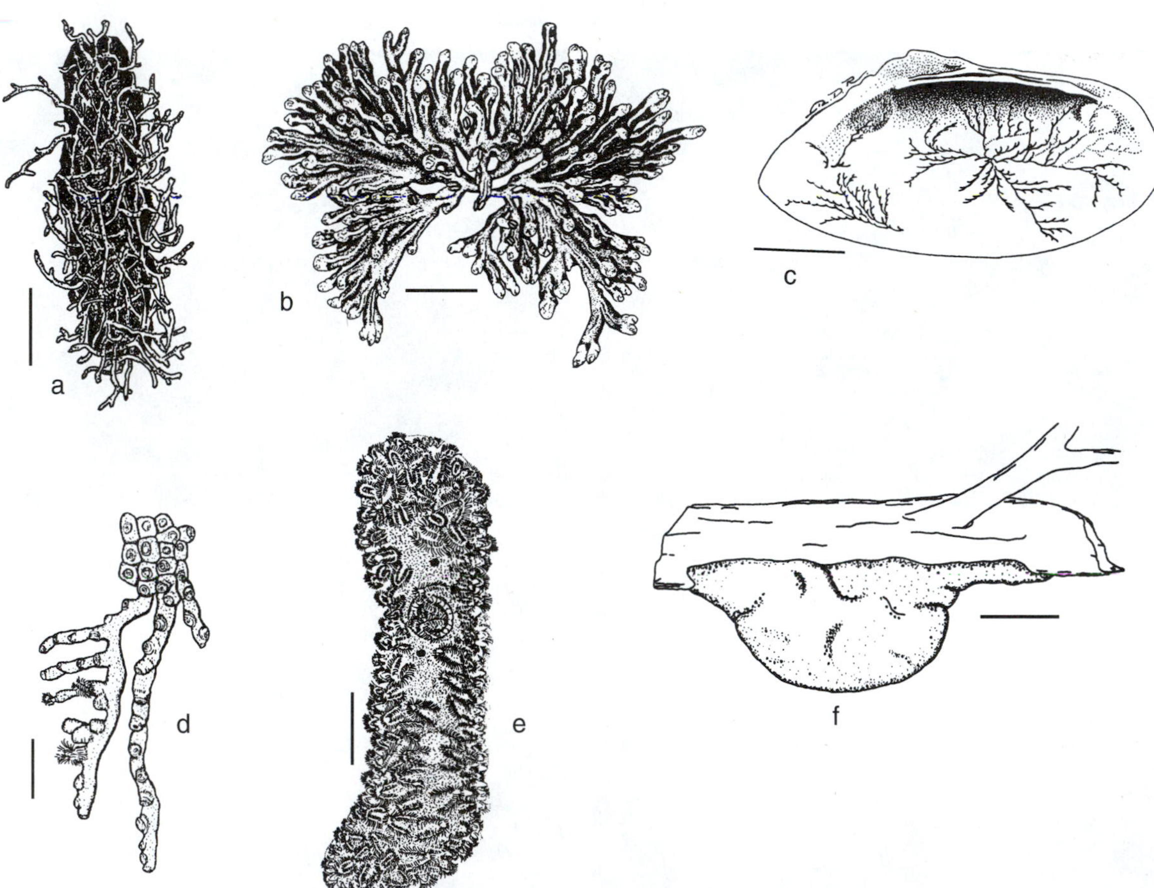

FIGURE 13 Various colony forms in Phylactolaemata. (a) *Fredericella indica* growing on a piece of wood, showing many free branches, scale bar = 5 cm; (b) *Plumatella casmiana*, with zooids attached throughout their length to the substrate or to each other, scale bar = 2 mm; c) *Plumatella rugosa* growing on the inner surface of a mussel valve, scale bar = 2 cm; (d) *Hyalinella punctata*, showing both densely packed and free-ranging growth, all zooids firmly attached to the substrate, scale bar = 5 mm; (e) *Cristatella mucedo*, showing a single statoblast inside, scale bar = 5 mm; (f) *Pectinatella magnifica* growing on the underside of a small log, scale bar = 2 cm [b from Rogick, 1941; c from Rogick and Brown, 1942; c–f from Wood, 1989].

25′ Mouth region without red pigmentation, lophophore lacking pair of white spots 26

26(25′) Colony not distinctly linear, statoblasts oblong, not round 27

26′ Colony linear, often longer than 2 cm (Fig. 13e); statoblasts round with wiry, hooked spines radiating beyond periphery from margin of fenestrae of both valves (Fig. 10b). Occuring mainly in oligotrophic waters. Cristatellidae. *Cristatella mucedo* Cuvier, 1798.

27(26) Statoblast with series of small hooks localized along the polar margins (Fig. 10c). Uncommon, but can be locally abundant (Fig. 4). *Lophopodella carteri* (Hyatt, 1866).

27′ Statoblast tapering to a single point at each pole (Fig. 10d); extremely rare worldwide, and possibly endangered. *Lophopus crystallinus* (Pallas, 1768).

* Species dropped from an earlier version of this key include *Stolella indica, S. evelinae*, and *Plumatella coralloides*. Recent work throws doubt on their reported occurrence in North America (Wood, 1998).

LITERATURE CITED

Backus, B. T. Mukai, H. 1987. Chromosomal heteromorphism in a Japanese population of *Pectinatella gelatinosa* and karyotypic comparison with some other phylactolaemate bryozoans. Genetica 73:189–196.

Bailey-Brock, J. H., Hayward, P. J. 1984. A freshwater bryozoan, *Hyalinella vaihiriae* Hastings (1929), from Hawaiian prawn ponds. Pacific Science 38:199–204.

Braem, F. 1896. Die geschlectliche Entwicklung von *Paludicella Ehrenbergii*. Zoologischer Anzeiger 19:54–57.

Brien, P. 1953. Etude sur les Phylactolémates. Annales de la Société Royale Zoologique de Belgique 84:301–444.

Brown, C. J. D. 1933. A limnological study of certain fresh-water Polyzoa with special reference to their statoblasts. Transactions of the american Microscopical Society 52:271–313.

Bullivant, J. S. 1968. The rate of feeding of the bryozoan, *Zoobotryon verticillatum*. New Zealand Journal of Marine and Freshwater Research 2:111–134.

Bushnell, J. H. 1965a. On the taxonomy and distribution of freshwater Ectoprocta in Michigan. Part I. Transactions of the American Microscopical Society 84:231–244.

Bushnell, J. H. 1965b. On the taxonomy and distribution of freshwater Ectoprocta in Michigan. Part II. Transactions of the American Microscopical Society 84:339–358.

Bushnell, J. H. 1965c. On the taxonomy and distribution of freshwater Ectoprocta in Michigan. Part III. Transactions of the American Microscopical Society 84:529–548.

Bushnell, J. H. 1966. Environmental relations of Michigan Ectoprocta, and dynamics of natural populations of *Plumatella repens*. Ecological Monographs 36:95–123.

Bushnell, J. H. 1974. Bryozoa (Ectoprocta), *in*: Hart, C. W. Fuller, S. L. H. Eds., Pollution ecology of freshwater invertebrates. Academic Press, New York.

Bushnell, J. H., Rao, K. S. 1974. Dormant or quiescent stages and structures among the Ectoprocta: physical and chemical factors affecting viability and germination of statoblasts. Transactions of the American Microscopical Society 93:524–543.

Canning, E. U., Okamura, B., Curry, A. 1996. Development of a myxozoan parasite *Tetracapsula bryozoides* gen. n. et sp. n. in *Cristatella mucedo* (Bryozoa: Phylactolaemata). Folia Parasitologica 43:249–261.

Canning, E. U., Okamura, B., Curry, A. 1997. A new microsporidium, *Nosema cristatellae* n. sp. in the bryozoan *Cristatella mucedo* (Bryozoa, Phylactolaemata). Journal of Invertebrate Pathology 70:177–183.

Cohen, B. L., Gawthrop, A. B. 1996. Brachiopod molecular phylogeny. Pages 73–80, *in*: Cooper, P., Jin, J. Eds. Brachiopods. Proceedings of the Third International Brachipod Congress. Balkema, Rotterdam.

Cooper, C. M., Burris, J. W. 1984. Bryozoans—possible indicators of environmental quality in Bear Creek, Mississippi. Journal of Environmental Quality 13(1):127–130.

Curry, M. G., Everitt, B., Vidrine, M. F. 1981. Haptobenthos on shells of living freshwater clams in Louisiana. The Wassman Journal of Biology 39:56–62.

Cusak, T. M., McCullough, J. D. 1985. *Urnatella gracilis* (Entoprocta) from Caddo Lake, Texas and Louisiana. The Texas Journal of Science 37:141–142.

Dehdashti, B., Blinn, D. W. 1986. A bryozoan from an unexplored cave at Montezuma Well, Arizona. The Southwestern Naturalist 31:557–558.

Dendy, J. S., 1963. Observations on bryozoan ecology in farm ponds. Limnology and Oceanography 8:478–482.

Dendy, J. S., Sublette, J. E. 1959. The Chironomidae (= Tendipedidae: Diptera) of Alabama with descriptions of six new species. Annals of the Entomological Society of America 52:506–519.

Eckman, J. E., Okamura, B. 1998. A model of particle capture by bryozoans in turbulent flow: significance of colony form. The American Naturalist 152(6):861–880.

Emig, C. 1984. On the origin of the Lophophorata. Zeitschrift für Zoologische Systematik und Evolutionforschung 22:91–94.

Emschermann, P. 1965. Über die dexualle Fortpflanzung und die Larve von *Urnatella gracilis* Leidy (Kamptozoa). Zeitshcrift für Morphologie und Ökologie der Tiere 55:859–914.

Eng, L. L. 1977. The freshwater entoproct, *Urnatella gracilis* Leidy, in the Delta–Mendota Canal, California. The Wassman Journal of Biology 35:196–202.

Everitt, B. C. 1975. Fresh-water Ectoprocta: distribution and ecology of five species in southeastern Louisiana. Transactions of the American Microscopical Society 94:130–134.

Franzén, A., Sensenbaugh, T. 1983. Fine structure of the apical plate in the larva of the freshwater bryozoan *Plumatella fungosa* (Pallas) (Bryozoa: Phylactolaemata). Zoomorphology 102:87–98.

Freeland, J. R., Jones, C. S., Noble, L. R., Okamura, B. 1999. Polymorphic microsatillite loci identified in the highly clonal freshwater bryozoan, *Cristatella mucedo*. Molecular Ecology 9:341–342.

Gilmour, T. H. J. 1978. Ciliation and function of the food-collecting and waste-rejecting organs of lophophorates. Canadian Journal of Zoology 56:2142–2155.

Goethals, B., Voss-Foucart, M. F., Goffinet, G. 1984. Composition chimique et ultrastructure de l'ectocyste et de la coque des statoblastes de *Plumatella repens* et *Plumatella fungosa* (bryozoaires phylactolémes). Annales des Sciences Naturelles, Zoologie, Paris 13e Serie 6:197–206.

Gutmann, W. F. 1978. Brachiopods: biomechanical interdependencies governing their origin and phylogeny. Science 100:890–893.

Halanych, K. M., Bacheller, J. D., Aguinaldo, A. M. A., Liva, S. M., Hills, D. M., Lake, J. A. 1995. Evidence from 18S ribosomal DNA that the lophophorates are protostome animals. Science 267:1641–1643.

Henry, V., Bussers, J. C., Bouquegneau, J. M., Thomé, J. P. 1989. Heavy metal and PCB contamination of bryozoan colonies in the River Meuse (Belgium). Hydrobiologia 202:147–152.

Hubschman, J. H. 1970. Substrate discrimination in *Pectinatella magnifica* Leidy (Bryozoa). Journal of Experimental Biology 52:603–607.

Hull, H. C., Bartos, L. F., Martz, R. A. 1980. Occurrence of *Urnatella gracilis* Leidy in the Tampa Bypass Canal, Florida. Florida Scientist 43:2–13.

Hyman, L. H. 1959. The lophophorate coelomates: Phylum Ectoprocta. The Invertebrates. McGraw–Hill, New York.

Jebram, D. 1973. The importance of different growth directions in the Phylactolaemata and Gymnolaemata for reconstructing the phylogeny of the Bryozoa, *in*: Larwood, G. P. Ed., Living and fossil Bryozoa. Academic Press, London.

Jones, C., Okamura, B., Noble, L. R. 1994. Parent and larval RAPD fingerprints reveal outcrossing in freshwater bryozoans. Molecular Ecology 3:193–199.

Jones, K., Marsh, T., Wood, T. 2000. Surveying for phylactolaemate bryozoans by seiving lentic sites for their statoblasts, *in*: Herrera Cubilla, A. and Jackson, J. B. C., Eds., Proceedings of the 11th International Bryozoology Association Conference, Smithsonian Tropical Research Institute, Panama, pp. 259–264.

Job, P. 1976. Intervention des populations de *Plumatella fungosa* (Pallas) (Bryozoaire phylactoléme) dans l'autoépuration des eaux d'unétang et d'un ruisseau. Hydrobiologica 48:257–261.

Kaminski, M. 1984. Food composition of three bryozoan species (Bryozoan Phylactolaemata) in a mesotrophic lake. Polskie Archiwum Hydrobiologii 31(1):45–53.

Kaminski, M. 1991a. Feeding of the freshwater bryozoan *Plumatella fungosa* (Pall.). 1. Food composition and particle size selection. Acta Hydrobiologica 33(3/4):229–239.

Kaminski, M. 1991b. Feeding of the freshwater bryozoan *Plumatella fungosa* (Pall.). 2. Filtration rate, food assimilation, and production of faeces. Acta Hydrobiologica 33(3/4):241–251.

King, D. K., King, R. H., Miller, A. C. 1988. Morphology and ecology of *Urnatella gracilis* Leidy, (Entoprocta), a freshwater macroinvertebrate from artificial riffles of the Tombigbee River, Mississippi. Journal of Freshwater Ecology 4(3):351–359.

Mackey, L. Y., Winnepenninckx, B., De Wachter, R., Backeljau, T., Emschermann, P., Garey, J. R. 1996. 18S rRNA suggests that Entoprocta are protostomes, unrelated to Ectoprocta. Journal of Molecular Evolution 42:552–559.

Marcus, E. 1926. Beobachtungen und Versuche an lebenden Süßwasserbryozoen. Zoologische Jahrbücher. Abteilung für Systematik, Ökologie und Geographie der Tiere 52:279–350.

Marcus, E. 1934. Uber *Lophopus crystallinus* (Pall.). Zoologische Jahrbücher. Abteilung für Anatomie und Ontogenie der Tiere 58:501–606.

Marcus, E. 1941. Sôbre Bryozoa do Brasil. Boletim da Faculdade de Filosopia, Ciências e Letras, Universidade de São Paulo 5:3–208.

Massard, J., Geimer, G. 1996. On the occurrence of *Fredericella indica* Annandale, 1909 (Phylactolaemata) in Europe, *in*: Gordon, D., Smith, A., Grant-Mackie, J. Eds., Bryozoans in space and time. National Institute of Water & Atmospheric Research, Wellington, New Zealand, pp. 187–192.

Massard, J. A., Geimer, G., Bromley, H. J., and Dimentman, C. 1992. Additional note on the fresh and brackish water Bryozoa of Israel (Phylactolaemata, Gymnolaemata). Bull. Soc. Nat. luxemb. 93: 199–214.

Meacham, R. H., Jr., Woolcott, W. S. 1968. Studies of the coelomic fluid and isotonic homogenates of the freshwater bryozoan *Lophopodella carteri* (Hyatt) on fish tissues. Virginia Journal of Science 19(3):143–146.

Morse, W. 1930. The chemical constitution of *Pectinatella*. Science 71:265.

Mukai, H. 1982. Development of freshwater bryozoans (Phylactolaemata), *in*: Harrison, F. W., Cowden, R. R. Eds. Developmental biology of freshwater invertebrates. Alan R. Liss, New York, pp. 535–576.

Mukai, H., Oda, S. 1980. Histological and histochemical studies on the epidermal system of higher phylactolaemate bryozoans. Annotationes Zoologicae Japonenses 53:1–17.

Mukai, H., Tsuchiya, M., Kimoto, K. 1984. Fusion of ancestrulae germinated from statoblasts in plumatellid freshwater bryozoans. Journal of Morphology 179:197–202.

Mukai, H., Fukushima, M., Jinbo, Y. 1987. Characterization of the form and growth pattern of colonies in several freshwater bryozoans. Journal of Morphology 192:161–179.

Mukai, H., Kobayashi, K. 1988. External observations on the formation of statoblasts in *Plumatella emarginata* (Bryozoa, Phylactolaemata). Journal of Morphology 196:205–216.

Neck, R., Fullington, R. 1983. New records of the freshwatyer ectoproct *Pectinatella magnifica* in eastern Texas. The Texas Journal of Science 35(3):269–271.

Nielsen, C. 1977. Phylogenetic considerations: the protostomian relationships, *in*: Woollacott R., Zimmer, R. Eds., Biology of Bryozoans. Academic Press, New York.

Oda, A. 1980. Effects of light on the germination of statoblasts in freshwater Bryozoa. Annotationes Zoologicae Japonenses 53:238–253.

Oda, S. 1982. *Urnatella gracilis*, a freshwater kamptozoan occurring in Japan. Annotationes Zoologicae Japonenses 55(3):151–166.

Okamura, B. 1996. Occurrence, prevalence, and effects of the myxozoan *Tetracapsula bryozoides* parasitic in the freshwater bryozoan *Cristatella mucedo* (Bryozoa: Phylactolaemata). Folia Parasitologica 43:262–266.

Okamura, B. 1997. The ecology of subdivided populations of a clonal freshwater bryozoan in southern England. Archiv für Hydrobiologie 141(1):13–34.

Okamura, B., Doolan, L. A. 1993. Patterns of suspension feeding in the freshwater bryozoan *Plumatella repens*. Biological Bulletin 184:52–56.

Okamura, B., C., Jones, S., Noble, L. R. 1993. Randomly amplified polymorphic DNA analysis of clonal population structure and geographic variation in a freshwater bryozoan. Proceedings of the Royal Society of London 253:147–154.

Protosov, A. A. 1980. On distribution of *Urnatella gracilis* (Kamptozoa) with reference to discharges of heated waters by thermal electrical power stations. Zoologicheskii Zhurnal 59:1569–1571 (in Russian).

Rao, K. S., Bushnell, J. H. 1979. New structures in binding designs of freshwater Ectoprocta dormant bodies (statoblasts). Acta Zoologica (Stockholm) 60:123–127.

Rao, K. S., Diwan, A. P., Shrivastava, P. 1978. Structure and environmental relations of sclerotized structures in fresh water Bryozoa. III. Observations on *Plumatella casmiana* (Ectoprocta: Phylactolaemata). Journal of Animal Morphology and Physiology 25:8–15.

Reynolds, J. D. 1976. Occurrence of the fresh-water Bryozoan, *Cristatella mucedo* Cuvier, in British Columbia. Syesis 9: 365–366.

Ricciardi, A., Reiswig, H. 1994. Taxonomy, distribution, and ecology of the freshwater bryozoans (Ectoprocta) of eastern Canada. Canadian Journal of Zoology 72:339–359.

Ricciardi, A. and Wood, T. S. 1992. Statoblast morphology and systematics of the freshwater bryozoan *Hyalinella orbisperma* (Kellicott, 1882). Can. J. Zool. 70:1536–1540.

Richelle, E., Moureau, Z., Van de Vyver, G. 1994. Bacterial feeding by the freshwater bryozoan *Plumatella fungosa* (Pallas, 1768). Hydrobiologia 291:193–199.

Rüsche, E. 1938. Nahrungsaufname und Nahrungsauswertung bei *Plumatella fungosa* (Pallas). (Ein Beitrag zur Frage des Stoffkreislaufs im Weiher). Archiv für Hydrobiolgie 33:271–293.

Schröder, O. 1912. Zur Kenntnis der *Buddenbrockia plumatellae* Ol. Schröder. Zeitschrift für wissenschafliche Zoologie 102: 79–91.

Shrivastava, P., Rao, K. S. 1985. Ecology of *Plumatella emarginata* (Ectoprocta: Phylactolaemata) in the surface waters of Madhya Pradesh with a note on its occurrence in the protected waterworks of Bhopal (India). Environmental Pollution (Series A) 39:123–130.

Silén, L. 1954. Developmental biology of Phoronidea of the Gullmar Fiordarea (west coast of Sweden). Acta Zoologica 31:215–257.

Sládeček, V. 1980. Indicator value of freshwater Bryozoa. Acta Hydrochimica and Hydrobiologica 8:273–276.

Smith, D. G. 1989a. On *Stephanella hina* Oka (Ectoprocta: Phylactolaemata) in North America, with notes on its morphology and systematics. Journal of the American Benthological Society 7:253–259.

Smith, D. G. 1989b. Keys to the freshwater macroinvertebrates of Massachusetts. No. 4: Benthic colonial phyla, including the Cnidaria, Entoprocta, and Ectoprocta (colonial hydrozoans, moss animals). Massachusetts Division of Water Pollution Control. Westborough, MA.

Smith, D.G. 1992. A new freshwater mos animal in the genus *Plumatella* (Ectoprocta: Phylactolaemata) from New England (U.S.A.). Canadian Journal of Zoology 70:2192–2201.

Tenney, W. R., Woolcott, W. S. 1966. The occurrence and ecology of freshwater bryozoans in the headwaters of the Tennessee, Savannah, and Saluda River systems. Transactions of the American Microscopical Association 85:241–245.

Vinogradov, A. V. 1996. New fossil freshwater bryozoans from the Asiatic part of Russia and Kazakhstan. Paleontological Journal 30(3):284–292.

Walburg et al. 1971. in Hall, Gordon E. Ed., Reservoir fisheries and limnology. Special Publication #8. American Fisheries Society, Washington, DC, pp. 449–467.

Wayss, K. 1968. Quantitative Untersuchungen über Wachstum und Regeneration bei *Plumatella repens* (L.). Zoologische Jahrbücher. Abteilung für Anatomie und Ontogenie der Tiere 85:1–50.

Weise, J. G. 1961. The ecology of *Urnatella gracilis* Leidy: Phylum Entoprocta. Limnology and Oceanography 6:228–230.

Wilde, G. A., Burke, T. A., Keenan, C. 1981. Bryozoa from Lake Mead and Lake Mohave, Nevada-Arizona. The Southwestern Naturalist 26:201.

Willmer, P. 1990. Invertebrate relationships. Cambridge University Press, Cambridge.

Wood, T. S. 1973. Colony development in species of *Plumatella* and *Fredericella* (Ectoprocta: Phylactolaemata), *in*: Boardman, R. S., Cheetham, A., Oliver, J. Eds., Development and function of animal colonies through time. Dowden, Hutchinson and Ross, Stroudsburg, PA, pp. 395–432.

Wood, T. S. 1989. Ectoproct bryozoans of Ohio. Bulletin of the Ohio Academy of Science, New Series 8:1–66.

Wood, T. S. 1996. Aquarium culture of freshwater invertebrates. The American Biology Teacher 58(1):46–50.

Wood, T. S. and Wood, L. J. 2000. Statoblast morphology in historical specimens of freshwater bryozoans, *in*: Herrera Cubilla, A. and Jackson, J. B. C., Eds., Proceedings of the 11th International Bryozoology Association Conference, Smithsonian Tropical Research Institute, Panama, pp. 421–430.

Wood, T. S., Marsh, T. G. 1996. *Sineportella forbesi*, a new victorellid bryozoan from Illinois (Ectoprocta: Ctenostomata). Journal of the North American Benthological Society 15(4):610–614.

Wood, T. S., Marsh, T. G. 1999. Biofouling of wastewater treatment plants by the freshwate bryozoan, *Plumatella vaihiriae* (Hastings, 1929). Water Research 33(3):609–614.

Wöss, E. 1996. Life history variation in freshwater bryozoans, *in*: Gordon, D., Smith, A., Grant-Mackie, J. Eds., Bryozoans in space and time. National Institute of Water & Atmospheric Research, Wellington, New Zealand, pp. 391–399.

Wöss, E. 1999. Colonization and development of freshwater bryozoan communities on artificial substrtes in the Laxenburg Pond (Lower Austria). Proceedings of the 11th International Bryozoology Association Conference, 1999 (in press).

Zimmer, R. L. 1973. Morphological and developmental affinities of the lophophorates. *in*: Larwood, G. P., Ed., Living and fossil Bryozoa. Academic Press, London, pp. 593–599.

Zimmerman, M. 1979. Larval release and settlement behavior in three species of plumatellid Bryozoa (Ectoprocta). M.S. Thesis, Wright State University, Dayton, OH.

15

TARDIGRADA

Diane R. Nelson

Department of Biological Sciences
East Tennessee State University
Johnson City, Tennessee 37614

I. INTRODUCTION

Since 1773, when J. A. E. Goeze reported the first observation of a "Kleiner Wasser Bär," tardigrades have commonly been called "water bears" (Goeze, 1773). Their bearlike appearance, legs with claws, and slow lumbering gait formed the basis for the descriptive name. In 1776, the naturalist Lazzaro Spallanzani introduced the name "il Tardigrado" (slow-stepper) to describe the slow, tortoiselike movement of the animal (Spallanzani, 1776). The systematic position of the Tardigrada, the taxon name assigned to the group by Doyère (1840), has ranged from inclusion within Infusoria to the Annelida to the Arthropoda (Acarina, Crustacea, Insecta). Tardigrada was acknowledged as a phylum by Ramazzotti (1962) in his first monograph, confirming the position of other workers (Richters, 1926). The phylogenetic affinities of the phylum, based on morphological studies, were clarified by recent molecular studies with 18S rRNA that showed tardigrades to be a sister group of the arthropods (Garey *et al.*, 1996; Giribet *et al.*, 1996; Garey *et al.*, 1999), although one study was inconclusive (Moon and Kim, 1996). The Ecdysozoa was recently established as a new clade containing the molting animals: arthropods, tardigrades, onychophorans, nematodes, nematomorphs, kinorhynchs, and priapulids (Aguinaldo *et al.*, 1997). This phylogenetic analysis, which showed no support for the Articulata (a clade of segmented animals uniting annelids and arthropods), offers an explanation for morphological similarities of tardigrades to both the onychophoran-arthropod and the aschelminth complex. Discussion on phylogenetic relationships of tardigrades and arthropods has been reviewed by Nielsen (1995) and Dewel and Dewel (1997). Thus far, about 800 species of tardigrades have been described from marine, freshwater, and terrestrial habitats; more species are expected as new areas are investigated (McInnes, 1994; Kinchin, 1994; McInnes and Pugh, 1998).

Tardigrades are hydrophilous micrometazoans with a bilaterally symmetrical body and four pairs of lobopodous legs usually terminating in claws (Figs. 1

Ecology and Classification of North American Freshwater Invertebrates. 2nd Edition

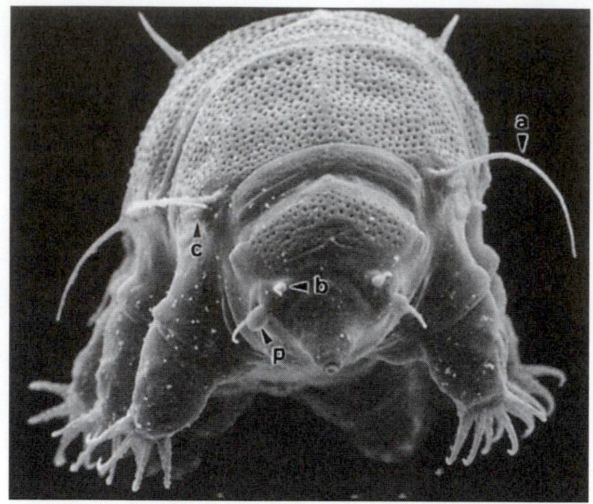

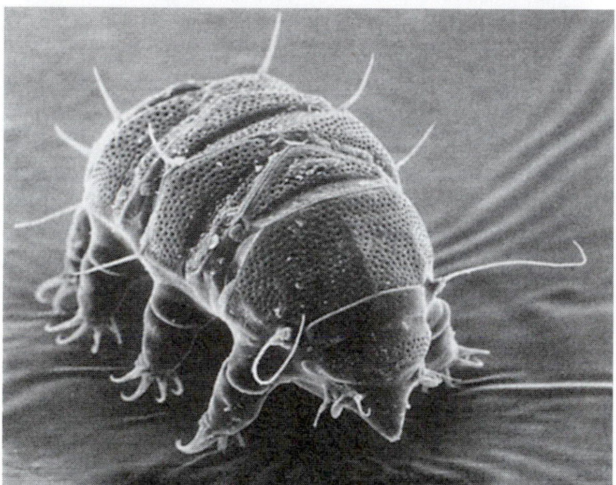

FIGURE 1 Scanning electron micrographs of a heterotardigrade. *Echiniscus spiniger.* a, Cirrus A; b, buccal cirrus; c, clava; p, cephalic papilla. [From Nelson, 1982a.]

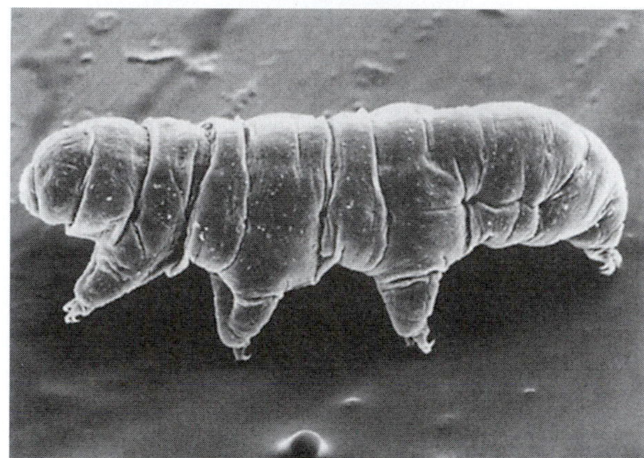

FIGURE 2 Scanning electron micrographs of a eutardigrade. *Macrobiotus tonollii.* [From Nelson, 1982a.]

and 2). Generally convex on the dorsal side and flattened on the ventral side, the body is indistinctly divided into a head (cephalic) segment, three trunk segments each bearing a pair of legs, and a caudal segment with the fourth pair of legs directed posteriorly. Although tardigrades range in body length from 50 μm in juveniles to 1200 μm in adults (both excluding the last pair of legs), mature adults average 250–500 μm with a few species exceeding 500 μm.

Despite their overall abundance and cosmopolitan distribution, the Tardigrada have been relatively neglected by invertebrate zoologists. Frequently categorized as one of the "minor phyla," they are more appropriately termed one of the "lesser-known phyla." Because of difficulties in collecting and culturing the organisms and their apparent lack of economic importance to humans, our knowledge of tardigrades has expanded slowly since their discovery over 225 years ago. The lack of basic biological information has hindered the growth of interdisciplinary studies, which are currently being developed.

Today, new insights into the biology of the Tardigrada are evolving from more comprehensive, contemporary investigations utilizing molecular techniques as well as transmission and scanning electron microscopy, phase- and interference-contrast microscopy, and histochemical and cytological techniques. Tardigrades provide a vast array of opportunities for classical and contemporary biological investigations. Within the last decade, interest and activity in tardigrade research has intensified.

II. ANATOMY

A. Cuticle

The cuticle, which is highly permeable to water, covers the body surface and lines the foregut and hindgut. [See Dewel *et al.,* 1993 for detailed review.]

TABLE I Subdivision of the Phylum Tardigrada with Habitat Classifications[a]

Taxon	*Habitat*
I. Class Heterotardigrada	
A. Order Arthrotardigrada	Marine, except *Styraconyx hallasi* (freshwater)
B. Order Echiniscoidea	Marine, freshwater, terrestrial
1. Family Echiniscoididae	Marine; genera *Echiniscoides* and *Anisonyches*
2. Family Oreellidae	Terrestrial; genus *Oreella*
3. Family Carphanidae	Freshwater; genus *Carphania*
4. Family Echiniscidae	Terrestrial, except for a few species of *Echiniscus, Hypechiniscus,* and *Pseudechiniscus* occasionally in freshwater; 12 genera
II. Class Mesotardigrada	
A. Order Thermozodia	
1. Family Thermozodiidae	Hot spring; genus *Thermozodium*
III. Class Eutardigrada	
A. Order Parachela	
1. Family Macrobiotidae	Terrestrial and freshwater; genera *Adorybiotus, Calcarobiotus, Dactylobiotus, Macrobiotus, Macroversum, Microbiotus, Minibiotus, Murrayon, Pseudodiphascon, Pseudohexapodibius, Richtersius, Xerobiotus*
2. Family Calohypsibiidae	Terrestrial; genera *Calohypsibius, Hexapodibius, Haplomacrobiotus, Parhexapodibius, Haplohexapodibius*
3. Family Eohypsibiidae	Terrestrial and freshwater; genera *Eohypsibius, Amphibolus*
4. Family Hypsibiidae	Marine; genus *Halobiotus* and a few *Isohypsibius, Ramajendas.* Terrestrial and freshwater; genera *Acutuneus Astatumen, Diphascon, Doryphoribius, Eremobiotus, Hebesuncus, Hypsibius, Isohypsibius, Itaquascon, Fujiscon, Mesocrista, Microhypsibius, Mixibius, Paradiphascon, Parascon, Platicrista, Pseudobiotus, Ramajendas, Ramazzottius, Thulinia*
5. Family Necopinatidae	Terrestrial; genus *Necopinatum*
6. *Incertae sedis*	Terrestrial; genus *Apodibius*
B. Order Apochela	
1. Family Milnesiidae	Terrestrial; *Milnesium* and *Limmenius*

[a]Additional references for table, not cited specifically in text: Binda, 1984; Binda and Pilato, 1986; Pilato and Beasley, 1987; Pilato and Binda, 1987a, b, 1989; Biserov, 1992; Dastych, 1992; Pilato and Binda, 1997.

Secreted by the underlying epidermis (= hypodermis), the cuticle consists of several layers which may be further divided into: (1) an outer complex epicuticle (= exocuticle); (2) an intracuticle (= mesocuticle), which is not present in all species; and (3) an inner procuticle (= endocuticle), which contains chitin (Greven and Peters, 1986; Greven and Greven, 1987; Dewel *et al.,* 1993). Although most freshwater tardigrades are white or colorless, some terrestrial species exhibit brown, yellow, orange, pink, red, or green coloration due to intestinal contents, body cavity cells and granules, or pigmentation in the epidermis and cuticle (Ramazzotti and Maucci, 1983). Based on cuticular structures, three classes, formerly considered orders, can be distinguished (Table I). The class Heterotardigrada (Figs. 1 and 3) includes mainly the marine and armored terrestrial tardigrades, while the class Eutardigrada (Figs. 2 and 4) primarily encompasses freshwater and other terrestrial species. A third class, Mesotardigrada (Rahm, 1937) is based on a single species, *Thermozodium esakii,* from a hot spring in Japan; it is difficult to evaluate the validity of this class due to the lack of type specimens and to the destruction of the type locality by an earthquake (Ramazzotti and Maucci, 1983).

Heterotardigrades (Fig. 1) are characterized by paired cephalic sensory appendages, which vary considerably in size and shape: internal buccal cirri, cephalic papillae, external buccal cirri, clavae, and lateral cirri (cirri A). Lateral cirri A mark the junction of the cephalic and scapular plates. Some marine species also have a median cirrus, particularly in the most primitive order, Arthrotardigrada. In the class Eutardigrada (Fig. 2), cephalic sensory structures, which may (Dewel *et al.,* 1993) or may not (Schuster *et al.,* 1980) be homologous to those in heterotardigrades, are present only in the order Apochela; *Milnesium* has two lateral papillae and six peribuccal (oral) papillae, and *Lemmenius* has two lateral papillae. Other eutardigrades (order Parachela) lack external sensory appendages but have sensory regions on the head (Walz, 1978).

The cuticle and its processes are taxonomically important in identifying genera and species and also form the basis for separating tardigrades into two large groups. The "armored" tardigrades are heterotardigrades

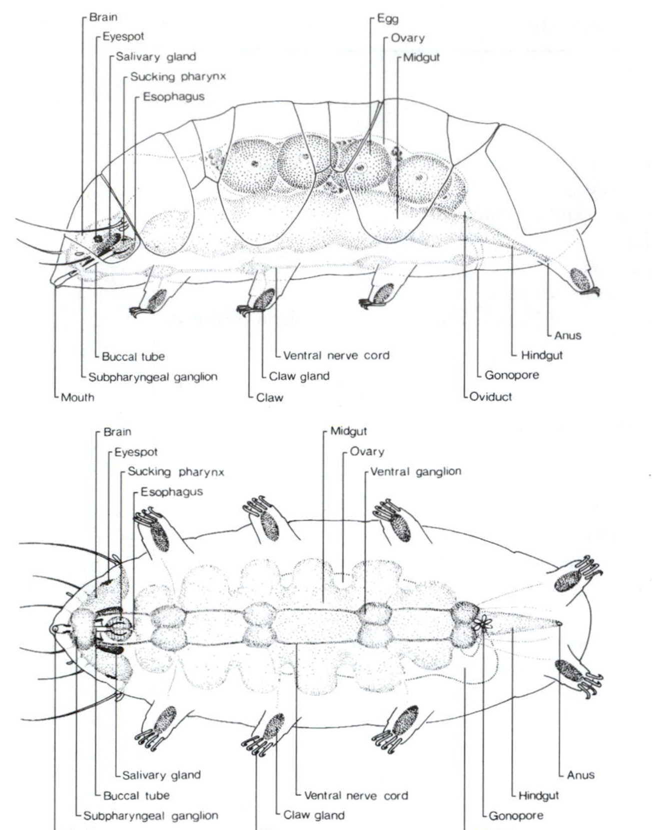

FIGURE 3 Heterotardigrade internal anatomy (*Bryodelphax parvulus,* female); lateral view (above) and ventral view (below). [From Nelson, 1982a; drawing by R. P. Higgins.]

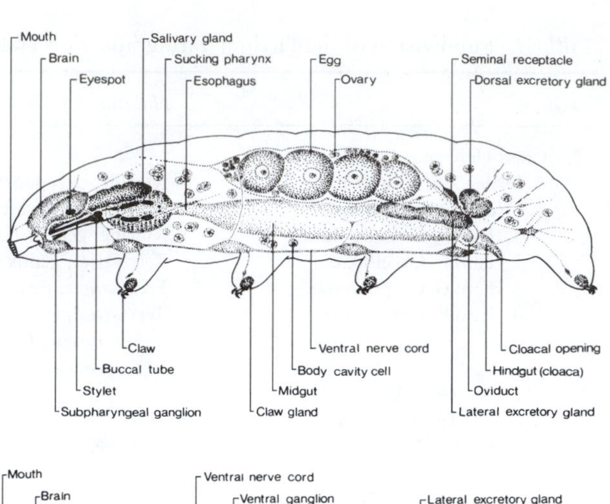

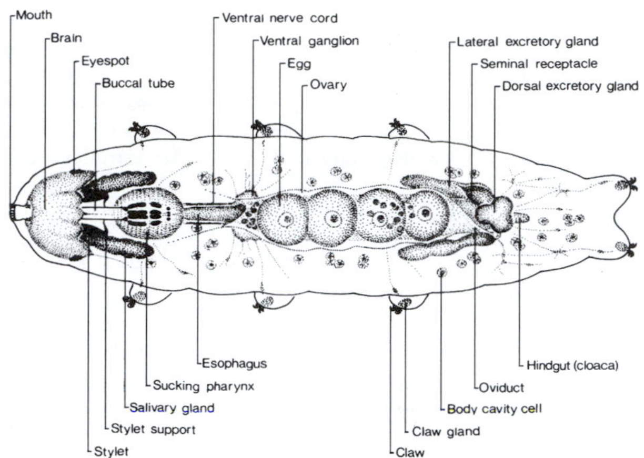

FIGURE 4 Eutardigrade internal anatomy (*Macrobiotus hufelandi,* female); lateral view (above) and dorsal view (below). [From Nelson, 1982a; drawing by R. P. Higgins.]

that have a thickened dorsal cuticle divided into individual plates, which have a species-specific pattern of sculpture. Armored tardigrades include some marine species (families Renaudarctidae and Stygarctidae, order Arthrotardigrada) and many terrestrial species (family Echiniscidae, order Echiniscoidea). Kristensen (1987) revised the genera of Echiniscidae, adding four new genera to the existing eight. Of these, some species in the genus *Pseudechiniscus* are found in freshwater, but the dorsal plates are nonsclerotized. The "naked" or unarmored tardigrades comprise the freshwater and terrestrial eutardigrades and most marine heterotardigrades, as well as the heterotardigrade families Oreellidae (terrestrial) and Carphanidae (freshwater) (Binda and Kristensen, 1986; Kristensen, 1987). Unarmored tardigrades have a thin, smooth cuticle or a sculptured one with pores, granulation, reticulation, tubercles, papillae, or spines, but lack plates.

B. Claws

In addition to cuticular structures, the size, shape and number of claws (Figs. 5 and 6) are important in tardigrade systematics and have been used for revision of the eutardigrades (Thulin, 1928; Marcus, 1929; Pilato, 1969, 1982; Bertolani and Kristensen, 1987; Kristensen and Higgins, 1984a, b; Ramazzotti and Maucci, 1983; Bertolani and Biserov, 1996). In the heterotardigrade family Echiniscidae, each leg terminates in four single claws in the adult and juvenile or two claws in the larva. Frequently, the two internal claws have a ventral spur, and occasionally one or more spurs may be present on the external claw near the base (Kristensen, 1987; Ramazzotti and Maucci, 1983). In the heterotardigrade family Carphanidae, the freshwater genus *Carphania* has two claws on legs I–III and only one claw on reduced leg IV; all claws are flexible and have two basal spurs (Binda and Kristensen, 1986).

Eutardigrades usually possess two double claws on each leg, arranged as external and internal claws (Schuster *et al.,* 1980; Bertolani, 1982a; Bertolani and Pilato, 1988; Bertolani and Biserov, 1996). Each double claw consists of a basal tract and a long primary branch with two accessory points and a secondary branch without accessory points. In the order Apochela

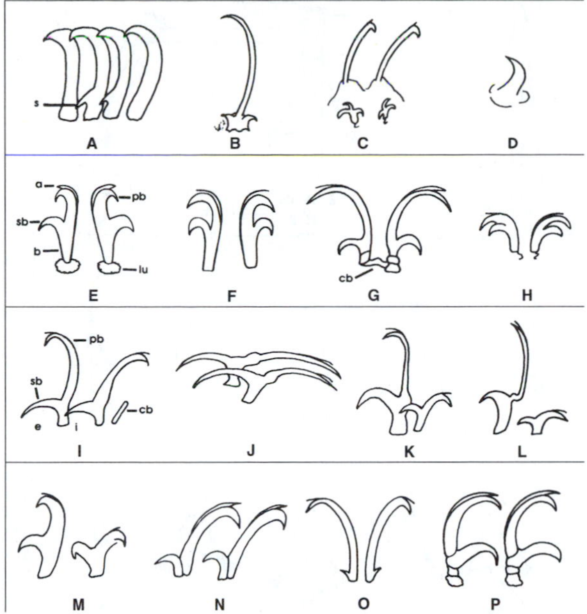

FIGURE 5 Claw types. All claws are shown in ventral view of the fourth right leg. (A) *Echiniscus*; (B) *Carphania*; (C) *Milnesium, Limmenius*; (D) *Necopinatum*; (E) *Macrobiotus, Macroversum, Murrayon, Pseudohexapodibius, Richtersius, Adorybiotus, Calcarobiotus, Xerobiotus*; (F) *Pseudodiphascon*; (G) *Dactylobiotus*; (H) *Minibiotus*; (I) *Isohypsibius, Thulinia, Doryphoribius*; (J) *Pseudobiotus*; (K) *Hypsibius, Mixibius, Platicrista, Mesocrista, Hebesuncus, Diphascon, Itaquascon, Astatumen, Fujiscon, Paradiphascon, Ramajendas*; (L) *Ramazzottius*; (M) *Calohypsibius, Microhypsibius*; (N) *Hexapodibius*; (O) *Haplomacrobiotus*; (P) *Eohypsibius, Amphibolus*. a, Accessory point; b, basal tract; cb, cuticular bar; e, external claw; i, internal claw; lu, lunule; pb, primary branch; s, spur; sb, secondary branch. [Redrawn from Nelson and Higgins, 1990.]

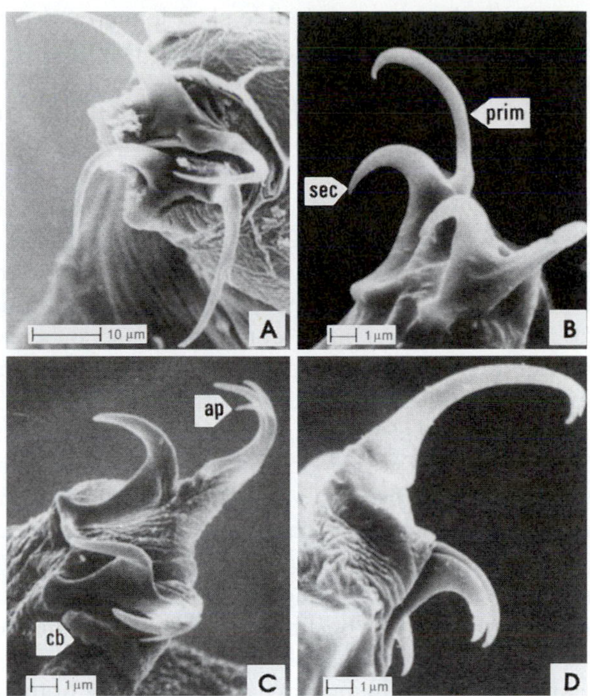

FIGURE 6 Scanning electron micrographs of eutardigrade claws. (A) *Pseudobiotus*; (B) *Hypsibius*; prim, primary branch; sec, secondary branch; (C) *Itaquascon*, ap, accessory point; cb, cuticular bar; (D) *Milnesium*. [From Schuster *et al.*, 1980.]

(family Milnesiidae), the primary branch is long, thin, and well separated from the stout, short secondary branch which bears hooks or spurs. In the order Parachela, the secondary branch and the primary branch arise from the common basal tract. The sequence of claw branches, with respect to the midline of extended legs, in some genera is alternate (2–1–2–1) (i.e., secondary–primary–secondary–primary). In other genera, the two primary branches are next to one another (2-1-1-2) (i.e., secondary–primary–primary–secondary). In some eutardigrades (e.g., *Macrobiotus*), a cuticular thickening called a lunule surrounds the base of each double claw. Lunules vary in size and shape; they may be smooth or dentate. In several genera, one or two cuticular bars may be present on the legs, separate from the claws but located between the double claws, just below the base of the claws, and/or along the sides of the internal claws.

Six major claw types have been identified based on the genera *Dactylobiotus, Macrobiotus, Calohypsibius, Hypsibius, Isohypsibius,* and *Eohypsibius*. Additional genera have been erected based on variations in these basic claw types. Claw reduction, particularly in the hind claws, has evolved independently in the eutardigrades, especially those dwelling in soil habitats (Bertolani and Biserov, 1996). In *Dactylobiotus*-type claws, the two large double claws on each leg are similar in size and shape, with a 2-1-1-2 sequence, lack lunules, and are connected at the base by a cuticular band. The secondary branch is much shorter than the primary and arises from a very short basal tract. The *Macrobiotus*-type claws are similar in size and shape, with a distinct peduncle (a septum separating the basal tract from the rest of the claw), and have a 2-1-1-2 sequence, but the claws usually possess lunules and are not connected at the base. A similar genus, *Calcarobiotus*, has spurs on the bases of the claws (Dastych, 1993a). In the genus *Murrayon*, separated from *Macrobiotus* by Bertolani and Pilato (1988), the claws are V-shaped and have a quadrate peduncle (Guidetti, 1998). *Macroversum* is separated from *Macrobiotus* by having secondary branch much shorter than the primary branch, and the lunules are connected by a cuticular bar (Pilato and Catanzaro, 1988; Bertolani and Pilato, 1988). The *Calohypsibius*-type claws are small and similar in size and shape but exhibit the alternate (2-1-2-1) sequence; the primary branch is rigidly attached to the secondary branch. The *Hypsibius*-type claws are usually dissimilar in size and shape, with the external double claw longer and more slender than the internal claw. The primary branch of the external double claw is inserted on the basal tract, which is continuous with the secondary branch, forming a flexible junction. The *Isohypsibius*-type claws resemble those of *Hypsibius*; however, they are usually similar in size and shape. To distinguish the two types, the secondary branch of the external claw from one of the first three pairs of legs should be observed in profile. In *Hypsibius*-type claws, the secondary branch and basal tract run in a continuous arc or curve; whereas, in *Isohypsibius*-type claws, the secondary branch forms a right angle with the basal tract. The genus *Ramajendas* was separated from *Hypsibius* by having very long primary branches (Pilato and Binda, 1990). The *Eohypsibius*-type claws are divided into three distinct parts: the base, the secondary branch, and the primary branch. The branches are usually arranged in a 2-1-2-1 sequence; however, the internal claw can rotate 180°, resulting in a 2-1-1-2 sequence. The external and internal claws are similar in size, but the angle between the primary and secondary branches is about 45° on the external claw and about 80° on the internal claw.

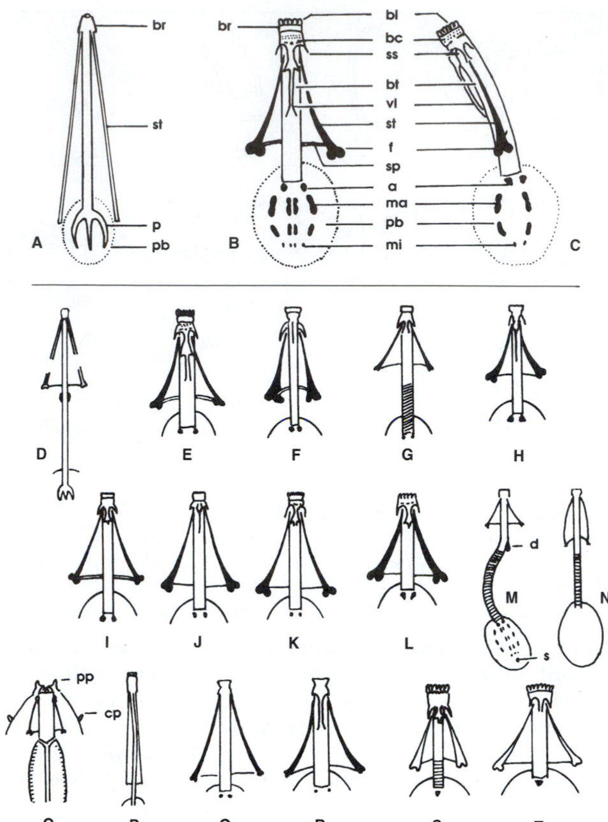

FIGURE 7 Buccal apparatus. (A) Heterotardigrada, *Echiniscus* (ventral); (B) Eutardigrada, *Macrobiotus* (ventral); (C) Eutardigrada, *Macrobiotus* (lateral); (D) *Carphania*; (E) *Macrobiotus, Macroversum, Murrayon, Dactylobiotus, Calcarobiotus, Pseudohexapodibius, Xerobiotus*; (F) *Adorybiotus, Richtersius*; (G) *Pseudodiphascon*; (H) *Minibiotus*; (I) *Hypsibius, Isohypsibius, Ramazzottius, Ramajendas, Mixibius, Necopinatum*; (J) *Doryphoribius, Apodibius*; (K) *Pseudobiotus*; (L) *Thulinia*; (M) *Diphascon, Platicrista, Mesocrista, Hebesuncus, Paradiphascon*; (N) *Itaquascon*; (O) *Milnesium*; (P) *Limmenius*; (Q) *Calohypsibius, Microhypsibius*; (R) *Hexapodibius, Haplomacrobiotus*, (S) *Eohypsibius*; and (T) *Amphibolus*. a, Apophysis; bc, buccal cavity; bl, buccal lamellae; br, buccal ring; bt, buccal tube; cp, cephalic papilla; d, droplike thickening; f, furca; ma, macroplacoid; mic, microplacoid; p, placoid; pb, pharyngeal bulb; pp, peribuccal papilla; s, septulum; sp, stylet support; ss, stylet sheath, st, stylet; and vl, ventral lamina. [Redrawn from Nelson and Higgins, 1990.]

C. Digestive System

1. General Anatomy

The digestive system consists of a foregut, a midgut, and a hindgut (Figs. 3 and 4). [For a detailed review, see Dewel *et al.*, 1993.] Both the foregut (buccal apparatus and esophagus) and the hindgut have a cuticular lining. The hindgut is divided into an anterior hindgut (rectum) and a true cloaca in eutardigrades; the opening is a ventral transverse slit just anterior to the fourth pair of legs. In heterotardigrades, however, reproductive and digestive systems are separate, the hindgut is a rectum, and the gonopore opens anterior to the anus, which is a longitudinal slit between or just anterior to the fourth pair of legs. In the heterotardigrade family Echiniscidae, first instar individuals have a mouth but no gonopore or anus; second instars have a mouth and anus but no gonopore; subsequent instars have all three openings.

2. Buccal Apparatus

The buccal apparatus is a complex structure with considerable taxonomic significance in the eutardigrades (Pilato, 1972, 1982; Schuster *et al.*, 1980). Basically, it consists of a buccal tube, a muscular sucking pharynx, and a pair of piercing stylets that can be extended through the mouth opening (Figs. 3,4,7,8). The position of the mouth may be terminal (as in *Macrobiotus* and *Milnesium*) or subterminal (as in *Hypsibius*). In some eutardigrades, cuticular structures surround the mouth such as elongate peribuccal papillae (6 in *Milnesium*), flattened peribuccal papulae (10 in *Minibiotus* and *Haplomacrobiotus* and 6 in *Calohypsibius*), or lobes (6 in *Isohypsibius* and others) (Schuster *et al.*, 1980).

The buccal tube begins with a buccal ring, which bears lamellae in some genera. The number of lamellae varies with the genus; e.g., there are 10 in *Macrobiotus*, 12 in *Thulinia*, 14 in *Amphibolus*, and about 30 in *Pseudobiotus*. In *Milnesium*, the mouth is surrounded

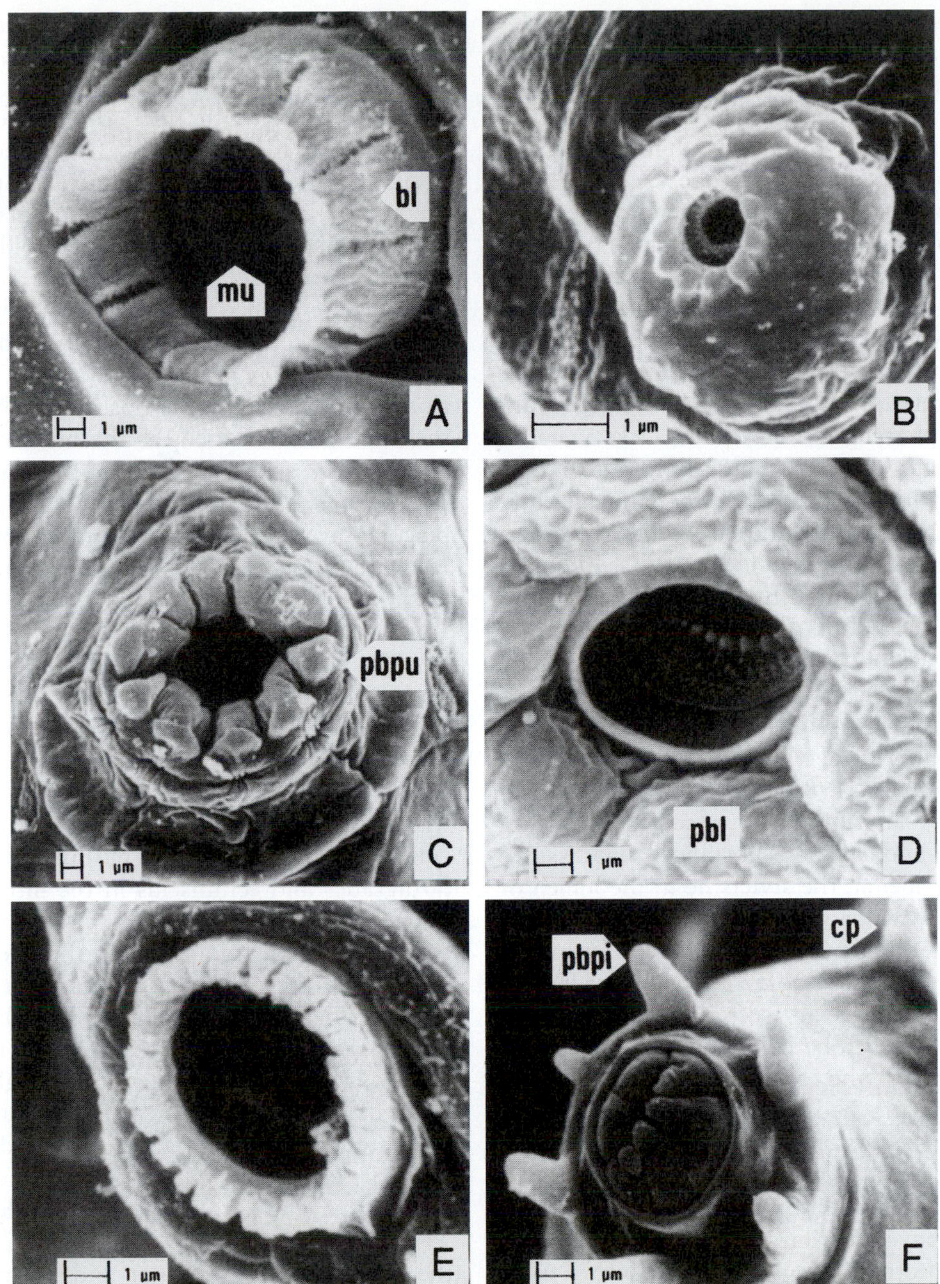

FIGURE 8 Scanning electron micrographs of eutardigrade buccal openings. (A) *Dactylobiotus,* mu, buccal mucrones; bl, buccal lamella; (B) *Minibiotus,* buccal opening and papulae; (C) *Haplomacrobiotus,* pbpu, peribuccal papulae; (D) *Isohypsibius,* pbl, peribuccal lobe; (E) *Pseudobiotus,* buccal opening and buccal lamellae; and (F) *Milnesium,* cp, cephalic (lateral) papilla; pbbl, peribuccal (oral) papilla. [From Schuster *et al.,* 1980.]

by six triangular lamellae, which close the opening. The interior of the buccal cavity may have anterior and posterior rows of small cuticular teeth (mucrones) followed by dorsal and ventral transverse ridges (also called crests or baffles), which have significant taxonomic value as species-specific characteristics (Pilato, 1972; Schuster *et al.,* 1980). Most eutardigrades have a completely rigid buccal tube. However, in some genera, (the *Diphascon* assemblage, *Pseudodiphascon, Itaquascon*), there is a rigid anterior buccal tube between the buccal opening and the stylet supports, and a flexible, spiral-walled posterior pharyngeal tube from just below the stylet supports to the pharynx. In a similar genus, *Astatumen,* the rigid portion of the buccal tube is only as long as the anterior apophyses, the pharyngeal tube is spiraled, and stylet supports are lacking (Pilato, 1997). In *Macrobiotus* and other genera, there is a median ventral lamina (= buccal tube support, = reinforcement rod) that extends from the buccal ring to the midregion of the tube.

Four major types of buccal apparatus have been described based on the genera *Macrobiotus, Hypsibius, Diphascon,* and *Pseudodiphascon.* The *Macrobiotus*-type has a rigid buccal tube with a ventral lamina (reinforcement rod), whereas the *Hypsibius*-type has a rigid buccal tube but lacks a ventral lamina. The *Diphascon*-type has a rigid anterior buccal tube and a flexible, annulated posterior pharyngeal tube; it also lacks a ventral lamina. The *Pseudodiphascon*-type has a buccal tube like *Diphascon,* but also has a ventral lamina. More types may be identified in the future as additional genera are described (Pilato and Binda, 1990, 1996). Although the buccal apparatus was used by Schuster *et al.* (1980) to revise the families of eutardigrades, Ramazzotti and Maucci (1983) and subsequent authors have basically followed Pilato's classification based on claw structure.

The stylet mechanism consists of two protrusible stylets, stylet sheaths, stylet supports, and associated muscles (Schuster *et al.,* 1980). The stylets are paired, piercing structures lateral to the buccal tube. Composed of cuticular and calcified substances that dissolve in some mounting media, the stylets lie in the lumen of the salivary glands, which are dorsolateral to the pharynx. The salivary epithelium secretes the new buccal tube, stylets, and stylet supports after the tardigrade molts. The stylets and salivary secretions enter the anterior end of the buccal tube through stylet sheaths on each side. Protractor and retractor muscles operate each stylet and are attached to the furca (condyle), the thickened base at the posterior end of each stylet. Protractor muscles extend from the furca to the buccal tube; retractor muscles insert on the pharynx.

Some genera have anterior apophyses for the insertion of the stylet protractor muscles on the buccal tube. In *Hypsibius,* the dorsal and ventral apophyses are hook-shaped; in *Isohypsibius* and other genera, they form dorsal and ventral crests. Pilato (1987) divided the genus *Diphascon* into four genera based on differences in the shape of the apophyses and other characteristics of the buccal–pharyngeal apparatus. The stylet support is a flexible lateral extension that attaches the furca of the stylet to the buccal tube, but it is poorly developed in the genus *Itaquascon* and absent in *Astatumen* (Pilato, 1997). In the heterotardigrades, the long, thin stylets may have delicate stylet supports or may be attached to the pharynx by a muscular band. In general, the buccal apparatus in heterotardigrades is of less taxonomic importance than in eutardigrades (Kristensen, 1987).

The buccal tube enters the muscular, tripartite pharyngeal bulb (= sucking pharynx, = myoepithelial bulb) (Eibye-Jacobsen, 1996), which, in most eutardigrades, is lined with three double rows of cuticular thickenings called placoids. The placoids alternate in position with three equal cuticular apophyses, which are posterior to the end of the buccal tube. The larger, anterior placoids, called macroplacoids, are present in either two or three transverse rows. Smaller, posterior placoids are termed microplacoids; when present, they are arranged in a single row. In some species of the genus *Diphascon,* a medial septulum is present posterior to and alternating in position with the placoids, in the same plane of focus as the buccal tube apophyses. The number, size, and shape of the placoids are important species characters in the eutardigrades; however, placoids are reduced or absent in some genera (e.g., *Itaquascon, Astatumen, Milnesium,* and *Limmenius).*

3. Esophagus, Midgut, and Hindgut

The pharyngeal bulb is followed by a short, tripartite esophagus, which is also lined with cuticle. A mucoid secretory substance is produced by esophageal cells, but its function is unknown.

The midgut, the largest section of the digestive tract, is derived from endoderm and is not lined with cuticle. There are significant differences between eutardigrades and echiniscids in the ultrastructure of the midgut epithelium (Dewel *et al.,* 1993). Eutardigrades have a straight midgut, and three excretory glands empty into the midgut–hindgut junction; in contrast, heterotardigrades have no excretory glands at the junction of the midgut and hindgut. Food is digested in the midgut, which has a convoluted single layer of large epithelial cells joined by deep continuous junctions. The intestinal content varies in color depending on the type of food consumed, the state of the digestive process, and the nutritional status of the animal.

The hindgut, which is lined with cuticle, receives the contents of the midgut and is specialized for osmoregulation. The anterior hindgut or rectum probably functions in modification of the urine from the excretory glands (= Malpighian tubules) as well as material from the midgut. In eutardigrades, the genital ducts empty into the posterior hindgut, which is therefore a true cloaca. In heterotardigrades, the anus and gonopore are separate openings.

D. Other Anatomical Systems

Reviews of the morphology of other tardigrade systems are found in Greven (1980), Bertolani (1982a), Nelson (1982a), Ramazzotti and Maucci (1983), Nelson and Higgins (1990), Dewel *et al.* (1993), and Kinchin (1994).

The nervous system consists of a dorsal lobed brain (= cerebral ganglion, = supraesophageal mass), a circumesophageal connective around the buccal tube (= ring neuropil), a "subesophageal ganglion", a circumpharyngeal connective around the pharynx (= posterior neuropil), and four ventral ganglia joined by paired nerve tracts (Figs. 3 and 4) (Dewel and Dewel, 1996, 1997; Wiederhöft and Greven, 1996). Many eutardigrades and some heterotardigrades have a pair of eye spots, cup-shaped structures with pigment granules, embedded in the dorsolateral lobes of the brain (Dewel *et al.,* 1993). Walz (1978) described four sensory areas that function as chemoreceptors or mechanoreceptors in the anterior head region of the eutardigrade *Macrobiotus*: the anteriolateral sensory field, the circumoral field, the suboral sensory region, and the pharyngeal organ. The cephalic appendages and buccal ventroposterior plates (third pair of clavae) in heterotardigrades also function as sense organs (Kristensen and Higgins, 1984a, 1989; Dewel *et al.* 1993).

Tardigrade musculature has been studied in detail in only a few species. [See Dewel *et al.,* 1993 for a detailed review.] The muscular system consists of dorsal and ventral longitudinal fibers, transverse muscles for movement of the legs, stylet muscles, pharyngeal and intestinal muscles, and muscles for defecation and deposition of eggs. Muscles are attached to the cuticle by protrusions (apodemes) that are reformed at each molt. The somatic muscles are single elongate cells that insert on the body wall. Those of eutardigrades are intermediate between smooth and obliquely striated muscles (Walz, 1974, 1975) or cross-striated in the pharyngeal bulb (Eibye-Jacobsen, 1996), whereas those of marine arthrotardigrades are cross-striated (Kristensen, 1978).

According to Ramazzotti and Maucci (1983), excretion in tardigrades probably occurs in four ways: (1) through the salivary glands at molting; (2) through shedding the cuticle containing accumulated excretory granules; (3) through the wall of the midgut; and (4) through excretory glands. Eutardigrades possess three excretory glands (= Malpighian tubules); one dorsal and two lateral glands empty into the junction between the midgut and hindgut. These function as excretory and/or osmoregulatory structures in addition to the hindgut (rectum). The glands of the marine eutardigrade *Halobiotus crispae* are unusually large, reflecting its secondary adaptation from freshwater to seawater, where it inhabits the subtidal zone and is subject to marked changes in salinity. In echiniscids, excretory and/or secretory organs lie above the ventro-medial body wall at the level of the second and third pair of legs (Dewel *et al.,* 1992). Each is comprised of one medial and two lateral cells enclosing a pair of convoluted ducts. They probably transport material in the body cavity fluid to the cuticle.

Associated with miniaturization, specialized respiratory and circulatory systems are absent. Respiration occurs through the cuticle. Circulation is accomplished by the movement of fluid and coelomocytes in the body cavity (hemocoel) when the animal moves.

III. PHYSIOLOGY (LATENT STATES)

All tardigrades are aquatic, regardless of their specific habitat, since they require a film of water surrounding the body to be active. They are capable, however, of entering a latent state (cryptobiosis) when environmental conditions are unfavorable. Crowe (1975) identified five types of latency: encystment, anoxybiosis, cryobiosis, osmobiosis, and anhydrobiosis. When a tardigrade is in a latent state, metabolism, growth, reproduction, and senescence are reduced or cease temporarily, and resistance to environmental extremes such as cold, heat, drought, chemicals, and ionizing radiation is increased. Latent states thus have a significant impact on the ecological role of the organism. (See review by Wright *et al.,* 1992; Kinchin, 1994.)

Encystment commonly occurs in freshwater tardigrades living in permanent pools, although cysts can also be formed by tardigrades inhabiting soil and moss. The conditions for encystment are not fully known; however, Węglarska (1957) determined that a gradual worsening of the environment is one inducing factor for encystment in the freshwater tardigrade *Dactylobiotus dispar.* Specimens encysted regularly within 12 h if placed in water containing cyanophytes or decomposing leaves, but a "swift worsening" of the environment caused the animals to go into an asphyxial state (anoxybiosis).

Before encystment, the tardigrade ingests large amounts of food for storage in body cavity cells (coelomocytes). Proceeding through stages similar to molting, the animal first ejects the buccal apparatus and enters the simplex stage. After this process, the gut is emptied by defecation, the hindgut lining is shed, and the organism contracts within the cuticle, becoming immobile and barrel-shaped. A cyst membrane is formed and gradually becomes smooth and dark, replacing the old cuticle which sloughs off. The cyst is encased in strong walls, often opaque and encrusted with debris. Metabolism is very reduced in the cysts. Before emergence from the cyst when environmental conditions improve, a new cuticle is produced, including buccal apparatus and claws. Encystment is not obligatory in the life cycle of a tardigrade, and reasons for emergence are not fully known. The cysts are not able to withstand high temperatures, as in anhydrobiosis, because of the high water content; however, the cysts can survive for over a year in nature without totally depleting food reserves.

Anoxybiosis, or asphyxia, is a cryptobiotic state induced by low oxygen tensions in the environmental water. The tardigrade becomes immobile, transparent, rigid, and extended due to water absorption resulting from loss of osmotic control. The asphyxial state is temperature dependent; the time required for inducement varies from a few hours to several days. Some species can survive up to 5 days in anoxybiosis; however, strictly aquatic species live only from a few hours to 3 days in the asphyxial state. Prior to this time limit, specimens can be revived by the addition of oxygen to the water; return to the active state requires a few minutes to several hours. The metabolic rate of tardigrades in anoxybiosis is unknown.

Induced by low temperatures, cryobiosis is a type of cryptobiosis that enables tardigrades to survive freezing and thawing, allowing limno-terrestrial tardigrades to be common in polar regions (Kristensen, 1982; De Smet and Van Rompu, 1994; Sømme, 1996). Slow cooling rates and the formation of barrel-shaped tuns enhance survival, and membrane protectants such as trehalose stabilize membrane structure (Rudolph and Crowe, 1985). [See Westh and Kristensen, 1992; Westh *et al.,* 1991; Ramløv and Westh, 1992, for detailed information.]

Osmobiosis is a form of cryptobiosis that results from elevated osmotic pressures and decreased water activity at the surface of the animal. Some species can tolerate variations in salinity, especially intertidal marine species and euryhaline limno-terrestrial species. However, placed in salt solutions of various ionic strengths, most freshwater and terrestrial tardigrades form contracted tuns (Wright *et al.,* 1992). The marine tardigrade *Echiniscoides sigismundi* becomes turgid in freshwater but can survive up to 3 days. Activity resumes when the animals are returned to water with the former osmotic concentrations.

Anhydrobiosis (formerly called anabiosis) is a type of cryptobiosis induced by evaporative water loss from terrestrial eutardigrades and echiniscids. [See Wright, 1989a, b; Wright *et al.,* 1992; Crowe *et al.,* 1992; Kinchin, 1994; Sømme, 1996 for reviews.] Nearly all marine and true freshwater tardigrades lack the ability to survive dehydration and undergo cryptobiosis, but experimental evidence is sparse. As the environmental water surrounding the animal evaporates, the tardigrade contracts, retracting the head and legs and becoming the characteristic barrel-shaped tun. The rate of desiccation must be slow to ensure survival and return to active life with the addition of water. Survival of dehydration is correlated with synthesis of trehalose, a disaccharide of glucose, during dehydration or its degradation following rehydration. Trehalose alters the physical properties of membrane phospholipids and stabilizes dry membranes in anhydrobiotic organisms (Crowe *et al.,* 1984; Westh and Ramløv, 1991). In the anhydrobiotic state, the tardigrade is highly resistant to extreme environmental factors (temperature, radiation, chemicals). Anhydrobiotic animals have remained viable on herbarium specimens for as long as 120 years. The eggs of terrestrial tardigrades can also withstand desiccation, but little is known about the eggs of the distinctly aquatic species. The life span of anhydrobiotic tardigrades can be greatly extended by going in and out of this cryptobiotic state, although frequent dehydration and rehydration results in reduced viability due to the depletion of lipid reserves, which can be converted to the membrane protectants glycerol and trehalose (Westh and Ramløv, 1991). In the anhydrobiotic state, tardigrades can also withstand extraordinarily high hydrostatic pressure, up to 600 MPa, in perfluorcarbon (Seki and Toyoshima, 1998).

IV. REPRODUCTION AND DEVELOPMENT

A. Reproductive Modes

Sexual reproduction and parthenogenesis are reproductive modes exhibited in the Tardigrada (Ramazzotti and Maucci, 1983; Nelson, 1982a; Bertolani, 1982b, 1983a, b, 1987a, 1990, 1992, 1994; Rebecchi and Bertolani, 1988, 1994; Kinchin, 1994; Bertolani and Rebecchi, 1999). Marine tardigrades are bisexual (gonochoristic) with sexual dimorphism evident in the gonopore structure; hermaphroditic species are very rare, and parthenogenesis is absent (Bertolani, 1994). Nonmarine species, especially the eutardigrades, are usually gonochoristic (dioecious) with unisexual or bisexual populations, but external sexual dimorphism is rarely observed. Males are generally smaller than females; however, mature males may be larger than immature females in the same population. Males may also exhibit modifications of the claws (Rebecchi and Nelson, 1998), especially on the first pair of legs, such as in *Milnesium* and the freshwater *Pseudobiotus megalonyx* and *P. kathmanae* (formerly *P. augusti,* in Kathman and Nelson, 1987, see revision in Nelson *et al.,* 1999). These secondary sexual characteristics arise during the molt preceding sexual maturity (Bertolani, 1992; Rebecchi and Nelson, 1998). In heterotardigrades, additional sexual dimorphism may be exhibited in the length of the cephalic appendages, especially the clavae (Kinchin, 1994; Claxton, 1996). Meiotic maturation of the gametes occurs in amphimictic tardigrades. In three species of gonochoristic freshwater and terrestrial tardigrades, all of the females were iteroparous, but males were either iteroparous or semelparous. In nonmarine tardigrades, unisexual populations are the most common type, and parthenogenesis is the mode of reproduction, often associated with polyploidy and with ameiotic maturation of the oocyte. Simultaneous hermaphroditism is sporadic but occurs in freshwater and terrestrial eutardigrades (Bertolani, 1994).

B. Reproductive Apparatus

A single (unpaired) gonad is present in both males and females (Figs. 3 and 4). This saclike structure is dorsal to the intestine and attached to the body wall by two anterior ligaments (small muscles, Dewel *et al.,* 1993) in adult eutardigrades or by a single median ligament (basement membrane, Kristensen, 1979) in heterotardigrades (Nelson 1982a, Dewel *et al.,* 1993). The morphology of the gonad varies with age, sex, species, and the reproductive activity of the tardigrade. The male tardigrade has two sperm ducts (vasa deferentia) that open into the hindgut (cloaca); the female has only one oviduct on either the right side or the left side of the intestine that opens into the hindgut. A small seminal receptacle (with internal epithelial spermatheca), which opens ventrally into the hindgut next to the oviduct opening, is present in only a few species of *Macrobiotus, Xerobiotus, Ramazzottius,* and *Hypsibius* (Marcus, 1929; Bertolani, 1983a; Rebecchi and Bertolani, 1988; Rebecchi, 1997). In eutardigrades, the cloacal opening is a ventral transverse slit anterior to the fourth pair of legs. In heterotardigrades, the preanal gonopore is rosette-shaped in females but is a protruding, round or oval-shaped aperture in males; only a few marine species have a postanal gonopore (Kinchin, 1994). Paired external cuticular seminal receptacles ("annex glands") lateral to the gonopore are present in females in many species of marine heterotardigrades, but in only a few eutardigrades. In some species of both heterotardigrades and eutardigrades, males have seminal vesicles. (See Dewel *et al.*, 1993 and Kinchin, 1994 for a review.)

C. Mating and Fertilization

Tardigrade spermatozoa have been described in some species (Kristensen, 1979; Baccetti, 1987; Bertolani, 1983b; Rebecchi and Guidi, 1991, 1995; Guidi and Rebecchi, 1996; Rebecchi, 1997; Rebecchi *et al.,* 1999). Henneke (1911) first observed the presence of two kinds of sperm in the freshwater tardigrade *Pseudobiotus macronyx.* The ultrastructure of the spermatozoa of *Macrobiotus hufelandi* was first investigated by Baccetti *et al.* (1971). Other workers described the morphology of spermatozoa and correlated morphology with a specialized mode of sperm transfer (Wolburg-Buchholz and Greven, 1979; Rebecchi and Guidi, 1991, 1995; Guidi and Rebecchi, 1996; Rebecchi, 1997). For example, in *Xerobiotus pseudohufelandi,* the nutcracker-shaped motile spermatozoa with a helical nucleus, kidney-shaped midpiece, and long tail are associated with internal fertilization; the female has a spermatheca that contains straight, nonmotile spermatozoa with a cylindrical midpiece and a very short tail (Rebecchi, 1997). In the freshwater eutardigrade *Isohypsibius granulifer,* transfer of modified sperm occurs by copulation, and spermatozoa are found in the ovary of the female (Wolburg-Buchholz and Greven, 1979). In some heterotardigrades, the spermatozoa are less specialized, the female has no seminal receptacle, and fertilization is external (Kristensen, 1979). Rebecchi *et al.,* (1999) reviewed ultrastructural differences in eutardigrade spermatozoa, which are considered highly specialized. Male gametes of eutardigrades are 25–100 μm in length, usually have an elongated head with a coiled piece (acrosome or nucleus), a midpiece, and a tail with a tuft of 9–11 fine filaments. Heterotardigrade spermatozoa, considered primitive, are shorter (14–20 μm), have a globose head, lack a well-defined midpiece, and have a tapering flagellum. In eutardigrades, spermatozoan morphology can be correlated with sclerified structure of adults and eggs, thus providing valuable taxonomic characters (Guidi and Rebecchi, 1996).

Mating and fertilization have been observed in only a few species of tardigrades (Ramazzotti, 1972; Nelson, 1982a; Bertolani, 1987a, 1992). In some freshwater species, one or more males deposit sperm into the cloacal opening of the old cuticle of the female prior to, or during, a molt. External fertilization occurs as the eggs are deposited in the old cuticle. In the eutardigrade *Pseudobiotus megalonyx,* diploid males were able to locate mates and to distinguish between diploid amphimictic individuals and triploid parthenogenetic ones (Bertolani, 1992). In heterotardigrades, cephalic sensory structures may play a role in mating.

D. Eggs

Eggs are often essential for the identification of freshwater and terrestrial species, especially in some genera of eutardigrades (Fig. 9). Ornamented eggs, such as those characteristic of *Macrobiotus, Murrayon, Minibiotus, Dactylobiotus, Amphibolus, Eohypsibius, Ramazzottius, Hebesuncus,* and some *Hypsibius* species, as well as the heterotardigrade *Oreella,* are usually deposited freely and singly or in small groups. The ornamentation may include pores, reticulations, and processes. The different types of egg ornamentations and their taxonomic significance have been discussed by various authors (Grigarick *et al.*, 1973; Toftner *et al.*, 1975; Greven, 1980; Ramazzotti and Maucci, 1983; Biserov, 1990a, b; Bertolani and Rebecchi, 1993; Bertolani *et al.*, 1996). In some genera,

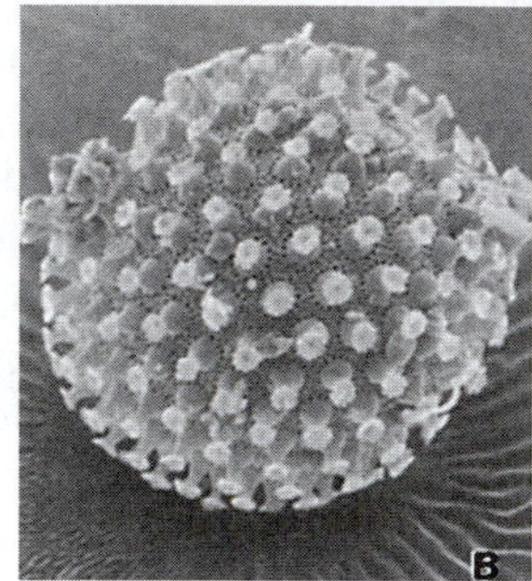

FIGURE 9 Scanning electron micrographs of *Macrobiotus* eggs. (A) *Macrobiotus areolatus*; (B) *Macrobiotus hufelandi* group; and (C) *Macrobiotus tonollii*.

smooth oval eggs with a thin shell are usually deposited in the molted cuticle. These smooth eggs are typical of the armored heterotardigrades, most eutardigrades in the genera *Pseudobiotus, Doryphoribius, Hypsibius, Isohypsibius, Diphascon, Platicrista, Itaquascon*, and *Milnesium*, and only a few of the *Macrobiotus* species. Some aquatic tardigrades deposit eggs inside the empty exoskeletons of cladocerans, ostracods, or insects. The aquatic eutardigrades *Isohypsibius annulatus* and two species of *Pseudobiotus* carry the exuvium containing eggs attached to the body for some time (Bertolani, 1982a; Rebecchi and Nelson, 1998). This is the only known example of parental care among tardigrades. The method of egg deposition is unknown for many species, especially in marine tardigrades.

E. Parthenogenesis

Parthenogenesis, a common mode of reproduction in nonmarine tardigrades, is correlated with the ability to undergo cryptobiosis (Bertolani, 1982b; 1994). According to Kristensen (1987), in echiniscid heterotardigrades, males had not been previously recorded in *Echiniscus, Bryochoerus, Parechiniscus*, and many *Pseudechiniscus* species; males were rare in *Bryodelphax, Testechiniscus, Cornechiniscus*, and *Mopsechiniscus* but common (50% of the population) in *Oreella, Hypechiniscus, Proechiniscus*, and *Antechiniscus*. In the absence of males in a tardigrade population, parthenogenesis was assumed to be the mode of reproduction. Dastych (1987a), however, reported the presence of males in four species of the genus *Echiniscus*. Claxton (1996) reported males to be common in a disproportionately large number of species (5 of the 8 species) of *Echiniscus* found in Australia; males and females were differentiated on the basis of gonopore morphology, the size of the males, and the length of the clavae.

Grigarick *et al.* (1975) reported two "types" of gonopores and published the first scanning electron micrograph of a male gonopore of a heterotardigrade. With regard to this report, Nelson (1982a) stated, "If males were indeed present, this would be the first observation of males in any *Echiniscus* species . . . " Based on the presence of ventral plates, Dastych (1987a) placed the species studied by Grigarick *et al.* (1975) in Kristensen's (1987) new genus *Testechiniscus*, in which males are rare.

Polyploidy is often associated with parthenogenesis and is frequently found in eutardigrades in various genera that inhabit different microhabitats (Bertolani, 1975, 1982b; Nelson, 1982a). Meiotic (automictic) parthenogenesis occurs in diploid biotypes of the freshwater eutardigrades *Dactylobiotus parthenogenesis* and *Hypsibius dujardini*. Ameiotic (apomictic) parthenogenesis occurs in triploid and tetraploid biotypes of some freshwater and terrestrial eutardigrades (*Macrobious, Xerobiotus, Pseudobiotus*). According to Pilato (1979), " . . . many species in which males do occur are actually an association of amphigonic with one or more parthenogenetic, generally polyploid and reproductively isolated strains, often distinguishable by their mechanism of egg deposition." Since parthenogenesis may be advantageous for invading new habitats, the evolution of parthenogenesis may be correlated with the evolution of cryptobiosis and the invasion of the terrestrial habitat (Pilato, 1979; Bertolani, 1987a; Rebecchi and Bertolani, 1988; Bertolani *et al.*, 1990; Wright *et al.*, 1992).

F. Hermaphroditism

Most eutardigrade populations are unisexual (thelytokus) although some are bisexual. Hermaphroditism is rare in eutardigrades and known only in one marine heterotardigrade (Bertolani, 1987a, 1994). In hermaphrodites, a single ovotestis with only one gonoduct is present. Both mature and almost mature spermatozoa and oocytes may be found in a single individual, indicating the possibility of self-fertilization. Cyclic maturation probably occurs, with the initial prevalence of one type of gamete and then another. Hermaphroditic eutardigrades include *Parhexapodibius pilatoi* from grasslands, *Macrobiotus joannae* from moss, *Amphibolus weglarskae* from leaf litter, and four species of *Isohypsibius* from freshwater (Bertolani and Manicardi, 1986).

G. Development

Developmental time, from deposition of the egg to hatching, varies considerably within and between species and with environmental conditions. A minimum of 5 to a maximum of 40 days have been required for egg development under experimental conditions (Marcus, 1928, 1929). Eggs can be found at any time of the year in mosses, lichens, and soil; however, few observations of eggs of freshwater tardigrades have been reported (Ramazzotti and Maucci, 1983). Recently the embryology of the marine eutardigrade *Halobiotus crispae* and the intertidal heterotardigrade *Echiniscoides sigismundi* were investigated (Eibye-Jacobsen, 1996, 1996/97, 1997). Cleavage was total, equal, and asynchronous; however, the cleavage pattern and mode of mesoderm formation could not be definitely determined. Normal development of *H. crispae* at 8 °C was estimated to be 15–16 days.

When embryonic development is complete, the immature tardigrade emerges from the egg by the action of stylets or hindlegs or by an increase in hydrostatic pressure created by pumping fluid into the gut with the pharynx (Ramazzotti and Maucci, 1983; Eibye-Jacobsen, 1996/97, 1997). Postembryonic development occurs through several successive molts, primarily by the growth of the individual cells rather than by cell division. Although tardigrades have been considered cell constant, Bertolani (1982b) observed mitosis in eutardigrades.

In eutardigrades, development is direct and the first-instar animals (juveniles) are similar to the adults but have immature gonads and slight differences in the claws and buccal apparatus. Bertolani *et al.* (1984) concluded that heterotardigrades undergo indirect development characterized by two larval stages. The first larval stage lacks an anus and a gonopore and has two claws less on each leg than the adult. The second larval stage (sometimes called a juvenile) has an anus and the same number of claws as the adult but the gonopore is rudimentary or lacking. The number of filamentous and spinous cuticular appendages generally increases with age (but decreases in *Mopsechiniscus*), although variations may also occur (Grigarick *et al.*, 1975; Ramazzotti and Maucci, 1983; Nelson 1982a; Claxton 1996).

V. ECOLOGY

A. Molting, Life History, and Cyclomorphosis

Molting, which usually requires 5–10 days, occurs periodically throughout the life of the tardigrade (Walz, 1982). The entire cuticular lining of the foregut, including the stylets and stylet supports, is ejected through the expanded buccal opening. The mouth opening closes and the animal cannot feed. This is the "simplex" stage, characterized by the absence of the buccal apparatus. The salivary glands reform the cuticular structures of the buccal tube, stylets, and stylet supports; the esophagus regenerates its own cuticular lining. Concomitantly, new body cuticle, including the hindgut lining, is synthesized by the underlying epidermis and the new claws are produced by claw glands in the legs (Walz, 1982). During the molt, the old body cuticle is shed, including the claws and the lining of the hindgut (Fig. 10). Generally, body length increases with each molt until maximum size is attained, although lack of food can result in a decrease in size. Although growth is more rapid during the earlier molts, molting and growth continue even after sexual maturity (Nelson 1982a). Defecation and oviposition may also be associated with molting and changes in the fluid pressure in the body cavity.

Life histories of certain tardigrade species have been reviewed by Walz (1982), Nelson (1982a), Ramazzotti and Maucci (1983), and Kinchin (1994). Based on frequency distributions of body length and

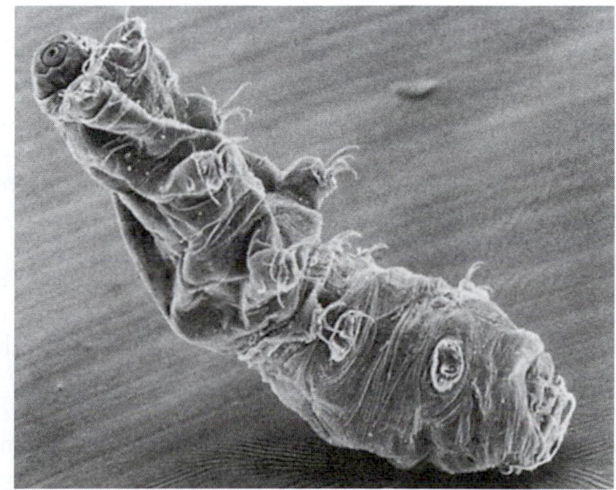

FIGURE 10 Scanning electron micrograph of *Pseudobiotus* Kathmanae undergoing a molt.

buccal length, the number of molts has been estimated to range from 4 to 12. Problems inherent in the method have been discussed by Baumann (1961), Morgan (1977), Ramazzotti and Maucci (1983), and Kinchin (1994). Considerable variation and overlapping of the stages may occur within a species. Sexual maturity is reached with the second or third molt, and egg production continues throughout life, although molting can occur without egg production.

The lifespan of active tardigrades (excluding cryptobiotic periods) has been estimated to be from 3–30 months (Higgins, 1959; Franceschi *et al.*, 1962/63; Pollock, 1970; Ramazzotti and Maucci, 1983). The total lifespan, from hatching until death, may be greatly extended by periods of latent life (encystment and anhydrobiosis). Since aquatic tardigrades have little or no ability to undergo anhydrobiosis, the period of active life generally corresponds to that of total life. Ramazzotti and Maucci (1983) concluded that freshwater species of *Macrobiotus* and *Hypsibius* live about 1–2 yr.; whereas moss-inhabiting species of the same genera average 4–12 yr. Shorter lifespans of 3–4 months were proposed by Pollock (1970) for marine species and Franceschi *et al.* (1962/63) for moss-inhabiting species. Morgan (1977) estimated a lifespan of 3–7 months for *Macrobiotus hufelandi* and up to 3 months for *Echiniscus testudo* in roof mosses.

Cyclomorphosis, an annual cycle of morphological change in individuals, has been documented in the marine eutardigrade *Halobiotus crispae* (Kristensen 1982). A sexually mature summer morph alternates with a sexually immature winter morph that tolerates freezing temperatures and low salinities. Cyclomorphosis may complicate identifications, especially since the distribution of the phenomenon is unknown.

B. Habitats

Although all active individuals require water, the environments in which tardigrades live are generally divided into: (1) marine and brackish water; (2) freshwater; and (3) terrestrial habitats. The distinction between freshwater and terrestrial species is sometimes unclear because some tardigrades can live in a wide range of habitats.

With few exceptions, marine tardigrades belong to the class Heterotardigrada, either in the order Arthrotardigrada or the order Echiniscoidea (Table I). In the class Eutardigrada, only five species of the genus *Halobiotus* (formerly *Isohypsibius*) and a few species of *Isohypsibius* and *Ramajendas* are marine (Crisp and Kristensen, 1983; Pilato and Binda, 1990). Although relatively few studies on marine tardigrades have been conducted, they indicate high morphological diversity, with most of the genera having one or two species (Renaud-Mornant, 1982, 1987; Grimaldi De Zio *et al.*, 1983; Villora-Moreno and De Zio Grimaldi, 1996). Marine tardigrades inhabit the intertidal zone and shallow waters off the continental shelf and as well as bathyl and abyssal depths, including manganese nodules in the deep sea of the eastern South Pacific (Bussau, 1992). Three major ecological groups of marine tardigrades are recognized:

1. a small number of species on intertidal algae or other substrates (e.g., *Echiniscoides* on barnacles), including facultative and obligatory invertebrate ectoparasites (*Tetrakentron* on a sea cucumber) and commensals (e.g., *Actinarctus, Pleocola*);
2. a larger group of species inhabiting the interstitial biotype, particularly in the top few centimeters of sand near the low tide mark (e.g., *Batillipes*); and
3. a highly diversified group of abyssal and deep-sea ooze dwellers (e.g., *Coronarctus*).

Hydrophilous tardigrades are "distinctly aquatic" species that live only in permanent freshwater habitats. Although they are occasionally found in plankton samples, they are benthic organisms that crawl on the surfaces of aquatic plants or in the interstitial spaces of sandy sediments in ponds, lakes, rivers, and streams (Schuster *et al.*, 1977; Wainberg and Hummon, 1981; Kathman and Nelson, 1987; Nelson *et al.*, 1987; Van Rompu and De Smet, 1988, 1991; Van Rompu *et al.*, 1991a, b, 1992; Strayer *et al.*, 1994). Most species live in the littoral zone; however, some individuals have been collected from lakes up to 150 m deep (Ramazzotti and Maucci, 1983). Benthic algal mats in maritime Antarctic lakes and pools are productive habitats for tardigrades (McInnes, 1995; Everitt, 1981; McInnes and Ellis-Evans, 1987, 1990). Cryoconite holes in glaciers, formed when heat is absorbed by surface accumulation of dark dust, also provide a specialized habitat for rotifers and tardigrades, which feed on the microflora living there (Dastych, 1993b; De Smet and Van Rompu, 1994; Sømme, 1996). Hygrophilous tardigrades, which inhabit moist mosses, and eurytopic species, which live in a wide range of moisture conditions, are often found in aquatic habitats.

In the Heterotardigrada, *Styraconyx hallasi* is the only freshwater arthrotardigrade. In the order Echiniscoidea, *Carphania fluviatilis* in the family Carphanidae (Binda and Kristensen, 1986), and a few species of *Echiniscus, Hypechiniscus,* and *Pseudechiniscus* in the family Echiniscidae are occasionally limnic (Kristensen, 1987). The single specimen of *Echinursellus longiunguis,* originally described as an arthrotardigrade from the shores of a Chilean lake, is a cyst or cyclomorphic

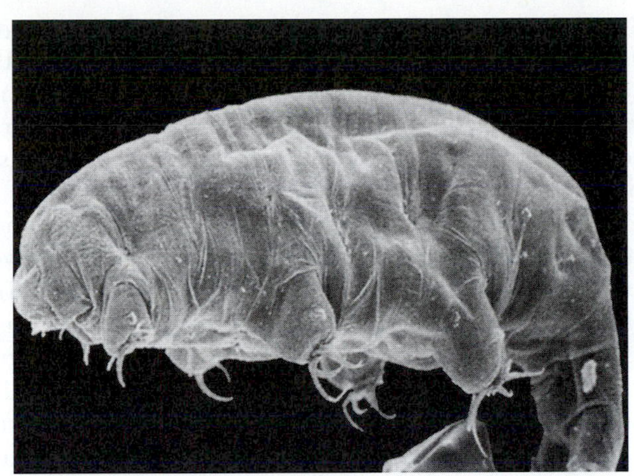

FIGURE 11 Scanning electron micrograph of an aquatic tardigrade, *Pseudobiotus* Kathmanae.

stage of a eutardigrade in the genus *Pseudobiotus*, and is not the missing link between Arthrotardigrada and Echiniscoidea as originally reported (Kristensen, 1987). The status of the mesotardigrade *Thermozodium esakii* from a Japanese hot spring is dubious.

The class Eutardigrada contains both terrestrial and freshwater species (Table I). Three genera *(Dactylobiotus, Pseudobiotus,* and *Thulinia)* are exclusively freshwater (Fig. 11). Other genera have one or more species that are typically aquatic, i.e., *Macrobiotus, Macroversum, Mixibius, Murrayon, Amphibolus, Doryphoribius, Ramajendas, Eohypsibius, Isohypsibius, Hypsibius,* and the *Diphascon* group (Bertolani, 1982a; Bertolani and Pilato, 1988; Pilato and Binda, 1990; Pilato, 1992). Several species of the genus *Hypsibius (* e.g., *H. dujardini* and *H. convergens)* are not obligate freshwater organisms but are often found in this habitat. *Milnesium tardigradum* and *Macrobiotus hufelandi* are eurytopic terrestrial species that may also inhabit freshwater. The tardigrade component of the psammon community in lakes and river banks is poorly known, but it consists primarily of hydrophilous and hygrophilous species that show a distinct seasonality in population densities, with peak periods mainly in the winter and spring (Ramazzotti and Maucci, 1983).

The largest number of tardigrade species are terrestrial and live in moist habitats, such as in soil and leaf litter or among mosses, lichens, liverworts, and cushion-shaped flowering plants (Ramazzotti and Maucci, 1983; Kinchin, 1994). Arctic and Antarctic species also live in terrestrial moss turfs (Dastych, 1984; Bertolani and Kristensen, 1987; Usher and Dastych, 1987; Maucci, 1996; Miller *et al.,* 1996). Most terrestrial tardigrades inhabit the moss environment, which can be divided into three groups: wet, moist, and dry mosses (Mihelčič, 1954/55; Ramazzotti and Maucci, 1983). Wet mosses such as submerged and emergent species in lakes, ponds, and streams never dehydrate completely. In contrast, moist and dry mosses living on tree trunks, rocks, walls, roofs, and soil are subject to total desiccation for various periods of time and are dependent on atmospheric precipitation for moisture. The typical terrestrial habitat of a tardigrade has: (1) sufficient aeration (tardigrades are hypersensitive to low oxygen levels); (2) alternate periods of wet and dry conditions; and (3) suitable and sufficient food. Tardigrades are less likely to inhabit impermeable clay soils or dense cushion mosses with thick cell walls. A distinct insular distribution of tardigrades within patches of mosses and soil has been frequently noted (Ramazzotti and Maucci, 1983). Lichens and liverworts are inhabited by a community usually similar to that in mosses.

High species diversity and large populations of eutardigrades are also common in soil and the upper layer of leaf litter; however, heterotardigrades are rare or absent (Fleeger and Hummon, 1975; Manicardi and Bertolani, 1987; Bertolani and Rebecchi, 1996; Guidetti, 1998; Guidetti *et al.,* in press). Interstitial eutardigrades in inorganic soils are characterized by very short legs with small claws and a reduction or absence of the hind claws (Bertolani and Biserov, 1996). The tardigrade community in leaf litter from beech forests contains some species found in freshwater, but not those characterized by very long claws (e.g., *Dactylobiotus, Pseudobiotus*). There is also some overlap with the turf community, except for turf species with very short claws. In contrast, the litter community is significantly different from that of mosses and lichens on beech trees in the same area (Guidetti *et al.,* 1999.)

C. Population Density

Population densities of tardigrades are highly variable; however, neither minimal nor optimal conditions for population growth are known. [See review in Kinchin, 1994.] Changes in tardigrade population densities have been correlated with a variety of environmental conditions, including temperature and moisture (Franceschi *et al.,* 1962/63; Morgan, 1977; Briones *et al.,* 1997), air pollution (Steiner, 1994a–c, 1995), and food availability (Hallas and Yeates, 1972). Other factors such as competition, predation, and parasitism may also play a role. Predators include nematodes, other tardigrades, mites, spiders, springtails, and insect larvae; parasitic protozoa and fungi often infect tardigrade populations (Ramazzotti and Maucci, 1983). Considerable variation in both population density and species diversity occurs in adjacent, seemingly identical, microhabitats. Usually 2–6 species are present, but 10 or more species have been found in a single sample.

In general, patchiness is a common characteristic of tardigrade populations.

Populations of tardigrades from soil and leaf litter habitats range from 5 to 40 individuals/cm^2. [See Nelson and Higgins, 1990 for review and Guidetti *et al.,* 1999.] Similarly, population densities in mosses and other terrestrial habitats vary widely, from 0 to 50–200 individuals/cm^2. (For review, see Ramazzotti and Maucci, 1983; Kinchin, 1994.) Large populations of both heterotardigrades and eutardigrades are often found in thin, spreading mosses growing in open, exposed places; in contrast, very few heterotardigrades, but many eutardigrades, can be collected in mosses from moist shady areas.

Studies of the population dynamics of freshwater tardigrades are less common than those of terrestrial habitats (Schuster *et al.,* 1977; Everitt, 1981; Wainberg and Hummon, 1981; Kathman and Nelson, 1987; Nelson *et al.,* 1987; Strayer *et al.,* 1994). Ramazzotti and Maucci (1983) and Nelson *et al.* (1987) reported that freshwater eutardigrades often reach a very high population density in the spring, whereas Schuster *et al.* (1977) and Wainberg and Hummon (1981) noted population peaks in the fall. Kathman and Nelson (1987) stated, "Several previous life-history studies indicated wide variation in total numbers of animals obtained from month to month, but few studies extended beyond 12 months or presented data on replicates, means, or standard errors for determining the validity of these variations. Unless there are recurring patterns over several years, abundance estimates may be invalid if adequate numbers of quantitative replicates are not obtained, as our data show that within-period variation was often greater than between-period variation."

D. Distribution

Due to the paucity of data and the uncertainty of identifications, the biogeography of the Tardigrada remains relatively unknown (McInnes, 1994). As a phylum, tardigrades have a worldwide distribution. Many species with broad ecological requirements are considered to be cosmopolitan, whereas others with more narrow tolerances are rare or endemic. Cluster analysis of the distribution of extant fauna showed that limnoterrestrial tardigrades have limited generic and familial distribution ranges defined by biogeographical provinces and major geological events (McInnes and Pugh, 1998; Pugh and McInnes, 1998). Distinct Laurasian and Gondwanan familial clusters correlate with the Triassic disintegration of Pangaea, while separate Antarctic, Australian and New Zealand familial/generic clusters relate to the subsequent Jurassic/Cretaceous disintegration of Gondwana. A "Nearctic-Arctic" biogeographic cluster is distinct from that of "Northern and Alpine Europe". In general, moss-inhabiting tardigrades are more common in polar and temperate zones than in the tropics. More data are required, however, before definitive biogeographical distributions of the phylum can be determined.

On a more localized scale, significant differences in altitudinal distributions of terrestrial tardigrades have been reported by Nelson (1975), Beasley (1988), Dastych (1987b, 1988), Manicardi and Bertolani (1987), Ramazzotti and Maucci (1983), and Guidetti *et al.* (1999). In contrast, Kathman and Cross (1991) found that the distribution of tardigrades at any altitude is primarily determined by abiotic factors, such as temperature and humidity, which affect the microhabitats at a particular altitude.

Physical and chemical monitoring of pollutants in environmental surveys have been used to indicate air and water quality; however, ecological monitoring has not been extensively developed. Terrestrial invertebrates in mosses and soils, as well as aquatic meiofauna, may be suitable as a monitoring system to indicate environmental quality. Steiner (1994a–c, 1995) investigated the influence of air pollution on moss-dwelling nematodes and tardigrades along an urban–rural gradient near Zürich, Switzerland. Although the abundance of tardigrades varied independently of SO_2 and NO_2, the number of tardigrade species decreased with increasing levels of SO_2. Fumigation experiments with SO_2 indicated that the effects were dose-dependent and taxon specific. The tardigrade *Macrobiotus persimilis* was typical of sites with relatively high levels of NO_2.

E. Dissemination

Eggs, cysts, and barrel-shaped anhydrobiotic specimens of terrestrial tardigrades are passively dispersed predominantly by wind, although rain, floodwaters, and melting snow can transport active as well as cryptobiotic forms (Ramazzotti and Maucci, 1983). Other animals, associated with terrestrial communities, e.g., birds, snails, isopods, mites, centipedes, millipedes, and insects, can also assist in the dispersal of tardigrades. Active dissemination by crawling is limited by the necessity for a constant film of water covering the body of the tardigrade, and by its size and speed of movement.

Little information is available on dispersal of the distinctly aquatic species, which probably undergo limited or no cryptobiosis. Transport of eggs or encysted forms from dry temporary ponds is possible. Active tardigrades may be distributed through river systems, particularly during periods of heavy rainfall and flooding. The mechanism for re-population of aquatic habitats after the summer disappearance of tardigrades or after habitat disturbances has not been investigated (Nelson *et al.*, 1987).

VI. TECHNIQUES FOR COLLECTION, EXTRACTION, AND MICROSCOPY

Qualitative collections of tardigrades from aquatic habitats with sandy bottoms can be made by stirring the sand in a container of water and decanting immediately after the sand settles (Schuster *et al.*, 1977). The decanted water is passed through a sieve (US Standard No. 325) with pores of about 44 μm, and the specimens are rinsed from the screen to a sample jar. Tardigrades in large volumes of water can be preserved by the addition of formalin (5%) or gluteraldehyde. A modified Boisseau apparatus has also been adapted for tardigrade extraction (Kinchin, 1994).

Modifications of techniques to extract soil nematodes by centrifugation in water, sucrose, or Ludox® have been used to remove tardigrades from aquatic and terrestrial habitats (Hallas, 1975; Hallas and Yeates, 1972; Bertolani, 1982a). A watery suspension of the sample (sediment or vegetation) is prepared and filtered through two stacked sieves with mesh openings of 500 μm or greater (top sieve) and 44 μm (bottom sieve). The sediment remaining on the fine mesh sieve is centrifuged with water at about 3000 rpm for 5 mins. The supernatant is then poured through the fine mesh sieve, and the remaining sediment is centrifuged with a sucrose solution (sp. gr. 1.18) at about 5000 rpm for 1 min. Ludox® (DuPont product, colloidal silica, Sigma–Aldrich Chemical Co.) can be used in place of sucrose; Ludox AM® is useful for extracting tardigrades from terrestrial/freshwater environment, whereas Ludox TM® is more effective with marine tardigrades (McInnes, personal communication). The supernatant is immediately rinsed through the fine mesh sieve with running tap water, and the residue is washed into a petri dish for examination under a dissecting microscope.

Collections of tardigrades from vegetation can be made by removing the plant (moss, lichen, liverwort, etc.) from the substrate (tree, rock, soil) and placing the sample in a small paper bag for dry storage until processing. The sample is then placed in a container of water for several hours to reverse anhydrobiosis. The sample is agitated and squeezed to remove specimens, eggs, and cysts. After the excess water is decanted, the remaining sediment containing the live tardigrades can be examined with a dissecting microscope at a magnification of 30× or greater. The sample may also be washed through sieves to remove particles smaller than 50 μm or larger than 1200 μm. Boiling water or boiling alcohol (85% or higher) added to specimens in small amounts of water is adequate for tardigrade fixation, but alcohol is recommended to preserve specimens to be mounted later. R. D. Kathman (personal communication) found that mounting the live tardigrade directly in Hoyer's medium was more efficient, however long-term preservation was not as satisfactory.

Core samples of soil and leaf litter can also be placed in paper bags for dry storage and processed according to the traditional procedure for vegetation. The sample is thoroughly soaked in water for several hours and examined in small quantities of thin aqueous suspensions since eggs and cysts are easily overlooked. The technique is extremely laborious and time-consuming; however, centrifugation with sucrose or Ludox®, as described above for aquatic samples, will reduce the time required for extraction of tardigrades from the substrate.

After fixation, individual specimens are mounted on microscope slides and observed with phase contrast and/or differential interference contrast (DIC) microscopy. Hoyer's mounting medium (distilled water, 50 ml; gum arabic, 30 g; chloral hydrate, 150 g; glycerol, 20 ml) produces cleared, distended specimens, but the coverslip must be sealed (e.g., with epoxy paint) to ensure permanence. The amount of chloral hydrate may be reduced to decrease the extent of clearing of specimens, and potassium iodide (KI, 1 g) may be added to stain specimens slightly to increase resolution of morphological structures. CMC® and CMCP® media are useful for temporary mounts; however, excessive stain may obscure important structures, and the specimens may clear completely after a period of time. Other media used include Faure's medium and polyvinyl lactophenol. More permanent mounts with balsam or synthetic resins are not as useful for phase microscopy. Kristensen and Higgins (1984a) recommend transferring animals with a drop of 2% formalin to slides and covering with a coverslip. The formalin preparation is infused with a 10% solution of glycerin and 90% ethyl alcohol and allowed to evaporate to glycerin over several days. Then the coverslip is sealed with Murrayite.

For scanning electron microscopy (SEM), fixed tardigrades are dehydrated to absolute alcohol, transferred to an intermediary liquid, such as amyl acetate, and dried by the critical point method (Grigarick *et al.*, 1975). Individuals are mounted on SEM stubs and coated with 200–300 nm of gold or gold–palladium.

Tardigrades can also be purchased commercially from Carolina Biological Supply Company (Burlington, NC) or Ward's (Rochester, NY). Specimens from Carolina, labeled "*Milnesium*" or "*Milnesium* and other species" and shipped in the anhydrobiotic state, may contain a mixture of *Macrobiotus, Milnesium,* and/or *Echiniscus*. Cultures of aquatic specimens shipped in the active state from Ward's, labeled

"*Hypsibius* (water bear)", were actually *Thulinia* in the samples personally examined. Caution should be exercised in relying on a company's identifications of specialized groups of invertebrates.

VII. IDENTIFICATION

Identification of species in the phylum Tardigrada is based primarily on the morphology of the cuticle, claws, buccal apparatus, and eggs, which have been discussed in previous sections. Early European workers described the majority of tardigrade species from temporary wet mounts of live specimens observed with the light microscope, often resulting in inadequate, incorrect, or incomplete descriptions and illustrations. Morphology, systematics, and natural history were focal points of their studies. Early monographs and other papers by Thulin (1928) and Marcus (1929, 1936) in German and Cuénot (1932) in French formed the basis of tardigrade systematics. Ramazzotti (1962, 1972) published a monograph, in Italian, which compiled keys and illustrated diagnoses of the species, thorough sections on morphology and biology, and an extensive bibliography; a supplement was produced in 1974. The last revision of this comprehensive monograph (Ramazzotti and Maucci, 1983) incorporated parts of the taxonomic revisions by Pilato (1969, 1982) and Schuster *et al.* (1980). An English translation of this important monograph is available from Dr. Clark Beasley, Department of Biology, McMurry University, Abilene, TX. Useful references in English include those by Morgan and King (1976), Nelson (1982a), Dewel *et al.* (1993), and Kinchin (1994) and the international symposia volumes (Higgins, 1975; Węglarska, 1979; Nelson, 1982b; Bertolani, 1987b; McInnes and Norman, 1996; and Greven, 1999). Kinchin (1994) reviewed the systematics, morphology, and ecology of tardigrades and presented a guide to common species. He synthesized classifications from various authors, dividing tardigrades artificially into "marine heterotardigrades, marine eutardigrades, limno-terrestrial heterotardigrades, and limno-terrestrial eutardigrades."

Currently, claw structure, considered more conservative than the bucco-pharyngeal apparatus, is the major character for distinguishing genera and families within the eutardigrades (Bertolani and Kristensen, 1987). Figures 5–8 illustrate the claw types and the buccal apparatus of the major genera of eutardigrades, basically according to the classification system of Ramazzotti and Maucci (1983), but with the inclusion of additional genera. The figures illustrate the range of variation in morphological structures that are considered of taxonomic importance in defining genera and also the need for further analysis.

The following key includes common genera previously found and likely to be found in freshwater and terrestrial habitats in North America. Positive identification with the key may be difficult because of inadequate and often vague generic descriptions, additions of new genera (often formed by splitting current genera), and the need for further revision of tardigrade systematics.

A. Taxonomic Key to Common Genera of Freshwater and Terrestrial Tardigrada

1a. Lateral cirri A present 2

1b. Lateral cirri A absent 4

2a(1a). Cuticle without dorsal plates 3

2b. Cuticle with complete or partial dorsal plates; terrestrial ...Echiniscidae [*Bryochoerus, Bryodelphax, Cornechiniscus, Echiniscus, Hypechiniscus, Mopsechiniscus, Parechiniscus, Pseudechiniscus* (see Kristensen, 1987 for revision).]

3a(2a). Body indistinctly divided into eight segments; clavae present; four claws on all legs in adults; terrestrial .... *Oreella*

3b. Body eutardigrade-like in shape; clavae absent; two claws on legs I–III, one claw on legs IV; freshwater .... *Carphania*

4a(1b). Head with cephalic papillae 5

4b. Head without cephalic papillae 6

5a(4a). Head with six peribuccal papillae and two lateral papillae present; terrestrial, occasionally in freshwater .... *Milnesium*

5b. Head without peribuccal papillae; two lateral papillae present; terrestrial .... *Limmenius*

6a(4b). Claws absent on all legs; legs very short; terrestrial (Binda, 1984) .... *Apodibius*

6b. Claws present on at least one pair of legs 7

7a(6b). Single unbranched claws without secondary branches 8

7b. Double claws, with primary and secondary branches 9

8a(7a). One unbranched conical claw on leg I, no claws on legs II–IV; terrestrial .... *Necopinatum*

8b. Two unbranched claws on legs I–III, small basal spur-like claws on leg IV; terrestrial .... *Haplomacrobiotus*

9a(7b). Two double claws on each leg, similar in size and shape and symmetrical with respect to the median plane of the leg; claw branch sequence is 2-1-1-2 (secondary, primary, secondary, primary) 10

9b. Two double claws on each leg usually dissimilar in size and shape and asymmetrical with respect to the median plane of the leg; claw branch sequence is 2-1-2-1 (secondary, primary, secondary, primary) 14

10a(9a). Buccal tube of distinctly greater length between stylet supports and placoids than between stylet supports and buccal opening; posterior part of tube of spiral composition; terrestrial .... *Pseudodiphascon*

10b. Buccal tube between stylet supports and placoids not longer than distance between stylet supports and buccal opening; tube without spiral composition 11

11a(10b). Stylet supports attach near middle of buccal tube; with 10 peribuccal papulae (not lamellae); terrestrial .... *Minibiotus*

11b. Stylet supports attach in posterior half of buccal tube 12

12a(11b). Branches of claws at right angles so that tips are remote; two claws structurally connected at base; claws without lunules; cuticle without pores; freshwater .... *Dactylobiotus*

12b. Both branches of claw follow similar axis so that tips are close together; claws not structurally connected at base; claws with lunules; cuticle with pores 13

13a(12b). Ventral lamina (reinforcement rod, buccal tube support) on buccal tube well developed; no anterior crest on buccal tube; 10 peribuccal lamellae; terrestrial or freshwater .... *Macrobiotus*

13b. Ventral lamina narrow or absent; unequal dorsal and ventral crests on buccal tube; lamellae unknown; terrestrial .... *Adorybiotus*

14a(9b). Buccal tube with median longitudinal anteroventral lamina (ventral lamina or buccal tube support) 15

14b. Buccal tube without median longitudinal anteroventral lamina (ventral lamina or buccal tube support) 16

15a(14a). Legs IV reduced or absent, with no claws or one or two minute double claws; terrestrial .... *Hexapodibius*

15b. Legs IV normally developed; terrestrial and freshwater .... *Doryphoribius*

16a(14b). Claw of *Calohypsibius*-type, primary and secondary branches rigidly connected, claws equal in size 17

16b. Claw with two or three distinct parts 18

17a(16a). Cuticle sculptured, two transverse rows of macroplacoids; terrestrial .... *Calohypsibius*

17b. Cuticle smooth; three transverse rows of macroplacoids; terrestrial .... *Microhypsibius*

18a(16b). Each double claw with three distinct parts 19

18b. Each double claw with two distinct parts 20

19a(18a). Posterior part of buccal tube of spiral composition; 14 (16?) lamellae present; terrestrial .... *Eohypsibius*

19b. Buccal tube rigid, without spiral composition; 14 lamellae present; terrestrial or freshwater .... *Amphibolus*

20a(18b). Buccal tube rigid, without posterior part of spiral composition 21

20b. Buccal tube with posterior part of spiral composition 25

21a(20a). Internal and external claws of dissimilar size and shape, with secondary branch continuous with the base; no lunules; no peribuccal lamellae 22

21b. Internal and external claws of dissimilar size and shape, with secondary branch forming right angle with base; lunules and/or peribuccal lamellae present or absent 23

22a(21a). Claws of *oberhaeuseri*-type: *Hypsibius*-type with internal and external claws of very dissimilar size and shape, with primary branch of external claw very long and narrow, with very thin flexible connection to upper basal portion; elliptical dorsolateral sense organs on head; terrestrial (Binda and Pilato, 1986) .... *Ramazzottius*

22b. Claws of *dujardini*-type: *Hypsibius*-type with external claw slightly larger than internal claw, with primary branch attached to median basal portion of claw; no sense organs on head; terrestrial or freshwater .... *Hypsibius*

23a(21b). Dorsal and ventral crests on buccal tube absent; peribuccal lunules rarely present; cuticle smooth or with granulations and/or gibbosites; terrestrial or freshwater .... *Isohypsibius*

23b. Dorsal and ventral crests on buccal tube present 24

24a(23b). With 12 peribuccal lamellae; freshwater .... *Thulinia*

24b. With 30 peribuccal lamellae; freshwater .... *Pseudobiotus*

25a(20b). Pharynx with placoids and stylet supports reduced or absent terrestrial, freshwater .... *Itaquascon*

25b. Pharynx with normal placoids; buccal tube with obvious stylet supports; terrestrial or freshwater; (formerly *Diphascon*; revised in Pilato, 1987) 26

26a(25b). Apophyses for insertion of stylet muscles in shape of a "hook" 27

26b. Apophyses for insertion of stylet muscles in shape of wide, flat "ridges" 28

27a(26a). Stylet apophyses in the shape of a semilunar hook; pharyngeal tube longer than the buccal tube; with (subgenus *Diphascon*) or without (subgenus *Adropion*) droplike thickening on buccal below insertion of stylet supports .... *Diphascon*

27b. Stylet apophyses in the shape of a blunt hook; pharyngeal tube shorter than the buccal tube .... *Hebesuncus*

28a(26b). Stylet furcae with posteriolateral processes thickened at apices .... *Mesocrista*

28b. Stylet furcae with posteriolateral processes spoon-like and tapering at their apices .... *Platicrista*

LITERATURE CITED

Aguinaldo, A. M., Turbeville, J. M., Linford, L. S., Rivera, M. C., Garey, J. R., Raff, R., Lake, J. A. 1997. Evidence for a clade of nematodes, arthropods, and other moulting animals. Nature 387:489–493.

Baccetti, B. 1987. The evolution of the sperm cell in the phylum Tardigrada (Electron microscopy of tardigrades. 5), *in:* Bertolani, R., Ed., Biology of tardigrades. Selected Symposia and Monographs U. Z. I., 1. Mucchi, Modena, Italy, pp. 87–91.

Baccetti, B., Rosati, F., Selmi, G. 1971. Electron microscopy of tardigrades. 4. The spermatozoon. Monitore Zoologico Italiano 5:231–240.

Baumann, H. 1961. Der Lebensablauf von *Hypsibius* (*H.*) *convergens* Urbanowicz (Tardigrada). Zoologischer Anzeiger 167:362–381.

Beasley, C. 1988. Altitudinal distribution of Tardigrada of New Mexico with the description of a new species. American Midland Naturalist 120:436–440.

Bertolani, R. 1975. Cytology and systematics in Tardigrada. Memorie dell'Istituto Italiano di Idrobiologia 32 (Suppl.)17–35.

Bertolani, R. 1982a. Tardigradi. Guide per il riconoscimento delle specie animali delle acque interne Italiane. Consiglio Nazionale Delle Ricerche, Verona, Italy. 104 pp.

Bertolani, R. 1982b. Cytology and reproductive mechanisms in tardigrades, *in*: Nelson, D. R., Ed., Proceedings of the third international symposium on the Tardigrada. East Tennessee State University Press, Johnson City, TN, pp. 93–114.

Bertolani, R. 1983a. 17. Tardigrada, *in*: Adiyodi, K. G., Adiyodi, R. G., Eds., Reproductive biology of invertebrates. Vol. I: Oogenesis, oviposition, and oosorption. Wiley, New York, pp. 431–441.

Bertolani, R. 1983b. 19. Tardigrada, *in*: Adiyodi, K. G., Adiyodi, R. G. Eds., Reproductive biology of invertebrates. Vol. II: Spermatogenesis and sperm function. Wiley, New York, pp. 387–396.

Bertolani, R. 1987a. Sexuality, reproduction, and propagation in tardigrades, *in*: Bertolani, R., Ed., Biology of tardigrades. Selected Symposia and Monographs U. Z. I., 1. Mucchi, Modena, Italy, pp. 93–101.

Bertolani, R., Ed., 1987b. Biology of tardigrades. Proceedings of the fourth international symposium on the Tardigrada, September 1985, Modena, Italy. Selected Symposia and Monographs U. Z. I., 1. Mucchi, Modena, Italy.

Bertolani, R. 1990. Tardigrada, *in*: Adiyodi, K. G., Adiyodi, R. G., Eds., Reproductive biology of invertebrates. Vol. 4B: Fertilization, development and parental care. Wiley, New York, pp. 49–60.

Bertolani, R. 1992. Tardigrada, *in*: Adiyodi, K. G., Adiyodi, R. G., Eds., Reproductive biology of invertebrates. Vol. 5B: Sexual differentiation and behaviour. Wiley, New York, pp. 255–266.

Bertolani, R. 1994. 2. Tardigrada, *in*: Adiyodi, K. G., Adiyodi, R. G., Eds., Reproductive biology of invertebrates. Vol. VI: Asexual propagation and reproductive strategies. Oxford and IBH Publ., New Delhi, pp. 25–37.

Bertolani, R., Biserov, V. 1996. Leg and claw adaptations in soil tardigrades, with erection of two new genera of Eutardigrada, Macrobiotidae: *Pseudohexapodibius* and *Xerobiotus*. Invertebrate Biology 115(4):299–304.

Bertolani, R., Kristensen, R. 1987. New records of *Eohypsibius nadjae* Kristensen, 1982, and revision of the taxonomic position of two genera of Eutardigrada (Tardigrada), *in*: Bertolani, R., Ed., Biology of tardigrades. Selected Symposia and Monographs U. Z. I., 1. Mucchi, Modena, Italy, pp. 359–372.

Bertolani, R., Manicardi, G. C. 1986. New cases of hermaphroditism in tardigrades. International Journal of Invertebrate Reproduction 9:363–366.

Bertolani, R., Pilato, G. 1988. Struttura delle unghie nei Macrobiotidae e descrizione di *Murrayon* n. gen. (Eutardigrada). Animalia 15:17–24.

Bertolani, R., Rebecchi, L. 1993. A revision of the *Macrobiotus hufelandi* group (Tardigrada, Macrobiotidae), with some observations on the taxonomic characters of eutardigrades. Zoologica Scripta 22:127–152.

Bertolani, R., Rebecchi, L. 1996. The tardigrades of Emilia (Italy). II. Monte Rondinaio. A multihabitat study on a high altitude valley of the northern Apennines. Zoological Journal of the Linnean Society 116:3–12.

Bertolani, R., Rebecchi, L. 1999. Tardigrada, *in*: Knobil, E., Neill, J. D., Eds., Encyclopedia of Reproduction, Vol. 4: Academic Press, San Diego, pp. 703–718.

Bertolani, R., Grimaldi De Zio, S., D'Addabbo Gallo, M., Morone de Lucia, M. R. 1984. Postembryonic development in heterotardigrades. Monitore Zoologico Italiano 18:307–320.

Bertolani, R., Rebecchi, L., Beccaccioli, G. 1990. Dispersal of *Ramazzottius* and other tardigrades in relation to type of reproduction. Invertebrate Reproduction and Development 18:153–157.

Bertolani, R., Rebecchi, L., Claxton, S. 1996. Phylogenetic significance of egg shell variation in tardigrades. Zoological Journal of the Linnean Society 116:139–148.

Binda, M. G. 1984. Notizie sui Tardigradi dell'Africa Meridonale con descrizione di una nuova specie di *Apodibius* (Eutardigrada). Animalia 11:5–15.

Binda, M. G., Kristensen, R. 1986. Notes on the genus *Oreella* (Oreellidae) and the systematic position of *Carphania fluviatilis* Binda, 1978 (Carphanidae fam. nov., Heterotardigrada). Animalia 13(1/3):9–20.

Binda, M. G., Pilato, G. 1986. *Ramazzottius,* nuovo genere di Eutardigrado (Hypsibiidae). Animalia 13:159–166.

Biserov, V. 1990a. On the revision of the genus *Macrobiotus.* The subgenus *Macrobiotus* s.st.: a new systematic status of the group *hufelandi* (Tardigrada, Macrobiotidae). Communication 1. Zoologicheskii Zhurnal 69:5–17 (in Russian, with English summary).

Biserov, V. 1990b. On the revision of the genus *Macrobiotus.* The subgenus *Macrobiotus* s.st. is a new systematic status of the group *hufelandi* (Tardigrada, Macrobiotidae). Communication 2. Zoologicheskii Zhurnal 69:38–50 (in Russian, with English summary).

Biserov, V. 1992. A new genus and three new species of tardigrades (Tardigrada: Eutardigrada) from the USSR. Bolletino di Zoologia 59:95–103.

Briones, M. J., Ineson, P., Piearce, T. G. 1997. Effects of climate change on soil fauna; responses of enchytraeids, Diptera larvae and tardigrades in a transplant experiment. Applied Soil Ecology 6:117–134.

Bussau, C. 1992. New deep-sea Tardigrada (Arthrotardigrada, Haleschiniscidae) from a manganese nodule area of the eastern South Pacific. Zoologica Scripta 21:79–91.

Claxton, S. 1996. Sexual dimorphism in Australian *Echiniscus* (Tardigrada, Echiniscidae) with descriptions of three new species. Zoological Journal of the Linnean Society 116:13–33.

Crisp, M., Kristensen, R. 1983. A new marine interstitial eutardigrade from East Greenland, with comments on habitat and ecology. Videnskabelige Meddeleser fra Dansk Naturhistorisk Foreng 144:99–114.

Crowe, J. 1975. The physiology of cryptobiosis in tardigrades. Memorie dell'Istituto Italiano di Idrobiologia 32(Suppl.):37–59.

Crowe, J., Crowe, L., Chapman, D. 1984. Preservation of membranes in anhydrobiotic organisms: the role of trehalose. Science 223:701–703.

Crowe, J., Hoekstra, F., Crowe, L. 1992. Anhydrobiosis. Annual Review of Physiology 54:579–599.

Cuénot, L. 1932. Tardigrades, *in*: Lechevalier, P., Ed., Faune de France 24:1–96.

Dastych, H. 1984. The Tardigrada from Antarctic with description of several new species. Acta Zoologica Cracoviensia 27:377–436.

Dastych, H. 1987a. Two new species of Tardigrada from the Canadian Subarctic with some notes on sexual dimorphism in the family Echiniscidae. Entomologische Mitteilungen aus dem Zoologischen Museum Hamburg 8 (129):319–334.

Dastych, H. 1987b. Altitudinal distribution of Tardigrada in Poland, *in*: Bertolani, R., Ed., Biology of tardigrades. Selected Symposia and Monographs U. Z. I., 1. Mucchi, Modena, Italy, pp. 169–176.

Dastych, H. 1988. The Tardigrada of Poland. Monografie Fauny Polski 16:1–255.

Dastych, H. 1992. *Paradiphascon manningi* gen. n. sp. n., a new water-bear from South Africa, with the erecting of a new subfamily Diphasconinae (Tardigrada). Mitteilungen aus dem Hamburgischen Zoologischen Museum und Institut 89:125–139.

Dastych, H. 1993a. A new genus and four new species of semiterrestrial water-bears from South Africa (Tardigrada). Mitteilungen aus dem Hamburgischen Zoologischen Museum und Institut 90:175–186.

Dastych, H. 1993b. Redescription of the cryoconital tardigrade *Hypsibius klebelsbergi* Mihelčič, 1959, with notes on the microslide collection of the late Dr. F. Mihelčič. Veröff. Museum Ferdinandeum 73:5–12.

De Smet, W., Van Rompu, E. A. 1994. Rotifera and Tardigrada from some cryoconite holes on a Spitsbergen (Svalbard) glacier. Belgian Journal of Zoology 124:27–37.

Dewel, R. A., Dewel, W. C. 1996. The brain of *Echiniscus viridissimus* Peterfi, 1956 (Heterotardigrada): a key to understanding the phylogenetic position of tardigrades and the evolution of the arthropod head. Zoological Journal of the Linnean Society 116:35–49.

Dewel, R. A., Dewel, W. C. 1997. The place of tardigrades in arthropod evolution, *in*: Fortey, R. A., Thomas, R. H., Eds., Arthropod relationships. Systematics Association Special Volume Series 55. Chapman & Hall, London, pp. 109–123.

Dewel, R. A., Dewel, W. C., Roush, B. G. 1992. Unusual cuticle-associated organs in the heterotardigrade, *Echiniscus viridissimus.* Journal of Morphology 212:123–140.

Dewel, R. A., Nelson, D. R., Dewel, W. C. 1993. Tardigrada, *in*: Harrison, F. W., Rice, M. E., Eds., Microscopic anatomy of invertebrates. Vol. 12: Onychophora, Chilopoda and lesser Protostomata. Wiley–Liss, Chichester, pp. 143–183.

Doyère, L. 1840. Memorie sur les Tardigrades. Annales des Sciences Naturelles, Zool., Paris, Series 2, 14:269–362.

Eibye-Jacobsen, J. 1996. On the nature of pharyngeal muscles cells in the Tardigrada, *In*: McInnes, S. J., Norman, D. B., Eds., Tardigrade biology. Zoological Journal of the Linnean Society 116:123–138.

Eibye-Jacobsen, J. 1996/97. New observations on the embryology of the Tardigrada. Zoologischer Anzeiger 235:201–216.

Eibye-Jacobsen, J. 1997. Development, ultrastructure and function of the pharynx of *Halobiotus crispae* Kristensen, 1982 (Eutardigrada). Acta Zoologica 78:329–347.

Everitt, D. A. 1981. An ecological study of an Antarctic freshwater pool with particular reference to Tardigrada and Rotifera. Hydrobiologia 83:225–237.

Fleeger, J. W., Hummon, W. D. 1975. Distribution and abundance of soil Tardigrada in cultivated and uncultivated plots of an old field pasture. Memorie dell'Istituto Italiano di Idrobiologia 32(Suppl.):93–112.

Franceschi, T., Loi, M. L., Pierantoni, R. 1962/63. Risultati di una prima indagine ecologica condotta su popolazioni di Tardigradi. Bollettino dei Musei e degli Istituti Biologici dell'Universita di Genova 32:69–93.

Garey, J., Nelson, D. R., Mackey, L.Y., Li, J. 1999. Tardigrade phylogeny: congruency of morphological and molecular evidence. Zoologischer Anzeiger 238(3–4): 205–210.

Garey, J., Krotec, M., Nelson, D. R., Brooks, J. 1996. Molecular analysis supports a tardigrade–arthropod association. Invertebrate Biology 115:79–88.

Giribet, G., Carranza, S., Baguñà, J., Riutort, M., Ribera, C. 1996. First molecular evidence for the existence of a Tardigrada + Arthropoda clade. Molecular Biology and Evolution 13:76–84.

Goeze, J. A. E. 1773. Uber den Kleinen Wasserbär, *in*: Bonnet, H. K., Ed., Abhandlungen aus der Insectologie, Ubers. Usw, 2. Beobachtg, pp. 67.

Greven, H. 1980. Die Bärtierchen. Die Neue Brehm-Bucherei, Vol. 537. Ziemsen Verlag, Wittenberg Lutherstradt, Germany.

Greven, H., Ed., 1999. Proceedings of the seventh international symposium on the Tardigrada, August 1997, Dusseldorf, Germany. Zoologischer Anzeiger 238(3–4): 1–346.

Greven, H., Greven, W. 1987. Observations on the permeability of the tardigrade cuticle using lead as an ionic tracer, *in*: Bertolani, R., Ed., Biology of tardigrades. Selected Symposia and Monographs U. Z . I., 1. Mucchi, Modena, Italy, pp. 35–43.

Greven, H., Peters, W. 1986. Localization of chitin in the cuticle of

Tardigrada using wheat germ agglutinin–gold conjugate as a specific electron dense marker. Tissue and Cell 18:297–304.

Grigarick, A. A., Schuster, R. O., Toftner, E. C. 1973. Descriptive morphology of eggs of some species in the *Macrobiotus hufelandi* group. The Pan-Pacific Entomologist 49(3):258–263.

Grigarick, A. A., Schuster, R. O., Toftner, E. C. 1975. Morphogenesis of two species of *Echiniscus*. Memorie dell'Istituto Italiano di Idrobiologia 32(Suppl.):133–151.

Grimaldi De Zio, S., D'Addabbo Gallo, M., Morone de Lucia, M. R. 1983. Marine tardigrades ecology. Oebalia 9:15–31.

Guidetti, R. 1998. Two new species of Macrobiotidae (Tardigrada, Eutardigrada) from the United States of America, and some taxonomic considerations of the genus *Murrayon*. Proceedings of the Biological Society of Washington 111(3):663–673.

Guidetti, R., Bertolani, R., Nelson, D. R. 1999. Ecological and faunistic studies on tardigrades in leaf litter of beech forests. Zoologischer Anzeiger 238(3–4): 215–223.

Guidi, A., Rebecchi, L. 1996. Spermatozoan morphology as a character for tardigrade systematics: comparisons with sclerified parts of animals and eggs in eutardigrades, *in*: McInnes, S. J., Norman, D. B., Eds., Tardigrade biology. Zoological Journal of the Linnean Society 116:101–113.

Hallas, T. E. 1975. A mechanical method for the extraction of Tardigrada. Memorie dell'Istituto Italiano di Idrobiologia 32(Suppl.):153–158.

Hallas, T. E., Yeates, G. W. 1972. Tardigrada of the soil and litter of a Danish beech forest. Pedobiologia 12:287–304.

Henneke, J. 1911. Beiträge zur Kenntnis der Biologie und Anatomie der Tardigraden (*Macrobiotus macronyx* Duj.). Zeitschrift fuer Wissenschaftliche Zoologie 97:721–752.

Higgins, R. P. 1959. Life history of *Macrobiotus islandicus* Richters with notes on other tardigrades from Colorado. Transactions of the American Microscopical Society 78:137–154.

Higgins, R. P., Ed., 1975. International symposium on tardigrades, Pallanza, Italy, 17–19, June 1974. Memorie dell'Istituto Italiano di Idrobiologia 32(Suppl.):1–469.

Kathman, R. D., Cross, S. F. 1991. Ecological distribution of moss-dwelling tardigrades on Vancouver Island, British Columbia, Canada. Canadian Journal of Zoology 69:122–129.

Kathman, R. D., Nelson, D. R. 1987. Population trends in the aquatic tardigrade *Pseudobiotus augusti* (Murray). *in*: Bertolani, R., Ed., Biology of tardigrades. Selected Symposia and Monographs U. Z. I., 1. Mucchi, Modena, Italy, pp. 155–168.

Kinchin, I. M. 1994. The biology of tardigrades. Portland Press, London, 186 p.

Kristensen, R. 1978. On the fine structure of *Batillipes noerrevangi* Kristensen 1978 (Heterotardigrada). 2. The muscle-attachments and the true cross-striated muscles. Zoologischer Anzeiger 200:173–184.

Kristensen, R. 1979. On the fine structure of *Batillipes noerrevangi* Kristensen 1978 (Heterotardigrada). 3. Spermiogenesis, *in*: Weglarska, B., Ed., Second international symposium on tardigrades, Kraków, Poland, 28–30, July 1977. Zeszyty Naukowe Uniwersytetu Jagiellonskiego, Prace Zoologiczne 25:97–105.

Kristensen, R. 1982. The first record of cyclomorphosis in Tardigrada based on a new genus and species from Arctic meiobenthos. Zeitschrift für Zoologische Systematik und Evolutionsforschung 20:249–270.

Kristensen, R. 1987. Generic revision of the Echiniscidae (Heterotardigrada), with a discussion of the origin of the family, *in*: Bertolani, R., Ed., Biology of tardigrades. Selected Symposia and Monographs U. Z. I., 1. Mucchi, Modena, Italy, pp. 261–335.

Kristensen, R., Higgins, R. P. 1984a. Revision of *Styraconyx* (Tardigrada: Halechiniscidae) with descriptions of two new species from Disko Bay, West Greenland. Smithsonian Contributions to Zoology 391:1–40.

Kristensen, R., Higgins, R. P. 1984b. A new family of Arthrotardigrada (Tardigrada: Heterotardigrada) from the Atlantic coast of Florida, U.S.A. Transactions of the American Microscopical Society 103:295–311.

Kristensen, R., Higgins, R. P. 1989. Marine Tardigrada from the southeastern United States coastal waters. 1. *Paradoxipus orzeliscoides* n. gen., n. sp. (Arthrotardigrada: Halechiniscidae). Transactions of the American Microscopical Society 108: 262–282.

Manicardi, C., Bertolani, R. 1987. First contribution to the knowledge of Alpine grassland tardigrades. *in*: Bertolani, R., Ed., Biology of tardigrades. Selected Symposia and Monographs U. Z. I., 1. Mucchi, Modena, Italy, pp. 177–185.

Marcus, E. 1928. Zur Embryologie der Tardigraden. Verhandlungen der Deutschen Zoologischen Gessellschaft 32:134–146.

Marcus, E. 1929. Tardigrada, *in*: Bronn, H. G., Ed., Klassen und Ordnungen des Tierreichs. Vol. 5. Akademische Verlagsgesellschaft, Leipzig, 608 pp.

Marcus, E. 1936. Tardigrada, *in*: Schultze, F., Ed., Das Tierreich. Walter de Gruyter, Berlin, 340 pp.

Maucci, W. 1996. Tardigrada of the Arctic tundra with description of two new species. Zoological Journal of the Linnean Society 116:185–204.

McInnes, S. 1994. Zoogeographic distribution of terrestrial/freshwater tardigrades from current literature. Journal of Natural History 28:257–352.

McInnes, S. 1995. Taxonomy and ecology of tardigrades from Antarctic lakes. M. Phil. Thesis, Open University. 248 p.

McInnes, S., Ellis-Evans, J. C. 1987. Tardigrades from maritime Antarctic freshwater lakes, *in*: Bertolani, R., Ed., Biology of tardigrades. Selected Symposia and Monographs U. Z. I., 1. Mucchi, Modena, Italy, pp. 111–123.

McInnes, S., Ellis-Evans, J. C. 1990. Micro-invertebrate community structure within a maritime Antarctic lake. Proceedings of the National Institute of Polar Research Symposium on Polar Biology 3:179–189.

McInnes, S., Norman, D., Eds., 1996. Tardigrade biology. Proceedings of the sixth international symposium on Tardigrada, August 1994, Cambridge, England. Zoological Journal of the Linnean Society 116:1–243.

McInnes, S., Pugh, P. 1998. Biogeography of limno-terrestrial Tardigrada, with particular reference to the Antarctic fauna. Journal of Biogeography 25:31–36.

Mihelčič, F. 1954/55. Zur Ökologie der Tardigraden. Zoologisher Anzeiger 153:250–257.

Miller, W. R., Miller, J. D., Heatwole, H. 1996. Tardigrades of the Australian Antarctic Territories: the Windmill Islands, East Antarctica. Zoological Journal of the Linnean Society 116:175–184.

Moon, S. Y., Kim, W. 1996. Phylogenetic position of the Tardigrada based on the 18S ribosomal RNA gene sequences. Zoological Journal of the Linnean Society 116:61–69.

Morgan, C. 1977. Population dynamics of two species of Tardigrada, *Macrobiotus hufelandi* (Schultze) and *Echiniscus (Echiniscus) testudo* (Doyère), in roof moss from Swansea. Journal of Animal Ecology 46:263–279.

Morgan, C., King, P. 1976. British Tardigrada. Tardigrada keys and notes for the identification of the species. Synopses of the British Fauna, No. 9. Academic Press, London, 132 p.

Nelson, D. R. 1975. Ecological distribution of tardigrades on Roan Mountain, Tennessee–North Carolina. Memorie dell'Istituto Italiano di Idrobiologia 32(Suppl.):225–276.

Nelson, D. R. 1982a. Developmental biology of the Tardigrada, *in*: Harrison, F., Cowden, R., Eds., Developmental biology of freshwater invertebrates. Alan R. Liss, New York, pp. 363–368.

Nelson, D. R., Ed., 1982b. Proceedings of the third international symposium on the Tardigrada, 3–6, August 1980, Johnson City, TN. East Tennessee State University Press, Johnson City, TN 236 pp.

Nelson, D. R., Higgins, R. P. 1990. Tardigrada, *in*: Dindal, D., Ed., Soil biology guide. Wiley, New York, pp. 393–419.

Nelson, D. R., Kincer, C. J., Williams, T. C. 1987. Effects of habitat disturbances on aquatic tardigrade populations, *in*: Bertolani, R., Ed., Biology of tardigrades. Selected Symposia and Monographs U. Z. I., 1. Mucchi, Modena, Italy, pp. 141–153.

Nelson, D. R., Marley, N., Bertolani, R. 1999. Re-description of the genus *Pseudobiotus* (Eutardigrada, Hypsibiidae) and of the new type species *Pseudobiotus kathmanae* sp. n., Zoologischer Anzeiger 238(3–4): 311–317.

Nielsen, C. 1995. Phylum Tardigrada, *in*: Animal evolution—Interrelationships of the living phyla. Oxford University Press, Oxford, pp. 176–181.

Pilato, G. 1969. Evoluzione e nuova sistemazione degli Eutardigrada. Bolletino di Zoologia 36:327–345.

Pilato, G. 1972. Structure, intraspecific variability and systematic value of the buccal armature of eutardigrades. Zeitschrift für Zoologische Systematik und Evolutionsforschung 10:65–78.

Pilato, G. 1979. Correlations between cryptobiosis and other biological characteristics in some soil animals. Bollettino di Zoologia 46:319–332.

Pilato, G. 1982. The systematics of Eutardigrada: A comment. Zeitschrift für Zoologische Systematik und Evolutionsforschung 20:271–284.

Pilato, G. 1987. Revision of the genus *Diphascon* Plate, 1889, with remarks on the subfamily Itaquasconinae (Eutardigrada, Hypsibiidae), *in*: Bertolani, R., Ed., Biology of tardigrades. Selected Symposia and Monographs U. Z. I., 1. Mucchi, Modena, Italy, pp. 337–357.

Pilato, G. 1992. *Mixibius,* nuovo genere di Hypsibiidae (Eutardigrada). Animalia 19:121–125.

Pilato, G. 1997. *Astatumen,* a new genus of the Eutardigrada (Hypsibiidae, Itaquasconinae). Entomologische Mitteilungen aus dem Zoologischen Museum Hamburg 12:205–208.

Pilato, G., Beasley, C. W. 1987. *Haplohexapodibius seductor* n. gen. n. sp. (Eutardigrada, Calohypsibiidae) with remarks on the systematic position of the new genus. Animalia 14:65–71.

Pilato, G., Binda, M. G. 1987a. *Parascon schusteri* n. gen. n. sp. (Eutardigrada, Hypsibiidae, Itaquasconinae). Animalia 14:91–97.

Pilato, G., Binda, M. G. 1987b. *Richtersia,* nuovo genere di Macrobiotidae, e nuova definizione di *Adorybiotus* Maucci & Ramazzotti 1981 (Eutardigrada). Animalia 14:147–152.

Pilato, G., Binda, M. G. 1989. *Richtersius,* nuovo nome generico in sostituzione di *Richtersia* Pilato e Binda 1987 (Eutardigrada). Animalia 16:147–148.

Pilato, G., Binda, M. G. 1990. Tardigradi dell'Antarctide. I. *Ramajendas,* nuovo genere di Eutardigrado. Nuova posizione sistematica di *Hypsibius renaudi* Ramazzotti, 1972 e descrizione di *Ramajendas frigidus* n. sp. Animalia 17:61–71.

Pilato, G., Binda, M. G. 1996. Additional remarks to the description of some genera of eutardigrades. Bollettino delle Sedute dell'Accademia Gioenia di Scienze Naturale, Catania 29:33–40.

Pilato, G., Binda, M. G. 1997. *Acutuncus* a new genus of Hypsibiidae (Eutardigrada). Entomologische Mitteilungen aus dem Zoologischen Museum Hamburg 12: 159–162.

Pilato, G., Catanzaro, R. 1988. *Macroversum mirum* n. gen. n. sp. nuovo Eutardigrado (Macrobiotidae) dei Monti Nebrodi (Sicilia). Animalia 15:175–180.

Pollock, L. W. 1970. Distribution and dynamics of interstitial Tardigrada at Woods Hole, Massachusetts, USA. Ophelia 7:145–165.

Pugh, P., McInnes, S. 1998. The origin of Arctic terrestrial and freshwater tardigrades. Polar Biology 19:177–182.

Rahm, G. 1937. A new ordo of tardigrades from the hot springs of Japan (Furu-Section, Unzen). Annotationes Zoologicae Japonenses 16:345–352.

Ramazzotti, G. 1962. Il Phylum Tardigrada. Memorie dell'Istituto Italiano di Idrobiologia 16:1–595.

Ramazzotti, G. 1972. II Phylum Tardigrada. (Seconda edizione aggiornata). Memorie dell'Istituto Italiano di Idrobiologia 28:1–732.

Ramazzotti, G. 1974. Supplement A, Il Phylum Tardigrada, Seconda Edizione, 1972. Memorie dell'Istituto Italiano di Idrobiologia 31:69–179.

Ramazzotti, G., Maucci, W. 1983. Il Phylum Tardigrada. III edizione riveduta e aggiornata. Memorie dell'Istituto Italiano di Idrobiologia 41:1–1012.

Ramløv, H., Westh, P. 1992. Survival of the cryptobiotic eutardigrade *Adorybiotus coronifer* during cooling to −196°C: Effect of cooling rate, trehalose level, and short term acclimation. Cryobiology 29:125–130.

Rebecchi, L. 1997. Ultrastructural study of spermiogenesis and the testicular and spermathecal spermatozoon of the gonochoristic tardigrade *Xerobiotus pseudohufelandi* (Eutardigrada, Macrobiotidae). Journal of Morphology 234:11–24.

Rebecchi, L., Bertolani, R. 1988. New cases of parthenogenesis and polyploidy in the genus *Ramazzottius* (Tardigrada, Hypsibiidae) and a hypothesis concerning their origin. Invertebrate Reproduction and Development 14:187–196.

Rebecchi, L., Bertolani, R. 1994. Maturative pattern of ovary and testis in eutardigrades of freshwater and terrestrial habitats. Invertebrate Reproduction and Development 26:107–117.

Rebecchi, L., Guidi, A. 1991. First SEM studies on tardigrade spermatozoa. Invertebrate Reproduction and Development 19:151–156.

Rebecchi, L., Guidi, A. 1995. Spermatozoon ultrastructure in two species of *Amphibolus* (Eutardigrada, Eohypsibiidae). Acta Zoologica 26(2):171–176.

Rebecchi, L., Nelson, D. R. 1998. Evaluation of a secondary sex character in eutardigrades. Invertebrate Biology 117(3): 194–198.

Rebecchi, L., Guidi, A., Bertolani, R. 1999. Tardigrada, *in*: Jamieson, B. G. M., Ed., Progress in male gamete biology. vol. 9, part B. Oxford and IBH, New Delhi.

Renaud-Mornant, J. 1982. Species diversity in marine Tardigrada, *in*: Nelson, D., Ed., Proceedings of the third international symposium on the Tardigrada. East Tennessee State University Press, Johnson City, TN., pp. 149–178.

Renaud-Mornant, J. 1987. Bathyl and abyssal Coronarctidae (Tardigrada), description of new species and phylogenetical significance, *in*: Bertolani, R., Ed., Biology of tardigrades. Selected Symposia and Monographs U. Z. I., 1. Mucchi, Modena, Italy, pp. 229–252.

Richters, F. 1926. Tardigrada, *in*: Kükenthal, W., Krumbach, T., Eds., Handbuch de Zoologie. Walter de Gruyter, Berlin, pp. 1–68.

Rudolph, A. S., Crowe, J. H. 1985. Membrane stabilization during freezing: the role of two natural cryoprotectants, trehalose and proline. Cryobiology 22:367–377.

Schuster, R. O., Toftner, E. C., Grigarick, A. A. 1977. Tardigrada of Pope Beach, Lake Tahoe, California. The Wasmann Journal of Biology 35:115–136.

Schuster, R. O., Nelson, D. R., Grigarick, A. A., Christenberry, D. 1980. Systematic criteria of the Eutardigrada. Transactions of the American Microscopical Society 99:284–303.

Seki, K., Toyoshima, M. 1998. Preserving tardigrades under pressure. Nature 395:853–854.

Sømme, L. 1996. Anhydrobiosis and cold tolerance in tardigrades. European Journal of Entomology 93:349–357.

Spallanzani, L. 1776. Opuscoli di fisica animale e vegetabile, Vol. 2, il Tardigrado etc. Opuscolo 4:222.

Steiner, W. 1994a. The influence of air pollution on moss-dwelling animals: 1. Methodology and composition of flora and fauna. Revue Suisse de Zoologie 101:533–556.

Steiner, W. 1994b. The influence of air pollution on moss-dwelling animals: 2. Aquatic fauna with emphasis on Nematoda and Tardigrada. Revue Suisse de Zoologie 101:699–724.

Steiner, W. 1994c. The influence of air pollution on moss-dwelling animals: 4. Seasonal and long-term fluctuations of rotifer, nematode and tardigrade populations. Revue Suisse de Zoologie 101:1017–1031.

Steiner, W. 1995. The influence of air pollution on moss-dwelling animals: 5. Fumigation experiments with SO_2 and exposure experiments. Revue Suisse de Zoologie 102:13–40.

Strayer, D., Nelson, D. R., O'Donnell, E. B. 1994. Tardigrades from shallow groundwaters in southeastern New York, with the first record of *Thulinia* from North America. Transactions of the American Microscopical Society 113:325–332.

Thulin, G. 1928. Über die Phylogenie und das System der Tardigraden. Hereditas 11:207–266.

Toftner, E. C., Grigarick, A. A., Schuster, R. O. 1975. Analysis of scanning electron microscope images of *Macrobiotus* eggs. Memorie dell'Istituto Italiano di Idrobiologia 32(Suppl.):393–411.

Usher, M. B., Dastych, H. 1987. Tardigrada from the Maritime Antarctic. British Antarctic Survey Bulletin 77:163–166.

Van Rompu, E. A., De Smet, W. H. 1988. Some aquatic Tardigrada from Bjørnøya (Svalbard). Fauna Norvegica, Series A, 9:31–36.

Van Rompu, E. A., De Smet, W. H. 1991. Contribution to the fresh water Tardigrada from Barentsøya, Svalbard (78°30′N). Fauna Norvegica, Series A, 12:29–39.

Van Rompu, E. A., De Smet, W. H., Bafort, J. M. 1991a. Some freshwater tardigrades from the Kilimanjaro. Natuurwetenschappelijk Tijdschrift 73:55–62.

Van Rompu, E. A., De Smet, W. H., Bafort, J. M. 1991b. Contributions to the Tardigrada of the Canadian High-Arctic. 2. Freshwater tardigrades from Little Cornwallis Island, N.W.T. Canada. Biologisch Jaarboek Dodonaea 59:132–140.

Van Rompu, E. A., De Smet, W. H., Beyens, L. 1992. Contributions to the Tardigrada of the Canadian High-Arctic 1. Freshwater tardigrades from Devon Island, Northwest Territories. The Canadian Field-Naturalist 106: 303–310.

Villora-Moreno, S., De Zio Grimaldi, S. 1996. New records of marine Tardigrada in the Mediterranean Sea. Zoological Journal of the Linnean Society 116:149–166.

Wainberg, R., Hummon, W. 1981. Morphological variability of the tardigrade *Isohypsibius saltursus*. Transactions of the American Microscopical Society 100:21–33.

Walz, B. 1974. The fine structure of somatic muscles of Tardigrada. Cell and Tissue Research 149:81–89.

Walz, B. 1975. Ultrastructure of muscle cells in *Macrobiotus hufelandi.* Memorie dell'Istituto Italiano di Idrobiologia 32(Suppl.): 425–443.

Walz, B. 1978. Electron microscopic investigation of cephalic sense organs of the tardigrade *Macrobiotus hufelandi* Schultze. Zoomorphologie 89:1–19.

Walz, B. 1982. Molting in Tardigrada. A review including new results on cuticle formation in *Macrobiotus hufelandi, in*: Nelson, D., Ed., Proceedings of the third international symposium on the Tardigrada. East Tennessee State University Press, Johnson City, TN, pp. 129–147.

Węglarska, B. 1957. On the encystation in Tardigrada. Zoologica Poloniae 8:315–325.

Węglarska, B., Ed. 1979. Second international symposium on tardigrades, Kraków, Poland, July 28–30, 1977, Zeszyty Naukowe Uniwersytetu Jagiellonskiego, Prace Zoologiczne 25:1–197.

Westh, P., Kristensen, R. M. 1992. Ice formation in the freeze-tolerant eutardigrades *Adorybiotus coronifer* and *Amphibolus nebulosus* studied by differential scanning calorimetry. Polar Biology 12:693–699.

Westh, P., Ramløv, H. 1991. Trehalose accumulation in the tardigrade *Adorybiotus coronifer* during anhydrobiosis. Journal of Experimental Zoology 258:303–311.

Westh, P., Kristiansen, J., Hvidt, A. 1991. Ice-nucleating activity in the freeze-tolerant tardigrade *Adorybiotus coronifer.* Comparative Biochemistry and Physiology [A] 73:621–626.

Wiederhöft, H., Greven, H. 1996. The cerebral ganglia of *Milnesium tardigradum* Doyère (Apochela, Tardigrada): three-dimensional reconstruction and notes on their ultrastructure. Zoological Journal of the Linnean Society 116:71–84.

Wolburg-Buchholz, K., Greven, H. 1979. On the fine structure of the spermatozoon of *Isohypsibius granulifer* Thulin 1928 (Eutardigrada) with reference to its differentiation, *in*: Węglarska, B., Ed., Second international symposium on tardigrades, Kraków, Poland, 28–30, July 1977. Zeszyty Naukowe Uniwersytetu Jagiellonskiego, Prace Zoologiczne 25:191–197.

Wright, J. C. 1989a. The tardigrade cuticle. 2. Evidence for a dehydration-dependent permeability barrier in the intracuticle. Tissue and Cell 21:263–279.

Wright, J. C. 1989b. Desiccation tolerance and water-retentive mechanisms in tardigrades. Journal of Experimental Biology 142: 267–292.

Wright, J. C., Westh, P., Ramløv, H. 1992. Cryptobiosis in Tardigrada. Biological Reviews of the Cambridge Philosophical Society 67:1–29.

16

WATER MITES (HYDRACHNIDA) AND OTHER ARACHNIDS

Ian M. Smith

Biodiversity Section
ECORC, Central Experimental Farm
Agriculture and Agri-Food Canada
Ottawa, Ontario, Canada K1A 0C6

Bruce P. Smith

Biology Department, Ithaca College
Ithaca, New York 14850

David R. Cook

7836 North Invergordon Place
Paradise Valley, Arizona 85253

I. INTRODUCTION TO ARACHNIDS

Members of the class Arachnida are the most familiar representatives of the arthropodan subphylum Chelicerata, an ancient lineage characterized by distinctive mouthparts with two pairs of appendages, the anterior chelicerae and posterior pedipalps, and four pairs of walking legs. Chelicerates first appear definitively in the fossil record as giant water scorpions (Merostomata, subclass Eurypterida) in freshwater deposits from the Silurian. By the Devonian, eurypterids had apparently disappeared but other chelicerate lineages were proliferating, such as horseshoe crabs (Merostomata, subclass Xiphosura) in marine environments, and early derivative arachnid groups, including mites, on land. Arachnids are the only group of chelicerates exhibiting extensive modern diversity. The higher classification is still being debated, but arachnids are usually considered to be a taxonomic class comprising 11 extant subclasses, mostly terrestrial groups of highly specialized predators of other arthropods. Only two subclasses of modern arachnids are represented in freshwater habitats, mites (Acari) which have diversified extensively to become arguably the most ubiquitous and adaptable clade of arthropods on Earth, and spiders (Araneae) which have maintained a conservative way of life as essentially terrestrial predators.

Mites belonging to several unrelated groups are commonly found in freshwater habitats. The true water mites comprise a taxon of actinedid acariform mites which we refer to here as Hydrachnida (Hydrachnellae, Hydracarina, or Hydrachnidia of other authors). Hydrachnida have flourished to become by far the most numerous, diverse, and ecologically important group of freshwater arachnids. Various families in certain suborders of Acariformes (namely, Actinedida,

Ecology and Classification of North American Freshwater Invertebrates. 2nd Edition

Oribatida and Acaridida) and Parasitiformes (namely, Gamasida) have independently invaded freshwater and become adapted for living there. Compared to Hydrachnida, these groups are less diverse taxonomically, usually less abundant, and relatively conservative in morphology and habits. We summarize diagnostic and ecological information for these mites and for the remarkable spiders of the genus *Dolomedes* at the end of this chapter.

II. WATER MITES (HYDRACHNIDA)

A. Introduction

Water mites are among the most abundant and diverse benthic arthropods in many habitats. One square meter area of substrate from littoral weed beds in eutrophic lakes may contain as many as 2000 deutonymphs and adults representing up to 75 species in 25 or more genera. Comparable samples from an equivalent area of substrate in rocky riffles of streams often yield over 5000 individuals of more than 50 species in over 30 genera (including both benthic and hyporheic forms). Water mites have co-evolved with some of the dominant insect groups in freshwater ecosystems, especially nematocerous Diptera, and typically interact intimately with these insects at all stages of their life histories.

We have made numerous additions and refinements to the information included in the first edition of this chapter (Smith and Cook, 1991), and some extensive revisions. We have tried to avoid unnecessary use of specialized acarological terms, and provide a glossary of essential terminology needed for describing morphological and anatomical features. Those wishing to consult a more comprehensive account of mite structure and function are referred to the appropriate sections of the books by Cook (1974) and Krantz (1978).

1. General Relationships

Hydrachnida, along with the enigmatic interstitial Stygothrombidioidea and terrestrial Calyptostomatoidea, Trombidioidea, and Erythraeoidea, belong to a remarkably diverse natural group of actinedid acariform mites, the Parasitengona. The complex and essentially holometabolous type of development of this group is unique among Acari. Typically, after emerging from the egg membrane, the hexapod larva seeks out an appropriate host, and becomes an ectoparasite which is passively transported while feeding on host fluids. When fully engorged the larva transforms to the quiescent protonymph (or nymphochrysalis). Radical structural reorganization occurs during this stage, giving rise to the active deutonymph which resembles the adult in being octopod and typically predaceous, but is sexually immature and exhibits incomplete sclerotization and chaetotaxy. The deutonymph feeds and grows in size before entering another quiescent stage, the tritonymph (or imagochrysalis). After completion of metamorphosis during this stage, the mature adult emerges (see Fig. 1).

Larval water mites can be distinguished morphologically from those of other parasitengones by having two setae, rather than one, on the genu of the pedipalp. Deutonymphal and adult water mites differ from all other Acari in having glandularia on the idiosoma.

2. Origin

Water mites evolved from terrestrial stock, and hypotheses on their origin usually presume an ancestral terrestrial parasitengone (Mitchell, 1957a; Davids and Belier, 1979). A recently proposed alternative hypothesis (Wiggins *et al.*, 1980; Smith and Oliver, 1986) suggests that the primitive parasitengone stock may have been water mites resembling certain extant Hydryphantoidea. According to this postulate, water mites diverged from terrestrial ancestors with direct development, perhaps resembling extant Anystoidea, while evolving the basic parasitengone life-history pattern as a set of adaptations for exploiting spatially and temporally intermittent aquatic habitats.

The fossil record provides little insight into the evolutionary history of water mites. The only reported fossils that undoubtedly are water mites (Cook, 1957; Poinar, 1985) are larval specimens from Tertiary deposits representing highly derived taxa. However, distributional data and host associations indicate that the group originated no later than the Triassic or Jurassic period.

Available morphological and behavioral data suggest that extant water mites are monophyletic (Mitchell, 1957a; Barr, 1972; Cook, 1974; Smith and Oliver, 1976, 1986), and that the major phyletic lineages of water mites, and possibly all Parasitengona, were derived from hydryphantoidlike ancestors (Mitchell, 1957a; Smith and Oliver, 1986).

3. Diversity and Classification

Well over 5000 species of water mites are currently recognized worldwide, representing more than 300 genera and subgenera in over 100 families and subfamilies (Viets, 1987). Acarologists usually consider water mites to be a taxon of intermediate rank between superfamily and suborder. Interestingly, the group appears to rival several orders of aquatic insects in diversity, and is comparable to some of them in age. Most of

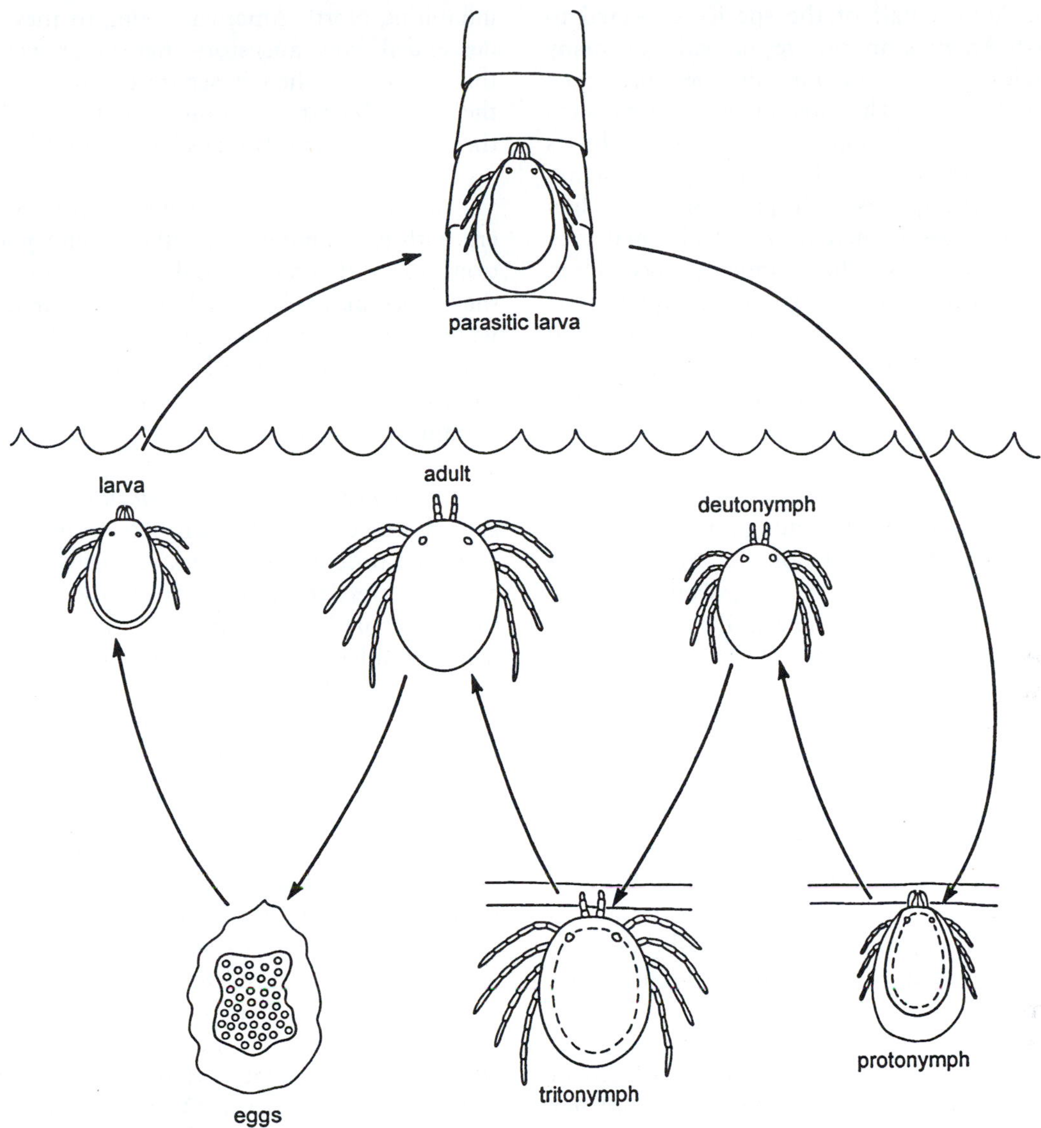

FIGURE 1 Diagrammatic illustration of a generalized water mite history.

the genera and families are reasonably stable and appear to approximate closely holophyletic lineages. The families are conservatively grouped into seven superfamilies. Three of these—Hydrovolzioidea, Hydrachnoidea, and Eylaoidea—probably represent natural groupings. The others—Hydryphantoidea, Lebertioidea, Hygrobatoidea and Arrenuroidea—are all either paraphyletic or polyphyletic assemblages subject to future revision. A comprehensive, detailed cladistic analysis of extant fauna is required to permit construction of more stable and informative superfamily groupings. The phylogenetic studies that prove most useful in understanding water mite evolution and developing a more natural classification provide syntheses of information based on the morphology of all instars, behavioral attributes, and life-history data (Mitchell, 1957a; Cook, 1974; Smith, 1976a; Smith and Oliver, 1976, 1986).

The over 1500 species currently estimated to occur in North America, north of Mexico, represent 136 genera in 69 subfamilies, 42 families, and all seven superfamilies. The fauna remains incompletely known at the generic level, and revision of the family level classification is ongoing (see Smith, 1990b; Cook *et al.,* 1998). Species of over 25 new or unreported genera have been discovered in North America since the publication of Cook's (1974) global review, an average of about one each year (see Smith, 1976b, 1979, 1983b, c, d, g, 1989a–c, 1990a, 1991b, d, 1992b, 1998; Smith and Cook, 1994, 1996, 1997, 1998a, b, 1999a, b; Gerecke *et al.,* 1999). Almost half of the species expected to occur in North America are not yet named, and many described species are known from only a few preserved specimens in collections. The most diverse genera, such as *Sperchon, Torrenticola* and *Aturus* in lotic habitats and *Piona* and *Arrenurus* in lentic habitats, contain 100 or more nearctic species each (probably more than 250 species in the case of *Arrenurus*). Additional taxa await discovery throughout the continent, especially in springs and interstitial habitats. The least explored areas, such as southern Appalachia and the boreal and arctic regions of northern Canada and Alaska, undoubtedly will yield some unexpected taxa when thoroughly studied.

4. Zoogeography

a. Global Patterns Water mites occur throughout the world, except Antarctica (Cook, 1974). All superfamilies, except Hydrovolzioidea, as well as many early derivative families and subfamilies, are richly represented in all zoogeographic regions. Knowledge of global distribution patterns of water mite taxa has been substantially improved by recent studies of the fauna of austral regions, especially those by K.O. Viets and D.R. Cook. The basic patterns can be explained as the result of vicariance due to plate tectonics (Cook, 1986, 1988; Smith and Cook, 1999c). Dispersal (along with hosts) between adjacent land masses (e.g., Southeast Asia and Australia) or across continents (e.g., Africa) significantly altered these patterns as more recent groups progressively displaced ancient ones.

b. Nearctic Patterns Four major groups of genera can be recognized among extant North American water mites based on their apparent origins, reflecting successive historical stages in the development of the modern fauna (see Table I).

i. Genera of Pangean origin Twenty-three early-derivative, cosmopolitan genera exhibiting substantial endemism in both the northern and southern hemispheres apparently originated in Pangea, and were represented in both Laurasia and Gondwanaland at the time of their separation during the Jurassic.

ii. Genera of Laurasian origin Seventy-two genera with predominantly northern hemispheric distributions, exhibiting comparable diversity and endemism in the Nearctic and Palearctic, probably originated in Laurasia.

The genera of Pangean and Laurasian origin probably were distributed throughout Laurasia during the late Cretaceous. The vast majority of species currently inhabiting North America belong to these genera, and descended from ancestors that either were present on this continent when it separated from Eurasia during the early Tertiary or immigrated from Eurasia with their hosts via land bridges later in that Period.

iii. Genera of Gondwanan origin Seventeen genera with predominantly southern hemispheric distributions, exhibiting substantial diversity and endemism in the Neotropics (Cook, 1980), are probably of Gondwanan or South American origin. Members of the Gondwanan taxa invaded southern North America from Central and South America after the Panamanian isthmus appeared during the Pliocene.

iv. Genera of North American origin Twenty-four genera are currently known only from North America. Some are probably endemics of recent origin while others have Tertiary relict distributions, suggesting that they originated in Laurasia and may have undiscovered members inhabiting poorly known areas of the Palearctic or Oriental Regions.

Distributions of water mite genera within North America were dramatically influenced by climatic cooling and glaciation during the late Pliocene and Pleistocene. Genera which show evidence of recent adaptive radiation have re-established pan-continental distributions, while those with relatively few species have remained within more limited areas. Many species groups came to be restricted to either temperate or boreal areas of the continent, as reflected to some extent in the distributions of genera (see Table I). As a result, typical "Tertiary-relict" distributions in, or adjacent to, unglaciated refugia such as coastal California and Oregon, the Ozark Plateau, and parts of the Appalachian and Rocky Mountains are exhibited by some genera (Smith, 1989d, e, 1991c, 1992a, b) . Other genera (e.g., *Teutonia, Midea, Acalyptonotus*) have characteristic "boreal" or "arctic-alpine" distributions at higher latitudes and elevations. Distributional limitations are frequently correlated with habitat specializations. For example, many interstitial genera have Tertiary-relict distributions because their habitats were destroyed in most northern regions of the continent by Pleistocene glaciation and will require a long period of climatic stability to become re-established there. The sister species of many nearctic Tertiary relicts now inhabit similar areas in temperate Asia, resulting in strikingly discontinuous distributions for certain genera. In contrast, species adapted to cold stenothermic pools inhabit subarctic areas, high mountains near permanent ice-fields, and temperate lowlands where groundwater springs create similar conditions. The sister-species of many stenophilic nearctic water mites live in northern Eurasia, yielding essentially circumpolar or boreal holarctic distribution patterns for these genera.

B. External Morphology and Internal Anatomy

Water mites exhibit the characteristic acarine body plan comprising the gnathosoma, or mouth region, and the idiosoma, or body proper, representing the fused cephalothorax and abdomen. The standard morphological terminology proposed by Greandjean (see Robaux, 1974) has been widely adopted by most authors dealing with prostigmatic mites, but until recently little effort was made to apply this terminology to water mites. The extreme plasticity of the adult exoskeleton during water mite evolution has obscured many homologies with other Prostigmata, and water mite workers have developed a highly useful set of peculiar morphological terms for external structures of this instar. Early taxonomic studies of larval water mites generated an additional set of special morphological terms for the exoskeleton of this strongly heteromorphic instar, and we used a modified version of this terminology in the 1st edition. However, it has become evident that the relatively plesiotypical body plan of larvae permits much of the standard idiosomal terminology of Grandjean to be used with confidence. We employ it here and indicate where appropriate the equivalence of the various terms in common use.

1. Morphology of Larvae

The short gnathosoma (Fig. 2) bears the stocky pedipalps which have five free segments (trochanter, femur, genu, tibia, tarsus) that flex ventrally. The tarsus of the pedipalp is relatively long and cylindrical in some early derivative genera (eg: *Wandesia, Thyas*) (Figs. 18, 22), but is typically reduced to a dome- or button-shaped pad in most derivative groups (Figs. 76, 182). A highly modified thick, curved seta is present dorsally at the end of the tibia (Fig. 2), the homolog of the tibial "claw" which characterizes the pedipalp of most terrestrial Anystoidea, Parasitengona and related groups. The paired chelicerae (Fig. 3), each consisting of a cylindrical basal segment and a movable terminal claw, lie between the pedipalps.

The idiosoma is plesiotypically mainly unsclerotized, as in extant Hydryphantoidea. The dorsal integument (Fig. 27) bears a medial eye-spot, two pairs of lenslike lateral eyes, four pairs of lyrifissures and the following complement of setae: four pairs on the propodosoma, two pairs of verticils, the internals (*vi*) or anteromedials (*AM*) and externals (*ve*) or anterolaterals (*AL*), and two pairs of scapulars, the internals (*si*) or sensillae (*SS*), and externals (*se*) or posterolaterals (*PL*); eight pairs on the hysterosoma, three pairs in the *c*-row (*c1*, *c2* and *c3*), two pairs in the *d*-row (*d1* and *d2*), two pairs in the *e*-row, (*e1* and *e2*) and one pair from the *f*-row (*f1*). Ventrally (Fig. 28) the integument bears the paired coxal plates, paired urstigmata (*Ur*) laterally between plates I and II, the excretory pore (*EP*), one pair of lyrifissures and the following complement of setae: three-to-six pairs on the coxal plates, two in the *1*-row (*1a* and *1b*), zero, one or two in the *2*-row (*2b* usually present, *2a* and *2c* usually absent) and one or two in the *3*-row (*3a* always present, *3b* usually absent), six pairs on the hysterosoma, one pair of preanals (*pa*), two pairs associated with the excretory pore (*ps1* and *ps2*), one pair from the *f*-row (*f1*) and two pairs in the *h*-row (*h1* and *h2*). The size, shape, and position of these dorsal and ventral structures, and the degree of fusion of the sclerites associated with them, provide useful taxonomic characters. The legs (Fig. 21) are inserted laterally on the coxal plates, and in the plesiotypical condition have six movable segments, namely trochanter (*Tr*), basifemur (*BFe*), telofemur (*TFe*), genu (*Ge*), tibia (*Ti*), and tarsus (*Ta*), that articulate to permit ventral flexion. The segments have characteristic complements of setae and solenidia (as in Figs. 6–8). The tarsi bear paired claws and clawlike empodia terminally.

The soft-bodied condition, retained in all known larval Hydryphantoidea, apparently provides adequate support for the muscles used in locomotion on the surface film. In all other major lineages, the dorsum (Figs. 3, 37, 56, 81, 144, 171) bears an extensive plate which incorporates the bases of the propodosomal setae and, in some cases, one or more hysterosomal pairs. Similarly, on the venter (Figs. 4, 38, 58, 83, 145, 172) the coxal plates are enlarged and variously fused, and the excretory pore plate (Figs. 5, 75, 109, 148, 176) is expanded to incorporate the bases of both pairs of excretory pore plate setae. The coxal plates and excretory pore plate may also bear various combinations of ventral setae.

Dorsal plates, and expanded coxal and excretory pore plates, developed independently in Hydrovolzioidea and in most Eylaoidea to provide support for muscles used in running and crawling on the surface film, and again in both Hydrachnoidea and the ancestral stock of the three "higher" superfamilies to anchor muscles used in swimming or crawling under water. In Hygrobatoidea, fused coxal plates II + III often have conspicuous lateral coxal apodemes (*LCA*), medial coxal apodemes (*MCA*), and transverse muscle attachment scars (*TMAS*) (Fig. 133).

Larvae of Hydryphantoidea and Eylaoidea retain six movable leg segments (Figs. 21, 40), but those of all other groups have the basifemoral and telofemoral segments fused (Figs. 6–8, 60). Larval Hydryphantoidea also retain the most plesiotypical complement of setae on the segments of the legs. Larvae of each of the more derivative groups exhibit characteristic leg chaetotaxies that reflect reductions from the plesiotypical complement. Certain ventral setae on the genua, tibiae and tarsi are elongate and plumose in larvae adapted for swimming (Fig. 149).

2. Morphology of Deutonymphs and Adults

The gnathosoma consists of the capitulum (or gnathosomal base) and associated appendages, namely the chelicerae and pedipalps (Fig. 200). The capitulum is plesiotypically a simple, short channel, derived from extensions of the pedipalpal coxae, leading to the oesophagus (Mitchell, 1962). A protrusible tube of integument connecting the capitulum to the idiosoma has developed independently in several distantly related genera (e.g., *Rhyncholimnochares, Clathrosperchon, Geayia*). The paired pedipalps (Fig. 200, 227), inserted on the capitulum, have both tactile and raptorial functions. In the plesiomorphic condition, the pedipalps have five movable segments, namely trochanter, femur, genu, tibia and tarsus, that are essentially cylindrical and articulate to allow ventral flexion. The tibia bears a thick, bladelike, dorsal seta distally in many ancient genera (e.g., *Hydryphantes, Tartarothyas, Pseudohydryphantes*). As in larvae, this seta is the homolog of the tibial "claw" of terrestrial relatives, and it often makes the pedipalps appear chelate. In derivative groups, other setae along with various denticles and tubercles may be elaborated to enhance the raptorial function of the pedipalps. Segmentation of the pedipalps is reduced by fusion in a few genera. A modification that has developed independently in various groups of Arrenuroidea is the so-called uncate condition. In these genera, the tibia is expanded and produced ventrally to oppose the tarsus, permitting the mites to grasp and hold slender appendages of prey organisms securely. The pedipalps are highly modified in some interstitial genera (e.g., Fig. 426), presumably to facilitate prey capture in confined hyporheic spaces. The paired chelicerae lie in longitudinal grooves between the pedipalps on the dorsal surface of the capitulum. Plesiotypically they consist of a cylindrical basal segment bearing a movable terminal claw. This cheliceral structure is designed for tearing the integument of prey organisms, and is retained in nearly all derivative groups. Hydrachnidae are unique in having unsegmented and stilletoform chelicerae, an obvious adaptation for piercing the insect eggs upon which they feed. The chelicerae are separate in all groups except Limnocharidae and Eylaidae where they are fused medially. The most comprehensive comparative studies of water mite mouthparts were published by Motas (1928) and Mitchell (1962).

The idiosoma, or body proper, is plesiotypically round or ovoid in outline, slightly flattened dorsoventrally, and mostly unsclerotized, as in certain extant members of ancient genera such as *Hydryphantes, Tartarothyas* and *Pseudohydryphantes*. The dorsal integument (Figs. 224, 256) bears an unpaired medial eye, paired lateral eyes that are usually enclosed in capsules, paired preocular and postocular setae, and longitudinal series of paired glandularia (six dorsoglandularia, five lateroglandularia), muscle attachment sites (five dorsocentralia, four dorsolateralia), and lyrifissures (five). Ventrally (Fig. 223) the integument bears the paired coxal plates (fused into anterior and posterior groups on each side), the genital field (comprising the gonopore, three pairs of acetabula, and paired genital valves), five pairs of ventroglandularia (including coxoglandularia I between the anterior and posterior coxal groups and coxoglandularia II behind the posterior groups) and the excretory pore. As in larvae, these idiosomal structures provide a wealth of useful taxonomic characters.

The legs (Figs. 200, 325) are inserted laterally on the coxae, and plesiotypically articulate on a vertical major axis and have six movable segments that articulate to permit ventral flexion. The segments are essentially cylindrical and have variable complements of setae. Though chaetotaxy of the legs provides a variety of taxonomic characters in deutonymphs and adults, the expression and position of individual setae are highly variable within taxa. Consequently, the rigorous analysis of chaetotactic patterns that proves so useful in the case of larvae is not practicable for later instars. The leg tarsi plesiotypically bear paired claws terminally.

The plesiotypical soft-bodied condition is retained in early derivative genera adapted for walking on the substrate (e.g., *Thyas*) or swimming in shallow habitats such as seepage pools or temporary pools (e.g., *Hydryphantes, Pseudohydryphantes*). Walking forms have relatively short, stocky leg segments and setae, whereas swimmers have longer segments bearing fringes of slender swimming setae. As pointed out by Mitchell (1957a, b, 1958, 1964b), the soft integument of these mites permits the highly precise local control of body shape and internal pressure that is needed to produce the walking and swimming movements of the legs.

3. Evolution of Adult Exoskeleton

The major lineages of water mites apparently differentiated from ancestral stock that resembled extant soft-bodied Hydryphantoidea (Mitchell, 1957a, b; Smith and Oliver, 1986), primarily through development of apotypical larval behavioral patterns and life histories. Hydrachnidae and Eylaidae became specialized for exploiting standing water habitats, especially temporary pools, whereas Hydryphantoidea, Lebertioidea, Hygrobatoidea, and Arrenuroidea ultimately diversified and radiated in a wide range of habitats. In each of these four groups, and presumably in Hydrovolzioidea, sclerotization of the integument came about as soft-bodied ancestral stock began to invade new habitats and diverge.

Adaptation to habitats such as seepage areas and springs or substrates in streams required a change in locomotor habits from walking or swimming. Mites adapting to moss mats and wet litter habitats developed hydrophilic integument which draws a film of water over the dorsal surface, creating sufficient downward force to press the body to the substrate and prevent efficient walking (Mitchell, 1960). Groups invading streams evolved a wedge-shaped body designed for negotiating confined spaces to avoid exposure to water turbulence and strong currents. Locomotion in both of these types of habitats necessitated evolution of a crawling gait. This change involved a shift in orientation of the major axis of the legs, shortening and thickening of leg segments and setae, enlargement of tarsal claws, and development of stronger, more massive muscles to control leg movements. Expansion of coxal plates and sclerites associated with glandularia, setal bases, and the genital field, along with sclerotization of the dorso—and laterocentralia, occurred to provide rigid exoskeletal support for these muscles. Fusion of these sclerotized areas led to development of complete dorsal and ventral shields in adults of certain groups. Multiple invasions of helocrene, rheocrene, and lotic habitats gave rise to a diverse array of dorsoventrally flattened, sclerotized, crawling groups of Hydryphantoidea, Lebertioidea, Hygrobatoidea and Arrenuroidea.

During the early evolution of each of these lineages certain groups became adapted for exploiting interstitial habitats in subterranean waters and the hyporheic zone of rheocrenes and streams. These mites tended to lose eyes and integumental pigmentation, and required further stream-lining of the body to facilitate locomotion in interstitial spaces. Consequently, soft-bodied forms adopted a vermiform shape, while sclerotized mites became extremely compressed laterally or dorsoventrally. The crawling mode of locomotion became secondarily modified in several groups of well-sclerotized interstitial genera (e.g., *Neomamersa, Frontipodopsis, Chappuisides*) to permit rapid and agile running in hyporheic and ground water habitats.

By late Pangean times, evolution within water mites apparently had produced basic communities of essentially soft-bodied species living in temporary and permanent standing water, and partly to fully sclerotized species living in emergent groundwater, lotic, and interstitial habitats. Subsequent phylogeny of Hydryphantoidea, Lebertioidea, Hygrobatoidea, and Arrenuroidea during late Mesozoic, Tertiary, and Quaternary times appears to have followed a recurring pattern of extended periods of gradual evolution leading to specialization within particular habitats, punctuated by episodes of relatively rapid and dramatic diversification. Divergence of many of the clades that originated during this time appears to have been correlated with successful invasion of new habitats. Repeated habitat diversification within the major clades of these four superfamilies, in response to opportunities provided by the explosive evolution of their primary host groups and by prolonged geological and climatic instability, resulted in several parallel and convergent trends in body sclerotization. Members of different clades developed superficially similar sclerite arrangements in adapting to lotic or interstitial habitats (e.g., species of *Diamphidaxona* and certain Axonopsinae); others underwent homoplastic reduction or loss of sclerites in secondarily invading lentic habitats (e.g., species of *Forelia* and *Piona*). Finally, some groups retained extensive sclerotization while becoming adapted for swimming in standing water (e.g., certain species of *Axonopsis* and *Mideopsis*). Consequently, modern communities are highly heterogeneous phylogenetically and morphologically, with each monophyletic component exhibiting unique exoskeletal adaptations for living in the particular habitat.

Evolutionary trends involving other readily observed and taxonomically useful characters—such as position and number of genital acetabula, positions of glandularia and lyrifissures, number and arrangement of setae on the body plates, and modification of claws on the tarsi of the legs—are still not adequately explained. For example, although the genital acetabula are plesiotypically borne in the gonopore in most Hydryphantoidea and Lebertioidea, they are either fused with the genital flaps or incorporated into acetabular plates flanking the gonopore in most taxa of other superfamilies. In addition, the number of acetabula has proliferated from the plesiotypical complement of three pairs several times independently in each of the six largest superfamilies. Alberti (1977, 1979) has recently proposed that acetabula in water mites may function as an osmoregulatory chloride epithelium, and Barr (1982) concluded that analysis of morphological data for early derivative taxa supports this hypothesis. However, until the function of acetabula is clearly understood on the basis of well-designed experiments, the significance of the trend to increasing number of acetabula will remain speculative.

The often striking and elaborate color patterns of adult water mites present an intriguing enigma. Members of most ancient clades (e.g., Hydrachnoidea, Eylaoidea, Hydryphantoidea) have colorless integument but are red, like terrestrial parasitengones, due to the presence of pigment granules which appear to be distributed throughout the body. Members of other superfamilies, except for taxa in interstitial habitats, exhibit highly distinctive patterns resulting from symmetrically arranged concentrations of pigment granules of various colors. In soft bodied taxa the pigments are located beneath the integument, but in sclerotized groups they are incorporated into the plates. Members of well-sclerotized taxa of Lebertioidea, Hygrobatoidea, and Arrenuroidea may be various shades of red, orange, yellow, green, or blue, and the dorsal shield frequently exhibits an intricate pattern combining several contrasting colors. The dorsal patterns of these mites can readily be interpreted as disruptive camouflage, but the adaptive value of their bright colors is more difficult to understand. The evolution of a wide range of highly colored pigments in these taxa certainly suggests that distinctive coloration confers some selective advantage. Though protection from predation may be part of the explanation, it is tempting to speculate that these mites, despite their apparently simple eye structure (see below), may detect and use color as one of several cues for recognizing conspecific individuals. Research is needed to determine the extent of their ability to detect and respond to the often subtle differences in color that distinguish closely related taxa.

4. Internal Anatomy

The organ systems of water mites occupy the hemocoel, bathed by hemolymph which is circulated by movements of the body musculature (Schmid, 1935; Bader, 1938; Mitchell, 1964b). As is typical for arthropods, there is no closed circulatory system.

a. Digestive System Digestion of food materials begins pre-orally, and only fluids are ingested. Food is drawn into the mouth (or buccal cavity) by the muscular pharynx, then passed through the tubular esophagus to the lobed midgut (Bader, 1938) where digestion and absorption occur. Undigested material accumulates as insoluble particles in a posterodorsal lobe of the midgut. All lobes of the midgut end blindly, and there is no connection between the gut and the excretory pore.

b. Excretory System This system consists of a large, thin-walled excretory tubule, apparently derived from the primitive hindgut, that lies dorsal to, and in close contact with, the midgut. Waste products are absorbed from hemolymph, and stored in the excretory tubule as insoluble, whitish or yellowish crystals of unknown chemical composition. When filled with crystalline material the excretory tubule may be visible through the dorsal integument as a "T"- or "Y"-shaped structure. The excretory tubule connects ventrally with the excretory pore, and is evacuated periodically by pressure generated through movements of the body muscles.

Osmoregulation is apparently accomplished by the urstigmata in larvae, and by the genital acetabula in deutonymphs and adults (Barr, 1982). Both of these structures have porous caps which apparently provide an ample surface area of chloride epithelium for maintaining water balance.

c. Respiratory System Many groups of water mites retain paired stigmata located between the bases of the chelicerae in all active instars. Plesiotypically, the stigmata lead to tracheal trunks which anastomose repeatedly into tracheolar tubules extending to all parts of the body. However, they appear to be nonfunctional in deutonymphs and adults, and respiration occurs by diffusion through the integument. In adults of many large species inhabiting standing water, a network of closed tubes of tracheolar dimensions lying beneath the integument transports gases to and from the tracheae that lead to internal organs (Mitchell, 1972). Heavily sclerotized mites have the body plates well supplied with regularly arranged "pores" that permit diffusion of gases between the tracheolar loops and surrounding water through areas of thin integument. In some species, there are regions of tracheal anastomosis lateral to the brain (Wiles, 1984).

d. Neural System The so-called brain, a fused, undifferentiated central ganglionic mass, surrounds the oesophagus. Nerve trunks dorsal to the oesophagus lead to the anterior sense organs and mouthparts, while ventral trunks lead to the legs and genital region.

The lateral eyes appear to be the primary light-sensing structures in water mites. There are typically two pairs, with the eyes of each side located close together and often enclosed in lenslike capsules that may lie above or beneath the integument. In postlarval instars of Eylaoidea, the eyes are borne on a medial sclerite. Many Hydryphantoidea also have a medial eye spot between the lateral eyes, but this structure is absent in most members of the other superfamilies. The eyes seem to function as ocelli, permitting the mites to detect the intensity and direction, and, at least in some taxa, the wavelength of incident light. Experiments with species of *Unionicola* have demonstrated that some species respond differentially to various wavelengths (Dimock and Davids, 1985) and that the response of the mites can be influenced by chemicals produced and released by their mollusc hosts (Roberts *et al.*, 1978). Retinal development, though rudimentary, has been reported for a variety of taxa (Lang, 1905).

Members of the various taxa have highly characteristic complements of setae on the idiosoma and appendages. Setae are the main tactile receptors, and many of them have also assumed secondary functions in locomotion, feeding, or mating. All water mites have specialized tactile organs, called glandularia, distributed in pairs on both the dorsum and venter of the body. Each glandularium consists of a small, goblet-shaped sac, or "gland", with a tiny external opening, and an associated seta. The "gland" contains a milky, viscous fluid which is ejected abruptly when the seta in stimulated. This substance quickly stiffens to a sticky gel after it comes in contact with water, and apparently evolved as a deterrent to attackers. In males of *Arrenurus,* however, material secreted by the glandularia seems to function also in cementing the body of the female in place during copulation.

The five pairs of lyrifissures, which are arranged in series on the dorsum and venter of the body are thought to be proprioceptors (Krantz, 1978). In addition, the distal segments of the pedipalps and legs are well supplied with a variety of specialized setiform structures, apparently derived from outgrowths of the integument, which function as chemoreceptors. The most conspicuous of these organs are the solenidia which adorn the dorsal surfaces of the genua, tibiae, and tarsi of the legs, and the tarsi of the pedipalps.

e. Reproductive System Males have paired testes and vasa deferentia which lead to an elaborate ejaculatory complex consisting of a series of membranous chambers attached to a sclerotized framework (Barr, 1972). Positioned immediately above the genital field, the ejaculatory complex functions as a syringelike organ for compacting masses of spermatozoa, assembling spermatophores, and expelling them from the genital tract through the gonopore.

Females have paired ovaries that are more or less fused, and paired oviducts that also fuse to form a single duct leading to the genital chamber within the gonopore. Paired spermathecae, for storing spermatozoa picked up in spermatophores, flank the genital chamber and are connected to it by short ducts.

C. Life History

The basic life-history pattern of water mites (Fig. 1) was first correctly inferred by Wesenburg-Lund (1918) and has been confirmed by subsequent studies on a wide range of taxa (see Smith and Oliver, 1986, for a summary of relevant literature). The first well-documented life-history studies were of palearctic species, especially members of the genus *Arrenurus,* investigated by Münchberg (1935a, b). Subsequently, Mitchell (1959, 1964a) dealt intensively with a number of nearctic species of *Arrenurus,* and Böttger (1962, 1972a, b) described life cycles of representative palearctic species of *Hydrachna, Limnochares, Eylais, Limnesia, Unionicola, Piona,* and *Arrenurus* in considerable detail. A meticulously comprehensive account of the life history of *Hydrodroma despiciens* (Müller) (family Hydrodromidae) was published by Meyer (1985).

Water mites in temperate latitudes typically live for just over one year, most of which is spent in the deutonymphal and adult stages. Most species are univoltine, with long-lived female adults producing multiple clutches of eggs. This tactic apparently minimizes the risks associated with the parasitic larval stage. In temperate latitudes, hosts are usually seasonally limited and parasitism occurs mainly in late spring and early summer. Larvae of most taxa spend no more than several days on their hosts and engorgement results in modest growth. Deutonymphal water mites typically feed and grow throughout the summer and adults appear in late summer or early fall and mate almost immediately. Inseminated females remain in obligative reproductive diapause until the following spring. This type of life history occurs in a wide range of water mite clades associated with host groups that are holometabolous and have aerial imagoes.

Several derivative life-history strategies occur among water mites, representing variations on this basic pattern (Smith and Oliver, 1986; Smith, 1988, 1999). One of these involves elimination of larval feeding and host association altogether, thus avoiding risk of not encountering a host but foregoing host-mediated dispersal (Smith, 1998). This has happened at least 20 different times in unrelated genera representing a wide diversity of families. Comparison of closely related species pairs with, and without, larval parasitism shows that loss of larval parasitism correlates with accelerated maturity and metamorphosis to adult at a smaller body size (Smith, 1998). Removal of the seasonal limitations of host association in species with nonparasitic larvae appears to favor a shift to a multivoltine life history. Several clades exhibit a third strategy by greatly extending the association with hosts, in some cases for several months, through the larval and protonymphal stages. Species in these groups typically undergo remarkable growth during the larvae stage, increasing in volume by up to 700 times their original size (Davids, 1973). Growth of these mites as deutonymphs is much more modest, and this stage often lasts only a few days or weeks. Extended larval association with hosts is exhibited by all species of Hydrachnoidea and Eylaioidea (excluding *Limnochares*), and also by atypical members of certain other taxa, such as *Hydryphantes tenuabilis* Marshall (Lanciani, 1971a). All known examples of extended association involve larval parasites of aquatic Hemiptera and Coleoptera, presumably reflecting the longevity, large size and mainly aquatic existence of these hosts. This type of life history allows at least some species to adopt multivoltine life cycles (Davids, 1973). There are interesting variations on these major life history patterns among species within genera (Lanciani, 1971b; Cook *et al.*, 1989).

1. Egg

Eggs are typically laid in masses in a gelatinous matrix and attached to plants, wood particles, or stones, but females of some species, such as *Midea expansa* Marshall, may scatter them individually on the substrate. Females of the genus *Hydrachna* use a uniquely developed, elongate ovipositor to lay eggs individually in stems of aquatic plants, and those of certain species of *Unionicola* use short stylets to oviposit in the tissues of sponges or mussels. The most comprehensive morphological study of water mite eggs was published by Sokolow (1977). Detailed accounts of oviposition behavior have been recorded by Crowell (1960), Ellis-Adam and Davids (1970) and Davids (1973).

Egg production varies greatly among species. For example, females of *Eylais discreta* Koenike may produce individual clutches of 1000–2000 eggs, with one individual able to produce over 13,000 eggs in a 3-month period (Davids, 1973). On the other extreme, species of *Thyas, Hygrobates, Piona* and *Arrenurus* that forego larval feeding typically produce 2–5 eggs per clutch (Smith, 1998). Most water mite species typically produce about a dozen-to-several hundred eggs in a clutch.

There is an evident trade-off between clutch size and egg size, probably related to risks to the larval instar and the proportion of total growth that occurs during this stage. An analysis of this trade-off in a group of 27 species of *Arrenurus* resulted in two evolutionary trajectories defined by egg size, clutch size, and female size: one representing species parasitic on adult nematocerous flies and the other species parasitizing adult Odonata (Cook *et al.*, 1989).

Large, communal clutches of eggs have been observed in several situations. Patches of eggs of *Eylais euryhalina* Smith in western Canada frequently cover areas of more than 100 cm^2 on plant or rock substrates. Females of *Arrenurus pseudosuperior* Cook produce communal egg masses entirely covering submerged oak leaves in central Canadian lakes. The stimulus to lay eggs communally may stem from a shortage of oviposition sites in nature, although there may be selective advantage if either eggs are distasteful or their sheer numbers exceed what local predators can consume.

Within the egg membrane, water mites pass through a transitory prelarval instar (Meyer, 1985) and then usually develop rapidly and directly into larvae. Fully formed individuals can be observed moving within the egg membrane just prior to emerging 1–3 weeks after oviposition. Arrested larval development has been reported in certain species of *Unionicola* (Mitchell, 1955), *Teutonia* (Smith, 1982), *Utaxatax* (Smith, 1982) and *Laversia* whose larvae do not become active and emerge until at least six months after eggs are laid.

Female mites presumably possess adaptations for selecting appropriate oviposition sites.

2. Larva

The vast majority of water mite species are parasitic on imaginal insects as larvae (reviewed by Smith and Oliver, 1976, 1986, and summarized in Table I), with the host providing nutrition as well as dispersal. Growth on the host is modest in many clades, but apparently essential for development as there are no records of purely phoretic associations. Mitchell (1966, 1969a) theorized that the probability of success for larval water mites can be calculated as the product of the probabilities of discovering a host, attaching at an appropriate site on the host, completing engorgement on the host and detaching from the host in a suitable habitat. The risk of failure can be high. Collins (1975) estimated that more than 75% of larval *Wandesia (Partnuniella) thermalis* Viets fail to locate a host in a system in which the distribution of hosts is extremely clustered and unpredictable (Collins *et al.*, 1976).

a. Evolution of Parasitic Associations

i. Host selection After emerging, larvae must quickly locate hosts that both provide adequate sites for attachment and feeding and regularly visit habitats that are suitable for postlarval development. Insect hosts provide water mites with both the source of nutrition necessary for larval growth, and their primary dispersal mechanism. Host associations of larval water mites were reviewed by Smith and Oliver (1976, 1986), and are summarized in Table I.

Larvae of *Hydrovolzia,* and most genera of Eylaoidea and Hydryphantoidea, rise to search for potential hosts on the surface film, exhibiting plesiotypical terrestrial behavior. In contrast, larvae of

Acherontacarus, Rhyncholimnochares, Wandesia, Hydrachna, Lebertioidea, Hygrobatoidea, and Arrenuroidea begin to swim in the water column or crawl on the substrate as fully adapted aquatic organisms. This apotypical behavior developed independently several times, within the superfamilies Hydrovolzioidea, Eylaioidea and Hydryphantoidea, in ancestral Hydrachnidae, and again in ancestors of the other three superfamilies. The development of fully aquatic larvae can best be understood as an adaptation enhancing the probability of success in locating hosts and, in the higher water mites, facilitating preparasitic attendance of hosts during their final preadult instar.

Surface dwelling larvae are essentially terrestrial, and typically can locate hosts only when they visit or pass through the surface film. Larvae of *Limnochares americana* Lundblad are unusual in being able to search for hosts up to 50 cm. away from the water surface. Opportunities of these mites to contact hosts are transitory events occurring irregularly in space and time, and are strongly influenced by specific aspects of behavior within a host population. For example, parasitic larvae of *Thyas barbigera* Viets are found only on parous female mosquitoes and those of *Limnochares americana* are found predominantly on territorial male dragonflies (Smith and Cook, 1991; Smith, 1999). Some water mites with terrestrial larvae overcome the risks of failure by utilizing a relatively broad range of host organisms and producing large numbers of eggs. For example, larvae of species of *Hydrovolzia* and certain genera of Hydryphantidae apparently exploit a wide range of hosts, including insects that are only casual visitors to the habitat of the mites. This strategy presumably results in considerable wastage, as larvae attaching to hosts that do not return to suitable mite habitats will die.

Larvae of most Eylaoidea (*Eylais, Piersigia, Neolimnochares,* some *Limnochares*) and various Hydryphantidae have developed strong specificity for particular groups of insect hosts that live on the surface film (e.g., Hydrometridae, Gerridae), regularly visit it to replenish air supplies (e.g., aquatic Hemiptera and Coleoptera), or pass through it just before ecdysis and regularly revisit it after emergence (e.g., Odonata, Trichoptera, Diptera). This strategy improves the probability that a larva finding a host will be returned to an appropriate habitat, but limits the availability of potential hosts. These mites apparently compensate for the high risk of failure in finding hosts by producing many hundred eggs per female.

Species with terrestrial larvae are generally limited to relatively shallow water microhabitats, given that these larvae must swim or crawl to the surface in order to search for potential hosts. Enclosure experiments by one of us (B.P. Smith) indicate that the likelihood of water striders of the genus *Gerris* being parasitized by larval *Limnochares aquatica* Linnaeus is strongly dependent on water depth. Differences in rates of infestation are substantial, even between insects caged over 0.5 m of water and those caged over 1.5 m.

In contrast, aquatic larvae have access to potential hosts throughout their aquatic existence and can actively seek them in the water column or substrate. Larvae of Hydrachnidae exploit virtually the same range of hosts as Eylaidae, but locate them opportunistically beneath the surface film with most species exhibiting no evidence of preparasitic attendance. Members of certain Anisitsiellidae and Teutoniidae (Lebertioidea), as well as Krendowskiidae and Arrenuridae, apparently also seek their hosts above the substrate, but have evolved preparsitic attendance of hosts during their penultimate instar prior to ecdysis. Their larvae parasitize tanypodine Chironomidae, Culicidae, Ceratopogonidae, Chaoboridae and Odonata, all groups that have the final aquatic instar active in the water column below the surface film.

Mites locating their hosts in the water column apparently exploit them more efficiently than do those with terrestrial larvae, as their females tend to lay smaller, though still substantial, numbers of eggs. For example, females of *Hydrachna conjecta* (Koenike) lay only about 10% as many eggs as those of *Eylais discreta* Koenike, yet these species co-occur in nature and parasitize the same species of corixids (Davids, 1973).

Larvae of *Rhyncholimnochares* and *Wandesia* are also aquatic and apparently independently evolved adaptations to locate their hosts in the substrate. Larvae of *Rhyncolimnochares* are subelytral parasites of adult elmids that remain in water. Those of *Wandesia* attend nymphal stoneflies before transferring to the emerging adults and becoming parasites. Interestingly, larval stygothrombidiids exhibit similar behavior in exploiting the same hosts. Larvae of the remaining families of Lebertioidea, Hygrobatoidea, and Arrenuroidea locate hosts as final instar larvae or pupae in cases or burrows in the substrate. The hosts, various Chironomidae or Trichoptera, construct cases or retreats where they remain during prepupal and pupal stages. These mites have developed suites of larval adaptations to exploit this sedentary behavior, greatly increasing their probability of successfully locating and associating with appropriate hosts. Larvae in all of these clades attend the host during the preparasitic phase, clinging to the integument of the pupa until it rises to the surface. They typically then transfer to the imago as it emerges, embed their chelicerae, and enter the parasitic phase. Individual females in these groups lay relatively small numbers of eggs (often less than 10), indicating that the larvae efficiently find hosts with a high probability of returning to a suitable habitat.

The development of fully aquatic larvae had important consequences each time it occurred during water mite evolution. In particular, it was a crucial preadaptation for ancestors of the various groups of Lebertioidea, Hygrobatoidea, and Arrenuroidea that became specialized in exploiting nematocerous Diptera as hosts. The origin and evolution of modern genera and families of these mites apparently followed the explosive radiation of the Nematocera, and especially Chironomidae, during the Jurassic and early Cretaceous. Refinement of their elegant adaptations for parasitizing these flies permitted the so-called higher water mites to co-evolve with them, and to exploit their potential for dispersal, invasion of new habitats, and rapid evolution.

ii. Site selection Larvae of many subfamilies of water mites exhibit strong selectivity for attaching to particular sites on the body of the host (Smith and Oliver, 1976, 1986; Oliver and Smith, 1980), as shown in Table I. These preferences appear to illustrate another adaptive process that has permitted mites to coevolve successfully with nematocerous flies by minimizing interference with the dispersal potential of the host.

Larvae of *Hydrovolzia* and certain genera of Hydryphantidae that parasitize Hemiptera, Plecoptera and Trichoptera (e.g., *Protzia, Wandesia*), seem to attach rather indiscriminately to various parts of the host's body. Those of all other groups are more selective. Larval *Eylais* are essentially terrestrial organisms and select attachment sites bathed by the host's air supplies, under the elytra of aquatic bugs and beetles (Fig. 427). Significantly, the fully aquatic larvae of *Hydrachna* are able to utilize a much wider range of sites on the same hosts (Fig. 428).

Larval hydryphantoid mites that parasitize Diptera tend to attach to thoracic sites, exhibiting apparently plesiotypical behavior. Larvae of many early derivative groups of Lebertioidea, and most Arrenuroidea parasitizing Diptera, also prefer thoracic sites. These larvae tend to be relatively large, and only a few individuals can share a single host. When more than one of these larvae attach to the same part of the host's thorax they are usually symmetrically arranged on either side of the body. Larvae of some species of *Sperchon, Lebertia, Oxus* and *Torrenticola* can use the anterior abdominal segments of hosts when thoracic sites are already occupied.

In virtually all Hygrobatoidea (except Limnesiidae which may not be closely related to other taxa in the superfamily) and a few families of Arrenuroidea (eg., Mideidae, Momoniidae) larvae attach exclusively to abdominal sites on their hosts, usually nematocerous flies but in some cases caddis flies. This apotypical behavior correlates with a marked trend to smaller size, allowing several larvae to share each attachment site on a host. These small larvae occupying abdominal attachment sites can exploit even the smallest species of Chironomidae as hosts without preventing them from flying and dispersing (Fig. 430). The relatively small larvae of various species of *Arrenurus (sensu stricto)* show preferences for either thoracic or abdominal sites on odonate hosts (Fig. 429).

The duration of the parasitic phase of larval development varies considerably. Larvae of certain species in early derivative genera such as *Limnochares* and *Hydryphantes* retain their mobility and can transfer from one instar to the next of the same host individual. Larval *Limnochares aquatica* engorge gradually by feeding on successive host instars, but larvae of *Hydryphantes tenuabilis* exhibit true preparasitic attendance and delay feeding until the host reaches adulthood. This stategy of tracking hemimetabolous hosts through several molts may reflect the plesiotypical condition in water mites. Larval *Hydrachna* and *Eylais* remain attached to the same adult bug or beetle throughout the parasitic phase, for several weeks or more. Larvae of species in these families that are adapted to temporary pools often remain on hosts throughout the dry phase of the habitat, and only engorge after a period of arrested development that may last as long as 10 months. Larvae of species parasitizing nematocerous Diptera or other aerial insects typically engorge rapidly and mature within several days. Fully fed larvae enter the quiescent protonymph stage.

The growth and dispersal functions of the larva have both strongly influenced the evolution of host associated behavior in water mites, with different results in the various major clades. Larvae of early derivative genera such as *Hydrachna, Eylais,* and certain Hydryphantidae grow substantially, increasing several hundred-fold in size in some cases, while feeding on their relatively long-lived hosts. Larvae of many relatively recently evolved groups associated with short-lived insects, especially nematocerous Diptera, still engorge on fluids from the host but increase more modestly in size in the process. The primary strategic role of the larval instar in the life history of most Hydryphantoidea, Lebertioidea, Hygrobatoidea and Arrenuroidea appears to be dispersal. Most feeding and growth occur during the deutonymph and adult stages in these mites.

The larva may be suppressed as an active instar in a few unrelated species in widely divergent genera (e.g., *Thyas, Lebertia, Limnesia, Hygrobates, Piona, Axonopsis* and *Arrenurus*), apparently permitting the mites to accelerate development and maximize exploitation of optimal conditions (see Smith and Oliver, 1986; Smith, 1998). In these cases, larvae either remain within egg membranes or emerge for a brief period of activity before entering the protonymph stage. Species known to forego a parasitic phase appear to be isolated and ephemeral evolutionary experiments (Smith and Oliver, 1986; Smith, 1998). Recent studies have detected relatively high levels of morphological divergence and lower levels of heterozygosity within populations of apparently recently diverged species lacking parasitism when compared with closely related species that retain parasitic larvae (Bohonak, 1998). We conclude that although elimination of parasitism confers short-term advantage in larval survival and potential for population growth, host-mediated dispersal is essential for long-term persistence of a lineage.

b. Ecological Aspects

i. Emergence of larvae In permanent water habitats at north temperate latitudes, most species of water mites overwinter primarily as inseminated females and there appears to be a necessary time interval in the spring for oviposition and development of larvae to proceed before parasitism can commence. Consequently, aquatic insect species whose adults emerge in early spring are rarely parasitized by water mites, even within groups that are otherwise suitable hosts. Interestingly, some of these early emerging species show high susceptibility to parasitism under laboratory conditions. For example, although only a small proportion of *Aedes communis* and *Aedes punctor* mosquitoes carry larval *Arrenurus angustilimbatus* and *Arrenurus kenki* in nature, they prove to be more susceptible to parasitism in choice experiments than the species typically utilized as hosts by these mites in the field (Smith and McIver, 1984a, b). A few mite species that apparently overwinter as eggs or larvae are adapted to exploit hosts that emerge in early spring. For example, larvae of *Huitfeldtia rectipes* Thor are found regularly on imagoes of early season species of *Chironomus* emerging from oligotrophic lakes in eastern Canada throughout the ice-free period beginning in mid-April. Available evidence suggests only loose synchrony between life cycles of mites and their host species, even when host specificity is strong (Smith and McIver, 1984a, c; Smith and Cook, 1991). In the case of *Coquillettidia perturbans* mosquitoes in Florida parasitized by larval *Arrenurus danbyensis* and *Arrenurus delawarensis,* peak infestation did not closely correlate with peak abundance of hosts and there was no close seasonal correspondence between population densities of the two mite species although both are dependent on the same host (Lanciani and McLaughlin, 1989).

ii. Seeking and locating hosts After emerging, larvae of species with a parasitic phase immediately begin host-seeking. The possible period of survival for free-living larvae in the preparasitic phase ranges from 4 days to 6 weeks (Smith, 1988), but is about 1 week for most species. Larval *Arrenurus* remain active and exhibit preparasitic attendance over a period of 2 weeks in the laboratory, but have the greatest chance to locate and infect a host during the first few days. Larvae older than 1 week rarely successfully attach to hosts, although those of some cold-adapted species apparently can remain infective for longer periods.

Some authors have suggested that water mite larvae find hosts by accidental contact, but close observation of both terrestrial and aquatic larvae reveals noticeable behavioral changes when in proximity to hosts. Terrestrial larvae seem to use visual, tactile and chemical cues to locate hosts at the surface film. Those of several Hydryphantidae have the hind legs modified for jumping and leap toward objects positioned above them, an obvious adaptation for encountering hosts on the surface film. Larval *Limnochares aquatica* exhibit questing behavior when within several millimeters of water striders, involving hesitation and reaching upward with the first pair of legs. Details of the sensory mechanisms used by aquatic larvae to locate hosts remain unclear, but there is evidence that both chemical and tactile cues are important (Böttger, 1972b). Larvae of *Hydrachna* often extend the gnathosoma when approaching a host before actual contact is made. Larval *Arrenurus* are attracted to mosquito pupae and slow their swimming just prior to contact (Smith and McIver, 1984b). These larvae swim in tight circles, as if attempting to orient to some cue, before making a direct approach (Page and B.P. Smith, unpublished data).

iii. Recognizing host Larvae of water mite species are typically able to exploit a range of host species. Some species with aerial larvae have remarkably broad host spectra encompassing several orders of insects, whereas species with preparasitic attendance will normally be limited to species within a subfamily or family of host insects. Commonly, one or two host species carry the majority of parasitic larvae of a given mite species in a particular habitat.

Although co-occurrence in space and time set the absolute boundaries for host utilization, larval mites also actively select among potential host species. In laboratory experiments, larval mites of diverse genera, such as *Hydrachna, Eylais, Limnochares* and *Arrenurus* exhibit preferences when exposed simultaneously to hosts of various species (Smith, 1977; Smith and McIver, 1984b). Choice of host by mite larvae appears to be influenced by differences in the intensity of cues, and larval mites favor certain individuals within a host population while rejecting others (Smith and McIver, 1984b). In some instances, hosts react strongly to the presence of mite larvae with avoidance behavior or vigorous grooming activity (Forbes and Baker, 1990; Smith and McIver, 1984b), and there may be differences in intensity of parasite avoidance among host species and individuals.

Strong bias for hosts of a specific gender or age class is evident in many water mite species with terrestrial larvae, apparently resulting from differences in the likelihood of exposure to mites of males and females of various ages within host populations that are determined by their behavior (Smith, 1999). Preparasitic attendance of hosts by aquatic larvae prevents gender based behavioral differences in exposure to parasitism and synchronizes initiation of larval engorgement with emergence of adult hosts. In these mites, infestation rates of male and female hosts of a given species at a particular time may differ in nature, but partitioning the data by date usually removes any apparent pattern. Gender-based selection of hosts among colonizing mites was demonstrated by Lanciani (1988). He found that larval *Arrenurus* exhibited significant but small preference for female pupae of *Anopheles crucians,* given equivalent exposure to both genders of host pupae, and adjusting for any sex-related differences in body size. Among mosquitoes, dragonflies and some other host groups, the likelihood of an adult parasitized by mite larvae returning to water is strongly influenced by its sex, and this should result in selective pressure for larval mites to recognize the gender of potential hosts.

iv. Infecting host After encountering a potential host, terrestrial larvae locate an appropriate attachment site and begin to feed. Species with larvae that exhibit preparasitic attendance must transfer from the penultimate instar of the host to the imago during ecdysis. Larval *Arrenurus* cling passively to the preimaginal host instar, becoming active only when it begins to molt. Interestingly, many larvae of species parasitizing mosquitoes appear to have difficulty in locating the split in the pupal exuvia and some that do find it are unable to penetrate the surface film quickly enough to reach the adult mosquito as it emerges. As a result, between 30 and 50% of preparasitic larvae fail to infect the adult host (Smith and McIver, 1984b). Larvae of *Arrenurus* species parasitic on odonates are carried out of the water on the penultimate instar of the host and almost all of them successfully move from the larval exuvia onto the adult odonate (Mitchell, 1969b). Larvae of species of the genera *Atractides, Hygrobates* and *Unionicola* pierce the pupal cuticle of their chironomid hosts and embed their chelicerae in the pharate adult. This behavior virtually ensures successful transfer to the adult because the larvae are passively pulled through the pupal exuvia as the host emerges (Böttger 1972b).

Little is known about the mechanisms by which mite larvae select sites on hosts, and virtually all available information pertains to a few species of the genus *Arrenurus.* Site selection of larval *Arrenurus* on odonates appears to be largely a function of timing (Mitchell, 1961). Progress of the host ecdysis at the point when mite larvae become active determines where the mites first come in contact with the host, and they typically attach close to that point. Larvae experimentally removed from the host and placed elsewhere on the body readily attached there. In contrast, there appears to be little correlation between the process of host ecdysis and site selection in larval *Arrenurus* parasitizing mosquitoes. For example, larval *Arrenurus kenki* almost always attach between the head and thorax when infesting adult *Aedes excrucians* but to the abdomen when attacking *Aedes cinereus,* without reference to the time taken by the host during ecdysis.

v. Attaching to host and engorgement Larval mites attach to hosts by using their pedipalps to stabilize the body and produce the leverage needed to pierce the cuticle of the host with their chelicerae. Larvae of some species of *Arrenurus* secrete a substance which solidifies and apparently helps to cement the mite to the host (Åbro, 1984), but this has not been observed in other cases (see Redmond and Lanciani, 1982).

After attachment, larvae of several genera (e.g., *Hydrachna, Rhyncholimnochares,* and *Arrenurus*) are known to form a feeding tube, or stylostome, in the tissues of the host (see Smith, 1988), and these structures are probably produced by larvae of species belonging to many other genera. Gross structure of stylostomes appears to be consistent for species within a genus, but differs greatly among genera. For example, stylostomes of larval *Hydrachna* (Davids, 1973) and *Rhyncholimnochares* are multibranched and open-ended, those of larval *Arrenurus* are single blind tubes (Åbro, 1979; Redmond and Hochberg, 1981) and the known example of a stylostome of a pionid appears to be a single, open-ended tube (Gledhill, 1985). Details of stylostome morphology appear to be constant for a given species of mite regardless of host but differ among mite species, leading to the conclusion that they are produced by larval mites rather than their hosts (Lanciani and Smith, 1989). Stylostomes are acellular, consisting of acid mucopolysaccharides, and have no perforations other than a terminal pore, when present. The mechanism of stylostome function is not well understood.

Davids (1973) argues that the stylostomes of larval *Hydrachna* are formed by coagulation of host hemolymph in reaction to mite saliva, and that the mites imbibe hemolymph from the host. Stylostomes of *Arrenurus* larvae are typically surrounded by a fluid-filled abscess, and it has been suggested that the mites feed upon liquefied tissues in much the same way as larval chiggers do (Pflügfelder, 1970; Åbro, 1982, 1984). It is possible that the stylostomes of larvae in these remotely related clades are non-homologous.

Duration of the period of larval engorgement varies considerably among water mite taxa. Larvae of *Limnochares aquatica* and *Hydryphantes tenuabilis* require 6–13 days to engorge (Böttger, 1972a; Lanciani, 1971a; Smith, 1989), but those of various species of *Hydrachna* and *Eylais* may spend from 2 weeks to 10 months on the host (e.g., Davids, 1973; Smith, 1977, 1988; Wiggins *et al.*, 1980). Larvae of some species of *Eylais* can complete engorgement within 2 weeks at the time of year when female hosts are producing eggs, but apparently diapause in a partially engorged state at other times of the year (Smith, 1988). Presumably, growth rate of larvae in these groups is generally dependent on physiological condition of the host. Larvae of more recently evolved groups associated with short-lived insects, especially nematocerous Diptera, require shorter times for engorgement. Those of pionid species inhabiting temporary pools, such as *Piona napio* and *Tiphys americanus,* can engorge in as little as 24 h on their chironomid hosts. Larval *Arrenurus* feeding on mosquitoes take approximately 6 days for full engorgement (Mullen, 1974).

Larval growth during engorgement is also highly variable. Larvae parasitic on nematocerous flies increase only modestly in size. For example, larvae of *Limnesia maculata* and *Unionicola crassipes* expand to between 3 and 4.3 times their original volume on their chironomid host and those of *Arrenurus* species feeding on mosquitoes increase in volume about 16-fold (Münchberg, 1938; Böttger, 1972b). Larvae of other species of *Arrenurus* parasitic on odonates augment their volume 80 to 90 times, and larval *Hydrachna* and *Eylais* grow substantially, increasing up to 600-fold in size in some cases (Davids, 1973; Fig. 427).

Larvae parasitizing long-lived hosts for extended periods risk premature detachment when the host molts. Caste exuvia of nepid and belastomatid bugs are commonly found with larval and protonymphal *Hydrachna* still attached. Larvae of *Limnochares aquatica* retain their mobility and can transfer from one host instar to the next to continue engorgement.

vi. Detaching from host Following engorgement, mite larvae must detach from the host and return to an aquatic habitat suitable for postlarval development. Larvae of various taxa may use host behavior, along with mechanical, visual, and chemical stimuli, as cues to initiate detachment (Böttger, 1972a, b). Environmental cues (such as moisture) may be sufficient to induce detachment in some species, but physiological cues from the host are essential in others (Smith and Laughland, 1990). Larvae of the Eurasian species *Arrenurus cuspidator* parasitic on the damselfly, *Coenagrion puella* require two simultaneous cues to stimulate detachment, high humidity and either mating or oviposition behavior by the host (Rolff and Martens, 1997). Presumably, the mite larvae can detect hormones or some other physiological correlate with reproductive behavior in the host.

Engorged larval mites risk detachment from hosts into unsuitable habitats. Fully fed larvae with limited mobility falling on dry ground quickly perish, and those introduced into unsuitable aquatic habitats may fail to survive or reproduce. At least some species of *Arrenurus* parasitic on odonates are regularly introduced into lakes that apparently cannot support their reproduction (Mitchell, 1964a, 1969b). A large proportion of *Cenocorixa bifida* waterboatmen living in inland saline lakes of central British Columbia are parasitized by nonresident species of *Eylais* and *Hydrachna* by late autumn (Smith, 1977).

Fully fed larvae are capable of only limited movement, and those successfully re-entering water typically seek out plant material or detritus, attach themselves by their chelicerae and become quiescent. Engorged larvae of *Piona* and *Arrenurus* may attempt to swim or crawl for up to 2 days before transforming to protonymphs .

3. Protonymph (Nymphochrysalis)

Engorged larvae of all groups of water mites enter a quiescent protonymph stage during which larval tissues are resorbed and reorganized and the deutonymph develops. After a few days fully formed deutonymphs emerge from the larval skins and become active.

Mites which parasitize large, long-lived insects, including *Hydrachna,* most Eylaoidea, and certain Hydryphantidae, apotypically pass through the protonymphal stage while still attached to the host, within the larval integument. This trait is correlated with extreme growth during the parasitic phase which renders engorged larvae incapable of locomotion. Along with other adaptations, extended attachment to the host allows many of these mites to exploit periodically temporary habitats efficiently.

Members of *Arrenurus planus* Marshall and *A. ventropetiolatus* Lavers survive for up to 10 months of the year in the dry basins of vernal temporary pools in eastern and western North America, respectively, as diapausing protonymphs and pharate deutonymphs. These mites require a period of desiccation, and probably cold temperatures, to release them from diapause (Münchberg, 1937; Wiggins *et al.*, 1980).

4. Deutonymph

In all known species the deutonymphal instar is active and predaceous and resembles the adult, but is sexually immature. Deutonymphs typically exhibit relatively undeveloped idiosomal sclerites and chaetotaxy compared to adults. They have only a rudimentary (or provisional) genital field bearing an incomplete complement of acetabula. They feed voraciously, often preying upon immature instars of the same taxa of insects they parasitized as larvae (see Table I). The deutonymphal stage varies in duration from a few days or weeks in early derivative groups such as Hydrachnidae, Eylaoidea and certain Hydryphantidae, to several months in many groups of Lebertioidea, Hygrobatoidea and Arrenuroidea. Deutonymphs of some hygrobatoid families are especially long-lived. Those of Pionidae typically live for many months through summer and the following winter, and this trait has pre-adapted certain species to exploit annually temporary pools. In these pionids, deutonymphs are able to endure the dry phase of the cycle in damp retreats in the substrate (Smith, 1976a; Wiggins *et al.*, 1980).

The deutonymph is the primary growth instar in most groups of water mites, and body size increases dramatically during this stage. On attaining adult size deutonymphs prepare to transform to tritonymphs by embedding their chelicerae in plant tissue or soft detritus and becoming inactive.

5. Tritonymph (Imagochrysalis)

During this stage further structural reorganization occurs to produce the adult. This final metamorphosis is rapid, and adults typically emerge within a few days.

Mature, fully fed deutonymphs anchor themselves to living or decaying plant material using the legs and mouthparts as they approach transformation to the tritonymph stage and become quiescent. Tritonymphs of several species of Pionidae inhabiting vernal temporary pools are regularly found attached to aquatic mosses in clusters, with each mite attached at a leaf axil. Conspicuous clumps of up to 500 tritonymphs comprising several species of the genera *Tiphys* and *Piona* can be observed in moss mats at the edges of these pools as the ice is melting in early spring. The significance of this clustering is not known. The tritonymphal instar remains enclosed by the deutonymphal exoskeleton, and the pharate adult may often be seen forming within it. Mites of various species of *Arrenurus* typically spend 4 to 5 days in the tritonymph stage under laboratory conditions at 20°C. Water mites appear to be particularly vulnerable to poor water quality during the tritonymph stage, perhaps because the metamorphosing adults must respire by diffusion across three levels of cuticle. For example, when dead and decaying prey are present in laboratory cultures, tritonymphs of *Arrenurus* may take up to 17 days to complete development and mortality rates of this instar may exceed 50%, despite minimal mortality among deutonymphs and adults living in the same container.

6. Adult

a. Sclerotization of Exoskeleton Adults emerge from the deutonymphal integument in teneral condition with soft, pliable, and colorless sclerites. They immediately become active, crawling or swimming about while sclerotization of the body is completed and distinctive color patterns become evident. Adults of the various groups display a wide range of body shapes and arrangements of idiosomal sclerites, primarily as adaptations for living in aquatic habitats of varying depth, turbulence, temperature, and substrate type. Shortly after emergence, both males and females mature sexually and begin to mate. In many taxa, males tend to emerge and become mature a few days earlier than conspecific females, and are ready to mate as soon as females become active in the habitat. Adults of certain *Limnochares* (Böttger, 1972a), and possibly some *Arrenurus* (Imamura, 1952), undergo supernumerary molts.

b. Mating and Spermatophore Transfer The adult is a highly specialized reproductive instar in water mites, as clearly evidenced by the remarkable behavioral modifications, and associated morphological adaptations, that have developed in various groups to facilitate successful spermatophore transfer (Smith and Oliver, 1986; Proctor, 1992a). Water mites exhibit a greater variety of spermatophore transfer methods than any other group of arachnids (Proctor, 1991b). Water mite genera can be grouped into categories based on degree of interaction between males and females during spermatophore transfer (Proctor, 1992b).

i. Dissociation during transfer The plesiotypical behavior pattern in water mites, complete dissociation involving no chemical communication or physical contact between the sexes, has been observed in species of a wide variety of early derivative genera (see Proctor, 1992a). In these species, males deposit stalked spermatophores on the substrate for subsequent retrieval by conspecific females, and those of some species have been observed to deposit large numbers when isolated in small containers of water. After finding the spermatophores, perhaps using pheromonal cues, females pick them up in the gonopore. An apotypical variant of this, incomplete dissociation requiring chemical or physical cues from females to that stimulate males for depositing spermatophores, occurs in several unrelated genera (Proctor, 1992a). Dissociation during spermatophore transfer is widespread in actinedid mites and probably occurs in most groups of water mites with unmodified males.

ii. Association during transfer A more interactive type of mating, pairing of males and females for indirect spermatophore transfer, has been observed in several species (see Proctor, 1992b). In these cases, males deposit spermatophores on the substrate, then actively assist females to find them and pick them up. Males of some *Eylais* lead receptive females in a circular dance, stopping repeatedly to deposit spermatophores at the same location on the circumference of the route. Halfway through each revolution of the dance, the female pauses with the male over his mass of spermatophores and picks up one or more of them in her gonopore (Lanciani, 1972). Males of *Neumania papillator* apparently exploit the hunting behavior of females to attract them to spermatophores (Proctor, 1991a, 1992a, c). Upon encountering a female, males vibrate their legs rapidly, simulating the stimuli produced by copepod prey, until she responds with characteristic ambush posture and aggression. The males then turn away, lay down a spermatophore net and rapidly fan water to create a current over the net toward the female, presumably directing pheromones toward her. A receptive female then ceases aggression, approaches the spermatophore net, and begins to collect spermatophores. Males of some species of *Unionicola* also use leg-trembling to generate vibrational cues to attract females to spermatophores (Proctor, 1992a, c).

iii. Copulation during transfer Pairing with direct spermatophore transfer has evolved independently in many water mite lineages. Active transfer of spermatophores to females is accomplished either by modified appendages or idiosomal sclerites of males or by direct gonopore-to-gonopore contact. Proctor (1992b), following Hinton (1964), considers both of these options as examples of copulation.

In *Unionicola intermedia,* a species that inhabits the mantle cavity of freshwater mussels, males retrieve their own spermatophores and carry them between the genua or tibiae of their hind legs while seeking receptive females. On finding a female, the male crawls beneath her and places his spermatophores in her gonopore (Hevers, 1978). Males of many groups of Hygrobatoidea produce unstalked spermatophores which they carry between modified claws on the tarsi of their flexed third pair of legs before transferring them directly into the gonopore of the female. This type of mating has been observed in various species of *Feltria* (Motas, 1928), *Forelia* (Uchida, 1940), *Tiphys* (Viets, 1914; Mitchell, 1957c), *Piona* (Koenike, 1891; Mitchell, 1957c; Böttger, 1962; Smith, 1976a) and *Brachypoda* (Motas, 1928; Halik, 1955). In each case, the male uses specially modified appendages, usually the fourth pair of legs, to hold the female in a characteristic posture so that he can transfer spermatophores to her gonopore by simply extending his third pair of legs. This mating behavior probably occurs in all water mites whose males have highly modified leg segments but no elaborate caudal development.

A variety of complex modes of direct spermatophore transfer occur in the genus *Arrenurus.* Males of the various species have the idiosoma modified posteriorly to form a characteristically shaped cauda (Figs. 419, 420). A receptive female mounts a male, positioning herself so that her gonopore rests over the back edge of his cauda. Secretions from glandularia of the male temporarily cement the pair together, and the male then deposits one or more stalked spermatophores on the substrate. In some species, the male then rocks his body down and forward to bring the gonopore of the female in contact with the spermatophore (Lundblad, 1929; Böttger and Schaller, 1961; Böttger, 1962). In others, the male uses his petiole, a cup-shaped posterior appendage of the cauda, to scoop up spermatophores into masses which he then inserts into the gonopore of the female (Böttger, 1965). Variations of this behavioral pattern occur throughout the many species of *Arrenurus,* along with the multitude of different modifications in caudal morphology that characterize the males. This type of mating probably can be inferred in all groups of water mites with elaborate caudal development in males.

Finally, direct transfer of spermatophores from gonopore to gonopore during copulation occurs in certain species of *Eylais* (Böttger, 1962; Lanciani, 1972) and in the arrenuroid genera *Midea* (Lundblad, 1929) and *Nudomideopsis* (I.M. Smith, unpublished data). In some of these cases, males either have the gonopore protruding from the venter on a cone-shaped projection (*Eylais*) or are equipped with winglike appendages flanking the gonopore (*Midea*) to facilitate transfer.

Clearly, copulation during spermatophore transfer has evolved many times from antecedent conditions involving association with complex sexual interaction. As defined by Proctor (1991b), copulation has developed independently no fewer than 91 times within the water mites, including at least three times within a single family, Pionidae (Mitchell, 1957c; Smith, 1976a). Direct transfer from gonopore to gonopore certainly developed independently in *Eylais* and Arrenuroidea.

The often profound modification of the male idiosoma and appendages that is associated with diverse modes of spermatophore transfer emphasizes the highly specialized mating function of the adult instar in water mites. Males in many groups mate and die within a few days of emerging. Those of some groups, such as *Arrenurus,* may live for several weeks. Males of some genera of Foreliinae and Tiphyinae are so modified morphologically that they are unable to move about easily, and are restricted to crawling slowly over the substrate using the anterior pairs of legs.

Sex pheromones appear to mediate pairing in at least some species of water mites. Males of *Limnesia undulata* increase spermatophore deposition when females are present, and produce comparable numbers when placed in containers from which females have recently been removed (Proctor, 1992a). The vigorous fanning of water over spermatophore nets toward females by males of *Neumania papillator* strongly suggests chemical communication (Proctor, 1991a). Females of various species of *Arrenurus* produce male-arrestant/attractant pheromones that can be extracted from water (B.P. Smith and Hagman, unpublished data). Unlike insect pheromones which are often species-specific, the pheromones of *Arrenurus* species often elicit response in several closely related and co-occurrent species (B.P. Smith and Florentino, unpublished data). There is also evidence that males of some species of *Arrenurus* produce pheromones attractive to females (Baker, 1996).

Males appear to be much less common than females in many genera of water mites (Meyer and Schwoerbel, 1981; Proctor, 1997). This may in part be due to the shorter life span of males. Biased sex ratios can also result from protandry which has been demonstrated in at least some water mite species (Proctor, 1989, 1992b). Females of *Arrenurus manubriator* produce strongly gender-biased clutches (Proctor, 1997; B.P. Smith, unpublished data). The bias for males or females appears to be consistent among successive clutches of eggs for a given female and appears to be a trait largely inherited from the maternal parent (B.P. Smith, unpublished data). Water mites appear to be diplo-diploid, with no recognizable sex chromosomes (Sokolow, 1954).

Mated females may live for many months, continuing to feed while they produce several clutches of eggs. In many groups, mating occurs in late summer and fertilization is delayed while females overwinter. These females lay their eggs after they have been fertilized early in the following spring or early summer. In contrast, females of species inhabiting temporary pools must lay their eggs within a few days of mating to ensure that their offspring reach a life-history stage capable of avoiding or withstanding the dry period before water disappears from the habitat (Smith, 1976a; Wiggins *et al.,* 1980).

Preliminary evidence suggests that the first male to mate with a female has almost total sperm precedence over subsequent mates. Given that sperm is regularly stored for many months over winter, there must be substantial selective pressure for behavioral and phenological adaptations to gain early access to females. Guarding of tritonymphs (presumably females) by young males has been observed in *Arrenurus planus.*

Males do not seem to discriminate between virginal and previously mated females. Males of *Neumania papillator* attempt to mate with all females they encounter. They eventually produce significantly greater numbers of spermatophores for virginal females, but this appears to be the result of avoidance behavior by inseminated females rather than discrimination by males (Proctor, 1991a). Males of *Arrenurus manubriator* choose larger females over smaller ones, perhaps selecting for greater egg production capacity (Proctor and Smith, 1994; Smith and Bovenzi, unpublished data).

D. Habitats and Communities

Throughout water mite evolution, successful invasion and exploitation of new habitats has depended on the development of compatible adaptive strategies for the various instars. Larval traits tend to promote parasitism and dispersal on hosts, while those of deutonymphs and adults favor feeding, growth and reproduction in water. Water mite species are regularly provided with opportunities to colonize new habitats by the passive transport of their larvae on hosts, and this has been a primary mechanism promoting speciation and divergence over time. Nevertheless, in modern communities most species and many monophyletic groups representing genus or family level taxa are restricted to one or a few similar types of habitats. The strong correlation between certain clades and habitats suggests that conservative factors such as adaptive requirements for locating hosts, prey, mates, and oviposition sites tend to constrain adaptive radiation. It also indicates that successful invasion of new habitats usually resulted in the rapid evolution of significant new life-history, behavioral and morphological traits.

The communities living in the major types of freshwater habitats in North America are outlined below and in Table I.

1. Springs (Including Seepage Areas)

Species of 47 genera live in moss and wet detritus associated with rheocrenes and helocrenes. Members of ancient clades that may well have originated in this type of habitat are soft-bodied (e.g., *Tartarothyas*), or partly to fully sclerotized (e.g., *Thyas, Panisopsis*). Adults of derivative groups that apparently invaded spring and seepage habitats from flowing water more recently have entire dorsal and ventral shields (e.g., *Paramideopsis*—Fig. 425, certain *Mideopsis, Laversia*). Most spring inhabiting species are cold-adapted stenophiles, but some *Wandesia* (subgenus *Partnuniella*) and all *Thermacarus* live only in hot springs. Over 115 species in 57 genera and 25 families have been identified from spring habitats in Canada including interstitial taxa (Smith, 1991a).

2. Riffle Habitats

Species of 49 genera live on the substrate in rapidly flowing areas of streams and rivers. Diverse communities from a single location may contain up to 50 species in 30 genera. Members of a few early derivative taxa are soft-bodied (e.g., *Protzia*), but adults of most groups are both strongly flattened and well sclerotized (e.g., *Kongsbergia*—Fig. 421, *Aturus*—Fig. 422). Most rheophilic species are cold- or cool-adapted stenophiles.

3. Interstitial Habitats

Species of 55 genera live in sand and gravel deposits to depths of 1 m or more, mainly in the hyporheic zone of streams and rheocrenes. Deutonymphs and adults crawl through the spaces between particles, often with considerable speed and agility. Members of most early derivative taxa are soft-bodied (e.g., *Rhyncholimnochares*) and some are also vermiform (e.g., *Wandesia*). In adults of more recently derived groups, the dorsum and venter are partly covered by various plates or small shields (e.g., *Clathrosperchon, Omartacarus, Atractides*), or are well sclerotized with either entire dorsal and ventral shields or greatly expanded coxal plates (e.g., *Frontipodopsis, Aturus*—Fig. 422, *Uchidastygacarus*—Fig. 424, *Arenohydracarus*). Mites adapted to interstitial habitats have reduced eyes and lack pigmentation of the integument.

4. Stenothermic Pools

Species of 47 genera live in silty substrates in spring-fed pools, fen pools and depositional areas of streams. Members of ancient taxa are soft-bodied (e.g., *Pseudohydryphantes*), while adults of early derivative clades exhibit various degrees of sclerotization ranging from slight enlargement of dorsal and ventral plates (e.g., *Teutonia, Wettina, Forelia*) to development of entire dorsal and ventral shields (e.g., *Momonia, Geayia*). Adults of certain taxa which invaded pools relatively recently from lotic and hyporheic habitats have retained dorsal and ventral shields, though they have become secondarily adapted for swimming (e.g., *Axonopsis, Mideopsis*). Certain groups inhabiting depositional areas of streams (e.g., *Teutonia, Oxus, Wettina*) include the fastest and most agile swimmers among the water mites. Most inhabitants of pools are cool-adapted stenophiles.

5. Lakes (Including Permanent Ponds, Marshes, Swamps and Bogs)

Species of 39 genera inhabit large bodies of standing water. Members of ancient and early derivative genera are generally soft-bodied (e.g., *Hydrodroma, Limnesia, Hygrobates, Huitfeldtia*). Adults of many genera in more recently evolved clades have extensive dorsal and ventral shields (e.g., *Koenikea*—Fig. 423), in at least some cases because the group has secondarily invaded standing water from lotic or hyporheic habitats (e.g., *Axonopsis, Mideopsis*). Most species are strong swimmers and have fringes of long, slender setae on the genua and tibiae of the second, third, and fourth pairs of legs (Fig. 290). These swimming setae help to propel the mites as they move about at the surface of the substrate or among plants. Some species of *Unionicola* and *Piona* have especially long swimming setae, and are essentially pelagic or planktonic (Riessen, 1982). A few species which lack swimming setae either walk or crawl actively on plants and detritus, reflecting recent rheobiontic ancestry (e.g., *Sperchon, Torrenticola*, certain *Mideopsis, Neoacarus*). Members of *Tyrrellia* are unusual in that they inhabit wet litter at the edges of permanent bodies of water, including lakes, where they crawl about at the surface film. Many species of *Unionicola* and the only known species of *Najadicola* are obligate parasites or commensals in the mantle cavity of molluscs in both lentic and lotic habitats (Mitchell, 1955; Simmons and Smith, 1984; Gledhill, 1985; Vidrine, 1986).

More than 25 species in each of the remarkable genera *Arrenurus* and *Piona* often occur together in a single small lake. Mitchell (1964a, 1969a) concluded that behavioral flexibility exhibited by larval *Arrenurus* in selecting and exploiting their odonate hosts is the main factor permitting such high levels of sympatry, and that competition between deutonymphs and adults of the various species is probably not significant in natural communities.

Most limnophilic taxa tolerate a broad range of temperatures which they encounter in weed beds and silty substrates in the littoral and sublittoral zones. However, a few are cold-adapted stenophiles restricted to either arctic-alpine glacial lakes (e.g., *Acalyptonotus*) or the profundal zone of oligotrophic lakes (e.g., *Huitfeldtia, Laversia*). Certain species of some genera (e.g., *Limnochares, Hydrodroma, Limnesia, Piona* and *Arrenurus*) are highly tolerant of extreme thermal and chemical regimes, and are able to inhabit desert sloughs, alkaline lakes or acid bogs. The water mite fauna of wetland habitats in Canada was treated by I.M. Smith (1987).

6. Temporary Pools

Species of 13 genera live in annually temporary pools. Members of ancient genera that may have originated in this type of habitat range from soft-bodied walkers (e.g., *Thyas*) and swimmers (e.g., *Eylais)* to mites with unique arrangements of integumental plates that either crawl (e.g., *Piersigia*) or swim awkwardly (e.g., *Hydrachna, Hydryphantes*). Species of derivative groups that apparently invaded temporary pools secondarily tend to be relatively good swimmers (eg:*Tiphys, Piona, Arrenurus*). Water mites inhabiting temporary pools are adapted either to avoid (e.g., species of *Hydrachna* and *Eylais*) or to endure the dry phase of the annual cycle (Wiggins *et al.,* 1980).

E. Ecology

The parasitic larvae and predaceous deutonymphs and adults of water mites have direct and almost certainly significant effects on the size and structure of insect populations in many habitats (Smith, 1983, 1988; Lanciani, 1983; Proctor and Pritchard, 1989). Unfortunately, their impact has rarely been measured accurately because of the routine neglect of mites in ecological studies of freshwater communities. Due to their small size and often cryptic habits, mites are usually absent or seriously under-represented in samples that are collected using standard techniques for capturing insects and crustaceans. Most entomologists are unfamiliar with mites, and tend either to disregard them as too poorly known ecologically and difficult taxonomically or to lump them together in a meaningless way, when conducting community studies. Failure to include realistic assessment of the roles played by water mites often results in seriously flawed analyses of the structure and dynamics of freshwater communities.

1. Impact as Parasites

a. Effect on Individual Hosts The impact of larval water mites on host insects was reviewed by Lanciani (1983) and B.P. Smith (1988). Numerous laboratory studies have demonstrated reduced survival or longevity of various insect hosts parasitized by larval water mites, including mosquitoes (Lanciani and Boyt, 1977; Lanciani, 1987), ceratopogonid midges (Lanciani, 1986a), chironomid midges (Weiberg and Edwards, 1997), juvenile water striders (Smith, 1989), backswimmers (Lanciani, 1982) and damselflies (Forbes and Baker, 1990, 1991; Leung and Forbes, 1997). Lanciani (1983) concluded that the ratio of larval mite weight-to-host weight is a good indicator of the probable impact of parasitism on the host. Studies of parasite-induced mortality in field populations have not been conclusive (Robinson, 1983; Åbro, 1990a; Smith, 1999). There is evidence of occasional crashes of host populations related to parasitism by mites (Smith, 1988).

Water mite parasitism reduces egg production of insect hosts. The most pronounced effects have been reported for Hemiptera infested with larval *Hydrachna* and *Eylais* (Davids, 1973; Smith, 1977), but parasitized nematocerous flies also show lowered fecundity (Smith and McIver, 1984c; Lanciani and Boyt, 1977). Development of juvenile instars is retarded in backswimmers (Notonectidae) parasitized by larvae of *Hydrachna virella* (Lanciani and May, 1982) and in water striders (Gerridae) attacked by *Limnochares aquatica* (Smith, 1989).

Mite parasitism influences frequency of foraging, intensity of territorial behavior and likelihood of mating in males of the damselfly *Enallagma ebrium* (Forbes, 1991). Mite parasitism apparently reduces the capacity of the damselfly *Nehalennia speciosa* to fly and disperse (Reinhardt, 1996). The corixid *Cymatia bonsdorfi* increases feeding activity when parasitized by larvae of *Hydrachna conjecta* (Reilly and McCarthy, 1991).

Some hosts exhibit strong avoidance behavior in the presence of larval water mites, involving grooming behavior to dislodge them. Larval *Ischnura verticalis* typically respond to fish by reducing movement, but in the presence of both fish and preparasitic larvae of *Arrenurus pseudosuperior* exhibit unrestrained grooming and mite-avoidance behavior despite the increased risk of detection by predators (Baker and Smith, 1997).

b. Effect on Population Structure of Hosts Larval water mites regularly parasitize 20–50% of adults in natural populations of aquatic insects in such diverse families as Corixidae (Hemiptera), Dytiscidae (Coleoptera), Libellulidae (Odonata), Culicidae, and Chironomidae (Diptera). Almost all individuals are parasitized in some host populations, and some may carry remarkably high mite loads. We have seen examples of over 1000 larvae of *Arrenurus* parasitizing a dragonfly (*Erythemus simplicicolis),* over 350 larvae of *Limnochares aquatica* on a water strider *(Gerris comatus),* over 50 larvae of *Arrenurus* on a ceratopogonid midge (*Bezzia* sp.) and over 40 larvae of *Arrenurus danbyensis* on a mosquito (*Coquillettidia perturbans*). Individual chironomid midges emerging from mesotrophic lakes in the Great Lakes Basin often carry 20 or more mite larvae representing as many as seven different genera.

The distribution of parasitic larval water mites typically follows a clustered pattern within a host population, with a few host individuals heavily infested and most carrying few or no mites (e.g., Smith and McIver, 1984a; Smith, 1988). Under natural conditions, clustering may reflect differences in the temporal and spatial distributions of host-seeking larval mites and their potential hosts. However, clustering also occurs in laboratory experiments involving synchronous exposure of insects to mites, where duration of exposure and spatial heterogeneity are controlled. This suggests either initial differential susceptibility to parasitism among host individuals or facilitation of additional loading on hosts that have become parasitized.

Intensity of infestation by larval *Arrenurus* appears to be negatively correlated with body size and condition (i.e., ratio of mass to body size) of their damselfly hosts in at least some cases (Forbes and Baker, 1991; Leung and Forbes, 1997). This may be a result of more vigorous avoidance behavior by larger and more robust potential hosts (Smith and McIver, 1984b; Forbes and Baker, 1990). Larval *Arrenurus* have greater intensity of parasitism on damselflies with asymmetry in forewing length, but it is not clear whether increased infestation is a response to or a cause of this asymmetry (Bonn *et al.*, 1996).

Behavioral subtleties can increase vulnerability of hosts of certain gender or age classes to colonizing larval mites (Smith and Cook, 1991), with resulting differentials in parasite loads (Smith, 1988, 1999). Observed differences in parasite loads among sex and age classes in natural populations can prove useful for inferring the behavior of the host and parasite and the age structure of the host population (Smith, 1999).

c. Effect on Other Mites Sharing the Same Host
Clustering of parasitic larvae on certain host individuals results in intraspecific competition between mites on those hosts that can affect their growth and survival (Lanciani, 1976, 1984, 1986b), and leaves many hosts underutilised. Consequently, clustering apparently both invokes density-dependent control of mite populations and keeps them below carrying capacity, effectively precluding interspecific competition.

Parasitism of a host individual by larvae of more than one species of water mite occurs more commonly than would be expected by chance (Stechmann, 1980; Smith and McIver, 1984a; Lanciani and McLaughlin, 1989). In addition, many species of water mites exploit a variety of related host species. As a result, parasitic associations often involve complex assemblages with considerable overlap in exploitation of host resources, including use of similar attachment sites (Kouwets and Davids, 1984). Interspecific competition appears to be negligible in these assemblages. Consequently, assuming that larval success is the limiting factor determining population size in water mite species, there is probably little competition for food resources in communities of deutonymphs and adults. This may help to explain the remarkably high levels of species richness in many water mite communities.

2. Impact as Predators

The feeding habits of deutonymphal and adult water mites were reviewed by Böttger (1970), Gledhill (1985) and Proctor and Pritchard (1989). Data for species inhabiting Canadian wetlands were reviewed by I.M. Smith (1987). Deutonymphs and adults of free-living species are voracious predators on a wide range of small aquatic organisms. Some species at least occasionally scavenge on recently killed organisms. Most species exhibit marked preferences for prey of a particular size with particular morphological and behavioral attributes. Species of *Hydrachna, Thyas, Hydryphantes* and *Hydrodroma* are specialized predators of arthropod eggs (Davids, 1973; Mullen, 1975; Lanciani, 1978, 1980; Wiles, 1982; Meyer, 1985). Many members of *Eylais* and *Piersigia* and the large genus *Arrenurus* appear to depend on ostracod crustaceans as prey (although some *Eylais* and *Arrenurus* feed on cladocerans), adults of *Neumania* ambush copepods, pelagic and planktonic species of *Unionicola* and *Piona* are highly efficient predators of cladoceran crustaceans and species of many genera of Lebertioidea, Hygrobatoidea and Arrenuroidea prey upon larval Chironomidae (Böttger, 1970; Proctor, 1991b; Proctor and Pritchard, 1989, 1990; also see Table I). Some species of *Limnesia* and *Piona* appear to be more opportunistic in their choice of prey, and frequently even attack other mites in crowded containers (Elton, 1922). Parasitism of freshwater molluscs during the deutonymphal and adult stages has evolved at least twice in the superfamily Hygrobatoidea. Many species of *Unionicola* inhabit the mantle cavity of mussels and snails (Mitchell, 1955; Vidrine, 1986) and the pionid *Najadicola ingens* has similar habits (Simmons and Smith, 1984).

Deutonymphs and adults frequently prey upon the immature instars of the same species of insects that they parasitize as larvae (Davids, 1973; Wiles, 1982; Proctor and Pritchard, 1989). Preliminary indications suggest that many species of mites are specialized to exploit one group of insects throughout their life history, a strategy that would presumably increase the probability of both spatial and temporal co-occurrence of the mites and their preferred hosts and prey.

Some water mites exhibit feeding rates in experimental studies that are high enough to be considered as potentially important influences on the size and structure of prey populations in nature (Paterson, 1970; Mullen, 1975; Lanciani, 1979; Wiles, 1982). Gliwicz and Biesiadka (1975) reported that individuals of *Piona limnetica* consume 10–20 cladocerans per day in a neotropical lake, and estimated that members of this species could reduce the average standing crop of prey by 50% each week. Adults of *Hygrobates nigromaculatus,* with reported population densities as high as 1000 individuals/m^2, apparently consumed an average of 14,500 larvae of the chironomid *Cladotanytarsus mancus* per square meter for two consecutive years in a European lake, accounting for a 50% mortality rate in the prey population (Ten Winkel *et al.,* 1989).

Tactile cues appear to be of primary importance to predatory water mites (Böttger, 1970; Davids *et al.,* 1981). Hunting mites often ignore stationary potential prey even at close range, but immediately attack and kill individuals of the same prey species after contact is made. Adults of species of *Piona* typically swim in wide arcs until colliding with potential prey. Upon contact, the mites attack aggressively, and if initially unsuccessful, swim in tight spirals for several seconds until the prey is encountered again (B.P. Smith, personal observation). Several species of *Neumania* and *Unionicola* orient toward vibrations in the water as a cue to locate potential prey (Proctor and Pritchard, 1990).

Water mites also use various cues to promote efficient prey capture. Adults of *Hydrodroma despiciens* seem to respond to chemicals in the gelatinous coating of the insect eggs on which they feed (Wiles, 1982). Adults of a number of species show behavioral responses to chemical cues from prey organisms, and Baker (1996) has demonstrated how the mites detect these stimuli. Adults of both free living and parasitic species of *Unionicola* react to host extracts suggesting that they use chemical cues to initiate regional searching (Proctor and Pritchard, 1989; Roberts *et al.,* 1978; Dimock and LaRochelle, 1980; LaRochelle and Dimock, 1981).

3. Importance as Prey

Water mites are typically under represented in the stomach contents of predatory freshwater vertebrates (Modlin and Gannon, 1973; Pieczynski, 1976; Eriksson *et al.,* 1980). Deutonymphs and adults at least occasionally form a significant part of the diet of fish and turtles (Marshall, 1933, 1940; Knight, 1989), and we have seen examples of stomach contents from individual brook trout and coregonids that consisted exclusively of several hundred adults of one species of *Piona* and *Hygrobates,* respectively. It appears that fish can develop narrowly focused search images for water mites in exceptional cases.

Adults of many species of water mites eject viscous, sticky fluid from the glandularia when handled roughly, apparently as a deterrent to predation. Experiments have shown that red-colored water mites are distasteful to fish which quickly learn to reject them as potential prey (Elton, 1922; Kerfoot, 1982). Kerfoot (1982) concluded that red pigmentation in water mites is aposematic coloration. However, most red water mites live in temporary pools and seepage areas where fish are absent. Terrestrial anystoid and parasitengone mites, and members of many early derivative water mite taxa, are predominantly red or orange suggesting that suffusion of the body with red carotenoid pigments is plesiotypical for the entire lineage. This type of pigmentation may have initially evolved in these mites as protection from ultraviolet wavelengths in sunlight (Garga, 1996). However, it is noteworthy that few members of the derivative water mite superfamilies Lebertioidea, Hygrobatoidea of Arrenuroidea have retained the plesiotypical condition. Most species in these groups are not red, and the relatively small number that are have acquired this coloration secondarily, as explained above. Interestingly, most red species in these groups also inhabit vernal temporary pools or springs.

Few studies have examined invertebrate predation on water mites. Elton (1922) reported an instance of predation by an adult dytiscid beetle on a species of *Hydrachna.* Two authors have reported positive associations between fish and the mite *Piona carnea* and suggested both that fish prefer insects to mites as prey and that predaceous insects, including hemipterans, larval odonates and larval chaoborid midges, feed on the mites (Eriksson *et al.,* 1980; Punčochér and Hrbáček 1991).

4. Potential as Indicators of Environmental Quality

Species of water mites are specialized to exploit narrow ranges of physical and chemical regimes, as well as the particular biological attributes (including physico-chemical constraints) of the organisms they parasitize and prey upon. Consequently, water mites should be exceptionally sensitive indicators of habitat conditions and the impact of environmental changes on freshwater communities. Preliminary studies of physico-chemical and pollution ecology of the relatively well-known fauna of Europe have demonstrated that water mites are excellent indicators of habitat quality (Schwoerbel, 1959; Pieczynski, 1976; Biesiadka, 1979; Kowalik and Biesiadka, 1981; Kowalik,

1981; Bagge and Merilainen, 1985; Cicolani and di Sabatino, 1991; Steenbergen, 1993). Gerecke and Schwoerbel (1991) analyzed changes in water mite communities in the Danube River between 1959 and 1984 and concluded that they were highly correlated with increased levels of organic pollution. One of the species they studied, *Hygrobates fluviatilis,* appears to be relatively resistant to pollution and increased in dominance during the period of declining water quality. The results of these studies, along with our own observations in sampling a wide variety of habitats in North America and elsewhere, lead us to conclude that water mite diversity is dramatically reduced in habitats that have been degraded by chemical pollution or physical disturbance.

There have been few laboratory studies of the tolerances of water mites to environmental variables. Deutonymphs of *Arrenurus manubriator* have proven to be more susceptible than other life-history stages to iron in solution, probably resulting from the major contribution of this instar to growth of the mites (Rousch *et al.,* 1997). Experiments to assess differential susceptibility among life-history instars may provide valuable insight into the mechanisms responsible for observed changes in species populations in stressed habitats.

Progress toward the establishment of essential baseline documentation on water mite communities in North America, however, continues to be hampered by lack of sufficient systematic and ecological information at the species level, and by inadequate sampling and extraction procedures. Application of potentially useful information on water mites in assessing and monitoring the well-being of freshwater habitats and communities demands clearer understanding of the individual niche requirements and trophic relationships of species. This can be achieved by improving systematic knowledge of the group, and by developing more appropriate collecting, observing, and rearing techniques.

F. Collecting, Rearing, and Preparation for Study

These subjects were treated comprehensively by Barr (1973), and we discuss here only those aspects needing further emphasis or refinement.

1. Collecting and Extracting Techniques

Thus far, field work on water mites in North America has focussed largely on qualitative sampling to obtain specimens required for systematic or life history studies, and the methods outlined below were developed primarily for these purposes. Efforts to modify these procedures for rigorous and cost-effective quantitative sampling are needed, particularly those aimed at capitalizing on behavioral traits to simplify extraction of mites from detritus and silt.

a. Deutonymphs and Adults The best results are achieved, in most habitats and for virtually all types of substrates, using a procedure that requires a strong net with a wide opening (25–40 cm diameter) and fine mesh size (250 μm), a supply of strong, clear polyethylene bags (4–5 kg capacity), a set of standard sieves of various mesh size (ranging from 250 μm to 2 mm), six large, leak proof polyethylene containers (1 L), an ice chest, six rectangular, white photographic trays (30 cm $\times$ 40 cm), a flashlight, a set of eye droppers with a range of different sized tip openings, and a supply of small, leak proof polyethylene containers (50 mL).

Once the desired habitat or microhabitat has been selected, the substrate must be thoroughly stirred. In flowing water, the net should be positioned and secured so that the current will carry dislodged organisms and particles of substrate into it. In lentic habitats, it must be swept back and forth through the column of mixed water and substrate to pick up the organisms.

In very shallow habitats, such as seepage areas, small rheocrenes, fens, and the edges of pools and ponds, the substrate must be gathered and stirred by hand. The net is worked beneath the edge of the wet moss, leaf litter, or detritus that is to be sampled so that mats of wet substrate can be separated and rinsed over the opening. Where there is sufficient surface water, successive handfuls of loose material are picked or scooped up with the aid of a small trowel, thoroughly picked apart and washed in the net. Larger pieces of detritus are discarded after careful washing. In seepage areas, spring edges, fens, and bog margin habitats where there is often an extensive carpet of wet moss and associated plants, mites can also be collected by methodically and vigorously treading down the plants while dragging the net along to scoop up the detritus, silt, and organisms that become temporarily dislodged as water is forced up through the disturbed layer of vegetation. In either case, when a mixture of fine silt, plant fragments and organisms begins to fill the bottom of the net, it is transferred to a plastic bag containing a small amount of water. This process is repeated until the bag is approximately half full of material. The bag is then nearly filled with clean water and the contents stirred gently but thoroughly to allow heavy inorganic material such as gravel and sand to settle to the bottom. The contents of the bag are then poured carefully through a set of sieves so that gravel and sand remain in the bag. This operation is repeated until all loose organic material is in the sieves. We find that a pair of upper and lower sieves with mesh sizes of 1.4 mm and 250 μm, respectively, yields good results. The coarse material in the upper sieve is thoroughly stirred and rinsed with clean water. The fine silt and small organisms accumulating in the lower sieve are regularly transferred to a large polyethylene container with enough water to just cover the surface of the silt. When the container is about half full of fine silt it is closed and placed in an ice chest to be transported back to the laboratory or sorting area. In situations where there is insufficient surface water to permit washing and rinsing of samples on site, accumulations of unsorted substrate material must be gathered, stored in plastic bags and transported on ice to a source of clean water for subsequent processing.

With few modifications, this procedure can be used effectively in other types of habitats. Where the water column is between 25 cm and 1 m in depth and the bottom is firm enough to permit wading, the feet, hands, and net can be used to agitate the substrate (with the aid of a spade when necessary). In stream riffles with a heterogeneous substrate of rocks, gravel, sand, and detritus, vigorous and persistent agitation is usually necessary to dislodge the large number and variety of mites that cling tenaciously to rocks or burrow in the silt under larger stones. We get good results by securing the net approximately 1–2 m downstream, and then systematically working through the substrate to loosen and thoroughly rinse all particles so that organisms are carried into the net. Where well-developed interstitial habitats occur in a stream bed, this approach can be extended to yield rich collections of mites including both surface and hyporheic species. This is accomplished by simply digging down through the gravel and sand beneath the riffle zone, removing all large stones, to allow the current to flush out the pockets of interstitial silty detritus and the organisms that live there. By digging out an area of substrate approximately 1 m^2 to a depth of 0.5 m, we are often able to recover good numbers of many hyporheic species. This technique can often be used to sample the hyporheic zone in habitats where the standard Karaman–Chappuis method (Chappuis, 1942; Gledhill, 1982) is not practicable.

Quantitative samples of hyporheic arthropods comprising large and diverse collections of water mites have recently been obtained in desert streams in Arizona by driving polyvinyl chloride tubing into the substrate to various depths and using a mechanical pump to evacuate the resulting wells periodically.

In littoral lentic habitats such as depositional pools in streams, ponds, and sheltered bays in lakes, the feet can often be used to thoroughly stir up the top few centimeters of sediment and any rooted aquatic plants that may be present. A net is then repeatedly swept back and forth through the entire volume of disturbed water column to ensure that both swimming and crawling species are captured. As some mites in lentic habitats are too large to pass readily through 1.4-mm mesh, we recommend use of an upper sieve with 2.0-mm mesh when sampling ponds and lakes.

The procedure outlined here can also be used by scuba divers and a boat operator to collect mites and insects from the sublittoral and profundal zones of lakes down to depths of 30 m or more, at a much higher level of efficiency than is possible with conventional grabs or dredges.

In the sorting area, a number of white photographic trays are set up on tables and filled to a depth of 2–4 cm with clean water. Once the water has settled, the contents of one polyethylene container are carefully decanted into each tray, so that the silt spreads out in the center of the tray but does not reach the edges. As the mites become active and begin to move around in the trays, they are picked up individually using an eye dropper and transferred to small polyethylene containers. Many swimming species are strongly positively phototropic and can be recovered efficiently from trays kept in bright light. However, many crawling mites and a considerable number of swimming species are moderately positively or negatively phototropic and become highly active only at low levels of ambient light. The effectiveness and rate of recovering mites from all types of habitats can be substantially increased by reducing ambient lighting to low levels indoors or waiting until after sunset outdoors, and then scanning the trays systematically and periodically with a flashlight beam. An experienced collector can then spot the mites as they move across a contrasting background provided by either the darkly colored silt or the white bottom of the tray. Mites remain active and continue to emerge from silt in the trays for up to 72 h, as deoxygenation of the water proceeds at room temperature. For this reason, it is suggested that, whenever possible, mites should be extracted from samples in situations that permit periodic observation of undisturbed trays over a period of at least 4–6 h with controlled lighting. Under these conditions, mites tend to congregate around the edges of the trays, especially in the corners, where they can be easily spotted and picked out.

Although water mites frequently are found in samples collected for quantitative analysis in ecological studies, they are almost always underrepresented in terms of both numbers and diversity. This is often because collecting devices such as dredges or sieves are designed to capture organisms that are larger or more sedentary than mites. Due to their small size and ability to cling tenaciously to substrates, mites in lotic habitats can often resist casual efforts to dislodge them or can simply pass through nets with coarse mesh. By running or swimming, mites in hyporheic or lentic habitats can avoid capture by traps that are designed to sample instantaneously small areas of superficial substrates. Mite specimens are frequently overlooked when preserved samples are sorted because extraction techniques concentrate on larger organisms or depend on flotation methods which we have found ill-suited for separating water mites efficiently from samples of detritus.

Water mites can be collected from deeper water with underwater light traps using either battery-powered incandescent bulbs (Hungerford *et al.,* 1955; Pieczynski, 1962, 1969; Pieczynski and Kajak, 1965) or chemoluminescent tubes (Barr, 1979) as light sources. Barr (1979) recommends use of small, portable light traps for qualitative sampling in lentic habitats, and provides a useful assessment of their utility for collecting water mites for various purposes, including quantitative analysis.

Recently, a promising technique has been developed for extracting water mites and other aquatic arthropods quantitatively from samples of substrate by inducing them to respond actively to a temperature gradient (Fairchild *et al.,* 1987). This method has yielded very good results for samples of baled sphagnum from bog ponds, and could probably be adapted for use with any type of substrate inhabited by mites.

b. Larvae Free-living larvae are often collected from aquatic habitats along with deutonymphs and adults, but these specimens must be preserved in order to be identified and often cannot be reliably associated with other instars of the same species. Parasitic larvae can be recovered from insect hosts collected at or near the parental habitats. Aquatic hosts are usually captured using a dip net, whereas aerial adults can be trapped as they emerge from water, netted while swarming (especially over water at dusk), swept from vegetation, or attracted to lights at night.

2. Rearing

Larvae needed for taxonomic or life history studies can be reared from females using the techniques described by Prasad and Cook (1972), Barr (1973), and Smith (1976a). Fertilized females of many species can be collected in spring and early summer. Females needed for rearing should be placed individually in small dishes or vials containing water and a few fragments of moss or wood as a substrate. The containers should then be covered and placed in a shaded, well-ventilated location where the temperature can be maintained near levels experienced in the natural habitat of the mites. Mites from cold, stenothermic habitats must be refrigerated. The rearing containers should be checked periodically for the presence of egg masses. Females should be preserved immediately after they have laid their eggs to ensure that good-quality specimens are available for confirming species identifications. Larvae should be either preserved or slide-mounted directly as they emerge. Species-level treatment of larvae in certain families can be found in Prasad and Cook (1972), Wainstein (1980) and various papers by Smith (1976a, 1978, 1979, 1982, 1983a–f, 1984, 1989d, 1992a).

3. Preservation and Preparation for Study

Deutonymphal and adult mites must be preserved in modified Koenike's solution (or GAW), consisting of 5 parts glycerin, 4 parts water, and 1 part glacial acetic acid, by volume (see Barr, 1973), so that they can be properly cleared, dissected, and slide-mounted for identification and study. Specimens preserved in alcohol or other hygroscopic agents become so distorted and brittle that subsequent preparation is difficult or impossible. Deutonymphs and adults must be cleared in either acetic corrosive or 10% KOH, dissected, and mounted in glycerine jelly, following the techniques described by Barr (1973) and Cook (1974).

Reared larvae are often slide-mounted directly, but can be preserved in 70% alcohol for subsequent mounting without being seriously damaged. Parasitic larvae should be preserved with the host in 70% alcohol, although larvae removed from hosts that have been pinned and dried often can be slide-mounted successfully. Before larvae are removed from hosts, the attachment sites should be noted and recorded. Larvae should be mounted whole in Hoyer's medium which also acts as an efficient clearing agent for these small specimens (see Barr, 1973).

G. Taxonomic Keys to Genera of Water Mites in North America

The following keys for larvae and adults permit identification of specimens to family, subfamily and genus levels. They are intended for use by nonspecialists and use characters that should be obvious on slide-mounted specimens to all careful observers. As a result, the keys are intentionally artificial in that genera do not necessarily key out with others belonging to the same family. The larval key does not include all taxa that appear in the adult key, as larvae of many genera remain unknown (see Table I). Knowledge of larvae in a number of other genera is based on only a few species. Consequently, the key for larvae will be subject to considerable revision and refinement as larvae of more species are reared and described. We include the genus *Stygothrombium* (superfamily Styothrombioidea) in the keys and in Table 1 because these mites are frequently collected with hydrachnids and closely resemble them in morphology, behavior and ecology.

The keys presented here are designed for use with a minimum of specimen preparation. Larvae should be whole-mounted on slides. The legs should be spread out from the body as much as possible by gently pressing down on the coverslip, as the chaetotaxy of leg segments is frequently used as a key character. Adults require clearing and some dissection prior to slide mounting. The mouth parts (capitulum and appendages) should be removed to permit examination of the pedipalps in medial and lateral view. Dorsal and ventral shields, when present, should be separated to allow clear viewing of certain key characters.

We follow the convention for numbering the setiform structures on the leg segments of larvae that was established by Prasad and Cook (1972) and modified slightly by I.M. Smith (1976a) (see Figs. 6–8). The setae are numbered starting proximally on the posterolateral surface, proceeding distally along the dorsal half of the segment to the end, then returning proximally along the ventral half of the segment. Exceptions are made in the cases of solenidia (setiform chemoreceptors, see Figs. 31, 32, arrows) which are numbered first, that is as "s1" on the genua, tibia III, and the tarsi, and as "s1" and "s2" on tibiae I and II, and tarsal eupathids (specialized setae, see Fig. 77, arrow) which are numbered second, that is as "2" on the tarsi. Where it has been necessary to use leg chaetotaxy in the key, the number of solenidia present on a segment is indicated in brackets (e.g. "+2s") immediately following the information on the number of setae, so that all setiform structures can be accounted for. We usually have omitted the empodia and claws from the illustrations of the legs of larvae so that the number and positions of distal setae on the tarsi can be seen clearly. The names of the segments of the appendages are consistently abbreviated as follows:trochanter, Tr; femur, Fe (basifemur, BFe; telofemur, TFe); genu, Ge; tibia, Ti; and tarsus, Ta.

1. Taxonomic Key to Genera for Known Larval Water Mites

1a. Legs with six movable segments, with basifemur (BFe) and telofemur (TFe) separated (Figs. 9, 21, 40, 48) 2

1b. Legs with five movable segments, with basifemur and telofemur fused (Figs. 6–8, 57, 60, 67) 15

2a(1a). Gnathosoma with elaborate camerostome enclosing chelicerae, with pedipalps inserted ventrally (Figs. 11, 13, 14) ; dorsal plate present and bearing only two pairs of setae (verticils—*ve* and *vi*) near anterior edge (Figs. 11 and 13); coxal plates III located posteriorly on idiosoma with insertions of legs III located at posterolateral edges (posterior to level of excretory pore) and legs III directed posteriorly (Figs. 12, 15) ....Hydrovolzioidea 3

2b. Gnathosoma lacking elaborate camerostome; dorsal plate absent (Figs. 30, 36), or present and bearing more than two pairs of setae with verticils (*ve* and *vi*) anteriorly and at least internal scapulars (*si*) in posterior half near edge (Figs. 27, 29, 33); coxal plates III located near midlength of idiosoma with insertions of legs III located anterolaterally (anterior to level of excretory pore) and legs III directed laterally (Fig. 28) 4

3a(2a). Dorsal plate relatively small (Fig. 11); venter with rows of numerous small urstigmata borne between coxal plates I and II (Fig. 12, arrow); coxal plates I and II separated from one another on each side and from opposite members medially (Fig. 12); pedipalps massive, with tibia (Ti) bearing four long, thick, curved setae (Fig. 10); setae *c3* often foliate (Fig. 11) ....Hydrovolziidae *Hydrovolzia*

3b. Dorsal plate relatively large (Fig. 13); venter with urstigmata either inconspicuous or absent (Fig. 15); coxal plates I and II fused together to form single anterior plate (Fig. 15); pedipalps moderate in size, with tibia (Ti) bearing two slightly thickened, straight setae (Fig. 14) ....Acherontacaridae *Acherontacarus*

4a(2b). Dorsal plate absent (Figs. 30, 36), or small, covering less than one-third length of idiosoma and usually with internal scapular setae (*si*) at posterior edge (Figs. 27, 29, 33); lateral eyes borne on separate platelets (Figs. 27, 29, 30, 36); leg tarsi with unmodified claws and empodium (emp) (Fig. 26).......................................Hydryphantoidea 5

4b. Dorsal plate large, covering well over one-third length of idiosoma and with internal scapular setae (*si*) near midlength (Figs. 37, 42, 46, 51); lateral eyes borne on single eye plates (Figs. 37, 42, 46); leg tarsi with claws modified or reduced (Figs. 41, 45, 50) ....Eylaoidea 11

5a(4a). Dorsal plate roughly quadrangular, bearing four pairs of setae (*ve, vi, se, si*) (Figs. 27, 33), or three pairs with external scapulars (*se*) borne on separate platelets near posterolateral angles of plate 6

5b. Dorsal plate triangular, fragmented, or absent, bearing fewer than three pairs of setae (Figs. 29, 30, 36).......................................9

6a(5a). Claw (movable digit) of chelicerae over one-half length of basal segment (Fig. 17); excretory pore setae (*ps1* and *ps2*) and their alveoli absent (Fig. 16) ....Thermacaridae *Thermacarus*

6b. Claw (movable digit) of chelicerae less than one-third length of basal segment (Fig. 25); excretory pore setae (*ps1* and *ps2*) or at least their alveoli, present (Figs. 24, 28, 34) ....Hydryphantidae (in part) 7

7a(6b). Coxal plates II lacking setae (*2b* absent) (Fig. 20); urstigmata relatively large (Fig. 20, arrow); pedipalp tibia with terminal clawlike seta deeply bifurcate (Fig. 18, arrow) and tarsus with only one thickened seta; excretory pore plate absent but setae *ps1* and *ps2* present (Fig. 20)....................................... Wandesiinae *Wandesia*

7b. Coxal plates II bearing setae *2b* (Fig. 28); urstigmata relatively small (Fig. 28, arrow); pedipalp tibia with terminal clawlike seta only slightly bifurcate or undivided, and tarsus with two thickened setae (Fig. 19, arrows); excretory pore plate present; setae *ps1* and *ps2* present or absent, but their alveoli always present (Fig. 28) 8

8a(7b). Medial eye present (Fig. 27, arrow) .... Thyadinae *Panisus, Panisopsis, Thyas, Thyasides, Euthyas, Zschokkea*

8b. Medial eye reduced (Fig. 33) .... Tartarothyadinae *Tartarothyas*

9a(5b). Dorsal plate triangular, bearing scapular setae (*se* and *si*), with verticil setae (*ve* and *vi*) borne on small platelets flanking apex of plate (Fig. 29); excretory pore setae present, with *ps1* borne on plate (Fig. 23) and *ps2* borne on small separate platelets; basal segment of chelicerae massive and longitudinally striate (Fig. 25) .... Hydryphantidae (in part) Hydryphantinae *Hydryphantes*

9b. Dorsal plate fragmented or absent, with verticil and scapular setae borne on paired platelets (Figs. 30, 36); excretory pore setae reduced to vestiges or absent, but their alveoli present (Figs. 24, 34); basal segment of chelicerae relatively slender and smooth (Fig. 27) 10

10a(9b). Medial eye present (Fig. 30, arrow); excretory pore plate well sclerotized and nearly triangular (Fig. 24); pedipalp tarsus with all setae slender; solenidia on leg tarsi slender (Fig. 31, arrow) .... Hydryphantidae (in part) Protziinae *Protzia*

10b. Medial eye absent (Fig. 36); excretory pore plate poorly sclerotized and oblong (Fig. 34); pedipalp tarsus with two thickened, bladelike setae (Fig. 35, arrows); solenidia on tarsi of legs I and II very thick (Fig. 32, arrow) .... Hydrodromidae *Hydrodroma*

11a(4b). Coxal plates I, II, and III on each side bearing two, two, and two setae, respectively (Figs. 38, 43); tarsi of legs bearing paired claws and clawlike empodium (emp)(Fig. 41) 12

11b. Coxal plates I, II, and III on each side bearing two or one, one, and one setae, respectively (Figs. 44, 53); tarsi of legs bearing two dissimilar clawlike structures terminally (Fig. 45, 50) .... Limnocharidae 13

12a(11a). Dorsum of idiosoma nearly covered by single, elongate dorsal plate in unengorged larvae, bearing seven pairs of setae (or their alveoli) (Fig. 37); coxal plates with setae all simple (Fig. 38); excretory pore plate elongate (Fig. 39) .... Eylaidae *Eylais*

12b. Dorsum of idiosoma with three separate plates bearing five, one, and one pairs of setae respectively (Fig. 42); coxal plates with setae *1a, 1b, 2b* and *3b* blunt and conical in shape (Fig. 43); excretory pore plate obcordate (Fig. 43) .... Piersigiidae *Piersigia*

13a(11b). Dorsal plate covering no more than anterior half of idiosomal dorsum, bearing four pairs of setae (verticils and scapulars) (Figs. 46, 51); coxal plates I bearing two pairs of setae, *1a* and *1b* (Fig. 44); excretory pore plate diamond-shaped (Fig. 47) or nearly oval (Fig. 49) .... Limnocharinae 14

13b. Dorsal plate covering entire idiosomal dorsum in unengorged larvae, bearing seven pairs of setae, including verticils, scapulars and three pairs of hysterosomal setae (Fig. 52); coxal plates I bearing one pair of setae (Fig. 53); excretory pore plate pyriform (Fig. 53) .... Rhyncholimnocharinae *Rhyncholimnochares*

14a(13a). Dorsal plate (Fig. 46) finely punctate; excretory pore plate relatively small, little larger than excretory pore (Fig. 47); leg tarsi bearing two similar clawlike structures with empodium and single ambulacral claw both slender (Fig. 45) .... *Limnochares*

14b. Dorsal plate (Fig. 51) strongly striate; excretory pore plate relatively large, much larger than excretory pore (Fig. 49); leg tarsi bearing two dissimilar clawlike structures, with empodium wider than single ambulacral claw (Fig. 50) .... *Neolimnochares*

15a(1b). Dorsal plate bearing three pairs of setae (external verticils, scapulars) and an unpaired anteromedial seta (probably representing internal verticils) (Fig. 59, arrow); urstigmata (Ur) stalked (Figs. 59, 61); pedipalp tibiae with terminal clawlike seta four-pronged (Fig. 63, arrow), leg tarsi bearing paired pectinate claws and simple empodium (Fig. 60) .... Stygothrombidioidea Stygothrombidiidae *Stygothrombium*

15b. Dorsal plate bearing at least four pairs of setae (verticils and scapulars), and in some cases one or more additional pairs of hysterosomal setae, but no unpaired setae (Figs. 56, 64, 68); urstigmata sessile (Figs. 58, 65); pedipalp tibiae with terminal clawlike seta undivided (Fig. 66, arrow), or deeply bisected (Figs. 55, 71, arrows); leg tarsi bearing paired simple claws (Fig. 175) or lacking claws (Fig. 54), but always bearing a simple empodium 16

16a(15b). Dorsal plate bearing eight pairs of setae, including verticils, scapulars, *c3*, and three additional pairs of hysterosomal setae (Fig. 56); leg tarsi lacking paired claws (Fig. 54); excretory pore plate tiny, and bearing neither setae nor their alveoli (Fig. 58, arrow) .... Hydrachnoidea Hydrachnidae *Hydrachna*

16b. Dorsal plate bearing four-to-six pairs of setae, including only the verticils and scapulars (Fig. 64) or, in some cases, the verticils, scapulars and one or two pairs of hysterosomal setae (Fig. 68); leg tarsi bearing paired claws (Fig. 143, 175); excretory pore plate small to large, always bearing setae *ps1* and *ps2* or at least their alveoli (Figs. 62, 80, 104, 109, 113–114, 172, 176, 185, 193) 17

17a(16b). Coxal plates I, II, and III on each side all separate (Figs. 65, 69, 73, 172, 179, 180, 186), *or* all fused (Fig. 185) *and* plates of two sides fused medially *and* dorsal plate round and bearing setae *c1* laterally (Fig. 184) 18

17b. Coxal plates I and II on each side separate and plates II and III fused at least medially (Figs. 83, 106, 145, 157), *or* all plates on each side fused (Figs. 92, 100) *but* plates of two sides separate medially *and* dorsal plate elliptical and not bearing setae *c1* (Figs. 91, 101, 144, 150, 153, 168) 34

18a(17a). Tarsi of legs III bearing 12 or more setae, including Ta8 (Fig. 8, arrow); dorsal plate usually elongate and elliptical and bearing only four pairs of setae (verticils and scapulars) (Figs. 3, 64, 74); *when* dorsal plate round and bearing five pairs of setae, including *c1* laterally (Fig. 68), *then* coxal plates III bearing setae *pa* and *h2* in addition to *3a* (Fig. 69) *and* all setae on pedipalp tarsi short and bladelike (Fig. 71) ....Lebertioidea (in part) 19

18b. Tarsi of legs III usually bearing 11 or fewer setae, lacking at least Ta8 (Figs. 174, 181, 196, 198); dorsal plate usually nearly round (Figs. 178, 184, 188, 190, 191), rarely elongate and elliptical (Fig. 171); *when* tarsi of legs III bearing 12 setae *then* dorsal plate round and bearing five pairs of setae, including *c1* laterally (Fig. 195) *and* coxal plates III bearing only setae *3a* (Fig. 193) *and* at least one seta on pedipalp tarsi long and whip-like (similar to Fig. 177) ....Arrenuroidea 23

19a(18a). Pedipalp tarsi with no setae as long as pedipalp (Fig. 71); tarsi of legs I, II, and III bearing 13 or 14 (+1s), 13 or 14 (+1s), and 12 setae respectively, lacking Ta15 ....Anisitsiellidae (in part) 20

19b. Pedipalp tarsi with at least one seta as long as pedipalp (Figs. 2, 66); tarsi of legs I, II, and III bearing 15 (+1s), 15 (+1s), and 13 or 14 setae respectively, including Ta15 (Figs. 6–8, 67)..21

20a(19a). Dorsal plate bearing six pairs of setae, including verticils, scapulars and both *c1* and *e1* on lateral edge, and with setae *si* long, thick, and located near midlength (Fig. 68); coxal plates III bearing three pairs of setae, including *3a* and both *pa* and *h2* near posterior edges (Fig. 69) ....*Bandakiopsis*

20b. Dorsal plate bearing four pairs of setae (verticils and scapulars), with setae *si* short, slender, and located in anterior third of plate (Fig. 70); coxal plates III bearing only setae *3a* (Fig. 72)..*Utaxatax*

21a(19b). Dorsal plate longitudinally striate (Figs. 3, 64); coxal plates III bearing only setae *3a* (Figs. 4, 65); excretory pore plate small and usually quadrangular (Figs. 5, 65) ....Sperchontidae 22

21b. Dorsal plate reticulate (Fig. 74); coxal plates III bearing two pairs of setae, *3a* and supernumerary setae *3b* (Fig. 73); excretory pore plate relatively large and diamond-shaped (Fig. 73) ....Teutoniidae *Teutonia*

22a(21a). Coxal plates II bearing one pair of setae, *2b,* posterolaterally (Fig. 65) ....*Sperchon*

22b. Coxal plates II lacking setae (Fig. 4)..*Sperchonopsis*

23a(18b). Coxal plates I, II, and III on each side all fused and coxal plates of two sides fused medially, with setae *1a, 2b* and *3a* reduced to vestiges (Fig. 185); pedipalp tarsi with no setae as long as pedipalp (Fig. 182); dorsal plate with setae *si* reduced to vestiges (Fig. 184) ....Acalyptonotidae *Acalyptonotus*

23b. Coxal plates I, II, and III on each side all separate and coxal plates of two sides separate (Figs. 172, 186, 189, 192, 193); pedipalp tarsi with at least one seta as long as pedipalp (Fig. 177) 24

24a(23b). Idiosoma moderately flattened dorsoventrally; gnathosoma projecting beyond anterior edge of dorsal plate, entirely exposed in dorsal view (Figs. 171, 195); excretory pore plate variously shaped (Figs. 172, 193, 197) but usually not attenuate anteriorly; pedipalp tarsi with long, thick, distal seta straight or only slightly bowed basally; leg genua with setae Ge5 borne on tubercles that usually are prominent (Fig. 173, arrow)..25

24b. Idiosoma extremely flattened dorsoventrally; gnathosoma recessed beneath anterior edge of dorsal plate, partially or entirely concealed in dorsal view (in unmounted specimens) (Fig. 188); excretory pore plate triangular with anterior apex attenuate (Figs. 176, 192); pedipalp tarsi with long, thick, distal seta strongly bowed, and usually lobed, basally (Figs. 177, 183, arrows); leg genua with setae Ge5 not borne on tubercles..29

25a(24a). Dorsal plate bearing four pairs of setae (verticils and scapulars), with setae *c1* on lateral membranous integument (Fig. 171); setae *c2* and *d2* thick and long relative to other hysterosomal setae (Fig. 171); excretory pore setae *ps1* and *ps2* absent, represented by their alveoli on excretory pore plate (Fig. 172); tibiae of legs I bearing eight setae (+2s), including Ti11; leg tarsi lacking setae Ta14 (Figs. 173, 175)..Momoniidae 26

25b. Dorsal plate bearing five pairs of setae, including the verticils, scapulars and setae *c1* laterally (Fig. 195); setae *c2* and *d2* similar to other hysterosomal setae (Fig. 195); excretory pore setae *ps1* and *ps2* present on plate (Fig. 193, 197), tibiae of legs I bearing seven setae (+2s), lacking Ti11; leg tarsi bearing setae Ta14 (Figs. 196, 198, arrows) 27

26a(25a). Legs with setae Tr1, Ge5, Ti9 and Ti11 much longer than respective segments; setae IITi9, IITi11, IIIGe5, IIITi9 and IIITi11 plumose (Fig. 173) ....Momoniinae *Momonia*

26b. Legs with setae Tr1, Ge5, Ti9 and Ti11 shorter than respective segments; setae IITi9, IITi11, IIIGe5, IIITi9 and IIITi11 bladelike (Fig. 175) ....Stygomomoniinae *Stygomomonia*

27a(25b). Tarsi of legs III bearing nine setae, lacking Ta12 (Fig. 198); tarsi of legs II and III with setae Ta4 and Ta6 long (usually as long as respective segments) (Figs. 198); coxal plates I and II with conspicuous denticulate projections on posterior edges (Figs. 197, 199, arrows), coxal plates III with transverse muscle attachment scars (TMAS), excretory pore plate subtriangular (Figs. 197, 199) ....Krendowskiidae 28

27b. Tarsi of legs III bearing at least 11 setae, including Ta12 (Fig. 196); tarsi of legs II and III with setae Ta4 and Ta6 usually short (conspicuously shorter than respective segments) (Fig. 196); coxal plates I and II lacking denticulate projections on posterior edges, coxal plates III lacking transverse muscle attachment scars (TMAS), excretory pore plate variously shaped, but rarely triangular (Fig. 193) .... Arrenuridae *Arrenurus*

28a(27a). Coxal plates I and II each with row of denticulate projections along whole posterior edge (Fig. 197, arrows); TMAS weakly developed and short (Fig. 197); tubercles bearing setae *f2* large, prominent, and medially attenuate (Fig. 197) .... *Krendowskia*

28b. Coxal plates I and II each bearing only few, denticulate projections in medial third of posterior edge (Fig. 199); TMAS strongly developed and long (Fig. 199); tubercles bearing setae *f2* not unusually prominent (Fig. 199) .... *Geayia*

29a(24b). Dorsal plate bearing five pairs of setae, including verticils, scapulars and setae *c1* laterally (Fig. 188) .... Mideopsidae *Mideopsis*

29b. Dorsal plate bearing four pairs of setae (verticils and scapulars), with setae *c1* on lateral membranous integument (Figs. 178, 190) 30

30a(29b). Tibiae of legs II and III bearing nine setae (+2s and +1s, respectively), including both Ti10 and Ti11 (Figs. 174, 181) 31

30b. Tibiae of legs II and III bearing fewer than nine setae (+2s and +1s, respectively), lacking either Ti10 or Ti11, or both (Fig. 187, 194) 32

31a(30a). Pedipalp tarsi with long, thick, distal seta bowed, but not deeply lobed or fringed basally, and with most medial seta moderately thick, but not fringed (Fig. 183); legs II and III with setae Ge5, Ti9 and Ti11 much longer than respective segments and plumose, and setae Ta4 and Ta6 much longer than respective segments (Fig. 181) .... Mideidae *Midea*

31b. Pedipalp tarsi with long, thick, distal seta bowed, deeply lobed and fringed basally, and most medial seta thick and fringed (Fig. 177); legs II and III with setae Ge5, Ti9 and Ti11 shorter than respective segments and simple, and setae Ta4 and Ta6 shorter than respective segments (Fig. 174) .... Nudomideopsidae *Nudomideopsis, Paramideopsis*

32a(30b). Tibiae of legs II bearing seven setae (+2s), lacking Ti10 and Ti11 (Fig. 194); tibiae of legs III bearing eight setae (+1s), lacking Ti10 .... Laversiidae *Laversia*

32b. Tibiae of legs II and III bearing eight setae (+2s and +1s respectively), including Ti10, but lacking Ti11 (Fig. 187) 33

33a(32b). Dorsum with setae *si* located well in anterior half of plate and setae *c1* located in lateral integument near midlength of plate, posterior to level of setae *si* (Fig. 191) .... Neoacaridae *Neoacarus, Volsellacarus*

33b. Dorsum with setae *si* near midlength of plate and setae *c1* located in lateral integument near setae *c3*, anterior to level of setae *si* (Fig. 190) .... Athienemanniidae Athienemanniinae *Chelomideopsis, Platyhydracarus*

34a(17b). Coxal plates III bearing setae *3a* and at least one other pair of setae (*pa* or *3b*) and excretory pore plate bearing only setae *ps1* and *ps2* (Figs. 83, 86, 90, 92) .... Lebertioidea (in part) 35

34b. Coxal plates III usually bearing only one pair of setae, *3a* (Figs. 80, 97, 100, 145, 152), *when* coxal plates III also bearing setae *pa*, *then* excretory pore plate bearing setae *h2* in addition to *ps1* and *ps2* (Figs. 104, 159–163, 166) 38

35a(34a). Coxal plates III truncate posteriorly, bearing three pairs of setae, including *3a* and both *pa* and *h2* on posterior edges (Fig. 83); tibiae of legs I, II, and III bearing nine setae (+2s, +2s, and +1s, respectively), including both Ti9 and Ti10 (Fig. 82); tarsi of legs I and II bearing 14 setae (+1s), including Ta14 and Ta15 (Fig. 82); cheliceral bases separate (Fig. 81, arrow) .... Lebertiidae *Lebertia, Estelloxus*

35b. Coxal plates III rounded or pointed posteriorly, bearing two pairs of setae, including setae *3a* and either *3b* laterally or *pa* posteromedially (Figs. 86, 90, 92); tibiae of legs I, II, and III bearing eight setae (+2s, +2s, and +1s, respectively), lacking Ti9 or Ti10 (Figs. 85, 88, 89); tarsi of legs I and II bearing 13 setae (+1s), lacking Ta15 (Fig. 88), or 12 setae (+1s), lacking both Ta15 and Ta14 (Fig. 85); cheliceral bases fused (Figs. 84, 87, 91) 36

36a(35b). Dorsal plate bearing five pairs of setae including verticils, scapulars and setae *c1* laterally; integument beneath posterior edge of dorsal plate intricately folded (Fig. 84, arrow); coxal plates I elongate, extending posteriorly nearly to level of excretory pore plate; coxal plates III bearing setae *3b* laterally (Fig. 86); tarsi of legs I and II bearing 12 setae (+1s), lacking Ta14 (Fig. 85) .... Oxidae *Oxus, Frontipoda*

36b. Dorsal plate bearing four pairs of setae (verticils and scapulars), or five pairs including verticils, scapulars and setae *d1* laterally; integument beneath posterior edge of dorsal plate not intricately folded (Figs. 87, 91); coxal plates I not elongate, extending posteriorly only to level of insertion of legs III (Fig. 90), or fused with plates II (Figs. 92); plates III bearing setae *pa* posteromedially (Figs. 90, 92); tarsi of legs I and II bearing 13 setae (+1s), including Ta14 (Fig. 88) .... Torrenticolidae 37

37a(36b). Dorsal plate bearing five pairs of setae including verticils, scapulars and setae *d1* laterally, and with setae *se* long and thick (Fig. 87); coxal plates I clearly delineated by complete suture lines, and plates II and III fused with suture line incomplete medially (Fig. 90); excretory pore plate with convex projection posteromedially (Fig. 90); tibiae of all legs lacking setae Ti10, genua of legs III bearing four setae (+1s) including Ge3, and tarsi of legs III bearing 11 setae, lacking Ta10 (Fig. 89) .... Testudacarinae *Testudacarus*

37b. Dorsal plate bearing only four pairs of setae (verticils and scapulars), and with setae *se* relatively short and slender (Fig. 91); coxal plates of each side all fused, with suture lines obliterated except laterally (Fig. 92); excretory pore plate without projection posteromedially (Fig. 92); tibiae of all legs lacking setae Ti9, genua of legs III bearing three setae (+1s), lacking Ge3, tarsi of legs III bearing 12 setae, including Ta10 (Fig. 93) .... Torrenticolinae *Torrenticola*

38a(34b). Cheliceral bases separate (Fig. 78, arrow); tarsi of legs III bearing 12 setae, including Ta9 (Fig. 79, arrow); excretory pore plate very large (equal in width to coxal plates of one side), bearing setae *ps1* and *ps2* anteromedially and occasionally also setae *h2* at posterolateral angles (Figs. 75, 80), *and* coxal plates I separate from posterior coxal groups, *and* coxal plates III bearing only setae *3a* (Fig. 80)................Lebertioidea (in part) Anisitsiellidae (in part) Anisitsiellinae (in part) *Bandakia*

38b. Cheliceral bases fused (Figs. 101, 107, 108, 144, 150); tarsi of legs III usually bearing 11 or fewer setae, lacking Ta9 (Fig. 132) (exception Limnesiidae); excretory pore plate small to very large, *when* equal in width to coxal plates of one side (measured from midline to insertion of leg III) (Figs. 100, 103, 104, 152, 164) *then either* excretory pore located near posterior edge of plate (Fig. 152, arrow), *or* coxal plates I fused to posterior coxal groups (Fig. 100), *or* coxal plates III bearing setae *pa* in addition to setae *3a* (Figs. 164)................Hygrobatoidea 39

39a(38b). Two pairs of urstigmata borne distally between coxal plates I and II (Figs. 97, arrows A; Fig. 98, arrows); tibiae of legs I and II bearing seven or eight setae (+2s), including only three ventral setae, lacking either Ti8 and Ti9 or Ti10 and Ti11 (Figs. 95, 99)................Limnesiidae 40

39b. One pair of urstigmata borne distally between coxal plates I and II (Figs. 100, arrow, 106, 111, 133, 164); tibiae of legs I and II usually bearing nine or more setae (+2s), including five ventral setae (Ti7, Ti8, Ti9, Ti10, and Ti11) (Figs. 102, 141, 149) (exception Feltriidae)................41

40a(39a). Coxal plates I and II completely separate (Fig. 98); tarsi of legs III bearing nine setae, lacking Ta9, Ta13 and Ta14 (Fig. 99); dorsal plate extending laterally well beyond level of eye plates (contrast with Fig. 94)................Tyrrelliinae *Tyrrellia*

40b. Coxal plates I and II on each side fused, with suture lines obliterated medially (Fig. 97, arrow B); tarsi of legs III bearing 12 setae, including Ta9, Ta13 and Ta14 (Fig. 96); dorsal plate narrow, usually confined to region between eye plates (Fig. 94)................Limnesiinae *Limnesia*

41a(39b). Coxal plates I, II, and III on each side all fused (Fig. 100)................Hygrobatidae 42

41b. Coxal plates I separate from posterior coxal group on each side (Figs. 106, 111, 145, 152, 169)................43

42a(41a). Excretory pore plate bearing three pairs of setae, including *ps1* and *ps2* medially and *h2* near anterolateral angles, with setae *pa* on posterior edges of coxal plates III (Fig. 104)................*Hygrobates*

42b. Excretory pore plate bearing four pairs of setae, including *ps1* and *ps2* medially, *pa* near anterior edge, and *h2* near lateral angles (Figs. 100, 103)................*Atractides*

43a(41b). Tarsi of legs I and II bearing 10 setae (+1s), lacking Ta14, and either Ta12 (Fig. 105) or Ta9 (Figs. 112, 116)................44

43b. Tarsi of legs I and II bearing 11 or more setae (+1s), including Ta9 (Figs. 128, 170, arrows; Figs. 147, 151, arrows A)................47

44a(43a). Dorsal and coxal plates, and leg sclerites, conspicuously longitudinally striate (as in Fig. 107); coxal plates III rounded posteriorly (Fig. 106); tibiae of legs I bearing seven setae (+2s), lacking Ti10 and Ti11 (Fig. 105)................Feltriidae *Feltria*

44b. Dorsal and coxal plates reticulate (Fig. 108); coxal plates III bearing pointed projections posteriorly (Fig. 111, arrow); tibiae of legs I bearing 9 setae (+2s), including Ti10 and Ti11 (Fig. 112, 116)................Unionicolidae 45

45a(44b). Tarsi of legs with empodium undivided (Fig. 115)................Pionatacinae 46

45b. Tarsi of legs with empodium bifurcate (Fig. 110, arrow)................Unionicolinae *Unionicola*

46a(45a). Excretory pore plate usually oval in shape with excretory pore borne near anterior edge (Fig. 114)................*Neumania*

46b. Excretory pore plate obcordate with excretory pore borne near middle of plate (Fig. 113)................*Koenikea*

47a(43b). Tibiae of all legs bearing seven setae (+2s, +2s, and +1s respectively), lacking Ti9 and Ti11 (Fig. 170)................Aturidae (in part) ....Aturinae *Aturus*

47b. Tibiae of all legs bearing eight or more setae (+2s, +2s, and +1s, respectively), including Ti9 and Ti11 (Fig. 151)................48

48a(47b). Coxal plates III with pointed or lobed projections posteriorly, bearing two pairs of setae, *3a* anteromedially and *pa* posteromedially (Figs. 156, 164); excretory pore plate large, bearing three pairs of setae, *ps1* and *ps2,* and *h2* posterolaterally (Figs. 159–163, 166)................Aturidae (in part) Axonopsinae (in part) 49

48b. Coxal plates III without, or with small, projections posteriorly, bearing only setae *3a* (Figs. 117, 145, 152, 154); excretory pore plate small or large, bearing only setae *ps1* and *ps2* (Figs. 118, 123, 135, 138, 142, 146)................52

49a(48a). Tarsi of legs I and II bearing 12 setae, including setae Ta8 (Fig. 158); excretory pore plate with setae *ps1* at or near anterior edge (Figs. 159, 160)................50

49b. Tarsi of legs I and II bearing 11 setae, lacking setae Ta8 (Fig. 167); excretory pore plate with setae *ps1* near anterior edge to posterior to midlength (Figs. 162, 163, 166)................51

50a(49a). Excretory pore plate subtriangular (Fig. 159) with acutely rounded posterolateral angles, with setae *ps2* anterior to midlength between setae *ps1* and *h2*; dorsal plate not reticulate; coxal plates I weakly fused medially with plates II, and with setae *1a* long, extending posteriorly beyond level of setae *3a* (Fig. 157)................*Estellacarus*

50b. Excretory pore plate obcordate (Fig. 160) or nearly quadrangular (Fig. 161), with broadly rounded posterolateral angles, and with setae *ps2* posterior to midlength, not between setae *ps1* and *h2*; dorsal plate reticulate; coxal plates I separate from plates II, and with setae *1a* relatively short, not extending posteriorly beyond level of setae *3a* (Fig. 156) .... *Woolastookia*

51a(49b). Suture line between coxal plates II and III long, extending medially to near base of setae *3a* (Fig. 164); excretory pore plate either moderately large and subtriangular (Fig. 162) or very large and subquadrangular (Fig. 163) .... *Brachypoda*

51b. Suture line between coxal plates II and III short, not extending medially to region of base of setae *3a* (Fig. 165); excretory pore plate moderately large and broadly elliptical or subquadrangular (Fig. 166) .... *Axonopsis*

52a(48b). Coxal plates III with small projections posteriorly (Fig. 117, arrow); excretory pore plate small, little larger than combined area of excretory pore and bases of setae *ps1* and *ps2* (Fig. 118).. Wettinidae *Wettina*

52b. Coxal plates III without projections posteriorly (Figs. 124, 133, 137, 140, 145, 152, 154); excretory pore plate considerably larger than combined area of excretory pore and bases of setae *ps1* and *ps2* (Figs. 123, 125, 129, 138, 142, 152) 53

53a(52b). Tarsi of legs I bearing 13 setae (+1s), including Ta8 (Figs. 147, 151, arrows B); excretory pore plate large, and usually triangular or obcordate (Figs. 138, 139, 148, 152) 54

53b. Tarsi of legs I bearing 12 setae (+1s), lacking Ta8 (Fig. 128); excretory pore plate small to large, and variously shaped (Figs. 119, 121, 123, 125, 134, 135) 58

54a(53a). Excretory pore plate with setae *ps1* close together adjacent to excretory pore in posterior half of plate (Figs. 138, 139, 142, 146, 148) .... Pionidae (in part) 55

54b. Excretory pore plate with setae *ps1* widely separated and removed from excretory pore in anterior half of plate (Fig. 154, arrow) .... Aturidae (in part) 57

55a(54a). Tarsi of legs II bearing 13 setae (+1s), including Ta8 (Fig. 141); tibiae of legs with setae Ti9 and Ti11 simple (Fig. 141) .... Najadicolinae *Najadicola*

55b. Tarsi of legs II usually bearing 12 setae (+1s), lacking Ta8 (Fig. 149), *when* bearing 13 setae *then* tibiae of legs with Ti9 and Ti11 long and plumose (as in Fig. 149) .... Pioninae 56

56a(55b). Femora of legs II with Fe5 shorter than segment; excretory pore plate usually much less than twice as wide as long, *when* nearly twice as wide as long *then* plate is concave or transverse anteromedially (Fig. 146) .... *Piona*

56b. Femora of legs II with Fe5 longer than segment; excretory pore plate twice as wide as long and convex anteromedially (Fig. 139) .... *Nautarachna*

57a(54b). Excretory pore plate broadly obcordate (Fig. 152).. Axonopsinae (in part) *Ljania*

57b. Excretory pore plate triangular (Fig. 154).. Albiinae *Albia*

58a(53b). Excretory pore plate small and subtriangular, with setae *ps2* borne near posterolateral angles of plate (Fig. 155); suture lines between coxal plates II and III parallel to anterior edge of plate II, coxal plates without distinct lateral coxal apodemes and coxal plates III lacking medial coxal apodemes and transverse muscle attachment scars (Fig. 155) .... Aturidae (in part) .... Axonopsinae (in part) *Neobrachypoda*

58b. Excretory pore plate variously shaped, *when* subtriangular with setae *ps2* borne near posterolateral angles of plate *then* suture lines between coxal plates II and III terminating in distinct lateral coxal apodemes (LCA) that are usually nearly transverse and coxal plates III bearing at least medial coxal apodemes (MCA) (Figs. 120, 124, 126, 129, 133, 136).. Pionidae (in part) 59

59a(58b). Suture line between coxal plates II and III on each side extending medially beyond lateral coxal apodeme nearly to midline (Fig. 120, arrow); excretory pore borne on elevated projection which extends to or beyond posterior edge of plate (Figs. 119, 121) .... Hydrochoreutinae *Hydrochoreutes*

59b. Suture line between coxal plates II and III on each side extending medially only as far as lateral coxal apodeme (Figs. 122, 124, 126, 129, 133, 136); excretory pore usually not borne on elevated projection, when on small projection this does not extend to posterior edge of plate 60

60a(59b). Trochanters of legs III bearing two setae, including supernumerary dorsal seta (Fig. 131, arrow), *or* tibiae of legs II and III bearing 11 or 12 setae (+2s and +1s, respectively), including two supernumerary posteroventral setae (Fig. 132, arrows), *or* coxal plates III lacking transverse muscle attachment scars *and* excretory pore plate with setae *ps2* near posterior edge of plate (Fig. 136) .... Foreliinae 61

60b. Trochanters of legs III bearing only one seta; tibiae of legs II and III bearing only nine setae (+2s and +1s, respectively) (as in Figs. 141, 149); coxal plates III usually with transverse muscle attachment scars (Figs. 124, 126, 133), *when* coxal plates lacking these scars (as in Fig. 122) *then* excretory pore plate with setae *ps2* removed from posterior edge of plate (Fig. 123).. 62

61a(60a). Trochanters of legs III bearing two setae, including supernumerary dorsal seta (Fig. 131, arrow) .... *Pseudofeltria*

61b. Trochanters of legs III bearing one seta.. *Forelia*

62a(60b). Excretory pore plate transversely elliptical or semicircular (Fig. 123) ....Huitfeldtiinae *Huitfeldtia*

62b. Excretory pore plate broadly elliptical, subcircular, or subtriangular ....Tiphyinae 63

63a(62b). Excretory pore plate three times as wide as long and broadly elliptical or dumbbell shaped (Figs. 124, 125)......................*Neotiphys*

63b. Excretory pore plate less than twice as wide as long and subcircular (Figs. 127, 130) to subtriangular....................*Pionopsis, Tiphys*

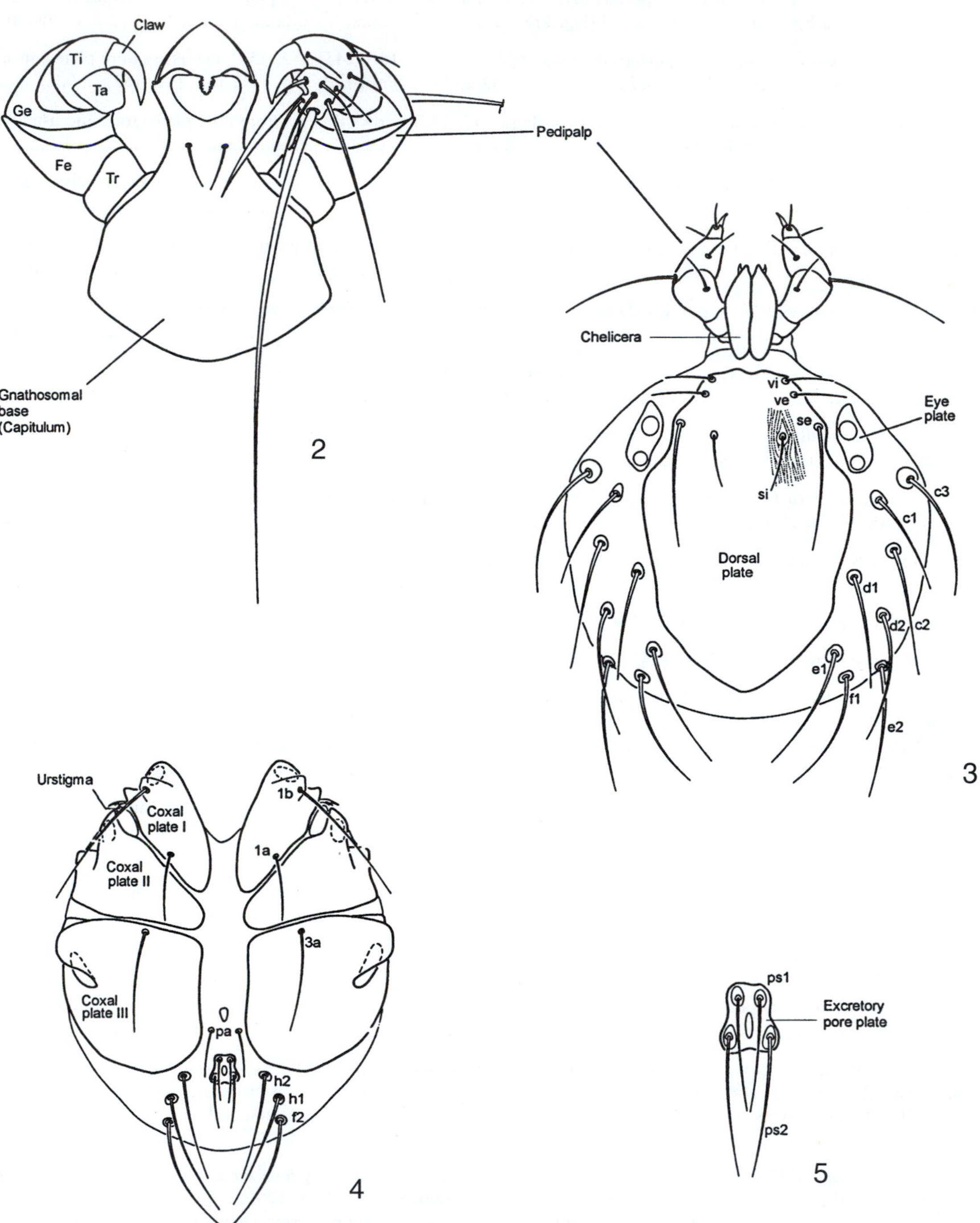

FIGURES 2–5 *Sperchonopsis ecphyma* Prasad and Cook (Sperchontidae), larva. Fig. 2, Venter of gnathosoma; Fig. 3, dorsum of idiosoma and gnathosoma; Fig. 4, venter of idiosoma; Fig. 5, excretory pore plate. See text for nomenclature and abbreviations of setae.

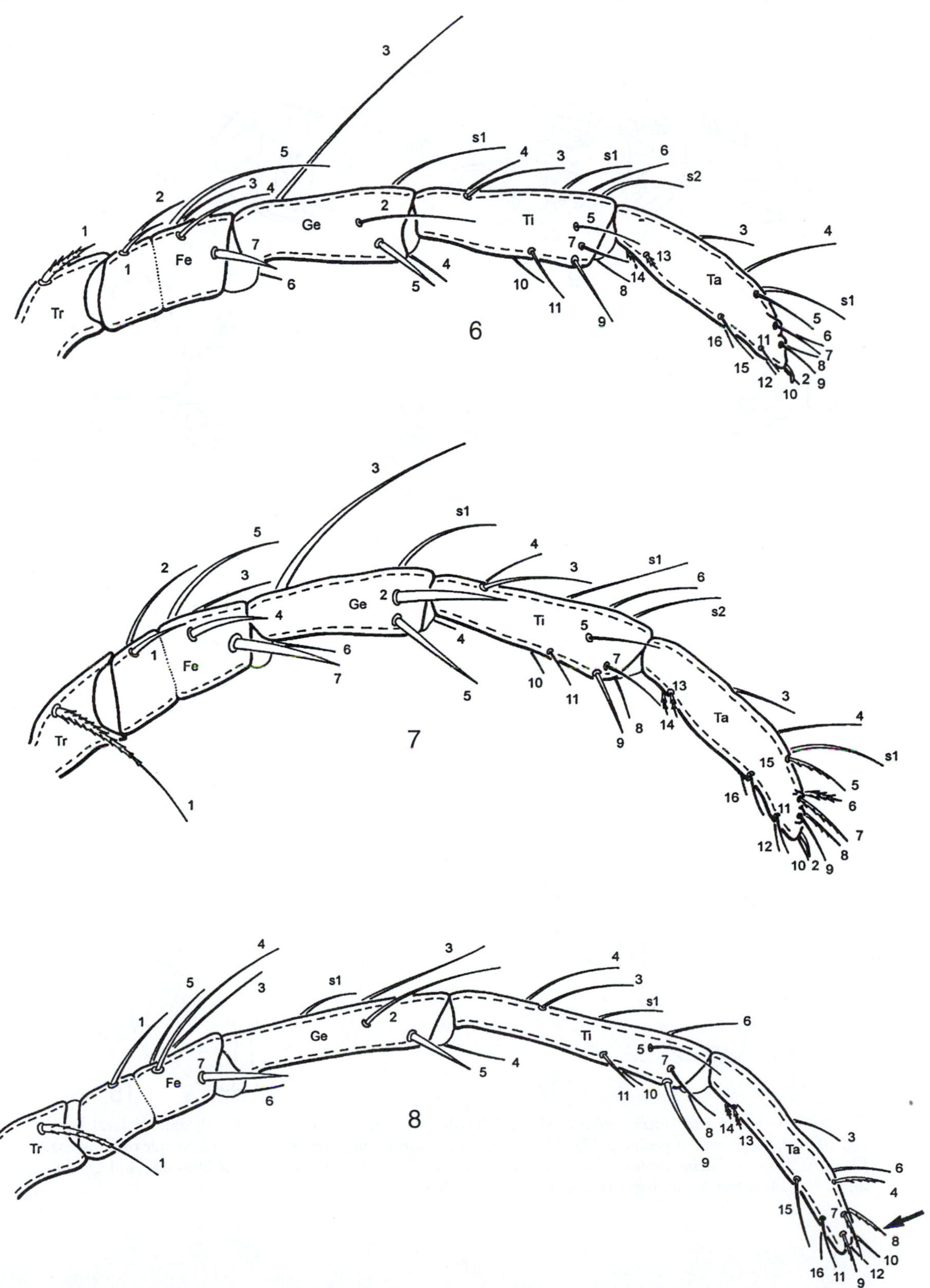

FIGURES 6–8 *Sperchonopsis ecphyma* Prasad and Cook (Sperchontidae), larva, anterolateral views of legs. Fig. 6, Leg I; Fig 7, Leg. II; and Fig. 8, Leg III. Fe, Femur; Ge, genu; Ta, tarsus; Ti, tibia; Tr, trochanter.

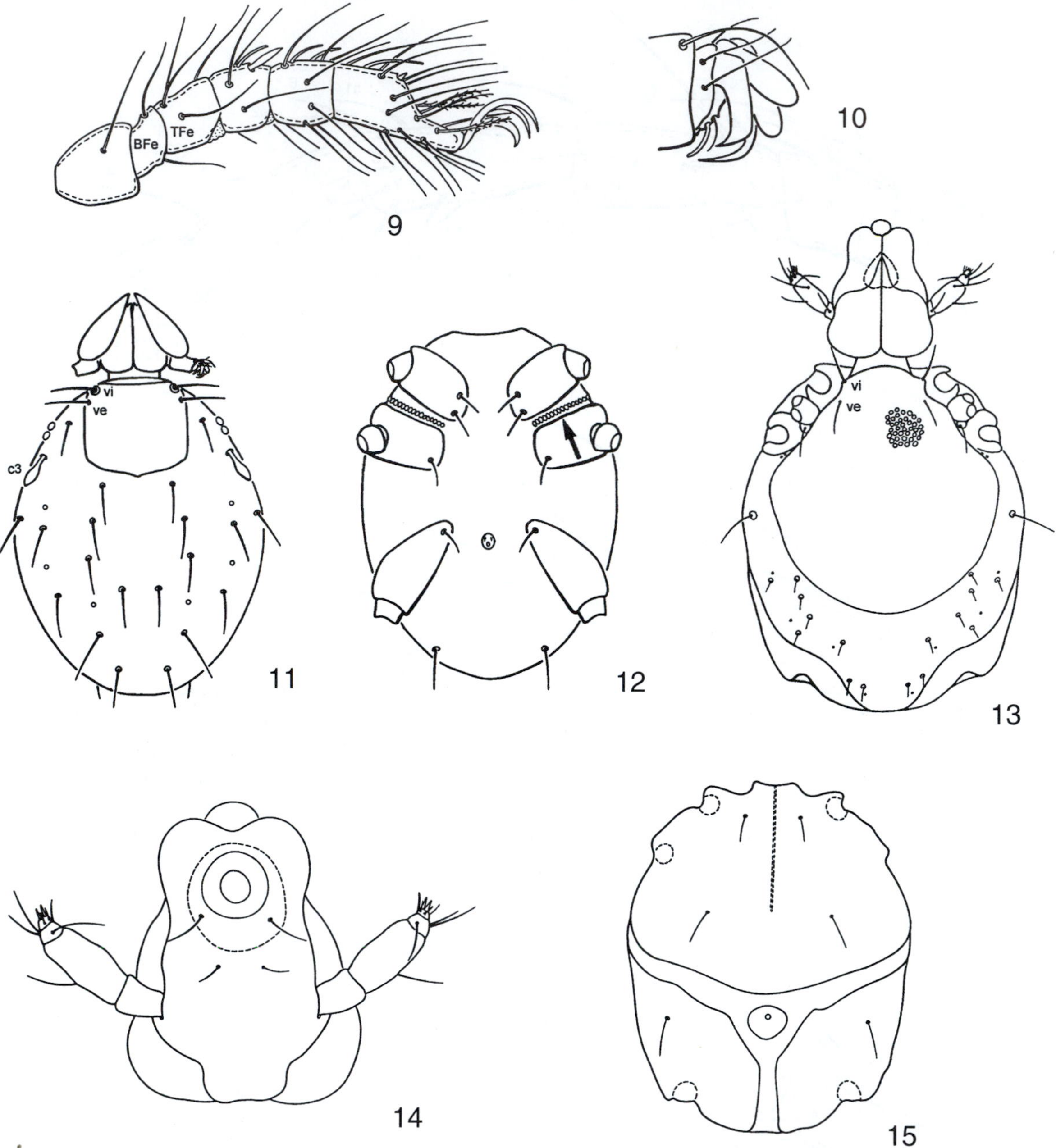

FIGURES 9–12 *Hydrovolzia gerhardi* Mitchell (Hydrovolziidae), larva. Fig. 9, Leg I, anterolateral view; Fig. 10, distal segments of pedipalp; Fig. 11, dorsum of idiosoma and gnathosoma; Fig 12, venter of idiosoma. *FIGURES 13–15* *Acherontacarus* sp. (Acherontacaridae), larva. Fig. 13, Dorsum of idiosoma and gnathosoma; Fig. 14, venter of gnathosoma; Fig. 15, venter of idiosoma.

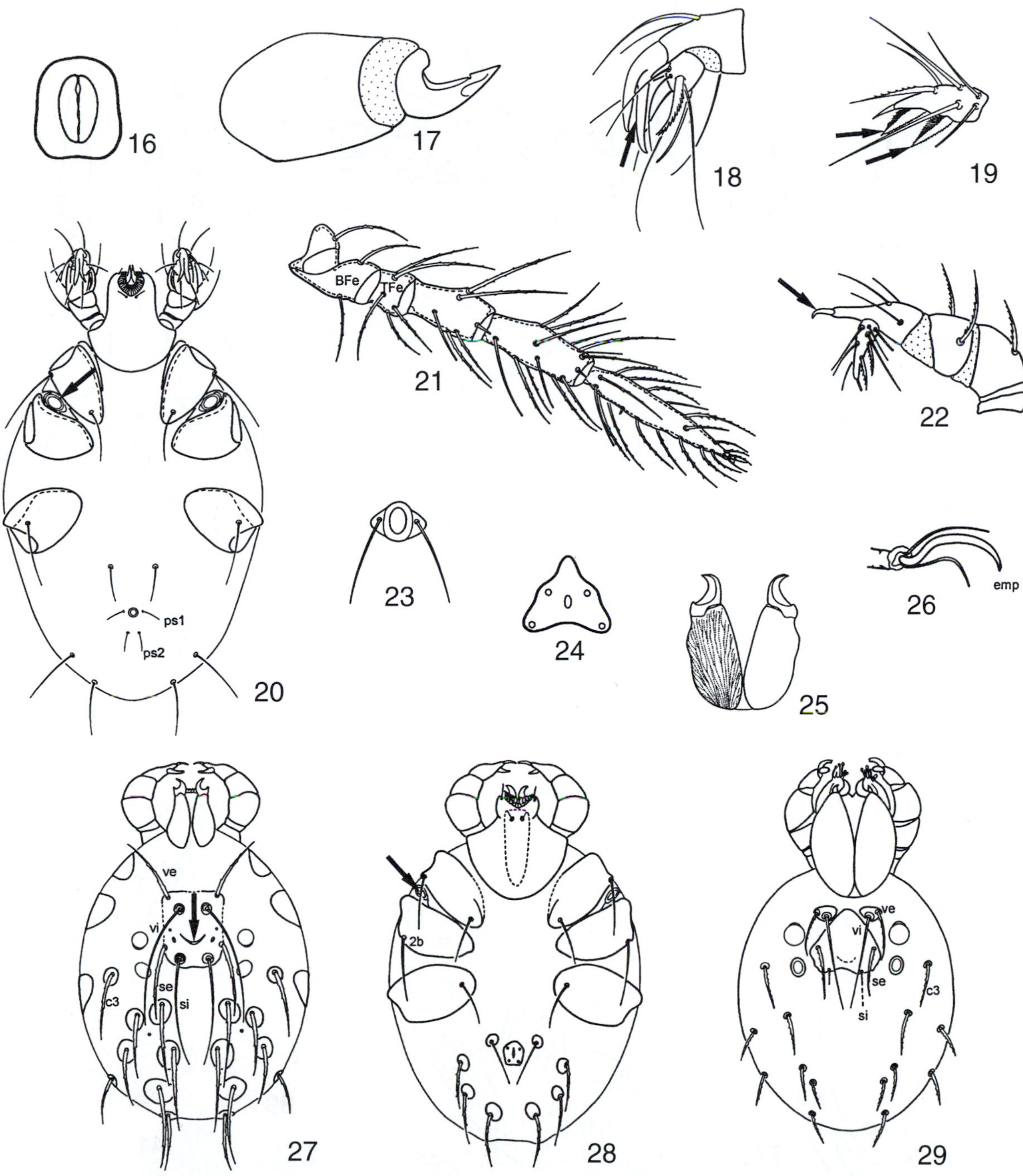

FIGURES 16–17 *Thermacarus nevadensis* Marshall (Thermacaridae), larva. Fig. 16, Excretory pore plate; Fig 17, chelicera. *FIGURE 18, 20* *Wandesia* sp. (Wandesiinae), larva. Fig. 18, Tibia and tarsus of pedipalp; Fig. 20, venter of idiosoma and gnathosoma. *FIGURES 19, 21–22, 26–28* *Thyas stolli* Koenike (Thyadinae), larva. Fig. 19, Tarsus of pedipalp; Fig. 21, leg I; Fig. 22, pedipalp; Fig. 26, claws and empodium; Fig. 27, dorsum of idiosoma and gnathosoma; Fig. 28, venter of idiosoma and gnathosoma. *FIGURES 23, 25, 29* *Hydryphantes ruber* (de Geer) (Hydryphantinae), larva. Fig. 23, Excretory pore plate; Fig. 25, chelicerae; Fig. 29, dorsum of idiosoma and gnathosoma. *FIGURE 24* *Protzia* sp. (Protziinae), larva, excretory pore plate.

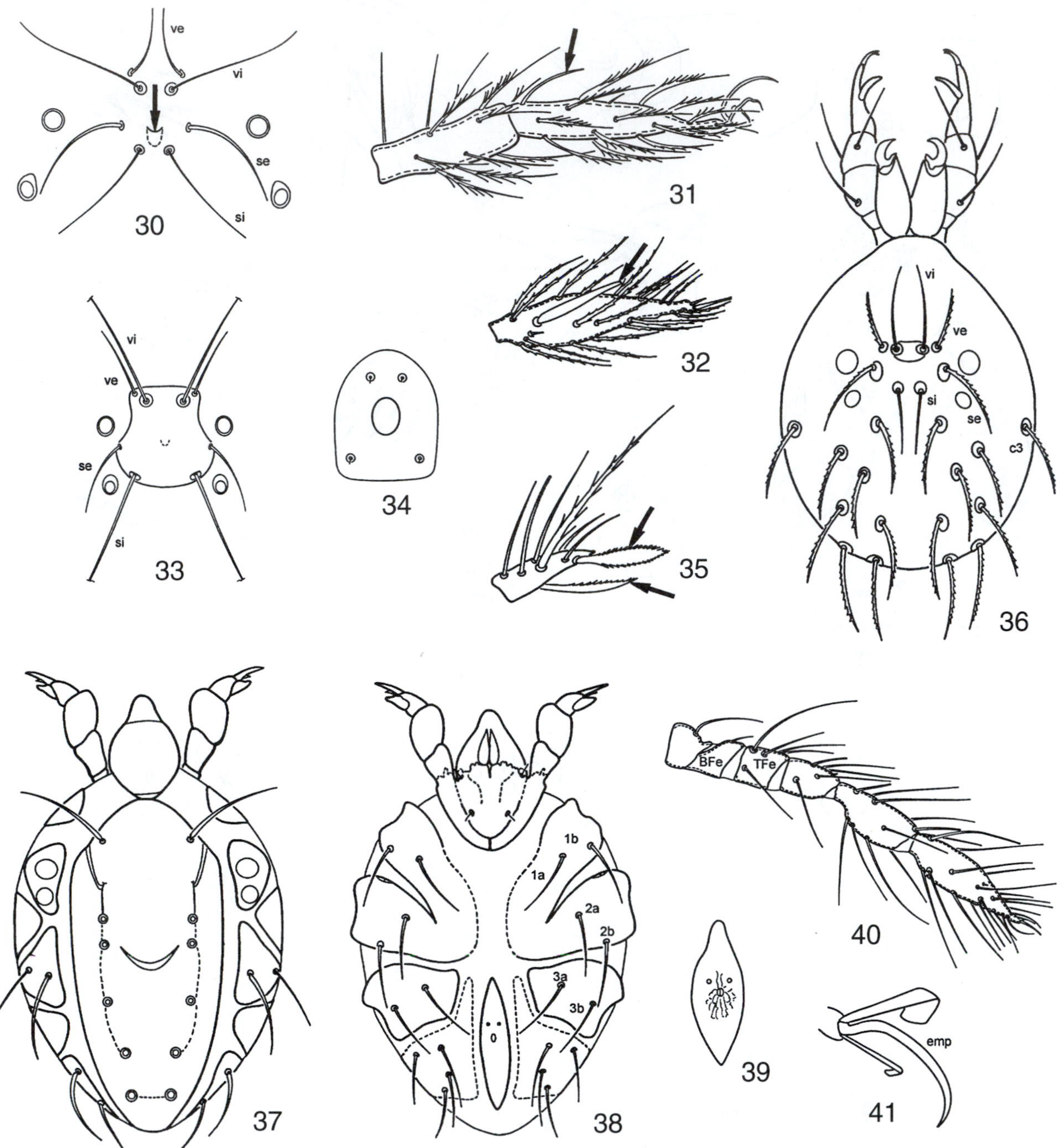

FIGURES 30–31 *Protzia* sp. (Protziinae), larva. Fig. 30, Prodorsal region of idiosoma; Fig. 31, tibia and tarsus of leg I. *FIGURES 32, 34–36* *Hydrodroma despiciens* (Müller) (Hydrodromidae), larva. Fig. 32, Tarsus of leg I; Fig. 34, excretory pore plate; Fig. 35, tarsus of pedipalp; Fig. 36, dorsum of idiosoma and gnathosoma. *FIGURE 33* *Tartarothyas* sp. (Tartarothyadinae), larva, dorsal plate. *FIGURES 37–41* *Eylais major* Lanciani (Eylaidae), larva. Fig. 37, Dorsum of idiosoma and gnathosoma; Fig. 38, venter of idiosoma and gnathosoma; Fig. 39, excretory pore plate; Fig. 40, leg I; Fig. 41, claws and empodium of leg I.

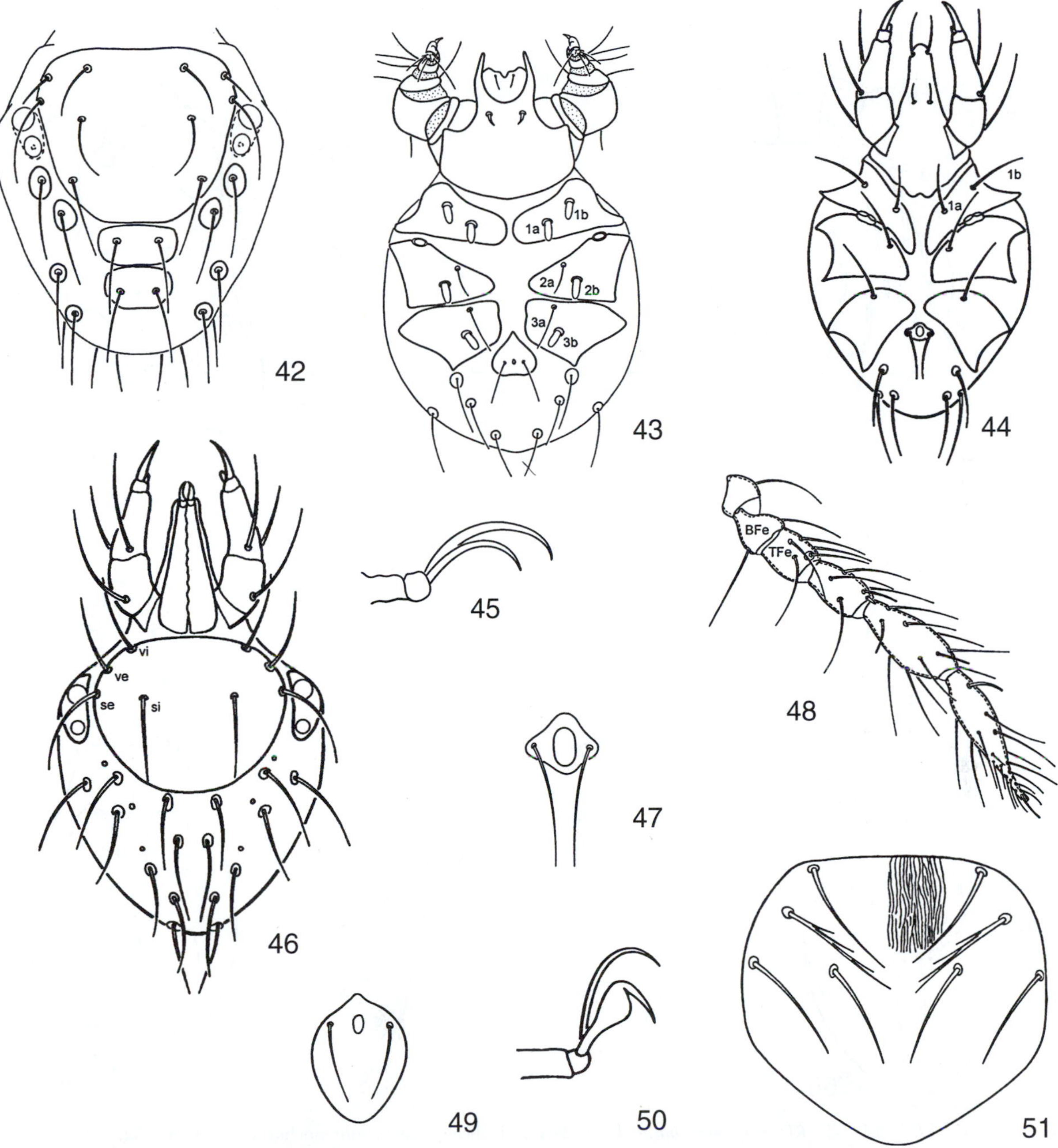

FIGURES 42–43 *Piersigia* sp. (Piersigiidae), larva. Fig. 42, Dorsum of idiosoma; Fig. 43, venter of idiosoma and gnathosoma. *FIGURES 44–48* *Limnochares americana* Lundblad (Limnocharinae), larva. Fig. 44, venter of idiosoma and gnathosoma; Fig. 45, claw and empodium of leg I; Fig. 46, dorsum of idiosoma and gnathosoma; Fig. 47, excretory pore plate; Fig. 48, leg I. *FIGURES 49–51* *Neolimnochares* sp. (Limnocharinae), larva. Fig. 49, Excretory pore plate; Fig. 50, claw and empodium of leg I; Fig. 51, dorsal plate.

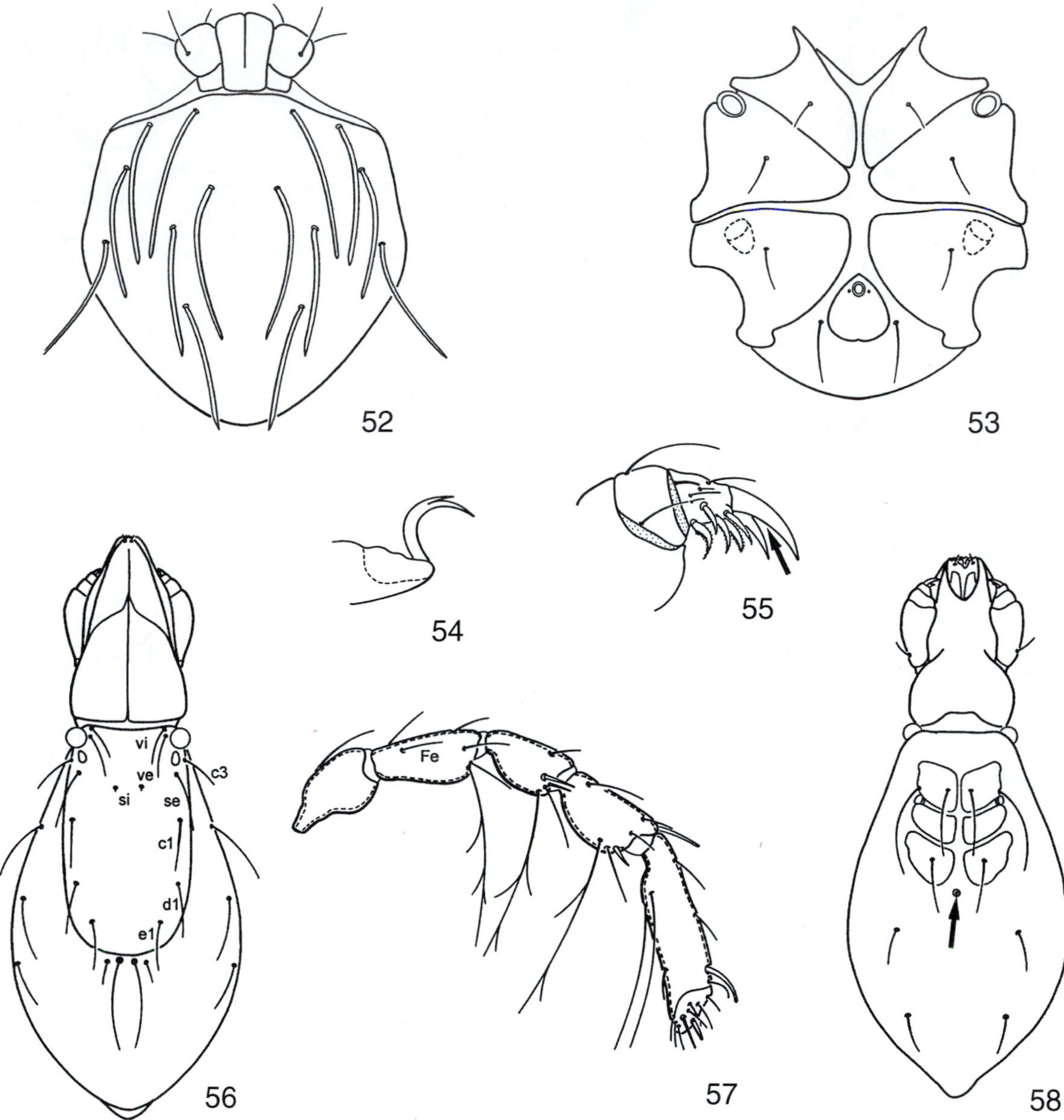

FIGURES 52–53 ***Rhyncholimnochares kittatinniana*** Habeeb (Rhyncholimnocharinae), larva. Fig. 52, Dorsum of idiosoma and gnathosoma; Fig. 53, venter of idiosoma. *FIGURES 54–58* ***Hydrachna magniscutata*** Marshall (Hydrachnidae), larva. Fig. 54, Empodium of leg I; Fig. 55, distal segment of pedipalp; Fig. 56, dorsum of idiosoma and gnathosoma; Fig. 57, leg I; Fig. 58, venter of idiosoma and gnathosoma.

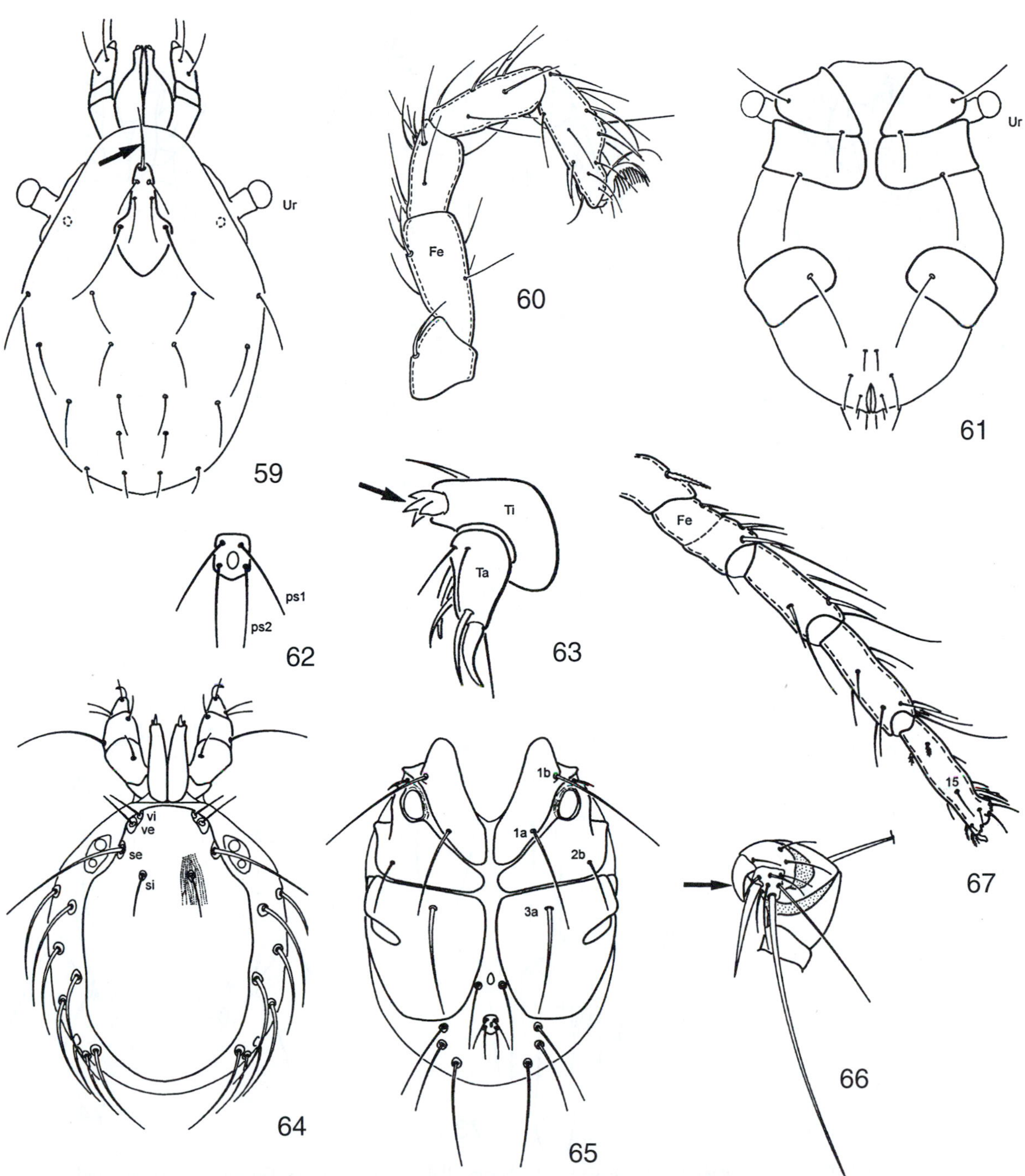

FIGURES 59–61, 63 ***Stygothrombium*** sp. (Stygothrombidiidae), larva. Fig. 59, Dorsum of idiosoma and gnathosoma; Fig. 60, leg I; Fig. 61, venter of idiosoma; Fig. 63, tibia and tarsus of pedipalp. *FIGURES 62, 64–67* ***Sperchon glandulosus*** Koenike (Sperchontidae), larva. Fig. 62, Excretory pore plate; Fig. 64, dorsum of idiosoma and gnathosoma; Fig. 65, venter of idiosoma; Fig. 66, pedipalp; Fig. 67, leg I.

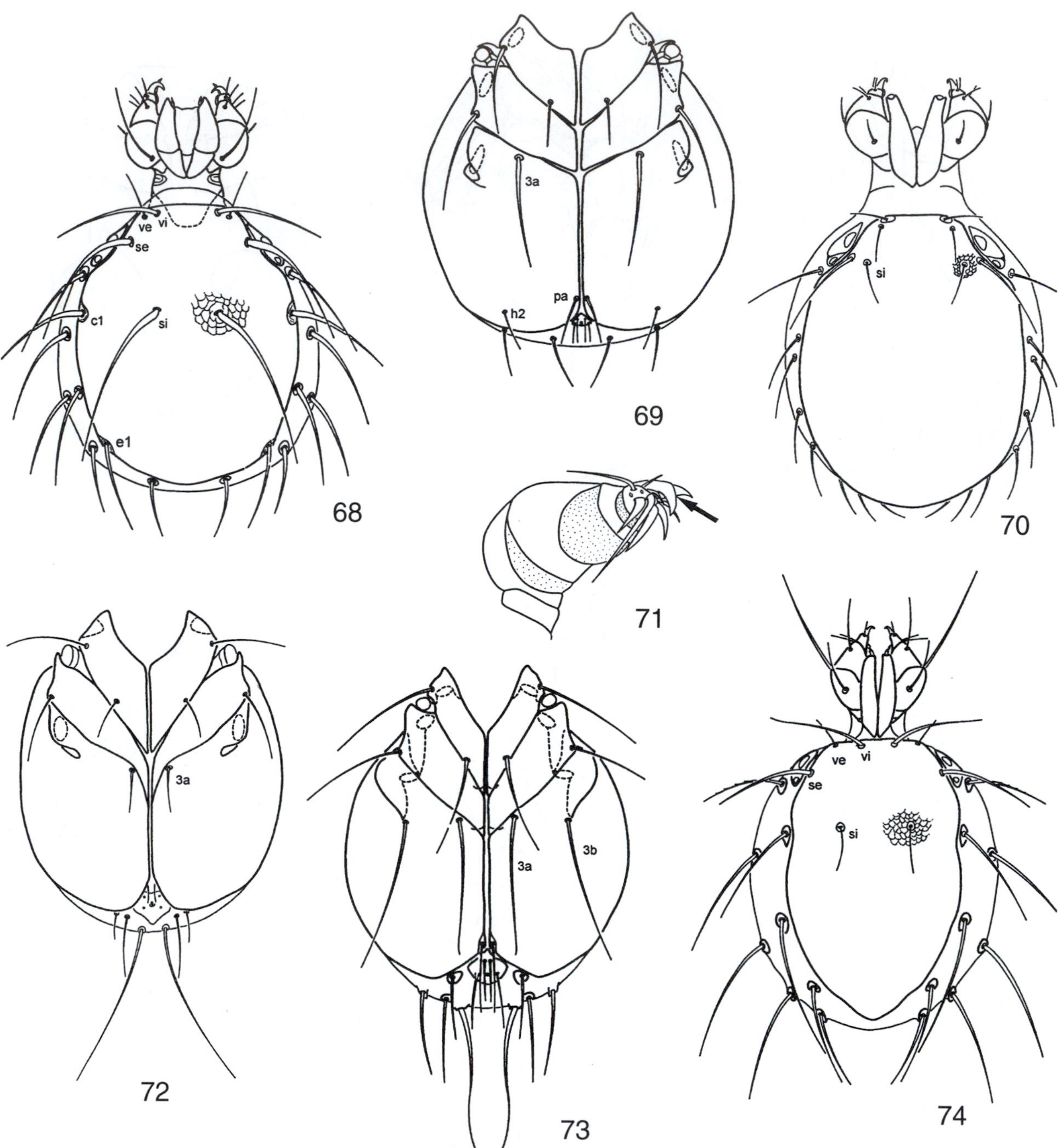

FIGURES 68–69, 71 *Bandakiopsis fonticola* Smith (Anisitsiellidae), larva. Fig. 68, Dorsum of idiosoma and gnathosoma; Fig. 69, venter of idiosoma; Fig. 71, pedipalp. *FIGURES 70,72* *Utaxatax newelli* Habeeb (Anisitsiellidae), larva. Fig. 70, dorsum of idiosoma and gnathosoma; Fig. 72, venter of idiosoma. *FIGURES 73, 74* *Teutonia lunata* Marshall (Teutoniidae), larva. Fig. 73, Venter of idiosoma; Fig. 74, dorsum of idiosoma and gnathosoma.

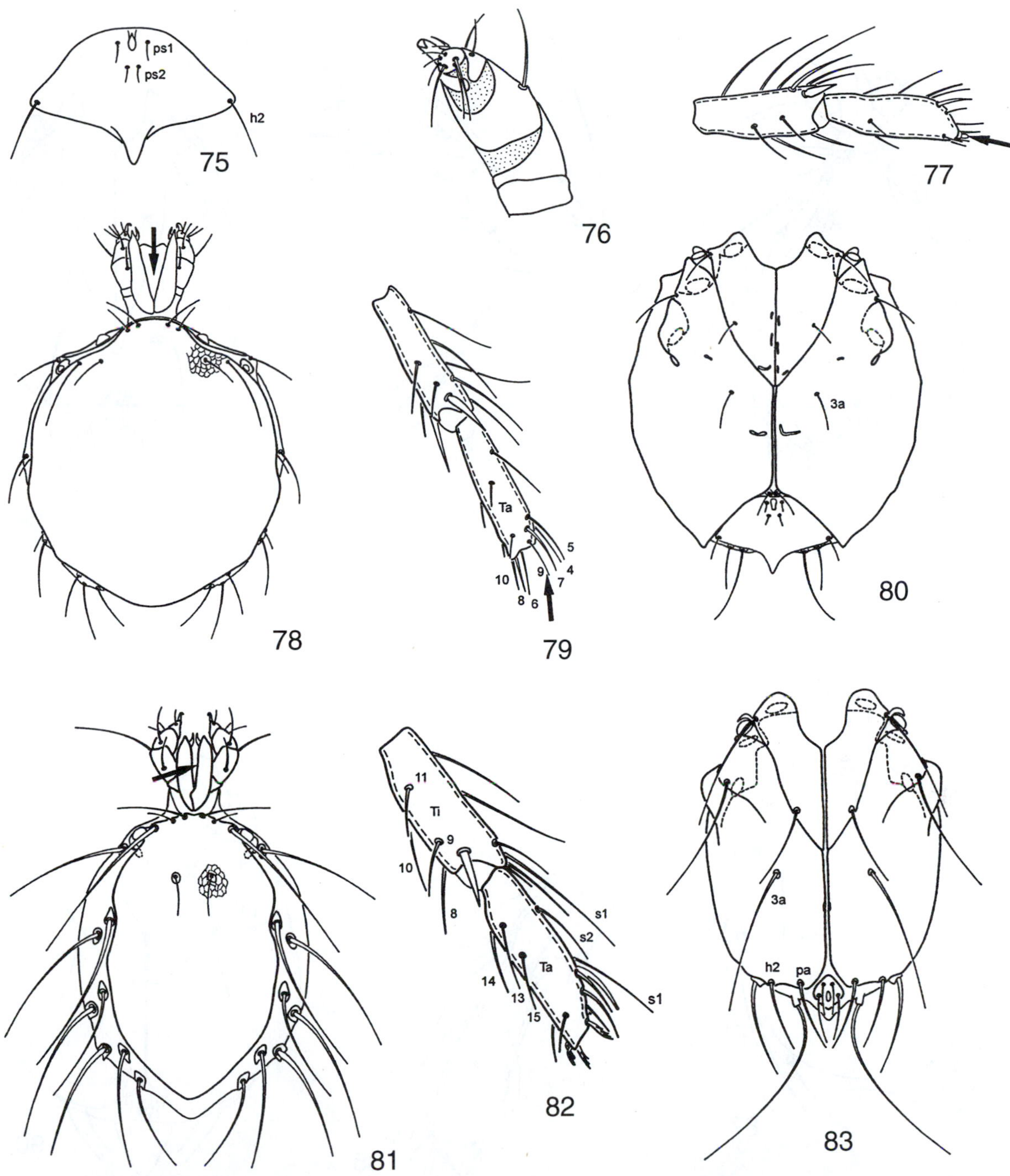

FIGURES 75–80 *Bandakia phreatica* Cook (Anisitsiellidae), larva. Fig. 75, Excretory pore plate; Fig. 76, pedipalp; Fig. 77, tibia and tarsus of leg I; Fig. 78, dorsum of idiosoma and gnathosoma; Fig. 79, tibia and tarsus of leg III. Fig. 80, venter of idiosoma. *FIGURES 81–83* *Lebertia* sp. (Lebertiidae), larva. Fig. 81, Dorsum of idiosoma and gnathosoma; Fig. 82, tibia and tarsus of leg I; Fig. 83, venter of idiosoma.

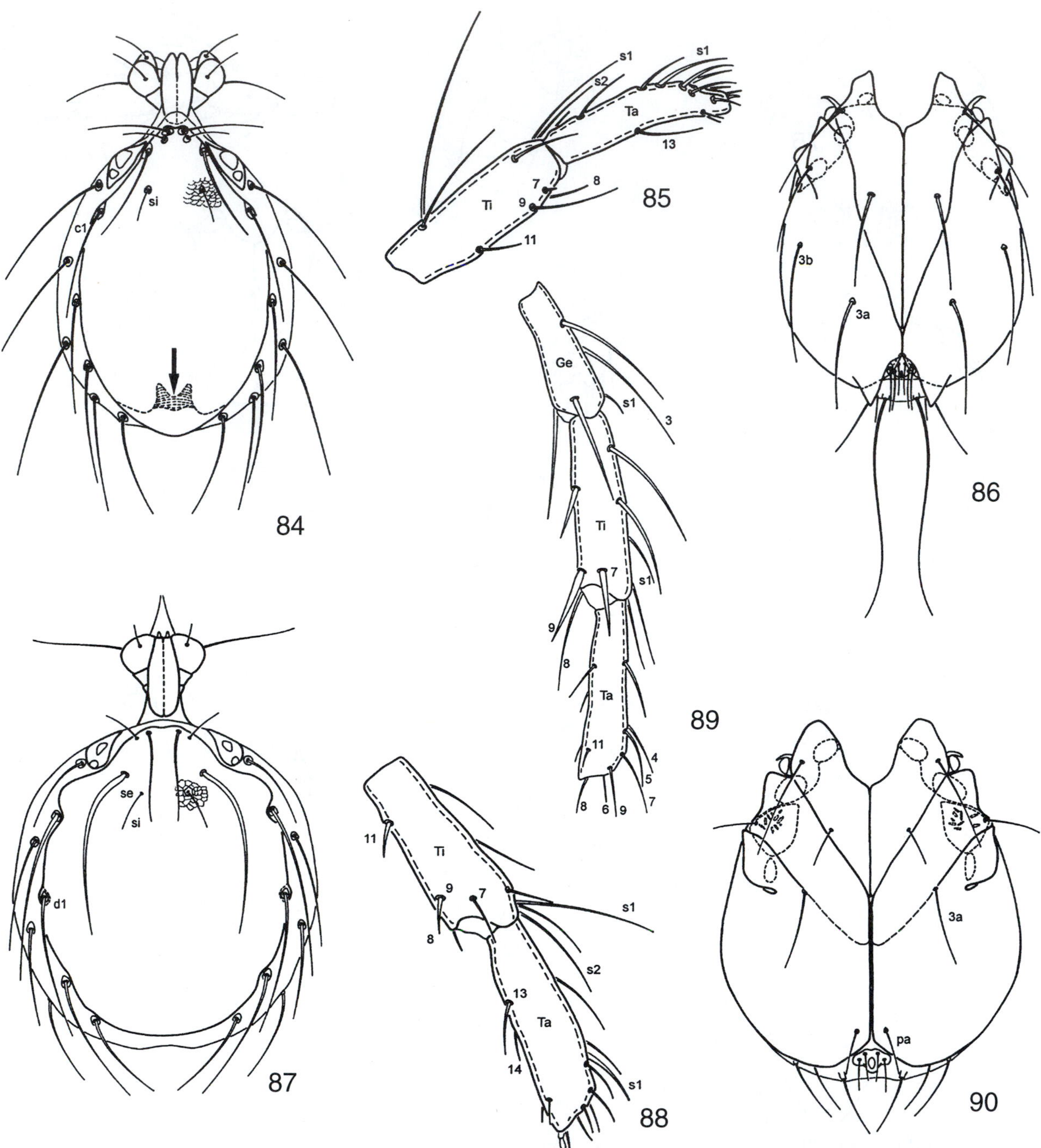

FIGURES 84–86 *Oxus elongatus* Marshall (Oxidae), larva. Fig. 84, Dorsum of idiosoma and gnathosoma; Fig. 85, tibia and tarsus of leg I; Fig. 86, venter of idiosoma. *FIGURES 87–90* *Testudacarus* sp. (Testudacarinae), larva. Fig. 87, Dorsum of idiosoma and gnathosoma; Fig. 88, tibia and tarsus of leg I; Fig. 89, distal segments of leg III; Fig. 90, venter of idiosoma.

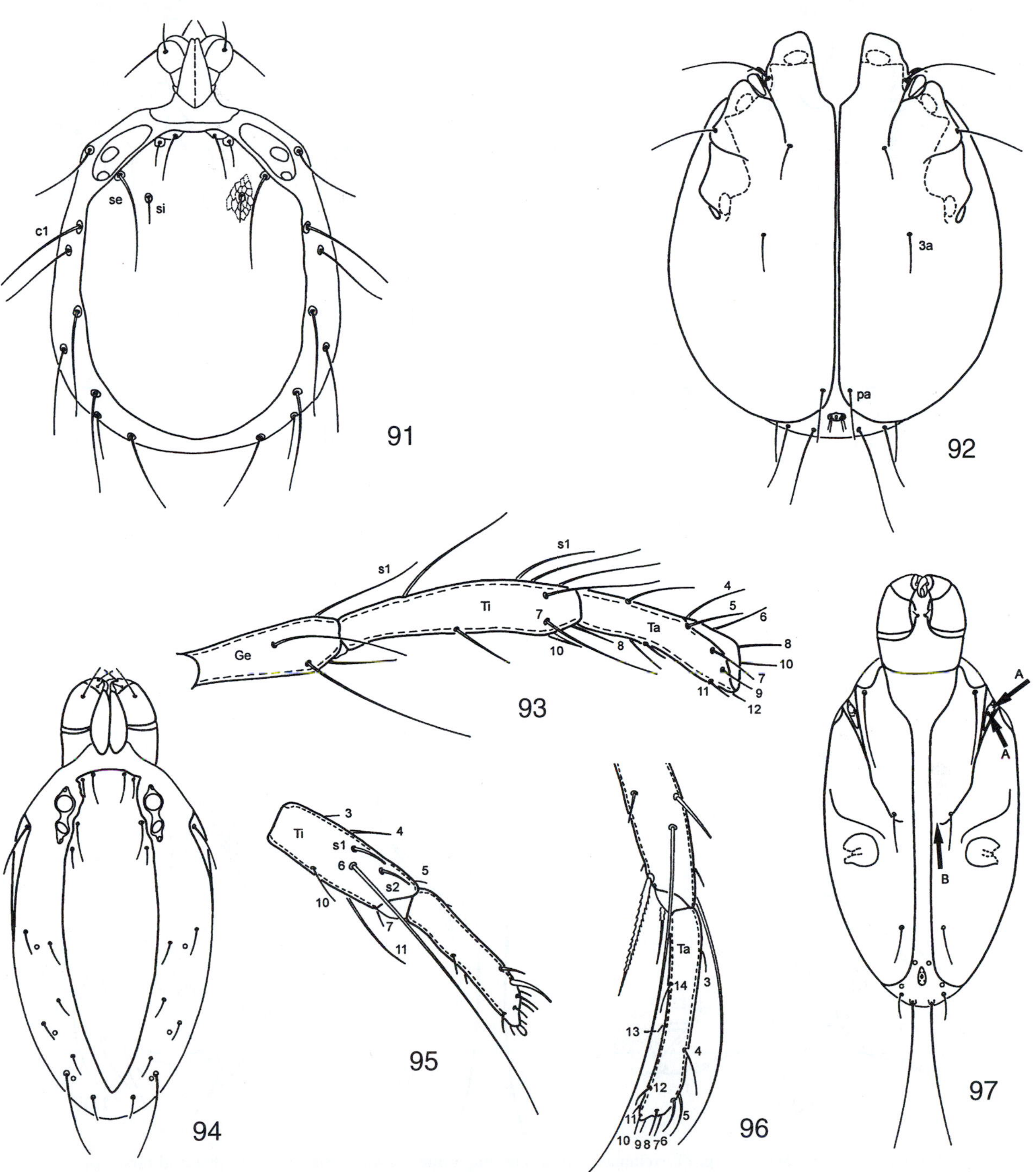

FIGURES 91–93 *Torrenticola* sp. (Torrenticolidae), larva. Fig. 91, dorsum of idiosoma and gnathosoma; Fig. 92, venter of idiosoma; Fig. 93, distal segments of leg III. *FIGURES 94–97* *Limnesia marshalliana* Lundblad (Limnesiinae), larva. Fig. 94, dorsum of idiosoma and gnathosoma; Fig. 95, tibia and tarsus of leg II; Fig. 96, tibia and tarsus of leg III; Fig. 97, venter of idiosoma and gnathosoma.

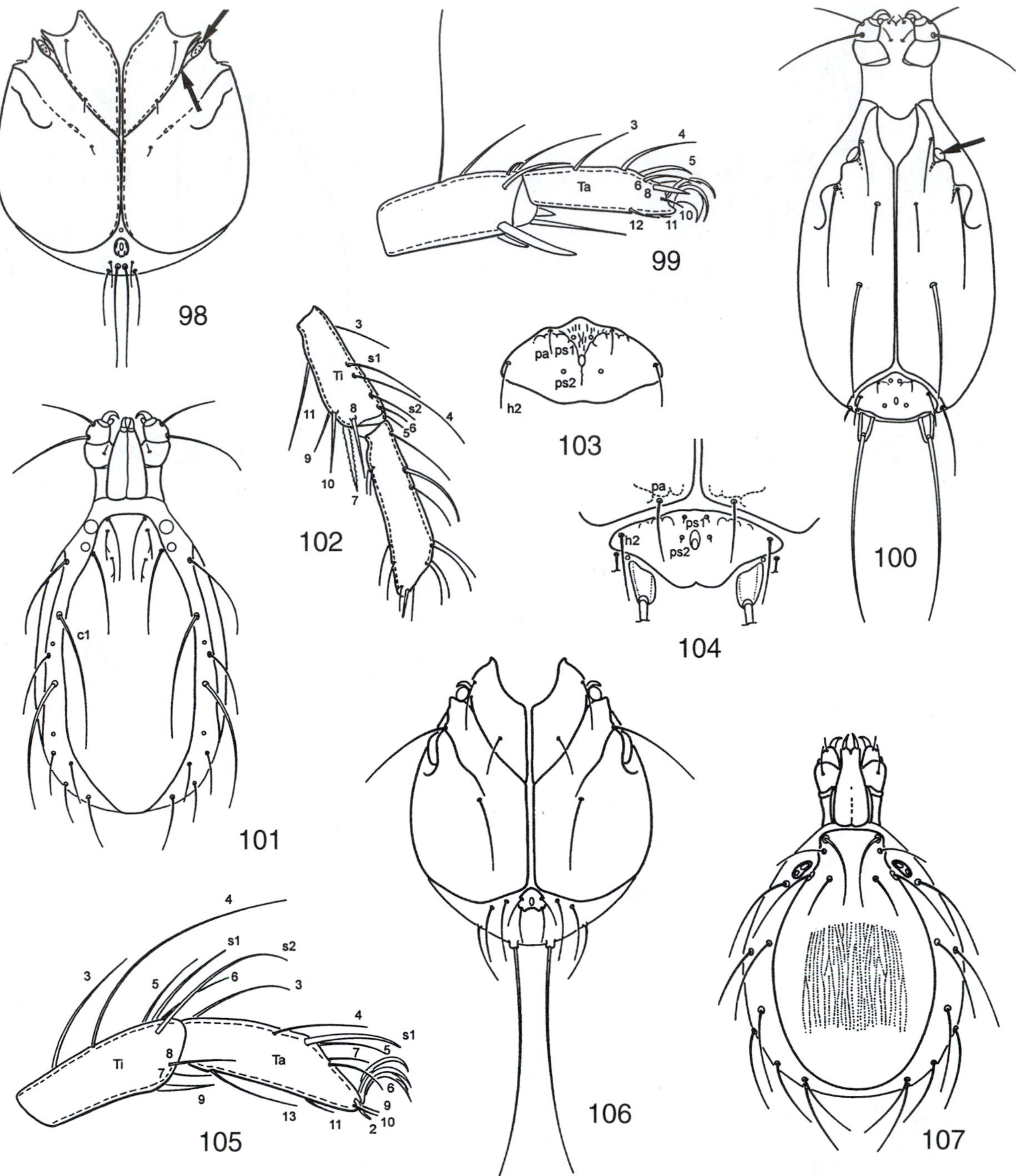

FIGURES 98–99 *Tyrrellia* sp. (Tyrrelliinae), larva. Fig. 98, Venter of idiosoma; Fig. 99, tibia and tarsus of leg III. *FIGURES 100–101, 103* *Atractides grouti* Habeeb (Hygrobatidae), larva. Fig. 100, Venter of idiosoma and gnathosoma; Fig. 101, dorsum of idiosoma and gnathosoma; Fig. 103, excretory pore plate. *FIGURE 102* *Hygrobates* sp. (Hygrobatidae), larva, tibia and tarsus of leg II. *FIGURE 104* *Hygrobates neocalliger* Habeeb (Hygrobatidae), larva, excretory pore plate region. *FIGURES 105–107* *Feltria* sp. (Feltriidae), larva. Fig. 105, Tibia and tarsus of leg I; Fig. 106, venter of idiosoma; Fig. 107, dorsum of idiosoma and gnathosoma.

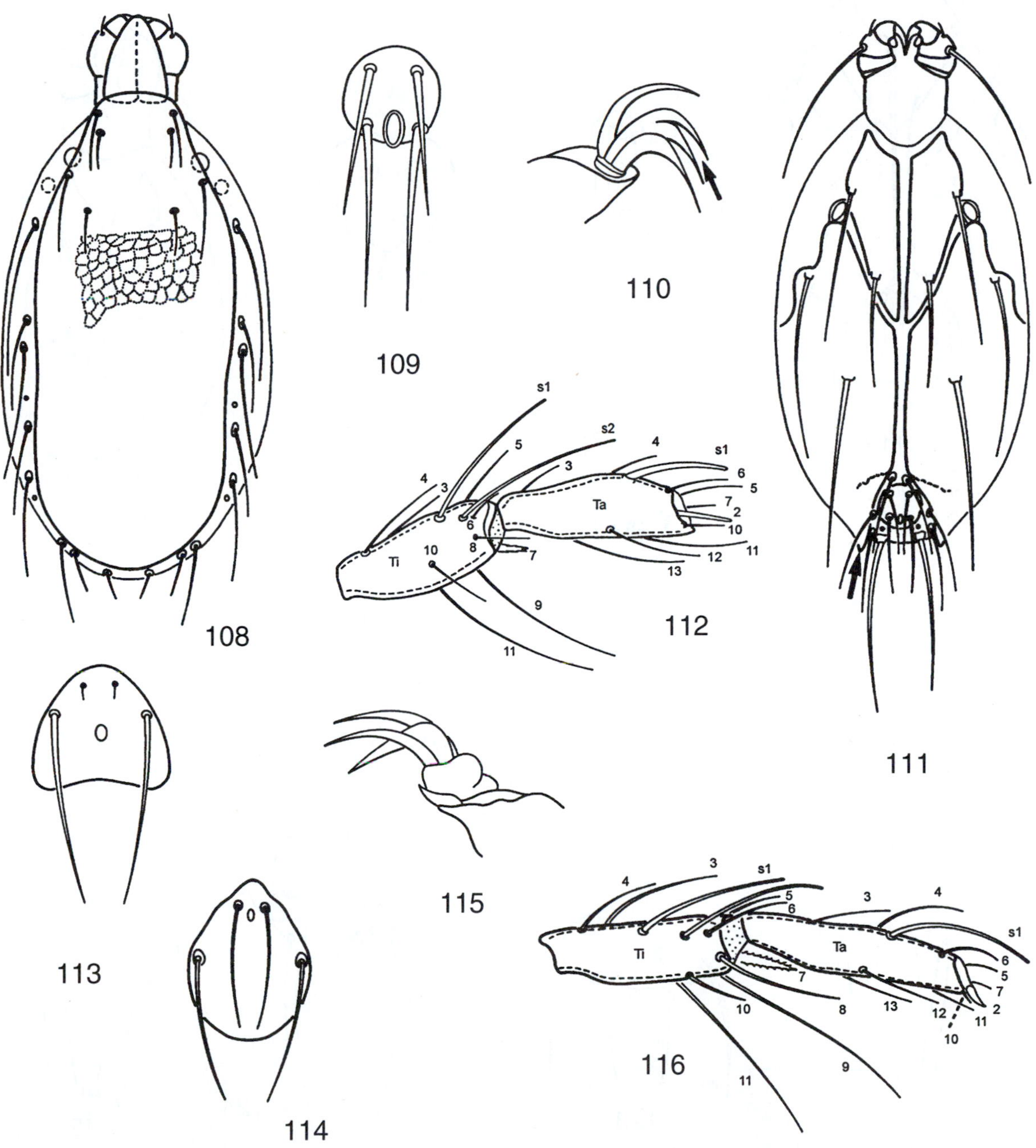

FIGURES 108–112 *Unionicola* sp. (Unionicolinae), larva. Fig. 108, Dorsum of idiosoma and gnathosoma; Fig. 109, excretory pore plate; Fig. 110, claws and empodium of leg I; Fig. 111, venter of idiosoma and gnathosoma; Fig. 112, tibia and tarsus of leg I. *FIGURE 113* *Koenikea marshallae* Viets (Pionatacinae), larva, excretory pore plate. *FIGURES 114–116* *Neumania punctata* Marshall (Pionatacinae), larva. Fig. 114, excretory pore plate; Fig. 115, claws and empodium of leg I; Fig. 116, tibia and tarsus of leg I.

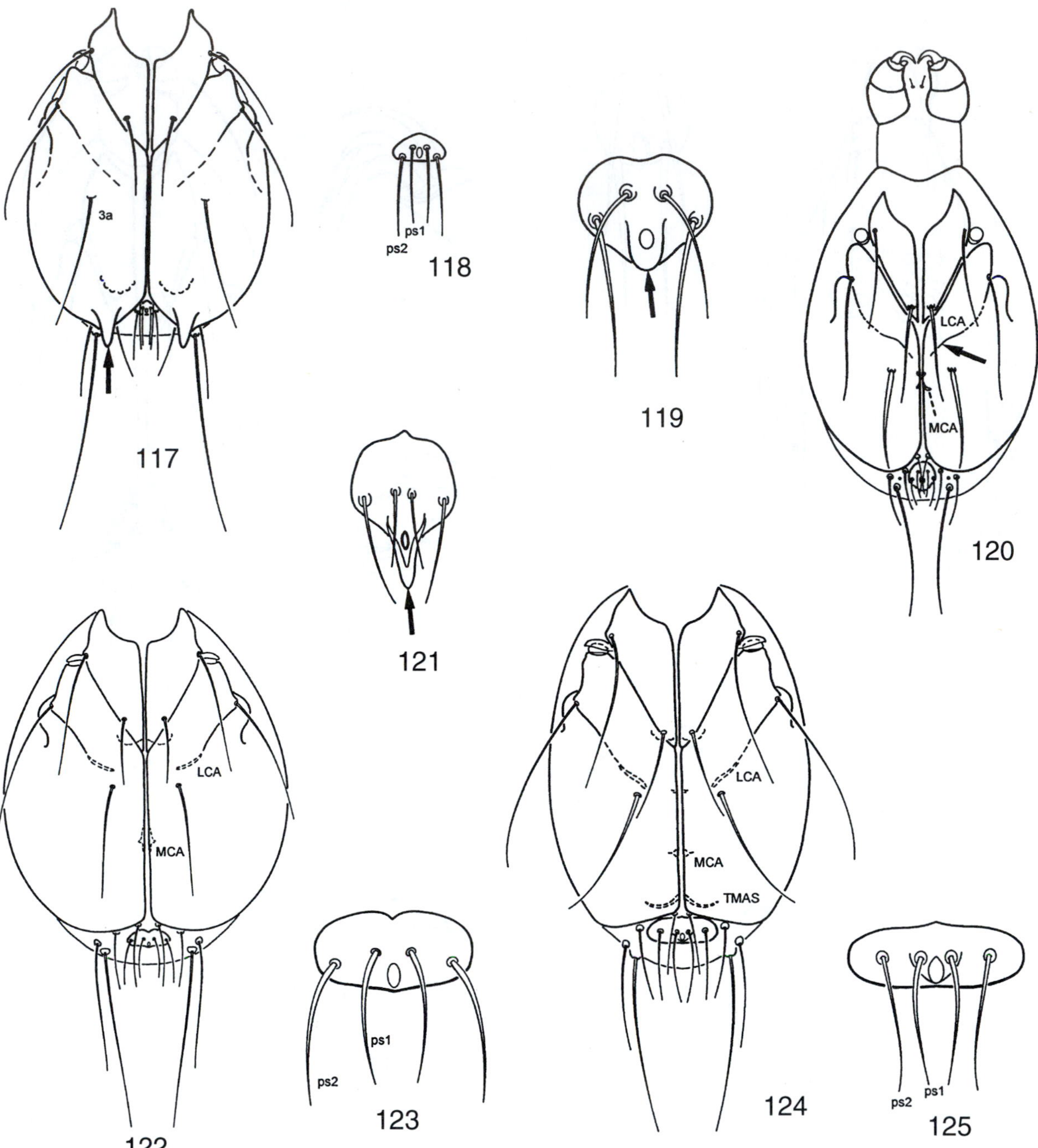

FIGURES 117–118 *Wettina ontario* Smith (Wettinidae), larva. Fig. 117, Venter of idiosoma; Fig. 118, excretory pore plate. *FIGURES 119–120* *Hydrochoreutes intermedius* Cook (Hydrochoreutinae), larva. Fig. 119, Excretory pore plate; Fig. 120, venter of idiosoma and gnathosoma. *FIGURE 121* *Hydrochoreutes minor* Cook (Hydrochoreutinae), larva, excretory pore plate. *FIGURES 122–123* *Huitfeldtia rectipes* Thor (Huitfeldtiinae), larva. Fig. 122, Venter of idiosoma; Fig. 123, excretory pore plate. *FIGURES 124–125* *Neotiphys pionoidellus* (Habeeb) (Tiphyinae), larva. Fig. 124, Venter of idiosoma; Fig. 125, excretory pore plate.

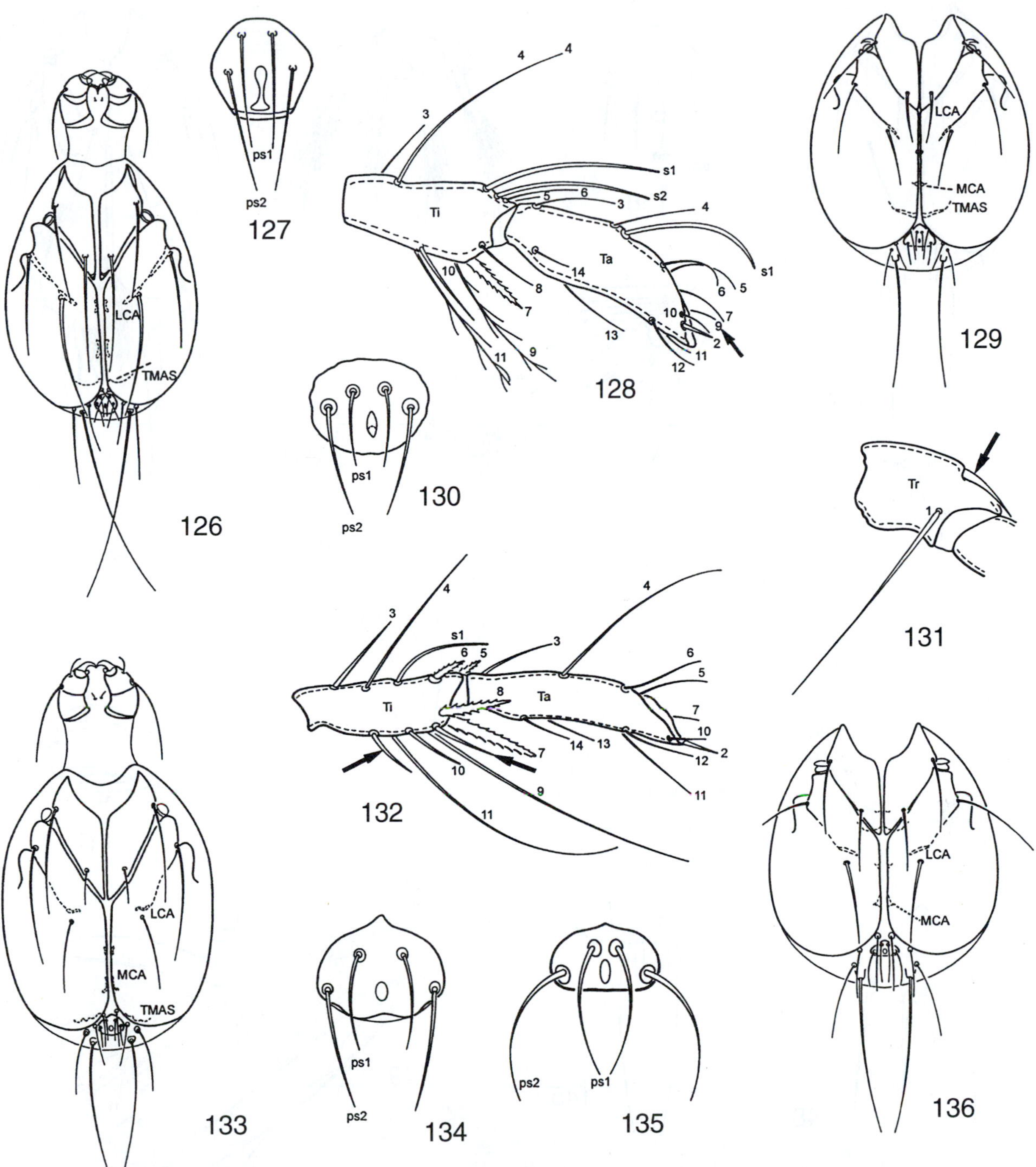

FIGURES 126–127 *Tiphys americanus* (Marshall) (Tiphyinae), larva. Fig. 126, Venter of idiosoma and gnathosoma; Fig. 127, excretory pore plate. *FIGURES 128, 130* *Tiphys ornatus* (Koch) (Tiphyinae), larva. Fig. 128, Tibia and tarsus of leg I; Fig. 130, excretory pore plate. *FIGURES 129, 131* *Pseudofeltria multipora* Cook (Foreliinae), larva. Fig. 129, Venter of idiosoma; Fig. 131, trochanter of leg III. *FIGURES 132–134* *Forelia ovalis* Marshall (Foreliinae), larva. Fig. 132, Tibia and tarsus of leg III; Fig. 133, venter of idiosoma and gnathosoma; Fig. 134, excretory pore plate. *FIGURES 135, 136* *Forelia onondaga* Habeeb (Foreliinae), larva. F ig. 135, excretory pore plate; Fig. 136, venter of idiosoma.

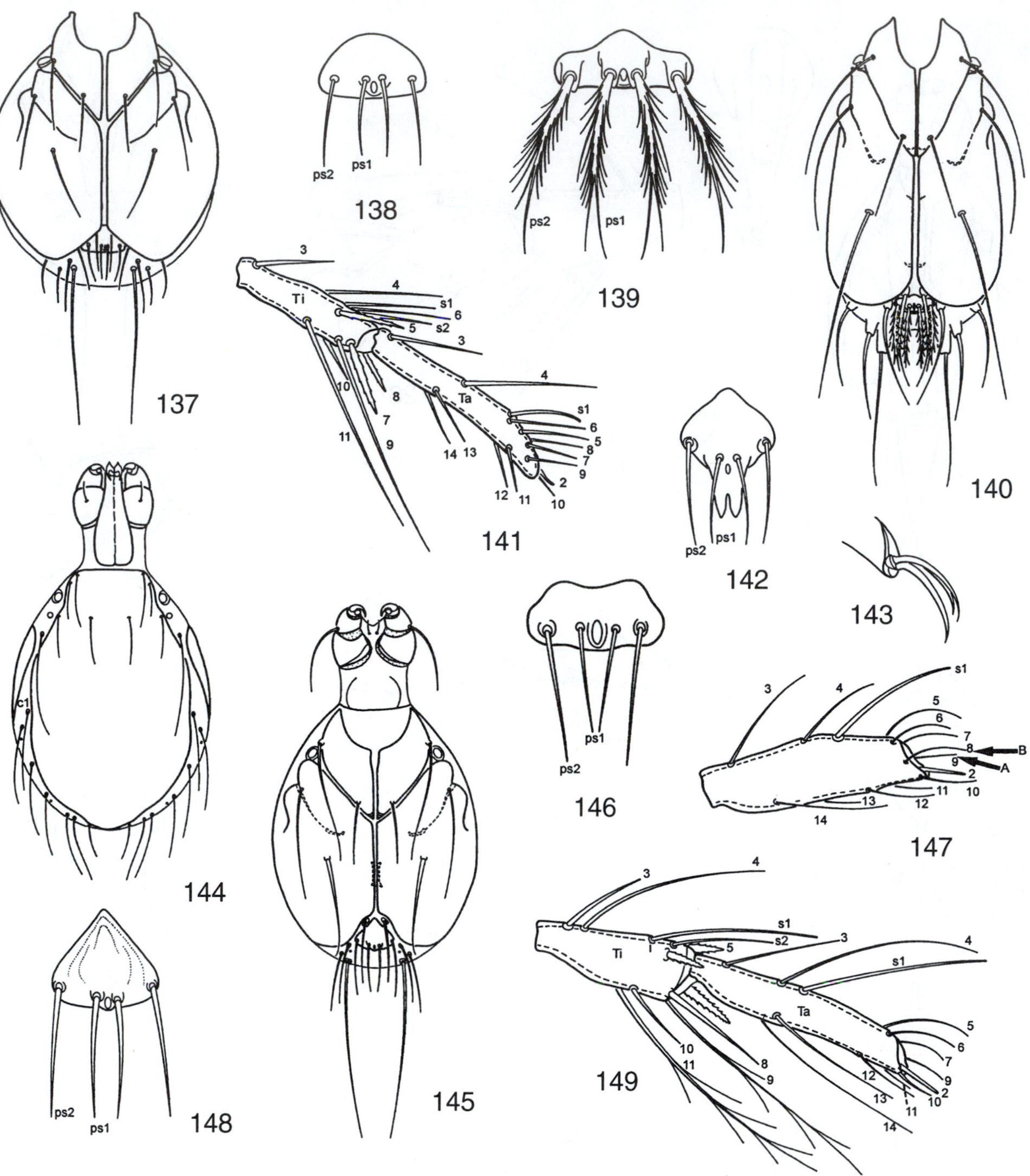

FIGURES 137–138, 141 *Najadicola ingens* Koenike (Najadicolinae), larva. Fig. 137, Venter of idiosoma; Fig. 138, excretory pore plate; Fig. 141, tibia and tarsus of leg II. *FIGURES 139, 140* *Nautarachna muskoka* Smith (Pioninae), larva. Fig. 139, Excretory pore plate; Fig. 140, venter of idiosoma. *FIGURES 142, 143* *Piona carnea* (Koch) (Pioninae), larva. Fig. 142, excretory pore plate; Fig. 143, claws and empodium of leg III. *FIGURES 144, 145, 148* *Piona interrupta* Marshall (Pioninae), larva. Fig. 144, dorsum of idiosoma and gnathosoma; Fig. 145, venter of idiosoma and gnathosoma; Fig. 148, excretory pore plate. *FIGURE 146* *Piona constricta* (Wolcott) (Pioninae), larva, excretory pore plate. *FIGURES 147, 149* *Piona mitchelli* Cook (Pioninae), larva. Fig. 147, tarsus of leg I; Fig. 149, tibia and tarsus of leg II.

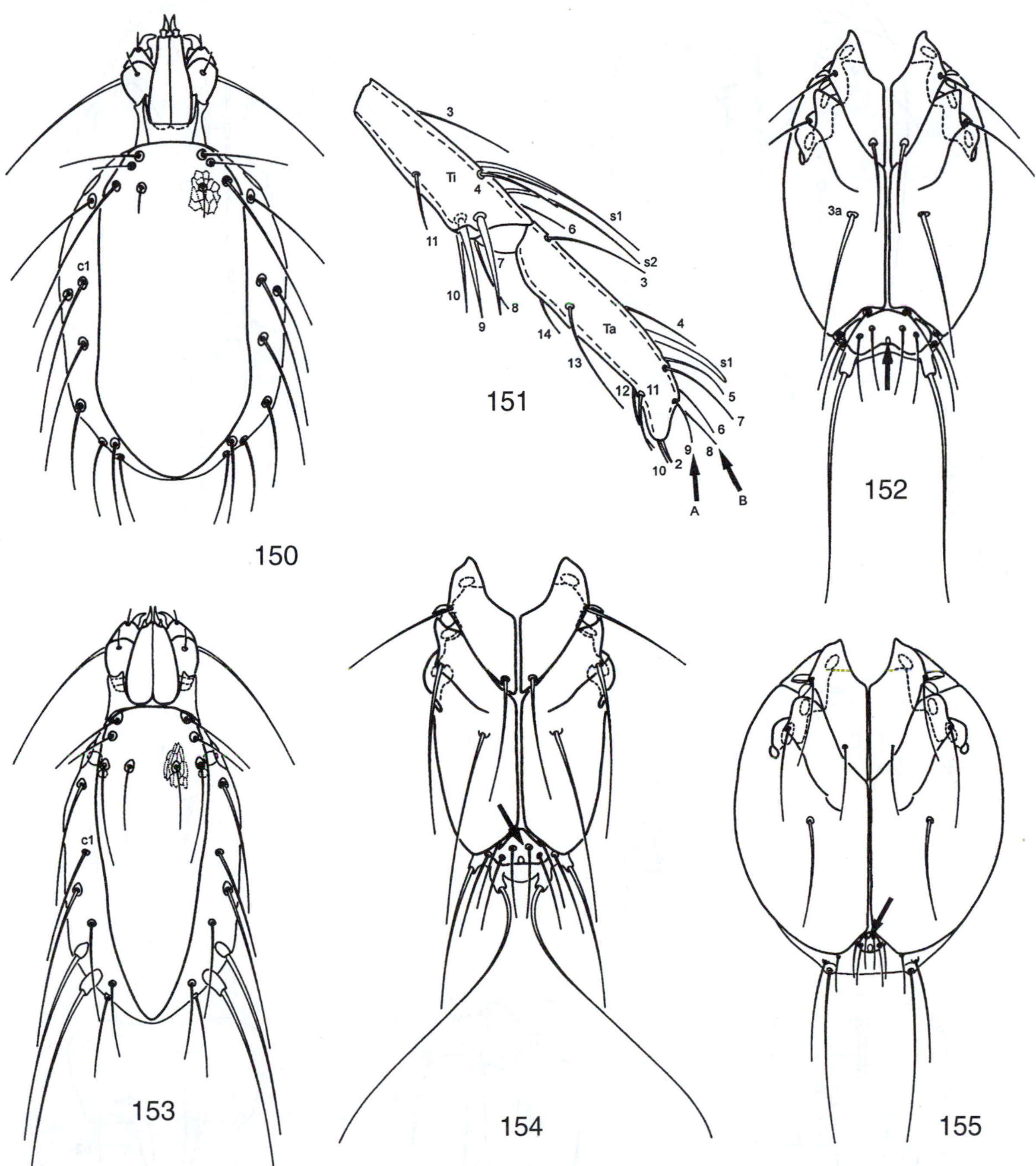

FIGURES 150–152 *Ljania bipapillata* Thor (Axonopsinae), larva. Fig. 150, Dorsum of idiosoma and gnathosoma; Fig. 151, tibia and tarsus of leg I; Fig. 152, venter of idiosoma. *FIGURES 153–154* *Albia neogaea* Habeeb (Albiinae), larva. Fig. 153, Dorsum of idiosoma and gnathosoma; Fig. 154, venter of idiosoma. *FIGURE 155* *Neobrachypoda ekmani* (Walter) (Axonopsinae), larva, venter of idiosoma.

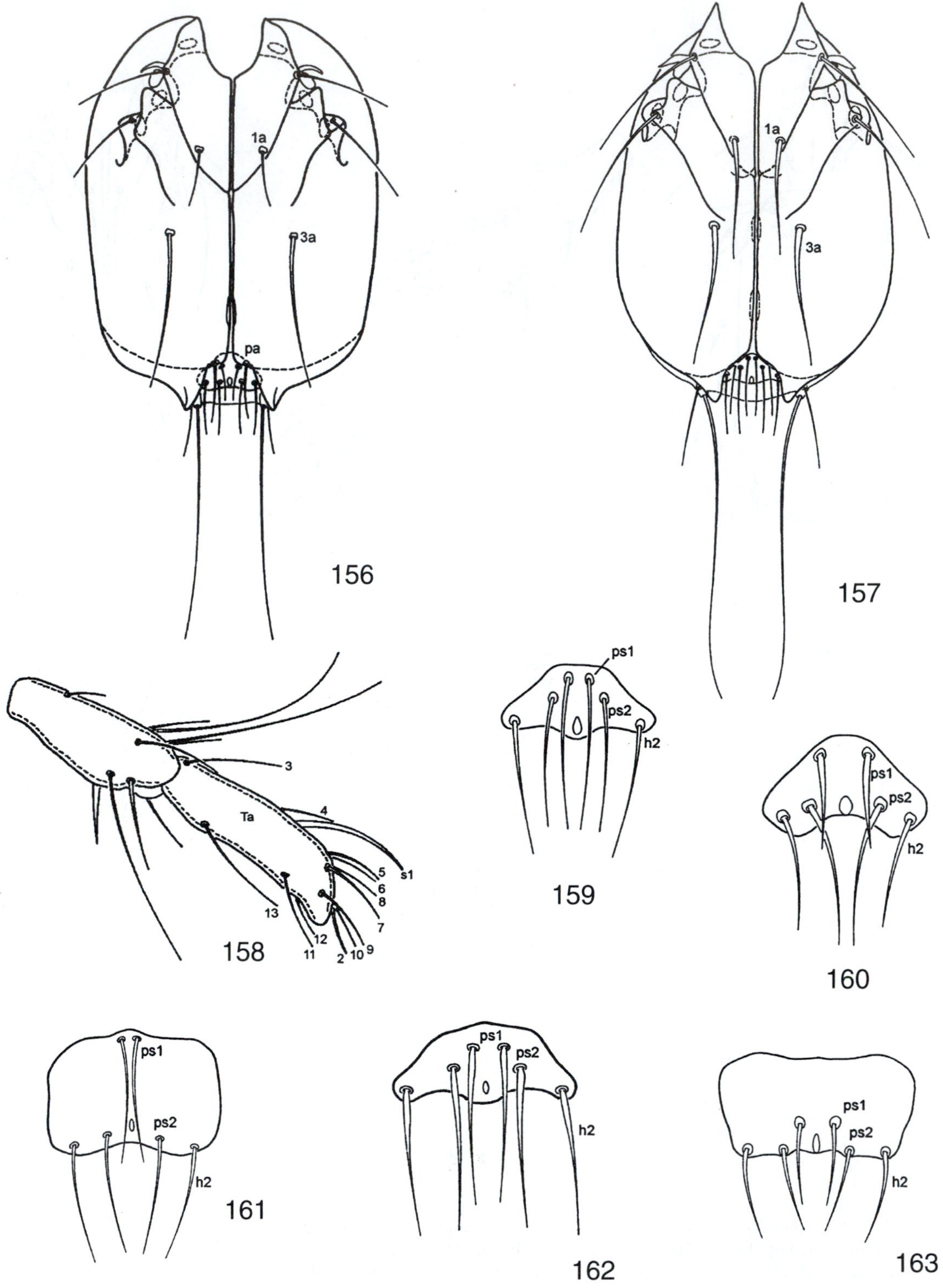

FIGURES 156,160 *Woolastookia setosipes* Habeeb (Axonopsinae), larva. Fig. 156, Venter of idiosoma; Fig. 160, excretory pore plate. *FIGURES 157–159* *Estellacarus unguitarsus* (Habeeb) (Axonopsinae), larva. Fig. 157, Venter of idiosoma, Fig. 158, tibia and tarsus of leg I; Fig. 159, excretory pore plate. *FIGURE 161* *Woolastookia pilositarsa* (Habeeb) (Axonopsinae), larva, excretory pore plate. *FIGURE 162* *Brachypoda cornipes* Habeeb (Axonopsinae), larva, excretory pore plate. *FIGURE 163* *Brachypoda setosicauda* Habeeb (Axonopsinae), larva, excretory pore plate.

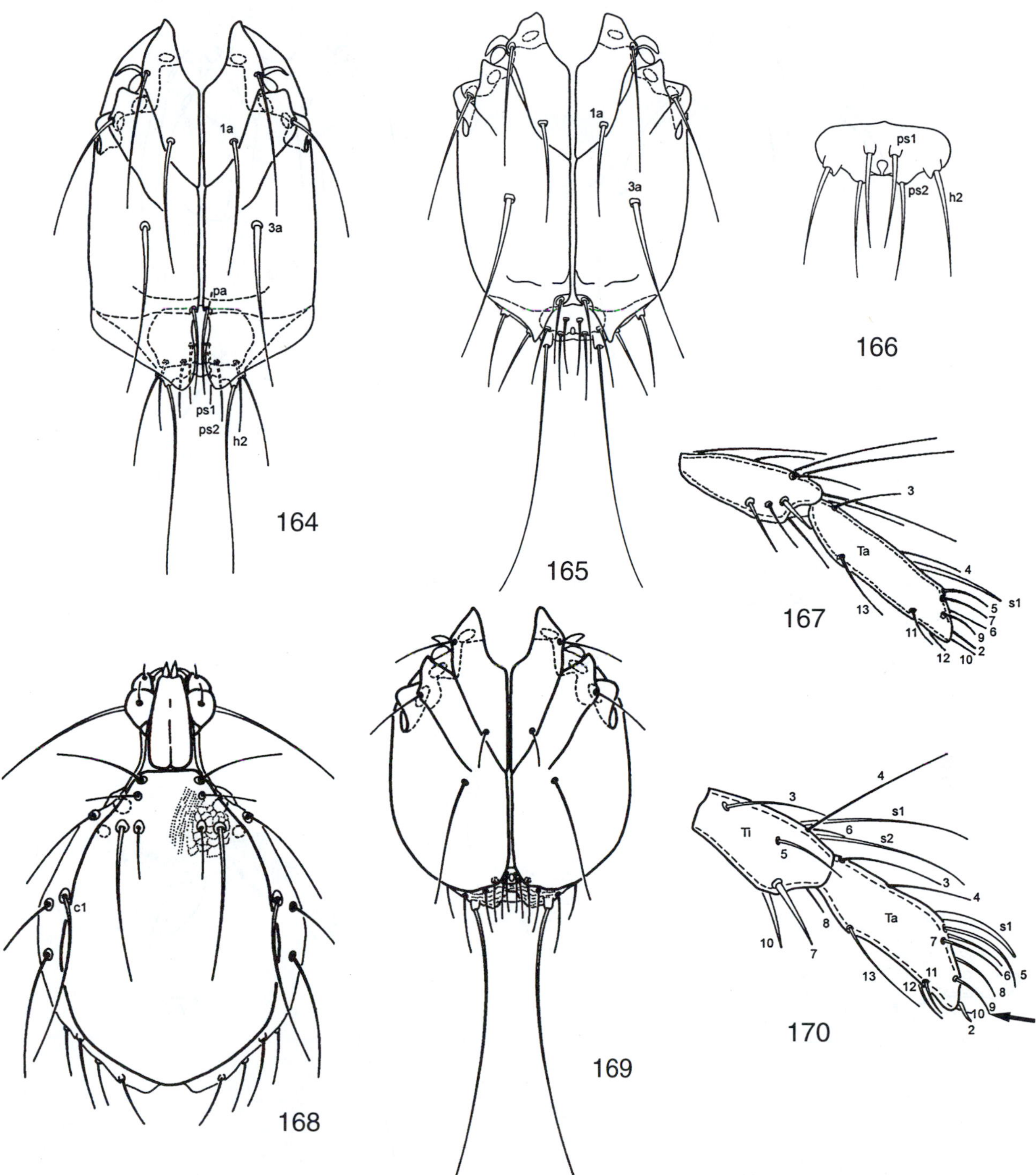

FIGURE 164 *Brachypoda setosicauda* Habeeb (Axonopsinae), larva, venter of idiosoma. *FIGURES 165–167* *Axonopsis setoniensis* Habeeb (Axonopsinae), larva. Fig. 165, venter of idiosoma; Fig. 166, excretory pore plate; Fig. 167, tibia and tarsus of leg I. *FIGURES 168–170* *Aturus* sp. (Aturinae), larva. Fig. 168, dorsum of idiosoma and gnathosoma; Fig. 169, venter of idiosoma; Fig. 170, tibia and tarsus of leg I.

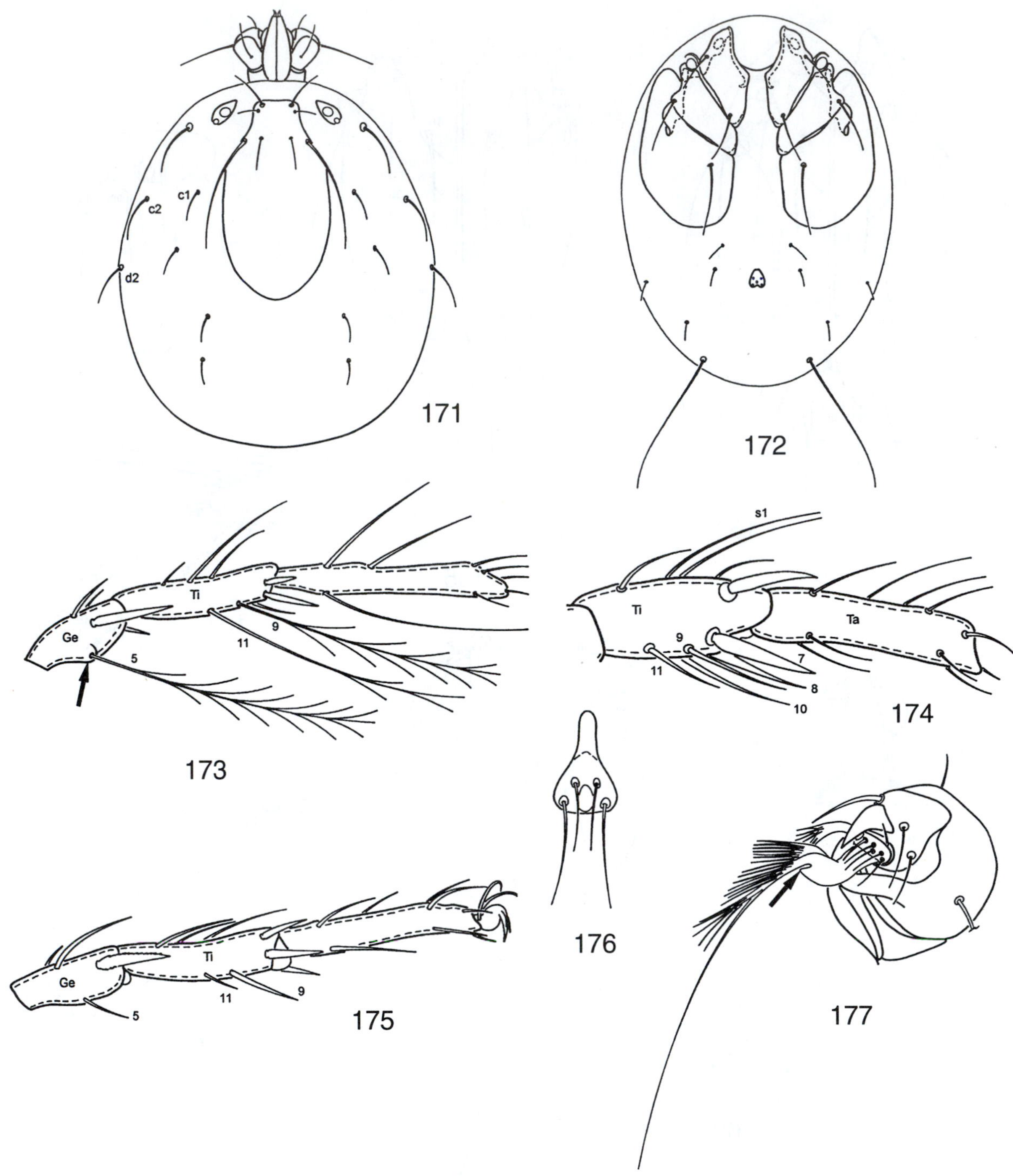

FIGURES 171–172, 175 *Stygomomonia mitchelli* Smith (Stygomomoniinae), larva. Fig. 171, dorsum of idiosoma and gnathosoma; Fig. 172, venter of idiosoma; Fig. 175, distal segments of leg III. *FIGURE 173* *Momonia campylotibia* Smith (Momoniinae), larva, distal segments of leg III. *FIGURES 174, 177* *Paramideopsis susanae* Smith (Nudomideopsidae), larva. Fig. 174, tibia and tarsus of leg III; Fig. 177, pedipalp. *FIGURE 176* *Nudomideopsis magnacetabula* (Smith), larva, excretory pore plate.

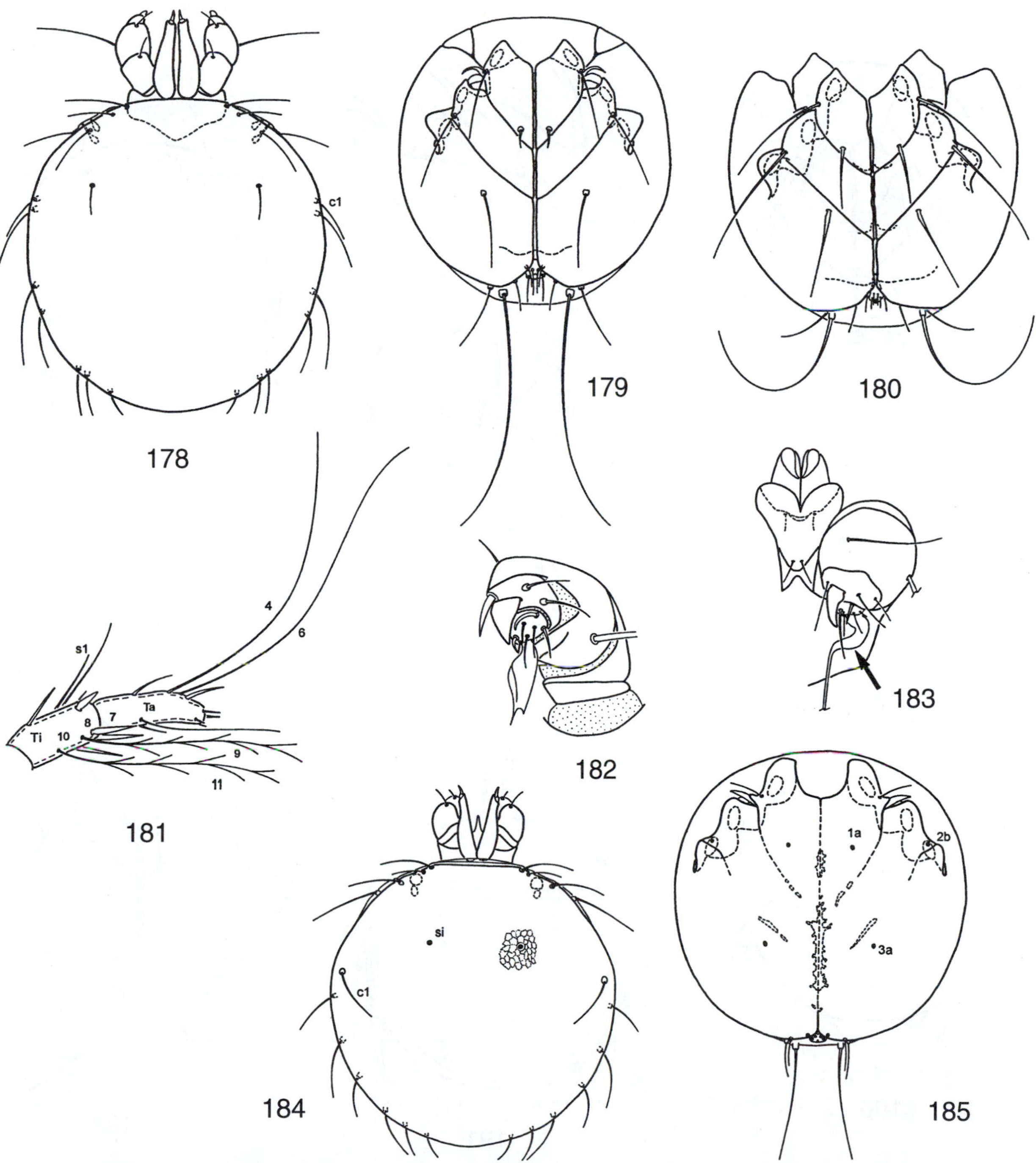

FFIGURES 178–179 Nudomideopsis magnacetabula (Smith) (Nudomideopsidae), larva. Fig. 178, dorsum of idiosoma and gnathosoma; Fig. 179, venter of idiosoma. *FIGURES 180–181, 183 Midea expansa* Marshall (Mideidae), larva. Fig. 180, venter of idiosoma; Fig. 181, tibia and tarsus of leg III; Fig. 183, venter of gnathosoma. *FIGURES 182, 184–185 Acalyptonotus neoviolaceus* Smith (Acalyptonotidae), larva. Fig. 182, pedipalp; Fig. 184, dorsum of idiosoma and gnathosoma; Fig. 185, venter of idiosoma.

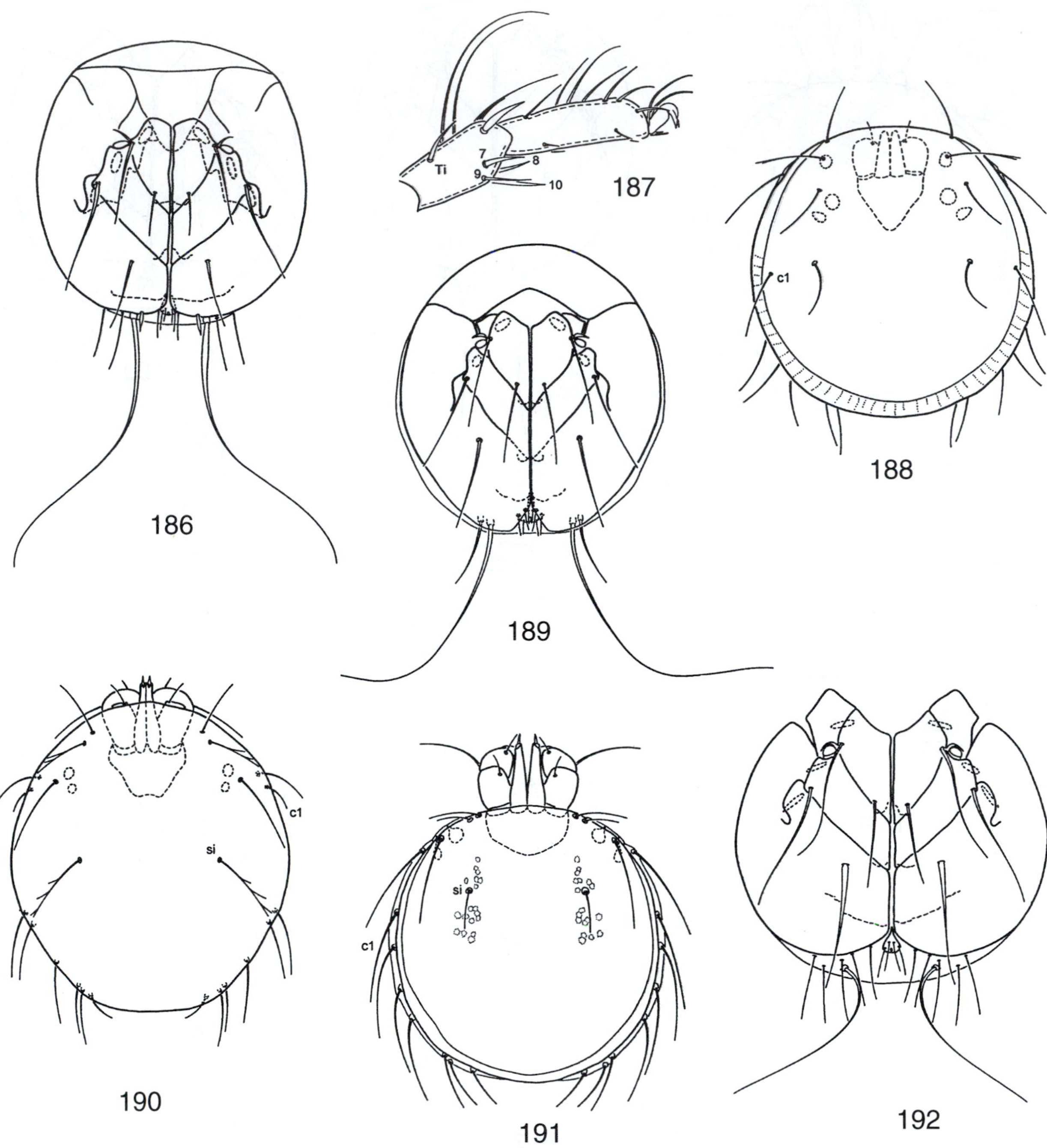

FIGURES 186, 188 *Mideopsis borealis* Habeeb (Mideopsinae), larva. Fig. 186, venter of idiosoma; Fig. 188, dorsum of idiosoma and gnathosoma. *Figures 187, 189–190* *Platyhydracarus juliani* Smith (Athienemanniinae), larva. Fig. 187, tibia and tarsus of leg II; Fig. 189, venter of idiosoma; Fig. 190, dorsum of idiosoma and gnathosoma. *FIGURE 191* *Neoacarus occidentalis* Cook (Neoacaridae), larva, dorsum of idiosoma and gnathosoma. *FIGURE 192* *Laversia berulophila* Cook (Larversiidae), larva, venter of idiosoma.

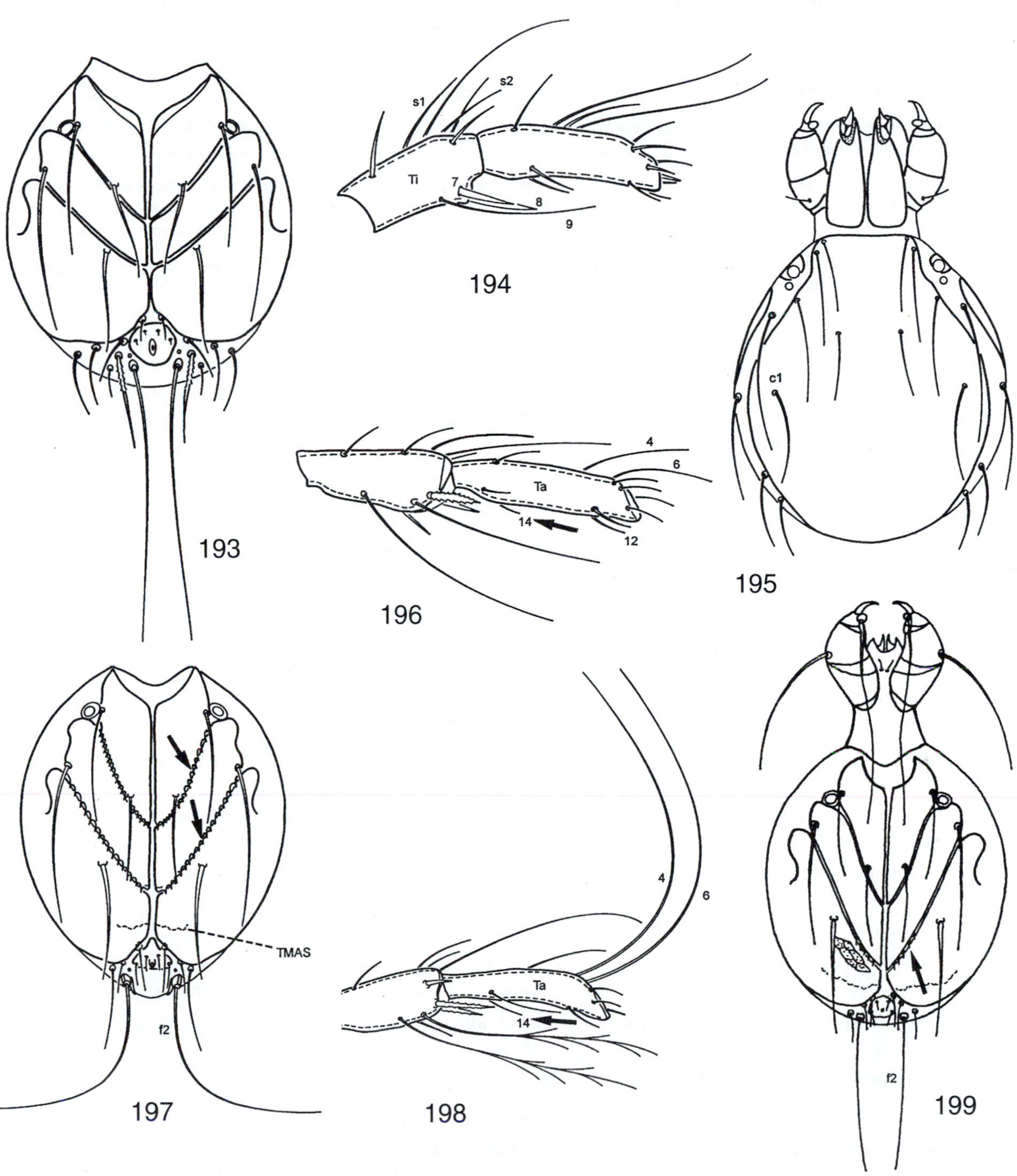

FIGURES 193, 195–196 Arrenurus planus Marshall (Arrenurinae), larva. Fig. 193, venter of idiosoma; Fig. 195, dorsum of idiosoma and gnathosoma; Fig. 196, tibia and tarsus of leg III. *FIGURE 194 Laversia berulophila* Cook (Laversiidae), larva, tibia and tarsus of leg II. *FIGURES 197–198 Krendowskia similis* Viets (Krendowskiidae), larva. Fig. 197, venter of idiosoma; Fig. 198, tibia and tarsus of leg III.

2. Taxonomic Key to Genera for Adult Water Mites

1a. Legs with tarsi bearing empodium and paired claws (Fig. 201); pedipalps with tarsus bearing single, long terminal seta (Fig. 202, arrow); idiosoma soft and elongate .... Stygothrombidioidea Stygothrombidiidae *Stygothrombium*

1b. Legs with tarsi bearing paired claws, but lacking empodium; pedipalps with tarsus lacking single long terminal seta as above; idiosoma various 2

2a(1b). Genital field with movable genital flaps, but with genital acetabula borne on coxal plates (Fig. 205, arrow) ......Hydrovolzioidea 3

2b. Genital field with or without movable genital flaps; when genital flaps present, genital acetabula borne on them (Fig. 200) or beneath them (Fig. 283) 4

3a(2a). Dorsal shield with large central plate surrounded by smaller unpaired plate anteriorly and numerous pairs of small similar platelets laterally and posteriorly (Fig. 204); glandularia represented by setae, with gland portions absent .... Acherontacaridae *Acherontacarus*

3b. Dorsal shield with large central plate surrounded by smaller unpaired plate anteriorly and several pairs of small disimilar plates laterally and posteriorly (Fig. 206); glandularia divided into gland and seta bearing sclerites (Fig. 203) .... Hydrovolziidae *Hydrovolzia*

4a(2b). Idiosoma nearly spherical (Fig. 207); pedipalps with tibia much shorter than genu and bearing a distidorsal projection extending well beyond insertion of tarsus (Fig. 209); chelicerae 1-segmented (Fig. 208) .... Hydrachnoidea Hydrachnidae *Hydrachna*

4b. Idiosoma variously shaped, but rarely spherical; pedipalps with tibia usually longer than genu, *when* shorter *then* lacking a distidorsal projection; chelicera 2-segmented (Fig. 316) 5

5a(4b). Lateral eye capsules borne on medial prodorsal plate (Figs. 210, 211, 212) or, in one family, on moderately large, rugose, paired platelets associated with complex of six prodorsal sclerites (Fig. 213) .... Eylaoidea 6

5b. Lateral eye capsules, when present, relatively small and not borne on prodorsal plate (Figs. 220, 224) 11

6a(5a). Dorsum with lateral eyes borne on lateral platelets separated by a medial triangular platelet (Fig. 213, arrows); idiosoma bearing 26 pairs of glandularia on large, rugose platelets (Figs. 213, 214); gnathosoma with capitulum wider than long and pedipalps with only one movable segment which flexes medially to oppose anetrior edge of capitulum (Fig. 214) .... Apheviderulicidae *Apheviderulix* (deutonymphs)

6b. Dorsum with lateral eyes borne on a medial prodorsal plate (Fig. 210, 211, 212); idiosoma bearing fewer than 26 pairs of glandularia on small sclerites; gnathosoma with capitulum longer than wide and pedipalps with 3 to 5 movable segments which flex ventrally 7

7a(6b). Lenses of lateral eyes well separated from each other on their respective sides and borne laterally on entire prodorsal plate bearing one pair of setae (Fig. 210) .... Eylaidae *Eylais*

7b. Lenses of lateral eyes relatively close together on their respective sides and borne laterally on smooth, entire prodorsal plate bearing four pairs of setae (Figs. 211, 212) 8

8a(7b). Prodorsal plate approximately as wide as long (Fig. 212); genital acetabula surrounded by acetabular plates .... Piersigiidae *Piersigia*

8b. Prodorsal plate much longer than wide (Fig. 211); genital acetabula scattered in ventral integument .... Limnocharidae 9

9a(8b). Gnathosoma protrusible; pedipalps three segmented, with tarsus very small and inserted dorsally on tibia (Fig. 215, arrow) .... Rhyncholimnocharinae *Rhyncholimnochares*

9b. Gnathosoma not protrusible; pedipalps 4- or 5-segmented, with tarsus inserted at distal end of tibia (Figs. 216, 217) .... Limnocharinae 10

10a(9b). Pedipalps 4-segmented (Fig. 217) .... *Neolimnochares*

10b. Pedipalps 5-segmented (Fig. 216) .... *Limnochares*

11a(5b). Dorsum with a series of reticulate platelets as shown in Figure 218 .... Rhynchohydracaridae Clathrosperchoninae *Clathrosperchon*

11b. Dorsum various, but not with an arrangement of platelets as shown in Figure 218 12

12a(11b). Pedipalps chelate, with dorsodistal portion of tibia extending well beyond insertion of tarsus (Figs. 221, 227, arrows); capitulum without anchoral process (as in Fig. 342, arrow) 13

12b. Pedipalps not chelate, *when* appearing chelate (as in some Tiphyinae) (Fig. 332) *then* capitulum with well-developed anchoral process (Fig. 330, arrow A) 38

13a(12a). Pedipalps with dorsodistal extension of tibia relatively long (Fig. 221, arrow) .... Hydrodromidae *Hydrodroma*

13b. Pedipalps with dorsodistal extension of tibia relatively short (Fig. 227, arrow) ....Hydryphantidae 14

14a(13b) Legs with swimming setae on distal segments....................15

14b. Legs without swimming setae 16

15a(14a). Dorsum with medial eye plate between lateral eyes bearing a pigmented medial eye and two pairs of setae (Figs. 219, 220) .... Hydryphantinae *Hydryphantes*

15b. Dorsum without medial eye plate between lateral eyes ....Pseudohydryphantinae *Pseudohydryphantes*

16a(14b). Third and fourth coxal plates with medial margins extensive and close together, not separated by genital field (Fig. 223); lateral eyes in capsules; body not elongated....................Cowichaniinae *Cowichania*

16b. Posterior coxal groups usually well separated from each other, with genital field extending between them (Fig. 237), *when* these coxal plates close together *then* lateral eyes not in capsules *and* body greatly elongated (Fig. 226) 17

17a(16b). Lateral eyes in capsules (Fig. 224, arrow)....................18

17b. Lateral eyes, when present, not in capsules....................37

18a(17a). Dorsum without dorsalia (Figs. 224, 230)....................19

18b. Dorsum with dorsalia of various sizes and arrangement (Figs. 232, 234, 245, 256) 22

19a(18a). Genital field bearing three pairs of acetabula ....Thyadinae (in part) 20

19b. Genital field bearing numerous acetabula (Fig. 228) ....Protziinae 21

20a(19a) Dorsum bearing well-developed medial eye containing two dots of pigment (Fig. 225); genital field with setiferous extensions of genital flaps extending anterior to first pair of acetabula (Fig. 229, arrow)....................*Notopanisus*

20b. Dorsum lacking medial eye (Fig. 230); genital field without setiferous extensions of genital flaps extending anterior to first pair of acetabula (Fig. 231)....................*Albertathyas*

21a(19b). Legs with tarsal claws bearing palmately arranged clawlets (Fig. 222) ....*Protzia*

21b. Legs with tarsal claws simple....................*Partnunia*

22a(18b). Dorsum with all dorsalia peripheral in position (Fig. 232, arrow) ....Cyclothyadinae *Cyclothyas*

22b. Dorsum with some dorsalia more medial in position (Figs. 245, 256)....................Thyadinae (in part) 23

23a(22b). Dorsum with single, large anterior plate, but without other dorsalia (Fig. 233) ....*Siskiyouthyas*

23b. Dorsum with or without single, large anterior plate, but always bearing other paired dorsalia 24

24a(23b). Dorsum with large anterior plate, and with two medial and four lateral pairs of platelets (Fig. 234)....................*Trichothyas*

24b. Dorsum with dorsalia variously developed and fused, but not as above 25

25a(24b). Dorsum with medial eye borne on spindle-shaped sclerite (Fig. 235, arrow); genital field with second pair of acetabula located about halfway between first and third pairs (Figs. 236, 237) ....*Euthyas*

25b. Dorsum with medial eye not borne on spindle-shaped sclerite; genital field with second pair of acetabula usually located in posterior half of gonopore 26

26a(25b). Dorsum with dorsalia fused into single dorsal shield (i.e., unsclerotized integument surrounds dorsoglandularia, but does not occur between dorsalia) (Figs. 238, 240, 241)....................27

26b. Dorsum with dorsalia not fused into single dorsal shield (i.e., unsclerotized integument separates dorsalia) (Figs. 242, 245).......29

27a(26a). Genital field with flaps not extending anterior to first pair of acetabula, and second pair of acetabula located beside third pair (Fig. 239) ....*Thyopsis*

27b. Genital field with flaps extending anterior to first pair of acetabula, and second pair of acetabula located well anterior to third pair (Figs. 244, 246)....................28

28a(27b). Dorsal shield well sclerotized, with some of thickened areas indicating medial dorsalia and adjacent lateral dorsalia fused on each side (Fig. 240, arrows) (edges of dorsalia obscured by secondary sclerotization in mature specimens); genital field with setiferous projections of genital flaps medial to third pair of acetabula relatively small (Fig. 244) ....*Thyopsella*

28b. Dorsal shield weakly sclerotized, with thickened areas indicating medial and lateral dorsalia separate (Fig. 241); genital field with setiferous projections of genital flaps medial to third pair of acetabula relatively large (Fig. 246)....................*Thyopsoides*

29a(26b). Dorsum with most posterior pair of dorsalia fused into single large plate (Figs. 242, 243, arrows)....................30

29b. Dorsum with most posterior pair of dorsalia separated medially (Figs. 245, 248, 250) 31

30a(29a). Dorsum with central region bearing three pairs of dorsalia that are separated medially (Fig. 242) .... *Panisus*

30b. Dorsum with central region bearing two pairs of dorsalia that are fused medially (Fig. 243) .... *Columbiathyas*

31a(29b). Genital field with well sclerotized extensions of genital flaps extending anterior to first pair of acetabula (Figs. 247, 249, arrows) 32

31b. Genital field without sclerotized extensions of genital flaps anterior to first pair of acetabula (Fig. 253) 34

32a(31a). Dorsum with lateral eye capsules raised and located dorsally on idiosoma (Fig. 245, arrow A) and with medial eye well developed (Fig. 245, arrow B); genital field with setae on anterior extensions of genital flaps relatively thick (Fig. 247, arrow) .... *Panisopsis*

32b. Dorsum with lateral eye capsules relatively flat and located at anterior edge of idiosoma (Fig. 248, arrow) and with medial eye indistinct or absent (Figs. 248); genital field with setae on anterior extension of genital flaps relatively slender (Fig. 249, arrow) 33

33a(32b). Dorsum with four pairs of medial dorsalia posterior to anteromedial plate (Fig. 248) .... *Parathyasella*

33b. Dorsum with five pairs of medial dorsalia posterior to anteromedial plate (Fig. 251) .... *Thyasella*

34a(31b). Dorsum with dorsalia large and occupying most of the dorsal area of the idiosoma (Fig. 250) .... *Amerothyasella*

34b. Dorsum with dorsalia small and occupying only a small part of the dorsal area of the idiosoma (Figs. 252, 256) 35

35a(34b). Dorsum with medial eye bearing two pigment dots and usually borne on small platelet formed by fused pre—and postocularia (Fig. 252, arrow) .... *Thyasides*

35b. Dorsum with medial eye unpigmented or with dispersed pigment, and not borne on small platelet (Fig. 256) 35

36a(35b). Genital field with posterior two pairs of acetabula not incorporated into edges of genital flaps (Fig. 253) .... *Thyas*

36b. Genital field with posterior two pairs of acetabula fused with edges of genital flaps (Fig. 257) .... *Zschokkea*

37a(17b). Body not greatly elongated; genital flaps well developed (Fig. 254, arrow) .... Tartarothyadinae *Tartarothyas*

37b. Body greatly elongated (Fig. 226); genital flaps poorly developed or absent .... Wandesiinae *Wandesia*

38a(12b). Femur of pedipalp with two long medial setae (Fig. 255, arrow); occuring only in hot springs .... Thermacaridae *Thermacarus*

38b. Femur of pedipalp usually without two medial setae and never as illustrated in Figure 255; not occurring in hot springs 39

39a(38b). Genital field with movable genital flaps flanking gonopore and either partially or completely covering gonopore when closed; genital acetabula lying free in gonopore (not on flaps or acetabular plates) (Figs. 259, 277, 283) .... Lebertioidea 40

39b. Genital field usually without movable genital flaps, *when* flaps present *then* acetabula lie on flaps rather than in gonopore (Fig. 200) 59

40a(39a). Fourth coxal plates with medial edges reduced to narrow angles and bearing a pair of glandularia (Fig. 264, arrow); tarsus of pedipalp bearing pad-shaped setae and appearing spatulate distally (Fig. 263, arrow) .... Rutripalpidae *Rutripalpus*

40b. Fourth coxal plates with medial edges usually extensive, *when* reduced to narrow angles *then* lacking glandularia as above 41

41a(40b). Dorsal and ventral shields present; dorsal shield usually comprising a large plate and a series of smaller platelets (Figs. 273, 276, 278); genital field with gonopore usually bearing six pairs of acetabula (Fig. 279), *when* only three pairs of genital acetabula present *then* dorsal shield with complete ring of marginal platelets (Fig. 273); one known North American species has platelets fused into an entire dorsal shield, but six pairs of genital acetabula are diagnostic .... Torrenticolidae 42

41b. Dorsal and ventral shields present or absent; when present, dorsal shield entire, without smaller platelets; genital field with gonopore bearing three pairs of acetabula 46

42a(41a). Dorsal shield with single anteromedial platelet and series of small marginal platelets (Fig. 273); genital field with gonopore bearing three pairs of genital acetabula .... Testudacarinae *Testudacarus*

42b. Dorsal shield with one or two pairs of anterior platelets (Figs. 276, 278) (in one species platelets are fused with large central plate); genital field with gonopore bearing six pairs of acetabula .... Torrenticolinae 43

43a(41b). Pedipalps 4-segmented (Fig. 275) .... *Neoatractides*

43b. Pedipalps 5-segmented (as in Fig. 274) 44

44a(43b). Gnathosoma protrusible, capitulum long and slender(Fig. 280) .... *Pseudotorrenticola*

44b. Gnathosoma not protrusible, capitulum relatively short and stocky (Fig. 281) 45

45a(44b). Capitulum with short proximodorsal projections (Fig. 281, arrow) .... *Torrenticola*

45b. Capitulum with long proximodorsal projections (Fig. 282, arrow) .... *Monoatractides*

46a(41b). Fourth coxal plates bearing glandularia (Fig. 265, arrow) .... Teutoniidae *Teutonia*

46b. Fourth coxal plates lacking glandularia (Figs. 271, 283) 47

47a(46b). Venter without medial suture line separating coxal plates (Fig. 289); all legs inserted near anterior end of idiosoma (Fig. 289) .... Oxidae 48

47b. Venter with medial suture line separating coxal plates of either side (Fig. 283, arrow A); fourth pair of legs inserted laterally near middle of idiosoma (Fig. 283, arrow C) 49

48a(47a). Idiosoma strongly compressed laterally with coxal plates extending onto dorsal surface leaving only narrow medial unsclerotized strip of soft integument that usually bears small sclerites (Fig. 291); sexual dimorphism not pronounced .... *Frontipoda*

48b. Idiosoma ovoid or moderately compressed laterally, without small sclerites in dorsal integument (Fig. 288); sexual dimorphism usually not pronounced, but in some species coxal plates extend more dorsally in males .... *Oxus*

49a(47b). Venter with suture lines between second and third coxal plates incomplete (Fig. 283, arrow B), but touching the genital field area posteriorly; genu of pedipalp usually with at least five long setae on medial surface (Fig. 284), but, in one known species, with four long setae on dorsal surface .... Lebertiidae 50

49b. Venter with suture lines between second and third coxal plates complete (Fig. 268) and these coxal plates often completely separated on their respective sides (Fig. 259); genu of pedipalp without long setae 52

50a(49a). Idiosoma strongly compressed laterally (Fig. 285) .... *Estelloxus*

50b. Idiosoma ovoid or dorsoventrally flattened 51

51a(50b). Dorsum covered by complete dorsal shield (Fig. 286); pedipalps with femur bearing several long setae dorsally and genu bearing four long setae dorsally (Fig. 287) .... *Scutolebertia*

51b. Dorsum usually covered by soft integument, rarely covered by dorsal shield; pedipalps with femur bearing few short setae and genu bearing five or six long setae on medial surface (Fig. 284) .... *Lebertia*

52a(49b). Dorsum *and* venter covered by complete shields .... Anisitsiellidae 53

52b. Dorsum and venter with sclerites variously developed, but complete dorsal *and* ventral shields never both present .... Sperchontidae 58

53a(52a). Tarsi of fourth pair of legs lacking claws (Fig. 267) .... *Mamersellides*

53b. Tarsi of fourth pair of legs bearing claws (Fig. 266) 54

54a(53b). Fourth pair of legs inserted at anterolateral angles of fourth coxal plates, just posterior to insertions of third pair of legs (Fig. 268, arrow) .... *Utaxatax*

54b. Fourth pair of legs inserted near posterolateral angles of fourth coxal plates, well posterior to insertions of third pair of legs (Fig. 271, arrow A) 55

55a(54b). Accessory glandularia ("Glandula Limnesiae") located in concavities of anteromedial edges of fourth coxal plates (Fig. 269, arrow) .... *Bandakiopsis*

55b. Accessory glandularia located on first or third coxal plates 56

56a(55b). Accessory glandularia located near anterior angles of first coxal plates (Fig. 270, arrow) .... *Cookacarus*

56b. Accessory glandularia located on third coxal plates 57

57a(56b). Accessory glandularia located near medial edges of third coxal plates (Fig. 271, arrow B); fourth coxal plates without or with only small projections associated with insertions of fourth pair of legs .... *Bandakia*

57b. Accessory glandularia located near lateral edges of third coxal plates (Fig. 272, arrow A); fourth coxal plates with large projections associated with insertions of fourth pair of legs (Fig. 272, arrow B) .... *Oregonacarus*

58a(52b). Idiosoma with glandularia borne on smooth platelets that are usually small and flat, rarely enlarged and raised (Fig. 258); pedipalps with tibia usually bearing two peglike setae ventrally (Fig. 260, arrows) .... *Sperchon*

58b. Idiosoma with glandularia borne on tuberculate or papillate platelets that are enlarged and raised (Fig. 261); pedipalps with tibia lacking peg-like setae ventrally (Fig. 262) .... *Sperchonopsis*

59a(39b). All coxal plates grouped together, with medial edges of anterior coxal groups noticeably longer than those of posterior coxal groups (Fig. 292); fourth coxal plates rounded posteriorly and suture lines between third and fourth plates extending posterolaterally (Fig. 292) .... Omartacaridae *Omartacarus*

59b. Coxal plates not as described and illustrated above 60

60a(59b). Pedipalps with femur usually bearing single ventral seta (Fig. 296, 306, arrows), *when* this seta absent *then* tarsi of pedipalps with curved terminal seta (Fig. 297, arrow); genital field with movable genital flaps which cover gonopore when closed in females (Fig. 200) and some males .... Limnesiidae 61

60b. Pedipalps with femur lacking ventral setae or, rarely, bearing three setae (Fig. 414, arrow); genital field without movable genital flaps which cover gonopore when closed 69

61a(60a). Pedipalps with tibia bearing five or more long, pointed ventral tubercles and a long medial seta (Fig. 296); idiosoma lacking dorsal and ventral shields, although dorsal plates present .... Kawamuracarinae *Kawamuracarus*

61b. Pedipalps with tibia bearing four long, pointed tubercles (Fig. 295) or lacking tubercles (Fig. 298), and lacking a long medial seta; idiosoma with dorsal and ventral shields 62

62a(61b). Fourth pair of legs with tarsus lacking claws 63

62b. Fourth pair of legs with tarsus bearing claws 67

63a(62a). Genital field with genital flaps on each side of gonopore divided into anterior and posterior sclerites (Fig. 299); dorsum with large, transversely divided dorsal shield (Fig. 294) .... Neomamersiinae 64

63b. Genital field with genital flaps of females (and acetabular plates of males) entire on each side (Fig. 200); dorsum usually without dorsal shield, but rarely with entire (undivided) shield .... Limnesiinae 66

64a(63a). Ventral shield with sclerotized bridge forming dorsal side of camerostome incomplete (Fig. 304, arrow A); venter with posterior region bearing several small platelets (Fig. 304, arrow B) or at least a small excretory pore platelet separate from ventral shield; genital field of both sexes with middle pair or group of acetabula borne on posterior sclerites of genital flaps (Fig. 300); pedipalps with femur and tibia relatively long and slender and tibia lacking ventral tubercles (Fig. 298) .... *Meramecia*

64b. Ventral shield with sclerotized bridge forming dorsal side of camerostome complete (Fig. 302, arrow A); ventral shield entire posteriorly (Figs. 302, 303); genital field of males with middle pair or group of acetabula borne on anterior sclerites of genital flaps (Fig. 301), or on pair of separate, small, medial sclerites (Fig. 299); pedipalps not as above 65

65a(64b). Pedipalps with femur bearing ventral seta, tibia bearing four long ventral tubercles and tarsus lacking long, curved distal seta (Fig. 295) .... *Neomamersa*

65b. Pedipalps with femur lacking ventral seta, tibia lacking ventral tubercles and tarsus bearing long, curved distal seta (Fig. 297, arrow) .... *Arizonacarus*

66a(65b). Genital field of males immediately flanked by pair of glandularia (Fig. 305) .... *Centrolimnesia*

66b. Genital field of males not immediately flanked by pair of glandularia .... *Limnesia*

67a(62b). Dorsum nearly covered by two large plates (Fig. 307); pedipalps with femur bearing ventral seta inserted directly on surface of segment (as in Fig. 296, arrow) .... Protolimnesiinae *Protolimnesia*

67b. Dorsum entirely covered by dorsal shield (Fig. 310) or less than half covered by small platelets (Fig. 311); pedipalps with femur bearing ventral seta inserted on prominent tubercle (Fig. 306, arrow) .... Tyrrelliinae 68

68a(67b). Third coxal plates with pair of accessory glandularia ("Glandula Limnesiae") (Fig. 308, arrow); genital field with numerous small acetabula (Fig. 308) .... *Neotyrrellia*

68b. Third coxal plates lacking glandularia (Fig. 309); genital field with 3–7 pairs of relatively large acetabula (Fig. 309) .... *Tyrrellia*

69a(60b). Integument between fourth coxal plates and genital field with two pairs of glandularia arranged more or less in a row, and either free (Fig. 312, arrows) or variously fused with other ventral sclerites; dorsal shield absent, or present and bearing fewer than five pairs of glandularia .... Feltriidae *Feltria*

69b. Integument between fourth coxal plates and genital field usually without two pairs of glandularia arranged in a row, *when* glandularia in this position (some *Koenikea*) *then* dorsal shield bearing six pairs of glandularia (Fig. 322) 70

70a(69b). Idiosoma strongly compressed laterally and much higher than wide (Fig. 314); fourth legs with all segments flattened (Fig. 313) .... Frontipodopsidae *Frontipodopsis*

70b. Idiosoma not strongly compressed laterally 71

71a(70b). Both dorsal and ventral shields present; first pair of legs with tibia lacking two thickened setae and a downturned seta distoventrally (ie. not as shown in Fig. 321) 72

71b. Either dorsal or ventral shield, or both, usually absent, *when* both shields present (males of some *Atractides*) *then* first pair of legs with tarsus bowed and tibia bearing two thickened setae and a downturned seta distoventrally (Fig. 321) 125

72a(71a). First leg with tarsus much shorter than tibia, and claws dorsal in position and flexing proximally (Fig. 381, arrow) .... Momoniidae 73

72b. First leg with distal segments not modified as described and illustrated above; claws terminal in position and flexing ventrally...75

73a(72a). Dorsal shield composed of large plate surrounded by numerous smaller platelets (Fig. 380) ......Cyclomomoniinae *Cyclomomonia*

73b. Dorsal shield entire or comprising large anterior and posterior plates, but not surrounded by smaller platelets (ie. not as in Fig. 380) 74

74a(73b). Venter with pair of glandularia located close to anterior end of genital field (Fig. 382, arrow); genital field of males with acetabula borne in gonopore ....Momoniinae *Momonia*

74b. Venter without glandularia near anterior end of genital field (Fig. 383); genital field of males with acetabula borne on plates flanking gonopore (Fig. 383)..Stygomomoniinae *Stygomomonia*

75a(72b). Dorsal shield large and transversely divided (somewhat as in Fig. 294); suture lines between third and fourth coxal plates noticeably looped around a pair of glandularia (Fig. 293, arrow) ....Hygrobatidae (in part) *Diamphidaxona*

75b. Dorsal shield usually entire, never transversely divided as above; suture lines between third and fourth coxal plates not looped as above 76

76a(75b). Venter with outer edges of coxal plates forming a smooth arc which does not extend beyond edge of idiosoma (Fig. 389, arrow); pedipalps highly modified (Fig. 390)..Chappuisididae (in part) Uchidastygacarinae *Uchidastygacarus*

76b. Venter with outer edges of coxal plates not forming a smooth arc; pedipalps not as described and illustrated above 77

77a(76b). Venter with pair of glandularia located near midline at junction of third and fourth coxal plates (Fig. 388, arrow)..Chappuisididae (in part) Chappuisidinae *Chappuisides*

77b. Venter without glandularia in position described and illustrated above..78

78a(77b). Ventral shield with suture lines between third and fourth coxal plates oblique, extending posteromedially to genital field region and well separated from each other medially (Fig. 395, arrow); genital acetabula borne in single rows on each side, either in gonopore (males) or on slender plates flanking gonopore (females) (Fig. 395) ....Neoacaridae 79

78b. Ventral shield not with above combination of characters 80

79a(78a). Pedipalps with tibia moderately produced distoventrally to oppose tarsus (Fig. 396) ....*Neoacarus*

79b. Pedipalps with tibia and tarsus highly modified to form chelate appendage (Fig. 397)..*Volsellacarus*

80a(78b). Ventral shield with coxoglandularia I shifted far forward in coxal plates II, near suture lines between first and second coxal plates (Figs. 384–387, arrows)..81

80b. Ventral shield with coxoglandularia I located between second and third coxal plates (Fig. 407, arrow) or in posterior edges of second coxal plates (Fig. 406, arrow A) 84

81a(80a). Ventral shield with genital field projecting between, and widely separating, fourth coxal plates (Fig. 384) ....Mideidae *Midea*

81b. Ventral shield with genital field located posterior to fourth coxal plates (Fig. 386) ....Nudomideopsidae 82

82a(81b). Genital field with three pairs of acetabula (Fig. 385)..*Nudomideopsis*

82b. Genital field with more than three pairs of acetabula 83

83a(82b). Genital field with four-to-six pairs of acetabula arranged in single rows (Fig. 386)..*Paramideopsis*

83b. Genital field with more than six pairs of acetabula arranged in several rows (Fig. 387) ....*Neomideopsis*

84a(80b). Pedipalps distinctly uncate (Fig. 401; Fig. 399, arrow A; Fig. 416, arrow) 85

84b. Pedipalps not distinctly uncate, with tibia bearing *either* slight distoventral projection (Fig. 417, arrow), *or* distinct ventral projection near middle of segment (Fig. 405), *or* no ventral projection..91

85a(84a). Ventral shield with fourth coxal plates bearing pair of glandularia (Fig. 413, arrow); genital field extending between fourth coxal plates, widely separating them medially, and usually bearing three or four pairs of genital acetabula (but with five or six pairs in a few species) ....Krendowskiidae 86

85b. Ventral shield with fourth coxal plates lacking glandularia; genital field extending at most only slightly between fourth coxal plates, not separating them medially, and bearing numerous acetabula (Fig. 402) 87

86a(85a). Gnathosoma protrusible..*Geayia*

86b. Gnathosoma not protrusible .... *Krendowskia*

87a(85b). Genital field with some or all acetabula lying free in gonopore (Fig. 402), or with acetabula lying on plates which flank gonopore, but are not fused with ventral shield (similar to Fig. 384); capitulum with a pair of long ventral setae (Fig. 401, arrow); tibia of pedipalp rotated relative to genu (up to 90°) .... Athienemanniidae 88

87b. Genital field with acetabula borne on acetabular plates incorporated into ventral shield, never in gonopore (Fig. 415); capitulum with all setae short; tibia of pedipalp not greatly rotated relative to genu (Fig. 416); posterior end of idiosoma variously (and often highly) modified in males (Figs. 419, 420) .... Arrenuridae Arrenurinae *Arrenurus*

88a(87a). Pedipalp with genu bearing pronounced ventral projection (Fig. 399, arrow B); genital field with acetabula in one or two rows on each side of gonopore (Fig. 398) .... Stygameracarinae *Stygameracarus*

88b. Pedipalp with genu lacking pronounced ventral projection (Figs. 401); genital field with acetabula in one to many rows on each side of gonopore (Figs. 402–404) .... Athienemanniinae 89

89a(88b). Venter with coxal plates I extending anteriorly beyond edge of ventral shield (Fig. 404); genital field of males with acetabula borne in one or two rows on each side of gonopore (Fig. 404) .... *Chelohydracarus*

89b. Venter with coxal plates I not extending anteriorly beyond edge of ventral shield (Fig. 402); genital field of males with acetabula borne in three or more rows on each side of gonopore (Figs. 400, 402) 90

90a(89b). Genital field of males with all acetabula borne in gonopore (Fig. 402) .... *Chelomideopsis*

90b. Genital field of males with acetabula borne both in gonopore and on acetabular plates flanking gonopore (Fig. 400) .... *Platyhydracarus*

91a(84b). Fourth pair of legs with tarsus indented or curved, but not greatly expanded, and bearing short peglike setae (Fig. 339) .... Pionidae (in part) Foreliinae (in part) 92

91b. Fourth pair of legs with tarsus not as described and illustrated above 94

92a(91a). Third pair of legs with tarsus and claws similar in shape to those of first two pairs of legs .... *Pseudofeltria* (males)

92b. Third pair of legs with tarsus and claws modified, different from those on first two pairs of legs (Fig. 340) 93

93a(92b). Genital field with three pairs of acetabula (Fig. 336) .... *Pionacercus* (males of some species)

93b. Genital field with more than three pairs of acetabula (Fig. 338) .... *Forelia* (males of some species)

94a(91b). Fourth pair of legs with genu notched dorsally and bearing short peglike setae in notch (Fig. 343) .... Pionidae (in part) Pioninae (in part) 95

94b. Fourth pair of legs with genu not notched as described and illustrated above 96

95a(94a). Capitulum with anchoral process (Fig. 330, arrow A) .... *Piona* (males of some species)

95b. Capitulum lacking anchoral process (Fig. 341 and as shown in Fig. 342, arrow) .... *Nautarachna* (males)

96a(94b). Pedipalps with segments short and partially fused, femur with three ventral setae (Fig. 414, arrow) .... Bogatiidae Horreolaninae *Horreolanus*

96b. Pedipalps not as described and illustrated above 97

97a(96b). Genital field with three or four pairs of acetabula 98

97b. Genital field with more than four pairs of acetabula 102

98a(97a). Ventral shield completely surrounding gonopore; genital field with acetabula borne in gonopore (Fig. 407) or in pair of straight or slightly curved rows on ventral shield flanking gonopore, *when* rows of acetabula slightly curved *then* genital field located well anterior to posterior edge of idiosoma (Fig. 394) 99

98b. Ventral shield not completely surrounding gonopore posteriorly (females of most genera) *or* completely surrounding gonopore and with genital acetabula usually arranged in pronounced triangle or arc (males of all genera and females of one genus) (Fig. 368), *when* acetabula arranged in weak arc *then* genital field located at extreme posterior edge of ventral shield 112

99a(98a). Dorsal shield with four pairs of glandularia including pair at anterior edge (Fig. 393, arrow); ventral shield with suture lines between third and fourth coxal plates oriented at right angles to long axis of body where they meet midline (Fig. 394, arrow) .... Chappuisididae (in part) Morimotacarinae 100

99b. Dorsal shield with three pairs of glandularia (as in Fig. 411); ventral shield with suture lines between third and fourth coxal plates oriented at oblique angle to long axis of body where they meet midline 101

100a(99a). Pedipalps with five segments (Fig. 391) .... *Morimotacarus*

100b. Pedipalps with four segments, including fused tibiotarsus (Fig. 392) .... *Yachatsia*

101a(99b). Pedipalps with base of tarsus only about half as high as distal end of tibia, and inserted distidorsally on tibia (Fig. 417); ventral shield with insertions of fourth pair of legs covered by projections of fourth coxal plates (Fig. 418, arrow) .... Acalyptonotidae (in part) *Paenecalyptonotus*

101b. Pedipalps with base of tarsus approximately equal in height to distal end of tibia, and inserted distally on tibia (Fig. 405); ventral shield with insertions of fourth pair of legs not covered by projections of fourth coxal plates (Fig. 407) .... Mideopsidae Mideopsinae *Mideopsis*

102a(97b). Dorsal shield bearing six pairs of glandularia, with three pairs grouped close together in an arc or triangle on each side (Fig. 322, circled area) .... Unionicolidae (in part) Pionatacinae (in part) *Koenikea*

102b. Dorsal shield not as described and illustrated above 103

103a(102b). Gonopore located near middle of ventral shield (Fig. 406) 104

103b. Gonopore located near posterior end of ventral shield 105

104a(103a). Ventral shield incorporating excretory pore (Fig. 406, arrow B) .... Laversiidae *Laversia*

104b. Ventral shield not incorporating excretory pore .... Pionidae (in part) Pioninae (in part) *Nautarachna* (females of some species)

105a(103b). Ventral shield with first coxal plates projecting beyond anterior edge of ventral shield (Figs. 377, 410, 412) 106

105b. Ventral shield with first coxal plates not projecting beyond anterior edge of ventral shield (Fig. 357) 123

106a(105a). Pedipalps with femur and tibia bearing distinct ventral projections (Fig. 409, arrows); third and fourth pair of legs with well-developed swimming setae (Fig. 408) .... Amoenacaridae *Amoenacarus*

106b. Pedipalps with femur and tibia usually lacking ventral projections, *when* bearing projections *then* third and fourth pair of legs lack swimming setae 107

107a(106b). Dorsal shield bearing three pairs of glandularia peripherally (Fig. 411) .... Arenohydracaridae *Arenohydracarus*

107b. Dorsal shield bearing one to several pairs of glandularia, with at least one pair located well medial to edge (Fig. 376) .... Aturidae (in part) Aturinae 108

108a(107b). Dorsal shield bearing only one pair of glandularia (Fig. 379); pedipalps with segments extremely long and slender (Fig. 378) .... *Bharatalbia*

108b. Dorsal shield bearing three or more pairs of glandularia (Fig. 371); pedipalps with segments relatively short and stocky (Fig. 373) 109

109a(108b). Ventral shield bearing pronounced projections associated with openings of insertions of fourth legs (Figs. 374, 375) 110

109b. Ventral shield lacking projections associated with openings of insertions of fourth legs (Fig. 372, 377) 111

110a(109a). Ventral shield with fourth coxal plates bearing pair of conspicuous glandularia medially (Fig. 374, arrow) .... *Neoaturus*

110b. Ventral shield with fourth coxal plates lacking glandularia (Fig. 375) .... *Kongsbergia*

111a(109b). Ventral shield with genital acetabula in more than one row and restricted to region immediately lateral to gonopore near posterior edge of idiosoma (Fig. 372) .... *Phreatobrachypoda*

111b. Ventral shield with genital acetabula in single row and distributed along posterolateral edge of shield between gonopore and insertions of fourth legs (Fig. 377) .... *Aturus*

112a(98b). Dorsal shield comprising large central plate completely surrounded by ring of small, paired platelets (Fig. 347) .... Lethaxonidae (in part) *Lethaxona*

112b. Dorsal shield comprising single large plate (although smaller posterior platelet bearing excretory pore may be present) (Fig. 365) .... Aturidae (in part) Axonopsinae (in part) 113

113a(112b). Ventral shield with posterior edges of fourth coxal plates clearly indicated by suture lines that are indented posteromedially to accommodate posterior coxoglandularia (Fig. 348, arrow) .... *Ljania*

113b. Ventral shield with posterior edges of fourth coxal plates usually faintly indicated or obliterated by fusion with shield (Fig. 368) and never indented posteromedially to accommodate posterior coxoglandularia 114

114a(113b). Fourth coxal plates with large projections associated with insertions of fourth pair of legs (Fig. 350, arrow A; Fig. 354, arrow) 115

114b. Fourth coxal plates with small projections or none associated with insertions of fourth pair of legs (Fig. 369) 116

115a(114a). Fourth coxal plates with pair of closely spaced glandularia in females (Fig. 349, arrow) and with looped suture lines extending anteriorly in males (Fig. 350, arrow B) ..... *Stygalbiella*

115b. Fourth coxal plates with glandularia widely spaced in females (Fig. 351) and without looped suture lines in males (Fig. 352) ..... *Axonopsella*

116a(114b). Ventral shield lacking ridges in region anterolateral to insertions of fourth pair of legs (Figs. 355, 356) 117

116b. Ventral shield bearing well-developed ridges extending anterolaterally from region of insertion of fourth pair of legs (Fig. 368, arrow) 118

117a(116a). Ventral shield with anterior edges of first coxal plates extending to or beyond anterior edge of ventral shield (Fig. 355); genital field with three pairs of acetabula (Fig. 355) ..... *Albaxona*

117b. Ventral shield with anterior edges of first coxal plates not extending to anterior edge of ventral shield (Fig. 356); genital field with four pairs of acetabula (Fig. 356) ..... *Javalbia*

118a(116b). Pedipalps with femur bearing spinelike projection in proximal three-quarters of ventral surface (Fig. 370); genital field of females separated from ventral shield (Fig. 358) 119

118b. Pedipalps with femur lacking spinelike projection ventrally, although rounded projection may be present at distiventral extremity (Fig. 366); genital field of females separated from or fused with ventral shield 121

119a(118a). Genital field bearing four pairs of acetabula (Fig. 369); fourth pair of legs of males not highly modified ..... *Neobrachypoda*

119b. Genital field bearing three pairs of acetabula (Fig. 358); fourth pair of legs of males highly modified 120

120a(119b). Fourth coxal plates bearing ridges extending posteriorly from region of insertions of fourth pair of legs (Fig. 358); fourth pair of legs of males with tarsus expanded distally and flattened (Fig. 362) ..... *Estellacarus*

120b. Fourth coxal plates lacking ridges extending posteriorly from region of insertions of fourth pair of legs (Fig. 359); fourth pair of legs of males with tibia expanded and curved and tarsus slightly bowed and with enlarged claws (Fig. 363) ..... *Brachypoda*

121a(118b). Fourth coxal plates of males bearing prominent ridges extending posteriorly from region of insertions of fourth pair of legs (Fig. 364, arrow); fourth pair of legs of males with tibia and tarsus thicker and stockier than those of females (Fig. 367); genital field of females separated from ventral shield ..... *Woolastookia*

121b. Fourth coxal plates of males lacking prominent ridges extending posteriorly from region of insertions of fourth pair of legs (as in Fig. 360), although short, faint ridges may be evident in some species (Fig. 368); fourth pair of legs of males similar to those of females, either flattened or unmodified; genital field of females fused with ventral shield (Fig. 360) 122

122a(121b). Fourth pair of legs with segments flattened and with telofemur relatively short and greatly expanded distally (Fig. 361) ..... *Erebaxonopsis*

122b. Fourth pair of legs with segments nearly cylindrical and with telofemur similar to other segments ..... *Axonopsis*

123a(105b). Fourth coxal plates usually with projections partially covering insertions of fourth pair of legs, *when* projections absent *then* genital field with acetabula arranged in single rows extending laterally from gonopore in ventral shield (Fig. 346) 124

123b. Fourth coxal plates lacking projections associated with insertions of fourth pair of legs *and* genital field with acetabula arranged in several rows on acetabular plates (Fig. 357) ..... Aturidae (in part) Albiinae *Albia*

124a(123a). Dorsal shield entire (Fig. 353); fourth coxal plates with projections associated with insertions of fourth pair of legs (Fig. 354, arrow) ..... Aturidae (in part) Axonopsinae (in part) *Submiraxona*

124b. Dorsal shield comprising large central plate completely surrounded by ring of small, paired platelets (Fig. 345); fourth coxal plates lacking projections associated with insertions of fourth pair of legs (Fig. 346) ..... Lethaxonidae (in part) *Lethaxonella*

125a(71b). Fourth coxal plates bearing glandularia (Fig. 319, arrow) ..... Hygrobatidae (in part) 126

125b. Fourth coxal plates lacking glandularia 129

126a(125a). Gnathosoma with capitulum broadly fused with first coxal plates (Fig. 315) 127

126b. Gnathosoma with capitulum separate from first coxal plates (Fig. 320) 128

127a(126a). First pair of legs with tarsus straight (Fig. 317) ..... *Hygrobates*

127b. First pair of legs with tarsus bowed (Fig. 318) ..... *Mesobates*

128a(126b). First coxal plates fused medially (Fig. 319); first pair of legs with distal segments modified as in Fig. 321 ..... *Atractides*

128b. First coxal plates separate medially (Fig. 320); first pair of legs with distal segments not modified .... *Corticacarus*

129a(125b). Venter with apodemes of anterior coxal groups extending to approximately middle of fourth coxal plates (Fig. 323, arrow) ....Unionicolidae (in part) Pionatacinae (in part) *Neumania*

129b. Venter with apodemes of anterior coxal groups not extending to middle of fourth coxal plates 130

130a(129b). Genital field surrounded by well developed ventral shield (Fig. 418) ....Acalyptonotidae (in part) *Acalyptonotus*

130b. Genital field not surrounded by ventral shield 131

131a(130b). Fourth coxal plates with medial edges reduced to angles (Fig. 338, arrow) 132

131b. Fourth coxal plates with medial edges not reduced to angles (Fig. 344, arrow) 135

132a(131a). First pair of legs with tarsal claw socket large, occupying much more than one half dorsal surface of segment (Fig. 326, arrow); genital field with three or four pairs of acetabula; fourth pair of legs with tarsus unmodified in males..............Wettinidae *Wettina*

132b. First pair of legs with tarsal claw socket smaller, occupying one half or less of dorsal surface of segment; genital field with three or more pairs of acetabula; fourth pair of legs with tarsus indented or bowed and bearing short peglike setae in males (Fig. 339) ....Pionidae (in part) Foreliinae (in part) 133

133a(132b). Genital field with three pairs of acetabula (Fig. 336) ....*Pionacercus*

133b. Genital field with more than three pairs of acetabula (Figs. 337, 338) 134

134a(133b). Legs with swimming setae on distal segments ....*Forelia* (females of all species, males of some species)

134b. Legs without swimming setae on distal segments ....*Pseudofeltria* (females)

135a(131b). Capitulum without anchoral process (Fig. 342, arrow); symbionts of clams................Pionidae (in part) Najadicolinae *Najadicola*

135b. Capitulum with anchoral process (Fig. 330, arrow A); symbionts of clams or free living 136

136a(135b). Pedipalps with segments relatively long and slender (Fig. 329); genital field with three pairs of acetabula; males with petiole posterior to genital field (Fig. 327); and with genu of third pair of legs modified (Fig. 328).. ....Pionidae (in part) Hydrochoreutinae *Hydrochoreutes*

136b. Pedipalps with segments relatively short and stocky; genital field with three to many pairs of acetabula; males lacking petiole posterior to genital field and with genu of third pair of legs unmodified..137

137a(136b). Venter with suture lines between third and fourth coxal plates extending only a short distance toward midline (Fig. 324, arrow A); with a pair of glandularia present at, and usually fused with, posteromedial angles of fourth coxal plates (Fig. 324, arrow B); first pair of legs with long movable setae in free-living species (Fig. 325); symbionts of clams or sponges, or free living ....Unionicolidae (in part) Unionicolinae *Unionicola*

137b. Venter with suture lines between third and fourth coxal plates extending to (or nearly to) medial margins of coxal plates (as in Fig. 330, usually longer than shown in Fig. 344); without glandularia at posteromedial angles of fourth coxal plates; first pair of legs lacking long, movable setae; free living...138

138a(137b). Genital field usually with three, but rarely up to six, pairs of acetabula ....Pionidae (in part) Tiphyinae 139

138b. Genital field with seven or more pairs of acetabula 142

139a(138a). Fourth pair of legs unmodified ....Tiphyinae (females) *Neotiphys, Pionopsis, Tiphys*

139b. Fourth pair of legs with segments modified..Tiphyinae (males) 140

140a(139b). Fourth pair of legs with tibia bearing long, thick, curved seta dorsally (Fig. 333, arrow) and with genu only slightly expanded ....*Neotiphys*

140b. Fourth pair of legs with tibia lacking long, thick, curved seta dorsally and with genu slightly or greatly expanded...................141

141a(140b). Fourth pair of legs with genu greatly expanded and bearing numerous long setae with enlarged bases (Fig. 335).................*Tiphys*

141b. Fourth pair of legs with genu slightly expanded and bearing short setae with normal bases (Fig. 334).............................*Pionopsis*

142a(138b). Pedipalps with genu bearing long seta medially (Fig. 331, arrow); venter with second and third coxal plates each with pair of thickened setae (Fig. 330, arrows B); legs of males unmodified....................................Pionidae (in part) Huitfeldtiinae *Huitfeldtia*

142b. Pedipalps with genu lacking long seta medially; venter with second and third coxal plates lacking thick setae (Fig. 344); fourth pair of legs in males with genu notched and with peglike setae associated with notch (Fig. 343)....................................... ....Pionidae (in part) Pioninae (in part) 143

143a(142b). Capitulum with anchoral process (as in Fig. 330, arrow A) ....*Piona* (females, males of most species)

143b. Capitulum without anchoral process (Fig. 341 and as in Fig. 342, arrow) ....*Nautarachna* (females of some species)

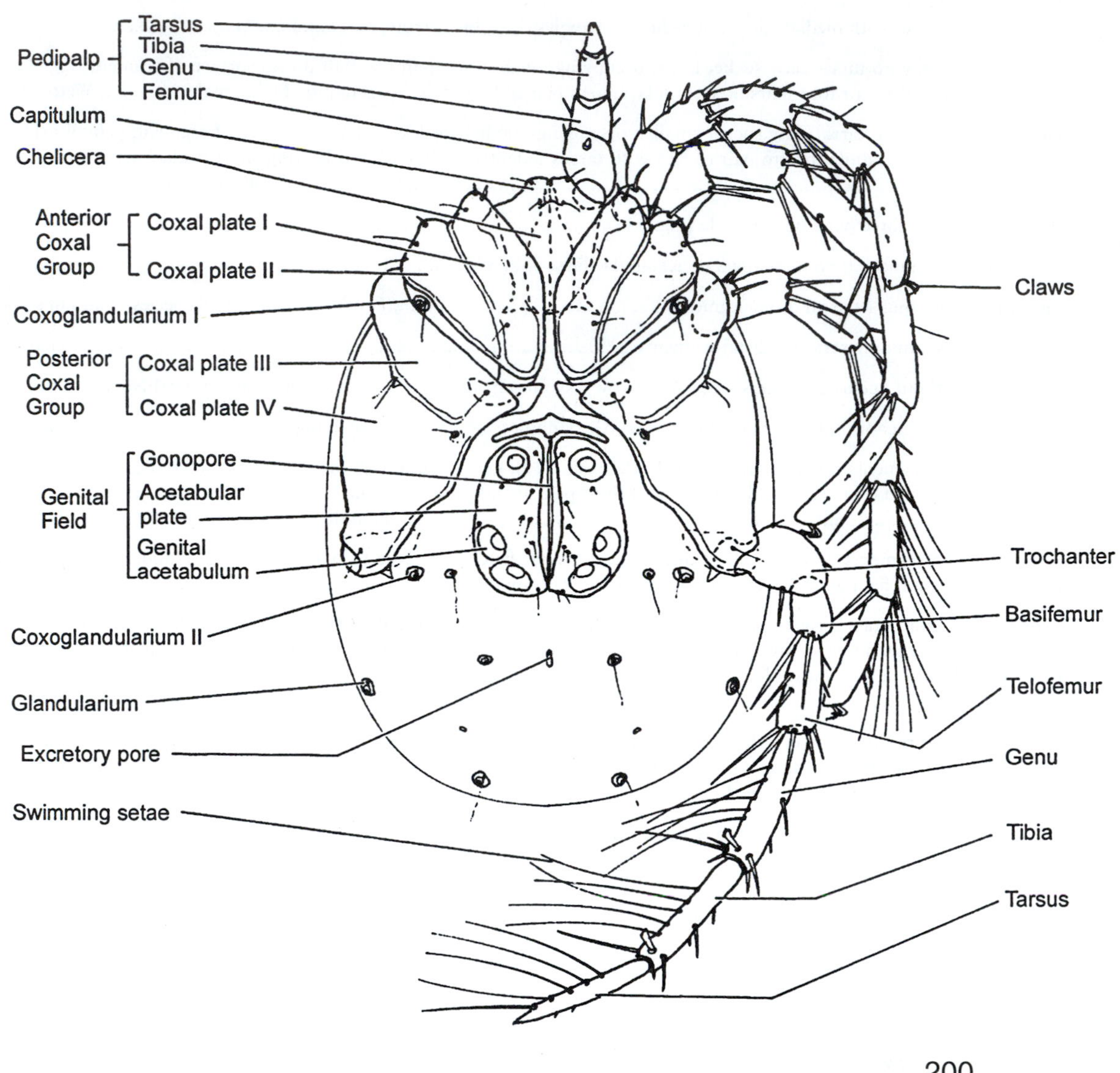

FIGURE 200 *Limnesia* sp. (Limnesiinae), adult female, ventral view.

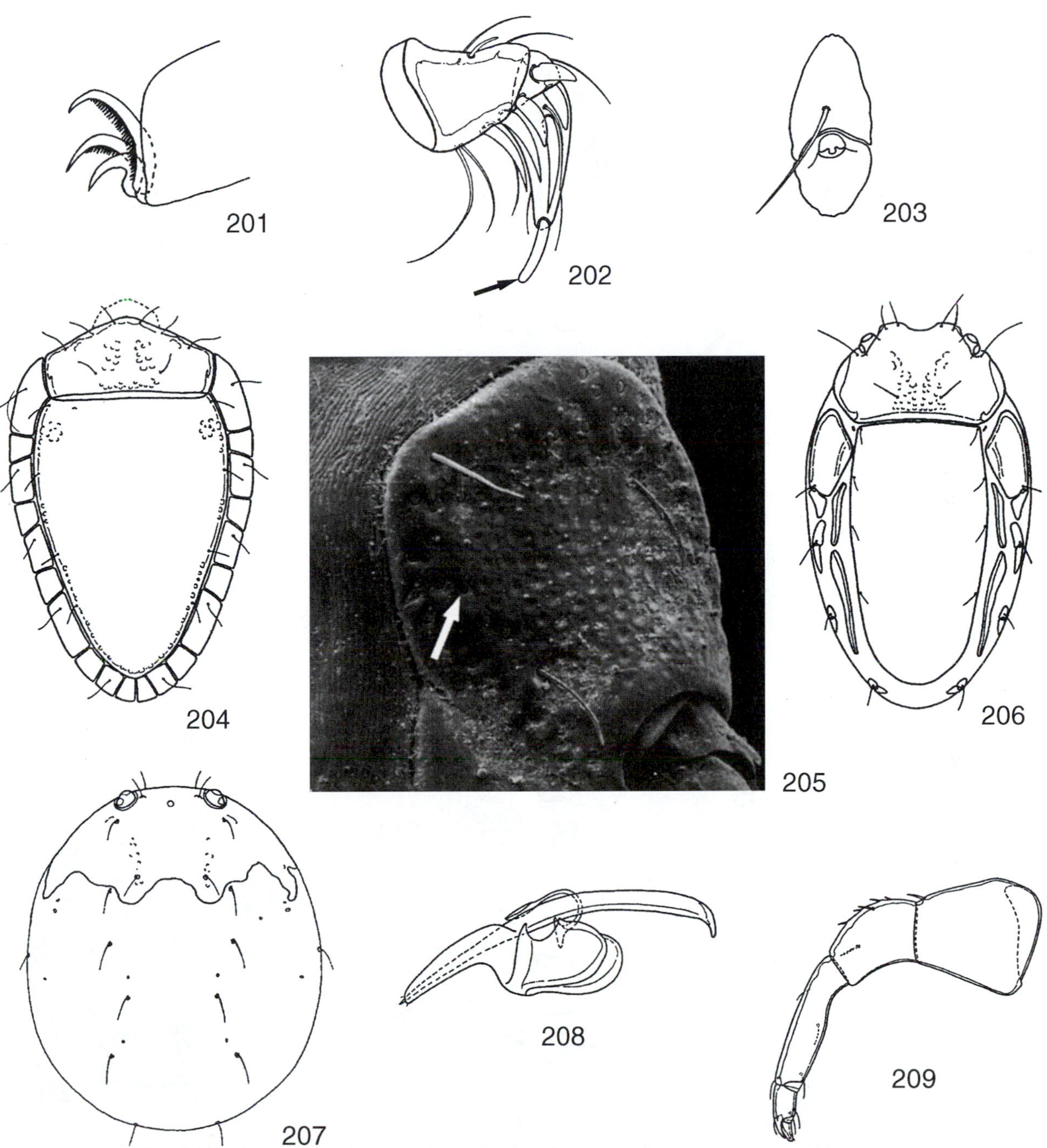

FIGURES 201, 202 ***Stygothrombium*** sp. (Stygothrombidiidae) adult. Fig. 201, claws and empodium of leg I; Fig. 202, pedipalp. *FIGURES 203, 205–206* ***Hydrovolzia marshallae*** Cook (Hydrovolziinae), adult. Fig. 203, glandularium; Fig. 205, coxal plate III ; Fig. 206, dorsum of idiosoma. *FIGURE 204* ***Acherontacarus*** sp. (Acherontacarinae), adult, dorsum of idiosoma. *FIGURES 207–209* ***Hydrachna*** sp. (Hydrachnidae), adult. Fig. 207, dorsal view of idiosoma; Fig. 208, capitulum and chelicera; Fig. 209, pedipalp.

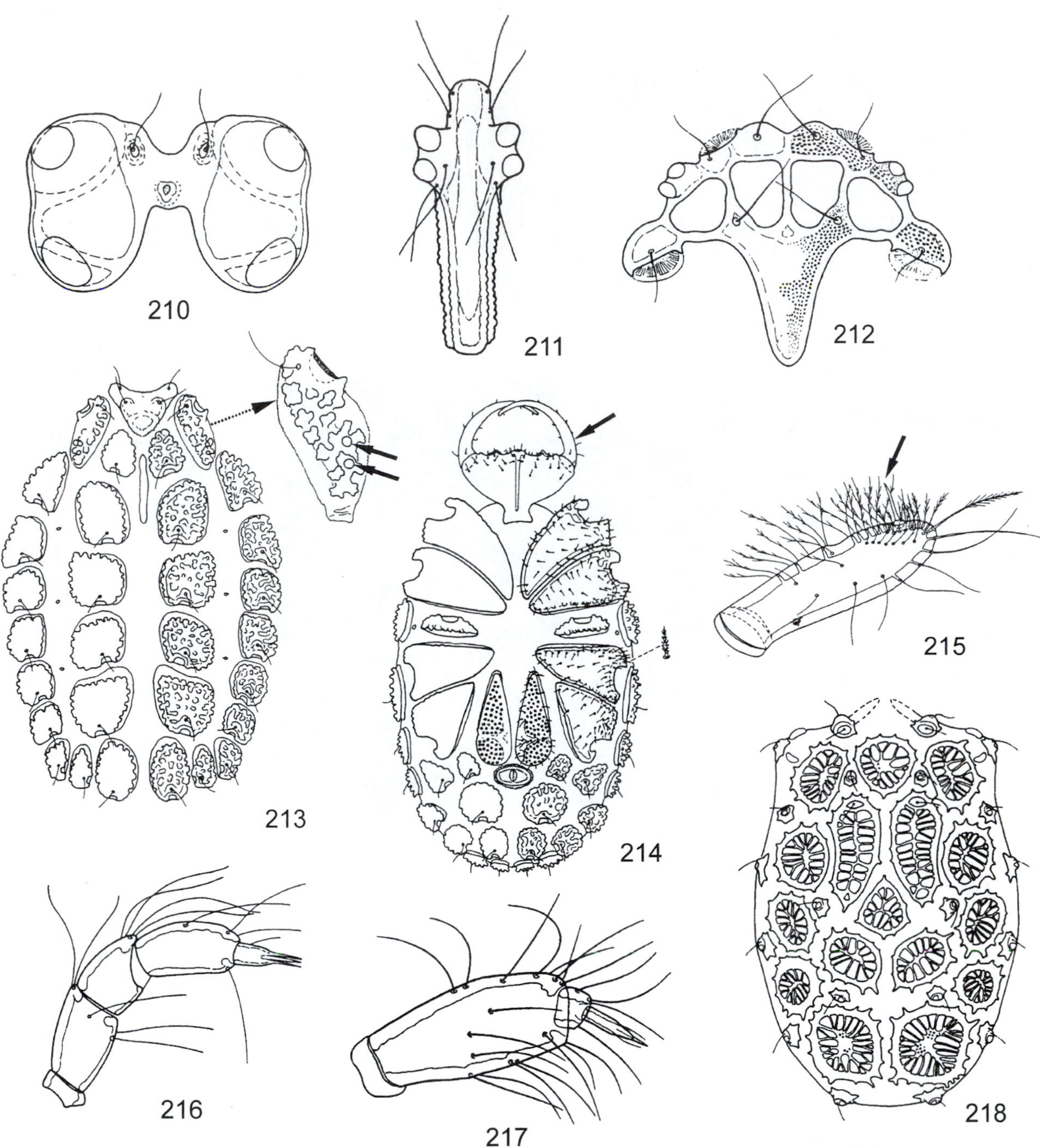

FIGURE 210 *Eylais* sp. (Eylaidae), adult, eye plate. *FIGURES 211, 216* *Limnochares* sp. (Limnocharinae), adult. Fig. 211, eye plate; Fig. 216, pedipalp. *FIGURE 212* *Piersigia limnophila* Protz (Protziidae), adult, eye plate. *FIGURES 213–214* *Apheviderulix santana* Gerecke, Smith and Cook. (Apheviderulicidae), deutonymph. Fig. 213 dorsum of idiosoma; Fig. 214, venter of idiosoma and gnathosoma. *FIGURE 215* *Rhyncholimnochares* sp. (Rhyncholimnocharinae), adult, pedipalp. *FIGURE 217* *Neolimnochares* sp. (Limnocharinae), adult, pedipalp. *FIGURE 218* *Clathrosperchon* sp. (Clathrosperchontinae), adult, dorsum of idiosoma.

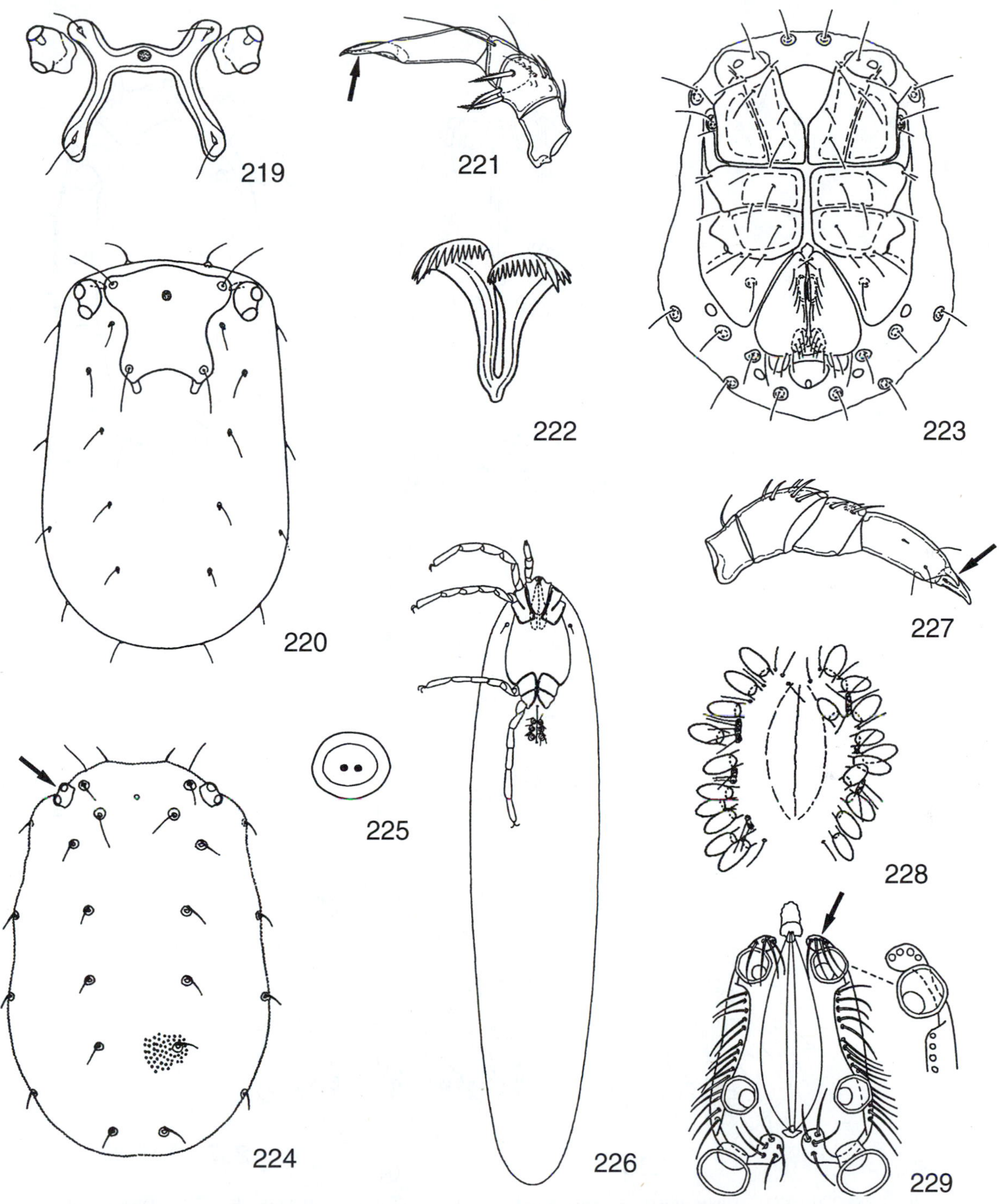

FIGURE 219 ***Hydryphantes*** sp. (Hydryphantinae), adult, eye plate. *FIGURE 220* ***Hydryphantes ruber*** (de Geer) (Hydryphantinae), adult, dorsum of idiosoma. *FIGURE 221* ***Hydrodroma*** sp. (Hydrodromidae), adult, pedipalp. *FIGURES 222, 228* ***Protzia*** sp. (Protziinae), adults. Fig. 222, claws of leg I; *FIG. 228,* genital field. *FIGURE 223* ***Cowichania interstitialis*** Smith (Cowichaniinae), adult female, venter of idiosoma. *FIGURE 224* ***Partnunia steinmanni*** Walter (Protziinae), adult, dorsum of idiosoma. *FIGURES 225, 229* ***Notopanisus*** sp. (Thyadinae), adult. Fig. 225, medial eye; Fig. 229, genital field. *FIGURES 226–227* ***Wandesia*** sp. (Wandesiinae), adult; Fig. 226, venter of idiosoma and gnathosoma; Fig. 227 pedipalp.

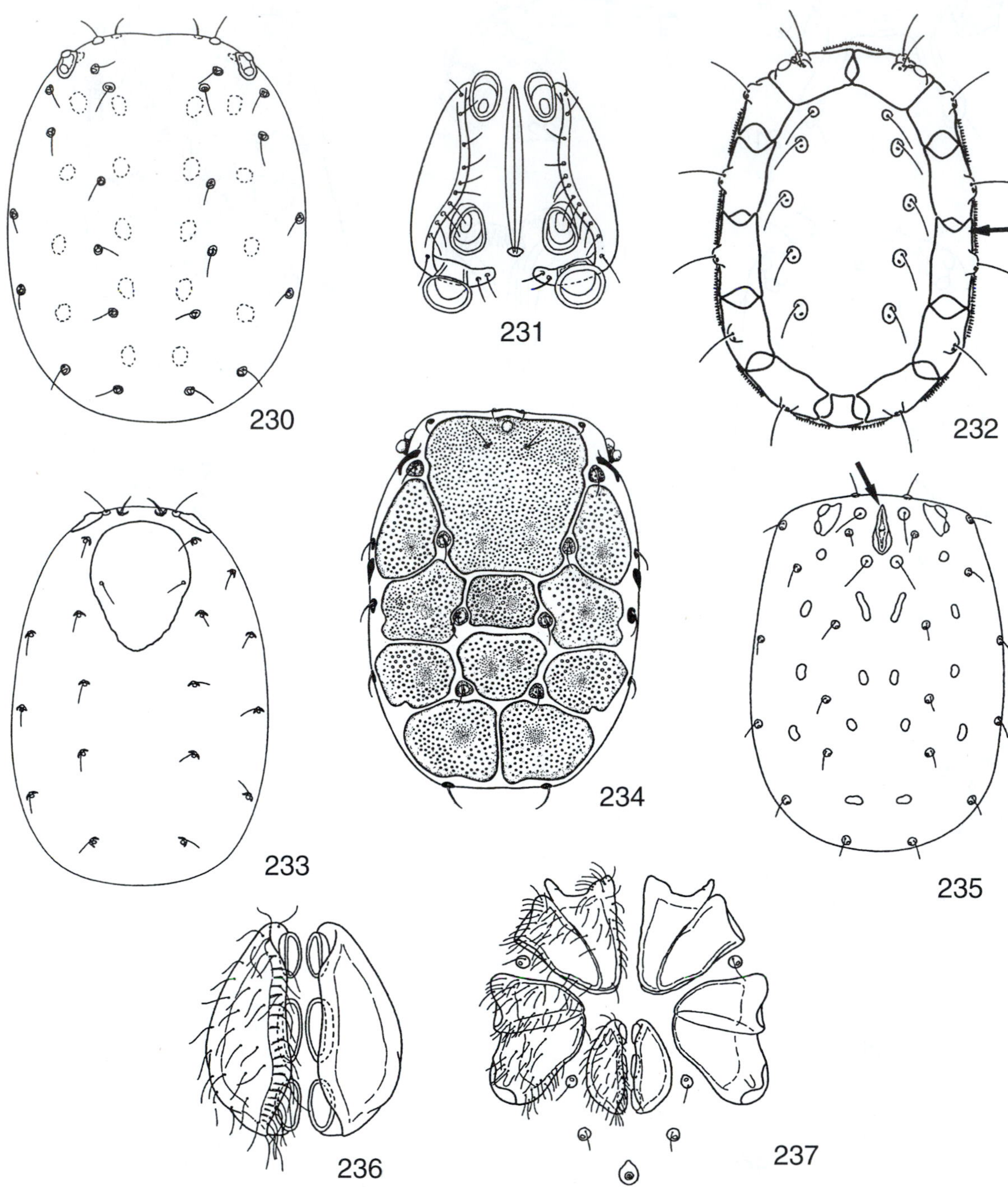

FIGURES 230, 231 *Albertathyas montana* Smith and Cook (Thyadinae), adult. Fig. 230, Dorsum of idiosoma; Fig. 231, genital field. *FIGURE 232* *Cyclothyas siskiyouensis* Smith (Cyclothyadinae), adult, dorsum of idiosoma. *FIGURE 233* *Siskiyouthyas rivularis* Smith and Cook (Thyadinae), adult, dorsum of idiosoma. *FIGURE 234* *Trichothyas muscicola* Mitchell (Thyadinae), adult, dorsum of idiosoma. *FIGURE 235* *Euthyas truncata* (Neumann) (Thyadinae), adult, dorsum of idiosoma. *FIGURE 236, 237* *Euthyas mitchelli* Cook (Thyadinae), adult. Fig. 236, Genital field; Fig. 237, venter of idiosoma.

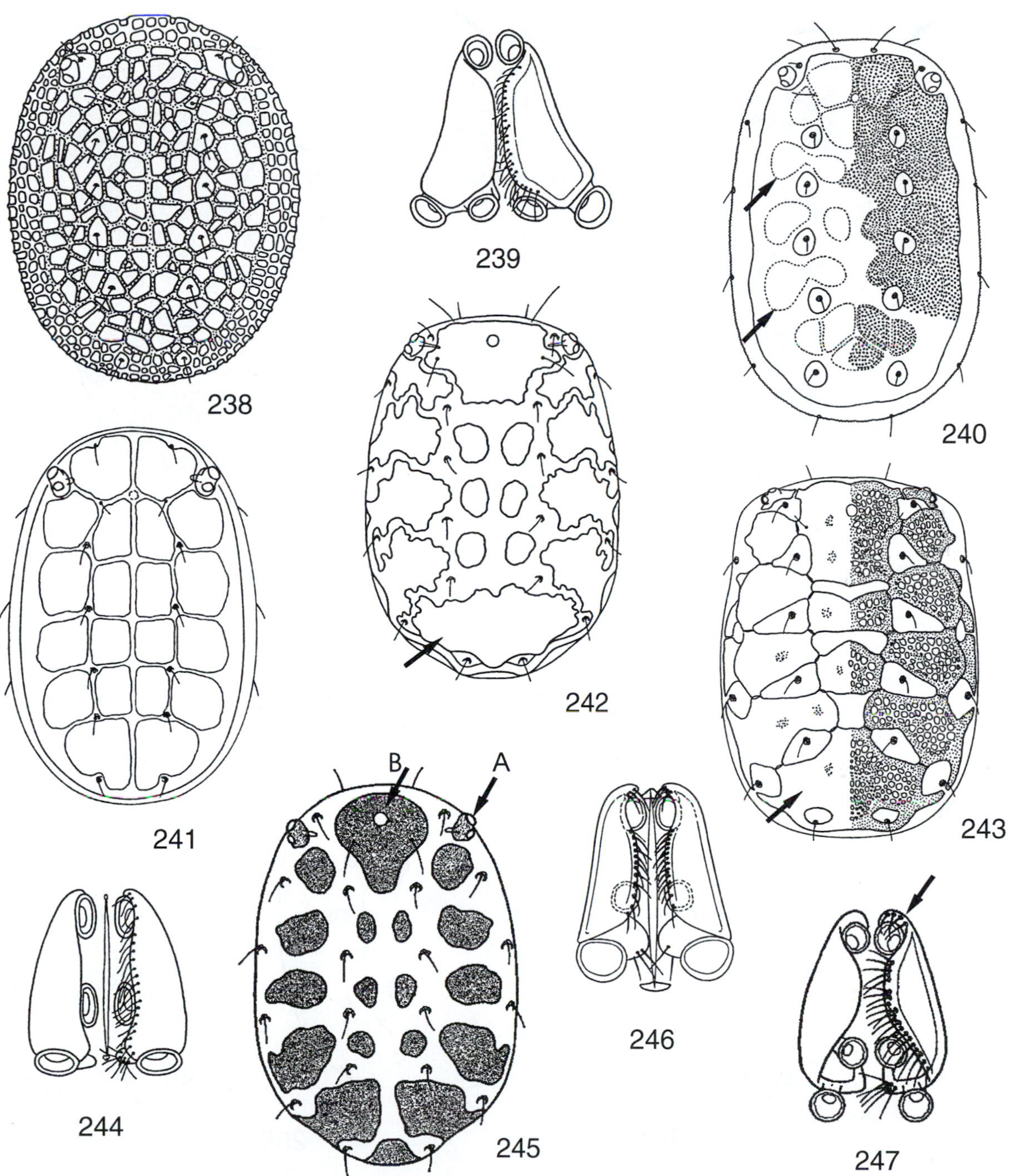

FIGURES 238, 239 *Thyopsis cancellata* (Protz) (Thyadinae), adult. Fig. 238, Dorsal shield; Fig. 239, genital field. *FIGURE 240* *Thyopsella occidentalis* Cook (Thyadinae), adult, dorsal shield. *FIGURES 241, 246* *Thyopsoides plana* (Habeeb) (Thyadinae), adult. Fig. 241, Dorsum of idiosoma; Fig. 246, genital field. *FIGURE 244* *Thyopsella dictyophora* Cook (Thyadinae), adult, genital field. *FIGURE 242* *Panisus condensatus* Habeeb (Thyadinae), adult, dorsum of idiosoma. *FIGURE 243* *Columbiathyas crenicola* Smith and Cook (Thyadinae), adult, dorsum of idiosoma. *FIGURES 245, 247* *Panisopsis setipes* (Viets) (Thyadinae), adult. Fig. 245, dorsum of idiosoma; Fig. 247, genital field.

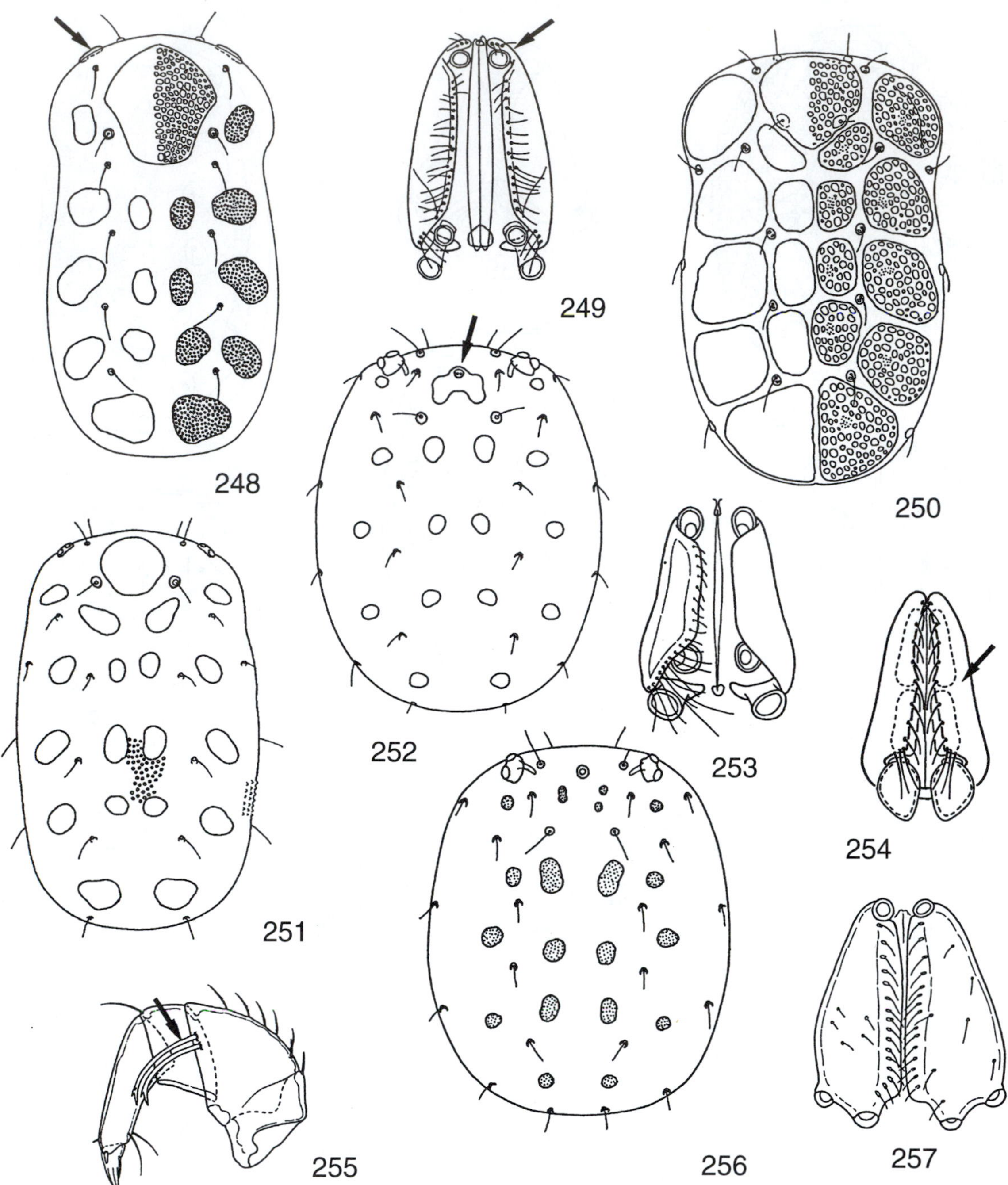

FIGURES 248, 249 *Parathyasella heatherensis* Smith and Cook (Thyadinae), adult. Fig. 248, Dorsum of idiosoma; Fig. 249, genital field. *FIGURE 250* *Amerothyasella appalachiana* Smith and Cook (Thyadinae), adult, dorsum of idiosoma. *FIGURE 251* *Thyasella* sp. (Thyadinae), adult, dorsum of idiosoma. *FIGURE 252* *Thyasides* sp. (Thyadinae), adult, dorsum of idiosoma. *FIGURE 253* *Thyas rivalis* Koenike (Thyadinae), adult, genital field. *FIGURE 254* *Tartarothyas* sp. (Tartarothyadinae), adult, genital field. *FIGURE 255* *Thermacarus nevadensis* Marshall (Thermacaridae), adult, pedipalp. *FIGURE 256* *Thyas stolli* Koenike (Thyadinae),adult, dorsum of idiosoma. *FIGURE 257* *Zschokkea* sp. (Thyadinae), adult, genital field.

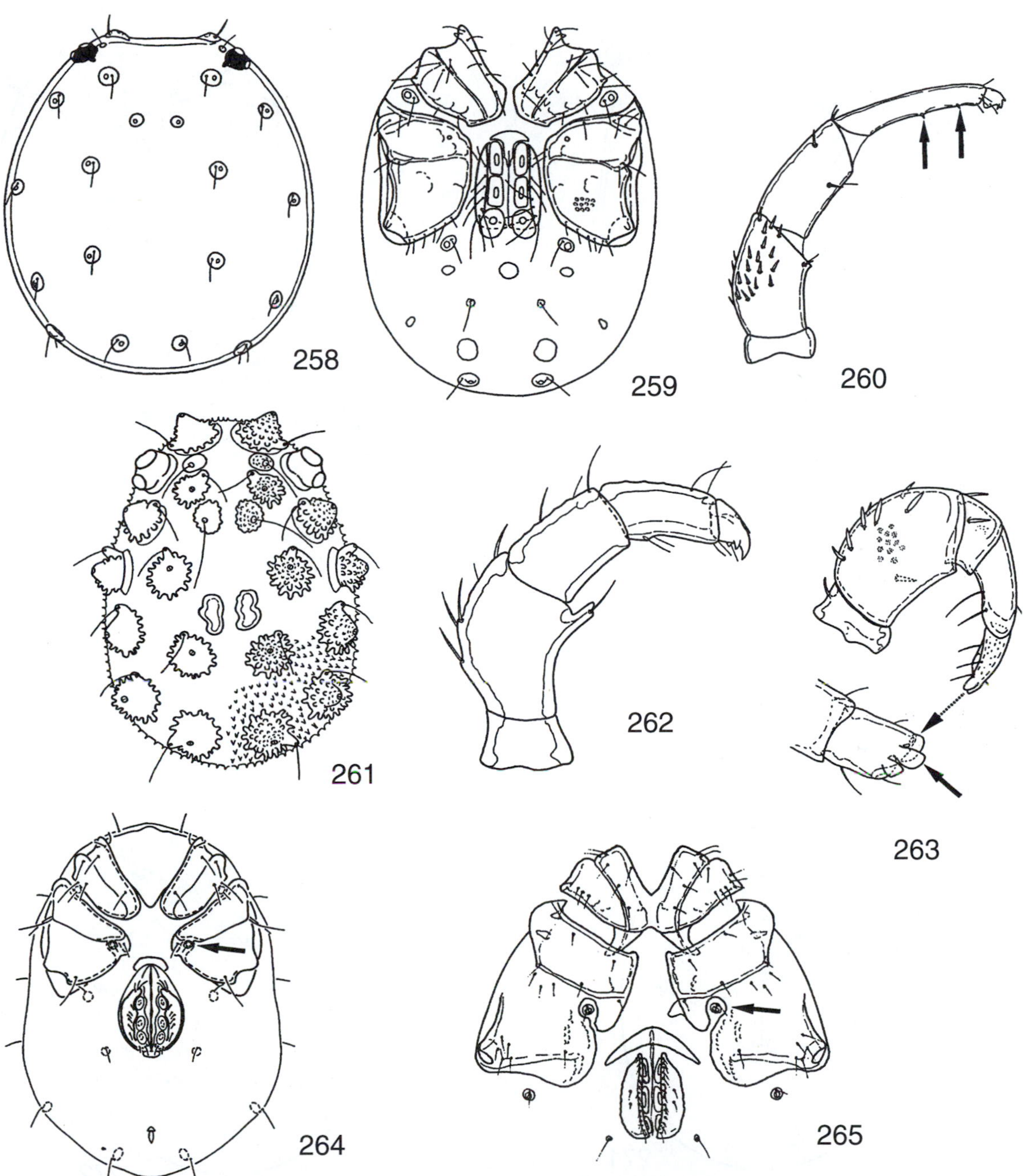

FIGURES 258–260 *Sperchon* spp. (Sperchontidae), adults. Fig. 258, Dorsum of idiosoma; Fig. 259, venter of idiosoma; Fig. 260, pedipalp. *FIGURES 261, 262* *Sperchonopsis nova* Prasad and Cook (Sperchontidae), adult. Fig. 261, Dorsum of idiosoma; Fig. 262, pedipalp. *FIGURES 263, 264* *Rutripalpus* spp. (Rutripalpidae), adult females. Fig. 263, Pedipalp; Fig. 264, venter of idiosoma. *FIGURE 265* *Teutonia* sp. (Teutoniidae), adult, venter of idiosoma.

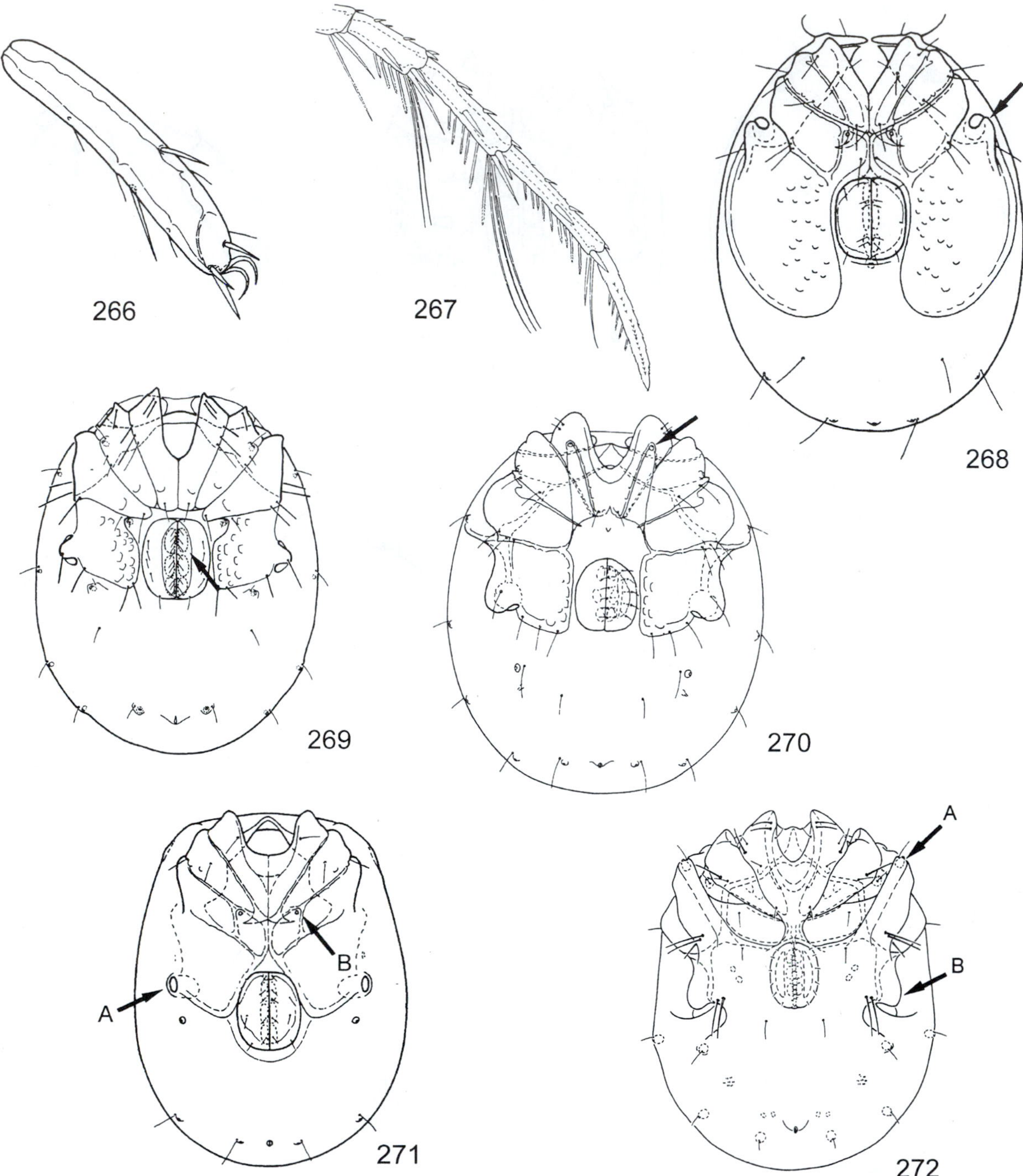

FIGURES 266, 268 *Utaxatax ovalis* Cook (Anisitsiellinae), adult. Fig. 266, Tarsus of leg IV; Fig. 268, male, ventral shield. *FIGURE 267* *Mamersellides* sp. (Anisitsiellinae), adult, leg. IV. *FIGURE 269* *Bandakiopsis fonticola* Smith (Anisitsiellinae), adult, ventral shield. *FIGURE 270* *Cookacarus columbiensis* Barr (Anisitsiellinae), adult male, ventral shield. *FIGURE 271* *Bandakia phreatica* Cook (Anisitsiellinae), adult male, ventral shield. *FIGURE 272* *Oregonacarus rivulicolus* Smith (Anisitsiellinae), adult male, ventral shield.

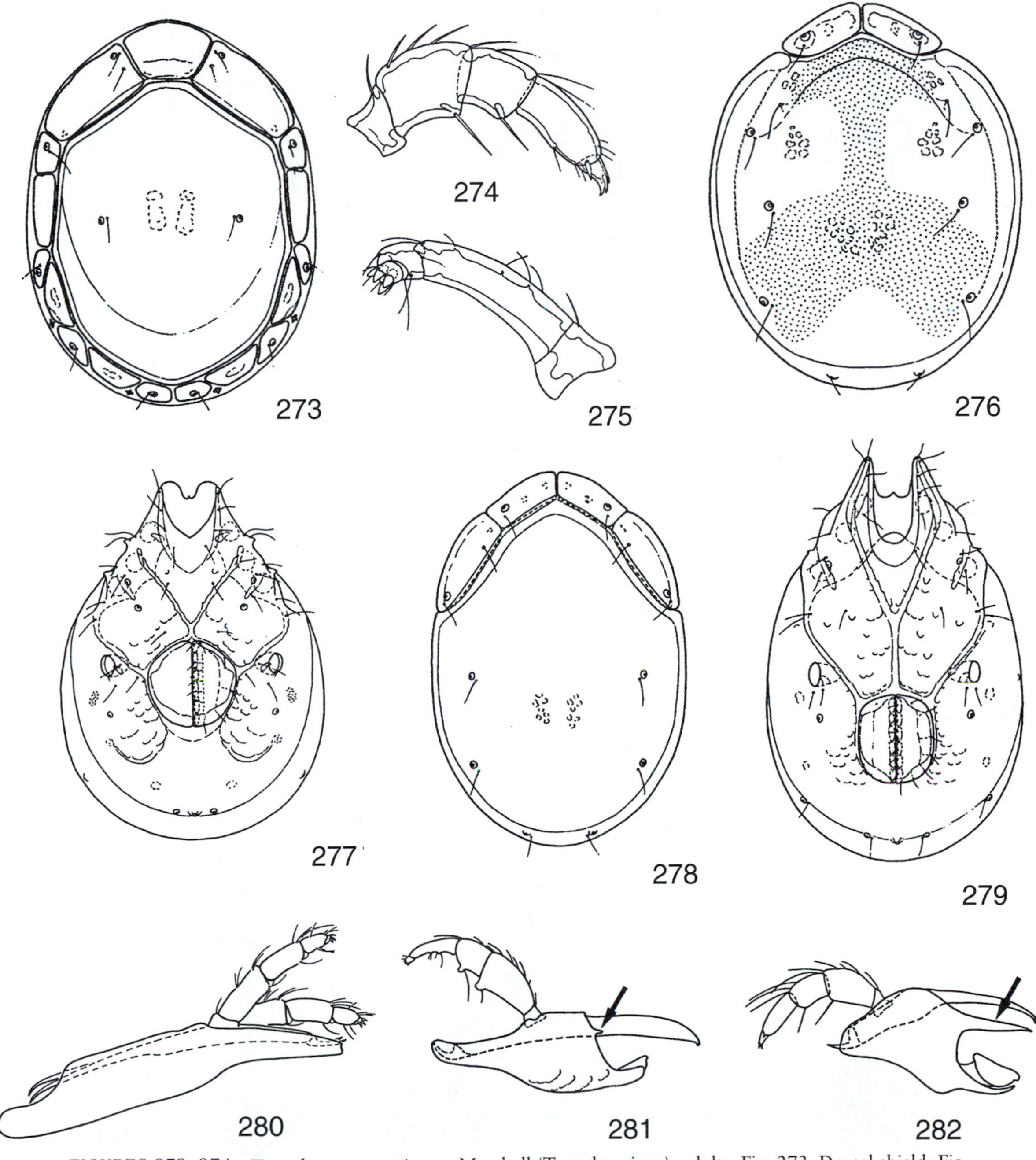

FIGURES 273, 274 *Testudacarus americanus* Marshall (Testudacarinae), adults. Fig. 273, Dorsal shield; Fig. 274, pedipalp. *FIGURES 275–277* *Neoatractides* sp. (Neoatractidinae), adults. Fig. 275, Pedipalp; Fig. 276, dorsal shield; Fig. 277, ventral shield. *FIGURES 278, 279, 281* *Torrenticola* sp. (Torrenticolinae), adults. Fig. 278, Dorsal shield; Fig. 279, ventral shield; Fig. 281, gnathosoma. *FIGURE 280* *Pseudotorrenticola chiricahua* Smith (Torrenticolinae), adult, gnathosoma. *FIGURE 282* *Monoatractides* sp. (Torrenticolinae), adult, gnathosoma.

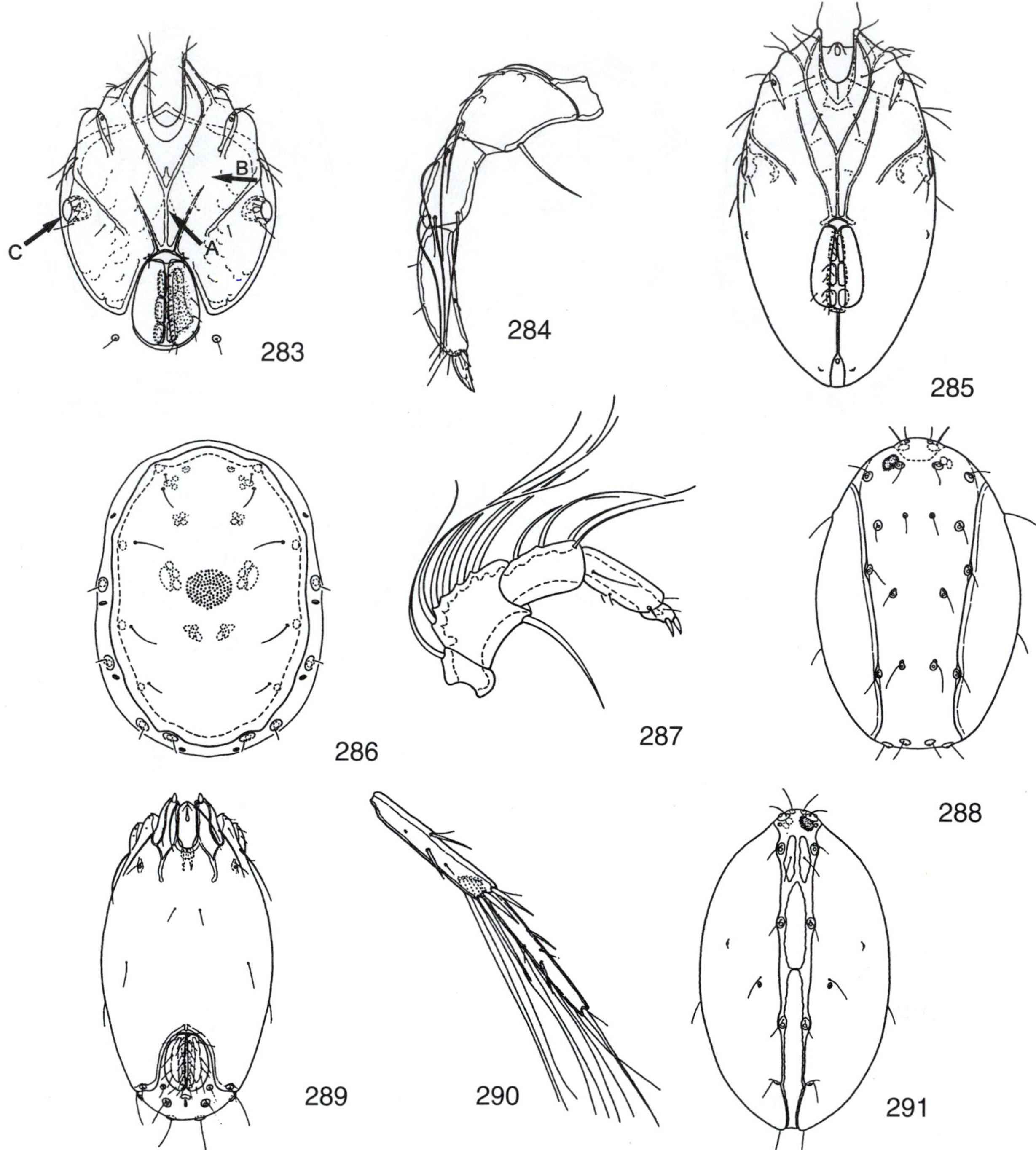

FIGURES 283, 284 *Lebertia* sp. (Lebertiidae), adults. Fig. 283, Venter of idiosoma; Fig. 284, pedipalp. *FIGURE 285* *Estelloxus montanus* Habeeb (Lebertiidae) adult, venter of idiosoma. *FIGURES 286, 287* *Scutolebertia trinitensis* Smith (Lebertiidae), adults. Fig. 286, Dorsum of idiosoma; Fig. 287, pedipalp. *FIGURES 288–290* *Oxus* spp. (Oxidae), adults. Fig. 288, Dorsum of idiosoma; Fig. 289, venter of idiosoma; Fig. 290, tibia and tarsus of leg IV. *FIGURE 291* *Frontipoda americana* Marshall (Oxidae), adult male, dorsum of idiosoma.

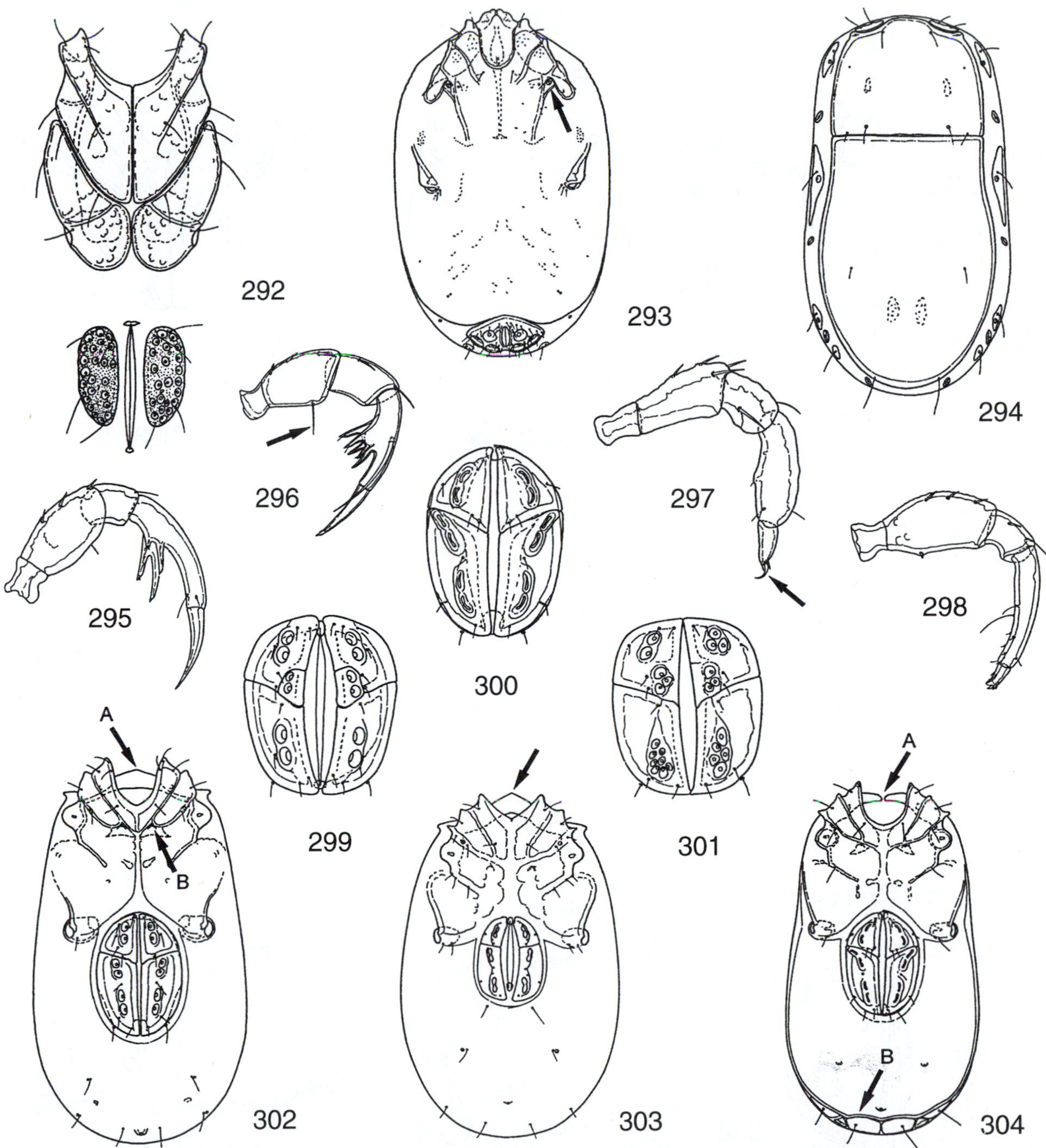

FIGURE 292 *Omartacarus* sp. (Omartacaridae), adult female, venter of idiosoma. *FIGURE 293* *Diamphidaxona* sp. (Hygrobatidae), adult male, venter of idiosoma. *FIGURE 294* *Neomamersa chihuahua* Smith and Cook (Neomamersinae), adult female, dorsum of idiosoma. *FIGURES 295, 299, 301* *Neomamersa* spp. (Neomamersinae), adult males. Fig. 295, Pedipalp; Figs, 299, 301, genital fields. *FIGURE 296* *Kawamuracarus* sp. (Kawamuracarinae), adult, pedipalp. *FIGURES 297, 303* *Arizonacarus chiricahuensis* Smith and Cook (Neomamersinae), adult females. Fig. 297, Pedipalp; Fig. 303, venter of idiosoma. *FIGURES 298, 300, 304* *Meramecia occidentalis* Smith and Cook (Neomamersinae), adult females. Fig. 298, Pedipalp; Fig. 300, genital field; Fig. 304, venter of idiosoma. *FIGURE 302* *Neomamersa californica* Smith and Cook (Neomamersinae), adult female, venter of idiosoma.

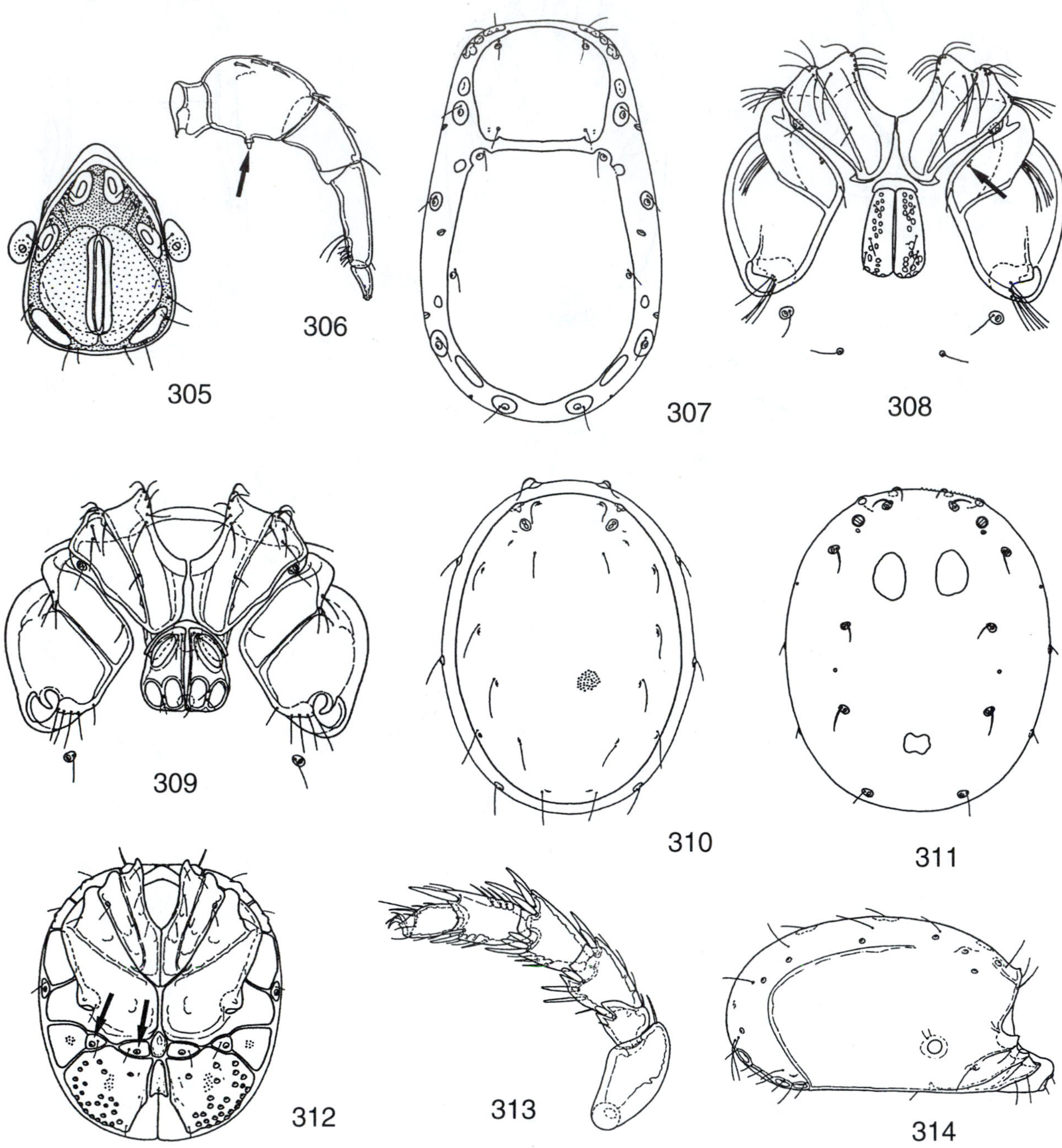

FIGURE 305 *Centrolimnesia bondi* Lundblad (Limnesiinae), adult male, genital field. *FIGURES 306, 309–311* *Tyrrellia* spp. (Tyrelliinae), adults. Fig. 306, Pedipalp; Fig. 309, male, venter of idiosoma; Fig. 310, male, dorsal shield; Fig. 311, male, dorsum of idiosoma. *FIGURE 307* *Protolimnesia* sp. (Protolimnesiinae), adult, dorsum of idiosoma. *FIGURE 308* *Neotyrrellia anitahoffmannae* Smith and Cook (Tyrrellinae), adult female, venter of idiosoma. *FIGURE 312* *Feltria exilis* Cook (Feltriidae), adult female, venter of idiosoma. *FIGURES 313, 314* *Frontipodopsis* sp. (Frontipodopsidae), adult. Fig. 313, Leg IV; Fig. 314, lateral view of idiosoma.

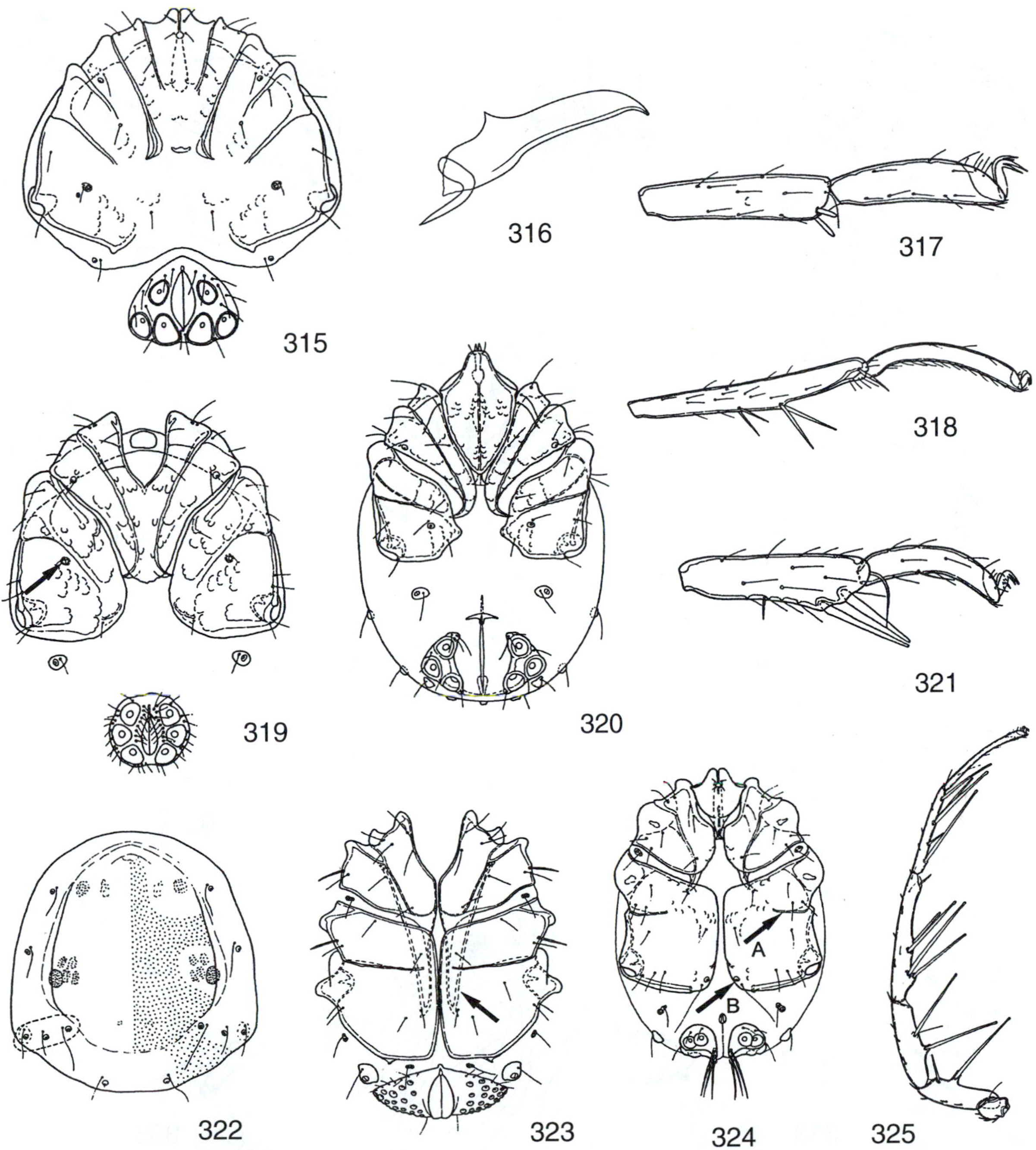

FIGURES 315–317 *Hygrobates* spp. (Hygrobatidae), adults. Fig. 315 male, venter of idiosoma; Fig. 316 , Chelicera; Fig. 317, tibia and tarsus of leg I. *FIGURE 318* *Mesobates* sp. (Hygrobatidae), adult, tibia and tarsus of leg I. *FIGURES 319, 321* *Atractides* sp. (Hygrobatidae), adults. Fig. 319, Male, venter of idiosoma; Fig. 321, tibia and tarsus of leg I. *FIGURE 320* *Corticacarus* sp. (Hygrobatidae), adult female, venter of idiosoma. *FIGURE 322* *Koenikea* sp. (Pionatacinae), adult dorsal shield. *FIGURE 323* *Neumania* sp. (Pionatacinae), adult male, venter of idiosoma. *FIGURES 324, 325* *Unionicola* sp. (Unionicolinae), adults. Fig. 324, female, venter of idiosoma; Fig. 325, leg I.

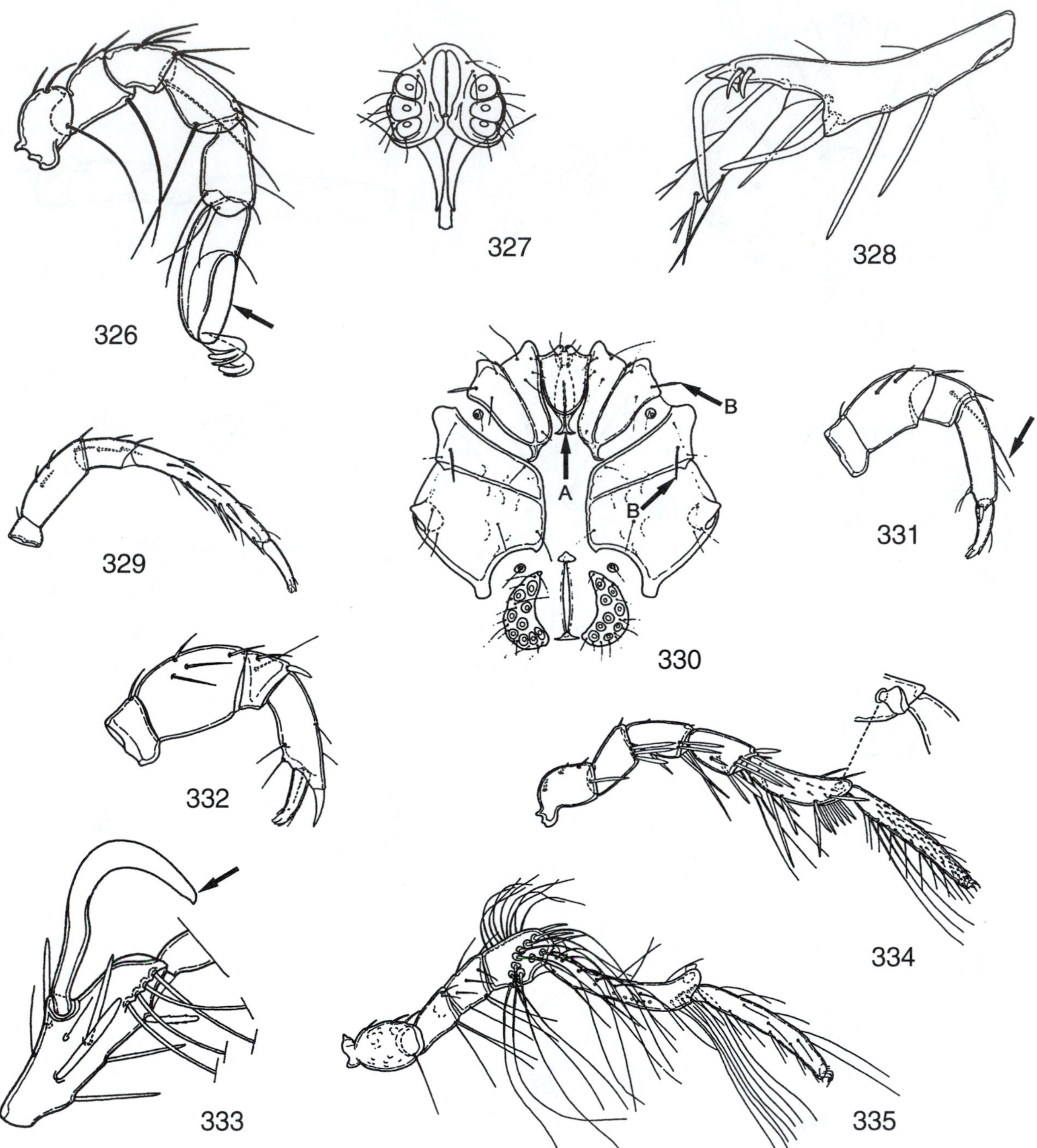

FIGURE 326 ***Wettina octopora*** Cook (Wettinidae), adult, leg I. *FIGURES 327–329* ***Hydrochoreutes intermedius*** Cook (Hydrochoreutinae), adults. Fig. 327, Male, genital field; Fig. 328, male, genu of leg III; Fig. 329, pedipalp. *FIGURES 330, 331* ***Huitfeldtia rectipes*** (Huitfeldtiinae), adults. Fig. 330, female, venter of idiosoma; Fig. 331, pedipalp. *FIGURES 332, 335* ***Tiphys*** spp. (Tiphyinae), adults. Fig. 332, pedipalp; Fig. 335, male, leg IV. *FIGURE 333* ***Neotiphys pionoidellus*** (Habeeb) (Tiphyinae), adult male, tibia of leg IV. *FIGURE 334* ***Pionopsis paludis*** Habeeb (Tiphyinae), adult male, leg IV.

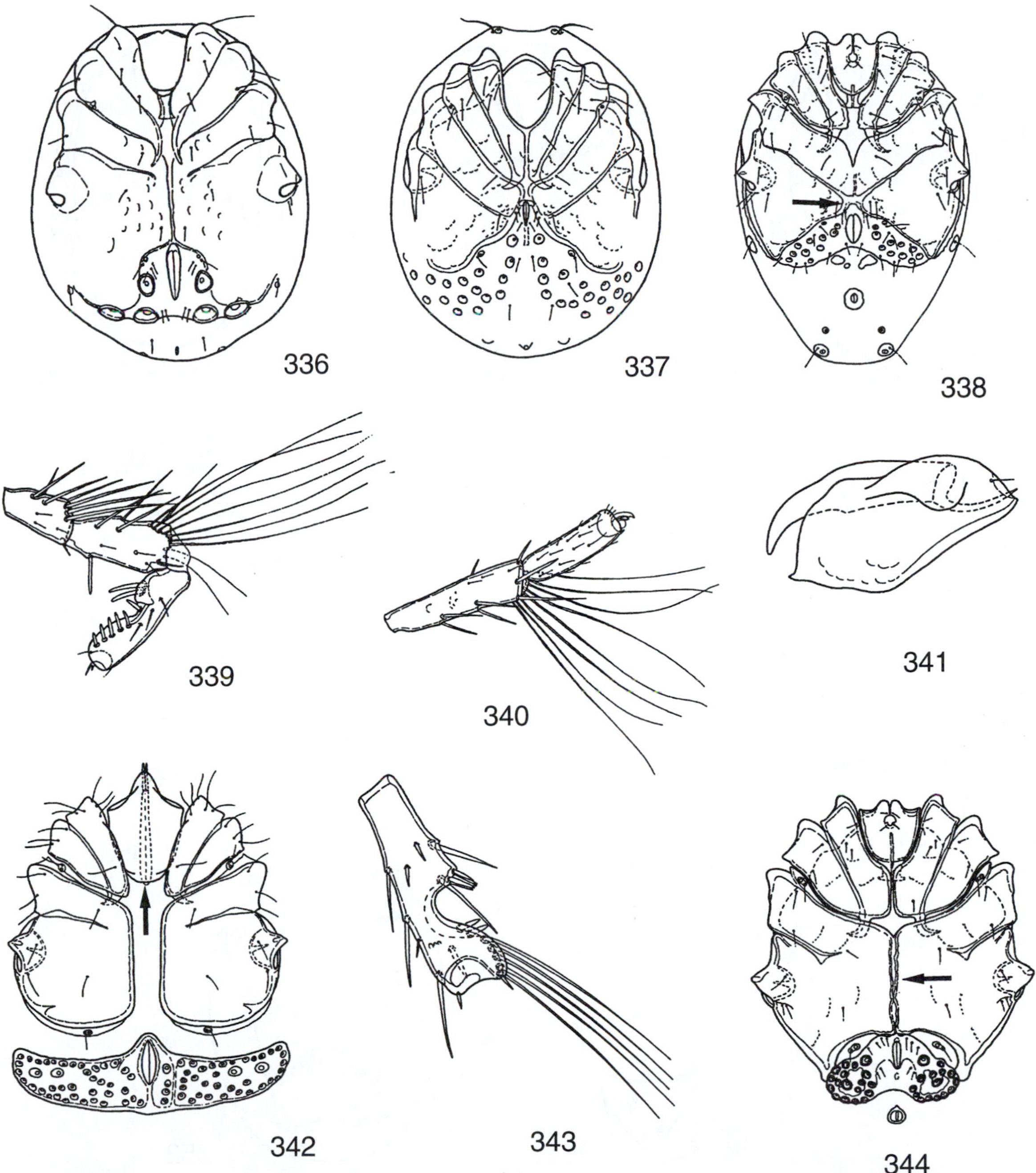

FIGURE 336 *Pionacercus* sp. (Foreliinae), adult male, venter of idiosoma. *FIGURE 337* *Pseudofeltria multipora* Cook (Foreliinae), adult male, ventral shield. *FIGURES 338–340* *Forelia floridensis* Cook (Foreliinae), adult male. Fig. 338, Venter of idiosoma; Fig. 339, distal segments of leg IV; Fig. 340, tibia and tarsus of leg III. *FIGURE 341* *Nautarachna queticoensis* Smith (Pioninae), adult female, gnathosoma. *FIGURE 342* *Najadicola ingens* Koenike (Najadicolinae), adult male, venter of idiosoma. *FIGURES 343, 344* *Piona* spp. (Pioninae), adult males. Fig. 343, Genu of leg IV; Fig. 344, venter of idiosoma.

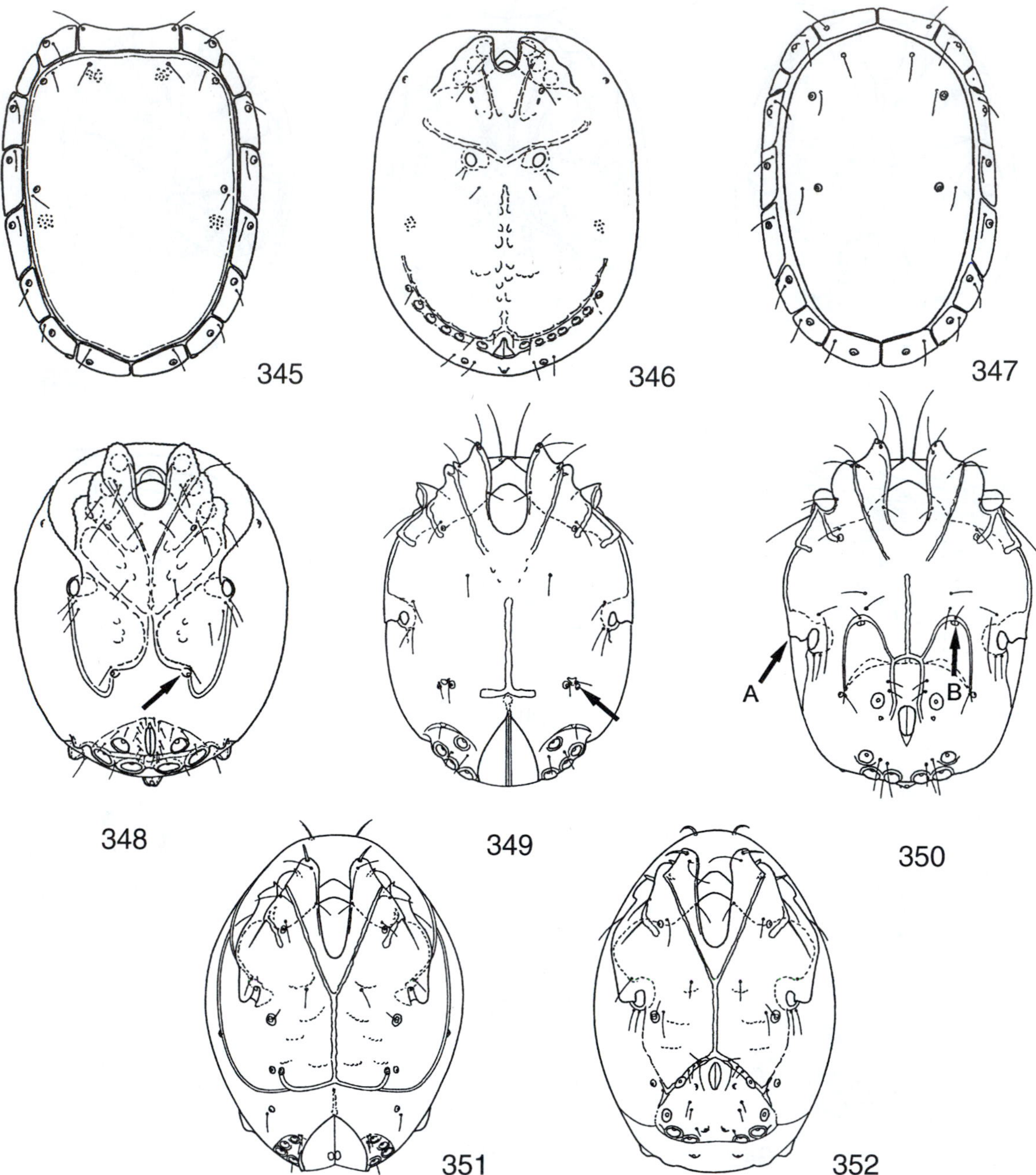

FIGURES 345, 346 ***Lethaxonella parva*** Cook (Lethaxonidae), adult female. Fig. 345, Dorsal plate; Fig. 346, ventral plate. *FIGURE 347* ***Lethaxona*** sp. (Lethaxonidae), adult, dorsal shield. *FIGURE 348* ***Ljania michiganensis*** Cook (Axonopsinae), adult male, ventral shield. *FIGURES 349, 350* ***Stygalbiella yavapai*** Smith and Cook (Axonopsinae), adults. Fig. 349, Female, ventral shield; Fig. 350, male, ventral shield. *FIGURES 351, 352* ***Axonopsella bakeri*** Smith and Cook (Axonopsinae), adults. Fig. 351, female, ventral shield; Fig. 352, male, ventral shield.

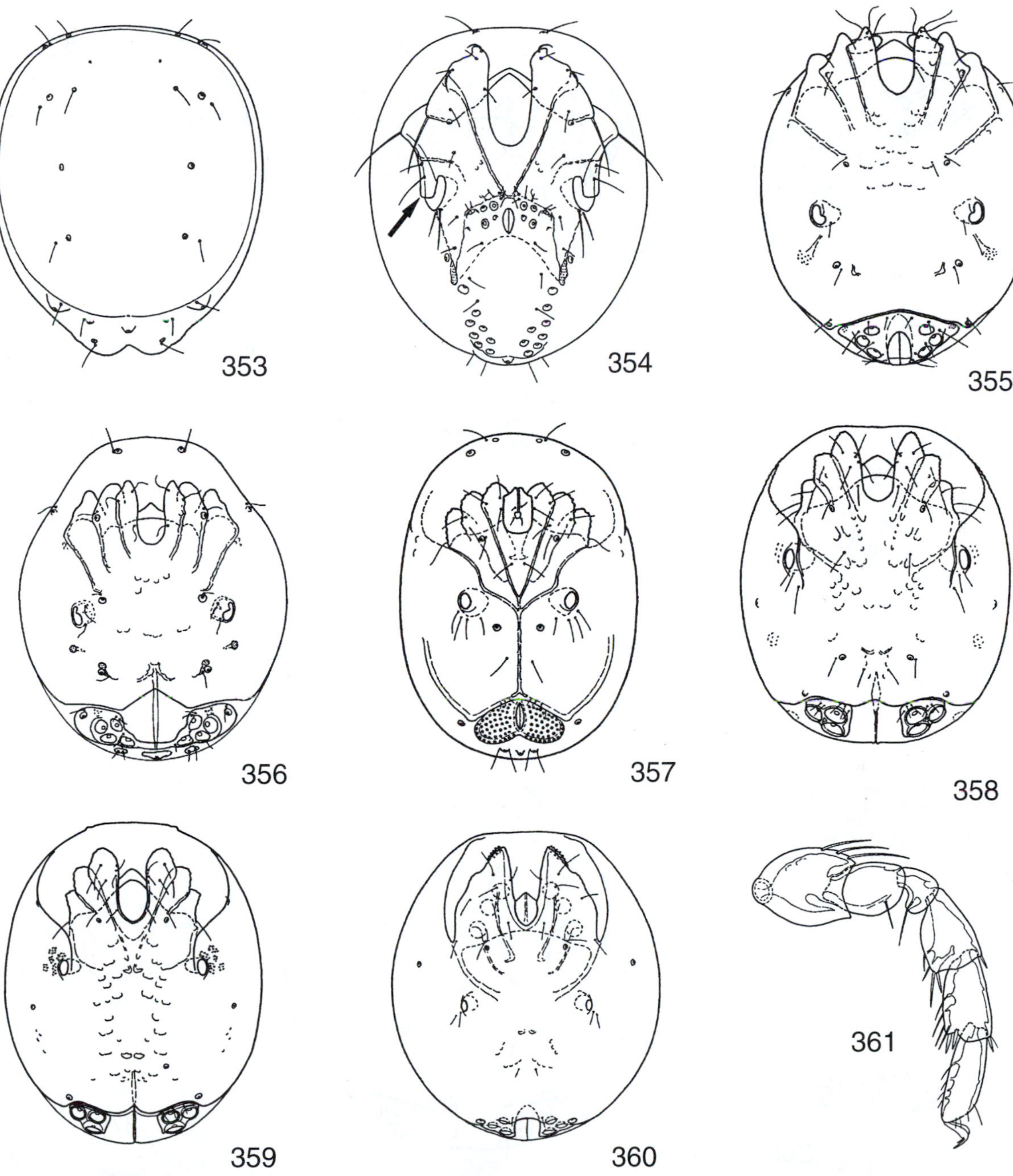

FIGURES 353, 354 ***Submiraxona gilana*** (Habeeb) (Axonopsinae), adults. Fig. 353, Dorsal shield; Fig. 354, male, ventral shield. *FIGURE 355* ***Albaxona nearctica*** Cook (Axonopsinae), adult male, ventral shield. *FIGURE 356* ***Javalbia*** sp. (Axonopsinae), adult female, ventral shield. *FIGURE 357* ***Albia*** sp. (Albiinae), adult male, ventral shield. *FIGURE 358* ***Estellacarus unguitarsus*** (Habeeb) (Axonopsinae), adult female, ventral shield. *FIGURE 359* ***Brachypoda oakcreekensis*** Habeeb (Axonopsinae), adult female, ventral shield. *FIGURES 360, 361* ***Erebaxonopsis nearctica*** Cook (Axonopsinae), adults. Fig. 360, female,ventral shield; Fig. 361, male, leg IV.

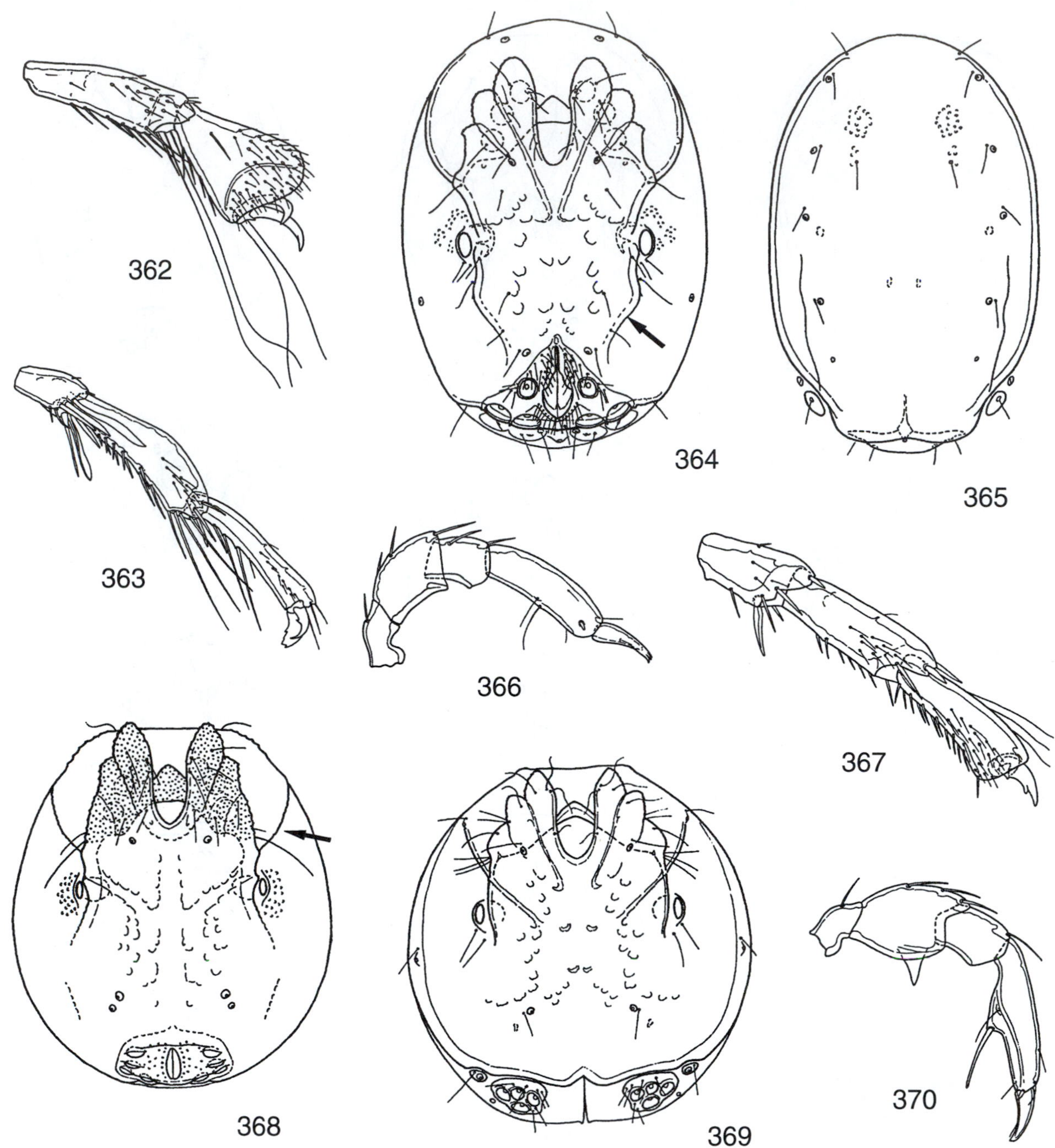

FIGURE 362 ***Estellacarus unguitarsus*** (Habeeb) (Axonopsinae), adult male, distal segments of leg IV. *FIGURE 363* ***Brachypoda oakcreekensis*** Habeeb (Axonopsinae), adult male, distal segments of leg IV. *FIGURES 364–367* ***Woolastookia pilositarsa*** (Habeeb) (Axonopsinae), adults. Fig. 364, Male, venter of idiosoma; Fig. 365, female, pedipalp; Fig. 366, female, dorsal shield; Fig. 367, male, distal segments of leg IV. *FIGURE 368* ***Axonopsis beltista*** Cook (Axonopsinae), adult male, ventral shield. *FIGURES 369, 370* ***Neobrachypoda ekmani*** (Walter) (Axonopsinae), adults. Fig. 369, Female, venter of idiosoma; Fig. 370, male, pedipalp.

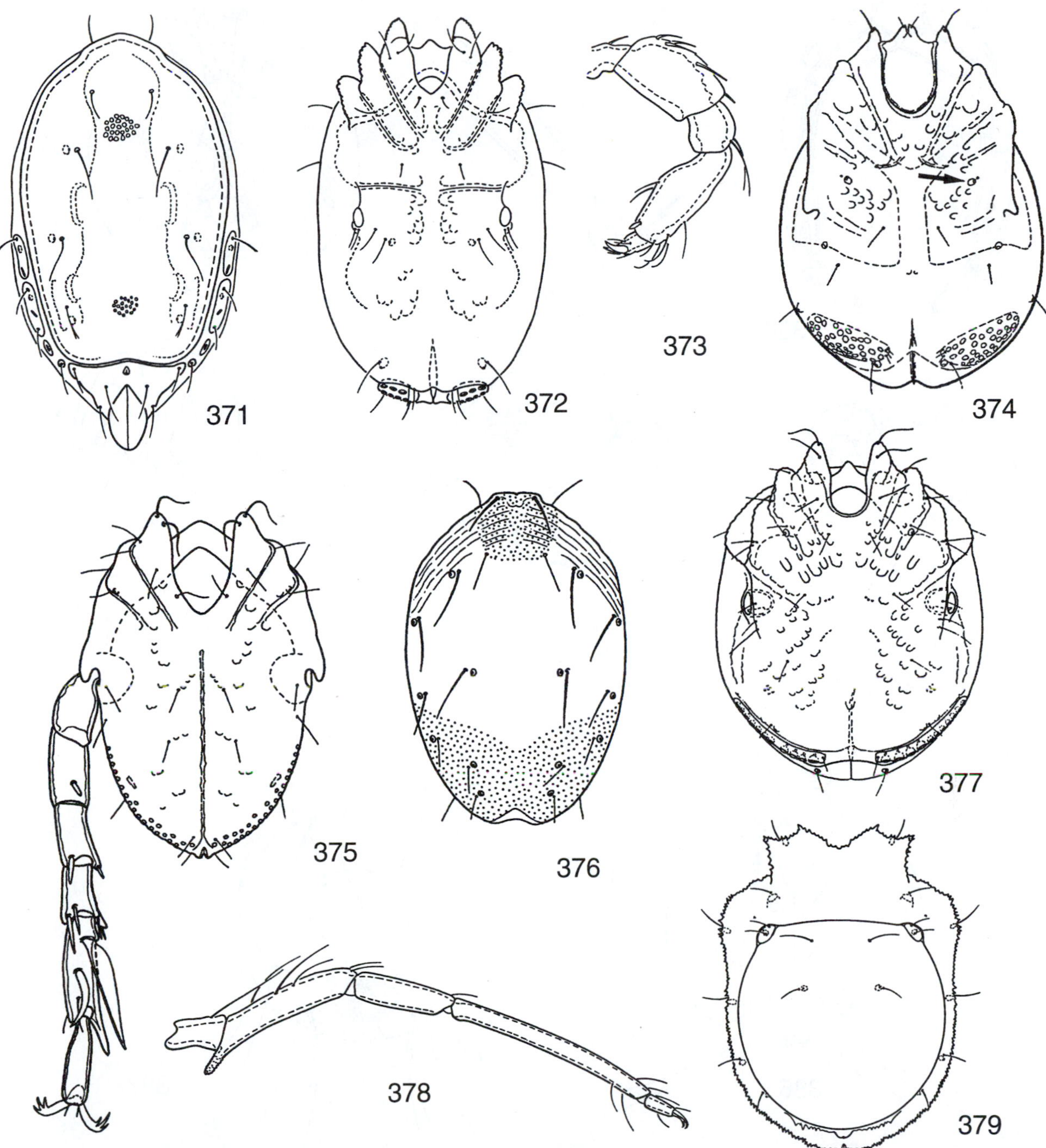

FIGURES 371, 372 *Phreatobrachypoda multipora* Cook (Aturinae), adult male. Fig. 371, Dorsum of idiosoma; Fig. 372, venter of idiosoma. *FIGURE 373* *Phreatobrachypoda oregonensis* Smith (Aturinae), adult male, pedipalp. *FIGURE 374* *Neoaturus* sp. (Aturinae), adult female, venter of idiosoma. *FIGURES 375, 376* *Kongsbergia* spp. (Aturinae), adult males. Fig. 375, Venter of idiosoma and fourth leg; Fig. 376, dorsal shield. *FIGURE 377* *Aturus* sp. (Aturinae), adult female, venter of idiosoma. *FIGURES 378, 379* *Bharatalbia* spp. (Aturinae), adults. Fig. 378, *B. cooki* Smith, female, dorsum of idiosoma; Fig. 379, *B. surensis* Smith, male, pedipalp.

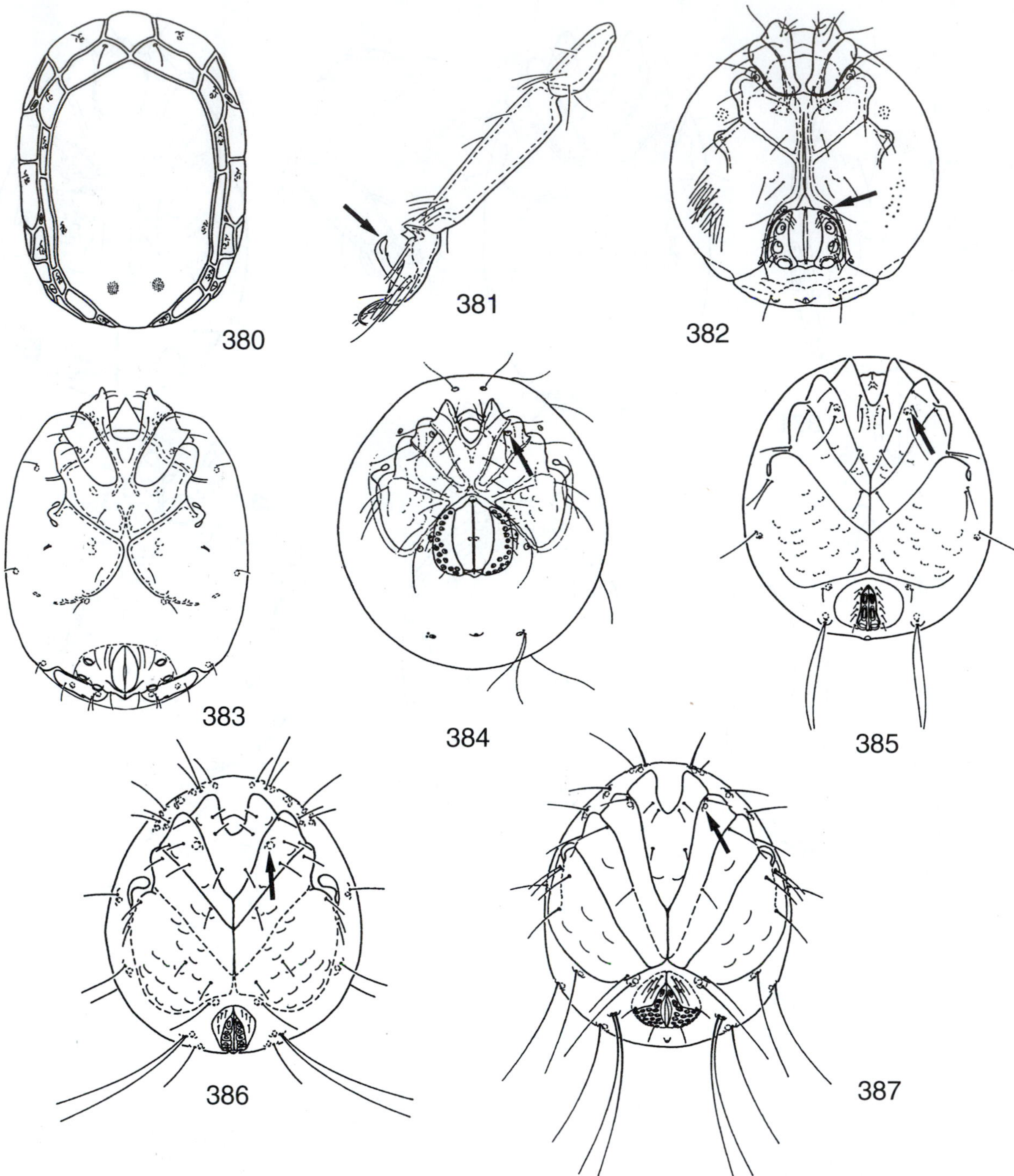

FIGURE 380 *Cyclomomonia andrewi* Smith (Cyclomomoniinae), adult, dorsal shield. *FIGURE 381* *Stygomomonia riparia* Habeeb (Stygomomoniinae), adult, distal segments of leg I. *FIGURE 382* *Momonia projecta* Cook (Momoniinae), adult female, ventral shield. *FIGURE 383* *Stygomomonia mitchelli* Smith (Stygomomoniinae), adult male, ventral shield. *FIGURE 384* *Midea expansa* Marshall (Mideidae), adult female, ventral shield. *FIGURE 385* *Nudomideopsis magnecetabula* (Smith) (Nudomideopsidae), adult male, ventral shield. *FIGURE 386* *Paramideopsis susanae* Smith (Nudomideopsidae), adult male, ventral shield. *FIGURE 387* *Neomideopsis suislawensis* Smith (Nudomideopsidae), adult male, ventral shield.

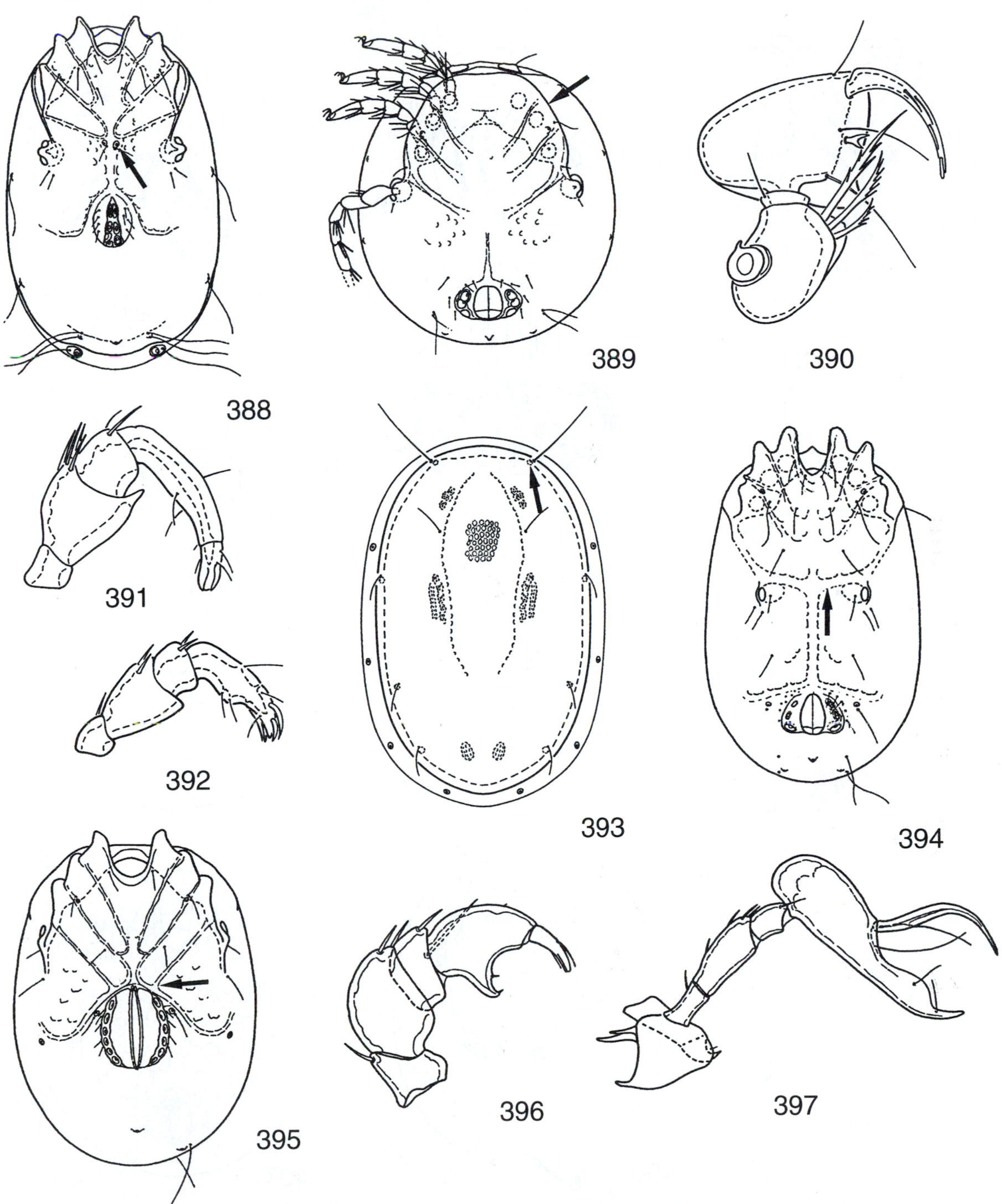

FIGURE 388 *Chappuisides eremitus* Cook (Chappuisidinae), adult male, ventral shield. *FIGURE 389* *Uchidastygacarus ovalis* Cook (Uchidastygacarinae), adult female, ventral shield and legs. *FIGURE 390* *Uchidastygacarus acadiensis* Smith (Uchidastygacarinae), adult, pedipalp. *FIGURE 391* *Morimotacarus nearcticus* Smith (Morimotacarinae), adult, pedipalp. *FIGURES 392–394* *Yachatsia mideopsoides* Cook (Morimotacarinae), adults. Fig. 392, Pedipalp; Fig. 393, dorsal shield; Fig. 394, female, ventral shield. *FIGURES 395, 397* *Volsellacarus sabulonus* Cook (Neoacaridae), adults. Fig. 395, female, ventral shield; Fig. 397, gnathosoma. *FIGURE 396* *Neoacarus minimus* Cook (Neoacaridae), adult, pedipalp.

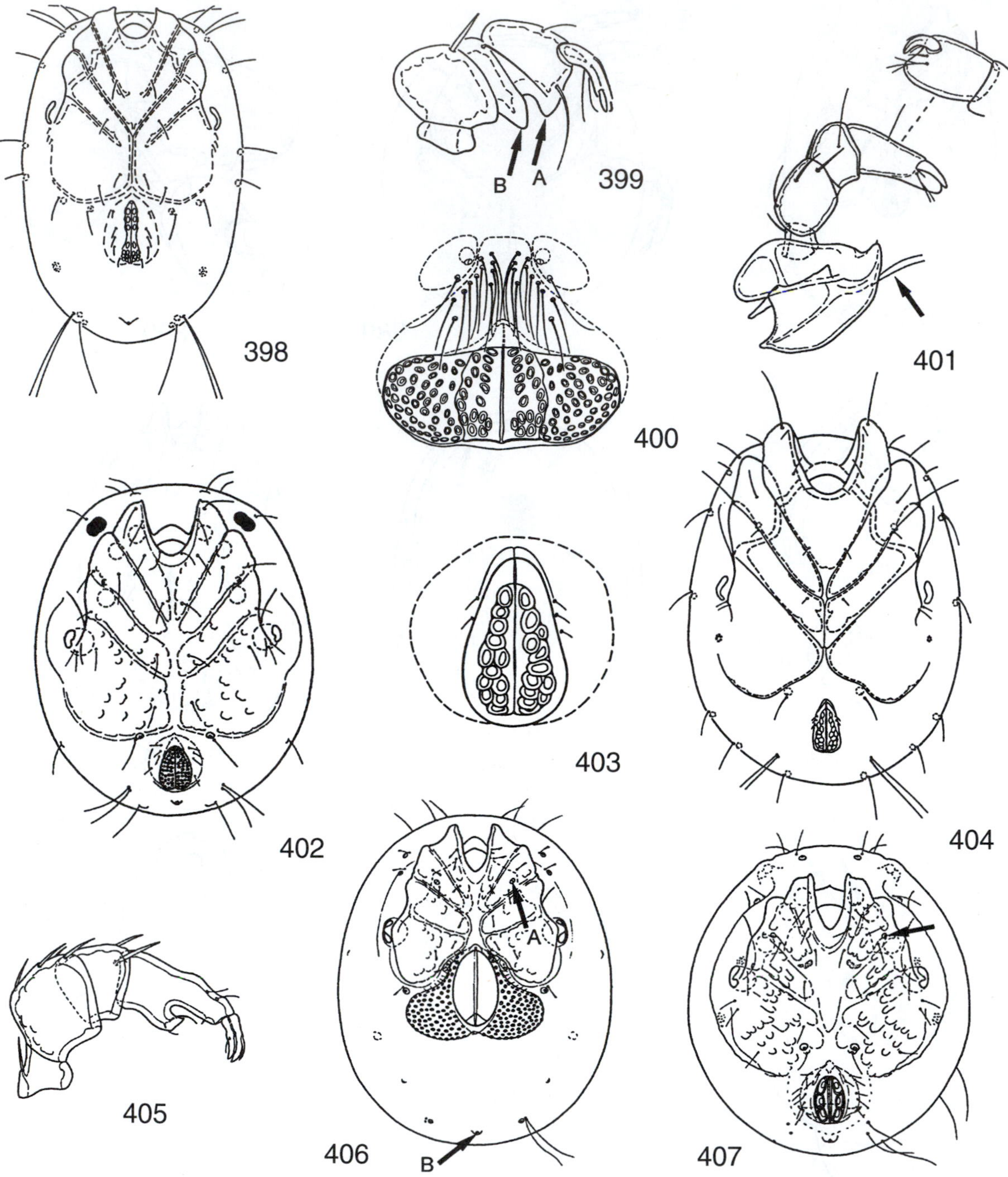

FIGURES 398, 399 ***Stygameracarus cooki*** Smith (Stygameracarinae), adults. Fig. 398, Male, ventral shield; Fig. 399, pedipalp. *FIGURE 400* ***Platyhydracarus juliani*** Smith (Athienemanniinae), adult male, genital field. *FIGURES 401, 402* ***Chelomideopsis besselingi*** (Cook) (Athienemanniinae), adults. Fig. 401, Gnathosoma; Fig. 402, adult male, ventral shield. *FIGURES 403, 404* ***Chelohydracarus navarrensis*** Smith (Athienemanniinae), adult male. Fig. 403, Genital field; Fig. 404, ventral shield. *FIGURES 405, 407* ***Mideopsis*** **sp.** (Mideopsinae), adults. Fig. 405, Pedipalp; Fig. 407, female, ventral shield. *FIGURE 406* ***Laversia berulophila*** Cook (Laversiidae), adult female, ventral shield.

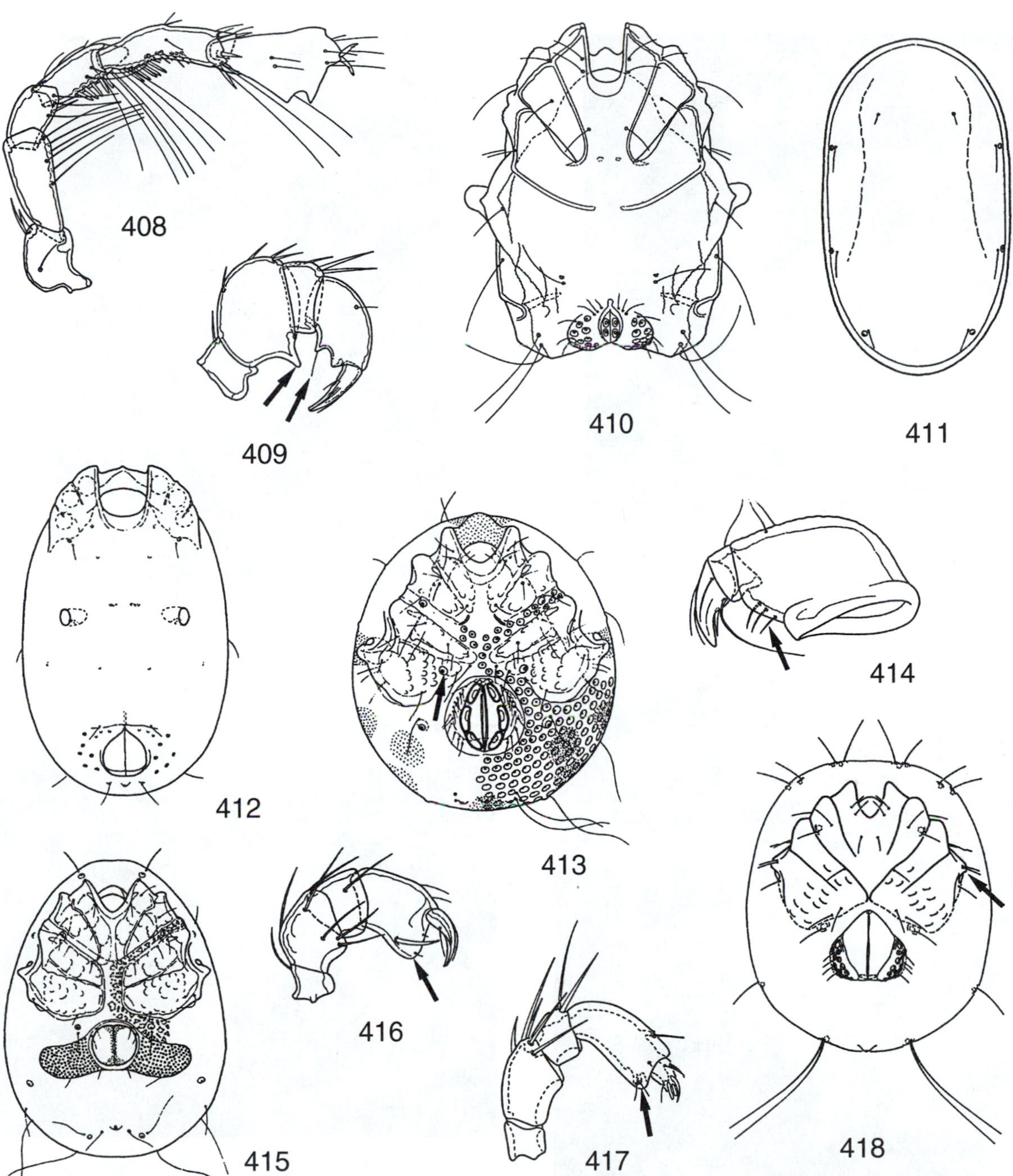

FIGURES 408–410 *Amoenacarus dixiensis* Smith and Cook (Amoenacaridae), adults. Fig. 408, Male, leg IV; Fig. 409, pedipalp; Fig. 410, male, ventral shield. *FIGURES 411, 412* *Arenohydracarus minimus* Cook (Arenohydracaridae), adults. Fig. 411, dorsal shield; Fig. 412, female, ventral shield. *FIGURE 413* *Geayia* sp. (Krendowskiidae), adult female, ventral shield. *FIGURE 414* *Horreolanus orphanus* Mitchell (Horreolaninae), adult, pedipalp. *FIGURES 415, 416* *Arrenurus* sp. (Arrenurinae), adults. Fig. 415, Female, ventral shield. Fig. 416, pedipalp. *FIGURES 417, 418* *Acalyptonotus neoviolaceus* Smith (Acalyptonotidae), adults. Fig. 417, Pedipalp; Fig. 418, female, ventral shield.

419

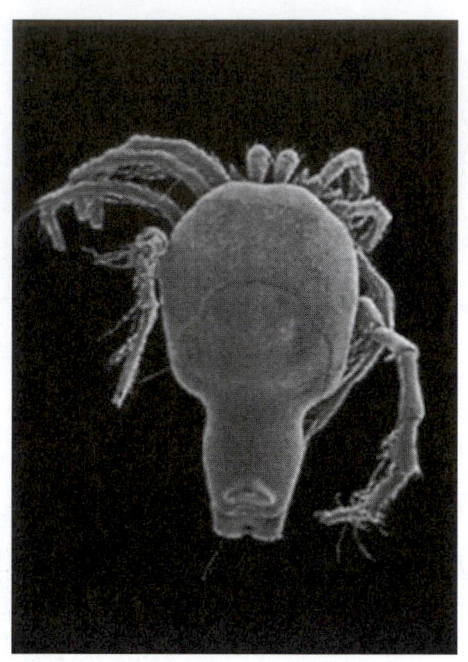

420

421

422

FIGURES 419–422 Scanning electron micrographs of adult water mites. Fig. 419 *Arrenurus fissicornis* Marshall (Arrenuruidae), male; Fig. 420 *Arrenurus (Megaluracarus) pseudocylindratus* Marshall (Arrenuridae), male; Fig. 421 *Kongsbergia* sp. (Aturidae), female; Fig. 422 *Aturus* sp. (Aturidae), male.

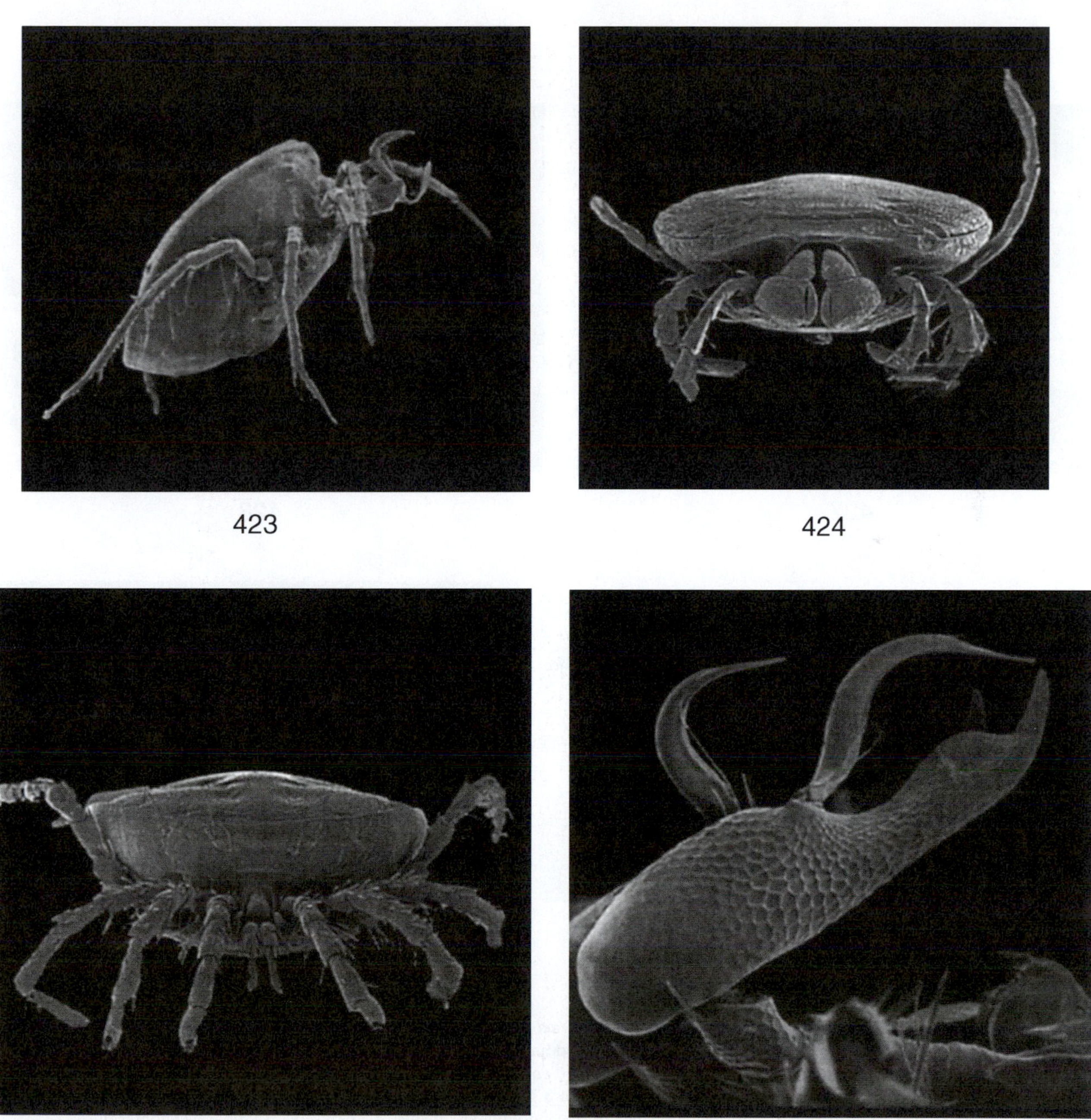

FIGURES 423–426 Scanning electron micrographs of adult water mites. FIGURE. 423 Koenikea wolcotti Viets (Unionicolidae), female; *FIGURE 424 Uchidastygacarus acadiensis* Smith (Chappuisididae), female; *FIGURE 425 Paramideopsis susanae* Smith (Nudomideopsidae), female; *FIGURE 426 Volsellacarus sabulonus* Cook (Neoacaridae), female.

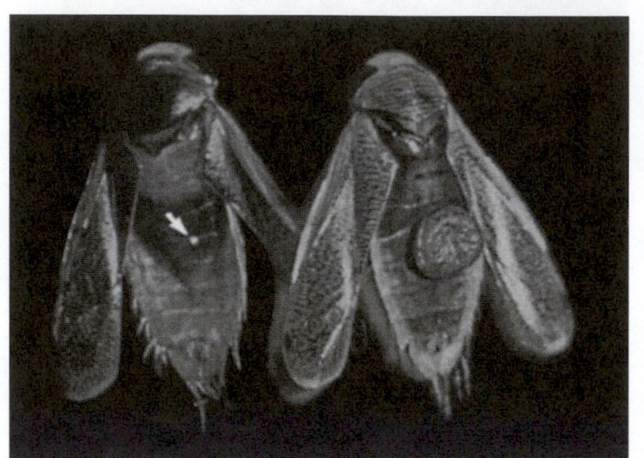

427

428

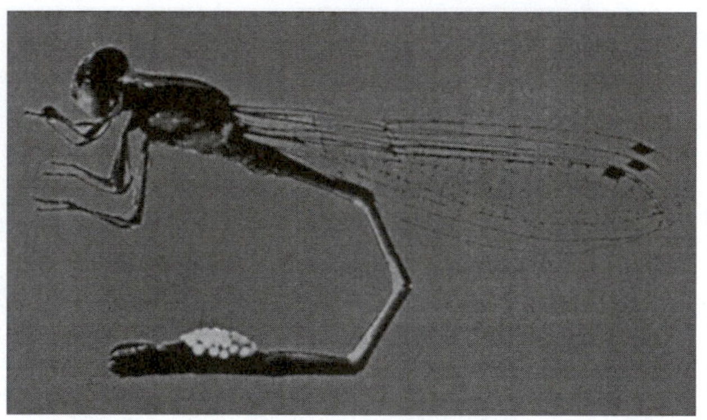

429

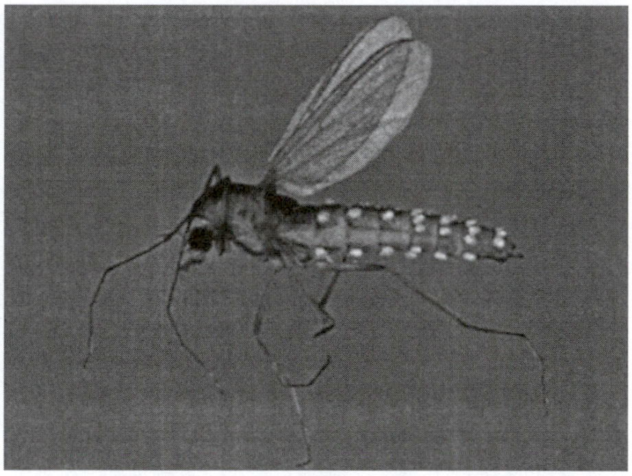

430

FIGURES 427–430 Photographs of parasitic larval water mites attached to host insects. FIGURE 427 Larvae of *Eylais* sp. attached to the subelytral abdominal integument of adult waterboatmen (Hemiptera: Corrixidae) illustrating larval size before and after engorgement; *FIGURE 428* larvae of *Hydrachna* sp. (Hydrachnidae) attached to the thorax and elytra of a giant water bug (Hemiptera:Belostomatidae); Fig. 429 larvae of *Arrenurus* sp. (Arrenuridae) attached to the abdominal segments of a damselfly (Odonata:Zygoptera); *FIGURES 430* larvae of *Feltria* sp. (Feltriidae) and *Aturus* sp. (Aturidae) attached to the abdominal segments of a chironomid midge (Diptera:Chironomidae).

TABLE I Summary of Some Important Zoogeographic and Ecological Characteristics of North American Water Mite Genera

	Zoogeography		Ecology						
			Larvae			Deutonymphs and Adults			Adults
Taxa	*Origin*	*North American distribution*	*Type*	*Order of hosts*	*Attachment site*	*Habitat type*	*Mode of locomotion*	*Prey*	*Spermatophore transfer mode*
Stygothrombidioidea									
Stygothrombidiidae									
Stygothrombium	Laurasian	Temperate, boreal	Aquatic	Plecoptera	Thorax, abdomen	Interstitial	Crawling	Unknown	Unknown
Hydrovolzioidea									
Hydrovolziidae									
Hydrovolzia	Laurasian	Temperate, boreal	Terrestrial	Hemiptera, Diptera	Thorax, abdomen	Springs, riffles	Crawling	Unknown	Unknown
Acherontacaridae									
Acherontacarus	Laurasian	Temperate	Aquatic	Unknown	Unknown	Interstitial	Crawling	Unknown	Unknown
Hydrachnoidea									
Hydrachnidae									
Hydrachna	Pangean	Throughout	Aquatic	Coleoptera, Hermiptera	Thorax, abdomen	Lakes, temporary pools	Swimming	Insect eggs	Dissociation
Eylaoidea									
Limnocharidae									
Limnocharinae									
Limnochares	Pangean	Temperate, boreal	Terrestrial	Hemiptera, Odonata	Thorax, abdomen	Riffles, pools, lakes	Crawling, swimming	Diptera larvae	UnKnown
Neolimnochares	Pangean	Temperate	Terrestrial	Hemiptera	Thorax, abdomen	Pools	Crawling	Unknown	Unknown
Rhyncholimnocharinae									
Rhyncholimnochares	Gondwanan	Temperate	Aquatic	Coleoptera	Abdomen	Riffles, interstitial	Crawling	Unknown	Unknown
Eylaidae									
Eylais	Pangean	Throughout	Terrestrial	Coleoptera, Hemiptera	Thorax, abdomen	Pools, lakes, temporary pools	Swimming	Ostracoda, Cladocera	Dissociation, association, copulation
Piersigiidae									
Piersigia	Laurasian	Temperate, boreal	Terrestrial	Coleoptera	Abdomen	Springs, interstitial, temporary pools	Crawling	Ostracoda	Unknown
Apheviderulicidae									
Apheviderulix	Laurasian	Temperate	Terrestrial	Coleoptera	Abdomen	Springs	Crawling	Unknown	Unknown
Hydryphantoidea									
Hydryphantidae									
Hydryphantinae									
Hydryphantes	Pangean	Throughout	Terrestrial	Hemiptera, Odonata, Diptera	Thorax, abdomen	Lakes, temporary pools	Swimming	Diptera eggs	Dissociation
Thyadinae									
Albertathyas	North American	Temperate	Unknown	Unknown	Unknown	Riffles	Crawling	Unknown	Unknown
Amerothyasella	North American	Temperate	Unknown	Unknown	Unknown	Riffles	Crawling	Unknown	Unknown
Columbiathyas	North American	Temperate	Unknown	Unknown	Unknown	Springs	Crawling	Unknown	Unknown

(continues)

TABLE I (Continued)

	Zoogeography		Ecology						
			Larvae			Deutonymphs and Adults			Adults
Taxa	Origin	North American distribution	Type	Order of hosts	Attachment site	Habitat type	Mode of locomotion	Prey	Spermatophore transfer mode
Euthyas	Laurasian	Temperate	Terrestrial	Diptera	Thorax	Riffles, temporary pools	Walking	Insect eggs	Unknown
Notopanisus	Pangean	Temperate	Unknown	Unknown	Unknown	Springs, riffles	Crawling	Unknown	Unknown
Panisus	Laurasian	Temperate, boreal	Terrestrial	Diptera, Hymenoptera	Thorax	Springs, riffles	Crawling	Unknown	Unknown
Panisopsis	Laurasian	Temperate, boreal	Terrestrial	Diptera	Unknown	Springs, riffles	Crawling	Unknown	Unknown
Parathyasella	Laurasian	Temperate, boreal	Unknown	Unknown	Unknown	Springs, riffles	Crawling	Unknown	Unknown
Siskiyouthyas	North American	Temperate	Unknown	Unknown	Unknown	Riffles	Unknown	Unknown	Unknown
Thyas	Laurasian	Temperate, boreal	Terrestrial	Collembola, Diptera, Trichoptera	Thorax	Springs, riffles, temporary pools	Walking	Insect eggs	Dissociation
Thyasella	Laurasian	Temperate	Unknown	Unknown	Unknown	Springs, interstitial	Walking	Unknown	Unknown
Thyasides	Laurasian	Boreal	Terrestrial	Diptera	Thorax	Temporary pools	Walking	Unknown	Unknown
Thyopsella	North American	Temperate, boreal	Terrestrial	Diptera, Trichoptera	Thorax, abdomen	Springs, riffles	Crawling	Unknown	Unknown
Thyopsis	Laurasian	Temperate	Terrestrial	Diptera	Thorax	Springs, temporary pools	Crawling	Unknown	Unknown
Thyopsoides	North American	Temperate	Unknown	Unknown	Unknown	Riffles	Crawling	Unknown	Unknown
Trichothyas	Laurasian	Temperate	Terrestrial	Diptera	Head, thorax, abdomen	Springs	Walking	Unknown	Unknown
Zschokkea	Laurasian	Boreal, arctic	Unknown	Unknown	Unknown	Temporary pools	Walking	Unknown	Unknown
Protziinae									
Partnunia	Laurasian	Temperate	Terrestrial	Plecoptera	Thorax, abdomen	Riffles	Walking	Unknown	Unknown
Protzia	Laurasian	Temperate, boreal	Terrestrial	Diptera, Trichoptera	Thorax, abdomen	Riffles	Walking	Unknown	Unknown
Wandesiinae									
Wandesia	Pangean	Temperate	Aquatic	Plecoptera	Thorax	Springs, interstitial	Crawling	Unknown	Unknown
Tartarothyadinae									
Tartarothyas	Pangean	Temperate	Terrestrial	Unknown	Unknown	Springs, riffles	Walking, crawling	Unknown	Unknown
Cyclothyadinae									
Cyclothyas	Laurasian	Temperate (W)	Unknown	Unknown	Unknown	Springs, interstitial	Crawling	Unknown	Unknown
Cowichaniinae									
Cowichania	North American	Temperate (W)	Unknown	Unknown	Unknown	Interstitial	Crawling	Unknown	Unknown
Hydrodromidae									
Hydrodroma	Pangean	Throughout	Terrestrial	Diptera	Thorax	Riffles, pools, lakes	Swimming	Insect eggs	Dissociation
Rhynchohydracaridae									
Clathrosperchoninae									
Clathrosperchon	Gondwanan	Temperate	Unknown	Unknown	Unknown	Riffles, interstitial	Crawling	Unknown	Unknown

Thermacaridae									
Thermacarus	Pangean	Temperate	Terrestrial	Anura (Amphibia)	Body dorsum	Hot springs	Crawling	Diptera larvae	Unknown
Lebertioidea									
Sperchontidae									
Sperchontinae									
Sperchon	Laurasian	Throughout	Aquatic	Diptera, Trichoptera	Thorax, abdomen	Springs, riffles, pools, lakes	Crawling	Diptera larvae	Unknown
Sperchonopsis	Laurasian	Temperate, boreal	Aquatic	Diptera	Thorax	Riffles, pools, lakes	Crawling	Unknown	Unknown
Teutoniidae									
Teutonia	Laurasian	Boreal	Aquatic	Diptera	Abdomen	Springs, pools	Crawling swimming	Diptera larvae	Unknown
Rutripalpidae									
Rutripalpus	Laurasian	Boreal (E)	Unknown	Unknown	Unknown	Springs, pools	Crawling	Unknown	Unknown
Anisitsiellidae									
Anisitsiellinae									
Bandakia	Laurasian	Temperate, boreal	Aquatic	Diptera	Thorax	Springs, riffles, interstitial	Crawling	Unknown	Unknown
Bandakiopsis	Laurasian	Temperate	Aquatic	Diptera	Thorax, abdomen	Springs	Crawling	Unknown	Unknown
Cookacarus	North American	Temperate	Unknown	Unknown	Unknown	Springs, interstitial	Crawling	Unknown	Unknown
Mamersellides	Gondwanan	Temperate (SE)	Unknown	Unknown	Unknown	Pools	Swimming	Unknown	Unknown
Oregonacarus	North American	Temperate	Unknown	Unknown	Unknown	Springs	Crawling	Unknown	Unknown
Utaxatax	Laurasian	Temperate	Aquatic	Diptera	Thorax, abdomen	Springs, interstitial	Crawling	Unknown	Unknown
Lebertiidae									
Estelloxus	North American	Temperate (W)	Aquatic	Diptera	Thorax	Springs, riffles	Walking	Unknown	Unknown
Lebertia	Laurasian	Throughout	Aquatic	Diptera	Thorax	Springs, riffles, pools, lakes	Walking, swimming	Diptera larvae	Dissociation
Scutolebertia	North American	Temperate (SW)	Unknown	Unknown	Unknown	Springs	Crawling	Unknown	Unknown
Oxidae									
Frontipoda	Pangean	Temperate, boreal	Aquatic	Diptera	Thorax	Pools, lakes	Swimming	Diptera larvae	Unknown
Oxus	Pangean	Temperate, boreal	Aquatic	Diptera	Thorax, abdomen	Pools, lakes	Swimming	Diptera larvae	Unknown
Torrenticolidae									
Neoatractidinae									
Neoatractides	Gondwanan	Temperate (SW)	Unknown	Unknown	Unknown	Riffles	Crawling	Unknown	Unknown
Testudacarinae									
Testudacarus	Laurasian	Temperate, boreal	Aquatic	Diptera	Thorax	Riffles, interstitial	Crawling	Diptera larvae	Unknown
Torrenticolinae									
Monatractides	Laurasian	Temperate	Aquatic	Diptera	Thorax	Riffles, pools, lakes	Crawling	Diptera larvae	Unknown
Pseudotorrenticola	Laurasian	Temperate	Unknown	Unknown	Unknown	Interstitial	Crawling	Unknown	Unknown
Torrenticola	Laurasian	Temperate, boreal	Aquatic	Diptera	Thorax	Riffles, interstitial, pools, lakes	Crawling	Diptera larvae	Unknown
Hygrobatoidea									
Limnesiidae									
Kawamuracarinae									
Kawamuracarus	Laurasian	Temperate (SW)	Unknown	Unknown	Unknown	Interstitial	Crawling, walking	Unknown	Unknown
Limnesiinae									
Centrolimnesia	South American	Temperate	Unknown	Unknown	Unknown	Pools	Swimming	Unknown	Unknown

(continues)

TABLE I (Continued)

	Zoogeography		Ecology						
			Larvae			Deutonymphs and Adults			Adults
Taxa	*Origin*	*North American distribution*	*Type*	*Order of hosts*	*Attachment site*	*Habitat type*	*Mode of locomotion*	*Prey*	*Spermatophore transfer mode*
Limnesia	Pangean	Temperate, boreal	Aquatic	Diptera	Thorax	Riffles, interstitial, pools, lakes	Crawling, swimming	Copepoda, cladocera, insect eggs and larvae	Unknown
Neomamersinae									
Arizonacarus	North American	Temperate (SW)	Unknown	Unknown	Unknown	Interstitial	Crawling, walking	Unknown	Unknown
Meramecia	South American	Temperate	Unknown	Unknown	Unknown	Interstitial	Crawling, walking	Unknown	Unknown
Neomamersa	South American	Temperate	Unknown	Unknown	Unknown	Interstitial	Crawling, walking	Unknown	Unknown
Protolimnesiinae									
Protolimnesia	South American	Temperate (SW)	Unknown	Unknown	Unknown	Interstitial	Walking	Unknown	Unknown
Tyrrelliinae									
Neotyrrellia	South American	Temperate	Unknown	Unknown	Unknown	Riffles	Crawling	Unknown	Unknown
Tyrrellia	South American	Temperate, boreal	Aquatic	Diptera	Abdomen	Springs, pools, lakes	Crawling	Diptera eggs and larvae	Unknown
Omartacaridae									
Omartacarinae									
Omartacarus	Gondwanan	Temperate (SW)	Unknown	Unknown	Unknown	Interstitial	Crawling	Unknown	Unknown
Hygrobatidae									
Atractides	Laurasian	Temperate, boreal	Aquatic	Diptera	Abdomen	Springs, riffles, interstitial, pools, lakes	Crawling, walking	Diptera larvae, Ostracoda, Cladocera	Unknown
Corticacarus	Gondwanan	Temperate	Unknown	Unknown	Unknown	Riffles	Crawling	Unknown	Unknown
Diamphidaxona	Gondwanan	Temperate	Unknown	Unknown	Unknown	Interstitial	Walking	Unknown	Unknown
Hygrobates	Laurasian	Boreal, temperate	Aquatic	Diptera, Trichoptera	Abdomen	Springs, riffles, pools, lakes	Crawling, swimming	Diptera larvae, Ostracoda	Dissociation
Mesobates	Laurasian	Boreal	Unknown	Unknown	Unknown	Riffles, pools	Crawling	Unknown	Unknown
Unionicolidae									
Pionatacinae									
Koenikea	Pangean	Temperate	Aquatic	Diptera, Trichoptera	Abdomen	Pools, lakes	Swimming	Diptera larvae	Unknown
Neumania	Pangean	Temperate, boreal	Aquatic	Diptera	Abdomen	Pools, lakes	Swimming	Diptera larvae, Copepoda	Dissociation, association
Unionicolinae									
Unionicola	Pangean	Temperate, boreal	Aquatic	Diptera, Trichoptera	Abdomen, legs	Pools, lakes, mollusc parasites	Crawling, swimming	Copepoda, Cladocera, Diptera larvae, mollusc tissue	Dissociation, association, copulation

Feltriidae									
Feltria	Laurasian	Temperate, boreal	Aquatic	Diptera	Abdomen	Springs, riffles, interstitial	Crawling	Diptera larvae	Copulation
Wettinidae									
Wettina	Laurasian	Temperate, boreal	Aquatic	Diptera	Abdomen	Pools, lakes	Swimming	Unknown	Dissociation
Lethaxonidae									
Lethaxona	Laurasian	Temperate (W)	Unknown	Unknown	Unknown	Interstitial	Walking	Unknown	Unknown
Lethaxonella	North American	Temperate (SW)	Unknown	Unknown	Unknown	Interstitial	Walking	Unknown	Unknown
Frontipodopsidae									
Frontipodopsis	Pangean	Temperate (W)	Unknown	Unknown	Unknown	Interstitial	Walking, swimming	Unknown	Unknown
Pionidae									
Foreliinae									
Forelia	Laurasian	Temperate, boreal	Aquatic	Diptera	Abdomen	Pools, lakes	Crawling, swimming	Diptera larvae	Copulation
Pionacercus	Laurasian	Boreal	Aquatic	Diptera	Abdomen	Springs, pools	Crawling, swimming	Unknown	Copulation
Pseudofeltria	Laurasian	Temperate	Aquatic	Diptera	Abdomen	Springs, interstitial	Crawling	Diptera larvae	Copulation
Huitfeldtiinae									
Huitfeldtia	Laurasian	Boreal	Aquatic	Diptera	Abdomen	Lakes (profundal)	Swimming	Cladocera	Unknown
Hydrochoreutinae									
Hydrochoreutes	Laurasian	Temperate, boreal	Aquatic	Diptera	Abdomen	Pools, lakes	Swimming	Cladocera	Unknown
Najadicolinae									
Najadicola	Laurasian	Temperate	Aquatic	Diptera	Abdomen	Lakes (mollusc parasite)	Crawling	Mollusc tissue	Unknown
Pioninae									
Nautarachna	Laurasian	Throughout	Aquatic	Diptera	Abdomen	Springs, riffles, pools	Crawling, swimming	Diptera larvae	Copulation
Piona	Laurasian	Throughout	Aquatic	Diptera	Abdomen	Pools, lakes, temporary pools	Crawling, swimming	Copepoda, Cladocera, Diptera larvae	Copulation
Tiphyinae									
Neotiphys	North American	Boreal, temperate	Aquatic	Diptera	Abdomen	Pools	Swimming	Unknown	Copulation
Pionopsis	Laurasian	Boreal	Aquatic	Diptera	Abdomen	Pools, temporary pools	Swimming	Ostracoda, Cladocera, Copepoda	Copulation
Tiphys	Laurasian	Throughout	Aquatic	Diptera	Abdomen	Pools, lakes, temporary pools	Swimming	Diptera larvae, Ostracoda, Cladocera, Copepoda	Copulation
Aturidae									
Albiinae									
Albia	Laurasian	Temperate	Aquatic	Trichoptera	Abdomen	Pools, lakes	Swimming	Unknown	Unknown
Aturinae									
Aturus	Laurasian	Temperate	Aquatic	Diptera	Abdomen	Springs, riffles, interstitial, pools	Crawling	Diptera larvae	Copulation
Bharatalbia	Laurasian	Temperate (W)	Unknown	Unknown	Unknown	Interstitial	Walking	Unknown	Unknown
Kongsbergia	Laurasian	Temperate	Unknown	Unknown	Unknown	Riffles	Crawling	Unknown	Copulation
Neoaturus	South American	Temperate	Unknown	Unknown	Unknown	Riffles	Crawling	Unknown	Unknown

(continues)

TABLE I (Continued)

	Zoogeography		Ecology						
			Larvae			Deutonymphs and Adults			Adults
Taxa	Origin	North American distribution	Type	Order of hosts	Attachment site	Habitat type	Mode of locomotion	Prey	Spermatophore transfer mode
Phreatobrachypoda	Laurasian	Temperate	Unknown	Unknown	Unknown	Interstitial	Walking	Unknown	Unknown
Axonopsinae									
Albaxona	Laurasian	Temperate (E)	Unknown	Unknown	Unknown	Interstitial	Walking	Unknown	Unknown
Axonopsella	Gondwanan	Temperate (E)	Unknown	Unknown	Unknown	Interstitial	Walking	Unknown	Unknown
Axonopsis	Laurasian	Temperate, boreal	Aquatic	Diptera	Abdomen	Riffles, interstitial, pools, lakes	Walking, swimming	Diptera larvae	Unknown
Brachypoda	Laurasian	Temperate, boreal	Aquatic	Diptera	Abdomen	Pools	Swimming	Diptera larvae	Copulation
Erebaxonopsis	Laurasian	Temperate	Unknown	Unknown	Unknown	Interstitial	Walking	Unknown	Unknown
Estellacarus	North American	Boreal	Aquatic	Diptera	Abdomen	Pools, lakes	Swimming	Diptera larvae	Copulation
Javalbia	Laurasian	Temperate	Unknown	Unknown	Unknown	Interstitial	Walking	Unknown	Unknown
Ljania	Laurasian	Temperate, boreal	Aquatic	Diptera	Abdomen	Riffles	Walking	Diptera larvae	Unknown
Neobrachypoda	Laurasian	Boreal	Aquatic	Unknown	Unknown	Springs, pools	Walking	Unknown	Unknown
Stygalbiella	Gondwanan	Temperate (W)	Unknown	Unknown	Unknown	Interstitial	Walking	Unknown	Unknown
Submiraxona	Gondwanan	Temperate (W)	Unknown	Unknown	Unknown	Riffles, pools	Walking, swimming	Unknown	Unknown
Woolastookia	Laurasian	Temperate, boreal	Aquatic	Diptera	Abdomen	Riffles, pools, lakes	Walking, swimming	Diptera larvae	Copulation
Arrenuroidea									
Mideidae									
Midea	Laurasian	Throughout	Aquatic	Diptera	Abdomen	Pools, lakes	Swimming	Diptera larvae	Copulation
Momoniidae									
Cyclomomoniinae									
Cyclomomonia	North American	Temperate (W)	Unknown	Unknown	Unknown	Interstitial	Walking	Unknown	Unknown
Momoniinae									
Momonia	Laurasian	Temperate	Aquatic	Trichoptera	Thorax, abdomen	Pools, lakes	Swimming	Diptera larvae	Unknown
Stygomomoniinae									
Stygomomonia	Laurasian	Temperate	Aquatic	Trichoptera	Abdomen	Interstitial	Walking	Unknown	Unknown
Nudomideopsidae									
Neomideopsis	North American	Temperate (W)	Unknown	Unknown	Unknown	Springs	Crawling	Unknown	Unknown
Nudomideopsis	Pangean	Temperate, boreal	Aquatic	Diptera	Thorax	Springs, interstitial	Crawling	Unknown	Unknown
Paramideopsis	Pangean	Temperate, boreal	Aquatic	Diptera	Thorax	Springs, interstitial	Crawling	Unknown	Unknown
Mideopsidae									
Mideopsinae									
Mideopsis	Laurasian	Temperate, boreal	Aquatic	Diptera	Thorax	Springs, riffles, interstitial, pools, lakes	Crawling, walking, swimming	Diptera larvae	Unknown

Chappuisididae									
Chappuisidinae									
Chappuisides	Laurasian	Temperate	Unknown	Unknown	Unknown	Interstitial	Walking	Unknown	Unknown
Morimotacarinae									
Morimotacarus	Laurasian	Temperate (W)	Unknown	Unknown	Unknown	Interstitial	Walking	Unknown	Unknown
Yachatsia	Laurasian	Temperate (W)	Unknown	Unknown	Unknown	Interstitial	Walking	Unknown	Unknown
Uchidastygacarinae									
Uchidastygacarus	Laurasian	Temperate	Unknown	Unknown	Unknown	Interstitial	Walking	Unknown	Unknown
Neoacaridae									
Neoacarus	Laurasian	Temperate, boreal	Aquatic	Diptera	Thorax	Riffles, interstitial, lakes	Walking	Unknown	Unknown
Volsellacarus	North American	Temperate	Aquatic	Diptera	Thorax	Riffles, interstitial	Walking	Unknown	Unknown
Bogatiidae									
Horreolaninae									
Horreolanus	North American	Temperate	Unknown	Unknown	Unknown	Interstitial	Walking	Unknown	Unknown
Acalyptonotidae									
Paenecalyptonotus	North American	Boreal	Aquatic	Unknown	Unknown	Springs	Crawling	Unknown	Unknown
Acalyptonotus	Laurasian	Boreal	Aquatic	Diptera	Thorax	Springs, pools, lakes	Walking	Unknown	Unknown
Athienemanniidae									
Athienemanniinae									
Chelohydracarus	Laurasian	Temperate (W)	Unknown	Unknown	Unknown	Interstitial	Crawling	Unknown	Unknown
Chelomideopsis	Laurasian	Temperate, boreal	Aquatic	Diptera	Thorax	Springs, riffles	Crawling	Diptera larvae	Unknown
Platyhydracarus	North American	Temperate (W)	Aquatic	Diptera	Thorax	Springs, riffles, interstitial, pools	Crawling	Unknown	Unknown
Stygameracarinae									
Stygameracarus	North American	Temperate	Unknown	Unknown	Unknown	Interstitial	Walking	Unknown	Unknown
Amoenacaridae									
Amoenacarus	North American	Temperate (E)	Unknown	Unknown	Unknown	Pools	Swimming	Unknown	Unknown
Arenohydracaridae									
Arenohydracarus	Gondwanan	Temperate (SW)	Unknown	Unknown	Unknown	Interstitial	Crawling	Unknown	Unknown
Laversiidae									
Laversia	North American	Temperate, boreal	Aquatic	Diptera	Thorax	Springs, riffles, lakes (profundal)	Crawling	Unknown	Unknown
Krendowskiidae									
Geayia	Gondwanan	Temperate	Aquatic	Diptera	Abdomen	Pools, lakes	Swimming	Unknown	Unknown
Krendowskia	Pangean	Temperate	Aquatic	Diptera	Abdomen	Pools, lakes	Swimming	Unknown	Unknown
Arrenuridae									
Arrenurinae									
Arrenurus	Pangean	Throughout	Aquatic	Diptera, odonata	Thorax, abdomen	Springs, riffles, interstitial, pools, lakes, temporary pools	Walking, swimming	Ostracoda, Cladocera, Copepoda, Insect larvae	Copulation

III. OTHER ARACHNIDS IN FRESHWATER HABITATS

A. Other Mites (ACARI)

1. Acariformes

a. Other Actinedida The mainly marine family Halacaridae is represented in both benthic and interstitial freshwater habitats in North America by species of the holarctic genera *Porohalacarus, Lobohalacarus, Stygohalacarus, Soldanellonyx* and *Porolohmanella* (Fig. 431). Halacarid mites have stilettoform chelicerae and 5-segmented pedipalps. Adults usually have four characteristic dorsal plates on the idiosoma and claws on the tarsi of the legs. All instars have a sprawling gait and crawl slowly on the substrate. Halacarid feeding habits are generally poorly understood, but predaceous, parasitic and algivorous species are known. Members of some species may be extremely abundant in mats of vegetation and detritus in ponds and small lakes. There are no comprehensive references for North American freshwater halacarid mites. All known taxa of freshwater Halacaridae were reviewed by K. H. Viets (1956) and K. O. Viets (1987) and species of the German fauna, including all of the holarctic genera, were treated by K.H. Viets (1936).

Homocaligidae is a small family of uncommon species inhabiting wet vegetation and detritus at the margin of ponds. Adults are well sclerotized and have prominent hemispherical dorsal shields and long, protruding cheliceral stylets. These mites frequently venture beneath the surface, apparently aided by accessory respiratory sacs and tubes. They have long legs and crawl slowly on the substrate, but are thought to be predaceous (Krantz 1978). Homocaligid taxa, including the North American species *Homocaligus muscorum* Habeeb, were reviewed by Wood (1969).

b. Oribatida The vast majority of oribatid mite taxa are strictly terrestrial, but members of certain genera of the families Camisiidae, Trhypochthoniidae, Malaconothridae, Hydrozetidae, Limnozetidae (Fig. 432), Ameronothridae, Tegeocranellidae and Zetomimidae are found only in freshwater habitats ranging from springs and streams to large lakes. Oribatid mites have chelate chelicerae and 5-segmented pedipalps. Adults are variously sclerotized, and usually have distinctively shaped prodorsal bothridia and claws on the tarsi of the legs. All active instars of oribatid mites crawl slowly on the substrate, feeding on both living and dead plant and fungal material. The feeding habits of a species of *Hydrozetes* (Hydrozetidae) were investigated by Baker (1985). Members of some species this genus may be very numerous in mats of plant material and detritus. Freshwater genera of oribatid mites can be identified using the keys published by Balogh and Balogh (1992a, b). Palearctic taxa, many of which are holarctic, were treated by Balogh and Mahunka (1983). There is no comprehensive reference for North American freshwater oribatid taxa, but systematic treatments have recently been published for the genera *Mucronothrus* (Trhypochthoniidae) (Norton, Behan-Pelletier and Wang, 1996), *Limnozetes* (Limnozetidae) (Behan-Pelletier, 1989) and *Tegeocranellus* (Tegeocranellidae) (Behan-Pelletier, 1997).

c. Acaridida Most acaridid mites are terrestrial saprophages, phytophages, fungivores or parasites, but certain free-living species of the families Histiostomatidae and Hyadesiidae inhabit freshwater. Adult acaridids are generally soft-bodied, have 2-segmented pedipalps and lack both prodorsal trichobothria and claws on the tarsi of the legs. Freshwater species are slow-moving crawlers typically associated with algal mats or decaying detritus, although some have been reported from water reservoirs in pitcher plants. Some species of histiostomatid mites are reportedly predaceous on microorganisms or oligochaete worms. Families of freshwater Acaridida can be identified using the key published by Krantz (1978).

2. Parasitiformes

Nearly all parasitiform taxa are strictly terrestrial, but members of a few genera of the gamasid family Ascidae, subfamily Platyseiinae (Fig. 433), live exclusively on wet plants and detritus and on the water surface of marginal freshwater habitats. Platyseiinae have chelate chelicerae and adults are long-legged mites with large dorsal and ventral shields and conspicuous peritremes laterally on the idiosoma. Active instars of these mites walk about on the surface film, apparently feeding on floating egg masses of nematocerous flies, especially mosquitoes. Adult females of Platyseiinae regularly disperse between aquatic habitats as phoretic associates of adult craneflies (Tipulidae).

B. Spiders

There are no truly aquatic spiders in North America, but many species of nurseryweb spiders (family Pisauridae) are associated with aquatic habitats (Carico, 1973). Species of *Dolomedes* (Fig. 434), the genus with the greatest affinity for water, are large, long-legged spiders with flattened, mottled bodies. They are often called fishing spiders, dock spiders or raft spiders because they venture onto water to hunt and capture prey at or just below the surface film, and can dive below the surface to escape potential predators (Dondale and Redner, 1990). Members of some species, such as *Dolomedes tenebrosus,* wander considerable distances from water (Carico, 1973; Dondale and Redner, 1990), and even those of the most aquatic North American species, *Dolomedes triton*, apparently spend a significant part of their lives away from water (Zimmermann and Spence, 1992, 1998).

The systematics and distribution of the nine described nearctic species of *Dolomedes* were thoroughly reviewed by Carico (1973) and the five species occurring in Canada and Alaska were treated in detail by Dondale and Redner (1990). Given the availability of these relatively recent and comprehensive taxonomic treatments, and in keeping with the level of coverage for other groups in this chapter, we do not include a key for the species of *Dolomedes* or the extensive morphological information that would be required to support it. The following brief summary of information is intended as an introduction to the literature on the biology and ecology of these spiders.

Members of *Dolomedes* are typically found on rocks or wood at the edge of water or on emergent or floating vegetation. They move across the surface film using a rowing gait and can dive beneath the surface to remain submerged for as long as 30 min (Dondale and Redner, 1990). When hunting, they typically remain motionless with the anterior 2–3 pairs of legs resting on the water surface. In this posture, they are able to detect vibrations and chemical stimuli indicating the direction and proximity of potential prey (Roland and Rovner, 1983). They apparently do not use visual stimuli to locate prey (Bleckmann and Lotz, 1987). Their primary prey are adults of terrestrial and aquatic insects that have fallen on the water, including Hemiptera, Diptera and Odonata (Zimmermann and Spence, 1989), but they also capture and consume small fish, tadpoles, and even small frogs, using their venom to immobilize prey as much as four times their own weight within a few minutes (Comstock, 1912; Bleckmann and Lotz, 1987; Dondale and Redner, 1990). Cannibalism apparently accounts for up to 5% of the prey of female *D. triton* (Zimmermann and Spence, 1989, 1992, 1998).

Certain species of *Dolomedes* show habitat preferences that seem to be based on exposure and vegetation type. For example, members of *D. scriptus* are most common along larger, fast-flowing streams, with exposed rocky shorelines, those of *D. vittatus* prefer edges of smaller and more shaded streams and individuals of *D. triton* and *D. striatus* frequent margins of ponds, lakes, and calm backwaters of streams, including floating and emergent vegetation. The various species have cryptic color patterns correlated with their habitat preferences:those commonly found on rocks or logs are predominately mottled with gray and brown while those frequenting emergent vegetation have conspicuous longitudinal stripes (Carico, 1973; Dondale and Redner, 1990).

Species of *Dolomedes* exhibit complex reproductive behavior. Females of *Dolomedes triton* mate only once, and because males of this species cannot distinguish between mated and unmated females they are at risk of sexual cannibalism. Males mature 5–10 days before females, increasing their probability of finding immature adult females and their likelihood of surviving multiple matings (Zimmermann and Spence, 1989, 1992). Males signal in the presence of females by waving the legs, drumming with the pedipalps, and twitching the body, to generate distinctive, regular, low-frequency waves in the water surface that differ from those produced by prey (Roland and Rovner, 1983; Bleckmann and Bender, 1987; Bleckmann and Lotz, 1987). They apparently also respond to pheromonal cues, because their courtship displays can be induced by water that has been in contact with conspecific females (Roland and Rovner, 1983). Sperm is transferred to receptive females by inserting the embolus, a specialized intromittent organ at the tip of the male's pedipalp, into a bursa copulatrix in the epigynum of the female (Carico, 1973; Roland and Rovner, 1983).

Gravid females of Dolomedes produce an egg sac consisting of a silk-wrapped cluster of between 250 and 1500 eggs (Comstock, 1912; Carico, 1973; Zimmermann and Spence, 1992; Spence et al., 1996), and carry it beneath their body using the chelicerae and a silken dragline until the eggs are ready to hatch (Dondale and Redner, 1990). This parental care apparently is necessary for successful hatching (Zimmermann and Spence, 1992). Shortly before the eggs begin to hatch, the female spins a nursery web, consisting of a tentlike structure with the egg sac suspended within it (Dondale and Redner, 1990). These webs are typically in low plant growth or among rocks and are vigorously defended by females against potential predators. Newly emerged juvenile spiders spend 3–7 days in the web, and disperse some time after their first molt either by ballooning or walking (Zimmermann and Spence, 1992, 1998).

In central Alberta, Canada, members of *Dolomedes triton* have approximately 12 juvenile instars but older individuals may experience supernumerary molts if exposed to warmer temperatures in autumn, possibly to delay maturation until spring. Duration of instars 1–9 is the same for both sexes, but duration of instars 10–12 is significantly longer for females, resulting in the protandry mentioned above (Zimmermann and Spence, 1998). At this latitude, *D. triton* is semivoltine, overwintering as instars 3–5 and

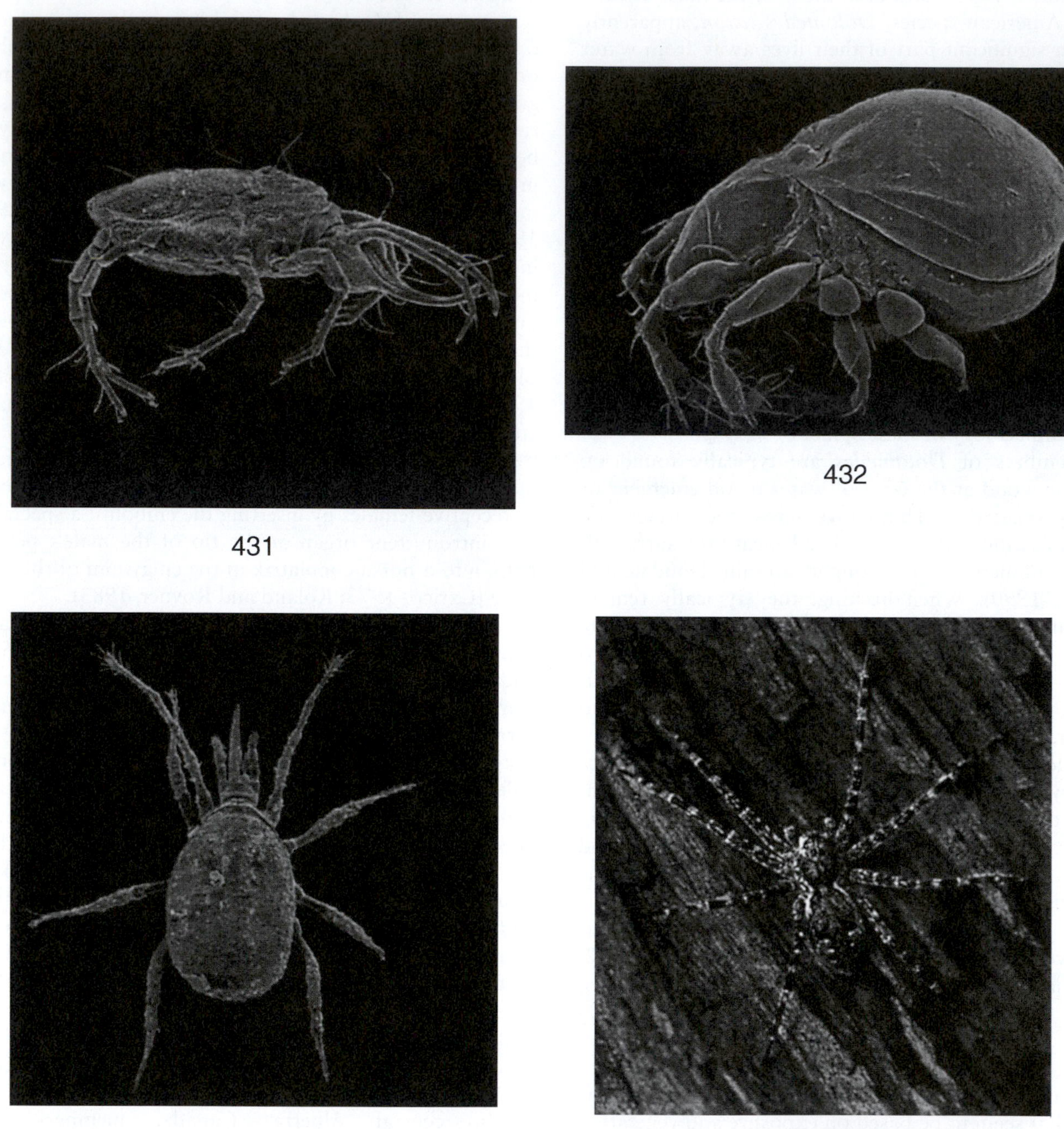

FIGURES 431–434 Scanning electron micrographs of adult mites. Fig. 431 *Porolohmannella* sp. (Halacaridae); Fig. 432 *Hydrozetes* sp. (Hydrozetidae); Fig. 433 *Cheiroseius* sp. (Ascidae). Fig. 434 Photograph of adult female *Dolomedes* sp. (Pisauridae).

9–11, with the first adults appearing in mid-May. Males have an estimated average adult survival time of 9–13 days and disappear by the end of June, female adults apparently live for 25–31 days and disappear by late August or early September (Zimmermann and Spence, 1992, 1998). Disappearance of males apparently reflects the influence of cannibalism by females (Zimmermann and Spence, 1992). Overwintering is facultative, with the later instars undergoing quiescence based upon temperature control but the earlier instars having dormancy determined by an interaction of photoperiod and temperature cues (Zimmermann and Spence, 1998).

Cannibalism is an important cause of mortality in *Dolomedes* populations. Interguild predation appears to be a strong selective force on juveniles of *D. triton,* and sexual cannibalism is correlated with the short seasonal occurrence of males of this species (Zimmermann and Spence, 1989, 1992, 1998). Members of *Dolomedes* are also eaten by birds (Carico, 1973) and attacked by predatory pompilid and sphecoid wasps (Carico, 1973; Dondale and Redner, 1990). Small- to intermediate-sized juveniles of these spiders are eaten by both water striders (Gerridae) and backswimmers (Notonectidae) (Zimmermann and Spence, 1998). The egg cocoons of *Dolomedes* are susceptible to infestation by hymenopterous parasitoids (Carico, 1973).

ACKNOWLEDGMENTS

We thank Michelle MacKenzie of the Biodiversity Section of Agriculture and Agri-food Canada for her expert help in assembling the plates of illustrations using computer-graphics software and for preparing many of the water mite specimens examined during preparation and testing of the keys. We also thank Drs. Valerie Behan-Pelletier, Evert Lindquist and Charles Dondale of the Biodiversity Section of Agriculture and Agri-food Canada for providing illustrations and helpful comments for the sections dealing with oribatid mites, ascid mites and spiders, respectively.

LITERATURE CITED

Åbro, A. 1979. Attachment and feeding devices of water-mite larvae (*Arrenurus* spp.) parasitic on damselflies (Odonata, Zygoptera). Zoologica Scripta 8:221–34.

Åbro, A. 1982. The effects of parasitic water mite larvae (*Arrenurus* spp.) on zygopteran imagoes (Odonata). Journal of Invertebrate Pathology 39:373–81.

Åbro, A. 1984. The initial stylostome formation by parasitic larvae of the watermite genus *Arrenurus* on zygopteran imagines. Acarologia (Paris) 25:33–46.

Åbro, A. 1990. The impact of parasites in adult populations of Zygoptera. Odonatologica 19:223–234.

Alberti, G. 1977. Zur Feinstruktur und Funktion der Genitalnäpfe von *Hydrodroma despiciens* (Hydrachellae, Acari). Zoomorphologie 87:155–164.

Alberti, G. 1979. Fine structure and probable function of genital papillae and claparede organs of Actinotrichida. 501–507, *in* Rodriguez, J. G., Ed.,. Recent Advances in Acarology, Vol. 2; 569 p.

Bader, C. 1938. Beitrag zur Kenntnis der Verdauungsvorgänge bei Hydracarinen. Revue Suisse de Zoologie 45:721–806.

Bagge, P., Merilainen, J. J. 1985. The occurrence of water mites (Acari: Hydrachnellae) in the estuary of the River Kyronjoki (Bothnian Bay). Annales Zoologici Fennici 22:123–127.

Baker, G. T. 1985. Feeding, moulting and the internal anatomy of *Hydrozetes* sp. (Oribatida: Hydrozetidae). Zoologische Jahrbücher Anatomie 113:77–83.

Baker, G. T. 1996. Chemoreception in four species of water mites (Acari: Hydrachnida): behavioural and morphological evidence. Experimental and Applied Acarology 20:445–455.

Baker, R. L., Smith, B. P. 1997. Conflict between antipredator and antiparasite behaviour in larval damselflies. Oecologia 109: 622–628.

Balogh, J., Balogh, P. 1992a. The oribatid mites genera of the World. Vol. 1. Hungarian National Museum Press, Budapest, 263 p.

Balogh, J., Balogh, P. 1992b. The oribatid mites genera of the World. Vol. 2. Hungarian National Museum Press, Budapest, 375 p.

Balogh, J., Mahunka, S. 1983. Primitive oribatids of the Palaearctic region. Elsevier Scientific Publication, Amsterdam. 372 p.

Barr, D. W. 1972. The ejaculatory complex in water mites (Acari:Parasitengona): Morphology and potential value for systematics. Life Sciences Contributions. Royal Ontario Museum 81; 87 p.

Barr, D. W. 1973. Methods for the collection, preservation, and study of water mites (Acari: Parasitengona). Life Sciences Miscellaneous Publication. Royal Ontario Museum, 28 p.

Barr, D. W. 1979. Water mites (Acari, Parasitengona) sampled with chemoluminescent bait in underwater traps. International Journal of Acarology, Vol. 5;187–194.

Barr, D. W. 1982. Comparative morphology of the genital acetabula of aquatic mites (Acari, Prostigmata): Hydrachnoidea, Eylaoidea, Hydryphantoidea, Lebertioidea. Journal of Natural History 16:147–160.

Behan-Pelletier, V. M. 1989. *Limnozetes* (Acari: Oribatida: Limnozetidae) of northeastern North America. Canadian Entomologist 121:453–506.

Behan-Pelletier, V. M. 1997. The semiaquatic genus *Tegeocranellus* (Acari: Oribatida: Ameronothroidea) of North and Central America. Canadian Entomologist 129:537–577.

Biesiadka, E. 1979. Wodopojki (Hydracarina) Pienin. Fragmenta Faunistica 24:97–173.

Bleckmann, H., Bender, M. 1987. Water surface waves generated by the male pisaurid spider *Dolomedes triton* Walckenaer during courtship behavior. Journal of Arachnology 15:363–370.

Bleckmann, H., Lotz, T. 1987. The vertebrate-catching behaviour of the fishing spider *Dolomedes triton*. Animal Behaviour 35:641–651.

Bohonak, A. J. 1998. Dispersal and gene flow in freshwater invertebrates. Ph.D. Thesis (unpublished), Cornell University, Ithaca, NY.

Bonn, A., Gasse, M., Rolff, J., Martens, A. 1996. Increased fluctuating asymmetry in the damselfly *Coenagrion puella* is correlated with ectoparasitic water mites: implications for fluctuating asymmetry theory. Oecologia 108:596–598.

Böttger, K. 1962. Zur Biologie und Ethologie der einheimischen Wassermilben *Arrenurus* (*Megaluracarus*) *globator* (Müll.),

1776, *Piona nodata nodata* (Müll.), 1776 und *Eylais infundibulifera meriodionalis* (Thon), 1989. (Hydrachnellae, Acari). Zoologische Jahrbücher Systematik 89:501–584.

Böttger, K. 1965. Zur Ökologie und Fortpflanzungsbiologie von *Arrenurus valdiviensis* Viets, K. O., 1964 (Hydrachnellae, Acari). Zeitschrift für Morphologie und Ökologie der Tiere 55:115–141.

Böttger, K. 1970. Die Ernährungsweise der Wassermilben (Hydrachnellae, Acari). Internationale Revue der gesamten Hydrobiologie 55:895–912.

Böttger, K. 1972a. Vergleichend biologisch-ökologische Studien zum Entwicklungszyklus der Süßwassermilben (Hydrachnellae, Acari). I. Der Entwicklungszyklus von *Hydrachna globosa* und *Limnochares aquatica*. Internationale Revue der gesamten Hydrobiologie 57:109–152.

Böttger, K. 1972b. Vergleichend biologisch-ökologische Studien zum Entwicklungszyklus der Süßwassermilben (Hydrachnellae, Acari). II. Der Entwicklungszyklus von *Limnesia maculata* und *Unionicola crassipes*. Internationale Revue der gesamten Hydrobiologie 57:263–319.

Böttger, K., Schaller, F. 1961. Biologische und ethologische Beobachtungen an einheimischen Wassermilben. Zoologischer Anzeiger 167:46–50.

Carico, J. E. 1973. The nearctic species of the genus *Dolomedes* (Araneae: Pisauridae). Bulletin of the Museum of Comparative Zoology 144:435–488.

Chappuis, P. A. 1942. Eine neue Methode zur Untersuchung der Grundwasserfauna. Acta Sciences, Mathematik und Natur, Kolozsvar 6:3–7.

Cicolani, B., di Sabatino, A. 1991. Sensitivity of water mites to water pollution. Modern Acarology 1:465–474.

Collins, N. C. 1975. Tactics of host exploitation by a thermophilic water mite. Miscellaneous Publications of the Entomological Society of America 9:359–370.

Collins, N. C., Mitchell, R. D., Wiegert, R. G. 1976. Functional analysis of a thermal spring ecosystem, with an evaluation of the role of consumers. Ecology 57:1221–1232.

Comstock, J. H. 1912. The spider book (revised and edited by Gertsch, W. J., for the 1940 edition). Cornell University Press, 5th printing (1980), 729 p.

Cook, D. R. 1957. Order Acarina. Suborder Hydracarina. Genus *Protoarrenurus* Cook, n. gen. *in:* Palmer, A. R., Ed. Miocene arthropods from the Mojave Desert, California. Geological Survey Professional Paper (U.S.) No. 294-G, pp. 248–249.

Cook, D. R. 1974. Water mite genera and subgenera. Memoirs of the American Entomological Institute 21;860 p.

Cook, D. R. 1980. Studies on Neotropical water mites. Memoirs of the American Entomological Institute 31;645 p.

Cook, D. R. 1986. Water mites from Australia. Memoirs of the American Entomological Institute 40;568 p.

Cook, D. R. 1988. Water mites from Chile. Memoirs of the American Entomological Institute 42;356 p.

Cook, D. R., Smith, I. M., Harvey, M. S. 2000. Assessment of lateral compression of the idiosoma in adult water mites as a taxonomic character and reclassification of *Frontipodopsis* Walter, *Wettina* Piersig and some other early derivative Hygrobatoidea (Acari:Hydrachnida). Invertebrate Taxonomy 14:433–448.

Cook, W. J., Smith, B. P., Brooks, R. J. 1989. Allocation of reproductive effort in female *Arrenurus* spp. water mites (Acari: Hydrachnidia; Arrenuridae). Oecologia 79:184–188.

Crowell, R. M. 1960. The taxonomy, distribution and developmental stages of Ohio water mites. Bulletin of the Ohio Biological Survey, (new series) 1;77 p.

Davids, C. 1973. The water mite *Hydrachna conjecta* Koenike, 1895 (Acari, Hydrachnellae), bionomics and relation to species of Corixidae (Hemiptera). Netherlands Journal of Zoology 23:363–429.

Davids, C., Belier, R. 1979. Spermatophores and sperm transfer in the water mite *Hydrachna conjecta* Koen. Reflections of the descent of water mites from terrestrial forms. Acarologia 21:84–90.

Davids, C., Heijnis, C. F., Weekenstroo, J. E. 1981. Habitat differentiation and feeding strategies in water mites in Lake Maarsseveen I. Hydrobiological Bulletin 15:87–91.

Dimock, R. V., Jr., LaRochelle, P. B. 1980. Chemically mediated host recognition: a behavioral basis for the specificity of water mite symbioses. American Zoologist 20:922.

Dimock, R. V., Jr., Davids, C. 1985. Spectral sensitivity and photobehavior of the water mite genus *Unionicola*. Journal of Experimental Biology 119:349–363.

Dondale, C. D., Redner, J. H. 1990. The wolf spiders, nurseryweb spiders, and lynx spiders of Canada and Alaska (Araneae: Lycosidae, Pisauridae, and Oxyopidae). The insects and arachnids of Canada; Part 17. Agriculture Canada. 383 pp.

Ellis-Adam, A. C., Davids, C. 1970. Oviposition and post-embryonic development of the watermite *Piona alpicola* (Neuman, 1880). Netherlands Journal of Zoology 20:122–137.

Elton, C. S. 1922. On the colours of water-mites. Proceedings of the Zoological Society of London 82:1231–1239.

Eriksson, M. O. G., Henrikson, L., Oscarson, H. G. 1980. Predator–prey relationships among water-mites (Hydracarina) and other freshwater organisms. Archiv für Hydrobiologie 88:146–154.

Fairchild, W. L., O'Neill, M. C. A., Rosenberg, D. M. l987. Quantitative evaluation of the behavioural extraction of aquatic invertebrates from samples of sphagnum moss. Journal of the North American Benthological Society 6:28l–287.

Forbes, M. R. L. 1991. Ectoparasites and mating success in male *Enallagma ebrium* damselflies (Odonata: Coenagrionidae). Oikos 60:336–342.

Forbes, M. R. L., Baker, R. L. 1990. Susceptibility to parasitism: experiments with the damselfly *Enallagma ebrium* (Odonata: Coenagrionidae) and larval water mites, *Arrenurus* spp. (Acari: Arrenuridae). Oikos 58:61–66.

Forbes, M. R. L., Baker, R. L. 1991. Condition and fecundity of the damselfly, *Enallagma ebrium* (Hagen): the importance of ectoparasites. Oecologia 86:335–341.

Garga, N. 1996. Aposematism in water mites (Acari: Hydracarina): a predator defense mechanism, a phylogenetic hold-over and protection from damaging light. M.Sc. Thesis (unpublished), Queens University, Kingston.

Gerecke, R., Schwoerbel, J. 1991. Water quality and water mites (Acari: Actinedida) in the upper Danube region, 1959–1984. Modern Acarology 1:483–491.

Gerecke, R., Smith, I. M., Cook, D. R. 1999. Three new species of *Apheviderulix gen. nov.* and proposal of Apheviderulicidae *fam. nov.* (Acari: Hydrachnida: Eylaoidea). Hydrobiologia 397:133–147.

Gledhill, T. 1982. Water-mites (Hydrachnellae, Limnohalacaridae, Acari) from the interstitial habitat of riverine deposits in Scotland. Polish Archives of Hydrobiology 29:439–451.

Gledhill, T. 1985. Water mites—predators and parasites. Freshwater Biological Association Annual Report 53:45–59.

Gliwicz, Z. M., Biesiadka, E. 1975. Pelagic water mites (Hydracarina) and their effect on the plankton community in a neotropical man-made lake. Archiv für Hydrobiologie 76:65–88.

Halik, L. 1955. O kopulach vodule *Brachypoda* versicolor (Müll.). Biologia (Bratislava) 10:464–474.

Hevers, J. 1978. Zur Sexualbiologie der Gattung *Unionicola* (Hydrachnellae, Acari). Zoologische Jahrbücher Systematik 105:33–64.

Hinton, H. E. 1964. Sperm transfer in insects and the evolution of haemocoelic insemination. *in* Higham, K. C., (Ed.), Insect Reproduction, Symposium No. 2, Royal Entomological Society, London, pp. 95–107.

Hungerford, H. B., Spangler, P. J., Walker, N. A. 1955. Subaquatic light traps for insects and other animal organisms. Transactions of the Kansas Academy of Sciences 58:387–407.

Imamura, T. 1952. Notes on the moulting of the adult of the water mite, *Arrenurus uchidai* n. sp. Annotationes Zoologicae Japonenses 25:447–451.

Kerfoot, W. C. 1982. A question of taste: crypsis and warning coloration in freshwater zooplankton communities. Ecology 63:538–554.

Knight, J. G. 1989. Investigation of reproductive biology, age, growth, and feeding habits of the bayou darter, *Ethiostoma rubrum* (Pisces, Percidae) from Bayou Pierre, Mississippi. M.Sc. Thesis (unpublished), University of Southern Mississippi, Hattiesburg.

Koenike, F. 1891. Seltsame Begattung unter der Hydrachniden. Zoologischer Anzeiger 14:253–256.

Kouwets, F. A. C., Davids, C. 1984. The occurrence of chironomid imagines in an area near Utrecht (the Netherlands), and their relations to water mite larvae. Archiv für Hydrobiologie 99:296–317.

Kowalik, W. 1981. Wodopojki (Hydracarina) rzek dorzecza Wieprza. Annales Universitatis Mariae Curie-Sklodowska, Section C 36:327–352.

Kowalik, W., Biesiadka, E. 1981. Occurrence of water mites (Hydracarina) in the River Wieprz polluted with domestic-industry sewage. Acta Hydrobiologica 23:331–347.

Krantz, G. W. 1978. A manual of Acarology. 2nd ed. Oregon State University Press, Corvallis. 335 p.

Lanciani, C. A. 1971a. Host related size of parasitic water mites of the genus *Eylais*. American Midland Naturalist 85:242–247.

Lanciani, C. A. 1971b. Host exploitation and synchronous development in a water mite parasite of the marsh treader *Hydrometra myrae* (Hemiptera: Hydrometridae). Annals of the Entomological Society of America 64:1254–1259.

Lanciani, C. A. 1972. Mating behavior of water mites of the genus *Eylais*. Acarologia 14:631–637.

Lanciani, C. A. 1976. Intraspecific competition in the parasitic water mite, *Hydryphantes tenuabilis*. American Midland Naturalist 96:210–214.

Lanciani, C. A. 1978. Parasitism of Ceratopogonidae (Diptera) by the water mite *Tyrrellia circularis*. Mosquito News 38:282–284.

Lanciani, C. A. 1979. The food of nymphal and adult water mites of the species *Hydryphantes tenuabilis*. Acarologia 20:563–565.

Lanciani, C. A. 1980. Parasitism of the backswimmer *Buenoa scimitra* (Hemiptera: Notonectidae) by the water mite *Hydrachna virella* (Acari:Hydrachnellae). Freshwater Biology 10:527–532.

Lanciani, C. A. 1982. Parasite-mediated reductions in the survival and reproduction of the backswimmer *Buenoa scimitra* (Hemiptera:Notonectidae). Parasitology 85:593–603.

Lanciani, C. A. 1983. Overview of the effects of water mite parasitism on aquatic insects. *in* Hoy, M. A., Cunningham, G. L., Knutson, L., Eds., Biological Control of Pests by Mites. Special Publications of the University of California Agricultural Experimental Station 3304, pp. 86–90.

Lanciani, C. A. 1984. Crowding in the parasitic stage of the water mite *Hydrachna virella* (Acari: Hydrachnidae). Journal of Parasitology 70:270–272.

Lanciani, C. A. 1986a. Competition among larval *Tyrrellia circularis* (Acariformes: Limnesiidae) on the host *Dasyhelea mutabilis* (Diptera: Ceratopogonidae). Florida Entomologist 69:438–439.

Lanciani, C. A. 1986b. Reduced survivorship in *Dasyhelea mutabilis* (Diptera: Ceratopogonidae) parasitized by the water mite *Tyrrellia circularis* (Acariformes: Limnesiidae). Journal of Parasitology 72:613–614.

Lanciani, C. A. 1987. Mortality in mite-infested, male *Anopheles crucians*. Journal of the American Mosquito Control Association 3:107–108.

Lanciani, C. A. 1988. Sexual bias in host selection by parasitic mites of the mosquito *Anopheles crucians* (Diptera: Culicidae). Journal of Parasitology 74:768–773.

Lanciani, C. A., Boyt, A. D. 1977. The effect of a parasitic water mite, *Arrenurus pseudotenuicollis* (Acari: Hydrachnellae), on the survival and reproduction of the mosquito *Anopheles crucians* (Diptera: Culicidae). Journal of Medical Entomology 14:10–15.

Lanciani, C. A., May, P. G. 1982. Parasite-mediated reductions in the growth of nymphal backswimmers. Parasitology 85:1–7.

Lanciani, C. A., McLaughlin, R. E. 1989. Parasitism of *Coquillettidia perturbans* by two water mite species (Acari: Arrenuridae) in Florida, U.S.A. Journal of the American Mosquito Control Association 5:428–431.

Lanciani, C. A., Smith, B. P. 1989. Constancy of stylostome form in two water mite species. Canadian Entomologist 121:439–443.

Lang, P. 1905. Über den Bau der Hydrachnidenaugen. Zoologische Jahrbücher Anatomie 21:453–494.

LaRochelle, P. B., Dimock, R. V. Jr. 1981. Behavioral aspects of host recognition by the symbiotic water mite *Unionicola formosa* (Acarina: Unionicolidae). Oecologia (Berlin) 48:257–259.

Leung, B., Forbes, M. R. L. 1997. Fluctuating asymmetry in relation to indices of quality and fitness in the damselfly, *Enallagma ebrium* (Hagen). Oecologia 110:472–477.

Lundblad, O. 1929. Über den Begattungsvorgang bei einigen *Arrhenurus*-Arten. Zeitschrift für Morphologie und Ökologie der Tiere 15:705–722.

Marshall, R. 1933. Water mites from Wyoming as fish food. Transactions of the American Microscopical Society. 52:34–41.

Marshall, R. 1940. On the occurrence of water mites in the food of turtles. American Midland Naturalist 24:361–364.

Meyer, E. 1985. Der Entwicklungszyklus von *Hydrodroma despiciens* (Müller, O. F. 1776) (Acari: Hydrodromidae). Archiv für Hydrobiologie, Supplement 66:321–453.

Meyer, E., Schwoerbel, J. 1981. Untersuchungen zur Phänologie der Wassermilben (Hydracarina) des Mindelsees. Archiv für Hydrobiologie, Supplement 59:192–251.

Mitchell, R. D. 1955. Anatomy, life history, and evolution of the mites parasitizing fresh-water mussels. Miscellaneous Publications of the Museum of Zoology, University of Michigan 89:1–41.

Mitchell, R. D. 1957a. Major evolutionary lines in water mites. Systematic Zoology 6:137–148.

Mitchell, R. D. 1957b. Locomotor adaptations of the family Hydryphantidae (Hydrachnellae, Acari). Abhandlungen herausgegeben vom naturwissenschaftlichen Verein zu Bremen 35:75–100.

Mitchell, R. D. 1957c. The mating behavior of pionid water mites. American Midland Naturalist 58:360–366.

Mitchell, R. D. 1958. Sperm transfer in the water mite *Hydryphantes ruber* Geer. American Midland Naturalist 60:156–158.

Mitchell, R. D. 1959. Life histories and larval behavior of arrenurid water-mites parasitizing Odonata. Journal of the New York Entomological Society 67:1–12.

Mitchell, R. D. 1960. The evolution of thermophilous water mites. Evolution 14:361–377.

Mitchell, R. D. 1961. Behavior of the larvae of *Arrenurus fissicornis* Marshall, a water-mite parasitic on dragonflies. Animal Behaviour 9:220–224.

Mitchell, R. D. 1962. The structure and evolution of water mite mouthparts. Journal of Morphology 110:41–59.

Mitchell, R. D. 1964a. A study of sympatry in the water mite genus *Arrenurus* (family Arrenuridae). Ecology 45:546–558.

Mitchell, R. D. 1964b. An approach to the classification of water mites. Acarologia 6:75–79.

Mitchell, R. D. 1966. Ecological aspects of sympatry in water mites. Proceedings of the Japanese Society of Systematic Zoology 2:28–31.

Mitchell, R. D. 1969a. A model accounting for sympatry in water mites. American Naturalist 103:331–346.

Mitchell, R. D. 1969b. Population regulation of larval water mites. Proceedings of the 2nd International Congress of Acarology, 1967:99–102.

Mitchell, R. D. 1972. The tracheae of water mites. Journal of Morphology 136:327–335.

Modlin, R. F., Gannon, J. E. 1973. A contribution to the ecology and distribution of aquatic Acari in the St. Lawrence Great Lakes. Transactions of the American Microscopical Society 92:217–224.

Motas, C. 1928. Contribution à la connaissance des Hydracariens français particilièrement du Sud-Est de la France. Travaux du Laboratoire d'Hydrobiologie et Pisciculture. Université de Grenoble 20:373 p.

Mullen, G. R. 1974. The taxonomy and bionomics of aquatic mites (Acarina: Hydrachnellae) parasitic on mosquitoes in North America. Ph.D. Thesis (unpublished), Cornell University, 263 p.

Mullen, G. R. 1975. Predation by water mites (Acarina: Hydrachnellae) on immature stages of mosquitoes. Mosquito News 35:168–171.

Münchberg, P. 1935a. Über die bisher bei einigen Nematocerenfamilien (Culicidae, Chironomidae, Tipulidae) beobachteten ektoparasitaren Hydracarinenlarven. Zeitschrift für Morphologie und Ökologie der Tiere 29:720–749.

Münchberg, P. 1935b. Zur Kenntnis der Odonatenparasiten, mit ganz besonderer Berucksichtigung der Ökologie der in Europa an Libellen schmarotzenden Wassermilbenlarven. Archiv für Hydrobiologie 29:1–122.

Münchberg, P. 1937. *Arrenurus planus* Marsh. in USA. und *A. papillator* (O. F. Müll.) in der alten Welt, zwei ökologisch und morphologisch einander entsprechende Arten (Ordnung: Hydracarina). Archiv für Hydrobiologie 31:209–228.

Münchberg, P. 1938. Dritter Beitrag über die an Stechmücken schmarotzenden *Arrenurus*-Larven (Ordn.: Hydracarina). Archiv für Hydrobiologie 33:99–116.

Norton, R. A., Behan-Pelletier, V. M., Wang, H-f. 1996. The aquatic oribatid mite genus *Mucronothrus* in Canada and western U.S.A. (Acari: Trhypochthoniidae). Canadian Journal of Zoology 74:926–949.

Oliver, D. R., Smith, I. M. 1980. Host association of some pionid water mite larvae (Acari: Prostigmata: Pionidae) parasitic on chironomid imagos (Diptera: Chironomidae). Acta Universitatis Carolinae, Biologica 1978:157–162.

Paterson, C. G. 1970. Water mites (Hydracarina) as predators of chironomid larvae (Insecta: Diptera). Canadian Journal of Zoology 48:610–614.

Pflügfelder, O. 1970. Schadwirkungen der *Arrenurus*-Larven (Acari, Hydrachnellae) am Flügel der Libelle *Sympetrum meridionale* Selys. Zeitschrift für Parasitenkunde 34:171–176.

Pieczynski, E. 1962. Notes on the use of light traps for water mites (Hydracarina). Bulletin de l'Académie Polonaise des Sciences, Classe II, Série des Sciences Biologiques 10:421–424.

Pieczynski, E. 1969. The trap method for ecological studies on water mites (Hydracarina) in lakes. Proceedings of the 2nd International Congress of Acarology, 1967; pp. 103–106.

Pieczynski, E. 1976. Ecology of water mites (Hydracarina) in lakes. Polish Ecological Studies 2:5–54.

Pieczynski, E., Kajak, Z. 1965. Investigations on the mobility of the bottom fauna in the lakes Taltowisko, Mikolajskie and Sniardwy. Bulletin de l'Académie Polonaise des Sciences, Classe II, Série des Sciences Biologiques 13:345–353.

Poinar, G. O. 1985. Fossil evidence of insect parasitism by mites. International Journal of Acarology 11:37–38.

Prasad, V., Cook, D. R. 1972. The taxonomy of water mite larvae. Memoirs of the American Entomological Institute 18:326 pp.

Proctor, H. C. 1989. Occurrence of protandry and a female-biased sex-ratio in a sponge-associated water mite (Acari: Unionicolidae). Experimental and Applied Acarology 7:289–297.

Proctor, H. C. 1991a. The evolution of copulation in water mites: a comparative test for nonreversing characters. Evolution 45: 558–567.

Proctor, H. C. 1991b. Courtship in the water mite *Neumania papillator*: males capitalize on female adaptations for predation. Animal Behaviour 42:589–598.

Proctor, H. C. 1992a. Sensory exploitation and the evolution of male mating behaviour: a cladistic test using water mites (Acari: Parasitengona). Animal Behaviour 44:745–752.

Proctor, H. C. 1992b. Mating and spermatophore morphology of water mites (Acari: Parsitengona). Zoological Journal of the Linnaean Society 106:341–384.

Proctor, H. C. 1992c. Discord between field and laboratory sex ratios of the water mite *Neumania papillator* Marshall (Acari: Unionicolidae). Canadian Journal of Zoology 70:2483–2486.

Proctor, H. C. 1997. Sex ratios and chromosomes in water mites (Acari: Parasitengona). Acarology IX 1:441–445.

Proctor, H. C., Pritchard, G. 1989. Neglected predators:water mites (Acari: Parasitengona: Hydrachnellae) in freshwater communities. Journal of the North American Benthological Society 8:100–111.

Proctor, H. C., Pritchard, G. 1990. Prey detection by the water mite *Unionicola crassipes* (Acari: Unionicolidae). Freshwater Biology 23:271–279.

Proctor, H. C., Smith, B. P. 1994. Mating behaviour of the water mite *Arrenurus manubriator* (Acari: Arrenuridae). Journal of Zoology, London 232:473–483.

Punčochář, P., Hrbáček, J. 1991. Water mites in the plankton of Hubenov reservoir and their relations to fish stock composition. Modern Acarology 1:449–457.

Redmond, B. L., Hochberg, J. 1981. The stylostome of *Arrenurus* spp. (Acari: Parasitengona) studied with the scanning electron microscope. Journal of Parasitology 67:308–313.

Redmond, B. L., Lanciani, C. A. 1982. Attachment and engorgement of a water mite, *Hydrachna virella* (Acari: Parasitengona), parasitic on *Buenoa scimitra* (Hemiptera: Notonectidae). Transactions of the American Microscopical Society 101:388–394.

Reilly, P., McCarthy, T. K. 1991. Watermite parasitism of Corixidae: infection parameters, larval mite growth, competitive interaction and host response. Oikos 60:137–148.

Reinhardt, K. 1996. Negative effects of *Arrenurus* water mites on the flight distances of the damselfly *Nehallenia speciosa* (Odonata: Coenagrionidae). Aquatic Insects 18:233–240.

Riessen, H. P. 1982. Pelagic water mites: Their life history and seasonal distribution in the zooplankton community of a Canadian lake. Archiv für Hydrobiologie, Supplement 62:410–439.

Robaux, P. 1974. Recherches sur le développement et la biologie des acarines. "Thrombidiidee." Mémories dee Museum National d' Histoire Naturelle, Série A, Zoologie 85:1–186.

Roberts, E. A., Dimock, R. V., Jr., Forward, R. B., Jr. 1978. Positive and host-induced negative phototaxis of the symbiotic water mite *Unionicola formosa*. Biological Bulletin of the Marine Laboratory (Woods Hole, Mass.) 155:599–607.

Robinson, J. V. 1983. Effects of water mite parasitism on the demographics of an adult population of *Ischnura posita* (Hagen) (Odonata: Coenagrionidae). American Midland Naturalist 109:169–174.

Roland, C., Rovner, J. S. 1983. Chemical and vibratory communication in the aquatic pisaurid spider *Dolomedes triton* (Araneae: Pisauridae). Journal of Arachnology 11:77–85.

Rolff, J., Martens, A. 1997. Completing the life cycle: detachment of an aquatic parasite (*Arrenurus cuspidator,* Hydrachnellae) from an aerial host (*Coenagrion puella,* Odonata). Canadian Journal of Zoology 75:655–659.

Rousch, J. M., Simmons, T. W., Kerans, B. L., Smith, B. P. 1997. Relative acute effects of low pH and high iron on the hatching and survival of the water mite, *Arrenurus manubriator* and aquatic insect, *Chironomus riparius.* Environmental Toxicology and Chemistry 16:2144–2150.

Schmidt, U. 1935. Beiträge zur Anatomie und Histologie der Hydracarinen, besonders von *Diplodontus despiciens* O. F. Müller. Zeitschrift für Morphogie und Ökologie der Tiere 30:99–175.

Schwoerbel, J. 1959. Ökologische und tiergeographische Untersuchungen über die Milben (Acari, Hydrachnellae) der Quellen und Bäche des südlichen Schwarzwaldes und seiner Randgebiete. Archiv für Hydrobiologie, Supplement 24:385–546.

Simmons, T. W., Smith, I. M. 1984. Morphology of larvae, deutonymphs, and adults of the water mite *Najadicola ingens* (Prostigmata: Parasitengona: Hygrobatoidea) with remarks on phylogenetic relationships and revision of taxonomic placement of Najadicolinae. Canadian Entomologist 116:691–701.

Smith, B. P. 1977. Water mite parasitism of water boatmen (Hemiptera: Corixidae). M.Sc. Thesis (unpublished), University of British Columbia, Vancouver, 117 pp.

Smith, B. P. 1983. The potential of mites as biological control agents of mosquitoes. *in* Hoy, M. A., Cunningham, G. L., Knutson, L. Eds. Biological Control of Pests by Mites. Special Publications of the University of California Agricultural Experimental Station. No. 3304, pp. 79–85.

Smith, B. P. 1988. Host–parasite interaction and impact of larval water mites on insects. Annual Review of Entomology 33:487–507.

Smith, B. P. 1989. Impact of parasitism by larval *Limnochares aquatica* (Acari: Hydrachnidia; Limnocharidae) on juvenile *Gerris comatus, Gerris alacris,* and *Gerris buenoi* (Insecta: Hemiptera; Gerridae). Canadian Journal of Zoology 67:2238–2243.

Smith, B. P. 1998. Loss of larval parasitism in parasitengone mites. Experimental and Applied Acarology 22:187–199.

Smith, B. P. 1999. Larval Hydrachnida and their hosts: biological inference and population structure. Pages 139–143 *in:* Needham, G. R., Mitchell, R., Horn, D. J., Welbourn, W. C., Eds. Acarology IX: Vol. 2, Symposia, Ohio Biological Survey, Columbus, OH, xvii + 507 pp.

Smith, B. P., Cook, W. J. 1991. Negative covariance between larval *Arrenurus* sp. and *Limnochares americana* (Acari: Hydrachnidia) on male *Leucorrhinia frigida* (Odonata: Libellulidae) and its relationship to the host's age. Canadian Journal of Zoology 69:226–231.

Smith, B. P., Laughland, L. A. 1990. Stimuli inducing detachment of larval *Arrenurus danbyensis* (Hydrachnidia: Arrenuridae) from adult *Coquillettidia perturbans* (Diptera: Culicidae). Experimental and Applied Acarology 9:51–62.

Smith, B. P., McIver, S. B. 1984a. The patterns of mosquito emergence (Diptera: Culicidae; *Aedes* spp.): their influence on host selection by parasitic mites (Acari: Arrenuridae; *Arrenurus* spp.). Canadian Journal of Zoology 62:1106–1113.

Smith, B. P., McIver, S. B. 1984b. Factors influencing host selection and successful parasitism of *Aedes* spp. mosquitoes by *Arrenurus* spp. mites. Canadian Journal of Zoology 62:1114–1120.

Smith, B. P., McIver, S. B. 1984c. The impact of *Arrenurus danbyensis* Mullen (Acari: Prostigmata; Arrenuridae) on a population of *Coquillettidia perturbans* (Walker) (Diptera: Cullcidae). Canadian Journal of Zoology 62:1121–1134.

Smith, I. M. 1976a. A study of the systematics of the water mite family Pionidae (Prostigmata: Parasitengona). Memoirs of the Entomological Society of Canada 98;249 pp.

Smith, I. M. 1976b. *Paenecalyptonotus fontinalis* n. gen., n. sp. with remarks on the family Acalyptonotidae (Acari: Parasitengona: Arrenuroidea). Canadian Entomologist 108:997–1000.

Smith, I. M. 1978. Descriptions and observations on host associations of some larval Arrenuroidea (Prostigmata: Parasitengona), with comments on phylogeny in the superfamily. Canadian Entomologist 110:957–1001.

Smith, I. M. 1979. A review of water mites of the family Anisitsiellidae (Prostigmata: Lebertioidea) from North America. Canadian Entomologist 111:529–550.

Smith, I. M. 1982. Larvae of water mites of the superfamily Lebertioidea (Prostigmata: Parasitengona) in North America with comments on phylogeny and higher classification of the superfamily. Canadian Entomologist 114:901–990.

Smith, I. M. 1983a. Description of larvae of *Neoacarus occidentalis* (Acari: Arrenuroidea: Neoacaridae). Canadian Entomologist 115:221–226.

Smith, I. M. 1983b. Description of adults of *Neomideopsis siuslawensis* n. gen., n. sp., with remarks on the family Mideopsidae (Acari: Parasitengona: Arrenuroidea). Canadian Entomologist 115:417–420.

Smith, I. M. 1983c. Description of *Cowichania interstitialis* n. gen., n. sp., and proposal of Cowichaniinae n. subfam., with remarks on phylogeny and classification of Hydryphantidae (Acari: Parasitengona: Hydryphantoidea). Canadian Entomologist 115: 523–527.

Smith, I. M. 1983d. Description of larvae and adults of *Paramideopsis susanae* n. gen., n. sp., with remarks on phylogeny and classification of Mideeopsidae (Acari: Parasitengona: Arrenuroidea). Canadian Entomologist 115:529–538.

Smith, I. M. 1983e. Description of larvae of *Nudomideopsis magnacetabula* (Acari: Arrenuroidea: Mideopsidae) with new distributional records for the species and remarks on the classification of *Nudomideopsis.* Canadian Entomologist 115:913–919.

Smith, I. M. 1983f. Descriptions of two new species of *Acalyptonotus* from western North America, with a new diagnosis of the genus based on larvae and adults, and comments on phylogeny and taxonomy of Acalyptonotidae (Acari: Parasitengona: Arrenuroidea). Canadian Entomologist 115:1395–1408.

Smith, I. M. 1983g. Descriptions of adults of a new species of *Cyclothyas* (Acari: Parasitengona: Hydryphantidae) from western North America, with comments on phylogeny and distribution of mites of the genus. Canadian Entomologist 115:1433–1436.

Smith, I. M. 1984. Larvae of water mites of some genera of Aturidae (Prostigmata: Hygrobatoidea) in North America with comments on phylogeny and classification of the family. Canadian Entomologist 116:307–374.

Smith, I. M. 1987. Water mites of peatlands and marshes in Canada. Pages 31–46 *in* Rosenberg, D. M., Danks, H. V., Eds. Aquatic insects of peatlands and marshes in Canada. Memoirs of the Entomological Society of Canada l40; l74 pp.

Smith, I. M. 1989a. Description of deutonymphs and adults of *Oregonacarus rivulicolus* gen. nov., sp. nov. (Acari: Lebertioidea: Anisitsiellidae). Canadian Entomologist 121:533–541.

Smith, I. M. 1989b. North American water mites of the family Momoniidae Viets (Acari: Arrenuroidea). I. Description of adults of *Cyclomomonia andrewi* gen. nov., sp. nov., and key to world genera and subgenera. Canadian Entomologist 121:543–549.

Smith, I. M. 1989c. Description of two new species of *Platyhydracarus* gen. nov. from western North America, with remarks on classification of Athienemanniidae (Acari: Parasitengona: Arrenuroidea). Canadian Entomologist 121:709–726.

Smith, I. M. 1989d. North American water mites of the family Momoniidae Viets (Acari: Arrenuroidea). II. Revision of species of *Momonia* Halbert, 1906. Canadian Entomologist 121:965–987.

Smith, I. M. 1989e. North American water mites of the family Momoniidae Viets (Acari: Arrenuroidea). III. Revision of species of *Stygomomonia* Szalay, 1943, subgenus *Allomomonia* Cook, 1968. Canadian Entomologist 121:989–1025.

Smith, I. M. 1990a. Description of two new species of *Stygameracarus* gen. nov. from North America, and proposal of Stygameracarinae subfam. nov. (Acari: Arrenuroidea: Athienemanniidae). Canadian Entomologist 122:181–190.

Smith, I. M. 1990b. Proposal of Nudomideopsidae fam. nov. (Acari: Arrenuroidea) with a review of North American taxa and description of a new subgenus and species of *Nudomideopsis* Szalay, 1945. Canadian Entomologist 122:229–252.

Smith, I. M. 1991a. Water mites (Acari: Parasitengona: Hydrachnida) of spring habitats in Canada. Pages 141–167 *in:* Williams, D. D., Danks, H. V., Eds. Arthropods of springs, with particular reference to Canada. Memoirs of the Entomological Society of Canada 155; 217 pp.

Smith, I. M. 1991b. North American water mites of the genera *Phreatobrachypoda* Cook and *Bharatalbia* Cook (Acari: Hygrobatoidea: Aturinae). Canadian Entomologist 123:465–499.

Smith, I. M. 1991c. North American water mites of the family Momoniidae Viets (Acari: Arrenuroidea). IV. Revision of species of *Stygomomonia (sensu stricto)* Szalay, 1943. Canadian Entomologist 123:501–558.

Smith, I. M. 1991d. Descriptions of new species representing new or unreported genera of Lebertioidea (Acari: Hydrachnida) from North America. Canadian Entomologist 123:811–825.

Smith, I. M. 1992a. North American species of the genus *Chelomideopsis* Romijn (Acari: Arrenuroidea: Athienemanniidae). Canadian Entomologist 124:451–490.

Smith, I. M. 1992b. North American water mites of the family Chappuisididae Motas and Tanasachi (Acari: Arrenuroidea). Canadian Entomologist 124:637–723.

Smith, I. M. 1998. *Chelohydracarus navarrensis* gen. nov., sp. nov. (Acari: Hydrachnida: Athienemanniidae). International Journal of Acarology 24:327–330.

Smith, I. M., Cook, D. R. 1991. Water mites. Chapter 16, Pages 523–592 *in* Thorp, J., Covich, A., Eds. Ecology and classification of North American freshwater invertebrates. Academic Press, San Diego, CA, 911 pp.

Smith, I. M., Cook, D. R. 1994. North American species of Neomamersinae Lundblad (Acari: Hydrachnida: Limnesiidae). Canadian Entomologist 126:1131–1184.

Smith, I. M., Cook, D. R. 1996. New and unreported species of *Neotyrrellia, Protolimnesia (Protolimnesella)* and *Centrolimnesia* (Acari: Hydrachnida: Limnesiidae) from the United States. Anales del Instituto de Biología, Universidad Nacional Autónoma de México, Serie Zoologia 67:265–277.

Smith, I. M., Cook, D. R. 1997. Description of *Amoenacarus dixiensis* gen. nov., sp. nov., and proposal of Amoenacaridae fam. nov. (Acari: Hydrachnida: Arrenuroidea). International Journal of Acarology 23:107–112.

Smith, I. M., Cook, D. R. 1998a. The *Axonopsella*-like water mites of the United States (Acari: Hydrachnida: Aturidae). International Journal of Acarology. 24:311–325.

Smith, I. M., Cook, D. R. 1998b. Three new species of water mites representing new genera of Thyadinae from North America (Acari: Hydrachnida: Hydryphantidae). International Journal of Acarology. 24:331–339.

Smith, I. M., Cook, D. R. 1999a. North American species of *Tartarothyas* Viets (Acari: Hydrachnida: Hydryphantidae). International Journal of Acarology. 25:37–42.

Smith, I. M., Cook, D. R. 1999b. North American species of *Parathyasella* stat. nov. and *Amerothyasella* gen. nov. (Acari: Hydrachnida: Hydryphantidae). International Journal of Acarology. 25:43–50.

Smith, I. M., Cook, D. R. 1999c. An assessment of global distribution patterns in water mites (Acari: Hydrachnida). Pages 109–124 *in:* Needham, G. R., Mitchell, R., Horn, D. J., Welbourn, W. C., Eds. Acarology IX: Volume 2, Symposia. Ohio Biological Survey, Columbus, OH, xvii + 507 pp.

Smith, I. M., Oliver, D. R. 1976. The parasitic associations of larval water mites with imaginal aquatic insects, especially Chironomidae. Canadian Entomologist 108:1427–1442.

Smith, I. M., Oliver, D. R. 1986. Review of parasitic associations of larval water mites (Acari: Parasitengona: Hydrachnida) with insect hosts. Canadian Entomologist 118:407–472.

Sokolow, I. I. 1954. Khromosomnye kompleksy kleshchei i ikh zhachenie dlia sistematiki i filogenii. Trudy Leningradskogo Obschchestva Estestvois-pytatelei 72:124–159.

Sokolow, I. I. 1977. The protective envelopes in the eggs of Hydrachnellae. Zoologischer Anzeiger 198:36–42.

Spence, J. R., Zimmermann, M., Wojcicki, J. P. 1996. Effects of food limitation and sexual cannibalism on reproductive output of the nursery web spider *Dolomedes triton* (Araneae: Pisauridae). Oikos 75:273–382.

Stechmann, D.-H. 1980. Zum Wirtskreis syntopischer *Arrenurus*-Arten (Hydrachnellae, Acari) mit parasitischer Entwicklung an Nematocera (Diptera). Zeitschrift für Parasitenkunde 62:267–283.

Steenbergen, H. A. 1993. Macrofauna-atlas of North-Holland: distribution maps and responses to environmental factors of aquatic invertebrates. Cip-gegevens Koninkliijke Bibliotheek, Den Haag, 651 pp.

Ten Winkel, E. H., Davids, C. 1985. Bioturbation by cyprinid fish affecting the food availability for predatory water mites. Oecologia 67:218–219.

Ten Winkel, E. H., Davids, C. 1987. Chironomid larvae and their food web relations in the littoral zone of Lake Maarsseveen. Kaal Boek, Amsterdam. 145 pp.

Ten Winkel, E. H., Davids, C., de Nobel, J. G. 1989. Food and feeding strategies of water mites of the genus *Hygrobates* and the impact of their predation on the larval population of the chironomid *Cladotanytarsus mancus* (Walker) in Lake Maarsseveen. Netherlands Journal of Zoology 39:246–263.

Uchida, T. 1940. Über die Begattungsakt bei *Forelia variegator.* Annotationes Zoologicae Japonenses 19:280–282.

Vidrine, M. F. 1986. Revision of the Unionicolinae (Acari: Unionicolidae). International Journal of Acarology 12:233–243.

Viets, K. H. 1914. Über die Begattungsvorgange bei *Acercus*-Arten. Internationale Revue gesamten Hydrobiologie und Hydrographie, Biological Supplement 6:1–10.

Viets, K. H. 1936. Wassermilben oder Hydracarina (Hydrachnellae und Halacaridae). Die Tierwelt Deutschlands 31:1–288. 32:289–574.

Viets, K. H. 1956. Die Milben des Sußwassers und des Meeres (Hydrachnellae und Halacaridae, Acari), Teil 2/3. Fischer Verlag, Jena. 870 pp.

Viets, K. O. 1987. Die Milben des Sußwassers (Hydrachnellae und Halacaridae [part.], Acari). 2. Katalog. Sonderbande Naturwissenschaftlichen Vereins Hamburg 8; 1012 pp.

Wainstein, B. A. 1980. Opredelitel lichinok vodjanych kleshchei. (Key to larval water mites). Institut Biologii Vnutrennikh Vod, Akademiia Nauk SSSR, 238 pp.

Weiberg, M, Edwards, D. D. 1997. Survival and reproductive output of *Chironomus tentans* (Diptera: Chironomidae) in response to parasitism by larval *Unionicola foili* (Acari: Unionicolidae). Journal of Parasitology 83:173–175.

Wesenburg-Lund, C. J. 1918. Contributions to the knowledge of the postembryonal development of the Hydracarina. Videnskabelige Meddelelser fra Dansk Naturhistorisk Forening 70:5–57.

Wiggins, G. B., Mackay, R. J., Smith, I. M. 1980. Evolutionary and ecological strategies of animals in annual temporary pools. Archiv für Hydrobiologie, Supplement 58:97–206.

Wiles, P. R. 1982. A note on the watermite *Hydrodroma despiciens* feeding on chironomid egg masses. Freshwater Biology 12:83–87.

Wiles, P. R. 1984. Watermite respiratory systems. Acarologia 25:27–31.

Wood, T. G. 1969. The Homocaligidae, a new family of mites (Acari: Raphignathoidea), including a description of a new species from Malaya and the British Solomon Islands. Acarologia 11:711–729.

Zimmermann, M., Spence, J. R. 1989. Prey use of the fishing spider *Dolomedes triton* (Pisuaridae: Araneae) an important predator of the neuston community. Oecologia 80:187–194.

Zimmermann, M., Spence, J. R. 1992. Adult population dynamics and reproductive effort of the fishing spider *Dolomedes triton* (Araneae, Pisauridae) in central Alberta. Canadian Journal of Zoology 70:2224–2233.

Zimmermann, M., Spence, J. R. 1998. Phenology and life-cycle regulation of the fishing spider *Dolomedes triton* Walckenaer (Araneae, Pisauridae) in central Alberta. Canadian Journal of Zoology 76:295–309.

17

DIVERSITY AND CLASSIFICATION OF INSECTS AND COLLEMBOLA[1]

William L. Hilsenhoff

Department of Entomology
University of Wisconsin
Madison, Wisconsin 53706

I. INTRODUCTION

This chapter includes orders and families in which one or more life stages are truly aquatic and adapted for survival under or on the water surface. Families that live on or burrow into emergent aquatic vegetation, internal parasites of aquatic animals, and riparian families that are closely associated with water but do not inhabit it are not included. Terrestrial stages of aquatic families are discussed briefly, but only as they relate to the general life history of the family. In addition, brief mention is made of another arthropod group, the semiaquatic springtails (class Entognatha, order Collembola) (see Sections IV and V.J.)

There are 10 orders of insects that contain aquatic species. Five of them (Ephemeroptera, Odonata, Plecoptera, Trichoptera, and Megaloptera) are "aquatic orders" in which almost all species have aquatic larvae. The remaining five orders (Heteroptera, Coleoptera, Diptera, Lepidoptera, and Neuroptera) are "partially

[1] Much of the material on Collembola was contributed by Barbara L. Peckarsky, Department of Entomology, Cornell University. The author and editors greatly appreciate her generous contribution of this material.

Ecology and Classification of North American Freshwater Invertebrates. 2nd Edition

aquatic orders" in which most species are terrestrial, but in which there are species or entire families that have one or more life stages adapted for living in an aquatic environment.

Three aquatic orders (Ephemeroptera, Odonata, and Plecoptera) have a hemimetabolous life cycle, which includes three developmental stages: egg; larva; and adult. The term "larva" is being used instead of "nymph" or "naiad," because in these orders the immature developmental stage is adapted for survival in aquatic environments and does not resemble the terrestrial adult, except in Plecoptera. The other two aquatic orders (Trichoptera and Megaloptera) have a holometabolous life cycle, which includes four developmental stages: egg; larva; pupa; and adult. In these orders the larva bears no resemblance to the adult.

Four of the five partially aquatic orders (Coleoptera, Diptera, Lepidoptera, and Neuroptera) also have holometabolous life cycles and larvae that do not resemble the adult. The fifth order, Heteroptera (=Hemiptera), has a paurometabolous life cycle, which includes three developmental stages: egg, nymph, and adult. In this order, nymphs inhabit the same habitat as adults and resemble them in many respects, differing mostly by being smaller and less sclerotized, having wing-pads instead of wings, and lacking genital structures

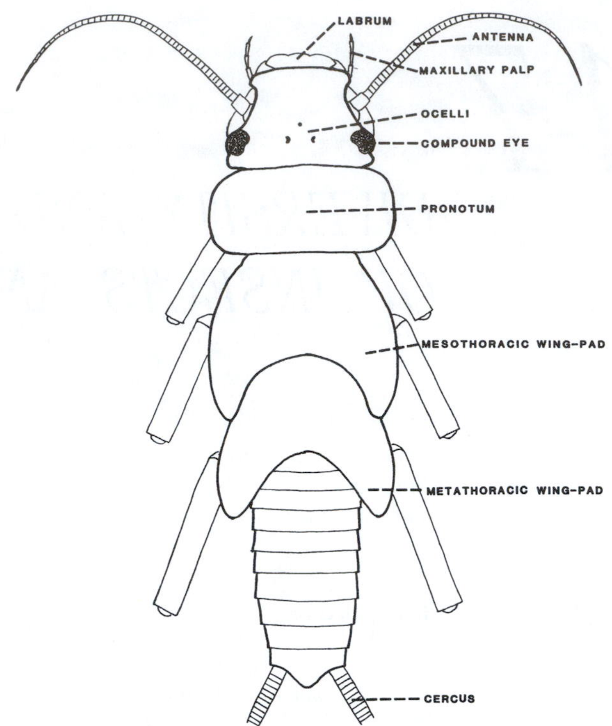

FIGURE 1 Plecoptera larva (Perlidae); *Acroneuria* larva (dorsal) showing various structures (thoracic gills and distal segments of legs and cerci not shown).

A. General Morphology of Aquatic Insects

Morphological terms used for identification of aquatic insects vary somewhat from order to order, but several terms are commonly used for all orders. A Plecoptera larva (Figs. 1 and 2) is used to illustrate these terms. Insects have three body regions: head, thorax, and abdomen. The anterior region is the head, which dorsally (Fig. 1) has a pair of antennae, a pair of compound eyes, and often three ocelli or simple eyes. On the underside of the head are several mouthparts (Fig. 2), including a labrum or upper lip, a pair of mandibles, a pair of maxillae, and a labium or lower lip. The mandibles usually are heavily sclerotized, toothlike structures. The maxillae and labium are usually multisegmented and frequently have a pair of elongate, segmented structures called palps.

The thorax is immediately posterior to the head and consists of three large segments (Fig. 1): a prothorax (anterior); a mesothorax; and a metathorax (posterior). The dorsal sclerite of each segment is referred to as a notum, and the ventral sclerite as a sternum. Each segment has a pair of legs that are divided into five parts (Fig. 2): a coxa, trochanter, femur, tibia, and tarsus. The femur and tibia are usually elongate and the tarsus has one-to-five segments, depending on the

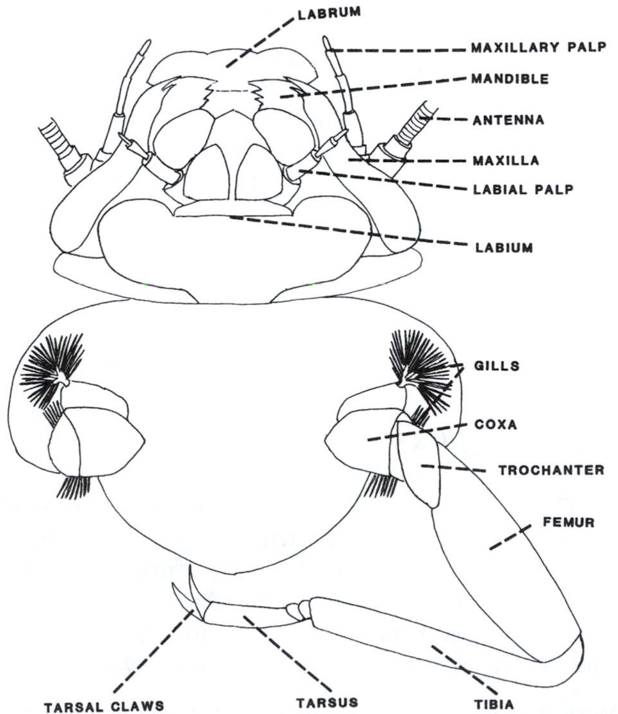

FIGURE 2 Plecoptera larva (Perlidae); head and prothorax (ventral) of *Acroneuria* larva showing mouthparts and structures of a leg (right leg and distal segments of antennae not shown).

order. Each tarsus has one or two tarsal claws. Prefixes pro-, meso- and meta-refer to the respective thoracic segments. Thus, a pronotum would be on the prothorax and a mesofemur would be on the mesothorax. In hemimetabolous and paurometabolous insect larvae or nymphs, wing-pads (developing wings) are usually present dorsally on the meso- and metathorax.

The abdomen is the posterior body region and may have as many as 10 visible segments, which are numbered posteriorly from the thorax. The dorsal sclerite of each segment is called a tergum, the ventral sclerite a sternum, and the lateral sclerite, when present, a pleuron. Terminal abdominal appendages vary widely or may be absent; cerci are present in Plecoptera and several other orders. In some orders, gills of various types may be present on the abdomen, thorax, or even on the head.

A taxonomic key to orders of aquatic insects is provided in Section V.A. Family level keys are also provided for aquatic larvae and aquatic adults in Sections V.B–I.; keys to eggs, Heteroptera nymphs, pupae, and terrestrial stages of aquatic species are not included. The most recent keys to genera of aquatic insects in North America appear in Merritt and Cummins (1996). Regional keys to genera or species of aquatic insects include those in Usinger (1956), Brigham *et al.* (1982), Peckarsky *et al.* (1990), Clifford (1991), and Hilsenhoff (1995c). The most recent keys to aquatic stages of North American species are listed along with regional species keys in tables that appear in the discussion of each order; reliable species keys do not exist for many families and genera. Regional keys are important because they are easier to use and may be more recent than North American keys. Subsequent descriptions of new species are not listed, but should be consulted when species are being identified. References to recent revisions and descriptions of new species may be found in Merritt and Cummins (1996), bibliographies such as those published annually by the North American Benthological Society, and through computer searches of data bases such as Zoological Record and Dissertation Abstracts Online.

B. Common Techniques for Studying Freshwater Insects

1. Qualitative Sampling and Rearing of Aquatic Insects

A D-frame or rectangular aquatic net with a mesh size of 1 mm or less is recommended for sampling lotic habitats. Shallow, rock or gravel riffles are sampled by placing the net firmly against the bottom and disturbing the substrate immediately upstream from the net with one's feet, allowing insects and debris washed into the water to be carried into the net. This is commonly called a "kick sample." The contents of the net are emptied into a shallow white pan containing 2 to 3 cm of water, taking care not to collect so much debris that it completely obscures the bottom of the pan. Insects that swim or crawl from the debris are most easily collected with a curved forceps and placed into a collecting jar containing 70% ethanol or Kahle's fluid (15 parts 95% ethanol, 30 parts distilled water, 6 parts formalin, and 1 part glacial acetic acid). The empty net should be examined for insects that remain clinging to it.

Additional samples can be collected with a net by pushing it under the stream bank or through silt and debris in pools and rinsing excess silt from the net before placing the sample in the pan. It is also important to remove pieces of decaying wood from the stream and to collect insects that crawl from the wood as it dries. Additional insects may be collected by examining stones that are temporarily removed from the stream bed. Deep areas of larger streams can be sampled by using grabs or dredges. Ponar or Ekman grabs are most commonly used, but the latter can be used only in soft sediments.

Shallow lentic habitats, such as ponds, marshes, swamps, and vegetated lake margins, are also best sampled with a flat-bottomed aquatic net. This is accomplished by pushing, pulling, or sweeping the net through the vegetation or debris, rinsing excess silt from the debris, and processing as described above. Additional larger collections can be processed by placing them on a half-inch (13 mm) mesh screen placed over a large white pan and allowing at least 15 min for insects to crawl from the debris and drop into the pan. This method is especially effective for collecting adult Coleoptera and Heteroptera, as is the use of bottle traps (Hilsenhoff, 1987). Wave-swept shorelines of lakes can be sampled by disturbing the sand or gravel bottom with one's feet and immediately collecting suspended material with a net. Removal and examination of decaying wood or stones from lake shores should not be overlooked, because these substrates often harbor species that normally will not be collected with a net. Deeper waters of lakes must be sampled with grabs, dredges, or corers. Special habitats such as tree holes or pitcher plants are best sampled with a large syringe or poultry baster. Appropriate sampling methods are discussed under each order.

Since immature stages in many families of aquatic insects cannot be identified to species, it is often necessary to rear them to the adult stage for identification. This can be readily accomplished in most orders, but the presence of a terrestrial pupal stage in Coleoptera, Megaloptera, Neuroptera, and some Diptera makes rearing of larvae in these orders more difficult. Chapters on qualitative and quantitative sampling techniques and rearing procedures appear in Merritt and Cummins (1996), Downing and Rigler (1984), McCafferty (1981), and Usinger (1956).

2. Preserving and Storing Aquatic Insects

Aquatic insects can most conveniently be killed and preserved by placing them in a jar containing a liquid preservative. Many preservatives have been recommended and used, with 70–80% ethanol or isopropanol, or Kahle's fluid being most popular. The latter penetrates insect larvae more rapidly and preserves some colors better, but should be replaced with 70% ethanol after one or two days.

For permanent storage, 70% ethanol containing 3% glycerine is recommended, especially for larvae. The glycerine serves to preserve larvae if the alcohol evaporates due to a poorly fitting stopper. Most aquatic entomologists store specimens in patent-lip vials with neoprene stoppers. Screw-cap vials and cork-stoppered shell vials should be avoided because of gradual evaporation of the alcohol, although screw-cap vials with bevelled polyethyene inserts are satisfactory. One-dram shell vials with polyethylene stoppers,—commonly used for liquid scintillation studies,—are an inexpensive substitute for neoprene stoppered vials. Initially, quality control of these polyethylene-stoppered vials was poor and many leaked, but those purchased in recent years do not leak and retain fluid as well as neoprene-stoppered vials. They have a wider opening than patent-lip or screw-cap vials, which facilitates removal of insects. Because they are optically clear, insects that are stored in these vials can be examined and often identified under a microscope while still in the vial, if the vial is only half-filled with alcohol so that a meniscus or bubble does not interfere with viewing. Adult aquatic Coleoptera and Heteroptera are normally preserved on pins, but study collections can be preserved in liquid, with several of the same species kept in a single vial to reduce storage space and curating time. Adult Odonata are preserved on pins or in envelopes. Adults of other orders also may be pinned, but are most frequently preserved in alcohol. Storage in liquid also facilitates removal of genitalia for study, which is important for identification of many species.

For examination under a dissecting microscope, specimens stored in liquid should be placed in a Syracuse watch glass or similar container and completely covered with 95% alcohol. If 70% alcohol is used, insects may float and the alcohol will become cloudy as it evaporates. If insects are not completely covered with alcohol, a meniscus will cause distortions. A layer of very small glass beads or silica sand in the bottom of the watch glass permits insects to be imbedded for viewing from any angle.

3. An Introduction to Biological Literature

A discussion of the general biology and habitat of insects in each order and family is provided in Section II, but it is often difficult to make generalizations at the family level. Literature on biology of the numerous genera and species is voluminous. References to biological literature pertaining to life histories and behavior of species in each genus can be found in Merritt and Cummins (1996) and in *Current and Selected Bibliographies on Benthic Biology* that is published each year by the North American Benthological Society. Several review articles treat specific aspects of biology such as drift of stream insects (Waters, 1972), predator–prey relationships (Bay, 1974), detritus processing (Anderson and Sedell, 1979), and filter feeding (Wallace and Merritt, 1980).

II. AQUATIC ORDERS OF INSECTS

A. Ephemeroptera—Mayflies

With only about 675 species in 20 families known from North America, this is a relatively small hemimetabolous order, but it is widespread and abundant in most regions. Larvae of all species are aquatic, and while the great majority inhabit streams, several inhabit a variety of permanent and temporary lentic habitats. Ephemeroptera larvae can be found in streams of all sizes and water temperatures, and even occur in temporary streams. In lakes, they occur among vegetation in littoral areas and sometimes also may be found on the bottom in deep water or along waveswept shores. Frequently, they occur in vernal or permanent ponds and marshes. Larvae of the various families often differ so greatly from one another that they can be easily identified to family in the field. They are frequently an important source of food for fish in streams. Because of the wide range of dissolved oxygen requirements among the species, they are very important in biological monitoring of streams. Some species that develop in lakes or large rivers may be sufficiently abundant that emerging subimagos (pre-adults) and adults create severe nuisance problems.

Except in warm waters of the southern United States, most species are univoltine, with eggs either hatching shortly after being laid and larvae developing slowly over a 1-year period (slow seasonal life cycle), or with eggs diapausing, followed by rapid development of the larvae after hatching several months later (fast seasonal life cycle). Other species are bivoltine, with rapid larval development and with eggs of the second generation diapausing over the winter. A few species are multivoltine or semivoltine, with multivoltinism being especially prevalent in the South. The number of larval instars varies widely among species, and even within a species, with 12 or more having been recorded for species that have been studied. After completion of larval development, the larva either crawls from the water or swims or floats to the surface where the subimago emerges. This winged subadult or "dun" stage is unique among insects. Subimagos of most species fly to nearby vegetation or supports where they spend about a day before molting into the adult (imago) stage. A few species molt almost immediately to the adult, sometimes while in flight. Since most subimagos are sexually mature, some females never molt to the adult.

Adult mayflies do not feed, and generally live only a few days. A few may live 2 weeks or more, and some live only an hour or two. Mating normally takes place when adult males form an aerial swarm to attract females and seize them when they enter the swarm. In some species, adult males mate with subimago females as they emerge from the water. Eggs are laid mostly on the water surface by females that repeatedly touch the water with their abdomen while in flight or by females alighting briefly on the surface to oviposit. Females of some species enter the water on a substrate to oviposit on substrates beneath the water. Most females die on the water surface upon completion of oviposition.

Mayfly larvae can be readily distinguished from other aquatic insects by their long, filamentous caudal filaments and the presence of lateral or ventrolateral gills on most of the first seven abdominal segments (Fig. 29). Usually three caudal filaments are present (a pair of cerci and a median filament); in a few species the median filament is so reduced that only cerci are visible. Larvae most closely resemble those of Plecoptera, which always have only cerci, lack gills on the middle abdominal segments, and have two tarsal claws (Figs. 44 and 45). Ephemeroptera larvae have compound eyes, ocelli, filamentous antennae, and 1-segmented tarsi with a single claw (Figs. 9 and 29). Subimagos and adults differ greatly from the larvae, having one or two pairs of wings that they hold vertically above the body when at rest, very large compound eyes, and no gills or functional mouthparts. In the subimago, unmarked areas of the wings are opaque; in adults, these areas are transparent.

While mayfly larvae mostly crawl on the substrate, many are able and rapid swimmers. They swim by moving their abdomen and caudal filaments up and down rather than horizontally as stonefly do. Their gills aid significantly in the uptake of oxygen, and larvae of most species are capable of moving their gills to increase water circulation. This enables larvae of some species to inhabit lentic habitats or lotic habitats that are less rich in oxygen. Almost all mayfly larvae are herbivores or detritivores; larvae of a few species are known to be predators on other invertebrates. Subimagos and adults do not feed.

Larvae may be readily collected with a net by techniques already described. Those that inhabit deep water must be collected with grabs, and those in shallow sand or silt areas can only be obtained by sieving these substrates. Subimagos and adults may be collected by sweeping stream-side vegetation, and large numbers are often collected at lights.

Since the first edition (Thorp and Covich, 1991), there have been several changes in the higher classification of Ephemeroptera, with the addition of five new families (McCafferty, 1991). The order was divided into three suborders by McCafferty (1991). Rectrachaeta, Setisura, and Pisciforma, which replace the former suborders Pannota and Schistonota. Although new species and genera are still being discovered, recent revisions have often synonymized species, sometimes resulting in a reduction of the number of known species. In general, larvae are poorly known and cannot be reliably identified to species in most families. Early books by Needham *et al.* (1935), Burks (1953), and Berner (1959) provided an essential background for more recent studies. The book by Edmunds *et al.* (1976) provides generic keys to adults and larvae as well as information on collecting, rearing, and mayfly biology; generic keys to adults and larvae were updated by G. F. Edmunds, Jr. and R. D. Waltz in Merritt and Cummins (1996). The biogeography and evolution of Ephemeroptera was reviewed by Edmunds (1972), and Brittain (1982) reviewed the biology of this order. A key to the families of freshwater Ephemeroptera larvae is presented in Section V.B. and references to species keys for identification of Ephemeroptera larvae are provided in Table I. Information on the biology, habitat, and identification of each family is presented next. Suborders and families are arranged alphabetically in this and subsequent orders.

1. Families of Ephemeroptera—Suborder Pisciforma

a. Acanthametropodidae (2 Genera, 2 Species) This family was formerly included in Siphlonuridae. The large larvae inhabit sandy bottoms of medium-to-large rivers where they feed primarily on larval Chironomidae. The rare and difficult-to-collect larvae are recognized by their bowed tibiae and tarsi and very long, slender claws. One species occurs from Saskatchewan

TABLE I Literature References to Species Keys for Identification of Ephemeroptera Larvae

Regional keys to larvae that include more than one family
- California—Day (1956)
- Florida—Berner and Pescador (1988)
- Illinois—Burks (1953)
- North and South Carolina—Unzicker and Carlson (1982)
- United States—(Ephemeroidea) McCafferty (1975)

Keys to larvae listed by family
- Ameletidae—(Alberta) Zloty and Pritchard (1997)
- Ametropodidae—Allen and Edmunds (1976)
- Baetidae
  - *Acentrella*—McCafferty *et al.* (1994)
  - *Americabaetis*—Waltz and McCafferty (1999)
  - *Baetis, Acerpenna, Diphetor, Labiobaetis*—Morihara and McCafferty (1979); (Wisconsin) Bergman and Hilsenhoff (1978)
  - *Baetodes*—McCafferty and Provonsha (1993)
  - *Callibaetis*—Check (1982)
  - *Camelobaetidius*—Lugo-Ortiz and McCafferty (1995); Wieresma (1998)
  - *Cloeodes*—Waltz and McCafferty (1987a)
  - *Fallceon*—Lugo-Ortiz *et al.* (1994)
  - *Labiobaetis*—McCafferty and Waltz (1995)
  - *Pseudocentroptiloides*—Waltz and McCafferty (1989); Wiersema and McCafferty (1998)
- Baetiscidae—Pescador and Berner (1981); (Wisconsin) Hilsenhoff (1984a)
- Behningiidae—Hubbard (1994)
- Caenidae
  - *Brachycercus*—(Northeast) Burian *et al.* (1997)
  - *Brachycercus* and *Cercobrachys*—Soldan (1984)
  - *Caenis*—Provonsha (1990)
- Ephemerellidae
  - *Attenella*—Allen and Edmunds (1961b)
  - *Caudatella*—Allen and Edmunds (1961a)
  - *Drunella*—Allen and Edmunds (1962b)
  - *Ephemerella*—Allen and Edmunds (1965)
  - *Eurylophella*—Allen and Edmunds (1963b); Funk and Sweeney (1994)
  - *Serratella*—Allen and Edmunds (1963a)
  - *Timpanoga*—Allen and Edmunds (1959); Allen and Edmunds (1962a); McCafferty (1977)
- Heptageniidae—(Wisconsin) Flowers and Hilsenhoff (1975)
  - *Heptagenia*—(Rocky Mountains) Bednarik and Edmunds (1980)
  - *Macdunnoa*—Flowers (1982)
  - *Stenonema*—Bednarik and McCafferty (1979)
- Isonychiidae—Kondratieff and Voshell (1984)
- Leptohyphidae
  - *Leptohyphes*—Allen (1978); (southwestern United States) Allen and Murvosh (1987)
- Leptophlebiidae
  - *Neochoroterpes*—Henry (1993)
  - *Paraleptophlebia*—(Oregon) Lehmkuhl and Anderson (1971)
  - *Thraulodes*—Allen and Brusca (1978)
  - *Traverella*—Allen (1973)
- Neoephemeridae—Berner (1956); Bae and McCafferty (1998)
- Oligoneuriidae
  - *Homoeoneuria*—Pescador and Peters (1980)
- Potamanthidae—Bae and McCafferty (1991)

to Utah, the other from Wisconsin to South Carolina. Life cycles are probably one year or longer.

b. Ameletidae (1 Genus, 32 Species) This family also was formerly included in Siphlonuridae. Larvae are excellent swimmers that inhabit small, rapid streams where they are mostly found clinging to vegetation or debris. They are recognized by their small, oval, lamellate gills that have a sclerotized mesal or lateral band (Fig. 30), short antennae, and lack of posterolateral spines on the apical abdominal segment. Life cycles are probably mostly univoltine, with larvae overwintering, and emergence from February to September, depending on latitude.

c. Ametropodidae (1 Genus, 3 Species) Larvae are relatively large and flat, with very long meso- and metathoracic claws and shorter prothoracic claws with long spines (Figs. 21 and 22). They partially burrow into sand in eddies of medium-to-large streams in western and north-central North America, and are generally rare. Larvae have a unique method for feeding on suspended algae and detritus, which they gather with their prothoracic legs from a vortex created in a pit that they construct in the sand. Their life cycle is 1 year, with a relatively long emergence period in late spring or early summer.

d. Baetidae (18 Genera, 146 Species) Larvae are relatively small, active swimmers, with lamellate abdominal gills (Fig. 29). They differ from larvae of other species with lamellate gills on most of the first seven abdominal segments in having relatively long antennae (mostly twice the width of head), usually no pronounced posterolateral spines on abdominal segment 9, and a median caudal filament that is often shorter than the cerci, or vestigial (Fig. 29). Baetidae is a widespread and abundant family that occurs in a variety of streams, and also in permanent and temporary ponds or littoral zones of lakes. Lotic species are sometimes univoltine, but more frequently bivoltine; most species overwinter as diapausing eggs. Emergence of various cohorts and species occurs from early spring to late autumn. Lentic species are often multivoltine. Recent studies of larvae (Waltz and McCafferty, 1987a–c, 1989; McCafferty *et al.*, 1994; McCafferty and Waltz, 1995; Lugo-Ortiz and McCafferty, 1998; Lugo-Ortiz *et al.*, 1999) have led to changes in the placement of species within genera. Northern populations of some species are parthenogenetic.

e. Metretopodidae (2 Genera, 9 Species) Larvae reach a relatively large size and have lamellate gills, very long meso- and metatarsal claws (Fig. 21), and shorter protarsal claws that are distinctively bifid (Fig. 20). They are excellent swimmers and may be found along banks and among vegetation in small-to-large streams and occasionally along shores of lakes. They occur in Canada, the upper Midwest, and in the East. Their life cycle is 1 year, with emergence in the spring or summer, depending on the species.

f. Pseudironidae (1 Genus, 1 Species) This family was formerly included in Heptageniidae. Larvae are apparently rare in sand bottoms of larger streams where they prey on larvae of Chironomidae. Larvae have a somewhat flattened head with dorsal eyes and antennae, as in Heptageniidae. They can be recognized by their long slender legs and claws and their unique gill lamellae (Fig. 19), which have a fibrilliform basal tuft and a ventral fingerlike projection. Life cycles are probably univoltine.

g. Siphlonuridae (4 Genera, 26 Species) Most larvae are relatively large with lamellate gills (Fig. 28). They most closely resemble larvae of Baetidae, but larvae are usually larger, have short antennae, and pronounced spines on the posterolateral angles of the ninth abdominal segment. Larvae are mostly herbivores–detritivores, but those of some species include insects in their diet. Larvae occur in vernal forest pools, stream-side pools, marshes, and swamps. Those of some species inhabit streams, but move into small pools just before emergence. Most, if not all, species are probably univoltine.

2. Families of Ephemeroptera—Suborder Rectrachaeta

a. Baetiscidae (1 Genus, 12 Species) Larvae are distinctive, having a mesonotal carapace that covers most of the abdomen and all abdominal gills (Fig. 9). They occur throughout eastern and central North America, and are rare or absent in the West. Larvae partially burrow into silty margins or eddies of streams and clean lakes, especially where the primary substrate is sand. All species probably are univoltine and overwinter as larvae. Larvae crawl from the water onto a substrate to emerge, with emergence occurring in spring or early summer.

b. Behningiidae (1 Genus, 1 Species) This family occurs mostly in the extreme southeastern United States, with a single record from northwestern Wisconsin. The unusual larvae burrow in the clean shifting sands of larger streams with a fairly rapid current. The relatively large larvae are unique, possessing crowns of bristles on the head and pronotum (Fig. 10), having ventrolateral filamentous gills, and lacking claws on their legs, which are highly modified for digging. The life cycle is 2 years, with a very long diapausing egg stage followed by slow larval growth and emergence in late April and early May. Unlike most other mayflies, larvae are predators, feeding mostly on larvae of Chironomidae.

c. Caenidae (4 Genera, 24 Species) The small larvae are easily recognized by their nearly square operculate gills that are not mesally fused (Fig. 14). They are widespread and common in a wide variety of lotic and lentic habitats, including streams of all sizes, spring seeps, marshes, swamps, ponds, and lakes. Here, they frequent the sediments and often are partially covered with silt. Larvae are generally more tolerant of low levels of dissolved oxygen than those in any other mayfly family. Life cycles are poorly known. In the North, most species are probably univoltine with overlapping cohorts; in the South, two or more generations per year are likely. Subimagos emerge at the surface, fly to a support, and molt immediately to adult. Adults, which live only a few hours, swarm and mate shortly after ecdysis.

d. Ephemerellidae (8 Genera, 98 Species) Larvae are recognized by the absence of gills on abdominal segment 2, and lamellate or operculate gills on segments 3- or 4–7 (Fig. 15). Larvae of most species inhabit clean streams, where they are often abundant in leaf litter, eddies, or near the banks. Those of a few species may persist in organically enriched streams, and others may be found on wave-swept lake shores. While most larvae are herbivores–detritivores, larvae of a few species are omnivorous, also feeding on small invertebrates. Most species are univoltine, with a spring emergence period and a summer egg diapause, or with emergence in summer and no egg diapause. Subimagos emerge from the surface or from protruding substrates.

e. Ephemeridae (4 Genera, 17 Species) Larvae are recognized by their relatively large size, filamentous gills that extend upward over their back (Fig. 4), mandibular tusks that curve upward and outward, and metatibiae with an acutely pointed apex (Fig. 8). Larvae burrow into sand or eddies in riffle areas of small-to-medium-sized streams or inhabit silt bottoms of medium-to-large streams; they also inhabit sandy or silt substrates in relatively clean lakes. In lakes and large rivers, they may occur in such tremendous numbers that synchronized emergences of subimagos create severe nuisance problems. Most species are apparently univoltine with emergence mostly in the summer, but 2-year life cycles are known to occur in Canada. Larvae are primarily filter feeders, which circulate water through their burrows; they also eat algae and detritus on the substrate surface. Species previously included in Palingeniidae are again included in this family.

f. Leptohyphidae (2 Genera, 44 Species) This family was previously named Tricorythidae. Larvae are recognized by their triangular (Fig. 11) or oval (Fig. 12) operculate gills on abdominal segment 2, which are well separated distally, and by the absence of fringed margins on other gills as found in Caenidae and Neoephemeridae. Larvae are widespread, inhabiting detritus, silt, and gravel in streams of all sizes; larvae of some species are quite resistant to lowered levels of dissolved oxygen. Life cycles are univoltine to multivoltine, with eggs often diapausing over winter and emergence occurring in spring and summer after rapid development of each generation.

g. Leptophlebiidae (10 Genera, 82 Species) The medium-sized larvae characteristically have gills on segments 2–6 that are forked, bilamellate and pointed apically, or terminating in filaments (Figs. 25–27). The gills differ markedly from the lamellate, operculate, or fringed gills found in other families. Larvae inhabit slow or fast currents in a variety of streams where they are found among debris, on rocks or wood, or in gravel. While predominantly lotic, larvae of at least one species migrate upstream in spring and leave the stream to enter adjacent ponds or marshes prior to emergence. Most species are univoltine, with eggs diapausing over winter or summer and adults emerging in spring, summer, or autumn; bivoltine and multivoltine life cycles occur in the South.

h. Neoephemeridae (1 Genus, 4 Species) This relatively uncommon family is confined to the East. Larvae have square operculate gills and resemble Caenidae, but the gills are fused mesally (Fig. 13), larvae grow to a larger size than Caenidae larvae, and have metathoracic wing-pads, which Caenidae larvae lack. Larvae cling to vegetation, debris, or undersides of rocks in slow to fairly rapid streams. They apparently are univoltine with a slow seasonal development and emergence in spring or early summer, depending on latitude.

i. Polymitarcyidae (4 Genera, 8 Species) Larvae in this small family of burrowing mayflies differ from other burrowing mayflies by having mandibular tusks that are curved inward and downward and are covered with small spines. Unlike Ephemeridae larvae, the metatibiae are rounded apically (Fig. 7). They inhabit medium-to-large streams and lakes, burrowing into silt, clay, or silt–gravel substrates; frequently they occur in riffles. Larvae circulate water through their burrows and filter out algae and detritus upon which they feed; they also graze on algae and detritus from the substrate. All species apparently are univoltine, with synchronized emergences during the summer months. In some species there is a prolonged winter egg diapause.

j. Potamanthidae (1 Genus, 5 Species) While having mandibular tusks and filamentous gills like other burrowing mayfly larvae, gills of Potamanthidae larvae are horizontal instead of arched over the back, and the legs also are spread more horizontally and not modified for digging (Fig. 6). They occur in medium-to-large streams where they most often sprawl on gravel and sand in shallow runs. They have a 1-year life cycle with emergence during the summer months.

3. Families of Ephemeroptera—Suborder Setisura

a. Arthropleidae (1 Genus, 1 Species) This family was formerly included in Heptageniidae. Larvae have dorsal antennae and eyes and were previously placed in the family Heptageniidae, but they differ in having extremely long maxillary palps that curve around both sides of the head and have long setae (Fig. 18). Larvae inhabit shallow ponds, bogs, or channels with almost no flow. Eggs hatch in spring and larvae develop rapidly with emergence in late spring. The life cycle is univoltine with a long egg diapause. Larvae use their long maxillary palps to strain plankton from the water for food. They are generally uncommon, occurring in northeastern North America.

b. Heptageniidae (14 Genera, 135 Species) The distinctive larvae are readily identified by their flattened appearance and the dorsal location of their eyes and antennae (Fig. 16). Larvae in some genera lack the median caudal filament. They are widespread and abundant in streams; some also occur on wave-swept shorelines of lakes. Larvae inhabit rocks, wood, debris, and other substrates to which they cling. Most species are probably univoltine, but a few may be bivoltine, especially in the South. Emergence occurs from spring to autumn, depending on the species, with most emerging in the summer.

c. Isonychiidae (1 Genus, 17 Species) This family was formerly included in Oligoneuriidae. The large larvae are strong swimmers and fairly common in rapidly flowing streams where they cling to rocks or debris. They are recognized by the double row of long setae on the inner margins of the profemurs and protibiae (Fig. 23), which they use to gather algae and small insects from the current. They also have gill tufts at the base of the procoxae (Fig. 23) and bilamellate gills with a fibrilliform tuft, which separate them from

Oligoneuriidae. Life cycles are mostly bivoltine or univoltine, depending on the latitude.

d. Oligoneuriidae 2 Genera, 8 Species) Larvae have two rows of long setae on the inner margin of the prothoracic legs, which distinguish them from larvae of other mayflies, except those of Isonychiidae. Gills are small and either rounded or elongate (Fig. 24), quite unlike in larvae of Isonychiidae. They use the setae on their prothoracic legs to filter diatoms and other algae from the current. Larvae are found either in sandy bottoms of streams or on sticks or rocks. Depending on latitude and species, life cycles are univoltine to multivoltine with emergence during the summer months.

B. Odonata—Dragonflies and Damselfies

This relatively small hemimetabolous order has two distinctive suborders in North America: Anisoptera (dragonflies), and Zygoptera (damselflies). Only about 434 species are known from North America. Larvae of all species are aquatic, with about two-thirds of them living in lentic habits and one-third in lotic environments. Larvae of lotic species occur in all types of permanent stream habitats, including gravel and rock riffles, debris along banks, bank vegetation, soft sediments, and sand. Those of lentic species inhabit permanent and temporary ponds, marshes, swamps, and littoral and shoreline areas of lakes. Because all larvae attain a relatively large size and are predators, mostly on other insects, they are often important apex predators in the invertebrate community. Adults are also predators, feeding on mosquitoes and other flying insects, and thus Odonata are regarded as highly beneficial insects.

Life cycles are relatively long, mostly 1 year in Zygoptera, and 1–6 years in Anisoptera. Slow seasonal developmental cycles predominate, but fast seasonal cycles occur in both suborders. There are normally 11 or 12 larval instars, but the number may vary somewhat between species or even within a species. When larval development is complete, the larva crawls from the water onto emergent vegetation or onto the bank, where ecdysis occurs. Larvae of some species may crawl several meters from the shore to emerge. Adults may live for several weeks. Newly emerged "teneral" adults are feeble fliers for a day or two and vulnerable to predation. After the teneral stage has passed, adults often disperse widely into fields and other habitats to feed for a week or more on flying insects. They then return to breeding sites where males patrol selected areas, chasing away rival males and other species. Before mating, the male transfers sperm from primary genitalia at the tip of the abdomen to a secondary genital organ on the venter of abdominal segment 2. Using claspers on the terminal segment, a male will seize a female of the same species that flies into his territory, grasping her either by the pronotum (Zygoptera) or by the head (Anisoptera). The female then loops the tip of her abdomen forward and upward to receive sperm from the penis on the second abdominal sternum of the male.

Eggs are laid singly in plant tissues (endophytic oviposition) above, at, or below the water surface, or they may be laid singly or in masses on plants, debris, or other substrates, or randomly on the water surface (exophytic oviposition). Oviposition is endophytic in all Zygoptera and in Aeshnidae and Petaluridae; it is exophytic in other families. In many species, especially species of Zygoptera and Libellulidae, the male continues to grasp the female while she is ovipositing, protecting her from other males. Eggs hatch after an incubation period of one to several weeks, or after a long diapause. The first larval instar is a prolarval stage that lasts only a few minutes to several hours. Prolarvae lack the well-developed appendages of other larval instars.

Odonata larvae can be readily distinguished from all other aquatic insects by their elongate, hinged labium that has been modified for seizing and grasping prey (Figs. 32–34). They have conspicuous compound eyes and relatively short filamentous antennae (Fig. 43). In Zygoptera, the abdomen is elongate and thin, terminating in three long, vertically oriented, caudal lamellae (Fig. 31). In Anisoptera, the abdomen is relatively broad and terminates in three triangular-shaped, pointed appendages (Fig. 44), the upper one being called the "epiproct" and the lower ones "paraprocts." In between them are triangular-shaped cerci, which usually are distinctly shorter than the epiproct or paraprocts. Adults do not resemble larvae, being much more elongate, having conspicuous genitalia, two pairs of elongate net-veined wings, and much larger compound eyes. Like Ephemeroptera, the primitive wings of Odonata cannot be folded. Wings of Anisoptera are held horizontally to the side when at rest; those of Zygoptera are held vertically just above or slightly to the side of the abdomen.

Odonata larvae move about on the substrates by crawling; some actively stalk prey. Zygoptera larvae may swim by moving their abdomen and caudal lamellae from side to side. Anisoptera larvae can move rapidly when disturbed by expelling water from a rectal chamber. There are tracheal gills within the rectal chamber, and movement of water in, and out of, the chamber aids in respiration of Anisoptera larvae. In Zygoptera, respiration is greatly enhanced by the caudal lamellae, enabling larvae of most species to live in water with relatively low levels of dissolved oxygen. Most larvae can be easily collected with an aquatic net using techniques described earlier. Adults are collected with large aerial nets after being sighted. Zygoptera and the Anisoptera adults that alight on vegetation are relatively easy to capture, but many large Anisoptera adults are extremely difficult to capture with a net.

Most larvae of North American Odonata can be identified to species by using keys and descriptions provided by Walker (1953, 1958) and Walker and Corbet (1975), Needham and Westfall (1955), and Westfall and May (1996). Most keys to Anisoptera larvae, however, are old, and problems often arise when using them. A revised edition of *Dragonflies of North America* (Needham *et al.*, 2000) will contain new larval and adult keys. Because Odonata are relatively homogeneous, differences between families, genera, and species are not as obvious as in most other orders. Larvae of Corduliidae cannot be readily separated from those of Libellulidae without identifying them to genus. There is also considerable disagreement about the status of various genera among odonatologists as well as some disagreement about family status. But, at the species level, North American Odonata are better known as larvae and adults than aquatic insects in any other order. This is perhaps because they have attracted the interest of many amateur odonatologists who have made significant contributions to our knowledge. Books by Corbet (1962) and Miller (1987), and a review article by Corbet (1980), summarize knowledge about the biology of dragonflies. References to species keys for the identification of Odonata larvae are provided in Table II, and a key to families of freshwater Odonata appears in Section V.C. Information on the biology, habitat, and identification of dragonfly and damselfly larvae in each family is presented in the following.

TABLE II Literature References to Species Keys for Identification of Odonata Larvae

Regional keys to larvae that include more than one family
Canada and Alaska—(Zygoptera) Walker (1953); (Anisoptera) Walker (1958); (Anisoptera) Walker and Corbet (1975)
California—Smith and Pritchard (1956)
Florida—(Zygoptera) Daigle (1991b); (Anisoptera) Daigle (1992)
North America—(Anisoptera) Needham and Westfall (1955); Needham *et al.* (2000); (Zygoptera) Westfall and May (1996)
North and South Carolina—Huggins and Brigham (1982)
Southeastern United States—(lotic Anisoptera) Louton (1982)
Texas—Young and Bayer (1979)
Utah—(Anisoptera) Musser (1962)
Keys to larvae listed by family
Aeshnidae
Gomphaeschna—Dunkle (1977)
Coenagrionidae
Enallagma—(West Coast) Garrison (1984)
Ischnura—(California) Garrison (1981)
Corduliidae
Somatochlora—Daigle (1991a)
Gomphidae
Dromogomphus—Westfall and Tennessen (1979)
Lanthus—Carle (1980)
Ophiogomphus—Carle (1992)
Progomphus—Tennesson (1993)
Libellulidae
Ladona—Bennefield (1965)
Sympetrum—Tai (1967)

1. Families of Odonata—Suborder Anisoptera (Dragonflies)

a. Aeshnidae (11 Genera, 39 Species) Larvae are large, with elongate tapered abdomens and can be distinguished from other Anisoptera by their elongate shape, flat prementum, and thin six- or seven-segmented antennae (Fig. 37). Most inhabit lentic habitats, especially weedy permanent ponds, marshes, and littoral areas of lakes. Here, they move about on the vegetation and stalk prey composed of invertebrates and small vertebrates. Larvae of a few species are lotic and inhabit either slower areas of streams or riffles. Life cycles of 2–4 years are likely for most species, especially in the North. One species, *Anax junius*, which does not overwinter in the North, migrates south in autumn and returns in early spring to oviposit. Larvae of this species develop more rapidly than other Aeshnidae and are bivoltine in the South, but rarely complete two generations in the northern United States. Flight periods of most species are during the summer months.

b. Cordulegastridae (1 Genus, 8 Species) The large, hairy, somewhat elongate larvae with small anterolateral eyes can be easily distinguished from other Anisoptera by the jagged teeth on the palpal lobes of their spoon-shaped prementum (Fig. 39). They inhabit small streams, where they lie partially concealed in silt to ambush prey, mostly at the upstream edge of pools. A 3–5-year life cycle is likely for most species, with adults flying from spring into early summer. Eggs are laid in the substrate of shallow riffles.

c. Corduliidae (7 Genera, 49 Species) Larvae are short and broad, with a spoon-shaped prementum, and cannot be readily separated from those of Libellulidae. In most, the palpal lobes are more deeply scalloped (Figs. 40 and 41) than in Libellulidae (Fig. 42), but larvae in one genus of Libellulidae also have deeply scalloped palpal lobes. Larvae are mostly lentic, with 2–4-year life cycles probably predominating in the North and shorter cycles in the South. They commonly occur in marshes, swamps, cool ponds, and littoral areas of lakes. However, several lotic species have larvae that inhabit debris in streams of all sizes. Synchronized emergences may lead to formation of exceptionally large swarms. Adult flight periods are mostly from May into early July.

d. Gomphidae (15 Genera, 97 Species) As in larvae of Aeshnidae and Petaluridae, the prementum is flat, but gomphid larvae have relatively shorter and broader abdomens. Their broad, four-segmented antennae are diagnostic (Fig. 36). Larvae of most species inhabit streams, where they may be found in riffles or partially buried in silt or sand in slower reaches. Those of other species inhabit lentic habitats, especially small permanent ponds and littoral areas of lakes, where they lie concealed in the substrate to ambush prey. Life cycles are long, usually 2–4 years, depending on water temperature and food. Most species have flight periods in late spring and summer; in some species eggs may diapause over the winter.

e. Libellulidae (26 Genera, 105 Species) The short, broad larvae closely resemble those of Corduliidae, but in almost all genera the crenulations on the palpal lobe of the mentum are extremely shallow (Fig. 42). Larvae occur in a variety of permanent and temporary lentic habitats where they crawl about in the vegetation and debris; occasionally they are found in vegetation along margins of streams. Littoral areas of lakes, permanent ponds, vernal ponds and marshes, cattail marshes, sphagnum swamps, and bogs all are habitats for larvae of a great variety of species. Many species are univoltine, with some having eggs that diapause. Other species may require two or more years to complete larval development. Flight periods range from spring through autumn, depending on the species.

f. Macromiidae (2 Genera, 9 Species) Larvae are readily recognized by their relatively broad abdomens, extremely long legs, and the presence of a frontal horn that projects forward between the eyes (Fig. 43). They are found among debris in the current of medium-to-large streams or in margins of lakes exposed to wave action. Life cycles of 2–4 years are likely. Flight periods extend from March through August, depending on the species.

g. Petaluridae (2 Genera, 2 Species) Both North American species are rare; neither has been found in the central part of the continent. Larvae of one semiterrestrial species inhabit bog seepages at high elevations in the western mountains; those of the other species have been found in bogs and seeps in eastern forests. The relatively broad larvae have a flat prementum, relatively broad six-or seven-segmented antennae that are hairy (Fig. 38), and patches of black bristles on the posterior abdominal terga. Because of their cold habitat, up to 6 years may be required to complete larval development. The flight period ranges from early spring in Florida to summer farther north.

2. Families of Odonata—Suborder Zygoptera (Damselflies)

a. Calopterygidae (2 Genera, 8 Species) The large larvae are readily distinguished from other Zygoptera by their very elongate first antennal segment (Fig. 31). All species are lotic, the larvae inhabiting permanent streams of all sizes, where they are found mostly among bank vegetation and accumulations of debris. Species are univoltine or semivoltine, with flight periods from late May into October.

b. Coenagrionidae (13 Genera, 96 Species) The relatively small larvae differ from those of Calopterygidae in having a short basal antennal segment, and differ from those of Lestidae in not having the base of the prementum greatly narrowed (Fig. 34). They are predominantly lentic, although species in one genus (*Argia*) inhabit riffles of streams and a few species in other genera occur along banks of streams. Larvae of lentic species are found mostly in permanent ponds, marshes, swamps, and littoral areas of lakes; occasionally, they occur among vegetation in parts of streams with little or no current. They crawl about on the vegetation and ambush or stalk prey, which consists mostly of small invertebrates. Most species are univoltine, some having eggs that diapause over the winter; a few species are bivoltine. Adults fly from spring to autumn, depending on the species.

c. Lestidae (2 Genera, 18 Species) The long, thin larvae with their long, parallel-sided caudal lamellae are distinctive. They differ from all other Zygoptera in having a prementum that is greatly narrowed and elongate basally (Fig. 33). Larvae of one genus (*Archilestes*) with 2 species are lotic. Larvae of other species (*Lestes*) are lentic, only occasionally being found among emergent vegetation along slow streams. They commonly inhabit vegetation in both permanent and vernal ponds and marshes and have a 1-year life cycle. Eggs are laid in vegetation above the water and often diapause until autumn or the following spring. Eggs hatch when the now dead vegetation in which they were laid falls into the water, or in vernal ponds, when water floods the dead vegetation. Adults fly from late spring through summer.

d. Protoneuridae (2 Genera, 2 Species) This tropical family is found only in Texas where larvae inhabit streams. Larvae are similar to those of Coenagrionidae, but their caudal lamellae are thick in the basal half and thin and leaflike in the distal half (Fig. 35).

C. Plecoptera—Stoneflies

In this relatively small hemimetabolous order at least 614 species are known from North America, and it is likely that many additional species will be described as the taxonomy of various families and genera is studied in greater detail and mountain streams are collected more thoroughly. The order is poorly represented in middle North America; species richness is greatest in mountain areas where fast, cold streams abound. Larvae of all species are aquatic. Almost all of them inhabit streams, but those of a few species have adapted to living in cold oligotrophic lakes. In the West, larvae of species in at least two genera inhabit the hyporheic zone of certain streams to depths of 10 m or more.

Stonefly larvae are an important part of the stream ecosystem. They provide food for fish and invertebrate predators, and often are top predators in the invertebrate food chain. They are not sufficiently abundant to create nuisance problems, and because larvae of most species are intolerant of lowered levels of dissolved oxygen they play an important role in biological monitoring of streams.

Most stoneflies are univoltine, but several larger species require 2 or 3 years to complete development. Species with 2 or 3 year life cycles in the North may be univoltine in the South. Univoltine species often have fast seasonal life cycles, with eggs or small larvae diapausing during the summer and rapidly developing in autumn, winter or spring. Many species, however, have a slow seasonal life cycle with eggs hatching within a few weeks and larval development progressing slowly throughout the year. Species that have a slow seasonal life cycle in the North may have a bivoltine, fast seasonal life cycle in the South.

The number of larval instars varies with the species and even among individuals within a species, with 12–23 or more instars having been recorded for species that have been studied. After completing development, larvae crawl from the water onto a convenient substrate where ecdysis occurs. Adults then crawl or fly to nearby vegetation or other structures to find a mate. In most species this is accomplished by drumming with their abdomen; males initiate the drumming signal and females respond. After repeated drumming and crawling, males find a female and mate. When eggs have matured, the female extrudes them into a mass on the tip of her abdomen. Females of some species fly over the water and land briefly on its surface to deposit their eggs; others jettison their egg mass while flying above the water. In still other species, females crawl into the water on a protruding substrate to oviposit beneath the surface. In most species the female will lay multiple egg masses.

Stonefly larvae can be distinguished from most other aquatic insects by their two long filamentous cerci and generally elongate or flattened appearance (Figs. 45, 48, 49, 54, 56, and 58). They have relatively long antennae, distinct compound eyes, two or three ocelli, chewing mouthparts, two pairs of thoracic wingpads, and three-segmented tarsi with two claws on each tarsus (Figs. 1 and 2). They are likely to be confused only with Ephemeroptera larvae, which also have long cerci, but Ephemeroptera larvae usually have three caudal filaments instead of two, have only one tarsal claw, and have gills on the middle abdominal segments. Gills on Plecoptera larvae, if present, are on the head, thorax, and/or basal or apical segments of the abdomen. Adult stoneflies resemble their larvae, the most noticeable difference being the presence of two pairs of net-veined wings that fold flat over the abdomen; a few species are brachypterous or apterous. Genitalia are well developed in adults and gills are reduced to remnants in species that have them as larvae.

Stonefly larvae generally crawl on the substrate; when forced to swim, most do so weakly by moving their abdomen from side to side. Oxygen for respiration is obtained from the water through the cuticle, with thoracic and, sometimes, abdominal gills aiding oxygen uptake in most larger species. Because they generally lack extensive gills, most Plecoptera larvae occur only in oxygen-rich water. Larval feeding habits vary among families and also among species within a family or genus. As in many other aquatic insects, larval feeding habits within a species will vary with developmental stage and availability of food.

Adult stoneflies are relatively short-lived; most live a few days to 2 weeks, some a little longer. Several species feed as adults, mostly on algae, lichens, and riparian leaves or flowers. Larvae can be readily collected with an aquatic net by using standard sampling techniques. Adults of most species can be collected by using a net to sweep bank vegetation or by examining rocks, tree trunks, and bridges in the vicinity of streams. Adults of some species, however, move high into trees and can be collected only by using nets with very long handles or by using sticky traps. Several species of adults are attracted to lights, while others are rarely found at lights.

Through the years there have been many changes in the higher classification of Plecoptera, but in the last two decades it has remained quite stable. The order is divided into two suborders, Euholognatha and Systellognatha. The former incudes four families (Capniidae,

Leuctridae, Nemouridae, and Taeniopterygidae) that previously were in the family Nemouridae; larvae in all of these families are primarily herbivores–detritivores. Larvae of two families of Systellognatha (Perlidae and Chloroperlidae) are predators, in two other families (Peltoperlidae and Pteronarcyidae) they are herbivores–detritivores, and in the remaining family (Perlodidae) feeding habits vary with the species and larval age.

The biology of Plecoptera was reviewed by Hynes (1976) and a recent book by Stewart and Stark (1988) discusses classification, phylogeny, biogeography, ecology, and behavior of Plecoptera. This book includes keys and illustrations for larvae in all North American genera, notes on their biology, and the known distribution of North American species. Species in some genera can be identified as mature larvae as well as adults; in other genera identification of larvae is not possible because larvae of several species remain unknown. About 55% of North American Plecoptera species are known in the larval stage. Information on the biology, habitat, and identification of larvae in each family is presented next. References to species keys for the identification of stonefly larvae are provided in Table III and a key to families of freshwater Plecoptera larvae appears in Section V.D.

1. Families of Plecoptera–Suborder Euholognatha

a. Capniidae (10 Genera, 151 Species) Capniidae are "winter stoneflies." Larvae are generally small, dark colored, and elongate (Fig. 56). They closely resemble larvae of Leuctridae, but can be recognized by their generally shorter and broader metathoracic wing-pads, which are nearly as widely separated as the mesothoracic wind-pads (Fig. 56), and by posterior abdominal terga that are usually widened and fringed posteriorly. Larvae inhabit streams of all sizes. They are especially abundant in smaller streams and spring seeps; some even inhabit streams that become dry in the summer.

Larvae in two genera occur in cold oligotrophic lakes. Life cycles are univoltine, with emergence in late winter or early spring. Eggs hatch a few weeks after being laid, and after some development, larvae of most species diapause. Larval development resumes in autumn, with some species developing most rapidly in late autumn and winter. A few species have a slow seasonal life cycle with steady development from late spring through autumn. Larvae are mostly detritivores that feed on allochthonous debris, especially decaying leaves. In the North, several species may emerge on warm days while streams are still ice covered; adults often can be found crawling on the snow. Adults of some species are apterous or brachypterous.

TABLE III Literature References to Species Keys for Identification of Plecoptera Larvae

Regional keys to larvae that include more than one family
- British Columbia (southwest)—Ricker (1943)
- Louisiana—Stewart *et al.* (1976)
- Minnesota—Harden and Mickel (1952)
- North and South Carolina—Unzicker and McCaskill (1982)
- Northeastern North America—Hitchcock (1974)
- Ozark and Ouachita mountains—Poulton and Stewart (1991)
- Pennsylvania—Surdick and Kim (1976)
- Rocky Mountains—Baumann *et al.* (1977)
- Saskatchewan—Dosdall and Lehmkuhl (1979)
- Texas—Szczytko and Stewart (1977)

Keys to larvae listed by family
- Capniidae—(eastern Canada) Harper and Hynes (1971b)
- Leuctridae—(eastern Canada) Harper and Hynes (1971a)
- Nemouridae—(eastern Canada) Harper and Hynes (1971d)
- Peltoperlidae
  - *Soliperla*—Stark (1983)
- Perlidae
  - *Agnetina*—Stark (1986)
  - *Hesperoperla*—Baumann and Stark (1980)
  - *Paragnetina*—Stark and Szczytko (1981)
  - *Perlinella*—Kondratieff *et al.* (1988)
- Perlodidae—(Wisconsin) Hilsenhoff and Billmyer (1973)
  - *Diploperla*—Kondratieff *et al.* (1981)
  - *Helopicus*—Stark and Ray (1983)
  - *Hydroperla*—Ray and Stark (1981); Ricker (1952)
  - *Isogenoides*—Ricker (1952)
  - *Isoperla*—(western North America) Szczytko and Stewart (1979)
  - *Malirekus*—Stark and Szczytko (1988)
  - *Setvena*—Stewart and Stanger (1985)
- Pteronarcyidae—(West Virginia) Tarter *et al.* (1975)
- Taeniopterygidae—Harper and Hynes (1971c)
  - *Taeniopteryx*—Fullington and Stewart (1980)

b. Leuctridae (7 Genera, 55 Species) The relatively small, elongate larvae are similar to those of Capniidae, but have metathoracic wing-pads that are similar in shape, but distinctly closer together than those on the mesothorax (Fig. 58). Larvae also resemble those of Chloroperlidae, which have shorter cerci and meso-and metathoracic wing-pads that are equally far apart (Fig. 49). Larvae are found mostly among gravel in small, cool, permanent streams where they feed on decaying allochthonous material; one genus occurs in temporary streams. Most species are univoltine with emergence from spring to autumn, depending on the species. A semivoltine life cycle occurs in at least one northern species.

c. Nemouridae (12 Genera, 72 Species) The small, relatively dark-colored larvae are short and broad, with divergent metathoracic wing-pads in more mature larvae and metathoracic legs that reach or exceed the apex of the abdomen (Fig. 54). Most nemourids are univoltine, with emergence from spring to autumn, depending on the species. Fast seasonal life cycles predominate, with an extended egg diapause in some species. At least one species is semivoltine. Larvae are detrital feeders that inhabit debris and soft sediments in smaller streams and spring seeps; they are often the only stonefly larvae in spring seeps and runs. Larvae in at least one genus inhabit temporary streams, and those of another species have been collected from Arctic lakes.

d. Taeniopterygidae (6 Genera, 35 Species) Like Capniidae, these are univoltine "winter stoneflies" that emerge in late winter or very early spring and probably spend the summer as diapausing larvae. In size and shape, larvae rsemble those of Perlodidae (Fig. 48), having divergent metathoracic wingpads and elongate abdomens, but they are not as heavily patterned as most Perlodidae, they move much more slowly, and they swim very inefficiently. Larvae are herbivores–detritivores that occur in permanent streams of all sizes. They are frequently found among bank vegetation.

2. Families of Plecoptera—Suborder Systellognatha

a. Chloroperlidae (13 Genera, 77 Species) Larvae are generally small, elongate, rounded in cross section, and have short cerci (Fig. 49). They are similar to larvae of Capniidae and Leuctridae and are difficult to distinguish from the larvae of Leuctridae in the field. They are not likely to be confused with Capniidae, however, because late-instar larvae of Chloroperlidae are present in late spring and summer, while those of Capniidae occur in late autumn, winter, and early spring. Larval mouthparts (Fig. 46) differ greatly from those of Capniidae and Leuctridae (Fig. 47), reflecting the carnivorous feeding habits of Chloroperlidae in later larval instars. Larvae of Chloroperlidae inhabit mostly gravel bottoms of cold, small- to medium-sized streams. Life histories are poorly known. Univoltine and semivoltine fast seasonal life cycles have been documented, with emergence in late spring or early summer.

b. Peltoperlidae (6 Genera, 20 Species) The broad, flat, roachlike appearance of the medium-sized larvae is distinctive (Fig. 45). They are often abundant in mountain streams of the eastern, southern, and western United States, but are absent from central North America. Larvae inhabit smaller streams, where they feed on decaying leaves, detritus, and associated microorganisms. Many species are semivoltine; some may be univoltine. Adults emerge in late spring or early summer.

c. Perlidae (15 Genera, 72 Species) Larvae are relatively large, broad, stoneflies (Fig. 1) that are readily recognized by the presence of branched filamentous gills at the base of each leg (Fig. 2) and the absence of gills on abdominal sterna. They are widespread and common, inhabiting permanent streams of all sizes where they prey on other macroinvertebrates. They are often the dominant insect predators, although early-instar larvae may feed on detritus and algae. Most larvae are found in fast water, especially among debris in riffles. Species in most genera have a 2 or 3 year slow seasonal life cycle in northern regions; univoltine life cycles may occur in the South. One common genus is univoltine with a fast seasonal life cycle. Emergence in most species is during the summer months.

d. Perlodidae (30 Genera, 122 Species) Larvae are mostly small- to medium-sized stoneflies, which usually have distinctive patterns dorsally. They lack the thoracic gills of the similarly shaped and patterned larvae of Perlidae. The metathoracic wing-pads are divergent and the abdomen is rather elongate (Fig. 48). While larvae of some species of Perlodidae are herbivores–detritivores throughout their larval development, most species are predominantly predators. They develop in all sizes and types of streams; some even develop in cold oligotrophic lakes. All species are apparently univoltine, with emergence in spring. Eggs of many species diapause during the summer and hatch in autumn, but in other species eggs hatch within a few weeks after being laid.

e. Pteronarcyidae (2 Genera, 10 Species) These large stoneflies require 1–4 years to complete development. They have a slow seasonal life cycle, with univoltine life cycles occurring only in the south. Larvae are elongate, and can be readily recognized by the branched filamentous gills at the base of thoracic legs and on the first two or three abdominal sterna. They feed mostly on coarse particulate organic matter and inhabit wood or accumulations of debris in small- to medium-sized permanent streams with moderate-to-rapid currents. Emergence occurs in spring or early summer, being earliest in the south.

D. Trichoptera–Caddisflies

Trichoptera is a rather large holometabolous order. Larvae and pupae of all species are aquatic, with one or two exceptions, and represent an important part of the insect fauna of most streams. About 1400 species are known from North America. Most larvae occur in lotic habitats, but those of some species in half of the families inhabit lentic habitats. Adults are terrestrial, mothlike insects, that sometimes are attracted to lights in large numbers. Most have very long filiform antennae, and their wings, which they fold rooflike over their body, are covered with hairlike setae. Eggs of most species are laid in masses in the water; adults of a few species lay their eggs on vegetation above the water or in moist soil of temporary ponds that will eventually be flooded. Most species are univoltine, often with rather synchronized emergences; however, in some species different cohorts emerge throughout the warm weather period. A few species are bivoltine, with the tendency for bivoltinism increasing farther south, and several species, especially those that develop in cold-water streams, are known to be semivoltine, especially in the northern part of their range.

Adult caddisflies can be identified to species by using the many keys and descriptions that have been published, but larval taxonomy lags, with about half of the described species still unknown in the larval stage. This makes accurate identification of larvae to species impossible in most genera, and larvae of four genera remain unknown. Wiggins (1996) published excellent keys, descriptions, and illustrations for larvae of known genera. Ross (1944) developed species keys for larvae in many genera that occur in the central United States as well as keys to adults, but in most of these genera there are species for which larvae are unknown, so his larval keys must be used with caution.

Trichoptera is divided into three suborders, Annulipalpia, Integripalpia, and Spicipalpia (Wiggins and Wichard, 1989; Frania and Wiggins, 1995). This parallels the division of the order into three superfamilies as used by Wiggins (1977) and others, and to some extent the classification proposed by Weaver and Morse (1986). In the Annulipalpia, larvae use silk from their labial glands to construct retreats and nets, which they use to filter or gather food such as algae, detritus, or macroinvertebrates. Larvae pupate within their retreat in a silk cocoon that has small openings at each end through which water circulates. In the suborder Spicipalpia, larvae in the four families are either free-living or construct portable cases that are barrel-shaped, purselike, or saddle-shaped. All larvae pupate in a completely closed cocoon of parchmentlike silk, differing from other caddisflies that have openings in their pupal case to allow circulation of water. Free-living larvae (Rhyacophilidae and Hydrobiosidae) are mostly predators on other arthropods. Larvae that build cases (Glossosomatiae and Hydroptilidae) are mostly herbivores that feed on periphyton. In the suborder Integripalpia, all larvae use silk from their labial glands to construct portable tubular cases of plant or mineral materials. Cases vary widely in size and shape, and are often sufficiently distinctive to be employed in generic and species identification. Integripalpia larvae are mostly herbivores or detritivores, but in some families (Brachycentridae, Leptoceridae, Limnephilidae, Molannidae, Odontoceridae, and Phryganeidae) omnivores or predators can be found. Larvae in eight families of Integripalpia have a prosternal horn, that larvae in other families lack. All larvae pupate within their last larval case after it has been attached to a substrate and closed at both ends. Small openings at each end permit water circulation.

Trichoptera larvae have a pair of simple eyes, chewing mouthparts, antennae that are so short that they often are difficult to see, three pairs of thoracic legs with a single tarsal segment and tarsal claw, and a pair of fleshy prolegs on the last abdominal segment (Figs. 66 and 67). Many larvae have single or branched gills on the abdominal segments. Most species have five larval instars, and after completion of larval development a pupal case is constructed, either by sealing off the open end of the portable case (Integripalpia) or by constructing a case and cementing it to the substrate (Annulipalpia and Spicipalpia). The larva then ceases feeding and becomes quiescent with its appendages pressed against its body. This is the prepupal stage, and may last a few days to several weeks. The pupa is of the exarate type, with appendages free from the body. Pupae in most families have well-developed mandibles that the pharate adults (adult within the pupal integument) use to cut their way out of the case just prior to emergence. After a pharate adult leaves the pupal case, it swims to the water surface where eclosion takes place or, more often, it swims to the surface and then rapidly to shore where it crawls out of the water onto some object to emerge. Eclosion is rapid.

Trichoptera larvae have many adaptations to their aquatic environment. Respiration in caddisfly larvae is through the integument, and in many families this is aided by the presence of abdominal gills. Additionally, especially in Integripalpia, larvae are able to increase respiration by undulating their body within their tubular case to increase the flow of water and dissolved oxygen through the case. Larval cases and retreats also serve to protect larvae from predators, acting either as a physical barrier or serving to camouflage the larva. Cases may also protect larvae from their environment. Hard cases protect those that live in sand and gravel from abrasion, and streamlined cases or nets allow others to inhabit substrates in very swift currents. Net-spinning larvae (Annulipalpia) use their nets to strain algae and other food material from the current in streams. Some caddisfly larvae (Leptoceridae) have fringes of setae on their legs that allow them to swim. Others (some Brachycentridae) have fringes of setae on their legs to permit them to filter food particles from the current.

Most caddisfly larvae can be readily collected from lotic and lentic habitats with an aquatic net by using methods already described. Debris in the pan should be examined carefully as many larvae have cases that blend in with the debris. In streams it is important to also examine large substrates such as wood and rocks, because larvae of many species attach themselves firmly to these substrates or inhabit moss or algae that grows on them. Species of larvae that burrow into decaying wood can be collected as they crawl from retreats when the wood is removed from a stream and allowed to dry. Larvae that inhabit sand or silt substrates are best collected by sieving these substrates through a net or strainer, or by closely watching sand substrates and collecting larvae that expose themselves by moving. Larvae of extremely small species are often overlooked because they remain attached to the substrate and are too small to see. Microscopic examination of macrophytes and other substrates returned to the laboratory may yield many small larvae, especially those of Hydroptilidae. In the field a hand lens is helpful in locating very small specimens.

Caddisflies are a very important part of the stream community; in some streams they dominate the insect biomass. Many species of fish feed on the larvae and emerging adults. Because of their general abundance and diversity in streams, and the wide variation in tolerance to pollution among larvae of various species, they are very important in biological monitoring. Occasionally, extremely large numbers of adults emerge, and because they are attracted to lights they may create severe problems. These include contamination of products in factories, annoyance of people in restaurants and elsewhere, and allergic reactions in people who are sensitive to them.

Larvae of many species have never been associated with adults. Larvae of these species must be reared, associated with adults, and described before we will be able to develop reliable keys to larvae at the species level. Such keys are needed to enable us to complete needed studies of the life history, ecology, and behavior of the various species. Relatively little is presently known about the life history and ecology of most North American species. Most of what is known has resulted from studies within a single stream where the fauna is limited and thoroughly known because larvae have been associated with adults. Wiggins (1996) summarized knowledge of the biology, ecology, morphology, and distribution of larvae in each genus. Information on the biology and habitat, and a brief description of the larva in each caddisfly family follows. A key to families of freshwater Trichoptera larvae appears in Section V.F, and references to species keys for the identification of larvae are provided in Table IV.

TABLE IV Literature References to Species Keys for Identification of Trichoptera Larvae

Regional keys to larvae that include more than one family
- Florida—Pescador *et al.* (1995)
- Illinois—(many incomplete) Ross (1944)
- North and South Carolina—Unzicker *et al.* (1982)

Keys to larvae listed by family
- Apataniidae
  - *Apatania*—Chen (1992)
- Brachycentridae—(Wisconsin) Hilsenhoff (1985)
  - *Brachycentrus*—Flint (1984)
  - *Micrasema*—Chapin (1978)
- Goeridae
  - *Goera*—(eastern United States) Flint (1960); Wiggins (1996)
  - *Goerita*—Parker (1998)
  - other genera—Wiggins (1973)
- Hydropsychidae—(Wisconsin) Schmude and Hilsenhoff (1986)
  - Arctopsychinae—Flint (1961); (western North America) Givens and Smith (1980)
  - *Ceratopsyche* (as *Hydropsyche)*—Schefter and Wiggins (1986)
  - *Hydropsyche* (including *Ceratopsyche)*—(eastern and central North America) Schuster and Etnier (1978)
- Lepidostomatidae—Weaver (1988)
- Leptoceridae
  - *Ceraclea*—Resh (1976)
  - *Mystacides*—Yamamoto and Wiggins (1964)
  - *Nectopsyche*—Haddock (1977)
  - *Oecetis*—Floyd (1995)
  - *Setodes*—Nations (1994)
  - *Triaenodes* and *Ylodes*—Glover (1996)
- Limnephilidae—(eastern United States) Flint (1960)
  - *Desmona*—Wiggins and Wisseman (1990)
  - *Dicosmoecus*—Wiggins and Richardson (1982)
  - *Hesperophylax*—Parker and Wiggins (1985)
  - *Pycnopshche*—Wojtowicz (1982)
- Molannidae
  - *Molanna*—Sherberger and Wallace (1971)
- Odontoceridae
  - *Psilotreta*—Parker and Wiggins (1987)
- Philopotamidae—Ross (1944)
- Phryganeidae—Wiggins (1960)
- Polycentropodidae
  - *Paranyctiophylax*—(as *Nyctiophylax*) Flint (1964); Wiggins (1996)
- Psychomyiidae
  - *Psychomyia*—Flint (1964)
- Rhyacophilidae—(eastern United States) Flint (1962); (Pennsylvania) Weaver and Sykora (1979)

1. *Families of Trichoptera—Suborder Annulipalpia*

a. Dipseudopsidae (1 Genus, 5 Species) The only genus *Phylocentropus*, was sometimes included in the generally predaceous family Polycentropodidae. Larvae differ from those of Polycentropodidae by having broad, flattened, setose tarsi (Fig. 72) and short, broad mandibles with mesal brushes. These are adaptations for their feeding on detritus, which they collect with a net placed within tubes that they construct in sand and silt along margins of sandy streams and lakes. The capture net and feeding habits were described by Wallace et al. (1976). All North American species occur east of the Great Plains. The life cycle is probably 1 year, with cohorts emerging from May through August.

b. Ecnomidae (1 Genus, 2 Species) Larvae of this Neotropical family were collected from a small, cold, rapid stream in the "hills" section of Texas (Waltz and McCafferty, 1983). A second species was described by Bowles (1995). These represent the only records of this family north of Mexico. While previously considered a subfamily of Polycentropodidae, larvae have sclerotized meso- and metanota as do larvae of Hydroptilidae and Hydropsychidae.

c. Hydropsychidae (11 Genera, 158 Species) Larvae in this very large and important family are easily recognized by the numerous, branched filamentous gills that occur ventrally on their abdomen. They frequently represent a large segment of the macroinvertebrate fauna in streams of all sizes, currents, and temperatures. Here, they build retreats on rocks and other larger substrates, in leafpacks, or in moss. Larvae of some species also tunnel into decaying wood and others may occur on wave-swept lake shores. Larvae of most species are omnivorous, feeding primarily on algae, crustacea, and insects collected by their capture nets; those of a few species tend to eat mostly algae and diatoms and others are mostly predaceous, especially in later instars. Life cycles vary from bivoltine to semivoltine, depending very much on stream temperature, with the majority of species probably being univoltine. Because of their widespread abundance and the great variation in the tolerance of species to organic pollution, larvae are very important in biological monitoring of streams.

The subfamily Arctopsychinae is generally recognized in Europe as a separate family, and in North America this family has been recognized by Schmid (1968) and Nimmo (1987), but Wiggins (1981) presented compelling arguments for retention of subfamily status. Capture nets of Arctopsychinae larvae are generally larger than those of other Hydropsychidae and larvae are predominantly predators in the later instars. Larvae differ from those of other Hydropsychidae by having an elongate, posteriorly narrowed gula that completely separates the genae.

d. Philopotamidae (3 Genera, 42 Species) Larvae live in silken retreats, usually on the bottom of rocks in warm-to-cold streams of all sizes. Their unmarked head with an unsclerotized T-shaped labrum is distinctive (Fig. 70) and readily separates them from other net spinning larvae with unsclerotized meso- and metanota. Larvae feed on algae and detritus that they capture in their silken nets, which have a smaller mesh size than those of other net spinning caddisflies. Most, if not all species, are probably univoltine, except in the south, where two or more generations may occur. Cohorts of most species emerge throughout the spring and summer months; one species also emerges in fall and winter.

e. Polycentropodidae (6 Genera, 71 Species) While larvae of most species live in streams, they inhabit a wide variety of other habitats, including littoral areas of lakes and temporary ponds. Larvae can be recognized by their pointed protrochantin (Fig. 71), unmodified tarsi, and the presence of dark or light spots (muscle scars) on the heads of most species. Reliable keys to larvae do not exist, because larvae of many species remain undescribed. Most larvae have a waxy cuticle that tends to repel water, giving the impression when collected that they are terrestrial. Larvae of a few species build trumpet-shaped capture nets similar to other Annulipalpia and are primarily herbivores, but those of most species are predators and have tubular retreats that they sometimes use to detect prey. These tubular retreats also function as an aid to respiration, with larvae circulating water through them by undulating their body. This is especially important for species that live in lentic habitats that occasionally may be deficient in oxygen. Life cycles are probably one year in most species, with emergence during the summer months.

f. Psychomyiidae (4 Genera, 17 Species) Larvae live in streams, frequently on or in wood; they also occur on rocks, which they cover with silken retreats that incorporate sand or detritus. Larvae can be readily separated from those of other families without sclerites on the meso- and metanotum by their hatchet-shaped protrochantin (Fig. 69). They graze food from the vicinity of their retreats, mostly consuming detritus, fungi, and periphyton. Most species are probably univoltine, with emergence in late spring and summer.

g. Xiphocentronidae (2 Genera, 2 Species) Sometimes considered as part of Psychomyiidae, the only described larvae of Xiphocentronidae north of Mexico differ from Psychomyiidae larvae and those in other families by having their tarsi and tibiae fused into a single segment on all legs (Fig. 68). Larvae are known in North America from southern Texas, where one species was collected from a spring run and described by Edwards (1961), and from northern Arizona, where larvae in a second genus were described by Moulton and Stewart (1997).

2. Families of Trichoptera—Suborder Integripalpia

a. Apataniidae (5 Genera, 33 Species) Until recently this family was included in Limnephilidae; the larvae are quite diverse morphologically. All are rather small (<12 mm) and live in cornucopia-shaped cases of small rock fragments. They occur in small cold-water streams or seeps, and occasionally in cold lakes where they feed mostly on algae. Most larvae differ from other Limnephilidae by having mandibles with uniform scraper blades; those with toothed mandibles have 25 or more setae on the anterior of the metanotum between the sclerites.

b. Beraeidae (1 Genus, 3 Species) The gently curved and tapered case is made of fine sand. Larvae are small and can be recognized by a transverse ridge on the pronotum that curves forward laterally to form rounded lobes (Fig. 77). Larvae, which are known for only two North American species (Wiggins, 1996), inhabit muck margins of spring seeps. Adults have been collected only in southeastern and northeastern North America, where they are rare. A species found in Ontario has been reported to be semivoltine.

c. Brachycentridae (5 Genera, 34 Species) The larval case is tapered, of vegetable matter or sand grains, and square or round in cross section. Larvae are distinctive, having their pronotum divided by a distinct furrow (Fig. 76), with the area in front of the furrow depressed. They have no dorsal or lateral humps on the first abdominal segment, which is unique among Integripalpia. Larvae in three of the genera are monotypic in North America; most others can be identified by keys listed in Table IV. Most species are univoltine, but some that inhabit cold northern streams are semivoltine. Larvae of all species inhabit streams, especially cold spring-fed streams, where they live in moss or other vegetation, or attach their cases to substrates in the current. The mostly herbivorous larvae graze on periphyton and other vegetation; some also consume small insects and crustaceans. Some are primarily filter feeders that use their elongate meso- and metathoracic legs to gather food from the current.

d. Calamoceratidae (3 Genera, 5 Species) Cases are of leaf pieces or hollowed sticks. Larvae, which may reach a relatively large size, can be recognized by the projecting, divergent, anterolateral corners of the pronotum and a row of about 16 long setae dorsally on the labrum (Fig. 79). Most species can be identified as larvae by using descriptions and distributions reported in Wiggins (1996). Larvae inhabit pool areas of cool streams in the west or east, where they feed mostly on detritus. Most species probably have a 2-year life cycle.

e. Goeridae (4 Genera, 11 Species) Larvae resemble those of Limnephilidae, but differ in their unique modification of the mesepisternum, which is formed into an anteriorly projecting elongate process (Fig. 83) or a rounded spiny prominece. Since there are only a limited number of species, it is often possible to identify larvae to species using publications listed in Table IV. The sand cases are either cornucopia-shaped or tubular with flanges of pebbles. Life cycles are probably univoltine or semivoltine. Larvae inhabit cobbles and other substrates in the current of small cold streams; larvae of one monotypic genus occur in the muck of spring seeps. They feed on algae, vascular plants, or detritus.

f. Helicopsychidae (1 Genus, 4 Species) The snail like case, which larvae never abandon, is unique (Fig. 59). Only the larva of *Helicopsyche borealis*, which is widespread in North America, is known; larvae of the three western species remain unknown. The normally lotic larvae may also be found along margins of lakes, often in rather deep water. Larvae inhabit rocks or wood and scrape periphyton and detritus from the substrate; early instars may burrow into the bed of sandy streams. They can tolerate very warm water, and have been collected from thermal springs with temperatures as high as 34°C. They are probably univoltine, with emergence of cohorts throughout the summer and a prolonged winter diapause of eggs in the North.

g. Lepidostomatidae (2 Genera, 82 Species) Cases are highly variable, being made of sand or bits of vegetation and square or round in cross section. The location of the antenna very close to the eye (Fig. 78) readily identifies all larvae. Larvae are primarily detritivores and are found among detritus on a variety of substrates. While larvae of most species inhabit cold streams or springs, some occur in lakes and may be found at a considerable distance from the shore. Species that have been studied are univoltine, with spatial or temporal separation of different species that occur in the same stream.

h. Leptoceridae (8 Genera, 113 Species) Larvae construct a wide variety of cases. In some genera they have elongate cases of vegetation, vegetation and sand, or of pure silk, and are adapted for swimming by having fringes of long setae on their legs. Larvae in other genera have shorter, stouter cases that are often cornucopia-shaped and constructed of vegetable or mineral material; these larvae are unable to swim. Most larvae can be recognized by their relatively long antennae (Fig. 63), which are much longer than those of other caddisfly larvae, and also by their elongate metathoracic legs, which are distinctly longer than their other legs. Larvae inhabit a variety of permanent lotic and lentic habitats; those in lotic habitats tend to be in slower water and those in lakes may be found far from shore as well as in the littoral zone. They are generally omnivorous in their feeding habits, but some are primarily predators and a few feed exclusively on freshwater sponges. Most species are probably univoltine.

i. Limnephilidae (39 Genera, 243 Species) Case construction in this large and diverse family varies widely, with vegetable or mineral material incorporated into a variety of shapes and sizes. Larvae also vary greatly in size and structure, making them difficult to characterize. All have a prosternal horn and antennae that are located about halfway between each eye and mandible (Fig. 80). Most have numerous setae on the first abdominal segment, and sclerites at SA-1 of the metanotum (Fig. 66). Many also have chloride epithelia ventrally or dorsally on abdominal segments. While species in some genera can be identified by keys listed below, accurate identification of larvae is not possible in most genera because larvae of too many species remain unknown. Most species are univoltine, but a few are semivoltine. Larvae of most species inhabit a wide range of lotic habitats, from springs and small streams to large rivers; others inhabit lakes, and several are found in permanent or temporary ponds. In these habitats larvae can be found on vegetation, rocks, or detritus, or in gravel, sand, or soft sediments. Most larvae are herbivores or detritivores, and are important in processing large particulate organic matter; larvae of one species are known to be carnivores.

j. Molannidae (2 Genera, 7 Species) Larvae construct characteristic sand cases with lateral flanges and a hood over the anterior opening. They also have long metathoracic legs with either very short or threadlike claws. Larvae live in sandy or silty substrates of lakes or streams where currents are reduced. Here they feed on algae, vascular plants, or even invertebrates. Most species are probably univoltine.

k. Odontoceridae (6 Genera, 13 Species) Cases are typically cornucopia-shaped and made of sand grains and tiny stones; they are hard and difficult to crush. Like Sericostomatidae, which they resemble, larvae lack a prosternal horn and have their eyes located close to the mandibles, but their protrochantin is small and not hook-shaped (Fig. 82). Species of larvae in the largest genus (*Psilotreta*) can be identified (Table IV); species in monotypic genera can be identified by using the key preposed by Wiggins (1996). Larvae inhabit small, cold streams or springs where they are found in sandy substrates. They are probably omnivorous, feeding on algae, vascular plants, and invertebrates. Most species are probably semivoltine in the north and univoltine in the south.

l. Phryganeidae (10 Genera, 28 Species) These large caddisflies have distinctive cases made mostly of pieces of vegetation that are spirally wound or in concentric rings. When disturbed, they readily abandon their cases, but may re-enter them. Larvae are distinctive, generally having a boldly striped head that is more prognathous than other Integripalpia (Fig. 65). They also have a prominent prosternal horn and lack significant sclerotization of the mesonotum. Identification of larvae to species is not possible in some genera. Larvae may be found among vegetation and detritus along streams of all sizes, in marshes, in temporary and permanent ponds, and even in lakes where they may occur far from shore. While many are mostly predators, vegetation is also consumed, especially by early instars. Life cycles are probably 1 year.

m. Rossianidae (2 Genera, 2 Species) This western North American family was erected for two genera previously placed in Limnephilidae and Goeridae. Larvae live in organic muck of spring seeps or in small, cold, mountain streams where they feed on plants and detritus. Their curved cases are made of tiny stones. Larvae are distinctive among Integripalpia because their head, pronotum, and mesepisternum are modified by prominent surface sculpturing.

n. Sericostomatidae (3 Genera, 15 Species) Cases are typically cornucopia-shaped and made of sand and small pebbles, but are not hard as in Odontoceridae. Larvae can be recognized by the pointed anterolateral angles of their pronotum, a hook-shaped protrochantin (Fig. 81), and broad, flat profemur. No key has been published for separating larvae of *Agarodes*, the largest genus. Larvae live in sandy substrates of streams, springs, and even lake margins, where they feed primarily on plant detritus. Most species are probably univoltine.

o. Uenoidae (5 Genera, 46 Species) The cornucopia-shaped cases of sand grains are either extremely long and thin or short and stout, depending on the genus. Larvae occur in streams and springs. Until recently they were considered part of Limnephilidae. Larvae differ from Limnephilidae by having a mesally notched anterior of the mesonotum (Fig. 84) and untoothed scraping mandibles, which they use to scrape diatoms and organic detritus from rock surfaces. There are several species in each of the genera, and keys to species of larvae have not yet been developed. Two genera (*Neothremma* and *Farula*) occur in the western mountains; *Neophylax*, a species widespread in North America, was recently added to this family (Vineyard and Wiggins, 1988).

3. Families of Trichoptera—Suborder Spicipalpia

a. Glossosomatidae (6 Genera, 76 Species) Larvae build unique turtlelike cases of sand and pebbles, but readily vacate them when collected. They can be separated from other Spicipalpia with unsclerotized meso- and metanota by their cases and their short anal prolegs with short claws (Fig. 73). Larvae can be identified to species only by associating them with adults. All species inhabit the upper surfaces of rocks or other large substrates in clean, cool streams or springs. Here they feed on diatoms and other algae that they scrape from the substrate. Most species probably have a one-year life cycle, often with many cohorts that emerge throughout the spring and summer.

b. Hydrobiosidae (1 Genus, 3 Species) Although sometimes included in Rhyacophilidae, the free-living larvae are unique in having chelate prothoracic legs (Fig. 75), which are suited to their predatory habits. They occur in cool streams in Texas, Arizona, and Nevada. The larva of one species was described by Edwards and Arnold (1961).

c. Hydroptilidae (16 Genera, 250 Species) Larvae are characterized by their completely sclerotized thoracic nota, lack of branched ventral gills, and very small size. They are frequently called "microcaddisflies." The first four larval instars are free-living, have no case, and have a normal-size abdomen; in the last instar the abdomen is greatly expanded and larvae have a case that is purse-shaped or barrel-shaped and sometimes attached to the substrate (Figs. 61, 62). Most species pass through the first four instars rapidly, and it is usually not possible to identify these early instars to genus. Larvae in the last instar can be identified to genus, but usually not to species. Larvae may be found in a wide variety of lotic and lentic habitats where they feed mostly on algae and other plant material. They are probably univoltine, or bivoltine in warmer habitats, but generally very little is known about life histories of the various species.

d. Rhyacophilidae (2 Genera, 128 Species) The larvae live in cool or cold streams, have no case, and do not build retreats. Most species occur in mountainous regions where cold, fast streams abound. They are recognized by their unsclerotized meso- and metanotum, sclerotized ninth abdominal tergum, and elongate anal prolegs (Fig. 74). Life cycles are probably univoltine for most species, but may be semivoltine in some. Larvae of most species are active predators, but those of a few species feed on living and dead vascular plant tissue and algae.

E. Megaloptera—Fishflies and Alderflies

This very small holometabolous order contains two aquatic families that once were included in the order Neuroptera. The aquatic larvae are distinctive, having seven or eight pairs of lateral filaments and large, conspicuous mandibles (Figs. 85, 86). They can be confused only with some aquatic Coleoptera larvae that have lateral filaments. Larvae often attain a very large size, and because all are predators, they may exert significant pressure on populations of macroinvertebrates that live in the same habitat. Studies indicate that there are usually 10 or 11 larval instars, and that life cycles range from 1–4 years, depending on the species, climate, and habitat. Larvae are aquatic; all other stages are terrestrial. Eggs are laid in masses on objects above the water, and pupation takes place in cells on land. The terrestrial adults are short-lived and generally secretive, but because of their large size and often conspicuous mandibles, they attract attention when discovered. Adults are weak fliers, but are often attracted to lights. They fold their net-veined wings rooflike over their abdomen when at rest.

Larvae of most species occur only in relatively oxygen-rich environments, because they usually obtain oxygen through their integument directly from the water in which they live. Their lateral filaments greatly increase surface area, allowing them to do this. Species that inhabit lentic habitats, which may at times be low in oxygen, have long caudal respiratory tubes that they use to obtain oxygen from air at the surface. All larvae have spiracles and are capable of living out of water in moist areas; larvae of some species survive in temporary streams.

Larvae can be readily collected with an aquatic net by using standard sampling procedures. However, greater numbers of stream species can often be collected by dislodging larger rocks. Rearing of larvae is difficult because of long life cycles, cannibalism, and the need to supply prey species and a terrestrial environment for pupation. Species of adults are well known, and as a result of recent studies larvae of most species can also be identified. References to species keys for the identification of larvae are provided in Table V. Information on the biology, habitat, and identification of larvae in the two families is presented next, and a

TABLE V Literature References to Species Keys for Identification of Megaloptera Larvae

Regional keys to larvae that include more than one family
North and South Carolina—Brigham (1982a)
Keys to larvae listed by family
Corydalidae—(Mississippi) Stark and Lago (1983)
Chauliodes—Cuyler (1958)
Nigronia—Neunzig (1966)
Sialidae—(eastern North America) Canterbury (1978)

key to families of freshwater Megaloptera larvae appears in Section V.F.

1. Families of Megaloptera

a. Corydalidae (7 Genera, 22 Species) The very large size (20–90 mm), lateral filaments, and large mandibles of late instar "fishfly" or "Dobsonfly" larvae are distinctive (Fig. 86), but smaller larvae resemble those of Gyrinidae and two other Coleoptera genera. Larvae can be distinguished from Coleoptera larvae by their pair of terminal prolegs, each with two hooks (Fig. 86). Larvae in most genera inhabit streams or spring seeps, which have a relatively high dissolved oxygen content. Because these habitats are generally cold, life cycles of two or more years are likely for most species, with different cohorts and extended emergence periods in some species. The habitat of *Chauliodes* larvae differs from that of other genera. They inhabit shallow lentic habitats or vegetated margins of streams and are univoltine, with emergence in late spring or early summer. They also are known to consume significant amounts of detritus in addition to prey, especially during the winter.

b. Sialidae (1 Genus, 24 Species) The lateral filaments and long, single tail filament at once separate "alderfly" larvae from all other aquatic insects (Fig. 85). Larvae live mostly in depositional sediments of streams, permanent ponds, and lakes. Species that live in lakes may inhabit areas that are several hundred meters from shore, necessitating a considerable migration to land for pupation. Most species are univoltine, but in northern regions species that live in streams or lakes are often semivoltine.

III. PARTIALLY AQUATIC ORDERS OF INSECTS

A. Aquatic Heteroptera—Aquatic and Semiaquatic Bugs

Because all aquatic and semiaquatic bugs are Heteroptera and none are Homoptera, the catalog by Henry and Froeschner (1988) will be followed and this order of insects will be referred to as Heteroptera rather than Hemiptera, or Hemiptera: suborder Heteroptera. The choice of names for this group of insects has created much controversy; pros and cons are discussed by Henry and Froeschner (1988).

Heteroptera is a medium-sized paurometabolous order that is primarily terrestrial, but about 8.5% of the species are aquatic, with adults and nymphs either living in water (220 species) or on its surface (106 species). It is divided into seven suborders, only two of which have aquatic species. Species with adults and nymphs that walk on the water surface are in the suborder Gerromorpha, and are referred to as "semiaquatic Heteroptera." Species with adults and nymphs that live in the water belong to the suborder Nepomorpha. These are referred to as "aquatic Heteroptera," because they spend their entire life under water, except during dispersal flights. Flightless species of Nepomorpha are the most aquatic of all insects because they rarely, if ever, leave the water. There are six families of Nepomorpha in which all species are aquatic, and five families of Gerromorpha in which all species are semiaquatic or occasionally riparian. There are also four families of Heteroptera with riparian adults and nymphs; these have been included in the key because they occasionally are collected along with aquatic species. These are Gelastocoridae and Ochteridae (Nepomorpha), Macroveliidae (Gerromorpha), and Saldidae (Leptopodomorpha).

Most species of aquatic and semiaquatic Heteroptera breed in lentic habitats, but there are also several lotic species, and adults of many lentic species fly to streams in the north to overwinter. Adults and nymphs of most species are predators, and their presence can significantly influence populations of other insects, especially in lentic habitats. They are known to have an impact on populations of mosquito larvae, and thus can benefit man. Adults of larger species, however, may prey on small fish and create problems in fish-rearing ponds. Those of some species also may bite people, and occasionally they become a nuisance when attracted by lights to swimming pools. The vast numbers of corixid adults that overwinter in northern streams are an important source of food for some fish, and dried corixids are sold as food for pet turtles and tropical fish. Most other Heteroptera are usually avoided by fish, probably because of secretions from their scent glands. Generally, however, aquatic and semiaquatic Heteroptera have little impact on humans.

Heteroptera adults can be distinguished from other aquatic insects by their mesothoracic wings, or hemelytra, which are hardened in the basal half and membranous apically (Fig. 87), and by sucking mouthparts that are formed into a tube or "rostrum" (Figs. 88, 90).

There are normally five nymphal instars; a few species have only four. Nymphs resemble adults, except that they lack wings and genitalia, and are distinctly less sclerotized. They inhabit the same habitat as adults and are often associated with them.

Aquatic and semiaquatic Heteroptera have been rather thoroughly studied, and adults of all North American species can be readily identified. Families and genera are quite heterogeneous, making it possible to recognize families, many genera, and some species in the field. The nymphs, however, have not been well-studied except for Gerridae, and few can be identified to species except regionally or through association with adults. Some cannot even be identified to genus. Since the pioneering works of H.B. Hungerford and others 45–80 years ago, there have been relatively few additions to the known fauna in North America. References to species keys to adults are provided in Table VI. Information on the biology, habitat, and identification of each family is presented next, along with information about suborders. A key to families of adults of aquatic, semiaquatic, and riparian Heteroptera appears in Section V.G.

TABLE VI Literature References to Species Keys for Identification of Heteroptera Adults

Regional keys to adults that include more than one family
Alberta, Saskatchewan, and Manitoba—Brooks and Kelton (1967)
California—Menke (1979a)
Florida—(northern) Herring (1951); Chapman (1958)
Minnesota—Bennett and Cook (1981)
Mississippi—Wilson (1958)
Missouri—Froeschner (1949, 1962)
Montana—Roemhild (1976)
North and South Carolina—Sanderson (1982)
Oklahoma—Schaefer and Drew (1968)
Virginia—Bobb (1974)
Wisconsin—Hilsenhoff (1984b, 1986)
Keys to adults listed by family
Belostomatidae–Menke (1979b); (Louisiana) Gonsoulin (1973c)
Corixidae–Hungerford (1948); (Oregon and Washington) Stonedahl and Lattin (1986)
Cenocorixa–Jansson (1972)
Hesperocorixa–Dunn (1979)
Trichocorixa–Sailer (1948)
Gerridae–(Arkansas) Kittle (1980); (British Columbia) Scudder (1971); (Louisiana) Gonsoulin (1974); (Oregon and Washington) Stonedahl and Lattin (1982); (Ontario) Cheng and Fernando (1970)
Aquarius–Andersen (1990)
Gerrinae–Drake and Harris (1934); Kuitert (1942)
Gerris–Andersen (1993); (Connecticut—also nymphs) Calabrese (1974)
Limnoporus–Andersen and Spence (1992)
Metrobates–Anderson (1932)
Rheumatobates–Hungerford (1954)
Trepobates–Kittle (1977)
Hebridae–Porter (1950)
Hydrometridae–Hungerford and Evans (1934); (Louisiana) Gonsoulin (1973a)
Mesoveliidae–Polhemus and Chapman (1979)
Naucoridae–(Texas) Davis (1996); (Louisiana) Gonsoulin (1973b)
Ambrysus–LaRivers (1951)
Pelocoris–LaRivers (1948); (Texas) Sites and Polhemus (1995)
Nepidae–Sites and Polhemus (1994); (Louisiana) Gonsoulin (1975)
Notonectidae–(Arizona) Zalom (1977); (British Columbia) Scudder (1965); (West Coast nymphs) Voigt and Garcia (1976)
Buenoa–Truxal (1953)
Notonecta–Hungerford (1933); (New England) Hutchinson (1945)
Veliidae–Smith and Polhemus (1978); Smith (1980)
Rhagovelia–Polhemus (1997)

1. Families of Semiaquatic Heteroptera—Suborder Gerromorpha

Adults of semiaquatic species reach peak abundance in September or October, depending on latitude, after which most of them find protected terrestrial sites in which to spend the winter; adults of species that overwinter as eggs die. Species that overwinter as adults are bivoltine or multivoltine; those that overwinter as eggs usually are univoltine, but may be bivoltine in the south. Eggs of semiaquatic species are laid on substrates at the water surface. Nymphs and adults feed mostly on invertebrates of an appropriate size in the surface film. Gerridae, and probably most other Gerromorpha, have sensillae on their femurs and trochanters that detect vibrations in the surface film, enabling them to find their prey. In most semiaquatic Heteroptera apterous and brachypterous adults are common. Apterous adults can be distinguished from nymphs by the presence of genitalia, greater sclerotization of the body, and by having two tarsal segments instead of one on most legs.

Semiaquatic Heteroptera adults and nymphs are adapted for walking on water by having hydrofuge hairs on their tarsi, and in the larger and heavier Gerridae and Veliidae, by additionally having preapical claws. They are most readily collected by sighting them on the water surface and sweeping them from the surface film with an aquatic net.

a. Gerridae (8 Genera, 44 Species) All of the largest adults of semiaquatic Heteroptera are species of Gerridae, but the family has many diverse sizes, forms, and life histories. Adult "water-striders" are distinguished by their relatively large size, preapical tarsal claws (Fig. 97), and elongate metafemurs that extend well past the tip of the short abdomen. Apterous and brachypterous forms are common. Although found in all types of aquatic habitats, adults and nymphs of most species prefer quiet areas where there is minimal wave action and current. Adults and nymphs of lotic species inhabit streams of all sizes; those of lentic species inhabit swamps, marshes, permanent and temporary ponds, littoral areas of lakes, and even the ocean. However, adults of most lotic species will occasionally inhabit lentic habitats and adults of lentic species may occasionally be found along margins of lotic habitats. Most are usually associated with emergent vegetation. Univoltine or bivoltine life cycles prevail, with some species overwintering as adults and some univoltine species overwintering as eggs. All water striders are predaceous. They feed on terrestrial insects and other invertebrates that become trapped in the surface film or aquatic insects that live just beneath the surface.

b. Hebridae (3 Genera, 15 Species) Because of their very small size (<2.5 mm), "velvet water bugs" are likely to be confused only with very small species of Veliidae. When on the water, Hebridae adults and nymphs move much more slowly than those of similar-sized Veliidae, and upon close examination Hebridae adults and nymphs have a rostral sulcus and apical claws, which those of Veliidae lack. Adults and nymphs inhabit very shallow, sheltered shoreline areas of lentic habitats, including swamps, marshes, ponds, lakes, and river sloughs. Those of some species are occasionally or even frequently riparian. Apterous adults are common in some species. Adults can be found from spring to autumn, suggesting that they are bivoltine or multivoltine and overwinter as adults.

c. Hydrometridae (1 Genus, 9 Species) The small (8–11 mm), delicate, sticklike "marsh-treaders" or "water-measurers" with their elongate head (Fig. 99) are unlike any other aquatic insect. Both macropterous and brachypterous adults are common. They walk on vegetation and the surface film of permanent habitats that lack wave action and current, including marshes, swamps, and vegetated margins of lakes, ponds, and slow streams. Here they feed by using their barbed rostrum to spear small macroinvertebrates in the surface film. They most often are found where fish are absent. Hydrometridae are multivoltine and overwinter as adults.

d. Mesoveliidae (1 Genus, 3 Species) "Water-treaders" become abundant in late summer in sheltered lentic habitats and margins of lotic habitats that have algae and duckweed along with emergent vegetation. Their relatively small size (2–4 mm) and the presence of black spines on the legs is distinctive. Apterous adults predominate, but winged adults are not uncommon, especially in autumn. They prey on small terrestrial and aquatic animals in the surface film. Adults and nymphs are killed by freezing temperatures, and in the north mesoveliids overwinter as eggs. All species are probably multivoltine.

e. Veliidae (5 Genera, 35 Species) "Broad-shouldered" or "short-legged water-striders" are often abundant in a wide variety of aquatic habitats. Some inhabit streams, living in the vicinity of riffles or under the banks; others inhabit spring ponds, swamps, marshes, ponds, and margins of lakes. Adults can be distinguished from other semiaquatic Heteroptera, except Gerridae, by their preapical tarsal claws. They differ from Gerridae in their generally smaller size and relatively shorter legs, the metafemurs not extending past the tip of the abdomen. Apterous adults predominate, with winged forms rare or absent in the various species. Most species are multivoltine and overwinter as adults, some along margins of streams or springs instead of in terrestrial habitals. A few species overwinter as eggs and are univoltine or bivoltine.

2. Families of Aquatic Heteroptera—Suborder Nepomorpha

Most species are univoltine in northern North America; a few are bivoltine. Farther south, bivoltine life cycles probably predominate. Adults reach peak abundance in October or November, and overwinter in aquatic habitats. In areas where ponds and lakes freeze, adults of species that typically breed in shallow lentic habitats usually fly to deep lentic habitats or larger streams to overwinter. In late April or early May, they return to their breeding sites to mate and oviposit. Eggs are laid on vegetation or other substrates, usually in the water. Some univoltine species diapause over the winter as eggs.

Adults and nymphs of most species are adapted for swimming by having fringes of long setae on their legs, especially the metathoracic legs. They obtain oxygen from air stores held under the hemelytra and from a bubble held ventrally on hydrofuge hairs, which may act as a physical gill or plastron. Air stores are renewed at the surface through air straps in Belostomatidae, air tubes in Nepidae, the pronotum in Corixidae, and the apex of the abdomen in Notonectidae and Naucoridae. In Corixidae, Notonectidae, Pleidae, and Naucoridae the ventral bubble is large in relation to their size and acts as a physical gill, with oxygen from the water diffusing into the bubble and carbon dioxide from respiration diffusing out of it. This allows prolonged submergence in oxygen-rich water. Adults and nymphs of almost all species are carnivores. They feed mostly on small invertebrates and occasionally on vertebrates such as minnows and tadpoles. The largest family, Corixidae, is an exception, with adults and nymphs of most species being primarily detritivores.

Nepomorpha adults and nymphs are best collected with an aquatic net along with debris. The contents of the net are then spread on a half-inch (13 mm) mesh screen over a large white pan, and insects are allowed to crawl from the debris and fall into the pan. Adults and nymphs of many species feign death and will crawl from the debris only if left undisturbed for several minutes. Additional specimens in all families can often be collected with bottle traps (Hilsenhoff, 1987). In northern regions, exceptionally large numbers of adults of both lotic and lentic species can be collected in autumn from along banks of large streams where the current is slow, but adults of some lentic species do not fly to streams and will be missed. Adults of many species are attracted to lights, and may be collected with light traps.

a. Belostomatidae (3 Genera, 19 Species) "Giant water bugs" are easily recognized by their large size and a pair of short, straplike, posterior respiratory appendages (Fig. 92) that they use to obtain air at the water surface. Adults and nymphs of two genera (*Belostoma* and *Lethocerus*) inhabit permanent lentic habitats, especially weedy ponds, margins of lakes, and marshes, and are important as apex invertebrate predators. Those of the third genus (*Abedus*) occur in streams among aquatic plants or under rocks in riffles. All adults are powerful predators that will capture and kill anything they can subdue, including fish, frogs, tadpoles, and other insects. Eggs are laid in large masses above the water on vegetation (*Lethocerus*) or on the backs of the males, which brood the eggs (*Abedus* and *Belostoma*). Belostomatids are univoltine or bivoltine, depending on the latitude, with adults of lentic species in northern regions flying in autumn to larger streams or deep lentic habitats where they overwinter. They return to their breeding sites in early May. Adults frequently fly to lights and are sometimes called "electric light bugs."

b. Corixidae (18 Genera, 129 Species) As the most abundant aquatic Heteroptera, "water boatmen" are found in most permanent aquatic habitats and frequently invade temporary ones as well. Adults most closely resemble those of Notonectidae, but are dorsoventrally flattened (Fig. 87) and have a short, broad, unsegmented, triangular rostrum and widened, scoop-shaped protarsi (palae) (Fig. 88). Unlike other aquatic and semiaquatic Heteroptera, which are predators, corixid adults and nymphs feed primarily on detritus, algae, protozoans, and other extremely small animals. Adults of a few species, however, will capture and eat larger insects, such as mosquito larvae. Adults and nymphs of most species are lentic, inhabiting permanent ponds and littoral areas of small lakes, but those of some species are strictly lotic and can be found only along margins of streams or in spring ponds and seeps. In northern latitudes, adults of most lentic species fly to larger streams to overwinter, and in October and November thousands can often be collected from slow, shallow areas of streams. Adults disperse widely, rapidly invading temporary ponds, and on warm nights they are readily attracted to lights. Eggs are laid on underwater structures in such great numbers that they are harvested for food in some tropical countries. Most northern species are probably univoltine, with overwintering adults returning to breeding sites in spring to mate and oviposit. In the south, two or more generations per year are likely. There is evidence that at least one species overwinters as an egg.

c. Naucoridae (4 Genera, 22 Species) "Creeping water bugs" are most common in the southern United States and are rarely found as far north as Canada. They are dorsoventrally flattened, have greatly expanded profemurs, and have a rounded appearance when viewed from above, with margins of the head, pronotum and elytra being continuous (Fig. 93). They are usually associated with lotic habitats, living in streams, spring ponds, or impoundments. Eggs are glued to vegetation or mineral substrates, depending on the species. Most species probably are univoltine and overwinter as adults in deeper areas of their breeding habitat. Because adults carry a large plastron and generally live in well-oxygenated water, they rarely need to surface for air. Although adults of many species have fully developed wings, flight rarely has been observed.

d. Nepidae (3 Genera, 13 Species) "Water scorpions" can be readily recognized by their long apical breathing tube (Fig. 91). They breed in permanent lentic habitats, especially shallow areas with much vegetation. Here they often remain with their breathing tube above the surface waiting to ambush invertebrates or small vertebrates, but they are capable of remaining completely submerged, using air stores under their hemelytra for respiration. Unlike other aquatic Heteroptera, adults and nymphs are very poor swimmers and usually cling to vegetation or other substrates. Most species are univoltine in northern latitudes and probably bivoltine in the south. They overwinter as adults, which in the north usually fly to deeper lentic or lotic habitats in autumn and return to breeding sites in late April or early May. Eggs that are laid on supports at the water surface have respiratory horns that protrude from the water. When flooded, the eggs respire through a plastron.

e. Notonectidae (3 Genera, 32 Species) "Backswimmers," as their name implies, swim ventral side up, but orient themselves dorsal side up when not in the water. They most resemble Corixidae, but are deep-bodied, not at all flattened (Fig. 95), and have an elongate, segmented rostrum (Fig. 90). Adults of most species breed in littoral areas of lakes and permanent ponds, others in stream pools; some species are transients in temporary ponds. Adults and nymphs of two genera, *Buenoa* and *Notonecta*, can inhabit relatively deep water and remain submerged for long periods because they have ventral channels to retain a plastron. *Notonecta* adults and nymphs typically hang from the surface film, while *Buenoa* adults and nymphs have hemoglobin, which enables them to maintain neutral buoyancy and remain planktonic. Adults and nymphs in both genera are active predators that pursue or ambush other invertebrates or small vertebrates. A third genus, *Martarega*, occurs only in extreme western United States, where adults and nymphs inhabit eddies of streams and ambush prey carried by the current. Several species overwinter as diapausing eggs and have a univoltine life cycle with nymphal development in late spring and summer. Others have univoltine or bivoltine life cycles, with overwintering adults that mate and lay eggs in the spring. In the north, adults of a few species may overwinter in larger streams and return to lentic habitats in spring.

f. Pleidae (2 Genera, 5 Species) "Pygmy backswimmers" are very small (<3 mm), convex insects that swim upside down. Adults do not have typical hemelytra, but instead their mesothoracic wings are shelllike (Fig. 94) and resemble elytra of Coleoptera. Their metathoracic wings are vestigial and they cannot fly. Adults and nymphs inhabit vegetation, mostly in permanent ponds, but also may be found in littoral areas of lakes, backwaters of streams, and swamps. If their habitat dries because of drought, they can survive for several months in the moist soil. Pleids are widespread and often abundant, except in the Pacific Coast region where they do not occur. Although adults and nymphs can swim quite well, they mostly crawl about on vegetation and feed on small invertebrates. Because of their small body size and relatively large plastron, they are able to remain submerged for long periods. Life cycles are bivoltine or multivoltine, but their life history is poorly known.

B. Aquatic Neuroptera—Spongillaflies

Neuroptera is a fairly large terrestrial order, but larvae in only one family are aquatic. They live in and feed upon several genera of freshwater sponges (Spongillidae) that occur in permanent lotic or lentic habitats. Taxonomic keys to species of larvae can be found in the following publications: Parfin and Gurney (1956), Poirrier and Arceneaux (1972), and for North and South Carolinas in Brigham (1982b). Information on the biology, habitat, and identification of larvae is presented next.

1. Families of Aquatic Neuroptera

a. Sisyridae (2 Genera, 6 Species) The small (<8 mm) aquatic larvae are distinctive. They have sucking mouthparts in the form of a pair of long stylets and are covered with long, spinose setae (Fig. 102). Most species are probably multivoltine, with larvae having only three instars. Other life stages are terrestrial. Eggs are usually laid in very small masses, mostly in crevices of vegetation or supports above aquatic habitats that support freshwater sponges. After an incubation period of about 1 week, the eggs hatch and larvae drop into the water where they swim or are carried by currents until they encounter a suitable species of sponge. After completing development on the sponge, they swim or crawl to shore, crawl out of the water, and find a suitable pupation site in a crack or crevice or on a flat surface that may be several meters from the water. Here, they spin a double-walled cocoon in which they pupate, the outer wall having a distinctive meshlike construction. After several days, pharate adults use pupal mandibles to cut their way out of the cocoon and emerge. Adults feed on nectar and apparently live no more than a few weeks. Most species overwinter as larvae on sponges, but some may overwinter as prepupae in the pupal cocoon.

Larvae have no special adaptations for the aquatic environment. They obtain oxygen from the water through their cuticle and swim by repeatedly arching their abdomen forward while in a vertical position and snapping it back. Larvae are rarely collected by standard sampling techniques. Substantial numbers can be collected only by collecting sponges and removing larvae that crawl from them as they dry. Rearing has been accomplished, but a suitable colony of freshwater sponges must be maintained and a terrestrial habitat for pupation must be provided.

C. Aquatic Lepidoptera—Aquatic Caterpillars

Larvae in several genera of this large, primarily terrestrial order, are associated with aquatic habitats. Some feed on emergent parts of aquatic macrophytes and others mine stems of these plants. Still others feed on submerged parts of plants attached to other substrates, and only these will be considered here. The 17 aquatic genera in which aquatic larvae occur are all in the family Pyralidae, and almost all are in the subfamily Nymphulinae.

Aquatic caterpillars have three pairs of thoracic legs and pairs of short prolegs ringed with hooklike crochets on abdominal segments 3–7 (Fig. 103). They are not likely to be confused with any other aquatic insect larvae, but often it is not possible to distinguish them from Lepidoptera that are inadvertently collected from emergent vegetation. Only those that have numerous filamentous gills covering their body and those that live in portable cases made from the aquatic vegetation that they inhabit, can be recognized with certainty as being aquatic. Larvae that live in cases are so well camouflaged that they often escape notice, even after capture with a net. Some species may become sufficiently abundant to have an impact on aquatic plant communities. While adults have been well studied and larvae of many of the species have been reared and described, reliable keys to species of larvae have not been constructed because larvae of many species remain unknown.

Larvae inhabit a wide variety of permanent aquatic habitats. Some inhabit rocks in rapid water of larger streams where they feed on diatoms and other algae. Others inhabit aquatic macrophytes or duckweed (*Lemna*) in ponds or along margins of streams and lakes. Here, they feed on the plants they inhabit and make cases from them. Respiration is cutaneous, and is enhanced by numerous filamentous gills in larvae that inhabit rocks in streams and also in some lentic casebuilders. Larvae of at least one species have hydrofuge hairs enabling them to maintain a plastron that is used as a physical gill. Pupation of aquatic species is in a cocoon, usually in the larval habitat. Emergence occurs after pupae swim or crawl to the surface of the water; legs of species that swim have long swimming hairs. Adults are generally nondescript small moths that hold their wings rooflike over their body and remain in the vicinity of their larval habitat. Adults of stream species crawl under the water to oviposit on rocks, while those of most lentic species deposit rows of eggs just below the water surface on the preferred food plant. Most species are univoltine or bivoltine and have five larval instars.

D. Aquatic Coleoptera—Water Beetles

Only about 3% of Coleoptera species have an aquatic stage, but because this is the largest insect order they represent a significant segment of the aquatic insect fauna. About 1450 aquatic species in 19 families have been found in North America. There are also several species of riparian Coleoptera that occasionally may be collected while sampling margins of aquatic habitats. In eight families both larvae and adults are aquatic in almost all species, and in nine others either all larvae or all adults are aquatic. A few families with mostly terrestrial species also have some species with larvae and/or adults that are adapted for living in the aquatic environment.

Three suborders of Coleoptera have aquatic representatives. Adephaga is represented by five families in which all species are aquatic both as larvae and adults. These families are often referred to as the "Hydradephaga". The suborder Myxophaga is represented in North America by a single aquatic species, *Hydroscapha natans*, in which both larvae and adults are aquatic. Several families of Polyphaga have some species with one or more aquatic stages, but only in a few families do most species have a tie to the aquatic environment.

Most aquatic beetles can be collected with an aquatic net by using standard collecting techniques. However, larger Dytiscidae and Hydrophilidae are ineffectively sampled with nets, and bottle traps must be relied on to capture substantial numbers of these larger beetles and also species that are nocturnal (Hilsenhoff, 1987, 1991). Adult Gyrinidae, which mostly swim on the surface in deeper water, are seldom captured without special effort. In late summer, large schools often appear in shallow bays of lakes and streams, and by herding them toward shore and sweeping through the school, large numbers often can be captured. In the north, where ponds and lakes freeze, lentic as well as lotic gyrinids congregate in streams in the fall to overwinter. Here, very large numbers of mixed species can be collected with a net from undercut banks where there is a current and the water is at least 0.5 m deep. Some overwintering lentic species of Dytiscidae, Haliplidae, and Hydrophilidae also may be found along stream banks in autumn and early spring. Large numbers of Hydrophilidae, Helphoridae, and Hydrochidae adults can often be collected by black-light traps on warm nights in spring and summer.

Adult water beetles range from 1–40 mm in length and vary greatly in structure. All are characterized by chewing or biting mouthparts and shell-like mesothoracic wings that are called elytra (Fig. 108). They have functional spiracles and rely on atmospheric oxygen for respiration. Adult Adephaga carry a bubble of air under their elytra, which is renewed at the surface by swimming or crawling (Amphizoidae) to the surface and breaking the surface film with the tip of the abdomen. This air supply must be frequently renewed, except when water temperatures are cold and metabolism is slowed, or when the water is well oxygenated. Hydroscaphidae adults (Myxophaga) crawl to the surface to renew their air supply. Elmidae and Dryopidae adults (Polyphaga) usually live in oxygen-rich water and maintain a plastron of air that is carried on hydrofuge hairs. The plastron acts as a physical gill, with oxygen for respiration diffusing into it from the water, and carbon dioxide from respiration diffusing out into the water. Other Polyphaga adults (Helophoridae, Hydrochidae, Hydrophilidae, Hydraenidae, and Curculionidae) also maintain an air supply on hydrofuge hairs, but must renew the air by swimming (most Hydrophilidae) or crawling (Curculionidae, Hydraenidae, Helophoridae, Hydrochidae, and some Hydrophilidae) to the surface when the water contains too little oxygen for plastron respiration. Unlike Adephaga, adult Polyphaga break the surface film with their antennae.

Larvae of aquatic Coleoptera vary greatly in size and general morphology. All have three pairs of thoracic legs, chewing or biting mouthparts, and are relatively well sclerotized, especially in the later instars. Some have lateral filaments and resemble larvae of Megaloptera. Most larvae of aquatic Coleoptera rely on cutaneous respiration, which may be enhanced by gills, although some larger Dytiscidae and Hydrophilidae larvae obtain oxygen at the water surface through functional posterior spiracles. Larvae of Chrysomellidae and Noteridae are able to utilize oxygen from plant tissues, and by this mechanism are able to pupate under water. It is also likely that larvae and adults of *Celina* (Dytiscidae) obtain oxygen from plants (Spangler 1973; Hilsenhoff, 1994). Most Coleoptera larvae lack spiracles in the early instars, but are equipped with a full complement of spiracles in the final instar, which allows them to leave the water and pupate in cells on land.

Adephaga adults (except Amphizoidae) and many Hydrophilidae adults are adapted for swimming by having long setae or swimming hairs on at least their metathoracic legs or by having greatly flattened legs (Gyrinidae). Some larvae of Dytiscidae are also adapted for swimming by having long setae on their legs, but larvae of most aquatic Coleoptera crawl on vegetation or other substrates. Gyrinidae larvae, which mostly crawl on vegetation, can swim rapidly with an undulating motion when disturbed.

Much of the taxonomy of North American water beetles was completed more than 50 years ago, but some genera need revision and several species remain undiscovered. Adults of almost all species can be reliably identified, but their distribution remains poorly defined in North America because in many areas aquatic beetles have not been adequately studied.

Unlike adults, very few larvae can be identified to species and a few cannot even be identified to genus. In most genera, larvae of too few species are known to permit development of reliable keys. Fortunately, the European fauna is much better known, and the European literature gives us insight into probable life cycles, larval and adult habitats, overwintering sites, feeding habits, and larval taxonomy. Development of keys and descriptions for aquatic Coleoptera larvae remains a pressing need. To accomplish this, larva–adult associations must be obtained through laboratory rearing. The most promising method for doing this is to capture mated females, or male–female pairs, allow females to oviposit, and then rear larvae that hatch from the eggs. By using this *ex-ovo* method (Alarie *et al.*, 1989), all larval instars can be associated with the adult.

Because of our inability to identify larvae, life histories of most North American species remain poorly known. This includes our knowledge of adult habits, especially such things as feeding, dispersal, overwintering, and estivation. From the occurrence of teneral adults and collection records for adults and larvae (if identified), life cycles and overwintering sites have been postulated for many species of Wisconsin Dytiscidae, Helophoridae, Hydrochidae, and Hydrophilidae (Hilsenhoff (1992, 1993 a-c, 1994, 1995a,b,d). Much research is needed before we will understand the bionomics of the many species of aquatic Coleoptera in North America.

References to species keys for the identification of Coleoptera adults and larvae are provided in Table VII. Keys to families of freshwater Coleoptera adults and

TABLE VII Literature References to Species Keys for Identification of Coleoptera Adults and Larvae

Regional keys to adults that include more than one family
- California–Leech and Chandler (1956)
- Florida–Young (1954); Epler (1996)
- Maine–(Adephaga and Hydrophilidae) Malcolm (1971)
- North Dakota–Gordon and Post (1965)
- North and South Carolina—Brigham (1982c)
- Pacific Northwest—Hatch (1953, 1965)

Keys to adults listed by family or superfamily
- Amphizoidae—Kavanaugh (1986)
- Dryopoidea—Brown (1972); (Louisiana) Barr and Chapin (1988); (Wisconsin) Hilsenhoff and Schmude (1992)
- Dytiscidae—(Alberta) Larson (1975); (Canada and Alaska) Larson *et al.* (2001); (Virginia) Michael and Matta (1977); (Utah) Anderson (1962); (Wisconsin) Hilsenhoff (1992, 1993a, b, c, 1994, 1995a)
  - *Acilius*—Hilsenhoff (1975)
  - *Agabus*—Fall (1922); Larson (1989, 1991, 1994, 1996, 1997); Larson and Wolfe (1998); (Pacific Coast) Leech (1938)
  - *Celina*—Young (1979a)
  - *Colymbetes*—Zimmerman (1981)
  - *Copelatus*—Young (1963a)
  - *Coptotomus*—Hilsenhoff (1980)
  - *Desmopachria*—Young (1981a, 1981b)
  - *Dytiscus*—Roughley (1990)
  - *Graphoderus*—Wallis (1939)
  - *Heterosternuta*—Matta and Wolfe (1981)
  - *Hydaticus*—Roughley and Pengelly (1981)
  - *Hydroporus*—Gordon (1969, 1981)

(Continues)

TABLE VII (Continued)

Hydrovatus—Bistrom (1996), Young (1956, 1963b)
Hygrotus—Anderson (1970, 1975, 1983)
Ilybius—Larson (1987)
Laccophilus—Zimmerman (1970)
Laccornis—Wolfe and Roughley (1990)
Liodessus—Larson and Roughley (1990); Miller (1998)
Lioporeus—(as *Falloporus*) Wolfe and Matta (1981)
Matus—Young (1953)
Nebrioporus—Shirt and Angus (1992); (as *Deronectes*) Zimmerman and A.H. Smith (1975)
Neoporus—(as *Hydroporus*) Fall (1923); Wolfe (1984)
Oreodytes—Zimmerman (1985); Alarie (1993)
Rhantus—Zimmerman and R. L. Smith (1975)
Sanfilippodytes—(as *Hydroporus*) Fall (1923)
Stictotarsus—(as *Deronectes*) Zimmerman and A. H. Smith (1975)
Thermonectus—McWilliams (1969)
Elmidae
Dubiraphia—Hilsenhoff (1973)
Optioservus—White (1978)
Stenelmis—Schmude (1992), (1999)
Gyrinidae—(Minnesota) Ferkinhoff and Gundersen (1983); (Quebec) Morrissette (1979); (Wisconsin) Hilsenhoff (1990)
Dineutus—Hatch (1930); Wood (1962)
Gyrinus—Oygur and Wolfe (1991)
Haliplidae—(Minnesota) Gundersen and Otremba (1988); (Virginia) Matta (1976); (Wisconsin) Hilsenhoff and Brigham (1978)
Haliplus—Wallis (1933)
Peltodytes—Roberts (1913)
Helophoridae—Smetana (1985); (Pacific Northwest) McCorkle (1965)
Hydraenidae—Perkins (1980)
Hydrochidae—Hellman (1975)
Hydrophilidae (includes Helophoridae and Hydrochidae)—(Canada and Alaska) Smetana (1988); (Illinois) Wooldridge (1967); (Michigan) Willson (1967); (Mississippi) Testa and Lago (1994); (Nevada) LaRivers (1954); (Pacific Northwest) Miller (1965); (Virginia) Matta (1974); (Wisconsin) Hilsenhoff (1995b, d)
Anacaena—(Europe) Berge Henegouwen (1986)
Berosus—Van Tassell (1966)
Crenitis—Winters (1926)
Cymbiodyta—Smetana (1974)
Enochrus—Gundersen (1978)
Hydrochara—Smetana (1980)
Laccobius—Cheary (1971); Gentili (1985, 1986)
Paracymus—Wooldridge (1966)
Tropisternus—Spangler (1960)
Noteridae
Hydrocanthus—Young (1985)
Suphisellus—Young (1979b)
Keys to larvae listed by family
Chrysomelidae—(Michigan) Hoffman (1939)
Dytiscidae—(Wisconsin) Hilsenhoff (1992, 1993a, b)
Heterosternuta—Alarie (1992)
Hydroporus—Alarie (1991)
Hygrotus—Alarie *et al.* (1990)
Oreodytes—Alarie (1997)
Ptilodactylidae
Anchytarsus—Stribling (1986)

larvae appear in Sections V.H.1 and V.H.2. Information on suborders and the biology, habitat, and identification of adults and larvae in each family is presented in the following.

1. Families of Aquatic Coleoptera—Suborder Adephaga

In the five aquatic families of this suborder (Hydradephaga), both larvae and adults of all species are aquatic. Eggs are laid singly or in masses, mostly on objects beneath the water surface. Larvae have three instars, and except for at least some Noteridae, pupation takes place in terrestrial cells near the larval development site. Almost all first and second instar larvae lack spiracles and rely on cutaneous respiration for oxygen. A few have functional posterior spiracles that they use to obtain oxygen at the water surface. Third instar larvae, which have functional spiracles on abdominal segments and the mesothorax, also rely mostly on cutaneous respiration until they leave the water to pupate. Life cycles are typically univoltine, although bivoltine and semivoltine life cycles have been recorded for some species. Some species complete larval development rapidly, requiring only a few weeks; others require several weeks, and those with semivoltine life cycles may need almost two years to complete larval development. Adults are long-lived, with records of unmated beetles living for several years, but typically most adults live several months and die after their final mating and oviposition, which is usually in spring.

a. Amphizoidae (1 Genus, 3 Species) "Trout-stream beetles" are restricted to mountain streams in northwestern North America. Here, both larvae and adults can be found near the water edge, crawling on plant roots, wood, or debris at or just beneath the water surface. Larvae are characterized by short, broad urogomphi and flattened lateral projections on each segment (Fig. 120). Adults are relatively large (12–14 mm long) and broad, with a narrow thorax and no swimming hairs on their legs (Fig. 108). Larvae cannot swim and adults swim very poorly; both usually remain in contact with the substrate. Third instar larvae frequently stay in the splash zone and enter the water only to capture prey. Both larvae and adults feed almost exclusively on Plecoptera larvae. Because their habitat is well-oxygenated, adults may remain submerged for prolonged periods, their air bubble acting as a physical gill. The life cycle is 1 year, with eggs laid on debris in the splash zone in late summer and pupation occurring the following summer, probably in cells in the stream bank.

b. Dytiscidae (48 Genera, 600 Species) The "predaceous diving beetles" are the largest family of water beetles, with more than 550 species distributed throughout North America. Adults range from 2 to 40 mm in length and are characterized by their ovate appearance, rounded sternum and dorsum, and by a usually spear-shaped prosternal process that projects back to the mesocoxae (Fig. 106). The beetles swim by moving both metathoracic legs backward simultaneously. Larvae are generally elongate, and range from 3 to 60 mm in length at maturity. Most have paired terminal appendages (urogomphi) of various lengths (Fig. 122). They have relatively elongate legs, which in several genera are modified by long hairs for swimming.

Adults and larvae of most species are lentic, inhabiting all types of permanent and temporary habitats, and showing a preference for the shallow, vegetated margins of ponds, marshes, bogs, and swamps. They are most abundant in habitats that do not contain significant numbers of insectivorous fish. Most species breed in well-defined habitats, but adults may fly to other habitats to feed or overwinter. Adults and larvae of lotic species are found mostly under stream banks and tend to remain in the habitat in which they breed.

Typically adults overwinter, mate in the spring, oviposit, and die. Eggs are laid in or on vegetation beneath the water surface, or on other objects below or just above the surface. The larvae develop over a period of a few to several weeks, then leave the aquatic habitat to pupate in cells they construct in the soil of protected areas nearby. The pupal stage lasts 5–14 days, after which the adult emerges and usually re-enters the aquatic habitat. Most species are univoltine, but bivoltine life cycles occur, especially in the south, and there is evidence northern species that breed in cold-water habitats are semivoltine. However, life histories of many species remain poorly known, and there are variations in the life cycle described above. There is ample evidence for estivation of adults during dry periods in the summer and also evidence that several northern species overwinter as diapausing eggs. Many lotic species and a few lentic species overwinter as larvae, even in northern areas that freeze. While most overwintering adults are found in ponds, lakes, or streams, some pass the winter in terrestrial habitats and enter their breeding sites late in the spring. Both adults and larvae are predators, feeding primarily on other invertebrates and small vertebrates. Adults of most species are also scavengers.

c. Gyrinidae (4 Genera, 60 Species) Adult "whirligig beetles" are small- to medium-sized beetles (3.5–14.0 mm long). They are widespread and often abundant. Most frequently, they are seen resting or swimming in groups of a few to several hundred on the water surface in areas protected from the wind, especially in late summer. However, they can, and often do, swim under the surface. Their greatly flattened legs and two pairs of eyes are distinctive. Larvae are also distinctive, having fringed projections on each abdominal segment posteriorly (Fig. 119). They swim with a vertical undulating motion when disturbed.

While adults and larvae of most species are lentic, those of several species are lotic, and adults of most lentic species frequent larger streams in autumn, or in late summer when it is dry. In streams, adult gyrinids will often escape the current by crawling onto vegetation just above the water. In northern regions, where ponds and lakes freeze, adults of most lentic species fly to larger streams or lakes to overwinter. They return to breeding sites in early spring, as soon as air temperatures have warmed enough to permit flight. All species tend to inhabit deeper water than other Hydradephaga, with larvae being found mostly among submerged vegetation in large ponds or rivers, or in littoral zones of lakes and impoundments. Larvae are predators, feeding mostly on other invertebrates, while adults are scavengers on dead animals or predators of small invertebrates in the surface film. Most species are probably univoltine, with adults overwintering and larvae developing during the summer. Eggs are laid mostly on submerged aquatic vegetation. Pupation is in cocoons on emergent vegetation, or on terrestrial vegetation adjacent to the larval habitat.

d. Haliplidae (4 Genera, 70 Species) "Crawling water beetles" are small (2.5–5.0 mm long) yellowish to orangish beetles with black blotches and spots, apically narrowed elytra, and a narrowed thorax. Their large metacoxal plates are distinctive (Fig. 105). They are often abundant in shallow lentic and lotic vegetation-choked habitats. In spite of their common name, they have fringed tarsi and tibiae and are able swimmers. The larvae, which crawl about on vegetation, are either elongate with a long terminal segment (Fig. 123), or have numerous long dorsal projections (Fig. 124). Life cycles of most species are probably univoltine, with eggs being deposited on or in filamentous algae, or other vegetation. Both larvae and adults are herbivores, feeding on algae or macrophytes. Pupation takes place in cells in moist soil near the larval development site. Most species overwinter in the water as adults, but some adults and larvae are known to overwinter in terrestrial sites adjacent to the water.

e. Noteridae (5 Genera, 18 Species) This small family, frequently referred to as "burrowing water beetles," is generally southern in its distribution and rarely found as far north as Canada or in the west. Adults are small (1.2–5.5 mm long) and resemble Dytiscidae, but are usually more attenuate apically than most Dytiscidae. They have a flat, V-shaped metasternal plate and a truncate, instead of a spear-shaped or rounded, prosternal process (Fig. 111). Like Dytiscidae, adults are mostly predators that live in close association with aquatic vegetation. The larvae, which are poorly known, have short legs, short urogomphi, and telescoping body segments (Fig. 121). Their morphology suggests that they are omnivores, which live among roots of aquatic vegetation. Larvae of a European species obtain oxygen from plant tissues and pupate in a cocoon on the plant roots, the pupae also obtaining oxygen from the plant. Life cycles are probably one year, with adults overwintering in the water.

2. Families of Aquatic Coleoptera—Suborder Myxophaga

a. Hydroscaphidae (1 Genus, 1 Species) The only North American species occurs from the western United States, north to Idaho. The fusiform shape, truncate elytra, and very small size (1.5 mm) of adult "skiff beetles" is distinctive (Fig. 118). Larvae and adults are found most commonly on algae over which a thin film of water flows, but adults have also been collected in deeper water. Both adults and larvae are herbivores, feeding on algae. They tolerate a wide range of water temperatures, and have been found in thermal springs as warm as 46°C and in cold streams that frequently freeze. Females lay single, well-developed eggs on algae, and pupation also takes place on algae at the air–water interface. Larval respiration is aided by distinctive pairs of balloon–like spiracular gills on the prothorax, and first and eighth abdominal segments. Adults, which may remain submerged for long periods, retain an air bubble that acts as a physical gill.

3. Families of Aquatic Coleoptera—Suborder Polyphaga

There are four families in which both aquatic adults and larvae are found. In Elmidae, adults and larvae of almost all species are aquatic, while in Hydrophilidae most species have aquatic adults and many also have aquatic larvae. Curculionidae has several species with aquatic adults and a few also with aquatic larvae, while Chrysomelidae has several species with aquatic larvae and two also with aquatic adults. In Dryopidae, Hydraenidae, Helophoridae, and Hydrochidae all adults are aquatic, and in Lutrochidae, Psephenidae, Ptilodactylidae, and Scirtidae almost all larvae are aquatic. Larvae in at least one species of Lampyridae are also aquatic. Additional families of Polyphaga have species that are associated with water, but not adapted for living in it, and these families will not be treated here.

Five families (Dryopidae, Elmidae, Lutrochidae, Psephenidae, and Ptilodactylidae) form a natural group of beetles in which almost all species have at least one aquatic stage. They formerly were placed in the superfamily Dryopoidea and will be discussed first (Section 3.a–e), although they now are included in a larger superfamily (Byrrhoidea) (Lawrence and Britton, 1991). They differ from most other aquatic beetles by generally having more larval instars, longer life cycles, and longer adult lives in species with aquatic adults. Aquatic adults and larvae are unable to swim, which confines them to well-oxygenated habitats. Aquatic larvae obtain oxygen from the water through their integument, usually with the aid of gills. Aquatic adults have a permanent air bubble (plastron) that is carried on hydrofuge hairs and acts as a physical gill. Terrestrial adults of species with aquatic larvae typically inhabit the splash zone on rocks that project from their aquatic habitats, and they also use a plastron to remain submerged for relatively long periods when ovipositing. In Dryopoidea that have been studied, five-to eight-larval instars have been recorded, but there is disagreement about the actual number of larval instars, with six being suggested for many species. In Adephaga, Helophoridae, Hydrochidae, Hydrophilidae, and Hydraenidae there are only three larval instars. Larval development probably takes about 2 years for most species of "Dryopoidea", with 1–3-year development times having been suggested for various species. The shorter times are for species in warmer and more southern streams. Aquatic adults typically live a year or more, and some may live several years; terrestrial adults of species with aquatic larvae typically live only a few days to a week. To date, studies of life histories have been mostly on Elmidae and Psephenidae; additional studies are needed. Brown (1987) reviewed the biology of these riffle beetles.

Larvae of "Dryopoidea" are poorly known, and identification beyond genus is usually not possible. Brown (1972) summarized what was known about the distribution and taxonomy of Dryopoidea in the United States, and provided species keys for adults and generic keys for larvae. Since his publication, the aquatic Limnichidae have been recognized as a separate family, Lutrochidae. Several new species have been discovered recently, and revisions are needed for some genera.

a. Dryopidae (5 Genera, 14 Species) In *Helichus*, the most common and widespread genus of "long-toed water beetles," only adults are aquatic, living among debris and on decaying wood in streams. Adults in other genera are also probably aquatic, with those in two genera being lotic and those in the other two genera lentic. Larvae and adults resemble those of other

Dryopoidea. Adults are larger than the more abundant Elmidae and have a pectinate antenna (Fig. 116). Larvae are elongate and lack gills in the operculate caudal chamber. Little is known about life histories of species in this family.

b. Elmidae 27 Genera, 101 Species) Commonly referred to as "riffle beetles," Elmidae are widespread and often abundant. Both larvae and adults are usually aquatic and often occur together; in a few species, adults are riparian. Adults are less than 4.5 mm long, smaller than Dryopidae adults, and have filiform or slightly clubbed antennae (Fig. 117). The sclerotized larvae are elongate, rounded in cross section, and have a ventral caudal operculum that closes a chamber containing hooks and numerous filamentous gills (Fig. 132). Both adults and larvae are found mostly in streams, where they inhabit a variety of substrates, including gravel riffles, algae laden rocks, aquatic macrophytes, and decaying wood. Adults and larvae of some species may also occur on these same substrates in spring ponds or along wave-swept shores of lakes. Larvae and adults are herbivores–detritivores, feeding on algae, decaying wood, and detritus. When larvae complete their development they leave the water and pupate in cells in protected areas on the adjacent shore. Upon emergence, adults disperse widely and frequently are captured in light-traps. Once they find a suitable aquatic habitat, they rarely if ever fly again, but stream species may move downstream by drifting in the current. Most adults probably live a year or more, and a semivoltine life cycle seems probable for most species in the north.

c. Lutrochidae (1 Genus, 3 Species) Larvae of Lutrochidae occur on calcareous encrustations in hardwater streams in eastern, central, and western United States. They closely resemble larvae of Elmidae, but the last abdominal tergum is rounded apically (Fig. 134) instead of being emarginate (Fig. 133), and thoracic sterna are membranous. They apparently feed on periphyton, but little is known about the bionomics of this family. Adults are riparian, inhabiting splash zones of larger rocks that project from the aquatic habitat.

d. Psephenidae (5 Genera, 14 Species) "Water pennies" are flat, rounded larvae (Fig. 128) that occur on rocks and wood in streams from Texas and Arizona north to British Columbia, and east of the Great Plains; they may also be found on wave-swept shores of lakes. In two genera (*Psephenus* and *Eubrianax*) larvae lack the ventral caudal operculum of other aquatic dryopoid larvae, but have ventral abdominal gills that aid in respiration. The remaining genera are in the subfamily Eubriinae, which is sometimes considered a separate family. Their larvae lack exposed ventral gills, possess a ventral caudal operculum, and have the lateral margins of each segment distinctly separated (Fig. 127). Pupation of most species occurs above the water line on rocks near the larval habitat, but at least one species is known to pupate under water. Adults are riparian, typically being found in splash areas of projecting rocks. Most species probably have life cycles of 1–2 years.

e. Ptilodactylidae (3 Genera, 3 Species) Larvae have been collected from streams and springs in California and Nevada, and from Georgia to New York, but not in central North America. They are elongate, lack a ventral caudal operculum, and have gills either on the first seven abdominal sterna or on the ninth sternum (Fig. 131). Very little is known about their life histories. Adults of aquatic species are riparian.

f. Chrysomellidae (6 Genera, about 70 Species) In this very large terrestrial family there are several species in which larvae and adults feed on emergent aquatic vegetation, but there is only one subfamily (Donaciinae) in which aquatic larvae feed on submerged portions of aquatic plants. They are adapted for underwater existence by having a pair of caudal spiracular spines that they insert into plant tissues to obtain oxygen, and they can be recognized by these conspicuous spines (Fig. 130). Larvae pupate on underwater portions of plants, with pupae also obtaining oxygen from plants for respiration. Adults of at least two species are also aquatic, feeding on plant stems.

g. Curculionidae (20 Genera, about 150 Species) In this very large terrestrial family, adults of species in several genera feed on aquatic plants, both above and below the water surface, and there is no easy way to distinguish truly aquatic species. Adult Curculionidae are readily distinguished by their unique head, which is elongated into a snout, and by their geniculate antennae (Fig. 104). Larvae are grublike, without distinct legs; most mine leaves and other plant tissues and are not aquatic. Aquatic adults crawl on underwater portions of plants and carry with them a plastron for respiration; a few species are capable of swimming. Most aquatic species are nocturnal, and difficult to find during the day, when they apparently hide in the bottom substrate. Life histories are poorly known.

h. Hydraenidae (3 Genera, 82 Species) Adult "minute moss beetles" are aquatic, living mostly in extremely shallow water along margins of streams, but also occurring along margins of lentic habitats. Adults resemble Hydrophilidae, and once were considered part of that family. They are much smaller (<2 mm long) than most Hydrophilidae and have five segments in the antennal club instead of three. Larvae are riparian, developing and pupating in moist soil adjacent to the water. Adults cannot swim, and float upside down in the surface film when dislodged from the substrate on which they live. Because of their very small size they are easily overlooked. Their life history and feeding habits are poorly known.

i. Helophoridae (1 Genus, 41 Species) ***Helophorus*** was previously included in the family Hydrophilidae, but a study by Archangelsky (1998) convinced me to include it in a separate family. Adults are aquatic, but are poorly adapted for swimming and mostly crawl about on aquatic vegetation in a variety of lentic habitats or along margins of streams. Larvae inhabit riparian habitats adjacent to aquatic habitats where adults are found. Adults are small, elongate beetles that are readily recognized by the seven longitudinal grooves and six raised areas on the pronotum (Fig. 114). The univoltine life cycles are variable, with adults or eggs overwintering in riparian habitats, depending on the species (Hilsenhoff, 1995b).

j. Hydrochidae (1 Genus, 19 Species) ***Hydrochus*** also was included in Hydrophilidae, but is now considered to be in a separate family (Archangelsky, 1998). Like Helophoridae, adults are aquatic, but swim poorly and mostly crawl about on the aquatic vegetation of shallow lentic habitats. Occasionally they occur along margins of streams. Life cycles are univoltine, with adults overwintering in terrestrial habitats, flying back to aquatic habitats in spring, and ovipositing in adjacent riparian habitats where the larvae develop.

Adults are readily identified by their small size and a very narrow pronotum that is covered with small granules and/or punctures (Fig. 115).

k. Hydrophilidae (19 Genera, 160 Species) Adults and larvae of this large and abundant family mostly inhabit shallow lentic habitats, although a few species are found in deep water and others are strictly lotic. In several genera only adults are aquatic, and in one subfamily (Sphaeridiinae) none of the life stages are aquatic. While commonly called "water scavenger beetles," all aquatic larvae are predators and adults feed on algae and perhaps small animals, as well as scavenging. Adults range in size from 1.5 to 40 mm and somewhat resemble Dytiscidae, but are flat ventrally, have club-shaped antennae (Fig. 113), and very long maxillary palps. The predatory larvae have large, conspicuous mandibles and relatively long antennae and maxillary palps (Fig. 125).

Adults in many genera mostly crawl on aquatic vegetation and are poorly adapted for swimming. In genera with adults that swim, beetles have fringes of long setae on their meso- and metathoracic legs, which they move alternately instead of in unison as in the Dytiscidae. While adults of most species can be readily collected with an aquatic net, those of larger species that swim rapidly are most effectively collected with bottle traps. Most species are univoltine, with adults being the normal overwintering stage. Adults of several species overwinter in terrestrial habitats, while others that inhabit shallow lentic habitats may fly to streams or larger lentic habitats in late summer and autumn to overwinter.

l. Lampyridae Larvae of at least one species of "firefly" have been collected from shallow lentic and semilotic habitats. They are readily recognized by their flat, platelike thoracic nota that mostly or completely conceal the head from above, and by gills ventrally on the last abdominal segment (Fig. 129). Larvae are predaceous, feeding on snails or other insects.

m. Scirtidae (8 Genera, 35 Species) Although adults are terrestrial, larvae of "marsh beetles" are aquatic and often common in marshes, swamps, margins of shallow ponds, tree holes, and a variety of other lentic habitats where they crawl about on the aquatic vegetation. Their elongate filiform antennae are unique among aquatic Coleoptera larvae (Fig. 126). They are apparently detritivores, with little being known about their life history.

E. Aquatic Diptera—Flies and Midges

Diptera is a large, mostly terrestrial order, but larvae and pupae of numerous species are aquatic, and it is the dominant order of insects in the aquatic environment. Diptera larvae inhabit all types of aquatic environments, and because of very short development periods in many species, they are able to utilize ephemeral aquatic habitats as well as permanent ones. About 40% of all species of aquatic insects belong to this order, and at least a third of the aquatic Diptera species are in one family, Chironomidae. It is impossible to accurately estimate the number of species of aquatic Diptera, because in most families only terrestrial adults can be identified and in many families it is not possible to know which species have aquatic larvae. It is also difficult to estimate the relatively large number of aquatic species that probably remain undiscovered and undescribed.

The order is divided into two suborders, Nematocera and Brachycera (Agriculture Canada, 1981), with

Nematocera dominating the aquatic fauna. Formerly a third suborder, Cyclorrhapha, was recognized; it is now included in Brachycera as the infraorder Muscomorpha. In Nematocera there are several families in which all or almost all species have aquatic larvae and pupae. In Brachycera most families that have species with aquatic larvae also have large numbers of terrestrial or riparian species. Larvae of the infraorder Muscomorpha are often referred to as cyclorrhaphous Brachycera, with the remaining Brachycera being called orthorrhaphous Brachycera.

Diptera larvae lack segmented thoracic legs, and thus can be readily distinguished from larvae of other aquatic insects. In Nematocera larvae, a completely sclerotized, relatively round head capsule is present (except in most Tipulidae), and the mandibles, which usually have subapical teeth, move laterally. In Brachycera larvae the head is reduced to an internal skeleton (cyclorrhaphous Brachycera) or poorly formed and not rounded (orthorrhaphous Brachycera), and the mandibles (mouth hooks) move vertically and lack subapical teeth.

Diptera larvae have many adaptations that allow them to live in a wide variety of environments. Those that live in oxygen-rich environments usually rely on cutaneous respiration, which may be aided by gills that increase the surface area. In Chironomidae, larvae of some species have a hemoglobinlike pigment that aids in oxygen storage. Larvae in many families obtain oxygen at the water surface, aided by short or elongate caudal respiratory siphons that reach to the surface and may be retractile. Other larvae have fringes of hairs surrounding caudal spiracles that allow them to float in the surface film with their caudal spiracles exposed to the air. Some larvae repeatedly swim to the surface to obtain oxygen (Culicidae). Other larval adaptations include papillae for regulation of salt concentration, sucker disks to anchor lotic forms in rapid currents, and prolegs, pseudopods, or creeping welts on various segments to aid in locomotion.

Pupae of Nematocera often have anterior respiratory horns that enable them to obtain oxygen at the surface; others rely on cutaneous respiration. A few swim actively to the surface (Culicidae, Chaoboridae) or pupate just above the water line (Dixidae). Most orthorrhaphous Brachycera pupate on land. Cyclorrhaphous Brachycera and Stratiomyidae pupate in a puparium, which may be terrestrial or float at the water surface where oxygen is plentiful.

Aquatic Diptera larvae can be readily captured with an aquatic net as described earlier, and when placed in a pan with some water will usually swim or crawl free of the debris, allowing them to be collected. Active Nematocera larvae are difficult to capture with forceps, and the use of a small loop (30-mm diameter) of nylon mesh is often advantageous. Such a loop can be made by bending the end of a piece of thin wire into a loop at a right angle to the remainder of the wire (the handle) and gluing a piece of nylon stocking to the loop. Diptera larvae are relatively easy to rear because most have short life cycles and pupate in the water. Lentic species, which require little or no aeration, are especially easy to rear.

Diptera larvae are not only a very important part of almost all aquatic ecosystems, but adults in many aquatic families are important because they transmit disease or create severe nuisance problems. Culicidae (mosquitoes) are the most important group from a public health standpoint because adults transmit diseases of humans, birds, and other animals. Adults of this family, along with those of Tabanidae (horseflies and deerflies), Simuliidae (blackflies), and Ceratopogonidae (biting midges) may create severe nuisance problems because they bite humans and other animals. Nonbiting midges in the families Chironomidae and Chaoboridae may emerge in such tremendous numbers from larger lakes that they damage property and create public health problems by causing severe allergic reactions in some people.

Much research is needed on aquatic Diptera in North America. Many species remain to be described, and larvae of most described species remain unknown. Only in families that are of public health importance (Culicidae, Tabanidae, Simuliidae) or useful as a biological control (Sciomyzidae) are larvae well enough known that reliable keys to most species have been developed. Even in these families there are taxonomic problems, especially with separation of sibling species. Great strides in larvae taxonomy have been made in recent years, especially in the largest family, Chironomidae, but until much more is accomplished it will not be possible for us to fully understand the role of Diptera larvae in aquatic ecosystems. The two-volume *Manual of Nearctic Diptera* (1981, 1987) by Agriculture Canada provides generic keys as well as information about morphology, biology, behavior, classification, and distribution.

References to species keys for the identification of larvae are provided in Table VIII and keys to families of freshwater Diptera larvae appear in Sections V.I. Information on suborders, and the biology, habitat, adaptations, and identification of larvae in each family is presented next.

1. Families of Aquatic Diptera–Suborder Nematocera

More than half of the families in this suborder have aquatic representatives, and in 12 families all or

TABLE VIII Literature References to Species Keys for Identification of Diptera Larvae

Regional keys to larvae that include more than one family
- North and South Carolina—Webb and Brigham (1982)

Keys to larvae listed by family
- Blephariceridae
  - *Blepharicera*—Hogue and Georgian (1986)
- Ceratopogonidae—Glukhova (1977)
- Chaoboridae—Cook (1956)
  - *Chaoborus*—Saether (1970)
  - *Chaoborus*—(*Schadonophasma*) Borkent (1979)
- Chironomidae—references since Simpson (1982) and Wiederholm (1983); (Florida) Epler (1995)
  - *Brillia*—Oliver and Rousell (1983)
  - *Cricotopus*—Simpson *et al.* (1983)
  - *Dicrotendipes*—Epler (1987)
  - *Endochironomus, Endotribelos, Synendotendipes, Tribelos*—Grodhaus (1987)
  - *Eukiefferiella, Tvetenia*—Bode (1983)
  - *Guttipelopia*—Bilyj (1988)
  - *Pagastia*—Oliver and Roussell (1982)
  - *Paraphaenocladius*—Saether and Wang (1995)
  - *Polypedilum*—Boesel (1985); Maschwitz and Cook (1999)
  - *Stenochironomus*—Borkent (1984)
- Tanypodinae—(eastern United States) Roback 1985, 1986a, 1986b, 1987)
- Corethrellidae—Cook (1956)
- Culicidae—Carpenter and LaCasse (1955); Darsie and Ward (1981); (Canada) Wood *et al.* (1979); several regional keys published prior to 1981; (Florida) Breeland and Loyless (1982)
- Deuterophlebiidae—Courtney (1990)
- Nymphomyiidae—Courtney (1994)
- Sciomyzidae
  - *Dictya*—Valley and Berg (1977)
  - *Elgiva*—Knutson and Berg (1964)
  - *Hoplodictya*—Neff and Berg (1962)
  - *Pherbella*—Bratt *et al.* (1969)
  - *Sciomyza*—Foote (1959)
  - *Sepedon*—Neff and Berg (1966)
  - *Tetanocera*—Foote (1961)
- Simuliidae—(Alabama) Stone and Snoddy (1969); (Alberta) Currie (1986); (Colorado) Peterson and Kondratieff (1995); (Michigan) Merritt *et al.* (1978); (New York) Stone and Jamnback (1955); (Ontario) Wood *et al.* (1963); (Pennsylvania) Adler and Kim (1986); (southeastern United States) Snoddy and Noblet (1976)
  - *Ectemnia*—Moulton and Adler (1997)
  - *Gymnopais, Twinnia*—Wood (1978)
  - *Prosimulium*—(Canada and Alaska) Peterson (1970)
  - *Simulium*—Moulton and Adler (1995); (southern California) Hall (1974)
- Tabanidae—(Arizona) Burger (1977); (eastern North America) Goodwin (1987); Teskey (1969); Teskey and Burger (1976); (Illinois) Pechuman *et al.* (1983); (Louisiana) Tidwell (1973)
  - *Atylotus*—Teskey (1983)
- Tanyderidae—(Alberta) Exner and Craig (1976)
- Thaumaleidae
  - *Thaumalea*—(Idaho and California) Gillespie *et al.* (1994)

almost all species have aquatic larvae and usually also aquatic pupae. Unless noted otherwise, larvae have four instars, respiration is through the integument, and pupation is in the larval habitat.

a. Blephariceridae (5 Genera, 28 Species) The distinctive flattened larvae of "net-winged midges" have seven apparent body segments and a sucker disk ventrally on the first six segments (Fig. 135). They use their sucker disks to cling to rocks in very fast water of western and eastern mountain streams, and also streams in the northern United States and Canada. Here, they feed on diatoms and other algae. Larvae move forward slowly with an undulating motion, or sideways by alternately moving anterior and posterior sucker disks. Pupation occurs in cracks and depressions of rocks near the water surface. Adults live 1–2 weeks, with some females being predators on other midges. Eggs are laid just above the water on rocks, usually when water levels are receding. They hatch when flooded by higher water levels.

b. Ceratopogonidae (20 Aquatic genera, 501 Species) Aquatic larvae of "biting midges" typically are small, elongate, and without prolegs (Fig. 151). While many species are riparian or live in moist terrestrial habitats, a large number of species occur in aquatic habitats that include tree holes, marshes, swamps, ponds, all areas of lakes, and streams. Most larvae are able swimmers and move about with a serpentine swimming motion. They are poorly known and most have not been associated with adults, making it difficult even to know which species develop in aquatic habitats. Reliable keys to larvae of North American genera have not been developed, although a key to larvae in the U.S.S.R. (Glukhova, 1977) includes most genera found in North America. Most larvae are carnivores; others are herbivores or detritivores. Pupae of aquatic species hang in the surface film by their respiratory horns. Adults of a few aquatic species bite people and may become annoying pests in some areas; most others feed on small insects. Kettle (1977) reviewed the biology and ecology of blood sucking Ceratopogonidae.

c. Chaoboridae (3 Genera, 14 Species) Most larvae are nearly transparent, except for hydrostatic organs, causing them to be called "phantom midge larvae." They are unique in that their enlarged antennae have been modified for capturing prey such as insect larvae and small crustaceans (Fig. 148). They occur in a wide variety of lentic habitats, and are especially abundant in the profundal or sublittoral zones of some lakes. In lakes, third and fourth instar larvae rest in the bottom mud during daylight hours and prey on zooplankton in the limnetic area at night; earlier instars remain limnetic. They are the only insects frequently found in the limnetic area of lakes. Larvae also occur in permanent ponds, spring ponds, vernal ponds, and margins of swamps. Adults do not feed, but their synchronized emergences may create severe nuisance problems around large lakes because adults are highly attracted to lights. Life cycles are univoltine to multivoltine, depending on species, climate, and habitat. Most larvae can be identified to species.

d. Chironomidae (208 Genera, 2000+ Species) Chironomidae is by far the largest family of aquatic insects, with the number of aquatic species surpassing that in any other order. The larvae, which are recognized because they usually have anterior and posterior pairs of prolegs (Fig. 158), are diverse in form and size. They inhabit all types of permanent and temporary aquatic habitats, and a few species inhabit semiaquatic or terrestrial habitats. Larvae are often the dominant insects in the profundal and sublittoral zones of lakes, and consequently adults are sometimes called "lake flies," although they are most often referred to as "midges." In larger lakes, adults of some species may emerge in such tremendous numbers that they create nuisance problems along their shores. The short-lived adults cause allergies in some people, invade factories, spot the paint on houses, and accumulate in large, odorous piles.

Larvae are an extremely important part of aquatic food chains, serving as prey for many other insects and food for most species of fish. Life cycles are highly variable, with many species being univoltine and many others being bivoltine or multivoltine. In the north, life cycles are longer, with some arctic species requiring several years to complete development. Feeding habits also vary widely, with herbivores, detritivores and carnivores all commonly represented. Many larvae, especially predators, are free living, but those of most species construct loose cases of substrate cemented together with salivary secretions. Most are herbivores and detritivores that graze on fine particles on the substrate, but some are filter feeders, which construct webs to filter water that they circulate through their case or retreat. Larvae of most species are quite tolerant of lowered levels of dissolved oxygen; some can survive in areas where oxygen levels are so low that oxygen cannot be detected. Such species are usually red and contain a hemoglobinlike pigment that retains oxygen. These "blood worms" may become abundant in sewage lagoons or organically polluted areas of lakes or streams. The pupal stage is short, with pupae of most species being quiescent. After completing development, the pupa swims to the surface of the water where emergence occurs. Adults do not feed, and generally live no more than 2 weeks. The biology of the family was recently reviewed by Pinder (1986).

In recent years many new species have been named and great progress has been made in rearing and identifying larvae and pupae, but larvae of too many species remain unknown to permit reliable species keys to be constructed for larvae in most genera. A bibliography of taxonomic literature was compiled by Simpson (1982), and in 1983 Wiederholm keyed larvae to genus and species groups, provided comments on identification, ecology, and distribution, and also listed known species keys. Only keys to species or species groups that were not included by Simpson or Wiederholm are listed in Table VIII.

e. Corethrellidae (2 Genera, 5 Species) Until recently, this was a subfamily of Chaoboridae. Larvae occur in small pools, tree holes, or artificial containers in the southeastern United States north to New Jersey and west to California. Like Chaoboridae larvae, they have prehensile antennae, but the antennae are very close together basally and not wide apart as in Chaoboridae. Little is known about their biology.

f. Culicidae (13 Genera, 172 Species) Because they bite people and transmit disease, "mosquitoes" have been studied more thoroughly than any other family of aquatic Diptera. All species can be identified as adults or larvae, and life histories and habitats of most species have been documented. Larvae are readily recognized by their swollen thoracic area (Fig. 147), the presence of a caudal respiratory tube (siphon) in most genera, and their characteristic flip-flop swimming motion. Because of this swimming motion larvae are commonly called "wrigglers." Larvae occur in a variety of shallow lentic habitats that include tree holes, artificial containers, catch basins, pitcher plants, swamps, shallow temporary or permanent ponds and marshes, and heavily vegetated margins of lakes and streams. They are not found in moving water or water subjected to wave action. Larvae obtain oxygen at the water surface through a caudal siphon, which is very abbreviated in *Anopheles* (Fig. 147). Larvae of *Coquillettidia* and *Mansonia* attach their spinelike siphon to aquatic plants to obtain oxygen. Most larvae are active swimmers, and feed on detritus and microorganisms; larvae in two genera are predators, mostly on other mosquito larvae. Pupae are also active swimmers, and obtain oxygen at the surface through thoracic respiratory horns.

Many species are univoltine, with eggs that diapause over the summer, autumn, and winter, and larval development and emergence occurring in the spring. Many others are bivoltine or multivoltine. Although most species overwinter as eggs, a few overwinter as larvae and several species overwinter as adults. Most female adults obtain blood from mammals, birds, reptiles, or amphibians, depending on the species, and may live several weeks or months. Male mosquitoes do not feed on vertebrates, instead they obtain nectar from plants. Culicidae larvae often dominate the insect fauna of temporary ponds and marshes, especially those that flood in spring and summer.

g. Deuterophlebiidae (1 Genus, 6 Species) The unique larvae of "mountain midges" are less than 6 mm long and readily recognized by their elongate, branched antennae and the presence of seven pairs of stout ventrolateral prolegs on the abdomen (Fig. 153). They are found in rapid currents of western mountain streams. Here, they inhabit the upper surface of light-colored, smooth rocks that have cracks or depressions, and are usually at or near the water surface. Pupation is in the same habitat, usually in depressions in darker-colored rocks. In cold water at higher elevations, there is one generation per year, while at lower elevations there may be several generations. Eggs within the pupa are mature, and when adults emerge, mating and oviposition occur almost immediately. Adults live only an hour or two.

h. Dixidae (3 Genera, 42 Species) Larvae of "dixid midges" are found at or just below the water surface in lentic habitats or densely vegetated margins of streams. Here, they characteristically lie bent in a U-shape, often resting on vegetation. They can be recognized by pairs of prolegs on the first one or two abdominal segments (Fig. 154). Larvae obtain oxygen from the air through caudal spiracles and feed on microorganisms and detritus in the surface film. Pupation occurs just above the water surface on vegetation. Life histories are poorly known; most species are probably multivoltine.

i. Nymphomyiidae (1 Genus, 2 Species) The small, slender larvae are recognized by pairs of ventrally projecting, slender, two segmented prolegs on abdominal segments 1–7 and 9 (Fig. 152). They are rare, and in North America have been found only in Quebec, New Brunswick, and Maine. Larvae occur mostly on moss-covered rocks in small, rapid, cold, spring-fed streams, where they apparently feed on diatoms and other algae. Species are probably bivoltine, with emergences in spring and late summer.

j. Psychodidae (6 Aquatic genera, 67 Species) Most species of "moth flies" develop in semiaquatic or moist terrestrial habitats, but larvae of several species occur in aquatic habitats. Aquatic larvae can be recognized by their small size and secondary annulations on thoracic and abdominal segments, which often also have sclerotized dorsal plates on many annuli (Fig. 150). In one aquatic genus (*Maruina*), the larvae have ventral suckerlike disks. Aquatic larvae occur among vegetation and debris in shallow lentic habitats and along margins of streams, and are frequently associated with organically polluted water. They even occur in sink and floor drains. Larvae feed near the surface on microorganisms and detritus. Life histories of aquatic species are not well known; most are probably multivoltine, especially in warm-water habitats. Larvae cannot be identified to species.

k. Ptychopteridae (3 Genera, 16 Species) Larvae of "phantom crane flies" are easily recognized by their elongate caudal respiratory siphon and the presence of pairs of prolegs on the first three abdominal segments (Fig. 155). They live in very shallow water along margins of lentic or lotic habitats where there is an accumulation of detritus. Here, they obtain oxygen at the surface through their caudal respiratory siphon and probably feed on detritus and associated microorganisms. Their life history is poorly known because they are generlly uncommon.

l. Simuliidae (11 Genera, 143 Species) "Blackfly" larvae are uniquely shaped, having a swollen abdomen that they attach to the substrate with a caudal sucker (Fig. 156). They also have a relatively large head, and a single ventral proleg on the thorax. They inhabit a wide variety of lotic habitats where they are often abundant on rocks, submerged wood, or vegetation in fast to slow currents, each species usually having rather specific ecological requirements. Most larvae are filter feeders, using labral fans to filter diatoms and other food from the current. Larvae without labral fans feed by grazing on the substrate around them. Many species are univoltine, while others are bivoltine or multivoltine. Species overwinter as either eggs or larvae, with larvae of some species growing significantly during the winter months. Unlike most other Nematocera, which have only four larval instars, simuliid larvae have six or sometimes seven instars. Pupation occurs at the site of larval attachment, with pupae having a pair of highly branched thoracic gills that aid in respiration. Females usually require a blood meal for maturation of eggs, and various species may be serious pests, biting humans, livestock, and poultry. They also transmit pathogenic organisms such as filarial nematodes, various viruses, and avian protozoan parasites.

m. Tanyderidae (2 Genera, 4 Species) The elongate larvae of "primitive crane flies", which may reach a length of 18 mm, are rare in eastern and western North America and have not been found in central North America. They resemble chironomid larvae, but lack anterior prolegs and have three pairs of long caudal filaments (Fig. 157). Larvae inhabit the shallow sand, silt, or gravel margins of large streams, but, because they are rare, their life history and feeding habits remain virtually unknown.

n. Thaumaleidae (2 Genera, 8 Species) The very rare "solitary midge" larvae resemble chironomid larvae and reach a length of 12 mm, but have a single, broad anterior and posterior proleg instead of paired prolegs (Fig. 159). They also have dorsally sclerotized segments and a pair of short, stalked spiracles dorsolaterally on the prothorax. Larvae inhabit vertical surfaces of rocks in cold, shady, mountain streams, where the water film is thin enough so as not to cover them completely. Here, they feed mostly on diatoms. Almost nothing is known about their life history.

o. Tipulidae (20 Aquatic Genera, about 240 Species) Larvae of "crane flies" differ from those of other Nematocera by having the posterior portion of the head capsule incompletely sclerotized (Figs. 136 and 137) and retracted into the thorax. In this very large family, the vast majority of species are not aquatic, but there are several genera with species that have aquatic larvae, and these species are often widespread and abundant. Because larvae of fewer than 10% of the species have been described, it is difficult to know which species have aquatic larvae. While most aquatic larvae are found in lotic habitats, larvae in a few genera inhabit shallow lentic habitats. Larvae of most aquatic species have functional caudal spiracles that they may utilize to obtain oxygen at the water surface, but those that live in well-aerated streams probably obtain most of their oxygen from the water through their cuticle. Larvae in some genera are very large. Feeding habits vary widely, with omnivores, herbivores, detritivores and carnivores all represented. Most species are univoltine or bivoltine, but some that live in very cold water may be semivoltine. Pupation most often occurs in riparian habitats adjacent to the larval habitat, but some species pupate in shallow water and have thoracic respiratory horns through which they breathe. Adults are short-lived. The biology of Tipulidae was reviewed by Pritchard (1983).

2. Families of Aquatic Diptera—Suborder Brachycera

In larvae of Brachycera, either the head and external head structures are replaced by an internal cephalopharyngeal skeleton (cyclorrhaphous Brachycera), or head structures are greatly reduced and not rounded and sclerotized as in Nematocera (orthorrhaphous Brachycera). Only in Stratiomyidae is there any significant head structure, but heads of Stratiomyidae larvae are angulate (Fig. 138), partially retracted into the thorax, and quite unlike heads of Nematocera. Mandibles of Brachycera larvae move vertically, not horizontally as in Nematocera. They lack subapical teeth and are usually referred to as "mouth hooks."

The majority of species in all families, except Athericidae and Sciomyzidae, have terrestrial or semiaquatic larvae, and it usually is not possible to know which species have aquatic larvae because larvae of most species of Brachycera remain unknown. Furthermore, many genera that are known to have aquatic larvae, also have terrestrial or semiaquatic larvae, so all species within a genus cannot be assumed to be aquatic because larvae of one or more species have been found to be aquatic. Because most larvae cannot be identified, life histories are poorly known.

a. Athericidae (2 Genera, 4 Species) Aquatic larvae of *Atherix* are distinctive, with pairs of ventral abdominal prolegs on the first seven abdominal segments and with a single ventral proleg and a pair of fringed caudal projections on the eighth segment (Fig. 144). Another genus (*Suragina*) occurs in Texas; larvae were described by Webb (1994) and are similar to those of *Atherix*. *Atherix* larvae commonly occur in riffles or among vegetation in streams where they feed mostly on insect larvae. They pupate in moist soil along the edge of the stream. Adults are known to drink water and perhaps feed on nectar; adult females of *Suragina* suck blood from humans and cattle. Eggs of *Atherix* are laid on vegetation or supports above the stream, with several females often contributing eggs to form a single, large mass. Larvae drop into the stream upon hatching. A 1-year life cycle is likely for most species.

b. Dolicopodidae (about 8 Aquatic Genera) The "long-legged flies" are mostly terrestrial, but larvae of a few species in eight genera are known to develop in a wide variety of lotic and lentic habitats. The taxonomy of larvae is so poorly known that identification to genus is not reliable and it is not possible to clearly define which species are aquatic. Aquatic larvae differ from other Brachycera by having a posterior spiracular pit that is surrounded by four lobes (Fig. 143). They are predaceous, feeding mostly on other insect larvae. Pupae differ from other orthorrhaphous Brachycera by having thoracic respiratory horns.

c. Empididae (about 8 Aquatic Genera) The small (<7 mm) larvae of "dance flies" are too poorly known to estimate the number of aquatic species. Larvae of most species are terrestrial or semiaquatic. Most aquatic larvae can be recognized because they have seven or eight pairs of abdominal prolegs and a pair of caudal respiratory appendages (Fig. 145), which are not elongate and fringed as in Athericidae. Larvae are predaceous on smaller animals, and most aquatic larvae live in rapid water of streams. Because larval taxonomy is so poorly known, almost nothing is known about the life history of aquatic species.

d. Ephydridae (68 Genera) Species in most genera have semiaquatic larvae. Some genera, however, have aquatic species with larvae that live in vegetation or detritus in shallow habitats near shore, in algal mats, or in unusual habitats such as hot springs and pools, and ponds containing oil, as well as saline and alkaline marshes, ponds and lakes, where larvae dominate the insect fauna. Larvae of other species are terrestrial or mine leaves of aquatic plants. "Shore fly" larvae are less than 12 mm long and usually have a pair of very short to elongate caudal respiratory tubes that frequently have a black sclerotized tip (Fig. 140). Many also have abdominal prolegs. Most larvae feed on blue-green and other algae, others feed on detritus, some are predators. Aquatic species pupate in a puparium that usually floats on the water surface. Eggs are laid singly at the surface of the aquatic habitat. Although the family has received a great deal of attention, larval taxonomy is still poorly known and identification of larvae, even at the generic level, is difficult or impossible. Foote (1995) reviewed their biology.

e. Muscidae (about 8 Aquatic Genera) In this extremely large terrestrial family, larvae of several aquatic species live in streams, marshes, or ponds; they are mostly predators. Larvae are poorly known, but most have a pair of very short caudal respiratory tubes, a pair of ventral prolegs on the last abdominal segment that are longer than the respiratory tubes, and often creeping welts ventrally on other abdominal segments (Fig. 146). Eggs are laid in algae or debris, and larvae usually remain near the surface. Pupation is in a puparium, which remains at or near the surface. A pair of respiratory siphons that are thrust through the fourth segment of the puparium penetrate the water surface.

f. Sciomyzidae (19 Genera, 175 Species) "Marsh fly" larvae are predators or parasitoids of snails, slugs, or fingernail clams, and only those species that attack land snails or slugs are not aquatic. They generally have a very wrinkled appearance, with girdles of pseudopods on abdominal segments. The abdomen terminates in a spiracular disk surrounded by eight to ten lobes and contains a pair of spiracles, each surrounded by numerous palmate hairs (Fig. 141). Larvae are found in lentic habitats and margins of lotic habitats inhabited by their prey. Here, they usually remain near the surface with their caudal spiracles at the surface. They swallow a large air bubble to keep themselves afloat, and are usually buoyant enough to keep their prey at the surface. Larvae pupate in a puparium that either remains in the snail shell or floats at the surface. Eggs are laid on emergent vegetation; newly hatched larvae drop into the water and actively search for prey. Adults of parasitoids oviposit directly on the shell of a host snail. Species may be univoltine or multivoltine. Because snails are intermediate hosts of liver flukes that infect cattle and of schistosomes that attack humans, sciomyzid larvae have been employed as a biological control for snails. For this reason they have been extensively studied and larvae in several genera can be identified to species. Berg and Knudson (1978) reviewed the biology and systematics.

g. Stratiomyidae (about 10 Aquatic Genera) Although most "soldier fly" larvae are terrestrial or semiaquatic, those of many species inhabit shallow, vegetated, lentic habitats. While often quite abundant, larvae are so poorly known that identification at even the generic level is not always reliable. The relatively flat larvae are easily recognized by their mostly visible truncate head (Fig. 138), caudal spiracles surrounded by long hydrofuge hairs, and calcium carbonate crystals that cover their integument. Larvae hang suspended in the surface film by hydrofuge hairs that surround the spiracular disk, and use their palps to move about. They may at times leave the water and inhabit moist shoreline debris or plants. They feed on detritus and also algae, differing from larvae of other orthorrhaphous Brachycera, which are predators. Pupation is in a puparium, which may float in the aquatic habitat. Adults feed mostly on flowers. Eggs are laid in masses, either on submerged or emergent aquatic plants, or on other objects in the water.

h. Syrphidae (about 7 Aquatic Genera) Most "rat-tailed maggosts" become fairly large and differ from other Brachycera by being broad and blunt anteriorly. Aquatic larvae are easily recognized by their very long, extensile, caudal breathing tube (siphon) (Fig. 139). Most also have ventral prolegs. Larvae inhabit shallow lentic habitats or margins of lotic habitats, especially areas high in decomposing organic matter. Here, they feed on detritus and microorganisms. Larvae of some species occur in tree holes. Because they obtain oxygen from the air through their long respiratory siphon, they frequently inhabit highly polluted areas such as sewage lagoons. Larvae pupate in a puparium, with the puparium of some species having anterior respiratory horns. Life histories are poorly known, but most species probably have more than one generation per year. Adults are called "flower flies" because they feed on nectar from flowers. Eggs are laid in masses on the debris or vegetation that the larvae inhabit.

i. Tabanidae (about 4 Aquatic Genera) Most "deerfly" or "horsefly" larvae develop in wet soil that may be closely associated with aquatic habitats. Some larvae in four genera are known to be aquatic, and a few in other genera also may be aquatic. Many North American species have been reared and described, and larval keys have been developed, but identification is still risky because larvae of some species remain unknown. Larvae are recognized by the absence of distinct prolegs and the presence of a girdle of six or more pseudopods on most abdominal segments (Fig. 142). Aquatic larvae have been collected from stream riffles, shallow stream margins, and shallow vegetated lentic habitats. Most species are probably univoltine, but bivoltine and semivoltine species likely occur. Larvae are predators on other aquatic macroinvertebrates; pupation of aquatic species takes place in semiaquatic areas. Most adult females feed on blood of humans, livestock, and other animals, and are often a persistent nuisance. Eggs are laid in masses on vegetation above the larval habitat.

IV. SEMIAQUATIC COLLEMBOLA—SPRINGTAILS

Until recently Collembola were regarded as insects, but most people no longer consider them to be in the class Insecta and place them in the class Entognatha or in a subclass of Hexopoda. The order Collembola is only partially aquatic, with the vast majority of the nearly 700 North American species inhabiting moist terrestrial habitats. Species that are collected from aquatic habitats are usually found on the water surface and are semiaquatic. Waltz and McCafferty (1979) reported that 96 species of springtails had been collected at least once from freshwater or coastal marine habitats in North America, but most of these records were believed to be incidental occurrences of terrestrial species. They regarded 10 species as semiaquatic and five as riparian with a propensity to venture onto the water surface. These 15 species exhibited various degrees of adaptation for the semiaquatic environment and are discussed in detail by Waltz and McCafferty, who also key species most likely to be found in freshwater habitats. Keys and descriptions for all North American species of Collembola are provided by Christiansen and Bellinger (1980).

Collembola are small, wingless arthropods, usually much less than 6 mm long. They resemble insects by having a distinct head with one pair of segmented antennae, a three-segmented thorax with three pairs of legs, and a segmented abdomen without legs (Figs. 160 and 161).

They differ from insects in having only six abdominal segments and in having their maxillae and mandibles concealed by the wall of the head capsule with which their labium is fused. They also have the tibia and tarsus of each leg fused into a single segment. Except for their smaller size and lack of genitalia, young Collembola closely resemble adults.

Springtails can be readily recognized by a cylindrical collophore or ventral tube (Fig. 160 and 161) on the first abdominal segment and by the furcula, a midventral appendage on the fourth abdominal segment (Fig. 160 and 161) that is present in most species and in all that are known to be semiaquatic. The furcula consists of a basal piece, the manubrium, and two arms (Fig. 162). The basal segment of each arm is the dens and the apical segment is the mucro. The furcula can be folded forward under the abdomen where it is held in place by the tenaculum, a midventral clasplike structure on the third abdominal segment. When the furcula is released, it springs backward and propels the Collembola several centimeters through the air, hence the common name "springtail." In most semiaquati species there is a widening of each mucro or lamellae on each mucro. This permits greater contact with the surface film, allowing individuals to spring farther when on the water surface.

The ecology and biology of Collembola remains very poorly known. Semiaquatic species are most often associated with lentic freshwater habitats; none occur on the ocean and reports from lotic habitats are rare. They feed primarily on algae, detritus, and other organic material in the surface film; some may eat bacteria. Adults do not copulate like insects. Instead males deposit stalked or sessile spermatophores, which are gathered by the females; usually no contact between sexes is involved. Eggs are laid in moist habitats; in at least one species they are laid beneath the water surface and newly hatched young are truly aquatic, but once they pass through the surface film and come in contact with the air they remain semiaquatic. Eggs usually hatch within a few weeks, except those that diapause over the winter or during periods of dryness. The young reach sexual maturity within a few weeks and molt from three to seven times. In some species the young may enter into diapause. Most adults live a few weeks to a few months, and unlike insects, they continue to molt throughout their adult life, with more than fifty molts reported for one species. Some species are cold-hardy and may be active at temperatures near freezing; however, most species overwinter as diapausing eggs.

Because of their springing ability Collembola are often difficult to capture. Semiaquatic species may be collected most easily by forcing them to jump into a white pan containing either 95% alcohol with 3% glacial acetic acid, or water with detergent to prevent them from jumping from the pan. Floating sticky traps may also be used to capture them. Christiansen and Bellinger (1980) recommend 70–95% isopropanol or 80–95% ethanol for preservation, with 1–2% glycerine added to prevent accidental desiccation. A taxonomic key to families of semiaquatic Collembola appears in Section V.J.

V. IDENTIFICATION OF THE FRESHWATER INSECTS AND COLLEMBOLA

The following taxonomic keys are separated by class (Entognatha and Insecta) and orders (Insecta only). Collembola, which usually are much less than 6 mm long, can be distinguished from small larval insects by the possession ventrally of a collophore (ventral tube) on the first abdominal segment and a furcula on the fourth (Figs. 158 and 159).

A. Taxonomic Key to Orders of Freshwater Insects

1a. Thorax with 3 pairs of segmented legs 3

1b. Thorax without segmented legs 2

2a(1b). Mummylike, with developing adult structures; in a case, often silk-cemented and contains vegetable or mineral matter .... pupae (not keyed)

2b. Not in a case; mobile larvae, mostly with prolegs, pseudopods, or creeping welts on one or more segments (Figs. 145–147 and 152–159) .... Diptera larvae

3a(1a). With large functional wings, which may be shelllike, or leatherlike at the base 4

3b. Wingless, or with developing wings (wing-pads) or brachypterous wings 6

4(3a). All wings completely membranous, with numerous veins .... ovipositing or terrestrial adults (not keyed)

4b. Mesothoracic wings hardened and shell-like (Figs. 108, 117, and 118), or leatherlike in basal half (Figs. 87 and 93–95) 5

5a(4b). Chewing mouthparts; mesothoracic wings hard, shell-like (Figs. 108, 117, 118) .... Coleoptera adults

5b. Sucking mouthparts formed into a broad or narrow tube (Figs. 88 and 90); mesothoracic wings hardened in basal half (Fig. 87); .... Heteroptera (= Hemiptera) adults

6a(3b). With 2 or 3 long, filamentous terminal appendages (Figs. 9, 29, 45, 48, 49, 54, 56, and 58) 7

6b. Terminal appendages absent, or with only 1 or 2 segments 8

7a(6a). Sides of abdomen with featherlike, platelike, or leaflike gills (Figs. 4, 11–15, and 25–30); usually with 3 tail filaments, occasionally only 2; tarsi with 1 claw .... Ephemeroptera larvae

7b. Gills absent from at least middle abdominal segments (Figs. 45, 48, 49, 54, 56, and 58); 2 tail filaments; tarsi with 2 claws .... Plecoptera larvae

8a(6b). Labium formed into an elbowed, extensile grasping organ (Figs. 32–34); abdomen terminating in 3 lamellae (Fig. 31) or 5 triangular points (Fig. 44) .... Odonata larvae

8b. With sucking or chewing mouthparts; abdomen not terminating in 3 lamellae or 5 triangular points 9

9a(8b). Mouthparts sucking, formed into a broad or narrow tube (Figs. 88 and 90) or a pair of long stylets (Fig. 102) 10

9b. Mouthparts not sucking, not formed into a tube or pair of stylets 11

10a(9a). Parasitic on sponges; mouthparts a pair of long stylets (Fig. 102); all tarsi with 1 claw (Fig. 102) .... Neuroptera; Sisyridae larvae

10b. Free-living; mouthparts a broad or narrow tube (Figs. 88 and 90); mesotarsi with 2 claws (Figs. 87 and 93–96) .... Heteroptera (= Hemiptera) adults or nymphs

11a(9b) Ventral prolegs on abdominal segments 3–6, each with a ring of fine hooks (Fig. 103) .... Lepidoptera; Pyralidae larvae

11b. Abdomen without ventral prolegs on abdominal segments 3–6 that have a ring of hooks 12

12a(11b). Antennae extremely small, inconspicuous, one-segmented (Figs. 63, 78, and 80) .... Trichoptera larvae

12b. Antennae elongate, with 3 or more segments 13

13a(12b). Without long lateral filaments (Figs. 120–130) ....Coleoptera larvae

13b. With long lateral filaments (Figs. 85, 86, and 119)....................14

14a(13b). A single claw on each tarsus (Fig. 121–124) ***or*** abdomen terminating in 2 slender filaments or a median proleg with 4 hooks (Fig. 119)....................Coleoptera larvae

14b. Each tarsus with 2 claws; abdomen terminating in a single slender filament (Fig. 85) or in 2 prolegs, each with 2 hooks (Fig. 86) ....Megaloptera larvae

B. Taxonomic Key to Families of Freshwater Ephemeroptera Larvae

1a. Mandibles with large forward-projecting tusks (Fig. 3); gills on abdominal segments 2–7 with fringed margins (Fig. 4) and projecting laterally or dorsally over abdomen 2

1b. Mandibles without large tusks; conspicuous fringed gills absent or projecting ventrolaterally....................4

2a(1a). Gills dorsal, curving up over abdomen; protibiae fossorial (Fig. 5)....................3

2b. Gills lateral, projecting from sides of abdomen; protibiae slender, subcylindrical (Fig. 6); East, Midwest....................Potamanthidae

3a(2a). Apex of metatibiae rounded (Fig. 7); mandibular tusks curved inward and downward apically....................Polymitarcyidae

3b. Apex of metatibiae projected into an acute point ventrally (Fig. 8); mandibular tusks curved upward or outward apically ....Ephemeridae

4a(1b). Mesonotum modified into a carapace-like structure that covers the gills on abdominal segments 1–6 (Fig. 9) ....Baetiscidae

4b. Mesonotum not modified into a carapace, gills exposed....................5

5a(4b). Head and pronotum with pads of long, dark setae on each side (Fig. 10); fringed gills projecting ventrolaterally; Southeast + WI ....Behningiidae

5b. Head and pronotum without pads of long, dark setae; gills not fringed and projecting ventrolaterally 6

6a(5b). Gills on abdominal segment 2 operculate or semi-operculate, covering or partially covering gills on succeeding segments (Figs. 11–14)....................7

6b. Gills on abdominal segment 2 similar to those on succeeding segments or absent 9

7a(6a). Operculate gills oval or subtriangular and well-separated from each other mesally (Figs. 11 and 12); gills on segments 3–6 without fringed margins ....Leptohyphidae

7b. Operculate gills quadrate and meeting along mesal edge (Figs. 13 and 14); gills on segments 3–6 with fringed margins 8

8a(7b). Operculate gills fused mesally (Fig. 13); East....................Neoephemeridae

8b. Operculate gills not fused mesally, but overlapping (Fig. 14) ....Caenidae

9a(6b). Gills absent from abdominal segment 2 and sometimes also from 1 and 3, gills on segments 3 or 4 may be operculate (Fig. 15) ....Ephemerellidae

9b. Gills present on abdominal segments 1 or 2 to 7 10

10a(9b). Head flattened dorsoventrally; eyes and antennae dorsal (Fig. 16); gills a single lamella, often with a fibrilliform tuft 11

10b. Head and body not dorsoventrally flattened; eyes, and usually also antennae, along lateral margin of head (Fig. 17) 13

11a(10a). Distal maxillary palpomere at least 4 times as long as galea-lacinia and visible from above (Fig. 18); Northeast....................Arthropleidae

11b. Distal maxillary palpomere much shorter, usually not visible from above 12

12a(11b). Gills with a ventral fingerlike projection on lamellae (Fig. 19); tarsal claws very long: West, Midwest, Southeast ....Pseudironidae

12b. Gill lamellae without such a projection; claws not exceptionally long....................Heptageniidae

13a(10b). Protarsal claws much shorter than tibiae and those on other tarsi (Figs. 20 and 22); claws on meso- and metatarsi long and slender, about as long as tibiae (Fig. 21)....................14

13b. Claws on all tarsi similar in structure and length....................15

14a(13a). Protarsal claws bifid (Fig. 20); no spinose pad on procoxae; Canada, midwest and, southeast United States ....Metretopodidae

14b. Protarsal claws simple, with long, slender denticles (Fig. 22); spinose pad present on procoxae; West, Upper MI....................Ametropodidae

15a(13b). Prothoracic legs with dense row of long setae along inner surface (Fig. 23)....................16

15b. Prothoracic legs without dense row of setae along inner surface 17

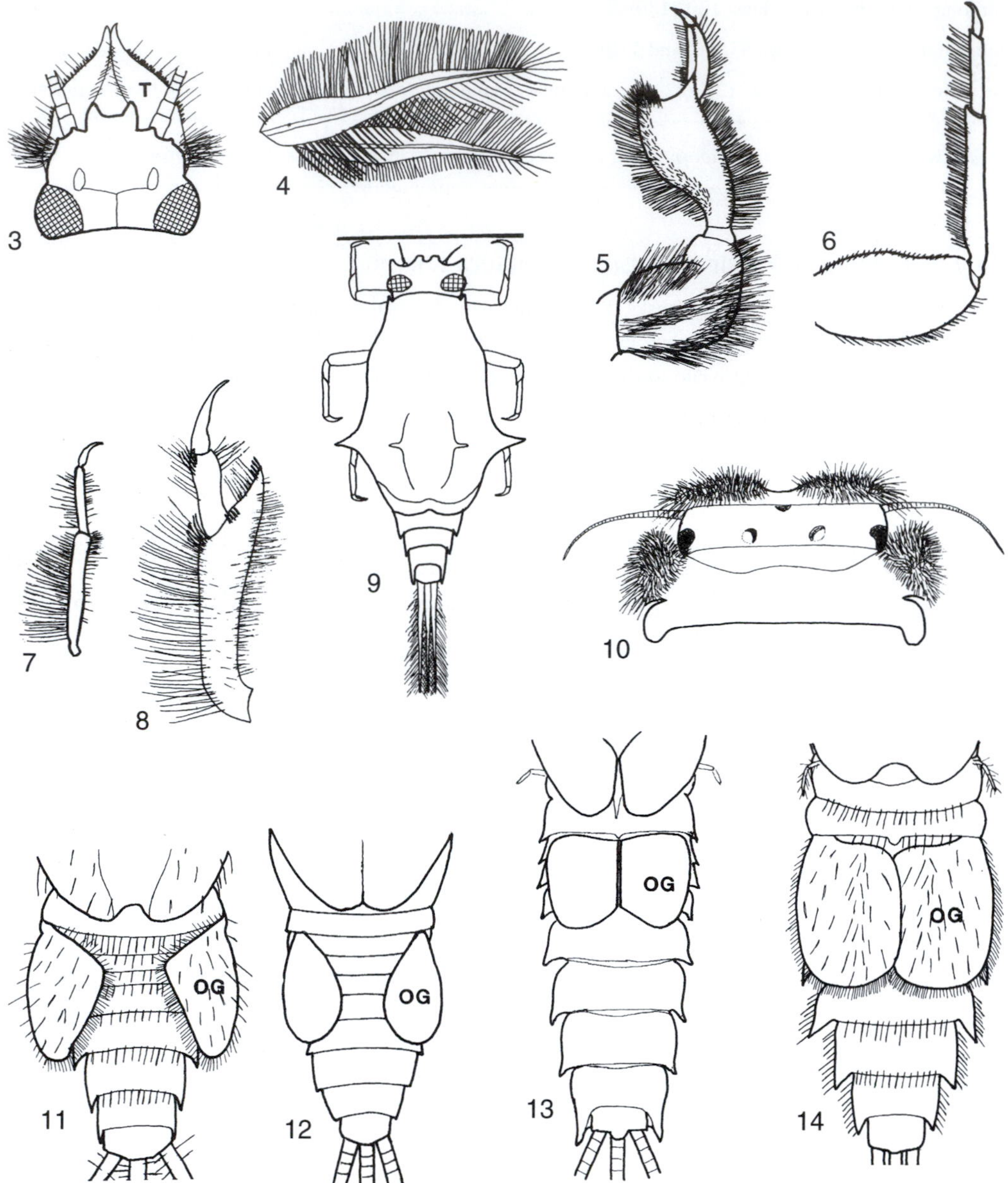

FIGURE 3 Ephemeridae; head of *Pentaqenia* (dorsal view) showing mandibular tusk (T). *FIGURE 4* Ephemeridae; gills of *Hexaqenia*. *FIGURE 5* Ephemeridae; prothoracic leg of *Hexaqenia*. *FIGURE 6* Potamanthidae; prothoracic leg of *Anthopotamus*. *FIGURE 7* Polymitarcyidae; metatibia and tarsus of *Ephoron*. *FIGURE 8* Ephemeridae; metatibia and tarsus of *Ephemera*. *FIGURE 9* Baetiscidae; *Baetisca* (dorsal view). *FIGURE 10* Behningiidae; head and pronotum of *Dolania*. *FIGURE 11* Leptohyphidae; abdomen of *Tricorythodes* (dorsal view) showing operculate gill (OG). *FIGURE 12* Leptohyphidae; abdomen of *Leptohyphes* (dorsal view) showing operculate gill (OG). *FIGURE 13* Neoephemeridae; abdomen of *Neoephemera* (dorsal view) showing operculate gill (OG). *FIGURE 14* Caenidae; abdomen of *Caenis* (dorsal view) showing operculate gill (OG).

16a(15a). Gills on abdominal segment 1 dorsolateral; gills lamellate, tapered apically, with a basal fibrilliform tuft; gill tuft at base of procoxae (Fig. 23) .... Isonychiidae

16b. Gills on abdominal segment 1 ventral; gills on abdominal segments 2–7 small, either lanceolate with a posterior fringe (Fig. 24), or rounded and platelike; no gill tuft at base of procoxae; West, midwest and and southeast United States .... Oligoneuriidae

17a(15b). Gills forked (Fig. 25), or bilamellate and terminating in a point (Fig. 26), or terminating in filaments (Fig. 27) .... Leptophlebiidae

17b. Gills single or double lamellae (Figs. 28 and 29) 18

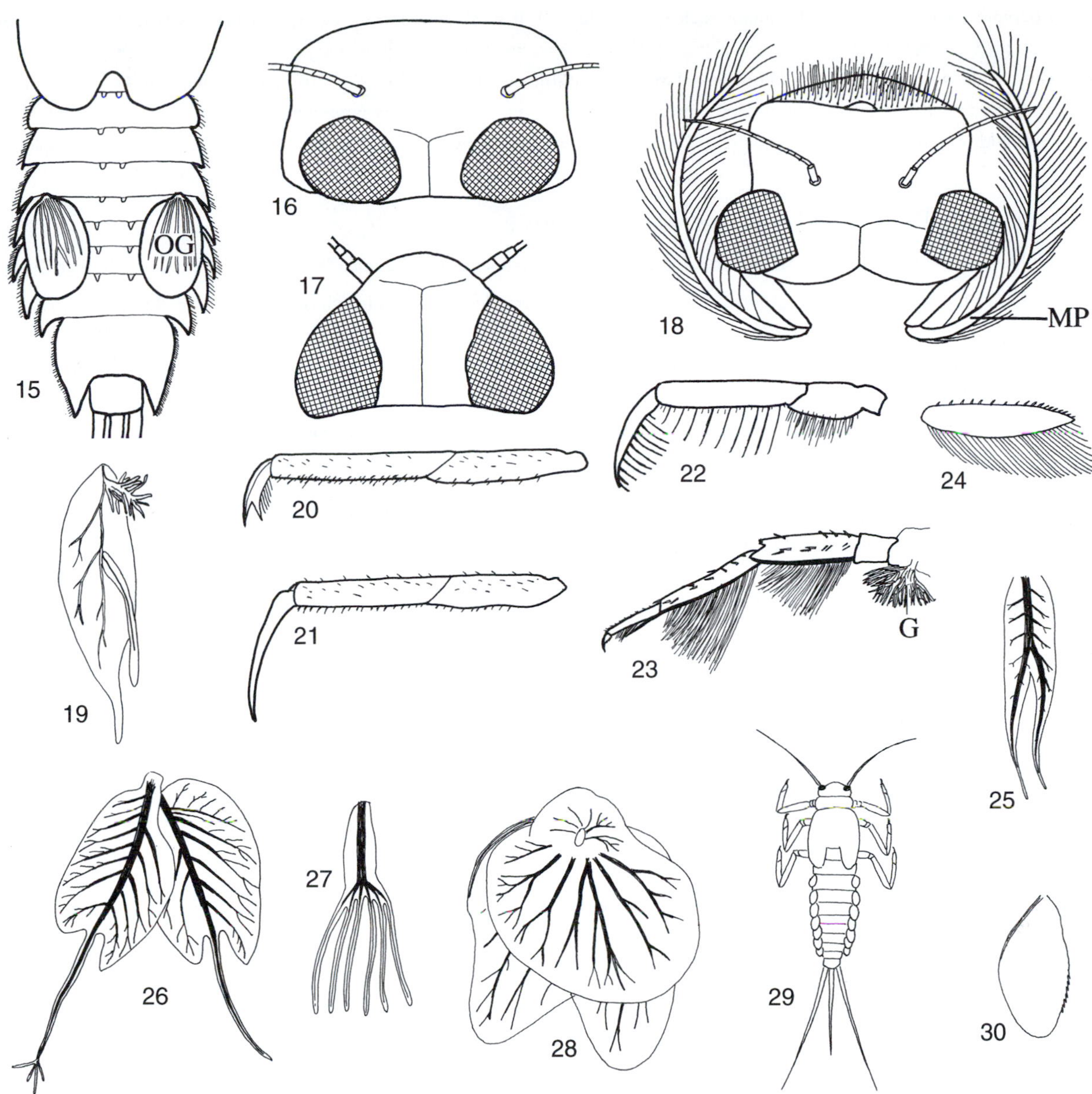

FIGURE 15 Ephemerellidae; abdomen of *Eurylophella* (dorsal view) showing operculate gill (OG). *FIGURE 16* Heptageniidae; head of *Stenonema* (dorsal view). *FIGURE 17* Siphlonuridae; head of *Siphlonurus* (dorsal view). *FIGURE 18* Arthropleidae; head of *Arthroplea* (dorsal view) showing maxillary palp (MP). *FIGURE 19* Pseudironidae; gill lamella of *Pseudiron* (ventral view). *FIGURE 20* Metretopodidae; protarsus and tibia of *Siphloplecton*. *FIGURE 21* Metretopodidae; metatarsus and tibia of *Siphloplecton*. *FIGURE 22* Ametropodidae; protarsus and tibia of *Ametropus*. *FIGURE 23* Isonychiidae; prothoracic leg of *Isonychia* showing gill (G). *FIGURE 24* Oligoneuriidae; gill on abdominal segment 3 of *Homoeoneuria*. *FIGURE 25* Leptophlebiidae; gill on abdominal segment 3 of *Paraleptophlebia*. *FIGURE 26* Leptophlebiidae; gills on abdominal segment 3 of *Leptophlebia*. *FIGURE 27* Leptophlebiidae; gill on abdominal segment 3 of *Habrophlebia*. *FIGURE 28* Siphlonuridae; gills on abdominal segment 2 of *Siphlonurus*. *FIGURE 29* Baetidae; *Baetis* (dorsal view). *FIGURE 30* Ameletidae; gill on abdominal segment 3 of *Ameletus*.

18a(17b). Tibiae and tarsi bowed; claws very long and slender; metatarsal claws about as long as tarsi; Illinois, Wisconsin, South Carolina, and Georgia .... Acanthametropodidae

18b. Tibiae and tarsi not bowed; claws not as above 19

19a(18b). Abdominal segments 8 and 9 produced posterolaterally into distinct, flattened spines (Figs. 13–15); if spines are weak, antenna more than twice width of head 20

19b. Abdominal segments 8 and 9 without such spines (Fig. 29), if weak spines are present, antenna less than twice width of head .... Baetidae

20a(19a). Gills single, with sclerotized band on ventral margin and little or no tracheation (Fig. 30); maxillae with crown of pectinate spines .... Ameletidae

20b. Gills with well-developed tracheation (Fig. 28); maxillae without a crown of pectinate spines .... Siphlonuridae

C. Taxonomic Key to Families of Freshwater Odonata Larvae

1a. Abdomen terminating in 3 caudal lamellae, longest more than half length of abdomen (Fig. 31) .... suborder Zygoptera 2

1b. Abdomen terminating in 3 stiff, pointed valves and 2 pointed cerci, longest <1/3 length of abdomen (Fig. 44) .... suborder Anisoptera 5

2a(1a). Zygoptera: first antennal segment as long as, or longer than, remaining segments combined (Fig. 31); prementum with deep, median cleft (Fig. 32) .... Calopterygidae

2b. First antennal segment much shorter than others combined; prementum with at most a very small median cleft (Figs. 33 and 34) 3

3a(2b). Basal half of prementum greatly narrowed and elongate (Fig. 33); prementum in repose extends back to or past mesocoxae . Lestidae

3b. Basal half of prementum not greatly narrowed (Fig. 34); prementum in repose extends only to procoxae 4

4a(3b). Caudal lamellae divided into a thick basal half and thin, lighter colored distal half (Fig. 35); one dorsal seta on each side of prementum; Texas .... Protoneuridae

4b. Caudal lamellae not distinctly divided as above; usually with 2 or more dorsal setae on each side of prementum .... Coenagrionidae

5a(1b). Anisoptera: Prementum flat or nearly so, without dorsal setae; palpal lobes also flat and usually without stout setae 6

5b. Prementum rounded, spoon-shaped, and usually with dorsal setae; palpal lobes also rounded, always with stout setae, and covering face to base of antennae 8

6a(5a). Antennae 4-segmented (Fig. 36); pro- and mesotarsi 2-segmented .... Gomphidae

6b. Antennae 6- or 7-segmented (Figs. 37 and 38); pro- and mesotarsi 3-segmented 7

7a(6b). Antennal segments slender, not hairy (Fig. 37) .... Aeshnidae

7b. Antennal segments short, thick, and hairy (Fig. 38); mountain seeps .... Petaluridae

8a(5b). Distal margin of palpal lobes with large, irregular teeth (Fig. 39) .... Cordulegastridae

8b. Distal margin of palpal lobes with small, even crenulations (Figs. 40 and 41) or nearly straight (Fig. 42) 9

9a(8b). Head with a prominent, almost erect, thick frontal process between bases of antennae (Fig. 43); legs very long, apex of each metafemur reaching to or beyond apex of abdominal segment 8; metasternum with broad, median tubercle .... Macromiidae

9b. Head without a prominent frontal process (except in *Neurocordulia molesta*); legs shorter, apex of metafemurs usually not reaching apex of abdominal segment 8; metasternum without a median tubercle 10

10a(9b). Crenulations on palpal lobes very shallow, 1/6 to <1/10 as long as wide (Fig. 42), or separated by minute notches bearing setae .... most Libellulidae

10b. Crenulations on palpal lobes large, at least 1/4 as long as wide (Figs. 40 and 41) 11

11(10b). Lateral spines on abdominal segment 8 as long or longer than middorsal length of segment 9 (Fig. 44) .... Libellulidae

Lateral spines on abdominal segment 8 shorter than middorsal length of segment 9 .... Corduliidae

D. Taxonomic Key to Families of Freshwater Plecoptera Larvae

1a. Conspicuous finely branched gills present ventrally or laterally on all thoracic segments (Fig. 2) 2

1b. Gills absent, confined to prosternum, or unbranched 3

2a(1a). Finely branched gills on abdominal sterna 1, 2 and sometimes 3 .... Pteronarcyidae

2b. Gills absent from first 3 visible abdominal sterna .... Perlidae

3a(1b). Thoracic sterna produced into plates that overlap succeeding segment (Fig. 45); single, double, or forked gills behind meso- and metacoxae; form roach-like; mountain streams .... Peltoperlidae

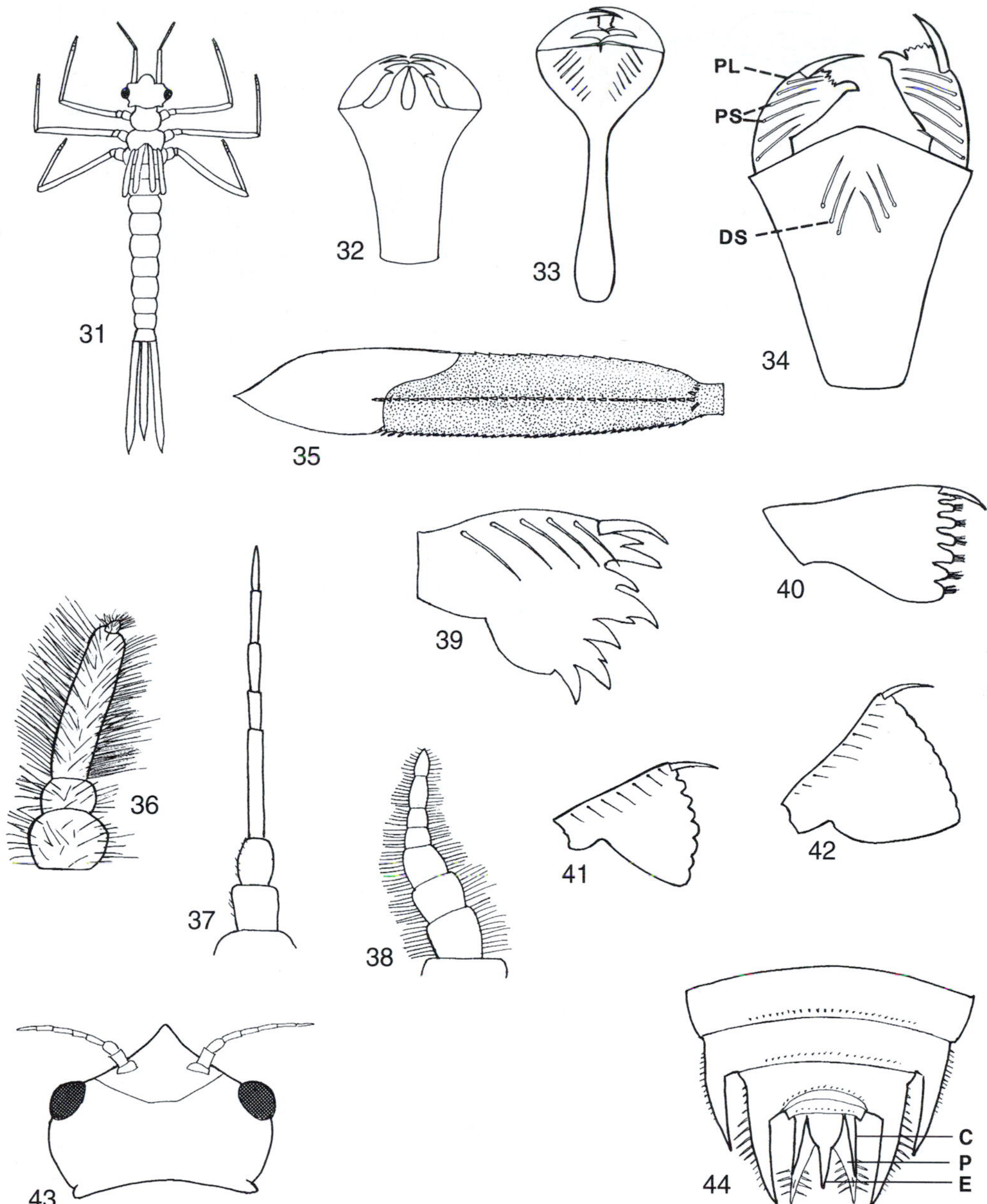

FIGURE 31 Calopterygidae; *Calopteryx* (dorsal view). *FIGURE 32* Calopterygidae; prementum of *Hetaerina* (dorsal view). *FIGURE 33* Lestidae; prementum of *Lestes* (dorsal view). *FIGURE 34* Coenagrionidae; prementum of *Enallagma* (dorsal view) showing dorsal setae (DS), palpal lobe (PL), and palpal setae (PS). *FIGURE 35* Protoneuridae; caudal lamella of *Protoneura* (lateral view). *FIGURE 36* Gomphidae; antenna of *Gomphus*. *FIGURE 37* Aeshnidae; antenna of *Anax*. *FIGURE 38* Petaluridae; right antenna of *Tachopteryx* (dorsal view). *FIGURE 39* Cordulegastridae; palpal lobe of *Cordulegaster* (dorsal view). *FIGURE 40* Corduliidae; palpal lobe of *Neurocordulia* (dorsal view). *FIGURE 41* Corduliidae; palpal lobe of *Cordulia* (dorsal view). *FIGURE 42* Libellulidae; palpal lobe of *Tramea* (dorsal view). *FIGURE 43* Macromiidae; head of *Macromia* (dorsal view). *FIGURE 44* Libellulidae; abdominal segments 7–10 of *Pantala* (dorsal view) showing epiproct (E), cerci (C), and paraprocts (P).

3b. Thoracic sterna not produced into overlapping plates; coxal gills, if present, segmented and not apically pointed; form usually elongate, not roachlike 4

4a(3b). Tips of glossae situated much behind tips of paraglossae (Fig. 46) 5

4b. Tips of glossae produced nearly as far forward as tips of paraglossae (Fig. 47) 6

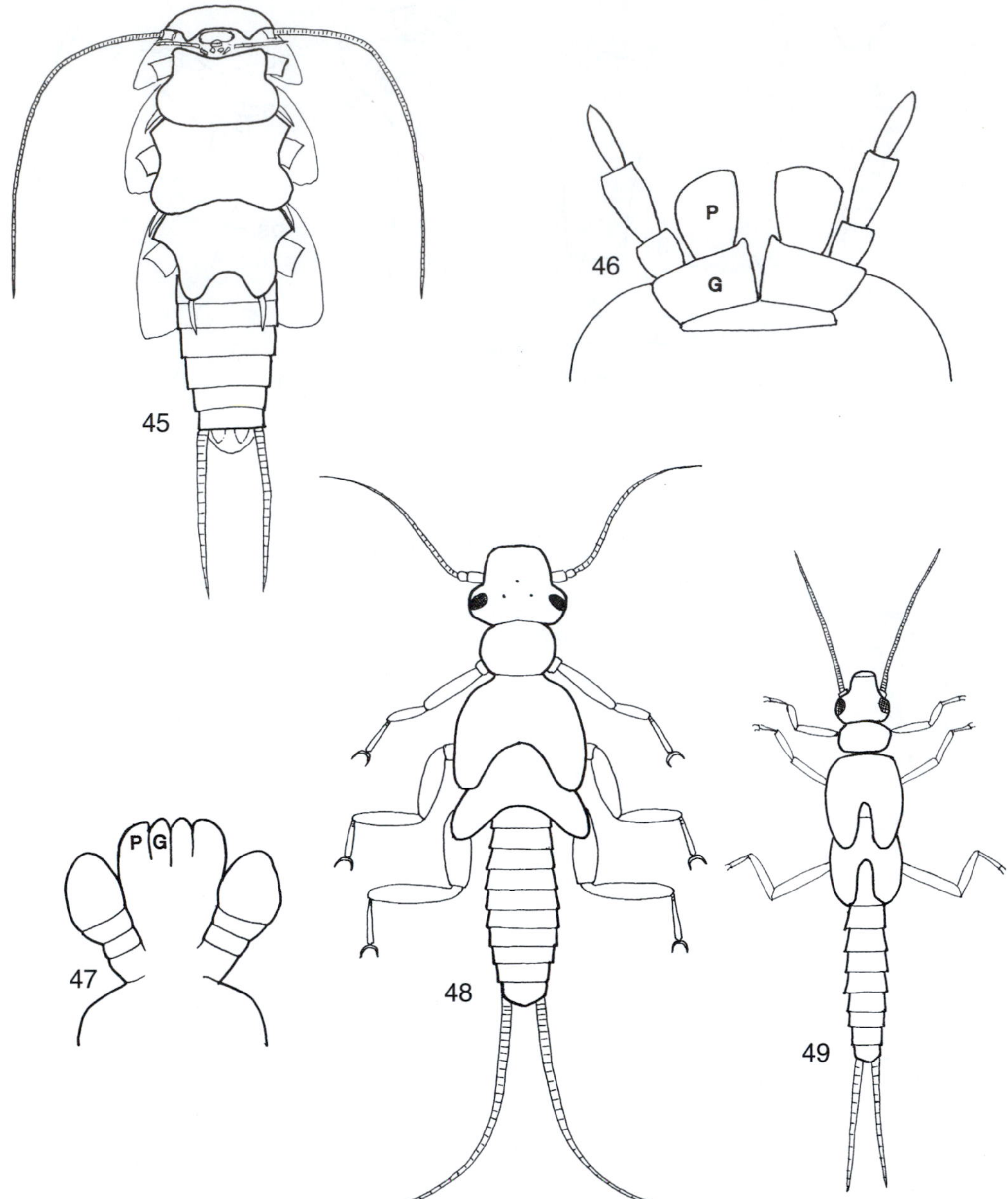

FIGURE 45 Peltoperlidae: *Peltoperla* (ventral view) with legs removed beyond coxae. *FIGURE 46* Chloroperlidae; labium of *Alloperla* (ventral view) showing glossae (G) and paraglossae (P). *FIGURE 47* Nemouridae; labium of *Nemoura* (ventral view) showing glossae (G) and paraglossae (P). *FIGURE 48* Perlodidae; *Isoperla* (dorsal view). *FIGURE 49* Chloroperlidae; *Haploperla* (dorsal view).

5a(4a). Cerci almost as long or longer than abdomen; metathoracic wing-pads with inner and outer margins strongly diverging from axis of body (Fig. 48); head and thorax usually with a distinct pattern .... Perlodidae

5b. Cerci distinctly shorter than abdomen; metathoracic wing-pads with inner and outer margins nearly parallel to axis of body (Fig. 49); head and thorax usually not patterned .... Chloroperlidae

6a(4b). Second tarsal segment (lateral view) about as long as, or longer than first (Fig. 50); single segmented gills on inner side of each coxa (Fig. 51), *or* ninth abdominal sternum greatly produced (Fig. 52) .... Taeniopterygidae

6b. Second tarsal segment much shorter than first (Fig. 53); gills absent from inner side of coxae and ninth sternum not produced 7

7a(6b). Robust larvae, with extended metathoracic legs reaching to or beyond tip of abdomen; metathoracic wing-pads with inner and outer margins strongly diverging from axis of body (Fig. 54) .... Nemouridae

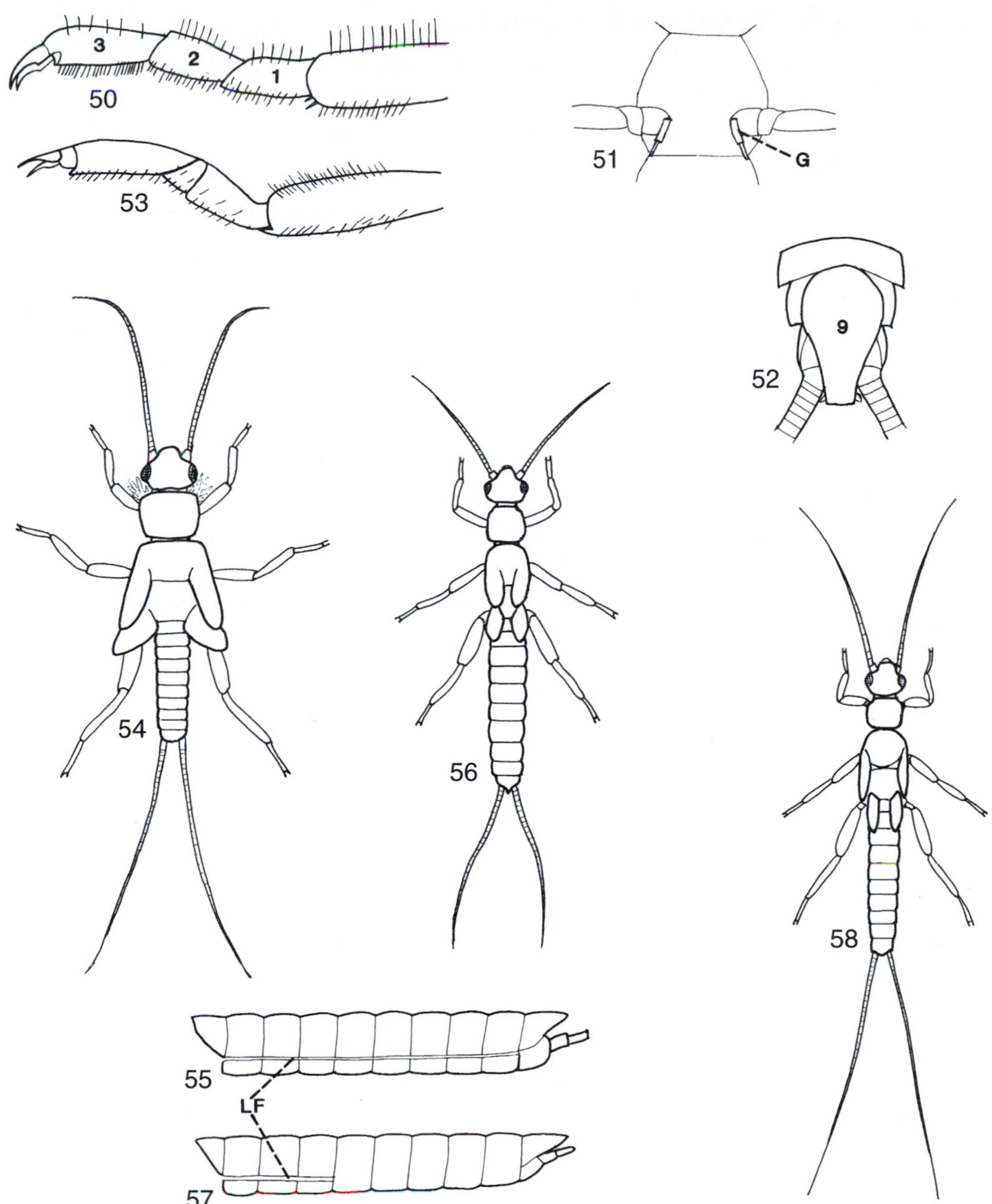

FIGURE 50 Taeniopterygidae; tarsal segments (1, 2, 3) of *Taeniopteryx* (lateral view). *FIGURE 51* Taeniopterygidae; mesosternum of *Taeniopteryx* showing gills (G) at base of coxae. *FIGURE 52* Taeniopterygidae; ninth abdominal sternum (9) of *Strophopteryx* (ventral view). *FIGURE 53* Nemouridae; tarsal segments of *Nemoura* (lateral view). *FIGURE 54* Nemouridae; *Amphinemura* (dorsal view). *FIGURE 55* Capniidae; abdomen of *Allocapnia* (lateral view) showing lateral fold (LF). *FIGURE 56* Capniidae; *Paracapnia* (dorsal view). *FIGURE 57* Leuctridae; abdomen of *Leuctra* (lateral view) showing lateral fold (LF). *FIGURE 58* Leuctridae; *Leuctra* (dorsal view).

7b. Elongate larvae, with extended metathoracic legs reaching well short of tip of abdomen; metathoracic wing-pads with inner and outer margins nearly parallel to axis of body (Figs. 56, and 58) 8

8a(7b). Abdominal segments 1–9 divided by a membranous ventrolateral fold (Fig. 55); metathoracic wing-pads, if present, about same distance apart as mesothoracic wing-pads and often short and broad (Fig. 56) .... Capniidae

8b. Abdominal segments 7–9 not divided by a membranous fold (Fig. 57); metathoracic wing-pads similar in shape to mesothoracic wing-pads, but distinctly closer together (Fig. 58) .... Leuctridae

E. Taxonomic Key to Families of Freshwater Trichoptera Larvae

1a. Larvae in spiral case of sand grains or tiny stones that resembles a snail shell (Fig. 59); anal claws with many short teeth forming a comb, not hook-shaped at apex (Fig. 60) .... Helicopsychidae

1b. Case not like a snail shell or absent; anal claws not forming a comb, and hook-shaped at apex 2

2a(1b). Each thoracic segment covered with a single dorsal plate, which may have a mesal or transverse fracture line 3

2b. Metanotum mostly membranous, having only scattered hairs or small plates, or with 2 or more sclerites 5

3a(2a). Abdomen with rows of branched gills ventrally; no portable case .... Hydropsychidae

3b. Abdomen without branched gills ventrally; often with a portable case 4

4a(3b). Anal prolegs very short and with a stout claw; abdominal tergum 9 sclerotized; fifth instar larvae with abdomen much enlarged; in a barrel- or purse like case (Figs. 61, and 62), or in a flat, silk case .... Hydroptilidae

4b. Anal prolegs elongate with a projecting claw; abdominal tergum 9 entirely membranous; abdomen not enlarged; larvae in sand retreats; southwestern United States .... Ecnomidae

5a(2b). Antennae long, at least 6 times as long as wide, and arising near base of mandibles (Fig. 63); *or* mesonotum membranous, except for a pair of sclerotized, narrow, curved or angled bars (Fig. 64); larvae in a case .... Leptoceridae

5b. Antennae very short, not more than 3 times as long as wide, often inconspicuous and arising at various points; mesonotum never with a pair of sclerotized, narrow, curved or angled bars; with or without case 6

6a(5b). Meso- and metanotum entirely membranous, or with only weak sclerites at SA-1 of mesonotum (Fig. 65); pronotum never with an anterolateral projection; larvae with or without a case 7

6b. Meso- and metanotum with some conspicuous sclerotized plates (Figs. 83 and 84); pronotum sometimes with an anterolateral projection; larvae in a case (Fig. 66) 15

7a(6a). Abdominal segment 9 with dorsum entirely membranous; no portable case (Fig. 67) 8

7b. Abdominal segment 9 bearing a sclerotized dorsal plate; with or without portable case 12

8a(7a). Tibia and tarsus fused to form a single segment on all legs; mesopleura with a forward projecting lobe (Fig. 68); southwestern United States .... Xiphocentronidae

8b. Tibia and tarsus distinct on all legs; mesopleura without a projecting lobe 9

9a(8b). Protrochantins broad, hatchet-shaped (Fig. 69) .... Psychomyiidae

9b. Protrochantins pointed or poorly developed 10

10a(9b). Protrochantins poorly developed; head without markings; labrum membranous and T-shaped (Fig. 70) .... Philopotamidae

10b. Protrochantins pointed anteriorly (Fig. 71); head usually with dark markings or dark or light muscle scars; labrum sclerotized and widest near base 11

11a(10b). Tarsi broad and densely pilose (Fig. 72); mandibles short and triangular, each with a large, thick mesal brush; East, Central .... Dipseudopsidae

11b. Tarsi with little or no pile and with a large claw (Fig. 67); mandibles elongate .... Polycentropodidae

12a(7b). Prosternal horn present (Fig. 80); SA-3 on meso- and metanotum with a small sclerite and a cluster of setae (Fig. 65); case of vegetation, spirally wound or a series of rings and readily abandoned .... Phryganeidae

12b. Prosternal horn absent; SA-3 on meso- and metanotum without a sclerite and usually with a single seta; case, if present, of sand and pebbles 13

13a(12b). Anal claws very small, much shorter than elongate sclerite on anal legs (Fig. 73); larvae with turtlelike case of sand and pebbles, which is readily abandoned .... Glossosomatidae

13b. Anal claws large, as long as elongate sclerite on anal legs (Fig. 74); larvae without a case 14

141(13b). Profemurs with ventral projection to form chelate legs (Fig. 75); southwestern United States .... Hydrobiosidae

14b. Prothoracic legs not chelate .... Rhyacophilidae

15a(6b). Claws of metathoracic legs very small or a slender filament, those of other legs long; case of sand, usually with lateral flanges .... Molannidae

15b. Claws of metathoracic legs as long as those of mesothoracic legs; case never of sand with lateral flanges 16

16a(15b). Pronotum divided by a sharp furrow across middle, area in front of furrow depressed (Fig. 76); no dorsal or lateral humps on abdominal segment 1 .... Brachycentridae

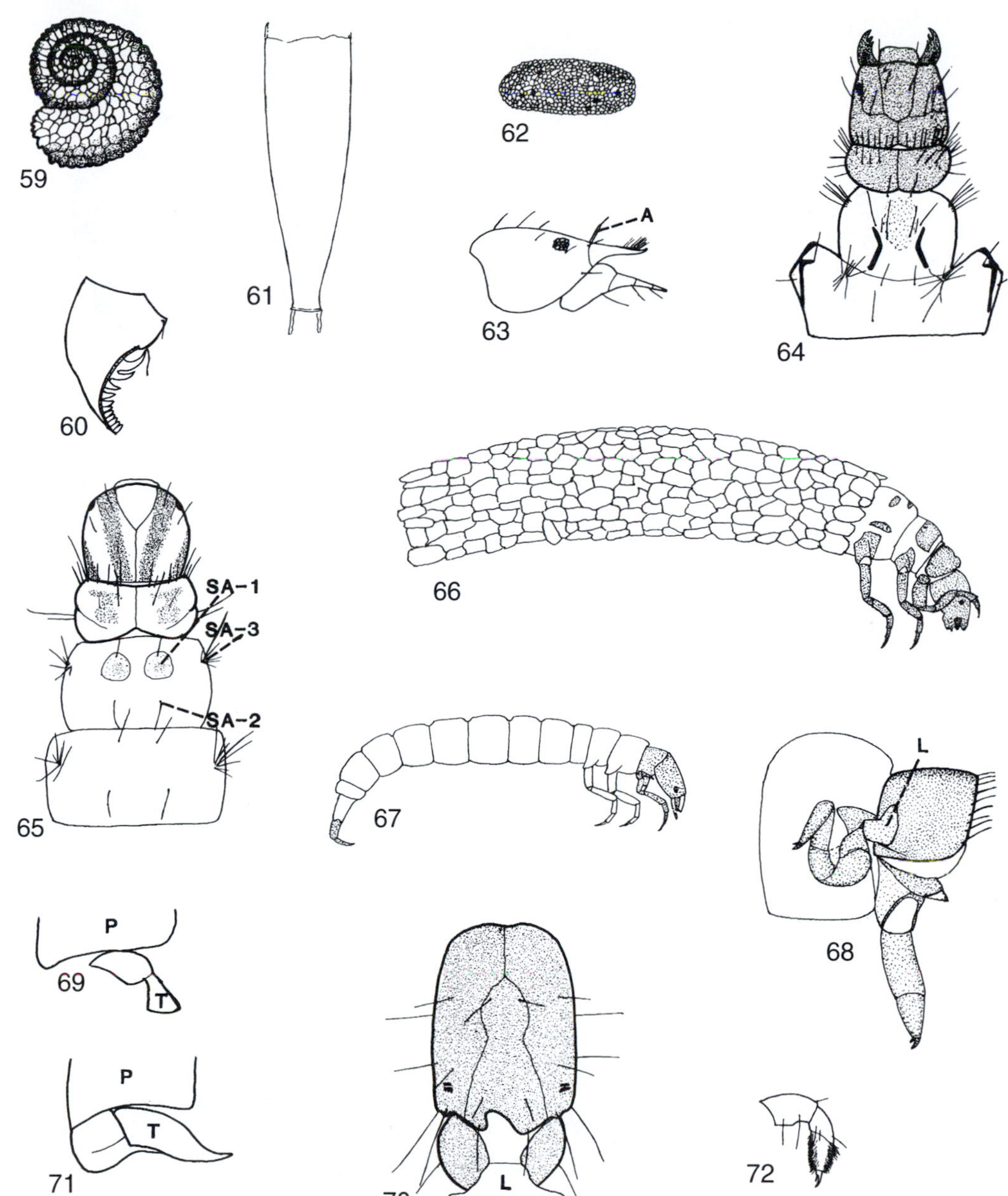

FIGURE 59 Helicopsychidae; case of *Helicopsyche* (dorsal view). *FIGURE 60* Helicopsychidae; claw on anal proleg of *Helicopsyche*. *FIGURE 61* Hydroptilidae; case of *Oxyethira* (lateral view). *FIGURE 62* Hydroptilidae; case of *Hydroptila* (lateral view). *FIGURE 63* Leptoceridae; head of *Oecetis* (lateral view) showing antenna (A). *FIGURE 64* Leptoceridae; head and thoracic terga of *Ceraclea*. *FIGURE 65* Phryganeidae; head and thoracic terga of *Oligostomis* showing location of setal areas (SA). *FIGURE 66* Limnephilidae; *Hesperophylax* (lateral view). *FIGURE 67* Polycentropodidae; *Polycentropus* (lateral view). *FIGURE 68* Xiphocentronidae; pro- and mesothorax of *Xiphocentron* (lateral view) showing projecting lobe (L). *FIGURE 69* Psychomyiidae; pronotum (P) and protrochantin (T) of *Psychomyia*. *FIGURE 70* Philopotamidae; head of *Chimarra* (dorsal view) showing labrum (L). *FIGURE 71* Polycentropodidae; pronotum (P) and protrochantin (T) of *Polycentropus*. *FIGURE 72* Dipseudopsidae; tarsus of *Phylocentropus*.

16b. Pronotum with at most a shallow furrow; lateral humps and usually dorsal hump present on abdominal segment 1 17

17a(16b). Pronotum divided obliquely by a sharp carina that terminates as a projecting, rounded lobe anterolaterally (Fig. 77); East, local ....Beraeidae

17b. Pronotum without a distinct carina terminating in a rounded anterolateral lobe 18

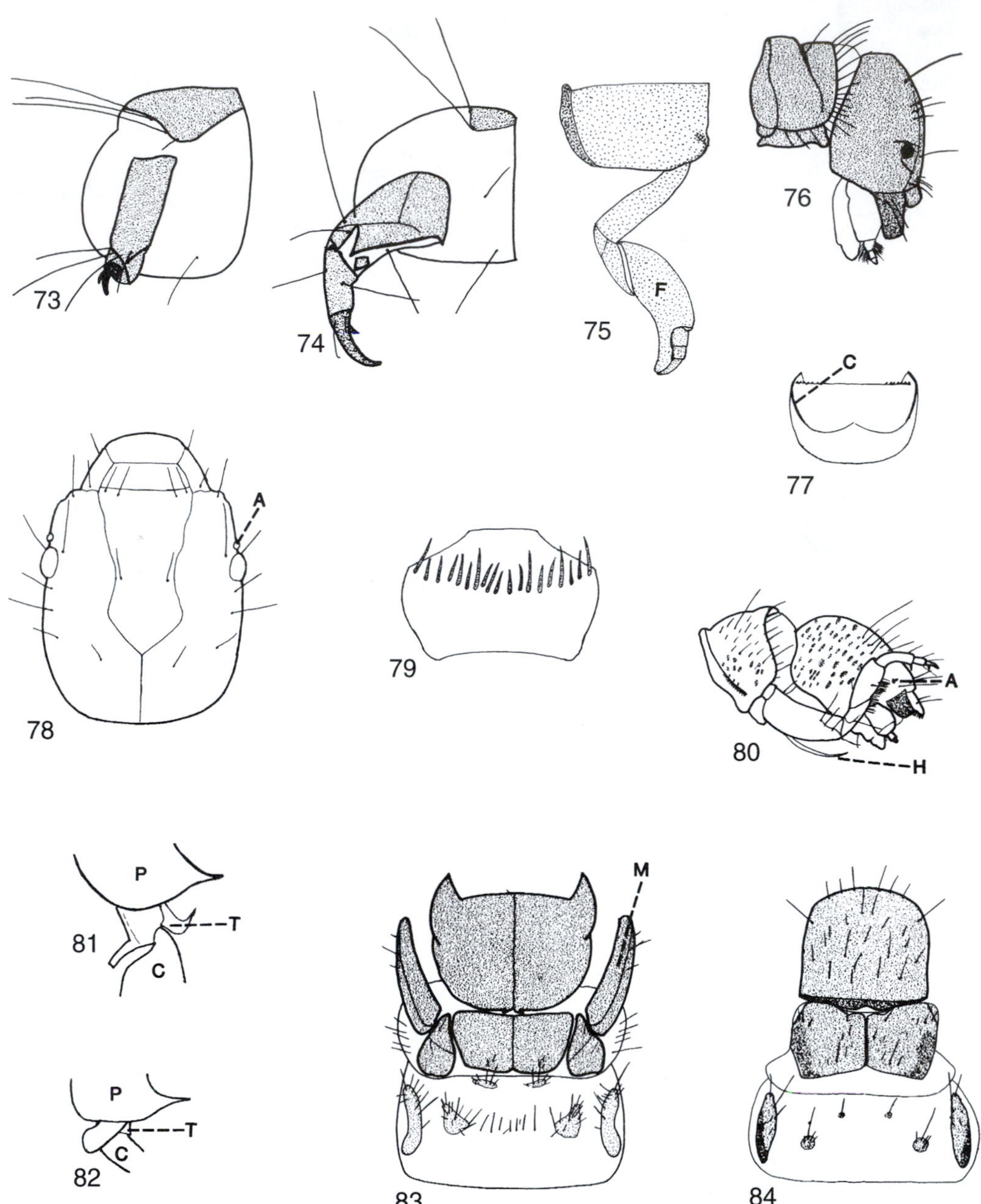

FIGURE 73 Glossosomatidae; last abdominal segment of *Glossosoma* (lateral view) showing anal proleg. *FIGURE 74* Rhyacophilidae; last abdominal segment of *Rhyacophila* (lateral view) showing anal proleg. *FIGURE 75* Hydrobiosidae; pronotum and prothoracic leg with chelate femur (F) of *Atopsyche* (posterolateral view). *FIGURE 76* Brachycentridae; pronotum and head of *Brachycentrus* (lateral view). *FIGURE 77* Beraeidae; pronotum of *Beraea* (dorsal view) showing carina (C). *FIGURE 78* Lepidostomatidae; head of *Lepidostoma* (dorsal view) showing antenna (A). *FIGURE 79* Calamoceratidae; labrum of *Heteroplectron* (dorsal view). *FIGURE 80* Limnephilidae; head and prothorax of *Platycentropus* (lateral view) showing antenna (A) and prosternal horn (H). *FIGURE 81* Sericostomatidae; pronotum (P), protrochantin (T), and coxa (C) of *Agarodes*. *FIGURE 82* Odontoceridae; pronotum (P), protrochantin (T), and coxa (C) of *Psilotreta*. *FIGURE 83* Goeridae; thoracic terga of *Goera* showing mesepisternum (M). *FIGURE 84* Uenoidae; thoracic terga of *Neophylax*.

18a(17b). Antennae located extremely close to eyes (Fig. 78); median dorsal hump absent from abdominal segment 1............Lepidostomatidae

18b. Antennae midway between eyes and base of mandibles or near base of mandibles; dorsal hump present on abdominal segment 1..........19

19a(18b). Anterolateral angles of pronotum produced and divergent; labrum with transverse dorsal row of about 18 long setae (Fig. 79); case of vegetation; East and West ....Calamoceratidae

19b. Anterolateral angles of pronotum if produced, not divergent; labrum without a dorsal row of about 18 long setae; case of various materials 20

20a(19b). Prosternal horn absent; antennae close to base of mandibles..........21

20b. Prosternal horn present (Fig. 80); antennae about midway between base of mandibles and eye (Fig. 80)..........22

21a(20a). Protrochantin large, hook-shaped (Fig. 81); anal prolegs each with about 30 long setae..........Sericostomatidae

21b. Protrochantins small, not hook-shaped (Fig. 82); anal prolegs each with about 5 long setae..........Odontoceridae

22a(20b). Mesepisternum formed anteriorly into a sharp, elongate process (Fig. 83) or a rounded spiny prominence..........Goeridae

22b, Mesepisternum not enlarged anteriorly as above..........23

23a(22b). Mesonotum with an anterior mesal notch (Fig. 84); SA-1 of metanotum unsclerotized and with only 1 or 2 setae (Fig. 84). ....Uenoidae

23b. Mesonotum without an anterior notch; SA-1 of metanotum with a sclerotized plate and/or more than 2 setae..........24

24a(23b). Head, pronotum, and mesepisternum with prominent surface sculpturing: northwestern United States..........Rossianidae

24b. Head, pronotum, and mesepisternum without prominent surface sculpturing..........25

25a(24b). Mandibles with uniform scraper blades *or* if toothed, SA-1 of metanotum with 25+ setae on membrane between sclerites; case cornucopia-shaped and mostly of sand grains; larvae <12 mm long ....Apataniidae

25b. Mandibles not modified into scraper blades; case of vegetation or mineral materials, larvae often >12 mm long and with a transverse depression of the pronotum in most species..........Limnephilidae

F. Taxonomic Key to Families of Freshwater Megaloptera Larvae

1a. Abdominal segments 1–7 with lateral filaments, last segment with a long, median filament (Fig. 85)..........Sialidae

1b. Abdominal segments 1–8 with lateral filaments, last segment without a median filament, but with a pair of prolegs, each with a pair of claws (Fig. 86) ....Corydalidae

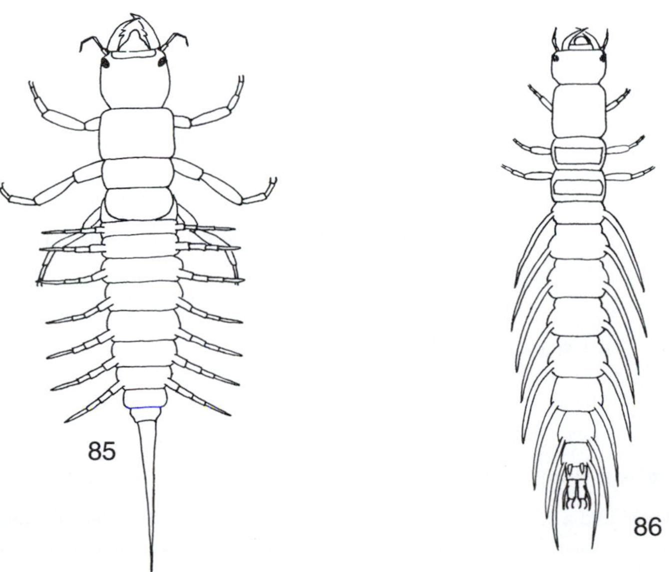

FIGURE 85 Sialidae; *Sialis* (dorsal view). *FIGURE 86* Corydalidae; *Niqronia* (dorsal view).

G. Taxonomic Key to Adults of Aquatic, Semiaquatic, and Riparian Heteroptera

1a. Antennae shorter than head, inserted beneath eyes and (except Ochteridae) not visible from above (Figs. 87, 88, 93, and 95) .... suborder Nepomorpha 2

1b. Antennae longer than head, inserted in front of eyes and visible from above (Figs. 96 and 99–101) 9

2a(1a). Nepomorpha: Rostrum broad, blunt, and triangular, not distinctly segmented (Fig. 88); each front tarsus a 1-segmented scoop fringed with setae (Fig. 89) .... Corixidae

2b. Rostrum cylindrical or cone-shaped, distinctly 3- or 4-segmented (Fig. 90); front tarsi not scooplike 3

3a(2b). Apex of abdomen with a long, slender, tubular respiratory appendage (Fig. 91) .... Nepidae

3b. Apical respiratory appendages absent, or if present, short and flat (Fig. 92) 4

4a(3b). Meso- and metathoracic legs with fringes of swimming hairs; ocelli absent; aquatic 5

4b. Meso- and metathoracic legs without fringes or swimming hairs; ocelli usually present; riparian 8

5a(4a). Dorsoventrally flattened, ovate insects; profemurs broad, raptorial (Fig. 93) 6

5b. Elongate or hemispherical insects, not flattened dorsoventrally (Figs. 94, and 95); profemurs slender, similar to other legs. 7

6a(5a). Length >18 mm; short, flat, straplike apical respiratory appendages present (Fig. 92); eyes protrude from margin of head .... Belostomatidae

6b. Length <16 mm; apical respiratory appendages absent; eyes do not protrude from margin of head (Fig. 93) .... Naucoridae

7a(5b). Hemispherical (Fig. 94); length <3 mm .... Pleidae

7b. Elongate (Fig. 95); length >5 mm .... Notonectidae

8a(4b). Profemurs broad, raptorial; antennae concealed from above .... Gelastocoridae

8b. Profemurs slender, similar to other legs; antennae visible from above .... Ochteridae

9a(1b). Membrane of hemelytra with 4 or 5 equal-sized cells (Fig. 96); metacoxae large, transverse; riparian ..... Leptopodomorpha, Saldidae

9b. Membrane of hemelytra without veins or with dissimilar sized cells; metacoxae small, conical; semiaquatic or riparian .... suborder Gerromorpha 10

10a(9b). Gerromorpha: Claws of at least protarsi inserted before apex (Fig. 97) 11

10b. Claws of all tarsi inserted at apex (Fig. 98) 12

11a(10a). Metafemurs very long, greatly surpassing apex of abdomen .... Gerridae

11b. Metafemurs short, not, or only slightly surpassing apex of abdomen. .... Veliidae

12a(10b). Head as long as entire thorax, very slender, with eyes set about halfway to base (Fig. 99) .... Hydrometridae

12b. Head short and stout, eyes near posterior margin 13

13a(12b). Head grooved ventrally to receive rostrum; tarsi 2-segmented; <2.5 mm long .... Hebridae

13b. Head not grooved ventrally; tarsi 3-segmented; >2.5 mm long 14

14a(13b). Inner margins of eyes converge anteriorly (Fig. 100); femurs with 1 or more dorsal black spines distally .... Mesoveliidae

14b. Inner margins of eyes rounded (Fig. 101); femurs without black spines; riparian; western United States .... Macroveliidae

H. Taxonomic Key to Families of Freshwater Coleoptera

1. Key to Adult Water Beetles

1a. Head formed into a snout or beak anteriorly; antennae geniculate (Fig. 104) .... Curculionidae

1b. Head not formed into a beak or snout; antennae not geniculate 2

2a(1b). Two pairs of eyes, a dorsal and a ventral pair divided by sides of head; meso- and metathoracic legs short, extremely flat, with tarsi folding fanlike .... Gyrinidae

2b. One pair of eyes; meso- and metathoracic legs not extremely flat, with tarsi not folded fanlike 3

3a(2b). Metacoxae expanded into large plates that cover 2 or 3 abdominal sterna and base of metafemurs (Fig. 105) .... Haliplidae

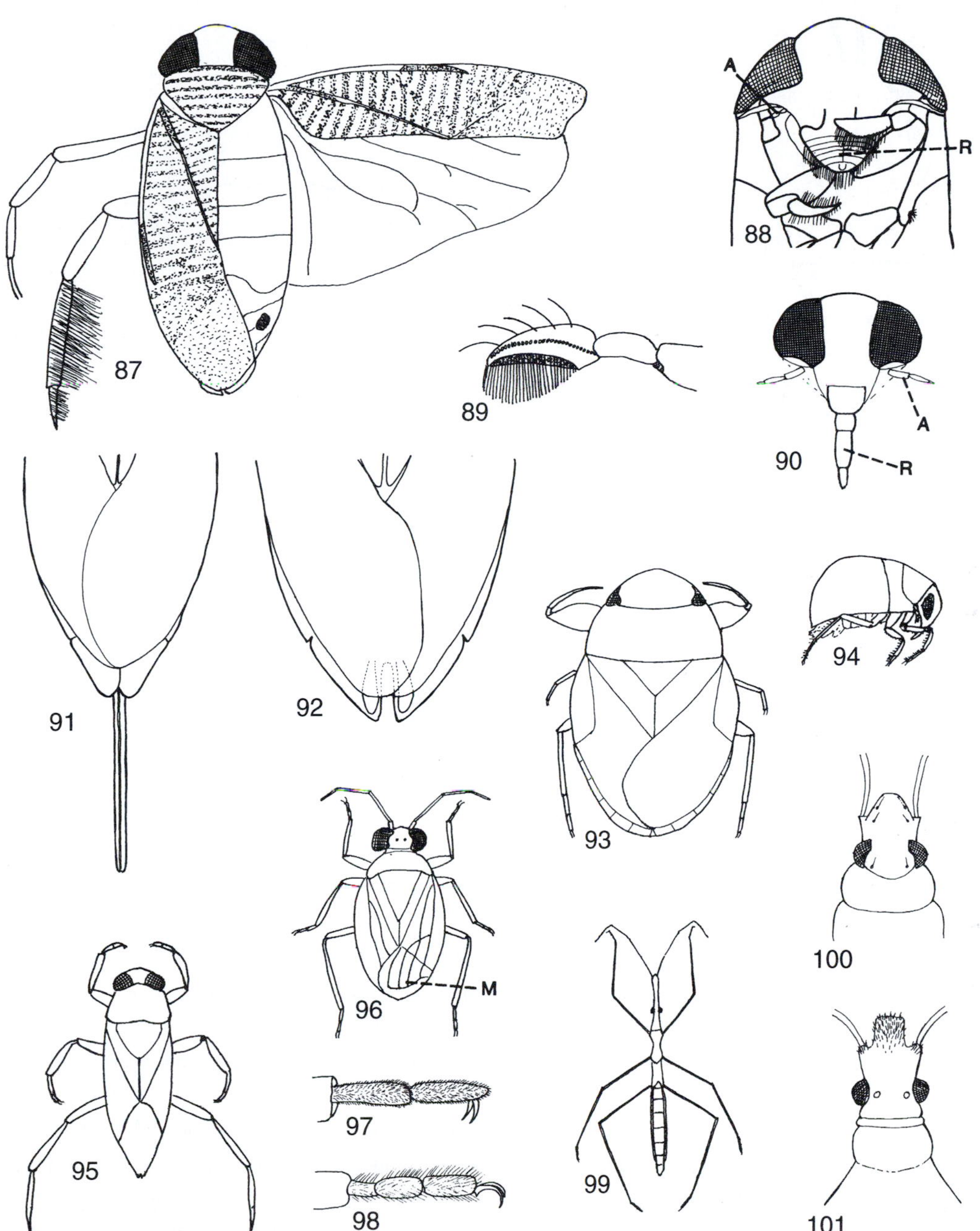

FIGURE 87 Corixidae; *Siqara* (dorsal view), with right wings extended laterally. *FIGURE 88* Corixidae; head and prothorax of *Siqara* (ventral view) showing rostrum (R) and antenna (A). *FIGURE 89* Corixidae; pala of male *Siqara*. *FIGURE 90* Notonectidae; head of *Notonecta* (ventral view) showing rostrum (R) and antenna (A). *FIGURE 91* Nepidae; apex of abdomen of *Nepa* (dorsal view). *FIGURE 92* Belostomatidae; apex of abdomen of *Belostoma* (dorsal view). *FIGURE 93* Naucoridae; *Pelocoris* (dorsal view). *FIGURE 94* Pleidae; *Neoplea* (lateral view). *FIGURE 95* Notonectidae; *Notonecta* (dorsal view). *FIGURE 96* Saldidae; *Salda* (dorsal view) showing membrane (M). *FIGURE 97* Gerridae: protarsus of *Gerris*. *FIGURE 98* Mesoveliidae; protarsus of *Mesovelia*. *FIGURE 99* Hydrometridae; *Hydrometra* (dorsal view). *FIGURE 100* Mesoveliidae; head of *Mesovelia* (dorsal view). *FIGURE 101* Macroveliidae; head of *Macrovelia* (dorsal view).

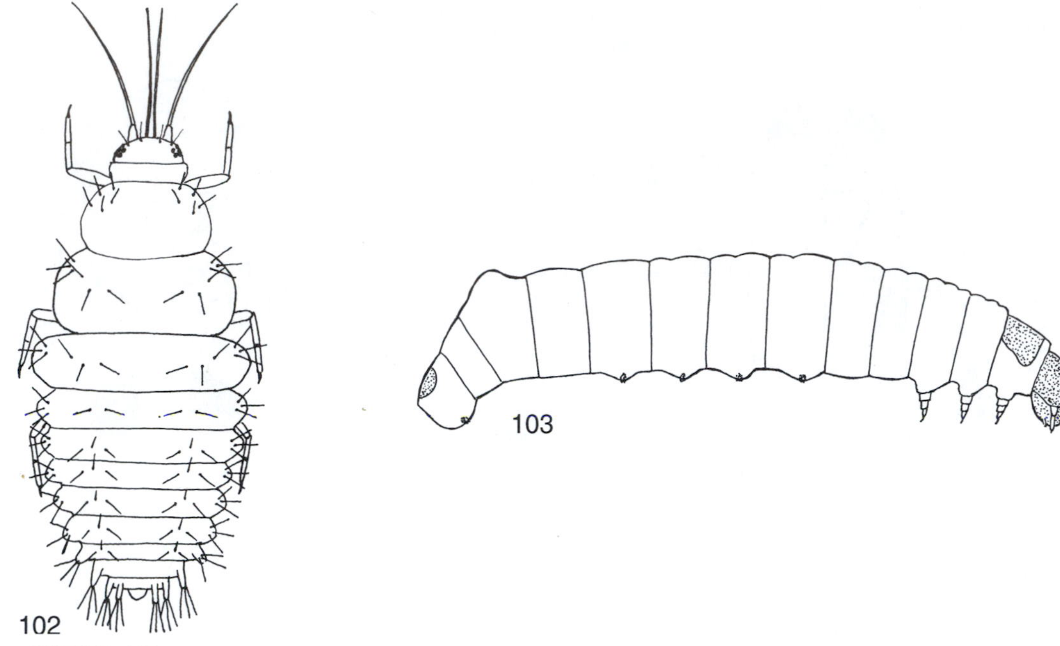

FIGURE 102 Neuroptera Larva (Sisyridae); *Climacia* (dorsal view). *FIGURE 103* Lepidoptera Larva (Pyralidae); *Nymphula* (lateral view).

3b. Metacoxae not expanded into large plates 4

4a(3b). Prosternum with a postcoxal process that extends posteriorly to mesocoxae (Fig. 106); first visible abdominal sternum completely divided by metacoxal process (Fig. 106) 5

4b. Prosternum with postcoxal process absent or short; first visible abdominal sternum extending for its entire breadth behind metacoxae (Fig. 107) 8

5a(4a). Base of pronotum much narrower than base of elytra (Fig. 108); metatarsi rounded and not fringed with long setae; large >10 mm long; western mountains of the United States .... Amphizoidae

5b. Base of pronotum subequal in width to base of elytra; metatarsi flattened and fringed with long setae *or* beetles <2 mm long 6

6a(5b). Anterior of prosternum, its postcoxal process and metasternum in same plane (Fig. 109); pro- and mesotarsi distinctly 5-segmented, segment 4 as long as 3 7

6b. Anterior of prosternum greatly depressed and not in same plane as its postcoxal process and metasternum (Fig. 110); pro- and mesotarsi appear to be 4-segmented .... Dytiscidae (in part)

7a(6a). Prosternal process pointed or nearly so (Figs. 106); no curved spur or hooked apex on protibiae; >4 mm long .... Dytiscidae (in part)

7b. Prosternal process truncate or rounded apically (Fig. 111); protibiae with curved spur or hooked apex (Fig. 112) or beetles <3 mm long; eastern, central United States .... Noteridae

8a(4b). Antennae short, club-shaped, with segment 4 or 6 modified to form a cupule (Fig. 113); maxillary palps usually longer than antennae 9

8b. Antennae pectinate or filiform (Figs. 116, 117), usually longer than maxillary palps 12

9a(8a). Antennae with 5 segments past cupule; less than 2.5 mm long .... Hydraenidae

9b. Antennae with 3 segments past cupule (Fig. 113); 1.5–40.0 mm long 10

10a(9b). Pronotum with 7 longitudinal grooves, including marginal grooves (Fig. 114); 2.0–7.1 mm long .... Helophoridae

10b. Pronotum without 7 longitudinal grooves 11

11a(10b). Pronotum much narrower than elytral base and scutellum very small (Fig. 115); pronotum granular or with large punctures; 2.2–5.3 mm long .... Hydrochidae

Pronotum not much narrower than base of elytra, or scutellum elongate; pronotum not granulate, large punctures lacking; 1.5–42.0 mm long .... Hydrophilidae

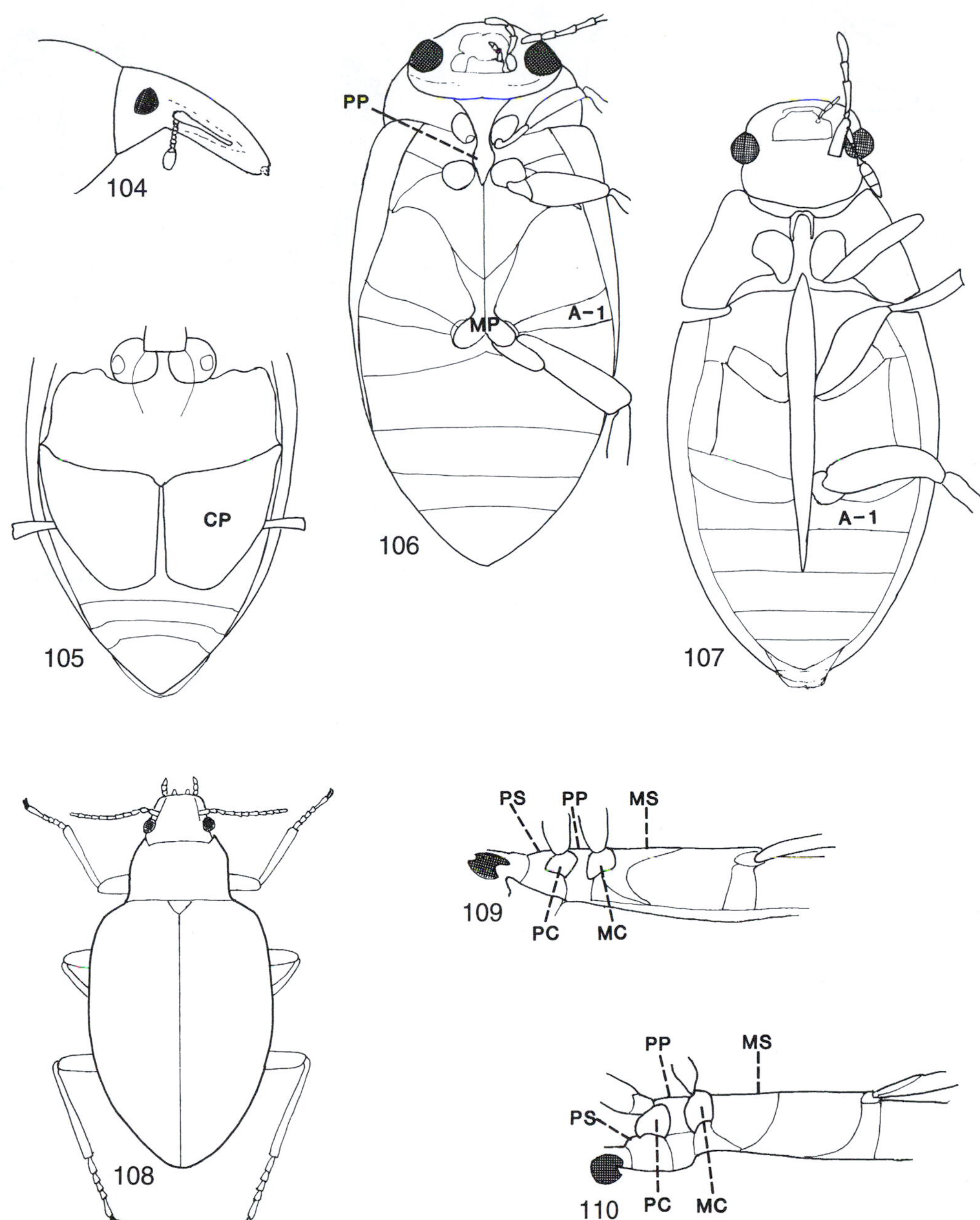

FIGURE 104 Curculionidae; head (lateral view). *FIGURE 105* Haliplidae; metathorax and abdomen of *Haliplus* (ventral view) showing metacoxal plate (CP). *FIGURE 106* Dytiscidae; *Aqabus* (ventral view) showing prosternal process (PP), first abdominal sternum (A-1), and metacoxal process (MP). *FIGURE 107* Hydrophilidae; *Tropisternus* (ventral view) showing first abdominal sternum (A-1). *FIGURE 108* Amphizoidae; *Amphizoa* (dorsal view). *FIGURE 109* Dytiscidae; *Aqabus* (lateral view) showing prosternum (PS), prosternal process (PP), metasternum (MS), procoxa (PC), and mesocoxa (MC). *FIGURE 110* Dytiscidae; *Hydroporus* (lateral view) showing prosternum (PS), prosternal process (PP), metasternum (MS), procoxa (PC), and mesocoxa (MC).

12a(8b). Antennae short with pectinate club (Fig. 116); > 5.0 mm long .... Dryopidae

12b. Antennae slender, filiform; (Fig. 117); < 4.5 mm long 13

13a(12b). Extremely small, < 1.5 mm long (Fig. 118); all tarsi 3-segmented; southwest United States, north to Idaho .... Hydroscaphidae

13b. Larger, > 1.7 mm long; all tarsi 5-segmented .... Elmidae

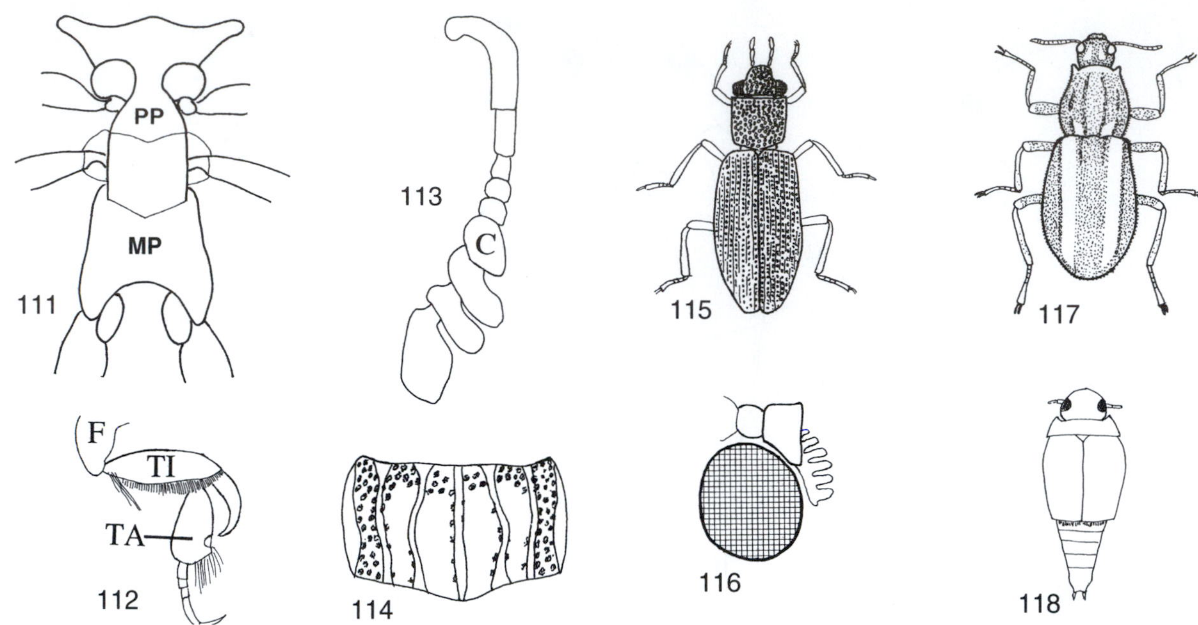

FIGURE 111 Noteridae; thorax of *Hydrocanthus* (ventral view) showing prosternal process (PP) and metasternal plate (MP). *FIGURE 112* Noteridae; prothoracic leg of *Hydrocanthus* showing profemur (F), tibia (TI) and tarsus (TA). *FIGURE 113* Hydrophilidae; antenna of *Tropisternus* showing cupule (C). *FIGURE 114* Helophoridae; pronotum of *Helophorus* showing longitudinal grooves. *FIGURE 115* Hydrochidae; *Hydrochus* (dorsal view). *FIGURE 116* Dryopidae; right antenna and eye of *Helichus* (dorsal view). *FIGURE 117* Elmidae; *Stenelmis* (dorsal view). *FIGURE 118* Hydroscaphidae; *Hydroscapha* (dorsal view).

2. Key to Larval Water Beetles

1a. Tarsi with 2 claws 2

1b. Tarsi with 1 claw 5

2a(1a). Abdomen with 4 conspicuous hooks on last segment; abdominal segments with at least 8 pairs of lateral filaments (Fig. 119) Gyrinidae

2b. No hooks on last abdominal segment; if lateral abdominal filaments are present, there are only 6 pairs 3

3a(2b). Abdominal and thoracic terga flattened and expanded laterally (Fig. 120); western mountains of the United States........Amphizoidae

3b. Abdominal and thoracic terga not flattened and expanded laterally 4

4a(3b). Urogomphi shorter than last abdominal segment; legs short, stout, adapted for digging (Fig. 121); mandibles short, adapted for chewing; eastern, central United States .... Noteridae

4b. Urogomphi usually longer than last abdominal segment (Fig. 122); *if shorter*, legs are elongate with setal fringe for swimming; mandibles elongate, pointed, for piercing .... Dytiscidae

5a(1b). Legs distinctly 5-segmented; abdomen terminating in 1 or 2 long filaments (Figs. 123, and 124) .... Haliplidae

5b. Legs apparently 4-segmented; abdomen not terminating in long filaments 6

6a(5b). Mandibles large, readily visible from above (Fig. 125) .... Hydrophilidae

6b. Mandibles not readily visible from above 7

7a(6b). Antennae long, filiform, as long as head and thorax combined (Fig. 126) .... Scirtidae

7b. Antennae much shorter than head and thorax combined 8

8a(7b). Body oval and extremely flat; head completely concealed from dorsal view (Figs. 127, and 128) .... Psephenidae

8b. Body elongate and round or triangular in cross section; head exposed, except mostly concealed in Lampyridae 9

9a(8b). Each thoracic and abdominal segment covered by a flat, platelike sclerite dorsally; prothoracic plate mostly or completely concealing head from above (Fig. 129) .... Lampyridae

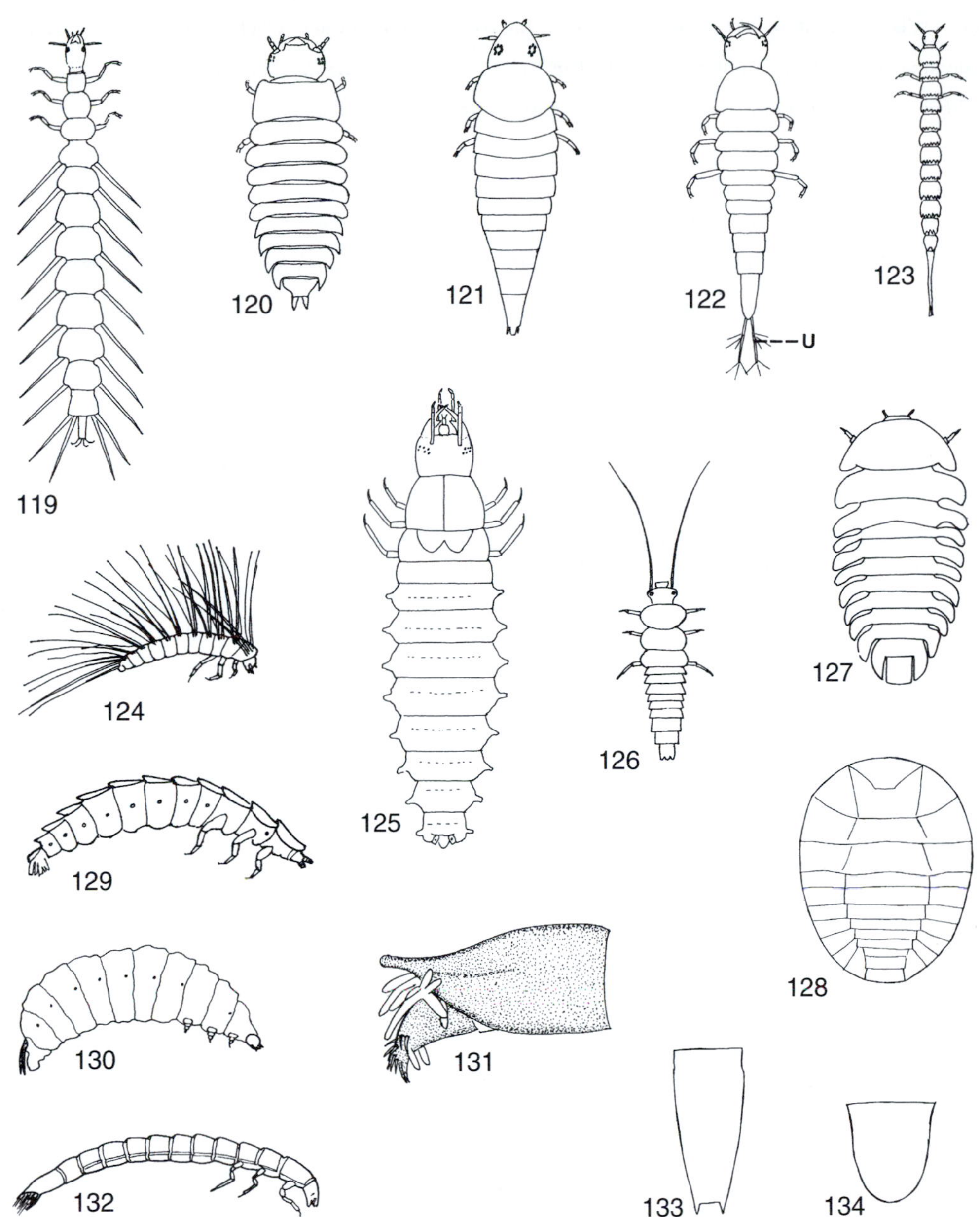

FIGURE 119 Gyrinidae; *Dineutus* (dorsal view). *FIGURE 120* Amphizoidae; *Amphizoa* (dorsal view). *FIGURE 121* Noteridae; *Hydrocanthus* (dorsal view). *FIGURE 122* Dytiscidae; *Aqabus* (dorsal view) showing urogomphi (U). *FIGURE 123* Haliplidae; *Haliplus* (dorsal view). *FIGURE 124* Haliplidae; *Peltodytes* (lateral view). *FIGURE 125* Hydrophilidae; *Tropisternus* (dorsal view). *FIGURE 126* Scirtidae; *Cyphon* (dorsal view). *FIGURE 127* Psephenidae; *Ectopria* (dorsal view). *FIGURE 128* Psephenidae; *Psephenus* (dorsal view). *FIGURE 129* Lampyridae (lateral view). *FIGURE 130* Chrysomelidae; *Donacia* (lateral view). *FIGURE 131* Ptilodactylidae; last abdominal segment of *Anchytarsus* (lateral view). *FIGURE 132* Elmidae; *Stenelmis* (lateral view). *FIGURE 133* Elmidae; last abdominal tergum of *Stenelmis*. *FIGURE 134* Lutrochidae; last abdominal tergum of *Lutrochus*.

9b.(8a). Thoracic and abdominal segments not covered by flat, platelike sclerites; head visible from above 10

10a(9b). All terga rounded and pale; grublike larvae with 2 spines on last abdominal tergum (Fig. 130) ....Chrysomelidae

10b. Body elongate and sclerotized; no spines on last abdominal tergum (Figs. 131, and 132)..11

11a(10b). Abdominal sterna with distinct tufts of gills, either on sterna 1–7 or on last segment and lacking and operculum (Fig. 131); eastern, United States and California..Ptilodactylidae

11b. Abdominal gills, if present, on last abdominal segment and covered by a ventral operculum (Fig. 132) 12

12a(11b). Last abdominal tergum bifid or notched apically (Fig. 133) ....Elmidae

12b. Last abdominal tergum rounded apically (Fig. 134); East, Southwest. ....Lutrochidae

I. Taxonomic Key to Families of Freshwater Diptera Larvae

1a. Larvae apparently 7-segmented; first 6 segments each with a prominent ventral sucker (Fig. 135) ....Nematocera, Blephariceridae

1b. Larvae with more than 7 apparent segments; without 6 ventral suckers 2

2a(1b). Head capsule completely sclerotized and fully visible (Figs. 147–159); mandibles opposed, moving in a horizontal plane, and usually with two or more apical teeth; body never flattened and with a posterior spiracular chamber margined with long, soft hairs. ....Nematocera 12

2b. Sclerotized head capsule absent (Figs 138–140, 142, 143, 145, and 146), or incomplete behind (Figs. 136 and 137), and retracted at least partially into thorax; mandibles parallel, without secondary apical teeth, and moving in a vertical plane *or* in a partially sclerotized retracted head may be opposed and moving in a horizontal plane 3

3a(2b). Mandibles opposed, moving in a horizontal plane, and with two or more apical teeth; larvae truncate anteriorly with head capsule retracted into thorax and incomplete posteriorly (Figs. 136 and 137)..Nematocera, Tipulidae

3b. Mandibles parallel, moving vertically, and without secondary apical teeth; larvae narrowed anteriorly with head capsule lacking or poorly developed and incompletely sclerotized *or* body terminating in a long respiratory tube (Fig. 139) or a posterior spiracular chamber margined with long, soft hairs ....Brachycera 4

4a(3b). Brachycera: Head mostly visible, truncate in shape (Fig. 138); body somewhat flattened; posterior spiracular chamber margined with long, soft hairs; integument covered with shiny calcium carbonate crystals ....Stratiomyidae

4b. Head mostly retracted into thorax and elongate, or indistinguishable; body nearly circular in cross section; without a posterior spiracular chamber margined with long, soft hairs; integument lacking calcium carbonate crystals..5

5a(4b). Larvae with a partially retractile caudal respiratory tube at least one-half as long as body (Fig. 139) ....Syrphidae

5b. Larvae without a long respiratory tube, if a short tube is present, it is divided apically 6

6a(5b). Body terminating in a short tube that is divided apically (Fig. 140), or in a pair of spines ....Ephydridae

6b. Body not terminating in a short tube or a pair of spines 7

7a(6b). Caudal spiracular disk with palmate hairs and surrounded by 8–10 lobes, some of which may be very short (Fig. 141); body wrinkled..Sciomyzidae

7b. Caudal spiracular disk without palmate hairs, if surrounded by lobes, body not wrinkled..8

8a(7b). Abdomen without distinct prolegs, paired ventral pseudopods, or terminal processes except a single spine, but with a girdle of 6 or more pseudopods on each segment (Fig. 142) ....Tabanidae

8b. Abdomen with distinct prolegs, paired pseudopods, or paired terminal processes present..9

9a(8b). Body terminating in a spiracular pit surrounded by 4 pointed lobes (Fig. 143)..Dolichopodidae

9b. Body not terminating in a spiracular pit with 4 lobes 10

10a(9b). Body terminating in a pair of ciliated, divergent processes (Fig. 144); abdomen with 8 pairs of ventral prolegs ....Athericidae

10b. Terminal processes, if present, not ciliated; abdomen with or without prolegs 11

11a(10b). Some external head structure visible, with palps and antennae usually present; abdomen usually with ventral prolegs and elongate, paired, terminal appendages (Fig. 145) or a bulbous segment..Empididae

11b. No visible external head structure; abdomen often with ventral pseudopods and usually short, paired, terminal appendages (Fig. 146) ....Muscidae

12a(2a). Nematocera: Prolegs absent (Figs. 147, 150, and 151) 13

12b. Prolegs present at one or both ends of body or on abdominal segments (Figs. 152–159)..17

13a(12a). Thoracic segments fused and distinctly thicker than abdomen (Fig. 147) 14

13b. Thoracic segments not fused and about equal to abdomen in diameter (Figs. 150 and 151)..16

14a(13a). Antennae prehensile, with long, strong apical spines (Fig. 148) 15

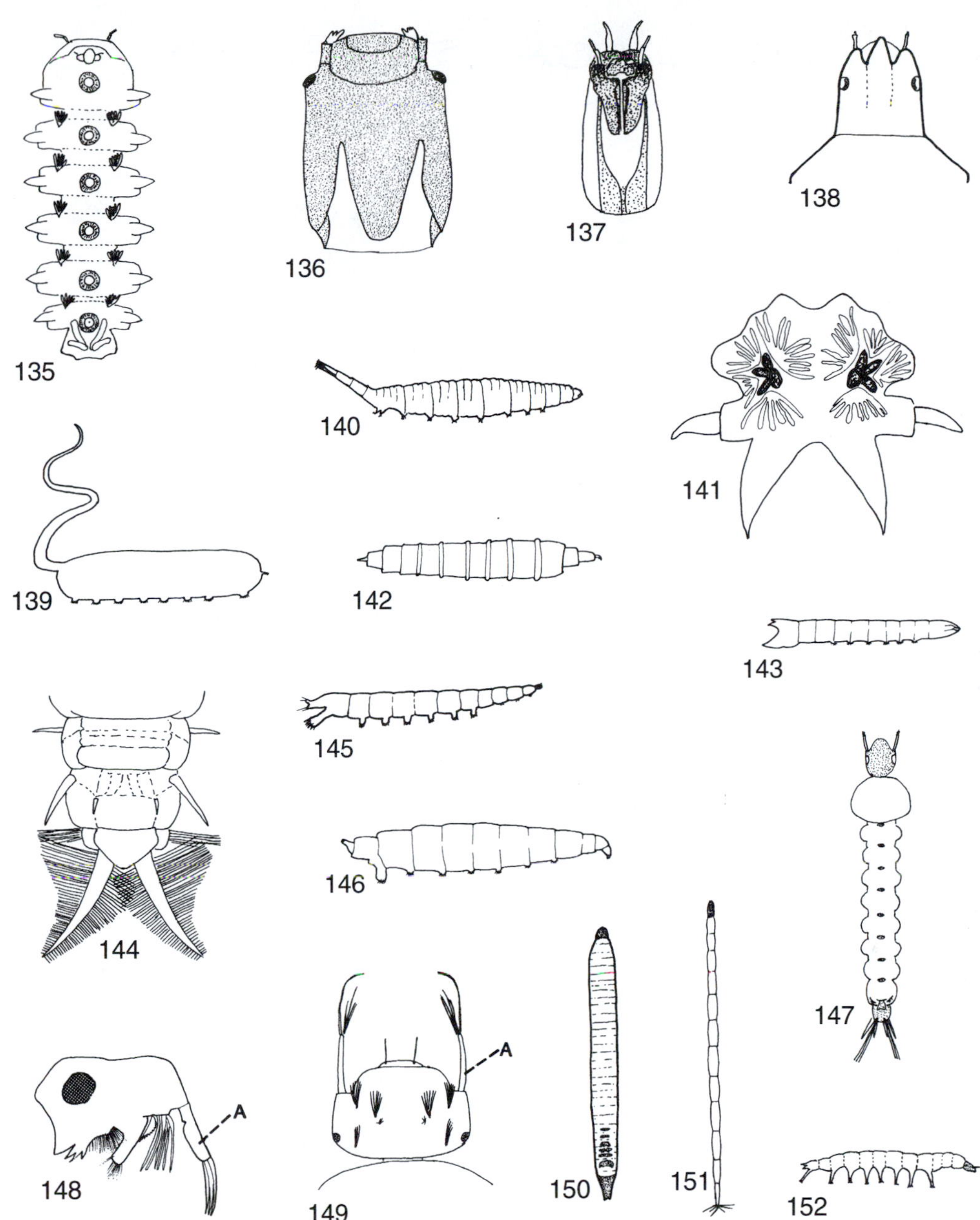

FIGURE 135 Blephariceridae; *Blepharicera* (ventral view). *FIGURE 136* Tipulidae; head of *Limonia* (dorsal view). *FIGURE 137* Tipulidae; head of *Hexatoma* (dorsal view). *FIGURE 138* Stratiomyidae; head of *Odontomyia* (dorsal view). *FIGURE 139* Syrphidae; *Eristalis* (lateral view). *FIGURE 140* Ephydridae (lateral view). *FIGURE 141* Sciomyzidae; spiracular disc of *Sepedon*. *FIGURE 142* Tabanidae; *Chrysops* (lateral view). *FIGURE 143* Dolicopodidae (lateral view). *FIGURE 144* Athericidae; terminal segments of *Atherix* (dorsal view). *FIGURE 145* Empididae (lateral view). *FIGURE 146* Muscidae; *Limnophora* (lateral view). *FIGURE 147* Culicidae; *Anopheles* (dorsal view). *FIGURE 148* Chaoboridae; head of *Chaoborus* (lateral view) showing antenna (A). *FIGURE 149* Culicidae; head of *Coquillettidia* (dorsal view) showing antenna (A). *FIGURE 150* Psychodidae; *Psychoda* (dorsal view). *FIGURE 151* Ceratopogonidae; *Palpomyia* (dorsal view). *FIGURE 152* Nymphomyiidae; *Palaeodipteron* (lateral view).

14b. Antennae not prehensile, lacking long apical spines (Fig. 149) .... Culicidae

15a(14a). Antennae close together at base; a row of spinose setae on dorsolateral margin of head .... Corethrellidae

15b. Antennae at anterolateral margins of head; head without spinose setae dorsolaterally .... Chaoboridae

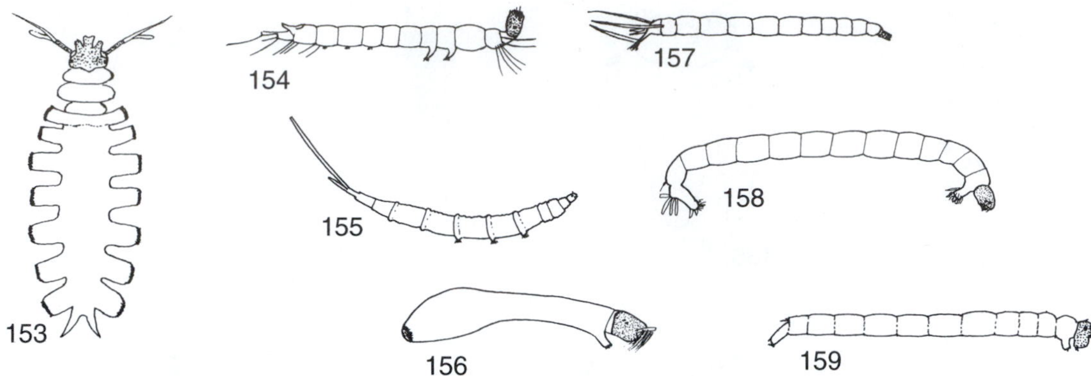

FIGURE 153 Deuterophlebiidae; ***Deuterophlebia*** (dorsal view). *FIGURE 154* Dixidae; ***Dixella*** (lateral view). *FIGURE 155* Ptychopteridae; ***Ptychoptera*** (lateral view). *FIGURE 156* Simuliidae; ***Simulium*** (lateral view). *FIGURE 157* Tanyderidae; ***Protoplasa*** (lateral view). *FIGURE 158* Chironomidae; ***Chironomus*** (lateral view). *FIGURE 159* Thaumaleidae; ***Thaumalea*** (lateral view).

16a(13b). Thoracic and abdominal segments each distinctly divided into 2 or 3 annuli, with sclerotized dorsal plates on some annuli (Fig. 150) Psychodidae

16b. No secondary annulations on abdominal segments (Fig. 151) ....Ceratopogonidae (in part)

17a(12b). Prolegs on intermediate body segments (Figs. 152–155) 18

17b. Prolegs on anterior and/or posterior ends of body only (Figs. 156–159) 21

18a(17a). Seven or eight pairs of distinct prolegs on abdomen (Figs. 152 and 153) 19

18b. Two or three pairs of weak prolegs on abdomen (Figs. 154 and 155) 20

19a(18a). Eight pairs of slender, ventrally projecting prolegs (Fig. 152); PQ, NB, ME ....Nymphomyiidae

19b. Seven pairs of stout, ventrolaterally projecting prolegs (Fig. 153); western mountains ....Deuterophlebiidae

20a(18b). Paired ventral prolegs on abdominal segment 1 and usually also on segment 2; posterior end of body with 2 pairs of fringed processes (Fig. 154) ....Dixidae

20b. Paired ventral prolegs on abdominal segments 1, 2, and 3; body terminating in a long respiratory tube (Fig. 155) ....Ptychopteridae

21a(17b). A single proleg present only on prothorax; posterior of abdomen swollen and terminating in a ring of numerous small hooks (Fig. 156) ....Simuliidae

21b. Posterior prolegs usually present; posterior of abdomen not swollen and terminating in a ring of hooks 22

22a(21b). Only posterior prolegs present (Fig. 157) 23

22b. Anterior and usually posterior prolegs present (Figs. 158 and 159) 24

23a(22a). Long filamentous processes arising from last two abdominal segments and from prolegs (Fig. 157); Eastern and Western United States ....Tanyderidae

23b. Last two abdominal segments without long filamentous processes ....Ceratopogonidae (in part)

24a(22b). Body covered with long, strong spines ....Ceratopogonidae (in part)

24b. Body at most covered with setae 25

25a(24b). At least one pair of prolegs separated distally (Fig. 158); stalked spiracles lacking ....Chironomidae

25b. Prolegs unpaired (Fig. 159); short, stalked spiracles dorsolaterally on the prothorax; mountain streams ....Thaumaleidae

J. Taxonomic Key to Families of Semiaquatic Collembola

1a. Body somewhat globular; thorax and abdomen indistinctly segmented (Fig. 160) ....Sminthuridae

1b. Body elongate; thorax and abdomen distinctly segmented (Fig. 161) 2

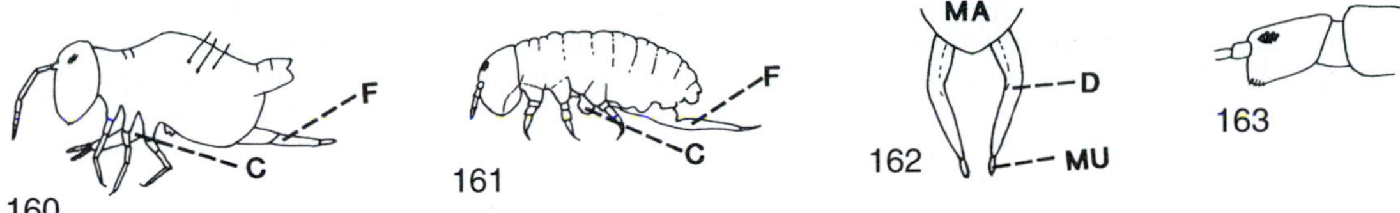

FIGURE 160 Sminthuridae (lateral view) showing furcula (F) and collophore (C). *FIGURE 161* Poduridae; *Podura* (lateral view) showing furcula (F) and collophore (C). *FIGURE 162* Poduridae; furcula of *Podura* (dorsal view) showing manubrium (MA), dens (D), and mucro (MU). *FIGURE 163* Isotomidae anterior portion (lateral view). [*Figures 158–161* redrawn from Peckarsky *et al.* (1990).]

2a(1b). Mouthparts directed downward (Fig. 161); furcula distinctly convergent apically (Fig. 162); first abdominal segment with dorsal setae .... Poduridae

2b. Mouthparts directed forward (Fig. 163); furcula not distinctly convergent apically; first abdominal segment lacking dorsal setae .... Isotomidae

LITERATURE CITED

Adler, P. H., Kim, K. C. 1986. The blackflies of Pennsylvania (Simuliidae, Diptera). Bionomics, taxonomy, and distribution. Pennsylvania State University Agricultural Experiment Station Bulletin 856, 88 pp.

Agriculture Canada, Research Branch. 1981. Manual of Nearctic Diptera Vol. 1. Monograph No. 27. Canadian Government Publishing Centre, Supply and Services Canada, Hull, Que., Canada, vi + 674 pp.

Agriculture Canada, Research Branch. 1987. Manual of Nearctic Diptera Vol. 2. Monograph No. 28. Canadian Government Publishing Centre, Supply and Services Canada, Hull, Que., Canada, vi + 658 pp.

Alarie, Y. 1991. Description of larvae of 17 Nearctic species of *Hydroporus* Clairville (Coleoptera: Dytiscidae: Hydroporinae) with an analysis of their phylogenetic relationships. Canadian Entomologist 123:627–704.

Alarie, Y. 1992. Descriptions of the larval stages of *Hydroporus (Heterosternuta) Cocheconis* Fall 1917 (Coleoptera: Dytiscidae: Hydroporinae) with a key to the known larvae of *Heterosternuta* Strand. Canadian Entomologist 124:827–840.

Alarie, Y. 1993. A systematic review of the North American species of the *Oreodytes alaskanus* clade (Coleoptera: Dytiscidae: Hydroporinae). Canadian Entomologist 125:847–867.

Alarie, Y. 1997. Taxonomic revision and phylogenetic analysis of the genus *Oreodytes* Seidlitz (Coleoptera: Dytiscidae: Hydroporinae) based on larval morphology. Canadian Entomologist 129: 399–503.

Alarie, Y., Harper, P. P., Maire, M. 1989. Rearing dytiscid beetles (Coleoptera: Dytiscidae). Entomologica Basiliensia 13:147–149.

Alarie, Y., Harper, P. P., Roughley, R. E. 1990. Description of the larvae of eleven Nearctic species of *Hygrotus* Stephens (Coleoptera: Dytiscidae: Hydroporinae) with an analysis of their phyletic relationships. Canadian Entomologist 122:985–1035.

Allen, R. K. 1973. Generic revisions of mayfly nymphs. 1. *Traverella* in North and Central America (Leptophlebiidae). Annals of the Entomological Society of America 66:1287–1295.

Allen, R. K. 1978. The nymphs of North and Central American *Leptohyphes* (Ephemeroptera: Tricorythidae). Annals of the Entomological Society of America 71:537–558.

Allen, R. K., Brusca, R. C. 1978. Generic revisions of mayfly nymphs II. *Thraulodes* in North and Central America (Leptophlebiidae). Canadian Entomologist 110:413–433.

Allen, R. K., Edmunds, G. F., Jr. 1959. A revision of the genus Ephemerella (Ephemeroptera: Ephemerellidae) I. The subgenus *Timpanoga*. Canadian Entomologist 91:51–58.

Allen, R. K., Edmunds, G. F., Jr. 1961a. A revision of the genus *Ephemerella* (Ephemeroptera: Ephemerellidae) II. The subgenus *Caudatella*. Annals of the Entomological Society of America 54:603–612.

Allen, R. K., Edmunds, G. F., Jr. 1961b. A revision of the genus *Ephemerella* (Ephemeroptera: Ephemerellidae) III. The subgenus *Attenuatella*. Journal of the Kansas Entomological Society 34:161–173.

Allen, R. K., Edmunds, G. F., Jr. 1962a. A revision of the genus *Ephemerella* (Ephemeroptera: Ephemerellidae) IV. The subgenus *Dannella*. Journal of the Kansas Entomological Society 35:333–338.

Allen, R. K., Edmunds, G. F., Jr. 1962b. A revision of the genus *Ephemerella* (Ephemeroptera: Ephemerellidae) V. The subgenus *Drunella* in North America. Miscellaneous Publications of the Entomological Society of America 3:147–187.

Allen, R. K., Edmunds, G. F., Jr. 1963a. A revision of the genus *Ephemerella* (Ephemeroptera: Ephemerellidae) VI. The subgenus *Serratella* in North America. Annals of the Entomological Society of America 56:583–600.

Allen, R. K., Edmunds, G. F., Jr. 1963b. A revision of the genus *Ephemerella* (Ephemeroptera: Ephemerellidae) VII. The subgenus *Eurylophella*. Canadian Entomologist 95:597–623.

Allen, R. K., Edmunds, G. F., Jr. 1965. A revision of the genus *Ephemerella* (Ephemeroptera: Ephemerellidae) VIII. The subgenus *Ephemerella* in North America. Miscellaneous Publications of the Entomological Society of America 4:244–282.

Allen, R. K., Edmunds, G. F., Jr. 1976. A revision of the genus *Ametropus* in North America (Ephemeroptera: Ametropodidae). Journal of the Kansas Entomological Society 49:625–635.

Allen, R. K., Murvosh, C. M. 1987. Mayflies (Ephemeroptera: Tricorythidae) of the southwestern United States and northern Mexico. Annals of the Entomological Society of America 80:35–40.

Andersen, N. M. 1990. Phylogeny and taxonomy of water striders, genus *Aquarius* Schellenberg (Insecta, Hemiptera, Gerridae), with a new species from Australia. Steenstrupia 16:37–81.

Andersen, N. M. 1993. Classification, phylogeny, and zoogeography of the pond skater genus *Gerris* Fabricius (Hemiptera: Gerridae). Canadian Journal of Zoology 71:2473–2508.

Andersen, N. M., Spence, J. R. 1992. Classification and phylogeny of the Holarctic water strider genus *Limnoporus* Stal (Hemiptera: Gerridae). Canadian Journal of Zoology 70:753–785.

Anderson, L. D. 1932. A monograph of the genus *Metrobates*. University of Kansas Science Bulletin 20:297–311.

Anderson, N. H., Sedell, J. R. 1979. Detritus processing by macroinvertebrates in stream ecosystems. Annual Review of Entomology 24:351–377.

Anderson, R. D. 1962. The Dytiscidae (Coleoptera) of Utah: keys, original citation, types, and Utah distribution. Great Basin Naturalist 22: 54–75.

Anderson, R. D. 1970. A revision of the Nearctic representatives of *Hygrotus* (Coleoptera: Dytiscidae). Annals of the Entomological Society of America 64:503–512.

Anderson, R. D. 1975. A revision of the Nearctic species of *Hygrotus* groups II and III (Coleoptera: Dytiscidae). Annals of the Entomological Society of America 69:577–584.

Anderson, R. D. 1983. Revision of the Nearctic species of *Hygrotus* groups IV, V, and VI (Coleoptera: Dytiscidae). Annals of the Entomological Society of America 76:173–196.

Archangelsky, M. 1998. Phylogeny of Hydrophiloidea (Coleoptera: Staphyliniformia) using characters from adult and preimaginal stages. Systematic Entomology 23:9–24.

Bae, Y. J., McCafferty, W. P. 1991. Phylogenetic systematics of the Potamanthidae (Ephemeroptera). Transactions of the American Entomological Society 117(3,4):1–143.

Bae, Y. J., McCafferty, W. P. 1998. Phylogenetic systematics and biogeography of the Neoephemeridae (Ephemeroptera: Pannota). Aquatic Insects 20:35–68.

Barr, C. B., Chapin, J. B. 1988. The aquatic Dryopoidea of Louisiana (Coleoptera: Psephenidae, Dryopidae, Elmidae). Tulane Studies in Zoology and Botany 26:89–164.

Baumann, R. W., Gaufin, A. R., Surdick, R. F. 1977. The stoneflies (Plecoptera) of the Rocky Mountains. Memoirs of the American Entomological Society 31, ii + 208 pp.

Baumann, R. W., Stark, B. P. 1980. *Hesperoperla hoguei*, a new species of stonefly from California (Plecoptera: Perlidae). Great Basin Naturalist 40:63–67.

Bay, E. C. 1974. Predator–prey relationships among aquatic insects. Annual Review of Entomology 19:441–453.

Bednarik, A. F., Edmunds, G. F. Jr. 1980. Descriptions of larval *Heptagenia* from the Rocky Mountain region. (Ephemeroptera: Heptageniidae). Pan-Pacific Entomologist 56:51–62.

Bednarik, A. F., McCafferty, W. P. 1979. Biosystematic revision of the genus *Stenonema* (Ephemeroptera: Heptageniidae). Canadian Bulletin of Fisheries and Aquatic Sciences 201, vi + 73 pp.

Bennefield, B. L. 1965. A taxonomic study of the subgenus *Ladona* (Odonata: Libellulidae). University of Kansas Science Bulletin 45:361–389.

Bennett, D. V., Cook, E. F. 1981. The semiaquatic Hemiptera of Minnesota (Hemiptera: Heteroptera). Technical Bulletin 332 Agricultural Experiment Station, University of Minnesota, 59 pp.

Berg, C. O., Knudson, L. 1978. Biology and systematics of the Sciomyzidae. Annual Review of Entomology 23:239–258.

Berge Henegouwen, van, A. 1986. Revision of the European species of *Anacaena* Thomson (Coleoptera: Hydrophilidae). Entomologica Scandinavica 17:393–407.

Bergman, E. A., Hilsenhoff, W. L. 1978. *Baetis* (Ephemeroptera: Baetidae) of Wisconsin. Great Lakes Entomologist 11:125–135.

Berner, L. 1956. The genus *Neoephemera* in North America (Ephemeroptera: Neoephemeridae). Annals of the Entomological Society of America 49:33–42.

Berner, L. 1959. A tabular summary of the biology of North American mayfly nymphs (Ephemeroptera). Bulletin of the Florida State Museum 4:1–58.

Berner, L., Pescador, M. L. 1988. The mayflies of Florida. Revised edition., Uni. Press of Florida, Gainesville, FL, 415 pp.

Bilyj, B. 1988. A taxonomic review of *Guttipelopia* (Diptera: Chironomidae). Entomologica Scandinavica 19:1–26.

Bistrom, O. 1996. Taxonomic revision of the genus *Hydrovatus* Motschulsky (Coleoptera, Dytiscidae). Entomologica Basiliensia 19:57–584.

Bobb, M. L. 1974. The insects of Virginia: No. 7. The aquatic and semi-aquatic Hemiptera of Virginia. Research Division Bulletin 87, Virginia Polytechnic Institute and State University, Blacksburg, iv + 196 pp.

Bode, R. W. 1983. Larvae of North American *Eukiefferiella* and *Tvetenia* (Diptera: Chironomidae). Bulletin 452, New York State Museum, v + 40 pp.

Boesel, M. W. 1985. A brief review of the genus *Polypedilum* in Ohio, USA with keys to known stages of species occurring in northeastern USA. Ohio Journal of Science 85:245–262.

Borkent, A. 1979. Systematics and bionomics of the species of the subgenus *Schadonophasma* Dyar and Shannon (*Chaoborus*, Chaoboridae, Diptera). Quaestiones Entomologicae 15:122–255.

Borkent, A. 1984. The systematics and phylogeny of the *Stenochironomus* complex (*Xestochironomus*, *Harrisius*, and *Stenochironomus*) (Diptera: Chironomidae). Memoirs of the Entomological Society of Canada 128, 269 pp.

Bowles, D. E. 1995. A new species of *Austrotinodes* (Trichoptera: Ecnomidae) from Texas. Journal of the New York Entomological Society 103:155–161.

Bratt, A. D., Knutson, L. V., Foote, B. A., Berg, C. O. 1969. Biology of *Pherbellia* (Diptera: Sciomyzidae). Memoirs of the Cornell University Agricultural Experiment Station 404, 247 pp.

Breeland, S. G., Loyless, T. M. 1982. Illustrated keys to the mosquitoes of Florida. Adult females and fourth stage larvae. Journal of the Florida Anti-mosquito Association 53:63–84.

Brigham, A. R., Brigham, W. U., Gnilka, A. Eds. 1982. The aquatic insects and oligochaetes of North and South Carolina. Midwest Aquatic Enterprises, Mahomet, IL, xi + 837 pp.

Brigham, W. U. 1982a. Megaloptera. *in*: Brigham, A. R., Brigham, W. U. Gnilka, A. Eds. The aquatic insects and oligochaetes of North and South Carolina. Midwest Aquatic Enterprises, Mahomet, IL, pp. 7.1–7.12.

Brigham, W. U. 1982b. Aquatic Neuroptera. *in*: Brigham, A. R., Brigham, W. U., Gnilka, A. Eds. The aquatic insects and oligochaetes of North and South Carolina. Midwest Aquatic Enterprises, Mahomet, IL, pp. 8.1–8.4.

Brigham, W. U. 1982c. Aquatic Coleoptera. *in* Brigham, A. R., Brigham, W. U., Gnilka, A. Eds. The aquatic insects and oligochaetes of North and South Carolina. Midwest Aquatic Enterprises, Mahomet, IL, pp. 10.1–10.136.

Brittain, J. E. 1982. Biology of mayflies. Annual Review of Entomology 27:119–147.

Brooks, A. R., Kelton, L. A. 1967. Aquatic and semiaquatic Heteroptera of Alberta, Saskatchewan, and Manitoba (Hemiptera). Memoirs of the Entomological Society of Canada 51. 92 pp.

Brown, H. P. 1972. Aquatic dryopoid beetles (Coleoptera) of the United States. Biota of Freshwater Ecosystems Identification Manual 6. U.S. Environmental Protection Agency Water Pollution Control Research Series 18050 ELDO4/72, ix + 82 pp.

Brown, H. P. 1987. Biology of riffle beetles. Annual Review of Entomology 32:253–273.

Burger, J. F. 1977. The biosystematics of immature Arizona Tabanidae (Diptera). Transactions of the American Entomological Society 103:145–268.

Burian, S. K., Novak, M. A., Bode, R. W., Abele, L. 1997. New record of *Brachycercus maculatus* Berner (Ephemeroptera:

Caenidae) from New York and a key to larvae of northeastern species. Great Lakes Entomologist 30:85–88.

Burks, B. D. 1953. The mayflies or Ephemeroptera of Illinois. Bulletin of the Illinois Natural History Survey 26:1–216.

Calabrese, D. 1974. Keys to the adults and nymphs of the species of *Gerris* Fabricius occurring in Connecticut. Memoirs, Connecticut Entomological Society 1974: 227–266.

Centerbury, L. E. 1978. Studies of the genus *Sialis* (Sialidae: Megaloptera) in eastern North America. Ph.D. Dissertation, University of Louisville, KY, x + 93 pp.

Carle, F. L. 1980. A new *Lanthus* (Odonata: Gomphidae) from eastern North America with adult and nymphal keys to American octogomphines. Annals of the Entomological Society of America 73: 172–179.

Carle, F. L. 1992. *Ophiogomphus (Ophionurus) australis* spec. nov. from the Gulf coast of Louisiana, with larval and adult keys to American *Ophiogomphus* (Anisoptera: Gomphidae). Odonatologica 21:141–152.

Carpenter, S. J., LaCasse, W. J. 1955. Mosquitoes of North America. Uni. of California Press, Berkeley, CA, vi + 360 pp.

Chapin, J. W. 1978. Systematics of Nearctic *Micrasema* (Trichoptera: Brachycentridae). Ph.D. Dissertation, Clemson University, Clemson, SC, xiii + 136 pp.

Chapman, H. C. 1958. Notes on the identity, habitat and distribution of some semiaquatic Hemiptera of Florida. Florida Entomologist 41:117–124.

Cheary, B. S. 1971. The biology, ecology, and systematics of the genus *Laccobius (Laccobius)* (Coleoptera: Hydrophilidae) of the new world. Ph.D. Dissertation, University of California, Riverside, xv + 178 pp. (privately printed).

Check, G. R. 1982. A revision of the North American species of *Callibaetis* (Ephemeroptera: Baetidae). Ph.D. Dissertation, University of Minnesota, Minneapolis, 164 pp.

Chen, Y. E. 1992. The larva and pupa of *Apatania praevolans* Morse (Trichoptera: Limnephilidae), with a key to described larvae of North American species of *Apatania*. Aquatic Insects 14:49–55.

Cheng, L., Fernando, C. H. 1970. The waterstriders of Ontario. Miscellaneous Life Science Publications of the Royal Ontario Museum, 23 pp.

Christiansen, K., Bellinger, P. 1980. The Collembola of North America north of the Rio Grande. A taxonomix analysis. Grinnell College, IA, iii + 1322 pp.

Clifford, H. F. 1991. Aquatic invertebrates of Alberta. Uni. of Alberta Press, Edmonton, 538 pp.

Cook, E. F. 1956. The Nearctic Chaoborinae (Diptera: Culicidae). Technical Bulletin 218, University of Minnesota Agricultural Experiment Station, 102 pp.

Corbet, P. S. 1962. A biology of dragonflies. H.F. & G. Witherby Ltd., London, xvi + 247 pp.

Corbet, P. S. 1980. Biology of Odonata. Annual Review of Entomology, 25:189–217.

Courtney, G. W. 1990. Revision of the Nearctic mountain midges (Diptera: Deuterophlebiidae). Journal of Natural History 2:81–118.

Courtney, G. W. 1994. Biosystematics of the Nymphomyiidae (Insecta: Diptera): life history, morphology and phylogenetic relationships. Smithsonian Contributions to Zoology 550:1–41.

Currie, D. C. 1986. An annotated list of and keys to the immature black flies of Alberta (Diptera: Simuliidae). Memoirs of the Entomological Society of Canada 134:1–90.

Cuyler, R. D. 1958. The larvae of *Chauliodes* Latreille (Megaloptera: Corydalidae). Annals of the Entomological Society of America 51:582–586.

Daigle, J. J. 1991a. A new key to the larvae of North American *Somatochlora*. Argia 3(3):9–10.

Daigle, J. J. 1991b. Florida damselflies (Zygoptera): A species key to the aquatic larval stages. Florida Department of Environmental Regulation Technical Series 11:1–12.

Daigle, J. J. 1992. Florida dragonflies (Anisoptera): A species key to the aquatic larval stages. Florida Department of Environmental Regulation Technical Series 12:1–29.

Darsie, R. F., Ward, R. A. 1981. Identification and geographical distribution of the mosquitoes of North America, north of Mexico. Mosquito Systematics Supplement 1, American Mosquito Control Association, 313 pp.

Davis, J. R. 1996. Naucoridae of Texas. Southwestern Naturalist 41:1–26.

Day, W. C. 1956. Ephemeroptera. *in*: Usinger, R. L., Ed. Aquatic insects of California with keys to North American genera and California species. Uni. of California Press, Berkeley., CA, pp. 79–105.

Dosdall, L., Lehmkuhl, D. M. 1979. Stoneflies (Plecoptera) of Saskatchewan. Quaestiones Entomologicae 15:3–116.

Downing, J. A., Rigler, F. H. 1984. A manual on methods for the assessment of secondary productivity in fresh waters. 2nd ed. IBP Handbook 17, Blackwell Scientific, Oxford, UK, xxiv + 501 pp.

Drake, C. J., Harris, H. M. 1934. The Gerrinae of the Western Hemisphere (Hemiptera). Annals of the Carnegie Museum 23:179–240.

Dunkle, S. W. 1977. Larvae of the genus *Gomphaeschna* (Odonata: Aeshnidae). Florida Entomologist 60:223–225.

Dunn, C. E. 1979. A revision and phylogenetic study of the genus *Hesperocorixa* Kirkaldy (Hemiptera: Corixidae). Proceedings of the Academy of Natural Sciences of Philadelphia 131:158–190.

Edmunds, G. F., Jr. 1972. Biogeography and evolution of Ephemeroptera. Annual Review of Entomology 17:21–42.

Edmunds, G. F., Jr., Jensen, S. L., Berner, L. 1976. The mayflies of North and Central America. Uni. of Minnesota Press, Minneapolis, x + 330 pp.

Edwards, S. W. 1961. The immature stages of *Xiphocentron mexico* (Trichoptera). Texas Journal of Science 13:51–56.

Edwards, S. W., Arnold, C. R. 1961. The caddisflies of the San Marcos River. Texas Journal of Science 13:398–415.

Epler, J. H. 1987. Revision of the Nearctic *Dicrotendipes* Kieffer, 1913 (Diptera: Chironomidae). Evolutionary Monographs 9, 102 pp. + 37 pl.

Epler, J. H. 1995. Identification manual for the larval Chironomidae (Diptera) of Florida. Florida Department of Environmental Protection, Tallahassee. 317 pp.

Epler, J. H. 1996. Identification manual for the water beetles of Florida. Florida Department of Environmental Protection, Tallahassee, iv + 253 pp.

Exner, K., Craig, D. A. 1976. Larvae of Alberta Tanyderidae (Diptera: Nematocera). Quaestiones Entomologicae 12:219–237.

Fall, H. C. 1922. A review of the North American species of *Agabus* together with a description of a new genus and species of the tribe Agabini. John D. Sherman, Jr., Mt. Vernon, NY, 36 pp.

Fall, H. C. 1923. A revision of the North American species of *Hydroporus* and *Agaporus*. John D. Sherman, Jr., Mt. Vernon, NY, 238 pp.

Ferkinhoff, W. D., Gundersen, R. W. 1983. A key to the whirligig beetles of Minnesota and adjacent states and Canadian provinces (Coleoptera: Gyrinidae). Scientific Publications of the Science Museum of Minnesota, New Series 5(3):1–55.

Flint, O. S., Jr. 1960. Taxonomy and biology of Nearctic limnephelid larvae (Trichoptera), with special reference to species in eastern United States. Entomologica Americana 40:1–120.

Flint, O. S., Jr. 1961. The immature stages of the Arctopsychinae occurring in eastern North America (Trichoptera: Hydropsychidae). Annals of the Entomological Society of America 54:5–11.

Flint, O. S., Jr. 1962. Larvae of the caddisfly genus *Rhyacophila* in eastern North America (Trichoptera: Rhyacophilidae). Proceedings of the United States National Museum 113:465–493

Flint, O. S., Jr. 1964. Note on some Nearctic Psychomyiidae with special reference to their larvae (Trichoptera). Proceedings of the United States National Museum 115:467–481.

Flint, O. S., Jr. 1984. The genus *Brachycentrus* in North America, with a proposed phylogeny of the genera of Brachycentridae (Trichoptera). Smithsonian Contributions to Zoology 398, 58 pp.

Flowers, R. W. 1982. Review of the genus *Macdunnoa* (Ephemeroptera: Heptageniidae) with description of a new species from Florida. Great Lakes Entomologist 15:25–30.

Flowers, R. W., Hilsenhoff, W. L. 1975. Heptageniidae (Ephemeroptera) of Wisconsin. Great Lakes Entomologist 8:201–218.

Floyd, M. A. 1995. Larvae of the caddisfly genus *Oecetis* (Trichoptera: Leptoceridae) in North America. Bulletin of the Ohio Biological Survey New Series 10(3), viii + 85 pp.

Foote, B. A. 1959. Biology and life history of the snail-killing flies belonging to the genus *Sciomyza* Fallen (Diptera: Sciomyzidae). Annals of the Entomological Society of America 52:31–43.

Foote, B. A. 1961. Biology and immature stages of the snail-killing flies belonging to the genus *Tetanocera* (Diptera: Sciomyzidae). Ph.D. Dissertation, Cornell University, Ithaca, NY, 190 pp.

Foote, B. A. 1995. Biology of shore flies. Annual Review of Entomology 40:417–442.

Frania, H. E., Wiggins, G. B. 1995. Analysis of morphological and behavioral evidence for the phylogeny and higher classification of Trichoptera. Royal Ontario Museum Life Science Contribution 160.

Froeschner, R. C. 1949. Contribution to a synopsis of the Hemiptera of Missouri. Part IV. Hebridae, Mesoveliidae, Cimicidae, Anthocoridae, Cryptostemmatidae, Isometopidae, Miridae. American Midland Naturalist 42:123–188.

Froeschner, R. C. 1962. Contribution to a synopsis of the Hemiptera of Missouri. Part V. Hydrometridae, Gerridae, Veliidae, Saldidae, Ochteridae, Gelastocoridae, Naucoridae, Belostomatidae, Nepidae, Notonectidae, Pleidae, Corixidae. American Midland Naturalist 67:208–240.

Fullington, K. E., Stewart, K. W. 1980. Nymphs of the stonefly genus *Taeniopteryx* (Plecoptera: Taeniopterygidae) of North America. Journal of the Kansas Entomological Society 53:237–259.

Funk, D. H., Sweeney, B. W. 1994. The larvae of eastern North American *Eurylophella* Tiensuu (Ephemeroptera: Ephemerellidae). Transactions of the American Entomological Society 120:209–286.

Garrison, R. W. 1981. Description of the larva of *Ischnura gemina* with a key and new characters for the separation of sympatric *Ischnura* larvae. Annals of the Entomological Society of America 56:356–364.

Garrison, R. W. 1984. Revision of the genus *Enallagma* of the United States west of the Rocky Mountains and identification of certain larvae by discriminant analysis (Odonata: Coenagrionidae). Uni. of California Publications in Entomology Vol. 105, ix + 129 pp.

Gentili, E. 1985. I *Laccobius americani* - I *Laccobius* del Canada. Osservatorio di Fisica Terrestre e Museo Antonio Stoppani del Seminario Arcivescovile di Milano 6 (1983):31–45.

Gentili, E. 1986. I *Laccobius americani* - II genere *Laccobius* a sud del Canada. Osservatorio di Fisica Terrestre e Museo Antonio Stoppani del Seminario Arcivescovile di Milano 7 (1984):31–40.

Gillespie, J. M., Barr, W. F., Elliott, S. T. 1994. Taxonomy and biology of the immature stages of species of *Thaumalea* occurring in Idaho and California (Diptera: Thaumaleidae). Myia 5:153–193.

Givens, D. R., Smith, S. D. 1980. A synopsis of the western Arctopsychinae (Trichoptera: Hydropsychidae). Melanderia 35:1–24.

Glover, J. B. 1996. Larvae of the caddisfly genera *Triaenodes* and *Ylodes* (Trichoptera: Leptoceridae) in North America. Bulletin of the Ohio Biological Survey New Series 11(2), vii + 89 pp.

Glukhova, V. M. 1977. Ceratopogonidae. Pages 431–457, *in* The identification of fresh water invertebrates of European USSR i.e. USSR west of the Urals (plankton and benthos). Zoological Institute., USSR Academy of Science. (translated from the Russian by D.S. Kettle).

Gonsoulin, G. J. 1973a. Seven families of aquatic and semiaquatic Hemiptera in Louisiana. Part I. Family Hydrometridae Billberg, 1820. "marsh treaders". Entomological News 84:9–16.

Gonsoulin, G. J. 1973b. Seven families of aquatic and semiaquatic Hemiptera in Louisiana. Part II. Family Naucoridae Fallen, 1814. "creeping water bugs". Entomological News 84:83–88.

Gonsoulin, G. J. 1973c. Seven families of aquatic and semiaquatic Hemiptera in Louisiana. Part III. Family Belostomatidae Leach, 1815. Entomological News 84:173–189.

Gonsoulin, G. J. 1974. Seven families of aquatic and semiaquatic Hemiptera in Louisiana. Part IV. Family Gerridae. Transactions of the American Entomological Society 100:513–546.

Gonsoulin, G. J. 1975. Seven families of aquatic and semiaquatic Hemiptera in Louisiana. Part V. Family Nepidae Latreille, 1802. Entomological News 86:23–32.

Goodwin, J. T. 1987. Immature stages of some eastern Nearctic Tabanidae (Diptera). VIII. Additional species of *Tabanus* Linnaeus and *Chrysops* Meigen. Florida Entomologist 70:268–277.

Gordon, R. D. 1969. A revision of the *niger-tenebrosus* group of *Hydroporus* (Coleoptera: Dytiscidae) in North America. Ph.D. Dissertation, North Dakota State University, Fargo., 311 pp. + 17 pl.

Gordon, R. D. 1981. New species of North American *Hydroporus, niger-tenebrosus* group (Coleoptera: Dytiscidae). Pan-Pacific Entomologist 57:105–123.

Gordon, R. D., Post, R. L. 1965. North Dakota water beetles. North Dakota Insects Publication No. 5. Department of Entomology, North Dakota State University, Fargo., 52 pp.

Grodhaus, G. 1987. *Endochironomus* Kieffer, *Tribelos* Townes, *Synendotendipes*, n. gen., and *Endotribelos*, n. gen. (Diptera: Chironomidae) of the Nearctic region. Journal of the Kansas Entomological Society 60:167–247.

Gundersen, R. W. 1978. Nearctic *Enochrus*. Biology, keys, descriptions and distribution (Coleoptera: Hydrophilidae). Department of Biological Science, St. Cloud State University, MN., 54 pp.

Gundersen, R. W., Otremba, C. 1988. Haliplidae of Minnesota. Scientific Publications of the Science Museum of Minnesota, New Series 6(3):1–43.

Haddock, J. D. 1977. The biosystematics of the caddisfly genus *Nectopsyche* in North America with emphasis on the aquatic stages. American Midland Naturalist 98:382–421.

Hall, F. 1974. A key to the *Simulium* larvae of southern California (Diptera: Simuliidae). California Vector Views 21:65–71.

Harden, P, Mickel, C. 1952. The stoneflies of Minnesota (Plecoptera). Technical Bulletin of the University of Minnesota Agricultural Experiment Station 210:1–84.

Harper, P. P., Hynes, H. B. N. 1971a. The Leuctridae of eastern Canada (Insecta; Plecoptera). Canadian Journal of Zoology 49:915–920.

Harper, P. P., Hynes, H. B. N. 1971b. The Capniidae of eastern

Canada (Insecta; Plecoptera). Canadian Journal of Zoology 49:921–940.

Harper, P. P., Hynes, H. B. N. 1971c. The nymphs of the Taeniopterygidae of eastern Canada (Insecta; Plecoptera). Canadian Journal of Zoology 49:941–947.

Harper, P. P., Hynes, H. B. N. 1971d. The nymphs of the Nemouridae of eastern Canada (Insecta: Plecoptera). Canadian Journal of Zoology 49:1129–1142.

Hatch, M. H. 1930. Records and new species of Coleoptera from Oklahoma and western Arkansas, with subsidiary studies. Publications of the University of Oklahoma Biological Survey 2:15–26.

Hatch, M. H. 1953. The beetles of the Pacific Northwest. Part I: Introduction and Adephaga. University of Washington Publications in Biology 16, 340 pp.

Hatch, M. H. 1965. The beetles of the Pacific Northwest. Part IV: Macrodactyles, Palpicornes, and Heteromera. Uni. of Washington Publications in Biology 16, 268 pp.

Hellman, J. L. 1975. A taxonomic revision of the genus *Hydrochus* of North America, Central America and West Indies. Ph.D. Dissertation, University of Maryland, College Park, 441 pp.

Henry, B. C. 1993. A revision of *Neochoroterpes* (Ephemeroptera: Leptophlebiidae) new status. Transactions of the American Entomological Society 199:317–333.

Henry, T. J., Froeschner, R. C. Eds. 1988. Catalog of the Heteroptera, or true bugs, of Canada and the continental United States. E.J. Brill, Publishers, Leiden, Netherlands, New York, NY, xx + 958 pp.

Herring, J. L. 1951. The aquatic and semiaquatic Hemiptera of northern Florida. Part IV. Classification of habitats and keys to the species. Florida Entomologist 34:146–161.

Hilsenhoff, W. L. 1973. Notes on *Dubiraphia* (Coleoptera: Elmidae) with descriptions of five new species. Annals of the Entomological Society of America 66:55–61.

Hilsenhoff, W. L. 1975. Notes on Nearctic *Acilius* (Dytiscidae), with the description of a new species. Annals of the Entomological Society of America 68:271–274.

Hilsenhoff, W. L. 1980. *Coptotomus* (Coleoptera: Dytiscidae) in eastern North America with descriptions of two new species. Transactions of the American Entomological Society 105:461–471.

Hilsenhoff, W. L. 1984a. Identification and distribution of *Baetisca* nymphs (Ephemeroptera: Baetiscidae) in Wisconsin. Great Lakes Entomologist 17:51–52.

Hilsenhoff, W. L. 1984b. Aquatic Hemiptera of Wisconsin. Great Lakes Entomologist 17:29–50.

Hilsenhoff, W. L. 1985. The Brachycentridae of Wisconsin (Trichoptera). Great Lakes Entomologist 18:149–154.

Hilsenhoff, W. L. 1986. Semiaquatic Hemiptera of Wisconsin. Great Lakes Entomologist 19:7–19.

Hilsenhoff, W. L. 1987. Effectiveness of bottle traps for collecting Dytiscidae (Coleoptera). Coleopterists Bulletin 41:377–380.

Hilsenhoff, W. L. 1990. Gyrinidae of Wisconsin, with a key to adults of both sexes and notes on distribution and habitat. Great Lakes Entomologist 23:77–91.

Hilsenhoff, W. L. 1991. Comparison of bottle traps with a D-frame net for collecting adults and larvae of Dytiscidae and Hydrophilidae (Coleoptera). Coleopterists Bulletin 45:143–146.

Hilsenhoff, W. L. 1992. Dytiscidae and Noteridae of Wisconsin (Coleoptera). I. Introduction, key to genera of adults, and distribution, habitat, life cycle, and identification of species of Agabetinae, Laccophilinae and Noteridae. Great Lakes Entomologist 25:57–69.

Hilsenhoff, W. L. 1993a. Dytiscidae and Noteridae of Wisconsin (Coleoptera). II. Distribution, habitat, life cycle, and identification of species of Dytiscinae. Great Lakes Entomologist 26:35–53.

Hilsenhoff, W. L. 1993b. Dytiscidae and Noteridae of Wisconsin (Coleoptera). III. Distribution, habitat, life cycle, and identification of species of Colymbetinae, except Agabini. Great Lakes Entomologist 26:121–136.

Hilsenhoff, W. L. 1993c. Dytiscidae and Noteridae of Wisconsin (Coleoptera). IV. Distribution, habitat, life cycle, and identification of species of Agabini (Colymbetinae). Great Lakes Entomologist 26:173–197.

Hilsenhoff, W. L. 1994. Dytiscidae and Noteridae of Wisconsin (Coleoptera). V. Distribution, habitat, life cycle, and identification of species of Hydroporinae, except *Hydroporus* Clairville sensu lato. Great Lakes Entomologist 26:275–295.

Hilsenhoff, W. L. 1995a. Dytiscidae and Noteridae of Wisconsin (Coleoptera). VI. Distribution, habitat, life cycle, and identification of species of *Hydroporus* Clairville sensu lato (Hydroporinae). Great Lakes Entomologist 28:1–23.

Hilsenhoff, W. L. 1995b Aquatic Hydrophilidae and Hydraenidae of Wisconsin (Coleoptera). I. Introduction, key to genera of adults, and distribution, habitat, life cycle, and identification of species of *Helophorus* Fabricius, *Hydrochus* Leach, and *Berosus* Leach (Hydrophilidae), and Hydraenidae. Great Lakes Entomologist 28:25–53.

Hilsenhoff, W. L. 1995c. Aquatic insects of Wisconsin. Keys to Wisconsin genera and notes on biology, habitat, distribution and species. Publication 3 of the Natural History Museums Council, University of Wisconsin-Madison, 79 pp.

Hilsenhoff, W. L. 1995d. Aquatic Hydrophilidae and Hydraenidae of Wisconsin (Coleoptera). II. Distribution, habitat, life cycle and identification of species of Hydrobiini and Hydrophilini (Hydrophilidae: Hydrophilinae). Great Lakes Entomologist 28:97–126.

Hilsenhoff, W. L., Billmyer, S. J. 1973. Perlodidae (Plecoptera) of Wisconsin. Great Lakes Entomologist 6:1–14.

Hilsenhoff, W. L., Brigham, W. U. 1978. Crawling water beetles of Wisconsin (Coleoptera: Haliplidae). Great Lakes Entomologist 11:11–22.

Hilsenhoff, W. L., Schmude, K. L. 1992. Riffle beetles (Coleoptera: Dryopidae, Elmidae, Lutrochidae, Psephenidae) of Wisconsin with notes on distribution, habitat, and identification. Great Lakes Entomologist 25:191–213.

Hitchcock, S. W. 1974. Guide to the Insects of Connecticut. Part VII. The Plecoptera or stoneflies of Connecticut. State Geological and Natural History Survey of Connecticut Bulletin 107, vi + 262 pp.

Hoffman, C. E. 1939. Morphology of the immature stages of some northern Michigan Donaciini (Chrysomelidae: Coleoptera). Papers of the Michigan Academy of Science, Arts and Letters 25:243–290 + 10 plates.

Hogue, C. L., Georgian, T. 1986. Recent discoveries in the *Blepharicera tenuipes* group, including descriptions of two new species from Apalachia (Diptera: Blephariceridae). Contributions to the Science and Natural History Museum, Los Angeles County 377:1–20.

Hubbard, M. D. 1994. The mayfly family Behningiidae (Ephemeroptera: Ephemeroidea): Keys to the recent species with a catalog of the family. Great Lakes Entomologist 27:161–168.

Huggins, D. G., Brigham, W. U. 1982. Odonata. *in*: Brigham, A. R., Brigham, W. U., Gnilka, A. Eds. 1981. The aquatic insects and oligochaetes of North and South Carolina. Midwest Aquatic Enterprises, Mahomet, IL, pp. 4.1–4.100.

Hungerford, H. B. 1933. The genus *Notonecta* of the world (Notonectidae-Hemiptera). University of Kansas Science Bulletin 21:5–193.

Hungerford, H. B. 1948. The Corixidae of the Western Hemisphere. University of Kansas Science Bulletin 32:1–288; 408–827.

Hungerford, H. B. 1954. The genus *Rheumatobates* Bergroth (Hemiptera-Gerridae). University of Kansas Science Bulletin 36:529–588.

Hungerford, H. B., Evans, N. W. 1934. The Hydrometridae of the Hungarian National Museum and other studies in the family. Annales historico-naturales Musei Nationalis Hungarici 28:31–112.

Hutchinson, G. E. 1945. On the species of *Notonecta* (Hemiptera-Heteroptera) inhabiting New England. Transactions of the Connecticut Academy of Arts and Sciences 36:599–605.

Hynes, H. B. N. 1976. The biology of Plecoptera. Annual Review of Entomology 21: 135–153.

Jansson, A. 1972. Systematic notes and new synonymy in the genus *Cenocorixa* (Hemiptera: Corixidae). Canadian Entomologist 104:449–459.

Kavanaugh, D. H. 1986. A systematic review of amphizoid beetles (Amphizoidae: Coleoptera) and their phylogenetic relationships to other Adephaga. Proceedings of the California Academy of Sciences 44:67–109.

Kettle, D. S. 1977. Biology and bionomics of bloodsucking ceratopogonids. Annual Review of Entomology 22:33–51.

Kittle, P. D. 1977. A revision of the genus *Trepobates* Uhler (Hemiptera: Gerridae). Ph.D. Dissertation, University of Arkansas, Fayetteville. 255 pp.

Kittle, P. D. 1980. The water striders (Hemiptera: Gerridae) of Arkansas. Arkansas Academy of Science Proceedings 34:68–71.

Knutson, L. V., Berg, C. O. 1964. Biology and immature stages of snail-killing flies: the genus *Elgiva* (Diptera: Sciomyzidae). Annals of the Entomological Society of America 57:173–192.

Kondratieff, B. C., Kirchner, R. F., Stewart, K. W. 1988. A review of *Perlinella* Banks (Plecoptera: Perlidae). Annals of the Entomological Society of America 81:19–27.

Kondratieff, B. C., Kirchner, R. F., Voshell, J. R., Jr. 1981. Nymphs of *Diploperla* (Plecoptera: Perlidae). Annals of the Entomological Society of America 74:428–430.

Kondratieff, B. C., Voshell, J. R., Jr. 1984. The North and Central American species of *Isonychia* (Ephemeroptera: Oligoneuriidae). Transactions of the American Entomological Society 110:129–244.

Kuitert, L. C. 1942. Gerrinae in the University of Kansas Collections. University of Kansas Science Bulletin 28:113–143.

LaRivers, I. 1948. A new species of *Pelocoris* from Nevada, with notes on the genus in the United States (Hemiptera: Naucoridae). Annals of the Entomological Society of America 41:371–376.

LaRivers, I. 1951. A revision of the genus *Ambrysus* in the United States. University of California Publications in Entomology 8:277–338.

LaRivers, I. 1954. Nevada Hydrophilidae (Coleoptera). The American Midland Naturalist 52:164–174.

Larson, D. J. 1975. The predaceous water beetles (Coleoptera: Dytiscidae) of Alberta: systematics, natural history and distribution. quaestiones Entomologicae 11:245–498.

Larson, D. J. 1987. Revision of North American species of *Ilybius* Erichson (Coleoptera: Dytiscidae), with systematic notes on Palaearctic species. Journal of the New York Entomological Society 95:341–413.

Larson, D. J. 1989. Revision of North American *Agabus* Leach (Coleoptera: Dytiscidae): introduction, key to species groups, and classification of the *ambiguus*-, *tristis*-, and *arcticus*-groups. Canadian Entomologist 121:861–919.

Larson, D. J. 1991. Revision of North American *Agabus* Leach (Coleoptera: Dytiscidae): *elongatus*-, *zetterstedti*-, and *confinis*-groups. Canadian Entomologist 123:1239–1317.

Larson, D. J. 1994. Revision of North American Agabus Leach (Coleoptera: Dytiscidae): *lutosus*-, *obsoletus*-, and *fuscipennis*-groups. Canadian Entomologist 126:135–181.

Larson, D. J. 1996. Revision of North American *Agabus* Leach (Coleoptera: Dytiscidae): The *opacus*-group. Canadian Entomologist 128:613–665.

Larson, D. J. 1997. Revision of North American *Agabus* Leach (Coleoptera: Dytiscidae): The *seriatus*-group. Canadian Entomologist 129:105–149.

Larson, D. J., Alarie, Y., Roughley, R. E. 2001. Dytiscidae of Canada and Alaska. NRC Research Press, Ottawa, Canada, about 850 pp.

Larson, D. J., Roughley, R. E. 1990. A review of the species of *Liodessus* Guignot of North America north of Mexico with the description of a new species (Coleoptera: Dytiscidae). Journal of the New York Entomological Society 98:233–245.

Larson, D. J., Wolfe, R. W. 1998. Revision of North American *Agabus* (Coleoptera: Dytiscidae): the *semivittatus*-group. Canadian Entomologist 130:27–54.

Lawrence, J. F., Britton, E. B. 1991. Colcoptera (beetles), *in*: CSIRO, Aquatic insects of Australia, 2nd edn. Melbourne Uni. Press, Australia, pp. 543–683.

Leech, H. B. 1938. A study of the Pacific Coast species of *Agabus* Leach, with a key to the Nearctic species. M.S. Thesis, University of California, Berkeley.

Leech, H. B., Chandler, H. P. 1956. Aquatic Coleoptera. *in*: Usinger, R. L., Edr. Aquatic insects of California with keys to North American genera and California species. Uni. of California Press, Berkeley, CA, pp. 293–371.

Lehmkuhl, D. M., Anderson, N. H. 1971. Contributions to the biology and taxonomy of the *Paraleptophlebia* of Oregon (Ephemeroptera: Leptophlebiidae). Pan-Pacific Entomologist 47:85–93.

Louton, J. A. 1982. Lotic dragonfly (Anisoptera: Odonata) nymphs of the southeastern United States: identification, distribution and historical biogeography. Ph.D. Dissertation, University of Tennessee, Knoxville, 357 pp.

Lugo-Ortiz, C. R., McCafferty, W. P. 1995. Taxonomy of the North and Central American species of *Camelobaetidius* (Ephemeroptera: Baetidae). Entomological News 106:178–192.

Lugo-Ortiz, C. R., McCafferty, W. P. Waltz, R. D. 1994. Contribution to the taxonomy of the Panamerican genus *Fallceon* (Ephemeroptera: Baetidae). Journal of the New York Entomological Society 102:460–475.

Lugo-Ortiz, C. R., McCafferty, W. P. 1998. A new North American genus of Baetidae (Ephemeroptera) and key to *Baetis* complex genera. Entomological News 109:345–353.

Lugo-Ortiz, C. R., McCafferty, W. P., Waltz, R. D. 1999. Definition and reorganization of the genus *Pseudocloeon* (Ephemeroptera: Baetidae) with new species descriptions and combinations. Transactions of the American Entomological Society 125:1–37.

Malcolm, S. E. 1971. The water beetles of Maine: including the families Gyrinidae, Haliplidae, Dytiscidae, Noteridae, and Hydrophilidae. Technical Bulletin 48, Life Sciences and Agricultural Experiment Station, University of Maine, Orono., 49 pp.

Maschwitz, D. E., Cook, E. F. 1999. Revision of the Nearctic species of *Polypedilum* (*Polypedilum*) and *Polypedilum* (*Urespedilum*) (Diptera: Chironomidae). Ohio Biological Survey Bulletin, New Series 12(3), vii + 136 pp.

Matta, J. F. 1974. The insects of Virginia: No. 8. The aquatic Hydrophilidae of Virginia (Coleoptera: Polyphaga). Research Division Bulletin 94, Virginia Polytechnic Institute and State University, Blacksburg., iv + 44 pp.

Matta, J. F. 1976. The insects of Virginia: No. 10. The Haliplidae of

Virginia (Coleoptera: Adephaga). Research Division Bulletin 109, Virginia Polytechnic Institute and State University, Blacksburg, vi + 26 pp.

Matta, J. F., Wolfe, G. W. 1981. A revision of the subgenus *Heterosternuta* Strand of *Hydroporus* Clairville (Coleoptera: Dytiscidae). Pan-Pacific Entomologist 57:176–219.

McCafferty, W. P. 1975. The burrowing mayflies (Ephemeroptera: Ephemeroidea) of the United States. Transactions of the American Entomological Society 101:447–504.

McCafferty, W. P. 1977. Biosystematics of *Dannella* and related subgenera of *Ephemerella* (Ephemeroptera: Ephemerellidae). Annals of the Entomological Society of America 70:881–889.

McCafferty, W. P. 1981. Aquatic entomology. The fishermen's and ecologists' illustrated guide to insects and their relatives. Science Books International, Boston, MA, xv + 448 pp.

McCafferty, W. P. 1991. Toward a phylogenetic classification of the Ephemeroptera (Insecta): a commentary on systematics. Annals of the Entomological Society of America 84:343–360.

McCafferty, W. P., Provonsha, A. V. 1993. New species, subspecies, and stage descriptions of Texas Baetidae (Ephemeroptera). Proceedings of the Entomological Society of Washington 95:59–69.

McCafferty, W. P., Waltz, R. D. 1995. *Labiobaetis* (Ephemeroptera: Baetidae): New status, new North American species, and related new genus. Entomological News 27:209–216.

McCafferty, W. P., Wigle, M. J., Waltz, R. D. 1994. Contributions to the taxonomy and biology of (*Acentrella turbida* (McDunnough) (Ephemeroptera: Baetidae). Pan-Pacific Entomology 70:301–308.

McCorkle, D. V. 1965. (Subfamily Elophorinae). *in*: Hatch, M. H., Edr. The beetles of the Pacific Northwest. Part IV. Macrodactyles, Palpicornes, and Heteromera. Uni. of Washington Publications in Biology 16, pp. 23–38.

McWilliams, K. L. 1969. A taxonomic revision of the North American species of the genus *Thermonectus* Dejean (Coleoptera: Dytiscidae). Ph.D. Dissertation, Indiana University, Bloomington., 220 pp.

Menke, A. S., Ed. 1979a. The semiaquatic and aquatic Hemiptera of California (Heteroptera: Hemiptera). Bulletin of the California Insect Survey 21, xi + 166 pp.

Menke, A. S., 1979b. Family Belostomatidae/giant water bugs, electric light bugs, toe biters, *in*: Menke, A. S., Ed. The semiaquatic and aquatic Hemiptera of California (Heteroptera: Hemiptera). Bulletin of the California Insect Survey Vol. 21, pp. 76–86.

Merritt, R. W., Cummins, K. W. Eds. 1996. An introduction to the aquatic insects of North America. 3 ed. Kendall/Hunt, Dubuque, IA., xiii + 862 p.

Merritt, R. W., Ross, D. H., Peterson, B. V. 1978. Larval ecology of some lower Michigan black flies (Diptera: Simuliidae) with keys to the immature stages. Great Lakes Entomologist 11;177–208.

Michael, A. G., Matta, J. F. 1977. The insects of Virginia: No. 12. The Dytiscidae of Virginia (Coleoptera: Adephaga) (subfamilies: Laccophilinae, Colymbetinae, Dytiscinae, Hydaticinae and Cybistrinae). Research Division Bulletin 124, Virginia Polytechnic Institute and State University, Blacksburg, vi + 53 pp.

Miller, D. C. 1965. Hydrophilidae, except Elophorinae and Sphaeridiinae. *in*: Hatch, M. H., Ed. The beetles of the Pacific Northwest. Part IV. Macrodactyles, Palpicornes, and Heteromera. Uni. of Washington Publications in Biology 16, pp. 21–23, 38–46.

Miller, K. B. 1998. Revision of the Nearctic *Liodessus affinis* (Say 1823) group (Coleoptera: Dytiscidae, Hydroporinae, Bidessini). Entomologica Scandinavica 29:281–314.

Miller, P. L. 1987. Dragonflies. (Naturalists' handbooks 7). Cambridge Uni. Press, Cambridge, UK. 84 pp.

Morihara, D. K., McCafferty, W. P. 1979. The *Baetis* larvae of North America (Ephemeroptera: Baetidae). Transactions of the American Entomological Society 105:139–221.

Morrissette, R. 1979. Les Gyrinidae (Coleoptera) du Quebec. Fabreries 5:51–58.

Moulton, J. K., Adler, P. H. 1995. Revision of the *Simulium jenningsi* species-group (Diptera: Simuliidae). Transactions of the American Entomological Society 121:1–57.

Moulton, J. K., Adler, P. H. 1997. The genus *Ectemnia* (Diptera: Simuliidae): taxonomy, polytene chromosomes, new species, and phylogeny. Canadian Journal of Zoology 75:1896–1915.

Moulton, S. R. II, Stewart, K. W. 1997. A new species of and first record of the caddisfly genus *Cnodocentron* Schmid (Trichoptera: Xiphocentronidae) north of Mexico. *in*: Holzenthal, R. W., Flint, O. S. Jr., Eds. Proceedings 8th International Symposium on Trichoptera, Ohio Biological Survey, Columbus, pp 343–347.

Musser, R. J. 1962. Dragonfly nymphs of Utah (Odonata: Anisoptera). Uni. of Utah Biological Series 12(6), viii + 66 pp.

Nations, V. L. 1994. A phylogenetic analysis of the North American species of *Setodes* (Trichoptera: Leptoceridae) with descriptions of the larvae and key to their identification. M.S. Thesis, University of Alabama, Tuscaloosa, 72 pp.

Needham, J. G., Traver, J. R., Hsu, Y-C. 1935. The biology of mayflies. Comstock Pub Co. Inc., New York, NY, xiv + 759 pp.

Needham, J. G., Westfall, M. J. Jr. 1955. A manual of the dragonflies of North America (Anisoptera). Uni. of California Press, Berkeley., xii + 615 pp.

Needham, J. G., Westfall, M. J., Jr., May, M. L. 2000. Dragonflies of North America. Scientific Publishers, Gainesville, FL. xv + 940 pp.

Neff, S. E., Berg, C. O. 1962. Biology and immature stages of *Hoplodictya spinicornis* and *H. setosa* (Diptera: Sciomyzidae). Transactions of the American Entomological Society 88:77–93.

Neff, S. E., Berg, C. O. 1966. Biology and immature stages of malacophagous Diptera of the genus *Sepedon* (Sciomyzidae). Bulletin of the Agricultural Experiment Station, Virginia Polytechnic Institute, Blacksburg 566:5–113.

Neunzig, H. H. 1966. Larvae of the genus *Nigronia* Banks (Neuroptera: Corydalidae). Proceedings of the Entomological Society of Washington 68:11–16.

Nimmo, A. P. 1987. The adult Arctopsychidae and Hydropsychidae (Trichoptera) of Canada and adjacent United States. Quaestiones Entomologicae 23:1–189.

Oliver, D. R., Roussel, M. E. 1982. The larvae of *Pagastia* Oliver (Diptera: Chironomidae) with descriptions of three Nearctic species. Canadian Entomologist 114:849–854.

Oliver, D. R., Roussel, M. E. 1983. Redescription of *Brillia* Kieffer (Diptera: Chironomidae) with descriptions of Nearctic species. Canadian Entomologist 115:257–279.

Oygur, S., Wolfe, G. W. 1991. Classification, distribution, and phylogeny of North American (north of Mexico) species of *Gyrinus* Muller (Coleoptera: Gyrinidae). Bulletin of the American Museum of Natural History, New York 207, 97 pp.

Parfin, S. I., Gurney, A. B. 1956. The spongilla-flies, with special reference to those of the Western Hemisphere (Sisyridae: Neuroptera). Proceedings of the United States National Museum 105:421–529.

Parker, C. R. 1998. A review of *Goerita* (Trichoptera: Goeridae), with description of a new species. Insecta Mundi 12:227–238.

Parker, C. R., Wiggins, G. B. 1985. The Nearctic caddisfly genus *Hesperophylax* Banks (Trichoptera: Limnephilidae). Canadian Journal of Zoology 63:2433–2472.

Parker, C. R., Wiggins, G. B. 1987. Revision of the caddisfly genus

Psilotreta (Trichoptera: Odontoceridae). Life Science Contributions Royal Ontario Museum 144, 55 pp.

Pechuman, L. L., Webb, D. W. Teskey, H. J. 1983. The Diptera or true flies of Illinois. I. Tabanidae. Bulletin of the Illinois Natural History Survey 33:1–122.

Peckarsky, B. L., Fraissinet, P., Penton, M. A., Conklin, D. J., Jr. 1990. Freshwater macroinvertebrates of northeastern North America. Cornell Uni. Press, Ithaca, NY, xi + 422 pp.

Perkins, P. D. 1980. Aquatic beetles of the family Hydraenidae in the Western Hemisphere: classification, biogeography and inferred phylogeny (Insects: Coleoptera). Quaestiones Entomologicae 16:1–554.

Pescador, M. L., Berner, L. 1981. The mayfly family Baetiscidae (Ephemeroptera). Part II Biosystematics of the genus *Baetisca*. Transactions of the American Entomological Society 107: 163–228.

Pescador, M. L., Peters, W. L. 1980. A revision of the genus *Homoeoneuria* (Ephemeroptera: Oligoneuriidae). Transactions of the American Entomological Society 106:357–393.

Pescador, M. L., Rassmussen, A. K., Harris, S. C. 1995. Identification manual for the caddisfly (Trichoptera) larvae of Florida. Florida Department of Environmental Protection, Tallahassee, iv + 132 pp.

Peterson, B. V. 1970. The *Prosimulium* of Canada and Alaska (Diptera: Simuliidae). Memoirs of the Entomological Society of Canada 69, 216 pp.

Peterson, B. V., Kondratieff, B. C. 1995. The black flies (Diptera: Simuliidae) of Colorado: An annotated list with keys, illustrations and descriptions of three new species. Memoirs of the American Entomological Society 42:1–121.

Pinder, L. C. V. 1986. Biology of freshwater Chironomidae. Annual Review of Entomology 31:1–23.

Poirrier, M. A., Arceneaux, Y. M. 1972. Studies on southern Sisyridae (spongillaflies) with a key to the third-instar larvae and additional sponge-host records. American Midland Naturalist 88:455–458.

Polhemus, D. A. 1997. Systematics of the genus *Rhagovelia* Mayr (Heteroptera: Veliidae) in the Western Hemisphere (exclusive of the *angustipes* complex). Monographs Thomas Say Publications in Entomology, ii + 386 pp.

Polhemus, J. T., Chapman, H. C. 1979. Family Mesoveliidae/water treaders. Pages 39–42, *in* Menke, A. S., Ed. The semiaquatic and aquatic Hemiptera of California (Heteroptera: Hemiptera). Bulletin of the California Insect Survey 21.

Porter, T. W. 1950. Taxonomy of the American Hebridae and the natural history of selected species. Ph.D. Dissertation, University of Kansas, Lawrence, 185 pp. + 12 pl.

Poulton, B. C., Stewart, K. W. 1991. The stoneflies of the Ozark and Quachita Mountains (Plecoptera). Memoirs of the American Entomological Society 38, 116 pp.

Pritchard, G. 1983. Biology of Tipulidae. Annual Review of Entomology 28:1–22.

Provonsha, A. V. 1990. A revision of the genus *Caenis* in North America (Ephemeroptera: Caenidae). Transactions of the American Entomological Society 116:801–884.

Ray, D. H., Stark, B. P. 1981. The Nearctic species of *Hydroperla* (Plecoptera: Perlodidae). Florida Entomologist 64:385–395.

Resh, V. H. 1976. The biology and immature stages of the caddisfly genus *Ceraclea* in eastern North America (Trichoptera: Leptoceridae). Annals of the Entomological Society of America 69:1039–1061.

Ricker, W. E. 1943. Stoneflies of southwestern British Columbia. Indiana University Publications, Science Series 12, 145 pp.

Ricker, W. E. 1952. Systematic studies in Plecoptera. Indiana University Publications, Science Series 18, 200 pp.

Roback, S. S. 1985. The immature chironomids of the eastern United States. VI. Pentaneurini - genus *Ablabesmyia*. Proceedings of the Academy of Natural Sciences of Philadelphia 137:73–128.

Roback, S. S. 1986a. The immature chironomids of the eastern United States. VII. Pentaneurini—genus *Monopelopia* new record with redescription of the male adults and description of some Neotropical material. Proceedings of the Academy of Natural Sciences of Philadelphia 138:350–365.

Roback, S. S. 1986b. The immature chironomids of the eastern United States. VIII. Pentaneurini—genus *Nilotanypus* with the description of a new species from Kansas. Proceedings of the Academy of Natural Sciences of Philadelphia 138:443–465.

Roback, S. S. 1987. The immature chironomids of the eastern United States. IX. Pentaneurini—genus *Labrundinia* with the description of some Neotropical material. Proceedings of the Academy of Natural Sciences of Philadelphia 139:159–209.

Roberts, C. H. 1913. Critical notes on the species of Haliplidae of America north of Mexico with descriptions of new species. Journal of the New York Entomological Society 21:91–123.

Roemhild, G. 1976. Aquatic Heteroptera (ture bugs) of Montana. Montana Agricultural Experiment Station Research Report 102, 70 pp.

Ross, H. H. 1944. The caddisflies or Trichoptera of Illinois. Bulletin of the Illinois Natural History Survey 23:1–326.

Roughley, R. E. 1990. A systematic revision of species of Dytiscus (Coleoptera: Dytiscidae). Part I. Classification based on adult stage. Quaestiones Entomologicae 26:383–557.

Roughley, R. E., Pengelly, D. H. 1981. Classification, phylogeny, and zoogeography of *Hydaticus* Leach (Coleoptera: Dytiscidae) of North America. Quaestiones Entomologicae 17:249–309.

Saether, O. A. 1970. Nearctic and Palaearctic *Chaoborus* (Diptera: Chaoboridae). Bulletin 174, Fisheries Research Board of Canada, vii + 57 pp.

Saether, O. A., Wang, X. 1995. Revision of the genus *Paraphaenocladius* Thienemann, 1924 of the world (Diptera: Chironomidae: Orthocladiinae). Entomologica Scandinavica Supplement 48:3–69.

Sailer, R. I. 1948. The genus *Trichocorixa* (Corixidae, Hemiptera). University of Kansas Science Bulletin 32:289–407.

Sanderson, M. W. 1982. Aquatic and semiaquatic Heteroptera, *in*: Brigham, A. R., Brigham, W. U., Gnilka, A. Eds. The aquatic insects and oligochaetes of North and South Carolina. Midwest Aquatic Enterprises, Mahomet, IL, pp. 6.1–6.94.

Schaefer, K. F., Drew, W. A. 1968. The aquatic and semiaquatic Hemiptera of Oklahoma. Proceedings of the Oklahoma Academy of Science 47:125–134.

Schefter, P. W., Wiggins, G. B. 1986. A systematic study of the Nearctic larvae of the *Hydropsyche morosa* group (Trichoptera: Hydropsychidae). Life Sciences Miscellaneous Publication Royal Ontario Museum, 94 pp.

Schmid, F. 1968. La famille des Arctopsychides. Memoires de la Societe Entomologique du Quebec 1, 84 pp

Schmude, K. L. 1992. Revision of the riffle beetle genus *Stenelmis* (Coleoptera: Elmidae) in North America, with notes on bionomics. Dissertation, Ph.D. University of Wisconsin–Madison, vi + 388 pp.

Schmude, K. L. 1999. Riffle beetles in the genus *Stenelmis* (Coleoptera: Elmidae) from warm springs in southern Nevada: new species, new status, and a key. Entomological News 110:1–12.

Schmude, K. L., Hilsenhoff, W. L. 1986. Biology, ecology, larval taxonomy, and distribution of Hydropsychidae (Trichoptera) in Wisconsin. Great Lakes Entomologist 19:123–145.

Schuster, G. A., Etnier, D. A. 1978. A manual for the identification of the larvae of the caddisfly genera *Hydropsyche* Pictet and

Symphitopsyche Ulmer in eastern and central North America (Trichoptera: Hydropsychidae). United States Environmental Protection Agency Environmental Monitoring and Support Laboratory 600/4-78-060, xii + 129 pp.

Scudder, G. G. E. 1965. The Notonectidae (Hemiptera) of British Columbia. Proceedings of the Entomological Society of British Columbia 62:38–41.

Scudder, G. G. E. 1971. The Gerridae (Hemiptera) of British Columbia. Journal of the Entomological Society of British Columbia 68:3–10.

Sherberger, F. F., Wallace, J. B. 1971. Larvae of the southeastern species of *Molanna*. Journal of the Kansas Entomological Society 44:217–224.

Shirt, D. B., Angus, R. B. 1992. A revision of the Nearctic water beetles related to *Potamonectes depressus* (Fabricius) (Coleoptera: Dytiscidae). Coleopterists Bulletin 46:109–141.

Simpson, K. W. 1982. A guide to basic taxonomic literature for the genera of North American Chironomidae (Diptera)—adults, pupae, and larvae. Bulletin of the New York State Museum 447, 43 pp.

Simpson, K. W., Bode, R. W., Albu, P. 1983. Keys for the genus *Cricotopus* adapted from "Revision der Gattung *Cricotopus* van der Wulp und ihrer Verwandten (Diptera, Chironomidae)" by Hirvenoja, M. Bulletin 450 New York State Museum, v + 133 pp.

Sites, R. W., Polhemus, J. T. 1994. The Nepidae (Hemiptera) of the United States and Canada. Annals of the Entomological Society of America 87:27–42.

Sites, R. W., Polhemus, J. T. 1995. The *Pelocoris* fauna of Texas. Southwestern Naturalist 40:249–254.

Smetana, A. 1974. Revision of the genus *Cymbiodyta* Bed. (Coleoptera: Hydrophilidae). Memoirs of the Entomological Society of Canada 93, iv + 113 pp.

Smetana, A. 1980. Revision of the genus *Hydrochara* Berth. (Coleoptera: Hydrophilidae). Memoirs of the Entomological Society of Canada 111, iv + 100 pp.

Smetana, A. 1985. Revision of the subfamily Helophorinae of the Nearctic region (Coleoptera: Hydrophilidae). Memoirs of the Entomological Society of Canada 131. 154 pp.

Smetana, A. 1988. Review of the family Hydrophilidae of Canada and Alaska. Memoirs of the Entomological Society of Canada 142, 316 pp.

Smith, C. L. 1980. A taxonomic revision of the genus *Microvelia* Westwood (Heteroptera: Veliidae) of North America including Mexico. Dissertation, Ph. D. University of Georgia, Athens, 388 pp.

Smith, C. L., Polhemus, J. T. 1978. The Veliidae (Heteroptera) of America north of Mexico—keys and check list. Proceedings of the Entomological Society of Washington 80:56-68.

Smith, R. F., Pritchard, A. E. 1956. Odonata. *in*: Usinger, R. L., Ed. Aquatic insects of California with keys to North American genera and California species. Uni. of California Press, Berkeley, pp. 106–153.

Snoddy, E. L., Noblet, R. 1976. Identification of the immature black flies (Diptera: Simuliidae) of the southeastern United States with some aspects of the adult role in transmission of *Leucocytozoon smithi* to turkeys. South Carolina Agricultural Experiment Station Technical Bulletin 1057, 58 pp.

Soldan, T. 1984. A revision of the Caenidae with ocellar tubercles in the nymphal stage (Ephemeroptera). Acta Universitatis Carolinae—Biologica 1982–1984:289–362.

Spangler, P. J. 1960. A revision of the genus *Tropisternus* (Coleoptera: Hydrophilidae). Dissertation, Ph. D. University of Missouri, Columbia, 364 pp.

Spangler, P. J. 1973. A description of the larva of *Celina angustata* Aube (Coleoptera: Dytiscidae). Journal of the Washington Academy of Science 63:165–168.

Stark, B. P. 1983. A review of the genus *Soliperla* (Plecoptera: Peltoperlidae). Great Basin Naturalist 43:30–44.

Stark, B. P. 1986. The Nearctic species of *Agnetina* (Plecoptera: Perlidae). Journal of the Kansas Entomological Society 59:437–445.

Stark, B. P., Lago, P. K. 1983. Studies of Mississippi fishflies (Megaloptera: Corydalidae: Chauliodinae). Journal of the Kansas Entomological Society 56:356–364.

Stark, B. P., Ray, D. H. 1983. A revision of the genus *Helopicus* (Plecoptera: Perlodidae). Freshwater Invertebrate Biology 2:16–27.

Stark, B. P., Szczytko, S. W. 1981. Contributions to the systematics of *Paragnetina* (Plecoptera: Perlidae). Journal of the Kansas Entomological Society 54:625–648.

Stark, B. P., Szczytko, S. W. 1988. A new *Malirekus* species from eastern North America (Plecoptera: Perlodidae). Journal of the Kansas Entomological Society. 61:195–199.

Stewart, K. W., Stanger, J. A. 1985. The nymphs, and a new species, of North American *Setvena* Illies (Plecoptera: Perlodidae). Pan-Pacific Entomologist 61:237–244.

Stewart, K. W., Stark, B. P. 1988. Nymphs of North American stonefly genera (Plecoptera). Thomas Say Foundation Vol. 12, xiii + 460 pp.

Stewart, K. W., Stark, B. P., Huggins, T. G. 1976. The stoneflies (Plecoptera) of Louisiana. Great Basin Naturalist 36:366–384.

Stone, A., Jamnback, H. A. 1955. The black flies of New York State (Diptera: Simuliidae). Bulletin of the New York State Museum 379. 144 pp.

Stone, A., Snoddy, E. L. 1969. The blackflies of Alabama (Diptera: Simuliidae). Bulletin of the Alabama Agricultural Experiment Station, Auburn University 390, 93 pp.

Stonedahl, G. M., Lattin, J. D. 1982. The Gerridae or water striders of Oregon and Washington (Hemiptera: Heteroptera). Oregon State University Agricultural Experiment Station Technical Bulletin 144, 36 pp.

Stonedahl, G. M., Lattin, J. D. 1986. The Corixidae of Oregon and Washington (Hemiptera: Heteroptera). Oregon State University Agricultural Experiment Station Technical Bulletin 150, 84 pp.

Stribling, J. B. 1986. Revision of *Anchytarsus* (Coleoptera: Dryopoidea) and a key to the new world genera of Ptilodactylidae. Annals of the Entomological Society of America 79:219–234.

Surdick, R. F., Kim, K. C. 1976. Stoneflies (Plecoptera) of Pennsylvania, a synopsis. Bulletin 808 The Pennsylvania State University Agricultural Experiment Station, University Park, 73 pp.

Szczytko, S. W., Stewart, K. W. 1977. The stoneflies (Plecoptera) of Texas. Transactions of the American Entomological Society 103:327–378.

Szczytko, S. W., Stewart, K. W. 1979. The genus *Isoperla* (Plecoptera) of western North America; holomorphology and systematics, and a new stonefly genus *Cascadoperla*. Memoirs of the American Entomological Society 32, i + 120 pp.

Tai, L. D. D. 1967. Biosystematic study of *Sympetrum* (Odonata: Libellulidae). Dissertation, Ph. D. Purdue University, West Lafayette, IN, xviii + 234 pp.

Tarter, D. C., Little, M. L., Kirchner, R. F., Watkins, W. D., Farmer, R. G., Steele, D. 1975. Distribution of pteronarcid stoneflies in West Virginia (Insecta: Plecoptera). Proceedings of the West Virginia Academy of Science 47:79–85.

Tennessen, K. J. 1993. The larva of *Progomphus bellei* Knopf and Tennessen (Anisoptera: Gomphidae). Odonatologica 22:373–378.

Teskey, H. J. 1969. Larvae and pupae of some eastern North American Tabanidae (Diptera). Memoirs of the Entomological Society of Canada 63, 147 pp.

Teskey, H. J. 1983. A revision of the North American species of *Atylotus* (Diptera: Tabanidae) with keys to adult and immature stages. Proceedings of the Entomological Society of Ontario 114:21–43.

Teskey, H. J., Burger, J. F. 1976. Further larvae and pupae of eastern North American Tabanidae (Diptera). Canadian Entomologist 108:1085–1096.

Testa, S., III, Lago, P. K. 1994. The aquatic Hydrophilidae (Coleoptera) of Mississippi. Mississippi Agricultural and Forestry Experiment Station, University of Mississippi Technical Bulletin 193, 71 pp.

Thorp, J. H., Covich, A. P., Eds. 1991. Ecology and classification of North American freshwater invertebrates. Academic Press, San Diego, CA, x + 911 pp.

Tidwell, M. A. 1973. The Tabanidae (Diptera) of Louisiana. Tulane Studies in Zoology and Botany 18:1–95.

Truxal, F. S. 1953. A revision of the genus *Buenoa* (Hemiptera: Notonectidae). University of Kansas Science Bulletin 35:1351–1517.

Unzicker, J. D., Carlson, P. H. 1982. Epemeroptera. *in*: Brigham, A. R., Brigham, W. U. Gnilka, A. Eds. The aquatic insects and oligochaetes of North and South Carolina. Midwest Aquatic Enterprises, Mahomet, IL, pp. 3.1–3.97.

Unzicker, J. D., McCaskill, V. H. 1982. Plecoptera. *in*: Brigham, A. R., Brigham, W. U. Gnilka, A. Eds. The aquatic insects and oligochaetes of North and South Carolina. Midwest Aquatic Enterprises, Mahomet, IL, pp. 5.1–5.50.

Unzicker, J. D., Resh, V. H. Morse, J. C. 1982. Trichoptera. *in*: Brigham, A. R., Brigham, W. U. Gnilka, A. Eds. The aquatic insects and oligochaetes of North and South Carolina. Midwest Aquatic Enterprises, Mahomet, IL, pp. 9.1–9.138.

Usinger, R. L., Ed. 1956. Aquatic insects of California with keys to North American genera and California species. Uni. of California Press, Berkeley, ix + 508 pp.

Valley, K. R., Berg, C. O. 1977. Biology, immature stages, and new species of snail-killing Diptera of the genus *Dictya* (Sciomyzidae). Search 7(2), 44 pp.

Van Tassell, E. 1966. Taxonomy and biology of the subfamily Berosinae of North and Central America and the West Indies (Coleoptera: Hydrophilidae). Dissertation, Ph. D. Catholic University, Washington, DC, v + 329 pp.

Vineyard, R. N., Wiggins, G. B. 1988. Further revision of the caddisfly family Uenoidae (Trichoptera): evidence for inclusion of Neophylacinae and Thremmatidae. Systematic Entomology 13:361–372.

Voigt, W. G., Garcia, R. 1976. Key to the *Notonecta* nymphs of the West Coast United States (Hemiptera: Notonectidae). Pan-Pacific Entomologist 52:172–176.

Walker, E. M. 1953. The Odonata of Canada and Alaska. Vol. 1. Part I: General. Part II: The Zygoptera—Damselflies. Uni. of Toronto Press, Toronto, Ont, xi + 292 pp.

Walker, E. M. 1958. The Odonata of Canada and Alaska. Vol. 2. Part III: The Anisoptera—four families. Uni. of Toronto Press, Toronto, Ont, xi + 318 pp.

Walker, E. M., Corbet, P. S. 1975. The Odonata of Canada and Alaska. Volume three. Part III: The Anisoptera—three families. Uni. of Toronto Press, Toronto, Ont. xvi + 308 pp.

Wallace, J. B., Merritt, R. W. 1980. Filter-feeding ecology of aquatic insects. Annual Review of Entomology 25:103–132.

Wallace, J. B., Woodall, W. R. Staats, A. A. 1976. The larval dwelling-tube, capture net and food of *Phylocentropus placidus* (Trichoptera: Polycentropdidae). Annals of the Entomological Society of America 69:149–154.

Wallis, J. B. 1933. Revision of the North American species, (north of Mexico), of the genus *Haliplus*, Latreille. Transactions of the Royal Canadian Institute 19:1–76.

Wallis, J. B. 1939. The genus *Graphoderus* Aube in North America (north of Mexico) (Coleoptera). Canadian Entomologist 71:128–130.

Waltz, R. D., McCafferty, W. P. 1979. Freshwater springtails (Hexopoda: Collembola) of North America. Purdue University Agricultural Experiment Station, West Lafayette, IN, 32 pp.

Waltz, R. D., McCafferty, W. P. 1983. *Austrotinodes* Schmid (Trichoptera: Psychomyiidae), a first U.S. record from Texas. Proceedings of the Entomological Society of Washington 85: 181–182.

Waltz, R. D., McCafferty, W. P. 1987a. Revision of the genus *Cloeodes* Traver (Ephemeroptera: Baetidae). Annals of the Entomological Society of America 80:191–207.

Waltz, R. D., McCafferty, W. P. 1987b. New genera of Baetidae for some Nearctic species previously included in *Baetis* Leach (Ephemeroptera). Annals of the Entomological Society of America 80:667–670.

Waltz, R. D., McCafferty, W. P. 1987c. Systematics of *Pseudocloeon, Acentrella, Baetiella,* and *Liebebiella,* new genus (Ephemeroptera: Baetidae). Journal of the New York Entomological Society 95:553–568.

Waltz, R. D., McCafferty, W. P. 1989. New species, redescriptions, and cladistics of the genus *Pseudocentroptiloides* (Ephemeroptera: Baetidae). Journal of the New York Entomological Society 97:151–158.

Waltz, R. D., McCafferty, W. P. 1999. Additions to the taxonomy of *Americabaetis* (Ephemeroptera: Baetidae): *A. lugoi*, n. sp., adult of *A. robacki*, and key to larvae. Entomological News 110:39–44.

Waters, T. F. 1972. The drift of stream insects. Annual Review of Entomology 17:253–272.

Weaver, J. S. III. 1988. A synopsis of the North American Lepidostomatidae (Trichoptera). Contributions of the American Entomological Institute 24(2), iv + 141 pp.

Weaver, J. S. III, Morse, J. C. 1986. Evolution of feeding and case-making behavior in Trichoptera. Journal of the North American Benthological Society 5:150–158.

Weaver, J. S., Sykora, J. L. 1979. The *Rhyacophila* of Pennsylvania, with larval descriptions of *R. banksi* and *R. carpenteri* (Trichoptera: Rhyacophilidae). Annals of the Carnegie Museum 48:403–423.

Webb, D. W. 1994. The immature stages of *Suragina concinna* (Williston) (Diptera: Athericidae). Journal of the Kansas Entomological Society 67:421–425.

Webb, D. W., Brigham, W. U. 1982. Aquatic Diptera. *in*: Brigham, A. R., Brigham, W. U. Gnilka, A. Eds. The aquatic insects and oligochaetes of North and South Carolina. Midwest Aquatic Enterprises, Mahomet, IL, pp. 11.1–11.111.

Westfall, M. J., May, M. L. 1996. Damselflies of North America. Scientific Publications, Gainesville, FL, 650 pp.

Westfall, M. J., Jr., Tennessen, K. J. 1979. Taxonomic clarification within the genus *Dromogomphus* Selys (Odonata: Gomphidae). Florida Entomologist 62:266–273.

White, D. S. 1978. A revision of the Nearctic *Optioservus* (Coleoptera: Elmidae), with descriptions of new species. Systematic Entomology 3:59–74.

Wiederholm, T., Ed. 1983. Chironomidae of the Holarctic region. Keys and diagnosis. Part I. Larvae. Entomologica Scandinavica Supplement 19, 457 pp.

Wiersema, N. A. 1998. *Camelobaetidius variabilis* (Ephemeroptera: Baetidae), a new species from Texas, Oklahoma and Mexico. Entomological News 109:21–26.

Wiersema, N. A., McCafferty, W. P. 1998. A new species of *Pseudocentroptiloides* (Ephemeroptera: Baetidae), with revisions to other previously unnamed baetid species from Texas. Entomological News 109:110–116.

Wiggins, G. B. 1960. A preliminary systematic study of the North American larvae of the caddisfly family Phryganeidae (Trichoptera). Canadian Journal of Zoology 38:1153–1170.

Wiggins, G. B. 1973. New systematic data for the North American caddisfly genera *Lepania, Goeracea* and *Goerita* (Trichoptera: Limnephilidae). Life Science Contributions Royal Ontario Museum 91, 33 pp.

Wiggins, G. B. 1977. Larvae of the North American caddisfly genera (Trichoptera). Uni. of Toronto Press, Toronto, Ont, xi + 401 pp.

Wiggins, G. B. 1981. Considerations on the relevance of immature stages to the systematics of Trichoptera. *in*: Moretti, G. P., Ed. Proceeding of the 3rd International Symposium on Trichoptera. Series Entomology 20. Junk, Dr. W. The Hague, Netherlands, pp. 395–407.

Wiggins, B. B. 1996. Larvae of the North American caddisfly genera (Trichoptera). 2nd ed. Uni. of Toronto Press, Toronto, Ont, Buffalo, NY, xiii + 457 pp.

Wiggins, G. B., Richardson, J. S. 1982. Revision and synopsis of the caddisfly genus *Dicosmoecus* (Trichoptera: Limnephilidae; Dicosmoecinae). Aquatic Insects 4:181–217.

Wiggins, G. B., Wichard, W. 1989. Phylogeny of pupation in Trichoptera, with proposals on the origin and higher classification of the order. Journal of the North American Benthological Society 8:260–276.

Wiggins, G. B., Wisseman, R. W. 1990. Revision of the North American caddisfly genus *Desmona* Trichoptera: Limnephilidae). Annals of the Entomological Society of America 82:155–161.

Willson, R. B. 1967. The Hydrophilidae of Michigan with keys to species of the Great Lakes Region. Thesis, M. S. Michigan State University, East Lansing, 100 pp.

Wilson, C. A. 1958. Aquatic and semiaquatic Hemiptera of Mississippi. Tulane Studies in Zoology 6:116–170.

Winters, F. C. 1926. Notes on the Hydrobiini (Coleoptera-Hydrophilidae) of boreal America. Pan-Pacific Entomologist 3:49–58.

Wojtowicz, J. A. 1982. A review of the adults and larvae of the genus *Pycnopsyche* (Trichoptera: Limnephilidae) with revision of the *Pycnopsyche scabripennis* (Rambur) and *Pycnopsyche lepida* (Hagen) complexes. Dissertation, Ph. D. University of Tennessee, 292 pp.

Wolfe, G. W. 1984. A revision of the *Hydroporus vittatipennis* species group of the subgenus *Neoporus* (Coleoptera: Dytiscidae). Transactions of the American Entomological Society 110:389–434.

Wolfe, G. W., Matta, J. F. 1981. Notes on nomenclature and classification of *Hydroporus* subgenera with the description of a new genus of Hydroporini (Coleoptera: Dytiscidae). Pan-Pacific Entomologist 57:149–175.

Wolfe, G. W., Roughley, R. E. 1990. A taxonomic, phylogenetic, and zoogeographical analysis of *Laccornis* Gozis (Coleoptera: Dytiscidae) with the description of Laccornini, a new tribe of Hydroporinae. Quaestiones Entomologicae 26:273–354.

Wood, D. M. 1978. Taxonomy of the Nearctic species of *Twinnia* and *Gymnopais* (Diptera: Simuliidae) and a discussion of the ancestry of the Simuliidae. Canadian Entomologist 110:1297–1337.

Wood, D. M., Dang, P. T., Ellis, R. A. 1979. The insects and arachnids of Canada. Part 6. The mosquitoes of Canada (Diptera: Culicidae). Agriculture Canada Publication 1686. 390 pp.

Wood, D. M., Peterson, B. V., Davies, D. M., Gyorkos, H. 1963. The black flies (Diptera: Simuliidae) of Ontario. Part II. Larval identification, with descriptions and illustrations. Proceedings of the Entomological Society of Ontario 93:99–129.

Wood, F. E. 1962. A synopsis of the genus *Dineutus* (Coleoptera: Gyrinidae) in the Western Hemisphere. M. S. Thesis, University of Missouri, Columbia, 99 pp.

Wooldridge, D. P. 1966. Notes on Nearctic *Paracymus* with descriptions of new species (Coleoptera: Hydrophilidae). Journal of the Kansas Entomological Society 39:712–725.

Wooldridge, D. P. 1967. The aquatic Hydrophilidae of Illinois. Illinois State Academy of Science 60:422–431.

Yamamoto, T. Wiggins, G. B. 1964. A comparative study of the North American species in the caddisfly genus *Mystacides* (Trichoptera: Leptoceridae). Canadian Journal of Zoology 42:1105–1126.

Young, F. N. 1953. Two new species of *Matus*, with a key to the known species and subspecies of the genus (Coleoptera: Dytiscidae). Annals of the Entomological Society of America 46:49–55.

Young, F. N. 1954. The water beetles of Florida. Uni. of Florida Press, Gainesville, FL, ix + 238 pp.

Young, F. N. 1956. A preliminary key to the species of *Hydrovatus* of the eastern United States (Coleoptera: Dytiscidae). Coleopterists Bulletin 10:53–54.

Young, F. N. 1963a. The Nearctic species of *Copelatus* Erichson (Coleoptera: Dytiscidae). Quarterly Journal of the Florida Academy of Science 26:56–77.

Young, F. N. 1963b. Two new North American species of *Hydrovatus*, with notes on other species (Coleoptera: Dytiscidae). Psyche 70:184–192.

Young, F. N. 1979a. A key to Nearctic species of *Celina* with descriptions of new species (Coleoptera: Dytiscidae). Journal of the Kansas Entomological Society 52:820–830.

Young, F. N. 1979b. Water beetles of the genus *Suphisellus* Crotch in the Americas north of Colombia (Coleoptera: Noteridae). Southwestern Naturalist 24:409–429.

Young, F. N. 1981a. Predaceous water beetles of the genus *Desmopachria*: the *convexa-grana* group (Coleoptera: Dytiscidae). Occasional Papers of the Florida Collection of Arthropods 2(III-IV):1–11.

Young, F. N. 1981b. Predaceous water beetles of the genus *Desmopachria* Babington: the *leechi-glabricula* group (Coleoptera: Dytiscidae). Pan-Pacific Entomologist 57:57–64.

Young, F. N. 1985. A key to the American species of *Hydrocanthus* Say, with descriptions of new taxa (Coleoptera: Noteridae). Proceedings of the Academy of Natural Sciences, Philadelphia 137:90–98.

Young, W. C., Bayer, C. W. 1979. The dragonfly nymphs (Odonata: Anisoptera) of the Guadelupe River Basin, Texas. Texas Journal of Science 31:85–98.

Zalom, F. G. 1977. The Notonectidae (Hemiptera) of Arizona. Southwestern Naturalist 22:327–336.

Zimmerman, J. R. 1970. A taxonomic revision of the aquatic beetle genus *Laccophilus* (Dytiscidae) of North America. Memoirs of the American Entomological Society 26., i + 275 pp.

Zimmerman, J. R. 1981. A revision of the *Colymbetes* of North America (Dytiscidae). Coleopterists Bulletin 35:1–52.

Zimmerman, J. R. 1985. A revision of the genus *Oreodytes* in North America (Coleoptera: Dytiscidae). Proceedings of the Academy of Natural Sciences, Philadelphia 137:99–127.

Zimmerman, J. R., Smith, A. H. 1975. A survey of the *Deronectes* (Coleoptera: Dytiscidae) of Canada, the United States, and northern Mexico. Transactions of the American Entomological Society 101:651–722.

Zimmerman, J. R., Smith, R. L. 1975. The genus *Rhantus* (Coleoptera: Dytiscidae) in North America. Part I. General account of the species. Transactions of the American Entomological Society 101:33–123.

Zloty, J., Pritchard G. 1997. Larvae and adults of *Ameletus* mayflies (Ephemeroptera: Ameletidae) from Alberta. Canadian Entomologist 129:251–289.

18 AQUATIC INSECT ECOLOGY

Anne E. Hershey
Department of Biology
University of North Carolina–Greensboro
Greensboro, North Carolina 27402

Gary A. Lamberti
Department of Biological Sciences
University of Notre Dame
Notre Dame, Indiana 46556

I. INTRODUCTION

Aquatic insects are abundant in most freshwater habitats and often exhibit high diversity. They play significant roles in freshwater ecosystems, including serving as food items for nearly the full range of vertebrate and invertebrate predators found in aquatic systems. The ecology of aquatic insects has been studied from many perspectives, including species diversity, life-history constraints, community structure, predator–prey interactions, detritivory, grazing, and implications for nutrient dynamics. Considerable information is available on insect responses to a variety of environmental conditions, including factors that operate on landscape-level scales. Thus, they often are used as indicators of water-quality conditions in both lentic and lotic systems. Longitudinal trends in insect functional feeding groups are an important component of the River Continuum Concept (see Fig. 6 in Chapter 2), a major paradigm in the discipline of stream ecology. Although the ecological literature on aquatic insects is extensive, most of it has been produced within the last 20 years, reflecting current interest in the topic. In spite of this extensive body of work, the enormous diversity of insects combined with their widespread distribution dictates that much remains to be learned.

This chapter provides a brief overview of aquatic insect communities in both lentic (standing water) and lotic (flowing water) habitats and describes the role of insects in ecological processes within these ecosystems. The discussion is placed in the context of how physical and life-history factors constrain the distribution and abundance of aquatic insects, thereby altering communities and ecosystem function.

II. AQUATIC INSECT COMMUNITIES

A. Insect Taxonomic Diversity

Insects are the most species-rich and often the most abundant group of substrate-dwelling macroinvertebrates, and have successfully invaded virtually all aquatic habitats. Although there are many specialized habitats, most aquatic insects are found either in lotic or lentic habitats (Table I), and are constrained to one or the other habitat by physiological adaptations (see below). We will follow the classification used

Ecology and Classification of North American Freshwater Invertebrates. 2nd Edition

TABLE I List of aquatic and partially aquatic orders of insects, with estimated number of North American species, preferred habitat, typical generation time, feeding mode, and examples of common genera.

Order	*Estimated no. of NA species*	*Primary (and 2°) habitat*	*Generation time* (yr)	*Primary feeding mode (FFG)*	*Examples of common NA genera*
Collembola	40	Water surface; semiaquatic	≤1	Collector-gatherers	*Podura, Sminthurides*
Ephemeroptera	600	Lotic (lentic)	≤1	Collector-gatherers, collector-filterers, scrapers	*Baetis, Hexagenia*
Odonata					
Anisoptera	300	Lentic + lotic	≥1	Predators	*Aeshna, Gomphus, Libellula*
Zygoptera	120	Lentic + lotic	≥1	Predators	*Argia, Lestes*
Orthoptera	40	Water margin; semiaquatic	≥1	Shredders	*Conocephalus*
Plecoptera					
6 families	330	Lotic (lentic)	≥1	Shredders	*Pteronarcys, Capnia*
3 families	250	Lotic (lentic)	≥1	Predators	*Acroneuria, Isoperla*
Heteroptera	410	Lentic + lotic	≤1	Predators, collector-gatherers	*Belostoma, Gerris, Trichocorixa*
Megaloptera	50	Lotic (lentic)	≥1	Predators	*Sialis, Corydalus*
Neuroptera	6	Lotic + lentic	≤1	Piercers	*Sisyra*
Trichoptera	1350	Lotic (lentic)	= 1 (some > or <1)	All major FFGs	*Rhyacophila, Hydropsyche, Glossosoma, Neophylax*
Lepidoptera	770	Lentic (lotic)	≤1	Shredders, scrapers	*Petrophila*
Coleoptera	5000	Lentic + lotic; water margin	≤1 (some >1)	All major FFGs	*Hydroporus, Gyrinus, Psephenus*
Diptera					
Tipulidae	1500	Lotic	≥1	Shredders, predators	*Tipula, Limonia*
Culicidae	175	Lentic	≤1	Collector filterers	*Aedes, Anopheles, Culex*
Simuliidae	150	Lotic	≤1	Collector-filterers, scrapers	*Prosimulium, Simulium*
Chironomidae	2000	Lotic + lentic	≤1	All major FFGs	*Chironomus, Polypedilum, Tanytarsus*
Others	3200	Lotic + lentic	≤1	All major FFGs	*Chaoborus, Chrysops, Ephydra*
Hymenoptera	60	Internal parasites	≤1	Parasites	*Apsilops, Trichogramma*

(NA, North American; FFG, functional feeding group.)

by Hilsenhoff in this volume (see Chapter 17). The orders Ephemeroptera (mayflies), Odonata (dragonflies and damselflies), Plecoptera (stoneflies), Trichoptera (caddisflies), and Megaloptera (fishflies and alderflies) are considered to be strictly aquatic in their immature stages. Other insect orders are partially aquatic (aquatic group listed in parentheses), including Collembola (semiaquatic springtails), Neuroptera (spongillaflies), Heteroptera (aquatic and semiaquatic bugs), Coleoptera (water beetles), Diptera (aquatic flies and midges, especially suborder Nematocera but some Brachycera), and Lepidoptera (aquatic caterpillars). All of these latter groups have important aquatic representatives as immatures. The Coleoptera and Heteroptera have groups with adult as well as larval stages that are aquatic. Even several species of Orthoptera (grasshoppers) are found in association with aquatic habitats (Cantrall and Brusven, 1996). These aquatic insect orders are treated taxonomically in Chapter 17.

Species diversity in many of the aquatic insect orders is very high. A pristine stream may contain well over one hundred species of insects, including many that are difficult or impossible to identify in the immature aquatic stages. Ponds, lakes, and wetlands can also exhibit very high diversity, depending on water-quality conditions, littoral development, and substrate characteristics. Due to taxonomic problems associated with this high diversity, the role of aquatic insects in aquatic ecosystems is often difficult to study. To minimize this problem, aquatic insects have been categorized into a smaller number of functional feeding groups (see Merritt and Cummins, 1996a, b), as discussed in more detail later in this chapter.

B. Physical Constraints on Aquatic Insects

The physical environment of aquatic ecosystems exerts substantial control over the population abundances and hence community composition of insects. Still, it is remarkable that virtually every North American freshwater environment has been colonized by aquatic insects, ranging from alpine lakes to desert hot springs, oligotrophic streams to sewage treatment lagoons, and treeholes to the extreme depths of large lakes. Within these habitats, physical factors of particular importance to aquatic insects include: (1) dissolved oxygen concentration; (2) water temperature; (3) water chemistry; (4) type of substrate; and (5) hydrodynamics.

1. Dissolved Oxygen

As aerobic organisms, all insects must obtain sufficient oxygen to drive their metabolic machinery (Eriksen *et al.*, 1996). This presents a particular challenge for aquatic insects because water, even when saturated, contains much less oxygen than do terrestrial environments (a maximum of about 15 ppm oxygen in water compared to over 200,000 ppm in the air). Dissolved oxygen concentrations are highly variable over time and space, and oxygen may be totally lacking from some aquatic habitats (termed "anoxia"). Furthermore, oxygen concentrations in lentic systems generally decrease with depth (Fig. 1A). Because the vast majority of aquatic insects are substrate dwellers, this presents a potential physiological constraint. Most cold-water oligotrophic lakes remain oxygenated throughout the open water season because solubility increases as water temperature decreases. However, even in cold climates, shallow ponds and wetlands can become anoxic due to respiratory oxygen demands under ice where no opportunity exists for oxygen replenishment. Deeper oligotrophic lakes typically have sufficient oxygen stores to meet under-ice metabolic demands. The likelihood of under-ice anoxia increases with productivity due to proportionally increasing respiratory requirements and decreases with depth due to high oxygen storage in the hypolimnion. Thus, seasonal and spatial variation in oxygen concentration greatly restricts the types and diversity of insects found in aquatic environments.

No insect lacking access to an oxygen supply can survive for long, but some can cope with short-term anoxia. For example, the chironomid midge *Chironomus* and related Chironomini (Diptera: Chironomidae) contain a hemoglobinlike pigment (hence the common name "bloodworm") that allows them to store oxygen and thus survive temporary anoxia. The phantom midge *Chaoborus* (Diptera: Chaoboridae) is also found in anoxic sediments, but uses a different strategy than do bloodworms. *Chaoborus* migrates vertically at night

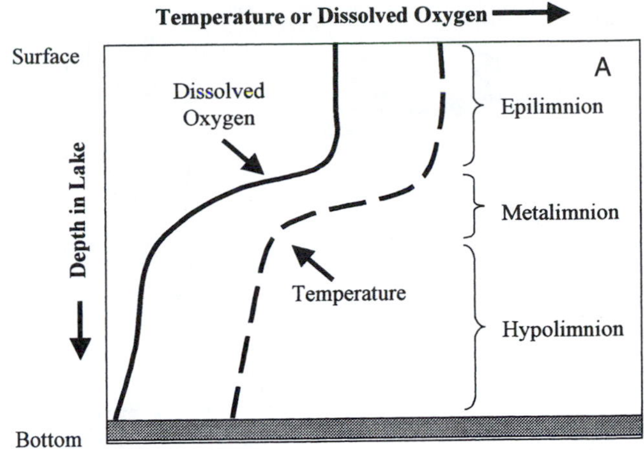

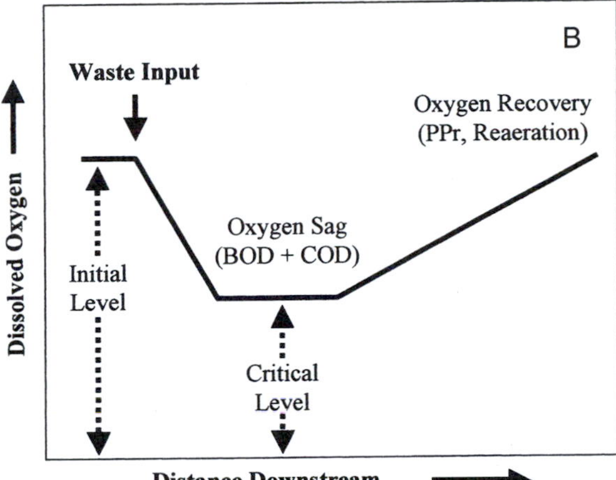

FIGURE 1 Oxygen features in lakes and streams. (A) Oxygen and temperature profiles in a temperate stratified lake. (B) Oxygen "sag" in a river receiving a point discharge of municipal waste. BOD, biological oxygen demand; COD, chemical oxygen demand; PPr, primary production.

toward the surface both to obtain oxygen and to hunt for prey. During the day, it returns to the deep, dark sediments to evade its own predators. Shallower environments (e.g., ponds and marshes) that experience oxygen depletion contain a more diverse array of insects because they can obtain oxygen in a myriad of ways. Some insects, including many heteropterans, carry an air store ("bubble") on their body that they occasionally replenish by swimming to the water surface (Fig. 2A). Others maintain a semipermanent connection to the atmosphere ("breathing tube") as exemplified by mosquito larvae (Diptera: Culicidae) and rat-tailed maggots (Diptera: Syrphidae). Yet other insects, such as some beetle larvae (Coleoptera) and a genus of mosquitoes (Fig. 2B), use specialized spiracles to pierce the vascular tissues of rooted aquatic plants and thus "steal" their oxygen (White and Brigham, 1996).

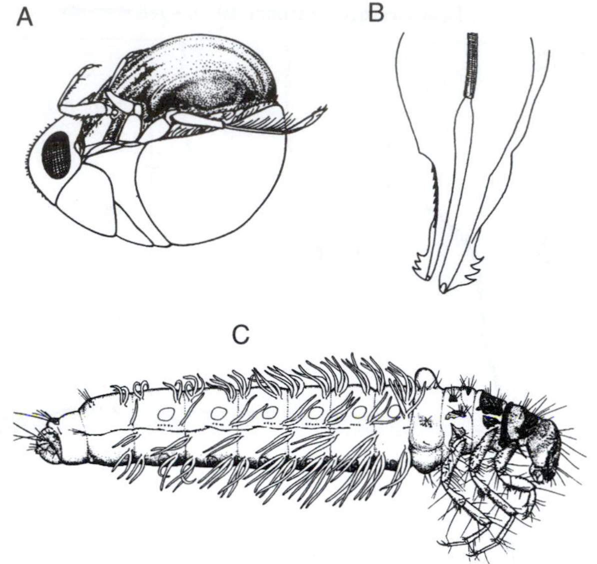

FIGURE 2 Respiratory structures of aquatic insects. (A) Ventral air bubble of the backswimmer *Neoplea* (Heteroptera: Pleidae). (B) Longitudinal section of postabdominal respiratory siphon of the mosquito larva *Taeniorhynchus* (Diptera: Culicidae), used to pierce plant air stores. (C) Abdominal filamentous tracheal gills of the caddisfly *Limnephilus* (Trichoptera: Limnephilidae). [Drawings (A) and (B) from Merritt and Cummins, 1996b; (C) from Wiggins, 1996.]

In most unimpacted streams, dissolved oxygen concentrations generally are near saturation and thus oxygen rarely limits insect diversity. We often see the greatest diversity of aquatic insects in small streams, due in part to the abundance of oxygen. However, insects requiring a connection to atmospheric air are typically in low numbers, because flow would quickly displace these taxa far downstream. Thus, most lotic insects obtain their oxygen by diffusion through the body wall, often aided by "tracheal gills," which are thinly sclerotized evaginations of the body wall designed to increase surface area (Fig. 2C). Most representatives of the insect orders that are wholly aquatic (e.g., Ephemeroptera, Plecoptera, and Trichoptera) possess such tracheal gills in the larval, aquatic state and achieve their highest diversity in flowing waters. Gills, however, rely on abundant dissolved oxygen for diffusion into the body, and thus gilled larvae can experience stress at low oxygen concentrations. Such oxygen "sags" can occur in larger rivers with low turbulence (and therefore low reaeration), especially at night when photosynthesis turns off or where municipal or industrial pollutants deplete oxygen due to high respiratory demand downstream of their input (Fig. 1B). In these cases, the aquatic insect fauna will more closely mimic that found in oxygen-poor standing waters (many dipterans, for example). Some behavioral adaptations allow insects to "weather" periods of low oxygen in flowing water. For example, perlid stoneflies perform "push-ups" when oxygen declines, presumably to move more oxygenated water across their ventral thoracic gills. Burrowing mayflies of the genus *Hexagenia* undulate their body to draw water, and thus food and oxygen, through their burrows.

2. Water Temperature

Water temperature also imposes constraints on aquatic insects. Temperature directly affects insects by regulating their metabolic rates and thus development from egg to adult, and indirectly by influencing such things as fluid dynamics, gas saturation constants, and primary production rates (Hauer and Hill, 1996). Fluctuation in water temperature is found both spatially and temporally. For example, stratified lakes show marked temperature gradients from top to bottom, while rivers can show horizontal (bank to bank) and longitudinal (source to estuary) gradients in temperature. Low- to midorder streams also may exhibit pronounced diel temperature fluctuations, especially on sunny days. Shallow unshaded wetlands may exhibit diel fluctuations of 20° (e.g., Hershey *et al.,* 1995a). Temporal fluctuations can range from subfreezing to tepid in northern wetlands. Thus, temperature variation imposes fundamental constraints on the types of insects that may be present in many aquatic habitats.

Some aquatic insects prefer a narrow range of temperature, usually cool water (cold "stenotherms"), whereas others can tolerate a broader range of temperatures ("eurytherms"). Consequently, temperature fluctuation can limit the types of insects found within a habitat. Any significant change in water temperature, as may result from thermal pollution or climate change, will likely alter the species composition of aquatic insects. Considerable variation in temperature tolerance occurs within the orders of aquatic insects, and so temperature preferences should be examined on a species-specific basis.

The upper lethal limit for all but the most specialized species is between 30 and 40°C (Pritchard, 1991; Wallace and Anderson, 1996), although most temperate aquatic habitats do not approach such temperatures. However, thermally destabilized habitats, such as cooling water reservoirs, have temperatures which may approach this limit intermittently or seasonally, and abundance and richness of most taxa are negatively affected (see Thorp and Diggins, 1982). Types of insects that can tolerate, and even thrive, in hot water include various beetles and true flies, although species diversity declines with increasing temperature. For example, in a survey of hot springs, Brock (1978) recorded 60 species of Coleoptera at 30°C; at 45°C only two species were found. At the other extreme, many species exhibit growth at temperatures only a few degrees above freezing in both lentic (e.g., Butler, 1982) and lotic systems (e.g., Rosenberg *et al.*, 1977).

Aquatic insects have several strategies for surviving subfreezing conditions. The antifreeze compounds produced by some species prevent freezing of bodily fluids to several degrees below 0°C (Danks, 1978; Duffy and Liston, 1985). Other species actually tolerate freezing of extracellular tissues, which provides some protection against intracellular freezing (Danks, 1978; Zachariassen, 1979; Ward and Stanford, 1982). Cocoons and diapausing eggs are important strategies to prevent ice damage (see Wallace and Anderson, 1996). Because water and snow cover provide some degree of thermal buffering, even streams and ponds in the arctic do not drop more than several degrees below freezing. In north temperate regions, where terrestrial temperatures may reach −35°C, stream temperatures typically do not drop below −1°C in pools and −2°C in riffles (Fig. 3; Hershey unpublished data). In arctic ponds that freeze solid, some insect species overwinter several times as larvae (Butler, 1982). Thus, aquatic insects only need mechanisms that provide them with a few degrees of thermal buffering. In some streams, avoidance of ice scour may be a more serious problem for benthic insects. Prior to the formation of surface ice, anchor ice attached to the stream bed may form at night due to cooling associated with back radiation (Barnes, 1928). As the stream warms during the day, anchor ice lifts off the bottom, scouring the stream bed and disturbing resident insects (Hynes, 1970).

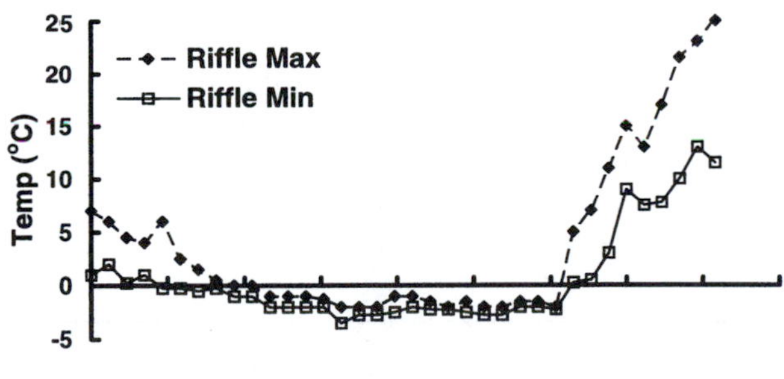

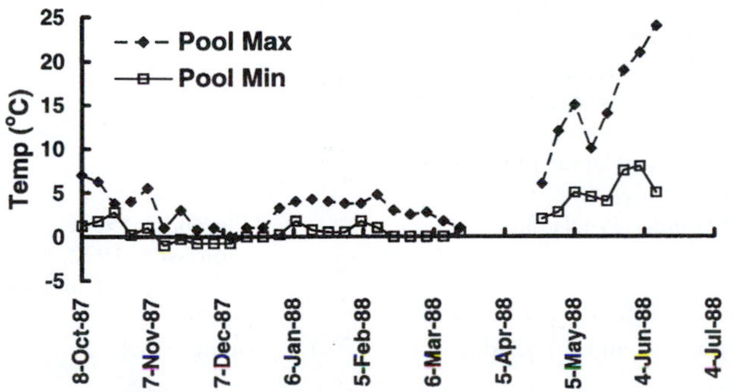

FIGURE 3 Pool and riffle temperature in the French River, Minnesota, which is strongly influenced by ice. Although air temperatures may drop as low as about −30°C, aquatic insects experience temperatures that are only a few degrees below freezing.

Temperature, and its variation, can influence aquatic insect diversity in any freshwater habitat. For streams, the River Continuum Concept (Vannote *et al.*, 1980; see also below) predicts that biological diversity should be highest in midorder (medium-sized) streams because temperature variation is highest, thereby providing the most temperature "niches" for insects. The literature on this topic, however, is mixed. Insect faunal diversity can be high in cold streams, which are thought to be the ancestral habitat for aquatic insects (Hynes, 1970). For example, Anderson and Hanson (1987) cataloged over 325 species of aquatic insects from a cool Oregon stream. On the other hand, Morse *et al.* (1983) found over 500 species of aquatic insects in a warmer coastal South Carolina stream. From wood snags in the warm Ogeechee River in Georgia, Benke and Jacobi (1994) collected over 20 species of mayflies alone, suggesting that aquatic insects can be diverse at a variety of temperatures.

3. Water Chemistry

Many aspects of water chemistry can restrict the occurrence or abundance of aquatic insects, including pH, salinity, and concentrations of specific ions or elements. Generally, it is the extremes in any of these parameters that result in change to aquatic insect communities, while levels around the "average" have less direct impact. Low pH, as is found in acidified lakes and streams of pH <5 (due to acid deposition, mine drainage, organic acids, or poor buffering) can alter community composition such that only acid-tolerant taxa are found (Hynes, 1970). However, it is unclear whether it is the low pH itself that adversely affects insects or some related factor. For example, iron can precipitate under acidic conditions with an associated reduction in oxygen. Acidic waters can also leach metals from soils and rock, especially aluminum, that can increase drift and possibly have toxic effects on aquatic insects (Hall *et al.*, 1987). Regardless, some aquatic insects thrive in acid streams and lakes (examples cited in Hynes, 1970). Salinity gradients that form in coastal estuaries, along saline lakes, and even from runoff after road salting, can affect insects, most of which are salt-intolerant. However, some insects such as brine flies (Diptera: Ephydridae) thrive in warm, saline water, where they have few competitors. Many other chemical features, such as calcium concentration and total ionic strength, have potential importance to aquatic insects, but little is known about the requirements of specific taxa.

4. Substrate and Flow

At the scale of local habitat, substrate and hydrodynamics are probably the most important factors determining the types and abundance of aquatic insects present. Swift currents in streams, wave action along shorelines, wind-generated turbulent mixing, and tidal action all present significant hydrodynamic challenges to insects. Most aquatic insects spend at least part, if not most, of their life cycle associated with the substrate, and hydrodynamic forces interact strongly with substrate type to produce the habitat experienced by the insect fauna. Hydrodynamic and substrate influences on aquatic insects are discussed in more detail in the sections below on lentic and lotic insect communities.

Cummins and Merritt (1996) recognized eight categories of aquatic insect habits, primarily based on the substrate or habitat they occupy. The first four groups remain associated with the substrate for most of their lives. "Burrowers" and "sprawlers" inhabit fine sediments where they either tunnel into the sediments or remain on top of the sediments, respectively. "Climbers" move along the stems of vascular macrophytes or pieces of detritus. "Clingers" attach themselves to the surfaces of substrates exposed to water movement, such as rocks in streams or along wave-swept lake shores. The final four groups spend part, or most, or their lives moving through or on the water. "Swimmers" periodically propel themselves through the water to change location, although they mostly remain attached to a substrate. "Divers" split their time among swimming to the surface, diving back through the water, and clinging to submerged objects. "Skaters" are adapted for life on top of the water surface, where they use hydrophobic body parts to move over water. "Planktonic" forms float or swim about in open water.

C. Lentic Insect Communities

Lentic environments, including lakes, ponds, and various types of wetlands offer a broad array of aquatic habitats that have been exploited by a wide variety of insects. As in any community, the insect species that are present reflect the physical, chemical, and biotic characteristics of the ecosystem, each of which places major physiological, morphological, and trophic constraints on aquatic insects. Insects exhibit a variety of adaptations to these habitat constraints, which combined have a large impact on the structure of lentic insect communities. In addition, numerous trophic and interspecific factors interact with the physical and chemical constraints to yield the dynamic and diverse insect communities that are present in lentic systems.

1. Littoral Communities

Littoral areas of ponds and lakes are typically better oxygenated, structurally more complex, and afford more abundant and diverse food resources than do profundal sediments of lakes. All of these factors lead to a high diversity of insects. Littoral communities generally have aquatic insect representatives from most of the aquatic orders (Table II), including nearly all of the

TABLE II Aquatic insect communities in lentic habitats

Habitat type	*Habitat characteristics*	*Typical insect community*	*References*
Littoral	High O_2, macrophytes, rock, sand, or silt substrates	Ephemeroptera (esp. Siphlonuridae, Baetidae, Heptageniidae, Caenidae, Leptophlebiidae), Odonata (representatives from most families), Diptera (many families, but esp. diverse Chironomidae), Trichoptera (esp. Phryganeidae, Limnephilidae, Leptoceridae), Plecoptera (some Perlidae and Perlodidae), Hemiptera (most aquatic families), Megaloptera (few genera), Neuroptera (both aquatic genera), Coleoptera (adults and larvae of many families), Hymenoptera (parasitoids only)	Merritt and Cummins, 1996a, b; Ward, 1992
Eutrophic profundal	Low O_2, soft mineral and organic sediments	Chironomidae (esp. *Chironomus, Procladius*)	Hilsenhoff, 1966; Jonasson, 1972; Merritt and Cummins, 1996a, b
Oligotrophic profundal	High O_2, soft mineral sediments	Chironomidae (esp. Tanytarsini and Orthocladiinae)	Saether,1979
Wetlands	Variable hydrology, variable O_2, organic sediments, macrophytes	High diversity of most lentic orders. Diptera and Coleoptera are especially well represented.	Rosenberg and Danks, 1987; Batzer and Wissinger, 1996; Hershey *et al.*, 1998

morphological and trophic types as well as the full size range of aquatic species. Large taxa such as dragonflies, damselflies, many of the lentic mayflies, aquatic beetles and bugs, and caddisflies are important members of littoral zone communities, and typically are uncommon in profundal zones (Table II). Aquatic insects of littoral zones also show considerable variation in habitat (see Ward, 1992). Many of these species use the macrophytes themselves as habitat, although others are characteristic inhabitants of inorganic and organic sediments and rocks. Macrophyte habitats usually have higher density and species diversity of insects than less structurally complex habitats, including many of the larger taxa mentioned above (Crowder and Cooper, 1982; Gilinsky, 1984; Hershey, 1985a). Many highly mobile species, particularly members of the Coleoptera and Heteroptera, swim in the water column (plankton), or use the upper or lower surface of the air–water interface (epineuston and hyponeuston, respectively). Gyrinid beetles, referred to as whirligig beetles because they swim in whirling swarms on the water surface, are especially well adapted for their habitat; they have dorsal and ventral compound eyes permitting them to see above and below the water surface simultaneously. The burrowing mayfly, *Hexagenia*, is often very abundant in littoral and sublittoral sediments of lakes, but is particularly sensitive to anoxia (Beeton, 1961; Rasmussen, 1988). A few Plecoptera (stoneflies) are found in erosional lentic habitats (White and Brigham, 1996), reflecting the high oxygen requirements of members of this order. Smaller species, especially representing the dipteran family Chironomidae, are usually very diverse and present in high densities (Thorp and Bergey, 1981; Crowder and Cooper, 1982; Gilinski, 1984; Hershey, 1985a).

2. Profundal Communities

Profundal zones of lakes have quite different communities than do littoral zones (Table II). In both oligotrophic and eutrophic lakes, chironomids most often dominate the insect fauna. However, in eutrophic lakes, where the profundal zone is often anoxic during much of the summer, *Chironomus* and *Procladius* may be the only insect genera present (Hilsenhoff, 1966; Saether, 1979). *Chironomus* feeds on suspended particles by pumping water into either U-shaped or J-shaped burrows, ingesting the particles that impinge on its silken funnel at the opening of the burrow (Walshe, 1947). Feces are deposited on the sediment surface forming mounds. At high density, this feeding mechanism can be important in returning phosphorus to the water column (Fig. 4; Gallepp, 1979). *Procladius* is a free-living chironomid, most often characterized as a predator (Kajak, 1980; Vodopich and Cowell, 1984), but it also ingests a variety of sediment dwelling diatoms and desmids (Hershey, 1986). Although ubiquitous in lentic habitats *Procladius* also is tolerant of low oxygen conditions. In oligotrophic lakes, the chironomid community in the profundal zone is more diverse, but still depauperate compared to littoral zones. The presence of *Chironomus* spp. versus *Tanytarsus* spp. has been considered indicative of lake trophic condition (e.g., Brinkhurst, 1974). An elaborate key to lake typology which recognizes several categories of trophic conditions has been developed based on the occurrence of chironomid genera and species (Saether, 1979), and lake classification has a long history based on chironomid assemblages (Brundin, 1949; Brinkhurst, 1974).

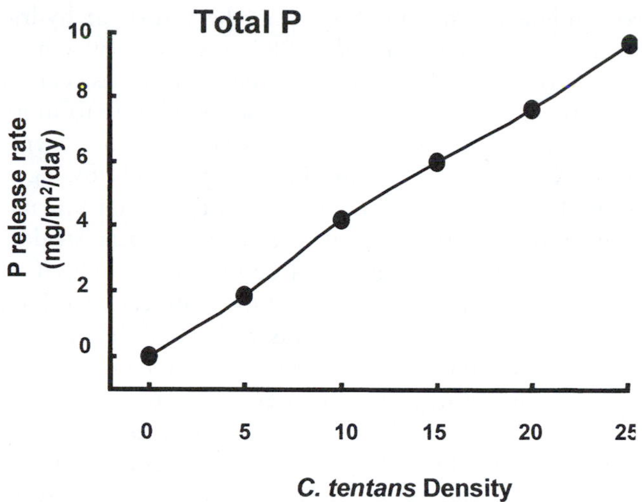

FIGURE 4 Feeding activity of *Chironomus tentans* returns P to the water column. At high density this can be a significant source of P to the water column. [From Gallepp, 1979.]

3. Wetland Communities

The various types of wetlands also afford a variety of habitats for aquatic insects. Water level is highly variable in wetlands, by definition, and many wetland types are subjected to periods of desiccation. Desiccation provides a variety of challenges for aquatic insects, and results in somewhat different communities than are found in more permanent lentic ecosystems. Hydroperiod is likely, therefore, to be the most important factor affecting wetland insect communities (Wiggins *et al.*, 1980; Batzer and Wissinger, 1996; Hershey *et al.*, 1998). Insect species in wetland habitats exhibit either desiccation-resistance or drought-avoidance strategies (Wiggins *et al.*, 1980). These strategies do not fall neatly along taxonomic lines. Desiccation resistant insects, especially many Diptera, and some odonates, caddisflies, and beetles, deposit drought-resistant eggs or lay eggs in plant stems, which then hatch to repopulate the habitat following the return of hydric conditions (Wiggins *et al.*, 1980, reviewed by Batzer and Wissinger, 1996). Insects that avoid desiccation have several different strategies. Some migrate to more permanent habitats to complete one or more generations. Wing polymorphism (i.e., winged and unwinged individuals) associated with this strategy is not uncommon and is particularly well-studied in water striders (Spence and Anderson, 1994) and some beetles (see Wallace and Anderson, 1996). Other insects have long flight periods as adults and thus will spend the drought cycle in the terrestrial environment (Wiggins *et al.*, 1980). This strategy is also used by chironomids and mosquitoes, as well as some dragonflies and limnephilid caddisflies (Wiggins *et al.*, 1980; Wissinger, 1988). These general strategies interact with life history (see below) such that wetland insect communities exhibit a succession following the drought cycle with dominance shifting from mosquitoes and chironomids to larger taxa such as beetles, odonates, heteropterans, and caddisflies (Batzer and Wissinger, 1996). The extreme of this cycle is that prolonged drought results in very low density of insects (Fig. 5; Hershey *et al.*, 1999). Under these conditions the wetland fauna is characterized by high abundance of molluscs and annelids and low abundance and species richness of insects, but it becomes more and more dominated by a rich insect fauna with prolonged inundation (Fig. 6; Hershey *et al.*, 1999).

D. Lotic Insect Communities

Lotic insect communities are functionally and structurally quite different than lentic communities due to the different physical and chemical challenges of the lotic environment. At a local scale in stream ecosystems, substrate and current velocity are probably the most important physical factors determining the community structure of aquatic insects. The stream substrate has obvious importance because the vast majority of stream insects spend most of their lives attached to substrates. Lotic substrates can be broadly divided into inorganic substrates (geologic material ranging from silt to boulders) and organic substrates (fine organic particles up to logs). The particle size of inorganic matter has a large influence on insect community

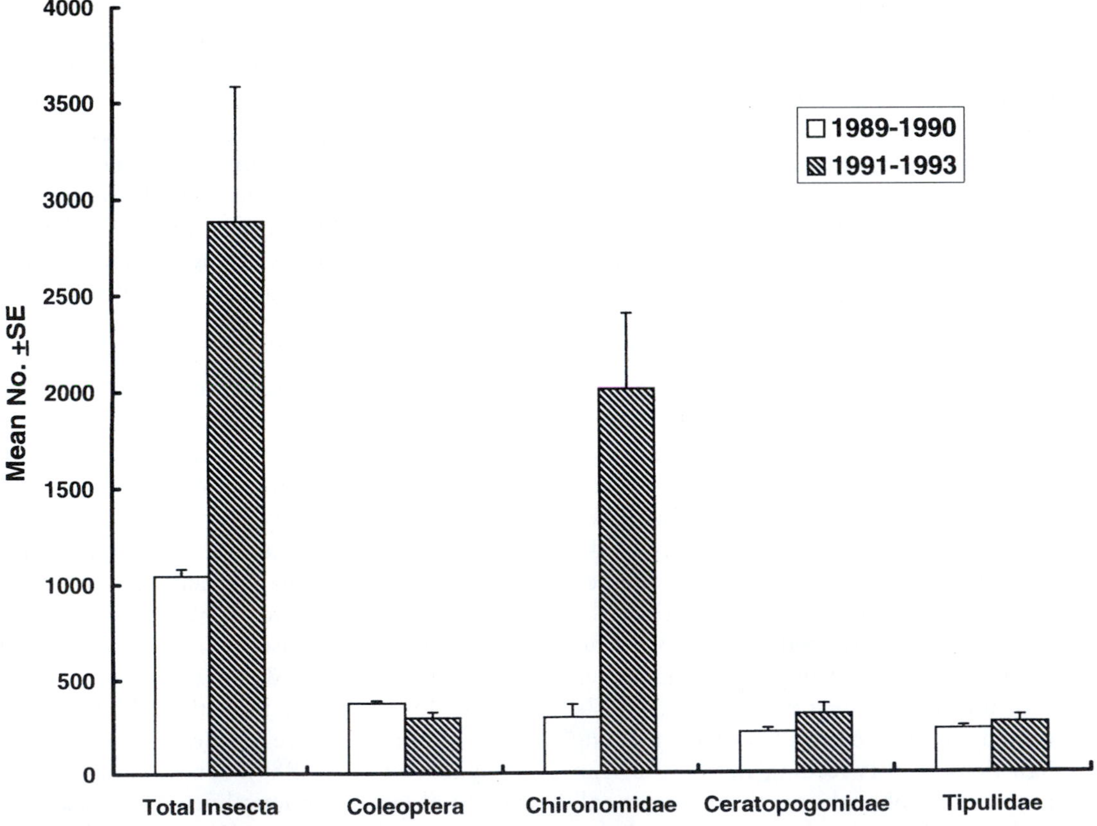

FIGURE 5 Prolonged drought in Minnesota prairie wetlands in 1987–1990 resulted in communities with significantly lower density of insects (measured in 1989–1990) than in wetter years of 1991–1993. Much of this difference was due to Chironomidae.

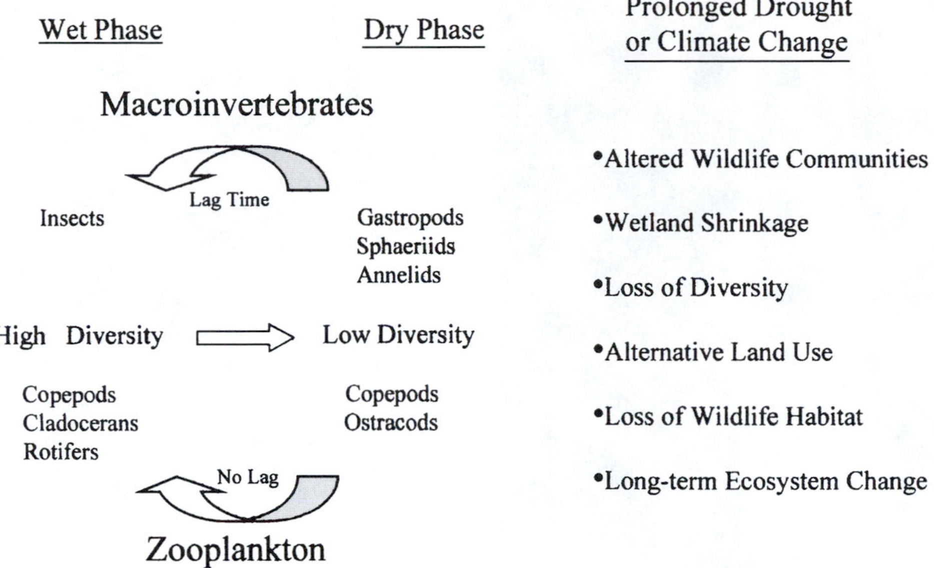

FIGURE 6 Patterns of insect and noninsect abundances in prairie wetlands over a hydrologic cycle. Insects and cladocerans are abundant during the wet phase, but molluscs and copepods dominate during drought. Because of their longer life cycles, the transition back to insects can be relatively long following the return of hydric conditions. [From Hershey *et al.*, 1999.]

structure. Coarser bed materials (e.g., gravel, cobble, and boulders) generally provide more interstitial habitat for insects than do finer sediments (e.g., sand and silt). For example, "water penny" beetles (family Psephenidae), hellgrammite larvae (Megaloptera: Corydalidae), and perlid stoneflies (Plecoptera) frequently are found in interstitial spaces on the undersides of rocks. Sedentary filter-feeding insects require the space between particles both to operate their filtration device (e.g., nets, specialized body parts) and to allow water flow to carry food items to them. Many case-building caddisflies (Trichoptera) pupate in dense aggregations on protected faces of cobbles and boulders where interstitial water flow carries the dissolved oxygen needed for metamorphosis (Lamberti and Resh, 1979; Resh *et al.*, 1981), but they are more dispersed in the larval stage. A lower diversity of insects is typically found in fine sediments because the tight packing of particles restricts physical habitat and the trapping of detritus, and can limit the availability of oxygen (Allan, 1995). However, taxa adapted to such habitats (e.g., chironomid midges, oligochaete worms, certain mayflies) may be abundant. For example, Soluk (1985) recorded densities of chironomid midges exceeding 80,000 individuals/m^2 in the sand substrate of an Alberta river. Organic substrates such as large woody debris or leaf packs often are "hot spots" for invertebrate activity because they provide both substrate and nutritional resources (Anderson and Sedell, 1979). In a survey of Oregon streams, Dudley and Anderson (1982) found 56 invertebrate taxa that were closely associated with wood, and another 129 taxa that were facultative associates of wood. The more specialized moss habitat typically supports a high density of chironomids and other flies (e.g., moth flies, family Psychodidae) and a different assemblage of species than found on adjacent mineral substrates (Glime and Clemons, 1972; Suren and Winterbourn, 1992).

Flow affects many aspects of insect biology, including body form, food acquisition, and movement. Taxa found in the swift currents of streams usually are streamlined (e.g., *Baetis* mayflies), dorsoventrally compressed (e.g., heptageniid mayflies and psephenid "water penny" beetles), possess suctorial disks (e.g., blephariceriid flies), use rock "ballast" in their cases to resist the flow (e.g., *Glossosoma* caddisflies), or are small enough to fit almost entirely within the sheltered "boundary layer" (e.g., chironomid midges). Many insects take advantage of swift flow by allowing current to transport food items to them. Examples include caddisflies of the family Hydropsychidae that construct silken nets to capture suspended food particles, black flies

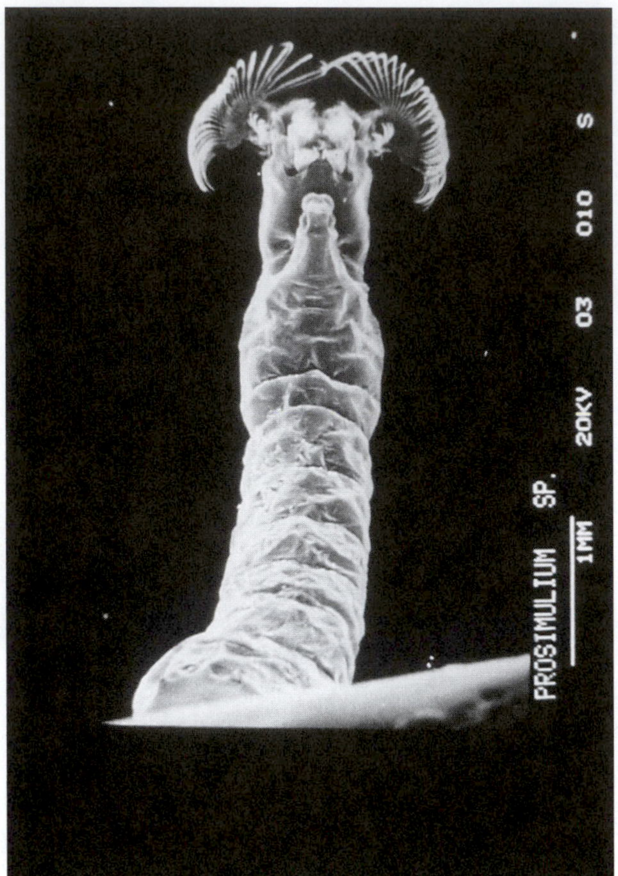

FIGURE 7 The larval blackfly is a common filter feeder in riffle habitats of North American streams. The fan-shaped objects, or cephalic fans, are a modification of the mouthparts which collect fine particulate organic matter (FPOM) from the water column. The larva orients itself in the current by forming a silk pad on the rock, then attaches itself to the pad using tiny hooks on the posterior portion of its abdomen. [Photograph courtesy of D. A. Craig.]

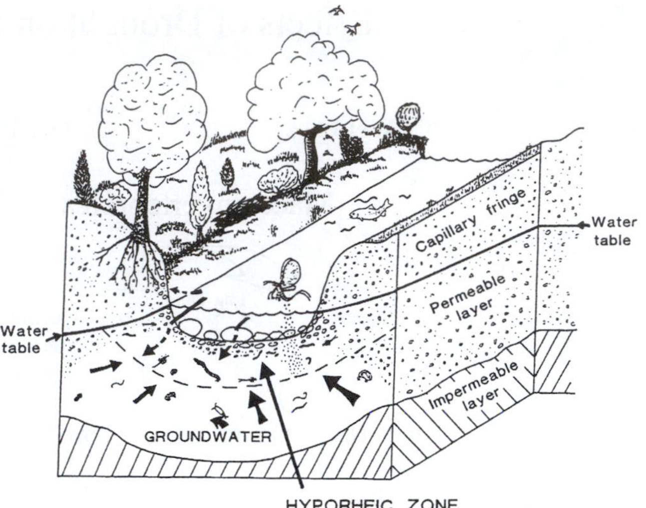

FIGURE 8 Channel cross section of a stream during summer low flow, showing position of hyporheic zone relative to surface water and groundwater. [Modified slightly from Williams, 1993].

(Simuliidae) that filter food with specialized mouthparts (Fig. 7), and *Brachycentrus* caddisflies that capture particles on the setae (hairs) of their front legs (Wallace and Merritt, 1980). Filter-feeding insects also can be found along the wave-swept shores of lakes or on stable substrates (e.g., wood debris) of large rivers, where similar hydrodynamic forces are experienced. Many insects prefer low water movement, such as found in quiescent stream pools and along stream margins. In these habitats, organic matter is deposited by gravity and thus can be "gathered" and ingested. These taxa often are more "wormlike" than swift-water forms.

Variation in flow (floods to desiccation) is the major cause of natural disturbance in streams (see section below on disturbance) and is responsible for large, usually temporary, reductions in macroinvertebrate abundance and diversity (Lamberti *et al.,* 1991). The hyporheic (i.e., interstitial) zone and the undersurfaces of rocks provide refugia during flood events (Fig. 8). During drought, the deep hyporheic zone is especially important, but is usually accessible only to smaller invertebrates (typically about 1–5 mm in length). The spatial extent of the hyporheic zone varies with local geomorphology. In bedrock-dominated reaches, it is virtually nonexistent; whereas in the Flathead River (Montana), the hyporheic zone may extend tens of meters vertically and up to three kilometers laterally from the river, supporting relatively large stoneflies several millimeters in length (Stanford and Ward, 1988). Finally, recolonization by egg-laying adults or by individuals drifting from upstream (Table III) serves to reintroduce taxa that become locally extinct during drought, flood, or other disturbance events (Fisher *et al.,* 1982; Smock, 1996).

1. Functional Feeding Groups

Rather that focusing on the species as the ecological unit in streams, aquatic insect ecologists often consider the insect community in terms of functional feeding groups (Cummins, 1973, 1974; Cummins and Klug, 1979; Merritt and Cummins, 1996a). The functional feeding group approach categorizes stream consumers according to their mode of feeding, or functional role, rather than taxonomic groups. From an ecosystem perspective, this approach reduces the number of groups to be considered by 1–2 orders of magnitude compared to the species approach. Thus, rather than studying dozens or hundreds of specific consumers, a smaller number of groups of organisms can be studied collectively from the perspective of their function in the stream ecosystem. This categorization also has utility in other aquatic ecosystems, and has recently been applied to all North American aquatic insects regardless of habitat (Merritt and Cummins, 1996b).

TABLE III Mechanisms of insect recolonization of stream reaches

Mechanism	*Time scales*	*Example*	*References*
Downstream drift	Nocturnal, constant, episodic, dispersal stages of some invertebrates	Many insects	Waters, 1972
Drift from tributaries	Nocturnal, constant, episodic, dispersal stages of some invertebrates	Many insects	Cairns, 1990
Upstream flight	Seasonal, depending on emergence period	*Baetis* mayflies	Hershey *et al.*, 1993
Flight from other watersheds	Seasonal, depending on emergence period	Many insects, *Baetis* mayflies	Wallace *et al.*, 1986; Schmidt *et al.*, 1995
Upstream swimming	Seasonal	*Leptophlebia cupida*, *Paraleptophlebia* mayflies	Hayden and Clifford, 1974; Bird and Hynes, 1981
Upstream crawling	Daily	*Dicosmoecus* caddisflies, *Juga* snails	Hart and Resh, 1980; L. M. Blair and G. A. Lamberti, unpublished data
Movement from hyporheic refuge	Episodic, seasonal	Many early instar insects; overwintering stages of many insects in cold climates; many residents of intermittent streams	Sedell *et al.*, 1990; Delucchi, 1989; Hershey unpublished data

[From Hershey and Lamberti, 1998]

The functional feeding groups recognized by Merritt and Cummins (1996a, b) are: (1) "scrapers" (= "grazers"), which remove and consume attached algae and associated periphytic material; (2) "shredders," which ingest coarse particulate organic matter (CPOM), as decomposing leaf litter, living macrophyte tissue, or dead wood; (3) "predators," which eat living animals; and (4) "collectors," which consume decomposing fine particulate organic matter (FPOM). The last group can be further broken down into "collector-gatherers," which collect FPOM from the sediments, and "collector-filterers," which collect FPOM from the water column (Fig. 9). Two other, less common functional feeding groups are the "macrophyte-piercers," which pierce the tissues of macroalgae and rooted hydrophytes, and "parasites," which develop on or in aquatic insects, generally killing them. Because groups of insect species can be studied collectively to unravel major avenues of organic matter processing, the functional feeding group approach greatly simplifies the study of insect function in aquatic ecosystems. It also provides a strong basis for comparative studies of the biomass and productivity of consumer groups across ecosystems, whereas it is much more difficult (and often less informative) to make such comparisons on a species-by-species basis.

Although implementation of this approach has been an important precedent for the development of other major paradigms of aquatic ecology (e.g., river continuum concept of Vannote *et al.*, 1980), the functional feeding group concept is not without limitations. First, assignment of individual organisms from aquatic samples to a functional feeding group generally requires identification of the organism to family (see Chapter 17 of this volume) or genus. This can be a very time-consuming task, although a shortcut approach using an illustrated key is effective for many taxa (see Merritt and Cummins, 1996a). Second, feeding ecology can vary within a genus, and even within a species depending on habitat or food availability. Thus, published functional group designations for the generic level are not always reliable (Mihuc, 1997). In fact, many aquatic insects are known to change feeding habit with growth and development (Chapman and Demory, 1963; Rader and Ward, 1987). For example, the large limnephilid caddisfly *Dicosmoecus gilvipes* begins larval life as a collector-gatherer, feeds primarily as a scraper of periphyton during the middle instars, and finally becomes partly predaceous in the final instar (Gotceitas and Clifford, 1983; Li and Gregory, 1989). Other insects are omnivorous throughout their lives. Filter-feeding caddisflies of the genus *Hydropsyche* will consume virtually anything that becomes entrapped in their nets, including FPOM, living algae, and small animals (Benke and Wallace, 1980), thus making them potential members of several functional feeding groups.

Some orders of aquatic insects, or the aquatic representatives within an order, perform mostly as a single functional feeding group, whereas others show great diversity in feeding (Table I). Generally, the more primitive orders of aquatic insects show more limited diversity in feeding mode. For example, aquatic collembolans are exclusively collector-gatherers, odonates are strictly predaceous, and stoneflies (Plecoptera) are either shredders or predators, generally depending on family. Less speciose orders also tend to show less feeding diversity, as one might expect. For example,

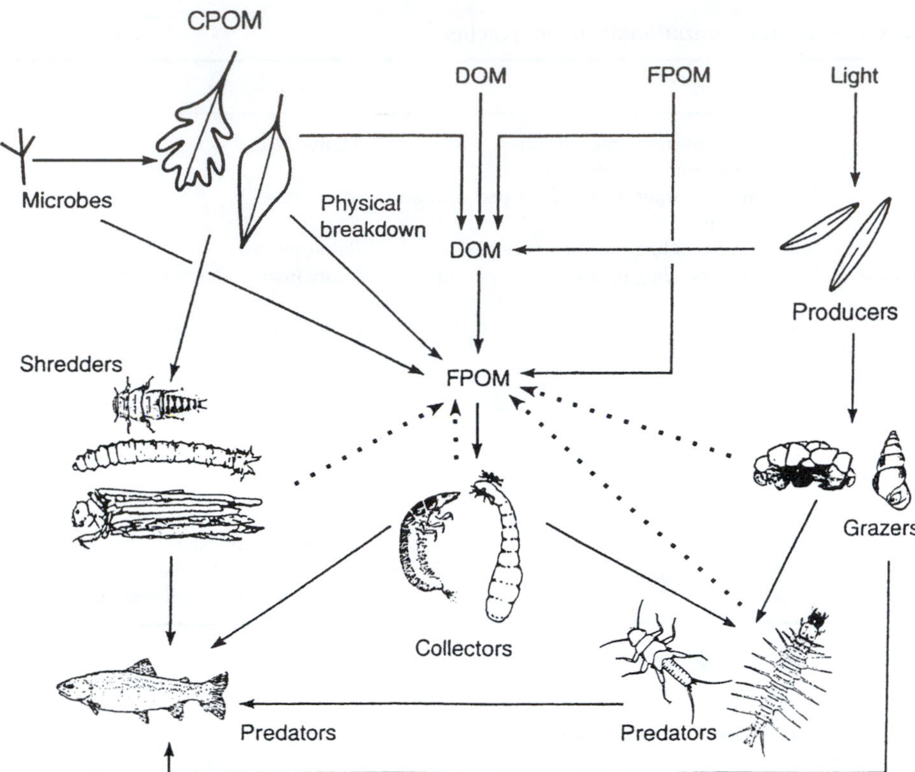

FIGURE 9 Simplified food web for a hypothetical North American forested stream, showing major energy inputs and use by different macroinvertebrate functional feeding groups [modified from Allan, 1995, after Cummins, 1973]. Macroinvertebrate examples are: shredders (*Peltoperla* stonefly, *Tipula* cranefly, *Hydatophylax* caddisfly); collector-filterers (*Hydropsyche* caddisfly, *Simulium* blackfly); grazers (*Glossosoma* caddisfly, *Juga* snail); predators (*Calineuria* stonefly, *Orohermes* hellgrammite). Broken lines indicate fate of fecal matter.

megalopterans are entirely predaceous, neuopterans pierce either animals or plants, and aquatic hymenopterans are mostly parasites. In contrast, more derived orders, such as Trichoptera, Diptera, and Coloptera, show considerable diversity in feeding mode, in part due to their large number of aquatic species. The Diptera and Trichoptera, in particular, have been enormously successful in freshwater ecosystems. Chironomid midges alone may equal the species diversity of all other insects combined in a typical aquatic ecosystem. In any specific stream, it is usually possible to find caddisflies represented in all functional feeding groups, and often dominating the biomass of those groups. Beetles possibly occupy the broadest range of aquatic habitats, ranging from ice fields to seashores, and can be very speciose.

2. The River Continuum Concept

The River Continuum Concept or "RCC" (Vannote *et al.*, 1980; see Fig. 6 in Chapter 2) is an important paradigm in stream ecology and functional feeding groups are a major component of that paradigm. The RCC predicts that small, heavily shaded streams will have high inputs of detritus ("allochthonous" CPOM) from adjacent riparian zones. Due to this abundant detritus, shredders will dominate the macroinvertebrate community. Both collector-filterers and collector-gatherers also will be abundant because high-quality FPOM will be produced as CPOM becomes fragmented. In medium-sized streams, light inputs increase and thus benthic algal production will increase. The shredders are replaced by scrapers, while collectors remain abundant. In large rivers, benthic algal production and direct riparian inputs decline. Food resources for macroinvertebrates are dominated by FPOM in suspension; collectors dominate the macroinvertebrate community. In all streams, predators comprise a small but fairly stable proportion of the fauna. These predictions were tested by Hawkins and Sedell (1981) for the McKenzie River drainage in Oregon, which consisted of streams ranging in size from the 1st- to 7th-order. They found that functional feeding group distributions were generally consistent with predictions of the RCC (Fig. 10). These longitudinal shifts were more pronounced

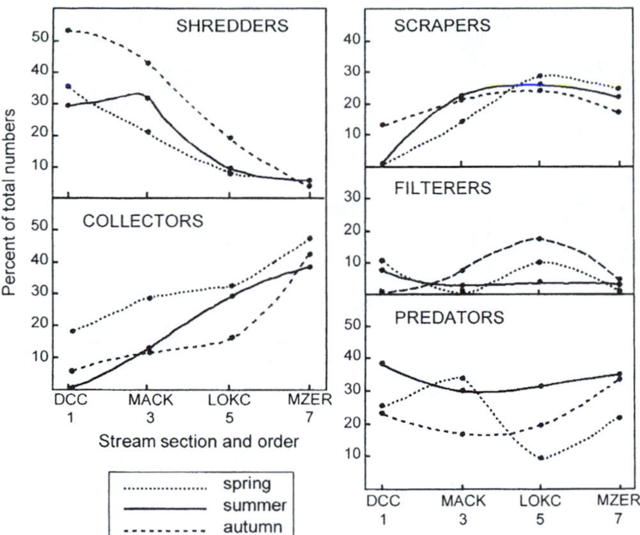

FIGURE 10 Longitudinal changes in relative abundance of macroinvertebrate functional feeding groups at different seasons within an Oregon, USA, river system. "Collectors" = collector-gatherers; "Filterers" = collector-filterers. Stream size increases along the *x*-axis; DCC, Devilsclub Creek; MACK, Mack Creek; LOKC, Lookout Creek; and MZER, McKenzie River. Note the general conformance to predictions of the River Continuum Concept (see Figure 2.6 in this volume). [Modified slightly from Hawkins and Sedell, 1981.]

than were seasonal fluctuations in community structure in the individual Oregon streams. Not all studies of functional feeding groups have shown the same degree of conformance to the RCC (e.g., Minshall *et al.*, 1985; Winterbourn *et al.*, 1981). Some groups, such as shredders, predictably decline with increasing stream size whereas other groups, such as scrapers, do not fit tightly with RCC expectations. In evaluating conformance to the RCC, abundance and biomass have often been considered as currency. However, some researchers have argued that production rather than biomass is a more appropriate currency as it is a better measure of transfer of energy between trophic levels (Benke, 1993, 1998; Grubaugh *et al.*, 1997). Large rivers, at least in their natural state, likely have more complex energy inputs than implied by the RCC. Thorp and Delong (1994) developed the riverine productivity model that suggests that riparian inputs and benthic primary production are quite important to insect production in large rivers, which is contrary to predictions of the RCC. Their preliminary test of the model in the Ohio River suggested that the shallow edges of large rivers function similarly (and have similar insect communities) as midorder streams, because they receive riparian litterfall and also have substantial periphyton production. Hawkins *et al.* (1982) found that riparian canopy highly influenced functional feeding group composition in six small streams in Oregon, but that absence of canopy increased the abundance of all functional feeding groups possibly because both autotrophic (on-site) and heterotrophic (from upstream) inputs were received. Hence, insect consumers responded to the abundance and composition of the available food resources. This notion of energy transfer via both autotrophic and heterotrophic food webs is an underpinning of the lotic ecosystem model of McIntire and Colby (1978), that postulates a hierarchical organization to stream food webs and a central role for aquatic insects.

3. Insect Drift

"Drift" is the downstream transport of oranisms in the current (usually expressed as number/per cubic meters), and it is a unique property of stream ecosystems. Because it provides insects with a unidirectional transportation system, it has the potential to be important to the distribution and ecology of stream insects. At any given time, the proportion of fauna in the drift, compared to that on the substrate, is very small (0.01–0.5%; Ulfstrand, 1968; Waters 1972). Some insect species and groups have a much greater propensity to drift than others. Mayflies, especially *Baetis*, and larval chironomids are perhaps the most common insect drifters.

It was once thought that when an insect entered the drift it was essentially lost from the system. Although estimates of drift distance vary widely, most experimental studies in small to mid-sized streams have shown that insects drift only a short distance (a few meters or less) during any drift event (Allan and Feifarek, 1989; Wilzbach and Cummins, 1989). However, working in a high-order blackwater river with an important but widely dispersed snag habitat, Benke *et al.* (1986) found that insects drifted between snags, a drift distance on the scale of kilometers.

Insects may drift passively or inadvertently if they are dislodged, such as might occur during a high discharge event or in response to an anthropogenic disturbance. However, most experimental studies have suggested that most drift is active, or a behavioral response to some stimulus. This can occur catastrophically due to chemical spills, or on a smaller scale in response to depleted food resources (Kohler, 1985) or the presence of a predator (Peckarsky, 1980). It was once thought that drift was a density-dependent response, such that drifting insects represented a portion of the population that exceeded the carrying capacity of the stream (Waters, 1972). However, empirical studies of drift have yielded little support for density dependent behavior (e.g., Hinterleitner-Anderson *et al.*, 1992).

In most streams, drift densities are much higher at night than during daylight hours. Because fish feed heavily on drifting invertebrates, nighttime drift behavior

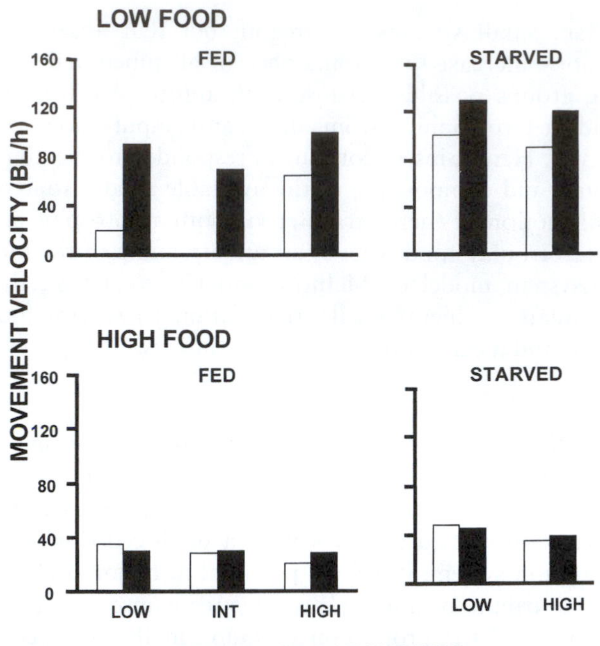

FIGURE 11 *Baetis* mayflies enter drift actively, not passively, as a result of greater activity on upper surfaces of substrates at night. Kohler (1985) showed that only food abundance and time of day affected drift entry, and only when no food was present was there any drift entry during the day. The figure shows the results from food abundance experiments. [Reprinted from Kohler, 1985.]

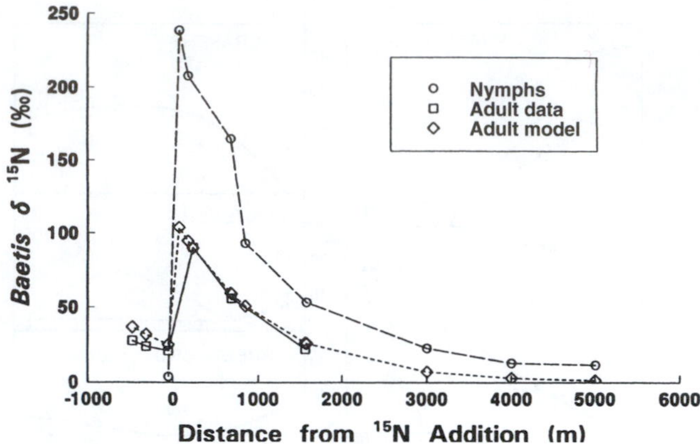

FIGURE 12 Using $\delta^{15}N$ as a tracer, adult *Baetis* mayflies were found to fly approximately 2 km upstream. The figure shows that although adults have a lower ^{15}N signature than nymphs, those captured from upstream of the ^{15}N dripper are considerably more labeled than background. An integral mixing model showed that this label reflected mayflies from 2 km downstream (hypothesized to be females) mixing with mayflies from upstream (hypothesized to be males). [From Hershey *et al.*, 1993.]

provides invertebrates with more protection from fish predation. Because many macroinvertebrates forage more on the upper surfaces of rocks at night, it was once thought that dislodgement was simply more likely at night. However, Kohler (1985) found that well-fed *Baetis* foraged only at night on tops of stones while starved nymphs foraged during both day and night, but drifted similarly to fed nymphs. Thus, activity during foraging was not related to drift (i.e., they were not accidentally dislodged) and drift entry was active (Fig. 11).

Drift clearly is an important mechanism for invertebrate dispersal, especially for early larval instars. It is also an important means for recolonizing reaches that are impacted by natural or anthropogenic disturbance (see Cairns, 1990). However, if there were no upstream movement of macroinvertebrates, downstream drift could result in a net displacement of populations, net depletion of upstream reaches, and loss of invertebrate-mediated ecosystem functions in depleted areas. The colonization cycle proposed by Müller (1954) states that insects displaced downstream as larvae later fly upstream as adults to oviposit, thereby repopulating depleted areas. Upstream flight has been observed for some species (e.g., Madsen *et al.*, 1973) and quantified for a *Baetis* population in arctic Alaska (Fig. 12; Hershey *et al.*, 1993).

E. Saline Habitats

Insects have enjoyed only limited success in colonizing the marine environment, and only water striders of the genus *Halobates*, which are surface dwelling, are truly oceanic (Spence and Andersen, 1994). The reasons for this are unclear, but have variously been attributed to physiological stresses associated with respiration and osmoregulation, physical stress of wave action, competition for habitat and resources with established species, and predation (see recent reviews by Ward, 1992; Wallace and Anderson, 1996). Because several species of beetles, true flies, and caddisflies have successfully colonized intertidal zones and brackish waters, it seems unlikely that physiological stresses alone have hindered radiation into the marine realm. Similarly, because many insects inhabit wave-swept lake shores and stream riffle habitats, and because there are many low energy marine habitats, physical stress also seems unlikely to be a major hindrance. These considerations suggest that competitive and predatory pressures from marine phyla, which have been established in marine habitats much longer, are the most likely mechanism that has precluded high insect diversity in marine habitats (Usinger, 1956; Ward, 1992; Wallace and Anderson, 1996). Alternatively, Williams and Feltmate (1992) suggested that insects may be in the early stages of invading marine systems, such that they may yet become established there after many more millennia.

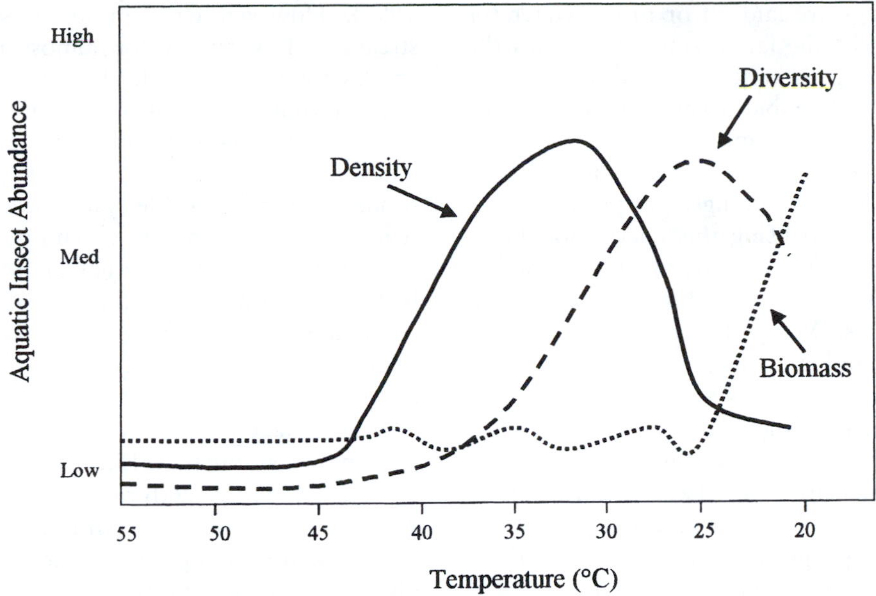

FIGURE 13 Characteristics of a stream insect community along a natural thermal gradient emanating from a California hot spring, depicting total density, species richness (diversity), and total biomass. [Modified from Lamberti and Resh, 1985.]

F. Specialized Habitats

Aquatic insects have invaded many specialized microhabitats. The species that comprise these communities often have morphological, behavioral, and life-history adaptations which are unique to those habitats and reflect the remarkable phenotypic plasticity of the Insecta. Some of the better studied examples of these habitats include treeholes, hot springs, pitcher plants, and caves and other subterranean environments (see Ward, 1992).

Mosquitoes are particularly well adapted to specialized ephemeral habitats such as treeholes, tires, rain gutters, and other containers, that catch rainwater (Merritt *et al.,* 1992). Filter feeding is the most common mechanism for acquiring food in such habitats, but detrital and bacterial particles may also be gleaned from leaf and bark surfaces, and a few species are predatory (Merritt *et al.,* 1992). In treehole habitats, a combination of leaf litter quality and quantity and stemflow nutrients influence mosquito productivity (Carpenter, 1982, 1983), but field experiments by Walker and Merritt (1988) found no evidence that quantity of leaf detritus was limiting.

Aquatic insects that inhabit subterranean habitats, including cave pools and streams, underground rivers, and extensive hyporheic zones, share some morphological characteristics. These insects, like other subterranean invertebrates, lack pigment, have reduced or absent compound eyes and ocelli, and possess reduced wings (Ward, 1992). Diptera and Coleoptera are perhaps the best represented among subterranean insects although representatives of other groups also have been reported (see Ward, 1992). The trophic ecology of these insect communities is dependent on organic matter imported from surface ecosystems. Most common among these nutritional sources are flocculated DOM (dissolved organic matter), entrained FPOM, bacteria (which utilize DOM), and bat guano.

Geothermal springs are another specialized habitat that has been exploited by some insects. Hot springs tend to have representatives from the Diptera, Coleoptera, and Odonata (see Ward, 1992). In addition to the high temperatures, these habitats also have high mineral content, which may pose toxicity problems for some species. However, Lamberti and Resh (1983a, 1985) showed that thermal effects were more important than chemistry in limiting insect diversity in a northern California stream receiving natural hot spring inputs (Fig. 13).

III. AQUATIC INSECT LIFE HISTORIES

Aquatic insects show tremendous diversity in their life-history patterns. This diversity includes variation in reproductive and dispersal strategies, life-cycle length, growth rate, developmental strategies, and phenology of the various life-history stages. This great variety serves to separate taxa seasonally and spatially, thus providing the basis for the varied and dynamic nature of aquatic insect community composition.

Virtually all aquatic insects have an aerial adult phase. Copulation usually occurs outside of the aquatic environment, and eggs are laid on or in the water for most species. Typically, the larval stage dominates the life cycle; the adult stage is very short and does not always involve feeding. Notable exceptions to this rule are the Odonata (dragonflies and damselflies), that are long-lived and voracious predators as adults. Dragonflies and damselflies also have long flight periods, feeding on insect swarms and using this energy for flight, food capture, and reproduction. Swarms of dragonflies often are thought to be important for control of mosquitoes and blackflies. Many dipterans also feed as adults. Mosquitoes and blackflies are notorius pests, but are also vectors of diseases such as malaria, encephalites, dengue, Rift Valley Fever, river blindness, and many others that infect and kill millions of people worldwide (e. g., Renshaw *et al.,* 1996; Jacobs-Lorena and Lemos, 1995; Kidwell and Wattam, 1998). With global warming, the geographic distribution of these diseases is expected to increase (Brown, 1996).

For many aquatic insects, the adult phase is often important to dispersal between lentic habitats, and permits upstream oviposition (see Müller, 1954; Hershey *et al.,* 1993), and oviposition between watersheds (Wallace *et al.,* 1986) in lotic habitats. Many aquatic beetles and bugs have extended adult life spans and are aquatic in both larval and adult stages of the life cycle, although dispersal still is accomplished by the adult stage. Dispersal varies tremendously among insects and has been studied thoroughly for only a few taxa. For example, *Baetis* mayflies in a tundra river in arctic Alaska have been shown to fly approximately 2 km upstream to lay eggs (Hershey *et al.,* 1993). Even larval insects can move substantial distances, aside from drift. For example, late-instar larvae of the grazing caddisfly *Dicosmoecus gilvipes* can crawl up to 25 m per day in search of algal food (Hart and Resh, 1980). Nymphs of the lotic mayfly, *Leptophlebia cupida,* move upstream at an average rate of 10 m "per hour" during seasonal migrations between habitats (Hayden and Clifford, 1974).

The frequency with which an organism completes its life cycle is referred to as "voltinism." Depending on environmental constraints, especially temperature and nutrition, as well as evolutionary factors, aquatic insects may be univoltine (one generation per year), multivoltine (more than one generation per year; e.g., bivoltine = two per year, trivoltine = three per year), semivoltine (2-year life cycle), or merovoltine (3- or more-year life cycle). Generally, larger insects will have longer life cycles than smaller species. For example, in the Kuparuk River in arctic Alaska, the relatively small *Baetis* mayflies are univoltine, but the larger *Brachycentrus* caddisflies are merovoltine (Hershey *et al.,* 1997). Univoltine development is by far the most common, especially in temperate regions (see Wallace and Anderson, 1996). However, even in arctic streams, where the stream bed is frozen for almost nine months, most species are univoltine (Hershey *et al.,* 1997). Arctic lakes have chironomid populations with life cycles ranging from 1 to 4 years, but univoltine development predominates (Hershey, 1985a). One arctic pond species of chironomid has a 7-year life cycle (Butler, 1982). Temperate regions display similar variation. The wood-boring aquatic beetle *Lara avara* has about a 5-year life cycle, likely due to the low nutritional value of wood (Anderson and Sedell, 1979). At the other extreme, some herbivorous chironomid midges in warm desert streams can complete their life cycle in as little as two weeks, a distinct advantage in desert streams where floods scour organisms and resources (Fisher and Gray, 1983).

Variation in growth rate is also important in distinguishing insect species, separating cohorts, and providing structure to aquatic insect communities. Within physiological constraints, temperature and food are the major factors affecting growth for most taxa. These factors vary widely within, and between, lentic and lotic ecosystems, as well as geographically.

In cool climates many aquatic insects have life cycles with synchronized development (Fig. 14). Environmental factors, especially temperature and photoperiod, as well as evolutionary factors, serve as proximal cues that synchronize life cycles (Butler, 1984). For example, many insects will overwinter in the egg stage, and hatch in the spring. Others may overwinter as larvae, emerging in the spring when the appropriate temperature and/or photoperiod cues are present. Although there may be several evolutionary as well as environmental mechanisms contributing to synchronous development, and contributing factors vary among species (see Butler, 1984), the result is that within a species, individuals in a "cohort," or age class, will emerge during a very short interval, enhancing the probability of mating success. Synchrony occurs through two general paths. Larval hatching and growth may be timed such that all individuals within a cohort are of identical or nearly identical size throughout their development and emerge together. This is common in many caddisflies, such as the widely distributed *Helicopsyche borealis* (Resh *et al.,* 1984) and the large limnephilid *Dicosmoecus gilvipes* (Lamberti and Resh, 1979). Alternatively, hatching and larval development may not be synchronous but instar-specific growth rates still may result in synchronized emergence (Lutz, 1974).

Within a community, co-occurring species typically exhibit a variety of patterns of voltinism and synchrony. For example, 14 species of dragonflies in a temperate pond were found to have several different developmental patterns (Wissinger, 1988). *Epitheca cynosura* was univotine with synchronous development (Fig. 15A), while one of its congeners, *E. princeps*, also

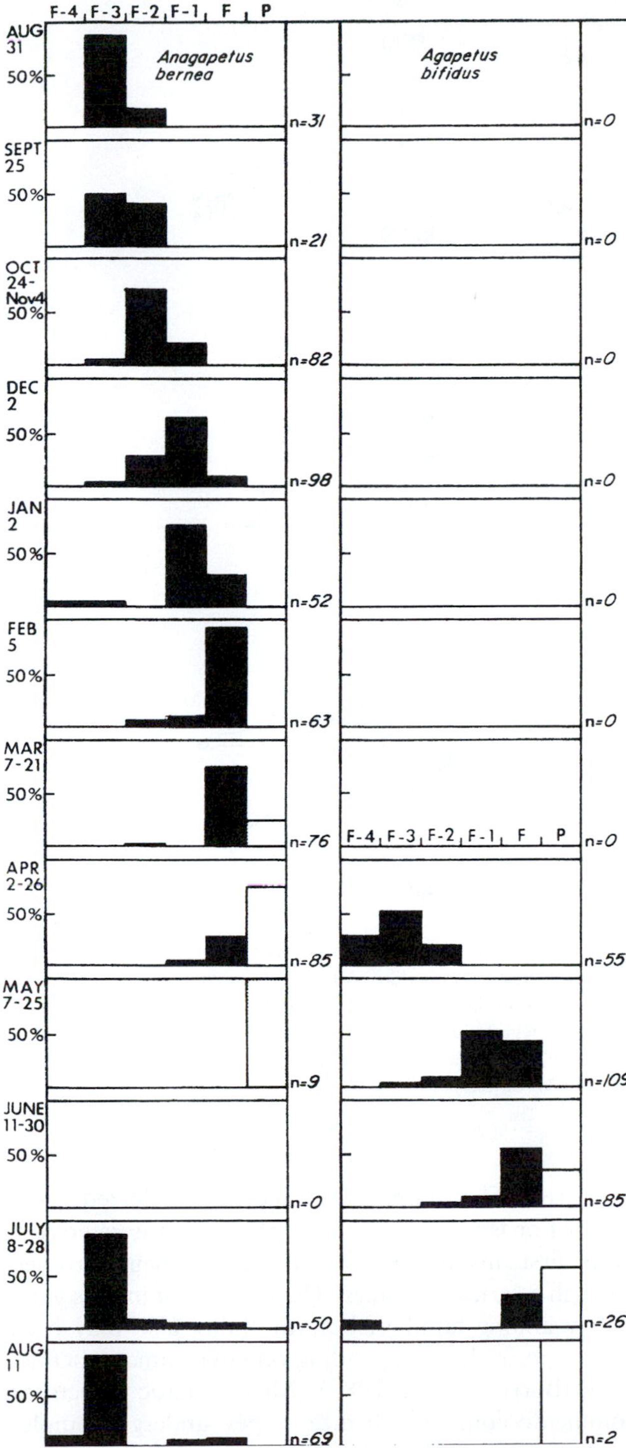

FIGURE 14 Examples of life cycle synchrony of stream macroinvertebrates in response to temperature and photoperiod cues. In temperate North America, many species overwinter as larvae (left). Larval growth and maturation occurs at approximately the same rate for the entire cohort, such that emergence will also be synchronized. Other species overwinter in the egg stage, then begin larval development in the spring (right). These developmental patterns increase the probability that adults will find mates, thus enhancing reproductive success. Larval instars indicated as: F, final instar; F-1, penultimate instar; etc; and P, pupae. [From Anderson and Bourne, 1975.]

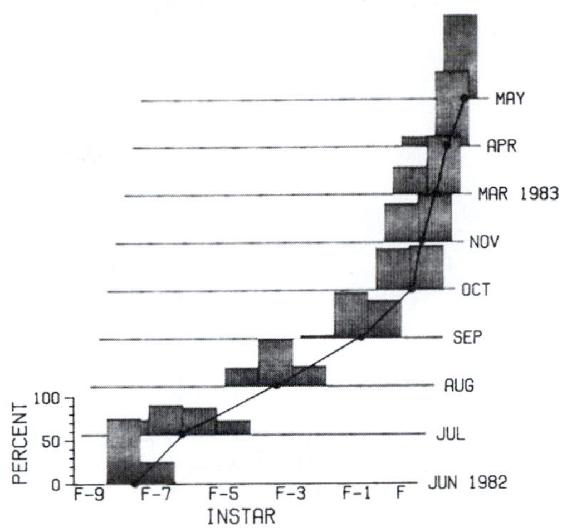

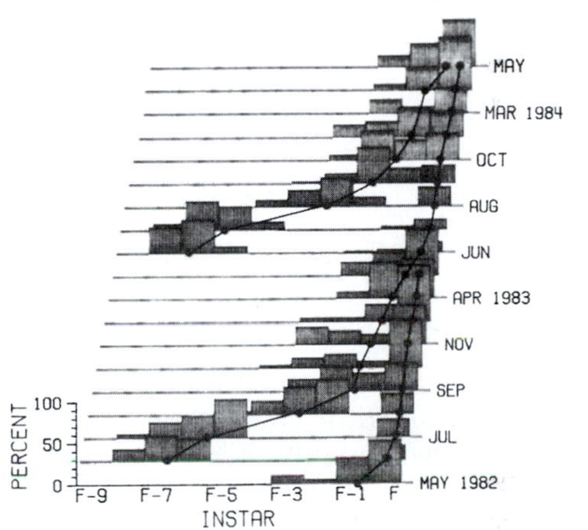

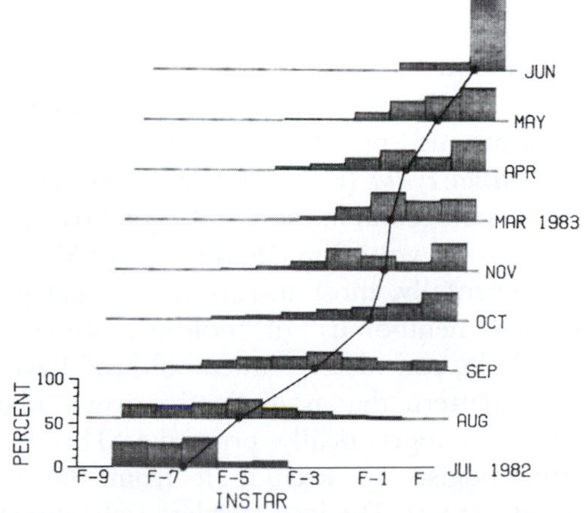

FIGURE 15 Developmental patterns of *Epitheca* spp and *Libellula lydia* illustrate that even closely related species from the same habitat can exhibit divergent life-history patterns. (A) Univoltine synchronized development. (B) Semivoltine synchronized development. (C) Semivoltine asynchronous development. [From Wissinger, 1988.]

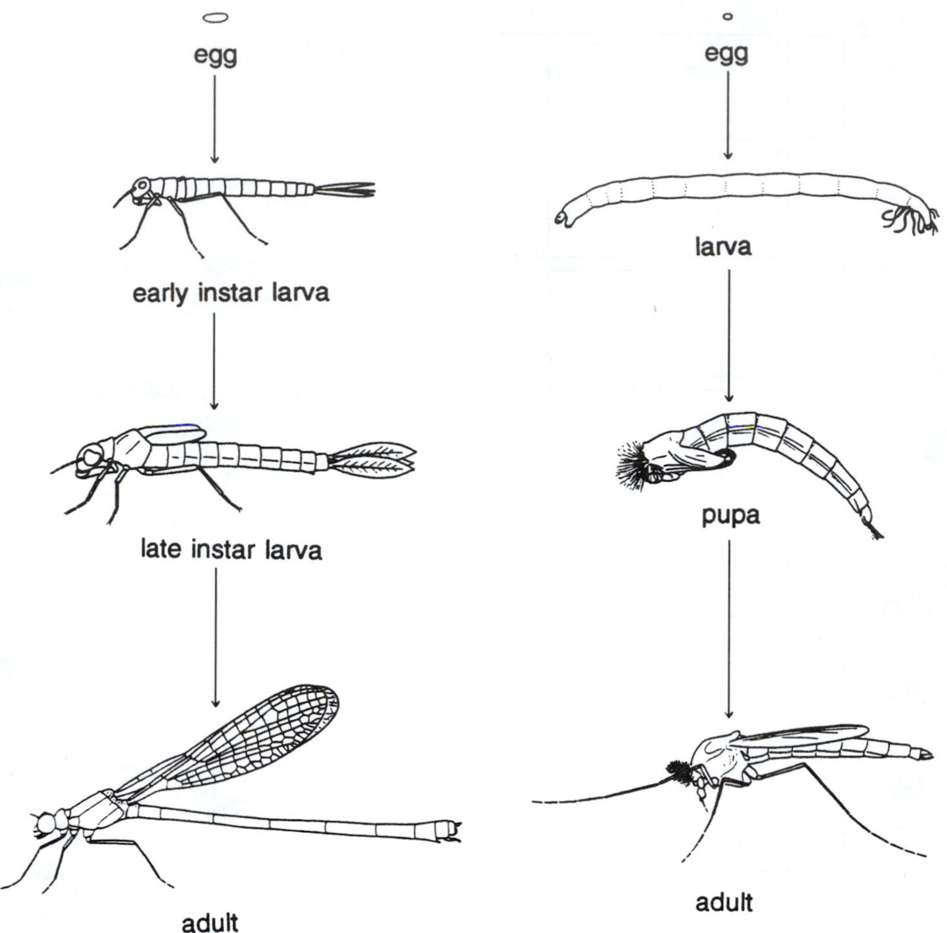

FIGURE 16 The two major forms of insect development seen in aquatic insects. In gradual metamorphosis (left), or hemimetabolous development, immature forms are insectlike. Adult characteristics develop gradually through the larval instars, and wings inflate in the final molt to the adult form. Insects demonstrating complete metamorphosis (right), or holometabolous development, have several larval instars that are less insectlike. Adult characteristics develop as a result of radical restructuring of the anatomy and physiology during the pupal instar. [From McCafferty, 1981.]

was synchronized but was semivoltine (Fig. 15B). Yet another congener, *E. simplicollis*, was sometimes bivoltine. Asynchronous univoltine development was exhibited by *Libellula lydia* (Fig. 15C), while other species showed different combinations of synchrony, asynchrony, and mixed voltinism (Wissinger, 1988).

Developmentally, most insects can be categorized as either "hemimetabolous" or "holometabolous" (Fig. 16). Collembola, considered "ametabolous," have a developmental pattern that is distinctive from other insects and is phylogenetically primitive. This has led some entomologists to exclude it from the Insecta (Borror *et al.*, 1989). The insect orders Ephemeroptera, Odonata, Heteroptera, and Plecoptera, are hemimetabolous. The aquatic immature forms, with few exceptions, respire using gills, and although they may differ considerably from adults, they are clearly insectlike (Fig. 16). The immature stages are referred to as nymphs or larvae. A newly hatched larva is referred to as a first instar, and larval development proceeds through a series of instars. The number of instars varies widely among families, but variability also may occur at the species level depending on environmental conditions (Borror *et al.*, 1989). Most aquatic insects are holometabolous, which means they undergo complete metamorphosis (Fig. 16). The immature instars are referred to as larvae and do not resemble the adults. Larval appearance is variable, but is usually not insectlike. Commonly larvae are described as wormlike, grublike, or caterpillarlike. A pupal stage that follows the last larval instar is characterized by development of the adult form in a protected habitat or cocoon. The insect does not feed during this stage, and appears to be quiescent. However, metabolic activity is very high as larval tissues are reorganized to form the morphologically and physiologically different adult form.

IV. INSECT-MEDIATED PROCESSES

Insects are important consumers and important prey items in virtually all aquatic habitats, thus are intimately involved in the flows of matter and energy that occur in these ecosystems. Insects also exert top–down controls on nutrient cycles, primary productivity, decomposition, and translocation of materials, and affect relative abundances of other consumers in lentic as well as lotic ecosystems. Specific relationships between insects and aquatic resources that are of particular interest include detritivory, grazing, and predation. Processes that integrate one or more of these specific relationships include secondary production and stream drift. All of these processes are shaped by physical habitat and life-history constraints.

A. Detritivory

Detritivory, the feeding on decaying organic matter, is a major insect-mediated process in both lentic and lotic systems, although it has received more attention in lotic studies. The relative importance of allochthonous versus autochthonous detritus in aquatic ecosystems is widely variable, and size of the water body is the major determining factor. Consumer resources in small lentic systems and low-order streams are typically dominated by allochthonous inputs, whereas lakes and middle-order streams (see Vannote *et al.,* 1980) are usually dominated by autochthonous production. When detritus is plentiful, however, detritivory by insects is usually important to ecosystem function.

Detritus forms a major habitat for aquatic insects in ponds and many wetlands. The coarseness of the detrital substrate provides habitat for large aquatic insects, such as odonate nymphs, which prey on small detritivorous chironomids, mayflies, and microcrustaceans. For some large insects, detritus serves as a food source, although availability of detritus seldom limits wetland insect abundance or production, and microbial decomposition appears to be more important than insect-mediated decomposition in many wetlands (Batzer and Wissinger, 1996). Detritus is divided into two major fractions: coarse particulate organic matter (CPOM—that portion ≥1 mm in size) and fine particulate organic matter (FPOM—the portion that is <1 mm).

Low-order streams are strongly influenced by the input of terrestrial organic matter from adjoining ecosystems. In temperate streams, much of this detritus enters as a seasonal (autumnal) pulse of leaves, which is then processed in the stream by microbial and macroinvertebrate communities, and serves as the major energy source for many stream consumers (Fig. 9). Life cycles of many detritivorous insects are cued to this seasonal pulse of leaves such that they develop through the fall and winter when their food resource also is plentiful. Microbes, which colonize the detritus, form an integral part of the CPOM complex and are nutritionally important to shredders (Anderson and Sedell, 1979; Cummins and Klug, 1979). Dominant shredder taxa vary according to stream type and biogeography. Shredders are found in many of the aquatic insect orders (Table I). Some of the more conspicuous insect shredders are pteronarcid and capniid stoneflies, larval craneflies, and several families of caddisflies. Low-order streams often have 30–50% shredders in their macroinvertebrate communities (Fig. 10; see also Fig. 6 in Chapter 2), as was found in Oregon streams (Hawkins and Sedell, 1981), by virtue of their closed canopy and high input of CPOM per unit of stream surface (discussed in Vannote *et al.,* 1980).

In a first-order mountain stream, Wallace *et al.* (1997) used an ecosystem-level experiment (litter exclusion by suspending a net over a small stream) to demonstrate that terrestrial litter supported the vast majority of insect secondary production in the stream. Interestingly, insects and other consumers occupying a moss microhabitat in the stream did not respond to litter exclusion, suggesting that moss inhabitants comprised a food web that was independent of that supported by terrestrial detritus (Wallace *et al.,* 1997). In a different study, experimental reduction of shredders in a stream using an insecticide resulted in lower CPOM decomposition (20–40% that of a reference stream) and 5- to 15-fold lower FPOM transport (Wallace *et al.,* 1982). Clearly, insects are of critical importance to CPOM processing in streams.

Formation of FPOM from CPOM by shredders during their feeding is an important process. FPOM is also formed by several other mechanisms, and is dynamic in stream food webs. All consumers produce feces, which are an important component of FPOM (Shepard and Minshall, 1981; Hershey *et al.,* 1996). Bacteria use DOM, thereby incorporating it into bacterial biomass, which itself constitutes a portion of the FPOM pool (e.g., Meyer, 1990). Bacteria produce extracellular materials, which are sloughed into the FPOM pool (e.g., Wotton, 1988). FPOM is produced by flocculation of DOM due to a variety of physical and chemical processes (Dahm, 1981; Petersen, 1986; Wotton, 1988; Ward *et al.,* 1994). The mass of FPOM usually exceeds that of nonwoody CPOM by about an order of magnitude (Wallace and Grubaugh, 1996) and turns over much more rapidly. Insect collector-filterers and collector-gatherers that use this resource are important components of the macroinvertebrate community in all streams (Fig. 9; Vannote *et al.*, 1980; Hawkins and Sedell, 1981). However, because FPOM is formed by many processes, it varies widely in quality; some FPOM is highly labile, but much is refractory (Newbold *et al.*, 1983). Its retention in streams is also high, but varies considerably due to physical and/or chemical characteristics of the particles (Miller *et al.*, 1998). The variety of collectors that use FPOM also is very high (Table I).

DOM is generally considered to be the exclusive resource of stream microbes, which assimilate labile components of DOM into microbial biomass (see Lock *et al.*, 1984; Meyer, 1990). However, larval blackflies also ingest DOM (Fig. 17, and see Wotton, 1976) and

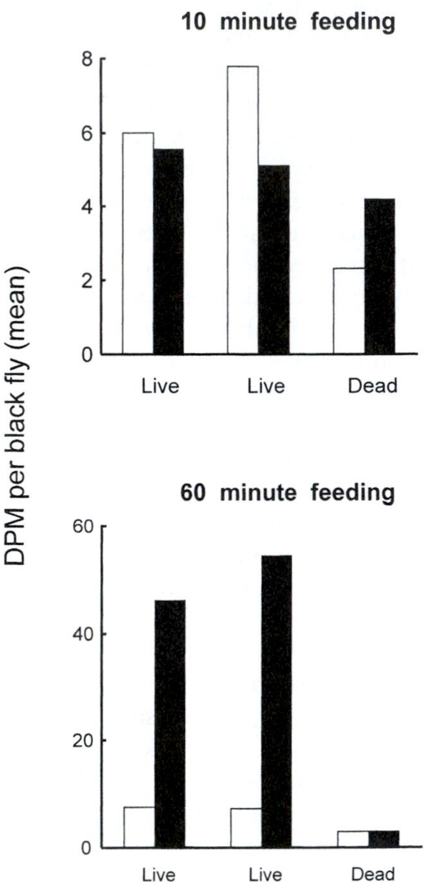

FIGURE 17 Ingestion of DOM by larval blackflies. Heads and bodies of black flies which had been permitted to feed for 10- or 60-min on 0.2-μm filtered ^{14}C-DOM were assayed separately for incorporation of labeled material. After 10 min, similar activity was seen on bodies of live and dead larvae, but higher activity was seen on heads of live larvae. This indicated that activity of live larvae enhanced accumulation of ^{14}C from DOM onto the head. After 60 min, considerably more material was present in the body, indicating that labeled DOM was being packed into the gut during feeding. Each set of bars shows results from a separate chamber (two live chambers and one dead chamber). [From Hershey *et al.*, 1996.]

this process can be important to the size and quality of the FPOM pool downstream of blackfly aggregations (Hershey *et al.*, 1996). DOM ingestion by blackflies includes some labile material, such as bacterial exopolymer (Couch *et al.*, 1996), and thus such feeding may be a nontrivial source of nutrition for these aquatic insects. Although direct DOM ingestion by stream insects is not thoroughly understood, its importance is likely limited to a few taxonomic groups.

Large woody debris (LWD) often forms an important pool of detritus in stream ecosystems (Anderson and Sedell, 1979; Bilby and Ward, 1989; Swanson *et al.*, 1982; Van Sickle and Gregory, 1990). Wood may cover over 25% of the beds of small streams in old-growth forests (Anderson and Sedell, 1979). However, only very specialized insects use this material as food, and these likely do not ingest quantities that are significant to wood decomposition (e.g., Wallace and Anderson, 1996). LWD is important as macroinvertebrate habitat. Benke *et al.* (1984) showed that much of the secondary production in a southeastern blackwater stream was associated with woody snags. In addition, the insect community on the snags was distinct from other substrates. LWD also serves as an important mechanism for retaining CPOM and FPOM resources that are used by insects (Benke *et al.*, 1984). In addition, LWD is a major factor defining pool and riffle habitat within a stream, which determines many aspects of macroinvertebrate community structure. Wood itself has low nutritional value, but the microorganisms that coat submerged wood (e.g., algae, bacteria, fungi, protozoans) provide a rich food resource for macroinvertebrates.

A few invertebrates ("wood gougers," a special group of shredders) are adapted to feed directly on the wood and include the cranefly larva *Lipsothrix* and the elmid beetle *Lara avara* (Anderson and Sedell, 1979; Wallace and Anderson, 1996). These taxa bore into and feed strictly on wood and have very long life cycles (3–6 years) as a result of the low nutritive value of their food. Examining a piece of water-logged wood will reveal the tunnels of these borers and often the insects themselves. Wood is also used by some larval caddisflies for case construction. *Heteroplecton californicum*, a caddisfly that can be found in shallow pools, hollow out twigs to construct portable cases (Fig. 18). When the larvae move, their cases appear to be moving twigs, and the larvae themselves are not visible.

B. Grazing

Benthic primary production is a major process in virtually all shallow aquatic ecosystems (Lamberti, 1996). As a consequence, grazing insects functioning as primary consumers can be found in nearly all of those

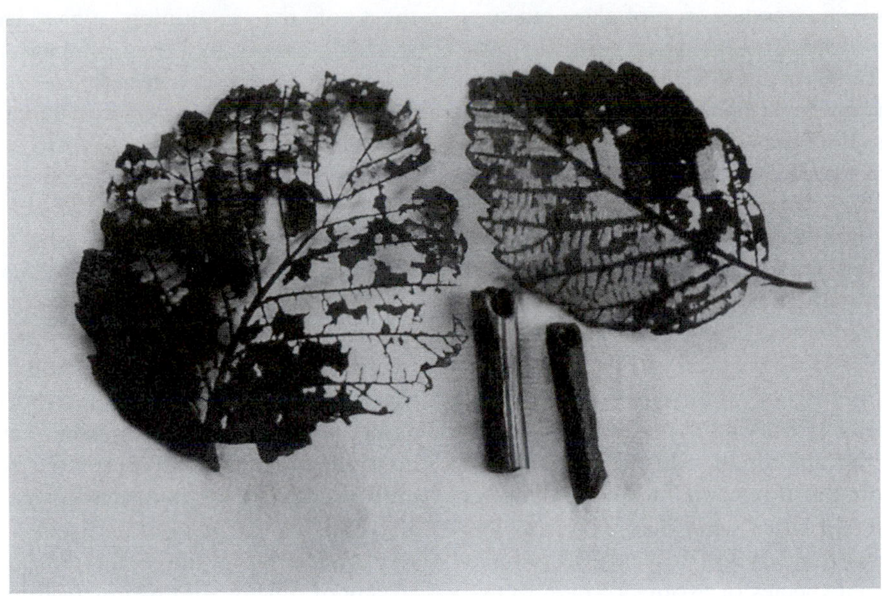

FIGURE 18 Wood cases of the caddisfly *Heteroplectron californicum* next to shredded alder leaves. Note opening chewed by caddisflies at upper end of twigs; larva resides inside the twig within a silk-lined tunnel. [Photograph by Jon Speir, reproduced from Anderson, 1976a.]

ecosystems except, perhaps, where extreme physical conditions predominate (Lamberti and Moore, 1984). Aquatic primary producers include algae, rooted macrophytes, and bryophytes (mosses and liverworts). Benthic algae frequently are the dominant plants in streams and along rocky, wave-swept littoral zones, whereas rooted macrophytes become more important in the finer sediments of lakes and large rivers where they are exploited for both habitat and food by insects. Fine sediments may also be covered with a layer of algal "epipelon," which is readily grazed by insects. Mosses can be locally important, especially in streams, and because of their fibrous texture and long-lived nature they provide excellent habitat for a variety of insects (Glime and Clemons, 1972; Suren and Winterbourn, 1992). Where mosses line the rock walls and spray zones of waterfalls, often a specialized insect fauna develops termed "torrenticolous" insects. In the discussion below, we will mostly concern ourselves with grazers of periphyton and macrophytes.

Scrapers can be found in six orders of aquatic insects (Table I), and for several of those orders (Ephemeroptera, Trichoptera, and Lepidoptera) algal consumption is a major way of life. Grazers often possess elaborate mechanisms to remove periphyton (attached algae, microbes, and associated organic matter) from surfaces (Lamberti *et al.*, 1987a). Many species of mayflies bear stout bristles on the labial or maxillary palps that are effective in removing periphyton from cracks and crevices. These bristles increase in length with successive instars, thereby allowing older larvae to remove larger particles while younger larvae remove smaller particles, effectively partitioning the resource by age. Many caddisfly larvae have robust or scoop-shaped mandibles for removing or "spooning" periphyton into their mouths. Some limnephilid caddisflies, such as *Dicosmoecus*, also use their tarsal claws to gather algae into small piles for ingestion (Lamberti *et al.*, 1987a). *Orthocladius* chironomids have stout mandibles lined with 4–5 teeth that wear down over time from scraping (Saponis, 1977); fortunately, molting periodically provides them with a new set of teeth. Even presumed shredders, such as cranefly larvae (Tipulidae), possess setae on the cutting surfaces of their mandibles that may help to collect algae from leaf surfaces, thereby allowing these insects to function as a facultative grazers (Pritchard, 1983). For shredders, the microorganisms, including algae, that grow on decaying leaves may be the most nutritional component of the leaves (Cummins *et al.*, 1989).

Grazing has been the focus of much study in streams (reviewed by Steinman, 1996, and Feminella and Hawkins, 1995, for field studies and by Lamberti, 1993, for laboratory stream studies) and to a lesser extent in lakes (reviewed by Lodge, 1991; Newman, 1991). In streams, benthic algae will grow on virtually any substrate exposed to light, including on benthic insects themselves or their cases (e.g., Pringle, 1985; Hershey *et al.*, 1988; Bergey and Resh, 1994). Many streams, however, have a low standing crop of benthic algae, sometimes barely perceptible to the naked eye. In comparison with obvious pools of dead organic matter (e.g., leaves and wood), it is tempting to dismiss the significance of algae and thereby grazing. However, a specific standing crop of algae can support a biomass of herbivores 10–20× greater than its own because of high algal turnover rates (McIntire, 1973), sometimes referred to as an "inverted trophic pyramid." For example, Mayer and Likens (1987) found that an abundant caddisfly (*Neophylax aniqua*) consumed mostly periphyton in a small, heavily shaded New Hampshire stream whose energy base was previously thought to be almost totally detrital (Fisher and Likens, 1973). Still, periphyton is often limiting to the growth of insect grazers, and competition for algal resources has been demonstrated (e.g., McAuliffe, 1984; Lamberti *et al.*, 1987b).

Among the stream insects, caddisflies (Trichoptera) and mayflies (Ephemeroptera) tend to be highly conspicuous grazers although small chironomid midges (Diptera) may be equally important herbivores due to their ubiquity, high densities, and short generation times (Berg and Hellenthal, 1992). Insect grazers have the ability as individual species or groups of taxa to exert large effects on the abundance, productivity, and community structure of benthic algae. Often, a single species can exert primary control over periphyton. For example, *Helicopsyche borealis* caddisflies consumed >95% of the algal standing crop in a northern California stream (Lamberti and Resh, 1983b). Experimental exclusion of these caddisflies produced a mat of filamentous algae almost 30-mm thick (Fig. 19). In another study, the large caddisfly *Dicosmoecus gilvipes* consumed about 60% of benthic algal production, and most of the remaining algae were dislodged and washed away by the grazing activities of these "sloppy" feeders (Lamberti *et al.*, 1987a). Furthermore, these large grazers can have negative effects on other benthic invertebrates, which they "bulldoze" and thereby displace as they harvest periphyton (Hawkins and Furnish, 1987; Lamberti *et al.*, 1992).

Tightly evolved associations between plants and herbivores are well documented in terrestrial ecosystems (Crawley, 1983). In aquatic ecosystems, however, there are few examples of host-specific associations between aquatic insects and benthic algae, for reasons that are still unclear (Gregory, 1983). One possible example is found in North American streams, where a mutualistic relationship may exist between the chironomid midge *Cricotopus nostocicola* and the blue-green alga *Nostoc parmelioides* (Brock, 1960). The first-instar midge larva finds and enters a small, globular colony of *Nostoc* and begins feeding on *Nostoc* cells, in the process changing the colony morphology to an ear-like form about 1 cm in diameter (Fig. 20). These dark-green colonies often are quite obvious on rocks, and the single larva can be seen clearly within the colony. The midge grows and pupates within the colony, finally emerging as an adult. The association also may benefit the colony, which has a higher photosynthetic rate when the fly is present (Ward *et al.*, 1985; Dodds, 1989). Amazingly, the midge will reattach the colony if it becomes dislodged from the substrate during a

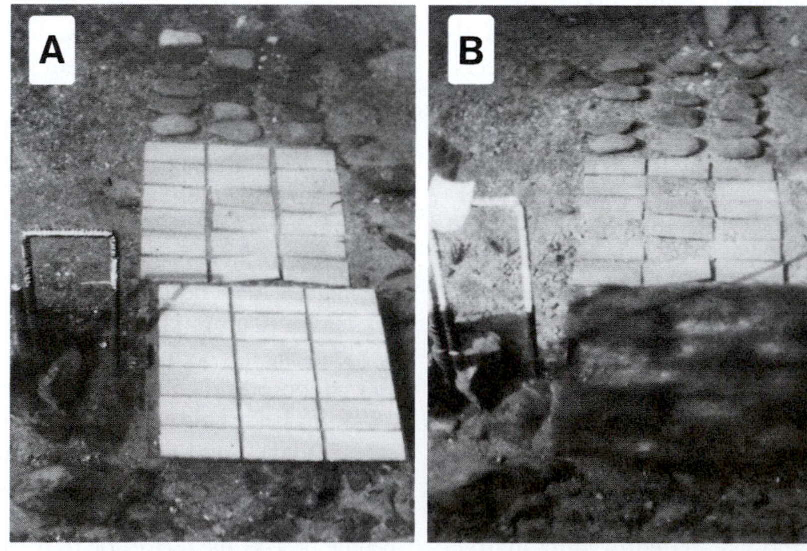

FIGURE 19 Effects of grazing by the caddisfly *Helicopsyche borealis* on algae in Big Sulphur Creek, California. Elevated but submersed platform in the foreground was used to exclude benthic caddisfly larvae; tiles in midground and rocks in background were freely grazed by caddisflies at densities $>1000/m^2$. Direction of water flow is from right to left. (A) Beginning of experiment. (B) After 5 weeks elapsed. Note dense growth of filamentous algae on ungrazed platform tiles. [From Lamberti and Resh, 1983b.]

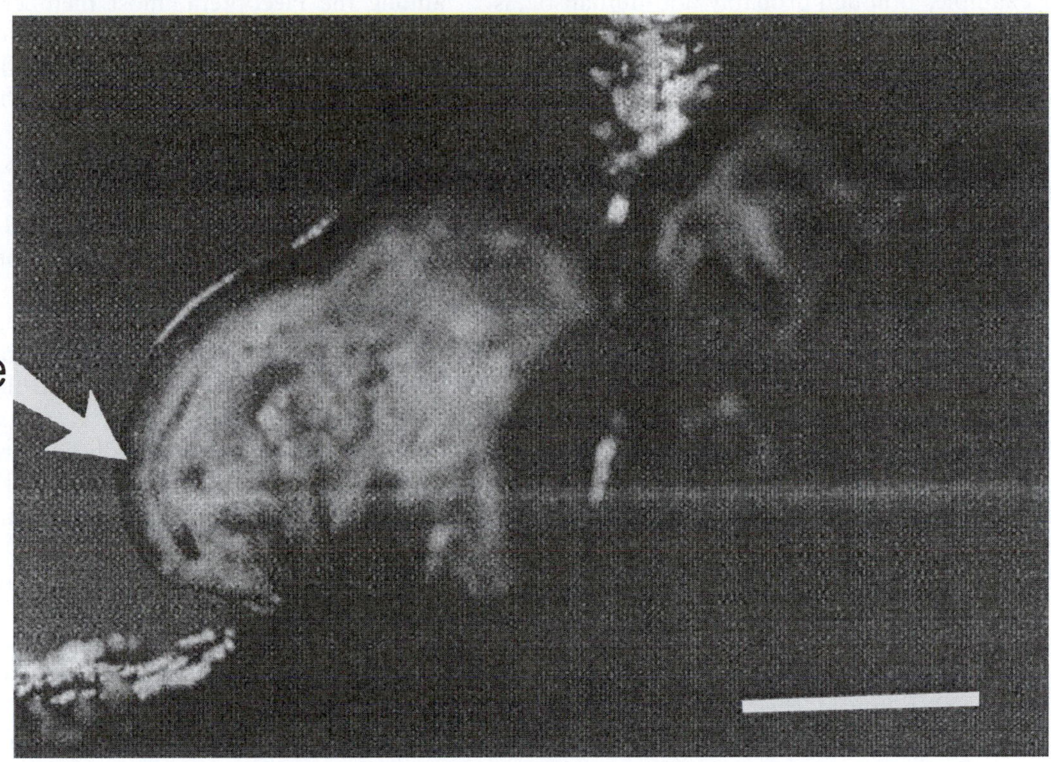

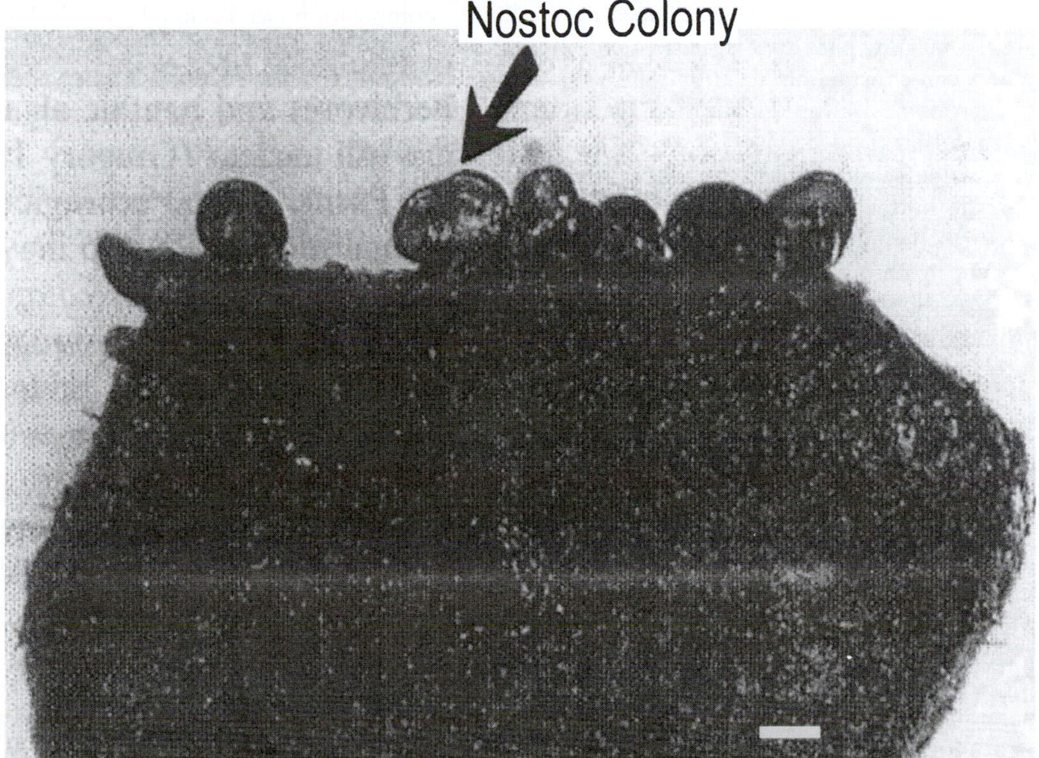

FIGURE 20 Colonies of the blue-green alga *Nostoc parmelioides* containing the midge larva *Cricotopus nostocicola*. Upper panel: *Nostoc* "ear" form with midge larva in left side of colony; Lower panel: *Nostoc* colonies attached to a rock. Bar = 5 mm. [Photographs by Louis Nelson, reproduced from Ward *et al.*, 1985.]

disturbance (Brock, 1960; Dodds and Marra, 1989), which doubtless benefits both the colony and the larva. Another example, although probably not mutualistic, is the association between the microcaddis fly *Dibusa angata* and the freshwater red alga *Lemanea australis* (Resh and Houp, 1986). Larval instars 1–4 of *Dibusa* are found among the basal holdfasts of *Lemanea*, where they consume epiphytic diatoms. Fifth-instar larvae construct cases made of *Lemanea* and consume only *Lemanea* tissue. There is no known benefit of this association to the alga.

Many insects consume living macrophyte tissue, a process that generally has been understudied and underappreciated (see review by Lodge, 1991). Unlike for algae, there appear to be a number of close associations between insects and aquatic macrophytes, perhaps because the large size of rooted plants makes specialization more feasible than for microscopic algae. The major insect orders involved in macrophyte grazing are the Trichoptera, Lepidoptera, Coleoptera, and Diptera. It is perhaps not surprising that rooted hydrophytes are eaten by the more advanced insect orders, as their terrestrial insect relatives are also mainly plant eaters. Indeed, it has been speculated that over evolutionary time terrestrial insects followed their host plants as they gradually invaded the water to eventually become true hydrophytes. Furthermore, parasitoids of those insects may have followed their hosts, as evidenced by the hymenopterans that now parasitize aquatic lepidopterans. Insect grazers can consume up to 100% of the standing stock of macrophytes, although the damage is usually much less than that (Lodge, 1991). It is more common to see the grazing scars of insects, such as beetles and moths, on the surfaces of floating leaves and other plant structures. Lepidopterans are known to feed on aquatic plants in numerous ways, including leaf mining, stem or root boring, foliage feeding, and flower or seed consumption (Lange, 1996). Coleopterans likewise exhibit a wide variety of feeding modes; leaf beetles (Chrysomelidae) are particularly common herbivores of water lillies, where their larval galleries are obvious on the floating leaves (Cronin *et al.*, 1998). Even though beetles may not consume large amounts of leaf tissue, their feeding activity can increase nutrient uptake and primary production by the plant and accelerate decoposition (Wallace and O'Hop, 1985).

C. Predator–Prey Interactions

Many aquatic insects function as predators, and predators are found in nearly all of the major groups of aquatic insects. A few groups, notably the Odonata, aquatic Heteroptera, and Megaloptera, are strictly predators, whereas other groups are more heterogeneous in their trophic affiliations, but may have important subdivisions that are predators. For example, among the Plecoptera, most members of the families Perlidae, Perlodidae, and Chloroperlidae are primarily predators (Stewart and Harper, 1996). Within the Trichoptera, members of the family Rhyacophilidae are free-roaming predators, which may inflict particular damage to larval black fly populations (Wheeler, 1994). Many of the sedentary, net-spinning Hydropsychidae are facultative predators (Benke and Wallace, 1980), and some larval Limnephilidae become opportunistic predators in later instars (Anderson, 1976a). Many Coleoptera and Heteroptera function as aquatic predators in both the immature and adult forms.

The role of predators in controlling the distribution and abundance of prey is a central question in aquatic ecology, and their role in determining aquatic insect communities also has received considerable attention. Predator and prey taxa are subjected to different physical constraints in various lentic and lotic ecosystems (see above), and these physical constraints interact with the predation process to produce different community types. Because predator–prey interactions do not always occur at the same spatial and temporal scales as abiotic factors that affect distribution and abundance of organisms, it is sometimes difficult to extrapolate predator–prey observations to community composition (see Peckarsky *et al.*, 1997).

Predators acquire prey using different behavioral strategies (see Peckarsky, 1982). Many predators, such as stoneflies, actively search for their prey ("hunters"). Ambush or "sit-and-wait" predators wait until a prey item is within range before striking. Odonate larvae often use this strategy. Once the prey is captured predators typically either engulf their prey intact, in bites, or may feed only on the internal fluids. Fluid feeders use specialized piercing (haustellate) mouthparts for injecting enzymes into their prey and sucking out the partially digested body. Clearly the relative sizes of predators and prey, as well as the mode of feeding used by the predator, are important mechanical constraints governing predator–prey interactions. Engulfing predators are limited to prey small enough to be subdued and swallowed intact (see Hershey, 1987), whereas piercing predators have access to a broader variety of prey sizes that are large enough to be handled and to accommodate the piercing mouthparts.

In a synthesis of predator effects in lentic communities, Wellborn *et al.* (1996) characterized control of community structure as falling along a gradient of physical and biotic control, with two major transitions (Fig. 21). Temporary ponds and lakes provide two extremes along a habitat size and permanence gradient, and two major transition thresholds occur along this perma-

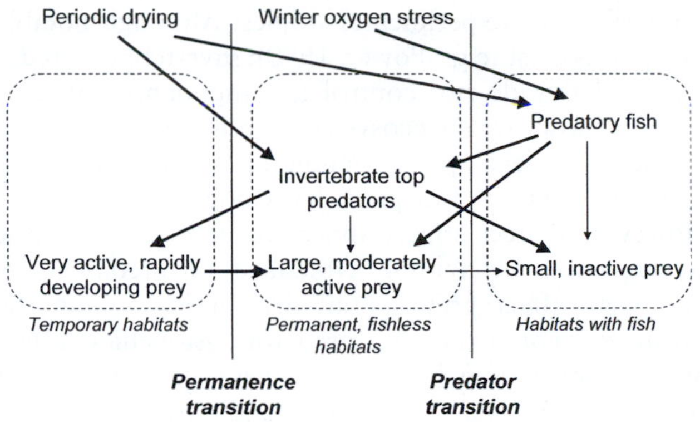

FIGURE 21 Aquatic communities are controlled by both biotic and abiotic factors. Two major transitions involving predator control of community structure are conceptualized to occur (Wellborn *et al.*, 1996). A permanence transition reflects when the wet phase of the hydroperiod is sufficiently long to support longer-lived predatory insects. A predator transition occurs when habitats contain fish, which then eliminate many predatory insects. [From Wellborn *et al.*, 1996.]

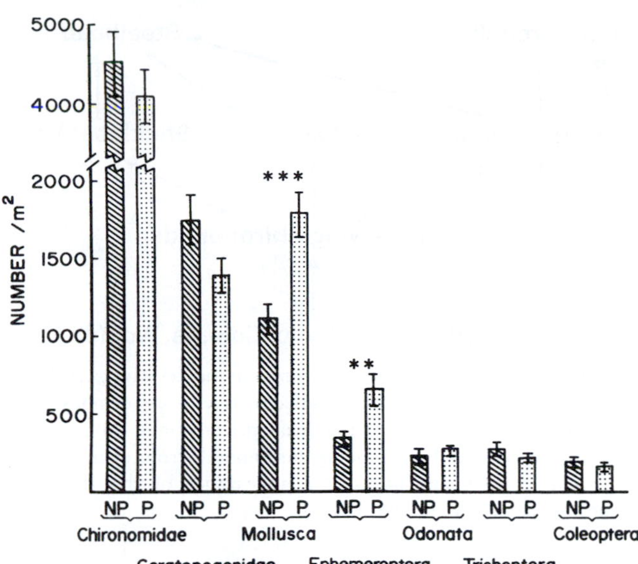

FIGURE 22 A fish exclusion experiment conducted in dense macrophytes of a cooling water reservoir in South Carolina did not show significant reduction of macroinvertebrates in the areas where no fish were present. NP, no predatory fish; and P, predatory fish present. [Reprinted from Thorp and Bergey, 1981.]

nence gradient, which provide the template for interaction of physical factors and predator control of insect communities (Fig. 21; Wellborn *et al.*, 1996). A permanence transition (Fig. 21) along the gradient controls predator presence or absence. In highly ephemeral habitats, most predatory insects are excluded (Batzer and Wissinger, 1996; Wellborn *et al.*, 1996, and see above). As a result, predator–prey interactions are relatively unimportant and the insect community is controlled by physical factors (Wellborn *et al.*, 1996), primarily drought cycles (Hershey *et al.*, 1998, 1999). Once the permanence transition is crossed, predatory insects become abundant and control various aspects of prey community composition (e.g., Batzer and Resh, 1996). These habitats tend to have the highest diversity of insect species (see Batzer and Wissinger, 1996). A predator transition occurs when habitats are large enough and deep enough to support fish (Fig. 21; Wellborn *et al.*, 1996). Fish are size-selective predators, typically eliminating large predatory taxa, which have a major impact on prey diversity and community structure. For example, in an experimental manipulation of bluegill and pumkinseed sunfish, mean invertebrate size decreased as fish density increased (Mittlebach, 1988). But fish may control insect prey density without dramatically affecting community composition (Bohanan and Johnson, 1983). Fish and predatory insects interact to increase the effect on prey compared to when either is present alone (Johnson *et al.*, 1996). But fish and large predatory insects such as dragonflies have different strategies for capturing prey (i.e., pursuit versus ambush). These alternative strategies provide contrasting selective environments for prey; McPeek *et al.* (1996) have shown that in lakes where dragonfly predation is important, *Enallagma* damselfies have larger caudal lamellae which enhance swimming speed, whereas in fish lakes, *Enallagma* caudal lamellae are smaller, associated with a cryptic strategy for predator avoidance.

Habitat heterogeneity moderates the effects of fish predation on insect communities (Gilinsky, 1984; Hershey, 1985); and in very dense macrophytes, fish are not important in controlling insect abundance or species composition (Fig. 22; Thorp and Bergey, 1981; Crowder and Cooper, 1982). Thorp (1986) concluded that although fish native to a system may control density or biomass of benthic insects, they have little impact on such community structure characteristics as diversity or species composition. However, historical presence or absence of fish within a community has a major impact on community structure (Fig. 21; Thorp, 1986; Wellborn *et al.*, 1996).

Similar to lentic studies of predation, predator–prey interactions involving stream insects also have been studied from two major perspectives: (1) the potential for their communities to be structured by fish predation; and (2) as predators interacting with, and possibly controlling, other macroinvertebrates. As in lentic systems, experimental studies of fish predation have not always led to the same conclusions. In an experimental removal of trout from a rocky mountain stream, Allan (1982) found no difference in macroinvertebrate density or diversity where fish were

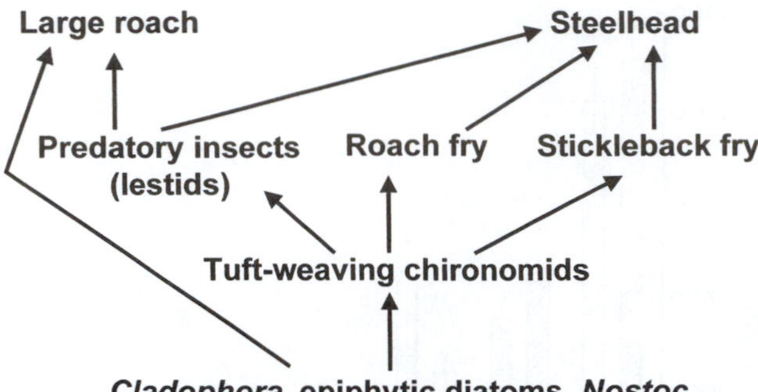

FIGURE 23 Simplified food web found in the Eel River, California (Power, 1990) where cascading effects have been demonstrated. Large fish control small fish that feed preferentially on predatory insects. This releases tube-building chironomids from control by insects, and dramatically affects the appearance of the algal mat. [Reprinted from Power, 1990.]

removed. Similarly, Reice (1983) found no evidence of trout control of any aspect of the stream insect communities that he studied. However, working in California's Eel River, Power (1990) has shown that predatory fish control predatory invertebrates (especially damselflies) which, in turn, control the abundance of larval chironomids, affecting both algal biomass and the physical appearance of the algal mat in the river (Fig. 23). Her study is an example of food web control via cascading trophic interactions (*sensu* Carpenter *et al.,* 1985), which have been well studied for pelagic systems. Although it is not clear that trophic cascades are common in stream ecosystems, it is clear that fish predation does not control insect abundance in all streams.

Regardless of their direct effects on prey density or community structure, fish have important indirect effects on prey in lotic ecosystems (e.g., Wooster, 1994; Wallace and Webster, 1996). Because of the threat of fish predation, many insects feed more actively on rock surfaces at night and are more prone to drift at night in search of better food patches (Fig. 11; Kohler, 1985). It appears that stream fishes are usually limited by their macroinvertebrate (especially insect) prey resources. This is suggested by Allen's paradox (Allen, 1951), but the strongest evidence for this comes from experimental fertilization studies that result in increased fish production (e.g., Stockner and Shortreed, 1978; Deegan and Peterson, 1992; Richardson, 1993).

The role of predatory macroinvertebrates in streams has been studied using a combination of gut analyses, laboratory stream microcosms, and in-stream experiments. Numerous studies of gut contents have yielded a wealth of information on which invertebrate species function as predators and which species or groups serve as their major prey. However, insight into their role in structuring stream communities has come primarily from experimental studies. Although notable exceptions exist (e.g., Power, 1990): invertebrate predators probably do not control the abundances of their prey in most stream ecosystems because stream communities are very open to immigration and emigration of both predators and prey (see Cooper *et al.,* 1990). However, indirect effects appear to be of considerable consequence. Behavioral avoidance of predators by prey may affect prey spatial distribution within the substrate. For example, predatory stoneflies affect mayfly distributions because mayflies will drift to avoid contact with them (Peckarsky and Dodson, 1980). When contact occurs, *Baetis* may further minimize the threat of predation by projecting their cerci forward in a "scorpionlike" posture that renders them more difficult for the stonefly to handle (Fig. 24; Peckarsky, 1980). Prey growth rates and/or fecundity can also be altered by the presence of predators, even if survivorship is not altered (Peckarsky *et al.,* 1993; Scrimgeour and Culp, 1994), and shredding activity can be reduced (Malmqvist, 1993). Alteration of growth rates may or may not have implications for prey demographics, depending on whether adult fecundity is a function of body size, as it is in *Baetis* mayflies but not in *Enallagma* damselflies (McPeek and Peckarsky, 1998). Prey also respond to predators through morphological adaptation. *Ephemerella* are armored against stonefly predators (Peckarsky, 1996). In stream habitats where *Hydra* are very abundant, long hairs on the midge larva *Cricotopus sylvestris* greatly decreases predation risk relative to short haired species, such as *C. bicinctus*, as well as to individuals of *C. sylvestris* whose hairs have been experimentally shortened (Fig. 25; Hershey and Dodson, 1987). A conceptual model by Peckarsky (1996) suggests that resource acquisition modes may account for investment in morphological versus behavioral strategies for predator avoidance. These examples show that predator–prey interactions involving insects are diverse, and variously may alter prey abundances, distributions, and behavior, and may have been important in selecting for morphological adaptations for predator defense.

D. Insects as Conduits for Energy Flow in Aquatic Food Webs

Insects provide a critical linkage in energy flow from microbial to vertebrate populations in aquatic ecosystems. Food web analyses have been used to understand these linkages, and thereby integrate organic-matter processing with community interactions (e.g., Closs and Lake, 1994; Benke and Wallace, 1997). The goals of food web studies for a particular system are to identify organic matter sources for the various

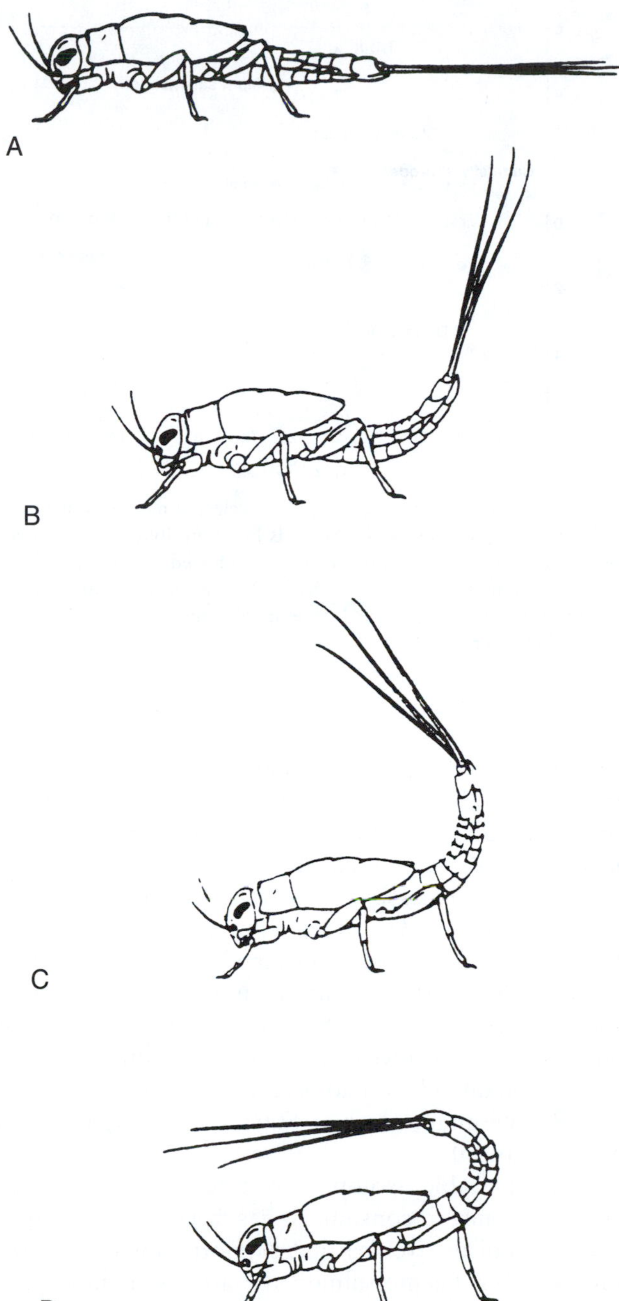

FIGURE 24 Mayflies exhibit a defensive "scorpion" posture when attacked by predatory stoneflies. [From Peckarsky, 1980.]

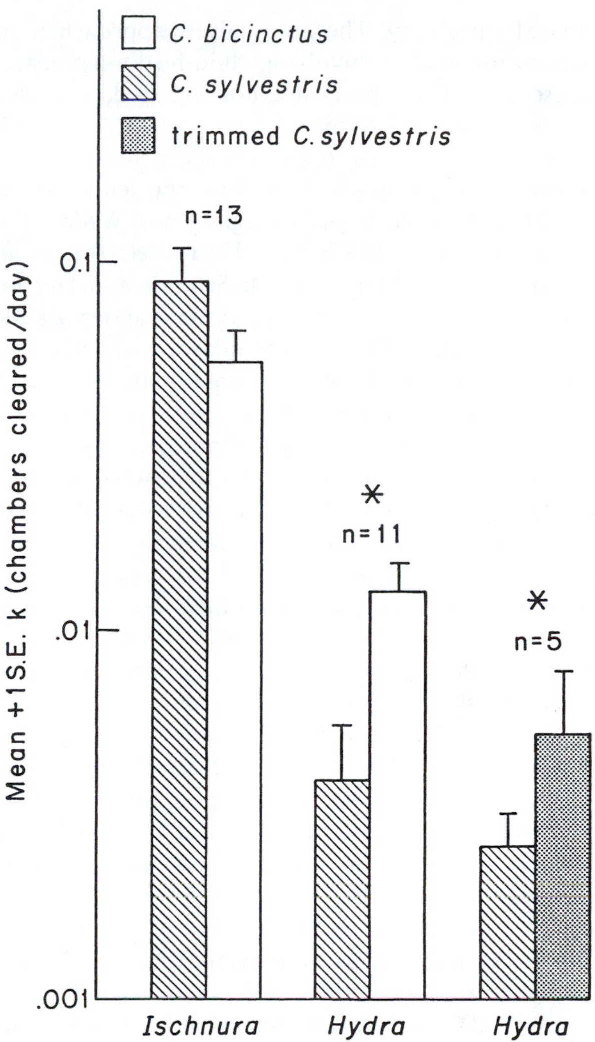

FIGURE 25 Long hairs provide *C. sylvestris* with protection against *Hydra*, compared to short haired *C. bicinctus* or *C. sylvestris* with experimentally shortened hairs. *Hydra* is a freshwater cnidarian that preys on chironomids and other aquatic invertebrates. [From Hershey and Dodson, 1987.]

consumers and to elucidate the trophic structure of the web. In most aquatic ecosystems, either three or four trophic levels are present including: (1) primary producers and detritus; (2) primary consumers, including detritivores (shredders and collectors) and grazers (scrapers); (3) secondary consumers (predators); and (4) tertiary consumers (vertebrate predators that consume invertebrate predators). These designations are oversimplifications because omnivory is common, such that many consumers occupy more than one trophic level (see Lamberti, 1996). In stream ecosystems, the relative importance of the detritivore food chain compared to the grazer food chain is a matter of some debate (Minshall, 1978), but this varies as a function of stream order and riparian conditions (Hawkins and Sedell, 1981; Cummins *et al.*, 1989; Gregory *et al.*, 1991). The food web in any particular aquatic ecosystem will reflect the combination of factors altering resource and invertebrate abundance and distribution. Thus, wide variation in food web structure occurs. Gut analyses are often used to delineate trophic structure (e.g., Hershey and Peterson, 1996). However, some food items are difficult or impossible to recognize, while others may be overlooked due to rarity or temporal variability. The gut analysis approach is impractical for studies involving fluid-feeding predators because sclerotized parts are not ingested, thus prey identity cannot be determined.

Recently, stable isotopes have been used to study food-web relationships in both lotic and lentic systems (e. g., Fry and Sherr, 1984; Minagawa and Wada, 1984; Peterson and Fry, 1987; Fry, 1991; Peterson *et al.*, 1993; Junger and Planas, 1994; Schuldt and Hershey, 1995; Whitledge and Rabeni, 1997; Vander Zanden and Rasmussen, 1999). One useful aspect of carbon stable isotopes is that the various organic matter sources for consumers often have different relative abundances of the two stable isotopes of carbon, ^{13}C and ^{12}C. Because the relative abundances of these isotopes change only slightly as the organic matter is processed by various consumers, the difference between the ratio ^{13}C to ^{12}C in a consumer and a standard (or value, expressed as parts per thousand, or per mil) can be used to infer the consumer's food source. Recently, ^{15}N has been used extensively in enrichment studies of lotic and lentic ecosystems to serve as a tracer of consumer food sources as well as nitrogen-distribution through the system (Kline *et al.*, 1990; Hershey *et al.*, 1993; Wolheim *et al.*, 1999). ^{15}N also is used in trophic studies. ^{15}N and ^{14}N change as food is processed by consumers because the various metabolic processes use, or fractionate, these isotopes differently. The net result is that with each trophic level, consumers become enriched (by about 3 per mil) in ^{15}N relative to ^{14}N, as ^{15}N is retained in the tissues and ^{14}N is excreted. Thus, consumers in a food web can generally be assigned to a trophic level even if their precise food resource is not known (see Fry, 1991; Peterson *et al.*, 1993). However, the del ^{15}N value of the base of the web may vary between systems, thus should be measured for each system where del ^{15}N is being studied in consumers (Vander Zanden and Rasmussen, 1999).

When organic matter sources in a stream are isotopically distinct, a combination of ^{13}C and ^{15}N analyses can yield considerable insight into a stream food web. Such an analysis was performed for the dominant consumers and their food resources of Lookout Creek in Oregon (Fig. 26; Fry, 1991). In Lookout Creek, the ^{15}N values revealed that macroinvertebrates were trophically distinct from two predatory vertebrates studied, whereas the ^{13}C values showed that consumers relied more heavily on stream algae than on terrestrial detritus. The macroinvertebrate scrapers studied (the snail *Juga* and the caddisfly *Dicosmoecus gilvipes*) appeared to consume algae, whereas the shredding caddisfly *Heteroplectron californicum* apparently ate terrestrial detritus. The stonefly *Calineuria californica*, a presumed predator, occupied a trophic level close to

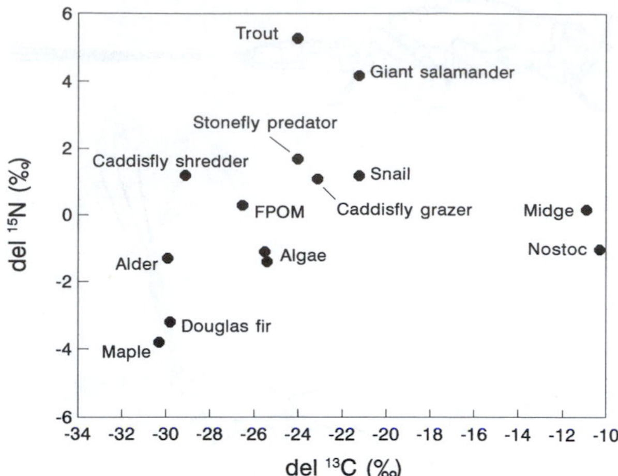

FIGURE 26 Stable isotope cross-plot of selected members of Lookout Creek, Oregon, food web. Animals horizontally positioned over other sources (i.e., have similar $\delta^{13}C$) are inferred to consume those materials. Animals, displaced vertically 2–4 $\delta^{15}N$ units upward are inferred to belong to the next higher trophic level in the food web. [Modified from Fry, 1991.]

primary consumers and thus may have consumed some plant matter along with small invertebrates that were not measured. The endosymbiotic midge larva *Cricotopus nostocicola* (described earlier) was positioned over its algal host *Nostoc parmeliodes*. Cutthroat trout, *Oncorhynchus clarki*, and Pacific giant salamanders, *Dicamptodon ensatus*, consumed invertebrates, as shown by their vertical positioning over the ^{13}C of their food and a 3–4 per mil difference in ^{15}N (Fig. 26). Gut contents of trout revealed a variety of invertebrates whereas salamanders had exclusively snails in their guts (R. Wildman, Oregon State University, personal communication).

Use of stable isotopes to trace organic matter sources for insect consumers has been a particularly powerful tool in streams that support runs of anadromous salmon. Pacific salmon die after spawning, and the carcasses provide nutrients that are used by stream algae (e.g., Brickell and Goering, 1970; Mathisen *et al.*, 1988), which are then fed upon by grazers (e.g., Richey *et al.*, 1975; Kline *et al.*, 1990). Salmon carcasses also serve as a high quality source of CPOM for stream shredders (Kline *et al.*, 1990). Stable isotope technology provides a means to quantify the importance of salmon carcasses to the stream food web. Salmon returning to streams to spawn have an isotopic signature that reflects their position in the marine food web. They are enriched in ^{15}N, reflecting their predation on marine fishes, and enriched in ^{13}C reflecting their dependence on marine sources of carbon. By comparing the isotopic signature of algae and insects in reaches

TABLE IV ^{15}N and ^{13}C values for food-web components and source materials in a Lake Superior tributary stream

	^{15}N		^{13}C	
Material	*Upstream*	*Downstream*	*Upstream*	*Downstream*
Salmon		8.9		−22.8
CPOM	−0.2		−28.5	
Hydropsyche	3.1	4.1	−29.1	−28.9
Heptageniidae	1.6	4.0	−29.2	−28.9

Salmon spawning occurs in the downstream reach of the river, but not the upstream reach. The stable isotope value shows that salmon-derived N, but not salmon-derived C, is incorporated into collector (*Hydropsyche*) and especially grazer (Heptageniidae) food-web components. [From Schuldt and Hershey, 1995.]

not used by salmon, to reaches where salmon spawn, the proportion of marine-derived N and C (from salmon) in the insect community can be traced. Thus, Kline *et al.* (1990) showed that at least 90% of the N in periphyton from Sashin Creek, in southeastern Alaska, was derived from spawning salmon or salmon eggs. Herbivores contained 50–100% salmon-derived N. In a Lake Superior tributary stream, the ^{15}N signature of heptageniid grazers suggested about a 30% reliance on salmon-derived N from Lake Superior (Table IV; Schuldt and Hershey, 1995).

V. SECONDARY PRODUCTION

Secondary production is the amount of animal biomass that is produced in a given area over a given time period, usually expressed in g/m^2 per year. For a cohort, this measurement integrates both individual growth rate and survivorship of individuals within the cohort (see Benke, 1984, 1996). In a food-web context, secondary production is the summation of the amount of biomass that is available to be consumed by the next trophic level over some period of time. Thus, an estimate of secondary production expresses energy or material flow through the consumer components of the ecosystem, and is useful for identifying major pathways of energy flow in an ecosystem (e.g., Smock and Roeding, 1986; Benke and Wallace, 1997).

Estimation of secondary production often provides insights that are not apparent from biomass or abundance measurements alone. This is especially true for organisms of small body size if they exhibit rapid growth rates. For example, Benke (1998) found extremely high secondary production of snag-dwelling chironomids in a high-order blackwater river despite their low biomass. This was due to very high biomass turnover rates of these small-bodied organisms (Benke, 1998). In Sycamore Creek, Arizona, macroinvertebrate secondary production, which was primarily insect production, was also extremely high despite fairly low biomass (Fisher and Gray, 1983). This high rate of production reflected a long growing season, small adult size, very rapid life cycles, and abundant (if low quality) FPOM (Fisher and Gray, 1983). But, even for large bodied insects measurement of production may provide surprising insights. O'Hop *et al.* (1984) showed that production of the detritivorous stonefly *Peltoperla maria* was similar in a disturbed and undisturbed stream despite more than twofold differences in abundances, reflecting faster growth and higher density of larger individuals in the disturbed stream. Although stream secondary production is very often dominated by insects, crustaceans and molluscs may also be quite important (see Waters and Hokenstrom, 1980; Fisher and Gray, 1983). In a recent study of production in a neotropical stream, Ramírez and Pringle (1998) show that insect production is low despite rapid growth rates. They attribute this low production to a combination of high abundance of non-insect macroconsumers and scouring of resources by flooding.

Because different habitats vary markedly in abundances and community composition of their insect fauna, calculation of secondary production on a habitat-specific basis is needed to reveal the spatial complexity of insect-mediated pathways of energy flow in aquatic ecosystems. In addition, failure to consider some habitats may grossly underestimate secondary production. For example, Smock *et al.* (1992) showed that debris dams were very important compared to other habitats in one low-gradient first-order stream, but that the floodplain was most productive in another stream. In addition, the vertical extent of the hyporheic zone in the two streams accounted for a large difference in secondary production between the two systems (Smock *et al.*, 1992). Snag habitats are very important sites of secondary production in higher order rivers, reflecting the success of both mayflies and chironomids in utilizing the amorphous detritus associated with that habitat (Benke and Jacobi, 1994; Benke, 1998).

Fisheries managers often use secondary production to assess the capacity of a stream reach to support fish. In many cases, there does not appear to be enough invertebrate production to support the observed fish production, a problem referred to as Allen's paradox (Allen, 1951). This may be due to underestimates of the production-to-biomass ratio (P/B) for individual species (Benke, 1996), or perhaps failure to consider the production of chironomid midges (Huryn and Wallace, 1986; Berg and Hellenthal, 1992). Recently, Huryn (1996) has shown that trout can consume a very large proportion of the benthic macroinvertebrates (mostly insects) in a stream.

Although conceptually very useful, secondary production is time-consuming and expensive to measure at the community level. The varied techniques for measuring secondary production, reviewed by Benke (1984, 1996), involve sampling efforts that quantify consumer standing stocks at several points in the growing season, at least a trophic level understanding of the various consumers, and some knowledge of life histories of the dominant consumer components (Table V). However, in the process of measurement, considerable insight is gained into the insect community, which sheds light on which consumers play key roles in the various aquatic ecosystem processes.

TABLE V Methods for measuring secondary production of stream macroinvertebrates. Modified from Benke, 1984, 1996, where secondary production methods are reviewed. [Modified from Hershey and Lamberti, 1998]

Method	*Data needed*	*Comments*	*Other references*
Allen curve	Density (No./m^2) for several dates spanning development of a cohort; mean individual mass for each date [often obtained from length × weight regressions (Smock, 1980)].	Graphical presentation of density versus mean individual mass. The area under the curve corresponds to the secondary production of a population. This method was developed prior to the widespread availability of electronic calculators and computers.	Allen, 1951
Increment-summation	Density (No./m^2) for several dates spanning development of a cohort; mean individual mass for each date [often obtained from length × weight regressions (Smock, 1980)].	Production (P) between sampling dates is calculated as the product of mean density (N) and change in mean individual mass (W): $P = NW$. Annual production is the summation of these incremental measurements for a year.	Waters, 1977; Waters and Crawford, 1973
Removal-summation	Density (No./m^2) for several dates spanning development of a cohort; mean individual biomass for each date [often obtained from length × weight regressions (Smock, 1980)].	Production lost between sampling dates is the product of the change in density and the mean individual mass: $P = WN$. Annual production is the sum of all production losses over a year.	Waters, 1977
Instantaneous growth	Growth rate; mean biomass	Production between sampling dates is the product of growth rate and mean biomass, where growth rate is determined in laboratory experiments, or from field samples of distinct cohorts, and biomass is determined from field samples: $P = g \times B$. Growth is temperature and size-dependent, so accuracy of this method depends on making measurements under the appropriate conditions. Annual production is: $P = \Sigma(g_i)(B_i)$, where the I's are the production intervals.	Ross and Wallace, 1981; Hauer and Benke, 1987
Size frequency	Mean annual density; size distribution of individuals throughout year; mean mass for each size category; cohort production interval (mean development time from hatching to emergence).	This method is widely used and does not depend on the presence of distinct cohorts in field populations. The mean size–frequency distribution, which is skewed toward smaller size categories due to mortality, represents an average cohort. For populations where distinct cohorts are known, this method gives a similar result as the increment summation method. $P = \Sigma W \Delta N_i$, where i = number of size classes. This estimate of secondary production needs to corrected for the CPI: (multiply P by 365/CPI)	Benke, 1979

VI. EFFECTS OF LAND USE ON AQUATIC INSECT COMMUNITIES

Because aquatic insects are sensitive to physical and chemical conditions in their environment as well as to top-down and bottom-up biotic controls, their communities also reflect changes in land use on both geological and ecological time scales and as well as on multiple spatial scales. For both lotic and lentic ecosystems, nutrient loading from either nonpoint or point sources, changes in hydrology, timber harvesting, construction, and riparian or shoreline development are among the most important land-use alterations that affect insects.

Lentic insect responses to land-use disturbances are seen through indirect effects of physical and chemical changes in littoral macrophyte zones and chemical changes in profundal zones. Nutrient loading results from many land-use changes, including nutrients associated with sediments derived from construction and tillage; fertilizer from agriculture, suburban and urban runoff; and wastewater and industrial inputs (Carpenter *et al.*, 1998). These inputs are derived from both inflow streams and land uses on lake shorelines themselves. Nutrient inputs enhance macrophyte growth and/or phytoplankton productivity. Since macrophytes support a high diversity of aquatic insects as well as affording refuge from fish predators (Thorp and Bergey, 1981; Crowder and Cooper, 1982; Gilinsky, 1984; Hershey, 1985, and see Fig. 22;), land-use changes altering macrophytes have strong indirect effects on littoral insects. However, if macrophyte growth becomes too dense, anoxia may be a problem, especially at the sediment/water interface. Low dissolved oxygen dramatically affects insect community structure (see Section II.B). Sedimentation may negatively affect macrophytes, thereby indirectly negatively affecting insect abundance and diversity. Nutrient inputs also enhance phytoplankton production (e. g., Schindler, 1977; Axler and Reuter, 1996), and at high level this often leads to bottom-water anoxia and associated low insect diversity and changes in insect community structure (see Hilsenhoff, 1966; Saether, 1979, and Section II.B).

Invertebrate populations in wetlands also are affected by vegetational and water management practices and by water quality and plant physical structure (see Ferenc *et al.*, 1999). However, Wolf *et al.* (1999) found little direct effect of land use on wetland invertebrates in urban playa wetlands, although they did find invertebrate abundances were positively associated with macrophyte growth. Similar to littoral areas of lakes, macrophyte beds respond to land use surrounding the wetlands. Water-level fluctuation in wetlands, often associated with land-use changes, has considerable influence on invertebrate community structure and the balance between insect and noninsect invertebrates (Anderson *et al.*, 1999; Hershey *et al.*, 1999). Wetland insects and other invertebrates are a major source of food for migratory and resident waterfowl (Swanson, 1984/1985; Martin *et al.*, 1951; Tester, 1995), thus wetland loss and associated loss of invertebrate food resources and habitat have had a major impact on wetland bird species (Richardson, 1979).

Hynes (1975) first articulated the importance of employing a landscape perspective for understanding stream ecosystem structure and function. The realization of this linkage between the stream and its terrestrial setting was a critical development in stream ecosystem theory (Minshall *et al.*, 1985) and aquatic insects have been central to this perspective because of their role in processing allochthonous inputs. This aquatic–terrestrial relationship is well appreciated qualitatively, but poorly understood quantitatively. Local geomorphology, vegetation, and climate determine the quality and quantity of organic matter entering a stream. These processes and disturbance history at a site determine the spatial and temporal patterns of riparian zone characteristics that dictate the physical characteristics of the stream habitat (Gregory *et al.*, 1991). Thus, riparian processes and natural disturbances, or disturbances due to changes in land use, will alter insect habitats in the stream channel. These disturbances operate over temporal scales ranging from geologic alterations of landform, to ecological succession of the riparian forest, as well as over spatial scales spanning watersheds, to local patchiness of algal growth on individual rocks within the channel (Gregory *et al.*, 1991). Aquatic insect abundance and diversity responds to all of these perturbations.

Aquatic insect communities reflect land use within the watershed because changes in land use alter stream habitat and water quality (Gregory *et al.*, 1987; Chauvet and Décamps, 1989; Merritt and Lawson, 1992). As discussed previously, insect communities are sensitive to predominant sources of organic matter, dictated largely by stream order, local geomorphology, and riparian vegetation (Vannote *et al.*, 1980; Naiman *et al.*, 1987; Naiman and Anderson, 1996). Biogeography (Resh, 1992; Hershey *et al.*, 1995b) and water quality factors (Richards and Minshall, 1992; Lamberti and Berg, 1995) also constrain insect communities.

Use of geographic information systems has facilitated landscape-level analyses and land-use studies of stream insect communities. In Minnesota north shore tributaries to Lake Superior, insect community structure and substrate characteristics within a stream are correlated with agricultural and residential land uses in the watersheds (Richards and Host, 1994). The types of alterations seen will depend on the nature and intensity

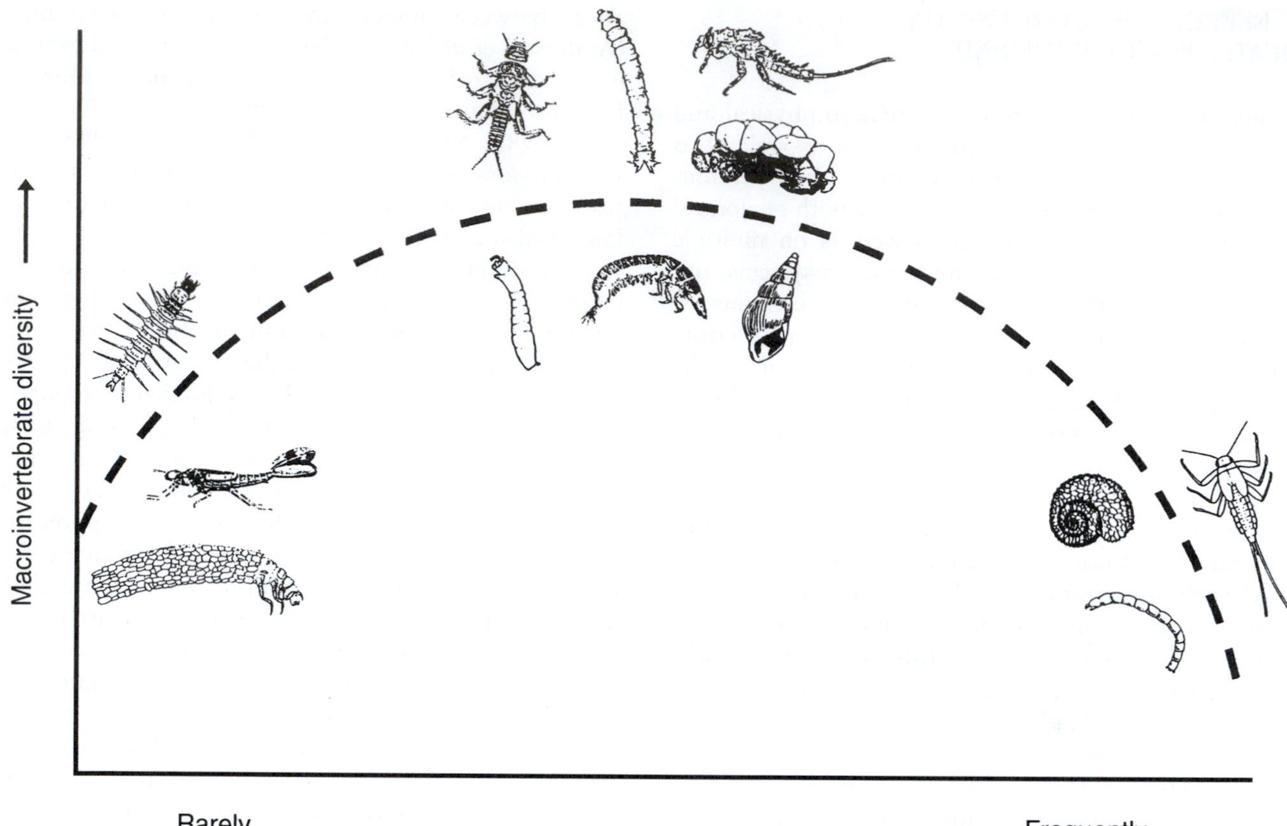

FIGURE 27 Theoretical relationship between macroinvertebrate species diversity and frequency of disturbance (e.g., floods) in streams ecosystems [redrawn from Ward and Stanford (1983) after Connell (1978)], with examples of macroinvertebrates that can be found under different conditions in North America. Rarely disturbed streams often contain large, long-lived taxa such as (top to bottom): *Neohermes* hellgrammite, *Argia* damselfly, and *Dicosmoecus* caddisfly. Streams with intermediate levels of disturbance usually contain a diverse fauna including (clockwise from upper left): *Calineuria* stonefly, *Tipula* cranefly, *Drunella* mayfly, *Glossosoma* caddisfly, *Juga* snail, *Hydropsyche* caddisfly, and *Simulium* black fly. Streams that are disturbed frequently may contain small-bodied taxa with short life cycles or high reproductive rates such as (clockwise from upper left): *Helicopsyche* caddisflies, *Baetis* mayflies, and chironomid midges.

of the land-use change (Cairns, 1990), but local geomorphology modifies some of the effects of land-use change, as well as constraining land cover and dictating what land uses might occur in a watershed (McIntire and Colby, 1978; Naiman *et al.*, 1993). In an extensive study of a 45-catchment river basin in central Michigan, Richards *et al.* (1996) found that geology and land use were approximately of equal importance in determining invertebrate communities in streams.

Land-use changes that alter large woody debris (LWD) have large impacts on insect communities (e.g., Fig. 27). Logging is one of the major such land uses. Logging has greatly altered the size and amount of LWD that enters stream channels (Sullivan *et al.*, 1987). Removal of riparian trees changes inputs of LWD for centuries, altering substrate stability and insect habitat (Naiman and Anderson, 1996); but recent changes in forest practices now offer some protection to the riparian zone from excessive harvest (Gregory and Ashkenas, 1990). Streams that lack LWD sources due to past logging are candidates for rehabilitation, which now emphasizes the input of logs in natural volumes and arrangements to improve stream habitat and to augment the food base. In addition to removing LWD and thereby altering detrital inputs and channel structure, logging also removes vegetative cover within the watershed. This results in increased sedimentation into the stream, reduced infiltration which decreases groundwater recharge, and an increased runoff which increases the magnitude of stream discharge change following storm events (e.g., Gregory *et al.*, 1991; Naiman and Anderson, 1996).

Other land uses, including urban, agricultural, and industrial development may follow timber harvest.

Agriculture has a large effect on sedimentation in a stream reach (Allan *et al.*, 1997), but stream buffers are more important than the whole catchment in determining sediment-related habitat variables in agricultural basins (Richards *et al.*, 1996). Road construction is usually associated with all types of human land use. Road construction often destabilizes hillsides, leading to landslides and sometimes debris flows in stream channels downslope (Lamberti *et al.*, 1991; Swanson *et al.*, 1987), which constitute a major disturbance to the stream ecosystem (see below). These changes in land use dramatically alter watershed hydrology, and hence stream hydrology. As discussed previously, insect communities are strongly affected by stream hydrology, but are also sensitive to the various forms of pollution that result from many land uses (see Richards *et al.*, 1993). Thus, insect communities are intimately linked to land use in the watershed as well as to local conditions in the stream channel.

VII. ROLE OF DISTURBANCE

A "disturbance" has been defined as any discrete event that disrupts population, community, or ecosystem structure, usually by changing resource abundance or the physical environment (Sousa, 1984; Resh *et al.*, 1988). Many types of natural and human-induced disturbance can disrupt the habitat, resources, or population densities of aquatic insects. Large-scale natural disturbances that can affect insects include: (1) hydrologic events such as floods, ice, or wave action that scour or remove substrates; (2) desiccation events such as droughts that lead to drying of aquatic habitats; (3) watershed disturbances, such as wildfire, that can disrupt nutrient and sediment inputs to streams and lakes for decades; and (4) seasonal events, such as summer or winter oxygen depletion, that can lead to insect mortality and major changes in community structure (see Fig. 27). Naturally, many small-scale disturbances can also affect local insect populations. Human-related disturbances of aquatic insect habitats are at least as important and include nominally: (1) point and nonpoint source pollutants that enter aquatic ecosystems from the land, connected water, or the atmosphere; (2) water withdrawal, diversion, and storage; (3) modifications of river channel geomorphology and destruction of wetlands; (4) watershed disruptions, such as logging and other land uses; and (5) introduction and establishment of exotic species. All of these mechanisms, singly and in combination, can alter population densities and community structure of insects, sometimes for decades or centuries as in the case of watershed land-use change. In general, however, natural disturbances tend to be episodic and, if the habitat returns to its orginal state, insect communities recover quickly. Human disturbance, by contrast, tends to be chronic and thus insect communities may not recover until the disturbance ceases or the habitat is restored (Cairns, 1990).

Variation in flow (floods to desiccation) is the major cause of natural disturbance in streams and is responsible for large, usually temporary, reductions in insect abundance and diversity (Fisher *et al.*, 1982; Lamberti *et al.*, 1991). Stream insects have evolved various mechanisms to deal with such flow variation, including life-history adjustments to minimize the presence of vulnerable stages during times of peak flows, behavioral movement into more protected hyporheic (interstitial) and lateral habitats during floods, and high reproductive rates to compensate for losses. Despite these adaptations, severe flood disturbance will still remove large proportions of the insect fauna. During drought, refuge provided by the deep hyporheic zone is especially important, but its accessibility to insects will depend on its grain size. Other insects may survive drought via resistant life stages, such as diapausing eggs or cocoons. Insects can recolonize streams from several sources: (1) egg laying by adults, usually aerial in the case of insects; (2) drift from upstream areas; and (3) movement along the bed from upstream, downstream, hyporheic, or lateral habitats (Smock, 1996). These behaviors are discussed in more detail above (see Section II. D. 3)

In general, aquatic insect communities recover rapidly from natural disturbance. Lamberti *et al.* (1991) studied the response and recovery of macroinvertebrates (mostly insects) in a Cascade Mountain stream that experienced an extreme disturbance. A debris flow, which is an episodic mass movement of sediment and debris through the stream channel (usually triggered by a landslide), devastated about a 1-km reach of Quartz Creek, Oregon, and partially disturbed areas downstream (Fig. 28). More than 99% of the macroinvertebrates were removed by the debris flow. An associated flood alone in an upstream reach removed >90% of the invertebrates, so that in-stream sources of insects clearly were low. Yet, in both the flood-impacted and debris flow-impacted reaches, macroinvertebrate species richness and total abundance recovered within one year of the disturbance to "normal" levels (Fig. 28). In another example, the eruption of Mt. St. Helens (southwest Washington) eliminated the entire biota from entire drainage systems, which included streams, lakes, and wetlands. Many stream channels were completely rerouted by topographic changes. Yet, in the first year after the eruption, 98 species of macroinvertebrates (mostly insects) were found in one devastated stream (Anderson, 1992). Five

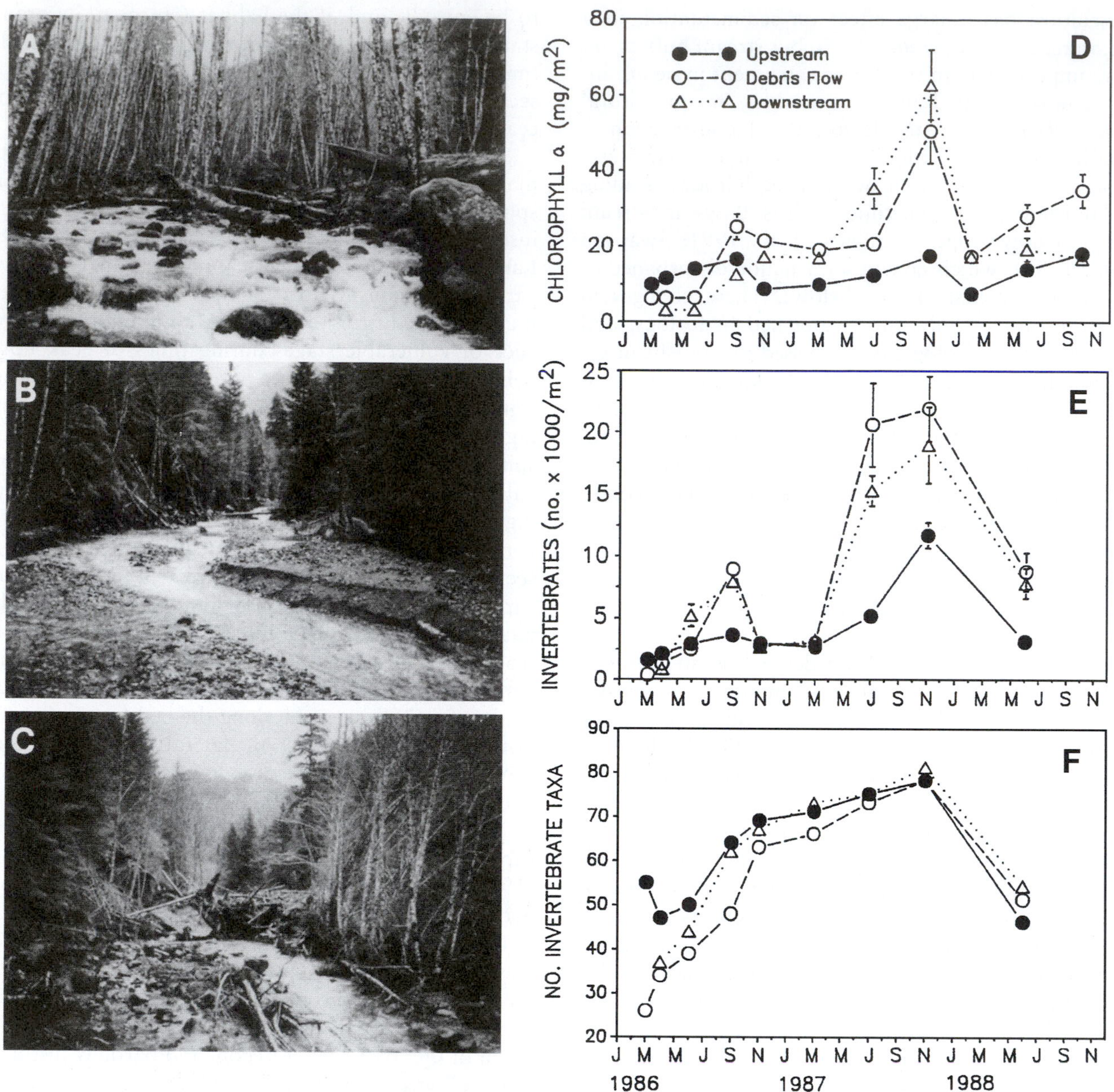

FIGURE 28 Debris flow and its effects in Quartz Creek, Oregon. (A) Upstream reach that was flooded only. (B) Reach most severely impacted by debris flow of February 1986. (C) Downstream of debris flow, showing debris dam and partial impact. (D) Recovery of periphyton chlorophyll *a* in the three reaches. (E) Recovery of macroinvertebrate density (mostly insects). (F) Recovery of macroinvertebrate species richness. [Modified from Lamberti *et al.*, 1991.]

years after the eruption, 141 species were found in the same stream, although nearly half of these were the ubiquitous chironomid midges. Both of these examples attest to the remarkable resilience of stream insects and other invertebrates to natural disturbance.

Episodic natural events contrast, however, with chronic anthropogenic disturbance, such as inputs of chemical and thermal pollutants or prolonged watershed disturbance. Prolonged point-source pollution can have devastating effects on aquatic insects. Wallace and Brady (1971) reported that low concentrations (0.2 ppm) of the insecticide dieldrin in a factory effluent reduced benthic insect densities by over 90% in a South Carolina stream, and that the remaining invertebrates had high tissue concentrations (up to 100 ppm) of dieldrin. Wallace and coworkers later explored the structural and functional significance of reduced insect populations in streams by conducting an experimental release of insecticide (methoxychlor) in an Appalachian Mountain stream. Insecticide application resulted in massive drift of insects, and reduction of benthic densities by over 85% (Wallace *et al.,* 1982, 1989). The remaining benthic invertebrates were composed of an impoverished fauna of mostly noninsects (e.g., oligochaetes) and chironomid midge larvae. Furthermore, important ecological processes in the stream were altered. For example, leaf litter decomposition declined drastically because an important shredding caddisfly, *Lepidostoma*, was eliminated by the insecticide treatment (Whiles *et al.,* 1993). When treatment ceased, *Lepidostoma* recolonized rapidly and leaf litter processing recommenced, again pointing to the resilience of aquatic insect communities. However, human chronic disturbance of aquatic ecosystems and their watersheds tends to preclude recovery of insect communities to predisturbance conditions. Watershed disturbances such as forest logging, agriculture, and urbanization can have long-term impacts on aquatic insect communities by altering energy inputs, thermal regimes, and sediment composition (Gregory *et al.,* 1987; Lamberti and Berg, 1995; Wallace *et al.,* 1988).

VIII. AQUATIC INSECTS IN BIOMONITORING STUDIES

Biomonitoring studies are used to measure response and recovery of aquatic communities to disturbances, protect biodiversity, evaluate compliance, and improve understanding of the relationship between physical, chemical, and biological components (Gurtz, 1994). Many biomonitoring programs incorporate fish, macroinvertebrate, and algae components in their studies, although macroinvertebrate studies, especially those focusing on aquatic insects, are widely used by many agencies. The National Water Quality Assessment (NAWQA) program of the USGS seeks to evaluate water quality at spatial scales ranging from local to national (Gurtz, 1994). Such biomonitoring studies are used to assess aquatic ecosystem health because organisms function as sensors of the quality of their environment in several ways that direct measurements of water quality cannot (Loeb, 1994). Insects have been particularly useful in biomonitoring of streams but also are used in lentic studies (Brinkhurst, 1974; Warwick and Tisdale, 1989) to assess changes in lake water quality and productivity. Warwick (1980) also has used insects in paleolimnological studies to reconstruct changes in land use surrounding a lake.

Stream macroinvertebrates, especially insects, offer distinct advantages in biomonitoring studies (Stewart and Loar, 1994). Their small size and limited mobility renders them relatively easy to sample. They also are known to be sensitive to a variety of pollutants. For example, insects absorb or transform chemical pollutants, such that direct chemical analyses may not be measuring toxicity in the same way as do insects. Insects also integrate environmental quality over a longer time interval than do static measures of water quality. Thus, biomonitoring incorporates and integrates ecology, life history, physiology, and taxonomy into a management and assessment tool. Although valuable and widely used, there are also a few disadvantages to using aquatic insects in these studies (Stewart and Loar, 1994). Spatial and temporal variation in abundance of aquatic life-history stages must be accounted for. Thus, an appropriate sampling design in a biomonitoring program must stratify this spatial and temporal variability to account for as much of it as possible (Hughes *et al.,* 1994). Stratification according to underlying geology is also of critical importance (Minshall *et al.,* 1985; Richards *et al.,* 1996). Furthermore, aquatic insect identification is time-consuming and requires trained personnel; errors in taxonomy can lead to erroneous conclusions (see Lenat and Barbour, 1994; Rosenberg and Resh, 1996).

Rosenberg and Resh (1996) provided a comprehensive review of current biomonitoring approaches that use aquatic insects. These approaches span a range of scales: (1) biochemical and physiological measurements, including metabolic studies and measures of enzyme activities; (2) individual attributes such as morphological, behavioral, and life-history parameters, or use of "sentinel organisms" that serve as bioaccumulators of toxic materials; (3) population- and assemblage-level responses, such as using occurrence or abundance of indicator species as a measure of sensitivity to a pollutant; (4) community-level approaches which synthesize many types of data into summary responses to a pollutant; and (5) ecosystem-level scales that assess effects of stressors on processes and function.

Of these, community-level approaches are probably in most widespread use, and these may be either qualitative or quantitative. Rapid assessment methods are often based on insect communities, are generally qualitative, but permit assessment of many streams with a reasonable degree of effort (see Lenat and Barbour, 1994). However, quantitative approaches are also very common, and the choice between qualitative and quantitative techniques depends on the objectives of the study. Several different metrics of aquatic insect communities are used (Rosenberg and Resh, 1996). Taxa richness is widely used, but is difficult to compare across studies because different levels of resolution tend to be used by different investigators and for different taxonomic groups. EPT richness, which is the combined

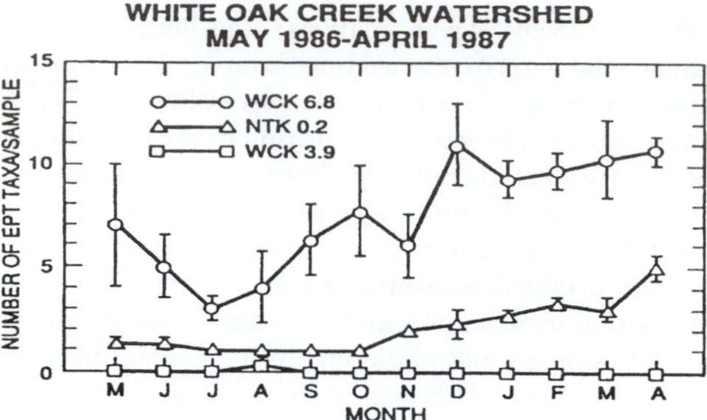

FIGURE 29 Richness of EPT (Ephemeroptera, Plecoptera, and Trichoptera) taxa is robust to seasonal variation within a site, although it varies longitudinally and between streams. [From Stewart and Loar, 1994.]

richness of Ephemeroptera, Plecoptera, and Trichoptera (EPT taxa) takes advantage of the fact that EPT taxa are relatively easy to identify and are pollution-sensitive (Lenat, 1988; Plafkin *et al.*, 1989; Resh *et al.*, 1996). This index varies longitudinally but is robust to seasonal and interannual fluctuations (Fig. 29, Stewart and Loar, 1994). Because chironomids as a group are considered pollution-tolerant and EPT taxa are not, the ratio of EPT/chironomids is another popular community-level metric. Diversity and similarity indexes afford the advantage of being readily comparable between sites, and biotic indexes offer a readily interpretable score (e.g., Hilsenhoff, 1982). Functional feeding group measures rely on functional rather than taxonomic information, and in this sense they bridge community- and ecosystem-level approaches to biomonitoring.

Rather than relying on a single metric for assessment in biomonitoring studies, multimetric and multivariate approaches often are recommended (ter Braak and Prentice, 1988; Karr, 1994; Lenat and Barbour, 1994; Richards *et al.*, 1996; Rosenberg and Resh, 1996). Multimetric approaches can combine the strengths of individual metrics to minimize the chance of an incorrect assessment, and often several metrics can be calculated from the same data set. Multivariate approaches permit partitioning of variance due to different factors, thus providing more useful information than a single measure. They also are less subjective than are univariate approaches. Resh *et al.* (1996) provide a useful exercise that demonstrates the utility of several different biomonitoring approaches that use benthic macroinvertebrates. They stress the importance of combining information on both physical habitat quality and macroinvertebrate composition when assessing water quality. A cornerstone of their approach is that macroinvertebrates, especially insects, provide an integrated picture of the recent history of an aquatic ecosystem.

IX. SUMMARY

The study of aquatic insect ecology is dynamic, with several hundred new papers being added to the literature each year. While it is not possible to provide a comprehensive treatment of aquatic insect ecology in a single chapter, we provide a broad overview of major topics in aquatic insect ecology. Our specific goals were to: (1) describe the basic life histories of aquatic insects and their general community structure; (2) outline their role in major ecological processes involving insects that occur in aquatic ecosystems; (3) discuss how the structure of insect communities and the services they perform are affected by disturbance to their aquatic ecosystems and the changing landscape in which they are found; and (4) summarize how their communities can be used by managers to assess aquatic ecosystem health. This synopsis should provide a useful starting point for those seeking an entry into the rich literature on aquatic insect ecology.

LITERATURE CITED

Allan, J. D. 1982. The effects of reduction in trout density on the invertebrate community of a mountain stream. Ecology 63:1444–1455.

Allan, J. D. 1995. Stream ecology, Chapman & Hall, London., 388 p.

Allan, J. D., Erickson, D. L., Fay, J. 1997. The influence of catchment land us on stram integrity across multiple spatial scales. Freshwater Biology 37:149–161.

Allan, J. D., Feifarek, B. P. 1989. Distances travelled by drifting mayfly nymphs: factors influencing return to the substrate. Journal of the North American Benthological Society 8:322–330.

Allen, K. R. 1951. The Horokiwi Stream. A study of a trout population. New Zealand Department of Fisheries Bulletin 10:1–238.

Anderson, N. H. 1976a. Carnivory by an aquatic detritivore, *Clistoronia magnifica* (Trichoptera: Limnephilidae). Ecology 57:1981–1085.

Anderson, N. H. 1976b. The distribution and biology of the Oregon Trichoptera. Oregon State University Agricultural Experiment Station Technical Bulletin 134, Corvallis, OR, 152 p.

Anderson, N. H. 1992. Influence of disturbance on insect communities in Pacific Northwest streams. Hydrobiologia 248:79–92.

Anderson, N. H., Bourne, J. R. 1974. Bionomics of three species of glossosomatid caddisflies (Trichoptera: Glossosomatidae) in Oregon. Canadian Journal of Zoology 52:405–411.

Anderson, N. H., Hanson, B. P. 1987. An annotated check list of aquatic insects collected at Berry Creek, Benton County, Oregon 1960–1984. Occasional Publication Number 2, Systematic Entomology Laboratory, Department of Entomology, Oregon State University.

Anderson, N. H., Sedell, J. R. 1979. Detritus processing by macroinvertebrates in stream ecosystems. Annual Review of Entomology 24:351–377.

Axler, R. P., Reuter, J. E. 1996. Nitrate uptake by phytoplankton and periphyton: Whole-lake enrichments and mesocosm ^{15}N experiments in an oligotrophic lake. Limnology and Oceanography 41:659–671.

Barnes, H. T. 1928. Ice engineering, Renouf Pub. Co., Montreal., 364 p.

Batzer, D. P., Resh, V. H. 1996. Trophic interactions among a beetle predator, a chironomid grazer, and periphyton in a seasonal wetland. Oikos 60:251–257.

Batzer, D. P., Wissinger, S. A. 1996. Ecology of insect communities in nontidal wetlands. Annual Review of Entomology 41:75–100.

Beeton, A. M. 1961. Environmental changes in Lake Erie. Transactions of the American Fisheries Society 87:73–79.

Benke, A. C. 1979. A modification of the Hynes method for estimating secondary production with particular significance for multivoltine populations. Limnology and Oceanography 24:168–174.

Benke, A. C. 1984. Secondary production of aquatic insects, *in* Resh, V. H., Rosenberg, D. M., Eds., The ecology of aquatic insects. Praeger, New York, pp. 289–322.

Benke, A. C. 1993. Concepts and patterns of invertebrate produciton in running waters. Internationale Verinigung fur theoretische und angewandte Limnologie, Verhandlungen 25:15–38.

Benke, A. C. 1996. Secondary production of macroinvertebrates, *in* Hauer, F. R., Lamberti, G. A., Eds., Methods in stream ecology. Academic Press, San Diego, CA, pp. 557–578.

Benke, A. C. 1998. Production dynamics of riverine chironomids: extremely high biomass turnover rates of primary consumers. Ecology 79:899–910.

Benke, A. C., Wallace, J. B. 1980. Trophic basis of production among net-spinning caddisflies in a southern Appalachian stream. Ecology 61:108–118.

Benke, A. C., Van Arsdall, T. C., Jr., Gillespie, D. M., Parrish, F. K. 1984. Invertebrate poductivity in a subtropical blackwater river: the importance of habitat and life hitory. Ecological Monographs 54:25–63.

Benke, A. C., Hunter, R. J., Parrish, F. K. 1986. Invertebrate drift dynamics in a subtropical blackwater river. Journal of the North American Benthological Society 5:173–190.

Benke, A. C., Jacobi, D. I. 1994. Production dynamics and resource utilization of snag-dwelling mayflies in a blackwater river. Ecology 75:1219–1232.

Benke, A. C., Wallace, J. B. 1997. Trophic basis of production among riverine caddisflies: implications for food web analyses. Ecology 78:1132–1145.

Berg, M. B., Hellenthal, R. A. 1992. Role of Chironomidae in energy flow of a lotic ecosystem. Netherlands Journal of Aquatic Ecology 26:471–476.

Bergey, E. A., Resh, V. H. 1994. Interactions between a stream caddisfly and the algae on its case: factors affecting algal quantity. Freshwater Biology 31:153–163.

Bilby, R. E., Ward, J. W. 1989. Changes in characteristics and function of woody debris with increasing stream size in western Washington. Transactions of American Fisheries Society 118:368–378.

Bird, G. A., Hynes, H. B. N. 1981. Movement of immature insects in a lotic habitat. Hydrobiologia 77:103–112.

Bisson, P. A., Bilby, R. E., Bryant, M. D., Dolloff, C. A., Brette, G. B., House, R. A., Murphy, M. L., Koski, K. V., Sedell, J. R. 1987. Large woody debris in forested streams in the Pacific Northwest: past, present, and future, *in* Salo, E. O., Cundy, T. W., Eds., Streamside management: forestry and fishery interactions. University of Washington Institute of Forest Resources Contribution No. 57, pp. 143–190.

Bohannan, R. E., Johnson, D. M. 1983. Response of littoral invertebrate populations to a spring fish exclusion experiment. Freshwater Invertebrate Biology 2:28–40.

Borror, D. J., Triplehorn, C. A., Johnson, N. F. 1989. An introduction to the study of insects, 6th ed., Saunders, Philadelphia, PA.

Brickell, D. C., Goering, J. J. 1970. Chemical effects of salmon decomposition on aquatic ecosystems, *in* Murphey, R. S., Ed. First International Symposium of Water Pollution Control in Cold Climates. U. S. Government Printing Office, Washington, DC, pp. 125–138.

Brinkhurst, R. O. 1974. The benthos of lakes. St. Martin's press, New York, NY.

Brock, E. M. 1960. Mutualism between the midge *Cricotopus* and the alga *Nostoc*. Ecology 41:474–483.

Brock, T. D. 1978. Thermophilic microorganisms and life at high temperatures, Springer, New York, NY. 465 p.

Brown, K. S. 1996. Do disease cycles follow changes in the weather? BioScience 46:479–481.

Brundin, L. 1949. Chironomide und andere Bodentiere der sudschwedischen Urgebirgsseen. Rep. Inst. Freshwat. Res. Drottningholm 30:1–94.

Butler, M. G. 1982. A 7-year life cycle for two *Chironomus* species in arctic Alaskan tundra ponds (Diptera: Chironomidae). Canadian Jounral of Zoology 60:58–70.

Butler, M. G. 1984. Life histories of aquatic insects, *in* Resh, V. H., Rosenberg, D. M., Eds., The ecology of aquatic insects. Praeger, New York, NY, pp. 24–55.

Cairns, J. Jr. 1990. Lack of theoretical basis for predicting rate and pathways of recovery. Environmental Management 14: 517–526.

Cantrall, I. J., Brusven, M. A. 1996. Semiaquatic Orthoptera, *in* Merritt, R. W., Cummins, K. W., Eds., 1996. An introduction to the aquatic insects of North America, 3rd ed. Kendall/Hunt, Dubuque, IA, pp. 212–216.

Carpenter, S. R. 1982. Stemflow chemistry: effects on population dynamics of detritivorous mosquitoes in tree-hole ecosystems. Oecologia 53:1–6.

Carpenter, S. R. 1983. Resource limitation of larval treehole mosquitoes subsisting on beech detritus. Ecology 64:219–223.

Carpenter, S. R., Kitchell, J. F., Hodgson, J. R. 1985. Cascading trophic interactions and lake productivity. BioScience 35: 634–639.

Carpenter, S. R., Caraco, N., Correll, D., Howarth, R., Sharpley, A., Smith, V. 1998. Nonpoint pollution of surface waters with phosphorus and nitrogen. Issues in Ecology No. 1. Ecological Society of America.

Chapman, D. W., Demory, R. L. 1963. Seasonal changes in the food ingested by aquatic insect larvae and nymphs in two Oregon streams. Ecology 44:140–146.

Chauvet, E., Décamps, H. 1989. Lateral interaction in a fluvial landscape: the River Garonne, France. Journal of the North American Benthological Society 8:9–17.

Closs, G. P., Lake, P. S. 1994. Spatial and temporal variation in the structure of an intermediate stream food web. Ecological Monographs 64:1–21.

Connell, J. H. 1978. Diversity in tropical rain forests and coral reefs. Science 199:1302–1310.

Cooper, S. D., Walde, S. J., Peckarsky, B. L. 1990. Prey exchange rates and the impact of predation in streams. Ecology 71: 1503–1514.

Couch, C. A., Meyer, J. L., Hall, R. O. 1996. Incorporation of bacterial extracellular polymer by blackfly larvae (Simuliidae). Journal of the North American Benthological Society 15:289–299.

Crawley, M. J. 1983. Herbivory. The dynamics of animal–plant interactions, Univ. of California Press, Berkeley, CA., 437 p.

Cronin, G., Wissing, K. D., Lodge, D. M. 1998. Comparative feeding selectivity of herbivorous insects on water lilies: aquatic vs semi-terrestrial insects and submersed vs floating leaves. Freshwater Biology 39:243–57.

Crowder, L. B., Cooper W. E. 1982. Habitat structural complexity and the interaction between bluegills and their prey. Ecology 63:1802–1813.

Cummins, K. W. 1973. Trophic relations of aquatic insects. Annual Review of Entomology 18:183–206.

Cummins, K. W. 1974. Structure and function of stream ecosystems. BioScience 24:631–641.

Cummins, K. W., Klug, M. J. 1979. Feeding ecology of stream invertebrates. Annual Review of Ecology and Systematics 10:147–172.

Cummins, K. W., Merritt, R. W. 1996. Ecology and distribution of aquatic insects, *in* Merritt, R. W., Cummins, K. W., Eds., An introduction to the aquatic insects of North America, 3rd ed. Kendall/Hunt, Dubuque, IA, pp. 74–86.

Cummins, K. W., Wilzbach, M. A., Gates, D. M., Perry, J. B., Taliaferro, W. B. 1989. Shredders and riparian vegetation. BioScience 39:24–30.

Dahm, C. N. 1981. Pathways and mechanisms for removal of dissolved organic carbon from leaf leachate in streams. Canadian Journal of Fisheries and Aquatic Sciences 38:68–76.

Danks, H. V. 1978. Modes of seasonal adaptations in the insects. I. Winter survival. Canadian Entomologist 110:1167–1205.

Deegan, L. A., Peterson, B. J. 1992. Whole river fertilization stimulates fish production in an arctic tundra river. Canadian Journal of Fisheries and Aquatic Sciences 54:269–283.

Delucchi, C. M. 1989. Movement patterns of invertebrates in temporary and permanent streams. Oecologia 78:199–207.

Dodds, W. K. 1989. Photosynthesis of two morphologies of *Nostoc parmelioides* (Cyanobacteria) as related to current velocities and diffusion patterns. Journal of Phycology 25:258–262.

Dodds, W. K., Marra, J. L. 1989. Behaviors of the midge, *Cricotopus* (Diptera: Chironomidae), related to mutualism with *Nostoc parmelioides* (Cyanobacteria). Aquatic Insects 11:201–208.

Dudley, T. L., Anderson, N. H. 1982. A survey of invertebrates associated with wood debris in aquatic habitats. Melanderia 39:1–21.

Duffy, W. G., Liston, C. R. 1985. Survival following exposure to subzero temperatures and respiration in cold acclimatized larvae of *Enallagma boreale* (Odonata: Zygoptera). Freshwater Invertebrate Biology 4:1–7.

Eriksen, C. H., Resh, V. H., Lamberti, G. A. 1996. Aquatic insect respiration. *in* Merritt, R. W., Cummins, K. W., Eds. An introduction to the aquatic insects of North America, 3rd ed. Kendall/Hunt, Dubuque, IA, pp. 29–40.

Feminella, J. W., Hawkins, C. P. 1995. Interactions between stream herbivores and periphyton: a quantitative analysis of past experiments. Journal of the North American Benthological Society 14:465–509.

Fisher, S. G., Gray, L. J. 1983. Secondary production and organic matter processing by collector macroinvertebrates in a desert stream. Ecology 64:1217–1224.

Fisher, S. G., Gray, L. J., Grimm, N. B., Busch, D. E. 1982. Temporal succession in a desert stream ecosystem following flash flooding. Ecological Monographs 52:93–110.

Fisher, S. G., Likens, G. E. 1973. Energy flow in Bear Brook, New Hampshire: an integrative approach to stream ecosystem metabolism. Ecological Monographs 43:421–439.

Fry, B. 1991. Stable isotope diagrams of freshwater food webs. Ecology 72:2293–2297.

Fry, B., Sherr, E. 1984. $\delta^{13}C$ measurements as indicators of carbon flow in marine and freshwater ecosystems. Contributions in Marine Science 27:13–47.

Gallepp, G. W. 1979. Chironomid influence on phosphorus release in sediment–water microcosms. Ecology 60:547–556.

Gilinsky, E. 1984. The role of predation and spatial heterogeneity in determining benthic community structure. Ecology 65: 455–468.

Glime, J. M., Clemons, R. M. 1972. Species diversity of stream insects on *Fontinalis* spp. compared to diversity on artificial substrates. Ecology 53:458–464.

Gotceitas, V., Clifford, H. F. 1983. The life history of *Dicosmoecus atripes* (Hagen) (Limnephilidae: Trichoptera) in a Rocky Mountain stream of Alberta, Canada. Canadian Journal of Zoology 61:586–596.

Gregory, S. V. 1983. Plant–herbivore interactions in stream systems. *in* Barnes, J. R., Minshall, G. W., Eds. Stream ecology. Plenum, New York, NY, pp. 157–189.

Gregory, S. V., Ashkenas, L. R. 1990. Field guide for riparian management. Willamette National Forest. U.S.D.A. Forest Service, Pacific Northwest Region, 65 pp.

Gregory, S. V., Lamberti, G. A., Erman, D. C., Koski, K. V., Murphy, M. L., Sedell, J. R. 1987. Influence of forest practices on aquatic production, *in* Salo, E. O., Cundy, T. W., Eds., Streamside management: forestry and fishery interactions. University of Washington Institute of Forest Resources Contribution No. 57, pp. 233–255.

Gregory, S. V., Swanson, F. J., McKee, W. A., Cummins, K. W. 1991. An ecosystem perspective of riparian zones. BioScience 41:540–552.

Grubaugh, J. W., Wallace, J. B., Houston, E. S. 1997. Production of benthic macroinvertebrate communities along a southern Appalachian river continuum. Freshwater Biology 37:581–96.

Gurtz, M. E. 1994. Design considerations for bioogical components of the National Water-Quality Assessment (NAWQA) Program, *in* Loeb, S. L., Spacie, A., Eds., Biological monitoring of aquatic systems. Lewis Pub., Boca Raton, FL, pp. 323–356.

Hall, R. J., Driscoll, C. T., Likens, G. E. 1987. Importance of hydrogen ions and aluminum in regulating the structure and function of stream ecosystems: an experimental test. Freshwater Biology 18:17–43.

Hart, D. D., Resh, V. H. 1980. Movement patterns and foraging ecology of a stream caddisfly larva. Canadian Journal of Zoology 58:1174–1185.

Hauer, F. R., Benke, A. C. 1987. Influence of temperature and river hydrograph on blackfly growth rates in a subtropical blackwater river. Journal of the North American Benthological Society 6:251–261.

Hauer, F. R., Hill, W. R. 1996. Temperature, light, and oxygen, *in* Hauer, F. R., Lamberti, G. A., Eds., Methods in stream ecology. Academic Press, San Diego, CA, pp. 93–106.

Hawkins, C. P., Furnish, J. K. 1987. Are snails important competitors in stream ecosystems? Oikos 49:209–220.

Hawkins, C. P., Murphy, M. L., Anderson, N. H. 1982. Effects of canopy, substrate composition, and gradient on the structure of macroinvertebrate communities in Cascade Range streams of Oregon. Ecology 63:1840–1856.

Hawkins, C. P., Sedell, J. R. 1981. Longitudinal and seasonal changes in functional organization of macroinvertebrate communities in four Oregon streams. Ecology 62:387–397.

Hayden, W., Clifford, H. F. 1974. Seasonal movements of the mayfly *Leptophlebia cupida* (Say) in a brown-water stream of Alberta, Canada. American Midland Naturalist 91:90–102.

Hershey, A. E. 1985. Effects of predatory sculpin on the chironomid communities in an arctic lake. Ecology 66:1131–1138.

Hershey, A. E. 1986. Selective predation by *Procladius* in an arctic Alaskan Lake. Canadian Journal of Fisheries and Aquatic Sciences. 43(12) 2523–2528.

Hershey, A. E. 1987. Tubes and foraging behavior in larval Chironomidae: implications for predator avoidance. Oecologia 73:236–241.

Hershey, A. E., Dodson, S. I. 1987. Predator avoidance by *Cricotopus*: cyclomorphosis and the importance of being big and hairy. Ecology 68:913–920.

Hershey, A. E, Hiltner, A. L., Hullar, M. A. J., Miller, M. C., Vestal, J. R., Lock, M. A., Rundle, S., Peterson, B. J. 1988. Nutrient influence on a stream grazer: *Orthocladius* microcommunities track nutrient input. Ecology 69:1383–1392.

Hershey, A. E., Pastor, J., Peterson, B. J., Kling, G. W. 1993. Stable isotopes resolve the drift paradox for *Baetis* mayflies in an arctic river. Ecology 74:2315–2325.

Hershey, A. E., Shannon, L., Axler, R., Ernst, C., Mickelson. P., 1995a. Effects of methoprene and *Bti* (*Bacillus thuringiensis* var *israelensis*) on non-target insects. Hydrobiologia 308:219–227.

Hershey, A. E., Merritt, R. W., Miller, M. C. 1995b. Trophic dynamics, diversity, and life history features of blackflies (Diptera: Simuliidae) in arctic Alaskan streams, *in* Chapin, S. F., Kroner, C., Eds., Arctic and alpine biodiversity. Springer, New York, pp. 283–295.

Hershey, A. E., Merritt, R. W., Miller, M. C., McCrea, J. S. 1996. Organic matter processing by larval black flies in a temperate woodland stream. Oikos 75:524–532.

Hershey, A. E., Peterson, B. J. 1996. Stream food webs, *in* Hauer, F. R., Lamberti, G. A., Eds., Methods in Stream Ecology. Academic Press, San Diego, CA, pp. 511–530.

Hershey, A. E., Bowden, W. B., Deegan, L. A., Hobbie, J. E., Peterson, B. J., Kipphut, G. W., Kling, G. W., Lock, M. A., Merritt, R. W., Miller, M. C., Vestal, J. R., Schuldt, J. A. 1997. The Kuparuk River: a long-term study of biological and chemical processes in an Arctic river, *in* Milner, A. M., Oswood, M. W., Eds., Freshwaters of Alaska. Springer, New York, pp. 107–129.

Hershey, A. E., Lamberti, G. A. 1998. Stream macroinvertebrate communities, *in* Bilby, R. E., Naiman, R. J., Eds., Ecology and management of streams and rivers in the Pacific Northwest coastal regions. Springer, New York, pp. 169–199.

Hershey, A. E., Lima, A. R., Niemi, G. J., Regal, R. R. 1998. Effects of methoprene and Bti (*Bacillus thuringiensis* var. *israelensis*) on non-target macroinvertebrates in Minnesota wetlands. Ecological Applications 8:41–60.

Hershey, A. E., Shannon, L., Regal, R., Lima, A., Niemi, G. 1999. Prairie wetlands of southcentral Minnesota: Effects of drought on invertebrate communities, *in* Batzer, D., Wissinger, S., Eds., Invertebrates in freshwater wetlands of North America: Ecology and management. John Wiley and Sons, pp. 515–541.

Hilsenhoff, W. L. 1966. The biology of *Chironomus plumosus* (Diptera: Chironomidae) in Lake Winnebago, Wisconsin. Annals of the Entomological Society of America 59:465–473.

Hilsenhoff, W. L. 1982. Using a biotic index to evaluate water quality in streams. Wisconsin Department of Natural Resources Technical Bulletin 132:1–22.

Hiltner, A. L., Hershey, A. E. 1992. Black fly (Diptera: Simuliidae) response to phosphorus enrichment in an arctic tundra stream. Hydrobiologia 240:259–265.

Hinterleitner-Anderson, D., Hershey, A. E., Schuldt, J. A. 1992. The effects of river fertilization on mayfly (*Baetis* sp.) drift patterns and population density in an arctic river. Hydrobiologia 240:247–258.

Hughes, R. M., Heiskary, S. A., Matthews, W. J., Yoder, C. O. 1994. Use of ecoregions in biological monitoring, *in* Loeb, S. L., Spacie, A., Eds., Biological monitoring of aquatic systems. Lewis Pub., Boca Raton, FL, pp. 125–154.

Huryn, A. D. 1996. An appraisal of the Allen paradox in a New Zealand trout stream. Limnology and Oceanography 41: 243–352.

Huryn, A. D., Wallace, J. B. 1986. A method for obtaining in situ growth rates of larval Chironomidae (Diptera) and its application to studies of secondary production. Limnology and Oceanography 31:216–222.

Hynes, H. B. N. 1970. The ecology of running waters. Univ. of Toronto Press, Toronto, Ontario. 555 p.

Hynes, H. B. N. 1975. The stream and its valley. Internationale verein für Theoretische und Angewandte Limnologie Verhandlungen 19:1–15.

Jacobs-Lorena, M., Lemos, F. J. A. 1995. Immunological strategies for control of insect disease vectors: a critical assessment. Parasitology Today 11:144–147.

Johnson, D. M., Martin, T. H., Crowley, P. H., Crowder, L. B. 1996. Link strength in lake littoral food webs: net effects of small sunfish and larval dragonflies. Journal of the North American Benthological Society 15:271–288.

Junger, M., Planas, D. 1994. Quantitative use of stable carbon isotope analysis to determine the trophic basis of invertebrate communities in a boreal forest lotic system. Canadian Journal of Fisheries and Aquatic Sciences 51:52–61.

Kajak, Z. 1980. Role of invertebrate predators (mainly *Procladius*) in benthos, *in* Murray, D., Ed. Chironomidae: ecology, systematics, cytology, and physiology. Pergamon, New York, pp. 339–348.

Karr, J. R. 1994. Biological monitoring: Challenges for the future, *in* Loeb, S. L., Spacie, A., Eds., Biological monitoring of aquatic systems. Lewis Pubs., Boca Raton, FL, pp. 357–373.

Kidwell, M. G., Wattam, A. R. 1998. An important step forward in genetic manipulation of mosquito vectors of human disease. Proceedings of the National Academy of Sciences 95:3349–3350.

Kline, T. C., Goering, J. J., Mathisen, O. A., Poe, P. H. 1990. Recycling of elements transported upstream by runs of pacific salmon: 1. ^{15}N and ^{13}C evidence in Sashin Creek, southeastern Alaska. Canadian Journal of Fisheries and Aquatic Sciences 47:136–144.

Kohler, S. L. 1985. Identification of stream drift mechanisms: an experimental and observational approach. Ecology 66:1749–1761.

Lamberti, G. A. 1993. Grazing experiments in artificial streams, *in* Lamberti, G. A., Steinman, A. D., Eds., Research in artificial streams: applications, uses, and abuses. Journal of the North American Benthological Society 12:313–384, pp. 337–342.

Lamberti, G. A. 1996. The role of periphyton in benthic food webs, *in* Stevenson, R. J., Bothwell, M. L., Lowe, R. L., Eds., Algal ecology: freshwater benthic ecosystems. Academic Press, San Diego, CA, pp. 533–572.

Lamberti, G. A., Ashkenas, L. R., Gregory, S. V., Steinman, A. D. 1987a. Effects of three herbivores on periphyton communities in laboratory streams. Journal of the North American Benthological Society 6:92–104.

Lamberti, G. A., Berg, M. B. 1995. Invertebrates and other benthic features as indicators of environmental change in Juday Creek, Indiana. Natural Areas Journal 15:249–258.

Lamberti, G. A., Feminella, J. W., Resh, V. H. 1987b. Herbivory and intraspecific competition in a stream caddisfly population. Oecologia 73:75–81.

Lamberti, G. A., Gregory, S. V., Ashkenas, L. R., Wildman, R. C., Moore, K. M. S. 1991. Stream ecosystem recovery following a catastrophic debris flow. Canadian Journal of Fisheries and Aquatic Sciences 48:196–208.

Lamberti, G. A., Gregory, S. V., Hawkins, C. P., Wildman, R. C., Ashkenas, L. R., DeNicola, D. M. 1992. Plant-herbivore interactions in streams near Mount St Helens. Freshwater Biology 27:237–247.

Lamberti, G. A., Moore, J. W. 1984. Aquatic insects as primary consumers, *in* Resh, V. H., Rosenberg, D. M., Eds., The ecology of aquatic insects. Praeger, New York, pp. 164–195.

Lamberti, G. A., Resh, V. H. 1979. Substrate relationships, spatial distribution patterns, and sampling variability in a stream caddisfly population. Environmental Entomology. 8:561–567.

Lamberti, G. A., Resh, V. H. 1983a. Geothermal effects on stream benthos: separate influences of thermal and chemical components on periphyton and macroinvertebrates. Canadian Journal of Fisheries and Aquatic Sciences 40:1995–2009.

Lamberti, G. A., Resh, V. H. 1983b. Stream periphyton and insect herbivores: an experimental study of grazing by a caddisfly population. Ecology 64:1124–1135.

Lamberti, G. A., Resh, V. H. 1985. Distribution of benthic algae and macroinvertebrates along a thermal stream gradient. Hydrobiologia 128:13–21.

Lange, W. H. 1996. Aquatic and semiaquatic Lepidoptera, *in* Merritt, R. W., Cummins, K. W., Eds., An introduction to the aquatic insects of North America, 3rd ed. Kendall/Hunt, Dubuque, IA, pp. 387–398.

Lenat, D. R. 1988. Water quality assessment of streams using a qualitative collection method for benthic macroinvertebrates. Journal of North American Benthological Society 7:222–233.

Lenat, D. R., Barbour, M. T. 1994. Using benthic macroinvertebrate community structure for rapid, cost effective, water quality monitoring: Rapid bioassessment, *in* Loeb, S. L., Spacie, A., Eds., Biological monitoring of aquatic systems. Lewis Pubs., Boca Raton, FL, pp. 187–215.

Li, J. L., Gregory, S. V. 1989. Behavioral changes in the herbivorous caddisfly *Dicosmoecus gilvipes* (Limnephilidae). Journal of the North American Benthological Society 8:250–259.

Lock, M. A., Wallace, R. R., Costerton, J. W., Ventullo, R. M., Charlton, S. E. 1984. River epilithon: toward a structural-functional model. Oikos 42:10–22.

Lodge, D. M. 1991. Herbivory on freshwater macrophytes. Aquatic Botany 41:195–224.

Loeb, S. L. 1994. An ecological context for biological monitoring, *in* Loeb, S. L., Spacie, A., Eds., Biological monitoring of aquatic systems. Lewis Pubs., Boca Raton, FL, pp. 3–7.

Lutz, P. E. 1974. Effects of temperature and photoperiod on larval development in *Tetragoneurai cynosura* (Odonata: Libellulidae). Ecology 55:370–377.

Madsen, B. L., Bengtson, J., Butz, I. 1973. Observations on upstream migration by imagines of some Plecoptera and Ephemeroptera. Limnology and Oceanography 18:678–681.

Malmqvist, B. 1993. Interactions in stream leaf packs: effects of a stonefly predator on detritivores and organic matter processing. Oikos 66:454–462.

Mathisen, O. A., Parker, P. L., Goering, J. J., Kline, T. C., Poe, P. H., Scalan, R. S. 1988. Recycling of marine elements transported into freshwater by anadromous salmon. Verhandlungen der Internationalen Vereinigung für Theoretische und Angewandte Limnologie 23:2249–2258.

Mayer, M. S., Likens, G. E. 1987. The importance of algae in a shaded headwater stream as food for an abundant caddisfly (Trichoptera). Journal of the North American Benthological Society 6:262–269.

McAuliffe, J. R. 1984. Competition for space, disturbance, and the structure of a benthic stream community. Ecology 65:894–908.

McCafferty, W. P. 1981. Aquatic entomology: the fisherman's and ecologist's illustrated guide to insects and their relatives, Jones and Bartlett, Boston, MA, 447 pp.

McIntire, C. D. 1973. Periphyton dynamics in laboratory streams: a simulation model and its implications. Ecological Monographs 43:399–420.

McIntire, C. D., Colby, J. A. 1978. A hierarchical model of lotic ecosystems. Ecological Monographs 48:167–190.

McPeek, M. A., Peckarsky, B. L. 1998. Life histories and the strengths of species interactions: combining mortality, growth, and fecundity effects. Ecology 79:867–879.

McPeek, M. A., Schrot, A. K., Brown, J. M. 1996. Adaptation to predators in a new community: swimming performance and predator avoidance in damselflies. Ecology 77:617–629.

Merritt, R. W., Cummins, K. W. 1996a. Trophic relations of macroinvertebrates, *in* Hauer, F. R., Lamberti, G. A., Eds., Methods in stream ecology. Academic Press, San Diego, CA, pp. 453–474.

Merritt, R. W., Cummins, K. W., Eds. 1996b. An introduction to the aquatic insects of North America, 3rd ed., Kendall/Hunt, Dubuque, IA, 862 p.

Merritt, R. W., Dadd, R. H., Walker, E. D. 1992. Feeding behavior, natural food, and nutritional relationships of larval mosquitoes. Annual Review of Entomology 37:349–376.

Merritt, R. W., Lawson, D. L. 1992. The role of macroinvertebrates in stream–floodplain dynamics. Hydrobiologia 248:65–77.

Meyer, J. L. 1990. A blackwater perspective on riverine ecosystems. BioScience 40:643–651.

Mihuc, T. B. 1997. The functional trophic role of lotic primary consumers: generalist versus specialist strategies. Freshwater Biology 37:455–462.

Miller, M. C., Hershey, A. E., Merritt, R. W. 1998. Feeding behavior of larval black flies and FPOM retention in a high gradient blackwater stream. Canadian Journal of Zoology 76:228–235.

Minagawa, M., Wada, E. 1984. Stepwise enrichment of ^{15}N along food chains: further evidence and the relation between $\delta^{15}N$ and animal age. Geochimica et Cosmochimica Acta 48:1135–1140.

Minshall, G. W. 1978. Autotrophy in stream ecosystems. Bioscience 28:767–771.

Minshall, G. W., Cummins, K. W., Petersen, R. C., Cushing, C. E., Bruns, D. A., Sedell, J. R., Vannote, R. L. 1985. Developments in stream ecosystem theory. Canadian Journal of Fisheries and Aquatic Sciences 42:1045–1055.

Minshall, G. W., Petersen, R. C., Cummins, K. W., Bott, T. L., Sedell, J. R., Cushing, C. E., Vannote, R. L. 1983. Interbiome comparison of stream ecosystem dynamics. Ecological Monographs 53:1–25.

Mittelbach, G. G. 1988. Competition among refuging sunfishes and effects of fish density on littoral zone invertebrates. Ecology 69:614–623.

Morse, J. C., Chapin, J. W., Herlong, D. D., Harvey, R. S. 1983. Aquatic insects of Upper Three Runs Creek, Savannah River Plant, South Carolina, Part II: Diptera. Journal of the Georgia Entomological Society 18:303–316.

Müller, K. 1954. Investigations on the organic drift in North Swedish streams. Institute of Freshwater Research, Drottingholm. Report No. 34:133–148.

Naiman, R. J., Anderson, E. C. 1996. Streams and rivers of the coastal temperate rain forest of North America: Physical and biological variability, *in* Schoomaker, P. K., von Hagen, B., Eds., The rain forests of home: an exploration of people and place. Island Press, Washington, DC, pp. 131–148.

Naiman, R. J., Beechie, T. J., Benda, L. E., Berg, D. R., Bisson, P. A., MacDonald, L. H., O'Connor, M. D., Olson, P. L., Steel, E. A. 1993. Fundamental elements of ecologically healthy watersheds in the Pacific Northwest coastal ecoregion, *in* Naiman, R. J., Ed., Watershed management. Springer, New York, pp. 127–188.

Naiman, R. J., Melillo, J. M., Lock, M. A., Ford, T. E., Reice, S. R. 1987. Longitudinal patterns of ecosystem processes and community structure in a subarctic river continuum. Ecology 68:1138–1156.

Newbold, J. D., Elwood, J. W., O'Neill, R. V., Sheldon, A. L. 1983. Phosphorus dynamics in a woodland stream ecosystem: a study of nutrient spiralling. Ecology 64:1249–1265.

Newman, R. M. 1991. Herbivory and detritivory on freshwater macrophytes by invertebrates: a review. Journal of North American Benthological Society 10:89–114.

O'Hop, J., Wallce, J. B., Haefner, J. D. 1984. Production of a stream shredder, *Peltoperla maria* (Plecoptora: Peltoperlidae) in disturbed and undisturbed hardwood catchments. Freshwater Biology 14:13–21.

Peckarsky, B. L. 1980. Predator-prey interactions between stoneflies and mayflies: behavioral observations. Ecology 61:932–943.

Peckarsky, B. L. 1982. Aquatic insect predator-prey relations. BioScience 32:261–266.

Peckarsky, B. L. 1996. Alternative predator avoidance syndromes of stream-dwelling mayfly larvae. Ecology 77:1888–1905.

Peckarsky, B. L., Cooper, S. D., McIntosh, A. R. 1997. Extrapolating from individual behavior to populations and communities in stream organisms. Journal of the North American Benthological Society 16:375–390.

Peckarsky, B. L., Cowan, C. A., Penton, M. A., Anderson, C. 1993. Sublethal consequences of stream-dwelling predatory stoneflies on mayfly growth and fecundity. Ecology 74:1836–1846.

Peckarsky, B. L., Dodson, S. I. 1980. Do stonefly predators influence benthic distributions in streams? Ecology 61:1275–1282.

Petersen, R. C. 1986. In situ particle generation in a southern Swedish stream. Limnology and Oceangraphy 31:432–437.

Peterson, B. J., Deegan, L., Helfrich, J., Hobbie, J., Hullar, M., Moller, B., Ford, T., Hershey, A., Hiltner, A., Kipphut, G., Lock, M. A., Fiebig, D. M., McKinley, V., Miller, M. C., Vestal, J. R., Ventullo, R., Volk, G. 1993. Biological responses of a tundra river to fertilization. Ecology 74:653–672.

Peterson, B. J., Fry, B. 1987. Stable isotopes in ecosystem studies. Annual Review of Ecology and Systematics 18:293–320.

Plafkin, J. L., Barbour, M. T., Porter, K. D., Gross, S. K., Hughes, R. M. 1989. Rapid bioassessment protocols for use in streams and rivers. U. S. EPA, Office of Water, EPA/444/4-89-001.

Pritchard, G. 1983. Biology of Tipulidae. Annual Review of Entomology 28:1–22.

Pritchard, G. 1991. Insects in thermal springs. Memoirs of the Entomological Society of Canada 155:89–106.

Pringle, C. M. 1985. Effects of chironomid (Insecta: Diptera) tube building activities on stream diatom communities. Journal of Phycology 21:185–194.

Power, M. E. 1990. Effects of fish in river food webs. Science 250:811–814.

Rader, R. B., Ward, J. V. 1987. Resource utilization, overlap and temporal dynamics in a guild of mountain stream insects. Freshwater Biology 18:521–528.

Ramírez, A., Pringle, C. M. 1998. Structure and production of a benthic insect assemblage in a neotropical stream. Journal of the North American Benthological Society 17:443–463.

Rasmussen, J. B. 1988. Habitat requirements of burrowing mayflies (Ephemeridae: *Hexagenia*) in lakes, with special reference to the effects of eutrophication. Journal of the North American Benthological Society 7:51–64.

Reice, S. R. 1983. Predation and substratum: factors in lotic community structure, *in* Fontaine, T. F. III, Bartell, S. M., Eds., Dynamics of lotic ecosystems. Ann Arbor Science, Ann Arbor, MI, pp. 325–345.

Renshaw, M., Elias, M., Maheswary, N. P., Hassan, M. M., Silver, J. B., Birley, M. H. 1996. A survey of larval and adult mosquitoes on the flood plains of Bangladesh, in relation to flood-control activities. Annals of Tropical Medicine and Parasitology 90:621–634.

Resh, V. H. 1992. Year-to-year changes in the age structure of a caddisfly population following loss and recovery of a springbrook habitat. Ecography 15:314–317.

Resh, V. H., Brown, A. V., Covich, A. P., Gurtz, M. E., Li, H. W., Minshall, G. W., Reice, S. R., Sheldon, A. L., Wallace, J. B., Wissmar, R. 1988. The role of disturbance in stream ecology. Journal of the North American Benthological Society 7:433–455.

Resh, V. H., Flynn, T. S., Lamberti, G. A., McElravy, E. P., Sorg, K. L., Wood, J. R. 1981. Responses of the sericostomatid caddisfly *Gumaga nigricula* (McL.) to environmental disruption. Series Entomologica 20:311–318.

Resh, V. H., Houp, R. E. 1986. Life history of the caddisfly *Dibusa angata* and its association with the red alga *Lemanea australis*. Journal of the North American Benthological Society 5:28–40.

Resh, V. H., Lamberti, G. A., Wood, J. R. 1984. Biology of the caddisfly *Helicopsyche borealis* (Hagen): a comparison of North American populations. Freshwater Invertebrate Biology 3:172–180.

Resh, V. H., Myers, M. J., Hannaford, M. J. 1996. Macroinvertebrates as biotic indicators of environmental quality, *in* Hauer, F. R., Lamberti, G. A., Eds., Methods in stream ecology. Academic Press, San Diego, CA, pp. 647–667.

Richards, C., Host, G. E. 1994. Examining land use influences on stream habitats and macroinvertebrates: a GIS approach. Water Resources Bulletin 30:729–738.

Richards, C., Host, G. E., Arthur, J. W. 1993. Identification of predominant environmental factors structuring stream macroinvertebrate communities within a large agricultural catchment. Freshwater Biology 29:285–294.

Richards, C., Johnson, L. B., Host, G. 1996. Landscape scale influences on stream habitats and biota. Canadian Journal of Fisheries and Aquatic Sciences 53:295–311.

Richards, C., Minshall, G. W. 1992. Spatial and temporal trends in stream macroinvertebrate species assemblages: the influence of watershed disturbance. Hydrobiologia 241:173–184.

Richardson, C. J. 1979. Primary productivity values in freshwater wetlands, *in* Greeson, P. E., Clark, J. R., Clark, J. E. Eds., Wetland function and values: the state of our understanding . American Water Resources Association, Minneapolis, MN, pp. 131–145.

Richardson, J. S. 1993. Limits to productivity in streams: evidence from studies of macroinvertebrates. Canadian Special Publications in Fisheries and Aquatic Sciences 118:9–15.

Richey, J. E., Perkins, M. A., Goldman, C. R. 1975. Effects of Kokanee salmon (*Oncorhynchus nerka*) decomposition on the ecology of a subalpine stream. Journal of the Fisheries Research Board of Canada 32:817–820.

Rosenberg, D. M., Danks, H. V. 1987. Aquatic insects of peatlands and marshes in Canada: introduction. Memoirs of the Entomological Society of Canada 140:1–4.

Rosenberg, D. M., Resh, V. H. 1996. Use of aquatic insects in biomonitoring. *in* Merritt, R. W., Cummins, K. W., Eds., An introduction to the aquatic insects of North America, 3rd ed. Kendall/Hunt, Dubuque, IA, pp. 87–97.

Rosenberg, D. M., Wiens, A. P., Saether, O. A. 1977. Life histories of *Cricotopus (Cricotopus) bicinctus* and *C. (C.) mckenziensis* (Diptera: Chironomidae) in the Fort Simpson area, Northwest Territories. Journal of the Fisheries Research Board of Canada 34:247–253.

Ross, D. H., Wallace, J. B. 1981. Production of *Brachycentrus spinae* Ross (Trichoptera: Brachycentridae) and its role in seston dynamics of a southern Applacahain stream (USA). Holarctic Ecology 6:270–284.

Saether, O. A. 1979. Chironomid communities as water quality indicators. Holarctic Ecology 2:65–74.

Saponis, A. R. 1977. A revision of the Nearctic species of *Orthcladius (Orthocladius)* van der Wulp (Diptera: Chironomidae). Memoirs of the Entomological Society of Canada 102:1–187.

Schindler, D. W. 1977. The evolution of phosphorus limitation in lakes. Science 195:260–262.

Schmidt, S. K., Huges, J. M., Bunn, S. E. 1995. Gene flow among conspecific populations of *Baetis* sp. (Ephemeroptera): adult flight and larval drift. Journal of the North American Benthological Society 14:147–157.

Schuldt, J. A., Hershey, A. E. 1995. Impact of salmon carcass decomposition on Lake Superior tributary streams. Journal of the North American Benthological Society 14:259–268.

Scrimgeour, G. J., Culp, J. M. 1994. Feeding while evading predators by a lotic mayfly: linking short-term foraging behavior to long-term fitness consequences. Oecologia 199:128–134.

Sedell, J. R., Reeves, G. H., Hauer, F. R., Standford, J. A., Hawkins, C. P. 1990. Role of refugia in recovery from disturbances: modern fragmentation and disconnected river systems. Environmental Management 14:711–724.

Shepard, R. B., Minshall, G. W. 1981. Nutritional value of lotic insect feces compared with allocthonous materials. Archiv fur Hydrobiologie 90:467–488.

Smock, L. A. 1996. Macroinvertebrate movements: drift, colonization, and emergence, *in* Hauer, F. R., Lamberti, G. A., Eds., Methods in Stream Ecology. Academic Press, San Diego, CA, pp. 371–390.

Smock, L. A., Roeding, C. E. 1986. The trophic basis of production of the macroinvertebrate community of a southeastern U.S.A. blackwater stream. Holarctic Ecology 9:165–174.

Smock, L. A., Gladden, J. E., Riekenberg, J. L., Smith, L. C., Black, C. R. 1992. Lotic macroinvertebrate production in three dimensions: channel surface, hyporheic, and floodplain environments. Ecology 73:876–886.

Soluk, D. A. 1985. Macroinvertebrate abundance and production of psammophilous Chironomidae in shifting sand areas of a lowland river. Canadian Journal of Fisheries and Aquatic Sciences 42:1296–1302.

Sousa, W. P. 1984. The role of disturbance in natural communities. Annual Review of Ecology and Systematics 15:353–391.

Spence, J. R., Andersen, N. M. 1994. Biology of water striders: interactions between systematics and ecology. Annual Review of Entomology 39:101–128.

Stanford, J. A., Ward, J. V. 1988. The hyporheic habitat of river ecosystems. Nature 335:64–66.

Steinman, A. D. 1996. Effects of grazers on freshwater benthic algae, *in* Stevenson, R. J., Bothwell, M. L., Lowe, R. L., Eds., Algal ecology: freshwater benthic ecosystems. Academic Press, San Diego, CA, pp. 341–373.

Stewart, A. J., Loar, J. M. 1994. Spatial and temporal variation in biological monitoring data, *in* Loeb, S. L., Spacie, A., Eds., Biological Monitoring of Aquatic Systems. Lewis Publishers, Boca Raton, FL, pp. 91–124.

Stewart, K. W., Harper, P. P. 1996. Plecoptera, *in* Merritt, R. W., Cummins, K. W., Eds., An introduction to the aquatic insects of North America, 3rd ed. Kendall/Hunt, Dubuque, IA, pp. 217–266

Stockner, J. G., Shortreed, K. R. S. 1978. Enhancement of autotrophic production by nutrient addition in a coastal rainforest stream on Vancouver Island. Journal of the Fisheries Research Board of Canada 35:28–34.

Sullivan, K., Lisle, T. E., Dolloff, C. A., Grant, G. E., Reid, L. M. 1987. Stream channels: the link between forests and fishes, *in* Salo, E. O., Cundy, T. W., Eds., Streamside management: forestry and fishery interactions. University of Washington Institute of Forest Resources Contribution No. 57, pp. 39–97.

Suren, A. M., Winterbourn, M. J. 1992. The influence of periphyton, detritus and shelter on invertebrate colonization of aquatic bryophytes. Freshwater Biology 27:327–339.

Swanson, F. J., Benda, L. E., Duncan, S. H., Grant, G. E., Megahan, W. F., Reid, L. M., Zeimer, R. R. 1987. Mass failures and other processes of sediment production in Pacific Northwest forest landscapes, *in* Salo, E. O., Cundy, T. W., Eds., Streamside management: forestry and fishery interactions. University of Washington Institute of Forest Resources Contribution No. 57, pp. 9–38.

Swanson, F. J., Gregory, S. V., Sedell, J. R., Campbell, A. G. 1982. Land–water interactions: the riparian zone, *in* Edmonds, R. L., Ed., Analysis of coniferous forest ecosystems in the western United States. US/IBP Synthesis Series 14. Hutchinson-Ross, Stroudsburg, PA, pp. 267–291.

Swanson, G. A. 1984/1985. Invertebrates consumed by dabbling ducks (Anatinae) on the breeding grounds. Jounnal of the Minnesota Academy of Science 50:37–40.

ter Braak, C. J. F., Prentice, I. C. 1988. A theory of gradient analysis. Advances in Ecological Research 18:271–317.

Tester, J. 1995. Minnesota's natural heritage: an ecological perspective. Univ. of Minnesota Press. Minneapolis, MN.

Thorp, J. H. 1986. Two distinct roles for predators in freshwater assemblages. Oikos 47:75–82.

Thorp, J. H., Bergey, E. A. 1981. Field experiments on responses of a freshwater, benthic macroinvertebrate community to vertebrate predators. Ecology 62:365–375.

Thorp, J. H., Delong, M. D. 1994. The riverine productivity model: an heuristic view of carbon sources and organic processing in large river ecosystems. Oikos 70:305–308.

Thorp, J. H., Diggins, M. R. 1982. Factors affecting depth distribution of dragonflies and other benhtic insects in a thermally destabilized reservoir. Hydrobiologia 87:33–44.

Ulfstrand, S. 1968. Benthic animals in Lapland streams. Oikos Supplement 10:1–120.

Usinger, R. L. 1956. Aquatic insects of California, Univ. of California Press, Berkeley, CA, 508 p.

Van Buskirk, J. 1989. Density-dependent cannibalism, and size class dominance in a dragonfly *Aeshna juncea*. Ecology 74: 1950–1958.

Vander Zanden, M. J., Rasmussen, J. B. 1999. Primary consumer $\delta^{13}C$ and $\delta^{15}N$ and the trophic position of aquatic consumers. Ecology 80:1242–1252.

Vannote, R. L., Minshall, G. W., Cummins, K. W., Sedell, J. R., Cushing, C. E. 1980. The river continuum concept. Canadian Journal of Fisheries and Aquatic Sciences 37:130–137.

Van Sickle, J., Gregory, S. V. 1990. Modeling inputs of coarse woody debris to streams from falling trees. Canadian Journal of Forest Research 20:1593–1601.

Vodopich, D. S., Cowell, B. C. 1984. Interaction of factors controlling the distribution of a predatory insect. Ecology 65:39–52.

Walker, E. D., Merritt, R. W. 1988. The significance of leaf detritus to mosquito (Diptera: Culicidae) productivity from treeholes. Environmental Entomology 17:199–206.

Wallace, J. B., Brady, U. E. 1971. Residue levels of dieldrin in aquatic invertebrates and effect of prolonged exposure on populations. Pesticides Monitoring Journal 5:295–300.

Wallace, J. B., Merritt, R. W. 1980. Filter-feeding ecology of aquatic insects. Annual Review of Entomology 25:103–132.

Wallace, J. B., Webster, J. R., Cuffney, T. F. 1982. Stream detritus dynamics: regulation by invertebrate consumers. Oecologia 53:197–200.

Wallace, J. B., O'Hop, J. 1985. Life on a fast pad: waterlily leaf beetle impact on water lilies. Ecology 66:1534–1544.

Wallace, J. B., Vogel, D. S., Cuffney, T. F. 1986. Recovery of a headwater stream from an insecticide-induced community disturbance. Journal of the North American Benthological Society 5:115–126.

Wallace, J. B., Gurtz, M. E., Smith-Cuffney, F. 1988. Long-term comparisons of insect abundances in disturbed and undisturbed Appalachian headwater streams. Verhandlungen Internationale Verein Limnologie 23:1224–1231.

Wallace, J. B., Lugthart, G. J., Cuffney, T. F., Schurr, G. A. 1989. The impact of repeated insecticide treatments on drift and benthos of a headwater stream. Hydrobiologia 179:135–147.

Wallace, J. B., Anderson, N. H. 1996. Habitat, life history, and behavioral adaptations of aquatic insects, *in* Merritt, R. W., Cummins, K. W., Eds., An Introduction to the Aquatic Insects of North America, 3rd ed. Kendall/Hunt, Dubuque, IA, pp. 41–73.

Wallace, J. B., Grubaugh, J. W. 1996. Transport and storage of FPOM, *in* Hauer, F. R., Lamberti, G. A., Eds., Methods in stream ecology, Academic Press, San Diego, CA, pp. 191–215.

Wallace, J. B., Webster, J. R. 1996. The role of macroinvertebrates in stream ecosystem function. Annual Review of Entomology 41:115–139.

Wallace, J. B., Eggert, S. L., Meyer, J. L., Webster, J. R. 1997. Multiple trophic levels of a forest stream linked to terrestrial litter inputs. Science 277:102–104.

Walshe, B. M. 1947. Feeding mechanisms of *Chironomus* larvae. Nature 160:474.

Ward, A. K., Dahm, C. N., Cummins, K. W. 1985. *Nostoc* (Cyanophyta) productivity in Oregon stream ecosystems: invertebrate influences and differences between morphological types. Journal of Phycology 21:223–227.

Ward, G. M., Ward, A. K., Dahm, C. N., Aumen, N. G. 1994. Origin and formation of organic and inorganic particles in aquatic systems, *in* Wotton, R. S., Ed., The biology of particles in aquatic systems, 2nd ed. Lewis Pubs., Boca Raton, FL, pp. 27–56.

Ward, J. V. 1992. Aquatic insect ecology. 1. Biology and habitat. Wiley, New York, NY.

Ward, J. V., Stanford, J. A. 1982. Thermal responses in the evolutionary ecology of aquatic insects. Annual Review of Entomology 27:97–117.

Ward, J. V., Stanford, J. A. 1983. The intermediate disturbance hypothesis: an explanation for biotic diversity patterns in lotic ecosystems, *in* Fontaine, T. D., Bartell, S. M., Eds., Dynamics of lotic ecosystems. Ann Arbor Science, Ann Arbor, MI, pp. 347–356.

Warwick, W. F. 1980. Paleobiology of the Bay of Quinte, Lake Ontario: 2800 years of cultural influence. Canadian Bulletin of Fisheries and Aquatic Sciences 206:1–117.

Warwick, W. F., Tisdale, N. A. 1989. Morphological deformities in *Chironomus*, *Cryptochironomus*, and *Procladius* larvae (Diptera: Chironomidae) from two differentially stressed sites in Tobin Lake, Saskatchewan. Canadian Journal of Fisheries and Aquatic Sciences 45:1123–1144.

Waters, T. F. 1972. The drift of stream insects. Annual Review of Entomology 17:253–272.

Waters, T. F. 1977. Secondary production in inland waters. Advances in Ecological Research 10:91–164.

Waters, T. F., Crawford, G. W. 1973. Annual production of a stream mayfly population: A comparison of methods. Limnology and Oceanography 18:286–296.

Waters, T. F., Hokenstrom, J. C. 1980. Annual production and drift of the stream amphipod *Gammarus pseudolimnaeus* in Valley Creek, Minnesota. Limnology and Oceanography 25:700–710.

Wellborn, G. A., Skelly, D. K., Werner, E. E. 1996. Mechanisms creating community structure across a freshwater habitat gradient. Annual Review of Ecology and Systematics 27:337–367.

Wheeler, J. R. 1994. Factors affecting black fly abundance and distribution in an arctic stream. M. S. Thesis, University of Minnesota, Duluth, MN.

Whiles, M. R., Wallace, J. B., Chung, K. 1993. The influence of Lepidostoma (Trichoptera: Lepidostomatidae) on recovery of leaf-litter processing in disturbed headwater streams. The American Midland Naturalist 130:356–363.

Whitledge, G. W., Rabeni, C. K. 1997. Energy sources and ecological role of crayfishes in an Ozark stream: insights from stable isotopes and gut analyses. Canadian Journal of Fisheries and Aquatic Sciences 54:2555–2563.

White, D. S., Brigham, W. U. 1996. Aquatic Coleoptera, *in* Merritt, R. W., Cummins, K. W., Eds., An introduction to the aquatic insects of North America, 3rd ed., Kendall/Hunt, Dubuque, IA, pp. 399–473.

Wiggins, G. B. 1996. Larvae of the North American caddisfly genera (Trichoptera), 2nd ed., Univ. of Toronto Press, Toronto, 457 p.

Wiggins, G. B., Mackay, R. J., Smith, I. M. 1980. Evolutionary and ecological strategies of animals in temporary pools. Archiv fur Hydrobiologie/Supplementum 58:97–206.

Williams, D. D. 1993. Nutrient and flow vector dynamics at the hyporheic/groundwater interface and their effects on the interstitial fauna. Hydrobiologia 251:185–198.

Williams, D. D., Feltmate, B. W. 1992. Aquatic insects. C.A.B. Int., Wallingford, Oxon, UK.

Wilzbach, M. A., Cummins, K. W. 1989. An assessment of short-term depletion of stream macroinvertebrate benthos by drift. Hydrobiologia 185:29–39.

Winterbourn, M. J., Rounick, J. S., Cowie, B. 1981. Are New Zealand stream really different? New Zealand Journal of Marine and Freshwater Research 15:321–328.

Wissinger, S. A. 1988. Life history and size structure of larval dragonfly populations. Journal of the North American Benthological Society 7:13–28.

Wolheim, W. M., Peterson, B. J., Deegan, L. A., Bahr, M., Jones, D., Bowden, W. B., Hershey, A. E., Kling, G. W., Miller, M. C. 1999. A coupled field and modeling approach for the analysis of nitrogen cycling in streams. Journal of the North American Benthological Society 18:199–221.

Wooster, D. 1994. Predator impacts on stream benthic prey. Oecologia 99:7–15.

Wotton, R. S. 1976. Evidence that blackfly larvae can feed on particles of colloidal size. Nature 261:697.

Wotton, R. S. 1988. Dissolved organic material and trophic dynamics. BioScience 38:172–177.

Zachariassen, K. E. 1979. The mechanism of the cryoprotective effect of glyceral in beetles tolerant to freezing. J. Insect Physiol. 25:29–32.

19

INTRODUCTION TO THE SUBPHYLUM CRUSTACEA

Alan P. Covich
Department of Fishery and Wildlife Biology
Colorado State University
Fort Collins, Colorado 80523

James H. Thorp
Department of Biology
Clarkson University
Potsdam, New York 13699

I. INTRODUCTION

Crustaceans are the most diverse of any group of arthropods and currently thrive in a wide array of habitats. Most species occur primarily in aquatic habitats but some are adapted to live on land. The fossil record extends back to the lower Cambrian and throughout most of this long period, crustacean fossils are associated with a broad range of aquatic habitats, especially marine waters (Schram, 1982). Groups of crustaceans adapted to freshwater habitats very early in their evolutionary history (Hutchinson, 1967; Abele, 1982). The fossil record for freshwater species is especially well preserved from the Tertiary for some groups that are clearly related to living forms. Although only about 10% of the nearly 40,000 extant species in the subphylum Crustacea (phylum Arthropoda) occur in the inland pools, lakes, and streams of the world (Table I) (Bowman and Abele, 1982; Banarescu, 1990; Barnes and Harrison, 1992), freshwater crustaceans are extremely important in many ecosystem processes of inland waters.

Crustaceans have evolved into several major groups (Fig. 1) that represent distinct modifications for exploiting resources and performing ecosystem services in many different surface waters and subsurface habitats. Amphipods and isopods, as well as mysids and branchiura are discussed in this chapter; ostracodes are reviewed in Chapter 20. The cladocerans and copepods together constitute the dominant component of most freshwater planktonic assemblages and are discussed in Chapters 21 and 22 respectively.

Crustacean zooplankton are major consumers of phytoplankton and they transfer energy from these primary producers up the food web to zooplanktivorous fish in lakes, ponds, and rivers. Some crustacean zooplankton also consume aggregations formed from dissolved organic matter and other organic particulates. Others are predatory on rotifers, protozoans, other crustaceans, aquatic insects, and fish eggs and larvae. Some zooplankton vertically migrate and thereby link deep, benthic habitats with open waters of lakes. These migratory species are important in nutrient cycling and transfers of toxins (such as heavy metals and pesticides) from sediments to pelagic foodwebs. Along with many other invertebrates, crustaceans provide hard surfaces as semipermanent, mobile substrates for diverse

TABLE I Classification of Crustacea and Estimates of Species Numbers in North American Fresh Waters[a,b]

Taxa	*Worldwide percentage found in Freshwaters*	*Approximate number of species in North America*	*Chapter(s) of this book*
Subphylum Crustacea	<10%	ca.1500	19–23
Class Cephalocarida	0%	0	—
Class Branchiopoda [eight orders, including cladocerans]	>95%	>200	21
Class Remipdia	0%	0	—
Class Maxillopoda	<15%		19, 22
subclass Branchiura	90%	23	19
subclass Copepoda	<15%	>200	22
Class Ostracoda	>50%	420	20
Class Malacostraca	10%		19, 23
Superorder Pancarida [order Thermosbaenacea]	50%	1	19
Superorder Peracarida [orders Mysidacea, Amphipoda, Isopoda]	<10%	>300	19
Superorder Eucarida [order Decapoda]	10%	334	23

[a] North of Mexico.
[b] Classification based in part on Bowman and Abele (1982).

epibiont species such as ciliated protozoa and sessile rotifers (chapters 3 and 8; Threlkeld *et al.,* 1993). Within inland hypersaline environments, brine shrimp (Chapter 21) are sometimes the only metazoan capable of surviving hyperosmotic conditions. As salinity fluctuates, the crustacean community can rapidly change its species composition and exploit different sources of energy (Galat *et al.,* 1988; Hammer *et al.,* 1990; Stephens, 1990; Wurtsbaugh and Berry, 1990; Williams, 1998).

Crustaceans also comprise a significant portion of the benthic biomass in many lakes and streams. Some species are abundant in the littoral zone and throughout the oxygenated deeper water where they feed on bacteria and fungi on decaying vegetation or live algae (Gee, 1988; Delong *et al.,* 1993). As with the zooplankton, these benthic consumers recycle nutrients that sustain high levels of microbial productivity. Some species horizontally migrate at night and link the littoral zone with pelagic water. This nocturnal mobility, like vertical migrations, can limit vulnerability of these species to visual predators. In streams, crustaceans also occupy many different habitats ranging from deep, subsurface waters to quiet pools and fast-flowing riffles where they actively consume a wide range of foods (Marchant and Hynes, 1981a, b; Boulton *et al.,* 1998). In caves, unique species of crustaceans have evolved that also feed on many types of resources (Culver *et al.,* 1995; Carpenter, 1999). Similarly, some species are found in wetlands and these feed on many types of detrital foods (Smock and Harlowe, 1983). These diverse benthic species function as herbivores, carnivores, and detritivores while also providing many ecosystem services (Palmer *et al.,* 1997; Boulton *et al.,* 1998; Covich *et al.,* 1999). As discussed below, some of these very productive benthic species live in extreme habitats, from cold arctic waters to hot springs and saline waters where they can dominate benthic communities and reach high densities.

One of the most ecologically and taxonomically diverse freshwater crustacean groups north of the Rio Grande are omnivorous crayfish (Decapoda; Chapter 23). Of the more than 500 species and subspecies currently known worldwide, 393 taxa (78%) occur in North America and 346 of these occur in the United States and Canada (Sket, 1999a; Taylor, 2000; see Hobbs, Chapter 23). River drainages in the southeastern United States have the highest diversity and highest concentrations of crayfish species of any region anywhere (Taylor, 2000). These diverse benthic consumers are not only important in nutrient cycling, but they also can change the life histories of some of their prey species that have evolved ways to avoid being eaten (Crowl and Covich, 1990). Crayfish maintain very high densities in many different habitats where they feed on a wide range of living and dead plants and animals. In addition to their crucial roles as consumers and secondary producers in natural aquatic ecosystems, some crustaceans are increasing in importance as a food source for humans. Crayfish have become widely available through aquacultural production, as have the larger "river shrimp" (*Macrobrachium*) in warm

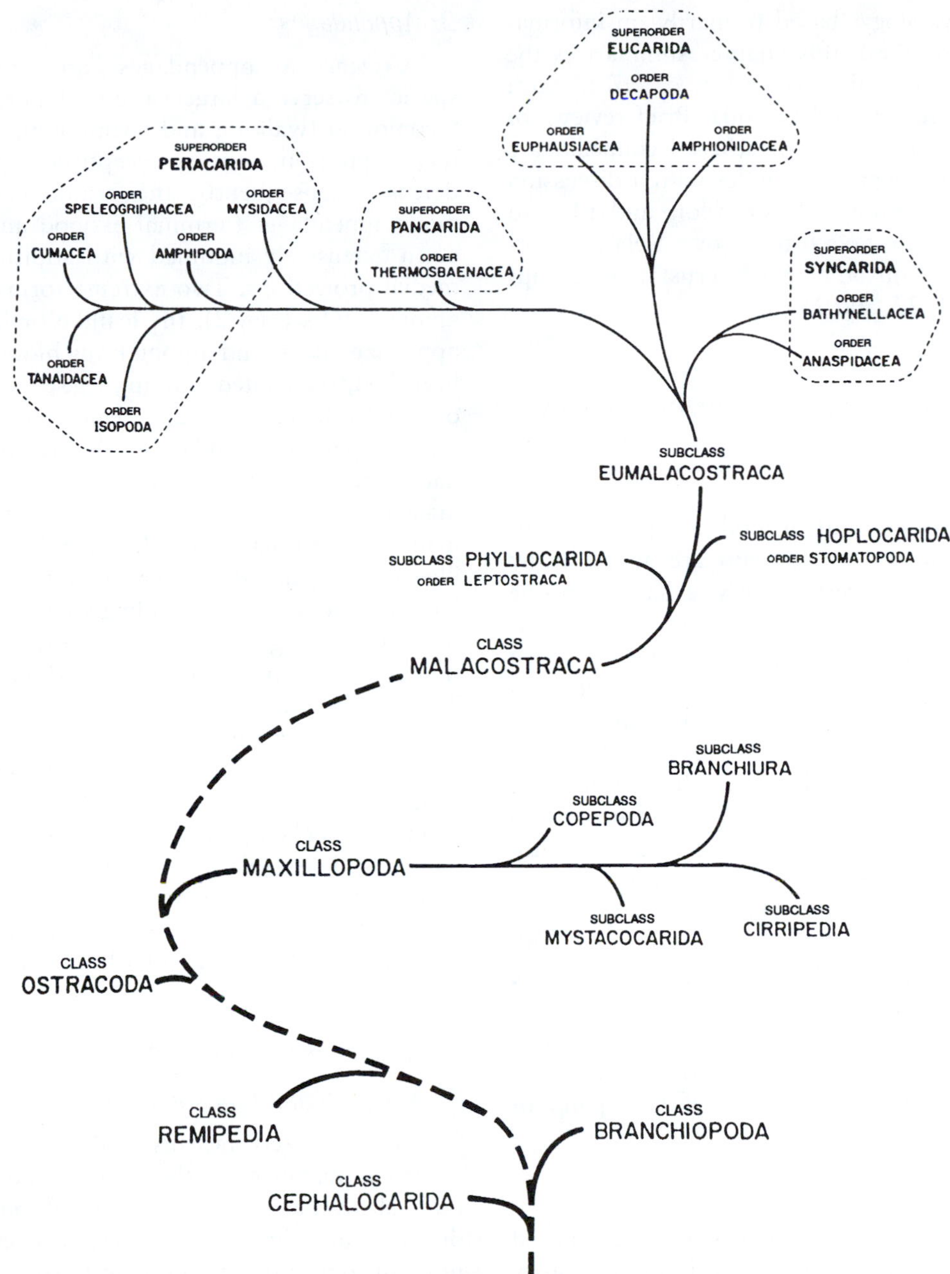

FIGURE 1 General evolutionary linkages of major groups of freshwater crustaceans. From Bowman and Abele (1982) who stated: "This [classification] is not to indicate phylogenetic relationships and should not be so interpreted. The dashed line at the base emphasizes uncertainties concerning the origins of the five classes and their relationships to each other."

temperate-zone and tropical locations. Benthic amphipods, isopods, ostracodes, and various littoral cladocerans and copepods consume a large amount of living organisms and particulate organic matter. All these crustaceans, in turn, can be significant constituents in the diets of many invertebrate predators and bottom-feeding fishes (Dahl, 1998; Wooster, 1998; MacNeil *et al.,* 1999a, b).

The present chapter provides a brief introduction to crustacean anatomy and physiology. More detailed information can be obtained from general invertebrate zoology texts (e.g., Barnes, 1987 and from several books on crustaceans, such as those by Schmitt (1965), Schram (1986), Fitzpatrick (1983), McLaughlin (1980), and the 10-volume set edited by Bliss (1982–1985) and several co-editors. Following an introduction to anatomy and physiology (based primarily on information from these sources), this chapter summarizes the ecology and evolution of three orders of crustaceans: Mysidacea, Amphipoda, and Isopoda. Brief reviews of the subclass Branchiura and the class Remipedia are also included. This chapter concludes with a discussion of crustacean biogeography, classification, and a key to the genera of the major taxa listed above. Related information is presented on other major crustacean groups in Chapters 20, 21, 22, and 23.

II. ANATOMY AND PHYSIOLOGY OF CRUSTACEA

A. External Morphology

Although many other arthropod groups use a single basic body plan, the crustaceans are anatomically very diverse (Schram, 1986). Once you have seen one insect or spider taxon, you have a good idea of the general body form. The crustaceans, however, have evolved a high diversity of body forms by fusing various segments or by developing highly specialized body segments and appendages. Crustaceans are basically metameric, protostomate coelomates whose bodies are covered by a hard exoskeleton and often divided into three regions, or tagmata: the head (cephalic area), thorax, and abdomen (sometimes the first two regions are combined as a cephalothorax). The jointed appendages are primitively biramous and may be present in all three body regions. Mandibles, two pairs of maxillae, and two pairs of antennae are present in all species during at least one life stage. Crustaceans are distinguished from the Insecta, one of the other two large arthropod groups, by the presence of two pairs of antennae.

1. Skeleton

All crustaceans have a hard exoskeleton composed of chitin that varies in rigidity at various life-history stages. The name Crustacea is derived from the Latin word for shell and refers to this exoskeleton, which in many taxa is hardened with very small inclusions of calcium carbonate. Although not a true "shell" as is found among the Mollusca, this exoskeleton is very durable. Crustaceans then, like other arthropods, are characterized by a chitinized exoskeleton consisting of a thin proteinaceous stiffened epicuticle and a thick, multilayer procuticle strengthened by calcium carbonate. The skeleton attains its maximum thickness and rigidity in the decapods. Projecting inward are chitinized struts, known as apodemes, which serve as sites for internal attachment of the muscles; they also lend partial protection to some organs.

2. Appendages

Crustacean appendages are modified among species to serve a large variety of purposes, including locomotion (walking and swimming), feeding, grooming, respiration, sensory reception, reproduction, and defense. Consequently, the primitive, generally biramous appendages (terminal exopod and endopod) are often extensively modified with additional lateral and medial projections. Two extreme forms are recognized among adults (Fig. 2): the leaflike or lobed phyllopod appendage (as found among branchiopods) and the unbranched, segmented walking leg, or stenopod (typical of crayfish). The cephalic region contains five basic types of paired appendages: (1) antennules (first antennae), which are uniramous in all crustaceans except the malacostracans; (2) antennae (second antennae); (3) mandibles; (4) maxillules (first maxillae); and (5) maxillae (second maxillae). The first two pairs generally have a sensory function (aiding some taxa in food location and filtering), whereas the last three pairs normally function in food acquisition, handling, or processing. The number of appendages on the thorax and abdomen vary greatly among large taxonomic groups. Malacostracans (such as decapods and amphipods) generally possess 5–8 pairs of thoracic appendages (sometimes called pereiopods) and six pairs of abdominal appendages (pleopods and terminal uropods). Primary abdominal appendages are absent in all nonmalacostracans except Notostraca, although comparable structures may have secondarily evolved.

B. Organ System Function

1. Nutrition and Digestion

Freshwater crustaceans predominantly employ their antennae, maxillae, or thoracic appendages to filter or grab algae, bacteria, microzooplankton, or suspended dead organic matter (detrital seston) from the water column. In some habitats, a relatively large number of taxa are primarily carnivores or scavengers (e.g., benthic decapods, certain copepods, and others). Parasitism is rare except among specialized copepods and branchiurans (Yamaguti, 1963; Margolis and Kabata, 1988) that attack fish and bopyrid isopods, such as *Probopyrus* which attack freshwater shrimps (Beck, 1980; Collart, 1990; Roman Contreras, 1996). Only one ostracod genus, *Entocythere,* is known to be parasitic. Cannibalism generally occurs when newly molted individuals are vulnerable, especially among the decapods (Goddard, 1988), amphipods (Dick, 1995; MacNeil *et al.,* 1997, 1999), and isopods (Jormalainen and Shuster, 1997).

Using their cephalic or thoracic appendages, crustaceans acquire food and pass it through a ventral

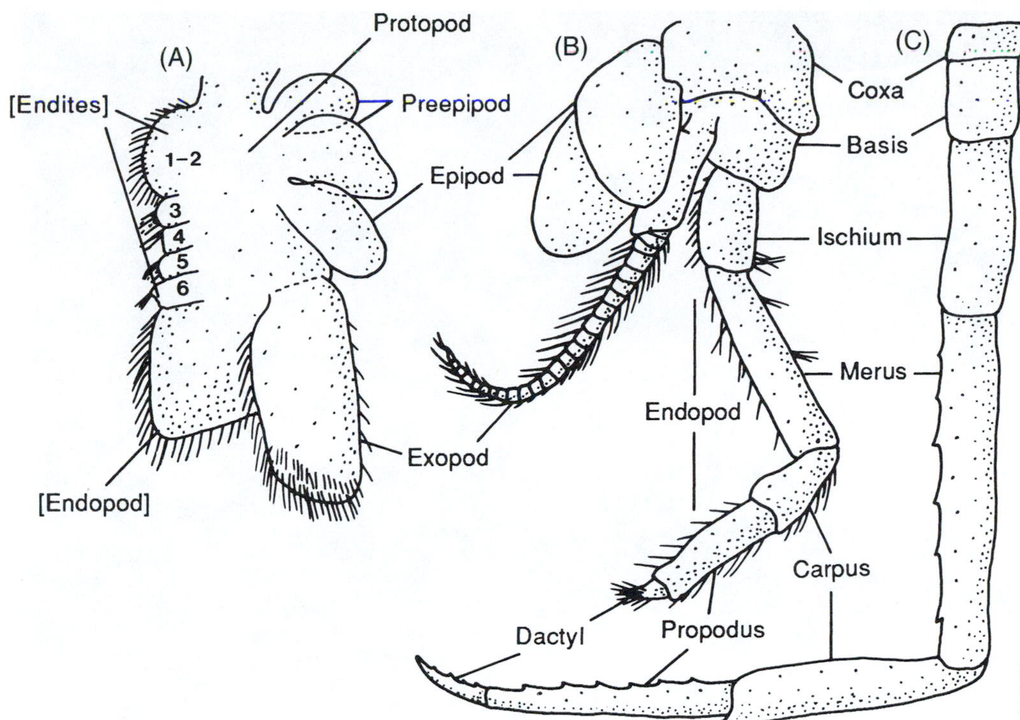

FIGURE 2 Representative crustacean appendanges: (A) a phyllopod appendage of Anostraca; (B) biramous appendage of Anaspidacea (superorder Syncarida); and (C) uniramous stenopod appendage of the Stenopodidea (Decapoda: Pleocyemata). [Redrawn from McLaughlin, 1982].

mouth into a tripartite alimentary tract (see Chapter 23, Fig. 4). Crayfish have sharp mandibles that assist in cutting and shredding their food. Crayfish chelae ("claws") can be used to grasp larger food items (although they are often used for defense) and their numerous sensilla (sensory hairs) may aid in chemically locating food (Fig. 3). In most crustaceans, food is ground into small, digestible particles by the gastric mill (the posterior end of the cardiac stomach) of the muscular foregut (e.g., Schmitz and Sherrey, 1983). Enzymes (primarily from the hepatopancreas) empty into the short midgut, where food is digested and assimilated. Wastes are conveyed by peristalsis through the long hindgut and out the terminal anus.

2. Circulation and Respiration

Crustaceans generally have an open circulatory system consisting of a dorsal, neurogenic heart with a single chamber, one or more arteries (in malacostracans only), and several sinuses for returning blood to the heart. Many noncalanoid copepods (and marine barnacles) lack a heart altogether; fluids are circulated as a result of general body movements. The heart is usually located in the thorax or cephalothorax, but is present in the abdomen in species with abdominal gills (such as isopods). The blood contains low amounts of a viscous, copper-based respiratory pigment (hemocyanin), which is dissolved in the hemolymph.

Respiration takes place entirely across the body integument in nonmalacostracans, but thoracic and abdominal gills are typical of other freshwater crustaceans. Peracaridans (discussed later in this chapter) feature rudimentary epipodal gills (flattened outgrowths of the coxa). Individual species differ in their rates of oxygen consumption and their ability to regulate their respiration as dissolved oxygen concentrations decrease (Toman and Dall, 1998). Freshwater decapods are distinct in having trichobranchiate gills composed of a central axis (with afferent and efferent blood vessels) bearing numerous filamentous branches (Fig. 4).

3. Fluid and Solute Balance

Paired maxillary and antennal glands (also called green glands) are the principal excretory organs in crustaceans. The "labyrinth" of the antennal gland is also involved in reabsorption of glucose, amino acids, and divalent ions from tubule fluids. Maxillary glands predominate within lower crustaceans, whereas the slightly more complex antennal glands characterize most malacostracans. More than 90% of the nitrogenous wastes are voided as ammonia.

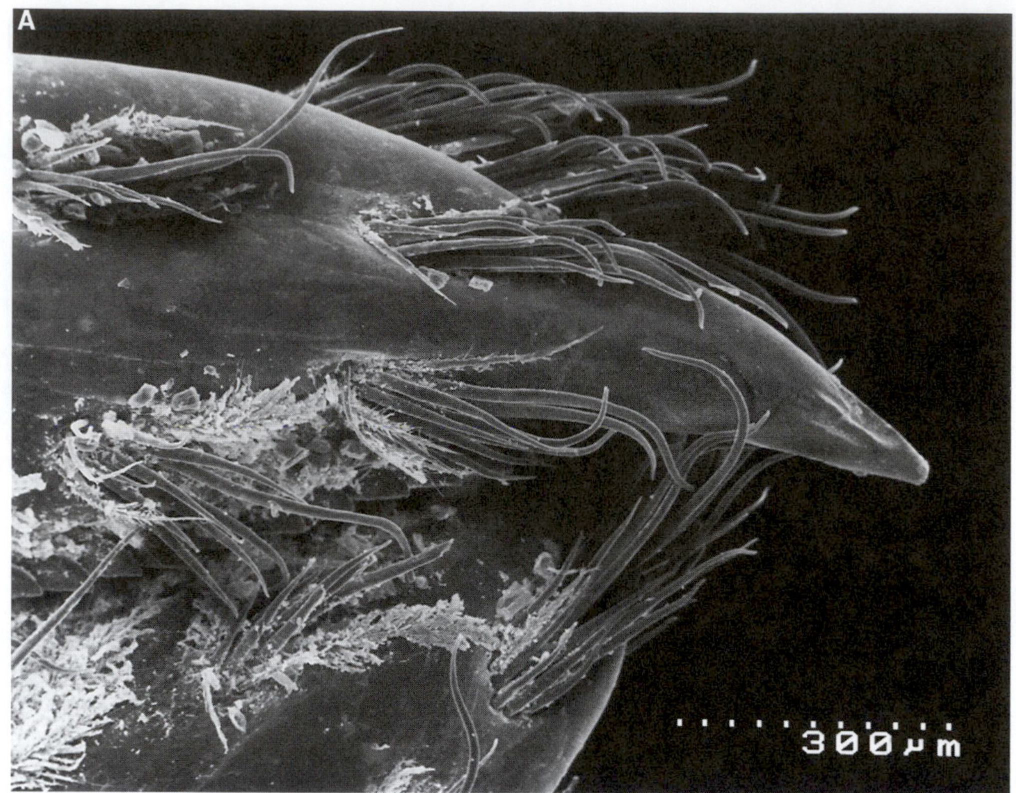

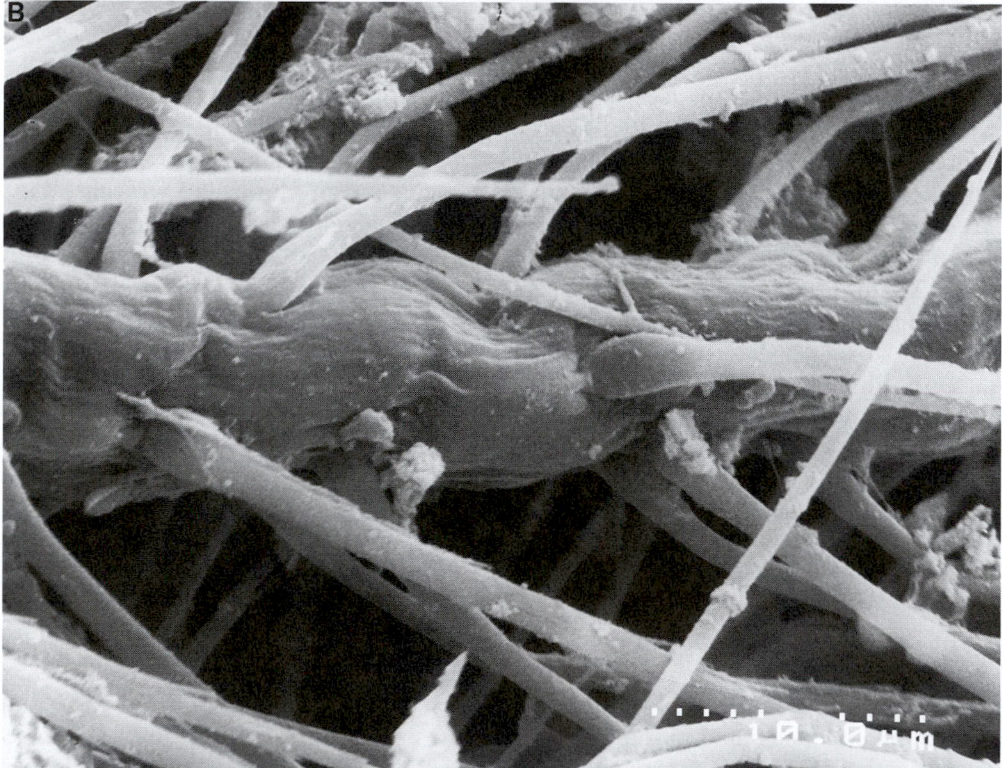

FIGURE 3 Scanning electron micrographs of crayfish chela: (A) view of distal portion of chela (*Orconectes propinquus,* female, 300× magnification); (B) Plumose sensillia (feathered hairs) located on the cutting edge of chela (*Orconectes virilis,* female, 6000× magnification). [Images provided by J. L. Borash, 1997. See Borash and Moore, 1997 for more information.]

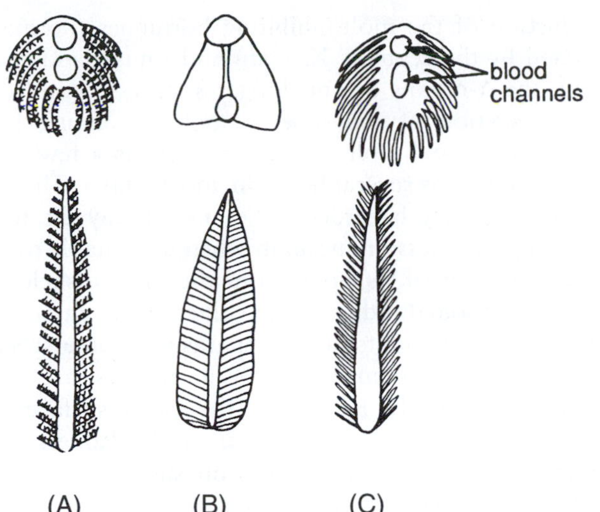

FIGURE 4 Structure of decapod gill branches, as shown in transverse (upper) and latteral (lower) views: (A) dendrobranchiate; (B) phyllobranchiate; and (C) trichobranchiate. [From McLaughlin, 1982].

Freshwater crustaceans are relatively competent osmoregulators. They face problems of water influx and salt loss (both primarily across the gills) and, therefore, must release copious quantities of urine that is isosmotic to the hemolymph (a few taxa produce urine that is hyposmotic to the blood). Freshwater branchiopods and some copepods maintain a dilute hemolymph equal to about 200 mOsm (20% of seawater), while the blood of crayfish is more salty (about 350 mOsm; see Mantel and Farmer, 1983). An opposite situation is faced by hyporegulating crustaceans living in inland salt lakes. The brine shrimp *Artemia salina,* tolerates salinities ranging from 10% seawater to crystallizing brine; it is hypertonic to the medium in dilute waters, but hypotonic in concentrated solutions.

4. Neural Systems

The crustacean neural system is comprised of a tripartite brain and paired ventral, ganglionated nerve cords linked by commissures in a ladderlike arrangement. The protocerebrum of the brain normally innervates the eye, sinus gland, frontal organs, and head muscles; the deutocerebrum controls the antennules; the tritocerebrum innervates the antennae and a portion of the alimentary tract. Crustaceans are generally sensitive to light, chemicals, temperature, touch, gravity, pressure, and sound. The eyes of adult malacostracans are compound and frequently mounted on a stalk, or peduncle; but the adult eye of many entomostracans (an older term inclusive of most nonmalacostracans) is not much advanced over the simple cluster of inverse pigment cup ocelli that characterizes the larval (naupliar) eye.

5. Reproduction, Development, and Growth

Crustaceans are primarily sexually reproducing, dioecious organisms, but hermaphroditism and parthenogenesis occur sporadically among entomostracans (especially in ostracodes and branchiopods) and a few marine malacostracans. Paired gonads lie above or lateral to the midgut in most species. In mysids, the ovaries are partially fused and linked by a cellular bridge. Paired, or rarely single, reproductive ducts open

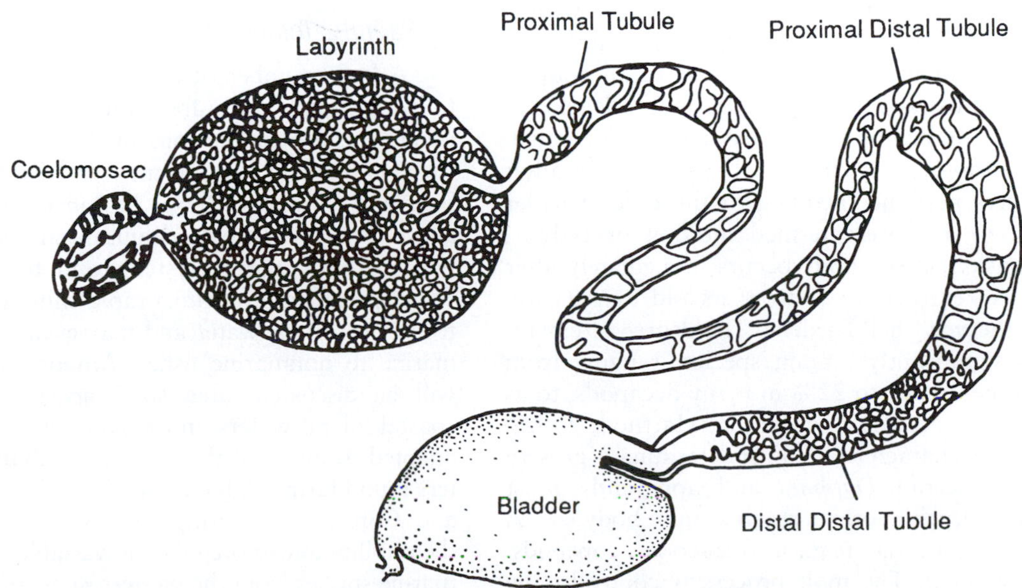

FIGURE 5 Structure of the antennal gland of a crayfish. [From Mantel and Farmer, 1983.]

ventrally through simple gonopores (paired or single), elevated papillae, or elaborate copulatory structures (as in decapods). The location of the gonopore varies greatly in entomostracans, whereas in malacostracans, it opens on the sternite or coxae of the sixth thoracic somite in females and at the eighth somite in males. Internal fertilization is the general rule, with males transferring sperm through a penis or specialized trunk appendages known as gonopods (in crayfish these are highly modified pleopods).

Females of most species protect their embryos either by retaining them in an internal brood chamber (e.g., in cladocerans) or external ovisac (e.g., in copepods) or by gluing them to certain appendages (decapods employ their abdominal pleopods). Mysids, isopods, and amphipods safeguard their young until an advanced stage in a ventral brooding shelf, the marsupium. Ovigerous female amphipods are readily recognizable when carrying eggs or young in late developmental stages because of their extended brood pouch and its yellowish to grey coloration. Stream-dwelling isopods apparently release their juveniles from the brood chamber when at risk from fish predation, but not when exposed to lower risk of predation by salamanders (Sparkes, 1996a,b).

The embryogeny of crustaceans includes modified spiral, total cleavage, and gastrulation by invagination. All taxa possess an initial nauplius stage in their development, but these larvae may be free swimming (e.g., in copepods) or enclosed in an egg (e.g., in decapods, cladocerans, and peracaridans). Direct development (i.e., without any external larval stages) characterizes taxa such as cladocerans and peracaridans. The young individuals appear morphologically identical to miniature adults. In contrast, indirect development with a free nauplius followed by either distinctive stages of metamorphosis or gradual development to an adult is typical of most ostracodes, copepods, decapods (except crayfish), and Branchiopoda (other than cladocerans).

Growth throughout the larval, juvenile, or adult stages requires periodic shedding of the older, smaller exoskeleton in a process termed molting or ecdysis. Rapid expansion of the body occurs immediately after the crustacean extracts itself from its old exoskeleton and before the new shell hardens. The degree of expansion varies significantly among species, ranging from 8–9% in some mysids to 22% in many decapods, to as high as 83% in certain cladocerans (Hartnoll, 1982). Some taxa are characterized by indeterminate growth (e.g., the cladoceran *Daphnia* and apparently most crayfish species), while others reach a finite body size in a fixed number of molts (e.g., in ostracodes, copepods, and some isopods). The molt process is controlled by hormones released by several organs. With a decline in production of the molt-inhibiting hormones (normally secreted by the cephalic X-organ and sinus gland complex), the Y-organs are no longer suppressed and can begin secretion of ecdysone, a molt-stimulating hormone. The actual molt requires as little as a few minutes to as long as several hours in aquatic taxa. This period is extremely hazardous; the animal may die from physiological stress or succumb when it is unable to extract a portion of its body from the old exoskeleton. The newly molted individual is vulnerable to rapid colonization by epibionts and ectoparasites. Some species of ciliated protozoans and sessile rotifers are well adapted to quickly recolonize the clean surfaces of newly molted isopods (Cook *et al.,* 1998; Roberts and Chubb, 1998). The temporarily soft shell of a recently molted individual is extremely susceptible to predatory attacks (and cannibalism). Some species may seek shelter prior to molting. The larger and older the individual becomes, the longer it takes for the molting process to be completed, and the longer is the exposure to possible cannibalism and predation. Various environmental factors influence susceptibility of crustaceans to predators after molting. For example, medium, and large-sized individuals experience lower risks of predation from salamanders when living in waters rich in calcium carbonate (tufa and travertine), because lime deposits on their bodies apparently make them less vulnerable (Ruff and Maier, 2000). Thus, although indeterminate growth typically has limitations imposed by physiological stress, parasite infections, epibiont loads, and predatory pressures, each habitat may vary in how environmental complexity influence rates of molting and survivorship.

C. Environmental Physiology

1. *Salinity, Temperature, and pH Tolerances*

A large number of crustacean species have adapted to living completely in freshwater habitats (Table I) while others spend only portions of their lives in these highly variable habitats (e.g., Mantel and Farmer, 1983; Vernberg and Vernberg, 1983; Holsinger, 1994a; Wollheim and Lovvorn, 1996). Branchiurans are distributed worldwide in both marine systems and freshwater habitats; some species can tolerate a rapid shifting of distributions from salt to freshwater and may even switch hosts from marine to nonmarine fishes. Among the Peracarida, as will be discussed later, some species are confined to coastal inland waters and apparently are not completely adapted to many of the relatively unbuffered, dilute waters found farther inland (Sutcliffe, 1971, 1974; Dormaar and Corey, 1978). Amphipod species differ greatly in their ability to osmoregulate in variable salinities. One estuarine species from the western coast of North America, *Corophium spinicorne,* occurs typically in tidal fresh waters (Hutchinson, 1967). This species is reported to have dispersed farther inland and now it occurs in fresh waters at two sites along the Snake River, upstream of Lewiston, Idaho (Lester and Clark, in press). This coastal species apparently was transported upriver by barges and may be widely distributed in the lower reaches of the Snake and Columbia rivers. Some species, such as *Hyalella azteca,* are found in saline lakes with salinities up to 25 parts per thousand (Hammer *et al.,* 1990). Other amphipod species are sensitive to direct or indirect effects of increased salinity (Wollheim and Lovvorn, 1996). Salt pollution from various sources can increase hypo-osmotic stress and create adverse effects on amphipod metabolism (Koop and Grieshaber, 2000).

Temperature is an important regulator of metabolism and growth rates as well as survival. The upper lethal temperature that limits survival varies greatly among different species. Some crustacean species have the highest known temperature tolerance among the aquatic metazoans and can live in hot springs. For example, the ostracode *Potamocypris,* lives on algal-coated substrates in habitats that range from 30 to 54°C (Wickstrom and Castenholz, 1973), the pancarids *Thermosbaena mirabilis* survives in waters of 45–48°C, and *Thermobatynella adami* lives in 55°C water (Capart, 1951). Some isopods are found in warm springs of the southwestern United States; for example, *Thermosphaeroma thermophilum,* an endangered species, and *T. subequalum* thrive in 34–35°C pools (Bowman, 1981; Cole and Bane, 1978; Shuster, 1981, Manuel Molles, personal communication). Other benthic crustaceans, such as the amphipod *Hyallela azteca,* also live in hot springs (Richard Wiegert and Dean Blinn, personal communication) with temperatures up to 33–34°C in Devils Hole, Nevada.

In general, individual growth rates of *Hyalella azteca* and other crustaceans are affected directly and indirectly by water temperatures. Variations in water temperatures often result in seasonal movements by differently aged individuals as they track changing temperatures at different depths. For example, field observations demonstrate that 20°C is a threshold for induction and termination of reproductive resting stages of *Hyalella azteca.* This species moves from warmer, shallow littoral zones into colder, deeper water as individuals mature (Panov and McQueen, 1998). If water depths decrease and water temperatures increase in response to both successional and climatic warming, it is anticipated that benthic species such as *Hyalella azteca* would likely have increased growth rates and earlier breeding in shallow lakes, ponds and intermittent streams. These shifts in water temperatures and habitat fragmentation could affect intra- and interspecific competition and gene flow among benthic species, especially during prolonged droughts (Hogg and Williams, 1996; Covich *et al.,* 1997; Hogg *et al.,* 1998a). Other amphipod species require colder, relatively constant waters and are often associated with cold springs. For example, *Gammarus minus* is rarely found in waters warmer than 15°C and is absent in waters above 20°C (Glazier *et al.,* 1992). Monitoring these cool-water habitats and their amphipod populations will provide important information about the long-term effects of climate on groundwater and surface-water temperatures (Wilhelm and Schindler 2000).

The widespread acidification of lakes and streams has drawn attention to the importance of pH tolerances of many organisms, especially the crustaceans. Because calcification of the carapace immediately following molting is highly sensitive to low pH, changes in regional distributions among crustaceans may reflect regional levels of acidification and natural differences in calcium concentrations (e.g., Nero and Schindler, 1983; Greenaway, 1985; Okland and Okland, 1985; Meyran, 1997, 1998; Meyran *et al.,* 1998). There are both lethal and sublethal effects of increased acidity that alter ion regulation, especially by juveniles in soft-water lakes. Although many studies have emphasized lentic species, there is evidence that lotic species may have a narrower range of tolerance to low pH and that egg mortality may be important for some species. Crustaceans generally are good indicators of changes in pH values in well-defined habitats. For example, a regional study documents that *Gammarus minus* distributions in 32 springs in central Pennsylvania are limited to waters with pH values of 6 and above, with most populations found in waters near pH 7 and where temporal variation was minimal (Glazier *et al.,* 1992, Glazier and Sparks, 1997).

Although crayfish are generally thought to require habitats with calcium concentrations in excess of 2 mg/L, low population densities of some species may be maintained by extracting sufficient calcium from their food. In adult *Orconectes virilis,* calcium uptake is impaired in habitats below pH 5; survivorship is greatly influenced by age and molt stage (Chapter 23).

2. Photoreception

Sensitivity to small differences in light intensities is extremely well developed in crustaceans. Many groups are able to locate microhabitats where they can optimize growth rates and reproduction (Hutchinson, 1967, 1993). Crustaceans have evolved receptors to detect even slight differences in light intensities, especially in deep, thermally stratified waters. Marine and freshwater zooplankton exploit thermal gradients by migrating daily and seasonally in both vertical and horizontal patterns to obtain their preferred conditions. The adaptive value of this migration for zooplankton is thought to be a combination of: (1) optimizing growth through reducing metabolic costs by remaining in cool waters; (2) maximizing growth through grazing at times when algae have the highest food value; and (3) avoiding exposure to visual predators. The relative importance of these variables can change seasonally and shift the exact timing of migration to maximize net energy gain under different conditions. When algal food resources are scarce, zooplankton may ascend "early" from deep cool waters and begin grazing 1–2 h before sunset, thereby gaining access to limited food supplies before competitors, but at a higher metabolic cost and risk of predation (Enright, 1977, 1979; Gliwicz, 1986). In arctic lakes and ponds, crustacean zooplankton are exposed to continuous daylight during most of their growing season and do not vertically migrate on a daily basis. However, they maintain phototactic reactions at approximately the same threshold values; their photosensitivity remains similar to temperate species (Buchanan and Haney, 1980). Controversy exists regarding how to evaluate the adaptive significance of variable migratory responses to different light intensities in the presence of predators (see chapters 21 and 22).

Many decapods are nocturnal and avoid sight-feeding predators by remaining inactive and undercover during the day. They have sensitive photoreception and synchonize their activity cycles within their population and thus increase their nighttime densities and reduce their individual vulnerability to predators. Many species of benthic amphipods and isopods seek dark microhabitats such as beneath leaf litter during daylight to minimize exposure to visual predators. This normal avoidance behavior of bright habitats by amphipods and isopods, however, can be affected by endoparasites. Infected crustacean hosts increase their daytime activity and can thereby increase their vulnerability to predators. This predation often continues the life cycle for the parasite with the predatory fish or bird serving as a second intermediate host (e.g., Helluy and Holmes, 1990; Maynard *et al.*, 1998). How photoreception and activity cycles of some crustacean species are altered by specific parasites is an active area of research.

3. Chemoreception

Crustaceans also have very well-developed chemosensory systems that allow individuals to locate food and mates while avoiding predators (Ache, 1982; Atema, 1988; Zimmer-Faust, 1989). Crustaceans demonstrate highly specific responses to pheromones (Dahl *et al*,. 1970; Dunham, 1978; Dahl *et al.*, 1998; Dahl, 1998; Wooster, 1998, see chapters 20–23). Hydrodynamics of stream flow and lake currents clearly mediate the effectiveness of diffuse chemicals, but the persistent unidirectional nature of many signals does provide a basis for orientation and communication. Laboratory and field studies demonstrate that chemical communication between crustacean predators such as crayfish and their gastropod prey can influence prey life histories (Crowl and Covich, 1990). Crustacean prey, such as *Gammarus minus*, respond differently to chemical cues released from injured conspecific individuals relative to chemical cues released from other species such as sympatric isopods, *Lirceus fontinalis* (Wisenden *et al.*,1999). *Gammarus lacustris* responds to alarm signals from fish predators (northern pike, *Esox lucius*), but not to dragonfly larvae (*Aeshna eremita*), apparently because fish are more effective predators (Wudkevich *et al.*, 1997). Similarly, in laboratory studies *Lirceus fontinalis* reduces their activity in the presence of alarm substances from fish mucus of predatory green sunfish (*Lepomis cyanellus*) but not from grazing fish, stonerollers (*Campostoma anomalum*) (Short and Holomuzki, 1992). Within the boundary layer of stream flow, chemoreception can be highly effective as a means for two-dimensional orientation. For organisms drifting downstream or migrating upstream, chemical cues may be extremely important. The relative size of the organism to the depth of the boundary layer is an important ratio (Dodds, 1990) that influences effectiveness of chemical communication.

During periods of stable-flow regimes in streams, crustaceans may depend on chemical cues as well as other sources of information to regulate daily and seasonal patterns of movement, to select foods or mates, and to avoid predators. As pointed out in chapters 21 and 23, there are recent studies on the effectiveness of chemical communication in crustaceans, and the situation is similar for other groups (Dodson *et al.*, 1994).

4. Mechanoreception

The ability to orient into currents (positive rheotaxis) is a common trait in many freshwater crustaceans. These physical, hydrodynamic cues are often associated with chemical cues that stimulate directed movement. Crustaceans have many types of setae, which function as mechanical and chemical sensors (Bush and Laverack, 1982). These external receptors are typically positioned near the antennae and the mandibles. However, there is a wide range of other locations so that information can be obtained from several directions simultaneously. The functional anatomy of these mechano-receptors is well studied in decapods. Behaviors associated with frequent cleaning activities are not as thoroughly studied in freshwater crustaceans as in marine taxa nor are there many studies on behavioral responses to changes in current speed (see Chapter 23 for further discussion). Behavioral studies of copepods have emphasized the importance of mechano-receptors for avoidance of predators (see review in Chapter 22), but relatively little is known for other groups.

5. Thermoreception

In warm waters rates of metabolism increase and the need for dissolved oxygen also increases. However, concentrations of dissolved oxygen decrease in warm waters and low-oxygen stress can cause mortality. This low-oxygen stress can be avoided if individuals move to certain microhabitats (e.g., groundwater seeps and deeper, stratified waters in summer). Thus, sensing differences in relative temperatures is an important ability for crustaceans to locate optimal microhabitats for growth and reproduction. Many species can orient in thermal gradients and combine information on temperature with simultaneous inputs from photo- and chemoreceptors (Ache, 1982). Freshwater crustaceans have distinct thermal preferences (e.g., Cheper, 1980; Oberlin and Blinn, 1997) and considerable information is available for crayfish (as discussed in Chapter 23). As mentioned previously, the interactions between thermal cues and light intensity are often associated with vertical migrations of zooplankton. In these behavioral responses to light and temperature gradients, there is growing evidence that zooplankton may also use chemical cues to orient themselves and to avoid predators (see chapters 21 and 22).

III. ECOLOGY AND EVOLUTION OF SELECTED CRUSTACEA

A. General Relationships

Evolution within the Crustacea has resulted in a very large number of different species. The fossil record demonstrates that both freshwater and marine crustaceans existed in the early Paleozoic and three main lines evolved from the monophyletic origin of arthropods (Schram, 1982; Emerson and Schram, 1990). Almost all of the crustaceans, that have adapted to freshwater environments, are thought to have evolved directly from marine ancestors without having gone through a phase of terrestrial adaptation (Hutchinson, 1967; Notenboom, 1991; Hutchinson, 1993; Holsinger, 1994a; Lee and Bell, 1999; Strayer, 1999).

The class Remipedia is a group of blind crustaceans that occur only in unique coastal (anchialine) cave environments (Yager, 1981, 1987; Schram, 1986; Schram *et al.*, 1986; Felgenhauer *et al.*, 1992; Carpenter, 1999). These crustaceans have a long trunk with 32 unfused segments and each segment bears a pair of similar biramous appendages. Remipedes can tolerate saline waters with very little dissolved oxygen and they apparently feed in overlying, well-oxygenated freshwater lens. This unique ecological salinity range gives them some advantages over other crustaceans living in marine polyhaline coastal cave environments. Remipedes are considered to be primitive crustaceans, and their evolution and biogeography have undergone active study since the relatively recent discovery of living representatives, primarily in the Caribbean region (Bahamas, West Indies, and in karst wells or "cenotes" of Yucatan, Mexico) but also from the Canary Islands. There are known to be fewer than a dozen species in two families. Their structural anatomy is better known (from transmission and scanning electron microscopic studies) than their functional anatomy, because they have not been successfully reared in the laboratory and studied experimentally (Felgenhauer *et al.*, 1992). Other freshwater crustaceans also occur in anchialine caves and other complex hypogean habitats as discussed below (Stock and Iliffe, 1990; Holsinger, 1993, 1994a; Holsinger *et al.*, 1994; Notenboom *et al.*, 1994; Sket, 1996; Danielopol *et al.*, 2000). Macrocrustaceans adapted to caves, artesian wells, and other groundwater habitats include a large number of amphipods, isopods, mysids, and decapods (Chapter 23) that inhabit these isolated inland waters (Holsinger, 1972; Holsinger and Longley, 1980; Abele, 1982; Bousfield, 1983; Barr and Holsinger, 1985; Holsinger, 1986, 1988, 1989, 1991, 1994a, b; Stock, 1995).

Although most benthic crustaceans are small and difficult to observe without intensive sampling, there are very large-sized crustaceans living in small and large rivers. *Macrobrachium* are the largest (>20 cm carapace length) freshwater crustaceans in North America and are distributed primarily in coastal rivers along the Gulf of Mexico (Chapter 23). They usually have amphidromous life cycles that limit their populations to coastal areas connected to brackish and marine waters where they spend part of their series of complex larval stages. Although more than 90 species occur in the tropics, there are only six species of these palaemonid shrimp in North America. However, their riverine migrations and long-term survival are vulnerable because of barriers created by impoundments, overharvesting, and water pollution. Only *M. ohione* is known from more interior rivers and its populations are thought to be declining (Bowles *et al.*, 2000).

1. Branchiura

The subclass Branchiura is widely distributed throughout the world and is comprised of about 150 species and four genera that are ectoparasites on fish. Although there are many freshwater species (Yamaguti, 1963; Cressey, 1972; Margolis and Kabata, 1988), only the genus *Argulus* is typically found in North American freshwater habitats (McLaughlin, 1980; Abele, 1982; Fitzpatrick, 1983; Schram, 1986). Some 23 North American species are encountered; three are found in western and 19 in eastern North America.

The largest number of species occurs along the Gulf of Mexico. Only a few are very restricted in their geographic ranges, while others are widely distributed; for example, *A. maculosus* occurs from New York to Michigan and down the length of the Mississippi River to Louisiana.

For many years, the Branchiura, or "fish lice," were classified as a group of Copepoda (Wilson, 1944) and earlier as Branchiopoda, but their separation as a distinct taxon is now widely accepted. Unlike copepods, the argulids have a pair of compound eyes. The Branchiura also differ from parasitic copepods in having several other unique structures that will be described. The branchiurans have undergone major modifications of the mandibles and associated feeding appendages. The mandibles typically form a pair of transversely toothed hooks between the labium and labrum, and in *Argulus,* the mandibles are incorporated into the specialized proboscis (McLaughlin, 1982). The maxillules are large suckers that are characteristic of adult branchiurans (Fig. 6). In *Argulus,* a sheathed, hollow spine is used to pierce the skin of the host. Some species are blood suckers while others feed on extracellular fluids or mucous. Once engorged, they can wait 2–3 weeks between meals.

Some reproductive and larval behaviors are unique to Branchiura. In contrast to copepods, female Branchiura do not carry eggs in external ovisacs but, instead, fasten them in rows to rocks and other objects. The actively swimming larvae attach themselves within the gill chambers, mouth, or on the outer surfaces of fishes using small, specialized antennal hooks and maxillules. After 4–5 weeks of development as ectoparasites, they become adults that again may actively swim (actually more like somersaulting through the water) to find mates and other hosts. Adult females are highly motile in seeking egg-laying sites. They apparently are opportunistic in selecting host species of fish (Lamarre and Cochran, 1992; Shafir and Oldewage, 1992). The mucous on fish surfaces does not interfere with attachment and low-level, short-term infection may not directly affect the fish host (Nolan *et al.,* 2000).

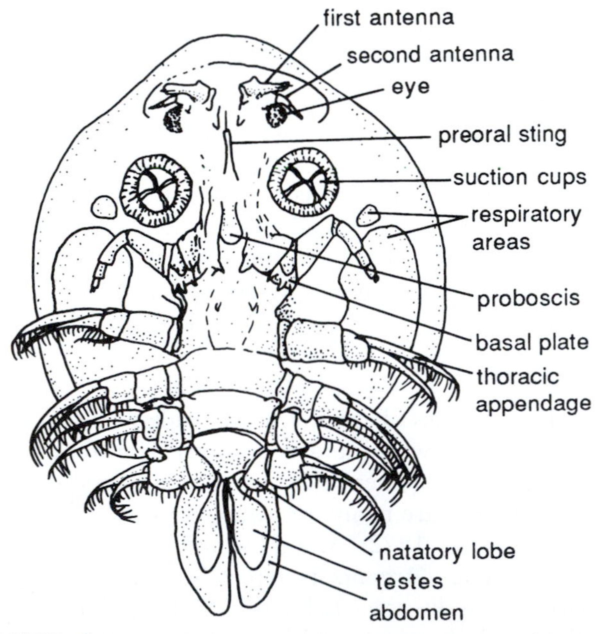

FIGURE 6 External anatomy of an ectoparasitic "fish lice" (Branchiura: *Argulus*).

2. Peracarida

The superorder Peracarida has a relatively low percentage of taxa living in inland waters. The three orders that comprise the Peracarida are the Amphipoda, Isopoda, and Mysidacea. These orders are commonly known as "scuds" (also "sideswimmers"), "sow bugs," and "opossum shrimp," respectively (Figs. 7–9). Members of a fourth order, the Thermosbaenacea, with some species living in thermal springs and others in fresh or saline groundwaters, have strong peracarid affinities but are generally considered to belong to the superorder Pancarida. Only one taxon, *Monodella texana,* lives in North American inland waters (Maguire, 1965; Stock and Longley, 1981). Most carcinologists exclude the Thermosbaenacea from the Peracarida because they brood their eggs dorsally under the carapace rather than in a brood pouch, a unique feature of the peracarids (Schram, 1982 1986). Within the Mysidacea, most of the 780 species are marine; worldwide, only 25 species occur in freshwater, and 18 additional species live in freshwater caves (Abele, 1982; Schmitz, 1992).

The Amphipoda is the largest peracarid order with some 6000 known species in more than 100 families (Schmitz, 1992). Amphipods are mostly small (5–15 mm in length), free-living benthic organisms although some are pelagic and a few are semiterrestrial. There are three large families, Gammaridae, Crangonyctidae and Hyalellidae and worldwide, there are approximately 800 gammarid amphipods (Abele, 1982; Pennak, 1989). In North America, these amphipods are represented mainly by six families (Table II) and about 175 species that occupy surface waters. Another 10 families of amphipods (composed of 28 genera and approximately 125–160 species) occupy subterranean waters (Holsinger, 1986, 1994a).

Compared to the Amphipoda, the Isopoda is a much less species-rich order but with endemic species often restricted to specific cave systems and other subterranean habitats. The Asellidae is the largest isopod family in North America with about 115 aquatic

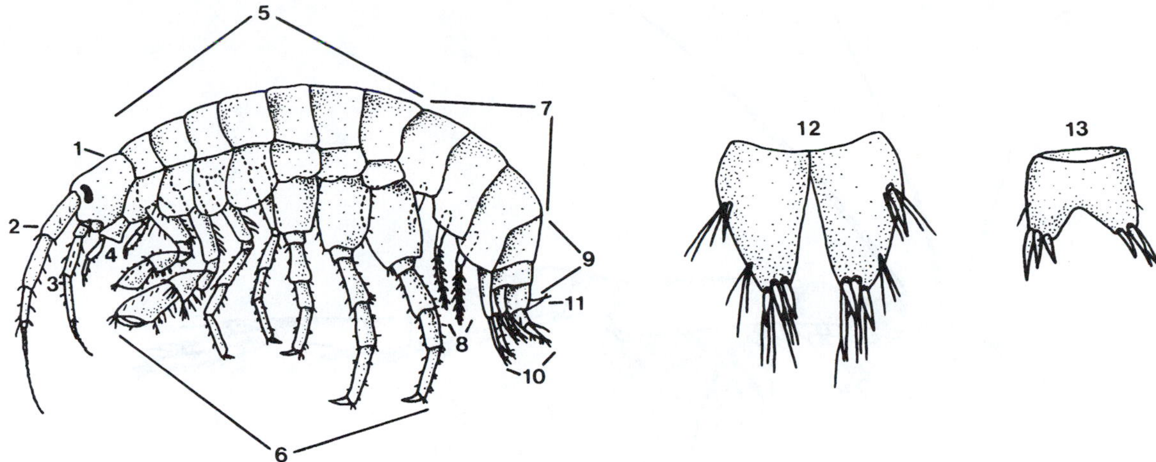

FIGURE 7 External side view of generalized freshwater gammarid amphipod: 1, head; 2, antenna 1; 3, antenna 2; 4, mouth parts; 5, pereonites 1–7; 6, pereopods 1–7; 7, pleonites 1–3; 8, pleopods 1–3; 9, uronites 1–3; 10, uropods 1–3; and 11, telson (redrawn from Holsinger, 1972). Top view of telsons: 12, *Gammarus*; and 13, *Crangonyx*.

species, of which approximately 65 are cave-dwelling species. The other families only contain 10 species in surface and six in subsurface waters of North America (Sket, 1999a). The relative scarcity of isopods in many freshwater habitats and their wider occurrence in hypogean waters and thermal springs may be related to their lack of competitive ability and more restricted timing of reproduction relative to amphipods and other benthic invertebrates (Bowman, 1981; Graça *et al.*, 1994a, b). In Europe, *Asellus aquaticus* generally occurs in the lower reachers of river drainages while *Gammarus pulex* is more common in the upper reaches, but their competitive interactions and susceptibility to predation are not completely documented. Isopods, especially juveniles, are sensitive to low oxygen concentrations, toxic metals, and organic enrichment of habitats. Ratios of widespread species of isopods and amphipods (such as *Asellus* and *Gammarus)* are used to compare regional water-quality conditions (Naylor *et al.*, 1990; Maltby, 1991, 1995; Whitehurst, 1991; Mullis *et al.*, 1996). Examples of widely distributed genera of isopods, such as *Caecidotea Lirceus,* and *Asellus,* are well studied and often compared with associated amphipod species in surface and sub-surface waters.

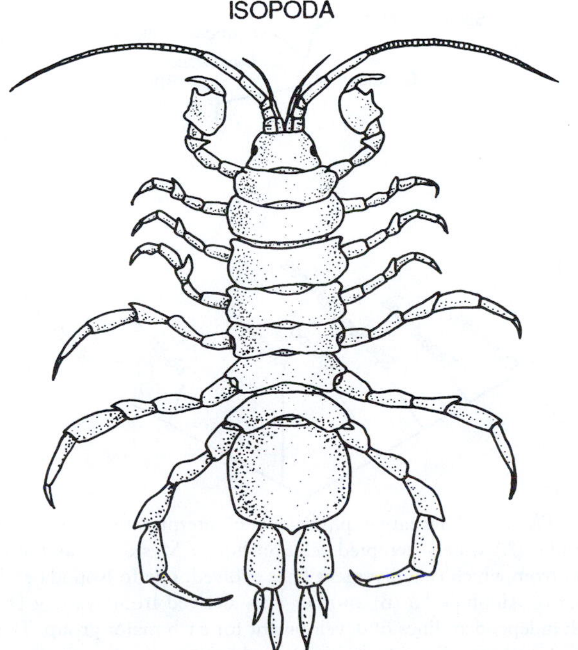

FIGURE 8 Diagram of top view of isopod *Caecidotea* (drawn without setation).

The Peracarida share several evolutionary similarities that demonstrate a common lineage. For example, all have modified the ancestral first thoracic leg into a mouth part (the maxilliped) and retain seven pairs of thoracic legs for movement. The first two pairs of these thoracic legs (the gnathopods) are specialized for grasping food while the remaining five pairs of legs (the pereiopods) lack any specialized grasping (chelate) appendages. As mentioned previously, peracarid females carry their eggs and young until a relatively advanced stage of development in a specialized ventral brooding chamber, the marsupium. Another character used for separating the peracarids from most other crustaceans is the highly developed *lacinia mobilis,* a small toothed process that articulates with the incisor process. A row of spines commonly separates the *lacinia mobilis* and the molar process (McLaughlin, 1982). Several phylogenetic relationships have been proposed for the

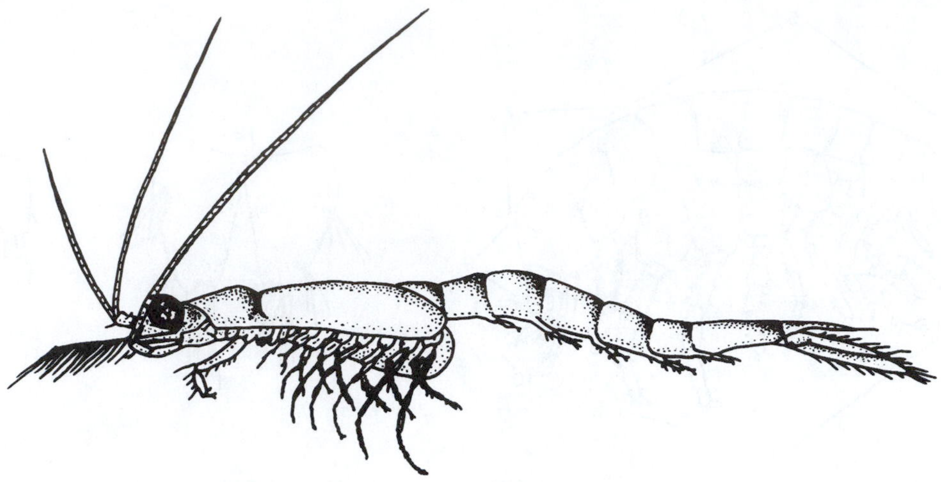

FIGURE 9 Diagram of side view of opossum shrimp, *Mysis*.

evolution of the Peracarida (see Schram, 1982, 1986, for review) based on how the fossil record is interpreted relative to modern morphological criteria. Many workers view the Isopoda and Amphipoda as derived from a common mysidlike stock. Watling (1981) proposed a different arrangement that is more consistent with the known, but incomplete, fossil record (Fig. 10). Most workers agree that the amphipods and mysids are more closely related to each other than to the isopods (Schram, 1982, 1986; Sieg, 1983).

Evolutionary relationships among species and subpopulations of widely distributed species are active areas of research (as discussed below) and new molecular techniques are providing insights regarding differences in rates of speciation. For example, studies using molecular time scales based on allozyme analysis suggest that the divergence of several amphipod lineages that gave rise to the three main species of *Pontoporeia* (*P. affinis, P. hoyi,* and *P. femorata*) is likely to have occurred over tens of millions of years. The divergence of *Mysis relicta* from its likely marine congeners apparently dates from the Tertiary (Vainola and Varvio, 1989; Vainola, 1990).

TABLE II Representative Taxa Found in North American Fresh Waters

Superorder Peracarida
- Order Amphipoda
  - Family Crangonyctidae
    - *Crangonyx, Stygobromus*
  - Family Gammaridae
    - *Gammarus minus, G. lacustris*
  - Family Hyalellidae
    - *Hyalella azteca, H. montezuma*
  - Family Pontoporeiidae
    - *Monoporeia affinis*
- Order Isopoda
  - Family Asellidae
    - *Asellus, Caecidotea, Lirceus*
- Order Mysidacea
  - Family Mysidae
    - *Mysis relicta, M. littoralis, Neomysis mercedis, Taphromysis louisiannae*

(A)

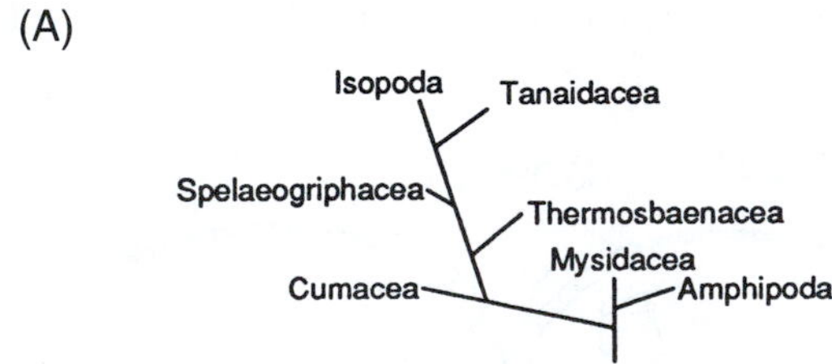

(B)

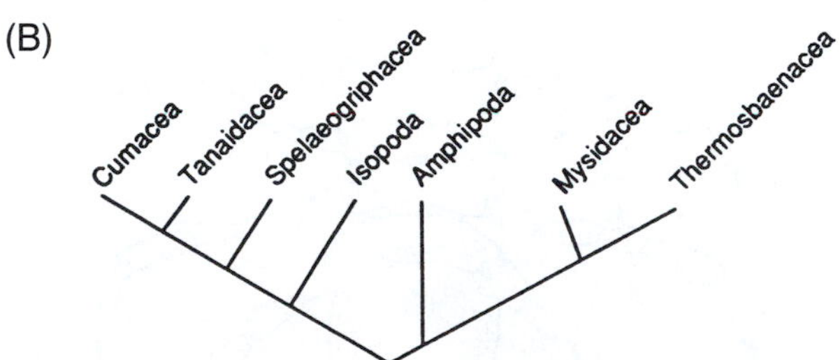

FIGURE 10 Alternative phylogenetic interpretations of the Peracarida: (A) widely accepted relationship of Mysidacea as the main stem from which two divergent lines evolved, one to Isopoda and the other to Amphipoda; (B) another view derived from Watling (1981) with independent lines of development for each major group. The orders Cumacea, Tanaidacea, and Spelaeogripacea are marine peracarids, while the order Thermosbaenacea is often considered a pancarid group closely related to the peracarids. [From Schram, 1982.]

B. Ecological Distributions and Interactions

1. Habitats

The widespread distribution of Branchiura suggests a lack of any specific pH or salinity requirements, but detailed studies of these ecological relationships are lacking. Temperature influences the rate of hatching for eggs of different species with a range of 12–30 days being required (Yamaguti, 1963). During the free-swimming stage *Argulus* may aggregate in a narrow zone of temperature and light conditions where encounters with host fishes can be increased. *Argulus* has no strict host specificity and is capable of attaching to both freshwater and marine telosts.

Most peracarid taxa that have fully adapted to the physiological and ecological conditions of inland waters share five ecological similarities in that they: (1) are typically restricted to permanent bodies of water that are relatively cool, clean, and well oxygenated; (2) have distinct behavioral patterns of vertical migration (e.g., among mysids and two species of amphipods; these are similar to daily migration in other groups of crustacean zooplankton, as discussed previously and in chapters 21 and 22); (3) have relatively limited ability to move upstream (positive rheotaxis is well studied in amphipods and isopods) or drift downstream in the current under specific ecological conditions (as do many of the lotic insects discussed in Chapter 18); (4) obtain much of their energy while feeding on the bottom substrates (even if they also feed while in the nekton on suspended algae and smaller taxa of zooplankton); and (5) serve as important prey to a large number of predatory fishes (also discussed in chapters 21–23). Some taxa also have various ways of inflicting damage to predatory populations either by being direct parasites on predators or by competing with young stages of those predators for the same prey (such as *Mysis* feeding on plankton) that would otherwise be potentially available to fish consumers.

In contrast to other crustaceans, most of these peracarid species typically lack a diapause phase and are missing any distinct adaptation for avoiding desiccation. Thus, their chances for passive dispersal (e.g., being carried from one habitat to another by ducks or other migratory animals or being carried by the wind during storms) are low in comparison with cladocerans, copepods, or ostracodes (Hairston and Caceres, 1996; Hairston and Bohonak, 1998; Bohonak, 1999). The peracarids also lack any strong active dispersal ability. They generally do not move upstream against strong current (Olyslager and Williams, 1993) so their migration throughout river-drainage networks is restricted, and populations apparently remain more or less in the same locality for long periods. Nonetheless, as discussed later, some taxa are very widely distributed and their patterns of distribution suggest a long history of slow dispersal and persistence.

There are a few very widely distributed amphipod species in surface waters. Some surface-dwelling and subsurface species have been well studied ecologically so that comparisons of population densities and life histories provide important information for their conservation and as indicators of water quality (e.g., Mullis *et al.*, 1996; Mosslacher, 1998; Knapp and Fong, 1999). The most widely distributed and diverse genera of cave-inhabiting amphipods are *Stygobromus* with approximiately 180 species and *Crangonyx* with 14 species (Peck, 1998). No doubt, many species remain undescribed in caves, hyporeheic zones, deep lakes, and even in well-studied, shallow-water habitats. The continued discovery of new species, and the recent applications of new techniques for genetic analysis of previously described taxa, suggest that many more genetically distinct populations occur than was first expected (Hogg *et al.*, 1998b). Many species of amphipods and isopods are exceptionally well adapted to live in surface waters as well as groundwater habitats and caves (Holsinger, 1972, 1986, 1988; Stanford and Ward, 1988; Fong, 1989; Culver *et al.*, 2000). In cave ecosystems, a single species of amphipod may dominate a relatively simple food web based on fine organic particulates (e.g., Drost and Blinn, 1997). The microbiogeographical distributions and evolution of relatively common species such as *Gammarus minus* are under intensive study (Kane *et al.*, 1992; Sarbu *et al.*, 1993; Culver *et al.*, 1994; Culver and Fong, 1994; Fong and Culver, 1994; Culver *et al.*, 1995; Kane *et al.*, 1995). As discussed below, other intensive studies have focused on the evolution of subpopulations of *Hyalella azteca*; some apparently have formed distinct species in certain surface-water habitats (Thomas *et al.*, 1997; McPeek and Wellborn, 1998).

Ecologically, most of the amphipods and isopods are photonegative, positively rheotactic, thigmotactic, cold stenotherms (i.e., restricted to relatively constant, cold waters where they avoid bright light by moving into the current and into crevices or under stones, leaves and roots). In complex substrates, they are less exposed to predators such as fish and crayfish. Typically, where refugia from predation exist, the peracarids occur at high densities in small, permanent, spring-fed streams, seeps, ponds, or sloughs (e.g., Allee, 1914, 1929; Juday and Birge, 1927). In isolated springs their populations sometimes genetically diverge from those in other springs (Gooch and Glazier, 1991; Glazier *et al.*, 1992; Culver *et al.*, 1994; Glazier, 1999). Some species of *Gammarus* reach densities of thousands of individuals per square meter where detrital food and cover are abundant and predators are few or absent.

A relatively smaller number of isopod species have apparently adapted to live in intermittent or permanent surface waters but distributional data are scarce (Richardson, 1905; Hubricht and Mackin, 1949; Williams, 1970; Ellis, 1971). Certain species are common in subsurface habitats. For example, the subterranean isopod *Caecidotea tridentata* is relatively abundant in groundwaters of the Konza prairie (Edler and Dodds, 1996). The most diverse genus of cave-inhabiting aquatic isopods is *Caecidotea* with more than 56 species (Peck, 1998). For most genera of isopods the number of described species found in groundwater habitats and caves of North America ranges from one to five (Holsinger, 1988; Fong and Culver, 1994; Culver *et al.*, 1995; Peck, 1998). Only a few known species tolerate the low oxygen concentrations found in polluted rivers and groundwaters (Williams, 1970; Strayer *et al.*, 1997; Henry and Danielopol, 1998; Mosslacher, 1998; Hervant *et al.*, 1999; Malard and Hervant, 1999) or the stress of high temperatures (Dadswell, 1974). For example, some isopods (*Caecidotea recurvata*) survived in cave pools only slightly polluted by septic tank effluents, but were absent from highly polluted pools (Simon and Buikema, 1997). There is concern that any level on increased food input by organic pollution could result in competitive displacement of subsurface species by surface-dwelling species that are better adapted to use higher concentrations of food (Sket, 1999b).

The order Mysidacea has only a single family and one very important species, *Mysis relicta* (Fig. 9), which occurs in cold, deep lakes as well as in shallow brackish ponds along the arctic coasts. A second species, *Neomysis mercedis*, has a more limited, coastal distribution but plays a significant role in some large lakes such as Lake Washington (Murtaugh, 1989). *Mysis relicta*, the "opossum shrimp," is not a true shrimp (or decapod as defined in Chapter 23), but is an active swimmer and can make long daily trips up and down the water column of deep, well-oxygenated lakes (Beeton and Bowers, 1982; Shea and Makarewicz, 1989). A third species, *Taphromysis louisianae*, lives in roadside pools in Louisiana and Texas (Banner, 1953; Pennak, 1989).

Mysids have a holarctic distribution; they occur in oligotrophic lakes of northern regions of North America and in northern Europe, but are known to tolerate relatively high temperatures and periods of low oxygen at least for short periods (Dadswell, 1974; Sherman *et al.*, 1987). As will be discussed later, their introduction into many western lakes has created several problems for fisheries managers (Martinez and Bergersen, 1989). Their natural limits of distribution apparently have been influenced by opportunities for migration during the Pleistocene and possibly by biotic interactions (Dadswell, 1974).

2. Food Resources

All three peracaridian orders are similar in the range of foraging behaviors used in obtaining food resources. Juveniles are typically dependent on microbial foods such as algae and bacteria associated with either periphyton or aquatic plants in brightly illuminated habitats. They also consume dead organic matter in forested streams, ponds, and lakes where bacteria and fungi living on the detritus provide essential protein for amphipods and isopods (Barlocher and Kendrick, 1973, 1975; Smock and Stoneburner, 1980; Smock and Harlowe, 1983; Findlay *et al.*, 1984, 1986; MacNeil *et al.*, 1997). Adults are not limited to grazing or detritivory because they broaden their food niche to include larger food items and are predatory (Schwartz, 1992; MacNeil *et al.*, 1999b). Some pelagic or nektoplanktonic amphipods in fishless lakes, ponds, and wetlands consume phytoplankton and zooplankton (Anderson and Raasveldt, 1974; Dehdashti and Blinn, 1991). Adults of well-studied species, such as *Gammarus minus*, are known to depend primarily on detritus when in surface waters but cave populations of this species are more flexible in their diet (Culver *et al.*, 1995). In addition to the many plant, animal, and detrital foods used by different species of *Gammarus*, intraguild predation and cannibalism are also thought to be relatively common in surface waters (Dick, 1995; MacNeil *et al.*, 1997, 1999b). Amphipods become opportunistic scavengers, predators, and omnivores, depending on food quality and which foods are most available (Delong *et al.*, 1993). Given their wide range of foods, it is clear that different modes of feeding are used to exploit these different resources (e.g., Crouau, 1989). These generalized consumers can reach high population densities by deriving their energy from many sources. Resource competition among guild members can be intense if access to suspended foods is space-limited, as may occur in soft-bottom streams among filter-feeding species that require hard surfaces for attachment or perching. Interference competition and aggressive behavior can also occur among species that differ in aggressiveness and mobility (e.g., Dick *et al.*, 1995; Haden *et al.*, 1999).

Laboratory and field studies of production have focused on juvenile and adult growth rates that result from feeding on different types and qualities of foods at widely different locations (Waters and Hokenstrom, 1980; Marchant and Hynes, 1981a; Sutcliffe *et al.*, 1981; Delong *et al.*, 1993; France, 1992, 1993). Only a few studies have considered the role of secondary plant chemicals as "defensive" compounds for aquatic plants to minimize effects of grazers. Concentrations of tannins, and other inhibitory chemicals are known to differ among various macrophytes, and some compounds influence crustacean grazing preferences (e.g., Newman *et al.,* 1990, 1992). Plant foods that contain a high percentage of cellulose are often considered to be of relatively low quality for grazers and detritivores (Newman, 1991). However, some amphipods (Monk, 1977) and mysids (Friesen *et al.,* 1986) have digestive enzymes that hydrolyze cellulose *in vitro.* These cellulases, however, are apparently confined to degrading small particles in the gut rather than whole plant cell walls that are ingested during grazing. Microbial breakdown of ingested cellulose within the gut is generally the mechanism used for digesting this refractory material and requires development of a complex community of various microorganisms within the gut. Given that crustaceans molt periodically and would need to re-establish this microbial community in their new gut lining, the evolution of cellulases would be an important adaptation for expanding use of detrital food resources.

3. Vertical Migrations and Feeding

Amphipods and isopods are often confined to burrowing and feeding on the bottoms of streams and lakes, although species like *Monoporeia affinis* (= *Pontoporeia affinis*) move from their benthic foraging areas during the day to nektonic, open-water foraging at night (Marzolf, 1965; Winnell and White, 1984; Donner *et al.,* 1987). These actively swimming amphipods consume a wide range of particles while on the sediment surface, although the exact nature of the food is not completely known for those species found in the Laurentian Great Lakes or other large basins (Dermott and Corning, 1988; Lopez and Elmgren, 1989; Nalepa *et al.,* 2000). The seasonal deposition of diatoms and other phytoplankton from the surface waters provides important food resources for bottom-dwelling amphipods (Johnson and Wiederholm, 1989; Hill *et al.,* 1992; Johnson and Wiederholm, 1992; Sly and Christie, 1992; Goedkoop and Johnson, 1994).

The vertical movements of amphipods, such as *Monoporeia affinis,* in deep lakes overlap those of mysids with whom they may share food resources. Mysids provide an intriguing example of vertical migration and complex foraging dynamics (Chess and Stanford, 1998, 1999; Rudstam *et al.,* 1999; Spencer *et al.,* 1999). *Mysis relicta* regulates nocturnal feeding patterns by responding to changing light conditions (Johannsson *et al.,* 1994). These very fast, active swimmers have well-developed eyes for precise photoreception. They rely on daily changes in intensity and quality of light (that penetrates through the photic zone) to determine when to begin swimming up through the water column from the dimly lit, deeper waters and to begin feeding on small zooplankton and phytoplankton (Morgan *et al.,* 1978; Cooper and Goldman, 1982; Folt *et al.,* 1982).

By precisely regulating their travel times, migrating crustaceans reduce their exposure to fish predators and still obtain the food they need to grow and reproduce. Both *Mysis* and *Monoporeia* have potential fish predators that feed on benthic and open-water habitats so that minimizing exposures to a wide range of predators, while also sometimes competing with them for zooplankton prey, is a complex process (McDonald *et al.,* 1990; Johannsson, 1992; Johannsson *et al.,* 1994; Johannsson, 1995; Wilhelm, 1996). As mentioned previously, those zooplankton that seek deeper, cooler waters during the day not only avoid brightly lit waters and exposure to visual predators, but they also lower metabolic rates while digesting the food that they obtained the previous night (Chipps, 1998; Rudstam *et al.,* 1999). Depending on the temperatures of the upper strata, *Mysis* migrates faster and closer to the surface than does *Monoporeia* (Wells, 1960). Because *Mysis* has less tolerance for low concentrations of dissolved oxygen in the deeper strata than does *Monoporeia,* it may not spend much time in those deep layers during the later portion of the growing season. Decomposition of organic matter in the deep hypolimnion depletes dissolved oxygen if the upper waters are highly productive and dead matter falls into these deep waters. Of considerable importance is the observation that *Mysis* can eat *Monoporeia* (Parker, 1980). The spatial refuge from mysid predation for amphipods then is dependent on the length of time that a lake remains thermally stratified, its level of productivity, and its bottom water temperature, because these factors determine both the concentration of dissolved oxygen in the deepest strata and the presence of various benthic predators.

Because mysids are consumed by a number of game fishes, there was a period of widespread, intentional introduction of *Mysis relicta* into many deep lakes in the western and northern United States (Martinez and Bergersen, 1989; Spencer *et al.,* 1999) and in western Canada (Lasenby *et al.,* 1986). As often happens with introductions of species into habitats where they do not naturally occur, the foodweb dynamics did not develop as expected. Of the 134 recorded introductions into the lakes of western North America, only 45 became directly established and eight others became indirectly established through passive migrations to other lakes (Martinez and Bergersen, 1989). In some lakes, instead of providing more food for predatory fishes such as lake trout and coregonids, the mysids ate many of the same zooplankton species that the juvenile fish had previously consumed (Lasenby *et al.*, 1986). Thus, the mysids began competing with fish for zooplankton prey (Cooper and Goldman, 1982; Spencer *et al.*, 1991, 1999). In these lakes, the mysids did not migrate high enough into the surface waters where the visually oriented fish could capture them. In the spring and summer, mysids avoided the warmer, upper, dimly lit surface waters and thus escaped predation; furthermore, they intercepted the upwardly moving zooplankton at midwater depths before the zooplankton could be consumed by fish. Even more unexpected was the predation by *Mysis relicta* on newly hatched larval fish (Seale and Binowski, 1988).

An unusual example of finding food occurs with the amphipod *Hyalella montezuma*. This species migrates to the surface waters of a large karst sink hole (Montezuma Well, Arizona) and filters out organic material entrapped on the surface film (Cole and Watkins, 1977; Blinn and Johnson, 1982; Blinn *et al.*, 1987; Wagner and Blinn, 1987). Its twilight migration is followed by a predatory leech, which swims up to feed on the dense concentration of amphipods and other prey (Blinn *et al.*, 1988, 1990; Blinn and Davies, 1990; McLoughlin *et al.*, 1999); no fish predators occur in this sink-hole lake. In contrast, *Hyalella azteca*, which coexists with *H. montezuma* in Montezuma Well, does not filter feed or migrate to the surface. In comparative studies of Devils Hole, Nevada, Dean Blinn and Kevin Wilson (personal communication) observed that *H. azteca* migrate horizontally from the shoreline to the open-water column at night, which is the reverse of what occurs in Montezuma Well (Blinn *et al.*,1987). In fishless lakes, ponds, and wetlands, other amphipods, such as *Gammarus lacutris* and *G. pulex*, are known to feed in the water column as well as on the bottom (Anderson and Raasveldt, 1974).

Other well-studied examples of complex vertical migrations of predatory macro-crustacean feeding on zooplankton occur among species living in very deep lakes and are similar to massive aggregations of *Mysis relicta* in North American lakes (Lasenby *et al.*, 1986). As discussed below, *Macrohectopus branickii* in ancient Lake Baikal has dynamics and functions similar to North American mysids (Rudstam*et al.*, 1992; Melnik *et al.*, 1993).

4. *Responses to Water Quality*

The general requirement for habitats with relatively high concentrations of dissolved oxygen by most peracarid crustaceans usually limits them to clean, cold waters especially, during their early life histories. Some species of amphipods are sensitive to changes in temperature, stream discharge, and predators in their choice of microhabitats through their onset of downstream drift responses and ability to move upstream in search of appropriate conditions (Waters and Hokenstrom, 1980; Williams and Moore, 1985). In productive habitats crustaceans can deposit excess energy in the form of lipid which they store internally (Hill *et al.*, 1992; Adare and Lasenby, 1994; Cavaletto *et al.*, 1996; Nalepa *et al.*, 2000). Because they feed on a variety of materials and incorporate chemical constituents from their food into their bodies (often in their lipids), these organisms are important in research on water-quality monitoring (Conlan, 1994; Landrum and Nalepa, 1998; Song and Breslin, 1998, 1999). Many species also move over relatively large areas and, thus, are exposed to a wide array of microhabitats that may contain specific toxins. The ease of collecting amphipods and isopods from a broad range of habitats and their hardiness in laboratory mesocosms make these crustaceans ideal candidates for monitoring pollution. They are especially sensitive to copper (even from copper pipes in laboratory plumbing) and a number of other toxic heavy metals (Thybaud and LeBrac, 1988; Borgmann and Munawar, 1989). They can be used as "sentinels" of chronic exposure to low concentrations of toxins or as indicators of acute episodes of toxic spills that might otherwise go unmeasured by routine sampling of streams and lakes.

Accidental spills of toxins can have major impacts on survival of crustaceans and their ecological roles in maintaining high water quality. Experimental studies demonstrate the importance of detrital processing by benthic crustaceans through comparisons of reference "control" streams with "treated" or disturbed streams (e.g., Newman *et al.*, 1987). Mysids also have become important organisms for monitoring the movement of toxins through foodwebs in large lakes (Evans *et al.*, 1982). The presence or absence of *Mysis relicta* in different lakes is important for understanding how toxins such as mercury accumulate in top fish predators because of the major role these crustaceans play in fish production (Cabana *et al.*, 1994; Branstrator *et al.*, 2000).

C. Life-History Traits

1. *Longevity*

Although most small species appear to be "annuals" and may complete their life cycle within a single year, larger amphipod species such as *Monoporeia* are thought to live for two years or more. Troglobitic species may live even longer (5–6 years) in stable habitats with low but relatively continuous inputs of organic detritus as food sources. Long-term data demonstrate cyclic population oscillations, although control mechanisms for the observed population growth cycles are not clear. Generally, effects of hydrographic and nutrient cycles on pelagic primary productivity and related changes in food quality and quantity on amphipod fecundity and secondary production are well documented (Hynes and Harper, 1972; Siegfried, 1985; Sarvala, 1986; Johnson and Wiederholm, 1989; Dehdashti and Blinn, 1991). Isopods inhabiting freshwater and saline ponds typically have an annual life cycle in contrast to the longer lifespan of four years for many terrestrial isopods (Sastry, 1983). *Caecidotea recurvata,* a cave-dwelling isopod, is known to live at least four years (D. C. Culver, personal communication).

2. Patterns of Reproduction

Generally, Peracarida mate at, or shortly after, the time of the female molt (Sastry, 1983). In most amphipods and isopods, females produce only a single brood during the annual life cycle. Seasonal peaks in annual reproduction occur in *Gammarus minus* populations in spring habitats but this seasonal pattern in less distinct in cave populations (Culver *et al.,* 1995). Some species are known to reproduce continuously and have multiple broods. For example, the subterranean isopod *Caecidotea tridentata* apparently reproduces throughout the year (Edler and Dodds, 1996). Amphipods, such as *Hyalella azteca,* can produce multiple broods during an extended breeding season (Cooper, 1965; Strong, 1972). Adult females can reproduce at every stage of molting. They carry their eggs in the marsupium and after fertilization continue to bear their young until they release the fully developed juveniles (ranging from a few to more than 20 per brood). Mate-guarding is known to occur in some species with time-limited reproductive cycles (e.g., Jormalainen and Shuster, 1999). Some families of amphipods differ in how mating occurs. The Pontoporeiidae have a pelagic mating system while the Gammaridae generally have a benthic mode (Strong, 1973; Naylor and Adams, 1987; Bousfield, 1989; Elwood and Dick, 1990; Wen, 1993).

Several types of broad life-history patterns have been documented for both North American and European species (e.g., Cooper, 1965; Strong, 1972; Gee, 1988; Tadini *et al.,* 1988). In general, the cold-spring faunas have longer life spans, produce fewer eggs, and lack marked seasonality in their reproduction compared to warm-water taxa, which only live one year and produce large numbers of young in the spring. A slowing down of reproduction during late summer is reported in some amphipod populations and may be related to food scarcity or shifts in breeding behavior controlled by other factors. In *Gammarus minus* the onset of amplexus is apparently induced by a rapid decline in temperature (D. C. Culver, personal communication). In widely distributed, warm-water species such as *Hyalella azteca,* day length has been ruled out as a controlling variable for triggering the onset of reproduction (Strong, 1972). For cold-water species such as *Gammarus lacustris,* a period of short days and long nights (typical of winter) is needed to induce reproduction (DeMarch, 1982). In *Monoporeia affinis,* the boreo-arctic amphipod, constant illumination inhibits gonadal development (Sergestrale, 1970).

Peracarid activity patterns on some substrates are also known to be influenced by the presence of predators (Newman and Waters, 1984; Malmqvist and Sjostrom, 1987; Holomuzki and Short, 1988, 1990; Holomuzki and Hoyle, 1990; Holomuzki and Hatchett, 1994; Wooster, 1998). These behavioral changes may influence life-history patterns in certain habitats. For example, different populations of the stream-dwelling isopod *Lirceus fontinalis* responded to different risks of predation from banded scuplins (*Cottus carolinae*) or streamside salamanders (*Ambystoma barbouri*) by maturing at larger sizes in those habitats where gape-limited salamanders were the dominant predators (Sparkes, 1996a, b). Endoparasites also influence isopod and amphipod behavior and may make some individuals more vulnerable to predators by increasing their tendency to drift (McCahon *et al.,* 1991; Hechtel *et al.,* 1993, Bakker *et al.,* 1997; Maynard *et al.,* 1998). In some cases the infected individuals are more active during daylight or more visible due to changed coloration. In these ways the parasite increases host vulnerability to predators and can enhance transmission of the parasite from the initial crustacean host to definitive vertebrate hosts such as fishes and birds that completes the parasite's life cycle (Bethel and Homes, 1977; Camp and Huizinga, 1979; Bakker *et al.,* 1997).

D. Biogeography

The distribution and abundance of any species reflects a large number of variables that may have interacted over very long periods. Climatic changes have clearly influenced the stability of aquatic habitats and their availability for colonization by crustaceans for millions of years. The most recent glacial advances and retreats were associated with not only major fluctuations in global temperatures, but also with changing water balances of lakes and river drainage ecosystems. When much of the earth's water was frozen in massive ice sheets, the levels of many lakes and rivers were much reduced in the temperate zones and throughout the Neotropics. As discussed later in this section, important refugia from thermal extremes and desiccation have persisted in below-ground habitats such as groundwater-fed caves.

The mechanisms of transport and available routes of dispersal for a species are primary factors in limiting access to new habitats and recolonization of old habitats that might have had major climatic disturbances. Physical and chemical barriers to entry can also limit range expansions, as may biological barriers such as the presence of predators or strong competitors already established in adjacent waters (MacNeil *et al.*, 1997; 1999b). Among the Peracarida, there is a pattern of restricted distribution for many species (sometimes to a single lake or spring-fed stream) that has interested biologists, geologists, and geographers. A few species are widespread and seem to tolerate a broad range of ecological parameters. For example, *Gammarus lacustris* occurs throughout most of the northern and parts of the western United States. *Hyalella azteca* is present from Canada to South America in the littoral zone of glacial lakes as well as in small ponds and streams. Several species of *Crangonyx* are frequently found in the southeastern United States as well as in Canada (Bousfield and Holsinger, 1989). Habitat modification has changed some amphipod distributions in historic times, such as the likely effects of dam building and regulation of flow regimes (e.g., Beckett *et al.*, 1998).

Many species are considered to be relatively rare. The isopod, *Thermosphaeroma theromophilum,* has been declared legally endangered; it is restricted to a currently protected hot-spring habitat near Socorro, New Mexico (Bowman, 1981). As discussed below some cave species are highly restricted in their distributions and some endemic species occur in only a single cave (Holsinger, 1991; Holsinger, 1994a, b). Recent reviews provide examples of threats to survival of crustacean species (Elliott, 2000; Ricciardi and Rasmussen, 1999; Culver *et al.,* 2000; Master *et al.,* 2000).

1. Groundwaters

The greatest number of amphipod species are found in subterranean interstitial habitats, and many are associated with caves or deep wells (Barr and Holsinger, 1985; Holsinger, 1993; 1994a). Bousfield (1983) suggested that more than 1000 hypogean amphipod species will eventually be identified. J. R. Holsinger (personal communication) completed a worldwide survey and found that 117 genera have subterranean species; approximately 615 species are fully adapted to live in groundwaters. There are about 116 species that live in groundwaters in North America, north of Mexico. Holsinger (1972) estimated that 65–70% of the North American amphipod fauna occurs in these groundwater habitats; such species are known generally as "stygobionts." He pointed out that not all of the stygobiont species are restricted to cave waters per se (true cave-dwelling species are termed "troglobites"), yet they may have similar morphological specializations associated with cave living (e.g., loss of eyes and pigmentation). Some cave-dwelling species are endemic to a single location within one county. Generally, the geographic locations of obligate cave-dwelling species are highly restricted and these subterranean communities appear regionally undersaturated. Culver *et al.* (2000) report that over 50% of all the aquatic obligate cave species and subspecies occur in only 19 counties (out of 3112) that comprise less than 1% of the land area in the 48 contiguous states. Most taxa are concentrated in Hays County, Texas or in a few counties in Florida, Oklahoma, Texas, Virgina, and West Viriginia. This high concentration of subterranean taxa in a relatively few locations makes these populations highly vulnerable to habitat destruction. For example, deforestation and dumping of sawdust (into sink-hole cave from a sawmill operation and leachate into groundwater from sawdust piles) in southwestern Virginia led to listing a cave stream isopod (*Lirceus usdagalun*) as an endangered species (Jacobs, 1991; Culver *et al.*, 1992). Although many subterranean populations are at risk, only a few have protection through federal status as rare or endangered.

Groundwaters are insulated from many of the environmental fluctuations that characterize surface waters, but prolonged droughts or massive ice formations associated with glaciation might eliminate these otherwise "constant" habitats. Until recently, it was thought that the few subterranean amphipods and isopods known from glaciated regions represented postglacial colonization from nonglaciated border regions to the south. Subglacial refugia are now thought likely to have persisted during the last glacial episode (known as the Wisconsin, which ended about 12,000 years ago). For example, Castleguard Cave in Alberta, Canada, is currently inhabited by the asellid isopod *Salmasellus steganothrix* and the crangonyctid amphipod *Stygobromus canadensis.* The amphipod is endemic to this single cave and was found about 2 km from the entrance. Eleven of the 100 species described in the genus *Stygobromus* are known to occur in various localities north of the southernmost boundary of the Wisconsin episode of Pleistocene glaciation (Holsinger *et al.,* 1983). Thus, the action of glacial scour of surface habitats did act as a barrier to northern distributions of many species, but others were adapted to subsurface waters and found refuge.

The fluctuations of sea levels associated with Pleistocene glaciations and earlier Tertiary events also created barriers to distributions of Peracarida. Colonization of coastal streams and subterranean habitats by marine-derived species has resulted in a complex mosaic of biogeographic boundaries. Initially, those organisms that adapted to brackish waters were able to escape most of their marine-based fish predators and to obtain abundant food resources in freshwater habitats. As shallow Cenozoic marine waters receded, the isolated populations of amphipods and isopods developed distinct adaptations; subsequent speciation produced many new taxa (Holsinger, 1986; Notenboom, 1991). In isolated refugia, these species diversified over a 70-million year period (possibly much longer) with apparently only slow and limited active dispersal. Because many of these underground habitats are deep aquifers, the faunal composition is difficult to sample. The Edwards Aquifer in south-central Texas has been studied intensively (Holsinger and Longley, 1980), and a unique ancient fauna of at least 22 species, including 10 amphipod species, is known to live in this stable groundwater habitat. Abele (1982) concluded that this artesian well in San Marcos, Texas, contains probably the richest cave crustacean fauna in North America.

2. Surface Waters

Today, the peracarids occupy many types of surface waters that range widely in physical and chemical characteristics. Historically, the chemical barrier of dilute water was a major isolating mechanism (Vernberg and Vernberg, 1983) in the evolution of crustaceans. In contrast to the frequently isolated subsurface populations and subpopulations, relatively wider geographic distributions and frequent gene flow can occur among surface-dwelling, interbreeding populations in large lakes and ponds interconnected by rivers. Long-distance transport by individuals adhering to mud carried by migratory birds, water fowl, and other animals is considered rare among amphipods but might infrequently provide wider dispersal by those individuals that do survive short-term drying.

Some species of isopods are reported to move through subterranean waters and into the surface waters for certain periods (e.g., Minckley, 1961). The importance of hyporheic habitats as spatial and temporal refugia for many types of benthic crustaceans and insects is now recognized as being of great importance, especially in previously glaciated regions where massive unconsolidated rock and sand deposits characterize the floodplains of large rivers (Stanford and Ward, 1988).

3. Barriers to Dispersal and Population Genetics

Neighboring populations can be genetically isolated from one another by many types of physical barriers. Dispersal of individuals to distant headwaters of many fast-flowing streams is limited within a drainage area unless there are subsurface connections among aquifers or many short tributary branches of streams that connect. Gooch (1989) quantified seven levels of isolation in his studies of *Gammarus minus* in the Appalachian Mountains of West Virginia and Pennsylvania. The larger the barrier, the more isolated were the populations. He has shown that there is a cline of lower genetic variability along a sequence of northern populations. These populations have apparently recolonized habitats since the last glacial advance and now occupy what was once a periglacial zone or perimeter around the ice sheets (Gooch and Glazier, 1986). The importance of the geomorphological setting in determining how gammarid populations are actively and passively distributed is also clear in other studies of karst populations (e.g., Gooch, 1990; Gooch and Glazier, 1991; Culver and Fong, 1994).

In addition to the studies on *Gammarus* mentioned above, other investigations have focused on *Hyalella* and these also provide important new information about adaptations among distinct populations and subpopulations. For example, in the glacial terrain of southeastern Michigan's there is evidence for recent divergence among populations of two ecotypes of *Hyalella azteca* (Saussure) that differ in terms of adult body size (McPeek and Wellborn, 1998). The large-bodied adults from different populations found in lakes interbreed as do the small-bodied adults found in marshes. In contrast, individual adults of different ecotypes do not interbreed. This type of recent divergence may have resulted from enhanced mating success favored through sexual selection and mediated by fish-predation selection in the lake populations of different *Hyalella* (Wellborn, 1994, 1995, 2000). In Arizona, there are distinct populations of *Hyalella azteca* living in different lake habitats. One group of populations lives in submerged vegetation while another group clings to roots of emergent vegetation (Thomas *et al.*, 1997). Studies have examined genetic differences and behavioral differences in swimming among different populations of *Hyalella azecta* from Oregon to Arizona (Thomas *et al.*, 1997, 1998) and suggest that xeric landscapes promote genetic and behavioral divergence among amphipods, presumably because of the long distances among freshwater habitats. Habitat preferences and possible segregation of populations likely occur in other amphipod species living in other structurally complex, stable habitats (e.g., Muller, 1998; Muller *et al.*, 2000) or habitats that vary temporally and spatially such as groundwater-fed cave streams and marshes.

Several other studies have considered the importance of habitat structure in the dispersal and population genetics of peracarid crustaceans (Hedgecock *et al.*, 1982; Meyran and Taberlet, 1998; Meyran *et al.*, 1997, 1998). The degree of genetic isolation among populations distributed within a drainage basin is usually related to topographic complexity and elevational gradients for many types of slow-moving stream invertebrates (Siegismund and Muller, 1991; Muller, 1998; Muller *et al.,* 2000). As discussed above, dispersal by downstream drift and infrequent upstream movements limits gene flow among subpopulations within river drainage networks. Long, high ridges dissect the landscape and form effective barriers to the mixing of genes within and among river networks and associated lake basins.

4. Comparisons Among Ancient Lakes

Isolation by topographic structure may also be important for benthic crustaceans in deeply dissected lake basins, as apparently has been the case in the ancient lakes such as Lake Titicaca in the Andes of Bolivia and Peru where some 11 species of *Hyalella* have evolved. One of the most interesting cases of speciation and adaptive radiation among freshwater invertebrates is the high species richness of amphipods in Lake Baikal, Siberia, the oldest and deepest lake in the world where nearly 2000 metazoan species occur (Brooks, 1950; Martens *et al.,* 1994; Yampolsky *et al.,* 1994; Kozhova and Izmest'va, 1998; Kamaltynov, 1999). Almost 25% of the total worldwide freshwater amphipod fauna known to exist today occurs in Baikal, and nearly all of these species are restricted to this single ancient lake (Kozhov, 1963; Martens, 1997; Martens and Schon, 1999; Martens and Danielopol, 1999). The progenitors of the modern fauna can be traced to much earlier Mesozoic brackish-water basins. Major climatic changes during the Tertiary are thought to have accelerated the amphipod species radiation within the Baikal basin during the last 25–30 million years (Kamalynov, 1999). Several endemic species also occur within the Baikal drainage in the Angara and Enisei Rivers (Kamaltynov, 1992). In some cases, the locations of the outlets and inlets of rivers apparently create barriers for dispersal along the lake shore and result in genetic separation of different localized populations (Mashiko *et al.,* 1997). There are at least 250 gammarid species in about 45 genera that have evolved specialized niches in this uniquely deep lake (maximal depth is over 1700 m). Some are armed with sharp cutaneous spines or ridges, and they range in size from 10 to 80 mm. Coloration is highly varied (red, green, yellow, violet, pink, and brown) with complex patterning, which is very unusual among freshwater invertebrates. A phylogenetic analysis of 18 of the endemic genera suggests that parallel evolution of body armament and large body size occurred among different genera during the history of the lake. The first lineage was apparently benthic and mostly unarmed; the second lineage was composed of mostly armed taxa and were primarily detritivorous and carnivorous (Sherbakov *et al.,* 1998). The evolutionary ages of the various taxa are being studied using molecular techniques such as analysis of mitrochondrial DNA and allozymes (Mashiko *et al.,* 1997; Ogarkov *et al.,* 1997; Vainola and Kamaltynov, 1999).

Because of the combination of low water temperatures and very deep circulation, there is sufficient dissolved oxygen in this lake to allow benthic gammarids to live in several depth zones. Deep-water (1400 m) forms have evolved different modes of surviving high hydrostatic pressure (Brauer *et al.,* 1980a; b). Littoral forms are extremely abundant, reaching densities of 30,000 or more per square meter. One endemic species of gammarid amphipod, *Macrohectopus branickii,* is a dominant pelagic omnivore. Stable isotopic analyses of N-15 demonstrate that individual consumers change their mode of feeding to consuming more zooplankton than phytoplankton as they grow larger (Yoshii *et al.,* 1999). This relatively large (maximum body length 38 mm) species is a major prey species of several fishes, especially omul, *Coregonus autumnalis migratorius* and pelagic sculpins. In the more productive bays in the middle basin of Lake Baikal, *Macrohectopus* forms dense aggregations and migrates vertically from 250 m depths to upper waters at night (Melnik *et al.,* 1993). Using vertical tows of especially designed closing nets and hydroacoustic equipment Rudstam *et al* (1992) were able to determine day and night distributions of individuals and biomass in Barguzin Bay. They found that these aggregations were dominated by immature females (<15 mm long) in late summer and contained some nektobenthic species of amphipods as well. The migratory patterns and apparent avoidance of fish predators by *Macrohectopus* in very similar to *Mysis* migrations in Lake Michigan and other North American deep lakes (Lasenby *et al.,* 1986; Rudstam *et al.,* 1992). Unlike the high species richness of amphipods, the isopods in Baikal are represented by a single genus, *Asellus,* with only five species. Generally, the diversification of crustaceans reflects differences in the response of isolated populations to local and regional conditions of long-term habitat stability, plasticity of life-history characteristics, predator–prey interactions, and access to limited food resources.

IV. COLLECTING, REARING, AND PREPARATION FOR IDENTIFICATION

Most amphipods and isopods can be collected by sweeping a dip net through the littoral zone of lakes and ponds or in vegetated reaches of streams. Mysids require a large diameter plankton net that can be used for deep vertical tows. Being fast swimmers, mysids can avoid small nets and are more difficult to collect. Quantitative sampling of amphipods and isopods has relied on drift nets and Surber samplers in fast-flowing streams and Ekman grab sampler in shallow ponds and lakes. Drift nets can entrap organisms that may be moving upstream as well as those drifting downstream (if the net is not positioned off the bottom of the stream), and, thus, interpretation of studies dealing with dispersal and migration can be complicated. The use of cores (e.g., Milstead and Threlkeld, 1986), can be very effective for intensive study of microhabitats. Considerable time can be spent in sieving or sorting specimens; elutriators increase the efficiency of separating specimens from the inorganic and organic matrix (Magdych, 1981). Baited traps with various mesh sizes also work well for quantitative sampling of isopods and gammarids that respond to current-dispersed chemicals given off by the bait (e.g., Fitzpatrick, 1983; Allan and Malmqvist, 1989). Sorting of specimens is relatively rapid, but construction of fine-meshed traps is necessary. Use of electrofishing equipment may also be useful in stunning larger decapod crustaceans so that they can be swept up more effectively with dip nets. Chemicals have been added to stun or poison animals, but these methods are not very effective and clearly can be very detrimental to the environment. As discussed previously, many species are extremely sensitive to even small concentrations of toxic chemicals and are used to monitor water quality.

Several papers provide details on care and feeding of laboratory stocks (e.g., Cooper, 1965; Strong, 1972; DeMarch, 1981a, b; Sutcliffe *et al.*, 1981; Barlocher *et al.*, 1989). Dead leaves conditioned with nutritious microbial growths of bacteria and fungi are sufficient food resources for most species of amphipods and isopods. These can be supplemented with cultures of green algae and diatoms or with boiled lettuce or spinach. As in rearing other taxa, the initial water used for setting up an aquarium must be well aerated and allowed to develop an adequate microbial community before the crustaceans are placed into the tank. Temperature fluctuations can be detrimental and some type of aquarium heater is useful to maintain optimal growing conditions. Parasites and diseases are, of course, widespread in natural populations (e.g., Pixell Goodrich, 1934; Brownell, 1970; Laberge and McLaughlin, 1989), and brood stock from previously reared laboratory cultures are less likely to suffer from these outbreaks as the complex life histories of many parasites require a variety of vertebrate and invertebrate hosts to maintain the life cycle.

Preparation of specimens for study usually requires rapid fixation and storage of field-collected organisms in vials of 60–70% ethyl alcohol. The alcohol will remove any pigment from the specimens, so careful notes or photographs will be useful if information on coloration or patterns of body markings are of interest. The long-term storage of specimens requires checking to ensure that the alcohol has not evaporated. Addition of a 5–20% solution of glycerin to the vials will protect the specimens from drying out. Glycerin jelly is a solid at typical room temperatures, and it is widely used as a permanent mounting medium. Thus, if the material is first stored in dilute glycerin, it is convenient to transfer specimens to slides for study and for preparing reference collections. These techniques are described in detail in several texts (e.g., Pennak, 1989; Peckarsky *et al.*, 1990).

V. CLASSIFICATION OF PERACARIDA AND BRANCHIURA

The taxonomic key for this chapter is intended to allow the reader to identify major taxa of amphipods, isopods, mysids, and Branchiura to the generic level. This key draws on characteristics used in previously published studies (Edmondson, 1959; Williams, 1970; Cressey, 1972; Holsinger, 1972, 1989, 1994b; Fitzpatrick, 1983; Pennak, 1989; Peckarsky *et al.*, 1990); some of these sources can be used to identify organisms to species level, others are useful for determining biogeographical distributions (e.g., Barnard and Barnard, 1983; Banarescu, 1990).

A. Taxonomic Key to Major Freshwater Genera of the Orders Mysidacea, Amphipoda, and Isopoda and the Subclass Branchiura

1a. Four pairs of legs and two prominent, highly modified suckers on upper ventral surface, specialized ectoparasite on fishes (Fig. 6)..... ....Subclass Branchiura, order Arguloida, family Arguilidae ....*Argulus*

1b More than four pairs of walking legs (pereiopods) prominent second antennae, compound eyes 2

2a(1b) Five pairs of legs, body flattened laterally (Fig. 7) ....order Amphipoda 4

2b Six or seven pairs of legs, body flattened dorsoventally 3

3a(2b) Seven pairs of legs, posterior legs longer than anterior legs, unstalked eyes (Fig.8) .... order Isopoda 8

3b Six pairs of legs, prominent stalked eyes (Fig. 9) .... order Mysidacea, family Mysidae .... *Mysis*

4a(2a) Amphipoda: antenna 1 shorter than antenna 2 (Fig. 7) 5

4b Antenna 1 longer than antenna 2 (Fig. 7) 6

5a (4a) Accessory flagellum present on antenna I with 3–5 segments (Fig. 7), telson deeply cleft but not to base .... family Pontoporeiidae .... *Monoporeia*

5b Accessory flagellum absent on antenna 1, telson entire .... family Hyalellidae .... *Hyalella*

6a (4b) Accessory flagellum of antenna 1 with 2–7 segments (usually 3 or more), telson typically cleft to base .... family Gammaridae .... *Gammarus*

6b Accessory flagellum of antenna 1 with never more than 2 segments, telson usually not deeply cleft but never to the base .... family Crangonyctidae 7

7a (6b) Eyes may be present and pigmented, apical margin of telson distinctly cleft .... *Crangonyx*

7b Eyes absent, apical margin of telson entire or with shallow cleft .... S*tygobromus*

8a (3a) Isopoda: margin of head with a pointed median protuberance (carina) between bases of first antennae (Fig. 8), eyes typically present .... family Asellidae .... *Lirceus*

8b Anterior margin of head without carina, eyes present or absent .... family Asellidae 9

9a (8b) Abdomen broader than long, up to 16 mm long .... *Asellus*

9b Abdomen not broader than long .... *Caecidotea*

ACKNOWLEDGMENTS

We greatly appreciate reviews and comments by David C. Culver and Dean W. Blinn as well as help from John R. Holsinger with several key references. Jennifer L. Borash provided scanning electron images of crayfish chelae. Rebecca Rudman assisted with a thorough proof-reading and many authors provided reprints of their studies.

LITERATURE CITED

Abele, L. G. 1982. Biogeography, *in:* Bliss, D. E., Ed., The biology of Crustacea. vol. 1. Systematics, the fossil record and biogeography. Academic Press. New York, pp. 242–304.

Ache, B. W. 1982. Chemoreception and thermoreception, *in:* Atwood, H. L., Sandeman, D. C., Eds., The biology of Crustacea. Vol. 3. Academic Press. New York, pp. 369–398.

Adare, K. I., Lasenby, D. C. 1994. Seasonal changes in the total lipid content of the opossum shrimp, *Mysis relicta* (Malacostraca, Mysideacea). Canadian Journal of Fisheries and Aquatic Sciences 51:1935–1941.

Allan, J. D., Malmqvist, B. 1989. Diel activity of *Gainmarus pulex* (Crustacca) in a South Swedish stream: comparison of drift catches vs baited traps. Hydrobiologia 179:73–80.

Allee, W. C. 1914. The ecological importance of the rheotactic reaction of stream isopods. Biological Bulletin 27:52–66.

Allee, W. C. 1929. Studies in animal aggregations: natural aggregations of the isopod, *Asellus communis.* Ecology 10:14–36.

Anderson, R. S., Raasveldt, L. G. 1974. *Gammarus* predation and the possible effects of *Gammarus* and *Chaoborus* feeding on the zooplankton composition in some small lakes and ponds in western Canada. Canadian Wildlife Service Occasional Papers 18:1–23.

Atema, J. 1988. Distribution of chemical stimuli, *in:* Atema, J., Ed. Sensory biology of aquatic animals. Springer, New York, pp. 29–56.

Bakker, T. C., Mazzi, D., Zala, S. 1997. Parasite-induced changes in behavior and color make *Gammarus pulex* more prone to fish predation. Ecology 78:1098–1104.

Banarescu, P. 1990. Zoogeography of fresh waters, Vol. 1. General distributions and dispersal of freshwater animals. AULA, Wiesbaden. 511 pp.

Banner, A. H. 1953. On a new genus and species of mysid from southern Louisiana. Tulane Studies in Zoology 1:1–8.

Barlocher, F., Kendrick, B. 1973. Fungi and food preferences of *Gammarus pseudolimnaeus.* Archiv fur Hydrobiologie 72:501–516.

Barlocher, F., Kendrick, B. 1975. Assimilation efficiency of *Gammarus pseudo! imnaeus* (Amphipoda). Oikos 24:295–300.

Barlocher, F., Tibbo, P. G., Christie, S. H. 1989. Formation of phenol-protein complexes and their use by two stream invertebrates. Hydrobiologia 173:243–249.

Barnard, J. L., Barnard, C. M. 1983. Freshwater Amphipoda of the world. 1. Evolutionary patterns. 2. Handbook and bibliography. Hayfield Associates, Mt. Vernon, VA.

Barnes, R. D. 1987. Invertebrate Zoology, 5th ed., Saunders, New York, 893 p.

Barnes, R. D., Harrison, F. W. 1992. Introduction, *in:* Harrison, F. W., Humes, A. G., Eds., Microscopic anatomy of invertebrates. Vol. 9. Crustacea. Wiley–Liss, New York, pp. 1–8.

Barr, T. C., Jr. Holsinger, J. R. 1985. Speciation in cave faunas. Annual Review of Ecology and Systematics 16:313–337.

Beck, J. T. 1980. Life history relationships between the bopyrid isopod *Probopyrus pandalicola* and one of its freshwater shrimp hosts *Palaemonetes paludosus.* American Midland Naturalist 104:135–154.

Beckett, D. C., Lewis, P. A., Green, J. H. 1998. Where have all the *Crangonyx* gone? The disappearance of the amphipod

Crangonyx pseudogracilis, and subsequent appearance of *Gammarus nr. fasciatus,* in the Ohio River. American Midland Naturalist 139:201–209.

Beeton, A. M., Bowers, J. A. 1982. Vertical migration of *Mysis relicta* Loven. Hydrobiologia 93:53–61.

Bethel, W. M., Holmes, J. C. 1977. Altered evasive behavior and responses to light in amphipods harboring acanthaocepthalan cystacanths. Journal of Parasitology 59:945–956.

Bohonak, A. J. 1999. Dispersal, gene flow, and population structure. Quarterly Review of Biology 74:21–45.

Blinn, D. W., Davies, R. W. 1990. Concomitant diel vertical migration of a predatory leech and its amphipod prey. Freshwater Biology 24:401–407.

Blinn, D. W., Dehdashti, B., Runck, C., Davies, R. W. 1990. The importance of prey size and density in an endemic predator prey couple (leech *Erpobdella montezuma,* amphipod *Hyalella montezuma*). Journal of Animal Ecology 59:187–192.

Blinn, D. W., Grossnickle, N. E., Dehdashti, B. 1987. Diet vertical migration of a pelagic amphipod in the absence of fish predation. Hydrobiologia 160:165–171.

Blinn, D. W., Johnson, D. B. 1982. Filter-feeding of *Hyalella montezuma,* an unusual behavior for a freshwater amphipod. Freshwater Invertebrate Biology 1:48–52.

Blinn, D. W., Pinney, C., Wagner, V. T. 1988. Intraspecific discrimination of amphipod prey by a freshwater leech through mechanoreception. Canadian Journal of Zoology 66:427–430.

Bliss, D. E., Editor-in-Chief, 1982–1985. The biology of Crustacea, Vols. 1–10, Academic Press, New York.

Borash, J. L., Moore, P. A. 1997. Ultrastructure, morphology and distribution of sensory structures on the chelae of the crayfish. Ohio Journal of Science 97(2): A-11.

Boulton, A. J., Findlay, S., Marmonier, P., Stanley, E. H., Valett, H. M. 1998. The functional significance of the hyporheic zone in streams and rivers. Annual Review of Ecology and Systematics 29:59–81.

Bousfield, E. L. 1983. An updated phyletic classification and paleohistory of the amphipods, *in:* Schram, F. R., Ed., Crustacean issues 1: Crustacean phylogeny. Balkema. Rotterdam, pp. 257–277.

Bousfield, E. L. 1989. Revised morphological relationships within the amphipod genera *Pontoporeia* and *Gammaracanthus* and the "glacial relict" significance of their postglacial distributions. Canadian Journal of Fisheries and Aquatic Sciences 46: 1714–1725.

Bousfield, E. L., Holsinger, J. R. 1989. A new crangonyctid amphipod crustacean from hypogean fresh waters of Oregon. Canadian Journal of Zoology 67:963–968.

Bowles, D. E., Aziz, K., Knight, C. L. 2000. *Macrobrachium* (Decapoda: Caridea: Palaemonidae) in the contiguous United States: a review of the species and an assessmet of threats to their survival. Journal of Crustacean Biology 20: 158–171.

Bowman, T. E. 1981. *Thermosphaeroma millerio* and *T. smithi,* new sphaeromatid isopod crustaceans from hot springs in Chihuahua, Mexico. with a review of the genus. Journal of Crustacean Biology 1:105–122.

Bowman, T. E., Abele, L. G. 1982. Classification of the recent Crustacea, *in:* Abele, L. G., Ed., The biology of Crustacea. Vol. 1. Systematics. the fossil record, and biogeography. Academic Press, New York, pp. 1–27.

Branstrator, D. K., Cabana, G., Mazumder, A., Rasmussen, J. B. 2000. Measuring life-history ominivory in the opossum shrimp, *Mysis relicta,* with stable nitrogen isotopes. Limnology and Oceanography 45:463–467.

Brauer, R. W., Bekman, M. Y., Keyser, J. B., Nesbit, D. L., Shvetzov, G. N., Wright, S. I. 1980a. Adaptation to high hydrostatic pressures of abyssal gammarids from Lake Baikal in eastern Siberia. Comparative Biochemistry and Physiology 65A:109–117.

Brauer, R. W., Bekman, M. Y., Keyser, J. B., Nesbit, D. L., Shvetzov, S. G., Sidelev, G. N., Wright, S. I. 1980b. Comparative studies of sodium transport and its relation to hydrostatic pressure in deep and shallow water gammarid crustaceans from Lake Baikal. Comparative Biochemistry and Physiology 65A: 119–127.

Brooks, J. L. 1950. Speciation in ancient lakes. Quarterly Review of Biology 25:30–60.

Brownell, W. N. 1970. Comparison of *Mysis relicta* and *Pontoporeia affinis* as possible intermediate hosts for the acanthocephalan *Echinorhynchus salmonis.* Journal of the Fisheries Research Board of Canada 27:1864–1866.

Buchanan, C., Haney, J. F. 1980. Vertical migrations of zooplankton in the arctic: a test of the environmental controls, *in:* Kerfoot, W. C., Ed., Evolution and ecology of zooplankton communities. Univ. Press of New England, Hanover, pp. 69–79.

Bush, B. M. H., Laverack, M. S. 1982. Mechanoreception, *in:* Atwood, H. L., Sandeman, D. C., Eds., The biology of Crustacea, Vol. 3. Academic Press, New York, pp. 399–468.

Cabana, G., Tremblay, A., Kalff, J., Rasmussen, J. B. 1994. Pelagic food chain structure in Ontario lakes: a determinant of mercury levels in lake trout (*Salvelinus namaycush*). Canadian Journal of Fisheries and Aquatic Sciences 51:381–389.

Camp, J. W., Huizinga, H. W. 1979. Altered color, behavior and predation susceptibility of the isopod *Asellus intermedius* infected with *Acanthocephalus dirus.* Journal of Parasitology 65:667–669.

Capart, A. 1951. *Thermobathynella adami,* gen. et spec. nov., Anaspidace du Congo Beige. Institut Royal des Sciences et Naturelles de Belgique, Bulletin 27:1–4.

Carpenter, J. H. 1999. Behavior and ecology of *Speleonectes epilimnius* (Remipedia, Speleonectidae) from surface water of an anchialine cave on San Salvador Island, Bahamas. Crustaceana 72:979–991.

Cavaletto, J. F., Nalepa, T. F., Dermott, R., Gardner, W. S., Quigley, M. A., Lang, G. A. 1996. Seasonal variation of lipid composition, weight, and length in juvenile *Diporeia* spp. (Amphipoda) form Lakes Michigan and Ontario. Canadian Journal of Fisheries and Aquatic Sciences 53:2044–2051.

Cheper, N. J. 1980. Thermal tolerance of the isopod *Lirceus brachyurus* (Crustacea: Isopoda). American Midland Naturalist 104:312–318.

Chess, D. W., Stanford, J. A. 1998. Comparative energetics and lifecycle of the opossum shrimp (*Mysis relicta*) in native and nonnative environments. Freshwater Biology 40:783–794.

Chess, D. W., Stanford, J. A. 1999. Experimental effects of temperature and prey assemblage on growth and lipid accumulation by *Mysis relicta* Loven. Hydrobiologia 412:155–164.

Chipps, S. R. 1998. Temperature-dependent consumption and gutresidence time in the opossum shrimp *Mysis relicta.* Journal of Plankton Research 20:2401–2411.

Cole, G. A., Bane, C. A. 1978. *Thermosphaeroma subequaluin,* n. gen., n. sp. (Crustacea: Isopoda) from Big Bend National Park, Texas. Hydrobiologia 59:223–228.

Cole, G. A., Watkins, R. L. 1977. *Hyalella montezuma,* a new species (Crustacea: Amphipoda) from Montezuma Well, Arizona. Hydrobiologia 52:175–184.

Collart, O. O. 1990. Interaction between the parasite *Probopyrus bithynis* (Isopoda, Bopyridae) and one of its hosts, the prawn *Macrobrachium amazonicum* (Decapoda, Palaemonidae). Crustaceana 58:258–269.

Conlan, K. E. 1994. Amphipod crustaceans and environmental disturbance- a review. Journal of Natural History 28:519–554.

Cook, J. A., Chubb, J. C., Veltkamp, C. J. 1998. Epibionts of *Asellus aquaticus* (L.) (Crustacea, Isopoda): a SEM study. Freshwater Biology 39:423–438.

Cooper, S. D., Goldman, C. R. 1982. Environmental factors affecting predation rates of *Mysis relicta*. Canadian Journal of Fisheries and Aquatic Sciences 39:203–208.

Cooper, W. E. 1965. Dynamics and production of a natural population of a freshwater amphipod, *Hyalella azteca*. Ecological Monographs 35:377–394.

Covich, A. P., Fritz, S. C., Lamb, P. J., Marzolf, R. D., Matthews, W. J., Poiani, K. A., Prepas, E. E., Richman, M. B., Winter, T. C. 1997. Potential effects of climate change on aquatic ecosystems of the Great Plains of North America.. Hydrological Processes 11:993–1021.

Covich, A. P., Palmer, M., Crowl, T. A. 1999. The role of benthic invertebrate species in freshwater ecosystems. BioScience 49: 119–127.

Cressey, R. F. 1972. The genus *Argulus* (Crustacea: Branchiura) of the United States. Biota of Freshwater Ecosystems, U.S. Environmental Protection Agency Identification Manual 2:1–14.

Crouau, Y. 1989. Feeding mechanisms of the Mysidacea, *in:* Felgenhauer, B. E., Watling, L., Thistle, A. B., Eds., Functional morphology of feeding and grooming in Crustacea. Balkema, Rotterdam, pp. 153–171.

Crowl, T. A., Covich, A. P. 1990. Predator-induced life-history shifts in a freshwater snail. Science 247:949–951.

Culver, D. C., Fong, D. W. 1994. Small-scale and large-scale biogeography of subterranean crustacean faunas of the Virginias. Hydrobiologia 287:3–9.

Culver, D. C., Jernigan, R. W., O'Connell, J., Kane, T. C. 1994 The geometry of natural selection in cave and spring populations of the amphipod *Gammarus minus* Say (Crustacea: Amphipoda). Biological Journal of the Linnean Society 52:49–67.

Culver, D. C., Jones, W. K., Holsinger, J. R. 1992. Biological and hydrological investigation of the Cedars, Lee County, Virginia, an ecologically significant and threatened karst area, *in*: Stanford, J. A., Simons, J. J., Eds., Proceedings of the first international conference on groundwater ecology. American Water Resources Association, Bethesda, MD, pp. 281–290.

Culver, D. C., Kane, T. C., Fong, D. W. 1995. Adaptation and natural selection in caves. The evolution of *Gammarus minus*. Harvard Univ. Press, Cambridge, MA.

Culver, D. C., Master, L. L., Christman, M. C., Hobbs, H. H., III. 2000. Obligate cave fauna of the 49 contiguous United States. Conservation Biology 14:386–401.

Dadswell, M. J. 1974. Distribution, ecology, and postglacial dispersal of certain crustaceans and fishes in eastern North America. Publications in Zoology No.11, National Museum of Natural Sciences, National Museums of Canada, Ottawa.

Dahl, E., Emanuelsson, H., von Mecklenburg, C. 1970. Pheromone transport and reception in an amphipod. Science 170:739–740.

Dahl, J. 1998. The impact of vertebrate and invertebrate predators on a stream benthic community. Oecologia 115:253–259.

Dahl, J., Nilsson, P. A., Pettersson, L. B. 1998. Against the flow: chemical detection of downstream predators in running waters. Proceedings of the Royal Society of London, series B- Biological Sciences 265:1339–1344.

Danielopol, D. L., Pospisil, P., Rouch, R. 2000. Biodiversity in groundwater: a large-scale view. Trends in Ecology and Evolution 15:223–224.

Dehdashti, B., Blinn, D. W. 1991. Population dynamics and production of the pelagic amphipod *Hyalella montezuma* in a thermally constant system. Freshwater Biology 25:131–141.

Delong, M. D., Summers, R. B., Thorp, J. H. 1993. Influence of food type on the growth of a riverine amphipod, *Gammarus fasciatus*. Canadian Journal of Fisheries and Aquatic Sciences 50: 1891–1896.

DeMarch, B. G. E. 1981a. *Hyalella azteca* (Saussure), *in:* Lawrence, S. G., Ed., Manual for the culture of selected freshwater invertebrates. Canadian Special Publication in Fisheries and Aquatic Sciences No. 54, pp. 61–77.

DeMarch, B. G. E. 1981b. *Gammarus lacustris lacustris* G. 0. Sars, *in:* Lawrence, S. G., Ed., Manual for the culture of selected freshwater invertebrates. Canadian Special Publication in Fisheries and Aquatic Sciences No. 54, pp. 79–94.

DeMarch, B. G. E. 1982. Decreased day length and light intensity as factors inducing reproduction in *Gammarus lacustris lacustris* Sars. Canadian Journal of Zoology 60:2962–2965.

Dermott, R. M., Corning, K. 1988. Seasonal ingestion rates of *Pontoporeia hoyi* (Amphipoda) in Lake Ontario. Canadian Journal of Fisheries and Aquatic Sciences 45:1886–1895.

Dick, J. T. A. 1995. The cannibalistic behaviour of two *Gammarus* species (Crustacea: Amphipoda). Journal of Zoology (London) 236:697–706.

Dick, J. T. A., Elwood, R. W., Montgomery, W. I. 1995. The behavioural basis of a species replacement: differential aggression and predation between the introduced *Gammarus pulex* and the native *G. duebeni celticus* (Amphipoda). Behavioral Ecology and Sociobiology 37:393–398.

Dodds, W. K. 1990. Hydrodynamic constraints on evolution of chemically mediated interactions between aquatic organisms in unidirectional flows. Journal of Chemical Ecology 16:1417–1430.

Dodson, S. I., Crowl, T. A., Peckarsky, B. L., Kats, L. B., Covich, A. P., Culp, J. M. 1994. Non-visual communication in freshwater benthos: an overview. Journal of the North American Benthological Society 13:268–282.

Donner, K. O., Lindstrom, A., Lindstrom, M. 1987. Seasonal variation in the vertical migration of *Pontoporeia affinis* (Crustacea, Amphipoda). Annales Zoologica Fennica 24:305–313.

Dormaar, K. A., Corey, S. 1978. Some aspects of osmoreculation in *Mysis relicta* Loven (Mysidacea). Crustaceana 34:90–93.

Drost, C. A., Blinn, D. W. 1997. Invertebrate community of Roaring Springs Cave, Grand Canyon National Park, Arizona. Southwestern Naturalist 42:497–500.

Dunham, P. 1978. Sex pheromones in Crustacea.. Biological Review 53:555–583.

Edler, C., Dodds, W. K. 1996. The ecology of a subterranean isopod, *Caecidotea tridentata*. Freshwater Biology 35:249–259.

Edmondson, W. T. Ed. 1959. Freshwater biology, 2nd ed., Wiley, New York.

Elliott, W. R. 1999. Conservation of the North American cave and karst biota, *in:* Wilken, H., Culver, D. C., Humphreys, W., Eds., Subterranean ecosystems. Elsevier, Oxford, UK, pp. 671–695.

Ellis, R. J. 1971. Notes on the biology of the isopod *Asellus tomalensis* Harford in an intermittent pond. Transactions of the American Microscopical Society 90:51–61.

Elwood, R. W., Dick, J. T. A. 1990. The amorous *Gammarus*: The relationship between precopula duration and size assortative mating in *G. pulex*. Animal Behaviour 39:828–833.

Emerson, M. J., Schram, F. R. 1990. The origin of crustacean biramous appendages and the evolution of Arthropoda. Science 250:667–669.

Enright, J. T. 1977. Diurnal vertical migration: adaptive significance and timing. Part I. Selective advantage: A metabolic model. Limnology and Oceanography 22:856–872.

Enright, J. T. 1979. The why and when of up and down. Limnology and Oceanography 24:788–791.

Evans, M. S., Bathelt, R. W., Rice, C. P. 1982. Polychlorinated biphenyls and other toxicants in *Mysis relicta*. Hydrobiologia 93:205–215.

Felgenhauer, B. E., Abele, L. G., Felder, D. L. 1992. Remipedia, *in:* Harrison, F. W., Humes, A. G., Eds., Microscopic anatomy of invertebrates, Vol. 9 Crustacea. Wiley–Liss, New York, pp. 225–247.

Findlay, S., Meyer, J. L., Smith, J. P. 1984. Significance of bacterial biomass in the nutrition of a freshwater isopod (*Lirceus* sp.). Oecologia 63:38–42.

Findlay, S., Meyer, J. L., Smith, P. J. 1986. Contribution of fungal biomass to the diet of a freshwater isopod (*Lirceus* sp.). Freshwater Biology 16:377–385.

Fitzpatrick, J. F., Jr. 1983. How to know the freshwater Crustacea. Brown. Dubuque, IA. 227 p.

Folt, C. L., Rybock, J. T., Goldman, C. R. 1982. The effect of prey composition and abundance on the predation rate and selectivity of *Mysis relicta*. Hydrobiologia 93:133–143.

Fong. D. W. 1989. Morphological evolution of the amphipod *Gammarus minus* in caves-quantitative genetic analysis. American Midland Naturalist 121:361–378.

Fong, D. W., Culver, D. C. 1994. Fine-scale biographic differences in the crustacean fauna of a cave system in West Virginia, USA. Hydrobiologia 287:29–37.

France, R. L. 1992. The North American latitudinal gradient in species richness and geographical range of freshwater crayfish and amphipods. American Naturalist 139:342–354.

France, R. L. 1993. Production and turnover of *Hyalella azteca* in central Ontario, Canada compared with other regions. Freshwater Biology 30:343–349.

Friesen, J. A., Mann, K. H., Novitsky, J. A. 1986. *Mysis* digests cellulose in the absence of a gut micro-flora. Canadian Journal of Zoology 64:442–446.

Galat, D. L., Coleman, M., Robinson, R. 1988. Experimental effects of elevated salinity on three benthic invertebrates in Pyramid Lake, Nevada. Hydrobiologia 158:133–144.

Gee, J. H. R. 1988. Population dynamics and morphometrics of *Gammarus pulex* L.: evidence of seasonal food limitation in a freshwater detritivore. Freshwater Biology 19:333–343.

Glazier, D. S. 1999. Variation in offspring investment within and among populations of *Gammarus minus* Say (Crustacea: Amphipoda) in ten mid-Appalachian springs (USA). Archiv fur Hydrobiologie 146:257–283.

Glazier, D. S., Horne, M. T., Lehman, M. E. 1992. Abundance, body composition and reproductive output of *Gammarus minus* (Crustacea: Amphipoda) in ten cold springs differing in pH and ionic content. Freshwater Biology 28:149–163.

Glazier, D. S., Sparks, B. L. 1997. Energetics of amphipods in ion-poor waters: Stress resistance is not invariably linked to low metabolic rates. Functional Ecology 11:126–128.

Gliwicz, M. Z. 1986. Predation and the evolution of vertical migration in zooplankton. Nature 320:746–748.

Goddard, J. S. 1988. Food and feeding, *in:* Holdich, D. M., Lowery, R. S., Eds., Freshwater crayfish. Biology, management and exploitation. Timber Press. Portland. OR, pp. 145–166.

Goedkoop, W., Johnson, R. K. 1994. Exploitation of sediment bacterial carbon by juveniles of the amphipod *Monoporeia affinis*. Freshwater Biology 32:553–563.

Gooch, J. L. 1989. Genetic differentiation in relation to stream distance in *Gammarus minus* (Crustacea, Amphipoda) in Appalachian watersheds. Archiv fur Hydrobiologie 114:505–519.

Gooch, J. L. 1990. Spatial genetic patterns in relation to regional history and structure—*Gammarus minus* (Amphipoda) in Appalachian watersheds. American Midland Naturalist 124:93–104.

Gooch, J. L., Glazier, D. S. 1986. Levels of heterozvgosity in the amphipod *Gammarus minus* in an area affected by Pleistocene glaciation. American Midland Naturalist 116:57–63.

Gooch, J. L., Glazier, D. S. 1991. Temporal and spatial patterns in mid-Appalachian springs. Memoirs of the Entomological Society of Canada 155:29–49.

Graca, M. A. S., Maltby, L., Calow, P. 1994a. Comparative ecology of *Gammarus pulex* (L.) and *Asellus aquaticus* (L.) I: population dynamics and microdistribution. Hydrobiologia 281:155–162.

Graca, M. A. S., Maltby, L., Calow, P. 1994b. Comparative ecology of *Gammarus pulex* (L.) and *Asellus aquaticus* (L.) II: Fungal preferences. Hydrobiologia 281:163–170.

Greenaway, P. 1985. Calcium balance and molting in the Crustacea. Biological Reviews of the Cambridge Philosophical Society 60:425–454.

Haden, A., Blinn, D. W., Shannon, J. P., Wilson, K. P. 1999. Interference competition between the net-building caddisfly *Ceratopsyche oslari* and the amphipod *Gammarus lacustris*. Journal of Freshwater Ecology 14:277–280.

Hairston, N. G., Bohonak, A. J. 1998. Copepod reproductive strategies: life-history theory, phylogenetic pattern and invasion of inland waters. Journal of Marine Systems 15:23–34.

Hairston, N. G., Caceres, C. E. 1996. Distribution of crustacean diapause: micro- and macroevolutionary pattern and process. Hydrobiologia 320:27–44.

Hammer, U. T., Sheard, J. S., Kranabetter, J. 1990. Distribution and abundance of littoral benthic fauna in Canadian prairie saline lakes. Hydrobiologia 197:173–192.

Hartnoll, R. G. 1982. Growth, *in:* Abele, L. G., Ed., The biology of Crustacea, Vol. 2. Academic Press, New York, pp. 111–196.

Hechtel, L. J., Johnson, C. L., Juliano, S. A. 1993. Modification of antipredator behavior of *Caecidotea intermedius* by its parasite *Acanthocephalus dirus*. Ecology 74:710–713.

Hedgecock, D., Tracey, M. L., Nelson, K. 1982. Genetics, *in:* Abele, L. G., Ed., The biology of Crustacea, Vol. 2. Academic Press, New York, pp. 284–403.

Helluy, S., Holmes, J. C. 1990. Serotonin, octopamine, and the clinging behavior induced by the parasite *Polymorphus paradoxus* (Acanthocephala) in *Gammarus lacustris* (Crustacea). Canadian Journal of Zoology 68:1214–1220.

Henry, K. S., Danielopol, D. L. 1998. Oxygen dependent habitat selection in surface and hyporheic environments by *Gammarus roeseli* Gervais (Crustacea, Amphipoda): experimental evidence. Hydrobiologia 390:51–60.

Hervant, F., Mathieu, J., Culver, D. C. 1999. Comparative responses to severe hypoxia and subsequent recovery in closely related amphipod populations (*Gammarus minus*) from cave and surface habitats. Hydrobiologia 392:197–204.

Hill, C., Quigley, M. A., Cavaletto, J. F., Gordon, W. 1992. Seasonal changes in lipid content and composition in the benthic amphipods *Monoporeia affinis* and *Pontoporeia femorata*. Limnology and Oceanography 37:1280–1289.

Hogg, I. D., Williams, D. D. 1996. Response of stream invertebrates to a global-warming thermal regime: An ecosystem-level manipulation. Ecology 77:395–407.

Hogg, I. D., Eadie, J. M., de Lafontaine, Y. 1998a. Atmospheric change and the diversity of aquatic invertebrates: Are we missing the boat? Environmental Monitoring and Assessment 49:291–301.

Hogg, I. D., Larose, C., de Lafontaine, Y., Doe, K. G. 1998b. Genetic evidence for a *Hyalella* species complex within the Great Lakes St. Lawrence River drainage basin: implication for ecotoxicology and conservation biology. Canadian Journal of Zoology 76:1134–1140.

Holomuzki, J. R., Hatchett, L. A. 1994. Predator avoidance costs and habituation to fish chemicals by a stream isopod. Freshwater Biology 32:585–592.

Holomuzki, J. R., Hoyle, J. D. 1990. Effect of predatory fish presence on habitat use and diel movement of the stream amphipod, *Gammarus minus*. Freshwater Biology 24:509–517.

Holomuzki, J. R., Short, T. M. 1988. Habitat use and fish avoidance behaviors by the stream-dwelling isopod *Lirceus fontinalis*. Oikos 52:79–86.

Holomuzki, J. R., Short, T. M. 1990. Ontogenetic shifts in habitat use and activity in a stream-dwelling isopod. Holarctic Ecology 13:300–307.

Holsinger, J. R. 1972. The freshwater amphipod crustaceans (Gammaridae) of North America. Biota of Freshwater Ecosystems, Identification Manual 5, U.S. Environmental Protection Agency.

Holsinger, J. R. 1986. Zoogeographic patterns of North American subterranean amphipod crustaceans, *in:* Gore, R. H., Heck, K. L., Eds., Crustacean biogeography. Balkema, Rotterdam, pp. 85–106.

Holsinger, J. R. 1988. Troglobites: the evolution of cave-dwelling organisms. American Scientist 76:146–153.

Holsinger, J. R. 1989. Allocrangonyctidae and Pseudocrangonyctidae, two new families of Holarctic subterranean amphipod crustaceans (Gammaridae), with comments on their phylogenetic and zoogeographic relationships. Proceedings of the Biological Society of Washington 102:947–959.

Holsinger, J. R. 1991. Species accounts: freshwater amphipods and freshwater isopods, *in* Hoffman, R. L., Ed., Arthropods in Virginia's endangered species. McDonald and Woodward, Blacksburg, VA, pp. 180–199.

Holsinger, J. R. 1993. Biodiversity of subterranean amphipod crustaceans: global patterns and zoogeographic implications. Journal of Natural History 27:821–835.

Holsinger, J. R. 1994a. Pattern and process in the biogeography of subterranean amphipods. Hydrobiologia 287:131–145.

Holsinger, J. R. 1994b. Amphipoda, *in:* Juberthie, C., Decu, V., Eds., Encyclopaedia Biospeologica, tome I, Societe de Biospeologie, Moulis-Bucarest, pp. 147–163.

Holsinger, J. R., Hubbard, D. A., Jr., Bowman, T. E. 1994. Biogeographic and ecological implications of newly discovered populations of the stygobiont isopod crustacean *Antrolana lira* Bowman (Cirolanidae). Journal of Natural History 28:1047–1058.

Holsinger, J. R., Longley, G. 1980. The subterranean amphipod crustacean fauna of an artesian well in Texas. Smithsonian Contributions to Zoology 308:1–62.

Holsinger, J. R., Mort, J. S., Reckles, A. D. 1983. The subterranean crustacean fauna of Castleguard Cave, Columbia Icefields, Alberta, Canada, and its zoogeographical significance. Arctic and Alpine Research 15:543–549.

Hubricht, L., Mackin, J. G. 1949. The American isopods of the genus *Lirceus* (Asellota, Asellidae). American Midland Naturalist 42:334–349.

Hutchinson, G. E. 1967. A treatise on limnology, Vol. 2. Wiley, New York. 1115 pp.

Hutchinson, G. E. 1993. A treatise on limnology, Vol. 4. Wiley, New York. 944 pp.

Hynes, H. B. N., Harper, F. 1972. The life histories of *Gammarus lacustris* and *G. pseudolimnacus* in southern Ontario. Crustaceana (Supplement) 3:329–341.

Jacobs, J. 1991. Endangered and threatened wildlife and plants; proposal to list the Lee County cave isopod (*Lirceus usdagalun*) as an endangered species. Federal Register 56 (221):58026–58029.

Johannsson, O. 1992. Life history and productivity of *Mysis relicta* in Lake Ontario. Journal of Great Lakes Research 18:154–168.

Johannsson, O. E. 1995. Response of *Mysis relicta* population dynamics and productivity to spatial and seasonal gradients in Lake Ontario. Canadian Journal of Fisheries and Aquatic Sciences 52:1509–1522.

Johannsson, O. E., Rudstam, L. G., Lasenby, D. C. 1994. *Mysis relicta* assessment of metalimnetic feeding and implications for competition with fish in lakes Ontario and Michigan. Canadian Journal of Fisheries and Aquatic Sciences 51: 2591–2602.

Johnson, R. K., Wiederholm, T. 1989. Long-term growth oscillations of *Pontoporeia affinis* Lindstrom (Crustacea: Amphipoda) in Lake Malaren. Hydrobiologia 175:183–194.

Johnson, R. K., Wiederholm, T. 1992. Pelagic–benthic coupling—the importance of diatom interannual variability for population oscillations of the amphipod *Monoporeia affinis* (Lindstrom). Limnology and Oceanography 37:1596–1607.

Jones, R., Culver, D. C. 1989. Evidence for selection on sensory structures in a cave population of *Gammarus minus* (Amphipoda). Evolution 43:688–693.

Jormalainen, V., Shuster, S. M. 1997. Microhabitat segregation and cannibalism in an endangered freshwater isopod, *Thermosphaeroma thermophilium*. Oecologia 111:271–279.

Jormalainen, V., Shuster, S. M. 1999. Female reproductive cycle and sexual conflict over precopulatory mate-guarding in *Thermosphaeroma* (Crustacea, Isopoda). Ethology 105:233–246.

Jormalainen, V., Shuster, S. M., Wildey, H. C. 1999. Reproductive anatomy, precopulatory mate guarding, and paternity in the Socorro isopod, *Thermosphaeroma thermophilum,* Marine and Freshwater Behaviour and Physiology 32:39–56.

Juday, C., Birge, E. A. 1927. *Pontoporeia* and *Alysis* in Wisconsin lakes. Ecology 7:445–452.

Kamaltynov, R. M. 1992. On the present state of systematics of the Lake Baikal amphipods (Crustacea, Amphipoda). Hydrobiological Journal 26:82–92.

Kamaltynov, R. M. 1999. On the evolution of Lake Baikal amphipods. Crustaceana 72:921–931.

Kane, T. C., Culver, D. C., Jones, R. T. 1992. Genetic structure of morphologically differentiated populations of the amphipod *Gammarus minus*. Evolution 46:272–278.

Kane, T. C., Mathieu, J., Culver, D. C. 1995. Biotic fluxes and gene flowages, *in:* Gibert, J., Danielopol, D. L., Stanford, J. A, Eds., Groundwater ecology. Academic Press, San Diego, CA, pp. 245–270.

Knapp, S. M., Fong, D. W. 1999. Estimates of population size of *Stygobromus emarginatus* (Amphipoda: Crangonyctidae) in a headwater stream in Organ Cave, West Virginia. Journal of Cave and Karst Studies 61:3–6.

Koop, J. H. E., Grieshaber, M. K. 2000. The role of ion regulation in the control of the distribution of *Gammarus tigrinus* (Sexton) in salt-polluted rivers. Journal of Comparative Physiology B. Biochemical Systematic and Environmental Physiology 170: 75–83.

Kozhov, N. I. 1963. Lake Baikal and its life, Dr. W. Junk, The Hague, 344 p.

Kozhova, O. M., Izmest'va, L. R. 1998. Lake Baikal: evolution and diversity, Backhuys Publishers, Leiden, The Netherlands.

Laberge, R. J. A., McLaughlin, J. D. 1989. *Hyalella azteca* (Amphipoda) as an intermediate host of the nematode *Streptocara crassicauda*. Canadian Journal of Zoology 67:2335–2340.

Lamarre, E., Cochran, P. A. 1992. Lack of host species slection by the exotic parasitic crustacean, *Argulus japonicus*. Journal of Freshwaer Ecology 7:77–80.

Landrum, P. F., Nalepa, T. F. 1998. A review of the factors affecting the ecotoxicology of *Diporeia* spp. Journal of Great Lakes Research 24:889–904.

Lasenby, D. C., Northcote, T. G., Furst, M. 1986. Theory, practice and effects of *Mysis relicta* introduction to North American and Scandinavian lakes. Canadian Journal of Fisheries and Aquatic Sciences 43:1277–1284.

Lee, C. E., Bell, M. A. 1999. Causes and consequences of recent freshwater invasions by saltwater animals. Trends in Ecology and Evoutlion 14:284–288.

Lester, G. T., Clark, W. H. Occurrence of *Corophium spinicore* Stimpson 1857 (Amphipoda: Corophiidae) in Idaho, U.S.A. Western North American Naturalist (in press).

Lopez, G., Elmgren, R. 1989. Feeding depths and organic absorption for the deposit-feeding benthic amphipods *Pontoporeia affinis* and *Pontoporcia femorata*. Limnology and Oceanography 34:982–991.

MacNeil, C., Dick, J. T., Elwood, R. W. 1997. The trophic ecology of freshwater *Gammarus* spp. (Crustacea: Amphipoda): Problems and perspectives concerning the functional feeding group concept. Biological Reviews of the Cambridge Philosophical Society 72:349–364.

MacNeil, C., Dick, J. T. A., Elwood, R. W. 1999. The dynamics of predation on *Gammarus* spp. (Crustacea: Amphipoda). Biological Reviews of the Cambridge Philosophical Society 74:375–395.

MacNeil, C., Elwood, R. W., Dick, J. T. A. 1999a. Predator-prey interactions between brown trout *Salmo trutta* and native and introduced amphipods: their implications for fish diets. Ecography 22:686–696.

MacNeil, C., Elwood, R. W., Dick, J. T. A. 1999b. Differential microdistributions and interspecific interactions in coexisting *Gammarus* and *Crangonyx* amphipods. Ecography 22:415–423.

Magdych, W. P. 1981. An efficient, inexpensive elutriator design for separating benthos from sediment samples. Hydrobiologia 85:157–159.

Maguire, B. 1965. *Monodella texana,* an extension of the range of the crustacean order Thermosbaenacea to the western hemisphere. Crustaceana 9:149–154.

Malard, F., Hervant, F. 1999. Oxygen supply and the adaptations of animals in groundwater. Freshwater Biology 41:1–30.

Malmqvist, B., Sjostrom, P. 1987. Stream drift as a consequence of disturbance by invertebrate predators. Field and laboratory experiments. Oecologia 74:396–403.

Maltby, L. 1991. Pollution as a probe of life-history adaptations in *Asellus aquaticus* (Isopoda). Oikos 61:11–18.

Maltby, L. 1995. Sensitivity of the crustaceans *Gammarus pulex* (L.) and *Asellus aquaticus* (L.) to short-term exposure to hypoxia and unionzed ammonia- observations and possible mechanisms. Water Research 29:781–787.

Mantel, L. H., Farmer, L. L. 1983. Osmotic and ionic regulation, *in:* Mantel, L. H., Ed., The biology of Crustacea, Vol. 5 Academic Press, New York, pp. 53–161.

Marchant, R., Hynes, H. B. N. 1981a. The distribution and production of *Gammarus pseudolimnaeus* (Crustacea: Amphipoda) along a reach of the Credit River, Ontario. Freshwater Biology 11:169–182.

Marchant, R., Hynes, H. B. N. 1981b. Field estimates of feeding rate for *Gammarus pseudolimnaeus* (Crustacea: Amphipoda) in the Credit River, Ontario. Freshwater Biology 11:27–36.

Margolis, L., Kabata, Z., Eds. 1988. Guide to the parasites of fishes of Canada, Part II. Crustacea. Canadian Special Publication of Fisheries and Aquatic Sciences 101. Department of Fisheries and Oceans, Ottawa. 184 p.

Martens, K. 1997. Speciation in ancient lakes. Trends in Ecology and Evolution 12:177–182.

Martens, K., Coulter, G., Goddeeris, B. 1994. Speciation in ancient lakes 40 years after Brooks. Advances in Limnology 44:75–96.

Martens, K., Danielopol, D. L. 1999. Concluding remarks- age and origin of crustacean deiversity in "extreme" environments. Crustaceana 72:1031–1037.

Martens, K., Schon, I. 1999. Crustacean biodiversity in ancient lakes: a review. Crustaceana 72:899–910.

Martinez, P. J., Bergersen, E. P. 1989. Proposed biological management of *Mysis relicta* in Colorado lakes and reservoirs. North American Journal of Fisheries Management 9:1–11.

Marzolf, G. R. 1965. Substrate relations of the burrowing amphipod *Pontoporeia affinis* in Lake Michigan. Ecology 46:579–592.

Mashiko, K., Kamaltynov, R. M., Shervakov, D. Y., Morino, H. 1997. Genetic separation of gammarid (*Eulimnogammarus cyaneus*) populations by localized topographic changes in ancient Lake Baikal. Archiv fur Hydrobiologie 139:379–387.

Master, L. L., Stein, B. A., Kutner, L. S., Hammerson, G. A. 2000. Vanishing assets: the conservation status of U.S. species, *in:* Stein, B. A., Kutner, L. S., Adams, J. S., Eds., Precious heritage: The status of biodiversity in the United States. Oxford Univ. Press, New York, pp. 93–118.

Maynard, B. J., Wellnitz, T. A., Wright, W. G., Dezfull, B. S. 1998. Parasite-altered behavior in a crustacean intermediate host: field and laboratory studies. Journal of Parasitology 84:1102–1106.

McCahon, C. P., Maund, S. J., Poulton, M. J. 1991. The effect of the acanthocephalan parasite (*Pomphorhynchus laevis*) on the drift of its intermediate host (*Gammarus pulex*). Freshwater Biology 25:507–513.

McDonald, M. E., Crowder, L. B., Brandt, S. B. 1990. Changes in *Mysis* and *Pontoporcia* populations in southeastern Lake Michigan: a response to shifts in the fish community. Limnology and Oceanography 35:220–227.

McLaughlin, P. A. 1980. Comparative morphology of recent Crustacea. Freeman. San Francisco. 177 p.

McLaughlin, P. A. 1982. Comparative morphology of crustacean appendages, *in:* Abele, L. G., Ed., The biology of Crustacea, Vol. 2. Academic Press, New York, pp. 197–256.

McLoughlin, N. J., Blinn, D. W., Davies, R. W. 1999. An energetic evaluation of a predator-prey (leech-amphipod) couple in Montezuma Well, Arizona, USA. Functional Ecology 13:45–50.

McPeek, M. A., Wellborn, G. A. 1998. Genetic variation and reproductive isolation among phenotypically divergent amphipod populations. Limnology and Oceanography 43:1162–1169.

Melnik, N. G., Timoshkin, O. A., Sideleva, V. G., Pushkin, S. V., Mamylov, V. S. 1993. Hydroacoustic measurement of the density of the Baikal macrozooplankter *Macrohectopus branickii* (Dyb.) (Amphipods, Gammaridae). Limnology and Oceanography 38:425–434.

Meyran, J-C. 1997. Impact of water calcium on the phenotypic diversity of alpine populations of *Gammarus fossarum*. Ecology 78:1579–1587.

Meyran, J-C. 1998. Ecophysiological diversity of alpine populations of *Gammarus lacustris* in relation to environmental calcium. Freshwater Biology 39:41–47.

Meyran, J-C., Gielly, L., Taberlet, P. 1998. Environmental calcium, and mitochondrial DNA polymorphism among local populations of *Gammarus fossarum* (Crustacea, Amphipoda). Molecular Ecology 7:1391–1400.

Meyran, J-C., Monnerot, M., Taberlet, P. 1997. Taxonomic status and phylogenetic relationships of some species of the genus *Gammarus* (Crustacea, Amphipoda) deduced from mitochrondrial DNA sequences. Molecular Phylogenetics and Evolution 8:1–10.

Meyran, J-C., Taberlet, P. 1998. Mitochondrial DNA polymorphism among alpine populations of *Gammarus lacustris* (Crustacea, Amphipoda). Freshwater Biology 39:259–265.

Milstead, B., Threlkeld, S. T. 1986. An experimental analysis of darter predation on *Hyalella azteca* using semipermeable enclosures. Journal of the North American Benthological Society 5:311–318.

Minckley, W. L. 1961. Occurrence of subterranean isopods in the epigean environment. American Midland Naturalist 63:452–455.

Monk, D. C. 1977. The digestion of cellulose and other dietary components, and pH of the gut in the amphipod *Gammarus pulex* (L.). Freshwater Biology 7:431–440.

Morgan, M. D., Threlkeld, S. T., Goldman, C. R. 1978. Impact of the introduction of kokanee *(Oncorhyncus nerka)* and opossum shrimp *(Mysis relicta)* on a subalpine lake. Journal of the Fisheries Research Board of Canada 35:1572–1579.

Mosslacher, F. 1998. Subsurface dwelling crustaceans as indicators of hydrological conditions, oxygen concentrations, and sediment structure in an alluvial aquifer. International Review of Hydrobiology 83:349–364.

Muller, J. 1998. Genetic population structure of two cryptic *Gammarus fossarum* types across a contact zone. Journal of Evolutionary Biology 11:79–101.

Muller, J., Partsch, E., Link, A. 2000. Differentiation in morphology and habitat partitioning of genetically characterized *Gammarus fossarum* forms (Amphipoda) across a contact zone. Biological Journal of the Linnean Society 69:41–53.

Mullis, R. M., Revitt, D. M., Shutes, R. E. 1996. A statistical approach for the assessment of the toxic influences on *Gammarus pulex* (Amphipoda) and *Asellus aquaticus* (Isopoda) exposed to urban aquatic discharges. Water Research 30:1237–1243.

Murtaugh, P. A. 1989. Fecundity of *Neomysis mercedis* Holmes in Lake Washington (Mysidacea). Crustaceana 57:194–200.

Naylor, C., Adams, J. 1987. Sexual dimorphism, drag constraints and male performance in *Gammarus duebeni* (Amphipoda). Oikos 48:23–27.

Naylor, C., Pindar, L., Calow, P. 1990. Inter and intraspecific variation in sensitivity to toxins: the effects of acidity and zinc on the freshwater crustaceans *Asellus aquaticus* (L.) and *Gammarus pulex* (L.). Water Research 24:757–762.

Nero, R. W., Schindler, D. W. 1983. Decline of *Mysis relicta* during acidification of Lake 223. Canadian Journal of Fisheries and Aquatic Sciences 40:1905–1911.

Newman, R. M. 1991. Herbivory and detritivory on freshwater macrophytes by invertebrates: A review. Journal of the North American Benthological Society 10:89–114.

Newman, R. M., Hanscom, Z., Kerfoot, W. C. 1992. The watercress glucosinolate-myrosinase system: a feeding deterrent to caddisflies, snails and amphipods. Oecologia 92:1–7.

Newman, R. M., Kerfoot, W. C., Hanscom, Z., III. 1990. Watercress and amphipods: potential chemical defense in a spring stream macrophyte. Journal of Chemical Ecology 16:245–259.

Newman, R. M., Perry, J. A., Tam, E., Crawford, R. L. 1987. Effects of chronic chlorine exposure on litter processing in outdoor experimental streams. Freshwater Biology 18:415–428.

Newman, R. M., Waters, T. F. 1984. Size-selective predation on *Gammarus pseudolimneaus* by trout and sculpins. Ecology 65:1535–1545.

Nolan, D. T., van der Salm, A. L., Bonga, S. E. W. 2000. The host-parasite relationship between the rainbow trout (*Oncorhynchus mykiss*) and the ectoparasite *Argulus foliaceus* (Crustacea: Branchiura): epithelial mucous cell response, corticol and factors which may influence parsite establishment. Contributions to Zoology 69:57–63.

Notenboom, 1991. Marine regression and the evolution of groundwater dwelling amphipods (Crustacea). Journal of Biogeography 18:437–454.

Notenboom, J., Plenet, S., Turquin, M. J. 1994. Groundwater contamination and its impact on groundwater animals and ecosystems, *in:* Gibert, J., Danielopol, D. L., Stanford, J. A., Eds., Groundwater ecology. Academic Press, San Diego, CA, pp. 477–504.

Oberlin, G. E., Blinn, D. W. 1997. The effect of temperature on the metabolism and behavior of an endemic amphipod, *Hyalella montezuma* from Montezuma Well, Arizona, USA. Freshwater Biology 37:55–59.

Ogarkov, O. B., Kamaltynov, R. M., Belikov, S. I., Shcherbakov, D. Y 1997. Phylogenetic relatedness of the Baikal Lake endemial amphipodes (Crustacea, Amphipoda) deduced from partial nucleotide sequences of the cyctochrome oxidase subunit III genes. Molecular Biology 31:24–29.

Okland, K. A., Okland, J. 1985. Factor interaction influencing the distribution of the fresh-water shrimp *Gammarus.* Oecologia 66:364–367.

Olyslager, N. J., Williams, D. D. 1993. Microhabitat selection by the lotic amphipod *Gammarus pseudolimneaus* Bousfield—mechanisms for evaluating local substrate and current suitability. Canadian Journal of Zoology 71:2401–2409.

Palmer, M., Covich, A. P., Finlay, B. J., Gibert, J., Hyde, K. D., Johnson, R. K., Kairesalo, T., Lake, S., Lovell, C. R., Naiman, R. J., Ricci, C., Sabater, F., Strayer, D. 1997. Biodiversity and ecosystem processes in freshwater sediments. Ambio 26:571–577.

Panov, V. E., McQueen, D. J. 1998. Effects of temperature on individual growth rate and body size of a freshwater amphipod. Canadian Journal of Zoology 76:1107–1116.

Parker, J. I. 1980. Predation by *Mysis relicta* on *Pontoporeia hoyi:* a food chain link of potential importance in the Great Lakes. Journal of Great Lakes Research 6:164–166.

Peck, S. B. 1998. A summary of diversity and distribution of the obligate cave-inhabiting fauna of the United States and Canada. Journal of Cave and Karst Studies 60:18–26.

Peckarsky, B. L., Frassinet, P. R., Penton, M. A., Conklin, D. J., Jr. 1990. Freshwater macroinvertebrates of northeastern North America. Cornell Univ. Press, Ithaca, New York. 422 pp.

Pennak, R. W. 1989. Freshwater invertebrates of the United States: protozoa to mollusca, 3rd ed., Wiley, New York. 628 p.

Pixell Goodrich, H. P. 1934. Reactions of *Gammarus* to injury and disease, with notes on microsporidial and fungoid disease. Quarterly Journal of the Microscopy Society 72:325–353.

Ricciardi, A., Rasmussen, J. B. 1999. Extinction rates of North American freshwater faunas. Conservation Biology 13:1220– 1222.

Richardson, H. 1905. A monograph of the isopods of North America. Bulletin of the U.S. National Museum 54:1–727.

Roberts, G. N., Chubb, J. C. 1998. The distribution and location of the symbiont *Lagenophyrs aselli* on the freshwater isopod *Asellus aquaticus.* Freshwater Biology 39:671–677.

Roman Contreras, R. 1996. A new species of *Probopyrus* (Isopoda, Bopyridae), parasite of *Macrobrachium americanum* Bate, 1868 (Decapoda, Palaemonidae). Crustaceana 69:204–210.

Rudstam, L. G., Melnik, N. G., Timoshkin, O. A., Hansson, S., Pushkin, S. V., Nemov, V. 1992. Diel dynamics of an aggregation of *Macrohectopus branickii* (Dyb) (Amphipoda, Gammaridae) in the Barguzin Bay, Lake Baikal, Russia. Journal of Great Lakes Research 18:286–297.

Rudstam, L. G., Hetherington, A. L., Mohammadian, A. M. 1999. Effect of temperature on feeding and survival of *Mysis relicta.* Journal of Great Lakes Research 25:363–371.

Ruff, H., Maier, G. 2000. Calcium carbonate deposits reduce predation on *Gammarus fossarum* from salamander larvae. Freshwater Biology 43:99–105.

Sarbu, S., Kane, T. C., Culver, D. C. 1993. Genetic structure and morphological differentiation: *Gammarus minus* (Amphipoda, Gammaridae) in Virginia. American Midland Naturalist 129:145–152.

Sarvala, J. 1986. Interannual variation of growth and recruitment in *Pontoporeia affinis* (Lindstrom) (Crustacea, Amphipoda) in relation to abundance fluctuations. Journal of Experimental Marine Ecology 101:41–59.

Sastry, A. N. 1983. Ecological aspects of reproduction, *in:* Vernberg, F. J., Vernberg, W. B., Eds., The biology of Crustacea, Vol. 8. Academic Press, New York, pp. 179–270.

Schmitt, W. L. 1965. Crustaceans, University of Michigan Press, Ann Arbor, 204 p.

Schmitz, E. H. 1992. Amphipoda, *in:* Harrison, F. W., Ed., Microscopic anatomy of invertebrates. Vol. 9 Crustacea. Wiley–Liss, New York, pp. 443–528.

Schmitz, E. H., Scherrey, P. M. 1983. Digestive anatomy of *Hyalella azteca* (Crustacea, Amphipoda). Journal of Morphology 175:91–100.

Schram, F. R. 1982. The fossil record and evolution of Crustacea, *in:* Abele, L. G., Ed., The biology of Crustacea, Vol. 1. Academic Press. New York, pp. 93–174.

Schram, F. R. 1986. Crustacea, Oxford University Press, New York, 606 p.

Schram, F. R., Yager, J., Emerson, M. J. 1986. Remipedia. Part I. Systematics. Memoirs of the San Diego Society of Natural History 15:1–60.

Schwartz, S. S. 1992. Benthic predators and zooplankton prey: predation by *Crangonyx shoemakeri* (Crustacea: Amphipoda) on *Daphnia obtusa* (Crustacea: Cladocera). Hydrobiologia 237: 25–30.

Seale, D. B., Binowski, F. P. 1988. Vulnerability of early life intervals of *Coregonus hoyi* to predation by a freshwater mysid. *Mysis relicta.* Environmental Biology of Fishes 21:117–126.

Sergestrale, S. G. 1970. Light control of the reproductive cycle of *Pontoporcia affinis* Lindstrom (Crustacea: Amphipoda). Journal of Experimental Marine Biology and Ecology 5:272–275.

Shafir, A., Oldewage, W. H. 1992. Dynamics of a fish ectoparasite population- opportunistic parasitism in *Argulus japonicus* (Branchiura). Crustaceana 62:50–64.

Shea, M. A., Makarewicz, J. C. 1989. Production, biomass, and trophic interactions of *Mysis relicta* in Lake Ontario. Journal of Great Lakes Research 15:223–232.

Sherbakov, D., R. M., Ogarkov, O. B., Verheyen, E. 1998. Patterns of evolutionary change in Baikalian gammarids inferred from DNA sequences (Crustacea, Amphipoda). Molecular Phylogenetics and Evolution 10:160–167.

Sherman, R. K., Lasenby, D. C., Hollet, L. 1987. Influence of oxygen concentration on the distribution of *Mysis relicta* Loven in a eutrophic temperate lake. Canadian Journal of Zoology 65: 2646–2650.

Short, T. M., Holomuzki, J. R. 1992. Indirect effects of fish on foraging behavior and leaf processing by the isopod *Lirceus fontinalis.* Freshwater Biology 27:91–97.

Shuster, S. M. 1981. Life history characteristics of Ther*mosphaeroma thermophi!um,* the Socorro isopod (Crustacea: Peracarida). Biological Bulletin 161:291–302.

Sieg, J. 1983. Evolution of the Tanaidacea, *in:* Schram, F. R., Ed., Crustacean phylogeny. Balkema, Rotterdam, pp. 229–256.

Siegfried, C. A. 1985. Life history, population dynamics and production of *Pontoporeia hoyi* (Crustacea, Amphipoda) in relation to the trophic gradient of Lake George, New York. Hydrobiologia 122:175–180.

Siegismund, H. R., Muller, J. 1991. Genetic structure of *Gammarus fossarum* populations. Heredity 66:419–436.

Simon, K. S., Buikema, A. L. 1997. Effects of organic pollution on an Appalachian cave: changes in macroinvertebrate populations. American Midland Naturalist 138:387–401.

Sket, B. 1996. The ecology of anchihaline caves. Trends in Ecology and Evolution 11:221–225.

Sket, B. 1999a. High biodiversity in hypogean waters and its endangerment—the situation in Slovenia, the Dinaric karst, and Europe. Crustaceana 72:767–779.

Sket, B. 1999b. The nature of biodiversity in hypogean waters and how it is endangered. Biodiversity and Conservation 8: 1319–1338.

Sly, P. G., Christie, W. J. 1992. Factors influencing densities and distributions of *Pontoporeia hoyi* in Lake Ontario. Hydrobiologia 235:321–352.

Smock, L. A., Harlowe, K. L. 1983. Utilization and processing of freshwater wetland macrophytes by the detritivore *Asellus forbesi.* Ecology 64:1556–1565.

Smock, L. A., Stoneburner, D. L. 1980. The response of macroinvertebrates to aquatic macrophyte decomposition. Oikos 33: 397–403.

Song, K. H., Breslin, V. T. 1998. Accumulation of contaminant metals in the amphipod *Diporeia* spp. in western Lake Ontario. Journal of Great Lakes Research 24:949–961.

Song, K. H., Breslin, V. T. 1999. Accumulation and transport of sediment metals by the vertically migrating opossum shrimp, *Mysis relicta.* Journal of Great Lakes Research 25:429–442.

Sparkes, T. C. 1996a.The effects of size-dependent predation risk on the interaction between behavioral and life history traits in a stream-dwelling isopod. Behavioral Ecology and Sociobiology 39:411–417.

Sparkes, T. C. 1996b. Effects of predation risk on population variation in adult size in a stream dwelling isopod. Oecologia 106:85–92.

Spencer, C. N., McClelland, B. R., Stanford, J. A. 1991. Shrimp introduction, salmon collapse, and eagle displacement: cascading interactions in the food web of a large aquatic ecosystem. BioScience 41:14–21.

Spencer, C. N., Potter, D. S., Bukantis, R. T., Stanford, J. A. 1999. Impact of predation by *Mysis relicta* on zooplankton in Flathead Lake, Montana, USA. Journal of Plankton Research 21:51–64.

Stanford, J. A., Ward, J. V. 1988. The hyporheic habitat of river ecosystems. Nature 335:64–66.

Strayer, D. 1999. Invasion of fresh waters by saltwater animals. Trends in Ecology and Evolution 14: 448–449.

Strayer, D. L., May, S. E., Nielsen, P., Wollheim, W., Hausam, S. 1997. Oxygen, organic matter, and sediment granulometry as controls on hyporheic animal communities. Archiv fur Hydrobiologie 140:131–144.

Stephens, D. W. 1990. Changes in lake levels, salinity and the biological community of Great Salt Lake (Utah, USA), 1847–1987. Hydrobiologia 197:139–146.

Stewart, B. A. 1992. Morphological and genetic differentiation between allopatric populations of a freshwater amphipod. Journal of Zoology (London) 228:287–305.

Stewart, B. A. 1993. The use of protein electrophoresis for determining species boudaries in amphipods. Crustaceana 65:265–277.

Stock, F. 1995. The ecological and historical determinants of crustacean diversity in groundwaters, or : why are these so many species? Memoirs of Biospeology 22:139–160.

Stock, J. H., Iliffe, T. M. 1990. Amphipod crustaceans from anchialine cave waters of the Galapagos Islands. Zoological Journal of the Linnean Society 98:141–160.

Stock, J. H., Longley, G. 1981. The generic status and distribution of *Monodella rexana,* Maguire, the only known North American thermosbaenacean. Proceedings of the Biological Society of Washington 94:569–578.

Strong, D. R. 1972. Life history variation among populations of an amphipod *(Hyalella azteca).* Ecology 53:1103–1111.

Strong, D. R., Jr. 1973. Amphipod amplexus, the significance of ecotypic variation. Ecology 54:1383–1388.

Sutcliffe, D. W. 1971. Regulation of water and some ions in gammarids (Amphipoda). I. *Gammarus duebeni* Lilljeborg from brackish water and fresh water. Journal of Experimental Biology 55:325–344.

Sutcliffe, D. W. 1974. Sodium regulation and adaptation to fresh water in the isopod genus *Asellus*. Journal of Experimental Biology 61:719–736.

Sutcliffe, D. W., Carrick, T. R., Willoughby, L. G. 1981. Effects of diet. body size, age and temperature on growth rates in the amphipod *Gammarus pulex*. Freshwater Biology 11:183–214.

Tadini, G. V., Tadini, E. A., Colangelo, M. 1988. The life history of *Asellus aquaticus* (L.) explains its geographical distribution. Verhandlungen Internationale Vereinigung fur Theoretische und Angewandte Limnologie 23:2099–2106.

Taylor, C. A. 2000. Preserving North America's unique crayfish fauna, *in:* Abell, R. A., Olson, D. M., Dinerstein, E., Hurley, P. T., Diggs, J. T., Eichbaum, W., Walters, S., Wettengel, W., Allnutt, T., Loucks, C. J., Hedau, P., Eds., Freshwater ecoregions of North America. A conservation assessment. Island Press, Washington, D C, pp. 36–37.

Thomas, P. E., Blinn, D. W., Keim, P. 1997. Genetic and behavioural divergence among desert spring amphipod populations. Freshwater Biology 38:137–143.

Thomas, P. E., Blinn, D. W., Keim, P. 1998. Do xeric landscapes increase genetic divergence in aquatic ecosystems? Freshwater Biology 40:587–593.

Threlkeld, S. T., Chiavelli, D. A., Willey, R. L. 1993. The organization of zooplankton epibiont communities. Trends in Ecology and Evolution 8:317–321.

Thybaud, E., LeBrac, S. 1988. Absorption and elimination by *Asellus aquaticus* (Crustacea, Isopoda). Bulletin of Environmental Contamination and Toxicity 40:731–735.

Toman, M. J., Dall, P. C. 1998. Respiratory levels and adaptations in four freshwater species *Gammarus* (Crustacea: Amphipoda). International Review of Hydrobiology 83:251–263.

Vainola, R. 1990. Molecular time scales for evolution in *Mysis* and *Pontoporeia*. Annales Zoologici Fennici 27:211–214.

Vainola, R., Kamaltynov, R. M. 1999. Species diversity and speciation in the endemic amphipods of Lake Baikal: molecular evidence. Crustaceana 72:945–956.

Vainola, R., Varvio, S. L. 1989. Molecular divergence and evolutionary relationships in *Pontoporeia* (Crustacea: Amphipoda). Canadian Journal of Fisheries and Aquatic Sciences 46: 1705–1713.

Vainio, J., Jazdzewski, K., Vainola, R. 1995. Biochemical systematic relationships among the freshwater amphipods *Gammarus varsoviensis, Gammarus lacustris* and *Gammarus pulex*. Crustaceana 68:687–694.

Vernberg, W. B., Vernberg, F. J. 1983. Freshwater adaptations, *in:* Vernberg, F. J., Vernberg, W. B., Eds., The biology of Crustacea, Vol 8. Academic Press, New York, pp. 335–363.

Wagner, V. T., Blinn, D. W. 1987. A comparative study of the maxillary setae for two coexisting species of *Hyalella* (Amphipoda), a filter feeder and a detritus feeder. Archiv fur Hydrobiologia 109:409–419.

Waters, T. F., Hokenstrom, J. C. 1980. Annual production and drift of the stream amphipod *Gammarus pseudolimnaeus* in Valley Creek, Minnesota. Limnology and Oceanography 25:700–710.

Watling, L. 1981. An alternative phylogeny of peracarid crustaceans. Journal of Crustacean Biology 1:201–210.

Wellborn, G. A. 1994. Size-biased predation and prey life histories: a comparative study of freshwater amphipod populations. Ecology 75:2104–2117.

Wellborn, G. A. 1995. Determinants of reproductive success in freshwater amphipod species that experience different mortality regimes. Animal Behaviour 50:353–363.

Wellborn, G. A. 2000. Selection on a sexually dimorphic trait in ecotypes within the *Hyalella azteca* complex (Amphipoda: Hyalellidae). American Midland Naturalist 143:212–225.

Wells, L. 1960. Seasonal abundance and vertical movements of planktonic Crustacea in Lake Michigan. U.S. Fish and Wildlife Service Fishery Bulletin 60:343–369.

Wen, Y. H. 1993. Sexual dimorphism and mate choice in *Hyalella azteca* (Amphipoda). American Midland Naturalist 129: 153–160.

Whitehurst, I. T. 1991.The *Gammarus–Asellus* ratio as an index of organic pollution. Water Research 25:333–339.

Wickstrom, C. E., Castenholz, R. W. 1973. Thermophilic ostracod: aquatic metazoan with the highest known temperature tolerance. Science 181:1063–1064.

Wilhelm, F. M. 1996. Predation on *Mysis relicta* by slimy sculpins (*Cottus cognatus*) in southern Lake Ontario—comment. Journal of Great Lakes Research 22:119–120.

Wilhelm, F. M., Schindler, D. W. 2000. Reproductive strategies of *Gammarus lacustris* (Crustacea: Amphipoda) along on elevational gradient. Functional Ecology 14: 413–422.

Williams, D. D. 1990. A field study of the effects of water temperature, discharge and trout odor on the drift of stream invertebrates. Archiv fur Hydrobiologie 119:167–181.

Williams, D. D., Moore, K. A. 1985. The role of semiochemicals to benthic community relationships of the lotic amphipod *Gammarus pseudolimnaeus*: a laboratory analysis. Oikos 44: 280–286.

Williams, W. D. 1970. A revision of North American epigean species of *Asellus* (Crustacea: Isopoda). Smithsonian Contributions to Zoology 49:1–80.

Williams, W. D. 1998. Salinity as a determinant of the structure of biological communities in salt lakes. Hydrobiologia 381: 191–201.

Willoughby, L. G., Sutcliffe, D. W. 1976. Experiments on feeding and growth of the amphipod *Gammarus pulex* (L.) related to its distribution in the River Duddon. Freshwater Biology 6:577–586.

Wilson, C. B. 1944. Parasitic copepods of the United States National Museum. Proceedings of the U.S. National Museum 94:529–582.

Winnell, M. H., White, D. S. 1984. Ecology of shallow and deep water populations of *Pontoporeia hoyi* (Smith) (Amphipoda) in Lake Michigan. Freshwater Invertebrate Biology 3:118–138.

Wisenden, B. D., Cline, A., Sparkes, T. C. 1999. Survival benefit to antipredator behavior in the amphipod *Gammarus minus* (Crustacea: Amphipoda) in response to injury-released chemical cues from conspecifics and heterospecifics. Ethology 105: 407–414.

Wollheim, W. M., Lovvorn, J. R. 1995. Salinity effects on macroinvertebrate assemblages and waterbird food webs in shallow lakes of the Wyoming high plains. Hydrobiologia 310:207–223.

Wollheim, W. M., Lovvorn, J. R. 1996. Effects of macrophyte growth forms on invertebrate communities in saline lakes of Wyoming high plains. Hydrobiologia 323:83–96.

Wooster, D. E. 1998. Amphipod (*Gammarus minus*) responses to predators and predator impact on amphipod density. Oecologia 115:253–259.

Wudkevich, K., Wisenden, B. D., Chivers, D. P., Smith, R. J. F. 1997. Reactions of *Gammarus lacustris* to chemical stimuli form natural predators and injured conspecifics. Journal of Chemical Ecology 23:1163–1173.

Wurtsbaugh, W. A., Berry, T. S. 1990. Cascading effects of decreased salinity on the plankton, chemistry, and physics of the Great Salt Lake (Utah). Canadian Journal of Fisheries and Aquatic Sciences 47:100–109.

Yager, J. 1981. Remipedia. a new class of Crustacea from a marine cave in the Bahamas. Journal of Crustacean Biology 1: 328–333.

Yager, J. 1987. *Cryptocorynectes haptodiscus* and *Speleonectes benjamini* from anchialine caves in the Bahamas, with remarks on distribution and ecology. Proceedings of the Biological Society of Washington 100:302–320.

Yamaguti, S. 1963. Parasitic Copepoda and Branchiura of fishes. Wiley Interscience, New York. 1104 pp.

Yampolsky, L. Y., Kamaltynov, R. M., Ebert, D., Filatov, D. A., Chernyck, V. I. 1994. Variation of allozyme loci in endemic gammarids in Lake Baikal. Biological Journal of the Linnean Society 53:309–323.

Yoshii, K., Melnik, N. G., Timoshkin, O. A., Bondarenko, N. A., Anoshko, P. N., Yoshioka, T., Wada, E. 1999. Stable isotope analyses of the pelagic food web in Lake Baikal. Limnology and Oceanography 44:502–511.

Zimmer-Faust, R. K. 1989. The relationship between chemoreception and foraging behavior in crustaceans. Limnology and Oceanography 34:1367–1374.

20

OSTRACODA

L. Denis Delorme[1]

National Water Research Institute
Canada Centre for Inland Waters
Burlington, Ontario, L7R 4A6

I. INTRODUCTION

The class Ostracoda in the Crustacea contains both marine and freshwater forms. The subclass Podocopa has one order, Podocopida in which two of the suborders (Metacopina, Podocopina) contain all of the freshwater ostracodes (Bowman and Abele, 1982; Schram, 1986).

Ostracodes are the oldest known microfauna. Their preservation as fossils is attributed to the calcitic shell present in most species. The first fossil representatives are reported from Cambrian marine sediments (Moore, 1961; Maddocks, 1982). The earliest freshwater ostracode species came from coal-forming swamps, ponds, and streams of early Pennsylvanian age (Benson, 1961).

[1] Current address: 621 Auburn Crescent, Burlington, Ontario, Canada L7L 5B3

Ecology and Classification of North American Freshwater Invertebrates. 2nd Edition

According to Moore (1961), the Podocopina evolved from Podocopida stock during the Ordovician period, and then branched out into the Cytheroidea and the Cypridoidea during the Devonian period (Maddocks, 1982). The Darwinulidae evolved during the Carboniferous period (Maddocks, 1982).

Freshwater ostracodes are free living except the Entocytheridae (Hoff, 1942) which are commensal upon crayfish. This chapter will concentrate on freshwater ostracode genera found in North America. To date there have been 56 genera and about 420 species identified for this continent.

Most references older than 1900 have been deleted from the second edition. Those interested in the older references are referred to the original edition.

II. ANATOMY AND PHYSIOLOGY

A. External Shell Morphology

The visceral mass is completely encased within a chitinous and a low-magnesium calcite exoskeleton (carapace). The only exception is the entocytherid group of ostracodes which lacks calcium in the exoskeleton. The exterior surface of the exoskeleton (cuticle) is made up of chitin (Fig. 1) (Harding, 1964). The outer chitinous coating of the epidermis is the only covering over the calcitic shell. The soft epidermal tissue lining the interior part of the carapace is composed of the inner and outer epidermis. The epidermal cells of the outer lamella (Fig. 1) are larger and more irregular in shape than the inner epidermal cells. Subdermal cells occur between the inner and outer epidermal cells as do the liver, ovaries, and testes. The epidermis is contained within a thin chitinous lining. When the ostracode molts, the complete exoskeleton is shed leaving only the visceral mass and the soft tissue or epidermis. The epidermis is continuous with the visceral mass in the dorsal region of the body.

When the ostracode molts, the space between the epidermal cells of the outer and inner layer is empty. The shells are secreted by the cells of the outer lamellae (Turpen and Angell, 1971). The immediate source of calcium and magnesium carbonate is thought to be stored in the body and not obtained directly from the water (Fassbinder, 1912). Fassbinder has also shown that ostracodes fed with crushed calcium carbonate from snail shells developed thicker carapaces with more pronounced marginal ornamentation. He concluded that food is an important source of carbonate. Turpen and Angell (1971), however, found that calcium dissolved in water is the source of calcium for the shell. They used dissolved ^{45}Ca as a tracer (pH 7.4, 23°C) to

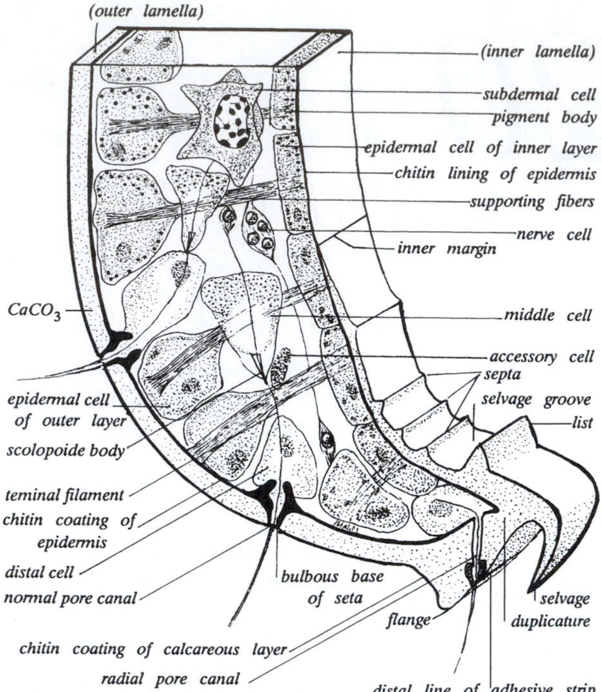

FIGURE 1 Diagrammatic cross section of the shell and the enclosing dermal layers, vestibule located between the outer lamella and the shell portion of the inner lamella in the anterior portion of the shell. [From Kesling, 1951a, b. Figure redrawn by J. L. Delorme.]

show that calcium is not resorbed from the shells prior to molting and that ostracodes do not store calcium in their bodies. Chivas *et al.*, (1985) ran laboratory experiments on the Australian *Mytilocypris henricae* and found that temperature influenced the rate of shell calcification. Roca and Wansard (1997) also found the rate of development to be linked to temperature. Using *Herpetocypris brevicaudata* they were able to show that below 15°C the Ca content of valves is strongly diminished and the length of time for calcification was longer (17 days). The optimum temperature was 23°C with the fastest rate of calcification (3–7 days). Bodergat (1978) demonstrated that phosphorus was an integral part of freshwater ostracode shells in all specimens investigated with a microprobe. She further noted a significant positive correlation between the amount of shell phosphorus and the habitat temperature of the ostracode.

The calcareous shell is the most obvious structure when the organism is viewed under a microscope. The shell is divided into two parts, the outer lamella and the duplicature (Fig. 1). The outer lamella is the major part of the shell; whereas the duplicature is the calcified part of the inner lamella bordering the posterior, ventral, and anterior margin. It forms a part of the free margin and projects toward the center of the carapace. The outer lamella may contain pores through which project setae (Fig. 2), that are sensitive to touch. The duplicature is welded to the outer lamella (Fig. 3) around the margins, by a strip of chitin, which joins with the chitin coating and the chitin lining. This juncture is referred to as the line of concrescence or the distal line of adhesive strip (Fig. 1). Within the chitinous strip, radial or marginal pore canals are found (Fig. 2). These are very fine tubes through which setae pass. The inner free margin of the duplicature is referred to as the inner margin (Fig. 1). When the duplicature extends beyond the line of concrescence toward the center of the shell, a shelf is formed (Fig. 3). The space between the shelf and the outer lamella is referred to as the vestibule (Fig. 1). Several structures are found on the duplicature which are very important in taxonomy down to the species level (Sylvester-Bradley, 1941).

Pustules, teeth, or crenulations (Fig. 3) may appear on the outer margin of the duplicature. A combination of septa, lists, and grooves are found on the duplicature. Some of these are shown in Fig. 1 (Kesling, 1951a, b). The shell may contain additional structures on the outer lamella such as lateral depressions called sulci (Fig. 30). Raised areas variously termed alae, knobs (Fig. 30), or pustules (Fig. 4), depending on their shape, size, and orientation, may also be found on the shell surface. The surface of the shell may also be pitted (Fig. 5), punctate, wrinkled (Fig. 6), or have a reticulate surface (Fig. 7). These ancillary structures of the shell are very often used for generic and specific identification (see Sylvester-Bradley and Benson, 1971, for terminology). Some authors, notably Benson (1974, 1981), have discussed the roles of ornamentation, form, function, and architecture of the ostracode shell.

Intraspecific morphological diversity of the shell may be pronounced. *Limnocythere itasca* has short recurved alae on its lateral surfaces. The individuals identified from Canada (Delorme, 1971a) and northern Minnesota (Forester *et al.,* 1987) appear to be characteristic of this species. R. M. Forester (personal communication) identified *L. itasca* from the Laguna de Texcoco, Mexico, which have very long and delicate recurved spines or alae. Intermediate forms have yet to be recovered. In another example, the caudal process of the posteroventral portion of the right valve of the female *Candona caudata* (may belong to *Fabaeformiscandona*; see Meisch, 1996, not confirmed for North America) may be blunt (Delorme, 1978) or sharp (= *C. novacaudata,* Benson and MacDonald, 1963). Both forms and intermediate morphs are found in Lake Erie.

B. Internal Shell Morphology

Structures on the internal surface of the shell may also be used in ostracode systematics. Many ostracode muscles are attached directly to the calcareous shells. The chitinous exoskeleton does not provide sufficient rigidity to anchor many of the muscles. The points of attachment of these muscles leave scars as raised or depressed areas on the interior of the shell (Figs. 8-11). The closing or adductor muscles form a distinctive pattern of scars as do the mandibular muscles (Benson, 1967; Benson and MacDonald, 1963; Smith, 1965). Traces of the ovaries and testes can also be seen in the posterior of the shell (see Section II.D).

C. Body and Appendage Morphology

Ostracodes have their body (visceral mass and appendages) suspended from the dorsal region in an elongate, chitinous pouch. The ostracode has a shortened body with a slight constriction in the midregion that separates the head from the thorax (Kesling, 1951a). There is no abdomen. Instead, the posterior of the body tapers off bluntly and ends in a pair of preanal furcae. The head region contains four pairs of appendages which are used for swimming, walking, and feeding. The thoracic region features two pairs (three pairs according to some authors) of appendages used or adapted for feeding, creeping, and cleaning of the shells (Broodbakker and Danielopol, 1982). There is some debate as to whether the maxilla are cephalic or thoracic appendage (Meisch, 1996) Ancillary cuticular, structures such as setae, claws, and pseudochaetae, found on most limbs, are recognized as important in functional morphology and systematics (Broodbakker and Danielopol, 1982). Appendages are shown in Figs. 9–12.

The head region is composed of the forehead, the upper lip, and the hypostome. The four pairs of appendages in the head region are the first antennae (antennules), second antennae (antennae), mandibles, and maxillae. When the distal four podomeres of the first antennae have very long, plumose setae, the ostracode is a good swimmer (e.g., *Cypridopsis vidua*). The first antennae are curved up and backward. If these setae are absent, as in the darwinulids (Fig. 10) and the limnocytherids (Fig. 11), the ostracode cannot swim. The second antennae extend down and curve backward. These appendages are robustly constructed and have a strong claw for walking and climbing. Swimming setae may be present on the first podomere of the endopodite. These structures are used for rowing when the ostracode is swimming. In most species, well-developed mandibles occur between the upper lip and the hypostome. The dorsal tips of the mandibular palps are attached to the interior of the shell by muscles. The ventral portion of the palps terminate in heavily toothed mandibles; these teeth grind the food before it

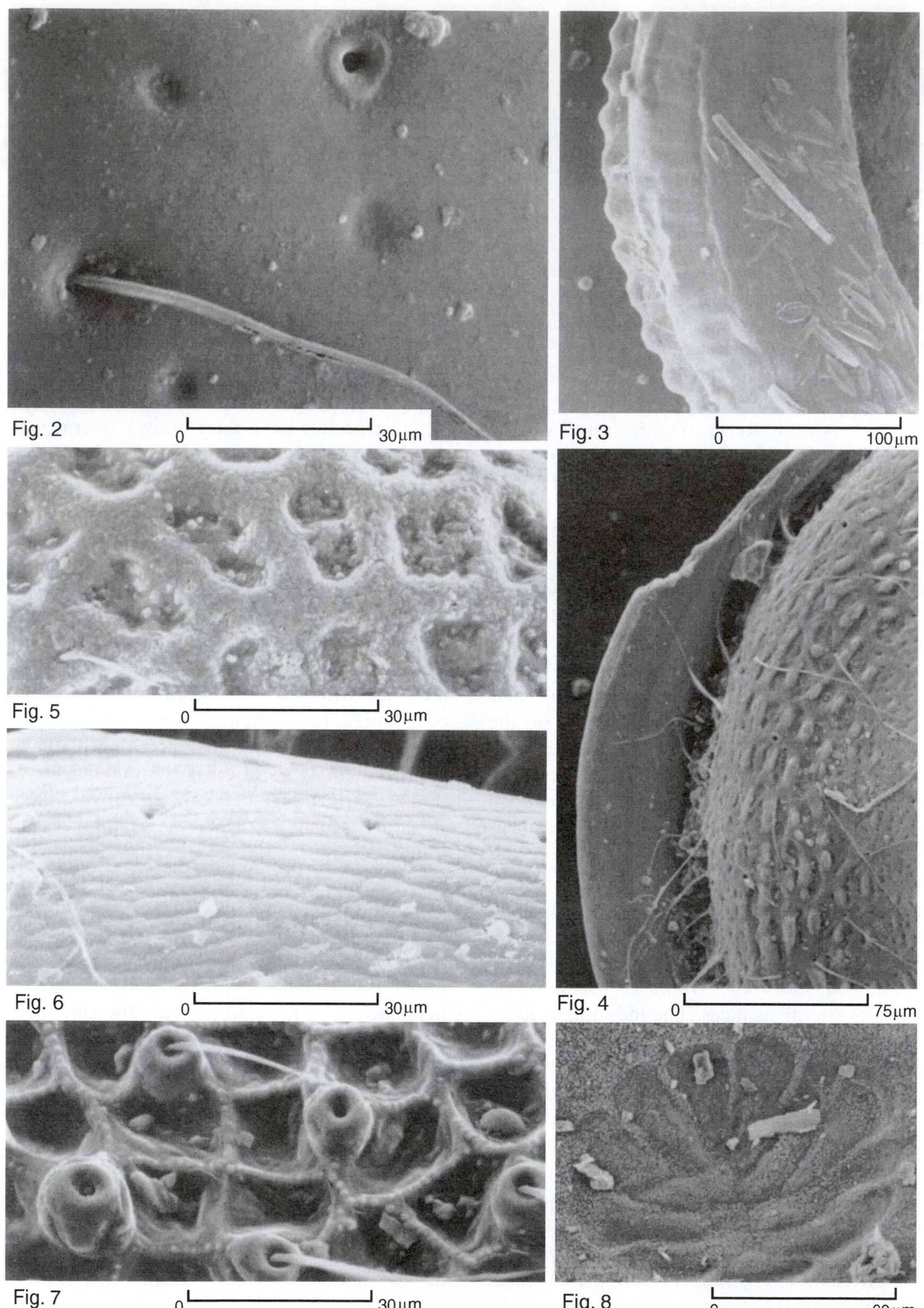

FIGURE 2 Shell surface of *Cyclocypris ampla* Furtos showing normal pore canals with funnels (without seta) and without funnels (with seta), and dimples. *FIGURE 3* Anterior margin of *Cyprinotus glaucus* Furtos showing crenulations, duplicature with position of radial pore canals, and calcified inner lamella forming a shelf. *FIGURE 4* Anterior margin of *Cyprinotus glaucus* Furtos with pustules on the right valve, left valve beneath. *FIGURE 5* Reticulate exterior structure on *Limnocythere itasca* Cole. *FIGURE 6* Anastomosing structure (lines) on the surface of *Cypria turneri*, note normal pore canals. *FIGURE 7* Reticulate structure with superimposed mamilliary pustules on *Paracandona euplectella* Brady and Norman. *FIGURE 8* Distinctive muscle scar pattern of *Darwinula*.

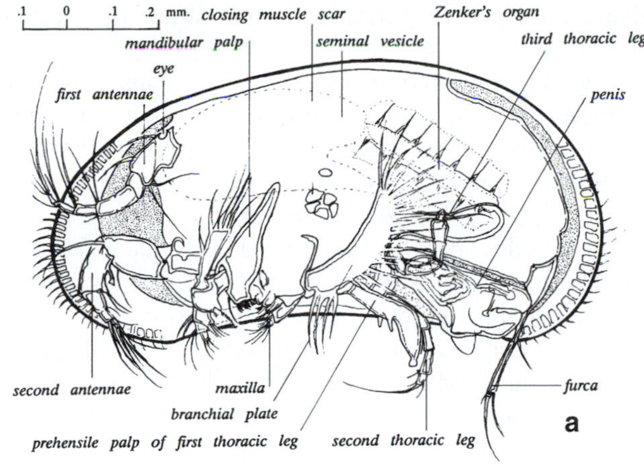

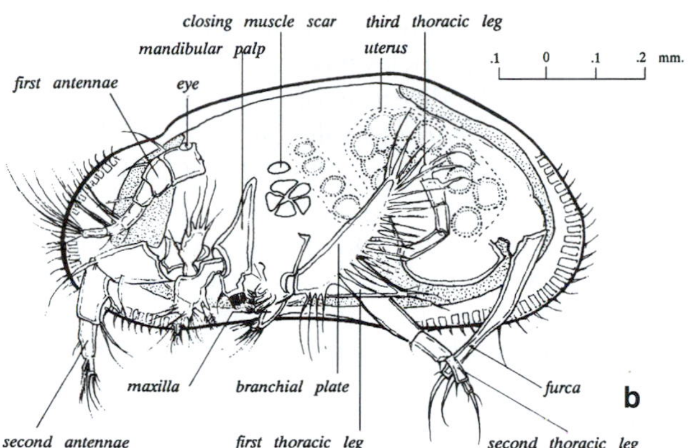

FIGURE 9 Sketch of the internal morphology of (a) male (b) female *Candona suburbana* Hoff (Cyprididae) from Kesling. First thoracic leg is referred to as the maxilla or maxillar endopodite and the second and third thoracic legs as the first and second thoracic legs in the text. [Figure redrawn by J. L. Delorme.]

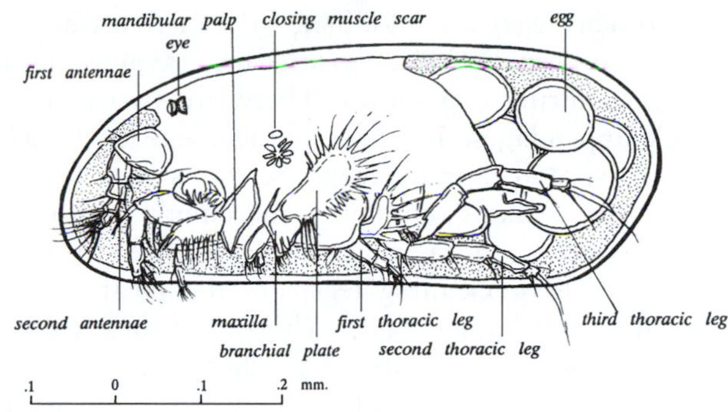

FIGURE 10 Sketch of the internal morphology of a female *Darwinula stevensoni* (Brady and Robertson) (Darwinulidae) from Kesling. First thoracic leg is referred to as the maxilla or maxillar endopodite and the second and third thoracic legs as the first and second thoracic legs in the text. [Figure redrawn by J. L. Delorme.]

is ingested. A respiratory (branchial) plate extends from the mandibular palp. On the ventral portion of the head is located a keel-shaped structure called the hypostome which forms the mouth. Rake-shaped organs are situated at the rear of the mouth. They consist of chitin shafts with terminal toothed structures and are used for straining and feeding. The fourth and final set of cephalic appendages are the maxillae. The maxillae, located on either side of the hypostome posterior of the mandibles, are made up of several cylindrical masticatory processes (maxillula). The maxillae pass the small particles toward the mouth. What were previously referred to as the first thoracic legs of most cyprids are now referred to as part of the maxillae. In other groups, this part of the maxillae (formerly the

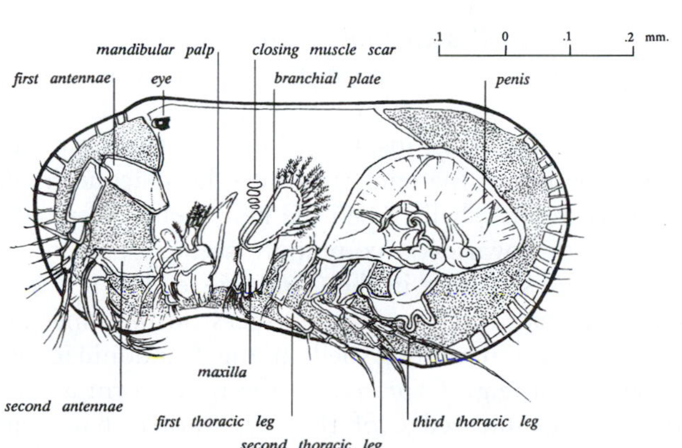

FIGURE 11 Sketch of the internal morphology of a male *Limnocythere sanctipatricii* (Brady and Robertson) (Cytheridae) from Kesling. First thoracic leg is referred to as the maxilla or maxillar endopodite and the second and third thoracic legs as the first and second thoracic legs in the text. [Figure redrawn by J. L. Delorme.]

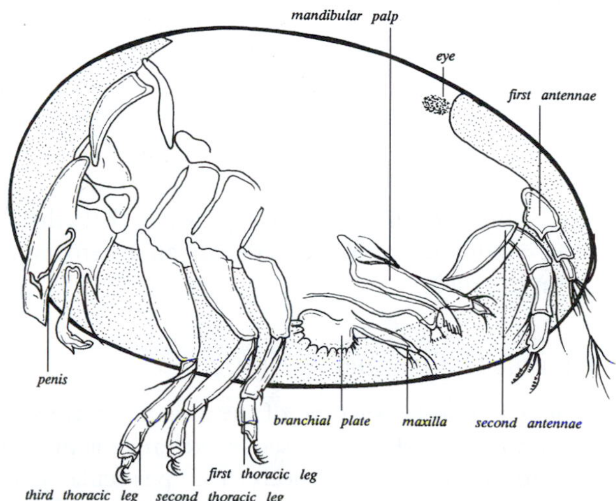

FIGURE 12 Generalized sketch of the internal morphology of a male entocytherid (Cytheridae) from Hart and Hart. First thoracic leg is referred to as the maxilla or maxillar endopodite and the second and third thoracic legs as the first and second thoracic legs in the text. [Figure redrawn by J. L. Delorme.]

first thoracic legs) are modified in sexual ostracodes into prehensile palps (Fig. 9a) used in grasping the female shell during copulation. There are two pairs of thoracic appendages. The first thoracic legs (previously called the second thoracic legs) are robustly developed and used for walking, climbing, and clinging onto surfaces. The second thoracic leg (formerly called the third thoracic legs), or cleaning legs, are the most flexible and well-muscled appendages. The end of these legs, in most ostracodes, are modified pinchers used to grasp and remove foreign objects from between the shells. In the entocytherid group, the modified maxillae and the two pairs of thoracic legs (Fig. 12) are modified terminally as hooks to hold onto the gills of the host crayfish (Hart and Hart, 1974).

Morphological diversity can be seen in the maxillae of ostracodes. Most notable are the first thoracic legs. These can be modified for feeding, for use during copulation (prehensile palps), or for walking. The commensal entocytherids have modified the maxillae and two thoracic legs for grasping onto the gills of the host.

Paired hemipenes are found in syngamic species behind the third thoracic legs on the ventral side of the thorax. Paired uterine openings lie behind and inside the vaginal openings of sexual and asexual females. The distal part of the thorax terminates in paired furcae. In some species, these are whip-shaped structures while in others they are well developed and armed at the end with long, serrated claws. Rome (1969) and Triebel (1953) consider the furcal attachment to be an important biocharacter to be used in systematics.

Sexual dimorphism is common in ostracodes. Determination of gender in most cases can be determined visually with a microscope. This is done by observing the presence/absence of testes and traces of the testes on the inner surface of the shell, Zenker's organs (see Section II.D), modification of the male maxillae (formerly called the first thoracic leg) into prehensile palps, hemipenes, and ovaries. For many species, it is necessary to separate the valves to view these organs. Except in cyprinotids, the male is usually larger than the female. In candonids, only the male has an anteroventral notch on the shell.

D. Internal Anatomy and Physiology

The circulatory system of freshwater ostracodes lacks both heart and gills. Gaseous exchange is through the entire surface of the body and particularly the membranous inner lamella of the exoskeleton. The respiratory plates of some appendages move and renew oxygenated water past the inner surfaces of the organisms. Large respiratory cells are known to occur in the inner lamella, forming a respiratory epithelium over the valve cavity which forms a blood sinus (Bernecker, 1909).

The digestive system of ostracodes consists of the mouth, esophagus, stomach, intestine, rear gut, and anus. The mouth is a large opening with the large teeth of the mandibular palps on either side. A gland, probably a salivary gland, opens into the mouth. The esophagus is very muscular and leads into the distended stomach. The hepatopancreas (liver) is found in the epidermal layers next to the shells and empties into the anterior part of the stomach. Most of the digestion takes place in the stomach where the food is formed into balls, passes from the stomach through the rear gut for absorption of the nutrients, and exits the anus as fecal pellets.

Approximately eight excretory glands, including the liver, have been identified in freshwater ostracodes. McGregor (1967) showed that for *Chlamydotheca arcuata* some food particles pass through the hepatopancreas, aiding in the digestion of food. Most glands have been studied in considerable detail (Bergold, 1910; Fassbinder, 1912; Schreiber, 1922; Cannon, 1925; Rome, 1947; Kesling, 1951a, 1965).

The nervous system of the ostracode is composed of a cerebrum (protocerebrum, deutocerebrum, and tritocerebrum; Turner, 1896; Hanström, 1924; Weygoldt, 1961), circumesophageal ganglion, and a ventral chain of fused ganglia (Kesling, 1951a). The first and second antennae, mandibular palps, and maxillae have small ganglia which connect to the central nervous system.

The ostracode has a single median eye made up of three optic cups each containing lenses. The number of cells composing a lens is variable (Novikoff, 1908; Rome, 1947). The black opaque eye lies in the anterior part of the body just below the rim of the shells. When the shells are open, the animal can detect shapes and motion; however, when the shells are closed it can only distinguish light intensity. Subterranean-interstitial (hypogean) ostracodes are blind (Danielopol, 1980a).

The ovaries are found in the epidermis next to the shell as a continuous sequence of gametes. The ova are fairly large in the last instar of many species, forcing the chitin coating of the epidermis into the space normally occupied by the shell. When the new shell is secreted, a trace of the ovaries is commonly seen as part of the shell's inner structure in the Candoninae. The testes are similarly placed and appear as traces on the inner and posterior region of the shell in the Candoninae. For some genera (e.g., *Cypricercus*) the testes form a spiral around the inner edge of the shell rather than being confined to the posterior area. The testes of Cyprididae change into the vasa deferentia as they leave the epidermis and enter the body cavity. The vas deferens joins up with the ejaculation ducts or Zenker's organs (Fig. 9a). These large organs are made up of longitudinal muscles in the shape of a tube on which are found radiating bristles and are covered in a chitinous sheath. The alternate contraction and expansion of these paired organs forces the sperm into the hemipenes during copulation. Zenker's organs lie in a nearly horizontal position in the posterior part of the body cavity and are visible in many live specimens. The hemipenes are very complex structurally. The ostracode must rotate them before copulation can take place (Kesling, 1957, 1965; Danielopol, 1969; McGregor and Kesling, 1969a, b). The testes and ovaries in Cytheridae are found next to the intestine, Zenker's organs are absent in this family. There are no males in the Darwinulidae. For further details on the anatomy and physiology, the reader is referred to Hoff (1942) and Kesling (1951a, 1965).

III. LIFE HISTORY

Ostracode eggs are constructed in a way that allows them to withstand physical and chemical extremes. The time of hatching may be soon after being laid or staggered over time indicating diapause. Once the egg is hatched, freshwater ostracodes will go through eight molt stages before becoming a mature adult. The length of time the organism lives depends on the species.

A. Egg

The freshwater ostracode egg is a double-walled sphere of chitin impregnated with calcium carbonate. The space between the two spheres is occupied by a fluid (Wohlgemuth, 1914). These two characteristics of the egg allow it to withstand desiccation and freezing.

Ostracodes develop from eggs which may or may not have been fertilized, depending on whether the species is bisexual or asexual. The eggs are laid singly or in masses, most often in the part of the habitat that will ensure an oxygen-rich environment when the eggs hatch (Wohlgemuth, 1914). Most ostracode eggs are white, but some are green (*Cypridopsis vidua*) or bright orange (*Cyprinotus incongruens, Potamocypris smaragdina*). All North American freshwater ostracodes are oviparous except several species of *Darwinula* and the marine brackish water *Cyprideis* which are ovoviviparous. These taxa brood their eggs in the posterior of the carapace from which the nauplii are released. Brood pouches are more common in marine ostracodes.

B. Reproduction

Reproduction in freshwater ostracodes may be either sexual or asexual. Depending on the gender, the ostracode has a pair of genital lobes or hemipenes.

Zissler (1969a, b, 1970) studied in detail the formation of sperm in *Notodromas monacha*. A single nucleus is formed by fusing of caryomeres; it is characterized by three granulated particles and by a double membrane-bounded region. The body then becomes spindle-shaped which includes the nucleus and the "Nebenkern" derivatives. Further development of the spindle-shaped cells takes place by the extension of two winglike structures, "Flügelstruckturen," on either side of the nucleus. The nucleus and the winglike structures extend throughout the sperm body, including the thread. The winglike structures are interpreted as providing motility to the sperm. Further differentiation of the sperm occurs between the testes and the vas deferens.

Lowndes (1935), following the research of other authors publishing between 1854 and 1889, reassessed the morphology and function of the sperm and concluded, "apparently the sperms are now functionless." Ostracode sperm have the dubious distinction of being the longest of any sperm relative to the host animal (Lowndes, 1935). Moore (1961) indicated that sperm do serve a function because females only lay eggs after copulation. However, Moore did not cite references to support his claim. Havel *et al.* (1990a) demonstrated that sperm have a function. Populations with males examined electrophoretically show genotypic frequencies in close agreement with those expected in a randomly mating sexual population.

Female-to-male sex ratios appear to vary depending on the species. Chaplin (1993) found that for *Candonocypris novaezelandiae* there are two morphs. For the "little brown morph" the sex ratio changes from 2.8:1 to 6.9:1 over a 27-month period. The conclusion was that the sex ratios show an increase in the abundance of parthenogenetic females. Dezfuli (1996) reports a sex ratio for *Cypria reptans* of 2:1. Sex ratios of modern North American ostracodes of 2:1 show a female bias (Chaplin *et al.*, 1994). These authors indicate there are no reports that male eggs are larger than female eggs so there is no logical explanation for the excess of females.

Based on earlier cytological studies, Bell (1982) concluded that all ostracodes are diploid. White (1973) indicated the diploid chromosome number varies between 12 and 19. Chaplin *et al.* (1994) provide an up-to-date review of the sex mechanisms for freshwater ostracodes. Although males are the heterogamic sex, some species have XO or XY sex determination systems. A case of multi-X systems is found in *Cyprinotus incongruens* where the females have six pairs of X chromosomes and the males have six X and a Y chromosome (Chaplin *et al.*, 1994). Further, they report that males of this species produce two types of sperms, one with five and the other with

10 chromosomes. Lécher *et al.* (1995) found some parthenogenetic ostracodes have "supernumerary chromosomes," and that they have complex sex chromosome mechanisms with XO, XY, and multiple X's and Y's. Three Canadian sexual ostracodes (*Cypricercus splendida, C. deltoidea,* and *C.tincta*) possess 22 chromosomes (Turgeon and Hebert, 1995). For *C. deltoidea* and *C. tincta* meiotic pairing was seen. These authors further indicate the asexuals, *Cypricercus horridus, C. reticulatus,* and *C. fuscatus,* had variable chromosome counts and no evidence of meiotic pairing was observed.

Electrophoretic studies by Havel *et al.* (1990b) show that 10 of 12 asexual ostracodes included clones which were polyploid. They also found that clonal diversity varied among species. For instance, *Cyprinotus incongruens* averaged 1.2 clones (1.3 clones; Havel and Hebert, 1993), whereas *Cypridopsis vidua* had 8.3 clones (9 clones; Havel and Hebert, 1993), *Cypricercus reticulatus* averaged 7.8 clones (Havel and Hebert, 1993) per pond. On a regional basis, high-diversity species indicate a range of 18–80 clones, while low diversity species show 1–7 clones (Havel and Hebert, 1993). Bell (1982) has indicated that asexuals are more abundant in modern habitats, whereas, sexuals are more abundant in habitats such as glacial lakes. The study of ostracodes from various habitats from the Canadian Arctic to the southern U.S.A. (Havel and Hebert, 1993) to tropical habitats in Jamaica (Little and Hebert, 1994) does not show an influence of latitude on the prevalence of asexual vs sexual reproduction. Rossi *et al.* (1998) reported similar results for Europe.

Hybridization has been documented between the asexual *Cyprinotus incongruens* and *C. glaucus* (Turgeon and Hebert, 1994; Chaplin and Hebert, 1997). Diploid asexual species when fertilized by a sexual species produce an asexual triploid. Suspected hybridization between *Cypricercus reticulatus* and other sexual cypricercids in Canada has been postulated (Turgeon and Hebert, 1995). Theoretical models predict sexually reproducing species should have selective advantage in unstable environments; however, a study by Chaplin and Ayre (1989) does not indicate a complete association of sexually derived recruitment in unstable environments.

Darwinula is thought to be the oldest asexual ostracode. Schön *et al.* (1998) believe that, because of this, accumulation of mutations should have occurred between alleles "with lineages and between lineages" through parthenogenesis. However, their data shows the opposite; there is "no variability in the nuclear ITSI region" despite analyzing individuals in the geographic range between Finland and South Africa.

C. Embryology

Embryology of ostracodes has been studied by Woltereck (1898), Schleip (1909), Müller-Cale (1913), and Weygoldt (1960). The small size, double-walled sphere, and textured outer surface of the egg does not allow for direct observation of the cleavage process. It is known, however, that after seven divisions, a 128-celled blastula is developed (Kesling, 1951a). Differentiation into ectoderm and endoderm (entoderm of Kesling) occurs after the eighth division.

D. Hatching and Seasonality

Most species begin hatching in the spring in the temperate zone. The timing of this process is controlled in part by temperature and geography (latitude). Species which live in temporary (vernal) ponds have a shorter life cycle. For example, *Cypridopsis vidua* develops in about one month (Kesling, 1951a) and during several months for *Cypria turneri* (Strayer, 1988). Some of these same species can also exist in permanent ponds where several generations per year may develop. For some species [e.g., *Candona caudata*; Delorme, 1978, 1982 (may belong to *Fabaeformiscandona,* see Meisch, 1996); not confirmed for North America, *Cyprinotus carolinensis* McLay, 1978c], hatching is delayed and spread over a much longer period. These apparent "resting eggs" allow the animal to sustain itself in the habitat in spite of stressful periods (such as anoxia and pollution; Delorme, 1978) when the nauplii cannot survive.

Eggs produced by some species can be either spring or summer forms. The ostracodes hatched from eggs laid in the spring do not tolerate warm summer temperatures. Ostracodes hatched from eggs laid in the summer will live only in warm water and expire in cold water (Schreiber, 1922). Martens *et al.* (1985) have noted that eggs of the Australian *Mytilocypris henricae* would not hatch below a certain temperature. Eggs of freshwater ostracodes can withstand freezing (Kesling, 1951a; Sohn and Kornicker, 1979). As early as 1915, Alm noted that temperatures affected the time of hatching. Ephemeral ponds in which the sediments stay frozen for up to 6 months during the winter produce many nauplii from eggs in the spring. Angell and Hancock (1989) found that eggs of *Heterocypris incongruens* could withstand freezing and remain viable down to −18°C when either wet or dry. Some authors have hatched ostracode eggs from dry lacustrine sediments kept in their laboratories for periods ranging from a few years to many decades (Sars, 1901; Sharpe, 1918). McLay (1978b) reported the eggs of *Cypricercus reticulatus* do not hatch until dried. Angell and Hancock

(1989) found that eggs of *Heterocypris incongruens* remained viable after drying at 22 and 40°C. The eggs did not remain viable if kept wet and heated to 40°C.

The eggs of some species (*Cypridopsis vidua, Cyprinotus incongruens, Physocypria* sp., and *Potamocypris* sp.) can pass through the lower intestinal tract of both wild and domestic ducks and remain viable (Proctor, 1964). Kornicker and Sohn (1971) reported that ostracode eggs were still viable after being egested by goldfish and swordtail fish. In an excellent paper detailing a meticulously carried out experiment, Rossi *et al.* (1996) have shown that, during its life time, the same individual of *Heterocypris incongruens* can produce diapausing eggs and eggs that hatch within a few days. Angell and Hancock (1989) categorize eggs that hatch within 10 days of being laid as being subitaneous. Rossi *et al.* (1991) and Rozzi *it al.* (1991)have shown the ratio of resting eggs depends on the genotype, temperature and photoperiod. Further they indicate the percentage of dormant eggs increases with temperature and daylight hours in a winter clone and decreases in the summer clone. Also their studies show egg hatching decreases at higher temperature, unaffected by photoperiod, in the winter clone and increases with both temperature and daylight hours in the summer clone. Rossi *et al.* (1996) found rather high number of first hatchlings (neonates) at the end of the experiment for the winter clone when they removed the algal mat to count cluster of eggs. They suggest that bioturbation may play a role in the hatching of eggs under field conditions. Rossi and Menozzi (1990) found from laboratory experiments that the dominant clone from the field have a better chance of survival at low temperatures and conversely the dominant summer clone survives better at high temperature. This indicates that temperature is most likely responsible for seasonal cycles.

Most ostracodes are controlled seasonally in their habitat. Research on seasonality has concentrated on those species living in temporary ponds. For temporary ponds in the temperate zone, early spring collections made after the basin receives water yield few live ostracodes [Hoff, 1943a (Illinois), Ferguson, 1958a (South Carolina)]. In these ponds a regular succession of juveniles can be observed until the pond dries, with some species exhibiting at least two generations per year. A temporary coastal habitat investigated by McLay (1978a, b, c) has ostracodes from the fall through the late spring. Ostracodes were absent through the summer because of habitat desiccation. Freezing had only a marginal effect on a part of the habitat. McLay (1978b) stated the maturation rate for *Herpetocypris reptans* was 190 days, and 45 days for *Cyprinotus carolinensis.* In permanent ponds, active species are absent for several months [Ferguson, 1944 (Missouri)]. Juveniles began appearing in early spring and continued until late spring, while adults were present until late fall to early winter. This pattern was repeated for the three species studied by Ferguson: *Cypridopsis vidua, Potamocypris smaragdina,* and *Physocypria pustulosa.* Ostracodes that live in ponds or ephemeral bodies of water have a short life cycle. Martens *et al.* (1985) noted that an Australian species took longer (4–5 months) to reach sexual maturity in the winter than in the summer (2–3.5 months). *Candona subtriangulata, Cytherissa lacustris,* and *Darwinula stevensoni,* that inhabit large lakes where the habitat is considered stable, will have a life cycle of 6–24 months (McGregor, 1969; Delorme, 1978; Ranta, 1979). The length of the life cycle depends on the species. Danielopol (1980b) observed that the length of the life cycle for five species ranged from 126 to 489 days, the latter being for a hypogean ostracode.

E. Molting

During the eight molt stages, the appendages change in size, shape, and function. A summary of these developments is found in Table I (after Kesling, 1951a).

Przibram (1931) determined that crustaceans doubled their weight from one molt to the next. He further concluded that because weight varies directly with volume, that volume would also increase by a factor of two. This led him to believe that a given linear dimension would increase by about the cube root of two (=1.26) for each molt. Based on this hypothesis, Kesling (1953) devised a circular slide rule which allows one to determine the variability of any linear dimension, area, or volume of any ostracode molt stage and adult. This slide rule is useful in determining which molt stage (shell or appendage) one is examining. Being able to identify an ostracode as an instar saves time because most instars cannot be identified to the species level.

TABLE I The Development of Appendages with Each Instar of the Freshwater Ostracode[a]

Instar	*1*	*2*	*3*	*4*	*5*	*6*	*7*	*8*	*9*
Antenna I	+	+	+	+	+	+	+	+	+
Antenna II	+	+	+	+	+	+	+	+	+
Mandible	+	+	+	+	+	+	+	+	+
Maxilla		−	+	+	+	+	+	+	+
Furca		−	−	−	+	+	+	+	+
Leg I				−	+	+	+	+	+
Leg II					−	+	+	+	+
Leg III						−	+	+	+
Ovaries							−	+	+
Testes								−	+

[a] (−) indicates the anlagen of a structure.

F. Competition, Co-existence, and Survival

The study of competition, coexistence, and survival by McLay (1978a–c) showed that, in a small temporary puddle, "competition is mediated via density-dependant mortality and maturation rates." He found that *Herpetocypris reptans* was a better competitor because of its longer maturation time (4419 degree days vs *Cyprinotus* with 2034 degree days) and could better withstand crowding and a shortage of food. His model shows that *Herpetocypris* was dominant over *Cyprinotus carolinensis* by a factor 7:1. He also found that competitive exclusion of *Cyprinotus* by *Herpetocypris* was prevented by periodic freezing. Freezing was also found to harm the population growth of *Herpetocypris* more than *Cyprinotus* because of the longer time required for maturation. Oertli (1995) studied the density of invertebrate of three substrates of a manmade pond. He found that life cycles and substrate controlled density fluctuations. High densities in the summer are related to life cycle and newly hatched individuals. He called the summer effect the "reproduction effect" with high mortality occurring in the winter and spring. For *Cypridopsis vidua* the highest densities occurred on *Chara* and *Typha,* with the lowest densities on terrestrial leaf litter. Rieradevall and Roca (1995) found that substrate and organic matter are important factors controlling the distribution of ostracode species in a karstic spanish lake. They collected high densities in the fall and spring.

IV. DISTRIBUTION

Ostracodes are found in nearly every conceivable aquatic habitat. These range from temporary and permanent ponds, lakes, intermittent, and permanent streams to ditches and irrigation canals, and fens. Caves and the interstices of aquifers are additional habitats. Some (referred to as terrestrial ostracodes) are found in moist organic mats of fens and in axial cups of certain plants such as bromeliads. North American freshwater and brackish water ostracodes thrive in a full range of salinities from 10 to 74,800 mg/L [high sulfate, low chloride waters] (personal files). Brackish water forms are common in high sulfate waters (*Limnocythere staplini*), as well as in chloride waters mixed with freshwater (*Cytheromorpha fuscata*) in inland lakes (Neale and Delorme, 1985). De Deckker (1981) lists an Australian species found in waters having salinitics greater than 170,000 mg/L. *Cyprideis,* also high salinity species, are common in southern coastal areas of North America (Benson, 1959), England and the Netherlands (van Harten, 1975; Sywula *et al.,* 1995). Jahn *et al.* (1996) have found that *Cyprideis torosa* can oxidize hydrogen sulfide to nontoxic thiosulfate and sulfite and rid itself of the oxidation products quickly.

A. Benthic Swimmers

Ostracodes are benthic organisms and are only rarely found in the plankton. They are divided into swimmers and nonswimmers. Most often they swim between aquatic plants. If they are disturbed while swimming, they often lose control and spiral down to the substrate. A few species of the genera *Cypria* and *Physocypria* are known to swim up into the water column, as they have been recovered from suspended sediment traps. Another swimmer, *Notodromas monacha,* has a peculiar habit of turning upside down when approaching the water surface. The venter of this species is flat, enabling the organism to adhere and hang beneath the water surface by surface tension. Marmonier *et al.* (1994) indicate that plant dwelling species, such as *Cyclocypris ovum* and *Cypridopsis vidua,* are more likely to be spherical in shape.

B. Benthic Nonswimmers

The nonswimmers, or true benthic forms, use their well-developed antennae (without setae) and first thoracic legs (previously called the maxillae) to crawl on the sediment–water interface. Some species, such as the limnocytherids, are infaunal, crawling beneath the sediment–water interface between the sediment particles (L. Delorme, personal observations; Benzie, 1989). Ostracodes live in the vents of warm and cold-water springs where groundwater is discharging (Forester, 1991; McKillop *et al.,* 1992). Särkkä *et al.* (1997) has reported *Potamocypris pallida, Candona candida,* and *Candona reducta* made up 21% of a spring meiofauna affected by road de-icing salt and gravel extraction. Ostracodes may live in the fractures of bedrock adjacent to streams, lakes, and pools in caves.

Hypogean (subterranean) ostracodes are being recovered more frequently from aquifers. These blind animals live in the interstices of sand aquifers. They are recovered by pumping the aquifer or trenching through it (Danielopol, 1980a). Marmonier *et al.* (1994) have shown that interstitial ostracodes have a geometric shape (triangular or trapezoidal) which is most suited to its habitat. The same can be said for most limnocytherids which live just below the sediment–water interface, crawling between the sediment interstices (personal observations). Ward *et al.* (1994) observed the highest concentration of hypogean ostracodes in a well close to the main channel of the Flathead River, Montana, and in wells farthest on the floodplain but near the Whitefish River. Rogulj *et al.* (1994) indicate those hypogean ostracodes (*Fabaeformiscandona wegelini*) that live in shallow interstitial habits are closely linked to running water rich in organic matter. This concept is supported by observations of Rouch and Danielopol (1997) and Marmonier and Creuzé des Châtelliers (1991). Creuzé des Châtelliers and Marmonier (1993) determined there were no meaningful correlations between abundance of ostracodes and particulate organic matter, alkalinity, calcium, or oxygen content of the interstitial water. Creuzé des Châtelliers and Marmonier (1990) found the interstitial fauna above and below a river riffle shows differing ostracode assemblages and abundances. Above the riffle, where the river bed was degrading, the water velocity was greater and the number of species and specimens was less than below the riffle. Below the riffle the river bed was aggrading and the velocity of the water was less, allowing more species to live in this area both in composition and numbers. Those living in deeper aquifers, prefer a more stable environment with less organic matter (Rogulj *et al.*, 1994). Along the length (320 km) of the Rhône River, Marmonier and Creuzé des Châtelliers (1992) found three different species assemblages tied to the ecological requirements of the member species. Two groups were restricted to the upper and lower ends of the river respectively and one was common to the whole length of the river.

V. CHEMICAL HABITATS

A. Chemical Water Types

Physical and chemical characteristics of the habitat are important factors in ostracode distribution and abundance. From Canadian habitat-waters, four chemical types are defined based on the predominant equivalent anion or oxyanion. These are bicarbonate, sulfate, chloride, and carbonate (personal files). A high percentage of genera (61%) are found only in bicarbonate waters. Bicarbonate, after being converted to carbonate, is a primary constituent of the shell. Only 9% of the known genera are found living in sulfate-rich waters. Several genera (30%) have species which have a preference for either of the two types of water. Chloride-dependent species, such as *Cytheromorpha fuscata* and species of *Cyprideis*, are more properly tolerant of marine brackish water. *C. fuscata* are occasionally found in inland lakes (Neale and Delorme, 1985) where brine seepages occur. Such species as *Candona rectangulata*, *Cyprinotus salinus*, and *Cypris bispinosa* are found in lake and pond habitats that may receive sea spray. These species are by no means restricted to chloride-enhanced waters.

B. Salinity and Solute Composition

Salinity and solute composition are of importance to freshwater ostracodes. Delorme (1989) has shown preliminary salinity tolerance distributions for some 43 species studied from Canadian habitats (Fig. 13). The sequence of species shown for salinity tolerance shows a gradual increase in the mean salinity value. Furthermore, the range, as shown by minimum and maximum values of salinity tolerance, is a function of the species. For example, *Cypridopsis vidua*, is a moderately saline-tolerant species able to survive in a very broad range of salt concentration. On the other hand, species such as *Candona subtriangulata* and *Cytheromorpha fuscata* are much more restricted in their chemical, and thus geographic range. The low end of the range, or pristine, low-salinity (SO_4) waters are inhabited by such species as *Candona subtriangulata* (23–93 mg/L),

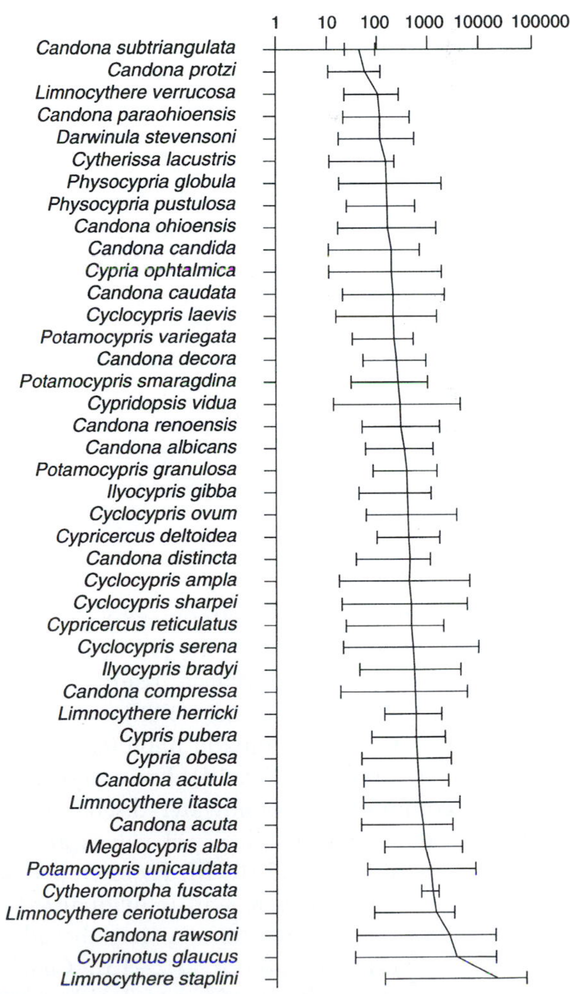

FIGURE 13 Total dissolved solids range for some Canadian freshwater ostracodes. The solid line represents the mean, the bars represent the minimum and maximum values. [Modified from Delorme, 1989.]

Limnocythere verrucosa (22–255 mg/L) from the Great Lakes area, and *Candona protzi* (may belong to *Fabaeformiscandona,* see Meisch, 1996, not confirmed for North America) (56–115 mg/L) of the high arctic. The high end of the range or high-salinity water bodies are inhabited by such species as *Limnocythere ceriotuberosa* (74–2780 mg/L) and *L. staplini* (122–74,840 mg/L). De Deckker (1983a) reported there are 14 North American ostracode species (one marine) that live in athalassic saline waters. Species such as *Cyprinotus glaucus* may be found in lakes where there is groundwater discharge of saline waters of intermediate flow systems (Delorme, 1972).

Forester (1983, 1986) and Forester and Brouwers (1985) have emphasized the importance of water composition over salinity. Examination of lake-water chemistry in North America, where *Limnocythere sappaensis* and *L. staplini* live, reveals that solute composition and not salinity is most important for their existence (Forester, 1983; Delorme personal files). Forester found that *L. sappaensis* lives in water enriched in Na^{+}–HCO_3–CO_3^{2-} and depleted in Ca^{2+}. *L. staplini* lives in water enriched in various combinations of Na^{2+}–Mg^{2+}–Ca^{2+}–SO_4^{2-}–$Cl^{-}$and depleted in HCO_3. Forester (1987) and De Deckker and Forester (1988) have expanded on the relationship between ostracode occurrences and major dissolved ion chemistry and salinity. They further suggested that relationships between species diversity, carbonate saturation, and relative abundance of dissolved calcium and carbonate ions exist, based on limited data from the United States. Most ostracode species may be placed in one of six or seven hydrochemical groups. The groups have well-defined boundaries based on carbonate saturation, major dissolved ion composition, and salinity.

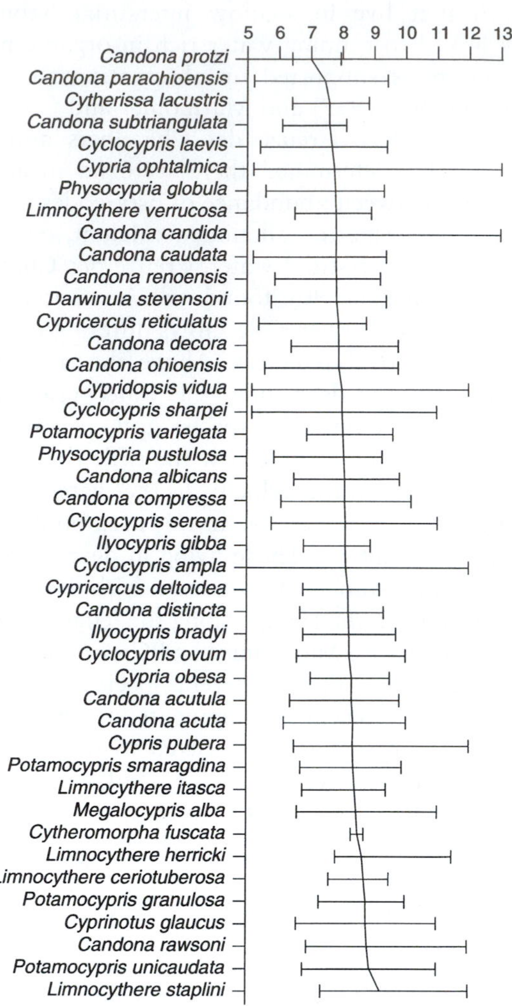

FIGURE 14 pH range for some Canadian freshwater ostracodes. The solid line represents the mean, the bars represent the minimum and maximum values. [Modified from Delorme 1989.]

C. Hydrogen Ion Concentration

The preservation of the calcareous exoskeleton for live ostracodes suggests that both the pH of the habitat must be circum-neutral and that a ready source of $CaCO_3$ must be available. Delorme (1989) indicated the "average" pH of Canadian aquatic habitats for ostracodes is between 7 and 9.2 (Fig. 14). For carbonate-enriched lakes, common to groundwater discharge areas of the United States and Mexico, pH values are typically in the range of 9.6–10.5 (R. M. Forester, personal communication). The species has an individualized pH range. Several of the species have a very broad pH range such as *Cypria ophtalmica* (5.2–13), *Candona candida* (5.4–13), *Cypridopsis vidua* (5.2–12), *Cyclocypris sharpei* (5.2–11), and *Cyclocypris ampla* (5–12). *Cytheromorpha fuscata* with its limited distribution in the Manitoba Lakes has a very restrictive range of 8.3 to 8.7 in inland lakes. This may be an artifact of the lakes from which the species was collected. The majority of ostracodes have a larger pH range than *C. fuscata.* When the epidermal lining covering the shell is ruptured and the water and sediments are poorly buffered, the calcareous shell will dissolve, destroying the animal. Larsen *et al.* (1996) used a number of statistical method to investigate the relationship between invertebrates to pH in Norwegian river systems. They indicate that a significant response occurs in the abundance of species when there is an increase in pH.

D. Dissolved Oxygen

Dissolved oxygen in the aquatic habitat is an important requirement for survival. The mean requisite

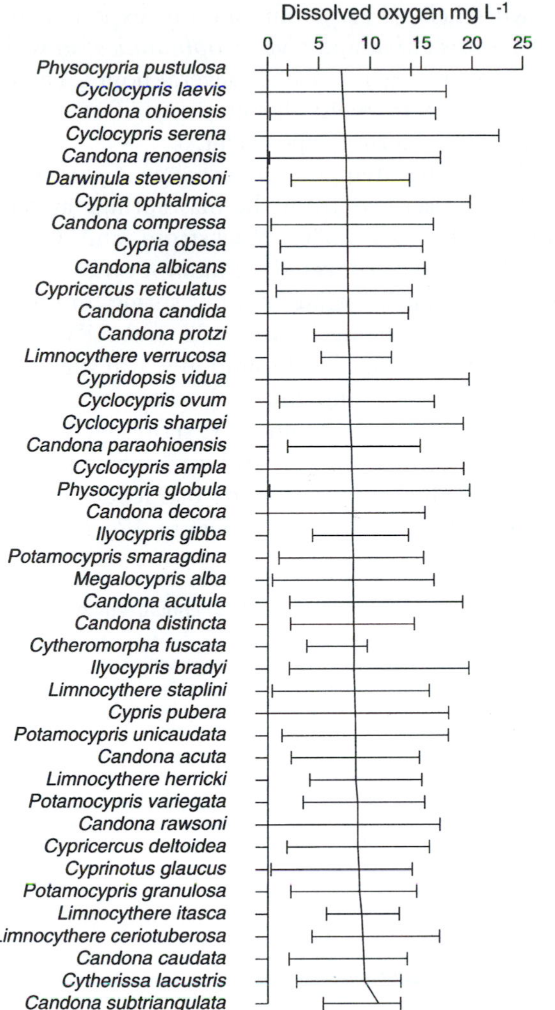

FIGURE 15 Dissolved oxygen range for some Canadian freshwater ostracodes. The solid line represents the mean, the bars represent the minimum and maximum values. [Modified from Delorme, 1991.]

for dissolved oxygen by ostracodes falls within a very narrow margin of 7.3 to 9.5 mg/L (Fig. 15). In general the concentration of dissolved oxygen in the water is broad. *Candona subtriangulata* has the highest minimum oxygen requirement of 5.6 mg/L. Most surprising in Figure 15 is the high number of species (34 out of 43, in Canadian habitats, L. D. Delorme, unpublished data) which can tolerate low dissolved oxygen concentrations. Those species which require a minimum 3 mg/L dissolved oxygen in the water are *Cytherissa lacustris, Cytheromorpha fuscata, Ilyocypris gibba, Limnocythere ceriotuberosa, L. herricki, L. itasca, L. verrucosa,* and *Potamocypris variegata.* Delorme (1978) found that *Candona caudata* (may belong to *Fabaeformiscandona,* see Meisch, 1996; not confirmed for North America) could survive in Lake Erie even though the hypolimnion has summer dissolved oxygen content below its lower tolerance limit of 2.8 mg/L. Its survival was accomplished by having a short life cycle allowing it to produce eggs before the onset of anoxia. This survival mechanism has been in place for some time in Lake Erie as indicated by the presence of fossil shells in the lake sediments (Delorme, 1982). Contrary to *C. caudata* (may belong to *Fabaeformiscandona*; see Meisch, 1996, not confirmed for North America), *C. subtriangulata* and *Cytherissa lacustris* have become locally extinct in the central basin of Lake Erie because they have a 1-year life cycle and cannot reach sexual maturity as they require a minimum dissolved oxygen content of 5.6 and 3.0 mg/L. *Cytherissa lacustris* is making a comeback with a reduction in anoxia. Newrkla (1985) found *C. lacustris* could survive at 1 mg O_2/L in the laboratory on a glass substrate and minimal food (20°C and 20-h exposure). Nine of the 43 species listed in Figure 15 can tolerate near-zero dissolved oxygen (below the detection limit of the Winkler azide method, APHA, 1965) for short periods. *Candona decora, C. candida, Cyclocypris ampla, C. sharpei,* and *Cypria ophtalmica* have been recovered from water of zero dissolved oxygen from a small river in early December (L. D. Delorme, unpublished data). The river had been covered by 1 ft of ice, suggesting these waters had been anoxic for some time. Fox and Taylor (1954, 1955) have found by experimentation that some ostracodes survive longer at oxygen levels below air saturation of 21%.

E. Carbonate

The effect of excess calcium carbonate and anoxia on *Cypridopsis vidua* was investigated by Reyment and Brännström (1962). Their results indicated the morphology of the shell is adversely affected (smaller shell dimensions) by excess calcium carbonate and anoxia in the environment. They reiterate the common belief that the major contributor to the size and shape of an ostracode is the environment, with a secondary contributor being the genome of the animal. Van Harten (1975) has shown the effect of salinity on shell size. The largest mean size of *Cyprideis torosa* was found in nearly freshwater, whereas the smallest mean size was found in waters where the salinity was very high.

VI. PHYSICAL HABITAT

A. Temperature

The water temperature of the aquatic habitat plays an important role in the habitat, seasonal, and geographical distribution of ostracodes. This is not to say that other factors such as availability of food, solute composition, turbidity, and energy levels are not important. The temperatures in shallow-water habitats can range from 0 to over 30°C, depending on the latitude and altitude. Species living in this niche must have the ability to tolerate this broad range in one or more life stages. Such species as *Candona acutula, C. candida, C. ohioensis, Cyclocypris ampla, C. sharpei, Cypria ophtalmica,* and *Limnocythere staplini* exhibit such a range for bottom water temperature (Fig. 16). The exception to the above are shallow ground water fed streams which may have a nearly constant temperature.

Candona subtriangulata has a low mean value of 5.5°C, with a range of 2.6 to 19.2°C (Delorme, 1978). This species is common in Lakes Superior, Huron, and Ontario at considerable depth, where the bottom-water temperature does not vary much.

Korschelt (1915) ran several experiments with *Cypridopsis vidua* and *Cypria ophtalmica,* in which the species were frozen in ice for several hours. The ice was then allowed to melt; the majority of the specimens survived the freezing experiment.

At the other temperature extreme, *Cypris balnearia* has been described from thermal springs at temperatures of 45 to 50.5°C. Wickstrom and Castenholz (1973) have recovered *Potamocypris* from an algal–bacterial substrate of a hot spring, in Oregon, at temperatures ranging between 30 and 54°C. *Chlamydotheca arcuata* has been found in warm springs (Utah, Nevada, Arizona, and Mexico) where water temperature varies between 24 and 39°C (Forester, 1991).

Mean annual air temperature assists in explaining the geographic distribution of ostracodes in habitats shallower than 25 m. From Figure 17, 20 out of 43

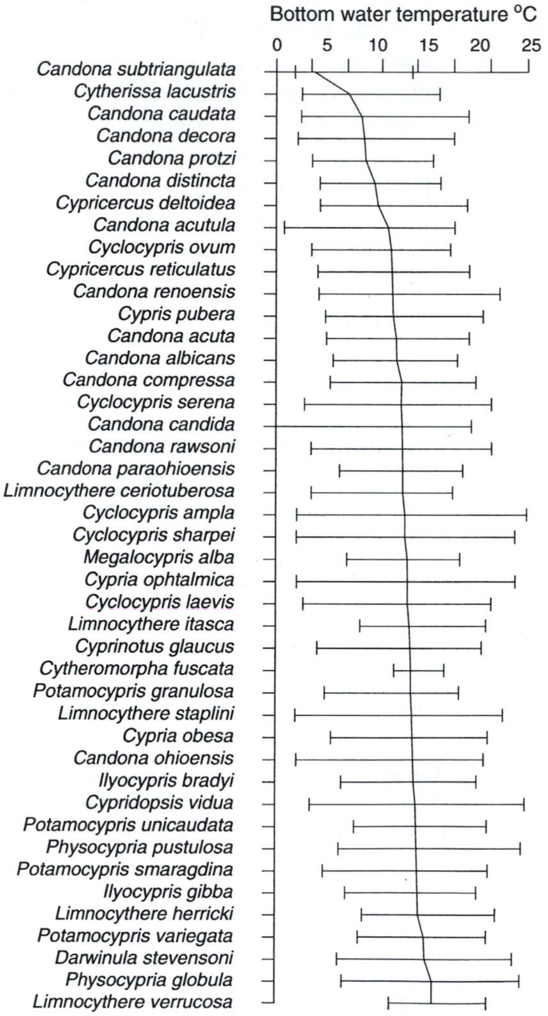

FIGURE 16 Bottom water or habitat temperature range (°C) for some Canadian freshwater ostracodes. The solid line represents the mean, the bars represent the minimum and maximum values. [Modified from Delorme, 1991.]

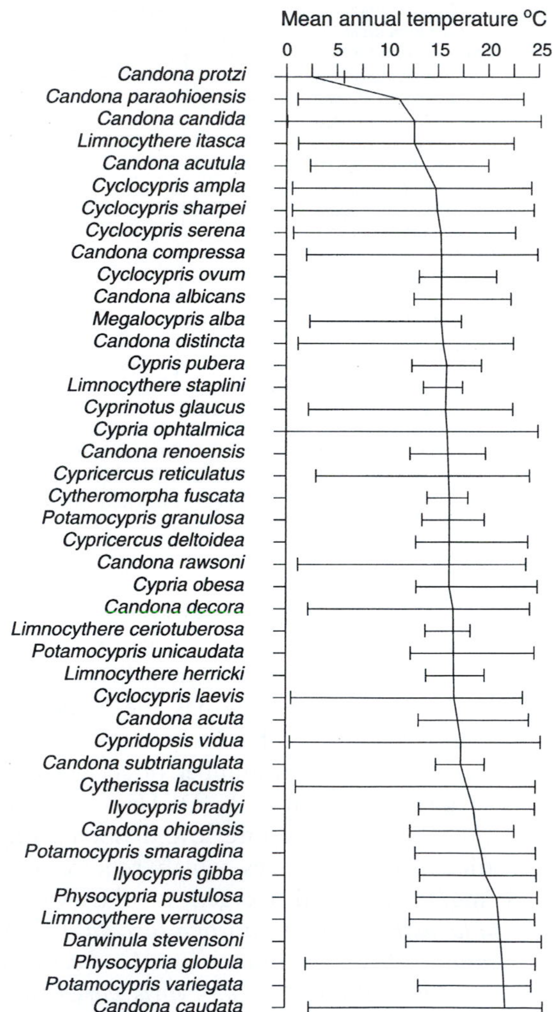

FIGURE 17 Mean annual temperature range (°C) for some Canadian freshwater ostracodes. The solid line represents the mean, the bars represent the minimum and maximum values. [Modified from Delorme, 1991.]

species can live in aquatic habitats north of latitude 60°. The minimum values of the range for mean annual air temperature for these species habitats are between −8.5 and −11°C. The species restricted to the arctic and low mean annual air temperatures are *Candona paralapponica, C. protzi* (may belong to *Fabaeformiscandona*; see Meisch, 1996, not confirmed for North America), *C. anceps, C. pedata, C. mülleri, C. ikpikpukensis, Cyclocypris globosa, Limnocythere liporeticulata,* and *Tonnacypris glacialis* . The habitats for those species living in the midlatitudes have a mean annual air temperature greater than −1.5°C.

B. Lakes, Ponds, and Streams

Most freshwater ostracode species may be found in more than one habitat (Fig. 18). The pond–lake habitats can be viewed as a continuum from shallow to deep water. Some ponds are temporary in nature and, therefore, harbor a specific faunal association such as *Candona renoensis, Cypricercus deltoidea, Megalocypris alba,* and *Cyclocypris laevis* (Delorme, 1989; King *et al.*, 1996; Wiggins *et al.*, 1980). At the opposite end of the scale are those species adapted or restricted to living in permanent lake habitats (*Candona subtriangulata, Cytherissa lacustris,* and *Cytheromorpha fuscata*). A faunal mix may occur between pond-lake and stream habitats. There may be a specific biotope developed at the mouth of a stream or river where it enters the lake or pond. The delta area is characterized by a combination of flowing and standing water. Many ostracode species are adapted but not restricted to living in this biotope such as *Candona acuta, Cyclocypris ampla, C. sharpei, Ilyocypris bradyi, I. gibba* and *Potamocypris variegata.* Large lakes have bottom currents which accommodate species such as *Candona caudata* (may belong to *Fabaeformiscandona*; see Meisch, 1996, not confirmed for North America) and *C. acuta.* None of the species listed by Delorme (1989) are completely restricted to flowing water. Those which are considered as lotic (fluvial) species are *Candona acuta, Cypria obesa, Ilyocypris bradyi, I. gibba,* and *Potamocypris variegata.* A special case is the fen. Up to 10 species have been recovered from fen vegetation at various levels down to 20 cm and varying levels of dissolved oxygen and redox potential (Douglas and Healy, 1991).

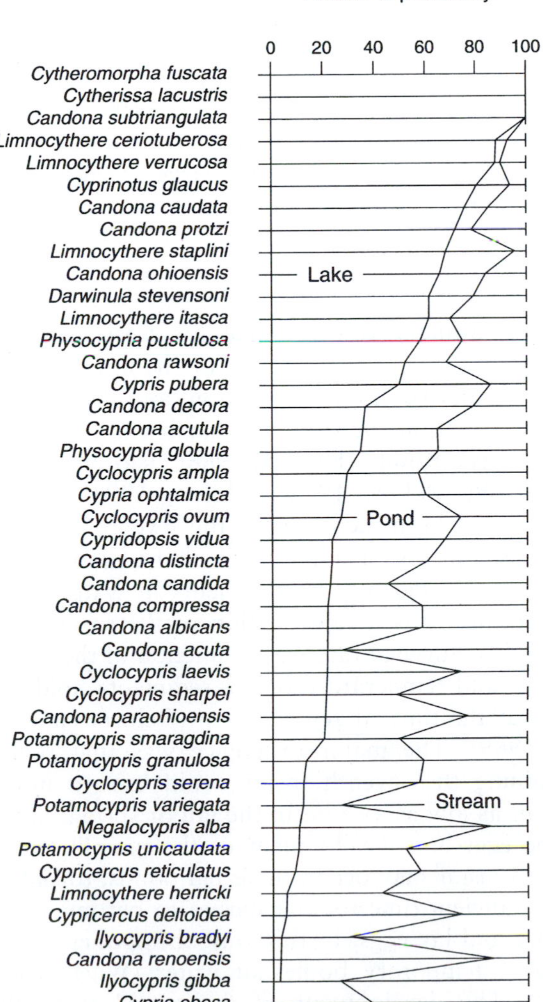

FIGURE 18 Percentage probability of finding a Canadian freshwater ostracode in a lake, pond, or stream. [Modified from Delorme, 1989.]

C. Bromeliads

Axial cups are formed by the spiral overlap of bromeliad leaves. These cups collect rain water and become inhabited by many kinds of invertebrates. Tressler (1956) described *Metacypris maracoensis,* recovered from the axial cups of Florida bromeliads. Ostracodes are more commonly found associated with bromeliads in Central and South America, and the Caribbean Islands (Little and Hebert, 1996).

D. Cave and Subterranean Habitats

Ostracodes have been recovered from several caves. Klie (1931) reported *Entocythere donnaldsonensis* commensal on the crayfish *Cambarus* from the Donnelson Cave of Indiana. From the Marengo Cave of Indiana, Klie named two new species:*Candona marengoensis* and *C. jeanneli.* Hart and Hobbs (1961) have described eight new troglobitic cave species from the eastern United States. Walton and Hobbs (1959) have also described troglobitic ostracodes from Florida caves.

Pools in deep sinkholes in limestone terrain (cenotes) of Yucatan, Mexico have a rich and varied ostracode fauna. Furtos (1936a) described 23 species from these habitats.

More research has been carried out on hypogean ostracodes. *Cavernocypris wardi* has been recovered from spring and interstitial water of the Flathead River, Montana (Ward *et al.*, 1994), from the South Platt River, Colorado (Danielopol *et al.*, 1994; Marmonier *et al.*, 1989; Marmonier and Ward, 1990 (*Candona candida, Cavernocypris wardi, Cypria* sp., *Eucypris* sp., *Fabaeformiscandona pennaki, F. wegelini, Ilyocypris* sp., *Nannocandona faba, Potamocypris* sp., *Pseudocandona* cf. *P. albicans, Strandesia canadensis*), and from springs in Wyoming, Nevada, and Colorado (Forester, 1991). *Cavernocypris subterranea* (common in Europe) has been found from a spring near Brush Creek, Idaho (Külköylüoĝlu and Vinyard, 1998). Not all ostracodes found near spring discharges are hypogean (Forester, 1991; McKillop *et al.*, 1992; Ward *et al.*, 1994).

E. Sediment–Water Interface, Particle Size, Food

The distribution of ostracodes at the sediment–water interface is a function of the availability of food, whether the substrate surface is clearly or poorly defined, the particle size distribution of the top centimeters of the substrate, as well as the time of year. The distribution of ostracodes is patchy. Where there is organic detritus, ostracodes will be abundant, particularly in shallow water. The presence of abundant floating organic debris just above an ill-defined sediment–water interface, discourages ostracodes from living in this part of the habitat. As detritivores and herbivores, ostracodes require a ready source of particulate organic matter that they can sweep into the mouth region. They also use their mandibular teeth to rasp or gnaw off small particles from large organic particles, living plants, or algae. On predominantly mineral substrates, ostracode densities will be low. As the sediment–water interface becomes deeper and further from the shoreline the number of species decrease. In this niche, the only food available will be the resistant cellulose raining down through the water column, bacteria, and fungi. In the deep troughs located in the eastern end of Lake Superior, only *Candona subtriangulata* has been recovered from depths of 45–305 m. Tressler (1957) identified *Candona decora* at a depth of 600 m from Great Slave Lake, Canada.

For many limnocytherids, especially *Limnocythere ceriotuberosa,* and *L. staplini,* the particle size of the substrate is very important. These ostracodes crawl through the interstices of very fine sand (0.0625 mm) to coarse sand (1.0 mm). In these niches, there must be sufficient food and oxygen for survival. Good porosity and permeability of these sediments allows a freeflow of oxygenated waters. Most clay-sized, organic-rich sediments are anoxic and reducing and not suitable for ostracodes to live in the interstices.

VII. PHYSIOLOGICAL AND MORPHOLOGICAL ADAPTATIONS

Ostracodes live in a wide variety of habitats, and within a given habitat physical and chemical conditions fluctuate. Physiological adaptations can be seen for some of these changes.

A. Eggs

Dispersal of freshwater ostracode eggs is thought to be passive (Neale and Delorme, 1985; Peck, 1994; Sywula *et al.*, 1995; Little and Hebert, 1996; Malmquist *et al.*, 1997) either in the gut or mud on the feet of birds, particularly for species living in the littoral habitat. Eggs that pass through the gut of fish and remain viable assist in passive dispersal (Kornicker and Sohn, 1971). For sampling sites that were up to 1000 km apart, Chaplin and Ayre (1997) found no evidence that stream flow was a "mediator of short or long distance gene flow in (the large green morph of) *Candonopsis novaezelandiae.*"

B. Resting Stage and Torpidity

The primary resting stage for freshwater ostracodes is the egg. The structure of the egg permits the yolk to survive both desiccation and freezing. The other adaptation for surviving harsh conditions in ostracodes is torpidity (Barclay, 1966; Delorme and Donald, 1969; McLay, 1978a; Horne, 1993). *Candona rawsoni* has been recovered in the seventh instar from frozen sediments of a pond which had gone dry the previous fall. Placing the sediments in water at room temperature caused rejuvenation of the seventh instar candonids. Survival rate was a function of the moisture content and temperature of the sediments holding the instar at the time of rejuvenation (Delorme and Donald, 1969). The major difference was that sediment containing the seventh instar could have a moisture content as low as 4–5%. In the fall or spring, hatching of the eggs and development of the nauplii from the seventh instar may occur. In vernal ponds, torpidity allows sufficient time for the species to become sexually mature and lay eggs. In this way, the species is propagated in temporary bodies of water (Wiggins *et al.*, 1980). The development of the torpid state allows ostracodes to survive the drying of vernal ponds. Horne (1993) found a similar situation for *Candona patzcuaro,* collected from playa lakes of the south plains of Texas.

C. Coloration

Coloration of ostracodes is a function of the pigment bodies located in epidermal cells of both the outer and inner layer. Ostracodes such as *Cypridopsis, Cypris, Cypricercus, Eucypris, Herpetocypris, Potamocypris,* and *Tonnacypris* are normally green, although *Cypridopsis vidua* and *Cypricercus* may be found in shades of brown and purple. These genera are all prevalent in littoral areas of lakes and in ponds where there is abundant vegetation. Green (1962) investigated the green coloration of *Eucypris virens* and found it to be a bile pigment, biladienes, derived from blue-green algae. Other species, such as *Cypricercus horridus* and *Megalocypris alba,* are purplish to reddish-brown in color which allows them to blend better with an organic substratum. Mbahinzireki *et al.* (1991) found that marked coloration increased predation risk by fish. The majority of the other species are a dirty white to yellowish-buff color which blends in with the mineral substrata. Candonids, cyprinotids, and limnocytherids fall into the last category.

D. Appendages

The first and second antennae are modified in swimming ostracodes. Finely feathered swimming setae on the first antennae are in constant motion and offer resistance to the water when the ostracode is swimming. Setae on the second antennae are used in a rowing motion. When the swimming setae are absent, the second antennae are used exclusively for walking and climbing.

In the male of some species, the maxillae (formerly referred to the first thoracic leg) is modified as a prehensile palp. This changes the appendage from a feeding function to a sexual function.

VIII. BEHAVIORAL ECOLOGY

Experiments were conducted by Towle (1900) on heliotropism using *Cypridopsis vidua.* She determined that ostracodes initially moved toward the light and then reversed directions.

Ostracodes that swim maintain a sustained motion. The swimming setae of the first antennae used in propelling the animal forward must be in constant motion or the animal will sink to the substrate. Observation of the swimming motion, using a binocular microscope, shows the swimming setae moving in a pattern similar to a hand-held egg beater directed upward. Because swimming is energetically quite expensive, ostracodes can only swim short distances (usually between adjacent plants).

The entry of a stimulus from a cyprinid fish tank into a microcosm of *Chara fragilis* (Uiblein *et al.* 1994) changed the behaviour pattern. The result was that *Cypridopsis vidua* responded by spending more time among the plant stems and leaves.

IX. FORAGING RELATIONSHIPS

The diet of most ostracodes is restricted to algae (phytophilic) and organic detritus. Klugh (1927) reported certain algae as being part of the food supply. Kesling (1951a) indicated diatoms are a part of the ostracode diet as did Strayer (1985) for *Cypria turneri.* Grant *et al.* (1983) reported feeding *Cyprinotus carolinensis* the blue-green algae *Nostoc* sp. Campbell (1995) calculated that tree pollen made up 8% and organic material 42% of the diet for *Australocypris insularis.* Tabacchi and Marmonier (1994) used *Chara* with alterable amounts of periphyton covering the stems as did *Roca et al.* (1993). When feeding, organic particles are swept into the mouth using the maxillae. The mandibular teeth grind large organic particles into a smaller size before they enter the mouth.

Although ostracodes are predominantly herbivores and detritivores, a few have shown carnivorous characteristics (Johansen, 1912). Ostracodes have been observed attacking and eating the soft tissue of certain snails (Deschiens *et al.,* 1953; Deschiens, 1954; Sohn and Kornicker, 1975). Some species (*Cypridopsis hartwigi, C. vidua, Cypretta kawatai,* and *Cyprinotus incongruens*) are known from experiments, by the above authors, to eat soft tissue of snails, such as the mantle and antennae, and to crawl into the respiratory organs. Have (1993) used *Paramecium* sp. as a source of food by *Eucypris* sp. in experiments on effects of area and patchiness on species richness. Campbell (1995) provides detailed information on the preying habits of *Australocypris insularis.* These large ostracodes feed on smaller ostracodes, *Diacypris compacta* and *D. dietzi.* She also found this large species grouped around drowned bee carcasses, presumably feeding on them, with terrestrial invertebrate detritus making up 8% and insects 3% of the diet.

The trophic position of most ostracodes is that of a herbivore and detritivore. This group of benthic animals scavenges and consumes organic matter primarily from the sediment–water interface. Some of this food is fresh, while the remainder is in the process of decaying. During the feeding process, nutrients and the more resistant cellulose and silica (diatom frustules) are moved back into the food chain.

Ostracodes are, in turn, consumed by higher animals. Predators include bottom dwelling fish. There are several references to the presence of ostracodes in the gut of fishes. Bigelow (1924), Harding (1962), and Ferguson (1964) found ostracodes in the stomachs of suckers and trout. Stomach contents of brook stickleback contained *Cyclocypris ampla,* that of yellow perch had *Candona ohioensis,* and bullheads had *Cypridopsis vidua* and *Physocypria globula* (L. D. Delorme, unpublished data). Griffiths *et al.* (1993) report the presence of Ostracoda in the guts of arctic charr, *Salvelinus alpinus,* and the three-spine stickleback, *Gasterosteus aculeatus* (Dezfuli, 1996). Bataille and Baldassarre (1993) investigated three prairie potholes (near the Delta Waterfowl Research Station, Manitoba) as to the availability of aquatic organisms (including ostracodes) as high protein food for ducks such as mallards and canvasbacks. They used activity traps to capture nektonic organisms during a 24-h period. The sample was then filtered through a #35 U.S. Standard sieve (500 μm). It should be noted that this sieve size would lose most ostracodes except *Cypris pubera.* Batzer *et al.* (1993) investigated the feeding habits of mallards and Green-Winged Teals on aquatic invertebrates. The contents of esophagi and habitat locations were studied using D-framed sweep nets (30-cm width, 1-mm mesh size) for nektonic and epiphytic invertebrates. For sediment sampling of the habitat, a bilge-pump intake was placed on the wetland bottom and four pumping repetitions of water and sediment were passed through a 0.3-mm sieve. This sieve size would lose most ostracodes. Ostracodes and other invertebrates were each consumed by only 3 or 4 of the 19 mallards. The relative spacing of bill lamellae of specific duck species influenced their selection of different orders of invertebrates (Nudds and Bowlby, 1984; Batzer *et al.,* 1993).

Green (1954) noted the oligochaete *Chaetogaster diaphanus* eats ostracodes as do cyclopoid copepods (Fryer, 1957) and the tanypodine midge (Roback, 1969). Benzie (1989) indicated "a negative association" between a predatory mite and *Herpetocypris reptans.* Swüste *et al.* (1973) showed experimentally that the phantom midge *Chaoborus flavicans* released *Cypria* immediately after capturing it. This suggests that this ostracode either tastes bad or releases a noxious chemical. Lancaster and Robertson (1995) looked at the preying habits of the polyphagous net-spinning caddisfly (*Plectrocnemia conspersa*) and the alderfly (*Sialis fuliginosa*). They found both flies had very high densities of ostracodes in their guts, probably because both groups are very abundant in leaf litter of slow moving streams. The introduction of fish into a pond, where larger predators have been absent, quickly reduces the invertebrate populations. Martens and de Moor (1995) found the invertebrate populations that lived in a large predator free habitat were adapted to living in such an environment. The consequence is these habitats will not be repopulated with the same invertebrate populations.

X. POPULATION REGULATION

A. Suitability of Habitat

Abundance of ostracodes at the sediment–water interface is a function of substrates, availability of food, season, and depth of water. Strayer (1985) has provided a useful and current summary of abundance and biomass measurements of ostracodes in lake environments. Bechara (1996) has indicated floodplain lakes with high floating plant biomass had a decrease in biomass during high-water periods, Ostracoda and Oligochaeta were dominant. The presence of food clearly has an impact on species richness and abundance. Long periods of warm temperatures down to the bottom of the euphotic zone favor ostracode production. The number of species and specimens recovered from the profundal sediment–water interface is small. For example, the ostracode assemblage in the deep troughs (190–365 m) of eastern Lake Superior is limited to *Candona subtriangulata.* At these depths, during the last week in May and the first week in June, the bottom water temperature varied between 2.6 and 4.0°C.

Based on Canadian habitats sampled for ostracodes, a maximum of nine genera have been counted from one sample. In these samples, between 13 and 19 species were represented with a maximum of nine species being alive at the time of collection (L. D. Delorme, unpublished data). Rouch and Danielopol (1997) defined high species diversity as "species collections of more than 25 species within an inventoried area, at the local scale (i.e., between 1 and 100-m length), low diversity is 1–5 species and medium level diversity is 6–24 species." King *et al.* (1996) found species richness varied up to 13 species per pond. Baltanás (1992) attempted to make a comparison between samples using species numbers referred to a fixed sampled area. He evaluated three methods for estimating species richness (the number of individual species per sample) using computer simulations. These were: Stout and Vandermeer method based on the species–area relation; Cohen's method which assumes the log-normal distribution of species abundances; and the jackknife method, a nonparametric procedure with no distributional assumptions. The author found the estimators tested showed a negative bias, with the second-order jackknife method giving the best results with the highest accuracy and with the best precision.

B. Toxicity of Herbicides and Pesticides

Sanders (1970) tested *Cypridopsis vidua* and five other crustaceans against 16 herbicides. Irritability and excitability were the first noticeable reactions. *Cypridopsis vidua* could only be tested at low concentrations because the animal would close its shell when it received the stimulus. This species was the second most sensitive to the herbicides. Landis *et al.* (1994) found a similar pattern when testing ecological risk assessment of *Cyprinotus* sp. and other organism in a microcosm against the water soluble fraction of jet fuel A. Dieter *et al.* (1996) tested several invertebrates in microcosms at the edge of lakes for toxicity to insecticides. They found ostracodes were resistant to phorate. Takamura and Yasuno (1986) found that ostracodes in rice fields decreased in numbers with a treatment of pesticide mixture of propoxur, thiobencarb, and simetryne. There were similar decreases in ostracodes with a treatment of bentazone. In another rice field, they used a mixture of methomyl, kasugamycin-hydrochloride, neo-asozin, thiram, and benomyl with some decrease in ostracode fauna.

C. Ultraviolet-B Radiation

Hurtubise *et al.* (1998) subjected a number of invertebrates to ultraviolet-B radiation. They found that *Cyprintous incongruens* was highly tolerant at a water depth of 4 cm when the UV-B irradiance was 56.1 μW/cm. They also contend the carapace protected the animal. Further that dissolved carbon (DOC) attenuated the irradiance.

D. Parasitism

More studies are now documenting the role freshwater ostracodes play in parasitism. Dezfuli (1996) has detailed the effects of the helminth *Neoechinorhynchus rutili* on the ostracode *Cypris reptans*. Between 8.6 and 14.5% of this species were infected, with up to three larvae found in one host (up to six worms; see references in Zelmer and Esch, 1998). The effect of this parasitism is a reduction in the number of eggs contained in females, 103 as compared to 207 in uninfected females. *Cypridopsis vidua* has been found to be the intermediate host of *Neoechinorhynchus prolixus* (B. B. Nickol, University of Nebraska, personal communication). *Cypridopsis* sp. acts as a second intermediate host for the parasite *Halipegus occidualis* (Zelmer and Esch, 1998). For the complete cycle, eggs from the worm pass through the feces of the green frog (*Rana clamitans*) and are ingested by the pulmonate snail, *Helisoma anceps*. While in *H. anceps* sporocysts and redia stages develop ending in the development of cercariocysts. Feeding activity of *Cypridopsis* sp. transfers the cercaria body into the hemocoel of the ostracode. Odonate naiiads eat the affected ostracodes and, in turn, are eaten by green frogs. Griffiths and Evans (1994) found the infestations of the peritrich *Nüchterleinella corneliae* on *Cypria ophtalmica* and *Cyclocypris ovum*. Infestations where more prevalent during winter and decreased during the summer. The parasite tended to be clustered around the bases of the appendages, particularly the genitalia and the furcae.

Bronnvall and Larsson (1994, 1995) described new microsporidian parasites, *Flabelliforma ostracodae* and *Binucleospora elongata* in the ostracode *Candona* sp. Their investigations, based on light microscopic and ultrastructural characteristics found the musculature, the frequent site for the development of microsporidia, was undisturbed. The infection was restricted to the adipose and connective tissues and to the haemocytes which appeared to be filled with spores. Diarra and Toguebaye (1996) examined sections of ostracodes infected by *Nosema stenocypris* which revealed that the microsporidium develops particulary in the muscles of *Stenocypris major* which are destroyed and replaced by the developmental stages of the parasite.

XI. COLLECTING AND REARING TECHNIQUES

A. Collecting and Preservation

Ostracodes have been collected using various kinds of gear such as a Birge cone net (Ferguson, 1958b). Using a net has certain disadvantages when gathering benthic organisms; many true benthic forms are not collected because the net usually does not thoroughly sample substrates which ostracodes live on and in. Bais and Agrawal (1995) collected zooplankton using a plankton net (0.25 mm); however, such an apparatus will only collect the larger ostracode species while missing a few smaller forms and the true benthic fauna. The use of an Ekman dredge or similar device is recommended for a quantitative sample of the benthos (Delorme, 1967). Care should be taken not to allow the dredge to settle too far into the substrate, 2 cm being optimal. The actual depth to which a sampling device should go may be checked out by pushing a short core tube into the sediments to observe the depth to which ostracodes have penetrated. Any dredge that is used should have flaps on the top so water will flow through the dredge when it is being lowered. This will prevent a bow wave from preceding the dredge and disturbing the sediment–water interface. When the dredge is retrieved, the flaps will close and prevent the interface and collected organisms from being disturbed and swept away by moving water.

The large quantity of sediment (>500 cm^3) retrieved using a dredge means that a method of concentrating the ostracodes must be used. The sample can be washed through a set of 8-in. sediment sieves (#20 mesh, 0.841 mm; #60 mesh, 0.25 mm; #100 mesh. 0.149 mm) using a gentle stream of water from a sprinkler head. This procedure removes particle size fractions (including instars) finer than fine sand (0.149 mm). The smallest freshwater ostracode is *Limnocythere friabilis* with a height of 0.23–0.31 mm. The adult shells of this species are readily retained by both the #60 and #100 mesh. Large cyprids, such as *Cypris pubera,* megalocyprids, herpetocyprids, and prionocyprids will be concentrated on the #20 mesh sieve. When the specimens are used to study appendages and other organs, ostracodes can be picked by hand from the washed residue by using a steromicroscope. The specimens should then be placed in 5–10% buffered formalin solution. Concentrated formalin, ethanol, and methanol should not be used, as these preservatives will decalcify the shell. It has been suggested that sodium tetraborate (250 cc saturated solution) be added to 20 L of 75% ethanol to prevent calcitic shells from being dissolved. Specimens may be mounted in Hoyer's solution, glycerin, or polyvinyl lactophenol. However, with the use of the scanning electron microscope (SEM) to study appendages, these mounting media are not necessary.

When preservation of the visceral mass or the appendages is not critical, freeze drying is suggested to preserve the specimens for easy storage. The residue from the sieves is transferred to a beaker using a spatula and a small amount of water from a wash bottle. The sediment is then frozen. After freezing, filter paper or a suitable tissue is placed on the beaker with an elastic band and set into a freeze-drying apparatus. In the freeze-dryer, ice is sublimed by vacuum from the specimens and sediments. This process takes 24–48 h, depending on the volume of ice being removed, the number of samples, and the size of the freeze-dryer. The specimens are then handpicked from the sample using a steromicroscope and placed in a separate vial or a micropaleontologic slide. This method has the added advantage of not requiring the more extensive labor and time curating wet samples. The dried specimens are also in excellent condition for further study with an SEM to view and photograph appendages.

When a large volume of sediment still remains in the sieves after washing and freeze drying, the ostracodes can be concentrated using acetone (Delorme, 1967). [Acetone should only be used under a safe operating fume hood.] The residue from the #60 and #100 mesh sieves is passed through a 3-in. filter sieve to ensure the sediment particles are disaggregated. The sieved residue is then sprinkled into a funnel (4-in.) containing acetone. The sediment particles sink and the ostracode carapaces float to the surface. Normally, the carapace will trap an air bubble which makes the shell buoyant. The lower tip of the funnel is equipped with a latex tube and a clamp. After the sediment has been sprinkled into the acetone, the clamp is carefully removed from the tube and the "sink" allowed to drain into another funnel lined with rapid-draining filter paper. Care must be taken not to drain all the acetone from the top funnel, otherwise the ostracodes will be removed. When the sink has been removed, another funnel lined with filter paper is placed beneath, the clamp removed, and the top funnel rinsed with acetone from a wash bottle. In this way, all supernatant particles (the float), including the ostracodes, will be transferred to the lower filter paper. Individual shells will also be a part of the float except *Cytherissa, Darwinula, Cytheromorpha,*and *Limnocythere* which do not have an anterior vestibule. The filter paper containing the ostracode residue is then air- or oven-dried, and the residue transferred to a vial. The shells and carapaces are picked with a very fine wetted sable hair brush and mounted onto a micropaleontological slide. The slide is preglued with the water-soluble gum Tragacanth. The ostracodes are then ready to be identified and counted.

For an accurate count of adult ostracodes, the foregoing method is adequate. If, however, all live instars and adults are required for biomass studies, then the complete sediment parcel will be required. Wet sieving would only assist in separating organisms and particles by size. For biomass studies, it might be preferable to use a short core tube pushed into the sediment some 10–15 cm. Nalepa and Robertson (1981) have recovered ostracodes using two mesh sizes, 0.595 and 0.106 mm, for biomass studies. The smallest mesh size required to collect all the ostracodes including instars is not indicated by them, as 0.106 mm will not retain all the instars. Hummon (1981) recommends a sieve size of 0.062 mm to retain all ostracodes for quantitative analyses. This sieve size would miss the first three instars of *Limnocythere friabilis.*

B. Rearing Techniques

Cosmopolitan ostracodes, such as *Cypridopsis vidua* and *Cyprinotus incongruens,* are relatively easy to rear in an aquarium. Collection of water, from the habitat associated with the species to be reared, along with some of the substrate will allow the population to develop. Natural water from the habitat is preferred to distilled or tap water. Habitat water contains the proper mix of major and minor ions for an established solute composition. The pH of the water should be checked periodically and adjusted when necessary. This type of aquarium will probably go "wild" and anoxic after several months. A more pristine culture can be set up by filtering the water from the habitat, placing it in an aquarium with sterilized silt, sand, and clay as a substrate. Cultured algae (Klugh, 1927; Kesling, 1951a; Sanders, 1970; Grant *et al.,* 1983; Havel and Talbot, 1995; Roca *et al.,* 1993; Roca and Wansard, 1997; Zelmer and Esch, 1998) have been used for food as well as fresh leaf lettuce and grated boiled egg which has been allowed to decompose bacterially for a month. Roca and Wansard (1997) also used benthic algae, gastropod pellets, and *Spirulina* with a high survival rate whereas chopped alfalfa as food with garden compost produced the lowest survival rate. Aquaria should not be placed on a window ledge where high temperatures may destroy the artificial habitat.

XII. USES OF FRESHWATER OSTRACODA

Ostracode shells are preserved as fossils. Autecological data bases are available for extant species, and so these crustaceans can be used to reconstruct past chemical and physical habitats.

With a suitable time frame, the paleolimnology of a pond or lake may be reconstructed from the fossils extracted from a sediment core. Initially, subjective interpretations were made of past conditions of the lake or pond (Gutentag and Benson, 1962; Benson and MacDonald, 1963; Staplin, 1963a, b; Neale, 1988). As autecological data bases were generated, paleointerpretive models were developed that used autecological data for objective reconstruction of the chemical and physical aspects of the habitat (Anderson *et al.,* 1985; Cvancara *et al.,* 1971; Delorme, 1968, 1969b, 1971b, c, 1982, 1996; Delorme and Zoltai, 1984; Delorme *et al.,* 1977, 1978; Forester *et al.,* 1987; Karrow *et al.,* 1975, 1995; Klassen *et al.,* 1967; McAllister and Harrington, 1969; Smith *et al.,* 1992; Westgate *et al.,* 1987; Westgate and Delorme, 1988; Porter *et al.,* 1999). In some instances, it is possible to use a number of different fossils (ostracodes, molluscs, pollen, and diatoms) to obtain comprehensive paleoenvironmental interpretations (Maher *et al.,* 1998). A summary of the use of Canadian ostracodes in paleolimnology and paleoclimatology is given by Delorme (1989).

Forester *et al.* (1994) have devised two interesting indices which estimate the "TDS ratio" of evaporation to outflow of a large lake and a "shore–zone ratio" which gives an indication of the proximity of the shore line to the core site. In each case, ostracode abundances are used in the ratios. Curry (1999) has devised two environmental tolerance indices (ETI) based on 341 modern aquatic environments in the United States. The indices are based on a limited set of autecological data and so at this time cannot be universally applied for North America.

Trace element analyses of Mg, Ca, and Sr in ostracode shells have been used for reconstructing past salinity changes in a variety of aquatic habitats (Chivas *et al.,* 1983, 1985, 1986a, b; De Deckker, 1983b, 1988; Lamb *et al.,* 1995). Similar studies are being carried out on North American ostracode shells. Engstrom and Nelson (1991) found that distribution coefficients (K_D) for $(Mg^{2+}/Ca^{2+})_{H2O}$ and $(Sr^{2+}/Ca^{2+})_{H2O}$ in shells of *Candona rawsoni* from cultured samples and from shells collected from Devils Lake, North Dakota have similar ratios. Analyses show excess Mg is incorporated into the shell during early calcification which is a confirmation of the work done by Chivas *et al.* (1986a). Hu *et al.* (1998) did trace-element analyses of ostracode shells from a sediment core of Farewell Lake, Alaska. From the Mg/Ca and Sr/Ca ratios they detailed the climatic reconstruction of the area indicating cold and warm climate periods. A 19-m core from Coldwater Lake, North Dakota documented several century-scale changes in the paleoenvironmental record using $\delta^{18}O$, $\delta^{13}C$, Mg/Ca, and Sr/Ca ratios (Xia *et al.,* 1997b). A total of seven phases of climate and salinity were identified. Yu and Ito (1999) using *Candona rawsoni* shells from a Rice Lake, North Dakota, core determined Mg/Ca ratios showing periods of salinity and aridity.

Oxygen isotope studies of calcitic shells are being used to develop paleoclimatic interpretations (von Grafenstein *et al.,* 1992, 1994). Middle Holocene ostracodes of Elk Lake, in Grant County, Minnesota, show an assemblage dominated by *Limnocythere staplini,* a halophyte. This species is a high-salinity form, yet the oxygen isotope values on shells of *Candona rawsoni* indicate a decrease by 2–3 ppt indicating that ground water mediated the paleoclimatic record. Xia *et al.* (1997a, c) have determined that calcitic shells of *Candona rawsoni* formed at temperatures of 15 and 25°C were not in isotopic equilibrium with water, but had a constant offset from equilibrium based on oxygen isotope fractionation of about 2 ppt. With the molting process occurring from spring to fall, the intraannual temperature change would exceed the interannual to century-scale variations expected during late-Quaternary climate shifts. The resulting intraannual signal may exceed the long-term paleoclimatic signals.

The authors suggest 10–15 shells per sample be analyzed to ameliorate the interannual noise.

XIII. CURRENT AND FUTURE RESEARCH PROBLEMS

The European literature on freshwater ostracodes continues to grow. Unfortunately, the North American literature is not doing as well. There is a real need for a comprehensive review and update of North American systematics. Detailed descriptions of soft and hard part morphology is desperately needed together with detailed drawings and SEM photographs. This information is critical if scientists hope to be able to continue to communicate about the biological units on which research is being conducted. Are we talking about the same species?

Specific research into ostracode systematics needs to be emphasized. This is particularly true of several generic pairs such as *Cypricercus* and *Eucypris, Cyprinotus* and *Heterocypris,* and *Megalocypris* and *Cypriconcha.* These groups require a detailed cytogenic and SEM study of shell and appendage morphology of species of the world (see Section XIV.A). Only in this way will we be able to resolve which biocharacters of a certain group should be used in defining a genus. The establishment of a universally acceptable systematic nomenclature is important for studies in ecology and paleolimnology.

Continuing studies are required in autecology of ostracodes. Data bases of chemical and physical parameters of ostracode habitats are being expanded (B. B. Curry, R. M. Forester, A. J. Smith, personal communication; L. D. Delorme unpublished). Analyses of these data bases are required to give a sense of the factors which control the species in its environment. The roles of solute composition and salinity of the habitat are an integral part of this analysis.

The role of ostracodes in the population dynamics of a pond or lake system is poorly known. Only a few studies have attempted to look at biomass (Bechara, 1996), productivity, densities in communities, positions in the aquatic food chain (as predators or prey), behavioral defense mechanisms, and host–parasite interactions. All these factors and others may affect the success of ostracodes in their habitat and need to be studied in more detail. More scientific papers have been written since the first edition of this book, but more are needed as many more questions are being raised.

XIV. CLASSIFICATION OF OSTRACODES

Ostracodes are typically identified from adult specimens. Insufficient information is obtained from the instars (juveniles) to make a proper identification to the species level unless a monoculture has been obtained and adults are present. In some instances, identification of the instar can be taken to the generic level. The most useful biocharacter in taxonomy and systematics is the shell. A field key is offered by Delorme (1967) for some of the North American ostracode species as well as his five-part series (Delorme, 1970a–d, 1971a). Other useful references are Furtos (1933) and Hoff (1942).

A. Taxonomic Controversies

A number of controversies exist around the naming of certain genera. These will be discussed separately. In order not to confuse the issue any further, a conservative approach will be taken for the use of generic names. Detailed systematics using the scanning electron microscope will be required, using specimens from the global arena rather than from local areas, in order to resolve the issues.

The names of two genera *Cypricercus* and *Eucypris* have been commonly used. Broodbakker (1983) has provided a literature review of the status for *Cypricercus* and *Strandesia.* There is no mention of the genus *Eucypris.* Sars (1928) differentiated between *Eucypris* and *Cypricercus* on the absence and presence of males respectively. He also noted the peculiar arrangement of the "spermatic vessels" (Sars, 1928, p. 117) in the males of *Cypricercus.* Furtos (1933) maintained the two genera based on reproduction and length of the furcal ramus (<1/2 the length of shell, *Eucypris*; >1/2 the length of shell, *Cypricercus*). Hoff (1942) noted the elevation of genera on the basis of propagation was not an acceptable practice. He opted for one genus based on an extremely long furcal ramus for *Cypricercus.* The species *Eucypris crassa* was originally assigned to *Cypris* by O. F. Müller and later to *Eucypris* by G. W. Müller (1912). This species has always been referred to the genus *Eucypris.* Martens (1989) has ascribed *Eucypris serrata* to the genus *Trajancypris,* but has not included the North American form (Delorme, 1970a) to the new genus. Broodbakker (1983) erected the genus *Bradleystrandesia* to include those species of *Cypricercus* which occur in Europe and North America. This author states "further research is needed to prove if the genus *Bradleystrandesia* is confined to Europe and North America" (Broodbakker, 1983, p. 329). Martens (in Turgeon and Hebert, 1995) continued to use *Cypricercus* for the North American species. *Bradleystrandesia* has not been confirmed for North America, but has been used by Rossi *et al.* (1998) in referencing work done by Turgeon and Hebert (1995). *Cypricercus* and *Eucypris* will be used in the key.

The genera *Cyprinotus* Brady and *Heterocypris* (Claus) have had a confused nomenclature. Purper and Würdig-Macial (1974) did a literature review of these two genera. They concluded the main difference between the two genera was the dorsal gibbosity of the right valve for *Cyprinotus*. Both genera are reported as having the anteroventral margin of the right valve with denticles or crenulations. Sars (1928) and Müller (1912) did not consider the gibbosity of the right valve as a valid generic biocharacter and, thus, did not recognize *Heterocypris*. Hoff (1942) also did not recognize *Heterocypris*. This morphological biocharacter in *Physocypria* (Delorme, 1970b) may or may not be present and is not used to split the genus. *Cyprinotus* is used in the key because it was published first in 1886.

The genus *Cypriconcha* was erected by Sars (1926) to incorporate the large cyprinids of North America. Sars (1898) had already established *Megalocypris* for the large cyprinids of Africa. Delorme (1969a) reviewed the two genera and concluded the similarities were many and discrepancies few. As a result he put the North American cypriconchids into the genus *Megalocypris*. Martens (1986) and McKenzie (1982, referenced in Martens, 1986) argued for the retention of *Cypriconcha* for the large cyprinids of North America. *Megalocypris* is used in the key.

A number of genera may be found in the North American ostracode literature, but are not included here. The following generic names are no longer considered valid:

Cyclocypria Dobbin, 1941—*Cyclocypris* Brady and Norman, 1888

Cypriconcha Sars, 1926—*Megalocypris* Sars, 1898 [see Delorme, 1969a]

Cytherites Sars, 1926—*Entocythere* Marshall, 1903 [see Hoff, 1943b]

Pionocypris Brady and Norman, 1896—*Cypridopsis* Brady, 1868

Pseudoilyocypris Ferguson, 1967—*Pelocypris* Klie, 1939 [see Delorme, 1970d]

Spirocypris Sharpe, 1903—*Cypricercus* Sars, 1895 [see Hoff, 1942]

Several species, as listed below, have been assigned to an incorrect genus; therefore, the generic names will not appear in the key:

Candonocypris deeveyi Tressler, 1954—*Megalocypris deeveyi* (Tressler) 1954

Candonocypris pugionis Furtos, 1936b—*Megalocypris*[?] *pugionis* (Furtos), 1936b [see Furtos, 1936b]

Candonocypris sarsi Danforth, 1948—*Megalocypris sarsi* (Danforth), 1948

Candonocypris serrato-marginata (Furtos), 1935—*Eucypris serrato-marginata* (Furtos), 1935

Erpetocypris barbatus Turner, 1899—*Megalocypris* sp. (Turner), 1899

Prionocypris canadensis Sars, 1926—*Strandesia canadensis* (Sars), 1926 [see Marmonier and Ward, 1990]

Prionocypris glacialis (Sars), 1890—*Tonnacypris glacialis* (Sars), [see Griffiths *et al.*, 1998]

Pseudocandona ikpikpukensis Swain, 1963—*Candona ikpikpukensis* (Swain), 1963 [see Delorme, 1970c]

Stenocypria longicomosa Furtos, 1933—*Isocypris longicomosa* (Furtos), 1933 [see Delorme, 1970a]

B. Classification of the Order Podocopida

The Podocopida[1] is the only order that contains both marine and freshwater ostracodes (Table II). Dorsal border of the valve is curved, or if straight, it is much shorter than the total length. No permanent anterior opening is present between the valves. There is no heart; two or three pairs of thoracic legs are present. Within Podocopida are two suborders: Metacopina and Podocopina. As will be described below, the suborder Metacopina contains only one freshwater genus in North America, *Darwinula* (superfamily Darwinuloidea, family Darwinulidae; Figs. 8 and 10). Its characteristics are given as follows[2].

Shell surface smooth, vestibules absent, distinctive adductor muscle scars (Figs. 8, 10, 19); first antenna composed of six podomeres with strong spikelike setae, swimming setae lacking on second antenna, mandibular exopodite with short palp of three podomeres of which the first is wide and has a row of long feathered setae, respiratory plate of mandible small, masticatory process of maxilla short and heavy, respiratory plate of maxilla with numerous feathered setae, maxillae with strong masticatory structure and leglike palp of three podomeres, first and second thoracic legs similar in structure and direction, furca lacking, ovaries do not originate between inner and outer epidermis, female carries eggs in posterior of carapace .... *Darwinula*

Freshwater and some marine ostracodes are included in the suborder Podocopina. Members of this taxon have two pairs of thoracic legs. The exopodite of the second antenna is represented by at most a scale and a few setae. The furcae are rodlike rather than lamelliform; eyes are present in most forms. Closing muscle scars are discrete and arranged in a distinctive group. Taxonomic keys to genera within two superfamilies of this suborder are given below.

C. Taxonomic Key to Genera of the Superfamily Cypridoidea[2]

Surface of the shell usually smooth, dorsal margin without interlocking teeth. Eyes developed to varying degrees, either separated or fused into a single median eye. The first antennae with basal portion of two or

[1] Modified after Kesling (1951a).

[2] Modified after Hoff (1942).

TABLE II Classification of the Class Ostracoda[a]

Class Ostracoda
Subclass Podocopa
Order Podocopida Sars, 1866
Suborder Metacopina Sylvester-Bradley, 1961
Superfamily Darwinuloidea Brady and Norman, 1888
Family Darwinulidae Brady and Norman, 1888
Genus *Darwinula* Brady and Norman
Suborder Podocopina Sars, 1866
Superfamily Cypridoidea Baird, 1845
Family Candoniidae Kaufmann, 1900
Subfamily Candoninae Daday, 1900
Genus *Candocyprinotus* Delorme
Genus *Candona* Baird
Genus *Fabaeformiscandona* Krstic
Genus *Nannocandona,* Eckman
Genus *Paracandona,* Hartwig
Genus *Pseudocandona,* Kaufmann
Family Cyprididae Baird, 1845
Subfamily Cypridinae Baird, 1845
Genus *Chlamydotheca* Sausseure
Genus *Cyprinotus* Brady
Genus *Cypris* Müller
Genus *Dolerocypris* Kaufmann
Genus *Herpetocypris* Brady and Norman
Genus *Ilyodromus* Sars
Genus *Isocypris* Müller
Genus *Megalocypris* Sars
Genus *Scottia* Brady and Norman
Genus *Stenocypris* Sars
Genus *Strandesia* Stuhlmann
Subfamily Eucypridinae Bronstein, 1947
Genus *Cypricercus* Sars
Genus *Eucypris* Vávra
Genus *Tonnacypris* Diebel and Pietrzeniuk
Genus *Trajancypris* Martens
Subfamily Cyclocyprinae Hoff, 1942
Genus *Candocypria* Furtos
Genus *Cyclocypris* Brady and Norman
Genus *Cypria* Zenker
Genus *Physocypria* Vávra
Subfamily Cypridopsinae Kaufmann, 1900
Genus *Cavernocypris* Hartmann
Genus *Cypretta* Vávra
Genus *Cypridopsis* Brady
Genus *Potamocypris* Brady
Genus *Sarscypridopsis* McKenzie
Family Ilyocyprididae Kaufmann, 1900
Subfamily Ilyocyprinae Kaufmann, 1900
Genus *Ilyocypris* Brady and Norman
Genus *Pelocypris* Klie
Family Notodromadidae Kaufmann, 1900
Subfamily Notodromadinae Kaufmann, 1900
Genus *Cyprois* Zenker
Genus *Notodromas* Lilljeborg
Superfamily Cytheroidea Baird, 1850
Family Cytheridae Baird, 1850
Subfamily Cytherideinae Sars, 1928
Genus *Cyprideis* Jones
Family Entocytheridae Hoff, 1942
Subfamily Entocytherinae Hoff, 1942
Genus *Ankylocythere* Hart
Genus *Ascetocythere* Hart
Genus *Cymocythere* Hart
Genus *Dactylocythere* Hart
Genus *Donnaldsoncythere* Hart
Genus *Entocythere* Marshall
Genus *Geocythere* Hart
Genus *Harpagocythere* Hobbs III
Genus *Hartocythere* Hobbs III
Genus *Litocythere* Hobbs and Walton
Genus *Lordocythere* Hobbs and Hobbs
Genus *Okriocythere* Hart
Genus *Ornithocythere* Hobbs
Genus *Phymocythere* Hobbs and Hart
Genus *Plectocythere* Hobbs III
Genus *Rhadinocythere* Hart
Genus *Saurocythere* Hobbs III
Genus *Sagittocythere* Hart
Genus *Thermastrocythere* Hobbs and Walton
Genus *Uncinocythere* Hart
Family Limnocytheridae Klie, 1938
Subfamily Limnocytherinae Klie, 1938
Genus *Limnocythere* Brady
Genus *Metacypris* Brady and Robertson
Family Loxoconchidae Sars, 1928
Subfamily Loxoconchinae Sars, 1928
Genus *Cytheromorpha* Hirschmann
Family Neocytherideidae Puri, 1957
Subfamily Neocytherideidinae Puri, 1957
Genus *Cytherissa* Sars

[a]The systematic list above is alphabetical (after Bowman and Abele, 1982), but the systematic treatment is not.

three podomeres and an endopodite of four or five podomeres, with swimming setae usually well developed. The second antennae with basal part of two podomeres and an endopodite of three or four podomeres. The exopodite is reduced to a small scale-like appendage bearing at most three setae. Maxillula not pediform, but modified as a mouth part, with the anterior margin of the base adapted for feeding. The maxillar endopodite forms a small palp in the female, but is enlarged to form prehensile palps in the male. The first thoracic leg has an endopodite of three or four podomeres and a strong distal claw. The second leg is bent dorsally and is probably used in cleaning the respiratory surfaces and other parts of the body. The second leg usually has three distal setae, but the distal end may be modified for grasping. The furca is typically well developed and rod-shaped, but may be reduced to a flagellum or whiplike structure. The gonads are located within the valves of the shell. In the male, a portion of the vas deferens is modified to form an ejaculatory duct.

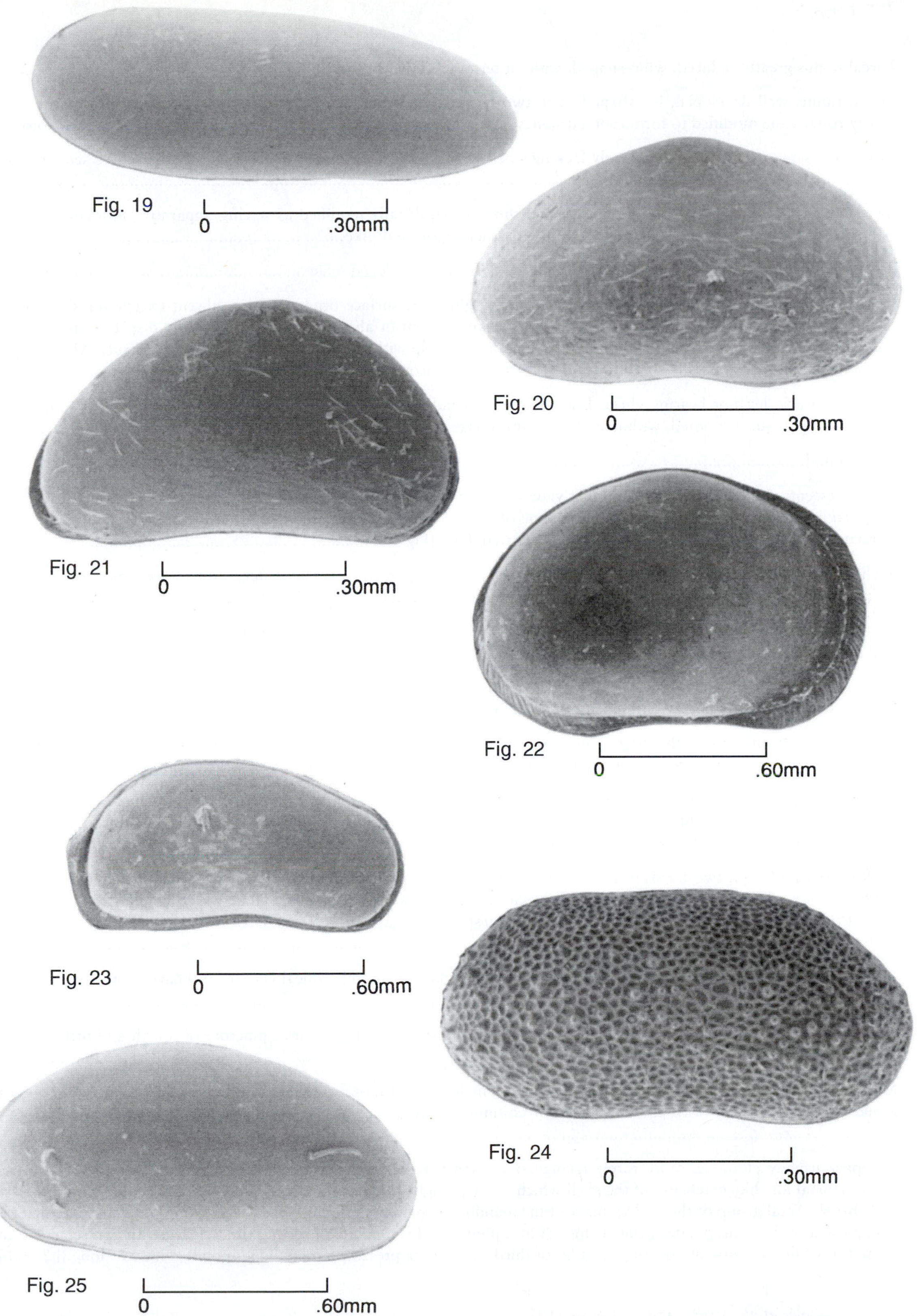

FIGURE 19 External view of *Darwinula stevensoni* (Brady and Robertson). *FIGURE 20* External view of *Cypridopsis vidua* (Müller). *FIGURE 21* External view of *Potamocypris smaragdina* (Vávra). *FIGURE 22* External view of *Cyprois marginata* Straus. *FIGURE 23* External view of a female *Candona rawsoni* Tressler, note the overlap of the right valve by the left valve. *FIGURE 24* External view of *Paracandona euplectella* Brady and Norman, note the reticulate structure and the pustules. *FIGURE 25* External view of *Cypricercus reticulatus* (Zaddach).

1a. Furcal ramus greatly reduced, whip-shaped, without terminal claw (difficult to observe)Cypridopsinae[3] 2

1b. Furcal ramus well developed, bar-shaped, with two terminal or subterminal claws; outer masticatory process of maxilla with six nearly equal setae modified to form toothed spines ....Notodromadinae[4] 4

1c. Swimming setae of antennae completely lacking, shell white; outer masticatory process with two or three setae modified as spines ....Candoninae[5] 5

1d. Swimming setae of first antenna present; second thoracic leg distally modified as seizing apparatus, ultimate podomere beak-like with two well developed bristle-like setae, third setae lacking or hook-like ....Cypridinae[6] 7

1e. Third thoracic leg with one long reflexed seta and two shorter unreflexed setae ....Cyclocyprinae[7] 10

1f. Carapace 1. 0–2.5 mm, elliptical to elongated, rarely globular; surface ornamentation absent except for shallow pits; marginal teeth possible on both valves; marginal zone with selvage present or absent, sometimes inwardly displaced in the right valve. First antenna with Rome organ small and undivided; mandibular palp with gamma-seta short, stout and hirsute. Maxillula with third endopodite carrying two, mostly smooth, teeth or bristles. The second process of the maxillula with "c"-seta ....Eucypridinae[8] 12

1g. Second thoracic leg not bearing chela, last podomere cylindrical and bearing three setae; shell oblong to subrectangular, dorsal margin straight, surface pitted, with one or more transverse sulci ....Ilyocyprinae[9] 4

2a(1a). Shells tumid..3

2b. Shells crescentic in shape, compressed, right valve higher than left and extending dorsally above left, $H/L < 0.5$, right valve overlaps left valve in venter, usually hairy; swimming setae of second antenna well developed, usually extending to tips of claws or beyond; ultimate podomere of maxillary palp distally wider than long (Fig. 21) ....*Potamocypris*

3a(2a). Shells subovate, $H/L > 0.5$, valves nearly equal, left valve overlaps right valve in venter; ultimate podomere of maxillary palp cylindrical, longer than wide (Fig. 20), swimming setae of second antennae well developed..*Cypridopsis*

3b. Shells subtriangular, $H/L > 0.5$, right valve overlaps left valve in venter; swimming setae of second antennae well developed ....*Sarscypridopsis*

3c. Shells elongate, <1 mm in length, left valve longer than right valve, $H/L = 0.5$, left valve overlaps right valve in venter; swimming setae of second antenna poorly developed, furcal rami absent in males..*Cavernocypris*

3d. Shells tumid, <1 mm in length, anterior margin of both valves with row of radiating septa; two maxillary spines may be smooth or toothed; ovaries or testes coiled in spiral pattern in posterior of carapace..*Cypretta*

4a(1b). Shell from side short and high, compressed; eyes not widely separated; first antenna of five podomeres; swimming setae almost reach tips of end claws; right and left prehensile palps of male maxillar endopodites differing little, with well-developed respiratory plate; setae on distal end of second thoracic leg modified for grasping; furca with two claws and two setae (Fig. 22).............*Cyprois*

4b. Shell short with height two-thirds of length, dorsal margin forming elevated arch, ventral surface flattened; eyes well separated; first antenna of six apparent podomeres, swimming setae extending beyond tips of terminal claws; palps of maxillae in males not similar, each formed with two podomeres; maxillae without respiratory plate; distal three setae of second thoracic leg nearly equal; furca with terminal seta lacking..*Notodromas*

5a(1c). Carapace with sides convex; outer masticatory process of maxilla smooth; apical claw of ultimate segment of second thoracic leg well developed, furcal claws weakly serrated..*Candocyprinotus*

5b. Shell subrectangular to subtriangular in shape, two special sensory setae present at juncture of fourth and fifth podomeres of male first antenna, respiratory plate of maxillae with two or three setae 6

6a(5b). Male carapace shape and size usually different than female; second thoracic leg with three unequally long setae on last podomere; Zenker's organs usually with seven wreaths of chitinous spines, openings of ejaculatory duct funnel-shaped (Figs. 9a, 9b, 23) ....*Candona*

6b. Carapace usually elongate, more rarely triangular in lateral view,distinctly compressed in dorsal view; left valve usually with a posterodorsal lobelike extension of the shell which overlaps right valve; male second antenna dimorphic, penultimate segment with male bristle. Setal group of the 2nd segment of the mandibular palp with three (fabaeformis group) or four (acuminata group) setae, gamma seta of the penultimate segment smooth (not plumose). Protopodite of second thoracic leg with two setae (d1 and dp), setae 'e' and 'f' of first endopodite missing, seta 'g' of third endopodite present, terminal endopodite with two long (h2 and h3) and one

[3] Modified after Marmonier *et al.* (1989), and McKenzie (1977).
[4] Modified after Hoff (1942).
[5] Modified after Hoff, 1942, Furtos (1933) and Meisch (1996).
[6] Modified after Hoff (1942) and Delorme (1970a).
[7] Modified after Hoff (1942) and Furtos (1933).
[8] Modified after Martens (1989).
[9] Modified after Hoff (1942) and Tressler (1949).

short (h1) setae. Female genital lobe usually with more or less well-developed process directed posteriorly. Hemipenis with 'M' process heavily sclerotized, proximal part slightly broader than the distal one .... *Fabaeformiscandona*

6c. Shell usually with small tubercles and reticulations on surface, shape subrectangular; penultimate segment of second thoracic leg divided, each division with strong distal seta; otherwise similar to *Candona* (Figs. 7, 24) .... *Paracandona*

6d. Carapace of medium length (0.8–1.2 mm), variously shaped, usually short and stout, more rarely elongate (subrectangular) or triangular in lateral view. Surface of adult valves smooth or pitted, usually with long stiff setae emanating through the shell, left valve overlaps the right valve ventrally. The penultimate endopodite of the second antennae subdivided or not, when subdivided with male bristles (t2 and t3) transformed. Setal group of the second endopodite of mandibular palp with three or five setae, distal seta of penultimate endopodite smooth (not plumose). Protopodite of second thoracic leg with three setae (d1-d2, dp), distal seta 'g' on the penultimate endopodite present, terminal endopodite with one short (h1), and two long setae (h2-3). Zenker's organ with 5 + 2 rings of spines. Hemipenis with 'M' process flat and proximally only weakly sclerotized .... *Pseudocandona*

7a(1d). Margins of valve usually smooth 8

7b. Margin of valve with spines or denticles 19

8a(7a). Shells <2.2 mm long 9

8b. Shells >2.2 mm long 16

9a(8a). Shell surface sculptured 10

9b. Shell surface smooth 13

10a(9a). Shell surface reticulate or scabrous especially in juveniles 11

10b. Shell surface punctate or striated 12

11a(10a). Outer masticatory process with two long, toothed spinelike setae; furcal ramus greater than half length of valve; testes coiled in anterior portion of each valve, males rare (Fig. 25) .... *Cypricercus*

12a. Outer masticatory process with two long, toothed spinelike setae; furcal ramus less than half length of valve, males rare (Fig. 26) .... *Eucypris*

12b. Swimming setae of second antenna rudimentary, length of gamma-seta of the mandibular palp is nearly five times its basal width, second segment of mandibular palp is trapezoidal and the terminal endopodite of the third endopodite bears two serrated teeth, first thoracic leg has the setae d_2 nearly two times d_1; shell elongate and moderately large, compressed and subrectangular to clavate, anterior inner lamellae of left valve with small peg tooth sometimes nearly invisible; furcal ramus smooth .... *Tonnacypris*

12c. Right valve with a frontal selvage inwardly displaced to a varying degree, but never marginal; left valve without selvage, but with large frontal inner list, situated about halfway on the calcified inner lamella .... *Trajancypris*

13a(10b). Shell elongate, narrow with scattered punctae; swimming seta of second antenna barely reaching tips of claws, second thoracic leg with well-developed apical claw; rami of furcae dissimilar in width, two end claws strongly pectinate .... *Stenocypris*

13b. Shells elongate and compressed, surface longitudinally striated; swimming setae of first and second antenna short, furcal ramus terminates in three short claws rather than two .... *Ilyodromus*

14. Swimming setae of second antenna strongly developed 15

15a(14b). Claws of furca finely denticulate; shell subrectangular, surface smooth and moderately hairy; characteristic marginal pore canals in anterior duplicature; outer masticatory process of maxilla with two smooth spines; second thoracic leg with well-developed apical claw .... *Isocypris*

15b. Claws of furca strongly developed 16

16a(15b). Shell reniform, densely hairy; maxillae bears well-developed respiratory plate; first thoracic leg with one long terminal seta and claw, furca with feathered dorsal seta .... *Scottia*

16b. Shells elongate, dorsum often with prominent dorsal flange, left valve always with conspicuous row of tuberclelike canals removed from the free margin; swimming setae of second antenna extend to tips of terminal claws, third masticatory process of maxilla with two strong spines, furcal ramus at least half as long as valve and with two claws and setae .... *Strandesia*

17a(8b). Shells with anterior flangelike projections which produces large vestibule; both antenna with well-developed swimming setae; outer masticatory process of maxilla with one toothed and one denticulate spine; second thoracic leg with well-developed apical claw; claws of furca finely denticulate .... *Chlamydotheca*

17b. Shells without anterior flangelike projection 18

18a(17b). Shells narrow and spindle-shaped, vestibules well developed at both extremities; both antennae with well-developed swimming setae; maxilla with outer masticatory process with two smooth spines; second thoracic leg with well-developed apical claw, claws of furca coarsely denticulate .... *Dolerocypris*

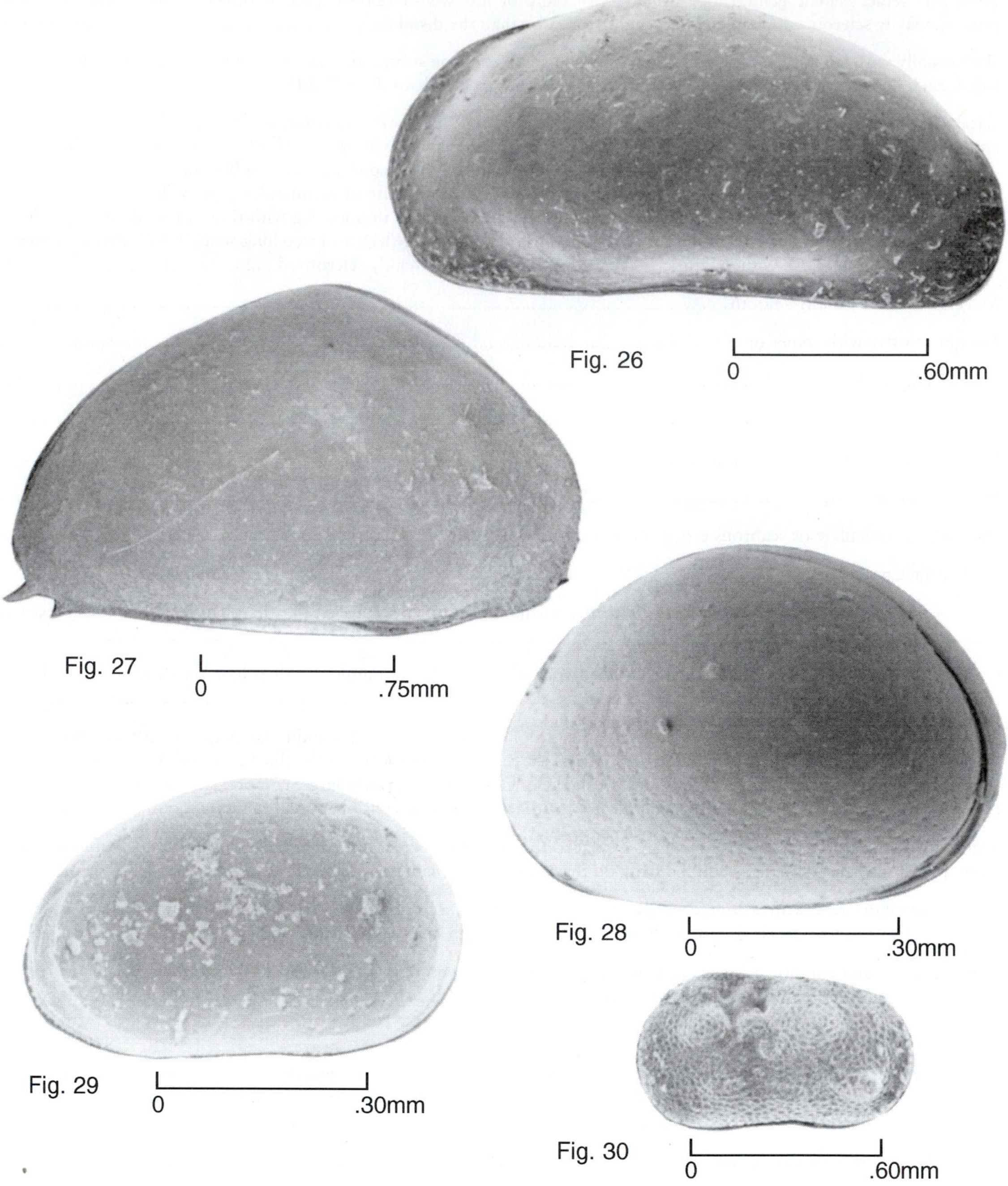

FIGURE 26 External view of *Eucypris crassa* (Müller). *FIGURE 27* External view of *Cypris pubera* Müller, note the posteroventral spines on the right valve. *FIGURE 28* External view of *Cyclocypris ampla* Furtos, note the pitted surface. *FIGURE 29* Internal view of *Physocypria pustulosa* (Sharpe) showing the denticles on the anteroventral margin. *FIGURE 30* Exterior view of *Pelocypris alatabulbosa* Delorme showing a reticulate surface, sulci, and nodes.

18b. Shells subrectangular 19

19a(18b). Shells elongate and large, compressed; swimming setae of second antenna not reaching mid position of terminal claw; spines of outer masticatory process of maxilla toothed; ventral edge of furca denticulate with combs of teeth .... *Herpetocypris*

19b. Shells subrectangular to subtrapezoidal; swimming setae on first antenna extend to end of ultimate podomere and not to tips of claws of second antenna; outer masticatory process made up of short, smooth bristles; maxillar endopodite of male developed into prehensile palp; second thoracic leg with apical claw not well developed; claws of furca denticulate........ *Megalocypris*

20a(7b). Shells <2 mm, margin of right or left valve tuberculate and margin of other valve smooth, dorsal margin may be arched or have an obtuse apex, valves commonly unequal, anterolateral surfaces pustulose; outer masticatory process of maxilla with two clawlike setae which may or may not be toothed; furcal claw longer than one-half length of ventral margin of furcal ramus, ramus with length less than twenty times least width (Figs. 3, 4) .... *Cyprinotus*

20b. Shells >2 mm, tumid, highly arched, venter flattened, anterior margins denticulate, posteroventral margin of right valve with two to four sharp marginal spines, surface wrinkled to minutely crenulate to punctate; second antenna with well-developed swimming setae; outermost masticatory process of maxilla with two denticulate spines, second thoracic leg with well-developed apical claw, claws of furca strongly denticulate (Fig. 27) .... *Cypris*

21a(1e). Swimming setae of second antenna lacking or rudimentary; shells ovoid, compressed; eyes well developed and fused; ultimate segment of second thoracic leg with two short and one long reflexed setae; tips of furcal claws serrate .... *Candocypria*

21b. Swimming setae of second antenna well developed 21

22a(21b). Shells tumid, height and width greater than one-half length, surface smooth and usually brown in color; eye well developed; respiratory plate of maxillae with six plumose setae; second thoracic leg consisting of four podomeres, ultimate podomere elongated and more than twice as wide and usually at least one-half as long as penultimate podomere; ultimate podomere of third thoracic leg with three distal setae, unequal, with outermost one very long and reflexed (Fig. 28) .... *Cyclocypris*

22b. Shells compressed 23

23a(22b). Shells short and high, occasionally elongate reniform, margins smooth not tuberculate; eyes well developed; second antenna of male with penultimate podomere divided and bearing specialized male setae; ultimate podomere of mandibular palp elongated three times as long as proximal width; palp of maxilla well developed, masticatory process weak; penultimate podomere of second thoracic leg undivided, ultimate podomere short scarcely longer than wide, longest distal seta reflexed; terminal and subterminal claws of furca strong, dorsal seta may be rudimentary; Zenker's organ with seven whorls of chitinous rays, proximal end much inflated; hemipene with two terminal lobes only, outer lobe lacking .... *Cypria*

23a(22b). Shells commonly unequal in height or length or both, at least anterior margin of one of the valves tuberculate; otherwise as in genus *Cypria* (Fig. 29) .... *Physocypria*

24a(1g). Shells may have large rounded humplike projections on lateral surface, marginal spines and pustules; first antenna with some swimming setae shortened and clawlike; swimming setae of second antenna may be shortened, male without special setae; maxillar endopodite of male transformed into prehensile palp, in female clearly leglike; ultimate podomere in second thoracic leg cylindrical with three setae, longest may be reflexed, claws well developed; Zenker's organs with numerous chitinous rods and with spherical inflated openings at each end; vasa deferentia twisted .... *Ilyocypris*

24b. Shells with pronounced mamilliarylike projections on lateral surface, marginal spines and pustules; second antenna with swimming setae which extend considerably beyond tips of terminal claws; maxillar endopodite with long curved terminal claw; first leg with one short and one long seta; furca well developed with dorsal seta longer than terminal claw (Fig. 30) .... *Pelocypris*

D. Taxonomic Key to Genera of the Superfamily Cytheroidea[10]

Shell variable in shape and sculpturing, seldom smooth, usually with reticulations, often with spines, furrows, or tubercles. Shells noncalcareous in the Entocytherinae. Valves nearly equal; often toothlike projections along the hinge. The first antennae consist of a base of two podomeres and an endopodite of three or four podomeres. The setae of the first antennae are short and stout, often clawlike. The exopodite or the flagellum of the second antennae is represented by a long hollow seta forming a duct carrying secretions from a gland. The endopodite of the second antennae of three podomeres, the long penultimate one may be divided. Swimming setae lacking. Maxillar endopodites and two pairs of thoracic legs similar and all adapted for crawling except in the Entocytherinae modified for grasping. Furca always greatly reduced. In the male, the hemipenes always present and well developed, the ejaculation duct is absent, a male accessory sense organ consisting of numerous setae on a short base is located between and somewhat medially to the bases of the first and second pairs of thoracic legs. The gonads do not lie between the inner and outer lamellae of the valves, but are in the body lateral to the intestine. In some species, the eggs are retained in the shell cavity during development.

1a. Free margins of valves without conspicuous pore canals, carapaces non calcareous, reniform or subelliptical in shape, furrows or protuberances may be present; respiratory plate of mandible reduced to two or three setae, masticatory lobes well developed, furca rudimentary, penis strongly curved; North American ostracodes commensal on crayfishes and a freshwater crab (Fig. 12) .... Subfamily Entocytherinae[11] 3

[10] Modified after Hoff (1942) and Sars (1928).
[11] Modified after Hoff (1942) and Hart and Hart (1974).

1b. Free margins of valves flattened, with many long marginal pore canals, subrectangular, often with protuberances or furrows, strongly reticulate; respiratory plate of mandible well developed, furca usually with two short setae .... Subfamily Limnocytherinae[12] 22

1c. Shell variable in shape and sculpturing, seldom smooth, usually with reticulations, spines, furrows, or tubercles, valves nearly equal, often with toothlike projections along hinge, vestibules absent; eyes distinctly separated, antennae well developed, exopodite of second antenna in form of long hollow seta carrying secretions from a gland near base of second antenna, setae of first antenna short and stout, often claw like, swimming setae on second antenna lacking, first thoracic leg of male prehensile, three pairs of thoracic leg similar, furca greatly reduced, hemipenes well developed, Zenker's organs absent, testes lying in body lateral to intestine, in some species eggs retained in body cavity during development .... Subfamily Cytherideinae[13]

[Shells distinctly sculptured, sometimes with lateral protuberances, right valve with spine in posteroventral corner] .... *Cyprideis*

1d. Thoracic legs successively increasing in length 2

2a(1d). Shell short and stout, hinge well developed; mandibular palp with terminal joint very small, respiratory plate well developed; furcal ramus edged at tip with two bristles .... Subfamily Loxoconchinae[13]

[Shell short and stout, surface pitted, marginal zones thickened, duplicature narrow; eyes distinctly separated; ultimate segment of first antenna articulated and with four robust spines; second antenna with two claw like spines inside penultimate joint; masticatory lobes of maxilla short] .... *Cytheromorpha*

2b. Shells plump, surface generally sculptured; hinge well defined, distinct closing teeth lacking in anterior and posterior of hinge line; mandibular palp slender, respiratory plate not well developed; furcal ramus extremely small, not well defined .... Neocytherideidinae[12]

[Shell club-shaped, surface rough and pitted; valves unequal with marginal zone thickened, duplicature narrow, vestibule absent, poorly developed teeth present on hinge; eyes well defined; antennae well developed, first antenna armed in front with three clawlike spines; respiratory plate of mandibular palp moderately well developed; maxilla with masticatory lobes short and stout; thoracic legs robustly developed; furcal ramus forming two oval-thickened pieces placed vertically, each provided behind with two very small bristles (Fig. 31)] .... *Cytherissa*

3a(1a). Penis with prostatic and spermatic elements widely separated along much of their lengths 4

3b. Penis simple; or if two elements recognizable, contiguous along their entire lengths 6

4a(3a). Ventral portion of peniferum tapering with tip of penis reaching, or almost reaching apex .... *Plectocythere*

4b. Ventral portion of peniferum usually rounded or with one or more prominences, seldom tapering; if tapering, tip of penis never approaching apex 5

5a(4b). Ventral portion of peniferum rounded, without prominences .... *Phymocythere*

5b. Ventral portion of peniferum with one or more prominences ventrally and/or anteriorly .... *Ascetocythere*

6a(3b). Penis directed posteroventrally from base .... *Lordocythere*

6b. Penis directed anteroventrally from base 7

7a(6b). Finger guard absent 8

7b. Finger guard present 18

8a(7a). Anteroventral portion of peniferum with acute beaklike projection .... *Ornithocythere*

8b. Anteroventral portion of peniferum never with beaklike projection 9

9a(8b). External border of horizontal ramus of clasping apparatus with one or more excrescence 10

9b. External border of horizontal ramus of clasping apparatus entire or with few shallow subapical grooves 14

10a(9a). Anteroventral portion of peniferum produced ventrally in rounded lobe 11

10b. Anteroventral portion of peniferum never produced ventrally in rounded lobe; or if produced, apex acute or truncate 12

11a(10a). Spermatic loop horizontal, peniferum distal to dorsal margin of spermatic loop at least twice as long as portion dorsal to loop, clasping apparatus with external border bearing single tubercle and terminating in fanlike cluster of serrations .... *Saurocythere*

11b. Spermatic loop vertical, peniferum distal to dorsal margin of spermatic loop much less than twice as long as portion dorsal to loop, clasping apparatus with external border broadly serrate and terminating in annulations .... *Okriocythere*

12a(10b). Anteroventral portion of peniferum never with conspicuous anterodorsally directed projection .... *Ankylocythere*

[12] Modified after Hoff (1942).

[13] Modified after Sars (1928).

12b. Anteroventral portion of peniferum with conspicuous anterodorsally directed projection 13

13a(12b). Anteroventral portion of peniferum entire .... *Geocythere*

13b. Anteroventral portion of peniferum bifid .... *Hartocythere*

14a(9b). Internal border of clasping apparatus with more than three teeth, apical cluster with more than two denticles, vertical ramus straight 15

14b. Internal border of clasping apparatus usually with no more than three teeth, if more than three, with only two apical denticles or vertical ramus strongly convex posteriorly 16

15a(14a). Ventral portion of peniferum tapering to slender tip .... *Rhadinocythere*

15b. Ventral portion of peniferum never slender nor tapering .... *Entocythere*

16a(14b). Clasping apparatus not clearly divisible into vertical and horizontal rami, extremities directed at angle of at least 100 degrees .... *Donnaldsoncythere*

16b. Clasping apparatus usually divisible into vertical and horizontal rami, extremities directed at angle of no more than 90 degrees 17

17a(16b). Penis large, S-shaped or sinuous and with curved posteroventral thickening of peniferum giving forcipate appearance to ventral portion of peniferum .... *Thermastrocythere*

17b. Penis of moderate size, L-shaped, and never so disposed as to contribute forcipate appearance to ventral portion of peniferum .... *Uncinocythere*

18a(7b). Peniferum with accessory groove (except in *Dactylocythere leptophylax* where finger guard always slender and trifid) 19

18b. Peniferum without accessory groove; finger guard never slender and trifid 20

19a(18a). Posteroventral portion of peniferum terminating in barbed point .... *Sagittocythere*

19b. Posteroventral portion of peniferum variable, but never ending in barbed point .... *Dactylocythere*

20a(18b). Ventral portion of peniferum bulbiform, clasping apparatus never extending so far ventrally as does peniferum .... *Cymocythere*

20b. Ventral portion of peniferum slender or strongly flattened; clasping apparatus extending ventrally to or beyond ventral extremity of peniferum 21

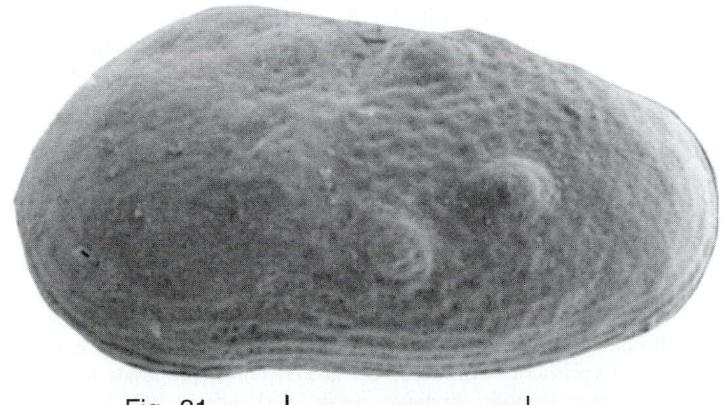

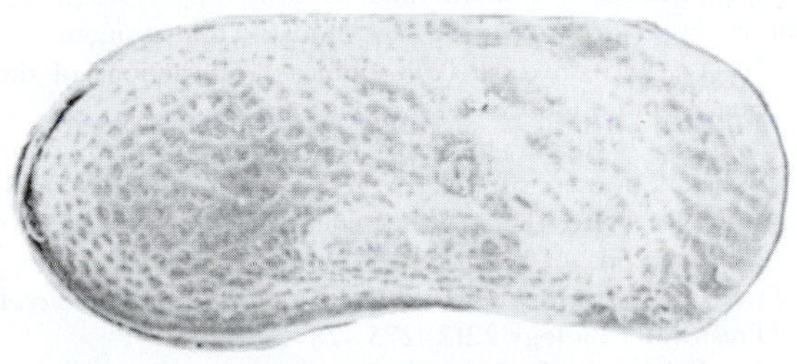

FIGURE 31 Exterior view of *Cytherissa lacustris* (Sars) showing coarse reticulate surface and nodes. *FIGURE 32* Exterior view of *Limnocythere itasca* Cole showing reticulate surface, sulci, and alae.

21a(20b). Ventral portion of peniferum slender, terminating in small recurved projection .... *Harpagocythere*

21b. Ventral portion of peniferum flattened and with concave border .... *Litocythere*

22a(1b). Shells subrectangular, reticulate, with sulci, alae or pustules on surface, margins may be denticulate, no vestibules developed; males longer, more inflated in posterior region than females; second antenna with three apical claws; respiratory plate of mandibular palp well developed; masticatory lobes of maxilla short; thoracic legs moderately slender; furcal ramus well defined, conical in shape with one terminal and one lateral seta; hemipene with basal part very large and protuberant in front (Figs. 11, 32 .... *Limnocythere*

22b. Shells short and broad, surface pitted, right valve with hinge line toothed; first antenna five or six segmented, second antenna four segmented, maxilla with three masticatory processes each longer than palp .... *Metacypris*

LITERATURE CITED

Alm, G. 1915. Monographie der Schwedischen SüsswasserOstracoden nebst systematischen Besprechungen der Tribus Podocopa. Zoologiska Bidrag från Uppsala 4:1–247.

Angell, R. W., Hancock, J. S. 1989. Response of eggs of *Heterocypris incongruens* (Ostracoda) to experimental stress. Journal of Crustacean Biology 9:381–386.

American Public Health Association Inc. (APHA). 1965. Standard methods for the examination of water and wastewater, including bottom sediments and sludges. American Public Health Association, Inc., New York. 769p.

Anderson, T. W., Mott, R. J., Delorme, L. D. 1985. Evidence for a pre-Champlain Sea glacial lake phase in Ottawa Valley, Ontario, and its implications, *in*: Current Research. Part A. Geological Survey of Canada, Paper 85-1A, pp. 239–245.

Bais, V. S., Agrawal, N. C. 1995. Comparative study of the zooplanktonic spectrum in the Sagar Lake and Military Engineering Lake. Journal of Environmental Biology 16(1):27–32.

Baltanás, A. 1992. On the use of some methods for the estimation of species richness. Oikos 65(3):484–492.

Barclay, M. H. 1966. An ecological study of a temporary pond near Auckland, New Zealand. Australian Journal of Marine and Freshwater Research 17:239–258.

Bataille, K. J., Baldassarre, G. A. 1993. Distribution and abundance of aquatic macroinvertebrates following drought in three prairie pothole wetlands. Wetlands 13(4):260–269.

Batzer, D. P., McGee, M., Resh, V. H., Smith, R. R. 1993. Characteristics of invertebrates consumed by mallards and prey response to wetland flooding schedules. Wetlands 13(1):41–49.

Bechara, J. A. 1996. The relative importance of water quality, sediment composition and floating vegetation in explaining the macrobenthic community structure of floodplain lakes (Parana River, Argentina). Hydrobiologia 333(2):95–109.

Bell, G. 1982. The masterpiece of nature, the evolution and genetics of sexuality. Univ. of California Press, Berkeley, 600 p.

Benson, R. H. 1959. Ecology of recent ostracodes of the Todos Santos Bay Region, Baja California, Mexico. University of Kansas Paleontological Contributions, Article 1:1–80.

Benson, R. H. 1961. Ecology of ostracode assemblages, *in*: Moore, R. D., Ed., Treatise on invertebrate paleontology. Part Q:Arthropoda 3, Crustacea, Ostracoda. Geological Society of America. Univ. of Kansas Press, Lawrence, pp. Q56–Q63.

Benson, R. H. 1967. Muscle-scar patterns of Pleistocene (Kansan) ostracodes; Essays in Paleontology and Stratigraphy. Raymond C. Moore Commemorative Volume. University Kansas Department of Geology Special Publication 2:211–241.

Benson, R. H. 1974. The role of ornamentation in the design and function of the ostracode carapace. Geoscience and Man 6:47–57.

Benson, R. H. 1981. Form, function, and architecture of ostracode shells. Annual Review of Earth and Planetary Science 9:59–80.

Benson, R. H., MacDonald, H. C. 1963. Postglacial (Holocene) ostracodes from Lake Erie. University of Kansas Paleontological Contributions 4:1–26.

Benzie, J. A. H. 1989. The distribution and habitat preference of ostracods (Crustacea:Ostracoda) in a coastal sand-dune lake, Loch of Strathbeg, north-east Scotland. Freshwater Biology 22(2):309–321.

Bergold, A. 1910. Beiträge zur Kenntnis des innern Baues der Süßasserostracoden. Zoologische Jahrbücher. Abteilung für Anatomie und Ontogenie der Tiere 30:1–42.

Bernecker, A. 1909. Zur Histologie der Respirationsorgane bei Crustaceen. Zoologischer Jahrbücher. Abteilung für Anatomie und Ontogenie der Tierecher Abteilung der Anatomie und Ontogenie der Tiere 27:583–630.

Bigelow, N. K. 1924. The food of young suckers in Lake Nipigon. University of Toronto Studies, Biological Series Number 24:81–116.

Bodergat, A. 1978. L'intensité lumineuse, son influence sur la teneur en phosphore des carapace d'ostracodes. Géobios 11:715–735.

Bowman, T. E., Abele, L. G. 1982. Classification of the recent Crustacea, *in*: Abele, L. G., Ed., The biology of Crustacea. Vol. 1:systematics, the fossil record, and biogeography. Academic Press, New York, pp. 1–27.

Bronnvall, A. M., Larsson, J. I. R. 1994. *Flabelliforma ostracodae* n. sp. (Microspora, Duboscqidae), a new microsporidian parasite of *Candona* sp. (Crustacea, Ostracoda) European Journal of Protistology 30(3):280–287.

Bronnvall, A. M., Larsson, J. I. R. 1995. Description of *Binucleospora elongata* gen. et sp. nov. (Microspora, Caudosporidae), a microsporidian parasite of ostracods of the genus *Candona* (Crustacea, Cyprididae) in Sweden. European Journal of Protistology 31(1):63–72.

Broodbakker, N. W. 1983. The genus *Strandesia* and other cypricercini (Crustacea, Ostracoda) in the West Indies. Part 1: Taxonomy. Bijdragen tot de Dierkunde 53:327–368.

Broodbakker, N. W., Danielopol, D. L. 1982. The chaetotaxy of Cypridacea (Crustacea, Ostracoda) limbs; proposals for a descriptive model. Bijdragen tot de Dierkunde 52:103–120.

Campbell, C. E. 1995. The influence of a predatory ostracod, *Australocypris insularis,* on zooplankton abundance and species composition in a saline lake. Hydrobiologia 302(3):229–239.

Cannon, H. G. 1925. On the segmental excretory organs of certain freshwater ostracods. Philosophical Transactions of the Royal Society of London 214:1–27.

Chaplin, J. A. 1993. The local displacement of a sexually reproducing ostracod by a conspecific parthenogen. Heredity 71(3):259–268.

Chaplin, J. A., Ayre, D. J. 1989. Genetic evidence of variation in the contributions of sexual and asexual reproduction to populations of the freshwater ostracod *Candonocypris novaezelandiae.* Freshwater Biology 22(2):275–284.

Chaplin, J. A., Ayre, D. J. 1997. Genetic evidence of widespread dispersal in a parthenogenetic freshwater ostracod. Heredity 78(1):57–67.

Chaplin, J. A., Hebert, P. 1997. *Cyprinotus incongruens* (Ostracoda):an ancient asexual? Molecular Ecology 6(2):155–168.

Chaplin, J. A., Havel, J. E., Hebert, P. D. N. 1994. Sex and ostracods (Review). Trends in Ecology and Evolution 9(11):435–439.

Chivas, A. R., DeDeckker, P., Shelley, J. M. G. 1983. Magnesium, strontium and barium partitioning in nonmarine ostracode shells and their use in paleoenvironmental reconstruction—a preliminary study, *in*: Maddocks, R. F., Ed., Applications of Ostracoda. University of Houston, Department of Geoscience, Houston, TX, pp. 238–249.

Chivas, A. R., DeDeckker, P., Shelley, J. M. G. 1985. Strontium content of ostracods indicate lacustrine paleosalinity. Nature (*London*) 316:251–253.

Chivas, A. R., DeDeckker, P., Shelley, J. M. G. 1986a. Magnesium and strontium in non-marine ostracod shells as indicators of paleosalinity and palaeotemperature. Hydrobiologia 143:135–142.

Chivas, A. R., DeDeckker, P., Shelley, J. M. G. 1986b. Magnesium content of non-marine ostracod shells:a new paleosalinometer and palaeothermometer. Palaeogeography, Palaeoclimatology, Palaeoecology 54:43–61.

Creuzé des Châtelliers, M., Marmonier, P. 1990. Macrodistribution of Ostracoda and Cladocera in a by-passed channel:exchange between superficial and interstitial layers. Stygologia 5(1): 17–24.

Creuzé des Châtelliers, M., Marmonier, P. 1993. Ecology of benthic and interstitial ostracods (Crustacea) of the Rhône River, France. Journal of Crustacean Biology 13(2):268–279.

Curry, B. B. 1999. An environmental tolerance index for ostracodes as indicators of physical and chemical factors in aquatic habitats. Paleogeography, Palaeoclimatology, Palaeoecology 148:51–63.

Cvancara, A. M., Clayton, L., Bickley, W. B., Jr., Jacob, A. F., Ashworth, A. C., Brophy, J. A., Shay, C. I., Delorme, L. D., Lammers, G. E. 1971. Paleolimnology of late Quaternary deposits, Seibold site, North Dakota. Science 171:172–174.

Danforth, W. A. 1948. A list of Iowa ostracods with descriptions of three new species. Proceedings of the Iowa Academy of Science 55:351–359.

Danielopol, D. L. 1969. Recherches sur la morphologie de l'organe copulateur male chez quelques ostracodes du genre *Candona* Baird (Fam. Cyprididae Baird), *in*: Neale, J. W., Ed., The taxonomy, morphology and ecology of Recent Ostracoda. Oliver & Boyd, Edinburgh, pp. 136–153.

Danielopol, D. L. 1980a. An essay to assess the age of the freshwater interstitial ostracods of Europe. Bijdragen tot de Dierkunde 50:243–291.

Danielopol, D. L. 1980b. Sur la biologie de quelques ostracodes Candoninae épigés et hypogés d'Europe. Bulletin du Musée National d'Histoire Naturelle (Paris) 2:471–506.

Danielopol, D. L., Marmonier, P., Boulton, A. J., Bonaduce, G. 1994. World subterranean ostracod biogeography; dispersal or vicariance. Hydrobiologia 287:119–129.

De Deckker, P. 1981. Ostracods of athalassic saline lakes. Hydrobiologia 81:131–144.

De Deckker, P. 1983a. Notes on the ecology and distribution of nonmarine ostracods in Australia. Hydrobiologia 106(3):223–234.

De Deckker, P. 1983b. Australian Salt Lakes; their history, chemistry, and biota. Hydrobiologia 105:231–244.

De Deckker, P. 1988. The use of ostracods in palaeolimnology in Australia. Palaeogeography, Palaeoclimatology, Palaeoecology 62:463–475.

De Deckker, P., Forester, R. M. 1988. The use of ostracods to reconstruct continental paleoenvironmental records, *in* DeDeckker, P., Collin, J.-P., Peypouquet, J.-P., Eds., Ostracoda in the earth sciences. Elsevier, Amsterdam, pp. 175–199.

Delorme, L. D. 1967. Field key and methods of collecting freshwater ostracodes in Canada. Canadian Journal of Zoology 45: 1275–1281.

Delorme, L. D. 1968. Pleistocene freshwater Ostracoda from Yukon, Canada. Canadian Journal of Zoology 46:859–876.

Delorme, L. D. 1969a. On the identity of the ostracode genera *Cypriconcha* and *Megalocypris*. Canadian Journal of Zoology 47:271–281.

Delorme, L. D. 1969b. Ostracodes as Quaternary paleoecological indicators. Canadian Journal of Earth Sciences 6:1471–1476.

Delorme, L. D. 1970a. Freshwater ostracodes of Canada. Part I:Subfamily Cypridinae. Canadian Journal of Zoology 48:153–169.

Delorme, L. D. 1970b. Freshwater ostracodes of Canada. Part II:Subfamilies Cypridopsinae, Herpetocypridinae, and family Cyclocyprididae. Canadian Journal of Zoology 48:253–266.

Delorme, L. D. 1970c. Freshwater ostracodes of Canada. Part III:Family Candonidae. Canadian Journal of Zoology 48:1099–1127.

Delorme, L. D. 1970d. Freshwater ostracodes of Canada. Part IV:Families Ilyocyprididae, Notodromadidae, Darwinulidae, Cytherideidae, and Entocytheridae. Canadian Journal of Zoology 48:1251–1259.

Delorme, L. D. 1971a. Freshwater ostracodes of Canada. Part V:Families Limnocytheridae, Loxoconchidae. Canadian Journal of Zoology 49:43–64.

Delorme, L. D. 1971b. Paleoecological determinations using Pleistocene freshwater ostracodes. Bulletin du Centre de Recherche de Pau-SNPA 5:341–347.

Delorme, L. D. 1971c. Paleoecology of Holocene sediments from Manitoba using freshwater ostracodes. Geological Association of Canada Symposium, Special Paper No. 9:301–304.

Delorme, L. D. 1972. Groundwater flow systems, past and present, *in* 24th International Geological Congress, Montreal. Section 11, pp. 222–226.

Delorme, L. D. 1978. Distribution of freshwater ostracodes in Lake Erie. Journal of Great Lakes Research 4:216–220.

Delorme, L. D. 1982. Lake Erie oxygen, the prehistoric record. Journal of Fisheries and Aquatic Sciences 39:1021–1029.

Delorme, L. D. 1989. Methods in Quaternary ecology No. 7:Freshwater Ostracoda. Geosciences Canada 16:85–90.

Delorme, L. D. 1996. Burlington Bay, Lake Ontario, its paleolimnology based on fossil ostracodes. Water Quality Research Journal of Canada 31:643–671.

Delorme, L. D., Donald, D. 1969. Torpidity of freshwater ostracodes. Canadian Journal of Zoology 47:997–999.

Delorme, L. D., Zoltai, S. C. 1984. Distribution of an arctic ostracod fauna in space and time. Quaternary Research 21:65–73.

Delorme, L. D., Zoltai, S. C., Kalas, L. L. 1977. Freshwater shelled invertebrate indicators of paleoclimate in Northwestern Canada during the late glacial times. Canadian Journal of Earth Sciences 14:2029–2046.

Delorme, L. D., Zoltai, S. C., Kalas, L. L. 1978. Freshwater shelled invertebrate indicators of paleoclimate in Northwestern Canada during late glacial times. Reply. Canadian Journal of Earth Sciences 15:462–463.

Deschiens, R. 1954. Mécansime de l'action léthale de *Cypridopsis hartwigi* sur les mollusques vecteurs des bilharzioses. Bulletin de la Société de Pathologie exotique 47:399–401.

Deschiens, R., Lamy, L., Lamy, H. 1953. Sur un ostracode prédateur de bulins et de planorbes. Bulletin de la Société de Pathologie exotique 46:956–958.

Dezfuli, B. S. 1996. *Cypria reptans* (Crustacea:Ostracoda) as an intermediate host of *Neoechinorhynchus rutili* (Acanthocephla: Eoacanthocephala) in Italy. Journal of Parasitology 82(3):503–505.

Diarra, K., Toguebaye, B. S. 1996. Ultrastructure of *Nosema stenocypris* Diarra and Toguebaye, 1994, a microsporidian parasite of *Stenocypris major* (Crustacea, Ostracoda, Cyprididae). Archiv für Protistenkunde 146(3–4):363–367.

Dieter, C. D., Duffy, W. G., Flake, L. D. 1996. The effect of phorate on wetland macroinvertebrates. Environmental Toxicology & Chemistry 15(3):308–312.

Dobbin, C. N. 1941. Fresh-water Ostracoda from Washington and other western localities. University of Washington Publications in Biology 4:175–246.

Douglas, D. J., Healy, B. 1991. The freshwater ostracods of two quagmires in Co. Louth, Ireland. Internationale Vereinigung für Theoretische und Angewandte Limnologie, Verhundlungen 24(3):1522–1525.

Engstrom, D. R., Nelson, S. R. 1991. Paleosalinity from trace metals in fossil ostracodes compared with observational records at Devils Lake, North Dakota, USA. Palaeogeography, Palaeoclimatology, Palaeoecology 83:295–312.

Fassbinder, K. 1912. Beiträge zur Kenntnis der Süsswasserostracoden. Zoologische Jahrbücher. Abteilung für Anatomie und Ontogenie der Tiere 32:533–576.

Ferguson, E., Jr. 1944. Studies of the seasonal life history of three species of fresh-water ostracodes. American Midland Naturalist 32:713–727.

Ferguson, E., Jr. 1958a. Seasonal life history studies of two species of freshwater ostracods. Anatomical Records 131:549–550.

Ferguson, E., Jr. 1958b. Freshwater ostracods from South Carolina. The American Midland Naturalist 59:111–119.

Ferguson, E., Jr. 1964. Stenocyprinae, a new subfamily of freshwater cyprid ostracods (Crustacea) with description of a new species from California. Proceedings of the Biological Society of Washington 77:17–24.

Ferguson, E., Jr. 1967. New ostracods from the playa lakes of eastern New Mexico and western Texas. Transactions of the American Microscopical Society 86:244–250.

Forester, R. M. 1983. The relationship of two lacustrine ostracode species to solute composition and salinity, implications for paleohydrochemistry. Geology 11:435–439.

Forester, R. M. 1986. Determination of the dissolved anion composition of ancient lakes from fossil ostracodes. Geology 14:796–799.

Forester, R. M. 1987. Late Quaternary paleoclimate records from lacustrine ostracodes, *in*: Ruddiman, W. E., Wright, H. E., Eds., North American and adjacent oceans during the last deglaciation. DNAG. Geological Society of America Boulder, CO, pp. 261–276.

Forester, R. M. 1991. Ostracode assemblages from springs in the western United States:implications for paleohydrology. Memoir of the Entomological Society of Canada 155:181–201.

Forester, R. M., Brouwers, E. M. 1985. Hydrochemical parameters governing the occurrence of estuarine and marginal estuarine ostracodes, an example from central Alaska. Journal of Paleontology 59:344–369.

Forester, R. M., Delorme, L. D., Bradbury, J. P. 1987. Mid-Holocene climate in Northern Minnesota. Quaternary Research 28: 263–273.

Forester, R. M., Colman, S. M., Reynolds, R. L., Keigwin, L. D. 1994. Lake Michigan's late Quaternary limnological and climate history from ostracode, oxygen isotope, and magnetic susceptibility. Journal of Great Lakes Research 20(1):93–107.

Forester, R. M., Delorme, L. D., Bradbury, J. P. 1987. Mid-Holocene climate in Northern Minnesota. Quaternary Research (N.Y.) 28:263–273.

Fox, H. M., Taylor, A. E. 1954. Injurious effect of air-saturated water on certain invertebrates. Nature (London) 174:312.

Fox, H. M., Taylor, A. E. 1955. The tolerance of oxygen by aquatic invertebrates. Proceedings of the Royal Society of London, Series B 143:214–225.

Fryer, G. 1957. The food of some freshwater cyclopoid copepods and its ecological significance. Journal of Animal Ecology 26: 263–286.

Furtos, N. C. 1933. The Ostracoda of Ohio. Ohio Biological Survey, Vol. 5, Bulletin 29:413–524.

Furtos, N. C. 1935. Fresh-water Ostracoda from Massachusetts. Journal of the Washington Academy of Science 25:530–544.

Furtos, N. C. 1936a. On the Ostracoda from the cenotes of Yucatan and vicinity. Carnegie Institution of Washington, Publ. Number 457:89–115.

Furtos, N. C. 1936b. Freshwater Ostracoda from Florida and North Carolina. American Midland Naturalist 17:491–522.

Grant, I. F., Egan, E. A., Alexander, M. 1983. Measurement of rate of grazing of the ostracod *Cyprinotus carolinensis* on blue-green algae. Hydrobiologia 106:199–208.

Green, J. 1954. A note on the food of *Chaetogaster diaphanus*. Annals of the Magazine of Natural History 12:842–844.

Green, J. 1962. Bile pigment in *Eucypris virens* (Jurine). Nature (London) 196:1318–1319.

Griffiths, H. I., Evans, J. G. 1994. Infestation of the freshwater ostracod *Cypria ophthalmica* [*sic*] (Jurine) by the Peritrich *Nüchterleinella corneliae* Matthes. Archiv für Protistenkunde 144(3):315–317.

Griffiths, H. I., Martin, D. S., Shine, A. J., Evans, J. G. 1993. The ostracod fauna (Crustacea, Ostracoda) of the profundal benthos of Loch Ness. Hydrobiologia 254(2):111–117.

Griffiths, H. I., Pietrzeniuk, E., Fuhrmann, R., Lennon, J. J., Martens, K., Evans, J. G. 1998. *Tonnacypris glacialis* (Ostracoda, Cyprididae):taxonomic position, (palaeo)ecology, and zoogeography. Journal of Biogeography 25(3):515–526.

Gutentag, E. D., Benson, R. H. 1962. Neogene (Plio-Pleistocene) fresh-water ostracodes from the central high plains. Bulletin of the Geological Survey of Kansas. Article 157:60 p.

Hanström, B. 1924. Beiträge zur Kenntnis des zentralen Nervensystem der Ostracoden und Copepoden. Zoologischer Anzeiger 61:31–38.

Harding, J. P. 1962. *Mungava munda* and four other new species of ostracod crustaceans from fish stomachs. Natural History of the Rennell Island, British Solomon Islands 4:51–62.

Harding, J. P. 1964. Crustacean cuticle with reference to the ostracod carapace. Pubblicazione della Stazione Zoologica di Napoli 33:9–31.

Hart, C. W., Jr., Hobbs, H. H., Jr. 1961. Eight new troglobitic ostracods of the genus *Entocythere* (Crustacea, Ostracoda) from the eastern United States. Proceedings of the Academy of Natural Sciences of Philadelphia 113:173–185.

Hart, D. G., Hart, C. W., Jr. 1974. The ostracod family Entocytheridae. Monograph 18. Academy of Natural Sciences of Philadelphia, 239 p.

Have, A. 1993. Effects of area and patchiness on species richness:an experimental archipelago of ciliate microcosms. Oikos 66(3): 493–500.

Havel, J. E., Hebert, P. D. N. 1993. Clonal diversity in parthenogenetic ostracods, *in*: McKenzie, K. G., Jones, P. J., Eds., Ostracoda in the Earth and Life Sciences. Proceedings of the 11th International, Symposium on Ostracoda. Balkema, Rotterdam, pp. 353–368.

Havel, J. E., Talbott, B. L. 1995. Life history characteristics of the freshwater ostracod *Cyprinotus incongruens* and their application to toxicity. Ecotoxicology 4:206–218.

Havel, J. E., Hebert, P. D. N., Delorme, L. D. 1990a. Genetics of sexual Ostracoda from a low arctic site. Journal of Evolutionary Biology 3:65–84.

Havel, J. E., Hebert, P. D. N., Delorme, L. D. 1990b. Genotypic diversity of asexual Ostracoda from a low arctic site. Journal of Evolutionary Biology 3:391–410.

Hoff, C. C. 1942. The ostracods of Illinois, their biology and taxonomy. Illinois Biological Monograph 19:196 p.

Hoff, C. C. 1943a. Seasonal changes in ostracod fauna of temporary ponds. Ecology 24:116–118.

Hoff, C. C. 1943b. Two new ostracods of the genus *Entocythere* and records of previously described species. Journal of the Washington Academy of Science 33:276–286.

Horne, F. R. 1993. Survival strategy to escape desiccation in a freshwater ostracod. Crustaceana 65(1):53–61.

Hu, F. S., Ito, E., Brubaker, L. B., Anderson, P. M. 1998. Ostracode geochemical record of Holocene climatic change and implications for vegetational response in the northwestern Alaska Range. Quaternary Research 49:86–95.

Hummon, W. D. 1981. Extraction by sieving: a biased procedure in studies of stream meibenthos. Transactions of the American Microscopical Society 100:278–284.

Hurtubise, R. D., Havel, J. E., Little, E. E. 1998. The effects of ultraviolet-B radiation on freshwater invertebrates:experiments with a solar simulator. Limnology & Oceanography 43(6):1082–1088.

Jahn, A., Gamenick, I., Theede, H. 1996. Physiological adaptations of *Cyprideis torosa* (Crustacea, Ostracoda) to hydrogen sulphide. Marine Ecology Progress Series 142(1–3):215–223.

Johansen, F. 1912. Freshwater life in north-east Greenland. Meddelelser om Gronland 45:321–337.

Karrow, P. F., Anderson, T. W., Delorme, L. D., Clarke, A. J., Jr. 1975. Stratigraphy, paleontology, and age of Lake Algonquin sediments in southwestern Ontario, Canada. Quaternary Research (*N.Y.*) 5:49–87.

Karrow, P. F., Anderson, T. W., Delorme, L. D., Miller, B. B., Chapman, L. J. 1995. Late-glacial paleoenvironment of Lake Algonquin sediments near Clarksburg Ontario. Journal of Paleolimnology 14:297–309.

Kesling, R. V. 1951a. The morphology of ostracod molt stages. Illinois Biological Monographs 21: 1–324.

Kesling, R. V. 1951b. Terminology of ostracod carapaces. Contribution from the Museum of Paleontology, University of Michigan 9:93–171.

Kesling, R. V. 1953. A slide rule for the determination of instars in ostracod species. Contributions from the Museum of Paleontology, University of Michigan 11:97–109.

Kesling, R. V. 1957. Notes on Zenker's organs in the ostracod *Candona*. American Midland Naturalist 57:175–182.

Kesling, R. V. 1965. Anatomy and dimorphism of adult *Candona suburbana* Hoff. Four Reports of ostracod Investigations. Report 1. Univ. of Michigan Press, Ann Arbor. 56 p.

King, J. L., Simovich, M. A., Brusca, R. C. 1996. Species richness, endemism and ecology of crustacean assemblages in northern California vernal ponds. Hydrobiologia 328:85–116.

Klassen, R. W., Delorme, L. D., Mott, R. J. 1967. Geology and paleontology of Pleistocene deposits in southwestern Manitoba. Canadian Journal of Earth Sciences 4:433–447.

Klie, W. 1931. Campagne spéologique de C. Bolivar et R. Jeannel dan L'Amérique du Nord (1928). Part 3: Crustacés Ostracodes. Archives Zoologie Expérimentale Générale 71:333–344.

Klie, W. 1939. Süsswasserostracoden aus Nordostbrasilien. Part 1. Zoologischer Anzeiger 128:84–91.

Klugh, A. B. 1927. The ecology, food-relations and culture of freshwater Entomostraca. Transactions of the Royal Canadian Institute 16:15–98.

Kornicker, L. S., Sohn, I. G. 1971. Viability of ostracode eggs egested by fish and effect of digestive fluids on ostracode shells; ecologic and paleoecologic implications, *in*: Oertli, H. J., Ed., Paleoecologie des ostracodes. Bulletin du Centre de Recherches de Pau-SNPA, Supplement 5:125–135.

Korschelt, E. 1915. Über das Verhalten verschiedener wirbelloser Tiere gegen niedere Temperaturen. Zoologischer Anzeiger 45:113–115.

Külköylüoĝlu, O., Vinyard, G. L. 1998. A new bisexual form of *Cavernocypris subterranea* (Wolf, 1920) (Crustacea, Ostracoda) from Idaho. Great Basin Naturalist. 58(4):380–385.

Lamb, H. F., Gasse, F., Benkaddour, A., El Hamouti, N., van der Kaars, S., Perkins, W. T., Pearce, N. J., Roberts, C. N. 1995. Relation between century-scale Holocene arid intervals in tropical and temperate zones. Nature 373(6510):134–137.

Lancaster, J., Robertson, A. L. 1995. Microcrustacean prey and macroinvertebrate predators in a stream food web. Freshwater Biology 34(1):123–134.

Landis, W. G., Matthews, G. B., Matthews, R. A., Sergeant, A. 1994. Application of multivariate techniques to endpoint determination, selection and evaluation in ecological assessment. Environmental Toxicology and Chemistry 13(2):1917–1927.

Larsen, J., Birks, H. J. H., Raddum, G. G., Fjellheim, A. 1996. Quantitative relationships of invertebrates to pH in Norwegian river systems. Hydrobiologia 328(1):57–74.

Lécher, P., Defaye, D., Noel, P. 1995. Chromosomes and nuclear DNA of Crustacea. Invertebrate Reproduction and Development 27(2):85–114.

Little, T. J., Hebert, P. D. N. 1994. Abundant asexuality in tropical freshwater ostracodes. Heredity 73:549–555.

Little, T. J., Hebert, P. D. N. 1996. Endemism and ecological islands: the ostracods from Jamaican bromeliads. Freshwater Biology 36(2):327–338.

Lowndes, A. G. 1935. The sperms of fresh-water ostracods. Proceedings of the Zoological Society of London. Part 2:35–48.

Maddocks, R. F. 1982. Ostracoda, *in*: Abele, L. G., Ed., The biology of Crustacea, Vol. 1:Systematics, the fossil record, and biogeography. Academic Press, New York, pp. 221–239.

Maher, L. J., Miller, N. G., Baker, R. G., Curry, B. B., Mickelson, D. M. 1998. Paleobiology of the sand beneath the Valders diamicton at Valders, Wisconsin. Quaternary Research 49:208–221.

Malmquist, B., Meisch, C., Nilsson, A. N. 1997. Distribution patterns of freshwater Ostracoda (Crustacea) in the Canary Islands with regards to habitat use and biogeography. Hydrobiologia 347:159–170.

Marmonier, P., Creuzé des Châtelliers, M. 1991. Effects of spates on interstitial assemblages of the Rhône River, importance of spatial heterogeneity. Hydrobiologia 210(3):243–251.

Marmonier, P., Creuzé des Châtelliers, M. 1992. Biogeography of the benthic and interstitial living ostracods (Crustacea) of the Rhône River (France). Journal of Biogeography 19:693–704.

Marmonier, P., Ward, J. V. 1990. Superficial and interstitial Ostracoda of the South Platt River (Colorado, USA)—systematics and biogeography. Stygologia 5(4):225–239.

Marmonier, P., Bodergat, A. M., Doledec, S. 1994. Theoretical habitat templets, species traits, and species richness:ostracods (Crustacea) in the upper Rhône River and its floodplain. Freshwater Biology 31(3):341–355.

Marmonier, P., Meisch, C., Danielopol, D. L. 1989. A review of the Genus *Cavernocypris* Hartmann (Ostracoda, Cypridopsinae): Systematics, Ecology and Biogeography. Bulletin de la Société de Luxembourg 89:221–278.

Marshall, W. S. 1903. *Entocythere cambaria* (nov. gen. et nov. spec.), a parasitic ostracod. Transactions of the Wisconsin Academy of Science, Arts, and Letters 14:117–144.

Martens, K. 1986. Taxonomic revision of the subfamily Megalocypridinae Rome, 1965. Verhandelingen van de Koninklijke Academie voor Wetenscappen, Letteren en schone Kunsten van Belgie 48:1–81.

Martens, K. 1989. On the systematic position of the *Eucypris clavata*—group, with a description of *Trajancypris* gen. nov.

(Crustacea, Ostracoda). Archiv für Hydrobiologie, Supplement 83(2):227–251.

Martens, K., de Moor, F. 1995. The fate of the rhino ridge pool at Thomas Baines Nature Reserve:cautionary tale for nature conservationists. South African Journal of Science 91(8): 385–387.

Martens, K., DeDeckker, P., Marples, T. G. 1985. Life history of *Mytilocypris henricae* (Chapman)(Crustacea:Ostracoda) in Lake Bathurst, New South Wales. Australian Journal of Marine and Freshwater Research 36:807–819.

Mbahinzireki, G., Uiblein, F., Winkler, H. 1991. Microhabitat selection of ostracods in relation to predation and food. Hydrobiologia 222(2):115–119.

McAllister, D. E., Harrington, C. R. 1969. Pleistocene grayling, *Thymallus,* from Yukon, Canada. Canadian Journal of Earth Sciences 6:1185–1190.

McGregor, D. L. 1967. Rhythmic pulsation of the hepatopancreas in freshwater ostracods. Transactions of the American Microscopical Society 86:166–169.

McGregor, D. L. 1969. The reproductive potential, life history and parasitism of the freshwater ostracod, *Darwinula stevensoni* (Brady & Roberstson), *in*: Neale, J. W., Ed., Taxonomy, morphology and ecology of recent Ostracoda. Oliver & Boyd, Edinburgh, pp. 194–221.

McGregor, D. L., Kesling, R. V. 1969a. Copulatory adaptations in ostracods. Part 1:Hemipenes of *Candona.* Contributions to the Museum of Paleontology University of Michigan 22:169–191.

McGregor, D. L., Kesling, R. V. 1969b. Copulatory adaptations in ostracods. Part 2:Adaptations in living ostracods. Contributions to the Museum of Paleontology University of Michigan 22:221–239.

McKenzie, K. G. 1977. Illustrated generic key to South African continental Ostracoda. Annals of the South African Museum 74(3):45–103.

McKillop, W. B., Patterson, R. T., Delorme, L. D., Norgrady, T. 1992. The origin, physico-chemistry and biotics of sodium chloride dominated saline waters on the western shore of Lake Winnipegosis, Manitoba. Canadian Field Naturalist 106: 454–473.

McLay, C. L. 1978a. Comparative observations on the ecology of four species of ostracods living in a temporary freshwater puddle. Canadian Journal of Zoology 56:663–675.

McLay, C. L. 1978b. The population biology of *Cyprinotus carolinensis* and *Herpetocypris reptans* (Crustacea, Ostracoda). Canadian Journal of Zoology 56:1170–1179.

McLay, C. L. 1978c. Competition, coexistence, and survival; a computer simulation study of ostracodes living in a temporary puddle. Canadian Journal of Zoology 56:1744–1758.

Meisch, C. 1996. Contribution to the taxonomy of *Pseudocandona* and four related genera, with the description of *Schellencandona* nov. gen., a list of the Candoninae genera, and a key to the European genera of the subfamily (Crustacea, Ostracoda). Bulletin de la Société des Naturalistes luxembourgeois 97:211–237.

Moore, R. C., Ed. 1961. Treatise on invertebrate paleontology. Part Q:Arthropoda 3, Crustacea, Ostracoda. Geological Society of America, Univ. of Kansas Press. Lawrence 442 pp.

Müller, G. W. 1912. Ostracoda, *in*: Schulze, F. E., Ed. Das Tierreich. Friedländer und Sohn, Lief, Berlin. 434 pp.

Müller-Cale, K. 1913. Über die Entwicklung von *Cypris incongruens.* Zoologische Jahrbücher. Abteilund für Anatomie und Ontogenie der Tiere 36:1–56.

Nalepa, T. F., Robertson, A. 1981. Screen mesh size affects estimates of macro- and meio-benthos abundance and biomass in the Great Lakes. Canadian Journal of Fisheries and Aquatic Sciences 38:1027–1034.

Neale, J. W. 1988. Ostracods and paleosalinity reconstruction, *in*: DeDeckker, P., Colin, J.-P., Peypouquet, J.-P., Eds., Ostracoda in the Earth Sciences. Elsevier, Amsterdam, pp. 125–155.

Neale, J. W., Delorme, L. D. 1985. *Cytheromorpha fuscata,* relict Holocene marine ostracode from freshwater inland lakes of Manitoba, Canada. Revista Espanola de Micropaleontologia 17:41–64.

Newrkla, P. 1985. Respiration of *Cytherissa lacustris* (Ostracoda) at different temperatures and its tolerance towards temperature and oxygen concentration. Oecologia 67:250–254.

Novikoff, M. 1908. Über den Bau des Midiamauges der Ostracoden. Zeitschrift für wissenschaftliche Zoologie 91:81–92.

Nudds, T. D., Bowlby, J. N. 1984. Predator-prey size relationships in North American dabbling ducks. Canadian Journal of Zoology 62(10):2002–2008.

Oertli, B. 1995. Spatial and temporal distribution of the zoobenthos community in a woodland pond (Switzerland). Hydrobiologia 300/301:195–204.

Peck, S. G. 1994. Diversity and zoogeography of the non-oceanic Crustacea of the Galápagos Islands, Ecuador (excluding terrestrial Isopoda). Canadian Journal of Zoology 72:54–69.

Porter, S. C., Sauchyn, D. J., Delorme, L. D. 1999. The ostracode record from Harris lake, southwestern Saskatchewan, 9200 years of local environmental change. Journal of Paleolimnology 21:35–44.

Proctor, V. W. 1964. Viability of crustacean eggs recovered from ducks. Ecology 45:656–658.

Przibram, H. 1931. Connecting laws in animal morphology. Four lectures held at the University of London, March, 1929. Univ. of London Press, London. 62 pp.

Purper, I., Würdig-Macial, N. I. 1974. Occurrence of *Heterocypris incongruens* (Ramdohr), 1808—Ostracoda—in Rio Grande do Sul, Brazil; discussion on the allied genera *Cyprinotus, Hemicypris, Homocypris* and *Eucypris.* Pesquisas 3:69–91.

Ranta, E. 1979, Population biology of *Darwinula stevensoni* (Crustacea, Ostracoda) in an oligotrophic lake. Annales Zoologici Fennici 16:28–35.

Reyment, R. A., Brännström, B. 1962. Certain aspects of the physiology of *Cypridopsis* (Ostracoda, Crustacea). Acta Universitatis Stockholmiensis 9:207–242.

Rieradevall, M., Roca, J. R. 1995. Distribution and population dynamics of ostracodes (Crustacea, Ostracoda) in a karstic lake: Lake Banyoles (Catalonia, Spain). Hydrobiologia 310(3): 189–196.

Roback, S. S. 1969. Notes on the food of Tanypodine larvae. Entomological News 80:13–19.

Roca, J. R., Wansard, G. 1997. Temperature influence on development and calcification of *Herpetocypris brevicaudata* Kaufmann, 1900 (Crustacea:Ostracoda) under experimental conditions. Hydrobiologia 347:91–95.

Roca, J. R., Baltanas, A., Uiblein, F. 1993. Adaptive responses in *Cypridopsis vidua* (Crustacea, Ostracoda) to food and shelter offered by a macrophyte (*Chara fragilis*). Hydrobiologia 262(2):127–131.

Rogulj, B., Marmonier, P., Lattinger, R., Danielopol, D. 1994. Fine-scale distribution of hypogean Ostracoda in the interstitial habitats of the Rivers Sava and Rhone. Hydrobiologia 287(1):19–28.

Rossi, V., Menozzi, P. 1990. The clonal ecology of *Heterocypris incongruens* (Ostracoda). Oikos 57(3):388–398.

Rossi, V., Gandolfi, A., Menozzi, P. 1996. Egg diapause and clonal structure in parthenogenetic populations of *Heterocypris incongruens* (Ostracoda). Hydrobiologia 320(1–3):45–54.

Rossi, V., Rozzi, M. C., Menozzi, P. 1991a. Life strategy differences among electrophoretic clones of *Heterocypris incongruens* (Crustacea, Ostracoda). Verhanlungen internationale Verheinlungen Limnologische 24:2416–2819.

Rossi, V., Schön, I., Butlin, R. K., Menozzi, P. 1998. Clonal genetic diversity, *in*: Martens, K., Ed., Sex and Parthenogenesis—evolutionary ecology of reproductive modes in non-marine ostracods. Backhuys Publ., Leiden, pp. 257–274.

Rozzi, M. C., Rossi, V., Benassi, G., Menozzi, P. 1991. Effetto della temperatura e del fotoperiodo sull'attivazione di uova durature di dloni elettroforetici di *Heterocypris incongruens* (Ostracoda). Società Italieana di Ecologia Atti 12:591–594.

Rome, D. R. 1947. *Herpetocypris reptans* (Ostracode) Étude morphologique et histologique I, morphologie externe et système nerveux. Cellule 51:51–152.

Rome, D. R. 1969. Morphologie de l'attache de la furca chez les Cyprididae et son utilisation en systematique, *in*: Neale, J. W., Ed., The taxonomy, morphology and ecology of recent Ostracoda. Oliver & Boyd, Edinburgh, pp. 168–193.

Rouch, R., Danielopol, D. L. 1997. Species richness of microcrustacea in subterranean freshwater habitats: comparative analysis and approximate evaluation. Internationale Revue der Gesamten Hydrobiologie 82(2):121–145.

Sanders, H. O. 1970. Toxicities of some herbicides to six species of freshwater crustaceans. Journal of Water Pollution Control Federation. 42:1544–1550.

Särkkä, J., Levonen, L., Mäkelä, J. 1997. Meiofauna of springs in Finland in relation to environmental factors. Hydrobiologia 347:139–150.

Sars, G. O. 1898. On *Megalocypris princeps,* a gigantic freshwater ostracod from South Africa. Archiv for Mathematik og Naturvidenskab 20:1–18.

Sars, G. O. 1901. Contributions to the knowledge of the fresh-water Entomostraca of South America, as shown by artificial hatching of dried material. Archiv for Mathematik og Naturvidenskab 24:16–52.

Sars, G. O. 1926. Freshwater Ostracoda from Canada and Alaska. Report of the Canadian Arctic Expedition 1913–1918 7:3–23.

Sars, G. O. 1928. An account of the Crustacea of Norway. Bergen Museum, Bergen. 277 pp.

Schleip, W. 1909. Vergleichende Untersuchung der Eireifung bei parthenogenetisch und bei geschlechtich sich fortplanzendan Ostracoden. Archiv für Zellforschung 2:390–431.

Schön, I., Butlin, R. K., Griffiths, H. I., Martens, K. 1998. Slow molecular evolution in an ancient asexual ostracod. Proceedings of the Royal Society, London B 265:235–242.

Schram, F. R. 1986. Crustacea. Oxford Univ. Press, New York. 606 pp.

Schreiber, E. 1922. Beiträge zur Kenntnis der Morphologie, Entwicklung und Lebensweise der Süsswasser-Ostracoden. Zoologische Jahrbücher. Abteilung für Anatomie und Ontogenie der Tiere. 43:485–539.

Sharpe, R. W. 1903. Report of the fresh-water Ostracoda of the United States National Museum including a revision of the subfamilies and genera of the family Cyprididae. Proceedings of the U.S. National Museum. 26:969–1001.

Sharpe, R. W. 1918. The Ostracoda, *in*: Ward, H. B., Whipple, G. C., Eds., Fresh-water biology. Wiley, New York, pp. 790–827.

Smith, A. J., Donovon, J. J., Ito, E., Engstrom, D. R. 1997. Groundwater processes controlling a prairie lake's response to middle Holocene drought. Geology 25(5):391–394.

Smith, A. J., Delorme, L. D., Forester, R. M. 1992. A lakes's solute history from ostracodes: comparison of methods, *in*: Kharaka, Y. K., Maest, A. S., Eds., Water-Rock Interaction, Balkema, Rotterdam, pp. 677–680.

Smith, R. N. 1965. Musculature and muscle scars of *Chlamydotheca arcuata* (Sars) and *Cypridopsis vidua* (O. F. Müller) (Ostracoda: Cyprididae), Vol. 3, Four Reports on ostracode investigations. University of Michigan Press, Ann Arbor, MI, pp. 1–40.

Sohn, I. G., Kornicker, L. S. 1975. Variation in predation behavior of ostracode species on *Schistosmoiasis* vector snails. Bulletin of American Paleontologist. 65:217–223.

Sohn, I. G., Kornicker, L. S. 1979. Viability of freeze-dried eggs of the freshwater *Heterocypris incongruens, in*: Proceedings of the Seventh International Symposium on Ostracoda, Belgrade, pp. 1–3.

Staplin, F. L. 1963a. Pleistocene Ostracoda of Illinois. Part 1: Subfamilies Candoninae, Cyprinae, general ecology, morphology. Journal of Paleontology 37:758–797.

Staplin, F. L. 1963b. Pleistocene Ostracoda of Illinois. Part 2: Subfamilies Cyclocyprinae, Cypridopsinae, Ilyocyprinae; families Darwinulidae and Cytheridae, stratigraphic ranges and assemblage patterns. Journal of Paleontology 37:1164–1203.

Strayer, D. 1985. The benthic micrometazoans of Mirror Lake, New Hampshire. Archiv für Hydrobiologie, Supplement 72: 287–426.

Strayer, D. 1988. Life history of a lacustrine ostracod. Hydrobiologia 160(2):189–191.

Swain, F. M. 1963. Pleistocene Ostracoda from the Gubik formation, Arctic Coastal Plain, Alaska. Journal of Paleontology 37: 798–834.

Swüste, H. F. J., Cremer, R., Parma, S. 1973. Selective predation by larvae of *Chaoborus flavicans* (Diptera, Chaoboridae). Internationale Vereinigung für Theoretische und Angewandte Limnologie. Verhundlungen 18:1559–1563.

Sylvester-Bradley, P. C. 1941. The shell structures of the Ostracoda and its application to their palaeontological investigation. The Annals and Magazine of Natural History Ser. 11:1–33.

Sylvester-Bradley, P. C., Benson, R. H. 1971. Terminology for surface features in ornate ostracodes. Lethaia 4:249–286.

Sywula, T., Glazewska, I., Whatley, R. C., Moguilevsky, A. 1995. Genetic differentiation in the brackish-water ostracod *Cyprideis torosa.* Marine Biology 121(4):647–653.

Tabacchi, E., Marmonier, P. 1994. Dynamics of the interstitial ostracod assemblage of a pond in the Adour alluvial plain. Archiv fur Hydrobiologie 131(3):321–340.

Takamura, K., Yasuno, M. 1986. Effects of pesticide application on chironomid larvae and ostracods in rice fields: Applied Entomology and Zoology 21(3):370–376.

Towle, E. W. 1900. A study in the heliotropism of *Cypridopsis.* The American Journal of Physiology 3:345–365.

Tressler, W. L. 1949. Freshwater Ostracoda from Brazil. Proceedings of the United States National Museum 100(3258):61–83.

Tressler, W. L. 1954. Fresh-water Ostracoda from Texas and Mexico. Journal of the Washington Academy of Science 44:138–149.

Tressler, W. L. 1956. Ostracoda from bromeliads in Jamaica and Florida. Journal of the Washington Academy of Science 16: 333–336.

Tressler, W. L. 1957. The Ostracoda of Great Slave Lake. Journal of the Washington Academy of Science 47:415–423.

Triebel, E. 1953. Genotypus und Schalen-Merkmale der Ostracoden-Gattung *Stenocypris.* Senckenbergiana 34:5–14.

Turgeon, J., Hebert, P. D. N. 1994. Evolutionary interactions between sexual and all-female taxa of *Cyprinotus* (Ostracoda, Cyprididae). Evolution 48(6):1855–1865.

Turgeon, J., Hebert, P. D. N. 1995. Genetic characterization of breeding systems, ploidy levels and species boundaries in *Cypricercus* (Ostracoda). Heredity 75(6):561–570.

Turner, C. H. 1896. Morphology of the nervous system of *Cypris.* Journal of Comparative Neurology 6:20–44.

Turpen, J. B., Angell, R. W. 1971. Aspects of molting and calcification in the ostracod *Heterocypris.* The Biological Bulletin (*Woods Hole, Mass.*) 140(2):331–338.

Uiblein, F., Roca, J. R., Danielopol, D. L. 1994. Experimental observations on the behavior of the ostracode *Cypridopsis vidua.*

Internationale Vereinigung für Theoretische und Angewandte Limnologie, Verhundlungen 25:2418–2420.

van Harten, D. 1975. Size and environmental salinity in the modern euryhaline ostracod *Cyprideis torosa* (Jones, 1850), a biometrical study. Palaeogeography, Palaeoclimatology, Palaeoecology 17(1):35–48.

von Grafenstein, V., Erlenkeuser, H., Müller, J., Kleinmann-Eisenmann, A. 1992. Oxygen isotope records of benthic ostracods in Bavarian Lake sediments. Naturwissenschaften 79:145–152.

von Grafenstein, V., Erlenkeuser, H., Kleinmann, A., Müller, J., Trimborn, P. 1994. High-frequency climatic oscillations during the last deglaciation as revealed by oxygen-isotope records of benthic organisms (Ammersee, southern Germany). Journal of Paleolimnology 11:349–357.

Walton, M., Hobbs, H. H., Jr. 1959. Two new eyeless ostracods of the genus *Entocythere* from Florida. Quarterly Journal of the Florida Academy of Science 22:114–120.

Ward, J. V., Stanford, J. A., Voelz, N. J. 1994. Spatial distribution patterns of Crustacea in the floodplain aquifer of an alluvial river. Hydrobiologia 287:11–17.

Westgate, J. A., Delorme, L. D. 1988. Lacustrine ostracodes in the late Pleistocene Sunnybrook diamicton of southern Ontario, Canada. Reply. Canadian Journal of Earth Sciences 25: 1717–1720.

Westgate, J. A., Chen, F.-J., Delorme, L. D. 1987. Lacustrine ostracodes in the late Pleistocene Sunnybrook diamicton of southern Ontario. Canadian Journal of Earth Sciences 24:2330–2335.

Weygoldt, P. 1960. Embryologische Untersuchungen an Ostrakoden; Die Entwicklung von *Cyprideis litoralis* (G. S. Brady) (Ostracoda, Podocopa, Cytheridae). Zoologische Jahrbücher. Abteilund für Anatomie und Ontogenie der Tiere 78:369–426.

Weygoldt, P. 1961. Zur Kenntnis der Sekretion im Zentralnervensystem der Ostrakoden *Cyprideis litoralis* (G. S. Brady) (Podocopa Cytheridae) und *Cypris pubera* (O.F.M.) (Podocopa Cypridae), Neurosekretion und Sekretionzellen im Perineurium. Zoologischer Anzeiger 166:69–70.

White, M. J. D. 1973. Animal cytology and evolution. Cambridge Univ. Press, London.

Wickstrom, C. E., Castenholz, R. W. 1973. Thermophilic ostracod: aquatic metazoan with the highest known temperature tolerance. Science 181:1063–1064.

Wiggins, G. B., Mackay, R. J., Smith, I. M. 1980. Evolutionary and ecological strategies of animals in annual temporary ponds. Archiv für Hydrobiologie Supplement 58:97–206.

Wohlgemuth, R. 1914. Beobactungen und Untersuchungen über die Biologie der Süsswasserostracoden; ihr Vorkommen in Sachsen und Böhmen, ihr Lebensweise und ihr Fortpflanzung. Biologisches Supplement zur Internationale Revue der Gesamten Hydrobiologie und Hydrobiologie 6:1–72.

Woltereck, R. 1898. Zur Bildung und Entwickelung des Ostracoden-Eies. Zeitschrift für Wissenschaftliche Zoologie 64:596–620.

Xia, J., Engstrom, D. R., Ito, E. 1997a. Geochemistry of ostracode calcite, part 2. The effects of water chemistry and seasonal temperature variation on *Candona rawsoni.* Geochimica et Cosmochimica Acta 61(2):383–391.

Xia, J., Haskell, B. J., Engstrom, D. R., Ito, E. 1997b. Holocene climate reconstruction from tandem trace-element and stable isotope composition of ostracodes from Coldwater Lake, North Dakota, USA. Journal of Paleolimnology 17:85–100.

Xia, J., Ito, E., Engstrom, D. R. 1997c. Geochemistry of ostracode calcite, part 1. An experimental determination of oxygen isotope fractionation. Geochimica et Cosmochimica Acta 61(2): 377–382.

Yu, Z., Ito, E. 1999. Possible solar forcing of century-scale drought frequency in the northern Great Plains. Geology 29(3):263–266.

Zelmer, D. A., Esch, G. E. 1998. Interactions between *Halipegus occidualiis* and its ostracod second intermediate host: evidence for castration? Journal of Parasitology 84(4):778–782.

Zissler, D. 1969a. Die Spermiohistogenese des Süsswasser-Ostracoden. Part 1: Die ovalen und spindelförmigen Spermatiden. Zeitschrift für Zellforschung und Mikroscopische Anatomie 96:87–105.

Zissler, D. 1969b. Die Spermiohistogenese SüsswasserOstracoden *Notodromas monacha* O. F. Müller. Part 2:Die spindelförmigen und schlauchürmigen Spermatiden. Zeitschrift für Zellforschung und Mikroscopische Anatomie 96:106–133.

Zissler, D. 1970. Zur Spermiohistogenese im Vas Deferens von Süsswasser-Ostracoden. Cytobiologie 2:83–86.

21

CLADOCERA AND OTHER BRANCHIOPODA

Stanley I. Dodson
Department of Zoology
University of Wisconsin
Madison, Wisconsin 53706

David G. Frey†

†Dr. David G. Frey's contribution to this chapter is greatly appreciated, and all of us regret his death prior to its publication.

Ecology and Classification of North American Freshwater Invertebrates. 2nd Edition

I. INTRODUCTION[1]

Branchiopods are small crustaceans that have flattened leaflike legs. They occur in all freshwater habitats and can be abundant enough to form conspicuous swarms. Branchiopods are useful for studies of community and population ecology, animal behavior, functional morphology, evolution of life history. They occupy a key position in aquatic communities, both as important herbivores eating algae and bacteria and as major prey items of fish, birds, backswimmers, and other aquatic predators. Their fossil remains open a window into the past climate and ecology of lakes.

The orders of branchiopods (Table I), including eight living and two extinct orders are related to each other only distantly (Fryer, 1987; Kerfoot and Lynch, 1987; Olesen, 1998). The Branchiopoda are a heterogeneous group of crustaceans that share a few characteristics, principally similar thoracic legs called phyllopods, which are flat, edged with setae, not distinctly segmented, and usually appear to be unbranched. While this leg architecture resembles that of some of the earliest crustacean fossils, it is probable that the similarities among the branchiopods is due at least as much to convergent evolution as to common ancestry. Branchiopods also have similar mouthparts. Mandibles are simple unsegmented rods with corrugated grinding inner surfaces. The first and second pairs of maxillae are reduced to small scalelike structures, or are absent. Branchiopods have on the last body segment a pair of spines or claws. A controversy exists concerning the evolutionary significance of these terminal spines (e.g., Bowman, 1971; Schminke, 1976).

Because branchiopods are such a diverse group, this chapter is divided into two major sections. The first focuses on the "cladocera," a group of four orders of small-sized branchiopods (Freyer 1987, Olesen 1998). The second section focuses on the "non-cladoceran orders," which are less diverse and typically somewhat larger than the cladocera.

TABLE I The Families and Present Number of Extant Genera of Class Branchiopoda

Order	*Family*	*Number of genera in world*
Anomopoda	Daphniidae[a]	8
	Moinidae[b]	2
	Bosminidae[c]	2
	Ilyocryptidae[d]	1
	Macrothricidae[e]	10
	Neothricidae[e]	2
	Acantholeberidae[e]	1
	Ophryoxidae[e]	2
	Chydoridae[f]	37
Ctenopoda	Sididae[g]	7
	Holopediidae[g]	1
Onychopoda	Podonidae[h]	7
	Polyphemidae[h]	1
	Cercopagidae[h]	2
Haplopoda	Leptodoridae[h]	1
Anostraca[i]	Artemiidae	1
	Branchinectidae	1
	Branchipodidae	6
	Chirocephalidae	7
	Linderiellidae	2
	Polyartemiidae	2
	Streptocephalidae	1
	Thamnocephalidae	1
Spinicaudata[i]	Cyclestheriidae	4
	Cyzcidae	4
	Leptestheriidae	3
	Limnadiidae	6
Laevicaudata[j]	Lynceidae	3
Notostraca[k]	Triopsidae	2

The "cladocera" include the first four orders listed in the table. The "conchostraca" include the orders Spinicaudata and Laevicaudata.

[a] Number of genera of Daphniidae, see Fryer (1991).

[b] For Moinidae, see Goulden (1968) and Fryer (1991).

[c] For Bosminidae, see DeMelo and Hebert (1994).

[d] For Ilyocryptidae, see Smirnov (1992).

[e] The families Macrothricidae, Neothricidae, Acantholeberidae, and Ophryoxidae are discussed in Smirmov (1992) and grouped into the superfamily Macrothricoidea by Dumont and Silva-Briano (1998).

[f] For family Chydoridae; see Smirnov (1974, 1996). Alonso (1996) suggests a name change to Eurycercidae. Because of long usage, I prefer the name "Chydoridae" for the family.

[g] For Sididae and Holopedidae, see Korovchinsky (1992).

[h] For the predaceous cladocerans, see Rivier (1998).

[i] Families and number of genera are based on an annotated update (personal communication) by D. Belk, based on Belk (1982).

[j] For Laevicaudata see Martin and Belk (1988).

[k] For Notostraca, see Fryer (1988).

CLADOCERA

Most plankton ecologists now agree that the term "cladocera" has no taxonomic significance (because the orders are probably not closely related), although the term is still used for convenience. The cladocera are grouped into four orders (Fryer, 1987), 15 families (including the revisions suggested by Dumont and Silva-Briano, 1998), about 80 genera, and roughly 400 species (Table I). There is some controversy over whether the orders Onychopoda and Haplopoda

[1] The references in this chapter are helpful starting places for futher reading and research. To find our more about a topic of interest, search electronic biological data bases, such as Biological Abstracts and the Zoalogical Record, which can be accessed at many university libraries. One typically finds a rich mine of publications. If your library lacks the publication, many are available through interlibrary loan programs

(which are related predaceous species; Olesen, 1998) are even members of the Branchiopoda (Starobogatov, 1986). The number of genera in each order (Table I) is only approximate, because new genera have been established recently in the families Sididae, Daphniidae, Macrothricidae, and Chydoridae, with certainly more to come.

Almost all members of the suborder Radopoda (Dumont and Silva-Briano, 1998) and some members of the Sididae (Table I) are benthic (bottom-dwelling), living on and in various surfaces, particularly aquatic plants (macrophytes), coarse plant detritus (decaying plant material), and organic sediments. They are called "meiobenthos" (Frey, 1988a) because of their small size (mostly <1 mm in total length) and because of their benthic habitat preference. Meiobenthos are often most abundant in the plants and associated sediments of the littoral (or shallow water) zone. Chydorid populations commonly have more than a million animals per square meter of bottom, and twenty or more species can coexist in the same small bit of habitat. Most species in the remaining eight families in Table I are primarily or totally planktonic (inhabiting the open water zone), where their facility for swimming makes them quite independent of surfaces.

Besides differing in habitat, planktonic and meiobenthic cladocerans differ also in ecological relationships and in their evolution. Hence, to imply, as students of plankton commonly do, that everything learned about *Daphnia* or *Bosmina* applies equally well to all cladocera is often misleading or incorrect. The present chapter seeks to present an even treatment of all cladoceran families.

II. ANATOMY AND PHYSIOLOGY

A. General Form and Function

1. General Female Form and Function

Martin (1992) provides a masterful and detailed description of branchiopod anatomy, with a wealth of drawings and photos. In general, cladocerans (water fleas) are a group of small animals belonging to one of four crustacean orders (Table I). Adults are 0.2–18.0 mm long. The genera *Daphnia* (Fig. 1A) and *Pleuroxus* (Fig. 1B) exemplify typical cladoceran morphology. Cladocerans do not have a clearly segmented body or appendages, although they do possess segmented second antennae. Obvious features of most cladocerans include a single central compound eye as adults and a transparent clear-to-yellow carapace that is used as a brood chamber. The carapace is attached to the back of the neck and acts like an overcoat, wrapping around part or all of the body, except for the head. In many species, the 4–6 pairs of thoracic legs are covered by carapace. Crustaceans differ from other arthropods in having two pairs of antennae. In Cladocera, the first pair of antennae (called antennules) are usually one-segmented and small, and have only a chemosensory function. The second pair of antennae (called antennae) are large and used for swimming.

Most cladocerans are small transparent animals, whose general shape and jerky swimming accounts for their common name, "water fleas." Although abundant in most standing freshwater, they are small enough to be easily overlooked. Nevertheless, it is possible to observe their behavior and identify at least the larger species using the unaided eye. For a clear view of their structures, however, it is necessary to use a dissecting microscope for gross features and a compound microscope for finer details.

Anatomy is best observed using living animals. At first, it is easiest to focus on large and transparent animals, such as an adult *Daphnia* or *Simocephalus*. Once structures are located in these large animals, they can also be found in small animals, such as chydorids and bosminids. Figure 1 indicates the appearance and relative arrangement of the major cladoceran structures.

For observation of a living cladoceran, one or more individuals can be put in a small volume of water in a Petri dish. If most of the water is then withdrawn, the animals will be held down by the surface tension and be unable to dart out of the microscope field of view. For more careful observation, an animal can be anchored to a small dab of clear grease. The grease can either be put on the bottom of a Petri dish or on the tip of a small glass rod.

Cladocerans typically are discus-shaped. When dead or trapped by a surface film of water, they will lie on their side. The body in side view (Fig. 1) is rounded, the legs are obscured by a transparent shelllike covering (the carapace), and the major pigmented parts are likely to be blue-to-yellow eggs in the brood chamber, the green or yellow gut, and the dark black eye.

The location of the head is marked by one or more eyes. The head is usually more or less dome-shaped but can have long, even pointed extensions (compare Fig. 2B and 2C) and an elongated beak may be present (Fig. 1). All species possess a single black compound eye except the chydorids *Monospilus* and *Bryospilus* (Fig. 14 and 15) and the blind cave-dwelling species of *Alona* and *Spinalona* (Dumont, 1995; Brancelj, 1990, 1992; Ciros-Perez and Gutierrez, 1997). The compound eye is single, the result of fusion of two eyes during embryonic development (Threlkeld, 1979). The black color of the cladoceran eye is due to accessory screening pigments, including melanins, ommochromes,

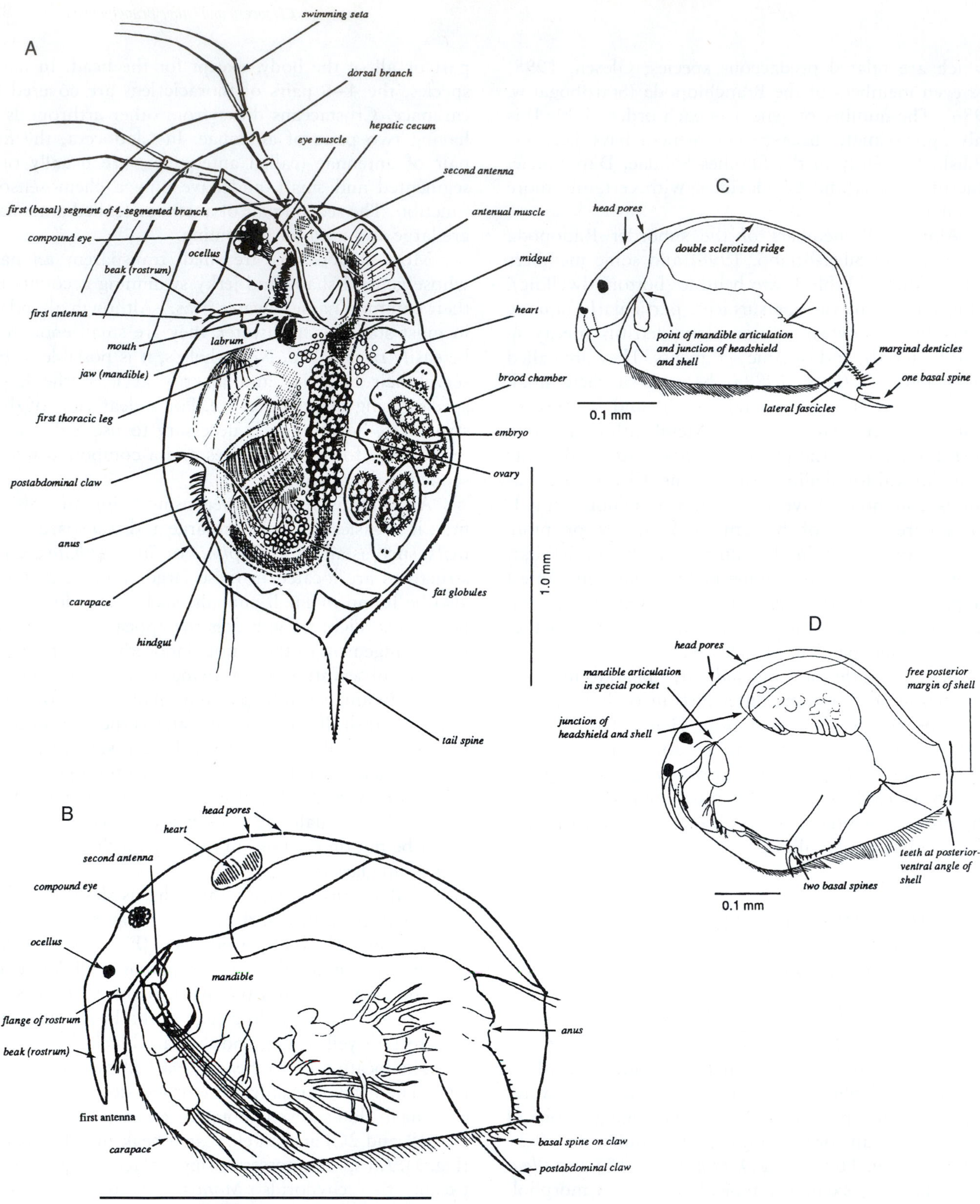

FIGURE 1 (A) The anatomy of a daphniid. *Daphnia pulicaria* Forbes, 1893, emend. Hrbácek, 1959. Nebish Lake, Vilas Co., Wisconsin, 6 June 1987. Figure drawn by Kandis Elliot (KE). (B) The whole-body anatomy of a chydorid *Pleuroxus trigonellus* (O.F. Mueller, 1776). Rybinsk Reservoir, USSR. 7 December 1962. Redrawn from Smirnov (1971) (KE); (C) Mandibular articulation in the chydorid subfamily Aloninae. *Oxyurella brevicaudis* Michael and Frey, 1983. D.G. Frey Sample 5019, Clearwater Lake, Putnam Co., Forida. 3 March 1979; (D) Mandibular articulation in the chydorid subfamily Chydorinae. *Pleuroxus aduncus* (Jurine, 1820). D. G. Frey sample 3004. Langemosa, Sealand, Denmark. 23 October 1972.

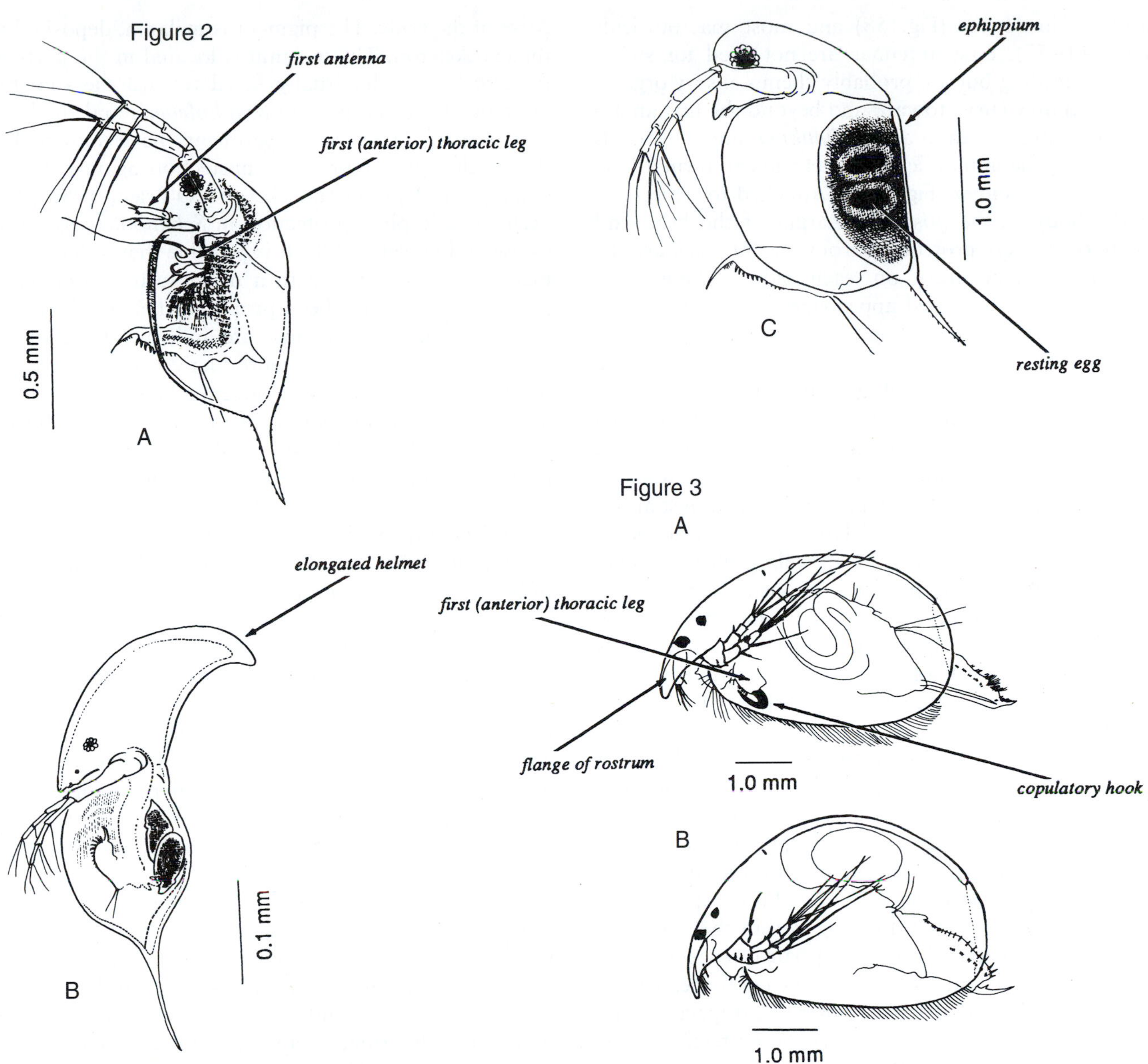

FIGURE 2 (A) Male *Daphnia pulicaria* Forbes, 1893, emend. Hrbáček, 1959. Nebish Lake, Vilas Co., WI, 6 June 1987; (B) *Daphnia retrocurva* Forbes, 1882. With elongated helmet. Lake Wingra, Dane Co., WI. 5 October 1978; (C) Ephippial female *Daphnia pulicaria*. Nebish Lake, Vilas Co. WI, 6 June 1997. (KE)
FIGURE 3 Male and female chydorid. *Alona bicolor* Frey, 1965. (A) Male redrawn from Frey (1965); (B) female redrawn from Frey (1965).

pteridines, or purines (Shaw and Stowe, 1982). Since the black color fades in several mounting media, such as Hoyer's, it is unlikely that much melanin is present. The actual photoreceptive pigment is a nearly colorless purple, masked by the accessory pigments. With magnification of 50×, it is possible to see nearly transparent muscles that are attached to the eye and presumably responsible for the motion. Once these muscles are seen it is easier to find other muscles, as for example, those attached between the basal segment of the swimming (second) antennae and the back of the thorax and head. Most cladocerans also have a small black ocellus (simple eye) just posterior to the compound eye. Like other crustaceans, cladocerans show evidence of five pairs of appendages on the head part of the body: two pairs of antennae, one pair of mandibles, and two pairs of maxillae. At the base of the head, near the carapace margin, is a pair of short cigarlike appendages—the first antennae (antennules). These are usually shorter than the head and inconspicuous but may be longer than the head, as in *Moina* (Fig. 58) and most macrothricids (Figs. 44–57). First antennae are not used for swimming or feeding but are probably chemosensory organs. The head may curve toward and beyond the first antennae, forming a beak, as in *Daphnia* and chydorids (Fig. 1), or the immobile first antennae can form a long tusk, as in *Bosmina* (Fig. 42B). Attached to either side of the body at the posterior margin of the head and partly or entirely protruding from the carapace are the second antennae (one on each side). These large branched and segmented appendages are typically used for swimming or crawling. The mandibles are unbranched and elongated, with a pivot point near the carapace and a darkened grinding surface near the mouth. The first pair of maxillae are typically reduced to a small flap with a few curved spines. The second pair of maxillae are missing or represented by a spine. In addition to these appendages on the head, just above the mouth is an upper lip (the labrum), which closes the mouth and secretes mucus. A maxillary gland (nuchal organ, or one or more head pores) opens on the dorsal margin of the head in conchostracans and the cladocerans (Olesen, 1996). It secretes a glue in some forms (*Sida*, *Simocephalus*), a large gelatinous capsule over the carapace (*Holopedium*), or salt in marine or hypersaline branchiopods.

The head smoothly joins the rest of the body which is often covered by a carapace. The transparent carapace is often ornamented with a faint geometric pattern in the planktonic species or by striae, spines, pits, or hexagonal meshes in many of the littoral forms. The cladoceran body posterior to the head is comprised of a thorax and an abdomen hanging within the overcoat-like carapace. The body ends in a pair of (postabdominal) claws which can reach out of the carapace. Several species have a stiff extension of the carapace: spines near the base of the second antennae or teeth, spines, or one long tail spine on the posterior end.

The chemistry of the carapace is important, since cladocerans tend to have a hydrophobic exoskeleton. Although surrounded by water, many cladocerans are unwetted. If brought into contact with the surface film, they become trapped. Their hydrophobic carapace is probably an adaptation which discourages growth of algae and protozoa on the carapace. Cladocerans cannot groom the outside of the carapace, and animals that have lived for a few weeks without molting often bear a thick load of encrusting organisms (Chiavelli *et al.*, 1993; Threlkeld and Willey, 1993) which may diminish their feeding and swimming efficiency, reduce their escape speed, and increase their sinking rate (Allen *et al.* 1993).

Some cladocerans, such as *Scapholeberis* (Fig. 40) and some forms of *Daphnia*, can be colored black over parts of the body. The pigment is melanin, deposited in the exoskeleton. The melanin is located in the parts of the exoskeleton habitually faced toward the sun: the margins of the carapace in *Scapholeberis* and the head and back in *Daphnia*. Experiments have shown that the black pigment provides protection against photodamage in *Daphnia* (Luecke and O'Brien, 1983), but perhaps little photoprotection in *Schapholeberis* (Hurtubise and Havel, 1998). It is possible that induction of melanin is associated with a decrease in reproductive potential. It has also been proposed that the dark pigment also increases heating rate and, therefore, allows cladocerans an increased reproductive rate in cold water. However, the small cladoceran body size and the high thermal conductivity of water suggest that this effect is insignificant. Finally, while melanin strengthens bird feathers and insect exoskeletons, it probably does not have this effect in cladocerans (Dodson, 1984). Melanistic animals brought into the lab lose their pigmentation at the first molt, suggesting that stimulation of bright sunlight is needed to induce melanin.

Scapholeberis contain the darkest pigmented cladoceran species. *Scapholeberis kingii* is the only known distasteful cladoceran; it is not even ingested by hydra (Schwartz *et al.*, 1983). *Scapholeberis* homogenate causes *Hydra* to writhe and contract, and it induces concerts of tentacular spasms. It is not yet known if *Scapholeberis* is distasteful to visual predators. If so, the black carapace pigmentation may be adaptive both as protection against photodamage and as a warning display.

The animal's thorax and abdomen can be seen moving within the carapace of living individuals. The thorax holds four or six pairs of legs. These flat leaf-shaped legs (called "phyllopods") are well supplied with rows of setae and spines and are used for handling food, filtering, scraping, pumping, creeping, or grasping females. The legs beat rhythmically back and forth at roughly five beats per second at room temperature. If a small amount of an alga suspension is placed with a pipette near the carapace margins, it is possible to follow the water current which brings the algae to the legs, and to watch algae collecting on the legs and being passed toward the mouth. The thoracic legs probably act as electrostatic filters (not sieves), collecting algae and other particles that stick to the flat surfaces and combs of setae (e.g., Gerritsen and Porter, 1982; Cheer and Koehl, 1987; Gerritsen *et al.*, 1988). Lampert (1994) showed that the filter comb size of *Daphnia* (7 species) increased by as much as 83% when the animals were grown at low food levels. Food caught on the thoracic legs is mixed with mucus to make a mass of food (the bolus), which is moved forward toward the mouth. The postabdominal claw is used to clean the filters or to reject a bolus containing distasteful algae. Cladocerans thus have the option of accepting or rejecting a mass of algae, but little control over the composition of the bolus (they have difficulty making decisions about whether to accept each individual algal or bacterial cell, although Strickler developed videos of *Daphnia* choosing individual particles). A bolus can be rejected if it contains a toxic alga or if its average "taste" is unacceptable (Richman and Dodson, 1983). If the food bolus is accepted, it is chewed by the mandibles and swallowed. The amount of food ingested depends on the algal concentration, until a concentration is reached at which the legs are operating at full capacity. Beyond this concentration, the amount of food ingested remains constant, or even declines, as the legs become clogged with food (Porter *et al.*, 1982).

The mouth is near the margin separating the carapace and the head. On either side of the mouth lies a pair of large but simple jaws (mandibles), each of which is a single rod, pointed toward the outer end. The inner dark chewing surfaces lie just in front of the mouth. The jaws are in nearly constant grinding motion. If the animals have been feeding, the gut will be easy to find. The gut runs from the mouth, loops through the head, continues along the body through the thorax and abdomen, and terminates at the anus near the posterior end of the animal. The gut can be divided into three regions. The foregut and hindgut are lined with cuticle, which is continuous with the cuticle that covers the outside of the animal. The midgut (in the thorax) is lined with epithelium elaborated into microvilli and is the site of absorption (Peters, 1987). In the head region of *Daphnia* are a couple of small sacs (hepatic caeca) attached to the gut. In some species, especially in the chydorids, there is a sac attached to the gut in the abdominal region and none in the head region. A green to brown, rather amorphous material typically fills the gut (Tappa, 1965) along with undigested algal cells and small animals (Porter, 1973, 1977). Waves of rhythmic (peristaltic) contraction move along the gut from the anus toward the midgut. By feeding a trapped animal two different kinds of food (yeast followed by algae), it is possible to measure the rate of movement of food particles through the gut.

Lying along the gut, in the thorax region of well-fed adults, are fat globules and a pair of gonads (Fig. 1). The fat globules are light yellow to orange, and are spread throughout the body cavity in greater or lesser abundance (Tessier and Goulden, 1982). The ovaries resemble a mass of large fat globules, but the maturing eggs are more often slightly green in color (the blue or green colors are due to carotenoids that are combined chemically with proteins). Each time an adult female molts, she extrudes a few new eggs into the brood chamber, as she releases the neonates developed over the previous between-molt period. Developing sperm are large round cells, similar in appearance to the fat globules. These are released through the anus, when the postabdomen is inserted into the female's brood chamber. When released, the spermatozoa are amoeboid, without a flagellum. In most species they are simple ovals 5–20 μm in diameter. Spermatozoa of *Diaphanosoma* are oval, about 60 μm in diameter, and those of *Sida* and *Moina* are spiny and nearly 100 μm long (Weismann, 1880; Wingstrand, 1978).

The heart (Fig. 1) is obvious because of its rapid beating in live animals. It is a clear muscular organ lying in the thorax above the gut and just behind the head. With the right light and at least 50× magnification, one can see the blood cells moving about, especially near the heart and in the bases of the second antennae, and in the beak of *Daphnia*.

The abdomen carries at its tip a pair of claws, which may appear to be single. These claws are used to groom the thoracic legs and can be seen to brush over the thoracic legs as the abdomen is swung forward. Some smaller species, especially those in the family Chydoridae, use the abdomen as a sort of foot to kick along surfaces.

The part of the body at the end of the abdomen that bears claws and is posterior to the anus is called the postabdomen. The anus lacks a muscle for closure (sphincter). In live animals, rhythmic (peristaltic) waves of contraction can be seen moving up the hindgut. Most, if not all cladocerans show this "anal drinking." The hindgut cause a current of water to flow into the anus; the water is absorbed by the columnar cells of the anterior gut. This means that dissolved organic compounds may stay in the gut much longer than the food from which they were derived (Fryer, 1970).

"Cyclomorphosis" (a type of phenotypic plasticity) is a change in body shape within a species caused by alternate developmental pathways induced by environmental factors (such as temperature and turbulence) and chemical communication (Dodson, 1989; Larsson and Dodson, 1993). For example, shown in Fig. 2b is a *Daphnia* with an elongated helmet. This same animal could have offspring with much lower helmets, depending on the chemical signals present during embryogenesis. Animals from the same clone mother, but influenced by different chemical signals, have such different forms that they have been described as different species (Krueger and Dodson, 1981). The changes in form include elongated heads, tail spines, or in *Bosmina*, elongated first antennae. These elongated body parts are sometimes accompanied by a reduced adult body size. Black and Slobodkin (1987) define cyclomorphosis as "temporal (seasonal or aseasonal) cyclic morphological changes that occur within a planktonic population." Animals generally lose elongation of body parts when cultured in the laboratory for one or two molts. For example, *Daphnia galeata mendotae* and *D. retrocurva* both produce higher helmets in warm turbulent laboratory environments. They also produce higher helmets in the presence of two invertebrate predators, *Chaoborus* larvae and *Notonecta* adults. Similarly, *Daphnia pulicaria* produced neck teeth the first summer that *Chaoborus* were present in Lake Lenore, Washington (Luecke and Litt, 1987). The induced morphological changes are generally thought to be advantageous for increasing survivorship in the face of specific predators: small size protects against fish and long spines protect against small invertebrate predators (Larsson and Dodson, 1993; Tollrian and Dodson, 1998).

2. Differences between the Sexes

The two sexes of cladocerans are morphologically similar. Because most cladocerans reproduce asexually at least part of the time, most individuals will be females. Males (Figs. 2A, 3B) resemble females but are smaller. Cladoceran males have a copulatory hook on the first thoracic leg, used for holding on to the female during mating. Macrothricidae and Chydoridae have the largest hooks, and *Polyphemus* and *Leptodora* have the smallest hooks (Lilljeborg, 1901). Some cladoceran males have longer first antennae than are present in the females (especially *Sida*, *Diaphanosoma*, *Daphnia*, *Ceriodaphnia*, *Moina*, *Polyphemus*, and *Leptodora*). *Daphnia* and *Ceriodaphnia* males have elongated setae on the anterior thoracic legs. Males of *Diaphanosoma*, *Latona*, *Bythothephes*, *Podon* (Lilljeborg, 1901) and *Penilia* (Della Croce and Gaino, 1970) have a pair of penes on the postabdomen, as do several species of *Alona* and *Leydigia*.

3. Life History and Development

Detailed information exists for the life histories of many cladocerans (Lynch, 1980; Frey; 1975, 1982c, 1987b, 1989) and especially for *Daphnia* (Schwartz, 1984; Lynch, 1989; Threlkeld, 1988). Cladoceran reproduce either sexually or asexually, depending on environmental conditions (Fig. 4). Most species reproduce parthenogenetically most of the time (Hebert, 1978, 1988), and most resting eggs develop into females. Species adapted to temporary habitats, such as *Moina* species, have resting eggs that can develop into either males or females.

Embryogenesis begins, and is usually completed, in the brood chamber, a space between the body and the carapace (Fig. 1, also see Kotov and Boikova, 1998). In the Onychopoda and Haplopoda the carapace does not

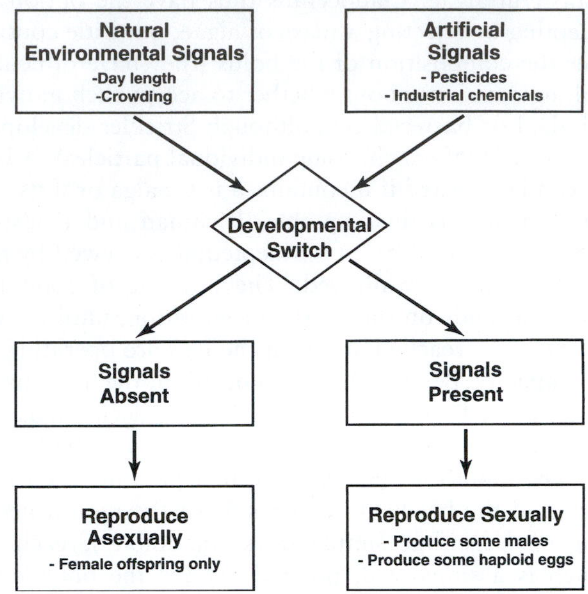

FIGURE 4 *Daphnia* reproductive strategy. Adult females can produce three different kinds of offspring, depending on environmental conditions. Diploid (subitaneous) eggs are produced asexually, and develop into females or males. Haploid eggs can also be produced, given the correct chemical signal from the environment.

cover the body, but is reduced to a brood chamber on the animal's back (e.g., Figs. 60 and 61). Cladoceran eggs (Figs. 1, 7, 9B, 13, 27, 56, and 61) are of three types:

- diploid "subitaneous" eggs, which develop immediately into young;
- resting "eggs," which come from haploid eggs that are fertilized, develop into early embryos, and then enter a diapause (state of physiological stasis) which is resistant to heating, drying, and freezing; and
- pseudosexual resting eggs, which are early diapausing embryos from asexually produced diploid eggs (no fertilization needed).

Why do cladoceran species often reproduce both asexually and sexually? Two adaptive advantages of asexual reproduction are that population growth rate is maximized, since all offspring are females, and genetic combinations particularly well adapted to present conditions are not disrupted by sexual recombination (Banta, 1939; Williams, 1975). Two disadvantages to asexual reproduction are that asexual clones are slow or unable to adjust to changing conditions by sexual recombination (and so are restricted to specific habitat conditions) and that asexual genotypes accumulate mutations (which are typically detrimental). Sexually derived resting eggs represent new genetic combinations and have a chance of both being pre-adapted to new environmental conditions and lacking detrimental mutations.

The different adaptive advantages of the sexual and asexual eggs would seem to dictate that only the sexual eggs should diapause, as is often the case. The exception to this is the production of "pseudosexual" resting eggs in obligate asexual clones. Pseudosexual resting eggs are produced without recombination but otherwise resemble sexual diapausing eggs and are enclosed in an ephippium. Pseudosexual eggs are probably a solution for the need to produce resting eggs in ponds (such as in the arctic or alpine zones) that are too cold to allow enough time for two generations of *Daphnia* (one generation of females out of the resting eggs, which produce males and females, which in turn produce new resting eggs). Pseudosexual eggs can be produced by the females that emerge from resting eggs.

"Asexual, immediate reproduction" is achieved via subitaneous eggs, which have a visible yolk center and are often pigmented yellow, blue, or green. Subitaneous eggs develop into neonates inside the mother's brood chamber. Egg developmental stages can be followed easily because the carapace is transparent (Threlkeld, 1979; Kotov and Boikova, 1998). The neonates are released from the brood chamber as the mother molts. After molting, she may extrude another set of eggs into the brood chamber, depending on her age, nutritional status, and the environmental conditions.

Except for *Eurycercus* and two other genera, the chydorids have a constant clutch size of two; whereas in other cladocera, clutch size increases with body size, so that the smallest species produce one or two eggs per clutch while the largest species can produce hundreds of eggs per clutch (Hann, 1985). Small individuals carry only a few eggs (which are huge relative to the adult, as in Fig. 1D). Reproduction is very expensive for small cladocerans. The small *Daphnia lumholtzi* typically carry two eggs in the first reproductive instar. These eggs account for 30–80% of the female's weight (King and Greenwood, 1992). Large *Daphnia* individuals (such as *Daphnia magna*) also allocate a significant proportion of their energy intake to reproduction; but because they are larger and produce relatively smaller eggs (not much larger than those of the small species), they can carry dozens or even hundreds of subitaneous eggs. These eggs begin developing as soon as they are extruded into the brood chamber. Egg development time and longevity are proportional to the inverse of temperature to an exponent of about 2.5 (Rigler and Downing, 1984).

"Delayed reproduction" is often associated with sexual reproduction (Hebert, 1978, 1988). Males are diploid and produced asexually, just as their sisters are. Production of males is induced by some environmental signal, such as a change in food concentration, a chemical signal produced by crowded *Daphnia*, or changing (usually decreasing) photoperiod (Ferrari and Hebert, 1982; Hobaek and Larsson, 1990; Kleiven *et al.*, 1992, see Fig. 4). When a population begins reproducing sexually, females produce mixed clutches of males and females. The male to female ratio varies greatly between populations and in different years. Highest sex ratios are typically seen in small ponds, and there is some concern that environmental contaminants may have reduced male production (Dodson and Hanazato, 1995).

Sexual reproduction begins when one or two haploid eggs are extruded into a modified (thickened) carapace (except in the Sididae and the predaceous species) called the ephippium (in Greek "saddle", see Fig. 2C). The changes in carapace development are induced by either environmental (chemical) signals from crowding or toxic food (Ferrari and Hebert, 1982; Carvalho and Hughes, 1983; Shei *et al.*, 1988; Kleiven *et al.*, 1992) or by environmental contamination (Van Der Hoeven, 1990; Shurin and Dodson, 1997). Ephippia are produced most often in temporary ponds, especially in harsh environments (Lynch, 1983; Weider and Hebert, 1987).

Upon fertilization of the haploid egg by the male (Jurine, 1820; Baird, 1850), the resulting zygote (or the diploid pseudosexual egg) undergoes several cell divisions (Banta, 1939; Ojima, 1958) and then enters diapause. These resting "eggs" (actually early embryos) are resistant to freezing and drying and are further protected by the thickened part of the carapace. The ephippium is darkly pigmented with melanin. The rough surface of the carapace may serve as a species-specific recognition cue for the male during mating. The ephippium is shed when the female molts. The ephippium closes around the resting eggs, and most of the rest of the exoskeleton detaches from the ephippium (except for the tail spine in *Daphnia*).

In some populations, resting eggs are not the result of sexual reproduction but are produced asexually (Hebert, 1987; Weider and Hebert, 1987; Innes and Hebert, 1988). These "pseudosexual" resting eggs are not produced by sexual reproduction, but are the result of a gene that suppresses meiosis. They act just like the sexually derived resting eggs, but are genetically identical to their mother.

Chydorid and macrothricid ephippia sink and are usually attached to the substrate (Bretschko, 1969; Fryer, 1972). *Daphnia* ephippia tend to float, because of their specific gravity and the hydrophobic character of the carapace. Dark pigmentation of ephippia may aid dispersal, since fish, amphibian, and bird predators can easily see the dark ephippium and the ephippium is not damaged by passage through a predator's gut (Mellors, 1975). The tail spine and, sometimes, a strip of chitin ventrally at the head end of the carapace (Fryer, 1972) may aid dispersal by allowing the ephippium to stick to fur or feathers of wading animals. Ephippia are small enough to blow about with the wind and resistant enough to remain alive on dry land for years (Moghraby, 1977).

"Egg banks" are composed of resting eggs stored in the sediments or watershed. The resting eggs may be years or decades old, and they appear to hatch leisurely over several years. Egg banks provide an important large-scale stabilizing influence on the genetic structure of a cladoceran population and a hedge against extirpation (Hairston, 1996). To continue development, the diapausing embryos in ephippia require stimuli such as water, long or increasing day lengths, and high oxygen concentrations (Schwartz and Hebert, 1987a). The resting eggs usually develop into females which produce more young asexually. Major exceptions are *Daphniopsis* and *Moina*, genera which specialize on short-lived temporary ponds and which produce both males and females from the ephippial eggs (Goulden, 1968; Schwartz and Hebert, 1987b).

Molting allows cladocerans to increase in size. As the embryos develop, they assume the appearance of the adult. After release from the brood chamber, the newborn (neonates) can grow only by molting. That is, when sufficient energy has been stored, the animals: (a) reabsorb the inner part of their exoskeleton; (b) begin to manufacture a new exoskeleton below the old; (c) absorb water; and (d) pull themselves out of the remaining outer layers of the old exoskeleton, exiting at the back of the neck. The old exoskeleton (the exuvia) is a transparent ghost image that retains the complexities of the outer surface of the animal, including all the fine setae of the thoracic legs. Exuviae have been used to measure changes in size of various structures as an animal grows from neonate to adult. The energetic cost of molting may be a major constraint on body size in *Daphnia* (Lynch, 1989). Given a limited rate of food intake, cladocerans may be faced with the choice of allocating energy to molting (to become larger) or allocating energy to reproduction.

The developmental stage between molts is called an instar. There is some variation in (and controversy about) how many times cladocerans molt. Neonates molt just twice in chydorids but up to seven times in daphnids before producing eggs of their own. The number of juvenile instars is more or less species-specific and depends on adult body size, with the smaller species having the smallest number of molts. The rate of molting (instar duration) in *Daphnia pulex* depends mostly on the temperature, less on food (Lynch, 1989). If animals are poorly fed, they may even lose weight when they molt. An instar lasts from about a day to weeks, depending on the temperature. In laboratory studies, cladocerans can live for months at room temperature.

Each time a well-fed juvenile molts, it approximately doubles its volume, and its length increases roughly by a factor of 1.26. Growth slows as cladocerans mature; successive adult instars grow slightly, but energy not used for maintenance is allocated to produce eggs instead of growth (Frey and Hann, 1985). Female cladocerans can molt several times after becoming adults. Chydorid males become mature in the third instar and stop molting, although *Daphnia* males continue to molt and grow after maturity.

4. Induced Changes in Life History

Many cladoceran life-history characteristics include instar duration, number of instars to first reproduction, time to first reproduction, primiparous and neonate size, and clutch size. These characters have a genetic basis (Spitze, 1995) and show trade offs and correlations that are suggestive of adaptations to the environment and to biological interactions, especially predation.

Life-history characteristics also show phenotypic plasticity (Hann, 1984; Schwartz, 1984; Larsson and Dodson, 1993). That is, the same genotype can develop differently, as it is influenced by temperature, food supply, and chemical signals produced by predators. The patterns of phenotypic response (reaction norms) appear to be, in general, adaptive responses to size-selective predation and food limitation.

B. Metabolism

The many physiological studies of *Daphnia* provide a reasonably complete picture of cladoceran metabolic rates (e.g., Peters, 1987). Metabolic rate is measured as the rate at which an animal uses oxygen or produces carbon dioxide. As long as oxygen is present, gas exchange (oxygen in and carbon dioxide out) is not a problem for small animals, because of their high surface-to-volume ratio. Gas exchange probably takes place across the entire surface of the animal, not just on the gill like thoracic legs. Porter *et al.* (1982) and Peters (1987) found no correlation between rate of leg movements and respiration rate. Animals collected in oxygen-poor habitats (sewage lagoons or temporary pools with decaying vegetation) will often have a general pink-to-red color (Weider and Lampert, 1985). This is due to hemoglobin dissolved in the blood. If a red animal is wounded, it will bleed red blood; and if the blood is examined at high magnification, the blood cells will appear less pigmented than the surrounding fluid. Animals become clear in a day or so if grown in well-oxygenated water, suggesting an energetic cost to maintenance of hemoglobin (Landon and Stasiak, 1983). Metabolic rate is influenced by a number of factors, including temperature, body size, and food concentration.

1. Temperature

Metabolic rate has a hump-shaped (inverted-U) relationship with temperature. At some low threshold temperature, metabolic rate is just fast enough to maintain growth and reproduction. A rise in temperature increases metabolic rate up to a maximum. Above the temperature of maximum metabolic rate, further increases cause death, either because critical metabolic enzymes are thermally damaged or because respiration rate exceeds the rate of energy intake (Moore *et al.*, 1996). Most cladocerans are unable to live at temperatures above 30°C, and animals native to cold waters may have much lower upper thermal limits. This sensitivity to high temperature can be increased if the cladocerans are feeding on toxic bluegreen algae (Threlkeld, 1986a). Both *Ceriodaphnia* and *Diaphanosoma* live naturally in water of 27–30°C. When bluegreen algae are present, *Ceriodaphnia* shows a decline in population growth rate. For short periods of time (15 min), *Daphnia* can tolerate temperatures up to about 35–38°C, depending on the temperature at which they were raised (MacIsaac *et al.*, 1985).

2. Body Size

Small animals, cladocerans have relatively fast rates of metabolism. Richman (1958) found that respiration rate of *Daphnia* is proportional to body length raised to the 2.14th power, a value similar to size-specific respiration rates reported for many invertebrates (Schmidt-Nielsen, 1984). The range of cladoceran body size is enough to ensure significant differences in metabolic rates between the smallest and largest cladocerans. While larger animals require more oxygen and energy in an absolute sense, large animals use less energy and oxygen per unit of weight.

3. Food Concentration

As food becomes more concentrated, the animals respire at an increasing rate, probably because of increased metabolic needs: assimilation rate also increases with food concentration (Peters, 1984; Lampert, 1986a). When food is superabundant, the grooming required to clean the filters on the thoracic legs may also increase the respiration rate (Porter *et al.*, 1982). This phenomenon can lead to starvation at concentrations of food high enough so that the increased respiration requirements exceed the maximum rate at which energy can be ingested. In the absence of food, *Daphnia* continue to move their legs, as if the thin gill-like legs were also being used as respiratory organs, requiring ventilation.

4. Interaction of Temperature, Body Size, and Food Concentration

Environmental temperature may be a causal factor for body size. For example, Stirling and McQueen (1986) found that the direction of change in environmental temperature was correlated with body size and reproductive strategy in *Daphniopsis ephereralis*, an inhabitant of cool forest ponds. Annual changes in average body size for the population was not directly related to predators and may have a physiological explanation.

Starvation is a problem faced by most cladoceran populations sometime during each year (Threlkeld, 1976, 1988). Because smaller animals have a higher metabolic rate per unit body weight, neonates are the more prone to starvation than adults, if no food is present (Tessier and Goulden, 1987). Eggs can receive a large store of energy-rich fat from the mother, if the mother is well fed. However, this energy store is reduced if the adult is faced with limited food. Similarly, if food is absent, the large species can persist for longer than small species. Even for large adults, a single day without food results in a drop in energy stores, body weight, and respiration rate.

At food concentrations above starvation levels, the relationship between food availability, energy allocation, and body size are complex. How do cladocerans allocate the energy gained from limited food input? Models of energy allocation based on physiological principles suggest that energy is used for different purposes at different life stages (for example, only for maintenance and growth in immature animals), and that adults have allocation priorities for maintenance, reproduction, and growth (Bradley *et al.*, 1991).

Large animals always have lower weight-specific respiration rates. However, smaller animals have the higher weight-specific assimilation at low food concentrations, and large animals have higher rates at high food concentrations (Tessier and Goulden, 1987). This suggests that net growth, estimated by the difference between assimilation rate and respiration rate, is highest for small animals at low food concentrations and highest for large animals at high food concentrations. In a study of three *Daphnia* species (Tillmann and Lampert, 1984), the large *D. magna* out-reproduced smaller *Daphnia*. When food was marginally limiting, the three species had similar reproductive rates; and when food was severely limiting, only the smallest species, *D. longispina* could still reproduce. Enserink *et al.* (1996) also found that larger individuals appeared to grow fastest at constant low food levels, whereas smaller individuals had the advantage with fluctuating food.

When a cladoceran is food limited, it tends to mature at a smaller size and produce smaller (but as many) offspring. The main response of *Daphnia pulex* to low food levels was a reduction in size-specific food intake and egg size (Lynch, 1989). Over a wide range, food concentration had no effect on length-weight relationship, instar duration, or weight-specific investment of energy in reproduction.

Population growth rate is linked to body growth rate. Competition among various species of cladocerans may largely depend on which species can grow fastest at a given food level. Thus, small species such as *Ceriodaphnia* and *Bosmina* may be able to out-compete, or at least out-grow, large species such as *Daphnia pulex* at low but not at high food concentrations (Tessier and Goulden, 1987).

III. DISTRIBUTION

A. Zoogeography

Cladocerans are a widespread group occurring in all but the most extreme freshwater habitats (Hutchinson, 1967). While they are more abundant in lakes, ponds, and slow-moving streams and rivers (Thorp *et al.*, 1994), they also occur in quiet water and marginal vegetation in fast-moving streams. Several species probably occur in groundwater, especially river gravels (Dumont, 1987). One chydorid cladoceran has been found in water trapped in mosses and bromeliads that drape trees in the cloud forest of El Yunque, Puerto Rico (Frey, 1980b). One species or another are found from sea level to ponds above tree line, and from arctic to tropical latitudes. In North America, the highest diversity of planktonic species is found in the glaciated midtemperate zone (Tappa, 1965), and the highest diversity of benthic species, especially small-bodied species, is found in the southeastern United States (Crisman, 1980; Kerfoot and Lynch, 1987).

B. Cladocera and the Species Concept

Our understanding of the distribution of cladoceran species depends on how well we can distinguish species. Cladoceran taxonomy is currently undergoing a reanalysis. Because of an almost complete disregard of fine but significant details of morphology and a consequent impression that taxa on different continents were the same, many investigators in the past (Forbes, 1925) and present have claimed that many species are cosmopolitan. But cosmopolitanism, at least among the Chydoridae, now seems to be mainly an illusory concept (Frey, 1987a), based on an incomplete appreciation of the morphology of seemingly the same taxa in different places and on an unquestioning acceptance of the efficiency of passive dispersal of these organisms via their resting eggs as carried by wind, water, birds, insects, and mammals (Frey, 1982a). Recent studies in North America (Frey, 1986a) are demonstrating that the geographical ranges of species of chydorids can be just as narrow and circumscribed as those of animals considered to be much less vagile. Careful genetic analysis (Hebert, 1987) has shown *Daphnia* species to be more diverse and to have distributions restricted to small portions of North America. In fact, those *Daphnia* species that are obligate parthenogens are genetically very diverse (Hebert, 1987), nearly as diverse as asexual ostracods (Havel and Hebert, 1989; Chaplin, 1994). Thus, the cladocera of different continents are probably different species (even though many now have the same specific names), and the distributions of species within a continent depend on the distribution of favorable environmental conditions.

Hybrids between cladoceran species are often reported (Hebert, 1985; Hann and Hebert, 1986; Hebert, 1987; Mort and Streit, 1992; Taylor and Hebert, 1993; Spaak, 1995; Colbourne *et al.*, 1997). The evidence has until recently been only morphological, or based on lab crosses. For example, *Pleuroxus denticulatus* and *P. procurvus* were crossed in the laboratory to produce one individual which could only reproduce asexually and was morphologically intermediate between its two parents (Shan and Frey, 1983). Recent studies based on molecular characters confirm that hybrids can indeed be formed between similar species. The persistence of clones in nature is due to their ability to reproduce asexually. Species remain distinct, in part, because the hybrids are not known to be able to reproduce sexually in nature. Thus, while cladoceran species are often difficult to distinguish using morphological characters, molecular characters show them to be clearly distinct.

C. Biodiversity

One measure of biodiversity is the number of species, also called "species richness." Cladoceran species richness in a lake depends on several factors, including water chemistry, lake size, productivity, the number of adjacent lakes, and biological interactions (Dodson *et al.*, 1999).

1. Water Chemistry

Carter *et al.* (1980) found that of 23 pelagic cladoceran species occurring in eastern Canada, eight may have distributions restricted by water chemistry: *Ceriodaphnia* and a species of *Eubosmina* tended to occur in hard water, and *Polyphemus*, *Sida*, two *Daphnia* species, and two other *Eubosmina* species were found mostly in soft water.

Salinity is a major factor that limits cladoceran distribution. Cladocerans as a group have a wide range of tolerance to salinity (Potts and Durning, 1980), but individual species are adapted to a relatively narrow range of salinity (Teschner, 1995). Particular species can be found from near-distilled water (Dodson, 1982) to the 3.2% salinity of oceans (Della Croce and Angelino, 1987), to water more saline than sea water. For example, *Moina* lived in Soap Lake Washington at salinities as high as 39 g/L (Edmondson, 1963). High salinity excludes most cladocerans while providing a refuge for physiologically specialized genera such as some *Moina*. The cladocerans of Pyramid Lake, Nevada, showed strong mortalities at salinities of 0.6% (*Diaphanosoma*), 0.7% (*Ceriodaphnia*), and 1.8% (*Moina*) (Galat and Robinson, 1983). The marine cladocerans *Evadne* and *Penilia* are restricted to coastal and open waters, while *Podon* is found in less saline coastal and estuarine water (Della Croce and Angelino, 1987).

Even within a species, there is evidence of adaptation to average salt conditions. For example, different clones of *Daphnia magna*, from fresh or brackish ponds, show differences in their level of salt tolerance and do poorly when raised at different salinity levels than their native pond (Weider and Hebert, 1987; Teschner, 1995).

Acidity also affects cladoceran distribution. Cladocerans are mostly found in neutral or alkaline water, although a few live in acidified lakes. Only a few small cladocerans such as *Bosmina*, *Chydorus*, *Diaphanosoma*, small *Daphnia*, *Holopedium*, and *Polyphemus* can tolerate extremely acidic water, occurring in lakes with pH values between 5.0 and 3.8 (Sprules, 1975; Frey, 1982c).

Turbidity affects cladoceran distribution, probably because clay particles interfere with cladoceran filters (Gerritsen and Porter, 1982). A combination of laboratory life table experiments and field observations suggested that *Moina* and *Diaphanosoma* are adapted to silt-laden water, while *Ceriodaphnia* and *Daphnia* (subgenus *Daphnia*) are not (Threlkeld, 1986b). Some species of *Daphnia* (*Ctenodaphnia*) (Dodson, 1985) and several chydorids (such as *Graptoleberis*) are found in very turbid water or in mud.

2. Lake Size, Productivity, and Adjacent Lakes

For pelagic cladoceran species, there are more species in larger lakes, with about 35 species in the largest freshwater lakes. Small ponds and pools (as small as a few liters) may have one or two species. As lake size increases, there are probably more distinct habitats for different species. The highest species richness occurs in mesotrophic lakes with average annual primary productivities of about 100 g C/m^2 per year. The mechanism by which productivity affects cladoceran species richness is still unclear. Lakes surrounded by lakes have a higher species richness than isolated lakes (such as Crater Lake in Oregon). Clustered lakes probably act as sources for each other for immigrants (usually via resting eggs) among lakes.

Whether a species occurs in a body of water depends partly on the: (a) zoogeographic region, that is on the probability of getting into the water; (b) chemical and physical requirements of the species; (c) food conditions within the water; and (d) predators living in the water. Although it is believed that some cladocerans are capable of long distance dispersal via resting eggs carried by the wind and by feathers and guts of water fowl, we still know little about actual rates of colonization. Inbreeding coefficients for small *Daphniopsis* populations suggest that populations receive an average of 0.3 migrants per generation (Schwartz and Hebert, 1987b). Studies of *Daphnia* indicate slightly higher rates of migration of about 0.5 migrants per generation. However, studies of several arctic cladocerans suggested that dispersal rates are quite low, and that pond populations, even when near each other, retain genetic differences (Boileau *et al.*, 1992).

3. Biological Interactions

Cladocerans compete for food with other cladocerans and with other herbivores (Threlkeld, 1988). In some cases, this competition is intense enough to result in marked reduction in abundance of other cladocerans and potentially to result in competitive exclusion. Vanni (1986) found that *Daphnia* could reduce algae concentrations to a level that caused *Bosmina* to decrease. He suggested competition for food is a cause of the often observed scarcity of small species in lakes inhabited by large cladocerans such as *Daphnia*. This effect may be especially strong in nutrient-rich lakes, where *Daphnia* populations can increase rapidly enough to produce a sudden and major reduction in algal abundance.

Size selective predation sometimes affects the distribution of cladocerans (Zaret, 1980; Kerfoot and Sih, 1987). In general, large-bodied species tend not to coexist with fish and small-bodied species can be excluded by small predators such as the larvae of the midge *Chaoborus* or the backswimmer *Notonecta*. Thorp (1986) analyzed 18 reports of predator–prey community experiments in freshwater zooplankton. He found that adding predators often reduced the number of individuals of cladocerans present, but that there are few studies showing an effect on the number or relative abundance of prey species. Over the long term, predators may influence cladoceran diversity via coevolution, size-selective predation, and restrictions on prey distributions.

While food limitation and selective predation are components determining the distribution of zooplankton, it is clear that the two factors are intertwined, that predators influence competition and *vice versa* via what are coming to be called "indirect effects" (Cooper and Smith, 1982; Lampert, 1987; Kerfoot and Sih, 1987; and Carpenter, 1988). Fryer (1985) concluded that diversity of chydorids within a lake depends only slightly on dispersal, competition or predation, and more on microhabitat preferences and presence of the appropriate specific food.

4. Littoral and Meiobenthic Species

We know much more about pelagic species richness and distribution than we know about the littoral (or meiobenthic) species, such as the chydorid, macrothricid, and sidid cladocerans. The meiobenthic cladocerans comprise a major part of the diverse littoral community of lakes, which includes such additional metazoans as cyclopoid and harpacticoid copepods, occasionally calanoids as well, ostracods, amphipods, rhizopods, hydra, bryozoans, rotifers, mites, nematodes, oligochaetes, and larval or nymphal stages of midges, corixids, notonectids, beetles, caddisflies, mayflies, damselflies, as well as some other groups. Besides this high diversity of animals, there are the various macrophytes, sessile and substrate-associated algae, and bacteria. Of all the groups of animals, the cladocerans typically are most abundant, at times reaching densities greater than 1 million per square meter of bottom, even under the ice in early winter. The composition of this littoral community and its importance relative to the offshore planktonic and benthic associations varies with the geographic location of the lake and its general characteristics of size, depth, and transparency.

There are typically about 8–15 species of meiobenthic cladocerans present in a given habitat, sometimes five or fewer and occasionally 25–30 or even more. As with pelagic cladoceran species, the species richness of littoral cladoceran species appears to be correlated with lake size. For example, in a study of 207 water bodies in Yorkshire, England, from small ponds (<5000 m^2) to small lakes ($>15,000$ m^2), Fryer (1985) found more littoral cladoceran species in the larger lakes. Small ponds had a median of about eight species; the larger lakes had about 30 species.

Some littoral species are associated only with the sediments and others with macrophytes. Some species occur only in acid, softwater habitats, others in quite strongly alkaline ones. Fryer (1968) showed that the species he studied are diversified as to food eaten and how this is obtained. One species of chydorid is a scavenger, another scrapes food from the bottom as a snail does, and the third an ectoparasite on hydra. Most of the remaining species use the same common food supply on substrates—organic detritus with its contained algae and bacteria. The structure of the five pairs of trunk limbs is remarkably uniform among all these species, suggesting little variation in method of getting the food. Most specializations within chydorids seem linked to obtaining food from slightly different microhabitats.

Meiobenthic cladocerans occur virtually wherever water is present, from water bodies of "normal" size down to marshes, swamps, pools, puddles, ditches, seasonal water bodies, groundwater, running water of all sizes and nearly all velocities, moist sphagnum, and even in overgrowths and cushions of leafy hepatics in rain and cloud forests, up to several meters above the forest floor. Most species are sensitive to salinity, although a few occur at concentrations up to 5–10% (g/L), sometimes even higher.

5. Conservation Issues

Unlike the large branchiopods (see below), conservation issues do not exist for cladocerans. Rare species probably exist, as do opportunities to conserve cladoceran biodiversity. However, rare cladoceran species are probably morphologically similar (or identical) to common closely related (sibling) species. Only a combination of morphological and molecular analysis is likely to identify rare cladoceran species (for an example of a fruitful protocol, see DeMelo and Hebert, 1994, and Hebert and Finston, 1996).

As in other areas of conservation, the invasion of exotic species is a concern for North American communities. For example, *Daphnia lumholtzi* was probably introduced into North America from Africa, along with Nile perch. It is now a major component of lake and reservoir plankton communities in southern North America (Havel, 1995) and occurs in some rivers, including the Ohio (Jack and Thorp, 1995). Besides being introduced along with fish, cladocerans invasions of new continents are either mysterious (Benzie and Hodges, 1996) or have been connected to heavy (military) equipment (Schrimpf and Steinberg, 1982).

IV. BEHAVIOR

Cladocerans display a number of behaviors relating to mating, escape from predators, feeding, and movements within lakes.

A. Mating Behavior

Jurine (1820) provides us with one of the few descriptions of mating behavior in cladocera. In November 1797, in the region of Geneva, he found a pond containing a dense population of *Daphnia*, which included ardent males. He observed numerous couplings in the laboratory, and reported (translated from the original French):

> "The male jumps onto the back of the female, which sometimes escapes, but if he is able to seize her with the long filaments of his first legs and secure her with his elongated first antennae, he is able to catch hold of her firmly. He soon scuttles quickly over the surface of her carapace until he has attained the lower margin. When he finds the position so that the openings of the two carapaces are opposed, he quickly introduces into her carapace the long setae of the first antennae and of the first thoracic legs, and thereby envelopes and binds, so to speak, the legs of the female. When he is established in this position, he curves his postabdomen forward, putting it out far enough to get that of the female. When she feels that part, she is greatly agitated, and fleeing, carries the male so fast it is hard to follow the amorous couple in the vase that contains them. Finally, this agitation ceases and the female in turn advances her postabdomen in order to touch that of the male. The postabdomens hardly touch before they separate. At the instant of this touch, the male is agitated by spasms which give his legs remarkable vibrations. It is during this contact that, in my opinion, the copulation occurs.
>
> The embrace is of variable duration; it rarely is sustained for more than eight or ten minutes. During this interval, the postabdomens were brought together more than once. Copulation terminated, the postabdomen of the male is slowly withdrawn into his carapace, but he is not yet able to withdraw the setae of his first antennae and first thoracic legs because of the spasms that persist in these parts. When these cease, the separation of the two animals takes place.
>
> This is not the only mode of embrace in these animals. I have several times seen the male introduce only one first antenna, one first leg seta, and one swimming (second) antenna with which he envelopes the setae of the females second and third legs. When the copulation is finished, he is not able to retrieve his swimming antenna before the spasms which he is experiencing are entirely gone.
>
> I noticed other animals that were embraced almost transversely, in such a way as to form a sort of cross.
>
> Sometimes other males arrived intending to unite with a female already in an embrace. One then sees them explore the female's body with a remarkable vivacity, trying, although futilely, to penetrate into the sanctuary of pleasure, by elongating their postabdomen as much as they can into the opening of the carapace.
>
> In all cases where copulation had taken place, the postabdomen of the male returned slowly and with shudders to its usual position, frequently interrupted by convulsive movements which stretched it out to re-approach the female, such that it took a minute or two for the postabdomen to be completely at rest again.
>
> Males attacked indiscriminately all females they encountered. Despite that, I assume that true copulation took place only with those females that had no eggs in the brood chamber, for with these the act was reiterated and prolonged. For the others, although the preliminaries were the same, they terminated very quickly. Ephippial females pushed aside the males and rendered their attacks in vain."

About 60 years later, Weismann (1880) gave cursory descriptions of mating behavior for *Moina*, *Daphnia*, and *Bythotrephes*, and conjectures on possible mating behavior, based on anatomy, for *Daphnella*, *Latona*, *Sida*, *Holopedium*, *Ceriodaphnia*, *Scapholeberis*, *Simocephalus*, *Bosmina*, *Pasithea*, *Macrothrix*, *Eurycercus*, *Pericantha* (now *Pleuroxus*) *truncata*, *Chydorus*, *Podon*, *Evadne*, and *Leptodora*. These mid-Victorian descriptions lack the enthusiastic attention to behavioral details shown by the pre-Victorian Jurine (1820), but do include a wealth of detail concerning cellular and tissue structure and ontogeny

> "Mating in *Daphnia* has been observed by many, although only the external association of the genders: the attachment of the male and the combined swimming about. Indeed, Jurine and after him many others noted that for *Daphnia pulex*, the ardent (begattungslustigen) males attacked nearly every female, whether they are carrying ephippia or embryos. This is indeed correct as far as it goes, but they always quickly let them go again and find a true mate—as far as I can judge—only succeeding with a female that carries a winter (haploid) egg in the oviduct. Sometimes the males are not so ardent, and wait for the correct moment. Thus, on the second of September 1875, I brought several *Daphnia pulex* males together with a female which carried ripe embryos in the brood chamber and a newly ripe haploid egg in each ovary. The males made no advances during the entire morning. First about 4:30 pm, after the female had delivered her young, a male attached himself on the back of the carapace, and quickly within a few seconds, hitched himself around in front. The female did not defend herself, but swam peacefully and slowly about, also sometimes lying still on the bottom. The male did not let go for more than a quarter of an hour. Sadly, I had to interrupt the observations, and when I was able to take them up an hour later, the mating was already consummated, the male had departed, and in the brood chamber, which had not yet developed an ephippium, lay two fertilized winter eggs . . . ." Weismann (1880) further argued that successful fertilization can occur in *Daphnia* only with the partners face to face and parallel.

The veil of modesty is further drawn over mating behavior by monographers of functional morphology, species, and evolution who largely ignored the topic of mating behavior (e.g., Banta, 1939; Brooks, 1957; Fryer, 1968; Goulden, 1968; Fryer, 1974). However, perhaps influenced by the renaissance of natural history in the last couple of decades and the general freeing-up of sexual attitudes of the late sixties, Shan (1969) and Smirnov (1971) have produced much more informative descriptions of chydorid mating behavior and functional morphology. The theme of female participation, and even perhaps female choice, introduced by Jurine (1820), is developed further by Shan (1969):

> "My observations on laboratory stocks of *P. denticulatus* do not agree with Weismann's (1880) description. At the

beginning, the male swims vigorously here and there, using his first antennae, trying to find a female that is receptive, usually an ephippial female. A receptive female often cooperates in the copulation process. Sometimes, the ephippial female he approaches may not be receptive and she may reject him using her postabdomen as he approaches her. At times a male may hook onto a parthenogenetic female and try to copulate with her, or a male may hook onto another male. In one occasion, I saw a parthenogenetic female that had three males attached in a series, one after another!

As the male hooks onto the edge of the female's carapace, he moves vigorously back and forth along the edge, often only on one side of the female. The contact points are the two copulatory hooks [on the first thoracic leg] and the tip of his rostrum. He jerks his postabdomen and moves his hooks vigorously while the female does not react to him in any way. This activity goes on for a while, usually a few minutes, possibly even longer than ten minutes. The female may become excited when the male reaches a position where his hooks are at the top of the free posterior edge of the female's carapace. In this position the male and the female are facing the same direction and have the same dorsal orientation. At this time the female may stop feeding, carry the male with her, and swim away. They swim together, both jerk their postabdomens and strike their antennae. They swim for some time, possibly ten minutes or longer. Finally, they fall onto the bottom, still hooked together, and lie on one side. Both of them stretch and jerk their postabdomens back and forth. At this time the female's postabdomen is in maximum extension downward and backward, while the male is flexing his forward trying to get as close as possible to the brood pouch of the female. Presumably at this time the sperm are ejected. The mechanism that allows the sperm to get into the brood pouch and fertilize the egg is unclear. It is quite possible that the jerks of the female's postabdomen create an inward current that helps bring the sperm into the brood pouch."

It is possible that the elongated first antennae of male daphniids and moinids are used to feel the roughened carapace of a female (Goulden, 1968; Kokkinn and Williams, 1987). The female's carapace ornamentation might tell the male whether she is the correct species and that she is producing haploid eggs. As with all aspects of behavior in *Daphnia*, we know few details of the individual behaviors that result in mating. Instead, cladoceran behavior has been studied mostly at the population level, using plankton nets in lakes.

B. Individual Swimming Behavior

Video and computer analysis techniques allow the analysis of three-dimensional swimming of individual pelagic cladocerans, especially *Daphnia* to measure swimming speeds, directions of movement (Dodson and Ramcharan, 1991). The characteristic hopping motion has not yet been satisfactorily characterized. Average swimming speed varies from just enough to maintain position (all pelagic cladocerans sink if quiescent), to sustained average speeds as high as about 25 mm/s (Dodson *et al.*, 1997b). Sinking rate is affected very little by temperature, less so than predicted by changes in water viscosity, suggesting *Daphnia* have behaviors to maintain constant sinking rate (Gorski and Dodson, 1996). Different clones swim differently, with the fastest swimming characteristic of clones from low-predation habitats. This suggests there is a benefit to fast swimming (perhaps increases filtering rate) that is sacrificed in communities with high predation rates (where encounter rates can be minimized by slow swimming). Fish predators pay attention to *Daphnia* swimming behavior, preferentially taking faster swimming individuals of real prey (O'Keefe *et al.*, 1998) or virtual (computer-simulated) prey (Brewer and Coughlin, 1996).

Predaceous cladocerans swim slowly through the water. The slow movement is probably a compromise between swimming fast enough to run into prey and slow enough to avoid attracting the attention of fish. Browman *et al.* (1989) describe the swimming movement of *Leptodora* (Fig. 60):

> "*Leptodora* swims randomly [at about 13 mm per sec] through the water column with all five pairs of thoracic appendages spread to form a 'feeding basket' and, seemingly by chance, encounters prey. Shortly after prey make contact with any part of *Leptodora*'s body, (usually ventral), the abdomen is rapidly pulled forward, clamping itself under the feeding basket so that the telson closes it at the posterior end. The duration of this movement is always the same and we conclude that it is an indiscriminate reflex. If the prey is encountered anywhere but a short distance directly in front and slightly below the *Leptodora*, it is not captured."

C. Diel Vertical Migration

One of the earliest observations of cladoceran behavior at the population level is Cuvier's, 1817, description of *Daphnia* vertical migration (translated by Bayly, 1986):

> "In the morning and evening, and even during the day when the sky is overcast, the daphnids usually stay at the surface. But during very hot weather, and when the sun beats fiercely down on the pools of stagnant water where they live, they sink down in the water, and stay at a depth of six or eight feet or more; often not a single one can be seen at the surface."

Cladoceran populations have been observed to migrate, on a daily cycle in both fresh and marine waters (Hutchinson, 1967; Bayly, 1986; Dodson, 1990). Distances migrated by pelagic species vary from less than a meter in a small rock pool to hundreds of meters in oceans. In freshwaters, the usual distance is 5–20 m. Because lake waters are vertically stratified, vertical migration takes the animals into very different habitats in terms of temperature, light, food, turbulence, oxygen and salts, and predators (Landon and Stasiak, 1983). The greatest depth of the migration is often set by the bottom of the lake, by the depth to which light penetrates, or by the limit of sufficient oxygen (about 1.0–0.5 mγ/L; Weider and Lampert, 1985; Nebeker *et al.*, 1992). Littoral organisms also have a pronounced diel up-and-down migration. This behavior allows >90% of the population below a sample area to be trapped over night in inverted funnels (Whiteside and Williams, 1975).

The daily pattern often seems to have the greatest amplitude during the summer and fall months. The larger developmental stages typically migrate the farthest. Populations usually migrate downwards at about dawn and do not return until after dusk.

Physiological response to light intensity and duration is a component of vertical migration behavior (e.g., Ringelberg, 1987). Directionality of light source is important for *Daphnia* to be able to assume normal swimming position; if the light comes equally from every direction, the animal is unable to swim normally. Cladocerans can probably orient in the water using polarized light (Ringelberg, 1987). There is also some evidence that cladoceran swimming speed and direction is influenced differently by red and blue light.

Many ecologists have concluded that, while response to light is an important element in vertical migration, the most important factor is escape from predators (Lampert, 1993; DeMeester *et al.*, 1998; Tollrian and Harvell, 1998). Cladocerans move downward during the day, into the darkness of deep water, where they are relatively safe from visual predators such as fish. However, deep water also tends to be cold. In order to grow and reproduce as fast as possible, the cladocerans return to warmer water near the surface at night. This story has been elegantly demonstrated by a series of experiments in the large plankton towers at the Max Planck Institut für Limnologie in Plön, Germany (Loose, 1993). When exposed to a combination of light and feeding predators, or even the water from cultures of predators, *Daphnia* rapidly swim downward. They accomplish the vertical migration not by sinking or gentle swimming, but by rapid directional swimming (Dodson *et al.*, 1997a). Because the prey respond to predator water, the behavior is clearly in response to a chemical signal released by the predator.

D. Swarming Behavior

Cladocerans sometimes form swarms. Baird (1850) reported:

> "I have, however, frequently seen large patches of water in different ponds assume a ruddy hue, like the red rust of iron, or as if blood had been mixed with it, and ascertained the cause to be an immense number of the *D. pulex*. The myriads necessary to produce this effect is really astonishing, and it is extremely interesting to watch their motions. On a sunshiny day, in a large pond, a streak of red, a foot broad, and ten or twelve yards in length, will suddenly appear in a particular spot, and this belt may be seen rapidly changing its position, and in a very short time wheel completely round the pond. Should the mass come near enough the edge to allow the shadow of the observer to fall upon them, or should a dark cloud suddenly obscure the sum, the whole body immediately disappear, rising to the surface again when they have reached beyond the shadow, or as soon as the cloud has passed over."

Swarms and movements similar to those described by Baird (1850), if not as colorful, have been reported for other planktonic and littoral cladocerans (Weismann, 1880). Tessier (1983) described swarms of *Holopedium*, and reviews reports of other cladocerans, including *Ceriodaphnia*, *Bosmina*, *Daphnia*, *Moina*, and *Sida*. *Daphnia* and *Polyphemus* sometimes form small swarms of a few hundred animals. Although earlier studies with *Ceriodaphnia*, *Daphnia*, and *Moina* suggested that females in swarms tended to produce ephippial eggs, Crease and Hebert (1983) failed to find any evidence for sexual pheromones produced by either sex of swarming *Daphnia*. Butorina (1986) observed *Polyphemus* swarms consisting of a few hundred animals in mushroom-shaped swarms about 50 cm deep, 6–30 cm in diameter, and with most of the animals in the top 5 cm of the swarm at the surface of the water (Butorina, 1986). The swarms were simple aggregations, with no difference in age structure from the surrounding nonswarming *Polyphemus*. The swarms are dynamic, with animals arriving and leaving as long as the sun shines. In addition to light, fish smell is also an environmental signal that appears to induce swarming behavior (Jensen *et al.*, 1998). Compared to controls, *Daphnia* in fish water swim more uniformly.

Ilyocryptus (Fig. 44) has a reputation for nesting (Williams, 1978). They are often found in dense aggregations, which have been interpreted as the result of young staying in the vicinity of their parent.

E. Escape Behavior

Cladocerans show various escape behavioral responses to predators (reviewed in Brewer *et al.*, 1998). At least some of the planktonic species have a fast-swimming response which takes them away from predators. This is effective against nonvisual predators such as *Leptodora* (Browman *et al.*, 1989), but may be useless against fish. Diel vertical migration or small body size are the most effective defenses against large predators, such as fish, that use vision to locate prey. When caught by a small predator, such as a copepod or insect larva, cladocerans struggle to break free. *Bosmina*, once it escapes a nonvisual predator, closes up its carapace and sinks without making detectable hydrodynamic pressure waves (noise) (Kerfoot *et al.*, 1980). Also, *Bosmina*, like chydorine chydorids, tucks the second antennae under the carapace for further protection when captured.

Brewer *et al.* (1998) measured the three-dimensional escape response of *Daphnia*. *Daphnia* of a given size always escaped at the same speed, regardless of origin of the clone, but individuals from lakes with fish tended to start escaping much sooner than those from ponds lacking fish.

V. WHAT CLADOCERANS EAT AND WHAT EATS CLADOCERANS

Planktonic cladocerans, such as *Daphnia*, are of great importance to the ecology of lakes, affecting the amount of algae, via grazing and nutrient regeneration, and providing food for young-of-the-year fish (Carpenter, 1987; Kitchell, 1992; Sommer, 1989; Lampert and Sommer, 1997).

A. Cladoceran Feeding

Cladocerans are discriminating in their choice of food (Porter, 1977). There are two major aspects of foraging relationships: what cladocerans eat, and how they get it. Techniques recently developed allow quantitative measurements of the various aspects of the feeding process: food collection, ingestion, assimilation, and defecation (Peters, 1984).

The so-called "herbivorous" cladocerans eat a variety of small particles, from bacteria less than a micrometer long (Porter *et al.*, 1983), through algae, up to ciliates, small rotifers, and copepod nauplii as large as 100 μm (Porter, 1973; Burns and Gilbert, 1986; Dodson, 1975; Pace *et al.*, 1994). The most important component of their diet is made up of small algae in the range of 1 to 25 μm. Algae larger than about 50 μm or algae with spines or in colonies are usually rejected. Also, cyanobacteria in the preferred size range are often toxic and are not eaten if abundant (Porter and Orcutt, 1980). The size of food particles is determined by more factors than the body size of the herbivore. For example, *Bosmina*, a small filter feeder, specializes on larger particles than do larger *Daphnia* (DeMott and Kerfoot, 1982).

Chydorids typically feed by crawling along surfaces or through mud and scraping up or filtering food (Fryer, 1968; Smirnov, 1971). Many cladocerans use the carapace to help filter feed or as a pair of cheeks which confine the feeding current. However, some chydorids also employ the carapace as a suction cup. Negative pressure can be created inside the carapace if water is pumped out, as long as there is a good seal between the carapace and substrate. Second antennae, which are used for swimming by pelagic cladocerans, are used for crawling and jumping in some chydorids. Thus, chydorids have a number of specialized feeding and related locomotion adaptions. For example, *Eurycercus*, the largest chydorid, climbs on upper surfaces of plants and is able to filter feed. *Alonopsis elongata* balances upright on its carapace margins on solid surfaces and scrabbles along using its second antennae, scraping up living and dead organic material with its thoracic legs. *Peracantha truncata* keeps its second antennae inside the carapace when it crawls and can crawl upside-down if the surface is rough enough to provide a hold for the claws on the first leg, aided by weak negative pressure. *Alonella exigua* can hang upside-down on aquatic vegetation solely by means of negative pressure. *Graptoleberis testudinaria* has the most effective pressure chamber and exhibits snail-like behavior. *Leydigia leydigi* has neither pressure chamber nor scraping spines on the thoracic legs. It pushes itself into soft ooze using its first antennae and large spiny postabdomen. Food is sieved from the mud using setae along the margin of the carapace and on the shortened thoracic legs. In *Leydigia* the respiratory current (which passes over the outside of the first two thoracic legs) is separated from the feeding current. While most chydorids are herbivorous, the scraping method of gathering food lends itself to carnivory, and *Anchistropus* is a predator on hydra, while *Pseudochydorus globosus* is a scavenger of dead animals.

Leptodora, *Polyphemus*, and *Bythotrephes* are predaceous cladocerans, eating protozoa, rotifers, planktonic stages of small chironomid larvae, and small crustaceans. *Leptodora* takes prey between about 0.5 and 1.5 mm and seems to prefer other cladocerans (Karabin, 1974; Browman *et al.*, 1989). The smaller polyphemids take smaller prey and prefer small cladocerans, rotifers and protozoans (Monakov, 1972; Lehman, 1987). Algae is a minor component of the diet of *Polyphemus* and *Bythotrephes*. Presumably, the metanauplii of *Leptodora* eat algae. Adult *Leptodora* and Onychopoda have a reputation of being vampire-like "fluid feeders" (Cummins, 1969; Karabin 1974). *Leptodora* (and other predaceous cladocerans) need actual contact with prey before they attempt to capture it with the outspread thoracic legs (Browman *et al.*, 1989). Once the prey is grasped, the *Leptodora* bites it with the mandibles, sucking up body fluids and perhaps small bits of tissue. Small prey, such as rotifers and protozoans are probably chewed and then swallowed. Edmondson and Litt (1987) report large numbers of whole rotifers (*Conochilus*) and even whole cyclopod copepods in the gut of *Leptodora*.

B. Predators on Cladocerans

Cladocerans are eaten by any freshwater predator that can either swallow them or chew pieces out of their small bodies. A few examples of predators include: common pond insects (Cooper, 1983), especially odonates (Hirvonen and Ranta, 1996) and larvae of the phantom midge *Chaoborus* (Swift and Federenko, 1975), flatworms (Schwartz *et al.*, 1983), hydra (Havel, 1985), salamanders (Dodson and Dodson, 1971), fish of all kinds but especially small pelagic forms (O'Brien, 1979) except for the smallest immature fish (Graham and Sprules, 1992), birds (Dodson and Egger, 1980), predaceous copepods (Dodson, 1974; 1984; Havel, 1985), other predaceous cladocerans (Havel, 1985), and even predaceous plants (Havel, 1985).

Examination of the gut contents of minnows, sticklebacks, and small fingerlings of other fishes shows a heavy uptake of littoral cladocerans. The young of percid and centrarchid fishes moving through the macrophyte beds in the littoral zone find an abundance of these animals to consume. Small catostomid fishes in streams near Bloomington, Indiana seem to feed almost exclusively on chydorids. Whiteside and his students believe that one of the possible causes of the midsummer population crash of chydorids in lakes of northern Minnesota is predation by early fingerling perch. The small size of the littoral cladocerans is suited to the feeding abilities of small fishes.

VI. POPULATION REGULATION

Cladoceran populations, especially of the planktonic species, are characterized by hundredfold annual variations in population size. The peak population generally occurs during an algae bloom, which occurs when lakes are mixed in the spring in northern areas or when the rainy season begins in southern areas. Cladoceran populations are typically at low points during cold weather or when edible algae are scarce. Littoral species show similar population dynamics (Whiteside, 1978).

What causes the large variations in *Daphnia* abundance? Until the work of Hall (1964), it was assumed that the only significant factor was the availability of edible algae. However, Hall showed that while the increase in *Daphnia* numbers was due to an algal bloom, the decrease was due to predation rather than starvation. Recent studies suggest that the usual cause of the decline in abundance is often a combination of predation and food limitation (Lampert, 1986b). *Daphnia* possess enough energy stores to carry them through short-term food shortages, such as the "clear-water phase" that sometimes follows peak *Daphnia* abundances. However, lowered fecundity during the summer, after the clear-water phase, indicates food limitation.

Mortality due to predation is generally size-specific, depending on the size of the prey and the size and hunting behavior of the predators (Zaret, 1980). Those cladocerans that can coexist with predators tend to have specialized life histories, morphologies, and behaviors that provide defenses against predation. Even so, mortality due to predation can be an important factor in determining the abundance of zooplankton.

Hall (1964) made a great contribution to aquatic ecology by showing that predation, as well as food limitation, limits zooplankton populations. He did this by separating the population growth rate coefficient (r) into a component of birth or recruitment (b) and a component of mortality (d). He used a life table technique which requires that individual animals be cultured in the laboratory to determine rates of mortality in the absence of predation and fecundity at various food levels. This important process of separating mortality and recruitment can be achieved much more easily using Edmondson's (1960) egg-ratio technique. This technique requires that one know the number of eggs per female and the developmental time of an egg in order to estimate b. Then, r can be estimated from changes in population size, and d is estimated by the difference between b and r. This technique provides good estimates of b when compared to life table estimates (Lynch, 1982), even when the assumptions of the egg-ratio technique (concerning age structure) are not met. The original Edmondson egg-ratio technique has been modified in various ways (Rigler and Downing, 1984). A comparison of these modifications with predictions based on life table studies (Gabriel, 1987) suggests that different modifications are most accurate depending on the population age structure and the frequency with which the population is sampled.

In general, a declining population with a high value of b indicates high predation, and a stable or declining population with a low value of d indicates food limitation. Studies on the effects of both food limitation and predation suggest that both regulate populations. The best competitors are generally the most susceptible to predators. In a lake dominated by *Holopedium* and *Daphnia*, the *Holopedium* were mainly limited by the food supply, while *Daphnia* showed significant mortality from fish during the summer (Tessier, 1986). A combination field and laboratory study showed that *Holopedium* required more food than *Daphnia*. Thus, when fish predation was intense and food was abundant, *Holopedium* increased; but whenever food was scarce, *Holopedium* declined relative to *Daphnia*; and when predation was weak and food abundant, *Daphnia* increased faster than *Holopedium*. Vanni (1986), using large plankton enclosures in a eutrophic lake, found that large *Daphnia* could reduce food levels sufficiently to reduce populations of *Bosmina* and several copepods. The *Daphnia* in the lake would have been kept in check by fish predation. Cooper and Smith (1982) found that when fish are absent, large invertebrate predators such as *Buenoa* could function in a size-selective manner like fish and remove the larger *Daphnia pulex*, which was able to live at lower food concentrations than the smaller *Daphnia laevis*. Kerfoot and Sih (1987), Carpenter (1987), Sommer (1989), Kitchell (1992), Lampert and Sommer (1997) provide further examples of direct and indirect effects of predators on cladoceran abundance.

VII. FUNCTIONAL ROLE IN THE ECOSYSTEM

Hrbácek (1958) first pointed out the importance of certain fish in excluding cladocerans from communities. These ideas were extended to the effects of selective grazing of cladocerans on algae to produce a model of a dynamic food chain, in which the number and abundance of species in a community is the result of the interaction of competition, selective predation, and nutrients (Fig. 6). For many aquatic predators, cladocerans are a major diet item.

While phosphate concentration in a lake is a major determinant of the kind and abundance of algae, and hence of "water quality," the phosphate concentration is not a particularly good predictor of water quality (Schindler, 1977). For any given level of phosphate, there is great variation in water quality. Several recent studies suggest that food chain dynamics also determine water quality (Shapiro, 1982; Carpenter, 1985; McQueen, 1986; Carpenter, 1988; Kitchell, 1992).

Large species of *Daphnia* appear to be the most efficient in cleaning up lakes made turbid by small algae. Filter-feeding cladocerans can remove significant fractions of the algae from a lake each day (Porter, 1977). Each animal, depending on its size, can clear ≥ 1 mL per animal per hour (Porter *et al.*, 1982). For example, large *Daphnia pulicaria* was scarce in Lake Washington until 1976, when it became common (Edmondson and Litt, 1982; Edmondson, 1991). At the same time the mean summer transparency of the lake doubled, due to grazing by *Daphnia* on diatoms and other algae. The large size and high feeding rate of the large *Daphnia* make them especially efficient at cleaning up lakes.

The hypothesis of food-chain dynamics is a central hypothesis in ecology, because of the perspective it provides and the potential it has for explaining the diversity of natural systems (Fretwell, 1987). The power of this hypothesis results from its inclusion of both direct effects, such as predation of fish on zooplankton, and indirect effects, such as the effects of fish predation on the abundance of algae eaten by zooplankton. In aquatic communities, this hypothesis predicts that the abundance of algae (especially nuisance algae, such as the scum-forming blue green algae) can be kept at low levels by maintaining dense zooplankton populations, which depend on the scarcity of plankton-eating fish, which in turn are controlled by an abundance of large fish-eating fish. Thus, the theory of food-chain dynamics predicts that management of game fish, such as bass or pike in a lake, will have effects on the abundance of algae and, consequently, various aspects of water quality depending on algae, such as clarity and odor. A key link in this chain are the large algae-eating zooplankton, especially *Daphnia*, which are common and often abundant, the favorite food of plankton-eating fish, and able to eat rapidly large quantities of algae (Carpenter, 1985). Smaller common cladocerans, such as *Bosmina* and *Ceriodaphnia*, while less susceptible to fish predation, are also less efficient at cleaning up algae.

Meiobenthic (littoral) cladocerans constitute one of the prime converters of organic detritus with its associated microorganisms into particles big enough to be handled by small fishes and predaceous invertebrates. Wetzel (1983) demonstrated that often the littoral community is more important than the open-water community in the annual production of a lake.

VIII. PALEOLIMNOLOGY AND MOLECULAR PHYLOGENY

A. Paleolimnology

In paleolimnology, we seek to interpret how a lake functioned in past time. The sediments accumulate chronologically; and hence from changing representation of kinds of sediments and their sorting coefficients, organic and inorganic chemicals, and morphological remains of plants and animals, we attempt to interpret past conditions in the water bodies and their watersheds. Sometimes the record is exceptional. Where varves (annual layers) are present, changes in response from year to year can be determined.

Lake sediments tell fascinating stories about the lake and the watershed. Lake sediments have been used to understand the past climate (from pollen from terrestrial plants in the watershed), lake productivity (from the kinds and amounts of microfossils of lake organisms), and the effects of species invasions into the lake. Lake sediments can also supply important archaeological insights. For example, a study of a small lake in central Italy (Lago di Monte Rosi; Hutchinson *et al.*, 1970) found a close correspondence between the ^{14}C date for a thin layer of charcoal and actual records of Roman road-building activity in the watershed.

Remains of animals that are preserved in lake sediments are mostly either silicious (sponges), calcareous (ostracods), or chitinous (cladocerans, midges). All groups of freshwater animals leave at least some remains in sediments (Frey, 1964), but certainly the cladocera are usually best represented, with quantities of remains generally between a few thousand and several hundred-thousand per cubic centimeter of fresh sediment. Once the species pool of a region is known, then virtually all the fragments can be identified positively as to producing species. Remains of the chydorids and bosminids are preserved best and nearly quantitatively, but all cladoceran taxa leave some remains. Those of the meiobenthic forms become integrated over time (at least a year) and habitat before being incorporated into deepwater sediments. The quantity of these remains varies from place to place, being greater toward shore than in the profundal zone, but the percentage composition by species is the same all over the bottom (Mueller, 1964). The best way to determine what species of cladocera occurs in a water body and in what relative abundance over time is from the animal remains in the sediments.

Close-interval stratigraphies are constructed of these remains in the same way as for pollen or diatoms, yielding changes in relative abundance over time of the various taxa (Frey, 1986b). Interpretation has been difficult because of an insufficient knowledge of the ecology of these forms. A few species, such as members of the *Chydorus sphaericus* complex, are considered to respond positively to eutrophication, sometimes comprising more than 90% of all remains. Others seem associated with bog conditions or changes resulting from acidification. *Daphnia* and *Bosmina* fossil remains record past predation regimes (Kerfoot, 1981; Kitchell and Carpenter, 1987).

What is really necessary is to analyze the community differences between time and place. Cotten (1985) studied cladoceran communities in 46 lakes in eastern Finland, using their remains in surficial sediments. She obtained about 70 different taxa. These and their relative abundances were then ordinated using measured chemical and physical characters of the water bodies and their watersheds. She recognized five kinds or groups of lakes that were almost completely concurrent with the five kinds revealed previously by diatom analysis. Thus, the cladocera are just as good responders as diatoms to conditions in a water body and to changes in these conditions over time naturally or in response to human activities.

A study of the cladocera present in a lake in Denmark during the last interglacial period (roughly 100,000 years ago) revealed 25 species of chydorids, all of which occur in Denmark today (Frey, 1962). No real differences in morphology were apparent; and moreover, the relative order of abundance of these 25 species in the interglacial was the same as in Denmark today, indicating that the ecology of these forms had not changed either.

Furthermore, evidence and suggestions from the meager remains of cladocera known from the present to the Tertiary, Cretaceous, and possibly even the Permian indicate that major evolution in the cladocera had occurred by the late Paleozoic or early Mesozoic, and that the species present today have been stabile for millions of years. Thus, knowledge of the present ecology of species will probably be usable in interpreting relationships in lakes even in the distant past.

The presence of a vast record of cladocera in lake sediments is valuable for more than just paleolimnology. Sometimes the presence of a particular species of cladoceran at a place is suspected to be the result of establishment by recent transport there, such as *Bythotrephes* in the Great Lakes (Sprules *et al.*, 1990). This hypothesis can be checked by sedimentary analysis if desired.

The development of the cladoceran community in a lake since late-glacial time, or even earlier in nonglacial lakes, can be followed easily (Goulden, 1966). The species diversity increases over time as the number of taxa in the community increases as conditions become stabilized, tending toward the limit for maximum diversity at any given number of taxa (Goulden, 1969). The distribution of species by their relative abundance in a stable community closely follows the MacArthur broken-stick model. Any major deviation in fit to the expected is considered to have arisen from changes in climate, agriculture, volcanism, or other controlling agents (Frey, 1974, 1976). In lakes with varved sediments one can follow changes in relative abundance of the total cladoceran community on a year-to-year basis and then relate these changes to potential causative agents (Boucherle and Zulig, 1983).

Access to the literature of paleolimnology can be obtained from the proceedings of the four international symposia on the subject (Frey, 1969; Klekowski, 1978; Löffler, 1987; Merilainen, 1983; North American Lake Management, 1995).

B. Cladoceran Fossil Record

Fossils of branchiopods and their relatives are known from the Cambrian Period, about 530 million years ago. Kerfoot and Lynch (1987) summarized data on the earliest fossils of Branchiopods. In the Cambrian

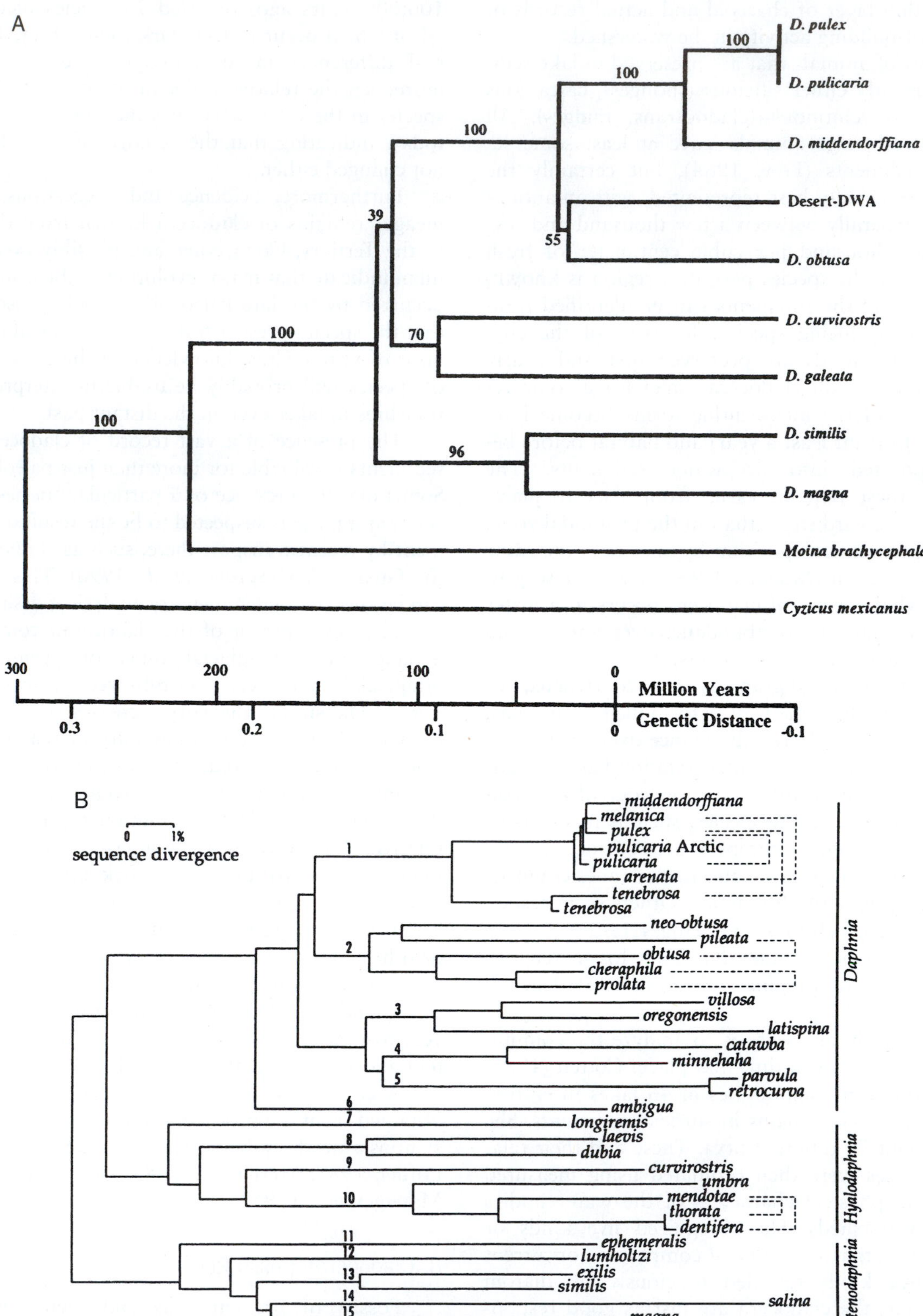

FIGURE 5 (A) Phylogenetic tree, showing the divergence of cladoceran genera during the Mesozoic, and differentiation of the subgroups of *Daphnia* during the Cretaceous (Lehman *et al.*, 1995); (B) phylogenetic tree showing the species in the three major species groups of the genus *Daphnia* (Colbourne *et al.*, 1997).

Period, animals first developed mineralized coverings about 540 million years before the present (mybp). There are branchiopodlike fossils from the Burgess Shale, such as *Protocaris* and *Branchiocaris*, but these animals are not considered branchiopods (Gould, 1989). A fossil of an animal that is probably an ancestor of the branchiopods (and distinct from other crustaceans) occurs in Upper Cambrian sediments of Sweden (Walossek, 1995).

By the lower Devonian (ca. 410 mybp, when fish also make their appearance), the larger Branchiopoda are well established as three recognizable modern groups: fairy, tadpole, and clam shrimp. The fairy shrimp are represented by *Lepidocaris*, an anostracan-like animal, but with a long segmented second antenna and more mouth parts than modern fairy shrimp. These animals are from sediments that indicate marine and estuarine, and perhaps freshwater sediments. Tadpole shrimp fossils are found in marine sediments, and a clam shrimp fossil also appears in marine sediments (this animal is possibly an ancestor of both modern clam shrimps and cladocera). By the Carboniferous (ca. 286–360 mybp), tadpole and clam shrimp species are diverse and occur in many habitats (marine, freshwater, streams). Smirnov (1970) described some possible *Daphnia*-like and chydoridlike fossil from the late Permian (ca. 240 mybp). The *Daphnia*-like animals are 5–8.5 mm long. Smirnov also has good specimens of *Daphnia*-like fossils from the Jurassic (ca. 225 mybp). By the early Triassic, most tadpole and clam shrimp groups are extinct, and cladocera appear, perhaps as neotenic conchostracans (as suggested by Brooks). The modern continental distribution of cladoceran orders suggests the orders were distinct when the large continent Pangea broke up into Laurasia and Gondwanaland, during the Cretaceous (approximately 100 mybp). Starting with the Oligocene (ca. 38 mybp), there are many good fossils of cladocerans, especially of ephippia from lake deposits such as the freshwater Florrisant beds in Colorado. Similarly, the modern distribution of cladoceran species in North America and Europe suggests that the major about of speciation associated with glaciation was associated with speciation which produced modern species (Brooks, 1957). Continental glaciers have been producing new lakes since about 1.8 mybp. For example, several species, such as *Daphnia galeata, retrocurva, catawba, longiremis*, are restricted to the glaciated portion of North America (or Europe: *D. cuculata*).

C. Molecular Phylogeny

Just as lake sediments contain a record of the history of the lake and its watershed, the nucleic acid (nuclear and mitochondrial) of cladocerans contains a record of the evolutionary history of living individuals. The molecular record is particularly valuable, because the fossil record of most zooplankton groups is poor.

Information about phylogeny exists in the nucleic acids of branchiopods. For example, analyses of the ribosomal RNA gene (encoded on mitochondrial DNA) suggest that *Daphnia* originated (differentiated from similar genera) in the Mesozoic (Lehman *et al.*, 1995; Colbourne *et al.* 1997; see Fig. 5A). Since then, the genus has evolved into at least three species complexes (Fig. 5B). A recent re-analysis of an Australian daphniid (*Daphnia jollyi*), suggests it may represent a fourth lineage. It is odd that there are only four complexes of *Daphnia* species, and that they are all rather similar morphologically, given 200–300 million years of evolution.

IX. TOXICOLOGY

Cladocerans, especially *Daphnia*, *Ceriodaphnia*, and *Moina* have long been used in toxicity tests. Cladocerans are relatively small, have short life spans, reproduce well in the laboratory, and are ecologically important—all desirable characteristics for a toxicity test. For a bioassay (a test using organisms), the animals are grown under optimal conditions in the laboratory (or less often, in enclosures in nature). Some animals are treated with known concentrations of chemicals, and their responses are compared to those of untreated animals. Survival and fecundity (number of offspring) are two standard bioassay endpoints (US EPA, 1994). Animals are exposed to test chemicals for a few hours or days as a test for acute toxicity, and for a generation or more as a test for long-term chronic toxicity.

Morphology, behavior, sex ratio, physiology, and developmental time are additional bioassay end points. These end points tend to be more sensitive than survivorship or fecundity, and they are of ecological importance. For example, the insecticide carbaryl induces morphological and swimming behavior changes in *Daphnia* (Hanazato and Dodson, 1993; Dodson *et al.*, 1995). The surfactant (nonionic detergent) and plasticizer nonylphenol interferes with normal development at levels found in streams and lakes, so that neonates are produced and they survive in the laboratory, but they are unable to swim (Shurin and Dodson, 1997).

Predator prey behavior is affected at low-dose and sublethal levels of many environmental pollutants. Carbaryl causes *Daphnia* to swim erratically and faster than normal (Dodson *et al.*, 1995), which in nature, would increase the predator–prey encounter rate, leading to an increased *Daphnia* mortality.

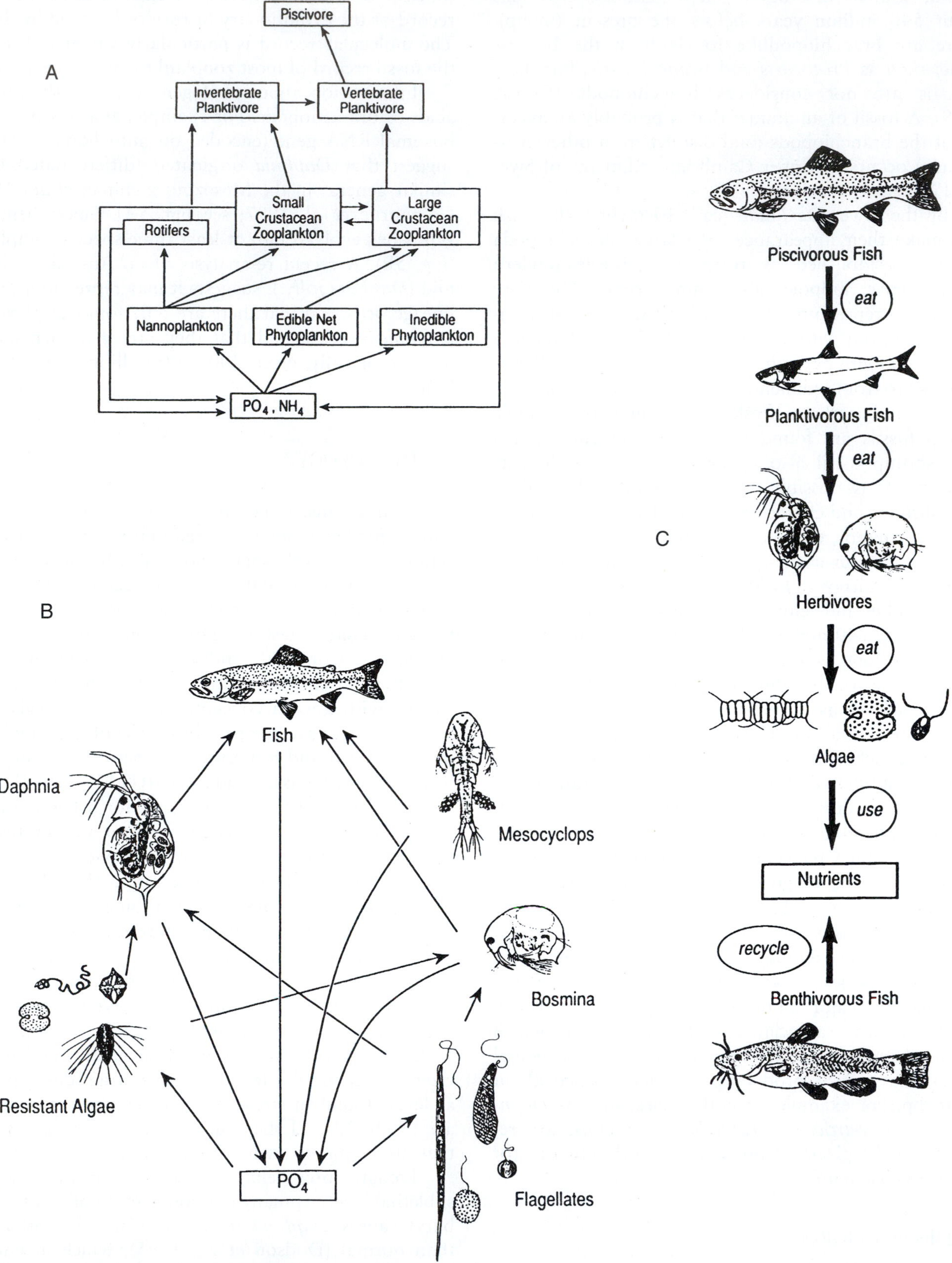

FIGURE 6 Food-chain models showing: (A) four trophic levels; (B) three trophic levels; and (C) three consumer trophic levels and one scavenger (decomposer) level.

Many aspects of animal development are controlled by hormones. There is concern that industrial and agricultural chemicals may be acting like or interfering with normal hormone systems in both wildlife and humans (Colborn *et al.*, 1996). The cladoceran sex ratio is an endpoint that can be used to test for these chemicals (called "endocrine disruptors). Chemicals that change sex ratio in cladocerans will have ecological consequences, and they may also act as endocrine disruptors for other animals, including humans. It is hypothesized that some chemicals that disrupt hormone systems in vertebrates can also interfere with the system regulating molting in arthropods (Zou and Fingerman, 1997).

X. CURRENT AND FUTURE RESEARCH

How many species are there of cladocerans? Do cladoceran species need conservation? How do you identify the different species? How closely related are the different morphologically similar species? Are cladocerans (or other branchiopods) even really crustaceans? There is a great deal of taxonomy and phylogeny that remains to be done. This is not just a matter of finding all the species and naming them, but more fundamentally, of deciding what sort of species concept makes sense with cladocerans. These animals are usually at least facultatively asexual; and in many groups, what seem to be good morphological species are made up of a combination of occasionally sexual and totally asexual populations. Descriptions of species before about 1950 are mostly very inadequate and hence cannot be used to compare populations on two different continents claimed to be the same taxon. To determine that any North American taxon is the same as or different than the taxon of the same name described from elsewhere is difficult. One first needs to compare both taxa closely, using large populations containing all instars and reproductive stages. Thus far about 16 pairs of taxa with the same name, from the New and Old World, have been compared (partial summary in Frey, 1986a), with the finding that almost all of them are full species. Sometimes, as in the *Chydorus sphaericus* complex, *Eurycercus*, *Ephemeroporus*, the *Chydorus reticulatus* group, the *Chydorus faviformis* group, and *Pleuroxus laevis*, the same specific name was being applied to two or more related but different species in the same geographical region. Consequently, for critical studies in ecology, one must be careful not to transfer the ecological requirements and indications of a species across an ocean or some other distributional barrier, on the assumption that all taxa presently bearing the same name are the same species. This question will be sorted out with a mixture of traditional morphological and new molecular and genetic techniques (e.g., DeMelo and Hebert, 1994; Colbourne *et al.*, 1997), and some new ideas!

The study of food-chain dynamics, using sets of whole lakes shows promise for interesting research into population and community ecology. Society has a great desire for clean lakes and lots of fish production. Cladocerans play crucial roles in both water quality and fish management, but we need to understand cladoceran biology better before we can meet society's desires. Food chain dynamics, or "biomanipulation" shows promise of being an inexpensive and safe method for improving both water quality and fish production.

Zooplankton live in a sea of chemical signals (Larsson and Dodson, 1993). The chemicals are produced by predators, food, competitors, and environmental sources. The signals induce changes in development, morphology, reproduction, and behavior. However, none of the chemical signals used between species have been characterized. When their chemical structure is known, it will be possible to do interesting experiments involving induction, and to understand how these signals affect development and behavior, which, in turn, affect the aquatic community and trophic dynamics.

Environmental toxicology has long depended on tests of cladoceran survival and fecundity. Future research will continue to use more ecologically relevant endpoints, such as behavior, developmental time, and sex ratio (Dodson *et al.*, 1999).

XI. COLLECTING AND REARING

A. Sampling Techniques

Cladocerans can be collected using a fine mesh net pulled through the water or a sieve through water is poured. A large number of specialized nets have been developed for specific applications (Bernardi, 1984). The mesh should be around No. 10, or have a pore size of about 90–150 μm. Nets are most useful for towing in open water. Weedy habitats can be sampled by a plankton net attached to a handle, and with a wide-mesh (1 cm or so) cover over the opening of the net. Weedy habitats or mud can be sampled by picking up water in a bucket and pouring it through a series of nets or sieves (Downing, 1984). In any case, keep mud out of the sample if possible. Large pieces of water weeds and detritus can be filtered out of a sample with screen having a mesh size of about 1 cm. Filters can be made by cutting the bottom off a plastic bottle. The center of the cap is cut out, and a piece of mesh glued over the hole. When the cap is screwed back onto the bottle, water can be poured into the inverted bottle, and cladocerans will be caught by the mesh. The captured cladocerans can then be washed into a jar of water previously filtered through the mesh.

Estimations of population dynamics or productivity require quantitative samples, which require special gear (Bernardi, 1984; Downing, 1984) and a careful experimental design and statistical analysis of samples (McCauley, 1984; Prepas, 1984).

If the sample organisms are to be killed, there are at least two ways to do it (the authors disagree on the best technique). Dodson prefers to wash the animals off the screen of a small sieve using a wash bottle filled with 95% ethanol, or to drop the sample into a large volume of 95% ethanol. This will kill the animals rapidly enough so there will be little distortion and will preserve the sample indefinitely. It will be necessary to keep the samples tightly capped and to add 70% alcohol from time to time. Although messy, a little glycerine will ensure that the sample never dries up. Frey preferred to kill and preserve animals by adding enough commercial formalin containing 40 g sucrose per liter to make a 5% formalin solution. He felt formalin was better for long-term museum preservation of cladocerans, because of lesser distortion of the specimens over time. For safety reasons, formalin samples are best examined in plain water, following filtration. Specimens can then be replaced in a formalin solution for storage.

B. Culture Methods

Animals will live in filtered pond water for a few days at least, especially if there are fewer than 50 animals/L. Animals can be transferred to new water using pipettes made of glass tubing and rubber bulbs, or for large quantities, turkey basters are useful. For long-term studies of cladocerans, it is possible to maintain cultures for years. Maintenance of cladoceran cultures can be divided into two processes: maintenance of food for the cladocerans and care of the cultures.

1. Cladoceran Food

The simplest to maintain and most reliable bulk food culture for cladocerans appears to be a mixture of green algae and bacteria maintained in a large aquarium. Keep the algae in a glass aquarium as large as possible; a 40-gal aquarium is sufficient. Place the aquarium in a south-facing window. Choose a place that can be kept at least 18–25°C; a greenhouse is preferable. Fill the aquarium with filtered lake or pond water, using a fine mesh (ca. 90 μm mesh) net, or use tap water aged for 1 day with a little additional filtered lake water as a seed source for the phytoplankton. Add a dozen or so small fish (bait fish, guppies, or gold fish) to the tank and feed the fish regularly. Aerate the tank. Keep the glass walls scraped clean and siphon off bottom detritus now and then. Regardless of what you start with, the algae species will probably eventually stabilize, and you will have a mixture of small greens (like *Chlorella*, *Selenastrum*, and *Scenedesmus*), along with a few other less common species such as *Ankistrodesmus*. Keep the algae in a healthy state by removing at least a quarter of the culture every week or so. Use a large-diameter siphon to remove the green algae.

Well-equipped laboratories can grow single-species algal cultures purchased from biological supply houses. Grow the algae either in a medium supplied by the supply house, or mix artificial lake water, such as "Combo" (Kilham *et al., 1998*). The advantage of Combo is that it has the same salinity as average lake water, and the algae can be added directly to branchiopod cultures. Grow the algae in 2-L clear plastic bottles (soda bottles). For fast growth, aerate the cultures and use bright light.

Culture Care: Branchiopods can be captured using plankton nets, dip nets, or by pouring water through filters of about 100-μm mesh. Wash the collected animals into some filtered water from the lake or pond. Bring the animals into the laboratory and let them equilibrate to the laboratory temperature. Many cladocerans tend to be captured by the surface film. This happens more often for the smaller lake animals, especially if they are initially in cold water with saturated oxygen, because bubbles form on the carapace as the water warms. In extreme cases of mortality, keep the surviving animals in a narrow-mouth bottle filled to the top and plugged with cotton pushed below the water surface. In the beginning, put the cultures on a dark substrate and keep them in a room with dim light. After a couple of days, it will be obvious which cultures are going to be vigorous. Be sure to label the cultures (with tape on the jar) as to the source and date of capture.

Feed the cultures algae suspension. If you cannot see large objects through the culture jar, the algae is probably too concentrated; dilute it. The mixture should be green but not opaque. As the water in each jar clears, add algal suspension. Do not let the water become clear. Even an overnight bout of clear water will cause some clones to starve. It is also possible to overfeed clones, but starvation seems a much more common occurrence. Continue adding algae suspension until the jar is full.

After the jar is full, there are two ways to feed the culture. First, if there are too many animals in the jar, pour out half or so of the culture and refill with algal suspension. Beware of cultures in which the animals concentrate in the top centimeter or so of the water; it is possible to pour out the entire culture! Second, if there are too few cladocerans in the jar, siphon off about half of the water and add algal suspension. Be sure to use fine enough mesh (about 90 μm) on the end of the siphon in the jar, so that the animals stay in the jar. Also, be sure to wash the siphon and dry it or dip it into hot water before using it on the next culture. The safest population size for a 1-L jar seems to be about 20–50 animals. Having some detritus on the bottom of the jar will not harm planktonic species and is beneficial to meiobenthic species. When there is a thick layer, decant off the culture and discard the detritus layer.

Clones, once they are established, can be kept in a cold room or refrigerator at about 4°C. At this temperature, the animals need to be fed only about once every two weeks. Beware—some species die when moved from 20 to 4°C, so do not put all your cultures in the cold at once.

XII. TAXONOMIC KEYS FOR CLADOCERAN GENERA

A. Available Keys and Specimen Preparation

Crustaceans are cladocerans if they have four to six pairs of (thoracic) legs, lack any paired eyes, swim with their second pair of antennae, and have at least the head not covered by a carapace.

The identification keys are based on a large number of sources, which are referenced in Table 1 and in the individual keys. Dodson is responsible for most of the keys. The key to the Chydoridae family was written by Frey.

These keys are designed to allow identification of adult females only. Dissection is seldom necessary, but it is a good idea to have several animals available for identification, because these small, delicate animals are easily crushed out of recognizable shape. Animals shorter than about 1 mm should be mounted on a slide for high-magnification viewing; the larger species can often be identified without being mounted, using a dissecting microscope at about 50 diameters magnification or less.

While live animals trapped in a thin film of water can usually be identified, it may be necessary to kill them with alcohol and mount them on slides. The best mounting medium is one which is more or less permanent and which is soluble in water and alcohol solutions so specimens can be added directly to the medium without tedious dehydration. If you have access to chloral hydrate (currently a "controlled substance"), a good medium is Hoyer's:

> Dissolve 30 g of ground Gum Arabic in 150 mL of hot distilled water. Add 200 g of chloral hydrate and 20 g of glycerine.

Divide the Hoyer's into two batches. Let one sit in a warm place until it is as thick as honey, and keep the second batch in a closed jar so it remains thin. You can always add more water to thin either batch. Keep at least some of the thick and thin Hoyer's in screw-top eye dropper bottles for easy use, and be careful to keep the Hoyer's off the threads of the bottles.

To make a slide, use the eye dropper to put a small streak of dilute Hoyer's (about a quarter of a drop) toward one end of the slide. Arrange the specimens in this streak. It is wise to place four or so of what you think are the same kind of animal on a slide to allow for comparison and viewing of difficult characters. Delicate specimens can be protected from compression by putting a few pieces of broken cover slips around the specimens. Let the Hoyer's streak dry on a slide warmer or in a warm place (below 100°C). When the streak is dry, add a drop or two of the thick Hoyer's and gently lower a cover glass over the Hoyer's. Put the slide back into a warm place to dry again. Label the slide as to the date and location of the collection! It is important to use thickened Hoyer's in the last step; otherwise, the Hoyer's will shrink as it dries, and produce large bubbles in the final preparation. On the other hand, if you use thickened Hoyer's, small bubbles produced when you lower the coverslip will disappear as the medium dries. These slides are semipermanent. If the climate is humid, the Hoyer's will thin and cover glasses will slip off the slide. If the climate is arid, the Hoyer's will desiccate enough so that the cover glasses pop off the slide. These problems may be avoided by storing the slides horizontally in a climate that has moderate humidity, or you can ring the slide with nail polish.

An alternate mounting method, especially useful for bosminids, chydorids, and macrothricids, is to use polyvinyl lactophenol stained with lignin pink. Temporary slides can be made using glycerol or glycerine jelly. Best of all for museum specimens is possibly Canada Balsam, although this requires the specimens be dehydrated through an alcohol series.

B. Taxonomic Key to Families of Freshwater Cladocerans

This key to families is followed by keys to genera in each family or group of families.

1a. Both branches (or branch) of second antenna ending in three setae (Fig. 1A) 2

1b. At least one of the second antenna branches ending in four or more setae (Fig. 8) 8

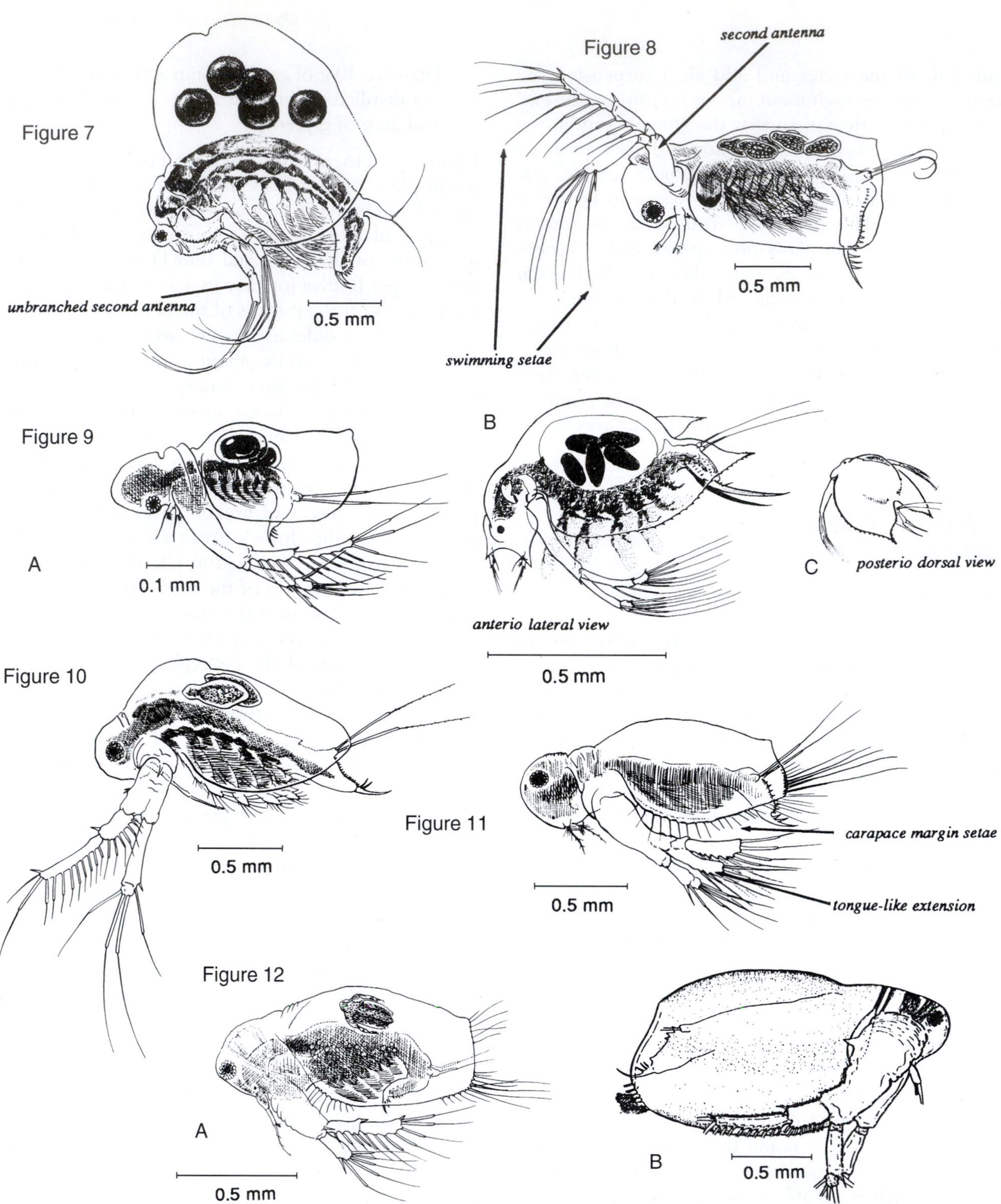

FIGURE 7 *Holopedium gibberum* Zaddach, 1855. Why Not Bog, Vilas Co., Wisconsin. 22 September 1973. (KE) *FIGURE 8* *Sida crystallina* (O.F. Mueller, 1776). Lake Mendota, Dane Co., Wisconsin. 11 September 1987. (KE) *FIGURE 9* (A) *Diaphanosoma birgei* (Kořínek, 1981). Lake Wingra, Dane Co., Wisconsin. 5 October 1978. (KE) (B) *Penilia avirostris* Dana, 1849. Anteriolateral view. Atlantic Ocean near Florida (29°55.8' N and 81°07.4°W). 27 October 1987. (KE); (C) *P. avirostris*. Posteriolateral view. (KE) *FIGURE 10* *Pseudosida bidentata* Herrick. 1884. Bull Frog Pond, S.R.P. area, South Carolina. 27 May 1985. (Coll. DGF #7483). (KE) *FIGURE 11* *Latona setifera* (O.F. Muelller, 1776). Lake Mendota, Dane Co., Wisconsin. 30 June 1987. (KE) *FIGURE 12* (A) *Latonopsis occidentalis* Birge, 1891. Dick's Pond, S.R.P. area, South Carolina. 28 May 1985. (Coll. DGF #7496). (KE); (B) *Sarsilatona serricauda* Sars, 1901. Itatiba, Brazil. Figure 91 from Korovchinsky 1992.

2a (1a). First antenna composed of two segments (one genus, *Ilyocryptus*, Fig. 44) .... Ilyocryptidae

2b. First antenna with one segment 3

3a(2a). Second antenna (the swimming antenna) with one branch (Fig. 7) .... Holopedidae

3b. Second antenna with two branches 4

4a(3b). First antennae fused with the rostrum to form two long and pointed tusklike structures (Figs. 41 and 42) .... Bosminidae

4b. First antennae not fused with the rostrum 5

5a(4b). One branch of the second antenna with three segments, the other with four (look for a small basal segment, as in Fig. 1) 6

5b. Both branches of the second antenna with three segments (Suborder Radopoda) 7

6a(5a). First antenna about as long as the width of the head; usually a tooth with two points (bifid) on the posterior angle of the postabdomen (near the claw), (Fig. 58) .... Moinidae

6b. First antenna less than half as long as the width of the head; teeth on postabdomen always single (Figs. 1A, 36–40) .... Daphniidae

7a(5b). First antenna is attached to the underside of the tip of the head, there is no rostrum (beak) (Figs. 44–57) .... Superfamily Macrothricoidea

7b. First antenna is attached to the ventral margin of the head, and is more or less covered with a rostrum (beak), (Figs. 13, 35) .... Chydoridae

8a(1b). Body (thorax and abdomen) covered by the two valves (shells) of the carapace (Figs. 8–12) .... Sididae

8b. Body not covered with a carapace (which is present in some forms as a saclike brood chamber attached to the back; legs are not covered with the carapace (Figs. 60–64) .... Orders Onychopoda and Haplopoda

C. Taxonomic Key to Genera of the Family Holopediidae

Hebert and Finston (1996) used results of allozyme analysis to conclude that there are two species in North America, *Holopedium gibberum* Zaddach, 1855 (Fig. 7) and *H. amazonicum* Stingelin, 1904, confirming the usage of Korovchinsky (1992). These are planktonic animals, which often occur in soft and even acidic water. *H. gibberum* is broadly distributed in the cool temperate regions of North America, while *H. amazonicum* occurs in the southern and eastern portions of the continent. These species are remarkable in that each individual is surrounded (out to about an additional body diameter) by a mass of clear jelly attached to the carapace (easily removed by rough handling). The jelly probably defends *Holopedium* against an array of invertebrate predators. Adult females are about 0.6–2.0 mm long.

D. Taxonomic Key to Genera of the Family Sididae

1a. Posterior margin of carapace flared out laterally; a point on either posterior-lateral margin of the carapace; marine (Figs. 9B, C) .... *Penilia*

1b. Posterior margin of carapace not flared laterally; postero-ventral corners of carapace rounded; freshwater 2

2a(1b). Swimming antenna appears to have three branches, the middle branch is an extension of the basal segment of the upper branch; postabdominal claw with two basal spines (Fig. 11) .... *Latona*

2b. Swimming antenna clearly has only two branches; postabdominal claw has 2–3 basal spines 3

3a(2b). Postabdominal claw with about 10 long teeth along the claw ("long" means longer than the width of the claw); a palearctic genus .... *Limnosida*

3b. Postabdominal claw with 2–4 long teeth near base 4

4a(3b). Postabdomen with a row of about 20 individual triangular spines along the margin (the anal margin); no clusters of spines on postabdomen; postabdominal claw with three long teeth and an occasional short fourth tooth near the base of the claw; both branches of swimming antennae with a total of 15 long swimming setae (Fig. 8) .... *Sida*

4b. Postabdomen with spines in clusters or patches, spines in rows are hairlike; swimming antennae with more than 15 long swimming setae 5

5a(4b). Posterior margin of carapace with setae about 20–25% as long as the carapace 4

5b. Posterior or ventral margin of carapace with setae shorter than 10% as long as the carapace 5

6a(5a). Swimming antennae branches with a total of 16 long swimming setae; postabdominal claw with two long basal spines (Fig. 11) .... *Latonopsis*

6b. Swimming antennae branches with a total of 24 long swimming setae; postabdominal claw with three long basal spines (Fig. 12B) .... *Sarsilatona*

7a.(5b). Swimming antennae with a total of 17 long swimming setae; postabdomen with rows of fine hairlike teeth; postabdominal claw with three long teeth (the most basal tooth may be only about as long as the width of the claw at the base (Fig. 9) *Diaphanosoma*

7b. Swimming antennae with a total of 20 long swimming setae, postabdomen with clusters of 4–6 spinelike teeth; postabdominal claw with two long teeth, and a short basal tooth less than half as long as the width of the claw base (10) .... *Pseudosida* [Korovchinsky (1992) has recently revised the Sididae, including the genus *Sarsilatona* and retaining the changes made by Kořínek (1981) for *Diaphanosoma. Diaphanosoma* and *Penilia* are planktonic and about 1.0 mm long; species in the other genera tend to be larger (2–4 mm) and are associated with littoral vegetation.]

E. Taxonomic Key to Genera of the Family Chydoridae (North of Mexico)

1a. Postabdomen strongly flattened, broad, with a single row of 80–150 marginal denticles, resulting in a saw like appearance; anus distal; subfamily Eurycercinae (Fig. 13) .... *Eurycerus*

[Genus is Holarctic except for isolated populations in southern Brazil–northern Argentina and in South Africa. At least five species in North America, distributed among three subgenera: one species in the subgenus *Eurycerus* only within 300 km of the Gulf of Mexico and the Atlantic Ocean as far north as the Pinelands of New Jersey (maximum length ca. 2 mm); two very common species in subgenus *Bullatifrons*, one in the northern tier of states and northward, the other in the south (length to 2.4 mm); and two species in the subgenus *Teretifrons* in Newfoundland and the Arctic (length to 6 mm). See Frey (1971, 1975, 1978, 1982d) and Hann (1982).]

1b. Postabdomen thicker, not so broad, and usually with two rows of marginal denticles, mostly fewer than 25 on each side; anus proximal 2

2a(1b). Proximal tip of mandible articulate where headshield and shell touch each other (Fig. 1C); usually two or three major headpores on midline, which usually are connected by a double sclerotized ridge, resembling a channel (Fig. 1C); occasionally only one median pore (*Monospilus, Euryalona*) or none (*Notoalona*); minor headpores lateral to these; free posterior margin of shell not greatly less than maximum height of animal; typically one, occasionally zero, basal spine(s) on postabdominal claw (Fig. 1C); postabdomen nearly always with lateral fascicles (Fig. 1C), sometimes without (*Monospilus*). .... Subfamily Aloninae 3

[In North America, Aloninae contains the genera *Acroperus, Alona, Alonopsis, Bryospilus, Camptocerus, Euryalona, Graptoleberis, Kurzia, Leydigia, Monospilus, Notoalona, Oxyurella*, and *Rhynchotalona*. Olesen (1998) presents evidence that Aloninae is a paraphyletic group.]

2b. Proximal tip of mandible articulates in a special pocket on the headshield, some distance in a dorsal direction from the point of contact between the headshield and shell (Fig. 1D); two major headpores on midline (Fig. 1B, D), completely separated from one another and with no sclerotized ridge connecting them, occasionally only one pore (*Dadaya*), or only one pore in first instar and none in later instars (*Ephemeroporus*); minor pores most commonly near midline between major pores; free posterior margin of shell usually considerably less than maximum height of animal; typically two basal spines on postabdominal claw (Fig. 1D); articulated lateral fascicles on postabdomen lacking, although *Pseudochydorus* has a row of fasciclelike groups of setae. .... Subfamily Chydorinae 15

[In North America, Chydorinae contains the genera *Alonella, Anchistropus, Chydorus, Dadaya, Disparalona, Dunhevedia, Ephemeroporus, Pleuroxus*, and *Pseudochydorus*.]

3a(2a). Only an ocellus present; no compound eye 4

3b. Compound eye and ocellus both present 5

4a(3a). Shells from previous instars firmly nested, so that instar number of specimen can be counted; headshield transversely trucate posteriorly; single median headpore (Fig. 14) .... *Monospilus*

[Supposedly *Monospilus* contains just one species in world, mainly Holarctic and Ethiopian in distribution, with several records from New Zealand as well. Length to 0.5 mm.]

4b. Shells lost molting, so that no nesting of shells occurs; two separated median pores without sclerotized connecting ridge; minor pores located far laterally; very short antennae (Fig. 15) .... *Bryospilus*

[Two species are known from New Zealand, Venezuela, and Puerto Rico, and hence this genus may possibly occur on north of Mexico. Length to 0.35 mm. Found among wet, leafy hepatics of rain forests and cloud forests (Frey, 1980b, d).]

5a(3b). Body strongly compressed from side to side, at times almost wafer thin; post-abdominal claw with a secondary spine midway along concave margin, and usually with a comb of fine setules decreasing in length proximally between it and the basal spine (comb also occurs in *Euryalona*) 6

5b. Body not so strongly compressed; postabdominal claw having only a basal spine of variable length (except in *Euryalona*); sometimes none at all (one species of *Leydigia*) 9

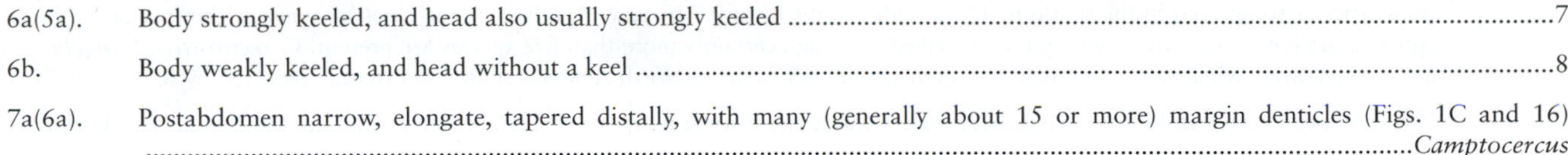

6a(5a). Body strongly keeled, and head also usually strongly keeled 7

6b. Body weakly keeled, and head without a keel 8

7a(6a). Postabdomen narrow, elongate, tapered distally, with many (generally about 15 or more) margin denticles (Figs. 1C and 16) ....*Camptocercus*

Figure 13

Figure 14

Figure 15

Figure 17

Figure 16

Figure 18

Figure 20

Figure 19

Figure 21

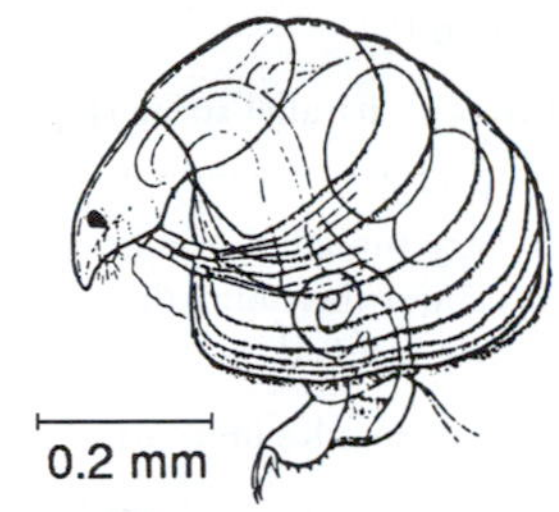

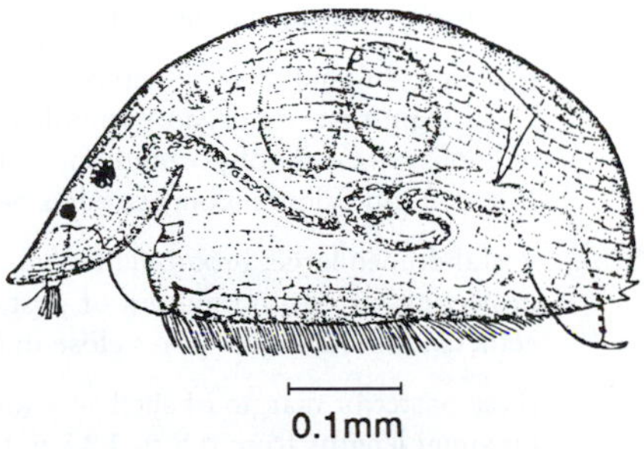

FIGURE 13 *Eurycercus (Eurycercus) lamellatus* (O.F. Mueller, 1776). Uppsala, Sweden. From Lilljeborg (1901). *FIGURE 14* *Monospilus dispar* Sars, 1862. Yddingesjon, Sweden. From Lilljeborg (1901). *FIGURE 15* Bryospilus *repens* Frey, 1980. El Yunque, Puerto Rico. From Frey (1980b). *FIGURE 16* *Camptocercus rectirostris* Schoedler, 1862. Kristianstad, Sweden. From Lilljeborg (1901). *FIGURE 17* *Acroperus* cf. *harpae* (Baird, 1834). Eastern North America. From Birge (1918). *FIGURE 18* Kurzia *longirostris* (Daday, 1898). Tammanna wewa, Sri Lanka. From Rajapaksa (1986). *FIGURE 19* *Alonopsis elongata Sars*, 1862. England. From Norman and Brady (1867). *FIGURE 20* *Rhynchotalona falcata* (Sars, 1861). Northern Michigan. From Birge. (1893). *FIGURE 21* *Graptoleberis testudinaria* (Fischer, 1851). Eastern North America. From Birge (1918).

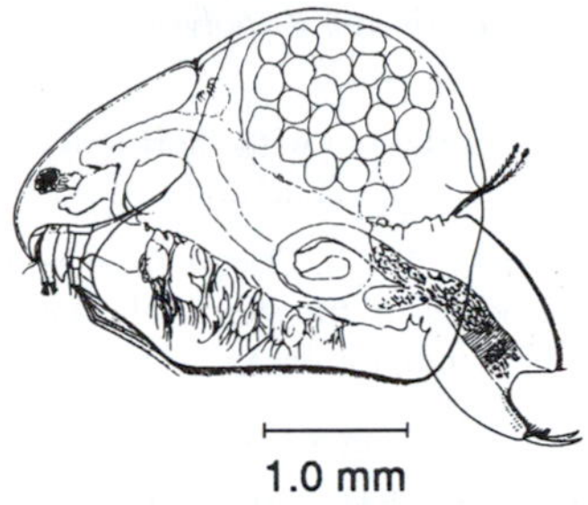

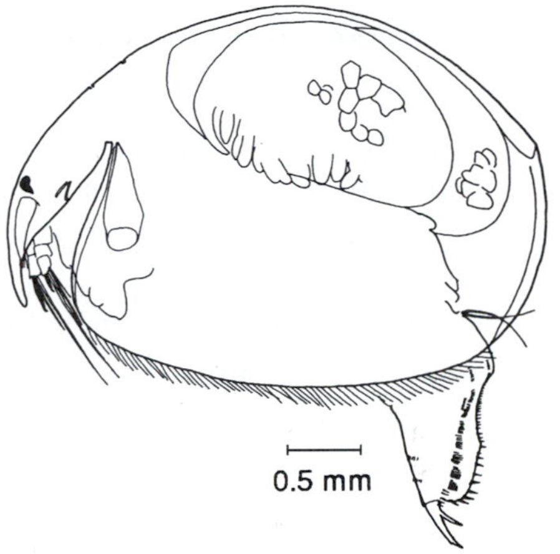

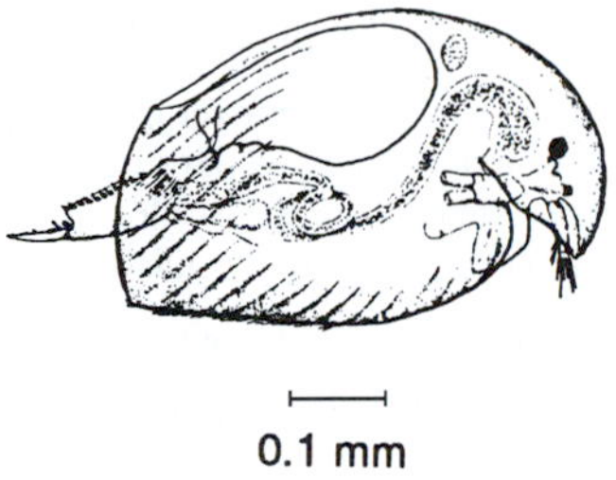

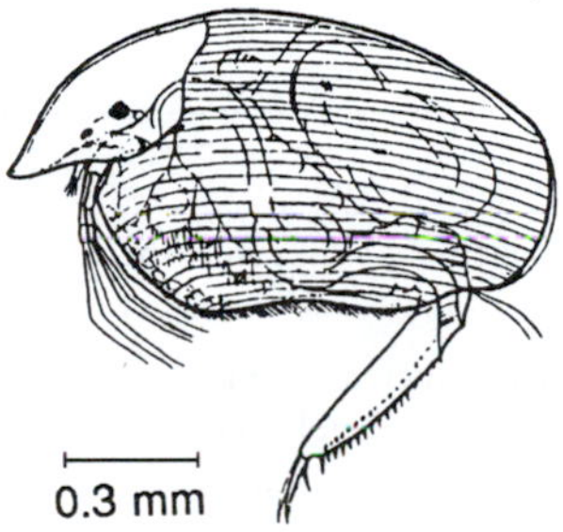

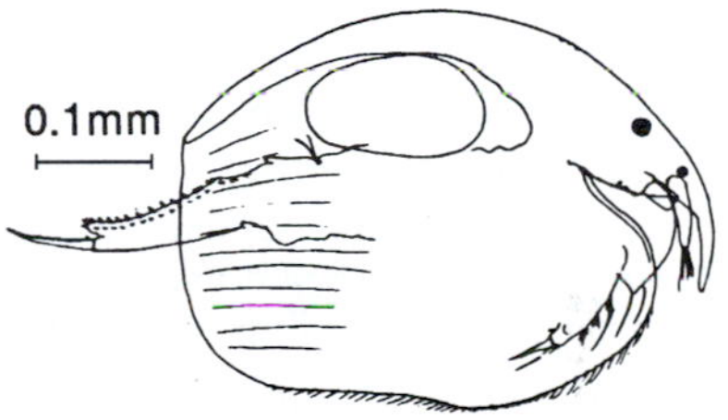

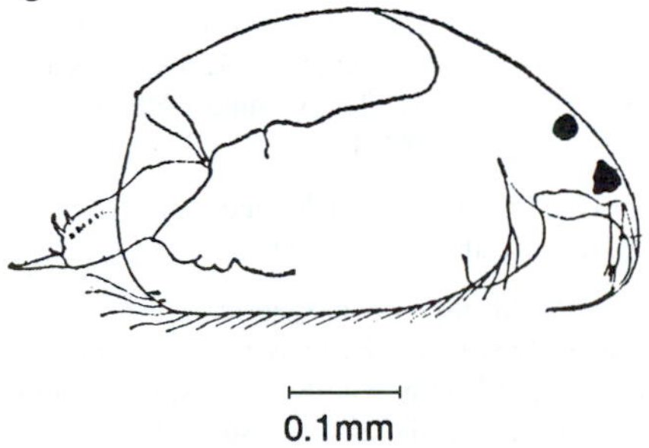

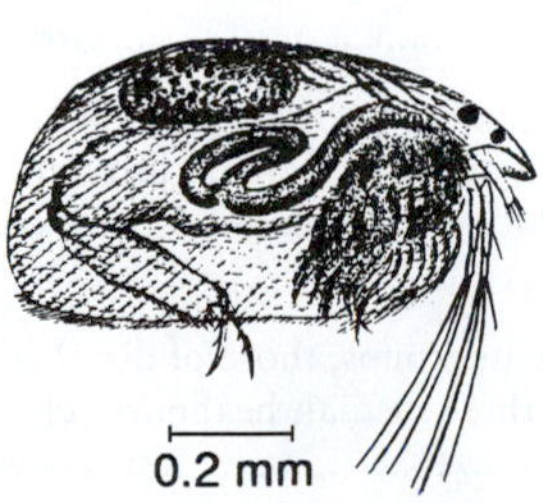

[Nine species recognized in the world in 1971, with lengths from 0.7 to 1.26 mm. Except for *C. oklahomensis*, the species in North America are completely unknown and undescribed, although certainly more than five species are present. *C. rectirostris*, *C. liljeborgi*, and *C. macrurus*, which were described from Europe, do not occur in North America at all, even though reported by Brooks (1959).]

7b. Postabdomen shorter and broader, parallel-sided, without any distinct marginal denticles (Fig. 17) .... *Acroperus*

[Seven species in the world in 1971, with maximal lengths from 0.59 to 0.85 mm. Like *Campocercus*, the species in *Acroperus* are not yet adequately described. There is marked conformity in body form and shape and in other morphologic characters, making it very difficult to sort out the species satisfactorily.]

8a(6b). Postabdomen rather slender, elongate, somewhat tapered distally; body high; shell widely open behind (Fig. 18) .... *Kurzia*

[Five species in world and three in North America, *K. latissima* being widespread, and *K. longirostris* and *K. polyspina* occurring only in the southernmost Gulf states. Length to 0.6 mm.]

8b. Postabdomen broader, elongate, parallel-sided; shell with parallel striae sloping downward posteriorly, and with short longitudinal scratch marks in between (Fig. 19) .... *Alonopsis*

[Two species in world, the one in North America very similar to the one in Europe (see Kubersky, 1977). Occurs from northern New England into the adjacent Canadian provinces and then into those farther East. Smirnov (1971) transferred the European species to the genus *Acroperus*, which is not adequately justified. Maximum length: American 0.91 mm, European 0.85 mm.]

9a(5b). Rostrum long, attenuate, recurved, greatly exceeding antennules in length (Fig. 20) .... *Rhynchotalona*

[Only one species listed in world, although there is a second species from North America not yet described. Postabdomen with 2–4 stout marginal denticles distally and with an almost continuous row of long, hairlike setae laterally. Length to 0.5 mm. A mud dweller.]

9b. Rostrum rounded, barely or only slightly exceeding antennules in length 10

10a(9b). Seen from above, rostrum very broad, semicircular, wider than body; shell and head strongly reticulated (Fig. 21) .... *Graptoleberis*

[Possibly only one species in world, although the taxonomy is not yet clear. Reported maximum length 0.7 mm. Ventral margin of shell provided with a dense fringe of setae, and generally with two large, sharp, triangular teeth at posterior-ventral corner of shell; postabdomen tapered distally, marginal denticles small, lateral fascicles minute; postabdominal claws small, with minute basal spine. This species functions much like a snail, scraping food off the substrate as it moves along. See Fryer (1968).]

10b. Rostrum more narrowly rounded; shell may be weakly reticulated, but head never is 11

11a(10b). Postabdomen elongate and rather narrow; with marginal denticles well developed 12

11b. Postabdomen much less elongate (except in *Leydigia*); marginal denticles usually well developed, but in some species greatly reduced 13

12a(11a). Dorsal margin of postabdomen straight or slightly convex; marginal denticles very short proximally, increasing in length distally to three very long curved denticles; four median headpores (middle one is divided into two), usually completely separated from one another, or with middle pores joined by a sclerotized ridge (Figs. 21. 1C and 22) .... *Oxyurella*

[Basal spine of postabdominal claw long, slender, attached some distance from base of claw; head and shell generally have a yellowish, hyaline appearance. Five species claimed to occur in the world in 1971. Two species in North America, one being confined to the Gulf states, the other more widely distributed in the East. The common North American species is now separated from the formerly cognate species in Eurasia. See Michael and Frey (1983).]

12b. Dorsal margin of postabdomen distinctly concave; marginal denticles increase in length distal, but never reach large size as in *Oxyurella*; one median headpore, sometimes completely absent (Fig. 23) .... *Euryalona*

[A tropical to subtropical genus. Three species in world, the one in America north of Mexico originally described from Sri Lanka. Head high; rostrum broadly rounded, with tip scarcely reaching halfway to ventral margin of shell; comb of setae on proximal half of concave margin of postabdominal claw, increasing in length distally; stout spine on inner distal lobe of trunklimb I, with coarse, rounded tubercles on concave edge. See Daday (1898) and Rajapaksa (1986).]

13a(11b). Postabdomen large, broad, flattened, almost semicircular; armed laterally with long, spinelike setae in groups, those of distal groups projecting far beyond margin of postabdomen; margin of postabdomen with short, fine spinules; three median headpores close together, and with minor pores close-in laterally (Fig. 24) .... *Leydigia*

[Free posterior margin of shell very long; ocellus as large as compound eye or larger. Twelve species claimed for world in 1971, with maximal lengths from 0.8 to 1.15 mm. Two species are widely distributed in United States, and one tropical species barely gets into Florida.]

13b. Postabdomen smaller, and with shorter setae in lateral fascicles; headpores variable 14

14a(13b). Shell strongly sculptured, with longitudinal striae in posterior part and about five vertical striae anteriorly, roughly parallel to the anterior margin; dorsal edge of postabdomen minutely serrate; about 14 fascicles laterally; no median headpores, but instead there are two comma-shaped thickenings, which presumably contain the minor pores (Fig. 25) .... *Notoalona*

[A genus containing two tropical to subtropical species, only one of which gets into the United States. The headpore configuration is unique among the *Chydoridae*. Length to 0.44 mm. See Rajapaksa and Fernando (1987a).]

14b. Shell usually not sculptured at all, or if so, then rather weakly; postabdomen usually with well-developed marginal denticles and lateral fascicles; two or three median headpores, most commonly united by a sclerotized ridge, although sometimes completely separated (Fig.26) .... *Alona*

[This is the largest genus in the *Chydoridae*, with 52 species claimed in the world in 1971 and with more than 20 species in the United States and Canada. *Alona* intuitively consists of a number of genera, but these have not yet been defined satisfactorily.

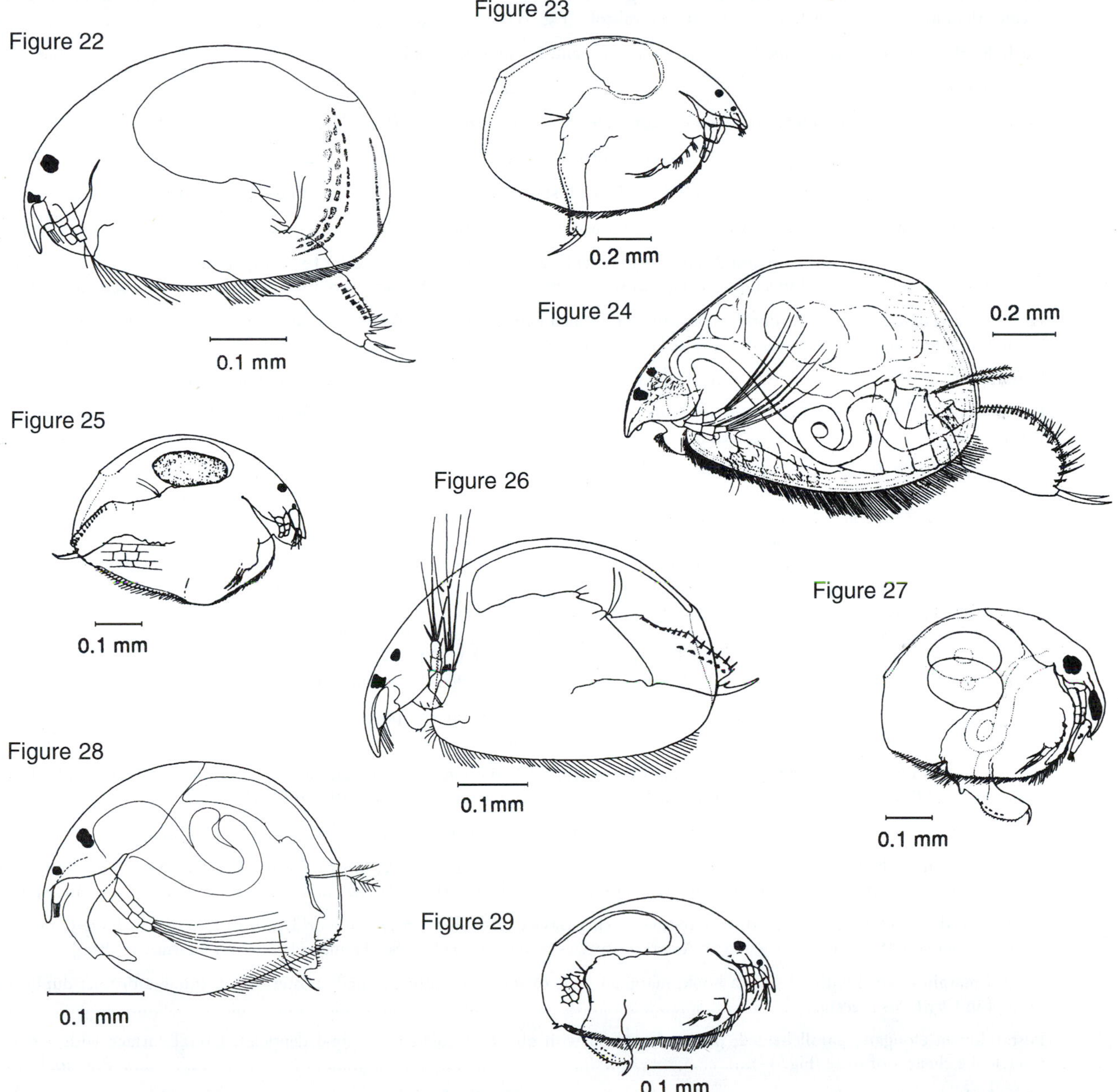

FIGURE 22 *Oxyruella brevicaudis* Michael and Frey, 1983. Skater's Pond, Monros Co., Indiana. From Michael and Frey (1983). *FIGURE 23* *Euryalona orientalis* (Daday, 1898). Ipiranga, Brazil. From Rajapaksa (1986). *FIGURE 24* *Leydigia acanthocercoides* (Fischer, 1854). Uplandia, Sweden. From Lilljeborg (1901). *FIGURE 25* *Notoalona freyi* Rajapaksa and Fernando, 1987. Homer Lake, Leon Co., Florida. From Rajapaksa and Fernando (1987a). *FIGURE 26* *Alona bicolor* Frey, 1965. Goodwil Pond, Barnstable Co., From Massachusetts. From Frey (1965). *FIGURE 27* *Dadaya macrops* (Daday, 1898). Polonnaruwa, Sri Lanka. From Rajapaksa and Fernando (1982). *FIGURE 28* *Ephermeroporus acanthodes* Frey, 1987. Audobon Park, New Orleans. From Frey (1982b). *FIGURE 29* *Dunhevedia americana* Rajapaksa and Fernando 1987. Glades Co., Florida. From Rajapaksa and Fernando (1987b).

Smirnov (1971) placed the species having just two median headpores in a separate genus, *Biapertura*. Three of the species in North America, currently named *A. affinis*, *A. intermedia*, and *A. verrucosa*, belong here. Smirnov and Timms (1983) found a much greater proportion of two-pored species in Australia, which do not seem to be closely enough related to be included in the same genus. Length of species in America north of Mexico: two-pored species to 1.05 mm; three-pored species 0.4–0.9 mm.]

15a(2b). Headshield not elongated posteriorly; in two-pored species, postpore distance usually considerably less than interpore distance.....16

[Includes the genera *Alonella, Dadaya, Disparalona, Dunhevedia*, and *Ephemeroporus*.]

15b. Headshield much elongated posteriorly, extending well beyond the heart to the middle of the back; postpore distance usually greater than interpore distance, sometimes by several-fold (Fig. 1D) 19

[Includes the genera *Anchistropus*, *Chydorus*, *Pleuroxus*, and *Pseudochydorus*.]

16a(15a). Only one median headpore, or none 17

16b. Two median headpores, well separated and not connected by a sclerotized ridge (Fig. 1D) 18

17a(16a). One median headpore in all instars (Fig 27); ocellus larger than eye ....*Dadaya*

[A single pantropical species occurring sparingly in the Gulf states. Color dark brown. Length to 0.41 mm.]

17b. One median headpore only in first instar; none in later instars (Fig 28); ocellus smaller than eye ....*Ephemeroporus*

[Postabdomen with long proximal and distal marginal denticles, and with shorter ones in between; labral keel large, with 1–4 teeth on anterior margin. A tropical to subtropical genus with many species, few of which have been described to date. See Frey (1982b).]

18a(16b). Postabdomen unique among chydorids, consisting of a much expanded postanal portion, having a series of short spines along dorsal margin and many small clusters of spinules laterally (Fig. 29) ....*Dunhevedia*

[Four species claimed for world, two in the United States, one of which is strictly southern. Length for 0.46–0.8 mm.]

18b. Postabdomen of more typical chydorid structure, with pre- and postanal angles and a well-developed series of postanal marginal denticles (Figs. 30 and 31) ....*Alonella* and *Disparalona*

[The taxon *Disparalona rostrata* of Europe was formerly in the genus *Alonella*. Fryer (1986) removed it from this genus because it had a long "sweeping" seta on trunklimb III, which the other species of *Alonella* were claimed not to posses. Fryer (1971) subsequently transferred *Alonella acutirostris* to *Disparalona*, and Smirnov (1971) did the same for *Alonella dadayi*. Michael and Frey (1984) closely examined a number of species of *Alonella* and of *Disparalona* and concluded that the sweeper seta is not an additional seta in a few species but rather is the proximal member of the gnathobasic filter comb, and that species still in *Alonella* have the same seta variably developed and not always distinctly separable from the *Disparalona* species. I know of no characters that will clearly separate the two groups of species, suggesting that the validity of *Disparalona* is questionable. *Disparalona* presently contains the species *D. acutirostris*, *D. dadayi*, *D. leei*, all occurring in North America, and *Alonella* the species *A. excisa*, *A. exigua*, *A. hamulata*, *A. nana*, and *A. pulchella*, out of a world total of 12. Maximal lengths of the North American taxa in *Disparalona* are 0.45–0.5 mm, and in *Alonella*, 0.25- 0.6 (Michael and Frey 1984, Hann and Chengalth 1981).]

19a(15b). Body elongate, somewhat flattened, usually with one or more teeth at the posterior-ventral angle of shell (Figs. 1D and 32) ....*Pleuroxus*

[Fifteen species claimed in world, with maximal lengths from 0.4–0.8 mm. Seven species in America north of Mexico, of which *P. procurvus* and *P. denticulatus* are most frequent and most abundant (Frey 1988c). Lengths from 0.5–0.8 mm.]

19b. Body globular, nearly round, seldom with any teeth at posteor-ventral angle 20

20a(19b). Ventral margin of shell anteriorly with hooklike development containing a groove, in which a stout seta with teeth on trunk limb I operates; postabdomen with a cluster of long, slender setae at distal angle (Fig. 33) ....*Anchistropus*

[Surface of shell and head reticulated, with meshes having wavy edges. Three species in world, at least two of which are parasitic on Hydra. Lengths 0.35–0.46 mm, the North American species being the smallest. See Hyman (1926) and Smirnov (1985).]

20b. Ventral margin of shell without such a hook; marginal setae of shell arise submarginally posteriorly, forming a distinct duplicature (except in *Chydorus piger*) 21

21a(20b). Postabdomen elongate, parallel-sided; postanal portion with about 15 slender marginal denticles; lateral surface with a row of fasciclelike clusters of setae (Fig. 34) ....*Pseudochydorus*

[*Pseudochydorus* presently contains a single species, distributed over much of the world, except possibly in South America. Length to 0.8 mm. It is a scavenger. See Fryer (1968).]

21b. Postabdomen shorter, with a smaller number of marginal denticles, and laterally with crescentic clusters of short spinules, which do not resemble fascicles (Fig. 35) ....*Chydorus*

[A highly complex genus, in which the species are difficult to separate, because they conform so closely to a common morphotype. At least ten species have been found in America north of Mexico. Smirnov (1971) recognized 18 species and many subspecies in the world, ranging in maximum length from 0.32–1 mm. The species occurring in North America range in length from 0.41 to 0.78 mm. The type of species of the family, *Chydorus sphaericus* (sens. str.), quite possibly does not occur in North America at all. See Frey (1980a, 1982a, c, 1987b).]

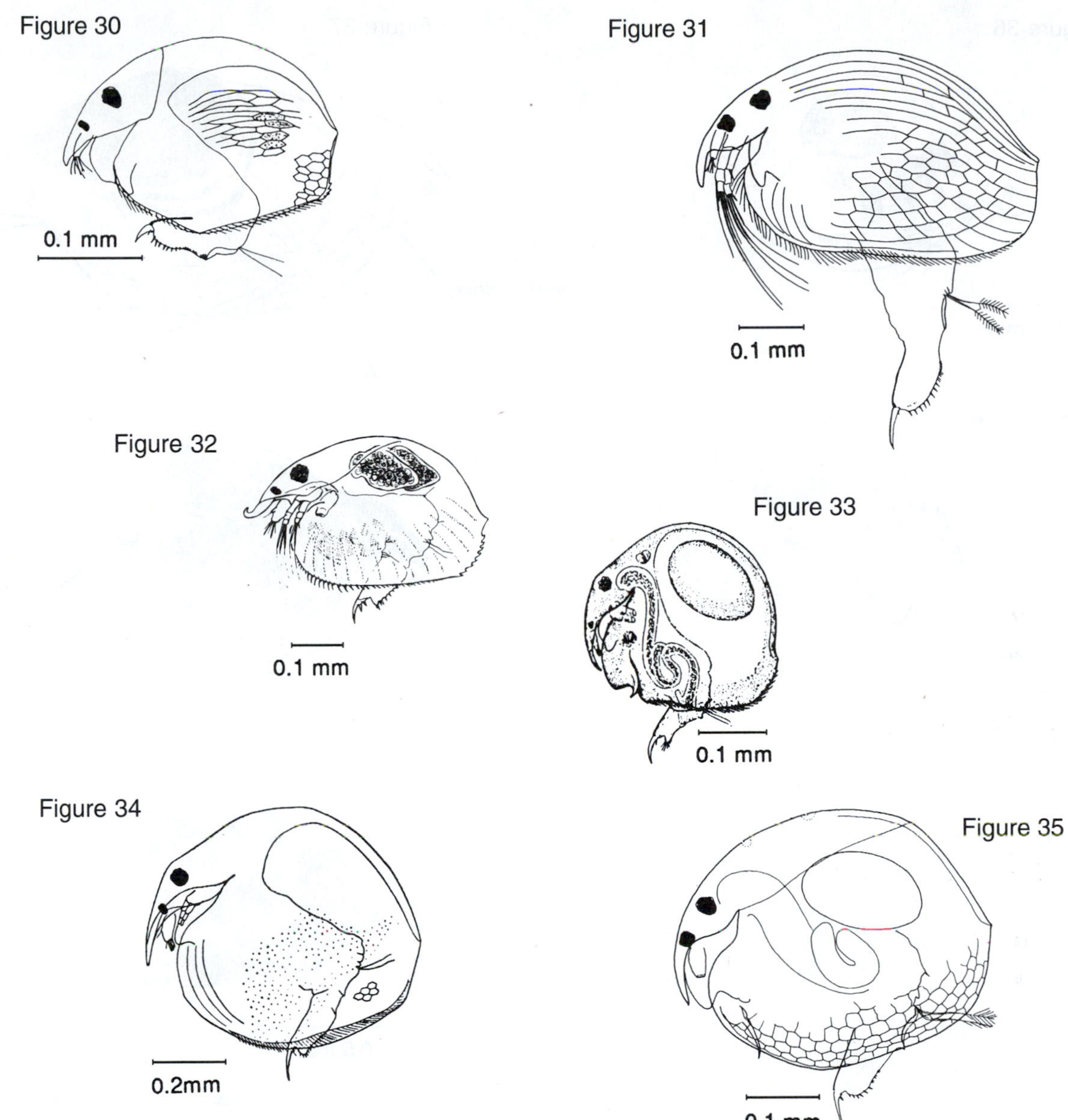

FIGURE 30 *Alonella pulchella* Herrick 1884. Kremer Lake, Minnesota. From Hann and Chengalath (1981). *FIGURE 31* *Disparalona leei* (Chien 1970) Griffey Lake, Monroe Co., Indiana. From Michael and Frey (1984). *FIGURE 32* *Pleuroxus procurvus* Birge, 1878. Lake Mendota, Dane Co., Wisconsin. 11 September 1987. (KE) *FIGURE 33* *Anchistropus minor* Birge, 1893. Eastern North America. From Birge (1918). *FIGURE 34* *Pseudochydorus globosus* (Baird, 1843). River Sutka, USSR. From Smirnov (1971). *FIGURE 35* *Chydorus brevlabris* Frey, 1980. Salmon Lake, Montana. From Frey (1980a).

F. Taxonomic Key to Genera of the Family Daphniidae and Moinidae

Fryer (1991) presents a complete and detailed analysis of the functional morphology of genera of the Daphniidae.

1a. First antenna about as long as the width of the head; usually a tooth with two points on the posterior angle of the Bostabdomen (near the claw), (Fig. 58) .... Family Moinidae 7

1b. First antenna less than half as long as the width of the head; teeth on postabdomen always single (Figs. 1A, 36–40). .... Family Daphniidae 2

2a. Ventral margin of carapace rounded and not pigmented 3

2b. Ventral margin of carapace straight and usually black 6

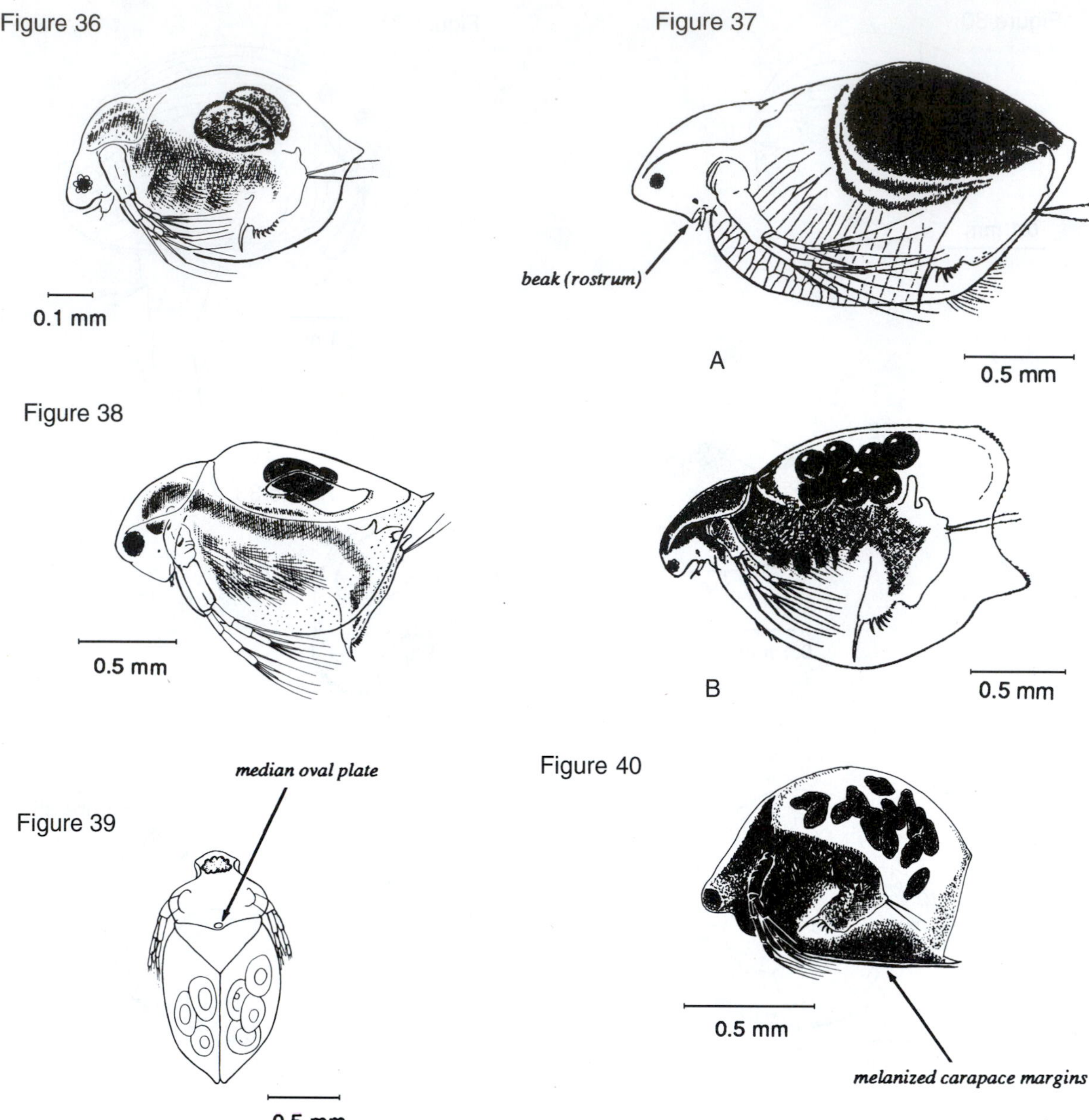

FIGURE 36 *Ceridodaphnia* cf. *quadrangula* (O.F. Mueller, 1785). Stewart Lake, Dane Co., Wisconsin. 5 September 1973. (KE) *FIGURE 37* *Simocephalus* (A) *S. vetulus* Schoedler, 1858. Gardener Pond, UW Arboretum, Dane Co., Wisconsin. 14 May 1987. (KE) (B) *S. serrulatus* (Koch, 1841). Coliseum Pond, Madison, Dane Co., Wisconsin. 15 May 1987. *FIGURE 38* *Daphniopsis ephemeralis* Schwartz and Hebert 1985. Baseline Pond, Ontario. 19 April 1988. (Coll. P.D.N. Hebert). (KE) *FIGURE 39* *Megafenestra nasuta* (Birge, 1879). Dorsal view. Elk Island National Park, Alberta. 5 August 1978. (KE). Redrawn from Dumont and Pensaert (1983). *FIGURE 40* *Scapholeberis rammneri* Dumont and Pensaert, 1983. Ojibwa, Ontario. 19 April 1988. (Coll. P.D.N. Hebert). (KE)

3a(2a). Adults with a tail spine, at least four times as long as broad and pointed (Figs. 1, 2, 4, 5) .... *Daphnia*

3b. Adults lack a tail spine, or if one is present, it is less than four times as long as broad and rounded 4

4a(3b). Fourth (distal) segment longer than the third segment on the four-segment branch of the second (swimming) antenna (Fig.38) .... *Daphniopsis*

[One species, *D. ephemeralis* in North America (Schwartz and Herbert 1987c); the genus is reviewed by Hann (1986).]

4b. Fourth segment shorter than the third segment on the four-segment branch of the second antenna 5

5a(4b). The second segment of the four-segment branch of the second antenna with a spine, the spine about 1/4 as long as the second segment (Fig 37) .... *Simocephalus*

[Allozyme studies of *Simocephalus* suggest there are at least five species in North America, two more than are recognized using morphological characters (Hann and Herbert 1986). Orlova-Bienkowskaja (1998) recognized at least eight North American species.]

5b. The second segment of the four-segment branch of the second antenna without an apical spine (Fig. 39) .... *Ceriodaphnia*

[Species of *Ceriodaphnia* were discussed by Brandlova *et al.* (1972), but are still poorly know and in need of taxonomic attention.]

6a(2b). Dorsum of headshield with a median oval plate about 20 μm in diameter, marked by slightly raised edges (Fig 39) .... *Megafenestra*

6b. Dorsum of headshield without such a plate (Fig 40) .... *Scapholeberis*

[With the exception of *Megafenestra* and *Scapholeberis* (Dumont and Pensaert 1983), and perhaps *Daphniopsis* (Hann, 1986; Schwartz and Hebert, 1987c), the daphniid genera are in need of revision. *Daphnia* are being carefully studied by Hebert (Hebert, 1995; Colbourne *et al.*, 1997), who recognizes about 34 species in North America. All of the daphniid genera tend toward the planktonic lifestyle. *Megafenestra* and *Scapholeberis* are the least planktonic; they are often found cruising along upside-down, at the surface of small pools. Their carapaces have flattened and turned-in margins (Dumont and Pensaert 1983), which may allow the carapace to serve as a suction cup, as in some chydorids. *Simocephalus* spends much of its time stuck to aquatic plants, using sticky mucous produced at the back of the neck. *Daphniopsis* has an extremely restricted habitat; it has been found in North America only in shallow temporary pools in maple forests. *Daphnia* and *Ceriodaphnia* are often restricted to open water of lakes and ponds. *Scapholeberis* and *Ceriodaphnia* adults are small, reaching about 1 mm, while the other genera produce adults in the range of 1 to 3 mm.]

7a.(1a). Branch of second antenna with four segments appears to have four terminal setae; ocellus (a small black spot just back of the eve) absent in North American species (Fig. 58) .... *Moina*

7b. Branch of second antenna with four segments has three setae and a short spine; an ocellus is present (Fig. 59)............ *Moinodaphnia*

[These two genera were monographed by Goulden (1968). The species are typically planktonic, but the majority are restricted to small temporary ponds, saline, or alkaline lakes. The habitat is often ephemeral, turbid, and warm. Adult females reach 1–2 mm, with one *Moina* species that is only about 0.5 mm long.]

G. Taxonomic Key to Genera of the Family Bosminidae

1a. First antennae attached separately to the head, immobile in female, usually nearly vertical and usually curving somewhat posteriorly (Fig. 42A) .... *Bosmina*
(with four subgenera) 2

1b. First antennae fused together in basal half, separated and curved outward in distal half (resembling a mermaid), firmly attached to head (Fig. 41) .... *Bosminopsis*

2a(1a). Lateral headpore located within five pore diameters of the edge of the headshield, just dorsal to the base of the second antenna and at the bottom of the reticulated part of the headshield (Figs. 42C, 43D); teeth of distal pecten (comb of spines) on postabdominal claw as wide as long and resemble an equilateral triangle 3

2b. Lateral headpore located more than 20 pore diameters from the edge of the headshield, near the point of articulation of the mandibles, in the reticulated region of the headshield; teeth of distal pecten much longer than wide and resemble needles................ 4

3a.(2a). Lateral head pore next to the basal striation at edge of the headshield (Fig. 42C); teeth of the proximal pecten of the postabdominal claw are hairlike, only the largest tooth is perceptibly triangular .... *Bosmina* (*Bosmina*) *longirostris*

3b. Lateral head pore between the two branches of the basal striation; teeth of the proximal pecten of the postabdominal claw are narrow triangles .... *Bosmina* (*Sinobosmina*)

4a(2b). Mucro (tail spine) absent in juvenile female (Fig. 43A), or if present, with minute incisions present only along the ventral margin; incisions may be lacking in the adult female; mostly in northern locations .... *Bosmina* (*Eubosmina*)

4b. Mucro present in juvenile female, with minute incisions or teeth present only on the dorsal margin (Fig. 42D); mostly in southern locations.. *Bosmina* (*Neobosmina*)

[There are just two genera in the family Bosminidae. The key to the subgenera of *Bosmina* is based on DeMelo and Hebert, (1994). The subgenus *B. Bosmina* contains the single species *B. (B.) longirostris*, which until DeMelo and Hebert (1994) had been considered to be very common. However, their revision indicated that *B. (B.) longirostris* is rare, and that instead most records probably should be for the subgenus *Sinobosmina*, either *B. (S.). leideri* or *B. (S.) freyi*. There are currently four North American species in the subgenus *Eubosmina* and three North American species of *Neobosmina*. The subgenera are separated using the shape of the carapace, spines on the postabdominal claw, and the position of the lateral head pores on the head shield (Fig. 42E–G). Bosminids are usually planktonic, with *Bosminopsis* the least so. Adult bosminids range in size from about 0.3–0.6 mm long. Because of their small size, bosminids often are abundant when fish predation is intense.]

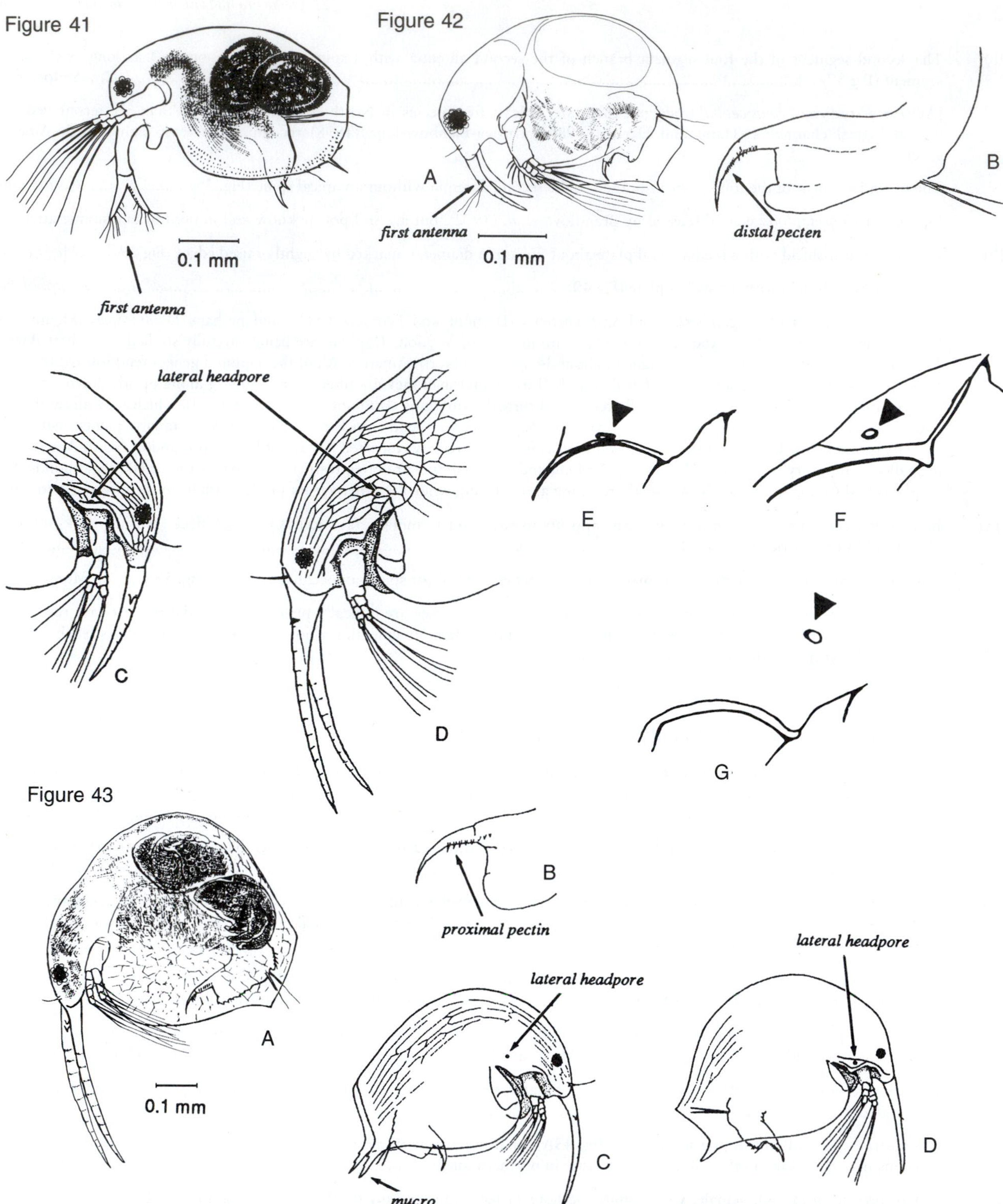

FIGURE 41 *Bosminopsis deitersi* Richard 1895. Singletary Lake, North Carolina. 8 August 1948. (coll. DGF #236. (KE) *FIGURE 42* (A) *Bosmina (Bosmina) longirostris* (O.F. Mueller, 1776). Lake Mendota, Dane Co., Wisconsin. 11 September 1987. (KE); (B) Postabdomen of A (KE); (C) Headshield of A, (KE); (D) Headshield of 20.43A. (KE); (E) Head pore position typical of *Bosmina* (Sinobosmina), (DeMelo and Hebert 1994); (F) Head pore position typical of *Bosmina (Bosmina)*, (DeMelo and Hebert 1994); (G) Head pore position typical of *Eubosmina (Bosmina)* and *Bosmina (Neobosmina)*, (DeMelo and Hebert 1994) *FIGURE 43* (A) *Bosmina (Eubosmina) coregoni* (Biard, 1850). Lake Waubesa, Dane Co., Wisconsin, 23 July 1988; (B) Postabdomen of A. (KE); (C) *Bosmina (Neobosmina) tubicen* Brehm, 1953. Presa Presidente Calles. Aquascalientes, Mexico. 8 October 1980. (KE); (D) *Bosmina (Sinobosmina) fatalis* Burchardt 1924. Redrawn from Kořínek (1971). (KE)

H. Key to the Genera of Superfamily Macrothricoidea

This key makes extensive use of setae on the second antenna (AII, or swimming antenna). A seta is longer than the segment it sits on, and it resembles a feather, having fine setules. A spine is usually shorter than the segment it sits on, and lacks the setules. The AII has two branches: the dorsal branch with four segments (look for a small segment at the base of one of the branches), and a ventral branch with three segments. For best results, dissect at least one AII off the specimen, and mount it so the branches and setae are spread out. This key is for adult females only, and animals are arranged to facilitate ease of identification, not to reflect phylogenetic relationships. For further details on morphology, see Smirnov (1992), Dumont and Silva-Briano (1998) and Silva-Briano (1998). For a careful morphological analysis of the macrothricids (and related taxa) see Olesen (1998).

1a. Carapace and head covered with fine hairlike setae (*Cactus* is South American, *Neothrix* is Australian; Family Neothricidae) (Fig. 56)........*Cactus* and *Neothrix*

1b. Carapace not covered with hairlike setae 2

2a.(1). AII dorsal (four-segmented) branch with no setae except for the three terminal setae........3

2b. At least one of the subterminal segments of the AII dorsal branch with a seta (Family Macrothricidae, in part) 8

3a.(2). Postabdominal claw with two basal spines, each longer than the width of the claw (Family Ophryoxidae) 4

3b. Postabdominal claw without basal spines, or with small spines much shorter than the width of the claw 5

4a.(3). The three-segmented branch of AII with a seta at the distal end of each of the two subapical segments (Family Ophryoxidae) (Fig. 45)........*Ophryoxus*

4b. The three-segmented branch of AII with no seta on the two subapical segments (Fig. 46)........*Parophryoxus*

5a.(3). The posterio-ventral corner of the carapace with setae about half as long as the carapace height (Family Acantholeberidae) (Fig. 47) ....*Acantholeberis*

5b. The posteroventral corner of the carapace with setae less than 0.1 times as long as the height of the carapace (Family Macrothricidae, in part)........6

6a.(5). Dorsal margin of the carapace with a small dorsal protrusion (about the size of the compound eye)........*Onchobunops*

6b. Dorsal margin of the carapace without a rounded protrusion 7

7a.(6). Dorsal margin of carapace with a backward-pointing tooth, dorsal margin of first antenna with teeth about 0.3 times as long as the width of the antenna a Palearctic genus) (Fig. 48)........*Drepanothrix*

7b. Dorsal margin of carapace without a backward-pointing tooth, margins of first antenna smooth or with small teeth less than 0.1 times as long as the width of the antenna (Fig. 53)........*Bunops*

8a.(2). Segments 2 and 3 (counting from the base) of the four-segmented branch of AII with setae........9

8b. Only segment 3 of the four-segmented branch of AII with a seta........11

9a.(8). Ventral margin of the carapace with flattened spines ("lanceolate") that are about twice as long as wide (Fig. 50) ....*Lathonura*

9b. Ventral margin of the carapace with cylindrical spines that are several times as long as wide........10

10a.(9). Postabdominal claw with a basal spine that is longer than the width of the claw (an Australasian genus) (Fig. 57)........*Pseudomoina*

10b. Postabdominal claw without a basal spine (a circum-tropical genus) (Fig. 52)........*Guernella*

11a.(8). Postabdomenal claw with a basal spine that is longer than the width of the claw (this character shown in the drawing in Silva-Briano, 1998, but not in Smirnov, 1992); a single large spine near the dorsal margin of the opening of the anus, the spine nearly as long as the width of the opening (a circum-tropical species) (Fig. 51)........*Grimaldina*

11b. Postabdominal claw without a basal spine, and anus opening without long spines 12

12a.(11). Postabdomen with large saw teeth along the margin; first antenna with a basal seta and four spinelike setae, all of which are longer than the width of the antenna (Fig. 49) ....*Streblocerus*

12b.. Postabdomen with fine hair-like teeth along the margin; first antenna with a basal seta and some other arrangement of setae toward the tip........13

13a.(12). First antenna with terminal seta that are less than 1/4 as long as the antenna, the setae are nearly the same length and thickness, the first antenna is cylindrical, not inflated toward tip (Fig. 55) ....*Wlassicsia*

13b. First antenna with at least two of the terminal setae that are at least 1/3 as long as the antenna, two of the terminal setae are at least 50% longer and thicker than the other setae; the tip of the first antenna may be wider at the tip than the base (Fig. 54) ....*Macrothrix*

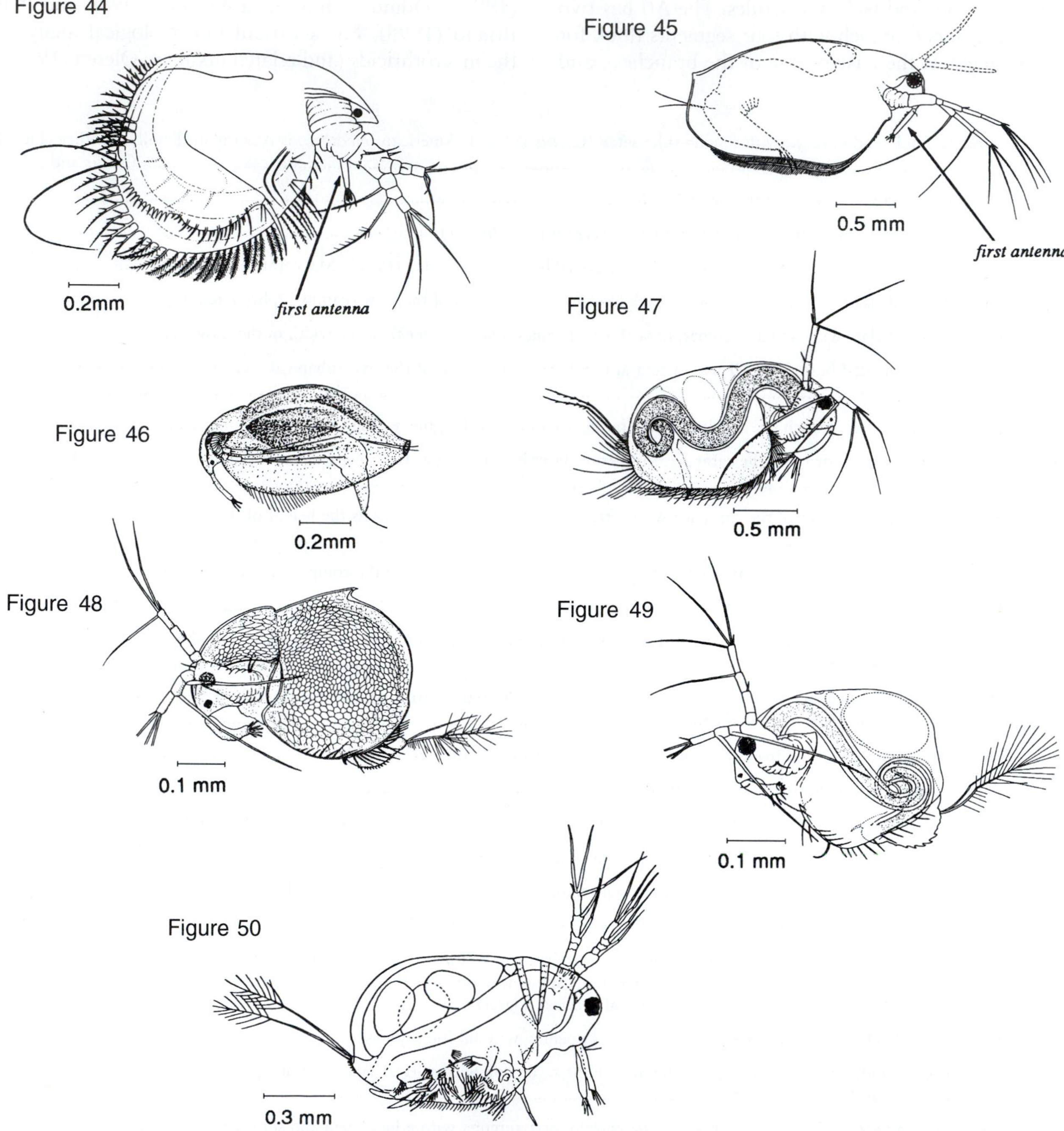

FIGURE 44 *Ilyocryptus sordidus* (Liévin 1848). English lake District. From Fryer (1974). *FIGURE 45* *Ofryoxus gracilis* Sars, 1861. English Lake District. From Fryer (1974). *FIGURE 46* *Paraphryoxus tubulatus* Doolittle, 1909. Anonymous Pond, Maine. From Doolittle (1911). *FIGURE 47* *Acnatholeberiscurvirostris* (O.F. Mueller, 1776). English Lake District. From Fryer (1974). *FIGURE 48* *Drepanothrix dentata* (Eurén, 1861). English Lake District. From Fryer (1974). *FIGURE 49* *Streblocerus serricaudatus* (Fischer 1849). English Lake District. From Fryer (1974). *FIGURE 50* *Lathonura rectirostris* (O. F. Mueller 1776). Near Leningrad, USSR. From Smirnov, after Sergeyev (1971).

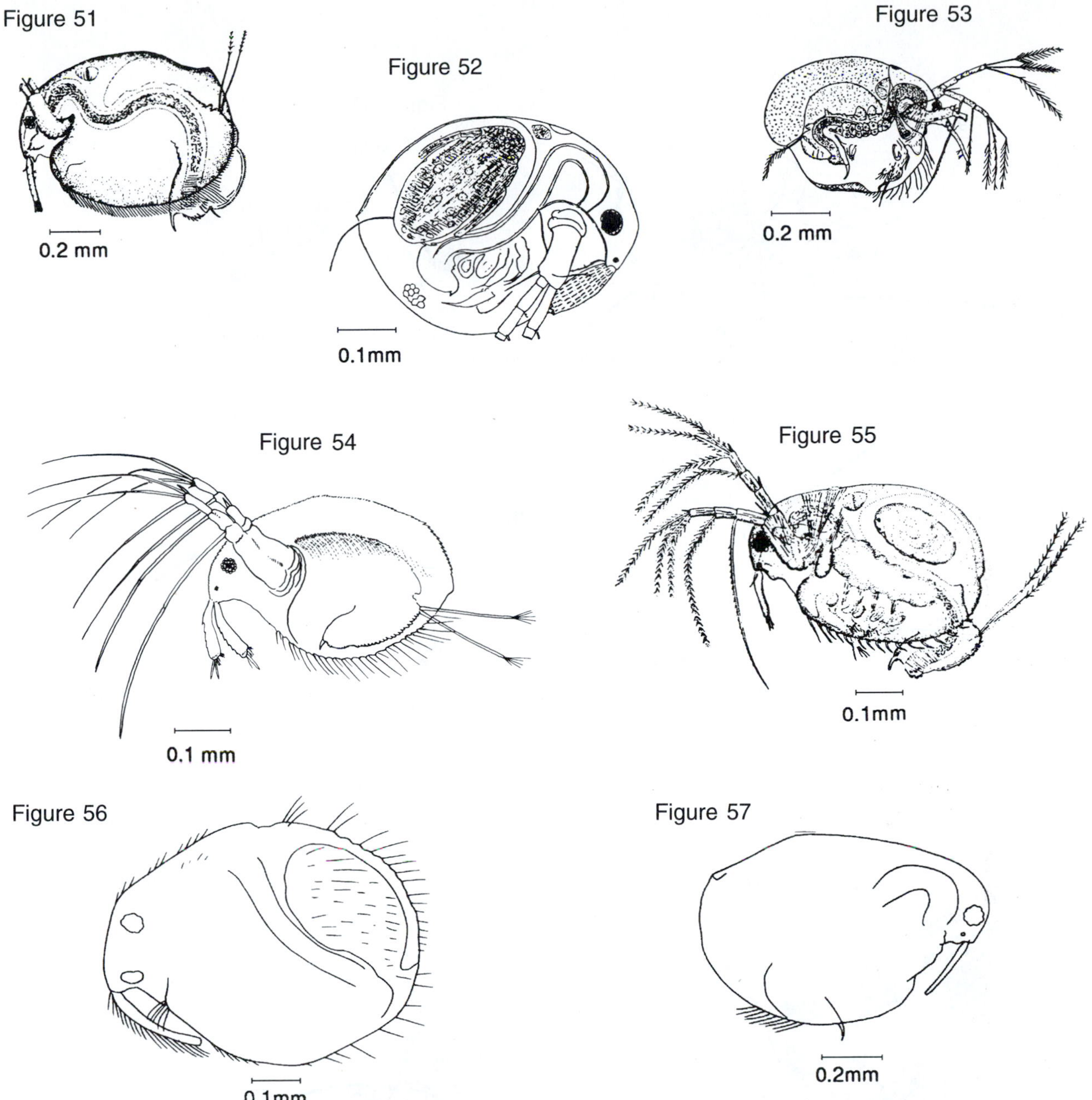

FIGURE 51 *Grimaldina brazzai* Richard, 1892. Louisiana. From Birge (1918). *FIGURE 52* *Guernella raphaelis* Richard, 1892. Mayumba, Gabon. From Richard (1892). *FIGURE 53* *Bunops acutifrons* Birge, 1893. Minocqua, Wisconsin. From Merrill (1893). *FIGURE 54* *Macrothrix rosea* (Jurine, 1820). Hook Lake, Dane Co., Wisconsin. 25 October 1972. (KE) *FIGURE 55* *Wlassicsia kinistinensis* Birge, 1910. Kinistino, Manitoba. From Birge (1910). *FIGURE 56* *Neothrix armata* Gurney, 1927. Fig. 491 from Smirnov (1992). *FIGURE 57* *Pseudomoina lemnae* (King, 1853). Fig. 475 from Smirnov (1992).

I. Key to the Orders Onychopoda and Haplopoda

See Rivier (1998) for details of these predatory cladocerans.

1a. Head cylindrical, more than twice as long as wide, swimming antennae with 36 setae (both branches) (Order Haplopoda, Fig. 60) .... *Leptodora*

[*Leptodora kindtii* (Fig. 60) is the sole species in this group. The adult females are up to 18 mm long. Despite their large size, the animals are so transparent as to be nearly invisible. They are common in the open water of large lakes, and occasionally are found in ponds. *L. kindtii* are voracious predators on smaller zooplankton.]

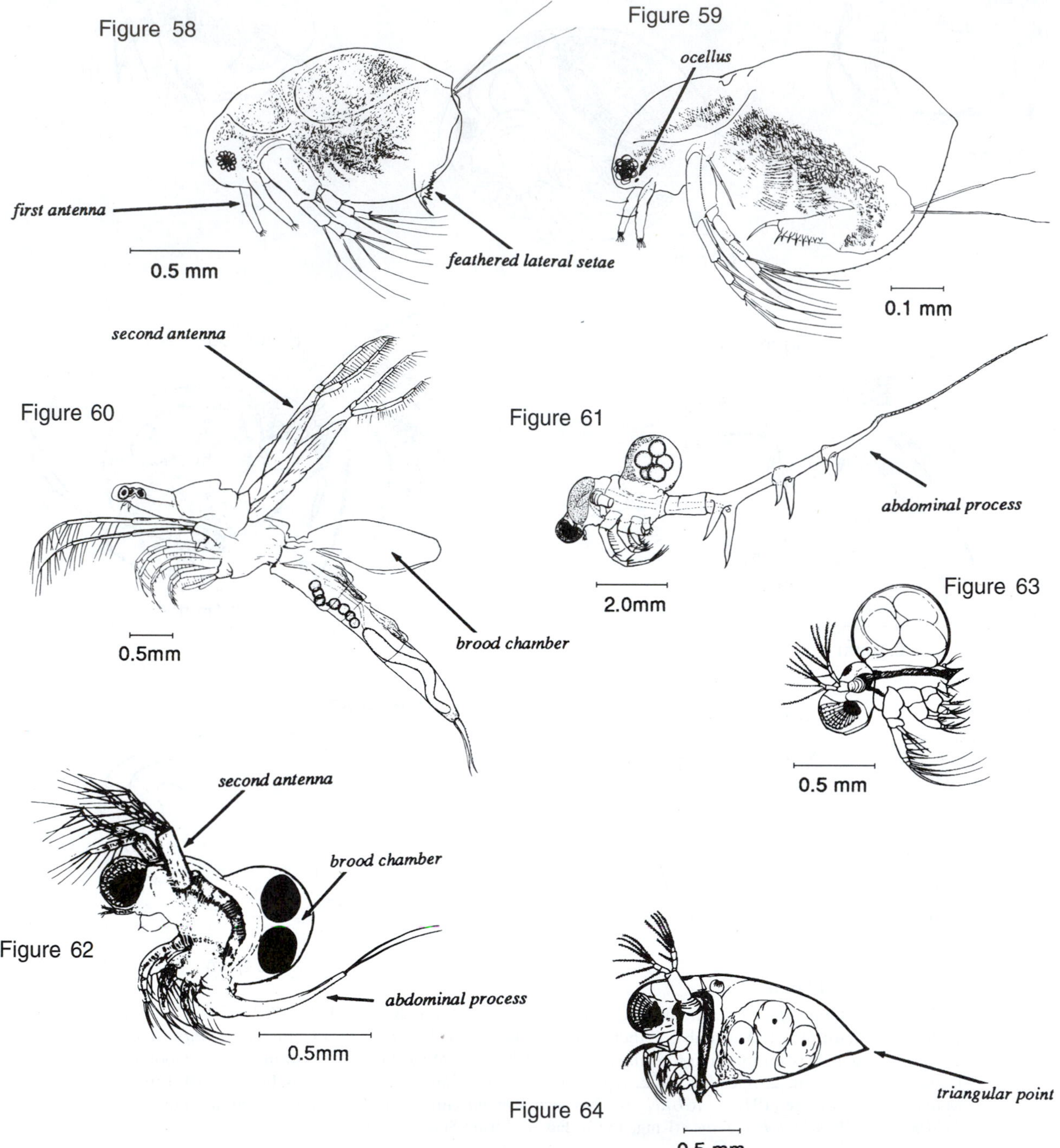

FIGURE 58 *Moina wierzejskii* Richard, 1895. 45 km NE of Las Cruces, New Mexico. 6 July 1984 (KE) *FIGURE 59* *Moinodaphnia macleayii* (King, 1853). Roadside canal, Highlands Co., Florida. 20 July 1960. (Coll. DGF #96). (KE) *FIGURE 60* *Leptodora kindti* (Focke, 1844). Lake Michigan, USA. From Balcer et al. (1984). Drawn by Nancy Korda. *FIGURE 61* *Bythotrephes cederstroemii* Schodler, 1877. Summer female from Lake Karesuando in Norrbotten. From Lilljeborg (1901). Redrawn by KE. *FIGURE 62* *Polyphemis pediculus* (Linnee, 1761). Lake Michigan, USA. From Balcer *et al.* (1984). Drawn by Nancy Korda. *FIGURE 63* *Evadne nordmanni* (LovÇn, 1836). Summer female from the sea near Dalaro. From Lilljeborg (1901). Redrawn by Cheryl Hughes. *FIGURE 64* *Podon leuckartii* Sars, 1862. Summer female from the sea near Bergen, Norway. From Lilljebory (1901). Redrawn by Cheryl Hughes.

1b. Head not obviously cylindrical, swimming antennae with 15 or fewer setae (both branches) (Order Onychopoda) 2

2a (1b). Adult female total length 1–2 mm, abdominal process shorter than the body 3

2a. Adult female total length greater than 2 mm, abdominal process longer than the body (Fig. 61, Family Cercopagidae) 5

3a (2a). Abdominal process slender and about half as long as the body, freshwater species in North America (Family Polyphemidae, Fig. 62) *Polyphemus*

[The single species in North America (*Polyphemus pediculus*) is often found along the margins of small bog lakes.]

3b. Abdominal process blunt and shorter than half of the body length, estuarine or marine species along the coasts of North America (Family Podonidae) 4

4a (3b). Body ends in a short extension of the abdomen, about twice as long as broad (Fig. 63) .... *Evadne*

4b. Body ends in a triangular point, but without a distinct abdominal process (Fig. 64) .... *Podon*

5a (2a). Swimming antennae with 15 setae (both branches) (Fig. 61) .... *Bythotrephes*

5b. Swimming antennae with 14 setae (both branches) .... *Cercopagis*

[*Bythotrephes longimanus* and *Cercopagis pengoi* are invaders of the Lawrentian Great Lakes, having come from Europe, probably in ballast water. See the Sea Grant web page at http://www.ansc.purdue.edu/sgnis/home.htm *B. longimanus* (probably from a Polish harbor) was first seen in Lake Huron in 1984 and *C. pengoi* (from the Caspian-Black Sea region) appeared in Lake Ontario in 1998.]

OTHER BRANCHIOPODS

XIII. ANATOMY AND PHYSIOLOGY OF THE OTHER BRANCHIOPODS

A. General External and Internal Anatomic Features

Martin (1992) provides a masterful and detailed description of branchiopod morphology, with many drawings and photographs. The non-cladoceran branchiopods are composed of four extant orders (Table I) They differ from cladocerans in a number of ways, including having more pairs of legs and paired compound eyes. The four orders are probably not closely related and are morphologically diverse (McLaughlin, 1980; Fryer, 1987).

Members of the first order, Anostraca (fairy shrimps), swim with their legs up (Fig. 65), have a pair of large compound eyes, and possess no shell-like covering. Anostracan morphology is described in detail by Linder (1941). Adults are 7–100 mm long and have 11, 17, or 19 pairs of thoracic legs along their rather delicate, transparent, and elongate body. The legs are all anterior to two fused genital segments (Fig. 65); hence, they can be called thoracic legs. There are no abdominal legs, but rather a pair of terminal lobes (cercopods, Fig. 67) on the last abdominal segment (telson). The genital segments contain the gonads (which extend into abdominal and thoracic segments). Males have a pair of ventral penes (Figs. 70C and 71). Females possess a medical brood pouch (Fig. 70B). Within the brood pouch are lateral oviducal pouches, a single median ovisac, and shell glands. Males have large and sometimes complex second antennae (Fig. 69A) and sometimes outgrowths from the head (Fig. 66C) or antennae (Fig. 70A). The second antennae are used for grasping females (Fig. 73B) but not for swimming (as in cladocerans and cochostracans).

Laevicaudata and Spinicaudata (the two conchostracan orders of clam shrimps, Figs. 75 and 76) swim with their legs down or forward, have a pair of close-set compound eyes (separated, except in *Cyclestheria* where they are fused), and a carapace that wraps around the entire body (Figs. 75 and 76). Adults are 2–16 mm long and have 10–32 pairs of thoracic legs covered by the transparent to brown carapace (Fig. 76). In species of Spinicaudata, the carapace closely resembles a clam shell, even having growth lines and umbo (Fig. 81). In Laevicaudata, the carapace is more globular, and only one Siberian) species has even a single growth line. The branched armlike appendages that can be extended from the carapace are the second antennae (Fig. 81); they are used for swimming, as in the cladocera.

Notostraca, or tadpole shrimps, are sometimes confused with trilobites by the unwary. These crustaceans swim with their legs down, have a pair of dorsal compound eyes, and possess a broad and flat carapace that covers the head and thorax (Fig. 82). Adults are 10–58 mm long and have 35–70 pairs of trunk legs partially hidden by their dark green or grown carapace. Immature notostracans have a transparent carapace that is folded along the dorsal midline and encloses the body and legs, making them appear somewhat like a conchostracan. Long filaments (endites) protruding from either side of the carapace are attached to the first pair of thoracic legs (Fig. 82).

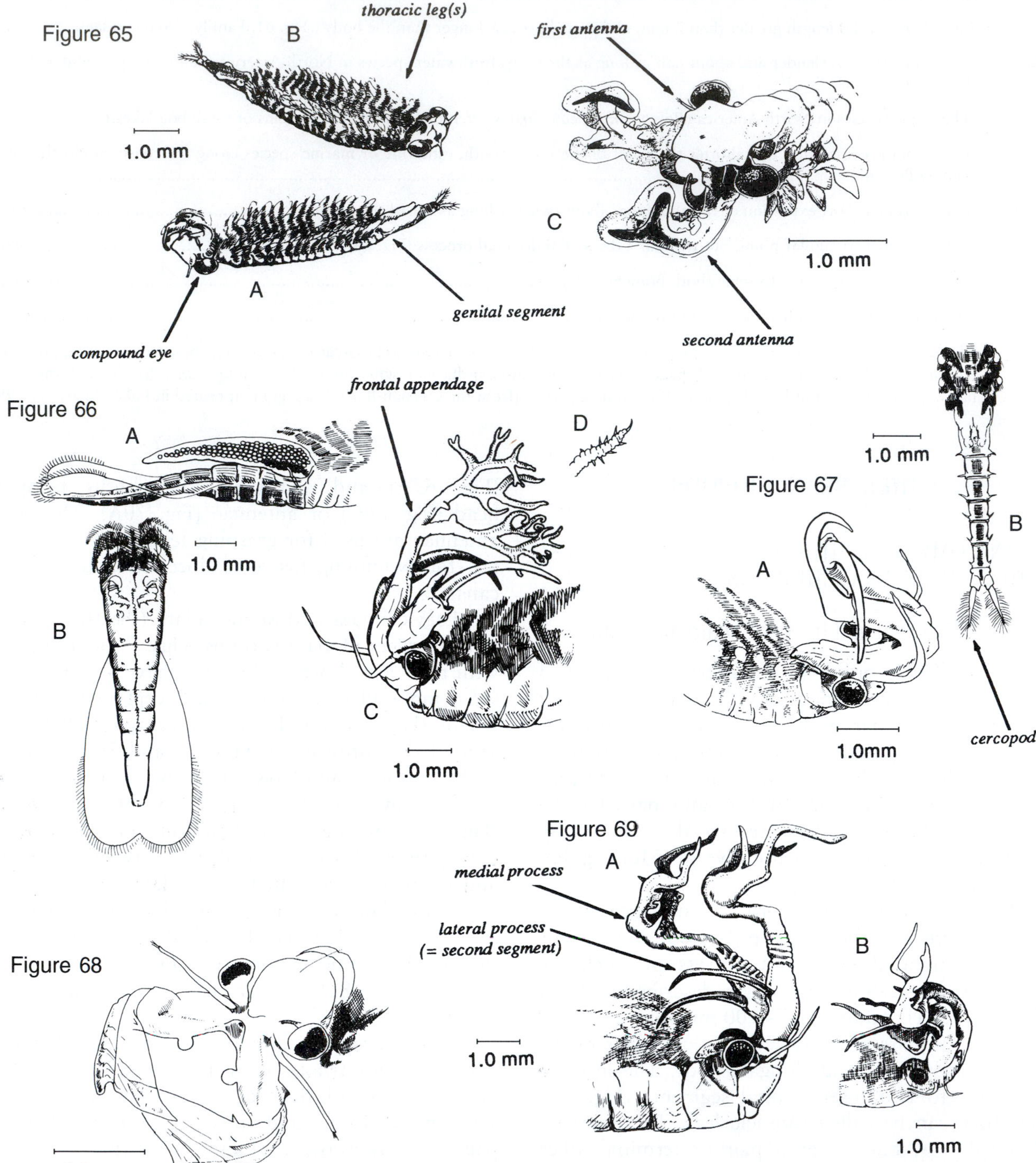

FIGURE 65 *Polyartemiella hazeni* (Murdoch, 1874). Pond #14, 18 July 1973, Point Barrow, Alaska. 18 July 1973. (A) female; (B) male; (C) head of male. (KE) *FIGURE 66* *Thamnocephalus platyrus* Packard, 1879. Pond near Ringold, McPherson Co., Nebraska. 28 June 1972. (Coll. S. Cooper). (KE) (A) Female abdomen from side; (B) male abdomen from below; (C) male head; (D) detail of spines on frontal process. *FIGURE 67* *Branchinella sublettei* Sissom, 1976. (Coll. D. Belk #717). (A) Head of male; (B) male abdomen. (KE) *FIGURE 68* *Artemia franciscana* (Kellogg 1906). Male head. Alkali lake, Owen's Valley, California. 2 July 1974. (KE) *FIGURE 69* *Streptocephalus texanus* Packard, 1971. Texas. (Coll. F. Wiman). (A) Head of male, second antenna unrolled; (B) Head of male, second antenna coiled. (KE)

Endites of the second pair of legs are less filamentous, and the succeeding legs lack filaments (Fryer 1988). The first pair of legs is used in swimming. Phyllopods posterior to the first pair are used for swimming and also walking, digging, and food handling (D. Belk, personal communication; Fryer, 1988).

Eggs are carried in brood pouches in the specialized eleventh pair of legs. The anterior lid of the concave brood pouch is derived from the exopodite of the leg and the bottom from the subapical lobe of the leg (Fryer, 1988). The eleventh pair of thoracic legs of males resembles the tenth and twelfth pairs. The gross morphological difference between males and females is that males lack the brood pouch (and ovaries) and tend to be slightly larger than the hermaphroditic "females" (Beaton and Hebert, 1988).

Maleless populations are usually (Akita, 1971; Beaton and Hebert, 1988; Longhurst, 1954; Fryer, 1988), but not always (Zaffagnini and Trentini, 1980) composed of hermaphrodites. Zaffagini and Trentini (1980) present evidence that reproduction in maleless populations is by automictic parthenogenesis (i.e., self-fertilizing hermaphrodites), not autogamic parthenogenesis as previously suggested (Longhurst, 1954). Beaton and Hebert (1988) found no electrophoretic variation in *Lepidurus arcticus* and were not able to clarify the role of males in reproduction.

The notostracan trunk is not properly divided into thorax and abdomen, because the legs continue past the genital opening on the eleventh segment and there are many more legs than "segments." The legs gradually decrease in size toward the tail. The body ends in a pair of long thin cercopods (Fig. 82); and in *Lepidurus*, a posterior extension of the telson (supra-anal plate, Fig. 82A).

B. Relevant Physiological Information

Branchiopods of the different orders are similar in regard to their physiological adaptations. Thus, the discussion of cladoceran physiology applies in general to the noncladoceran orders as well. *Artemia* is a model animal for physiologists (e.g., Sorgeloos *et al.*, 1987).

Conchostracans, notostracans, and anostracans probably depend more on gills for respiration than do the cladoceran orders (Eriksen and Brown, 1980a). Notostracans and anostracans show the usual correlation between surface area and the rate of oxygen consumption, although in conchostracans, small juveniles have an unusually high VO_2 (rate of oxygen uptake), perhaps due to the well-developed gills on the trunk legs.

Osmoregulation (maintenance of proper salt concentration in the blood and tissues) is important to these animals, which often live in conditions of extreme or variable salinity. The general mechanisms of osmoregulation are probably similar in all branchiopods (Peters, 1987), but there are many variations on the common theme. Different species, even in the same genus, show differences in internal salt concentrations, degree of osmoregulation, and tolerance of extremes (e.g., Horne, 1966; Broch, 1988).

The hemoglobins of *Moina* (a cladoceran), *Cyzicus* (Spinicaudata), and *Triops* (Notostraca) are electrophoretically similar (Horne and Beyenbach, 1974).

Further physiological topics, including diapause, respiration, and ion balance are discussed in Section VII. C.

XIV. ECOLOGY OF THE OTHER BRANCHIOPODS

A. Life History

Notostracans begin life as nauplii or metanauplii which hatch out of eggs, often resting eggs. Non-cladoceran branchiopods use a variety of signals to begin development from the egg, including day length, salinity level, temperature, and oxygen concentration. Depending on the species, survival through suboptimal conditions is accomplished by either (or both) diapause (an internal physiological state) or quiescence (slow development resulting from low temperature; Brendonck 1996). There are a dozen or so molts before maturity, and the adults continue molting throughout their life (Fryer. 1988).

Anostracans hatch from resting eggs as nauplii or metanauplii. Some species have the egg shell as nauplii (*Artemia*, Anderson, 1967; *Branchinecta*, Daborn, 1977a), and some hatch as advanced metanauplii with as many as ten pairs of swimming legs (*Eubranchipus*, Daborn, 1976). As the metanauplii molt, they add segments and apendages, gradually developing into adults (Fryer, 1983). Adults molt several times. Fertilization normally takes place just after the adult female molts. The brood pouch has a pore through which sperm are injected by one or the other of the penes of a male. [It is not clear how the eggs of *Artemiopsis*, which lacks a brood-pouch pore, are fertilized.] Fairy shrimp typically carry eggs until they molt. New eggs are released into the ovisac only after mating in all genera studied, except *Artemia* (Munuswamy and Subramoniam, 1985). Anostracans probably do not store sperm (Wiman, 1979).

Most conchostracans hatch from resting eggs as nauplii (Anderson, 1967). The carapace appears in about the third naupliar stage. By successive molts, the juveniles gradually come to resemble adults.

The eggs of most species of large branchiopods enter diapause soon after development commences and become resistant resting eggs (Belk and Cole, 1975). Exceptions include *Artemia*, in which development may continue in the ovisac and free-swimming nauplii may pass out the brood pouch, and perhaps *Cyclestheria hislopi* (Spinicaudata), which may exhibit direct larval development similar to cladocera (D. Belk, personal communication). Eggs produced parthenogenetically or sexually have similar diapausing characteristics (*Eulimnadia antlei*; Belk, 1972). Anostracans, conchostracans, and notostracans of temporary ponds typically have only one generation per wet episode and produce resting eggs. There is some evidence that diapausing eggs of different anotracan species specialize on hatching under specific rainfall patterns, allowing different species to take advantage of variations in climatic conditions from year to year in the same temporary pond (Donald, 1982; Gallagher, 1996).

Thiel (1963) and Belk (1972) raised *Triops* (Notostraca) and *Eulimnadia* (Spinicaudata), respectively, in the laboratory without drying the eggs. The eggs of *Triops* overwinter best if exposed to low oxygen. The eggs of *Streptocephalus seali* show sensitivity to strong drying conditions. This many limit their distribution, excluding them from ephemeral (temporary) ponds in highly desiccating environments (Belk and Cole, 1975). *Artemia* can develop directly in permanent lakes. *Artemia* of Mono Lake, a permanent saline lake in the Great Basin desert of mideastern California, hatch from resting eggs in the spring, resulting in adults that produce a second generation from eggs carried and matured in their brood sacs (Lenz, 1984).

The large size achieved by most noncladocerans requires a longer time to first reproduction, which is crucial in ephemeral ponds (Loring *et al.*, 1988). Notostracans (1–2 cm long) require 2–3 weeks to develop from the egg to mature adults (Rzoska, 1961). Anostracans (1–2 cm long) mature in 3 weeks at 15°C (Mossin, 1986) and about 45 days at 4–17°C (Modlin, 1982). Successive generations of *Artemia* are separated by a month or so in permanent lakes (Lenz, 1984). Depending on temperature, *Eulimnadia diversa* (3–4 mm) took 4–11 days to develop from the egg to an adult (Belk, 1972). Mature, 1 to 2 mm long cladocerans, which live in ephemeral ponds (e.g., *Daphnia, Moina*), could produce a clutch of eggs in as few as 2 days at high temperatures (Rzoska, 1961). Thus, cladocerans can mature in the most temporary of ponds, while the larger noncladocerans require ponds that retain water long enough to allow completion of their longer life cyles. These large branchiopods may then disappear as slower-developing predators and competitors become abundant (Sublette and Sublette, 1967; Dodson, 1987; Loring *et al.*, 1988).

B. Distribution and Biogeography

Because of their vulnerability to fish predation, the noncladoceran branchiopods typically occur in the absence of fish (Hartland-Rowe, 1972; Kerfoot and Lynch, 1987). However, in each group, there are a few reports of occurrences in lakes with fish. Anderson (1974) gives examples of all three groups in lakes. Some anostracans co-occur with fish in large, deep freshwater lakes where fish predation is low (*Branchinecta* in Canada, Anderson 1974; *Polyartemia* in Sweden, Nilsson and Pejler, 1973).

The modern distribution of noncladoceran branchiopods is based on 500 million years of history (Walossek, 1995). A branchiopodlike animal first appears in the Upper Cambrian. The major groups of branchiopods date from the Paleozoic, with the exception of cladocerans, which probably diverged from conchostracans in the Mesozoic or Cretaceous, perhaps barely 100 million years ago.

Conchostracans are rare north of southern Canada, while anostracans and notostracans are found from arctic islands to Central America. All three groups appear to be especially diverse in the arid southwestern United States. Anostracan diversity in Arizona depends mainly on two general factors: (1) chemical heterogeneity among (mostly temporary) habitats; and (2) thermal variation resulting both from ponds filling at different seasons and from altitudinal and latitudinal effects (Belk, 1977). It is uncommon to find congeneric assemblages within the same pond, although associations of different genera are common (Sublette and Sublette, 1967; Dodson, 1979, 1987, Donald, 1982; Kerfoot and Lynch, 1987); however, the two genera of tadpole shrimp, *Triops* and *Lepidurus*, do not appear to co-occur.

Conservation ecology is an important issue for the larger branchiopods; hence, a conservation-oriented newsletter, *Anostracan News*, is devoted entirely to them. As its name suggests, the focus is on fairy shrimp, but it often includes information about large branchiopods in general. Conservation of large branchiopods is an issue because many of the species have restricted ranges with only a few small populations in fragile habitats (such as rock pools in deserts) in danger of loss or modification. Several large branchiopod species are listed as endangered species by the U.S. federal government. As careful studies of species continue (for example, Torrentera and Dodson, 1995) additional species with restricted distribution are being discovered.

C. Physiological Adaptations

Ponds without fish usually exhibit characteristics that discourage fish; that is, the ponds tend to be small, isolated, and often have extreme environments that can be tolerated by branchiopods but not fish. Physiological factors important to the ecology of large branchiopods include high salinity and low oxygen. Because fishless ponds are typically temporary, large branchiopods also need a strategy, such as diapause, to live through periods when the pond is dry.

1. Salinity and Temperature

Some of these habitats are saline, and some species within the anostracan and conchostracan orders show a high tolerance to salinity (Hartland-Rowe, 1966; 1972; D'Agostino, 1980). *Artemia salina* and *Branchinecta campestris* are the most tolerant, surviving in waters several times as saline as seawater. Most fairy shrimp, conchostracans, and notostracans tolerate a wide range of salinities; but a few taxa, probably most species of *Eubranchipus* and some of *Streptocephalus* and *Branchinecta*, are restricted to low-salinity water. On the other hand, *Artemia* are rarely found in water as dilute a seawater (D'Agostino, 1980; Eriksen and Brown, 1980b). The greater salt tolerance of *Branchinecta mackini* (Broch, 1988) may provide an occasional refuge in time or space from its co-occurring congeneric predator *B. gigas*.

The distributions of *Artemia* species and populations within species can be limited by their salinity tolerances (Hartland-Rowe, 1972; Bowen *et al.*, 1985; Abreu-Grobois, 1987). The viability of several populations in the *A. franciscana* (Kellogg) superspecies depended on salinity, the concentration of bicarbonate, and the chloride/sulfate ratio. These differences in salinity requirements between populations may be a factor in the process of speciation. Populations in different types of saline lakes are genetically isolated in nature, because nauplii from one source will die in water from a different type of saline lake (Bowen *et al.*, 1985).

Because temporary ponds can often get quite warm, these crustaceans must frequently tolerate high temperatures (Horne, 1971; Hartland-Rowe, 1972; Eriksen and Brown, 1980a–c). Conchostracans do not perform well below 10°C, but they can live for at least a few hours at 30–35°C. Fairy shrimp from desert areas tolerate for several hours temperatures as high as 40°C. Several species that tolerate the highest temperatures are eurythermal, in that they can also swim at temperatures near 0°C. Notostracans have thermal tolerances similar to those of fairy shrimp, but there is significant variation among taxa. For example, the higher temperature tolerance of *Branchinecta mackini* provides a temporal refuge from its predator *Lepidurus lemmoni* (Eriksen and Brown, 1980b).

2. Oxygen

In warm climates, the combination of high temperatures and high productivities can lead to low oxygen concentrations during the night, or in the deeper levels of stratified ponds at any time (Horne, 1971; Loring *et al.*, 1988). Functional hemoglobin has been observed in some notostracans (*Triops*), conchostracans (*Cyzicus*), and anostracans (*Artemia*, Horne and Beyenbach, 1974; *Streptocephalus mackini*, D. Belk, personal communication). All of the larger branchiopods can regulate their oxygen consumption and live at low oxygen concentrations. In the laboratory, adult conchostracans (*Cyzicus californicus*) showed the greatest tolerance to low oxygen, surviving down to levels of about 0.6–0.7 mg/L (roughly 7–9% saturation at 20°C) (Eriksen and Brown, 1980a). Horne (1971) found a natural population of the conchostracan *Cyzicus setosa* able to tolerate oxygen concentrations of 0.14 mg/L for 2 h with no significant mortality. Anostracans (*Branchinecta mackini*) tolerated oxygen concentrations down to about 1.4–2.8 mg/L (roughly 15–30% saturation at 20°C) (Eriksen and Brown, 1980b). Notostracans (*Lepidurus lemmoni*) have a minimum oxygen tolerance similar to anostracans, and show the least ability to regulate oxygen consumption at low oxygen concentrations (Eriksen and Brown, 1980c). Oddly enough, a field study of lethal oxygen thresholds of these animals revealed that conchostracans (*Eulimnadia inflecta*) had a lower tolerance than two anostracan species (*Eubranchipus moorei* and *Streptocephalus seali*) (Moore and Burn, 1968). The apparently greater tolerance of the anostracans may have been due to the behavioral tendency of anostracans (but not *E. inflecta*) to seek the oxygenated surface of the pond. All three groups show a maximum rate of weight-specific oxygen consumption at about 25°C (Eriksen and Brown, 1980a–c).

The saturation level of oxygen concentration is reduced exponentially by salinity. Seawater holds about 20% less oxygen than freshwater at corresponding temperatures and pressures. More saline waters hold even less oxygen. Broch (1969) suggested that *Branchinecta campestris* may be limited to waters of lower salinity than *Artemia salina*, both because of salt tolerances and because *Artemia* produces hemoglobin, which may be necessary at high concentrations of salt and the correlated low oxygen concentrations.

3. Diapause

In Anostraca, Conchostraca, and Notostraca, the eggs are carried for at least a short period by the adult female. After a few cell divisions, the egg typically enters diapause. This resting egg (actually a developing embryo) has a dark covering (tertiary envelope or shell) and is prepared to survive drying, heat, freezing, and ingestion by birds (Belk and Cole, 1975). Belk (1970) removed the covering of diapausing conchostracan embryos to demonstrate that the shell does not reduce desiccation effects, but may reduce mortality due to abrasion or intense (ultraviolet) sunlight.

Signals that break diapause include temperature and oxygen concentration (Hart-Rowe, 1972; Belk and Cole, 1975; Clegge and Conte, 1980; Wiggins *et al.*, 1980; Mossin, 1986). In some anostracans, the final signal that breaks diapause may be a combination of high oxygen and an increase in CO_2 concentration (Mossin, 1986). In fact, the fastest average hatching time of the European anostracan *Siphonophanes grubei* occurred at pH of 5.5 (associated with high CO_2 concentration). At least one conchostracan, *Caenestheria setosa*, has a requirement for high temperatures to break diapause; its resting eggs hatch between 20–35°C. Hatching in *Eulimnadia antelei* (Spinicaudata) was inhibited by high salinity and darkness (Belk, 1972).

Resistant eggs appear to be an adaptation to temporary habitats and long-distance dispersal (Belk and Cole, 1975). The habit of carrying diapausing eggs probably has two major advantages (Daborn, 1977b). First, the eggs are protected from numerous egg predators, such as other crustaceans and chironomid larvae. Second, the eggs are more likely to be ingested by a bird for long-range dispersal. As with cladocerans and copepods, ingestion of dipausing eggs may be an important means of dispersal. Proctor (1964) raised *Artemia, Streptocephalus, Triops*, and *Cyzicus* from feces of mallard ducks, *Anas platyrhynchos*. Yet, curiously, anostracans show subtle genetic differences among populations, even when close to each other, suggesting that the dispersal rate is actually very low (Boileau *et al.*, 1992).

D. Behavior

1. Swimming Behavior

Notostracans are detritus feeders and predators, plowing through the superficial sediments to capture anostracans and benthic invertebrates. When feeding, the tadpole shrimp is face down and its many legs moving rhythmically, digging into the sediments, and throwing up a plume of sediment behind the animal. Most anostracans feed on their backs, filtering with rhythmic movements of their legs as they swim through the water. *Artemiopsis* and *Branchinecta* also roll over and scrape surfaces with their thoracic legs (Daborn, 1977b). *Branchinecta gigas* catches and holds smaller anostracans with its spiny phyllopods (Fryer, 1966; Daborn, 1975).

Conchostracans swim slowly, spending most of their time skimming along the pond bottom in aquatic vegetation (Martin *et al.*, 1986), or above algal mats (e.g., Sissom, 1980). Notostracans and anostracans will swim away from a person walking up to the edge of a pond; conchostracans show no such response.

2. Mating Behavior

Tadpole shrimp are generally considered to be hermaphrodites that need to find a mate for exchange of sperm. The immediate goal of notostracan mating behavior is for two individuals to join their eleventh pair of thoracic legs. Male and female gonopores are in the same position at the base of the first endopodite on these legs (Akita, 1971), so some bodily contortion is required to deposit spermatozoa into the brood pouch of the partner. Engelmann *et al.* (1996) reported several populations composed of males and females and suggested that sexual reproduction may be common in notostracans.

Most conchostracan species have even numbers of males and females, which are morphologically distinct (Sassaman, 1995). Males have specialized hooks on the first (Laevicaudata) or first and second (Spinicaudata) pairs of thoracic legs. These hooks are probably used in a similar manner to those of cladocerans; the hooks hold onto the female and position the animals for ventral apposition copulation (Mathias, 1937; Martin and Belk, 1988). In laboratory cultures and in nature, it is common to see a female swimming about with a male holding on to the lower (posterior) margins of her carapace. In some species, females can fertilize themselves.

Anostracans typically have two sexes, although in some populations the females reproduce asexually (Browne, 1993; Badaracco *et al.*, 1995). The sexes are dimorphic. Males possess enlarged second antennae, often with long and complex fingerlike appendages. These antennae are used to grasp the female just anterior to her genital segments (just behind the last thoracic leg). Only the second antennae are prehensile. Belk (1984) offered evidence indicating no "holding" role for antennal appendages and suggested none for frontal appendages. Anostracans may swim in tandem for days, especially *Artemia* and *Artemiopsis* (Daborn, 1977b). Belk (1984) described the mating behavior of *Eubranchipus serratus*:

> "Once a sexually active male *Eubranchipus* locates another fairy shrimp, he positions himself with his head below the dorsal surface of the other shrimp's genital segments. From this station-taking posture (Moore and Ogren, 1962), the male may be able

to assess suitability of mates (Wiman, 1981). Assuming the station-taking male is below a female, his next action is to clasp her rapidly just anterior to her genital region and his second antennae. As he grabs her, he extends his antennal appendages, laying them along her back. Holding his body stiff with his phyllopods pressed against the ventral surface of his trunk and flexing only at the neck region, he bats at the female several times with quick ventral movements. As soon as she is grabbed, the female rolls the anterior half of her body into a ball with her head pressed ventrally against her genital region. Her abdomen remains straight. This position is usually held only a second or so. Then the female will either struggle violently and dislodge the male, or she will relax and begin swimming slowly. If the female accepts the male, he too relaxes and begins swimming with her while still clasping her with his second antennae. His antennal appendages remain stretched along her back. The male now attempts intromission, trying from first one side, then the other. Once one of his penes is inserted into the female's ovisac, the male assumes an S-posture. He ceases swimming and his phyllopods make only slight vibration movements. The female may continue to swim or she may lie on the bottom. Disengagement is effected by the female with one or more rapid jerking movements. As soon as copulation terminates, the male resumes normal swimming." (p. 69)

Mating in *Streptocephalus mackini* was similar to that of *E. serratus*, except that males took station above rather than below females (Wiman, 1981). Male *E. serratus* were successful in grasping a female in only about 20% of their attempts (Belk, 1984), whereas *S. mackini* males mate with "undiscriminating eagerness," approaching males as well as females and also orienting toward fairy shrimp of different genera (Wiman, 1981). Copulation is terminated by the female, possibly when she releases his penis from her gonopore (Wiman, 1981). Males are sometimes observed being towed along by their penis and one occasionally finds males with only one functional penis, the other being everted permanently. The spines on anostracan penes may be used to hold the penis inside the gonopore during swimming (Wiman, 1981). The pattern of spines on the penis is a useful taxonomic character (Brendonck, 1995, 1997).

Eubranchipus bundyi occurs in April in roadside ditches. Dodson (unpublished) watched four adult males and one adult female that were swimming in several adjacent depressions in a marshy area near Madison, Wisconsin. Each depression was about 20 cm deep, with dead grasses at the bottom. One or two males swam around the edge of each depression, making circles about 20 cm in diameter and keeping at middepth, or about 10 cm most of the time. Although both sexes beat their legs 8–10 times/s, males swam at about 1 cm/s, females at 0.1 cm/s. Females swam in long straight lines, changing direction every minute or so with a quick burst of speed. They emerged from grass cover for a few minutes every few hours. A female that came up to the midwater level was followed by a male, who adjusted his swimming rate to hers, and swam with his head a millimeter or so below her uropods. He followed for about 10 s, in, then moved closer. The female escaped with a quick burst of speed, about 10 cm/s for 4 cm. The male showed no further interest. The impression these observations give is that males patrol bits of open water, and females spend most of their time in concealment. When the females desire to mate, they swim upward, find a male, mate, and then return to the protection of the bottom. The leklike pattern of male and female fairy shrimp mating behavior may result from their reproductive strategy (Wiman, 1981). Because females product large clutches of eggs while males spend little of their energy budget on gametes, life-history theory predicts that females should show discrimination in mating while males should be active and mate as many females as possible. As a consequence of higher activity levels, males are more vulnerable than females to predators.

E. Foraging Relationships

In general, anostracans are free-swimming filter feeders, conchostracans process detritus from surfaces or collect plankton, and notostracans are benthic detritus feeders tending toward carnivory (Kaestner, 1970).

There are many records of anostracans eating algae. The filtration mechanisms of the thoracic legs is described by Fryer (1983). Cannon (1933) analyzed the feeding movements of fairy shrimp legs, and assumed they lived mainly on algae. However, Bernice (1971) reported that *Streptocephalus dichotomus* ate a variety of foods, including large algae (e.g., diatoms and blue-green bacteria), ciliates, small invertebrates (e.g., rotifers, nematodes, and cladocerans), and crustacean eggs. The diet of *Artemia* is well known both for natural populations and for laboratory, cultures (D'Agostino, 1980). *Artemia* feed on coccoid green algae in Mono Lake (Winkler, 1977), and Baird (1850) relayed an account of *Artemia* catching and eating their own nauplii. *Branchinecta gigas*, the largest extant anostracan, catches and eats other smaller fairy shrimp, and possibly other small zooplankton and benthos (Fryer, 1966; Daborn, 1975). Several species of *Branchinecta* and *Artemiopsis* (Fryer, 1966; Daborn, 1977b) have the habit of rolling over and scraping the pond substrate with their thoracic legs. Several species of *Branchinecta* have been seen scooping up detritus and then sorting through it. *Branchinecta* species have a row of short curved spines along the inner distal edge of the legs which are probably used for scraping. Even congeneric species show differences in their feeding behavior. *Eubranchipus holmni* has spinier legs than *E. vernalis* and shows more scraping behavior (Modlin, 1982). This scraping action seems a preadaption for feeding on benthos and catching larger planktonic prey, methods employed by the predaceous *B. gigas*.

Conchostracans feed somewhat like cladocerans, by drawing water into the carapace to remove food particles using the phyllopods (Kaestner, 1970). They probably can remove large particles, such as large algal colonies, small crustaceans, and rotifers. They are able to feed in soft mud.

Notostracans feed by plowing along the surface or through muddy sediments. They appear to pump mud posteriorly and probably extract any organic particles. Their diet includes algae, amphibian eggs, smaller crustaceans, insect larvae, anostracans, and tadpoles (Kaestner, 1970; Dodson, 1987). They may be significant predators on resting eggs of other aquatic invertebrates such as fairy shrimp and clam shrimp (Daborn, 1977b). Notostracans capture mosquito larvae, which is seen as a benefit to humans, but they are also occasionally pests in rice fields, eating the young plants off at mud level (Dodson, 1987).

Tadpole shrimp, available commercially as dried eggs, can be easily cultured in the laboratory. Horne (1966) cultured *Triops* in the lab using live fairy shrimp. In the lab, notostracans have been observed to kill and eat goldfish (C. Sassaman, personal communication).

F. Population Regulation

Cladoceran populations are often controlled by a combination of physical and biological factors, with food limitation and predation being important for several generations. Anostracans, conchostracans, and notostracans typically have one generation each time their habitat appears, in some cases briefly once a year (Belk and Cole, 1975; Loring *et al.*, 1988). A major strategy is to make as many small resistant eggs as possible in the shortest amount of time. These eggs then hatch when the pond fills again, and the animals typically show rapid growth due to a combination of high temperatures and physiological capabilities specialized for rapid growth (Loring *et al.*, 1988). Population regulation can be accomplished by physiologic factors (e.g., salinity, Broch, 1988).

In the laboratory when food is not limiting, *Artemia* females produced the same total number of eggs with little regard as to whether they are mated early or late in life (Browne, 1982). They accomplished this by having larger clutches closer together when mated late in life. Only at limiting food levels was there a trade-off between adult life span and number of eggs produced.

Populations of *Artemia* in permanent lakes can be reduced by bird predation. Mono Lake, a large saline lake in mideastern California, hosts a dense population of *Artemia*. The *Artemia* occupy the epilimnion (about 15 m deep) from early spring to December. Cooper *et al.* (1984) found that the *Artemia* population declined rapidly in response to feeding by eared grebes using the lake as a migratory stop in August and September. The *Artemia* in Mono Lake typically have two generations per year, with a significant fraction of the second generation being consumed by grebes. The *Artemia* population of Mono Lake lives at great enough depths to be protected from predation by most waterfowl, which are surface feeders or shallow divers. However, waterfowl can be important predators of large branchiopods in shallow ponds (Dodson and Egger, 1980), and flamingoes are a major predator of *Artemia* in shallow tropical ponds and lakes (Hulbert *et al.*, 1986). Anostracans are an important part of the diet of many waterfowl, especially dabbling ducks (Swanson *et al.*, 1985) feeding in midwestern prairie ponds. *Artemia* in a permanent lake are a significant source of food for California gull chicks (Lenz, 1984).

G. Functional Role in the Ecosystem

Because of their large size and abundance, the large branchiopods living in ephemeral ponds are a major link between primary production and detritus on the one hand and predators (often terrestrial) on the other hand (Loring *et al.*, 1988). Temporary ponds in southwestern North America, especially in physically simple playas or rock pools, provide community ecologists with simple, highly replicated, and easily manipulated natural communities (Dodson, 1987; Loring *et al.*, 1988). These natural ponds are easily mimicked by constructing artificial ponds for community studies of ephemeral pond branchiopods (Tribbey, 1965).

Ephemeral ponds in arid areas are closely linked to the surrounding terrestrial landscape (Loring *et al.*, 1988). The ponds are an important water source for terrestrial animals, which then act as dispersal agents for crustaceans. Primary production both in the ponds and on the pond bottoms when they dry are an important food source for terrestrial animals. The position of *Artemia* in the food web of a permanent saline lake is discussed by Winkler (1977).

H. Toxicology

Anostracans, especially species of *Artemia* and *Streptocephalus*, are frequently used in tests of the acute or chronic effects of environmental contaminants. The *Artemia* Reference Center of Ghent is a huge data base, especially for the biology, aquaculture, and toxicology of *Artemia*. Endpoints of branchiopod bioassays are typically survivorship or fecundity (for example, see Centeno *et al.*, 1993), although more subtle endpoints, such as hatching of cysts (resting eggs) is also used (Crisinel *et al.*, 1994).

XV. CURRENT AND FUTURE RESEARCH PROBLEMS

Temporary ponds are a chancy habitat to study, because they do not always contain water. However, biologists who are willing to work around the vagaries of these animals and their habitats have an almost unlimited field for genetic, ecological, and behavioral studies (e.g., Loring *et al.*, 1988). Areas of great potential include application of molecular techniques to questions of phylogeny and speciation, field studies of patterns of life-history evolution, grazing and predator–prey relationships, individual behavior, population dynamics, community structure, and landscape ecology.

XVI. COLLECTING AND REARING TECHNIQUES

Collecting these animals is easy enough when they are in their adult forms. A large-mesh plankton net, a D-frame aquatic dip net, or even a large-bore pipette is sufficient to catch them. The *Eubranchipus* illustrated in this chapter were collected with a spoon. A traditional technique is to collect mud from a dried habitat, add water, and rear the animals that hatch from the resistant eggs. Rearing is easy. Anostracans and conchostracans eat commercial tropical fish food. Notostracans can be fed anostracans, earthworms, or dried cladoceran fish food. *Artemia* diets are myriad, as are recipes for culture media. The most common culture medium is artificial seawater, although more specialized media may give better viability (Bowen *et al.*, 1985).

XVII. TAXONOMIC KEYS FOR NON-CLADOCERAN GENERA OF BRANCHIOPODS

A. Available Keys and Specimen Preparation

General keys to North American branchiopod genera can be found in Pennak (1989). More recent revisions of large branchiopod groups include keys for Streptocephalidae (fairy shrimp, Maeda-Martinez *et al.*, 1995) and Lynceidae (clam shrimp, Martin and Belk, 1988).

This key applies to those crustaceans that have more than ten pairs of thoracic (or trunk) leaflike legs and paired compound eyes. As in the cladoceran orders, the taxonomy of the larger branchiopods in undergoing revision based on studies of morphology (e.g., Brendonck, 1995, 1997, Maeda-Martinez, 1995) and molecular studies (e.g., Wiman, 1979; Browne and Sallee, 1984). The animals in this key can be identified to genus with a good dissecting microscope, without making dissections. Male or female conchostracans and notostracans may be identified to genus; males only are keyed for Anostraca.

B. Taxonomic Key to Genera of Non-Cladoceran Freshwater Branchiopoda

1a. Body soft and flexible, not covered by a shell (order Anostraca) (for a key to species see Belk, 1975) 3

[Belk and Brtek (1995) prepared a checklist of the global anostracan fauna. They provide a strong argument for conservation of these animals and their habitat.]

1b. Body covered at least in part by a hard shell (Figs. 75, and 82) 2

2a(1b). Head and body covered with a clamlike shell that wraps around the legs; the abdomen is enclosed inside the shell (conchostracan orders Laevicaudata and Spinicaudata) (Figs. 75 and 76) 13

2b. Head and body covered dorsally with a flattened shell; the abdomen is straight and trails out behind the shell (order Nostraca) (Fig. 82) 20

[Fryer (1988) is an excellent source for notostracan biology and taxonomy.]

3a(1a). Seventeen pairs of thoracic legs (Fig. 65) .... *Polyartemiella*

[Belk and Brtek (1995) list a single species for North America.]

3b. Eleven pairs of thoracic legs 4

4a(3b). Abdomen ends in and is bordered by a flattened blade (Fig. 66A, B) .... *Thamnocephalus*

4b. Abdomen ends in two blade-shaped appendages (cercopods) (Fig. 67) 5

5a(4b). Large branched *frontal appendage* growing from between the second antennae (similar to that of *Thamnocephalus*, Fig. 66C); when unrolled, frontal appendage is longer than the second antennae .... *Branchinella* (in part)

[Belk and Sissom (1992) report four species of *Branchinella* from North America. They suggest that at least two of these species qualify as endangered species, because of habitat destruction.]

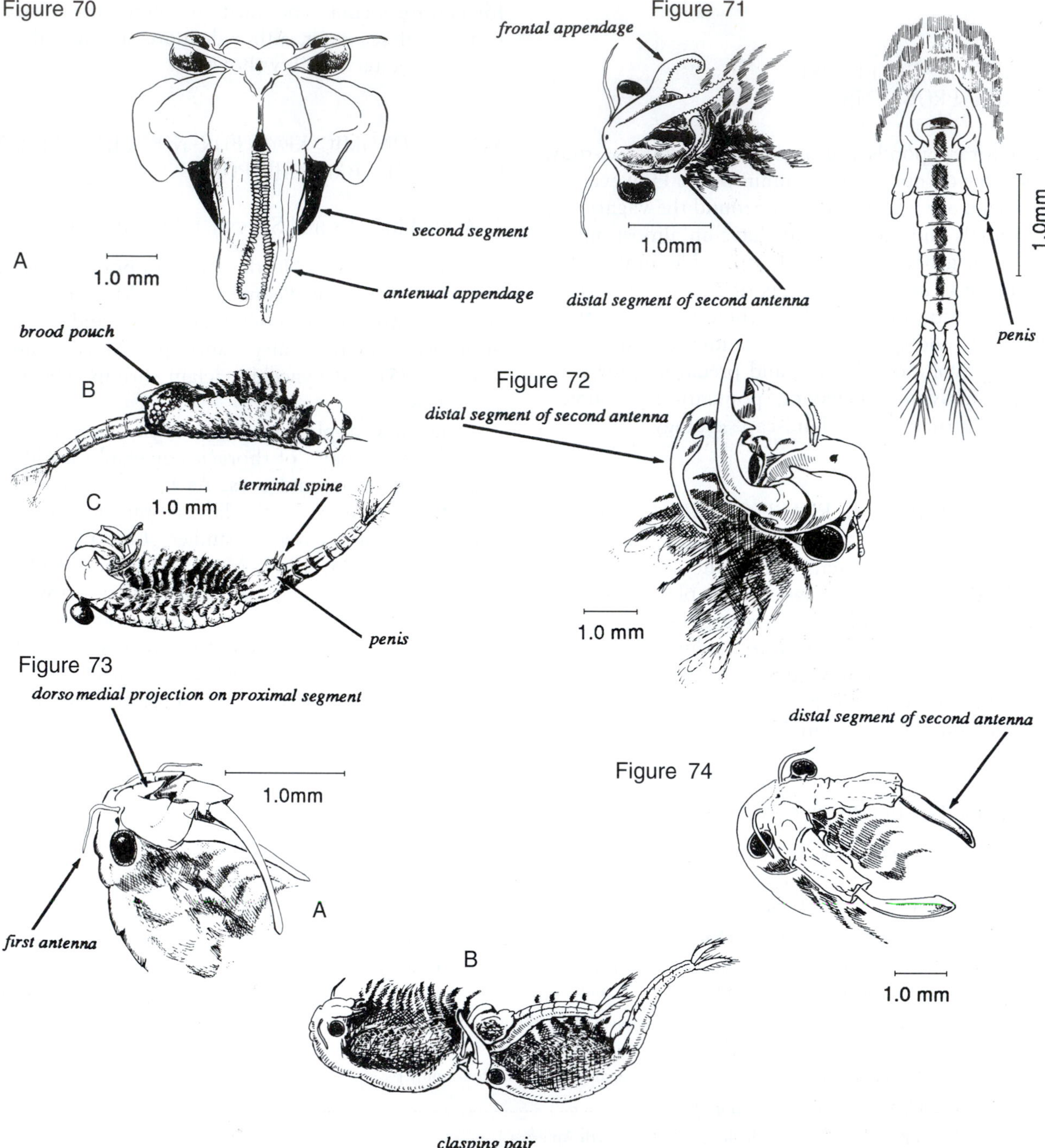

FIGURE 70 *Eubranchipus bundyi* Forbes, 1876. AB Woods Pond, Dane Co., Wisconsin. 27 April 1986. (Coll. K. Parejko). (A) Head of male, antennal appendage unrolled; (B) female; (C) male. (KE) *FIGURE 71* *Dexteria floridanus* (Dexter, 1953). Alachua Co., Florida. 7 March 1939. From paratypes in US National Museum. (Collection #93537). (KE) *FIGURE 72* *Artemiopsis stephanssoni* (Johansen, 1922). Head of male. (Coll. D. Belk #506). (KE) *FIGURE 73* *Linderiella occidentalis* (Dobbs, 1923). Sulfur Mtn. Pond, Venture Co., California. 2 March 1986. (Coll. S. Copper). (A) Head of male; (B) Male clasping female. (KE) *FIGURE 74* *Branchinecta coloradensis* Packard, 1874. Head of male. Stock Pond in Rock Garden, Grand Co., Utah. 29 October 1986. (KE)

5b. Frontal appendage absent, or if present, unbranched and less than half as long as the basal segment or the second antennae 6

6a(5b). Lateral spines at the posterior margin of each abdominal segment, spines about 1/3 as long as the width of the segment (Fig. 67B) .... *Branchinella* (in part)

6b. Abdomen without lateral spines 7

7a(6b). Distal (second) segment of the second antennae a flat triangular blade, the base of the triangle about half of the length (Fig. 68A) .... *Artemia*

[Belk and Brtek (1995) list three species for North America. Hontoria and Amat (1992) used a multivariate analysis of 25 different North American populations, and found evidence for only two species. However, studies in the Caribbean (Robert Mayer, personal communication) and Yucatan (Torrentera and Dodson, 1995) suggest the existence of several additional species.]

7b. Distal segment not triangular and thin, instead more or less rounded, but may be flattened toward the tip (do not confuse the distal antennal appendage, which is attached on the basal-medial margin of the second antenna and often carried folded or rolled up) (Fig. 70A) 8

8a(7b). Antennal appendages present and longer than the basal (first) segment of the second antennae 9

8b. Antennal appendages (projections) on the basal segment, if present, no longer than half of the length of the basal segment of the second antennae 10

9a(8a). Complex (branched) medial process attached at the inner corner of the apex of the first segment of the second antennae. The second segment is also attached at the apex of the basal segment, but it is outside (lateral to) the medial process and unbranched (Fig. 69) .... *Streptocephalus*

[Maeda-Martinez *et al.* (1995) give a diagnosis and phylogeny of the thirteen North American species of *Streptocephalus*.]

9b. Antennal appendage unbranched but with a series of lobes and teeth along at least one margin; antennal appendage attached near the proximal end of the basal segment of the second antenna 11

10a(8b). Penis with a stout terminal spine about as long as the eversible (distal) part of the penis; several species have a second antenna with branch-like processes on the second segment (Fig. 70) .... *Eubranchipus*

[Belk and Brtek (1995) list eight species for North America.]

10b. Penis without a terminal spine, distal segment of second antenna without branch-like processes (Fig. 71) .... *Dexteria*

[According to Belk and Brtek (1995), there is one species in this genus which is closely related to *Eubranchipus*. The species is known only from its type locality near Gainesville, FL.]

11a(9b). Second segment of the second antenna with one or two sharp projections from the medial surface, about halfway along the segment, and approximately doubling the width of the segment at that point; tip of the second segment tapers to a blunt point (Fig. 72) *Artemiopsis*

[Belk and Brtek (1995) list a single species for North America, restricted to arctic ponds.]

11b. Second segment of second antenna with no projections at middle, but sometimes flattened, curved, or with a knob at the tip......... 12

12a(11b). First segment of second antenna with a dorsal-medial projection from the base of the segment about half as long as segment; this is the only projection on the basal segment (Fig 73) .... *Linderiella*

[One species is recognized in North America (Belk and Brtek, 1995), and it is a candidate for endangered species status.]

12b. First segment of the second antenna may be without ornamentation, however, usually there is a projection, bump or patch or row of spines on the medial surface of the segment from about the middle of the segment, or a ventral-medial projection (curving upward) from the base of the segment, or some combination of the above; none of the projections more than half as long a the basal segment (Fig. 74) .... *Branchinecta*

[Belk and Brtek (1995) list 14 species for northern North America, and two from Mexico. Three of the species found in California are candidates for listing as endangered species.]

13a(2a). Dorsal union of the carapace valves a true hinge with a groove (Fig. 75A) (order Laevicaudata) 19

[Martin and Belk, 1988) provide keys for the family Lynceidae in the Americas.]

13b. Dorsal union of the valves a simple fold (order Spinicaudata) 14

14a(13b). Dorsal margin of head with a stalked rounded organ (Fig. 76), and an occipital crest at the posterior end of the dorsal margin of the head 15

14b. Dorsal margin of head with occipital crest only (Fig. 78), may be indistinct........................ 16

15a(14a). Ventral surface of telson at point of articulation of terminal claws with a pine (Fig. 76) .... *Eulimnadia*

[Belk (1989) reviews the North American species of *Eulimnadia*, recognizing five of the twelve described species as valid, based on traditional morphologic characters and new information on egg-shell morphology.]

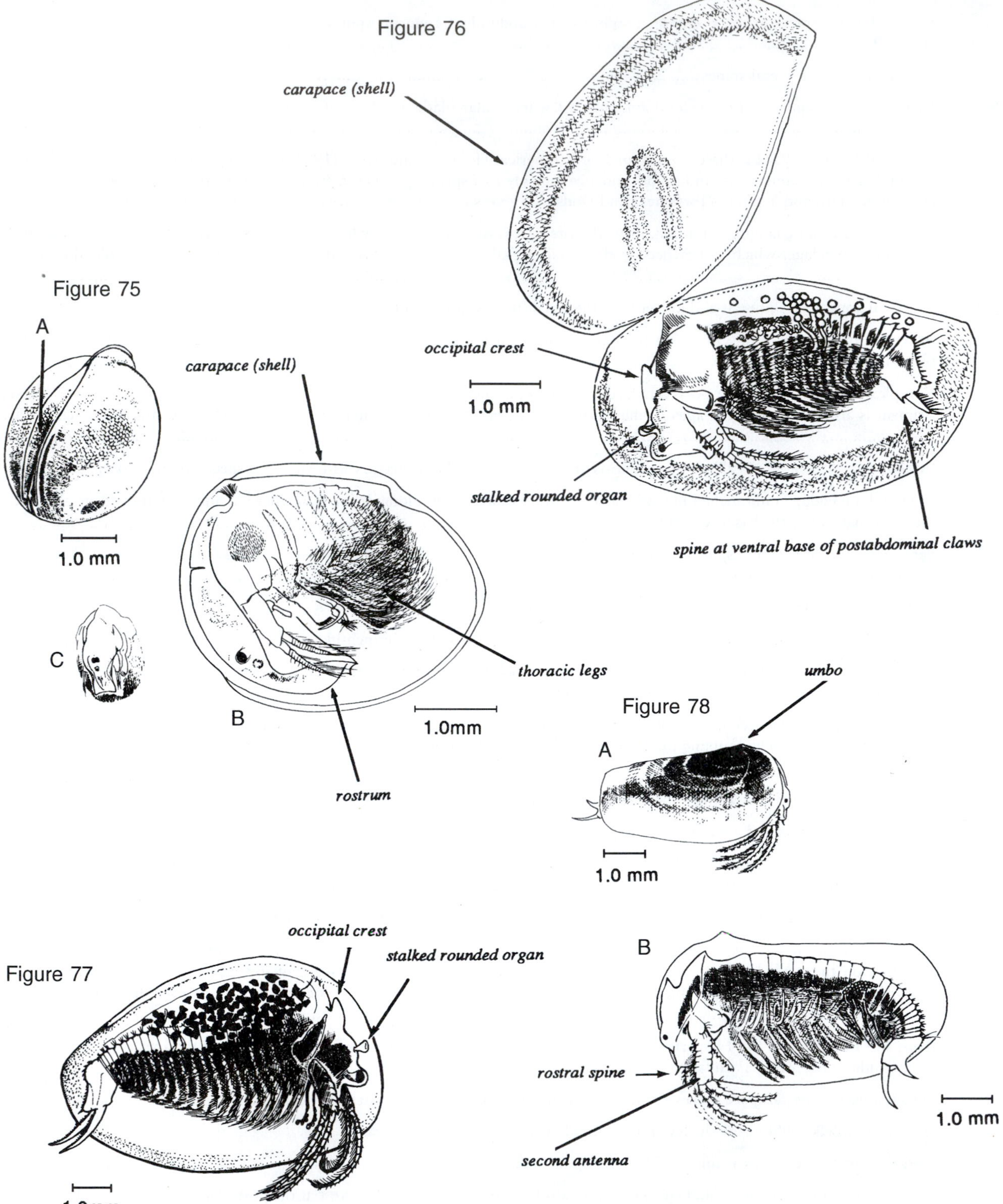

FIGURE 75 *Lycneus brachyurus* O.F. Mueller, 1785. Dane Co., Wisconsin. (A) Dorsal groove and hinge; (B) lateral view; (C) anterior view showing rostrum shape. (KE) *FIGURE 76* *Eulimnadia* cf. *agassizzii* Packard, 1874. Hwy 190, Oxaca, Mexico. 12 July 1976. (Coll. P. Mündel #235). (KE) *FIGURE 77* *Limnadia lenticularis* (Linnacus, 1761). Female within shell. Woods Hole, MA. 27 August 1887. (USNM coll. Smith). (KE) *FIGURE 78* *Leptisthera compleximanus* (Packard, 1877). Mexico? (A) Female within shell; (B) Female with one valve of shell removed. (KE)

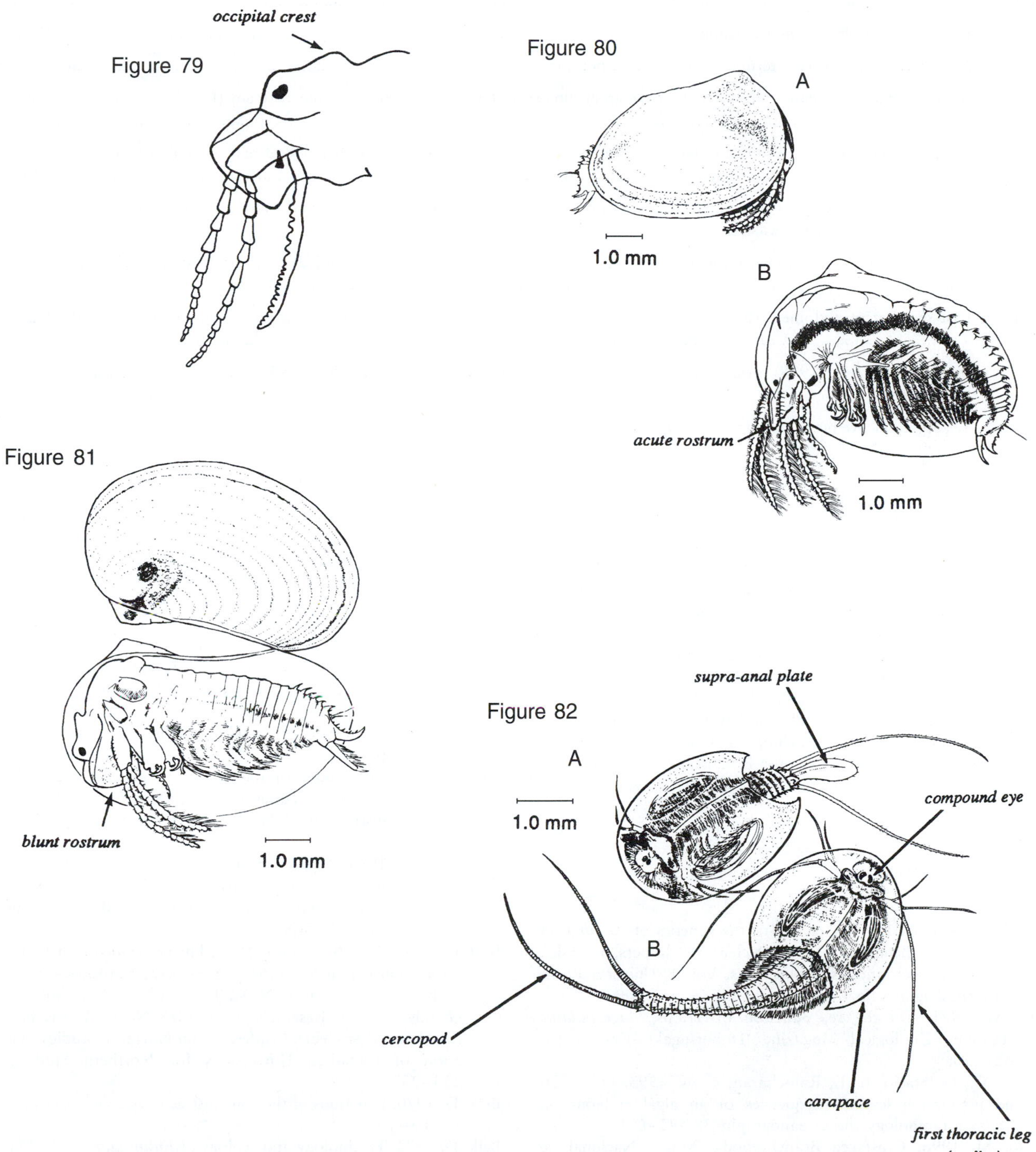

FIGURE 79 *Eocyzicus concavus* (Mackin, 1939). Profile view of male head. Texas. Copied from Mattox, 1950. *FIGURE 80* *Caenestheriella* cf. *setosa* (Pearse, 1912). (A) Male within shell; (B) male with one valve removed. (KE) *FIGURE 81* *Cyzicus californicus* (Packard, 1874) Male. Collected east of lordsburg, Hidalgo Co., New Mexico. 13 August 1955. (Coll Lynch). (KE) *FIGURE 82* (A) *Lepidurus couessii* (Packard, 1875). Fish hatchery at Valentine, Cherry Co., Nebraska, Spring 1977. (Coll. D.C. Ashley). (KE) (B) *Triops longicaudatus* LeConte, 1846. Hwy 385, Philips Co., Colorado. 13 June 1972. (Coll. F. Wiman). (KE)

15b. Ventral surface telson at point of articulation of terminal claws without a spine (Fig. 77) .... *Limnadia*

16a(14b). Rostrum (snout or beak) with a sharp spine (Fig. 78) .... *Leptistheria*

16b. Rostrum with a spine 17

17a(16b). Occipital crest acute, conspicuous 18

17b. Occipital crest obtuse, inconspicuous (Fig. 79) .... *Eocyzicus*

18a(17a). Male and female rostrum terminates in an acute point (Fig. 80) .... *Caenestheriella*

18b. Male rostrum terminates bluntly, anterior margin almost as high as the length of the rostrum (Fig. 81), female rostrum acute .... *Cyzicus*

19a(13a). Ridge running from top of head down center of rostrum unbranched; second thoracic leg of male is similar to the more posterior legs .... *Lynceus*

[Martin *et al.* (1986) provide comparative descriptions of the North American species, and Martin and Belk (1988) review the family Lynceidae in the Americas.]

19b. Head ridge branched toward the rostrum; second thoracic leg of male is modified different from the more posterior legs .... *Paralimnetis*

20a(2b). Last segment of abdomen (the telson) has a medial extension, the supra-anal plate, between the two long cercopods (Fig. 82A) .... *Lepidurus*

[King and Hannert (1998) present molecular evidence of at least five species of Lepidurus in North America; only four of these have been named.]

20b. Telson is concave between the two cercopods (Fig. 82B) .... *Triops*

ACKNOWLEDGMENTS

Professor Frey, a meiobenthic specialist, contributed the keys for the Bosminidae, Chydoridae, and Macrothricidae and discussion of these families, plus the section on paleolimnology. Professor Dodson wrote the remaining keys and recently revised the key to the macrothricids and related families. Each author modified the other's contribution. We are grateful for helpful comments from reviewers, especially James Thorp, Alan Tessier, Brenda Hann, Denton Belk, Carlos Santos-Flores, Henri Dumont, Professor Smirnov, and John Havel. The section on noncladoceran branchiopods could have been written only with the extensive advice of Denton Belk. Many of the figures for this chapter were drawn by Kandis Elliot ("KE" in the figure legends).

LITERATURE CITED

Abreu-Grobois, F. A. 1987. A review of the genetics of *Artemia*, *in* Sorgeloos, P., Bengtson, D. A., Decleir, W., Jaspers, E., Eds., *Artemia* research and its applications. Vol. 1. Universa Press, Wetteren, Belgium, pp. 61–73.

Akita, M. 1971. On the reproduction of *Triops longicaudatus* (LeConte). Zoological Magazine (Dobutsugaku Zasshi) 80: 242–250.

Allen, Y. C., De Stasio, B. T., Ramcharan, C. W. 1993. Individual and population level consequences of an algal epibiont on *Daphnia*. Limnology and Oceanography 38:592–601.

Alonso, M. 1996. Crustacea Branchiopoda. Museo Nacional de Ciencias Naturales, Consejo Superior de Investigaciones Cientificas, Madrid. Fauna Iberica 7:486.

Anderson, D. T. 1967. Larval development and segment formation in the branchiopod crustaceans *Limnadia stanleyana* and *Artemia salina* (L.) (Anostraca). Australian Journal of Zoology 15:47–91.

Anderson, R. S. 1974. Crustacean plankton communities of 340 lakes and ponds in and near the National Parks of the Canadian Rocky Mountains. Journal of the Fisheries Research Board of Canada 31:855–869.

Badaracco, G., Bellorini, M., Landsberger, N. 1995. Phylogenetic study of bisexual *Artemia* using random amplified polymorphic DNA. Journal of Molecular Evolution 41:150–154.

Baird, W. 1850. The natural history of the British Entomostraca. Printed for the Ray Society, London. reprinted by Johnson Reprint Corporation, New York. 1968. 364 p + 36 tables.

Balcer, M. D., Korda, N. L., Dodson, S. I. 1984. Zooplankton of the Great Lakes: a guide to the identification and ecology of the common crustacean species. Uni. of Wisconsin Press, Madison, WI. 175 pp.

Banta, A. M. 1939. Studies on the physiology, genetics and evolution of some Cladocera. Paper No. 39, Department of Genetics, Carnegie Institution of Washington Publication No. 513, pp. iii–x, 1–285.

Bayly, I. A. E. 1986. Aspects of diel vertical migration in zooplankton, and its enigma variations. *in*: De Deckker P., Williams, W. D. Eds., Limnology in Australia. CSIRO Melbourne, Australia, pp. 349–368.

Beaton, M. J., P. D. N. Hebert. 1988. Further evidence of hermaphroditism in *Lepidurus arcticus* (Crustacea, Notostraca) from the Melville Peninsula area, N. W. T., *in*: Adams W. P., Johnson, P. G. Eds., Student Research in Canada's North: Proceedings of the National Student's Conference on Northern Studies. Association of Canadian Universities for Northern Studies, pp 253–257.

Belk, D. 1970. Functions of the conchostracan egg shell. Crustaceana 19:105–106.

Belk, D. 1972. The biology and ecology of *Eulimnadia antlei* Mackin (Conchostraca). The Southwestern Naturalist 16:297–305.

Belk, D. 1975. Key to the Anostraca (fairy shrimps) of North America. The Southwestern Naturalist 20:91–103.

Belk. D. 1977. Evolution of egg size strategies in fairy shrimps. The Southwestern Naturalist 22:99–105.

Belk, D. 1984. Antennal appendages and reproductive success in the Anostraca. Journal of Crustacean Biology 4:66–71.

Belk, D. 1989. Identification of species in the conchostracan genus *Eulimnadia* by egg shell morphology. Journal of Crustacean Biology 9:115–125.

Belk, D., Cole, G. A. 1975. Adaptational biology of desert temporary-pond inhabitants. Pages 207–226, *in*: Hadley, N. F., Ed. Environmental physiology of desert organisms. Dowden, Hutchinson and Ross, Stroudsburg, Pa.

Belk, D., Brtek, J. 1995. Checklist of the Anostraca. Hydrobiologia 298:315–353.

Belk, D., Sissom, S. L. 1992. New *Branchinella* (Anostraca) from Texas, USA, and the problem of antenna-like processes. Journal of Crustacean Biology 12:312–316.

Bernardi, R. de. 1984. Methods for the estimation of zooplankton abundance. Pages 59–86, *in*: Downing, J. A., Rigler, F. H. Eds. A Manual on methods for the assessment of secondary productivity in fresh waters, 2nd ed. Blackwell Sci., Oxford.

Berner, D. B. 1982. Key to the cladocera of Par Pond on the Savannah River Plant. A publication of the Savannah River Plant, National Environment Research Park Program, U.S. Department of Energy. NERP-SRO–11. Savanna River Ecology Laboratory, Aiken, SC.

Bernice, R. 1971. Food, feeding, and digestion in *Streptocephalus dichotomus* Baird (Crustacea: Anostraca). Hydrobiologia 38: 507–520.

Benzie, J. A. H., Hodges, A. M. A. 1996. *Daphnia obtusa* Kurz, 1874 ememd. Scourfield, 1942 from Australia. Hydrobiologia 333:195–199.

Birge, E. A. 1893. Notes on Cladocera, III. Transactions of the Wisconsin Academy of Sciences, Arts and Letters 9:275–317.

Birge, E. A. 1910. Notes on Cladocera, IV. Transactions of the Wisconsin Academy of Sciences, Arts and Letters 16:1017–1066.

Birge, E. A. 1918. The water fleas (Cladocera), Pages 676–750, *in*: H. B. Ward, Whipple, G.C. Eds. Fresh-water Biology, Wiley, New York.

Black, R. W. III, Slobodkin, L. B. 1987. What is cyclomorphosis? Freshwater Biology (1987) 18:373–378.

Boileau, M. G., Hebert, P. D. N., Schwartz, S. S. 1992. Non-equilibrium gene frequency divergency. Persistent founder effects in natural populations. Journal of Evolutionary Biology 5:25–40.

Boucherle, M. M., Zullig, H 1983. Cladoceran remains as evidence of change in trophic state in three Swill lakes. Hydrobiologia 103:141–146.

Bowen, S. T., Buoncristiani, M. R., Carl, J. R. 1988. *Artemia* habitats: ion concentrations tolerated by one superspecies. Hydrobiologia 158:201–214.

Bowen, S. T., Fogarino, E. A., Hitchner, K. N., Dana, G. L., Chow, V. H. S., Buoncristiani, M. R., Carl, J. R. 1985. Ecological isolation in *Artemia*: population differences in tolerance of anion concentrations. Journal of Crustacean Biology 5:106–129.

Bowman, T. E. 1971. The case of the nonubiquitous telson and the fradulent furca. Crustaceana 21:165–175.

Bradley, M. C., Perrin, N., Calow, P. 1991. Energy allocation in the cladoceran *Daphnia magna* Straus, under starvation and refeeding. Oecologia (Heidelberg) 86:414–418.

Brancelj, A. 1990. *Alona hercegovinae*, new species (Cladocera: Chydoridae) a blind cave-inhabiting Cladoceran. Hydrobiologia 199:7–16.

Brancelj, A. 1992. *Alona sketi*, new species (Cladocera: Chydoridae) the second cave-inhabiting cladoceran from former Yugoslavia. Hydrobiologia 248:105–114.

Brandlova, J., Brandl, Z., Fernando, C. H. 1972. The cladocera of Ontario with remarks on some species and distribution. Canadian Journal of Zoology. 50:1373–1403.

Brendonck, L. 1995. An updated diagnosis of the branchipodid genera (Branchiopoda: Anostraca: Branchipodidae) with reflections on the genus concept by Dubois (1988) and the importance of genital morphology in anostracan taxonomy. Archiv Fuer Hydrobiologie Supplementband 107:149–186.

Brendonck, L. 1996. Diapause, quiescence, hatching requirements: What we can learn from large freshwater branchipods (Crustacea: Branchiopoda: Anostraca, Notostraca, Conchostraca). Hydrobiologia 320:85–97.

Brendonck, L. 1997. The anostracan genus *Branchinella* (Crustacea: Branchiopoda), in need of a taxonomic revision: Evidence from penile morphology. Zoological Journal of the Linnean Society 119:447–455.

Bretschko, G. 1969. Zur Ephippienablage bei Chydoridae (Crustacea, Cladocera). Zool. Anz. Suppl. 33, Vehr. Zool. Ges. 1969:95–97.

Brewer, M. C., Coughlin, J. 1996. Virtual plankton: A novel approach to the investigation of aquatic predator-prey interactions. Pages 425–434, *in*: Lenz, P.H., Hartline, D.K., Purcell, J.E., Macmillan, D.L. editors. Zooplankton: Sensory Ecology and Physiology. Gordon & Breach Amsterdam.

Brewer, M. C., Dawidowicz, P., Dodson, S.I. 1998. Interactive effects of fish kairomone and light on behavior in *Daphnia*. **Ecology**

Broch, E. S. 1969. The osmotic adaptation of the fairy shrimp *Branchinecta campestris* Lynch to saline astatic waters. Limnology and Oceanography 14:485–492.

Broch, E. S. 1988. Osmoregulatory patterns of adaptation to inland astatic waters by two species of fairy shrimps, *Branchinecta gigas* Lynch and *Branchinecta mackini* Dexter. Journal of Crustacean Biology 8:383–391.

Brooks, J. L. 1957. The systematics of North American *Daphnia*. Memoirs of the Connecticut Academy of Arts and Sciences 13:1–180.

Brooks, J. L. 1959. Cladocera. Pages 587–656, *in*: W.T. Edmondson, Editor. Fresh-water Biology, 2nd ed., Wiley, New York.

Browman, H. L., Kruse, S., O'Brien, W. J. 1989. Foraging behavior of the predaceous cladoceran, *Leptodora kindti* and escape responses of their prey. Journal of Plankton Research. 11: 1075–1088.

Browne, R. A. 1982. The costs of reproduction in brine shrimp. Ecology 63:43–47.

Browne, R. A. 1993. Sex and the single brine shrimp. Natural History 1993 (5):35–38.

Browne, R. A., Sallee, S. E. 1984. Partitioning genetic and environmental components of reproduction and lifespan in *Artemia*. Ecology 65:949–960.

Burns, C. W., Gilbert, J. J. 1986. Effects of daphnid size and density on interference between *Daphnia* and *Keratella cochlearis*. Limnology and Oceanography 31:848–858.

Butorina, L. G. 1986. On the problem of aggregation of planktonic crustaceans *[Polyphemus pediculus* (L.), Cladocera]. Archiv fur Hydrobiologie 105:355–386.

Cannon, H. G. 1933. On the feeding mechanism of the branchiopods. Philosophical Transactions of the Royal Society of London 222:267–352

Carpenter, S. (Ed.) 1987. Complex interactions in lake communities. Springer.

Carpenter, R., Kitchell, J. F., Hodgson, J. R. 1985. Cascading trophic interactions and lake productivity. BioScience 35:634–639.

Carter, J. C. H., Dadswell, M. J., Roff, J. C., Sprules, W. G. 1980. Distribution and zoogeography of planktonic crustaceans and dipterans in glaciated eastern North America. Canadian Journal of Zoology 58:1355–1387.

Carvalho, G. R., Hughes, R. N. 1983. The effect of food availability, female culture-density and photoperiod on ephippia production in *Daphnia magna* (Crustacea: Cladocera). Freshwater Biology 13:37–46.

Centeno, M. D. F., Brendonck, L., Persoone, G. 1993. Acute toxicity tests with *Streptocephalus proboscideus* (Crustacea: Branchiopoda: Anostraca): influence of selected environmental conditions. Chemosphere 27:2213–2224.

Chaplin, J. A., Havel, J. E., Hebert, P. D. N. 1994. Sex and ostracods. Trends in Ecology and Evolution 9:435–439.

Cheer, A. Y. L., Koehl, M. A. R. 1987. Paddles and rakes: fluid flow through bristled appendages of small organisms. Journal of Theoretical Biology 129:17–39.

Chiavelli, D. A., Mills, E. L., Threlkeld, S. T. 1993. Host preference, seasonality, and community interactions of zooplankton epibionts. Limnology and Oceanography 38:574–583.

Ciros-Perez, J., Elias-Gutierrez. M. 1997. *Spinalona anophtalma*, n. gen. n. sp. (Anomopoda, Chydoridae) a blind epigean cladoceran from the Neovolcanic Provence of Mexico. Hydrobiologia 353:19–28.

Clegge, J. S., Conte, F.P. 1980. A review of the cellular and developmental biology of *Artemia*, Pages 11–54, *in*: Persoone, G., Sorgeloos, P., Roels, O., Jaspers, E. Eds. The Brine shrimp *Artemia*, Vol. 2: Physiology, Biochemistry, Molecular Biology, Universa Press, Wettern, Belgium.

Colborn, T., Dumanoski, D., Myers, J. P. 1996. Our stolen future. Dutton. New York. 306 pp.

Colbourne, J. K., Hebert, P. D. N., Taylor, D. J. 1997. Evolutionary origins of phenotypic diversity in *Daphnia*. Pages 163–188, *in*: T. J. Givnish and K. J. Sytsma (Eds.), Molecular evolution and adaptive radiation. Cambridge Uni. Press.

Cooper, S. D. 1983. Selective predation on cladocerans by common pond insects. Canadian Journal of Zoology 61:879–886.

Cooper, S. D., Winkler, D. W., Lenz, P. H. 1984. The effect of grebe predation on a brine shrimp population. Journal of Animal Ecology 53:51–64.

Cooper, S. C., Smith, D. W. 1982. Competition, predation and the relative abundances of two species of *Daphnia*. Journal of Plankton Research 4:859–879.

Cotten, C. A. 1985. Cladoceran assemblages related to lake conditions in eastern Finland. Ph.D. dissertation, Indiana University, Bloomington, viii, 96 pp.

Crease, T. J., Hebert, P. D. N. 1983. A test for the production of sexual pheromones by *Daphnia magna* (Crustacea: Cladocera) Freshwater Biology 13:491–496.

Crisinel, A., Delaunay, L., Rossel, D., Tarradellas, J., Meyer, H., Saiah, H., Vogel, P., Delisle, C., Blaise, C. 1994. Cyst-based ecotoxicological tests using anostracans: Comparison of two species of *Streptocephalus*. Environmental Toxicology and Water Quality 9:317–326.

Crisman, T. L. 1980. Chydorid cladoceran assemblages from subtropical Florida. American Society of Limnology and Oceanography Special Symposium 3:657–668.

Cummins, K. W., Costa, R. R., Rowe, R. E., Moshiri, G. A., Scanlon, R. M., Zajdel, R. K. 1969. Ecological energetics of a natural population of the predaceous zooplankter *Leptodora kindtii* Focke (Cladocera). Oikos 20:189–223.

Daborn, G. R. 1975. Life history and energy relations of the giant fairy shrimp *Branchinecta gigas* Lynch 1937 (Crustacea: Anostraca). Ecology 56:1025–1039.

Daborn, G. R. 1976. The life cycle of *Eubranchipus bundyi* (Forbes) (Crustacea: Anostraca) in a temporary vernal pond of Alberta. Canadian Journal of Zoology 54:193–201.

Daborn, G. R. 1977a. The life history of *Branchinecta mackini* Dexter (Crustacea: Anostraca) in an agrillotrophic lake of Alberta. Canadian Journal of Zoology 55:161–168.

Daborn, G. R. 1977b. On the distribution and biology of an arctic fairy shrimp *Artemiopsis stefanssoni* Johansen, 1921 (Crustacea: Anostraca). Canadian Journal of Zoology. 55:280–287.

Daday, E. 1898. Mikroskopische Susswasserthiere aus Ceylon. Terme's. Fuzetek, Anhangsheft 21:1–123.

D'Agusto, A. 1980. The vital requirements of *Artemia*: physiology and nutrition. Pages 56–82, *in*: Persoone, G., Sorgeloos, P., Roels, O., Jaspers, E. (Eds.) 1980. The brine shrimp Artemia. Vol. 2. Physiology, Biochemistry, Molecular Biology. Universa Press: Wetteren, Belgium.

Deevey, E. S. Jr., Deevey, G.B. 1971. The American species of *Eubosmina* Seligo (Crustacea, Cladocera). Limnology and Oceanography 16:201–218.

Della Croce, N., Angelino, M. 1987. Marine Cladocera in the Gulf of Mexico and the Caribbean Sea. Cahiers de Biologie Marine 28:263–268.

Della Croce, N., Gaino, E. 1970. Osservazioni sulla biologia del maschio di *Penilia avirostris* Dana. Cahiers de Biologie Marine 11:361–365.

DeMeester, L., Dawidowicz, P., van Gool, E., Loose, C. J. 1998. Ecology and evolution of predator-induced behavior of zooplankton: Depth selection behavior and diel vertical migration, *in*: Tollrian, F., Harvell, C. D., Eds. The ecology and evolution of inducible defenses. Princeton University Press, Princeton, NJ [in press].

DeMelo, R., Hebert, P. D. N. 1994. Allozyme variation and species diversity in North American Bosminidae. Canadian Journal of Fisheries and Aquatic Science 51:873–880.

Demott, W. R., Kerfoot, W. C. 1982. Competition among cladocerans: nature of the interaction between *Bosmina* and *Daphnia*. Ecology 63:1949–1966.

Dodson, S. I. 1974. Zooplankton competition and predation: An experimental test of the size-efficiency hypothesis. Ecology 55:605–613.

Dodson, S. I. 1975. Predation rates of zooplankton in arctic ponds. Limnology and Oceanography 20:426–433.

Dodson, S. I. 1979. Body size patterns in arctic and temperate zooplankton. Limnology and Oceanography 24:940–949.

Dodson, S. I. 1982. Chemical and biological limnology of six west-central Colorado mountain ponds and their susceptibility to acid rain. American Midland Naturalist 107:173–179.

Dodson, S. I. 1984. Predation of *Heterocope septentrionalis* on two species of *Daphnia*: Morphological defenses and their cost. Ecology 65:1249–1257.

Dodson, S. I. 1985. *Daphnia* (*Ctenodaphnia*) *brooksi* (Crustacea: Cladocera), a new species from eastern Utah. Hydrobiologia 126:75–79.

Dodson, S. I. 1987. Animal assemblages in temporary desert rock pools: aspects of the ecology of *Dasyhelea sublettei* (Diptera: Ceratopogonidae). Journal of the North American Benthological Society 6:65–71.

Dodson, S. I. 1989. Predator-induced reaction norms. Bioscience 39:447–452.

Dodson, S. 1990. Predicting diel vertical migration of zooplankton. Limnology and Oceanography 35:1195–1200.

Dodson, S., Ramcharan, C. 1991. Size-specific swimming behavior of *Daphnia pulex*. Journal of Plankton Research. 13:1367–1379.

Dodson, S. I., Egger, D. L. 1980. Selective feeding of red phalaropes on zooplankton of arctic ponds. Ecology 61:755–763.

Dodson, S. I., Dodson, V. E. 1971. The diet of *Ambystoma tigrinum* larvae from Western Colorado. Copeia 1971:614–624.

Dodson, S. I., Merritt, C. M., Shannahan, J.-P., Shults, C. M. 1999. Low doses of Atrazine increase male production in *Daphnia pulicaria*. Environmental Toxicology and Chemistry 18: [in press].

Dodson, S. I, Tollrian, R., Lampert, W. 1997a. *Daphnia* swimming behavior during vertical migration. Journal of Plankton Research. 19:969–978.

Dodson, S. I., Arnot, S. E., Cottingham, K. L. 1999. The relationship in lake communities between primary productivity and species richness. Ecology [in review]

Dodson, S. I., Ryan, S., Tollrian, R., Lampert, W. 1997b. Individual swimming behavior of *Daphnia*: effects of food, light, and container size on four clones. Journal of Plankton Research. 19:1537–1552.

Dodson, S. I., Hanazato, T., Gorski, P. R. 1995. Behavioral responses of *Daphnia pulex* exposed to carbaryl and *Chaoborus* kairomone. Environmental Toxicology and Chemistry 14:43–50.

Dodson, S. I., Hanazato, T. 1995. Commentary on effects of anthropogenic and natural organic chemicals on development, swimming behavior, and reproduction of *Daphnia*, a key member of aquatic ecosystems. Environmental Health Perspectives 103 (Suppl. 4):7–11.

Donald, D. B. 1982. Erratic occurrence of anostracans in a temporary pond: colonization and extinction or adaptation to variations in annual weather. Canadian Journal of Zoology 61:1492–1498.

Doolittle, A. A. 1911. Descriptions of recently discovered cladocera from New England. Proceedings of the U.S. National Museums. 41:161–170.

Downing, J. A. 1984. Sampling the benthos of standing waters. Pages 87–130, *in*: Downing, J. A., Rigler, F. H., Eds. A Manual on Methods for the Assessment of Secondary Productivity in Fresh Waters, 2nd ed. Blackwell Sci., Oxford. 501 pp.

Dumont, H. J. 1987. Groundwater Cladocera: A synopsis. Hydrobiologia 145:169–173.

Dumont, H. J., Pensaert, J. 1983. A revision of the Scapholeberinae (Crustacea: Cladocera). Hydrobiologia 100:3–45.

Dumont, H. J. 1995. The evolution of groundwater Cladocera. Hydrobiologia 307:69–74.

Dumont, H.J., Silva-Briano, M. 1998. A reclassification of the anomopod families Macrothricidae and Chydoridae, with the creation of a new suborder, the Radopoda (Crustacea: Branchiopoda). Hydrobiologia 384:119–149.

Edmondson, W. T. 1960. Reproductive rates of rotifers in natural populations. Memorie dell'Istituto Italiano di Idrobiologia 12:21–77.

Edmondson, W. T. 1963. Pacific Coast and Great Basin. Pages 371–392 *in*: D. G. Frey, Ed. Limnology in North America,. Univ. of Wisconsin Press. Madison, WI.

Edmondson, W. T. 1991. The uses of ecology: Lake Washington and beyond. Uni. of Washington Press, Seattle, WA.

Edmondson, W. T., Litt, A. H. 1982. *Daphnia* in Lake Washington. Limnology and Oceanography 27:272–293.

Edmondson, W. T., Litt, A. H. 1987. *Conochilus* in Lake Washington. Hydrobiologia 147:157–162.

Engelmann, M. Hoheisel, G., Hahn, T., Joost, W., Vieweg, J., Naumann, W. 1996. Populations of *Triops cancriformis* (Bosc) (Notostraca) in Germany north of 50 degrees N are not clonal and at best facultatively hermaphroditic. Crustaceana (Leiden) 69:755–768.

Engle, D. L. 1985. The production of hemoglobin by small pond *Daphnia pulex*: Intraspecific variation and its relation to habitat. Freshwater Biology 15:631–638.

Enserink, E. L., Van Der Hoeven, N., Smith, M., Van Der Klis, C. M., Van Der Gaag, M. A. 1996. Competition between cohorts of juvenile *Daphnia* magna: A new experimental model. Archiv für Hydrobiologie 136:433–454.

Eriksen, C. H., Brown, R. J. 1980a. Comparative respiratory physiology and ecology of phyllopod crustacea. I. Conchostraca. Crustaceana 39:1–10.

Ericksen, C. H., Brown, R. J. 1980b. Comparative respiratory physiology and ecology of phyllopod crustacea. II. Anostraca. Crustaceana 39:11–21.

Eriksen, C. H., Brown, R. J. 1980c. Comparative respiratory physiology and ecology of phyllopod crustacea. III. Notostraca. Crustaceana 39:22–32.

Ferrari, D. C., Hebert, P. D. N. 1982. The induction of sexual reproduction in Daphnia magna: genetic differences between arctic and temperate populations. Canadian Journal of Zoology 60:2143–2148.

Forbes, S. A. 1925. The lake as a microcosm. Bulletin of the Illinois State Laboratory of Natural History (Survey) 15:537–550.

Fretwell, S. J. 1987. Food chain dynamics: the central theory of ecology? Oikos 50:291–301.

Frey, D. G. 1962. Cladocera from the Eemian Interglacial of Denmark. Journal of Paleontology 36:1133–1154.

Frey, D. G. 1964. Remains of animals in Quaternary lake and bog sediments and their interpretation. Archiv für Hydroliologie, Supplement Ergebnisse der Limnologie 2:1–116.

Frey, D. G. 1965. Differentiation of *Alona costata* Sars from two related species (Cladocera, Chydoridae). Crustaceana 8:159–173.

Frey, D.G. 1966. Phylogenetic relationships in the family Chydoridae. Marine Biology Association India, Proceedings of a Symposium on Crustacea, Part 1:29–37.

Frey, D. G. 1969. Symposium on paleolimnology. International Association of Theoretical and Applied Limnology, Mitteilungen. 17:1–448.

Frey, D. G. 1971. Worldwide distribution and ecology of *Eurycercus* and *Saycia* (Cladocera). Limnology and Oceanoraphy. 16: 254–308.

Frey, D. G. 1974. Paleolimnology. 95–123 *in* Jubilee Symposium: 50 years of limnological research. W. Rodhe (Ed.). International Association of Theoretical and Applied Limnology, Mitteilungen. 20:1–402.

Frey, D. G. 1975. Subgeneric differentiation within Eurycercus (Cladocera, Chydoridae) and a new species from northern Sweden. Hydrobiologia 46:263–300.

Frey, D. G. 1976. Interpretation of Quaternary paleoecology from Cladocera and midges, and prognosis regarding usability of other organisms. Canadian Journal of Zoology 54:2208–2226.

Frey, D. G. 1978. A new species of *Eurycercus* (Cladocera, Chydoridae) from the southern United States. Tulane Studies Zoology & Botany 20:1–26.

Frey, D. G. 1980a. On the plurality of *Chydorus sphaericus* (O. F. Muller)(Cladocera, Chydoridae) and designation of a neotype from Sjaelso, Denmark. Hydrobiologia 69:83–123.

Frey, D. G. 1980b. The non-swimming chydorid Cladocera of wet forests, with descriptions of a new genus and two new species. Internationale Revue der gesamten. Hydrobiologie 65:613–641.

Frey, D. G. 1982. Cladocera. Pages177–185, *in*: Hulbert, S. H., Villalobos-Figueroa, A. Eds. Aquatic Biota of Mexico, Central America and the West Indies. San Diego State University, San Diego, CA.

Frey, D. G. 1982a. Questions concerning cosmopolitanism in Cladocera. Archiv für. Hydrobiol. 93:484–502.

Frey, D. G. 1982b. Relocation of *Chydorus barroisi* and related species (Cladocera, Chydoridae) to a new genus and descriptions of two new species. Hydrobiologia 86:231–269.

Frey, D. G. 1982c. The reticulated species of *Chydorus* (Cladocera, Chydoridae): two new species with suggestions of convergence. Hydrobiologia 93:255–279.

Frey, D. G. 1982d, Cladocera. Pages 177–185, *in*: Hulbert, S. H., Villalogos-Fiqueroa, A. Eds. Aquatic Bioat of Mexico, Central America, and the West Indies. SanDiego State Uni. Press, San Diego, CA.

Frey, D. G. 1985. Cladocera. Proceedings of the Cladocera Symposium, Budapest 1985, Forro', L., Frey, D. G., Eds. Junk: Dordrecht.

Frey, D. G. 1986a. The non-cosmopolitanism of chydorid Cladocera: implications for biogeography and evolution, Pages 237–256, *in*: Gore, R. H., Heck, K. L., Eds. Crustacean Biogeography. Balkema: Rotterdam.

Frey, D. G. 1986b. Cladocera analysis, *in*: Berglund, B. E. Ed. Handbook of Holocene Palaeoecology and Palaeohydrology. Wiley, Chicester. pp. 667–692,

Frey, D. G. 1987a. The taxonomy and biogeography of the Cladocera. Hydrobiologia 145:5–17.

Frey, D. G. 1987b. The North American *Chydorus faviformis* (Cladocera, Chydoridae) and the honeycombed taxa of other continents. Philosophical Transactions of the Royal Society of London B315:353–402.

Frey, D. G. 1988a. Cladocera. Chapter, *in*: Higgins, R.P., Thiel, Hjalmar. Eds. Introduction to the study of meiofauna. Smithsonian Press. Washington, DC.

Frey, D. G. 1988b. Are there tropicopolitan macrothricid Cladocera? Acta Limnologica Brasiliensia 2:513–525.

Frey, D. G. 1988c. Separation of *Pleuroxus laevis* Sars, 1961, from two resembling species in North America: *Pleuroxus straminius* Birge, 1879 and *P. chiangi* n. sp. (Cladocera, Chydoridae). Canadian Journal of Zoology 66:2534–2563.

Frey, D. G. 1989. Separation of *Pleuroxus laevis* Sars, 1961, from two resembling species in North America—*P. straminius* Birge, 1879, and *P. chiangi* n. sp. (Cladocera, Chydoridae). Canadian Journal of Zoology. 66:2534–2563.

Frey, D. G., Hann, B. J. 1985. Growth in cladocera. Pages 315–335, *in*: A. W. Wenner, Ed. Factors in adult growth. Balkema, Boston.

Fryer, G. 1966. *Branchinecta gigas* Lynch, a non-filter-feeding raptatory anostracan, with notes on the feeding habits of certain other anostracans. Proceedings of the Linnean Society of London 177:19–34.

Fryer, G. 1968. Evolution and adaptive radiation in the Chydoridae (Crustacea: Cladocera): a study in comparative functional morphology and ecology. Philosophical Transactions of the Royal Society of London, Series B, Biological Sciences 254:221–385.

Fryer, G. 1970. Defecation in some macrothricid and chydorid cladocerans. Zoological Journal of the Linnean Society 49:255–269.

Fryer, G. 1971. Allocation of *Alonella acutirostris* (Birge) (Cladocera, Chydoridae) to the genus *Disparalona*. Crustaceana 21: 221–222.

Fryer, G. 1972. Observations on the ephippia of certain macrothricid cladocerans. Zoological Journal of the Linnean Society 51: 79–96.

Fryer, G. 1974. Evolution and adaptive radiation in the Macrothricidae (Crustacea: Cladocera): a study in comparative functional morphology and ecology. Philosophical Transactions of the Royal Society of London, Series B, Biological Sciences 269:137–274.

Fryer, G. 1983. Functional ontogenetic changes in *Branchinecta ferox* (Milne-Edwards) (Crustacea:Anostraca). Philosophical Transactions of the Royal Society of London, Series B. 303:229–243.

Fryer, G. 1985. Crustacean diversity in relation to the size of water bodies: some facts and problems. Freshwater Biology 15: 347–361.

Fryer, G. 1987. A new classification of the branchiopod Crustacea. Zoological Journal of the Linnean Society 91:357–383.

Fryer, G. 1988. Studies on the functional morphology and biology of the Notostraca. Philosophical Transactions of the Royal Society of London. B. 321(1203):27–124.

Fryer, G. 1991. Functional morphology and the adaptive radiation of the Daphniidae (Branchiopoda: Anomopoda). Philosophical Transactions of the Royal Society of London. Series B. 331:1–99.

Gabriel, W., Taylor, B. E., Kirsch-Prokosch, S. 1987. Cladoceran birth and death rates estimates: experimental comparisons of egg-ratio methods. Freshwater Biology 18:361–372.

Galat, D. L., Robinson, R. 1983. Predicted effects on increasing salinity on the crustacean zooplankton community of Pyramid Lake, Nevada. Hydrobiologia 105:115–131.

Gallagher, S. P. 1996. Seasonal occurrence and habitat characteristics of some vernal pool branchiopoda in northern California, U.S.A. Journal of Crustacean Biology 16:323–329.

Gerritsen, J., Porter, K. G. 1982. The role of surface-chemistry in filter feeding by zooplankton. Science 216:1225–1227.

Gerritsen, J., Porter, K. G., Strickler, J. R. 1988. Not by sieving alone: Suspension feeding in *Daphnia*. Bulletin of Marine Science 43:366–376.

Gorski, P. R., Dodson, S. I. 1996. Free-swimming *Daphnia* can avoid following Stokes' law. Limnology and Oceanography 41: 1815–1821.

Gould, S. J. 1989. Wonderful life. Norton, NY.

Gophen, M., Geller, W. 1984. Filter mesh size and food particle uptake by *Daphnia*. Oecologia (Berlin) 64:408–412.

Goulden, C. E. 1966. The animal microfossils. Pages 84–120 *in*: Cogill, U. M., Goulden, C. E., Hutchinson, G. E., Patrick, R., Racek, A. A., Tsukada, M. (Eds), The history of Laguna de Petenxil. Memoirs of the Connecticut Academy of Arts and Sciences Vol. 17.

Goulden, C. E. 1968. The systematics and evolution of the Moinidae. Transactions of the American Philosophical Society, Vol. 58, New Series, Part 6. pp. 1–101.

Goulden, C. E. 1969. Developmental phases of the biocenosis. Proceedings of the National Academy of Sciences 62:1066–1073.

Graham, D. M., Sprules, W. G. 1992. Size and species selection of zooplankton by larval and juvenile walleye (*Stizostedion vitreum vitreum*) in Oneida Lake, New York. Canadian Journal of Zoology 7-:2059–2067.

Hairston, N. G., Jr. 1996. Zooplankton egg banks as biotic reservoirs in changing environments. Limnology and Oceanography 41:1087–1092.

Hall, D. J. 1964. An experimental approach to the dynamics of a natural population of *Daphnia galeata mendotae*. Ecology 45:94–112.

Hanazato, T., Dodson, S. I. 1993. Morphological responses of four species of cyclomorphic *Daphnia* to a short-term exposure to the insecticide carbaryl. Journal of Plankton Research 15: 1087–1095.

Hann, B. J. 1982. Two new species of *Eurycercus* (*Bullatifrons*) from Eastern North America (Chydoridae, Cladocera). Taxonomy, ontogeny, and biology. Int. Rev. ges. Hydrobiology. 67:585–610.

Hann, B. J. 1984. Influence of temperature on life-history characteristics of two sibling species of *Eurycercus* (Cladocera, Chydoridae) Canadian Journal of Zoology 63:891–898.

Hann, B. J. 1985. Influence of temperature on life-history characteristics of two sibling species of Eurycercus (Cladocera, Chydoridae). Canadian Journal of Zoology 63:891–898.

Hann, B. J. 1986. Revision of the genus *Daphniopsis* Sars, 1903 (Cladocera: Daphniidae) and a description of *Daphniopsis chilensis*, new species, from South America. Journal of Crustacean Biology 6:246–263.

Hann, B. J., Chengalath, R. 1981. Redescription of *Alonella pulchella* Herrick, 1884 (Cladocera, Chydoridae), and a description of the male. Crustaceana 41:249–262.

Hann, B. J., Hebert, P. D. N. 1986. Genetic variation and population differentiation in species of *Simocephalus* (Cladocera, Daphniidae). Canadian Journal of Zoology 64:2246–2256.

Hartland-Rowe, R. 1966. The fauna and ecology of temporary pools in Western Canada. Verhandlungen. Internationale Vereinigung für Theoretische und Angewandte Limnologie 16:577–584.

Hartland-Rowe, R. 1972. The limnology of temporary waters and the ecology of Euphyllopoda. Pages 15–30, *in*: Clark, R.B., Wootton, R.J., Eds. Essays in Hydrobiology presented to Leslie Harvey. University of Exeter.

Havel, J. E. 1985. Predaton of common invertebrate predators on long- and short-featured *Daphnia retrocurva*. Hydrobiologia 124:141–149.

Havel, J. E., Hebert, P. D. N. 1989. Apomictic parthenogenesis and genotypic diversity in Cypridopsis vidua (Ostracoda: Cyprididae). Heredity 62:383–392.

Havel, J. E., Mabee, W. R., Jones, J. R. 1995. Invasion of the exotic cladoceran *Daphnia lumholtzi* into North American reservoirs. Canadian Journal of Fisheries and Aquatic Sciences 52:151–160.

Hebert, P. D. N. 1978. The population biology of *Daphnia* (Crustacea, Daphnidae). Biological Review 53:387–426.

Hebert, P. D. N. 1985. Interspecific hybridization between cyclic parthenogens. Evolution 39:216–220.

Hebert, P. D. N. 1987. Genotypic characteristics of the Cladocera. Hydrobiologia 145:183–193.

Hebert, P. D. N. 1988. The comparative evidence. Pages 175–196, *in*: S. C. Stearns, ed. The evolution of sex and its consequences. Birkhauser, Boston.

Hebert, P. D. N. 1995. The *Daphnia* of North America—An illustrated fauna. Version 1. CD ROM.

Hebert, P. D. N., Finston, T. L. 1996. A taxonomic reevaluation of North American *Daphnia* (Crustacea: Cladocera). II. New species in the *Daphnia pulex* group from the south-central United States and Mexico. Canadian Journal of Zoology 74:632–653.

Hirvonen, H., Ranta, E. 1996. Prey to predator size ratio influences foraging efficiency of larval *Aeshna juncea* dragonflies. Oecologia (Berlin) 106:407–415.

Hontoria, F., Amat, F. Morphological characterization of adult *Artemia* (Crustacea, Branchiopoda) from different geographical origins: American populations. Journal of Plankton Research 14: 1461–1471.

Horne, F. R. 1966. Some aspects of ionic regulation in the tadpole shrimp *Triops longicaudatus*. Comparative Biochemistry and Physiology 19:313–316.

Horne, F. R. 1971. Some effects of temperature and oxygen concentration on phyllopod ecology. Ecology 52:343–347.

Horne, F. R., Beyenbach, K. W. 1974. Physio-chemical features of hemoglobin of the Crustacean *Triops longicaudatus*. Archives of Biochemistry and Biophysics 161:369–374.

Hobaek, A., Larsson, P. 1990. Sex determination in *Daphnia magna*. Ecology 71:2255–2268.

Hrbácek, J. 1958. Density of the fish population as a factor influencing the distribution and speciation of the species of *Daphnia*. 15th Proceedings of the International Congress of Zoology, Section 10:794–796.

Hurlbert, S. H., Loayza, W., Moreno. T. 1986. Fish–flamingo–plankton interactions in the Peruvian Andes. Limnology and Oceanography 31:457–468.

Hurtubise, R. D., Havel, J. E. The effects of ultraviolet-B radiation on freshwater invertebrates: Experiments with a solar simulator. Limnology and Oceanography 43:1082–1088.

Hutchinson, G. E. 1967. A treatise on limnology, Vol. II, Introduction to lake biology and the limnoplankton. Wiley, New York. 1115 pp.

Hutchinson, G. E., Bonatti, E. Cogill, U. M., Goulden, C. E., Leventhall, E. A., Mallett, M. E. Margaritora, F., Patrick, R., Racek, A., Roback, S. A., Stella, E., Ward-Perkins, J. B., Wellman, T. 1970. Ianula: An account of the history and development of the Lago di Monterosi, Latium, Italy. Transactions of the American Philosophical Society. 60 (New Series): pt. 4:3–178.

Hyman, L. H. 1926. Note on the destruction of *Hydra* by a chydorid cladoceran, *Anchistropus minor* Birge. Transactions of the American Microscopical Society 45:298–301.

Innes, D. J., Hebert, P. D. N. 1988. The origin and genetic basis of obligate parthenogenesis in *Daphnia pulex*. Evolution 42: 1024–1035.

Jack, J. D., Thorp, J. H. 1995. *Daphnia lumholtzi*: appearance and likely impacts of an exotic cladoceran in the Ohio River. Transactions of the Kentucky Academy of Science. 56(3/4):101–103.

Jurine, L. 1820. Histoire des Monocles, qui se trouvent aux environs de Geneve. J. J. Paschoud, Paris, 260 pp., 22 plates.

Jensen, K. H., Jakobsen, P. J., Kleiven, O. T. 1998. Fish kairomone regulation of internal swarm structure in *Daphnia pulex* (Cladocera: Crustacea). Hydrobiologia [In Review].

Karabin, A. 1974. Studies on the predatory role of the cladoceran, *Leptodora kindtii* (Focke), in secondary production of two lakes with different trophy. Ekologia Polska 22:295–310.

Kaestner, A. 1970. Invertebrate Zoology. Vol. 3. Crustacea [English edition] Wiley Interscience, New York. 523 pp.

Kerfoot, W. C. 1981. Long-term replacement cycles in cladoceran communities: a history of predation. Ecology 62:216–233.

Kerfoot, W. C., Lynch, M. 1987. Branchiopod Communities: Associations with planktivorous fish in time and space. Pages 367–378, *in*: Kerfoot, W. C., Sih, A., Eds. Predation: Direct and indirect impacts on aquatic communities. Uni. Press of New England. Hanover.

Kerfoot, W. K., Sih, A. (Eds.) 1987. Predation: direct and indirect impacts on aquatic communities. Uni. Press of New England. Hanover. 386 pp.

Kerfoot, W. C., Kellogg, D. L., Jr. Strickler, J. R. 1980. Visual observations of live zooplankters: Evasion, escape, and chemical defenses. American Society of Limnology and Oceanography Special Symposium 3:10–27.

Kilham S. S., Kreeger D. A., Lynn S. G., Goulden C. E., Herrera L. 1998. COMBO: A defined freshwater culture medium for algae and zooplankton. Hydrobiologia 377:147–159.

King, C. R., Greenwood, J. G. 1992. The productivity and carbon budget of a natural population of *Daphnia lumholtzi* Sars. Hydrobiologia 231:197–207.

King, J. L., Hannert, F. 1998. Cryptic species in a "living fossil" lineage: Taxonomic and phylogenetic relationships within the genus *Lepidurus* (Crustacea: Notostraca) in North America. Molecular Phylogenetics and Evolution 10:23–36.

Kitchell, J. F. 1992. Food Web Management: A case study of Lake Mendota. Springer, New York. 553 pp.

Kitchell, J. F., Carpenter, S. A. 1987. Piscivores, planktivores, fossils, and phorbins. Pages 132–146, *in*: W. C. Kerfoot, Ed. Predation: Direct and indirect impacts on aquatic communities. New England Press, Hanover, N.H. 386 pp.

Kleiven, O. T., Larsson, P., Hobaek, A. 1992. Sexual reproduction in *Daphnia magna* requires three stimuli. Oikos 65:197–206.

Klekowski, R. Z. (Ed.) 1978. (Proceedings) Polskie Archiwum Hydrobiologii 25:1–501.

Kokkinn, M. J., Williams, W. D. 1987. Is ephippial morphology a useful taxonomic descriptor in the Cladocera? An examination based on a study of *Daphniopsis* (Daphniidae) from Australian salt lakes. Hydrobiologia 145:67–73.

Kořínek, B. 1981. *Diaphanosoma birgei* n.sp. (Crustacea, Cladocera). A new species from America and its widely distributed subspecies *Diaphanosoma birgei* ssp. *lacustris* n. ssp. Canadian Journal of Zoology 59: 1115–1121.

Korovchinsky, N. M. 1992. Sididae and Holopediidae. Guides to the identification of the macroinvertebrates of the continental waters of the world, No. 3. SPB Academic Publishing. The Hague. 77 p.

Kotov, A. A., Boikova, O. S. 1998. Comparative analysis of the late embryogenesis of *Sida crystallina* (O.F. Muller, 1776) and *Diaphanosoma brachyurum* (Lievin, 1848) (Crustacea: Branchiopoda: Ctenopoda) Hydrobiologia [In Press].

Kubersky, E. S. 1977. Worldwide distribution and ecology of *Alonopsis* (Cladocera: Chydoridae) with a description of *Alonopsis americana* sp. nov. Int. Rev. ges. Hydrobiol. 62:649–685.

Krueger, D. A., Dodson, S. I. 1981. Embryological induction and predation ecology in *Daphnia pulex*. Limnology and Oceanography 26:219–223.

Lampert, W. 1984. The measurement of respiration. Pages 413–468, *in*: Downing, J. A., Rigler, F. H. Eds. 1984. A manual on methods for the assessment of secondary productivity in fresh waters, 2nd edn. Blackwell Sci., Oxford.

Lampert, W. 1986a. Response of the respiratory rate of *Daphnia magna* to changing food conditions. Oecologia (Berlin) 70:495–501.

Lampert, W. 1986b. Phytoplankton control of grazing zooplankton: A study on the spring clear-water phase. Limnology and Oceanography 31:478–490.

Lampert, W. 1987. Vertical migration of freshwater zooplankton: Indirect effects of vertebrate predators on algal communities. Pages 291–299, *in*: Kerfoot, W.C. Ed. Predation: direct and indirect impacts on aquatic communities. New England Press, Hanover, NH.

Lampert, W. 1993. Ultimate causes of diel vertical migration of zooplankton: new evidence for the predator-avoidance hypothesis. Archiv für Hydrobiologie. Beih. Ergebn. Limnol. 39:79–88.

Lampert, W., Sommer, U. 1997. Limnoecology: the ecology of lakes and streams, (Translated by J. Haney) Oxford Uni. Press.

Landon, M. S., Stasiak, R. H. 1983. *Daphnia* hemoglobin concentration as a function of depth and oxygen availability in Arco Lake, Minnesota. Limnology and Oceanography 28:731–737.

Larsson P., Dodson, S. I. 1993. Invited review. Chemical communication in planktonic animals. Archiv für Hydrobiologie 129: 129–155.

Lehman, J. T. 1987. Palearctic predator invades North American Great Lakes. Oecologia (Berlin) 74:478–480.

Lehman, N., Pfrender, M. E., Morin, P. A. Crease, T. J, Lynch, M. 1995. A hierarchical molecular phylogeny within the genus *Daphnia*. Molecular Phylogenetics and Evolution 4:395–407.

Lenz, P. H. 1984. Life-history analysis of an *Artemia* population in a changing environment. Journal of Plankton Research 6:967–983.

Lilljeborg, W. 1901. Cladocera Sueciae, oder Beitraage zur Kenntniss der in Schweden lebenden Krebsthiere von der Ordnung der Branchiopoden und der Unterordnung der Cladoceren. Nova Acta reg. Soc. Sci. Upps. Ser. 3, pp. i-vi, 1–701.

Linder, F. 1941. Contributions to the morphology and the taxonomy of the Branchiopoda Anostraca. Zoologiska Bidrag fran Uppsala 10:101–302, 1 plate.

Löffler, H., Ed. 1987. Paleolimnology IV. Proceedings of the Fourth International Symposium on Paleolimnology, held at Ossiach, Carinthia, Austria. Hydrobiologia Vol. 143, 431 pp.

Longhurst, A. R. 1954. Reproduction in Notostraca (Crustacea). Nature 173:781–782.

Loose, C. 1993. Lack of endogenous rhythmicity in *Daphnia* diel vertical migration. Limnology and Oceanography 38:1837–1841.

Loring, S. J., MacKay, W. P., Whitford, W. G. 1988. Ecology of small desert playas. Pages 89–113, *in*: Thame, J.L., Ziebel, C. D. Eds. Small water impoundments in semi-arid regions. Uni. of New Mexico Press, Albuquerque.

Luecke, C., Litt, A. H. 1987. Effects of predation by *Chaoborus flavicans* of Lake Lenore, Washington. Freshwater Biology 18: 185–192.

Luecke, C., O'Brien, W. J. 1983. Photoprotective pigments in a pond morph of *Daphnia middendorffiana*. Arctic 36:365–368.

Lynch, M. 1980. The evolution of cladoceran life histories. The Quarterly Review of Biology 55:23–42.

Lynch, M. 1982. How well does the Edmondson–Paloheimo model approximate instantaneous birth rates? Ecology 63:12–18.

Lynch, M. 1983. Ecological genetics of *Daphnia pulex*. Evolution 37: 358–374.

Lynch, M. 1989. The life history consequences of resource depression in *Daphnia pulex*. Ecology 70: 246–256.

MacIsaac, H. J., Hebert, P. D. N., Schwartz, S.S. 1985. Inter- and intraspecific variation in acute thermal tolerance of *Daphnia*. Physiological Zoology 58:350–355.

Maeda-Martinez, A. M. 1991. Distribution of the species of Anostraca, Notostraca, Spinicaudata, and Laevicaudata in Mexico. Hydrobiologia 212: 209–220.

Maeda-Martinez, A. M., Belk, D., Obregon-Barboza, H., Dumont, H. J. 1995. Diagnosis and phylogeny of the New World Streptocephalidae (Branchiopoda: Anostraca). Hydrobiologia 298: 15–44.

Martin, J. W. 1992. Branchiopoda. Pages 25–244 in Harrison, F. W., Humes, A. G. Eds. Microscopic Anatomy of Invertebrates. Vol. 9: Crustacea. Wiley– Liss, New York, 625 pp.

Martin, J. W., Felgenhauer, B. E., Abele, L. G. 1986. Redescription of the clam shrimp *Lynceus gracilicornis* (Packard) (Branchiopoda, Conchostraca, Lunceidae) from Florida, with notes on its biology. Zoologica Scripta 15:221–232.

Martin, J. W., Belk, D. 1988. Review of the clam shrimp family Lynceidae Stebbing, 1902 (Branchiopoda: Conchostraca), in the Americas. Journal of Crustacean Biology 8:451–482.

Mathias, P. 1937. Biologie des Crustaces Phyllopodes. Actualites Scientifiques et Industrielles 447:1–107.

McCauley, E. 1984. The estimation of the abundance and biomass of zooplankton in samples. Pages 228–265, *in*: Downing, J. A., Rigler F. H. (Eds.) 1984. A Manual on Methods for the Assessment of Secondary Productivity in Fresh Waters, 2nd edn. Blackwell Sci., Oxford. 501 pp.

McLaughlin, P. A. 1980. Comparative morphology of recent Crustacea. Freeman, San Francisco, CA.

McQueen, D. J, Post, J. R., Mills, E. L. 1986. Trophic relationships in freshwater pelagic ecosystems. Canadian Journal of Fisheries and Aquatic Science. 43:1571–1581.

Mellors, W. K. 1975. Selective predation of ephippial *Daphnia* and the resistance of ephippial eggs to digestion. Ecology 56: 974–980.

Merilainen, J., Huttunen, P., Battarbee, R. W. (Eds.) 1983. Paleolimnology. Proceedings of the Third International Symposium on Paleolimnology, held at Joensuu, Finland. Hydrobiologia Vol. 103, 318 p.

Merrill, H. B. 1893. The structure and affinities of *Bunops scutifrons*, Birge. Transactions of the Wisconsin Academy of Sciences Arts and Letters. 9:319–342.

Michael, R. G., Frey, D. G. 1983. Assumed Amphi-Atlantic distribution of *Oxyurella tenuicaudis* (Cladocera, Chydoridae) denied by a new species from North America. Hydrobiologia 106:3–35.

Michael, R. G., Frey, D. G. 1984. Separation of *Disparalona leei* (Chien, 1970) in North America from *D. rostrata* (Koch, 1841) in Europe (Cladocera, Chydoridae). Hydrobiologia 114:81–108.

Modlin, R. F. 1982. A comparison of two *Eubranchipus* species (Crustacea: Anostraca). American Midland Naturalist 107:107–113.

Moghraby, A. E. el. 1977. A study on diapause of zooplankton in a tropical river—the Blue Nile. Freshwater Biology 7:207–212.

Monakov, A. V. 1972. Review of studies on feeding of aquatic invertebrates conducted at the Institute of Biology of Inland Waters,

Academy of Science, USSR. Journal of the Fisheries Research Board of Canada 29:363–383.

Moore, M. V., Folt, C. L., Stemberger, R.S. 1996. Consequences of elevated temperatures for zooplankton assemblages in temperate lakes. Archiv für Hydrobiologie 135:289–319.

Moore, W. G., Burn, A. 1968. Lethal oxygen thresholds for certain temporary pond invertebrates and their applicability to field situations. Ecology 49:349–351.

Moore, W. G., Ogren, L. H. 1962. Notes on the breeding behavior of *Eubranchipus holmani* (Ryder). Tulane Studies in Zoology 9:315–318.

Mort, M. A., Streit, B. 1992. Measuring molecular variation in zooplankton populations: DNA extraction from small *Daphnia* species. Aquatic Sciences 54: 77–84.

Mossin, J. 1986. Physiochemical factors inducing embryonic development and spring hatching of the European fairy shrimp *Siphonophanes grubei* (Dybowsky) (Crustacea: Anostraca). Journal of Crustacean Biology 6:693–704.

Mueller, W. P. 1964. The distribution of cladoceran remains in surficial sediments from three northern Indiana lakes. Invest. Indiana Lakes and Streams 6:1–63.

Munuswamy, N., Subramoniam, T. 1985. Influence of mating on ovarian and shell gland activity in a freshwater fairy shrimp *Streptocephalus dichotomus* (Anostraca). Crustaceana 49:225–232.

Nebeker, A. V., Onjukka, S. T., Stevens, D. G., Chapman, G. A., Dominguez, S.E. 1992. Environmental Toxicology and Chemistry 11: 373–379.

Nilsson, N., Pejler, B. 1973. On the relation between fish fauna and zooplankton composition in north Swedish lakes. Institute of Freshwater Research, Drottningholm. Report 53:51–77.

Norman, A. M., Brady, G. S. 1867. A monograph of the British Entomostraca belonging to the families Bosminidae, Macrothricidae, and Lynceidae. Natural History Transaction of Northumberland and Durham 1:354–408.

North American Lake Management. 1995. 15th International Symposium of the North American Lake Management Society, Toronto, Ontario, Canada, 6–11 November. Lake and Reservoir Management 11:113–206.

O'Keefe, T. C., Brewer, M. C., Dodson, S.I. 1998. Swimming behavior of *Daphnia*: Its role in determining predation risk. Journal of Plankton Research [in press].

O'Brien. W. J. 1979. The predator–prey interaction of planktivorous fish and zooplankton. American Scientist 67:572–581.

Ojima, Y. 1958. A cytological study on the development and maturation of the parthenogenetic and sexual eggs of *Daphnia pulex* (Crustacea–Cladocera). Kwansei Gakuin University Annual Studies 6:123–176, 28 plates.

Olesen, J. 1996. External morphology and phylogenetic significance of the dorsal/neck organ in the Conchostraca and the head pores of the cladoceran family Chydoridae (Crustacea, Branchiopoda). Hydrobiologia 330:213–226.

Olesen, J. 1998. A phylogenetic analysis of the Conchostraca and Cladocera (Crustacea, Branchiopoda, Diplostraca) Zoological Journal of the Linnean Society 122:491–536.

Orlova-Bienkowskaja, M. J. 1998. A revision of the cladoceran genus Simocephalus (Crustacea, Daphniidae). Bulletin of the Natural History Museum of London (Zoology) 64:1–62.

Pace, M. L., Vaque, D. 1994. The importance of *Daphnia* in determining mortality rates of protozoans and rotifers in lakes. Limnol. Oceanogr. 39:985–996.

Persoone, G., Sorgeloos, P. Roels, O., Jaspers, E. (Eds.) 1985. The brine shrimp *Artemia*. Vols. 1–3. Universa Press: Wettere, Belgium.

Peters, R. H. 1984. Methods for the study of feeding, grazing, and assimilation by zooplankton. pp. 336–412, *in*: Downing, J. A., Rigler, F. H. Eds. A Manual on Methods for the Assessment of Secondary Productivity in Fresh Waters, 2nd edn. Blackwell Sci., Oxford.

Peters, R. H. 1987. Metabolism in *Daphnia*. *in*: Peters, R. H., de Bernardi, R. Eds. "*Daphnia*" Memorie dell'Istituto Italiano di Idrobiologia Vol. 45, pp. 193–243.

Porter, K. G. 1973. Selective grazing and differential digestion of algae by zooplankton. Nature 244:179–180.

Porter, K. G. 1977. The plant-animal interface in freshwater ecosystems. American Scientist 65:159–170.

Porter, K. G., Feig, Y. S., Vetter, E. F. 1983. Morphology, flow regimes, and filtering rates of *Daphnia*, *Ceriodaphnia*, and *Bosmina* fed natural bacteria. Oecologia 58:156–163.

Porter, K.G., Gerritsen, J., Orcutt, Jr., J. D. 1982. The effect of food concentration on swimming patterns, feeding behavior, ingestion, assimilation, and respiration by *Daphnia*. Limnology and Oceanography 27:935–949.

Porter, K. G., Orcutt, J. D. 1980. Nutritional adequacy, manageability, and toxicity as factors that determine the food quality of green and blue-green algae for *Daphnia*. American Society of Limnology and Oceanography Special Symposium 3:268–281.

Potts, W. T. W., Durning, C. T. 1980. Physiological evolution in the Branchiopods. Comparative Biochemistry and Physiology 67B:475–484.

Prepas, E. E. 1984. Some statistical methods for the design of experiments and analysis of samples. Pages 266–335, *in*: Downing, J. A., Rigler, F. H. Eds. A manual on methods for the assessment of secondary productivity in fresh waters, 2nd edition. Blackwell Sci., Oxford.

Proctor, V. W. 1964. Viability of crustacean eggs recovered from ducks. Ecology 45:656–658.

Rajapaksa, R. 1986. A contribution to the taxonomy, biogeography and gamogenesis of freshwater Cladocera, with special reference to the tropical region. Ph.D. dissertation, University of Waterloo, Ontario. xiv, 221, & 21 Appendix pp.

Rajapaksa, R., Fernando, C.H. 1982. The first description of the male and ephippial female of Dadaya macrops (Daday, 1898) (Cladocera, Chydoridae), with additional notes on this common tropical species. Canadian Journal of Zoology 60:1841–1850.

Rajapaksa, R., Fernando, C. H. 1987a. Redescription and assignment of *Alona globulosa* Daday, 1898 to a new genus *Notoalona* and a description of *Notoalona freyi* sp. nov. Hydrobiologia 144:131–153.

Rajapaksa, R., Fernando, C. H. 1987b. Redescription of *Dunhevedia serrata* Daday, 1898) (Cladocera, Chydoridae) and a description of *Dunhevedia americana* sp. nov. from America. Canadian Journal of Zoology 65:432–440.

Richard, J. 1892. Cladoceres nouveaux du Congo. *Grimaldina Brazzai, Guernella Raphaelis, Moinodaphinia Macquerysi.* Mem. Soc. Zool. Fr. 5:213–226.

Richman, S. 1958. The transformation of energy by *Daphnia pulex*. Ecological Monographs 28:273–291.

Richman, S. E., Dodson, S. I. 1983. The effect of food quality on feeding and respiration by *Daphnia* and *Diaptomus*. Limnol. Oceanogr. 28(5):948–956.

Rigler, F. H., Downing, J. A. 1984. The calculation of secondary productivity. Pages 19–58, *in*: Downing, J. A., Rigler, F. H., Eds. A Manual on Methods for the Assessment of Secondary Productivity in Fresh Waters, 2nd edn. Blackwell Sci., oxford.

Ringelberg, J. 1987. Light induced behavior in *Daphnia*. Pages 285–323, *in*: Peters, R. H., de Bernardi, R., Eds. "*Daphnia*." Memorie dell'Istituto Italiano di Idrobiologia, Vol. 45.

Rivier, I. K. 1998. The predatory Cladocera and Leptodorida of the world. Backhuys Publishing, Leiden, The Netherlands. 213 pp.

Rzòska, J. 1961. Observations on tropical rainpools and general remarks on temporary waters. Hydrobiologia 17:265–286.

Sars, G. O. 1900. Description of *Iheringula paulensis* G. O. Sars, a new generic type of Macrothricidae from Brazil. Arch. Math. Naturvidensk. 22(6):1–27.

Sars, G. O. 1904. Pacifische Plankton-Crustaceen. (Ergebnisse einer Reise nach dem Pacific. Schauinsland 1896/97). Zoologische jahrbücher. Abteilung für Systematik, Geographie und Biologie der Tiere. 19:629–646.

Sars, G. O. 1916. The fresh-water Entomostraca of Cape Province (Union of South Africa). Part I. Cladocera. Annals of the South African Museums 15:303–351.

Sassaman, C. 1995. Sex determination and evolution of unisexuality in the conchostraca. Hydrobiologia 298:45–65.

Schindler, D. 1977. Evolution of phosphorus limitation in lakes. Science 195:260–262.

Schmidt-Nielsen, K. S. 1984. Scaling: Why is animal size so important. Cambridge Uni. Press. 241 p.

Schminke, H. K. 1976. The ubiquitous telson and the deceptive furca. Crustaceana 30: 292–300.

Schrimpf, A., Steinberg. C. 1982. Further recordings of the newly observed cladoceran *Daphnia* parvula, new record in southern Germany. Archiv für Hydrobiologie 94: 372–381.

Schwartz, S. S., Hann, B. J., Hebert, P.D.N. 1983. The feeding ecology of *Hydra* and possible implications in the structuring of pond zooplankton communities. Biological Bulletin 164: 136–142.

Schwartz, S. S. 1984. Life history strategies in *Daphnia*: a review and predicitons. Oikos 42:114–122.

Schwartz, S. S., Hebert, P.D.N. 1987a. Methods for the activation of the resting eggs of *Daphnia*. Freshwater Biology 17:373–379.

Schwartz, S. S., Hebert, P. D. N. 1987b. Breeding system of *Daphniopsis ephemeralis*: adaptations to a transient environment. Hydrobiologia 145:195–200.

Schwartz, S. S., Hebert, P. D. N. 1987c. *Daphniopsis ephereralis* sp. n. (Cladocera, Daphniidae): a new genus for North America. Canadian Journal of Zoology 63: 2689–2693.

Sergeyev, V. N. 1971. Povedeniye i mekhanizm pitaniya *Lathonura rectirostris* (Cladocera, Macrothricidae). Zoologicheskii Zhurnal 50: 1002–1010.

Shan, R. K., Frey, D. G. 1983. *Pleuroxus denticulatus* and *P. procurvus*(Cladocera, Chydoridae) in North America: distribution, experimental hybridization, and the possiblity of natural hybridization. Canadian Journal of Zoology 61:1605–1617.

Shapiro, J., Forsberg, B., Lamarra, V., Lynch, M., Smeltzer, E., Zoto, G. 1982. Experiments and experiences in biomanipulation: Studies of biological ways to reduce algal abundance and eliminate blue-greens. Interim Report No. 19 of the Limnological Research Center, University of Minnesota, Minneapolis Minnesota. 251pp.

Shaw, S. R., Stone, S. 1982. Photoreception. Pages 291–367 *in*: Atwood H.L., Sandeman, D. C., Eds. Neurobiology: Structure and Function. 479 pp. Vol. III, D. E. Bliss, Editor-in-Chief. The Biology of Crustacea. Academic Press. New York.

Shei, P., Iwakuma, T., Fujii, K. 1988. Population dynamics of *Daphnia rosea* in a small eutrophic pond. Ecological Research 3:291–304.

Shen, C.-j., Tai, A.-y., Chiang, S.-c. 1966. (On the cladoceran fauna of Hsi-song-pang-na and vicinity, Yunnan Province). Acta Zootaxonomica Sinica 3:29–423. [In Chinese, with English summary.]

Shurin, J. B., Dodson, S.I. 1997. Sublethal toxic effects of cyanobacteria and nonylphenol on environmental sex determination and development in *Daphnia*. Environmental Toxicology and Chemistry 16:1269–1276.

Sissom, S. L. 1980. An occurrence of *Cycloestheria hislopi* in North America. Texas Journal of Science 32:175–176.

Smirnov, N. N. 1970. Cladocera (Crustacea) from the Permian of Eastern Kazakhstan. Paleontological Journal 3:1–37 [In Russian.]

Smirnov, N. N. 1971. Chydoridae Fauny Mira. Fauna SSSR, Nov. Ser. No. 101. Rakoobraznyye, T. 1, vyp. 2. 531 p. [Available in English from Israel Program for Scientific Translations, Jerusalem 1974.]

Smirnov, N. N. 1985. *Anchistropus ominosus* sp. n. (Cladocera, Chydoridae) iz Reki Shingy (pritok Amazonki). Zoologicheskii Zhurnal 64:137–139.

Smirnov, N. N. 1992. The Macrothricidae of the world. Guides to the identification of the microinvertebrates of the Continental Waters of the World. SPB Academic Publishing.

Smirnov, N. N. 1996. Cladocera: The Chydorinae and Sayciinae (Chydoridae) of the world. Guides to the identification of the microinvertebrates of the continental waters of the world, No. 11. SPB Academic Publishing. The Hague. 197 p.

Smirnov, N. N., Timms, B. V. 1983. A revision of the Australian Cladocera (Crustacea). Records of Australian Museums, Supplement 1:1–132.

Sommer, U. 1989. Plankton ecology: succession in plankton communities. Springer, Berlin.

Spaak, P. 1995. Sexual reproduction in *Daphnia*: Interspecific differences in a hybrid species complex. Oecologia. 104:501–507.

Spitze, K. 1995. The quantitative genetics of zooplankton life histories. Experientia 51:454–464.

Sprules, W. G. 1975. Midsummer crustacean zooplankton communities in acid-stressed lakes. Journal of the Fisheries Research Board of Canada 32:389–385.

Sprules, W. G., Riessen, H. P., Jin, E. H. 1990. Dynamics of the Bythotrephes invasion of the St. Lawrence Great Lakes. Journal of Great Lakes Research 16:346–351.

Starobogatov, Ja.I. 1989. Sistema Crustacea. Zoologicheskii Zhurnal 65:1769–1781.

Stirling, G., McQueen, D. J. 1986. The influence of changing temperature on the life history of *Daphniopsis ephemeralis*. Journal of Plankton Research 8:583–595.

Sublette, J. E., Sublette, M. S. 1967. The limnology of playa lakes on the Llano Estacado, New Mexico and Texas. The Southwestern Naturalist 12:369–406.

Swanson, G. A., Meyer, M. I., Adomaitis, V. O. 1985. Foods consumed by breeding mallards on wetlands of south-central North Dakota. Journal of Wildlife Management 49:197–203.

Swift, M. C., Federenko, A. Y. 1975. Some aspects of prey capture by *Chaoborus* larvae. Limnology and Oceanography 20: 418–425.

Tappa, D. W. 1965. The dynamics of the association of six limnetic species of *Daphnia* in Aziscoos lake, Maine. Ecological Monographs 35:395–423.

Taylor, D. J., Hebert, P.D.N. 1993. Habitat-dependent hybrid parentage and differential introgression between neighboring sympatric *Daphnia* species. Proceedings of the National Academy of Sciences of the United States of America 90:7079–7083.

Teschner, M. 1995. Effects of salinity on the life history and fitness of *Daphnia magna*: Variability within and between populations. Hydrobiologia 307:33–41.

Tessier, A. J. 1983. Coherence and horizontal movements of patch of *Holopedium gibberum* (Cladocera). Oecologia 60:71–75.

Tessier, A. J. 1986. Comparative population regulation of two planktonic cladocera (*Holopedium gibberum* and *Daphnia catawba*). Ecology 67:285–302.

Tessier, A. J., Goulden, C. E. 1982. Estimating food limitation in cladoceran populations. Limnology and Oceanography 27:707–717.

Tessier, A. J., Goulden, C. E. 1987. Cladoceran juvenile growth: Implications for competitive ability. Limnology and Oceanography 32:680–685.

Thiel, H. 1963. Zur Entwicklung von *Triops cancriformis* Bosc. Zoologischer Anzeiger 170:62–68.

Thorp, J. H. 1986. Two distinct roles for predators in freshwater assemblages. Oikos 47:75–82.

Thorp, J. H., Black, A. R., Haag, K. H., Wehr, J. D. 1994. Zooplankton assemblages in the Ohio River: Seasonal, tributary, and navigational dam effects. Canadian Journal of Fisheries and Aquatic Sciences 51:1634–1643.

Threlkeld, S. T. 1976. Starvation and the size structure of zooplankton communities. Freshwater Biology 6:489–496.

Threlkeld, S. A. 1979. Estimating cladoceran birth rates: the importance of egg mortality and the egg age distribution. Limnology and Oceanography 24:601–612.

Threlkeld, S. T. 1986a. Differential temperature sensitivity of two cladoceran species to resource variation during a blue-green algal bloom. Canadian Journal of Zoology 64:1739–1744.

Threlkeld, S. T. 1986b. Life table responses and population dynamics of four cladoceran zooplankton during a reservoir flood. Journal of Plankton Research 8:639–647.

Threlkeld, S. T. 1988. *Daphnia* population fluctuations: patterns and mechanisms. Pages 367–388 *in*: Peters, R., deBernardi, R. editors. *Daphnia*. Memorie Istituto Italiano di Idrobiologia Vol. 45.

Threlkeld, S. T. Willey, R. L. 1993. Colonization, interaction, and organization of cladoceran epibiont communities. Limnology and Oceanography 38:584–591.

Tillmann, U., Lampert, W. 1984. Competitive ability of differently sized *Daphnia* species: An experimental test. Journal of Freshwater Ecology 2:311–323.

Tollrian, R., Dodson, S.I. 1998. Predator induced defenses in cladocerans. Pages 177–202, *in*: Tollrian, R., Harvell, C. D. Eds. The ecology and evolution of inducible defenses. Princeton Uni. Press, Princeton, NJ.

Tollrian, R., Harvell, C. D. 1998. The evolution of inducible defenses: current ideas. *in*: Tollrian R., Harvell, C. D., Eds. The ecology and evolution of inducible defenses. Princeton Uni. Press, Princeton, NJ, pp. 306–322

Torrentera, L., Dodson, S. I. 1995. Morphological diversity in populations of *Artemia* in Yucatan (Branchiopoda). Journal of Crustacean Biology 15:86–102.

Tribbey, B. A. 1965. A field and laboratory study of ecological succession in temporary ponds. Ph.D. Thesis, University of Texas, Austin.

EPA, U. S. 1992. Environmental monitoring and assessment program: Great Lakes monitoring and research strategy. Office of Research and Development, Environmental Research Laboratory-Duluth, Duluth, MN 55804.

Van Der Hoeven, N. 1990. Effect of 3,4-dichloroaniline and metavanadate on *Daphnia* populations. Ecotoxicology and Environmental Safety 20:53–70.

Vanni, M. J. 1986. Competition in zooplankton communities: Suppression of small species by *Daphnia pulex*. Limnology and Oceanography 31:1039–1056.

Walossek, D. 1995. The Upper Cambrian Rehbachiella, its larval development, morphology, and significance for the phylogeny of Branchiopoda and Crustacea. Hydrobiologia 298:1–13.

Weider, L. J., Hebert, P. D. N. 1987. Ecological and physiological differentiation among low-arctic clones of *Daphnia pulex*. Ecology 68:188–198.

Weider, L. J., Lampert, W. 1985. Differential response of *Daphnia* genotypes to oxygen stress: respiration rates, hemoglobin content and low-oxygen tolerance. Oecologia (Berlin) 65: 487–491.

Weismann, A. 1880. Beitrage zur Naturgeschichte der Daphnoiden. VI. Samen und Begattung der Daphnoiden. Zeitschrift fur wissenschaftliche Zoologie 33:55–256.

Wetzel, R. G. 1983. Limnology, 2nd ed., Saunders, Philadelphia, 753 p.

Whiteside, M. C. 1970. Danish chydorid Cladocera: modern ecology and core studies. Ecological monographs 40:79–118.

Whiteside, M. D., Williams, J. B. 1975. A new sampling technique for aquatic ecologists. International Association of Theoretical and Applied Limnology. Proceedings. 19: 1534–1539.

Whiteside, M. D., Williams, J. B., White, C. P. 1978. Seasonal abundance and pattern of chydorid Cladocera in mud and vegetative habitats. Ecology 59:1177–1188.

Wiggins, G. B., Mackay, R. J., Smith, I. M. 1980. Evolutionary and ecological strategies of animals in annual temporary pools. Archiv für Hydrobiologie Supplement. 58:97–206.

Williams, G. C. 1975. Sex and evolution. Monographs in Population Biology No. 8. Princeton, Princeton. 200 pp.

Williams, J. L. 1978. *Ilyocryptus gouldeni*, a new species of water flea, and the first American record of *I. agilis* Kurz (Crustacea: Cladocera: Macrothricidae). Proceding of the Biological Society of Washington. 91:666–680.

Wiman, F. H. 1979. Mating patterns and speciation in the fairy shrimp genus *Streptocephalus*. Evolution 33:172–181.

Wiman, F. H. 1981. Mating behavior in the *Streptocephalus* fairy shrimps (Crustacea: Anostraca). The Southwestern Naturalist 25:541–546.

Wingstrand, K. G. 1978. Comparative spermatology of the Crustacea Entomostraca. 1. Subclass Branchiopoda. Det Kongelige Danske Videnskabernes Selskab Biologiske Skrifter 22:1–67+20pl.

Winkler, D. W. (Ed.) 1977. An ecological study of Mono Lake, California. Institute of Ecology Publication Number 12, University of California, Davis.

Wolvecamp, H. P., Waterman, T. Respiration. Pages 35–100, *in:* Waterman, T. Ed. The Physiology of Crustacea. Vol. 1. Academic Press: New York. 670 p.

Zaret, T. M. 1980. Predation and freshwater communities. Yale University Press. New Haven. Zou, E.M, Fingerman, M. 1997. Effects of estrogenic xenobiotics on molting of the water flea, *Daphnia magna*. Ecotoxicology and Environmental Safety 38: 281–285.

Zaffagnini, R., Trentini, R. 1980. The distribution and reproduction of *Triops cancriformis* (Bosc) in Europe. Monitore Zoologico italiano nuova serie 14:1–8.

22 COPEPODA

Craig E. Williamson
Department of Earth and Environmental Sciences
Lehigh University
Bethlehem, Pennsylvania 18015

Janet W. Reid
Department of Invertebrate Zoology
National Museum of Natural History
Smithsonian Institution
Washington, DC 20560

I. INTRODUCTION

The subclass Copepoda H. Milne Edwards, 1840, in the class Crustacea is the largest of the entomostracan classes with over 10,000 known species. Due largely to their high densities in the world oceans, they are estimated to be some of the most abundant metazoans on earth (Hardy, 1970). Copepods are also abundant in terrestrial systems where they may reach densities in excess of 100,000 individuals/m^2 in wet organic soils (Reid, 1986). Although of marine origin, copepods may also dominate the fauna of some freshwater systems. For example, in Lake Baikal, which is both the deepest lake in the world and one which contains 20% of the unfrozen freshwater on earth, a single species (*Epischura baikalensis*) contributes up to 96% of the zooplankton (Huys and Boxshall, 1991).

Free-living freshwater copepods can be distinguished from other small aquatic invertebrates by a variety of morphological characteristics. They have a somewhat cylindrical, segmented body with numerous segmented appendages on the head and thorax, and two setose caudal rami on the posterior end of the abdomen. They possess an exoskeleton, conspicuous first antennae, and a single, simple, anterior eye. The defining apomorphy of the Copepoda is the structure of their swimming legs, each pair of which is connected at the base by a "coupler" or "intercoxal sclerite". The name of the class is from the Greek words *kope* for oar and *podos* for foot, derived from the coupler, which apparently increases the energetic efficiency of the legs.

Here, we follow Huys and Boxshall (1991) in recognizing two infraclasses and 10 orders of copepods. The infraclass Progymnoplea includes the order Platycopioida, while the infraclass Neocopepoda contains the other nine orders in two superorders. The superorder Gymnoplea includes the Calanoida, while the superorder Podoplea includes the remaining eight orders. Three of the 10 orders (Monstrilloida, Siphonostomatoida, and Poecilostomatoida) are primarily parasitic on a variety of fish and invertebrates, and the great majority are marine. Some poecilostomatoids, such as *Ergasilus* may have free-living planktonic males and copepodids while the adult females are ectoparasites on fish. Three other orders (Platycopioida, Misophrioida, and Mormonilloida) are primarily marine free-living

Ecology and Classification of North American Freshwater Invertebrates. 2nd Edition

species of the benthos or plankton. This chapter will focus on the three orders that contain the widespread, abundant, and primarily free-living copepods (Calanoida, Cyclopoida, Harpacticoida), of which there are over 7750 species. The harpacticoids include approximately 3000 species, 10% of which inhabit freshwaters; calanoids include about 2300 species, 25% of which are freshwater; while there are about 2450 species of cyclopoids, 20% of which are freshwater (Bowman and Abele, 1982). Many cyclopoids are parasitic. The remaining order, Gelyelloida, includes two recently discovered free-living species in the genus *Gelyella* from subterranean habitats in Europe. Another species has been reported from a single location in North America (Reid, unpublished).

Free-living freshwater copepods generally range in size from less than 0.5 to 2.0 mm in length, although some species such as the cyclopoids *Macrocylops fuscus* and *Megacyclops gigas,* and calanoids in several genera including *Heterocope, Epischura, Limnocalanus,* and *Hesperodiaptomus* can reach lengths of 3–5 mm. While the vast majority of freshwater copepods are transparent or a pale gray or brown, colors may range from black to red, orange, pink, purple, green and blue. The brighter colors are often caused by plant pigments such as carotenoids in oil droplets of the copepod.

Copepods are common in a variety of aquatic and semiaquatic habitats ranging from moist soils, leafpacks, groundwater, wetlands, and phytotelmata, to lakes, estuaries, and open oceans. Most species are omnivorous to some extent, with foods ranging from detritus and pollen, to phytoplankton, other invertebrates, and even larval fish. Copepods frequently comprise a major portion of the consumer biomass in these habitats. They play a pivotal role in aquatic food webs both as primary and secondary consumers, and as a major source of food for many larger invertebrates and vertebrates.

II. ANATOMY AND PHYSIOLOGY

A. External Morphology

The generalized copepod body in freshwater systems consists of an elongate, segmented body with an exoskeleton. There is usually a single major articulation that divides the body functionally into an anterior portion, the prosome, and a posterior portion, the urosome (Figs. 1 and 2). In the superorder Podoplea which includes the cyclopoids, harpacticoids, and gelyelloids, the major body articulation (not distinct in the gelyelloids) is between the fifth and sixth thoracic segments (somites of the fourth and fifth legs); while in the superorder Gymnoplea, which includes the calanoids, it is between the sixth thoracic segment (somite of the fifth leg) and the genital segment. The head is fused with the first and sometimes second thoracic segment to form the cephalosome. The thorax consists of seven segments, starting with the segment that bears the maxillipeds and ending with the genital segment. The number of segments in the urosome is variable, but usually between three and five. The prosome thus includes the head and most of the thorax, while the urosome includes the last one or two segments of the thorax and the entire abdomen, except for the terminal caudal rami.

Five pairs of jointed appendages occur on the head: the first antennae (antennules), second antennae (antennae), mandibles, first maxillae (maxillules), and second maxillae (Figs. 1–3). The first antennae serve many important functions related to reproduction, locomotion, and feeding. They are armed with both chemoreceptors and mechanoreceptors that may aid in the discrimination between mates, prey, and potential predators. The first antennae of male copepods are geniculate (have a joint that can bend abruptly) and modified for grasping the female during copulation (Figs. 1 and 3). In male harpacticoids, cyclopoids, and gelyelloids, both first antennae are geniculate, while in male calanoids only the right antenna is geniculate. The second antennae of calanoids, harpacticoids, and gelyelloids are biramous, while those of most freshwater cyclopoids lack an exopod and are thus uniramous.

Five or six pairs of well-developed thoracic appendages are generally present. The first thoracic segment bears the maxillipeds, while the second through the fifth bear four pairs of morphologically similar, biramous swimming legs which are joined by a coupler. In most freshwater copepods, the somite (or segment) bearing the first pair of swimming legs is fused with the head and segment bearing the maxillipeds, forming the cephalosome (Fig. 2). The swimming legs consist of two broad basal segments (the coxa, or coxopod and the basis, or basipod), to which are attached inner (endopod) and outer (exopod) rami of 1–3 segments each (Fig. 3). In cyclopoids and harpacticoids the fifth pair of legs is highly reduced. In calanoids the fifth legs are well developed and symmetrical in the female and highly asymmetrical and modified for grasping females during copulation in the male (Fig. 3). Cyclopoids and harpacticoids have vestigial sixth legs, which are larger in males. In adult gelyelloids the first three or four pairs of swimming legs have fused coxae and are highly reduced; the fourth legs are absent in the two European species. The fifth legs are absent, and male gelyelloids have vestigial sixth legs in the form of flaps that cover the genital aperture, while females have no sixth legs. Abdominal appendages are absent in copepods,

FIGURE 1 Major body types of freshwater copepods: (A) a calanoid, *Arctodiaptomus dorsalis* male; (B) a cyclopoid *Mesocyclops americanus* female, grasping a chironomid; (C) epibenthic harpacticoids, *Attheyella spinipes,* with the male guarding the female; (D) an interstitial harpacticoid, *Parastenocaris palmerae,* female with egg sac; (E) an undescribed species of gelyelloid, female; and (F) a first stage nauplius (NI) of *Macrocyclops fuscus* [redrawn from Dahms and Fernando, 1994].

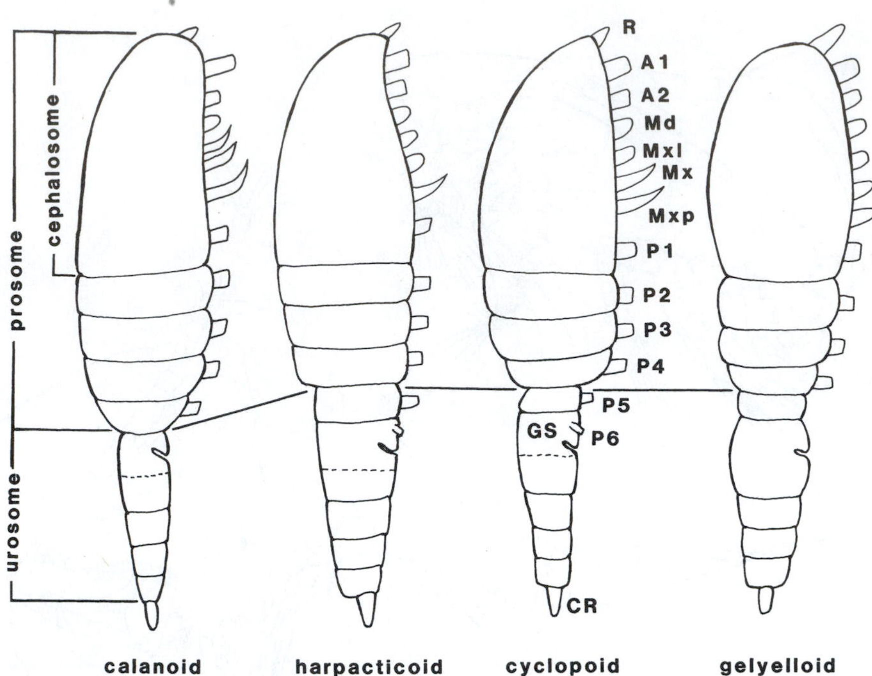

FIGURE 2 Generalized body plan of free-living freshwater copepods. Abbreviations: R, rostrum; A1, first antenna; A2, second antenna; Md, mandible; Mxl, first maxilla, Mx, second maxilla; Mxp, maxilliped; P1–P4, first through fourth swimming legs; P5, fifth leg; P6, sixth leg; GS, genital segment; CR, caudal ramus. The fifth and sixth legs in Gelyelloida, and the sixth legs in Calanoida, are lacking.

although the abdomen terminates in two caudal rami that are variously armed with setae and spines.

B. Internal Anatomy and Physiology

The circulatory system of cyclopoids and harpacticoids consists of a hemocoel with no heart or blood vessels. Blood is circulated by body and gut movements, and respiration occurs through the general body surfaces. Copepods have no gills or other respiratory organs. Calanoids have a more developed circulatory system that includes a dorsal heart with one pair of lateral ostia and one ventral ostium. Blood is carried anteriorly by a short anterior aorta; no other blood vessels are present.

The digestive system of copepods includes a mouth, esophagus (foregut), midgut, hindgut, and anus. The midgut may have a saclike diverticulum or cecum, and the anus is located at the posterior end of the urosome. The esophagus and hindgut are chitinized. A variable number of labral glands are located around the mouth. They function in a manner analogous to salivary glands by secreting mucus that mixes with food before entering the esophagus. The overall structure of the alimentary canal and the numbers and types of associated cells may vary with the feeding habits of the species (Musko, 1983). Egested fecal material is surrounded by a peritrophic membrane secreted by cells in the midgut of at least some calanoid and harpacticoid copepods, forming pellets (Gould, 1957; Fahrenbach, 1962).

Nitrogenous excretion occurs through antennal glands in naupliar (juvenile) stages, and through maxillary glands in adults. Osmoregulation is accomplished largely by low osmolality of the hemolymph ($<$200 mOsm, Mantel and Farmer, 1983). Some harpacticoids have integumental windows on their dorsal cephalothorax that serve as sites of ion exchange, and aid in osmoregulation (Hosfeld and Schminke, 1997). These windows are fairly common in most freshwater and some estuarine canthocamptids and parastenocaridids, as well as in some species of *Halicyclops*. They do not appear in marine copepods.

With the exception of a few species of harpacticoids (see Sarvala, 1979), reproduction in copepods is sexual. Copulation between male and female followed by union of haploid gametes is required for the formation of a fertile diploid zygote. Females have a paired or unpaired ovary dorsal to the midgut, with paired oviducts extending anteriorly. Diverticuli extend from the oviducts posteriorly to the gonopore on the genital segment. In calanoids the diverticuli join just before entering an atrium. Female calanoids also have a pair of

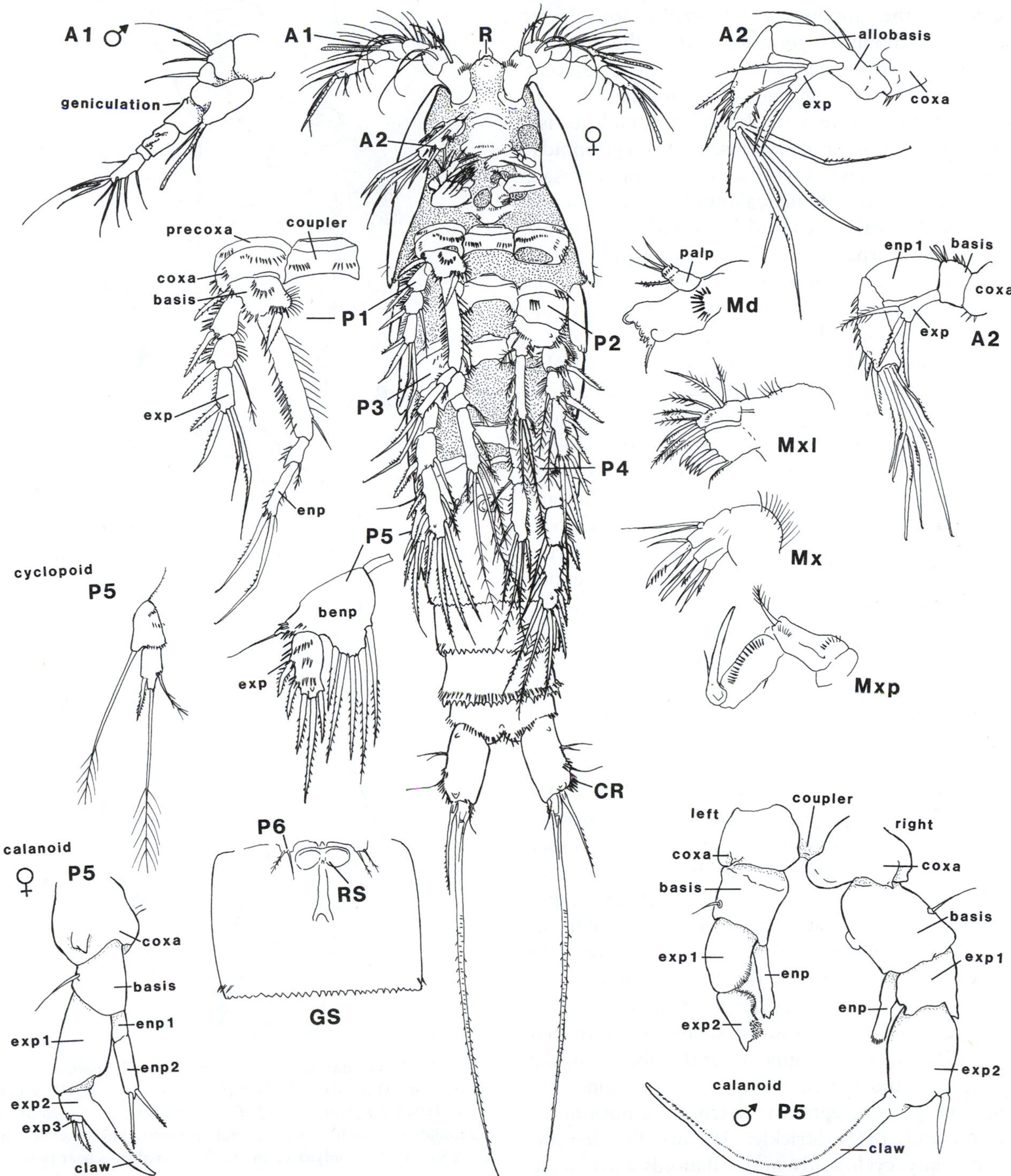

FIGURE 3 Copepod body parts and appendages, mainly of a harpacticoid, from anterior to posterior order: R, rostrum; A1, first antenna; A2, second antenna; Md, mandible; Mxl, first maxilla; Mx, second maxilla; Mxp, maxilliped; P1–4, first through fourth swimming legs; P5, fifth leg; P6, sixth leg; GS, genital segment; RS, seminal receptacle; and CR, caudal ramus. Other abbreviations: exp, exopod; enp, endopod; and benp, baseopod. Note differences in the structure of the fifth legs of male and female calanoids (lower right and lower left in figure) showing the modification of the male fifth leg for grasping the female during mating. Cyclopoid and harpacticoid fifth legs are smaller, and the basal segment of the harpacticoid P5 is enlarged on the inner margin (middle left of figure). In the upper right of the figure the second antennae of two different harpacticoids illustrate the difference in A2 structure with basis and allobasis.

seminal receptacles that open into the atrium. Female cyclopoids have a single seminal receptacle which is connected to the gonopores by fertilization ducts. Glandular cells along the terminal end of the oviducts secrete material for the egg shells. Skin glands adjacent to the gonopore secrete the egg-sac membrane.

Male copepods have a single dorsal gonad (testis). The vas deferens (one in calanoids, two in cyclopoids) is composed of the seminal vesicle, spermatophoric sac, and ejaculatory duct. Male calanoids have a single gonopore, female calanoids and both male and female cyclopoids and harpacticoids have paired gonopores. Spermatophores are elongate and cylindrical in calanoids and harpacticoids, and more compact and often kidney-shaped in cyclopoids (Fig. 4).

The nervous system of copepods is composed of a supraesophageal ganglion (brain) connected to a ventral nerve cord by two large circumesophageal connectives. The ventral nerve cord has several thoracic ganglia. Copepods have a variety of sensory receptors. Freshwater copepods have a simple eye spot that cannot form images; they lack the highly developed compound eyes with lenses characteristic of some marine pontellids (Gophen and Harris, 1981). Most copepods have a frontal organ consisting of a complex of fine sensory hairs and pores located on the anterior tip of their rostrum. The organ, referred to as the sensory pore X organ, includes the organ of Belloncі (Hosfeld, 1996), is connected to the brain by paired nerves and probably functions in chemoreception (Elofsson, 1971; Strickler and Bal, 1973).

The structures of several other types of sensory receptors have been investigated in freshwater copepods, and their function established by analogy with other known arthropod receptors. The body surfaces of calanoids and cyclopoids are covered with two types of integumental sensilla:small pegs (sensilla basiconica) and fine hairs (sensilla trichodea). The pegs are thought to be chemoreceptors, while the hairs may serve as both chemoreceptors and mechanoreceptors (Strickler, 1975a). Cyclopoids have rows of small spines located at the joints between the segments of the abdomen, the swimming legs, and the caudal rami. These spines are innervated by proprioceptors that sense the position of the relevant body parts (Strickler, 1975b). The first antennae of many cyclopoids and calanoids have setae that are modified ciliary processes that may serve in mechanoreception (Strickler and Bal, 1973; Friedman, 1980). The mouthparts of diaptomids have numerous contact chemoreceptors, but no mechanoreceptors (Friedman and Strickler, 1975; Friedman, 1980). Experiments in which copepods were fed flavored and unflavored polystyrene beads confirm the chemoreceptive abilities of copepods (DeMott, 1986).

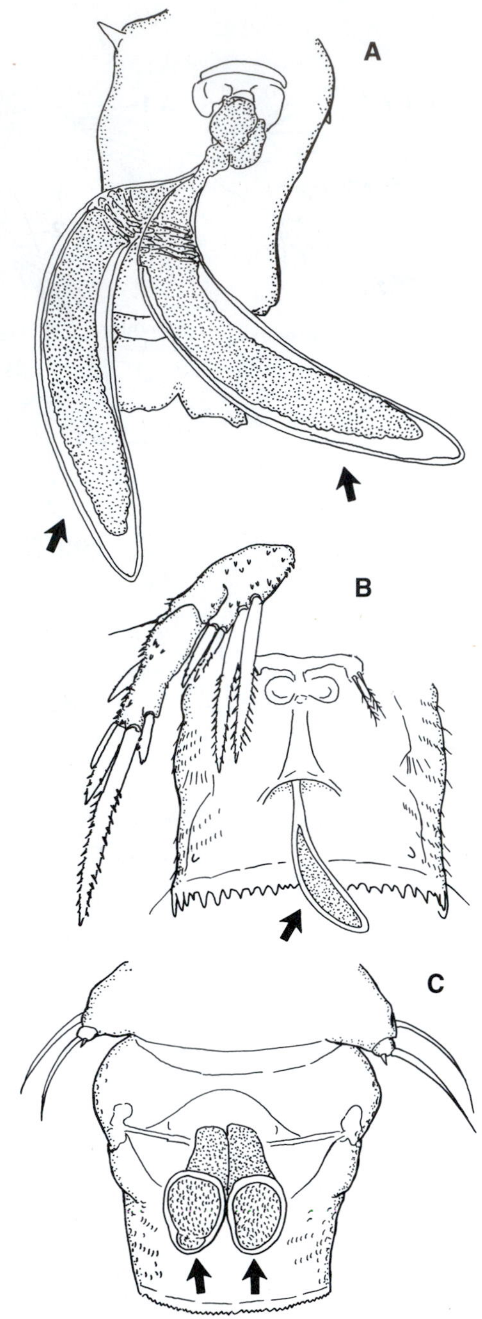

FIGURE 4 Genital segments with attached spermatophores on females of (A) a calanoid, *Arctodiaptomus dorsalis,* (B) a harpacticoid, *Attheyella spinipes,* and (C) a cyclopoid, *Metacyclops cushae.* Calanoids may bear one to several spermatophores, while harpacticoids usually bear only one, and cyclopoids always bear two.

III. ECOLOGY AND EVOLUTION

A. Reproduction and Life History

Copepods develop from fertilized eggs which hatch into a larval stage called a nauplius (Fig. 1F). There are six naupliar stages (N1–N6) followed by six copepodid stages (C1–C6), the last of which is the adult. [Some authors use the term "copepodite" instead of copepodid, but the "ite" ending is more properly used to refer to a part of a structure.] Growth is determinate, and adults do not continue to molt as in cladocerans. The nauplius larva starts out with a somewhat rounded body with three pairs of appendages: the first antennae, the second antennae, and the mandibles. During development the body of the nauplius becomes more elongate, additional appendages appear, and existing appendages become more elongate and specialized. A pronounced metamorphosis occurs between the last naupliar and the first copepodid instar. Copepodids are morphologically much more similar to the adults, as the body is clearly divided into a prosome and a urosome with caudal rami (Fig. 5).

The sexes of adult copepods are dimorphic. The mechanism of sex determination is genetic in some species but not known in others (Wyngaard and Chinappa, 1982). Sexual dimorphism is characterized by differences in the structure of the first antennae and the fifth and sixth legs, as well as by the number of urosomal segments and the generally larger size of the females versus the males (see taxonomy section). Sexual dimorphism in body size is more pronounced in the cyclopoids than in the calanoids or harpacticoids. Cyclopoids have a mean female:male body length ratio of 1.44; while for calanoids, this ratio is 1.13 (Gilbert and Williamson, 1983). In some cyclopoids the third and fourth swimming legs are dimorphic, while sexual dimorphism of the swimming legs is most pronounced in harpacticoids where some or all of the five pairs of swimming legs may be dimorphic.

Morphological variation brought about by allometric growth of certain body parts (cyclomorphosis) is subtle and not well understood in copepods. Seasonal variation in adult body size is common, with body size generally being smaller during the warmer summer months; both temperature and food may play a role in this seasonal variation (Reed, 1986; Reed and Aronson, 1989; Lescher-Moutoué, 1996). Some cyclopoids also may have more plumose setae on their caudal rami and swimming legs during the summer (Lescher-Moutoué, 1996). In a study where 22 different body dimensions were measured, allometric variation was established in *Tropocyclops prasinus* (Riera and Estrada, 1985).

The development time from extrusion to hatching for subitaneous (nondiapausing) eggs is usually between 1 and 5 days. Development time from egg to adult is normally on the order of 1–3 weeks, while the adult life span is generally from one to several months. Development times and life spans are much longer at lower temperatures and may vary considerably with species. Some cyclopoids have periods of diapause that may extend their life spans up to 1–2 years. Males generally mature more rapidly and have shorter life spans than females.

After reaching maturity copepods begin to produce gametes and can mate and produce viable embryos (usually called eggs) within a few days of maturation. Under favorable conditions, a single adult female produces a new clutch of eggs every few days and several hundred eggs over her lifetime. Clutches of 2–50 or more eggs may be carried laterally in two egg sacs (cyclopoids), ventrally in a single egg sac (most calanoids and harpacticoids), or scattered freely in the water (certain calanoids such as *Limnocalanus* and *Senecella*). Diel periodicities in egg laying have been reported for *Mesocyclops ogunnus* (Gophen, 1978; the original species determination of *M. leuckarti* was later corrected to *M. ogunnus*).

Calanoids and harpacticoids may produce either of two types of eggs. The normal mode of reproduction is through subitaneous eggs which hatch within a few days of being extruded. Resting eggs (diapausing eggs) that are capable of extended periods of dormancy may also be produced in many freshwater as well as coastal marine calanoids (Hairston, 1996; Hairston and Van Brunt, 1994). Initial studies on the "egg banks" of diaptomid copepods suggested that these resting eggs could survive several years (De Stasio, 1990), but more recent studies have demonstrated that mean ages of the resting eggs in these egg banks may be several decades, and that some eggs may even remain viable for up to several hundred years (Hairston *et al.,* 1995). Some calanoids, such as *Limnocalanus macrurus,* may produce only resting eggs in some systems but both resting eggs and subitaneous eggs in other systems (Roff, 1972). Cyclopoids produce only subitaneous eggs, but may enter diapause as copepodid stage II—adult, including fertilized females (Dobrzykowski and Wyngaard, 1993; Naess and Nilssen, 1991). These resting stages may vary from a simple developmental arrest to a fully encysted diapausing stage. In temporary ponds, cyclopoids such as *Diacyclops* can emerge from diapause within a single day of the appearance of water (Wyngaard *et al.,* 1991). Cyclopoid diapause is considered a true diapause similar to that observed in insects (Elgmork and Nilssen, 1978; Naess and Nilssen, 1991). A few species of harpacticoids may encyst in response to adverse environmental conditions (Sarvala, 1979; Frenzel, 1980).

Diapause can be initiated during either the spring and summer or the autumn. The proximate environmental cues that trigger the onset of diapause stages (eggs or copepodids) include overcrowding, photoperiod, and possibly age (Walton, 1985). The ultimate factors that

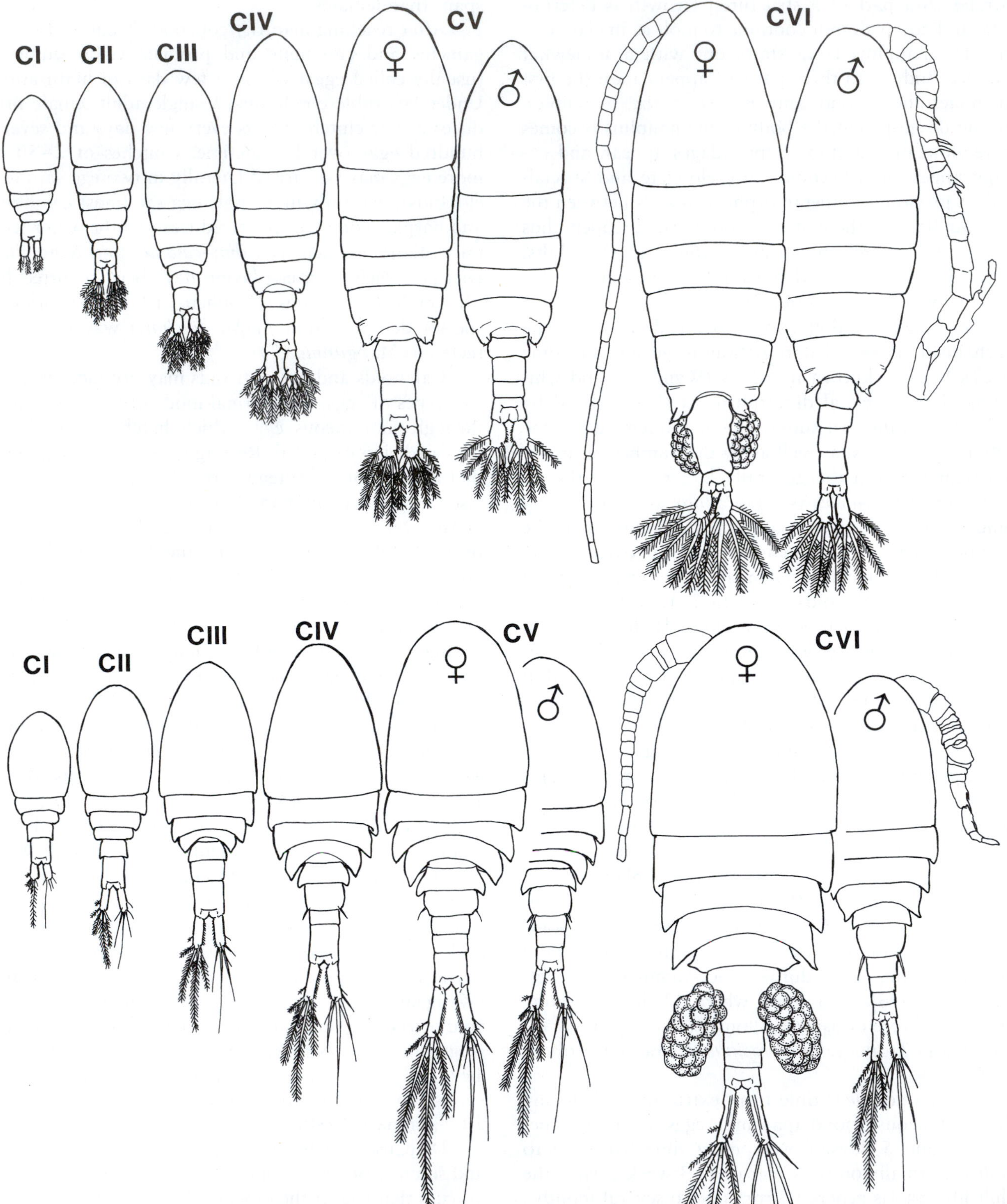

FIGURE 5 Copepodid and adult stages of calanoid (upper) and cyclopoid (lower) copepods. Secondary sex characteristics are apparent in the preadult stage (fifth copepodid, CV).

confer a selective advantage on diapausing individuals may be related to the avoidance of adverse environmental conditions, including an increase in predation pressures (Hairston and Olds, 1984; Hairston and Walton, 1986). The amphipod *Gammarus lacustris* preys on copepod resting eggs in the sediments and may deplete the "egg bank" to the extent that later recovery from adverse conditions such as the presence of fish predators is not possible (Parker *et al.*, 1996). The termination of diapause may be caused by either endogenous mechanisms, exogenous factors (such as low oxygen concentrations, changes in temperature, light, or physical disturbance), or some combination of these (Elgmork, 1967).

Female cyclopoids and harpacticoids can store sperm and thus produce fertile clutches of eggs for extended periods of time in the absence of males (Whitehouse and Lewis, 1973; Hicks and Coull, 1983). Most calanoid copepods, on the other hand, must mate repeatedly in order to continue clutch production. Diaptomids must mate before each clutch is produced (Watras and Haney, 1980; Williamson and Butler, 1987), while *Epischura* can produce multiple clutches from a single mating (Chow-Fraser and Maly, 1988). Under favorable environmental conditions female diaptomids alternate between gravid (oviducts filled with mature ova) and nongravid reproductive phases. Females must mate when they are gravid in order to produce viable eggs (Watras and Haney, 1980; Watras, 1983a, b; Buskey, 1998; Lonsdale *et al.*, 1998).

Mating behavior is similar for freshwater and marine calanoids (Blades and Youngbluth, 1980; Watras, 1983a; Lonsdale *et al.*, 1998). However, while male marine calanoids can locate females at distances of up to 20 cm with the aid of sex pheromones, diaptomids seem to respond to mates at distances of only a few body lengths (Watras, 1983a). Evidence for sex pheromones in freshwater calanoids is scarce (van Leeuwen and Maly, 1991). Watras (1983a) has described a behavioral sequence for mating in diaptomids wherein the male actively pursues the female (pursuit), grasps the female with his geniculate first antenna (precopula 1), grasps the female's urosome with his chelate right fifth leg (precopula 2), and uses the left fifth leg to attach a cigar-shaped spermatophore to the genital segment of the female (copula). The spermatophore is affixed with the aid of a cementlike secretion. During both precopula 2 and copula, the male and female are aligned with their urosomes adjacent and their metasomes extending in opposite directions. It is likely that precopula is accompanied by stroking behavior, as occurs in marine calanoids (Blades and Youngbluth, 1980). The entire mating sequence lasts from one to several minutes. Variations in sexual size dimorphism in calanoids may be important to mating success in some species, but not in others (Grad and Maly, 1988, 1992). Eggs are fertilized upon extrusion from the genital segment. In diaptomids, eggs that are not fertilized are released into the environment and disintegrate rapidly (Watras and Haney, 1980; Williamson and Butler, 1987).

Mating behavior is similar in harpacticoid and cyclopoid copepods, but the behaviors of both groups differ somewhat from the reproductive activities of calanoids (Gophen, 1979; Glatzel and Schminke, 1996; Dürbaum, 1995, 1997). Male harpacticoids and cyclopoids are more active swimmers than females, and mates are encountered randomly, with no evidence of sex pheromones. When encountering a female, a male grasps her with both his geniculate antennae; usually her third or fourth legs or urosome are held, but any part of her body will do. A period of high activity follows, during which the pair may jump and turn through the water. While still grasping the female with his first antennae, the male works himself into the final mating position with his ventral metasome facing the ventral surface of the female's genital segment so that the two bodies are parallel but facing in opposite directions. The male strokes this area of the female with his swimming legs and, with sharp flexing of the abdomen, deposits a compact spermatophore on her genital segment. Both individuals may remain passive for 5–10 s after spermatophore transfer. The entire mating process usually takes from 5 to 20 min (in some cases 2 h or more), after which the male releases the female. In some harpacticoids the male may retain his grasp on the female for up to several hours or even days in what is known as postcopulatory mate guarding (Dürbaum, 1995). During this period the female swims around with the passively attached male presumably preventing access to the female by any other males. Fertilization occurs upon extrusion of the eggs.

In some harpacticoids precopulatory mate-guarding has also been observed. This involves the male attaching to copepodid stage females (as early as CI in some species) so that they maximize their chances of being able to mate with a virgin female (Dürbaum, 1997). Interspecific mating has been observed in some cyclopoid species (Maier, 1995), and hybridization has been reported between two species of *Leptodiaptomus* (Chen *et al.*, 1997).

Feifarek *et al.* (1983) have shown that there may be a cost to mating for cyclopoid copepods. In a laboratory experiment, female *Mesocyclops edax* that mated and reproduced throughout their lives exhibited lower survivorship than unmated females. Within the group of females which mated, however, there was no relationship between reproductive output and survival. This suggests that the cost is associated with the mating itself rather than with allocation of resources to reproduction versus survival.

Environmental factors have a great influence on the life-history characteristics of copepods. Temperature, food availability, and predation are the three factors most often implicated as causes of variations in life-history traits. Adult female body size and interclutch duration are generally inversely related to temperature, while clutch size shows no consistent relationship to temperature (but see Chow-Fraser and Maly, 1991). Clutch size is often positively correlated with female body size (Maly, 1973; Hebert, 1985; Chow-Fraser and Maly, 1991). Food availability has also been demonstrated to be an important factor limiting rate of egg production and body size both in the laboratory and in the field (Edmondson, 1964; Elmore, 1983; Jersabek and Schabetsberger, 1995). Some species of diaptomids exhibit high reproduction during winter algal blooms (Vanderploeg *et al.,* 1992a). In some marine copepods ingestion of certain species of dinoflagellates can cause production of low-quality sperm that results in poor fertilization and hatching failure of eggs (Ianora *et al.,* 1999).

Food limitation can decrease fecundity by either decreasing clutch size, increasing interclutch duration, or both. The highly transparent bodies of many copepods allow one to monitor the presence of ova in the oviducts as well as external eggs carried by females. The proportion of a copepod population that occurs in each of these reproductive phases at one point in time can be used to quantify the relative importance of food and (in calanoids) mate limitation in nature (Fig. 6; Williamson and Butler, 1987). The presence or absence of ova in the oviducts alone are not by themselves, however, a good indicator of reproductive activity (Watras and Haney, 1980). This may be due at least in part to mate limitation and to an increase in the percentage of gravid individuals caused by the need for diaptomids to re-mate before each clutch is produced.

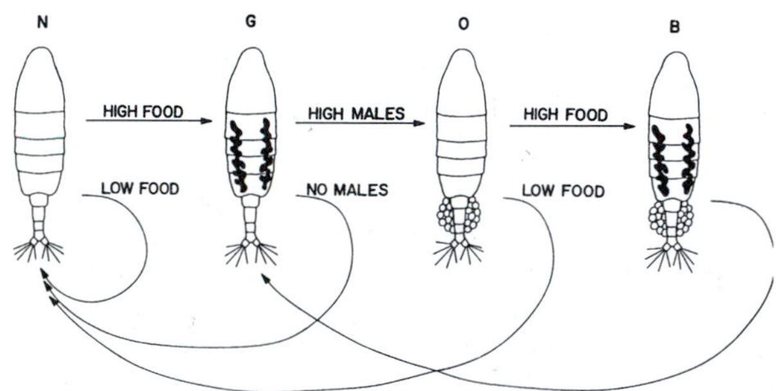

FIGURE 6 Four major reproductive phases of clutch-bearing female diaptomid copepods: G, gravid; O, ovigerous; N, neither; B, both. When abiotic environmental conditions are favorable for reproduction, transitions between phases are influenced by the availability of males and food. Phase B may be absent in species where embryos are released before oocyte maturation. [From Williamson and Butler, 1987.]

Predation may influence adult body size (Carter *et al.,* 1983), sex ratio (Hairston *et al.,* 1983; Svensson, 1997), body pigmentation, or the timing of diapause events in copepods (Hairston and Olds, 1984; Hairston and Walton, 1986). Some of these life history variations have a definite genetic component to them and, therefore, represent evolutionary adjustments rather than just simple acclimatory responses (Hairston and Walton, 1986; Wyngaard, 1986). When the invertebrate predator *Chaoborus* preys upon clutch-carrying diaptomids, the clutches may become detached and thus not be ingested; larger clutches are spared more often than smaller clutches (Svensson, 1996).

Life-history events of copepods may also differ among populations. Florida populations of *Mesocyclops edax* have shorter maturation times than populations of conspecifics in Michigan (Wyngaard, 1986). These differences seem to be largely genotypic, as evidenced by their persistence through two generations in the laboratory under the same controlled environmental conditions, as well as their persistence during a reciprocal transfer experiment carried out with the two different populations (Wyngaard, 1998).

B. Distribution

Copepods are found in a wide variety of aquatic environments, ranging from the benthic, littoral, and pelagic waters of lakes and oceans, to swamps, wetlands, marshes, large rivers, temporary ponds, and small puddles. Numerous species are common in interstitial and subterranean systems as well. Copepods are often common in phytotelmata, the small volumes of water that collect in the various parts of plants (Janetzky *et al.,* 1996; Reid and Janetzky, 1996). Copepods are present but less abundant in the flowing waters of streams and rivers. Although most prevalent in aquatic habitats, copepods have been collected from many semiterrestrial habitats such as moist forest soil, leaf litter, and even arboreal mosses (Reid, 1986). Calanoids are primarily planktonic, while cyclopoids and harpacticoids are generally associated with substrates in littoral or benthic habitats, although a few species of cyclopoids are planktonic and may contribute substantially to the zooplankton biomass in many lakes and ponds. Benthic copepods generally do not have pelagic larvae, and dispersal may occur in all developmental stages (Hicks and Coull, 1983).

The copepod component of the zooplankton is usually dominated by one or two species of calanoid or cyclopoid copepods. The most common and widely distributed planktonic cyclopoid species in North America are probably *Mesocyclops edax, Diacyclops thomasi,* and *Acanthocyclops* spp. Species richness in benthic copepods generally shows little or no clear pattern with latitude in the Americas, but the number of genera of Canthocamptidae correlates with latitude (Reid, 1992).

The distribution of many copepods seems to be related to temperature. For example, *Mesocyclops* often dominates in the south and in the warmer months; while in the north and in the cooler months in intermediate climates, *Mesocyclops* is often replaced by either *Cyclops, Diacyclops,* or *Acanthocyclops.* In systems where *Mesocyclops* does not reach high densities (e.g., Lake Michigan), these other cyclopoids may persist through the summer. The only calanoid genus endemic to North America is *Osphranticum.* The distribution of planktonic copepods in the northeastern United States indicates that historically some species depended on surface waters along retreating glacial ice fronts for postglacial dispersal; while other species, commonly found in smaller ponds and at high elevations less accessible to glaciers, may have been aided in their dispersal by their capability for diapause (Stemberger, 1995). Glaciation has strong residual effects on the present-day distribution of hyporheic (streambed) cyclopoids. Previously glaciated sites contain fewer species of interstitial specialists and fewer narrowly endemic species, but generalist species have apparently easily reinvaded these areas (Strayer and Reid, 1999).

The pH of a system may also influence the distribution and abundance of copepods, but the effects vary with species. In a study of 20 small lakes in the northeastern United States ranging in pH from 4.5 to 7.2, *Mesocyclops edax* and *Leptodiaptomus minutus* were both observed across the full range of pH values, but the former was more abundant at higher pH values while the latter was more abundant at lower pH values (Confer *et al.,* 1983). In this same study *Epischura lacustris* and *Cyclops scutifer* were found only at pH values of 5.3 or above. Bioassay experiments showed 100% mortality within 8.5 h for *Mesocyclops leuckarti* at pH values of <5, or of >9.2 (Ramalingam and Raghunathan, 1982). The distribution and abundance of various copepod species may also vary with lake size, depth, water transparency, color, and the presence of vertebrate and invertebrate predators. The horizontal distribution of copepods may be influenced by aquatic macrophytes (Gehrs, 1974) or fish (Carter and Goudie, 1986).

In benthic habitats, copepods occur primarily in the top 1–2 cm of the sediments, although diapausing animals may be found up to depths of 10–20 cm or more. The species richness of copepods in lake sediments is usually between 7 and 25 species, with about equal numbers of harpacticoids and cyclopoids (Strayer, 1985). The Laurentian Great Lakes together harbor 30, mostly benthic species of cyclopoids and 34 harpacticoid species (Hudson *et al.,* 1998). The vertical distribution of copepods within the sediments is primarily dependent upon redox potential, and only a few species can survive anaerobic sediments (Hicks and Coull, 1983).

The abundance of harpacticoids in benthic communities tends to increase with larger sediment particle size (Hicks and Coull, 1983). The most common species of harpacticoids are in the cosmopolitan family Canthocamptidae, with the most common North American genera being *Attheyella, Bryocamptus,* and *Canthocamptus.* Harpacticoids in the family Parastenocarididae are also abundant in well-oxygenated groundwater and interstitial environments. Common littoral–benthic cyclopoids include *Acanthocyclops, Diacyclops, Eucyclops, Macrocyclops,* and *Paracyclops.*

C. Physiological Adaptations

Copepods face a wide range of fluctuating environmental conditions in littoral, benthic, pelagic, and semiterrestrial habitats that require physiological adaptation. Perhaps the most widespread physiological adaptation of copepods is their ability to respond to seasonally unfavorable environmental conditions by reducing their metabolic rate and entering diapause in either the egg or late copepodid stages (see Section IIIA). Diapause may be induced by changes in temperature (Smyly, 1962) or oxygen concentration (Wierzbicka, 1962). These resting stages are highly resistant to temperature extremes and desiccation.

Physiological responses to temperature may vary among species and even sex of the copepod. Surface-dwelling harpacticoid species that are exposed to a greater seasonal variability in food availability and temperature can also better regulate respiration rates with temperature than can species living deeper in the sediments (Gee and Warwick, 1984). The longevity of adult male *Megacyclops viridis* may be extended under fluctuating versus constant temperature regimes, whereas female longevity is reduced under similar conditions (Smyly, 1980).

Copepods are frequently found in the benthic sediments of productive lakes and ponds which become hypoxic or anoxic. While low oxygen concentrations may induce diapause, many copepods still remain active under these rigorous oxygen conditions. A few planktonic species (*Mesocyclops edax, Microcyclops varicans*) vertically migrate into anoxic hypolimnetic waters for several hours on a daily basis (Williamson and Magnien, 1982; Threlkeld and Dirnbirger, 1986).

Laboratory experiments have demonstrated that adult *M. edax* can tolerate anoxia for 6–12 h at 24.6°C (Woodmansee and Grantham, 1961), while adult *M. varicans* live for at least 36 h. Cyclopoid nauplii can survive anoxia for less than 1 h at unspecified temperatures (Chaston, 1969). Benthic cyclopoids may persist for at least four days, and probably indefinitely under hypoxic (25% saturation, 8°C) conditions, but die within 2–5 h under anoxic conditions at 10°C (Tinson and Laybourn-Parry, 1985). Females are more tolerant of anoxia than males, and smaller species resist anoxia longer than larger species. A few species of harpacticoids are also known to tolerate anoxia (Hicks and Coull, 1983).

In surface waters where oxygen is more plentiful, solar radiation may have lethal effects on copepods (Hairston, 1976; Ringelberg *et al.*, 1984; Williamson *et al.*, 1994). The presence of plant-derived carotenoid pigments such as astaxanthin in the body, low temperature conditions, or nutritional state may reduce these lethal light effects (Hairston, 1976, 1979a–c; Ringelberg, 1980; Ringelberg *et al.*, 1984; Luecke and O'Brien, 1981). The darker color of the pigmented copepods may, however, make them more vulnerable to visually feeding vertebrate predators. Some species seem to have overcome these conflicting selective pressures by producing two separate morphs: one which has dark red pigments that increase survivorship under high light conditions, and one with a carotenoid–protein complex that is a pale green in systems where vertebrate predation pressures are high (Hairston, 1976; Luecke and O'Brien, 1981).

D. Behavioral Ecology

Copepods are active and highly proficient swimmers that exhibit a diversity of behavioral responses to environmental stimuli, ranging from the intensity and wavelength of light to the prospective reward or danger of other approaching organisms. The swimming mode of copepods varies with the species and habitat. Benthic harpacticoids and cyclopoids swim and crawl over substrates in the littoral, benthic, and interstitial habitats. The swimming behavior of planktonic copepods has been investigated in some detail (Rosenthal, 1972; Strickler, 1982) and has been reviewed for cyclopoids (Williamson, 1986).

Planktonic cyclopoids use their swimming legs, first antennae, and urosome to swim through the water, alternating between an active hop and a passive sink phase. The hop phase lasts about 20 ms and consists of a posteriorly directed thrust of the first antennae and a metachronal (sequential, one pair at a time, back to front) thrust of each of the pairs of swimming legs. The hops occur about once every second and during normal swimming will accelerate the copepod to a maximum velocity of about 80 mm/s. The sinking phase is the time between hops during which the copepod remains passive.

Many variations on this generalized swimming behavior occur. For example, the hops may have a reduced intensity and increased frequency and appear to be more of a shuttle behavior than a hop and sink. In addition, most cyclopoids swim right-side up, while others such as the small, more herbivorous *Tropocyclops,* swim upside down. When startled, cyclopoids may exhibit escape responses that consist of continuous, more vigorous hops which may accelerate the copepod up to a velocity of 350 mm/s. *Cyclops singularis* adults, which are on the order of 2 mm in body length, can move over 10 cm in a single jump—a behavior so pronounced that it allows this copepod to be separated from other species (Einsle, 1996b).

The swimming behavior of calanoids is distinctly different from that of cyclopoids. Calanoids spend a majority of their time gliding slowly through the water, propelled by the rapid (20–80 Hz) vibration of their feeding appendages. The more carnivorous species, such as *Epischura* and *Limnocalanus,* may actively cruise through the water in an upright position (dorsal surface up), while the more herbivorous species tend to hang vertically or at a slight angle in the water with their anterior end up. Every few seconds the gliding motion may be interrupted by a hop that involves the same appendages and motions as observed in cyclopoids. Less frequently, calanoids stop vibrating their appendages and sink passively through the water.

Nauplii spend a larger proportion of their time in a stationary position than do adults and copepodids; calanoid nauplii vibrate their appendages for only short periods between passive phases, while cyclopoid nauplii hop less frequently than do copepodids or adults (Gerritsen, 1978). Nauplii sink very slowly, if at all, but can exhibit escape responses which are the fastest recorded for any metazoan relative to body size (364 body lengths/s, = 55 mm/s, Williamson and Vanderploeg, 1988).

Copepods alter their swimming behavior in response to many diverse environmental stimuli, including the presence of food (Williamson, 1981) and predators (De Stasio, 1993; Ramcharan and Sprules, 1991). Light seems to be a particularly important stimulus in regulating the distribution of copepods. For example, *Heterocope septentrionalis* exhibits a swarming behavior in which it preferentially aggregates over light-colored substrates (Hebert *et al.*, 1980), and *Leptodiaptomus tyrrelli* may swarm to densities exceeding 11,000 per liter in response to contrasting substrates (Byron *et al.*, 1983). Copepods may also respond to different wavelengths of light in different ways. For example, *Hesperodiaptomus nevadensis* swims faster in blue light than in red, and individuals that contain large amounts of red carotenoid pigments swim faster than those that do not (Hairston, 1976). Light is also responsible for the distinct "avoidance of shore" ("Uferflucht") exhibited by pelagic copepods (Siebeck, 1969, 1980).

The diel vertical migration of planktonic copepods through the water column seems to be driven largely by responses to light, predation, and food limitation (Hutchinson, 1967; Williamson and Magnien, 1982; Neill, 1990). The most commonly observed migration patterns consist of downward migrations of several meters into the deeper, darker strata at dawn, and a similar distance upward into the surface waters at dusk. Reverse migrations downward at dusk have also been noted, particularly in response to invertebrate predation (Neill, 1990; Neill, 1992). Adult females often migrate farther than do either males or juveniles (Burgi *et al.*, 1993; Williamson and Magnien, 1982).

The adaptive significance of these vertical migrations is thought to be related to the reduction of predation pressures by predators (Zaret and Suffern, 1976; Wright *et al.*, 1980; Neill, 1990, 1992), but vertical migration may also influence food availability to the predators (Williamson *et al.*, 1989; Williamson, 1993; Makino and Ban, 1998). Changes in vertical overlap between predatory copepods and their prey populations can be quantified with a predation risk model that couples behavioral responses with field distribution and migration patterns of predator and prey (Williamson, 1993). The damaging effects of visible radiation cannot, however, be ruled out as an additional ultimate cause of these migrations. Damaging solar ultraviolet radiation may penetrate to depths of 10 m or more in clear lakes with low dissolved organic carbon concentrations (Morris *et al.*, 1995; Williamson *et al.*, 1996) and thus can cause substantial mortality in copepods that remain in the surface waters throughout the day (Williamson *et al.*, 1994). Interestingly, sublethal concentrations of free cupric ions can reduce the phototactic sensitivity of some estuarine and nearshore marine copepods such as *Acartia* and *Temora* (Stearns and Sharp, 1994).

Copepods may also respond to changes in the density of food in the environment. Cyclopoid copepods exhibit two distinct types of looping behavior: horizontal and vertical looping (Williamson, 1981). Horizontal loops consist of the copepod swimming in horizontal circles in a normal hop and sink swimming mode. When food densities are high, the frequency of horizontal looping behavior increases and serves to maintain the copepod in the vicinity of high food densities. Predatory cyclopoids may swim in rapid, tight, vertical loops after contacting individual prey organisms. This behavior aids the copepods in capturing prey and in recovering prey that attempt to escape after an initial attack (Williamson, 1981). Calanoids also alter their swimming behavior in response to the presence of food. *Senecella calanoides* and *Epischura lacustris* both increase the frequency of their passive sinking phase in the presence of prey. This behavior may reduce the "noise" levels of the predators and thereby permit them to approach and capture their prey by surprise (Wong and Sprules, 1986).

In addition to behavioral responses to food, copepods may also react to predators by altering their behavior. *Leptodiaptomus tyrrelli* reduces its filtering rate (Folt and Goldman, 1981), and *Leptodiaptomus minutus* decreases the frequency of its hops (Wong *et al.*, 1986) in the presence of predatory copepods. Similarly, *Cyclops vicinus* reduces its activity levels when the vertebrate predator *Abramis brama* (bream) is near (Winfield and Townsend, 1983). These diminished activity levels decrease the ability of both types of predators to detect these small copepods.

E. Foraging Relationships

The foraging relationships of copepods have received a great deal of attention over the years because of the important intermediate position of copepods between the phytoplankton and fish in aquatic food chains, and the fact that copepods are so abundant in a wide variety of aquatic ecosystems. Copepods contribute a major portion of the biomass and secondary production in a wide variety of aquatic communities. The importance of copepods in these communities derives largely from their impact on energy and material processing, prey mortality rates, and depression of available resources resulting from foraging relationships.

1. Diet

Copepods feed on a wide variety of foods including algae, pollen, detritus, bacteria, rotifers, crustaceans, dipteran larvae including mosquitos, chironomids, and chaoborids, and even larval fish (Fryer, 1957a, b; Monakov, 1976; Fischer and Moore, 1993; Hartig *et al.*, 1982; Reid, 1989b; Brandl, 1998). Therefore, they occupy three of the four major trophic positions in the food chain: detritivore; herbivore; and carnivore. Most species of calanoids and cyclopoids are omnivorous but selective in their feeding (Adrian, 1991; Williamson and Butler, 1986). Calanoids ingest particles ranging in size from small (a few micrometers) algae and bacteria to larger algae, microzooplankton, and macrozooplankton (>1 mm). Cyclopoids are grasping feeders that generally eat larger foods than calanoids. Their diet may include algae, rotifers, copepods, other crustaceans, oligochaetes, chironomids, nematodes, larval fish, and a variety of other foods. This creates an interesting situation where cyclopoids can prey on the larvae of their own predators, and may cause significant declines in the predator populations (Fischer and Frost, 1997; Fischer and Moore, 1993). Smaller cyclopoids such as *Tropocyclops prasinus mexicanus* tend to be more herbivorous than larger cyclopoid species (Adrian and Frost, 1992). Gelatinous chlorophytes are usually considered inedible or of low food value for zooplankton, but they can support survival and reproduction in some diaptomids (Stutzman, 1995).

Harpacticoids consume microalgae, fungi, protozoa, bacteria, and detritus, and some take larger prey such as nematodes. The algae may include phytoflagellates, cyanobacteria, and both epipelic and epiphytic diatoms. Harpacticoids also appear to use dissolved organic matter as a food source (Hicks and Coull, 1983). The microbial communities associated with dead foods may be more important than the foods themselves in the nutrition of harpacticoids. Although carnivory has been reported in harpacticoids, predation seems to be relatively unimportant as a mechanism for obtaining food (Hicks and Coull, 1983). However, *Phyllognathopus viguieri* is an active predator on soil and plant parasitic nematodes (Lehman and Reid, 1993).

The structure of the feeding appendages often correlates with dietary preferences, and both chemoreception and mechanoreception are important in selection of different types of food (Price *et al.,* 1983; DeMott, 1986; Legier-Visser *et al.,* 1986). Diaptomids are highly selective in their feeding and tend to prefer larger, more nutritious algae, or even small invertebrates such as rotifers (Williamson and Butler, 1986; DeMott, 1995a; Vanderploeg, 1994). Selectivity in diaptomids varies with food concentration in a way that is consistent with optimal foraging models (DeMott, 1995b). While diaptomids are capable of ingesting filamentous and colonial cyanobacteria at substantial rates (Schaffner *et al.,* 1994), they exhibit a distinct avoidance of toxic strains (DeMott and Moxter, 1991); existing data suggest that diaptomids are more sensitive than *Daphnia* to cyanobacterial toxicity (DeMott *et al.,* 1991). When feeding on filamentous diatoms or cyanobacteria, diaptomids may bite the ends off, a phenomenon known as "filament clipping" which may in turn reduce filament length in natural phytoplankton populations (DeMott and Moxter, 1991; Schaffner *et al.,* 1994; Vanderploeg *et al.,* 1988).

Both calanoid and cyclopoid copepods feed on ciliates, although many ciliates have behavioral, morphological, or chemical defenses that reduce their vulnerability to attack, capture, or ingestion (Williamson, 1980; Burns and Gilbert, 1993; Hartmann *et al.,* 1993; Wickham, 1995; Burns and Schallenberg, 1996). Reproduction is enhanced in diaptomids when ciliates are added to algal foods, indicating that the ciliate biomass is assimilated and utilized rather than just ingested (Sanders *et al.,* 1996). Thus diaptomids, which were once thought to use only algal biomass, can also use carbon sources from larger heterotrophs such as rotifers, and from protozoan carbon derived from the microbial component of the foodweb.

Diet may vary with the developmental stage, especially in the more carnivorous species. Nauplii and early copepodids are generally more herbivorous. The transition to more predatory feeding occurs in the late copepodid stages in calanoids (Maly and Maly, 1974; Chow-Fraser and Wong, 1986), while in cyclopoids even the earlier copepodid stages may be predatory (Brandl and Fernando, 1986; Williamson, 1986). Adult females are generally larger in size and correspondingly more voracious than adult males or juveniles. Many species are cannibalistic.

2. Feeding Mechanisms

The feeding mechanisms of suspension-feeding calanoids appear to be similar for marine and freshwater species. Food is brought in toward the copepod on microcurrents created by rapid vibrations of the second antennae, mandibular palps, first maxillae, and maxillipeds (Koehl and Strickler, 1981; Vanderploeg and Paffenhöfer, 1985). Smaller food particles are captured passively and are funneled into the mouth through the setae of the second maxillae (Vanderploeg and Paffenhöfer, 1985; Price and Paffenhöfer, 1986). During passive captures, the vibrations of the feeding appendages continue uninterrupted. Larger food particles are captured actively with an outward fling of the second maxillae followed by an inward squeeze to remove the excess water surrounding the food particle.

When microzooplankton such as rotifers and nauplii are brought in on the feeding currents of suspension-feeding diaptomids, the copepods will often attack these small animal prey from a distance with an active thrust response in which the antennae and swimming legs are used to orient and pounce toward the prey (Williamson and Butler, 1986; Williamson, 1987; Williamson and Vanderploeg, 1988). Particles less than 5 μm are generally captured passively by diaptomids, while those greater than 50 μm are generally captured actively or by attack; in intermediate size ranges, the frequency of active captures increases with increasing particle size (Vanderploeg and Paffenhöfer, 1985; Williamson and Vanderploeg, 1988).

Many of the larger calanoid species such as *Heterocope, Epischura, Limnocalanus,* and some of the large species of diaptomids have predatory tendencies and feed on other zooplankton as well as algae. These predators cruise through the water, attack their prey with a pounce, and grasp them with their first and second maxillae. If a capture attempt fails, the copepod may swim in a vertical loop to try again to capture the prey (Kerfoot, 1978).

Cyclopoid copepods do not create currents to aid their feeding but instead grasp their food directly with their first maxillae or, to a lesser extent, with their second maxillae and maxillipeds. These appendages push food between the mandibles. By oscillating rapidly during feeding bouts, the mandibles tear prey into pieces and stuff them into the esophagus (Fryer, 1957a).

Cyclopoids detect their prey with the help of mechanoreceptors on their first antennae. Prey which generate much disturbance as they swim can be detected at distances of several millimeters. Cyclopoids actively orient and attack prey with considerable precision (Kerfoot, 1978). Larger prey such as cladocerans, copepods, and chironomids may be handled for 30 min or more before ingestion (Fryer, 1957a; Gilbert and Williamson, 1978; Williamson, 1980). Small, generally less active prey such as algae, protozoans, rotifers, and nauplii may be detected only after contact, and are generally eaten whole. Cyclopoid nauplii use their mandibles and second antennae to capture and ingest food (Monakov, 1976).

Harpacticoid copepods are primarily surface-feeders that use their mouthparts for scraping food from a diversity of substrates. Suspension-feeding has been reported in two primitive marine families, the Longipediidae and the Canuellidae (Hicks and Coull, 1983). In surface-feeding, the second maxillae and second antennae grasp the prey while the maxillipeds aid in anchoring the copepod to the substrate. Once the food is grasped, the first maxillae shred and stuff the food into the vibrating mandibles. The second maxillae and maxillipeds are not used in food handling. Harpacticoid nauplii have modified second antennae that are employed to grasp, masticate, and push food into the mouth (Fahrenbach, 1962). The mandibles of nauplii serve only for locomotion and not for feeding.

Suspension-feeding harpacticoids feed by lying on their dorsal or lateral sides and vibrating their first and second maxillae and mandibles. These vibrations create feeding currents that bring water and food in toward the mouth and out posteriorly along the ventral surface of the copepod. Some harpacticoids can use both surface-feeding and suspension-feeding mechanisms (Hicks and Coull, 1983).

3. Feeding Rates

Feeding rates are usually expressed quantitatively as either clearance rates (expressed as volume of water cleared of food per unit time) or ingestion rates (food items ingested per copepod per unit time). Ingestion rates generally depend upon prey density below a saturation food density referred to as the incipient limiting concentration; above this density there is no further increase in ingestion rate. Clearance rates are density-independent below this saturation food concentration and decrease above it. Feeding rates are calculated based on the exponential equations first applied to zooplankton feeding by Gauld (1951). The standard experimental design is to fill a number of experimental vessels with either food alone (controls) or food and copepods (experimentals) and then incubate them for a period of time on a rotating wheel to prevent settling and minimize patchiness of the food and copepods. Feeding rates can then be calculated as follows:

$$\text{Ingestion rate } (I) \ 5 \ \frac{D_{ct} \ 2 \ D_{et}}{TN} \tag{1}$$

$$\text{Clearance rate } (F) = \frac{V(ln\ D_{ct} - ln\ D_{et})}{TN} \tag{2}$$

where D_{ct} and D_{et} are the food densities in the control and experimental vessels at the end of the incubation, respectively, T the duration of the experiment, and N the number of copepods per experimental vessel. The average food density to which the copepods are exposed during the experiment is then:

$$d = \frac{D_{et} - D_{eo}}{V(ln\ D_{et} - ln\ D_{eo})} \tag{3}$$

where d is the average food density, D_{et} and D_{eo} are the final and initial food densities in the experimental vessels, respectively, V the volume of the experimental vessel, and $I = Fd$.

An alternative method for estimating feeding rates is to add different numbers of predators to the experimental vessels and use regression analysis to determine clearance rates (Lehman, 1980; Landry and Hassett, 1982). This method assumes that changes in the experimental vessels are proportional to the number of predators per vessel and to the incubation time according to the following equation:

$$ln\ D_{et} = D_{ct} + FNT \tag{4}$$

This relationship is derived from the conventional population growth rate equation (see Eq. 5). If D_{et}/T is plotted against N, the slope of the regression relationship is the clearance rate (F), the y intercept is an estimate of ln D_{eo}, and standard regression statistics can be used to determine the statistical significance of the regression relationship as well as the percentage of the variation in prey densities that can be attributed to predation (or, how good was the experiment?). An assumption of both of the above methods of estimating feeding rates is that $D_{ct} = D_{eo}$, or, in other words, that prey birth rates as well as nonpredatory death rates were minimal.

Gut content analyses can also be helpful in distinguishing the diet of copepods. Copepods can be squashed between a coverslip and a microscope slide for a crude assessment of gut contents. A more quantitative analysis can be obtained by dissection. A pair of fine minuten needles (insect pins) secured to the tips of applicator sticks or glass pipets can be used to tease the metasomal gut tube out of preserved copepods. The gut is placed in a very small drop of water, covered with a coverslip, and examined under a compound microscope before and after the addition of 5% sodium hypochlorite (commercial bleach) (Williamson, 1984).

Clearance rates of predatory calanoids are generally in the range of 80 to 180, but values have been reported as high as 630 mL per predator per day, for *Hesperodiaptomus shoshone* feeding on calanoid nauplii at 20°C (Maly, 1976) and over 1,000 mL per predator per day, for *Heterocope septentrionalis* feeding on *Daphnia pulex* at 15°C (Luecke and O'Brien, 1983). Ingestion rates averaging as high as 18.2 prey per predator per day have been reported for *Hesperodiaptomus shoshone* feeding on various crustacean prey (Anderson, 1970); while rates of up to 54 prey per predator per day have been observed for *Heterocope septentrionalis* feeding on *Bosmina longirostris* at 15°C (O'Brien and Schmidt, 1979). The benthic harpacticoid *Attheyella* reduced bacterial densities in an Appalachian headwater stream by up to 58%; the greater reductions were observed under lower flow conditions (Perlmutter and Meyer, 1991).

The feeding rates of suspension-feeding calanoids vary widely and probably depend largely on dietary preferences and available food. Many investigators have obtained clearance rates of only a few milliliters per copepod per day or less for certain algae, and calanoids are thus often thought to have clearance rates that are much lower than similarly sized planktonic cladocerans (Wetzel, 1983). However, when offered high quality algae as food, diaptomids filter 19–25 milliliters per copepod per day, and weight-specific rates are comparable to or greater than those of cladocerans (Bogdan and Gilbert, 1984; Vanderploeg *et al.*, 1984; Williamson and Butler, 1986). Although most of the smaller diaptomids have historically been considered to be herbivorous, clearance and ingestion rates of diaptomids on rotifers may be five to six times greater than those for algae (Williamson and Butler, 1986). The presence of algae reduces the predation rate of suspension-feeding diaptomids (Williamson and Butler, 1986), but does not similarly affect more predatory species such as *Epischura* (Wong, 1981). In general, diaptomids feed more selectively than *Daphnia* (Richman and Dodson, 1983; DeMott, 1988, 1989; Vanderploeg, 1994).

Cyclopoid ingestion rates are density-dependent and may reach 42 prey per predator per day at prey densities approaching 1000 prey/L (Williamson, 1984, 1986). Clearance rates may exceed 150 mL per predator per day (Williamson, 1983b). Ingestion rates, clearance rates, and prey selectivities may be influenced by hunger levels of the predator, water temperature, and the size, morphology, behavior, and density of the prey (Williamson, 1980, 1986; Stemberger, 1986; Roche, 1987).

F. Population Regulation

1. Factors Controlling Population Density

Estimates of the per capita growth rates (r) of copepod populations vary from less than 0.1 per day to as high as 0.4 per day (Allan, 1976; Hicks and Coull, 1983). Doubling times ($=\ln 2/r$) are generally on the order of 1 to 2 weeks but may be less than 2 days under optimal conditions. These high reproductive rates generate pronounced density oscillations in nature. These oscillations, in conjunction with the 12 distinct life-history stages, a small body size, and short generation times, make copepods good subjects for population studies.

Several hypotheses have been proposed to explain the observed oscillations in copepod populations. The most compelling of these include food limitation, temperature, and predation. Mate limitation may also be important for diaptomids that must mate before the production of each clutch (Williamson and Butler, 1987). The relative contributions of each of these factors vary between systems and are difficult to separate in nature.

The physiological importance of temperature in controlling rates of growth and development has been demonstrated in numerous copepod species. Gamete maturation, egg development, clutch production, and instar development rates are all temperature-dependent (Watras, 1983b; Hicks and Coull, 1983). In addition, temperature may determine the seasonal timing of reproductive activity. A recent review of temperature responses of freshwater zooplankton indicates that many species of copepods are cold-water species potentially sensitive to climate warming (Moore *et al.*, 1996). Within the temperature range that permits reproductive activity, food availability may limit the rates of reproduction and development of copepods by prolonging interclutch duration and reducing clutch size (Woodward and White, 1981; Elmore, 1982, 1983). Response to food limitation may be quite rapid in some species (Williamson, *et al.*, 1985).

The differential response of the reproductive phases of copepods to food density, mate density, and temperature may permit the analytical separation of food and mate limitation from temperature effects in some diaptomid copepods (Williamson and Butler, 1987). The transitions between reproductive phases are largely regulated by food and mate availability and are independent of temperature; hence, the proportion of adult females in the different phases will reflect which factors are limiting (Fig. 6). Food limitation can also be examined with an index that quantifies both lipid storage and reproductive condition (Vanderploeg *et al.*, 1992b). This latter index is similar to that developed for marine copepods (Marshall and Orr, 1952) and freshwater cladocerans (Tessier and Goulden, 1982). Optical–digital methods have also been used to quantify seasonal changes in lipid reserves in diaptomids (Arts and Evans, 1991). They are much simpler than, and compare favorably with, more expensive chemical methods of analysis. The response of copepods to food and mate limitation will vary with species and environmental conditions, and all of these indices must therefore be applied only with an adequate understanding of the reproductive physiology of each species (Jersabek and Schabetsberger, 1995).

In many populations the nauplii are the "bottleneck" stage that exhibits the highest mortality rate in nature (Confer and Cooley, 1977; Feller, 1980). The small size of the nauplii relative to copepodids may result in an increased vulnerability to both starvation (due to higher weight-specific metabolic rates) and invertebrate predation, and consequently higher mortality rates for the nauplii (Williamson *et al.*, 1985).

The relative abundance of calanoid versus cyclopoid copepods may change with the trophic status of an ecosystem. Several investigators found that as system productivity rose, the abundance of cyclopoid copepods and cladocerans increased while calanoid abundance either decreased or varied independently (Byron *et al.*, 1984; Pace, 1986; Stemberger and Lazorchak, 1994). Feeding and respiration rate experiments with *Daphnia* and diaptomids indicated that food quality may also play an important role in the relative abundance of calanoids versus cladocerans (Richman and Dodson, 1983; Lampert and Muck, 1985). Richman and Dodson (1983) suggested that calanoids will predominate when either food quality is low or when food density is either extremely high or extremely low. Lampert and Muck (1985) suggested that diaptomids are better adapted to low, fluctuating food densities than is *Daphnia,* because the former allocates more energy to storage for survival while the latter puts its resources into reproduction under these food-limiting conditions. There is some evidence that the threshold food level for survival is lower for calanoid than for cyclopoid nauplii, and that this "bottleneck" stage may be important in regulating the relative abundance of these two taxa (Santer, 1994). Enclosure experiments in 450 L tanks indicate that the interactions between cyclopoids and calanoids are complex and may involve predation on calanoids by cyclopoids, and depression of food resources for cyclopoid nauplii by calanoids (Soto and Hurlbert, 1991).

Predation may also influence copepod population densities. Cannibalism is common among the more carnivorous species (Gabriel, 1985), and both vertebrate and invertebrate predators prey on copepods. The inconsistent relationship between calanoid abundance and lake trophic status discussed above may be due to invertebrate predation on diaptomids (Edmondson, 1985), or possibly predation by the more carnivorous cyclopoids on calanoids (Soto and Hurlbert, 1991; Adrian, 1997). Other invertebrate predators, including predatory copepods and larvae of the phantom midge *Chaoborus,* may inflict substantial mortality on copepod populations (Luecke and Litt, 1987). The predatory cladoceran *Bythotrephes cederstroemi* was introduced into the Laurentian Great Lakes from Europe in the 1980s, and has spread to nearby smaller lakes as well. This exotic species preys on both adult copepods and nauplii; and, while copepods seem to be some of the less-preferred prey (Vanderploeg *et al.*, 1993), some cyclopoids such as *Mesocyclops edax* and *Tropocyclops prasinus mexicanus* have either declined or disappeared in lakes when *Bythotrephes* invades (Yan and Pawson, 1997). Vertebrate predators, including fish, birds, and larval and adult salamanders, may influence the distribution and abundance of copepods in lakes (Dodson and Egger, 1980; Zaret, 1980; Carter and Goudie, 1986). Larger species such as *Hesperodiaptomus* may be excluded from lakes when fish are introduced (Donald *et al.,* 1994). Vertebrate predators may also be instrumental in the induction of diapause (Hairston and Walton, 1986) or diel vertical migrations (Williamson and Magnien, 1982), and both vertebrate and invertebrate predators may contribute to skewed sex ratios in copepod populations on which they prey (Maly, 1970; Hairston *et al.,* 1983; Blais and Maly, 1993).

Heavy organic and nutrient loading and associated reductions in dissolved oxygen may differentially reduce populations of different species of benthic harpacticoids, as well as reduce the abundance of harpacticoids relative to other meiobenthos (Särkkä, 1992).

2. Methods for Analysis of Populations

The instantaneous per capita birth rates, death rates, and growth rates of copepods can be estimated using the egg ratio technique developed by Edmondson (1960) and Edmondson *et al.* (1962), as modified by Paloheimo (1974). The growth rate of the population (r) is estimated from the population densities of adult females at time zero (N_0) and time t (N_t) with the following equation:

$$r = \frac{\ln N_t - \ln N_0}{t} \tag{5}$$

The birth rate (b) can be estimated from the egg ratio (E) and the development time (D) from egg to adult as:

$$b = \frac{\ln (E + 1)}{D} \tag{6}$$

where E is the number of female eggs in the population (assume a 1:1 sex ratio) divided by the number of adult females in the population. The death rate (d) can then be estimated from:

$$d = b - r \tag{7}$$

These estimates of r are for the midpoint between the two sampling dates while the estimates of b are for the given sample date. Therefore, when estimating d from Eq. 6, the values of r should be linearly interpolated for the given date (Wyngaard, 1983). These methods assume a constant age structure and continuous egg laying in the populations. Estimates of r with this egg-ratio method are best applied to populations with shorter developmental stages such as those found at higher temperatures. An alternative approach which requires much greater resolution of the developmental stage durations is cohort analysis (Hairston and Twombly, 1985).

A careful analysis of the allocation of effort between and within sampling stations as well as between subsamples is critical in order to maximize the accuracy of population estimates (Nie and Vijverberg, 1985). The variance associated with estimates of per capita birth, death, and growth rates can be obtained with either jackknife or bootstrap techniques (Meyer *et al.*, 1986).

G. Functional Role in the Ecosystem

Copepods make up a major portion of the biomass and productivity of freshwater systems. In Mirror Lake, New Hampshire, copepods contributed about 55% of the total zooplankton biomass over a three year period (Makarewicz and Likens, 1979). Cyclopoid copepods may contribute up to 77% and calanoid copepods up to 49% of the macrozooplankton biomass in some lakes (Pace, 1986). Benthic samples in a second-order stream in North Carolina revealed biomass to be on the order of 22 mg dry mass/m^2, equivalent to densities of 3000–18,000 individuals/m^2, and cyclopoid densities of about 100–1600 individuals/m^2 (O'Doherty, 1985, 1988). In the benthic communities of lakes, it is not uncommon to find densities of copepods in excess of 10,000 and, occasionally, up to even 70,000 individuals/m^2, with a biomass of 50–70 mg dry weight individuals/m^2 (Strayer, 1985).

In addition to their numerical contribution, copepods occupy an important intermediate position in aquatic food chains (Kerfoot and DeMott, 1984). Their generally omnivorous habits make them important in energy transfer up the food chain; they also have the potential to regulate the structure of phytoplankton and zooplankton prey assemblages. By packaging bacterioplankton, phytoplankton, and microzooplankton prey into larger particles (their own bodies) copepods may actually increase food-chain efficiency in spite of the energy lost by the additional intermediate trophic level (Confer and Blades, 1975). Copepods may also contribute to phosphorus regeneration in lakes (Bowers, 1986).

As predators, copepods may have a substantial impact on their prey populations. Predatory calanoids and cyclopoids are intense, selective predators that can alter the distribution and morphology of their prey. *Mesocyclops edax,* an important cyclopoid predator in the plankton, may crop up to 24% of its prey populations per day (Brandl and Fernando, 1979; Williamson, 1984). Populations of even some of the smaller suspension-feeding diaptomids have the potential to clear the rotifers from a volume of water equivalent to the entire volume of the lake in a single day (Williamson and Butler, 1986). *Cyclops vicinus* can prey on the rotifer *Synchaeta* at rates of up to 37 *Synchaeta* per day, rates that suggest that this cyclopoid may be responsible for declines in this rotifer in Lake Constance in the spring (Plassmann *et al.,* 1997). Cyclopoid copepods prey selectively on, and may selectively inhibit the growth and reproductive rates of, smaller cladoceran species as well as juveniles within species (Gliwicz, 1994; Gliwicz and Umana, 1994; Santer, 1993). In contrast, only very large *Daphnia* appear to be susceptible to brood predation by copepods. *Eudiaptomus gracilis* and *Acanthocyclops robustus* copepodids have both been observed to enter the brood cavities of *Daphnia* over 2.25 mm long, where they prey on the eggs (Gliwicz and

Lampert, 1994; Gliwicz and Stibor, 1993). Some of the large species of *Hesperodiaptomus* may regulate the relative abundance of different rotifer species and the clonal composition of *Daphnia* in arctic and alpine systems (Paul *et al.,* 1995; Paul and Schindler, 1994; Wilson and Hebert, 1993). Several calanoid species in the Great Lakes can feed on the larvae of the zebra mussel, *Dreissena polymorpha* (Liebig and Vanderploeg, 1995).

Although smaller prey are generally preferred by predatory copepods, the coincident vertical migration of the copepods with some of the larger cladocerans may cause ingestion rates to be disproportionally greater on these otherwise undesirable prey types (Williamson *et al.,* 1989). Many prey species exhibit elaborate morphological and behavioral defense mechanisms that suggest a strong coevolutionary relationship between copepod predators and their prey. The hard loricas, spines, gelatinous sheaths, and rapid escape responses of many prey species are all effective at reducing predation by copepods (Williamson, 1983a, 1986, 1987; Stemberger, 1985; Roche, 1987). Some of the morphological defenses are nongenetic polymorphisms that are induced by chemical substances released by the copepods (Stemberger and Gilbert, 1984).

Copepods are themselves important prey for other copepods (Peacock and Smyly, 1983), *Chaoborus* larvae (Peacock, 1982; Wyngaard, 1983), and fish. Adult copepods are preyed on by many planktivorous and benthic-feeding fish, and the nauplii are particularly important prey for larval fish (Guma'a, 1978; Hicks and Coull, 1983).

Some of the larger cyclopoid copepods such as *Mesocyclops* and *Macrocyclops* may serve as important agents of biological control of pest insects such as mosquitos (Marten *et al.,* 1994; Vu *et al.,* 1998). Predation rates on early instar mosquito larvae may be quite high (Brown *et al.,* 1991; Reid, 1989b), and cyclopoids can be cultivated in large quantities for this purpose (Marten *et al.,* 1994; Suárez *et al.,* 1992).

Cyclopoids and calanoids are also important intermediate hosts for many parasites, including flukes, nematodes, and tapeworms of fish, amphibians, birds, and mammals. Some cyclopoids such as *Mesocyclops* and *Thermocyclops* may serve as intermediate hosts of the human parasite *Dracunculus medinensis* (guinea worm) in the tropics. Ingestion of the larvae of this nematode in drinking water can cause serious illness or death in humans in areas of equatorial Africa and India (Steib and Mayer, 1988). The bodies of planktonic and epibenthic copepods provide a substrate for the *Vibrio cholerae* bacterium that carries cholera (Huq *et al.,* 1983).

Recent studies on the stoichiometric relationships in planktonic populations have revealed that nutrient ratios in copepods differ from other crustacean plankton. Since high growth rates require high ribosomal RNA, high specific growth rates should be tied to a higher percentage of phosphorus in zooplankton (Main *et al.,* 1997). Copepods tend to have a lower percentage of phosphorus in their bodies than many other zooplankters (Sterner and Hessen, 1994). For example, the mean atomic ratios of C:N:P are on the order of 212:39:1 for *Acanthodiaptomus* and 85:14:1 for *Daphnia* (Andersen and Hessen, 1991). The implications are that these nutrient ratios will give diaptomids a competitive advantage over cladocerans such as *Daphnia* during periods of phosphorus limitation, and also influence the importance of copepods in nutrient cycling as well as a variety of other ecosystem processes (Sterner and Hessen, 1994). Although seasonal changes in zooplankton communities may contribute to temporal variations in levels of toxic trace metals such as cadmium, concentrations are much lower in copepods than in cladoceran zooplankton (Yan *et al.,* 1990).

IV. CURRENT AND FUTURE RESEARCH PROBLEMS

One of the most striking things about freshwater copepods is how little is known about them in comparison with either cladocerans or the marine copepods. Basic information on things as elementary as morphological characterization, life cycles, reproductive biology, feeding ecology, physiology, and genetics is lacking for most species. This is particularly true for the harpacticoids. Outside of Europe, taxonomic knowledge is inadequate for many species, especially for the non-planktonic species. Information of this sort will increase our understanding of copepods as vectors of diseases such as cholera and dracunculiasis, as well as help us to realize their full potential in biological control of pest insects such as mosquitos (discussed above).

In addition to expanding our knowledge of copepods themselves, research on copepods can contribute to our knowledge of more general ecological questions. The small size, amenability to culture, discrete instars, existence of diapause stages, and variability in life-history patterns permit many interesting questions in population biology to be addressed. The intense and selective feeding of copepods, as well as their importance as food for higher trophic levels, open the way for studies in foraging behavior, community ecology, and systems ecology.

While we are aware that food may limit copepod populations, we know very little about the relative importance of food in controlling the presence or abundance of different copepod species in nature. We are particularly naive about the interactive effects of environmental variables such as temperature, pH, solar ultraviolet radiation, the presence of alternate food, the hunger level of the predator, predation risk, and various forms of pollution and environmental manipulation on copepods. With ecosystem disturbance being the norm rather than the exception now across the globe (Vitousek, 1994), it is critical that we gain a better understanding of how many of these disturbance-related abiotic and biotic variables affect key organisms such as copepods.

Another interesting area of research to pursue concerns the relative abundance of cyclopoid versus calanoid copepods across lakes of different trophic status (discussed in Section III. F.1). The densities of cyclopoid copepods and most other zooplankton groups (ciliates, rotifers, cladocerans) are positively correlated with the primary productivity of lakes. This does not, however, seem to be universally true for calanoid copepods (Pace, 1986). What are the characteristics of calanoids that permit them to respond differently from most other groups to variations in primary productivity?

Ecological genetics is another area that is advancing rapidly and is ripe for future study. The presence or absence and timing of chromatin diminution varies among species and may be useful in deciphering phylogenetic and evolutionary relationships among taxa (Einsle, 1996; Leech and Wyngaard, 1996; Dorward and Wyngaard, 1997). Little is known about the endogenous versus exogenous control of the wide range of variations in life-history characteristics observed for copepods in different systems or through different seasons. Many interesting evolutionary questions can be asked about the timing and control of life-history events of copepods due to the genetic isolation of populations and variability in selective pressures acting on these populations among lakes. Of particular interest is the recent finding that resting eggs may contribute substantially to the evolution and persistence of copepod populations over long periods of time (Hairston *et al.*, 1995; Hairston, 1996), thus potentially permitting copepods to survive bottleneck periods related to both anthropogenic and natural disturbances.

Recent finds in North America of groundwater-related copepod species that are closely related to Eurasian forms (e.g., Reid, 1998; Reid *et al.*, 1999) are throwing new light on biogeographical relationships and the nature of evolution in copepods. While some groups seem to be extremely conservative, apparently having diverged little since the pre-Tethyan continental separation, other species evolved more rapidly to take advantage of new niches such as the cenotes of the Yucatan Peninsula (Fiers *et al.*, 1996).

V. COLLECTING AND REARING TECHNIQUES

A. Collection and Preservation

Planktonic copepods can be collected with either integrating samplers, such as a plankton net, or tube samplers (Lewis and Saunders, 1979), or with a variety of point samplers such as Schindler traps and Van Dorn bottles (Schindler, 1969). Sampling devices with narrow openings and pump samplers may elicit avoidance responses in copepods and should be tested for sampling efficiency before use. Quantitative samples must be taken at night for species that exhibit nocturnal vertical migrations into the sediments. Benthic copepods and the resting stages of planktonic species can be collected with either Ekman grab or core samplers. Separation of the smaller benthic copepods from the sediments presents difficulties. Sieving and sorting are tedious and not very efficient for copepods. Density gradient centrifugation seems to be a more promising method (Strayer, 1985).

Copepods can be preserved in a 10% solution of formalin (=3.7% formaldehyde) or 70% ethanol. Formalin is less expensive but more toxic, and tends to cause specimens to get brittle over time due to the low pH. Buffering formalin with a saturated solution of borax (sodium tetraborate, $Na_2B_{4.1}H_2O$) will improve preservation quality by making the specimens less brittle, but the high pH may also lead to more rapid color loss. A 70% solution of ethanol is actually a better preservative but is also more expensive, bulky, and may create difficult eddies in open counting chambers, making quantification extremely difficult. Prior narcotization of copepods in carbonated water may enhance the retention of the gut contents of some species when they are placed in formalin (Gannon and Gannon, 1975) but is not always necessary (Williamson, 1984). The retention of egg clutches may be enhanced by prior narcotization in a 1:10 solution of chloroform: ethanol (Williamson and Butler, 1987). The chilling and addition of 40 g of sucrose/L of formalin commonly employed in the preservation of cladocerans (Prepas, 1978) is not required, but it also has no adverse effect on copepods. The adult stages of copepods can be counted in a Bogorov chamber under a dissecting microscope (Gannon, 1971), while the nauplii are more easily counted in a Sedgwick–Rafter chamber under a compound microscope.

Several techniques have been developed for the separation of copepods in samples. Many live copepods can be separated from cladocerans with siphon tubes that remove the slower cladocerans but not the evasive copepods; or, one can take advantage of the greater resistance of copepods to pH shock (Bulkowski *et al.*, 1985).

Live copepods can be separated from dead copepods by staining with neutral red (Dressel *et al.*, 1972; Flemming and Coughlan, 1978). A stock solution is made with 0.5 g of neutral red powder per liter of distilled water. This stock solution is then added to the sample containing the copepods in a 2:100 ratio for a staining period of up to 1 h before preservation in formalin. The prior addition of 4 g of NaOH per liter of formalin reduces the leaching of the stain out of the copepods for up to 28 weeks. The high pH makes the stain much less visible, however, and glacial acetic acid must be used to titrate the sample back down to an acid pH to regenerate the deep red color before sorting.

Techniques have also been developed for the critical examination of the anatomy of copepods. The external anatomy, including sensilla and other delicate integumental organs, can be prepared by critical point drying and plating for examination under a scanning electron microscope, or cleared, stained, and examined under a light microscope (Fleminger, 1973; Friedman and Strickler, 1975). The gonads of some copepods can be selectively stained with fast green for easier visualization (Batchelder, 1986), and techniques are available for the histopathological examination of copepods (Barszcz and Yevich, 1976).

B. Rearing Techniques

Inorganic medium, spring water, or filtered lake water can be used in conjunction with a variety of foods to culture copepods. The utmost care should be taken to avoid glassware contaminated with soaps, acids, formaldehyde, or other toxins. Certain plastic containers may also be unsuitable. An easily rinsed detergent such as 7X®, should be used to clean all glassware thoroughly. [The source of 7X® is ICN Biomedicals, Inc., Costa Mesa, CA; phone:800-854-0530; www.icnpharm.com.] Standard detergents may cause high mortalities. Dilute acids may also be satisfactory for washing glassware (Wyngaard and Chinnappa, 1982), but thorough soaking or extra rinsing is suggested. Water quality may also be enhanced by running distilled water through a mixed-bed resin before use. All glassware should be sterilized between uses to avoid contaminant buildup.

Harpacticoids have been cultured on a wide range of artificial and natural foods, including baker's yeast, dried fish food, algae, bacteria, and dried, mashed, or rotting animals, eggs, and plants (Hicks and Coull, 1983). Multiple foods are often more successful than a single food type.

Freshwater diaptomids ranging from the smaller *Leptodiaptomus minutus* to the larger more omnivorous *Hesperodiaptomus shoshone* can be raised through multiple generations on a diet of *Cryptomonas* alone. Cyclopoid copepods are generally best reared on a combination of plant and animal foods. The nauplii will survive and grow well on a variety of algae including *Cryptomonas* and *Scenedesmus,* while animal food is usually necessary for the growth and survival of late copepodids and adults and for successful reproduction of some species (Jamieson, 1980b; Wyngaard and Chinnappa, 1982; Hansen and Santer, 1995; Hart and Santer, 1994). Algae as well as protozoa and invertebrates may still be an important food source for these older predatory stages (Santer, 1996). In culture, the dinoflagellate *Ceratium furcoides* supports the growth and reproduction of several genera of cyclopoid copepods but not those of the calanoid *Eudiaptomus gracilis* (Santer, 1996; Hopp *et al.,* 1997). The early copepodid stages may need a mixture of plant and animal foods. *Artemia* nauplii are the most common and convenient source of animal food for cyclopoids, but calanoid copepods or soft-bodied cladocerans and rotifers may also be used. Protozoans are a good food source, but they may lead to bacterial contamination unless care is taken to rinse and remove the organic medium or use an inorganic medium for culture. Mass rearing of cyclopoids is also possible (Suárez *et al.,* 1992).

VI. IDENTIFICATION TECHNIQUES

A. Dissection Techniques for Identification

Dissection is usually required for positive identification of copepods. Small insect pins (minuten nadeln, available from entomological supply houses) can be mounted on the end of wood applicator sticks or melted glass pipets for use in dissection. The dissection should be performed on a glass microscope slide so that a coverslip can be placed over the final preparation for examination under a compound microscope. The dissection is most easily performed by placing the copepod on its side in a drop of medium, and placing one pin in the dorsal part of the thorax to hold it down. A second pin is then used to tease off first the urosome, then each thoracic segment and its associated pair of legs—one segment at a time, and finally the first antennae. The chance of confusing legs is lessened if the first cut is made between the second and third legs, and the posterior part of the animal transferred to a second drop of medium on the same slide. Dissection and handling of the pieces is easier when performed in a drop of glycerin or other viscous, water-miscible mounting medium such as Hoyer's or CMC-9. Placing a copepod directly in full strength medium may cause osmotic stress and deformation. This problem can be avoided by placing the copepod in a well slide in a dilute mixture of about one drop of the medium to 5–10 drops of water and letting the water evaporate. Gentle warming will speed evaporation.

Semi-permanent mounts can be made by sealing a glycerin coverslip with clear fingernail polish. One of several standard mounting media can be used for more permanent mounts. A small amount of neutral red dye in glycerin or a mounting medium with a dye will color the dissected parts so that they are easier to locate after dissection. The dissection can be performed right in the viscous medium, the dissected appendages appropriately arranged in order, and the coverslip put in place. When using a mounting medium other than glycerin it is useful to let the mounting medium dry somewhat after dissection before adding more medium and adding a coverslip. This helps the dissected parts stay put. If the medium is allowed to dry for a week and the edges sealed with fingernail polish or spar varnish, the mount will last indefinitely.

For more details on techniques for copepod preparation, dissection, staining, mounting, and preservation, the reader is referred to an unpublished pamphlet prepared by J. W. Reid and available in the C. B. Wilson Copepod Library at the National Museum of Natural History, Smithsonian Institution, Washington, DC.

B. Characteristics for Distinguishing the Free-Living Freshwater Orders

Copepods are most easily identified from adult specimens, although information is available for the separation of the immature stages of some species (Comita and Tommerdahl, 1960; Czaika and Robertson, 1968; Katona, 1971; Comita and McNett, 1976; Shih and Maclellan, 1977; Bourguet, 1986; Pinel-Alloul and Lamoureux, 1988a, b). Growth rates and some external morphological characteristics of copepods can be analyzed by examining the highly transparent molted exoskeletons (exuviae) (Twombly and Burns, 1996). In freshwater habitats, the three major copepod orders can be distinguished by a number of characteristics of the adults; the length of the first antennae and taper of the body are perhaps the most useful of these (Fig. 1). The first antennae of harpacticoids are very short (5–9 segments in the female) and rarely extend past the posterior end of the cephalothorax. In cyclopoids the first antennae are intermediate in length (6–17 segments in the female), and rarely extend beyond the posterior end of the prosome. In calanoids the first antennae have 23–25 segments in the female and generally extend well beyond the end of the prosome to the urosome or even the caudal setae. The body of harpacticoids generally shows only a slight taper from the anterior to the posterior, so that the metasome and urosome are of similar widths. In cyclopoids and calanoids the metasome is substantially wider than the urosome. In cyclopoids there is a strong but gradual taper from front to back; while in calanoids there is a more abrupt change in body width where the metasome joins the urosome. The body shape of harpacticoids is quite diverse and varies greatly with their habitat. For example, interstitial species are smaller and more elongate, burrowing species are larger and broader, while epibenthic species are large and variable in their body shape (Hicks and Coull, 1983). The distinguishing features of the gelyelloids (all three species of which are under 0.5 mm in body length) are the absence of the fourth (in the European species) and fifth swimming legs, and the fusion of the coxae of the first three pairs of swimming legs; their general body shape is narrow near the midsection (Fig. 1E).

Other factors that distinguish the four free-living orders include general habitat preferences and the number of egg sacs. Calanoids are almost exclusively planktonic and either carry a single egg sac medially or release their eggs directly into the water. Cyclopoids are primarily benthic, although several planktonic species are widespread and abundant, and carry two egg sacs attached laterally to their genital segment. Harpacticoids are generally smaller than either calanoids or cyclopoids, are generally benthic or associated with other substrates or particles, and usually carry a single egg sac medially. Wilson and Yeatman (1959) provided a useful table that summarizes other differences among these three orders. Smallest of all are the gelyelloids, which occur exclusively in subterranean karst or stream hyporheic sediments; the nature of their egg sacs is unknown.

Adult copepods are sexually dimorphic and sexes can be distinguished on the basis of their relative size, as well as certain primary or secondary sexual characteristics. Females are generally larger than males, and mature gonads, spermatophores, or externally-carried egg sacs are often visible. The subitaneous and resting eggs of diaptomids can be distinguished from each other in preserved samples by the presence of a space between the chorion and the vitelline membrane in the subitaneous eggs, and its absence in resting eggs (Lohner *et al.*, 1990). Adult male copepods have one (calanoids) or two (harpacticoids, cyclopoids, and gelyelloids) geniculate first antennae that are modified for grasping the females during mating (Fig. 1 and 3). After preservation the geniculate antennae of males often become sharply bent, or elbowed, so sexes are quite easy to distinguish. In harpacticoids, males can be separated from females by their first antennae, which have fewer and broader segments, and often more densely clumped setae. Other characteristics such as morphological differences in the caudal rami and legs 2–4 can be used to distinguish male and female harpacticoids in many genera. In addition, female harpacticoids generally have a more highly developed fifth leg, and the sixth leg tends to be tiny. Characteristics that distinguish sex are generally not visible in any of the earlier larval stages and are less conspicuous in stage C5.

Adult copepods can be distinguished from immature copepodids either by size, or by the primary or secondary sexual characteristics described above. In adult cyclopoids and harpacticoids the last segment of the urosome is shorter than, or similar in length to, the next to last (penultimate) segment, while in immatures the last urosomal segment has not yet divided and is about twice as long as it is wide (Fig. 5). Characteristics of the genital segment such as a ventral swelling or more angular side profile in the adults may also be helpful in separating adult females from immatures. The larval stages within each order can be differentiated on the basis of body shape and the number and structure of the appendages. Several individuals of each sex and stage should be compared to establish valid criteria for distinguishing the sexes and stages of each species examined.

C. Identification of the Continental Free-Living Copepoda

1. Introduction

Here we provide keys to the orders and genera, followed by tables that list the families, genera, and species of copepods recorded from continental and nearshore oligohaline habitats in North America. Keys and revisions for individual genera are cited in the tables, while some of the broader, more comprehensive keys and taxonomic references are listed in the paragraphs below. The taxonomy and systematics of copepods are active areas of research. In North America, the faunas of ephemeral, subterranean, interstitial, and semiterrestrial habitats are still poorly known. Moreover, refinements of taxonomic methods have led to redefinition of many taxa (Reid, 1998). Investigators should consult with appropriate specialists and archive reference specimens.

Dussart and Defaye (1995) provided an introduction to the taxonomy and systematics of the continental Copepoda, with diagnoses and keys to most genera. For North American free-living continental copepods, the most complete and best-illustrated keys are those by Wilson and Yeatman (1959). The keys by Pennak (1989) include most species of Calanoida and some Cyclopoida. Basic general taxonomic references for marine harpacticoids include the works of Coull (1977), Lang (1948, 1965), and Huys *et al.* (1996).

There are several useful keys and checklists for different regions of North America. Works treating calanoids, cyclopoids, and harpacticoids include those of Cole (1959) for Kentucky, Robertson and Gannon (1981) for the Great Lakes, Clamp *et al.* (1999) for North Carolina, and Chengalath and Shih (1994) for northwestern North America. Lists for calanoids and cyclopoids include those of Harris (1978) for northern Mississippi, Reed (1963) for northern Canada, and Smith and Fernando (1978) for Ontario. Cole (1961, 1963, 1966) summarized records of Diaptomidae (Calanoida) from Arizona, and Robertson (1970, 1972, 1975) listed diaptomids from Oklahoma. Czaika (1982) provided keys and tables for adults and developmental stages of planktonic calanoids and cyclopoids of the Great Lakes. Regional works on Cyclopoida include the key by Torke (1976) for Wisconsin and the checklists by Bunting (1973) for Tennessee and Reid and Marten (1995) for Louisiana and Mississippi. Hudson *et al.* (1998) provided a checklist of cyclopoids and harpacticoids of the Great Lakes and a key for the cyclopoids.

Checklists for Mexico were furnished by Suárez-Morales and Reid (1998) and by Reid (1990) for Mexico, Central America, and the Antilles. Suárez-Morales *et al.* (1996) provided keys and taxonomic discussion for species of the Yucatan Peninsula, as well as a Spanish-language general introduction to the study of continental copepods.

Two public resources for copepod studies are the C. B. Wilson Copepod Library at the National Museum of Natural History, Smithsonian Institution, Washington DC., and the Monoculus Library at Oldenburg University in Germany. The Wilson Library contains some 10,000 works, indexed by author and species. Its website (http://www.nmnh.si.edu/iz/copepod) contains five databases: (1) a bibliography of all known literature on Copepoda and Branchiura; (2) taxonomic lists of families, genera, and species; (3) a world list of researchers; and (4) the taxonomic "types" of Copepoda and Branchiura held in the National Museum of Natural History. Contact F. D. Ferrari, T. C. Walter, or J. W. Reid for more information. The Monoculus Bibliography Project also holds about 10,000 works, many entered in its database; contact H. K. Schminke, Fachbereich 7-Biologie, Universität Oldenburg, Postfach 2503, D-26111 Oldenburg, Germany for details.

The true continental Calanoida are dominated by the families Centropagidae, Diaptomidae, and Temoridae, and the Cyclopoida by the family Cyclopidae. The continental harpacticoid fauna consists mainly of the families Canthocamptidae, Parastenocarididae, and Phyllognathopodidae, but the primarily marine families Ameiridae, Huntemanniidae, and Laophontidae also contain some true freshwater species. Several additional calanoid, cyclopoid, and harpacticoid families include euryhaline species that may wander into coastal freshwaters, but these are not listed in the tables. Useful taxonomic references for euryhaline species include C. B. Wilson (1932) for calanoids and cyclopoids; Walter (1989) for *Pseudodiaptomus*; Rocha (1991), Rocha and Hakenkamp (1993), Rocha *et al.* (1998), and M. S. Wilson (1958) for *Halicyclops*; and Coull (1977), Lang (1965), and Huys *et al.* (1996) for harpacticoids.

2. Key to Orders and Families of the Free-Living Continental Copepoda of North America

[Note that there are 3 or 4 couplets in some parts of this taxonomic key.]

1a. Body constricted, or major body joint between somites bearing 4th and 5th legs; first antenna ♀ of 6–17 segments 2

1b. Body constricted between somite bearing 5th leg and genital segment; first antenna ♀ long, of 23–25 segments (Figs. 1A, 2, 5) Order Calanoida, ... 4

1c. No distinct body articulation between prosome and urosome; first antenna ♀ with 14 poorly defined segments; swimming legs without couplers (intercoxal sclerites), and joined at base (Fig. 1E) .... Order Gelyelloida

2a(1a). Second antenna of 3 or 4 segments, not prehensile; first antenna of ♀ with 6–17 segments 3

2b. Second antenna of 3 segments, with large prehensile terminal claw; first antenna of ♀ with 6 segments .... Order Poecilostomatoida (Ergasilidae) ....

3a(2a). Fifth leg small, of 1–3 short segments, or represented only by setae; segment 1 not enlarged on inner margin (Figs. 1B, 2, 5) .... Order Cyclopoida 24

3b. Fifth leg broad, of 1–2 segments, segment 1 (baseoendopod) enlarged on inner margin (Figs. 1C,D, 2, 3) .... Order Harpacticoida, ... 44

ORDER CALANOIDA

4a(1b). Caudal ramus ♀ ♂ with 3 or 5 well-developed setae (plus 1 or 2 shorter, slender setae) 5

4b. Caudal ramus ♀ ♂ with 4 well-developed setae (plus shorter, slender outer and inner setae); legs 1–4 endopods of 1, 2, 3, and 3 segments respectively; first antenna ♂ not geniculate; leg 5 ♀ lacking........................ Aetideidae *Senecella*

5a(4b). Caudal ramus ♀ ♂ with 3 well-developed terminal setae (plus reduced or spiniform outer seta and slender, dorsally placed inner seta) 6

5b. Caudal ramus ♀ ♂ with 5 well-developed setae (plus slender, dorsally placed inner seta) 7

6a(5a). Caudal ramus ♀ ♂, outer seta slender, about as long as ramus; leg 5 ♀ and left leg 5 ♂, last segment with long apical spine; urosome ♂ symmetrical .... Temoridae *Heterocope*

6b. Caudal ramus ♀ ♂, outer seta shorter than ramus, and or spiniform; leg 5 ♀ and left leg 5 ♂, last segment without long apical spine; urosome ♂ asymmetrical, with various processes on right side .... Temoridae *Epischura*

7a(5b) Caudal ramus ♀ ♂ long (more than 3 times longer than wide) 8

7b. Caudal ramus ♀ ♂ short (3 times or less than 3 times longer than wide) 10

8a(7a). Maxillipeds elongate (about 2 times body width) in lateral view; legs 1–4, endopods all of 3 segments 9

8b. Maxillipeds not elongate (subequal to body width) in lateral view; legs 1–4, endopods of 1, 2, 2, and 2 segments, respectively; leg 5 ♀ ♂ lacking endopods .... Temoridae *Eurytemora*

8c. Maxillipeds reduced; legs 1–4, endopodites all of 2 segments; leg 5 ♀ ♂ lacking endopodite........................ Acartiidae .... *Acartia sinensis*

8d. Maxillipeds not elongate; legs 1–4, endopodites all of 3 segments........................ Pseudodiaptomidae *Pseudodiaptomus*

9a(8a). Caudal ramus with tiny spines on surface; body somite bearing leg 5 with no spines; right leg 5 ♂, exopod segment 2, coxa with, basis without medial process; leg 5 ♀, exopod segment 2 with outer spine, and endopod segment 1 with inner seta .... Centropagidae........................ *Limnocalanus*

9b. Caudal ramus with no surface spines; body somite bearing leg 5 with 1 spine on each posterior corner; right leg 5 ♂, coxa without medial process, and basis with medial process; leg 5 ♀, exopod segment 2 with no outer spine, and endopod segment 1 with no inner seta .... Centropagidae *Sinocalanus*

10a(7b). Caudal ramus ♀ ♂, lengths of setae unequal, fourth seta from outer margin longer and stouter than others; legs 1–4, endopods all with 3 segments; leg 5 ♀ ♂, endopods 3-segmented, last segments with 6 long setae .... Centropagidae .... *Osphranticum*

10b. Caudal ramus ♀ ♂, lengths of setae nearly equal; leg 1 endopod of 2 segments, legs 2–4 endopods of 3 segments; leg 5 ♀ ♂, endopods 1- or 2-segmented, last segments with 0–2 short setae .... Diaptomidae 11

11a(10b). Last thoracic segment ♀ with rounded or spiniform dorsal process 12

11b. Last thoracic segment of female with no dorsal process 13

12a(11a). Right leg 5 ♂, terminal claw of exopod segment 2 slender, at least 1.5 times longer than segment; leg 5 ♀, endopod with row of fine terminal hairs and 2 long terminal spines .... *Mastigodiaptomus*

12b. Right leg 5 ♂, terminal claw of exopod segment 2 slender, usually twice or more length of segment; leg 5 ♀, endopod usually with fine terminal hairs only .... *Arctodiaptomus* (in part)

12c. Right leg 5 ♂, terminal claw of exopod segment 2 stout and about as long as segment; left leg 5 ♂, terminal process strongly corrugated; leg 5 ♀, endopod with row of terminal hairs and 1 short spine .... *Sinodiaptomus*

13a(11b). Right first antenna ♂, terminal segment with end rounded, bearing only setae 14

13b. Right first antenna ♂, terminal segment with small clawlike process on distal end .... *Acanthodiaptomus*

14a(13a). Left leg 5 ♂, distal process of exopod not distinct, i.e., exopod of 2 segments 16

14b. Left leg 5 ♂, distal process of exopod distinct, i.e., exopod of 3 segments 15

15a(14b). Right first antenna ♂, antepenultimate segment with short stout curved spinous process on distal end; leg 5 ♀, exopod segment 3 absent (with 2 remnant spines inserted directly on segment 2) .... *Onychodiaptomus*

15b. Right first antenna ♂, antepenultimate segment with no process; leg 5 ♀, exopod segment 3 present, distinct .... *Nordodiaptomus*

16a(14a). Right first antenna ♂, antepenultimate segment with distal spinous process 18

16b. Right first antenna ♂, antepenultimate segment with at most narrow hyaline lamella 17

17a(16b). Both first antennae ♀ and left first antenna ♂, with 2 setae on segment 11 and 2 setae on segments 15, 16, and 17 .... *Mixodiaptomus*

17b. Both first antennae ♀ and left first antenna ♂, with 1 seta on each of segments 11 and 15–17 .... *Skistodiaptomus*

18a(16a). Right first antenna ♂ with simple, rodlike or dentiform process on antepenultimate segment 19

18b. Right first antenna ♂ with comb-shaped (corrugated) process on antepenultimate segment; leg 5 ♂ with segment 2 shorter than segment 1 and with 1 rounded, haired pad; leg 5 ♀, endopod usually with terminal hairs only .... *Arctodiaptomus* (in part)

19a(18a). Right first antenna ♂, antepenultimate segment with short dentiform process (usually shorter than penultimate segment) 21

19b. Right first antenna ♂, antepenultimate segment with long dentiform process (usually longer than penultimate segment) 20

20a(19b). Left leg 5 ♂, segment 2 usually longer than segment 1, bearing two distinctly divided, more or less hairy pads; leg 5 ♀, exopod segment 3 present, distinct .... *Hesperodiaptomus*

20b. Left leg 5 ♂, segment 2 shorter than segment 1, bearing 2 indistinctly divided pads; leg 5 ♀, exopod segment 3 absent, with only 2 remnant spines, inserted directly on segment .... *Leptodiaptomus* (in part)

21a(19a). Both first antennae ♀ and left first antenna ♂, setae on segments 17, 19, 20, and 22 with normal tapered ends (not hooked); left leg 5 ♂, exopod segment 2 with lateral seta short, not extending past end of segment 22

21b. Both first antennae ♀ and left first antenna ♂, setae on segments 17, 19, 20, and 22 with ends stiffly hooked; left leg 5 ♂, lateral seta of exopod 2 very long .... *Aglaodiaptomus*

22a(21a). Both first antennae ♀ and left first antenna ♂, segment 11 with 1 seta 23

22b. Both first antennae ♀ and left first antenna ♂, segment 11 with 2 setae .... *Diaptomus*

23a(22a). Left leg 5 ♂, exopod segment 2 with both terminal process and inner spine short and stout; leg 5 ♀, exopod segment 3 absent, with only 2 remnant spines, inserted directly on exopod 2 .... *Leptodiaptomus* (in part)

23b. Left leg 5 ♂, exopod segment 2 with both terminal process and inner spine relatively long and slender; leg 5 ♀, exopod segment 3 present and distinct .... *Eudiaptomus*

ORDER CYCLOPOIDA

24a(3a). Mandibular palp consisting of tiny segment bearing 1–3 setae, or mandibular palp entirely absent ....Cyclopidae 25

24b. Mandibular palp large, complex, with many plumose setae ....Oithonidae *Limnoithona*

25a(24a). Leg 5 completely fused to body somite and bearing 3 setae and or spines; first antenna ♀ of 9–11 (usually 11) segments 26

25b. Leg 5 consisting of 1 or 2 distinct segments; first antenna ♀ of 6–17 segments 28

25c. Leg 5 consisting of 3 distinct segments; first antenna ♀ of 16 segments ....*Orthocyclops*

26a(25a). Leg 5 reduced to narrow plate or knobs, bearing 3 tiny setae; legs 1–4, endopods and exopods of 2 or fewer segments 27

26b. Leg 5 consisting of broad plate, bearing 2 strong inner spines and 1 long outer seta; legs 1–4, endopods and exopods all of 3 segments ....*Ectocyclops*

27a(26a). Anal operculum large, triangular, with dentate margin; leg 4 coxa, inner corner lacking seta; leg 3 ♂, endopod terminal segment with modified apical spine *Bryocyclops* ....

27b. Anal operculum small, quadrate, with smooth margin; leg 4 coxa with seta on inner corner; leg 3 ♂, endopod terminal segment with apical spine simple *Stolonicyclops* ....

28a(25b). Leg 5 consisting of 1 distinct, broad segment and bearing 1 inner spine and 2 outer setae 29

28b. Leg 5 consisting of 1 or 2 distinct, usually narrow segments (proximal segment may be fused to body somite, but remnant seta of this segment always present), segment 2 with 1–3 spines and setae (if 3, there is 1 middle spine between 2 setae) 33

29a(28a). First antenna ♀ of 8, 11, or 12 segments; small species (usually under 1.4 mm long) 30

29b. First antenna ♀ of 17 segments; large robust species (1.7–2.9 mm long) ....*Homocyclops*

30a(29a). First antenna ♀ of 12 segments 31

30b. First antenna ♀ of 8 or 11 segments ....*Paracyclops* (in part)

31a(30a). Caudal ramus ♀, with tiny spines next to lateral and outermost terminal setae only 32

31b. Caudal ramus ♀ with longitudinal row of tiny spines on outer margin; caudal ramus ♀ ♂ at least 4 times longer than wide ....*Eucyclops*

32a(31a). First antenna ♀, last 2 segments not more that 2 times longer than wide; caudal ramus ♀ about 2.5 times longer than wide ....*Paracyclops* (in part)

32b. First antenna ♀, last 2 segments more than 3 times longer than wide; caudal ramus ♀ ♂ about 3 times longer than wide ....*Tropocyclops*

33a(28b). Leg 5 with 2 distinct segments 34

33b. Leg 5 with 1 distinct segment 40

34a(32a). Leg 5, distal segment narrow, bearing 1–2 spines or setae 35

34b. Leg 5, distal segment broad, bearing 3 spines and setae ....*Macrocyclops*

35a(34a). Distal segment of leg 5 bearing 1 apical seta and 1 long inner marginal or subapical spine (length of spine more than 4 times length of segment); first antenna always of 17 segments 36

35b. Distal segment of leg 5 bearing 1 apical seta and (usually) 1 rather short inner marginal or subapical spine (length of spine not more than twice length of segment); first antenna of 17 or fewer segments 37

36a(35a). Leg 5, inner spine of distal segment inserted about midlength of inner margin of segment; first antenna, last segment usually 3 or more times longer than wide ....*Mesocyclops*

36b. Leg 5, inner spine of distal segment inserted subterminally on segment; first antenna, last segment about twice longer than wide ....*Thermocyclops*

37a(35b). Leg 5, distal segment with apical seta and (usually) small slender subapical spine on inner margin; caudal ramus without longitudinal dorsal ridge, and inner surface with or without hairs 38

37b. Leg 5, distal segment with apical seta and large spine (about equal in length to distal segment) attached at middle of inner margin of segment; caudal ramus usually with longitudinal dorsal ridge, and inner margin hairy; first antenna of 14–17 segments........*Cyclops*

38a(37a). Leg 5, segment 2 with apical seta and small spine or spur on middle of inner surface, or somewhat longer subapical spine (usually shorter than segment 2); caudal ramus with or without hairs on inner surface 39

38b. Leg 5, segment 2 with inner subapical spine longer than segment 2; caudal ramus usually without hairs on inner surface ....*Diacyclops*

39a(38a). Leg 5, segment 2 with apical seta and small spine or spur (usually a spur, continuous with segment) at about middle of inner surface; caudal ramus, inner surface hairy; first antenna of 17 segments .... *Megacyclops*

39b. Leg 5, segment 2 with apical seta and small spine slightly distal to middle of segment, or almost apical; caudal ramus with or without hairs on inner margin; first antenna of 11–17 segments .... *Acanthocyclops*

40a(37b). Leg 5, free segment narrow and cylindrical, or only slightly longer than wide, with inner apical spine or seta (if present) inserted near outer apical seta 41

40b. Leg 5, free segment very broad, with inner apical spine inserted far from outer apical seta .... *Apocyclops*

41a(40a). Leg 5, free segment narrow, cylindrical, with inner spine (if present) tiny and inserted subterminally or on inner surface of segment; proximal segment completely absent, except for remnant seta 42

41b. Leg 5, free segment only slightly longer than wide, with slender inner terminal seta inserted next to longer outer terminal seta; proximal segment obvious although continuous with body somite 43

41c. Leg 5, free segment only slightly longer than wide, with stout inner apical spine and outer apical seta both inserted terminally; proximal segment completely absent, except for remnant seta .... *Metacyclops*

42a(41a). Legs 1–4, couplers and coxa-basis very wide (about 2.5 times wider than long); legs 2–4, basis with rows of tiny spines on inner expansion; leg 4 endopod, outer terminal spine tiny, less than 1/6 length of inner terminal spine .... *Cryptocyclops*

42b. Legs 1–4, couplers and coxa-basis less wide (about 2 times wider than long); legs 2–4, inner expansion of basis with fine hairs; leg 4 endopod, outer terminal spine at least 1/2 length of inner terminal spine .... *Microcyclops*

43a(41b). Proximal segment of leg 5 present as tiny knob set close to free segment; legs 1–4 with 3- or 2-segmented rami; if exopodites of 2,2,3,3 segments, then exopodite segment 1 with medial (inner) seta .... *Rheocyclops*

43b. Proximal segment of leg 5 present as large knob set apart from free segment; legs 1–4 with endopodites 2-segmented and exopodites of 2,2,3,3 segments; exopodite segment 1 without medial (inner) seta .... *Itocyclops*

ORDER HARPACTICOIDA

44a(3b). Somite bearing leg 1 completely fused with cephalothorax; maxilliped prehensile (ending in long claw), never flat and leaflike 45

44b. Somite bearing leg 1 not fused to cephalothorax; maxilliped flat and leaflike, with several setae .... Phyllognathopodidae *Phyllognathopus* ....

45a(44a). Leg 1 endopod prehensile (with dist almost segment long) or not, with 2 or more terminal setae (which may be stout) 46

45b. Leg 1 endopod prehensile, much longer than exopod, ending in 1 stout claw .... Laophontidae *Onychocamptus*

46a(43a). Leg 1 exopod of 3 segments, middle segment with no spine on outer margin; body vermiform (narrow, cylindrical) 47

46b. Leg 1 exopod of 3 segments, middle segment with a spine on outer margin, *or* leg 1 exopod of only 1 or 2 segments; body stout, usually not vermiform 48

47a(46a). Legs 1 and 2, endopods of 2 and 1 segment, respectively .... Parastenocarididae *Parastenocaris*

47b. Legs 1 and 2, endopods of 3 and 2 segments, respectively .... Ameiridae *Psammonitocrella*

48a(46b) Second antenna with basis (Fig. 3) 49

48b. Second antenna with allobasis (Fig. 3) 51

49a(48a). Leg 5 ♀ ♂ unsegmented .... Tachidiidae *Tachidius*

49b. Leg 5 ♀ ♂ of 2 distinct segments, i.e. exopod segment separated from baseoendopod .... Ameiridae 50

50a(49b). Caudal rami short, 1-2 times as broad as long; legs 2–4, 1 or more endopod 3-segmented 65

50b. Caudal rami about 5 times longer than wide; legs 2–4, endopod with 2, 2, and 1 segments, respectively .... *Stygonitocrella*

51a(48b). Leg 1 with endopod of 2 or 3 segments 52

51b. Leg 1 with endopod of 1 short segment .... Huntemanniidae *Huntemannia*

52a(51a). First antenna ♀ of 6–9 segments; leg 1 ♀ ♂, endopod segment 1 with inner seta 53

52b. First antenna ♀ of 5 segments; leg 1 ♀ ♂, endopod segment 1 without inner seta .... Huntemanniidae *Nannopus*

53a(52a). First antenna ♀ of 6–9 (usually 8) segments; leg 5 ♀ ♂ of 2 distinct segments, i.e., exopod separated from baseoendopod .... Canthocamptidae 54

53b. First antenna ♀ of 6 segments; leg 5 ♀ ♂ unsegmented .... Canthocamptidae incertae sedis *Cletocamptus*

54a(53). Leg 1 ♀ ♂, exopod of 3 segments 55

54b. Leg 1 ♀ ♂, exopod of 2 segments .... *Maraenobiotus*

55a(54a). Leg 1 ♀ ♂, exopod segment 2 without inner seta 56

55b. Leg 1 ♀ ♂, exopod segment 2 with inner seta 60

56a(55a). Leg 1 ♀ ♂, endopod of 2 segments 57

56b. Leg 1 ♀ ♂, endopod of 3 segments .... *Bryocamptus* (in part)

57a(56a). Leg 1 ♀ ♂, endopod reaches from about midlength to slightly past end of exopod segment 3 58

57b. Leg 1 ♀ ♂, endopod does not reach end of exopod segment 2 .... *Epactophanes*

57c. Leg 1 ♀ ♂, endopod longer than exopod by half of endopod segment 3 or more .... *Paracamptus*

58a(57a). Legs 2–4 ♀ ♂, exopod 2 without inner seta 59

58b. Legs 2–4 ♀ ♂, exopod 2 with inner seta .... *Bryocamptus* (in part)

59a(58a). Leg 3 ♀ ♂, exopod segment 3 with 1 inner seta; legs 2–4 ♀ ♂, exopod segment 3 with all normal-sized setae and spines (i.e., inner terminal seta not reduced) .... *Moraria*

59b. Leg 3 ♀ ♂, exopod segment 3 with 2 inner setae; legs 2–4 ♀ ♂, exopod segment 3, inner terminal seta tiny .... *Gulcamptus*

60a(55b). Leg 1 ♀ ♂, endopod of 3 segments and rostrum small to moderate in size, i.e., not extending past segment 1 of first antenna; first antenna ♀ of 7–9 (usually 8) segments 61

60b. Leg 1 ♀ ♂, endopod of 2 segments and rostrum small, not extending past segment 1 of first antenna; first antenna ♀ of 7 or 8 segments .... *Bryocamptus* (in part)

60c. Leg 1 ♀ ♂, endopod of 2 or 3 segments and rostrum large, extending past end of segment 1 of first antenna; first antenna ♀ of 6 or 7 segments .... *Mesochra*

61a(60a). Leg 5 ♀, next outermost seta of baseoendopod not reduced, usually equal or longer than outermost seta; leg 4 ♂, outer corner of last segment of endopod with distinct spine or seta 62

61b. Leg 5 ♀, next outermost seta of baseoendopod tiny; leg 4 ♂, outer corner of last segment of endopod with spinous process .... *Canthocamptus*

62a(61a). Leg 1 ♀ ♂, endopod segment 1 not reaching beyond end of exopod segment 2 63

62b. Leg 1 ♀ ♂, endopod segment 1 reaching midlength of exopod segment 3, or longer .... *Attheyella* (in part)

63a(62a). Legs 2–4 ♀ ♂, exopod segment 3 with 3 outer spines (total number of spines and setae 6, 7, and 6 or 7, respectively); leg 5 ♀, baseoendopod with 5 or 6 (rarely 4) setae; leg 5 ♂, baseoendopod usually with 2 setae 64

63b. Legs 2–4 ♀ ♂, exopod segment 3 with 2 outer spines (total number of spines and setae 5, 6, and 6, respectively); leg 5 ♀, baseoendopod with 3–4 setae; leg 5 ♂, baseoendopod with no setae .... *Elaphoidella*

64a(63). Legs 2 and 3 ♀, endopods of 3 segments; *or* of 2 segments and leg 5 ♀ with 5 setae on baseoendopod; leg 2 ♂, endopod segment 2 modified, bearing 1 or 2 knobs, spinous processes, or notches .... *Bryocamptus* (in part)

64b Legs 2 and 3 ♀, endopods of 2 segments; leg 5 ♀ with 6 setae on baseoendopod; leg 2 ♂, endopod unmodified (may have spines on outer margin) .... *Attheyella* (in part)

65a(50a). Leg 2-4 ♀ ♂, all endopods of 3 segments .... *Nitokra*

65b. Leg 2-4 ♀ ♂, endopods of 3, 3 and 2 segments respectively .... *Nitocrellopsis*

LITERATURE CITED

Adrian, R. 1991. The feeding behavior of *Cyclops kolensis* and *C. vicinus* (Crustacea, Copepoda). Verhandlungen der Internationale Vereinigung für Theoretische und Angewandte Limnologie 24:2852–2863.

Adrian, R. 1997. Calanoid–cyclopoid interactions: evidence from an 11-year field study in a eutrophic lake. Freshwater Biology 38:315–326.

Adrian, R., Frost, T. M. 1992. Comparative feeding ecology of *Tropocyclops prasinus mexicanus* (Copepoda, Cyclopoida). Journal of Plankton Research 14:1369–1382.

Allan, J. D. 1976. Life history patterns in zooplankton. American Naturalist 110:165–180.

Andersen, T., Hessen, D. O. 1991. Carbon, nitrogen, and phosphorus content of freshwater zooplankton. Limnology and Oceanography 36:807–814.

Anderson, R. S. 1970. Predator–prey relationships and predation rates for crustacean zooplankters from some lakes in western Canada. Canadian Journal of Zoology 48:1229–1240.

Arts, M. T., Evans, M. S. 1991. Optical–digital measurements of energy reserves in calanoid copepods: Intersegmental distribution and seasonal patterns. Limnology and Oceanography 36:289–298.

Barszcz, C. A., Yevich, P. P. 1976. Preparation of copepods for histopathological examinations. Transactions of the American Microscopical Society 95:104–108.

Batchelder, H. P. 1986. A staining technique for determining copepod gonad maturation: application to *Metridia pacifica* from the northern Pacific Ocean. Journal of Crustacean Biology 6: 227–231.

Blades, P. I., Youngbluth, M. J. 1980. Morphological, physiological, and behavioral aspects of mating in calanoid copepods. *In*: Kerfoot, W. C., Ed., Evolution and ecology of zooplankton communities. Univ. Press of New England, Hanover, NH, pp. 39–51.

Blais, J. M., Maly, E. J. 1993. Differential predation by *Chaoborus americanus* on males and females of two species of *Diaptomus*. Canadian Journal of Fisheries and Aquatic Sciences 50:410–415.

Bogdan, K.G., Gilbert, J.J. 1984. Body size and food size in freshwater zooplankton. Proceedings of the National Academy of Sciences 81:6427–6431.

Bourguet, J.-P. 1986. Contribution àl'étude de *Cletocamptus retrogressus* Schmankewitch, 1875 (Copepoda, Harpacticoida) II. Développement larvaire—stades naupliens. Crustaceana 51:113–122

Bowers, J. A. 1986. Phosphorus regeneration by the predatory copepod *Diacyclops thomasi*. Canadian Journal of Fisheries and Aquatic Sciences 43:361–365.

Bowman, T. E., Abele, L. G. 1982. Classification of the recent Crustacea. Pages 1–25, *in*: Bliss, D. E., Ed. The biology of Crustacea. Vol. 1, Abele, L. G., Ed., Systematics, the fossil record, and biogeography, Academic Press, New York.

Brandl, Z. 1998. Feeding strategies of planktonic cyclopoids in lacustrine ecosystems. Journal of Marine Systems 15:87–95.

Brandl, Z., Fernando, C. H. 1979. The impact of predation by the copepod *Mesocyclops edax* (Forbes) on zooplankton in three lakes in Ontario, Canada. Canadian Journal of Zoology 57:940–942.

Brandl, Z., Fernando, C. H. 1986. Feeding and food consumption by *Mesocyclops edax*. Syllogeus 58:254–258.

Brown, M. D., Kay, B. H., Greenwood, J. G. 1991. The predation efficiency of north-eastern Australian *Mesocyclops* (Copepoda: Cyclopoida) on mosquito larvae. Bulletin of the Plankton Society of Japan, Special Vol.:329–338.

Bulkowski, L., Krise, W. F., Kraus, K. A. 1985. Purification of *Cyclops* cultures by pH shock (Copepoda). Crustaceana 48:179–182.

Buskey, E. J. 1998. Components of mating behavior in planktonic copepods. Journal of Marine Systems. 15:13–21.

Bunting, D. L. 1973. The Cladocera and Copepoda of Tennessee. II. Cyclopoid copepods. Journal of the Tennessee Academy of Science 48:138–141.

Burgi, H.-R., Elser, J. J., Richards, R. C., Goldman, C. R. 1993. Zooplankton patchiness in Lake Tahoe and Castle Lake U.S.A. Verhandlungen der Internationale Vereinigung für Theoretische und Angewandte Limnologie 25:378–382.

Burns, C. W., Gilbert, J. J. 1993. Predation on ciliates by freshwater calanoid copepods: rates of predation and relative vulnerabilities of prey. Freshwater Biology 30:377–393.

Burns, C. W., Schallenberg, M. 1996. Relative impacts of copepods, cladocerans and nutrients on the microbial food web of a mesotrophic lake. Journal of Plankton Research 18:683–714.

Byron, E. R., Folt, C. L., Goldman, C. R. 1984. Copepod and cladoceran success in an oligotrophic lake. Journal of Plankton Research 6:45–65.

Byron, E. R., Whitman, P. T., Goldman, C. R. 1983. Observations of copepod swarms in Lake Tahoe. Limnology and Oceanography 28:378–382.

Carter, J. C. H., Goudie, K. A. 1986. Diel vertical migrations and horizontal distributions of *Limnocalanus macrurus* and *Senecella calanoides* (Copepoda, Calanoida) in lakes of southern Ontario in relation to planktivorous fish. Canadian Journal of Fisheries and Aquatic Sciences 43:2508–2514.

Carter, J. C. H., Sprules, W. G., Dadswell, M. J., Roff, J. C. 1983. Factors governing geographical variation in body size of *Diaptomus minutus* (Copepoda, Calanoida). Canadian Journal of Fisheries and Aquatic Sciences 40:1303–1307.

Chaston, I. 1969. Anaerobiosis in *Cyclops varicans*. Limnology and Oceanography 14:298–301.

Chen, C. Y., Folt, C. L., Cook, S. 1997. The potential for hybridization in freshwater copepods. Oecologia 111:557–564.

Chengalath, R., Shih, C-t. 1994. Littoral freshwater copepods of northwestern North America: northern British Columbia. Verhandlungen der Internationale Vereinigung für Theoretische und Angewandte Limnologie 25:2421–2431.

Chow-Fraser, P., Maly, E. J. 1988. Aspects of mating, reproduction, and co-occurrence in three freshwater calanoid copepods. Freshwater Biology 19:95–108.

Chow-Fraser, P., Maly, E. J. 1991. Factors governing clutch size in two species of *Diaptomus* (Copepoda:Calanoida). Canadian Journal of Fisheries and Aquatic Sciences 48:364–370.

Chow-Fraser, P., Wong, C. K. 1986. Dietary change during development in the freshwater calanoid copepod *Epischura lacustris* Forbes. Canadian Journal of Fisheries and Aquatic Sciences 43:938–944.

Clamp, J. C. (Compiler), Adams, W. F., Reid, J. W., Taylor, A. Y., Cooper, J. E., McGrath, C., Williams, D. J., DeMont, D. J., McLarney, W. O., Mottesi, G., Alderman, J. 1999. A report on the conservation status of North Carolina's freshwater and terrestrial crustacean fauna. Scientific Council on Freshwater and Terrestrial Crustaceans, North Carolina Wildlife Resources Commission, Raleigh. 92 p.

Cole, G. A. 1959. A summary of our knowledge of Kentucky crustaceans. Transactions of the Kentucky Academy of Science 20:66–81.

Cole, G. A. 1961. Some calanoid copepods from Arizona with notes on congeneric occurrences of *Diaptomus* species. Limnology and Oceanography 6:432–442.

Cole, G. A. 1963. Calanoid copepods from some old Arizona collections. Journal of the Arizona Academy of Science 2:176–183.

Cole, G.A. 1966. Contrasts among calanoid copepods from permanent and temporary ponds in Arizona. American Midland Naturalist 76:351–368.

Comita, G. W., McNett, S. J. 1976. The postembryonic developmental instars of *Diaptomus oregonensis* Lilljeborg, 1889 (Copepoda). Crustaceana 30:123–163.

Comita, G. W., Tommerdahl, D. M. 1960. The postembryonic developmental instars of *Diaptomus siciloides* Lilljeborg. Journal of Morphology 107:297–355.

Confer, J. L., Blades, P. I. 1975. Omnivorous zooplankton and planktivorous fish. Limnology and Oceanography 20:571–579.

Confer, J. L., Cooley, J. M. 1977. Copepod instar survival and predation by zooplankton. Journal of the Fisheries Research Board of Canada 34:703–706.

Confer, J. L., Kaaret, T., Likens, G. E. 1983. Zooplankton diversity and biomass in recently acidified lakes. Canadian Journal of Fisheries and Aquatic Sciences 40:36–42.

Coull, B. C. 1977. Marine flora and fauna of the northeastern United States. Copepoda: Harpacticoida. NOAA Technical Report NMFS Circular 399, 48 pp.

Czaika, S. C. 1982. Identification of nauplii N1–N6 and copepodids CI–CVI of the Great Lakes calanoid and cyclopoid copepods (Calanoida, Cyclopoida, Copepoda). Journal of Great Lakes Research 8:439–469.

Czaika, S. C., Robertson, A. 1968. Identification of the copepodids of the Great Lakes species of *Diaptomus* (Calanoida, Copepoda). Proceedings of the 11th Conference on Great Lakes Research 17:39–60.

DeMott, W. R. 1986. The role of taste in food selection by freshwater zooplankton. Oecologia 69:334–340.

DeMott, W. R. 1988. Discrimination between algae and artificial particles by freshwater and marine copepods. Limnology and Oceanography 33:397–408.

DeMott, W. R. 1989. Discrimination between algae and detritus by freshwater and marine zooplankton. Bulletin of Marine Science 43:486–499.

DeMott, W. R. 1995a. Food selection by calanoid copepods in response to between-lake variation in food abundance. Freshwater Biology 33:171–180.

DeMott, W. R. 1995b. Optimal foraging by a suspension-feeding copepod: responses to short-term and seasonal variation in food resources. Oecologia 103:230–240.

DeMott, W. R., Moxter, F. 1991. Foraging on cyanobacteria by copepods: responses to chemical defenses and resource abundance. Ecology 72:1820–1834.

DeMott, W. R., Zhang, Q.-X., Carmichael, W. W. 1991. Effects of toxic cyanobacteria and purified toxins on the survival and feeding of a copepod and three species of *Daphnia*. Limnology and Oceanography 36:1346–1357.

De Stasio, B. T. Jr. 1990. The role of dormancy and emergence patterns in the dynamics of a freshwater zooplankton community. Limnology and Oceanography 35:1079–1090.

De Stasio, B. T. Jr. 1993. Diel vertical and horizontal migration by zooplankton:population budgets and the diurnal deficit. Bulletin of Marine Science 53:44–64.

Dobrzykowski, A. E., Wyngaard, G. A. 1993. Phenology of dormancy in a Virginia population of *Mesocyclops edax* (Crustacea: Copepoda). Hydrobiologia 250:167–171.

Dodson, S. I., Egger, D. L. 1980. Selective feeding of red phalaropes on zooplankton of arctic ponds. Ecology 61:755–763.

Donald, D. B., Anderson, R. S., Mayhood, D. W. 1994. Coexistence of fish and large *Hesperodiaptomus* species (Crustacea: Calanoida) in subalpine and alpine lakes. Canadian Journal of Zoology 72:259–261.

Dorward, H. M., Wyngaard, G. A. 1997. Variability and pattern of chromatin diminution in the freshwater Cyclopidae (Crustacea: Copepoda). Archiv für Hydrobiologie, Supplement 107: 447–465.

Dressel, D. M., Heinle, D. R., Grote, M. C. 1972. Vital staining to sort dead and live copepods. Chesapeake Science 13:156–159.

Dürbaum, J. 1995. Discovery of postcopulatory mate guarding in Copepoda Harpacticoida (Crustacea). Marine Biology 123:81–88.

Dürbaum, J. 1997. Precopulatory mate guarding and mating in *Tachidius discipes* (Copepoda:Harpacticoida). Contributions to Zoology 66:201–214.

Dussart, B. H., Defaye, D. 1995. Introduction to the Copepoda. Guides to the Identification of the Microinvertebrates of the Continental Waters of the World. Dumont, H. J. (Ed.) 7:1–277.

Edmondson, W. T. 1960. Reproductive rates of rotifers in natural populations. Memorie dell'Istituto Italiano di Idrobiologia 12:21–77.

Edmondson, W. T. 1964. The rate of egg production by rotifers and copepods in natural populations as controlled by food and temperature. Verhandlungen der Internationale Vereinigung für Theoretische und Angewandte Limnologie 15:673–675.

Edmondson, W. T. 1985. Reciprocal changes in abundance of *Diaptomus* and *Daphnia* in Lake Washington. Archiv für Hydrobiologie, Beihefte Ergebnisse der Limnologie 21:475–481.

Edmondson, W. T., Comita, G. W., Anderson, G. C. 1962. Reproductive rate of copepods in nature and its relation to phytoplankton population. Ecology 43:625–634.

Einsle, U. 1996. Copepoda:Cyclopoida. Genera *Cyclops, Megacyclops, Acanthocyclops.* Guides to the Identification of the Microinvertebrates of the Continental Waters of the World. Dumont, H.J. (Ed.) 10:1–83.

Einsle, U. 1996. *Cyclops heberti* n. sp. and *Cyclops singularis* n. sp., two new species within the genus *Cyclops* ('*strenuus*-subgroup') (Crustacea: Copepoda) from ephemeral ponds in southern Germany. Hydrobiologia 319:167–177.

Elgmork, K. 1967. Ecological aspects of diapause in copepods. Symposium Series of the Marine Biological Association of India 2(3): 947–954 and 1 plate.

Elgmork, K., Nilssen, J. P. 1978. Equivalence of copepod and insect diapause. Verhandlungen der Internationale Vereinigung für Theoretische und Angewandte Limnologie 20:2511–2517.

Elmore, J. L. 1982. The influence of food concentration and container volume on life history parameters of *Diaptomus dorsalis* Marsh from subtropical Florida. Hydrobiologia 89:215–223.

Elmore, J. L. 1983. Factors influencing *Diaptomus* distributions: An experimental study in subtropical Florida. Limnology and Oceanography 28:522–532.

Elofsson, R. 1971. The ultrastructure of a chemoreceptor organ in the head of copepod crustaceans. Acta Zoologica 52:299–315.

Fahrenbach, W. H. 1962. The biology of a harpacticoid copepod. Cellule 62:303–376.

Feifarek, B. P., Wyngaard, G. A., Allan, J. D. 1983. The cost of reproduction in a freshwater copepod. Oecologia 56:166–168.

Feller, R. J. 1980. Development of the sand-dwelling meiobenthic harpacticoid copepod *Huntemannia jadensis* Poppe in the laboratory. Journal of Experimental Marine Biology and Ecology 46:1–15.

Fiers, F., Reid, J. W., Iliffe, T. M., Suárez-Morales, E. 1996. New hypogean cyclopoid copepods (Crustacea) from the Yucatan Peninsula, Mexico. Contributions to Zoology 66:65–102.

Fischer, J. M., Frost, T. M. 1997. Indirect effects of lake acidification on *Chaoborus* population dynamics: the role of food limitation and predation. Canadian Journal of Fisheries and Aquatic Sciences 54:637–646.

Fischer, J. M., Moore, M. V. 1993. Juvenile survival of a planktonic insect:effects of food limitation and predation. Freshwater Biology 30:35–45.

Fleminger, A. 1973. Pattern, number, variability, and taxonomic significance of integumental organs (sensilla and glandular pores) in the genus *Eucalanus* (Copepoda, Calanoida). Fishery Bulletin 71:965–1010.

Flemming, J. M., Coughlan, J. 1978. Preservation of vitally stained zooplankton for live/dead sorting. Estuaries 1:135–137.

Folt, C. L., Goldman, C. R. 1981. Allelopathy between zooplankton: A mechanism for interference competition. Science 213:1133–1135.

Frenzel, P. 1980. Die Populationsdynamik von *Canthocamptus staphylinus* (Jurine) (Copepoda, Harpacticoida) im Litoral des Bodensees. Crustaceana 39:282–286.

Friedman, M. M. 1980. Comparative morphology and functional significance of copepod receptors and oral structures. *in*: Kerfoot, W. C. (Ed.), Evolution and ecology of zooplankton communities. Uni. Press of New England, Hanover, NH, Pages 185–197.

Friedman, M. M., Strickler, J. R. 1975. Chemoreceptors and feeding in calanoid copepods (Arthropoda:Crustacea). Proceedings of the National Academy of Sciences of the U.S.A. 72:4185–4188.

Fryer, G. 1957a. The feeding mechanism of some freshwater cyclopoid copepods. Proceedings of the Zoological Society of London 129:1–25.

Fryer, G. 1957b. The food of some freshwater cyclopoid copepods and its ecological significance. Journal of Animal Ecology 26:263–286.

Gabriel, W. 1985. Overcoming food limitation by cannibalism: a model study on cyclopoids. Archiv für Hydrobiologie, Beihefte Ergebnisse der Limnologie 21:373–381.

Gannon, J. E. 1971. Two counting cells for the enumeration of zooplankton micro-Crustacea. Transactions of the American Microscopical Society 90:486–490.

Gannon, J. E., Gannon, S. A. 1975. Observations on the narcotization of crustacean zooplankton. Crustaceana 28:220–224.

Gauld, D. T. 1951. The grazing rate of planktonic copepods. Journal of the Marine Biological Association of the United Kingdom 29:695–706.

Gee, J. M., Warwick, R. M. 1984. Preliminary observations on the metabolic and reproductive strategies of harpacticoid copepods from an intertidal sandflat. Hydrobiologia 118:29–37.

Gehrs, C. W. 1974. Horizontal distribution and abundance of *Diaptomus clavipes* Schacht in relation to *Potamogeton foliosus* in a pond and under experimental conditions. Limnology and Oceanography 19:100–104.

Gerritsen, J. 1978. Instar-specific swimming patterns and predation of planktonic copepods. Verhandlungen der Internationale Vereinigung für Theoretische und Angewandte Limnologie 20:2531–2536.

Gilbert, J. J., Williamson, C. E. 1978. Predator–prey behavior and its effect on rotifer survival in associations of *Mesocyclops edax, Asplanchna girodi, Polyarthra vulgaris,* and *Keratella cochlearis.* Oecologia 37:13–22.

Gilbert, J. J., Williamson, C. E. 1983. Sexual dimorphism in zooplankton (Copepoda, Cladocera, and Rotifera). Annual Review of Ecology and Systematics 14:1–33.

Glatzel, T., Schminke, H. K. 1996. Mating behaviour of the groundwater copepod *Parastenocaris phyllura* Kiefer, 1938 (Copepoda:Harpacticoida). Contributions to Zoology 66:103–108.

Gliwicz, Z. M. 1994. Retarded growth of cladoceran zooplankton in the presence of a copepod predator. Oecologia 97:458–461.

Gliwicz, Z. M., Lampert, W. 1994. Clutch-size variability in *Daphnia*: Body-size related effects of egg predation by cyclopoid copepods. Limnology and Oceanography 39:479–485.

Gliwicz, Z. M., Stibor, H. 1993. Egg predation by copepods in *Daphnia* brood cavities. Oecologia 95:295–298.

Gliwicz, Z. M., Umana, G. 1994. Cladoceran body size and vulnerability to copepod predation. Limnology and Oceanography 39:419–424.

Gophen, M. 1978. Errors in the estimation of recruitment of early stages of *Mesocyclops leuckarti* (Claus) caused by the diurnal periodicity of egg-production. Hydrobiologia 57:59–64.

Gophen, M. 1979. Mating process in *Mesocyclops leuckarti* (Crustacea:Copepoda). Israel Journal of Zoology 28:163–166.

Gophen, M., Harris, R. P. 1981. Visual predation by a marine cyclopoid copepod, *Corycaeus anglicus.* Journal of the Marine Biological Association of the United Kingdom 61:391–399.

Gould, D. T. 1957. A peritrophic membrane in calanoid copepods. Nature 179:325–326.

Grad, G., Maly, E. J. 1988. Sex size ratios and their influence on mating success in a calanoid copepod. Limnology and Oceanography 33:1629–1634.

Grad, G., Maly, E. J. 1992. Further observations relating sex size ratios to mating success in calanoid copepods. Journal of Plankon Research 14: 903–913

Guma'a, S. A. 1978. The food and feeding habits of young perch, *Perca fluviatilis,* in Windermere. Freshwater Biology 8: 177–187.

Hairston, N. G. Jr. 1976. Photoprotection by carotenoid pigments in the copepod *Diaptomus nevadensis.* Proceedings of the National Academy of Sciences of the U.S.A. 73:971–974.

Hairston, N. G. Jr. 1979a. The adaptive significance of color polymorphism in two species of *Diaptomus* (Copepoda). Limnology and Oceanography 24:15–37.

Hairston, N. G. Jr. 1979b. The relationship between pigmentation and reproduction in two species of *Diaptomus* (Copepoda). Limnology and Oceanography 24:38–44.

Hairston, N. G. Jr. 1979c. The effect of temperature on carotenoid photoprotection in the copepod *Diaptomus nevadensis.* Comparative Biochemistry and Physiology 62:445–448.

Hairston, N. G. Jr. 1996. Zooplankton egg banks as biotic reservoirs in changing environments. Limnology and Oceanography 41:1087–1092.

Hairston, N. G. Jr., Olds, E. J. 1984. Population differences in the timing of diapause: adaptation in a spatially heterogeneous environment. Oecologiaphy 61:42–48.

Hairston, N. G., Jr., Twombly, S. 1985. Obtaining life table data from cohort analyses: A critique of current methods. Límnology and Oceanography. 30:886–893.

Hairston, N. G. Jr., Van Brunt, R. A. 1994. Diapause dynamics of two diaptomid copepod species in a large lake. Hydrobiologia 292/293:209–218.

Hairston, N. G. Jr., Van Brunt, R. A., Kearns, C. M., Engstrom, D. R. 1995. Age and survivorship of diapausing eggs in a sediment egg bank. Ecology 76:1706–1711.

Hairston, N. G. Jr., Walton, W. E. 1986. Rapid evolution of a life history trait. Proceedings of the National Academy of Sciences of the U.S.A. 83:4831–4833.

Hairston, N. G. Jr., Walton, W. E., Li, K. T. 1983. The causes and consequences of sex-specific mortality in a freshwater copepod. Limnology and Oceanography 28:935–947.

Hansen, A.-M., Santer, B. 1995. The influence of food resources on the development, survival and reproduction of the two cyclopoid copepods: *Cyclops vicinus* and *Mesocyclops leuckarti.* Journal of Plankton Research 17:631–646.

Hardy, A. 1970. The Open Sea. The World of Plankton. Collins, London, 335 p.

Harris, M. J. 1978. Copepoda of northern Mississippi with a description of a new subspecies. Tulane Studies in Zoology and Botany 20:27–34.

Hart, R. C., Santer, B. 1994. Nutritional suitability of some uni-algal diets for freshwater calanoids: unexpected inadequacies of commonly used edible greens and others. Freshwater Biology 31:109–116.

Hartig, J. H., Jude, D. J., Evans, M. S. 1982. Cyclopoid predation on Lake Michigan fish larvae. Canadian Journal of Fisheries and Aquatic Sciences 39:1563–1568.

Hartmann, H. J., Taleb, H., Aleya, L., Lair, N. 1993. Predation on ciliates by the suspension-feeding calanoid copepod *Acanthodiaptomus denticornis.* Canadian Journal of Fisheries and Aquatic Sciences 50:1382–1393.

Hebert, P. D. N. 1985. Ecology of the dominant copepod species at a Low Arctic site. Canadian Journal of Zoology 63:1138–1147.

Hebert, P. D. N., Good, A. G., Mort, M. A. 1980. Induced swarming in the predatory copepod *Heterocope septentrionalis.* Limnology and Oceanography 25:747–750.

Herbst, H.-V. 1986. Beschreibung des *Thermocyclops hastatus antillensis* n. ssp. mit einem Bestimmungschlüssel für die Gattung *Thermocyclops* Kiefer, 1927. Bijdragen tot de Dierkunde 56:165–180.

Hicks, G. R. F., Coull, B. C. 1983. The ecology of marine meiobenthic harpacticoid copepods. Annual Review of Oceanography and Marine Biology 21:67–175.

Hopp, U., Maier, G., Bleher, R. 1997. Reproduction and adult longevity of five species of planktonic cyclopoid copepods reared on different diets: a comparative study. Freshwater Biology 38:289–300.

Hosfeld, B. 1996. The relationship between the rostrum and the organ of Bellonci in copepods: an ultrastructural study of the rostrum of *Canuella perplexa*. Zoologischer Anzeiger 234:175–190.

Hosfeld, B., Schminke, H. K. 1997. The ultrastructure of ionocytes from osmoregulatory integumental windows of *Parastenocaris vicesima* (Crustacea, Copepoda, Harpacticoida). Archiv für Hydrobiologie 139:389–400.

Hudson, P. L., Reid, J. W., Lesko, L. T., Selgeby, J. H. 1998. Cyclopoid and harpacticoid copepods of the Laurentian Great Lakes. Bulletin of the Ohio Biological Survey, New Series. 12: vii, 1–50 pp.

Huq, A., Small, E. B., West, P. A., Huq, M. I., Rahman, R., Colwell, R. R. 1983. Ecological relationships between *Vibrio cholerae* and planktonic crustacean copepods. Applied and Environmental Microbiology 45:275–283.

Hutchinson, G. E. 1967. A Treatise on Limnology, Vol. 2. Wiley, New York.

Huys, R., Boxshall, G. A. 1991. Copepod Evolution. The Ray Society, London.

Huys, R., Gee, J. M., Moore, C. G., Hamond, R. 1996. Marine and brackish water harpacticoid copepods. Part 1. Keys and notes for identification of the species. Synopses of the British Fauna, New Series 51:1–352.

Ianora, A., Miralto, A., Buttino, I., Romano, G. 1999. First evidence of some dinoflagellates reducing male copepod fertilization capacity. Limnology and Oceanography 44:147–153.

Jamieson, C. D. 1980. Observations on the effect of diet and temperature on rate of development of *Mesocyclops leuckarti* (Claus) (Copepoda, Cyclopoida). Crustaceana 38:145–154.

Janetzky, W., Martínez Arbizu, P., Reid, J. W. 1996. *Attheyella (Canthosella) mervini* sp. n. (Canthocamptidae, Harpacticoida) from Jamaican bromeliads. Hydrobiologia 339:123–135.

Jersabek, C. D., Schabetsberger, R. 1995. Resting egg production and oviducal cycling in two sympatric species of alpine diaptomids (Copepoda: Calanoida) in relation to temperature and food availability. Journal of Plankton Research 17:2049–2078.

Karaytug, S. 1999. Cyclopoida—Genera Paracyclops, Ochridacyclops and key to the Eucyclopina. Guides to the Identification of the Microinvertebrates of the Continental Waters of the World. Dumont, H. J. (Ed.) 14: 1–217

Katona, S. K. 1971. The developmental stages of *Eurytemora affinis* (Poppe, 1880) (Copepoda, Calanoida) raised in laboratory cultures, including a comparison with the larvae of *Eurytemora americana* Williams, 1906, and *Eurytemora herdmani* Thompson and Scott, 1897. Crustaceana 21:5–20.

Kerfoot, W. C. 1978. Combat between predatory copepods and their prey: *Cyclops, Epischura,* and *Bosmina.* Limnology and Oceanography 23:1089–1102.

Kerfoot, W. C., DeMott, W. R. 1984. Food web dynamics: Dependent chains and vaulting. Pages 347–382 *in*: Meyers, D. G., Strickler, J. R. Ed., Trophic interactions within aquatic ecosystems. American Association for the Advancement of Science Selected Symposium 85. Westview Press Inc., Boulder, CO.

Koehl, M. A. R., Strickler, J. R. 1981. Copepod feeding currents: Food capture at low Reynolds number. Limnology and Oceanography 26:1062–1073.

Lampert, W., Muck, P. 1985. Multiple aspects of food limitation in zooplankton communities: the *Daphnia-Eudiaptomus* example. Archiv für Hydrobiologie, Beihefte Ergebnisse der Limnologie 21:311–322.

Landry, M. R., Hassett, R. P. 1982. Estimating the grazing impact of marine micro-zooplankton. Marine Biology 67:283–288.

Lang, K. 1948. Monographie der Harpacticiden. Vols. I, II. H. Ohlsson, Lund.

Lang, K. 1965. Copepoda Harpacticoida from the Californian Pacific coast. Kunglige Svenska Vetenskapsakademiens Handlingar, N.S., 10(2):1–560 + 5 plates.

Leech, D. M., Wyngaard, G. A. 1996. Timing of chromatin diminution in the free-living, freshwater Cyclopidae (Crustacea:Copepoda). Journal of Crustacean Biology 16:496–500.

Legier-Visser, M. F., Mitchell, J. G., Okubo, A., Fuhrman, J. A. 1986. Mechanoreception in calanoid copepods. A mechanism for prey detection. Marine Biology 90:529–535.

Lehman, J. T. 1980. Release and cycling of nutrients between planktonic algae and herbivores. Limnology and Oceanography 25:620–632.

Lehman, P. S., Reid, J. W. 1993. *Phyllognathopus viguieri* (Crustacea: Harpacticoida), a predaceous copepod of phytoparasitic, entomopathogenic, and free-living nematodes. Soil and Crop Science Society of Florida Proceedings 52:78–82.

Lescher-Moutoué, F. 1996. Seasonal variations in size and morphology of *Acanthocyclops robustus* (Copepoda Cyclopoida). Journal of Plankton Research 18:907–922.

Lewis, W. M. Jr., Saunders, J. F. III. 1979. Two new integrating samplers for zooplankton, phytoplankton, and water chemistry. Archiv für Hydrobiologie 85:244–249.

Liebig, J. R., Vanderploeg, H. A. 1995. Vulnerability of *Dreissena polymorpha* larvae to predation by Great Lakes calanoid copepods: The importance of the bivalve shell. Journal of Great Lakes Research 21:353–358.

Lohner, L. M., Hairston, Jr., N. G., Schaffner, W. R. 1990. A method for distinguishing subitaneous and diapausing eggs in preserved samples of the calanoid copepod genus *Diaptomus.* Limnology and Oceanography 35:763–767.

Lonsdale, D. J., Frey, M. A., Snell, T. W. 1998. The role of chemical signals in copepod reproduction. Journal of Marine Systems 15:1–12.

Luecke, C., Litt, A. H. 1987. Effects of predation by *Chaoborus flavicans* on crustacean zooplankton of Lake Lenore, Washington. Freshwater Biology 18:185–192.

Luecke, C., O'Brien, W. J. 1981. Phototoxicity and fish predation: Selective factors in color morphs in *Heterocope.* Limnology and Oceanography 26:454–460.

Luecke, C., O'Brien, W. J. 1983. The effect of *Heterocope* predation on zooplankton communities in arctic ponds. Limnology and Oceanography 28:367–377.

Maier, G. 1995. Mating frequency and interspecific matings in some freshwater cyclopoid copepods. Oecologia 101:245–250.

Main, T. M., Dobberfuhl, D. R., Elser, J. J. 1997. N:P stoichiometry and ontogeny of crustacean zooplankton: a test of the growth rate hypothesis. Limnology and Oceanography 42:1474–1478.

Makarewicz, J. C., Likens, G. E. 1979. Structure and function of the zooplankton community of Mirror Lake, New Hampshire. Ecological Monographs 49:109–127.

Maly, E. J. 1970. The influence of predation on the adult sex ratios of two copepod species. Limnology and Oceanography 15:566–573.

Maly, E. J. 1973. Density, size, and clutch of two high altitude diaptomid copepods. Limnology and Oceanography 18:840–848.

Maly, E. J. 1976. Resource overlap between co-occurring copepods: effects of predation and environmental fluctuation. Canadian Journal of Zoology 54:933–940.

Maly, E. J., Maly, M. P. 1974. Dietary differences between two co-occurring calanoid copepod species. Oecologia 17:325–333.

Mantel, L. H., Farmer, L. L. 1983. Osmotic and ionic regulation. *In*: Bliss, D. E. Ed., The Biology of Crustacea, Vol.5: Mantel, L. H. Ed., Internal anatomy and physiological regulation. Academic Press, New York, Pages 53–162.

Marshall, S. M., Orr, A. P. 1952. On the biology of *Calanus finmarchicus*. VII. Factors affecting egg production. Journal of the Marine Biological Association of the United Kingdom 30:527–547.

Marten, G. G., Bordes, E. S., Nguyen, M. 1994. Use of cyclopoid copepods for mosquito control. Hydrobiologia 292/293: 491–496.

Meyer, J. S., Ingersoll, C. G., McDonald, L. L., Boyce, M. S. 1986. Estimating uncertainty in population growth rates: jackknife vs bootstrap techniques. Ecology 67:1156–1166.

Monakov, A. V. 1976. Feeding and food interrelationships in freshwater copepods. Nauka Press, Leningrad, 170 p. [in Russian].

Moore, M. V., Folt, C. L., Stemberger, R. S. 1996. Consequences of elevated temperatures for zooplankton assemblages in temperate lakes. Archiv für Hydrobiologie 135:289–319.

Morris, D. P., Zagarese, H., Williamson, C. E., Balseiro, E. G., Hargreaves, B. R., Modenutti, B., Moeller, R. E., Queimalinos, C. 1995. The attenuation of solar UV radiation in lakes and the role of dissolved organic carbon. Limnology and Oceanography 40:1381–1391.

Musko, I. B. 1983. The structure of the alimentary canal of two freshwater copepods of different feeding habits studied by light microscope. Crustaceana 45:38–47.

Naess, T., Nilssen, J. P. 1991. Diapausing fertilized adults: A new pattern of copepod life cycle. Oecologia 86:368–371.

Neill, W. E. 1990. Induced vertical migration in copepods as a defence against invertebrate predation. Nature 345:524–526.

Neill, W. E. 1992. Population variation in the ontogeny of predator induced vertical migration of copepods. Nature 356:54–57.

Nie, H. W. de, Vijverberg, J. 1985. The accuracy of population density estimates of copepods and cladocerans, using data from Tjeukemeer (The Netherlands) as an example. Hydrobiologia 124:3–11.

O'Brien, W. J., Schmidt, D. 1979. Arctic *Bosmina* morphology and copepod predation. Limnology and Oceanography 24:564–568.

O'Doherty, E. C. 1985. Stream-dwelling copepods: Their life history and ecological significance. Limnology and Oceanography 30:554–564.

O'Doherty, E. C. 1988. The ecology of meiofauna in an Appalachian headwater stream. Ph.D. Dissertation, University of Georgia, Athens, 113 p.

Orsi, J. J., Bowman, T. E., Marelli, D. C., Hutchinson, A. 1983. Recent introduction of the planktonic calanoid copepod *Sinocalanus doerrii* (Centropagidae) from mainland China to the Sacramento-San Joaquin Estuary of California. Journal of Plankton Research 5:357–375.

Pace, M. L. 1986. An empirical analysis of zooplankton community size structure across lake trophic gradients. Limnology and Oceanography 31:45–55.

Paloheimo, J. 1974. Calculation of instantaneous birth rate. Limnology and Oceanography 19:692–694.

Parker, B. R., Wilhelm, F. M., Schindler, D. W. 1996. Recovery of *Hesperodiaptomus arcticus* populations from diapausing eggs following elimination by stocked salmonids. Canadian Journal of Zoology 74:1292–1297.

Paul, A. J., Leavitt, P. R., Schindler, D. W., Hardie, A. K. 1995. Direct and indirect effects of predation by a calanoid copepod (subgenus:*Hesperodiaptomus*) and of nutrients in a fishless alpine lake. Canadian Journal of Fisheries and Aquatic Sciences 52:2628–2638.

Paul, A. J., Schindler, D. W. 1994. Regulation of rotifers by predatory calanoid copepods (subgenus *Hesperodiaptomus*) in lakes of the Canadian Rocky Mountains. Canadian Journal of Fisheries and Aquatic Sciences 51:2520–2528.

Peacock, A. H. 1982. Responses of *Cyclops bicuspidatus thomasi* to alterations in food and predators. Canadian Journal of Zoology 60:1446–1462.

Peacock, A. H., Smyly, W. J. P. 1983. Experimental studies on the factors limiting *Tropocyclops prasinus* (Fischer) 1860 in an oligotrophic lake. Canadian Journal of Zoology 61:250–265.

Pennak, R. W. 1989. Fresh-water Invertebrates of the United States. Protozoa to Mollusca. 3rd ed. Wiley, New York.

Perlmutter, D. G., Meyer, J. L. 1991. The impact of a stream-dwelling harpacticoid copepod upon detritally associated bacteria. Ecology 72:2170–2180.

Pinel-Alloul, B., Lamoureux, J. 1988a. Développement post-embryonnaire du copépode calanoïde *Diaptomus* (*Aglaodiaptomus*) *leptopus* S. A. Forbes, 1882. I. Phase nauplienne. Crustaceana 54:69–84.

Pinel-Alloul, B., Lamoureux, J. 1988b. Développement post-embryonnaire du copépode calanoïde *Diaptomus* (*Aglaodiaptomus*) *leptopus* S. A. Forbes, 1882. II. Phase copepodite et adulte. Crustaceana 54:171–195.

Plassmann, T., Maier, G., Stich, H. B. 1997. Predation impact of *Cyclops vicinus* on the rotifer community in Lake Constance in spring. Journal of Plankton Research 19:1069–1079.

Prepas, E. 1978. Sugar-frosted *Daphnia*: An improved fixation technique for Cladocera. Limnology and Oceanography 23:557–559.

Price, H. J., Paffenhöfer, G. -A. 1986. Effects of concentration on the feeding of a marine copepod in algal monocultures and mixtures. Journal of Plankton Research 8:119–128.

Price, H. J., Paffenhöfer, G. -A., Strickler, J. R. 1983. Modes of cell capture in calanoid copepods. Limnology and Oceanography 28:116–123.

Ramalingam, K., Raghunathan, M. B. 1982. A study on the pH tolerance and survival of a cyclopoid copepod *Mesocyclops leucarti* (Claus). Comparative Physiology and Ecology 7:188–190.

Ramcharan, C. W., Sprules, W. G. 1991. Predator-induced behavioral defense and its ecological consequences for two calanoid copepods. Oecologia 86:276–286.

Reddy, Y. R. 1994. Copepoda: Calanoida: Diaptomidae, Key to the genera Heliodiaptomus, Allodiaptomus, Neodiaptomus, Phyllodiaptomus, Eodiaptomus, Arctodiaptomus and Sinodiaptomus. Guides to the Identification of the Microinvertebrates of the Continental Waters of the World. Dumont, H. J. (Ed.) 5:1–221.

Reed, E. B. 1963. Records of freshwater Crustacea from Arctic and Subarctic Canada. National Museum of Canada Bulletin No. 199, Contributions to Zoology 199:29–62.

Reed, E. B. 1986. Esteval phenology of an *Acanthocyclops* (Crustacea, Copepoda) in a Colorado tarn with remarks on the *vernalis–robustus* complex. Hydrobiologia 139:127–133.

Reed, E. B. 1994. *Arctodiaptomus novosibiricus* Kiefer, 1971 in Alaska and Northwest Territories, with notes on *A. arapahoensis* (Dodds, 1915) and a key to New World species of *Arctodiaptomus* (Copepoda:Calanoida). Proceedings of the Biological Society of Washington 107:666–679.

Reed, E. B. 1995. *Cyclops kolensis alaskaensis* Lindberg, 1956, revisited (Copepoda: Cyclopoida). Journal of Crustacean Biology 15:365–375.

Reed, E. B., Aronson, J. G. 1989. Seasonal variation in length of copepodids and adults of *Diacyclops thomasi* (Forbes) in two Colorado montane reservoirs (Copepoda). Journal of Crustacean Biology 9:67–76.

Reed, E. B., McIntyre, N. E. 1995. *Cyclops strenuus* (Fischer, 1851) sensu lato in Alaska and Canada, with new records of occurrence. Canadian Journal of Zoology 73:1699–1711.

Reid, J. W. 1986. Some usually overlooked cryptic copepod habitats. Syllogeus 58:594–598.

Reid, J. W. 1988. Copepoda (Crustacea) from a seasonally flooded marsh in Rock Creek Stream Valley Park, Maryland. Proceedings of the Biological Society of Washington 101:31–38.

Reid, J. W. 1989a. The distribution of species of the genus *Thermocyclops* (Copepoda, Cyclopoida) in the western hemisphere, with description of *T. parvus,* new species. Hydrobiologia 175:149–174.

Reid, J. W. 1989b. RE:"Infection of a field population of *Aedes cantator* with a polymorphic microsporidium, *Amplyospora connecticus* via release of the intermediate copepod host, *Acanthocyclops vernalis.*" Journal of the American Mosquito Control Association 5:616–617.

Reid, J. W. 1990. Continental and coastal free-living Copepoda (Crustacea) of Mexico, Central America and the Caribbean region. *In*: Navarro L. D. Robinson, J. G. (Eds.), Diversidad Biologica en la Reserva de la Biosfera de Sian Ka'an, Quintana Roo, México. Centro de Investigaciones de Quintana Roo (CIQRO) and Program of Studies in Tropical Conservation, University of Florida; Chetumal, Quintana Roo. Pp. 175–213.

Reid, J. W. 1991a. The genus *Metacyclops* (Copepoda:Cyclopoida) present in North America: *M. cushae,* new species, from Louisiana. Journal of Crustacean Biology 11:639–646.

Reid, J. W. 1991b. Some species of *Tropocyclops* (Crustacea, Copepoda) from Brazil, with a key to the American species. Bijdragen tot de Dierkunde 61:3–15.

Reid, J. W. 1992. Taxonomic problems: A serious impediment to groundwater ecological research in North America. *In*: Stanford J. A., Simons, J. J. Eds., First International Conference on Groundwater Ecology. American Water Resources Association, Bethesda, MD. Pages 133–142.

Reid, J. W. 1995. Redescription of *Parastenocaris brevipes* Kessler and description of a new species of *Parastenocaris* (Copepoda:Harpacticoida:Parastenocarididae) from the U.S.A. Canadian Journal of Zoology 73:173–187.

Reid, J. W. 1998. How "cosmopolitan" are the continental cyclopoid copepods? Comparison of North American and Eurasian faunas, with description of *Acanthocyclops parasensitivus* sp.n. (Copepoda:Cyclopoida) from the U.S.A. Zoologischer Anzeiger 236(1997):109–118.

Reid, J. W., Hare, S. G. F., Nasci, R. S. 1989. *Diacyclops navus* (Crustacea:Copepoda) redescribed from Louisiana, U.S.A. Transactions of the American Microscopical Society 108:332–344.

Reid, J. W., Hunt, G. W., Stanley, E. H. 2001. A new species of *Stygonitocrella* (Crustacea:Copepoda:Ameiridae) from North America. Proceedings of the Biological Society of Washington 114 [in press]

Reid, J. W., Ishida, T. 1993. New species and new records of the genus *Elaphoidella* (Crustacea:Copepoda:Harpacticoida) from the United States. Proceedings of the Biological Society of Washington 106:137–146.

Reid, J. W., Ishida, T. 1996. Two new species of *Gulcamptus* (Crustacea:Copepoda:Harpacticoida) from North America. Japanese Journal of Limnology 57:133–144.

Reid, J. W., Ishida, T. 2000. *Itocyclops,* a new genus proposed for *Speocyclops yezoensis* Ito (Copepoda:Cyclopoida). Journal of Crustacean Biology 20:589–596.

Reid, J. W., Janetzky, W. 1996. Colonization of Jamaican bromeliads by *Tropocyclops jamaicensis* n. sp. (Crustacea:Copepoda:Cyclopoida). Invertebrate Biology 115:305–320.

Reid, J. W., Marten, G. G. 1995. The cyclopoid copepod (Crustacea) fauna of non-planktonic continental habitats in Louisiana and Mississippi. Tulane Studies in Zoology and Botany 30:39–45.

Reid, J. W., Reed, E. B. 1994. First records of two neotropical species of *Mesocyclops* (Copepoda) from Yukon Territory: cases of passive dispersal? Arctic 47:80–87.

Reid, J. W., Strayer, D. L., McArthur, J. V., Stibbe, S. 1999. *Rheocyclops,* a new genus of copepods from the southeastern and central U.S.A. (Copepoda:Cyclopoida). Journal of Crustacean Biology 19:384–396.

Richman, S., Dodson, S. I. 1983. The effect of food quality on feeding and respiration by *Daphnia* and *Diaptomus.* Limnology and Oceanography 28:948–956.

Riera, T., Estrada, M. 1985. Dimensions and allometry in *Tropocyclops prasinus.* Empirical relationships with environmental temperature. Verhandlungen der Internationale Vereinigung für Theoretische und Angewandte Limnologie 22:3159–3163.

Ringelberg, J. 1980. Aspects of red pigmentation in zooplankton, especially copepods. *In*: Kerfoot, W. C. (Ed.), Evolution and ecology of zooplankton communities. Uni. Press of New England, Hanover, NH. Pages 91–97.

Ringelberg, J., Keyser, A. L., Flik, B. J. G. 1984. The mortality effect of ultraviolet radiation in a translucent and in a red morph of *Acanthodiaptomus denticornis* (Crustacea, Copepoda) and its possible ecological relevance. Hydrobiologia 112:217–222.

Robertson, A. 1970. Distribution of calanoid copepods (Calanoida, Copepoda) in Oklahoma. Proceedings of the Oklahoma Academy of Science 50:98–103.

Robertson, A. 1972. Calanoid copepods: new records from Oklahoma. Southwestern Naturalist 17:201–203.

Robertson, A. 1975. A new species of *Diaptomus* (Copepoda, Calanoida) from Oklahoma and Texas. American Midland Naturalist 93:206–214.

Robertson, A., Gannon, J. E. 1981. Annotated checklist of the free-living copepods of the Great Lakes. Journal of Great Lakes Research 7:382–393.

Rocha, C. E. F. 1991. A new species of *Halicyclops* (Copepoda, Cyclopidae) from California, and a revision of some *Halicyclops* material in the collections of the US Museum of Natural History. Hydrobiologia 226:29–37.

Rocha, C. E. F., Hakenkamp, C. C. 1993. New species of *Halicyclops* (Copepoda Cyclopidae) from the United States of America. Hydrobiologia 259:145–156.

Rocha, C. E. F., Iliffe, T. M., Reid, J. W, Suárez-Morales, E. 1998. A new species of *Halicyclops* (Copepoda, Cyclopoida) from cenotes of the Yucatán Peninsula, Mexico, with an identification key for the species of the genus from the Caribbean region and adjacent areas. Sarsia 83:387–399.

Roche, K. F. 1987. Post-encounter vulnerability of some rotifer prey types to predation by the copepod *Acanthocyclops robustus.* Hydrobiologia 147:229–233.

Roff, J. C. 1972. Aspects of the reproductive biology of the planktonic copepod *Limnocalanus macrurus* Sars, 1863. Crustaceana 22:155–160.

Rosenthal, H. 1972. Über die Geschwindigkeit der Sprungbewegungen bei *Cyclops strenuus* (Copepoda). Internationale Revue der gesamten Hydrobiologie 57:157–167.

Sanders, R. W., Williamson, C. E., Stutzman, P. L., Moeller, R. E., Goulden, C. E., Aoki-Goldsmith, R. 1996. Reproductive success of "herbivorous" zooplankton fed algal and nonalgal food resources. Limnology and Oceanography 41:1295–1305.

Santer, B. 1993. Do cyclopoid copepods control *Daphnia* populations in early spring, thereby protecting their juvenile instar stages from food limitation? Verhandlungen der Internationale

Vereinigung für Theoretische und Angewandte Limnologie 25:634–637.

Santer, B. 1994. Influences of food type and concentration on the development of *Eudiaptomus gracilis* and implications for interactions between calanoid and cyclopoid copepods. Archiv für Hydrobiologie 131:141–159.

Santer, B. 1996. Nutritional suitability of the dinoflagellate *Ceratium furcoides* for four copepod species. Journal of Plankton Research 18:323–333.

Särkkä, J. 1992. Effects of eutrophic and organic loading on the occurrence of profundal harpacticoids in a lake in southern Finland. Environmental Monitoring and Assessment 21: 211–223.

Sarvala, J. 1979. A parthenogenetic life cycle in a population of *Canthocamptus staphylinus* (Copepoda, Harpacticoida). Hydrobiologia 62:113–129.

Schaffner, W. R., Hairston Jr., N. G., Howarth, R. W. 1994. Feeding rates and filament clipping by crustacean zooplankton consuming cyanobacteria. Verhandlungen der Internationale Vereinigung für Theoretische und Angewandte Limnologie 25:2375–2381.

Schindler, D. W. 1969. Two useful devices for vertical plankton and water sampling. Journal of the Fisheries Research Board of Canada 26:1948–1955.

Shen, C.-j., Lee, F.-s. 1963. The estuarine Copepoda of Chiekong and Zaikong Rivers, Kwangtung Province, China. Acta Zoologica Sinica 15:571–596. [In Chinese with English summary].

Shih, C.-T., Maclellan, D. C. 1977. Descriptions of copepodite stages of *Diaptomus* (*Leptodiaptomus*) *nudus* Marsh 1904 (Crustacea:Copepoda). Canadian Journal of Zoology 55:912–921.

Siebeck, O. 1969. Spatial orientation of planktonic crustaceans. I. The swimming behaviour in a horizontal plane. Verhandlungen der Internationale Vereinigung für Theoretische und Angewandte Limnologie 17:831–840.

Siebeck, O. 1980. Optical orientation of pelagic crustaceans and its consequences in the pelagic and littoral zones. *in*: Kerfoot, W. C. (Ed.), Evolution and ecology of zooplankton communities. Uni. Press of New England, Hanover, NH. Pages 28–38.

Smith, K. E., Fernando, C. H. 1978. A guide to the freshwater calanoid and cyclopoid Copepoda Crustacea of Ontario. University of Waterloo Biology Series 18:1–76.

Smyly, W. J. P. 1962. Laboratory experiments with stage V copepodites of the freshwater copepod, *Cyclops leuckarti* Claus, from Windermere and Esthwaite Water. Crustaceana 4:273–280.

Smyly, W. J. P. 1980. Effect of constant and alternating temperatures on adult longevity of the freshwater cyclopoid copepod, *Acanthocyclops viridis* (Jurine). Archiv für Hydrobiologie 89:353–362.

Soto, D., Hurlbert, S. H. 1991. Long-term experiments on calanoid-cyclopoid interactions. Ecological Monographs 61:245–265.

Stearns, D. E., Sharp, A. A. 1994. Sublethal effects of cupric ion activity on the phototaxis of three calanoid copepods. Hydrobiologia 292/293:505–511.

Steib, K., Mayer, P. 1988. Epidemiology and vectors of *Dracunculus medinensis* in northwest Burkina Faso, West Africa. Annals of Tropical Medicine and Parasitology 82:189–199.

Stemberger, R. S. 1985. Prey selection by the copepod *Diacyclops thomasi.* Oecologia 65:492–497.

Stemberger, R. S. 1986. The effects of food deprivation, prey density and volume on clearance rates and ingestion rates of *Diacyclops thomasi.* Journal of Plankton Research 8:243–251.

Stemberger, R. S. 1995. Pleistocene refuge areas and postglacial dispersal of copepods of the northeastern United States. Canadian Journal of Fisheries and Aquatic Sciences 52:2197–2210.

Stemberger, R. S., Gilbert, J. J. 1984. Spine development in the rotifer *Keratella cochlearis*: induction by cyclopoid copepods and *Asplanchna.* Freshwater Biology 14:639–647.

Stemberger, R. S., Lazorchak, J. M. 1994. Zooplankton assemblage responses to disturbance gradients. Canadian Journal of Fisheries and Aquatic Sciences 51:2435–2447.

Sterner, R. W., Hessen, D. O. 1994. Algal nutrient limitation and the nutrition of aquatic herbivores. Annual Review of Ecology and Systematics 25:1–29.

Strayer, D. L. 1985. The benthic micrometazoans of Mirror Lake, New Hampshire. Archiv für Hydrobiologie, Supplement 72:287–426.

Strayer, D. L., Reid, J. W. 1999. Distribution of hyporheic cyclopoids (Crustacea: Copepoda) in the eastern United States. Archiv für Hydrobiologie 145:79–92.

Strickler, J. R. 1975a. Intra- and interspecific information flow among planktonic copepods: Receptors. Verhandlungen der Internationale Vereinigung für Theoretische und Angewandte Limnologie 19:2951–2958.

Strickler, J. R. 1975b. Swimming of planktonic *Cyclops* species (Copepoda, Crustacea): Pattern, movements and their control. *In*: Wu, T. Y., Brokaw, C. J., Brennan, C. (Eds.), Swimming and flying in nature. Vol. 2. Plenum Press, New York. Pages 599–613.

Strickler, J. R. 1982. Calanoid copepods, feeding currents, and the role of gravity. Science 218:158–160.

Strickler, J. R., Bal, A. K. 1973. Setae of the first antennae of the copepod *Cyclops scutifer* (Sars): Their structure and importance. Proceedings of the National Academy of Science of the U.S.A. 70:2656–2659.

Stutzman, P. 1995. Food quality of gelatinous colonial chlorophytes to the freshwater zooplankters *Daphnia pulicaria* and *Diaptomus oregonensis.* Freshwater Biology 34:149–153.

Suárez, M. F., Marten, G. G., Clark, G. G. 1992. A simple method for cultivating freshwater copepods used in biological control of *Aedes aegypti.* Journal of the American Mosquito Control Association 8:409–412.

Suárez-Morales, E., Elías-Gutiérrez, M. 2000. Two new Mastigodiaptomus (Copepoda, Diaptomidae) from southeastern Mexico, with a key for the identification of the know on species of the genus. Journal of Natural History 34:693–708.

Suárez-Morales, E., Reid, J. W. 1998. An updated list of the free-living freshwater copepods (Crustacea) of Mexico. Southwestern Naturalist 43:256–265.

Suárez-Morales, E., Reid, J. W., Iliffe, T. M., Fiers, F. 1996. Catálogo de los Copépodos (Crustacea) Continentales de la Península de Yucatán, México. Comisión Nacional para el Conocimiento y Uso de la Biodiversidad (CONABIO) and El Colegio de la Frontera Sur (ECOSUR), Mexico City. 296 p.

Svensson, J.-E. 1997. *Chaoborus* predation and sex-specific mortality in a copepod. Limnology and Oceanography 42:572–577.

Tessier, A. J., Goulden, C. E. 1982. Estimating food limitation in cladoceran populations. Limnology and Oceanography 27:707–717.

Threlkeld, S. T., Dirnberger, J. M. 1986. Benthic distributions of planktonic copepods, especially *Mesocyclops edax.* Syllogeus 58:481–486.

Tinson, S., Laybourn-Parry, J. 1985. The behavioural responses and tolerance of freshwater benthic cyclopoid copepods to hypoxia and anoxia. Hydrobiologia 127:257–263.

Torke, B. G. 1976. A key to the identification of the cyclopoid copepods of Wisconsin, with notes on their distribution and ecology. Wisconsin Department of Natural Resources Research Report 88:1–15.

Twombly, S., Burns, C. W. 1996. Exuvium analysis: A nondestructive method of analyzing copepod growth and development. Limnology and Oceanography 41:1324–1329.

Vanderploeg, H. A. 1994. Zooplankton particle selection and feeding mechanisms. *In* R. S. Wotton (Ed.), The biology of particles in aquatic systems. Lewis, Ann Arbor, MI. Pages 205–234.

Vanderploeg, H. A., Bolsenga, S. J., Fahnenstiel, G. L., Liebig, J. R., Gardner, W. S. 1992a. Plankton ecology in an ice-covered bay of Lake Michigan: utilization of a winter phytoplankton bloom by reproducing copepods. Hydrobiologia 243/244:175–183.

Vanderploeg, H. A., Gardner, W. S., Parrish, C. C., Liebig, J. R., Cavaletto, J. F. 1992b. Lipids and life-cycle strategy of a hypolimnetic copepod in Lake Michigan. Limnology and Oceanography 37:413–424.

Vanderploeg, H. A., Liebig, J. R., Omair, M. 1993. *Bythotrephes* predation on Great Lakes' zooplankton measured by an *in situ* method: implications for zooplankton community structure. Archiv für Hydrobiologie 127:1–8.

Vanderploeg, H. A., Paffenhöfer, G.-A. 1985. Modes of algal capture by the freshwater copepod *Diaptomus sicilis* and their relation to food-size selection. Limnology and Oceanography 30: 871–885.

Vanderploeg, H. A., Paffenhöfer, G.-A., Liebig, J. R. 1988. *Diaptomus* vs net phytoplankton: effects of algal size and morphology on selectivity of a behaviorally flexible, omnivorous copepod. Bulletin of Marine Science 43:377–394.

Vanderploeg, H. A., Scavia, D., Liebig, J. R. 1984. Feeding rate of *Diaptomus sicilis* and its relation to selectivity and effective food concentration in algal mixtures and in Lake Michigan. Journal of Plankton Research 6:919–941.

van Leeuwen, H. C., Maly, E. J. 1991. Changes in swimming behavior of male *Diaptomus leptopus* (Copepoda:Calanoida) in response to gravid females. Limnology and Oceanography 36:1188–1195.

Vitousek, P. M. 1994. Beyond global warming: ecology and global change. Ecology 75:1861–1876.

Vu, S. N., Nguyen, T. Y., Kay, B. H., Marten, G. G., Reid, J. W. 1998. Eradication of *Aedes aegypti* from a village in Vietnam, using copepods and community participation. American Journal of Tropical Medicine and Hygiene 59:657–660.

Walter, T. C. 1989. Review of the New World species of *Pseudodiaptomus* (Copepoda: Calanoida), with a key to the species. Bulletin of Marine Science 45:590–628.

Walton, W. E. 1985. Factors regulating the reproductive phenology of *Onychodiaptomus birgei* (Copepoda: Calanoida). Limnology and Oceanography 30:167–179.

Watras, C. J. 1983a. Mate location by diaptomid copepods. Journal of Plankton Research 5:417–423.

Watras, C. J. 1983b. Reproductive cycles in diaptomid copepods: Effects of temperature, photocycle, and species on reproductive potential. Canadian Journal of Fisheries and Aquatic Sciences 40:1607–1613.

Watras, C. J., Haney, J. F. 1980. Oscillations in the reproductive condition of *Diaptomus leptopus* (Copepoda: Calanoida) and their relation to rates of egg-clutch production. Oecologia 45:94–103.

Wetzel, R. G. 1983. Limnology, 2nd ed. Saunders, Philadelphia, PA. 767 p.

Whitehouse, J. W., Lewis, B. G. 1973. The effect of diet and density on development, size and egg production in *Cyclops abyssorum* Sars, 1863 (Copepoda, Cyclopoida). Crustaceana 25:225–236.

Wickham, S. A. 1995. *Cyclops* predation on ciliates: species-specific differences and functional responses. Journal of Plankton Research 17:1633–1646.

Wierzbicka, M., 1962. On the resting stage and mode of life of some species of Cyclopoida. Polish Archives of Hydrobiology (Polskie Archiwum Hydrobiologii) 10:215–229.

Williamson, C. E. 1980. The predatory behavior of *Mesocyclops edax*: Predator preferences, prey defenses, and starvation-induced changes. Limnology and Oceanography 25:903–909.

Williamson, C. E. 1981. Foraging behavior of a freshwater copepod: Frequency changes in looping behavior at high and low prey densities. Oecologia 50:332–336.

Williamson, C. E. 1983a. Behavioural interactions between a cyclopoid copepod predator and its prey. Journal of Plankton Research 5:701–711.

Williamson, C. E. 1983b. Invertebrate predation on planktonic rotifers. Hydrobiologia 104:385–396.

Williamson, C. E. 1984. Laboratory and field experiments on the feeding ecology of the freshwater cyclopoid copepod, *Mesocyclops edax*. Freshwater Biology 14:575–585.

Williamson, C. E. 1986. The swimming and feeding behavior of *Mesocyclops*. Hydrobiologia 134:11–19.

Williamson, C. E. 1987. Predator–prey interactions between omnivorous diaptomid copepods and rotifers: The role of prey morphology and behavior. Limnology and Oceanography 32:167–177.

Williamson, C. E. 1993. Linking predation risk models with behavioral mechanisms: identifying population bottlenecks. Ecology 74:320–331.

Williamson, C. E., Butler, N. M. 1986. Predation on rotifers by the suspension-feeding calanoid copepod *Diaptomus pallidus*. Limnology and Oceanography 31:393–402.

Williamson, C. E., Butler, N. M. 1987. Temperature, food, and mate limitation of copepod reproductive rates: separating the effects of multiple hypotheses. Journal of Plankton Research 9: 821–836.

Williamson, C. E., Butler, N. M., Forcina, L. 1985. Food limitation in naupliar and adult *Diaptomus pallidus*. Limnology and Oceanography 30:1283–1290.

Williamson, C. E., Magnien, R. E. 1982. Diel vertical migration in *Mesocyclops edax*: Implications for predation rate estimates. Journal of Plankton Research 4:329–339.

Williamson, C. E., Stemberger, R. S., Morris, D. P., Frost, T. M., Paulsen, S. G. 1996. Ultraviolet radiation in North American lakes: attenuation estimates from DOC measurements and implications for plankton communities. Limnology and Oceanography 41:1024–1034.

Williamson, C. E., Stoeckel, M. E., Schoeneck, L. J. 1989. Predation risk and the structure of freshwater zooplankton communities. Oecologia 79:76–82.

Williamson, C. E., Vanderploeg, H. A. 1988. Predatory suspension-feeding in *Diaptomus*: prey defenses and the avoidance of cannibalism. Bulletin of Marine Science 43:561–572.

Williamson, C. E., Zagarese, H. E., Schulze, P. C., Hargreaves, B. R., Seva, J. 1994. The impact of short-term exposure to UV-B radiation on zooplankton communities in north temperate lakes. Journal of Plankton Research 16:205–218.

Wilson, C. B. 1932. The copepods of the Woods Hole region, Massachusetts. Bulletin of the United States National Museum 158:1–635 and plates 1–41.

Wilson, C. C., Hebert, P. D. N. 1993. Impact of copepod predation on distribution patterns of *Daphnia pulex* clones. Limnology and Oceanography 38:1304–1310.

Wilson, M. S. 1958. The copepod genus *Halicyclops* in North America, with description of a new species from Lake Pontchartrain, Louisiana, and the Texas coast. Tulane Studies in Zoology 6:176–189.

Wilson, M. S., Yeatman, H. C. 1959. Free-living Copepoda. *In* Edmondson, W. T. (Ed.), Ward and Whipple's Fresh-water biology. J Wiley, New York, Pages 735–868.

Winfield, I. J., Townsend, C. R. 1983. The cost of copepod reproduction: increased susceptibility to fish predation. Oecologia 60:406–411.

Wong, C. K. 1981. Predatory feeding behavior of *Epischura lacustris* (Copepoda, Calanoida) and prey defense. Canadian Journal of Fisheries and Aquatic Sciences 38:275–279.

Wong, C. K., Sprules, W. G. 1986. The swimming behavior of the freshwater calanoid copepods *Limnocalanus macrurus* Sars, *Senecella calanoides* Juday and *Epischura lacustris* Forbes. Journal of Plankton Research 8:79–90.

Woodmansee, R. A., Grantham, B. J. 1961. Diel vertical migrations of two zooplankters (*Mesocyclops* and *Chaoborus*) in a Mississippi lake. Ecology 42:619–628.

Woodward, I. O., White, R. W. G. 1981. Effects of temperature and food on the fecundity and egg development rates of *Boeckella symmetrica* Sars (Copepoda: Calanoida). Australian Journal of Marine and Freshwater Research 32:997–1002.

Wright, D., O'Brien, W. J., Vinyard, G. L. 1980. Adaptive value of vertical migration: A simulation model argument for the predation hypothesis. *In*: W. C. Kerfoot (Ed.), Evolution and ecology of zooplankton communities. Uni. Press of New England, Hanover, NH. Pages 138–147.

Wyngaard, G. A. 1983. *In situ* life table of a subtropical copepod. Freshwater Biology 13:275–281.

Wyngaard, G. A. 1986. Genetic differentiation of life history traits in populations of *Mesocyclops edax* (Crustacea: Copepoda). Biological Bulletin 170:279–295.

Wyngaard, G. A. 1998. Reciprocal transfer study of north temperate and subtropical populations of *Mesocyclops edax* (Copepoda: Cyclopoida). Journal of Marine Systems 15:163–169.

Wyngaard, G. A., Chinnappa, C. C. 1982. General biology and cytology of cyclopoids. *In*: Harrison, F. W., Cowden, R. R. (Eds.), Developmental biology of freshwater invertebrates. A. R. Liss, New York. Pages 485–533.

Wyngaard, G. A., Taylor, B. E., Mahoney, D. L. 1991. Emergence and dynamics of cyclopoid copepods in an unpredictable environment. Freshwater Biology 25:219–232.

Yan, N. D., Mackie, G. L., Dillon, P. J. 1990. Cadmium concentrations of crustacean zooplankton of acidified and nonacidified Canadian Shield lakes. Environmental Science and Technology 24:1367–1372.

Yan, N. D., Pawson, T. W. 1997. Changes in the crustacean zooplankton community of Harp Lake, Canada, following invasion by *Bythotrephes cederstroemi*. Freshwater Biology 37:409–425.

Zaret, T. M. 1980. Predation in freshwater communities. Yale Uni. Press, New Haven, CT, 187 pp.

Zaret, T. M., Suffern, J. S. 1976. Vertical migration in zooplankton as a predator avoidance mechanism. Limnology and Oceanography 21:804–813.

TABLE I Orders and Superorders of Copepods

Platycopioida Fosshagen 1985—Small, hyperbenthic marine species in shallow seas and anchialine caves.

Calanoida Sars 1903—Freshwater and marine, free-swimming.

Misophrioida Gurney 1933—shallow coastal waters, deep sea, deep-water plankton, and anchialine caves.

Harpacticoida Sars 1903— Freshwater and marine, free-living, primarily benthic or attached.

Monstrilloida Sars 1903—nauplii and copepodids are endoparasitic on polychaetes and prosobranch molluscs. Adults are free-swimming, but nonfeeding. Adult females carry eggs on spines.

Mormonilloida Boxshall 1979—only two species known, both widely distributed and abundant in deep oceans (400–1500 m depth). They are free-swimming particle-feeders.

Gelyelloida Huys 1988—only three species known, two from Europe (France and Switzerland) and one from the United States (South Carolina). All are freshwater and benthic/subterranean.

Cyclopoida Burmeister 1835—Freshwater and marine, benthic and planktonic.

Siphonostomatoida Thorell 1859—parasitic or associated with fish or invertebrate hosts; primarily marine, but some freshwater species exist. Body often highly reduced, lacking appendages.

Poecilostomatoida Thorell 1859—Mostly parasitic on, or associated with, a wide variety of fish and invertebrates. Morphologically the most diverse copepod order; mostly marine. Four families are often abundant in the marine plankton. Some feed visually (Corycaeidae and Sapphirinidae). Includes the family Ergasilidae.

Superorder Gymnoplea—Articulation between prosome and urosone is between fifth pedigerous and genital somites. This is the ancestral tagmosis that includes the Platycopioida and Calanoida.

Superorder Podoplea—Articulation between prosome and urosome is between fourth and fifth pedigerous somites. This is the derived tagmosis, and it includes the other eight orders.

From Huys and Boxshall, 1991.

TABLE II North American Copepods of the Order Calanoida G. O. Sars, 1903

Family Acartiidae G. O. Sars, 1903
Acartia: sinensis Shen and Lee, 1963 (I)
Family Aetideidae Giesbrecht, 1892
Senecella: calanoides Juday, 1923
Family Centropagidae Giesbrecht, 1892
Limnocalanus: *johanseni* Marsh, 1920; *macrurus* G. O. Sars, 1863.
Osphranticum: *labronectum* S. A. Forbes, 1882.
Sinocalanus: *doerri* (Brehm, 1909) (I) (see Orsi *et al.*, 1983).
Family Diaptomidae G. O. Sars, 1903
Acanthodiaptomus: *denticornis* (Wierzejski, 1887).
Aglaodiaptomus: *atomicus* DeBiase and Taylor, 1997; *clavipes* (Schacht, 1897); *clavipoides* M. S. Wilson, 1955; *conipedatus* (Marsh, 1907); *dilobatus* M. S. Wilson, 1958; *forbesi* Light, 1938; *kingsburyae* Robertson, 1975; *leptopus* (S. A. Forbes, 1882); *lintoni* (S. A. Forbes, 1893); *marshianus* M. S. Wilson, 1953; *pseudosanguineus* (Turner, 1921); *saskatchewanensis* M. S. Wilson, 1958; *savagei* DeBiase and Taylor, 2000; *spatulocrenatus* (Pearse, 1906); *stagnalis* (S. A. Forbes, 1882).
Arctodiaptomus: *arapahoensis* (Dodds, 1915); *bacillifer* (Koelbel, 1884); *dorsalis* (Marsh, 1907); *floridanus* (Marsh, 1926); *kurilensis* Kiefer, 1937; *saltillinus* (Brewer, 1898). Key: Reddy, 1994, Reed, 1994.
Diaptomus: *glacialis* Lilljeborg, 1889.
Eudiaptomus: *gracilis* (G. O. Sars, 1863); *yukonensis* Reed, 1991.
Hesperodiaptomus: *arcticus* (Marsh, 1920); *augustaensis* (Turner, 1910); *breweri* M. S. Wilson, 1958; *caducus* Light, 1938; *californiensis* Scanlin and Reid, 1996; *eiseni* (Lilljeborg, 1889); *franciscanus* (Lilljeborg, 1889); *hirsutus* M. S. Wilson, 1953; *kenai* M. S. Wilson, 1953; *kiseri* (Kincaid, 1953); *nevadensis* Light 1938; *novemdecimus* M. S. Wilson, 1953; *schefferi* M. S. Wilson, 1953; *shoshone* (S. A. Forbes, 1893); *victoriaensis* Reed, 1958; *wardi* (Pearse, 1905); *wilsonae* Reed, 1958.
Leptodiaptomus: *ashlandi* (Marsh, 1893); *assiniboiaensis* (Anderson, 1970); *coloradensis* (Marsh, 1911); *connexus* Light, 1938; *cuauhtemoci* (Osorio-Tafall, 1941; synonym assiniboiaensis Anderson, 1970); *insularis* Kincaid, 1956; *judayi* (Marsh, 1907); *minutus* (Lilljeborg, 1889); *moorei* M. S. Wilson, 1954; *novamexicanus* (Herrick, 1895); *nudus* (Marsh, 1904); *pribilofensis* (Juday and Muttkowski, 1915); *sicilis* (S. A. Forbes, 1882); *siciloides* (Lilljeborg, 1889); *signicauda* (Lilljeborg, 1889); *spinicornis* Light, 1938; *trybomi* (Lilljeborg, 1889); *tyrrelli* (Poppe, 1888).
Mastigodiaptomus: *albuquerquensis* (Herrick, 1895); *texensis* M. S. Wilson, 1953.Key: Suárez-Morales and Elías-Gutiérrez, 2000.
Mixodiaptomus: *theeli* (Lilljeborg, 1889).
Nordodiaptomus: *alaskaensis* M. S. Wilson, 1951.
Onychodiaptomus: *birgei* (Marsh, 1894); *hesperus* M. S. Wilson and Light, 1951; *louisianensis* M. S. Wilson and Moore, 1953; *sanguineus* (S. A. Forbes, 1876); *virginiensis* (Marsh, 1915).
Sinodiaptomus: *sarsi* (Rylov, 1923)(I). Key: Reddy, 1994.
Skistodiaptomus: *bogalusensis* M. S. Wilson and Moore, 1953; *bogalusensis marii* Harris, 1978; *carolinensis* Yeatman, 1986; *mississippiensis* (Marsh, 1894); *oregonensis* (Lilljeborg, 1889); *pallidus* (Herrick, 1879); *pygmaeus* (Pearse, 1906); *reighardi* (Marsh, 1895); *sinuatus* Kincaid, 1953.
Family Temoridae Giesbrecht, 1892
Epischura: *fluviatilis* Herrick, 1883; *lacustris* S. A. Forbes, 1882; *massachusettsensis* Pearse, 1906; *nevadensis* Lilljeborg, 1889; *nordenskioldi* Lilljeborg, 1889.
Eurytemora: *affinis* (Poppe, 1880); *americana* Williams, 1906; *arctica* M. S. Wilson and Tash, 1966; *bilobata* Akatova, 1949 (synonym *yukonensis* M. S. Wilson, 1953); *canadensis* Marsh, 1920; *composita* Keiser, 1929; *gracilicauda* Akatova, 1949.
Heterocope: *septentrionalis* Juday and Muttkowski, 1915.

Identification is based primarily on males. Species primarily occurring in brackish coastal waters, some of which may occasionally enter fresh water, are not included. Introduced species are indicated by (I). Keys and taxonomic revisions published after 1959 are noted.

TABLE III North American Copepods of the Order Cyclopoida Burmeister, 1834

Family Cyclopidae Dana, 1853

Acanthocyclops: *brevispinosus* (Herrick, 1884); *capillatus* (G. O. Sars, 1863); *carolinianus* Yeatman, 1944; *columbiensis* Reid, 1990; *exilis* (Coker, 1934); *montana* Reid and Reed *in* Reid *et al.*, 1991; *parasensitivus* Reid, 1998; *parvulus* Strayer, 1989; *pennaki* Reid, 1992; *robustus* (G. O. Sars, 1863); *venustoides* (Coker, 1934); *vernalis* (Fischer, 1853). Key: Einsle, 1996a.

Apocyclops: *dimorphus* (Kiefer, 1931); *panamensis* (Marsh, 1913); *spartinus* Ruber, 1968.

Bryocyclops: *muscicola* (Menzel, 1926). (I)

Cryptocyclops: *bicolor* (G. O. Sars, 1862).

Cyclops: *canadensis* Einsle, 1988; *columbianus* Lindberg, 1956 (inadequately described species); *furcifer* (Claus, 1857); *kolensis alaskaensis* Lindberg, 1956; *laurenticus* Lindberg, 1956 (inadequately described); *scutifer* G. O. Sars, 1863; *strenuus* Fischer, 1851; *vicinus* Uljanin, 1875. Revisions: Reed (1995), Reed and McIntyre (1995); Key: Einsle, 1996a.

Diacyclops: *alabamensis* Reid, 1992; *albus* Reid, 1992; *bernardi* (Petkovski, 1986); *bicuspidatus* (Claus, 1857); *bicuspidatus lubbocki* (Brady, 1868); *bisetosus* (Rehberg, 1880); *chrisae* Reid, 1992; *crassicaudis* (G. O. Sars, 1863); *crassicaudis* var. *brachycercus* (Kiefer, 1927); *dimorphus* Reid and Strayer, 1994; *harryi* Reid, 1992; *haueri* (Kiefer, 1931); *hypnicola* (Gurney, 1927); *jeanneli* (Chappuis, 1929); *jeanneli putei* (Yeatman, 1943); *languidoides* (Lilljeborg, 1901) s.l.; *languidus* (G. O. Sars, 1863); *nanus* (G. O. Sars, 1863); *navus* (Herrick, 1882); *nearcticus* Kiefer, 1934; *palustris* Reid, 1988; *sororum* Reid, 1992; *thomasi* (S. A. Forbes, 1882); *yeatmani* Reid, 1988; Key: Reid, 1988, emend. Reid *et al.*, 1989.

Ectocyclops: *phaleratus* (Koch, 1838); *polyspinosus* Harada, 1931; *rubescens* Brady, 1904.

Eucyclops: *agilis* (Koch, 1838); *agilis montanus* (Brady, 1878); *bondi* Kiefer, 1934; *conrowae* Reid, 1992; *elegans* (Herrick, 1884; synonym *E. neomacruroides* Dussart and Fernando, 1990); *macruroides denticulatus* (Graeter, 1903); *prionophorus* Kiefer, 1931.

Homocyclops: *ater* (Herrick, 1882).

Itocyclops: *yezoensis* (Ito, 1954) (see Reid and Ishida, 2000).

Macrocyclops: *albidus* (Jurine, 1820); *fuscus* (Jurine, 1820).

Megacyclops: *donnaldsoni* (Chappuis, 1929); *gigas* (Claus, 1857); *latipes* (Lowndes, 1927); *magnus* (Marsh, 1920); *viridis* (Jurine, 1820)(I). Key: Einsle, 1996a.

Mesocyclops: *americanus* Dussart, 1985; *edax* (S. A. Forbes, 1891); *longisetus* (Thiébaud, 1912); *longisetus* var. *curvatus* Dussart, 1987; *ruttneri* Kiefer, 1981 (I); *venezolanus* Dussart, 1987. Key: Reid and Reed, 1994.

Metacyclops: *cushae* Reid, 1991; in Reid, 1991a.

Microcyclops: *pumilis* Pennak and Ward, 1985; *rubellus* (Lilljeborg, 1901); *varicans* (G. O. Sars, 1862).

Orthocyclops: *modestus* (Herrick, 1883).

Paracyclops: *canadensis* (Willey, 1934); *chiltoni* (Thomson, 1882); *fimbriatus* (Fischer, 1853); *poppei* (Rehberg, 1880); *smileyi* Strayer, 1989; *yeatmani* Daggett and Davis, 1974. Key: Karaytug, 1999.

Rheocyclops: *carolinianus* Reid, *in* Reid *et al.*, 1999; *hatchiensis* Reid and Strayer *in* Reid *et al.*, 1999; *talladega* Reid and Strayer *in* Reid *et al.*, 1999; *virginianus* (Reid, 1993). Key: Reid *et al.*, 1999.

Stolonicyclops: *heggiensis* Reid and Spooner, 1998.

Thermocyclops: *crassus* (Fischer, 1853) (I); *inversus* Kiefer, 1936; *parvus* Reid, 1989; *tenuis* (Marsh, 1909). Revision: Reid (1989a); Key: Herbst, 1986.

Tropocyclops: *extensus* Kiefer, 1931; *extensus* f. *longispina* Kiefer, 1931; *jerseyensis* Kiefer, 1931; *prasinus* (Fischer, 1860); *prasinus mexicanus* Kiefer, 1938. Key: Reid, 1991b.

Family Oithonidae Dana, 1853

Limnoithona: *sinensis* (Burckhardt, 1912) (I).

Identification is based primarily on females. Species primarily occurring in brackish coastal waters, some of which may occasionally enter freshwater, are not included. Introduced species are indicated by (I). Keys and taxonomic revisions published after 1959 are noted.

TABLE IV North American Copepods of the Order Harpacticoida G. O. Sars, 1903

Family Ameiridae Monard, 1927
Nitocrellopsis: texana Fiers and lliffe, 2000
Nitokra: hibernica (Brady, 1880) (I); *lacustris* (Shmankevich, 1875).
Psammonitocrella: boultoni Rouch, 1992; *longifurcata* Rouch, 1992.
Stygonitocrella sequoyahi Reid, Hunt, and Stanley, 2001.
Family Canthocamptidae Brady, 1880
Attheyella: americana (Herrick, 1884); *carolinensis* Chappuis, 1932; *dentata* (Poggenpol, 1874); *dogieli* (Rylov, 1923); *idahoensis* (Marsh, 1903); *illinoisensis* (S. A. Forbes, 1882); *nordenskioldi* (Lilljeborg, 1902); *obatogamensis* (Willey, 1925); *pilosa* Chappuis, 1929; *spinipes* Reid, 1987; *trispinosa* (Brady, 1880); *ussuriensis* Rylov, 1933.
Bryocamptus: arcticus (Lilljeborg, 1902); *calvus* (Brehm, 1927); *cuspidatus* (Schmeil, 1893); *douwei* (Willey, 1925); *hiatus* (Willey, 1925); *hiemalis* (Pearse, 1905); *hutchinsoni* Kiefer, 1929; *minnesotensis* (Herrick, 1884; inadequately described species); *minusculus* (Willey, 1925); *minutus* (Claus, 1863); *morrisoni* (Chappuis, 1929); *morrisoni elegans* (Chappuis, 1929); *newyorkensis* (Chappuis, 1927); *nivalis* (Willey, 1925); *pilosus* Flössner, 1989; *pygmaeus* (G. O. Sars, 1862); *subarcticus* (Willey, 1925); *tikchikensis* M. S. Wilson, 1958; *umiatensis* M. S. Wilson, 1958; *vejdovskyi* (Mrázek, 1893); *vejdovskyi* f. *minutiformis* Kiefer, 1934; *washingtonensis* M. S. Wilson, 1958; *zschokkei* (Schmeil, 1893); *zschokkei alleganiensis* Coker, 1934.
Canthocamptus: assimilis Kiefer, 1931; *oregonensis* M. S. Wilson, 1956; *robertcokeri* M. S. Wilson, 1958; *sinuus* Coker, 1934; *staphylinoides* Pearse, 1905; *staphylinus* (Jurine, 1820) (records of *staphylinus* in North America are doubtful); *vagus* Coker and Morgan, 1940.
Elaphoidella: amabilis Ishida *in*: Reid and Ishida, 1993; *bidens* (Schmeil, 1894); *californica* M. S. Wilson, 1975; *carterae* Reid *in*: Reid and Ishida, 1993; *fluviusherbae* Bruno and Reid, *in* Bruno et al., 2000); *kodiakensis* M. S. Wilson, 1975; *marjoryae* Bruno and Reid, *in* Bruno et al., 2000); *reedi* M. S. Wilson, 1975; *shawangunkensis* Strayer, 1989; *subgracilis* (Willey, 1934); *tenuicaudis* (Herrick, 1884; inadequately described species); *wilsonae* Hunt, 1979. Key: Reid and Ishida, 1993.
Epactophanes: richardi Mrázek, 1893.
Gulcamptus: alaskaensis Ishida, 1996 *in*: Reid and Ishida, 1996; *huronensis* Reid, 1996 *in* Reid and Ishida, 1996; *laurentiacus* (Flössner, 1992). Revision: Reid and Ishida, 1996.
Maraenobiotus: brucei (Richard, 1898); *canadensis* Flössner, 1992; *insignipes* (Lilljeborg, 1902).
Mesochra: alaskana M. S. Wilson, 1958.
Moraria: affinis Chappuis, 1927; *arctica* Flössner, 1989; *cristata* Chappuis, 1929; *duthiei* (T. and A. Scott, 1896); *laurentica* Willey, 1927; *mrazeki* T. Scott, 1902; *virginiana* Carter, 1944.
Paracamptus: reductus M. S. Wilson, 1956; *reggiae* M. S. Wilson, 1958.
Family Canthocamptidae incertae sedis:
Cletocamptus: albuquerquensis (Herrick, 1895); *brevicaudata* (Herrick, 1895); *deitersi* (Richard, 1897).
Family Huntemanniidae Por, 1986
Huntemannia: lacustris M. S. Wilson, 1958.
Nannopus: palustris Brady, 1880.
Family Laophontidae T. Scott, 1904
Onychocamptus: mohammed (Blanchard and Richard, 1891).
Family Parastenocarididae Chappuis, 1940
Parastenocaris: brevipes Kessler, 1913; *delamarei* Chappuis *in* Chappuis and Delamare Deboutteville, 1958; *lacustris* Chappuis *in* Chappuis and Delamare Deboutteville, 1958; *palmerae* Reid, 1992; *texana* Whitman, 1984; *trichelata* Reid, 1995.
Family Phyllognathopodidae Gurney, 1932
Phyllognathopus: viguieri (Maupas, 1892).

Identification is based on either sex. Species primarily occurring in brackish coastal waters, some of which may occasionally enter freshwater, are not included. Introduced species are indicated by (I). Keys and taxonomic revisions published after 1959 are noted.

TABLE V North American Copepods of the Order Poecilostomatoida Thorell, 1859

Family Ergasilidae Nordmann, 1832
Ergasilus: arthrosis Roberts, 1969; *auritus* Markevich, 1940; *caeruleus* C. B. Wilson, 1911; *celestis* Mueller, 1937; *centrarchidarum* Wright, 1882; *cerastes* Roberts, 1969; *chautauquaensis* Fellows, 1887; *clupeidarum* Johnson and Rogers, 1972; *cotti* Kellicott, 1892; *cyprinaceus* Rogers, 1969; *elongatus* C. B. Wilson, 1916; *japonicus* Harada, 1930 (I); *lanceolatus* C. B. Wilson, 1916; *luciopercarum* Henderson, 1926; *megaceros* C. B. Wilson, 1916; *nerkae* Roberts, 1963; *rhinos* Burris and Miller, 1972; *sieboldi* Nordmann, 1832; *tenax* Roberts, 1965; *versicolor* C. B. Wilson, 1911; *wareaglei* Johnson, 1971.

Adult females are ectoparasites of fishes, but males and copepodids are planktonic. Identification is based on females. Species primarily occurring in brackish coastal waters, some of which may occasionally enter freshwater, are not included. Introduced species are indicated by (I). No keys or taxonomic revisions have been published since 1959.

23 DECAPODA

H. H. Hobbs III

Department of Biology
Wittenberg University
Springfield, Ohio 45501

I. INTRODUCTION

The Decapoda (Latreille) represent a very significant order that is assigned to the class Malacostraca and encompasses an immense diversity of marine, freshwater, and semiterrestrial crustaceans, with some 10,000 species having been described. The two infraorders treated in this chapter, Caridea and Astacidea, are nearly worldwide in distribution and for purposes of this discussion are represented by freshwater shrimps and crayfishes (see Bowman and Abele, 1982 for a detailed classification; see also Hart, 1994). These aquatic arthropods have successfully invaded a wide variety of aquatic and semiaquatic habitats, occurring as obligate cave-dwellers [stygobites (= stygobionts)], as stream, lake, pond, and swamp dwellers, and as primary burrowers; a few even invade saline environments. The crayfishes and one genus of shrimps (*Macrobrachium*), in particular, attain the greatest size among the freshwater crustaceans in North America.

Summarized in this chapter are the ecology and distribution of shrimps and crayfishes in North American freshwaters (treatment generally restricted to the United States). Species richness and niche diversification are particularly well demonstrated in aquatic ecosystems of the southeastern United States. That these crustaceans have successfully colonized such diverse habitats is reflected in various morphological and physiological adaptations discussed. Life-history patterns are reviewed, particularly with reference to how they are influenced by environmental conditions. An examination is made of the role of these decapods in the functioning of various aquatic communities. As might be anticipated, a much larger body of data is available for crayfishes than for shrimps and is reflected in the treatment of these groups. Under each major section of the chapter, a discussion of shrimps appears first, followed by that of crayfishes. Where data are somewhat limited (particularly with reference to shrimps), both groups are discussed in the same section.

II. ANATOMY AND PHYSIOLOGY

A. External Morphology

Although decapods represent the greatest diversity among orders of crustaceans, all possess a number of common characteristics, including a carapace that encloses the bronchial chamber. In addition, the first three

Ecology and Classification of North American Freshwater Invertebrates. 2nd Edition

pairs of thoracic appendages are modified in all species as maxillipeds. Carideans (Fig. 1) are easily distinguished from other shrimp groups (e.g., penaeids, stenopodids) by the large second abdominal pleura that overlap those of both the first and third somites; moreover, all carideans lack terminal chelae on the third pereiopods (Fig. 1a). Astacideans (Fig. 2) include the crayfishes of the northern and southern hemispheres and the marine lobsters; only North American crayfishes are treated here. In contrast to the carideans, crayfishes are rarely compressed laterally and their first three pairs of pereiopods are always chelate. Extensive body segmentation and the presence of jointed appendages on all metameres give most decapods a primitive appearance; yet the nervous, sensory, circulatory, and digestive systems are complex, a necessary requirement for animals of large body size. The body of these freshwater decapods is encased in an exoskeleton consisting of complex polysaccharides hardened with inorganic salts (except at joints, where the cuticle is thin and pliable). The head and thoracic segments are fused to form a large cephalothorax covered by a single shield, the carapace (Figs. 1 and 2), but the six abdominal segments are individually distinct. The anterior portion of the cephalothorax contains a pair of large, stalked eyes, a median rostrum, two pairs of antennae, and a pair of mandibles. The body is composed of somites, each with a pair of ventral, jointed appendages that are serially homologous and basically biramous (Fig. 3), but are modified for various functions (e.g., sensory, food handling, cleaning, gill-bailing, pinching, walking, copulation, egg attachment and incubation, and swimming). The typical biramous appendage is Y-shaped and the base of the Y is attached to the somite. The base is the protopodite (consisting of two joints—coxopodite and basiopodite) that bears a mesial endopodite, typically of five podomeres, and a lateral exopodite, with few to many segments.

B. Organ System Function

The digestive system of crayfishes is simple and includes the mouth, a short tubular esophagus, the stomach, two large hepatopancreatic glands, a short midgut, and a long tubular intestine extending dorsally through the abdomen to the anus (Fig. 4). The relatively simple, straight gastrointestinal tract has few significant areas for storage or microbial degradation although the stomach functions as a storage compartment (cardiac portion), as a masticating structure (gastric mill), and as a filter for gathering digestible material (pyloric portion). Morphological differences among species in the grinding surfaces of the gastric mill relate to diet (Caine, 1975). For example, the surface is reduced in

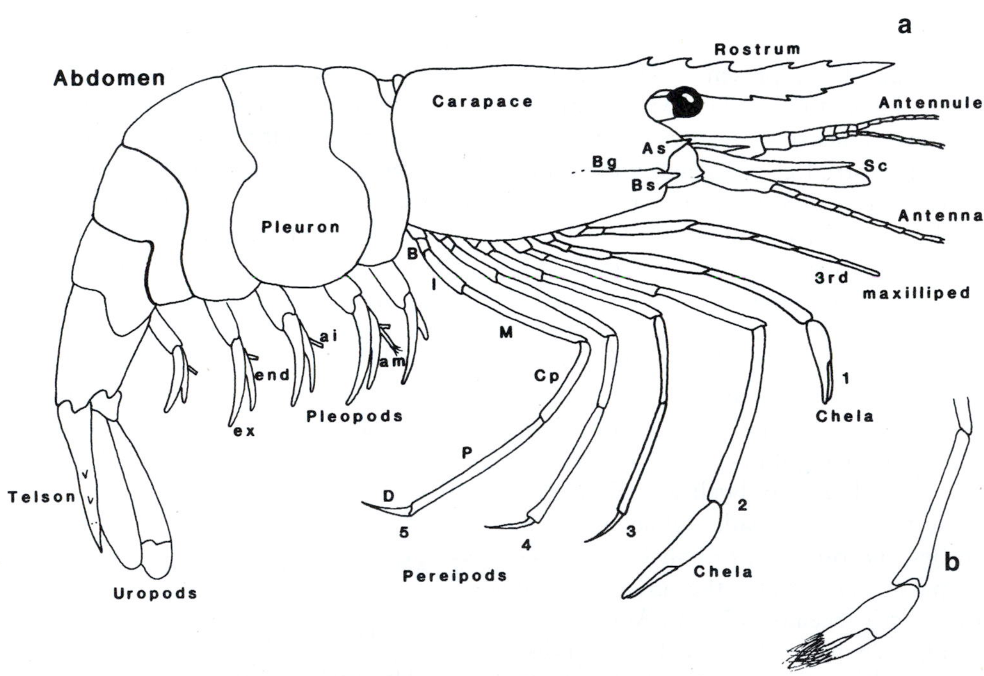

FIGURE 1 (A) Lateral view of generalized shrimp (after Hobbs and Jass, 1988). Ai, appendix interna; Am, appendix masculina; As, antennal spine; B, basis; Bg, branchiostegal groove; Bs, branchiostegal spine; Cp, carpus; D, dactyl; end, endopod; I. ischium; M, merus; P, propodus; and Sc, scaphocerite. (B) Chela of second pereiopod with apical tufts of setae; *Palaemonias ganteri* (after Hobbs *et al.*, 1977).

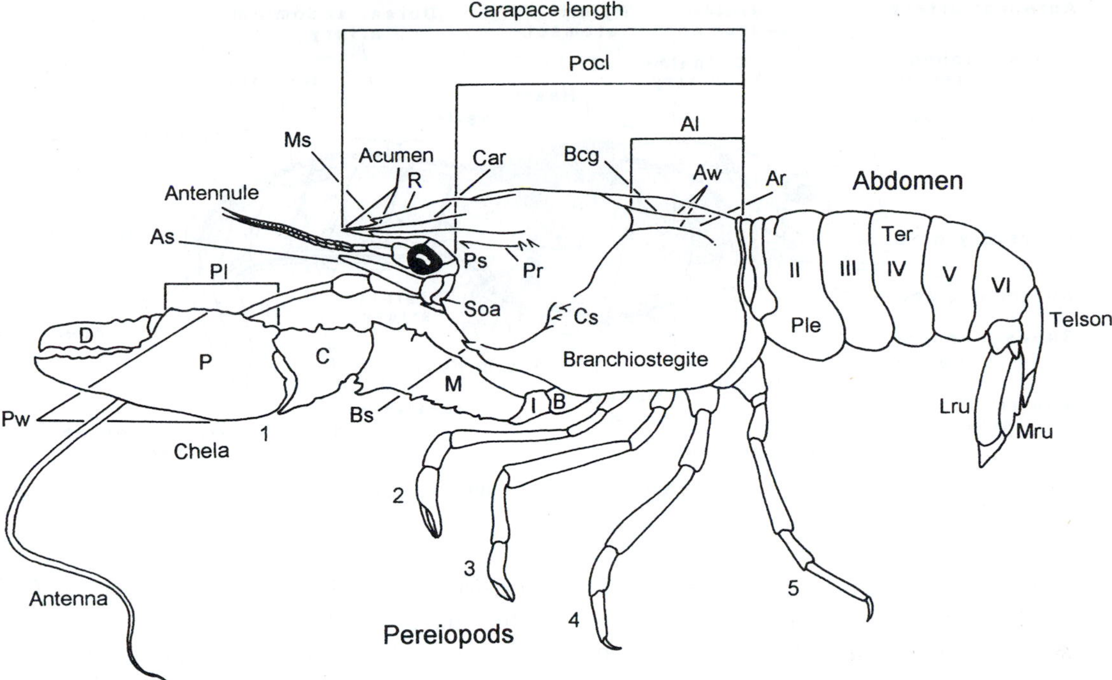

FIGURE 2 Dorsolateral view of generalized crayfish. Al, areola length; Ar, areola; As, antennal scale; Aw, areola width; B, basis; Bcg, branchiocardiac groove; Bs, branchiostegal spine; C,carpus; Car, rostral carina; Cs, cervical spine; D, dactyl of chela; I, ischium; Lru, lateralramus of uropod; M, merus; Mru, mesial ramus of uropod; Ms, marginal spine; P, propodus-palm of chela; Pl, palm length; Ple, pleuron; Pocl, postorbital carapace length; Pr, postorbital ridge; Ps, postorbital spine; Pw, width of palm; R, rostrum; Soa, suborbital angle; Ter, tergum; and pleopods on ventral side of abdomen.

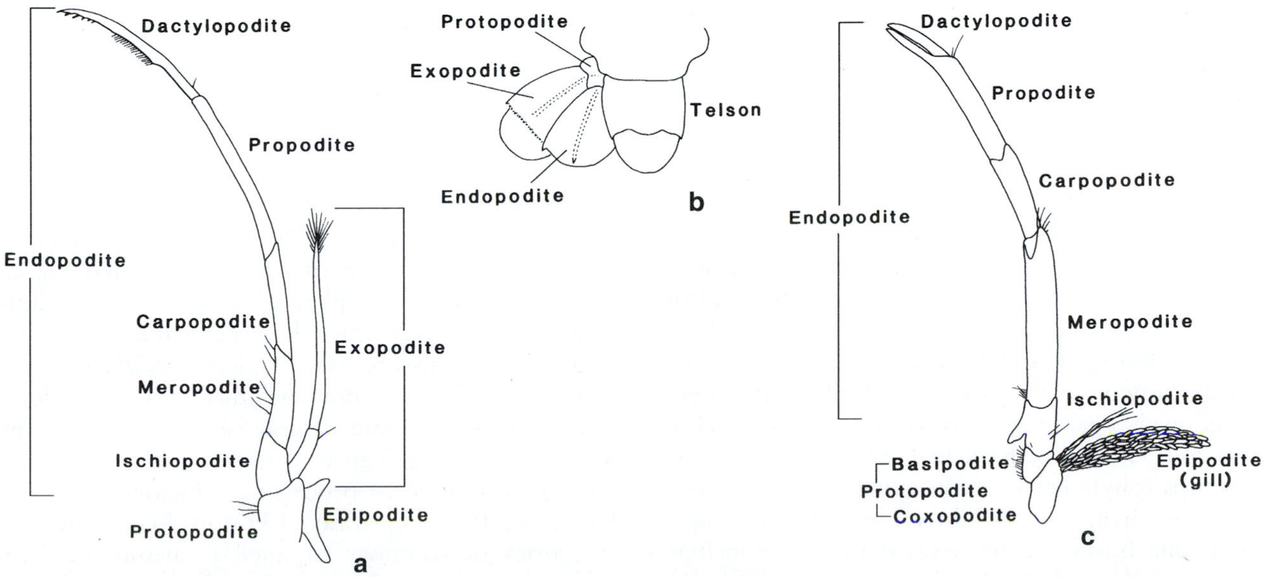

FIGURE 3 Biramous appendages:(a) *Palaemonias alabamae,* third maxilliiped (after Smalley, 1961); (b) *Cambarus* (*Depressicambarus*) *strigosus Hobbs,* dorsal view of telson and biramous uropod (after Hobbs, 1981); and (c) *Procambarus* (*Ortmannicus*) *lunzi* (Hobbs), third pereiopod.

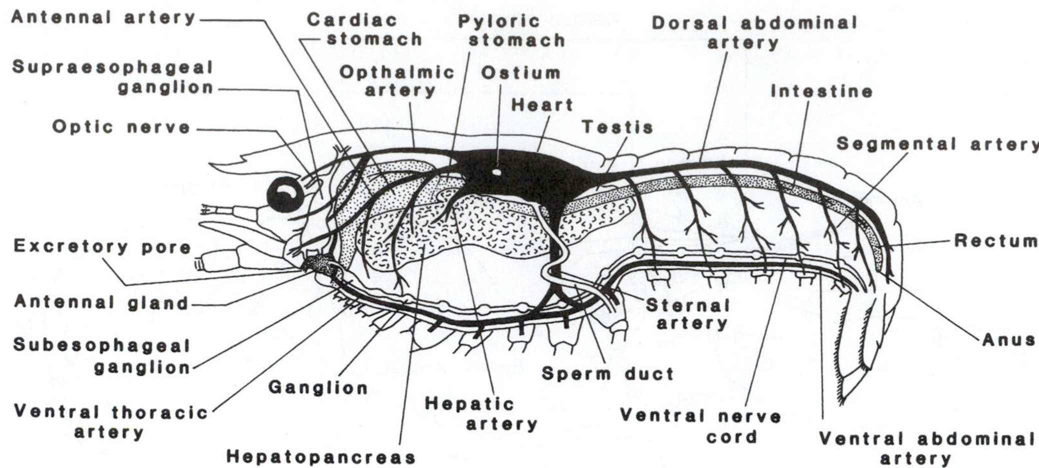

FIGURE 4 Generalized diagram of crayfish, demonstrating internal organization.

some cave crayfishes that feed on organic silt and expanded in epigean species that utilize macromaterials. The pH of the stomach is relatively alkaline and the enzymes of the gastrointestinal tract and hepatopancreas are particularly effective in the processing of a wide spectrum of food materials. Chymotrypsin and pepsin are lacking, but an alkaline protease, trypsin, cellulase (hepatopancrease origin), lysozyme (muramidase), and probably chitinase and chitobiase are present (Brown, 1995).

The circulatory system is an open or lacunar system and consists of a heart, arteries (no veins), and sinuses. The heart is situated in a large middorsal pericardial sinus, and blood in the sinus enters the heart via three pairs of ostia (valves). The nearly colorless blood is pumped into arteries that distribute it to various organs and afferent sinuses leading to the gills, and the blood returns through a series of efferent sinuses that ultimately join to the pericardial sinus. Hemocyanin is the principal blood pigment transporting oxygen.

The gills are attached to the thoracic appendages (maxillipeds and pereiopods) and are situated along either side of the thorax in the bronchial (gill) chamber. Lateral extensions of the carapace (branchiostegites) cover the gills (Figs. la, 2, and 7). In crayfishes, anterior to the gills, paddlelike second maxillae (scaphognathites, see Burggren and McMahon, 1983) beat back and forth, drawing water over the gill filaments to ensure gas exchange. The gills of crayfishes (trichobranchs composed of unbranched, fingerlike filaments) and shrimps (phyllobranchs consisting of platelike elements) are arranged in three series: the single podobranchia borne on the coxa, the arthrobranchiae on the articular membrane between the coxa and basis, and the pleurobranchia borne on the pleura. The pleurobranch series is absent in all of the North American crayfishes except in members of the genus *Pacifastacus*, which have a single pair.

Considerable water constantly diffuses into the blood through the gill surfaces of freshwater decapods. Osmotic control and excretion are maintained by two large antennal (green) glands situated ventral to the subesophageal ganglia and anterior to the esophagus. These glands have an intimate association with the circulatory system and thus maintain water balance (Flynn and Holliday, 1994). The green gland consists of a complexly folded end sac, a convoluted labyrinth, a nephridial canal, a bladder, and an excretory pore situated at the coxae of the second antennae. Crayfishes excrete a quantity of hypotonic urine (some ammonia, urea, uric acid, and amines), which is generally isosmotic with the blood. High ammonia excretion rates have been correlated with low pH values (pH 4.6; Daveikis and Alikhan, 1996).

The central nervous system functions as the destination for sensory input, the center of synaptic integration, and the source of motor control and consists of paired supraesophageal and subesophageal ganglia, which are joined, forming the "brain." This juncture is accomplished by a pair of connectives extending caudally from the subesophageal ganglia to the double ventral nerve cord which has segmental ganglia connected by commissures (see Sandeman *et al.*, 1992 for a proposed common nomenclature for homologous structures of the brain of crayfishes, crabs, and spiny lobsters). Of significance, 30–40% of crayfish brain volume is devoted to processing olfactory input (Mellon *et al.*, 1992). Cuadras (1993) evaluated the eleven categories of secretory organelles (membrane bound carriers) present in the central nervous system of crayfishes. The peripheral nervous system consists of nerves and receptor cells that function primarily as conducting pathways (see Swerup and Rydqvist, 1992). Atwood and Nguyen (1995) reviewed neural adaptations of crayfishes to different activity levels, describing how various species are equipped for a variety of environmental changes.

The basal segment of each first antenna supports an organ of balance (statocyst). Chemoreceptors are abundant on mouth parts, pereiopods, and antennae; these last appendages also function as tactile organs. Lateral rami of the antennules bear chemoreceptive sensilla called aesthetascs that are innervated by sensory neurons. Tierney and Dunham (1984) and Tierney *et al.* (1984) suggested that these structures may be receptive to pheromones although there have been conflicting results from various studies concerning under what conditions and the location of receptor sites where chemical sensation takes place (see Oh and Dunham, 1991; Dunham and Oh, 1992).

Rutherford *et al.* (1996) reported that in agonistic encounters of conspecific Form I males (breeding males—see below) of *Orconectes* (*Procericambarus*) *rusticus* (Girard), winners had higher rates of antennal movements than losers. Winners also demonstrated longer periods of "large amplitude depression" (LAD—holding antennules approximately 45° below the horizontal plane) which is used often in aggressive encounters. The researchers postulated that the increased use of crayfish LAD movements may be in response to some "stress pheromone" delivered to them by their opponent's maxillipeds. Pheromones and other biochemicals are used by crayfishes to recognize species, sex, food, the presence of animals that are stressed, disturbed, or physically damaged. These alarm substances are widespread and the detection of these will elicit distinct responses by crayfishes (Hazlett, 1994b).

Dunham *et al.* (1997) examined the use of the inner and outer rami of the antennules in mediation of feeding behavior of *Cambarus* (*Cambarus*) *bartonii bartonii* (Fabricius) and found that the several behavior patterns used were dependent on the presence of either inner or outer antennular rami, but not on both. Many setae on the body surface contain sensory neurons and are mechanoreceptors of one kind or another and at least three sensory structures on the chelipeds appear to have chemo- and mechanosensory function (Borash and Moore, 1997). A pair of large compound eyes characterizes most decapods [excluding stygobites (obligate, aquatic cave dwellers)], and reduced eye size in most primary burrowing species); each eye is composed of a number of virtually identical, discrete optical units called ommatidia ("mosaic eyes"—good for detecting movement, but limited resolving power) that are arranged in geometrical array. In addition to compound eyes, a pair of neurons originating in the sixth abdominal ganglion function as extraocular photoreceptors and are responsible for photonegative behavior, but also receive information from mechanoreceptors in the abdomen (see Fernández-de-Miguel and Aréchiga, 1992; Pel and Moss, 1996).

Further details of the morphology of decapods, in general, can be examined in numerous invertebrate zoology textbooks (see also Holdich and Reeve, 1988), and the external anatomy of crayfishes specifically is treated well by Crocker and Barr (1968).

C. Aspects of Decapod Physiology

1. Respiration

The adaptation of organisms to various habitats usually involves physiological, biochemical, and behavioral adjustments. Obviously a dynamic interaction exists between the external environment of an organism and the functioning of its internal machinery. For example, reduced metabolic rates in cave decapods may stem from biochemical and physiological adaptations to greater environmental stability, to low number of predators, or to a severely constrained resource base in terms of quantity and variety (Huppop, 1985), as demonstrated by whole animal respiration studies (see Weingartner, 1977). Also, the reduction of oxygen consumption and energy turnover of gill tissues reported by Dickson and Franz (1980) further substantiate the highly specialized nature of physiological and biochemical adaptations in stygobitic animals.

In order for any life form to survive, the variety of factors constituting the ambient environmental complex must not exceed genetically determined tolerance limits for that species. That crayfishes and shrimps have radiated into nearly every type of aquatic habitat suggests that these crustaceans have been tolerant and highly adaptable, particularly with regard to respiratory physiology. For example, Wiens and Armitage (1961) demonstrated that, for the crayfishes *Orconectes* (*Trisellescens*) *immunis* (Hagen) and *O.* (*Gremicambarus*) *nais* (Faxon), the effect of body size on oxygen consumption and the effects of both temperature and oxygen saturation on oxygen consumption by them are highly significant (inverse relationship between body weight and metabolic rate; direct relationship between temperature and oxygen consumption). They also showed that the cumulative effect of increased temperature and lowered oxygen concentration produces greater stress than either parameter acting independently. *O. immunis* has a daily metabolic rate 10% lower than *O. nais,* and *O. nais* does not regulate as well as does *O. immunis* when under stressed conditions. Therefore, the inability of *O. nais* to maintain a higher and more nearly regulated metabolic rate (as

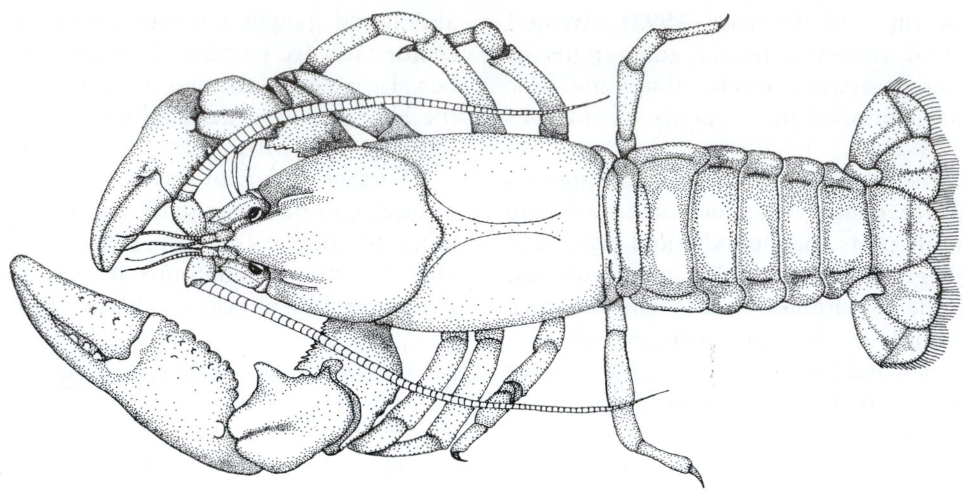

FIGURE 5 *Cambarus tenebrosus* from epigean and hypogean (stygophile) lotic habitats in the midwestern and southeastern United States.

does *O. immunis)* under such conditions, probably excludes it from roadside ditches where *O. immunis* is found. Hence, *O. immunis* is better adapted physiologically to live under environmental conditions of periodic high temperatures and low oxygen saturations (see Reiber, 1997). Daveikis and Alikhan (1996) found that *Cambarus (Puncticambarus) robustus* Girard from an acidic, metal-contaminated lake in Ontario, Canada, had lower oxygen consumption rates than conspecific crayfishes from a circumneutral, uncontaminated, high velocity stream. Nelson and Hooper (1982) showed that the shrimp, *Palaemonetes kadiakensis* Rathbun, exhibited an acute thermal preference within a temperature range of 6–10°C below their maximum tolerance limits of 37–40°C. This shrimp was unable to tolerate sustained exposure to temperatures greater than 32°C. For additional treatment of metabolism, temperature acclimation, preference, and avoidance in these two groups of crustaceans, see McWhinnie and O'Connor (1967), Crawshaw (1974), Becker *et al.* (1975), Loring and Hill (1978), Mathur *et al.* (1982), Layne *et al.* (1985), and Reiber (1995).

Burrowing crayfishes are often observed above the water table; many species are capable of remaining out of water for extended periods in the humid environment of their tunnels (Hobbs, 1981; McMahon and Hankinson, 1993, and McMahon and Stuart, 1995). The same is true for many cavernicolous crayfishes (stygophiles–Fig. 5, and stygobites–Fig. 6), which, in the saturated air, can move from one pool to another or remain out of the wa-

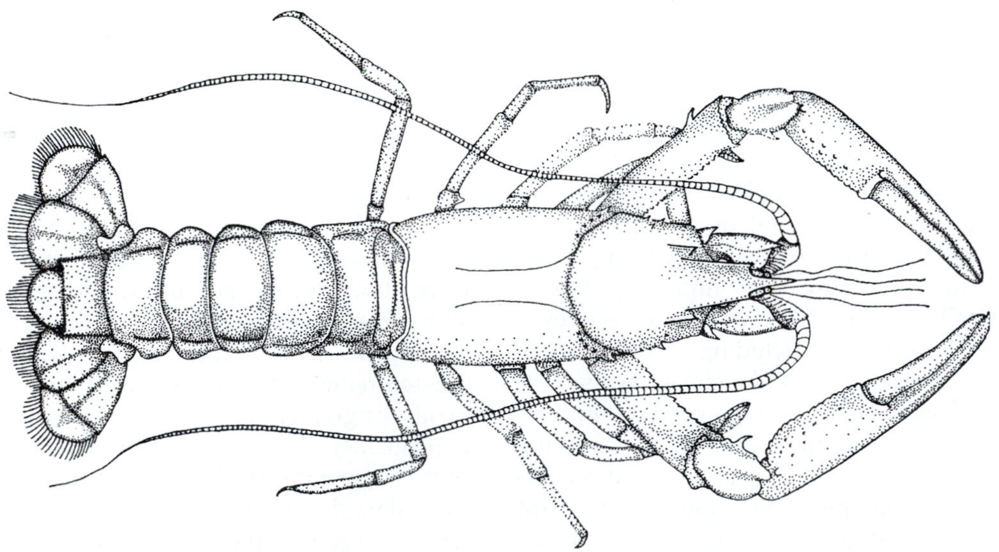

FIGURE 6 *Orconectes pellucidus,* a stygobitic crayfish from Kentucky and Tennessee.

ter for hours. Burrowers and surface-dwelling crayfishes are commonly noted at the air–water interface, aerating in response to low oxygen levels; the oxygen concentration in burrow waters is frequently below 2 mg/L (Hobbs, 1981), and pool and ditches often become oxygen-depleted. These species that inhabit such oxygen-poor habitats have evolved a greater branchial volume by adopting several characteristics (Figs. 7 and 8), such as a vaulted and elongated carapace as well as branchiocardiac grooves that are closer together thus narrowing the areola and raising the height of the chamber. Kushlan and Kushlan (1980) observed the shrimp *Palaemonetes paludosus* (Gibbes) swimming at the surface using oxygen that diffused across it as a means of surviving low oxygen tension in the Everglades.

Many species (e.g., the crayfish *Procambarus* (*Girardiella*) *simulans* (Faxon)) are oxygen regulators and increase their ventilation rates in response to reduced oxygen or increased carbon dioxide concentrations. Others [e.g., *Pacifastacus* (*Pacifastacus*) *l. leniusculus* (Dana)] are oxygen conformers, lacking specific adaptations for regulation of oxygen uptake at low oxygen tensions (see Vernberg, 1983). For further discussion of respiratory physiology in crayfishes and shrimps, see Moshiri *et al.* (1970), Cox and Beauchamp (1982), Nelson and Hooper (1982), and Reiber (1995).

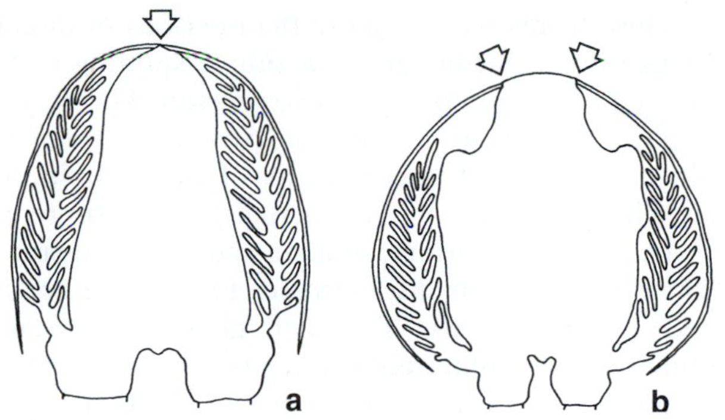

FIGURE 7 Semi-diagrammatic cross section through the thoracic region of (a) *Cambarus diogenes*, a primary burrower living in oxygen-poor groundwaters and (b) *Procambarus*(*Pennides*) *versutus* (Hagen), dwelling in the oxygen-rich lotic environment (after Hobbs, 1976); arrows point to branchiocardiac grooves.

2. Ecdysis

Molting (ecdysis), a process critical for growth and seasonal changes in form (dimorphic only in male cambarines), is a recurring crisis in the lives of shrimps and

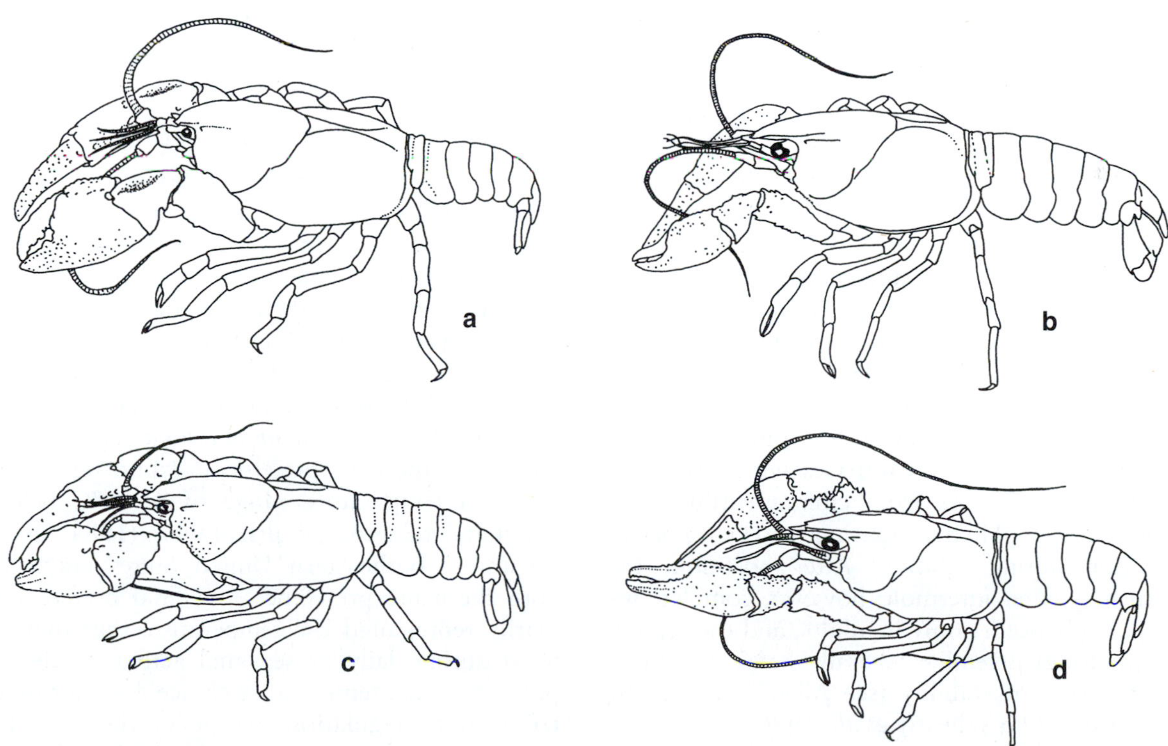

FIGURE 8 Dorsolateral views of (a) *Cambarus* (*Jugicambarus*) *gentryi* Hobbs, a primary burrower (Cheatham County, TN); (b) *Procambarus lunzi*, found in roadside ditch lentic environments (Hampton County, SC); (c) *Cambarus* (*Jugicambarus*) *distans* Rhoades, a dweller of swift, surface streams (Dade County, GA); and (d) *Procambarus* (*Ortmannicus*) *enoplosternum* Hobbs, occupies sluggish surface streams (Washington County, GA).

crayfishes. It involves far more than periods of discontinuous growth facilitated by a simple splitting of the cuticle and secretion of a new exoskeleton. The life of a decapod consists of alternating periods of premolt, molt, postmolt, and intermolt; all of these phases are controlled by hormones (endocrines). Neurosecretory cells occur in the sensory papillae and in the X-organs of the medulla terminalis; both secretary masses are situated in the eye stalk. The sinus gland, also located within the eye stalk, receives secretions from the X-organs and, in turn, produces hormones that inhibit ecdysis. The Y-organs, a pair of glands in the maxillary somites, secrete hormones derived from dietary cholesterol (three ecdysteroids: ecdysone; 25-deoxyecdysone; and 3-dehydroecdysone) that stimulate molting (under the control of the X-organ). During postmolt and intermolt, the X-organ/sinus gland complex liberates a neuropeptide molt-inhibiting hormone, thus regulating the length of the intermolt period (see Lachaise *et al.*, 1993). Under appropriate external or internal conditions (light, temperature, loss of limbs—see Stoffel and Hubschman, 1974), sinus gland hormones are not released and the Y-organ, no longer suppressed, secretes the molt-initiating hormone. This acts on the epidermis to initiate premolt activities that affect most body parts. Glycogen reserves are increased and minerals (e.g., calcium) are resorbed from the exoskeleton and stored. With the secretion of an enzyme that softens the cuticle at its base, it pulls away from the epidermal cells (apolysis), stimulating the formation of a new epicuticle that is impervious to the molting enzyme. At the termination of premolt, the old cuticle splits, the animal emerges from the old exoskeleton (exuvium) including shedding the cuticle covering the gills (see Andrews and Dillaman, 1993), and the soft body within retains the maximum volume due to imbibing of water by the decapod. Tissues grow rapidly; and during the short postmolt period, the new layer becomes rigid as the stored minerals are deposited in the hardening cuticle. Bendell-Young and Harvey (1991) found that crayfishes from acid lakes have carapaces with lower Mg concentrations than those from circumneutral lakes, suggesting that low pH conditions may prevent normal mineralization during molting. After postmolt, the crustacean may enter intermolt; however, complete intermolt probably occurs only in adults, and the rapidly growing (molting) juveniles generally have, at best, a very short period of stability (see Aiken and Waddy, 1992; Wheatly, 1995; Wheatly *et al.*, 1996).

3. Reproductive Physiology

The development and function of gonads and the ontogeny of secondary sexual characters in the dioecious decapods are regulated by hormones. The Y-organ influences the development and maturation of gonads in juveniles. In females, the sinus gland/X-organ complex is important in controlling the ovary. During nonbreeding periods, the sinus gland produces a hormone that inhibits egg development. However, during the breeding season, the blood level of this hormone declines and egg development is subsequently initiated. Development of the testes and male sexual characteristics is controlled by hormones produced in the androgenic gland, a small mass of secretory tissue near the vas deferens (Taketomi *et al.*, 1996). Additional information concerning the reproductive cycle for shrimps and crayfishes is presented below. Even though the source and nature of interspecific and intraspecific chemicals (pheromones) are unknown for crayfishes, studies have shown that these chemical signals play an integral role in the behavior of these decapods, although other studies have not demonstrated this (see review by Bechler, 1995). The role of chemoreception in mediating agonistic behavior, species recognition, and mate location in crayfishes has been documented by Hazlett (1985). Discrimination between male and female conspecifics using chemical cues has been reported by some authors; these pheromones may promote sex recognition and induce mating behavior (see Hazlett, 1985; Dunham and Oh, 1996). Tierney *et al.* (1984) determined that the site of pheromone reception in *Orconectes* (*Crockerinus*) *propinquus* (Girard) is the lateral antennular flagellum (exopod) bearing the chemoreceptive sensilla (aesthetascs).

4. Other Physiological Adaptations

As ectotherms, these decapods are subjected to temperature changes that directly affect metabolic rate, and most species demonstrate various thermal tolerances (most upper tolerances range from 32 to 36°C) and preferences for their aquatic medium. Unfortunately, upper lethal temperatures are known only for a few species (see Cox and Beauchamp, 1982, and references therein). Nelson and Hooper (1982) showed experimentally that *Palaemonetes kadiakensis* exhibited an acute thermal preference (short-term response) within a temperature range of 6–10°C below their maximum tolerance limits. Taylor (1984) determined for three southeastern United States crayfishes that both the acute preference and final preferendum primarily represented the temperature that they encountered during daily or seasonal activity cycles. He hypothesized that temperature choice does not function in temperature regulation except in the very broadest sense; instead, it serves as an "error detection system," and cues for behavioral arousal from either seasonal hibernation or from their daily refugia. Clearly, water temperature plays an important role in the ecology of these crustaceans (see Mundahl and Benton, 1990).

Biological rhythms are recognized in decapods and reviewed by DeCoursey (1983). Not only are organismic cycles pronounced (e.g., crayfishes have daily locomotor rhythms) but cellular and physiological rhythms are prominently exhibited (e.g., circadian rhythmicity of heart rate, Pollard and Larimer, 1977). Most seem to be in response to environmental cues (e.g., light, temperature); however, even where rhythmic environmental input may be lacking, circadian clock regulation can be maintained. For instance, Weingartner (1977) demonstrated a circadian activity pattern in the cave crayfish *Orconectes* (*Orconectes*) *i. inermis* Cope from Indiana. In a similar way, the Kentucky stygobite *0.* (*Orconectes*) *pellucidus* (Tellkampf) has retained a circadian clock regulation of locomotor activity and oxygen consumption in spite of isolation from day–night cycles.

Crayfishes and shrimps are faced with the problems of constant influx of water by osmosis and loss of salts by diffusion (Flynn and Holliday, 1994). Maintenance of constant body fluid composition is accomplished by active sodium uptake via the gills, production of a copious hypotonic (dilute) urine, and by some reduction in boundary membrane permeability (see Newsom and Davis, 1991; Dickson *et al.*, 1991; Sarver *et al.*, 1994). As indicated above, the green gland is intimately associated with the circulatory system.

The chromatophore is an integument cell with branched, radiating, noncontractile processes that are associated with one or more types of pigment granules. These white, red, yellow, blue, brown, or black pigments can flow into the processes or may be confined to the center of the cell. A single chromatophore may possess any number of pigments and is accordingly classified as mono-, bi-, di-, or polychromatic (see Fig. 9). In shrimps, the number of pigments and the number of chromatophores can change (morphological color change). However, physiological color change (rapid color adaptation to background resulting from dispersal and concentration of pigments within the chromatophores) is the more common type of color alteration and is controlled by hormones and the central nervous system. Chromatophorotropins are released from both the sinus gland and the central nervous system. The latter elaborates several chromatophorotropins, resulting in pigments either dispersing or concentrating in the chromatophores (darkening or blanching of body color).

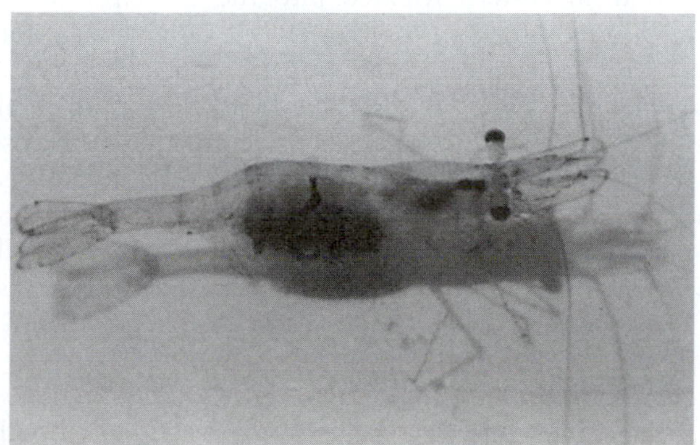

FIGURE 9 Female *Palaemonetes kadiakensis* with eggs (note chromatophores, e.g., small speckles on abdomen).

Color morphs of shrimps and crayfishes have been documented in a number of species, such as *Macrobrachium rosenbergii* (deMan), *Orconectes propinquus*, *Procambarus* (*Ortmannicus*) *a. acutus* (Girard), *Procambarus* (*Scapulicambarus*) *paeninsulanus* (Faxon). Studies have demonstrated that color variations may be due to genetic differences (e.g., Black, 1975) or they can be environmentally induced (see Fitzpatrick, 1987b; Thacker *et al.*, 1993).

Increasing awareness of human impact on the environment has directed attention to the effects of various pollutants on the physiological ecology of crustaceans, including shrimps and crayfishes. Accumulation of trace metals occurs through runoff or direct dumping of sewage or other effluents into aquatic systems. Lead has become particularly important because of its relative toxicity and increased environmental contamination via automobile exhaust and highway runoff. Anderson (1978) demonstrated for *0.* (*Gremicambarus*) *virilis* (Hagen) that exposure to lead resulted in damage to gills, which caused a decrease in the uptake of oxygen; ventilation rates increased as ambient lead concentrations increased. Hubschman (1967) noted that the toxic effect of copper on crayfishes is dependent not only on concentration and duration of exposure but also on the age (or size) of the crustacean (see Maranhão *et al.*, 1995). At concentrations above 1 mg/L, respiratory enzymes are inhibited quite rapidly as the mechanism of detoxification is apparently overwhelmed. At concentrations below 1 mg/L, copper has a chronic effect on cell maintenance. For example, actively secreting cells such as those of the antennal gland apparently are destroyed because cell repair cannot keep pace with cellular activity (see also Taylor *et al.*, 1995). Cadmium, a highly toxic metal, has been implicated as both a carcinogen and a mutagen and causes a number of cardiovascular problems (all resulting from enzyme dysfunction). Cadmium is bioaccumulated and thus when incorporated in tissues of freshwater decapods can have far-reaching effects within aquatic ecosystems as well as in humans (e.g., Dickson *et al.*, 1982; Thorp and Gloss, 1986; Bendell-Young and Harvey, 1991; Naqvi and Howell, 1993a, b). Vermeer (1972) noted high mercury levels in muscle tissue of *0. virilis* from several localities in Manitoba and Ontario and suggested that this species is a good indicator of mercury contamination in aquatic ecosystems (see also Wright and Welbourn, 1993). Insecticides (Albaugh, 1972), herbicides (Naqvi and Leung, 1983), lampricides, and numerous other chemicals have been shown to have various detrimental effects on freshwater decapods.

The bioaccumulation of metals and other toxic or persistent chemicals poses particular problems for long-lived cavernicoles (stygobites)(see below, Cooper and Cooper, 1978). Dickson *et al.* (1979) showed higher concentrations of cadmium and lead in tissues of *Orconectes* (*Orconectes*) *a. australis* (Rhoades) (a stygobite) than in those of *Cambarus* (*Erebicambarus*) *tenebrosus* [a stygophile—this species now inclusive of *C. cahni* (see Hobbs, 1989), *C. laevis,* and *C. ornatus* (see Taylor, 1997)], crayfishes occurring syntopically in a Tennessee cave. It is possible that some stygobitic decapod populations could be extirpated not from a single, massive dose of a chemical (though this has occurred; see Bechler, 1983; Hobbs III, 1987), but slowly from long-term exposure to low levels of these metals or other toxic substances. Because the longevity of primary burrowing species is apparently greater than that for nonburrowing taxa living in surface waters, a similar threat from chronic pollution exists for these crustaceans. This is particularly true for crayfishes living in roadside ditches that are periodically sprayed with various pesticides (Hubschman, 1967; Dimond *et al.*, 1968).

III. ECOLOGY AND EVOLUTION

A. Diversity, Distribution, and Evolutionary Relationships

Most freshwater decapods of North America occur in one or more of the following habitats: shallow (1–2 m depths, although this is probably a function of sampling bias) lentic and lotic waters, epigean and hypogean habitats, oligotrophic to hypereutrophic lakes, ponds, marshes, ditches, steep-gradient headwater streams, low-gradient large rivers, springs, stygean streams and pools, and terrestrial burrows leading to groundwater. Individuals have been found in lakes or springs at depths of 30–90 m (stygobitic crayfishes in submerged caves in north-central Florida are observed by divers at depths greater than 30 m) but crayfishes certainly thrive in shallow, littoral areas. They are most abundant in the central, eastern, and southern regions of the United States, while a few species are found in the Pacific slope drainages; they are rare or absent in the large area of the western Great Plains and the Rocky Mountains (see Taylor *et al.*, 1996). This simply means that American crayfishes are crustaceans possessing biological plasticity. They have become established in nearly every type of freshwater habitat, except glacial and thermal effluents, and some can tolerate brackish water. Considerable data suggest that habitat has a channelizing influence (convergence) on certain morphological aspects of crayfishes (Hobbs, 1976). For example, stygobites (a polyphyletic group) are characterized by an albino-sightless combination and attenuated appendages; they are classic K-strategists and demonstrate extreme resource efficiency. In contrast, "burrowers" (also polyphyletic) possess short, broad, flattened chelipeds, lack carapace spines, have a reduced abdomen and tail fan (length, width, and height), and possess an enlarged gill chamber (an adaptation to the oxygen-poor burrow habitat—see above) (Fig 7). Water currents have certainly influenced the body form of crayfishes, with those species dwelling in fastest flowing water having depressed (dorsoventrally flattened—most common) or compressed (laterally deflated) bodies; these include, for example, *Orconectes* (*Gremicambarus*) *compressus* (Faxon) and *Procambarus* (*Tenuicambarus*) *tenuis* Hobbs. Spination is also typically reduced in such species. Those crayfishes inhabiting riffle areas of streams and residing in pebble and cobble substrates (rubble dwellers) are markedly depressed dorsoventrally; examples of such taxa are *Cambarus* (*Jugicambarus*) *friaufi* Hobbs and *Cambarus* (*J.*) *brachydactylus* Hobbs.

Several decapods (Infraorder Brachyura—crabs) occur sporadically (naturally or introduced) in coastal brackish or freshwater systems, but are not included in the following discussions, numerical summaries, or key. They are represented by the portunid blue crab, *Callinectes sapidus* Rathbun, a xanthid mud crab, *Rhithropanopeus harrisii* (Gould) (found in streams of the Atlantic and Gulf coasts and introduced along the coasts of California and Oregon), and a grapsid river crab, *Platychirograpsus spectabilis* Rathbun, introduced from eastern Mexico into the Hillsboro River near Tampa, Florida (Powers, 1977). Of curious interest is the occurrence of the grapsid crab, *Hemigrapsis estellinensis* Creel, in Estelline Salt Spring in Hall County, Texas. This representative of the otherwise marine family Grapsidae is endemic (probably a Pleistocene relict) to the spring which is located 800 km from the sea at an elevation of 531 m (see Reddell, 1994). For further information on cavernicolous crabs, refer to Guinot (1994).

1. Shrimps

The atyid shrimps (fingers of chelate pereiopods with apical tufts of setae, Fig. 1B) of North America are restricted to two genera, one with two southeastern, obligate cave species and the other represented by two species found in coastal streams of California. Freshwater palaemonids (fingers of chelate pereiopods lacking apical tufts of setae, Fig. lA) are a very successful family of shrimps; the genus *Palaemonetes* contains approximately 17 species in the western hemisphere, and *Macrobrachium* is assigned about 100 species worldwide (35 in the western hemisphere). These genera are represented in North America by 10 and 12 species, respectively. Shrimps are generally found in the macrophyte-rich littoral zone of lakes or in sluggish, vegetation-choked streams. Within these two families residing in the confines of the United States, five shrimps are stygobitic and are very restricted in distribution (two species known to occur only in a single karst locality).

The North American stygobitic decapods (inclusive of Bahamas, Barbuda, Belize, Bermuda, Caicos Islands, Canada, Cuba, Dominican Republic, Guatamala, Hispaniola, Honduras, Isla de Pinos, Isla Mona, Jamaica, Mexico, Puerto Rico, and the United States) are represented by 36 shrimps assigned to five families and 16 genera:Agostocarididae [*Agostocaris* (2)], Alpheidae [*Automate* (1), *Potamalpheops* (1), *Yagerocaris* (1)], Atyidae [*Palaemonias* (2), *Typhlatya* (8)], Hippolytidae [*Barbouria* (1), *Calliasmata* (1), *Janicea* (1), *Somersiella* (1)], and Palaemonidae [*Cryphiops* (2), *Creaseria* (1), *Macrobrachium* (3), *Neopalaemonon* (1), *Palaemonetes* (3), *Troglocubanus* (6)], Procarididae [*Procaris* (1)] (Hobbs III, 1998).

a. Atyidae North American freshwaters are inhabited by three federally endangered and one probable extinct atyid species:*Palaemonias ganteri* Hay, *P. alabamae* Smalley, *Snycaris pacifica* (Holmes), and *S. pasadenae* (Kingsley), respectively (Hedgpeth, 1968). The first two taxa are stygobitic shrimps, the former restricted to Mammoth Cave in Kentucky, and the latter is known only from five caves (three groundwater basins) in Madison County, northeastern Alabama (Jacobson and Hartfield, 1997; McGregor *et al.*, 1997). Because most atyids currently inhabit warmer southern aquatic ecosystems, it is probable that these two stygobites are thermophilic relicts of a former, more widely distributed common ancestor. *Syncaris pacifica* and S. *pasadenae* are found in small coastal streams of California. The former is reported from Marin, Napa, and Sonoma counties and the latter is previously known from Los Angeles, San Bernardino, and San Diego (may be erroneous report) counties but is likely extinct due to habitat destruction and over-collecting.

b. Palaemonidae Those North American palaemonids that occur within the United States and that

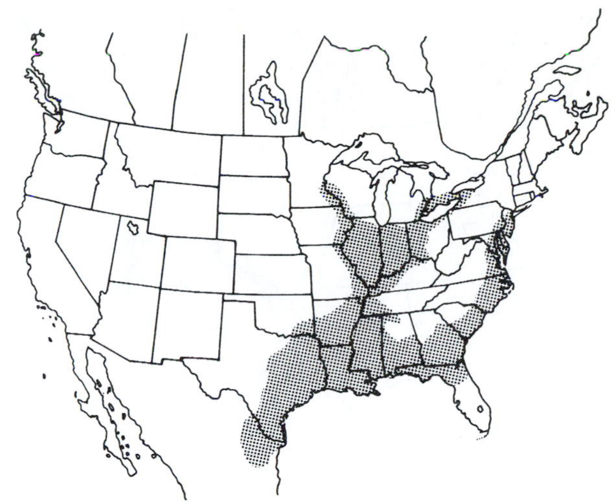

FIGURE 10 Geographical distribution of *Palaemonetes kadiakensis* and *P. paludosus*.

are strictly freshwater inhabitants are represented by 10 species belonging to the genus *Palaemonetes* (three are stygobitic): *Palaemonetes antrorum* (Benedict) from wells and caves in Hays and Uvalde counties, Texas (federally endangered), *P. holthuisi* (Strenth) from a cave in Hays County, Texas (the only known site in the United States where two species of stygobitic shrimps are found together), and *P. cummingi* Chace from a cave in Alachua County, Florida. *Palaemonetes cummingi* is listed as "vulnerable" (threatened) on the 1996 IUCN Red List of Threatened Animals (Baillie and Groombridge, 1996). Seven are inhabitants of lentic and sluggish lotic systems, including springs (see Strenth, 1976, 1994; Wu and Brown, 1980) (Fig. 10—distribution map does not include disjunct populations such as *P. paludosus* in Tygart Valley River, Randolph County, West Virginia—population probably introduced). In addition, five species assigned to the widespread genus *Macrobrachium* (North, Central, and South America, and the West Indies) are found in freshwaters in the middle and lower Mississippi River drainage and in low-lying streams from New Jersey to Texas (Figs. 11 and 12). *Macrobrachium acanthurus* (Wiegmann) ranges from Georgia to Brazil (also Bahamas) and is found in the United States from Georgia and Florida to Texas along the Gulf coast. *Macrobrachium carcinus* (Linnaeus) is the largest and most spectacular species and is found from Florida to Texas and south to Brazil and the West Indies. *Macrobrachium ohione* (Smith) is the only endemic species of the genus in North America and is found from Virginia to Texas and in the lower and middle Mississippi River drainage (Fig. 11). The natural range of *M. olfersii* (Wiegmann) is from Mexico to Brazil (Chace

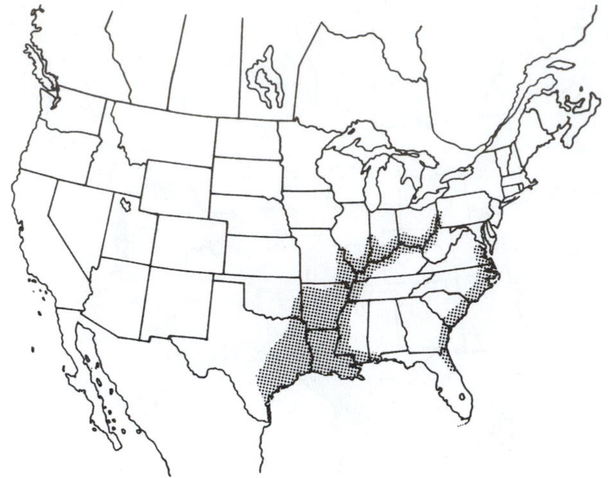

FIGURE 11 Geographical distribution of *Macrobrachium ohione.* Gross reduction in populations throughout its range is attributable to loss of suitable habitat.

and Hobbs, 1969); however, it likely has been introduced into the St. Augustine, Florida area (Hedgpeth, 1949). As referred to above, the giant Malaysian prawn, *M. rosenbergii,* has been introduced into ponds of central Florida, South Carolina, and numerous other parts of the world (e.g., Israel, Jamaica) (Smith *et al.,* 1978).

Strenth (1976), in a hypothetical discussion of the origin and dispersal of *Palaemonetes* and *Macrobrachium,* proposed that freshwater *Palaemonetes* may have arisen during the late Mesozoic or early Cenozoic and that the widely disjunct species of the genus are largely of monophyletic or of limited polyphyletic origin. He presented salinity tolerance data from laboratory experiments supporting the premise that worldwide dispersal of freshwater *Palaemonetes* over large expanses of open ocean would have been unlikely. Results of his studies on *P. kadiakensis,* coupled with those of others for *P. paludosus,* demonstrate that this group of shrimps cannot osmoregulate in salinities greater than 25 ppt. Thus, the distributional patterns of *Palaemonetes* and *Macrobrachium* appear consistent with the theories of plate tectonics and continental drift. He proposed that *Palaemonetes* preceded *Marcobrachium* into the Gulf of Mexico area and that *Macrobrachium* (primarily juveniles), through competitive exclusion, barred *Palaemonetes* from the middle latitudes. Most of the freshwater species of *Palaemonetes* are found above or below the 30° latitudes; or, if found between, they always occur at higher elevations or are

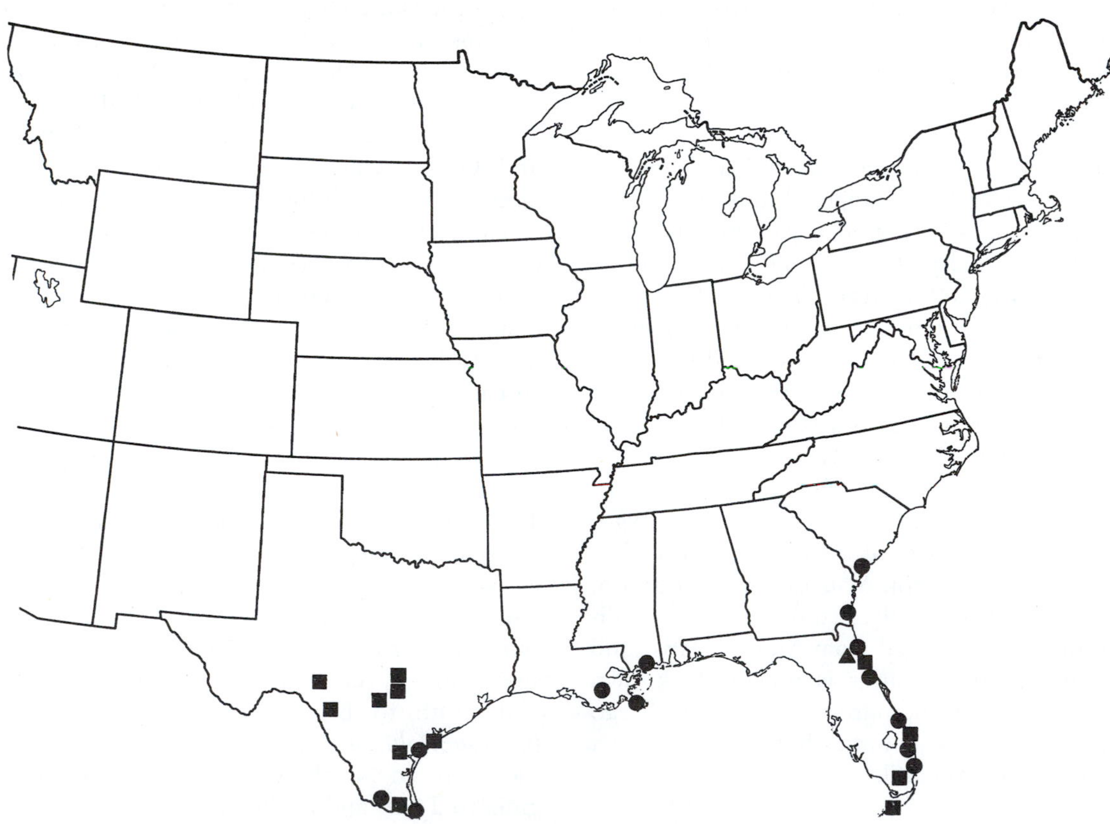

FIGURE 12 Geographical distribution of *Macrobrachium acanthurus* (filled circles), *M .carcinus* (filled squares), and *M. olfersi* (filled triangles) (after Hedgpeth, 1949).

located at considerable distances from the coast. Parameters that appear to limit the further spread of *Macrobrachium* are current velocity, salinity, temperature tolerances, and anthropogenic perturbations.

2. Crayfishes

The freshwater astacideans (superfamily Astacoidea) constitute a very successful group and currently are represented by 393 described species and subspecies (4 extinct species), 342 species and subspecies which are restricted (introductions ignored) to North America, north of Mexico (see Hobbs, 1986 for lineages of the major groups in North America). These are assigned to twelve genera and two families (Table I), of which species richness sustains a progressive decline with increasing latitude from 30–60°N (France, 1992). The families, Astacidae and Cambaridae, have historically occupied allopatric ranges. The former inhabits the Pacific slope drainages, except for *Pacifastacus* (*Hobbsastacus*) *gambelii* (Girard) which has crossed the divide into the upper Missouri; and the Cambaridae occurs in some Hudson Bay watersheds, the Atlantic, and Gulf drainages from southern Canada to Honduras, certain Pacific slope basins in Mexico, and Cuba (Fig. 13). The three most successful genera of crayfishes are *Procambarus, Cambarus,* and *Orconectes* (166, 86, 81 species and subspecies, respectively—see below). The coastal plain of the southern states is dominated by *Procambarus,* whereas members of the genus *Cambarus* (Fig. 13) are major components of aquatic environments from the Cumberland Plateau to the North and East. West and northwest of the Cumberland Plateau, species of the genus *Orconectes* are the most commonly encountered crayfishes. Hobbs (1989) produced a useful listing of the crayfish faunas by country and state and included references treating species within each state. Two recent publications that discuss decapods in the northeastern United States are those of Smith (1988) and Peckarsky *et al.* (1990), and the following contemporary publications present data for decapods within various states and Canada:Guiasu *et al.* (1996)—Ontario, Canada; Deyrup and Franz

TABLE I List of Freshwater Decapod Genera and Subgenera Occurring in the United States

Infraorder	*Family*	*Subfamily*	*Genus*	*Subgenus*	*N. A.*	*U. S.*
Caridea (Shrimps)	Atyidae	———	*Palaemonias**	———	4	4
			Syncaris	———	2	2
	Palaemonidae	———	*Palaemonetes**	———	10	6
		———	*Macrobrachium*	———	12	5
Shrimp Totals					28	17
Astacidea (Crayfishes)	Astacidae	———	*Pacifastacus*	*Hobbseus*	5	5
		———	*Pacifastacus*	*Pacifastacus*	3	3
	Cambaridae	Cambarellinae	*Cambarellus*	*Cambarellus*	9	0
			Cambarellus	*Dirigicambarus*	1	1
			Cambarellus	*Pandicambarus*	7	7
		Cambarinae	*Barbicambarus*	———	1	1
			Bouchardina	———	1	1
			*Cambarus**	*Aviticambarus**	3	3
				Cambarus	7	7
				Depressicambarus	16	16
				*Erebicambarus**	5	5
				Exilicambarus	1	1
				*Glareocola***	1	1
				Hiaticambarus	9	9
				*Jugicambarus**	23	23
				Lacunicambarus	4	4
				*Puncticambarus**	15	15
				Tubericambarus	1	1
				Veticambarus	1	1
			Distocambarus	*Distocambarus*	2	2
				Fitzcambarus	3	3
			Fallicambarus	*Creaserinus*	9	9
				Fallicambarus	7	7
			Faxonella	———	4	4
			Hobbseus	———	7	7

(Continues)

TABLE I (Continued)

Infraorder	*Family*	*Subfamily*	*Genus*	*Subgenus*	*N. A.*	*U. S.*
			*Orconectes**	*Billecambarus*	1	1
				Buannulifictus	6	6
				Crockerinus	14	14
				Faxonius	3	3
				Gremicambarus	5	5
				Hespericambarus	7	7
				*Orconectes**	8	8
				Procericambarus	25	25
				Rhoadesius	2	2
				Tragulicambarus	1	1
				*Trisellescens***	9	9
			*Procambarus**	*Acucauda*	1	1
				*Austrocambarus**	20	1
				Capillicambarus	3	3
				Girardiella	19	18
				Hagenides	10	10
				*Leconticambarus**	14	14
				*Lonnbergius**	2	2
				Mexicambarus	1	0
				*Ortmannicus**	56	50
				Paracambarus	2	0
				Pennides	19	18
				Procambarus	1	0
				*Remoticambarus**	1	1
				Scapulicambarus	6	5
				Tenuicambarus	1	1
				Villalobosus	10	0
			*Troglocambarus**	———	1	1
Crayfish Totals					393	342

Summary listing: [refer also to Hedgpeth (1949), Strenth (1976), and Hobbs (1989)].

Infraorder Caridea (shrimps)
- Family Atyidae
  - Genus *Palaemonias*—4;
  - Genus *Syncaris*—2;
- Family Palaemonidae
  - Genus *Palaemonetes*—10;
  - Genus *Macrobrachium*—12;

Infraorder Astacidea (crayfishes)
- Family Astacidae
  - Genus *Pacifastacus*—2 subgenera and 8 species;
- Family Cambaridae
  - Subfamily Cambarellinae
    - Genus *Cambarellus*—3 subgenera, 17 species;
  - Subfamily Cambarinae
    - Genus *Barbicambarus*—monotypic;
    - Genus *Bouchardina*—monotypic;
    - Genus *Cambarus*—12 subgenera and 86 species;
    - Genus *Distocambarus*—2 subgenera and 5 species;
    - Genus *Fallicambarus*—2 subgenera and 16 species;
    - Genus *Faxonella*—4 species;
    - Genus *Hobbseus*—7 species;
    - Genus *Orconectes*—11 subgenera and 81 species and subspecies;
    - Genus *Procambarus*—16 subgenera and 166 species and subspecies; only 12 subgenera and 124 species and subspecies in North America north of Mexico [*Procambarus* (*Austrocambarus*) *primaevus* (Packard) is an extinct species known only from tertiary deposits in Wyoming and is not included in these figures];
    - Genus *Troglocambarus*—monotypic.

Respective numbers of species in North America (inclusive of Belize, Canada, Cuba, Guatemala, Honduras, Isla de Pinos, Mexico, and the United States) are included. Data are presented only for those genera that are assigned U. S. species (* denotes genera/subgenera having stygobitic species; ** see Bouchard and Bouchard, 1995).

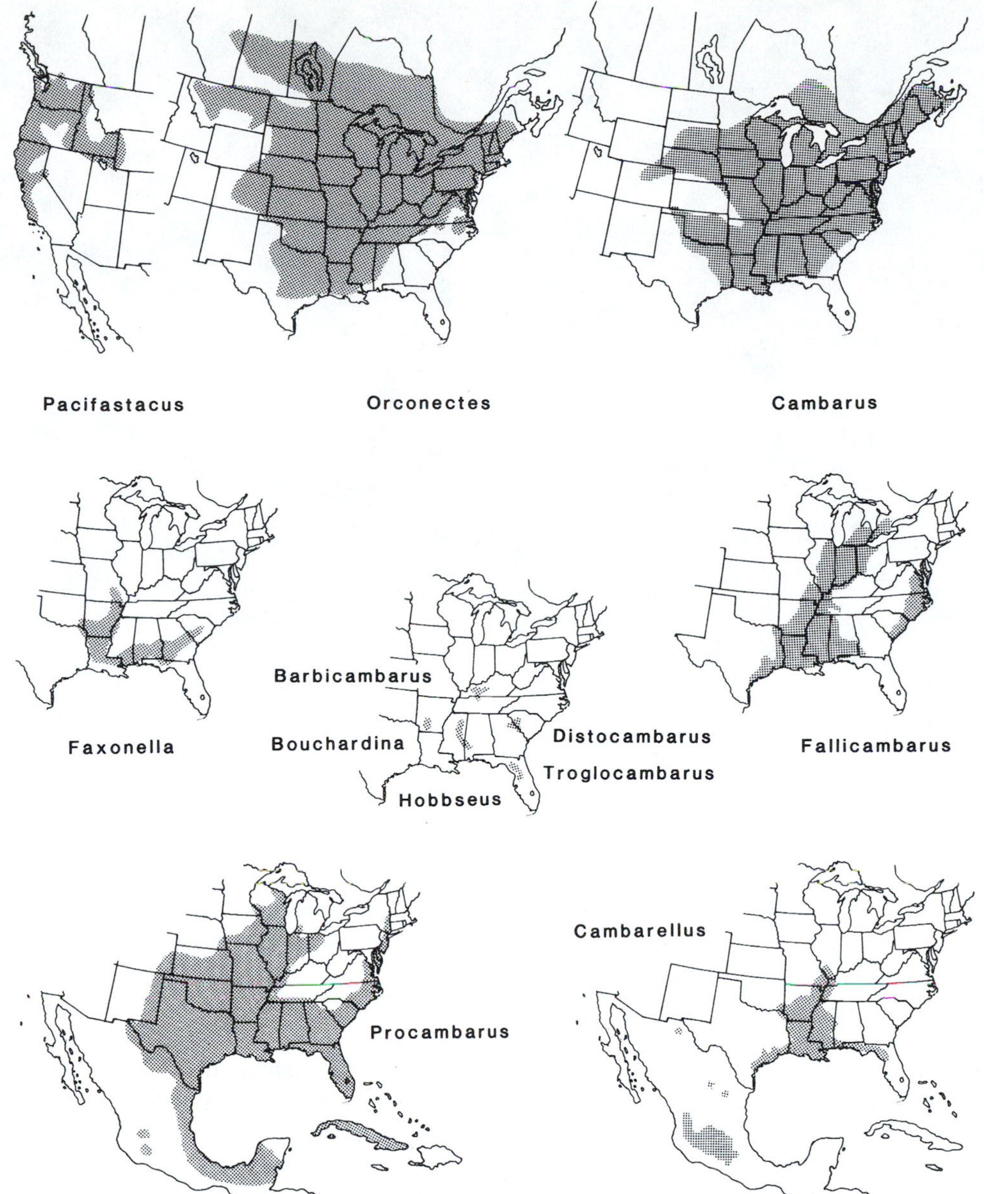

FIGURE 13 Geographical distribution of North American crayfish genera (not inclusive of introductions) (modified from Hobbs, 1988).

(1994) and Franz *et al.* (1994)—Florida, Georgia; Page (1985)—Illinois; Page and Mottesi (1995)—Indiana; Martin (1997)—Maine; Pflieger (1996)—Missouri; Cooper and Braswell (1995)—North Carolina; Eversole (1995)—South Carolina; Jezerinac *et al.* (1995)—West Virginia; and Hobbs III and Jass (1988)—Wisconsin. The reader should refer to Taylor *et al.* (1995) for a more complete listing of publications having state crayfish data.

Many epigean crayfishes occur abundantly in large interstices that afford protection and concealment (dark thigmotactic cover; Alberstadt *et al.*, 1995), particularly during the daytime hours, such as among stones, weed beds, leaf litter, brush piles, logjams, roots of riparian trees; and they often find shelter in cans, tires, or among other anthropogenic debris. The genus *Cambarellus* in North America is distributed below the Fall Line zone in the Gulf Coastal Plain. Whereas a few species are known only from lotic habitats, most members of this genus, as well as those of *Faxonella* and *Bouchardina,* frequent lentic habitats such as ponds, backwaters of streams, and roadside ditches. These small crayfishes are successful in littoral waters subject to low oxygen concentrations and elevated temperatures. The

FIGURE 14 Pool habitat for the stygobitic crayfish *Orconectes a. australis* in Russell Cave, Jackson County, AL.

members of the genus *Hobbseus* also occupy such habitats, but they are more common in very small, shallow streams. The monotypic *Bouchardina* probably lives in similar habitats since it has been collected from leaf litter and aquatic vegetation in a single stream backwater ditch. The monotypic *Barbicambarus* is found only under large limestone rocks in the swift currents of large streams in the Green and Barren river basins of the Highland Rim. Of note, introductions of nonnative species into various parts of the world are leading to species displacements among crayfishes (see below; Hobbs III *et al.,* 1989; Olson *et al.,* 1991; Clancy, 1997).

Hypogean species are represented by burrowing and cave-dwelling crayfishes. Primary burrowers, which are found in the genera *Cambarus, Distocambarus, Fallicambarus,* and *Procambarus* (Fig. 13), construct tunnels (Fig. 22a, d) in such habitats as seepage areas, along stream banks, in open fields, and in other low-lying areas. Among the stygobitic species are members of the genera *Cambarus, Orconectes, Procambarus,* and *Troglocambarus;* some of these cavernicolous crayfishes occupy lotic habitats while others appear to have become adapted to lentic situations. Those species inhabiting caves with active streams generally are found in pooled areas where they have abandoned their preferred, and now less abundant, cover of cobble-gravel in favor of accumulated allochthonous, food-rich, silty substrates (Hobbs III, 1976) (Fig. 14). Some stygobitic crayfishes living in sluggish lotic/lentic systems (e.g., caves developed in the porous Ocala limestone of northcentral Florida) are found near cave entrances where organic debris is available; they are generally positioned on the walls and silty substrates of these submerged solution passages (Hobbs *et al.,* 1977; Hobbs and Franz, 1986). Franz and Lee (1982) showed that the geographical and ecological distributions of the cavernicolous crayfishes of Florida are correlated with low or high energy (food) cave systems. One species, *Troglocambarus maclanei* Hobbs, is often found on walls or clinging pendant to ceilings of submerged caves. The stygobitic crayfishes are derived from surface ancestors that entered caves during late Tertiary (Hobbs *et al.,* 1977; see also Hobbs and Franz, 1986; Holsinger, 1988; Hobbs III, 1994, 1999).

a. Astacidae A single genus, *Pacifastacus,* is represented by eight species and subspecies in the Pacific drainages of western North America and headwaters of the Missouri River in Wyoming (Fig. 13). One species, *P.* (*Hobbsastacus*) *chenoderma* (Cope), is extinct, known only from Miocene and Pliocene (?) deposits, and is not included in Table I. *P.* (*H.*) *nigrescens* (Stimpson) is probably extinct and *P.* (*H.*) *fortis* (Faxon) populations have declined significantly (Light *et al.,* 1995) and it is listed as a Federal Endangered Species.

These crayfishes are found in lentic and lotic habitats ranging from alpine lakes and streams to streams crossing cold deserts. They occur in California, Idaho, Montana, Nevada, Oregon, Utah, Washington, Wyoming, and British Columbia (Fig. 13). Hobbs (1974) suggested that two evolutionary lines arose from a Mesozoic marine nephropoid ancestral stock. One line (represented today by the genera *Astacus, Austropotamobius,* and *Pacifastacus*) moved into the freshwater habitat and maintained the primitive characteristic of absence of cyclic dimorphism in males, and females exhibited a sclerite (annular plate) lacking both a sinus and fossa. A second, less conservative stock also made the transition to freshwaters (represented by extant members of the family Cambaridae). These males evolved cyclic dimorphism and associated characters (e.g., hooks on ischia of certain pereiopods in males), while the females developed an annular sclerite with a sinus or fossa for sperm storage (the annulus ventralis).

b. Cambaridae As previously stated, the ancestral stock, in which cyclic dimorphism became established, moved into freshwater habitats and gave rise to the Cambaridae. This family consists of three subfamilies, two of which (Cambarellinae and Cambarinae) occur in North America (Fig. 13). The third, Cambaroidinae, is restricted to eastern Asia-Amur Basin, Korea, and Japan, and is treated only briefly in further discussions.

The following scenario is an abbreviation of arguments presented by Hobbs (1969, 1981, 1989) concerning the origin and dispersal of cambarid decapods. Whether or not the Asian Cambaroidinae and the American subfamilies shared a common freshwater ancestor or whether there were separate Asiatic and American invasions of freshwaters may never be known. Hobbs, however, suggested that the ancestral cambarine stock entered freshwaters of the southeastern portion of the North American continent during the late Cretaceous or early Cenozoic eras. Through adaptive radiation, this ancestral *Procambarus* stock moved through estuarine environments and populated available habitats in many of the streams flowing from the southern and southeastern slopes of the existing continent. Except for rare passive dispersal resulting from stream piracy or from crayfishes moving during periods of low salinity from one stream mouth to another, stocks became isolated in their respective drainage systems.

Before the close of the Miocene (at least 25 million years ago), some stream-dwellers were able to exist in backwaters of lotic habitats; these forms gave rise to the ancestors of some of the lentic species of *Procambarus* (e.g., subgenera *Ortmannicus* and *Scapulicambarus*). These crayfishes moved into an ecological vacuum and were no longer confined to flowing water systems. Pools may have dried, or oxygen concentrations became low, or perhaps individuals simply moved across land from one aquatic habitat to another. Whatever the reason, additional ponds, lakes, ditches, streams, etc., were occupied and thus dispersal into lower elevation habitats occurred; also several stocks penetrated the water table by burrowing. *Distocambarus* occupied the groundwater niche and probably represents a remnant of a much more widely distributed Piedmont stock. The monotypic *Troglocambarus* represents an offshoot of the *Procambarus* stock that moved into the karst groundwaters in northcentral Florida [see below, and Hobbs *et al.* (1977) for further discussion of the evolution of stygobitic decapods]. The genus *Cambarellus* probably arose from a primitive ancestral *Procambarus* stock in the Gulf coastal plain and occupied ephemeral swamps, ponds, and ditches.

One of the lotic ancestral procambarid stocks moved upstream, reaching the vicinity of the Cumberland Plateau, where ancestors to two very successful genera, *Cambarus* and *Orconectes,* became established. The major evolutionary lines were probably established by the Miocene Epoch, and the final stage was set for the refinement of evolutionary patterns begun during the early Cenozoic. Such major features as large rivers, the Appalachian and Ozark mountains, the coastal plain, the Cumberland Plateau, the Nashville Basin, and the Highland Rim influenced the dispersal of these stocks. Superimposed on these were the effects of Pleistocene glaciation.

Representatives of the ancestral *Cambarus* stock were successful in various habitats and spread through lowland areas, giving rise to the species currently assigned to *Fallicambarus.* This genus, which probably arose on the west Gulf coastal plain, developed into primary burrowers. *Barbicambarus* is another descendant of the early *Cambarus;* it moved into sizeable streams with large substrates, rapid currents, and high oxygen concentrations. *Hobbseus* also may have evolved from the cambaroid line (Fitzpatrick, 1977), but it is more likely that it arose from the orconectoid line [Hobbs (1969)].

The genus *Orconectes* probably evolved in the Interior Lowland Plateau province (Kentucky/Tennessee) and radiated virtually in every direction (see Fitzpatrick, 1986, 1987a; Fetzner, 1996). Some orconectoid stock successfully moved into the lowland regions where species arose that are now assigned to the genus *Faxonella.* This genus also seems to have evolved on the west Gulf coastal plain and today these small, tertiary burrowers are found in lentic situations (ponds, ditches). They occur in flood plains and, when the water level rises, are fairly common in the shallow flowing water. *Bouchardina* also is derived from the orconectoid stock and occupies backwater areas. Ancestors of the members of these two genera likely arrived along the margins of the retreating Mississippi embayment during the early Cenozoic.

Thirty-eight stygobite crayfish species and subspecies (refer to Table I) belonging to the family Cambaridae are assigned to the four genera:*Cambarus* (11 species), *Orconectes* (7 species and subspecies), *Procambarus* (19 species and subspecies), and the monotypic *Troglocambarus*. These obligate cave shrimps and crayfishes were derived from surface ancestors that made numerous colonizations of karst waters during late Tertiary. Fluvial barriers would generally have had minimal effects on the dispersal of stygobites (see Barr and Holsinger, 1985, and Holsinger, 1988; also see above and Hobbs *et al.*, 1977 and Hobbs III, 1994 for a more detailed discussion of their evolution and dispersal).

The profound effects of Pleistocene glaciation on the distribution of organisms should be mentioned. The obliteration of immense drainage systems (e.g., Teays) resulted in total habitat destruction for virtually all organisms. The principal zoogeographical effects were the elimination of populations and entire species from the northern portions of North America, with displacements of species farther south than they had occurred preglacially. Retreat of the glaciers resulted in the establishment of new drainage systems (e.g., the Ohio) which undoubtedly were rapidly invaded (re-invaded) primarily from the south and southeast, though some species may have entered from more than one refugium.

(1). Cambarellinae The subfamily Cambarellinae is confined to North and middle America where its members are found in the coastal plain from northern Florida to southern Illinois and Texas (Fig. 13) and, disjunctly, on the Pacific slope and Central Plateau of Mexico. Seventeen species are assigned to three subgenera in the single genus *Cambarellus* (Fitzpatrick, 1983); however, only eight of them are found north of the Rio Grande (Table I). In the United States, eight species are known from the following states:Alabama, Arkansas, Florida, Illinois, Kentucky, Louisiana, Mississippi, Missouri, Tennessee, and Texas (Fig 13); they may have been introduced by humans into Georgia.

(2). Cambarinae The subfamily Cambarinae, which is comprised of 10 genera and 368 species, ranges from New Brunswick across southern Canada to Mexico, Guatemala, Honduras, and Cuba (Fig. 13). Ten genera represented by 326 species occur in North America, north of Mexico (Table I; see Hobbs, 1989 for detailed presentation of genera, subgenera, species, and subspecies of American crayfishes).

B. Reproduction and Life History

1. Shrimps

Like all astacideans, freshwater shrimps have external fertilization, with the ova being fertilized as they are extruded and then attached to the pleopods of females. The two common epigean, freshwater shrimps, *Palaemonetes kadiakensis* (Fig. 9) and *P. paludosus*, have similar life histories, both species living approximately one year (Fig. 15). Although many individuals reproduce only once (semelparous), Nielsen and Reynolds (1977) observed ovigerous females of *P. kadiakensis* from Missouri also bearing mature eggs, indicating at least two broods of young per female (iteroparous); Beck and Cowell (1976) also noted that *P. paludosus* spawned twice. The length of the breeding season varies with latitude and is generally longer in southern localities. For instance, ovigerous females occur in populations of *P. kadiakensis* in Illinois and Michigan from April until August, but are present in Louisiana populations from February to October; ovigerous females of *P. paludosus* occur throughout the year in the Florida Everglades. As a general rule, breeding does not begin until late spring or early summer in more northern areas. Only large females breed early in the season, and they are replaced by smaller females as the reproductive season progresses. Females are larger and usually more abundant than males. Mature (egg-bearing) females range from 20–49 mm in total length and produce one or two broods (two is particularly representative of populations of *P. paludosus* in Florida where the growing season is longer). The number of eggs produced per female is highly variable, ranging from 8 to 160, with fecundity generally a linear function of total length. The incubation period depends on water temperature and varies from 12 to 24 days; zoeae larval lengths are approximately 4 mm at hatching and these free-swimming larvae pass through six stages (varies from 3 to 8) in about three weeks. Accounts of larval development and postembryonic growth in these two species are summarized by Hubschman and Rose (1969), Beck and Cowell (1976), Beck (1980), Kushlan and Kushlan (1980), Page (1985), and Hobbs III and Jass (1988). Shrimp usually mature when they attain a total length of 20 mm although this is not necessarily the case. Most shrimp die after reproducing, but some females molt within three days (generally within 24 h) after young are released and can later produce another brood. Generally, adults disappear from the population in late summer or early

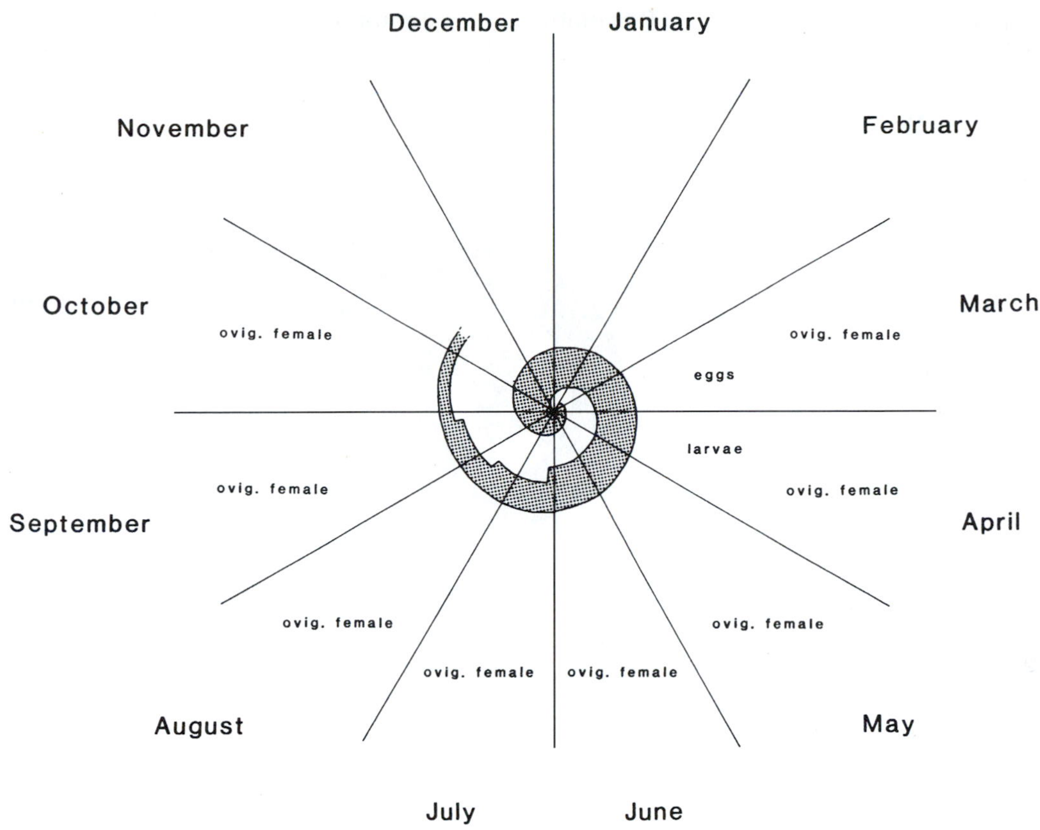

FIGURE 15 Generalized life history of *Palaemonetes* (see text for explanation).

fall, having a life span of approximately one year. Marine and brackish water *Palaemonetes* (e.g., *P. intermedius* Holthuis, *P. pugio* Holthuis) usually have smaller and more numerous eggs per female than do freshwater species. The production of fewer but larger eggs may enable the freshwater larvae to hatch at an advanced stage with a corresponding reduction in the number of free-swimming larval stages.

Dobkin (1971) described three larval stages in the stygobite, *P. cummingi,* and studies of atyids consist of observations made on *Palaemonias ganteri* in Mammoth Cave, Kentucky (Hobbs *et al.,* 1977; Leitheuser *et al.,* 1985); the Alabama cave shrimp, *P. alabamae,* conducted by Cooper (1975) and McGregor *et al.* (1997); and *Syncaris pacifica* (Eng, 1981). The study of the life history of *Macrobrachium ohione* has not received as much attention as have the two taxa of *Palaemonetes* previously discussed. This species is larger (total length of ovigerous females ranging from 27 to 93 mm) and it lives a maximum of two years (Fig. 16). [Huner (1977) determined that *M. ohione* at Port Allen, Louisiana lived up to two years.] Fecundity is relatively enormous, the number of eggs varying from 6273–24,000 per female (Truesdale and Mermilliod, 1979). Ovigerous females are known only during May in Illinois, from March through September in Louisiana, and during March in Texas. Generally, females are larger than males and only the largest individuals in the population bear eggs. Additional ecological data can be obtained from Truesdale and Mermilliod (1979), Page (1985), and references therein.

2. Crayfishes

Surprisingly few crayfishes have been studied carefully for extended periods. Although considerable data are available for many species (e.g., Penn, 1943; Fielder, 1972; Hamr and Berrill, 1985; Corey, 1990; Mitchell and Smock, 1991; Norrocky, 1991; Johnston and Figiel, 1997; see also Hart and Clark, 1987), most are observations from isolated collections. Astacid life-history investigations are limited to *Pacifastacus leniusculus leniusculus* and *Pacifastacus* (*Pacifastacus*) *leniusculus trowbridgii* (Stimpson) (see references in Mason, 1970a, b; McGriff, 1983a). Life-history studies of the more diverse cambarids have been conducted on only about 20 species.

In a general sense, the life history of a crayfish can be divided into several phases (Momot, 1984), each having a slightly different component important to the completion of the life history of a given cohort as well

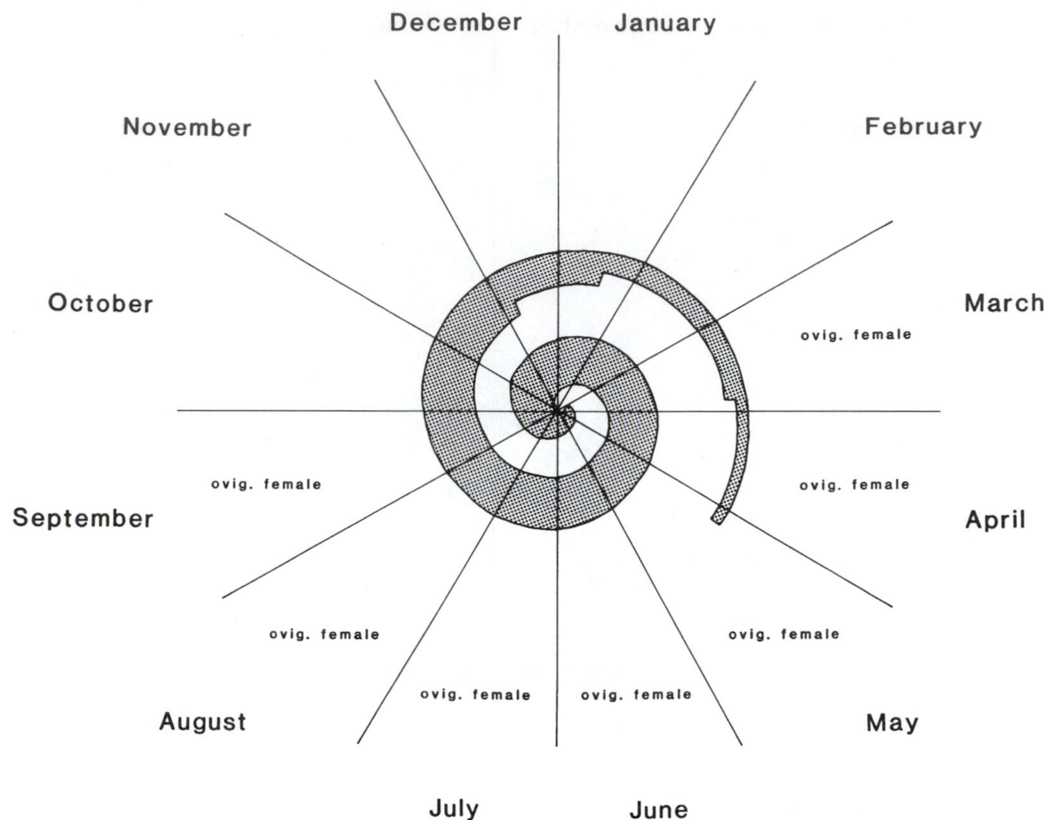

FIGURE 16 Generalized life history of *Macrobrachium ohione* (see text for explanation).

as to the long-term success of the species. Recently hatched juveniles search lake littoral and stream riffle areas for the best food and shelter available (see Figiel *et al.,* 1991). Rapid growth of young-of-the-year (YOY) allows a partial escape from the devastating effects of predation (Gowing and Momot, 1979). As juveniles grow larger, they abandon littoral and riffle areas for deeper waters, only to return after reaching adult size; this habitat segregation is probably related, in part, to such factors as competition with adults and vulnerability to predators. During postembryonic development, food conversion is expressed in maximum growth. Adulthood represents a conservative phase in which energy is shunted from rapid growth toward a periodic maximization of egg output among several cohorts per year. Rapid growth and shorter lifespan are reported to be more common in crayfishes in lower latitudes (Momot, 1984; Hazlett and Rittschof, 1985); however, many "northern" species demonstrate these same characteristics.

Unlike the cambarids, astacid males do not exhibit sexual dimorphism. Copulation in *Pacifastacus l. trowbridgii* coincides with the autumn decline of water temperature and is soon followed by oviposition (late October and early November) (Fig. 17). Females carry 90–251 eggs (number directly proportional to mass) on their abdomens, and hatching generally occurs after 7–8 months, during April to June. After the second molt, the young leave the mother as 3- or 4-week old juveniles. During the first year, individuals may undergo as many as 11 molts and nearly triple in length. Some individuals live for eight years, and a few become sexually mature at age three; the majority, however, become part of the breeding population during the fall of their fourth year. This certainly follows the trend suggested by Momot (1984) that crayfishes occurring at higher latitudes and in colder environments usually live longer and mature later. The smallest ovigerous female observed by Mason (1974) was 60 mm total length and the largest 102 mm. This suggests that female astacids lay eggs more than once (iteroparous) and, indeed, may spawn as many as three or four times (Fig. 17). Refer to McGriff (1983a, b) for life-history data for *Pacifastacus l. leniusculus.*

The life history of cambarid crayfishes (Fig. 18) is somewhat more variable and complex than is typical of the astacid crayfishes. The sexually dimorphic males exhibit two distinct and usually alternating body forms (Taylor, 1985). The change from one form to another occurs among mature males during the semiyearly

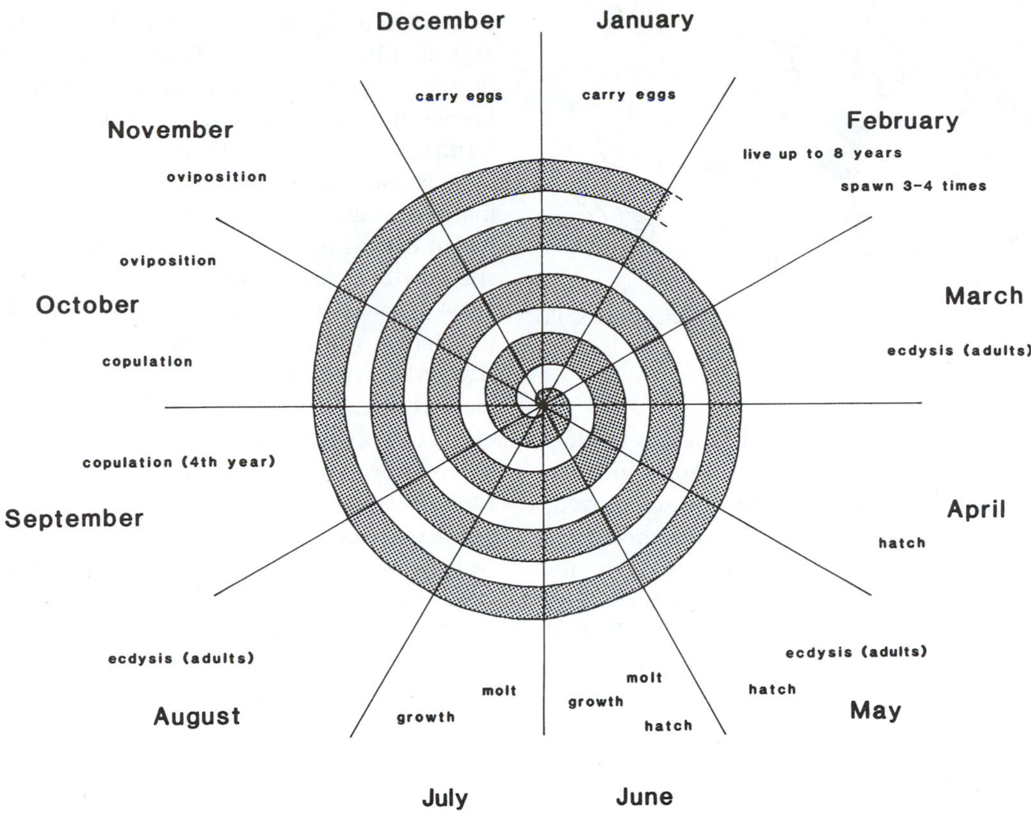

FIGURE 17 Generalized life history of astacid crayfishes (see text for explanation).

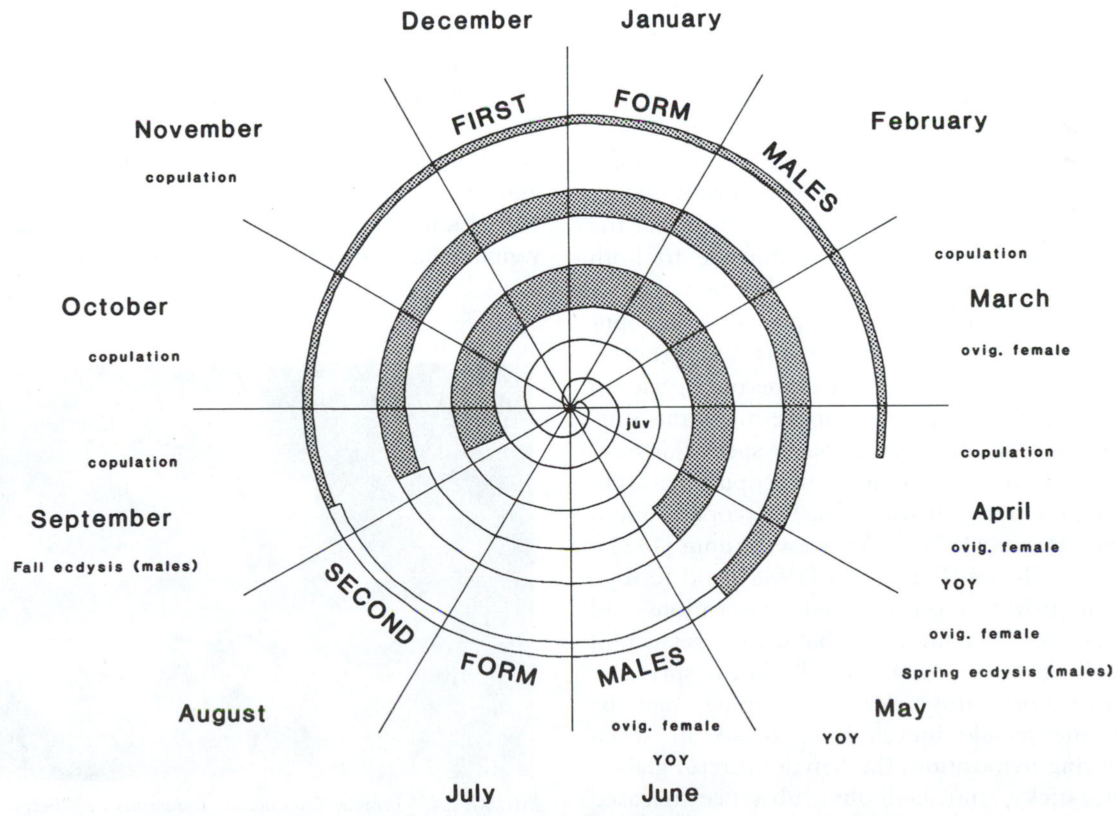

FIGURE 18 Generalized life history of cambarid crayfishes (see text for explanation).

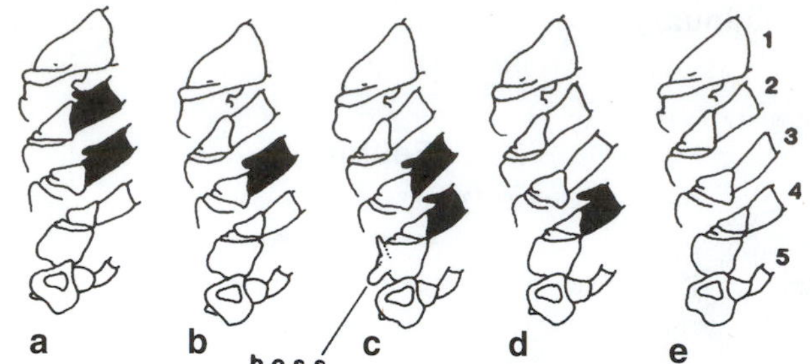

FIGURE 19 Ventral view of basal portions of left pereiopods with ischia bearing hooks darkened (after Hobbs, 1972).

molts. The sexually competent stage (Form I) is first attained with the last juvenile molt. This more aggressive breeding form can be distinguished by the sclerotized, amber-colored, and lengthened condition of the terminal elements of at least one of the first pleopods. In addition, the ischial spines are more pronounced (Fig. 19), the first chelipeds are enlarged (Stein, 1975), and the sperm ducts are full of recently formed spermatids (Word and Hobbs, 1958). In Form II males (the nonbreeding form), however, the terminal elements of the first pleopod are not as well differentiated (and never corneous), the ischial hooks are shorter and weaker, and the chelipeds are less robust.

The ovarian cycle of female cambarids is only partly known and from only a handful of species but endocrine regulation of ovarian development coupled with abiotic influences such as temperature and photoperiod are certainly paramount (Kulkarni *et al.*, 1991; Dubé and Portelance, 1992; Daniels *et al.*, 1994). Kulkarni *et al.* (1991) reviewed the studies devoted to this process and, working with *Procambarus* (*Scapulicambarus*) *clarkii* (Girard), determined that the ovary of this species is directly responsive to both ovary-inhibiting and ovary-stimulating hormones.

Although Form I males may be present in a population year round (e.g., in *Cambarellus* and *Procambarus;* see also life-history data for numerous species in Hobbs, 1981), generally they are more numerous during the fall and/or spring. Hence, many species demonstrate one or two peak periods of copulation (fall and/or spring) although mating clearly occurs throughout the year. [Mason (1970a), Ameyaw-Akumfi (1981), Bechler (1981), Hobbs III and Jass (1988), and Kasuya *et al.* (1996) provide more detailed observations and literature review of the mating behavior in crayfishes.]

Because oviposition occurs usually in the spring or early summer, spermatophores (sperm plug) may be carried by the female for as long as six or seven months. During oviposition, the female secretes glair, a translucent, sticky, mucilaginous substance released from "cement glands" located on the ventral side of the abdomen. After eggs are released from the genital pore and simultaneously fertilized, they attach to the abdominal pleopods by a short stalk. Amazingly, it is not known how nonmotile sperm inside a hard, waxy plug fertilize an egg mass! Once the glair hardens, the female is considered to be "in berry." Typically, larger females of a given species carry more eggs, yet the number of eggs is highly variable within as well as between species (epigean forms carry as many as 600–700 eggs). For example, Hobbs (1981, p. 415) reported that females of *Procambarus* (*Ortmannicus*) *pubescens* (Faxon) carried between 89 and 443 eggs. He cited numerous other examples of variation in egg number for Georgia crayfishes. Corey (1987a, b) reported on the egg number versus individual size for three species of *Orconectes* as well as for *Cambarus robustus.* She indicated that there was a positive linear relationship between the total number of eggs carried per female (fecundity) and the size of the female. The total number of eggs for each was:*O. rusticus* ranged from 75 to 351, mean 161.8; *O. virilis* from 29 to 195, mean 139.1; *O. propinquus* from 26 to 195, mean 68.5; and *C. robustus* ranged from 25 to 128, mean 64.9. See Stechey and Somers (1995) for fecundity data on *O. immunis.*

Little is known about the fecundity of stygobitic species because ovigerous females or those carrying young have not been observed in most species. For the few taxa studied, however, the eggs were larger (more yolk), but less numerous (27–80) (Fig. 20), reflecting the constraints of life for these K-strategists in such energy-poor systems. Noblitt and Payne (1995) compared water, lipid, and protein content of the eggs of *P. clarkii* and *P.* (*Ortmannicus*) *zonangulus* Hobbs and Hobbs and found that water contributes significantly to weight changes. They also determined that *P. clarkii*

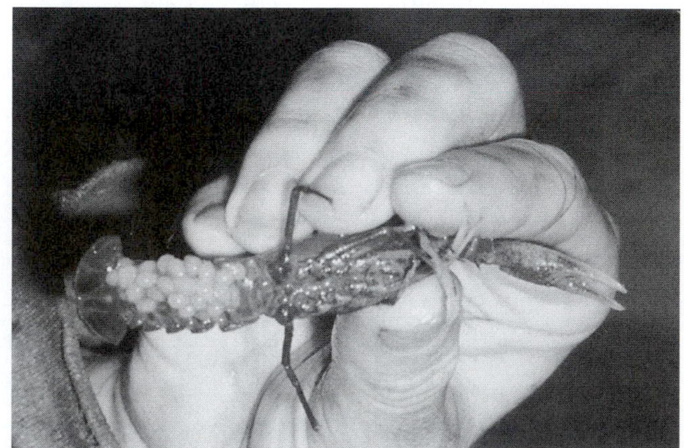

FIGURE 20 Female *Orconectes i. inermis* in "berry," Marengo Cave, Crawford County, IN (note size and number of eggs).

eggs have significantly higher percentages of proteins while *P. zonangulus* eggs have significantly higher percentages of lipids. *Procambarus clarkii* produces relatively large numbers of small eggs (among the highest recorded for crayfishes), whereas *P. zonangulus* generates relatively few but larger eggs with greater size variation. *P. clarkii* exhibits a profligate reproductive strategy (well adapted to warm water habitats where nutrient input is continuous) whereas *P. zonangulus* demonstrates a prudential reproductive strategy, being better adapted to cool water environments where nutrient flow is pulsed (Noblitt and Payne, 1995; Noblitt *et al.*, 1995).

Depending on the season of laying and the water temperature, eggs are carried by females for 2 to 20 weeks, during which time rhythmic movements of the pleopods effectively aerate the developing eggs. Corey (1987a, b) noted an inverse relationship between the number of newly released eggs and the current velocity. Incomplete extrusion, failure to attach to pleopods, and fertilization failure are three other factors leading to egg loss at oviposition. Additional losses of eggs and young while being carried on the female result from abrasion, predation, high current velocity, low food availability, high density of females, and insufficient cover for females (see also Figler *et al.*, 1995). After hatching, the first juvenile instar attaches to the mother with hooked chelae and a telson "thread" anchored by specialized hammate setae (Price and Payne, 1984). The cephalothorax of this instar is disproportionately large, with immense eyes, a yolk-laden carapace, and an incompletely developed abdomen. Generally after 2–7 days of attachment, juveniles molt and grow to a form more closely approximating the adult appearance. This instar loses its posterior connection to the mother and uses only chelae to cling to her abdomen. An additional molt usually occurs within 4–12 days, producing a large, third instar crayfish. Although this juvenile hangs onto the mother with its chelae and pereiopods, it may leave the parental pleopods intermittently; subsequent instars are free living. [Refer to Price and Payne (1984) for more details of postembryonic ontogeny and to Mason (1970b) for maternal–offspring behavior.] The adult female commonly molts 2–3 weeks after all young have left. By fall, 6–10 molts have occurred in the developing juvenile, resulting in a considerably larger and often sexually mature adult. Mating among mature yearlings frequently takes place; however, many individuals do not become sexually active until the following late summer or fall. Most adults can live 2.5–3 years, and the majority breed more than once. The longest life span reported for noncave dwelling cambarids is 6 or 7 years for the burrowing crayfish *Procambarus* (*Girardiella*) *h. hagenianus* (Faxon) in eastern Mississippi and western Alabama (Lyle, 1938). Longevity in stygobitic crayfishes has been demonstrated to be considerably greater than for epigean forms (Cooper and Cooper, 1978; Hobbs III, 1980; Streever, 1996).

C. Ecological Interactions

1. Functional Role in the Ecosystem

Crayfishes have successfully invaded a wide variety of habitats and play important roles in processing organic matter and in the transformation and flow of energy. They are the largest and longest lived crustaceans in freshwater ecosystems of North America and process spatially and temporally large quantities of organic matter (see Huryn and Wallace, 1987) in most lotic and lentic systems. Although they can function as food specialists, most are opportunistic generalists, feeding virtually at all trophic levels (polytrophic) and serve as efficient shredders and macrograzers (see Huryn and Wallace, 1987; France and Welbourn, 1992). Shrimps, on the other hand, are generally restricted to large, sluggish backwaters of rivers where aquatic vegetation (cover) is dense. They function primarily as grazers and, like crayfishes, shrimps are eaten by a myriad of predators (Hobbs III, 1993). Indeed, crayfishes are almost the sole source of food for some predators, such as the striped swamp snake, *Regina alleni* ((Franz, 1977; Fotenot *et al.*, 1993). Naturally, shrimps do not play as important a role as do crayfishes in energy flow since they are much more limited ecologically (largely restricted to slowly moving streams or lakes and ponds with dense vegetation). For additional information on food and feeding by these decapods, see below.

Clearly crayfishes are important ecological constituents in numerous freshwater ecosystems yet few studies have defined the habitat requirements for these organisms. Taylor (1983) determined that the distribution of *Procambarus* (*Pennides*) *spiculifer* (LeConte) was influenced by water depth. Rabeni (1985) ascertained that the sympatric species *O. luteus* and *O. punctimanus* in south-central Missouri co-exist by partitioning the physical environment. *Orconectes punctimanus* tends to be more restricted in its use of habitats, occurring as YOY in shallow water with macrophyte cover. Both YOY and adults forage on larger-sized substrates and both life stages prefer slow current velocities. This species maintains a size advantage over *O. luteus* (larger individuals are dominant and occupy their preferred microhabitat). *O. luteus* tolerates a wider range of environmental conditions and exploits habitats unavailable or undesirable to its congener. Additionally, Rabeni (1992) determined that these crayfishes are the most energetically important component in the diets of rock bass and small-mouth bass and that predation accounts for a significant percentage of total mortality in these crayfishes. Eversole and Foltz (1993) demonstrated that abundance and standing crop biomass of *Cambarus* (*Puncticambarus*) *chaugaensis* Prins and Hobbs and *Cambarus* (*Depressicambarus*) *latimanus* (LeConte) were altered by substrate type, velocity, cover, and water depth. For *Procambarus* (*Leconticambarus*) *alleni* (Faxon) occupying the freshwater marsh habitat mosaic of southern Florida, differences in relative risk of predation, food availability, or a combination of these factors are probably influencing variances in habitat occupation (Jordan *et al.*, 1996).

Symbiotic and parasitic relationships involving crayfishes and shrimps have been summarized by Johnson (1977), Beck (1980), France and Graham (1985), Lichtwardt (1986), Keller (1992), Hobbs and Peters (1993), Holt and Opell (1993), Font (1994), and Gelder (1996). These decapods serve as intermediate or definitive hosts and/or substrates to a wide variety of bacteria, algae, protozoa, fungi, worms, and crustaceans. For instance, the metacercariae of the lung fluke *(Paragonimus* spp.*)* invade crayfishes and are transmitted to another host when the crustacean is eaten. *Paragonimus westermani* is an important parasite of humans in Asia; fortunately for North Americans, few human infestations have been noted for *P. kellicoti.* Ecdysis provides periodic relief from the buildup of various ectocommensals.

Crayfishes serve as a food resource as well as a pest for humans. A multimillion dollar business centers around "crawfish" production in Louisiana and other parts of the world (Huner and Barr, 1984; Avault, 1993). In addition, certain crayfishes are economically important as local seasonal nuisances for farming. Burrowing species can be extremely abundant:Hobbs and Whiteman (1991) reported that *Fallicambarus* (*Fallicambarus*) *devastator* Hobbs and Whiteman (Fig. 21) was responsible for constructing immense burrows with as many as 62,676 mounds/ha in the prairie section of eastern Texas! Farm equipment can obviously be damaged, crops destroyed, and earthen dams can be weakened by the activities of these fossorial decapods.

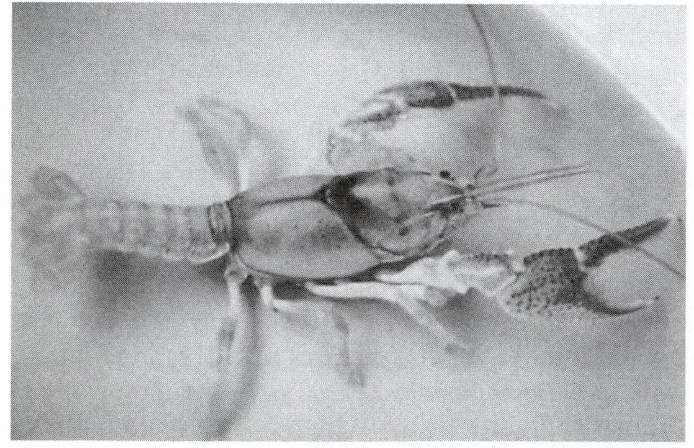

FIGURE 21 Form I male (breeding stage) of *Fallicambarus devastator,* a primary burrowing crayfish from Texas.

2. Foraging Relationships

Shrimps are somewhat opportunistic feeders, foraging on a variety of plant and animal material. The shrimp *Palaemonetes paludosus* feeds heavily on epiphytic algae on vascular plants. Grazing on diatoms *(Fragilaria, Navicula, Stephanodiscus, Gomphonema, Synedra,* and *Cymbella)* appears to be the primary method of food intake for populations in Hillsboro County, Florida (Beck and Cowell, 1976); however, they also forage for immature insects.

Barr and Kuehne (1971) reported that the stygobitic shrimp *Palaemonias ganteri* (Mammoth Cave, Kentucky) strained and filtered pool sediments with mouth parts, suggesting that its food consisted of microorganisms in the silt (sediments contained protozoa; Barr, 1968). However, Lisowski (1983) observed this shrimp exploiting another resource; an individual was observed foraging upside-down on the surface film of a pool apparently feeding on floating material. This foraging behavior was similarly noted by Cooper (1975) for *Palaemonias alabamae* in Shelta Cave, Alabama, (he also observed the sediment-straining behavior as has this author in Bobcat Cave, Madison County, Alabama).

Crayfishes play a polytrophic role in most ecosystems and, as such, contribute much to the complexities of energy flow in aquatic systems (see above; see also Lorman and Magnuson, 1978; Momot *et al.*, 1978; Momot, 1984; Hobbs III, 1993). By feeding directly on carrion and organic debris of aquatic and terrestrial origin, these two decapods contribute to the processing of organic matter in aquatic ecosystems. They serve in part as decomposers by breaking down particulate matter, altering the chemical composition of detritus, and releasing fecal material back into the system. Budd *et al.* (1979) demonstrated that juvenile *Cambarus robustus* and *Orconectes propinquus* function as collectors by filter feeding on algae. In Missouri, YOY populations of O. (*Procericambarus*) *punctimanus* (Creaser) possessed a greater percentage of plant detritus, O. (*Procericambarus*) *luteus* (Creaser) had a higher percentage of filamentous algae, and both species had a higher percentage of plant rather than animal matter in their stomachs (Rabeni *et al.*, 1995). Lorman (1980) and Lorman and Magnuson (1978) showed that males of *Orconectes rusticus* in Wisconsin are more carnivorous than females yet are less carnivorous than juveniles. Others (e.g., Momot *et al.*, 1978) found that all age classes tend to utilize vegetation as a food source.

They are capable of switching roles from herbivore/carnivore to scavenger/detritivore merely in response to food availability. It is important to recognize that gut contents of crayfishes typically reflect availability of food type much more readily than they indicate preference for a specific food item. Food materials constituting the majority of gut contents are not necessarily what crayfishes prefer to consume. Although individuals of some species and sizes do forage selectively, many species are quite opportunistic and catholic in their diets. For example, Horns and Magnuson (1981) found that crayfishes consume significant numbers of Lake Trout eggs in northern Wisconsin when this seasonally restricted food item is available. Savino and Miller (1991) determined that significant impact on lake trout egg survival would occur only in lake trout spawning grounds with relatively low egg densities where there was a large crayfish population, and where cobble or large rock substrate was available. Juvenile crayfishes are somewhat limited to detritivory and herbivory, as they filter suspended particulates and grasp coarse particles. Adults actively prey and graze on larger items, yet they too scrape microbes from hard substrates and shred vegetation. The total litter processing by these crustacea was estimated to range from 4–6% of the annual litter input in a West Virginia headwater stream (Huryn and Wallace, 1987). Because of this flexibility, crayfishes can maintain large population densities despite fluctuations in food resources. Although crayfishes forage nearly continuously, the peaks of this behavior occur shortly after sunset and during the night. In addition, for example, this activity for *Procambarus clarkii* decreased as water temperatures began to cool to around 14°C in Oklahoma (Covich, 1978). In contrast, Price (1981), working in Arkansas, found that both diversity in the diet and length of foraging increased for *0.* (*Procericambarus*) *neglectus chaenodactylus* Williams as food abundance decreased during the winter months. Another factor significantly affecting foraging is the interaction of sympatric species, since dominance could be critical in determining both foraging and reproductive successes.

Crayfishes feed readily on macrophytes, and they can significantly reduce species richness and biomass of aquatic vegetation by grazing selectively on plant species (see Lorman and Magnuson, 1978; Lodge and Lorman, 1987; Feminella and Resh, 1989; Lodge, 1991; Creed, 1994; Lodge *et al.*, 1994; Momot, 1995; Nystrom and Strand, 1996). By moving through the shallow littoral zone of lakes and cutting stems near the bottom, they can wreak havoc even though they actually ingest very little of the vascular plant "harvest" (an example of consumptive and nonconsumptive destruction). In contrast, Saffran and Barton (1993) found that *O. propinquus* in Georgian Bay (Ontario, Canada) had little impact on vascular plants, but that they influenced primary productivity of the littoral zone by the consumption of periphyton, charophytes, and invertebrates.

Crayfishes, as predators (Lorman and Magnuson, 1978; Covich *et al.*, 1981; Covich, 1982; Lodge *et al.*, 1987; Hobbs III, 1993; Lodge and Hill, 1994), apparently attempt to maximize energy gain with minimal energy expenditure. For instance, rates of predation by *Procambarus clarkii* on thin-shelled snails (*Physa* sp.) were higher than those on thicker-shelled *Helisoma* sp. (Covich, 1978). Yet, MacIsaac (1994) showed that *O. propinquus* attacked medium and large zebra mussels (*Dreissena polymorpha*) more often than small individuals but that predation was limited primarily to small- and medium-sized mussels. Also, *O. rusticus* damaged more snails and had a greater weight-specific ingestion rate on snails than did its congeners in Wisconsin (Olsen *et al.*, 1991). Hazlett (1994a) found that crayfishes feeding on zebra mussels did not respond to food odors unless they had prior experiences with that protein source. He also determined that food odors are not produced in sufficient quantities to stimulate feeding by experienced individuals unless the food has been degraded by microbial activity. Results of other studies suggest that crayfish predation reduces snail populations in lakes and streams (e.g., Lodge and Lorman, 1987; Hanson *et al.*, 1990; Weber and Lodge, 1990; Olsen *et al.*, 1991; Lodge *et al.*, 1994; Perry *et al.*, 1997) and also induces shifts in snail life history characteristics (Crowl and Covich, 1990). Clearly these crustaceans have ramifying impacts on the structure and dynamics of aquatic communities (e.g., Hanson *et al.*, 1990). That is, crayfishes prey upon organisms occupying various trophic levels and compete for food and other resources at the intra- and interspecific levels. Thus they are positioned high enough in the trophic structure to influence numerous interactions among several subsets of the community foodwebs (Rabeni *et al.*, 1995 found that *Orconectes* spp. accounted for more than half of all consumption by the invertebrate community of Missouri Ozark streams).

Foraging behavior is altered in the presence of a predator. Stein (1977) showed that the susceptibilities of the crayfish *0. propinquus* to fish predators reflected the size and condition of the potential prey. Life stages with the greatest reproductive potential were those least vulnerable to small-mouth bass predation (berried females and Form I males). Young-of-the-year are particularly susceptible to predation; yet they, like other size classes, must balance predation risk with the benefits of foraging. Thus, juveniles, Form II males, females, and recently molted individuals modify their behaviors (reducing activity) and microdistribution (seeking substrates affording maximum protection) to minimize risk of predation. The degree of behavioral response appears to correlate positively with vulnerability.

The importance of top-down forces on community composition and productivity (e.g., the trophic cascade model—see Carpenter and Kitchell, 1992) has received much attention (e.g., Power, 1992; Rabeni, 1992; Strong, 1992; Carpenter and Kitchell, 1993; Lodge *et al.,* 1994). It is apparent that in lakes, the presence of predatory fishes may decrease crayfish impact on lower trophic levels by reducing crayfish abundance (Lorman and Magnuson, 1978; Covich, 1982; Lodge *et al.,* 1987; DiDonato and Lodge, 1993; Garvey *et al.,* 1994), by nonconsumptively increasing crayfish mortality (Hill and Lodge 1995), and by reducing foraging activity by surviving crayfish (Hill and Lodge, 1994). Hill and Lodge (1995) further suggest that a significant component of the impact of top predators on food webs may be the result of sublethal effects.

Although several studies have determined the thresholds of detection by crayfishes for various stimulatory substances (e.g., Hatt, 1984; Tierney and Atema, 1986; Hazlett, 1990; Ciruna *et al.,* 1995), little effort has been made to examine the role or influence of chemical stimuli and chemoreception on the foraging behavior of crayfishes (see Blake and Hart, 1993; Willman *et al.,* 1994). Studies of the hierarchial structure of food search and feeding in the spiny lobster (e.g., Zimmer-Faust and Case, 1983) suggest that chemosensory-induced foraging in crayfishes may also be geared primarily to obtaining nearby rather than distant food sources. Johannes and Webb (1970) have demonstrated that several species of invertebrates release free amino acids into the environment, and Rittschof (1980) showed that amino acids diffuse rapidly from carrion. Thus, amino acids are ubiquitous in aquatic ecosystems and are readily available for use as chemical feeding cues for crayfishes. Of particular note, Hatt (1984) demonstrated that amino acid receptors are present on the pereiopods and antennules of crayfishes. Tierney and Atema (1988) suggested that the responsiveness of *Orconectes rusticus* and *0. virilis* to a variety of chemicals (e.g., amino acids) may reflect the omnivorous foraging habits of these crustaceans.

Stygobitic crayfishes and shrimps live in heterotrophic environments where the primary source of energy is from the surface in the form of allochthonous materials carried or washed into the cave, such as vegetative debris and bat guano. In addition, the stygobitic crayfish, *Troglocambarus maclanei,* also is presumed to be a filter feeder; however, cannibalism by this obligate cave dweller has been noted (Hobbs *et al.,*1977). In such resource-depressed systems, these crustaceans tend to congregate in regions of the cave where food is spatially or temporally abundant, and thus are commonly observed in pooled areas of streams among accumulated vegetation. Crayfish densities are particularly high in these microhabitats, and these crustaceans often can be observed foraging over the substrate and feeding on microbe-encrusted debris. Their particularly enhanced foraging efficiency reflects their improved sensory perception, reduced random movements, diminished body size, and increased metabolic efficiencies.

Foraging behavior of burrowing crayfishes is variable; but generally on warm, humid nights, individuals leave their burrows presumably in search of food (vegetative material). A peculiar phenomenon is noted regularly in Louisiana in which large numbers of crayfishes (primarily males) leave their marsh habitat, move over land, and die during the fall. They are generally in poor physical condition, and the cause of this mass "death-migration" is unknown (Penn, 1943, p. 15). See the following references for additional information concerning burrowing crayfishes: Penn (1943), Hobbs (1981), Hobbs III and Jass (1988), and Hobbs and Whiteman (1991).

3. Behavioral Ecology

The behavioral and distributional responses of organisms to features of their physical and biotic environment is encompassed within the discipline of behavioral ecology. Behavioral responses to drought (burrowing), orientation to current (positive rheotaxis), light (negative phototrophism), reaction to cover, and substrate preference (cobble–pebble), are examples of responses by decapods to the physical environment (see Gore and Bryant, 1990). These factors, to a large extent, determine the nature of selected habitats. Habitat use is influenced by food, feeding rate, predators, competitors, and other sources of mortality (see Mather and Stein, 1993). Aggressive behavior, maternal-offspring relationships, and behavioral mechanisms resulting in reproductive isolation of a shrimp or crayfish species are representative patterns related to the biotic features of the environment.

a. Feeding See Section III.C. 2.

b. Burrowing Atkinson and Taylor (1968) suggest that burrowing is a strategy for minimizing adverse extremes of humidity and temperature and probably for protection against predators; this behavior is certainly well developed among the cambarids. Hobbs (1981) recognized three categories of burrowing crayfishes. "Primary burrowers" (e.g., *Fallicambarus devastator,* Fig. 21) spend most of their lives in burrows, occasionally exiting to forage or to mate. Their burrows consist

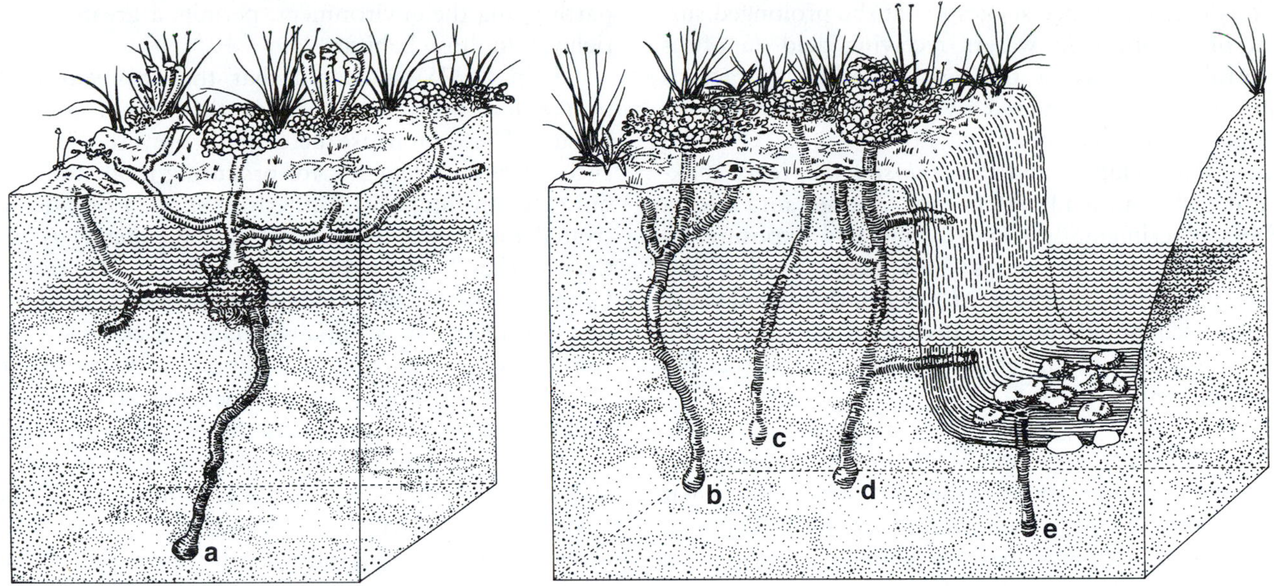

FIGURE 22 Generalized crayfish burrows:(a, d) those of primary burrowers; (b) that of secondary burrower; and (c, e) those of tertiary burrowers (after Hobbs, 1981, p. 32; see also Hobbs and Whiteman, 1991, Fig. 2).

of complex tunnels that are often quite removed from open bodies of water (Fig. 22a, d) and some species (e.g., *Fallicambarus* (*Creaserinus*) *gordoni* Fitzpatrick) are found only in pitcher-plant bogs (Johnston and Figiel, 1997). Crayfishes tend to be attracted out of their burrows by low light intensities, and this may be considered as an initial response that ultimately gives rise to more complex behavior patterns involving food searching, reproduction, and other social interactions. Higher levels of light initiate a withdrawal response, one that protects them from predators (see Fernández-de-Miguel and Aréchiga 1992).

"Secondary burrowers" [e.g., *Fallicambarus* (*Creaserinus*) *fodiens* (Cottle)—see Norrocky 1991] spend much of their life history in tunnels but frequent open water during rainy seasons (Fig. 22b). Their burrows usually feature a single subvertical tube that either slopes gently or descends in an irregular spiral (Hobbs, 1981, p. 32). "Tertiary burrowers" (e.g., *Procambarus a. acutus*) live in open waters and move into burrows only at specific periods (Fig. 22c, e). For example, they burrow to brood eggs, to move below the frost line during winter, and to avoid desiccation. Burrows also are refugia and can be defended, and juveniles can exhibit communal living within them (see Hasiotis, 1995). Nonburrowing crayfishes will occasionally construct burrows when surface waters disappear (Berrill and Chenoweth, 1982) or burrow into the hyporheos of streams. For further details concerning the burrowing behavior of crayfishes, see Grow (1981) and Rogers and Huner (1985).

c. Water Currents Water currents certainly influence life processes of lotic organisms and bar many species from rapidly flowing streams. Maude and Williams (1983) demonstrated that when crayfishes are exposed to increases in current velocity, they alter their body posture to counteract the effects of drag and to maintain position. Not only is this a behavioral adaptation resulting in certain species optimally occurring in fast-flowing streams, but it may very well play a role in the widespread distribution of species as well as in the expansion of exotics, such as *O. rusticus,* into a variety of habitats following their introduction.

d. pH Many temperate crayfishes inhabit shallow littoral waters during early spring. In some regions, this time period coincides with the spring pulses of acidic water from melting snow. Because crayfishes are capable of extensive movement and can avoid streams acidified by industrial sulfuric acid or mine drainage (pH 4.0–4.5; France, 1985b), they should be able to elude environments acidified by spring melt pulses. France (1985b) determined that *Orconectes virilis* can avoid lethally acidic waters, implying that during the spring melt, they could escape mass mortality by seeking refuge in the profundal zone of lakes. Yet, no avoidance was demonstrated to lake water with pH values as low as 5.6, levels that are reported to cause serious reproductive difficulties [from decreased aeration of eggs, leading to fungal infections and egg mortality (France, 1985a)]. The failure of this crayfish to avoid sublethal acid waters could, therefore, interfere with recruitment;

for this reason, France suggests that the prolonged survival of *O. virilis* in waters receiving acid runoff is doubtful. This may be the case for other crustaceans living in poorly buffered systems but, surprisingly enough, some crayfishes [e.g., *Procambarus* (*Ortmannicus*) *fallax* (Hagen) and *P. paeninsulanus*] reach maximal densities in acid waters. Yet, Tierney and Atema (1986) experimentally showed that acid exposure inhibits behavioral responses (specifically feeding and grooming; see also Uiska *et al.*, 1994); this could cause acid-stressed populations to decline at pH levels well above those at which lethal physiological effects become apparent.

e. Shelter and Aggression Obviously important aspects of the ecology of crayfishes and shrimps are their intra- and interspecific behavioral responses. Although they are nonterritorial, they are aggressive and rely on agonism to procure shelter, food, and mates (Stein, 1975). Most crayfishes tend to be solitary and occupy and defend crevices in a coarse substrate, both of which are of survival value. Crevices not only provide shelter from pounding shore waves or from high stream velocities, but they also furnish a refuge that can be defended against predators and competitors (see Blank and Figler, 1996). Considering the generalistic feeding habits of crayfishes and the apparent absence of food limitation in most systems, the principal resource bottleneck may be crevice availability. Hobbs III (1976) demonstrated a positive relationship between crayfish abundance and availability and size of cobble substrates. Caine (1978) observed that competition among Florida epigean crayfishes is primarily for shelter, and individuals unable to obtain cover succumb to predators. A specific home range has been demonstrated for a number of species (e.g., Hazlett *et al.*, 1974; Hobbs III, 1980), and dispersion and dispersal often appear related not to population density but to water depth, number of crevices, predators, temperature, etc. Experiments examining intraspecific aggression of resident maternal (carrying eggs and/or hatchlings) *P. clarkii* against intrusions by Form I males and nonmaternal female conspecifics showed that maternal residents won a significantly higher proportion of encounters with either sex, thus demonstrating maternal aggression in crayfishes (Figler *et al.*, 1995). Kershner and Lodge (1995) showed that *O. rusticus*' use of cobble habitat in northern Wisconsin lakes was significantly and positively correlated with lakewide predator density, reflecting the potential importance of predation risk in establishing crayfish distribution patterns. They suggested that the availability of cobble habitat may influence the success of the invasions of this crayfish. Additionally, crevices serves as a feature, which, by further partitioning the environment, permits a greater species richness in those habitats.

Stein (1975) indicated that the chelipeds of *O. propinquus* make up 40% of their adult total dry weight. These immense pereiopods are used in sexual encounters with females, for procuring food, and in agonistic confrontations. Rutherford *et al.* (1995) demonstrated that chela length (but not symmetry) is positively related to fighting ability of *O. rusticus.* Levenach and Hazlett (1996) found that if the display function of chelae is impaired then the crayfish is prevented from competing successfully with conspecifics (*O. virilis*), yet impairing the mechanical function of chelae does not reduce the crayfishes' competitive ability. Since crayfishes can autotomize their chelae, Figiel and Miller (1995) investigated the influence of chela autotomy on growth and survival of *P. clarkii.* They showed that as population density increased, the frequency of chela loss increased and that injured individuals demonstrated reduced growth and survival rates.

Hayes (1975) demonstrated eight innate components of social interaction in the primary burrowing crayfish, *Procambarus* (*Girardiella*) *gracilis* (Bundy): alert, approach, threat, combat, submission, avoidance, escape, and courtship. These also have been observed in other species and Ameyaw-Akumfi (1979) summarized the sequence of behaviors demonstrated by interacting crayfishes during aggressive encounters (Fig. 23). He proposed that movement of gnathal appendages and pereiopods as well as repeated antennal waving occurs prior to the termination of an agonistic encounter and this behavior may in fact constitute an appeasement display. He further stated that "the use of appeasement signals in mating behavior is, probably, a capitalization by the male as this behavior appears to reduce aggressive tendencies" (Ameyaw-Akumfi 1979, p. 41). Copulatory behavior (reviewed in Bechler, 1981) is generally seasonal with the pair assuming several positions, the male mounting the female most commonly (Fig. 24). Mating generally lasts from 10 min to 10 h (e.g., Andrews, 1895; Mason, 1970a). After oviposition, egg fanning, and hatching, maternal–offspring behavior is mediated by pheromones released into the water by the female. These cues are species-specific but not brood-specific, since juveniles are attracted to conspecific brooding females. Also, Peck (1985) showed that social interaction is an important factor affecting temperature selection by *O. virilis.* Individuals demonstrated behavioral mechanisms, not only to find favorable temperature ranges but also to segregate according to social rank.

f. Predation Avoidance The presence of predators elicits various behavioral response in crayfishes. Young-

FIGURE 23 Agonistic behavior between two male *Orconectes propinquus* in the "body up" position; crayfish on right performing "antenna tap" (Bruski and Dunham, 1987) (photograph courtesy of Ann Jane Tierney).

of-the-year are particularly susceptible to predation and must balance predation risks with foraging benefits (Butler and Stein, 1985). Stein and Magnuson (1976) and Stein (1977) showed that in the presence of smallmouth bass *(Micropterus dolomieui), 0. propinquus* foraged less, adopted defensive postures, and sought protective substrates. Ideally, crayfishes use crypsis to avoid detection by predators. However, when discovered, crayfishes typically backswim seeking shelter in vegetation, beneath rocks or debris, or in burrows. If

FIGURE 24 Copulating pair of *Orconectes propinquus,* Form I male on top (photograph courtesy of Ann Jane Tierney).

escape is not possible, individuals assume and intensify a species specific defensive posture, the predator response posture (PRP). Hayes (1977) described PRP of crayfishes and discussed the evolutionary significance of these behavioral patterns in three species of the genus *Procambarus*. The crayfishes studied have similar lateral merus displays which result in a lateral presentation of the chelae, allowing for defensive coverage of the entire carapace and for quick flight by tail-flipping. Also, elongated chelae of these species enable them to sense tactilely the approach of a predator and to keep it at a greater distance. The PRP of the terrestrial (burrowing) species *P. gracilis* appears to have been modified for threat from terrestrial predators with attack coming primarily from above. Chelae are held shield-like in front of the carapace, and the tail is folded tightly beneath the body, allowing for rapid reorientation to the attacks of a shifting predator. This posture frees the tail from use as an escape organ on land. Stein and Magnuson (1976) also suggested that, in addition to their effects on the distribution and behavior of crayfishes, predators have an influence beyond simple predator–prey interactions because of their impact on several trophic levels.

4. Population Regulation

The population dynamics of shrimps and crayfishes are not well studied except for a few species. Traditionally, the four major features of any demographic model are summarized as natality, mortality, immigration, and emigration. Each of these clearly affects the size and growth of natural populations and all are under the constraints of density-dependent (D-D) and/or density-independent (D-I) factors. It should be emphasized, however, that conventional D-D and D-I regulatory factors, while present, are too often superceded in importance by the effects on density of human activities (Hobbs and Hall, 1974). Rather than a single factor, it should be emphasized that any of these elements may interact strongly in affecting shrimp and crayfish community composition, population sizes, and production.

a. Shrimps A 1-year life history is characteristic of epigean shrimps, whereas cave forms probably live longer. Food resources tend not to be limiting for epigean shrimps, but fish predation can have significant impact on their densities (Nielsen and Reynolds, 1977). The same investigators noted that fluctuating water conditions (seasonal drying and extended high water) also elicit population changes. Fecundity appears to be related to size (length) of female and to time of year.

In cave environments of the southeast, amblyopsid fishes (and crayfishes) are predators of stygobitic shrimps and crayfishes. In these food-limited systems, where shrimp and some crayfish populations are very small, loss of ovigerous females (and thus loss of recruitment of young) can be very significant to immediate and particularly long-term stability of population densities. Lisowski (1983) attributed the extreme decrease in density of *Palaemonias ganteri* from base-level streams in the Mammoth Cave System, Kentucky, to pollution of local groundwater. The sources of contaminants were human sewage and water from the Green River, which had received inputs of oil brines and hydrocarbons during the 1960s. He also indicated that modifications of habitat and flooding regimes caused by dams within the Green River drainage adversely affected shrimp by reducing the reproductive success and by increasing predation pressure on them.

b. Crayfishes Lodge and Hill (1994) reviewed the variables regulating population size and the importance of both D-D and D-I factors for some cool-water crayfishes. They summarized the D-I parameters as temperature, calcium concentration, pH, dissolved oxygen, salinity, habitat (substrate particle size and food availability), and water velocity and indicated that the D-D factors are diet, competition (intra- and interspecific), predation, molting, and ectosymbionts and disease. Momot and others, in a variety of population studies (Momot and Gowing, 1977a, b; Gowing and Momot, 1979; France, 1985a), have shown that various populations of *O. virilis* are regulated by several D-D and D-I factors, including cannibalism, natural mortality at molting (physiological and mechanical problems), predation by snakes (Godley *et al.*, 1984), fishes (e.g., trout, bass) and some insects (e.g., dragonfly) on yearlings, predation on adults by various terrestrial vertebrates (Lodge and Hill, 1994), flooding, temperature, pH (Davies, 1984; Siewert and Buck, 1991; France and Collins, 1993), seiches (Emery, 1970), and other factors. Crayfishes are generally characterized by high juvenile mortality and reduced adult mortality (concave survivorship curve). It has been suggested (Momot and Gowing, 1977b) that overall, most changes in biomass and productivity of these populations result from internal density-dependent processes operating on fecundity and mortality rates rather than on growth. In several Michigan lakes during most years studied by Momot, the breeding success of *O. virilis* depended mainly on the survival and subsequent fecundity of the 1.5-year age group females. At 2.5 years, females are considered an important strategic reserve.

Where natality, mortality, emigration, and immigration rates vary, population sizes will fluctuate, especially where birth recruitment is inadequate. As populations expand, they face potential resource depletion and the possibility of attracting more predators—both factors that can limit population size. Predation pressure may cause prey to modify their behavior (e.g., increased burrowing and chelae displays, suppressed foraging, and reduced overall activity). In turn, these antipredatory behavioral patterns act as selective (coevolutionary) forces on the predator, thus operating to modify predator strategies. Stein and Magnuson (1976) and Stein (1977, 1979) have demonstrated that juvenile *O. propinquus* are the preferred crayfish prey of smallmouth bass and, as such, are rarely found on open, sandy substrates. Adult crayfish (least preferred prey) are abundant in these areas, demonstrating both that preferred prey make distributional shifts to protective habitats and that use of space by nonpreferred prey is regulated much less by predators.

Inter- and intraspecific competition for habitat (refuge) and nutrients (food) affects crayfish populations in a variety of ways. When food or shelter are limiting crayfishes interact agressively, suggesting that interference competition may be common (Capelli and Munjal, 1982; Capelli and Hamilton, 1984). Bovbjerg (1956) noted that larger size entrusts a competitive advantage; and during intraspecific competitive encounters, male crayfish usually dominate female crayfish. Also, others (e.g., Flynn and Hobbs III, 1984; Hazlett *et al.*, 1992; Lodge and Hill, 1994) have suggested that interspecific competition for habitat and food results in competitive exclusion.

Length–frequency distributions have traditionally been used by investigators to show age and growth of individuals in populations (see France *et al.*, 1991 for conditions required to produce reliable results). Growth of crayfishes is related directly to the number of molts during a growing season and the growth increment per molt; growth is stepwise rather than continuous. Average growth increments range from 0.41 mm for the dwarf crayfish *Faxonella clypeata* (Hay)(Black, 1958) to 2.6 mm for *Procambarus clarkii* (Penn, 1943). The difference in yearly size increment between males and females is due primarily to the frequency of molting and not necessarily to the growth increment at each molt (Hazlett and Rittschof, 1985). However, the increment for individuals living in optimum conditions may well have as much impact as does the number of molts. Female growth is depressed, probably in response to the effects of carrying eggs and young, resulting overall in fewer molts.

Conditions that promote rapid juvenile growth rather than adjustments in egg production per se modify any stock recruitment interaction. In more northern climates, the temperature of the water in the shallows, especially in "nursery areas," is an important determinant of the molting rate in juveniles. Flint (1975) found that a coastal stream population of *Pacifastacus leniusculus* had a slightly faster overall growth rate than a subalpine lake population; he ascribed this to higher water temperatures and a longer growing season. Because crayfishes molt more frequently as juveniles, any parameter reducing juvenile molt frequency also reduces growth rate; this, in turn, may determine size at maturity, although there is no absolute size at which a crayfish becomes mature. Because fecundity is a function of female size, populations with smaller females should have lower recruitment of YOY. Obviously, a cohort that grows faster and reaches a larger overall size will have a better chance of ensuring the replacement of its population. Huryn and Wallace (1987) demonstrated that long life span (five years) and slow growth of *Cambarus bartonii* in a West Virginia headwater stream resulted in low production (see also Roell and Orth, 1992); biomass exceeded production and transpired in a production/biomass ratio (P/B) of 0.58, a value similar to those reported from other temperate crayfish populations (e.g., Momot, 1984). Working with long-term data sets for *Orconectes virilis* in two Ontario, Canada lakes, Momot and Hauta (1995) suggested that crayfishes can have cohort P/B ratios ranging from 1 to 8 and that fluctuations in growth and mortality can alter these ratios significantly.

Crayfishes commonly establish sole occupancy of protective crevices in coarse substrates (when available) which they defend agonistically. A study conducted by Aiken (1968) in Alberta found that both male and female *Orconectes virilis* move to deeper water in late summer, females preceding males. Severe winter conditions would eliminate a population if it remained in the littoral zone where water freezes to the bottom. Rather than burrow to avoid ice, individuals escape by migrating to deeper water. Young-of-the-year appear to have higher mortality rates during the winter than adults. At spring thaw, both males and females return to shallow water. In streams, crayfishes generally burrow into the hyporheic zone to overwinter.

Periods of low water obviously affect all members of the aquatic community, yet crayfishes can escape desiccation by burrowing into the gravel armor of stream channel beds. Prior to complete loss of an aquatic habitat, other responses by crayfishes have been noted. Taylor (1983), studying a lotic population of *P. spiculifer* during a drought in northeastern Georgia, suggested that population shifts (decrease in adult population densities, change in reproductive timing, increase in number of juveniles) resulted from loss of preferred deep-water sites, emigration, and subsequent increase in predation of larger adults.

Population regulation among sympatric species has received considerable attention in the last 15–20 years (e.g., Lodge *et al.*, 1986; Lodge and Hill, 1994). Many

North American crayfishes occur sympatrically with close relatives, yet hybridization appears to occur relatively infrequently although it may be much more common than previously recognized (see Roush, 1997). One mechanism that may be operating to ensure reproductive isolation is chemical or visual recognition of species during mate selection (Smith, 1988). Chemoethological isolating mechanisms could be effective for sympatric but not allopatric species. In consequence, hybridization of a resident and an exotic species might be more likely than between two distinct but geographically overlapping resident species. However, morphometric analyses indicate a strong likelihood that hybridization does occur between *O. rusticus* and *O. propinquus* in northern Wisconsin lakes where the former is an exotic. Lodge and Hill (1994) report that putative hybrids comprise up to 27% of mixed-species populations, and laboratory crosses have shown that viable offspring from matings of these two species are produced but in numbers that are 60% lower than in conspecific matings (Berrill and Arsenault, 1985). Also, Cesaroni *et al.* (1992) report a narrow hybrid zone maintained by two closely related undescribed crayfishes (*Procambarus* spp.) cohabiting a cave in northern Chiapas, Mexico.

Sequential invaders (natural or introduced) can elicit drastic changes in community structure (Hobbs III *et al.*, 1989), including replacement of native species (Hill *et al.*, 1993; Kershner and Lodge, 1995; Gamradt and Kats, 1996). In recent years, biological invasions have been considered such a major environmental problem that many volumes have been devoted to their study (see Lodge, 1993a). Of particular importance to crayfish biology is the apparent replacement of native crayfishes in various parts of North America (and world, Hobbs III *et al.*, 1989) by introductions of nonnative species (e.g., interactions of *0. virilis, 0. propinquus,* and *0. rusticus* in Wisconsin, Lodge; 1993b; Capelli, 1982; Capelli and Magnuson, 1983; Page, 1985; Hobbs III and Jass, 1988; Momot, 1992; Charlebois and Lamberti, 1996; Taylor and Redmer, 1996). Displacement mechanisms are complex, and the mechanisms regulating shifts and/or replacements of species are not well understood in most instances (Flynn and Hobbs III, 1984). However, field and laboratory results indicate that reproductive interference, growth, predation, competition, aggressive dominance, mortality, and interspecific morphological differences probably serve as mechanisms regulating replacement (Capelli, 1982; Capelli and Munjal, 1982; Butler and Stein, 1985; Lodge *et al.*, 1986; Mather and Stein, 1993, Hill *et al.*, 1993; Hill and Lodge, 1994; Garvey and Stein, 1993; Garvey *et al.*, 1994). Based on data from various lakes in northern Wisconsin, Lodge *et al.* (1986) and Lodge (1993b) stated that outcomes of such invasions are not necessarily predictable and Lodge *et al.* (1998) suggested that "best-guess" models of invasions, short of experimental work before or during an invasion, is the best approach. This unpredictability is due to a complex suite of factors such as reproductive success, competition for resources, diel and year to year fluctuations in predation risk, and parasitism. A variety of disturbances are involved in the "final" community structure (see Olsen *et al.*, 1991; DiDonato and Lodge, 1993; Garvey and Stein, 1993; Hill and Lodge, 1994).

IV. CURRENT AND FUTURE RESEARCH PROBLEMS

Even though considerable effort has been directed toward the study of these decapods, with particular emphasis placed historically on taxonomy, systematics, evolution, anatomy, and some areas of physiology, satisfying answers only to a few questions concerning them have been obtained. Barely a handful of functional morphology studies have been conducted. Recent realization of the impact of exotic species introductions (e.g., *Orconectes rusticus,* Capelli, 1982; Magnuson and Beckel, 1985; Hobbs III and Jass, 1988; Hobbs III *et al.*, 1989; Lodge, 1993a) has initiated numerous studies to determine the underlying mechanisms controlling crayfish species replacements and community changes (Butler and Stein, 1985; France and Welbourn, 1992; Mather and Stein, 1993; Garvey *et al.*, 1994).

Our knowledge of life histories of shrimps and crayfishes is incomplete, at best. As noted above, only about 20 species have been studied in any detail, and varying quantities of data exist for the remaining taxa. Although much time and energy are required for these research projects, the need is paramount. Detailed studies of ovarian development and oogenesis as well as testis development and production of spermatozoa are sorely needed. Also, much is lacking in our understanding of the dynamics of populations for virtually all species of freshwater decapods. Best known are factors influencing population dynamics in the crayfish *0. virilis.*

The Fish and Wildlife Service of the U.S. Department of Interior periodically issues a List of Endangered and Threatened Wildlife. Currently there are seven decapod species placed on the Endangered list, four of which are stygobites. *Palaemonias alabamae* and *P. ganteri* are stygobitic shrimps in Alabama and Kentucky, respectively; *Cambarus* (*Jugicambarus*) *aculabrum* Hobbs and Brown is known only from two caves in northwest Arkansas and is the "newest addition" to the

Federal List (Jacobson, 1996; Elliott, 1997; Boyd, 1997); and *Cambarus* (*Jugicambarus*) *zophonastes* Hobbs and Bedinger is found in an Arkansas cave. The crayfishes *Orconectes* (*Crockerinus*) *shoupi* Hobbs and *Pacifasticus fortis* are restricted to streams in western Tennessee and Shasta County, northeastern California, respectively, and the shrimp *Snycaris pacifica* inhabits streams in central California. Additionally, the Fish and Wildlife Service listed the stygobitic shrimp, *Palaemonetes cummingi* Chace, as a Threatened species. The 1996 IUCN Red List of Threatened Animals (Baillie and Groombridge, 1996) cited three atyid (2 *Palaemonias,* 1 *Syncaris*) and two palaemonid shrimps (*Palaemonetes*); two astacid (*Pacifastacus*) and 118 cambarid (3 *Camabrellus,* 34 *Cambarus,* 2 *Distocambarus,* 8 *Fallicambarus,* 6 *Hobbseus,* 24 *Orconectes,* and 41 *Procambarus*) crayfishes as Threatened; and the atyid shrimp *Syncaris pasadenae* is listed as Extinct.

The American Fisheries Society Endangered Species Committee listed all crayfishes occurring in the United States and Canada and reviewed the conservation status of all taxa (Taylor *et al.,* 1996). There are state lists as well (e.g., Florida—Deyrup and Franz, 1994). Clearly, it is important that these species and their critical habitats be monitored.

The ecology of decapods inhabiting resource limited cave habitats is poorly understood (see Hobbs III, 1992; Brown *et al.,* 1994; Streever, 1996). Further investigations should be conducted to determine the physiological mechanisms that allow the reduction of energy utilization in tissues of stygobites. These should be run in conjunction with studies delineating the levels of environmental flexibility of these forms.

Brown (1981), using electrophoretic analysis of 19 biochemical loci of six crayfishes, substantiated previous reports of low genetic variability in decapod crustaceans [(also demonstrated for crayfishes by Fuller *et al.* (1989), Crandall (1993), Koppelman and Figg (1995), and Fetzner (1996)]. She attributed the low heterozygosity in these crayfishes from the southeastern United States to their limited geographical distribution in a low number of drainage systems and to their high mobility within those watersheds—both are factors allowing crayfishes to perceive their habitats as fine-grained. Nevo (1978) proposed that generalists are more genetically variable than habitat specialists among all major taxonomic groups and Brown, on the basis of her findings, suggested that crayfishes should be classified as habitat specialists. Biochemical investigations of other crayfishes occupying a variety of habitats and from different phylogenetic positions should be undertaken to test the latter hypothesis. Also, additional allozyme studies as well as nucleotide sequence data from the amplified mitochondrial DNA from the 16S ribosomal subunit may aid in determining more precisely the relationships of currently recognized genera, subgenera, species groups, species, and subspecies (see Fetzner, 1996; Crandall and Fitzpatrick, 1996). Current hypotheses of taxonomic relationships may not reflect actual phylogenetic relationships.

Recently, considerable effort has been directed toward the study of pheromones and other chemical substances released into the environment by various decapods. Much is to be learned about the importance of chemical communication in the behavior of aquatic animals and about the impact of these substances on the structure of communities. Basic questions remain unanswered concerning the roles of pheromones and chemical cues in population and community dynamics. Obviously, a tremendous variety of behavioral studies await the ethologist.

Although James (1979, see his bibliography) has shown that benthic macro invertebrates are good indicators of various forms of pollution and are useful as test organisms for determining the presence of toxic substances, the importance of the role of freshwater decapods has not been completely demonstrated (Vermeer, 1972; France, 1986). Recent technical advances and syntheses have brought together existing data on the effects and measurement of toxicity in decapods (Hobbs and Hall, 1974; Doughtie and Rao, 1984; Roldman and Shivers, 1987; Wright and Welbourn, 1993). Acidification due to acid precipitation and loss of buffering capability in freshwater systems in the northeastern United States and Canada have stimulated research on the effects of increased acidity on organisms (France and Graham, 1985; Tierney and Atema, 1986; Wheatly *et al.,* 1996). Crayfishes are highly sensitive to an increase in the hydrogen ion concentration (Wood and Rogano, 1986; Bendell-Young and Harvey, 1991) and although there are inter- and intraspecific variations in the degree of tolerance (Berrill *et al.,*1985), France (1984, 1985a) has shown that *Orconectes virilis* juveniles are particularly sensitive. DiStefano *et al.* (1991) determined that adult *C. b. bartonii* and *C. robustus* are among the most acid-tolerant freshwater organisms they tested. This implies that if populations of these two species would begin to be affected negatively by increasing acidification, then it may already be too late for the recovery of most other species in those lotic or lentic habitats (see Clancy, 1997). Additionally, studies are needed in order to determine the impact of toxic materials on spermatozoa, ova, developing embryos, young, and individuals in various molt stages (premolt, postmolt, molt, intermolt). For those species that regularly experience dilute seawater in some portions of their ranges (e.g., *Pacifastacus leniusculus* and *P. clarkii*), relatively little is

FIGURE 25 Practice of "ditching" to lower local water table significantly (note person for scale) is a common practice (Hillsborough County, FL) with numerous negative environmental impacts.

known concerning the identity and character of short and long-term control of the sodium pump in the crustacean gill and antennal gland (Sarver *et al.,* 1994). Much is yet to be discovered about the physiology of the gut, including enzyme systems, digestion and absorption of nutrients, and metabolic fate of absorbed nutrients.

As a result of increasing demands by humans for energy, aquatic ecosystems undoubtedly will be subjected to considerable additional waste heat. Because crayfishes and shrimps occupy important positions in complex food webs of lentic and lotic ecosystems, precise data concerning temperature preferences, heat tolerances, and thermal responses (e.g., McWhinnie and O'Connor, 1967; Claussen, 1980; Mathur *et al.,* 1982; Taylor, 1984) of these and other organisms will be invaluable in establishing temperature standards such that aquatic communities will be protected.

The following is only a partial list of threats to freshwater ecosystems. A thorough understanding of these and other potential sources of perturbations is paramount if we are to attempt to protect freshwater organisms, including shrimps and crayfishes. Point and nonpoint agricultural (e.g., pesticides, sediments) and other sources (e.g., municipal runoff and discharge) of pollutants; urban and residential development; heavy recreational use of private and public lands (e.g., horse camps); gem, feldspar, coal, and mica mining; logging and chip mills; paper mills; industry (particularly chemical); impoundment structures (dams) and their operations; highway expansion; dewatering of stream sections (e.g., Hiwassee River), including "ditching" (Fig. 25); disturbances (chemical, physical) to hydrological systems in karst areas; sport fishery (stocking, bait-bucket introductions); pig lots; disruption of riparian vegetation; acid rain; and catastrophic spills (highways, trains). Sadly, this is only a brief start to a lengthy list of potential threats to freshwater ecosystems.

Much effort is and will continue to be required in order to recover disturbed lotic and lentic habitats and to conserve and preserve existing natural ecosystems (Richter *et al.,* 1997); this includes surface and subsurface water resources. Aside from basic research conducted to recognize contamination and to treat perturbed areas, implementation of new laws, education, and stricter enforcement of existing legislature are necessary strategies of the future. In some parts of the region treated in this book (e.g., Louisiana), shrimp and crayfish production are important economic activities. Considerable work has been directed toward increasing the yields of crayfish farmers and this effort should be continued (Avault, 1993; see also Daniels *et al.,* 1994). As marine shellfish populations are depleted, humans will undoubtedly turn to alternate food sources, including crayfishes. In various countries in Europe, these crustaceans are valued as food and extensive stocking of exotic species (extreme caution advised!), and rehabilitation programs for native species are ongoing. Most of the interest in North America is centered on development of an export trade for Europe, yet there is potential for an expanded domestic market. Those crayfish species with known economic importance in North America, such as the burrowers *Cambarus (Lacunicambarus) diogenes* Girard and *Failicambarus devastator,* the stream and lake-dweller *Orconectes rusticus,* and the tasty *Procambarus clarkii,* particularly should be studied in detail.

V. COLLECTING AND REARING TECHNIQUES

Because decapods are found in such a large diversity of habitats, a variety of techniques should be used in procuring representatives.

A. Shrimps

1. Atyidae

Collections should not be made for any of the atyid shrimps in North American freshwaters. Populations of the two species of *Syncaris* and the two cave shrimps assigned to *Palaemonias* are very small (two may already be extirpated) and should not be disturbed!

2. Palaemonidae

Freshwater shrimps occupy various sluggish lotic or lentic environments and are most readily collected with the aid of a sturdy, long-handled, fine-meshed dip net or a small-meshed seine. When a particular locality is choked with vegetation or if the water is greater than 1.5 m deep, the use of fine-meshed wire traps with inverted cones baited with meat can be employed. In clear to low turbidity waters, they may be collected at night with the aid of a headlight and dip net (the ruby-colored reflection of their eyes makes for easy location of individuals). Shrimps can be maintained in well-aerated aquaria with considerable vegetation; the water should be changed weekly.

B. Crayfishes: Astacidae and Cambaridae

The North American astacids are known primarily from lakes and streams west of the Continental Divide, and methods for collecting these crayfishes are identical to those for the cambarid stream/lake-dwellers. In shallow streams that have little vegetation, a small-meshed seine (2–5 m long) can be most useful. Deep sections or pooled areas of streams as well as shallow ponds and the littoral zone of lakes may be sampled by pulling the seine (5–8 m length) through the water, taking care to ensure that the weighted margin of the seine remains in contact with the substrates. When electrofishing equipment is used, small-meshed seines can collect any organism that is stunned and carried downstream. Use of various wire traps is recommended for vegetation-choked streams, swamps, marshes, ponds, or lakes or when sampling deep waters of lakes. Inverted-cone minnow (funnel) traps with enlarged openings (4–5 cm diameter) have proven to be quite effective. A residence time of 1–2 nights, using chicken and fish scraps, liver, canned cat food, etc. as bait, is an adequate and productive duration for this sampling technique but this method suffers from sampling bias (Stuecheli, 1991).

A sturdy dip net is very helpful in sampling ditches, mud-bottomed pools, overhanging vegetation, and areas choked with vegetation. This type of net (delta net preferred) can be used to collect individuals from rocky substrates in lakes and streams. Probably most crayfishes burrow at least occasionally for one or more reasons. The diameter of the burrow tube and chimney pellet size provide good indicators of the size of the crayfish inhabiting the burrow (Hobbs III and Jass, 1988). One may examine the burrow aperture and assume that the smaller openings contain juveniles; to sample the adult population, concentrate on burrows with larger diameters. [This would not be an appropriate procedure for small burrowing species, such as *Cambarellus* (*Pandicambarus*) *puer* Hobbs, *Faxonella clypeata,* and *Procambarus* (*Hagenides*) *pygmaeus* Hobbs]. On warm, rainy, humid nights, burrowing crayfishes can be collected fairly easily. These crayfishes come to the mouths of their burrows and often leave them to forage for food. With the aid of a headlight, they can occasionally be collected in large numbers (see also Norrocky, 1984). Use of scuba equipment to observe and/or collect crayfishes in deep streams, pools, ponds, and lakes also is a very useful technique [see France *et al.* (1991) and Lamontagne and Rasmussen (1993)].

During population studies where long-term data are required for individual crayfish, various marking (tagging) techniques are employed. Because these crustaceans periodically molt and thus lose their exoskeleton, external tags are not used since they interfere with casting off the exuvium or they are lost when the exoskeleton is shed. Consequently, internal tags are recommended [see review in Hobbs III (1980); Weingartner (1977) for colored filiaments; Wiles and Guan (1993) for microchip implants].

Stygobitic decapods generally lack pigments and thus appear white or translucent. In subterranean streams, use of a small aquarium net is generally adequate to capture the readily visible organisms. Hand-grabbing also is a fairly productive means of collecting them once they have been disturbed and are "swimming" in open water. Wire minnow traps can be used if checked often. [Note: cave decapod populations are usually small and because of reduced biotic potentials in stygobitic forms, yearly recruitment is very low. Thus, care must be taken not to collect many (if any) individuals from any cave locality—at any time!]

Because crayfishes are aggressive, antisocial, and cannibalistic, rearing and maintaining them can be problematic. On the other hand, because they are scavengers and detritus feeders, they will feed on almost anything organic. Ideally, individuals should be maintained separately, each crayfish in a single container (e.g., 20 × 8 cm stackable glass bowl) and at room temperature. No aeration is required, provided the water does not become fouled with food, wastes, etc. The bowls should have small gravel substrates, and the water should be sufficiently deep to cover the crayfish. If larger containers are used, small cobbles limestone fragments, or broken pieces of flower pots, bricks, or PVC pipe cut to varying sizes can be added for cover (see Alberstadt *et al.,* 1995). Individuals should be fed 2–3 times per week, and the water should be changed approximately every other week. Food can range from various aquatic plants *(Potamogeton, Elodea),* scraps of meat, and earthworms to dried cat food. Continuous illumination will ensure a growth of algae but is not necessary.

Shrimps can be raised in much the same manner as described for crayfishes except more vegetation and deeper water are required for these crustaceans. Females with eggs should definitely be kept isolated. When young hatch, they will remain attached to the pleopods of the female for up to several weeks. Most of the young will then leave the mother and the female should be separated from them to prevent parent/offspring cannibalism. As the young undergo molting and growth, some sibling cannibalism will probably occur. By providing cover for the molting young or by dividing them among several containers, this can be reduced significantly.

On a much grander scale, large tanks or ponds can be utilized for raising crayfishes or shrimps. These should be designed for complete draining with sluice gates or standpipes and should be long and narrow rather than square (Figure 26). Water depth should be maintained at about 1–1.5 m and food can be variable, but catfish feed or poultry starter are good. For additional information on various aspects of crayfish aquaculture, see Huner and Barr (1984), Avault (1993), and Huner (1995).

Upon collection, crayfishes and shrimps should be killed in 5% neutral formalin and kept in the solution from 12 h to a week, depending on the size and number of individuals. Specimens should then be washed in running tap water for several hours and transferred to 70–75% ethanol. If epizooites (e.g., protozoans, copepods, entocytherids, branchiobdellids) are to be saved,

FIGURE 26 Aquaculture ponds for raising crayfish (the blue color morph, *Procambarus paeninsulanus*) in Hillsborough County, FL.

the formalin in which the decapods were killed should be filtered, the exoskeleton rinsed, and the rinse water poured through a fine seive. Any symbionts should be preserved in 75% ethanol and carefully labeled (Hart and Hart, 1974). Where possible, crayfishes and shrimps should be stored in clamp-top glass jars with gaskets. The organisms should be placed in the jar "anterior end down" with room for adequate 70% ethanol preservative. The specimens should be accompanied by a 100% rag paper label on which data are clearly printed using insoluble black ink.

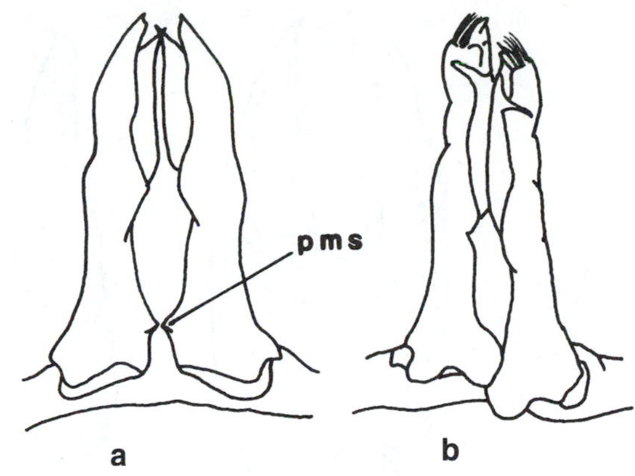

FIGURE 27 Caudal view of first pleopod (gonopod) of first-form males (after Hobbs, 1972). (a) *Bouchardina robinsoni* Hobbs (symmetrical); and (b) *Procambarus* (*Ortmannicus*) *acutissimus* (Girard) (asymmetrical); (pms, proximomesial spur).

VI. IDENTIFICATION

A. Preparation of Specimens

Reliable identification of freshwater shrimps can be made only if appendages are removed, cleared, and mounted on a microscope slide. The second pleopod of adult males should be extirpated and then heated in a lactic acid–chlorazol Black E stain solution [3–5 drops in 1% ethanol (95%) stain solution per 20 mL of lactic acid; a 1% solution of fast green can be substituted] at 150°C for 15 min. The translucent appendage should be transferred to glycerine and then examined.

Crayfishes can be identified with the aid of a hand lens or a stereoscope. First-form males are required for positive identification, and a crayfish's left pleopod should be removed and used for working with the key.

B. Taxonomic Key to Genera of Freshwater Decapoda

Morphological characteristics used in the following key are illustrated in Figures 1, 2, 19, and 27–29. First-form males are required to identify crayfishes at the generic level.

la. Rostrum and abdomen compressed laterally; third pair of pereiopods never bearing chelae .... infraorder Caridea 2

lb. Rostrum and abdomen compressed dorsoventrally; third pair of pereiopods bearing chelae .... infraorder Astacidea 5

2a(la). Fingers of chelae of first and second pereiopods with apical tufts of long setae (Fig.lb); supraorbital spine on either side of base of rostrum; some pereiopods with exopods .... family Atyidae 3

2b. Fingers of chelae of first and second pereiopods without tufts; no supraorbital spines; all pereiopods lacking exopods ....family Palaemonidae 4

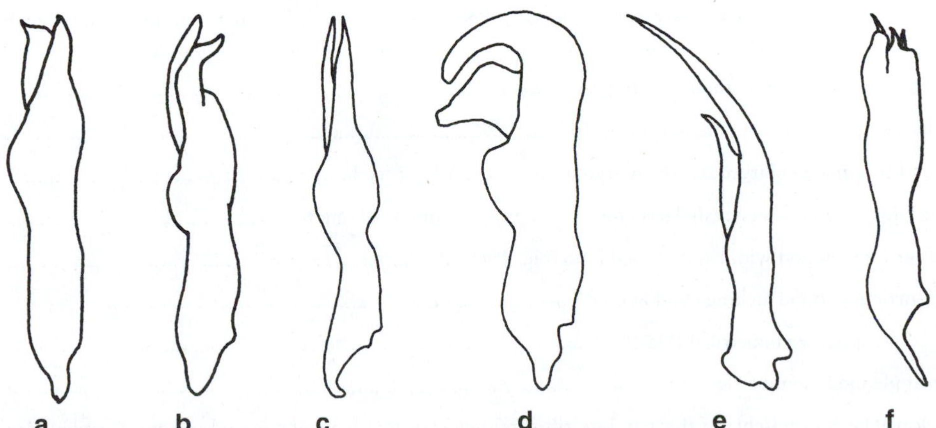

FIGURE 28 (a–d, f) Lateral view of left first pleopods of first-form males (after Hobbs, 1972); (e) caudal view of same. a, *Orconectes i. inermis*; b, *Procambarus* (*Remoticambarus*) *pecki* Hobbs; c, *Orconectes propinquus*; d, *Cambarus latimanus*; e, *Faxonella creaseri* Walls; and f, *Procambarus gracilis*.

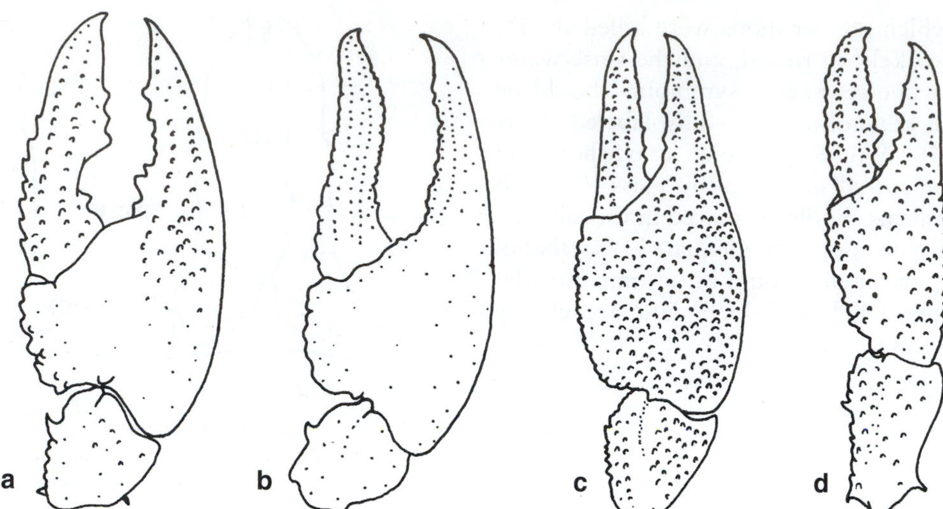

FIGURE 29 Dorsal view of chelae and carpus of first form males (after Hobbs, 1972). a, *Fallicambarus fodiens*; b, *Orconectes rusticus*; c, *Hobbseus cristatus* (Hobbs); *Distocambarus* (*Distocambarus*) *crockeri* Hobbs and Carlson.

3a(2a). Eyes reduced, without pigment; all pereiopods with exopods; stygobitic .... *Palaemonias*

3b. Eyes well developed, pigmented; fifth pereiopod without exopod; epigean .... *Syncaris*

4a(2b). Second pereiopod distinctly longer than first; mandibular palp present; branchiostegal spine on carapace situated distant to anterior margin; epigean .... *Macrobrachium*

4b. Second pereiopod only slightly longer than first; mandibular palp absent; branchiostegal spine on carapace situated near or on anterior margin; epigean or hypogean .... *Palaemonetes*

5a(lb). Males lacking hooks on ischia of all pereiopods; first pleopod of male with distal portion rolled to form cylinder; male never demonstrating cyclic dimorphism; female lacking annulus ventralis .... family Astacidae *Pacifastacus*

5b. Males with ischial hooks on one or more of second through fourth pereiopods (Fig. 19); first pleopod of male complexly folded with sperm tube opening on one terminal element; male always demonstrating cyclic dimorphism; female with annulus ventralis .... family Cambaridae 6

6a(5b). Males with hooks on ischia of second and third pereiopods; very small (15–33 mm total length) .... subfamily Cambarellinae *Cambarellus*

6b. Males lacking hooks on ischia of second pereiopods; hooks present on ischia of third, third and fourth, or fourth pereiopod .... subfamily Cambarinae 7

7a(6b). Mesial surface of flagellum of antennae fringed .... *Barbicambarus*

7b. Mesial surface of flagellum of antennae never fringed 8

8a(7b). Third maxilliped much enlarged; mesial margin of ischium without teeth .... *Troglocambarus*

8b. Third maxilliped not conspicuously large; mesial margin of ischium bearing teeth 9

9a(8b). Coxa of fourth pereiopod with caudomesial boss (Fig. 19c) 10

9b. Coxa of fourth pereiopod lacking caudomesial boss 15

10a(9a). First pair of pleopods asymmetrical (Fig. 27b) .... *Procambarus* (in part)

10b. First pair of pleopods symmetrical (Fig. 27a) 11

11a(10b). Well-developed hooks on ischia of third and fourth pereiopods, or terminals of pleopods disposed cephalodistally 12

11b. Well-developed hooks on ischia of third pereiopods only, terminals of pleopods never disposed cephalodistally 13

12a(11 a). Shaft of first pleopods straight and terminating in two very short elements directed caudodistally or distally (Fig. 28a) .... *Orconectes* (in part)

12b. Shaft of first pleopods straight or bent caudally; if straight, never terminating other than in two very short elements directed caudodistally or distally (Fig. 28b) .... *Procambarus* (in part)

13a(11 b). Opposable margin of dactyl of chela with angular excision in proximal half (Fig. 29a) .... *Fallicambarus*

13b. Opposable margin of dactyl of chela lacking angular excision in proximal half (Fig 29b) 14

14a(13b). First pleopod with terminal elements directed distally (Fig. 28c) .... *Orconectes* (in part)

14b. First pleopod with terminal elements directed caudally or, rarely, caudodistally (Fig. 28d) .... *Cambarus*

15a(9b). First pleopod with two terminal elements, one of which at least twice as long as other (Fig. 28e) .... *Faxonella*

15b. First pleopod with two or more terminal elements, never two with one at least as long as other 16

16a(15b). First pleopod terminating in at least three distinct elements (Fig. 28f) .... Procambarus (in part)

16b. First pleopod terminating in only two distinct elements 17

17a(16b). Dorsal surface of chela studded with crowded small tubercles (Fig. 29c .... *Hobbseus*

17b. Dorsal surface of chela with tubercles over lateral half widely scattered at most 18

18a(17b). First pleopod with proximomesial spur (Fig. 27a) .... *Bouchardina*

18b. First pleopod lacking proximomesial spur 19

19a(18b). Carpus of cheliped slender and longer than mesial margin of palm of chela (Fig 29d); rostrum without marginal spines .... *Distocambarus*

19b. Carpus of cheliped never conspicuously slender and seldom longer than mesial margin of palm (Fig. 29b); if so, rostrum with marginal spines .... *Orconectes* (in part)

LITERATURE CITED

Aiken, D. E. 1968. The crayfish *Orconectes virilis:* survival in a region with severe winter conditions. Canadian Journal of Zoology 46:207–211.

Aiken, D. E., Waddy, S. L. 1992. The growth process in crayfish. Reviews in Aquatic Sciences 6(3,4):335–381.

Albaugh, D. W. 1972. Insectidice tolerances of two crayfish populations (*Procambarus acutus*) in southcentral Texas. Bulletin of Environmental Contamination and Toxicology 8:334–338.

Alberstadt, P. J., Steele, C. W., Skinner, C. 1995. Cover-seeking behavior in juvenile and adult crayfish, *Orconectes rusticus*: effects of darkness and thigmotactic cues. Journal of Crustacean Biology 15(3):537–541.

Ameyaw-Akumfi, C. E. 1979. Appeasement displays in cambarid crayfish (Decapoda, Astacoidea). Crustaceana, Suppl. 5:136–141.

Ameyaw-Akumfl, C. E. 1981. Courtship in the crayfish *Procambarus clarkii* (Girard) (Decapoda, Astacidea). Crustaceana 40:57–64.

Anderson, R. V. 1978. The effects of lead on oxygen uptake in the crayfish, *Orconectes virilis* (Hagen). Bulletin of Environmental Contamination and Toxicology 20:394–400.

Andrews, E. A. 1895. Conjugation in an American crayfish. American Naturalist 29:867–873.

Andrews, S. C., Dillaman, R. M. 1993. Ultrastructure of the gill epithelia in the crayfish *Procambarus clarkii* at different stages of the molt cycle. Journal of Crustacean Biology 13(1):77–86.

Atkinson, R. J. A., Taylor, A. C. 1968. Physiological ecology of burrowing decapods. Symposia of the Zoological Society of London 59:201–226.

Atwood, H. L., Nguyen, P. V. 1995. Neural adaptation in crayfish. American Zoologist 35:28–36.

Avault, J. W., Jr. 1993. A review of world crustacean aquaculture with special reference to crayfish. Freshwater Crayfish 9:1–12.

Baillie, J., Groombridge, B., (Eds.). 1996. 1996 IUCN Red List of Threatened Animals. The International Union for Conservation of Nature and Natural Resources, Kelvyn Press, U.S.A.

Barr, T. C. 1968. Cave ecology and the evolution of troglobites, *in*: Dobzhansky, T., Hecht, M. K., Steere, W. C., Eds. Evolutionary Biology 2:35–102.

Barr, T. C., Holsinger, J. R. 1985. Speciation in cave faunas. Annual Review of Ecology and Systematics 16:313–337.

Barr, T. C., Kuehne, R. A. 1971. Ecological studies in the Mammoth Cave system of Kentucky. II. The ecosystem. Annales de Speleologie Revue Trimestrielle, 26:47–96.

Bechler, D. L. 1981. Copulatory and maternal–offspring behavior in the hypogean crayfish, *Orconectes inermis inermis* Cope and *Orconectes pellucidus* (Tellkampf) (Decapoda, Astacidae). Crustaceana 40:136–143.

Bechler, D. L. 1983. Contamination of Maramec Spring. North American Biospeleology Newsletter 29:5–6.

Bechler, D. L. 1995. A review and prospectus of sexual and interspecific pheromonal communication in crayfish. Freshwater Crayfish 8:657–667.

Beck, J. T. 1980. Life history relationships between the bopyrid isopod *Probopyrus pandalicola* and one of its freshwater shrimp hosts *Palaemonetes paludosus*. American Midland Naturalist 104:135–154.

Beck, J. T., Cowell, B. C. 1976. Life history and ecology of the freshwater caridean shrimp, *Palaemonetes paludosus* (Gibbes). American Midland Naturalist 96:52–65.

Becker, C. D., Genoway, R. G., Merrill, J. A. 1975. Resistance of a northwestern crayfish, *Pacifastacus leniusculus* (Dana), to elevated temperatures. Transactions of the American Fisheries Society 104:374–387.

Bendell-Young, L., Harvey, H. H. 1991. Metal concentrations in crayfish tissues in relation to lake pH and metal concentrations in water and sediments. Canadian Journal of Zoology 69:1076–1082.

Berrill, M., Arsenault, M. 1985. Laboratory induced hybridization of two crayfish species, *Orconectes rusticus* and *O. propinquus*. Journal of Crustacean Biology 5:346–349.

Berrill, M., Hollett, L., Margosian, A., Hudson, J. 1985. Variation in tolerance to low environment pH by the crayfish *Orconectes rusticus, 0. propinquus,* and *Cambarus robustus.* Canadian Journal of Zoology 63:2586–2589.

Black, J. B. 1958. Ontogeny of the first and second pleopods of the male crayfish *Orconectes clypeatus* (Hay). Tulane Studies in Zoology 6:190–203.

Black, J. B. 1975. Inheritance of the blue color mutation in the crawfish *Procambarus acutus acutus* (Girard). Proceedings of the Louisiana Academy of Science 38:25–27.

Blake, M. A., Hart, P. J. B. 1993. The behavioral responses of juvenile signal crayfish *Pacifastacus leniusculus* to stimuli from perch and eels. Freshwater Biology 29(1):89–97.

Blank, G. S., Figler, M. H. 1996. Interspecific shelter competition between the sympatric crayfish species *Procambarus clarkii* (Girard) and *Procambarus zonangulus* (Hobbs and Hobbs). Journal of Crustacean Biology 16(2):300–309.

Borash, J. L., Moore, P. A. 1997. Ultrastructure, morphology and distribution of sensory structures on the chelae of the crayfish, *Orconectes rusticus, Orconectes virilis* and *Orconectes propinquus.* Ohio Journal of Science 97(2-Program Abstracts):A-11.

Bouchard, R. W., Bouchard, J. W. 1995. Two new species and subgenera (*Cambarus* and *Orconectes*) of crayfishes (Decapoda: Cambaridae) from the Eastern United States. Notulae Naturae 471:1–21.

Bovbjerg, R. V. 1956. Some factors affecting aggressive behavior in crayfish. Physiological Zoology 29(2):127–136.

Bowman, T. E., Abele, L. G. 1982. Classification of the recent Crustacea, *in:* Bliss, D. E. Ed., The biology of Crustacea. Vol. 1:Systematics, the fossil record, and biogeography. Academic Press, New York, pp. 1–27.

Boyd, G. L. 1997. Metabolic rates and life history of aquatic organisms inhabiting Logan Cave stream in northwest Arkansas. Coop Unit Publication No. 29, Arkansas Cooperative Fish and Wildlife Research Unit, U. S. Geological Survey, Fayetteville, Arkansas, 107 pp.

Brown, A. V., Pierson, W. K., Brown, K. B. 1994. Organic carbon resources and the payoff-risk relationship in cave ecosystems. Proceedings of the Second International Conference on Ground Water Ecology, United States Environmental Protection Agency, Atlanta, Georgia, pp. 67–76.

Brown, K. 1981. Low genetic variability and high similarities in the crayfish genera *Cambarus* and *Procambarus.* American Midland Naturalist 105:225–232.

Brown, P. B. 1995. Physiological adaptations in the gastrointestinal tract of crayfish. American Zoologist 35:20–27.

Bruski, C. A., Dunham, D. W. 1987. The importance of vision in agonistic communication of the crayfish *Orconectes rusticus.* I: An analysis of bout dynamics. Behavior 103:83–107.

Budd, T. W., Lewis, J. C., Tracey, M. L. 1979. Filtration feeding in *Orconectes propinquus* and *Cambarus robustus* (Decapoda, Cambaridae). Crustaceana, Supple. 5:131–134.

Burggren, W. W., McMahon, B. R. 1983. An analysis of scaphognathite pumping performance in the crayfish *Orconectes virilis:* compensatory changes to acute and chronic hypoxic exposure. Physiological Zoology 56:309–318.

Butler, M. J., Stein, R. A. 1985. An analysis of the mechanisms governing species replacements in crayfish. Oecologia 66:168–177.

Caine, E. A. 1975. Feeding and masticatory structures of six species of the crayfish genus *Procambarus* (Decapoda, Astacidae). Forma et Functio 8:49–66.

Caine, E. A. 1978. Comparative ecology of epigean and hypogean crayfish (Crustacea: Cambaridae) from northwestern Florida. American Midland Naturalist 99:315–329.

Capelli, G. M. 1982. Displacement of northern Wisconsin crayfish by *Orconectes rusticus* (Girard). Limnology and Oceanography 27:741–745.

Capelli, G. M., Hamilton, P. A. 1984. Effects of food and shelter on aggressive activity in the crayfish *Orconectes rusticus* (Girard). Journal of Crustacean Biology 4:252–260.

Capelli, G. M., Magnuson, J. J. 1983. Morphoedaphic and biogeographic analyses of crayfish distribution in northern Wisconsin. Journal of Crustacean Biology 3:548–564.

Capelli, G. M., Munjal, B. M. 1982. Aggressive interactions and resource competition in relation to species displacement among crayfish of the genus *Orconectes.* Journal of Crustacean Biology 2:486–492.

Carpenter, S. R., Kitchell, J. F. 1992. Trophic cascade and biomanipulation: interface of research and management. Limnology and Oceanography 37:208–213.

Carpenter, S. R., Kitchell, J. F. 1993. The trophic cascade in lakes, Cambridge University Press, Cambridge, United Kingdom.

Cesaroni, D., Allegrucci, G., Sbordoni, V. 1992. A narrow hybrid zone between two crayfish species from a Mexican cave. Journal of Evolutionary Biology 5:643–659.

Chace, F. A., Jr., Hobbs, H. H., Jr. 1969. The freshwater and terrestrial decopod crustaceans of the West Indies with special reference to Dominica. U. S. National Museum Bulletin 292:1–258.

Charlebois, P. M., Lamberti, G. A. 1996. Invading crayfish in a Michigan stream: direct and indirect effects on periphyton and macroinvertebrates. Journal of the North American Benthological Society 15(4):551–563.

Ciruna, K. A., Dunham, D. W., Harvey, H. H. 1995. Detection and response to food versus conspecific tissue in the crayfish *Cambarus bartonii* (Fabricius, 1978) (Decapoda, Cambaridae). Crustaceana 68(6):782–788.

Clancy, P. 1997. Feeling the pinch: the troubled plight of America's crayfish. Nature Conservancy 47(3):10–15.

Claussen, D. L. 1980. Thermal acclimation in the crayfish, *Orconectes rusticus* and *0. virilis.* Comparative Biochemistry and Physiology 66A:377–384.

Cooper, J. E. 1975. Ecological and behavioral studies in Shelta Cave, Alabama, with emphasis on decapod crustaceans. Ph.D. Thesis, University of Kentucky, Lexington, 364 pp.

Cooper, J. E., Braswell, A. L. 1995. Observations on North Carolina crayfishes (Decapoda: Cambaridae). Brimleyana 22:87–132.

Cooper, J. E., Cooper, M. R. 1978. Growth, longevity, and reproductive strategies in Shelta Cave crayfishes. National Speleological Society Bulletin 40:97.

Corey, S. 1987a. Comparative fecundity of four species of crayfish in southwestern Ontario, Canada (Decapoda, Astacidea). Crustaceana 53:276–286.

Corey, S. 1987b. Intraspecific diferences in reproductive potential, realized reproduction and actual production in the crayfish *Orconectes propinquus* (Girard 1852) in Ontario. American Midland Naturalist 118:424–432.

Corey, S. 1990. Life history of *Cambarus robustus* Girard in the Eramosa-Speed river system of south-western Ontario, Canada (Decapoda, Astacidae). Crustaceana 59:225–230.

Covich, A. P. 1978. Spatial and temporal patterns of foraging activity among radio-monitored crayfish. Page 13, *in:* 26th Annual Meeting of the North American Benthological Society, Titles and Abstracts.

Covich, A. P. 1982. Crayfish foraging and ecosystem control by selective omnivory. Page 14, *in:* Payne, J. F., Ed. Crayfish Distribution Patterns. Symposium, American Society of Zoologists, Louisville, KY.

Covich, A. P., Dye, L. L., Mattice, J. S. 1981. Crayfish predation on *Corbicula* under laboratory conditions. American Midland Naturalist 105:181–188.

Cox, D. K., Beauchamp, J. J. 1982. Thermal resistance of juvenile crayfish, *Cambarus bartoni* [sic] (Fabricius): experiment and model. American Midland Naturalist 108:187–193.

Crandall, K. A. 1993. Molecular systematics and evolutionary biology of the crayfish subgenus *Procericambarus* (Decapoda: Cambaridae). Ph.D. Thesis, Washington University, St. Louis, MO.

Crandall, K. A., Fitzpatrick, J. F., Jr. 1996. Crayfish molecular systematics: using a combination of procedures to estimate phylogeny. Systematic Biology 45(1):1–26.

Crawshaw, L. I. 1974. Temperature selection and activity in the crayfish *Orconectes immunis*. Journal of Comparative Physiology 95:315–322.

Creed, R. P., Jr. 1994. Direct and indirect effects of crayfish grazing in a stream community. Ecology 75(7):2091–2103.

Crocker, D. W., Barr, D. W. 1968. Handbook of the crayfishes of Ontario. Miscellaneous Publication of the Royal Ontario Museum. Uni. of Toronto Press, Toronto, 158 pp.

Crowl, T. A., Covich, A. P. 1990. Predator-induced life-history shifts in a freshwater snail. Science 247:949–951.

Cuadras, J. 1993. Secretory organelles in the crayfish nervous system. Comparative Biochemistry and Physiology 104A(3):419–422.

Daniels, W. H., D'Abramo, L. R., Graves, K. F. 1994. Ovarian development of female Red Swamp Crayfish (*Procambarus clarkii*) as influenced by temperature and photoperiod. Journal of Crustacean Biology 14(3):530–537.

Daveikis, V. F., Alikhan, M. A. 1996. Comparative body measurements, fecundity, oxygen uptake, and ammonia excretion in *Cambarus robustus* (Astacidae, Crustacea) from an acidic and a neutral site in northeastern Ontario, Canada. Canadian Journal of Zoology 74:1196–1203.

Davies, I. J. 1984. Effects of on experimental whole-lake acidification on a population of the cray fish *Orconectes virilis*. North American Benthological Society Abstracts, 32:35.

DeCoursey, P. J. 1983. Biological timing. *in*: Bliss, D. E., Ed. The biology of Crustacea. Vol. 7, Behavior and Ecology. Academic Press, New York, pp. 107–162.

Deyrup, M., Franz, R. (Eds.). 1994. Rare and Endangered Biota of Florida. Vol. IV. Invertebrates. Uni. Press of Florida, Gainesville, 798 pp.

Dickson, G. W., Franz, R. 1980. Respiration rates, ATP turnover and adenylate energy change in excised gills of surface and cave crayfish. Comparative Biochemistry and Physiology 65A:375–379.

Dickson, G. W., Briese, L. A., Giesey, J. P., Jr. 1979. Tissue metal concentrations in two crayfish species cohabiting a Tennessee cave stream. Oecologia 44:8–12.

Dickson, G. W., Giesy, J. P. Jr., Briese, L. A. 1982. The effect of chronic cadmium exposure on phosphoadenylate concentrations and adenylate energy charge of gills and dorsal muscle tissue of crayfish. Environmental Toxicology and Chemistry 1:147–156.

Dickson, J. S., Dillaman, R. M., Roer, R. D., Roye, D. B. 1991. Distribution and characterization of ion transporting and respiratory filaments in the gills of *Procambarus clarkii*. Biological Bulletin 180:154–166.

DiDonato, G. T., Lodge, D. M. 1993. Species replacements among *Orconectes* crayfishes in Wisconsin lakes: the role of predation by fish. Canadian Journal of Fisheries and Aquatic Sciences 50:1484–1488.

Dimond, J. B., Kadunce, R. E., Gelchell, A. S., Blease, J. A. 1968. Persistence of DDT in crayfish in a natural environment. Ecology 49:759–762.

DiStefano, R. J., Neves, R. J., Helfrich, L. A., Lewis, M. C. 1991. Response of the crayfish *Cambarus bartonii bartonii* to acid rain exposure in southern Appalachian streams. Canadian Journal of Zoology 69:1585–1591.

Dobkin, S. 1971. The larval development of *Palaemonetes cummingi* Chace, 1954 (Decapoda, Palaemonidae), reared in the laboratory. Crustaceana 20:285–297.

Doughtie, D. G., Rao, K. R. 1984. Histopathological and ultrastructural changes in the antennal gland, midgut, hepatopancreas, and gill of grass shrimp following exposure to hexavalent chromium. Journal of Invertebrate Pathology 43:89–108.

Dubé, P., Portelance, B. 1992. Temperature and photoperiod effects on ovarian maturation and egg laying of the crayfish, *Orconectes limosus*. Aquaculture 102:161–168.

Dunham, D. W., Ciruna, K. A., Harvey, H. H. 1997. Chemosensory role of antennules in the behavioral integration of feeding by the crayfish *Cambarus bartonii*. Journal of Crustacean Biology 17(1):27–32.

Dunham, D. W., Oh, J. W. 1992. Chemical sex discrimination in the crayfish *Procambarus clarkii*: role of antennules. Journal of Chemical Ecology 18:2362–2372.

Dunham, D. W., Oh, J. W. 1996. Sex discrimination by female *Procambarus clarkii* (Girard, 1852) (Decapoda, Cambaridae): use of chemical and visual stimluli. Crustaceana 69(4):534–542.

Elliott, W. R. 1997. Conservation of the North American cave and karst biota. Ecosystems of the World, Subterranean Biota, Elsevier Science, pp. 1–24.

Emery, A. R. 1970. Fish and crayfish mortalities due to an internal seiche in Georgian Bay, Lake Huron. Journal of the Fisheries Research Board of Canada, 27:1165–1168.

Eng, L. L. 1981. Distribution, life history, and status of the California freshwater shrimp, *Syncaris pacifica* (Holmes). Inland Fish. Endangered Species Program, Special Publication 81–1:1–27.

Eversole, A. G. 1995. Distribution of three rare crayfish species in South Carolina. Freshwater Crayfish 8:113–120.

Eversole, A. G., Foltz, J. W. 1993. Habitat relationships of two crayfish species in a mountain stream. Freshwater Crayfish 9:300–310.

Feminella, J. W., Resh, V. R. 1989. Submersed macrophytes and grazing crayfish:an experimental study of herbivory in a Californian freshwater marsh. Holarctic Ecology 12:1–8.

Fernández-de-Miguel, F., Aréchiga, H. 1992. Sensory inputs mediating two opposite behavioural responses to light in the crayfish *Procambarus clarkii*. Journal of Experimental Biology 164:153–169.

Fetzner, J. W., Jr. 1996. Biochemical systematics and evolution of the crayfish genus *Orconectes* (Decapoda: Cambaridae). Journal of Crustacean Biology 16(1):111–141.

Fielder, D. D. 1972. Some aspects of the life histories of three closely related crayfish species, *Orconectes obscurus, O. s. sanborni*, and *O. propinquus*. Ohio Journal of Science 72(3):129–145.

Figiel, C. R., Jr., Babb, J. G., Payne, J. F. 1991. Population regulation in young of the year crayfish, *Procambarus clarkii* (Girard, 1852) (Decapoda, Cambaridae). Crustaceana 61(3):301–307.

Figiel, C. R., Jr., Miller, G. L. 1995. The frequency of chela autotomy and its influence on the growth and survival of the crayfish *Procambarus clarkii* (Girard, 1852) (Decapoda, Cambaridae). Crustaceana 68(4):472–483.

Figler, M. H., Twum, M., Finkelstein, J. E., Peeke, H. V. S. 1995. Maternal aggression in red swamp crayfish (*Procambarus clarkii*, Girard): the relation between reproductive status and outcome of aggressive encounters with male and female conspecifics. Behaviour 132(1,2):107–125.

Fitzpatrick, J. F., Jr. 1977. A new crawfish of the genus *Hobbseus* from northeast Mississippi, with notes on the origin of the genus (Decapoda, Cambaridae). Proceedings of the Biological Society of Washington 90:367–374.

Fitzpatrick, J. F., Jr. 1983. A revision of the dwarf crawfishes. Journal of Crustacean Biology 3:266–277.

Fitzpatrick, J. F., Jr. 1986. The pre-Pliocene Tennessee River and its bearing on crawfish distribution (Decapoda: Cambaridae). Brimleyana 12:123–146.

Fitzpatrick, J. F., Jr. 1987a. The subgenera of the crawfish genus *Orconectes* (Decapoda: Cambaridae). Proceedings of the Biological Society of Washington 100:44–74.

Fitzpatrick, J. F., Jr. 1987b. Notes on the so-called "blue color phase" in North American cambarid crawfishes (Decapoda, Astacoidea). Crustaceana 52(3):316–319.

Flint, R. W. 1975. Growth in a population of the crayfish *Pacifastacus leniusculus* from a subalpine lacustrine environment. Journal of the Fisheries Research Board of Canada 32:2433–2440.

Flynn, M. A., Holliday, C. W. 1994. Renal Na, K-ATPase and osmoregulation in the crayfish, *Procambarus clarkii*. Comparative Biochemistry and Physiology 107(2):349–356.

Flynn, M. F., Hobbs, H. H. III. 1984. Parapatric crayfishes in southern Ohio: evidence of competitive exclusion? Journal of Crustacean Biology 4:382–389.

Font, W. F. 1994. *Alloglossidium greeri* n. sp. (Digenea:Macroderoididae) from the Cajun Dwarf crayfish, *Cambarellus schufeldti*, in Louisiana, U.S.A. Transactions of the American Microscopical Society 113(1):86–89.

Fotenot, L. W., Platt, S. G., Dwyer, C. M. 1993. Observation on crayfish predation by water snakes, *Nerodia* (Reptilia, Colubridae). Brimleyana 19:77–82.

France, R. L. 1984. Comparative tolerance to low pH of three life stages of the crayfish *Orconectes virilis*. Canadian Journal of Zoology 62:2360–2363.

France, R. L. 1985a. Preliminary investigations of effects of sublethal acid exposure on maternal behavior in the crayfish *Orconectes virilis*. Bulletin of Environmental Contamination and Toxicology 35:641–645.

France, R. L. 1985b. Low pH avoidance by crayfish (*Orconectes virilis):* evidence for sensory conditioning. Canadian Journal Zoology 63:258–262.

France, R. L. 1986. Current status of methods of toxicological research on freshwater crayfish. Canadian Technical Report of Fisheries and Aquatic Sciences No. 1404:1–20.

France, R. L. 1992. The North American latitudinal gradient in species richness and geographical range of freshwater crayfish and amphipods. The American Naturalist 139(2):342–354.

France, R. L., Collins, N. C. 1993. Extirpation of crayfish in a lake affected by long-range anthropogenic acidification. Conservation Biology 7(1):184–188.

France, R. L., Holmes, J., Lynch, A. 1991. Use of size-frequency data to estimate the age composition of crayfish populations. Canadian Journal of Fisheries and Aquatic Sciences 48(12): 2324–2332.

France, R. L., Graham, L. 1985. Increased microsporidian parasitism of the crayfish *Orconectes virilis* in an experimentally acidified lake. Water Air and Soil Pollution 26:129–136.

France, R. L., Welbourn, P. M. 1992. Influence of lake pH and macrograzers on the distribution and abundance of nuisance metaphytic algae in Ontario, Canada. Canadian Journal of Fisheries and Aquatic Sciences 49(1):185–195.

Franz, R. 1977. Observations on the food, feeding behavior, and parasites of the striped swamp snake, *Regina alleni*. Herpetologica 33:91–94.

Franz, R., Bauer, J., Morris, T. 1994. Review of biologically significant caves and their faunas in Florida and south Georgia. Brimleyana 20:1–109.

Franz, R., Lee, D. S. 1982. Distribution and evolution of Florida's troglobitic crayfishes. Bulletin of the Florida State Museum, Biological Sciences 28:53–78.

Fuller, S. J., Mellen, A. J., Mosher, M. W., Parker, K. E. 1989. Enzymatic comparison of three species of crayfish by cellulose acetate electrophoresis. Transactions Missouri Academy of Science 23:7–11.

Gamradt, S. C., Kats, L. B. 1996. Effect of introduced crayfish and mosquitofish on California newts. Conservation Biology 10(4):1155–1162.

Garvey, J. E., Stein, R. A. 1993. Evaluating how chela size influences the invasion potential of an introduced crayfish (*Orconectes rusticus*). American Midland Naturalist 129:172–181.

Garvey, J. E., Stein, R. A., Thomas, H. M. 1994. Assessing how fish predation and interspecific prey competition influence a crayfish assemblage. Ecology 75(2):532–547.

Gelder, S. R. 1996. A review of the taxonomic nomenclature and a checklist of the species of the Branchiobdellae (Annelida: Clitellata). Proceedings of the Biological Society of Washington 109(4):653–663.

Godley, J. S., McDiarmid, R. W., Rojas, N. N. 1984. Estimating prey size and number in crayfish-eating snakes, genus *Regina*. Herpetologica, 40:82–88.

Gore, J. A., Bryant, R. M., Jr. 1990. Temporal shifts in physical habitat of the crayfish, *Orconectes neglectus* (Faxon). Hydrobiologia 199:131–142.

Gowing, H., Momot, W. T. 1979. Impact of brooktrout (*Salvelinus fontinalis*) predation on the crayfish *Orconectes virilis* in three Michigan lakes. Journal of the Fisheries Research Board of Canada 36:1191–1196.

Grow, L. 1981. Burrowing behaviour in the crayfish, *Cambarus diogenes diogenes* Girard. Animal Behaviour 29:351–356.

Guiasu, R. C., Barr, D. W., Dunham, D. W. 1996. Distribution and status of crayfishes of the genera *Cambarus* and *Fallicambarus* (Decapoda: Cambaridae) in Ontario, Canada. Journal of Crustacean Biology 16(2):373–383.

Guinot, D. 1994. Decapoda Brachyura. *in*: Juberthie C., Decu, V., Eds. Encyclopedaedia Biospeologica, Tome I, Société de Biospéologie, Moulis-Bucarest, pp. 165–179.

Hamr, P., Berrill, M. 1985. The life histories of north temperate populations of the crayfish *Cambarus robustus* and *Cambarus bartoni* [sic]. Canadian Journal of Zoology 63:2313–2322.

Hanson, J. M., Chambers, P. A., Prepas, E. E. 1990. Selective foraging by the crayfish *Orconectes virilis* and its impact on macroinvertebrates. Freshwater Biology 24:69–80.

Hart, C. W., Jr. 1994. A dictionary of non-scientific names of freshwater crayfishes (Astacoidea and Parastacoidea), including other words and phrases incorporating crayfish names. Smithsonian Contributions to Anthropology No. 38, 127 pp.

Hart, C. W., Jr., Clark, J. 1987. An interdisciplinary bibliography of freshwater crayfishes (Astacoidea and Parastacoidea) from Aristotle through 1985. Smithsonian Contributions to Zoology No. 455, 437 pp.

Hart, D. G., Hart, C. W., Jr. 1974. The ostracod family Entocytheridae. Academy of Natural Sciences of Philadelphia, Monograph 18, 239 pp.

Hasiotis, S. T. 1995. Notes on the burrow morphologies and nesting behaviors of adults and juveniles of *Procambarus clarkii* and *Procambarus acutus* acutus (Decapoda: Cambaridae). Freshwater Crayfish VIII:623–634.

Hatt, H. 1984. Structural requirements of amino acids and related compounds for stimulation of receptors in crayfish walking leg. Journal of Comparative Physiology 155A:219–231.

Hayes, W. A., II. 1975. Behavioral components of social interactions in the crayfish *Procambarus gracilis* (Bundy) (Decapoda, Cambaridae). Proceedings of the Oklahoma Academy of Science 55:1–5.

Hayes, W. A., II. 1977. Predator response postures of crayfish. I. The genus *Procambarus* (Decapoda, Cambaridae). Southwestern Naturalist 21:443–449.

Hazlett, B. A. 1985. Chemical detection of sex and condition in the crayfish *Orconectes virilis*. Journal of Chemical Ecology 2:181–189.

Hazlett, B. A. 1990. Source and nature of the disturbance–chemical system in crayfish. Journal of Chemical Ecology 16(7): 2263–2275.

Hazlett, B. A. 1994a. Crayfish feeding responses to zebra mussels depend on microorganisms and learning. Journal of Chemical Ecology 20(10):2623–2630.

Hazlett, B. A. 1994b. Alarm responses in the crayfish *Orconectes virilis* and *Orconectes propinquus*. Journal of Chemical Ecology 20(7):1525–1535.

Hazlett, B. A., Anderson, F. E., Esman, L. A., Stafford, C., Munro, E. 1992. Interspecific behavioral ecology of the crayfish *Orconectes rusticus*. Journal of Freshwater Ecology 7:69–76.

Hazlett, B. A., Rittschof, D. 1985. Variation in rate of growth in the crayfish *Orconectes virilis*. Journal of Crustacean Biology 5:341–346.

Hazlett, B. A., Rittschof, D., Rubenstein, D. 1974. Behavioral biology of the crayfish *Orconectes virilis*. Home range. American Midland Naturalist 92:301–319.

Hedgpeth, J. W. 1949. The North American species of *Macrobrachium* (River Shrimp). Texas Journal of Science 1(3):28–38.

Hedgpeth, J. W. 1968. The atyid shrimp of the genus *Syncaris* in California. Internationale Revue der Gesamten Hydrobiologie und Hydrographie 53(4):511–524.

Hill, A. M., Lodge, D. M. 1994. Diel changes in resource demand: competition and predation in species replacement among crayfishes. Ecology 75(7):2118–2126.

Hill, A. M., Lodge, D. M. 1995. Multi-trophic-level impact of sublethal interactions between bass and omnivorous crayfish. Journal of the North American Benthological Society 14(2):306–314.

Hill, A. M., Sinars, D. M., Lodge, D. M. 1993. Invasion of an occupied niche by the crayfish *Orconectes rusticus*: potential importance of growth and mortality. Oecologia 94:303–306.

Hobbs, H. H., Jr. 1969. On the distribution and phylogeny of the crayfish genus *Cambarus*. *in:* Holt, P. C., Hoffman, R., Hart, C. W., Jr, Eds. The distributional history of the biota of the southern Appalachians. Part I: Invertebrates. Research Division Monograph 1. Virginia Polytechnic Institute, Blacksburg, VA, pp. 93–178.

Hobbs, H. H., Jr. 1972. Crayfishes (Astacidae) of North and Middle America. Biota of Freshwater Ecosystems. U.S. Environmental Protection Agency, Water Pollution Control Research Service Identification Manual 9, 173 pp.

Hobbs, H. H., Jr. 1974. Synopsis of the families and genera of crayfishes (Crustacea: Decapoda). Smithsonian Contributions to Zoology No. 164, 32 pp.

Hobbs, H. H., Jr. 1976. Adaptations and convergence in North American crayfishes. Freshwater Crayfish 2:541–551.

Hobbs, H. H., Jr. 1981. The crayfishes of Georgia. Smithsonian Contributions to Zoology No. 318, 549 pp.

Hobbs, H. H., Jr. 1986. Highlights of a half century of crayfishing. Freshwater Crayfish 6:12–23.

Hobbs, H. H., Jr. 1988. Crayfish distribution, adaptive radiation, and evolution. *in:* Holdich, D. M., Lowery, R. S., Eds. Freshwater crayfish biology: Management and exploitation. Croom Helm, London, pp. 52–82.

Hobbs, H. H., Jr. 1989. An illustrated checklist of American crayfishes (Decapoda: Astacidae, Cambaridae, and Parastacidae). Smithsonian Contributions to Zoology No. 480, 236 pp.

Hobbs, H. H., Jr., Franz, R. 1986. New troglobitic crayfish with comments on its relationship to epigean and other hypogean crayfishes of Florida. Journal of Crustacean Biology 6:509–519.

Hobbs, H. H., Jr., Hall, E. T. Jr. 1974. Crayfishes (Decapoda: Astacidae). *in:* Hart, C. W., Jr. Fuller, S. L. H., Eds. Pollution ecology of freshwater invertebrates. Academic Press, New York, pp. 195–241.

Hobbs, H. H., Jr., Peters, D. J. 1993. New records of entocytherid ostracods infesting burrowing and cave-dwelling crayfishes, with descriptions of two new species. Proceedings of the Biological Society of Washington 106(3):455–466.

Hobbs, H. H., Jr., Whiteman, M. 1991. Notes on the burrows, behavior, and color of the crayfish *Fallicambarus (F.) devastator* (Decapoda: Cambaridae). Southwestern Naturalist 36(1):127–135.

Hobbs, H. H., Jr., Hobbs, H. H. III, Daniel, M. A. 1977. A review of the troglobitic decapod crustaceans of the Americas. Smithsonian Contributions to Zoology No. 244, 183 pp.

Hobbs, H. H., III. 1976. Observations on the cave-dwelling crayfishes of Indiana. Freshwater Crayfish 2:405–414.

Hobbs, H. H., III. 1980. Studies of the cave crayfish, *Orconectes inermis inermis* Cope (Decapoda, Cambaridae). Part IV: Mark-recapture procedures for estimating population size and movements of individuals. International Journal of Speleology 10:303–322.

Hobbs, H. H., III. 1987. Gasoline pollution in an Indiana cave. *in*: Program of the 1987 National Speleological Society Convention, p. 33.

Hobbs, H. H., III. 1992. Caves and Springs, *in*: Hackney, C. T., Adams, S. M., Martin, W. H. Eds., Biotic communities of the southeastern United States, Chapter 3: Wiley, New York pp. 59–131.

Hobbs, H. H., III. 1993. Trophic relationships of North American freshwater crayfishes and shrimps. Contributions in Biology and Geology, Milwaukee Public Museum. Milwaukee, Wisconsin, 110 pp.

Hobbs, H. H., III. 1994. Biogeography of subterranean decapods in North and Central America and the Caribbean region (Caridea, Astacidea, Brachyura). Hydrobiologia 287:95–104.

Hobbs, H. H., III. 1998. Decapoda (Caridea, Astacidea, Anomura). *in*: Juberthie, C., Decu, V. Eds. Encyclopædia Biospeologica. Société de Biospélogie, Moulis-Bucarest, Pages 891–911.

Hobbs, H. H., III. 2000 Subterranean Exosystems: Crustacea, *in*: Wilkens, H., Culver, D. C., Humphreys, W. F., Eds., Ecosystems of the World—Subterranean Biota, Chapter 5: Elsevier Science, The Netherlands pp. 95–107.

Hobbs, H. H., III, Jass, J. P. 1988. The crayfishes and shrimp of Wisconsin (Decapoda: Palaemonidae, Cambaridae). Special Publications in Biology and Geology, No. 5. Milwaukee Public Museum. Milwaukee, WI. 177 p.

Hobbs. H. H., III, Jass, J. P., Huner, J. V. 1989. A review of global crayfish introductions with particular emphasis on two North American species (Decapoda:Cambaridae). Crustaceana 56:299–316.

Holdich, D. M., Reeve, I. D. 1988. Functional morphology and anatomy. *in*: Holdich, D. M., Lowery, R. R. S., Eds. Freshwater Crayfish:Biology, Management and Exploitation. Timber Press, Portland, OR, Pages 11–51.

Holsinger, J. R. 1988. Troglobites: the evolution of cave-dwelling organisms. American Scientist 76:146–153.

Holt, P. C., Opell, B. D. 1993. A checklist of and illustrated key to the genera and species of the Central and North American Cambarincolidae (Clitellata:Branchiobdellida). Proceedings of the Biological Society of Washington 106(2):251–295.

Horns, W. H., Magnuson, J. J. 1981. Crayfish predation on lake trout eggs in Trout Lake, Wisconsin. Rapports et Proces- Verbaux des Reunions, Conseil International pour Exploration de la Mer 178:299–303.

Hubschman, J. H. 1967. Effects of copper on the crayfish *Orconectes rusticus* (Girard). II. Mode of toxic action. Crustaceana 12: 141–150.

Hubschman, J. H., Rose, J. A. 1969. *Palaemonetes kadiakensis* Rathbun: post embryonic growth in the laboratory (Decapoda, Palaemonidae). Crustaceana 16:81–87.

Huner, J. V. 1977. Observations on the biology of the river shrimp from a commercial bait fishery near Port Allen, Louisiana. *in:* 31st Annual Conference of the Southeastern Association of Fish and Wildlife Agencies, Proceedings, Pages 380–386.

Huner, J. V. 1995. Ecological observations of red swamp crayfish, *Procambarus clarkii* (Girard, 1852), and white river crayfish, *Procambarus zonangulus* Hobbs and Hobbs 1990, as regards their cultivation in earthen ponds. Freshwater Crayfish 10:456–468.

Huner, J. V., Barr, J. E. 1984. Red Swamp Crawfish:Biology and Exploitation. Louisiana Sea Grant College Program. Louisiana State University, Baton Rouge, 136 pp.

Huppop, K. 1985. The role of metabolism in the evolution of cave animals. National Speleological Society Bulletin 47:136–146.

Huryn, A. D., Wallace, J. B. 1987. Production and litter processing by crayfish in an Appalachian mountain stream. Freshwater Biology 18:277–286.

Jacobson, R. R. 1996. Cave crayfish *Cambarus aculabrum* recovery plan. U. S. Fish and Wildlife Service, Southeast Region, Atlanta, 37 pp.

Jacobson, T. R., Hartfield, P. 1997. Alabama cave shrimp *Palaemonias alabamae* recovery plan. U. S. Fish and Wildlife Service, Southeast Region, Atlanta, 46 pp.

James, A. 1979. The value of biological indicators in relation to other parameters of water quality. *in:* James, A., Evison, L. Eds. Biological Indicators of water quality. Wiley, Chichester, U K, Pages 1–16.

Jezerinac, R. F., Stocker, G. W., Tarter, D. C. 1995. The crayfishes (Decapoda:Cambaridae) of West Virginia. Bulletin of the Ohio Biological Survey, New Series 10(1):1–193.

Johannes, R. E., Webb, K. L. 1970. Release of dissolved organic compounds by marine and freshwater invertebrates. *in:* Hood, D. W. Eds. Symposium on organic matter in natural waters. Institute of Marine Sciences. University of Alaska, Fairbanks, Pages 257–273.

Johnson, S. K. 1977. Crawfish and freshwater shrimp diseases. Texas A&M University, Agricultural Extension Service, College Station, 19 pp.

Johnston, C. E., Figiel, C. 1997. Microhabitat parameters and life-history characteristics of *Fallicambarus gordoni* Fitzpatrick, a crayfish associated with pitcher-plant bogs in southern Mississippi. Journal of Crustacean Biology 17(4):687–691.

Jordan, F., Babbitt, K. J., McIvor, C. C., Miller, S. J. 1996. Spatial ecology of the crayfish *Procambarus alleni* in a Florida wetland mosaic. Wetlands 16(2):134–142.

Kasuya, E., Tsurumaki, S., Kanie, M. 1996. Reversal of sex roles in the copulatory behavior of the imported crayfish *Procambarus clarkii*. Journal of Crustacean Biology 16(3):469–471.

Keller, T. A. 1992. The effects of the branchiobdellid annelid *Cambarincola fallax* on the growth rate and condition of the crayfish *Orconectes rusticus*. Journal of Freshwater Ecology 7(2):165–171.

Kershner, M. W., Lodge, D. M. 1995. Effects of littoral habitat and fish predation on the distribution of an exotic crayfish, *Orconectes rusticus*. Journal of the North American Benthological Society 14(3):414–422.

Koppelman, J. B., Figg, D. E. 1995. Genetic estimates of variability and relatedness for conservation of an Ozark cave crayfish species complex. Conservation Biology 9(5):1288–1294.

Kulkarni, G. K., Glade, L., Fingerman, M. 1991. Oogenesis and effects of neuroendocrine tissues on *in vitro* systhesis of protein by the ovary of the Red Swamp Crayfish *Procambarus clarkii* (Girard). Journal of Crustacean Biology 11(4):513–522.

Kushlan, J. A., Kushlan, M. S. 1980. Population fluctuations of the prawn, *Palaemonetes paludosus*, in the Everglades. American Midland Naturalist 103:401–403.

Lachaise, F., Le Roux, A., Hubert, M., Lafont, R. 1993. The molting gland of crustaceans: localization, activity, and endocrine control (a review). Journal of Crustacean Biology 13(2):198–234.

Lamontagne, S., Rasmussen, J. B. 1993. Estimating crayfish density in lakes using quadrats: maximizing precision and efficiency. Canadian Journal of Fisheries and Aquatic Sciences 50: 623–626.

Layne, J. R., Manis, M. L., Claussen, D. L. 1985. Seasonal variation in the time course of thermal acclimation in the crayfish *Orconectes rusticus*. Freshwater Invertebrate Biology 4:98–104.

Leitheuser, A. T., Holsinger, J. R., Olson, R., Pace, N. R., Whitman, R. L., Whitmore, T. 1985. Ecological analysis of the Kentucky cave shrimp, *Palaemonias ganteri* Hay, at Mammoth Cave National Park (Phase V). Old Dominion University Research Foundation, Norfolk, 102 pp.

Levenbach, S., Hazlett, B. A. 1996. Habitat displacement and the mechanical and display functions of chelae in crayfish. Journal of Freshwater Ecology 11(4):485–492.

Lichtwardt, R. W. 1986. The Trichomycetes: Fungal Associates of arthropods. Springer, New York, 343 pp.

Light, T., Erman, D. C., Myric, C., Clarke, J. C. 1995. Decline of the Shasta crayfish (*Pacifastacus fortis* Faxon) of northeastern California. Conservation Biology 9(6):1567–1577.

Lisowski, E. A. 1983. Distribution, habitat, and behavior of the Kentucky cave shrimp, *Palaemonias ganteri* Hay. Journal of Crustacean Biology 3:88–92.

Lodge, D. M. 1991. Herbivory on freshwater macrophytes. Aquatic Botany 41:195–224.

Lodge, D. M. 1993a. Species invasions and deletions: community effects and responses to climate and habitat change. *in:* Kareiva, P. M., Kingsolver, J. G., Huey, R. B. Eds. Biotic interactions and global change. Sinauer Associates, In., Sunderland, MA, Pages 367–387.

Lodge, D. M. 1993b. Biological invasions: lessons for ecology. Trends in Ecology and Evolution 8:133–137.

Lodge, D. M., Brown, K. M., Klosiewski, S. P., Stein, R. A., Covich, A. P., Leathers, B. K., Bronmark, C. 1987. Distribution of freshwater snails: spatial scale and the relative importance of physicochemical and biotic factors. American Malacological Bulletin 5:73–84.

Lodge, D. M., Hill, A. M. 1994. Factors governing species composition, population size, and productivity of cool-water crayfishes. Nordic Journal of Freshwater Research 69:111–136.

Lodge, D. M., Kershner, M. W., Aloi, J. E., Covich, A. P. 1994. Effects of an omnivorous crayfish (*Orconectes rusticus*) on a freshwater littoral food web. Ecology 75:1265–1281.

Lodge, D. M., Kratz, T. K., Capelli, G. M. 1986. Long-term dynamics of three crayfish species in Trout Lake, Wisconsin. Canadian Journal of Fisheries and Aquatic Sciences 43:993–998.

Lodge, D. M., Lorman, J. G. 1987. Reductions in submersed macrophyte biomass and species richness by the crayfish *Orconectes rusticus*. Canadian Journal of Fisheries and Aquatic Sciences 44:591–597.

Lodge, D. M., Stein, R. A., Brown, K. M., Covich, A. P., Brönmark, C., Garvey, J. E., Klosiewski, S. P. 1998. Predicting impact of freshwater exotic species on native biodiversity: Challenges in spatial scaling. Australian Journal of Ecology 23:53–67.

Loring, M. W., Hill, L. G. 1978. Temperature selection and shelter utilization of the crayfish *Orconectes causeyi*. Southwestern Naturalist 21:219–226.

Lorman, J. G. 1980. Ecology of the crayfish *Orconectes rusticus* in northern Wisconsin. Ph.D. Thesis, University of Wisconsin, Madison, WI, 227 pp.

Lorman, J. G., Magnuson, J. J. 1978. The role of crayfishes in aquatic ecosystems. Fisheries, 3:8–10.

Lyle, C. 1938. The crawfishes of Mississippi, with special reference to the biology and control of destructive species. Iowa State College Journal of Science 13:75–77.

MacIsaac, H. J. 1994. Size-selective predation on zebra mussels (*Dreissena polymorpha*) by crayfish (*Orconectes propinquus*). Journal of the North American Benthological Society 13(2):206–216.

Magnuson, J. J., Beckel, A. L. 1985. Exotic species:A case of biological pollution. Wisconsin Academy Review 32:8–10.

Maranhão, P. J. C. Marques, Madeira, V. 1995. Copper concentrations in soft tissues of the red swamp crayfish *Procambarus clarkii* (Girard, 1852), after exposure to a range of dissolved copper concentrations. Freshwater Crayfish 10:282–286.

Martin, S. M. 1997. Crayfishes (Crustacea: Decapoda) of Maine. Northeastern Naturalist 4(3):165–188.

Mason, J. C. 1970a. Copulatory behavior of the crayfish, *Pacifastacus trowbridgii* (Stimpson). Canadian Journal of Zoology 48:969–976.

Mason, J. C. 1970b. Maternal–offspring behavior of the crayfish, *Pacifastacus trowbridgi* (Stimpson). American Midland Naturalist 84:463–473.

Mason, J. C. 1974. Crayfish production in a small woodland stream. Department of Environmental Fisheries and Marine Sciences, Pacific Biological Station, Nanaimo, British Columbia. 30 pp.

Mather, M. E., Stein, R. A. 1993. Using growth/mortality trade-offs to explore a crayfish species replacement in stream riffles and pools. Canadian Journal of Fisheries and Aquatic Sciences 50:88–96.

Mathur, D., Schutsky, R. M., Purdy, E. J. Jr. 1982. Temperature preference and avoidance responses of the crayfish, *Orconectes obscurus*, and associated statistical problems. Canadian Journal of Fisheries and Aquatic Sciences 39:548–553.

Maude, S. H., Williams, D. D. 1983. Behavior of crayfish in water currents: hydrodynamics of eight species with reference to their distribution patterns in southern Ontario. Canadian Journal of Fisheries and Aquatic Sciences 40:68–77.

McGregor, S. W., O'Neil, P. E., Rheams, K. F., Moser, P. H., Blackwood, R. 1997. Biological, geological, and hydrological investigations in Bobcat, Matthews, and Shelta caves and other selected caves in north Alabama. Geological Survey of Alabama Bulletin 166. 198pp.

McGriff, D. 1983a. The commercial fishery for *Pacifastacus leniusculus*, from the Sacramento-San Joaquin Delta. Freshwater Crayfish 5:403–417.

McGriff, D. 1983b. Growth, maturity, and fecundity of the crayfish *Pacifastacus leniusculus* from the Sacromento–San Joaquin Delta. California Fish and Game 69:227–242.

McMahon, B. R., Hankinson, J. J. 1993. Respiratory adaptations in burrowing crayfish. Freshwater Crayfish 9:174–182.

McMahon, B. R., Stuart, S. A. 1995. Simulating the crayfish burrow environment: air exposure and recovery in *Procambarus clarkii*. Freshwater Crayfish 8:451–461.

McWhinnie, M. A., O'Connor, J. D. 1967. Metabolism and low temperature acclimation in the temperate crayfish, *Orconectes virilis*. Comparative Biochemistry and Physiology 20:131–145.

Mellon, D., Jr., Sandeman, C., Sandeman, R. E. 1992. Characterization of oscillatory olfactory interneurons in the protocerebrum of the crayfish. Journal of Experimental Biology 167:15–38.

Mitchell, D. J., Smock, L. A. 1991. Distribution, life history and production of crayfish in the James River, Virginia. American Midland Naturalist 126(2):353–363.

Momot, W. T. 1984. Crayfish production: a reflection of community energetics. Journal of Crustacean Biology 4:35–54.

Momot, W. T. 1992. Further range extensions of the crayfish *Orconectes rusticus* in the Lake Superior basin of northwestern Ontario. Canadian Field-Naturalist 106(3):397–399.

Momot, W. T. 1995. Redefining the role of crayfish in aquatic ecosystems. Reviews in Fisheries Science 3:33–63.

Momot, W. T., Gowing, H. 1977a. Response of the crayfish *Orconectes virilis* to exploitation. Journal of the Fisheries Research Board of Canada 34:1212–1219.

Momot, W. T., Gowing, H. 1977b. Production and population dynamics of the crayfish *Orconectes virilis* in three Michigan lakes. Journal of the Fisheries Research Board of Canada 34:2041–2055.

Momot, W. T., Gowing, H., Jones, P. D. 1978. The dynamics of crayfish and their role in ecosystems. American Midland Naturalist 99:10–35.

Momot, W. T., Hauta, P. L. 1995. Effects of growth and mortality phenology on the cohort P/B of the crayfish, *Orconectes virilis*. Freshwater Crayfish 8:265–275.

Moshiri, G. A., Goldman, C. R., Godshalk, G. L., Mull, D. R. 1970. The effects of variations in oxygen tension on certain aspects of respiratory metabolism in *Pacifastacus leniusculus* (Dana) (Crustacea:Decapoda). Physiological Zoology 43:23–29.

Mundahl, N. D., Benton, M. J. 1990. Aspects of the thermal ecology of the rusty crayfish *Orconectes rusticus* (Girard). Oecologia 82:210–216.

Naqvi, S. M., Howell, R. D. 1993a. Cadmium and lead uptake by red swamp crayfish (*Procambarus clarkii*) of Louisiana. Bulletin of Environmental Contamination and Toxicology 51:296–302.

Naqvi, S. M., Howell, R. D. 1993b. Toxicity of cadmium and lead to juvenile red swamp crayfish, *Procambarus clarkii*, and effects on fecundity of adults. Bulletin of Environmental Contamination and Toxicology 51:303–308.

Naqvi, S. M., Leung, T. 1983. Trifluralin and Oryzalin herbicide toxicities to juvenile crawfish (*Procambarus clarkii*) and mosquitofish (*Gambusia affinis*). Bulletin of Environmental Contamination and Toxicology 31:304–308.

Nelson, D. H., Hooper, D. K. 1982. Thermal tolerance and preference of the freshwater shrimp *Palaemonetes kadiakensis*. Journal Thermal Biology 7:183–187.

Newsom, J. E., Davis, K. B. 1991. Ionic responses of white river crayfish, *Procambarus zonangulus*, and red swamp crayfish, *Procambarus clarkii*, to changes in temperature and salinity. American Zoologist 31:43A.

Nevo, E. 1978. Genetic variation in natural populations; pattern and theory. Theoretical Population Biology 13:121–177.

Nielsen, L. A., Reynolds, J. B. 1977. Population characteristics of a freshwater shrimp, *Palaemonetes kadiakensis* Rathbun. Transactions of the Missouri Academy Science 10/11:44–57.

Noblitt, S. B., Payne, J. F. 1995. A comparative study of selected chemical aspects of the eggs of the crayfish *Procambarus clarkii* (Girard, 1852) and *P. zonangulus* Hobbs & Hobbs, 1990 (Decapoda, Cambaridae). Crustaceana 68(6):695–704.

Noblitt, S. B., Payne, J. F., Delong, M. 1995. A comparative study of selected physical aspects of the eggs of the crayfish *Procambarus clarkii* (Girard, 1852) and *P. zonangulus* Hobbs & Hobbs, 1990 (Decapoda, Cambaridae). Crustaceana 68(5):575–582.

Norrocky, M. J. 1984. Burrowing crayfish trap. Ohio Journal of Science 84:65–66.

Norrocky, M. J. 1991. Observations on the ecology, reproduction and growth of the burrowing crayfish *Fallicambarus* (*Creaserinus*) *fodiens* (Decapoda:Cambaridae) in north-central Ohio. American Midland Naturalist 125(1):75–86.

Nystrom, P., Strand, J. A. 1996. Grazing by a native and an exotic crayfish on aquatic macrophytes. Freshwater Biology 36(3):673–682.

Oh, J. W., Dunham, D. W. 1991. Chemical detection of conspecifics in the crayfish *Procambarus clarkii*: role of antennules. Journal of Chemical Ecology 17:161–166.

Olsen, T. M., Lodge, D. M., Capelli, G. M., Houlihan, R. J. 1991. Mechanisms of impact of an introduced crayfish (*Orconectes rusticus*) on littoral congeners, snails, and macrophytes. Canadian Journal of Fisheries and Aquatic Sciences 48(10):1853–1861.

Page, L. M. 1985. The crayfishes and shrimps (Decapoda) of Illinois. Illinois Natutal History Survey Bulletin 33:335–448.

Page, L. M., Mottesi, G. B. 1995. The distribution and status of the Indiana crayfish, *Orconectes indianensis*, with comments on the crayfishes of Indiana. Proceedings of the Indiana Academy of Science 104:103–111.

Peck, S. K. 1985. Effects of aggressive interactions on temperature selection by the crayfish, *Orconectes virilis*. American Midland Naturalist 114:159–167.

Peckarsky, B. L., Fraissinet, P. R., Penton, M. A., Conklin, D. J. Jr. 1990. Freshwater macroinvertebrates of northeastern North America. Cornell Uni. Press, Ithaca, NY, 422 pp.

Pel, X., Moss, F. 1996. Characterization of low-dimensional dynamics in the crayfish caudal photoreceptor. Nature 379(6566): 618–621.

Penn, G. H., Jr. 1943. A study of the life history of the Louisiana red-crawfish, *Cambarus clarkii* Girard. Ecology 24:1–18.

Perry, W. L., Lodge, D. M., Lamberti, G. A. 1997. Impact of crayfish predation on exotic zebra mussels and native invertebrates in a lake-outlet stream. Canadian Journal of Fisheries and Aquatic Sciences 54(1):120–125.

Pflieger, W. L. 1996. The crayfishes of Missouri. Missouri Department of Conservation, Jefferson City, 152 pp.

Pollard, T. G., Larimer, J. L. 1977. Circadian rhythmicity of heart rate in the crayfish, *Procambarus clarkii*. Comparative Biochemistry and Physiology 57A:221–226.

Power, M. E. 1992. Top-down and bottom-up forces in food webs: do plants have primacy? Ecology 73:733–746.

Powers, L. W. 1977. A catalogue and bibliography to the crabs (Brachyura) of the Gulf of Mexico. Contribution to Marine Science, 20(Suppl.):1–190.

Price, J. O. 1981. Observations on the trophic ecology of the crayfish *Orconectes neglectus chaenodactylus* (Decapoda:Astacidae). ASB Bulletin 28:91–92.

Price, J. O., Payne, J. F. 1984. Postembryonic to adult growth and development in the crayfish *Orconectes neglectus chaenodactylus* Williams, 1952 (Decapoda, Astacidea). Crustaceana, 46:176–194.

Rabeni, C. F. 1985. Resource partitioning by stream dwelling crayfish: The influence of body size. American Midland Naturalist 113:20–29.

Rabeni, C. F. 1992. Trophic linkage between stream centrarchids and their crayfish prey. Canadian Journal of Fisheries and Aquatic Sciences 49:1714–1721.

Rabeni, C. F., Gossett, M., McClendon, D. D. 1995. Contribution of crayfish to benthic invertebrate production and trophic ecology of an Ozark stream. Freshwater Crayfish 10:163–173.

Reddell, J. R. 1994. The cave fauna of Texas with special reference to the western Edwards Plateau. *in*: Elliott, W. R., Veni, G. Eds. The caves and karst of Texas. A guide-book for the 1994 convention of the National Speleological Society with emphasis on the south-western Edwards Plateau. National Speleological Society, Brackettville, Texas, Pages 31–48.

Reiber, C. L. 1995. Physiological adaptations of crayfish to the hypoxic environment. American Zoologist 35:1–11.

Reiber, C. L. 1997. Oxygen sensitivity in the crayfish *Procambarus clarkii*: peripheral O_2 receptors and their effect on cardiorespiratory functions. Journal of Crustacean Biology 17(2): 197–206.

Richter, B. D., Braun, D. P., Mendelson, M. A., Master, L. L. 1997. Threats to imperiled freshwater fauna. Conservation Biology 11(5):1081–1093.

Rittschof, D. 1980. Chemical attraction of hermit crabs and other attendants to simulated gastropod predation sites. Journal of Chemical Ecology 6:103–118.

Roell, M. J., Orth, D. J. 1992. Production of three crayfish populations in the New River of West Virginia, USA. Hydrobiologia 228:185–194.

Rogers, R., Huner, J. V. 1985. Comparison of burrows and burrowing behavior of five species of cambarid crawfish (Crustacea, Decapoda) from the Southern University campus, Baton Rouge, Louisiana. Proceedings of the Louisiana Academy of Science 48:23–29.

Roldman, B. M., Shivers, R. R. 1987. The uptake and storage of iron and lead in cells of the crayfish (*Orconectes propinquus*) hepatopancreas and antennae gland. Comparative Biochemistry and Physiology 86C:201–214.

Roush, W. 1997. Hybrids consummate species invasion. Science 227:316.

Rutherford, P. L., Dunham, D. W., Allison, V. 1995. Winning agonistic encounters by male crayfish *Orconectes rusticus* (Girard) (Decapoda, Cambaridae): chela size matters but chela symmetry does not. Crustaceana 68(4):526–529.

Rutherford, P. L., Dunham, D. W., Allison, V. 1996. Antennule use and agonistic success in the crayfish *Orconectes rusticus* (Girard, 1852) (Decapoda, Cambaridae). Crustaceana 69(1): 117–122.

Saffran, K. A., Barton, D. R. 1993. Trophic ecology of *Orconectes propinquus* (Girard) in Georgian Bay (Ontario, Canada). Freshwater Crayfish 9:350–358.

Sandeman, D., Sandeman, R., Derby, C., Schmidt, M. 1992. Morphology of the brain of crayfish, crabs, and spiny lobsters: a common nomenclature for homologous structures. Biological Bulletin 183:304–326.

Sarver, R. G., Flynn, M. A., Holliday, C. W. 1994. Renal Na, K-ATPase and osmoregulation in the crayfish, *Procambarus clarkii*. Comparative Biochemistry and Physiology 107A(2):349–356.

Savino, J. F., Miller, J. E. 1991. Crayfish (*Orconectes virilis*) feeding on young lake trout (*Salvelinus namaycush*): effect of rock size. Journal of Freshwater Ecology 6(2):161–170.

Siewert, H. F., Buck, J. P. 1991. Effects of low pH on survival of crayfish (*Orconectes virilis*). Journal of Freshwater Ecology 6(1):87–91.

Smalley, A. E. 1961. A new cave shrimp from southeastern United States (Decapoda, Atyidae). Crustaceana 3:127–130.

Smith, D. G. 1988. Keys to the freshwater macroinvertebrates of Massachusetts. No. 3: Crustacea Malacostraca (Crayfish, isopods, amphipods). Publ. No. 15, 236–61–250–2–88-CR. Division of Water Pollution Control, Westboro, MA, 58 pp.

Smith, T. I, Sandifer, P. A., Smith, M. H. 1978. Population structure of malaysian prawns, *Macrobrachium rosenbergii* (deMan), reared in earthen ponds in South Carolina, 1974–1976. Proceedings of world Mariculture Society 9:21–38.

Stechey, D. P. M., Somers, K. M. 1995. Potential, realized, and actual fecundity in the crayfish *Orconectes immunis* from southwestern Ontario. Canadian Journal of Zoology 73:672–677.

Stein, R. A. 1975. Sexual dimorphism in crayfish chelae: functional significance linked to reproductive activities. Canadian Journal of Zoology 54:220–227.

Stein, R. A. 1977. Selective predation, optimal foraging, and the predator–prey interaction between fish and crayfish. Ecology 58:1237–1253.

Stein, R. A. 1979. Behavioral response of prey to fish predators. *in:* Stroud, R. H., Clepper, H. Eds. Predator–prey systems in fisheries management. Sport Fisheries Institute, Washington, DC 343–353 pp.

Stein, R. A., Magnuson, J. J. 1976. Behavior response of crayfish to a fish predator. Ecology 57:751–761.

Stoffel, L. A., Hubschman, J. H. 1974. Limb loss and the molt cycle in the freshwater shrimp, *Palaemonetes kadiakensis*. Biological Bulletin (Woods Hole, Mass.) 147:203–212.

Streever, W. J. 1996. Energy economy hypothesis and the troglobitic crayfish *Procambarus erythrops* in Sim's Sink Cave, Florida. American Midland Naturalist 135(2):357–366.

Strenth, N. E. 1976. A review of the systematics and zoogeography of the freshwater species of *Palaemonetes* Heller of North America (Crustacea:Decapoda). Smithsonian Contributions to Zoology No. 228: 27 pp.

Strenth, N. E. 1994. A new species of *Palaemonetes* (Crustacea: Decapoda:Palaemonidae) from northeastern Mexico. Proceedings of the Biological Society of Washington, 107(2):291–295.

Strong, D. R. 1992. Are trophic cascades all wet? Differentiation and donor-control in speciose ecosystems. Ecology 73:747–754.

Stuecheli, K. 1991. Trapping bias in sampling crayfish with baited funnel traps. North American Journal of Fisheries Management 11:236–239.

Swerup, C., Rydzvist, B. 1992. The abdominal stretch receptor organ of the crayfish. Comparative Biochemistry and Physiology 103A(3):423–431.

Taketomi, Y., Nishikawa, S., Koga, S. 1996. Testis and androgenic gland during development of external sexual characteristics of the crayfish *Procambarus clarkii*. Journal of Crustacean Biology 16(1):24–34.

Taylor, C. A. 1997. Taxonomic status of members of the subgenus *Erebicambarus*, Genus *Cambarus* (Decapoda:Cambaridae) east of the Mississippi River. Journal of Crustacean Biology 17(2): 352–360.

Taylor, C. A., Redmer, M. 1996. Dispersal of the crayfish *Orconectes rusticus* in Illinois, with notes on species displacement and habitat preference. Journal of Crustacean Biology 16(3):547–551.

Taylor, C. A., Warren, M. L. Jr., Fitzpatrick, J. F. Jr., Hobbs, H. H. III, Jezerinac, R. F., Pflieger, W. L., Robison, H. W. 1996. Conservation status of crayfishes of the United States and Canada. Fisheries 21(4):25–38.

Taylor, R. C. 1983. Drought-induced changes in crayfish populations along a stream continuum. American Midland Naturalist 110:286–298.

Taylor, R. C. 1984. Thermal preference and temporal distribution in three crayfish species. Comparative Biochemistry and Physiology 77A:513–517.

Taylor, R. C. 1985. Absence of Form I to Form II alternation in male *Procamkerus spiculifer* (Cambaridae). American Midland Naturalist 114:145–151.

Taylor, R. M., Watson, G. D., Alikhan, M. A. 1995. Comparative sublethal acute toxicity of copper to the freshwater crayfish *Cambarus robustus* Astacidae, Decapoda, Crustacea from an acidic metal contaminated lake and circumneutral uncontaminated stream. Water Research 29(2):401–408.

Thacker, R. W., Hazlett, B. A., Esman, L. A., Stafford, C. P., Keller, T. 1993. Color morphs of the crayfish *Orconectes virilis*. American Midland Naturalist 129:182–199.

Thorp, J. H., Gloss, S. P. 1986. Field and laboratory tests on acute toxicity of cadmium to freshwater crayfish. Bulletin of Environmental Contamination and Toxicology 37:355–361.

Tierney, A. J., Atema, J. 1986. Effects of acidification on the behavioral responses of crayfishes *(Orconectes virilis* and *Procambarus acutus)* to chemical stimuli. Aquatic Toxicology 9:1–11.

Tierney, A. J., Atema, J. 1988. Behavioral responses of crayfish *(Orconectes virilis* and (*Orconectes rusticus*) to chemical feeding stimulants. Journal of Chemical Ecology 14:123–133.

Tierney, A. J., Dunham, D. W. 1984. Morphology of aesthetasc sensilla in surface and cave dwelling crayfishes. American Zoologist 24:67A.

Tierney, A. J., Thompson, C, S., Dunham, D. W. 1984. Site of pheromone reception in the crayfish *Orconectes propinquus* (Decapoda, Cambaridae). Journal of Crustacean Biology 4:554–559.

Truesdale, F. M., Mermilliod, W. J. 1979. The river shrimp *Macrobrachium ohione* (Smith)(Decapoda, Palaemonidae): its abundance, reproduction, and growth in the Atchafalaya River basin of Louisiana, USA. Crustaceana 36:61–73.

Uiska, E., Dunham, D. W., Harvey, H. H. 1994. Cumulative pattern in pH change alters response to food in the crayfish *Cambarus bartoni* [sic]. Canadian Journal of Zoology 72:187–190.

Vermeer, K. 1972. The crayfish, *Orconectes virilis,* as an indicator of mercury contamination. The Canadian Field-Naturalist 86:123–125.

Vernberg, F. J. 1983. Respiratory adaptations. *in:* Bliss, D. E., Ed. The biology of crustacea. Vol. 8. Environmental adaptations. Academic Press, New York, 1–42 pp.

Weber, L. M., Lodge, D. M. 1990. Periphytic food and predatory crayfish: relative roles in determining snail distribution. Oecologia 82:33–39.

Weingartner, D. L. 1977. Production and trophic ecology of two crayfish species cohabiting an Indiana cave. Ph.D. Thesis, Michigan State University, East Lansing, 323 pp.

Wheatly, M. G. 1995. An overview of electrolyte regulation in the freshwater crayfish throughout the molting cycle. Freshwater Crayfish 8:420–436.

Wheatly, M. G., de Souza, S. C. R., Hart, M. K. 1996. Related changes in hemolymph acid–base status, electrolytes, and ecdysone in intermolt crayfish (*Procambarus clarkii*) at 23°C during extracellular acidosis induced by exposure to air, hyperoxia, or acid. Journal of Crustacean Biology 16(2):267–277.

Wiens, A. W., Armitage, K. B. 1961. The oxygen consumption of the crayfish *Orconectes immunis* and *Orconectes nais* in response to temperature and to oxygen saturation. Physiological Zoology 34:39–54.

Wiles, P. R., Guan, R. Z. 1993. Studies on a new method for permanently tagging crayfish with microchip implants. Freshwater Crayfish 9:419–425.

Willman, E. J., Hill, A. M., Lodge, D. M. 1994. Response of three crayfish congeners (*Orconectes* spp.) to odors of fish carrion and live predatory fish. American Midland Naturalist 132(1):44–51.

Wood, C. M., Rogano, M. 1986. Physiological responses to acid stress in crayfish (*Orconectes*): haemolymph ions, acid–base status, and exchanges with the environment. Canadian Journal of Fisheries and Aquatic Sciences 43:1017–1026.

Word, B. H., Jr., Hobbs, H. H. Jr. 1958. Observations on the testis of the crayfish *Cambarus montanus acuminatus* Faxon. Transactions of the American Microscopical Society 77:435–450.

Wright, D. A., Welbourn, P. M. 1993. Effects of mercury exposure on ionic regulation in the crayfish *Orconectes propinquus*. Environmental Pollution 82:139–142.

Wu, S., Brown, S. E. S. 1980. The occurrence of a freshwater shrimp *Palaemonetes paludosus* (Gibbes, 1850) (Crustacea:Palaemonidae) in a warm spring of Wellsville, Colorado. Natural History Inventory of Colorado 5:1–10.

Glossary

acclimation process of adjustment of a rate function (i.e., respiratory rate) to a change in ambient environmental conditions.

acetabular plates sclerites associated with the genital field in adults (provisional genital field in deutonymphs) of certain mite taxa bearing the genital acetabula.

acetabulum (acetabula) knob-like structure with a porous cap associated with the genital field in deutonymph and adult mites, thought to function as an osmoregulatory chloride epithelium. Acetabula are strictly or roughly paired, and usually are borne on acetabular plates, but may lie in the gonopore or scattered in the integument flanking the gonopore.

acidosis decline in the pH of the blood through accumulation of carbon monoxide or anaerobic endproducts.

acrosome a cytoplasmic inclusion in the cell body of a sperm cell.

Note: Glossary terms were defined by individual authors in reference to their chapter. Consequently, terms may actually apply to other taxa and have broader definitions than indicated here.

acute sharp angle at the end.

adductor muscles in bivalve molluscs, the muscles extending between shell valves, closing the valves on contraction.

adductor muscle scars raised areas or scars left on the interior of the ostracode shell by the adductor muscles, generally leave a distinctive pattern of scars.

adenal produced in the parenchyma (rhabdoids).

adenodactyl auxiliary glandular bulb in the male system of some turbellarian taxa.

adhesive strip thin layer of chitin between the duplicature and the outer lamella of the ostracode valve.

adnate closely adherent, joined together.

adoral zone of membranelles orderly arrangement of three or more compound ciliary organelles (membranelles) that are serially arranged along the left side of the oral area of a ciliate.

afferent branchial vessel in Mollusca, the blood vessels bringing deoxygenated blood from the visceral mass, foot, and mantle to the distal ends of the gill filaments.

ala(e) a coarse or fine, wing-like spine, as in the posteriorly directed alae on the lateral surface of the ostracode carapace.

allelochemic biochemical substance produced by one species and released into the environment that subsequently affects the behavior or physiology of another species.

allochthonous substances (or sometimes species) originating outside the immediate habitats (e.g., carbon produced by plants located upstream or in the riparian zone); opposite of autochthonous.

allopatric referring to multiple species whose habitat distributions do not overlap.

allorecruitive type of growth of colonial rotifers whereby young recruited into the colony come predominantly from other colonies; results in higher genetic diversity of the colony. Originally defined as Type I colony formation; see autorecruitive.

allozymes enzyme having a specific biochemical function, but which differs slightly in its primary structure and in its electrophoretic mobility, from other enzymes coded by different alleles of this gene

amictic phase in the life cycle of monogonont rotifers in which reproduction is parthenogenetic only; see mictic and mixis.

amoeboid amoeba-like, have no fixed shape, and creeping on surfaces.

amphidelphic with two opposite ovaries that normally join at the common vagina.

amphids paired sensory organs located on opposite lateral surfaces of nematodes; variable in position (lip or neck region) and shape (pore, slit, circular, spiral or cub-shaped).

amphimixis union of egg and sperm through ordinary sexual reproduction, with the gametes derived from separate individuals.

amphoteric type of female rotifers of the class Monogononta that can produce two types of eggs in the same clutch, one developing into female (diploid) and the other into male (haploid) offspring.

anastomosing interconnecting.

ancestrula in ectoproct bryozoans, the original, single zooid that emerges from a statoblast, differing from subsequent zooids in the colony by its smaller number of tentacles and inability to form statoblasts.

anchoral processes posteriorly directed apodemes of the gnathosoma (capitulum) in deutonymphs and adult mites.

angulate having an angle.

anhydrobiosis ability of bdelloid rotifers to undergo desiccation and then be revived at some later time; a cryptobiotic (latent) state in tardigrades induced by loss of water through evaporation.

annual occurring once per year.

annulate ringed; surrounded by a ring of a different color; formed into ring-like segments.

annulations transverse striations (trenches) in the cuticle of nematodes.

annulus ring; in ectoproct bryozoans, the ring of enlarged chambers, usually gas-filled, on the periphery of floatoblast valves.

annulus ventralis seminal (spermatophore) receptacle of female crayfish.

anoxia lack of dissolved oxygen.

anoxybiosis cryptobiotic (latent) state in tardigrades induced by low oxygen tensions.

antenna whiplike, paired, generally elongate sensory appendages arising from anterior area of the head (or cephalothorax); 1-2 pairs of antennae are present in arthropods.

antennal gland one of pair of complex excretory glands in many decapods with duct opening on antenna; green gland.

antennule paired appendage of first cephalic somite of crayfish; sensory, often bearing aesthetascs.

aperture opening to an anterior cavity, as in the opening of a snail shell from which foot and body protrude.

apex that part of any structure opposite the base by which it is attached.

apical pertaining to the apex.

apodeme internal extension of thickened cuticle or exoskeleton in some arthropods, often functioning as muscle attachment site.

apomixis form of parthenogenesis in which meiosis is suppressed and offspring are genetically identical to their mother.

apophyses see pharyngeal or buccal apophyses.

aposymbiotic describing an organism that would normally have symbionts but the symbionts are not present.

apotypical (apomorphic) derived, or modified, state of a biologic attribute or morphologic character.

apterous without wings.

areoles segment of hairworm cuticle separated from adjacent segments (areoles) by longitudinal and transverse furrows; may bear granules, bristles, and contain pores.

articulation a joint.

ascus glandular pocket in some turbellarians (Typhloplanidae).

atrium area where two or more passages meet, as in the space in flatworms where male and female systems open.

attenuate tapering rapidly or abruptly to a point.

aufwuchs German term whose broader North American usage refers to all small, attached (except macrophytes) and free-moving organisms forming a living film on aquatic substrates such as rocks, snags, and plants; included are some microinvertebrates, fungi, algae, and bacteria.

autochthonous substances (or sometimes species) originating within the immediate habitat, for example, carbon produced by resident phytoplankton; opposite from allochthonous.

autogamy self-fertilization phenomenon (not true sexual reproduction) in ciliates involving a single individual; haploid gametic nuclei are formed and two fuse with each other in a single individual; see cytogamy.

autorecruitive type of growth of colonial rotifers whereby young produced by a colony tend to remain in that colony; characterized by low genetic diversity of the colony; seeallorecruitive and geminative.

autotoky type of reproduction where progeny are produced by a single parent (i.e., hermaphroditism or parthenogenesis).

autotrophy synthesis of organic compounds from carbon dioxide using light as the energy source (in photosynthesis) or inorganic chemical compounds (in chemoautotrophy); contrasts with heterotrophy.

axe-head glochidium unionacean (Mollusca) glochidium larva characterized by a distinctly quadrate shell form.

axenic culture a bacteria-free culture containing only one species of organism.

axoneme longitudinal bundle of microtubular fibers in flagella, cilia, and axopodia.

axopodium characteristic feeding pseudopodium of Actinopoda, distinguished by needlelike shape, axonemes, and bidirectional streaming of cytoplasm.

baffles dorsal and/or transverse ridges posterior to mucrones on interior of the buccal cavity in some eutardigrades.

basal at or pertaining to the base or point of attachment to, or nearest the main body.

basal bulb swollen, set-off, basal portion of the pharynx (esphagus) of a nematode.

basipod (basipodite) second segment of an arthropod appendage.

basis second segment from proximal end of segmented appendage.

beak raised portion of dorsal margin of the bivalve shell; oldest portion of the shell.

benthos animals living on or near the bottom of an aquatic habitat.

bident having two points.

bifid cleft, or divided into two parts; forked.

bilamellate divided into two lamellae or plates.

bioturbation disturbance of sediments by an animal.

bivoltine having two broods or generations per year.

bolus a ball.

boss expanded protuberance, as on caudomesial surface of coxa of fourth pereiopod of male crayfishes.

brachypterous with short or abbreviated wings.

bristles variously shaped cuticular projections arising from cuticle.

bromeliads common name of a family of plants with cup-shaped leaves that can hold water at the base (mostly tropical species, e.g. the pineapple); protozoa, ostracodes, and some other invertebrates regularly inhabit these specialized aquatic microhabitats.

brood chamber body space in parent where eggs or embryos develop; space between the body and carapace of a cladoceran, where eggs develop.

brosse distinctive "brush" of cilia arising from specialized short kineties which is present on the anterodorsal surface of nondividing individuals of certain gymnostome ciliates.

buccal apparatus anterior part of tardigrade foregut, consisting of buccal tube, muscular pharynx, and pair of piercing stylets.

buccal apophyses cuticular thickenings at junction of buccal tube for insertion of stylet protractor muscles.

buccal ciliature compound ciliary organelles, the bases of which are associated with the oral area of some ciliates.

buccal cirri in heterotardigrades, paired sensory appendages, usually short and hair-like, on either side of each cephalic papilla on the cephalic plate.

buccal tube support median ventral lamina, also called a "reinforcement rod," from the buccal ring to the midregion of the buccal tube in some eutardigrades.

bulbous see pharynx.

bursa folds of cuticle on the male nematode tail that assist in attaching males to females during mating.

bursa (copulatrix) copulatory bursa; receptacle in some female reproductive systems into which donor spermatozoa are deposited.

byssus proteinaceous threads produced from a byssal gland at the base of the foot, used to attach juvenile or adult molluscan bivalves to hard surfaces.

calcium phosphate concretions small extracellular spherules found in gills of unionacean mussels made of calcium phosphate deposited in consecutive lamellae on a proteinaceous matrix.

calotte head region of a Nematomorpha, usually lighter in color than the adjoining body.

capitulum sclerotized base of a mite's gnathosoma.

carapace expanded, hard covering of a major body region (usually head and/or thorax of some arthropods).

cardate type of trophi (jaw structure) found in rotifers which functions by creating a sucking action.

cardinal teeth massive conical projections formed at the center of the shell-hinge plate which interdigitate to form the fulcrum on which the valves open and close.

carina sharp, spiral edge or keel on a snail's shell, usually along the center of the whorl.

carpus fifth segment from proximal end of segmented appendage.

cauda posterior extension of the idiosoma in adult males of certain mite taxa, functioning as an copulatory adaptation during spermatophore transfer.

caudal toward the posterior of an organism.

caudal glands glands found in the tail region, as in those of nematodes that secrete mucus.

caudal rami bifurcated terminal extension of the abdomen in some arthropods.

cellulase enzyme that degrades cellulose of plant cell walls.

cement glands glands that secrete material to assist fixation of egg capsules to a substrate, often near female gonopore in turbellarians.

centroplast central structure in some heliozoa from which axonemes radiate; synonym: central granule, axoplast; a type of microtubule organizing center.

cephalic pertaining to, or toward, the head.

cephalic papillae large papillae in head region of nematodes; also, paired sensory appendages on cephalic plate of heterotardigrades, located between internal and external cirri.

cercus (cerci) paired appendage of the last abdominal segment. In dragonflies, one of two pairs of lateral, terminal, triangular sclerites.

cerebropleural ganglion mass of nerve cell bodies near the mouth.

chaetae see setae.

chaetotaxy number and arrangement of setae on the body and appendages.

chela claw or pincer; in crayfish, two opposed distal podomeres (dactyl and propodus) of certain pereiopods.

chelate presence of pincer-like, exoskeletal or cuticular structures at the end of an appendage. In mites, a condition of the pedipalp in deutonymphs and adults of certain taxa in which tibia bear a dorsodistal seta, often associated with a projecting tubercle that opposes the dorsal surface of the tarsus to form a grasping organ.

chelicera in mites, a two-segmented, paired feeding appendage located dorsomedially on the gnathosoma; usually two-segmented, but one-segmented in deutonymphs and adults of Hydrachnidae; the distal segment, or "claw," functions in piercing and tearing the integument of hosts or prey.

chitin a long chain, polymeric glucosamine containing 6% nitrogen; used in exoskeletons of many arthropods and often complexed with protein or infiltrated with calcium for strength.

chloride epithelia oval patches (as in insects) that differ from normal epithelia and are used for osmoregulation.

chrysolaminarin liquid carbohydrate storage form, β-1:3-linked glucan; characteristic of flagellates in the order Chrysomonadida; usually in one or more large refractile vesicles at the posterior of the cell, optically bluish under strong light; synonym is leucosin.

cingulum one of two ciliated rings present in the typical rotiferan corona (apical end); seetrochus.

circadian rhythm diurnal or daily activity rhythms.

cirrus A in heterotardigrades, paired filamentous sensory appendage at the anterior lateral margin of the scapular plate, situated slightly dorsal to the clava.

cirrus eversible, spiny male duct; may contain ejaculatory, prostatic, or accessory ducts and be enclosed in a bulb or special propulsory sac.

cladogram classification, usually in the form of a branching diagram, of a group of organisms based on cladistic principles (i.e., tracing the evolution of shared, derived characters from ancestral to derived forms).

clasping apparatus in ostracodes, the heavily sclerotized, apparently moveable sclerite articulating with midportion of the peniferum in a socket near the ventral cardo and loop of the spermatic tube of entocytherid hemipenes; includes vertical and horizontal rami.

clava in heterotardigrades, a paired sensory appendage, usually short and broad, at the anterior lateral margin of the scapular plate, situated slightly ventral to lateral cirrus A.

claw raptorial structures; in all mite instars, true claws are paired, curved structures located terminally on tarsi of the legs.

clear-water phase period, usually in spring, when lake water becomes unusually clear due to grazing of herbivorous zooplankton.

clutch group of eggs.

cocoon structure formed by some species to enclose many egg capsules or embryos.

cohort group of equal-aged individuals whose survivorship and fecundity are to be followed throughout the lifespan of the entire group.

columellar strong central support around which a snail shell coils.

commensal animal or plant that lives on, in, or with another, sharing its resources but neither parasitic on it nor injured by it.

concentric in Mollusca, describing the growth lines of a snail operculum that lie entirely within each other (not forming a spiral).

conglutinate mass of unionacean (Mollusca) glochidia larval bound by mucus; often resembling prey items of unionacean glochidial fish hosts (increasing chances of host contact).

conic shaped like a cone.

conjugation in ciliates, a reciprocal fertilization type of sexual phenomenon occuring only between members of differing mating types and involving temporary or total fusion of the pair.

contractile vacuole liquid-filled organelle which functions as an osmoregulator in the cytoplasm of many protozoa.

copepodid larval stage of copepods that follows the first stage (the nauplius).

copulatory bulb swollen muscular and connective tissue region containing the male copulatory organ and often also the seminal vesicle and/or prostate.

copulatory organ terminus of male reproductive systems which delivers spermatozoa into the body of a mate, i.e. an intromittent organ. When not armed with hard structures, yet capable of protrusion, it is simply called a penis (or penis papilla). Copulatory organs may be equipped with hard (sclerotized) structures such as a stylet or cirrus spines.

corneous slightly hardened but still pliable (cornified or "horny") proteinaceous material, as in the operculum of a snail or the modified terminal elements of first pleopods of first-form male crayfish.

corona apical region of rotifers; usually a ciliated band around the anterior end, characteristically composed of two ciliated rings called the trochus and cingulum; in some forms, ciliation is absent and may be replaced by long setae which surround the rim of a funnel-shaped structure called the infundibulum.

cosmopolitanism idea that the same species are found on different continents.

costa rib or ridge on a snail shell that lies transverse to (at right angles to) the coiling of the whorl.

coxa proximal (closest to body) segment of a segmented appendage.

coxal plate in mites, a sclerite representing the modified basal leg segment, attached to the venter of the idiosoma in all instars. Coxal plates may be expanded and fused with one another to cover large areas of idiosomal integument.

coxoglandularia series of paired glandularia associated with the coxal plates of mites.

coxopod proximal segment of an arthropod appendage.

creeping welt slightly raised, often darkened structure on dipteran larvae.

crenulate evenly rounded and often rather deeply curved; having an irregularly wavy or serrate outline; possessing a scalloped margin or rounded tooth.

crescent (postanal crescent) crescent-shaped fold between the cloaca and terminus in male hairworms; usually ornamented with setae or bristles.

cross fibers fine criss-cross fibers in the upper cuticular layers in mermithid nematodes.

cryobiosis cryptobiotic (latent) state in tardigrades induced by low temperatures.

cryptobiosis latent state in which metabolism comes to a reversible halt; also known as "anabiosis" or "abiosis."

crystalline style mucopolysaccharide rod containing digestive enzymes projecting dorsally from the style sac in the floor of the bivalve stomach where it is secreted; the crystalline style is rotated by cilia lining the style sac against the gastric shield on the dorsal side of the stomach.

ctenidium molluscan gill, in bivalves consisting of two demibranchs formed from the reflection of right and left gill filaments upon themselves.

cupule cup-shaped segment at the base of the club on some antennae.

cuticular bar external cuticular thickening on the leg, near or between the claws in some eutardigrades.

cyclomorphosis regular (often seasonal or annual) phenotypic change in body size, spine length, etc., found in successive generations of zooplankton within a single habitat (does not apply to morphological changes shown by a single animal).

cyrtos basket-like structure supporting the cytopharynx of Ciliophora in the protistan subphylum Cyrtophora.

cytogamy occurrence of autogamy in each member of a pair of temporarily fused ciliates.

cytolosomes digestive organelles of protozoa, analogous with food vacuoles, but containing the organelles of the cell, which are being broken down under conditions of starvation; synonymous with autophagic vacuole or autophagic vesicle.

cytopharynx cell pharynx, leading from the cytostome to site of food vacuole formation in protozoa.

cytoproct cell anus; permanent site for phagocytosis in some protozoa.

cytostome cell mouth; permanent site for phagocytosis in some protozoa.

dactyl most distal segment of the endopod of segmented appendages; smaller, mesially situated, movable finger of chela.

demibranch portion of the bivalve ctenidium (gill) formed from a set of filaments reflecting upon themselves; in each ctenidium, the inner lateral filaments form the inner demibranch lying next to the visceral mass and the outer lateral filaments form the outer demibranch lying next to the mantle.

denticles small, delicate, spinelike projections, usually of shell material in numerous taxa; also the small teeth-like processes lining the stoma of some nematodes; small teeth.

depressed spire a spire that is not raised above the body whorl of a snail shell.

determinate growth describing an organism that ceases to grow in size after reaching a certain size or stage; contrasts with indeterminate growth.

detritus decaying organic material.

deutocerebrum portion of a crustacean's brain controlling antennules and innervating antennae and portions of the alimentary tract.

deutonymph second nymphal instar in the generalized mite life cycle and the only active, free-living nymphal instar in the water mite life cycle.

dextral snail shells with the aperture on right (when viewed with the apex pointed away from the observer and with the aperture fully visible).

diapause period of arrested growth and development; often used to survive harsh environmental conditions.

diel daily

digestive diverticulum in bivalve molluscs, a digestive organ surrounding the stomach, area in which the final intracellular phase of digestion takes place.

digitate many finger-like extensions of a mollusc's mantle.

dioecious individual possessing either a male or a female reproductive system; reproductive systems contained in separate individuals.

diploid having twice the number of chromosomes normally occurring in a germ cell, most somatic cells are diploid.

discobolocyst extrusome of some flagellated protozoa, especially Chrysomonadida.

discus-shaped round and flat disk with a thin edge and a thicker middle.

distal toward the end of a structure and away from the central body opposite of proximal.

diurnal recurring every day and/or relating to the daytime.

diverticulum sac-like structure branching from an internal organ such as the intestine.

doliiformis see pharynx.

dormancy state of minimal metabolic activity when growth ceases, primarily enabling organisms to survive periods of adverse conditions.

dorsal furrow strip of soft integument separating dorsal and ventral shields in highly sclerotized adult mites.

dorsalia small platelets on the dorsum of the idiosoma in deutonymphs and adult mites that function as muscle attachment sites.

dorsal plate large sclerite covering the prodorsal region of the idiosoma in any mite instar, often expanded to cover part of the opisthosomal dorsum as well.

dorsal shield single large sclerite or a series of closely fitting platelets covering the entire dorsum of the idiosoma in deutonymphs and adults of certain mite taxa.

dorsocentralia series of paired, medially located dorsalia in mites.

dorsoglandularia series of paired glandularia located on the dorsum of the idiosoma in mites.

dorsolateralia series of paired, laterally located dorsalia in mites.

dorsum flattened area adjacent to the hinge line and set off from the lateral surface of the ostracode carapace.

ductus communis in turbellarians, the terminal (or common) part of the female canal distal to the joining of vitelline duct(s) and oviduct(s) in ectolecithal systems.

duplicature narrow band of shell material around outer edge of the proximal face of the epidermis, composed of the same three layers as the outer lamella; only calcified layer of the proximal covering, lining, or epidermis separated into the list strip, selvage strip, and flange strip of the ostracode shell.

ecdysis shedding or molting of an older, smaller exoskeleton in arthropods.

ecdysone the principal molt-stimulating hormone (in several forms) of arthropods.

ecophenotypic plasticity variation in a species induced by nongenetic environmental factors.

ectocyst nonliving outer layer of the body wall of bryozoans; may be sclerotized or gelatinous.

ectolecithal pertains to systems in which the yolk is not incorporated into the oocyte, i.e., part of the female gonad is a yolk gland (as in higher Turbellaria).

efferent branchial vessel blood vessel in the bivalve gill axis carrying oxygenated hemolymph from gill filaments to the longitudinal vessel of the kidney and eventually to the heart.

ejaculatory complex series of membranous chambers and associated sclerotized framework located distally in the reproductive tract of male adult mites, functioning as a syringe-like organ for compacting masses of spermatozoa, assembling spermatophores, and expelling them from the genital tract through the gonopore.

ejaculatory duct (ductus ejaculatorius) terminal part of the male duct in Turbellaria, sometimes eversible; carries spermatozoa and usually prostatic secretions through the copulatory organ.

ejectisome extrusome of some flagellated protozoa, especially in Cryptophyta.

electromorph organism possessing a set of enzymes with a specific migration pattern as determined by electrophoretic techniques.

electrostatic (biotic) filter surface of an animal that attracts charged particles, because of static-electricity-like charges on the surface.

elytra hardened, shell-like mesothoracic wings of Coleoptera.

emarginate notched; with an obtuse, rounded, or quadrate section cut from the margin.

empodium medial unpaired, claw-like structure located terminally between the true claws on the tarsi of legs in larvae of all mite taxa, and in deutonymphs and adults of Stygothrombidiidae.

encystment latent state in aquatic tardigrades in which resistant cysts are produced under deteriorating environmental conditions.

endemic restricted to a particular geographic region.

endites projections or processes from the inner margin of the inner branch of a branched appendage.

endocrine relating to the system of glands that produce chemical signals (hormones) in the body.

endocyst in ectoproct bryozoans, the inner living tissues of the body wall, including epidermis, muscle layers, basement membrane, and peritoneum.

endocytosis process by which minute food particles are engulfed by cells into food vacuoles for the final stages of digestion, as occurs in terminal tubules of molluscan digestive diverticula.

endogenous initiated from within an individual, such as nervous or hormonal cues controlling diurnal rhythmicity in the absence of external cues.

endolecithal (entolecithal) pertaining to those systems in which the yolk is incorporated into the oocyte (as occurs in more primitive Turbellaria).

endopod mesial, or inner branch of an appendage; mesial ramus of biramous appendage, originating on second segment from base.

endosymbiotic describing an organism which currently has internal symbionts.

epibranchial cavity in bivalve molluscs, the exhalant mantle cavity formed dorsal to the ctenidium, which carries water leaving the gill to the exhalant siphon; it also receives the openings of the gonad, kidney, and anus.

epigean referring to surface (above ground) habitats as opposed to hypogean.

epilimnion portion of a lake above the thermocline and located in the limnetic zone.

epineuston organisms living on or slightly immersed in the air/water interface; see neuston.

ephippium darkened thickened carapace that protects resting eggs in cladocera.

estivation general term for a physiologic process whereby animals reduce metabolism and activity levels to survive harsh conditions, such as pond drying for freshwater bivalves.

eulaterofrontal cilia in bivalve molluscs, stiffened cilia extending laterally from the inhalant side of a gill filament to form a mesh with eulaterofrontal cilia of adjacent gill filaments on which particles are filtered.

eupathid specialized, thickened seta of mites located on distal, leg segments in all active instars.

euryoecic pertaining to species tolerant of a wide range of habitats or environmental conditions.

exchange diffusion transport of ions across epithelia such that diffusion of one ion species down its concentration gradient is coupled with transport of a second ion in the opposite direction.

excrescences teeth on internal margins, denticles on the tips, and talons on the external margins of the horizontal ramus of the entocytherid ostracode hemipenes.

excretophore large vacuolated cells of the gut epithelium of turbellarians.

excretory pore plate in mites, a sclerite bearing the excretory pore in larvae, usually expanded to incorporate the bases of the setae associated with the pore.

exogenous initiated from outside individuals, such as changes in light intensity controlling valve-gaping activity of a bivalve.

exopod outer branch of a crustacean appendage; lateral ramus of biramous appendage, originating on the second segment from the base.

exoskeleton hard external supporting structure, often composed of chitin (with or without calcium carbonate); as in covering of crustaceans.

exploitative competition form of competition in which the outcome is determined by differential abilities to harvest a resource and/or by differences in reproductive rates; competition in which one organism does not directly inhibit another's access to a limiting resource by behavioral means; compare with interference competition.

extrapallial fluid in bivalve molluscs, the fluid filling the space (extrapallial space) between the mantle and the shell.

extrapallial space area between the mantle and shell in molluscs.

extrusome organelle of the pellicle of protozoa, which extrudes a substance or structure, usually in response to a stimulus.

eye plate in mites, a sclerite bearing one or eyes.

facilitation indirect food web effects which span two trophic levels; for example, an increase in the abundance of algae because a predator removes a snail herbivore.

facultative an optional process.

fascicles small tooth-like bundle of fibers.

fenestra in ectoproct bryozoans, the clear central portion of a floatoblast valve.

file row, as in file of cilia or setae.

filiform thread-like; slender and of equal diameter.

filopodium long, filamentous pseudopodium of certain Rhizopoda (amoeba), sometimes branching out without anastomoses; may function in both food capture and locomotion.

finger guard extension of the ventral cardo along the side of the dorsal and ventral fingers of the entocytherid ostracode hemipenes.

first-form male one of two morphological forms of cambarine crayfish; sexually functional male; at least one terminal element of first pleopod usually corneous.

flagellum multiarticulate, endopodite of antennae in some arthropods; whiplike locomotory structure of many protozoa.

flange distal ridge of the contact margin of the duplicature between the list and groove, it may be a part of the outer lamella of the ostracode shell.

flange groove that part of the duplicature surface of the ostracode valve between the selvage and flange.

flange strip part of the duplicature that forms the flange groove and the flange of the ostracode valve.

floatoblast in ectoproct bryozoans, a type of statoblast having a ring of gas-filled chambers; in some cases, the ring becomes buoyant only after being dried.

follicular organ arranged in several small compartments (follicles), i.e., neither diffuse nor compact.

foot a highly muscular locomotory or digging structure.

forcipate type of trophi (jaw structure) found in rotifers having an action like forceps in which the trophi are projected from the mouth to grasp prey.

fossorial associated with digging or burrowing.

fragmentation spherules in Mollusca, the shed tips of digestive cells lining the terminal tubules of digestive diverticulum which are released after intracellular digestion has been completed; they contain undigested material remaining in food vacuoles and are carried on ciliary rejection tracts in the digestive diverticulum back into the stomach where their breakdown releases the undigested contents of their food vacuoles to be egested.

frontal glands glands at anterior terminal region of turbellarians (frontal organ in Acoela).

fulcrate type of aberrant trophi (jaw structure) found in rotifers in the order Seisonidae.

fulcrum one of seven pieces comprising trophi of rotifers; together with the paired rami, they form the incus subunit.

funiculus in ectoproct bryozoans, a tubular strand of tissue joining the blind end of the gut to the inner colony wall; the site of statoblast and sperm formation

furca forked posterior end of some gastrotrichs. Also, an appendage-like structure that is the termination of the body proper in cypridid and cytherid ostracodes; it is articulated to the body and may assist in locomotion but is not considered a true thoracic leg.

fusiform spindle-shaped; tapering at each end.

gastric mill posterior end of the cardiac stomach of crustaceans and some other arthropods where food is ground for later digestion.

gastric shield chitinous plate on the dorsal side of the stomach of bivalves (Mollusca) against which the distal tip of the crystalline style rotates.

geminative type of growth of colonial rotifers whereby young produced by a colony on any day are all born within a span of a few hours and leave the parent colony together as a free-swimming larval colony, attaching to a new substrate in concert. As a consequence of this phenomenon, the genetic diversity of the colony is low; see autorecruitive and allorecruitive.

gena (genae) the cheek; part of the head on each side below the eyes, extending ventrally to the gular suture.

genet genetically distinct individual; often used to describe individuals that commonly grow as colonies in which the individual is not easily distinguished as a separate genetic entity; contrasts with ramet.

geniculate jointed to permit bending at an abrupt angle; distinctly elbowed.

genital bay area between the posterior coxal groups that accommodates the genital field in adult mites.

genital field area of the idiosoma occupied by the gonopore and genital acetabula in adult mites.

genital flaps paired movable flaps which close to cover the gonopore in adults of certain mite taxa; they typically also cover the genital acetabula, but in some derivative groups, the acetabula lie on the flaps.

genital papillae papillae situated on the surface of the cuticle or on elongate finger-like projections on the tail of male nematodes.

germarium an ovary.

germovitellarium organ containing both oogonia and oocytes, i.e., ovary and vitelline cells.

gibbosity hump, marked by convexity or swelling, protuberant structure normally found on the dorsal part of an ostracode carapace.

gill axis in Mollusca, portion of the ctenidium from which gill filaments extend and which is attached to the dorsal mantle wall.

gill filaments in bivalve molluscs, long, thin, fused lateral extensions from the gill axis which extend ventrally and then reflect dorsally to form the inner and outer demibranchs.

glandularium (glandularia) specialized organ located in the integument of deutonymphs and adult mites, consisting of a small, goblet-shaped sac, or "gland," with a tiny external opening, and an associated seta. Glandularia have both a sensory and a secretory function.

globose globe-shaped.

glochidium bivalved larval stage of unionacean bivalves (Mollusca), generally parasitic on fish.

glossa median terminal (or subterminal) lobe of the labium in some arthropods.

glycocalyx "hairy" secreted layer covering the outer cell membrane of many of the "naked" rhizopod amoebae.

gnathosoma in mites, anterior region of the body comprising the mouth and associated appendages, the chelicerae and pedipalps.

gonochoristic two sexes; males and females in the same population reproducing by cross-fertilization; contrasts with hermaphroditism.

gonopore external opening of the reproductive system in adults.

green glands paired excretory glands in crustaceans; also known as antennal glands.

gubernaculum sclerotized structure that supports and guides the spicules out of the cloacal opening during mating.

gula throat sclerite, forming the central part of the head beneath the genae.

gullet in flagellated protozoa, anterior invagination from which the flagella emerge.

gyttja watery, organic sediment that consists mainly of excretory material from benthic animals.

haploid having the number of chromosomes characteristic of the mature germ cell (half the number of the somatic cell).

haptocyst extrusome of some predatory Ciliophora (Sectoria) used to capture prey.

heliotropism tendency of certain plants or other organisms to turn or bend under the influence of light, especially sunlight.

helocrene seepage area where spring water percolates through a substratum of mosses, leaf litter, and detritus.

hemelytra mesothoracic wings of heteropteran insects.

hemipenes paired penes.

hemocoel internal body cavity formed by the expansion of the vascular system.

hemocyanin copper-based respiratory pigment which is dissolved within the hemolymph of crustaceans.

hemoglobin iron-containing blood respiratory pigment; found in a few invertebrates living in aquatic habitats with low oxygen levels.

hemolymph the circulatory fluid or blood of animals with open circulatory systems.

hemolymph sinuses or hemocoels open cavities in which the blood or hemolymph bathes the tissues directly, allowing exchange of gases, ions, nutrients, and wastes.

hermaphroditic individual containing both male and female reproductive organs.

heterotrophy consuming, rather than synthesizing, organic compounds for metabolic activities; contrasts with autotrophy.

hibernaculum thick-walled resting bud occurring in certain gymnolaemate bryozoans species.

hindgut portion of intestine lying between midgut and rectum, primarily involved with consolidation of undigested particles into feces.

hinge section of dorsal shell valve in bivalve molluscs containing the teeth and forming the fulcrum on which valves open and close.

hinge ligament elastic proteinaceous portion of the shell of bivalve molluscs secreted by the mantle isthmus which connects the mineralized shell valves; it is flexed or compressed when the valves are closed and when adductor muscles relax and functions to open the valves by returning to an unflexed state.

hinge teeth interdigitating mineralized projections of shell in bivalve molluscs on the dorsal margins of the valves that hold the valves in position and serve as the fulcrum on which they open and close.

histophages protozoa that feed on the tissue of dead metazoa, as opposed to a parasite feeding on live hosts.

holophyletic group monophyletic clade including all species descended from a common ancestor.

homoplasy parallel adaptive modification of homologous attributes or structures in species of different clades due to similar selective pressures.

hooked glochidia glochidia larvae in bivalve molluscs with spiny hooks on their ventral shell margins.

hormone chemical signal produced and used inside an individual's body.

hyaline margin clear ectoplasmic material surrounding and/or preceding the granular body mass of some amoebae during locomotion.

hybrid result of a mating between organisms with different genotypes; used also for matings between species.

hypersaline term generally referring to a water body whose salinity is higher than seawater (35 mg/L).

hypodermal cords expansions of the hypodermis between muscle fields of nematodes.

hypogean subterranean; burrow, cave, or groundwater environment.

hypolimnion portion of a lake below the thermocline and located in the limnetic zone.

hyponeuston organisms living just below the air/water interface of a water body; see neuston.

hyporheic zone groundwater present below and to the side of a stream and located close enough to be affected by the movement of stream surface waters (unlike the phreatic zone).

hypostome pertaining to the ventral portion of the ostracode head is located a keel-shaped structure which forms the mouth.

hypoxia condition of environmental oxygen concentration below air oxygen saturation levels but not necessarily anoxic.

hysterosoma region of idiosoma of mites posterior to level of primitive sejugal furrow.

idiosoma body proper of mites, comprising the fused cephalothorax and abdomen.

imagochrysalis quiescent transitional instar between the deutonymph and adult in the water mite life cycle, representing the suppressed tritonymph.

incudate type of trophi (jaw structure) found in rotifers, which functions by grasping prey with a forceps-like action.

incus subunit of a rotifer trophi composed of fulcrum and paired rami.

induction change in development or behavior due to receiving an internal or external signal (such as a hormone or kairomone)

infauna sediment-dwelling animals; also, but more rarely, animals that burrow into and live inside another organism.

infraciliature total assemblage of all kinetosomes and the associated subpellicular microfibrular and microtubular structures in ciliate protozoa.

infundibulum corona of rotifers of the order Collothecacea.

inner lamella thin layer of chitin, which, together with the duplicature, forms a proximal covering of the epidermis in ostracodes; at inner limit, it is folded back on itself to form the body wall of the pouch-shaped body of the animal.

instar any developmental stage between molts.

interareolar furrows network of transverse and longitudinal trenches separating areoles in hairworms; may bear pores or support bristles and warts.

interference competition form of competition in which one organism directly or indirectly interferes with another organism (e.g., by aggression), thus preventing it from obtaining a limiting resource; forms of interference include physical combat, the threat of combat, noxious chemical compounds, etc.

interlamellar space cavity formed between ascending and descending gill filaments in Mollusca; water flowing through the ostia of gill filaments enters the interlamellar space to be carried dorsally to the epibranchial cavity; also called the water tube or water channel.

interstitial microhabitat comprising spaces between particles of submerged sand and gravel in groundwater and hyporheic habitats.

intracytoplasmic lamina filament layer of varying thickness present within the syncytial integument of rotifers and acanthocephalans.

ischiopodite third segment from the base of a segmented appendage.

isthmus portion of nematode pharynx between the metacorpus and basal bulb, usually surrounded by the nerve ring. In bivalve molluscs, the constricted dorsal section of the mantle separating the mantle into two lateral shell-secreting halves; the isthmus secretes the proteinaceous hinge ligament connecting the valves.

iteroparous reproducing repeatedly at multiple times during an organism's life (cf., semelparous).

kairomone chemical signal produced by a predator that affects behavior, development, or life history of another species so as to result in higher relative survival of individuals of that species that react to the signal.

karst terrain underlain by fractured and extensively dissolved carbonate rock (e.g., limestone) and typified by numerous sinkholes, springs, caves, and few surface streams.

kinetid kinetosome and its associated pellicular structures in Ciliophora, including the cilium, the elementary repeating structure of the ciliate cortex; the polykinetid is an organellar complex composed of two or more kinetids.

kinetocyst extrusome of Actinopoda, which secretes materials that probably serve to make the surface of an axopod sticky.

kinetoplast DNA-rich organelle near the anterior of some flagellated protozoa (Kinetoplastida), associated with their unique large mitchondrion.

kinetosome subpellicular structure forming the base of a cilium, but may be without a cilium; it is characterized by nine triplets of microtubules at right angles to the plasma membrane arranged in a circle.

kinety single, structurally and functionally integrated (somatic) row of kinetosomes along with their cilia (cilia not necessarily associated with each inetosome), i.e., a line of kinetids.

knob in bivalve molluscs, a high rounded major protuberance with the sides joining the rest of the valve at a distinct line and at a steep angle.

K-selection selection for life-history characteristics that increase fitness in stable environments where populations reach densities near environmental carrying capacity; associated with high levels of intraspecific competition.

labial palps pair of thin flattened structures extending from either side of the mouth region of bivalve molluscs to lie against the outer surface of the outer demibranch and the inner surface of the inner demibranch; they function to receive filtered particles from the ctenidia and sort them into those accepted for ingestion and those rejected as pseudofeces.

lacustrine see lentic.

lamellae layers; flattened, leaf-like structures in several phyla, such as the gills of bivalve molluscs and the buccal plate-like projections on the mouth (buccal) ring of some eutardigrades.

Lansing effect putative negative effect of elevated calcium on survivorship in rotifers in which calcium accumulation in older mothers is passed on to their offspring; seeorthoclone.

lateral coxal apodeme conspicuous apodeme indicating the line of fusion between the second and third coxal plates in larvae of certain mite taxa.

lateral field track usually containing several longitudinal striations running on both sides of the nematode.

lateral teeth elongated lamellar hinge teeth in bivalve molluscs forming anterior and posterior to the cardinal teeth or posterior to the pseudocardinal teeth in unionacean bivalves.

LC_{50} concentration of a toxic agent at which 50% of a cohort will perish within a certain predetermined period; also called the median lethal concentration and LD_{50}.

lentic standing water environments (e.g., lakes, ponds, wetlands, and both permanent and temporary pools).

limnetic open water, deeper areas of a lake or pond and away from the shoreline littoral zone.

line of concrescence proximal line of junction of the duplicature and outer lamella of the ostracode valve, the inner border of the chitin adhesive strip.

lira rib or ridge running along the whorl of a snail shell in the direction of the coiling.

list proximal ridge on the contact margin of the duplicature of the ostracode valve, not always present.

list strip part of the duplicature from the inner margin to and including the list of the ostracode valve.

littoral zone or body of water shallow enough for growth of rooted aquatic vegetation.

lobopodium blunt pseudopodium used in feeding and/or locomotion.

lophophore organized structure of ciliated tentacles used for capturing suspended food particles from the water, and probably also providing an important surface for gas exchange; occurring in ectoprocts and certain other phyla.

lorica thickened body wall of rotifers, organized as a syncytial integument possessing two keratin-like proteins which compose a layer called the intracytoplasmic lamina. A loose-fitting surface covering secreted and/or assembled by protozoa; generally synonymous with test.

lotic running water environments (e.g., creeks, rivers, and some springs).

lunule crescent or half-moon shaped structure at the base of each double claw in some eutardigrades.

lyrifissure (lyriform fissure) slender sac-like structures in mites opening through slits in the integument, apparently functioning as proprioceptors, in all active instars.

lysosome membrane-bound organelle containing hydrolytic enzymes; in protozoa, they are involved in intracellular digestion.

macroinvertebrate organisms larger than microinvertebrates and meiobenthos; retained on coarse sieves with a mesh ≥ 2 mm.

macronucleus in ciliates, the nucleus active in transcription and thus responsible for the phenotype and regulation of metabolism; see micronucleus.

macrophagous in protozoa, a species that consumes large items singly by phagocytosis.

macrophyte macroscopic photosynthetic organism growing submersed, floating, or emergent in water; most are angiosperms but also includes some nonvascular plants and macroalgae.

macropterous with fully formed wings.

malleate type of trophi (jaw structure) found in certain monogonont rotifers in which the rami are massive and may possess teeth along the inner margin.

malleoramate type of trophi (jaw structure) found in monogonont rotifers of the order Flosculariacea, resembling the malleate form, except in number of teeth on the unci.

malleus subunit of rotifer trophi composed of an uncus and manubrium.

mandible hardened mouthparts (cephalic appendages) in many arthropods that are used to grasp, tear, and push food into the mouth; most anterior gnathal appendage and just behind the antennae.

mandibular palp endopodite segment of the mandible.

mantle thin extension of the dorsal body wall of molluscs underlying the shell and responsible for its secretion.

mantle cavity space formed between the body and surrounding mantle tissues in molluscs; completely surrounds the body in bivalves.

manubrium one of seven pieces comprising the trophi of rotifers; a manubrium together with an uncus form the malleus subunit.

marsupium space formed for incubation of embryos and larval stages.

mastax muscular pharynx of rotifers containing jaws called trophi.

maxillae first or second pair of mouthparts (cephalic appendages) posterior to the mandibles; crustaceans have two pairs of maxillae, insects have one pair.

maxillary glands paired excretory glands of crustaceans.

maxilliped one of pair of three sets of gnathal appendages situated immediately posterior to second pair of maxillae.

maxillule first maxilla.

medial coxal apodemes series of short apodemes at the medial edges of coxal plates in some larval mites.

medial eye plate (frontal plate) sclerite bearing medial eye in early derivative mite taxa; the medial eye may be absent in highly derived groups.

meiobenthos small benthic organisms that pass through a 2 mm mesh but are retained on a 40–200 μm mesh.

melanin black, brown, or red pigment made of polymerized amino acids (tyrosine); used for photoprotection and making structures stronger.

merus fourth segment from proximal end of segmented appendage.

mesial (mesal) pertaining to the middle; toward the middle.

metachronal occurring one after another, in succession.

metacorpus swollen median portion of nematode pharynx, located between the procarpus and isthmus.

metamere homologous body segment, a somite.

metasome portion of the copepod body anterior to the major body articulation.

microaerophilic protozoa with aerobic metabolism living in microhabitats with very low oxygen concentrations.

micronucleus nuclei of ciliates mediating sexual recombination and reproduction, typically much smaller than the macronucleus; diploid; often multiple.

microphagous protozoa that consume relatively small items by phagocytosis, collecting many into a single food vacuole.

microplankton plankton that are 20–200 μm in size.

microsome simple membrane-bound organelle containing oxidative enzymes; synonymous with microbody.

microvilli microscopic projections from a cell surface that increase the surface area of the cell.

mictic (mixis) phase in the life cycle of monogonont rotifers in which reproduction is sexual, resulting in a diapausing embryo called a resting egg; seeamictic.

midden refuse heap; commonly, shells of unionacean mussels that had been consumed by humans or other mammals and then left at one site.

middle piece structure in sperm cells lying between the head and flagellum containing mitochondria.

midgut portion of the intestine between stomach and hindgut; in molluscs, it is characterized by the presence of the typhlosole, the ciliated surfaces of which sort particles into those returned to the stomach and those passed to the hindgut for consolidation into feces and egestion.

mixotroph organism driving energy from both autotrophy and heterotrophy, including those whose autotrophy is via endosymbionts.

molting process of shedding the cuticle at periodic intervals to allow expansion of the body in a newly developed and larger exoskeleton.

monoecious individual possessing both male and female reproductive systems, at least during part of its life cycle.

monophyletic group phylogenetic grouping of species descended from a common ancestor and which are classified together within a single taxonomic unit; compare with polyphyletic.

monopodial describing an unbranched, cylindrical type of pseudopodium.

morph variant form within a species.

morphotype morphologically distinct form; in rotifers, a form present within a single population whose polymorphic feature(s) (e.g., spines, body size) undergo a seasonal phenotypic change.

mouth hook vertically-oriented, mandible-like structure in dipteran larvae.

muciferous body variety of mucocyst found in some Phytomonadida.

mucocyst extrusome of protozoa responsible for secretion of mucous-like material onto the cell surface.

mucrones small cuticular teeth on interior of the buccal cavity in some eutardigrades.

multispiral growth rings of an operculum of a snail comprised of many, slowly enlarging spirals.

multivoltine having more than two broods or generations per year.

muscle scars area marking the former position of attachment of a muscle on the interior of a shell or other support structure, distinguished from the rest of the shell by a surrounding groove or by being higher or depressed.

mutualism association between two organisms in which both benefit.

nacre or nacreous layer inner shell layer of most molluscs, secreted by the mantle as successive layers of calcium carbonate crystals (absent in the Sphaeriidae).

nannoplankton plankton that range from 2–20 μm in size.

nauplius earliest larval stage of many crustaceans.

neonate newly-born animal.

net growth efficiency percentage of assimilated energy represented by that allocated to tissue growth and gamete production.

neuston organisms living at or very near the air/water interface of a water body; see hyponeuston, epineuston, and pleuston.

niche partitioning differences in use of resources that alleviate interspecific competition.

node on the ostracode carapace, a protuberance smaller than a lobe or knob but larger than a tubercle, may have the form of a half sphere.

nonrespired assimilation portion of assimilated energy or food not catabolized for maintenance energy and, therefore, available for production of new tissue or gametes.

nymphochrysalis quiescent transitional instar between the larva and deutonymph in the water mite life cycle, representing the suppressed protonymph.

oblique slanting; neither parallel nor perpendicular to a reference point.

obtuse blunt angle at the end.

ocellus simple eye consisting of a single, bead-like lens, occurring singly or in small groups; distinct from a compound eye.

omnivore consuming food at more than one trophic level, e.g., crayfish eat live animals (carnivory) and plants (herbivory) and will also consume dead animals and plants (detritivory).

open circulatory system circulatory system characterized by blood not being continually maintained in vessels; rather, the blood bathes the tissues directly in large hemocoel sinuses before returning to the heart.

operculate formed as a cover.

operculum lid or covering structure, as in some insects. Horny or calcareous disk on the foot of a snail that covers the aperture when foot is withdrawn; used as protection for predators.

opsiblastic egg type of resting egg produced by gastrotrichs; see tachyblastic egg.

orifice opening; in ectoprocts, the terminal opening in a zooid through which the polypide extends.

orthoclone clone of a rotifer species established after many generations of culturing the ith offspring of every generation (seeLansing effect).

osmobiosis cryptobiotic (latent) state in tardigrades induced by elevated osmotic pressures.

ostia pores; openings penetrating fused gill filaments in bivalve molluscs with eulamellibranch ctenidia which allow passage of water across the filaments into the interlamellar space.

outer lamella in ostracodes, a hard shell material that covers the outside of the epidermis that secreted it, composed of a thick layer of calcite enclosed between two layers of chitin.

ovate (ovoid) somewhat oval in shape.

oviduct duct from the ovary, through which ova pass.

oviparous females that produce eggs that undergo embryonic development and hatch outside the body of the female; see viviparous and ovoviviparous.

ovoviviparous females which produce eggs that undergo embryonic development inside the female and that hatch just before or soon after being released; see oviparous and viviparous.

paddles feather-shaped locomotory structures attached anteriorly in *Polyarthra* rotifers.

pallial pertaining to the pallium or mantle cavity of a mollusc.

pallial line line of muscle scars on the outer perimeter of the inside of the bivalve mollusc shell, marking points at which pallial muscles attach the mantle to the shell.

pallial sinus posterior indentation of the pallial line on the inside of the bivalve mollusc shell, marking the space into which the siphons are withdrawn on valve closure.

palmate like the palm of the hand, with fingerlike processes.

palpal lobes broad, paired, movable lobes on the distal end of the prementum in odonate insects.

palustrine relating to wetland habitats.

papilla one of many small discrete protuberances (as on the ostracode shell surface) each with steep sides; smaller than tubercles.

papillate many lobe-like extensions.

paraglossa lateral terminal lobe of the labium.

paramylon reserve food material characteristic of euglenid flagellates; occurs in a variety of shapes, usually with a laminated structure.

paraphyletic group monophyletic grouping including only some of the species descended from a common ancestor.

paratene repeated kinetidal patterns at right angles to the longitudinal axis of a ciliate's body; superficially gives impression that the ciliary rows run circumferentially rather than longitudinally in the affected part of the body.

parietal inside wall of a snail aperture.

paroral membrane synonymous with undulating membrane.

parthenogenesis sexual reproduction in which egg development occurs without fertilization. Amictic parthenogenesis is sexual reproduction without recombination of genetic material.

paucispiral growth lines of a snail operculum, present as a few, rapidly enlarging spirals.

pectinate having narrow parallel projections suggestive of the teeth of a comb.

pedal referring to the foot.

pedal feeding use of the foot in bivalve molluscs to feed on organic detrital deposits in or on sediments; generally involves ciliary tracts on the foot to bring detritus to the palps or gill food grooves.

pedal gland cement glands in the foot region of rotifers used to achieve temporary (in pelagic or littoral zooplankton) or permanent (in sessile forms) attachment to a surface.

pedicles lateral extensions of the corona of certain bdelloid rotifers.

pedipalp feeding appendage; in mites, the pedipalp typically has five movable segments, located laterally on the gnathosoma in all instars. Pedipalps have both sensory and raptorial functions.

peduncle pseudopodium-like cytoplasmic extension emerging from the sulcus in some dinoflagellates and used in feeding. Also, general name for a holdfast organelle in diverse groups of free-living protozoa.

pelagic see limnetic.

pellicle surface layer of protozoa, including the plasma membrane and associated membranous, microfibrillar and microtubular structures and organelles, but excluding nonliving external coverings such as tests or loricas.

peniferum portion of the entocytherid ostracode copulatory apparatus that bears the penis (usually its distal portion), and its proximal portion forms a hinge on which the entire apparatus may be turned through an arc of 180 degrees about the zygum; includes the dorsal and ventral cardo, costa, spermatic tube, and penis.

penultimate next to last.

pereiopods thoracic appendages of many crustaceans; usually having a prime locomotory function (walking or swimming) and one or more secondary roles.

perennial living for several years.

periblast part of a statoblast valve of ecotoprocts that excludes the capsule; in floatoblasts, the periblast includes both annulus and fenestra.

peribuccal lobes flattened cuticular projections surrounding the mouth of some eutardigrades.

peribuccal (oral) papillae elongated cuticular projections surrounding the mouth of some eutardigrades.

peribuccal papulae short, distinct cuticular projections surrounding the mouth of some eutardigrades.

pericardial cavity remnant of the coelomic cavity surrounding the heart ventricle in molluscs.

pericardial gland specialized tissue in the wall of the heart auricles of molluscs through which hemolymph fluid is ultrafiltered into the pericardial space from which it enters the kidney.

pericardium epithelial layer lining the pericardial cavity.

periostracum proteinaceous outer layer of mollusc shells secreted by the mantle edge.

periphyton microalgae growing in a thin layer on mostly vascular plants; sometimes used loosely for microalgae growing on other substances.

peristalsis rhythmic waves of contraction which moves material through the gut.

peristome area surrounding the cytostome of ciliates, especially where it is specialized relative to the rest of the body surface.

peristomium first functional segment, usually containing the mouth, in segmented worms.

phagocytosis process by which a cell engulfs a particle by extending its plasma membrane to form a vacuole or invagination.

phagotrophy nutrition by phagocytosis.

pharyngeal apophyses cuticular thickenings at the junction of the buccal tube and lumen of the pharynx, alternating in position with the placoids.

pharyngeal intestinal junction; in nematodes, a cellular valve located between the pharynx and intestine; may protrude down into the intestine.

pharynx portion of the alimentary tract located between the stoma and intestine, often called the esophagus. In turbellarians, "pharynx simplex" is a mostly short, unfolded tube; "pharynx plicatus" is a protrusible, more or less tubular fold enclosed in the pharynx cavity; "pharynx bulbosus" is a bulb separated from the surrounding tissue by a muscular septum. Pharynx bulbosus is represented by three types: rosulatus, doliiformis, and variabilis. Pharynx rosulatus is mostly round with a more or less vertical axis; pharynx doliiformis is barrel-shaped, anteriorly situated and with a horizontal axis; pharynx variabilis is weakly muscular with variable shape and weakly differentiated septum.

phasmids paired sensory organs located laterally in the caudal area of nematodes.

phenotypic plasticity expression of different morphologies depending on environmental influences or signals and expressed by a single genotype.

pheromone chemical messenger substance secreted by one individual into the external environment, which elicits a specific response in another individual of the same species.

photodamage damage caused to molecules due to ultraviolet radiation in sunlight.

photoperiod length of the light period during a 24 hour light-dark cycle.

photoprotection protection from ultraviolet light, as by a dark pigment that absorbs light.

phototaxis locomotion directed by light.

phreatic zone flowing groundwater located far enough from surface streams so that water movement in the streams does not influence movement of the phreatic waters; see-hyporheic zone.

phyllopod a leaf-like or lobed, biramous appendage in some non-malacostracan crustaceans.

phytophilic aquatic herbivores with a preference for consuming algae; also called algivores.

phytoplanktivorous pertaining to feeding on phytoplankton (i.e., suspended algae).

phytoplankton unicellular and multicellular algae and cyanobacteria suspended in the water column.

picoplankton plankton 0.2–2 μm in size, primarily bacteria.

pinocytosis process by which cells ingest dissolved or colloidal material, involving attraction of these substances to cell surfaces and then internalizing them through invagination of the plasma membrane.

piptoblast type of statoblast in bryozoans that has no annulus and is not cemented to the colony substrate; formed only in the family Fredericellidae.

placoids cuticular thickenings in the lumen of the tardigrade pharynx (pharyngeal bulb); large anterior ones are macroplacoids, small posterior row are microplacoids.

plankton eukaryotic and prokaryotic organisms living all or part of their lives in the water column of aquatic ecosystems and at the mercy of water currents.

planospiral having shell whorls confined to a single plane; coiled but without an elevated spire.

plasma membrane unit or cell which is the outer-most living layer of protozoa.

plasmodium multinucleate mass of protoplasm enclosed by a plasma membrane; nonhomologous trophic stage in the life cycle of some protozoa.

plastron semipermanent bubble of air through which carbon dioxide from respiration is exchanged with oxygen from the water.

pleopod one of five pairs of appendages on first five segments of a decapod's abdomen; also called "swimmerets," or modified into male gonopods.

plesiotypical (plesiomorphic) conservative, or unmodified, state of a biologic attribute or morphologic character.

podomere leg segment of an arthropod.

polykinetid see kinetid.

polykinety compound ciliary organelle composed of several kineties, e.g., a cirrus.

polyphyletic group taxonomic assemblage of species that are not all descended from a common ancestor.

polypide retractable portion of a bryozoan zooid comprising the central ganglion and all structures related to feeding and digestion.

polyploid having more than twice the number of chromosomes in somatic cells.

polytrophic see omnivore.

polyvoltine with several generations during the season; contrasts with univoltine.

pore canals in nematomorphs, pores connected to fine tubes in the cuticle; in ostracodes, a passage through the entire valve in which are located sensory hairs to the nerves of the hypodermis.

postcloacal crescent see crescent.

posterior ridge external ridge on unionacean (Mollusca) shells extending from the umbos posterioventrally to the shell margin.

postocularia pair of prominent setae located on the dorsum of the idiosoma posterior to the level of the eyes in deutonymphs and adult mites.

prehensile fitted or adapted for grasping, holding, or seizing.

prehensile palp appendage modified to grasp, hold, or seize; generally formed of two podomeres, the proximal one referred to as the prodopus, distal one referred to as the dactylus of some crustaceans such as ostracodes.

prementum distal segment of the labium in odonate insects, which bears the palpal lobes.

preocularia pair of prominent setae located on the dorsum of the idiosoma anterior to the level of the eyes in deutonymphs and adult mites.

preparasitic attendance phoretic association by larvae of some species of water mites with the preimaginal stage of their insect host.

prismatic layer molluscan shell layer composed of elongated calcium carbonate crystals oriented at a 90° angle to the surface plane of the shell and lying between the outer periostracum and inner nacreous layer.

procorpus anterior portion of the nematode pharynx between the stoma and metacorpus.

prodelphic with a single anterior-directed ovary in nematodes.

profundal deep zone of a lake below the littoral and open water photic zone; often restricted to benthic habitats but may include pelagic zone.

proleg process or appendage that serves the purpose of a leg but is not a true leg.

promitosis form of nuclear division wherein the endosomal body in the nucleus divides and forms two polar masses with a spindle between them; characteristic of some naked amoeba of the class Heterolobosea.

propodosoma region of a mite's idiosoma posterior to the level of the primitive sejugal furrow.

propodus penultimate segment (sixth from base) of a segmented appendage.

proprioceptor sensory receptor that responds to physical or chemical stimuli originating from the organism.

prostate cells, glands, or organs in male reproductive systems, releases secretions that mix with sperm.

prostaglandins local chemical mediators inducing short-lasting effects in adjacent tissues.

prostomium region anterior to mouth in some oligochaete worms.

protandry sequential hermaphroditism with the male stage preceding the female stage.

protocerebrum part of a crustacean's brain that normally innervates eyes, sinus gland, frontal organs, and head muscles.

protoconch shell of newborn snail or "spat."

protonephridia excretory-osmoregulatory system of certain pseudocoelomates, including gastrotrichs and rotifers, comprised of flame cells and tubules.

protonymph first nymphal instar in the generalized mite life cycle.

protopodite basal part of a typical crustacean limb consisting of two, more-or-less consolidated segments and bearing at its distal extremity an exopodite, endopodite, or both.

proventriculus large cavity in collothecid rotifers that is an extension of the mastax in which prey are held prior to being passed into the stomach.

provisional genital field region of the idiosoma occupied by the rudimentary gonopore and genital acetabula in mite deutonymphs.

proximal toward the median plane of a body.

pseudobranch "false gill" present in ancylid and planorbid snails; conical extension of epithelium used in respiration, analogous to ctenidium of prosobranchs.

pseudocardinal teeth in bivalve molluscs, compact shell lamellae on the anterior portion of the hinge plate of unionacean mussel shells which performs function of cardinal teeth.

pseudocoelom body cavity not lined with mesodermal tissue, as is a true coelom, and found in gastrotrichs, nematodes, and rotifers.

pseudofeces masses of mucous-bound particles rejected from the labial palps of bivalve molluscs before ingestion and release into the mantle cavity to be carried out the siphon on water currents induced by valve clapping (i.e., rapid valve adduction).

pseudopodium in protozoa, a dynamic extension of the cytoplasm used in feeding or locomotion; in insects, a soft, foot-like appendage that is less prominent than a proleg.

pseudosexual eggs in cladocera, asexually produced eggs that resemble those produced by a sexual process. In rotifers, diploid resting egg supposedly produced in rotifers by parthenogenesis in the absence of males; seemixis.

pseudostome aperature of a test or shell of a testate amoeba; pseudopodia protrude from it during locomotion and/or feeding.

punctae small pit-like opening of a pore canal on the ostracode shell. Microscopic pores traversing the mineral portion of the shells of sphaeriid bivalve molluscs to open beneath the periostracum; contain extensions of the pyramidal cells.

pustule protuberance with a pore in the middle, similar to a volcano with a crater, about the size of a tubercle on the ostracode carapace.

pyramidal cells in bivalve molluscs, mantle epithelial cells that extend through shell punctae (pores) to lie just below the periostracum.

Q_{10} difference in metabolic rate over a 10°C temperature range (e.g., Q_{10} would equal 2 if the respiration rate doubled between 10 and 20°C).

Q_{10} **(acc.)** factor by which a biologic rate function changes over a 10°C increase in acclimation temperature.

quadrate square or rectangular in shape.

radial pore canal passage through the adhesive strip, between the duplicature and the outer lamella of the proximal valve edge of an ostracode.

ramate type of trophi (jaw structure) found in bdelloid rotifers in which the rami are large and semicircular in shape and unci possess many teeth.

ramus one of the seven pieces comprising rotifer trophi; paired rami together with the fulcrum form the incus subunit.

receptaculum seminis part of the female reproductive system of turbellarians in which recipient spermatozoa are stored.

reniform suggesting a kidney in shape.

respiratory plate also referred to as branchial plate, modified from the exopodite and may contain numerous feathered setae.

resting egg thick-walled cyst enclosing a diapausing, diploid embryo, usually resistant to extreme environmental conditions. In rotifers these "eggs" are produced by sexual reproduction.

reticulopodium threadlike, repeatedly branching and anastomosing pseudopodium, functioning more often in food gathering than locomotion.

retrocerebral organ structure of unknown function in rotifers composed of the unpaired retrocerebral sac and paired subcerebral glands together; possible function as an exocrine gland lubricating the anterior part of the body.

retrocerebral sac unpaired glandular structure near the subcerebral gland in the head region of rotifers possessing a forked duct that opens into the corona.

rhabdions cuticular segments of the nematode stoma, may be fused and produce an entire stomal wall or separate and exhibit a segmented wall.

rhabdos organelle supporting the cytopharynx of some ciliates, especially in the subphylum Rhabdophora; although like a cyrtos, rhabdos is considered evolutionarily the more primitive organelle.

rheocrene flowing spring with a substratum of sand and gravel.

rheotaxis orientation to current (can be positive or negative).

riparian living on the bank of a lake, pond, or stream.

rostrum any extended portion of the head end of an invertebrate.

r-selection selection for life-history traits increasing fitness in an unstable habitat where catastrophic population reductions prevent high levels of intraspecific competition.

scapular setae second most anterior series of sensory setae on the dorsal surface of the idiosoma of water mites, consisting of an internal pair (si) and an external pair (se).

sclerite hardened area of the insect body wall bounded by sutures or membranes.

semivoltine having one brood or generation every two years.

sanguivorous feeding on blood.

scabrous having small, round, raised scales or cones.

second-form male one of two morphological forms of male cambarine crayfishes; sexually nonfunctional male lacking corneous terminal elements on first pleopod (gonopod).

sediment focusing uneven deposition of sediments in a basin so that certain portions of the basin accumulate sediments at a greater rate; usually this occurs in the deepest parts of the basin.

sejugal furrow primitive line of demarcation on idiosoma at level between insertions of second and third pairs of legs in all mite instars, but expressed only in certain early derivative taxa.

selvage middle and principal ridge of the contact margin of the duplicature of the ostracode valve.

selvage groove part of the duplicature surface between the list and selvage of ostracode valve.

selvage strip part of the duplicature forming the selvage and selvage groove of the ostracode valve.

semelparous breeding only once during an organism's life; one massive reproductive effort.

seminal receptacle (spermatheca) sac-like structure in the female reproductive system that receives and stores spermatozoa.

seminal vesicle saclike structure in the male reproductive system that stores spermatozoa.

sensillum epithelial sense organ.

septulum cuticular thickening in the lumen of the pharynx, posterior to the placoids, in the same plane as the pharyngeal apophyses in some eutardigrades.

septum dividing wall or membrane; in ostracodes, a small ridge on the list strip on the duplicature of the valve.

serial homologies structures having the same embryonic and evolutionary origin but which differ segmentally (e.g., all appendages of a crayfish, except the first antennae, arose from similar embryonic primordia and from a series of similarly formed appendages in an ancient ancestral stock).

serotonin chemical messenger stimulating active transport of sodium ions from the medium in freshwater bivalve molluscs.

serrate notched or toothed on the edge.

sessoblast relatively large statoblast cemented firmly through the body wall to the substratum; formed only in the bryozoan family Plumatellidae.

setae (or chaetae) epidermal, hair-like extensions of the cuticle.

sexual dimorphism differences in shape and/or size between females and males of the same species.

sheath in nematodes, a cuticle retained from the last molt which may surround the entire body or head region.

simplex stage molting stage in tardigrades characterized by the absence of the buccal apparatus.

sinistral snail shells with the aperture opening toward the left (when viewed with the apex pointed away from the observer and with the aperture fully visible).

siphon in bivalve molluscs, tubular structure formed by fusion of opposite mantle margins directing inhalant (inhalant siphon) and exhalant (exhalant siphon) respiratory and feeding water currents into and out of the mantle cavity. An elongated breathing tube in some insects.

solenidion (solenidia) specialized setiform chemosensory structure in mites; solenidia are located on distal segments of the pedipalps and legs in all active instars; though seta-like in appearance, solenidia differ structurally and ontogenetically from setae and are usually given distinctive designations in chaetotactic formulae.

somatic cell cell that become differentiated into tissues, organs; opposed to embryonic germ cell.

somite homologous body segment, a metamere.

spear protrusible pointed, solid rod-like structure in the mouth region of some nematodes.

species richness number of species in a habitat.

spermatogenesis formation of sperm.

spermatophore packet of spermatozoa enclosed in a membranous capsule, produced by adult male mites and either deposited on the substratum or transferred to the female gonopore.

spermatozoan individual haploid cell produced by males for sexual reproduction.

sphincter circular muscle forming a ring around a tube, such as the gut, that closes the tube by contracting.

spicules internal support and/or protective structures of sponges composed principally of silica or calcium salts. Also, paired sclerotized structures in the tail of the male nematode.

spindle snail shell with a thick middle and a tapered point at both the apex and the aperture ends.

spinneret in nematodes, a terminal pore out of which passes secretions of the caudal glands.

spiracle respiratory opening connected to the tracheal (respiratory) system of insects.

standing crop biomass amount of tissue produced per unit area at a set point in time, usually measured as g carbon, Kcal, or ash free dry mass (AFDM) per m^2.

statoblast in phylactolaemate bryozoans, an asexually produced capsule composed of two convex valves enclosing a mass of yolky material and undifferentiated tissue; capable of germinating a new zooid after obligate dormancy.

statoconia in bivalve molluscs, many small mineral inclusions in the cavity of a statocyst (gravity orientation organ).

statocyst gravity-orientation organ containing mineral inclusions (statolith or statoconia) in a vesicle lined with sensory cilia.

statolith mineral inclusion in the cavity of a statocyst.

stenopod unbranched, segmented walking leg in some crustaceans, particularly decapods.

stenothermal tolerating little variation in temperature.

stigmata paired external openings of the tracheal system, located between bases of the chelicerae in all active instars of mites.

stolon in certain ectoproct bryozoans, a cylindrical stem-like structure from which individual zooids grow at intervals.

stoma portion of the alimentary tract between the oral opening (mouth) and pharynx.

stream order method for classifying streams in which the smallest permanent streams is designated as a "first order" stream; two first orders streams combine to form a second order stream; a first and second in combination are still considered a second order, but when two second order tributaries join, a third order stream results.

striations (striae) parallel grooves; also, the spiral, microscopic grooves along the whorls of a snail shell (sometimes incorrectly used to refer to spiral ridges or lirae).

stygobitic species occur exclusively in hypogean habitat.

stygophilous species occurring in both surface and subterranean waters but with a preference for the latter habitat.

style sac evagination of the stomach floor of molluscs, which secretes the crystalline style.

stylet sclerotized tube or channel of variable shape and complexity; a hollow, sclerotized, hard, rod-like structure in the mouth region of some nematodes.

stylet knobs enlarged protuberances (usually three) at the base of the nematode stylet, usually serve for attachment of muscles that aid in stylet extension.

stylostome mucopolysaccharide feeding tube secreted by parasitic larval water mites into the body of their host.

subcerebral glands paired glands near the retrocerebral sac in the head region of rotifers with ducts that lead to the corona; seeretrocerebral organ.

subitaneous egg one developing and hatching with no arrest in development.

sulcus depression or furrow on the lateral surface of the organisms in several phyla (e.g., Ostracoda and Mollusca); in protozoa, a longitudinal groove on the surface of dinoflagellates from which the flagella and, if present, the peduncle or other feeding structures emerge.

suture region where two structures (e.g., plates or valves) are joined.

symbiont one member of a symbiotic relationship.

sympatric two or more species whose habitat distributions overlap.

syngamy fusion of haploid gametes or gametic cells in sexual reproduction to form a diploid zygote; phenomenon exhibited by some members of all major protozoan taxa except ciliates.

tachyblastic egg type of egg produced by gastrotrichs in which development occurs immediately; see opsiblastic egg.

tagma or tagmata body regions (especially in arthropods) consisting of metameres grouped or fused to perform similar functions, e.g., head, thorax, and abdomen of a crustacean or insect.

test loose surface covering of protozoa, created by the inhabitant with secreted and, sometimes, gathered materials; generally synonymous with lorica, but test is normally applied to Rhizopoda.

thermocline zone in a lake of relatively rapid temperature change over a small range of depth; commonly defined as 1°C change per meter of depth.

tocopherol (*a*-tocopherol or vitamin E) organic compound that has been shown to induce mixis (sexual reproduction) in some rotifers and lengthen the lifespan in others.

torpidity condition or quality of being dormant or inactive, a temporary loss of all or part of the power of sensation or motion.

transverse muscle attachment scar conspicuous muscle attachment scar in the posterior region of the third coxal plates in larvae of certain mite taxa.

trichocyst type of extrusome emitting a spindle-shaped, nontoxic projectile.

triploid having three times the haploid number of chromosomes.

tritocerebrum portion of a crustacean's brain that principally innervates the antennae and a section of the alimentary tract.

tritonymph third nymphal instar in the generalized mite life cycle.

trochantin small, forward projecting sclerite at the base of the trochanter.

trochus one of two ciliated rings of the typical rotiferan corona; seecingulum.

troglobite obligate cave-dweller, often characterized in crustaceans by a reduction in eye structure and a lack of pigments; also, frequently demonstrating K-strategies.

troglophile organism living in a cave in a nonobligate relationship; such organisms usually complete their life history in the cave and commonly have conspecifics surviving outside the cave; contrasts with troglobite.

trophi complex set of hard jaws characteristically found in the pharynx (mastax) of rotifers.

trophozoite mature, vegetative, feeding, adult stage in the protozoan life cycle; used primarily with reference to parasitic species of nonciliate taxa; synonymous with trophont.

tubercles small, rounded knobs on the body surface of an animal, larger than papillae, smaller than nodes.

tun barrel-shaped form of a tardigrade in a cryptobiotic (latent) state.

turnover time amount of time necessary for the standing crop biomass to be replaced completely in a species.

type taxon (or specimen) original individual of a species on which the species name is based; type specimens are kept in museums for comparative purposes and to reduce confusion about scientific nomenclature.

typhlosole invaginated, cilated, particle-sorting structure running the length of the midgut and extending into the ciliated sorting surfaces of the floor of the stomach and digestive diverticulum in bivalve molluscs.

umbo in bivalve molluscs, portion of the shell raised above the dorsal shell margin near the hinge (also called the beak).

uncate condition of the pedipalp in deutonymphs and adults of some mites in which the ventral surface of the tibia is expanded and produced distally to oppose the tarsus, forming an efficient grasping organ.

uncinate type of trophi (jaw structure) found in rotifers characterized by unci possessing few teeth, usually with one large one and a few small ones.

uncus one of the seven pieces comprising rotifer trophi; an uncus together with a manubrium form the malleus subunit.

undulating membrane compound ciliary apparatus typically lying on the right side of the buccal cavity of some protozoa; synonomous with paroral membrane.

undulipodium swimming organelle of eukaryotic cells, including protozoa, includes both the flagellum and the cilium.

univoltine having one brood or generation per year.

urogomphus paired, unsegmented terminal appendage of coleopteran larvae.

uroid posterior part of moving "naked" lobose amoeba.

urosome portion of the copepod body posterior to the major body articulation.

urstigma (urstigmata) knob-like structure with a porous cap located between the first and second coxal plates in larval mites; the paired urstigmata are apparently homologous with the genital acetabula of deutonymphs and adults and probably function as an osmoregulatory chloride epithelium.

uterus region in female reproductive system of invertebrates that stores egg capsules prior to release.

vagina portion of the female reproductive tract connecting the vulva with the uterus or uteri; a female tract used in copulation, distal to entrance of the oviducts.

valve one of either the left or right shell of a seed shrimp (Ostracoda) or mussel (Mollusca), generally referred to as left or right valve. In nematodes, the disc or nipple-like structure which may occur in the metacorpus or basal bulb of the pharynx.

varve layer of sediment representing one year's accumulation.

vas deferens duct from the testes, through which spermatozoa pass.

veliger free-swimming larval stage of molluscs characterized by a ciliated velum used for swimming and feeding; in North American freshwater bivalves, veligers occur only in exotic *Dreissena* spp.

ventral shield single large sclerite or series of closely fitting plates and platelets completely covering the venter of the idiosoma in deutonymphs and adults of certain mite taxa.

verge penis or organ that bears the penis.

verticil setae most anterior series of sensory setae on the dorsal surface of the idiosoma, consisting of and internal pair (vi) and an external pair (ve).

vesicula seminalis enlarged part of the sperm duct or ejaculatory duct holding sperm.

vestibule space between the outer lamella and the duplicature in the proximal portion of an ostracode valve.

virgate type of trophi (jaw structure) found in rotifers that are modified for piercing and pumping; generally can be recognized by the long fulcrum and manubria; some are asymmetrically shaped.

vitellarium yolk gland associated with the ovary in bdelloid and monogonont rotifers.

vitelline membrane most external membrane of a molluscan egg.

viviparous type of reproduction in which the young develop internally and are maintained and nourished by the mother before birth; see oviparous and ovoviviparous.

VO_2 abbreviation for volume of oxygen consumed per unit time.

voltine refers to the number of generations during the year or during the reproductive season (e.g., univoltine, bivoltine, polyvoltine).

vulva ventrally located genital opening of the female nematode, usually located in the midbody region.

water tube in bivalve molluscs, the channels formed between the descending and ascending limbs of lamellibranch gill filaments which carry water past gill ostia dorsally to the epibranchial cavity; also form gill marsupial chambers in which eggs and larval stages are incubated through development to juveniles or glochidia.

wetlands ecosystems whose soil is saturated for long periods seasonally or continuously; include marshes, swamps, ephemeral ponds, etc.

wing in bivalve molluscs, a thin, flattened, dorsally extended portion of the posterior slope occurring on the shell of some unionacean species.

YOY (young-of-the-year) juveniles during the period from the last larval stage to adulthood, or one year of age-whichever comes sooner.

Zenker's organ ejaculation ducts, composed of longitudinal muscles in the shape of a tube on which occur radiating bristles and are covered in a chitinous sheath in freshwater ostracodes.

zooid one of the physically connected, asexually replicated, morphologic units that comprise a colony; in bryozoans, the zooid includes the polypide and associated musculature as well as all parts of the adjacent body wall.

Subject Index

Taxonomic Index